STABILITY OF STRUCTURES

THE OXFORD ENGINEERING SCIENCE SERIES

1. D. R. Rhodes: *Synthesis of Planar Antenna Sources* (1974)
2. L. C. Woods: *Thermodynamics of Fluid Systems* (1975)
3. R. N. Franklin: *Plasma Phenomena in Gas Discharges* (1976)
4. G. A. Bird: *Molecular Gas Dynamics* (1976)
5. H. G. Unger: *Optical Waveguides and Fibres* (1977)
6. J. Heyman: *Equilibrium of Shell Structures* (1977)
7. K. H. Hunt: *Kinematic Geometry of Mechanisms* (1978)
8. D. S. Jones: *Methods in Electromagnetic Wave Propagation. 2 Volumes* (1979)
9. W. D. Marsh: *Economics of Electric Utility Power Generation* (1980)
10. P. Hagedorn: *Non-linear Oscillations* (*2nd edn*) (1988)
11. R. Hill: *Mathematical Theory of Plasticity* (1950)
12. D. J. Dawe: *Matrix and Finite Element Displacement Analysis of Structures* (1984)
13. N. W. Murray: *Introduction to the Theory of Thin-walled Structures* (1984)
14. R. I. Tanner: *Engineering Rheology* (1985)
15. M. F. Kanninen and C. H. Popelar: *Advanced Fracture Mechanics* (1985)
16. R. H. T. Bates and M. J. McDonnel: *Image Restoration and Reconstruction* (1986)
18. K. Huseyin: *Multiple-parameter Stability Theory and its Applications* (1986)
19. R. N. Bracewell: *The Hartley Transform* (1986)
20. J. Wesson: *Tokamaks* (1987)
21. P. B. Whalley: *Boiling, Condensation, and Gas–liquid Flow* (1987)
22. C. Samson, M. Le Borgnc, and B. Espiau: *Robot Control* (1989)
23. H. J. Ramm: *Fluid Dynamics for the Study of Transonic Flow* (1990)
24. R. R. A. Syms: *Practical Volume Holography* (1990)
25. W. D. McComb: *Physics of Fluid Turbulence* (1990)
26. Z. P. Bažant and L. Cedolin: *Stability of Structures* (1991)

STABILITY OF STRUCTURES

Elastic, Inelastic, Fracture, and Damage Theories

Zdeněk P. Bažant
Walter P. Murphy Professor
of Civil Engineering
Northwestern University
Evanston, Illinois, USA

and

Luigi Cedolin
Professor of Structural Engineering
Politecnico di Milano
Milano, Italy

New York Oxford
Oxford University Press
1991

Oxford University Press

Oxford New York Toronto
Delhi Bombay Calcutta Madras Karachi
Petaling Jaya Singapore Hong Kong Tokyo
Nairobi Dar es Salaam Cape Town
Melbourne Auckland

and associated companies in
Berlin Ibadan

Published by Oxford University Press, Inc.,
200 Madison Avenue, New York, New York 10016

Library of Congress Cataloging-in-Publication Data
Bažant, Z. P.
Stability of structures: elastic, inelastic, fracture, and
damage theories / by Zdeněk, P. Bažant and Luigi Cedolin.
p. cm.— (The Oxford engineering science series; 26)
Includes index.
ISBN 0-19-505529-2
1. Structural stability. 2. Structural analysis (Engineering)
I. Cedolin, Luigi, II. Title. III. Series.
TA656.B39 1991 623.1'7—dc20 90-7291
CIP

Printing (last digit): 9 8 7 6 5 4 3 2 1

Printed in the United States of America
on acid-free paper

Dedicated to our beloved and esteemed fathers
Zdeněk J. Bažant
and
Lionello Cedolin

Dr. ZDENĚK J. BAŽANT, one of the two persons to whom this book is dedicated, was born on June 11, 1908, in Nové Město in Moravia, Czechoslovakia. He was Professor of Foundation Engineering at the Czech Technical University at Prague during 1947–1975. Prior to that he had been the chief design engineer of Lanna Co. in Prague (1932–1947). Since 1975, he has been Professor Emeritus and Consultant to Geoindustria Co. in Prague. He has been a fellow of ASCE without interruption since 1947. He has authored a number of books and numerous research articles.

Preface

It is our hope that this book will serve both as a textbook for graduate courses on stability of structures and a reference volume for engineers and scientists. We assume the student has a background in mathematics and mechanics only at the level of the B.S. degree in civil or mechanical engineering, though in the last four chapters we assume a more advanced background. We cover subjects relevant to civil, structural, mechanical, aerospace, and nuclear engineering, as well as materials science, although in the first half of the book we place somewhat more emphasis on the civil engineering applications than on others. We include many original derivations as well as some new research results not yet published in periodicals.

Our desire is to achieve understanding rather than just knowledge. We try to proceed in each problem from special to general, from simple to complex, treating each subject as concisely as we can and at the lowest possible level of mathematical apparatus we know, but not so low as to sacrifice efficiency of presentation. We include a large number (almost 700) of exercise problems. Solving many of them is, in our experience, essential for the student to master the subject.

In some curricula, the teaching of stability is fragmented into courses on structural mechanics, design of steel structures, design of concrete structures, structural dynamics, plates and shells, finite elements, plasticity, viscoelasticity, and continuum mechanics. Stability theory, however, stands at the heart of structural and continuum mechanics. Whoever understands it understands mechanics. The methods of stability analysis in various applications are similar, resting on the same principles. A fundamental understanding of these principles, which is not easy to acquire, is likely to be sacrificed when stability is taught by bits, in various courses. Therefore, in our opinion, it is preferable to teach stability in a single course, which should represent the core of the mechanics program in civil, mechanical, and aerospace engineering.

Existing textbooks of structural stability, except for touching on elastoplastic columns, deal almost exclusively with elastic stability. The modern stability problems of fracture and damage, as well as the thermodynamic principles of stability of irreversible systems, have not been covered in textbooks. Even the catastrophe theory, as general is it purports to be, has been limited to systems that possess a potential, which implies elastic behavior. Reflecting recent research results, we depart from tradition, devoting about half of the book to nonelastic stability.

Various kinds of graduate courses can be fashioned from this book. The first-year quarter-length course for structural engineering students may, for example, consist of Sections 1.2–1.7, 2.1–2.4, 2.8, 3.1, 3.2, 3.5, 3.6, 4.2–4.6,

5.1–5.4, 6.1–6.3, 7.1–7.3, 7.5, 7.8, 8.1, 8.3, and 8.4, although about one-third of these sections can be covered in one quarter only partly. A semester-length course can cover them fully and may be expanded by Sections 1.8, 1.9, 2.7, 3.3, 4.5, 4.6, 5.5, 7.4, 7.8, 8.2, and 8.6. The first-year course for mechanical and aerospace engineers may, for example, be composed of Sections 1.1–1.5, 1.7, 1.9, 2.1–2.3, 3.1–3.7, 4.2–4.6, 5.1–5.4, 6.1–6.3, 7.1–7.3, 7.5, 7.8, 8.1–8.3, and 9.1–9.3, again with some sections covered only partly. A second-year sequel for structural engineering students, dealing with inelastic structural stability, can, for example, consist of Sections 8.1–8.6, 9.1–9.6, 10.1–10.4, 13.2–13.4, and 13.6, preceded as necessary by a review of some highlights from the first course. Another possible second-year sequel, suitable for students in theoretical and applied mechanics, is a course on material modeling and stability, which can be set up from Sections 11.1–11.7, 10.1–10.6, 13.1–13.4, 13.8–13.10, and 12.1–12.5 supplemented by a detailed explanation of a few of the constitutive models mentioned in Section 13.11. A course on Stability of Thin-Wall Structures (including plates and shells) can consist of a review of Sections 1.1–1.8 and detailed presentation of Chapters 6 and 7. A course on Inelastic Columns can be based on a review of Sections 1.1–1.8 and detailed presentation of Chapters 8 and 9. A course on Stability of Multidimensional Structures can be based on a review of Sections 1.1–1.9 and detailed presentation of Chapters 7 and 11. A course on Energy Approach to Structural Stability can be based on a review of Sections 1.1–1.8 and detailed presentation of Chapters 4, 5, and 10. A course on Buckling of Frames can be based on Chapters 1, 2, and 3. Chapter 3, along with Section 8.6, can serve as the basis for a large part of a course on Dynamic Stability.

The present book grew out of lecture notes for a course on stability of structures that Professor Bažant has been teaching at Northwestern University every year since 1969. An initial version of these notes was completed during Bažant's Guggenheim fellowship in 1978, spent partly at Stanford and Caltech. Most of the final version of the book was written during Professor Cedolin's visiting appointment at Northwestern between 1986 and 1988, when he enriched the text with his experience from teaching a course on structural analysis at Politecnico di Milano. Most of the last six chapters are based on Bažant's lecture notes for second-year graduate courses on inelastic structural stability, on material modeling principles, and on fracture of concrete, rock, and ceramics. Various drafts of the last chapters were finalized in connection with Bažant's stay as NATO Senior Guest Scientist at the Ecole Normale Supérieure, Cachan, France, and various sections of the book were initially presented by Bažant during specialized intensive courses and guest seminars at the Royal Institute of Technology (Cement och Betonginstitutet, CBI), Stockholm; Ecole des Ponts et Chaussées, Paris; Politecnico di Milano; University of Cape Town; University of Adelaide; University of Tokyo; and Swiss Federal Institute of Technology. Thanks go to Northwestern University and the Politecnico di Milano for providing environments conducive to scholarly pursuits. Professor Bažant had the good fortune to receive financial support from the U.S. National Science Foundation and the Air Force Office of Scientific Research, through grants to Northwestern University; this funding supported research on which the last six chapters are partly based. Professor Bažant wishes to express his thanks to his father, Zdeněk J. Bažant, Professor Emeritus of Foundation Engineering at the

Czech Technical University (ČVUT) in Prague and to his grandfather, Zdeněk Bažant, late Professor of Structural Mechanics at ČVUT, for having introduced him to certain stability problems of structural and geotechnical engineering.

We are indebted for many detailed and very useful comments to Leone Corradi and Giulio Maier, and for further useful comments to several colleagues who read parts of the text: Professors J. P. Cordebois, S. Dei Poli, Eduardo Dvorkin, Theodore V. Galambos, Richard Kohoutek, Franco Mola, Brian Moran, and Jaime Planas. Finally, we extend our thanks to M. Tabbara, R. Gettu, and M. T. Kazemi, graduate research assistants at Northwestern University, for checking some parts of the manuscript and giving various useful comments, to Vera Fisher for her expert typing of the manuscript, and to Giuseppe Martinelli for his impeccable drawings.

Evanston, Ill. Z. P. B. and L. C.
October, 1989

Contents

Introduction

One of the principal objectives of theoretical research in any department of knowledge is to find the point of view from which the subject appears in its greatest simplicity.

—J. Willard Gibbs
(acceptance letter of Rumford Medal, 1881)

Failures of many engineering structures fall into one of two simple categories: (1) material failure and (2) structural instability. The first type of failure, treated in introductory courses on the strength of materials and structural mechanics, can usually be adequately predicted by analyzing the structure on the basis of equilibrium conditions or equations of motion that are written for the initial, undeformed configuration of the structure. By contrast, the prediction of failures due to structural instability requires equations of equilibrium or motion to be formulated on the basis of the deformed configuration of the structure. Since the deformed configuration is not known in advance but depends on the deflections to be solved, the problem is in principle nonlinear, although frequently it can be linearized in order to facilitate analysis.

Structural failures caused by failure of the material are governed, in the simplest approach, by the value of the material strength or yield limit, which is independent of structural geometry and size. By contrast, the load at which a structure becomes unstable can be, in the simplest approach, regarded as independent of the material strength or yield limit; it depends on structural geometry and size, especially slenderness, and is governed primarily by the stiffness of the material, characterized, for example, by the elastic modulus. Failures of elastic structures due to structural instability have their primary cause in geometric effects: the geometry of deformation introduces nonlinearities that amplify the stresses calculated on the basis of the initial undeformed configuration of the structure.

The stability of elastic structures is a classical problem which forms the primary content of most existing textbooks. We will devote about half the present treatise to this topic (Part I, Chapters 1–7).

We begin our study of structural stability with the analysis of buckling of elastic columns and frames, a bread-and-butter problem for structural engineers. Although this is a classical research field, we cover in some detail various recent advances dealing with the analysis of very large regular frames with many members, which are finding increasing applications in tall buildings as well as certain designs for space structures.

The study of structural stability is often confusing because the definition of structural stability itself is unstable. Various definitions may serve a useful purpose for different problems. However, one definition of stability—the dynamic definition—is fundamental and applicable to all structural stability problems. Dynamic stability analysis is essential for structures subjected to nonconservative loads, such as wind or pulsating forces. Structures loaded in this manner may falsely appear to be stable according to static analysis while in reality they fail through vibrations of ever increasing amplitude or some other accelerated motion. Because of the importance of this problem in modern structural engineering we will include a thorough treatment of the dynamic approach to stability in Chapter 3. We will see that the static approach yields correct critical loads only for conservative structural systems, but even for these it cannot answer the question of stability completely.

The question of stability may be most effectively answered on the basis of the energy criterion of stability, which follows from the dynamic definition if the system is conservative. We will treat the energy methods for discrete and discretized systems in Chapter 4 and those for continuous structures in Chapter 5, in which we will also focus on the approximate energy methods that simplify the stability analysis of continuous structures.

In Chapters 6 and 7 we will apply the equilibrium and energy methods to stability analysis of more complicated thin structures such as thin-wall beams, the analysis of which can still be made one-dimensionally, and of two-dimensional structures such as plates and shells. Because many excellent detailed books deal with these problems, and also because the solution of these problems is tedious, requiring lengthy derivations and mathematical exercises that add little to the basic understanding of the behavior of the structure, we limit the treatment of these complex problems to the basic, prototype situations. At the same time we emphasize special features and approaches, including an explanation of the direct and indirect variational methods, the effect of imperfections, the postcritical behavior, and load capacity. In our computer era, the value of the complicated analytical solutions of shells and other thin-wall structures is diminishing, since the solutions can be obtained by finite elements, the treatment of which is outside the scope of the present treatise.

While the first half of the book (Part I, Chaps. 1–7) represents a fairly classical choice of topics and coverage for a textbook on structural stability, the second half of the book (Part II, Chaps. 8–13), devoted to inelastic and damage theories of structural stability, attempts to synthesize the latest trends in research. Inelastic behavior comprises not only plasticity (or elastoplasticity), treated in Chapters 8 and 10, but also creep (viscoelastic as well as viscoplastic), treated in in Chapter 9, while damage comprises not only strain-softening damage, treated in Chapter 13, but also fracture, which represents the special or limiting case of localized damage, treated in Chapter 12. Whereas the chapters dealing with plasticity and creep present for the most part relatively well-established theories, Chapters 10–13, dealing with thermodynamic concepts and finite strain effects in three dimensions, as well as fracture, damage, and friction, present mostly fresh results of recent researches that might in the future be subject to reinterpretations and updates.

Inelastic behavior tends to destabilize structures and generally blurs the aforementioned distinction between material failures and stability failures. Its effect can be twofold: (1) it can merely reduce the critical load, while instability is still caused by nonlinear geometric effects and cannot occur in their absence—this is typical of plasticity and creep (with no softening or damage); or (2) it can cause instability by itself, even in the absence of nonlinear geometric effects in the structure—this is typical of fracture, strain-softening damage, and friction and currently represents a hot research subject. An example of this behavior is fracture mechanics. In this theory (outlined in Chapter 12), structural failure is treated as a consequence of unstable crack propagation, the instability being caused by the global structural action (in which the cause of instability is the release of energy from the structure into the crack front) rather than the nonlinear geometric effects.

Stability analysis of structures that are not elastic is complicated by the fact that the principle of minimum potential energy, the basic tool for elastic structures, is inapplicable. Stability can, of course, be analyzed dynamically, but that too is complicated, especially for inelastic behavior. However, as we will see in Chapter 10, energy analysis of stability is possible on the basis of the second law of thermodynamics. To aid the reader, we will include in Chapter 10 a thorough discussion of the necessary thermodynamic principles and will then apply them in a number of examples.

Irreversibility, which is the salient characteristic of nonelastic behavior, produces a new phenomenon: the bifurcation of equilibrium path need not be associated with stability loss but can typically occur in a stable manner and at a load that is substantially smaller than the stability limit. This phenomenon, which is not found in elastic structures, will come to light in Chapter 8 (dealing with elastoplastic columns) and will reappear in Chapters 12 and 13 in various problems of damage and fracture. A surprising feature of such bifurcations is that the states on more than one postbifurcation branch of the equilibrium path can be stable, which is impossible for elastic structures or reversible systems in general. To determine the postbifurcation path that will actually be followed by the structure, we will need to introduce in Chapter 10 a new concept of stable path, which, as it turns out, must be distinct from the concept of stable state. We will present a general thermodynamic criterion that makes it possible to identify the stable path.

The stability implications of the time-dependent material behavior, broadly termed creep, also include some characteristic phenomena, which will be explained in Chapter 9. In dealing with imperfect viscoelastic structures under permanent loads, we will have to take into account the asymptotic deflections as the time tends to infinity, and we will see that the long-time (asymptotic) critical load is less than the instantaneous (elastic) critical load. In imperfect viscoelastic structures, the deflections can approach infinity at a finite critical time, and can again do so under a load that is less than the instantaneous critical load. For creep buckling of concrete structures, we will further have to take into account the profound effect of age on creep exhibited by this complex material.

The most important consequence of the instabilities caused by fracture or damage rather than by geometric effects is that they produce size effect, that is, the structure size affects the nominal stress at failure. By contrast, no size effect exists according to the traditional concepts of strength, yield limit, and yield surface in the stress or strain space. Neither does it according to elastic stability theory. The most severe and also the simplest size effect is caused by failures due to propagation of a sharp fracture where the fracture process happens at a point. A less severe size effect, which represents a transition from failures governed by strength or yield criteria to failures governed by instability of sharp fractures, is produced by instability modes consisting either of propagation of a fracture with a large fracture process zone (Chap. 12) or of damage localization (Chap. 13). As a special highlight of the present treatise, these modern problems are treated in detail in the last two chapters.

The practical design of metallic or concrete columns and other structures is an important topic in any stability course. In this text, the code specifications and design approaches are dispersed through a number of chapters instead of being presented compactly in one place. This presentation is motivated by an effort to avoid a cookbook style and present each aspect of design only after the pertinent theory has been thoroughly explained, but not later than that. It is for this reason, and also because fundamental understanding of inelastic behavior is important, that the exposition of column design is not completed until Chapters 8 and 9, which also include detailed critical discussions of the current practice.

The guiding principle in the presentation that follows is to advance by induction, from special and simple to general and complex. This is one reason why we choose not to start the book with general differential equations in three dimensions and thermodynamic principles, which would then be reduced to special cases. (The general three-dimensional differential equations governing stability with respect to nonlinear geometric effects do not appear in the book until Chap. 11.) There is also another reason—the three-dimensional analysis of stability is not necessary for slender or thin structures made of structural materials such as steel or concrete, which are relatively stiff. It is only necessary for dealing with incremental deformations of massive inelastic structures or structures made of highly anisotropic or composite materials which can be strained to such a high level that some of the tangential moduli of the material are reduced to values that are of the same order of magnitude as the stresses.

As another interesting phenomenon, which we will see in Chapter 11, various possible choices of the finite-strain tensor lead to different expressions for the critical loads of massive bodies. It turns out that the stability formulations corresponding to different choices of the finite-strain tensor are all equivalent, but for each such formulation the tangential moduli tensor of the material has a different physical meaning and must be determined from experimental data in a different manner. If this is not done, then three-dimensional finite-strain stability analysis makes no sense.

As we live in the new era of computers, stability of almost any given structure could, at least in principle, be analyzed by geometrically nonlinear finite element codes with incremental loading. This could be done in the presence of complex nonlinear behavior of the material as well. Powerful though this approach is, the value of simple analytical solutions that can be worked out by hand must not be underestimated. This book attempts to concentrate on such solutions. It is these solutions that enhance our understanding and also must be used as test cases for the finite element programs.

I

ELASTIC THEORIES

1

Buckling of Elastic Columns by Equilibrium Analysis

Under an axial compressive load, a column that is sufficiently slender will fail due to deflection to the side rather than crushing of the material. This phenomenon, called buckling, is the simplest prototype of structural stability problems, and it is also the stability problem that was historically the first to be solved (cf. Timoshenko, 1953).

The essential characteristic of the buckling failures is that the failure load depends primarily on the elastic modulus and the cross-section stiffness, and it is almost independent of the material strength or yield limit. It is quite possible that a doubling of the material strength will achieve a less than 1 percent increase in the failure load, all other properties of the column being the same.

After a brief recall of the elements of the theory of bending, we analyze, as an introductory problem, a simply supported (pin-ended) column, first solved by Euler as early as 1744. Next we generalize our solution to arbitrary columns treated as beams with general end-support conditions and possible elastic restraints at ends. We seek to determine critical loads as the loads for which deflected equilibrium positions of the column are possible.

Subsequently we examine the effect of inevitable imperfections, such as initial curvature, load eccentricity, or small disturbing loads. We discover that even if they are extremely small, they still cause failure since they produce very large destructive deflections when the critical load is approached. Taking advantage of our solution of the behavior of columns with imperfections, we further discuss the method of experimental determination of critical loads. We also, of course, explain the corresponding code specifications for the design of columns, although we avoid dwelling on numerous practical details that belong to a course on the design of concrete or steel structures rather than a course on stability. However, those code specifications that are based on inelastic stability analysis will have to be postponed until Chapter 8. Furthermore, we show that for certain columns the usual bending theory is insufficient, and a generalized theory that takes into account the effect of shear must be used. We conclude the first chapter by analysis of large deflections for which, in contrast to all the preceding analysis, a nonlinear theory is required.

In stability analysis of elastic structures, the equilibrium conditions must be formulated on the basis of the final deformed shape of the structure. When this is done under the assumption of small deflections and small rotations, we speak of second-order theory, while the usual analysis of stresses and deformations in which the equilibrium conditions are formulated on the basis of the initial, undeformed shape of the structure represents the first-order theory. The first-order theory is linear, whereas the second-order theory takes into account the leading nonlinear terms due to changes of structural geometry at the start of buckling. All the critical load analyses of elastic structures that follow fall into the category of the second-order theory.

1.1 THEORY OF BENDING

Because of its practical importance, stability of beam structures subjected to bending will occupy a major part of this text. In the theory of bending we consider beams that are sufficiently slender, that is, the ratio of their length to the cross-section dimensions is sufficiently large (for practical purposes over about 10:1). For such slender beams, the theory of bending represents a very good approximation to the exact solution according to three-dimensional elasticity. This theory, first suggested by Bernoulli in 1705 and systematically developed by Navier in 1826 is based on the following fundamental hypothesis:

During deflection, the plane normal cross sections of the beam remain (1) plane and (2) normal to the deflected centroidal axis of the beam, and (3) the transverse normal stresses are negligible.

The Bernoulli–Navier hypothesis, which is applicable not only to elastic beams but also to inelastic beams, provided that they are sufficiently slender, implies that the axial normal strains are $\varepsilon = -z/\rho$, where z = transverse coordinate measured from the centroid of the cross section (Fig. 1.1), and ρ = curvature radius of the deflected centroidal axis of the beam, called the deflection curve. At the beginning we will deal only with linearly elastic beams; then the Bernoulli–Navier hypothesis implies that the axial normal stress is $\sigma = E\varepsilon = -Ez/\rho$ and E = Young's elastic modulus.

Substituting this into the expression for the bending moment (Fig. 1.1)

$$M = -\int_A \sigma z \, dA$$

where A = cross-section area (and the moment is taken about the centroid) one gets $M = E\int z^2 \, dA/\rho$ or

$$M = EI/\rho \tag{1.1.1}$$

where $I = \int z^2 \, dA$ = centroidal moment of inertia of the cross section. (For more details, see, e.g., Popov, 1968, or Crandall and Dahl, 1972.) In terms of

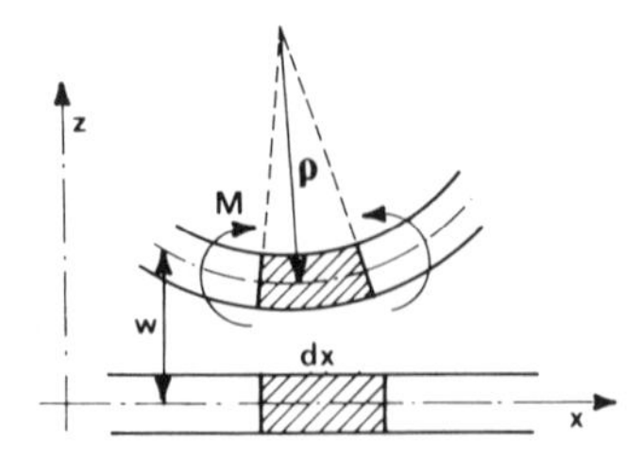

Figure 1.1 Bending of a straight bar according to Bernoulli-Navier hypothesis.

deflection w (transverse displacement of the cross section), the curvature may be expressed as

$$\frac{1}{\rho} = \frac{w''}{(1 + w'^2)^{3/2}} = w''\left(1 - \frac{3}{2}w'^2 + \frac{(3)(5)}{(2)(4)}w'^4 - \cdots\right) \tag{1.1.2}$$

in which the primes denote derivatives with respect to the length coordinate x of the beam, that is, $w'' = d^2w/dx^2$. In most of our considerations we shall assume that the slope of the deflection curve $w(x)$ is small, and then we may use the linearized approximation

$$\frac{1}{\rho} \simeq w'' \tag{1.1.3}$$

If $|w'|$ is less than 0.08, then the error in curvature is within about 1 percent. Equation 1.1.1 then becomes

$$M = EIw'' \tag{1.1.4}$$

which is the well-known differential equation of bending for small deflections. In writing the equilibrium equations for the purpose of buckling analysis, however, displacements w, even if considered small, cannot be neglected, that is, M must be calculated with respect to the deformed configuration.

Part 2 of the Bernoulli–Navier hypothesis implies that the shear deformations are neglected. Sometimes this may be unacceptable, and we will analyze the effect of shear later.

Problems

1.1.1 For a simply supported beam of length l with sinusoidal $w(x)$, calculate the percentage of error in $1/\rho$ and M at midspan if $w_{max}/l = 0.001, 0.01, 0.1, 0.3$ (subscript "max" labels the maximum value).

1.1.2 Derive Equations 1.1.1 and 1.1.2.

1.2 EULER LOAD, ADJACENT EQUILIBRIUM, AND BIFURCATION

Let us now consider a pin-ended column of length l shown in Figure 1.2 (also called the hinged or simply supported column). The column is loaded by axial load P, considered positive when it causes compression. We assume the column to be perfect, which means that it is perfectly straight before the load is applied,

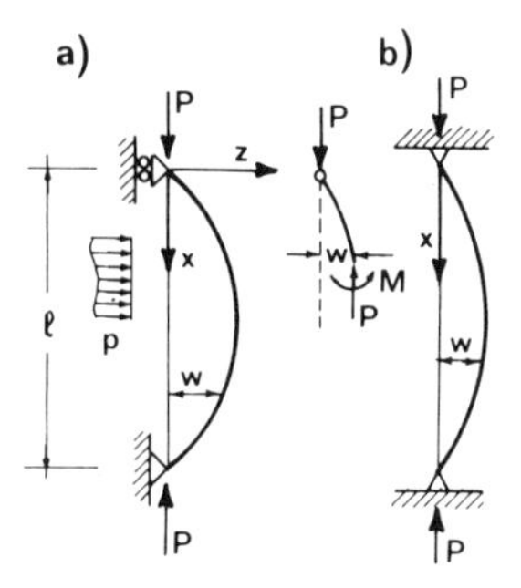

Figure 1.2 Euler column under (a) imposed axial load and (b) imposed axial displacement.

and the load is perfectly centric. For the sake of generality, we also consider a lateral distributed load $p(x)$, although for deriving the critical loads we will set $p = 0$.

The undeflected column is obviously in equilibrium for any load P. For large enough load, however, the equilibrium is unstable, that is, it cannot last. We now seek conditions under which the column can deflect from its initial straight position and still remain in equilibrium. If the load keeps constant direction during deflection, as is true of gravity loads, the bending moment on the deflected column (Fig. 1.2a) is $M = -Pw + M_0(x)$ where $M_0(x)$ is the bending moment caused by a lateral load $p(x)$, which may be calculated in the same way as for a simply supported beam without an axial load, and $-Pw$ is an additional moment that is due to deflection and is the source of buckling. Substituting Equations 1.1.1 and 1.1.3 into the preceding moment expression, we obtain $EIw'' = -Pw + M_0$ or

$$w'' + k^2 w = \frac{M_0}{EI} \qquad \text{with } k^2 = \frac{P}{EI} \tag{1.2.1}$$

This is an ordinary linear differential equation. Its boundary conditions are

$$w = 0 \quad \text{at} \quad x = 0 \qquad w = 0 \quad \text{at} \quad x = l \tag{1.2.2}$$

Consider now that $p = 0$ or $M_0 = 0$. Equations 1.2.1 and 1.2.2 then define a linear eigenvalue problem (or characteristic value problem). To permit a simple solution, assume that the bending rigidity EI is constant. The general solution for $P > 0$ (compression) is

$$w = A \sin kx + B \cos kx \tag{1.2.3}$$

in which A and B are arbitrary constants. The boundary conditions in Equation 1.2.2 require that

$$B = 0 \qquad A \sin kl = 0 \tag{1.2.4}$$

Now we observe that the last equation allows a nonzero deflection (at $P > 0$) if and only if $kl = \pi, 2\pi, 3\pi, \ldots$. Substituting for k we have $Pl^2/EI = \pi^2, 4\pi^2, 9\pi^2, \ldots$, from which

$$P_{\mathrm{cr}_n} = \frac{n^2\pi^2}{l^2} EI \qquad (n = 1, 2, 3, \ldots) \tag{1.2.5}$$

The eigenvalues P_{cr_n}, called critical loads, denote the values of load P for which a nonzero deflection of the perfect column is possible. The deflection shapes at critical loads, representing the eigenmodes, are given by

$$w = q_n \sin \frac{n\pi x}{l} \tag{1.2.6}$$

where q_n are arbitrary constants. The lowest critical load is the first eigenvalue $(n = 1)$; it is denoted as $P_{\mathrm{cr}_1} = P_{\mathrm{E}}$ and is given by

$$P_{\mathrm{E}} = \frac{\pi^2}{l^2} EI \tag{1.2.7}$$

It is also called the Euler load, after the Swiss mathematician Leonhard Euler, who obtained this formula in 1744. It is interesting that in his time it was only known that M is proportional to curvature; the fact that the proportionality constant is EI was established much later (by Navier). The Euler load represents the failure load only for perfect elastic columns. As we shall see later, for real columns that are imperfect, the Euler load is a load at which deflections become very large.

Note that at P_{cr_n} the solution is not unique. This seems at odds with the well-known result that the solutions to problems of classical linear elasticity are unique. However, the proof of uniqueness in linear elasticity is contingent upon the assumption that the initial state is stress-free, which is not true in our case, and that the conditions of equilibrium are written on the basis of the geometry of the undeformed structure, whereas we determined the bending moment taking the deflection into account. The theory that takes into account the effect of deflections (i.e., change of geometry) on the equilibrium conditions is called the *second-order theory,* as already said.

At critical loads, the straight shape of the column, which always represents an equilibrium state for any load, has adjacent equilibrium states, the deflected shapes. The method of determining the critical loads in this manner is sometimes called the method of adjacent equilibrium. Note that at critical loads the column is in equilibrium for any q_n value, as illustrated by the equilibrium load-deflection diagram of P versus w in Figure 1.3. The column in neutral equilibrium behaves the same way as a ball lying on a horizontal plane (Fig. 1.3).

In reality, of course, the deflection cannot become arbitrarily large because we initially assumed small deflections. When finite deflections of the column are solved, it is found that the branch of the P-w diagram emanating from the critical load point is curved upward and has a horizontal tangent at the critical load (Sec. 1.9).

At critical loads, the primary equilibrium path (vertical) reaches a bifurcation point (branching point) and branches (bifurcates) into secondary (horizontal) equilibrium paths. This type of behavior, found in analyzing a perfect beam, is called the buckling of bifurcation type. Not all static buckling problems are of this type, as we shall see, and bifurcation buckling is not necessarily implied by the existence of adjacent equilibrium.

It is interesting to note that axial displacements do not enter the solution.

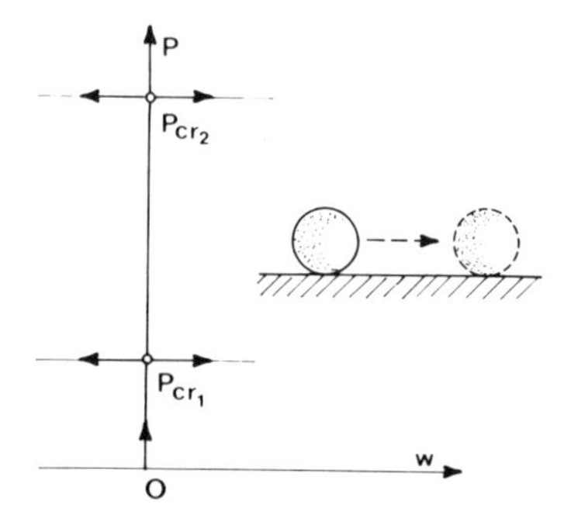

Figure 1.3 Neutral equilibrium at critical load.

Thus, the critical load is the same for columns whose hinges slide axially during buckling (Fig. 1.2a) or are fixed (Fig. 1.2b). (This will be further clarified by the complete diagram of load vs. load-point displacement in Sec. 1.9.)

Dividing Equation 1.2.7 by the cross-section area A, we obtain the critical stress

$$\sigma_E = \frac{\pi^2}{(l/r)^2} E \tag{1.2.8}$$

in which $r = \sqrt{I/A}$ = radius of inertia (or radius of gyration) of the cross section. The ratio l/r is called the slenderness ratio of the column. It represents the basic nondimensional parameter for the buckling behavior of the column. The plot of σ_E versus (l/r) is called Euler's hyperbola (Fig. 1.4). This plot is obviously meaningful only as long as the values of σ_E do not exceed the compression strength or yield limit of the material, f_y. Thus, buckling can actually occur only for columns whose slenderness ratio exceeds a certain minimum value:

$$\frac{l}{r} = \pi\sqrt{\frac{E}{f_y}} \tag{1.2.9}$$

Substituting the typical values of Young's modulus and the compression strength or the yield limit for various materials, we obtain the minimal slenderness ratios as 86 for steel, 90 for concrete, 50 for aluminum, and 13 for glass fiber-reinforced plastics. Indeed, for the last type of material, buckling failures are the predominant ones.

For real columns, due to inelastic effects and imperfections, the plot of the axial stress versus the slenderness ratio exhibits a gradual transition from the horizontal line to the Euler hyperbola, as indicated by the dashed line in Figure 1.4. Consequently, buckling phenomena already become practically noticeable at about one-half of the aforementioned slenderness ratios (see Chap. 8).

Up to this point we have not reached any conclusion about stability. However, the fact that for critical loads the deflection is indeterminate, and can obviously become large, is certainly unacceptable to a designer. A certain degree of safety against reaching the lowest critical load must evidently be assured.

Problems

1.2.1 Sketch the load-deflection diagram of the column considered and the critical state; explain neutral equilibrium and adjacent equilibrium.

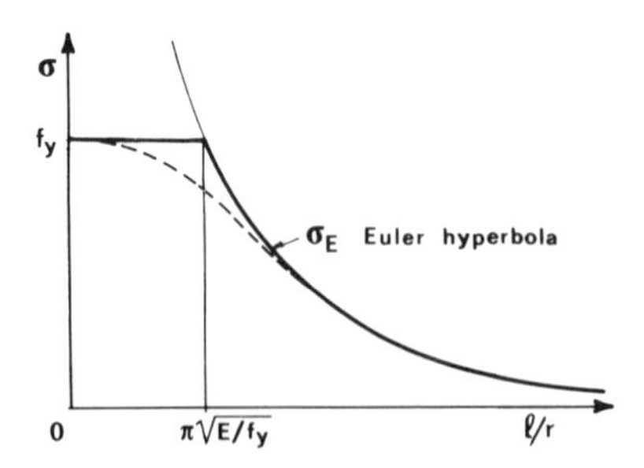

Figure 1.4 Buckling stress as a function of slenderness.

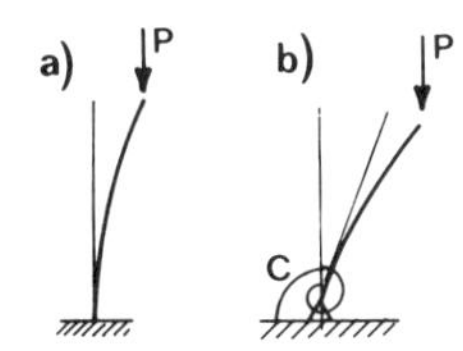

Figure 1.5 Exercise problems on critical loads of free-standing columns.

1.2.2 If a timber column has slenderness 15:1, compression strength $f_c' = 1000$ psi, and $E = 300{,}000$ psi, does it fail by buckling or by material failure?
1.2.3 Solve for the critical load of a free-standing column of height l (Fig. 1.5a).
1.2.4 Do the same for a free-standing column with a hinge and rotational spring of stiffness C at the base (Fig. 1.5b). Study the dependence of P_{cr} on EI/Cl and consider the limit case $EI/Cl \rightarrow 0$.

1.3 DIFFERENTIAL EQUATIONS OF BEAM-COLUMNS

In the foregoing solution we determined the bending moment directly from the applied load. However, this is impossible in general, for example, if the column has one or both ends fixed. Unknown bending moment M_1 and shear force V_1 then occur at the fixed end, and the bending moments in the column cannot be determined from equilibrium. Therefore, we need to establish a general differential equation for beams subjected to axial compression, called beam-columns.

Consider the equilibrium of a segment of the deflected column shown in Figure 1.6a. The conditions of force equilibrium and of moment equilibrium about the top support point are

$$
\begin{gathered}
V(x) - V_1 + \int_0^x p(x^*)\, dx^* = 0 \\
M(x) + Pw(x) - M_1 + Vx + \int_0^x p(x^*)x^*\, dx^* = 0
\end{gathered}
\tag{1.3.1}
$$

Differentiating Equations 1.3.1, we obtain $V' + p = 0$ and $M' + Pw' + V + V'x + px = 0$ from which

$$
V' = -p \qquad M' + Pw' = -V \tag{1.3.2}
$$

These relations represent the differential equations of equilibrium of beam-columns in terms of internal forces M and V. The term Pw' can be neglected only if $P \ll P_{cr_1}$, which is the case of the classical (first-order) theory. The quantities w', M', V, and p are small (infinitesimal), and so Pw' is of the same order of magnitude as M' and V unless $P \ll P_{cr_1}$.

In the special case of a negligible axial force ($P = 0$), Equations 1.3.2 reduce to the relations well known from bending theory.

Alternatively, the differential equations of equilibrium can be derived by considering an infinitesimal segment of the beam, of length dx (Fig. 1.6b). Taking into account the increments dM and dV over the segment length dx, we get the

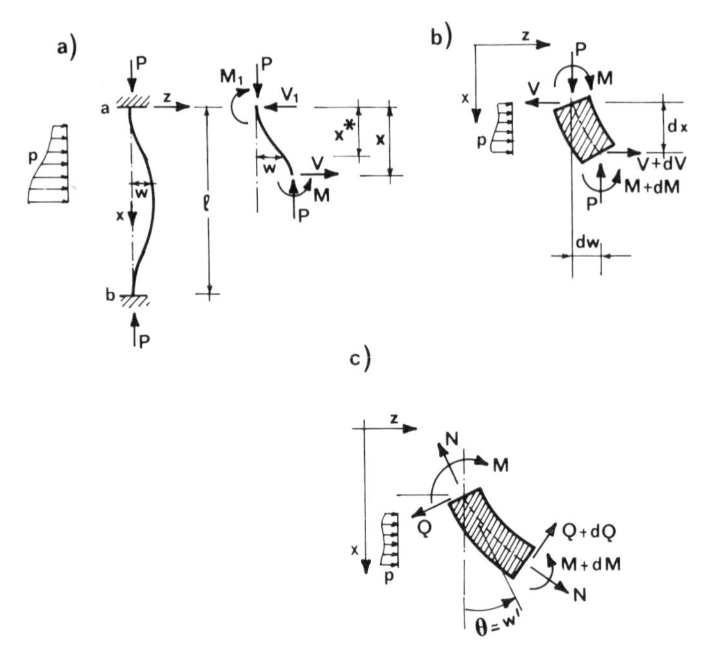

Figure 1.6 Equilibrium of (a) a segment of a statically indeterminate column and (b, c) of an infinitesimal element.

following conditions of equilibrium of transverse forces and of moments about the centroid of the cross section at the end of the deflected segment:

$$(V + dV) - V + p\,dx = 0$$
$$(M + dM) - M + V\,dx + P\,dw - (p\,dx)\left(\frac{dx}{2}\right) = 0 \tag{1.3.3}$$

Dividing these equations by dx, and considering that $dx \to 0$, we obtain again Equations 1.3.2.

Note that the shear force V is defined in such a way that its direction remains normal to the initial beam axis, that is, does not rotate during deflection (and remains horizontal in Fig. 1.6). Sometimes a shear force Q that remains normal to the deflected beam axis is introduced (Fig. 1.6c). For this case the conditions of equilibrium yield $Q = V\cos\theta + P\sin\theta \simeq V + Pw'$ and $-N = P\cos\theta - V\sin\theta \simeq P - Vw'$ (where N = axial force, positive if tensile and θ = slope of the deflected beam axis). Substituting the expression for Q in the second of Equations 1.3.2 one obtains

$$M' = -Q \tag{1.3.4}$$

which has the same form as in the first-order theory. Equation 1.3.4 could be derived directly from the moment equilibrium condition of the forces shown in Figure 1.6c.

Differentiating the second of Equations 1.3.2 and substituting in the first one, we eliminate V; $M'' + (Pw')' = p$. Furthermore, substituting $M = EIw''$ we get

$$(EIw'')'' + (Pw')' = p \tag{1.3.5}$$

This is the general differential equation for the deflections of a beam-column. It is an ordinary linear differential equation of the fourth order.

Alternatively, one could write the force equilibrium condition with the cross-section resultants shown in Figure 1.6c, which gives $p\,dx + dQ + d(Nw') = 0$. Substituting Equation 1.3.4 as well as the relations $M = EIw''$ and $-N = P$ (valid for small rotations), one obtains again Equation 1.3.5. A generalization of N is customarily used for plates (Sec. 7.2).

Note that two integrations of Equation 1.3.5 for $P = \text{const.}$ yield $EIw'' + Pw = Cx + D + \int\int p\,dx\,dx$, where C, D are integration constants. The right-hand side represents the bending moment $M_0(x)$, and so this equation is equivalent to the second-order differential (Eq. 1.2.1). However, if the column is not statically determinate, then M_0 is not known.

The term that makes Equations 1.3.2 and 1.3.5 the second-order theory is the term Pw', which is caused by formulating equilibrium on the deflected column. If this term is deleted, one obtains the familiar equations of the classical (first-order) theory, that is, $M' = -V$ and $(EIw'')'' = p$. This theory is valid only if $P \ll P_{cr_1}$ (this will be more generally proven by Fig. 2.2, which shows that the column stiffness coefficients are not significantly affected by P as long as $|P| \leq 0.1P_E$). When $P \ll P_{cr_1}$, the term Pw' is second-order small compared with M, w', and V, and therefore negligible (this is how the term "second-order theory" originated).

The terms Pw' and $(Pw')'$ would be nonlinear for a column that forms part of a larger structure, for which not only w but both w and P are unknown. The problem becomes linear when P is given, which is the case for columns statically determinate with regard to axial force. However, even if P is unknown, the problem may be treated as approximately linear provided that the deflections and rotations are so small that the variation of P during deflection is negligible (see Sec. 1.9). P may then be considered as constant during the deflection. In this manner we then get a linearized formulation.

To make simple solutions possible, consider that the axial (normal) force P and the bending rigidity EI are constant along the beam. The differential equation then has constant coefficients, and the fundamental solutions of the associated homogeneous equation ($p = 0$) may be sought in the form $w = e^{\lambda x}$. Upon substitution into Equations 1.3.5 we see that $e^{\lambda x}$ cancels out and we obtain the characteristic equation $EI\lambda^4 + P\lambda^2 = 0$, or $\lambda^2(\lambda^2 + k^2) = 0$. The roots are $\lambda = ik, -ik, 0, 0$ provided that $P > 0$ (compression). Since $\sin kx$ and $\cos kx$ are linear combinations of e^{ikx} and e^{-ikx}, the general solution of Equations 1.3.5 for constant EI and P is

$$w(x) = A \sin kx + B \cos kx + Cx + D + w_p(x) \qquad (P > 0) \qquad (1.3.6)$$

in which A, B, C, and D are arbitrary constants and $w_p(x)$ is a particular solution corresponding to the transverse distributed loads $p(x)$.

For columns in structures, it is sometimes also necessary to take into account the effect of an axial tensile force $P < 0$ on the deflections. In that case the characteristic equation is $\lambda^2(\lambda^2 - k^2) = 0$, with $k^2 = |p|/EI$. The general solution then is

$$w(x) = A \sinh kx + B \cosh kx + Cx + D + w_p(x) \qquad (P < 0) \qquad (1.3.7)$$

The basic types of boundary conditions are

$$
\begin{aligned}
&\text{Fixed end:} && w=0 && w'=0 && && \\
&\text{Hinge:} && w=0 && M=0 && \text{or} && w''=0 \\
&\text{Free end:} && M=0 && V=0 && \text{or} && (EIw'')'+Pw'=0 \\
&\text{Sliding restraint:} && w'=0 && V=0 && &&
\end{aligned}
\qquad (1.3.8)
$$

When we seek critical loads, we set $p=0$. The boundary conditions are homogeneous, and the boundary-value problem defined by Equations 1.3.5 and 1.3.8 becomes an eigenvalue problem.

Generally, there exists (for $p=0$) an infinite series of critical loads, and the load-deflection diagram is of the kind shown in Figure 1.3. At critical loads, one has bifurcation of the equilibrium path and neutral equilibrium.

When the bending rigidity EI or the axial force P vary along the beam, approximate solutions are in general necessary. This may be accomplished, for example, by the finite difference or the finite element method, which leads to an algebraic eigenvalue problem for a system of homogeneous algebraic equations. Solutions in terms of orthogonal series expansions are also possible, and normally very efficient, especially for hand calculations.

Problems

1.3.1 Without referring to the text, derive the differential equations of equilibrium of a beam-column and the general solution for $w(x)$, using the cross-section resultants the directions of which do not rotate during deflection (Fig. 1.6b).

1.3.2 Do the same as in Problem 1.3.1 but use the cross-section resultants the directions of which rotate during deflection (Fig. 1.6c).

1.3.3 Explain why, for small rotations w' and large P, $Q \neq V$ while $(-N) \simeq P$. *Hint:* Is the $Q-V$ first-order small in w', that is, proportional to w', and $|N|-P$ second-order small, that is, proportional to w'^2? Note that w, w', M, V, and Q are considered small (infinitesimal) while P is finite.

1.3.4 Write the finite difference equations that approximate Equations 1.3.5 and 1.3.8 for case of variable $EI(x)$.

1.3.5 Find the general solution of the beam-column equation for variable EI and P such that $EI=a+bx$, $P=a+b(x+\tan x)$ where $a, b=$ constants. *Hint:* Is $w=\sin x$ a solution?

1.4 CRITICAL LOADS OF PERFECT COLUMNS WITH VARIOUS END RESTRAINTS

Let us now examine the solution of critical loads for perfect columns with various simple end restraints shown in Figure 1.7. As an example, consider a column with one end fixed (restrained, built-in) and one end hinged (pinned), sometimes called propped-end column, (Fig. 1.7a). Let $p=0$. Because the general solution has four arbitrary constants, four boundary conditions are needed. They are of two kinds, kinematic and static. The kinematic ones are $w=0$ and $w'=0$ at $x=l$,

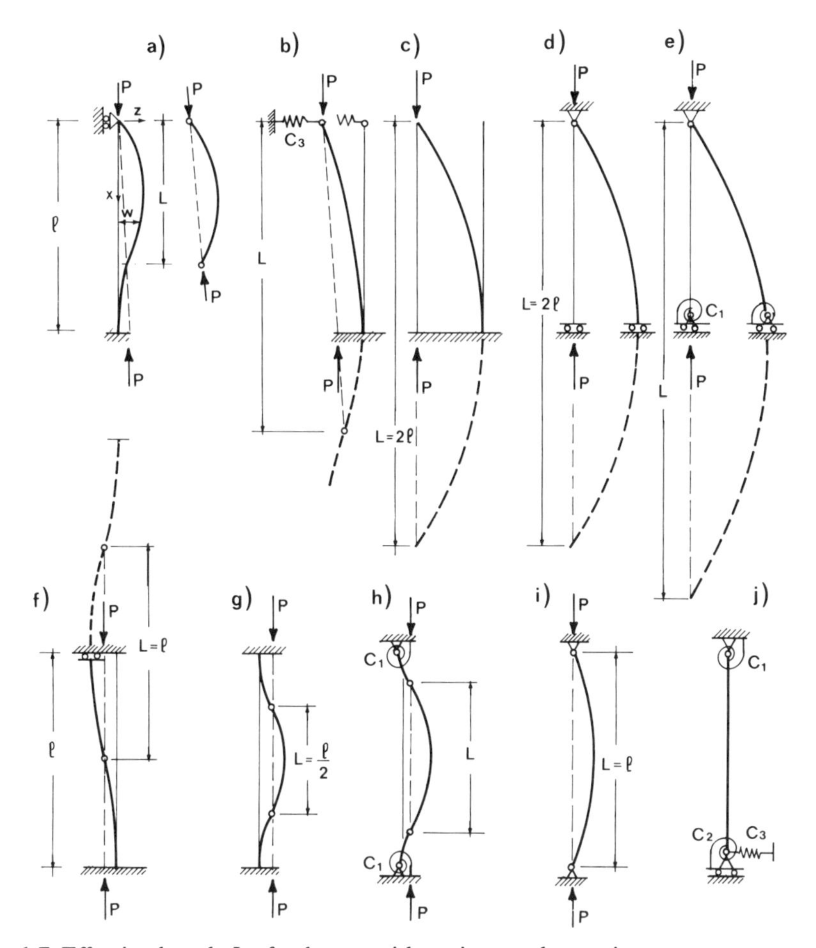

Figure 1.7 Effective length L of columns with various end restraints.

and $w=0$ at $x=0$. The remaining boundary condition is static: $M=0$ or $EIw''=0$ or $w''=0$ at $x=0$ (axial coordinate x is measured from the free end; see Fig. 1.7). In terms of Equation 1.3.6 (with $w_p=0$), these boundary conditions are

$$
\begin{aligned}
&\text{For } x=0: && B+D=0 \\
& && -Bk^2=0 \\
&\text{For } x=l: && A\sin kl + B\cos kl + Cl + D = 0 \\
& && Ak\cos kl - Bk\sin kl + C = 0
\end{aligned}
\tag{1.4.1}
$$

This is a system of four linear homogeneous algebraic equations for the unknowns A, B, C, and D. This system, representing an algebraic eigenvalue problem, can have a nonzero solution only when the determinant of the equation system vanishes. This condition may be reduced, after some algebraic rearrangements, to the equation $\sin kl - kl\cos kl = 0$. Because $\cos kl = 0$ does not solve this equation, we may divide it by $\cos kl$ and get

$$\tan kl = kl \tag{1.4.2}$$

This is a transcendental algebraic equation. The approximate values of the roots may be located graphically as the intersection points of curves $y = u$ and $y = \tan u$ ($u = kl$) (see Fig. 1.8). Using the iterative Newton method, one can then determine the roots with any desired accuracy. The smallest positive root (Fig. 1.8) is $u = kl = 4.4934$, and noting that $k = \sqrt{P/EI}$, we find that the smallest critical load is

$$P_{cr} = \frac{\pi^2}{(0.699l)^2} EI \tag{1.4.3}$$

The shape of the buckled column may be obtained by eliminating B, C, and D from Equation 1.4.1. This yields

$$w = A(\sin kx - kx \cos kl) \tag{1.4.4}$$

where A is an arbitrary constant (limited, of course, by the range of small deflections). Solving x for which $w'' = 0$, we find that there is an inflection point at $x = 0.699l$. Note that, after buckling, the reaction at the base is no longer aligned with the beam axis (Fig. 1.7a), but has the eccentricity $(M)_{x=l}/P_{cr} = (EIw'')_{x=l}/P_{cr} = -A \sin kl = 0.976A$, and its line of action runs through the deflected inflection point, as well as the column top, as expected.

Columns with other end restraints shown in Figure 1.7 can be solved in a similar manner. Aside from the simple boundary conditions listed in Equations 1.3.8 and illustrated by Figure 1.7c, d, f, g, one can have elastically restrained ends. Such restraints can sometimes be used as approximations for the behavior of columns as parts of larger structures, the action of the rest of the structure upon the column being replaced by an equivalent spring. An end supported elastically in the transverse direction (Fig. 1.7b) is described by the boundary conditions $M = 0$ and $V = -C_3 w$. A hinge that slides freely but is elastically restrained against rotation (Fig. 1.7e) is characterized by the boundary conditions $V = 0$ and $M = C_1 w'$. A hinged end elastically restrained against rotation (Fig. 1.7h) is characterized by the boundary conditions $w = 0$ and $M = C_1 w'$, where C_1 is the spring stiffness.

The buckling modes of a column under different boundary conditions are demonstrated in Figure 1.9, which portrays one of the teaching models developed at Northwestern University (1969). Intermediate supports are used to obtain higher buckling modes.

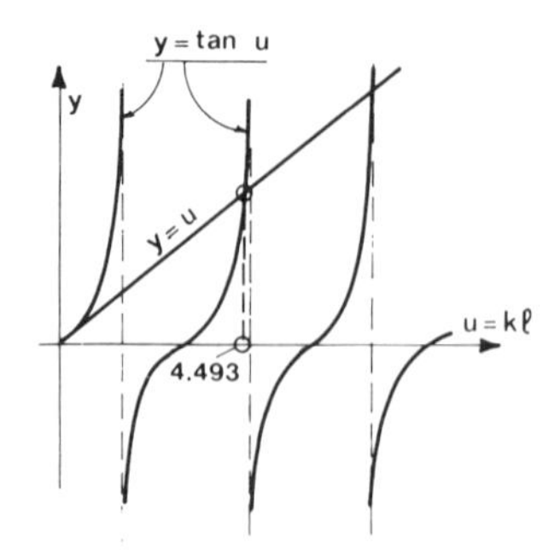

Figure 1.8 Determination of critical states for fixed-hinged columns.

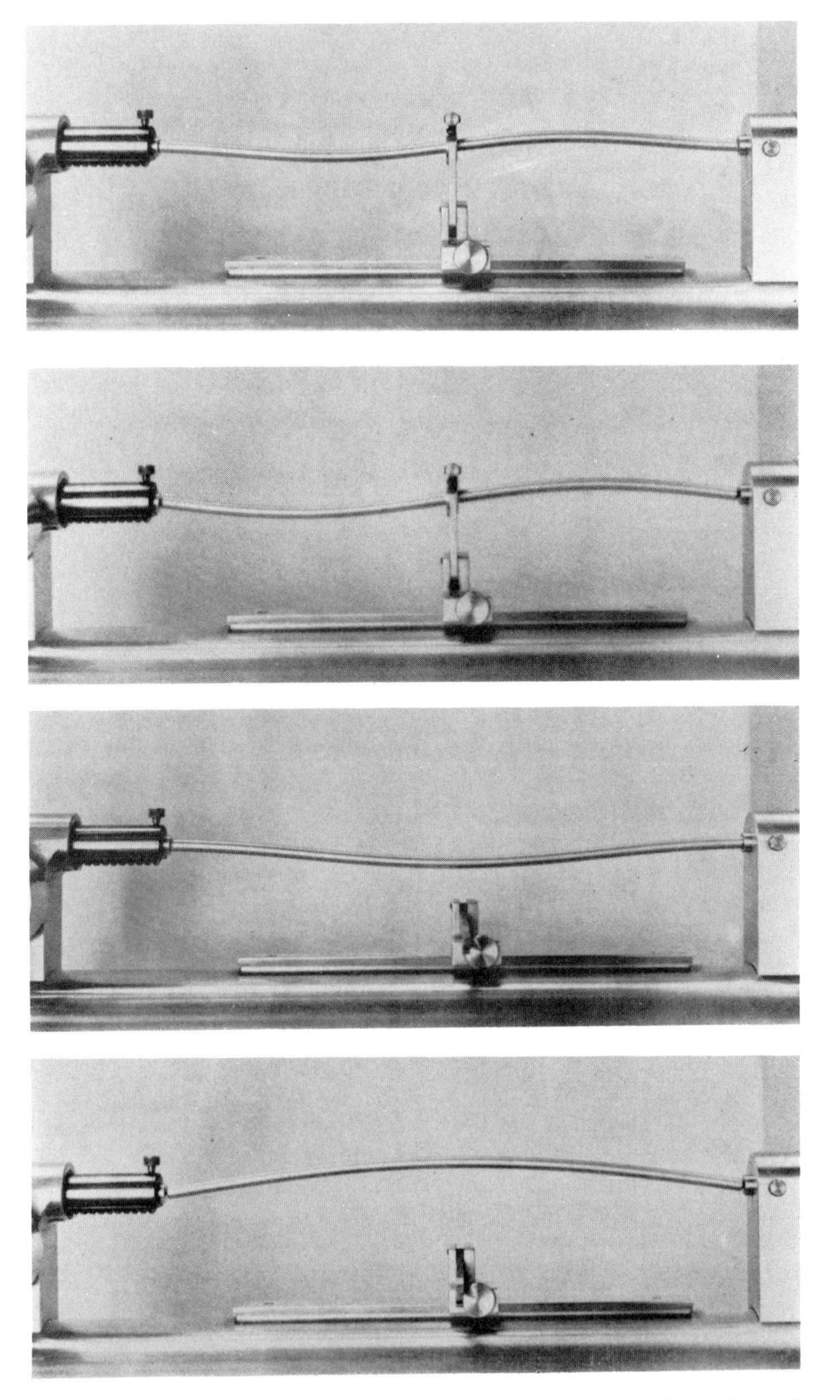

Figure 1.9 Northwestern University (1969) teaching models: buckling of axially compressed columns with various end restraints.

Instead of solving for constants A, B, C, and D from a system of four algebraic equations for each particular type of column, one can determine the critical loads more expediently using the idea of the so-called effective length, L (also called the free length, the reduced length, or the buckling length). This approach is based on the fact that any segment of the column between two adjacent inflection points of the deflection curve is equivalent to a pin-ended column of length L equal to the distance between these two inflection points. This is because $M=0$ ($w''=0$) at the inflection points. Since the expression in Equation 1.3.6 for the general solution can be extended beyond the length domain of the column, one can also consider inflection points of this extended deflection curve that lie outside the column. This needs to be done when two inflection points within the actual column length do not exist. The critical load of any column may now be written in the form

$$P_{cr} = \frac{\pi^2}{L^2} EI \tag{1.4.5}$$

As an example, comparison with Equation. 1.4.3 suggests that the effective length for the fixed–hinge column (Fig. 1.7a) should be $L = 0.699l$ where l is the actual length of the column. Differentiating Equation 1.4.4, we find $w'' = -Ak^2 \sin kx$, which becomes 0 for $x = L$ when $kL = \pi$. From this $L/l = \pi/kl = \pi/4.4934 = 0.699$. This confirms the effective length approach.

Equation 1.3.6 (with $w_p = 0$) represents a transversely shifted and rotated sine curve. Thus the effective length L can be intuitively figured out by trying to sketch a sine curve that fulfills the given boundary conditions. This is illustrated for various columns in Figure 1.7. For the fixed–fixed column (Fig. 1.7g), the inflection points are obviously at quarter-length points. Therefore, $L = l/2$, from which $P_{cr_1} = 4\pi^2 EI/l^2$. For the fixed–free column (Fig. 1.7c), one inflection point is at the free end ($M=0$), and the other one may be located by extending the deflection curve downward; $L = 2l$, from which $P_{cr_1} = \pi^2 EI/4l^2$. Similarly, for the column with a fixed end and a sliding restraint (Fig. 1.7f), extension of the deflection curve shows that $L = l$, and so the first critical load is the Euler load. For a column with one hinge and one sliding restraint (Fig. 1.7d), extension of the deflection curve shows that the effective length is $L = 2l$, and so $P_{cr_1} = \pi^2 EI/4l^2$. For the column with rotational springs (Fig. 1.7h), the bending moments at the ends oppose rotation, which causes the inflection point to shift away from the ends, as compared with a pin-ended column; consequently, $l/2 < L < l$, which means that $P_E < P_{cr_1} < 4P_E$. Similarly, for the column in Figure 1.7e, we conclude that $P_{cr_1} < P_E/4$. For the column in Figure 1.7b, we conclude that $P_{cr_1} > P_E/4$.

A useful approximate formula for columns whose ends do not move and are restrained against rotation by springs of spring constants C_1 and C_2 was developed by Newmark (1949):

$$P_{cr_1} = \frac{\pi^2 EI}{l^2}\left[\frac{(0.4+\lambda_a)(0.4+\lambda_b)}{(0.2+\lambda_a)(0.2+\lambda_b)}\right] \qquad \lambda_a = \frac{EI}{C_1 l} \qquad \lambda_b = \frac{EI}{C_2 l} \tag{1.4.6}$$

The error of this formula is generally less than 4 percent.

The most general supports of a column are obtained when one end is supported by a hinge and is restrained by a rotational spring of spring constant C_1, and the other end can rotate and move laterally, being restrained by a rotational spring and a lateral spring of spring constants C_2 and C_3 (Fig. 1.7j). One extreme case of this column is the fixed-end column, which is obtained when C_1, C_2, and C_3 all become infinite. Another extreme case is obtained when C_1, C_2, and C_3 all tend to zero. In this case the column becomes a mechanism, and its critical load P_{cr_1} vanishes. Therefore, the first (smallest) critical load of a column of constant cross section is bounded as

$$0 \leq P_{cr_1} \leq 4P_E \tag{1.4.7}$$

This inequality also applies for a column as part of a frame. The reason is that the replacement of the action of the rest of a frame onto the column by elastic springs cannot increase the critical load, as we will explain later (Sec. 2.4), although usually such a replacement leads to a higher critical load than the actual one.

In some structural systems the axial load may rotate during buckling. Some problems of this type may be nonconservative, which we will discuss later. However, even for conservative problems of this type one must be careful to give proper consideration to the lateral force component of an inclined load P, since such a component generally affects buckling and may cause a significant reduction of the critical load. Consider, for example, the case that the load P passes through a fixed point C (Fig. 1.10a) at a distance c from the free end of a cantilever, as would happen, for example, if a cable were stretched between the free end of the cantilever and the fixed point. The boundary condition for shear at the free end ($x = 0$) becomes $V = -EIw''' - Pw' = P\Delta/c$. The other boundary conditions are for $x = 0$: $M = EIw'' = 0$; and for $x = l$: $w = 0$ and $w' = 0$. Substituting Equation 1.3.6 and requiring that, for $x = 0$, $w = \Delta$, one obtains $\tan kl = kl(1 - c/l)$ as the condition for the critical load. The solution is tabulated in Timoshenko and Gere (1961, p. 57) and gives a critical load that is higher than

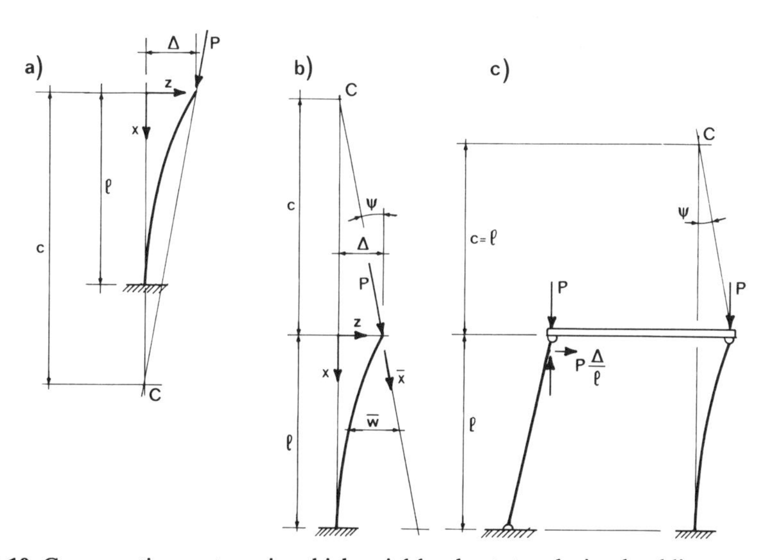

Figure 1.10 Conservative systems in which axial load rotates during buckling.

$P_E/4$. It is also interesting to note that, in the case $c = l$, the critical load becomes equal to P_E, because the moment at the base becomes zero, and the conditions are the same as for a beam with hinged ends.

In the case that the fixed point C is situated above the free end (Fig. 1.10b), the boundary condition on shear becomes $V = -P\Delta/c$ and the critical load is given by $\tan kl = kl(1 + c/l)$. In this case the critical load becomes smaller than $P_E/4$, and for the case $c = l$ we find $P_{cr} = 0.138P_E$.

The solution can be found more directly if one uses a coordinate axis $\bar{x}$, which coincides with the direction of the applied force P, that is, is inclined by $\psi = \Delta/c$ (Fig. 1.10b). Relative to axis $\bar{x}$, the deflection curve is $\bar{w} = A \sin k\bar{x}$, and for $\bar{x} = l$ (base), one has $\bar{x} = A \sin kl = \Delta + \Delta l/c$, from which $A = \Delta(1 + l/c)/\sin kl$. The slope at the base relative to axis $\bar{x}$ is $w' = \Delta/c$, which yields $Ak \cos kl = \Delta/c$, and substituting the value found for A one obtains again $kl(1 + c/l) = \tan kl$.

A situation in which a free-standing column is subjected to this second type of load is depicted in Figure 1.10c and was analyzed by Lind (1977) (see also Prob. 1.4.5). One may be tempted in this case to say that P_{cr_1} equals $P_E/4$, but this would neglect the fact that the reaction of the beam upon the fixed-base leg is inclined rather than vertical because of the horizontal component of the axial force in the pin-ended column (Fig. 1.10c).

Problems

1.4.1 For the fixed–hinged column, a lateral reaction component develops at the sliding hinge and causes the line of the force-resultant at the hinge to pass through the inflection point of the deflection curve. Prove it.

1.4.2 Find the critical load of the systems in Figure 1.7b, e, h, j with the following values of the spring constants $C_1 = EI/l$, $C_2 = EI/2l$, $C_3 = 12EI/l^3$. Find also the inflection points of the deflection curves and verify the validity of Equation 1.4.5.

1.4.3 Find the critical load of the system in Figure 1.11a in which the load is applied through a frictionless plunger. Note that the load point is fixed, that

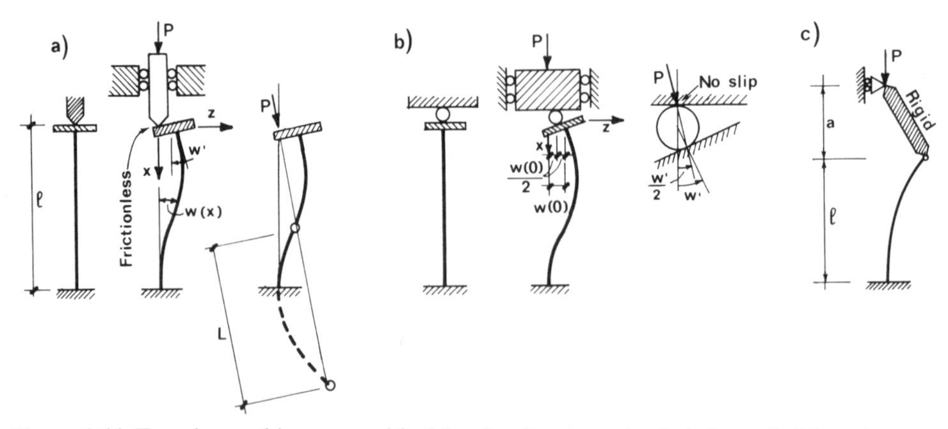

Figure 1.11 Exercise problems on critical loads of systems loaded through (a) a plunger, (b) a roller, and (c) a rigid member.

is, does not move with the column top, but the load direction rotates with the column top. So the boundary conditions are $M(0) = -Pw(0)$ and $V(0) = -Pw'(0)$, which is equivalent to $Q(0) = 0$. The load resultant P is a follower force, since its direction follows the rotation of the column top. Yet load P is conservative because the lateral displacement is zero.

1.4.4 Find the critical load of the system in Figure 1.11b, assuming that the roller on top rolls perfectly without friction and without slip. Note that the load P acting on the column top moves by distance $w(0)/2$ if the deflection on top is $w(0)$. So one boundary condition on top is $M(0) = -Pw(0)/2$. Since the load P transmitted by the roller has inclination $w'(0)/2$, the second boundary condition on top is $V(0) = -Pw'(0)/2$.

1.4.5 Find the critical load of the system in Figure 1.11c.

1.4.6 Find the value of c for the frame of Figure 1.10c in the case that the pin-ended leg is shorter than l. Does the critical load increase or decrease in this case? Show that if the pin-ended leg is placed above the horizontal beam, the situation for the free-standing column becomes that of Figure 1.10a.

1.4.7 Find the critical load of the systems in Figure 1.12a–e.

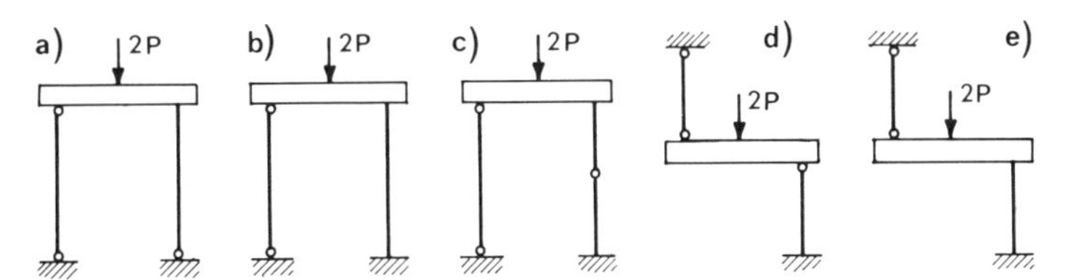

Figure 1.12 Exercise problems on critical loads of structures with rigid members.

1.4.8 Solve the critical loads of the beams in Figure 1.13 in which the rigid bars are attached to the beam ends (see Ziegler, 1968). Note that the columns are under tension or, if loaded by a couple, under no axial force.

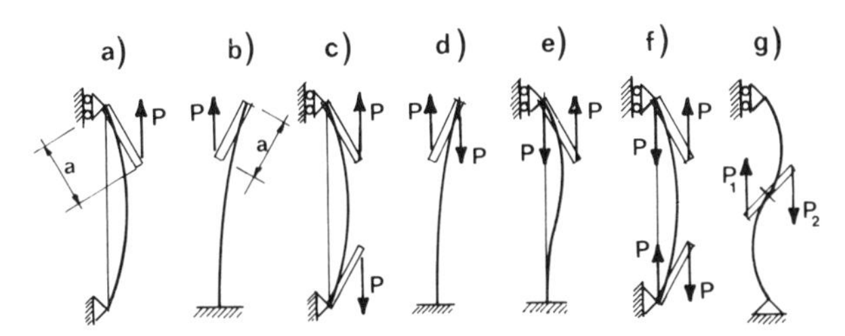

Figure 1.13 Exercise problems on critical loads of beams loaded through rigid bars at ends.

1.5 IMPERFECT COLUMNS AND THE SOUTHWELL PLOT

The perfect column which we have studied so far is an idealized model. In reality, several kinds of inevitable imperfections must be considered. For example, columns may be subjected to unintended small lateral loads; or they may be initially curved rather than perfectly straight; or the axial load may be slightly eccentric; or disturbing moments and shear forces may be applied at column ends, which is essentially equivalent to an eccentric or inclined axial load. Unlike beams subjected to transverse loads and small axial forces, columns are quite sensitive to imperfections, although not as much as shells.

Lateral Disturbing Load

As the simplest prototype case, consider again the pin-ended column (Fig. 1.14a). We assume the column to be perfectly straight in the initial stress-free state but subjected to lateral distributed loads $p(x)$ and to end moments M_1 and M_2. We imagine these loads to be applied first, as the first stage of loading. As the second stage of loading, we then add the axial load P.

The coordinates of the deformed center line of the column under the action of p, M_1, and M_2 are denoted as $z_0(x)$, and the associated bending moments as $M_0(x)$ (Fig. 1.14a). Subsequent application of the axial load P increases the deflection ordinates to $z(x)$ (see Fig. 1.14a); this causes the bending moments to change to $M(x) = M_0(x) - Pz(x)$. Now we may substitute $M = EIz''$, and this yields the differential equation

$$z'' + k^2 z = \frac{M_0}{EI} \tag{1.5.1}$$

in which $k^2 = P/EI$. Note that if load P were applied first, and loads $p(x)$, M_1, and M_2 subsequently, the differential equation as well as the final deflection $z(x)$ would be the same because elastic behavior is path-independent.

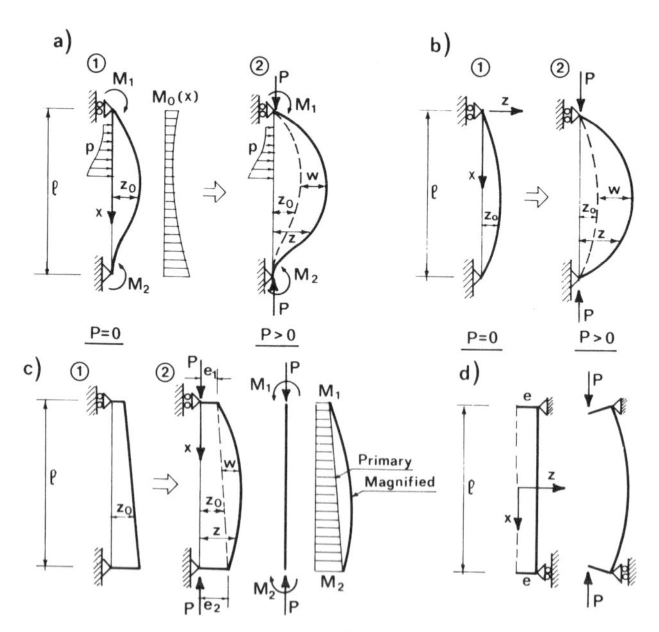

Figure 1.14 Buckling of imperfect columns: (a) lateral loads, (b) initial curvature, (c, d) axial load eccentricity.

Let us assume again that k is constant and expand the initial bending moments in a Fourier sine series:

$$M_0(x) = \sum_{n=1}^{\infty} Q_{0_n} \sin \frac{n\pi x}{l} \tag{1.5.2}$$

in which l is the length of the column, and $Q_{0_n} = 2\int_0^l M_0(x) \sin(n\pi x/l)\, dx/l$ are Fourier coefficients, which can be determined from the given distribution $M_0(x)$ (see, e.g., Rektorys, 1969; Pearson, 1974; Churchill, 1963). We may then seek the solution also in the form of a Fourier sine series:

$$z(x) = \sum_{n=1}^{\infty} q_n \sin \frac{n\pi x}{l} \tag{1.5.3}$$

in which q_n are unknown coefficients. Note that each term of the last equation satisfies the boundary conditions $z(0) = z(l) = 0$. Substituting Equations 1.5.2 and 1.5.3 into Equation 1.5.1, we obtain

$$\sum_{n=1}^{\infty} \left[k^2 q_n - \left(\frac{n\pi}{l}\right)^2 q_n - \frac{Q_{0_n}}{EI} \right] \sin \frac{n\pi x}{l} = 0 \tag{1.5.4}$$

Since this equation must be satisfied for any value of x, and since the functions $\sin(n\pi x/l)$ are linearly independent, the bracketed terms must vanish. This provides

$$q_n = \frac{Q_{0_n}}{P - n^2 P_E} \tag{1.5.5}$$

Let us now relate this result to the deflection $z_0(x)$ existing before the application of the axial load P. The initial deflection $z_0(x)$ may be expanded, similarly to Equation 1.5.3, in Fourier sine series with coefficients q_{0_n}, which are obtained from Equation 1.5.5 by setting $P = 0$, that is, $q_{0_n} = -Q_{0_n}/(n^2 P_E)$. From this relation and Equation 1.5.5 we conclude that

$$q_n = q_{0_n} \frac{1}{1 - (P/P_{cr_n})} \tag{1.5.6}$$

in which $P_{cr_n} = \pi^2 n^2 EI/l^2 =$ the nth critical load of the perfect column.

Initial Curvature or Load Eccentricity

Consider now another type of imperfection—the initial curvature (crookedness) characterized by the initial shape $z_0(x)$ (Fig. 1.14b). Application of axial load P produces deflections w, and the ordinates of the deflection curve become $z = z_0 + w$ (Fig. 1.14b). For a pin-ended column, we have, from the equilibrium conditions, $M = -Pz$. At the same time, the bending moment is produced by a change in curvature, which equals w'' (not z''), and so $M = EIw'' = EI(z'' - z_0'') = -Pz$. This leads to the differential equation

$$z'' + k^2 z = z_0'' \tag{1.5.7}$$

We see that the previously solved Equation 1.5.1 is identical to this equation if we set $M_0 = EIz_0''$. Therefore, there is no need to carry out a separate analysis for the case of initial curvature. The results are the same as those for lateral disturbing loads, which we have already discussed.

Equation 1.5.7 was derived by Thomas Young (1807), who was the first to take into account the presence of imperfections.

Note that in formulating the conditions of equilibrium for the deformed structure, we use displacements with regard to the initial undeflected (but axially deformed) state; that is., $w = z - z_0$, where z and z_0 are the initial and final coordinates of a material point, both z and z_0 being expressed in the initial coordinate system. In the general theory of finite strain this is called the Lagrangian coordinate description, which contrasts with the Eulerian description in which the local coordinates move with the points of the structure. Except for some instances in Chapter 11, in this book we use exclusively the Lagrangian coordinate description, which is more convenient for stability analysis than the Eulerial description.

Another type of imperfection is the eccentricity of the axial load applied at the end. This case may be treated as a special case of initial curvature such that the beam axis is straight between the ends and has right-angle bends at the ends (Fig. 1.14c). Expanding this distribution of z_0 into a Fourier sine series, the previous solution may be applied. Alternatively, an eccentrically loaded column is equivalent to a centrically loaded perfect column with disturbing moments $M_1 = M_0(0) = -P/e_1$ and $M_2 = M_0(l) = -P/e_2$ applied at the ends, which is a case we have already solved (Eq. 1.5.1) (e_1 and e_2 are the end eccentricities).

Behavior Near the Critical Load

Figure 1.15a illustrates the diagrams of load P versus maximum deflection w for various values of q_{0_1} characterizing the imperfection (only the first sinusoidal component is considered in the calculation). As the initial imperfection tends to zero, or $q_{0_1} \rightarrow 0$, the load-deflection diagram asymptotically approaches that of a perfect column with a bifurcation point at the first critical load.

Since a small imperfection of the most general type ($Q_{0_i} \neq 0$ or $q_{0_i} \neq 0$ for $i = 1, 2, \ldots$) can never be excluded, we must conclude from Equation 1.5.5 or 1.5.6 that the deflection tends to infinity at $P \rightarrow P_{cr_1}$ no matter how small the initial imperfection is (it would be unreasonable to assume that q_{0_1} is exactly zero). This leads us to conclude that columns must fail at the first critical load and that load values higher than the first critical load cannot be reached.

We must also conclude that in a static experiment one can never realize the

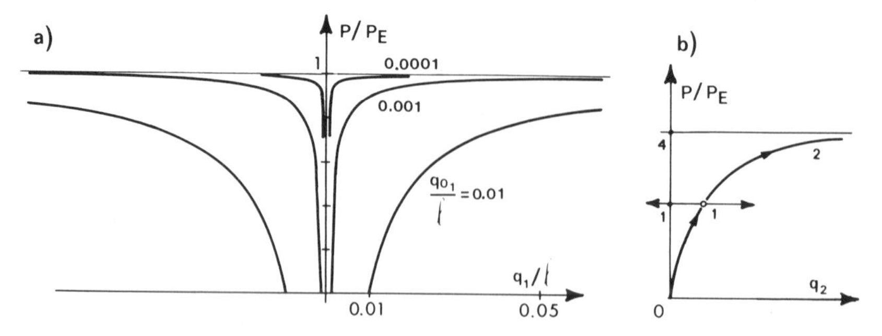

Figure 1.15 Behavior of column with initial curvature given by (a) first sinusoidal component and (b) second sinusoidal component.

critical load. However, it should not be concluded that the imperfection analysis always yields the same maximum load as the bifurcation analysis (critical state). Counterexamples will be given in Section 4.5. Note also that if lateral disturbing loads are large, the column material will fail before the critical load is reached.

When the amplitude q_{0_1} of the first sinusoidal component of the imperfection is thought to be zero, the dominant term in Equation 1.5.6 is $q_2 = q_{0_2}/[1-(P/4P_E)]$, as plotted in Figure 1.15b (curve 012). Although the assumption that any imperfection is exactly zero is physically unreasonable, one might think that for this assumption the column would not fail for $P < 4P_E$. Not so, however. At $P = P_E = P_{cr_1}$, the solution of the homogeneous differential equation for homogeneous boundary conditions may be superimposed, and this indicates that at $P = P_E$ the equilibrium path bifurcates at point 1 shown in Figure 1.15b. Thus arbitrary deflection becomes possible at $P = P_E$, and $P < P_E$ must be required to prevent it.

These conclusions are, of course, limited to the small-deflection theory. If a nonlinear, finite-deflection theory is used, the deflection does not approach infinity at $P \rightarrow P_{cr_1}$. Rather, it simply becomes large, and from a practical viewpoint usually unacceptably large.

The foregoing analysis brings to light the importance of the first critical load. We are not sure what are the type and value of the initial imperfection; however, as long as they are small we do not need to know their precise values since they make little difference for the load at which the column fails. To illustrate this point better, consider some value of P near the first critical load, say $P = 0.95P_{cr_1}$. Then Equation 1.5.6 becomes

$$q_1 = 20q_{0_1} \qquad q_2 = 1.31q_{0_2} \qquad q_3 = 1.12q_{0_3} \qquad q_4 = 1.06q_{0_4} \cdots \tag{1.5.8}$$

We see that the first term of the Fourier series expansion dominates, and the other terms are unimportant. So the precise shape of the initial bending moment distribution or curvature need not be known closely if the behaviour under overload is of interest. This is a comforting finding since the initial imperfections are not well known in practice. We may also conclude that for practical purposes it is sufficient to consider only the first Fourier series component, that is, an initial bending moment or curvature distributed according to a sinusoidal half-wave. It then follows that, near the first critical load,

$$\max z \simeq \mu \max z_0 \qquad \max M \simeq \mu \max M_0 \tag{1.5.9}$$

in which

$$\mu = \frac{1}{1-(P/P_{cr_1})} \tag{1.5.10}$$

Factor μ is called the magnification factor because it indicates the ratio by which the initial deflections due to initial bending moments are magnified by the axial load. Factor μ was discovered by Thomas Young (1807). The same magnification rule applies to initial curvature, in which case it is called Perry's rule (Timoshenko and Gere, 1961, p. 32).

The magnification factor (Eqs. 1.5.9 and 1.5.10) is utilized in the current structural engineering code specifications, including those of the American Institute of Steel Construction (AISC) and the American Concrete Institute (ACI).

Another simple formula was used in codes for the case of equal eccentricities $e_1 = e_2 = e$ at both ends. In this case it is convenient to measure the axial coordinate x from the midlength point (Fig. 1.14d). Deflection ordinates $z(x)$ are measured from the line of the axial load P. Due to symmetry, $A = C = 0$ in Equation 1.3.6, and $z(x) = B \cos kx + D$. The boundary conditions at $x = \pm l/2$ require that $z = e$ and $M = -Pe$. This yields two equations for B and D: $B \cos (kl/2) + D = e$ and $-EIk^2 B \cos (kl/2) = -Pe$. Their solution yields $D = 0$ and $B = e \sec (kl/2)$. Noting that $kl = \pi\sqrt{P/P_E}$, we thus obtain the well-known secant formula

$$\max z = e \sec \left(\frac{\pi}{2} \sqrt{\frac{P}{P_E}} \right) \tag{1.5.11}$$

It is interesting to check that the asymptotic form of this formula at $P \to P_E$ is equivalent to the use of the magnification factor in Equation 1.5.9 (except for the factor $\pi/4$). Denoting $\xi = 1 - P/P_E$, we confirm it by the following approximations:

$$\begin{aligned} \frac{e}{\max z} &= \cos \left(\frac{\pi}{2} \sqrt{\frac{P}{P_E}} \right) = \sin \left(\frac{\pi}{2} - \frac{\pi}{2} \sqrt{\frac{P}{P_E}} \right) \simeq \frac{\pi}{2} - \frac{\pi}{2} \sqrt{\frac{P}{P_E}} \\ &= \frac{\pi}{2}(1 - \sqrt{1 - \xi}) \simeq \frac{\pi}{2}\left(1 - 1 + \frac{\xi}{2}\right) = \frac{\pi}{4}\left(1 - \frac{P}{P_E}\right) \end{aligned} \tag{1.5.12}$$

in which we exploit the fact that $\xi \to 0$ and that the argument of the sine is very small.

Although our analysis has been limited to pin-ended columns, similar conclusions may be obtained for all types of columns with various end restraints.

Southwell Plot

Unavoidable imperfections must be taken into account in evaluating the results of buckling tests. Every column has some initial curvature or eccentricity, and so $z = z_0 + w$. The initial imperfection is, however, difficult to measure, and one needs a method of determining the critical load independently of z_0. Now, which quantity is easier to measure, deflection w or ordinate z? Deflection w, since it suffices to install a deflection gauge and its change of reading gives w. Determination of the deflection ordinate z would require measuring the distance from the point on the abstract line of the load axis, which is impractical.

To determine the critical load, we measure deflection w at various values of load P close to P_{cr_1}. For these values the magnification factor, Equations 1.5.9 and 1.5.10, should be applicable as a good approximation, regardless of the shape of initial imperfections. So we have $z = z_0 + w = z_0/[1 - (P/P_{cr_1})]$. Rearranging, we get $w = (z_0 + w)P/P_{cr_1}$, from which

$$\frac{w}{P} = \frac{1}{P_{cr_1}} w + \frac{z_0}{P_{cr_1}} \tag{1.5.13}$$

Denoting $Y = w/P$, $A = 1/P_{cr_1}$, $B = z_0/P_{cr_1}$, we see that this equation can be written as $Y = Aw + B$, which represents the equation of a straight line of slope A

and Y-intercept B in the plot of Y versus w. This plot is called the *Southwell plot*; see Figure 1.16 (Southwell, 1932). We measure deflections w at various values of load P and plot the values of w/P versus w. We obtain a series of data points whose middle portion is essentially straight, except for the scatter of measurements. To eliminate the statistical scatter, we pass a straight regression line (visually, or better, by using the method of least squares) through the middle portion of the data points. The inverse slope of the regression line is the first critical load, and the Y-intercept gives information on the magnitude of imperfections.

The data points at the lower left of the Southwell plot normally deviate from the regression line upward, since for small P/P_{cr_1} the higher harmonics in Equation 1.5.3 have a nonnegligible effect (note the calculation in Eq. 1.5.8). The upper right range of data points normally deviates from the regression line also upward, which is usually caused by large deflections at which Equation 1.5.13 does not apply, as it is limited to the linear small-deflection theory. The smaller the imperfection, the longer is the linear range (and the better is the estimate of P_{cr_1}).

The foregoing observations underscore for the experimentalist the importance of minimizing imperfections of test columns. If the imperfections are larger, P_{cr_1} will not be approached closely before the deflection becomes so large that either the material breaks or the linearized small-deflection theory, on which the Southwell plot is based, is no longer valid.

The Southwell plot is applicable not only to columns, but also to many other buckling problems, which we will study later. However, for structures such as plates or shells, as well as some frames, significant deviation from the Southwell plot is caused by postcritical behavior, which we will study in Chapters 4 and 7. The load may start to either increase or decrease soon after the first significant lateral deflections near the critical load take place. In these cases of postcritical reserve or postcritical loss of carrying capacity, which are especially marked for plates and shells (Chap. 7), the Southwell plot deviates from the straight line quite significantly at the upper right end of the plot. Such deviations make the Southwell plot useless for plates. A modification of the Southwell plot that can take these postcritical deflections approximately into account was proposed by Spencer and Walker (1975) (see Eq. 7.4.18). For a discussion of other deviations,

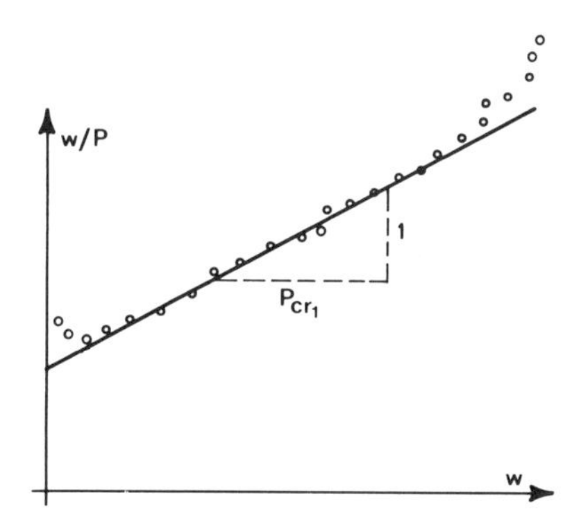

Figure 1.16 Southwell plot for the evaluation of critical loads of imperfect columns.

see Roorda (1967). To mitigate the effect of higher harmonics, and thus to extend the linear range, Lundquist (1938) proposed a modified plot of $(w - w')/(P - P')$ versus $(w - w')$ where (P', w') is a certain suitably chosen "pivot" state (see also Taylor and Hirst, 1989, with extensions to imperfection increase due to cyclic loading). For further discussions, see also Leicester (1970) (extension to lateral buckling of beams, Chap. 6) and Allen and Bulson (1980).

Problems

1.5.1 Consider a pin-ended column with initial curvature expressed by the first term of a Fourier series $z_0(x) = q_{0_1} \sin(\pi x/l)$. Find the values of shear forces V and Q at the ends (according to their definition in Sec. 1.3).

1.5.2 Consider a pin-ended column with a lateral load $p = \text{const.}$, and find the deflection through the first five components of the Fourier series. Do the same for the case $p = 0$ but the axial loads P at the ends have an eccentricity e on the same side on the beam. In both cases construct the diagram of P/P_{cr_1} versus the ratio $w(l/2)/w_0(l/2)$ where $w_0(l/2)$ is the deflection at midlength if the axial load P were absent.

1.5.3 Solve columns whose one end is supported on a hinge that slides on a plane of small inclination β, as shown in Figure 1.17a. An interesting aspect of this problem is the postcritical behavior, which is characterized by a decrease of the maximum load with increasing imperfection (see Sec. 2.6, Prob. 2.6.7, and Sec. 4.5).

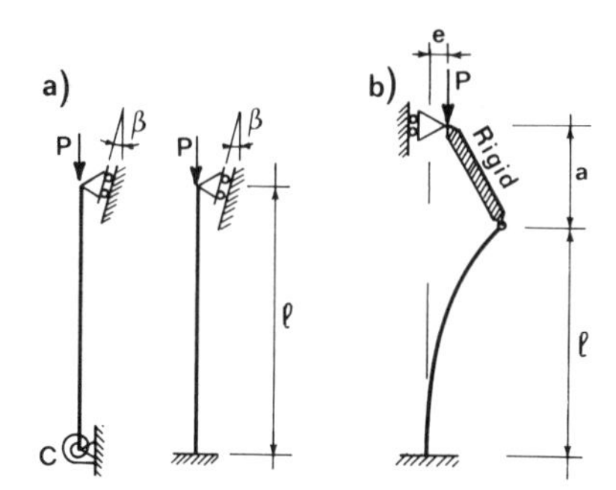

Figure 1.17 Exercise problems on critical loads (a) of columns with a hinge sliding on an inclined support, and (b) of a system with a rigid member.

1.5.4 Solve P_{cr_1} for the column in Figure 1.17b, with $a/l = 0.05$, 0.3, 1, 3, 10, 1000.

1.5.5 Consider a free-standing column of height h.

(a) Solve deflections in the presence of $M_0(x)$, first sinusoidal $M_0(x)$, then general.

(b) Do the same for $w_0(x)$.

(c) Verify the magnification factor for this column.

1.5.6 Stochastic imperfection. A hinged column has a sinusoidal initial shape whose amplitude a is uncertain and a normal probability distribution with mean $\bar{a}$ and standard deviation s_a. For $P = 0.5P_E$, $0.7P_E$, and $0.9P_E$, determine the mean and standard deviation of max M.

1.5.7 Do the same as Problem 1.5.6; however, not only a but also E, P, and minimum allowable compressive stress σ_{al} (<0) are uncertain, all with normal distributions. The means and standard deviations of a, E, P, and σ_{al} are $\bar{a}$, $\bar{E}$, $\bar{P}$, $\bar{\sigma}_{al}$, s_a, s_E, s_p, s_σ. Formulate the problem and discuss the numerical method of calculation of the probability of failure.

1.6 CODE SPECIFICATIONS FOR BEAM-COLUMNS

Before World War II, most code specifications were based on the secant formula (Eq. 1.5.11). Presently, most codes, including that of the American Institute for Steel Construction (AISC) for steel structures and that of the American Concrete Institute (ACI) for concrete structures, are based on the magnification factor (Eq. 1.5.10). A typical problem is the buckling analysis of columns as parts of frame structures. The analysis is carried out approximately in two stages. In the first stage, called first-order analysis, one solves the frame in the usual manner, without regard to buckling. This means that the equilibrium conditions are written for the undeflected structure, on the basis of its initial geometry (and the so-called P–Δ effects are ignored, Δ being deflection w). The first-order analysis yields bending moments called the primary bending moments. In the second stage of analysis, called the second-order analysis, the deflections (i.e., the P-Δ effects) are taken into account when the equilibrium conditions are written. This second-order analysis may be iterated to improve accuracy. In regular practice, however, the second-order analysis is replaced, in an approximate sense, by the use of the magnification factor in the following form:

$$M_{\max} \simeq \frac{C_m}{1-(P/P_{\mathrm{cr}_1})} M_{0_{\max}} \tag{1.6.1}$$

in which $M_{0_{\max}}$ is the maximum primary moment within the column and C_m is a coefficient that takes into account the distribution of primary moments along the column. For the constant loading moment we already found in Equation 1.5.12 that $C_m = \pi/4$. If the calculated primary moments happen to be negligible, a certain minimum value of $M_{0_{\max}}$, corresponding to minimum imperfections, is prescribed by some codes (e.g., ACI).

In most cases, the distribution of bending moments in a column in a frame is linear according to the first-order analysis (Fig. 1.14c). We restrict our attention to braced frames in which column ends do not move. So we may consider a pin-ended column with moments M_1 and M_2 applied at the ends. The larger end moment we denote as M_2, that is, $M_2 \geq M_1$. According to Equation 1.3.6, the distribution of bending moments after the axial load P is superimposed on the initial bending moment has the form

$$M(x) = EIw''(x) = C_1 \sin kx + C_2 \cos kx \tag{1.6.2}$$

in which C_1 and C_2 are arbitrary constants. The boundary conditions are $M = M_1$ at $x = 0$, and $M = M_2$ at $x = l$, from which

$$C_2 = M_1 \qquad C_1 = \frac{M_2 - M_1 \cos kl}{\sin kl} \tag{1.6.3}$$

We need to find the maximum moment M_{max}. To this end, we evaluate $dM/dx = C_1 k \cos kx - C_2 k \sin kx$, and set it equal to 0. This yields $\tan kx = C_1/C_2$, from which

$$\sin kx = \frac{C_1}{\sqrt{C_1^2 + C_2^2}} \qquad \cos kx = \frac{C_2}{\sqrt{C_1^2 + C_2^2}} \tag{1.6.4}$$

Now, M_{max} can occur either within the column length, in which case the above equations apply, or at the end, in which case $M_{max} = M_2$. Therefore, substituting Equation 1.6.4 into 1.6.2,

$$M_{max} = \max(M_2, \sqrt{C_1^2 + C_2^2}) \tag{1.6.5}$$

This result should be approximately equivalent to Equation 1.6.1, from which we may obtain the expression

$$C_m = \left(1 - \frac{P}{P_{cr_1}}\right)\left(\frac{\max(M_2, \sqrt{C_1^2 + C_2^2})}{M_2}\right) \tag{1.6.6}$$

This is a nondimensional expression which depends on M_1/M_2 and P/P_{cr_1}.

The curves of coefficient C_m versus ratio M_1/M_2 have been plotted for various values of the ratio of P/P_{cr_1}. This plot, taken from the work of Wang and Salmon (1979), is shown in Figure 1.18. This figure also shows the plot of the expression $C_m = [0.3(M_1/M_2)^2 + 0.4(M_1/M_2) + 0.3]^{1/2}$, proposed by Massonet (1959). In order to simplify practical analysis, a line approximately representing the bundle of the C_m curves has been drawn (see Fig, 1.18). This line is described by the following simple formula:

$$C_m = 0.6 + 0.4\frac{M_1}{M_2} \quad \begin{cases} \leq 1.0 \\ \geq 0.4 \end{cases} \tag{1.6.7}$$

in which the inequalities mean that the value of C_m should be taken as 0.4 if it comes out to be less than 0.4, and as 1.0 if it comes out to be greater than 1.0. This last equation is now used in the current AISC (1978, -1986) and ACI (1977) codes for columns braced against lateral sway. It should be noted that the

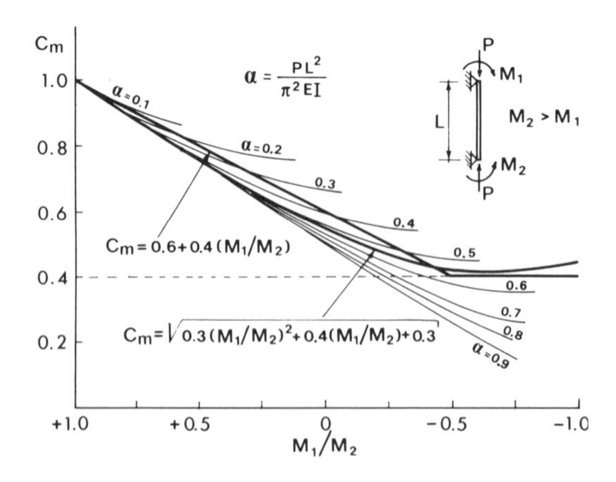

Figure 1.18 Equivalent-moment factor C_m for beam with unequal end moments. (*After C. Massonet, 1959; see also Wang and Salmon, 1979.*)

limitation $C_m \geq 0.4$ applies for a range of values M_1/M_2 in which the initial double curvature of the beam reverses on approach to critical load to a single curvature (Fig. 1.15b). In that case the concept of magnification factor does not give a good approximation.

For the relatively infrequent cases when primary bending moments $M_0(x)$ are caused by a lateral load on the column, the AISC and ACI codes prescribe $C_m = 1$.

Equation 1.6.7 is permitted only for braced frames, since we assumed the column ends do not move. Consider now that one column end moves (Fig. 1.19a); this is typical of unbraced frames, which buckle with a sidesway. For the column base, consider a hinged support, which is the most unfavorable assumption yielding the maximum effective length L. The primary bending moment before applying the axial load P is $M_0(x)$ (Fig. 1.19a). Application of axial load P would cause an additional (second-order) bending moment Pw so that the total moment is $M = M_0 + Pw$. According to the magnification factor, $w = w_0/[1 - (P/P_{cr_1})]$, in which w_0 is the deflection due to applied lateral load alone, that is, for $P = 0$ (primary deflection). We want to express the final moment in the form of Equation 1.6.1, and so we have the condition

$$M_0 + \frac{Pw_0}{1 - P/P_{cr_1}} = M_0\left(\frac{C_m}{1 - P/P_{cr_1}}\right) \tag{1.6.8}$$

Solving this equation we get

$$C_m = 1 - \left(1 - P_{cr_1}\frac{w_0}{M_0}\right)\left(\frac{P}{P_{cr_1}}\right) \tag{1.6.9}$$

which is valid for all columns with one end hinged and an arbitrary support at the other end. Equation 1.6.9 is generally applicable only for columns which initially

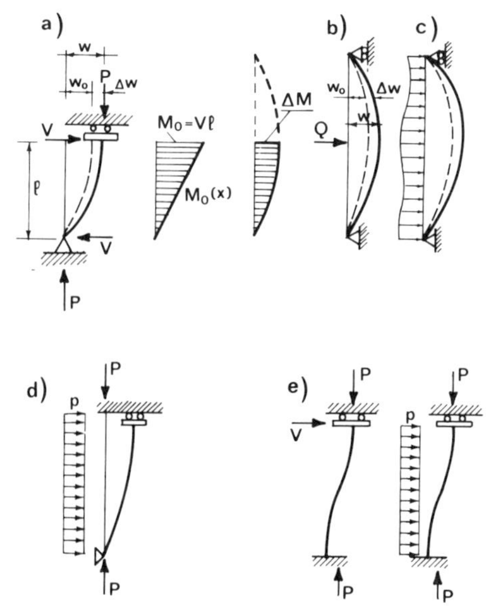

Figure 1.19 Primary bending caused by lateral load.

bend in a single curvature. For initial double curvature the magnification factor does not work well.

Consider now in particular that there is no load between column ends, and primary bending is caused by lateral load V applied at the column top (Fig. 1.19a). Then the maximum of M_0 occurs at the column top, $M_0 = Vl$. The primary deflection at the column top is the deflection corresponding to the triangular bending moment distribution shown in Figure 1.19a, and by virtual work or the moment-area theorem, we have $w_0 = Vl^3/3EI$. Substitution into Equation 1.6.9 then yields the equation

$$C_m = 1 - \left(1 - \frac{\pi^2}{12}\right)\left(\frac{P}{P_{cr_1}}\right) = 1 - 0.18\frac{P}{P_{cr_1}} \tag{1.6.10}$$

which is suggested in sec. 1.6.1 of the AISC Commentary. ACI is more conservative, requiring $C_m = 1$ for any load P on sway columns. On the other hand, AISC specification 1.6.1 permits using $C_m = 0.85$, which is unconservative except for the case when $P/P_{cr_1} > \frac{5}{6}$, that is, the range in which buckling is important.

As for columns subjected to lateral loads (Fig. 1.19b, c), one may assume the most unfavorable case of hinged supports. Then our starting equation $M = M_0 + Pw$ again applies, and with an identical derivation C_m is again found to be given by Equation 1.6.10.

To sum up, the practical analysis procedure consists of the following steps: (1) analyze the primary bending moments M_0 by the first-order theory, (2) find the first critical load P_{cr_1} for the column in the perfect structure (we shall complete the discussion of this in the next chapter), and (3) apply the magnification factor with the correction coefficient C_m.

Many aspects of code specifications have to do with inelastic material behavior. It is not possible to do them justice until the basic theory of inelastic buckling is presented. See Sections 8.4, 8.5, and 10.3.

Problems

1.6.1 Solve the cases of Figure 1.19a–c directly from the differential equation (for the distributed load assume $p = \text{const.}$).

1.6.2 Derive the exact value of C_m for a pinned column with a parabolic $p(x)$ and various values of P/P_{cr}.

1.6.3 Do the same as Problem 1.6.2 but for a fixed–fixed column.

1.6.4 Do the same as Problem 1.6.2 but for a uniform load eccentricity e.

1.6.5 Derive expressions similar to Equations 1.6.8 and 1.6.9 for the column of Figure 1.19a for a distributed lateral load $p = \text{const.}$ (Fig. 1.19d).

1.6.6 Consider the case of a column with nonrotating ends but sidesway (Fig. 1.19e). Derive expressions similar to Equations 1.6.8 and 1.6.9 for the cases $V = \text{const.}$ and $p = \text{const.}$ Solve directly from the differential equations.

1.7 EFFECT OF SHEAR AND SANDWICH BEAMS

When a column buckles, the axial load causes not only bending moments in the cross sections, but also shear forces. The deformations due to shear forces are neglected in the classical bending theory, since the cross sections are assumed to remain normal to the deflected beam axis. This assumption is usually adequate; however, there are some special cases when it is not.

Pin-Ended Columns

The shear deformations can be taken into account in a generalization of the classical bending theory called Timoshenko beam theory. In this theory (whose generalizations for plates and sandwich plates were made by Reissner, Mindlin, and others), the assumption that the plane cross sections remain normal to the deflected beam axis is relaxed, that is, the slope θ of the deflected beam axis is no longer required to be equal to the rotation ψ of the cross section (see Fig. 1.20a). The difference of these two rotations is the shear angle γ, which may be expressed as

$$\gamma = \theta - \psi = \frac{Q}{GA_0} \tag{1.7.1}$$

in which Q represents the shear force that is normal to the deflected beam axis and rotates as the beam deflects (introduced in Sec. 1.3).

Furthermore, G is the elastic shear modulus, and $A_0 = A/m$, where A is the area of the cross section and m is the shear correction coefficient. This coefficient takes into account the nonuniform distribution of the shear stresses throughout the cross section (m would equal 1 if this distribution were uniform). As shown in mechanics of materials textbooks, for a rectangular cross section $m \simeq 1.2$, for a solid circular cross section $m \simeq 1.11$, for a thin-walled tube $m \simeq 1.65$, and for an I-beam bent in the plane of the web $m \simeq A/A_w$, where A_w is the area of the web (Timoshenko and Goodier, 1973; Dym and Shames, 1973).

The axial normal strains are given as $\varepsilon = -z\psi'$ since $\psi'\,dx$ represents a relative rotation of two cross sections lying a distance dx apart. Substituting this

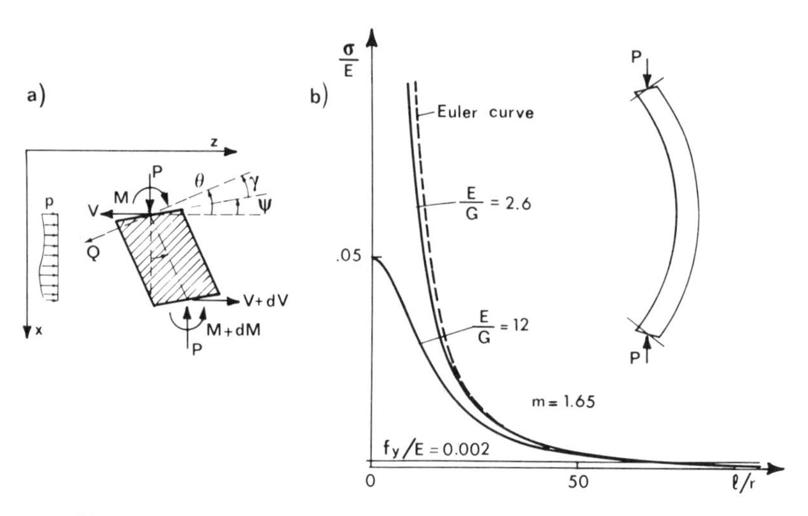

Figure 1.20 (a) Beam element with shear deformation, (b) Influence of shear deformation on critical stress.

into the bending moment expression $M = -\int E\varepsilon z\, dA$, one obtains

$$\psi' = \frac{M}{EI} \tag{1.7.2}$$

We have already shown (Eq. 1.3.4) that $Q = -M'$. Inserting this and $\theta = w'$ into Equation 1.7.1, we have

$$w' - \psi = -\frac{M'}{GA_0} \tag{1.7.3}$$

This equation and Equation 1.7.2 represent two basic differential equations of the problem. Differentiating the last relation and substituting Equation 1.7.2, variable ψ can be eliminated and it follows that

$$w'' = \frac{M}{EI} + \left(\frac{-M'}{GA_0}\right)' \tag{1.7.4}$$

This equation means that the total curvature w'' is a sum of the flexural curvature and the curvature due to shear. This fact may, alternatively, be introduced as the starting assumption.

Consider now a pin-ended column. In this case we have $M = -Pw$, and Equation 1.7.4 becomes

$$w'' = \left(\frac{Pw'}{GA_0}\right)' + \frac{-Pw}{EI} \tag{1.7.5}$$

Furthermore, if the bending rigidity EI, the shear rigidity GA_0, and the axial load are constant, Equation 1.7.5 yields

$$w'' + k^2 w = 0 \qquad \text{where } k^2 = \frac{P}{EI(1 - P/GA_0)} \tag{1.7.6}$$

The boundary conditions are $w = 0$ at $x = 0$ and $x = l$. We see that the differential equation and the boundary conditions are the same as for buckling without shear (Sec. 1.2). The general solution is $w = A \sin kx + B \cos kx$, and the boundary conditions require that $B = 0$ and $kl = n\pi$ $(n = 1, 2, \ldots)$ if A should be nonzero. Thus $k^2 = (n\pi/l)^2$. If this is substituted into Equation 1.7.6, the resulting equation may be solved for P, which furnishes the formula

$$P_{\mathrm{cr}_n} = \frac{P^0_{\mathrm{cr}_n}}{1 + (P^0_{\mathrm{cr}_n}/GA_0)} = \frac{P^0_{\mathrm{cr}_n}}{1 + n^2 m[\pi/(l/r)]^2 (E/G)} \tag{1.7.7}$$

due to Engesser (1889, 1891). Here $P^0_{\mathrm{cr}_n} = n^2 P_{\mathrm{E}}$ = critical load when bending is neglected. We see that again the smallest critical load occurs for $n = 1$. When slenderness, l/r, is sufficiently large, the shear strains become negligible, and the classical Euler solution is recovered.

From Equation 1.7.7 we observe that shear strains decrease the critical load. The smaller the ratio G/mE, the larger is the shear effect in buckling. Also, the smaller the slenderness, l/r, the larger is the shear effect. Therefore, the shear correction becomes significant for short columns (small l/r), but only if the yield stress f_y is so high that the short column still fails due to buckling rather than yield. The critical stress for $n = 1$ is, from Equation 1.7.7, $\sigma_{\mathrm{cr}_1} = \sigma^0_{\mathrm{cr}_1}/(1 + \sigma^0_{\mathrm{cr}_1} m/G)$ in which $\sigma^0_{\mathrm{cr}_1} = P^0_{\mathrm{cr}_1}/A = \pi^2 E/(l/r)^2$ = critical stress without the effect of shear. This equation yields the plot in Figure 1.20b. We note that as the slenderness tends to zero, the critical stress tends to a finite value while with the neglect of shear it tends to infinity.

Evaluating the ratio f_y/E for typical metals, (e.g., 0.002 for structural steel), and noting that the largest possible value of σ_E is f_y, we find that the correction due to shear is generally negligible (<1 percent). So it is for reinforced concrete columns. However, for compression members made of an orthotropic material that has a high elastic modulus in the axial direction and a low shear modulus, the shear correction can be quite large. This is often the case for fiber composites based on a polymer matrix. Other practical cases in which the shear correction is important are the built-up columns and tall building frames that will be discussed later (Secs. 2.7 and 2.9).

Generalization

Our solution in Equation 1.7.5 is based on introducing $M = -Pw$, which is valid only for pin-ended columns. In general, we must express M from Equation 1.7.2. Then, substituting this into the differential equilibrium equations (Eq. 1.3.3) and into Equation 1.7.3, we obtain the following equilibrium equations:

$$\begin{aligned} (EI\psi')'' + (Pw')' &= p \\ (EI\psi')' + GA_0(w' - \psi) &= 0 \end{aligned} \tag{1.7.8}$$

which correspond to what is called Timoshenko beam theory (Timoshenko, 1921). Equation 1.7.8 represents a system of two simultaneous linear ordinary differential equations for functions $w(x)$ and $\psi(x)$. If EI, GA_0, and P are all constant, the general solution for $p = 0$ can be sought in the form $w = Ce^{\lambda x}$, $\psi = De^{\lambda x}$. Substitution into Equation 1.7.8 yields a system of two linear equations for C and D, which are homogeneous if $p = 0$. They have a solution if and only if their determinant vanishes. This represents a characteristic equation which yields $\lambda^2 = -k^2$ where k^2 is again given by Equation 1.7.6. The general solution for $p = 0$ now is

$$w = C_1 + C_2 x + C_3 \cos kx + C_4 \sin kx \qquad \psi = C_2 - \beta C_3 k \sin kx + \beta C_4 k \cos kx \tag{1.7.9}$$

with $\beta = 1 - P/GA_0$. The boundary conditions are $w = 0$, $\psi = 0$ for a fixed end, $w = 0$, $\psi' = 0$ for a hinged end, and $\psi' = 0$, $\psi'' + \beta k^2 w' = 0$ for a free end. The solutions show that Equation 1.7.7 for $n = 1$ (in which $P^0_{\text{cr}_n}$ is no longer equal to $n^2 P_E$) is still valid for all the situations in which $V = 0$ for all x (Corradi, 1978; see also Ziegler, 1968, p. 51). In particular this is so for fixed and free standing columns (as well as columns with sliding restraints; Fig. 1.7f). For a fixed–hinged column, however, the critical load is given by the equation $\beta kl = \tan kl$ (with k defined by Eq. 1.7.6). Since $\beta < 1$, the critical load is smaller than that predicted by Equation 1.7.7, and the difference from Equation 1.7.7 is about the same as that between Equation 1.7.7 and P_{cr_1} for no shear (see Prob. 1.7.3).

All the foregoing formulations rest on Equation 1.7.1 in which the shear angle γ is intuitively written as a function of the variable direction shear force Q rather than the fixed shear direction force V. That this must indeed be so can be rigorously shown on the basis of an energy formulation with finite strain, utilizing the calculus of variations (an approach which will be discussed in Sec. 11.6). In such an analysis, it can be shown that our use of shear force Q normal to the deflection curve is associated with the use of the classical Lagrangian (or Green's) finite-strain tensor. Other definitions of the finite-strain tensor are possible, and if they are used to calculate the strain energy, the resulting differential equations differ somewhat from Equation 1.7.8 (see, e.g. Eq. 11.6.11 proposed by Haringx, 1942, for helical springs), and Q has a different meaning (e.g., it could represent a shear force inclined at angle ψ, i.e., in the direction of the rotated cross-section plane). Careful analysis shows, however, that the various possible definitions of finite strain as well as Q correspond to different definitions of elastic moduli E and G in finite strain, and that all these formulations are physically equivalent (as proven in Bažant, 1971). Distinctions among such approaches are unimportant if the strains are small.

Sandwich Beams and Panels

The shear beam theory can also be used to analyze cylindrical bending of sandwich beams or panels, whose applications are especially important in the aerospace industry, and are presently growing in structural engineering as well. Sandwich beams are composite beams, which consist of a soft core of thickness c (Fig. 1.21), for example, a hardened polymeric foam or honeycomb, bonded to stiff faces (skins) of thicknesses f. The contribution of the longitudinal normal stresses in the core is negligible compared with those in the skins. Consequently, the shear stress is nearly uniform through the thickness of the core. Since the skins are thin, the shear stresses in the core carry nearly all the shear force, and the shear deformation of the core is very important. The skins alone behave as ordinary beams whose cross sections remain normal, but the cross section of the core does not remain normal.

Rotation ψ (Fig. 1.21) is defined by longitudinal displacements of the skin centroids, which slightly differs from the rotation ψ_c of the core cross section. Writing $\frac{1}{2}(c+f)\gamma = \frac{1}{2}c\gamma_c$ where γ_c = shear strain in the core and γ = average shear strain, we have, per unit width of plate (see Fig. 1.21), $Q/G_c c = \gamma_c = (1 + f/c)\gamma = (1+f/c)(\theta - \psi)$, that is,

$$\theta - \psi = \frac{Q}{G_c A_1} \qquad G_c A_1 = G_c(f+c) \tag{1.7.10}$$

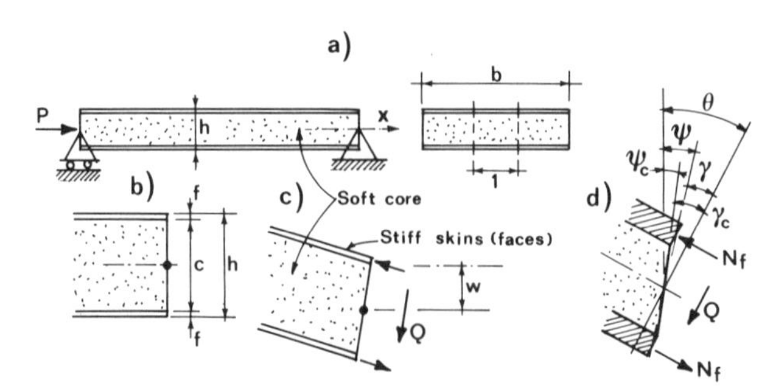

Figure 1.21 Sandwich-beam deformation under bending and shear.

where G_c = elastic shear modulus of the core. The bending moment per unit width of the plate is $M = N_f(c+f) + 2M_f$ where $N_f = (E'_f f)\frac{1}{2}(c+f)\psi'$ and $M_f = \frac{1}{12}f^3 E'_f w''$. So

$$M = E'_f I_1 \psi' + E'_f I_f w'' \qquad I_1 = \frac{f}{2}(c+f)^2 \qquad I_f = \frac{f^3}{12} \tag{1.7.11}$$

where $E'_f = E_f/(1-\nu_f^2)$ = elastic modulus of the faces (skins) for uniaxial strain (ν_f = Poisson ratio). E'_f must be used instead of E_f since in a wide panel the lateral strains ε_y must be zero, or else the bending could not be cylindrical, and curvature would arise also in the lateral direction y (cf. Chap. 7).

The differential equation for $w(x)$ is obtained by differentiating Equation 1.7.10 where $Q = -M'$ and $\theta = w'$, expressing from this ψ' and substituting it into Equation 1.7.11. The result, for constant cross section, is identical to Equation 1.7.4 in which $EI = E_f(I_1 + I_f)$ and $GA_0 = G_c A_1(1 + I_f/I_1)$. So the solution is similar. If the faces are thin, that is, $f \ll c (I_f \to 0)$, Equations 1.7.1 and 1.7.2 apply directly, with $GA_0 = G_c c$, $EI = E'_f f c^2/2$.

Problems

1.7.1 Derive in detail Equation 1.7.9 giving the general solution of Equations 1.7.8 (with $p = 0$).

1.7.2 Using Equation 1.7.9, show that Equation 1.7.7 still gives correct P_{cr_1} for (a) a fixed column, (b) a free-standing column, and (c) a column with a sliding restraint (Fig. 1.7f).

1.7.3 For a fixed–hinged beam, the expression for P_{cr_1} that takes shear into account differs from Equation 1.7.7. Calculate P_{cr_1} for $\beta = 1 - P/GA_0 = 0.9$, 0.8, 0.7, 0.6, 0.5, using the result of Problem 1.7.1. *Hint:* The graphic solution illustrated in Figure 1.8 is still valid, although the straight line now has slope β.

1.7.4 Solve the critical axial load (per unit width) of a simply supported sandwich panel in cylindrical bending (Fig. 1.21a). Express the results in a nondimensional form and discuss the effects of G_c/E_f and c/f.

1.8 PRESSURIZED PIPES AND PRESTRESSED COLUMNS

Pressurized Pipes

It is instructive to consider a variant to the column problem—a pressurized pipe. We consider, for example, the steel pipes shown in Figure 1.22a, b filled with water and loaded axially by a frictionless piston, with applied force, $P = p_h A$ where p_h = hydrostatic pressure in water and A = area of the interior cross section of the pipe. At first thought, one is tempted to say that this column will never buckle since the pipe carries no axial stresses due to P. Indeed, patent applications for such columns have been submitted. How wonderful would that be! We could make this column, say 10 cm in diameter, 100 m long, support on it a building, and it would never buckle.

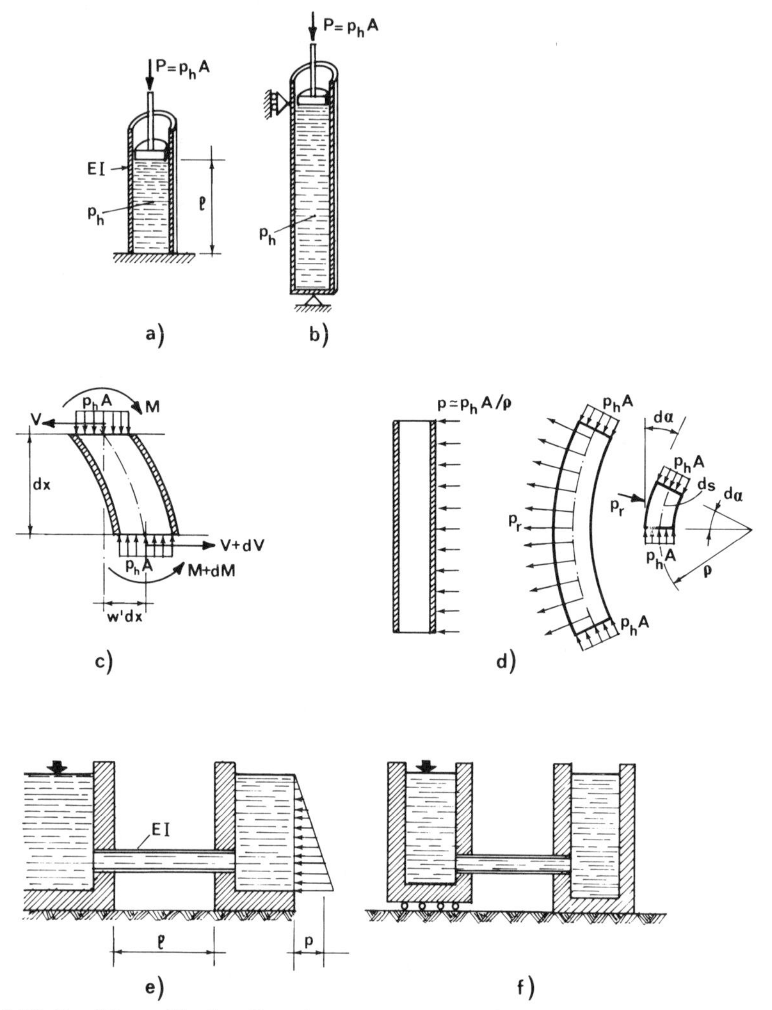

Figure 1.22 Buckling of hydraulic column supports and pressurized pipes.

In reality these columns do buckle, and the critical value of the axial force on the piston is exactly the same as it is when the pipe is empty and an axial force is applied to the pipe. This fact can be demonstrated in various ways.

The easiest demonstration is to consider the differential equilibrium conditions for the composite of the pipe and the fluid. The forces acting on an element of this composite of length dx are sketched in Figure 1.22c. Taking the conditions of equilibrium of horizontal forces and of moments above the center of the lower end cross section, we obtain the differential equilibrium equations $V' = 0$, $M' + p_h A w' = -V$, which are the same as Equations 1.3.2 we deduced before provided we set $P = p_h A$. Then, introducing $M = EIw''$ and eliminating V, we obtain the differential equation $EIw^{\text{IV}} + p_h A w'' = 0$, which implies that for a pipe with pin-ended supports (Fig. 1.22b) the critical pressure p_{cr} in water is given by $p_{\text{cr}} A = EI\pi^2/l^2$.

As another demonstration, we may consider the pipe separately from water

(Fig. 1.22d). The pipe has no axial load, but it is loaded laterally by the pressure of water. Now, if the pipe deflects as shown in Figure 1.22d the area of the internal wall of the pipe on the left of the deflected axis becomes larger than the area on the right, causing a net resultant p per unit length applied radially from water to the pipe. The value of this resultant may be calculated most easily by considering equilibrium of a slice of the water column (without steel walls) of length ds (Fig. 1.22d). The condition of equilibrium of radial forces is $p_r\, ds - p_h\, dA\, d\alpha = 0$, with $d\alpha = ds/\rho$, in which ρ = curvature radius of the center line of the column of water. This yields the well-known expression for the radial force $p_r = p_h A/\rho$, which is, for small rotations, equal in magnitude to the lateral distributed force applied on the steel pipe of bending stiffness EI. The bending equation for the pipe, without water, is $EIw^{\text{IV}} = p$, and substituting $p = -p_h A/\rho$, where $1/\rho \simeq w''$ (for a positive curvature, p is negative), we obtain again the governing equation $EIw^{\text{IV}} + p_h A w'' = 0$, which is identical to the beam column equation if $p_h A = P$.

Based on these considerations, it is clear that, for example, the fixed-end pipe of length l, connecting two rigid and rigidly supported tanks (Fig. 1.22e), can buckle, and if the Poisson ratio ν of pipe material is 0, the buckling pressure p_{cr} at the level of the pipe satisfies the equation $p_{\text{cr}}A = 4EI\pi^2/l^2$. For nonzero ν, one must note that pressurization of the pipe generates axial force $N_t = 2\nu pA$ in the pipe. Thus the critical pressure p must be solved from the relation $p_{\text{cr}}A - 2\nu p_{\text{cr}}A = 4EI\pi^2/l^2$.

By contrast, if the system of the two tanks with the pipe is statically determinate, for example, when one of the tanks can freely displace laterally on rollers, no pressure can cause the pipe to buckle. The reason is that in this case the steel pipe does not have zero axial force but transmits an axial tensile force that is exactly equal in magnitude and opposite to the force transmitted by the column of water inside the pipe, with the consequence that the net resulting axial force in the composite of pipe and water is exactly zero.

Column supports of the type shown in Figure 1.22a, b are actually used in mining, particularly, in very deep gold mines in the Transvaal. (They are very stocky, operating far below the critical load, and are provided with valves which leak liquid when a certain limit pressure is exceeded, thus endowing the column with an infinite yield plateau.)

Prestressed Columns

A situation that is analogous and mathematically equivalent to a pressurized pipe arises in a prestressed column (Fig. 1.23). The prestressed tendon in a prestressed concrete column is analogous to the pressurized column of water in a pipe, except that the force in it is tensile rather than compressive, and the reaction of the tendon is always applied to the concrete part of the column. At first it might seem that the axial compressive force F applied on the concrete part of the column by the prestressing anchors at the end of the column would cause buckling when $F = P_{\text{cr}_1}$, for example, when $F = EI\pi^2/l^2$ for a pin-ended column. Not so, however. The prestressing force F has in fact no effect on the column buckling, and the prestressed column buckles only due to its externally applied axial load P (however, this is not true for inelastic columns; see Sec. 8.5). This is obvious if

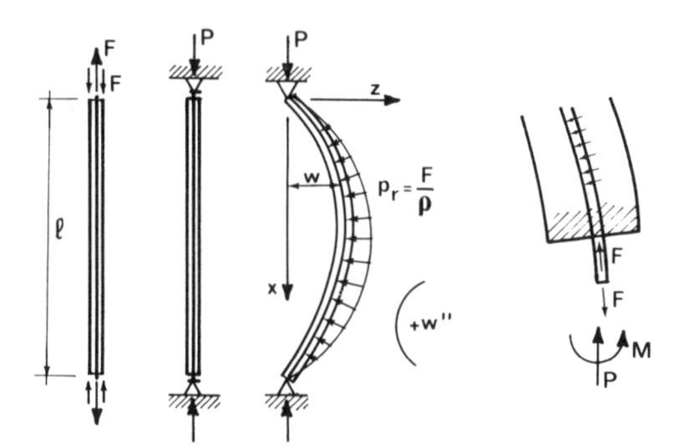

Figure 1.23 Buckling of prestressed columns.

one realizes that in the composite cross section of tendon and concrete, the tensile force in the tendon, F, is exactly canceled by the compressive force $-F$ in the concrete part of the cross section, with the result that the net axial force in the composite cross section is zero.

Another way to look at the prestressed column problem is to consider the concrete part of the column separately from the tendon, in which case the concrete part is loaded by axial force $P + F$ at the ends, and also by the radial distributed forces $p_r = F/\rho$ applied from the tendon onto the concrete due to column curvature $1/\rho = w''(x)$ as the column deflects (Fig. 1.23). The differential equation for the deflections $w(x)$ of the concrete part of the column is $(EIw'')'' + (P + F)w'' = p$ in which P is the externally applied axial force. Now, for small rotations, $p = -p_r = Fw''$ (since for a negative curvature p is negative). Upon substitution into the preceding differential equation one thus obtains $(EIw'')'' + Pw'' = 0$, which is the same as the differential equation of a column loaded only by axial load P, and is independent of the prestressing force F.

Our preceding analysis is contingent upon the assumption that the tendon is snugly embedded in concrete without any free play. In post-tensioning construction this is achieved by grouting. If a post-tensioned tendon is left ungrouted, then the axial reaction from the tendon anchors may cause small buckling deflections. However, as the deflection grows, the tendon comes in contact with concrete, and a radial distributed force develops, limiting further increase of deflections.

Problems

1.8.1 Discuss the behavior of a concrete hinged column which is prestressed by a straight tendon of constant eccentricity e. The prestress force is F, and a center axial load P is applied on the column.

1.8.2 Discuss the behavior of a simply supported pipe with initial curvature $w_0(x)$. Axial load P is applied on a piston that has constant frictional force P_f.

1.8.3 Analyze a prestressed free-standing column—calculate the lateral reactions of the tendon, and from them show that the column cannot buckle.

1.9 LARGE DEFLECTIONS

All our considerations so far have been limited to the linearized, small-deflection theory (second-order theory), which applies only for infinitely small deformations from an initial stressed state of the structure. This is sufficient for most practical purposes in structural engineering; nevertheless, a full understanding of the column behavior calls for a nonlinear finite-deflection theory. Although the initial nonlinear behavior is more easily determined by an approximate energy approach (see Sec. 5.9), an exact equilibrium solution is possible for some columns, and we will demonstrate it now.

Solution of Rotations by Elliptic Integrals (Elastica)

Consider again the pin-ended column shown in Figure 1.24. We assume that the length of the deflected centroidal axis of the column remains the same, l. Instead of deflection w it is more convenient to work with the slope angle θ. Since $ds = \rho\, d\theta$ where s is the length coordinate measured from the column end along the deflection curve and ρ is the curvature radius, the curvature is given as $1/\rho = d\theta/ds$. Now using primes to denote derivatives with regard to s rather than x, we have $M = EI\theta'$. This must equal $-P(w+e)$ according to the equilibrium condition, e representing the load eccentricity at the end (Fig. 1.24). Differentiating with regard to s and noting that $dw/ds = \sin\theta$, we obtain for the function $\theta(s)$ the differential equation

$$EI\theta'' = -P\sin\theta \tag{1.9.1}$$

It so happens that this nonlinear second-order differential equation can be easily solved. The solution we are going to present was given in 1859 by Kirchhoff after he obtained Equation 1.9.1 and noticed that it is mathematically identical to the equation that describes large oscillations of a pendulum, which was solved earlier by Lagrange (this was called a kinetic analogy of columns; Love, 1927). We assume EI to be constant. It is expedient to set $\sin\theta = 2\sin(\theta/2)\cos(\theta/2)$, multiply Equation 1.9.1 by θ', and then, upon noting that $\theta'\theta'' = \frac{1}{2}(\theta'^2)'$, and that $[\sin^2(\theta/2)]' = \theta'\sin(\theta/2)\cos(\theta/2)$, we may integrate Equation 1.9.1 to get

$$\frac{1}{4k^2}\theta'^2 = -\sin^2\frac{\theta}{2} + c^2 \qquad \left(k^2 = \frac{P}{EI}\right) \tag{1.9.2}$$

in which c is an undetermined integration constant. Denoting as θ_0 the slope at the end of the beam ($s = 0$), and noting that $\theta_0' = M/EI = -Pe/EI$, we obtain by

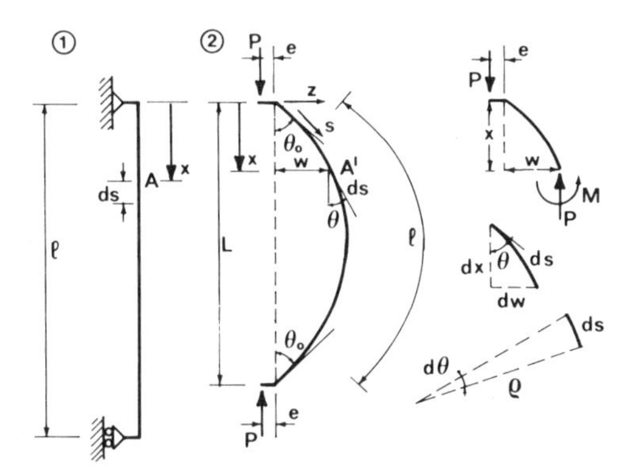

Figure 1.24 Beam-column under large deflections.

substitution into Equation 1.9.2

$$c^2 = \frac{1}{4k^2}\left(\frac{Pe}{EI}\right)^2 + \sin^2\frac{\theta_0}{2} = \frac{\pi^2}{4}\left(\frac{e}{l}\right)^2\left(\frac{P}{P_E}\right) + \sin^2\frac{\theta_0}{2} \tag{1.9.3}$$

Taking a square root of Equation 1.9.2 and separating the variables, we get

$$\frac{\pm d\theta}{\sqrt{c^2 - \sin^2(\theta/2)}} = 2k\,ds = \frac{2\pi}{l}\sqrt{\frac{P}{P_E}}\,ds \tag{1.9.4}$$

The plus and minus signs distinguish between buckling to the left and to the right. We will consider only the minus sign. It is convenient to introduce in Equation 1.9.4 the substitution

$$\sin\frac{\theta}{2} = c\sin\phi \qquad d\theta = \frac{2c\cos\phi\,d\phi}{\cos(\theta/2)} = \frac{2c\cos\phi\,d\phi}{\sqrt{1 - c^2\sin^2\phi}} \tag{1.9.5}$$

The integral of Equation 1.9.4 yields a functional relation between the slope of the deflected column and the length coordinate s, which completely defines the shape of the column, a curve known as the *elastica.* We are interested only in the end rotation, and we consider from now on only the special case of centric load, $e = 0$ (even though the solution is possible for any e). Then $c = \sin(\theta_0/2)$, which yields $\sin\phi_0 = 1$ or $\phi_0 = \pi/2$ as the integration limit corresponding to the end of the beam. Thus, integration of Equation 1.9.4 from the end of the beam ($s = 0$, $\phi = \phi_0$) to midlength ($s = l/2$, $\theta = \phi = 0$) yields

$$\frac{\pi}{2}\sqrt{\frac{P}{P_E}} = \int_0^{\pi/2}\frac{d\phi}{\sqrt{1 - c^2\sin^2\phi}} \qquad \left(c = \sin\frac{\theta_0}{2}\right) \tag{1.9.6}$$

This integral is recognized as the complete elliptic integral of the first kind, tables of which are available. Equation 1.9.6 represents a relation between P and θ_0.

Consider now the case of small end rotation θ_0. Then $c^2\sin^2\phi \ll 1$, and so $(1 - c^2\sin^2\phi)^{-1/2} \simeq 1 + \frac{1}{2}c^2\sin^2\phi$. Upon substitution in Equation 1.9.6 we can easily integrate, obtaining for the integral the value $(1 + c^2/4)\pi/2$. Substituting this into Equation 1.9.6, setting $c = \sin(\theta_0/2) \simeq \theta_0/2$, and neglecting higher-order terms in $\theta_0/2$, we obtain from Equation 1.9.6 the asymptotic relation

$$P = P_E\left(1 + \frac{\theta_0^2}{8}\right) \tag{1.9.7}$$

which is valid when θ_0 is not too large (for $\theta_0 < 55°$ the error in P is <1 percent). For $\theta_0 \to 0$ we have $P = P_E$. This means that deflected equilibrium shapes at very small θ_0 are possible if $P = P_E$. This agrees with the results of linearized theory (Eq. 1.3.5). The error in P for the linearized theory is less than 1 percent if $\theta_0 < 16°$.

We conclude further from Equation 1.9.7 that with increasing deflection the load P increases. This is an important feature, which contrasts with some other buckling problems (especially shells) for which the load may decrease with increasing buckling deflections (see Secs. 2.6, 4.5, and 7.7). (The latter behavior requires the use of larger safety factors in design.)

Deflections and Shortening

Let us now calculate deflections w and the axial coordinates x' of points whose coordinate was x in the initial undeflected (straight) state. To this end, we multiply Equation 1.9.4 either by $\sin\theta$ or by $(1+\cos\theta)$. Then, noting that $\sin\theta\,ds = dw$, $(1+\cos\theta)\,ds = ds + dx'$, $\sin\theta = 2\sin(\theta/2)\cos(\theta/2)$, and $1+\cos\theta = 2[1-\sin^2(\theta/2)]$, and taking only the minus sign in Equation 1.9.4, we get

$$-\frac{2\sin(\theta/2)\cos(\theta/2)\,d\theta}{\sqrt{c^2-\sin^2(\theta/2)}} = 2k\,dw \qquad -\frac{2[1-\sin^2(\theta/2)]\,d\theta}{\sqrt{c^2-\sin^2(\theta/2)}} = 2k(ds+dx') \tag{1.9.8}$$

Making again the substitution according to Equation 1.9.5, we obtain

$$dw = -\frac{2c}{k}\sin\phi\,d\phi \qquad -\frac{2}{k}\sqrt{1-c^2\sin^2\phi} = ds + dx' \tag{1.9.9}$$

Integration between the limits ϕ_0 and ϕ, which correspond to θ_0 and θ, yields w and x'at any point of the column as functions of s and θ. Integration between the limits $\phi = 0$ and $\phi_0 = \pi/2$ (for $e = 0$) then gives

$$w_{\max} = \frac{2cl}{\pi}\sqrt{\frac{P_E}{P}} \qquad l' = l\left(\frac{4}{\pi}\sqrt{\frac{P_E}{P}}\int_0^{\pi/2}\sqrt{1-c^2\sin^2\phi}\,d\phi - 1\right) \tag{1.9.10}$$

in which $w_{\max}$ is the maximum column deflection, which occurs at midlength, and l' is the length of the chord of the deflected column, l being the length of the deflection curve. We recognize the integral in Equation 1.9.10 to be the complete elliptic integral of the second kind.

Consider again that the end slope θ_0 is small. Then, noting that $(1-c^2\sin^2\phi)^{1/2} \simeq 1 - \frac{1}{2}c^2\sin^2\phi$, we get from Equations 1.9.10 the approximations

$$w_{\max} = \theta_0\frac{l}{\pi}\sqrt{\frac{P_E}{P}} \qquad u = l - l' = 2l\left[1-\left(1-\frac{1}{16}\theta_0^2\right)\sqrt{\frac{P_E}{P}}\right] \tag{1.9.11}$$

in which u represents the magnitude of axial displacement under the load P. From Equations 1.9.11, $\theta_0 \simeq \pi w_{\max}/l$, and substituting this into Equation 1.9.7, we finally obtain the asymptotic relation

$$P = P_E\left(1+\frac{\pi^2}{8l^2}w_{\max}^2\right) \tag{1.9.12}$$

The diagram of axial load versus maximum lateral displacement according to Equation 1.9.12 is plotted in Figure 1.25a. The figure shows also the exact solution obtained with the help of elliptic integrals. The approximate solution is good for maximum deflections up to 0.2 of column length, which is sufficient for all practical purposes. Furthermore, we note that the increase of axial load over P_{cr_1} is less than 2 percent for deflections up to $0.1l$, and this is why the axial load after the critical state can be considered approximately constant for most practical purposes. Moreover, substituting Equation 1.9.12 into the second equation in Equations 1.9.11 along with $\theta_0 \simeq \pi w_{\max}/l$, we obtain

$$u = w_{\max}^2\frac{\pi^2}{4l} \tag{1.9.13}$$

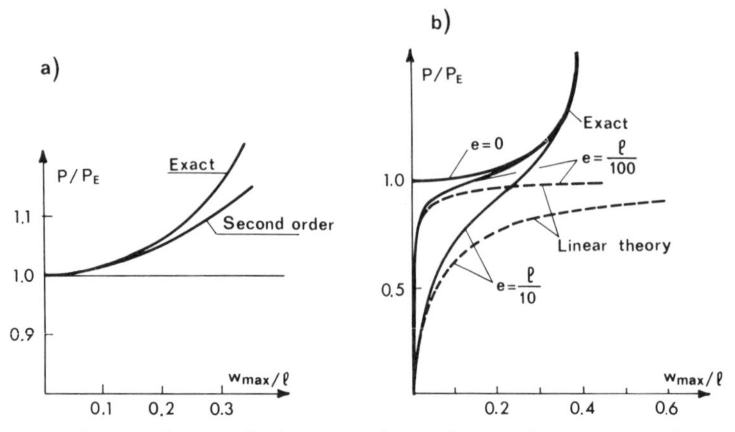

Figure 1.25 Load versus lateral displacement for perfect and imperfect columns.

It is interesting to note that this asymptotic relation for not too large deflections is exactly the same as if the deflection curve were sinusoidal, with a constant length of arc. Furthermore, expressing w^2_{max} and substituting it in Equation 1.9.12, we obtain for not too large deflections of an inextensible column the asymptotic relation (Fig. 1.26a)

$$P \simeq P_{cr}\left(1 + \frac{u}{2l}\right) \qquad \text{or} \qquad \sigma = \sigma_{cr} + \frac{\sigma_{cr}}{2}\bar{\varepsilon} \qquad (P \geq P_{cr}) \tag{1.9.14}$$

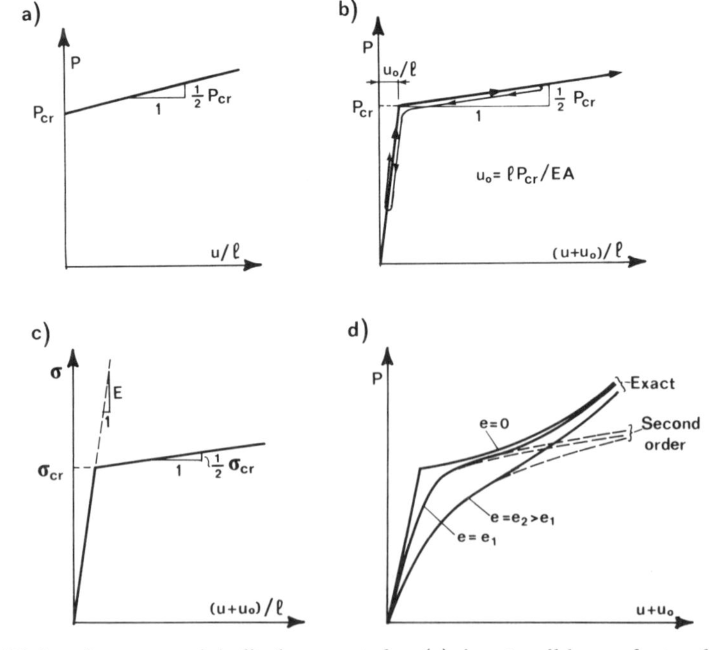

Figure 1.26 Load versus axial displacement for (a) inextensible perfect columns, (b, c) extensible perfect columns, and (d) extensible imperfect columns.

where the stresses refer to the centroid at column midlength and $\bar{\varepsilon} = u/l$. So we see that after the bifurcation of equilibrium path, the axial load is approximately proportional to the axial displacement. (The present derivation of Eq. 1.9.14 was given by Bažant, 1985.) In Section 5.9, Equation 1.9.14 as well as Equation 1.9.7 will be generalized to imperfect columns.

Equation 1.9.14 holds true under the assumption of an inextensible column. To take into account the elastic strain increment ε at neutral axis, we need to superimpose $\varepsilon = (\sigma - \sigma_{cr})/E$ upon the strain $\bar{\varepsilon} = u/l \simeq 2(P/P_{cr} - 1)$, obtained by solving Equation 1.9.14. This yields $u \simeq 2l\,\Delta P/P + l\,\Delta P/EA$, with $\Delta P = P - P_{cr}$. Setting $\Delta P/P \simeq \Delta P/P_{cr}$ (for small $\Delta P \ll P$), and solving this equation for ΔP, we then obtain the approximate load-shortening relation

$$P \simeq P_{cr} + \frac{P_{cr}}{2 + P_{cr}/EA}\left(\frac{u}{l}\right) \quad \text{or} \quad \sigma = \sigma_{cr} + \frac{\sigma_{cr}}{2 + (\sigma_{cr}/E)}\,\bar{\varepsilon} \qquad (P \geq P_{cr}) \tag{1.9.15}$$

which is valid for not too large deflections. In this relation, the axial load-point displacement u is measured from the critical state (that is, $u = 0$ at $P = P_{cr}$). The total axial displacement is $\bar{u} = u + u_0$, where $u_0 = P_{cr}l/EA$.

According to Equation 1.9.15, the diagram of axial load versus load-point axial displacement, as well as the axial stress–strain diagram, is bilinear, as shown in Figure 1.26b, c.

The axial stress–strain diagram of a perfect column, as shown in Figure 1.26c, looks the same as the stress–strain diagram for an elastic–plastic material. However, there is one crucial difference. For unloading, the elastic column retraces back the same path as it traced on previous loading (Fig. 1.26b), while an elastic–plastic material unloads along a different path. Of course, in an elastic material the deformations must be perfectly reversible even if the response diagram of the structure is nonlinear.

Thus the stress–strain diagram of a column that buckles is approximately bilinear, with slope E (Young's modulus) for $\sigma \leq \sigma_{cr}$, and slope $\frac{1}{2}\sigma_{cr}$ for $P > P_{cr}$ and the load-axial displacement diagram slope $\frac{1}{2}P_{cr}/l = \frac{1}{2}\pi^2 EI/l^3$. These slopes represent the incremental (tangential) stiffnesses of the buckling column.

It may seem surprising that the postcritical dependence of the axial load on the axial displacement may be linear. That this is indeed possible can be intuitively understood by noting that both the axial load change and the axial displacement are proportional to the square of the maximum deflection.

Discussion of Results

For normal columns, the decrease of the incremental stiffness after the critical state is tremendous. The ratio of the stiffness before and after buckling is $2E/\sigma_{cr}$, which is always larger than $2E/f_u$ where f_u = strength (ultimate stress); thus, for steels with yield limit $f_y = f_u = 40{,}000$ psi and Young's modulus $E = 30 \times 10^6$ psi, the stiffness decreases over 1500 times as the critical state is exceeded, while for a concrete column with $f_u = 5000$ psi and $E = 4 \times 10^6$, the stiffness decreases over 1600 times. For a fiber-reinforced polymer, with $f_u = 40{,}000$ psi and $E =$ 700,000 psi, the stiffness decreases over 35 times. Nevertheless, the incremental stiffness after the critical state remains positive, which contrasts with some other structures in which the incremental stiffness after the critical state may become negative (e.g., shells; Secs. 7.7, 2.6, and 4.5). This indicates that the pin-ended column has a postcritical reserve strength.

A similar solution is possible for columns with other end supports. However, the solutions for other types of supports can be obtained directly from the present solution by a procedure similar to the effective length approach as described for small deflections (Sec. 1.4). Thus the elastica curve for a free-standing column (Fig. 1.27b) is one-half of the curve for the pin-ended column (Fig. 1.27a) with the same load. The curve for a fixed–fixed column is obtained by adding to segment 12 segments 41 and 25 which are identical to 13 or 32 but inverted. The curve for a column with sliding ends restrained against rotation (curve 325 in Fig. 1.27d) is obtained as one-half of curve 435 (Fig. 1.27c) for a fixed–fixed column. The curve for a fixed–hinged column (126 in Fig. 1.27e) is obtained by finding point 6 at which the tangent 16 passes through point 1 in Figure 1.27c. The present solution, however, cannot be applied to certain elastically restrained columns of columns within a frame.

Real columns are always imperfect. Based on Equations 1.9.3 and 1.9.4, one can obtain an exact solution for nonzero e or an approximate (asymptotic) second-order solution by a similar procedure as just shown (Renzulli, 1961). The curves of load versus maximum lateral deflection or load versus axial displacement, which are obtained for various initial load eccentricities e, are illustrated in Figures 1.25b and 1.26d.

The approximate large-deflection behavior can also be solved by integrating a differential equation in terms of w rather than θ. To this end the curvature expansion from Equation 1.1.2 may be introduced (Thompson and Hunt, 1973; Dym and Shames, 1973). However, this approach is considerably more involved.

Our analysis leads to several important observations:

1. In contrast to the linearized small-deflection theory, the deflection does not tend to infinity at P_{cr} but remains bounded at any load.
2. After the critical state, the load increases as the deflection increases. Thus columns possess a postcritical reserve. This contrasts with buckling of some other structures, especially shells, for which the load decreases after the critical state.
3. The deviations from the linearized small-deflection theory do not become significant until the deflections become quite large.

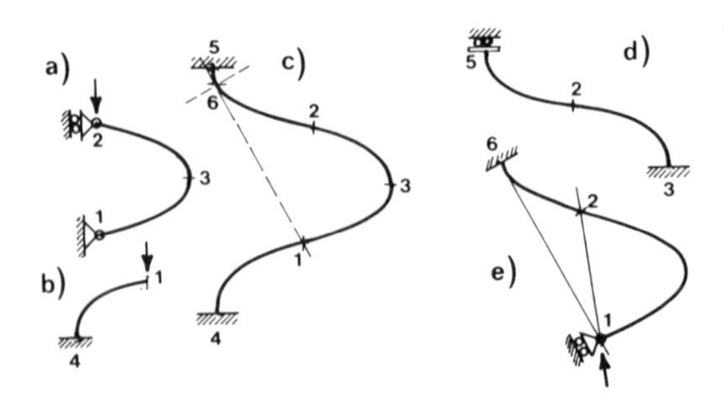

Figure 1.27 Elastica curves for beams with various end supports.

4. Normal structural columns would suffer material yielding or fracture well before the finite-deflection theory becomes important, and so the linearized small-deflection theory suffices for design.

Knowledge of postcritical behavior is important for evaluating the effect of buckling of a member within a statically indeterminate truss (see Sec. 2.5). It is also useful for various applications outside structural engineering. As one such example, the elastica solution has been used in a method of measurement of stress relaxation of highly flexible specimens made of a linearly viscoelastic material (Bažant and Skupin, 1965, 1967). For such a material, one can show that the specimen deflected as shown in Figure 1.28 keeps a constant shape, and so it undergoes stress relaxation at constant strain. The relaxation ratio for stress is proportional to the relaxation ratio for the support reaction, and this reaction can be easily measured. The main advantage is that the reaction corresponding to large strains in bending (Fig. 1.28) is far smaller than the reaction corresponding to large axial strains without bending, so that the frame that provides the reaction is cheap to make. Thus one can economically produce many loading frames to test long-time relaxation (as well as strength degradation) of many specimens. The elastica solution is necessary to determine the stresses from the measured reaction (Bažant and Skupin, 1967).

Analytical solution for nonlinear finite-deflection theory of columns, such as outlined above, are generally much more difficult than the linearized small-deflection solutions. Today, however, all these problems can be solved easily with a computer using finite elements. The usefulness of the analytical nonlinear solutions now lies mainly in the understanding that they provide us, as well as in serving as a check on accuracy of finite element solutions.

For a generalization of the solution of initial postcritical behavior to imperfect columns, see Sec. 5.9.

Problems

1.9.1 Obtain the P-w_{max} and P-u relations for (a) a free-standing column and (b) a fixed–fixed column.

1.9.2 Study the approximate large-deflection behavior for the case that the load has a finite eccentricity e.

1.9.3 Repeat the preceding solution for a free-standing column.

1.9.4 Consider a pin-ended column whose ends cannot slide, under the effect of an axial thermal expansion due to temperature change ΔT. Construct the diagrams of P versus ΔT and P versus midspan deflection w.

1.9.5 Use the $P(u)$ diagram to obtain the critical load and postcritical behavior of a column with nonsliding hinges, heated from temperature T_0 to temperature T. Plot $P(T)$, $w(T)$.

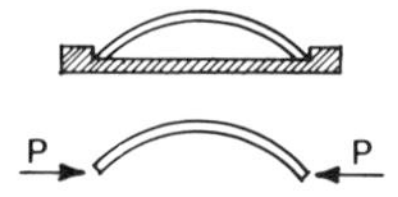

Figure 1.28 Device for long-time relaxation tests. (*After Bažant and Skupin, 1965, 1967.*)

1.10 SPATIAL BUCKLING OF BEAMS UNDER TORQUE AND AXIAL FORCE

All the problems we have analyzed so far have been two-dimensional (planar). In the presence of torque, the analysis has to be generalized to three dimensions. Spatial buckling of twisted and compressed shafts is important for the design of rotors of turbines, generators, and other rotating machinery (Ziegler, 1968). Spatial buckling may also be important for frames. Recently, design of latticed struts that can be collapsed for transport by means of torsion became of interest for construction of an orbiting space station.

Buckling of beams due to torque and axial load exhibits critical load combinations that represent equilibrium path bifurcations at which a deflection of a perfect shaft or beam becomes possible. The method of solution of these problems is a generalization of the method in Sections 1.2–1.4.

Consider a geometrically perfect beam or shaft supported on two spherical hinges, loaded by axial force P and torque M_t, which is assumed to keep its direction during buckling; see Figure 1.29b, where the axial vector of torque is represented by a double arrow. The deflection must be expected to be a spatial curve characterized by displacements $v(x)$ and $w(x)$ in directions y and z (x, y, and z are Cartesian coordinates). In a cross section at x, the torque vector $\mathbf{M}_t$ has bending components of magnitudes $M^T = M_t v'$ and $-M_t w'$ in the z and y directions. Adding these components to the expressions $EIw'' = -Pw$ and $EIv'' = -Pv$ without the torque (Sec. 1.3), we get the differential equations

$$EIw'' - M_t v' + Pw = 0$$
$$EIv'' + M_t w' + Pv = 0 \tag{1.10.1}$$

We assume here that the bending stiffness EI is the same for both y and z directions (the cross section need not be circular, though). Note that the change of torque about the deflected beam axis due to deflections $v(x)$ and $w(x)$ is second order small and therefore negligible, since it has the magnitude $\Delta T = M_t(1 - \cos\theta) \simeq M_t\theta^2/2$ where θ = slope of the deflection curve. For this reason the torsional stiffness of the beam is, curiously, irrelevant for buckling of shafts, and only the bending stiffness matters (this, however, contrasts with torsional buckling of thin-wall beams; see Chap. 6).

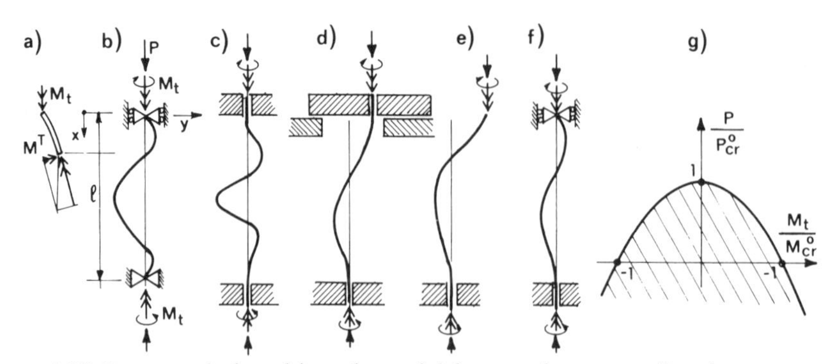

Figure 1.29 Beam or shafts subjected to axial force and constant-direction torque.

The general solution of Equations 1.10.1 may be sought in the form

$$v = Ae^{i\omega x} \qquad w = Be^{i\omega x} \tag{1.10.2}$$

where $i = \sqrt{-1}$. Substituting this into Equations 1.10.1, we can eliminate $e^{i\omega x}$ and obtain a system of two homogeneous linear equations

$$\begin{bmatrix} P - EI\omega^2 & iM_t\omega \\ -iM_t\omega & P - EI\omega^2 \end{bmatrix} \begin{Bmatrix} A \\ B \end{Bmatrix} = \mathbf{0} \tag{1.10.3}$$

Deflection is possible if the determinant vanishes, which yields the characteristic equation $EI\omega^2 \pm M_t\omega - P = 0$. Considering only $+M_t\omega$ (torque of one sign), we have the roots

$$\omega_{1,2} = \frac{1}{2EI}\left(-M_t \pm \sqrt{M_t^2 + 4EIP}\right) \tag{1.10.4}$$

The general solution is the real or imaginary part of

$$v = A_1 e^{i\omega_1 x} + A_2 e^{i\omega_2 x} \qquad w = B_1 e^{i\omega_1 x} + B_2 e^{i\omega_2 x} \tag{1.10.5}$$

where A_1, A_2, B_1, and B_2 are constants, which may be complex. The boundary conditions for the hinged beam are $v = w = 0$ for both $x = 0$ and $x = l$. This yields for A_1, A_2 the conditions

$$A_1 + A_2 = 0 \qquad A_1 e^{i\omega_1 l} + A_2 e^{i\omega_2 l} = 0 \tag{1.10.6}$$

and the conditions for B_1, B_2 happen to be identical. Equations 1.10.6 have a nonzero solution only if $e^{i\omega_2 l} = e^{i\omega_1 l}$, that is, if $\omega_2 l = \omega_1 l + 2\pi n$, where n is an integer. The case $n = 1$, that is, $\omega_2 = \omega_1$, would require that $M_t = 0$ (if $P > 0$), and so the first critical state is obtained for $n = 1$, that is, $\omega_2 l = \omega_1 l + 2\pi$ or $\omega_2 = \omega_1 + 2\pi/l$. According to Equation 1.10.4, this requires that $(M_t^2 + 4EIP)^{1/2}/2EI = \pi/l$, which can be rearranged as

$$\frac{P}{P_{\text{cr}}^0} + \left(\frac{M_t}{M_{\text{cr}}^0}\right)^2 = 1 \tag{1.10.7}$$

where

$$P_{\text{cr}}^0 = \frac{\pi^2}{l^2} EI \qquad M_{\text{cr}}^0 = k\frac{\pi EI}{l} \qquad k = 2 \tag{1.10.8}$$

P_{cr}^0 is the critical load for buckling without torque, and M_{cr}^0 is the critical torque for buckling without axial force. This result was obtained by Greenhill (1883); see also Love (1927), Timoshenko and Gere (1961), and Ziegler (1968). Equation 1.10.7 is plotted in Figure 1.29g.

To determine the deflection curve, one obtains from Equation 1.10.3 the eigenvector $B_1 = iA_1c_1$ and $B_2 = iA_2c_2$ where $c_j = (P - EI\omega_j)/M_t\omega_j$ $(j = 1, 2)$. Assuming A_1 to be real, we then have from Equations 1.10.6 $A_2 = -A_1$, $B_2 = -B_1$, and the solution is obtained as the real part of Equation 1.10.5:

$$v = A_1(\cos\omega_1 x - \cos\omega_2 x) \qquad w = -A_1 c_1(\sin\omega_1 x - \sin\omega_2 x) \tag{1.10.9}$$

This is obviously a spatial (nonplanar) curve. For a more detailed analysis, see Konopasek (1968), Grammel (1949), and Herrmann (1967).

The foregoing problem, however, is not as simple as it seems. The work of a torque vector $\mathbf{M}_t$ which keeps its direction and is applied at either a hinged end or a free end is path-dependent. To understand why, imagine that we reach the slope $(v'^2 + w'^2)^{1/2}$ at the beam end by rotating the beam end tangent first by angle w' about axis y and then by angle v' about axis z. During these rotations $\mathbf{M}_t$ does no work. However, we can reach the same final angle in other ways. For example, we can rotate the beam tangent by the angle $(v'^2 + w'^2)^{1/2}$ about axis y and bring then the beam end to the final position by rotating it about axis x. In that case, $\mathbf{M}_t$ does work. Therefore, the loading is path-dependent, nonconservative. When we deal with such problems in Chapter 3, we will see that static analysis is a priori illegitimate, and dynamic analysis is required. By chance, however, it so happens that for the present problem the dynamic analysis yields the same result as the static analysis, that is, Equation 1.10.7 happens to be correct.

A shaft (Fig. 1.29c) with fixed slopes at both ends obviously represents a conservative problem, in which the work of $\mathbf{M}_t$ due to $v(x)$ and $w(x)$ is always zero. In this case, there are at the ends redundant bending moments M_z about axis y and M_y about axis z, as well as horizontal reactions Q_x and Q_y. Consequently, moments $(M_z - Q_z x)$ and $(M_y - Q_y x)$ must be added to the right-hand sides of Equations 1.10.1. To get rid of the unknowns M_z, M_y, Q_z, Q_y, one may differentiate these equations twice, which yields (for constant EI and P)

$$
\begin{aligned}
EIw^{\mathrm{IV}} - M_t v''' + Pw'' &= 0 \\
EIv^{\mathrm{IV}} + M_t w''' + Pv'' &= 0
\end{aligned}
\qquad (1.10.10)
$$

These differential equations have a general applicability. They are of the fourth order, as we encountered it in Sections 1.3 and 1.4 for the fixed beam without torque, while the hinged beam as well as the cantilever can be handled by second-order equations, as in Section 1.2.

The solution may again be considered in the form of Equations 1.10.2. Substitution into Equations 1.10.10 yields a characteristic equation that can be reduced for $P = 0$ to the condition $\tan(\omega l/2) = \omega l/2$. The smallest root is $\omega l/2 = 4.494$. The corresponding critical torque for buckling without axial force is found to be again given by M_{cr}^0 in Equation 1.10.8, with $k = 2.861$. For a detailed solution, see Ziegler (1968).

Another conservative problem is the shaft in Figure 1.29d. The solution yields again Equations 1.10.8, but with $k = 2$. However, for the shafts in Figure 1.29e, f, which are nonconservative, there is a surprise. The static analysis yields $k = \infty$, that is, the column appears to be unconditionally stable, for any load. Dynamic stability analysis (see Chap. 3), on the other hand, indicates that the (undamped) column buckles under an arbitrarily small torque (Ziegler, 1951). This raises the question whether it is actually possible to apply on such columns a torque $\mathbf{M}_t$ whose vector keeps its directions, and whose magnitude is constant in time. This question becomes rather complex if one tries to consider the hydrodynamic forces that produce the torque, for example, those acting on the blades of a turbine rotor.

A conservative torque can be applied at free or pinned ends of a shaft by long cables as shown in Figure 1.30. In case (a), called semitangential torque, the torque vector $\mathbf{M}_t$ tilts due to deflection angle v' but not w', and in case (b), called the quasitangential torque, $\mathbf{M}_t$ tilts due to both v' and w'. Beck (1955) and

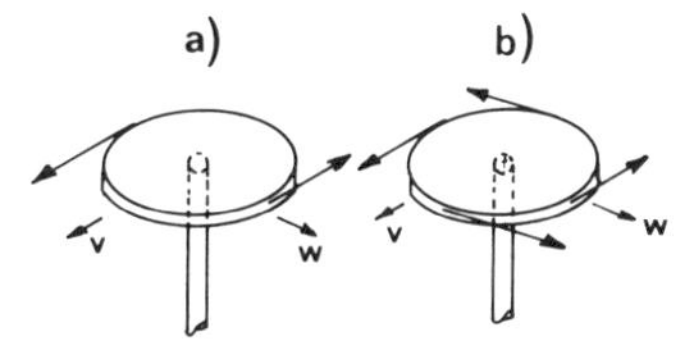

Figure 1.30 (a) Semitangential and (b) quasitangential conservative torque.

Ziegler (1968) show that for semitangential torque, $k = 1.564$, 1, and 2.168, and for quasi-tangential torque, $k = 1$, 0.5, and 1.576 for the cases in Figure 1.29b, e, f, respectively. This illustrates that the precise manner in which the torque is applied has a great influence and must be studied carefully.

Problems

1.10.1 Derive Equations 1.10.10 in detail, with sketches.

1.10.2 Carry out the detailed solutions of the shafts in Figure 1.29c, d.

1.10.3 Solve the column in Figure 1.29b for (a) semitangential or (b) quasi-tangential torque at $P = 0$.

1.10.4 Equations 1.10.10 can be combined into one differential equation

$$EIY^{IV} - iM_tY''' + PY'' = 0 \tag{1.10.11}$$

in which $Y = v + iw$ (which is a complex variable). Indeed, the real and imaginary parts of this equation yield Equations 1.10.10. The boundary conditions can be written completely in terms of Y; for a hinged end $Y = 0$ and $Y'' = 0$, and for a fixed end $Y = 0$, $Y' = 0$. Equations 1.10.11 can be solved by $Y = Ae^{i\omega x}$ where A is complex. Carry out this solution for the problems in the text (for $P = 0$ this method was shown by Ziegler, 1968).

1.10.5 P and M_t are increased proportionally, at a fixed P/M_t ratio. Suppose the beam has initial imperfection $v_0(x) = \alpha v_1(x)$ and $w_0(x) = \alpha w_1(x)$ where α = imperfection parameter and $v_1(x)$, $w_1(x)$ = first buckling mode of perfect beam for this P/M_t ratio. Calculate the diagrams of max v and max w versus M_t and the magnification factor. Also discuss the behavior near the critical state.

1.10.6 Formulate the differential equations for initial postcritical deflections of a beam under P and M_t that is correct only up to second-order terms in v and w. Does P produce a second-order torque? Does M_t? Do you need to introduce rotation $\theta(x)$ about the beam axis as a third unknown? Discuss how you would solve the problem by finite differences.

References and Bibliography[1]

Allen, H. G., and Bulson, P. S. (1980), *Background to Buckling,* McGraw-Hill, New York, pp. 81–85 (Sec. 1.5).

Bažant, Z. (1944), *Strength of Materials,* 3rd ed. (in Czech), Česká matice technická, Prague.

1. Not all references are cited in the text. For entries cited, the sections where they are first quoted are indicated.

Bažant, Z. P. (1971), "A Correlation Study of Formulations of Incremental Deformation and Stability of Continuous Bodies," *J. Appl. Mech.* (ASME), 38:919–27 (Sec. 1.7).

Bažant, Z. P. (1985), *Lectures on Stability of Structures, Course 720–D24*, Northwestern University, Evanston, Ill. (Sec. 1.9).

Bažant, Z. P., and Skupin, L. (1965), "Measurement Method of Relaxation and Corrosion of Stressed Specimens" (in Czech), Czechoslovak Patent No. 126345, Oct. 21 (Sec. 1.9).

Bažant, Z. P., and Skupin, L. (1967), "Méthode d'essai de viellissement des plastiques renforcés sous contrainte," *Verre Textile-Plastiques Renforcés (Paris)*, 5:27–30 (Sec. 1.9).

Beck, M. (1955), "Knickung gerader Stäbe durch Druck und konservative Torsion," *Ing. Arch.*, 23:231–53 (Sec. 1.10).

Bleich, F. (1952), *Buckling Strength of Metal Structures,* McGraw-Hill, New York.

Bognár, L., and Strauber, B. G. (1989), "Load-Shortening Relationships for Bars," *J. Struct. Eng.* (ASCE), 115(7):1711–25.

Brèzina, V. (1962), *Buckling Strength of Metallic Columns and Beams* (in Czech), Czechosl. Academy of Sciences Publ. House, Prague.

Britvec, S. J. (1965), "The Theory of Elastica in the Nonlinear Behavior of Plane Frame-Works," *Int. J. Mech. Sci.*, 7:661.

Calcote, L. R. (1969), *The Analysis of Laminated Composite Structures,* Van Nostrand Reinhold, New York.

Chen, Y. Z., Cheung, Y. K., and Xie, J. R. (1989), "Buckling Loads of Columns with Varying Cross Sections," *J. Eng. Mech.* (ASCE), 115:662–67.

Churchill, R. V. (1963) *Fourier Series and Boundary Value Problems,* 2d ed., MacGraw-Hill, New York (Sec. 1.5).

Corradi, L. (1978), *Instabilità delle strutture,* CLUP, Milano (Sec. 1.7).

Crandall, S. H., and Dahl, N. C. (1972), *An Introduction to the Mechanics of Solids,* McGraw-Hill, New York (Sec. 1.1).

Dym, C. L., and Shames, I. H. (1973), *Solid Mechanics: A Variational Approach,* McGraw-Hill, New York (Sec. 1.7).

Engesser, F. (1889), "Die Knickfestigkeit gerader Stäbe," *Z. Arch. Ing. Vereins Hannover* 35:455 (Sec. 1.7).

Engesser, F. (1891), "Die Knickfestigkeit gerader Stäbe," *Zentralb. Bauverwaltung,* 11:483 (Sec. 1.7).

Euler, L. (1744), *Methodus Inveniendi Lineas Curvas Maximi Minimive Proprietate Gaudentes* (Appendix, De Curvis Elasticis), Marcum Michaelem Bousquet, Lausanne (Sec. 1.2).

Grammel, R. (1949), "Scherprobleme", *Ing. Arch.*, 17:107–118 (Sec. 1.10).

Greenhill, A. G. (1883), "On the Strength of Shafting When Exposed Both to Torsion and End Thrust," *Proc. Inst. Mech. Eng.*, p. 182 (Sec. 1.10).

Haringx, J. A. (1942), "On the Buckling and Lateral Rigidity of Helical Springs," *Proc. Konink. Ned. Akad. Wetenschap.* 45:533 (cf. also Timoshenko and Gere, 1961, p. 142) (Sec. 1.7).

Hegedus, I., and Kollar, L. P. (1984a), "Buckling of Sandwich Columns with Thin Faces under Distributed Normal Loads," *Acta Tech.*, Budapest, 97(1–4):111–122.

Hegedus, I., and Kollar, L. P. (1984b), "Buckling of Sandwich Columns with Thick Faces Subjected to Axial Loads of Arbitrary Distribution," *Acta Tech.*, Budapest, 97(1–4):123–131.

Herrmann, G. (1967), "Stability of the Equilibrium of Elastic Systems Subjected to Nonconservative Forces." *Appl. Mech. Rev.* 20:103 (Sec. 1.10).

Kirchhoff, G. R. (1859), "Über des Gleichgewicht und die Bewegung eines unendlich dunnen elastischen Stabes," *J. Math.* (Crelle), 56:285–313 (Sec. 1.9).

Kollbrunner, C. F., and Meister, M. (1955), *Knicken,* Springer-Verlag, West Berlin.
Konopasek, H. (1968) *Classical Elastic Theory and Its Generalizations.* NATO Adv. Study Inst. Ser. E. 38:255–74 (Sec. 1.10).
Leicester, R. H. (1970), "Southwell Plot for Beam Columns," *J. Eng. Mech.* (ASCE), 96(6):945–66 (Sec. 1.5).
Lind, N. C. (1977), "Simple Illustrations of Frame Instability," *J. Struct. Eng.* (ASCE) 103:1–8 (Sec. 1.4).
Love, A. E. H. (1927), *A Treatise of the Mathematical Theory of Elasticity,* 4th ed., Cambridge University Press, Cambridge (also Dover, New York, 1944), p. 418 (Sec. 1.10).
Lundquist, E. E. (1938), "Generalized Analysis of Experimental Observations in Problems of Elastic Stability," NACA TN658, Washington, D.C. (Sec. 1.5).
Maier, G. (1962), "Sul calcolo del carico critico di aste a sezione variabile" (in Italian), *Costruzioni Metalliche,* 14(6):295–314.
Massonet, C. (1959), "Stability Considerations in the Design of Steel Columns," *J. Struct. Eng.* (ASCE), 85:75–111 (Sec. 1.6).
Michiharu, O. (1976), "Antisymmetric and Symmetric Buckling of Sandwich Columns under Compressive Load," *Trans. Jpn. Soc. Aeron. Space Sci.,* 19(46):163–78.
Newmark, N. M. (1949), "A Simple Approximate Formula for Effective End-Fixity of Columns," *J. Aero. Sci.,* 16:116 (Sec. 1.4).
Northwestern University (1969) *Models for Demonstrating Buckling of Structures.* Prepared by Parmelee, R. A., Herrmann, G., Fleming, J. F., and Schmidt, J. under NSF grant GE-1705 (Sec. 1.4).
Pearson, C. E. (1974), *Handbook of Applied Mathematics,* Van Nostrand Reinhold, New York (Sec. 1.5).
Pflüger, A. (1964), *Stabilitätsprobleme der Elastostatik,* 2nd ed., Springer-Verlag, West Berlin.
Popov, E. P. (1968), *Introduction to Mechanics of Solids.* Prentice Hall, Englewood Cliffs, N.J. (Sec. 1.1).
Rektorys, K. (1969), *Survey of Applicable Mathematics.* Iliffe Books, London (Sec. 1.5).
Renzulli, T. (1961), "Configurazioni di Equilibrio di una Trave Caricata di Punta," (in Italian). L'Ingegnere, Roma, pp. 613–20 (Sec. 1.9).
Roorda, J. (1967) "Some Thoughts on the Southwell Plot," *J. Eng. Mech.* (ASCE), 93:37–48 (Sec. 1.5).
Rutenberg, A., Leviathan, I., and Decalo, M. (1988), "Stability of Shear-Wall Structures," *J. Eng. Mech.* (ASCE), 114(3):707–61.
Sheinman, I. (1989), "Cylindrical Buckling Load of Laminated Columns," *J. Eng. Mech.* (ASCE) 115(3):659–61.
Sheinman, I., and Adan, M. (1987), "The Effect of Shear Deformation on Post-Buckling Behavior of Laminated Beams," *J. Appl. Mech.* (ASME), 54:558.
Southwell, R. V. (1932), "On the Analysis of Experimental Observations in Problems of Elastic Stability," *Proc. R. Soc., Lond.* [Series A], 135:601–16 (Sec. 1.5).
Spencer, H. H., and Walker, A. C. (1975), "A Critique of Southwell Plots with Proposals for Alternative Methods," *Experimental Mechanics,* 15:303–10 (Sec. 1.5).
Taylor, N., and Hirst, P. (1989), "Southwell and Lundquist Plots for Struts Suffering Subbuckling Cyclic Excursions," *Experimental Mechanics,* 29(4):392–8 (Sec. 1.5).
Thompson, J. M. T., and Hunt, G. W. (1973) *A General Theory of Elastic Stability,* John Wiley & Sons, New York (Sec. 1.9).
Timoshenko, S. P. (1921), "On the Correction for Shear of the Differential Equation of Transverse Vibrations of Prismatic Bars," *Philos. Mag.* [Ser. 6], 21:747 (Sec. 1.7).
Timoshenko, S. P. (1953), *History of Strength of Materials,* McGraw-Hill, New York; also, Dover Publications, New York, 1983 (Sec. 1.0).

Timoshenko, S. P., and Gere, J. M. (1961), *Theory of Elastic Stability*. McGraw-Hill, New York (Sec. 1.5).

Timoshenko, S. P., and Goodier, J. N. (1973), *Theory of Elasticity*, 3rd ed. McGraw-Hill, New, York (Sec. 1.7).

Van den Broek, J. A. (1947), "Euler's Classic Paper 'On the Strength of Columns'," *J. Physics*, 15:309.

Vinson, J. R ., and Sierakowski, R. L. (1986), *The Behavior of Structures Composed of Composite Materials*. Martinus Nijhoff, Dordrecht, Netherlands.

Wang, C. Y. (1987a), "Approximate Formulas for Buckling of Heavy Columns with End Load," *J. Eng. Mech.* (ASCE), 113(10):2316–20.

Wang, C. Y. (1987b), "Buckling and Postbuckling of Heavy Columns," *J. Eng. Mech.* (ASCE), 113(8):1229–33; discussion and closure, 115(8):1840–42, 1989.

Wang, C. K., and Salmon, C. G. (1979), *Reinforced Concrete Design*, Intext Educational Publishers, New York (Sec. 1.6).

Yang, Y.-B., and Yau, J.-D. (1989), "Stability of Pretwisted Bars with Various End Torques," *J. Eng. Mech.* (ASCE) 115(4): 671–88.

Young, T. (1807), *A Course of Lectures on Natural Philosophy and the Mechanical Arts*, 2 vols., London (Sec. 1.5).

Ziegler, H. (1951), "Stabilitätsprobleme bei geraden Stäben und Wellen," *Z. Angew. Math. Phys. (ZAMP)*, 2:265 (Sec. 1.10).

Ziegler, H. (1968) *Principles of Structural Stability*. Blaisdell Publishing Co., Waltham, Mass. (Sec. 1.7).

2

Buckling of Elastic Frames by Equilibrium Analysis

Buckling of framed structures is no doubt the most important buckling problem for civil and structural engineers, and it is often encountered by mechanical and aerospace engineers as well. Obviously, buckling of frames involves buckling of their individual members, and so the results of our analysis in the preceding chapter must be applicable. In particular, buckling of a beam-column with arbitrary spring supports at both ends has some essential characteristics of the buckling of members of a frame. Not all the essential characteristics, though.

In a frame, the individual members interact and buckle simultaneously, the axial load of one member influencing not only the critical load of that member but also the critical loads of the adjacent members. This fact is often overlooked in practice and may sometimes lead to serious underestimations of the critical load of a frame. The problem is that a member of a frame does not quite behave as a column with spring supports at both ends, because the spring stiffnesses, which are equivalent to the elastic reaction of the rest of the frame on the member under consideration, are not constant but depend on the unknown critical load of the frame. In fact, the equivalent spring stiffnesses for the effect of the rest of the frame upon a given member are generally smaller than those calculated with the neglect of the dependence of these spring stiffnesses on the critical load.

In contrast to columns, the initial axial force P in the column of a frame before buckling usually is not known exactly. To circumvent this difficulty, the linearized second-order theory is used in an approximate form. From the first-order theory, in which the equilibrium conditions are formulated for the initial undeformed shape of the structure, one obtains by the usual methods of frame analysis the axial forces in the columns (recall that the distribution of floor loads into the columns depends on the bending of floor beams). These axial forces in the columns are then treated as known in a subsequent second-order analysis. By virtue of this fact, the incremental second-order formulation becomes linear. In reality, of course, the distribution of the floor loads into the column is not constant and varies as the column buckles; however, it may be considered as approximately constant if the buckling deflections are small, the error being higher-order small.

The analysis of buckling of frames begins with the determination of the stiffness or flexibility matrices of its members. Although the method of determination of these matrices is well known for arbitrary beam-columns of variable cross section, we will confine our attention mostly to members of constant cross section since their solution is easy and sufficient for instructional purposes. Subsequently we will apply the stiffness or flexibility matrices to solve various typical problems for frames and continuous beams appearing in practice, and we will also generalize our analysis to a form suitable for computer programming. Next we shall study large regular frames or lattices, which are used in tall buildings or planned for certain structures in space. Due to the large number of unknown joint rotations and displacements, we will emphasize simplified and approximate solutions, among which the use of finite difference calculus and of continuum approximations is particularly interesting.

Furthermore, we will also discuss how the frame behaves after the critical load is almost reached, and we will briefly point out the important aspects of redundancy in frames, which lend to many structures a large postcritical reserve and prevent a failure of the frame when only a limited number of the members has buckled. We will also briefly investigate built-up columns, in which the effect of shear can be very important, and similar shear effects in column-type approximations of tall regular frames. Finally, we will present the elements of calculation of the critical loads of arches, a problem which is important for bridge design and also has some analogies in buckling of cylindrical shells.

2.1 STIFFNESS AND FLEXIBILITY MATRICES OF BEAM-COLUMNS

Stiffness Matrix for End Rotations

To determine the critical loads of continuous beams or framed structures, one needs the stiffness or flexibility matrix of the beam-column. The stiffness coefficients of a linearly elastic structural element represent the forces produced by a unit displacement. To determine the stiffness coefficients, we consider a perfectly straight elastic beam which is fixed at end b, is loaded by axial compressive force P, and has length l (while under load P); see Figure 2.1.

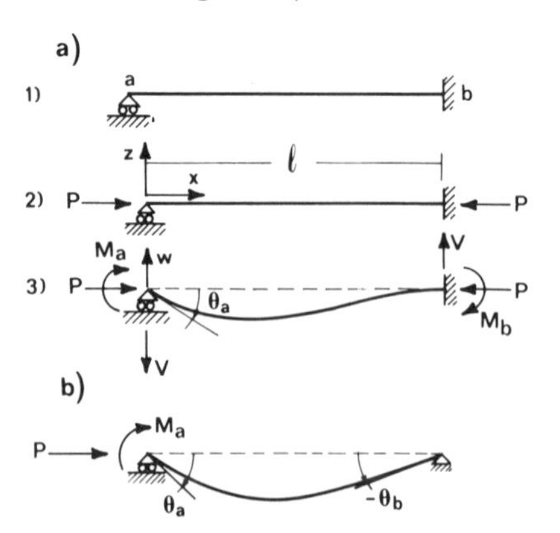

Figure 2.1 (a) Fixed–hinged beam (1), subjected to axial force P (2), and subsequently to end rotation (3); (b) hinged–hinged beam subjected to axial force and end rotation.

Subsequently, let a small rotation θ_a be imposed at end a. We want to calculate the end moment M_a that has to be applied in order to produce end rotation θ_a. We also need to calculate the corresponding reaction moment M_b that arises at the fixed end b.

In accordance with the customary sign convention in frame analysis, the end moments applied on the beam and the end rotations are considered positive when turning clockwise. This means that the moments acting upon a joint are positive when turning counterclockwise. As for the bending moments within the beam, we retain the previous sign convention according to which a positive bending moment is that which causes positive curvature. For the sake of simplicity, EI is considered as constant. The deflection curve of the beam is then given by Equation 1.3.6 (in which $w_p = 0$ because $p = 0$). The boundary conditions are $w = 0$ and $w' = -\theta_a$ at $x = 0$, and $w = 0$ and $w' = 0$ at $x = l$, which yields the conditions

$$\begin{aligned} B + D &= 0 \\ Ak + C &= -\theta_a \\ A \sin kl + B \cos kl + Cl + D &= 0 \\ Ak \cos kl - Bk \sin kl + C &= 0 \end{aligned} \tag{2.1.1}$$

For the sake of brevity we denote

$$\lambda = kl = \sqrt{\frac{P}{EI}}\, l = \pi \sqrt{\frac{P}{P_{\mathrm{E}}}} = \pi \frac{l}{l^*} \tag{2.1.2}$$

where l^* = half-wavelength of the deflection curve. Eliminating C and D from Equations 2.1.1 we obtain the equations

$$\begin{aligned} A(\sin \lambda - \lambda) + B(\cos \lambda - 1) &= \theta_a l \\ A(\cos \lambda - 1) - B\lambda \sin \lambda &= \theta_a l \end{aligned} \tag{2.1.3}$$

and by subtracting these two equations we get

$$A = B\left(\frac{1 - \cos \lambda - \lambda \sin \lambda}{\sin \lambda - \lambda \cos \lambda}\right) \tag{2.1.4}$$

Substituting this into the second of Equations 2.1.3, we obtain an expression for B. Now we may note that $M_a = M(0) = EIw''(0) = -EIk^2 B$. After some trigonometric manipulations we then obtain

$$M_a = K\theta_a \tag{2.1.5}$$

in which $K = sEI/l$ = stiffness coefficient, and

$$s = \frac{\lambda(\sin \lambda - \lambda \cos \lambda)}{2 - 2 \cos \lambda - \lambda \sin \lambda} \tag{2.1.6}$$

The bending moment at the opposite end of the beam may be obtained as $M_b = -M(l)$ (note the minus sign!). This yields $M_b = -EIw''(l) = EIk^2(A \sin \lambda + B \cos \lambda)$. Dividing this expression by $M_a = -EIk^2 B$ and utilizing Equation 2.1.4, we obtain after some further trigonometric transformations

$$c = \frac{M_b}{M_a} = \frac{\lambda - \sin\lambda}{\sin\lambda - \lambda\cos\lambda} \tag{2.1.7}$$

Parameter c is termed the carry-over factor, same as in the slope-deflection method.

Having solved the bending moments caused by rotating end a while end b is fixed, we automatically know the answer when end b is rotated while keeping end a fixed. Due to symmetry, the resulting moments are $M_b = K\theta_b$ and $M_a = cM_b$. The case when both ends are rotated may be obtained by superposition of the two cases in which only one end is rotated and the other end is kept fixed. Superposition applies because the boundary-value problem at hand is linear. Thus we find that $M_a = K\theta_a + Kc\theta_b$, $M_b = K\theta_b + Kc\theta_a$, which may be written in matrix form as

$$\begin{Bmatrix} M_a \\ M_b \end{Bmatrix} = \frac{EI}{l} \begin{bmatrix} s & sc \\ sc & s \end{bmatrix} \begin{Bmatrix} \theta_a \\ \theta_b \end{Bmatrix}. \tag{2.1.8}$$

The square matrix in this equation, together with the factor EI/l, is called the stiffness matrix of beam-columns. This matrix was introduced by James (1935) in a work dealing with the moment-distribution method, Goldberg (1954), and Livesley and Chandler (1956).

Considering a pin-ended column subjected to moment M_a at one end (Fig. 2.1b), we may use a similar procedure to show that the rotations at the ends are given by $\theta_a = (l/EI)\psi_S M_a$, $\theta_b = -(l/EI)\phi_S M_a$, in which

$$\psi_S = \frac{1}{\lambda}\left(\frac{1}{\lambda} - \cot\lambda\right) \qquad \phi_S = \frac{1}{\lambda}\left(\frac{1}{\sin\lambda} - \frac{1}{\lambda}\right) \tag{2.1.9}$$

When moments are applied at both ends of a simply supported column, symmetry and superposition arguments yield

$$\begin{Bmatrix} \theta_a \\ \theta_b \end{Bmatrix} = \frac{l}{EI} \begin{bmatrix} \psi_S & -\phi_S \\ -\phi_S & \psi_S \end{bmatrix} \begin{Bmatrix} M_a \\ M_b \end{Bmatrix} \tag{2.1.10}$$

Here the square matrix, together with the factor l/EI, represents the flexibility matrix of beam-columns. The inverse of this matrix is the stiffness matrix in Equation 2.1.8. The flexibility coefficients were introduced in stability analysis of frames by von Mises and Ratzersdorfer (1926); see also Timoshenko and Gere (1961), Chwalla (1928), and Bažant (1943). Because functions s, c, ψ_S, and ϕ_S serve as the basis of stability analysis of frames, they are also called stability functions.

Equations 2.1.6, 2.1.7, and 2.1.9 are limited to axial compression ($P > 0$). In calculations of buckling of frame structures it is sometimes necessary also to take into account the effect of axial tension on the stiffness of a beam ($P < 0$). In this case the general solution of the differential equation for $w(x)$ is given in terms of hyperbolic functions; see Equation 1.3.7. Using this expression, and proceeding similarly as before, one can show that for tension ($P < 0$, $\lambda = \pi\sqrt{(-P)/P_E}$) the parameters s and c in Equation 2.1.8 and ψ_S and ϕ_S in Equation 2.1.9 have the form

$$s = \frac{\lambda(\lambda \cosh \lambda - \sinh \lambda)}{2 - 2\cosh \lambda + \lambda \sinh \lambda} \qquad c = \frac{\sinh \lambda - \lambda}{\lambda \cosh \lambda - \sinh \lambda} \tag{2.1.11a}$$

$$\psi_S = \frac{1}{\lambda}\left(\frac{1}{\tanh \lambda} - \frac{1}{\lambda}\right) \qquad \phi_S = \frac{1}{\lambda}\left(\frac{1}{\lambda} - \frac{1}{\sinh \lambda}\right) \tag{2.1.11b}$$

If there is no axial load, $P = 0$, the deflection curve is given neither by trigonometric nor by hyperbolic functions, but is a cubic polynomial. The values of c and s must be in this case $\frac{1}{2}$ and 4, respectively, as is known from the slope-deflection method. Since s and c must be continuous functions of λ, one may check that correct limits of s and c are obtained as $\lambda \to 0$. In determining these limits, one repeatedly obtains expressions of the type $\frac{0}{0}$ and one needs to use L'Hospital's rule several times to verify that the correct limits are indeed obtained.

Calculating the derivatives of s and sc (Eqs. 2.1.6 and 2.1.7) at $\lambda = 0$, one may obtain the following Taylor series expansion (e.g., Dean and Ugarte, 1968):

$$\begin{aligned} s &= 4 - \frac{2\pi^2}{15}\left(\frac{P}{P_E}\right) - \frac{11\pi^4}{6300}\left(\frac{P}{P_E}\right)^2 - \cdots \\ sc &= 2 + \frac{\pi^2}{30}\left(\frac{P}{P_E}\right) + \frac{13\pi^4}{12{,}600}\left(\frac{P}{P_E}\right)^2 + \cdots \end{aligned} \tag{2.1.12}$$

which ensue by introducing $\lambda = \pi\sqrt{P/P_E}$ (Eq. 2.1.2). The same Taylor series expansion about $\lambda = 0$ is obtained from Equations 2.1.11a for $P < 0$. The series expansions in Equations 2.1.12 need to be used for very small values of P/P_E, in order to avoid problems of numerical accuracy. The exact expressions are then ratios of very small numbers.

The diagrams of parameters s, c, and sc are shown in Figure 2.2a as functions

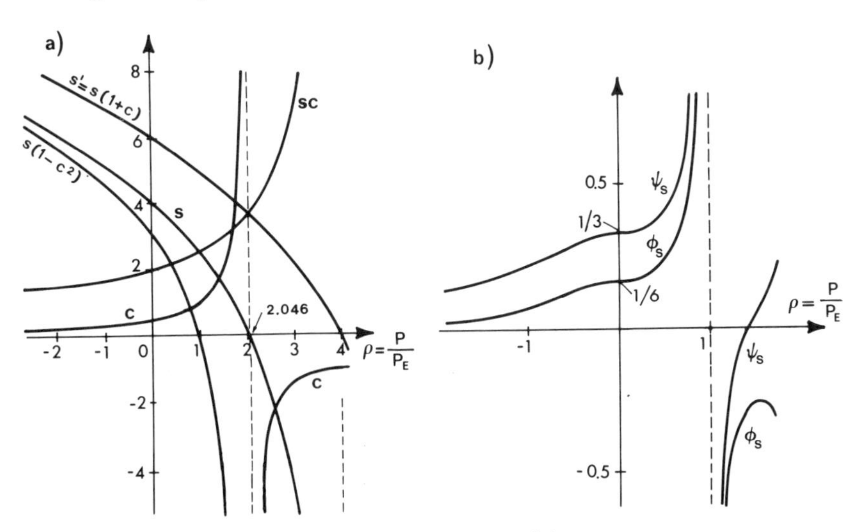

Figure 2.2 Variation of (a) stiffness coefficients and (b) flexibility coefficients with axial force P.

of $\rho = P/P_E$. Coefficients s, c, and sc are also tabulated in Horne and Merchant (1965). The basic property that should be emphasized is that the bending stiffnesses of a beam-column depend on its axial load. With increasing compression, s decreases, while the carry-over factor c increases. For $\rho = 2.046$, we have $s = 0$ and $c \to \infty$, with sc finite. The zero value of s indicates that a zero moment applied to the hinged end of a hinged–fixed beam can produce a finite rotation of the hinge, that is, an infinite flexibility. The value $P = 2.046P_E$ was already recognized (see Eq. 1.4.3) to be the first critical load for a fixed–hinged beam.

The diagrams of parameters ψ_S and ϕ_S are shown in Figure 2.2b as functions of $\rho = P/P_E$. For $P = P_E$ (critical load for a hinged–hinged beam) ψ_S and ϕ_S tend to infinity, thus indicating zero stiffness.

The stability functions for beam-columns with shear deformations (Sec. 1.7) have been derived by Absi (1967) and used by Bažant and Christensen (1972b). They are given by

$$s = \frac{\lambda \sin \lambda - \lambda^2 \beta \cos \lambda}{2 - 2\cos \lambda - \lambda\beta \sin \lambda} \qquad c = \frac{\lambda\beta - \sin \lambda}{\sin \lambda - \lambda\beta \cos \lambda} \qquad \text{with } \beta = 1 - \frac{P}{GA_0} \tag{2.1.13}$$

Stiffness Matrix for End Rotations and Relative Lateral Displacement

For general frame analysis we need the stiffness matrix of a beam-column whose ends are subjected not only to rotation but also to lateral translation. Consider a beam which is initially parallel to the axis x and is subjected to axial load P under which it has length l (Fig. 2.3). Subsequently, small lateral displacements w_a and w_b along with small rotations θ_a and θ_b are imposed at the ends of the beam (Fig. 2.3). To accomplish the transformation from the initial position to the final position, the beam may be first deformed by imposing at its ends the rotations

$$\phi_a = \theta_a + \frac{\Delta}{l} \qquad \phi_b = \theta_b + \frac{\Delta}{l} \tag{2.1.14}$$

while keeping the positions of the ends fixed, and then a small rigid body rotation of the whole beam by angle $-\Delta/l$ (together with a translation) may be carried out to obtain the final position. Here, $\Delta = w_b - w_a$ = small relative lateral displacement of beam ends. The rigid body translation and rotation do not affect the

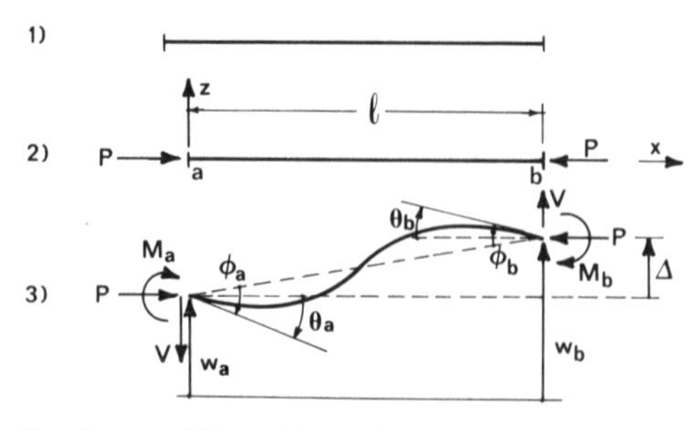

Figure 2.3 Member of a frame (1), subjected to axial force P (2), and subsequently to general end displacements (3).

internal forces in the beam, and so according to Equation 2.1.8,

$$M_a = \frac{EI}{l}(s\phi_a + sc\phi_b) = \frac{EI}{l}\left[s\theta_a + sc\theta_b + \frac{s(1+c)}{l}\Delta\right]$$
$$M_b = \frac{EI}{l}(sc\phi_a + s\phi_b) = \frac{EI}{l}\left[sc\theta_a + s\theta_b + \frac{s(1+c)}{l}\Delta\right] \qquad (2.1.15)$$

In these relations, there are three independent kinematic variables θ_a, θ_b, and Δ, but only two force variables, M_a and M_b. Thus, we need to introduce one additional internal force that is associated with Δ. This additional internal force is properly the shear force V since it is the shear force that does work on Δ. Writing the moment equilibrium condition of the whole beam in the deformed state (Fig. 2.3), we have $M_a + M_b - Vl - P\Delta = 0$, and substituting for M_a and M_b from Equations 2.1.14, we find

$$V = \frac{M_a + M_b}{l} - P\frac{\Delta}{l} = \frac{EI}{l}\left\{\frac{s(1+c)}{l}\theta_a + \frac{s(1+c)}{l}\theta_b + \left[\frac{2s(1+c)}{l^2} - \frac{\pi^2}{l^2}\left(\frac{P}{P_E}\right)\right]\Delta\right\} \qquad (2.1.16)$$

So far we have considered loads applied at the ends of the beam. If a lateral load is applied between the ends of the beam, we may utilize the principle of superposition. First we consider the ends of the beam to be rotated and displaced, and subsequently, holding the ends fixed, we apply the lateral load. This load produces fixed-end moments M_a^L and M_b^L and fixed-end shear force V^L. Their values can be calculated from the general solution (Eq. 1.3.6) by imposing the boundary conditions $w = w' = 0$ at both ends. The line of reasoning is the same as in the slope-deflection method; however, M_a^L, M_b^L, and V^L depend on the ratio P/P_E. A table of fixed-end moments and forces is given by Horne and Merchant (1965).

Combining Equations 2.1.15 and 2.1.16, we acquire the following matrix equilibrium equation of a beam-column:

$$\begin{Bmatrix} M_a \\ M_b \\ V \end{Bmatrix} = \frac{EI}{l}\begin{bmatrix} s & sc & \frac{\bar{s}}{l} \\ sc & s & \frac{\bar{s}}{l} \\ \frac{\bar{s}}{l} & \frac{\bar{s}}{l} & \frac{s^*}{l^2} \end{bmatrix}\begin{Bmatrix} \theta_a \\ \theta_b \\ \Delta \end{Bmatrix} + \begin{Bmatrix} M_a^L \\ M_b^L \\ V^L \end{Bmatrix} \qquad (2.1.17)$$

in which we introduce the notations

$$\bar{s} = s(1+c) \qquad s^* = 2\bar{s} - \pi^2\frac{P}{P_E} \qquad (2.1.18)$$

The square matrix (with the factor EI/l) represents the general stiffness matrix for small deformations of a beam-column. Note that this matrix is symmetric. Symmetry is a required property since the material behavior is assumed to be elastic. Note also that θ_a, θ_b, and Δ are here considered incremental displacements starting from the initial state of equilibrium under axial force P and possible lateral loads.

A broad spectre of stiffness matrices of beam-columns is presented in Tuma's torsion (simple torsion, Chap. 6), the planar and spatial stiffness matrices of a beam-column encased in an elastic foundation resisting deflection as well as twisting rotation (cf. Sec. 5.2), the load terms of the matrix stiffness equation of equilibrium for various typical cases of transverse loads, the modified stiffness matrices and load terms for various end supports (both planar and spatial), the critical load for many typical variable-section columns (tapered symmetrically or asymmetrically, or stepped), and various elastically restrained columns. Moreover, this monumental handbook presents the transfer matrices (or transport matrices) that relate the column matrix of forces and displacements for one cross section to that for another cross section.

A more detailed discussion of the stability functions and their use in frame stability analysis is given by Horne and Merchant (1965). The stability functions s and c can be generalized to dynamics, to include the effect of the axial force on vibration frequencies. A unified approach for matrix analysis of buckling and vibration problems is given by Williams (1981), and Williams and Wittrick (1983); also Livesley and Chandler (1956).

Note that the theory presented so far is limited to negligible axial extensions. Normally this is acceptable, but not for very tall frames (Sec. 2.9).

Problems

2.1.1 Derive the expressions for s (Eq. 2.1.6), c (Eq. 2.1.7), ψ_S, and ϕ_S (Eqs. 2.1.9).

2.1.2 Invert the relation of Equation 2.1.10 and show that it becomes equivalent to Equation 2.1.8.

2.1.3 Derive the expressions of parameters s, c, ψ_S, ϕ_S for the case that the beam is under axial tension.

2.1.4 Find the expressions of parameters s, c, ψ_S, ϕ_S, taking into account the effect of shear deformation (Eqs. 2.1.3).

2.1.5 For a fixed-end beam under axial compression, find the end moments and shear force due to a constant lateral load.

2.1.6 Using Equation 1.7.4, try to derive Absi's stability functions for beam-columns with shear.

2.1.7 Calculate M_a, M_b due to $\theta_a = 1$ for a free-standing column (Fig. 2.4a). Plot the associated stiffness as a function of P/P_E.

2.1.8 Calculate lateral displacement and rotation at the end due to $V = 1$ (Fig. 2.4b).

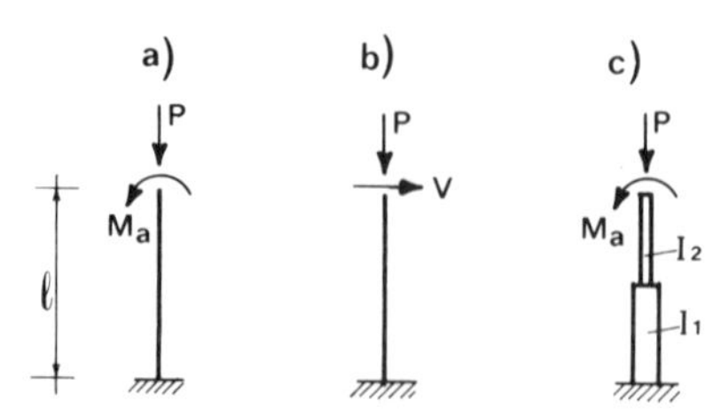

Figure 2.4 Exercise problems on stiffness and flexibility coefficients of free-standing columns.

2.1.9 Calculate the stiffness for rotation on top of a free-standing column (Fig. 2.4a) in terms of s and c functions. Then compare to the result of Problem 2.1.7.

2.1.10 Same as above, but for the stepwise variable I (I_1 and I_2) (Fig. 2.4c).

2.1.11 The so-called transfer (or transport) matrix (Pestel and Leckie, 1963) relates the column matrix (N, V, M, u, w, θ) at one cross section to that at another cross section (N = axial force, positive if tensile). (a) Derive it by rearrangement of the stiffness matrix equation of equilibrium, that is, of $\mathbf{Kq} = \mathbf{f}$; (b) also derive some of the transfer matrix elements by integration of the beam-column differential equation (check: Tuma, 1988).

2.1.12 Use the transfer matrix obtained above to solve the critical load of a simply supported continuous beam of two equal spans.

2.1.13 Derive the modified stiffness matrix relating all the end forces and displacements of columns with nonzero F, V, M at one end and at the other end (a) $M = 0$ (hinge), (b) $M = V = 0$ (free end), (c) $V = 0$ (sliding restraint); see Tuma (1988, p. 220).

2.1.14 Integrating the beam-column differential equation, derive the load term in the matrix stiffness equation of equilibrium for (a) a distributed uniform load, (b) a uniform load over a half-span, (c) two concentrated loads at the third points, (d) a moment applied at the quarter-span (cf. Tuma 1988).

2.2 CRITICAL LOADS OF FRAMES AND CONTINUOUS BEAMS

We are now ready to demonstrate calculations of the critical loads of continuous beams and simple frames.

Simple Structures

Example 1. Consider a continuous beam of two spans, shown in Figure 2.5a, under the axial load μF, where F is a reference (design) load and μ is a safety factor (load multiplier) for which we want to calculate the critical value μ_{cr}. Consider that, in general, small moments m_2 and m_3 may be applied at joints 2 and 3.

There are two unknown generalized displacements, rotations θ_2 and θ_3 above the supports. The associated conditions of equilibrium of the joints at these supports are $M_{21} + M_{23} = m_2$ and $M_{32} = m_3$, where M_{ij} is now used to denote the end moment in beam ij at end i. Substituting from Equation 2.1.8, we obtain

$$
\begin{aligned}
M_{21} + M_{23} &= \frac{EI_{12}}{l_{12}}(s_{12}\theta_2) + \frac{EI_{23}}{l_{23}}(s_{23}\theta_2 + s_{23}c_{23}\theta_3) = m_2 \\
M_{32} &= \frac{EI_{23}}{l_{23}}(s_{23}c_{23}\theta_2 + s_{23}\theta_3) = m_3
\end{aligned}
\tag{2.2.1}
$$

in which the subscripts 21 and 23 refer to beams 21 and 23.

Let us consider first the case $m_2 = m_3 = 0$. We have a system of two algebraic linear homogeneous equations for the unknown rotations θ_2 and θ_3 above

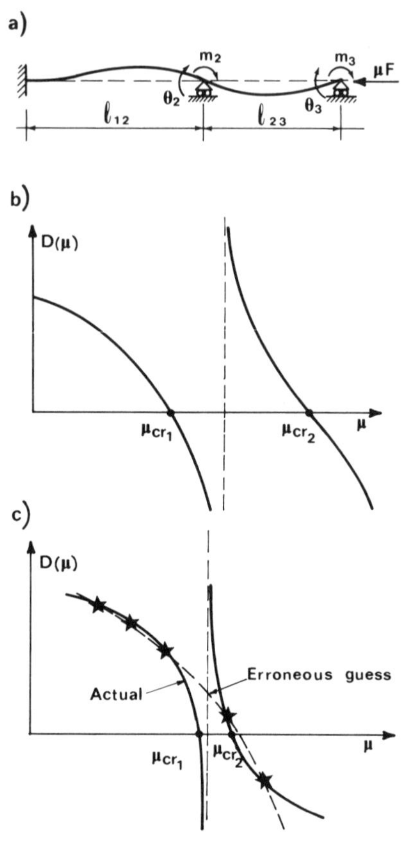

Figure 2.5 (a) Continuous beam; (b) plot of determinant of stiffness matrix as function of load multiplier; (c) possible source of error in the determination of the critical value of the load multiplier.

supports 2 and 3. Nonzero solutions exist if and only if

$$D(\mu) = \det \mathbf{K}(\mu) = \left[s_{12}\left(\frac{l_{23}}{l_{12}}\right) + s_{23}\right]s_{23} - (s_{23}c_{23})^2 = 0 \qquad (2.2.2)$$

where $\mathbf{K}(\mu)$ = stiffness matrix = matrix of the coefficients at θ_1 and θ_2 in these two equations. The determinant is a function of μ since s_{12} is a function of $\rho_{12} = \mu F/P_{E_{12}}$ and s_{23} and c_{23} are functions of $\rho_{23} = \mu F/P_{E_{23}}$.

To obtain the smallest value of μ which satisfies Equation 2.2.2, one may construct the plot of $D(\mu)$ versus μ to find the approximate location of the roots (Fig. 2.5b). Then an accurate value of μ may be obtained by the Newton iterative method. Assuming a uniform moment of inertia, $I_{12} = I_{23}$, and $l_{12} = 1.3l_{23}$, one finds $\mu_{cr}F = 1.088P_{E_{23}}$.

If disturbing moments m_2 and m_3 are present, Equation 2.2.1 is a system of nonhomogeneous linear equations for unknowns θ_2 and θ_3. According to Kramer's rule, its solutions are $\theta_i = D_i/D(\mu)$ $(i = 2, 3)$ where $D(\mu)$ is the determinant of the left-hand-side matrix, same as before, and D_i is the determinant of the matrix obtained when the right-hand side replaces the column corresponding to θ_i. The critical state is obtained when $\theta_i \to \infty$, and this occurs when $D(\mu) = 0$. So the critical state condition is the same as before.

A warning of a possible pitfall may be in order. The plot of the critical state determinant $D(\mu)$ versus μ often has vertical asymptotes, as illustrated in Figure 2.5b. If the subdivisions of μ at which $D(\mu)$ is evaluated are too coarse, one may succeed to pass through the points a smooth curve such as the dashed curve in Figure 2.5c, and thus miss the root lying to the left of the vertical asymptote, on the correct solid curve (Fig. 2.5c). But this could fool the analyst only if the roots μ_{cr_1} and μ_{cr_2} are very close.

Example 2. Consider now a two-span continuous beam on simple supports (Fig. 2.6a). For this structure, the flexibility method is more efficient. While with the displacement method there are three unknown joint rotations θ_1, θ_2, θ_3 (Fig. 2.6a), with the flexibility method there is only one unknown, the moment M_2 at the intermediate support (Fig. 2.6b). The primary structure is obtained by inserting a hinge at this support. The condition of compatibility of the rotations at this support reads $M_2(\psi_{S_{12}} l_{12}/EI_{12} + \psi_{S_{23}} l_{23}/EI_{23}) = 0$, in which the subscripts 12 and 23 refer to beams 12 and 23 and ψ_S is the flexibility coefficient given by Equations 2.1.9. A nonzero value of M_2 is possible if and only if the coefficient of M_2 is zero, that is,

$$D(\mu) = \left[\psi_{S_{12}} \left(\frac{l_{12}}{EI_{12}} \right) + \psi_{S_{23}} \left(\frac{l_{23}}{EI_{23}} \right) \right] = 0 \tag{2.2.3}$$

Note that in the case of a single unknown, $D(\mu)$ is the determinant of a 1×1 flexibility matrix. For $l_{23} = 1.3 l_{12}$ and $EI_{12} = EI_{23}$, Equation 2.2.3 indicates the lowest critical value of the load multiplier to be $\mu_{cr} F = 1.24 P_{E_{23}}$. The plot of $D(\mu)$ versus μ is shown in Figure 2.6f. We see that the curve representing $D(\mu)$ jumps from ∞ to $-\infty$ before the critical load is reached. This is due to the fact that the flexibility coefficient $\psi_{S_{23}}$ of the longer beam of the primary structure jumps from ∞ to $-\infty$.

This example at the same time illustrates one limitation of the stiffness and flexibility methods. If the spans are equal, that is, $l_{12} = l_{23} = l$ and $P_{E_{12}} = P_{E_{23}} = P_E$,

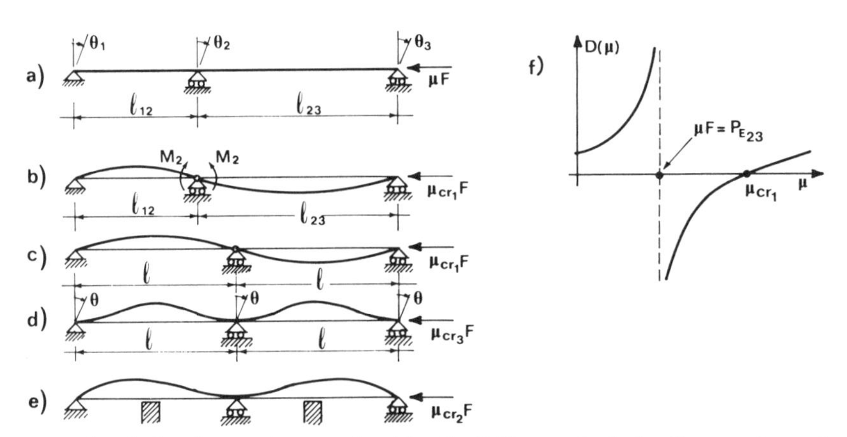

Figure 2.6 (a, b) Continuous beam solved by flexibility method; (c, d) case of equal spans; (e) presence of one-sided constraints; (f) plot of the determinant of the flexibility matrix.

then $\mu F = P_E$ is obviously the lowest critical load, $\mu_{cr_1}F$. In this case, however, the ends of the beams meeting at joint 2 can rotate freely (Fig. 2.6c), and so compatibility is obtained for $M_2 = 0$. But the critical load $\mu_{cr_1}F = P_E$ cannot be detected by the flexibility method since the determinant of the equation system (only one equation in this case) does not have to vanish when all unknown moments are zero.

The stiffness method has an analogous limitation. It cannot detect the critical load $\mu_{cr_3}F = 4P_E$ corresponding to a fixed-end beam (Fig. 2.6d). Indeed, in this case the end rotations and displacements are zero, and so the associated stiffness matrix does not have to vanish to satisfy the equilibrium equations.

Generally, in passing from a direct solution of the differential equations to a discretized formulation, we can find only the eigenstates that correspond to a nonzero solution of the discretized problem. So it is necessary to check the possibility that the first critical load might correspond to a homogeneous (zero) solution according to the flexibility or stiffness method. In this regard, though, the use of the displacement method is safer since the shape of such eigenstates is more easily detected with the stiffness method.

The presence of one-sided constraints on deformation may change the buckling load. For example, for the case of Figure 2.6e, the lowest critical load is $\mu_{cr_2}F = 2.046P_E$, which corresponds to a fixed–hinged beam.

Example 3. Next consider an unbraced portal frame as shown in Figure 2.7a. The frame is loaded by vertical loads μF and also by a small lateral disturbing load f. Let the bending rigidity EI be the same for all members, and the span l be equal to the height. There are, in general, three unknown generalized displacements, Δ, θ_2, and θ_3. Due to symmetry of the frame, one may expect that buckling would occur either in a symmetric mode or in an antisymmetric mode. The general buckling mode can be obtained as a superposition of these two.

Consider first the antisymmetric buckling mode, $\theta_2 = \theta_3 = \theta$. As known from the slope-deflection method for frames, the conditions of equilibrium may be written for moments acting on the corner joint, in which case the same equation ensues for joints 2 and 3, and for the horizontal forces acting on the horizontal

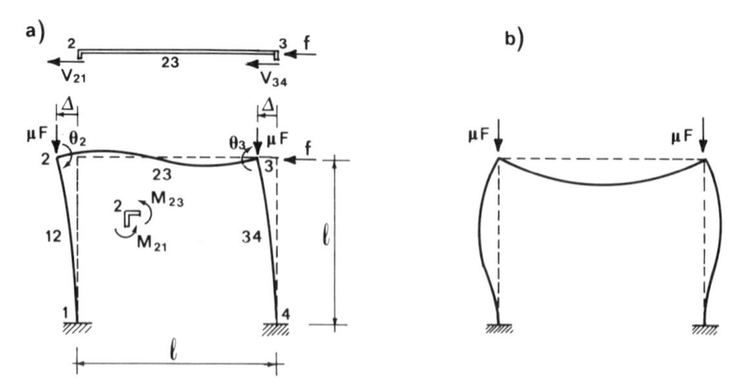

Figure 2.7 Portal frame: (a) antisymmetric buckling mode (unbraced frame), (b) symmetric buckling mode (braced frame).

beam separated from the columns. These two conditions are:

$$M_{23} + M_{21} = \frac{EI}{l}\left(s_{12}\theta + s_{23}\theta + s_{23}c_{23}\theta + \frac{1}{l}\bar{s}_{12}\Delta\right) = 0$$
$$V_{21} + V_{34} = \frac{EI}{l}\left[\frac{\bar{s}_{12}}{l}\theta + \frac{\bar{s}_{34}}{l}\theta + \frac{1}{l^2}(s^*_{12} + s^*_{34})\Delta\right] = f \qquad (2.2.4)$$

Due to the consideration of the lateral disturbing force f, this system of two linear algebraic equations is nonhomogeneous. The stiffness coefficients that appear in Equations 2.2.4 are functions of the axial forces in beams 12, 23, and 34. For finite deformations, these forces would, in general, depend not only on the externally applied load μF, but also on f and on the moments at the joints, which in turn are functions of Δ, θ_1, and θ_2. We consider, however, only very small (infinitesimal) deviations from the initial state at which $f = \Delta = \theta_1 = \theta_2 = 0$. Consequently, the stiffness coefficients are functions of μF alone, and are independent of the additional axial forces produced by f, Δ, θ_1, and θ_2, which are all infinitesimal. (The shear forces V_{21}, V_{34}, and the moments M_{23} and M_{21} are also infinitesimal.) Thus, for a given load multiplier μ, all the coefficients of these equations can be evaluated, and the frame rotations and displacements can be solved, unless the determinant of the equation is 0. If this happens, the solution tends to infinity (just as it does for a column under lateral load as the axial load approaches the critical value). Thus, the vanishing of the determinant is the condition of critical load.

If there is no lateral disturbing load f, the reasoning is different but the conclusion is the same. In this case, the equation system is homogeneous ($f = 0$), and we seek the value of μ for which a nonzero solution exists: again, this happens when the determinant vanishes.

Evaluating the determinant of Equations 2.2.4, and noting that $\bar{s}_{23} = 4(1 + \frac{1}{2}) = 6$ (since the initial axial force in the horizontal beam is zero), we find $D(\mu) = (s_{12} + 6)l^2 s^*_{12} - (l\bar{s}_{12})^2 = 0$. Choosing various values for the load multiplier μ and applying the iterative Newton's method, we find that the smallest critical load is $\mu_{\mathrm{cr}_1} F = 0.744 P_{\mathrm{E}}$, where P_{E} is the Euler load of length l. Note that for $\mu = \mu_{\mathrm{cr}}$ none of the stiffness coefficients becomes zero, which means that the critical load of the structure does not coincide with the critical load of any single member.

The symmetric buckling mode ($\theta_2 = -\theta_3 = \theta$, $\Delta = 0$) occurs only if $f = 0$. In this case one gets from moment equilibrium of the corner joint a single equation: $(EI/l)(s_{12} + s_{23} - s_{23}c_{23})\theta = 0$, which is satisfied (at $\theta \neq 0$) for $\mu_{\mathrm{cr}} F = 2.55 P_{\mathrm{E}}$. The buckling mode is shown in Figure 2.7b. The corresponding critical load governs only if the frame is braced against lateral sway. Otherwise the nonsymmetric mode occurs, since its critical load is smaller.

We have now seen an example that a symmetric and symmetrically loaded structure can buckle nonsymmetrically. This contrasts with first-order theory (linear elasticity), for which the symmetry of structure and loads always implies symmetry of deformation.

Figure 2.8 shows a teaching model of a symmetric portal frame developed at Northwestern University (1969). The model demonstrates the buckling modes

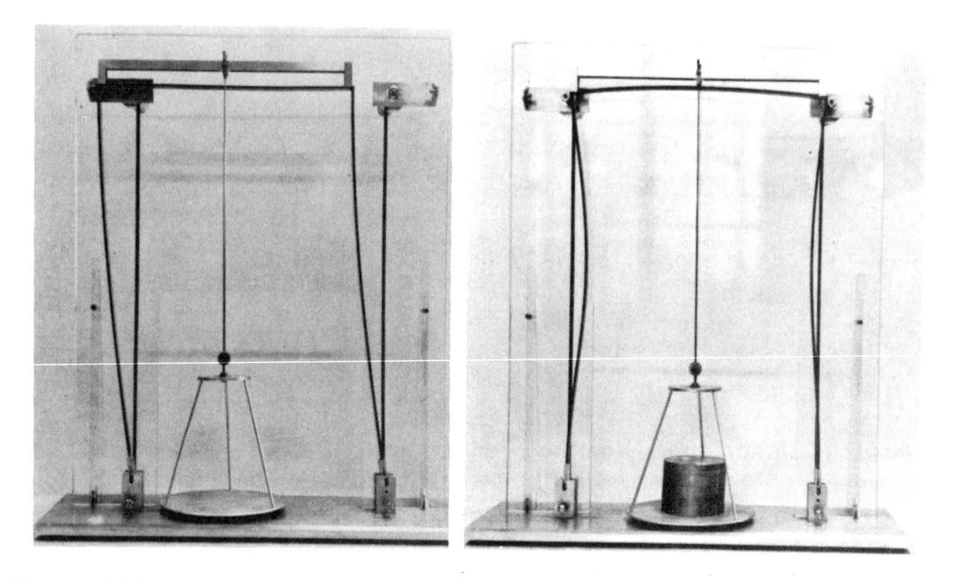

Figure 2.8 Northwestern University (1969) teaching models: sway buckling and symmetric buckling of portal frame with hinged based or fixed base.

under equal vertical loads centered on the columns. The sway (asymmetric) buckling is shown for the case of a hinged base. The symmetric buckling is obtained, for the case of a fixed base, by preventing translation of the horizontal beam.

When there are several independent loads, such as F_1 and F_2 in Figure 2.9a, one normally considers them fixed and seeks the common multiplier μ that produces buckling. In some design problems, though, the ratio of these loads can be arbitrary, and then it is of interest to solve the critical load value of multiplier μ for various ratios F_1/F_2. This then allows constructing the surface of critical states, as illustrated in Figure 2.9b.

A remark on the concept of "perfect." A frame may be defined to be perfect if it is described by a homogeneous system of equations, that is, yields an eigenvalue problem. Otherwise the frame is imperfect.

Example 4. Consider the triangular frame shown in Figure 2.9c. Let the bending rigidity EI and the length l be the same for all members and assume that the members are very slender so that their axial stiffnesses are much higher than their bending rigidities. If this is so, then the shear forces V_{21}, V_{23}, . . . (Fig. 2.9d) in all members are negligible in comparison with the axial forces, and the condition of equilibrium at the corner joint is determinate, yielding the following axial forces (Fig. 2.9e): $P_{23} = \mu F \tan 30° = \mu F/\sqrt{3}$ (compression), $P_{21} = -2\mu F/\sqrt{3}$ (tension). The neglect of V_{21} and V_{23} is only approximate if the frame is imperfect, but it is correct in the limit of very small imperfections or very slender bars. For not too slender members, of course, the diagram with shear forces (Fig. 2.9d) needs to be used. The fact that the inclined member is under tension is important, since it significantly increases the stiffness of this member (Eq. 2.1.11 must be used). For a symmetric buckling mode, for which the top joint does not rotate, the

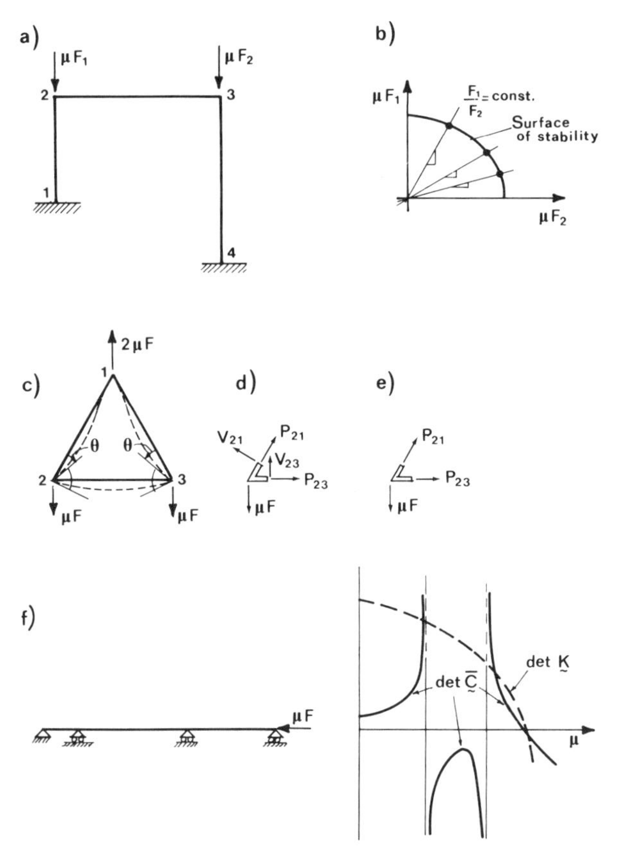

Figure 2.9 (a) Unsymmetric portal frame subjected to independent loads F_1 and F_2; (b) surface of critical states; (c, d, e) triangular frame with members 12 and 13 under tension; (f) variation of determinant of stiffness and flexibility matrix for continuous beam.

conditions of equilibrium of the moments acting at the top joint and at either one of the lower joints yield

$$M_{23} + M_{21} = \frac{EI}{l}[s_{23}\theta + c_{23}s_{23}(-\theta)] + \frac{EI}{l}(s_{21}\theta) = 0$$

or

$$f(\mu) = s_{23}(1 - c_{23}) + s_{21} = 0 \tag{2.2.5}$$

in which again s_{23} and c_{23} are functions of $\rho_{23} = \mu F/\sqrt{3}\,P_E$ and s_{21} is a function of $\rho_{21} = -2\mu F/\sqrt{3}\,P_E$. Using the iterative Newton's method, we obtain $\mu_{cr_1}F = 4.76P_E$.

Difficulties with the Flexibility Method

As illustrated by Example 1, the elements k_{ij} of the stiffness matrix $\mathbf{K}$ can never be infinite and its determinant gradually decreases as the load multiplier is increased. On the contrary, Example 2 has shown us that elements of the flexibility matrix of the primary structure $\bar{\mathbf{C}}$ typically become infinite before the load multiplier becomes critical. The variation of $\det \bar{\mathbf{C}}$ may even be rather complex; see, for example, Figure 2.9f for a three-span beam with one span considerably shorter than the other two. These examples reveal that the flexibility matrix $\bar{\mathbf{C}}$ need not be positive definite for stable states (i.e., below P_{cr_1}). Consequently various iteration methods cannot be based on flexibility, because they would not be guaranteed to converge.

There are also other reasons which favor the stiffness method. One is that the equations of this method can be set up more directly without searching first for a statically determinate primary structure that is required by the flexibility method. The other is that the second-order effects are functions of the displacements and, consequently, more readily accounted for if the displacements are chosen as unknowns. For all these reasons, computer analysis is usually done on the basis of the stiffness method.

General Approach for Computer Analysis

For computer analysis, it is more convenient to set up the equilibrium equations in terms of independent lateral end displacements w_a and w_b rather than their difference Δ, and to introduce separate quantities for the shear forces acting at either end. In some cases, such as very tall building frames, it is also necessary to consider axial shortening of the members, due to increments of the axial force P, but even if the problem does not require it, programming is easier if the axial displacements at member ends are treated as unequal. Thus we consider, for a member, six independent displacement parameters $u_1, u_2, \ldots, u_6$, forming a column matrix $\mathbf{u}$; the associated forces acting upon the members are $F_1, \ldots, F_6$, forming column matrix $\mathbf{F}$; see Figure 2.10, which shows the usual matrix analysis convention for positive signs. Now substitute $\Delta = u_5 - u_2$, $\theta_a = u_3$, and $\theta_b = u_6$ into Equation 2.1.17; express $F_5 = V$, $F_2 = -V$, $F_3 = M_a$, and $F_6 = M_b$; and further substitute $F_4 = (u_4 - u_1)EA/l = (u_4 - u_1)EI/r^2 l$; $F_1 = -(u_4 - u_1)EA/l = -(u_4 - u_1)EI/r^2 l$. This yields, in matrix notation,

$$\mathbf{F} = \mathbf{K}\mathbf{u} + \mathbf{F}^L \tag{2.2.6}$$

Column matrix $\mathbf{F}^L$ consists of components F_i^L that represent the fixed-end forces produced by the loads on the member and depend on the ratio P/P_E. Matrix $\mathbf{K}$ is a symmetric (6×6) stiffness matrix of the member:

$$\mathbf{K} = \frac{EI}{l} \begin{bmatrix} r^{-2} & 0 & 0 & -r^{-2} & 0 & 0 \\ & s^* l^{-2} & -\bar{s} l^{-1} & 0 & -s^* l^{-2} & -\bar{s} l^{-1} \\ & & s & 0 & \bar{s} l^{-1} & sc \\ & \textit{sym.} & & r^{-2} & 0 & 0 \\ & & & & s^* l^{-2} & \bar{s} l^{-1} \\ & & & & & s \end{bmatrix} \tag{2.2.7}$$

To explain the meaning of the stiffness coefficients in this matrix, note that coefficient K_{23} that represents the V_a value when $\theta_a = 1$ is the only nonzero displacement, and, according to Equation 2.1.17, this must be equal to $-\bar{s}EI/l^2$ (the negative sign is due to the difference in the sign convention for the shear force at end a). The term K_{32} that represents the M_a value when $w_a = 1$ is the only

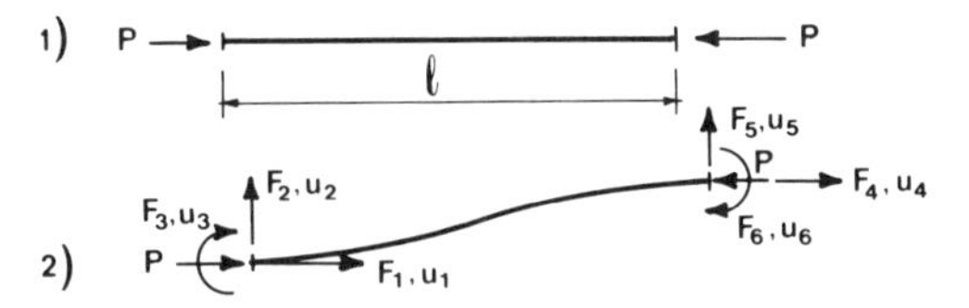

Figure 2.10 Member of a frame in the initial state (1), and after a general infinitesimal displacement at the ends (2).

nonzero displacement, and, according to Equation 2.1.17, this must be equal to $-\bar{s}EI/l^2$ because $\Delta = -w_a$.

Note that the 6×6 stiffness matrix **K** of the beam-column is singular and cannot be inverted. A corresponding flexibility matrix does not exist. The reason is that the end displacements $u_1, \ldots, u_6$ include rigid body translations and rotation (which cannot be determined from the forces because the forces do no work on rigid body translations and rotations). On the other hand, the 3×3 stiffness matrix of beam-columns is not singular and can be inverted, and so the corresponding 3×3 flexibility matrix exists (in this case the displacements are relative and do not include any rigid body translations or rotation). The singularity of the 6×6 stiffness matrix nevertheless causes no singularity of the assembled structural stiffness matrix that takes into account the boundary conditions.

The initial axial forces P in frame members that are to be used in critical state buckling analysis, particularly the distribution of floor loads, have to be calculated according to the first-order theory. Note that the use of the 6×6 full stiffness matrix of a column (Eq. 2.2.6) makes it possible to calculate changes N of the axial loads that occur during buckling. It must be emphasized that the theory is valid only for very small (infinitesimal) values of N compared to the initial axial forces $N^0 = -P$, that is, for $|N| \ll P$. If we wanted to calculate large changes of the axial load, we would need to use nonlinear analysis with incremental loading.

Matrix **K** of Equation 2.2.7 also can be used for incremental loading analysis of load-deflection behavior of the structure. In this case the increments of member forces are added to the initial values, and the coordinates of member ends are updated after each loading step. For better accuracy, each loading step may be iterated, determining the stiffness coefficients in the second and further iterations from the midstep forces for the preceding iteration.

Problems

2.2.1 Same as Example 1, but $l_{12} = 1.4 l_{23}$.
2.2.2 Same as Example 2, but $l_{23} = 1.5 l_{12}$.
2.2.3 Same as Example 3, but span $l = 1.2$ times height h.
2.2.4 Same as Example 4, but change the angle at the top corner from 60° to 90°.
2.2.5 For the continuous beam in Figure 2.5a, determine μ_{cr_1} from Equation 2.2.2 (equilibrium equations). Assume that $I_{12} = I_{23}$, $l_{12} = 1.3 l_{23}$.
2.2.6 For the continuous beam in Figure 2.5a, determine μ_{cr_1} using flexibility coefficients (compatibility equations). Assume $I_{12} = I_{23}$, $l_{12} = 1.3 l_{23}$.

2.2.7 For the frame in Figure 2.7, determine μ_{cr_1}, μ_{cr_2}, and the corresponding buckling modes. Repeat the analysis for $I_{23} = 4I_{12}$. Repeat the calculation for the frame in Figure 2.9a assuming $l_{12} = l_{34}/2$, $I_{23} = 2I_{14} = 4I_{12}$, and $F_1/F_2 = 2$.

2.2.8 Verify the solution of Example 4 and find the second critical value of the load multiplier (it corresponds to an antisymmetric mode: $\theta_2 = \theta_3$, $\theta_1 \neq 0$).

2.2.9 Suppose that the determinant characterizing the critical state gives $s(2 + c) = 0$. Does $s = 0$ give the critical load? (Only if simultaneously $c \neq \infty$.)

2.2.10 Find the first buckling mode and critical load of the continuous beam with fixed ends shown in Figure 2.11a.

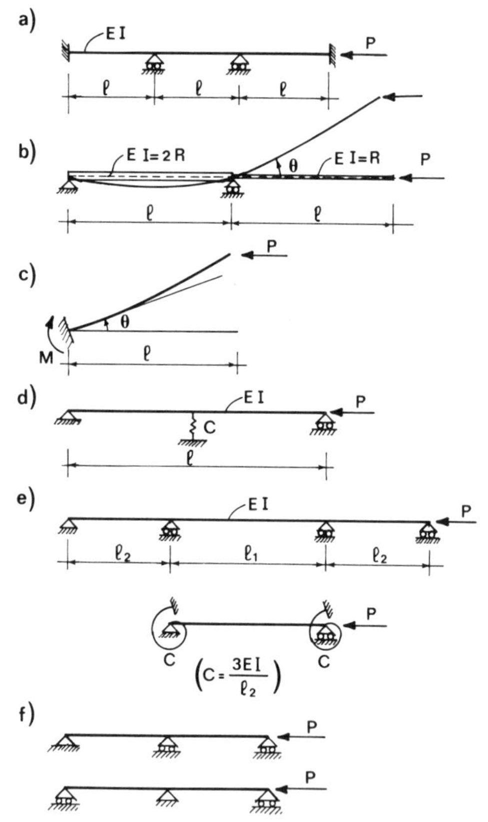

Figure 2.11 Exercise problems on critical loads of continuous beams.

2.2.11 Using the stiffness method, find the first critical load of the beam with a cantilever shown in Figure 2.11b. (Find first the fixed end moment due to rotation θ of the fixed end of the cantilever, Fig. 2.11c. The result is $P_{cr_1} = 0.1813\pi^2 R/l^2$.)

2.2.12 Find the first critical load of the beam with an intermediate spring support shown in Figure 2.11d. Assume $C = 20EI/l^3$ (the result is $P_{cr_1} = 13.892\pi^2 EI/l^2 = 1.4076P_E$).

2.2.13 Solve P_{cr} for a three-span continuous beam (Fig. 2.11e), and then compare it with the P_{cr} value that is obtained when the central span is treated as an isolated column with rotational end springs whose stiffness constants equal the stiffness of the side span at zero axial force. Then find how the results change with the ratio l_2/l_1 in the range (0, 1).

Note: One may be tempted to analyze a column in a frame by replacing the action of the rest of the frame upon the column by equivalent springs. This practice, however, is incorrect and may cause a large error on the unsafe side, as this problem demonstrates. The reason is that the spring stiffnesses, which are equivalent to the rest of the frame, are actually not constant but depend on the unknown value of the critical load. The only exception is when the rest of the frame has no axial forces, as in the case of the L-frame in Figure 2.12b (Prob. 2.2.15).

2.2.14 Find the critical load of the two beams shown in Figure 2.11f. (Note that the solutions are not the same.)

2.2.15 Find the critical load of the L-frames in Figure 2.12a and b, assuming axial inextensibility for beams 12 and 23. ($P_{cr_1} = 1.406P_E$ for Fig. 2.12a and $P_{cr_1} = 0.144P_E$ for Fig. 2.12b.) Compare the results with that obtained for the columns in Figure 2.12c and d with a spring on top whose spring stiffness is taken the same as the stiffness of the horizontal beam ($C = 3EI/l$).

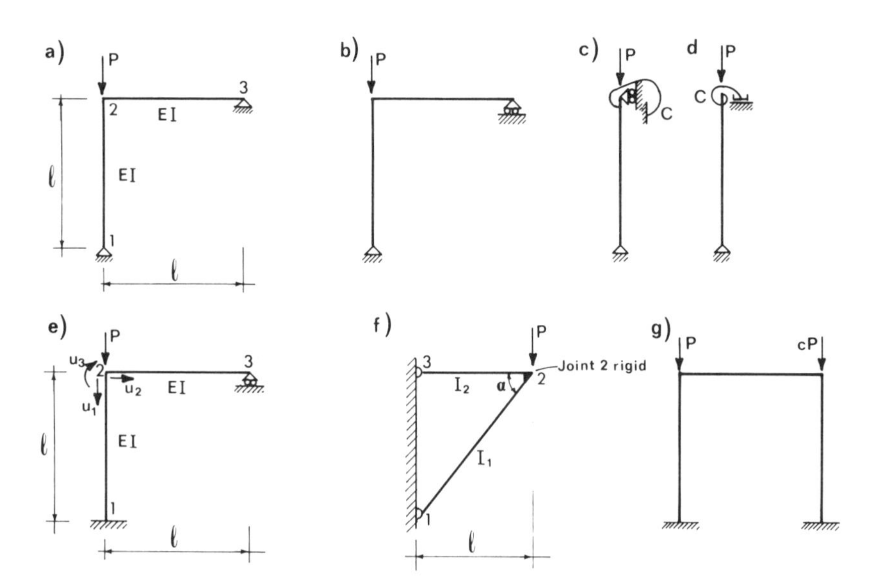

Figure 2.12 Exercise problems on critical loads of frames.

2.2.16 Using the stiffness matrix of Equation 2.2.7 write the incremental equilibrium equations for the L-frame in Figure 2.12e taking into account the elastic shortening of beam 12, and find the critical load P_{cr_1}. Compare the result with that obtained assuming axial inextensibility.

Note: In the presence of axial shortening in the column, the L-frame in Figure 2.12a does not exhibit neutral equilibrium at critical load and behaves as imperfect since the equation system is not homogeneous. The reason is that support 3 resists the axial shortening of column 12.

2.2.17 Consider the truss of Figure 2.12f with a rigid joint at 2.

(a) Solve P_{cr} for $\alpha = 45°$, $I_2 = I_1$.
(b) Same as (a) but limit $I_2/I_1 \to 0$.
(c) Same as (a) but limit $I_1/I_2 \to 0$.
(d) Let $I_1 = I_2$. Find angle α for which P_{cr_1} = maximum.
(e) $\alpha = 45°$. Is the cost of the truss lower if joint 2 is made a pin (a single bolt instead of a welded connection)? Assume (1) making the joints rigid doubles the cost of the truss; (2) the cost of a pin-jointed truss is proportional to its

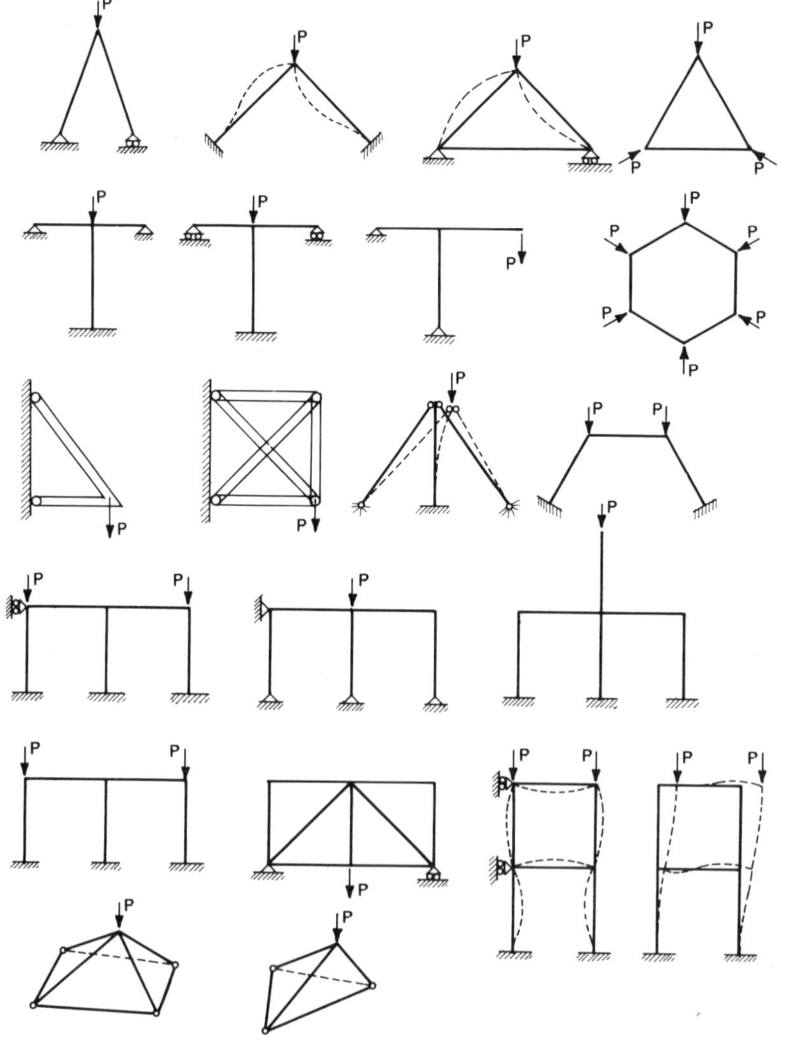

Figure 2.13 Further exercise problems on critical loads of frames.

weight; (3) the cross sections are geometrically similar in which case the weight of the bars per unit length is proportional to $\sqrt{I}$. (Demonstrate this, too.)

2.2.18 Solve P_{cr} for the frame in Figure 2.12g, with $c = 1$, 0.98, 0.90, 0.30, 0, assuming the frame to be (a) braced, (b) unbraced (constant EI).

2.2.19 Solve P_{cr} for the frames in Figure 2.13.

2.2.20 Solve the critical load of the structures with rigid members in Figure 2.14a, b. *Hint:* Assume axial inextensibility and include in the translational equilibrium equation the component $P\Delta/a$ due to the inclination of the rigid member.

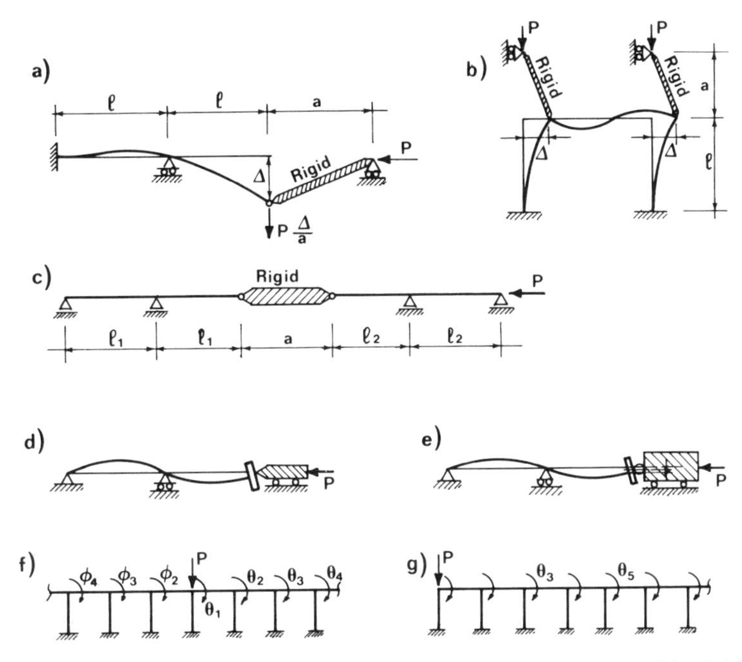

Figure 2.14 Exercise problems on critical loads (a–c) of structures with rigid members, (d, e) of continuous beams loaded through a frictionless plunger or roller, and (f, g) of infinite one-story frames.

2.2.21 Solve the structure in Figure 2.14c for (a) $l_1 = l_2$ and (b) $l_1 = 2l_2$.

2.2.22 Solve the critical load of the structures in Figure 2.14d (frictionless contact) and in Figure 2.14e (no-slip rolling support). *Hint:* See Probs. 1.4.3 and 1.4.4.

2.2.23 Consider the infinite one-story frame in Figure 2.14f. Calculate the ratios θ_{i+1}/θ_i and ϕ_{i+1}/ϕ_i, which must be constant ($i > 1$).

2.2.24 Solve P_{cr} for the infinite one-story frame in Figure 2.14g.

2.2.25 Calculate the critical load P_{cr} of the arch frame in Figure 2.15a for the antisymmetric buckling shown, which satisfies (to the first-order) the condition of axial inextensibility of members ($EI = \text{const.}$). *Note:* There are two unknowns: θ and v. One equation for θ and v is obtained as the condition of

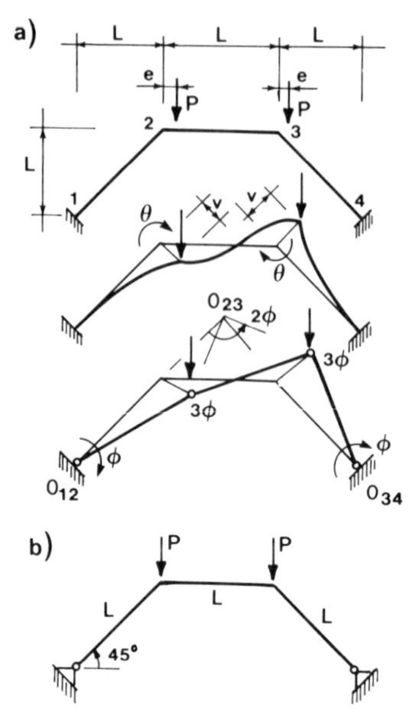

Figure 2.15 Exercise problems on critical loads of arch frames.

equilibrium of the moments acting on joint 2 or 3. To obtain a second equation for θ and v it is convenient to consider a single-degree-of-freedom mechanism as shown, in which members 12, 23, and 34 have centers of rotation O_{12}, O_{23}, and O_{34}, and the rotations in the joints 1, 2, 3, and 4 are obtained as ϕ, -3ϕ, 3ϕ, and $-\phi$. According to the principle of virtual work, equilibrium requires that $M_{12}\phi - 3M_{21}\phi + 3M_{34}\phi - M_{43}\phi = 0$ in which expressions for end moments M_{12}, M_{21}, M_{34}, and M_{43} in terms of θ and v need to be substituted (ϕ cancels out). Also note that this problem has much similarity with the buckling of a fixed arch (Sec. 2.8).

2.2.26 Solve P_{cr} for the frame in Figure 2.15b.

2.3 BUCKLING AS A MATRIX EIGENVALUE PROBLEM AND USE OF FINITE ELEMENTS

From the examples presented we see the analysis of frame buckling by the stiffness method generally reduces to a linear matrix equation:

$$\mathbf{K}(\mu)\mathbf{u} = \mathbf{f} \tag{2.3.1}$$

in which $\mathbf{u}$ is the column matrix of small generalized displacement increments u_i from the initial state, $\mathbf{f}$ is the column matrix of the associated small generalized force increments f_i (which include resultants of distributed loads such as the fixed end forces), $\mathbf{K}$ is a matrix of incremental stiffness coefficients K_{ij} $(i, j = 1, 2, \ldots, n)$ and μ is the parameter of initial loads that are finite rather than infinitely small and are independent of $\mathbf{f}$. The analysis by the flexibility method also reduces to a linear matrix equation, but from now on we will consider only the stiffness method since in the general case it is more easily programmed for a computer, as discussed in the preceding section.

If the determinant of matrix $\mathbf{K}$ approaches zero, displacements u_i increase to infinity and any f_i, no matter how small, causes infinitely large displacements (according to the linearized theory). This represents instability. The behavior is the same as that obtained for columns. Obviously, the vanishing of the determinant is the condition of critical load. When $f_i = 0$, the frame in the critical state is in neutral equilibrium, that is, is in equilibrium for any magnitude of deflection in the buckling mode. In rectangular frames with negligible axial extensions, the case $f_i = 0$ occurs when all the loads are vertical and are applied at the joints (thus causing zero-fixed-end moments), and when no disturbing loads act at the joints.

Equation 2.3.1 represents a matrix eigenvalue problem, but not of the standard type because the coefficients K_{ij} depend on μ nonlinearly, and μ appears not only in the diagonal terms. The formulation can be modified in various ways to get rid of the nonlinearity. One way is incremental linearization of $\mathbf{K}(\mu)$. We select a certain value μ_0 of load parameter μ and expand the coefficients $K_{ij}(\mu)$ in Taylor series about μ_0. Truncating the series after the linear terms, we have

$$\mathbf{K}(\mu)\mathbf{u} = [\mathbf{K}_0 + (\mu - \mu_0)\mathbf{K}_1]\mathbf{u} = \mathbf{0} \tag{2.3.2}$$

where $\mathbf{K}_0$ and $\mathbf{K}_1$ are constant matrices. Premultiplying this equation by the inverse matrix $\mathbf{K}_1^{-1}$ we get

$$(\mathbf{C} - \mu\mathbf{I})\mathbf{u} = \mathbf{0} \tag{2.3.3}$$

in which $\mathbf{C} = \mu_0\mathbf{I} - \mathbf{K}_1^{-1}\mathbf{K}_0$ and $\mathbf{I}$ = identity matrix. Thus we have acquired a standard matrix eigenvalue problem, for whose solution efficient computer library subroutines are available. After solving the smallest root $\mu = \mu_1$, we expand $K_{ij}(\mu)$ again around point μ_1 and repeat the entire computation until the results differ negligibly. This process converges very rapidly.

Another way to get rid of nonlinearity, which does not require iterations but increases the number of unknowns and leads to a linear eigenvalue problem that is not of the standard form, consists in subdividing the columns of the frame into three or more short elements (beam segments). Then the Euler buckling load of these short elements, P_E, is at least nine-times that of the whole column, P_E^c, while the buckling load of the frame is often $\leq P_E^c$ and always $\leq 4P_E^c$. Thus the ratio $\rho = P_{cr}/P_E$ for each member may be expected to be small (always less that $\frac{4}{9}$ and typically about 0.1). In such a range the graphs of s, sc, etc. (see Fig. 2.2a) are nearly linear and a linearized expression, corresponding to the linear part of the Taylor series expansion about $\rho = 0$, may be used for functions s, sc, s^*, and $\bar{s}$. From Equations 2.1.12 and 2.1.18, we have

$$\begin{aligned} s &\simeq 4 - \frac{2\pi^2}{15}\left(\frac{P}{P_E}\right) \qquad & sc &\simeq 2 + \frac{\pi^2}{30}\left(\frac{P}{P_E}\right) \\ \bar{s} &\simeq 6 - \frac{\pi^2}{10}\left(\frac{P}{P_E}\right) \qquad & s^* &\simeq 12 - \frac{6}{5}\pi^2\left(\frac{P}{P_E}\right) \end{aligned} \tag{2.3.4}$$

Substituting the values from Equations 2.3.4 into Equation 2.2.7, we obtain

$$\mathbf{K} = \mathbf{K}^e + \mathbf{K}^\sigma \tag{2.3.5}$$

where $\mathbf{K}^e$ is the linear elastic stiffness matrix of the beam at no axial force and $\mathbf{K}^\sigma$

represents the so-called "geometric stiffness matrix," which is given by

$$\mathbf{K}^{\sigma} = -\frac{P}{l}\begin{bmatrix} 0 & 0 & 0 & 0 & 0 & 0 \\ & \frac{6}{5} & -\frac{l}{10} & 0 & -\frac{6}{5} & -\frac{l}{10} \\ & & \frac{2l^2}{15} & 0 & \frac{l}{10} & -\frac{l^2}{30} \\ & & & 0 & 0 & 0 \\ & \text{sym.} & & & \frac{6}{5} & \frac{l}{10} \\ & & & & & \frac{2l^2}{15} \end{bmatrix} \qquad (2.3.6)$$

The geometric stiffness matrix is also called the "initial stress matrix" or the "load-geometric matrix," because it depends both on the initial load and on the geometry of the structure. See also Chapter 10, Equations 10.1.32 and 10.1.33 and Chapter 11, Equation 11.8.8.

Note that the subdivision of beams into a number of shorter elements provides also one way of treating a beam of variable EI or variable P (P can vary due to own weight). For a sufficient number of elements, a smooth variation of EI (or P) may be replaced by a stepwise variation of EI (or P) (Fig. 2.16), thus enabling the use of beam elements of constant EI and constant P, as developed before.

Still another effective method of solving Equation 2.3.1 is as follows. We select one of the displacement parameters, u_k, such that we do not expect u_k to be negligible as compared with max $|u_i|$ in the buckling mode. Then we replace the kth equation by the equation $u_k = 1$ (which can be visualized as imposing a unit displacement upon the frame). Thus, the system of equations becomes nonhomogeneous and, choosing a certain value of μ, it can be solved. After solving it, we evaluate $Q_k = K_{kj}u_j$, which represents a force needed to produce displacement u_k. Now, Q_k can be regarded as a function of μ, and the problem is to find the smallest μ which yields $Q_k = 0$ (i.e., the smallest μ for which the force needed to produce the displacement is zero and the frame is in neutral equilibrium). This can be accomplished by choosing various values of μ and applying Newton's *regula falsi* method. A subroutine which implements the above algorithm is given in Bažant (1974); see also Bažant and Estenssoro (1979). Using this method, buckling of frames with as many as 2000 displacement unknowns has

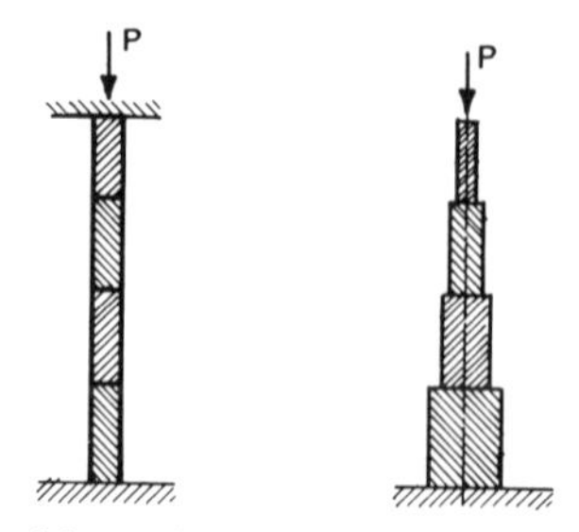

Figure 2.16 Column subdivided into elements of shorter length.

use a nonlinear optimization subroutine, such as that for the Levenberg–Marquardt algorithm, to minimize the sum of Q_k^2 as a function of μ.

As we might expect, Equation 2.3.5 is the same as that used in the finite element method, and we now outline its well-known derivation. Since, as it appears from Equation 2.3.6, the terms which refer to the axial deformation are not affected by the second-order theory, we will consider only lateral displacements w and rotations $\phi = w'$ of the axis of the beam. In the finite element method the stiffness matrix is normally derived from the principle of virtual work, whose statement can be obtained directly from the differential equations of equilibrium (Eqs. 1.3.2) by integration over the length of the beam

$$\int_0^l (V' + p)\, \delta w\, dx + \int_0^l (M' + V + Pw')\, \delta w'\, dx = 0 \tag{2.3.7}$$

where $\delta w(x)$ is any continuous and differentiable function representing virtual displacements. One may now substitute $\delta\phi = \delta w'$, integrate by parts the terms which contain V' and M', and denote as R_k the values of V and M at the two ends of the beam and as δr_k the corresponding virtual (generalized) displacements. This yields the following relation, representing a statement of the principle of virtual work:

$$\int_0^l M\, \delta\phi'\, dx - \int_0^l Pw'\, \delta\phi\, dx = \int_0^l p\, \delta w\, dx + \sum_k R_k\, \delta r_k \tag{2.3.8}$$

Let us now introduce the approximation $w = \mathbf{N}^T\mathbf{u}$ where $\mathbf{N}^T$ is a column matrix of interpolation functions and $\mathbf{u}$ is a column matrix of joint displacements (see, e.g., Gallagher, 1975). We then have $w' = \mathbf{C}^T\mathbf{u}$, where $C_i = dN_i/dx$. Assuming the same approximation for δw, we also have $\delta\phi = \mathbf{C}^T\, \delta\mathbf{u}$ and $\delta\phi' = \mathbf{B}^T\, \delta\mathbf{u}$ where $B_i = d^2N_i/dx^2$. Substituting these relations together with the constitutive relation $M \simeq EIw'' = EI\mathbf{B}^T\mathbf{u}$ into the left-hand side of Equation 2.3.8, one obtains the expression $\delta\mathbf{u}^T\mathbf{K}\mathbf{u}$ in which

$$\mathbf{K} = \int_l \mathbf{B}EI\mathbf{B}^T\, dx - P\int_l \mathbf{C}\mathbf{C}^T\, dx = \mathbf{K}^e + \mathbf{K}^\sigma. \tag{2.3.9}$$

Using a vector of shape functions $\mathbf{N}$ corresponding to a cubic approximation of the displacement distribution, $w = a_0 + a_1x + a_2x^2 + a_3x^3$ (see, e.g. Gallagher, 1975), one obtains for $\mathbf{K}^e$ and $\mathbf{K}^\sigma$ the same matrices already introduced in Equation 2.3.5. So we verify that both approaches are equivalent.

The fact that a cubic parabola yields exactly the same s and c when P/P_E is small can be verified by developing in Taylor series the sine and cosine functions which appear in the exact deflection curve (Eq. 1.3.6). One has $\sin x = x - x^3/6 + \cdots$, $\cos x = 1 - x^2/2 + \cdots$, and we therefore see that a cubic polynomial approximates the deflection curve as closely as desired if P is small enough.

Problems

2.3.1 Solve the critical load of the beams in Figure 2.17a, b, c, d using one beam finite element (with stiffness matrix given by Eq. 2.3.5). Calculate the percentage error with respect to the exact solution.

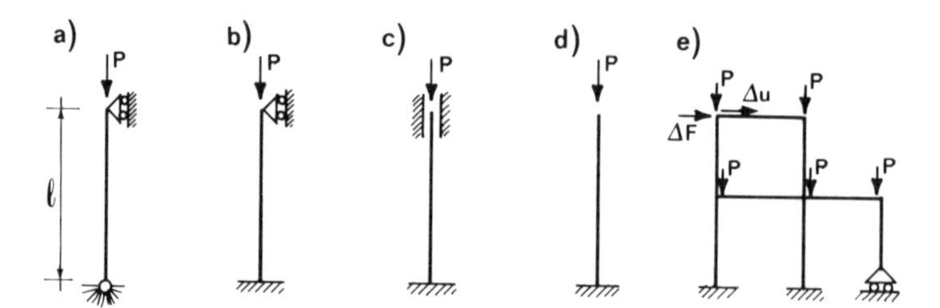

Figure 2.17 Exercise problems on finite element solutions of critical loads of structures.

2.3.2 Same as Problem 2.3.1, but use two finite elements of equal length.

2.3.3 Solve the critical load of the continuous beam of Example 1 in Section 2.2 (Fig. 2.5a) using one finite element for each span. Compare the result with the exact solution.

2.3.4 Same as Problem 2.3.3, but use two finite elements for each span.

2.3.5 Solve the same example as in Problem 2.3.3, using incremental linearization of **K** (Eq. 2.3.2).

2.3.6 Same as Problem 2.3.3, imposing $\theta_3 = 1$ to one end of the beam and searching for the value of P that yields a moment $M_3 \simeq 0$ as the reaction at the same end.

2.3.7 Solve by finite elements the critical load of the beams in Figure 2.11.

2.3.8 Solve the critical load of the frame in Figure 2.17e: (a) use the matrix in Equation 2.2.7, plotting the reaction ΔF caused by imposed displacement $\Delta u = 1$ as a function of P; (b) use the linearized matrix in Equation 2.3.5, short beam elements, and a computer library subroutine for eigenvalues.

2.4 LARGE REGULAR FRAMES

Large frames give rise to a large number of unknowns, and this may engender computational problems. However, they are, in general, regular, and this regularity may be exploited to greatly reduce the number of unknowns. Consider a large rectangular frame shown in Figure 2.18. The vertical axial forces in the columns vary from floor to floor; however, if the frame is very tall, then the change of axial force from one floor to the next is small, and locally a constant value of axial load P may be assumed. Similarly, the changes in column

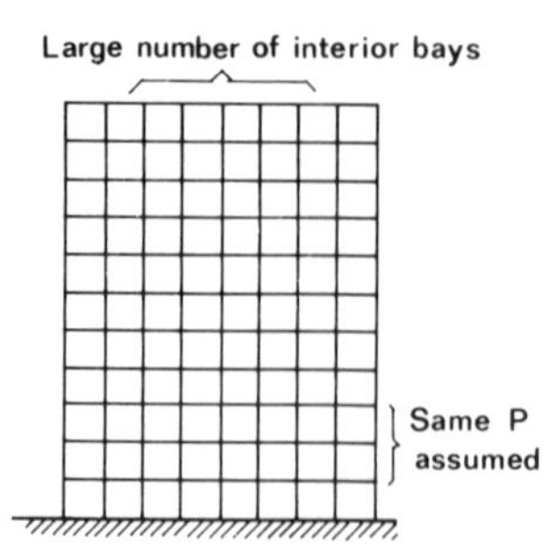

Figure 2.18 Large rectangular frame.

dimensions from floor to floor may be neglected. As an approximation the frame may be assumed to extend in both the vertical and horizontal directions to infinity.

Consider first buckling of rectangular frames that are braced against lateral sway. We consider columns that are remote from the bracing, and we neglect the axial deformations of all members. Among various possible periodic patterns of joint locations, the smallest critical value of P will be obtained for the pattern for which the curvature of the columns is the smallest, that is, the distance between adjacent inflection points (effective length) is the largest. The pattern of joint rotations that satisfies this condition is sketched in Figure 2.19a. In this pattern, the joint rotations, denoted as θ, are all equal and their sign alternates from joint to joint in both directions. The displacements of the joints are zero. So there is only one unknown displacement, θ. We need one equilibrium equation, which is provided by the condition of equilibrium of all four moments acting on one joint (Fig. 2.19a); $M_{12} + M_{13} + M_{14} + M_{15} = 0$, in which the numerical subscripts refer to the joints as numbered in Figure 2.19a. Expressing these moments according to Equation 2.1.5, and noting that $s_{12} = s_{14}$, $s_{13} = s_{15}$, we have

$$m_{12}\theta + m_{13}\theta = 0 \qquad m_{12} = \frac{2EI_{12}}{l_{12}}(s_{12} - s_{12}c_{12}) \qquad m_{13} = \frac{2EI_{13}}{l_{13}}(s_{13} - s_{13}c_{13}) \tag{2.4.1}$$

We may set $s_{13} = 4$ and $c_{13} = 0.5$ because the axial load is zero in the horizontal members. Thus, if we assume that the bending rigidity as well as the length of the

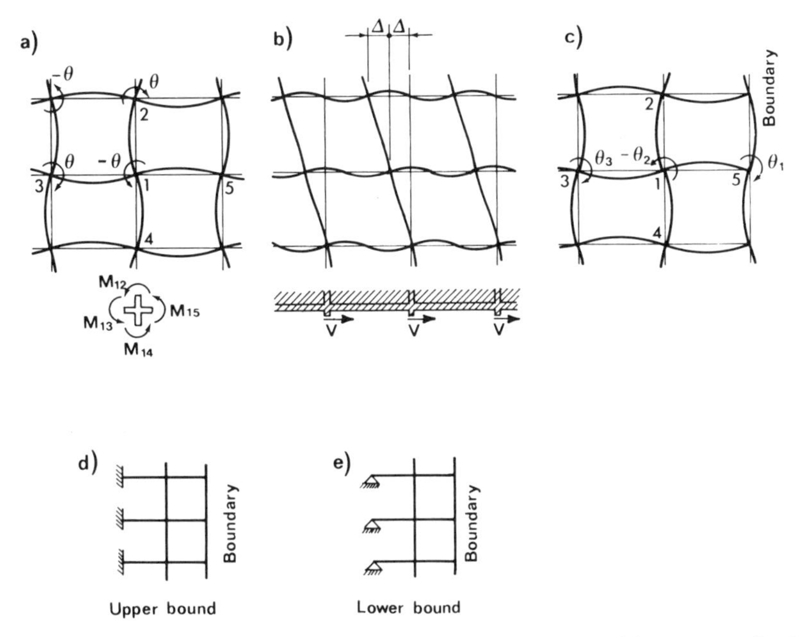

Figure 2.19 Regular buckling modes of large regular frames: (a) nonsway buckling of interior members (braced frame), (b) sway buckling of interior members (unbraced frame), (c) nonsway buckling of boundary members (braced frame), (d, e) approximate models for boundary column buckling.

horizontal and vertical members is the same, Equations 2.4.1 reduce to

$$D(P) = s_{12}(1 - c_{12}) + 2 = 0 \tag{2.4.2}$$

Using the iterative Newton method, we find the solution $P_{\mathrm{cr}_1} = 1.6681 P_{\mathrm{E}}$.

The foregoing solution applies only if the frame is braced. In an unbraced frame, a horizontal relative displacement Δ is possible between adjacent floors. The smallest curvature of columns and the longest effective length is obtained when all the joint rotations θ are equal (which also implies the same sign); see Figure 2.19b. Assuming that the relative horizontal displacements between two adjacent floors are the same, we have only two unknowns, Δ and θ. So we need two equations of equilibrium. One of them is the condition of equilibrium of all the four bending moments acting on one joint. The other one is a condition of equilibrium of the entire upper part of the frame above a certain floor, separated from the lower part of the frame by cutting the columns just below the floor (Fig. 2.19b). Since the deformations at each point are assumed to be the same, the horizontal shear forces V applied from the columns onto the bottom of the floor are all the same. At the same time, the sum must be zero since no lateral load is assumed to be applied. Therefore, their horizontal shear force in each column must be zero. So we have the following two conditions of equilibrium:

$$\begin{aligned} \sum M &= m_{12}\theta + m_{13}\theta + m_{12}^{s}\Delta = 0 \\ \sum V &= k_{12}\theta + k_{12}^{s}\Delta = 0 \end{aligned} \tag{2.4.3}$$

where $m_{12} = (2EI_{12}/l_{12})(s_{12} + s_{12}c_{12})$, $m_{13} = (2EI_{13}/l_{13})(s_{13} + s_{13}c_{13})$, $m_{12}^{s} = -(2EI_{12}/l_{12}^{2})\bar{s}_{12}$, $k_{12} = -(2EI_{12}/l_{12}^{2})\bar{s}_{12}$, and $k_{12}^{s} = (EI/l_{12}^{3})s_{12}^{*}$. Since there are no axial forces in the horizontal beams, we may substitute $s_{13} = 4$ and $c_{13} = 0.5$. For a nonzero deformation to exist, the determinant of the foregoing two equations must vanish, which yields the condition

$$D(P) = 2\bar{s}_{12}^{2} - s_{12}^{*}[6 + s_{12}(1 + c_{12})] = 0 \tag{2.4.4}$$

provided that the bending rigidity and the length are the same for the columns and the beams. The solution by the iterative Newton method gives the first critical load, $P_{\mathrm{cr}_1} = 0.577 P_{\mathrm{E}}$.

It is interesting to check the limits of the critical loads for the sway and nonsway modes as the horizontal beams become either infinitely rigid or infinitely flexible. If they are infinitely rigid, $I_{13} \rightarrow \infty$, then Eq. 2.4.1 for the nonsway frame yields $P_{\mathrm{cr}_1} = 4P_{\mathrm{E}}$. This is the critical load for a fixed-end column, as we expect. For the sway mode, Equation 2.4.4 yields the critical load $P_{\mathrm{cr}_1} = P_{\mathrm{E}}$, which is the critical load of a column with sliding rotational restraints at the end. If the beam is infinitely flexible, $I_{13} \rightarrow 0$, Equations 2.4.1 for the nonsway mode yield $P_{\mathrm{cr}_1} = P_{\mathrm{E}}$, and Equation 2.4.4 for the sway mode yields $P_{\mathrm{cr}_1} \rightarrow 0$.

We conclude, therefore, that the critical load of a column in a braced frame is never less than the Euler load, that is, the effective length is never longer than the floor height. On the other hand, for an unbraced frame the critical load of the column is never larger than the Euler load, that is, the effective length is never shorter than the floor height.

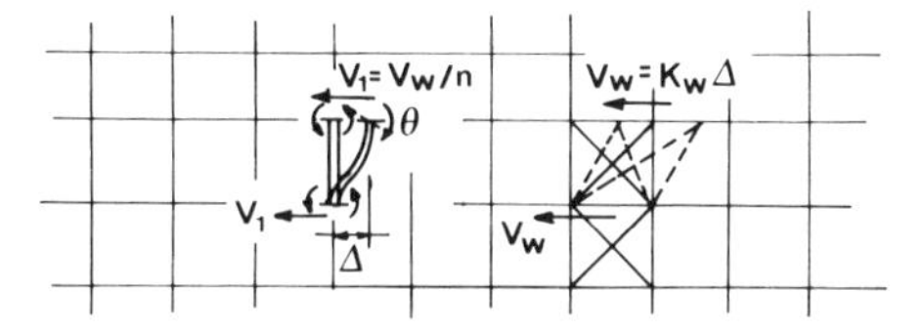

Figure 2.20 Frame with elastically restrained sway.

In the preceding examples we assumed the lateral sway of the frame to be either perfectly free or perfectly prevented. In most practical frames, though, we have an intermediate situation in which lateral sway is possible but is elastically resisted by a shear wall (Fig. 2.20). Denote the effective shear stiffness of such a wall as K_w, and the number of all columns in a floor as n. Assume that the shear force $V_w = K_w\Delta$ of the shear wall is balanced approximately uniformly by all the columns. Then the resisting shear force per column is $V_1 = K_w\Delta/n$, and this force must be included in the horizontal condition of equilibrium (adding it to the left-hand side of the second of Eqs. 2.4.3). (In practice the stiffness of exterior walls and partition walls may also contribute to the horizontal shear force V_1, but it is on the safe side to neglect this contribution.) This yields for the critical load (again with the assumption that $I_{12} = I_{13} = I$, $l_{12} = l_{13} = l$) the characteristic equation $D(P) = (s_{12} + s_{12}c_{12} + 6)(s_{12}^* + K_w l^3/nEI) - 2\bar{s}_{12}^2 = 0$. As an example, assuming that $K_w/n = 3EI/l^3$, the solution of the characteristic equation yields $P_{\text{cr}_1} = 0.862P_E$.

The preceding solution for braced and unbraced frames may be easily generalized to slab buildings in which the horizontal beams are replaced by a flat slab or a slab stiffened by horizontal beams. The stiffness of the slab, just like the stiffness of the horizontal beam, is not affected by the critical load P. All that is needed to apply our previous approach is to calculate the rotations ϕ at column ends when moments $M = 1$ are imposed at all column ends. For a braced building, the unit moments are applied in an alternating pattern (Fig. 2.21a), and for the sway buckling of a building that is not braced the applied unit moments are all of the same sign (Fig. 2.21b). Then one must calculate the rotation ϕ

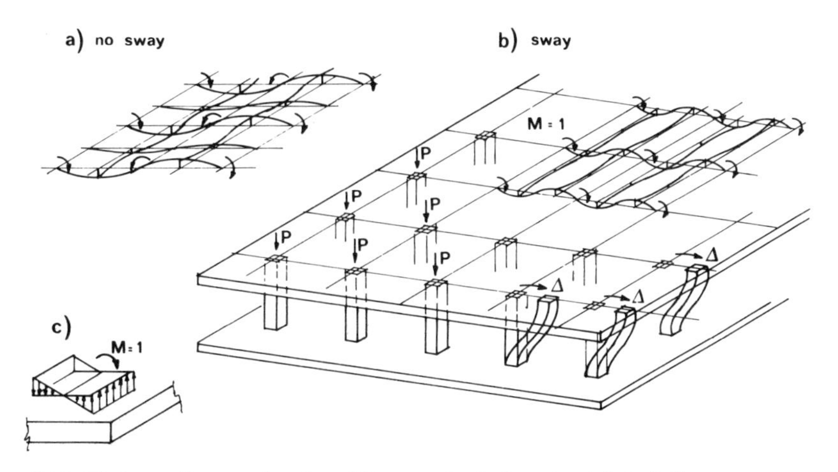

Figure 2.21 Deformation modes for stiffness calculations in slab buildings.

caused by these unit applied moments, and the effective stiffness of the rotational elastic restraint provided to each column by the floor slab is $m_{13} = 1/\phi$. This value must then be substituted into Equation 2.4.1 or 2.4.3. Note that moments M cannot be applied as a concentrated moment applied to a point, because the plate-bending equation would have a singularity and would yield an infinite rotation ϕ. Therefore M must be applied by an equivalent distributed load over area A (Fig. 2.21c). This area must be the larger of (1) the cross section of the column at its end and (2) the area h^2 where h = slab thickness (this is because the plate-bending equation does not apply at a scale $<h$).

The foregoing solutions are applicable only to columns in the interior of a large regular framework. A different solution is required for the columns at or near the boundary of the regular framework (Fig. 2.19c). This is because the boundary joints receive only three instead of four moments, and also because the cross section and load of the boundary columns usually differ from those of the interior columns.

A simple solution for the boundary columns is possible if a finite number of bays near the boundary, say, two, are isolated from the rest of the frame. This may be achieved by inserting either hinges or fixed supports at the joints lying on the third column line, as shown in Figure 2.19d, e. When hinges are used, the frame is obviously weakened, and the critical load will be lower than the actual one, that is, we obtain a lower bound. When fixed ends are inserted, the frame will obviously be stiffer and the critical load will be obtained larger than the actual one, that is, we obtain an upper bound. These two bounds are usually close enough for practical purposes. If closer bounds are desired, the hinges and fixed ends are inserted at the joints lying on the third rather than the second column line, or even the fourth one, etc.

Let us assume for all the solutions of boundary column buckling that the frame is braced. Note also that the axial force in the boundary columns normally differs from that in the interior columns and may be denoted as kP where k is some constant give a priori. In the present solutions we assume that $k = 1$.

Consider first that fixed ends are inserted in the joints on the third column line (Fig. 2.19d). We denote by θ_1 the joint rotations of the boundary column line and by θ_2 those on the second column line. The moment conditions of equilibrium of the joints of the first and second column lines are

$$\begin{aligned} \frac{EI_{12}}{l_{12}}(2s_{12} - 2s_{12}c_{12})\theta_1 + \frac{EI_{13}}{l_{13}}(s_{13}\theta_1 + s_{13}c_{13}\theta_2) = 0 \\ \frac{EI_{12}}{l_{12}}(2s_{12} - 2s_{12}c_{12})\theta_2 + \frac{EI_{13}}{l_{13}}(2s_{13}\theta_2 + s_{13}c_{13}\theta_1) = 0 \end{aligned} \tag{2.4.5}$$

Then, assuming that the bending rigidities EI and the lengths l of all members are the same, the condition of a vanishing determinant of these two equations is

$$D(P) = (2s_{12} - 2s_{12}c_{12} + s_{13})(2s_{12} - 2s_{12}c_{12} + 2s_{13}) - (s_{13}c_{13})^2 = 0 \tag{2.4.6}$$

where, for the horizontal beams, we have $s_{13} = 4$ and $c_{13} = 0.5$, because they carry no axial load. From this condition, $P_{\text{cr}_1} = 1.550P_{\text{E}}$, which represents an upper bound on the actual critical load, as already explained.

To get the lower bound, hinges may be inserted just to the left of the joints at the third column line (Fig. 2.19e). So we have three unknowns, the joint rotations, θ_1, θ_2, and θ_3 at the first, second, and third column line. The moment conditions of equilibrium of the joints on the first, second, and third column lines, and the condition of zero moments at the hinge, yield a system of three linear homogeneous equations for θ_1, θ_2, and θ_3. The determinant of these equations vanishes if

$$D(P) = (2s_{12} - 2s_{12}c_{12} + 4)(2s_{12} - 2s_{12}c_{12} + 7) - (2)^2 = 0 \tag{2.4.7}$$

Equation 2.4.7, which has been derived assuming that the bending rigidities and lengths of all members are the same and introducing the condition that horizontal members do not carry axial loads, yields $P_{cr_1} = 1.525P_E$. This is a lower bound on the actual critical load. We see that the upper and lower bounds are sufficiently close for practical purposes.

An exact solution to boundary column buckling can be obtained easily if the frame is assumed to extend to infinity away from the boundary, and if the joint equilibrium equations for all joints $j \geq 2$ are the same. In this case, the ratios of rotations of adjacent hinges are constant, that is, $\theta_2/\theta_1 = \theta_3/\theta_2 = \theta_4/\theta_3 = \cdots = a = \text{constant}$, or $\theta_2 = a\theta_1$, $\theta_3 = a^2\theta_1$. The reason is that for joints $j \geq 2$ the joint equilibrium equations represent difference equations with constant coefficients, the solutions of which have the general form $\theta_j = Ae^{\lambda j}$ where $j = 2, 3, 4, \ldots$, and λ, A are, in general, complex constants (e.g., Wah and Calcote, 1970; and also Sec. 2.9). Consequently, the moment conditions of equilibrium of the boundary joints and of the first joint behind the boundary have the form:

$$A_1\theta_1 + A_2(a\theta_1) = 0 \qquad B_1\theta_1 + B_2(a\theta_1) + B_3(a^2\theta_1) = 0 \tag{2.4.8}$$

in which A_1, A_2, B_1, B_2, and B_3 are expressed in terms of stability functions s and c of the members and depend on axial load P, as well as kP. Canceling θ_1 and eliminating a from these equations we get the condition

$$D(P) = B_1 - B_2\frac{A_1}{A_2} + B_3\left(\frac{A_1}{A_2}\right)^2 = 0 \tag{2.4.9}$$

To determine the critical load, we need to find the value of P for which this expression vanishes. If the bending stiffnesses and lengths are the same for all members and the boundary column is subjected to the same axial load as the other columns, we find in this manner that $P_{cr_1} = 1.525P_E$, a value which happens to coincide with the lower bound that we obtained before (but this would not happen, e.g., if the boundary column had a different bending rigidity or were subjected to a different axial force).

By similar reasoning one can find the critical load for a boundary column in an unbraced frame.

One might be tempted to assume, for the sake of simplicity, that the buckling of a regular frame (Fig. 2.22a) is equivalent to the buckling of a single column elastically supported at the ends by springs of constant stiffness (Fig. 2.22c), the springs modeling the resistance of both the adjacent columns and beams. However, this is incorrect, since one ignores the fact that the adjacent columns are in fact axially loaded and, consequently, their stiffness is a function of the unknown axial load and becomes zero for $P = P_{cr}$. Springs can replace only the action of the adjacent beams without axial forces.

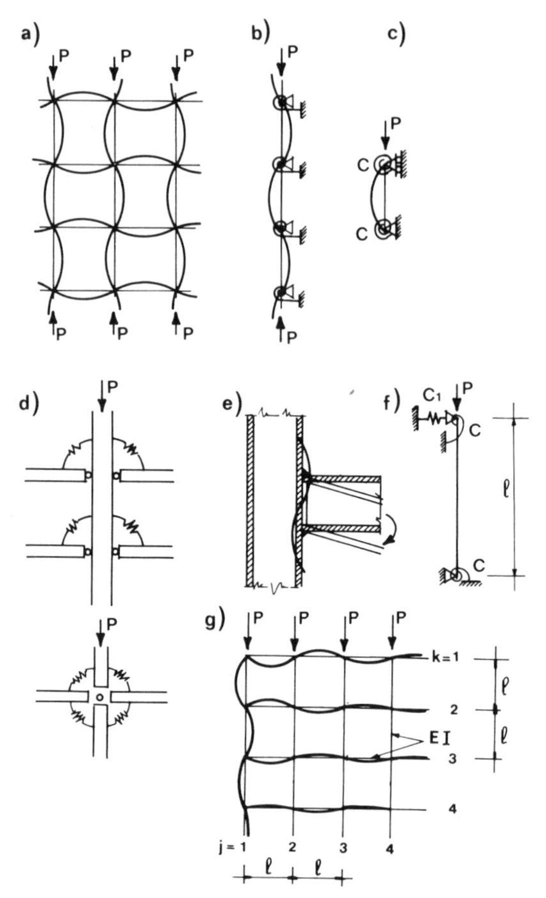

Figure 2.22 (a) Buckling of a regular frame; (b, c, f) schemes for approximate buckling analysis; (d) flexible joints; (e) inelastic deformation of joint; (g) corner column buckling.

On the other hand, it is certainly possible to replace the action of those parts of the frame that are not subjected to axial loads by equivalent springs. Thus it is possible to replace the horizontal beams in the regular frame (Fig. 2.22a) by rotational springs as shown in Figure 2.22b. This means that the problem of buckling of braced frames is equivalent to the problem of buckling of an infinite continuous beam. (See Probs. 2.2.13 and 2.4.6.)

One aspect that we have so far neglected is extremely important for practical applications. In most frames, the joints are not perfectly rigid but flexible, that is, the ends of the beam column can rotate with regard to the joint (Fig. 2.22d). This means that one must introduce an effective rotational spring at each end of the beam-column to take the deformability of the connection into account (and possibly a shear spring needs to be introduced as well). One must then distinguish between the rotations of the ends of the beam-column and of the joint, and add to the system of equations the spring deformation relations involving these rotations. Otherwise the formulation of frame-buckling problems is the same and may be based on the stability functions s and c. Many practical examples of the buckling of frames with elastic connections, as well as tables of critical loads, have been worked out by Fiřt (1964, 1974), whose work was mainly intended for frames assembled from precast concrete members.

A further complication, however, often arises, due to inelastic behavior of the connection. All joints, including welded ones, undergo some limited plastic deformations (Fig. 2.22e) and in riveted or bolted joints frictional slip may also take place. One consequence is that cyclic lateral loads of steel building frames cause moment redistributions such that the bending moments in beams at the column joints due to vertical loads become almost zero. The problem then is how to determine for buckling analysis the initial bending moments and the effective elastic constants relating the rotations of the beam and of the column at the joint. For detailed analysis, see, for example, Ackroyd and Gerstle (1982, 1983), and Moncarz and Gerstle (1981).

Problems

2.4.1 Calculate P_{cr} for a regular frame in which the sway is resisted by an elastic shear wall of shear stiffness K_w (see Fig. 2.20). Do the calculations for various values of K_w such that $K_w l^3/nEI = 0.1, 0.5, 1, 2, 4$. Check that $0.577P_E < P_{cr_1}$. (Note that the bound $1.668P_E$ will be exceeded for sufficiently large K_w because the buckling mode assumed does not approach the buckling mode for no-sway frames, causing the present critical load to become the second rather than the first critical load. Also note that for plotting P_{cr} versus K_w it is easier to choose various values of P/P_E and calculate the corresponding K_w than vice versa.)

2.4.2 Solve P_{cr_1} from Equation 2.4.6. (Answer: $P_{cr_1} = 1.55P_E$.)

2.4.3 Find the lower bound on the critical load at the boundary of a regular frame, assuming the same given data as in the text but the hinge is inserted at the joints of the third column line (see Fig. 2.19e). (Answer: $P_{cr_1} = 1.52P_E$.)

2.4.4 Same as Problem 2.4.3, but assume the frame to extend to infinity to the right and obtain a solution of the type $\theta_j = Ae^{\lambda j}$ for $j = 2, 3, 4, \ldots$ ($A, \lambda =$ complex) using the difference calculus. Compare the result with Problem 2.4.3.

2.4.5 Solve P_{cr_1} from Equation 2.4.9.

2.4.6 Isolate an interior column from a regular frame by replacing with springs of constant C the restraint against rotation furnished at the joint by the adjacent floor beams (in which the axial force may be assumed to be negligible) (Fig. 2.22f).

(a) Consider the case of no sway ($C_1 = \infty$), What would you assume for the value of C? Then, assuming $C = 2(2EI)/l$, would you find the same value as P_{cr} given by Equation 2.4.2 (for the case when the bending stiffnesses and lengths of the horizontal and vertical members are the same)? Explain.

(b) Consider the case of an elastic restraint against sway ($C_1 l^3/EI = 0.1, 0.5, 1, 2, 4$). What value would you assume for C? Compare with the results of Problem 2.4.1.

2.4.7 Solve P_{cr} for the corner column buckling in Figure 2.22g, in which the joint rotation decays both in the horizontal and vertical directions. Formulate the equilibrium condition for interior joints relating $\theta_{k,j}$ to $\theta_{k-1,j}$, $\theta_{k+1,j}$, $\theta_{k,j-1}$, and $\theta_{k,j+1}$, as well as the conditions for the boundary joints $k = 1$, $j \geq 2$, and

$k \geq 2$, $j = 1$, and the corner joint $k = 1$, $j = 1$. For $j \geq 2$, $k \geq 2$ they represent partial difference equations with constant coefficients and may be solved by $\theta_{k,j} = Ce^{aj}e^{bk}$ where a, b = complex constants (these difference equations are a special case of Eqs. 2.9.1 for zero axial extensions); for $j = 1$, $k \geq 2$ the solution has the form $\theta_{k,j} = Ae^{\lambda j}$, and for $k = 1$, $j = \geq 2$, $\theta_{k,j} = Be^{\kappa k}$ where λ, κ = complex constants (A, B, C = arbitrary complex constants).

2.5 POSTCRITICAL RESERVE IN REDUNDANT TRUSSES

In statically indeterminate structures, buckling failure of one member does not usually mean collapse. The axial forces redistribute, and the loads may then be further increased until another member fails by buckling, and so forth until eventually so many members buckle that the whole structure collapses. Thus, statical indeterminacy generally endows structures with an additional reserve strength that is not available in isolated columns.

Example of a Statically Indeterminate Truss

The simplest case to analyze is a statically indeterminate truss. Consider the example of the truss in Figure 2.23a. All the members are assumed to be pin-ended and of the same cross section. If we analyze the truss by the force method, considering, for example, the force in member 23 as redundant, we note that members 23 and 36 have the highest axial compression, $P_{23} = kP$, where $k = (2 + \sqrt{2})/(2 + 2\sqrt{2})$. When the load P reaches the value $P_1 = P_{E_{23}}/k$, $P_{E_{23}}$ being the Euler load of member 23, the member buckles. In the postcritical behavior after buckling, the deflection of member 23 increases at approximately constant axial force (Fig. 2.24a), and so the length of member 23 can decrease quite substantially without any significant change in the axial force P_{23} carried by

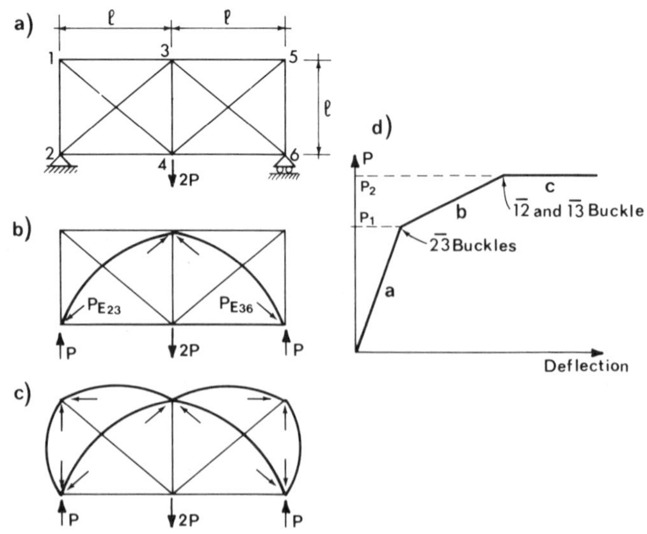

Figure 2.23 Postcritical behavior of a truss (a, b, c) and its load-deflection diagram (d).

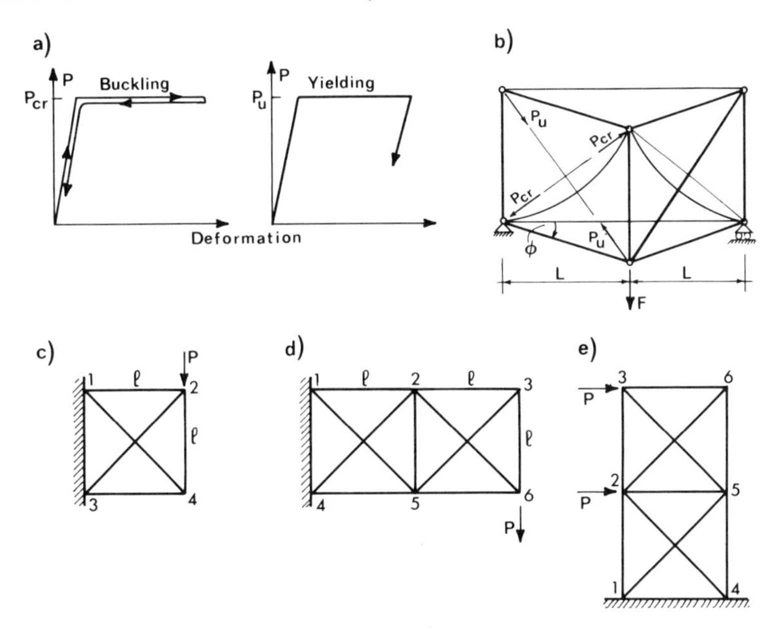

Figure 2.24 (a) Limited analogy between yielding and buckling of a column; (b) buckling mechanism; (c, d, e) truss examples for problems.

this member (Fig. 2.23b). This behavior is quite similar to plastic yielding (Fig. 2.24b) of the member in compression, $P_{E_{23}}$ being imagined as the yield value of the normal force. After members 23 and 36 buckle, which occurs simultaneously, the truss can be treated as a statically determinate structure since the forces in the two buckled members are constant and known. Stress analysis of this statically determinate truss shows that an equal axial compression next occurs in members 13, 35, 12, and 56; its value is $P_{12} = (1 - k/\sqrt{2})P_1 + (P - P_1)$. The frame then becomes a mechanism. It continues to deflect at constant load and its stress state remains the same. Since, due to symmetry, the mechanism has four degrees of freedom, the ratios of the shortenings of members 13, 35, 12, and 56 are indeterminate and are left to chance. But that has not effect on the load-deflection diagram. (In reality, though, only one member is likely to shorten, due to inevitable small random differences between these members.)

The load-deflection diagram of the truss is sketched in Figure 2.23d. It consists of straight-line segments of different slope, and the changes in slope correspond to the attainment of critical loads in individual columns.

Generalization and Limit Analysis Method

From this example, it is evident that significant redistributions of internal forces can take place in statically indeterminate structures. This problem has been studied in detail by Masur (1954). Since the behavior of the buckled column is about the same as axial yielding of the column, one may use the methods of limit analysis known from plasticity. In particular, the lower and upper bound theorems can be applied. It should be pointed, however, that the shake-down theorem for cyclic loading does not apply because the deformation of a buckled elastic column upon loading is reversible (as long as no inelastic deformations take place), while the deformation of a yielded column is irreversible (see Fig. 2.24a).

Buckling of some compressed members may be combined with tensile (or compression) yielding of other members. Elaborating further on our preceding example, suppose that while the compressed diagonal buckles, the tensile diagonal fails by yielding (Fig. 2.24b). Then, by equating the work of external forces and the work of internal reactions for the limit load F ($W_{ext} = W_{int}$), one obtains $FL\phi = P_u L\phi/\sqrt{2} + P_{cr} L\phi/\sqrt{2}$, where P_{cr} and P_u are the critical load of member 23 and the yielding axial force in member 14, respectively. The result is $F = (P_u + P_{cr})\sqrt{2}/2$.

In practice, the designer cannot always take advantage of the plastic-like force redistributions due to buckling of individual members. Often the deflections associated with such buckling may be too large. Or for some structures such as bridges the major loads are repetitive, in which case repeated buckling would arise, which cannot be permitted because of fatigue of the material. Frequently, too, the member may enter a softening regime, in which the axial load descends at increasing shortening (see Secs. 8.2 and 8.7).

It must be emphasized, however, that the foregoing solution becomes insufficient for large deflections for which the member can develop force P significantly higher than P_E. See the discussion of postcritical reserve in Sections 1.9 and 5.9. Then a nonlinear incremental loading analysis of the truss is required. This case can be practically important only for very flexible (slender) members.

Order of Approximation

It is helpful to realize what is the order of approximation in this analysis. The classical bifurcation buckling of columns at small deflections (Sec. 1.1–1.8, 2.1–2.4) is called the second-order theory because it takes into account bending moments $M = P_{cr} w$. In terms of displacements w, this represents the first order of approximation. For large-deflection buckling of the members of a pin-jointed truss, we have $P = P_{cr}(1 + cw^2)$ where $c =$ constant, $w =$ deflection (Eq. 1.9.12). The bending moments are $M = P_{cr}(1 + cw^2)w$, and because of the term $P_{cr} w^3$ this formulation may be called the fourth-order theory, although in terms of w the formulation is of the third order.

One is now tempted to ask: Would it suffice to use a third-order theory, that is, a second-order approximation in terms of w^2? The answer is no. The second-order terms cancel out for a symmetric problem such as a pin-jointed column. In frames, however, this need not happen since nonsymmetry is possible. Then it suffices to use a third-order theory (i.e., a second-order approximation in w). Such an analysis is quite different, as we see next.

Problems

2.5.1 With reference to the example of Figure 2.23, find the values P_1 and P_2 of the truss load that correspond to the buckling of members 23 and 21.

2.5.2 Solve the limit load of the trusses in Figure 2.24c, d, and e assuming the same P_E for all members but no yielding. Determine the sequence of buckling of the members.

2.5.3 Solve the truss in Figure 2.23 assuming that the height of the truss is not l but the diagonal has a 30° slope. (In that case the truss becomes a two-degree-of-freedom mechanism when member 12 buckles.)

2.5.4 Solve the maximum load of the truss in Figure 2.23 assuming that $P_{E_{13}} = P_{E_{35}} = P_{E_{56}} = 1.01 P_{E_{12}}$. (In that case the truss becomes a single-degree-of-freedom mechanism when member 12 buckles.)

2.6 POSTCRITICAL BEHAVIOR OF FRAMES

The limit analysis as we described it in the preceding section applies only to pin-jointed trusses, in which there are no bending interactions among the members of the truss and each member buckles independently as the Euler column. Rigidly jointed frames, however, exhibit more intricate behavior. It appears that for large deflections the equilibrium value of a load is affected by the incremental shear forces and by the axial shortenings of members resulting from their deflections. The incremental shear forces induce axial force changes in the adjoining members meeting at an angle, thereby modifying their bending stiffness. These effects are second-order small if the deflection is small. Yet they are of a lower order than the large-deflection effects in columns, and therefore more important.

This behavior leads to a new phenomenon we have not yet discussed, called asymmetric bifurcation. In such a bifurcation, the initial postcritical response is characterized by a plot of load versus deflection of a perfect frame and does not have a horizontal slope; rather it has a finite inclination. This phenomenon distinguishes frames in general (as well as shells) from typical columns, continuous beams, and pin-jointed trusses. It causes sensitivity to imperfections, manifested by the fact that an imperfect frame has a smaller maximum load than the perfect frame.

L-Frame of Koiter and Roorda

For the sake of illustration, consider the L-frame shown in Figure 2.25. In their famous, and by now classical, papers, Koiter (1967) used this example to illustrate asymmetric bifurcation and Roorda (1965a, b) confirmed Koiter's theoretical predictions by experiment. Koiter's analysis was based on power series expansion of the potential-energy expression for the structure, which provides complete information for both equilibrium states and their stability. Roorda and Chilver (1970) showed a different method of analysis, which was based solely on equilibrium equations and employed the perturbation method with power series expansions. Although their method does not deal with stability of equilibrium states, it yields all the information needed for practical purposes, and was shown by Roorda and Chilver to agree with Koiter's previous solution. The previous solutions, however, still have certain shortcomings; (1) they rely solely on mathematical manipulations and do not provide much insight as to the source and

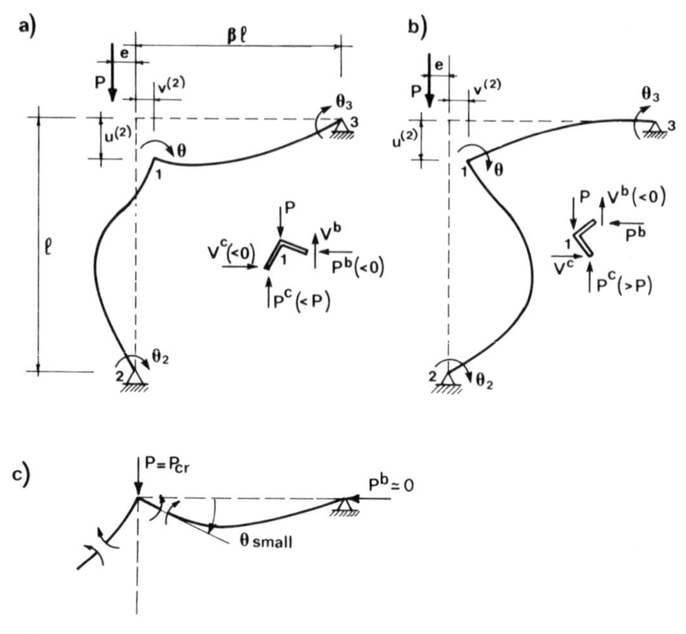

Figure 2.25 Koiter-Roorda's L-frame: (a) buckling to the left; (b) buckling to the right; (c) neutral equilibrium at $P = P_{cr}$ for perfect frame (and small θ).

mechanism of the imperfection sensitivity; and (2) they are relatively complicated because the nonlinear differential equations are integrated directly, without exploiting the existing powerful matrix method for frames.

A simplified method for nonlinear postcritical analysis of frames has recently been presented by Kounadis (1985) and illustrated by the example of Koiter-Roorda's L-frame. This method relies on direct integration of the differential equations for the deflection curves of members and is approximate since it achieves simplification by relaxing in a certain intuitive manner the compatibility conditions at the joints. We will present here a relatively simple method that, by contrast, does not require integration of differential equations as it utilizes the stiffness matrices with stability functions, is asymptotically exact, and gives information on the magnitude of contributions to imperfection sensitivity from various mechanisms (Bažant and Cedolin, 1989).

In the frame in Figure 2.25, the bars have equal uniform bending rigidities EI, and the ratio of their lengths is β. The vertical load P is applied at the corner with a small eccentricity e. The deformation of the frame is characterized by rotations θ, θ_2, and θ_3 (Fig. 2.25a, b). The column and beam are assumed to be so slender that their first-order axial shortenings due to axial forces are negligible.

The deflections w produce second-order axial shortenings of the column and the beam. They are second-order small in terms of w or θ and cause joint displacements $u^{(2)}$ downward and $v^{(2)}$ to the right (Fig. 2.25a, b).

Due to these displacements, buckling of the column to the right (Fig. 2.25b) produces an incremental shear force V^b in the beam which tends to make the axial compression force P^c in the column larger than the applied load P. On the other hand, for buckling to the left the shear force V^b is of opposite sign and tends to make P^c less than P (Fig. 2.25a). This favors buckling to the right. Second, buckling of the column to the right (Fig. 2.25b) produces a smaller curvature of the beam than buckling to the left (Fig. 2.25a) (since for both cases the joint moves down and to the right). This, too, favors column buckling to the right (Fig. 2.25b). Third, buckling of the column to the right (Fig. 2.25b) produces an incremental shear force in the column which translates into a compressive force in the beam, lowering its stiffness, while buckling to the left causes a tensile force in the beam, increasing its stiffness. This again favors buckling to the right.

So we see that the response does not exhibit the same symmetry that is found for centrically loaded perfect columns, which are equally likely to buckle left or right. This asymmetry is manifested in the load-deflection diagram of imperfect columns and has an important consequence: It causes the maximum load to become less than P_{cr}. We will demonstrate it.

Taking $u^{(2)}$ and $v^{(2)}$ into account but still considering small deflections, we obtain from Equation 2.1.17 the following moment equilibrium condition of the joint (Fig. 2.25):

$$\frac{EI}{l}\left[s_c\theta + s_c c_c \theta_2 - \bar{s}_c\left(\frac{v^{(2)}}{l}\right)\right] + \frac{EI}{\beta l}\left[s_b\theta + s_b c_b \theta_3 + \bar{s}_b\left(\frac{u^{(2)}}{\beta l}\right)\right] = -Pe \quad (2.6.1)$$

where s_c, c_c, and $\bar{s}_c$ are functions s, c and $\bar{s}$ for the column, depending on $\rho^c = P^c/P_E^c$ (P^c = axial compressive force in the column) and P_E^c = Euler load of the column; s_b, c_b, and $\bar{s}_b$ are functions s, c, and $\bar{s}$ for the beam, depending on $\rho^b = P^b/P_E^b$ (P^b = axial load of the beam and P_E^b = Euler load of the beam).

The moment equilibrium conditions at hinges 2 and 3 yield $s_c\theta_2 + s_c c_c\theta - \bar{s}_c v^{(2)}/l = 0$ and $s_b\theta_3 + s_b c_b\theta + \bar{s}_b u^{(2)}/\beta l = 0$. Expressing θ_2 and θ_3 from these conditions and substituting them into Equation 2.6.1, one obtains

$$\left[s_c(1 - c_c^2) + \frac{s_b}{\beta}(1 - c_b^2)\right]\theta + s_c(c_c^2 - 1)\left(\frac{v^{(2)}}{l}\right) + \frac{s_b}{\beta}(1 - c_b^2)\left(\frac{u^{(2)}}{\beta l}\right) = -\frac{Pel}{EI} \quad (2.6.2)$$

The horizontal and vertical equilibrium conditions for the joint (Fig. 2.25) can be written as

$$P^b = V^c = \frac{EI}{l^2} s_c(1 - c_c^2)\theta \quad (2.6.3)$$

and

$$P^c = P - V^b \qquad V^b = \frac{EI}{(\beta l)^2} s_b(1 - c_b^2)\theta \quad (2.6.4)$$

where V^c and V^b are the shear forces in the column and the beam (Fig. 2.25). In the calculation of V^c and V^b, higher-order terms have been omitted since the validity of our theory is limited to small displacements.

Let us now determine $u^{(2)}$ and $v^{(2)}$. Small deflections of the members of the frame may be expressed as $w = \theta f_c(x)$ for the column and $w = \theta f_b(x)$ for the beam, where x is the axial coordinate of the column or beam measured from the joint, $f_c(x) = A_c \sin kx + B_c \cos kx + C_c x + D_c$, $k = \lambda_c/l$, $\lambda_c = \pi\sqrt{\rho^c}$, and $f_b(x) = A_b x^3 + B_b x^2 + C_b x + D_b$. The expression adopted for $f_b(x)$ is a cubic parabola, which is exact for $P^b = 0$ and is always a sufficient approximation for the present since P^b (unlike P^c) is small (this will be confirmed later). The constants A_c, $B_c, \ldots, D_b$ are determined from the conditions $w = 0$ and $w' = 1$ for $x = 0$, and $w = 0$, $w'' = 0$ for $x = l$ or $x = \beta l$, where the primes denote derivatives (here with respect to x). This yields $A_c = l/(\sin \lambda_c - \lambda_c \cos \lambda_c)$, $C_c = -(A_c/l) \sin \lambda_c$, $B_c = D_c = 0$, $A_b = -C_b/l^2$, $C_b = \frac{1}{2}$, $B_b = D_b = 0$. Since the beam axis may be considered to be inextensible during buckling (as the axial force change is negligible) we have $u^{(2)}$ or $v^{(2)} = \int (1 - \cos w')\, dx = \int [1 - (1 - \frac{1}{2}w'^2)]\, dx = \int \frac{1}{2}w'^2\, dx$. Consequently, for small θ,

$$u^{(2)} = \eta_c \theta^2 \qquad \eta_c = \int_0^l \tfrac{1}{2}[f'_c(x)]^2\, dx \tag{2.6.5a}$$

$$v^{(2)} = \eta_b \theta^2 \qquad \eta_b = \int_0^{\beta l} \tfrac{1}{2}[f'_b(x)]^2\, dx \tag{2.6.5b}$$

where coefficients η_b and η_c are positive. Their values are found to be $\eta_b = 0.1\beta l$ and $\eta_c = A_c^2(2\lambda_c^2 + \lambda_c \sin 2\lambda_c - 4 \sin^2 \lambda_c)/8l$. Substituting Equations 2.6.5a, b into Equation 2.6.2 one gets

$$\left[s_c(1 - c_c^2) + \frac{s_b}{\beta}(1 - c_b^2)\right]\theta + \left[s_c(c_c^2 - 1)\left(\frac{\eta_b}{l}\right) + s_b(1 - c_b^2)\left(\frac{\eta_c}{\beta^2 l}\right)\right]\theta^2 = -\frac{Pel}{EI} \tag{2.6.6}$$

It is clear that if we would correct the deflection shapes $f_c(x)$ and $f_b(x)$ by taking $u^{(2)}$, $v^{(2)}$ into account, we would introduce into Equation 2.6.6 only terms whose order is higher than θ^2 but would not change the terms with θ and θ^2. Also note that since $P^b = 0$ for $\theta = 0$, the consideration of $P^b \neq 0$ in the expression of $f_b(x)$ would add a term proportional to θ in the expression of η_b (Eq. 2.6.5b), which would translate into a higher-order term in Equation 2.6.6.

Equations 2.6.3, 2.6.4, and 2.6.6 represent a system of five equations relating θ, P, P^b, P^c, V^b, V^c. If P^b, P^c, V^b, V^c are eliminated, one gets the relation of P to θ. For a convenient calculation of the curve $P(\theta)$, one may choose a series of closely spaced increasing values of P^c. For each P^c, one evaluates s_c and c_c. Then using approximately the previous rather than current value of θ in Equation 2.6.3, one solves from it P^b and then evaluates s_b and c_b, upon which one solves two values from Equation 2.6.6 (a quadratic equation), giving different portions of the $P(\theta)$ curve. Accuracy could be improved by iterating the procedure with the latest value of θ used in Equation 2.6.3 but this is not necessary if the chosen P^c values are very closely spaced.

The curves $P(\theta)$ are plotted in Figure 2.26a for various values of relative eccentricity e/l. As expected, for $e > 0$ the rotation θ is negative. A crucial fact to note is that, for $e > 0$, the curve $P(\theta)$ has a maximum, while the curves that are obtained upon neglecting $u^{(2)}$ and $v^{(2)}$ (dashed in Fig. 2.26a) do not. As a limit case for $e = 0$, we obtain the initial postcritical response of the perfect system.

Second-Order Solution of the L-Frame

As already shown, the foregoing solution has only second-order (quadratic) accuracy in θ. Therefore, any simplifications that preserve the second-order

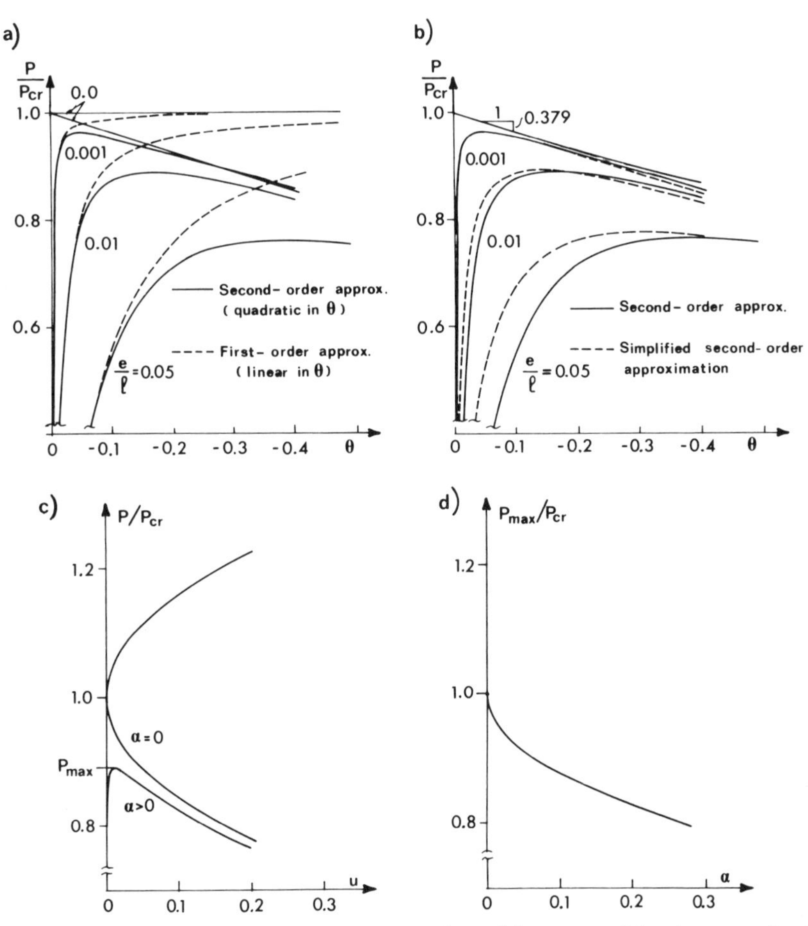

Figure 2.26 (a, b) Curves of load versus rotation; (c) curve of load versus load-point displacement; (d) imperfection sensitivity diagram.

accuracy are admissible and cause no error. To this end, we expand all the variables in Equation 2.6.6 into a power series with respect to θ about the critical state ($e = 0$). Then we discard all the terms of powers higher than θ^2, that is, we solve Equation 2.6.6 with second-order accuracy in θ. Since we are expanding all the coefficients about the critical state, we evaluate them for the critical state (onset of buckling, $\theta = 0$). At that state, $u^{(2)} = v^{(2)} = 0$, $P = P_{\text{cr}} = P^c$ and $V^b = V^c = 0$. Using power series expansions for functions s and c (see Dean and Ugarte, 1968, and Eq. 2.1.12), the second-order approximations for s_b and c_b near the critical state are

$$s_b \simeq 4 - \frac{2\pi^2}{15}\left(\frac{P^b}{P_E^b}\right) \qquad c_b \simeq \frac{1}{2} + \frac{3}{8}\left(\frac{\pi^2}{15}\right)\left(\frac{P^b}{P_E^b}\right) \tag{2.6.7}$$

because $P^b = 0$ at the critical state.

The approximation for P^b as a function of θ can be obtained from Equation 2.6.3 but it can be more directly reasoned from the fact that, at the onset of buckling, the joint is in a state of neutral equilibrium (Fig. 2.25c). Therefore, the moment acting on the column must be equal to the moment acting on the beam, which equals $3EI/\beta l$ because initially $P^b = 0$. Then, using Equation 2.6.3 one may calculate the shear force in the column:

$$P^b = V^c = -\frac{3EI}{\beta l^2}\theta, \qquad \rho^b = \frac{P^b}{P^b_{\mathrm{E}}} = -\frac{3\beta}{\pi^2}\theta \tag{2.6.8}$$

From this it appears that $P^b > 0$ (compression) if $\theta < 0$. So we conclude that the change of beam stiffness also contributes to the initial downward slope for $\theta < 0$.

Substituting ρ^b from Equations 2.6.8 into Equations 2.6.7 and then s_b and c_b into Equation 2.6.6, and neglecting terms that contain powers higher than θ^2, we obtain

$$\chi\theta + \xi\theta^2 = 0 \tag{2.6.9}$$

where $\chi = s_c(1 - c_c^2) + 3/\beta$ and $\xi = s_c(c_c^2 - 1)\eta_b/l + 3\eta_c/\beta^2 l + 3\beta/5$. Coefficient χ depends on $\rho^c = P^c/P^c_{\mathrm{E}}$. For P close to P_{cr}

$$\chi \simeq \chi_{\mathrm{cr}} + (P^c - P_{\mathrm{cr}})\chi' \qquad \text{with } \chi' = d\chi/dP^c \tag{2.6.10}$$

Now $\chi = 0$ for $P = P_{\mathrm{cr}}$, and so $\chi_{\mathrm{cr}} = 0$. (The reason is that at the onset of buckling all the second-order terms in Eq. 2.6.6 or 2.6.9 disappear.) This condition also gives the critical load of the perfect frame: $P_{\mathrm{cr}} = \rho^c_{\mathrm{cr}} P^c_{\mathrm{E}}$, in which $\rho^c_{\mathrm{cr}} = 1.407$ for $\beta = 1$. The value of $\chi' = (d\chi/d\rho^c)/P^c_{\mathrm{E}}$ is evaluated at $\rho^c = P_{\mathrm{cr}}/P^c_{\mathrm{E}}$, and it may be noted that $\chi' < 0$. In view of Equation 2.6.4 we have $\chi = \chi'(P - V^b - P_{\mathrm{cr}})$. Taking into account Equations 2.6.7 and neglecting third-order terms in θ, Equation 2.6.8 becomes

$$P = P_{\mathrm{cr}}(1 + a\theta) \tag{2.6.11}$$

in which

$$a = a_1 + a_2 + a_3 \qquad a_1 = \frac{3EI}{(\beta l)^2 P_{\mathrm{cr}}} \qquad a_2 = \frac{3\beta}{5(-\chi')P_{\mathrm{cr}}} \qquad a_3 = \frac{\xi_{\mathrm{cr}}}{(-\chi')P_{\mathrm{cr}}} \tag{2.6.12}$$

Note that in writing Equation 2.6.11 we neglected the increment $\Delta\xi = (P^c - P_{\mathrm{cr}})\xi'$ because $(P^c - P_{\mathrm{cr}})$ is a small quantity, which would be multiplied by θ^2 when substituted in Equation 2.6.9, and thus it would yield a third-order small term. In Figure 2.26b, Equation 2.6.11 is represented by a straight line of slope a, which is very close to the solution we previously obtained from the full equation system for $e = 0$ (and is asymptotically, i.e., for $\theta \to 0$, the same).

There are three terms that contribute to the slope $dP/d\theta$: (1) the stiffness change of the column caused by its axial force change due to the vertical shear force transmitted to it from the beam (term a_1); (2) the stiffness change of the beam caused by its axial force change due to the horizontal shear force transmitted to it from the column (term a_2); and (3) the displacement of the corner due to axial shortenings of the beam and column caused by their deflections (term a_3). Note that in a symmetric frame the shear forces represented by a_1 as well as a_2 would be canceled by the shear force from the opposite member, and term a_3 would vanish also if the joint displacement is precluded by symmetry.

The diagram of load versus load-point displacement u ($u = u^{(2)}$) at $e = 0$ (Fig. 2.25) is obtained, according to Equation 2.6.5a by substituting $\theta = \pm\sqrt{u/\eta_c}$ (with

η_c evaluated for $P^c = P_{cr}$), that is,

$$P = P_{cr}\left(1 \pm \frac{a}{\sqrt{\eta_c}}\sqrt{u}\right) \qquad (2.6.13)$$

Equation 2.6.13 is plotted in Figure 2.26c.

For small values of e and θ, and for values of P^c not too different from P_{cr}, introduction of the approximations in Equations 2.6.7, 2.6.8, and 2.6.10 into Equation 2.6.6 and elimination of higher-order terms in θ provides

$$P = P_{cr}\left[\frac{1 + a\theta}{1 - (\alpha/\theta)}\right] \qquad \text{with } \alpha = \frac{le}{EI(-\chi')} = \frac{\pi^2}{(-d\chi/d\rho^c)}\left(\frac{e}{l}\right) \qquad (2.6.14)$$

where α represents a nondimensional imperfection. Figure 2.26b compares the solution of the full second-order nonlinear system of Equations 2.6.3–2.6.6 to the simplified second-order solution (Eq. 2.6.14). We may observe that for very small imperfections ($e/l = 0.0001$, $e/l = 0.001$) these two second-order solutions agree closely. For stronger imperfections this is not so, mainly because P is no longer close to P_{cr}.

Imperfection Sensitivity

In Equation 2.6.14 we may now introduce the expansion $1/(1 - \alpha/\theta) = 1 + (\alpha/\theta) + (\alpha/\theta)^2 + \cdots$ and assume that $\theta \gg \alpha$ at maximum load. Then, neglecting in the resulting expression all higher than linear terms in θ, we get

$$\frac{P}{P_{cr}} \simeq 1 + \frac{\alpha}{\theta} + a(\theta + \alpha) \qquad (2.6.15)$$

Setting $dP/d\theta = 0$, the value of θ at maximum P is obtained as $\theta_m = -\sqrt{\alpha/a}$ (only the negative root is of interest because we know the column deflects to the right). Substituting this into Equation 2.6.15 we obtain $P_{max}/P_{cr} = 1 - 2\sqrt{a\alpha} - a\alpha$, and for small imperfections α we have

$$P_{max} = P_{cr}(1 - 2\sqrt{a\alpha}) \qquad (2.6.16)$$

Equation 2.6.16 is plotted in Figure 2.26d.

It is important to note that after reaching the critical state the load declines with increasing deflections, that is, the structure exhibits softening. The diagram of load P versus rotation $(-\theta)$ begins to descend with a finite slope $P_{cr}a$ (Fig. 2.26a). On the other hand, the diagram of the load versus the associated displacement, that is, the axial load-point displacement u, begins to descend with a vertical slope (Fig. 2.26c).

The postcritical behavior we just illustrated is generally called asymmetric bifurcation, since the equilibrium path $P(\theta)$ or $P(u)$ at the critical point bifurcates in an asymmetric manner (symmetry would require a horizontal slope at P_{cr}). An important consequence is that the imperfect column has a maximum load P_{max} that is less than P_{cr}. The larger the imperfection, the smaller is P_{max} (Fig. 2.26d). (Note that ACI requires the columns to be designed for approximately $e > 0.01l$ even if the load is supposed to be centric.) Applied loads in buildings can often cause e to be as large as $0.3l$ or more.

The validity of Equation 2.6.16 where a is the initial slope, and especially the fact that P_{max} declines in proportion to $\sqrt{\alpha}$ (α = imperfection), is not limited to this example. It represents the famous half-power law of Koiter (1945) (Sec. 4.6), which applies generally to all asymmetric bifurcations. This law implies a rather severe sensitivity of the maximum load to the magnitude of imperfection, the severity being manifested by the fact that the curves in Figure 2.26c, d start to descend with a vertical tangent (for imperfection-sensitive structures for which the bifurcation is symmetric, Koiter derived a $\frac{2}{3}$-power law, $P_{max} - P_{cr} \simeq \alpha^{2/3}$, which is less severe than $\alpha^{1/2}$).

The frame we analyzed was tested by Roorda (1965a, b). Figure 2.27 shows his test results for two different eccentricities e, which are above and below the value e_0 that offsets the geometric imperfections of the model (adjustment of the theoretical curve by horizontal shift in Fig. 2.27b is necessary due to inevitable imperfections of the experiment). The same figure also illustrates the results of the theoretical calculations of Roorda and Chilver (1970). They found the initial slope of the load-rotation curve to be $(dP/d\theta)/P_{cr} = 0.381$ (at $\theta = 0$), which agreed with Roorda's experimental results (see Fig. 2.27a). The present calculation gives $a = 0.379$, which might be more accurate since the present method is more direct.

If, for example, $e = 0.01l$, then $\alpha = 0.00871$, and Equation 2.6.16 with $a = 0.379$ yields $P_{max} = 0.885P_{cr}$, that is, P drops by about 12 percent below P_{cr}. Calculation of this kind of drop in P_{max} has not been the practice in the design of frames.

Generalizations and Implications

To generalize the foregoing procedure to arbitrary frames one needs to consider second-order joint displacements due to lateral deflections and the second-order

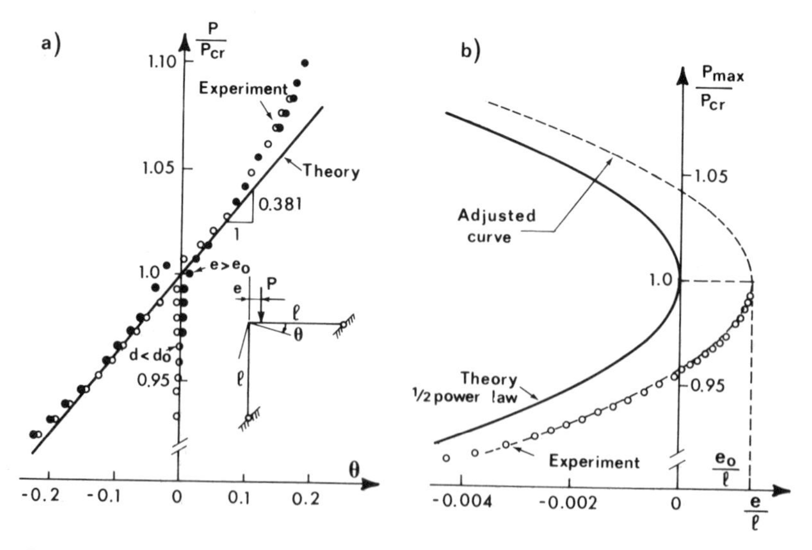

Figure 2.27 Experimental results of Roorda compared to theory.

changes of the stiffness coefficients due to second-order changes of the axial forces in members. The solution may be expected to lead in general to a system of quadratic equations.

In some frames such as the symmetric portal frame in Figure 2.28c, the axial shortening of columns meets with no resistance and the incremental shear force for asymmetric buckling in the horizontal beam has no effect on the sum of vertical loads. The bifurcation is, therefore, symmetric, and neutral equilibrium exists at the critical state (for small deflections), just as it does for columns and, obviously, also for continuous beams. However, if the portal frame columns have different lengths or different stiffnesses (Fig. 2.28d, e) or different axial loads, their axial shortenings u_1 and u_2 are different and cannot be freely accommodated, and the incremental shear force in the horizontal beam affects the sum of the loads. Then the bifurcation is asymmetric. For frames the symmetric bifurcation is in fact an exception rather than a rule.

The redundancy of the frame per se is not the source of asymmetric bifurcation. For example, the two-bar frame in Figure 2.28a, b, which is statically determinate and is obtained from our previous frame by replacing the upper hinge support with a simple support, must also exhibit asymmetric bifurcation. Indeed, the column buckling to the left (Fig. 2.28a) meets in this frame also with less resistance than buckling to the right (Fig. 2.28b) since it produces less curvature in the horizontal beam, due to the fact that the vertical column shortens in proportion to θ^2.

Asymmetric bifurcation is also exhibited by columns if the sliding plane of a simple support is misaligned with the beam axis (see Probs. 1.5.3 and 2.6.7). Since very small misalignments are inevitable, a small bifurcation asymmetry must be always present.

Order of Approximation

We should be aware of the order of approximation in the foregoing solution. The highest-order terms that are contained accurately in the equilibrium equation (Eq. 2.6.2) are of the second-order in rotation θ. Together with the load, the highest-order term in Equation 2.6.2 is $P\theta^2$, which is proportional to Pw^2. So, in view of the fact that $M = P_{cr}w$ is called the second-order theory, the present

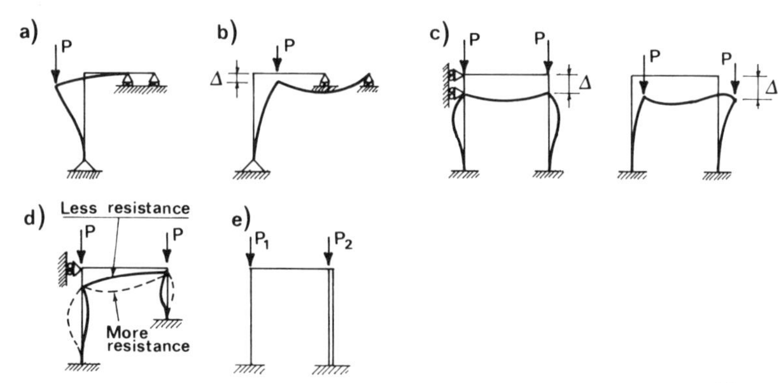

Figure 2.28 Frames exhibiting (a, b, d, e) asymmetric bifurcation, (c) symmetric bifurcation.

approximation (which is of the second order in deflections) can be called the third-order theory.

Note that the power series expansion of beam curvature, which reads $1/\rho = w''(1+w'^2)^{-3/2} = w''(1 - \frac{3}{2}w'^2 + \frac{15}{4}w'^4 - \cdots)$, lacks the second-order term $w''w'$. The first term beyond the linear term w'' is the term $w''w'^2$, which is of the third order, and so it affects in the joint equilibrium condition only terms of order θ^3 but not of order θ^2. It is for this reason that the second-order axial shortening can be calculated from the first-order deflection solution based on the linear curvature expression $1/\rho \simeq w''$ and the linear stability functions s and c. For the same reason, functions s and c (linear stiffness) are insufficient to determine initial postcritical behavior in symmetric bifurcations. Thus the symmetric bifurcation is harder to analyze than the asymmetric one. A higher-order accuracy is needed.

We may recall from the comments in Section 2.5, that the postbuckling solution for columns (Sec. 1.9) as well as pin-jointed trusses can be considered as a fourth-order theory (or a third-order approximation in w). This is the lowest-order theory it can be since, due to symmetry of the column-buckling problem, the second-order terms in w cancel out. For the present L-frame, they do not cancel out due to nonsymmetry, and consequently the third-order terms in w are not needed for the initial postcritical behavior. That is why our solution method for asymmetric bifurcation of the L-frame is rather different and could not be applied in the same manner to postcritical behavior of a column. It is for this reason that one can get correct results calculating the shortening of the members (Eqs. 2.6.5a, b) on the basis of sinusoidal deflection curves, while for columns a higher-order approximation of the deflection curve must be used.

For symmetric frame-buckling problems, by contrast, a fourth-order theory, based on the moment-curvature relation from Section 1.9, is needed to determine the postcritical behavior.

Postcritical Reserve Due to Redundancy

Similar to redundant pin-jointed trusses, which we studied in the previous sections, frames, of course, also exhibit postcritical reserve due to redundancy. For example, when a single column of a large frame buckles, the entire frame need not collapse, since the axial force can be transferred from this column to the adjacent columns.

As long as the bifurcation asymmetry effects we just illustrated are absent (or at least unimportant), the postcritical behavior may again be analyzed by methods of elastoplastic limit analysis, similar to the procedure we demonstrated for pin-jointed trusses. Such analysis can indicate that significant redistributions of internal forces occur due to large-deflection buckling in frames. The increase of load capacity due to these redistributions may overshadow the effects of bifurcation asymmetry due to displacements $u_i^{(2)}$ and $v_i^{(2)}$ and the incremental shear forces. The load of the frame may still increase even if the axial force in one column decreases. This was demonstrated for continuous beams by Masur and Milbradt (1957), who showed that they can possess a much larger postcritical reserve than columns (see also Masur, 1970, and Powell and Klingner, 1970). If, however, the column force decreases with deflection significantly, then the limit analysis methods are inapplicable and step-by-step loading needs to be used.

The existence of either postcritical reserve or postcritical softening is important for probabilistic analysis and safety of frames. Obviously, if only one out of many columns in the frame fails, the adverse consequences may be quite limited and much less severe than the consequences of overall failure of the entire frame due, for example, to long-wave extensional buckling. Probabilistic analysis of these problems is needed.

Finite Element Computational Procedure

Large-deflection buckling of frames can be calculated with the help of finite element programs, using the technique of incremental loading (see, e.g., Bažant and El Nimeiri, 1973). Each beam of the structure is subdivided into beam elements that are so short that their curvature will remain negligible even for the large deflections anticipated. Even after large deflections, the incremental stiffness matrix of such short elements can be assumed to be the same as that for small deformations as indicated in the preceding text (Eq. 2.3.5). For each load increment the computation proceeds as follows (Bažant and El Nimeiri, 1973):

1. Using the current axial force P in each element and current x and y coordinates of the end nodes of the element, generate the linearized 6×6 stiffness matrix $\mathbf{k}$ for each beam element (Eq. 2.3.5). Then assemble these element stiffness matrices into the structural stiffness matrix $\mathbf{K}$. Solve the displacement increment vector $\Delta\mathbf{u} = (\Delta u_1, \Delta u_2, \ldots)^T$ from the linear equation system $\mathbf{K}\,\Delta\mathbf{u} = \Delta\mathbf{f}$ in which $\Delta\mathbf{f}$ is the vector of load increments for the current loading step. [To obtain the postpeak response, one must prescribe here the load-point displacement increments, rather than the load increments; if matrix $\mathbf{K}$ happens to be singular, it means that a critical state (failure) has been reached.]
2. Adding $\Delta\mathbf{u}$ to the old nodal coordinates, generate new coordinates of the nodes, and adding calculated $\Delta\mathbf{f}$ to the old $\mathbf{f}$, obtain new reaction values. Using $\Delta\mathbf{u}$, calculate from $\mathbf{K}\,\Delta\mathbf{u}$ the increments $\Delta\mathbf{P}$ of the axial forces $\mathbf{P}$ in all finite elements, and add them to the previous values of $\mathbf{P}$. Unless the given termination condition is satisfied, return to step 1 and start computations for the next loading increment.

We will return to large-deflection buckling in Section 4.9 where we will solve both perfect and imperfect columns by the energy approach.

Often the response of frames at large deflections becomes inelastic, and then a different solution is required; see Chapters 8 and 10.

Problems

2.6.1 For the frame in Figure 2.25 (with $\beta = 1$) solve the critical load and then calculate η_c, η_b (Eqs. 2.6.5), a (Eqs. 2.6.12), and α (in terms of e). What are the relative magnitudes of η_c and η_b? Compare with the value $a = 0.381$ given by Roorda and Chilver. (The results are $\eta_b/l = 0.1$, $\eta_c/l = 0.621$, $\rho_{cr} = 1.407$, $a = 0.379$.) Determine $P_{cr} - P_{max}$ for $e = 0.01l$.

2.6.2 In Equations 2.6.11 and 2.6.12, a_1 and a_2 represent the effect of the incremental shear forces in the beam and in the column, respectively, and a_3 represents the effect of the second-order displacements of the joint. Verify that $a_3 = 0.2160$, $a_2 = 0.0372$, and $a_3 = 0.1255$.

2.6.3 Same as Problem 2.6.1 but (a) the column and beam are fixed at the ends, (b) the column base is hinged and the beam end is fixed, (c) vice versa.

2.6.4 Same as Problem 2.6.1, but for the frames in Figure 2.29a, b, c.

2.6.5 Discuss the postbuckling stiffness change for the asymmetric bifurcation of the frame in Figure 2.29d.

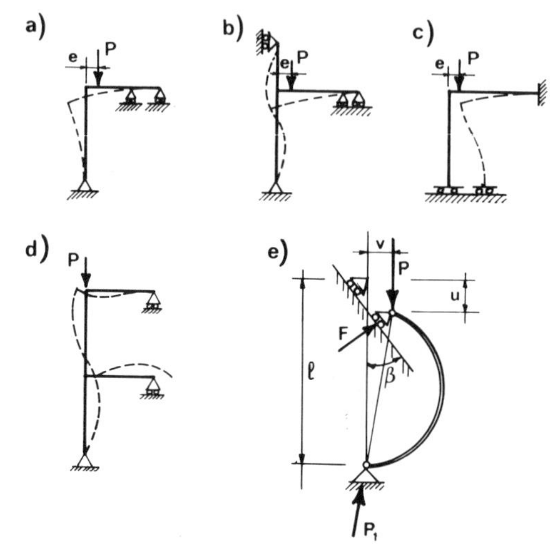

Figure 2.29 Exercise problems on postcritical behavior (a–d) of frames and (e) of hinged column with inclined support.

2.6.6 Calculate the plot of curves $P(\theta)$ for $e = \pm 0.001l$, $e = \pm 0.01l$, $e = \pm 0.1l$ corresponding to Problem 2.6.1.

2.6.7 Asymmetric bifurcation is also exhibited by columns if their symmetry is broken by a support. Show that this is the case for a hinged column in Figure 2.29e in which the support slides on a plane of inclination β with respect to the column axis (cf. Prob. 1.5.3 and Sec. 4.5). *Hint:* Force equilibrium of the slider along the plane of sliding requires that $P \cos \beta = P_1 \cos(\beta + v/l)$, and sliding can occur only if $P_1 = P_E$.

2.6.8 Asymmetric bifurcation is also exhibited by asymmetric or asymmetrically loaded rigidly jointed trusses. Analyze the sensitivity to the eccentricity of the load for the trusses in Figure 2.30a, b, c. Such trusses were tested and analyzed by Roorda.

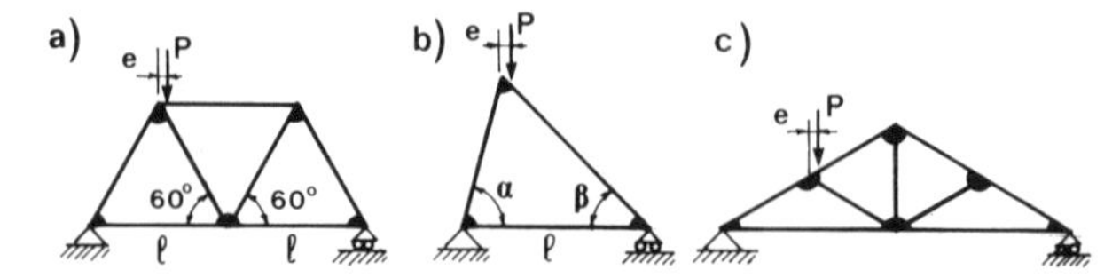

Figure 2.30 Exercise problems on postcritical behavior of trusses.

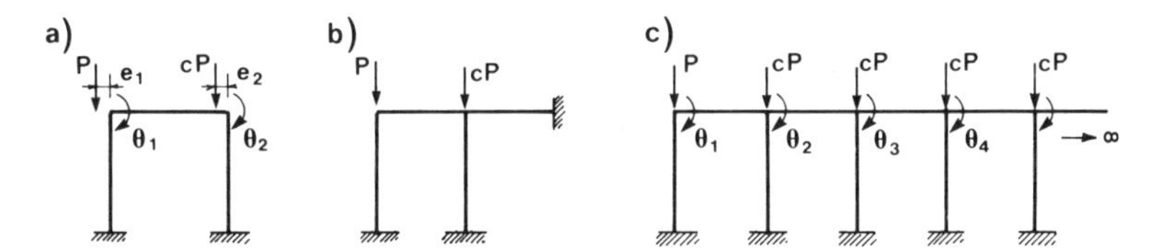

Figure 2.31 Further exercise problems on postcritical behavior of frames.

2.6.9 Show that for $c \neq 1$ the frame in Figure 2.31a exhibits asymmetric bifurcation. Assume the frame to be (a) braced, $e_1 = e_2 = e$, (b) unbraced, $e_1 = -e_2 = e$; and for $c = 0.98$, 0.95, 0.90, 0.70, 0.30, 0.00 solve (1) P_{cr} (see Prob. 2.2.18), (2) $\partial P_{cr}/\partial \theta_1$ for the initial postbuckling behavior, (3) $P_{cr} - P_{max}$ for $0 = 0.01l$.

2.6.10 Same as above, but for the frame in Figure 2.31b.

2.6.11 Same as above, but for frame in Figure 2.31c. *Hint:* First determine the ratio $\theta_{1+1}/\theta_i = r$ which must be constant for $i > 1$.

2.6.12 Again solve Problem 2.2.25 taking now, however, axial shortenings of the members due to θ and u into account. *Note*: This arch frame exhibits asymmetric bifurcation. Based on the first-order solution from Problem 2.2.25, one may calculate the second-order approximations to axial shortenings $u_a^{(2)}$ of members $\overline{12}$ and $\overline{34}$, and $u_b^{(2)}$ of member $\overline{23}$. They represent quadratic forms in θ and v. The effect of $u_a^{(2)}$ and $u_b^{(2)}$ is to cause the displacement v at joint 2 to increase to $v + \Delta v$, and displacement v at joint 3 to decrease to $v - \Delta v$. The magnitude of Δv is determined from the condition that the horizontal projections of all the second-order displacements (Fig. 2.32a) must have a zero sum, as dictated by the support conditions. This yields the condition $-u_a^{(2)}/\sqrt{2} + \Delta v/\sqrt{2} - u_b^{(2)} + \Delta v/\sqrt{2} - u_a^{(2)}/\sqrt{2} = 0$, from which $\Delta v = u_a^{(2)} + u_b^{(2)}/\sqrt{2}$. Replacing v with $v + \Delta v$ and $v - \Delta v$ for joints 2 and 3, respectively, in the equilibrium equations of Problem 2.2.25, we then get two quadratic equations relating P, θ, and v, from which the initial postcritical response can be calculated.

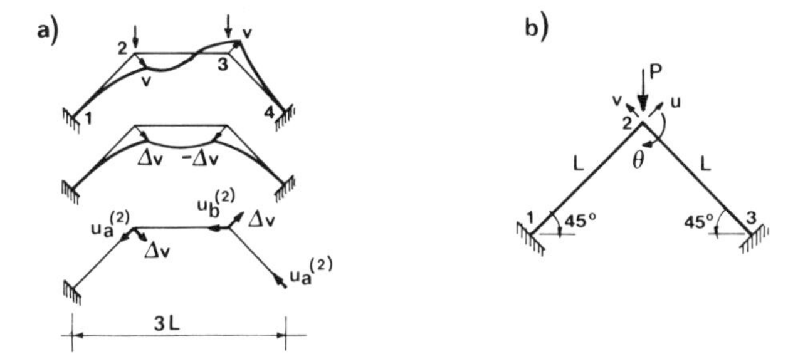

Figure 2.32 Further exercise problems on postcritical behavior of frames.

2.6.13 Consider the L-frame in Figure 2.32b, which represents a symmetric structure-load system. Applying the same method as we did for the Koiter-Roorda frame, verify that terms proportional to θ^2 in the joint equilibrium equation cancel out. Consequently, this frame does not exhibit asymmetric bifurcation. Rather, it must exhibit symmetric bifurcation, for which the initial slope of load versus deflection is horizontal. At larger deflections the load can either increase (as in Sec. 1.9) or decrease. To decide it, higher-order terms must be included (see Sec. 4.9).

2.6.14 The distribution of P in the systems of Figure 2.33a, b, c, d varies with deflection. Is this sufficient to produce asymmetric bifurcation? Is the postbuckling behavior symmetric?

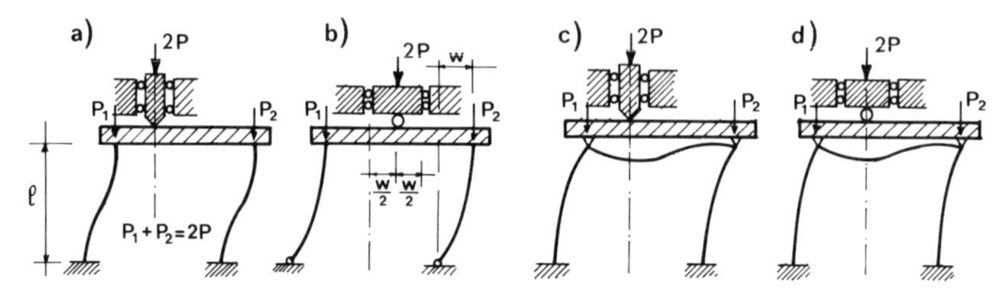

Figure 2.33 Exercise problems on postcritical behavior of frames with load distribution varying with deflection.

2.6.15 Stochastic eccentricity. Load P on the L-frame of Koiter and Roorda, which was studied in the text, has an uncertain eccentricity e that has a normal probability distribution with mean $\bar{e}$ and standard deviation s_e. Using the final simplified Equation 2.6.16 determine the mean and standard deviation of $\max P$ (it is the problem of statistics of a function a of a random variable; see, e.g., Benjamin and Cornell, 1970).

2.6.16 Same as Problem 2.6.15, however not only the eccentricity but also the load P are uncertain, with log-normal distributions. The means and standard deviations of $\log e$ and $\log P$ are μ_e, μ_p, s_e, and s_p. Formulate the problem and discuss the numerical method of calculation of the probability that P would exceed $P_{\max}$.

2.7 BUILT-UP COLUMNS AND REGULAR FRAMES AS COLUMNS WITH SHEAR

In Section 1.7 we analyzed the buckling of columns that exhibit significant shear deformations. Such deformations happen to be very important when built-up columns are approximately analyzed as homogeneous columns. Disregard of the shear effect proved tragic, causing one of the greatest disasters in the history of bridge building. As a world record span of over 500 m was being completed over the St. Lawrence River in Quebec at the beginning of this century (year 1907), one of the compression diagonals of this truss railroad bridge (a cantilever system with inserted simple spans) buckled and caused total structural collapse sending many workers to their deaths. The critical diagonal was a built-up member and the source of failure was traced to a very low shear stiffness of this diagonal. Although Engesser's formula for the shear effect had been presented about two decades before, it passed unnoticed by the bridge designers. Learning from this experience, the bridge was then successfully rebuilt in the 1920s and by virtue of this disaster it became generally accepted that built-up columns must be designed for buckling with shear.

Let us now illustrate how a built-up column may be approximated by a homogeneous column. Consider the column shown in Figure 2.34a, consisting of two channel beams connected by welded batten plates. Although such a column would be more accurately treated as a frame, it suffices to treat it as a single homogeneous column if the number of panels is large. The distribution of the bending moment along the channel beams is then approximately periodic, repeated in each panel. This requires the bending moment in the channel to have a zero value at middistance between the batten plates. Thus, we may imagine an H-shaped element to be cut by cross sections at middistance between the batten plates (Fig. 2.34b), and we may consider that the ends of this element are subjected only to shear forces. Using the principle of virtual work and the bending moment distributions along the channel beams and the batten plates as shown in Figure 2.34c, one may calculate that the deflection δ (Fig. 2.34c) caused by shear force Q is

$$\delta = \gamma a \qquad \gamma = \frac{Q}{GA_0} \qquad \frac{1}{GA_0} = \frac{ab}{12EI_b} + \frac{a^2}{24EI_c} \tag{2.7.1}$$

in which γ is the average shear angle, a is the distance between the centroids of the batten plates, b is the distance between the centroids of the channel beams, and I_c and I_b are the centroidal cross-section moment of inertia of the channel beams and of the batten plates, respectively. The expression for GA_0 in Equations 2.7.1 represents the equivalent (approximate) shear stiffness of the built-up column. Based on this stiffness, one may then use Engesser's formula (Eq. 1.7.7 of Chap. 1) to obtain the critical load.

Consider now the column shown in Figure 2.35a. Since the bending moments on the individual members are negligible, we may assume hinges at each node

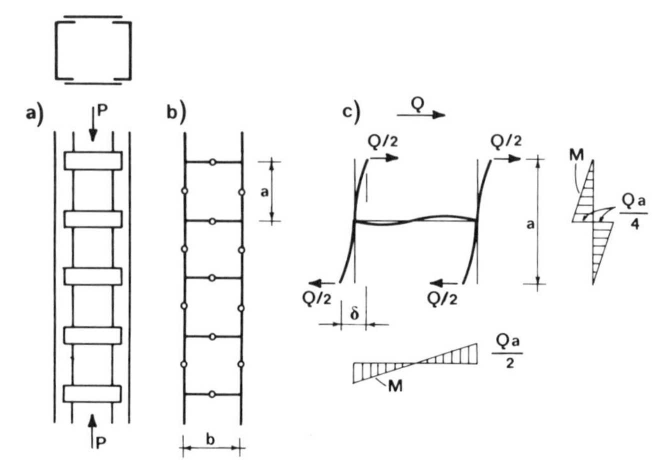

Figure 2.34 (a) Battened column, and (b, c) subdivision in cells for approximate calculation of shear deformation.

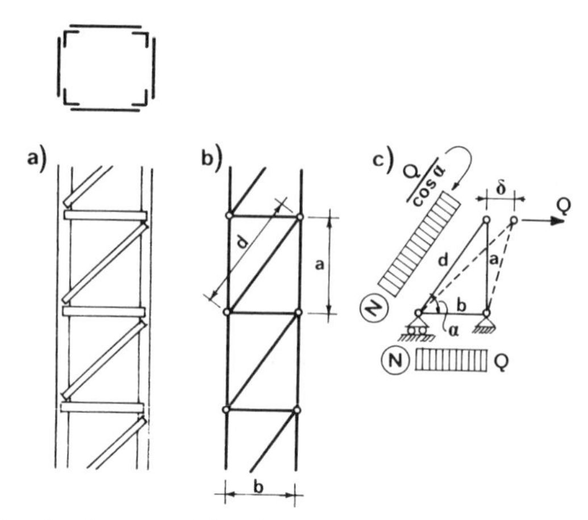

Figure 2.35 (a) Latticed column and (b, c) subdivision in cells for approximate calculation of shear deformation.

(Fig. 2.35b) and treat the column as a truss. To calculate the equivalent shear deformation we isolate one cell of the truss whose length a represents the distance between two adjacent lacing bars, and we subject the cell to a shear force Q (Fig. 2.35c). The relative lateral displacement δ may be calculated from the principle of virtual work on the basis of the axial forces in the lacing bars shown in Figure 2.35c. Neglecting the axial deformation of the vertical members, the result is

$$\delta = \frac{Q}{GA_0} a \qquad \frac{1}{GA_0} = \frac{1}{ab^2}\left(\frac{d^3}{EA_d} + \frac{b^3}{EA_t}\right) \tag{2.7.2}$$

in which b is the distance between the centroids of the vertical members, d is the length of the diagonal measured to the centroids of the vertical members, and A_d and A_t are the cross-sectional areas of the diagonal and transversal lacing bars.

Similar expressions can be derived for other types of built-up columns or latticed columns; see, for example Bleich (1952) and Timoshenko and Gere (1961). A similar consideration of shear is also possible for columns with webs that are weakened by openings.

The column with shear can also be used as a model that greatly simplifies the analysis of long-wave extensional buckling of large regular frames. If the width-to-height ratio (Fig. 2.36a) is not too high, this approach gives results that are very close (errors of 5 to 10 percent) to the exact solutions obtained by finite difference calculus or by computer analysis based on the assembly of all members (Bažant and Christensen, 1972b).

Consider a planar regular rectangular frame (Fig. 2.36b) in equilibrium under axial forces P_y in columns and P_x in beams (usually $P_x = 0$). Let L_x and L_y be the length of horizontal and vertical members; A_x and A_y their cross-sectional area, and I_x, I_y their cross-sectional moments of inertia (subscripts x and y refer to members in the horizontal and vertical directions).

In analogy with the well-known portal method of approximate frame analysis, it will be assumed that the bending moments at the midlength points of all columns and beams are zero. This is equivalent to the assumption that the rotations of all joints in a given floor and in the two adjacent floors are equal.

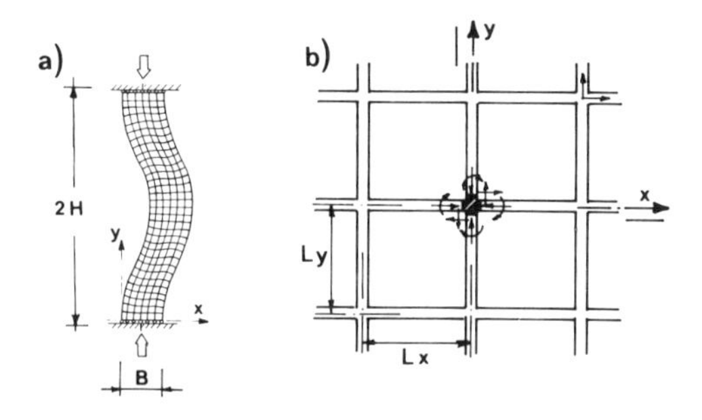

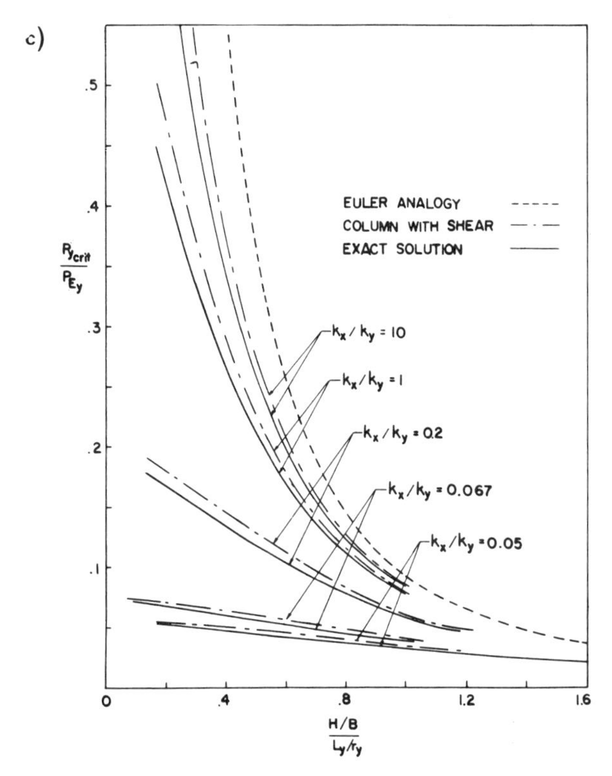

Figure 2.36 (a) Frame geometry; (b) equilibrium at a typical joint; (c) comparison of frame critical loads obtained by various methods. (*After Bažant and Christensen, 1972b.*)

Consider an imposed relative horizontal displacement γL_y between the adjacent floors. Taking the initial forces, P_y, into account and neglecting P_x, we obtain the moment equilibrium equations for the interior joints and the boundary joints: $6\beta k_x \phi + 2k_y \bar{s}_y \phi - 2k_y \bar{s}_y \gamma = 0$ in which $\beta = 2$ for an interior joint and $\beta = 1$ for a boundary joint, and $k_x = EI_x/L_x$, $k_y = EI_y/L_y$. Then we calculate the sum $\sum V$ of the corresponding shear forces V in all the columns of the floor, each of which is expressed (after substitution for ϕ) as follows: $V/\gamma = (s_y^* \gamma - 2\bar{s}_y \phi)k_y/L_y = \{12(k_y/L_y)/[(6/\bar{s}_y) + 2k_y/(k_x \beta)]\} - P_y$. From this, the shear rigidity $\tilde{R}$ of the cross section of the whole frame is obtained as

$$\tilde{R} = \frac{\sum V}{\gamma} = 12 \frac{m\kappa k_y}{L_y} \tag{2.7.3}$$

$$\kappa = \frac{1 - 2/m}{(6/\bar{s}_y) + (k_y/k_x)} + \frac{2/m}{(6/\bar{s}_y) + (2k_y/k_x)} - \frac{\pi^2}{12}\left(\frac{P_y}{P_{E_y}}\right) \tag{2.7.4}$$

in which m = the number of columns in the floor. The moment of inertia of the horizontal cross section of the whole building (neglecting the contribution of the values of I_y) is

$$\tilde{I} = A_y L_x^2[1^2 + 2^2 + \cdots + (m-1)^2] - mA_y\left[\frac{L_x(m-1)}{2}\right]^2 = \frac{m}{12}(m^2 - 1)A_y L_x^2 \tag{2.7.5}$$

According to Engesser's formula (Eq. 1.7.7), the critical value $\tilde{P}_{cr}$ of the resultant of initial axial forces in all columns is

$$\tilde{P}_{cr} = mP_{y_{cr}} \simeq \frac{\tilde{P}_E}{1 + \tilde{P}_E/\tilde{R}} \tag{2.7.6}$$

or

$$\frac{P_{y_{cr}}}{P_{E_y}} = \left[12\left(\frac{m-1}{m+1}\right)\left(\frac{r_y H}{L_y B}\right)^2 + \frac{\pi^2}{12\kappa}\right]^{-1} \tag{2.7.7}$$

in which $\tilde{P}_E = \pi^2 E\tilde{I}/H^2$ = Euler load for the frame taken as a single column without shear, B = width of frame, H = height corresponding to half-wavelength (Fig. 2.36a), and $r_y = \sqrt{I_y/A_y}$.

Equation 2.7.7 is not an explicit formula for $P_{y_{cr}}$, since $P_{y_{cr}}$ appears in Equation 2.7.4 for κ. In the cases of interest for long-wave buckling, however, $P_{y_{cr}} \ll P_{E_y}$ and so an approximate value of $P_{y_{cr}}$ may be obtained by setting $P_y \simeq 0$ and $\bar{s}_y \simeq 6$ in Equation 2.7.4.

The results of this approximate analysis are shown by the dash-dot lines in Figure 2.36c in comparison with the exact solutions for a typical frame. It is seen that the predictions are surprisingly accurate and quite satisfactory for most design purposes, even for the cases where the actual distribution of vertical displacements along the floor is far from a linear one (although the derivation of Engesser's formula assumed the cross section to remain plane).

Still more accurate values can be obtained for high values of $(H/B)/(L_y/r_y)$ by solving Equations 2.7.4 and 2.7.7 for $P_{y_{cr}}$ exactly, for example, by the *regula falsi* method. Curiously, though, for low values of $(H/B)/(L_y/r_y)$, solving Equations 2.7.7 and 2.7.4 by an iterative procedure leads to a value for $P_{y_{cr}}/P_E$ that has a greater error than the first approximation.

The approximation by a column with shear is particularly useful for practical problems that are difficult to solve exactly by finite difference calculus, such as a free-standing frame in which the column cross sections and the axial forces vary from floor to floor.

The consideration of shear in regular frames exemplified in the preceding analysis may be used also in more complex structural systems. An example is the buckling of a framed tube, a term used for a stiff-perimeter frame frequently used in modern skyscrapers (e.g., the Sears Tower, Chicago and the World Trade Center, New York, the two tallest buildings in the world in the 1980's) or the buckling of planar frames or frame tubes that are coupled with a stiffening truss (e.g., John Hancock Building, Chicago) or with strong shear walls in the core of the building (e.g. Lake Point Tower, Chicago). In that case, one needs to write the differential equation for a beam-column with shear using an unknown distributed reaction from the stiffening shear wall, or unknown concentrated reaction forces from the stiffening shear wall, or unknown concentrated reaction forces from the large stiffening truss. These unknowns are then determined from the conditions of compatibility of deflections of the regular frame and the shear wall of the stiffening truss.

Since even in the tallest contemporary buildings the overall critical loads are generally much higher than the actual loads, of main practical interest is the extension of the foregoing calculations to dynamics, particularly to the determination of free vibration frequencies as influenced by the axial loads, in the presence of shear deformations in the beam-column approximating the frame. Accurate knowledge of the vibration frequencies is obviously important for the consideration of seismic loads or wind loads.

Problems

2.7.1 Calculate the equivalent shear stiffness for the trusses in Figure 2.37a, b.

2.7.2 Calculate the equivalent shear stiffness and the critical load for the structures in Figure 2.37c, d, e.

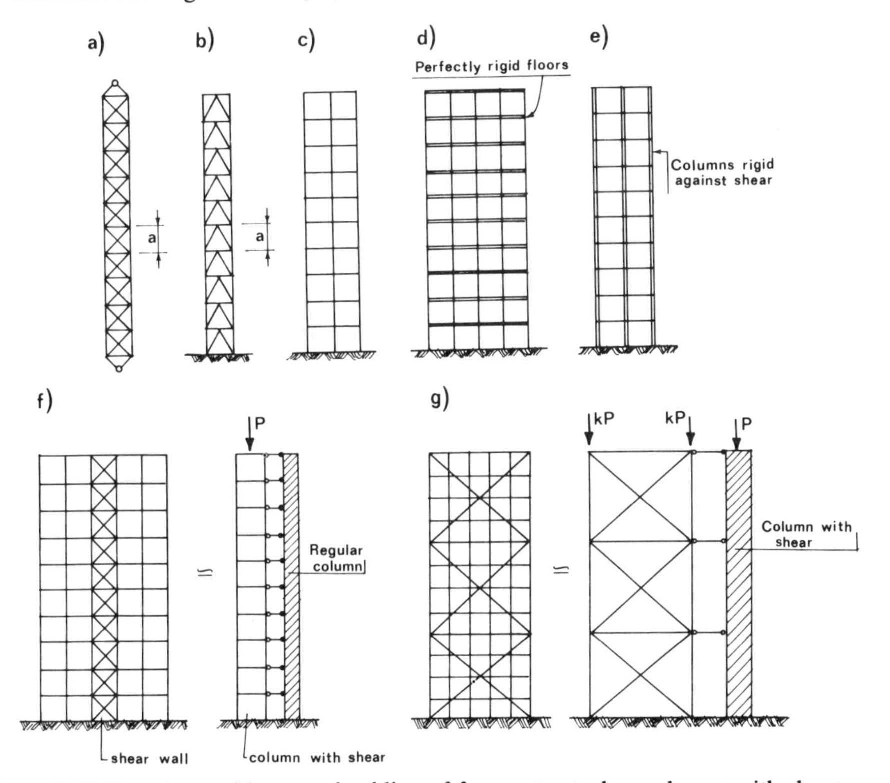

Figure 2.37 Exercise problems on buckling of frames treated as columns with shear.

2.7.3 Consider a regular frame (see Fig. 2.37f) the sway of which is opposed by a shear wall. Analyze it as a column with shear coupled continuously to an ordinary column of bending stiffness EI and the same height, and impose the entire load on the top of the column with shear. Express the critical load.

2.7.4 Consider a regular frame connected at discrete points to a stiffening truss as shown in Figure 2.37g. Express the critical load.

2.8 HIGH ARCHES

From the viewpoint of buckling analysis, it is useful to distinguish two basic types of arches: high arches and flat arches. High arches are those for which the center line of the arch may be considered incompressible, and flat (or shallow) arches are those for which its shortening is important. We will now analyze high arches from the viewpoint of differential equations of equilibrium, and shallow arches we will treat from the viewpoint of energy in Section 4.4. Later, in Section 5.5, we will return to high arches, applying approximate energy methods. Note also that in Problems 2.2.25 and 2.6.12 we have already analyzed a polygonal frame that behaves essentially the same as an arch.

Curvature Change

First we must describe the geometry of deformation. Let R be the curvature radius of the arch, s the length coordinate measured along the curved center line, and w the deflection normal to the center line, positive if toward the center of curvature. The bending moment M is proportional to the curvature change κ, and if the arch is sufficiently slender, the proportionality constant is EI (like for beams). Therefore

$$M = EI\kappa \qquad \kappa = \frac{1}{R^*} - \frac{1}{R} \tag{2.8.1}$$

in which R^* is the curvature radius of the center line after the deflection. As an approximation, we can imagine that the curvature change (Fig. 2.38a) is a sum of the curvature change at constant deflection, which is $1/(R-w) - 1/R$, and the

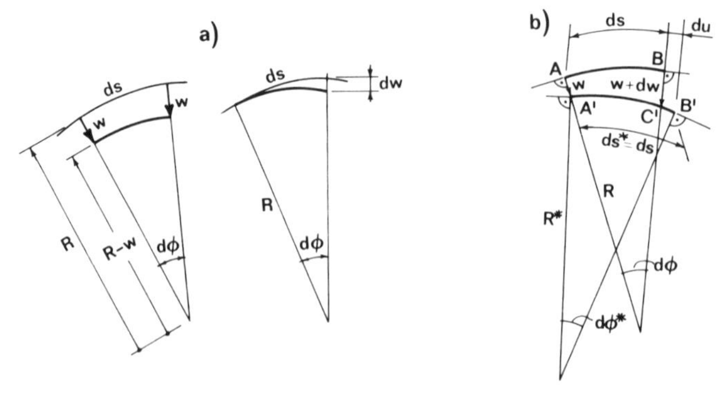

Figure 2.38 Geometry of arch deformation.

curvature change due to deflection variation, which is d^2w/ds^2. So we have

$$\kappa = \frac{d^2w}{ds^2} + \left(\frac{1}{R-w} - \frac{1}{R}\right) = \frac{d^2w}{ds^2} + \frac{w}{R(R-w)} \simeq \frac{d^2w}{ds^2} + \frac{w}{R^2} \tag{2.8.2}$$

where s = length coordinate of arch center line.

However, calculating the curvature change due to w under the assumption of constant w has been an intuitive step. We need a more rigorous justification by calculating the curvature change at variable $w(s)$. To this end, we note (Fig. 2.38b) that because the center line of the arch is inextensible, $ds = R\,d\phi = R^*\,d\phi^* = ds^*$. The angle subtended by the tangents of the center line at the ends A' and B' of the deformed element ds^* (Fig. 2.38b) is $d\phi^* = (w' + w''\,ds) + d\phi - w' + du/R$, in which the last term represents the additional relative rotation between the tangents at B' and C' that is caused by axial displacement u. This displacement is required to satisfy the inextensibility assumption, that is, $du = w\,d\phi$; see Figure 2.38b. Therefore we have

$$\kappa = \frac{1}{R^*} - \frac{1}{R} = \frac{d\phi^*}{ds^*} - \frac{1}{R} = \frac{d\phi^*}{ds} - \frac{1}{R} = \frac{1}{ds}\left[d\phi + \left(\frac{dw}{ds} + \frac{d^2w}{ds^2}ds\right) - \frac{dw}{ds} + \left(w\frac{d\phi}{R}\right)\right] - \frac{1}{R}$$

$$= \frac{d^2w}{ds^2} + \frac{d\phi}{ds}\left(1 + \frac{w}{R}\right) - \frac{1}{R} = \frac{d^2w}{ds^2} + \frac{w}{R^2} \tag{2.8.3}$$

This is the same as Equation 2.8.2 because $d\phi/ds \simeq 1/R$.

Substituting the linearized curvature change expression in Equation 2.8.2 into Equation 2.8.1, we obtain the following governing differential equation for the buckling of high arches:

$$\frac{d^2w}{ds^2} + \frac{w}{R^2} = \frac{M}{EI} \tag{2.8.4}$$

Approximate Theory for Perfect Arches with a Fixed Compression Line

It is expedient to analyze first a perfect system. A perfect arch is an arch in which the center line before buckling coincides with the compression line, which represents the locus of the points of the normal force resultant within the cross sections (and is also called the funicular line). Perfect arches are, for example, circular arches under uniform radial pressure, parabolic arches under a vertical distributed load that is uniform on a horizontal projection, and catenary arches under dead weight that is uniform along the arch, provided that the boundary conditions do not introduce significant bending moments. (For example, the two-hinge arch or fixed arch would need to have radially sliding supports in order to achieve zero bending moments upon application of uniform radial pressure; see Fig. 2.39a, b). Although such boundary conditions are not normally used, if the arch is not too slender the bending moments produced by boundary conditions due to center-line compression are small and can be neglected. Anyhow, the assumption of incompressibility of the center line implies these bending moments to be zero (in Fig. 2.39a, b it does not permit radial sliding of supports).

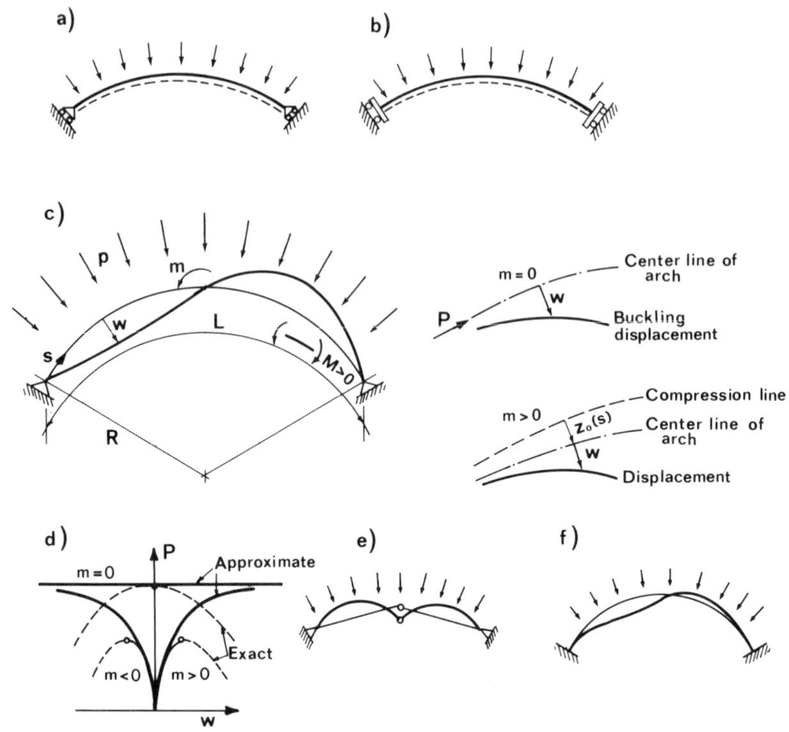

Figure 2.39 (a, b) Boundary conditions for perfect arches; (c) circular arch subjected to uniform radial pressure and disturbing moment m; (d) approximate and exact behavior; (e, f) buckling modes of hinged and fixed arches.

A useful simplification is due to the fact that, for hinged high arches, the reaction resultants of the arch do not move significantly during buckling. Then the compression line remains fixed, that is, does not move with the cross sections during buckling. Consequently, the bending moment M due to buckling deflections w simply is $M = -Pw$ (Fig. 2.39c), and Equation 2.8.4 takes the form:

$$\frac{d^2w}{ds^2} + k^2 w = 0 \qquad k^2 = \frac{1}{R^2} + \frac{P}{EI} \tag{2.8.5}$$

Together with the boundary conditions for $w(s)$, Equation 2.8.5 defines a one-dimensional boundary-value problem for $w(x)$.

The problem may be readily solved if k is constant, which occurs for a circular arch ($R = \text{const.}$) if the axial force P is constant. This case is obtained if the loading consists of a uniformly distributed radial load p, and then $P = pR =$ const., according to the differential equation of equilibrium. For the other types of arches mentioned above, the axial force increases from the crown to the support.

Consider now a two-hinged circular arch with incompressible center line and no imperfection, subjected to uniform radial pressure p (Fig. 2.39c). The differential equation (Eq. 2.8.5) as well as the boundary conditions at hinges ($w = 0$) may be satisfied (for $P = \text{const.}$) by the function $w = A \sin(n\pi s/L)$ where

L is the length of the arch center line from one hinge to another. Upon substitution in Equation 2.8.5 it then follows $n^2\pi^2/L^2 = 1/R^2 + pR/EI$, which yields for the critical value of p the formula, due to Hurlbrink (1908),

$$p_{\mathrm{cr}_n} = \frac{EI}{R^3}\left[\frac{n^2\pi^2}{(L/R)^2} - 1\right] \qquad (n = 2, 4, 6, \ldots) \tag{2.8.6}$$

Now there exists an interesting point particular to arches. Not every sine curve is admissible if the center line is assumed inextensible. This inextensibility condition could in no way be satisfied if the number of half-sine waves from support to support were odd. The number n of half-sine waves must be even, because deflections outward tend to extend the arch while those downward tend to shorten the arch, and one must exactly compensate for the other. Therefore, the smallest admissible value of n in Equation 2.8.6 is not $n = 1$ but $n = 2$. This was noted by Hurlbrink (1908), although the governing differential equation (Eq. 2.8.5) was obtained long before him by Boussinesq (1883).

The formula also applies for a full circular ring, in which case, however, it yields $p_{\mathrm{cr}_2} = 0$ because $L/R = 2\pi$. This is not surprising since, in this case, the two support hinges coincide and the arch no longer has a statically determinate support. Rather, it can freely rotate as a rigid body about the hinge; see Timoshenko and Gere (1961, p. 298). The lowest load that causes buckling is $p_{\mathrm{cr}_4} = 3EI/R^3$ (Bresse formula, cf. Timoshenko and Gere, p. 291).

The foregoing solution may also be used for cylindrical shells, which buckle in a mode that is translationally symmetric along the axis of the cylinder (this can be assumed only if the load does not change in the direction of the axis). In this case the bending stiffness EI must be replaced by the cylindrical stiffness of the shell (Chap. 7) $B = Eh^3/12(1 - \nu^2)$ where h = shell thickness and ν = Poisson ratio.

Various Types of Arches and the Effect of Imperfections

Often the cross sections of the arch are not, before buckling, loaded centrically, that is, there is an initial bending moment $M_0(s) = -Pz_0(s)$ where $z_0(s)$ is the distance of the center line of the undeflected (stress-free) arch from the compression line (Fig. 2.39c). In such a case we have on the right-hand side of Equation 2.8.5 the term $M_0(s)/EI$ instead of 0. This is the case, for example, when moment $m \neq 0$ is applied as in Figure 2.39c. Instead of bifurcation-type buckling with a critical load given by Equation 2.8.6, we obtain a load-deflection curve that is similar to the linearized solution of imperfect columns, with the deflections approaching infinity as the critical load in Equation 2.8.6 is approached (Fig. 2.39d). Note, however, that such behavior is obtained only if the arch is loaded asymmetrically. The symmetric components of loads can produce only finite deflections but cannot excite the first critical mode since n must be even (Eq. 2.8.6).

Equation 2.8.6 can be put into the form (for $n = 2$)

$$p_{\mathrm{cr}}R = P_{\mathrm{cr}} = \frac{\pi^2 EI}{(\beta L/2)^2} \tag{2.8.7}$$

in which βL = effective length, $\beta = 2/[4 - (L^2/\pi^2 R^2)]^{1/2}$, and P_{cr} = critical axial thrust. For $L/R \to 0$ we have $\beta = 1$, which corresponds to an arch of a very low rise (for which, however, Eq. 2.8.7 is invalid since the effect of axial extensibility becomes important). For $L/R = \pi$, we have $\beta = 1.15$, which corresponds to a semicircular arch. In view of the fact that the buckling shape (Fig. 2.39c) has an inflection point at the crown, it is interesting that the value of β, lying between 1 and 1.15, is not much different from that for a hinged column of length $L/2$ ($\beta = 1$).

A three-hinged arch of a sufficiently high rise has an antisymmetric buckling shape similar to that of a two-hinged arch. As calculated by Austin (1971), $\beta = 1.14$ to 1.15. It is interesting that this is not very different from the β value for a semicircular two-hinged arch (see also Timoshenko and Gere, 1961, p. 301, in which the results obtained by Dinnik, 1934, are reported). For a not too deep three-hinged arch, however, the buckling becomes symmetric, according to the shape illustrated in Figure 2.39e.

A fixed arch cannot be solved by the present procedure since the compression line moves during buckling. It is then best to solve the arch by approximate energy methods, which will be shown in Section 5.5. Timoshenko and Gere (1961, p. 299) present a solution due to Nicolai (1918), from which it appears that $\beta = 0.70$ to 0.71 (see also Austin, 1971). Note that the effective length factor β is rather close to the value for a fixed–hinged column of length $L/2$. This is not surprising, since this type of arch buckles with an inflection point ($M = 0$) at the crown (Fig. 2.39f).

The data presented in Timoshenko and Gere (1961) for parabolic arches under a vertical load uniformly distributed over the horizontal projection, as well as catenary arches subjected to own weight, show that the value of the effective length factor is very close to that which we have mentioned for circular arches subjected to normal load (Austin, 1971). For typical arch shapes, and for P defined as the critical axial thrust at quarter-points of the span, it has been found that

$$\begin{aligned} \beta &= 0.68 \text{ to } 0.73 \qquad &&\text{(fixed arch)} \\ \beta &= 1.10 \text{ to } 1.24 \qquad &&\text{(two-hinged arch)} \\ \beta &= 1.10 \text{ to } 1.15 \qquad &&\text{(three-hinged arch, except catenary)} \end{aligned} \tag{2.8.8}$$

For the case of own weight, DaDeppo and Schmidt (1971) have shown that two-hinged circular arches have a critical value of the axial thrust significantly smaller than under normal loading.

The case of circular arches loaded by a concentrated load at the crown has been analyzed by DaDeppo and Schmidt (1969). The critical thrust at the quarter-points, expressed by Equation 2.8.7, corresponds in this case to the values $\beta = 1.0$ to 1.14, which is about the same as for a uniformly distributed load. The prebuckling deformation due to the primary bending moment obviously cannot be neglected in this case. It leads, for a two-hinged arch, to a load-deflection curve of the type illustrated in Figure 2.40a. At the critical load the equilibrium path intersects another path that corresponds to an antisymmetric mode with sidesway and has a negative slope. The arch follows this second path after the bifurcation point, that is, the deformation ceases to be symmetric. This causes sensitivity to imperfections (of a similar type as illustrated in Sec. 2.6). As is clear from Figure 2.40a, the presence of inevitable imperfections causes the

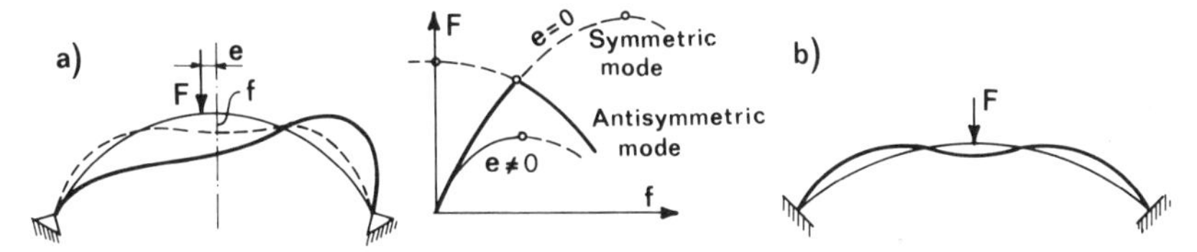

Figure 2.40 (a) Hinged circular arch subjected to a concentrated load and (b) the load-displacement curve; fixed arch subjected to concentrated load.

load-deflection curve to have a maximum load that is less than the critical load for the bifurcation point.

On the other hand, fixed arches under concentrated load at the crown buckle symmetrically (Fig. 2.40b). For this case the value of the effective length factor is $\beta = 0.77$ to 0.83, which is considerably larger than for the case of distributed normal loading (Schmidt and DaDeppo, 1972).

An important fact about slender arches is that they are imperfection-sensitive. This means that, in contrast to columns, imperfections such as asymmetry of the applied load cause a decrease of the maximum load compared to the critical load of the perfect arch. This is due to appearance of an asymmetric deformation mode. The type of bifurcation in buckling of arches is called unstable symmetric (Fig. 2.39d). We will discuss such behavior in more detail in Section 4.5 (see also Thompson, 1982, p. 57).

The symmetric buckling modes of a circular arch under radial or vertical loads are illustrated by the teaching model in Figure 2.41a, b developed at Northwestern University (1969). The results of the experiments of Roorda (1965a, b) are shown in Figure 2.42a, b. Figure 2.42a gives the load-rotation diagram for a certain value of the eccentricity. The critical load of the imperfect arch decreases with the eccentricity of the load, as shown by the imperfection sensitivity diagram in Figure 2.42b. The milder effect of the imperfection with respect to asymmetric bifurcation (approximating the $\frac{2}{3}$-power law) is evident. We will return to this example in Section 4.5.

Arches can also buckle laterally, in a bending-torsional mode. This will be touched on in Section 6.3; see also Timoshenko and Gere (1961), where many other interesting results about buckling of arches and rings are given.

General Linearized Theory for Uniformly Compressed Circular Arches

The equilibrium relation $M = -Pw$, which we used to derive Equation 2.8.5, is not applicable in general. Let us now outline a linearized formulation based on more general equilibrium relations. We restrict attention to the case of an inextensible circular arch of radius R (Fig. 2.43a) that is in equilibrium under a uniformly distributed radial load p, and is characterized by the absence of bending moments, shear forces, and displacements in the prebuckled state. Buckling produces the increments M, V, and N of bending moment, shear force, and axial force (Fig. 2.43b). For the case of radial loads whose direction remains constant during deflection (dead pressure), the differential equilibrium conditions

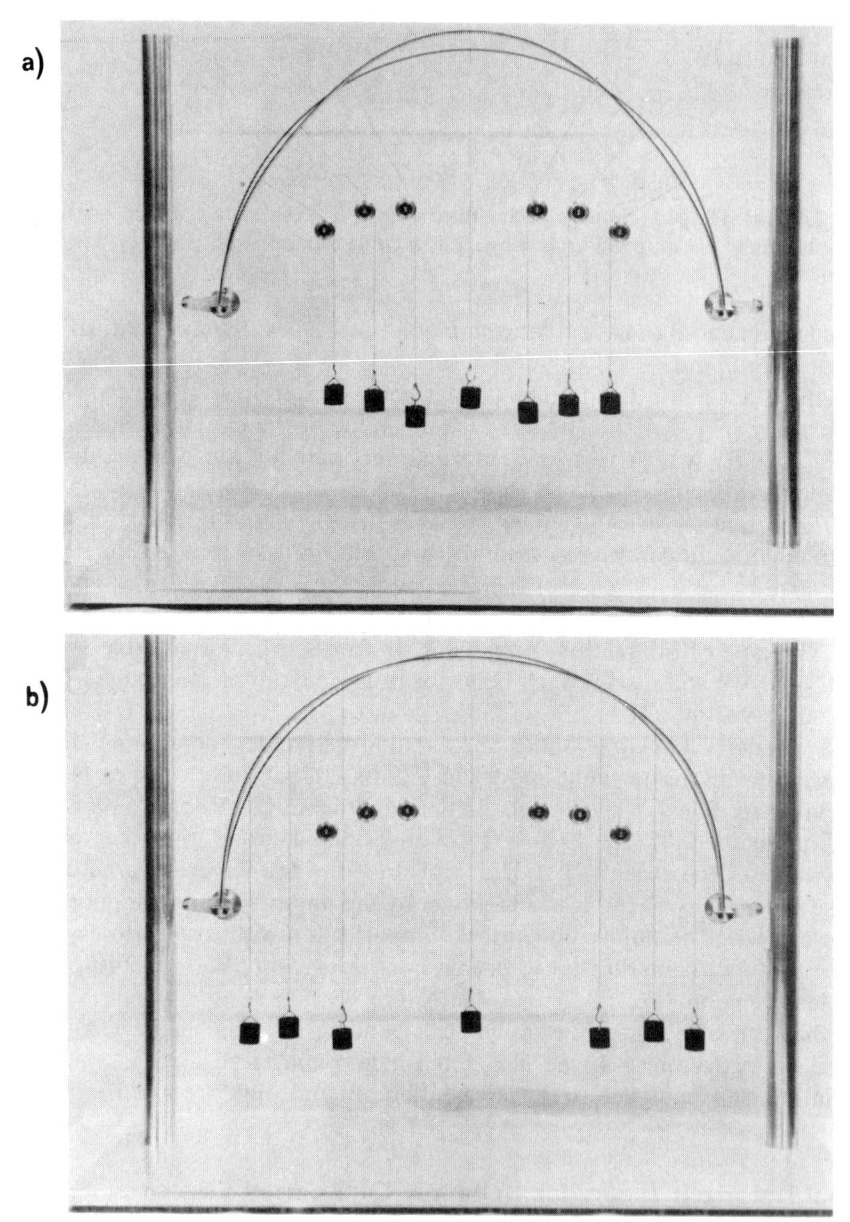

Figure 2.41 Northwestern University (1969) teaching models: in-plane buckling of high arch under normal and vertical loading.

for the radial and tangential directions and for moments are (Oran and Reagan, 1969)

$$V' + N = 0 \qquad N' - V = 0 \tag{2.8.9}$$

$$M' - RV - pR^2\theta = 0 \tag{2.8.10}$$

in which the prime indicates differentiation with respect to angle ϕ, and $\theta = (u + w')/R$ = rotation of the arch cross section (Fig. 2.43b). Equations 2.8.9 are the same as in the first-order theory. The rotational equilibrium equation (Eq.

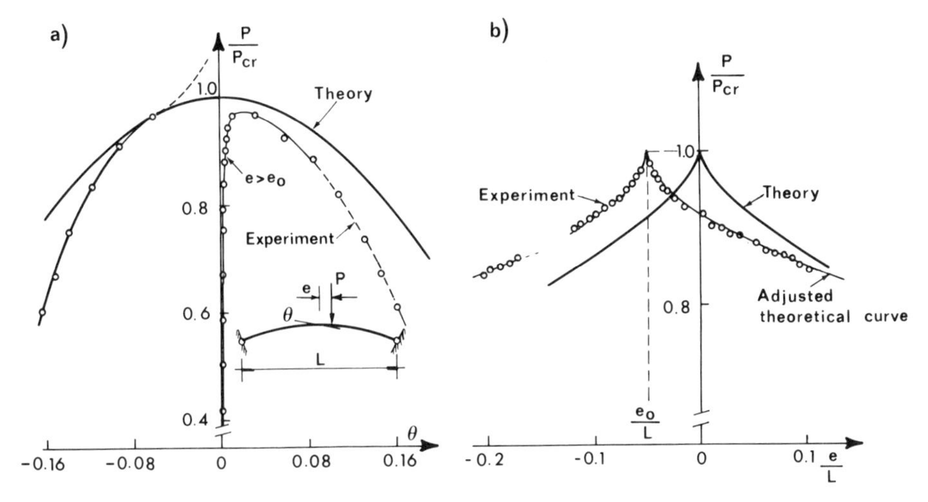

Figure 2.42 Experimental results of Roorda (1965a, b) compared to theory.

2.8.10), however, has an additional term which results from the fact that the arm of the axial force pR (Fig. 2.43b) is proportional to $\theta\, ds$.

We may now express V from Equation 2.8.10, in which $M = EI(w'' + w)/R^2$ from Equation 2.8.4. Then we substitute this into $V' + N = 0$ and $N' - V = 0$. Realizing that the inextensibility condition requires that $u' = w$, and denoting $k^2 = pR^3/EI$, we thus obtain

$$(u^{\mathrm{IV}} + 2u'' + u)'' + k^2(u^{\mathrm{IV}} + 2u'' + u) = 0 \tag{2.8.11}$$

This is a homogeneous sixth-order differential equation for the tangential displacement $u(\phi)$. Its general solution can be expressed (if $k^2 \neq 1$) as $u(\phi) = u^S(\phi) + u^A(\phi)$ in which $u^S(\phi)$ and $u^A(\phi)$ are the symmetric and antisymmetric parts, given by

$$u^S(\phi) = A_1 \cos k\phi + A_2 \cos \phi + A_3 \phi \sin \phi \tag{2.8.12}$$

$$u^A(\phi) = A_4 \sin k\phi + A_5 \sin \phi + A_6 \phi \cos \phi \tag{2.8.13}$$

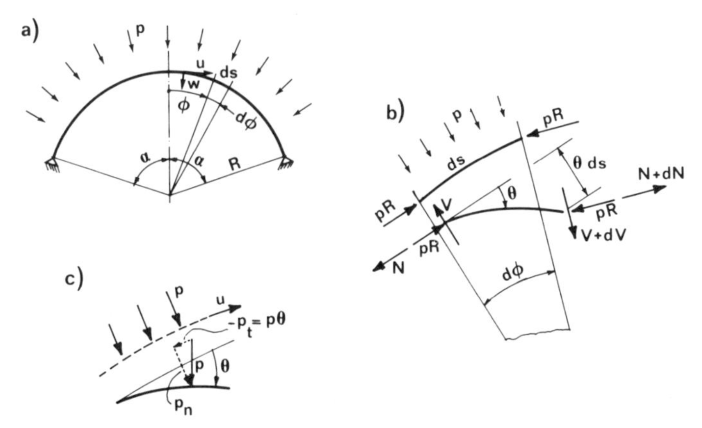

Figure 2.43 (a) Inextensible circular arch and (b, c) equilibrium of arch element.

Considering the case of symmetric boundary conditions, Equations 2.8.12 and 2.8.13 yield the antisymmetric and symmetric buckling modes, respectively (note that antisymmetric buckling corresponds to a symmetric solution for the tangential displacement u). For hinged ends, the boundary conditions are

$$\text{At } \phi = \pm\alpha: \quad u = 0 \quad w = u' = 0 \quad \kappa = w'' + \frac{w}{R^2} = u''' = 0 \tag{2.8.14}$$

For clamped ends, they are

$$\text{At } \phi = \pm\alpha: \quad u = 0 \quad w = u' = 0 \quad \theta = \frac{u + w'}{R} = u'' = 0 \tag{2.8.15}$$

Consider now antisymmetric buckling. The system of three homogeneous equations that express the boundary conditions has a nonzero solution if, for hinged ends,

$$2(k \sin k\alpha \cos \alpha - \sin \alpha \cos k\alpha) \sin \alpha - k(k^2 - 1)(\alpha + \sin \alpha \cos \alpha) \sin k\alpha = 0 \tag{2.8.16}$$

and for clamped ends,

$$2(k \sin k\alpha \cos \alpha - \sin \alpha \cos k\alpha) \cos \alpha - (k^2 - 1)(\alpha + \sin \alpha \cos \alpha) \cos k\alpha = 0 \tag{2.8.17}$$

Expressing the critical value of the radial loads through Equation 2.8.7, one obtains for a semicircular arch $\beta = 1.1078$ if the ends are hinged and $\beta = 0.6667$ if the ends are clamped. Comparing this with the values given by Timoshenko and Gere ($\beta = 1.15$ and $\beta = 0.70$), we see that their approximate solution in these cases is not too inaccurate and is also on the safe side.

The effect of elastic restraints at the ends was analyzed for circular arches by Oran and Reagan (1969). They also studied the effect of pressure loading, which differs from the loadings we examined before in that the load direction changes as the arch deflects in such a manner that the load vector remains normal to the deflected shape of the arch (normal pressure). Therefore, the load acquires in the buckled configuration (Fig. 2.43c) a component, $p_t = -p\theta$, in the direction tangent to the initial undeformed center line of the arch while the normal component $p_n = p \cos \theta \simeq p$ remains approximately constant. The equilibrium equation in the tangential direction, which replaces the second of Equations 2.8.9, takes the form $N' - V + p_t R = N' - V - pR\theta = 0$, and the governing differential equation is found to be

$$u^{\mathrm{VI}} + (1 + k_1^2)u^{\mathrm{IV}} + k_1^2 u'' = 0 \tag{2.8.18}$$

in which $k_1 = 1 + pR^3/EI$. For $k_1 \neq 0$ and $k_1 \neq 1$ the solution can be expressed as (Oran and Reagan, 1969)

$$u^S(\phi) = B_1 + B_2 \cos \phi + B_3 \cos k_1\phi \tag{2.8.19}$$

$$u^A(\phi) = B_4\phi + B_5 \sin \phi + B_6 \sin k_1\phi \tag{2.8.20}$$

Considering again the case of a semicircular arch ($\alpha = \pi/2$), and imposing the boundary conditions expressed by Equations 2.8.14 and 2.8.15 for the cases of hinged and clamped ends, respectively, one finds $\beta = 1.1547$ for a hinged arch and $\beta = 0.7071$ for a clamped arch. This is to be compared with the values $\beta = 1.1078$ and $\beta = 0.6667$ that we found before for the case that the loads do not rotate as the arch deflects. We see that the rotation of the loading produces a decrease of the critical load.

We will return to this point in Section 5.5 where we will apply energy methods to arches.

Problems

2.8.1 Without using the text, derive Equation 2.8.6.

2.8.2 Solve p_{cr} for a circular arch whose footings slide radially but cannot rotate (Fig. 2.39b).

2.8.3 Assume that the compression line (obtained from first-order analysis) is a circle of radius r passing through the hinge supports, such that $r \neq R$. Assume that P, p, and r are constant during buckling, and solve the load-deflection diagram from Equation 2.8.4, and plot it.

2.8.4 Sketch the forces acting on an infinitesimal element $ds = R\,d\phi$ of the arch and give a detailed derivation of Equations 2.8.9 and 2.8.10, as well as Equation 2.8.11.

2.8.5 Derive Equations 2.8.16 and 2.8.17 by applying the proper boundary conditions.

2.8.6 Solve antisymmetric buckling of a hinged semicircular arch under pressure loading, taking into account the rotation of the load vector as the arch deflects. *Hint*: Apply to the solution given by Equation 2.8.19 the boundary conditions given by Equations 2.8.14. The result is $k_1 = 2$, $p_{cr}R = 3EI/R^2$, $B_2 = 0$, $B_3 = B_1 = B$, and $u = B(1 + \cos 2\phi)$.

2.8.7 Same as Problem 2.8.6 but for a clamped arch.

2.8.8 Express the boundary conditions for the case of a hinged arch with rotational springs attached at the ends. Find, for a semicircular arch, the critical value of p as a function of the spring rigidity.

2.8.9 A stiffened arch (Fig. 2.44a) is an efficient structure, which was a trend in bridge design during the 1930s. Assuming that the arch as well as the vertical supports of the beam (the roadway) are so flexible that they can transmit only axial forces and are at the same time axially inextensible, set up the differential equations of equilibrium of an element ds of the arch coupled with

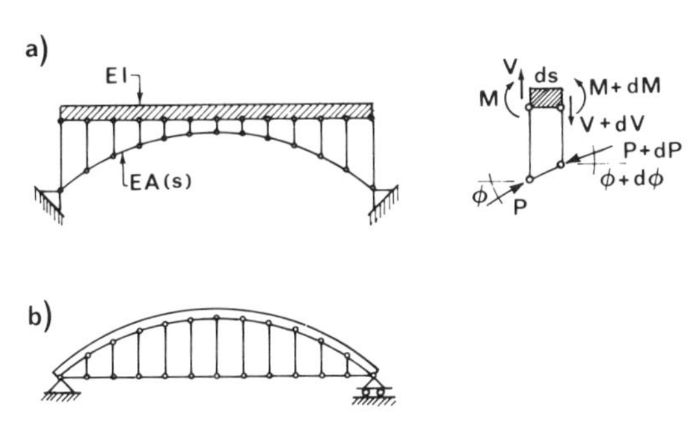

Figure 2.44 Exercise problems on buckling of (a) stiffened and (b) tied arches.

an element ds of the beam. Assume the vertical supports to be infinitely densely distributed. [The critical loads were obtained by Dischinger (1937, 1939); see also Oran and Reagan (1969).]

2.8.10 Discuss buckling of a tied arch (Fig. 2.44b), in which the suspenders and the roadway beam are perfectly flexible, the arch is axially inextensible, and the beam is axially extensible (cf. Dischinger, 1937, 1939; and Austin, 1971).

2.9 LONG-WAVE BUCKLING OF REGULAR FRAMES

In Section 2.4, we did not find the complete solution of regular frames but used intuition and experience to pick the lowest buckling modes. Such an approach might cause one to assume an incorrect mode for the lowest critical load if the frame is more complicated. However, for a regular frame, in which the member properties and loading are repeated from joint to joint (or more generally from a group of joints to the next group of joints), a complete and mathematically rigorous solution is possible with the help of finite difference calculus; see Bleich and Melan (1927), Bleich (1952), Chwalla and Jokisch (1941), Dean and Ugarte (1968), Omid'varan (1968), Tsang (1963), Gutkowski (1963), Jordan (1965), Bažant and Christensen (1972b), and Wah and Calcote (1970). Interest in regular frames or lattices has surged recently, stimulated by some prospective applications proposed for space stations.

System of Difference Equations

A planar regular frame (Fig. 2.45) is considered to be initially in equilibrium under axial forces P_y in all columns and P_x in all beams (although in buildings usually $P_x = 0$). Subsequently the initial equilibrium is disturbed by infinitely small load increments f^x and f^y and moments m applied at the joint, causing the joints to undergo infinitely small displacements u and v in the horizontal and vertical directions x and y, and small rotations ϕ, positive if counterclockwise; see Figure 2.45. If all the members in each direction have the same properties, the conditions of equilibrium of incremental horizontal forces, vertical forces, and moments acting on interior joint r, s lead to the equations (Bažant and

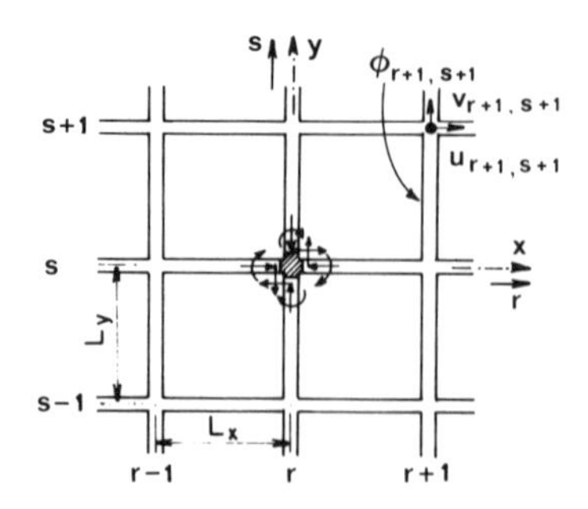

Figure 2.45 Section of large regular frame with joint and generalized displacements numbering.

Christensen, 1972b):

$$E'_x(u_{r+1,s} - 2u_{r,s} + u_{r-1,s}) + \frac{k_y\bar{s}_y}{L_y}(\phi_{r,s+1} - \phi_{r,s-1})$$
$$+ \frac{k_y s_y^*}{L_y^2}(u_{r,s+1} - 2u_{r,s} + u_{r,s-1}) + f^x_{r,s} = 0$$
$$E'_y(v_{r,s+1} - 2v_{r,s} + v_{r,s-1}) - \frac{k_x\bar{s}_x}{L_x}(\phi_{r+1,s} - \phi_{r-1,s})$$
$$+ \frac{k_x s_x^*}{L_x^2}(v_{r+1,s} - 2v_{r,s} + v_{r-1,s}) + f^y_{r,s} = 0 \qquad (2.9.1)$$
$$k_x s_x c_x(\phi_{r+1,s} - 2\phi_{r,s} + \phi_{r-1,s}) + 2k_x\bar{s}_x\phi_{r,s}$$
$$- \frac{k_x\bar{s}_x}{L_x}(v_{r+1,s} - v_{r-1,s}) + k_y s_y c_y(\phi_{r,s+1} - 2\phi_{r,s} + \phi_{r,s-1})$$
$$+ 2k_y\bar{s}_y\phi_{r,s} + \frac{k_y\bar{s}_y}{L_y}(u_{r,s+1} - u_{r,s-1}) - m_{r,s} = 0$$

in which subscripts x and y refer to members in the horizontal and vertical directions, and subscripts r and s represent the joint numbers in the horizontal and vertical directions, $r = 1, 2, \ldots,$ $s = 1, 2, \ldots;$ L_x, L_y = length of horizontal and vertical members; $E'_x = EA_x/L_x$, $E'_y = EA_y/L_y$ in which A_x, A_y = cross-sectional areas; $k_x = EI_x/l_x$, $k_y = EI_y/l_y$ in which I_x, I_y = cross-sectional moments of inertia; s_x, c_x and s_y, c_y are the stability functions of P_x, P_y (Eqs. 2.1.6, 2.1.7 for axial compression, and Eqs. 2.1.11a for axial tension), and $\bar{s}_x$, s_x^* or $\bar{s}_y$, s_y^* the expressions given in Equation 2.1.18. Equations 2.9.1 have the general form:

$$a_0^\nu u_{r,s} + a_1^\nu u_{r+1,s} + a_2^\nu u_{r,s+1} + a_3^\nu u_{r-1,s} + a_4^\nu u_{r,s-1}$$
$$+ b_0^\nu v_{r,s} + b_1^\nu v_{r+1,s} + b_2^\nu v_{r,s+1} + b_3^\nu v_{r-1,s} + b_4^\nu v_{r,s-1}$$
$$+ c_0^\nu \phi_{r,s} + c_1^\nu \phi_{r+1,s} + c_2^\nu \phi_{r,s+1} + c_3^\nu \phi_{r-1,s} + c_4^\nu \phi_{r,s-1} = -f^\nu_{r,s}$$
$$(\nu = 1, 2, 3 \text{ with } f^1_{r,s} = f^x_{r,s},\ f^2_{r,s} = f^y_{r,s},\ f^3_{r,s} = m_{r,s}) \qquad (2.9.2)$$

These equations are two-dimensional, linear, second-order partial difference equations with constant coefficients. (The fact that the coefficients are constant means that they do not depend on r and s, which is a property facilitating the solution.) The solution is entirely analogous to the solution of linear differential equations with constant coefficients. If the equations are nonhomogeneous, that is, have nonzero right-hand sides, then the general solution of these equations is the sum of some particular solution and a general solution of the corresponding homogeneous linear difference equations, for which the right-hand sides are zero, that is, $f^x_{r,s} = f^y_{r,s} = m_{r,s} = 0$. If one wishes to analyze the deflections of regular frames with imperfections, nonzero disturbing loads at the joints are considered. However, we will be interested only in the critical loads, and for that purpose we may set the disturbing joint loads equal to zero, that is, we only consider the homogeneous linear difference equations. Similar to the method of solution of linear differential equations, Equation 2.9.2 is satisfied by solutions of the type:

$$u_{r,s} = U_r e^{i\gamma s} \qquad v_{r,s} = V_r e^{i\gamma s} \qquad \phi_{r,s} = R_r e^{i\gamma s} \qquad (2.9.3)$$

where $i^2 = -1$; γ = constant (in general complex); and U_r, V_r, R_r = functions (in general complex) of the integer subscript $r = 1, 2, 3, \ldots$. Indeed, noting that $e^{i\gamma(s+1)} = e^{i\gamma}e^{i\gamma s}$, $e^{i\gamma(s-1)} = e^{-i\gamma}e^{i\gamma s}$, one finds that upon substitution of Equations 2.9.3 into Equation 2.9.2 the terms $e^{i\gamma s}$ cancel out, that is, the dependence on s is eliminated, and ordinary linear homogeneous difference equations for integer functions U_r, V_r, R_r are obtained (in general they can have complex coefficients). For the case of infinitely extending frames we considered before, Equation 2.9.2 may always be solved by Equations 2.9.3 with

$$U_r = Ae^{ikr} \qquad V_r = Be^{ikr} \qquad R_r = Ce^{ikr} \tag{2.9.4}$$

in which k, A, B, C are, in general, complex constants. Upon substitution into Equations 2.9.3, the dependence on r disappears and one obtains a system of three homogeneous linear algebraic equations for the unknowns A, B, and C. The condition that the determinant of these equations must vanish to make nonzero deflections possible then yields an algebraic equation with coefficients in general complex, depending on k as well as the axial loads P_x and P_y. Upon choosing some values for P_x and P_y, we can solve this equation to obtain the corresponding k. Our objective is to find, for a given ratio P_y/P_x, the smallest value P_x for which a solution k exists. This solution then represents the lowest critical load of the infinitely extending frame. The general solution of an infinitely extending frame under any loads is a linear combination of all possible solutions of the type of Equations 2.9.4 plus some particular solution.

In this manner we could rigorously obtain all the buckling modes, including the solutions for the critical loads of the interior columns in large regular frames, which we previously obtained by an intuitive, semiempirical choice of the decisive buckling mode (Sec. 2.4).

Solution for Tall Building Frames

Following Bažant and Christensen (1972a, b and 1973), we will now seek buckling solutions of wide and tall regular frames depicted in Figure 2.46a, in which the

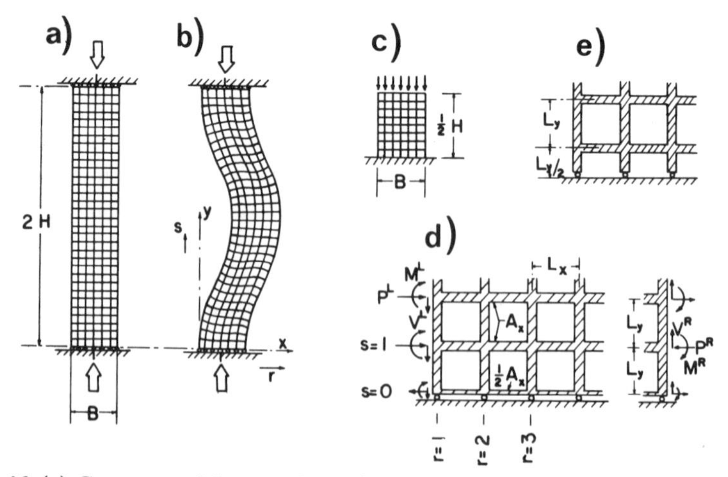

Figure 2.46 (a) Geometry of frame solved; (b) long-wave buckling mode; (c) free-standing frame of approximately the same initial load; (d) detail of boundaries of frame solved; (e) alternative boundary support which could be analyzed similarly.

vertical members may undergo significant axial extensions or shortenings, thus permitting the entire building frame to buckle as a whole. Such solutions are of interest for the tallest skyscraper frames presently built, although further generalizations would be needed to take into account the spatial action, interaction with flow slabs, in-fill panels or walls, bracing or shear walls, etc. Moreover, a generalization of the solution that we will now describe can be made for overall long-wave vibrations of large regular frames with columns under significant axial forces; such solutions are of interest for the response to earthquake or wind loads.

The buckling mode shown in Figure 2.46b can be characterized by the following simple expressions:

$$u_{r,s} = U_c - U_r \cos \gamma s \qquad v_{r,s} = V_r \sin \gamma s \qquad \phi_{r,s} = R_r \sin \gamma s \tag{2.9.5}$$

which may be regarded as the real part of a special case of Equations 2.9.2 and 2.9.3. U_r, V_r, R_r are discrete real functions of one integer subscript, r, and U_c is a constant. Using Equations 2.9.5 instead of Equations 2.9.3 means that we expect the solution to be periodic vertically. Substituting Equations 2.9.5 into Equations 2.9.1, we find that the trigonometric functions of s cancel out and the equations reduce to a system of three second-order simultaneous linear ordinary homogeneous difference equations with constant coefficients:

$$\begin{aligned}
&-E_x'(U_{r+1} - 2U_r + U_{r-1}) + \frac{k_y \bar{s}_y}{L_y}(2 \sin \gamma) R_r - \frac{k_y s_y^*}{L_y} 2(\cos \gamma - 1) U_r = 0 \\
&E_y' 2(\cos \gamma - 1) V_r - \frac{k_x \bar{s}_x}{L_x}(R_{r+1} - R_{r-1}) + \frac{k_x s_x^*}{L_x}(V_{r+1} - 2V_r + V_{r-1}) = 0 \\
&k_x s_x c_x (R_{r+1} - 2R_r + R_{r-1}) + (2k_x \bar{s}_x + 2k_y \bar{s}_y) R_r \\
&\qquad + k_y s_y c_y 2(\cos \gamma - 1) R_r + \frac{k_y \bar{s}_y}{L_y}(2 \sin \gamma) U_r - \frac{k_x \bar{s}_x}{L_x}(V_{r+1} - V_{r-1}) = 0
\end{aligned} \tag{2.9.6}$$

These equations have the advantage that their coefficients are real, in contrast to those arising from the more general substitution in Equations 2.9.3.

Consider now the boundary conditions for the long-wave extensional buckling mode in Figure 2.46. The kinematic boundary conditions for the top and bottom boundaries of the rectangular frame may be written as

$$\begin{aligned}
&v = \phi = 0 \qquad \text{for } s = 0 \text{ or } s = 2n_y, \text{ and all } r \\
&u = 0 \qquad \text{for } s = 0 \text{ or } s = 2n_y, \text{ and } r = c
\end{aligned} \tag{2.9.7}$$

$n_y = \pi/\gamma$ is assumed to be an integer denoting the number of floors contained within the half-wavelength, H (Fig. 2.46a), and c is the subscript for the column line that is held against lateral sliding at the base and top of the frame, while the bases and tops of all other column lines are allowed to slide freely in the horizontal direction at bottom and top boundaries. Furthermore, Equations 2.9.5 satisfy the condition:

$$\bar{V}_{y_{r,s}} = \tfrac{1}{2}(V_{y_{r,s-1}} + V_{y_{r,s}}) = 0 \tag{2.9.8}$$

in which

$$V_{y_{r,s}} = \frac{k_y s_y^*}{L_y^2}(u_{r,s+1} + u_{r,s}) + \frac{k_y \bar{s}_y}{L_y}(\phi_{r,s} + \phi_{r,s+1}) \tag{2.9.9}$$

$\bar{V}_y$ represents the shear force per length L_x transmitted along a horizontal section of the frame passed through the neutral axis of one line of horizontal members. This implies that the total shear force in the horizontal cross sections of the entire frame is zero (this shear force corresponds to the fixed direction shear force V, and not to the rotating shear force Q we previously introduced for columns with shear).

The boundary conditions in Equations 2.9.7 imply that the rectangular frame in Figure 2.46 is supported rigidly in the vertical direction at the joints of the top and bottom boundaries, while horizontally the joints are allowed to slide freely, with no friction. Furthermore, since $\bar{V}_y$ represents the shear force along the neutral axis of the horizontal beams, the beams at the top and bottom boundaries are implied to have one-half of the cross-sectional area of the other members. Although practical frames do not normally satisfy this condition, the difference is unimportant because the horizontal members at the top and bottom boundaries are not subjected to bending (since the joint rotations on top and bottom are zero). So the axial strains of these horizontal members are negligible. These boundary conditions are illustrated in Figure 2.46d.

It may be noted that a frame with the boundary conditions on top and bottom as illustrated in Figure 2.46e also could be solved exactly in a similar manner.

The simple solution according to Equations 2.9.5 also gives approximately correct results for the critical load of the free-standing rectangular frame in Figure 2.46c. This was verified by comparisons with exact solutions of frames up to 52 stories high (Bažant, Christensen, 1972a). For the free-standing frame, its height is taken as $H/2$ and represents the quarter-wavelength of the buckling mode shown in Figure 2.46b. The boundary conditions for the free-standing frame are not fulfilled on the top boundary exactly. However, they are fulfilled in the integral sense. The vertical force resultant and the moment resultant above the vertical center line of all axial forces of columns on top are zero for the solution according to Equations 2.9.5 at $y = H/2$. The deviation from the exact boundary conditions consists of a self-equilibrated system of vertical column forces on the top boundary, and according to the Saint-Venant principle this system should have only a local effect and should decay rapidly away from the top of the free-standing frame.

The boundary conditions on the left and right sides of the rectangular frame are most conveniently formulated if the frame is imagined to extend beyond the actual boundary, adding vertical rows of fictitious joints $r = 0$ and $r = m + 1$ outside the boundary (m is the number of column lines). This has the advantage that the boundary joints $r = 1$ and $r = m = n_x + 1$ (n_x = number of bays) can be treated as interior joints of the frame, for which the discrete field equations (Eqs. 2.9.6) apply. The internal forces transmitted into the joints at $r = 1$ from the left and the joints at $r = m$ from the right, that is, from the fictitious outside extension of the frame, must be made equal to the prescribed incremental loads applied at the left and right boundaries. This condition yields

$$E'_x(u_{0,s} - u_{1,s}) = P^L \qquad E'_x(u_{m,s} - u_{m+1,s}) = P^R$$

$$k_x s_x(c_x \phi_{0,s} + \phi_{1,s}) - \frac{k_x \bar{s}_x}{L_x}(v_{1,s} - v_{0,s}) = M^L$$

$$
\begin{aligned}
&k_x s_x(\phi_{m,s} + c_x\phi_{m+1,s}) - \frac{k_x \bar{s}_x}{L_x}(v_{m+1,s} - v_{m,s}) = M^R \\
&\frac{k_x s_x^*}{L_x^2}(v_{1,s} - v_{0,s}) - \frac{k_x \bar{s}_x}{L_x}(\phi_{0,s} + \phi_{1,s}) = V^L \\
&\frac{k_x s_x^*}{L_x^2}(v_{m+1,s} - v_{m,s}) - \frac{k_x \bar{s}_x}{L_x}(\phi_{m,s} + \phi_{m+1,s}) = V^R
\end{aligned}
\tag{2.9.10}
$$

in which P^L, V^L, M^L, P^R, V^R, and M^R are the incremental horizontal force, vertical force, and moment applied at the left or the right boundary joint (Fig. 2.46d). For the calculation of critical loads we of course consider $P^L = V^L = \cdots = 0$.

The difference equations (Eqs. 2.9.6) can be satisfied if U_r, V_r, and R_r are expressed according to Equations 2.9.4, or alternatively, $U_r = Ae^{\beta r}$, $U_s = Be^{\beta r}$, and $R_r = Ce^{\beta r}$. Considering then the most general linear combination of all possible solutions of this type, one can satisfy all the boundary conditions. However, it is more convenient to first reduce the difference equations from the second order to the first order. This may be accomplished by introducing twice as many unknowns, $F_{1_r}, F_{2_r}, \ldots, F_{6_r}$, defined as

$$
F_{1_r} = U_{r-1} \qquad F_{3_r} = V_{r-1} \qquad F_{5_r} = R_{r-1} \qquad F_{2_r} = U_r \qquad F_{4_r} = V_r \qquad F_{6_r} = R_r
\tag{2.9.11}
$$

Substituting this into Eqs. 2.9.6 and appending three additional finite difference equations $F_{1_{r+1}} = F_{2_r}$, $F_{3_{r+1}} = F_{4_r}$, $F_{5_{r+1}} = F_{6_r}$, we then obtain the following first-order matrix difference equation in the canonical (standard) form:

$$
\begin{Bmatrix} F_{1_{r+1}} \\ F_{2_{r+1}} \\ F_{3_{r+1}} \\ F_{4_{r+1}} \\ F_{5_{r+1}} \\ F_{6_{r+1}} \end{Bmatrix}
=
\begin{bmatrix}
0 & 1 & 0 & 0 & 0 & 0 \\
-1 & a_{22} & 0 & 0 & 0 & a_{26} \\
0 & 0 & 0 & 1 & 0 & 0 \\
0 & a_{42} & a_{43} & a_{44} & a_{45} & a_{46} \\
0 & 0 & 0 & 0 & 0 & 1 \\
0 & a_{62} & a_{63} & a_{64} & a_{65} & a_{66}
\end{bmatrix}
\begin{Bmatrix} F_{1_r} \\ F_{2_r} \\ F_{3_r} \\ F_{4_r} \\ F_{5_r} \\ F_{6_r} \end{Bmatrix}
\tag{2.9.12}
$$

in which

$$
\begin{aligned}
&a_{22} = 2\left[\frac{k_y s_y^*}{L_y^2 E_x'}(1 - \cos\gamma) + 1\right] \qquad a_{26} = \frac{2k_y \bar{s}_y}{L_y E_x'} \sin\gamma \\
&a_{42} = \frac{2L_x \bar{s}_x k_y \bar{s}_y \sin\gamma}{L_y k_x(\bar{s}_x^2 - s_x^* s_x c_x)} \qquad a_{43} = \frac{\bar{s}_x^2 + s_x^* s_x c_x}{\bar{s}_x^2 - s_x^* s_x c_x} \\
&a_{44} = \frac{-2s_x^* s_x c_x}{\bar{s}_x^2 - s_x^* s_x c_x}\left[\frac{E_y' L_x^2}{k_x s_x^*}(1 - \cos\gamma) + 1\right] \\
&a_{45} = \frac{2L_x \bar{s}_x s_x c_x}{\bar{s}_x^2 - s_x^* s_x c_x} \qquad a_{46} = \frac{2L_x \bar{s}_x(k_x s_x + k_y s_y + k_y s_y c_y \cos\gamma)}{k_x(\bar{s}_x^2 - s_x^* s_x c_x)} \\
&a_{62} = \frac{2k_y \bar{s}_y s_x^* \sin\gamma}{L_y k_x(\bar{s}_x^2 - s_x^* s_x c_x)} \qquad a_{63} = \frac{2s_x^* \bar{s}_x}{L_x(\bar{s}_x^2 - s_x^* s_x c_x)}
\end{aligned}
\tag{2.9.13}
$$

$$a_{64} = -\frac{2\bar{s}_x}{L_x k_x}\left[\frac{E'_y L_x^2(1-\cos\gamma) + k_x s_x^*}{\bar{s}_x^2 - s_x^* s_x c_x}\right] \qquad a_{65} = \frac{\bar{s}_x^2 + s_x^* s_x c_x}{\bar{s}_x^2 - s_x^* s_x c_x}$$

$$a_{66} = \frac{2s_x^*(k_x s_x + k_y s_y + k_y s_y c_y \cos\gamma)}{k_x(\bar{s}_x^2 - s_x^* s_x c_x)} \tag{2.9.13}$$

The solution of these equations may be found in the form

$$F_1(r) = K_1\rho^r, \quad F_2(r) = K_2\rho^r, \ldots, F_6(r) = K_6\rho^r \tag{2.9.14}$$

in which $\rho, K_1, \ldots, K_6$ are arbitrary constants. Note that one can equivalently replace ρ^r with $e^{\beta r}$ where $\beta = \ln\rho$. Substituting this in Equations 2.9.13 we obtain a system of linear homogeneous algebraic equations for $K_1, K_2, \ldots, K_6$. Their determinant must vanish to make a nonzero solution possible, that is,

$$\begin{vmatrix} -\rho & 1 & 0 & 0 & 0 & 0 \\ -1 & a_{22}-\rho & 0 & 0 & 0 & a_{26} \\ 0 & 0 & -\rho & 1 & 0 & 0 \\ 0 & a_{42} & a_{43} & a_{44}-\rho & a_{45} & a_{46} \\ 0 & 0 & 0 & 0 & -\rho & 1 \\ 0 & a_{62} & a_{63} & a_{64} & a_{65} & a_{66}-\rho \end{vmatrix} = 0 \tag{2.9.15}$$

This equation represents a standard linear eigenvalue problem. Computational experience has shown that the roots are normally real, and usually also distinct. (To avoid the need of programming for the case of double roots, the value of the axial load, which influences the values of the coefficients of the determinant, was slightly adjusted so as to avoid the occurrence of either double roots of two roots that are too close.) The general solution of the discrete field equation (Eq. 2.9.12) may then be expressed as

$$F_i(r) = C_1K_i^1\rho_1^r + C_2K_i^2\rho_2^r + \cdots + C_6K_i^6\rho_6^r \qquad (i = 1, 2, \ldots, 6) \tag{2.9.16}$$

in which $\rho_1, \rho_2, \ldots, \rho_6$ are the roots of the determinant (Eq. 2.9.15); $C_1, C_2, \ldots, C_6$ are arbitrary constants to be determined from the boundary conditions; and $(K_1^j, K_2^j, \ldots, K_6^j)$ is the jth eigenvector associated with root ρ_j. Always

$$K_1^j\rho_j = K_2^j \qquad K_3^j\rho_j = K_4^j \qquad K_5^j\rho_j = K_6^j \tag{2.9.17}$$

Substitution of the solution in Equation 2.9.16 into the boundary conditions in Equations 2.9.10 for the left and right sides of the plane (with $P^L = P^R = \cdots = M^R = 0$) yields a system of six linear algebraic homogeneous equations for C_j:

$$\sum_{j=1}^{6} b_{ij}(P_y)C_j = 0 \qquad (i = 1, \ldots, 6) \tag{2.9.18}$$

in which coefficients b_{ij} depend on the value of the axial load P_y, both directly (through s_y, c_y, s_x, etc.) as well as indirectly through the values of the roots $\rho_1, \ldots, \rho_6$ (we assume that the ratio P_x/P_y is prescribed, usually as zero). The expressions for these coefficients are found to be

$$
\begin{aligned}
b_{1j} &= \left[c_x K_5^j + K_6^j - (K_4^j - K_3^j) \frac{\bar{s}_x}{s_x L_x} \right] \rho_j \\
b_{2j} &= \left[K_5^j + c_x K_6^j - (K_4^j - K_3^j) \frac{\bar{s}_x}{s_x L_x} \right] \rho_j^{m+1} \\
b_{3j} &= \left[K_4^j - K_3^j - (K_5^j - K_6^j) \frac{\bar{s}_x L_x}{s_x''} \right] \rho_j \\
b_{4j} &= \left[K_4^j - K_3^j - (K_5^j + K_6^j) \frac{\bar{s}_x L_x}{s_x''} \right] \rho_j^{m+1} \\
b_{5j} &= (K_2^j - K_1^j)(\rho_j + \rho_j^{m+1}) \\
b_{6j} &= \begin{cases} K_2^j \rho_j^{(m+1)/2} & \text{for } m \text{ odd} \\ \frac{1}{2}(K_1^j + K_2^j)\rho_j^{1+m/2} & \text{for } m \text{ even} \end{cases}
\end{aligned}
\qquad (2.9.19)
$$

The computation of the lowest critical load for a given frame may now proceed as follows. We choose a series of values of P_y; for each of them we solve the roots $\rho_2, \ldots, \rho_6$, the corresponding eigenvectors, and the coefficients b_{ij}. Equations 2.9.18 can have a nonzero solution only if the determinant vanishes. Therefore, we calculate the value of the determinant corresponding to each value P_y and construct the plot of the determinant value versus P_y (actually, it is computationally more effective to plot the $\frac{1}{6}$ power of the determinant). Then we search for the zero values of this plot. First one locates the approximate values of the roots graphically, and then one may find the accurate value of $P_{y_{cr}}$ according to Newton's iterative procedure.

Numerical Results and Their Discussion

It may be shown that (for $P_x = 0$) the critical load, normalized with respect to the Euler load P_{E_y} of the columns, is in general a function of five nondimensional parameters:

$$
\frac{P_{y_{cr}}}{P_{E_y}} \simeq f\left(\frac{k_x}{k_y}, \frac{Hr_y}{BL_y}, \frac{L_y}{r_y}, n_x, n_y\right) \qquad (2.9.20)
$$

where r_y^2 = radius of gyration of the column cross section. Numerical results characterizing the effects of various nondimensional parameters were obtained by Bažant and Christensen (1972b) and some of them are plotted in Figures 2.47–2.49.

The weakening influence of the number of floors is seen from the lines for H/B = constant in Figure 2.47. The effect of n_x, n_y is negligible if, roughly, n_x or $n_y > 10$ for $H/B > 2$, or $n_x > 5$, $n_y > 10$ for $H/B > 4$. That the effect of the third parameter in Equation 2.9.20 is indeed rather small can be seen from Equations 2.9.5.

The value of the critical axial force for short-wave buckling modes in which axial extensions are negligible lies, for typical building frames without diagonal bracing, above $0.8P_{E_y}$. The critical forces for the long-wave extensional buckling

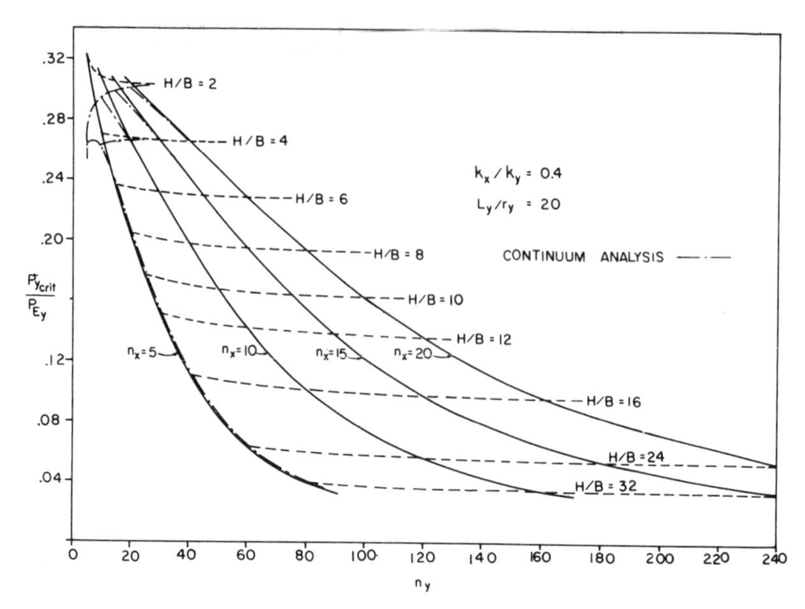

Figure 2.47 Critical loads per column for typical frame properties. (Solid lines for constant number of bays, n_x; dashed lines for constant height-to-width ratio.) Note the small effect of n_x, n_y at H/B constant. (*After Bažant and Christensen, 1972b.*) Dash-dot lines show solution according to continuous approximation. (*From Bažant and Christensen, 1973.*)

modes are thus of no interest when $P_{y_{cr}}$ is in this range or above it. Such cases have not been included in Figures 2.47–2.49. (It was shown, however, that even in such cases the presence of axial force P_y may considerably reduce the frame stiffness in long-wave deformation modes; this is of importance especially for vibration analysis.

It is found that frames for which long-wave buckling needs to be considered have either small k_x/k_y or are very high and slender, with members so stiff that short-wave buckling is precluded, as is typical for stiffening frames of modern tall buildings. In practice, these frames are arranged as a framed tube to provide also torsional rigidity.

The critical load values are also irrelevant when they exceed the plastic yield load of the columns, P_u. Lines indicating the P_u value for the frame with various L_y/r_y values have been included in Figure 2.49 for structural steels with 60,000 psi yield limit. Obviously, for high-strength steels the range in which long-wave buckling modes are of importance is considerably wider than for low-strength steels.

It is also of interest to determine the relationship of the present solutions to the exact solutions for buckling of free-standing frames, as in Figure 2.46. The critical axial loads for long-wave buckling of such frames has been solved by Bažant and Christensen (1972a) exactly for a few cases (frames 52 stories high). Comparison of numerical results indicates that the critical load of the free-standing frames about equals the critical load of the frames solved previously (Fig. 2.46a) if the height of the free-standing frame equals the quarter-wavelength of the buckling modes solved previously, that is, $H/2$. This agrees with intuition.

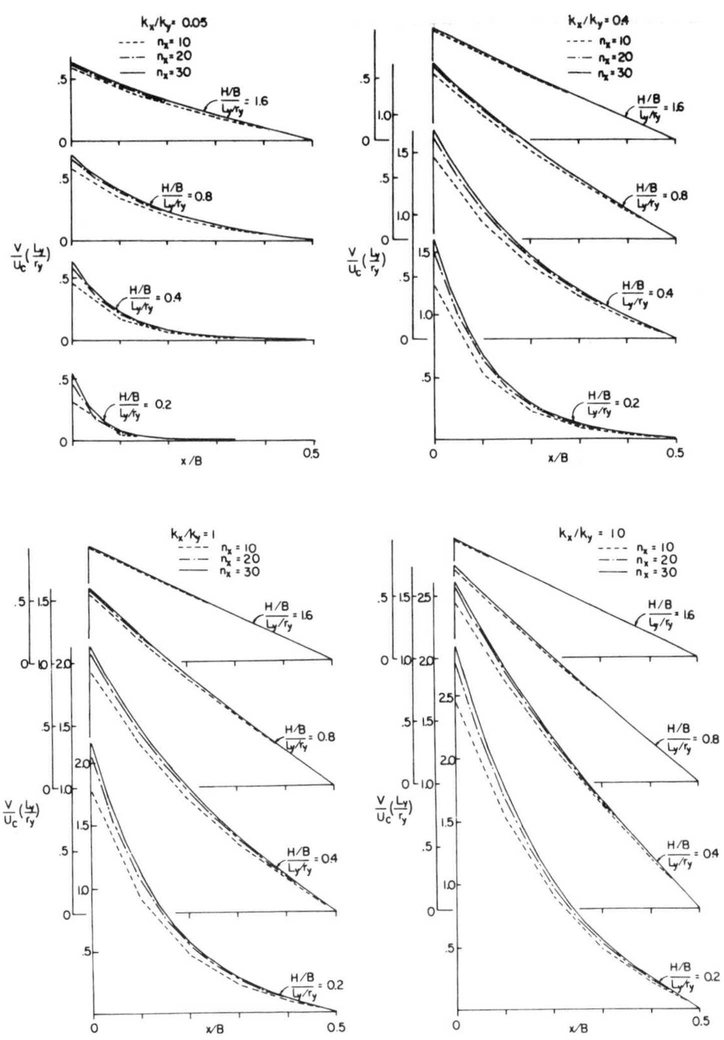

Figure 2.48 Distribution of amplitude of vertical joint displacements across frame width at buckling for various beam-to-column stiffness ratios k_x/k_y, frame height-to-width ratios H/B, and column slenderness ratios L_y/r_y. (Distributions of column extensions are the same. For P_y less than the critical load, distributions are appreciably different only if P_y/P_{E_y} is not small.) (*After Bažant and Christensen, 1972b.*)

To sum up, the critical loads of large regular frames with uniformly distributed axial loads can be effectively calculated by the methods of finite difference calculus. This approach offers the possibility of easily treating long-wave extensional buckling modes in which the buckling is not local, confined to a few members, but the frame buckles as a whole.

Problems

2.9.1 Verify that Equations 2.9.3 and 2.9.4 indeed satisfy Equations 2.9.2. Also verify that Equation 2.9.14 satisfies Equation 2.9.12, yielding Equation 2.9.15.

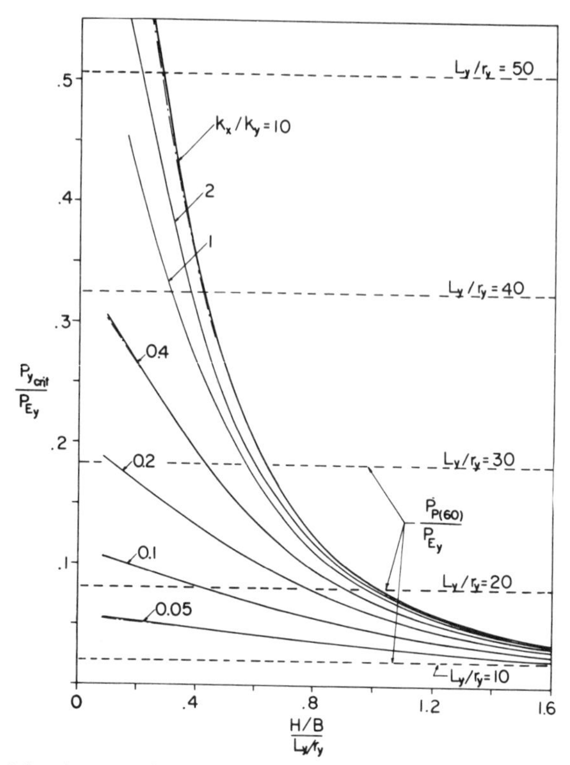

Figure 2.49 Critical load per column as function of ratio of frame slenderness H/B to column slenderness L_y/r_y (solid lines) (*after Bažant and Christensen, 1972b*). [Dashed lines indicate plastic yield loads for columns with various L_y/r_y having yield limit of 60,000 psi. Dash-dot lines show results according to continuous approximation (*from Bažant and Christensen, 1973*).]

2.9.2 Write the finite difference equations for a free-standing tall regular single-bay frame (Fig. 2.50a) and solve P_{cr} by the methods of finite difference calculus (cf. Bleich, 1952). Assume the horizontal beams to be infinitely stiff axially but not the columns. Compare the result with the solution of a built-up column as a column with shear (Sec. 2.7).

2.9.3 Same as Problem 2.9.2, but both ends fixed (Fig. 2.50b).

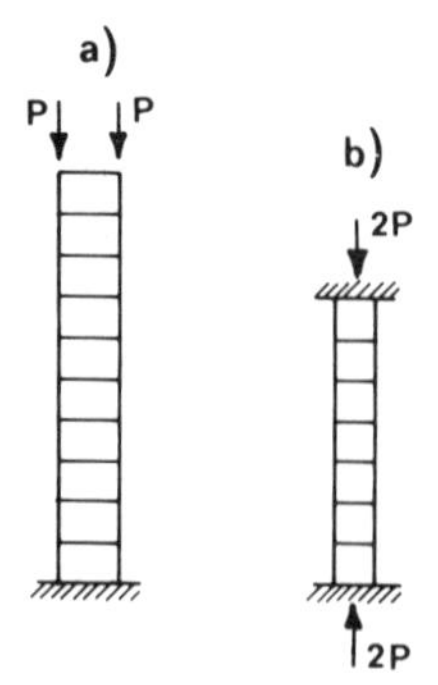

Figure 2.50 Exercise problems on long-wave buckling of regular frames.

2.9.4 Same as Problem 2.9.3, but the top joints slide vertically.
2.9.5 Derive the difference equations for a regular three-dimensional frame.

2.10 CONTINUUM APPROXIMATION FOR LARGE REGULAR FRAMES

Apart from the methods of finite difference calculus, the long-wave buckling and generally all large-scale overall behavior of very large regular frames, trusses, and lattices can be analyzed by approximating the structure as a continuum in which the joint displacements and rotations are represented by the values of some continuous functions evaluated at the joint locations. The smoothing of the structure with a macroscopic continuum always involves some error, and, therefore, in contrast to the solution of finite difference equations, this approach is never exact, although the approximation error often is negligible (Bažant and Christensen, 1972a).

The replacement of the discrete structure with a continuum can also be advantageous in that it makes it possible to utilize various methods of numerical solution of partial differential equations, which are generally more developed than those for difference equations (e.g., when the coefficients of the difference equations are not constant but variable). Aside from the classical methods such as the trigonometric series expansion for structures with rectangular boundaries, one can utilize the finite element method, in which the finite elements can be quite large, with dimensions corresponding to many bays of a frame or many cells of a lattice. With this approach, the variation of member properties or the initial axial forces along the frame poses no particular problem.

Transition from Difference to Differential Equations

Although the method of continuum approximation can be generally applied to all types of large regular frames, trusses, and lattices, we will demonstrate it only for the case of a large two-dimensional regular rectangular grame, considered before. We consider again the frame shown in Figure 2.45. However, we now allow the member stiffnesses to vary from one cell of the frame to the next, either because of a change in the cross section of the members or change in the initial axial force. As a generalization of Equations 2.9.1, the conditions of equilibrium of the horizontal forces, vertical forces, and moments acting on joint r, s read:

$$
\begin{aligned}
&E'_x(u_{r+1,s} - 2u_{r,s} + u_{r-1,s}) + \frac{k_y \bar{s}_y}{L_y}(\phi_{r,s+1} - \phi_{r,s-1}) \\
&\quad + \frac{k_y s_y^*}{L_y^2}(u_{r,s+1} - 2u_{r,s} + u_{r,s-1}) + \frac{1}{2}\Delta_x(E'_x)(u_{r+1,s} - u_{r-1,s}) \\
&\quad + \frac{1}{2}\Delta_y\left(\frac{k_y \bar{s}_y}{L_y}\right)(\phi_{r,s+1} - 2\phi_{r,s} + \phi_{r,s-1} + 4\phi_{r,s}) \\
&\quad + \frac{1}{2}\Delta_y\left(\frac{k_y s_y^*}{L_y^2}\right)(u_{r,s+1} - u_{r,s-1}) + f_{x_{r,s}} = 0
\end{aligned}
$$

$$
\begin{aligned}
&E'_y(v_{r,s+1}-2v_{r,s}+v_{r,s-1})-\frac{k_x\bar{s}_x}{L_x}(\phi_{r+1,s}-\phi_{r-1,s})\\
&\quad+\frac{k_xs_x^*}{L_x^2}(v_{r+1,s}-2v_{r,s}+v_{r-1,s})+\tfrac{1}{2}\Delta_y(E'_y)(v_{r,s+1}-v_{r,s-1})\\
&\quad-\frac{1}{2}\Delta_x\left(\frac{k_x\bar{s}_x}{L_x}\right)(\phi_{r+1,s}-2\phi_{r,s}+\phi_{r-1,s}+4\phi_{r,s})\\
&\quad+\frac{1}{2}\Delta_x\left(\frac{k_xs_x^*}{L_x^2}\right)(v_{r+1,s}-v_{r-1,s})+f_{y_{r,s}}=0
\end{aligned}
\tag{2.10.1}
$$

$$
\begin{aligned}
&k_xs_xc_x(\phi_{r+1,s}-2\phi_{r,s}+\phi_{r-i,s}+2\phi_{r,s})+2k_xs_x\phi_{r,s}-\frac{k_x\bar{s}_x}{L_x}(v_{r+1,s}-v_{r-1,s})\\
&\quad+k_ys_yc_y(\phi_{r,s+1}-2\phi_{r,s}+\phi_{r,s-1}+2\phi_{r,s})+2k_ys_y\phi_{r,s}+\frac{k_y\bar{s}_y}{L_y}(u_{r,s+1}-u_{r,s-1})\\
&\quad+\frac{1}{2}\Delta_x(k_xs_xc_x)(\phi_{r+i,s}-\phi_{r-1,s})-\frac{1}{2}\Delta_x\left(\frac{k_x\bar{s}_x}{L_x}\right)(v_{r+1,s}-2v_{r,s}+v_{r-1,s})\\
&\quad+\frac{1}{2}\Delta_y(k_ys_yc_y)(\phi_{r,s+1}-\phi_{r,s-1})+\frac{1}{2}\Delta_y\left(\frac{k_y\bar{s}_y}{L_y}\right)(u_{r,s+1}-2u_{r,s}+u_{r,s-1})-m_{r,s}=0
\end{aligned}
$$

in which all notations are the same as before except that for the case of nonuniform properties we introduce the difference operators Δ_x and Δ_y denoting the differences in the quantity to which they apply between two adjacent members. Also, in the case of nonuniform properties, k_x, s_x, k_y, etc., are defined as the average values (attributed to joint r, s) of the corresponding quantities in the adjacent members.

The finite difference expressions in Equations 2.10.1 are obviously the well-known second-order finite difference approximations to the first and second derivatives. Therefore, Equations 2.10.1 may be approximated by the partial differential equations

$$
\begin{aligned}
&L_x^2(E'_xu_{,x})_{,x}+2(k_y\bar{s}_y\phi)_{,y}+(k_ys_y^*u_{,y})_{,y}+\frac{L_y^2}{2}(k_y\bar{s}_y)_{,y}\phi_{,yy}+f_x=0\\
&L_y^2(E'_yv_{,y})_{,y}-2(k_x\bar{s}_x\phi)_{,x}+(k_xs_x^*v_{,x})_{,x}-\frac{L_x^2}{2}(k_x\bar{s}_x)_{,x}\phi_{,xx}+f_y=0\\
&2k_x\bar{s}_x(v_{,x}-\phi)-2k_y\bar{s}_y(u_{,y}+\phi)-L_x^2(k_xs_xc_x\phi_{,x})_{,x}-L_y^2(k_ys_yc_y\phi_{,y})_{,y}\\
&\qquad+\frac{L_x^2}{2}(k_x\bar{s}_x)_{,x}v_{,xx}-\frac{L_y^2}{2}(k_y\bar{s}_y)_{,y}u_{,yy}+m=0
\end{aligned}
\tag{2.10.2}
$$

Here u, v, ϕ, f_x, f_y, and m represent continuous and sufficiently smooth functions of x and y whose values at points (x_i, y_j) approximate the values of $u_{i,j}$, $v_{i,j}$, etc.; subscript x or y following a comma denotes partial derivatives, for example, $v_{,x}=\partial v/\partial x$, $\phi_{,xx}=\partial^2\phi/\partial x^2$; k_x, s_x, $\bar{s}_x$, c_x, E'_x, etc. are also understood as continuous and smooth functions whose values at the midspans of the members approximate the actual member properties. (Introduction of these continuous functions is meaningful only if member properties and initial axial forces vary sufficiently regularly from member to member.)

Continuum Approximation Based on Potential Energy

An alternative approach to obtaining Equations 2.10.2 can be based on potential energy. The expression for the second-order approximation to the incremental strain energy U_1 for a single member is

$$U_1 = \frac{1}{2}[M_a(\phi_a - \psi) + M_b(\phi_b - \psi) + P(u_a - u_b)] - P^0\left(\frac{L\psi^2}{2}\right) \quad (2.10.3)$$

where P^0 is the initial axial force, subscripts a and b refer to member ends, and ψ is the rotation of the line joining the member ends, plus linear terms $-P^0(u_b - u_a)$, $M_a^0(\phi_a - \psi)$, and $M_b^0(\phi_b - \psi)$. However, as is well known, the linear terms need not be taken into account since they would yield the conditions of equilibrium for the initial state which are assumed to be satisfied. The value $(L\psi^2/2)$ represents, with an error $0(\psi^4)$, the axial extension of the member due to small lateral displacements v_a, v_b. If M_a, M_b, and P are expressed according to the stiffness matrix in Equation 2.2.7, it follows on rearrangement that

$$U_1 = \frac{1}{2}E'(u_b - u_a)^2 + \frac{1}{2}ks(\phi_b - \phi_a)^2 + k\bar{s}(\phi_a - \psi)(\phi_b - \psi) - P^0\frac{L\psi^2}{2} \quad (2.10.4)$$

The incremental strain energy U_x contained in a pair of horizontal members at joint (r, s) is a sum of two expressions of the form of Equation 2.10.4. For the sake of simplicity, the properties and initial axial forces of members whose longitudinal axes lie on the same line will now be considered as constant. Then, expanding the value of u, v, and ϕ in joints $(r+1, s)$ and $(r-1, s)$ in a Taylor's series about the point (r, s) leads to the continuum approximation

$$U_x = L_x^2 E_x' u_{,x}^2 + L_x^2 k_x s_x \phi_{,x}^2 + L_x^2 k_x \bar{s}_x \phi\phi_{,xx} + 2k_x\bar{s}_x(\phi - v_{,x})^2 - P_x^0 L_x v_{,x}^2 \quad (2.10.5)$$

In Equation 2.10.5 we retain only the terms with first derivatives of u, v, and ϕ, but with the notable exception of those higher-order terms that can be converted on integration by parts to terms containing only the first-order derivatives. The term $\phi\phi_{,xx}$ is of this type since, in the expression for the total energy Π of the structure, it can be converted, upon integration by parts, to the term $-\phi_{,x}^2$.

The incremental strain energy U_y stored in a pair of vertical members meeting in the joint (r, s) can be obtained similarly. The element of the continuum approximation as shown in Figure 2.51 is periodically repeated in both directions. From this it can be seen that the strain energy corresponding to the area $L_x L_y$ of the frame is $\frac{1}{2}(U_x + U_y)$.

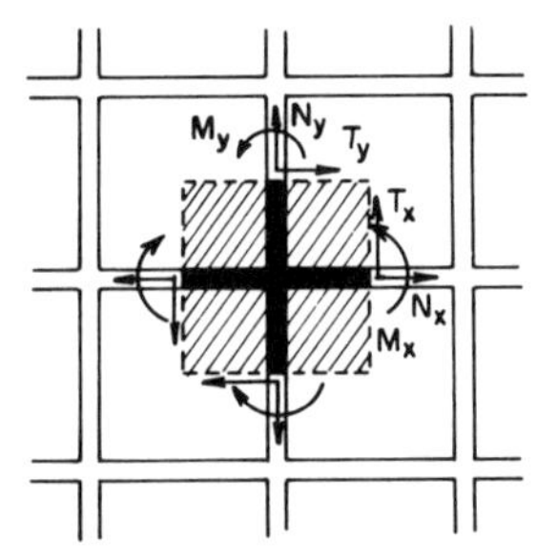

Figure 2.51 Internal forces at the member midspans and their intuitive analogy with the stresses acting on an element of micropolar continuum. (*After Bažant and Christensen, 1972a.*)

The total incremental potential energy of the structure, Π, is approximated by the expression

$$\Pi = \int_{(x)} \int_{(y)} \left[\frac{1}{2}(U_x + U_y) - f_x u - f_y v - m\phi \right] \frac{dx\,dy}{L_x L_y} - W_B \tag{2.10.6}$$

where W_B = the work of the incremental loads applied at the boundary of the structured body. Integrating the terms involving the products $\phi\phi_{,xx}$ and $\phi\phi_{,yy}$ by parts, or applying Green's theorem (e.g. Pearson, 1974), the integral in Equation 2.10.6 takes the form:

$$\Pi = \int_{(x)} \int_{(y)} U\,dx\,dy - \int_{(x)} \int_{(y)} (f_x u + f_y v + m\phi) \frac{dx\,dy}{L_x L_y} - W'_B \tag{2.10.7}$$

for which

$$\begin{aligned} U = {} & [L_x^2 E'_x u_{,x}^2 + L_y^2 E'_y v_{,y}^2 - L_x^2 k_x s_x c_x \phi_{,x}^2 - L_y^2 k_y s_y c_y \phi_{,y}^2 + 2k_x \bar{s}_x (\phi - v_{,x})^2 \\ & + 2k_y \bar{s}_y (\phi + u_{,y})^2 - P_x^0 L_x v_{,x}^2 - P_y^0 L_y u_{,y}^2](2L_x L_y)^{-1} \end{aligned} \tag{2.10.8}$$

and $W'_B = W_B$ plus a certain contour integral of terms involving products $\phi\phi_{,x}$ and $\phi\phi_{,y}$. Obviously, u can be regarded as the second-order terms of the specific incremental strain energy of the continuum approximating the frame. The presence of negative terms involving $-\phi_{,x}^2$ and $-\phi_{,y}^2$ in Equation 2.10.8 or terms such as $\phi\phi_{,xx}$ and $\phi\phi_{,yy}$ in Equation 2.10.5, which can also become negative, raises questions regarding the conditions for positive definiteness of the total potential energy. Such conditions must always be met if $P_x^0 = P_y^0 = 0$. That this is indeed so follows from the positive definiteness of the expression in Eq. 2.10.4 for $P_x^0 = P_y^0 = 0$.

The differential equilibrium equations may be derived from the first variation of the incremental potential Π. If the first-order terms are omitted, this variation is

$$\begin{aligned} \delta\Pi = \int_x \int_y \{ & L_x^2 E'_x u_{,x}\,\delta u_{,x} + L_y^2 E'_y v_{,y}\,\delta v_{,y} - L_x^2 k_x s_x c_x \phi_{,x}\,\delta\phi_{,x} \\ & - L_y^2 k_y s_y c_y \phi_{,y}\,\delta\phi_{,y} + 2k_x \bar{s}_x (\phi - v_{,x})(\delta\phi - \delta v_{,x}) \\ & + 2k_y \bar{s}_y (\phi + u_{,y})(\delta\phi + \delta u_{,y}) - P_x^0 L_x v_{,x}\,\delta v_{,x} - P_y^0 L_y u_{,y}\,\delta u_{,y} \\ & - f_x\,\delta u - f_y\,\delta v - m\,\delta\phi \} \frac{dx\,dy}{L_x L_y} \end{aligned} \tag{2.10.9}$$

plus a certain contour integral that is relevant only for the boundary conditions. If the terms containing derivatives of the variations are integrated by parts (or if Green's theorem is used), and if the equilibrium condition $\delta\Pi = 0$ is applied, one obtains the differential equations of equilibrium that are identical with Equations 2.10.2 for constant member properties, as expected.

If the incremental properties of members and initial axial forces vary from member to member, additional terms must be included in the strain energy density expression. Proceeding in the same manner as we did from Equation

2.10.7 to Equations 2.10.9 and 2.10.2, it can be verified that in order to obtain the equilibrium equations (Eqs. 2.10.2) for the case of variable properties, the strain energy density must be given by Equation 2.10.8 plus the term

$$\frac{L_x^2(k_x\bar{s}_x)_{,x}\phi_{,x}v_{,x} - L_y^2(k_y\bar{s}_y)_{,y}\phi_{,y}u_{,y}}{2L_xL_y} \tag{2.10.10}$$

Micropolar Continuum and Couple Stresses

It is now interesting to observe that the continuum described by Equations 2.10.2 is a special case of the so-called micropolar continuum proposed by Eringen (1966). This continuum is defined as a continuum that is characterized not only by displacements $u(x, y)$ and $v(x, y)$ but also by material rotations $\phi(x, y)$, which are independent of the rotations ω based on u and v, that is, of $\omega = \frac{1}{2}(v_{,x} - u_{,y})$. Thus, the potential energy of a micropolar continuum depends not only on the strains $u_{,x}$, $v_{,y}$, and γ where $\gamma = 2\varepsilon_{xy} = v_{,x} + u_{,y}$, but also on three additional quantities: $\omega - \phi$, r_x, and r_y where $r_x = \frac{1}{2}\gamma + (\omega - \phi) = v_{,x} - \phi$ and $r_y = \frac{1}{2}\gamma - (\omega - \phi) = u_{,y} + \phi$. The fact that the potential-energy density in Equation 2.10.7 is indeed of the form $U(u_{,x}, v_{,y}, r_x, r_y, \phi_{,x}, \phi_{,y})$ means that the continuum that approximates a rectangular frame is a special case of a micropolar continuum. Obviously, this continuum is orthotropic.

The micropolar continuum is a special case of continua with couple stresses m_{xz}, m_{yz} that exist because of the fact that the shear stresses are not symmetric, that is, $\sigma_{xy} \neq \sigma_{yx}$. (The fact that $\sigma_{xy} = \sigma_{yx}$ for the usual continua is derived from the condition of moment equilibrium of an infinitesimal element under the condition that the moments applied to its faces are zero.)

The stresses σ_{xx}, σ_{yy}, σ_{xy}, σ_{yx} and the couple stresses m_{xz}, m_{yx} in the micropolar continuum of constant properties are defined and expressed from Equation 2.10.8 as follows:

$$\begin{aligned}
\sigma_{xx}^0 + \sigma_{xx} &= \frac{\partial \bar{U}}{\partial u_{,x}} = \sigma_{xx}^0 + \frac{E_x' u_{,x} L_x}{L_y} \\
\sigma_{yy}^0 + \sigma_{yy} &= \frac{\partial \bar{U}}{\partial v_{,y}} = \sigma_{yy}^0 + \frac{E_y' v_{,y} L_y}{L_x} \\
\sigma_{xy} &= \frac{\partial \bar{U}}{\partial r_x} = \frac{\partial \bar{U}}{\partial v_{,x}} = \frac{k_x s_x^* v_{,x} - 2k_x\bar{s}_x\phi}{L_xL_y} \\
\sigma_{yx} &= \frac{\partial \bar{U}}{\partial r_y} = \frac{\partial \bar{U}}{\partial u_{,y}} = \frac{k_y s_y^* u_{,y} + 2k_y\bar{s}_y\phi}{L_xL_y} \\
m_{xz} &= \frac{\partial \bar{U}}{\partial \phi_{,x}} = \frac{-k_x s_x c_x \varphi_{,x} L_x}{L_y} \\
m_{yz} &= \frac{\partial \bar{U}}{\partial \phi_{,y}} = \frac{-k_y s_y c_y \phi_{,y} L_y}{L_x}
\end{aligned} \tag{2.10.11}$$

where $\bar{U} = U + \sigma_{xx}^0 u_{,x} + \sigma_{yy}^0 v_{,y}$ = the total strain energy density and $\sigma_{xx}^0 = -P_x^0/L_y$, $\sigma_{yy}^0 = -P_y^0/L_x$ are the initial stresses. Note that the symmetric part $\sigma_{(xy)}$ and the antisymmetric part $\sigma_{[xy]}$ of shear stress σ_{xy} can also be expressed directly as $\sigma_{(xy)} = \frac{1}{2}\,\partial U/\partial\varepsilon_{xy}$, $\sigma_{[xy]} = \frac{1}{2}\,\partial U/\partial(\omega - \phi)$.

Stresses and Boundary Conditions

It is now interesting to define the continuum counterparts of the internal forces in the regular frame. These forces are associated with the incremental internal forces in the frame members at midspan (Fig. 2.51) and include axial force N, shear force T, and bending moment M, which will be referred for reasons of convenience to the point on the straight line connecting the ends of the member in the deformed position (and not to the point on the deformed neutral axis). Therefore,

$$N = E'(u_b - u_a) = -P$$
$$V = \frac{k[s^*(v_b - v_a)/L - \bar{s}(\phi_a + \phi_b)]}{L} \tag{2.10.12}$$
$$M = \frac{M_b - M_a}{2} = \frac{1}{2} ks(1-c)(\phi_b - \phi_a)$$

The end moments M_a and M_b are related to M, N, and V by the following equations:

$$M_a = -M - \frac{(V + P^0\psi)L}{2} \qquad M_b = M - \frac{(V + P^0\psi)L}{2} \tag{2.10.13}$$

In terms of the member properties within the framework, the continuum approximations of Equations 2.10.12 in the x- and y-directions are straightforward and are given by

$$N_x = L_x E'_x u_{,x} \qquad N_y = L_y E'_y v_{,y}$$
$$T_x = \frac{k_x s_x^* v_{,x} - 2k_x \bar{s}_x \phi}{L_x} \qquad T_y = \frac{k_y s_y^* u_{,y} + 2k_y \bar{s}_y \phi}{L_y} \tag{2.10.14}$$
$$M_x = \tfrac{1}{2} L_x k_x s_x (1 - c_x)\phi_{,x} \qquad M_y = \tfrac{1}{2} L_y k_y s_y (1 - c_y)\phi_{,y}$$

where N_x, N_y, T_x, T_y, M_x, and M_y are continuous functions whose values at midspan approximate the internal forces in Equations 2.10.12, and where T_y is taken as $-V_y$ so that its positive direction would correspond to the usual continuum convention.

Equations 2.10.11 for the stresses and couple stresses are simply related to the continuous approximations in Equations 2.10.14 for the internal forces at the midspans

$$\sigma_{xx} = \frac{N_x}{L_y} \qquad \sigma_{yy} = \frac{N_y}{L_x}$$
$$\sigma_{xy} = \frac{T_x}{L_y} \qquad \sigma_{yx} = \frac{T_y}{L_x} \tag{2.10.15}$$
$$m_{xz} = -\frac{2c_x M_x}{L_y(1 - c_x)} \qquad m_{yz} = -\frac{2c_y M_y}{L_x(1 - c_y)}$$

It is interesting to note that while the normal and shear forces are expressed as the resultants of the normal and shear stresses over lengths L_y and L_x, the bending moments are not equal to the resultants $L_y m_{xz}$, $L_x m_{yz}$ of the couple stresses m_{xz} and m_{yz}. (For a zero axial force, $M_x = -L_y m_{xz}/2$ and $M_y = -L_x m_{yz}/2$, so that even the sign of M_x and m_{xz} is opposite.) Thus, the analogy that some authors based intuitively on the assumption that the midspan bending moments may be expressed as the resultants of couple stresses, that is, $M_x = L_y m_{xz}$ and $M_y = L_x m_{yz}$, is not completely correct. The reason for the inequality of M_x and $L_y m_{xz}$ consists, roughly speaking, in the fact that the bending moment varies along the member, while N_x and T_x are constant along the member. Also, the strain energy of the member due to $\phi_{,x}$ equals $\frac{1}{2} m_{xz} \phi_{,x}$ rather than $\frac{1}{2}(M_x/L_y)\phi_{,x}$, whereas the energies due to N_x and T_x equal $\frac{1}{2}(N_x/L_y)u_{,x}$ and $\frac{1}{2}(T_x/L_y)(v_{,x} - \phi)$.

After defining the internal forces, we can consider the boundary conditions. At each boundary joint of the framework, one quantity of each of the pairs (u, f_x), (v, f_y), and (ϕ, m) must be given. In the case of a continuum approximation, Equations 2.10.14 have an error of an order higher than second only if the functions u, v, ϕ, and their derivatives are evaluated at the midspan point of the beam. Therefore, when the applied loads f_x, f_y, or m at the boundary joint are prescribed, the simplest formulation of the boundary conditions may be achieved in a manner similar to that used in the solution of continuous boundary-value problems by the finite difference method. The gridwork is imagined to be extended beyond the boundary, and the hypothetical values for the nodes outside the physical boundary are used in such a way that the internal forces at the midspan points of the imagined members crossing the boundary transmit the prescribed forces into the actual boundary joints (Fig. 2.52). Thus, using Equations 2.10.12 and 2.10.13, the conditions on the left vertical and top horizontal boundaries of the framework are

$$
\begin{aligned}
N_x &= -P_x^B & N_y &= -P_y^B \\
T_x &= V_x^B & T_y &= -V_y^B \\
M_x &= M_x^B + (V_x^B + P_x^0 \psi_x)\frac{L_x}{2} & M_y &= -M_y^B - (V_y^B + P_y^0 \psi_y)\frac{L_y}{2}
\end{aligned}
\qquad (2.10.16)
$$

where P^B, V^B, and M^B denote the prescribed incremental normal load, tangential load, and moment, respectively, at the actual boundary joint of the frame. By substitution of Equations 2.10.14, the respective conditions for the left vertical

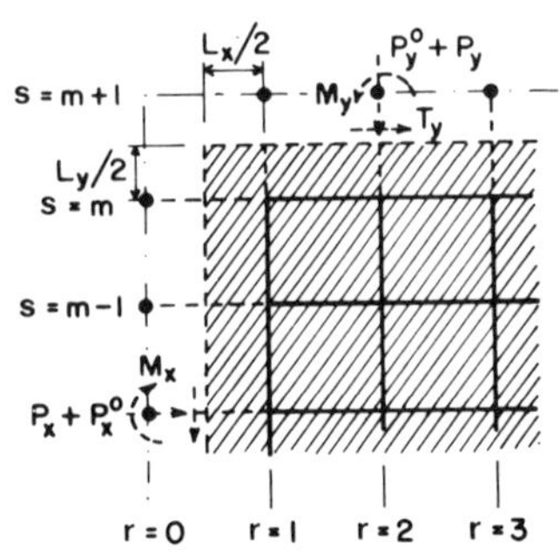

Figure 2.52 Fictitious extension of the grid beyond its boundary for the formulation of the free boundary condition. (*After Bažant and Christensen, 1972a.*)

and top horizontal boundaries of the continuum are obtained as follows:

$$
\begin{aligned}
L_x E'_x u_{,x} &= -P^B_x \\
k_x s^*_x v_{,x} - 2k_x \bar{s}_x \phi &= V^B_x L_x \\
L_x k_x s_x (1 - c_x)\phi_{,x} &= 2M^B_x + (V^B_x + P^0_x v_{,x})L_x \\
L_y E'_y v_{,y} &= -P^B_y \\
k_y s^*_y u_{,y} + 2k_y \bar{s}_y \phi &= -V^B_y L_y \\
L_y k_y s_y (1 - c_y)\phi_{,y} &= -2M^B_y - (V^B_y - P^0_y u_{,y})L_y
\end{aligned}
\tag{2.10.17}
$$

These conditions are imposed along the line connecting the midspan points of the imaginary members crossing the boundary. The continuum boundary can thus be imagined to be located at distances $L_x/2$ and $L_y/2$ beyond the boundary joints of the frame.

As an interesting digression, we may note one consequence for the problem of a substitute frame. Structural analysts have often been seeking a substitute frame that would behave approximately in the same manner as a given large regular frame yet would have much greater distances between the joints, thus permitting simpler analysis. The substitute frame was usually determined on the basis of some limited conditions of equivalence, but this could not guarantee the same behavior under all types of loading. From the viewpoint of our formulation, it is clear that a rational definition of a substitute frame can be based on the requirement that the continuum approximation of the substitute frame be the same as for the actual given frame. Labeling the quantities for the substitute frame with subscript S the condition that the coefficients of all the terms in Equations 2.10.2 and in the finite difference approximations of these differential equations based on grid steps $\Delta x = L_{x_S}$, $\Delta y = L_{y_S}$ be the same (or proportional), we find that the conditions of equivalence of a given frame and a substitute frame with members of greater lengths L_{x_S} and L_{y_S} are

$$
\begin{aligned}
E'_x L^2_x &= \beta E'_{x_S} L^2_{x_S} & \qquad E'_y L^2_y &= \beta E'_{y_S} L^2_{y_S} \\
k_x \bar{s}_x &= \beta k_{x_S} \bar{s}_{x_S} & \qquad k_y \bar{s}_y &= \beta k_{y_S} \bar{s}_{y_S} \\
k_x s^*_x &= \beta k_{x_S} s^*_{x_S} & \qquad k_y s^*_y &= \beta k_{y_S} s^*_{y_S} \\
k_x s_x c_x L^2_x &= \beta k_{x_S} s_{x_S} c_{x_S} L^2_{x_S} & \qquad k_y s_y c_y L^2_y &= \beta k_{y_S} s_{y_S} c_{y_S} L^2_{y_S}
\end{aligned}
\tag{2.10.18}
$$

in which β is an arbitrary parameter. This is a system of eight equations, which however involve only five unknowns, namely k_{x_S}, E'_{x_S}, k_{y_S}, E'_{y_S}, and β. Therefore we must conclude that a substitute frame in general does not exist.

Obviously, the length of the members of the given frame translates into a certain characteristic length of the equivalent micropolar continuum, and this characteristic length is a fundamental property of the continuum that cannot be altered. In a limited sense, though, a substitute frame can be defined for those frames that are adequately described as a column with shear; see Section 2.7. It suffices to ensure that the substitute frame with greater floor heights and wider bays has the same moment of inertia in its horizontal cross section and also the same overall horizontal shear stiffness. These two conditions can only be satisfied by altering the ratio of elastic constants, that is, by using $E_S \neq E$.

While the micropolar continuum can describe the overall behavior of large regular frames very closely, it is incapable of capturing local disturbances. Such disturbances are usually induced at the boundaries due to deviations of the boundary conditions from those required by the continuous approximation. It can be shown, however, that such disturbances decay rapidly with the distance from the boundary (Bažant and Christensen, 1972a).

Numerical Results and Discussion

The exact solutions of buckling for the continuum approximation of the large regular frame were obtained by an exact integration of the partial differential equations of the problem, in a manner that is analogous to that demonstrated in the preceding section for the solution by means of finite difference calculus (Bažant and Christensen, 1973).

Extensive numerical comparisons of the deformations of frames and of their continuum approximations were given by Bažant and Christensen (1972a, b). Further results for the nonlocal continuum approximation are plotted in the figures of the previous section. The dash-dot line of Figure 2.47 represents the plot of the critical load for long-wave buckling versus the number of floors, n_y. It is seen that this curve is very close to the solid curve for the exact solution, except when the frame height-to-width ratio is very small. The results for the continuum approximation are also plotted in Figure 2.49 showing the dependence of the long-wave critical load on the height-to-width ratio of the frame; it is seen that in most cases the continuum solution is graphically indistinguishable, and when a difference is seen, it is still practically negligible.

Bažant and Christensen's (1972a) original formulation of the continuum approximation to large regular frames with columns under axial forces was later extended by other authors to vibrations as well as to certain problems of more general lattices. Recently there has been an increasing interest in these formulations with regard to the analysis of large space lattices for certain special structures to be put in orbit in space.

As a historical note, the usefulness of continuum approximations of frames was recognized already during the 1960s. The possibility of applying the couple stress medium to lattice structures and frames was discussed by Wozniak (1966) in rather general terms. A more specific model was presented by Banks and Sokolowski (1968); however, their model was not quite satisfactory because it was based on a classical type of couple stress continuum (Cosserat's continuum) in which the microrotation ϕ and the macrorotation ω are equal, whereas for a frame the continuum counterparts of these quantities (the joint rotation and the member chord rotation) are in general unequal. The micropolar continuum as a model for a rectangular gridwork with diagonals was proposed by Askar and Cakmak (1968). However, this model was also not quite satisfactory because certain essential terms were missing from the expression for the potential-energy density (the missing terms were $\phi\phi_{,xx}$, which are essential because they transform, upon integration by parts, to the term $-\phi_{,x}^2$). A similar model for a spatial cubic gridwork with diagonals was investigated by Tauchert (1969) who assumed, similarly to Banks and Sokolowski (1968), that for a unit element of the continuous medium the couple stresses approximate the moments transmitted by the lattice bars at their midlength. Plausible though this assumption might seem at first, it is nevertheless incorrect as Equations 2.10.15 demonstrate.

To sum up, the continuum approximation by a micropolar continuum is an effective means to solve buckling (as well as vibrations at initial forces) of very large regular frames or lattices. Some problems can be solved exactly on the basis of the partial differential equations, while in general one can discretize the continuum approximation with a mesh whose step is larger than the cells of the frame or lattice, which enables a considerable reduction of the number of unknowns.

Problems

2.10.1 Find the trigonometric functions which solve Equations 2.10.2.

2.10.2 Based on the finite difference equations for Problem 2.9.2, deduce the differential equations for the smoothing continuum for the single-bay frame. Then solve the critical load from this differential equation and compare it to the result of Problem 2.9.2.

2.10.3 Do the same as above for Problem 2.9.3 or 2.9.4.

2.10.4 Generalize the present formulation to three dimensions.

References and Bibliography

Absi, E. (1967), "Équations intrinsèques d'une poutre droite à section constante," *Annales de l'Institut Technique du Bâtiment et des Travaux Publics,* 21(229):151–67 (Sec. 2.1).

Ackroyd, M. H., and Gerstle, K. H. (1982), "Behavior of Type 2 Steel Frames," *J. Struct. Eng.* (ASCE), 108(7):1541–56 (Sec. 2.4).

Ackroyd, M. H., and Gerstle, K. H. (1983), "Elastic Stability of Flexibility Connected Frames," *J. Struct. Eng.* (ASCE), 109(1):241–45 (Sec. 2.4).

Askar, A., and Cakmak, A. S. (1968), "A Structural Model of a Micropolar Continuum," *Int. J. Eng. Sci.,* 6:583–89 (Sec. 2.10).

Austin, W. J. (1971), "In-Plane Bending and Buckling of Arches," *J. Struct. Eng.* (ASCE), 97(5):1575–92 (Sec. 2.8).

Banks, C. B., and Sokolowski, U. (1968), "On Certain Two-Dimensional Applications of the Couple-Stress Theory," *Int. J. Solids and Structures,* 4:15–29 (Sec. 2.10).

Bathe, K. J., and Bolourchi, S. (1979). "Large Displacement Analysis of Three-Dimensional Beam Structures," *Int. J. Numer. Methods Eng.,* 14:961–86.

Bažant, Z. (1943), "Buckling Strength of Frame Structures" (in Czech), *Technický obzor* (Prague), 51(7, 8, and 16) (Sec. 2.1).

Bažant, Z. P. (1966), "Analysis of Framed Structures, Part II," in *Applied Mechanical Surveys,* ed. by Abramson *et al.,* (App. Mech. Reviews), Spartan Books, Washington, D.C., pp. 453–64.

Bažant, Z, P. (1974), "Three-Dimensional Harmonic Functions near Termination or Intersection of Singularity Lines: A General Numerical Method," *Int. J. Eng. Sci.,* 12:221–43 (Sec. 2.3).

Bažant, Z. P. (1985), *Lectures on Stability of Structures,* Course 720-D24, Northwestern University, Evanston, Illinois.

Bažant, Z. P., and Cedolin, L. (1989), "Initial Postcritical Analysis of Asymmetric Bifurcation in Frames," *J. Struct. Eng.* (ASCE), 115(11):2845–57 (Sec. 2.6).

Bažant, Z. P., and Christensen, M. (1972a), "Analogy Between Micropolar Continuum and Grid Frameworks Under Initial Stress," *Int. J. Solids and Structures,* 8:327–46 (Sec. 2.9).

Bažant, Z. P., and Christensen, M. (1972b), "Long-Wave Extensional Buckling of Large Regular Frames," *J. Struct. Eng.* (ASCE), 98(10):2269–88 (Sec. 2.1).

Bažant, Z. P., and Christensen, M. (1973), "Continuum Solutions for Long-Wave Extensional Buckling of Regular Frames," *Int. J. Eng. Sci.,* 11:1255–63 (Sec. 2.9).

Bažant, Z. P., and El Nimeiri, M. E. (1973), "Large-Deflection Spatial Buckling of Thin-Walled Beams and Frames," *J. Eng. Mech.* (ASCE), 99(6):1259–81 (Sec. 2.6).

Bažant, Z. P., and Estenssoro, L. F. (1979). "Surface Singularity and Crack Propagation," *Int. J. Solids and Structures,* 15:405–26. "Addendum," 16:479–81 (Sec. 2.3).

Benjamin, J. R., and Cornell, C. A. (1970), *Probability, Statistics and Decision for Civil Engineers,* McGraw-Hill, New York (Sec. 2.6).

Blaszkowiak, S., and Kaczkowski, Z. (1966), *Iterative Methods in Structural Analysis,* Pergamon Press, Oxford, England.

Bleich, F. (1952), *Buckling Strength of Metal Structures,* McGraw-Hill Book Co., New York (Sec. 2.7).

Bleich, F., and Melan, E. (1927), *Die gewöhnlichen und partiellen Differenzengleichungen der Baustatik,* Springer Verlag, Berlin (Sec. 2.9).

Boussinesq, J. (1883), "Résistance d'un anneau à la flexion, quand sa surface extérieure supporte une pression normale," *Compt. rend.,* 97:843 (Sec. 2.8).

Bridge, R. Q., and Fraser, D. J. (1987), "Improved G-Factor Method for Evaluating Effective Lengths of Columns," *J. Struct. Eng.* (ASCE), 113(6):1341–56. See also discussion by Koo (1988), 114(12):2828–30.

Britvec, S. J., and Chilver, A. H. (1964), "Elastic Buckling of Rigidly-Jointed Braced Frames," *J. Eng. Mech.* (ASCE), 89(6):217–55.

Bryant, A. H. (1987), "Built-Up Wood Columns," *J. Struct. Eng.* (ASCE), 113(1):107–21. See also discussion, 114(6):1441–43.

Budiansky, B., Frauenthal, J. C., and Hutchinson, J. W. (1969), "On Optimal Arches," *J. Appl. Mech.* (ASME), 36:880–82.

Chajes, A., and Churchill, J. E. (1987), "Non-Linear Frame Analysis by Finite Element Methods," *J. Struct. Eng.* (ASCE), 113(6):1221–35.

Chen, W. F., and Liu, E. M. (1987), *Structural Stability: Theory and Implementation.* Elsevier, New York.

Chini, S. A., and Wolde-Tinsae, A. M. (1988a), "Buckling Test of Prestressed Arches in Centrifuge," *J. Eng. Mech.* (ASCE), 114(6):1063–75.

Chini, S. A., and Wolde-Tinsae, A. M. (1988b), "Critical Load and Post-Buckling of Arch Frameworks," *J. Eng. Mech.* (ASCE), 114(9):1435–53.

Chini, S. A., and Wolde-Tinsae, A. M. (1988c), "Effect of Prestressing of Elastica Arches," *J. Eng. Mech.* (ASCE), 114(10):1791–1800.

Chugh, A. K. (1977), "Stiffness Matrix for a Beam Element Including Transverse Shear and Axial Force Effects," *Int. J. Numer. Methods Eng.,* 11:1681–97.

Chwalla, E. (1928), "Die Stäbilität zentrisch und exzentrisch gedrückter Stäbe aus Baustahl," Sitzungsberichte der Akademie der Wissenschaften in Wien, Part IIa, p. 469 (Sec. 2.1).

Chwalla, E., and Jokisch, F. (1941), "Über das ebene Knickproblem des Stockwerkrahmens," *Der Stahlbau,* 14:33 (Sec. 2.9).

Corradi, L. (1978), *Instabilità delle strutture,* CLUP, Milano.

Courbon, J. (1962), "Déversement d'un arc circulaire de section constante," *Annales des Ponts et Chaussées,* 132(1):225–44.

DaDeppo, D. A., and Schmidt, R. (1969), "Sidesway Buckling of Deep Circular Arches under a Concentrated Load," *J. Appl. Mech.* (ASME), 36:325–27 (Sec. 2.8).

DaDeppo, D. A., and Schmidt, R. (1971), "Stability of Heavy Circular Arches with Hinged Ends," *AIAA Journal*, 9(6):1200–1 (Sec. 2.8).

Dean, D. L., and Jetter, F. R. (1972), "Analysis for Truss Buckling," *J. Struct. Eng.* (ASCE), 98(8):1893–7.

Dean, D. L., and Ugarte, C. P. (1968), "Field Solutions for Two-Dimensional Frameworks," *Int. J. Mech. Sci.*, 10:315–39 (Sec. 2.6).

Dinnik, A. N. (1934), *Vestnik Inzhenerov*, No. 6 (Sec. 2.8).

Dischinger, F. (1937), "Untersuchung über die Knicksicherheit, die elastiche Verformung und das Kriechen des Betons bei Bogenbrücken," *Bauingenieur*, 18:487–520 (*Sec.* 2.8).

Dischinger, F. (1939), "Elastische und plastische Verformungen der Eisenbetontragwerke und insbesondere der Bogenbrücken," *Bauingenieur*, 20:290 (Sec. 2.8).

Duan, L., and Chen, W. F. (1988), "Effective Length Factor for Columns in Braced Frames," *J. Struct. Eng.*, (ASCE), 114(10):2357–70.

Duan, L., and Chen, W. F. (1989), "Effective Length Factor for Columns in Unbraced Frames," *J. Struct. Eng.*, (ASCE), 115(1):149–65.

Eisenberger, M., and Gorbonos, V. (1985), "Stability of Plane Frames Omitting Axial Strains," *J. Struct. Eng.* (ASCE), 111(10):2261–5. See also discussion and closure, 113(9):2080–3.

Ekhande, S. G., Selvappalam, M., and Madugula, M. K. S. (1989), "Stability Functions for Three-Dimensional Beam-Columns," *J. Struct. Eng.* (ASCE), 115(2):467–79.

Eringen, C. E. (1966), "Linear Theory of Micropolar Elasticity," *J. Math. Mech.*, 15:909–23.

Fiřt, V. (1964), *Stabilita montovanyćh konstruckí* (*Stability of Precast Structures*), SNTL (State Publishers of Technical Literature), Prague (Sec. 2.4).

Fiřt, V. (1974), *Stabilita a kmitání konstruckí s netuhými spoji* (*Stability of Frames with Deformable Joints*), Academia, Prague (Sec. 2.4).

Gallagher, R. H. (1975), *Finite Element Analysis—Fundamentals*, Prentice-Hall, Englewood Cliffs, N.J. (Sec. 2.3).

Goldberg, J. E. (1954), "Stiffness Charts for Gusseted Members under Axial Load," *Trans. ASCE*, 119(Paper No. 2657):43–54 (Sec. 2.1).

Goldberg, J. E. (1965), "Buckling of Multistory Buildings," *J. Eng. Mech.* (ASCE), 91(1):51–70.

Goldberg, J. E. (1968), "Lateral Buckling of Braced Multistory Frames," *J. Struct. Eng.* (ASCE), 94(12):2963–83.

Goto, Y., and Chen, W. F. (1987), "Second-Order Elastic Analysis for Frame Design," *J. Struct. Eng.*, (ASCE), 113(7):1501–19.

Gutkowski, W. (1963), "The Stability of Lattice Struts," *Zeitschrift für Angew. Math. und Mech.* (*ZAMM*), 43:284 (Sec. 2.9).

Harrison, H. B. (1982), "In-Plane Stability of Parabolic Arches," *J. Struct. Eng.* (ASCE), 108(1):195–205.

Horne, M. Z., and Merchant, W. (1965), *The Stability of Frames*, Pergamon Press, New York (Sec. 2.1).

Hu, K. K., and Lai D. C. (1986), "Effective Length Factor for Restrained Beam-Column," *J. Struct. Eng.* (ASCE), 112(2):241–56.

Hurlbrink, E. (1908), *Schiffbau*, 9:517 (Sec. 2.8).

James, B. W. (1935), "Principal Effects of Axial Load by Moment Distribution Analysis of Rigid Structures," NACA Tech. Note No. 534 (Sec. 2.1).

Jordan, C. (1965), *Calculus of Finite Differences*, 3rd ed., Chelsea, New York (Sec. 2.9).

Kitipornchai, S., and Finch, D. L. (1986), "Stiffness Requirements for Cross Bracing," *J. Struct. Eng.*, (ASCE), 112(12):2702–10.

Koiter, W. T. (1945), "Over de stabiliteit van het Elastische Evenwicht," Dissertation, Delft, Holland. Translation "On the stability of Elastic Equilibrium," NASA TT-F-10833, 1967 and AFFDL-TR-70-25, 1970 (Sec. 2.6).

Koiter, W. T. (1967), "Post-Buckling Analysis of Simple Two-Bar Frame," in *Recent Progress in Applied Mechanics,* ed. by B. Broberg et al. (Folke Odqvist Volume), Almqvist and Wiksell, Sweden, p. 337 (Sec. 2.6).

Kounadis, A. N. (1985), "An Efficient Simplified Approach for the Non-Linear Buckling Analysis of Frames," *AIAA Journal,* 23(8): 1254–59 (Sec. 2.6).

Kuranishi, S., and Yabuki, T. (1984), "Lateral Load Effect on Arch Bridge Design," *J. Struct. Eng.* (ASCE), 110(9): 2263–74.

Lightfoot, E. (1980), "Exact Straight-Line Elements," *J. Strain Analysis,* 15(2): 89–96.

Lightfoot, E., McPharlin, R. D., and Le Messurier, A. P. (1979), "Framework Instability Analysis using Blaszkowiak's Stiffness Functions," *Int. J. Mech. Sci.,* 21: 547–55.

Lightfoot, E., and Oliveto, G. (1977), "The Collapse Strength of Tubular Steel Scaffold Assemblies," *Proc. Instn. Civ. Engrs.,* 63(2): 311–29.

Lin, F. J., Glauser, E. C., and Johnston, B. G. (1970), "Behavior of Laced and Battened Structural Members," *J. Struct. Eng.* (ASCE), 96(7): 1377–1401.

Livesley, R. K. (1954), *Matrix Methods in Structural Analysis,* Pergamon Press, New York.

Livesley, R. K., and Chandler, D. B. (1956), *Stability Functions for Structural Frameworks,* Manchester University Press, Manchester (Sec. 2.1).

Lui, E. M., and Chen, W.-F. (1988), "Behavior of Braced and Unbraced Semi-Rigid Frames," *Int. J. Solids Struct.,* 24(9): 893–913.

Maier, G. (1966), "Behavior of Elastic-Plastic Trusses with Unstable Bars," *J. Eng. Mech.* (ASCE), 92(3): 67–91.

Martin, H. C. (1965), "On the Derivation of Stiffness Matrices for the Analysis of Large Deflection and Stability Problems," *Proc. Conf. Matrix Methods Struct. Mech.,* AFFDL TR 66-80, Wright-Patterson AFB, Ohio.

Masur, E. F. (1954), "Post-Buckling Strength of Redundant Trusses," *Trans. ASCE,* 119: 699 (Sec. 2.5).

Masur, E. F. (1970), "Buckling, Post-Buckling and Limit Analysis of Completely Symmetric Structures," *Int. J. Solids and Structures,* 6: 587–604 (Sec. 2.6).

Masur, E. F., and Milbradt, K. P. (1957), "Collapse Strength of Redundant Beams after Lateral Buckling," *J. Appl. Mech.* (ASME), 24: 283–88 (Sec. 2.6).

Moncarz, P. D., and Gerstle, K. H. (1981), "Steel Frames with Nonlinear Connections," *J. Struct. Eng.* (ASCE), 107(8): 1427–41. (Sec. 2.4).

Nicolai, E. L. (1918), *Bull. Polytech. Inst.* (St. Petersburg), Vol. 27 (Sec. 2.8).

Noor, A. K. (1981), "Survey of Computer Programs for Solution of Non-Linear Structural and Solid Mechanics Problems," *Comput. Struct.,* 13: 425–65.

Noor, A. K., and Nemeth, M. P. (1980), "Analysis of Spatial Beamlike Lattices with Rigid Joints," *Comput. Methods Appl. Mech. Eng.,* 24(1): 35–39.

Noor, A. K., and Weisstein, L. S. (1981), "Stability of Beamlike Lattice Trusses," *Comput. Methods Appl. Mech. Eng.,* 25(2): 179–93.

Northwestern University (1969), "Report on 'Models for Demonstrating Buckling of Structures'," prepared by R. A. Parmelee, G. Herrmann, J. F. Fleming, and J. Schmidt under NSF Grant GE-1705, (Sec. 2.2).

Omid'varan, C. (1968), "Discrete Analysis of Latticed Columns," *J. Struct. Eng.* (ASCE), 94(1): 119–30 (Sec. 2.9).

Oran, C. (1980), "General Imperfection Analysis in Shallow Arches," *J. Eng. Mech.* (ASCE), 106(6): 1175–93.

Oran, C., and Bayazid, H. (1978), "Another Look at Buckling of Circular Arches," *J. Eng. Mech.* (ASCE), 104(6): 1417–32.

Oran, C., and Reagan, R. S. (1969), "Buckling of Uniformly Compressed Circular Arches," *J. Eng. Mech.* (ASCE), 95(4):879–95 (Sec. 2.8).
Pearson, C. E. (1974), *Handbook of Applied Mathematics,* Van Nostrand Reinhold, New York (Sec. 2.9).
Pestel, E. C., and Leckie, F. A. (1963), *Matrix Methods in Elastostatics,* McGraw-Hill Book Co., New York.
Powell, G. H. (1969), "Theory of Non-Linear Elastic Structures," *J. Struct. Eng.* (ASCE), 95(12), 2687–2701.
Powell, G., and Klingner, R. (1970), "Elastic Lateral Buckling of Steel Beams," *J. Struct. Eng.* (ASCE), 96(9):1919–32 (Sec. 2.6).
Powell, G. H., and Simons, J. (1981), "Improved Iteration Strategy for Non-Linear Structures," *Int. J. Numer. Methods Eng.*, 17:1455–67.
Roorda, J. (1965a), "Stability of Structures with Small Imperfections," *J. Eng. Mech.* (ASCE), 91(1):87–106 (Sec. 2.6).
Roorda, J. (1965b), "The Instability of Imperfect Elastic Structures," Ph.D. Dissertation, University of London, England (Sec. 2.6).
Roorda, J. (1965c), "The Buckling Behavior of Imperfect Structural Systems," *J. Mech. Phys. Solids,* 13:267–80.
Roorda, J. (1968), "On the Buckling of Symmetric Structural Systems with First and Second Order Imperfections," *Int. J. Solids and Structures,* 4:1137–48.
Roorda, J., and Chilver, A. M. (1970), "Frame Buckling: An Illustration of the Perturbation Technique," *Int. J. Non-Linear Mechanics,* 5:235–46 (Sec. 2.6).
Rutenberg, A. (1982), "Simplified P-Delta analysis for Asymmetric Structures," *J. Struct. Eng.* (ASCE), 108(9):1995–2013.
Sakimoto, T., and Komatsu, S. (1983), "Ultimate Strength Formula for Steel Arches," *J. Struct. Eng.* (ASCE), 109(3):613–27.
Schmidt, R., and DaDeppo, D. A. (1972), "Discussion of the paper by Austin (1971)," *J. Struct. Eng.* (ASCE), 98(1):373–78 (Sec. 2.8).
Seide, P. (1987), "Buckling of a Rectangular Frame Revisited," *J. Appl. Mech.* (ASME), 54:617.
Stelmack, T. W., Marley, M. J., and Gerstle, K. H. (1986), "Experimental Verification of Behavior of Flexibly-Connected Steel Frames," *J. Struct. Eng.* (ASCE), 112(7):1573–88.
Tauchert, T. R. (1969), "A Lattice Theory for Representation of Thermoelastic Composite Materials," in *Recent Advances in Engineering Science,* ed. by A. C. Eringen, vol. 5, pt. I, pp. 325–45, Gordon and Breach, New York (Sec. 2.10).
Thompson, J. M. T. (1982), *Instabilities and Catastrophes in Science and Engineering,* John Wiley and Sons, New York (Sec. 2.8).
Thompson, J. M. T., and Hunt, G. W. (1973), *A General Theory of Elastic Stability,* John Wiley and Sons, New York.
Timoshenko, S. P., and Gere, J. M. (1961), *Theory of Elastic Stability,* 2nd ed., McGraw-Hill Book Co., New York (Sec. 2.1).
Tsang, H. S. (1963), "Analysis of Rigid Frames by Difference Equations," *J. Struct. Eng.* (ASCE), 89(2):127–59 (Sec. 2.9).
Tuma, J. (1988), *Handbook of Structural and Mechanical Matrices,* McGraw-Hill Book Co., New York (Sec. 2.1).
Von Mises, R., and Ratzersdorfer, J. (1926), Die Knicksicherheit von Rahmentragwerken, vol. 6, *Zeitschrift für Angew. Math. und Mech.* (*ZAMM*) p. 181 (Sec. 2.1).
Wah, T. (1965), "The Buckling of Gridworks," *J. Mech. Phys. Solids,* 13:1–16.
Wah, T., and Calcote, L. R. (1970), *Structural Analysis by Finite Difference Calculus,* Van Nostrand-Reinhold, New York (Sec. 2.4).

Williams, F. W. (1981), "Stability Functions and Frame Instability—A Fresh Approach," *Int. J. Mech. Sci.*, 23(12):715–22 (Sec. 2.1).

Williams, F. W. and Wittrick, W. H. (1983), "Exact Buckling and Frequency Calculations Surveyed," *J. Struct. Eng.*, 109(1):169–87 (Sec. 2.1).

Wolde-Tinsae, A. M., and Assaad, M. C. (1984), "Non-Linear Stability of Prebuckled Tapered Arches," *J. Eng. Mech.* (ASCE), 110(1):84–94.

Wolde-Tinsae, A. M., and Huddleston, J. V. (1977), "Three-Dimensional Stability of Prestressed Arches," *J. Eng. Mech.* (ASCE), 103(5):855–67.

Woolcock, S. T., and Trahair, S. N. (1975), "Post-Buckling of Redundant Rectangular Beams," *J. Eng. Mech.* (ASCE), 101(4):301–16. See also discussion, 102(3):585–87.

Wozniak, C. (1966), "Load-Carrying Structures of the Dense Lattice Type. The Plane Problem," *Archwm. Mech. Stosow.*, 18:581–97; see also "Bending and Stability Problems with Lattice Structures," *Archwm. Mech. Stosow.*, 18:781–96 (Sec. 2.10).

Wu, C. H. (1968), "The Strongest Circular Arch—A Perturbation Solution," *J. Appl. Mech.* (ASME), 35:476–80.

Yabuki, T., and Vinnakota, S. (1984), "Stability of Steel Arch-Bridges, A State-of-the-Art Report," *Solid Mech. Arch.*, Vol. 9, No. 2, Noordhoff International Publishers, Leyden, Netherlands.

Yang, Y. B., and McGuire, W. (1986), "Stiffness Matrix for Geometric Non-Linear Analysis," *J. Struct. Eng.* (ASCE), 112(4):853–87. See also discussion, 113(7):1632–37.

3

Dynamic Analysis of Stability

Failure of structures is a dynamic process, and so it is obviously more realistic to approach buckling and stability from the dynamic point of view. At the same time, it appears that a dynamic approach is necessary to define the concept of stability precisely. So far we have used the term stability only in a loose, intuitive way.

In this chapter, we begin with the analysis of the stability of vibrations of structures subjected to conservative loads, such as gravity loads. Then we proceed to show that a completely different behavior may be encountered when the structure is subjected to nonconservative loads, such as wind or hydrodynamic forces in general, or pulsating loads produced by rotating machinery or moving vehicles, etc. This will lead us to state the precise fundamental definition of stability of motion and to consider when equivalent results for stability can be achieved by static analysis or energy methods, which are generally much simpler. Finally, we will touch the topic of nonlinear dynamic systems. They can exhibit a complex, apparently random response called chaos, which nevertheless shows some degree of order.

3.1 VIBRATION OF COLUMNS OR FRAMES AND DIVERGENCE

According to the D'Alembert principle, the differential equation of motion of a column may be obtained from the differential equation of equilibrium by including the transverse inertia force per unit length of column in the applied transverse load p. We will also consider that there may be damping that is linear and velocity-dependent. We will neglect the effect of rotational inertia, which may be appreciable only at high frequencies (Raju and Rao, 1986). Thus, we may set $p = -\mu\, \partial^2 w/\partial t^2 - \beta\, \partial w/\partial t$, where w = deflection, μ = mass of the column per unit length, β = damping coefficient, and t = time. Then, substituting this into Equation 1.3.5, we get

$$\frac{\partial^2}{\partial x^2}\left(EI\frac{\partial^2 w}{\partial x^2}\right) + \frac{\partial}{\partial x}\left(P\frac{\partial w}{\partial x}\right) = -\mu\frac{\partial^2 w}{\partial t^2} - \beta\frac{\partial w}{\partial t} \tag{3.1.1}$$

where x = length coordinate, EI = bending stiffness, and P = axial load (positive if compression). When forced vibrations are studied, then, of course, the prescribed distributed load $p(t)$ must be added on the right-hand side of this equation.

In our calculations we will assume, for the sake of simplicity, that P, EI, μ, and β are constants, in which case this equation becomes

$$EI\frac{\partial^4 w}{\partial x^4}+P\frac{\partial^2 w}{\partial x^2}=-\mu\left(\frac{\partial^2 w}{\partial t^2}+2b\frac{\partial w}{\partial t}\right) \tag{3.1.2}$$

where $b=\beta/2\mu=$ damping parameter. Equation 3.1.1 or 3.1.2 is a linear partial differential equation. Together with the appropriate boundary conditions at column ends $x=0$ and $x=l$, and initial conditions at $t=0$, we have an initial boundary-value problem.

Columns

For the sake of simple illustration let us study first the free vibration of a pin-ended column (Fig. 3.1a) loaded by axial force P, whose magnitude and direction remain constant as the column moves (this is an example of a conservative load). Since the boundaries of the domain in the xt plane are rectangular (Fig. 3.1b), the solutions of this initial boundary-value problem may be expected to be separable in the form $w(x, t)=Y(x)f(t)=$ product of a function of x that gives the modal shape (or mode) of the vibration of the structure and a function of t. As the coefficients of Equation 3.1.2 are constant, we may seek the general solution in the form:

$$w(x, t)=\sum_{n=1}^{\infty} f_n(t)\sin\frac{n\pi x}{l} \tag{3.1.3}$$

Note that the sine functions automatically satisfy the bondary conditions for a pin-ended column ($w=w''=0$ at $x=0$ and $x=l$). Note also that Equation 3.1.3 for $t=0$ can represent any initial deformation of the beam and any initial velocity distribution. The initial values $f_n(0)$ and $\dot{f}_n(0)$ can be found through Fourier sine series expansion of given $w(x, 0)$ and $\dot{w}(x, 0)$. The Fourier cosine series is not included in Equation 3.1.3 since its terms do not satisfy the boundary conditions.

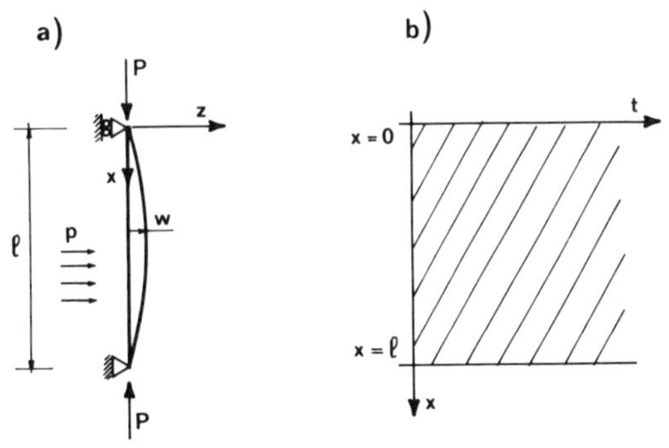

Figure 3.1 (a) Vibration mode of a pin-ended column; (b) region of (x, t) plane in which the solution is sought.

For general boundary conditions, such as fixed-end columns, one would have to replace Equation 3.1.3 with $w = \sum a_n \exp c_n x$ where a_n and c_n are complex numbers. This would have to be done also if both even and odd derivatives with respect to x were present in the differential equation, but fortunately this does not happen for bending theory (Eq. 3.1.2). Complex-valued a_n and c_n also need to be used to derive the generalized s and c functions and the stiffness matrix for free vibrations of beam columns under axial load. Then w is also complex, and the actual deflection is represented by the real part of w (or alternatively the imaginary part of w).

Let us try the D'Alembert substitution $f_n(t) = A_n e^{\lambda_n t}$ in Equation 3.1.3. Substitution of Equation 3.1.3 into Equation 3.1.2 furnishes

$$\sum_{n=1}^{\infty} \left[\left(EI \frac{n^2\pi^2}{l^2} - P \right) \frac{n^2\pi^2}{l^2} + \mu\lambda_n^2 + 2b\mu\lambda_n \right] A_n \sin \frac{n\pi x}{l} e^{\lambda_n t} = 0 \qquad (3.1.4)$$

To satisfy this equation for all x and all t, the bracketed terms must vanish (otherwise this equation could be satisfied only for some x and some t). Noting that $n^2 EI\pi^2/l^2 = n^2 P_E = P_{cr_n} = n$th critical load, we thus get the condition

$$\lambda_n^2 + 2b\lambda_n + \frac{n^4\pi^4}{l^4} \left(\frac{EI}{\mu} \right) \left(1 - \frac{P}{P_{cr_n}} \right) = 0 \qquad (3.1.5)$$

Consider now that there is no damping ($b = 0$). The foregoing quadratic equation for λ_n then has the roots:

$$\lambda_n = \pm i\omega_n \qquad \omega_n = \frac{n^2\pi^2}{l^2} \sqrt{\frac{EI}{\mu}} \sqrt{1 - \frac{P}{P_{cr_n}}} \qquad (3.1.6)$$

The corresponding general solutions for $f_n(t)$ are the linear combinations of all the fundamental solutions associated with ω_n or λ_n:

$$\text{For } P < P_{cr_n}: \quad f_n(t) = A_n \cos \omega_n t + B_n \sin \omega_n t \qquad (3.1.7)$$

$$P = P_{cr_n}: \quad f_n(t) = A_n + B_n t \qquad (3.1.8)$$

$$P > P_{cr_n}: \quad f_n(t) = A_n e^{-\lambda_n t} + B_n e^{-\lambda_n t} \qquad (3.1.9)$$

where A_n and B_n are constants to be determined from the initial conditions at $t = 0$, that is, from the values $f_n(0)$, $\dot{f}_n(0)$. [These values can be found by Fourier series expansion of given $w(x, 0)$ and $\dot{w}(x, 0)$.]

Types of Motion and Dependence of Natural Frequency on Load

Of prime interest is the first natural frequency ω_1. Let us examine its dependence on the axial load P. For $P = 0$ we have the first natural frequency of a load-free beam $\omega_1^0 = (EI/\mu)^{1/2}(\pi^2/l^2)$. As the magnitude of the axial compressive load P increases, the frequency ω_1 according to Equation 3.1.6 decreases. The plot of P versus ω_1^2 is a straight line (Fig. 3.2a); and $f_1(t)$ is given by Equation 3.1.7. For $P = P_{cr_1}$ = first critical load, the first vibration frequency ω_1 vanishes (Fig. 3.2a), and in this case $\lambda_1 = 0$ is a double root of Equation 3.1.5. This means that the corresponding fundamental solutions are $e^{0 \cdot t}$ and $te^{0 \cdot t}$, or 1 and t, which yields the general solution in Equation 3.1.8. For $P > P_{cr_1}$, ω_1 becomes an imaginary

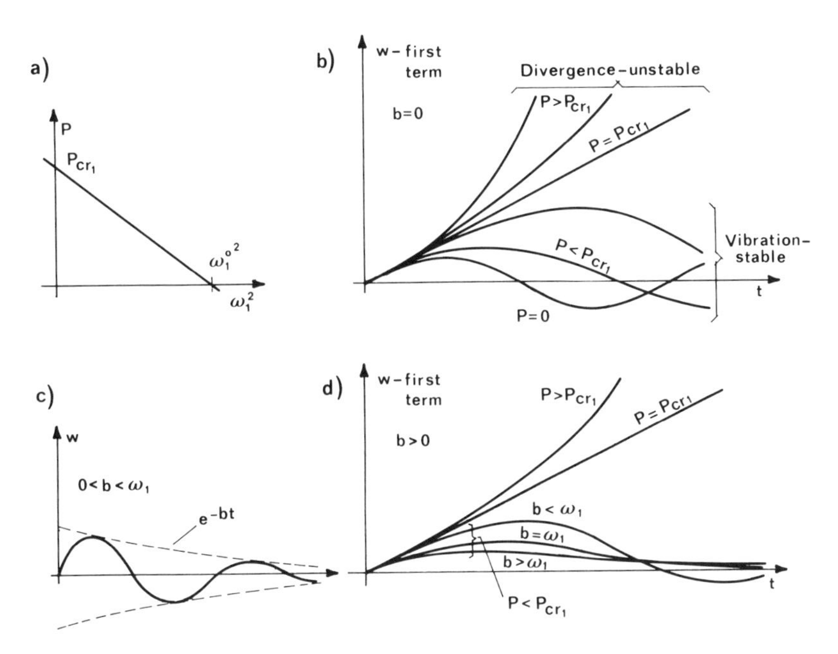

Figure 3.2 (a) Dependence of frequency on axial load P; time variations of deflection at various P for (b) $b=0$ and (c, d) $b>0$ (b = damping coefficient).

number, which means that λ_1 becomes real, and so the solution consists of exponentials (Eq. 3.1.9).

The types of motion obtained for various values of P are plotted in Figure 3.2b. We see that there is a qualitative change in the type of solution as the axial load exceeds the first critical load. For small $P<P_{cr_1}$, the solutions are bounded, and if the initial velocity (or the initial deflection) is very small, the deflection remains very small for all times. However, when $P \geq P_{cr_1}$, the deflections will become very large no matter how small the initial velocity (or the initial deflection) is. Since arbitrarily small initial velocities or initial deflections (analogous to imperfections, with arbitrary distributions along x) cannot, in reality, be, prevented, that is, $f_1(0) \neq 0$, $\dot{f}_1(0) \neq 0$, the deflection for $P \geq P_{cr_1}$ will increase beyond any bounds, according to our linearized theory. This may be interpreted as failure.

Therefore, the response of the column for $P \geq P_{cr}$ is termed unstable, and the condition of stability of the column is

$$P < P_{cr_1} \tag{3.1.10}$$

The type of instability exhibited by our column, and illustrated in Figure 3.2b by the solution for $P=P_{cr_1}$, is called divergence, because the column is incapable of vibrating and diverges to one side. The typical feature of this type of instability is that it occurs at $\omega=0$. The motion is at constant velocity, at which the inertia forces vanish. Such motion implies the existence of neutral equilibrium at $x=0$ when $f_1(0)=0$, $\dot{f}_1(0)=0$. This explains why the static analysis gave the same critical load. For $P>P_{cr_1}$, the motion is accelerated.

The complete solution is given by the infinite sum expressed by Equation

3.1.3. This means that, for example, the solution for $P_{cr_1} \le P < P_{cr_2}$ consists of a diverging motion according to Equation 3.1.8 or 3.1.9 for the first mode ($n = 1$) and superimposed oscillations according to Equation 3.1.7 (for $n \ge 2$). These oscillations are of limited magnitude if the corresponding initial perturbation is small, but this does change the diverging character of the response caused by the first mode.

Effect of Damping

If damping is present ($b > 0$), the solution is different, but the stability condition (Eq. 3.1.10) remains the same. In this case, the solution of Equation 3.1.5 for $b < \omega_1$ is

$$\lambda_n = -b \pm i\omega'_n \qquad \text{with } \omega'_n = \sqrt{\omega_n^2 - b^2} \tag{3.1.11}$$

in which ω'_n = vibration frequency of the damped column. The corresponding fundamental solutions are of the type e^{-bt} and $e^{\pm i\omega'_n t}$, that is,

$$w \sim e^{-bt} \cos \omega'_n t \qquad \text{or} \qquad e^{-bt} \sin \omega'_n t \tag{3.1.12}$$

These motions represent damped vibrations with envelope e^{-bt} (Fig. 3.2c). The deflections obviously remain very small if the initial velocity or the initial deflection is very small, provided that ω'_1 is real or $\omega_1'^2$ is positive, which occurs when $\omega_1 > b$ (and requires that $P < P_{cr_1}$ because ω_1 is real if $P < P_{cr_1}$).

When $\omega'_1 = 0$ or $\omega_1 = b$ ($P < P_{cr_1}$), the solutions for $n = 1$ are of the type e^{-bt} and te^{-bt}, and so they are also bounded (Fig. 3.2d). When $\omega_1'^2 < 0$ or $\omega_1 < b$, the solutions for $n = 1$ are of the type $e^{-(b+\omega''_1)t}$ and $e^{-(b-\omega''_1)t}$ with $\omega''_1 = i\omega'_1$ (ω''_1 is real because in this case ω'_1 is imaginary); these solutions remain bounded only as long as $\omega''_1 < b$ or $-\omega_1'^2 < b^2$, or $b^2 - \omega_1^2 < b^2$, that is, $\omega_1^2 > 0$ or $P < P_{cr_1}$ (Fig. 3.2d).

So we see that the stability condition of the column is not affected by linear damping, that is, damping has no stabilizing (or destabilizing) influence. This is generally true for linear systems with conservative loads, but not always true for systems with nonconservative loads (such as circulatory loads, as corroborated by certain examples).

Thus the main effect of damping is to prevent an oscillating motion (a motion with alternating sign of the deflection) at loads sufficiently close to critical. In dynamics, the value of the damping coefficient b at which the oscillatory motion changes to nonoscillatory is usually called critical damping (in our case, $b_{cr} = \omega_1$). Here the use of the term critical has nothing to do with stability.

Frames and Other Generalizations

The foregoing analysis can be generalized to beams with arbitrary end supports as well as to frames. In the linearized, small-deflection theory, the motion can be sought in the form

$$w(x, t) = A\, \bar{w}(x) \sin \omega t \tag{3.1.13}$$

in which $\bar{w}(x)$ represents what is called the modal shape or mode of the structure. In contrast to our solution for a simply supported column, this shape is generally not sinusoidal. By a procedure similar to our derivation of the stiffness matrix for a beam-column, one can derive a generalized stiffness matrix for vibrations, in which the coefficients s and c are functions of not only $\rho = P/P_E$ but also of the vibration frequencies ω; see, for example, Mohsin and Sadek (1968), Howson and Williams (1973), Åkesson (1976), and Kohoutek (1985) (along with a discussion of some pitfalls). Thus, the problem of free vibrations of arbitrary beams or frames with axial forces in the beams generally leads to the following matrix eigenvalue problem:

$$\mathbf{K}(\rho, \omega)\bar{\mathbf{u}} = \mathbf{0} \tag{3.1.14}$$

in which $\bar{\mathbf{u}}$ is the column matrix of the modal amplitudes of generalized displacements (displacements and rotations at the joints of members), and $\mathbf{K}$ is a square matrix whose coefficients are generally nonlinear functions of ρ and ω. Free vibrations are possible if $\det \mathbf{K} = 0$, and from this condition for each chosen ρ one can find the corresponding vibration frequencies ω, especially the first frequency ω_1. The value of ρ for which ω_1 ceases to be real is then the first critical load. The condition $\det \mathbf{K} = 0$ must in general be solved numerically and the same procedure as already expounded for static buckling of frames may be used.

When the beam has a variable stiffness, or the axial load is variable, a solution may be obtained by splitting the beam into a number of short beam elements, each with different I or different P. This is the approach used in the finite element method, and the solution based on s and c functions then becomes asymptotically identical to the solution based on beam finite elements with cubic distribution functions. As before, one advantage of the finite element approach is that it yields a linear eigenvalue problem (i.e., matrix $\mathbf{K}$ that is linearly dependent on small ρ and ω). This is achieved, though, at the cost of a greatly increased number of unknowns.

Forced vibrations of structures with axial forces generally lead to the matrix equation of motion:

$$\mathbf{K}(\rho)\mathbf{u}(t) = \mathbf{f}(t) \tag{3.1.15}$$

where ρ = parameter of axial forces and $\mathbf{u}, \mathbf{f}$ = column matrices of joint (or node) displacements and applied forces. The general solution can be composed from the individual modal responses based on Equation 3.1.14.

Note that in our example we can determine stability without obtaining the complete solutions. The precise initial conditions are irrelevant for stability. We need to determine only the type of solution, that is, solve the problem qualitatively. This is the basic objective in most dynamic stability analyses. We are interested in qualitative methods for dynamics, which give only the stability information we need.

The divergence type of instability that we illustrated for our column is typical of all conservative elastic systems. Next we will see that nonconservative systems may lose stability in a different manner—dynamically.

Problems

3.1.1 Solve free vibrations of a massless pin-ended column with point mass m attached at midspan (Fig. 3.3a). Determine ω_1, P_{cr_1}.

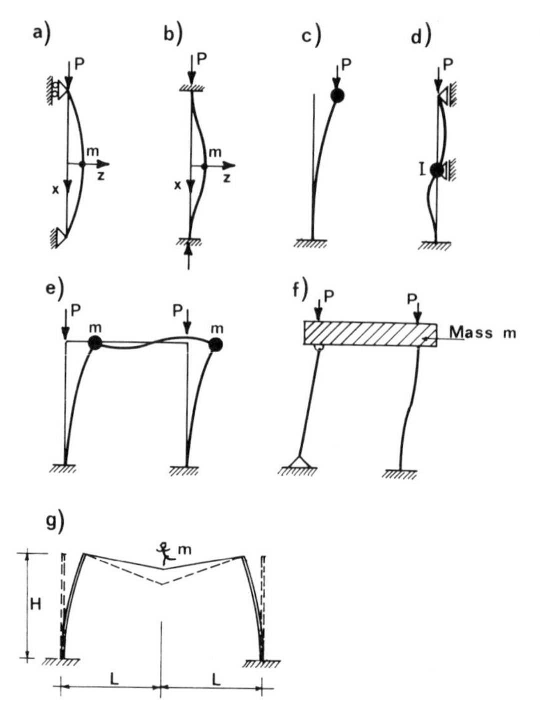

Figure 3.3 Exercise problems on free vibrations of beams and frames under conservative loads.

3.1.2 Same for a fixed-end column (Fig. 3.3b).

3.1.3 Same for a free-standing column, mass m on top (Fig. 3.3c).

3.1.4 Solve sway vibrations of a massless portal frame, with mass m at each node (Fig. 3.3e). Find ω_1, P_{cr_1} (use s and c functions).

3.1.5 Solve vibrations of a massless two-span continuous beam in Figure 3.3d with a mass of rotational inertia moment I attached at the middle support (use s and c functions).

3.1.6 Solve free vibrations of the two-column structure in Figure 3.3f, and P_{cr_1}.

3.1.7 Using Equation 3.1.14, solve the free vibration frequencies and the first critical load of a fixed-end beam. *Hint:* Start with $w = \sin \omega t(e^{\beta x})$ with β = complex constant. After finding roots β, use $w = \sin \omega t(A \sin \beta_1 x + B \cos \beta_1 x + Ce^{\beta_2 x} + De^{-\beta_2 x})$. [*Note:* Instead of the exponentials, one can also use $C \sinh \beta_2 x + D \cosh \beta_2 x$, but then the resulting stiffness matrix loses accuracy when x is large; see Kohoutek (1985).]

3.1.8 Same for a free-standing column of height l (load P travels with column top, remaining vertical).

3.1.9 Using $w(x, t) = \sum f_n(t)e^{c_n x}$ find the critical load of a pin-ended beam.

3.1.10 Using $w(x, t) = \sum a_n e^{c_n x}$ find the generalized s and c functions (cf. Howson and Williams, 1973, and Åkesson, 1976).

3.1.11 A rope dancer (Fig. 3.3g) of mass m performs on an inextensible rope of axial stiffness EA attached to the tops of poles (free-standing columns) of distributed mass μ and stiffness EI, receiving substantial axial force due to the dancer's weight mg. Calculate the natural frequency ω which needs to be outside the frequency range of the dancer's motions.

3.2 NONCONSERVATIVE LOADS AND FLUTTER

We will now show that a qualitatively different loss of stability can occur under nonconservative loads. In the preceding problem, it was stipulated that the direction of the load does not change as the column deflects. By contrast, in some cases the load direction may vary depending on the deformation of the structure. Let us study a simple example.

Massless Column under Follower Load

Consider a free-standing column that is loaded by a so-called follower force, P, a force that turns its direction so as to always remain tangential to the deflection curve at the column top (Fig. 3.4a). Follower loading is typical of wind and may

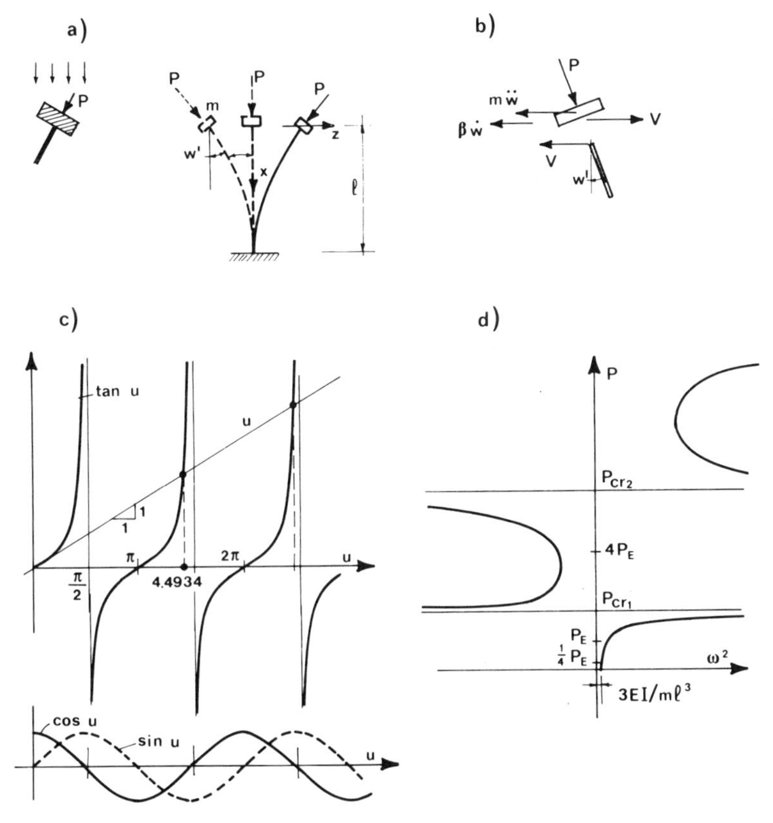

Figure 3.4 (a) Column loaded by a constant follower force (massless column, point mass at end); (b) diagram of forces at column top; (c) condition for critical load; (d) dependence of frequency of free vibrations upon load.

be generated by attaching to the column end a disk normal to the column and placing the column in a wind tunnel with an airstream in the x-direction. The same loading also may be approximately obtained by striking the disk with a water jet in the x-direction, or by attaching a rocket engine of thrust P at the end of the column.

To make a simple solution possible, we consider that the column in Figure 3.4a is massless while the disk at the end has mass m and may be treated as a mass point (i.e., its rotational inertia is negligible). We consider the linearized small-deflection theory and constant EI. The absence of distributed mass and of damping along the column makes the static solution (Eq. 1.3.6) valid for the shape of the deflection curve, that is,

$$w(x, t) = f(t)(A \sin kx + B \cos kx + Cx + D) \tag{3.2.1}$$

in which $f(t)$ is a function of time t to be determined and $k = (P/EI)^{1/2}$.

To formulate the boundary condition on top, we may recall that the shear force is expressed as $V = -M' - Pw'$ (Eq. 1.3.2); w now denotes the deflection on top, that is, $w = w(0, t)$. Referring to the diagram of forces, including the inertia and damping forces, as sketched in Figure 3.4b, we obtain for mass m the equation of motion: $m\ddot{w} = V + Pw' - \beta\dot{w}$ or $m\ddot{w} + \beta\dot{w} = -M' - Pw' + Pw' = -M'$ where $\ddot{w} = d^2w/dt^2$, $\dot{w} = dw/dt$, and $\beta =$ damping coefficient. Substituting $M = EIw''$, and writing also the remaining three boundary conditions (the second boundary condition at the top is $M = 0$ for $x = 0$), we have the conditions:

$$\begin{aligned} &\text{For } x = 0: \quad && w'' = 0 \quad && -(EIw'')' = m\ddot{w} + \beta\dot{w} \\ &\text{For } x = l: \quad && w = 0 \quad && w' = 0 \end{aligned} \tag{3.2.2}$$

It is interesting to examine the problem first by the static approach, assuming that $\ddot{w} = 0$ and $\dot{w} = 0$. Then, for $EI =$ constant, substitution of Equation 3.2.1 into Equations 3.2.2 yields the boundary conditions,

$$A = 0 \qquad B = 0 \qquad Cl + D = 0 \qquad C = 0 \tag{3.2.3}$$

This further implies that $D = 0$. Thus, the column loaded by a follower force has no deflected equilibrium position, that is, it cannot buckle in a static manner. However, this does not mean that the column cannot buckle at all—a striking conclusion, which was incorrectly drawn in one of the first studies of this problem. We cannot exclude the possibility that the column could lose stability while in motion.

Turning our attention to the dynamic problem, we obtain from the boundary conditions (Eqs. 3.2.2)

$$\begin{aligned} &B = 0 \qquad && -AEIk^3 f = -D(m\ddot{f} + \beta\dot{f}) \\ &A \sin kl + Cl + D = 0 \qquad && Ak \cos kl + C = 0 \end{aligned} \tag{3.2.4}$$

Eliminating the unknown constants A and C, we get for functions $f(t)$ the second-order ordinary linear differential equation

$$\ddot{f} + 2b\dot{f} + \omega^2 f = 0 \tag{3.2.5}$$

in which we introduce the notations

$$\omega^2 = \frac{EI}{m}\left(\frac{k^3}{\sin u - u\cos u}\right) \qquad b = \frac{\beta}{2m} \qquad (u = kl) \tag{3.2.6}$$

Consider first that damping is negligible, that is, $b = 0$. A solution of Equation 3.2.5 may be sought in the form $f = e^{i\omega t}$, which satisfies automatically Equation 3.2.5 (ω is the vibration frequency). As long as Equations 3.2.6 yield $\omega^2 > 0$, the column is stable because the deflections remain very small for all time if the initial velocity or the initial deflection is very small. However, when $\omega^2 < 0$, the solutions are of the type $f = e^{\lambda t}$ or $e^{-\lambda t}$ where $\lambda = i\omega$ real (ω = imaginary); one of these solutions leads to arbitrarily large deflections no matter how small the initial velocities or initial deflections are.

According to Equations 3.2.6, ω^2 changes its sign when $\sin u - u \cos u = 0$ or $\tan u = u$. As shown in Figure 3.4c, the smallest positive load value for which this condition can be satisfied is $u = 4.4934$, which yields the first critical load

$$P_{cr_1} = 20.19\frac{EI}{l^2} = 8.183\frac{P_E}{4} \tag{3.2.7}$$

The fact that P_{cr_1} is much larger than the critical load $P_E/4$ for the dead load is due to the horizontal component of P, which is opposite to the deflection if P is not too large.

Furthermore, as is clear from the graphs of functions $\tan u$, $\cos u$, and u in Figure 3.4c, ω^2 is positive, and the column is stable, when $P < P_{cr_1}$, or when $P_{cr_2} < P < P_{cr_3}$, etc, where P_{cr_2}, $P_{cr_3}, \ldots$ are the load values corresponding to successive change of sign of ω^2. On the other hand, for $P_{cr_1} \le P \le P_{cr_2}$, $\omega^2 \le 0$, which means instability.

The plot of P versus ω^2, calculated according to Equations 3.2.6, is shown in Figure 3.4d. It is noteworthy that ω^2 does not become negative by passing through a zero value, as it did in our previous study of column vibrations under dead load, but it does so by jumping from $+\infty$ to $-\infty$. This means that neutral equilibrium does not exist at the critical load. The loss of stability in which the structure is oscillating at the critical load is called the oscillatory instability or flutter, and the critical load is also called the flutter load. The present example is a limiting case of flutter in which the frequency tends to infinity at the loss of stability. (The infinite value of the frequency is due to our neglect of the mass between the column ends.)

It is interesting but not surprising to note that the value of P_{cr_1} coincides with the value that would be found by static analysis for the case in which the load passes through a fixed point at distance c from the free end (Sec. 1.4), with $c \to 0$. Note also that, for $P = 0$, $\omega^2 = 3EI/ml^3$ (i.e., the limit of Eqs. 3.2.6 for $y \to 0$) can be calculated as the frequency of a single-degree-of-freedom elastic system of stiffness $3EI/l^3$ and mass m.

Consider now that damping is present, that is, $b > 0$. Substituting $f = e^{\lambda t}$ into Equation 3.2.5, one obtains the characteristic equation $\lambda^2 + 2b\lambda + \omega^2 = 0$, whose roots yield the fundamental solutions:

$$e^{-bt}\sin\omega' t \qquad e^{-bt}\cos\omega' t \qquad \text{where } \omega' = \sqrt{\omega^2 - b^2} \tag{3.2.8}$$

or

$$e^{-(b+\omega'')t} \qquad e^{-(b-\omega'')t} \qquad \text{where } \omega'' = \sqrt{b^2 - \omega^2} \tag{3.2.9}$$

Obviously, deflections that are initially small remain small if $b > \omega''$. This means that $b^2 > \omega''^2$ and finally leads to the stability condition $\omega^2 > 0$, which is the same as before. So we see again that linear damping has no effect on the critical loads.

Effect of Distributed Mass

A free-standing column that has a distributed mass and is loaded on top by a follower force (called Beck's column) also loses stability by means of flutter rather than divergence (Beck, 1952). The solution is considerably more involved; see, for example, Timoshenko and Gere (1961, p. 152). The loss of stability, however, is different in that the vibration frequency at the critical state is finite, not infinite. We illustrate such a type of stability loss with another, simpler example.

Consider the free-standing column shown in Figure 3.5a, which consists of two rigid bars of length l connected by hinges with rotational springs of stiffness C. The rigid bars are assumed to be massless, however masses m (considered as point masses) are attached on top and at the middle hinge. The column is loaded on top by a follower force P.

The horizontal inertia forces acting at the upper and lower point masses are, for small deflections, $ml(\ddot{\theta}_1 + \ddot{\theta}_2)$ and $ml\ddot{\theta}_1$. Writing the moment equations of equilibrium of the entire column as a free body about its base, and of the top part as a free body about the middle hinge, we obtain the following system of two linear ordinary differential equations for θ_1 and θ_2 as functions of time t:

$$\begin{aligned} ml(\ddot{\theta}_1 + \ddot{\theta}_2)2l + ml\ddot{\theta}_1 l + Pl(\theta_2 - \theta_1) + C\theta_1 &= 0 \\ ml(\ddot{\theta}_1 + \ddot{\theta}_2)l + C(\theta_2 - \theta_1) &= 0 \end{aligned} \tag{3.2.10}$$

If the terms with $\ddot{\theta}_1$ and $\ddot{\theta}_2$ are deleted, we have a system of two equilibrium equations. They are homogeneous and have a nonzero solution only if their determinant D vanishes. However, one finds that $D = C^2$, which cannot vanish. Therefore, this column again cannot lose stability in a static manner, that is, by neutral equilibrium. It may also be noted that the system of equilibrium equations

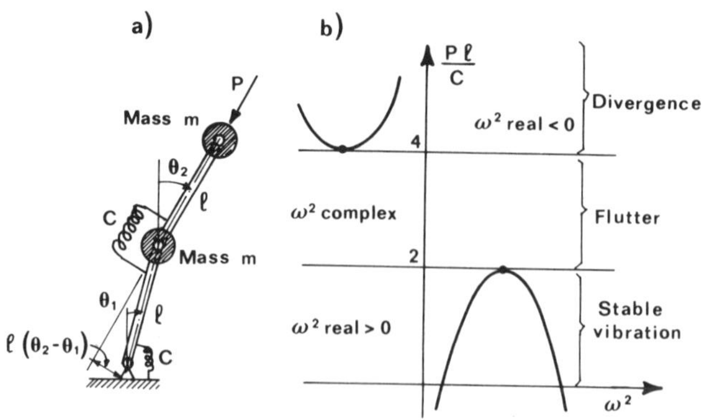

Figure 3.5 (a) Free-standing two-bar column loaded by a constant follower force (massless bars, point masses on top and at middle hinge); (b) dependence of frequency of free vibrations upon load.

has a nonsymmetric matrix (stiffness matrix); this means that a potential energy does not exist, that is, the column is nonconservative. Later we will discuss it more.

The dynamic solution of Equations 3.2.10 may be sought in the form $\theta_1 = q_1 e^{i\omega t}$, $\theta_2 = q_2 e^{i\omega t}$. Indeed, upon substitution in Equations 3.2.10 the term $e^{i\omega t}$ cancels out. Then a system of two homogeneous linear algebraic equations for q_1 and q_2 results:

$$\begin{bmatrix} (C - Pl) - 3ml^2\omega^2 & Pl - 2ml^2\omega^2 \\ -C - ml^2\omega^2 & C - ml^2\omega^2 \end{bmatrix} \begin{Bmatrix} q_1 \\ q_2 \end{Bmatrix} = \begin{Bmatrix} 0 \\ 0 \end{Bmatrix} \tag{3.2.11}$$

Nonzero amplitudes q_1 and q_2 are possible only if the determinant of this equation system vanishes. This leads to a quadratic equation for ω^2:

$$(ml^2\omega^2)^2 - 2(3C - Pl)(ml^2\omega^2) + C^2 = 0 \tag{3.2.12}$$

the solution of which is

$$ml^2\omega^2 = (3C - Pl) \pm \sqrt{(3C - Pl)^2 - C^2} = (3C - Pl) \pm \sqrt{(4C - Pl)(2C - Pl)} \tag{3.2.13}$$

Obviously, the frequency ω might not only become imaginary, as we saw it in our preceding examples, but it could also become complex. What does that mean for stability? Consider that $\omega = \alpha \pm i\beta$ where α and β are real numbers. Since ω is a root of an algebraic equation with real coefficients, the complex frequency must come in conjugate pairs, that is, it $\alpha + i\beta$ is a root of the characteristic equation, then $\alpha - i\beta$ is also a root. Then $e^{i\omega t} = e^{i\alpha t} e^{\mp \beta t}$, which means that the general solution is a linear combination of the following four solutions:

$$e^{\mp \beta t} \cos \alpha t \qquad e^{\mp \beta t} \sin \alpha t \tag{3.2.14}$$

Due to the fact that the complex roots come in conjugate pairs, we have both $+\beta$ and $-\beta$ in these solutions. This means that one of the exponents must be positive regardless of the sign of β. Two of these solutions represent then oscillations with an amplitude growing beyond any bounds, and the response can become very large no matter how small the initial velocity or the initial deflection may be. Hence, when the vibration frequency becomes a complex number, the structure loses stability. This is called Borchardt's criterion.

Let us now finish the problem in Figure 3.5a. For $P < 2C/l$, there are four real distinct frequencies ω, which implies stability. For $P = 2C/l$, all frequencies ω are still real but there are double roots, which means that solutions of the type $t \sin \omega t$ and $t \cos \omega t$ are present: they are unbounded, which means instability. For $2C/l < P < 4C/l$, the expression under the square root in Equation 3.2.13 is negative, and so the roots ω are complex: this implies instability of the oscillatory type, that is, flutter. For $P \geq 4C/l$, ω^2 becomes again real, but negative, and so ω is purely imaginary. This means that the deflection grows exponentially, diverging at monotonically increasing velocity and increasing acceleration to one side. This is an unstable situation. Note that a solution proportional to t is not present, unlike the case of a column under a dead load, and divergence does not begin with a state of neutral equilibrium, that is, with $\omega = 0$. The regions of instability are graphically represented in Figure 3.5b.

It may also be noted that if the same column were loaded by a vertical force P

(which corresponds to gravity load and is conservative), then $P_{cr_1} = 0.3820C/l$ (see Eq. 4.3.13). Our lowest critical load is $P_{cr_1} = 2C/l$, which is 5.23 times larger. Obviously, the directional behavior of the load has a tremendous effect.

The last problem was studied by Herrmann and Jong (1965) and Herrmann (1967b). They demonstrated with this problem that in a nonconservative system a small internal (viscous) damping can have a destabilizing influence, that is, it can lower the critical load (Thompson, 1982, p. 32).

Elastically Supported Rigid Plate under Aerodynamic Forces

As another example consider the stability of a massive rigid plate of unit width and specific mass μ per unit length, suspended on springs of stiffness C_1 and C_2 as shown in Figure 3.6. The plate, initially in a horizontal position of static equilibrium, is loaded by wind of velocity v, characterized by wind force resultant $F = kv^2\theta$ acting at a distance a ahead of the downwind end of the plate; k = constant, θ = rotation of the plate. (The foregoing definition of F is valid only for very slow steady oscillations, such that $\omega b/v \ll 1$; see Bisplinghoff, Ashley and Halfman, 1955.) The location of the resultant of the aerodynamic forces on the plate is called in aeroelasticity the aerodynamic center. Its location does not depend on angle θ. For two-dimensional incompressible flow this center is located at $a = 3b/4$, while for supersonic flow it is located at $a = b/2$ (Dowell et al., 1978, p. 4). Denoting the deflection from the static equilibrium position at midpoint as w (Fig. 3.6), the inertia effects are characterized by vertical inertia force $\mu b\ddot{w}$ applied at the center of the plate, and inertia moment $\mu b^3\ddot{\theta}/12$. The equations of motion may be obtained as the equations of dynamic equilibrium of vertical forces and moments about the center of the plate, and they yield the following system of two linear ordinary differential equations for w and θ:

$$\ddot{w} + a_{11}w + a_{12}\theta = 0 \qquad \ddot{\theta} + a_{21}w + a_{22}\theta = 0 \tag{3.2.15}$$

in which

$$a_{11} = \frac{C_1 + C_2}{\mu b} \qquad a_{12} = \frac{C_1 - C_2}{2\mu} - \frac{kv^2}{\mu b}$$
$$a_{21} = \frac{6}{b^2\mu}(C_1 - C_2) \qquad a_{22} = \frac{3}{b\mu}(C_1 + C_2) - \frac{12kv^2}{\mu b^3}\left(a - \frac{b}{2}\right) \tag{3.2.16}$$

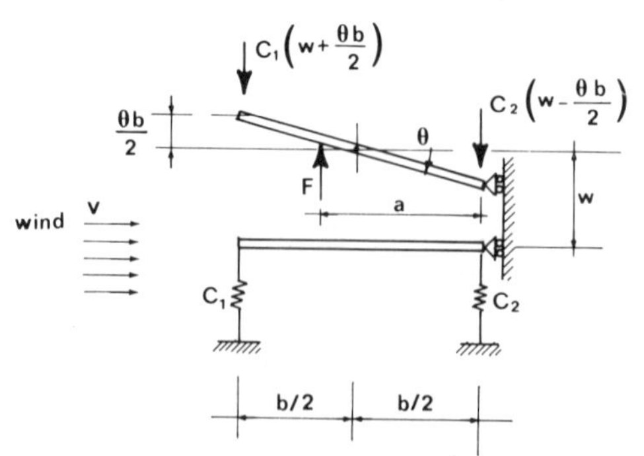

Figure 3.6 Rigid plate suspended on springs and loaded by wind of constant velocity v.

The loss of stability due to divergence (equivalent to static buckling) may be determined by setting $\ddot{w} = \ddot{\theta} = 0$. Equations 3.2.15 then becomes a system of two linear algebraic homogeneous equations, and so the loss of stability occurs when $\det a_{ij} = 0$. From this condition, the critical wind velocity for divergence may be solved, and it is found that (since v is real) divergence is possible only if $C_2 > C_1(b/a - 1)$.

The dynamic solutions are found by substituting $w = q_1 e^{i\omega t}$ and $\theta = q_2 e^{i\omega t}$. This reduces Equations 3.2.15 to the following homogeneous linear algebraic equation for the amplitudes q_1, q_2:

$$\begin{bmatrix} a_{11} - \omega^2 & a_{12} \\ a_{21} & a_{22} - \omega^2 \end{bmatrix} \begin{Bmatrix} q_1 \\ q_2 \end{Bmatrix} = \begin{Bmatrix} 0 \\ 0 \end{Bmatrix} \tag{3.2.17}$$

Nonzero amplitudes are possible only if the determinant of this equation system vanishes, which yields the condition

$$\omega^2 = \frac{a_{11} + a_{22}}{2} \pm \sqrt{\left(\frac{a_{11} + a_{22}}{2}\right)^2 - a_{11}a_{22} + a_{12}a_{21}} \tag{3.2.18}$$

Assuming that $a_{11} + a_{22} > 0$, the condition that ω^2 be real and positive requires that

$$0 < a_{11}a_{22} - a_{12}a_{21} < \left(\frac{a_{11} + a_{22}}{2}\right)^2 \tag{3.2.19}$$

The second of these inequalities reduces to $(a_{11} - a_{22})^2 + 4a_{12}a_{21} > 0$, and it is found that condition 3.2.19 is always satisfied if $C_1 < C_2$ (because $a_{12}a_{21} > 0$); in this case no flutter is possible. If $C_1 > C_2$, then this condition can be violated and flutter is possible for a sufficiently high wind speed v, which is solved from the foregoing condition.

The condition $a_{11}a_{22} - a_{12}a_{21} = 0$ yields the critical velocities for divergence ($\omega = 0$), that is, static buckling, the case we have already solved.

It may be noted that the condition for critical velocity is not changed if the initial angle of attack of the plate is different from zero. In this case θ and w must be interpreted as incremental generalized displacements due to elastic deformations of the springs; the problem is analogous to that of lateral deflection of an imperfect column (see Bisplinghoff et al., 1955, p. 424).

The last problem illustrates some but not all of the essential aspects of instability of aircraft wings as well as suspension bridges. In wing design it is desirable to make flutter impossible at any velocity. This is achieved with $C_2 > C_1$. It can be shown that, more generally, stability of a plate whose center of gravity is not at midlength is achieved if the center of the spring stiffness of the links is behind the center of gravity.

In relation to the suspension bridge problem, the plate in our example may be imagined as a section of the roadway and the springs model the suspension cables. A realistic analysis of a suspension bridge, however, must also take into account the effect of motions $\theta(t)$ and $w(t)$ on the aerodynamic forces, as well as the inertia of the added mass of air that is forced to move with the plate.

The understanding of dynamic stability of suspension bridges was greatly advanced by the collapse of the Tacoma bridge near Seattle in 1940 (Bowers, 1940; Simiu and Scanlan, 1986). Shortly after the erection of the bridge, which was of a much lighter design than the preceding large suspension bridges (the

Figure 3.7 Oscillations and collapse of Tacoma bridge. (*Reprinted by permission of the Centro di Cinematografia Scientifica del Politecnico di Milano.*)

Golden Gate bridge in San Francisco, built in 1937, and the George Washington bridge, built earlier over the Hudson River in New York), a strong gale caused large torsional–flexural oscillations of the center span that produced collapse within a few hours. These oscillations as well as the subsequent collapse were recorded on a film (Fig. 3.7). Analysis of this disaster (e.g., T. von Kármán, 1940; F. B. Farquarson, 1954; Simiu and Scanlan, 1986) eventually led to a much better understanding of the behavior of suspension bridges and made it possible to design light large-span suspension bridges that are safe against aerodynamic instability. An example is the Humber bridge over the mouth of the Humber River in England, whose span (currently a world record) is 1410 m (compared with 853 m for the Tacoma bridge and 1280 m for the Golden Gate bridge, which held the world record until the Verrazzano Narrows bridge in New York, also of classical design, surpassed it in the 1960s; see, e.g., Leonhardt, 1982, which gives a good discussion of practical design aspects).

Conservative and Nonconservative Forces

Let us now consider the nature of the forces that produce dynamic instability. When a structure oscillates, its amplitude as well as its kinetic energy increases beyond any bounds. This energy must come from somewhere. Clearly, it is extracted from the load. However, if the loads are conservative and have a potential, the energy supplied to the structure cannot exceed the potential-energy loss of the applied loads due to deflection, and it is therefore bounded. Hence, dynamic instability (flutter) can be caused only by nonconservative loads.

The loads on a structure are generally defined as a force field $\mathbf{P} = \mathbf{P}(\mathbf{x})$, where $\mathbf{x} = (x, y, z)$ = position vector in Cartesian coordinates x, y, z. During the deflection of the structure, the load vector **P** moves along path s consisting of displacement vectors $d\mathbf{s}$. A load is said to be conservative if the work $W = \int_A^B \mathbf{P}\, d\mathbf{s}$ from position A to position B is independent of the path taken and depends only on these final positions. An equivalent condition is that the closed path integral $\oint \mathbf{P}\, d\mathbf{s} = \oint (P_x\, dx + P_y\, dy + P_z\, dz)$ must be zero for all closed paths. This is always true if a potential-energy function $\Pi = \Pi(x, y, z)$ exists such that

$$P_x = \frac{\partial \Pi}{\partial x} \qquad P_y = \frac{\partial \Pi}{\partial y} \qquad P_z = \frac{\partial \Pi}{\partial z} \tag{3.2.20}$$

The potential-energy function always exists if the following integrability conditions are satisfied:

$$\frac{\partial P_x}{\partial y} = \frac{\partial P_y}{\partial x} \qquad \frac{\partial P_x}{\partial z} = \frac{\partial P_z}{\partial x} \qquad \frac{\partial P_y}{\partial z} = \frac{\partial P_z}{\partial y} \tag{3.2.21}$$

The force field is conservative if it possesses a potential. However, it can be conservative even if it does not have a potential, as we will see from an example of a rotating shaft in Section 3.4.

For the follower loads on a column, as well as the wind force on the plate in the preceding examples, a potential cannot be defined and does not exist because the load direction depends on the deformation of the structure. This may be illustrated by the sketch in Figure 3.8. Depending on the trajectory of the column top and its rotation history, the disk can move from its original position 1 to its final position 3 in infinitely many different ways, for example, those illustrated by a, b, and c in Figure 3.8. In the first case the disk moves first laterally and then it rotates, in which case no work is done. If the disk first rotates clockwise and then moves to the right, the work done by the load is negative, and if it first rotates counterclockwise, then moves to the right and again rotates clockwise, the work done by the load is positive. Different amounts of work are done for different paths leading to the same final state.

We should caution, however, that the dependence of the load direction on the structure deformation does not in itself cause the load to be nonconservative. This is, for example, demonstrated by considering a shell under hydrostatic pressure. Hydrostatic pressure is a special case of follower loads always oriented orthogonally to the deformed surface of the shell. The work done is nevertheless proportional to the cross-hatched area in Figure 3.9a, which is determined fully by the final deflected position and does not depend on the way in which this final position has been reached.

Another example of a load whose direction depends on the deformation of the structure is a spring load on the column shown in Figure 3.9b. This load is also conservative because it is a force produced by an elastic spring. Such a load, as

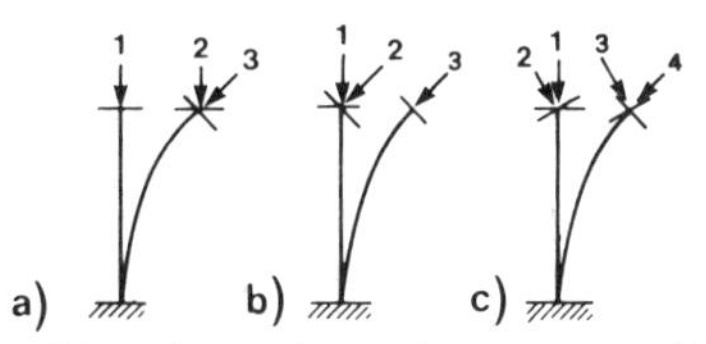

Figure 3.8 Different rotation histories of a column top which correspond to different amounts of work done by the follower force (for the same initial and final state).

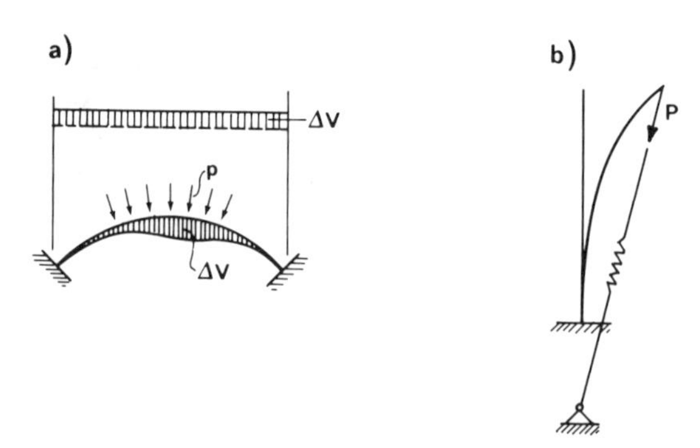

Figure 3.9 Conservative systems with load direction dependent on deformation.

well as the hydrostatic pressure load, cannot supply unlimited kinetic energy to the structure. Still other examples were shown in Figure 1.10.

The fact that the loads are constant in magnitude and keep constant direction does not in itself guarantee that they are conservative. This may be illustrated by the example (presented in Sec. 1.10 and Fig. 1.29) of a shaft supported at the ends by pins and loaded by axial force P and torque M_t. Such a loading is nonconservative, since the work of torque M_t depends on the deflection path (Timoshenko and Gere, 1961, p. 156). Therefore, dynamic analysis of stability is required. This example, however, is somewhat artificial since no mechanical way to apply M_t in this manner is known.

The basic lesson from our preceding examples is that examination of equilibrium states (i.e., static analysis) is in general insufficient to determine stability of an elastic structure subjected to nonconservative loads.

Equations Governing Flutter of Suspension Bridges

Finally, to supplement the preceding discussion of the behavior of suspension bridges, let us state the full equations that are adequate for most practical purposes. The equations of motion of the bridge deck beam (Fig. 3.10) that vibrates flexurally and torsionally in a steady flow of air and does not exhibit significant warping torsion or cross-section distortion (Chap. 6) are partial differential equations of the form:

$$\begin{aligned}(EIw_{,xx})_{,xx} + \mu w_{,tt} + \mu y_e \theta_{,tt} + C_w w_{,t} + L = 0 \\ -(GJ\theta_{,xx})_{,x} + \mu y_e w_{,tt} + I_m \theta_{,tt} + C_t \theta_{,t} + M = 0\end{aligned} \qquad (3.2.22)$$

in which subscripts following a comma denote partial derivatives (e.g., $w_{,tt} = \ddot{w} = \partial^2 w/\partial t^2$), θ = rotation about beam axis; GJ = torsional stiffness (for simple or Saint Venant's torsion, Chap. 6), μ = beam mass per unit length, I_m = mass moment of inertia about the elastic axis (neutral axis for bending), y_e = distance

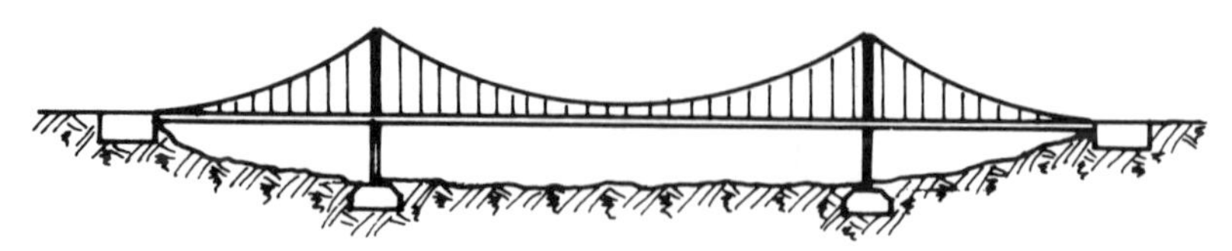

Figure 3.10 Suspension bridge.

of the elastic axis from the inertia axis; C_W, C_T = damping coefficients for bending and torsional motions; and L, M = aerodynamic lift force and torque. According to Simiu and Scanlan (1986) (see also Meirovitch and Ghosh, 1987):

$$L = -\frac{1}{2}\rho v^2 2b\left[KH_1(K)\left(\frac{\dot{w}}{v}\right) + KH_2(K)\left(\frac{b\dot{\theta}}{v}\right) + K^2H_3(K)\theta\right]$$
$$M = -\frac{1}{2}\rho v^2(2b)^2\left[KA_1(K)\left(\frac{\dot{w}}{v}\right) + KA_2(K)\left(\frac{b\dot{\theta}}{v}\right) + K^2A_3(K)\theta\right] \tag{3.2.23}$$

in which b = width of the deck; $K = b\omega/v$-reduced frequency; ω = frequency of oscillations; v = wind velocity; and H_1, H_2, H_3, A_1, A_2, A_3 = aerodynamic coefficients depending on K. Further one must add the boundary condition of the beam; for example, $w = w_{,x} = \theta = 0$ at $x = 0$ and $x = L$ for fixed ends.

Problems

3.2.1 Solve the problem of the free-standing two-bar massless column in Figure 3.5a for the case that the intermediate mass is $2m$.

3.2.2 Same as above, but assume that the lower and upper bars have lengths $2l$ and l.

3.2.3 Same for a free-standing two-bar massless column with only one mass m on top (Fig. 3.11a).

3.2.4 Same as Problem 3.2.1, but the load is vertical.

3.2.5 Same as Problem 3.2.1, but the load passes through a fixed point at a distance a from the base, Figure 3.11b. *Hint:* Is P a conservative load? What is the limit case for $a \to \infty$?

3.2.6 Same as Problem 3.2.1, but the load has an angle $\phi = \theta/2$ (partial follower load) (Fig. 3.11c).

3.2.7 Show that the system of Figure 3.9b is conservative.

3.2.8 Is the column in Figure 3.11e conservative? *Hint:* Refer to Problem 1.4.3.

3.2.9 Frequency ω is given by the equation $\omega^4 - 2(2+P)\omega^2 + P(P+6) = 0$. Find P_{cr_1}. What type of instability occurs?

3.2.10 Show that the work done by hydrostatic pressure p (Fig. 3.9a) is equal to $p\,\Delta V$, where ΔV is the volume between the initial and final deflection surface. This volume is obviously independent of the path in which the structure got from the initial to the final state. Why is this sufficient to conclude that the load is conservative?

3.2.11 Show that the column in Figure 3.11d loaded by a follower force P exhibits no flutter and $\omega = 0$ at the critical state (so that P_{cr} can be obtained by static analysis).

3.2.12 Solve the system in Figure 3.11f for the cases: (a) $\phi = \theta$ (follower load); (b) $\phi = \theta/2$ (partial follower load); (c) $\phi = \theta$ fixed direction load (dead load).

3.2.13 Solve the system in Figure 3.11g.

3.2.14 Consider the equations of motion of a suspension bridge deck. For the purpose of this exercise, assume that $H_1 = -0.1(v/fB)$; $H_2 = 0$; $H_3 = 0$; $A_1 = 0$; $A_2 = -0.025(v/fB)(2.0 - v/fB)$; $A_3 = 0$; in which $f = \omega/2\pi$, and analyze dynamic stability.

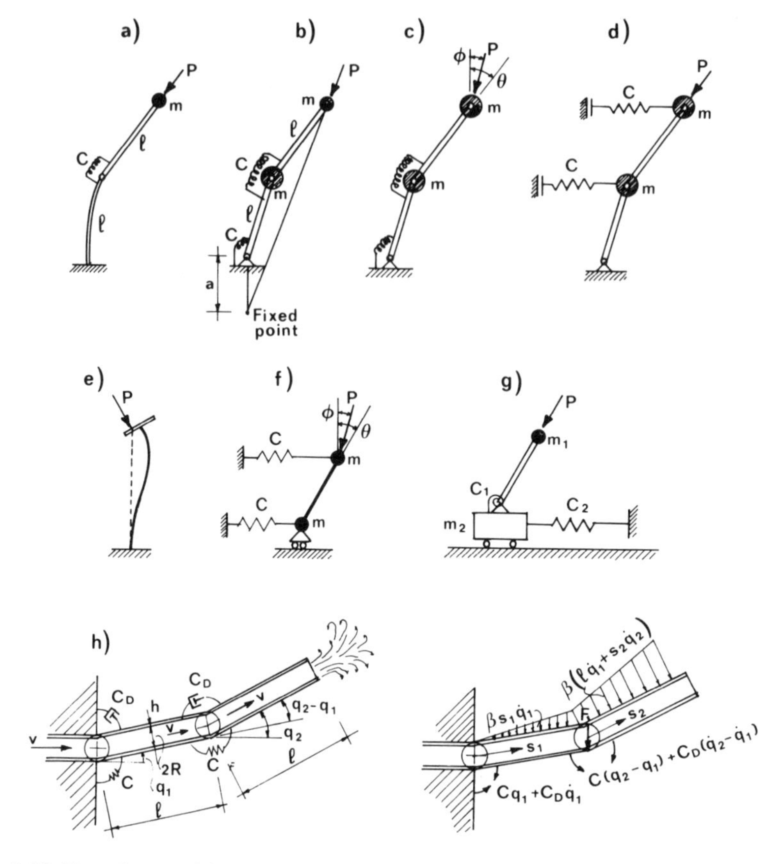

Figure 3.11 Exercise problems on (a–g) structures loaded by follower forces, and (h) flexible pipes with high-speed fluid flow.

3.2.15 Flexible pipes can fail by flutter due to high-speed flow of the fluid they convey. As a simple prototype problem consider the pipe in Figure 3.11h, which can flex only at two points (by angles q_1 and q_2), restrained by springs of constant C and dampers of coefficient C_D. Let ρ be the fluid mass density (constant, incompressible fluid), v the fluid velocity, R the inner radius of the pipe, and h the wall thickness. The pipes are subjected to external linear viscous damping β. Due to inertia there is a lateral force F exerted by the fluid on the pipe at the pipe bend. It is proportional to $q_2 - q_1$ and can be calculated as the rate of change of the momentum vector of the fluid as it passes through the bend. (a) Derive the differential equations of motion for q_1 and q_2, (b) discuss the solution method, and (*c*) determine the stability limits. For detailed analysis see Sugiyama (1987).

3.3 PULSATING LOADS AND PARAMETRIC RESONANCE

Another type of nonconservative load is pulsating loads. Such loads often approximate reasonably well the action of rotating machinery such as turbines and power generators on the columns of structures, or the action of moving vehicles on the piers of bridges.

Axial Pulsating Load on a Column

Consider again a pin ended column of length l (Fig. 3.12a) that has a uniform distributed mass μ, a constant bending rigidity EI, and is subjected to axial compressive load $P(t) = P_0 + P_t \cos \Omega t$ in which Ω is a given forcing frequency, and P_0, P_t = constants. Since no restriction on constancy of axial load P needed to be made in deriving the differential equation of beam-columns, substitution of inertia force and of $P(t)$ into Equation 3.1.2 (with no damping) provides the differential equation

$$EI \frac{\partial^4 w}{\partial x^4} + (P_0 + P_t \cos \Omega t) \frac{\partial^2 w}{\partial x^2} = -\mu \frac{\partial^2 w}{\partial t^2} \tag{3.3.1}$$

For the boundary conditions of a pin-ended column, we may seek the solution in the form

$$w(x, t) = \sum_{n=1}^{\infty} f_n(t) \sin \frac{n\pi x}{l} \tag{3.3.2}$$

in which $f_n(t)$ are unknown functions to be found. Substitution into Equation

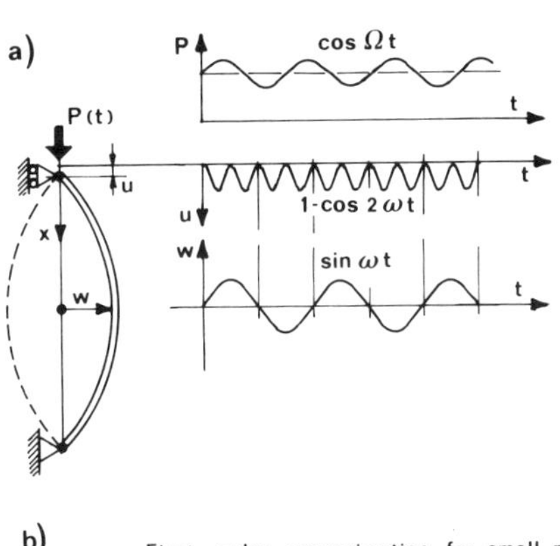

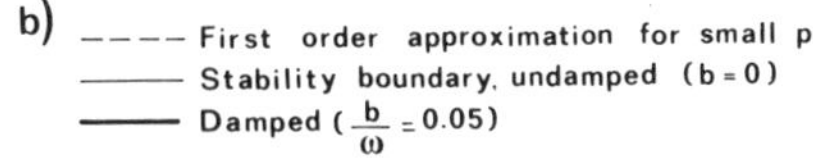

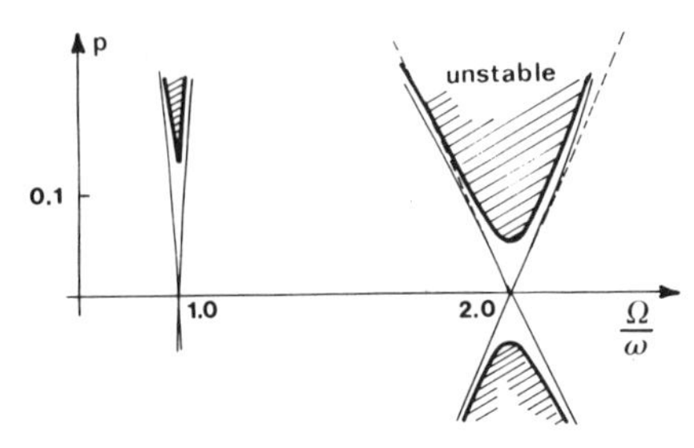

Figure 3.12 (a) Pin-ended column under pulsating axial force; (b) limits of stability if damping is considered (heavy lines) or not (light lines).

3.3.1 yields the condition

$$\sum_{n=1}^{\infty} \sin \frac{n\pi x}{l} \left\{ \ddot{f}_n + \frac{EI}{\mu}\left(\frac{n^4\pi^4}{l^4}\right)\left[1 - \frac{l^2}{n^2\pi^2 EI}(P_0 + P_t \cos \Omega t)\right] f_n \right\} = 0 \quad (3.3.3)$$

For this equation to be satisfied at any value of x, the expression in braces must vanish for every n. This yields the differential equation

$$\frac{d^2f}{dt^2} + \omega^2(1 - 2p \cos \Omega t)f = 0 \quad (3.3.4)$$

in which the following notations are made:

$$\omega^2 = \omega^{0^2}\left(1 - \frac{P_0}{P_{\text{cr}_n}}\right) \qquad \omega^{0^2} = \frac{n^4\pi^4}{l^4}\left(\frac{EI}{\mu}\right) \qquad p = \frac{P_t}{2(P_{\text{cr}_n} - P_0)} \qquad P_{\text{cr}_n} = \frac{n^2\pi^2}{l^2} EI$$

(3.3.5)

ω^0 = free vibration frequency of the column for no axial load, ω = free vibration frequency under static load P_0, and p = excitation parameter. We have discarded subscripts n, that is, replaced ω_n, ω_n^0, p_n, and f_n by ω, ω^0, p, and f, since Equation 3.3.4 has the same form for all n.

Equation 3.3.4 is an ordinary linear differential equation of second order, with one variable coefficient. It is known as the Mathieu differential equation, and its solutions are Mathieu functions. We will now indicate an approximate solution of the limiting response that was given by Rayleigh and can be obtained without relying on the theory of Mathieu functions.

Undamped Vibration

The motion of the column produced by the periodic load may be expected to be quasi-periodic, and when the amplitude increases in time, the column is unstable. At the limit of stability, a periodic solution may be expected:

$$f(t) = \sum_{k=1}^{\infty} (A_k \cos k\gamma t + B_k \sin k\gamma t) \quad (3.3.6)$$

in which γ = constant. Substituting this into Equation 3.3.4 and utilizing the relations $2 \cos \Omega t \cos k\gamma t = \cos(\Omega + k\gamma)t + \cos(\Omega - k\gamma)t$, $2 \cos \Omega t \sin k\gamma t = \sin(\Omega + k\gamma)t - \sin(\Omega - k\gamma)t$, we obtain

$$\sum_{k=1}^{\infty} \{(\omega^2 - k^2\gamma^2)(A_k \cos k\gamma t + B_k \sin k\gamma t)$$
$$- p\omega^2[A_k \cos(\Omega - k\gamma)t - B_k \sin(\Omega - k\gamma)t]$$
$$+ p\omega^2[A_k \cos(\Omega + k\gamma)t + B_k \sin(\Omega + k\gamma)t]\} = 0 \quad (3.3.7)$$

There are six different trigonometric functions in these equations for each k. To satisfy these equations identically for all t, the coefficient at each trigonometric function would have to vanish. However, this condition would, in general, yield six equations for each k, while there are only two unknowns A_k and B_k. So, in general, we would have more unknowns than equations, which is an unsolvable problem. To satisfy these equations for all t (at nonzero A_k, B_k), the sine functions as well as the cosine functions must be identical, that is, their arguments must coincide. Therefore, we must choose γ such that the numbers $\Omega - k\gamma$ and $\Omega + k\gamma$ are all contained among the numbers $k\gamma = 0, \pm\gamma, \pm 2\gamma, \pm 3\gamma, \ldots$.

One way to achieve it is to set $\Omega - \gamma = 0$, that is, $\gamma = \Omega$. Then the coefficient of $\cos \gamma t$ must vanish, that is, $\omega^2 - \gamma^2 = 0$, which implies that $\Omega = \omega$ represents a critical load at which stability is lost. This critical load is the same as in static buckling ($P_t = 0$).

Another way to achieve it is to set $\Omega - \gamma = \gamma$, that is, $\gamma = \Omega/2$, and it can be verified that this yields the smallest excitation frequency Ω that causes instability. Then we have in Equation 3.3.7 only two types of trigonometric functions: $\cos(k\Omega t/2)$ and $\sin(k\Omega t/2)$; indeed, arguments $k\gamma t$ and $(\Omega - n\gamma)t$ are equal when $n = 2 - k$. But this means that identical sine or cosine functions occur in different brackets of Equation 3.3.7 for different k values. Substituting $\gamma = \Omega/2$ into Equation 3.3.7, writing out all terms for $k = 1, 2, 3$ in detail (without Σ), and collecting the trigonometric functions of the same argument, we obtain a condition of the form

$$-p\omega^2 A_2 + a_1 \cos\frac{\Omega t}{2} + b_1 \sin\frac{\Omega t}{2} + a_2 \cos \Omega t + b_2 \sin \Omega t + a_3 \cos\frac{3\Omega t}{2} + \cdots = 0 \tag{3.3.8}$$

in which

$$a_1 = [(1-p)\omega^2 - \tfrac{1}{4}\Omega^2]A_1 - p\omega^2 A_3 \qquad b_1 = [(1+p)\omega^2 - \tfrac{1}{4}\Omega^2]B_1 - p\omega^2 B_3 \qquad \cdots \tag{3.3.9}$$

Now, to satisfy Equation 3.3.8 identically for all times t, it is necessary that A_2 as well as the coefficient of each trigonometric function vanish. We get an infinite system of equations for $A_1, B_1, A_2, B_2, A_3, B_3, \ldots$, with the same number of equations as unknowns (Rayleigh, 1894 and 1945, pp. 82–83).

We are not really interested in a complete solution, however. If we find some nonzero solution, it represents a critical state. One such solution can readily be obtained by setting $A_2 = 0$ and $A_3 = B_3 = 0$, which has the advantage that the equations for A_1 and B_1 become uncoupled from the rest. Note also that even if we considered those solutions in which A_3 and B_3 are present, then the terms with A_3 and B_3 for $p \ll 1$ are higher-order smaller than the terms with A_1 and B_1. This justifies omission of A_3 and B_3 in more general circumstances.

The system of equations $a_1 = 0$ and $b_1 = 0$ in this case can be satisfied for nonzero A_1 and B_1 (of arbitrary magnitude) if its determinant is zero, and this occurs if the excitation frequency is

$$\Omega = 2\omega\sqrt{1 \pm p} \simeq 2\omega\left(1 \pm \frac{p}{2}\right) \tag{3.3.10}$$

The limits of stability according to Equation 3.3.10 are plotted in Figure 3.12b as continuous light curves for $2\omega\sqrt{1 \pm p}$ and as dashed straight lines for $2\omega(1 \pm p/2)$. For a vanishing amplitude of the pulsating load, instability is obtained for excitation frequencies $\Omega = 2\omega$ where ω is the free vibration frequency corresponding to the mean load P_0.

The stability loss that occurs at other than the natural frequencies for $P_0 = 0$ and is determined by further parameters, such as those in Equations 3.3.5, is called the parametric resonance. For columns, it occurs at double the natural frequency for $P_0 = 0$. It can be any natural frequency, not just the first one. In general, for each natural frequency one can find infinitely many parametric resonance frequencies, all except one being lower than the natural frequency. It may be checked that $\Omega = 2\omega$ is the only destabilizing frequency of a column that is associated with the first natural frequency and is higher than this frequency. It may be interesting to note that a pulsating tensile force ($P < 0$) can also produce instability (Kohoutek, 1984).

The diagram in Figure 3.12b is called a Strutt diagram; for its complete picture see, for example, Timoshenko and Gere (1961, p. 160). A more complete investigation can be found in Rayleigh (1894 and 1945).

Damped Vibration

For design engineers, it may be disturbing that according to the previous solution an infinitely small amplitude of the pulsating loads may destabilize the column at a certain frequency that is less than the static critical load. In reality, however, there is always damping. It is an essential feature of parametric resonance that damping has a strong stabilizing effect, since it keeps energy from flowing from the excitation into the system (Rayleigh, 1894 and 1945, p. 84).

For linear damping, the solution may be obtained easily in the same manner as we demonstrated. It leads again to an equation of the type of Equation 3.3.8 in which, however, each of the coefficients a_1, a_2 depends on both A_1 and B_1. Instability occurs when the determinant of the equations $a_1 = 0$ and $b_1 = 0$ vanishes, which yields (for $b \ll \omega$) the condition

$$\Omega = 2\omega \sqrt{1 \pm \sqrt{p^2 - \frac{4b^2}{\omega^2}}} \tag{3.3.11}$$

The limit of stability given by this equation is approximately (for $p \to 0$) a hyperbola, sketched with a bold line in Figure 3.12b. The lowest magnitude of p is obtained for the case of parametric resonance, $\Omega = 2\omega$, for which $p = 2b/\omega$ and so we see that a sufficiently small excitation amplitude cannot cause instability. In practice, a relatively small damping usually suffices to stabilize columns under normal excitation amplitudes. At parametric resonance, the damping is least effective. For the highest parametric resonant frequency, damping has a much stronger stabilizing influence than at lower ones, that is, it pushes the unstable region much farther away from the horizontal axis in the Strutt diagram of Figure 3.12b (see Timoshenko and Gere, 1961).

More accurate solutions can be obtained considering the coefficients of further trigonometric terms in Equation 3.3.8. However, if $p < 0.6$, the present approximate solution lies within about 1 percent of the exact one.

Another way to obtain an approximate solution of the Mathieu differential equation is by the perturbation method. For the general theory of Mathieu equations see Whittaker and Watson (1969).

Simple Energy Analysis of Parametric Resonance

From the energy viewpoint it is important to note that the lateral deflections cause second-order small axial shortening $u(t)$ of the column. Consequently, the axial load can do unbounded work on the axial shortenings if its oscillations are synchronous with those of $u(t)$. For the sake of simplicity (Fig. 3.12a), let us now assume that

$$w(x, t) = a \sin \omega t \sin \frac{\pi x}{l} \tag{3.3.12}$$

Then

$$w' = \frac{\partial w}{\partial x} = a \frac{\pi}{l} \sin \omega t \cos \frac{\pi x}{l} \qquad \dot{w} = \frac{\partial w}{\partial t} = a\omega \cos \omega t \sin \frac{\pi x}{l} \tag{3.3.13}$$

The axial shortening may be calculated as $u = \int_0^l (ds - dx)$ where $ds^2 = dx^2 + dw^2$. Setting $dw = w'\,dx$ and noting that $(1 + w'^2)^{1/2} \simeq \frac{1}{2}w'^2$ if w' is small, we have $u = \frac{1}{2}\int_0^l w'^2\,dx$, which yields

$$u = \frac{\pi^2 a^2}{8l}(1 - \cos 2\omega t) \qquad \dot{u} = \frac{\pi^2 a^2 \omega}{4l} \sin 2\omega t \tag{3.3.14}$$

Because this equation involves $\sin 2\omega t$ rather than $\cos 2\omega t$, it is now more convenient to assume that $P = P_0 + P_t \sin \Omega t$. Then

$$W(t) = \int P\dot{u}\,dt = \frac{\pi^2 a^2 \omega}{4l} \int (P_0 + P_t \sin \Omega t) \sin 2\omega t\, \mathrm{d}t$$

$$= \frac{\pi^2 a^2 \omega}{8l} \left\{ P_t \int_0^t [\cos(2\omega - \Omega)t - \cos(2\omega + \Omega)t]\,dt + \frac{P_0}{\omega}(1 - \cos 2\omega t) \right\}. \tag{3.3.15}$$

When $\Omega \neq 2\omega$, $W(t)$ is obviously a periodic function and therefore the energy supplied to the structure is bounded. That must be a stable situation. Not so, however, when $\Omega = 2\omega$.. In this case Equation 3.3.15 yields for $t = t_n = \pi n/\omega$ $(n = 1, 2, 3, \ldots)$

$$W(t_n) = \frac{\pi^2 a^2 \omega}{8l} P_t t_n \tag{3.3.16}$$

So, if no energy is absorbed by damping, the energy of the structure grows beyond any bound. This proves, in a much simpler manner than before, that $\Omega = 2\omega$ is an unstable situation (i.e., parametric resonance).

Consider now damping. The energy dissipated by damping is $D = \iint p_D \dot{w}\,dx\,dt$ where $p_D = \beta\dot{w} = 2b\mu\dot{w}$ = damping force per unit length of column. Hence

$$D(t) = \int_0^t \int_0^l 2b\mu a^2 \omega^2 \cos^2 \omega t \sin^2 \frac{\pi x}{l}\,dx\,dt = b\mu a^2 \omega^2 \left(\frac{l}{2}\right)\left(t + \frac{\sin 2\omega t}{2\omega}\right) \tag{3.3.17}$$

The energy that can be stored in the column at times t_n is $U(t_n) = W(t_n) - D(t_n)$. Obviously, if $W(t_n) < D(t_n)$, the column cannot acquire unbounded energy, that is, it must be stable. This yields the stability condition $|P_t| < 4b\mu\omega^2 l^2/\pi^2\omega$. Substituting $\omega^2 = (\pi^4 EI/l^4\mu)(1 - P_0/P_{cr})$ according to Equations 3.3.5 for $n = 1$, and noting that $EI\pi^2/l^2 = P_{cr}$, the stability condition becomes

$$\frac{|P_t|}{2(P_{cr} - P_0)} < \frac{2b}{\omega} \tag{3.3.18}$$

(in which, of course, $P_0 < P_{cr}$, according to the static stability condition). When this inequality is reversed, one finds that the value of $U(t_n)$ will increase beyond any bound, which is an unstable situation.

Note that when $\Omega = 2\omega$, Equation 3.3.11 for the critical state reduces to $p = \pm 2b/\omega$ where $p = P_t/2(P_{cr} - P_0)$. This confirms the correctness of Equation 3.3.18. Also note that the larger P_0 is, the smaller is the amplitude $|P_t|$ that can be stably sustained by the column.

The foregoing energy approach (Bažant, 1985) does not seem to be effective for $\Omega \neq 2\omega$ because the time history of motion is more complex. Other methods, as already shown, are necessary for a complete solution of the problem. Nevertheless, the foregoing energy method for parametric resonance with or without damping yields all the information needed for practical design purposes. This method can be extended easily to arbitrary columns, frames, arches, plates, shells, thin-wall beams, etc. (see Prob. 5.2.14, 6.2.5, 6.3.11, 6.4.6, 7.3.19 and 7.5.7).

To sum up:

1. The frequency of the applied axial load has a great influence on stability.
2. A large destabilizing effect, called parametric resonance, can be obtained when the frequency of the load is other than the natural vibration frequencies.
3. Typically, as we saw it for a column, the strongest parametric resonance is experienced at double the first natural frequency.
4. Damping changes the stability limits, significantly enlarging the stable domain (this contrasts with the examples in Secs. 3.1 and 3.2).

Problems

3.3.1 Define and describe the phenomenon of parametric resonance.

3.3.2 Without referring to the text, derive the differential equation governing the column response to the pulsating load.

3.3.3 Using the energy method, show that parametric resonance occurs also when $\Omega = 2\omega_n$ when ω_n ($n = 1, 2, \ldots$) are all the higher natural frequencies of a column under constant axial load P_0. Also calculate the maximum stable amplitude $|P_t|$ for $\Omega = 2\omega_n$.

3.3.4 Using the energy method, estimate the frequency Ω for parametric resonance and the corresponding maximum stable $|P_t|$ in the presence of damping for (a) a fixed–fixed column, (b) a fixed–hinged column, and (c) a portal frame. Assume approximate deflection shapes for this purpose.

3.3.5 Calculate the lowest natural vibration frequency of a circular hinged arch under radial pressure p.

3.3.6 Consider the axial pressure $p(t)$ to be pulsating, and extend the preceding problem to parametric resonance, with and without damping. *Hint:* Calculate the second-order shortening of the arch center line due to $w(s, t)$ and the work done on it by the axial force N^0 due to p ($N^0 = -pR$). Compare this to the energy dissipated by damping due to $\dot{w}(s, t)$.

3.3.7 The simply supported column in Figure 3.13a, having uniform EI and uniform distributed mass μ, is loaded by the weight of mass m through a

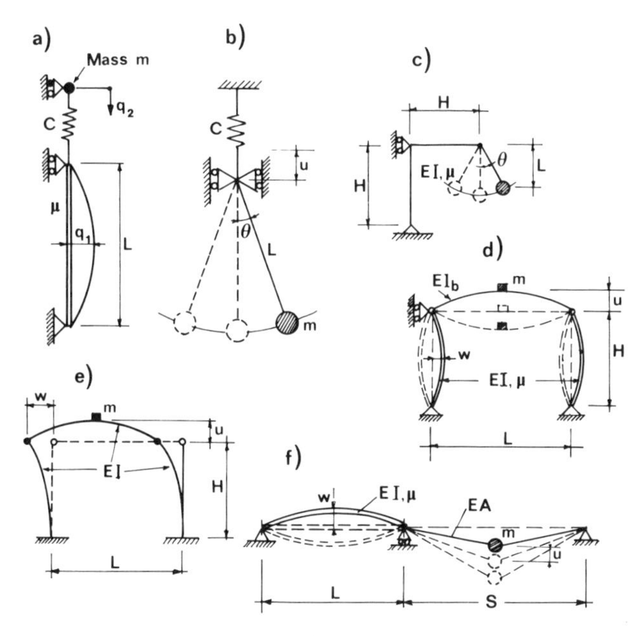

Figure 3.13 Exercise problems on parametric resonance.

spring of stiffness C as shown. Neglect the first-order axial shortening of the column ($EA \rightarrow \infty$). Calculate (a) the first frequency ω_1 of flexural vibrations characterized by midspan deflection q_1, and (b) the frequency ω_1 of axial vibration of mass m characterized by q_2 (at $q_1 = 0$). Determine the value of C for which parametric resonance occurs ($\omega_2 = 2\omega_1$). Then show that a disturbance initially exciting only axial vibrations q_2 must eventually also produce flexural vibrations q_1 and vice versa.

3.3.8 (a) The spring C in Figure 3.13b supports an inextensible massless pendulum of length L with point mass m at the end. Determine the length L for which lateral oscillations will arise if mass m is initially displaced purely vertically. (b) Do the same for the pendulum in Figure 3.13c suspended on a massless frame.

3.3.9 (a) Point mass m in Figure 3.13d is fixed at midspan to a massless beam of stiffness EI_b, supported on columns of uniform distributed mass μ and stiffness EI as shown. Calculate span L for which an initial vertical displacement of mass m will lead to lateral oscillations of the columns. (b) Do the same for the point mass m in Figure 3.13e, where both the beam and columns are massless (and inextensible). (c) Do the same for point mass m in Figure 3.13f fixed at midlength to a massless string of stiffness EA (the beam, with uniform distributed mass μ, receives significant axial force due to weight mg).

3.3.10 The massless rigid bar column in Figure 3.14a has point mass m_2 attached at the joint, and while $\theta = 0$, mass m_1 is excited to oscillate vertically. Oscillations u at $\theta = 0$ is one possibility. Find the conditions that this produces oscillations in θ.

3.3.11 (a) The rigid block of mass m_1 in Figure 3.14b is elastically supported on a rigid plate of mass m_2 that is supported by two massless columns. While

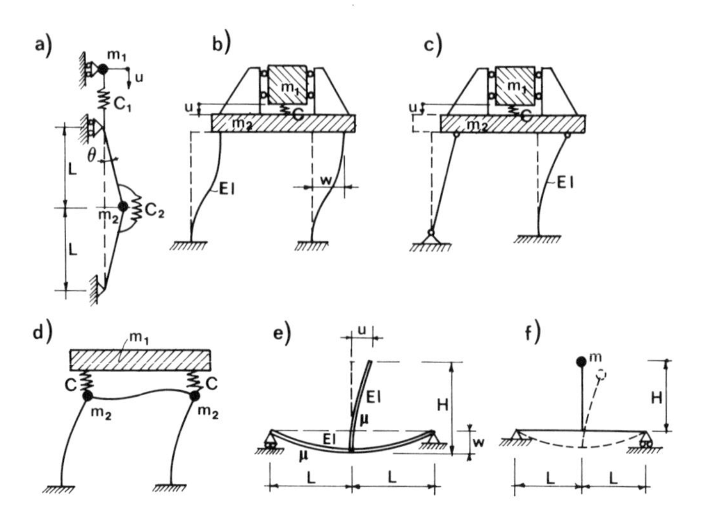

Figure 3.14 Further exercise problems on parametric resonance.

$w = 0$, mass m_1 is excited to oscillate vertically. Oscillations u at $w = 0$ are one possibility. Find the condition when this produces lateral oscillations w. (d) Do the same for the rigid block in Figure 3.14c. (c) Do the same, when mass m_1 is supported elastically (Fig. 3.14d) on a massless rigid-jointed elastic portal frame with point masses m_2 attached at the joint.

3.3.12 A column of height H is rigidly joined with a beam of span L. Both have uniform distributed mass μ and bending stiffness EI, and are inextensible. While $u = 0$, the beam deflection $w = 0$ is initially excited to oscillate. Oscillations at $u = 0$ are one possibility. Find the ratio H/L for which this produces simultaneous lateral vibrations u of the column.

3.3.13 Do the same as Problem 3.3.12, but the beam and column are massless, and point mass m is attached at the column top (Fig. 3.14f).

3.4 OTHER TYPES OF DYNAMIC LOADS

Other interesting, qualitatively different aspects of dynamic stability are revealed in problems of rotating machinery, whose vibrations must be sustained by structures and their foundations. Following Ziegler (1968), consider the example of a rigid circular disk of mass m mounted on a rotating massless elastic shaft, as shown in Figure 3.15. It is convenient to use rectangular coordinates x, y that rotate with the disk at its angular velocity ω. The deflections in the directions x and y are denoted as u and v, and the spring constants of the shaft in these directions as C_1 and C_2 (we will see that even a very small difference between C_1 and C_2 due to inevitable imperfections is of interest for the solution).

The spring and inertia forces acting on mass m are sketched in Figure 3.15b. Since we cannot assume the deflection to be constant, we must also include the inertia forces due to relative accelerations, in particular the centrifugal force and the Coriolis force. The centrifugal force components in the x- and y-directions are $m\omega^2 u$ and $m\omega^2 v$. The Coriolis force components are $2m\omega\dot{v}$ and $-2m\omega\dot{u}$ in the x-

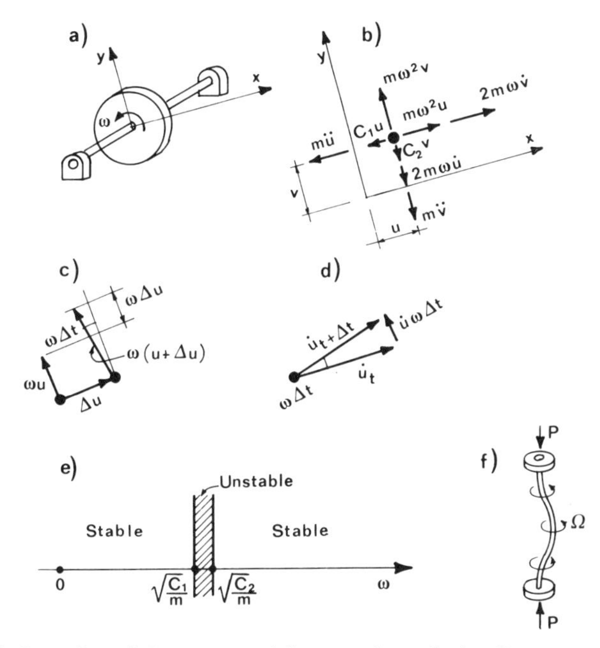

Figure 3.15 (a) Rotating disk supported by massless shaft; diagrams of (b) forces, (c) velocities, (d) accelerations; (e) limits of stability.

and y-directions, as shown in Figure 3.15b. These expressions can be derived either formally, by vector differentiation, or simply from the velocity vector diagrams in Figure 3.15c, d. As is apparent from these figures, the radial displacement $\Delta u = \dot{u}\,\Delta t$ during the time interval Δt in the x-direction causes the magnitude of the circumferential velocity ωx to increase by $\omega\,\Delta u$ or $\omega\dot{u}\,\Delta t$. In addition to this, the radial velocity vector changes direction by angle $\omega\,\Delta t$, and so we have an additional velocity increment $\dot{u}(\omega\,\Delta t)$ in the tangential direction. This combines to give the total velocity increment $2\dot{u}\omega\,\Delta t$ during Δt. Therefore, the acceleration in the y-direction is $2\omega\dot{u}$. By similar reasoning (Fig. 3.15c, d), the radial velocity $\dot{v}$ leads to acceleration $-2\omega\dot{v}$ in the x-direction.

Thus it is found that the equations of motion of the mass disk in the rotating coordinates are

$$-C_1 u + m\omega^2 u + 2m\omega\dot{v} - m\ddot{u} = 0 \qquad -C_2 v + m\omega^2 v - 2m\omega\dot{u} - m\ddot{v} = 0 \quad (3.4.1)$$

This is a system of two linear ordinary differential equations with constant coefficients. Seeking the solution in the form $u = q_1 e^{\lambda t}$, $v = q_2 e^{\lambda t}$, we get for the unknown coefficients q_1 and q_2 the following system of homogeneous linear algebraic equations:

$$\begin{aligned} \left(\lambda^2 - \frac{C_1}{m} - \omega^2\right)q_1 - 2\omega\lambda q_2 &= 0 \\ 2\omega\lambda q_1 + \left(\lambda^2 + \frac{C_2}{m} - \omega^2\right)q_2 &= 0 \end{aligned} \qquad (3.4.2)$$

A nonzero solution is possible only if the determinant vanishes. This yields the condition $\lambda^4 + 2a\lambda^2 + c = 0$, in which we introduce the notations

$$a = \frac{C_1 + C_2}{2m} + \omega^2 \qquad c = \left(\frac{C_1}{m} - \omega^2\right)\left(\frac{C_2}{m} - \omega^2\right) \tag{3.4.3}$$

This biquadratic equation has four roots:

$$\lambda = \pm\sqrt{-a \pm \sqrt{a^2 - c}} \tag{3.4.4}$$

There is always at least a very small unavoidable difference in the values of the spring constants, and the smaller one may be denoted as C_1, that is, $C_1 < C_2$. For $\omega^2 < C_1/m$, as well as for $\omega^2 > C_2/m$, c is positive, $a^2 - c < a^2$, and so all the roots are imaginary; this means that the shaft oscillates with a constant amplitude and is stable. For $\omega^2 = C_1/m$ or $\omega^2 = C_2/m$, we have $c = 0$ and $\lambda = 0$, which is a double root; this means that $u, v \sim t$ is one fundamental solution, and so the shaft is unstable. For $C_1/m < \omega^2 < C_2/m$, we have $c < 0$ and two of the roots λ are complex; then one root has a positive real part, and so the shaft is unstable. Thus, the region

$$C_1/m \leq \omega^2 \leq C_2/m \tag{3.4.5}$$

is the region of instability (Fig. 3.15e) and $\omega_1 = (C_1/m)^{1/2}$ and $\omega_2 = (C_2/m)^{1/2}$ are the critical angular velocities.

One interesting feature of this problem is that the shaft becomes stable at supercritical angular velocities, which exceed the angular velocities, for which the shaft is unstable. This contrasts with the buckling of columns under axial dead loads, for which all loads exceeding the first critical one cause instability.

The fact that $u, v \sim t$ at the critical states, that is, the stability limit corresponding to translation at constant velocity (no acceleration) implies that it must be possible to obtain the critical velocities by static analysis. Indeed, using the static approach and assuming the disk to be mounted with unavoidable eccentricities e_1, e_2, one gets the equilibrium equatons $m\omega^2(e_1 + u) - C_1 u = 0$, $m\omega^2(e_2 + v) - C_2 v = 0$, from which

$$e_1 + u = e_1 \frac{C_1}{C_1 - m\omega^2} \qquad e_2 + v = e_2 \frac{C_2}{C_2 - m\omega^2} \tag{3.4.6}$$

This gives infinite deflections at the critical velocities $\omega = (C_1/m)^{1/2}$ and $\omega = (C_2/m)^{1/2}$. However, for all other angular velocities ω the deflection is small and the shaft appears to be stable, which is false, so we have an example of a problem where the static method fails to give the complete answer, even though the instability is of static (divergence) type. The reason for the failure of the static approach is the neglect of the Coriolis force. Instability occurs as a translation at constant velocity in the rotating system, but this implies that there must be acceleration relative to fixed coordinates, and this acceleration is the Coriolis acceleration.

With reference to the energy approach to be studied later (Chapter. 4), it may be noted that the present problem possesses potential energy, $\Pi = \frac{1}{2}(C_1 - m\omega^2)u^2 + \frac{1}{2}(C_2 - m\omega^2)v^2$, and indeed the derivatives $-\partial\Pi/\partial u$ and $-\partial\Pi/\partial v$ yield the correct spring forces and centrifugal forces. The existence of potential energy implies that the system is conservative. Π ceases to be positive definite when

$\omega^2 > C_1/m$, and thus we are led to conclude from the energy analysis that all angular velocities exceeding the first critical one would be unstable, which is incorrect.

It may be checked that the same incorrect result would have been obtained from the dynamic method if the Coriolis force were omitted. Thus, the Coriolis force has a stabilizing effect on the shaft.

The failure of the energy approach is due to the fact that the Coriolis force, while being conservative, does not possess a potential because it always does zero work (as it is always oriented normal to the direction of motion). There are other examples of forces that are conservative (since they do no work), yet possess no potential; for example e.g, the gyroscopic moments. Generally, such forces may have a stabilizing influence.

Another striking feature of the rotating shaft problem, discovered by Ziegler (1968), is the possibility that damping may have a destabilizing influence on a system stabilized by the Coriolis force (or gyroscopic forces in general). Consider velocity-dependent linear damping forces $-2mb\dot{u}$ and $-2mb\dot{v}$ in Figure 3.15, and carry out the same type of analysis as before. Assuming $C_1 = C_2$, one finds that the shaft is stable for $\omega^2 < C_1/m$ and unstable for $\omega^2 > C_1/m$. So, we see that damping cancels the stabilizing effect of the Coriolis force. However, although this result is mathematically correct, it contradicts experiments that all show that $\omega = (C_1/m)^{1/2}$ is the only unstable angular velocity. The explanation is that there always exist further damping forces such as air drag and bearing friction, or nonlinear damping, and according to Ziegler (1968) their consideration removes the discrepancy with experiment.

As is apparent from the preceding examples, it is important to distinguish various types of forces. Following Ziegler (1968), the forces on dynamic systems may be classified as follows:

1. *Nonstationary* (or *heteronomous*) loads, which are specified functions of time. They are obviously nonconservative and always require a kinetic approach. Pulsating loads are an example.
2. *Stationary* loads, which do not depend directly on time.
   a. *Velocity-dependent* loads.
      (1) *Dissipative* loads, which are nonconservative; they do work (dissipate energy) as the structure moves.
      (2) *Gyroscopic* loads (such as the Coriolis force or gyroscopic moments), which are *conservative* but have no potential since they do no work.
   b. *Velocity-independent* loads.
      (1) Loads having *no potential,* which may be termed *circulatory* (borrowing this expression from hydrodynamics). They are *nonconservative.*
      (2) Loads having a *potential,* which may be termed *noncirculatory.* They are *conservative.*

Note that the class of conservative loads includes two categories: (1) velocity-independent loads having a potential; (2) velocity-dependent gyroscopic loads. All other loads are nonconservative. The last class of loads occurs in the classical stability problems, in which static (as well as energy methods) are always applicable.

Among *reactions*, one may distinguish (1) *nonworking* reactions (i.e., reactions of rigid supports) and (2) *dissipative* reactions (e.g., reaction of a viscous dashpot or frictional slider). The work of reactions can never be positive in scleronomic systems. *Scleronomic* systems are systems in which the constraints do not depend explicitly on time (Gantmacher, 1970, p. 11), that is, the position vector of the reaction point may depend on a set of Lagrangian coordinates but not explicitly on time (Ziegler, 1968, p. 27).

Problems

3.4.1 Work out the solution for the rotating shaft in the presence of linear damping.

3.4.2 If a shaft is subjected to axial compression force P, its critical angular velocity ω_{cr} is, of course, reduced. The value of ω_{cr} may be calculated from $\omega = (C_1/m)^{1/2}$ provided that the spring constant is evaluated taking P (and the support conditions) into account. This may be done in the manner shown in Section 2.1. Solve ω_{cr} for a simply supported shaft of length l, uniform bending stiffness EI, one central disk, and $P = 0.5P_E$. (For a detailed analysis of many cases, see Ziegler, 1968.)

3.4.3 Analyze stability of a massive beam-column (shaft) that has fixed ends, carries axial load P, and rotates with angular velocity Ω (Fig. 3.15f). The column has uniform distributed mass μ. {Note that the centrifugal force $\mu\Omega^2 w$ contributes to buckling [cf. Watson and Wang, 1983; also *J. Eng. Mech.* (ASCE), 1990 in press].}

3.5 DEFINITION OF STABILITY

The examples in the preceding sections demonstrate that stability of structures in general must be defined in the dynamic sense. A static definition is in general insufficient.

We consider structures with a finite number of degrees of freedom, characterized by generalized displacements $q_i(t)$. Except for some pathological cases, the behavior of continuous structures may be approximated in this manner by some discretization procedure, such as the finite element method, the finite difference method, or the truncated Fourier series expansion. The equations of motion of the structure, which in general may be derived as the Lagrange equations of motion (see, e.g., Flügge, 1962), have the form

$$[M_{ij}]\{\ddot{q}_j\} + [C_{ij}]\{\dot{q}_j\} + [K_{ij}]\{q_j\} = \{f_i\} \tag{3.5.1}$$

in which $[M_{ij}]$ = mass matrix, $[C_{ij}]$ = damping matrix, $[K_{ij}]$ = stiffness matrix, and $\{f_i\}$ = column matrix of applied forces ($i, j = 1, 2, \ldots, n$). For reasons of convenience, mathematicians prefer to convert Equation 3.5.1 to a system of first order differential equations. This is accomplished by introducing new variables $Y_1, Y_2, \ldots, Y_N$ with $N = 2n$, such that $Y_{2i-1} = q_i$ and $Y_{2i} = \dot{q}_i$. The latter equations, together with Equation 3.5.1, yield a system of $2n$ first-order differential equations for the unknowns $Y_K(t)$:

$$[M_{ij}]\{\dot{Y}_{2j}\} + [C_{ij}]\{Y_{2j}\} + [K_{ij}]\{Y_{2j-1}\} = \{f_i\} \qquad \{\dot{Y}_{2i-1}\} = \{Y_{2i}\} \tag{3.5.2}$$

Multiplying by the inverse $[M_{ij}]^{-1}$, one can obtain a system of $2n$ first-order differential equations in the standard (canonical) form:

$$\dot{Y}_k = F_k(Y_1, Y_2, \ldots, Y_N, t) \qquad k = 1, 2, \ldots, N \tag{3.5.3}$$

Since the stiffness and damping coefficients may depend on displacements, functions F_k can, in general, be nonlinear. The N-dimensional space of variables $Y_1, \ldots, Y_N$ is called the *phase space*.

In a general sense, one needs to decide whether a certain solution $Y_k^0(t)$, which corresponds to certain given initial values $Y_k^0(t_0) = Y_{k_0}$ at time $t = t_0$ $(k = 1, \ldots, N)$, is stable. In the static problems as well as in the problems of Sections 3.1 and 3.2, the solution $Y_k^0(t)$ whose stability is to be examined consisted of a state at rest, that is, $Y_k^0 =$ constant, although in general Y_k^0 may be functions of time. As illustrated by the preceding examples, the crucial question with regard to stability is what happens when the system is disturbed, for example, when the initial values Y_k^0 are changed to slightly different initial values $Y_k^0 + v_k^0$ where v_k^0 are some small initial perturbations. The solution corresponding to these initial values may be written as

$$Y_k(t) = Y_t^0(t) + v_k(t) \tag{3.5.4}$$

in which functions $v_k(t)$ represent the change of the solution caused by the change in the initial conditions. For most practical problems, functions F_k are smooth, and so they may be expanded in Taylor series about Y_k^0. This yields

$$\dot{Y}_k = \dot{Y}_k^0 + \dot{v}_k = F_k(Y_1^0, \ldots, Y_N^0, t) + \sum_{m=1}^{N} a_{km} v_m + \frac{1}{2!} \sum_{m=1}^{N} \sum_{p=1}^{N} \frac{\partial^2 F_k}{\partial Y_m \, \partial Y_p} v_m v_p + \cdots \tag{3.5.5}$$

in which $a_{km} = \partial F_k / \partial Y_m$. At the same time, by definition, we have $\dot{Y}_k^0 = F_k(Y_1^0, \ldots, Y_N^0, t)$, and introducing this into Equation 3.5.5, we obtain the differential equations

$$\dot{v}_k = \sum_{m=1}^{N} a_{km} v_m + \psi_k(v_1, \ldots, v_N, t) \qquad k = 1, 2, \ldots, N \tag{3.5.6}$$

in which ψ_k are functions representing the second and higher terms of the Taylor series expansion. They have the important special property that their Taylor series expansions about $v_k = 0$ contain no constant terms and no linear terms in v_k. Equations 3.5.6, introduced by Poincaré, reduce the investigation of stability of any solution to the investigation of stability of a zero (trivial) solution, $v_k = 0$. In the problems we have considered so far, we obtained equations of the type of Equations 3.5.6 directly, without having to carry out the foregoing reduction.

Definition of Stability. If for an arbitrary positive number ε there exists a positive number δ such that every solution with initial values

$$|v_k^0| \le \delta \qquad k = 1, \ldots, N \tag{3.5.7}$$

satisfies the inequalities

$$|v_k(t)| < \varepsilon \qquad k = 1, \ldots, N \tag{3.5.8}$$

for all times $t > t_0$, the solution $Y_k^0(t)$ is said to be *stable* (in the sense of Liapunov). When, in addition, $\lim v_k = 0$ for $t \to \infty$, the solution is said to be *asymptotically* stable.

Simply stated, *a structure* (or any system) *is stable if a small change in the initial conditions* (input) *leads to a small change in the solution* (output, response).

The foregoing definition of stability, due to Liapunov (1892), is generally used in all fields—not only structural mechanics but also biology, economics, etc.

The meaning of the stability definition is graphically illustrated in Figure 3.16a. For any given band of width 2ε (cross-hatched), if one can find a nonzero positive initial deviation δ such that the corresponding solution curve (dashed) remains within the band 2ε, then the solution (the solid curve) is stable. On the other hand, if for all initial deviations δ (≥ 0) the solution deviates outside the band 2ε (the dash-dot curve), the solution will be unstable.

If the system is linear, then the solution is proportional to the initial conditions. Then stability may also be defined as follows: the structure (system) is stable if a finite change in initial conditions (input) does not cause an infinite change in the solution (output, response).

The conditions in Equations 3.5.7 and 3.5.8 may be slightly modified, for example, by requiring that the vectors v_k^0 and $v_k(t)$ be contained within hyperspheres of radii δ and ε in the phase space, that is, that

$$v_1^{0^2} + v_2^{0^2} + \cdots + v_N^{0^2} \leq \delta \qquad \text{and} \qquad v_1^2 + v_2^2 + \cdots + v_N^2 < \varepsilon \tag{3.5.9}$$

instead of the conditions in Equations 3.5.7 and 3.5.8 in the foregoing definition. The inequalities in the definition stated may be regarded as describing hypercubes in N-dimensional phase space. The prefix *hyper* refers to the fact that the phase space has generally more than three dimensions.

An example illustrating the stability definition is given in Figure 3.16b. Let $a > 0$, that is, a is finite, even though it could be very small. If one gives $0 < \varepsilon < a$, then the ball will move beyond position ε (all the way to position a) no matter how small δ is ($\delta > 0$).

As another example, consider a linear system (an oscillator) for which q = deviation from the static equilibrium position and $\dot{q}$ = velocity. If the system is damped, the motion after an imposed displacement δ brings the system back to

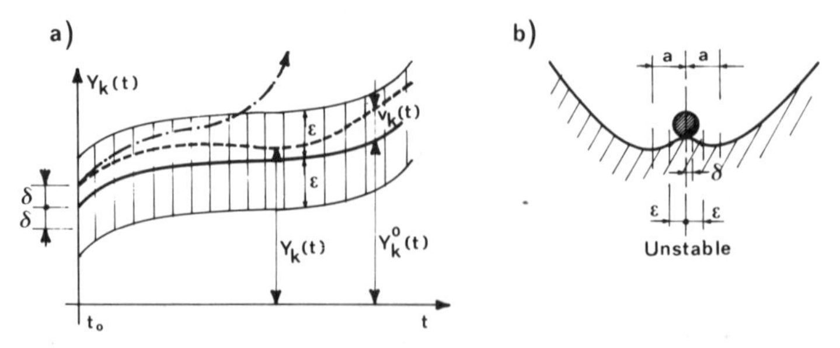

Figure 3.16 (a) Responses of perturbed systems: stable (dashed curve) or unstable (dash-dot curve); (b) example of unstable system.

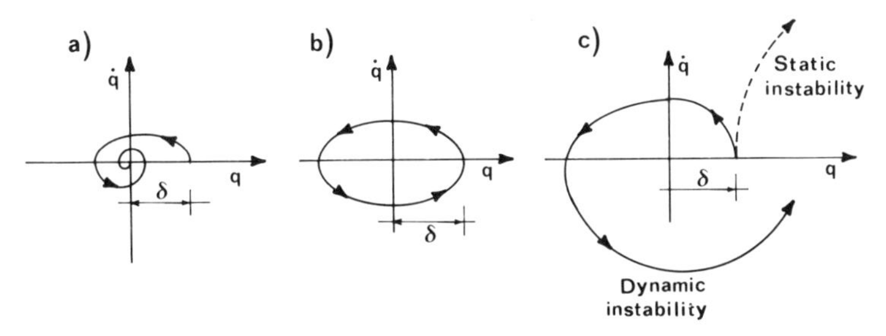

Figure 3.17 Stable (a, b) and unstable (c) responses in the phase space.

the original position of equilibrium (Fig. 3.17a). If the system is not damped, a motion in which q and $\dot{q}$ are always limited ensues; see the diagram in the phase space in Figure 3.17b. If the stiffness of the system becomes negative, a static instability is produced (Fig. 3.17c) but if the damping becomes negative a dynamic instability is produced (Fig. 3.17c).

As is now obvious from our discussion, we are normally not interested in obtaining complete solutions of the dynamic problem. We are only interested in the *qualitative* nature of the solutions for various values of the load, to the extent that we can judge stability. The qualitative analysis of dynamic systems owes its basic development to Poincaré (1885, 1892) who introduced this approach in analyzing stability of bodies in orbit and stability of rotating fluid bodies in a gravity field.

Problems

3.5.1 On the basis of the Liapunov definition of stability, determine whether the balls in Figure 3.18a are stable.

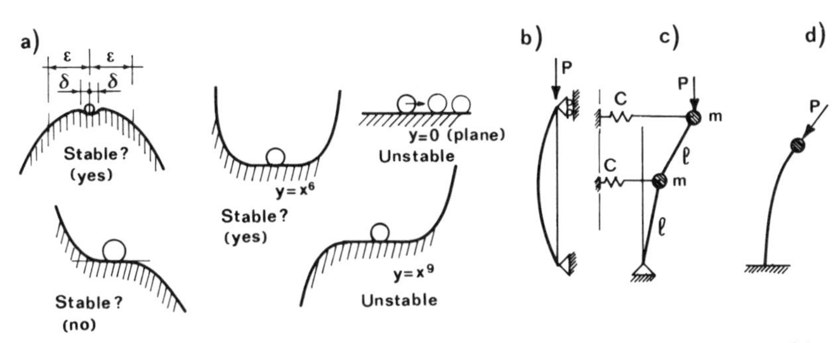

Figure 3.18 (a) Stable and unstable ball supports, and (b, c) exercise problems on Liapunov's definition of stability.

3.5.2 Consider a single-degree-of-freedom system for which the (equilibrium) solution is $q_1 = 0$. Let the motion be defined by $q_1 = a(t+1)^r + be^{-t}$, where a and b characterize the initial deviations, and $r = P^2 - 3P + 2$. Find the stability condition for P. Then for (a) $\varepsilon = 10^{-30}$, $P = 0.5$ and (b) $\varepsilon = 10^{-30}$, $P = 1.5$, try to find δ ($\delta > 0$).

3.5.3 Analyze stability of the solution $q_1 = t$ for a single-degree-of-freedom system. Let the perturbed motion of the system be defined by $q_1 = t + a \operatorname{Re} e^{i\omega t}$, where $\omega^2 = -3 + \sqrt{1-P}$, $a = \text{const}$. Consider
(a) $P = 0.5$, $\varepsilon = 0.1$—find $\delta > 0$
$P = 0.6$, $\varepsilon = 10^{-3}$—find $\delta > 0$
$P = 0.9$, $\varepsilon = 10^{-50}$—find $\delta > 0$
(b) $P = 1.1$, $\varepsilon = 10^{-18}$—can you find $\delta > 0$?
$P = 1.4$, $\varepsilon = 10^{-9}$—can you find $\delta > 0$?

3.5.4 For the hinged column in Figure 3.18b try to find δ ($\delta > 0$) if (a) $P = 0.5P_E$; (b) $P = 0.999P_E$; (c) $P = P_E$; (d) $P = 1.1P_E$.

3.5.5 For the column in Figure 3.18c, try to find δ ($\delta > 0$) for (a) $P = C/8l$; (b) $P = 18C/l$; (c) $P = 2C/l$.

3.5.6 Do the same as in Problem 3.5.5 for the column in Figure 3.18d.

3.6 THEOREMS OF LAGRANGE–DIRICHLET AND OF LIAPUNOV

As we have seen, dynamics permits a more general and more fundamental approach to stability problems than statics. The dynamic solutions, however, are also much more difficult than the analysis of equilibrium states. At the same time, the complete dynamic solution is usually not needed for stability analysis. The magnitudes of the initial disturbances are normally unknown anyway. What is needed are only the limits of stability, that is, the critical states.

Much effort has therefore been devoted in stability theory to finding criteria that make it possible to determine the limits of stability without actually having to solve the motion of the system. The simplest and oldest of such criteria is an energy criterion of stability, which was in its essence known to Torricelli in 1644 for mechanical systems subjected to gravity loads, was presented by Lagrange in his *Mécanique Analytique* (1788), and was rigorously proven by Dirichlet in the early nineteenth century.

Theorem 3.6.1 Lagrange–Dirichlet Theorem. Assuming the total energy to be continuous, the equilibrium of a system containing only conservative and dissipative forces is stable if the potential energy of the system has a strict minimum (i.e., is positive definite).

Proof The state of the system may be imagined as a point in a $2n$-dimensional space with coordinates $v_1 = q_1$, $v_2 = \dot{q}_1$, $v_3 = q_2$, $v_4 = \dot{q}_2, \ldots, v_N = v_{2n} = \dot{q}_n$, called the *phase space*. The energy of the system is $E = \Pi + T$ where Π = potential energy and T = kinetic energy. As shown before, we may assume without any loss of generality that the equilibrium state whose stability is investigated is $q_1 = \dot{q}_1 = q_2 = \dot{q}_2 = \cdots = \dot{q}_{2n} = 0$, and the corresponding energy value is $E = 0$. The condition of strict minimum means that $\Pi \geq 0$ for all points of the phase space sufficiently close to the origin. Figure 3.19a illustrates that a surface $\Pi(q_1, \ldots, q_n)$ has a strict local minimum at the origin $q_1 = \cdots = q_n = 0$ if and only if the intersections with horizontal planes $\Pi = \text{const.}$ result in closed contours around the origin. For a surface that does not have a strict minimum, Figure 3.19b illustrates that the intersections are open curves such as hyperbolas. Since the kinetic energy T is always nonnegative, $\Pi > 0$ implies $E > 0$ for all

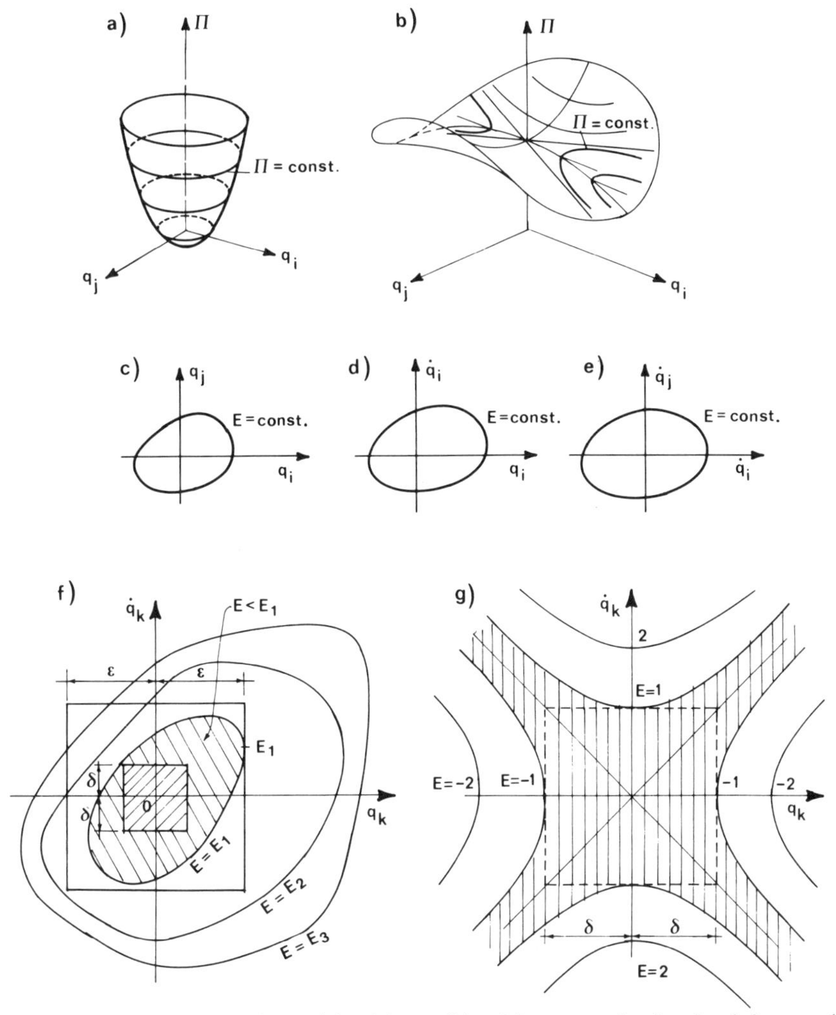

Figure 3.19 Total energy surfaces (a) with or (b) without a strict local minimum; (c–e) typical two-dimensional cross sections of total energy surface for positive-definite Π; cross sections corresponding to increasing energy values for (f) positive-definite Π and (g) nonpositive-definite Π.

points of the phase space close to the origin. Note that the position in the phase space defines the kinetic energy completely since the phase space coordinates include the velocities. Typical two-dimensional cross sections of E are sketched in Figure 3.19c, d, e. In Figure 3.19c, the contour $E = \text{const.}$ is closed because Π is positive definite. In Figure 3.19d it is closed because E is positive definite, and in Figure 3.19e it is closed because $T > 0$ always.

Consider now a finite hypercube of size 2ε centered in the phase space around the origin, that is, $|v_k| < \varepsilon$ for all k, ε being any given positive number. In any two-dimensional cross section, this hypercube appears as a square of side 2ε (Fig. 3.19f). Among all the values of E on the boundary of the hypercube, there exists one minimum value $E_1 > 0$ (this follows, e.g., from the Weierstrass theorem, which states that every continuous function may be uniformly approximated within a finite region with any desired accuracy by a polynomial). Because E has a strict minimum, the hypersurfaces of constant E value must be closed and contain the origin. In the two-dimensional cross section, these hypersurfaces appear as the closed contours shown in Figure 3.19f. Also, the subsequent hypersurfaces (the subsequent contours in Fig. 3.19f) correspond to progressively increasing energy values $E_1, E_2, E_3, \ldots$. Consequently the hypersurface $E = E_1$ is contained entirely within the 2ε hypercube (within the 2ε square in Fig. 3.19f). Therefore, one can choose a 2δ hypercube (2δ square in Fig. 3.19f) in such a manner that it is contained entirely within the hypersurface $E = E_1$. For all points in the interior of the 2δ hypercube (i.e., for $|v_k| < \delta$ for all k), we have $E < E_1$.

Next we note that for a system containing only conservative and dissipative forces, energy E cannot increase. Therefore, if we choose an initial state (i.e., the initial deviations from the equilibrium position and initial velocities) within the 2δ hypercube (2δ square in Fig. 3.19f), the initial energy E_0 is such that $0 < E_0 < E_1$. So the energy E for all subsequent times will be bounded as $0 < E \leq E_0 < E_1$, which implies that the state of the system will always remain within the hypersurface $E = E_1$ (contour $E = E_1$ in Fig. 3.19f). Since the surface is entirely contained within the 2ε hypercube, the state of the system always remains within this hypercube (i.e., within the 2ε square in Fig. 3.19f).

For further discussion see e.g. Leipholz (1970), Ziegler (1968), and Gantmacher (1970). Equivalently, one can use in this proof hyperspheres instead of hypercubes (Ziegler, 1968). Note that the proof depends on the fact that dissipative forces cannot increase the value of E. This follows from the second law of thermodynamics (or the dissipation inequality), which in fact serves as the most fundamental criterion of stability (see Sec. 10.2). In the absence of dissipation, the constancy of E follows simply from the principle of conservation of mechanical energy.

If the dissipative forces are absent, the energy of the system will remain constant ($\dot{E} = 0$) and equal to its initial value E_0 which corresponds to the initial deviations from the equilibrium state and the initial velocities. In the phase space, the state will remain on the hypersurface $E = E_0$ (and in a two-dimensional cross section, on the closed contour $E = E_0$); see Figure 3.20a.

When dissipative forces are present we have $dE/dt \leq 0$. Assuming that the

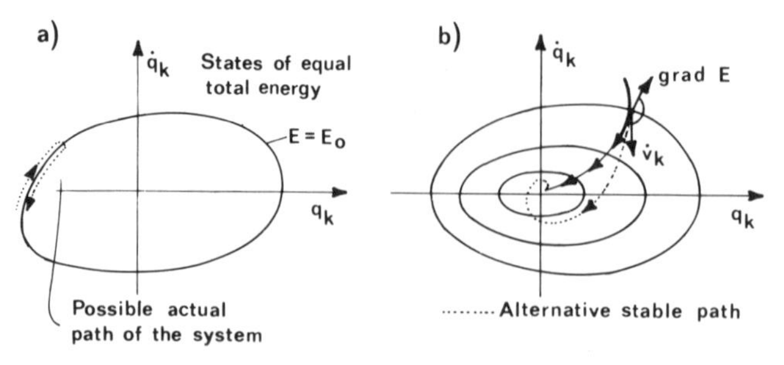

Figure 3.20 Stable paths in the phase space if dissipative forces are (a) absent or (b) present.

potential energy and the nonpotential forces do not depend explicitly on time

$$\frac{dE}{dt} = \sum_k \left(\frac{\partial E}{\partial v_k}\right) \dot{v}_k \tag{3.6.1}$$

that is, the derivative with respect to time represents in the phase space the projection of the vector $\dot{v}_k$ of velocities and accelerations onto the normal to the surface $E = \text{constant}$. Thus the path of the system must either cross the surfaces $E = \text{constant}$ in the direction toward the origin (Fig. 3.20b) or run along them. In the case that $dE/dt = 0$ only for $\dot{q}_k = 0$ (definite dissipative system) and that the equilibrium position is isolated (i.e., there are no other infinitely close equilibrium positions) the equilibrium position is asymptotically stable, that is, all the deviations and velocities tend to zero as the time increases (Gantmacher, 1970, p. 176).

Figure 3.19g illustrates that if the potential energy is not positive definite, stability cannot be proven. In this case the contours $E = \text{constant}$ are not closed curves, and thus if the initial state is within the 2ε square in Figure 3.19g, it only guarantees that a subsequent state will remain in the infinite cross-hatched domain, which, of course, cannot fit within any finite 2δ square.

Simple examples of positive-definite potential-energy functions are

$$\begin{gathered} 3q_1^2 + 5q_2^2 \qquad q_1^2 + (q_2 - 2q_1)^2 \qquad (q_1 - q_3)^4 + q_2^6 + q_3^2 \qquad q_1^2 - 9q_1^3 \\ q_1^2 + q_2^2 - q_2^3 \qquad q_1^2 + q_2^2 - q_1^4 \end{gathered} \tag{3.6.2}$$

Note that there exists a number $\varepsilon > 0$ such that, for $|q_1| < \varepsilon$, one always has $q_1^2 - 9q_1^3 > 0$ (because q_1^3 is higher-order small). One can denote $\delta^2\Pi = q_1^2 + q_2^2$, $\delta^3\Pi = -q_2^3$, and then, for all q_1 and q_2, $\delta^2\Pi > 0$; see for more detail Section 4.1. On the other hand, the expressions

$$\begin{gathered} q_1^2 - 4q_2^2 \qquad q_1^2 - q_2^3 \qquad q_1^2 - (q_3 - 3q_2)^2 + q_3^2 \qquad q_1^2 + q_2^3 \\ q_1 + q_2^4 \qquad q_1^2 + (q_2 + q_3)^9 \end{gathered} \tag{3.6.3}$$

are not positive definite.

As revealed by our previous dynamic solution of a rotating shaft (Sec. 3.4) as well as other examples, the gyroscopic forces, which are conservative but not derivable from a potential, can stabilize a system that would otherwise be unstable. However, they cannot destabilize a stable system, as implied by the Lagrange–Dirichlet theorem.

The minimum of potential energy defines an equilibrium state. Therefore, the limits of stability of conservative systems can be analyzed statically. This is true whether or not dissipative forces are present.

In the Lagrange–Dirichlet theorem, positive definiteness is a sufficient condition, not a necessary one. When the potential energy is not positive definite, the system may or may not be stable. The absence of the local minimum of the potential energy (absence of positive definiteness) does not necessarily imply instability, and examples to this effect are known. However, for two important cases Liapunov proved instability. This is stated by the following two Liapunov instability theorems (Gantmacher, 1970, pp. 174–176; Rektorys, 1969, p. 827; Kamke, 1956, p. 61).

Theorem 3.6.2 Liapunov's First Theorem. If the potential energy in the equilibrium state is not minimum and if the absence of a minimum is caused by the second-order terms in the Taylor series expansion of the potential energy then the system is unstable. (Thus, a saddle point, $\Pi = q_1^2 - q_2^2 + q_1^4$, implies instability, but $\Pi = q_1^2 + q_1^4 - q_2^4$ does not necessarily imply instability, while not guaranteeing stability; for ramifications see Leipholz, 1970).

Theorem 3.6.3 Liapunov's Second Theorem. If the potential energy is maximum with respect to all adjacent states (local maximum) and the maximum is characterized by the terms of the lowest order (not necessarily the second order) in the Taylor series expansion of potential energy, then the system is unstable. (For example, $\Pi = -q_1^2 - q_2^2 + q_1^4 + q_2^4$ and $\Pi = -q_1^4 - q_2^4$ each implies instability, but $\Pi = -q_1^2 + q_2^4$ or $\Pi = -q_1^2 + q_1^4 + q_2^6$ does not necessarily imply instability.)

The crucial point in the proof of the Lagrange–Dirichlet theorem is the existence of the energy function E, serving as the test function for stability. As noted by Liapunov, it is possible to restate the proof of the Lagrange–Dirichlet theorem, using instead of the function E a continuous function Ψ that has a strict minimum in the equilibrium state, and that, during any motion of the system, does not increase, $d\Psi/dt \leq 0$. If, again, $d\Psi/dt = 0$ only for $\dot{q}_k = 0$ the solution is asymptotically stable.

Is it possible to find more general test functions that we could apply at least to some nonconservative system? Some such test functions, which are positive definite and are called the Liapunov functions, have been found. One of them was given by Liapunov in the following theorem (Liapunov, 1892):

Theorem 3.6.4 If one can find for a given system a positive-definite function $\Psi(t, v_k)$ such that its total derivative

$$\frac{d\Psi}{dt} = \frac{\partial \Psi}{\partial t} + \sum_k \frac{\partial \Psi}{\partial v_k} \dot{v}_k \tag{3.6.4}$$

is either identically zero or negative semidefinite ($d\Psi/dt \leq 0$), then the state $v_k = 0$ is stable. Furthermore, if $d\Psi/dt$ is negative definite ($d\Psi/dt < 0$) and Ψ can be made less than any positive number, (i.e., $\Psi \to 0$ at the origin), then the solution is asymptotically stable.

Proof The proof is analogous to that of the Lagrange–Dirichlet theorem (Leipholz, 1970, p. 77) and its main point is illustrated in Figure 3.21. When Ψ is positive definite then the surfaces $\Psi = \text{constant}$ are closed around the origin (Fig. 3.21) and for $\Psi_1 < \Psi_2$ the surface $\Psi = \Psi_1$ lies entirely within the surface $\Psi = \Psi_2$. Obviously $\sum_k (\partial\Psi/\partial v_k)\dot{v}_k = \dot{\mathbf{v}} \cdot \text{grad}\,\Psi$, which represents in the phase space the projection of the vector $\dot{\mathbf{v}}$ onto the normal of the surface $\Psi = \text{constant}$ (Fig. 3.21). Thus, the path of the system (curve p in Fig. 3.21) must either cross all the surfaces $\Psi = \text{constant}$ in the direction toward the origin, in which case $\dot{\mathbf{v}} \cdot \text{grad}\,\Psi$ must be negative, or it must run along one of such surfaces, in which case $\dot{\mathbf{v}} \cdot \text{grad}\,\Psi = 0$. For the special case that $\partial\Psi/\partial t = 0$ (i.e., surfaces $\Psi = \text{const.}$ do not vary in time) it is then clear that the trajectory of subsequent states will keep approaching the origin or remain at a fixed surface $\Psi = \text{const.}$ Thus, for any given 2ε square in Figure 3.21, one can find a contour $\Psi = \Psi_3$ lying entirely within the

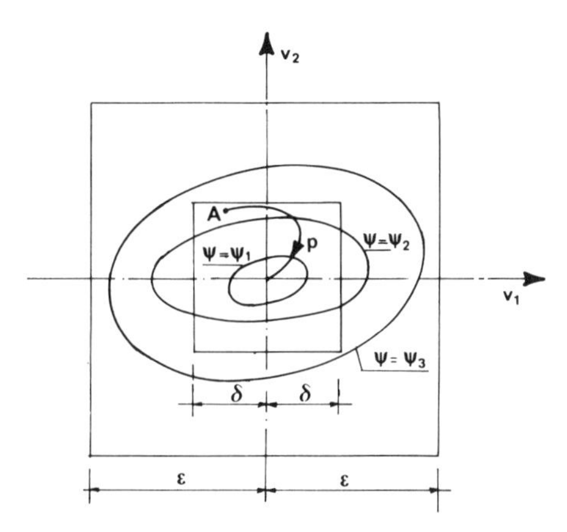

Figure 3.21 Cross sections of positive-definite test function Ψ and path of stable system.

2ε square, and so by choosing the initial state A within a 2δ square that fits within the contour $\Psi = \Psi_3$, it is ensured that the state of the system cannot get outside the region $\Psi = \Psi_3$, and therefore not outside the 2ε square. If $\partial\Psi/\partial t < 0$, the surfaces $\Psi =$ const. shrink in time, and so the movement toward the origin cannot be slower than for $\partial\Psi/\partial t =$ const., which implies stability. If $\partial\Psi/\partial t > 0$, the surfaces $\Psi =$ const. expand in time, however by requiring that $d\Psi/dt \leq 0$ one ensures that the movement toward the origin relative to the expanding surfaces prevails, thus ensuring stability. Finally, the asymptotic stability is ensured if Ψ is diminishing during the motion of the system ($d\Psi/dt \to 0$) and $\Psi \to 0$ implies reaching the origin in the limit.

Evidently, the Lagrange–Dirichlet theorem is a special case of Liapunov's first theorem such that $\Psi = E =$ total energy.

In conclusion we should emphasize that the Lagrange–Dirichlet theorem permits only dissipative forces that do not destroy the existence of the potential-energy function $\Pi\ (q_1, \ldots, q_n)$ from which all conservative forces are derived by differentiation. Dissipative phenomena such as material creep, plasticity, damage, or fracture generally make potential energy nonexistent, and thus the Lagrange–Dirichlet theorem is not applicable. However, as we will see in Chapter 10, static stability analysis can then be conducted on the basis of thermodynamic criteria.

Problems

3.6.1 In two dimensions sketch typical contours of surfaces $\Pi =$ const. when Π is (a) positive definite, (b) negative definite, (c) positive semidefinite, and (d) indefinite.

3.6.2 State and prove (with a sketch) the Lagrange–Dirichlet theorem without referring to the text.

3.6.3 Indicate the type of definiteness of the following expressions: $2q_1^2 + 3q_2^2$, $(q_1 - 2q_2^2) + q_2^2$, $q_1^2 - 4q_1q_2 + 3q_2^2$, $(2q_1 - q_3)^4 - q_2^8 + q_3^2$, $(q_1 - q_2)^2 + q_2^7 + q_3^4$, $q_1^3 + 4q_1^2$, $q_1^2 - q_2^2 + q_2^4$, $2q_1^2 + q_2^2 + q_2^5$, $q_1^2 - 2q_2^2 - (q_1 - q_2)^4$.

3.7 STABILITY CRITERIA FOR DYNAMIC SYSTEMS

Practically all problems of stability of structures are nonlinear. To make the solution feasible, they usually need to be linearized. For a certain class of problems, specified by the following theorem due to Poincaré and Liapunov, the linearization is legitimate and is certain to give correct results as far as the stability limits are concerned.

Theorem 3.7.1 The nonlinear system (Eq. 3.5.6) in which the nonlinear terms ψ_k for sufficiently small v_k admit a Taylor series expansion that contains nonlinear terms is stable if the linearized system ($\psi_k = 0$) is stable.

For a proof, see Leipholz (1970, p. 81) or Gantmacher (1970, p. 193). The converse of this theorem is not necessarily true.

The method of stability analysis of a linear system has already been illustrated by several examples (Secs. 3.1–3.4). Generally, the solution is sought in the form $q_k = Q_k e^{\lambda t}$ where Q_k are constants (or $v_k = A_k e^{\lambda t}$), and substitution into the equations of motion (Eq. 3.5.6 with $\psi_k = 0$) then yields the characteristic equation, an algebraic equation in λ. If the problem is formulated in terms of the first-order differential equations in the standard (canonical) form (Eq. 3.5.6 with $\psi_k = 0$), the characteristic equation defines a standard linear matrix eigenvalue problem for a real matrix:

$$\begin{vmatrix} a_{11}-\lambda & a_{12} & a_{13} & \cdots & a_{1N} \\ a_{21} & a_{22}-\lambda & a_{23} & \cdots & a_{2N} \\ a_{N1} & a_{N2} & a_{N3} & \cdots & a_{NN}-\lambda \end{vmatrix} = 0 \qquad (N = 2n) \tag{3.7.1}$$

The roots are in general complex, appearing in conjugate pairs $\lambda = a \pm ib$, and the corresponding fundamental solutions then are of the form $e^{at}\cos bt$ and $e^{at}\sin bt$ if the roots are distnct. In the case of multiple roots, of multiplicity r, solutions $te^{at}\cos bt, \ldots, t^{r-1}e^{at}\sin bt$ are also present. It has been made clear by the preceding examples that a change in initial conditions will become arbitrarily magnified with time when $a > 0$. Hence, the stability condition is

$$Re\,\lambda < 0 \qquad \text{or} \qquad \text{Im}\,\omega < 0 \tag{3.7.2}$$

in which $\omega = i\lambda = \mp b + ia$ = frequency. In the case of multiple roots, the case $a = 0$ is unstable. Equation 3.7.2 must hold for all roots to ensure stability.

According to the theorem of Poincaré and Liapunov, Equation 3.7.2 is a sufficient stability condition for the actual, nonlinear system. Furthermore, if $\text{Re}\,\lambda > 0$ for at least one root, the actual, nonlinear system is unstable.

No general theorem is available for the case when the critical state is characterized by $\lambda = 0$ ($\omega = 0$) for some root and $\text{Re}\,\lambda \le 0$ for all roots. The question of stability of the actual nonlinear system is then decided by the nonlinear terms ψ_k in Equation 3.5.6.

Figure 3.22 illustrates the behavior of the roots in the Gauss plane when the load P is varied. When the root passes into the right half-plane $\text{Re}\,\lambda > 0$ through the origin ($\lambda = 0$), stability is lost by divergence (i.e., by buckling), and the critical load then may be obtained by static analysis because the inertia forces at

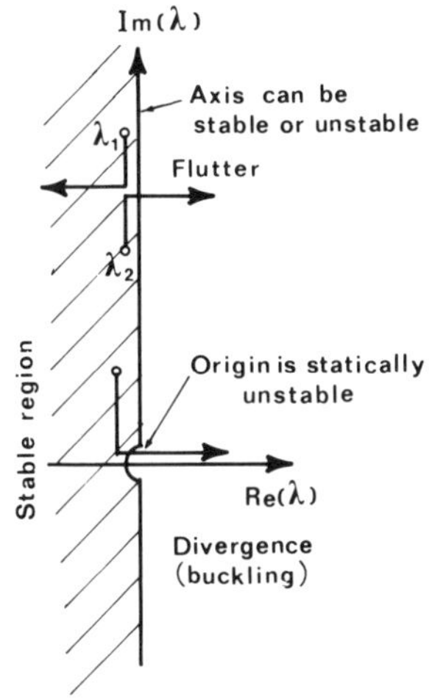

Figure 3.22 Stability conditions expressed through the behavior of roots λ in the Gauss plane.

incipient instability vanish. When the roots leave the stable half-plane (left half-plane) through the imaginary axis, that is, at a point other than the origin, stability is lost by flutter, and then static analysis of the critical loads is generally inapplicable.

Every conservative system (with forces derivable from a potential) exhibits static instability (buckling, divergence). This is because the condition of minimum potential energy defines an equilibrium state (Leipholz, 1970).

By expanding the determinant in Equation 3.7.1 one may bring the characteristic equation to the form

$$f(\lambda) = p_0\lambda^N + p_1\lambda^{N-1} + \cdots + p_{N-1}\lambda + p_N = 0 \tag{3.7.3}$$

whose coefficients $p_0, \ldots, p_N$ are not all zero. If the characteristic equation for λ is neither quadratic nor biquadratic, simple explicit expressions for the roots are unavailable. However, the precise values of the roots are not needed; one only needs to know the sign of the real parts of all roots. There exist certain methods that can decide this question without actually solving the roots. One such method utilizes the following matrix, called the Hurwitz matrix:

$$\mathbf{H} = \begin{vmatrix} p_1 & p_0 & 0 & 0 & 0 & 0 & \cdots & 0 \\ p_3 & p_2 & p_1 & p_0 & 0 & 0 & \cdots & 0 \\ p_5 & p_4 & p_3 & p_2 & p_1 & p_0 & \cdots & 0 \\ \vdots & \vdots & & & \ddots & & & \\ 0 & 0 & & & & & & p_N \end{vmatrix} \tag{3.7.4}$$

in which 0 is substituted for all the terms whose subscript exceeds N.

Theorem 3.7.2 Hurwitz Theorem. The necessary and sufficient condition for all the roots of Equation 3.7.1 to have negative real parts is that all the principal minors of the Hurwitz matrix (Eq. 3.7.4), that is,

$$\Delta_1 = p_1 \qquad \Delta_2 = \begin{vmatrix} p_1 & p_0 \\ p_3 & p_2 \end{vmatrix} \qquad \Delta_3 = \begin{vmatrix} p_1 & p_0 & 0 \\ p_3 & p_2 & p_1 \\ p_5 & p_4 & p_3 \end{vmatrix} \qquad \cdots \qquad \Delta_N = \det \mathbf{H} \quad (3.7.5)$$

must be positive. (For the proof, see Leipholz, 1970, pp. 33 and 37.) An equivalent criterion was derived earlier by Routh (Gantmacher, 1970, p. 199).

As an example, in the case of a fourth-degree polynomial one has

$$\mathbf{H} = \begin{bmatrix} p_1 & p_0 & 0 & 0 \\ p_3 & p_2 & p_1 & p_0 \\ 0 & p_4 & p_3 & p_2 \\ 0 & 0 & 0 & p_4 \end{bmatrix} \qquad (3.7.6)$$

and the conditions of stability are $p_1 > 0$, $p_1 p_2 - p_0 p_3 > 0$, $(p_1 p_2 - p_0 p_3) p_3 - p_1^2 p_4 > 0$, $p_4 > 0$. Note also that for the evaluation of the determinants in Equations 3.7.5, it is best to expand them always with respect to the last row, because then the preceding determinant appears as one of the minors used in the expansion.

Another useful method of examining stability is due to Nyquist and Michailov (Leipholz, 1970, p. 27). One considers the mapping of the complex plane (Gauss plane Re λ, Im λ) into the complex plane (Re f, Im f) given by the polynomial $f(\lambda)$ in Equation 3.7.3. In the mapped plane, the image of all roots coincides with the origin. Therefore, if in the complex plane of λ all the roots lie to the left of the imaginary axis, in the complex f plane the origin must also lie to the left of the image of the imaginary axis. Choosing various y values in an increasing sequence, and evaluating $f(\lambda)$ from Equation 3.7.3 for all values $\lambda = iy$, one can trace from individual points (such as points 1 through 9 illustrated in Fig. 3.23) the curve that is the image of the imaginary axis. If the origin remains to the left of this curve (proceeding from $y = -\infty$ to $y = +\infty$), the system is stable.

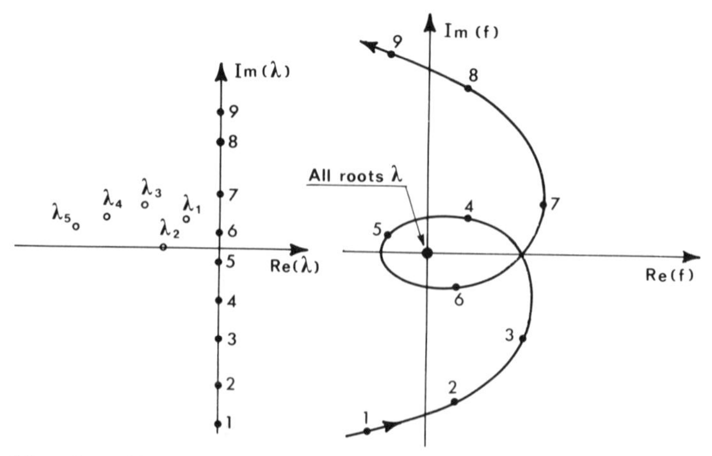

Figure 3.23 Mapping of imaginary axis of Gauss plane and of roots λ for stable system.

A caveat with respect to linearization needs to be mentioned. In some stability problems, such as shell buckling, it is insufficient to investigate infinitely small variations in the initial conditions, and small but *finite* imperfections of the system must be taken into account, in order to obtain physically relevant results. The linearized stability analysis still yields the correct critical loads, but they are insufficient to predict the actual behavior which is dominated by finite imperfections, as it will be shown in Chapter 7.

Problems

3.7.1 Verify the stability of the solution of the example presented in Section 3.4 (disk mounted on shaft) using the Routh-Hurwitz condition.

3.7.2 Analyze stability of the nonlinear dynamic system:

$$2ml^2(\ddot{q}_1 + \ddot{q}_2) + ml^2\ddot{q}_1 + Pl(q_2 - q_1) + Cq_1(1 + kq_1) = 0$$
$$ml^2(\ddot{q}_1 + \ddot{q}_2) + C(q_2 - q_1) + Ck(q_2 - q_1)^2 = 0$$

in which m, l, k are positive constants. (*Note:* This system gives the motion of the column with two rigid bars and two point masses that was described before by Eqs. 3.2.10, but the springs are now nonlinear.)

3.7.3 Analyze stability of a massless free-standing column with a point mass on top loaded by a follower force P that varies with the slope θ as $P = P_0 \cos\theta$ ($P_0 = \text{const.}$).

3.7.4 Same P as above, but for a free-standing column consisting of two rigid bars connected by springs, with point masses at the joint and on top.

3.7.5 Can a conservative system with forces derivable from a potential lose stability by flutter? Is dynamic analysis necessary?

3.8 STABILITY OF CONTINUOUS ELASTIC SYSTEMS

In generalizing Liapunov's stability definition to continuous structures it would be unreasonable to require that the displacements, strains, and strain rates caused by an infinitely small disturbance (less than δ) remain infinitely small (less than ε) at all points of the structure. This is clarified, for example, by the example (due to J. D. Achenbach of Northwestern University) of an elastic sphere that is subjected on its surface to a disturbance (pressure jump) generating a radial pressure wave (Fig. 3.24a). When this wave reaches the center of the sphere, a finite strain is

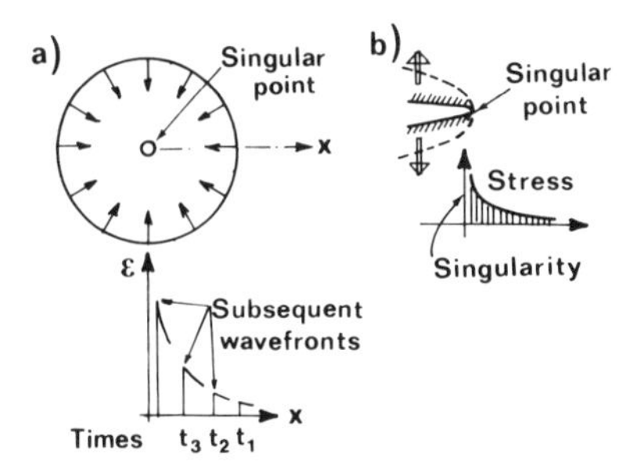

Figure 3.24 Pressure wave (a) in elastic sphere and (b) in elastic solid with sharp crack or sharp notch.

produced even if the disturbance is infinitely small. Similarly, when the tip of a sharp crack (or a sharp acute corner) (Fig. 3.24b) in an elastic solid is hit by a plane wave of an infinitely small amplitude, finite displacements, strains, and strain rates are produced within an infinitely small neighborhood of the crack tip (or corner). If the body is in its natural (unstressed) state prior to the disturbance, obviously it would be senseless to consider these states as unstable, just because of the response in one isolated, singular point. Therefore, stability of continuous bodies can be meaningfully defined only in a certain average sense. Thus, the definition of stability must employ a certain metric (or norm) of a function space, applied to the displacements as continuous functions of location.

Consider a structure with one continuous displacement variable $w = w(x, y; t)$ that describes a deviation from a certain basic state whose stability is to be examined. Following Movchan (1956) and others (e.g., Herrmann and Nemat-Nasser, 1966; Fu and Nemat-Nasser 1970) one defines a certain metric $\rho(w, t)$ such that $\rho(0, t) = 0$, and ρ is a positive-definite continuous function of time t (w being continuous). For $t = t_0$ one may in general consider a different metric $\rho_0(w, t)$ of the same properties as ρ, and such that ρ continuously depends on ρ_0. An example of an acceptable metric is $\rho = \int_V w^2(\mathbf{x}, t)\, dV$ *and* $\rho_0 = \int_V w^2(\mathbf{x}, t_0)\, dV$. Liapunov's definition of stability may now be extended as follows:

The state $w = 0$ is stable if, for an arbitrary positive ε, there exists a positive δ such that $\rho(w, t) < \varepsilon$ for all $t > t_0$ if $\rho_0(w, t_0) < \delta$.

Using this definition, Movchan proved a theorem which states the condition whose fulfillment guarantees stability:

Theorem 3.8.1 The state $w = 0$ is *stable* (with respect to metrics ρ_0 and ρ) if there exists some positive-definite functional $\Phi(w, t)$ (called the Liapunov functional) such that $\dot{\Phi} \le 0$ for all $t \le t_0$ and Φ tends to zero as ρ_0 tends to zero, that is, if for any positive ε_1 one can find a positive δ_1 such that $\Phi(w, t_0) < \varepsilon_1$ if $\delta_0(w, t_0) < \delta_1$. (The proof is analogous to that of Lagrange–Dirichlet theorem.)

Similar to the Lagrange–Dirichlet stability theorem, a physically meaningful choice of the Liapunov functional is energy. For a structure with nonconservative loads, such as follower forces (Sec. 3.2), one cannot choose the total energy, $E = \Pi + T$, because nonconservative loads may supply further energy W to the structure, that is, E may increase. However, because of the law of conservation of mechanical energy, $\Pi + T$ must increase (in absence of dissipation) precisely by the amount W. So the energy functional

$$\Phi = \Pi + T - W \qquad \text{with } W = \int_V p\dot{w}\, dV\, dt \tag{3.8.1}$$

(where V denotes the volume of the structure and p is the distributed applied load) remains constant, that is, $\dot{\Phi} = 0$. Furthermore, Φ obviously tends to zero as w tends to zero, and so the assumptions of Movchan's theorem are satisfied. Since the kinetic energy T is always positive definite, one may exclude it from Φ, setting

$$\Phi^* = \Pi - W \tag{3.8.2}$$

and state the *energy criterion of stability* as follows:

If Φ^* is a *positive-definite* functional, the stability of an elastic system with nonconservative loads is ensured.

A special case of this theorem for the case when only conservative loads are present ($W = 0$) represents an extension of the Lagrange–Dirichlet theorem to continuous structures. For nonconservative loads a positive-definite Π does not ensure stability, as has been shown by Ziegler (1952) via an example.

Note that the foregoing energy criterion states only a sufficient condition of stability, and not the necessary one. Stability might exist even when Φ^* is not positive definite, and even when it is outright negative definite. For example, the gyroscopic loads, which are conservative but do no work, can stabilize a structure that would otherwise be unstable (Sec. 3.4), and the foregoing energy criterion says nothing about this possibility. Nevertheless, the criterion does imply that the gyroscopic forces cannot destabilize a structure that is otherwise stable, same as implied by the Lagrange–Dirichlet theorem for discrete systems.

Problems

3.8.1 State various other possible metrics (norms) that can be used to define stability of a continuous structure.

3.8.2 Is $\rho = \int w^3 \, dV$ an acceptable metric?

3.8.3 Is $\rho = \max |w|$ an acceptable metric? If not, give some examples which demonstrate it.

3.9 NONLINEAR OSCILLATIONS AND CHAOS

Our previous dynamic analysis has been confined to linear or linearized systems. Nonlinear structural behavior may arise due to geometric nonlinearity of large deflections as well as nonlinear material behavior (Chaps. 8–13). Nonlinear dynamic systems show a much richer and intricate specter of behavior. During the last few decades it became apparent that nonlinear dynamic systems can exhibit not only a simple dynamic response such as periodic oscillation or divergence but also a complex response which is nonperiodic and appears to be random, despite the deterministic nature of the system with its loads. Such a response, called chaos, nevertheless shows a certain degree of order, and so it would be incorrect to treat it by methods of random dynamics. Therefore, considerable attention has recently been devoted to the study of the order in chaos (see, e.g., the books by Thompson and Stewart, 1986; Thompson, 1982; Moon, 1986; Schuster, 1989; Guckenheimer and Holmes, 1983; and the papers by Moon and Holmes, 1983; and by Thompson, 1989).

The typical responses of a single-degree-of-freedom oscillator are illustrated in Figure 3.25, which shows the histories of deflection $q(t)$ and the trajectories in the phase space $(q, \dot{q})$. The motion of an undamped, unforced, linear oscillator is of the type $q = A \sin \omega t$, $\dot{q} = A\omega \cos \omega t$; from this we see that $q^2 + (\dot{q}/\omega)^2 = A^2$, and so the trajectory in the phase space is an ellipse (Fig. 3.25a). A typical property of damped, unforced, linear oscillators is that the trajectories are attracted to the point of equilibrium, provided the equilibrium is stable. This point in the phase

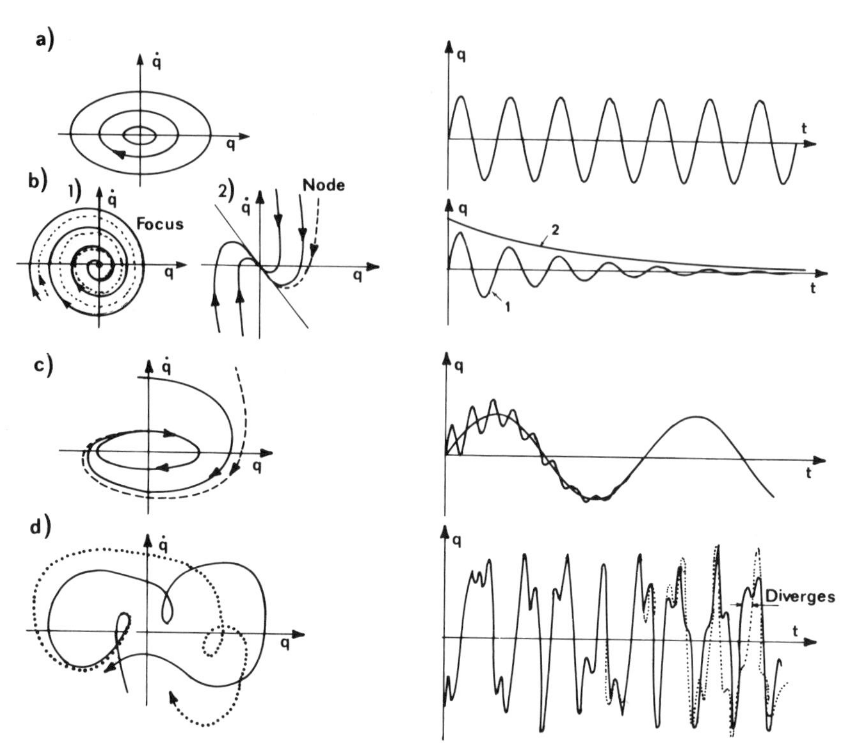

Figure 3.25 Typical responses of a single-degree-of-freedom oscillator: (a) linear, undamped, and unforced; (b) linear, damped, and unforced; (c) linear, damped, and forced; (d) nonlinear, damped, and forced (*reproduced, with permission, from Thompson and Stewart, 1986*).

space is then called a point attractor. The attractor can be a focus (Fig. 3.25b-1) if damping is light, subcritical, or a node (Fig. 3.25b-2) if the damping is strong, supercritical. The trajectories for the damped, forced oscillator are attracted to the trajectory of the steady-state cyclic motion, which is called a cycle attractor (Fig. 3.25c).

For a nonlinear oscillator, by contrast, the time history as well as the phase-space trajectory may happen to appear chaotic, attracted to no simple trajectory. Nevertheless, in some cases the trajectory is not completely random but upon closer scrutiny is found to be attracted to something, called the *strange* (or *chaotic*) attractor. (More often, though, the response is periodic, with a period equal to that of the forcing function or its multiple.) The strange attractor describes a hidden order in the chaotic response. It is typical of chaotic response that a very small change in the initial conditions produces a trajectory that exponentially diverges from the original trajectory (Fig. 3.25d); this means that the system is unstable and the response in time is unpredictable.

Chaotic response may be experimentally demonstrated, for example, on a buckled beam loaded by a constant axial force $P \geq P_{cr}$ and magnetically excited by a sinusoidal lateral force; see Figure 3.26 (Moon and Holmes, 1979). The nonlinearity stems from large deflections that cause the restoring force to be equivalent to a spring of a quadratically varying secant stiffness, $\bar{C} = C_1 + C_2q^2$

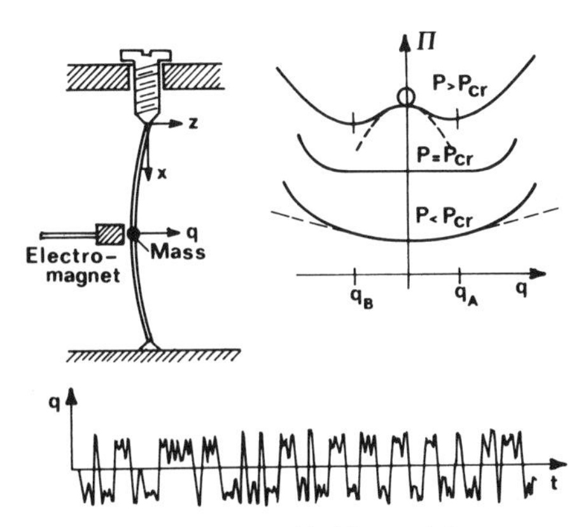

Figure 3.26 Buckled beam excited by sinusoidal lateral force.

where C_1, C_2 = given coefficients, depending on P. Thus the equation of motion has the form

$$\ddot{q} + k\dot{q} + C_1 q + C_2 q^3 = \beta \cos \Omega t \tag{3.9.1}$$

or

$$\ddot{q} + k\dot{q} + C_1 q^3 = \beta \cos \Omega t \tag{3.9.2}$$

where k = damping coefficient, Equation 3.9.2, called the Duffing equation, is the special case for which there is no linear stiffness, that is, $C_1 = 0$. This happens to the column at $P = P_{cr}$. Equation 3.9.1, with $C_1 < 0$, applies for $P > P_{cr}$. The corresponding surfaces of potential energy Π have the shapes shown in Figure 3.26, described as $\Pi = a_1 q^2 + a_2 q^4$ where $a_2 > 0$ while $a_1 = 0$ at $P = P_{cr}$ and $a_1 < 0$ at $P > P_{cr}$. This means that the surface $\Pi(q)$ for $P = P_{cr}$ has a nearly flat portion at the origin, which agrees with the fact that linear analysis for small deflections yields a zero natural frequency (Sec. 3.1). However, taking into account the quartic term in $\Pi(q)$ one finds that bounded finite oscillations at $P = P_{cr}$ are possible. Understanding is helped by noting that a ball forced to roll on these surfaces oscillates in the same manner as the column.)

For Equations 3.9.1 and 3.9.2, as well as in general, the order in a chaotic response becomes apparent by plotting in the phase plane a map of the discrete states of the system at periodic instants $t = T, 2T, 3T, \ldots$ where T = period of the forcing action. This map, called Poincaré map, was introduced already at the turn of the century by Poincaré who studied the subsequent positions of an orbiting body at intersections with one plane.

To construct a Poincaré map of this or other systems, one needs to calculate the mapping $\mathbf{Y}_{i+1} = f(\mathbf{Y}_i)$ which yields the new intersection point $\mathbf{Y}_{i+1}$ in terms of the previous one, $\mathbf{Y}_i$ ($i = 1, 2, 3, \ldots$ = subscript denoting here subsequent discrete times) ($\mathbf{Y}$ = phase space vector of displacements $\mathbf{q}$ and their rates $\dot{\mathbf{q}}$). This mapping is generally nonlinear, exhibiting linear and quadratic terms. Quadratic mappings $f(\mathbf{Y}_i)$, also called iterates of quadratic polynomials, have been studied intensely. An important mapping is Hénon's mapping: $x_{i+1} = y_i + 1 - ax_i^2$, $y_{i+1} = bx_i$ (where $x_i = Y_i$, $y_i = \dot{Y}_i$), which may be shown to include transformations by folding, contracting, and rotating.

After the initial transients are damped out, the Poincaré maps of chaotic

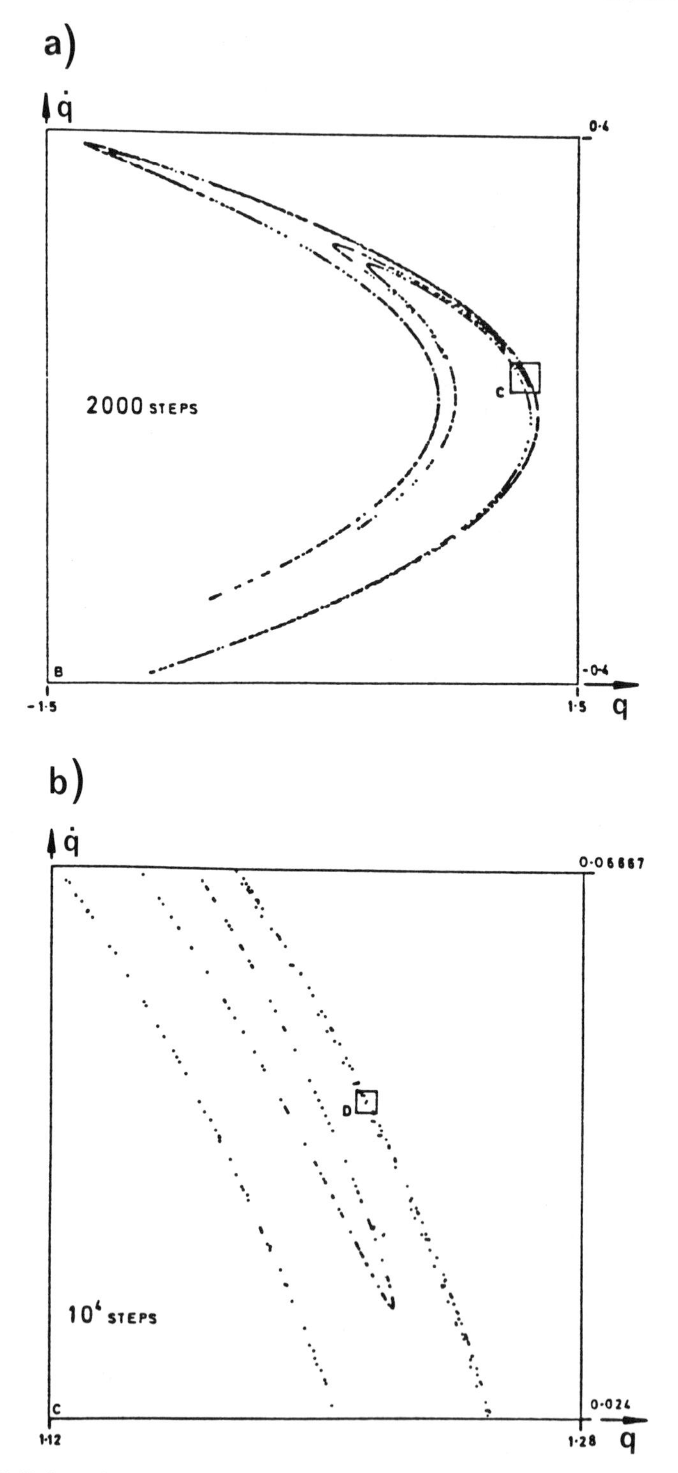

Figure 3.27 Poincaré maps of chaotic systems (*from Instabilities and Catastrophies in Science and Engineering, by J. M. T. Thompson, Copyright* 1982, *John Wiley & Sons. Reprinted by permission of John Wiley & Sons, Ltd.*).

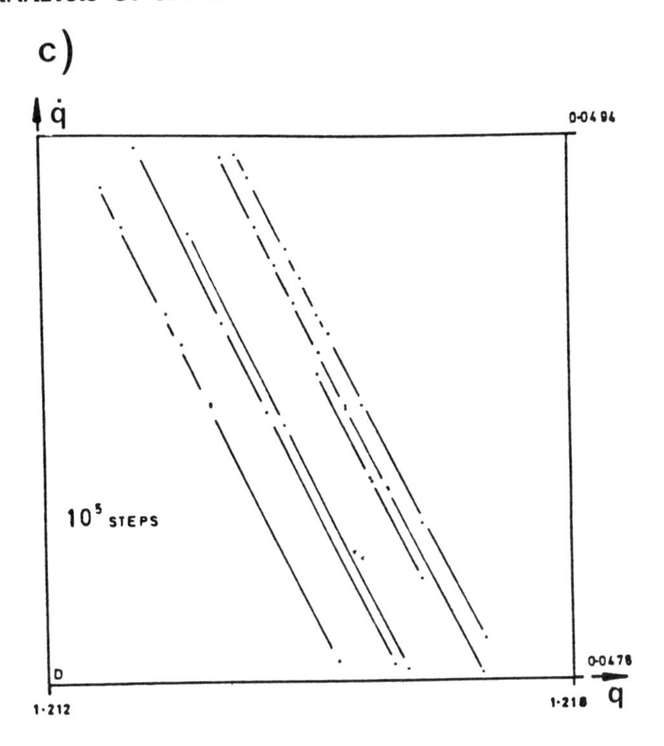

Figure 3.27 (cont.)

systems (or quadratic iterates) typically consist of discrete points all of which are located on regular and repetitive patterns. An example of such a pattern, obtained for Hénon mappings by Thompson (1982), is shown in Figure 3.27a, b, c.

The subsequent iterates do not lie next to each other but jump randomly over the patterns in this map. Part (b) of Figure 3.27 is an enlargement of the rectangular region C from part (a), and part (c) is an enlargement of the rectangular region D from part (b). Upon subsequent refinements, finer and finer self-similar patterns emerge, *ad infinitum*. This is called a fractal structure. (An example of such maps are Cantor sets.) The fractal structure does not have to consist just of lines as in Figure 3.27. Beautiful intricate multidimensional patterns, reproduced with self-similarity at subsequent refinements of scale, have been demonstrated for various chaotic systems or quadratic mappings.

Solutions that start at an arbitrary state may be attracted to a point attractor or a cycle attractor as illustrated for linear systems, or to a strange attractor, such as exemplified in Figure 3.27. There can be a multitude of attractors for a given nonlinear dynamic system, and then one may construct regions in the mapping space, called the attractor basins. Each basin includes all the starting states for which the solution is attracted to the same attractor.

To sum up, nonlinear deterministic systems under periodic excitation may exhibit chaotic response that appears to be random but in reality has a certain orderly structure. Determination of this structure is the objective of the theory of chaos.

Problems

3.9.1 Explain the meaning of a strange attractor.

3.9.2 Construct the Poincaré map for the steady-state motion of a single-degree-of-freedom oscillator such as a massless cantilever column with mass m on top, loaded on top by a constant vertical force, and periodic lateral force $f = f_0 \sin \omega t$ on top. Consider $P < P_{cr}$, and also discuss the cases $P = P_{cr}$ and $P > P_{cr}$. Finally, referring to Section 1.9, discuss the change in the governing equation due to large deflections.

3.9.3 For Hénon mapping with $a = 1.4$ and $b = 0.3$, the point $q = 0.63135448$ and $\dot{q} = 0.18940634$ was found to correspond to a state at which the initial transients are already damped out. Using your computer calculate the subsequent iterates and plot them on the Poincaré map. In this manner, a plot of the strange attractor is obtained (cf. Thompson, 1982, p. 146).

References and Bibliography

Ahmadian, M., and Inman, D. J. (1986), "Some Stability Results for General Linear Lumped-Parameter Dynamic Systems," *J. Appl. Mech.* (ASME), 53:10–13.

Akesson, B. A. (1976), "PFVIBAT—A Computer Program for Plane Frame Vibration Analysis by an Exact Method," *Int. J. Numer. Methods Eng.*, 10(6):1221–31 (Sec. 3.1).

Bahar, L. Y., and Law, G. E. (1981), "Dynamic Response by Means of Function of Matrices," *Comput. Struct.*, 14:173–78.

Bažant, Z. P. (1985), *Lectures on Stability of Structures*, Course 720–D24, Northwestern University, Evanston, Ill. (Sec. 3.3).

Beck, M. (1952), "Die Knicklast des einseitig eingespannten, tangential gedrückten Stabes," *Z. Angew. Math. Phys.*, 3:225–(Sec. 3.2).

Beliveau, J. G., Vaicaitis, R., and Shinozuka, M. (1977), "Motion of Suspension Bridge Subject to Wind Loads," *J. Struct. Eng.* (ASCE), 103(6):1189–1205.

Benjamin, T. B. (1961), "Dynamics of a System of Articulated Pipes Conveying Fluid, I. Theory, II. Experiment," *Proc. R. Soc., Lond.*, A261:457–86, 487–99.

Bisplinghoff, R. L., Ashley, H., and Halfman, R. L. (1955), *Aeroelasticity*, Addison Wesley, Reading, Mass. (Sec. 3.2).

Bolotin, V. V. (1963), *Non-conservative Problems of the Theory of Elastic Stability*, Pergamon Press, London.

Bolotin, V. V. (1964), *The Dynamic Stability of Elastic Systems*, Holden-Day, San Francisco.

Bowers, N. A. (1940), "Tacoma Narrows Bridge Wrecked by Wind," *Engineering News Record*, Nov. 14, pp. 647 and 656 (Sec. 3.2).

Brownjohn, J. M., Dumanoglu, A. A., Severn, R. T., and Taylor, C. A. (1987), "Ambient Vibration Measurements of the Humber Suspension Bridge and Comparison with Calculated Characteristics," *Proc. Instn. Civ. Engrs.*, Part 2, 83:561–600.

Bucher, C. G., and Lin, Y. K. (1988). "Stochastic Stability of Bridges Considering Coupled Modes," *J. Eng. Mech.* (ASCE), 114(12):2055–70.

Bucher, C. G., and Lin, Y. K. (1989), "Stochastic Stability of Bridges Considering Coupled Modes: II," *J. Eng. Mech.* (ASCE), 115(2):384–400.

Capron, M. D., and Williams, F. W. (1988), "Exact Dynamic Stiffnesses for an Axially Loaded Uniform Timoshenko Member Embedded in an Elastic Medium," *J. Sound and Vibration*, 124(3):435–66.

Cesari, L. (1971), *Asymptotic Behavior and Stability Problems in Ordinary Differential Equations*, 3rd ed., Springer, Berlin.

Farquarson, F. B. (1954), "Aerodynamic Stability of Suspension Bridges with Special Reference to the Tacoma Narrows Bridge," University of Washington Engineering Experimental Station, Bullettin No. 116 (Sec. 3.2).

Flügge, W. (1962), *Handbook of Engineering Mechanics,* McGraw-Hill, New York, chap. 62 (Sec. 3.5).

Fu, F. C. L., and Nemat-Nasser, S. (1970), "Stability of Dynamic Systems Subjected to Nonconservative Harmonic Forces," *AIAA Journal,* 8:1174–76 (Sec. 3.8).

Fung, Y. C. (1969), *An Introduction to the Theory of Aeroelasticity,* Dover Publications, Inc., New York.

Gantmacher, F. (1970), *Lectures in Analytical Mechanics,* MIR Publishers, Moscow (Sec. 3.4).

Guckenheimer, J., and Homes, P. (1983), *Nonlinear Oscillations, Dynamical Systems and Bifurcations of Vector Fields,* Springer-Verlag, New York (Sec. 3.9).

Gurgoze, M. (1985), "On the Dynamic Stability of a Pre-twisted Beam Subjected to a Pulsating Axial Load," *J. Sound and Vibration,* 102(3):415–22.

Halanay, A. (1966), *Differential Equations, Stability, Oscillations, Time Lags,* Academic Press, New York.

Harrison, H. B. (1973), *Computer Methods in Structural Analysis,* Prentice-Hall, Englewood Cliffs, N.J.

Hayashi, C. (1964), *Non-linear Oscillations in Physical Systems,* McGraw-Hill, New York.

Herrmann, G. (1967a), *Dynamic Stability of Structures,* Pergamon Press, New York.

Herrmann, G. (1967b), "Stability of Equilibrium of Elastic Systems Subjected to Non-conservative Forces," *Appl. Mech. Rev.,* 20:103–108 (Sec. 3.2).

Herrmann, G., and Jong, I. (1965), "On the Destabilizing Effect of Damping in Non-conservative Elastic Systems," *J. Appl. Mech.* (ASME), 32:592 (Sec. 3.2).

Herrmann, G., and Nemat-Nasser, S. (1966), "Energy Considerations in the Analysis of Stability of Nonconservative Structural Systems," in *Dynamic Stability of Structures,* Pergamon Press, New York, pp. 299–308 (Sec. 3.8).

Hill, J. L., and Davis, C. G. (1974), "The Effect of Initial Forces on the Hydroelastic Vibration and Stability of Planar Curved Tubes," *J. Appl. Mech.* (ASME), 41:355–9.

Holmes, P. J. (1978), "Pipes Supported at Both Ends Cannot Flutter," *J. Appl. Mech.* (ASME), 45:619–22.

Holmes, P. J. (1980), "Averaging and Chaotic Motions in Forced Oscillations," *J. Appl. Math.* (SIAM) 38(1):65–80.

Holmes, P. J., and Moon, F. C. (1983), "Strange Attractors and Chaos in Nonlinear Mechanics," *J. Appl. Mech.* (ASME), 50:1021–32.

Housner, J. M., and Knight, N. F., Jr., (1983), "The Dynamic Collapse of a Column Impacting a Rigid Surface," *AIAA Journal,* 21(8):1187–95.

Howson, W. P., and Williams, F. W. (1973), "Natural Frequencies of Frames with Axially Loaded Tomoshenko Members," *J. Sound and Vibration,* 26(4):503–15 (Sec. 3.1).

Huston, D. R. (1987), "Flutter Derivatives Extracted from Fourteen Generic Deck Sections," in *Bridges and Transmission Line Structures,* ed. by L. Tall, ASCE, pp. 281–91.

Irwin, P. A. (1987), "Wind Buffeting of Cable-Stayed Bridges During Construction," in *Bridges and Transmission Line Structures,* ed. by L. Tall, ASCE, pp. 164–77.

Kohli, A. K., and Nakra, B. C. (1984), "Vibration Analysis of Straight and Curved Tubes Conveying Fluids by Means of Straight Beam Finite Elements," *J. Sound Vibration,* 93:307–11.

Kamke, E. (1956), *Differentialgleichungen, Lösungsmethoden und Lösungen, 3. verbesserte Aufl.,* Akademische Verlagsgesellschaft, Leipzig, p. 61 (Sec. 3.6).

Irwin, P. A. (1987), "Wind Buffeting of Cable-Stayed Bridges During Construction," in *Bridges and Transmission Line Structures,* ed. by L. Tall, ASCE, pp. 164–77.

Kohli, A. K., and Nakra, B. C. (1984), "Vibration Analysis of Straight and Curved Tubes Conveying Fluids by Means of Straight Beam Finite Elements," *J. Sound Vibration,* 93:307–11.

Kamke, E. (1956), *Differentialgleichungen, Lösungsmethoden und Lösungen, 3. verbesserte Aufl.,* Akademische Verlagsgesellschaft, Leipzig, p. 61 (Sec. 3.6).

Kohoutek, R. (1984), "The Dynamic Instability of the Bar Under Tensile Periodic Force," *Trans. Canadian Society Mech. Engineers,* 8(1):1–5 (Sec. 3.3).

Kohoutek, R. (1985), "Analysis of Continuous Beams and Frames," Chap 4, in *Analysis and Design of Foundations for Vibrations,* ed. by P. J. Moore, A. A. Balkema, Rotterdam (Sec. 3.1).

Kounadis, A. N., and Belbas, S. (1977), "On the Parametric Resonance of Columns Carrying Concentrated Masses," *J. Struct. Mech.,* 5:383–94.

Krajcinovic, D., and Herrmann, G. (1968), "Parametric Resonance of Straight Bars Subjected to Repeated Impulsive Compression," *AIAA Journal,* 6:2025–27.

Lagrange, J. L. (1788), *Mécanique Analytique,* Courier, Paris (Sec. 3.6).

Landau, L., and Lifchitz, E. (1969), *Physique Theorique,* Tome I, MIR, Moscow.

La Salle, J., and Lefschetz, S. (1961), *Stability by Liapunov's Direct Method with Applications,* Academic Press, New York.

Leipholz, H. (1970), *Stability Theory,* Academic Press, New York. See also 2nd ed., John Wiley & Sons, Chichester (1987) (Sec. 3.6).

Leipholz, H. H. (1980), *Stability of Elastic Systems,* Sijthoff and Noordhoff, Alphen aan den Rijn.

Leonhardt, F. (1982), *Bridges,* Deutsche Verlag-Anstalt, Stuttgart (Sec. 3.2).

Liapunov, A. (1892), "The General Problem of Stability of Motion," Dissertation, Kharkov, French translation in *Ann. Fac. Sci. Univ. Toulouse,* 9(1907):203–475. Reprinted in *Ann. Math. Studies,* vol. 17, 1949, Princeton University Press, Princeton, N.J. (Sec. 3.5).

Liapunov, A. M. (1893), "Issledovanic odnogo iz ossobenykh sluchaev zadachi ob ustoichivosti dvizhenia," *Matem. Sbornik,* 17:253–333 (also Liapunov's 1982 dissertation).

Liapunov, A. M. (1966), *Stability of Motion,* Academic Press, New York (collected works of Liapunov).

Lin, Y. K. (1987), "Structural Stability in Turbulent Flow," in *Recent Trends in Aeroelasticity, Structures and Structural Dynamics,* ed. by P. Hajela, papers from the Professor R. L. Bisplinghoff Memorial Symposium, University of Florida, Feb. 6–7, 1986, Univ. of Florida Press, Gainesville.

Lin, Y. K., and Yang, J. N. (1983), "Multimode Bridge Response to Wind Excitations," *J. Eng. Mech.* (ASCE), 109(2):586–603.

Lo, D. L. C., and Masur, E. F. (1976), "Dynamic Buckling of Shallow Arches," *J. Eng. Mech. Div.,* 102:901–17.

Meirovitch, L., and Ghosh, D. (1987), "Control of Flutter in Bridges," *J. Eng. Mech.* (ASCE), 113(5):720–36 (Sec. 3.2).

Mettler, E. (1967), "Stability and Vibration Problems of Mechanical Systems under Harmonic Excitation," in *Dynamic Stability of Structures,* ed. by G. Herrmann, Pergamon Press, New York.

Mohsin, M. E., and Sadek, E. A. (1968), "The Distributed Mass-Stiffness Technique for the Dynamical Analysis of Complex Frameworks," *The Structural Engineer,* 46(11):345–51 (Sec. 3.1).

Moon, F. C. (1980), "Experiments on Chaotic Motions of a Forced Non-linear Oscillator: Strange Attractors," *J. Appl. Mech.* (ASME), 47:638–44.

Moon, F. C. (1986), *Chaotic Vibrations,* John Wiley and Sons, New York (Sec. 3.9).
Moon, F. C., and Holmes, P. J. (1979), "A Magnetoelastic Strange Attractor," *J. Sound and Vibration,* 65(2):275–96 (Sec. 3.9).
Moon, F. C., and Holmes, P. (1983), "Strange Attractors and Chaos in Nonlinear Mechanics," *J. Appl. Mech.* (ASME), 50:1021–32 (Sec. 3.9).
Moon, F. C., and Li, G. X. (1985), "Fractal Basin Boundaries and Homoclinic Orbits for Periodic Motion in a Two-Well Potential," *Phys. Rev. Letters,* 55(14):1439–42.
Movchan, A. A. (1956), "On the Oscillations of a Plate Moving in Gas" (in Russian), *Prikladnaya Matematika ì Mekhanika (PMM),* 20(2):211 (Sec. 3.8).
Nemat-Nasser, S. (1972), "On Elastic Stability Under Non-conservative Loads," in *Stability,* ed. by H. H. Leipholz, Study No. 6, University of Waterloo, Canada.
Nemat-Nasser, S., and Herrmann, G. (1966), "Some General Cosiderations Concerning the Destabilizing Effect in Non-conservative Systems," *Z. Angew. Math. Phys. (ZAMP),* 17:305–13.
Paidoussis, M. P. (1987), "Flow-Induced Instabilities of Cylindrical Structures," *Appl. Mech. Rev.,* 40(2):163–75.
Paidoussis, M. P., and Issid, N. T. (1974), "Dynamic Stability of Pipes Conveying Fluids," *J. Sound and Vibration,* 33:267–94.
Panovko, Y. G. (1971), *Elements of the Applied Theory of Elastic Vibration,* Mir Publ., Moscow.
Panovko, Y. G., and Gubanova, I. I. (1965), *Stability and Oscillations of Elastic Systems—Paradoxes, Fallacies and New Concepts,* Consultants Bureau, New York (transl. from Russian book publ. by Nauk. A., Moscow, 1964).
Poincaré, H. (1885), "Sur l'equilibre d'une masse fluide animée d'un mouvement de rotation," *Acta Math.,* 7:259 (Sec. 3.5).
Poincaré, H. (1892), *Méthodes nouvelles de la mécanique céleste,* vols. I, III, Gauthier-Villars & Cie, Paris (Sec. 3.5).
Raju, K. K., and Rao , G. V. (1986), "Free Vibration Behavior of Prestressed Beams," *J. Struct. Eng.* (ASCE), 112(2):433–57. See also discussion by Bažant and Cedolin, 113(9):2087 (Sec. 3.1).
Rayleigh, J. W. S. (1894), *The Theory of Sound,* vol. 1, Macmillan, New York; also Dover Publications, New York, 1945 (Sec. 3.3).
Rektorys, K. (1969), *Survey of Applicable Mathematics,* Iliffe Books, London, p. 827; see also Perron, *Math. Zeitschr.,* 1929, p. 129 (Sec. 3.6).
Richardson, A. S. (1988a), "Predicting Galloping Amplitudes: I," *J. Eng. Mech.* (ASCE), 114(4):716–23.
Richardson, A. S. (1988b), "Predicting Galloping Amplitudes: II," *J. Eng. Mech.* (ASCE), 114(11):1945–52.
Roth, V. N. (1964), "Instabilität durchstromter Rohre," *Ingenieur Archiv.,* 33(4):236–63.
Rousselet, R., and Herrmann, G. (1977), "Flutter of Articulated Pipes at Finite Amplitudes," *J. Appl. Mech.* (ASME), 44:154–58.
Scanlan, R. H. (1978a), "The Action of Flexible Bridges Under Wind: I. Flutter Theory," *J. Sound and Vibration,* 60(2):187–99.
Scanlan, R. H. (1978b), "The Action of Flexible Bridges Under Wind: II. Buffeting Theory," *J. Sound and Vibration,* 60(2):201–11.
Scanlan, R. H. (1981), "State-of-the-Art Methods for Calculating Flutter, Vortex-Induced, and Buffeting Response of Bridge Structures," Rep. No. FHWA/RD-80-050, Federal Highway Administration, Washington, D.C.
Scanlan, R. H. (1987), "Interpreting Aerolastic Models of Cable-Stayed Bridges," *J. Eng. Mech.* (ASCE), 113(4):555–75.
Scanlan, R. H. (1988), "On Flutter and Buffeting Mechanism in Long-Span Bridges," *Prob. Eng. Mech.,* 3(1):22–27.

Schuster, H. G. (1989), *Deterministic Chaos,* 2nd rev. ed., Physik-Verlag, Weinheim, (Sec. 3.9).

Shimegatsu, T., Hara, T., and Ohga, M. (1987), "Dynamic Stability Analysis by Matrix Function," *J. Eng. Mech.* (ASCE), 113(7):1085–1100. See also discussion, 115:881–83.

Simitses, G. J. (1983), "Effect of Static Preloading on the Dynamic Stability of Structures," *AIAA Journal,* 21:1174–80.

Simitses, G. J. (1984), "Suddenly-Loaded Structural Configurations," *J. Eng. Mech.* (ASCE), 110:1320–34.

Simitses, G. J. (1987), "Instability of Dynamically-Loaded Structures," *Appl. Mech. Rev.,* 40(10):1403–08.

Simiu, E., and Scanlan, R. H. (1986), *Wind Effects on Structures: An Introduction to Wind Engineering,* 2nd ed., John Wiley and Sons, New York (Sec. 3.2).

Stein, R. A., and Torbriner, M. W. (1970), "Vibrations of Pipes Containing Flowing Fluid," *J. Appl. Mech* (ASME), 37:906–16.

Stoker J. J. (1950), *Non-linear Vibrations in Mechanical Electrical Systems, II,* Interscience, London.

Sugiyama, Y. (1987), "Experiments on the Non-conservative Problems of Elastic Stability," 24th Annual Technical Meeting, Society of Engineering Science, University of Utah, Sept. 20–23 (Sec. 3.2).

Thompson, J. M. T. (1982), *Instabilities and Catastrophes in Science and Engineering,* John Wiley and Sons, Chichester–New York (Sec. 3.2).

Thompson, J. M. T. (1989), "Chaotic Dynamics and the Newtonian Legacy," *Appl. Mech. Rev.,* 42(1):15–25 (Sec. 3.9).

Thompson, J. M. T., and Stewart, H. B. (1986), *Nonlinear Dynamics and Chaos,* John Wiley and Sons, Chichester–New York (Sec. 3.9).

Timoshenko, S. P., and Gere, J. M. (1961), *Theory of Elastic Stability,* McGraw-Hill, New York (Sec. 3.2).

Vol'mir, A. S. (1963), *Stability of Elastic Systems,* (in Russian), Fizmatgiz, Moscow; English translation, Wright-Patterson Air Force Base, FTD-MT 64–335 (1964) and NASA, AD 628 508 (1965).

von Kármán, T. (1940), "Aerodynamic Stability of Suspension Bridges," *Engineering News Record,* 125(Nov.21):670 (Sec. 3.2).

Watson, L. T., and Wang, C. Y. (1983), "The Equilibrium States of a Heavy Rotating Column," *Int. J. Solids Structures,* 19(7):653–8 (Sec. 3.4).

Whittaker, E. T., and Watson, G. N. (1969), *A Course of Modern Analysis,* Cambridge University Press, Cambridge (Sec. 3.3).

Williams, F. E. (1979), "Consistent, Exact, Wind and Stability Calculations for Substitute Sway Frames with Cladding," *Proceedings of the Institution of Civil Engineers,* London, England, vol. 67, part. 2, pp. 355–67.

Williams, F. W. and Wittrik, W. H. (1983), "Exact Buckling and Frequency Calculations Surveyed," *J. Struct. Eng.,* (ASCE), 109(1):169–87.

Wittrick, W. H., and Williams, F. W. (1974), "Buckling and Vibration of Anisotropic or Isotropic Plate Assemblies under Combined Loadings," *Int. J. Mech. Sci.,* 16(4):209–39.

Ziegler, H. (1952), "Die Stabilitätskriterien der Elastomechanik," *Ingenieur Archiv,* 20:49–56 (Sec. 3.8).

Ziegler, H. (1953), "Linear Elastic Stability," *Z. Angew. Math. Phys. (ZAMP),* 4:89–121, 168–85.

Ziegler, H. (1956), "On the Concept of Elastic Stability," *Advn. Appl. Mech.,* 4:351–403.

Ziegler, H. (1968), *Principles of Structural Stability,* Blaisdell Publishing Company, Waltham, Mass. (Sec. 3.4).

4

Energy Methods

In this chapter we analyze conservative structural systems, which constitute the majority of applications in structural engineering practice. According to the Lagrange-Dirichlet theorem, which we proved in the preceding chapter, stability of these systems can be determined by energy methods. Static analysis, which we exclusively used in Chapters 1 and 2, represents a part of the energy approach, however a part that cannot answer the question of stability. Statics can only yield equilibrium states, which may be stable or unstable. Of course the question of failure of conservative and dissipative systems can be answered by the static method provided that imperfections are taken into account. However, solutions of perfect structures, for example, structures with no initial curvature, are generally simpler than the analysis of structures with imperfections, and the fact that energy analysis can decide stability of a perfect system represents one important advantage of the energy approach. Compared to the dynamic analysis, which serves as the fundamental test of stability, the static energy analysis generally brings about a great simplification.

First we will use the energy approach to analyze the stability of discretized elastic systems. Then we will study stability of equilibrium states of the structure after the first critical state is passed (postcritical behavior). We shall see that there exist diverse types of postcritical behavior, which leads to an important classification of conservative stability problems.

4.1 POSITIVE-DEFINITE MATRICES, EIGENVALUES, AND EIGENVECTORS

For the reader's convenience, this section reviews the basic results that we will need from the algebra of quadratic forms (e.g., Franklin, 1968, Rektorys, 1969; Korn and Korn, 1968; Courant and Hilbert, 1962; Zurmühl, 1958; Pearson, 1974; Smirnov, 1970; Hohn, 1958; Shilov, 1977; Gantmacher, 1959; Faddeev and Faddeeva, 1963). A general quadratic form with coefficients K_{ij} is written as

$$2\Pi = \sum_{i=1}^{n} \sum_{j=1}^{n} K_{ij} q_i q_j = \mathbf{q}^T \mathbf{K} \mathbf{q} \tag{4.1.1}$$

in which $\mathbf{q}$ = column matrix (vector) of the generalized displacements q_i, superscript T is used to denote the transpose of a matrix, $\mathbf{K}$ = matrix of the quadratic form = $(n \times n)$ square matrix with elements K_{ij}. In application to elastic structures, matrix $\mathbf{K}$ is real and represents the stiffness matrix of the structure with regard to its generalized displacements, and Π is the potential energy.

Matrix $\mathbf{K}$ is said to be nonsingular (or regular, invertible) if a matrix $\mathbf{C}$ exists such that $\mathbf{CK} = \mathbf{KC} = \mathbf{I}$, where $\mathbf{I}$ = identity matrix (that is, a matrix with 1's on the main diagonal, and 0's everywhere else). $\mathbf{C}$ is said to be the inverse of $\mathbf{K}$, and it is denoted by $\mathbf{K}^{-1}$. If $\mathbf{K}$ has no inverse, $\mathbf{K}$ is said to be singular. A matrix is singular if and only if $\det \mathbf{K} = 0$.

In a conservative system (see Sec. 3.2), Π must be equal to the work $\int \sum_i f_i \, dq_i$ done by the loads to reach the current state (cf. Eq. 3.2.20). This work expression is path-independent if and only if it represents a total differential, that is, $\partial f_i / \partial q_j = \partial f_j / \partial q_i$ (see also the integrability conditions in Eq. 3.2.21). Substituting the stiffness definition $K_{ij} = \partial f_i / \partial q_j$, we conclude that the potential energy exists if and only if

$$K_{ij} = \frac{\partial^2 \Pi}{\partial q_i \, \partial q_j} = \frac{\partial^2 \Pi}{\partial q_j \, \partial q_i} = K_{ji} \tag{4.1.2}$$

that is, the stiffness matrix is symmetric.

In Chapters 10 and 13 we will see that friction or damage can make the stiffness matrix $\mathbf{K}$ nonsymmetric. Then the potential energy does not exist and the value of the quadratic form is independent of the antisymmetric part of $\mathbf{K}$ because $\mathbf{q}^T\mathbf{K}\mathbf{q} = \mathbf{q}^T\hat{\mathbf{K}}\mathbf{q}$ where $\hat{\mathbf{K}} = \frac{1}{2}(\mathbf{K} + \mathbf{K}^T)$ = symmetric part of matrix $\mathbf{K}$. In this chapter we will consider only symmetric stiffness matrices.

Matrix $\mathbf{K}$ is said to be positive definite when the associated quadratic form Π is positive definite. A quadratic form is said to be positive definite when it has a strict minimum, that is, when $\Pi > 0$ for all $q_1, q_2, \ldots, q_n$ except $q_1 = q_2 = \cdots = q_n = 0$. An example is $\Pi = q_1^2 + 5q_2^2$. Another example is $\Pi = q_1^2 - 2q_1q_2 + 2q_2^2 = (q_1 - q_2)^2 + q_2^2$. A quadratic form is positive semidefinite when $\Pi \geq 0$ for all q_i while $\Pi = 0$ for some nonzero q_i. The simplest example is $\Pi = q_1^2$ provided that we deal with a two-dimensional space, $\Pi = \Pi(q_1, q_2)$. Another example is $\Pi = q_1^2 + 4q_1q_2 + 4q_2^2 = (q_1 + 2q_2)^2$, which vanishes for $q_1 = -2q_2$. A quadratic form is indefinite when $\Pi > 0$ for some q_i and $\Pi < 0$ for some other q_i (then of course there also exist nonzero q_i for which $\Pi = 0$, because $\Pi(q)$ is a continuous function). An example is $\Pi = q_1^2 - 3q_2^2$. Other examples are $\Pi = q_1^2 + 6q_1q_2 + 4q_2^2 = (q_1 + 3q_2)^2 - 5q_2^2$ both in the (q_1, q_2) and (q_1, q_2, q_3) spaces, and $\Pi = q_1^2 + q_2^2 + q_3^2 + 2q_1q_2 - 2q_1q_3 = (q_1 + q_2)^2 + (q_1 - q_3)^2 - q_1^2$. Π is negative definite when $-\Pi$ is positive definite, and it is negative semidefinite when $-\Pi$ is positive semidefinite. As one can check from the previous examples, the matrices of positive-semidefinite and negative-semidefinite quadratic forms are singular, while the matrices of positive-definite and negative-definite quadratic forms are regular (nonsingular). The matrices of indefinite forms can be regular or singular. Examples of quadratic surfaces $\Pi = \Pi(q_1, q_2)$ of various types are shown in Figure 4.1.

To decide whether $\Pi = q_1^2 - 4q_1q_2 + 5q_2^2$ is positive definite, one may realize that this can be written as $\Pi = y_1^2 + y_2^2$ where $y_1 = q_1 - 2q_2$ and $y_2 = q_2$; obviously, this expression is positive definite. On the other hand, $\Pi = q_1^2 - 4q_1q_2 + 3q_2^2$ is not positive definite because $\Pi = (q_1 - 2q_2)^2 - q_2^2$. So one can decide positive

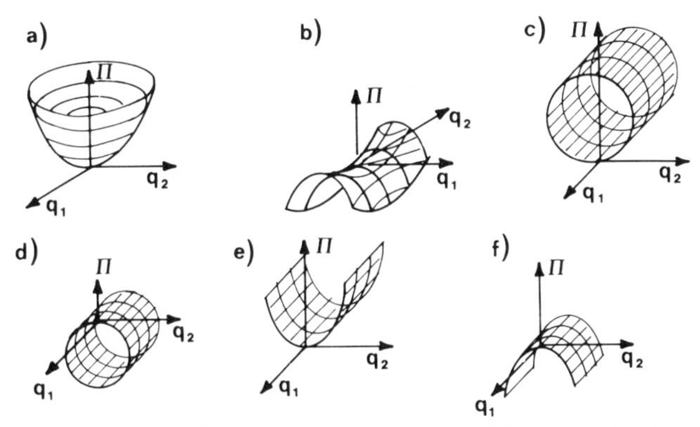

Figure 4.1 Potential-energy surfaces in two dimensions: (a) stable; (b–f) unstable.

definiteness by introducing new variables $y_i = \sum_j B_{ij} q_j$ or $\mathbf{y} = \mathbf{Bq}$. If $\mathbf{B}$ is a nonsingular matrix, the inverse $\mathbf{q} = \mathbf{Ty}$ where $\mathbf{T} = \mathbf{B}^{-1}$ exists; this is called a regular linear substitution (or linear transformation). The quadratic form transforms as

$$2\Pi = \mathbf{q}^T\mathbf{K}\mathbf{q} = (\mathbf{Ty})^T\mathbf{K}\mathbf{T}\mathbf{y} = \mathbf{y}^T\mathbf{K}^*\mathbf{y} \tag{4.1.3}$$

where

$$\mathbf{K}^* = \mathbf{T}^T\mathbf{K}\mathbf{T} \tag{4.1.4}$$

is the transformed matrix. Equation 4.1.3 leads to the following theorem:

Theorem 4.1.1 Matrix $\mathbf{K}$ is positive definite (or positive semidefinite, indefinite, singular), if and only if matrix $\mathbf{K}^*$ is.

Proof According to Equation 4.1.3, the quadratic forms of $\mathbf{K}$ and $\mathbf{K}^*$ are equal, and if $\mathbf{T}$ is nonsingular, then for every $\mathbf{q}$ one can find a corresponding $\mathbf{y}$, and for every $\mathbf{y}$ one can find a corresponding $\mathbf{q}$. So, if among $\mathbf{K}$ and $\mathbf{K}^*$ one is always positive, the other must be, too.

Theorem 4.1.1 permits us to choose any independent linear combinations of displacements as the independent variables in stability analysis.

Positive definiteness of a matrix can be decided by its *eigenvalues* (also called *proper values* or *characteristic values*). The homogeneous equation $(\mathbf{K} - \lambda\mathbf{I})\mathbf{q} = \mathbf{0}$ has a nonzero solution if and only if $\det(\mathbf{K} - \lambda\mathbf{I}) = 0$. The roots of this equation are the eigenvalues $\lambda_1, \lambda_2, \ldots, \lambda_n$ of matrix $\mathbf{K}$. Normally we number them so that $\lambda_1 \leq \lambda_2 \leq \cdots \leq \lambda_n$. In general, $\mathbf{q}$ and λ may be complex numbers. Since the determinant is an nth-degree polynomial in λ with roots λ_r, it must have the form $\det(\mathbf{K} - \lambda\mathbf{I}) = (\lambda_1 - \lambda)(\lambda_2 - \lambda) \cdots (\lambda_n - \lambda)$. Substituting in this equation $\lambda = 0$, we get Vieta's rule:

$$\det \mathbf{K} = \lambda_1 \lambda_2 \cdots \lambda_n \tag{4.1.5}$$

and comparing the coefficients at λ^{n-1} on both sides of the equation, we also get the relation $\lambda_1 + \lambda_2 + \cdots + \lambda_n = K_{11} + K_{22} + \cdots + K_{nn} = \operatorname{tr} \mathbf{K}$. From Equation 4.1.5 it follows that $\mathbf{K}$ is singular if and only if at least one eigenvalue is zero.

Theorem 4.1.2 All eigenvalues of a real symmetric matrix are real.

Theorem 4.1.3 A real symmetric matrix is positive definite (or positive semidefinite, indefinite) if and only if all its eigenvalues are positive (or nonnegative, or of different signs).

Proof If $\mathbf{K}$ is real symmetric positive definite, then $\mathbf{q}^T\mathbf{K}\mathbf{q} > 0$ for any vector $\mathbf{q}$. If $\mathbf{q}$ is an eigenvector, we may substitute $\mathbf{K}\mathbf{q} = \lambda\mathbf{q}$, which yields $\mathbf{q}^T(\lambda\mathbf{q}) > 0$ or $\lambda(\mathbf{q}^T\mathbf{q}) > 0$. Because $\mathbf{q}^T\mathbf{q} = |q_1|^2 + |q_2|^2 + \cdots + |q_n|^2 > 0$ (which is valid even if $\mathbf{q}$ is complex), we conclude that $\lambda > 0$. This proves that the condition in Theorem 4.1.3 is necessary. To prove that it is also sufficient, consider any vector $\mathbf{q}$, which may be represented as a linear combination of all eigenvectors $\mathbf{q}^{(i)}$, that is, $\mathbf{q} = c_1\mathbf{q}^{(1)} + \cdots + c_n\mathbf{q}^{(n)}$ (because $\mathbf{q}^{(i)}$ are linearly independent, cf. Theorem 4.1.4). Multiplying by $\mathbf{K}$, we get $\mathbf{K}\mathbf{q} = c_1\mathbf{K}\mathbf{q}^{(1)} + \cdots + c_n\mathbf{K}\mathbf{q}^{(n)} = c_1\lambda_1\mathbf{q}^{(1)} + \cdots + c_n\lambda_n\mathbf{q}^{(n)}$, and $\mathbf{q}^T\mathbf{K}\mathbf{q} = (c_1\mathbf{q}^{(1)^T} + \cdots + c_n\mathbf{q}^{(n)^T})(c_1\lambda_1\mathbf{q}^{(1)} + \cdots + c_n\lambda_n\mathbf{q}^{(n)}) = c_1^2\lambda_1\mathbf{q}^{(1)^T}\mathbf{q}^{(1)} + \cdots + c_n^2\lambda_n\mathbf{q}^{(n)^T}\mathbf{q}^{(n)}$ because $\mathbf{q}^{(i)^T}\mathbf{q}^{(j)} = 0$ when $i \neq j$ (according to Theorem 4.1.4). As $\mathbf{q}^{(i)^T}\mathbf{q}^{(i)} > 0$ for all i, positiveness of all λ_i suffices for the last sum to be positive.

Corollary From Theorem 4.1.3 and Equation 4.1.5 we know that $\det \mathbf{K} > 0$ is a necessary condition of positive definiteness. But it is not sufficient.

As revealed by Equations 2.3.2 and 2.3.5, the stiffness matrix of an elastic structure linearized with respect to load parameter λ has the form

$$\mathbf{K} = \mathbf{M} - \lambda\mathbf{N} \tag{4.1.6}$$

where $\mathbf{M}$ and $\mathbf{N}$ are constant square real symmetric nonsingular matrices. The (generalized) eigenvalues $\lambda = \lambda_r$ $(r = 1, \ldots, n)$ are those for which the equation $\mathbf{K}\mathbf{q} = 0$ has a nonzero solution. The (generalized) eigenvector $\mathbf{q}^{(r)}$ associated with λ_r is the nonzero solution of $\mathbf{K}^{(r)}\mathbf{q}^{(r)} = 0$ where $\mathbf{K}^{(r)} = \mathbf{K}(\lambda_r) = \mathbf{M} - \lambda_r\mathbf{N}$. The generalized linear eigenvalue problem $(\mathbf{M} - \lambda\mathbf{N})\mathbf{q} = 0$ can be transformed to the standard eigenvalue problem by multiplying this equation by $\mathbf{N}^{-1}$ from the left: this yields $(\mathbf{D} - \lambda\mathbf{I})\mathbf{q} = 0$ where $\mathbf{D} = \mathbf{N}^{-1}\mathbf{M}$. Obviously the eigenvalues λ remain the same, and so do the eigenvectors. [An alternative conversion to a standard eigenvalue problem is to multiply by $\mathbf{M}^{-1}$, which yields $(\mathbf{M}^{-1}\mathbf{N} - \mu\mathbf{I})\mathbf{q} = 0$ where $\mu = 1/\lambda$.]

Theorem 4.1.4 The eigenvectors corresponding to two different eigenvalues are mutually orthogonal.

Proof The eigenvectors $\mathbf{q}^{(r)}$ and $\mathbf{q}^{(s)}$ of the generalized eigenvalue problem are the nonzero solutions of the equations $(\mathbf{M} - \lambda_r\mathbf{N})\mathbf{q}^{(r)} = 0$ and $(\mathbf{M} - \lambda_s\mathbf{N})\mathbf{q}^{(s)} = 0$. We multiply these equations from the left by $\mathbf{q}^{(s)T}\mathbf{N}^{-1}$ and $\mathbf{q}^{(r)T}\mathbf{N}^{-1}$, respectively, and subtracting them we get $\mathbf{q}^{(s)T}\mathbf{D}\mathbf{q}^{(r)} - \mathbf{q}^{(r)T}\mathbf{D}\mathbf{q}^{(s)} = \lambda_r\mathbf{q}^{(r)T}\mathbf{q}^{(s)} - \lambda_s\mathbf{q}^{(s)T}\mathbf{q}^{(r)}$ where $\mathbf{D} = \mathbf{N}^{-1}\mathbf{M}$. Both terms on the left-hand side are equal because $\mathbf{q}^{(s)T}\mathbf{D}\mathbf{q}^{(r)} = (\mathbf{q}^{(s)T}\mathbf{D}\mathbf{q}^{(r)})^T = \mathbf{q}^{(r)T}\mathbf{D}^T\mathbf{q}^{(s)}$ and $\mathbf{D}^T = \mathbf{D}$. Also $\mathbf{q}^{(r)T}\mathbf{q}^{(s)} = \mathbf{q}^{(s)T}\mathbf{q}^{(r)}$. So we conclude that $(\lambda_r - \lambda_s)\mathbf{q}^{(r)T}\mathbf{q}^{(s)} = 0$. Thus, if $\lambda_r - \lambda_s \neq 0$, we must have $\mathbf{q}^{(r)T}\mathbf{q}^{(s)} = 0$, that is, vectors $\mathbf{q}^{(r)}$ and $\mathbf{q}^{(s)}$ are orthogonal. (Note that this is true not only for the standard eigenvalue problem, but also for the generalized one.)

Consider now the linear transformation $\mathbf{q} = \mathbf{Q}\mathbf{y}$ (that is, $\mathbf{T} = \mathbf{Q}$) such that $Q_{ir} = q_i^{(r)}$, that is $\mathbf{Q} = [\mathbf{q}^{(1)}, \ldots, \mathbf{q}^{(n)}] =$ square matrix whose columns are the eigenvectors $\mathbf{q}^{(r)}$ represented as column matrices. The eigenvectors satisfy the equations $\mathbf{K}^{(r)}\mathbf{q}^{(r)} = \mathbf{0}$ in which the right-hand side is a zero column matrix. Combining similar equations for all the eigenvectors, we have $[\mathbf{K}^{(1)}\mathbf{q}^{(1)}, \ldots, \mathbf{K}^{(n)}\mathbf{q}^{(n)}] = [\mathbf{0}, \ldots, \mathbf{0}]$, and substituting $\mathbf{K}^{(r)} = \mathbf{M} - \lambda_r\mathbf{N}$ we get

$$[\mathbf{M}\mathbf{q}^{(1)}, \ldots, \mathbf{M}\mathbf{q}^{(n)}] = [\lambda_1\mathbf{N}\mathbf{q}^{(1)}, \ldots, \lambda_n\mathbf{N}\mathbf{q}^{(n)}] \tag{4.1.7}$$

Multiplying this equation from the left by $\mathbf{Q}^T\mathbf{N}^{-1}$, we get

$$\mathbf{Q}^T\mathbf{N}^{-1}\mathbf{M}\mathbf{Q} = [\mathbf{q}^{(1)}, \ldots, \mathbf{q}^{(n)}]^T[\lambda_1\mathbf{q}^{(1)}, \ldots, \lambda_n\mathbf{q}^{(n)}] \tag{4.1.8}$$

Now we may note that $\mathbf{q}^{(r)T}\mathbf{q}^{(s)} = 0$ if $\lambda_r \neq \lambda_s$ and 1 if $\lambda_r = \lambda_s$ (here we assume the eigenvectors to be normalized). It follows that

$$\mathbf{Q}^T\mathbf{D}\mathbf{Q} = \begin{bmatrix} \lambda_1 & 0 & \cdots & 0 \\ 0 & \lambda_2 & \cdots & 0 \\ 0 & 0 & \cdots & \lambda_n \end{bmatrix} \qquad \mathbf{D} = \mathbf{N}^{-1}\mathbf{M} \tag{4.1.9}$$

For the standard eigenvalue problem we have $\mathbf{N} = \mathbf{N}^{-1} = \mathbf{I}$, $\mathbf{M} = \mathbf{D} = \mathbf{K}$, and $\mathbf{Q}^T\mathbf{D}\mathbf{Q} = \mathbf{Q}^T\mathbf{K}\mathbf{Q} = \mathbf{K}^*$ (Eq. 4.1.4), and thus the following theorem ensues:

Theorem 4.1.5 The linear substitution $\mathbf{q} = \mathbf{Q}\mathbf{y}$ in which the columns of matrix $\mathbf{Q}$ are the normalized eigenvectors of $\mathbf{K}$ transforms a real symmetric quadratic form to

$$2\Pi = \mathbf{q}^T\mathbf{K}\mathbf{q} = \lambda_1 y_1^2 + \lambda_2 y_2^2 + \cdots + \lambda_n y_n^2 \tag{4.1.10}$$

Corollary For the generalized eigenvalue problem ($\mathbf{N} \neq \mathbf{I}$), the matrix $\mathbf{Q}^T\mathbf{K}\mathbf{Q}$ is not diagonal.

Note that Equation 4.1.10 can also be used as an alternative proof of Theorem 4.1.3.

Furthermore, one can introduce new variables z_i by substituting $y_i = z_i/\sqrt{|\lambda_i|}$ if $\lambda_i \neq 0$ and $y_i = z_i$ if $\lambda_i = 0$. This leads to the following theorem:

Theorem 4.1.6 Any real symmetric quadratic form can be transformed to the form

$$2\Pi = -z_1^2 - \cdots - z_k^2 + z_m^2 + \cdots + z_n^2 \tag{4.1.11}$$

(called the canonical form) where $0 \leq k \leq m \leq n$.

The transformation $\mathbf{q} = \mathbf{T}\mathbf{z}$, which leads from Equation 4.1.1 to Equation 4.1.11, is not unique, but the numbers of plus and minus signs in Equation 4.1.11 are (Sylvester's law of inertia).

It may be useful to note a subtle difference between (1) looking for the (generalized) eigenvalues and (2) expressing Π as a sum of squares. In the former problem we look for the eigenvalues λ_r for which the equation $\mathbf{K}\mathbf{q} = 0$ or $(\mathbf{M} - \lambda\mathbf{N})\mathbf{q} = 0$ admits nonzero $\mathbf{q}$. In the latter problem, Π is to be expressed for a given load parameter $\lambda = \lambda_0$; to this end, we need to find the eigenvalues $\lambda_0 = \lambda_{0_r}$ and eigenvectors $\mathbf{q} = \mathbf{q}_0^{(r)}$ for which the equation $(\mathbf{K}_0 - \lambda_0\mathbf{I})\mathbf{q} = 0$ with $\mathbf{K}_0 = \mathbf{M} - \lambda_0\mathbf{N}$ admits nonzero $\mathbf{q}$. Then $2\Pi = \sum_r \lambda_{0_r} y_r^2$ in which $\mathbf{y} = \mathbf{Q}_0^{-1}\mathbf{q}$ and $\mathbf{Q}_0$ is the matrix of eigenvectors $\mathbf{q}_0^{(r)}$ of matrix $\mathbf{K}_0$. For the special case where $\mathbf{N} = \mathbf{I}$, we have $\lambda_{0_r} = \lambda_r - \lambda_0$, but not for general $\mathbf{N}$.

The magnitudes of vectors $\mathbf{y}$ and $\mathbf{q} = \mathbf{R}\mathbf{y}$ are the same if and only if $\mathbf{R}^T = \mathbf{R}^{-1}$, because $\mathbf{q}^T\mathbf{q} = (\mathbf{R}\mathbf{y})^T(\mathbf{R}\mathbf{y}) = \mathbf{y}^T\mathbf{R}^T\mathbf{R}\mathbf{y} = \mathbf{y}^T\mathbf{R}^{-1}\mathbf{R}\mathbf{y} = \mathbf{y}^T\mathbf{I}\mathbf{y} = \mathbf{y}^T\mathbf{y}$. In that case matrix $\mathbf{R}$ is said to be unitary and the transformation orthogonal (it can be geometrically interpreted as a rotation in an n-dimensional space). The transformed matrix $\mathbf{K}' = \mathbf{R}^{-1}\mathbf{K}\mathbf{R}$ is said to be similar to $\mathbf{K}$.

Theorem 4.1.7 Similar matrices have the same eigenvalues, that is, the eigenvalues are invariant during orthogonal transformations (rotations).

Proof Let λ be an eigenvalue of $\mathbf{K}'$, that is $\mathbf{K}'\mathbf{y} = \lambda\mathbf{y}$ or $\mathbf{R}^{-1}\mathbf{KRy} = \lambda\mathbf{y}$. Multiply this by $\mathbf{R}$ from the left: $(\mathbf{RR}^{-1})\mathbf{K}(\mathbf{Ry}) = \lambda(\mathbf{Ry})$. Denoting $\mathbf{Ry} = \mathbf{q}$, we then have $\mathbf{Kq} = \lambda\mathbf{q}$, which means that if λ is an eigenvalue of $\mathbf{K}'$ it is also an eigenvalue of $\mathbf{K}$. The reverse holds true as well, since the foregoing procedure can be reversed.

For students of mechanics of materials it may be interesting to note that invariance of the principal stresses during coordinate rotations, and the fact that their values are real, are special cases of Theorems 4.1.7 and 4.1.2.

Corollary The determinant and trace of a matrix, or any other function of the eigenvalues, are invariant.

Theorem 4.1.8 The inverse of a positive-definite real symmetric matrix is also positive definite.

Proof If a real symmetric matrix $\mathbf{K}$ is positive definite, then every eigenvalue λ satisfying the equation $(\mathbf{K} - \lambda\mathbf{I})\mathbf{q} = 0$ is positive. Also, the inverse matrix $\mathbf{K}^{-1}$ must exist. Premultiplying this equation by $-\lambda^{-1}\mathbf{K}^{-1}$ from the left, we get $(-\lambda^{-1}\mathbf{K}^{-1}\mathbf{K} + \mathbf{K}^{-1}\mathbf{I})\mathbf{q} = 0$, and noting that $\mathbf{K}^{-1}\mathbf{K} = \mathbf{I}$, $\mathbf{K}^{-1}\mathbf{I} = \mathbf{K}^{-1}$, we have $(\mathbf{K}^{-1} - \lambda^{-1}\mathbf{I})\mathbf{q} = 0$. Now, from the fact that all the possible values of λ^{-1} are positive, it follows $\mathbf{K}^{-1}$ is positive definite.

Corollary If λ is an eigenvalue of stiffness matrix $\mathbf{K}$, then λ^{-1} is an eigenvalue of the corresponding compliance matrix $\mathbf{C} = \mathbf{K}^{-1}$ provided that $\mathbf{K}$ is nonsingular. Thus, if $\det \mathbf{K} \to 0$ $(\lambda \to 0)$, then $\det \mathbf{C} \to \infty$ (because $1/\lambda \to \infty$).

From the relation $(\mathbf{K}^{-1} - \lambda^{-1}\mathbf{I})\mathbf{q} = 0$ and Eq. 4.1.5 it further follows that $\det \mathbf{K}^{-1} = \lambda_1^{-1}\lambda_2^{-1} \cdots \lambda_n^{-1} = 1/(\lambda_1\lambda_2 \cdots \lambda_n)$, that is, $\det \mathbf{K}^{-1} = 1/\det \mathbf{K}$.

The consequence for structural mechanics is that if a quadratic form describing the potential energy is positive definite, so is the quadratic form for the associated complementary energy. Or, if the stiffness matrix is positive definite, so is the associated flexibility matrix (Hohn, 1958, p. 257).

Theorem 4.1.9 If both $\mathbf{A}$ and $\mathbf{K}$ are positive definite, so are $\mathbf{AK}$ and $\mathbf{A}^{-1}\mathbf{K}$.

Proof If $\mathbf{K}$ is positive definite, then all the eigenvalues λ for which the equation $\mathbf{Kq} = \lambda\mathbf{q}$ has a nonzero solution are positive. Multiplying this equation by $\mathbf{q}^T\mathbf{A}$ from the left we get $\mathbf{q}^T(\mathbf{AK})\mathbf{q} = \mathbf{q}^T\mathbf{A}\lambda\mathbf{q} = \lambda(\mathbf{q}^T\mathbf{Aq})$. Here $\lambda > 0$ and also, if $\mathbf{A}$ is positive definite, $\mathbf{q}^T\mathbf{Aq} > 0$ for any $\mathbf{q} \neq 0$. Therefore $\mathbf{q}^T(\mathbf{AK})\mathbf{q} > 0$ for any $\mathbf{q} \neq 0$. Furthermore, according to Theorem 4.1.8, $\mathbf{A}^{-1}$ is also positive definite, and replacing $\mathbf{A}$ by $\mathbf{A}^{-1}$ in the foregoing argument we find that $\mathbf{q}^T(\mathbf{A}^{-1}\mathbf{K})\mathbf{q} > 0$ for any $\mathbf{q} \neq 0$.

A convenient test of positive definiteness is given by the following theorem:

Theorem 4.1.10 Sylvester's Criterion. A real symmetric matrix K_{ij} is positive definite if and only if all its principal minors are positive, that is,

$$D_1 = K_{11} > 0 \qquad D_2 = \begin{bmatrix} K_{11} & K_{12} \\ K_{21} & K_{22} \end{bmatrix} > 0 \qquad \cdots \qquad D_n = \begin{bmatrix} K_{11} & \cdots & K_{1n} \\ & \cdots & \\ K_{n1} & \cdots & K_{nn} \end{bmatrix} > 0 \tag{4.1.12}$$

Corollary If $\mathbf{K}$ is positive definite, so is any submatrix with coincident diagonal.

This means that if the structure is stable (i.e., its potential energy has a strict minimum) and we fix some of the displacements, then the structure remains stable (i.e., its potential energy will still have a strict minimum; see Sec. 4.2). This observation is, of course, clear on physical grounds. It can in fact be used to prove Theorem 4.1.10 physically (Bažant, 1985), as follows:

Proof of Theorem 4.1.10 We imagine that we fix $q_n = 0$. The stiffness matrix of the structure is thereby reduced from $n \times n$ to $(n-1) \times (n-1)$, and its determinant changes from D_n to D_{n-1}. Now, because the structure must remain stable (i.e., cannot be destabilized), if we fix one displacement we must have $D_{n-1} > 0$ on the basis of Equation 4.1.5 and Theorem 4.1.3. (Conversely, if $D_{n-1} \leq 0$, the structure with q_n fixed is unstable.) Furthermore, if we fix also $q_{n-1} = 0$, by the same physical argument we conclude that $D_{n-2} > 0$, etc., until we get $D_1 > 0$, which completes the physical proof of Theorem 4.1.10. (A purely mathematical version of this proof, without mechanical interpretation, is also possible.)

Now the algebraic proof. That $D_n > 0$ is a necessary condition we already know. For $n = 1$, Theorem 4.1.10 is obvious. For $n = 2$, the proof is easy. We have $2\Pi = K_{11}q_1^2 + 2K_{12}q_1q_2 + K_{22}q_2^2$, which can be transformed to

$$2\Pi = K_{11}\left(q_1 + \frac{K_{12}}{K_{11}} q_2\right)^2 + \frac{K_{11}K_{22} - K_{12}^2}{K_{11}} q_2^2 \tag{4.1.13}$$

Obviously, Π is positive definite if and only if $K_{11} > 0$ and $K_{11}K_{22} - K_{12}^2 > 0$.

It is also easy to verify the fact that K_{ij} cannot be positive definite if any of the principal minors D_k is zero. Let $y_1, y_2, \ldots, y_k$ be one eigenvector of the minor D_k $(k \leq n)$, and set $q_1 = y_1, q_2 = y_2, \ldots, q_k = y_k, q_{k+1} = q_{k+2} = \cdots = q_n = 0$. If the determinant D_k vanishes, then $\sum_{j=1}^{k} K_{ij}q_j = 0$ for $i = 1, 2, \ldots, k$ and $\sum_{j=1}^{n} K_{ij}q_j = 0$ for $i = 1, 2, \ldots, n$. Hence $2\Pi = \sum_{i=1}^{n} \sum_{j=1}^{n} K_{ij}q_iq_j = \sum_{i=1}^{n} q_i(\sum_{j=1}^{n} K_{ij}q_j) = 0$. This means that K_{ij} is not positive definite.

For $n > 2$, the algebraic proof of Theorem 4.1.10 is more involved and can be made on the basis of Jacobi's theorem (Zurmühl, 1958) or by induction on the basis of Courant's minimax principle (Franklin, 1968, p. 152). In a more elementary way Theorem 4.1.10 for a general n can be algebraically proven as follows (e.g., Simitses, 1976, pp. 11–12). Introduce a new variable:

$$y_1 = q_1 + \sum_{i=2}^{n} \frac{K_{1i}}{K_{11}} q_i = q_1 + \sum_{i=2}^{n} \frac{K_{1i}^{(1)}}{K_{11}^{(1)}} q_i \qquad \text{with } K_{ij}^{(1)} = K_{ij} \tag{4.1.14}$$

Then it is easy to check that

$$\Pi = \frac{1}{2} K_{11}^{(1)} y_1^2 + \frac{1}{2} \sum_{i=2}^{n} \sum_{j=2}^{n} K_{ij}^{(2)} q_i q_j \tag{4.1.15}$$

in which

$$K_{ij}^{(2)} = \frac{K_{11}^{(1)} K_{ij}^{(1)} - K_{1i}^{(1)} K_{1j}^{(1)}}{K_{11}^{(1)}} \tag{4.1.16}$$

Now, by choosing $q_2 = 1$ and $y_1 = q_3 = \cdots = q_n = 0$, we have $\Pi = K_{22}^{(2)}/2$, which must be positive. So the numerator of Equation 4.1.16 must be positive (since $K_{11}^{(1)} > 0$). This means that $D_2 = K_{11}K_{22} - K_{12}^2 > 0$, that is, the second principal minor D_2 of the $n \times n$ matrix K_{ij} must be positive. Subsequently, one may introduce consecutive substitutions

$$y_k = q_k + \sum_{i=k+1}^{n} q_i(K_{ki}^{(k)}/K_{kk}^{(k)}) \tag{4.1.17}$$

obtaining

$$\Pi = \frac{1}{2} K_{11}^{(1)} y_1^2 + \cdots + \frac{1}{2} K_{kk}^{(k)} y_k^2 + \frac{1}{2} \sum_{i=k+1}^{n} \sum_{j=k+1}^{n} K_{ij}^{(k+1)} q_i q_j \tag{4.1.18}$$

with

$$K_{ij}^{(k)} = \frac{K_{11}^{(k-1)} K_{ij}^{(k-1)} - K_{1i}^{(k-1)} K_{1j}^{(k-1)}}{K_{11}^{(k-1)}} \tag{4.1.19}$$

and, in the same manner as before, one obtains

$$D_k = K_{11}^{(k+1)} K_{22}^{(k+1)} - K_{12}^{(k+1)^2} > 0 \tag{4.1.20}$$

D_k defined in this manner represents the kth principal minor since it consists of a sum of products of k elements of a $k \times k$ matrix such that each product has one and only one element from each row and each column, and involves all possible such terms with the signs corresponding to the definition of a determinant.

Theorem 4.1.11 For symmetric matrices $\mathbf{K}$, the number of positive (or negative) pivots is equal to the number of positive (or negative) eigenvalues (see, e.g., Strang, 1980, 1986).

Problems

4.1.1 Is the form $3q_1^2 - 12q_1q_2 + 9q_2^2$ positive definite? (see also Problem 3.6.3).

4.1.2 (a) Determine for which P-values the structure with the stiffness matrix $K_{11} = 2 - P$, $K_{22} = 3 - P$, $K_{12} = K_{21} = \sqrt{2}$ is stable. (b) Do the same for $K_{11} = 9 - P$, $K_{22} = 6 - P$, $K_{12} = K_{21} = 3\sqrt{2}$.

4.1.3 (a) Prove that all the diagonal elements K_{ii} of a positive-definite matrix $\mathbf{K}$ are positive. *Hint:* Consider the sign of Π for the case when $q_i = 1$ with all other $q_j = 0$. (b) Prove Theorems 4.1.2 and 4.1.3.

4.1.4 A nonsymmetric matrix $\mathbf{K}$ can be decomposed as $\mathbf{K} = \hat{\mathbf{K}} + \mathbf{A}$ where matrix $\hat{\mathbf{K}}$ is symmetric ($\hat{\mathbf{K}}^T = \hat{\mathbf{K}}$) and matrix $\mathbf{A}$ is antisymmetric ($\mathbf{A}^T = -\mathbf{A}$). Noting that $\mathbf{q}^T\mathbf{A}\mathbf{q} = \frac{1}{2}[\mathbf{q}^T\mathbf{A}\mathbf{q} + (\mathbf{q}^T\mathbf{A}\mathbf{q})^T]$, prove that $\mathbf{q}^T\mathbf{A}\mathbf{q}$ is always zero. This, of course, means that $\mathbf{q}^T\mathbf{K}\mathbf{q} = \mathbf{q}^T\hat{\mathbf{K}}\mathbf{q}$, as already said.

4.1.5 For systems with friction or damage, the stiffness matrix $\mathbf{K}$ can be nonsymmetric (see Sec. 10.4 and Chap. 13). As an example, consider a system with two displacements q_1, q_2 such that $K_{11} = 2 - P$, $K_{12} = 3$, $K_{21} = 1$, $K_{22} = 4$. Let $\hat{\mathbf{K}}$ be the symmetric part of $\mathbf{K}$. Show that the state of neutral equilibrium (for which there exists an eigenvector q_j satisfying the equations $f_i = \sum_j K_{ij} q_j = 0$, $i = 1, 2$), occurs for $P_{cr} = \frac{5}{4}$, while the critical state of stability limit (for which $\hat{\mathbf{K}}$, rather then $\mathbf{K}$, becomes singular) occurs for $\hat{P}_{cr} = 1$, which is less than P_{cr}. Further show that the eigenvectors of $\mathbf{K}$ and $\hat{\mathbf{K}}$, corresponding to P_{cr} and $\hat{P}_{cr}$, respectively, are (4, −1) and (2, −1) (they differ from each other). Then, noting that the energy function $2\hat{\Pi} = \sum f_i q_i = \sum_i \sum_j K_{ij} q_i q_j = \sum_i \sum_j \hat{K}_{ij} q_i q_j = (q_1 + 2q_2)^2 + (1 - P)q_1^2$ (representing the work done on the system but not the potential energy), find one nonzero vector (q_1, q_2) such that $2\hat{\Pi} < 0$ at $P = \frac{9}{8}$ (which is smaller than P_{cr}). Finally verify that for $P = \hat{P}_{cr} = 1$, one can find a nonzero vector (q_1, q_2) (eigenvector of $\hat{\mathbf{K}}$) that is orthogonal to the corresponding force vector $f_i = \sum_j K_{ij} q_j$, that is, $\sum_i k_i q_i = 0$, thus causing $\hat{\Pi}$ to vanish even though the vector (f_1, f_2) is always nonzero at $P=1$.

4.1.6 Determine for which P-values the structure with the stiffness matrix $K_{11}=4-P$, $K_{22}=6-P$, $K_{12}=3$, $K_{21}=5$ is stable.

4.1.7 In the foregoing problems the eigenvalues of $\mathbf{K}$ were real, but for a nonsymmetric matrix $\mathbf{K}$ they can also be complex. Show that this is the case for $K_{11}=1-P$, $K_{22}=2-P$, $K_{12}=-K_{21}=1$. Show that the limit of stability is $P=\hat{P}_{cr}=1$, and that there is no state of neutral equilibrium. (See also Prob. 10.4.4.)

4.2 POTENTIAL ENERGY FOR DISCRETE ELASTIC SYSTEMS

A discrete system is a system with a finite number, n, of degrees of freedom, characterized by the generalized displacements $q_1, \ldots, q_n$ (kinematic variables) that may represent actual displacements or rotations, or parameters of some deformation modes. Continuous structures such as beams, which do not represent a discrete system, may usually be approximated by a discrete system using some discretization procedure, such as the finite element method or expansion of the displacement distribution into a Fourier series. So the theory of discrete systems has a rather general applicability.

The Lagrange-Dirichlet theorem reduces the stability analysis to a test of positive definiteness (or existence of a strict minimum) of the energy function of the structure-load system, called the potential energy, Π. The dissipative forces (permitted by the Lagrange-Dirichlet theorem) must be such that Π is a function of the generalized displacements $q_1, \ldots, q_n$ in the vicinity of the equilibrium state under consideration, such that the variation of Π with $q_1, \ldots, q_n$ be path-independent, that is, reversible. Consequently, the loads must have a potential (i.e., must be conservative), at least in the vicinity of the equilibrium state, and dissipative forces, if present, may depend only on velocities $\dot{q}_1, \ldots, \dot{q}_n$, which is the case for damping; they vanish when static deformations are analyzed.

Structure-Load System

The potential energy Π consists of the (elastic) strain energy U and the work of loads W. If we consider that an equilibrium state of the structure is changed to another, adjacent equilibrium state in response to a change in the given loads (e.g., a change of the weight that is carried by the structure), then we must have $\Delta U = \Delta W$, or $\Delta U - \Delta W = 0$, according to the principle of conservation of energy. It follows that if the loads are not changed (e.g., the weights carried by the structure remain constant), then the net change of energy of the structure-load system is $\Delta U - \Delta W$. Therefore, the potential energy is defined as

$$\Pi = U - W \tag{4.2.1}$$

For the case of constant loads, called dead loads, we have $W = \sum_k P_k q_k$ where $P_k =$ load associated with q_k $(k = 1, 2, \ldots, n)$. For variable loads, $W = \sum_k \int P_k(q_k)\, dq_k$ where $P_k(q_k)$ are functions specified independently of the properties of the structure (see also Eq. 4.3.20 and the discussion below it). An example of a variable conservative load is the attractive or repulsive force between two electric charges or between two magnets, for which $P_k = a_k(r_k - q)^{-2}$ where a_k, r_k = constants, or the force produced by hydrostatic pressure of a heavy liquid, for which $P_k = a_k(r_k - q)$. In the following we will tacitly assume dead loads unless specified otherwise.

The fact that positive definiteness of $\Delta\Pi$ guarantees stability can also be proven [independently of the Lagrange-Dirichlet theorem (Theorem 3.6.1)] on the basis of the second law of thermodynamics. If the change of state is isothermal, $\Delta\Pi$ represents the Helmholtz free energy of the structure-load system, and if the change of state is isentropic, $\Delta\Pi$ represents the total energy of the structure-load system. For details see Chapter 10.

Second Variation of Potential Energy

The loading may in general be considered to change as a function of some control parameter λ, which may represent the load ($\lambda = P$), or the parameter of a system of loads, or the prescribed displacement, or the parameter of a system of prescribed displacements. Let $\delta q_1, \ldots, \delta q_n$ be small variations of the generalized displacements from the equilibrium state assumed to occur at constant λ. Assuming Π to be a smooth function (i.e., continuous and with continuous derivatives up to a sufficient order), function Π may be expanded into a Taylor series about the equilibrium state; this provides

$$\begin{aligned}\Delta\Pi &= \Pi(q_1 + \delta q_1, \ldots, q_n + \delta q_n; \lambda) - \Pi(q_1, \ldots, q_n; \lambda) \\ &= \delta\Pi + \delta^2\Pi + \delta^3\Pi + \cdots \end{aligned} \tag{4.2.2}$$

in which

$$\delta\Pi = \frac{1}{1!}\sum_{i=1}^{n} \frac{\partial\Pi(q_1, \ldots, q_n; \lambda)}{\partial q_i}\,\delta q_i \qquad \delta^2\Pi = \frac{1}{2!}\sum_{i=1}^{n}\sum_{j=1}^{n} \frac{\partial^2\Pi(q_1, \ldots, q_n; \lambda)}{\partial q_i\,\partial q_j}\,\delta q_i\,\delta q_j$$

$$\delta^3\Pi = \frac{1}{3!}\sum_{i=1}^{n}\sum_{j=1}^{n}\sum_{k=1}^{n} \frac{\partial^3\Pi(q_1, \ldots, q_n; \lambda)}{\partial q_i\,\partial q_j\,\partial q_k}\,\delta q_i \delta q_j \delta q_k \tag{4.2.3}$$

$\delta\Pi$, $\delta^2\Pi$, $\delta^3\Pi, \ldots$ are called the first, second, third, etc., variations of the potential energy. The conditions of equilibrium are

$$\delta\Pi = 0 \quad \text{for any } \delta q_i \qquad \text{or} \qquad \partial\Pi/\partial q_i = 0 \quad \text{for each } i \tag{4.2.4}$$

According to the Lagrange-Dirichlet theorem, the equilibrium state is stable for those values of the control parameter λ for which

$$\delta^2\Pi > 0 \qquad \text{for any } \delta q_i, \delta q_j \tag{4.2.5}$$

When, for some λ value, $\delta^2\Pi = 0$ for some δq_i, δq_j and $\delta^2\Pi > 0$ for some other δq_i, δq_j, the system may be, but need not be, unstable for that λ value, depending on the higher-order variations of Π. When $\delta^2\Pi = 0$ identically for all δq_i, δq_j the system might or might not be stable, and it will be stable if also $\delta^3\Pi = 0$ and $\delta^4\Pi > 0$ for all δq_i, δq_j. When, for some λ value, $\delta^2\Pi < 0$ for some δq_i, δq_j, the system is unstable for that λ value. This follows from Liapunov's stability theorem (Theorem 3.6.2), which states that the system is unstable if the potential energy is not positive definite, and the lack of positive definiteness is indicated by the second-order terms in the expansion of the function Π (Sec. 3.6).

In one dimension q_1, the condition of stability is illustrated in Figure 4.2. The potential energy of a ball in a gravity field is proportional to its vertical

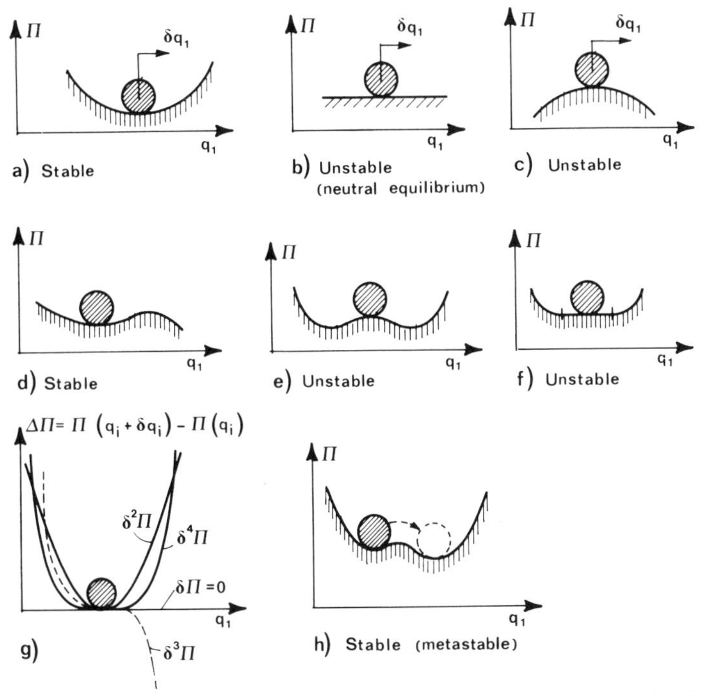

Figure 4.2 Ball in a gravity field with single degree of freedom: (a–f, h) different conditions of stability of equilibrium position; (g) variations of potential energy for stable system represented in (d).

coordinate. It should be also noted that the foregoing conditions of stability (Eq. 4.2.5) are based only on the variations of potential energy Π in an infinitely small neighborhood of the equilibrium state. In Figure 4.2g the various surfaces $\delta\Pi$, $\delta^2\Pi$, $\delta^3\Pi$, $\delta^4\Pi$ are shown for the case of Figure 4.2d. $\delta^2\Pi$ describes the actual potential surface only asymptotically, in an infinitely small neighborhood of the equilibrium state.

Of course stability can also be defined in the large. For example, the unstable state in Figure 4.2e may be characterized as stable in the large. The stable state in Figure 4.2h is called metastable if a stable state of lower Π exists at a finite but small distance away (a finite but small disturbance then changes the state of the system).

Critical State

At the limit of stability, the second variation ceases to be positive definite. This requires that the first variation of $\delta^2\Pi$ must turn to zero, and so the condition for the stability limit, called the critical state, may be written as

$$\delta[\delta^2\Pi(q_1, \ldots, q_n; \lambda)] = 0 \qquad \text{or} \qquad \sum_i \frac{\partial(\delta^2\Pi)}{\partial(\delta q_i)}\,\delta q_i = 0 \qquad \text{for all } \delta q_i$$

or

$$\frac{\partial(\delta^2\Pi)}{\partial(\delta q_i)} = 0 \qquad \text{for all } i \tag{4.2.6}$$

This is called the Trefftz criterion. [Note that $\delta(\delta^2\Pi)$ is not the third variation, $\delta^3\Pi$.]

If the structural system is linear, then the potential energy is a quadratic form, and if $q_1 = \cdots = q_n = 0$ is the equilibrium state, then $\Pi = \delta^2\Pi$, that is, the second variation of the potential energy coincides with the potential energy itself. Thus, for a structural system that is linear and for which $q_i = 0$ represents the equilibrium state, the critical state condition in Equation 4.2.6 reduces to

$$\delta\Pi = 0 \qquad \text{or} \qquad \sum_i \frac{\partial \Pi}{\partial q_i}\,\delta q_i = 0 \quad \text{for all } \delta q_i \qquad \text{or} \qquad \frac{\partial \Pi}{\partial q_i} = 0 \quad \text{for all } i \tag{4.2.7}$$

This condition has the appearance of an equilibrium condition, although it is, in fact, an implementation of the Trefftz criterion (Eq. 4.2.6). Calculation of critical states according to Equation 4.2.7 is also known as the method of adjacent equilibrium.

The equation $\delta\Pi(q_1, \ldots, q_n; \lambda) = 0$ is equivalent to the equilibrium conditions formulated directly as we did it in Chapters 1 and 2. The solution of this equation is the equilibrium state $q_1(\lambda), \ldots, q_n(\lambda)$ as a function of the control parameter λ. The critical value λ_{cr} is solved from the eigenvalue problem defined by the equation $\delta[\delta^2\Pi(q_1, \ldots, q_n; \lambda)] = 0$. This equation, however, coincides with the equation $\delta\Pi(q_1, \ldots, q_n; \lambda) = 0$ if the equilibrium state is $q_1 = \cdots = q_n = 0$ because then $\delta q_1 = q_1$ and $\delta^2\Pi = \Pi$. The direction of motion at the critical state is characterized by the eigenvector $q_1^{(1)}, \ldots, q_n^{(1)}$ obtained by solving the eigenvalue problem.

It should be noted that when, for certain λ, $\delta^2\Pi < 0$ occurs for some q_i, the direction of dynamic motion is not determined by the ratio $q_1 : q_2 : \cdots : q_n$ for which $\delta^2\Pi$ is minimized (at fixed λ) with respect to $q_1, q_2, \ldots, q_n$; rather, it can be determined only by dynamic analysis that includes inertia forces, viscous forces, etc. This direction is generally not the same as the static instability mode for λ_{cr}.

An Example

Consider now a simple example: a rigid bar of length l restrained at one end with a spring of stiffness C, loaded by force P of constant vertical direction (Fig. 4.3a) ($\lambda = P$). The angle q may be taken as the generalized displacement, and if for

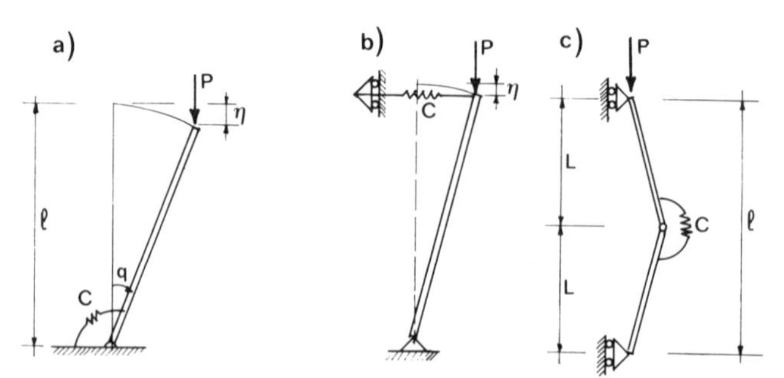

Figure 4.3 Single-degree-of-freedom systems.

$q=0$ the spring is free of stress, $q=0$ is an equilibrium position, which we want to investigate for stability. The change of the potential energy of the system from the position $q=0$ (Fig. 4.3a) is $\Delta\Pi=\Pi=\frac{1}{2}C\delta q^2-P\delta\eta=\frac{1}{2}Cq^2-Pl(1-\cos q)$ $(\delta q=q)$ where η is the vertical displacement of the load. Expanding in a Taylor series, one obtains $\delta\eta=l(1-\cos\delta q)=l(\frac{1}{2}\delta q^2-\frac{1}{24}\delta q^4+\cdots)$. So we have $\delta\Pi=0$ (as required if $q=0$ is an equilibrium position), $\delta^2\Pi=\frac{1}{2}(C-Pl)\delta q^2$, $\delta^3\Pi=0$, $\delta^4\Pi=\frac{1}{24}Pl\delta q^4,\ldots$. The condition of stability is $\delta^2\Pi>0$, that is, $P<C/l$, and the critical load $P_{cr}=C/l$. The same answer would be obtained from the Trefftz criterion, since the condition $\delta(\delta^2\Pi)=(C-Pl)\delta q=0$ yields $P_{cr}=C/l$ (because $\delta q\neq 0$).

Alternatively, one may set $\Pi=0$ for $q=0$ and express the potential energy of the system as $\Pi=\frac{1}{2}Cq^2-Pl(1-\cos q)=\frac{1}{2}Cq^2-Pl[(q^2/2)+\cdots]$. Then the equilibrium condition is $\delta\Pi=(C-Pl)q\delta q=0$ for any δq, which gives $q=0$ for $P\neq C/l$. The second variation $\delta^2\Pi=(C-Pl)\delta q^2$ gives the same stability condition as before, that is, $P<C/l$.

Note that if one would ignore the geometric effect of lateral deflection on the work of axial load, we would get the incorrect equilibrium condition $Cq\delta q=0$. This is equivalent to neglecting in the expression of Π the term $Pl(1-\cos q)\approx Plq^2/2$, which is quadratic and of the same order as the strain energy $\frac{1}{2}Cq^2$. Obviously, the second-order theory must include in the potential-energy expression all the terms up to the second order in displacements.

Effect of Higher-Order Derivatives of Π

For nonlinear systems, the determination of stability gets more complicated as it depends on higher than second derivatives of Π. When $\delta^2\Pi=0$ for some path and $\delta^2\Pi>0$ for all others, the system can be stable or unstable depending on higher variations. For polynomial expressions without cross terms, such as q_1q_2 (exemplified in Eqs. 3.6.3), the assessment of stability is straightforward, but not when cross terms are present. While for a quadratic surface it suffices to check the change of Π for all the radial paths away from the origin (constant ratio of q_i), paths of curved projection need to be also considered for a higher-order surface.

Figure 4.4 shows an example (adapted from Thompson and Hunt, 1984) of a quartic polynomial for Π, which is curious by the fact that Π is not positive definite even though Π initially rises along every radial path (constant q_1/q_2) from the origin. The reason is that there exist paths with a curved projection onto the q_1q_2 plane (of the type $q_2=-kq_1^2$, $k=\text{constant}>0$) for which a negative Π is reached. Therefore the condition $\Pi_{,1111}>0$ does not guarantee stability in this case (the subscripts preceded by a comma denote partial differentiation). It turns out that stability requires that $\Pi_{,1111}-3(\Pi_{,112})^2/\Pi_{,22}>0$ (see Thompson and Hunt, 1984, p. 23).

Difficulties with Complementary Energy

For any stiffness matrix $\mathbf{K}$ corresponding to displacements q_i of a kinematically determinate structure, one can define the associated flexibility matrix $\mathbf{C}$ corresponding to the associated forces f_i, which are such that $\sum f_i\,dq_i$ is the work

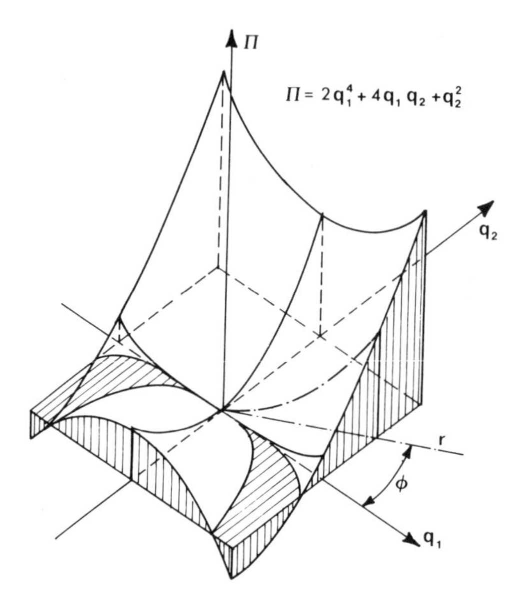

Figure 4.4 Higher-order surface that represents unstable behavior due to motion along curved nonradial path.

(the meaning of stiffness coefficients K_{ij} and the associated flexibility coefficients C_{ij} is illustrated in Fig. 4.5). Matrix $\mathbf{C}$ defines the quadratic form of complementary energy of the structure, $\Pi^* = \frac{1}{2}\mathbf{f}^T\mathbf{C}\mathbf{f}$.

Since $\mathbf{f} = \mathbf{K}\mathbf{q}$ and $\mathbf{q} = \mathbf{C}\mathbf{f}$, we have $\mathbf{C} = \mathbf{K}^{-1}$. Because the inverse of a real symmetric positive-definite matrix is also positive definite (Theorem 4.1.8), positive definiteness of either $\delta^2\Pi$ or $\delta^2\Pi^*$ $(= \frac{1}{2}\delta\mathbf{f}^T\mathbf{C}\delta\mathbf{f})$ implies stability. Then $\delta^2\Pi^*$ as a function of f_i (not q_i) has a strict minimum. In this sense, the methods of potential energy and complementary energy appear as two dual approaches, as do the associated principles of virtual displacements $(\delta\Pi = 0)$ or virtual forces $(\delta\Pi^* = 0)$; see, for example, Washizu (1975), or Argyris and Kelsey (1960).

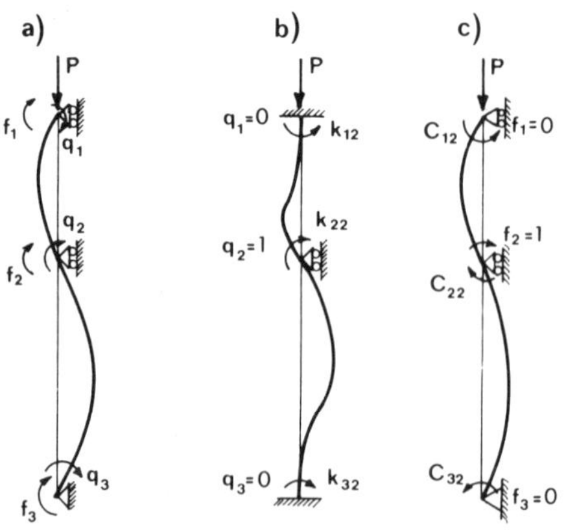

Figure 4.5 Definition of stiffness coefficients (b) and flexibility coefficients (c) for a continuous beam (a).

This duality, however, does not apply for the flexibility method of calculation of critical loads of redundant (statically indeterminate) beams or frames. In this method, one uses another flexibility matrix $\bar{\mathbf{C}}$, which represents the flexibility matrix of the primary structure, that is, a structure obtained by releasing the redundant forces X_i (through imaginary cuts). For the primary structure, X_i may be regarded as applied (external) forces, and so one can also define $\bar{\Pi}^* = \frac{1}{2}\mathbf{X}^T\bar{\mathbf{C}}^T\mathbf{X}$ = complementary energy of the primary structure with respect to X_i. Now it is important to realize that $\bar{\Pi}^*$ is not associated with $\mathbf{K}$ (because X_i and q_i are not associated by work). Therefore, $\bar{\mathbf{C}} \neq \mathbf{K}^{-1}$.

When the load (or load parameter) P is varied, $\det \mathbf{K}$, $\det \mathbf{C}$, and $\det \bar{\mathbf{C}}$ change. Det $\mathbf{K}$ can never become infinite before reaching the first critical load, because it would imply rigid response. It can become infinite only at critical loads for which the buckling mode involves some zero displacements, e.g., the stiffness matrix for end rotations of a beam when the load corresponds to buckling of a fixed-fixed beam; see Figure 2.5. Det $\mathbf{C}$ first becomes infinite at the first critical load. On the other hand, $\det \bar{\mathbf{C}} = 0$ at the first critical load because at neutral equilibrium the redundants X_i can change at no change of loads. However, det $\bar{\mathbf{C}}$ typically becomes infinite and jumps from ∞ to $-\infty$ at some load (or loads) $\bar{P}_1 < P_{cr_1}$ (= first critical load of the structure), see Figures 2.6f and 2.9 (cf. Prob. 4.2.7). This is caused by the fact that parts of the primary structure reach and exceed their own critical load $\bar{P}_1$ before P_{cr_1} is reached. Indeed, in that case the stiffness matrix of that part becomes singular, and therefore the determinant of the compliance matrix of that part (as well as some flexibility coefficient of that part) becomes infinite. This must make det $\bar{\mathbf{C}}$ also infinite because the $\bar{C}_{ij}$ are the sums of the flexibilities of all the parts of the primary structure.

Consequently, it is possible (and in fact typical) that $\bar{\Pi}^*$ loses positive definiteness before P_{cr_1} is reached (see the profiles of $\Delta\Pi$ and $\Delta\bar{\Pi}^*$ in Fig. 4.6).

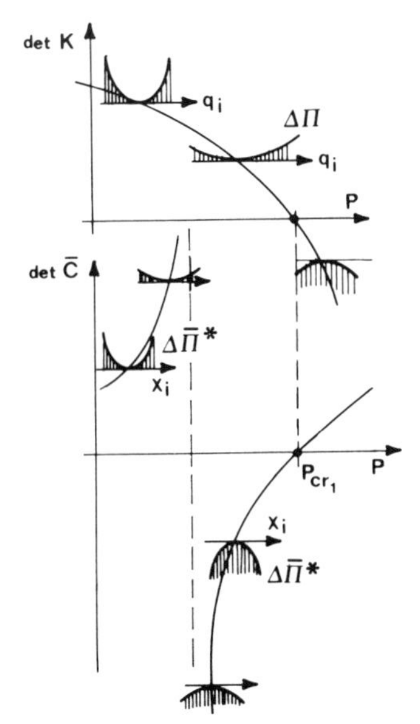

Figure 4.6 Typical profiles of potential energy Π and of complementary energy $\bar{\Pi}^*$ of the primary structure.

So we conclude that the flexibility method, which is based on the flexibility matrix $\bar{\mathbf{C}}$ of the primary structure, cannot be used to determine stability.

That positive definiteness of $\bar{\Pi}^*$ cannot decide stability of the actual structure is actually also a direct consequence of the fact that $\bar{\Pi}^*$ represents the complementary energy of the primary structure rather than the actual structure. This is further reflected in the fact that complementary energy of the primary structure is used to enforce conditions of compatibility rather than equilibrium.

The condition $\det \bar{\mathbf{C}} = 0$ can of course be used to determine P_{cr_1} (see Sec. 2.2). However, this approach is not advisable for any structure with more than a few unknowns, because the graph of $\det \bar{\mathbf{C}}$ versus P typically has many jumps between ∞ and $-\infty$ (Fig. 2.9). By contrast, $\det \mathbf{K}$ cannot exhibit such behavior (as we illustrated in more detail in Sec. 2.2).

To sum up, the complementary energy of the primary structure as a function of the redundants can, and typically does, lose positive definiteness while the actual structure is still stable.

Overturning Instability of a Block: Discontinuous Π'

The preceding stability conditions apply only if function $\Pi(q_1, \ldots, q_n)$ is continuous and has a continuous derivative. This is true for most elastic structures, but not for some systems of bodies in contact that can separate (lose contact). For example, consider stability of a rigid block (e.g., a retaining wall or a building) of weight P that rests on a rigid base and is loaded by a constant horizontal force H (e.g., dead load exerted by earth pressure or water pressure) applied at height a as shown in Figure 4.7a. We assume the block cannot slip; but it can lift from the base and thus it can lose stability by overturning. The work done on the block during overturning is $W = Mq$ where $M = Ha - Pb$ and $q = dq =$ small rotation about corner O. So the potential energy of the block is

$$\Pi = \begin{cases} -Mq = (Pb - Ha)q & \text{for } q \geq 0 \\ \infty & \text{for } q < 0 \end{cases} \tag{4.2.8}$$

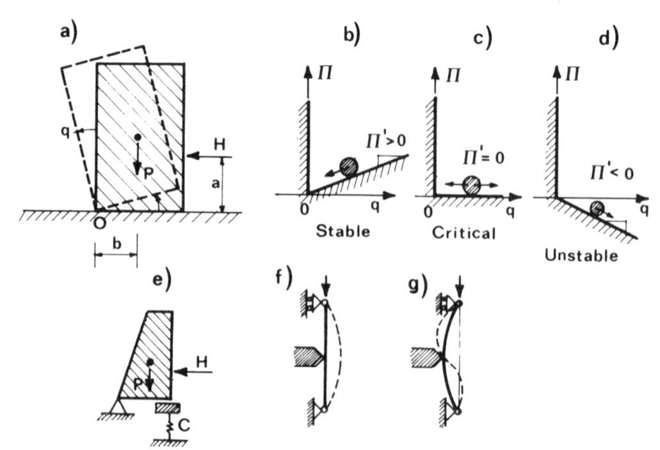

Figure 4.7 (a–d) Stability of rigid block; (e) rigid block on elastic base, and (f, g) buckling of column with one-sided constraints (exercise problems).

Here we considered the rigid base to be the limiting case of an elastic base (Fig. 4.7e) with stiffness approaching infinity, in which case Π for negative rotation immediately becomes infinite. According to the Lagrange-Dirichlet theorem (Theorem 3.6.1), the block is stable if $\Pi > 0$ for any possible $\delta q \neq 0$. This means that the block is stable if $Pb - Ha > 0$, critical if $Pb - Ha = 0$, and unstable if $Pb - Ha < 0$. The system behaves the same way as a ball rolling on the potential surfaces shown in Figure 4.7b, c, d. Conceptually it is interesting to note that (1) equilibrium exists without $\delta\Pi = 0$; (2) the condition of stability is $\delta\Pi > 0$ (or $\Pi' = \partial\Pi/\partial q > 0$) for $dq > 0$, but not $\delta^2\Pi > 0$ (or $\partial^2\Pi/\partial q^2 > 0$); and (3) Π, Π', and $\partial^2\Pi/\partial q^2$ are discontinuous at $q = 0$, and the derivatives Π' and $\partial^2\Pi/\partial q^2$ are one-sided, defined only for $q \to 0^+$ (that is, $q \to 0$ at $q \geq 0$); see Figure 4.7b, c, d.

Problems

4.2.1 Similar to the example in Figure 4.3a, determine the equilibrium and stability conditions for the rigid-bar column in Figure 4.3b. Also apply the Trefftz criterion.

4.2.2 Same as Problem 4.2.1 for the two-bar column in Figure 4.3c.

4.2.3 Without referring to the text explain why the complementary energy of the primary structure does not decide stability.

4.2.4 Discuss stability against overturning of the elastically supported block in Figure 4.7e, which can lift from the springs. Plot Π versus q.

4.2.5 Figure 4.7f shows a perfect Euler column for which buckling to the left is prevented by a rigid block. Plot Π versus deflection and show that the limit of stability is P_E, that is, the same without the block, despite discontinuity of $\Pi(q)$.

4.2.6 However, the behavior is different if the column has an initial crookedness toward the block (Fig. 4.7g). Show that in that case the column buckles in two half-waves and can support loads up to $4P_E$.

4.2.7 A fixed-free column, whose first critical load is $P_{cr_1} = EI\pi^2/(0.699l)^2$, may be considered as a system with one degree of freedom—the end rotation ϕ_1. So $\det \mathbf{K} = K_{11} = sEI/l$ (Sec. 2.1), $\det \mathbf{C} = 1/K_{11}$. The statically determinate primary system is a hinged column, with end rotations ϕ_1, ϕ_2, and $\bar{\mathbf{C}}$ is a 2×2 matrix given by Equation 2.1.10. Using the graphs of stability functions s, ϕ_s, ψ_s in Figure 2.2, show that (a) $\det \mathbf{K}$ first becomes infinite at the critical load $P^*_{cr} = 4EI\pi^2/l^2$ of a fixed-fixed column (i.e., $P^*_{cr} > P_{cr_1}$), (b) $\det \mathbf{C}$ first becomes infinite at P_{cr_1}, and (c) $\det \bar{\mathbf{C}}$ first becomes infinite at the critical load of a hinged column, $\bar{P}_1 = P_E = EI\pi^2/l^2$ (i.e., $\bar{P}_1 < P_{cr_1}$). Also plot $\det \mathbf{C}$, $\det \mathbf{K}$, and $\det \bar{\mathbf{C}}$ as functions of P.

4.2.8 Similar to Problem 4.2.7, plot $\det \mathbf{K}$, $\det \mathbf{C}$, and $\det \bar{\mathbf{C}}$ as a function of P for a (a) fixed column whose primary system is taken as a free-standing column, (b) fixed-hinged column whose primary system is taken as (1) a hinged column or (2) a free-standing column, and (c) simply supported continuous beam-column of two equal spans whose primary system consists of two simply supported columns.

4.3 BIFURCATION BUCKLING AT SMALL DEFLECTIONS

Consider again the prototype buckling problem of the hinged elastic column, with a sliding support at one end (Fig. 4.8). Let the column be perfect (which implies it is perfectly straight), and let EI be constant. In Chapter 1 we saw (Fig. 1.3) that the equilibrium path of this column in the plot of P versus max w bifurcates, and for this reason we speak of bifurcation buckling to distinguish this type of buckling from other types at which there is no bifurcation (see Sec. 4.4).

Calculation of Potential Energy of Beam-Columns

Let us now calculate the potential energy of a perfectly straight column, possibly with a lateral disturbing load $p(x)$. At zero load, the initial length of the column is l_0. Then we increase the axial compressive load until (at column length l, state 2 in Fig. 4.8) we reach the value P, at which the column instantly (but statically) buckles. The load remains constant during buckling (according to the linearized small-deflection theory), that is, the load in the buckled state is $P' = P$ (state 3 in Fig. 4.8). This implies that (for small deflections) the axial strain at neutral axis remains constant during buckling, that is, the length of the arc of the deflected neutral axis does not change. Therefore, the axial load P moves during buckling axially by the distance

$$\Delta l = \int_0^l (ds - dx) = \int_0^l \{[(dx)^2 + (dw)^2]^{1/2} - dx\}$$

$$= \int_0^l [(1 + w'^2)^{1/2} - 1]\, dx \simeq \frac{1}{2}\int_0^l w'^2\, dx \qquad (4.3.1)$$

in which we introduce the approximation $(1 + w'^2)^{1/2} \simeq 1 + \frac{1}{2}w'^2$, since we assume w' to be small (linearized bending theory). The strain energy of the column per unit length (i.e., the bending energy) is $\frac{1}{2}Mw''$, in which $M = EIw''$. Thus the potential energy of the column relative to the state just before buckling (state 2 in

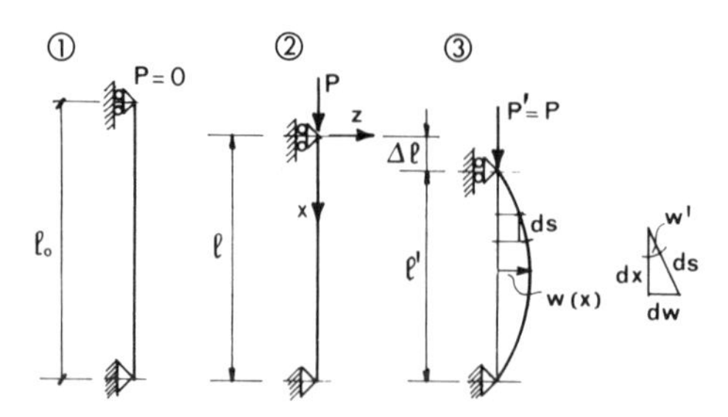

Figure 4.8 Pin-ended column (1) subjected to axial force (2), and lateral deflection due to buckling (3).

Fig. 4.8) is

$$\Pi = U - W \qquad \text{with } U = \int_0^l \frac{1}{2} EIw''^2 \, dx \quad W = P\,\Delta l + \int_0^l pw\, dx \tag{4.3.2}$$

in which W includes the potential energy of constant lateral disturbing loads $p(x)$, which might be present. Expand now the deflection curve $w(x)$ and the lateral disturbing load $p(x)$:

$$w(x) = \sum_{n=1}^{\infty} q_n \sin \frac{n\pi x}{l} \qquad p(x) = \sum_{n=1}^{\infty} p_n \sin \frac{n\pi x}{l} \tag{4.3.3}$$

in which p_n and q_n are Fourier coefficients. Substituting this into Equation 4.3.2 with Δl according to Equation 4.3.1, we have

$$\Pi = \frac{EI}{2} \int_0^l \left(- \sum_n q_n \frac{n^2\pi^2}{l^2} \sin \frac{n\pi x}{l} \right)^2 dx - \frac{P}{2} \int_0^l \left(\sum_n q_n \frac{n\pi}{l} \cos \frac{n\pi x}{l} \right)^2 dx$$
$$- \int_0^l \left(\sum_n p_n \sin \frac{n\pi x}{l} \right) \left(\sum_n q_n \sin \frac{n\pi x}{l} \right) dx \tag{4.3.4}$$

Squaring the sums, we obtain terms of the type $\sin(n\pi x/l)\sin(m\pi x/l)$, which are to be integrated from 0 to l ($m, n = 1, 2, 3, \ldots$). The integration, however, is greatly simplified by the fact that the integral of all terms for which $m \neq n$ vanishes (which is a general property of orthogonal functions). Then, noting that $\int_0^l \sin^2(n\pi x/l)\, dx = l/2$, and that $n^2\pi^2 EI/l^2 = P_{\text{cr}_n} = n$th critical load, we may reduce Equation 4.3.4 to the form:

$$\Pi(q_1, \ldots, q_n) = \frac{EIl}{4} \sum_n \frac{n^4\pi^4}{l^4} q_n^2 - \frac{Pl}{4} \sum_n \frac{n^2\pi^2}{l^2} q_n^2 - \frac{l}{2} \sum_n p_n q_n$$

or

$$\Pi(q_1, \ldots, q_n) = \sum_{n=1}^{\infty} \frac{n^2\pi^2}{4l} (P_{\text{cr}_n} - P) q_n^2 - \frac{l}{2} \sum_n p_n q_n \tag{4.3.5}$$

Equilibrium and Stability

The first variation of Equation 4.3.5 yields the equilibrium state, and by setting $\partial\Pi/\partial q_n = 0$ we obtain $\delta\Pi = \sum_n [2q_n n^2\pi^2(P_{\text{cr}_n} - P)/4l - lp_n/2]\,\delta q_n = 0$ for any δq_n, from which

$$q_n = q_{0_n} \frac{1}{1 - P/P_{\text{cr}_n}} \qquad q_{0_n} = \frac{l^2 p_n}{n^2\pi^2 P_{\text{cr}_n}} \tag{4.3.6}$$

This is the same as obtained before by the static method (Eq. 1.5.6), and involves the previously derived magnification factor (Eq. 1.5.10).

Stability is determined by the second variation of potential energy $\delta^2\Pi$, which is represented by the first term in Equation 4.3.5. Note that in this problem the lateral disturbing load $p(x)$ has no effect on the stability limits, because we use here a linearized theory for which Π is a quadratic function, and the lateral load affects only the linear terms. However, $\delta^2\Pi$ coincides with Π only if the system is perfect, that is, if $p_1 = \cdots = p_n = 0$. The second variation consists of a sum of squares, and represents a quadratic form in its standard (canonical) form. Thus the column is stable if the coefficients at all q_n^2 are positive and is unstable if at least one of these coefficients is negative or zero. Obviously, for $P < P_{cr_1}$, all the coefficients are positive. So the column is stable below the first critical load, while for $P \geq P_{cr_1}$, the column is unstable, the case $P = P_{cr_1}$ representing the loss of stability. Thus we confirm by energy analysis the result which we already established by dynamic analysis—the most general and fundamental approach.

The fact that the potential energy is obtained in the form of a sum of squares, that is, that the matrix of the quadratic form is diagonal, is a consequence of the fact that the deflection expansion in Equations 4.3.3 happens to represent a linear combination of the fundamental buckling modes. (This would not be so if EI were variable.) The buckling modes, representing the eigenstates of the eigenvalue problem of buckling, always represent a system of orthogonal functions, whether one deals with a column of arbitrary end restraints, a column of variable cross section, or a continuous beam or frame. Therefore, it is generally advantageous to represent the deflection curve as a linear combination of the fundamental buckling modes $w^{(n)}(x)$, and if this is done, then the potential energy is always obtained in the form of a sum of squares, which makes the stability analysis trivial. In this manner, the foregoing stability analysis could be generalized to arbitrary elastic structures.

Role of Axial Strain and Shortening Due to Deflections

We may now recall that in the previous chapters we formulated the equilibrium conditions without taking into account the axial shortening Δl (e.g., in deriving Eq. 2.1.16 for V). Why is it, then, that now we must take Δl into account? The reason is that Δl is quadratic in w. This makes it possible to neglect Δl in the equilibrium equations since they are linear in w, provided the deflections are small. The incremental expression for $\Delta\Pi$, however, has no linear terms. Its lowest-order terms are quadratic, and so the term $P\,\Delta l$, which is neglected in the first-order theory, must be taken into consideration here.

In certain problems, the finite strain ε at beam axis during buckling can be nonzero. With second-order accuracy, we may write

$$\varepsilon = e + \varepsilon^{(2)} \simeq u' + \tfrac{1}{2}w'^2 \qquad (4.3.7)$$

in which $e = u' = \partial u/\partial x$ = first-order linearized strain at beam axis (u = axial displacement), and $\varepsilon^{(2)} = \frac{1}{2}w'^2$ = the second-order strain at beam axis. This strain represents the approximation $(ds - dx)/dx = [(1 + w'^2)^{1/2}\,dx - dx]/dx \simeq \frac{1}{2}w'^2$ (see Fig. 4.8) where ds is the length of the arc of the deflected beam axis whose projection onto the original (undeflected, straight) beam axis is dx. The approximation is sufficient for small enough rotation w'. In the previously considered problems, we had $\varepsilon = 0$ at beam axis during buckling.

So far we solved column buckling assuming the column top to be axially sliding during buckling, so that $\varepsilon = 0$. However, if the column is axially restrained, then $\varepsilon \neq 0$ during buckling. To solve such problems, it is important to realize that the term $-P\int_0^l \frac{1}{2}w'^2\,dx$ no longer represents the work of the load. However, it still represents a part of potential energy that depends on P and is neglected in the first-order theory (a similar situation typically occurs for plates, see Sec. 7.2). To illustrate it, consider the pin-ended (hinged) column in Figure 4.9a where the axial force is produced before buckling (i.e., at $w = 0$) by an

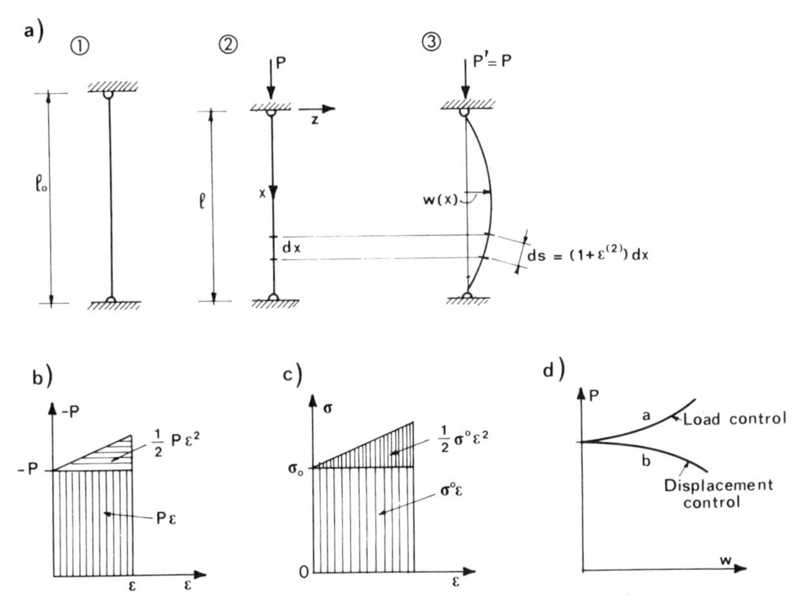

Figure 4.9 (a) Pin-ended column (1) subjected to axial (controlled) displacement (2) and lateral deflection due to buckling (3); (b, c) first- and second-order variations of strain energy and strain energy density due to elongation at beam axis; (d) difference in postbuckling behavior for load and displacement controls.

enforced axial displacement, but during buckling the axial displacement at the end is prevented, that is, both end hinges are fixed axially (Fig. 4.9a). This situation may be imagined to characterize an ideal buckling test in which the crosshead of the testing machine is moved at a small velocity; the end of the column is displaced slowly, the axial force P grows, and then the column suddenly buckles during a time interval that is so short that the axial displacement of the crosshead during the buckling is negligible. In this case the incremental strain energy density during buckling, $\frac{1}{2}EIw''^2$, must be augmented by the term $(N\varepsilon + \frac{1}{2}N\varepsilon^2)$ where ε is the incremental axial strain during buckling at column axis, and $N = -P =$ axial force, positive for tension (Fig. 4.9b). The first-order strain, $e = du/dx$, is zero in this case, same as for axially sliding hinges. But the second-order strain is nonzero, that is, $\varepsilon = \varepsilon^{(2)} = \frac{1}{2}w'^2 \neq 0$. Since ε is second-order small the term $\frac{1}{2}P\varepsilon^2$ is fourth-order small and negligible compared to $P\varepsilon$. So the incremental strain energy of the column for small deflections is

$$\Delta\Pi = \int_0^l \left[\tfrac{1}{2}EIw''^2 + (-P)(\tfrac{1}{2}w'^2)\right] dx = \Pi \tag{4.3.8}$$

But this at the same time represents the total energy Π since the incremental work W of the end load P is in this case *zero*. So we conclude that the potential-energy expression is the same as for the Euler column (an axially sliding hinge at the end, i.e., simple support), and so the stability limits (as well as the differential equation for small deflections w) for the columns in Figures 4.8 and 4.9 are exactly the same. A difference is, of course, encountered for finite postbuckling deflections (Fig. 4.9d) (e.g., because of the term $\frac{1}{2}N\varepsilon^2$, that we neglected).

Calculation of Π from the Work of Initial and Incremental Stresses

Instead of expressing directly the energy density ($\frac{1}{2}EIw''^2 + N\varepsilon$) per unit length of column, one can begin with the energy density per unit volume. Let us illustrate this alternative but equivalent procedure for the case of a column with axially fixed hinges and a lateral distributed load $p(x)$. The column may be either perfect or imperfect. If imperfect w represents the increment of $z(x)$ from the initial ordinate $z^0(x)$ (which is assumed to be negligibly small), and the initial bending moment M^0 is nonzero (superscript 0 is used throughout the text, e.g., in Sec. 6.1, as a label for initial equilibrium states). We have

$$\Delta\Pi = \int_0^l \int_A \left(\sigma^0 + \frac{1}{2}\sigma\right)\varepsilon_1 \, dA\, dx - \int_0^l pw\, dx = \int_0^l \int_A \left(\sigma^0 \varepsilon_1 + \frac{1}{2}E\varepsilon_1^2\right) dA\, dx$$
$$- \int_0^l pw\, dx \qquad \sigma^0 = -\frac{P}{A} - \frac{M^0}{I} z \tag{4.3.9}$$

where (Fig. 4.9c) σ^0 = initial axial stresses before buckling, M^0 = bending moment in the initial equilibrium state caused by p, by the applied end moments (also considered to be dead loads), and by P, $\varepsilon_1 = \varepsilon - \partial(-w'z)/\partial x = \varepsilon - w''z$ = increment of normal strain from the initial equilibrium state for a generic point of the cross section, and $\sigma = E\varepsilon$ = the corresponding stress increment. P must now be understood as the initial value of the axial compression force (reaction at column top). During buckling the axial force is now changing. Its change is $-EA\varepsilon$, and the corresponding energy is taken into account by the term $\frac{1}{2}E\varepsilon^2$. With regard to variational analysis, the new equilibrium state represents a state to which the adjacent non-equilibrium states are compared in terms of minimization of Π. Using for ε the expression given by Equation 4.3.7, assuming p to be a dead load (i.e., constant), and integrating, we get from Equation 4.3.9:

$$\Delta\Pi = \int_0^l \int_A \left[-\left(\frac{P}{A} + \frac{M^0}{I}z\right)\left(-w''z + \frac{1}{2}w'^2\right) + \frac{1}{2}Ew''^2 z^2\right] dA\, dx - \int_0^l pw\, dx$$
$$= \int_0^l \left[\frac{1}{2}Ew''^2 \int_A z^2\, dA + \frac{P}{A}w'' \int_A z\, dA - \frac{P}{2A}w'^2 \int_A dA\right.$$
$$\left. + \frac{M^0}{I}w'' \int_A z^2\, dA - \frac{M^0}{2I}w'^2 \int_A z\, dA\right] dx - \int_0^l pw\, dx$$
$$= \int_0^l \left(\tfrac{1}{2}EIw''^2 - \tfrac{1}{2}Pw'^2 + M^0 w''\right) dx - \int_0^l pw\, dx = \int_0^l \tfrac{1}{2}\left(EIw''^2 - Pw'^2\right) dx = \delta^2\Pi \tag{4.3.10}$$

Here we used the relations $\int z^2\, dA = I$, $\int z\, dA = 0$, and $\int dA = A$, assuming axis x to be centroidal. The last expression in Equation 4.3.10 is obtained upon noting that, due to initial equilibrium, $\int M^0 w''\, dx = -\int M^{0\prime} w'\, dx \simeq \int V^0 w'\, dx = -\int V^{0\prime} w\, dx = \int pw\, dx$ because either M^0 or w' and either V^0 or w are zero at the ends, and $M^{0\prime} \simeq V^0$. We neglect here the contribution to M^0 due to Pz^0 as higher-order small because for the initial state z_0 is assumed to be small.

This last derivation demonstrated a procedure that has the most general applicability. We will use it again, for example, for thin-walled beams (Sec. 6.1).

Example with Two Degrees of Freedom

As another example, consider the column shown in Figure 4.10a consisting of two perfectly rigid bars of length l, connected at hinges by rotational springs of spring stiffness C. The deflected state of the column is defined by the inclination angles of the bars, q_1 and q_2. These angles are assumed to be small. In general the column may be imperfect, with the initial state at load $P = 0$ given by angles $q_1 = \alpha_1$ and $q_2 = \alpha_2$. The same column was studied in Section 3.2; here, however, the load remains vertical, that is, is conservative. The potential energy $\Pi(q_1, q_2)$ may be expressed as $\Pi = U - W$ where

$$\begin{aligned} U(q_1, q_2) &= \tfrac{1}{2}C(q_1 - \alpha_1)^2 + \tfrac{1}{2}C(q_2 - q_1 - \alpha_2 + \alpha_1)^2 \\ W(q_1, q_2) &= P(l\cos\alpha_2 - l\cos q_2 + l\cos\alpha_1 - l\cos q_1) \\ &\simeq \frac{Pl}{2}(q_1^2 + q_2^2 - \alpha_1^2 - \alpha_2^2) \end{aligned} \tag{4.3.11}$$

(U = strain energy, W = work of load P).

In the last equation we introduced the approximation $\cos q_1 = 1 - \frac{1}{2}q_1^2$ and $\cos\alpha_2 = 1 - \frac{1}{2}\alpha_2^2$, which is acceptable for small angles. Note that, even though the angles are small, we cannot substitute $\cos q_1 = 1$ and $\cos\alpha_2 = 1$, because the energy is quadratic and must be approximated correctly at least up to the terms of second order in q_1. By differentiating, we obtain the conditions of equilibrium:

$$\begin{aligned} \frac{1}{C}\frac{\partial\Pi}{\partial q_1} &= \left(2 - \frac{Pl}{C}\right)q_1 - q_2 - 2\alpha_1 + \alpha_2 = 0 \\ \frac{1}{C}\frac{\partial\Pi}{\partial q_2} &= -q_1 + \left(1 - \frac{Pl}{C}\right)q_2 - \alpha_2 + \alpha_1 = 0 \end{aligned} \tag{4.3.12}$$

When the structure is perfect, that is, $\alpha_1 = \alpha_2 = 0$, a deflected equilibrium state is possible if and only if the determinant of the equation system vanishes. This

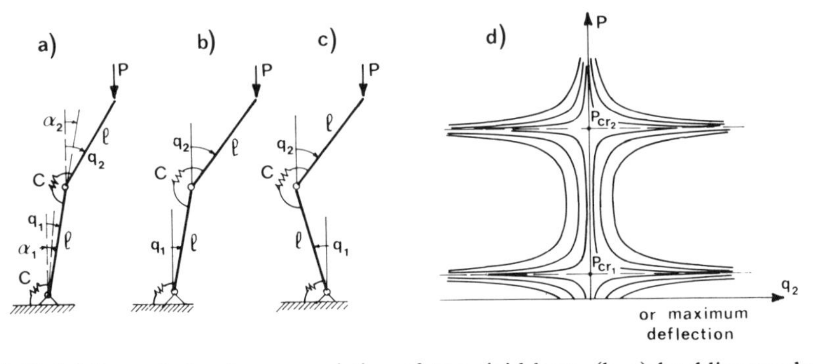

Figure 4.10 (a) Imperfect column consisting of two rigid bars, (b, c) buckling modes, and (d) equilibrium paths.

yields the condition $(Pl/C)^2 - 3(Pl/C) + 1 = 0$, from which one may solve:

$$P_{\mathrm{cr}_{1,2}} = \frac{3 \mp \sqrt{5}}{2}\left(\frac{C}{l}\right) = \begin{cases} 0.3820C/l \\ 2.6180C/l \end{cases} \tag{4.3.13}$$

By substituting these critical values into either the first or the second equation in Equations 4.3.12, one obtains for the eigenstates, that is, the buckling modes,

$$\left(\frac{q_1}{q_2}\right)_{\mathrm{cr}_{1,2}} = \frac{(-1 \pm \sqrt{5})}{2} = 0.6180 \quad \text{or} \quad -1.6180 \tag{4.3.14}$$

The buckling modes defined by these ratios are sketched in Figure 4.10b, c.

By eliminating q_1 from Equations 4.3.12, one can solve P as a function of q_2, for various values of the initial imperfection. This yields the equilibrium path shown in Figure 4.10d. Only the equilibrium path of the perfect system exhibits bifurcation. For the imperfect system, the equilibrium paths do not reach the critical load value at any finite deflection.

It may be checked that the same equations of equilibrium and the same critical loads are obtained by either the virtual work method or by writing the equilibrium conditions on the basis of free-body diagrams. However, this equilibrium approach does not answer the question of stability. To answer it, we must consider the second variation $\delta^2\Pi$, which happens to coincide in the present case with Π, provided the structure is perfect. The matrix of the second variation $2\,\delta^2\Pi = K_{11}\,\delta q_1^2 + 2K_{12}\,\delta q_1\,\delta q_2 + K_{22}\,\delta q_2^2$ is obtained by differentiating Equations 4.3.12:

$$\frac{K_{11}}{C} = \frac{1}{C}\frac{\partial^2\Pi}{\partial q_1^2} = 2 - \frac{Pl}{C} \qquad \frac{K_{22}}{C} = \frac{1}{C}\frac{\partial^2\Pi}{\partial q_2^2} = 1 - \frac{Pl}{C} \qquad \frac{K_{12}}{C} = \frac{1}{C}\frac{\partial^2\Pi}{\partial q_1\,\partial q_2} = -1 \tag{4.3.15}$$

Note that for our structure these coefficients give the second variation for both the perfect and imperfect cases, and so the stability regions of the perfect and imperfect structures are the same (this is generally true for linear theory, for which the energy is quadratic, but it is not generally true for nonlinear theory; see Secs. 4.4–4.6). According to Theorem 4.1.10, the structure is stable if and only if

$$\begin{vmatrix} 2 - \dfrac{Pl}{C} & -1 \\ -1 & 1 - \dfrac{Pl}{C} \end{vmatrix} > 0 \quad \textit{and} \quad 2 - \frac{Pl}{C} > 0 \tag{4.3.16}$$

which may be rewritten as

$$(P - P_{\mathrm{cr}_1})(P - P_{\mathrm{cr}_2}) > 0 \quad \textit{and} \quad 2 - \frac{Pl}{C} > 0 \tag{4.3.17}$$

We may now distinguish three cases:

1. $P < P_{\mathrm{cr}_1}$, in which case both conditions in Equations 4.3.17 are satisfied.
2. $P_{\mathrm{cr}_1} \le P \le P_{\mathrm{cr}_2}$, in which case the first condition in Equations 4.3.17 is violated, and so the system is unstable.
3. $P > P_{\mathrm{cr}_2}$ in which case the first condition in Equations 4.3.17 is fulfilled, but the second one is violated, and so the system is unstable.

Overall, the system is stable if and only if $P < P_{cr_1}$, whether it is perfect or imperfect.

There are certain advantages to stability analysis in which the buckled shape is expressed as a linear combination of the buckling modes (eigenmodes) $q_i^{(k)}$ $(k = 1, \ldots, n)$ whose amplitudes y_k are taken as the kinematic variables. Using again the example in Figure 4.10a, and expressing q_1 and q_2 as a linear combination of the buckling modes (Eq. 4.3.14), we have

$$\begin{aligned} q_1 &= y_1 q_1^{(1)} + y_2 q_1^{(2)} = y_1(-1+\sqrt{5}) + y_2(-1-\sqrt{5}) \\ q_2 &= y_1 q_2^{(1)} + y_2 q_2^{(2)} = y_1 \cdot 2 + y_2 \cdot 2 \end{aligned} \tag{4.3.18}$$

which is in fact the orthogonal transformation $\mathbf{q} = \mathbf{T}\mathbf{y}$ or $q_i = \sum_r y_r q_i^{(r)}$ introduced in Section 4.1. Substituting into Equations 4.3.11 (with $\alpha_1 = \alpha_2 = 0$), we get

$$\begin{aligned} U &= \frac{C}{2}[4(5-2\sqrt{5})y_1^2 + 4(5+2\sqrt{5})y_2^2] \\ W &= \frac{Pl}{2}[2(5-\sqrt{5})y_1^2 + 2(5+\sqrt{5})y_2^2] \end{aligned} \tag{4.3.19}$$

Note that the cross product $y_1 y_2$ is absent from these expressions. The absence of cross products from the expression for Π is, of course, a general property of eigenmodes, and their amplitudes y_k are therefore called the orthogonal coordinates. The potential energy becomes a sum of squares, and thus the question of stability can be decided easily. We have $\Pi = U - W = B_1 y_1^2 + B_2 y_2^2$, and so the stability conditions are $B_1 > 0$ and $B_2 > 0$. This immediately yields the critical values in Equation 4.3.13 with the same stable domain as we obtained it from Equations 4.3.17.

In the plane (q_1, q_2) (Fig. 4.11a) the coordinates y_1 and y_2 can be represented by observing that $y_1 = 0$ (axis y_2) corresponds to the line $q_1/q_2 = (-1-\sqrt{5})/2$ and $y_2 = 0$ (axis y_1) corresponds to the line $q_1/q_2 = (-1+\sqrt{5})/2$. The axes y_1 and y_2 are the principal axes of the quadratic surface of potential energy, as illustrated in Figure 4.11b–f for various load intervals. For a system with two degrees of freedom (Fig. 4.10a), these surfaces are a convex elliptic paraboloid for $P < P_{cr_1}$ (Fig. 4.11b), a cylinder with horizontal axis for $P = P_{cr_1}$ and $P = P_{cr_2}$ (Fig. 4.11c, e), a hyperbolic paraboloid (saddle surface) for $P_{cr_1} < P < P_{cr_2}$ (Fig. 4.11d), and a concave elliptic paraboloid for $P > P_{cr_2}$ (Fig. 4.11f).

The expression for $\Pi = U - W$ according to Equation 4.3.19 can, of course, be obtained from the eigenvalues P_{cr_1} and P_{cr_2} and eigenvectors according to Equation 4.1.3. The reason that this is a sum of squares is that the stiffness matrix in Equation 4.3.15 happens to involve P only in the diagonal terms. In other problems this is not the case in general, and then the use of eigenvectors in substitution $\mathbf{q} = \mathbf{T}\mathbf{y}$ does not yield Π as a sum of squares. However, if the eigenvalue problem is first converted to the standard form, a sum of squares always results (see Sec. 4.1).

A salient property of bifurcation buckling of symmetric structures, as illustrated by our preceding examples of perfect structures, is that the symmetry of response breaks down. In physics and other fields of science, the breakdown of symmetry of response typically leads to instability, and the bifurcation theory may be developed from this viewpoint.

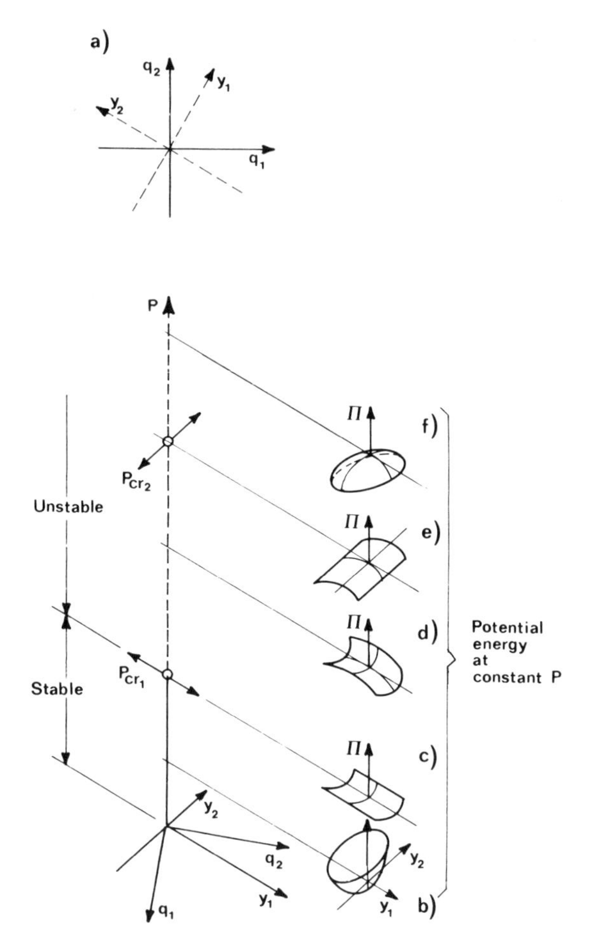

Figure 4.11 (a) Orthogonal coordinates and (b–f) surfaces of potential energy for various values of P.

Some Fundamental Aspects of Potential Energy

As pointed out in Section 4.2, the derivation of the Lagrange-Dirichlet theorem shows that Π represents the energy of the entire structure-load system (in thermodynamics, Π is the Helmholtz free energy if the conditions are isothermal and the total energy if the conditions are isentropic; see Sec. 10.1). As stated in Equation 4.2.1 the principle of conservation of energy implies that $\Pi = U - W$ (where we write Π, U, W, instead of $\Delta\Pi$, ΔU, ΔW because Π, U, W is considered to be zero in the initial state); U = strain energy of the structure, and W = work of the loads, P_k $(k = 1, 2, \ldots, n)$, that is,

$$W = \sum_n \int P_k(w_k)\, dw_k \tag{4.3.20}$$

where w_k = load-point displacements and P_k = given functions of w_k. Obviously, $P_k = \partial W/\partial w_k = -\partial\Phi/\partial w_k$ where $\Phi = -W$ = potential energy of the loads. Now a crucial point is that Φ is defined independently of the structure and is a function of w_k (considered as load-point coordinates in space). Potential Φ exists only if $\partial P_k/\partial w_j = \partial P_j/\partial w_k$ for all k, j; this is necessary to ensure that forces P_k are conservative, as required by the Lagrange-Dirichlet theorem (Theorem 3.6.1).

For the special case of dead loads (P_k = const.), $W = \sum P_k w_k$. In that case the load is exerted by gravity and may be imagined to be applied by an object whose weight is P_k (as in Fig. 4.12). Then $\Phi = -W$ = change of gravitational potential energy of the object when it is lowered by distance w_k.

For the sake of simple illustration, the potential energy of a rigid ball of mass m moving on a rigid curved surface (Fig. 4.12a) is $\Pi = -Pw$ (not $-\frac{1}{2}Pw$!), w = displacement of the ball in the direction of force P, $P = mg$, g = acceleration of gravity (Fig. 4.12a); indeed $P = -\partial\Pi/\partial w$. The potential energy of a weightless block resting on a spring of stiffness C is $\Pi = \frac{1}{2}Cu^2$ where u = shortening of the spring (Fig. 4.12b); this satisfies the condition that $F = Cu = \partial\Pi/\partial u$ = force in the spring. The potential energy of a system of mass m moving on massless curved block that is supported by a spring of stiffness C is $\Pi = -Pw + \frac{1}{2}Cu^2$ where $P = mg$ (Fig. 4.12c); this satisfies the conditions $P = -\partial\Pi/\partial w$, $F = Cu = \partial\Pi/\partial u$. If a weightless block supported on a spring of stiffness C is loaded by a force $P(w)$ that is defined as a function of displacement w (Fig. 4.12d), the potential energy of the system is $\Pi = -\int P(w)\,dw + \frac{1}{2}Cu^2$; again $P(w) = -\partial\Pi/\partial w$, $F = Cu = \partial\Pi/\partial u$. For the special case of dead load, that is, for P = constant, this last case reduces to the previous one (Fig. 4.12c). Finally, if the spring is nonlinear, then $\Pi = -\int P(w)\,dw + \int F(u)\,du$ where $F(u)$ = force in the spring.

The strain energy is generally calculated as

$$U = \int_V \int \boldsymbol{\sigma} : d\boldsymbol{\varepsilon}\, dV \tag{4.3.21}$$

where $\boldsymbol{\sigma}$ = stress tensor, $\boldsymbol{\varepsilon}$ = strain tensor, V = volume of structure (and : denotes a tensor product contracted on two indices $\boldsymbol{\sigma}:d\boldsymbol{\varepsilon} = \sigma_{ij}\, d\varepsilon_{ij}$). For the special case of uniaxial linear elasticity ($\sigma = E\varepsilon$) this yields

$$U = \int_V \frac{1}{2}\sigma\varepsilon\, dV = \int_V \frac{1}{2}E\varepsilon^2\, dV = \int_V \frac{\sigma^2}{2E}\, dV \tag{4.3.22}$$

where $\int \sigma\, d\varepsilon = \frac{1}{2}\sigma\varepsilon$ represents the triangular area under the stress-strain diagram.

There is another viewpoint that often makes things puzzling to inquisitive students. Does not the reaction from the loading device increase gradually from zero during the load application? Yes, it does. So, should not the work of load be

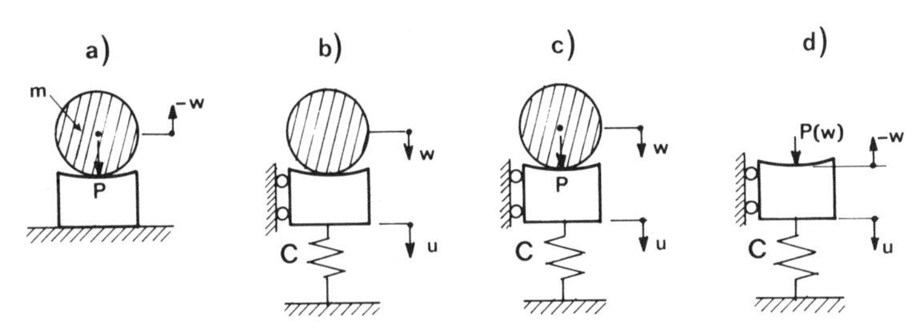

Figure 4.12 Ball on (a) rigid block and (b, c) block resting on springs; (d) weightless block resting on springs loaded by a force.

$\frac{1}{2}P_k w_k$ instead of $P_k w_k$ for a linear increase of reaction, or $\int \bar{P}_k(u)\,du$ in general? [u is the axial displacement increasing from 0 to w_k, and $\bar{P}_k(u)$ are the reaction values increasing from 0 to the final value P_k as the load is applied.]

No, it should not. As we said, a constant load may always be imagined to be exerted by some object of weight P_k. If the object is positioned in contact with the undeflected structure and then suddenly released, the reaction it receives from the structure does not, of course, become immediately equal to its weight, but increases gradually as the structure deflects. The difference between this reaction and the weight accelerates the object and is manifested as the inertia force. Clearly, the analysis would have to be done dynamically if the process of development of the reaction from the structure onto the object were to be taken into account. The rapid application of a mass onto an elastic structure is analyzed in a course on dynamics, and it is found that, at the deflection w_k equal to the static deflection, the structure also has kinetic energy, which is exactly equal to $P_k w_k - \frac{1}{2}P_k w_k$. So there is no contradiction with the previous viewpoint.

Problems

4.3.1 Write the expression of the potential energy Π for (a) a free-standing column, (b) a fixed-end column, (c) a two-span continuous beam of equal spans; and expand the deflection curve and lateral disturbing load in a Fourier series. Consider also an eccentricity in the applied axial load. Write the expressions for the first and second variations of Π and find the critical loads. Determine the deflected shapes. Verify that the principle of virtual work gives the same equations as the first variation.

4.3.2 Solve the stability conditions of the same column as in Figure 4.10a except that (a) the top bar length is $2l$, (b) the spring constant of the base spring is $2C$.

4.3.3 Do the same for the column in Figure 4.13a.

4.3.4 Do the same for the column in Figure 4.13b.

4.3.5 Do the same for the column in Figure 4.13c.

4.3.6 Do the same for the column in Figure 4.13d.

4.3.7 Do the same for the column in Figure 4.13e (where P is a spring force).

4.3.8 Do the same for the system in Figure 4.13f.

4.3.9 Do the same for the system in Figure 4.13g.

4.3.10 Draw Figure 4.10d to scale.

4.3.11 Show that the potential energy $\Pi = U - W$ calculated from Equations 4.3.19 is equivalent to the form $\Pi = \lambda_1 y_1^2 + \lambda_2 y_2^2$ in which λ_1 and λ_2 are the eigenvalues of the matrix $\mathbf{K}$ of coefficients of the quadratic form Π (Theorem 4.1.5, Sec. 4.1).

4.3.12 Draw Figure 4.11 to scale.

4.3.13 *Thermal buckling.* An elastic column with two hinges that cannot slide axially (Fig. 4.13h) is heated by ΔT. Express its potential energy and show that the stability limit is the same as for a column on which the thermal force is applied externally. Express the stability limit in terms of ΔT. See also Problem 1.9.5. *Note:* $\sigma = E(\varepsilon - \varepsilon'')$ where $\varepsilon'' = \alpha\,\Delta T$, α = thermal expansion coefficient; $W = 0$, $\Pi = U = \int (\sigma^2/2E)\,dV = \int \frac{1}{2}E(\varepsilon - \varepsilon'')^2\,dV$ and substitute $\varepsilon = -zw'(x)$. (The postcritical behavior is, however, very different.)

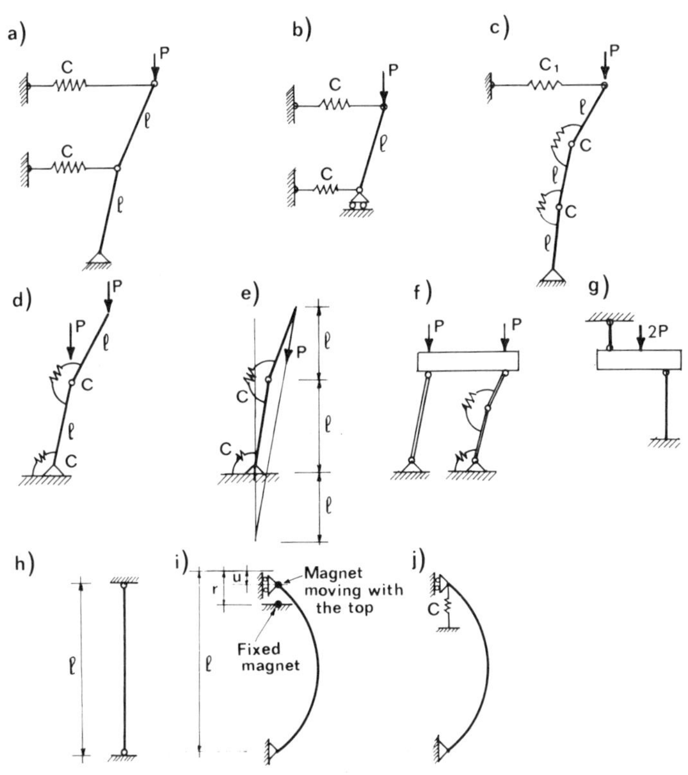

Figure 4.13 Exercise problems on bifurcation buckling at small deflections.

4.3.14 *Magnetic forces* (*variable load*). A hinged column is loaded by a magnet that is attached to its top and is attracted to another fixed magnet at initial distance r (Fig. 4.13i), such that the attractive force, directed toward the opposite end, is $P = a(r-u)^{-2}$ where a, r = constants and u = axial displacement. Calculate P_{cr_1} from potential energy and also verify P_{cr_1} by equilibrium analysis using $P(u)$ from Equation 1.9.14. Show that $P_{cr_1} \simeq P_E$ unless r is very small. How small must r be to make $P_{cr_1} \leq 0.8 P_E$? (*Hint*: The second derivative of the potential energy of the load is nonzero and equal to $-dP/du$.)

4.3.15 Do the same, but force P is exerted by a spring of stiffness C whose lower end is moved away until P_{cr_1} is reached but is fixed during buckling (Fig. 4.13j). This case can be solved in two ways: (1) as a column loaded by a variable force, or (2) as a single elastic system comprising both the column and the spring. Show that both ways yield the same result.

4.3.16 Derive Equations 4.3.19 on the basis of eigenvalues and eigenvectors as presented in Section 4.1.

4.4 SNAPTHROUGH AND FLAT ARCHES

So far we have dealt with problems that can be cast in a linear form. However, for some problems, such as stability of flat arches, linearization would deprive the formulation of certain essential features that lead to instability and failure. We first illustrate these different features with an example of a truss, then we extend the analysis to flat arches and finally analyze the effects of imperfections and of interaction with bifurcation buckling.

Von Mises Truss

As the first illustration, consider the simple two-bar truss (von Mises truss) shown in Figure 4.14a. The bars are elastic, characterized by axial stiffness $EA/(L/\cos\alpha)$, and the Euler load of each bar is assumed to be so large that the bars never buckle. The initial length of each bar is $L/\cos\alpha$, with L being the half-span and α the initial inclination angle of the bars. Noting that the axial strain of the bars is $\varepsilon = (L\cos\alpha/\cos q - L)/L$, we find the potential energy:

$$\Pi(q) = U - Pw = \frac{EAL}{\cos q}\left(\frac{\cos\alpha}{\cos q} - 1\right)^2 - PL(\tan\alpha - \tan q) \qquad (4.4.1)$$

By differentiation, we obtain the equilibrium condition:

$$\frac{\partial\Pi}{\partial q} = \frac{L}{\cos^2 q}\left[2EA(\cos\alpha\tan q - \sin q) + P\right] = 0 \qquad (4.4.2)$$

from which we can calculate

$$P = 2EA(\sin q - \cos\alpha\tan q) \qquad (4.4.3)$$

The equilibrium path according to this equation is plotted in Figure 4.14b.

To decide the question of stability of equilibrium states, we need to calculate the second derivative of Π. Evaluating $\partial^2\Pi/\partial q^2$ (at constant P) from Equation 4.4.2 and substituting Equation 4.4.3 for P, we obtain

$$\frac{\partial^2\Pi}{\partial q^2} = \frac{2EAL}{\cos^4 q}(\cos\alpha - \cos^3 q) \qquad \text{(for equilibrium states)} \qquad (4.4.4)$$

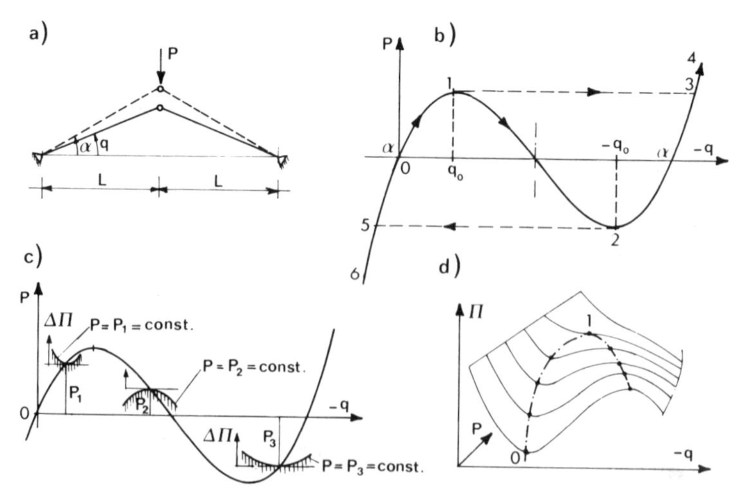

Figure 4.14 (a) Two-bar truss, (b) equilibrium path, (c) variation of potential energy at various points on equilibrium path, and (d) potential-energy curves for various constant values of P.

The truss is stable if and only if this expression is positive. This yields the condition that $q > q_0$ or $q \leq -q_0$ where

$$\cos q_0 = (\cos \alpha)^{1/3} \tag{4.4.5}$$

For $-q_0 \leq q \leq q_0$ the truss is unstable.

It is interesting to compare the critical state values q_0 with the q value corresponding to the limit points (the maximum and minimum points) of the equilibrium diagram in Figure 4.14b. Calculating the derivative $\partial P/\partial q$ from Equation 4.4.3 and setting $\partial P/\partial q = 0$, we find that the q values for the limit points are $-q_0$ and $+q_0$, with q_0 given by Equation 4.4.5. Thus we see that the stable equilibrium states lie on the equilibrium diagram in Figure 4.14b on the segments of positive slope (segments $\overline{501}$ and $\overline{234}$), while the unstable equilibrium states lie on the segment of negative slope (segment $\overline{12}$). This result is not by chance. Later we will see that generally, for single-degree-of-freedom systems, a positive slope of the equilibrium path implies stability, and a negative slope instability. Figure 4.14c illustrates, for various points on the equilibrium path, the variation of the potential energy with q at $P =$ constant.

The equilibrium path can be followed also in the plot of the potential energy Π versus the generalized displacement q (Fig. 4.14d), in which load P is plotted in the third spatial direction. The resulting surface is represented through the curves of constant load, which have horizontal tangents for the equilibrium states. These states are stable if the curve is convex (i.e., positive second derivative of Π) and unstable if not convex. The limit point (a critical state) is, on these curves, the inflection point with a horizontal tangent.

Let us now discuss what happens when the von Mises truss is loaded in a load-controlled manner. First the stable equilibrium path $\overline{01}$ is followed, until the critical state point 1 is reached. When the load is increased an infinitesimal amount above the peak point value, there is suddenly no adjacent equilibrium state and the only equilibrium state possible is a finite distance apart; it is the state corresponding to point 3 in Figure 4.14b. Therefore, the structure snaps through from point 1 to point 3, as indicated by the dashed horizontal line in Figure 4.14b. So an infinitely small change of load causes a finite change of deflection, which is defined as an unstable situation. The snapthrough path $\overline{13}$ does not represent equilibrium states, and therefore it happens dynamically, in the presence of inertia forces which are equal to the vertical deviation from the equilibrium path in Figure 4.14b. The work done by the inertia forces is equal to the area 1231 in Figure 4.14b (see also Sec. 4.8). This work goes into kinetic energy, causing the movement along the horizontal path $\overline{13}$ to be accelerated, with the kinetic energy at point 3 equal to the area 1231. If there were no damping, the structure would then oscillate about point 3 indefinitely, but in reality these oscillations will be damped and the structure will ultimately come to equilibrium at point 3. When the load is further increased, the stable path $\overline{34}$ is followed.

When the load P is decreased, the structure follows path $\overline{432}$ in Figure 4.14b, until a critical state is reached at point 2. If the load is further decreased by an infinitesimal amount, there is suddenly no adjacent equilibrium state and the structure will snap through to point 5. This snapthrough will again be dynamic, and the work corresponding to area 2152 will be converted to kinetic energy and then dissipated as heat. Due to damping, the structure will eventually come to equilibrium at point 5. If the load is further decreased, the stable segment $\overline{56}$ is followed.

From this discussion we see that segment $\overline{12}$, which has a negative slope, can never be reached under load control. Nevertheless, it can be reached in a stable manner under displacement control. We will discuss it in more detail later (Sec. 4.8).

Even though the structure is perfectly elastic, it exhibits hysteresis. The energy lost during a complete load–unload cycle, which goes first into kinetic energy of vibrations and is ultimately dissipated as heat due to ever-present damping, is given by the area 132501.

Note that in contrast to the previous examples, the snapthrough instability does not involve any bifurcation of the equilibrium path.

The snapthrough instability is also called the *limit-point instability,* or snap instability. This term, however, does not make a clear distinction from the snapdown instability that occurs under displacement control and will be examined later (Sec. 4.8). Structural behavior with a negative tangential stiffness, that is, declining load-deflection diagram, is also called *softening.*

Consider now what simplification is possible when the angles q and α are small. Then one may set $\cos\alpha \simeq 1-\alpha^2/2$ and $1/\cos q \simeq 1+q^2/2$, $\tan q \simeq q$ and $\tan\alpha \simeq \alpha$. Equation 4.4.1 reduces to

$$\Pi = \frac{EAL}{4}(q^2-\alpha^2)^2 - PL(\alpha - q) \tag{4.4.6}$$

which is a fourth-degree polynomial. A quadratic approximation, however, would lose the essential feature of this problem. The equilibrium condition $\partial\Pi/\partial q = 0$ yields

$$P = EAq(\alpha^2 - q^2) \tag{4.4.7}$$

The stability condition $\partial^2\Pi/\partial q^2 > 0$, which may be again shown to be equivalent to the condition $\partial P/\partial q > 0$, yields the regions of stable equilibrium:

$$q > \frac{\alpha}{\sqrt{3}} \quad \text{or} \quad q < \frac{-\alpha}{\sqrt{3}} \quad \text{(stable)} \tag{4.4.8}$$

The behavior obtained by this small-angle approximation is obviously similar, and for very small angles asymptotically equivalent, to our previous exact result.

The essential aspect of snapthrough buckling is that it leads to a nonlinear problem that cannot be meaningfully linearized. For very small angles q and α, all the quartic and cubic terms in Equation 4.4.6 could be dropped, making Π quadratic. But then the equilibrium condition $\partial\Pi/\partial q = 0$ would yield a linear equation between P and q that does not formulate an eigenvalue problem and yields no critical load. This linear equation would describe the tangent of the equilibrium path at $q = 0$, which is obviously useless for determining failure. Likewise, linearization could be carried out about the initial state $q = \alpha_0$, but this would yield a linear equation that describes the tangent to the equilibrium path at the initial state and is, therefore, useless for predicting failure. So we see that there are buckling problems that are inherently nonlinear and cannot be linearized even if the deflections are very small. We will see more problems of this kind in this and the following sections. The nonlinearity aspect is essential in the buckling of axially loaded cylindrical shells (see Sec. 7.5).

Flat Arches

The snapthrough instability is exhibited by flat (or shallow) arches, that is, arches whose rise is small compared to the span. Timoshenko (1935) and Biezeno (1938) presented solutions for the cases of distributed and concentrated load, respectively, and Marguerre (1938) discussed some implications for the theory of buckling. Fung and Kaplan (1952) considered various types of arches and of lateral load. Hoff and Bruce (1953) analyzed the problem dynamically, and Masur and Lo (1972) gave a general discussion of the problem including imperfection analysis. Nonlinear material behavior was considered by Franciosi, Augusti, and Sparacio (1964) and loading through an elastic foundation was studied by Simitses (1973). Experimental results were reported by Roorda (1965).

As experiments as well as theoretical studies confirm, flat arches may be assumed to fail in a symmetric mode, whose basic characteristic is the shortening of the center line of the arch. This contrasts with the behavior of high arches (also called deep arches, see Sec. 2.8), for which the center-line shortening is negligible, making the asymmetry of bending the paramount feature. Analysis of high arches, on the other hand, must include the curvature term w/R^2 (Eq. 2.8.2), which is negligible for flat arches (since R is large).

Consider a two-hinge arch whose initial shape before loading is

$$z_0(x) = a \sin \frac{\pi x}{l} \tag{4.4.9}$$

where l = length of the span (Fig. 4.15a) and a = rise of the arch; $a \ll l$ since the arch is flat. The arch is loaded by a vertical distributed load p. The solution is

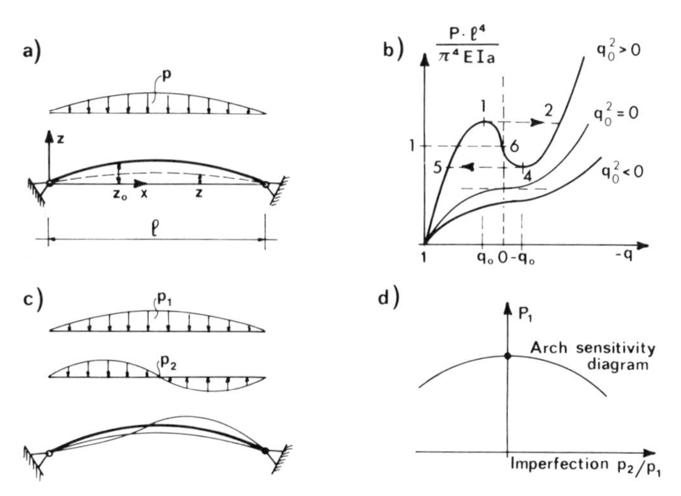

Figure 4.15 (a) Flat arch subjected to symmetric loading, (b) equilibrium paths, (c) symmetric and antisymmetric disturbances, and (d) imperfection sensitivity.

simple if the distributed load is sinusoidal:

$$p = P \sin \frac{\pi x}{l} \tag{4.4.10}$$

The deflection ordinate may be sought in the form:

$$z = qa \sin \frac{\pi x}{l} \tag{4.4.11}$$

The potential energy of the flat arch may be expressed as

$$\Pi = \int_0^l \frac{1}{2} EI(\Delta k)^2 \, dx + \frac{1}{2} EA\left(\frac{\Delta l}{l}\right)^2 l - \int_0^l p(z_0 - z) \, dx \tag{4.4.12}$$

in which EI and EA are considered to be constant along x, and

$$\Delta k \simeq z'' - z_0'' = \frac{\pi^2}{l^2} a(1-q) \sin \frac{\pi x}{l} \tag{4.4.13a}$$

$$\Delta l = \int_0^l \sqrt{dx^2 + dz^2} - \int_0^l \sqrt{dx^2 + dz_0^2} \simeq \frac{1}{2} \int_0^l (z'^2 - z_0'^2) \, dx = \frac{\pi^2 a^2}{4l} (q^2 - 1) \tag{4.4.13b}$$

Δk represents the change of curvature of the arch, and Δl the change of length of the center line of the arch (for the case of nonuniform EI, see Simitses and Rapp, 1977). (We neglect the term w/R^2 that we included in the curvature change expression for high arches, Eq. 2.8.2; the reason is that for flat arches R^2 may be assumed to be very large.) Substituting Equations 4.4.9 to 4.4.11 and 4.4.13 into Equation 4.4.12, and integrating, one obtains

$$\begin{aligned}
\Pi(q) &= \int_0^l \left[\frac{EI}{2} a^2\left(\frac{\pi^4}{l^4}\right)(1-q)^2\left(\sin\frac{\pi x}{l}\right)^2 - Pa(1-q)\left(\sin\frac{\pi x}{l}\right)^2\right] dx \\
&\quad + \frac{EA}{2l}\left(\frac{\pi^4}{16l^2}\right)a^2(1-q^2)^2 \\
&= \frac{EI}{2} a^2\left(\frac{\pi^4}{l^4}\right)(1-q)^2\left(\frac{l}{2}\right) + \frac{EA}{32}\left(\frac{\pi^4}{l^3}\right)a^4(1-q^2)^2 - Pa(1-q)\left(\frac{l}{2}\right) \\
&= \frac{\pi^4 EI}{2l^3} a^2\left[\frac{1}{2}(1-q)^2 + \frac{n}{4}(1-q^2)^2\right] - \frac{a}{2} Pl(1-q)
\end{aligned} \tag{4.4.14}$$

in which we introduce the notation $n = Aa^2/4I = a^2/4r^2$ = nondimensional parameter of arch slenderness, r = radius of gyration. Differentiating, we obtain for the equilibium states the condition:

$$\frac{\partial \Pi}{\partial q} = -\frac{\pi^4 EI}{2l^3} a^2[1 + (n-1)q - nq^3] + \frac{a}{2} Pl = 0 \tag{4.4.15}$$

which furnishes

$$P = \frac{\pi^4}{l^4} EIa[1 + (n-1)q - nq^3] \tag{4.4.16}$$

The diagram of $P(q)$ may, but need not, have limit points (Fig. 4.15b). The condition of the limit points is $\partial P/\partial q = 0$, which yields $n - 1 - 3nq^2 = 0$ or

$$q = \pm q_0 = \pm\sqrt{\frac{n-1}{3n}} \tag{4.4.17}$$

The equilibrium diagram has limit points only if $n > 1$, that is, if the ratio of the rise of the arch to the radius of gyration is not too small.

The equilibrium is stable if

$$\frac{\partial^2 \Pi}{\partial q^2} = -\frac{EI\pi^4}{2l^3} a^2(n - 1 - 3nq^2) > 0 \tag{4.4.18}$$

This is satisfied if $n - 1 - 3nq^2 < 0$ or $q_0^2 - q^2 > 0$, or $(q_0 - q)(q_0 + q) < 0$. So we get the stability condition:

$$\text{Either} \quad q > q_0 \qquad \text{or} \qquad q < -q_0 \tag{4.4.19}$$

For $-q_0 \le q \le q_0$, the arch is unstable. It is again easy to check that the states $q = \pm q_0$ are identical to the maximum and minimum points of the equilibrium curve of P versus q.

The equilibrium path of the system can be obtained also from the principle of virtual work. Another method, used by Timoshenko and Gere (1961), is to apply equilibrium conditions to a primary system in which one hinge is allowed to slide horizontally and then restore compatibility by introducing the statically indeterminate horizontal thrust in the arch.

When the distributed load p is uniform, $p = p_0$, Equation 4.4.10 may be approximately replaced by the first term of the Fourier sine expansion of uniform load, which is $p = [4P \sin(\pi x/l)]/\pi$. The rest of the analysis is similar. It is found that the resulting solution is only 0.38 percent less than an accurate solution for the critical value of a uniform load. More sinusoidal Fourier components may be considered to obtain more accurate solutions, for any $p(x)$. But the first term has dominant influence.

Except for the fact that for $n < 1$ there are no limit points (no critical states) and the response is always stable (Fig. 4.15b), the behavior of the present system is entirely analogous to the von Mises truss. Especially, the flat arch exhibits snapthrough buckling (Fig. 4.15b, path $\overline{12}$), and the snapthrough leads to hysteretic energy dissipation due to spontaneous conversions of potential energy into kinetic energy and then into heat.

Effect of Imperfections

The main difference from high arches is that shortening of the arch center line is important while antisymmetric bending is not. This may be verified by replacing Equations 4.4.10 and 4.4.11 by more general expressions:

$$p = P_1 \sin\frac{\pi x}{l} + P_2 \sin\frac{2\pi x}{l} \tag{4.4.20a}$$

$$z(x) = q_1 a \sin\frac{\pi x}{l} + q_2 a \sin\frac{2\pi x}{l} \tag{4.4.20b}$$

in which the added (antisymmetric) sine terms may be regarded as imperfections perturbing symmetry (see Fig. 4.15c); P_1 and P_2 are given parameters for symmetric and antisymmetric loading (in general, the first two components of a Fourier series expansion), and q_1, q_2 are the generalized displacements characterizing the symmetric and antisymmetric deformation. Substituting Equations 4.4.20a and 4.4.20b, along with Equations 4.4.9, 4.4.13a, and 4.4.13b, into Equation 4.4.12 and integrating, one gets the potential-energy expression:

$$\Pi = \frac{Bl}{2}\left[\frac{1}{2}(q_1-1)^2 + \frac{n}{4}(1-q_1^2)^2\right] + P_1\left(\frac{al}{2}\right)(q_1-1) + P_2\left(\frac{al}{2}\right)q_2 + 4Blq_2^2$$
$$+ 2nBlq_2^4 + Bln(q_1^2-1)q_2^2 \qquad (4.4.21)$$

in which $B = \pi^4 EIa^2/l^4$. The equilibrium conditions then are

$$\frac{2}{Bl}\frac{\partial\Pi}{\partial q_1} = q_1 - 1 - n(1-q_1^2)q_1 + P_1\left(\frac{a}{B}\right) + 4nq_1q_2^2 = 0 \qquad (4.4.22a)$$

$$\frac{2}{Bl}\frac{\partial\Pi}{\partial q_2} = P_2\left(\frac{a}{B}\right) + 4(4 - n + 4nq_2^2 + nq_1^2)q_2 = 0 \qquad (4.4.22b)$$

Solution of these equations yields the equilibrium paths $P_1(q_1, q_2)$ and $P_2(q_1, q_2)$.

A question of concern is whether a small inevitable imperfection of load represented by the load asymmetry parameter P_2 can greatly reduce the critical load of the arch. Consider, therefore, that P_2 and q_2 are very small while P_1 and q_1 are finite. In Equation 4.4.22b the term $4nq_2^2$ may then be neglected, and so this equation yields $q_2 = -f_0(q_1)P_2$ where $f_0(q_1)$ is a function of q_1. Substituting this into Equation 4.4.22a, we find that $P_1 = f_1(q_1) - f_2(q_1)P_2^2$, where $f_1(q_1)$ and $f_2(q_1)$ are positive functions and $f_1(q_1)$ represents the previous solution for symmetric load ($P_2 = 0$), that is, Equation 4.4.16. This is valid for all q_1, including the critical value. Thus we may conclude that if P_2 is small, then the change in load-carrying capacity for the symmetric load component is second-order small (Fig. 4.15d). This is a rather weak imperfection sensitivity. It means that, in stability analysis of flat arches, one does not need to worry about inevitable small asymmetric load components. (By contrast, later we will see buckling problems in which a small imperfection of magnitude ξ causes a critical load reduction that is proportional to $\xi^{1/2}$ or $\xi^{2/3}$, which represents a much more severe sensitivity to imperfections.)

Another type of imperfection of interest for snapthrough is an initial dynamic disturbance. Imagine that kinetic energy ΔT equal to the cross-hatched area in Figure 4.16a is initially imparted to the flat arch. Then the arch can snap through along the horizontal path $\overline{AB}$ at load $P_{cr} - \Delta P$, which is lower than the static critical load P_{cr}. Near the critical points, the equilibrium curve may be approximated as a parabola. For a parabola the area $A1C$, and thus also ΔT, is proportional to $\Delta P^{3/2}$, and so the lowering of the critical load, ΔP, due to initial dynamic disturbance is proportional to $\Delta T^{2/3}$. This means it is proportional to $v^{4/3}$, v being the initial downward velocity of the arch at the apex. This is not a very strong imperfection sensitivity; if v is very small, the load reduction is higher-order small, that is, $\sim v^{4/3}$).

There is one particular dynamic path (Fig. 4.16b) for which the cross-hatched areas $A1C$ and $C2B$ are exactly equal. For that path there is no energy loss.

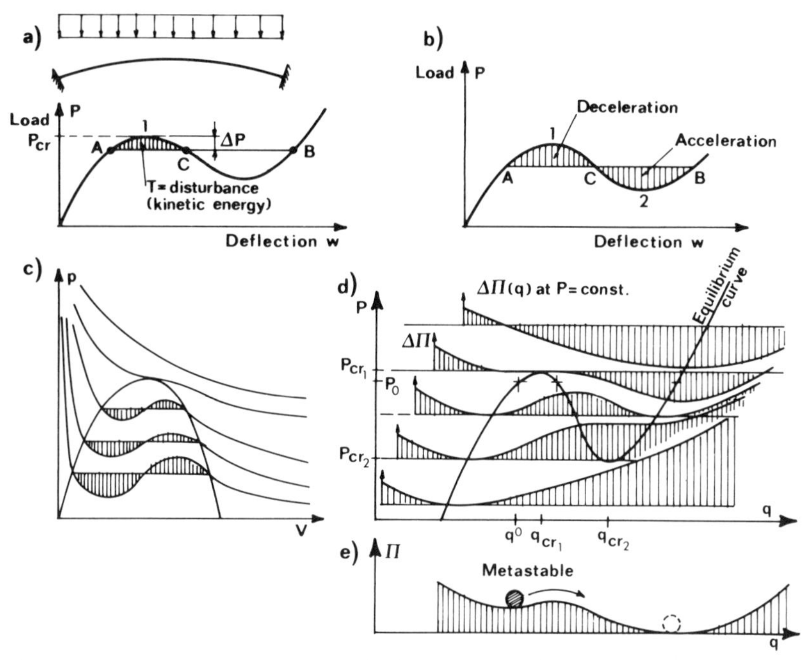

Figure 4.16 (a, b) Horizontal snapthrough due to dynamic disturbance; (c) similarity to pressure-volume diagram for real gases; (d, e) variation of potential energy at constant P.

The horizontal snapthrough has an analogy in the pressure-volume diagram according to the van der Waals equation for real gases (Fig. 4.16c; see, e.g., "Thermodynamics," in Guggenheim, *Enc. Brit.*, 1980). In this diagram, due to inevitable disturbances in a gas, the real behavior usually consists of a horizontal snapthrough which cuts equal areas below and above as shown in Figure 4.16c, and is therefore equivalent in energy to the curved diagram. (Note that this pressure-volume diagram is also analogous to a strain-softening material; Chap. 13.)

For a state (q_0, P_0) that occurs just before reaching P_{cr} (Fig. 4.16d) the potential-energy curve (at $P = \text{const.}$) has the shape shown in Figure 4.16e, and the structure behaves as a ball rolling on this curve. In physics and chemistry, such a situation is called *metastable*. It is a special case of stability in which a small disturbance can lead to another stable state that exists nearby, at a finite distance away, and has a lower potential energy (or, more generally, a lower value of some thermodynamic potential).

Other Examples of Snapthrough

Snapthrough buckling is exhibited by other types of arches or trusses; for example, the tied arch or tied truss (Fig. 4.17a, b) with simple supports, or a flat fixed arch (Fig. 4.17c), for which the choice of the deflected shape must respect the condition of zero rotation at the supports. The long strip of a flat cylindrical shell, uniformly loaded, also exhibits snapthrough buckling and our solution for the flat arch may be directly applied. Snapthrough buckling is also exhibited by a

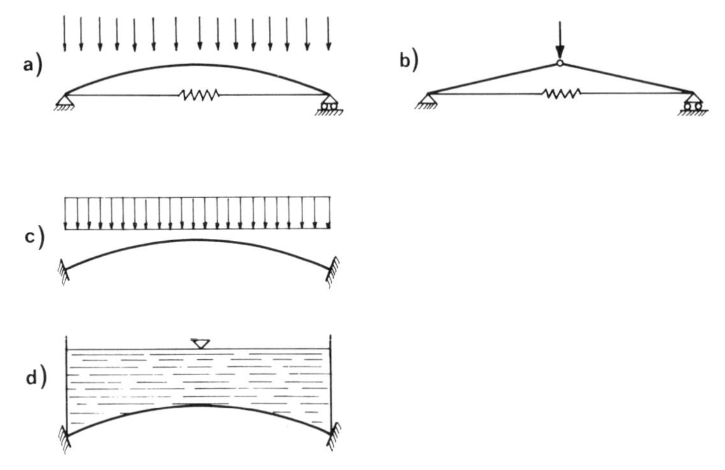

Figure 4.17 Examples of systems which exhibit snapthrough buckling: (a) tied arch, (b) tied truss, (c) flat fixed arch, and (d) shell under hydrostatic pressure.

flat spherical dome shell under hydrostatic pressure (Fig. 4.17d). Snapthrough buckling of rectangular panels of cylindrical shells of finite length requires a rather sophisticated analysis, but the behavior is similar.

Other examples of snapthrough instability are found in various switches (e.g., the standard wall switch for electric lights). Safety ski bindings involve a mechanism that behaves in the same manner as snapthrough buckling (e.g., Bažant, 1960; and Bažant's Patent No. 97175 of safety ski binding, Czechoslovakia, 1959).

Problems

4.4.1 Find the critical load of the structural systems in Figure 4.18a, b, c assuming that the axial force in the bars is less than the Euler load. For Figure 4.18b solve for (a) $C > 0$, EA finite ($<\infty$); and (b) $C = 0$, EA finite. For Figure 4.18c solve (a) $\Pi = \Pi(\theta)$ and $P = P(\theta)$; (b) critical loads; (c) stable regions, (d) limit cases: (1) $C \to 0$, $A_c > 0$; (2) $C > 0$, $A_c \to 0$.

4.4.2 Find the critical load of a flat tied bridge arch.

4.4.3 Find the value of arch rise a such that critical loads from flat arch theory and high arch theory become the same. Do this for (a) a hinged arch (Fig. 4.18d) and (b) a tied arch (Fig. 4.18e).

4.4.4 For the low portal frame of a large span in Figure 4.18f, calculate $\Pi = \Pi(w)$, $P = P(w)$ by virtual work. *Hint:* Solve the stiffness of the L-shape segment first. Plot the result and discuss stability.

4.4.5 Do a similar analysis for the overpass-type frame in Figure 4.18g. *Hint:* Solve stiffnesses, then proceed as in Problem 4.4.4.

4.4.6 Solve $\Pi(w)$, $P(w)$ for the four-bar pyramidal roof truss in Figure 4.18h.

4.4.7 Do the same for the three-bar pyramidal roof truss in Figure 4.18i.

4.4.8 Solve the critical load of a fixed cylindrical shell panel that is shallow (i.e., its rise above the chord of the arc is not high) and is loaded by water pressure, as shown in Figure 4.18j. Assume $z_0 \simeq ax(l - x)$ and $w \simeq bx^2(l - x)^2$ where

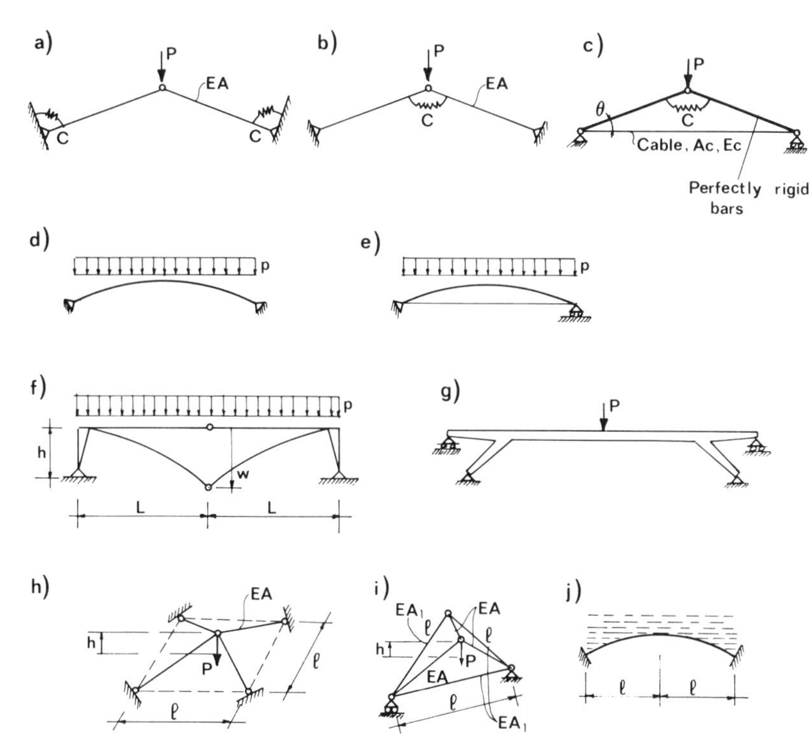

Figure 4.18 Exercise problems on snapthrough buckling.

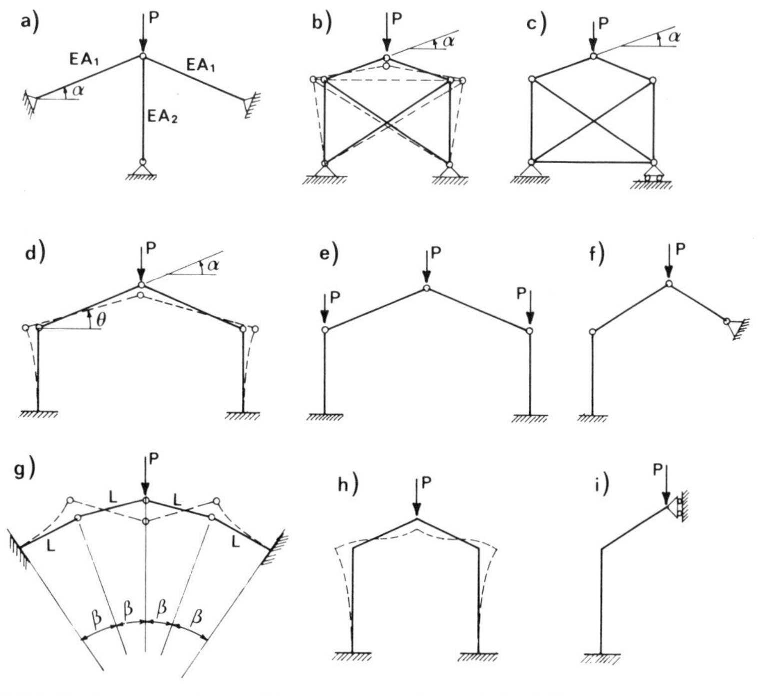

Figure 4.19 Further exercise problems on snapthrough buckling.

$a, b =$ constants. (Note that the work of load is $p\,\Delta V$, where $p =$ water pressure and $\Delta V =$ volume of water displaced by deflection.)

4.4.9 For what values of A_2/A_1 and α does the system in Figure 4.19a exhibit snapthrough?

4.4.10 For the systems in Figure 4.19b, c find for which α the critical load for snapthrough becomes equal to the critical load for bifurcation buckling of this frame. (*Note:* Mode interaction then produces intricate postcritical behavior.)

4.4.11 The systems in Figure 4.19d, e, f, g consist of axially inextensible bars in which the axial force is less than the Euler load. Solve the $P(\theta)$ diagram. Find the critical load. Discuss stability.

4.4.12 Find the critical load for snapthrough for the rigidly jointed frames in Figure 4.19h, i. *Hint:* Snapthrough analysis requires taking into account the effect of axial shortening of the members on their rotation.

4.5 LARGE-DEFLECTION POSTCRITICAL BEHAVIOR AND TYPES OF BIFURCATION

Our analysis of bifurcating systems (Sec. 4.3) has so far been limited to the linearized formulation for small deflections. Under that simplification, the stable regions of the perfect and imperfect systems are the same because the second variations $\delta^2\Pi$ for both systems are given by identical expressions. Not so, however, for the nonlinear, finite-deflection theory. The stable regions of the imperfect and perfect systems then need not be the same.

Symmetric Stable Bifurcation: Example

Consider first the column shown in Figure 4.20a, consisting of a perfectly rigid bar of length L held upright by means of a rotational spring of stiffness C (analyzed by Augusti, 1961). The load P remains constant and vertical, and the column initially has a small inclination angle α (for $\alpha = 0$ the column is perfect, and we studied this case in Sec. 4.2). Let $q =$ inclination angle of the bar. The potential energy is

$$\Pi = U - W = \tfrac{1}{2}C(q-\alpha)^2 - PL(\cos\alpha - \cos q) \tag{4.5.1}$$

Its derivatives are

$$\frac{\partial\Pi}{\partial q} = C(q-\alpha) - PL\sin q \qquad \frac{\partial^2\Pi}{\partial q^2} = C - PL\cos q \tag{4.5.2}$$

Setting $\partial\Pi/\partial q = 0$, we have the equilibrium condition:

$$P = \frac{C}{L}\left(\frac{q-\alpha}{\sin q}\right) \tag{4.5.3}$$

The equilibrium diagrams $P(\theta)$ are plotted (as the solid and dashed curves) in Figure 4.20b for various values of the initial imperfection, α. Setting $\partial^2\Pi/\partial q^2 = 0$, we find that the critical states lie on the curve (dash-dot curve in Fig. 4.20b):

$$P_{\text{cr}} = \frac{C}{L\cos q} \tag{4.5.4}$$

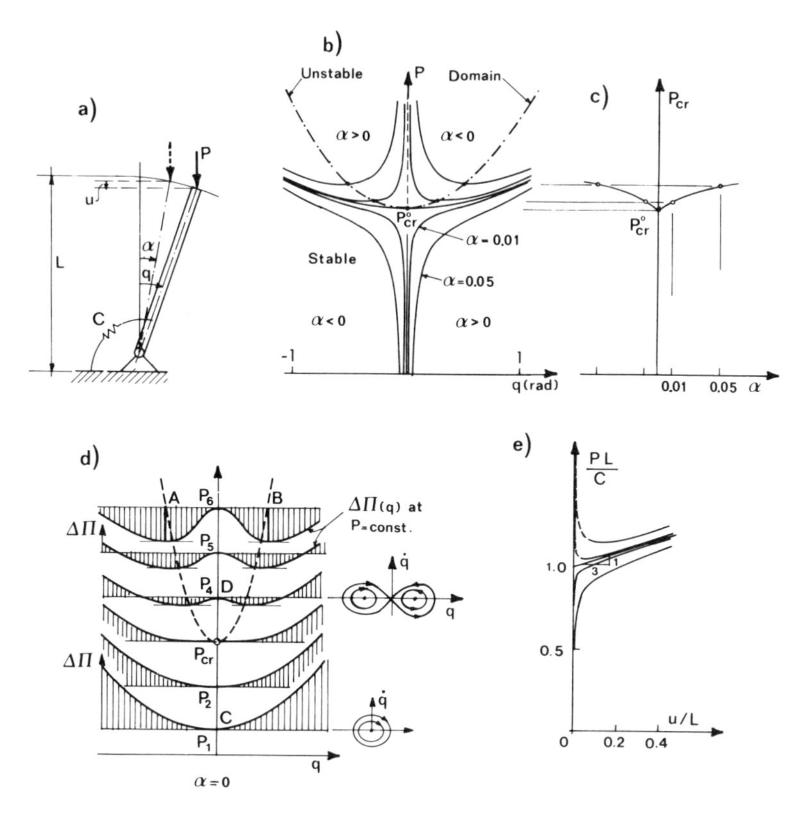

Figure 4.20 (a) Rigid-bar column with rotational spring at base, (b) equilibrium path, (c) imperfection sensitivity diagram, (d) variation of potential energy at constant P, and (e) load-displacement diagrams.

The critical load of a perfect column ($\alpha = 0$) is $P^0_{cr} = C/L = \lim P_{cr}$ for $q \to 0$, or $\lim P$ (Eq. 4.5.3) for $q \to 0$.

Examine now stability of the equilibrium states. Substituting the equilibrium value of P (Eq. 4.5.3) into Equations 4.5.2, we get

$$\frac{\partial^2 \Pi}{\partial q^2} = C[1 - (q - \alpha)\cot q] \tag{4.5.5}$$

The column is stable if $\partial^2\Pi/\partial q^2 > 0$. So the stability condition is $\tan q > q - \alpha$ if $\alpha < q < \pi/2$ and $\tan q < q - \alpha$ if $-\pi/2 < q < 0$. This may be related to the slope of the equilibrium diagram, which is given by

$$\frac{\partial P}{\partial q} = \frac{C}{L}\left(\frac{1}{\sin q}\right)[1 - (q - \alpha)\cot q] \tag{4.5.6}$$

Consider now that $\alpha > 0$. The equilibrium curves are then the curve families at bottom right and top left in Figure 4.20b. From Equation 4.5.6 we conclude that, for $q > \alpha$, the equilibrium curves for $\alpha > 0$ in Figure 4.20b are stable when their slope is positive, while for $q < 0$ they are stable when their slope is negative.

The point of zero slope on each curve represents the critical state (Eq. 4.5.4). According to Equation 4.5.5, the value of q at the critical state is related to α by the transcendental equation $q - \tan q = \alpha$. Note that this column can be stable for loads higher than the critical load P^0_{cr} of the perfect column. Also, critical states exist for $P > P^0_{cr}$ even if the system is imperfect (see the dash-dot curve connecting the critical states in Fig. 4.20b).

Repeating the same derivation for $\alpha < 0$ one gets symmetrical results—the curve families at bottom left and top right in Figure 4.20b. It may be concluded that for $q > 0$ stability is always associated with positive slopes, and for $q < 0$ with negative ones.

Although, for this column, the critical states other than the one for the perfect structure cannot be reached under load control, it is interesting to determine the dependence of the critical load on the initial imperfection α (Fig. 4.20c). The diagram for this dependence is generally called the imperfection sensitivity diagram. According to the condition $\partial^2\Pi/\partial q^2 = 0$ and Equation 4.5.5, the critical states are characterized by $q - \alpha = \tan q$. For small values of q (and α), we have $\tan q \simeq q + \frac{1}{3}q^3$, $\cos q \simeq 1 - \frac{1}{2}q^2$, and substituting these expressions into Equation 4.5.4 we obtain

$$P_{cr} = \frac{C}{L}\left[1 + \frac{3^{2/3}}{2}(\alpha^{2/3})\right] \qquad \text{(for small } \alpha\text{)} \tag{4.5.7}$$

The perfect column has a point of bifurcation at $P = P^0_{cr}$, and for $P > P^0_{cr}$ it follows a rising postbuckling path. Therefore this type of bifurcation is called *stable*. It is also *symmetric*, because the postbuckling path is symmetric with q. The critical state for the perfect column is in this case stable since for $P = C/L = P^0_{cr}$ and $\alpha = 0$ we have $\Pi = C(\frac{1}{2}q^2 - 1 + \cos q) \simeq q^4 C/4! > 0$. [This illustrates that stability at the critical state is decided by fourth-order terms in the expansion of Π. However, if the solution were done only with second-order accuracy, as is usually the case for more complex structures, then the critical state appears to be a neutral equilibrium state ($\Pi = 0$ for any q).]

The shape of the potential-energy function near the equilibrium state at various stages of loading is sketched for the case $\alpha = 0$ in Figure 4.20d. Positive curvature indicates stability.

With regard to Chapter 3 on dynamics (Secs. 3.5–3.6), Figure 4.20d shows for various levels of load P the diagrams of constant total energy $E = \Pi + T$ (where T = kinetic energy) plotted in the phase space $(q, \dot{q})$. These diagrams show the trajectories followed by the motion of the column if it is undamped. A set of closed trajectories shrinking into a point indicates stability (points A, B, C in Fig. 4.20d), and the lack of closed contours around a point indicates instability (point D in Fig. 4.20d).

It is also important to understand the diagram of load P versus load-point displacement u. We have (Fig. 4.20a) $u = L(\cos\alpha - \cos q)$. Restricting attention to small angles q and α, we may set $\cos q = 1 - q^2/2$, $\cos\alpha = 1 - \alpha^2/2$, which yields $q = \pm(\alpha^2 + 2u/L)^{1/2}$. Substituting this approximation, which is accurate up to terms of third order in q, and also the approximation $\sin q \simeq q - \frac{1}{6}q^3$, we obtain from Equation 4.5.3

$$\frac{PL}{C} = \frac{1 \mp \alpha(\alpha^2 + 2u/L)^{-1/2}}{1 - \frac{1}{6}(\alpha^2 + 2u/L)} \tag{4.5.8}$$

For the perfect column we have

$$\frac{PL}{C} = \frac{1}{1 - u/3L} \simeq 1 + \frac{1}{3}\frac{u}{L} \tag{4.5.9}$$

The diagrams of these relations are plotted in Figure 4.20e. We see that in the plot of load versus load-point displacement, the initial postcritical response is given by a straight line of slope $K_p = \partial P/\partial u = \partial^2\Pi/\partial u^2 > 0$. On the other hand, the force associated with q by work is not P, but a disturbing applied moment m (since $m\,dq$ = work). The tangential stiffness associated with q is $K_m = \partial m/\partial q = \partial^2\Pi/\partial q^2$. Note that $K_m = 0$ at the critical state ($q = 0$) (this is a special case of the condition $\det \mathbf{K} = 0$). This means that there is a neutral equilibrium at the critical state. This further implies that applying an arbitrarily small disturbance m causes at critical state a finite rotation q. However, this is detectable neither in the $P(q)$ nor in the $P(u)$ plot because $m = 0$ for these plots. The fact that $\partial P/\partial q = 0$ for $q = 0$ (and $m = 0$) is *not* the cause of neutral equilibrium because P and q are not associated by work.

To obtain only the behavior near the bifurcation point, the potential energy in Equation 4.5.1 may be approximated by a polynomial. It is interesting to note that it does not suffice to substitute in Equation 4.5.1 $\cos q \simeq 1 - \frac{1}{2}q^2$, $\cos\alpha = 1 - \frac{1}{2}\alpha^2$; one can check that such simplification yields only the critical load $P^0_{cr} = C/L$ of the perfect system (from $\partial^2\Pi/\partial q^2 = 0$) but nothing about the postcritical behavior. One must include at least the next higher-order term in the approximation of Π, that is, set $\cos q \simeq 1 - \frac{1}{2}q^2 + \frac{1}{24}q^4$, $\cos\alpha \simeq 1 - \frac{1}{2}\alpha^2 + \frac{1}{24}\alpha^4$ in Equation 4.5.1, and then

$$\Pi = \tfrac{1}{2}C(q - \alpha)^2 + \tfrac{1}{24}PL(q^4 - 12q^2 - \alpha^4 + 12\alpha^2) \tag{4.5.10}$$

One may now follow the same procedure as before to obtain the asymptotic approximations for equilibrium paths $P(q, \alpha)$ and critical states $P_{cr}(\alpha)$ (see problem assignment). A salient property of the stable symmetric bifurcation is that the polynomial expansion of Π contains no cubic term and has a positive quartic term.

The behavior of the rigid-bar column supported by a rotational spring (Fig. 4.20a) is entirely analogous to the large-deflection behavior of an elastic column as solved in Section 1.9. The equilibrium diagrams obtained (Fig. 1.25) were similar to those in Figure 4.20b, and it can be shown in general that the rising portions of these diagrams are stable, the declining portions are unstable, and the critical states are characterized by a horizontal tangent (see our later general discussion of equilibrium curves). The load-point displacement diagram (Fig. 1.26) was also similar to the one in Figure 4.20e.

Symmetric Unstable Bifurcation: Example

The fact that the critical states of the imperfect column occur at loads higher than P^0_{cr} for the perfect system is particular to this example and is not verified in general. To illustrate it, consider a slightly different column, as shown in Figure 4.21a, in which the rotational spring at the base is replaced by a horizontal, vertically sliding spring of stiffness C. Unlike before, the resisting moment of the

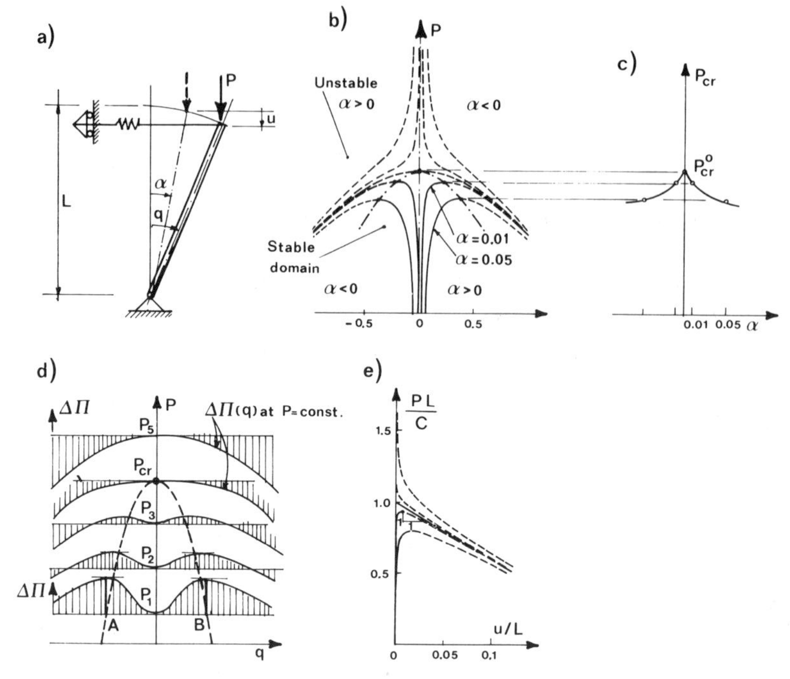

Figure 4.21 (a) Rigid-bar column with horizontal spring at top, (b) equilibrium path, (c) imperfection sensitivity diagram, (d) variation of potential energy at constant P, and (e) load-displacement diagrams.

spring force about the hinge at the base is not proportional to the rotation at the base, because the force arm of the spring decreases as the column deflects. The potential energy is

$$\Pi = \tfrac{1}{2}CL^2(\sin q - \sin \alpha)^2 - PL(\cos \alpha - \cos q) \tag{4.5.11}$$

By differentiation:

$$\begin{aligned} \frac{\partial \Pi}{\partial q} &= CL^2(\sin q - \sin \alpha)\cos q - PL \sin q \\ \frac{\partial^2 \Pi}{\partial q^2} &= CL^2(\cos 2q + \sin \alpha \sin q) - PL \cos q \end{aligned} \tag{4.5.12}$$

Setting $\partial \Pi / \partial q = 0$, we obtain the equilibrium condition:

$$P = CL\left(1 - \frac{\sin \alpha}{\sin q}\right)\cos q \tag{4.5.13}$$

The equilibrium diagrams $P(\theta)$ for various values of the initial imperfection α are plotted in Figure 4.21b. Setting $\partial^2\Pi/\partial q^2 = 0$, we find that the critical states lie on the curve (dash-dot curve in Fig. 4.21b):

$$P_{cr} = CL\frac{\cos 2q + \sin \alpha \sin q}{\cos q} \tag{4.5.14}$$

The critical load of a perfect column ($\alpha = 0$) is $P^0_{cr} = CL = \lim P_{cr}$ for $q \to 0$, or $\lim P$ (Eq. 4.5.13) for $q \to 0$.

Examine now stability of the equilibrium states. Substituting P from Equation 4.5.13 into Equations 4.5.12 we get

$$\frac{\partial^2 \Pi}{\partial q^2} = \frac{CL^2}{\sin q}(\sin \alpha - \sin^3 q) \qquad (4.5.15)$$

It may be noted that the condition of positiveness of this expression is identical (for $0 \leq q \leq 90°$) to the condition $\partial P/\partial q > 0$. Thus, as is generally true, the rising portions of the equilibrium curves in Figure 4.21b are stable, the declining portions are unstable, and the critical states occur at the limit points, that is, the points with a horizontal tangent. From Equation 4.5.15 the critical states are characterized by the condition $\sin q_{cr} = (\sin \alpha)^{1/3}$.

In one respect the behavior of the present column is markedly different from that in Figure 4.20a. The critical states of the imperfect system occur at loads less than the critical load of the perfect system. The larger the imperfection, the smaller is the critical load. For such systems, it is important to quantify the dependence of the critical load P_{cr} of the imperfect system on the magnitude α of the imperfection. Substituting $\sin q = \sin q_{cr} = (\sin \alpha)^{1/3}$ and $\cos q = \cos q_{cr} = [1 - (\sin \alpha)^{2/3}]^{1/2}$ into Equation 4.5.13, we obtain for the critical load of the imperfect column the equation

$$P_{cr} = CL[1 - (\sin \alpha)^{2/3}][1 - (\sin \alpha)^{2/3}]^{1/2} \simeq CL(1 - \tfrac{3}{2}\alpha^{2/3}) \qquad (4.5.16)$$

in which the last expression applies to sufficiently small imperfections α. Note that the critical load decrease is proportional to the $\frac{2}{3}$ power of the imperfection, a behavior which is typical of many systems (see Sec. 4.6). Therefore, the tangent of the imperfection sensitivity diagram $P_{cr}(\alpha)$ (Fig. 4.21c) has a vertical downward slope at $\alpha = 0$. This means that the critical load decreases very rapidly with only a very small imperfection.

Structures for which the critical load (i.e., maximum load) decreases at increasing imperfection are called *imperfection sensitive* (Fig. 4.21c) while those with the opposite behavior (Fig. 4.20c) are called *imperfection insensitive*. The imperfection sensitivity diagram $P_{cr}(\alpha)$ always begins with a vertical downward slope (see Eq. 4.6.6). If the drop of P_{cr} is large for typical imperfection values in practice, the structure is said to be *strongly imperfection sensitive*. For columns and frames, the drop of P_{cr} due to actual imperfections is usually quite small, even though the imperfection sensitivity diagram may be starting with a vertical tangent. For example, a typical unavoidable imperfection for a column may be $\alpha = 0.05$, for which Equation 4.5.16 indicates that the critical load is reduced by 4.6 percent, not a very large reduction. For shells, however, a typical inevitable imperfection may cause a 60 to 80 percent reduction of the critical load (Chap. 7).

The equilibrium curve emanating from the bifurcation decreases with displacement symmetrically and for this reason the bifurcation is termed *unstable symmetric*. Unlike the previous example, the bifurcation state itself is unstable, according to the exact solution. The reason: For $P = CL = P^0_{cr}$ and $\alpha = 0$ we have $\Pi = CL^2(\frac{1}{2}\sin^2 q + \cos q - 1)$, and upon expanding $\sin q$ and $\cos q$ in a power series we get $\Pi \simeq -\frac{1}{8}CL^2 q^4$, which is not positive definite. On the other hand, at

$P = P^0_{cr}$ the tangential stiffness associated with q is $K_m = \partial^2\Pi/\partial q^2 = 0$; this means that one finds neutral equilibrium at the critical state. The shape of the potential-energy function near the equilibrium state at various stages of loading is sketched for the case $\alpha = 0$ in Figure 4.21d.

It is interesting to calculate again the diagram of load P versus load-point displacement w. Considering q and α to be small, noting that $u = L(\cos\alpha - \cos q)$, and introducing the approximation $(\sin q)^{-1} \simeq (1 + \frac{1}{6}q^2)/q$, which is accurate up to the third-order terms in q, we obtain from Equation 4.5.13

$$\frac{P}{CL} \simeq \left(\cos\alpha - \frac{u}{L}\right)\left[1 - \frac{\sin\alpha}{q}\left(1 + \frac{q^2}{6}\right)\right] \quad \text{with } q = \pm\left(\alpha^2 + \frac{2u}{L}\right)^{1/2} \tag{4.5.17}$$

For the perfect column ($\alpha = 0$), this reduces to

$$\frac{P}{CL} = 1 - \frac{u}{L} \tag{4.5.18}$$

These equations are plotted in Figure 4.21e. It is noteworthy that the initial postbuckling response of the perfect column consists of an inclined straight line with negative slope. This contrasts with the previous case of a rigid bar supported by a rotational spring, for which the postbuckling slope was positive.

To determine only the behavior near the bifurcation, Π can be approximated by a polynomial. As before, it does not suffice to use a quadratic approximation for $\cos q$ and $\sin q$ in Equation 4.5.11 (this would yield only P^0_{cr} but no information on the postcritical behavior). Rather, the approximation must be at least of the fourth degree, and Equation 4.5.11 yields

$$\begin{aligned}\Pi &= \frac{CL^2}{2}\left[\left(q - \frac{q^3}{6}\right) - \left(\alpha - \frac{\alpha^3}{6}\right)\right]^2 - PL\left[\left(1 - \frac{\alpha^2}{2} + \frac{\alpha^4}{24}\right) - \left(1 - \frac{q^2}{2} + \frac{q^4}{24}\right)\right] \\ &\simeq \frac{CL^2}{6}[3(q-\alpha)^2 + \alpha q(\alpha^2 + q^2) - \alpha^4 - q^4] \\ &\quad + \frac{PL}{24}(q^4 - 12q^2 - \alpha^4 + 12\alpha^2)\end{aligned} \tag{4.5.19}$$

Asymptotically, this is the same type of behavior as is obtained by the preceding procedure (see also problem assignment). For obtaining unstable symmetric bifurcation, it is essential that the cubic term (q^3) vanishes when $\alpha = 0$, and that the quartic term (q^4) is negative near the critical load; this is so because $\frac{1}{24}PLq^4 - \frac{1}{6}CL^2q^4 = \frac{1}{24}L(P - 4P^0_{cr})q^4$ and $CL = P^0_{cr}$. Note that even though Equation 4.5.19 contains the cubic term αq^3, this term has a negligible effect here because the linear term αq (which is much larger if q is small) is also present.

Asymmetric Bifurcation: Example

In the preceding examples the equilibrium curves were symmetric with regard to the undeflected equilibrium state at which bifurcation takes place. However, asymmetric equilibrium curves near the bifurcation point are also possible and quite frequent for frames (see Sec. 2.6) as well as shells. Consider the column shown in Figure 4.22a, in which a rigid bar supported by a hinge is held upright

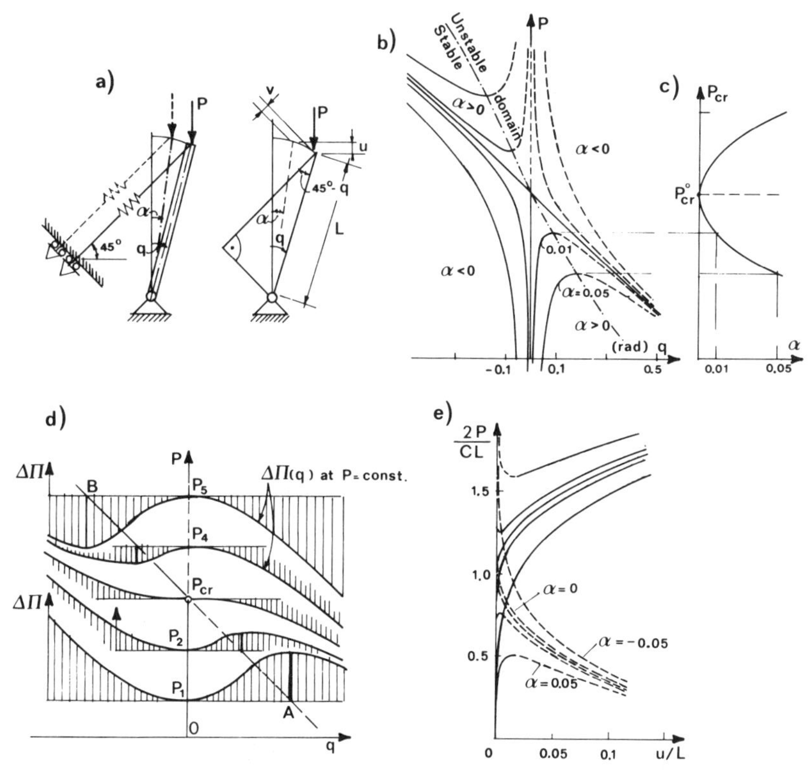

Figure 4.22 (a) Rigid-bar column with inclined spring, (b) equilibrium path, (c) imperfection sensitivity diagram, (d) variation of potential energy at constant P, and (e) load-displacement diagrams.

by a laterally sliding spring of stiffness C and constant inclination angle 45°. The column is generally imperfect, with initial inclination angle α. The extension of the spring is $v = L\cos(45° - \alpha) - L\cos(45° - q)$, and so the potential-energy expression is

$$\Pi = \frac{C}{2} L^2[\cos(45° - q) - \cos(45° - \alpha)]^2 - PL(\cos\alpha - \cos q) \qquad (4.5.20)$$

from which

$$\frac{\partial \Pi}{\partial q} = CL^2[\tfrac{1}{2}\cos 2q - \sin(45° - q)\cos(45° - \alpha)] - PL\sin q$$
$$\frac{\partial^2 \Pi}{\partial q^2} = CL^2[\cos(45° - \alpha)\cos(45° - q) - \sin 2q] - PL\cos q \qquad (4.5.21)$$

Setting $\partial\Pi/\partial q = 0$, we obtain the equilibrium curves:

$$P = \frac{CL}{\sin q}[\tfrac{1}{2}\cos 2q - \sin(45° - q)\cos(45° - \alpha)] \qquad (4.5.22)$$

These equilibrium curves are plotted for various values of the initial imperfection angle α in Figure 4.22b, and in view of our previous experience one may be stricken by the fact that these curves are asymmetric. The curve on which the initial states lie is obtained from Equations 4.5.21 by setting $\partial\Pi^2/\partial q^2 = 0$, which yields (dash-dot curve in Fig. 4.22b):

$$P_{cr} = CL\left[\frac{\cos(45° - \alpha)\cos(45° - q)}{\cos q} - 2\sin q\right] \qquad (4.5.23)$$

The critical load of perfect column ($\alpha = 0$) is $P^0_{cr} = CL/2 = \lim P_{cr}$ for $q \to 0$, or $\lim P$ (Eq. 4.5.22) for $q \to 0$.

Stability is characterized by the condition

$$\frac{\partial^2\Pi}{\partial q^2} = CL^2[\cos(45° - q)\cos(45° - \alpha) - \sin 2q] - PL\cos q > 0 \qquad (4.5.24)$$

and substituting here for P from Equation 4.5.22 we can check that the condition $\partial^2\Pi/\partial q^2 > 0$ is (for positive q) equivalent to the condition $\partial P/\partial q > 0$. Thus, as is generally true, the rising portions of the equilibrium curves in Figure 4.22b are stable, the declining portions are unstable, and for $\alpha \neq 0$ the points with horizontal tangents represent the critical states. For $\alpha = 0$ (perfect column), the bifurcation point, representing the critical state, is characterized by vanishing tangential stiffness $K_m = \partial m/\partial q = 0$, despite the fact there is no horizontal equilibrium path in the $P(q)$ plot. Indeed, $K_m = \partial^2\Pi/\partial q^2 = CL^2(\cos 45° \cos 45° - 0) - P^0_{cr}L = 0$. Consequently, an arbitrarily small disturbing moment m produces at the critical state a finite rotation q, but this is manifested neither in $P(u)$ nor in $P(q)$ plots (Fig. 4.22b, e) because $m = 0$ for these plots.

Note that an initial inclination of the column to the right causes a decrease in the critical load, while to the left it causes an increase in the critical load, compared to P^0_{cr}. This type of behavior is called asymmetric bifurcation. While symmetric bifurcations are divided into imperfection insensitive and imperfection sensitive, asymmetric bifurcation is obviously always imperfection sensitive.

The reason for the asymmetry of this bifurcation is the fact that the arm of the spring reaction force increases as the column deflects to the left, and decreases as it deflects to the right. An equivalent behavior can be obtained with a rotational spring (Fig. 4.20) with nonlinear properties, such that the incremental spring stiffness C varies with the rotation.

Let us now examine the imperfection sensitivity of the asymmetric bifurcation of our column. Restricting our attention to small angles q and α, we may substitute $\sin(45° - q) = \sin 45° \cos q - \cos 45° \sin q \simeq (1 - q - \frac{1}{2}q^2)/\sqrt{2}$, $\cos 2q \simeq 1 - 2q^2$, etc., into Equation 4.5.22. Then, after rearranging and discarding all terms of higher than second order in α and q, we obtain the approximate equilibrium diagram for small deflections

$$P = \frac{CL}{2}\left(1 - \frac{3}{2}q - \frac{\alpha}{q}\right) \qquad (4.5.25)$$

Then, setting $dP/dq = 0$ we get for the critical states the relation $q_{cr} = (2\alpha/3)^{1/2}$, and substituting this into Equation 4.5.25, we obtain the relation

$$P_{cr} = \frac{CL}{2}(1 - \sqrt{6}\sqrt{\alpha}) = P^0_{cr} - P_1\alpha^{1/2} \qquad (4.5.26)$$

Equation 4.5.26 yields the imperfection sensitivity diagram plotted in Figure 4.22c. Again, the initial slope of this diagram is vertical. The fact that Equation 4.5.26 involves a $\frac{1}{2}$-power law rather than a $\frac{2}{3}$-power law, which we found before, reveals that the imperfection sensitivity of this asymmetric bifurcation is stronger than in our previously analyzed symmetric unstable bifurcation. For imperfection angle $\alpha = 0.01$, Equation 4.5.26 indicates a drop of maximum load by 24.5 percent and, for $\alpha = 0.001$, by 7.8 percent, which is certainly quite significant for design. (A significant drop was also demonstrated for a certain frame in Sec. 2.6.) These effects are in the current practice lumped into a uniform safety factor. But obviously distinction properly should be made between imperfection-sensitive and imperfection-insensitive frames.

The shape of the potential-energy function near the equilibrium states at various stages of loading is sketched for the case $\alpha = 0$ in Figure 4.22d.

To calculate the diagram of load P versus load-point displacement u, we restrict our attention to small angles q and the perfect column $(\alpha = 0)$. Introducing the approximations $\cos q = 1 - q^2/2$, $\sin q = q(1 - q^2/6)$, $\cot q = (1 - q^2/3)/q$, we acquire from Equation 4.5.22 the relation $P = (CL/2)(1 - 3q/2)$. Then, by substituting $q \simeq \pm(2u/L)^{1/2}$ according to Equation 4.5.17, we obtain

$$\frac{P}{CL/2} \simeq 1 \mp \frac{3}{2}\left(\frac{2u}{L}\right)^{1/2} \tag{4.5.27}$$

This equation is plotted in Figure 4.22e. We see that the initial postcritical response of the perfect column is not a straight inclined line, but a curve that starts downward with a vertical slope if the column deflects to one side $(q > 0)$, and upward with a vertical slope if the column deflects to the other side. This behavior obviously produces, for the relationship of load versus load-point displacement, an extreme imperfection sensitivity.

To determine the initial postcritical behavior near bifurcation, we may write $\cos(45° - q) = (\cos q + \sin q)/\sqrt{2}$ and approximate $\cos q$, $\sin q$, $\cos\alpha$, $\sin\alpha$ in Equation 4.5.20 for Π by fourth-order polynomials:

$$\begin{aligned}\Pi = {} & \frac{CL^2}{4}\left[\left(1 - \frac{q^2}{2} + \frac{q^4}{24}\right) - \left(1 - \frac{\alpha^2}{2} + \frac{\alpha^4}{24}\right) + \left(q - \frac{q^3}{6}\right) - \left(\alpha - \frac{\alpha^3}{6}\right)\right]^2 \\ & - PL\left[\left(1 - \frac{\alpha^2}{2} + \frac{\alpha^4}{24}\right) - \left(1 - \frac{q^2}{2} + \frac{q^4}{24}\right)\right] \\ \simeq {} & \frac{CL^2}{4}(q^2 + \alpha^2 - q^3 + q^2\alpha + q\alpha^2 - 2q\alpha - \alpha^3) + \frac{PL}{2}(\alpha^2 - q^2)\end{aligned} \tag{4.5.28}$$

The essential property of this polynomial approximation that causes the bifurcation to be asymmetric is that is contains the *cubic* term q^3. All terms of degree higher than cubic in q and α have been dropped from Equation 4.5.28, since they are not essential.

L-Shaped Rigid-Bar Frame

Asymmetric bifurcation is exhibited by many types of frames. For example the L-shaped frame analyzed by Koiter (1967) and already solved in Section 2.6

behaves in the same manner as the column in Figure 4.22. This kind of behavior can be easily illustrated with a simple model consisting of two rigid segments joined by elastic springs of stiffness C_1 and C_2 (Fig. 4.23a). The frame may be imperfect, due to initial inclination $q = \alpha$. To obtain the postcritical behavior, the equilibrium conditions must take into account (similar to Sec. 2.6) the load-point displacement u due to lateral deflection v. Same as in Equation 4.5.17 we have $u = H(q^2 - \alpha^2)/2$, where H = frame height. This displacement affects the relative rotation $(\theta - \theta_0)$ in the hinge restrained by spring C_2; $(\theta - \theta_0) = (q - \alpha) + (\phi - \phi_0) = 2(q - \alpha) + u/(L/2)$. The moment equilibrium condition at this hinge can be written as $C_2(\theta - \theta_0) = V(L/2)\cos\phi$, in which V is the vertical reaction at the sliding support and $\cos\phi \simeq 1 - q^2/2$. The moment equilibrium condition at the base gives $C_1(q - \alpha) - PqH + V[qH + L(1 - q^2/2)] = 0$. By substitution for V and elimination of $\theta - \theta_0$ with the help of the preceding relation, one obtains the load-rotation relation:

$$P = (qH)^{-1}\left\{C_1(q - \alpha) + \frac{C_2}{L}\left(1 - \frac{q_2^2}{2}\right)^{-1}\left[4(q - \alpha) + 2\left(\frac{H}{L}\right)(q^2 - \alpha^2)\right] \times \left[qH + L\left(1 - \frac{q^2}{2}\right)\right]\right\} \tag{4.5.29}$$

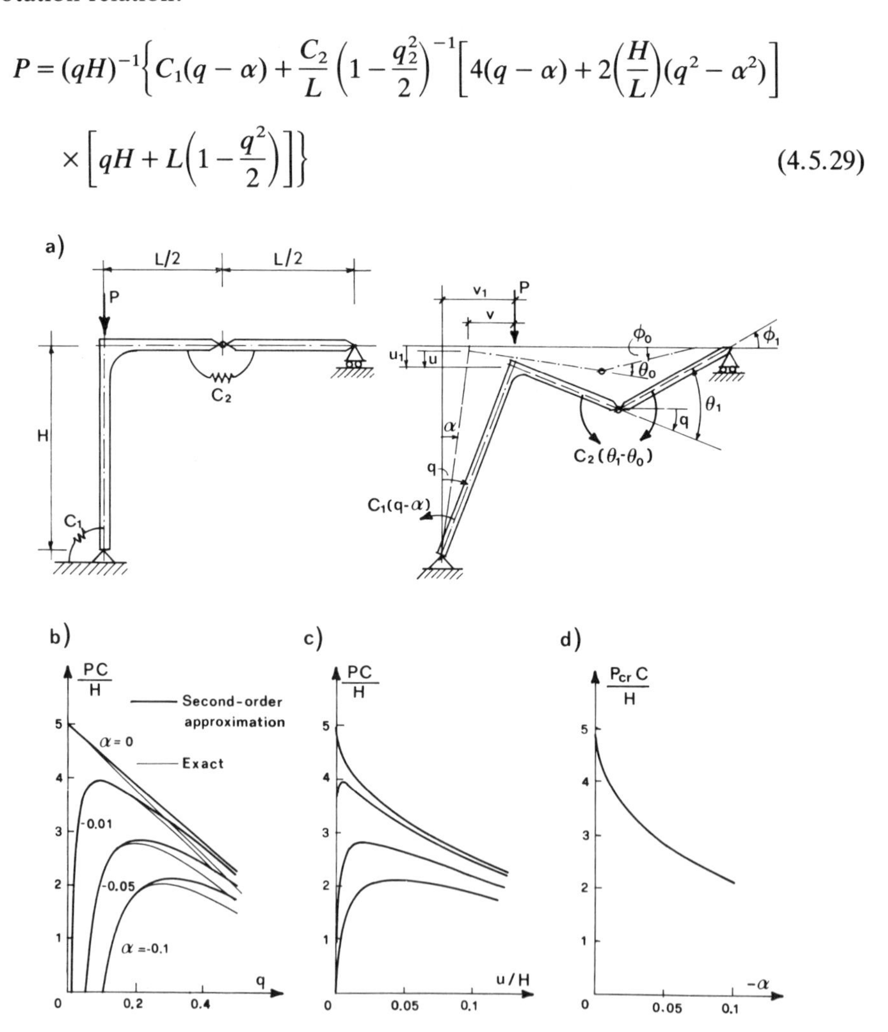

Figure 4.23 (a) L-shaped rigid-bar frame, (b) equilibrium paths, (c) load-displacement diagrams, and (d) imperfection sensitivity diagram.

This relation is plotted in Figure 4.23b for various values of the imperfection α. Figure 4.23c shows the curves $P(u)$. Figure 4.23d shows the imperfection sensitivity diagram obtained from the maximum points of the $P(\theta)$ curves, and it may be checked that this diagram follows the $\alpha^{1/2}$-power law, that is, the reduction in the maximum load is quite significant, even for a small imperfection such as $\alpha = 0.01$ rad. The calculations presented are second-order accurate. This problem, however, is easy to solve exactly even for finite (large) deflections, in which case $\cos\gamma$, $\tan\gamma$, and $\sin\gamma$ in the foregoing relations are not replaced by approximations. This solution yields the light curves in Figure 4.23b.

Rigid-Bar Arch

An important example of symmetric bifurcation is the inextensional buckling of arches, analyzed already is Section 2.8. In this case a good model can again be

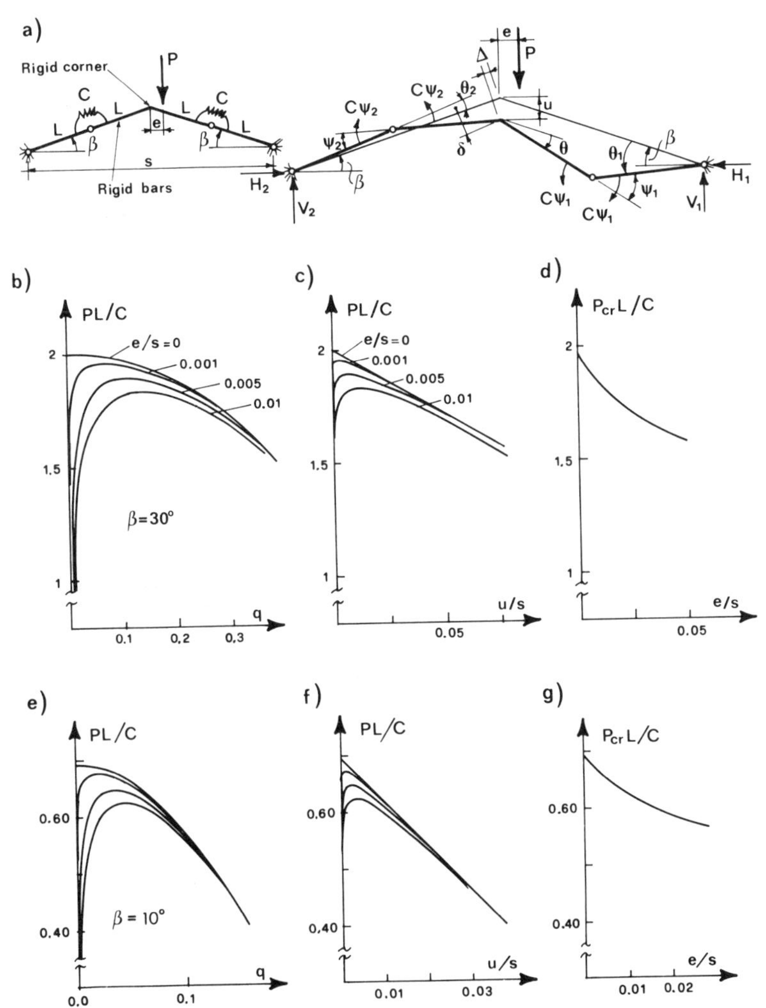

Figure 4.24 (a) Rigid-bar arch; (b, e) equilibrium paths; (c, f) load-deflection diagrams; and (d, g) imperfection sensitivity diagram.

obtained by considering an angle frame made of rigid bars, in which hinges are inserted at quarterpoints; see Figure 4.24a. The relative rotation of the hinges is restrained by springs. To obtain the postcritical behavior, one must consider the load-point displacement u due to the lateral deflection caused by rotation θ (Fig. 4.24a). If one neglects the terms of higher than second order, the displacement of the rigid joint (vertex) may be assumed to be vertical. Its projections parallel and normal to the initial direction of the bars (which form an angle β with the horizontal) are, respectively, $\Delta = 2L(1-\cos\theta) \simeq L\theta^2$ and $\delta = \Delta/\tan\beta = L\theta^2/\tan\beta$. If θ is the rotation of the rigid joint, the rotations at the supporting hinges are $\theta_1 \simeq \theta + \delta/L = \theta + \theta^2/\tan\beta$ and $\theta_2 \simeq \theta - \delta/L = \theta - \theta^2/\tan\beta$. The relative rotations at the hinges with springs are $\Psi_1 = \theta + \theta_1$ and $\Psi_2 = \theta + \theta_2$. Global equilibrium requires that $V_1 + V_2 = P$, $H_1 = H_2$ and $4V_1 L\cos\beta = P(2L\cos\beta + e)$. Then one substitutes these equations into the moment equilibrium equations for the hinges with springs, which are $C\Psi_1 = V_1 L\cos\beta_1 - H_1 L\sin\beta_1$, $C\Psi_2 = -V_2 L\cos\beta_2 + H_1 L\sin\beta_2$, in which $\beta_1 = \beta - \theta_1$, $\beta_2 = \beta + \theta_2$. This provides the following equation for the relation of P and θ:

$$P = \frac{C}{L}\left(\frac{f_3 g_1 - f_1 g_3}{f_2 g_3 + f_3 g_2}\right) \tag{4.5.30}$$

in which

$$f_1 = -\frac{2\theta^2}{\tan\beta} \qquad f_2 = \left(1 + \frac{\theta^2}{2}\right)\cos\beta + \frac{e\theta}{2L}\tan\beta$$

$$f_3 = \left[1 - \theta^2\left(\frac{1}{2} + \frac{1}{\tan^2\beta}\right)\right]\sin\beta \qquad g_1 = 4\theta \tag{4.5.31}$$

$$g_2 = \frac{\theta}{2L}\left(1 - \frac{\theta^2}{2}\right) + \theta\sin\beta \qquad g_3 = \theta\cos\beta$$

The results are plotted in Figure 4.24b, for the frame with $\beta = 30°$, for various values of the ratio e/s where e is the eccentricity of the load and s is the span. Figure 4.24c shows the curves of the load versus the load-point displacement. Figure 4.24d shows the imperfection sensitivity diagram, whose asymptotic form for small e agrees with the $\frac{2}{3}$-power law (as it must; see Sec. 4.6). For a more shallow angle frame ($\beta = 10°$), for which the assumption of inextensibility still holds, the curves $P(q)$, $P_{cr}(e)$ are shown in Figure 4.24e, g. They are seen to be similar to those measured and analyzed for an angle frame by Roorda (1965). They are also similar to those measured and analyzed by Roorda (1965) for a shallow arch, already shown in Section 2.8. The postcritical $P(u)$ diagram for the perfect angle frame, shown in Figure 4.24f, is a straight line, which agrees with that measured and analyzed by Roorda (1971) for a shallow prestressed arch.

Nonlinear Springs and Polynomial Approximation of Potential Energy

The physical property that causes asymmetric bifurcation in Figure 4.22 is (as already pointed out) the fact that the linearly elastic extensional spring installed nonhorizontally behaves, relative to the buckling mode, as a nonlinear rotational spring. Indeed, the elastic restraint against rotation of the rigid column about the base can be characterized by $dM = \tilde{C}\,dq$, where M is the moment of the spring force about the pivot and $\tilde{C}$ is the incremental rotational spring stiffness. From the geometry in Figure 4.25a (perfect system, $\alpha = 0$) we have $M = CL^2[\cos(\gamma - q) - \cos\gamma]\sin(\gamma - q)$, $dM/dq = \tilde{C} = (CL^2/2)[2\cos^2\gamma\cos q - 2\cos 2\gamma\cos 2q + \sin 2\gamma(\sin q - 2\sin 2q)]$, and, for small rotations q, this can be written as

$$\tilde{C} = C_1 - C_2 q \tag{4.5.32}$$

in which C_1, C_2 = constants; $C_1 = CL^2(\cos^2\gamma - \cos 2\gamma)$, $C_2 = (3CL^2/2)\sin 2\gamma$, constants for fixed γ. For the imagined nonlinear rotational spring, we have, by integration, $M = \int \tilde{C}\,dq$ or

$$M = C_1 q - \tfrac{1}{2}C_2 q^2 \tag{4.5.33}$$

Now the potential energy of the system is calculated as $\Pi = \int M\,dq - PL(1 - \cos q)$, which yields

$$\Pi = \frac{1}{2}\left(1 - \frac{P}{P_{cr}}\right)C_1 q^2 - \frac{1}{6}C_2 q^3 \qquad P_{cr} = \frac{C_1}{L} \tag{4.5.34}$$

The salient property of this polynomial expression for potential energy is that, unlike linear systems, it is not quadratic but cubic in the deflection parameter q. The cubic form of the potential-energy expression, which is typical for certain shells, always leads to asymmetric bifurcation and the associated strong imperfection sensitivity.

Note that the tangential stiffness $K_m = \partial m/\partial q = \partial^2\Pi/\partial q^2 = (1 - P/P_{cr})C_1 -$

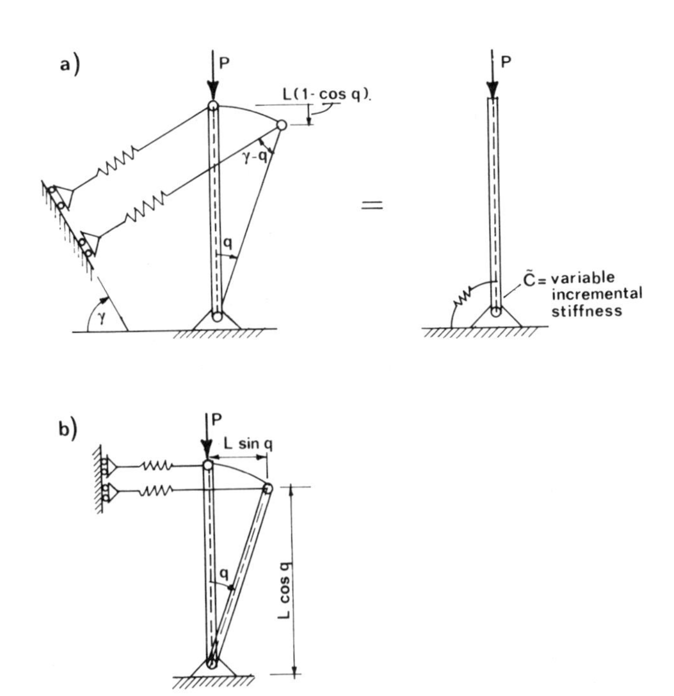

Figure 4.25 Equivalence between linear extensional springs and nonlinear rotational springs.

$C_2 q$ (where m = disturbing applied moment) vanishes at the bifurcation point, which means that one has neutral equilibrium at $q=0$ (for perfect column).

For the case of symmetric bifurcations, $K_m = \partial m/\partial q = \partial^2\Pi/\partial q^2 = (1 - P/P_{cr})C_1 + C_3 q^2$ (where C_1, C_3 are constants) and positiveness (or negativeness) of C_3 means stable (or unstable) bifurcation.

The existence of neutral equilibrium means that an arbitrarily small applied moment m causes a finite rotation q. But this is detectable in the plots of neither $P(q)$ nor $P(u)$ because these are taken at $m=0$.

For the special case when the linear spring is horizontal (Fig. 4.25b) we have for the equivalent rotational spring $C_2 = 0$. This means that the cubic term in the potential-energy expression vanishes. Then one must include the higher-order term (quartic form). From the geometry in Figure 4.25b, we have $M = CL^2 \sin q \cos q$, $dM/dq = \bar{C} = CL^2 \cos 2q$, and, for a small rotation q,

$$\bar{C} = C_1 - C_3 q^2 \tag{4.5.35}$$

where $C_1 = CL^2$ and $C_3 = 2CL^2$. We see that the nonlinearity of the effective rotational spring lacks the linear term, and the leading term is quadratic. By integration, the effective rotational spring is characterized by

$$M = C_1 q - \tfrac{1}{3} C_3 q^3 \tag{4.5.36}$$

By further integration, the potential-energy expression is expressed by the polynomial:

$$\Pi = \tfrac{1}{2}(C_1 - PL)q^2 - \tfrac{1}{12} C_3 q^4 \tag{4.5.37}$$

We see that in this case the potential-energy polynomial lacks the cubic term and the leading term, which causes nonlinear postcritical behavior, is quartic. This property, already shown in Equations 4.5.10 and 4.5.19 for rigid columns with a linear spring at the base or on top, is typical of all symmetric bifurcations.

If the quartic term is negative ($C_3 < 0$), the symmetric bifurcation is imperfection sensitive, as is true of this example. If it is positive, the symmetric bifurcation is imperfection insensitive and exhibits a postcritical reserve. The foregoing properties, resulting from cubic or quartic terms in Π, are applicable generally. They apply to all kinds of phenomena in physics and science where instabilities can be characterized by a potential (Thompson and Hunt, 1984). Classification and qualitative analysis of various types of instabilities on the basis of the potential polynomial is the subject of catastrophy theory (Sec. 4.7).

Two Degrees of Freedom: Example

Consider now an example of a two-degree-of-freedom system (Fig. 4.26a). The rigid-bar column loaded vertically by constant load P is supported on a horizontally sliding hinge and is held upright by two horizontal springs of stiffness C. The inclination angle of the bar is q_1, and the horizontal displacement of the hinge is q_2. The initial imperfection is characterized by the initial angle of

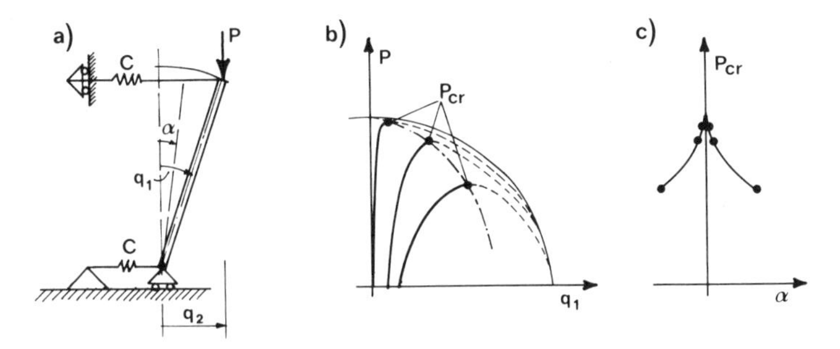

Figure 4.26 (a) Sliding rigid-bar column, (b) equilibrium paths, and (c) imperfection sensitivity.

inclination, α (Fig. 4.26a). The potential energy of this column is

$$\Pi = \frac{C}{2}(q_2 + L\sin q_1 - L\sin\alpha)^2 + \frac{C}{2}q_2^2 + PL(\cos q_1 - \cos\alpha) \tag{4.5.38}$$

The equilibrium conditions are given by

$$\begin{aligned} \Pi_{,1} &= CL\cos q_1(q_2 + L\sin q_1 - L\sin\alpha) - PL\sin q_1 = 0 \\ \Pi_{,2} &= C(q_2 + L\sin q_1 - L\sin\alpha) + Cq_2 = 0 \end{aligned} \tag{4.5.39}$$

Eliminating q_2, we obtain for the equilibrium path the equation

$$P = \frac{CL}{2}\cos q_1\left(1 - \frac{\sin\alpha}{\sin q_1}\right) \tag{4.5.40}$$

These equilibrium paths are plotted for various values of initial imperfection α in Figure 4.26b.

To analyze stability, we calculate the second partial derivatives of Π:

$$\begin{aligned} &\Pi_{,11} = C(L\cos q_1)^2 - C(q_2 + L\sin q_1 - L\sin\alpha)L\sin q_1 - PL\cos q_1 \\ &\Pi_{,12} = CL\cos q_1 \qquad \Pi_{,22} = 2C \end{aligned} \tag{4.5.41}$$

where $\Pi_{,ij} = \partial^2\Pi/\partial q_1\,\partial q_j = K_{ij}$ = tangential stiffness matrix. These expressions form the matrix of the second variation $\delta^2\Pi = \frac{1}{2}(\Pi_{,11}\,\delta q_1^2 + 2\Pi_{,12}\,\delta q_1\,\delta q_2 + \Pi_{,22}\,\delta q_2^2)$.

For this matrix to be positive, one diagonal coefficient must be positive, which is guaranteed by $\Pi_{,22} = 2C$ (Eqs. 4.5.41), and the determinant D of the matrix must be positive (Theorem 4.1.10). The latter condition reduces, after some algebraic manipulations, to the condition $(\sin\alpha - \sin^3 q_1)/\sin q_1 > 0$. For $\sin q_1 > 0$ this yields the stability condition:

$$\sin q_1 < (\sin\alpha)^{1/3} \qquad (0 < q_1 < \pi/2) \tag{4.5.42}$$

which means that the critical states $q_{1_{cr}}$ are given by the condition $\sin q_1 = (\sin\alpha)^{1/3}$. Substitution into Equation 4.5.40 yields the critical load as a function of the imperfection α:

$$P_{cr} = \frac{CL}{2}[1 - (\sin\alpha)^{2/3}]^{3/2} \tag{4.5.43}$$

A plot of this equation is the imperfection sensitivity diagram; see Figure 4.26c. For small imperfections, $P_{cr} \simeq \frac{1}{2}CL(1 - \frac{3}{2}\alpha^{2/3})$. Note again that for this structure the critical load markedly decreases with imperfection α; for $\alpha = 0.01$, the maximum load drops by 7.0 percent compared with the perfect system. Also note that at the bifurcation point ($P = P_{cr}$, $\alpha = 0$, $q = 0$), $\det \mathbf{K} = \det [\Pi_{,ij}] = 0$, which is in fact a necessary condition of all bifurcations (see also Sec. 10.4).

The initial postbuckling behavior could be obtained upon approximating Equation 4.5.38 with a fourth-degree polynomial in q_1, q_2, and α.

One could construct other examples of two-degree-of-freedom systems that exhibit symmetric imperfection-insensitive bifurcation, asymmetric bifurcation, and snapthrough.

Limit Points of Equilibrium Paths

Our foregoing determination of stability limits based on calculation of the determinant D of the second variation can be quite tedious. A shortcut is possible by realizing that at the maximum points of the equilibrium path $P(q_1)$ the determinant D must vanish or else these points would not be critical states (in view of the fact that $\partial^2\Pi/\partial q_2^2 > 0$, Eqs. 4.5.41).

Another reason for the limit point to be a critical state is that the vector $(\delta q_1, \delta q_2)$ can be nonzero at $\delta P = 0$. Thus, by differentiating Equation 4.5.40 we have

$$\frac{dP}{dq_1} = \frac{CL}{2}\left(-\sin q_1 + \sin \alpha \frac{1}{\sin^2 q_1}\right) \tag{4.5.44}$$

Setting this equal to zero, we indeed get the same conditions for the critical states as before, namely, $\sin q_{1_{cr}} = (\sin \alpha)^{1/3}$ (Eq. 4.5.42). At all other points of each equilibrium path $P(q_1)$ at fixed imperfection α, determinant D is nonzero, and so it must be positive everywhere on one side of the critical point and negative everywhere on the other side of the critical point. Then, from the fact that the determinant of $\delta^2\Pi$ is positive in the initial state $P = 0$, it follows that the states $0 < q_1 < q_{1_{cr}}$ are stable, and the states $q_1 > q_{1_{cr}}$ are unstable.

We saw in the previous example that at the critical states the determinant of the matrix of the second variation vanishes. This property is true in general, that is, there cannot be a critical state in which the determinant does not change sign. To show it, the potential energy may be written in the form $\Pi = U - P\bar{W}$ where P = load or load parameter and $\bar{W}$ = work per unit load. Equilibrium requires that $\Pi_{,i} = U_{,i} - P\bar{W}_{,i} = 0$ (subscripts preceded by a comma denote partial derivatives). From this, the equilibrium load-deflection diagram is defined by $P = U_{,i}/\bar{W}_{,i}$. The load variation along the equilibrium diagram may be expressed as

$$\begin{aligned}\delta P &= \sum_j P_{,j}\,\delta q_j = \bar{W}_{,i}^{-2}\Big(\bar{W}_{,i} \sum_j U_{,ij}\,\delta q_j - P\bar{W}_{,i} \sum_j \bar{W}_{,ij}\,\delta q_j\Big) \\ &= \bar{W}_{,i}^{-1} \sum_j \Pi_{,ij}\,\delta q_j \end{aligned} \tag{4.5.45}$$

Now it is obvious that if $\delta P = 0$ while at least one among δq_j vectors is nonzero, then at least one of the principal minors of the matrix $\Pi_{,ij}$ must vanish or else the right-hand side of Equation 4.5.45 could never vanish. Similarly, if $\delta P \neq 0$ for all δq_j, then all principal minors of matrix $\Pi_{,ij}$ must be nonzero.

In view of the foregoing analysis, the critical states on an equilibrium path without bifurcation can be determined as the maximum points of the equilibrium paths of load versus deflection. These points may be obtained by static analysis, for example, from the principle of virtual work.

Bifurcation Criterion in Terms of the Tangential Stiffness Matrix

As illustrated for several preceding examples, the tangential stiffness vanishes at the bifurcation point regardless of whether the bifurcation is symmetric or asymmetric. For systems with many degrees of freedom, one can similarly show that for bifurcation points always $\det \mathbf{K} = \mathbf{0}$ where $\mathbf{K}$ = tangential stiffness matrix (see also Sec. 10.4). This means that $\mathbf{K}\,\delta\mathbf{q} = \mathbf{0}$ for some admissible eigenvector $\delta\mathbf{q}^* = \delta\mathbf{q}$, that is, the bifurcation point of elastic structures always represents a state of neutral equilibrium (this is not generally true for inelastic structures because the eigenvector $\delta\mathbf{q}^*$ may be inadmissible according to the unloading criteria, see Sec. 10.4). Thus, the condition

$$\det \mathbf{K} = \det [\Pi_{,ij}] = 0 \qquad (\text{at } \mathbf{q} = \mathbf{0}) \tag{4.5.46}$$

in general may be used to detect the bifurcation point, whether symmetric or asymmetric.

For a general derivation of Equation 4.5.46, we observe that the fundamental (primary, main) path at given increments δP is characterized by $\delta\mathbf{q} = \mathbf{0}$. At the bifurcation point suddenly a path with $\delta\mathbf{q} \neq \mathbf{0}$ becomes possible for a given increment δP. Generally $\mathbf{K}\,\delta\mathbf{q} = \delta\mathbf{f}$. Now we have $\delta f_1 = \delta f_2 = \cdots = \delta f_n = 0$ (that is, $\delta\mathbf{f} = \mathbf{0}$), since δP does not represent any of the loads δf_i (i.e., is not associated by work with any q_i); δP is merely a parameter on which $\mathbf{K}$ depends. Therefore the matrix equation $\mathbf{K}\,\delta\mathbf{q} = \mathbf{0}$ must have a nonzero solution at the bifurcation point. Hence $\det \mathbf{K} = 0$ (i.e., matrix $\mathbf{K}$ is singular, its lowest eigenvalue is zero). The eigenvector $\delta\mathbf{q}^*$ of matrix $\mathbf{K}$ determines the initial path direction (Fig. 4.27)—not in any Pq_i plane (i.e., not the slope dP/dq_i) but in the n-dimensional space of $q_1, \ldots, q_n$ (in all but the last of our examples this was a one-dimensional space). To determine the increment δP associated with $\delta\mathbf{q} = \delta\mathbf{q}^*$,

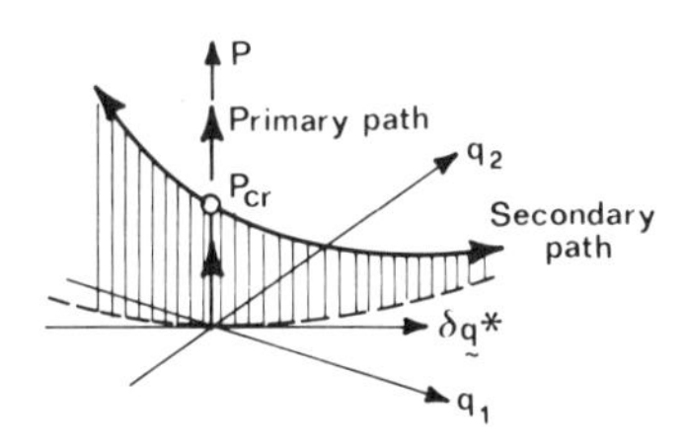

Figure 4.27 Initial path direction at bifurcation.

one must of course consider equilibrium in the adjacent states, as documented by our examples.

As will be seen in Section 10.3, for inelastic columns an important difference is that the axial shortening does not depend on deflection, but is independent, and must be taken as one of the δq_i. Thus δP does work on one of the δq_i. Consequently, neutral equilibrium need not exist at bifurcation.

Classification of Elementary Instabilities of Elastic Structures

We have encountered in this book examples of various elementary types of instabilities of elastic structures. They may be classified as follows:

I. Dynamic (energy approach is insufficient)
II. Static (Π exists, energy approach is sufficient)
  A. Snapthrough (limit point)
  B. Bifurcation
    1. Asymmetric bifurcation
    2. Symmetric bifurcation
      a. Stable (imperfection insensitive)
      b. Unstable (imperfection sensitive)

Problems

4.5.1 Calculate equilibrium paths $P(\theta, \alpha)$, critical loads $P_{cr}(\alpha)$, function $P(u)$, and $P_{cr}(\alpha)$ for the polynomial approximation of Π in Equation 4.5.10 for a rigid column with a rotational spring at the base (Fig. 4.20a) and analyze stability.

4.5.2 Do the same as above but for Equaton 4.5.19 for a rigid column with a horizontal spring on top (Fig. 4.21a)

4.5.3 Do the same as above, but for Equation 4.5.28 for a rigid column with an inclined spring on top (Fig. 4.22a)

4.5.4 Do the same as above, but for the two-degree-of-freedom column in Figure 4.26a (approximate Eq. 4.5.38 by a fourth-degree polynomial).

4.5.5 Solve $P(q)$, $P_{cr}(\alpha)$, $P(u)$, and analyze stability for the structures in Figure 4.28a, b, c, d, e, f.

4.5.6 For the inextensible columns in Figure 4.28g, h solve $P(u)$. *Hint:* $u = \int_0^L (w'^2/2)\,dx$.

4.5.7 Consider the rigid column in Figure 4.28i with both a rotational spring C_1 at the base and a horizontal spring C_2 on top. Analyze stability. For a certain combination of C_1 and C_2, the fourth-order terms in the polynomial expansion of Π vanish. Find this combination and analyze stability. (*Note:* A polynomial approximation of Π must in this case include all terms up to sixth degree in q.)

4.5.8 Do the same for a column with an initially horizontal spring on top attached to a fixed point at a distance a (Fig. 4.28j). For a certain a value, the fourth-degree terms in Π vanish. Find such a and analyze stability.

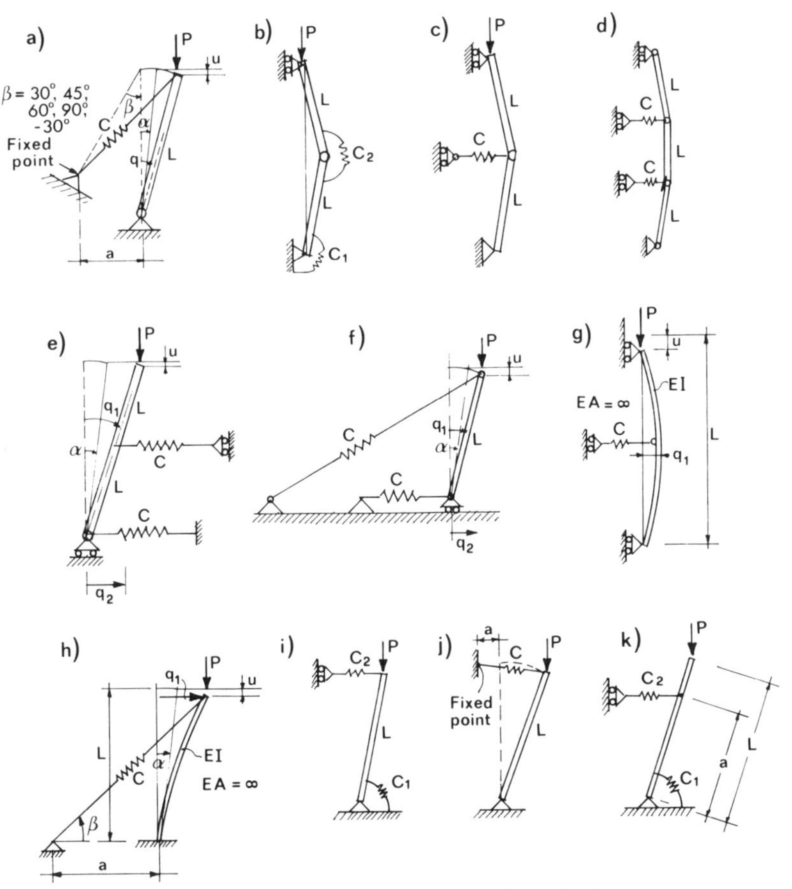

Figure 4.28 Exercise problems on large-deflection buckling of columns.

4.5.9 Do the same but for a column with a sliding horizontal spring attached at height a below the top (Fig. 4.28k). For a certain a, the fourth-degree terms in Π vanish. Find such a and analyze stability.

4.5.10 For the column in Figure 4.28a with $\beta = 90°$, solve $P(q)$ and $P(u)$ for any a and L. For which ratio a/L is the slope $dP/du = 0$ (if any)? Correlate the answer to the polynomial expansion in Π.

4.5.11 For the (nonrigid) elastic column in Figure 4.28h with $\beta = 0°$, (a) calculate dP/du. (b) For which ξ and η do you get $dP/du = 0$? ($\xi = a/L$, $\eta = EI/CaL^2$).

4.5.12 Using a computer graphics package, plot the surfaces $\Pi = \Pi(q_1, P)$ for the one-degree-of-freedom structures in Figures 4.14a, 4.20a, 4.21a, 4.22a, 4.28a and $\Pi = \Pi(q_1, q_2)$ for the two-degree-of-freedom structures in Figures 4.26a and 4.28e, f. Discuss instability on the basis of the shape of these surfaces.

4.5.13 Determine by the principle of virtual work the equilibrium path of (a) a flat arch; (b) an imperfect Euler column; (c) a rigid bar with elastic restraint. Compare it with potential-energy analysis.

4.5.14 The column in Figure 4.29a has a pin and a support sliding on a plane of inclination β. The column is imperfect, with initial curvature $z_0 =$

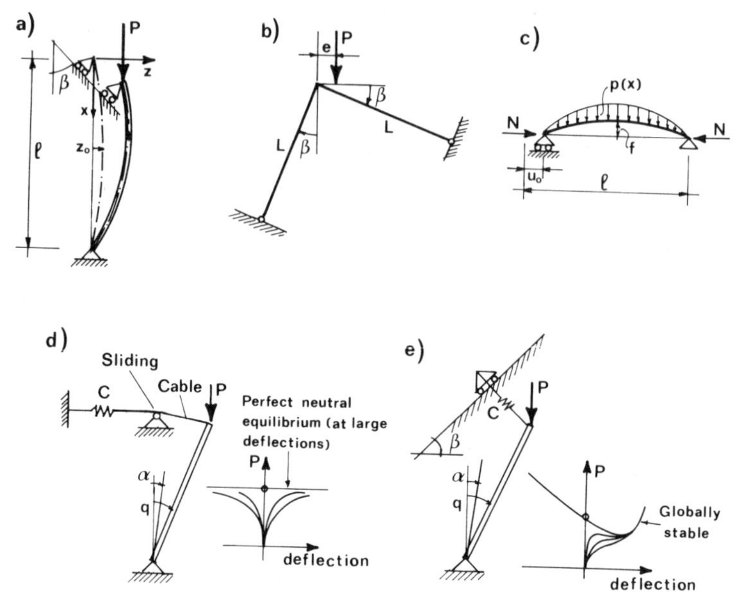

Figure 4.29 Exercise problems on large-deflection buckling.

$\alpha l \sin(\pi x/l)(\alpha \ll 1)$; determine $P_{\max}$ and the power law for sensitivity to imperfection α. Calculate $P_{\max}/P_E$ for $\alpha = 0.01$ if $\beta = 1°$, $5°$, $45°$. (See first Prob. 2.6.7, which deals with the postcritical behavior of the perfect column.)

4.5.15 Determine the postcritical response and the imperfection sensitivity diagram of the inclined L-frame in Figure 4.29b. Compare the results with Roorda's experimental results (Roorda, 1971).

4.5.16 Precompressed arch. A straight beam is shortened by u_0 to form an arch of height f. (Fig. 4.29c). Then load $p(x) = p_0 \sin(\pi x/l)$ is applied. Calculate the diagram of p_0 versus $\max w$ at constant u_0 (the base held fixed during buckling).

4.5.17 Solve the $P(q)$ diagrams for the structures in Figure 4.29d, e. *Note:* The shapes of the diagrams that should be obtained are also shown in these figures; the column in Figure 4.29d exhibits perfect neutral equilibrium [a horizontal $P(q)$ diagram] at large deflections, and the column in Figure 4.29e exhibits rehardening and may therefore be said to be globally stable, while being locally unstable near the critical point.

4.5.18 Solve exactly (for large deflections) the $P(q)$ and $P(u)$ diagrams for the structure in Figure 4.23a.

4.5.19 (a) Derive Equations 4.5.30 and 4.5.31 in detail and plot $P(\theta)$ and $P_{\max}$ versus e/L for $\beta = 45°$, $60°$. (b) Using the same procedure, analyze the rigid-bar L-frame in Figure 4.30a. This structure is actually the same as the rigid-bar arch in Figure 4.24, except that the direction of load P is now different. Check that the critical load (for $e = 0$) is $P_{cr} = 4C/L$. This value can be found by writing the condition of neutral equilibrium in the deformed configuration for small θ. (*Caution:* Even for small θ there are small additional horizontal reactions at the supports, proportional to θ, because the structure is redundant.) Figure 4.30b shows the results.

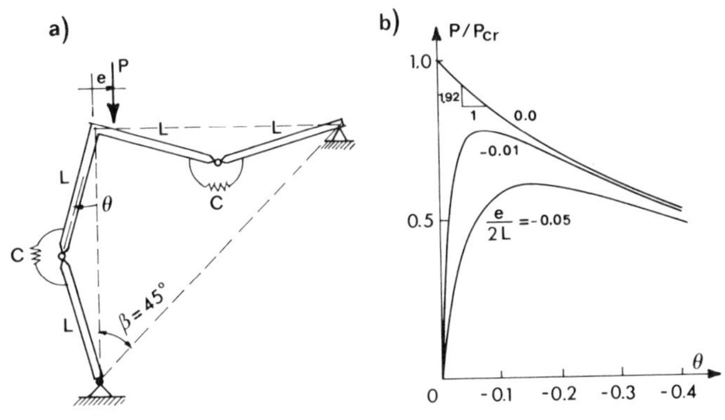

Figure 4.30 Equilibrium paths of rigid-bar L-frame (exercise problem).

4.5.20 Solve $P(q)$, $P(u)$, and the imperfection sensitivity for the L-frame in Figure 4.31a in which the load has eccentricity e and (a) $\alpha=0$, $C_1=0$, $C_2>0$, $H=L$; (b) $\alpha=0$, $C_2=2C_1$, $H=L$; (c) $\alpha=0$, $C_2=C_1$, $H=2L$. Solve the problem also for $\alpha=\alpha_1$ (small imperfection).

4.5.21 Same as Problem 4.5.20 for the L-frames in Figure 4.31b, c, e.

4.5.22 Same as Problem 4.5.20 for the asymmetric portal frame made of rigid segments shown in Figure 4.31d (consider either $C_1=C_2=C_3$ or $C_1=C_3=0$).

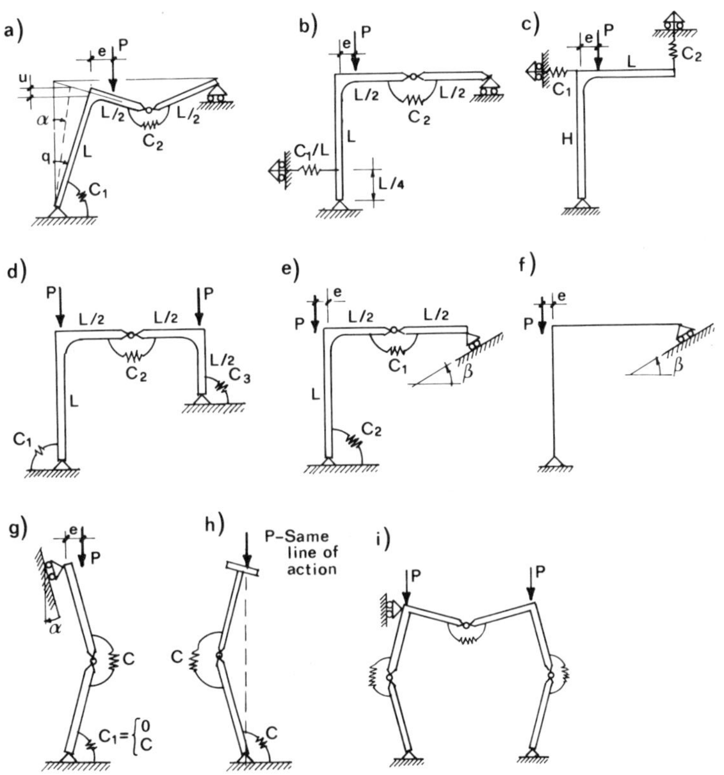

Figure 4.31 Exercise problems on equilibrium paths (a–e, g–i) of rigid-bar frames and (f) of two-bar frame.

4.5.23 Solve the postcritical behavior of the frame in Figure 4.31f, in which one support slides on a plane of inclination β.

4.5.24 Solve the postcritical behavior of the two-bar columns in Figure 4.31g, h.

4.5.25 Analyze the postcritical behavior of the model frames in Figure 4.32 made of rigid bars that are rigid in flexure but elastic axially (with stiffness EA). At the joints the bars are connected by pins and are restrained by springs.

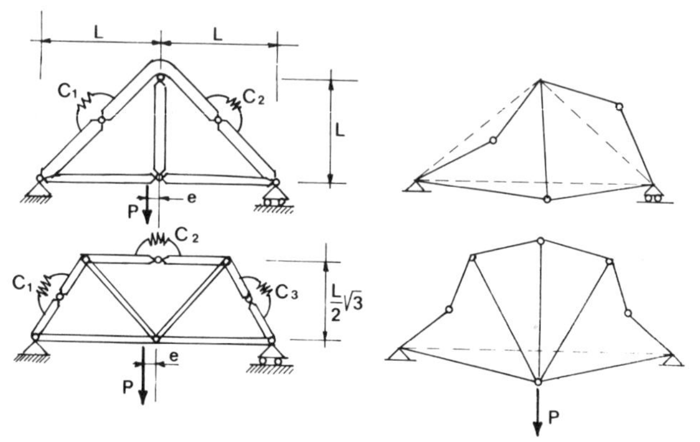

Figure 4.32 Exercise problems on large-deflection buckling of frames with infinite bending rigidity of members.

4.5.26 Determine the $P(u)$ and $P(q)$ diagrams for various imperfections α, the types of bifurcation and imperfection sensitivity of the rigid columns in Figure 4.33a–e ($\beta = 45°$ or $90°$); in c, d, e the spring C_2 is precompressed exactly to the critical load but during subsequent buckling its end is held fixed. Consider $l = 2L$. Then consider $l = -L$. Also check the limit cases $l \to \infty$ and $l \to -\infty$, which are equivalent for gravity load.

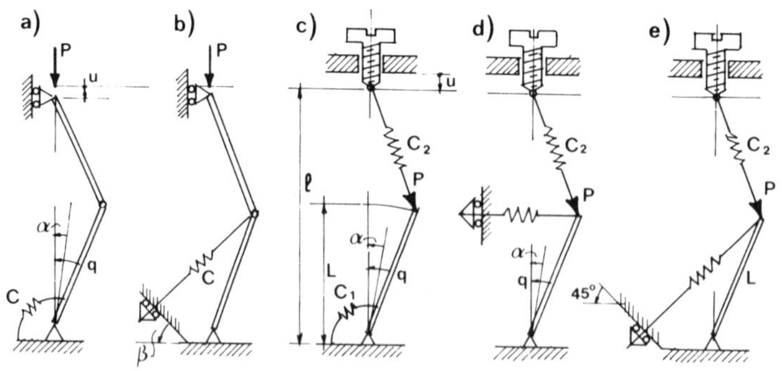

Figure 4.33 Exercise problems on equilibrium paths of rigid-bar columns.

4.5.27 Using the results from Section 1.9, determine the $P(u)$ and $P(w_{max})$ diagrams and the type of bifurcation for the column in Figure 4.34a, considering $z_0 = 0$ (perfect column). (Later, after studying Sec. 5.9, we will be

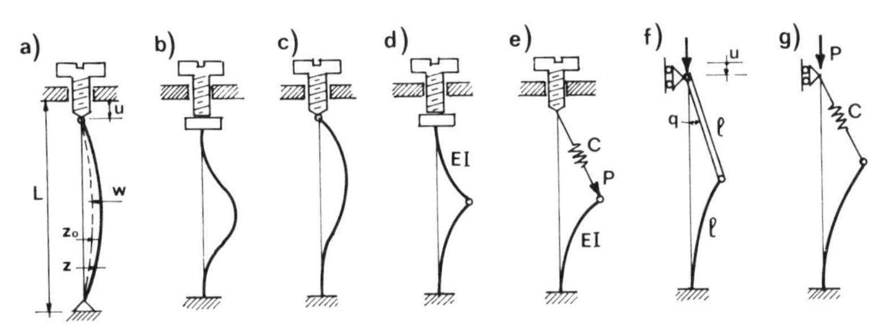

Figure 4.34 Exercise problems on equilibrium paths of beam-columns.

able to consider also $z_0 \neq 0$.) Repeat for the columns in Figure 4.34b. c, d, e, f, g.

4.5.28 Equation 4.5.9 may be written as $P = P_{cr}(1 + u/3L)$. The same equation applies to the simply supported rigid-bar column in Figure 4.3c with a spring-stiffened hinge at midspan. A similar equation, namely $P = P_{cr}(1 + u/2l)$, was derived before for an elastic beam-column (Eq. 1.9.14). The relation between end rotation θ and axial displacement u for the rigid-bar column is $u = L(1 - \cos\theta) \simeq L\theta^2/2$, and for the elastic sinusoidally deflected column it is $u = l\theta^2/4$, which is again similar except for a factor. Considering, for example, a hinged symmetric portal frame, check whether these similarities could be exploited for approximate analysis of the postcritical behavior of complex trusses or frames. For example, an elastic frame may be replaced by an assembly of rigid segments with hinges placed at the estimated max M-points, as shown in Figure 4.31i. Obviously, the postcritical behavior of such an assembly of rigid segments could be analyzed more easily, and if one would assume that $P = P_{cr}(1 + u/2L)$ and $u = L\theta^2/4$ instead of the correct relations for rigid-bar members, the results are likely to be very close to those for the exact solution of the elastic frame. Could this similarity be further extended for the effect of imperfections?

4.5.29 Check for the L-shaped rigid-bar frame (Fig. 4.23) that the tangential stiffness $K = \partial^2\Pi/\partial q^2$ at the point of asymmetric bifurcation vanishes.

4.5.30 (a) For the perfect column with nonlinear spring (Fig. 4.25) calculate rotations q due to vertical load P and disturbing moment m. Then evaluate $K = \partial m/\partial q = \partial^2\Pi/\partial q^2$, $K_p = \partial P/\partial u = \partial^2\Pi/\partial u^2$ (where u = load-point displacement) and discuss the question of neutral equilibrium with its relation to the $P(u)$, $P(q)$, $m(q)$ diagrams. (b) Repeat for the columns in Figure 4.28a, b, c.

4.5.31 For the column in Figure 4.26a (Eq. 4.5.38), obtain the eigenvectors and express Π as a sum of squares of orthogonal coordinates y_k according to Equations 4.1.3.

4.6 KOITER'S THEORY, IMPERFECTION SENSITIVITY, AND INTERACTION OF MODES

As we have seen, structures exhibit various types of bifurcation and imperfection sensitivity. A completely general theory of the initial (linearized) postbuckling behavior in bifurcation-type problems was formulated by Koiter (1945) in Holland during the second world war. His famous doctoral thesis in Dutch had not become widely known and appreciated until its English translation appeared in 1967 (Koiter, 1967). By examining all possible forms of the potential-energy surface near a bifurcation point, and exploiting the physically required smoothness properties, Koiter generally proved that

1. The equilibrium at the critical state is stable if load P for the adjacent postcritical equilibrium states is higher than the bifurcation load P_{cr}. The postcritical states are stable and the structure is imperfection insensitive (Fig. 4.20b).
2. The equilibrium at the critical state of the perfect structure is unstable if there exist adjacent postcritical equilibrium states for which load P is lower than the bifurcation load P_{cr} of the perfect structure. Imperfections cause the load at which the structure becomes unstable to be smaller than P_{cr} (Figs. 4.21b, 4.22b).

Koiter also proved in general that, among the unstable bifurcations, the asymmetric ones have a much higher imperfection sensitivity than the unstable symmetric ones. The behavior can be characterized by the angle β of the initial postbifurcation load-deflection diagram shown in Figure 4.35. If angle β is positive, the bifurcation is stable and the structure is imperfection insensitive. If β is negative, the bifurcation is unstable, the structure is imperfection sensitive, and the larger the magnitude of the negative angle β, the higher is the imperfection sensitivity. The imperfection sensitivity becomes very strong when, after bifurcation, not only load P but both the load and the deflection decrease. We will see examples of such behavior, now called snapback, in Section 4.8. This type of postbifurcation behavior is typical of axially compressed or bent cylindrical shells as well as compressed spherical domes under transverse distributed loading.

In his general theory, Koiter (1945, 1967) also studied interaction of various buckling modes occurring at the same load, which can occur in frames and is particularly important for cylindrical shells. He conceived the notion of a multidimensional space with one axis for the load and one axis for the amplitude of each possible buckling mode. He found how to determine in this space the equilibrium paths on the basis of the first derivatives of the total potential-energy function, as well as how to determine the stability of the states on these paths on the basis of the second derivatives of the potential-energy function. As one powerful general result, Koiter established that at least one stable equilibrium

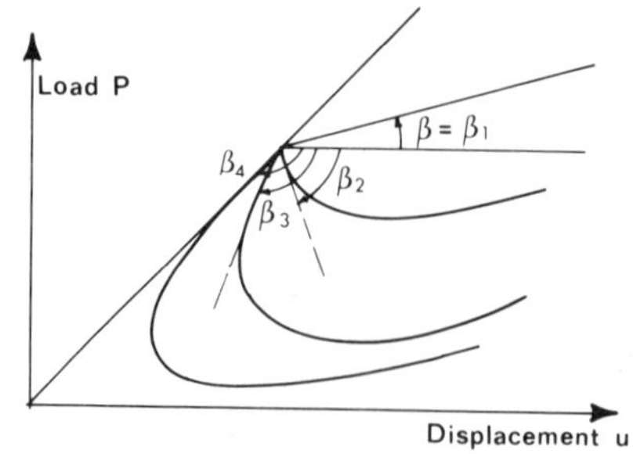

Figure 4.35 Postbifurcation load-deflection diagram.

path must emanate from a bifurcation point. Koiter's theory was later refined by Hutchinson and others, particularly with reference to shell buckling; see Hutchinson (1967), Hutchinson and Koiter (1970), Stein (1968), Budiansky (1974), Thompson and Hunt (1973), and Croll and Walker (1972).

General Validity of Koiter's $\frac{1}{2}$-Power and $\frac{2}{3}$-Power Laws

To analyze imperfection sensitivity in full generality, Koiter studied the asymptotic form of a smooth potential-energy function $\Pi(q, \lambda, \alpha)$ where α = small imperfection parameter and λ = load parameter. This parameter may characterize either a single load ($\lambda = P/P^0_{cr}$) or a system of loads. As exemplified by Equations 4.5.19, 4.5.28, 4.5.34, and 4.5.37, the polynomial approximation of function Π for small q and α may in general be written in the form (Bažant and Cedolin, 1989):

$$\Pi(q, \lambda, \alpha) = (c_1 - a_1\alpha)q + (c_2 - a_2\alpha)q^2 + (c_n - a_n\alpha)q^n - \lambda u$$
$$u = b_2(q^2 - \alpha^2) + b_4(q^4 - \alpha^4) \tag{4.6.1}$$

where c_1, a_1, c_2, a_2, c_n, a_n, b_2, b_4, n = constants, and u = generalized displacement defined so that λu would represent the work of loading. In our previous examples we saw that $n = 3$ for the asymmetric bifurcation and $n = 4$ for the symmetric bifurcation. Generally we will assume that $n = 3$ or 4. It may be checked by generalization of what follows that inclusion of terms with $\alpha^2 q$, $\alpha^2 q^2, \ldots$ would not change the result. The reason is that near the bifurcation $\alpha \ll q$, so that $\alpha^2 q \ll \alpha q$, $\alpha^2 q^2 \ll \alpha q^2$, etc.

The condition of equilibrium is

$$\frac{\partial \Pi}{\partial q} = c_1 - a_1\alpha + 2(c_2 - a_2\alpha - b_2\lambda)q + n(c_n - a_n\alpha)q^{n-1} - 4b_4\lambda q^3 = 0 \tag{4.6.2}$$

and (according to Eq. 4.5.45) the condition of maximum load on the equilibrium path (i.e., critical load of an imperfect system) is

$$\frac{\partial^2 \Pi}{\partial q^2} = 2(c_2 - a_2\alpha - b_2\lambda) + n(n-1)(c_n - a_n\alpha)q^{n-2} - 12b_4\lambda q^2 = 0 \tag{4.6.3}$$

Since $\alpha = 0$, $q = 0$ is an equilibrium state ($\partial\Pi/\partial q = 0$), we must have $c_1 = 0$, and if this state is also the bifurcation point, the condition $\partial^2\Pi/\partial q^2 = 0$ (at $\alpha = q = 0$) must reduce to $\lambda = \lambda^0_{cr}$ for the perfect system. This yields $\lambda^0_{cr} b_2 = c_2$. We may always define λ so that $\lambda^0_{cr} = 1$; then $b_2 = c_2$ and $1 - \lambda$ is small near the critical state of a perfect structure. For a single load, $\lambda = P/P^0_{cr}$.

Consequently, Equations 4.6.2 and 4.6.3 can be put in the form:

$$(1 + k_3 q^2)^{-1}\left(1 - k_1 \frac{\alpha}{q} - k_2 q^{n-2}\right) = \lambda$$
$$(1 + 3k_3 q^2)^{-1}[1 - (n-1)k_2 q^{n-2}] = \lambda \tag{4.6.4}$$

in which $k_1 = a_1/2c_2$, $k_2 = nc_n/2c_2$, $k_3 = 2b_4/c_2 \neq 0$ if $n = 4$ and $k_3 = 0$ if $n = 3$. The case $n = 3$ follows from the fact that the terms in q^2 are of higher order. Since the terms $k_3 q^2$ are small with respect to 1, one can set $(1 + k_3 q^2)^{-1} \simeq [1 - k_3 q^2 + (k_3 q^2)^2 - \cdots]$, $(1 + 3k_3 q^2)^{-1} \simeq [1 - 3k_3 q^2 + (3k_3 q^2)^2 - \cdots]$ in Equations 4.6.4. Then, multiplying the expressions on the left-hand sides, and neglecting higher-order terms, one gets

$$\lambda = 1 - k_1\left(\frac{\alpha}{q}\right) - (k_2 + k_3)q^{n-2}$$
$$\lambda = 1 - [(n-1)k_2 + 3k_3]q^{n-2} \qquad (4.6.5)$$

where $k_3 = 0$ if $n = 3$ and $k_3 \neq 0$ if not. The derivation of Equations 4.6.5 took into account the fact that, if $n = 4$, the terms with q^{n-2} and q^2 in the original equations may be combined. Eliminating λ from Equations 4.6.5, one finds the values of q for the critical points of the imperfect column to be given by $q_{cr} = (\{k_1/[(n-2)k_2 + 2k_3]\}\alpha)^{1/(n-1)}$. Then substituting this into the second of Equations 4.6.5, one gets the following basic result:

$$\lambda = 1 - \left(\frac{\alpha}{\alpha_0}\right)^m \qquad m = \frac{n-2}{n-1} \qquad (4.6.6)$$

where $\alpha_0 = \text{const.} = [(n-2)k_2 + 2k_3]k_1^{-1}[(n-1)k_2 + 3k_3]^{-1/m}$ and $\lambda = P_{max}/P^0_{cr}$ for the case of a single load.

The smaller the exponent m, the stronger is the imperfection sensitivity. In particular, for $n = 3$ we have $m = \frac{1}{2}$ (asymmetric bifurcation), and for $n = 4$ we have $m = \frac{2}{3}$ (unstable symmetric bifurcation). These are the famous Koiter's $\frac{1}{2}$-power and $\frac{2}{3}$-power laws, which we already illustrated and discussed in the preceding section (Eqs. 4.5.16 and 4.5.26; Figs. 4.21b, c, 4.22b, c, 4.23b, d, 4.24b, d, and 4.26b, c). For the $\frac{1}{2}$-power law the imperfection sensitivity is stronger than for the $\frac{2}{3}$-power law.

Equation 4.6.6, of course, makes sense only if α_0 exists, that is, if $k_1 + k_3 > 0$ for $n = 4$ and if $k_1 k_2 > 0$ for $n = 3$. If not, it means that there is no solution for P_{max}; this occurs either for the stable symmetric bifurcation (if $n = 4$) or for buckling to one of the sides (right or left) in the case of asymmetric bifurcation ($n = 3$).

Equation 4.6.6 could also be extended to structures for which $n > 4$.

Equation 4.6.6 represents the celebrated results of Koiter's dissertation, indicating the type of imperfection sensitivity. These results (which were derived here by a considerably simpler procedure than that presented before) apply to any elastic structure, not just frames. Briefly, these results may be summarized as follows (Koiter, 1945, sec. 4.5.5).

If the equilibrium of the perfect system at the critical state is unstable and the instability is caused by an odd power n of q in the potential-energy expression, then the structure exhibits asymmetric bifurcation. If the instability is caused by an even power n of q, then the structure exhibits unstable symmetric bifurcation. Both cases involve imperfection sensitivity in which the maximum load rapidly decreases with increasing imperfection parameter α. The decrease of maximum load is stronger, the smaller the value of n; it happens only for one sign of α if n is odd and for both signs of α if n is even.

Interaction of Buckling Modes

Optimizing the weight of a structure from the viewpoint of the first critical load of the perfect structure usually produces a design for which the critical loads for several buckling modes coincide. This causes interaction of buckling modes that leads to increased imperfection sensitivity, in which either the coefficient that multiplies α^m is increased or the exponent m is decreased. We will illustrate it by two examples. (For a general analysis see, e.g., Thompson and Hunt, 1973).

Consider first Augusti's (1964) column shown in Figure 4.36a; see also Thompson and Hunt (1973, p. 242), Thompson and Supple (1973), Supple (1967), and Chilver (1967). It is a rigid free-standing column of length L, which can deflect in any spatial direction characterized by angles q_1, q_2 with the vertical. It is supported by springs of stiffness C_1 and C_2. Initial imperfections are given by angles $q_1 = \alpha_1$ and $q_2 = \alpha_2$. Consider q_1, q_2, α_1, α_2 to be small. Then the angles of the bar with x and y axes are approximately $90° - q_1$ and $90° - q_2$, respectively, and denoting as ϕ the angle of the bar with the z axis we have $\cos^2(90° - q_1) + \cos^2(90° - q_2) + \cos^2\phi \simeq 1$. From this, $\cos\phi = (1 - \sin^2 q_1 - \sin^2 q_2)^{1/2}$. The vertical displacement of the load is $L(1 - \cos\phi)$. Therefore, the potential energy is

$$\Pi(q_1, q_2, P, \alpha) = \tfrac{1}{2}C_1(q_1 - \alpha_1)^2 + \tfrac{1}{2}C_2(q_2 - \alpha_2)^2 - PL[1 - (1 - \sin^2 q_1 - \sin^2 q_2)^{1/2}] \tag{4.6.7}$$

from which, for the case $\alpha_1 = \alpha_2 = \alpha$,

$$\Pi \simeq \tfrac{1}{2}[C_1 q_1^2 + C_2 q_2^2 - PL(q_1^2 + q_2^2 - \tfrac{1}{12}q_1^4 - \tfrac{1}{12}q_2^4 + \tfrac{1}{2}q_1^2 q_2^2)] - \alpha(C_1 q_1 + C_2 q_2) \tag{4.6.8}$$

because $\sin q \simeq q - q^3/6$ and $(1 + \delta)^{1/2} \simeq 1 + \frac{1}{2}\delta - \frac{1}{8}\delta^2$ if q and δ are small. Setting $\Pi_{,1} = 0$, $\Pi_{,2} = 0$, and $\alpha = 0$, we get the critical loads $P_{cr_1} = C_1/L$ and $P_{cr_2} = C_2/L$. If one critical load, say P_{cr_1}, is much less than the other, one may assume $q_2 = 0$, and the problem becomes identical to the one we already solved in Section 4.2 (Fig. 4.3a). This problem is imperfection insensitive.

The optimum design that takes into account only the critical load is obtained for $C_1 = C_2 = C$, $P_{cr_1} = P_{cr_2} = P_{cr}$. Consider now this special case, in which the buckling modes q_1 and q_2 interact. The equilibrium equations are $\Pi_{,1} = 0$ and $\Pi_{,2} = 0$, and for the perfect column ($\alpha = 0$) the condition of existence of a

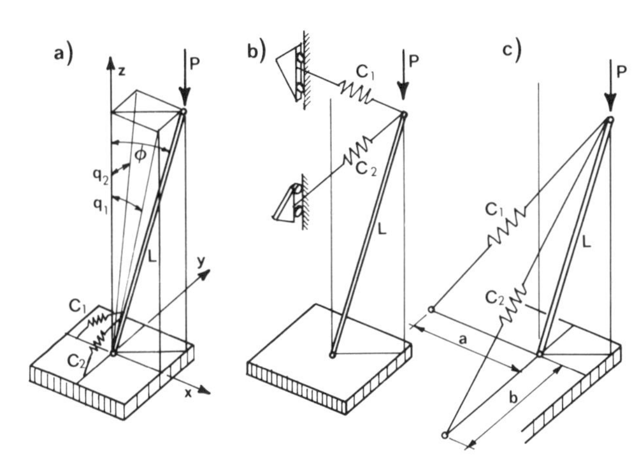

Figure 4.36 (a) Rigid free-standing column in space (Augusti's column, 1964), and (b, c) its modifications.

nonzero solution of these homogeneous equations is found to require that $q_2 = \pm q_1$. This means that the column buckles along the plane of symmetry, as might have been expected in view of the symmetry of the problem when $C_1 = C_2$.

When $\alpha_1 = \alpha_2 = \alpha \neq 0$, the buckling direction may be expected to be very close to or on the plane of symmetry ($q_2 = q_1$). Moreover, by writing out in detail the equations $\Pi_{,1} = 0$ and $\Pi_{,2} = 0$, which are nonlinear, we find they are symmetric and are satisfied by $q_2 = q_1 = q$, and one finds $PL = C(1 - \alpha/q)(1 + q^3/3)$. By the physical nature of the problem, this solution should be unique. This justifies that, for $\alpha \neq 0$, we may set $q_1 = q_2 = q$ in Equation 4.6.8, which furnishes

$$\Pi = Cq^2 - PL(q^2 + \tfrac{1}{6}q^4) - 2C\alpha q \tag{4.6.9}$$

At maximum load we must have $\Pi_{,qq} = 0$, which yields $q^2 = P_{cr}/P - 1$ where $P_{cr} = C/L$. Equilibrium requires that $\Pi_{,q} = 0$, which yields the relation $P_{cr}(q - \alpha) = P(q + q^3/3)$. Substituting $P_{cr} = P(1 + q^2)$, this relation becomes $q^3 - \alpha - \alpha q^2 = q^3/3$. Here the term αq^2 may be neglected, being higher-order small, and so $q^2 = (3\alpha/2)^{2/3}$ at $P = P_{max}$. Recalling that $q^2 = P_{cr}/P - 1$, we thus obtain the result

$$\frac{P_{max}}{P_{cr}} = 1 - \left(\frac{3}{2}\alpha\right)^{2/3} \tag{4.6.10}$$

This is again the $\frac{2}{3}$-power law. For initial angle $\alpha = 0.01$, the drop of maximum load is by 6.1 percent.

Now the important phenomenon to note is that interaction of buckling modes q_1 and q_2 conspires to produce imperfection sensitivity although each mode taking place alone exhibits no imperfection sensitivity, as we recall from Equation 4.5.7 and Figure 4.20c. Such an increase of sensitivity to imperfections is typical when critical loads coincide and modes interact.

As another example, let us look at the reticulated (lattice) strut (built-up column) shown in Figure 4.37 (Koiter and Kuiken, 1971; Thompson and Hunt, 1973, p. 277). The longitudinal bars (flanges), distance $2b$ apart, have axial

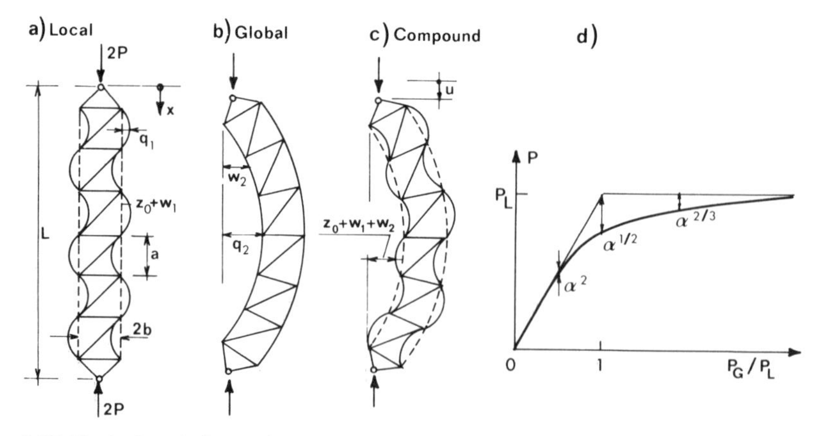

Figure 4.37 Reticulated (lattice) strut buckling according to (a) local, (b) global, and (c) compound modes; (d) plot of maximum load P versus ratio between global and local critical loads.

stiffness EA_f and bending stiffness EI_f. The bracing is pin-jointed but the flanges are continuous through the joints. We may distinguish two basic buckling modes: local and global; see Figure 4.37. The flange deflection for these modes and the initial imperfection may be assumed as

$$w_1 = q_1 \sin\frac{\pi x}{a} \qquad w_2 = q_2 \sin\frac{\pi x}{L} \qquad z_0 = \alpha a \sin\frac{\pi x}{a} \tag{4.6.11}$$

where q_1, q_2 = modal amplitudes, α = imperfection parameter, L = strut length, and a = length of each bracing cell. The load on the strut is $2P$, causing axial force P in each flange. In the local mode (Fig. 4.37a) the flange may be considered to buckle as an infinite continuous beam of spans a, in a sinusoidal curve with zero bending moments (inflection points) at the joints. Thus, the local and global critical loads are

$$P_L = \frac{\pi^2}{a^2} EI_f \qquad P_G = \frac{\pi^2}{L^2} EA_f b^2 \tag{4.6.12}$$

Consider first the local buckling alone ($q_2 = 0$), which is equivalent to an Euler column of length a. According to the magnification factor (Eq. 1.5.10), $\alpha a + q_1 = \alpha a/(1 - P/P_L)$, that is, $q_1 = \alpha a P/(P_L - P)$, and so the axial displacement u_f per length a is

$$u_f = \frac{1}{2}\int_0^a [(z_0' + w_1')^2 - z_0'^2]\,dx = \frac{\pi^2}{4a}(q_1^2 + 2q_1\alpha a) = \frac{\pi^2\alpha^2 a P(2P_L - P)}{4(P_L - P)^2} \tag{4.6.13}$$

The lateral deflection w_1 increases the axial compliance of the flange to the following value:

$$\frac{1}{EA_f^*} = \frac{1}{EA_f} + \frac{1}{a}\frac{\partial u_f}{\partial P} = \frac{\pi^2 b^2}{L^2 P_G} + \frac{\pi^2\alpha^2 P_L^2}{2(P_L - P)^3} \tag{4.6.14}$$

Now, taking simultaneous global buckling into account, load $2P$ on the strut with nonzero w_2 can be approximately taken as the Euler load of a column with bending stiffness $2b^2EA_f^*$. Therefore $2P = 2b^2EA_f^*\pi^2/L^2$. Substituting Equation 4.6.14 we get the relation

$$\frac{1}{\lambda} - \frac{P_L}{P_G} = \frac{L^2\alpha^2}{2b^2(1-\lambda)^3} \qquad \lambda = \frac{P}{P_L} \tag{4.6.15}$$

If $P_L < P_G$ (but close to P_G) and λ is close to 1, the left-hand side of this equation is approximately $1 - P_L/P_G$, and solving for $1 - \lambda$ we get

$$\frac{P}{P_L} = 1 - \left(\frac{\alpha}{\alpha_0}\right)^{2/3} \qquad \text{where } \alpha_0 = \sqrt{2}\,k_1\left(1 - \frac{P_L}{P_G}\right)^{1/2}, \qquad k_1 = \frac{b}{L} \simeq \frac{r_f}{a} \tag{4.6.16}$$

where we introduced the radius of gyration of the flange $r_f = (I_f/A_f)^{1/2}$, and took into account the fact that the present analysis is made only for P_L close to P_G for which $r_f^2A_f/a^2 \simeq A_fb^2/L^2$, that is, $r_f/a \simeq b/L$ (which means the slenderness of the flange and of the strut are nearly equal). If $P_L = P_G$, and λ is close to 1, the left-hand side of Equation 4.6.15 is approximately $1 - \lambda$, and so we get

$$\frac{P}{P_L} = 1 - \left(\frac{\alpha}{\alpha_1}\right)^{1/2} \qquad \text{where } \alpha_1 = \sqrt{2}\,k_1 \tag{4.6.17}$$

If $P_L > P_G$ (but close to P_G) and P is close to P_G, we may substitute $1-\lambda \simeq 1-P_G/P_L$ and obtain from Equation 4.6.15

$$\frac{P}{P_G} = 1 - \left(\frac{\alpha}{\alpha_2}\right)^2 \qquad \text{where } \alpha_2 = \sqrt{2}\, k_1 \left[\frac{(P_L - P_G)^{3/2}}{P_L P_G^{1/2}}\right] \tag{4.6.18}$$

The typical plot of maximum load P versus P_G/P_L for a strut of the same cross section is shown in Figure 4.37d. For $P_L > P_G$ (but close to P_G), the reticulated strut has a mild imperfection sensitivity, characterized by exponent 2. For $P_L < P_G$ (but close to P_G), the strut has a strong imperfection sensitivity, characterized by exponent $\frac{2}{3}$. For $P_L = P_G$, that is, when the local and global critical loads coincide, the strut has a severe imperfection sensitivity, characterized by exponent $\frac{1}{2}$. In this case, which represents the optimum design based on critical load alone, the drop of maximum load for $\alpha = 0.01$ and $L/2b = 20$ is 53 percent. This is a very large loss of capacity indeed!

From the foregoing analysis it is, of course, not clear whether the approximations made have been of a sufficiently high order, in particular, whether the uses of sinusoidal deflection curves in Equations 4.6.11 of w_2, of the magnification factor in Equation 4.6.13, and of the global critical load based on EA_f^* have been sufficiently accurate. Comparison with the complete analysis made by Koiter and Kuiken (1971), however, indicates that the results are correct, and therefore the accuracy of these approximations has been sufficient. In other problems, if there is uncertainty about the sufficiency of the order of approximation, one should preferably use Koiter's general method for compound bifurcation (branching) points. This method is based on the potential-energy expression and utilizes orthogonalized buckling modes (cf. Eqs. 4.3.18 and 4.3.19), but is considerably more involved than the analysis just presented.

In the present context, it may be noted that some codes (e.g., Italian) impose a maximum slenderness λ_m of the members in the built-up columns; for example, $\lambda_m \leq 50$ for Figure 4.40. It so happens that this is enough to avoid the drop of critical load due to interactive buckling.

Problems of mode interaction abound for thin-wall structures. Some are not imperfection sensitive, for example, the buckling modes of web-stiffening ribs and the rib-stiffened plate as a whole (Fig. 4.38). Usually however, mode interaction causes severe imperfection sensitivity. In most cases, it is naive to optimize the design of such structures so that the local and global critical loads are the same or close (this is called naive optimization). Such situations were identified to have been significant contributing factors of several collapses in steel box-girder bridges (see, e.g., Dowling, 1975).

Koiter's theory is further discussed in Section 7.7.

Problems

4.6.1 Without referring to the text, discuss imperfection sensitivity diagrams for various types of buckling.

4.6.2 Analyze imperfection sensitivity for Π according to Equation 4.6.1 with $n = 5$.

4.6.3 Knowing that $q_2 = q_1 = q$, Augusti's column for large deflection can easily be solved exactly. In that case it is better to use the angle ϕ of the bar with the

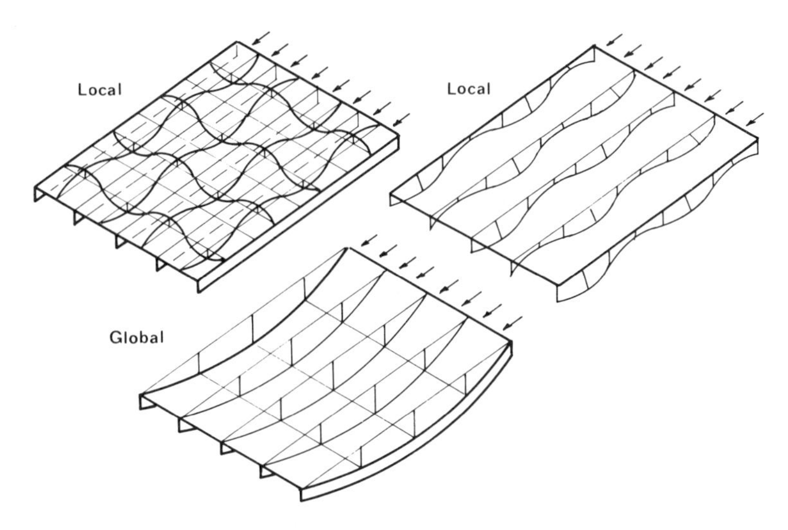

Figure 4.38 Local and global buckling modes of web-stiffening ribs and rib-stiffened plates.

vertical axis, and, instead of Equation 4.6.7, we may write (exactly) $\Pi = C(q-\alpha)^2 - PL(1-\cos\phi)$, in which $\tan\phi = \sqrt{2}\tan q$.

4.6.4 Analyze interaction of modes q_1, q_2 and imperfection sensitivity for the three-dimensional buckling of the columns in Figure 4.36b, c, which represent generalizations of the two-dimensional columns in Figures 4.21 and 4.22 and are modifications of Augusti's column.

4.6.5 Analyze buckling mode interaction and imperfection sensitivity for the reticulated struts in Figure 4.39a, b for $P_G < P_L$, $P_G = P_L$, and $P_G > P_L$.

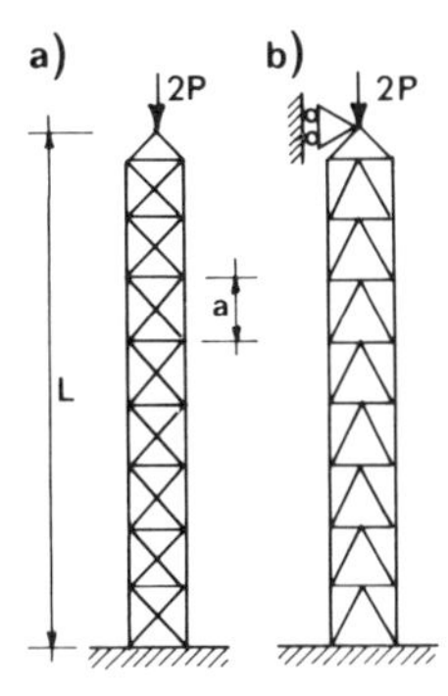

Figure 4.39 Reticulated trusses; Exercise problems on interaction of buckling modes.

4.6.6 Analyze interaction of the shear and bending modes for Figure 4.40. This structure approximates either a built-up column with plates or a tall building frame. Pay particular attention to the case when the critical loads for the shear and bending modes coincide (which would represent a naive optimum design). Solve it by a smeared, continuum approach (valid for large L/a), which is

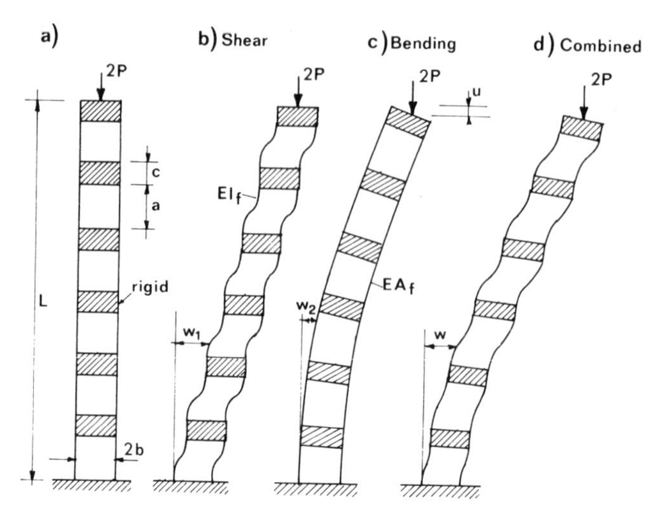

Figure 4.40 Battened column: Exercise problem on interaction of buckling modes.

equivalent to a column with shear or a sandwich plate, which we treated in Section 1.7, but at small deflections only.

4.6.7 Calculate the potential-energy polynomial $\Pi(q_1, q_2, P, \alpha)$ for the reticulated strut in Figure 4.37 making the same order of approximations as we did in Equations 4.6.13 to 4.6.15. Repeat it for Figures 4.39 and 4.40.

4.6.8 Analyze load-deflection paths and stability of Bergan's truss (Bergan, 1979, 1982); see Figure 4.41. This structure, which has two degrees of freedom, exhibits both bifurcation and snapthrough, and for a certain combination of spring stiffnesses both can occur simultaneously. As a generalization, the lateral spring may be nonlinear.

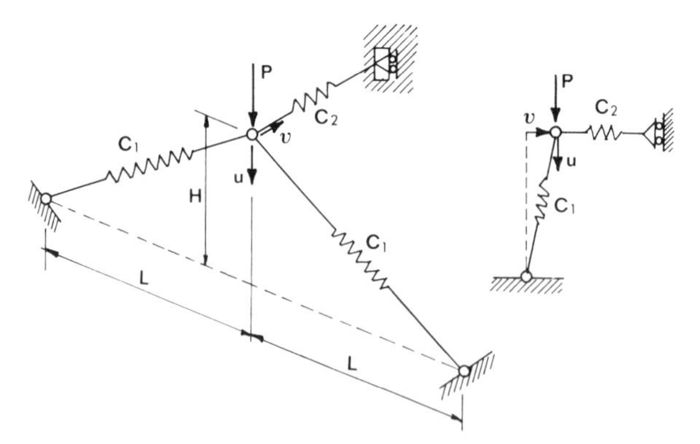

Figure 4.41 Exercise problem on Bergan's truss.

4.7 CATASTROPHE THEORY AND BREAKDOWN OF SYMMETRY

In the preceding sections we have seen that the type of elastic instability is completely determined by the form of the potential-energy function. In fact, what matters are the basic topological characteristics of the potential surface. This is the viewpoint taken in catastrophe theory.

This theory, developed by Thom (1975), Zeeman (1977), Poston and Stewart (1978), Smale (1967), Arnold (1963, 1972), Andronov and Pontryagin (1937), and others, classifies instabilities, called more generally catastrophes, from a strictly qualitative viewpoint. It seeks to identify properties that are common to various catastrophes known in the fields of structural mechanics, astrophysics, atomic lattice theory, hydrodynamics, phase transitions, biological reactions, psychology of aggression, spacecraft control, population dynamics, prey–predator ecology, neural activity of brain, economics, etc. Simply, the theory deals with the basic mathematical aspects common to all these problems.

As shown by Thom (1975), the number of possible types of instability is determined by the number of essential parameters that can be independently controlled, such as the loads (called the active control parameters) or imperfection magnitudes (called the passive control parameters). If there is only a single control parameter λ, we can observe only one type of catastrophe called the fold. In buckling it can take the form of either snapthrough (limit-point instability) or asymmetric bifurcation, for which the typical potential surfaces for problems with one generalized displacement q are shown in Figure 4.42a, b and the equilibrium curves are shown in Figure 4.44. If there are two independent control parameters λ_1 and λ_2, for example, the load P and the imperfection magnitude α, we can observe not only the fold catastrophe (Fig. 4.44), but also another type of catastrophe called the cusp, which corresponds to the symmetric bifurcations (see

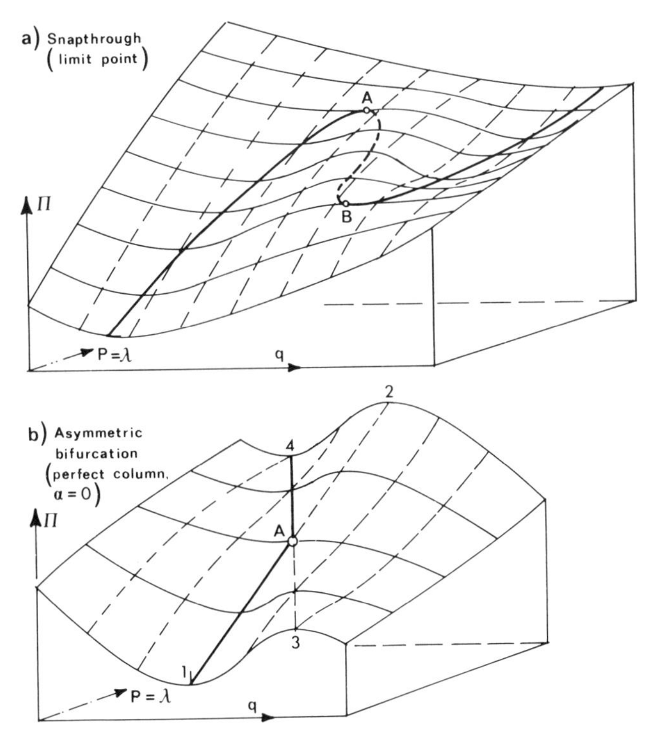

Figure 4.42 Surface $\Pi(q, \lambda)$ with fold catastrophes.

Fig. 4.43 and 4.45). If there are three independent control parameters, we can, in addition, observe three more catastrophes called the swallowtail, the hyperbolic umbilic, and the elliptic umbilic.

Using topological concepts, Thom (1975) proved a remarkable property: For systems with only one control parameter, one can observe only one type of catastrophe; for systems with up to two control parameters, one can observe only two types of catastrophes; for systems with up to three control parameters, one can observe only five types of catastrophes; and for systems with up to four control parameters, one can observe at most seven types of catastrophes. These catastrophes, listed in Table 4.7.1, are called elementary. Those with up to three control parameters have already been mentioned, and those with four independent control parameters include the butterfly and the parabolic umbilic.

The potential functions for the elementary catastrophes can be reduced by coordinate transformations to the basic forms listed in Table 4.7.1. This table also lists one additional catastrophe, the double cusp, which has eight control parameters and has been identified in various buckling problems with interacting modes. The practical advantage of the classification of instabilities is that their analysis can be systematized and the entire behavior near the critical state understood as soon as the potential-energy function is formulated.

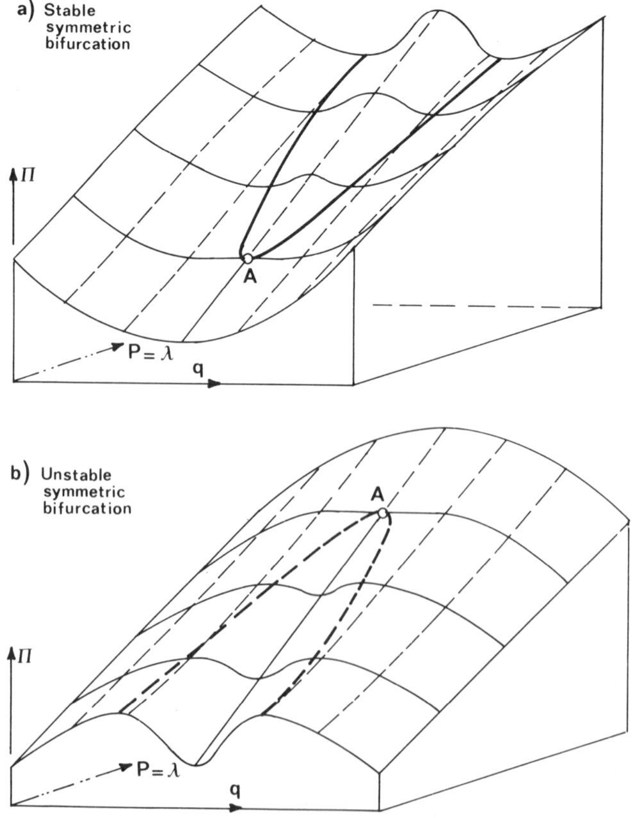

Figure 4.43 Surfaces $\Pi(q, \lambda)$ with cusp catastrophes at point A.

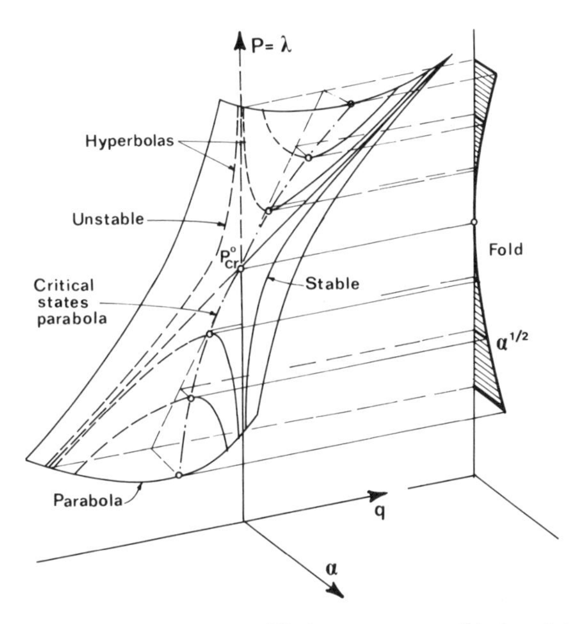

Figure 4.44 Asymmetric bifurcation: equilibrium curves $q(\lambda)$ for fold catastrophes at various values of imperfection α (the surface is quadratic near P_{cr}—a hyperbolic paraboloid).

Note that the first four catastrophes in Table 4.7.1 involve only one variable q, representing the amplitude of the buckling mode. The remaining three elementary catastrophes have two independent variables q_1 and q_2 (generalized coordinates), representing the amplitudes of two interacting buckling modes. Generally the greater the number of independently controlled parameters, the more pathologic the type of instability can get. The equilibrium surfaces for the higher-order catastrophes are quite complicated, as we see from the example of a hyperbolic umbilic in Figure 4.46 adapted from Thompson and Hunt (1984, p. 129).

Interaction of simultaneous buckling modes is the feature that produces the more complicated types of catastrophes (instabilities) numbered 5–8 in Table 4.7.1. Two examples of such interactions have already been given in Section 4.6. As other examples, mode interaction arises when the flanges of a column buckle at the same critical load as the column as a whole; or when the critical loads for the overall buckling of a rib-stiffened panel and for the local buckling of either the plate panels or the ribs happen to coincide (Byskov and Hutchinson, 1977; Thompson and Hunt 1977, 1984; Thompson, 1982; Hunt 1983), see Figure 4.38; or when the stiffness of a beam on an elastic foundation and the foundation modulus are such that the critical loads for such buckling modes coincide (Hansen, 1977; Wicks, 1987); or when the web of a box girder buckles at the same critical load as the top plate or the box overall.

As already noted in Section 4.6, optimum design in which the structure has the minimum possible weight to resist the lowest critical load typically leads to interaction of simultaneous buckling modes. Such optimum designs were labeled naive (Koiter and Škaloud, 1963; Thompson, 1969; Tvergaard, 1973; Thompson

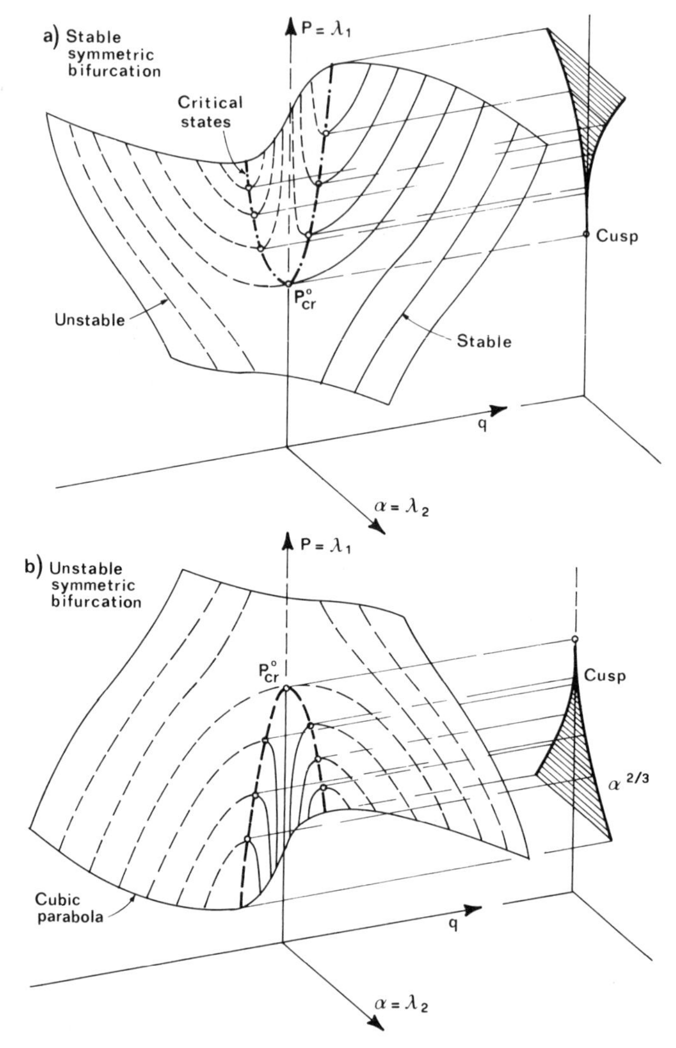

Figure 4.45 Equilibrium surfaces $q(\lambda_1, \lambda_2)$ for cusp catastrophes showing contours of constant imperfection α and lines of critical states (the surface is cubic near P^0_{cr}).

and Hunt, 1973) because they are not really optimum as the interaction of simultaneous buckling modes is found to increase the imperfection sensitivity, often to a high degree. For the double cusp catastrophe, this was documented by Hui's (1986) analysis of beams on nonlinear elastic foundations (see also Wicks, 1987). The optimum design should have sufficiently different critical loads for various modes. Otherwise, optimization produces severe imperfection sensitivity, which has been a contributing factor in some failures of box girder bridges.

Interaction of many simultaneous buckling modes that have the same or nearly the same critical load is typical for compressed cylindrical or spherical shells and is the source of their enormous imperfection sensitivity. In some structures, however, for example, rib-stiffened plates, interaction of buckling modes need not cause imperfection sensitivity (Magnus and Poston 1977).

Table 4.7.1 Seven Elementary Catastrophes (Thom, 1975; Zeeman, 1977) Having Up to Four Control Parameters, and One Catastrophe with More Parameters

Type of Catastrophe	No. of Control Parameters	No. of Variables	Potential-Energy Expression	Structural Instability	
(1) Fold	1	1	$q^3 + \lambda q$	Limit point (snapthrough), asymmetric bifurcation	
(2) Cusp	2	1	$q^4 + \lambda_2 q^2 + \lambda_1 q$	Stable symmetric bifurcation, unstable symmetric bifurcation	
(3) Swallowtail	3	1	$q^5 + \lambda_3 q^3 + \lambda_2 q^2 + \lambda_1 q$	Beams on nonlinear elastic foundation[c] (for (3) and (4))	
(4) Butterfly	4	1	$q^6 + \lambda_4 q^4 + \lambda_2 q^3 + \lambda_2 q^2 + \lambda_1 q$		
(5) Hyperbolic umbilic	3	2	$q_2^3 + q_1^3 + \lambda_1 q_2 q_1 - \lambda_2 q_2 - \lambda_3 q_1$	Rib-stiffened plate,[a] beam on elastic foundation[c]	Interaction of simultaneous buckling modes q_1 and q_2 (for (5)–(8))
(6) Elliptic umbilic	3	2	$q_2^3 - 3q_2 q_1^2 + \lambda_1(q_2^2 + q_1^2) - \lambda_2 q_2 - \lambda_3 q_1$	Beam on quadratic foundation[c]	
(7) Parabolic umbilic	4	2	$q_2^2 q_1 + q_1^4 + \lambda_1 q_2^2 + \lambda_2 q_1^2 - \lambda_3 q_2 - \lambda_4 q_1$	Pressurized shallow spherical shell,[b] plate on elastic foundation[b]	
(8) Double cusp	8	2	$q_1^4 + q_2^4 + \lambda_1 q_1^2 q_2^2 + \lambda_2(q_1^3 + 2q_1^2 q_2) + \lambda_3(q_2^3 + 2q_1 q_2^2) + \lambda_4 q_1^3 + \lambda_5 q_2^2 + \lambda_6 q_1 q_2 + \lambda_7 q_1 + \lambda_8 q_2$	Nonlinear beams on elastic foundation,[i] stiffened cylindrical shells,[d] compressed plates,[e,f] thin-wall angle columns[h]	

[a] Thompson and Hunt (1977, 1984), Thompson (1982); Hunt (1983).
[b] Hui and Hansen (1980 and 1981).
[c] Hansen (1977).
[d] Byskov and Hutchinson (1977).
[e] Magnus and Poston (1977).
[f] Poston and Stewart (1978).
[g] Sridharan (1983).
[h] Hui (1984).
[i] Hui (1986).

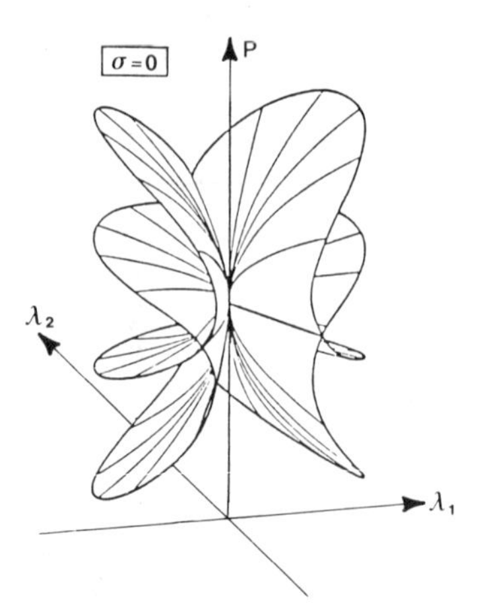

Figure 4.46 Equilibrium surface $P(\lambda_1, \lambda_2)$ for hyperbolic umbilic catastrophe. (From *Elastic Instability Phenomena, by J. M. T. Thompson and G. W. Hunt, copyright* 1984, *John Wiley and Sons. Reprinted by permission of John Wiley and Sons, Ltd.*).

The term elementary means that the potential function has been reduced to its most elementary form that retains the essential qualitative features of the catastrophe. Thus, merely adding various higher-order terms in q, q_1, q_2, λ, or λ_i to the expressions in Table 4.7.1 does not cause different behavior. For example, it has been shown that the control parameters need to appear in the potential function only linearly. This tells us, for example, that it was legitimate to delete the terms with α^2 and α^4 from Equations 4.6.1. Likewise, for $n = 3$ Table 4.7.1 (line 1) indicates the fold catastrophe and shows that it would have been legitimate to set $c_1 = c_2 = a_2 = a_3 = b_2 = b_4 = 0$. For $n = 4$, Table 4.7.1 (line 2) indicates the cusp catastrophe and shows that it would have been legitimate to set $c_1 = a_2 = a_4 = 0$. The fact that these parameters are nonessential can be proven by certain topological concepts called determinacy and unfolding (Poston and Stewart, 1978), applied to the surface of equilibrium states in the space of control parameters $(\lambda_1, \lambda_2, \ldots)$. Considerations of determinacy decide which terms in q, q_1, q_2 are needed and which can be omitted. Considerations of unfolding do the same for the control parameters, thus reducing the stability problem to its simplest form, with the minimum possible number of control parameters.

As an elementary illustration of topological arguments in stability, let us show a topological proof of the following theorem (Thompson, 1970): An initially stable equilibrium path $P(q)$ rising monotonically with the load becomes unstable if and only if it (1) intersects another equilibrium path (bifurcation) or (2) attains a horizontal tangent (limit point, snapthrough). Graphically, this theorem prohibits the situation shown in Figure 4.47c.

To give a proof, following Thompson (1970) we consider the plane (P, q) where, by definition, the values of $\Pi_{,q}$ are zero for the points on the equilibrium path. If the states on this path are stable, $\Pi_{,q}$ is negative to the left of the path and positive to the right of the path, as shown in Figure 4.47. First assume that curve $P(q)$ is always rising (Fig. 4.47a). If the path becomes unstable at some point such as B, and $P(q)$ is still rising, then $\Pi_{,q}$ must be positive on the left and

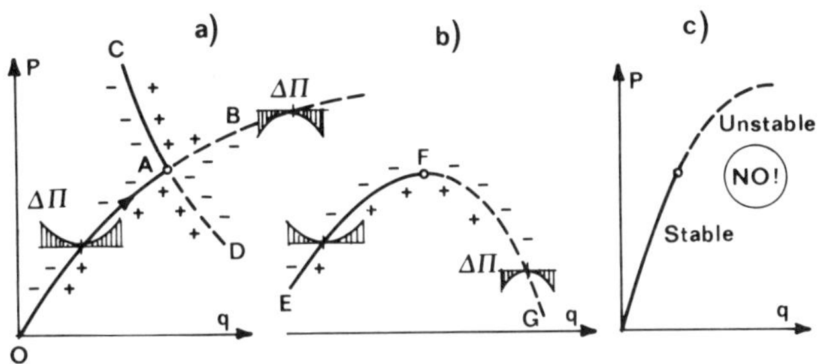

Figure 4.47 (a, b) Possible and (c) impossible equilibrium paths.

negative on the right of the path. Obviously, between 0 and B there must be a point A, where the $\Pi_{,q}$ values on the left switch from negative to positive and those on the right switch from positive to negative. Since $\Pi_{,q}$ changes continuously, we must have $\Pi_{,q} = 0$ at all points where $\Pi_{,q}$ changes its sign, that is, those points must be equilibrium states. Planar topology now requires that the regions of positive and negative signs on each side must be separated by a curve, such as CAD in Figure 4.47, at which the sign switch takes place. This curve is an interacting equilibrium path.

If curve $P(q)$ is not always rising but reaches a limit point F, after which it descends, it is clear from Figure 4.47 that $\Pi_{,q}$ must be positive on the left of the descending segment FG and negative on the right. Indeed, if the opposite were true, that is, the signs of $\Pi_{,q}$ on each side of the curve switched, the regions of positive and negative signs on each side would have to be separated by a curve where $\Pi_{,q} = 0$, that is, another equilibrium path would have to intersect, which is the case we already examined.

From a mathematical viewpoint, the most fundamental characteristic of bifurcation of solutions is the breakdown of symmetry. For example, the perfect pin-ended column is in a symmetric state since it can deflect either left or right. If, after bifurcation, the column is deflected to the right, it can no longer deflect to the left and so its symmetry has broken down.

Since bifurcation-type stability problems always represent some type of a breakdown of symmetry, bifurcation and stability analysis may be avoided if the symmetry is deliberately destroyed by introducing prescribed imperfections. This approach is applicable not only to elastic systems but also to inelastic systems (see Secs. 8.2, 8.4, and 8.5). The advantage is that for a slightly imperfect system one has no bifurcation, and if the imperfection tends to zero the path of the imperfect (nonsymmetric) system approaches the stable path of the perfect (symmetric) system. The disadvantage is that imperfect (nonsymmetric) systems are usually much harder to analyze (e.g., for elastic shells, or for inelastic structures—see Secs. 8.2, 8.4, and 8.5) and their analysis does not yield simple basic properties of the structural system, such as the bifurcation load.

Despite the intent of catastrophe theory to be most general, its formulation has so far been limited to reversible systems for which a smooth potential-energy surface exists. To be really general, the catastrophe theory needs to be extended to irreversible systems, such as inelastic structures (Chaps. 8–13), for which one must apply more general thermodynamic stability criteria. Such criteria correspond to maximization on surfaces with curvature discontinuities, which represent the increment of internal entropy but do not constitute a potential in the mathematical sense (i.e., with the attribute of path independence except in the small).

Problems

4.7.1 Without referring to the text, indicate the forms of potential-energy function that govern the field and cusp catastrophies and give examples for elastic structures.

4.7.2 Set up the potential-energy expression for Π for the column with a horizontal spring on top, Figure 4.21. If imperfection α is not a control parameter but is given, say $\alpha = 0.01$, what type of catastrophe occurs, and at what values of q and P?

4.7.3 Setting up the potential-energy expression for the rigid column with a horizontal spring on top shown Figure 4.21, decide which type of catastrophe takes place. Alternatively, do it for the columns in Figures 4.20 and 4.22.

4.7.4 Formulate the potential-energy expression up to quartic terms for the reticulated strut in Figure 4.37, and comparing it to the expressions in Table 4.7.1 decide which type of catastrophe takes place. Also discuss which, if any, terms can be omitted.

4.8 SNAPDOWN AT DISPLACEMENT-CONTROLLED LOADING

Loads are often applied on structures as reactions to prescribed displacements. Such types of loading may be called the displacement-controlled loading. An example is the loading of a specimen in a typical laboratory testing machine. If the machine is much stiffer than the specimen, and if the movement of the loading crosshead is prescribed, the specimen is loaded essentially in a displacement-controlled manner.

It seems plausible that the structure is always stable when the displacements are prescribed. This is certainly true for the von Mises truss (Fig. 4.14a), as well as all structures with a single degree of freedom. However, for a structure with n degrees of freedom, stability is guaranteed only if all the displacements $q_1, \ldots, q_n$ are prescribed. If only one of them is prescribed, a state that would be unstable under load control may or may not be stabilized.

Structures with Two Degrees of Freedom

To illustrate the problem (following Bažant, 1985), consider again the von Mises truss. However, the truss is not loaded directly but through a spring of stiffness C, as shown in Figure 4.48. The load-point displacement, denoted as q_2, is assumed to be small, and so is the initial angle α. Extension of the spring then is $q_2 - L(q_1 - \alpha)$. The strain energy of the spring must now be added to that we previously figured out for the von Mises truss (Eq. 4.4.6). Thus, the potential energy of the entire structure is

$$\Pi = \frac{EAL}{4}(q_1^2 - \alpha^2)^2 + \frac{C}{2}[q_2 - L(\alpha - q_1)]^2 - Pq_2 \tag{4.8.1}$$

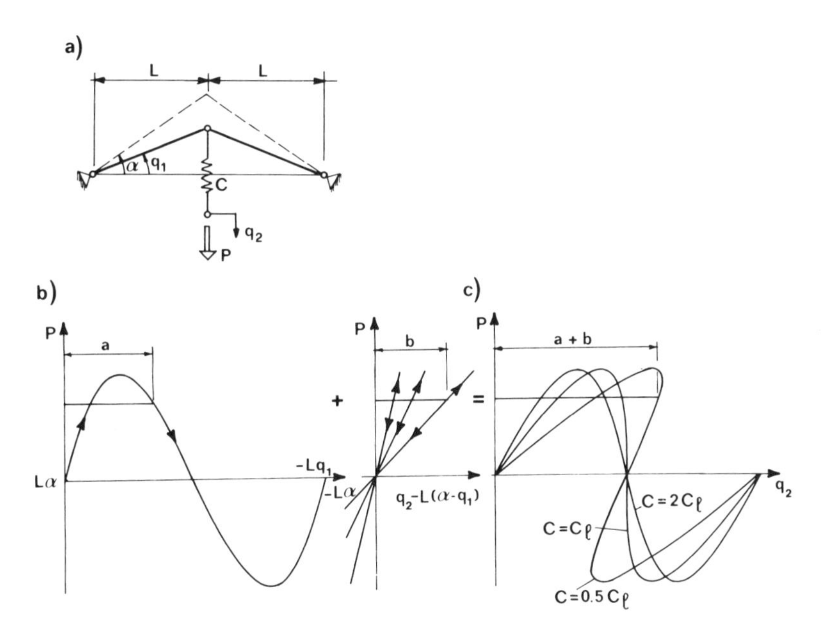

Figure 4.48 (a) Spring-loaded von Mises truss, and (b, c) graphic construction of load-displacement curve.

The equilibrium conditions are

$$\Pi_{,1} = EAL(q_1^2 - \alpha^2)q_1 + CL[q_2 - L(\alpha - q_1)] = 0 \tag{4.8.2}$$

$$\Pi_{,2} = C[q_2 - L(\alpha - q_1)] - P = 0 \tag{4.8.3}$$

Multiplying Equation 4.8.3 by L and adding it to Equation 4.8.2, we find that the equilibrium path may be described as a function of only one variable, q_1:

$$P = EA(\alpha^2 - q_1^2)q_1 \tag{4.8.4}$$

This equation is the same as before (Eq. 4.4.7). However, q_1 is not the work-associated quantity for P. Rather, the work-associated quantity is q_2, whose relation to q_1 (resulting from Eq. 4.8.3) is $q_2 = P/C + L(\alpha - q_1)$. Substituting in Equation 4.8.4, we find that the dependence of load P on load-point displacement q_2 is

$$P = EA\left[\alpha^2 - \left(\alpha - \frac{q_2}{L} + \frac{P}{CL}\right)^2\right]\left(\alpha - \frac{q_2}{L} + \frac{P}{CL}\right) \tag{4.8.5}$$

This equation represents a cubic equation in both P and q_2. The plots of the function $P(q_2)$ are shown in Figure 4.48c for various values of spring stiffness C at fixed EA, α, and L. These curves look different from what we have seen so far—for small enough values of C, the displacement growth reverses and the curve turns back with a positive slope—a phenomenon known as the *snapback*.

The shapes of the equilibrium curves $P(q_2)$ may be more easily figured out graphically. Since the truss and the spring are coupled in series, they are both under the same load P and their deformations are added. Therefore, for any given P value one nceds to add the corresponding q_1 value for the truss and the spring displacement P/C. As illustrated in Figure 4.48b, c, this amounts to adding the horizontal segments a and b on the same horizontal line. This construction displaces the peak to the right but the points of intersection with the q_2 axis do not displace. So the slopes of both the rising segment and the snapback become smaller, while the softening segment becomes steeper. The snapback happens only if C is smaller than a certain limit value C_l that corresponds to a vertical tangent at $q_1 = 0$. From the condition $(\partial P/\partial q_2) \to \infty$ or $(\partial q_2/\partial P) = 0$ at $q_1 = 0$ we get $C = C_l = \alpha^2 EA/L$. (The reversal of softening to a positive slope is typical of many stability problems, especially in the mechanics of distributed cracking; Chap. 13.)

Let us now examine stability of the equilibrium paths. First consider that the load P is controlled, in which case we have two independent kinematic variables, q_1 and q_2. Since $\Pi_{,22} = C > 0$, stability is completely decided by the determinant $D = \Pi_{,11}\Pi_{,22} - \Pi_{,12}^2$ (the subscripts preceded by a comma denote derivatives, e.g., $\Pi_{,12} = \partial^2\Pi/\partial q_1\, \partial q_2$). From Equations 4.8.2 and 4.8.3 we calculate

$$D = CEAL(3q_1^2 - \alpha^2)\begin{cases} >0 & \text{(stable)} \\ =0 & \text{(critical)} \\ <0 & \text{(unstable)} \end{cases} \tag{4.8.6}$$

This is the same condition we found before for small α (see Eq. 4.4.8), and so, same as before, the limit points (the maximum and minimum points) of the curves $P(q_1)$ as well as $P(q_2)$ in Figure 4.48b, c represent the critical states, and the segment of each path between the maximum and the minimum (having a negative slope) is unstable while the rest of each path is stable. See also Figure 4.49a.

Second, consider that displacement q_2 is controlled. Then q_2 becomes a parameter, P is the reaction, and we are left with only *one* kinematic variable —displacement q_1, which represents an internal displacement (internal kinematic variable) of the system. So the stability condition now is $\Pi_{,11} > 0$.

Looking at it in another way, we may consider the general expression for the second variation $2\,\delta^2\Pi = \Pi_{,11}\,\delta q_1^2 + 2\Pi_{,12}\,\delta q_1\,\delta q_2 + \Pi_{,22}\,\delta q_2^2$ and substitute in it $\delta q_2 = 0$ (since δq_2 cannot be arbitrarily varied if the displacement q_2 is controlled). This yields $2\,\delta^2\Pi = \Pi_{,11}\,\delta q_1^2$, showing again that the condition $\delta^2\Pi > 0$ reduces to $\Pi_{,11} > 0$.

From Equation 4.8.2 we get

$$\frac{\partial^2\Pi}{\partial q_1^2} = EAL(3q_1^2 - \alpha^2) + CL^2\begin{cases} >0 & \text{(stable)} \\ =0 & \text{(critical)} \\ <0 & \text{(unstable)} \end{cases} \tag{4.8.7}$$

It is interesting to compare this condition with the condition $\partial q_2/\partial P = 0$. Differentiating Equation 4.8.5 with respect to P, and then solving for $\partial q_2/\partial P$, we obtain exactly the same condition as Equations 4.8.7 for the critical state. Therefore, the critical states are the points of vertical tangent on the load-deflection path in Figure 4.48b.

In Figure 4.49c, these critical points are projected onto the curve $P(q_1)$. We see that the critical points are pushed away from the limit points of the curve $P(q_1)$, and the unstable portion of the equilibrium path becomes smaller. Thus,

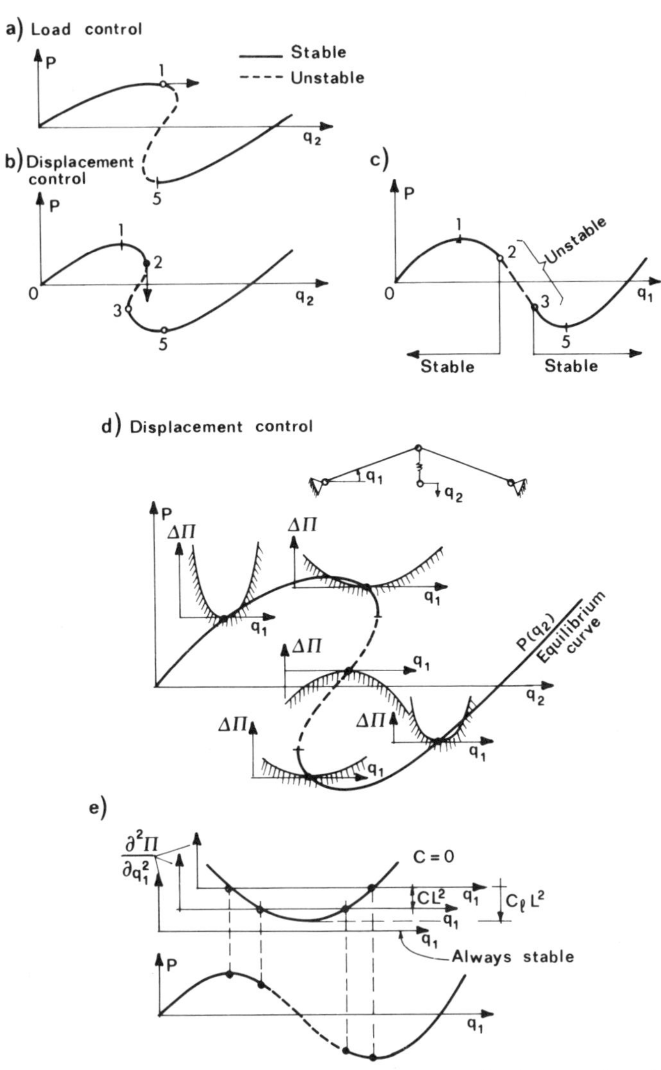

Figure 4.49 (a–c) Stable and unstable segments of equilibrium paths; (d) variation of potential energy at constant P; (e) graphic construction of the critical points.

displacement control has a stabilizing influence, as naturally expected. This example nevertheless teaches us that a displacement-controlled elastic system does not always have to be stable.

The fact that the points of vertical tangent of the diagram $P(q_2)$ represent critical states is obvious even without consideration of energy. These points must be at the stability limit because there exist adjacent (infinitely close) equilibrium states with the same value of the control variable q_2. However, the fact that the portion of the equilibrium path between the critical points is unstable and the rest of it is stable is fundamentally justified on the basis of potential energy. This fact is not obvious intuitively, since segment $\overline{23}$ of the path in Figure 4.49b has a positive slope. We see that under controlled displacement the equilibrium path of positive slope may be either stable or unstable—a point to note.

Figure 4.49d illustrates how the potential-energy change $\Delta\Pi$ with respect to the equilibrium state varies in the vicinity of the equilibrium state. Since q_2 is controlled, it cannot vary freely, and so the nonequilibrium positions adjacent to those on the equilibrium curve can differ only by displacement q_1. Therefore, $\Delta\Pi$ can vary only as a function of q_1, at fixed q_2. The $\Delta\Pi$-plots with convex curvature are characteristic of the stable points on the equilibrium states, and those with concave curvature are characteristic of the unstable points (Fig. 4.49d). Realize that on all these plots of $\Delta\Pi$ versus q_1 only one point, namely that on the equilibrium curve, is an equilibrium state, while the other ones are not. It should be kept in mind that the principle of minimum potential energy Π compares states of different deflection at the same load, and the equilibrium (stable) is that state for which Π is stationary (minimum). Therefore, determination of min Π does not involve any differentiation of P ($P = \text{const.}$).

When we are calculating the equilibrium path $P(q_2)$, we are looking for equilibrium states at various *fixed* values of P; but again minimization of Π at *constant* P is implied at each load level (load value). Thus we have an infinite number of minimization problems for Π for all values of P.

Figure 4.49e illustrates a graphic construction of the critical points. The plot of the second derivative $\Pi_{,11}$ versus q_1 is a parabola, and the critical points are obtained, according to Equations 4.8.7, as the intersections with the horizontal line of ordinate CL^2. Where the parabola is below this line, the system is unstable. If $C = 0$ (the limiting case of no spring), the equilibrium path reduces to that which we found before for the von Mises truss. If $C > C_l = EA\alpha^2/L$, the parabola is always above the line, and the system is always stable (under displacement control). For $C \rightarrow \infty$, that is, an infinitely stiff spring, the system is always stable. This is also evident from the fact that $L(\alpha - q_1) = q_2$, indicating that, in effect, displacement q_1 is controlled.

As the structure is loaded under controlled q_2, the load P (actually a reaction) varies in a smooth equilibrium manner until point 2 (Fig. 4.50a) of the vertical tangent is reached. As the displacement q_2 is further increased by an infinitesimal amount, there is no equilibrium state possible in the immediate vicinity of point 2. So, rather than pursuing the snapback curve, the structure must snap rapidly down to point 4 on the second stable branch—a phenomenon that may be generally called the *snapdown*.

The vertical downward path from point 2 to point 4 (Fig. 4.50a), of course, cannot be an equilibrium path, and inertia forces will be present. Imagine that the mass m of the system is concentrated just under the hinge of the truss (Fig. 4.50d). During the snapdown, the force $P_2 = P$ in the spring, which is related to relative spring displacement linearly (Fig. 4.50c), is also represented by the dotted line in the diagram (Fig. 4.50b) of the force on the hinge of the von Mises truss, denoted as P_1, versus hinge displacement u_1. The force difference $P - P_1$ will act on the mass m, imparting to it the kinetic energy $W = \int (P - P_1)\, du_1 = \int m\ddot{u}_1\, du_1 = \int m\ddot{u}_1\dot{u}_1\, dt = \int (\frac{1}{2}m\dot{u}_1^2)^{\cdot}\, dt = \frac{1}{2}m\dot{u}_1^2$. At point 4, W is exactly equal to the cross-hatched area 2342 in Figure 4.50b. It is also equal, due to the way the diagram in Figure 4.50a is constructed, to the cross-hatched area 2342 in Figure 4.50a. During snapdown along path $\overline{24}$ the velocity of mass m increases; then the mass swings beyond point 4, is decelerated, and continues to oscillate about point 4 at constant q_2. Due to inevitable damping, the oscillation eventually ceases, and

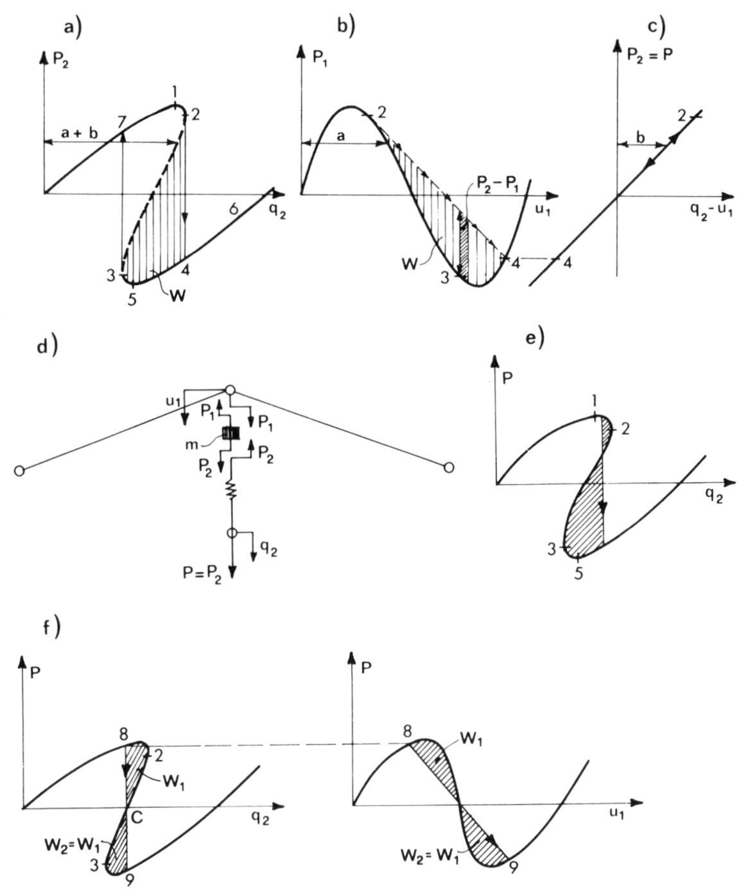

Figure 4.50 (a) Snapdown at controlled displacement q_2; (b–c) force-displacement diagrams for von Mises truss and spring; (d) system with concentrated point mass; (e) snapdown path; (f) snapdown path with no energy change.

the system finds new equilibrium at point 4. After point 4 in Figure 4.50a, the system follows the stable path $\overline{46}$ with increasing displacement q_2.

On subsequent unloading, the system returns along the stable path $\overline{6453}$. As the displacement is decreased below that for point 3, there is no longer any equilibrium state adjacent to 3, and the system snaps dynamically up to point 7 on the first stable branch. After oscillations about point 7 cease due to damping, further unloading follows the stable equilibrium along path $\overline{70}$.

Similar to snapthrough, the system exhibits hysteresis even though it is perfectly elastic . The energy dissipated during a cycle is equal to the area 72437 and is dissipated due to inevitable damping of the dynamic motion. Note also that the softening segment $\overline{12}$ of the stable path cannot be reached at unloading from the opposite branch. However, if the loading reverses to unloading at a point of the segment $\overline{12}$, then, of course, unloading follows the softening segment $\overline{12}$. (This contrasts with the behavior of strain-softening materials, in which unloading is always irreversible. Otherwise, though, the behavior of such a structure is analogous to the stress-strain diagrams of strain-softening materials.)

If our system, under displacement control, receives an initial disturbance, for example, initial velocity, it may follow other snapdown paths as shown in Figure 4.50e. There is one particular path for which there is no net energy loss; it is path $\overline{89}$ in Figure 4.50f, for which the cross-hatched areas 8C2 and C93 are exactly equal. Along path $\overline{8C9}$ the structure first decelerates from its initial velocity at point 8, thus reducing its initial kinetic energy by an amount equal to area 8C2. Along path $\overline{C9}$ the structure accelerates and ends at point 9 with exactly the same velocity as the initial velocity at point 8, and the same final kinetic energy (because areas 82C and C39 are equal). When a structure exhibiting snapdown instability constitutes a part of a larger structural system and when the analysis is static, then the path of the snapdown subsystem must be assumed to follow the straight vertical path $\overline{89}$ with equal areas on both sides, since for this path there is no energy change. Any other snapdown paths, and particularly path $\overline{24}$ of Figure 4.50a, would not be admissible for static elastic analysis because energy dissipation would be implied while elastic analysis presupposes no energy dissipation.

Softening Specimen in a Testing Machine

Consider now the stability of a softening specimen or structure loaded in a testing machine. Assume that displacement q_m is controlled (Fig. 4.51a) and that the

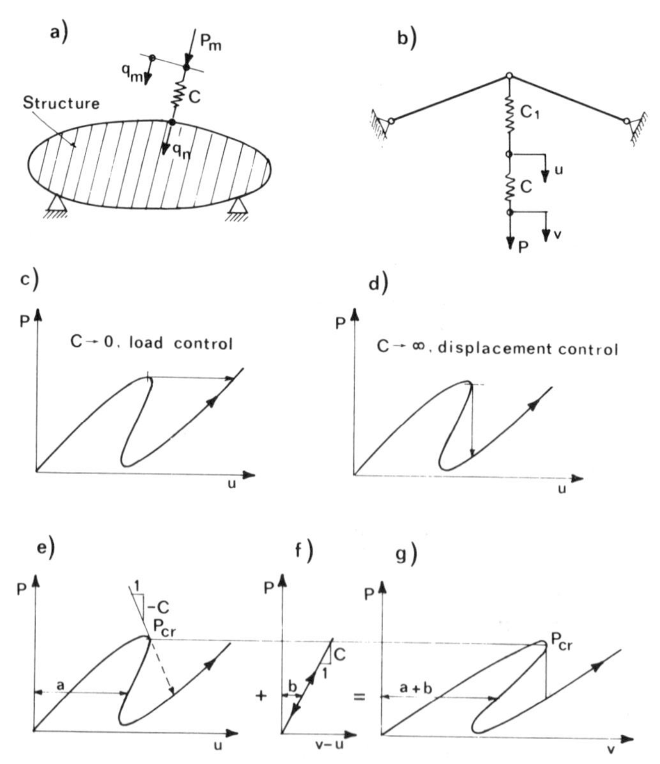

Figure 4.51 (a) Softening structure and, as an example, (b) von Mises truss with added spring C_1 loaded in a testing machine of stiffness C; responses under (c) load control and (d) displacement control; (e–g) graphical construction of load-displacement diagram.

spring stiffness C characterizes the stiffness of the machine frame. If the structure to be tested has a softening response (negative incremental stiffness) with regard to its displacement q_n (Fig. 4.51a), a certain minimum stiffness of the machine frame is necessary for stability. As an example, assume that the specimen is a von Mises truss with a spring C_1 (Fig. 4.51b). When $C \to 0$ we have load control (Fig. 4.51c), and when $C \to \infty$ we have displacement control (Fig. 4.51d).

To treat the intermediate case of finite C (that is, $0 < C < \infty$), we can have either of two approaches. The basic one is to include spring C with the original system, that is, consider the combined system of a von Mises truss with both springs as one system, in which case the two springs have an equivalent stiffness $C_{eq} = 1/(1/C_1 + 1/C)$. In this case we have displacement control for the combined system, see Figure 4.51g. Another approach is to consider only the original system. In this case the critical state arises when the unloading displacement rate du/dP of the original system offsets the unloading displacement rate of the spring $d(v-u)/dP$, because then $dv/dP = 0$ for the combined system. Since $d(v-u)/dP = 1/C$, the critical state condition for the original system is $du/dP = -1/C$, or

$$\frac{dP}{du} = -C \tag{4.8.8}$$

Thus the critical state may be found by drawing a tangent of slope $-C$ as shown in Figure 4.51e. As is clear from the graphic construction, the critical load P_{cr} obtained in this manner is of course the same as that obtained by drawing a vertical tangent to the diagram in Figure 4.51g for the combined system.

The condition $dP/du = -C$, however, is correct only if the softening specimen cannot be further decomposed into a system consisting of a softening element plus a spring (cf. Sec. 13.2).

Generalization of Snapdown Analysis

Consider first an elastic structure with a single degree of freedom, q_1. The potential energy may be written as $\Pi(q_1) = U(q_1) - P\bar{W}(q_1)$, where P is the load, U = strain energy, and $\bar{W}$ = work per unit load; U and $\bar{W}$ are functions of q_1, independent of P. Equilibrium is characterized by the condition $\Pi_{,1} = U_{,1} - P\bar{W}_{,1} = 0$. Since this condition must be satisfied for all points of the equilibrium path $\Pi(q_1)$, the total derivative of $\Pi_{,1}$ along the path must be zero, that is, $d\Pi_{,1}/dq_1 = 0$. Hence, $U_{,11} - P\bar{W}_{,11} - P_{,1}\bar{W}_{,1} = 0$, or $\Pi_{,11} - P_{,1}\bar{W}_{,1} = 0$ (notation: $\Pi_{,11} = \partial^2\Pi/\partial q_1^2$). Thus we conclude that

$$\frac{dP}{dq_1} = \frac{\partial^2 \Pi/\partial q_1^2}{\partial \bar{W}/\partial q_1} \tag{4.8.9}$$

Normally $\partial\bar{W}/\partial q_1$ does not change sign during the deformation process, and so q_1 may always be defined so that $\partial\bar{W}/\partial q_1$ is positive. Then stability is decided by the sign of $\partial^2\Pi/\partial q_1^2$. So a single-degree-of-freedom structure is stable if and only if the slope of the equilibrium curve $P(q_1)$ is positive. If the slope is negative, the system is unstable. A horizontal slope indicates the critical state.

Let us now generalize the analysis to an arbitrary elastic structure with two generalized displacements q_1, q_2. If there is a single load P (or if there are several loads that depend on a single parameter P), then the potential energy may be written as

$$\Pi(q_1, q_2) = U(q_1, q_2) - P\bar{W}(q_1, q_2) \tag{4.8.10}$$

Differentiating with respect to q_1 and q_2, we obtain the following equations for the equilibrium path:

$$\Pi_{,1} = U_{,1} - P\bar{W}_{,1} = 0 \qquad \Pi_{,2} = U_{,2} - P\bar{W}_{,2} = 0 \tag{4.8.11}$$

Let us now relate these conditions to the equilibrium path $P(q_2)$, which we assume to be continuous and smooth, that is, without bifurcation. The partial derivatives $\Pi_{,12}$ and $\Pi_{,22}$ are generally nonzero; however the total derivatives $d\Pi_{,1}/dq_2$ and $d\Pi_{,2}/dq_2$ along the equilibrium path must be zero since Equations 4.8.11 continue to be satisfied for all points of the path $P(q_2)$. Calculating the total derivative of a function according to the chain rule, that is, according to the rule $dF(q_1, q_2, P)/dq_2 = \partial F/\partial q_2 + (\partial F/\partial q_1)(dq_1/dq_2) + (\partial F/\partial P)(dP/dq_2)$, we obtain from Equations 4.8.11:

$$\begin{aligned} \frac{d\Pi_{,1}}{dq_2} &= \Pi_{,12} + \Pi_{,11} q_{1,2} - \bar{W}_{,1} P_{,2} = 0 \\ \frac{d\Pi_{,2}}{dq_2} &= \Pi_{,22} + \Pi_{,21} q_{1,2} - \bar{W}_{,2} P_{,2} = 0 \end{aligned} \tag{4.8.12}$$

in which $q_{1,2} = dq_1/dq_2$ = derivative of q_1 with respect to q_2 along the equilibrium path. Eliminating the derivative from the last two equations, we obtain the relation $\Pi_{,11}\Pi_{,22} - \Pi_{,12}^2 = \mathrm{P}_{,2}(\bar{W}_{,2}\Pi_{,11} - \bar{W}_{,1}\Pi_{,21})$, that is,

$$\frac{dP}{dq_2} = \frac{\det \Pi_{,ij}}{\bar{W}_{,2}\Pi_{,11} - \bar{W}_{,1}\Pi_{,21}} \tag{4.8.13}$$

Under *load control,* a necessary and sufficient condition of stability (Theorem 4.1.10) is that both $\det \Pi_{,ij} > 0$ $(i, j = 1, 2)$ and $\Pi_{,11} > 0$. Either $\det \Pi_{,ij} = 0$ or $\Pi_{,11} = 0$ indicates the critical state. Thus we conclude that if the load is controlled (i.e., if q_2 can vary independently), then a stationary point on the equilibrium curve $P(q_2)$ (a point of horizontal tangent, $dP/dq_2 = 0$) always indicates a critical state, and this critical state is associated with $\det \Pi_{,ij} = 0$. (This condition was already demonstrated for a structure with two degrees of freedom treated in Sec. 4.5.)

Let us further assume P to be defined so that it is associated (or conjugate) with q_2. This means that the work of load P (at constant P) is given by Pq_2, that is, $\bar{W} = q_2$, as was the case in our preceding example. Then $\bar{W}_{,2} = 1$ and $\bar{W}_{,1} = 0$, and Equation 4.8.13 reduces to

$$\frac{dP}{dq_2} = \frac{\det \Pi_{,ij}}{\Pi_{,11}} \tag{4.8.14}$$

The stability conditions $\det \Pi_{,ij} > 0$ and $\Pi_{,11} > 0$ in this case imply that if a load-controlled system with two degrees of freedom is stable, then $dP/dq_2 > 0$. However, the condition $dP/dq_2 > 0$ implies stability only if $\Pi_{,11} > 0$ also. When does $\Pi_{,11}$ become negative?

To answer this question, consider that displacement q_2 is controlled. So $\Pi_{,11} > 0$ is then both necessary and sufficient for stability. The displacement-controlled system becomes critical when $\Pi_{,11} = 0$, and according to Equation 4.8.14 this happens when $dP/dq_2 \rightarrow \infty$, that is, at the snapdown point of the equilibrium curve $P(q_2)$ (point 2 in Fig. 4.50a), provided that $\det \Pi_{,ij}$ does not vanish simultaneously. Thus, $\Pi_{,11}$ can change sign only at the snapdown point. Also $\Pi_{,11}$ must vary continuously and be positive at the start of loading ($P = q_2 = 0$). So $\Pi_{,11}$ must be positive up to the snapdown point and negative beyond it. Also, since the snapdown point cannot occur before the peak point, $\Pi_{,11}$ must be positive from the origin up to the peak point.

So we may conclude that for a two-degree-of-freedom structure a smooth equilibrium path without bifurcation is (1) stable up to the peak point (maximum point) and unstable after it if the load is controlled, and (2) stable up to the snapdown point and unstable after it if the displacement is controlled. Thus, the conditions of horizontal or vertical tangents may be used for determining the critical states under load or displacement control. This is shorter for calculation than the second derivatives of Π.

Let us now extend our analysis to an arbitrary system with n degrees of freedom and one load (or load parameter) P. The potential energy may be written as

$$\Pi(q_1, \ldots, q_n) = U(q_1, \ldots, q_n) - P\bar{W}(q_1, \ldots, q_n) \qquad (4.8.15)$$

The equilibrium conditions are

$$\Pi_{,i} = U_{,i} - P\bar{W}_{,i} = 0 \qquad (i = 1, \ldots, n) \qquad (4.8.16)$$

If displacement q_n is controlled, the equilibrium path may be described as $P(q_n)$ and $q_i(q_n)$. If load P is controlled, then the equilibrium path may be described as $q_n(P)$ and $q_i(q_n)$. Although the derivatives $\Pi_{,in}$ are generally nonzero, the total derivative along the equilibrium path, $d\Pi_{,i}/dq_n$, must vanish since Equation 4.8.16 must apply for all the values q_n. Hence, along the equilibrium path,

$$\Pi_{,in} = \sum_{j=1}^{n} \Pi_{,ij} q_{j,n} - \bar{W}_{,i} P_{,n} = 0 \qquad (i = 1, \ldots, n) \qquad (4.8.17)$$

Recognizing that $q_{n,n} = 1$, these equations may be rewritten as

$$\sum_{j=1}^{n-1} \Pi_{,ij} q_{j,n} = P_{,n} \bar{W}_{,i} - \Pi_{,in} \qquad (i = 1, \ldots, n-1) \qquad (4.8.18)$$

from which we have deleted the last equation for $i = n$. These equations represent a system of $n-1$ linear algebraic equations for the unknowns $q_{1,n}$, $q_{2,n}, \ldots, q_{n-1,n}$. By Kramer's rule, the solution of these equations may be expressed as

$$q_{j,n} = \frac{1}{D_{(n-1)}} \sum_{i=1}^{n-1} (-1)^{i+j} \Delta_{ij} (P_{,n} \bar{W}_{,i} - \Pi_{,in}) \qquad (j = 1, \ldots, n-1) \quad (4.8.19)$$

in which $D_{(n-1)}$ is the determinant of the system of equations (Eq. 4.8.18) and represents the principal minor of the $n \times n$ matrix $\Pi_{,ij}$ (illustrated by the cross hatching in Fig. 4.52a). The sum represents the determinant obtained when the

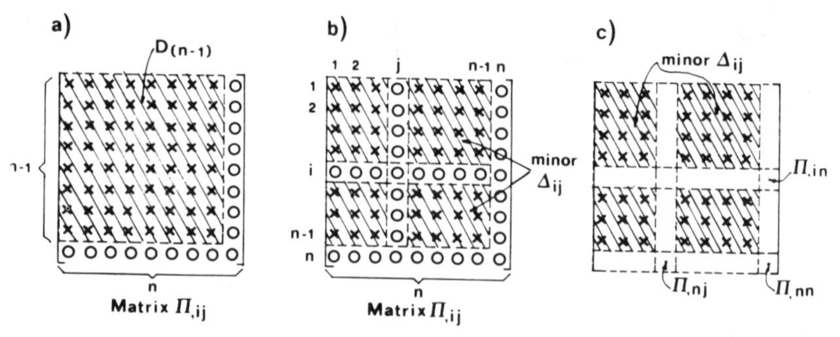

Figure 4.52 Determinant expansions into minors.

jth column of matrix $\Pi_{,ij}$ is replaced by the right-hand sides of Equation 4.8.18. This determinant is expressed in Equation 4.8.19 by its expansion into minors Δ_{ij} obtained from matrix $\Pi_{,ij}$ when the jth column and ith row are deleted (the cross-hatched subdeterminant pictured in Fig. 4.52b).

Substituting now Equation 4.8.19 into Equation 4.8.18 written for $i=n$ (i.e., the last of Eqs. 4.8.17) and solving the ensuing equation for $P_{,n}$ we obtain

$$P_{,n}A_{(n)} = \Pi_{,nn}D_{(n-1)} - \sum_{j=1}^{n-1}\sum_{i=1}^{n-1}(-1)^{i+j}\Delta_{ij}\Pi_{,in}\Pi_{,jn} \tag{4.8.20}$$

in which we denoted

$$A_{(n)} = W_{,n}D_{(n-1)} - \sum_{j=1}^{n-1}\sum_{i=1}^{n}(-1)^{i+j}\Delta_{ij}\Pi_{,nj}\bar{W}_{,i} \tag{4.8.21}$$

Now the right-hand side of Equation 4.8.20 is equal to the determinant $D_{(n)} = \det \Pi_{,ij}$ since it represents the expansion of this determinant into minors according to the nth row (Fig. 4.52c). Hence,

$$\frac{dP}{dq_n} = \frac{D_{(n)}}{A_{(n)}} \tag{4.8.22}$$

If P and q_n are associated (conjugate), then $\bar{W} = q_n$, and $W_{,n} = 1$, $W_{,i} = 0$ for all $i \neq n$, and also $A_{(n)} = D_{(n-1)} =$ principal minor of matrix $\Pi_{,ij}$ (Fig. 4.52a). Then, along the equilibrium path,

$$\frac{dP}{dq_n} = \frac{D_{(n)}}{D_{(n-1)}} \tag{4.8.23}$$

Under *load control,* $D_{(n)} > 0$ and $D_{(n-1)} > 0$ represent necessary (albeit insufficient) conditions of stability (sufficient if $n \leq 2$). Therefore, the initial branch of the equilibrium path has a positive slope as long as the system is stable (Fig. 4.49a). The maximum point (with a horizontal tangent) represents the critical state. After the maximum point, the sign of either $D_{(n)}$ or $D_{(n-1)}$ must change, and so the negative slope of the equilibrium path $P(q_n)$ always implies instability.

Under *displacement control,* however, $D_{(n)}$ is irrelevant for stability. Stability requires that only the principal minors $D_{(1)}, D_{(2)}, \ldots, D_{(n-1)}$ be positive. A vertical tangent of the equilibrium path implies that $D_{(n-1)} = 0$ (assuming that

$D_{(n)} \neq 0$), and so under displacement control the equilibrium path does not become unstable until a point of vertical downward slope is reached (Fig. 4.49b). This conclusion in general legitimizes the determination of stability from the equilibrium path and makes it unnecessary to calculate the matrix of second derivatives of the potential energy.

Note that $D_{(n)}$ becomes negative at the maximum point, and it cannot change sign again because it would imply a horizontal tangent (provided $D_{(n)}$ varies continuously). Also note that $D_{(n-1)}$ becomes negative at the point of vertical downward slope. For these reasons the snapback segment that follows the point of vertical downward slope is unstable, even though the slope of the load-deflection curve is positive. This is an important conclusion, which is not intuitively obvious.

To sum up, instability can occur not only under load control but also under displacement control when there are other uncontrolled displacements that can cause instability. Calculation of the matrix of the second variation of the potential energy can be bypassed by determining stability from the slope and shape of the equilibrium path. If there are no more than two kinematic variables, the equilibrium path gives complete information on stability.

Equilibrium Paths with Bifurcations, Snapthrough, and Snapdown

Our foregoing analysis of the stability of equilibrium paths is limited to the paths which do not bifurcate. We will not carry out a complete analysis of equilibrium paths with bifurcations, but we will at least give some illustrative examples.

Consider a rigid bar with a hinge at the base, supported by a laterally sliding inclined spring, of stiffness C. We investigated this structure before (Sec. 4.5, Fig. 4.22). However, in contrast to our previous study, we consider that load P is applied through a spring of spring constant C_2 (Fig. 4.53a). The initial imperfection is characterized by the initial inclination angle of the bar, α. For $\alpha = 0$, the structure is perfect and exhibits asymmetric bifurcation.

The diagrams of load P versus the associated displacement q_2 can be deduced either by simple calculations, or graphically. We pursue the latter, which is more instructive. We have already derived, for the perfect column, the diagram (Fig. 4.22e, Eq. 4.5.27) of load P versus the vertical displacement u on top of the bar. We show this diagram again in Figure 4.53b. Now, since $q_2 = u + P/C_2 =$ sum of the vertical displacements due to bar inclination and to spring C_2, we may obtain the diagram $P(q_2)$ by adding the displacements of the bar top and the spring as shown graphically by the addition of segments a and b in Figure 4.53b and c. The resulting diagram of load P versus load-point displacement q_2 are shown in Figure 4.53d. The diagram $P(q_2)$ in Figure 4.53d exhibits snapback soon after its peak point, and if the displacement q_2 is controlled the perfect structure will experience snapdown right from the bifurcation point, along the dashed line $\overline{12}$ in Figure 4.53d. In this case the bifurcation point is a point of instability under both load control and displacement control.

The diagram $P(q_2)$ for the imperfect structure can be constructed in a similar way. The diagrams of load versus load-point displacement can easily be calculated exactly, by choosing a series of values of q_1 and evaluating explicitly the value of P (load) and of the load-point displacement q_2. The resulting diagrams are

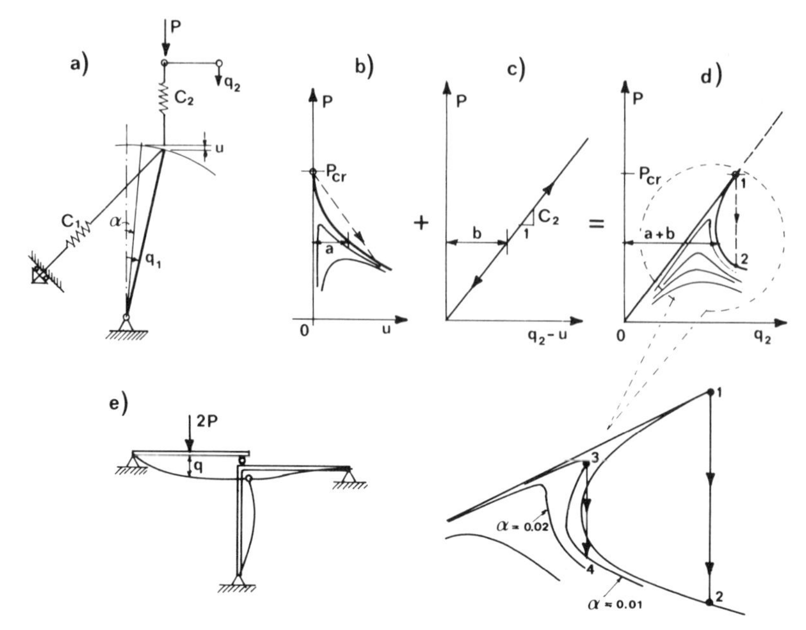

Figure 4.53 (a) Spring-loaded rigid bar with inclined spring; (b–d) graphic construction of load-displacement response; and (e) Roorda frame loaded through a beam.

sketched in Figure 4.53d. Depending on the value of the imperfection and on the stiffness C_2 of the spring, these diagrams may exhibit snapdown paths, such as $\overline{34}$ in Figure 4.53d.

The load-displacement diagram shown in Figure 4.53d is typical for axially compressed or bent cylindrical shells or compressed spherical shells. For such shells, a significant load reduction of maximum load occurs at rather small displacements.

Another example that would give rise to similar diagrams is the Roorda frame, studied in Sections 2.6 and 4.5, if the load is applied through a spring or another flexible element such as a beam (Fig. 4.53e). The load versus load-point displacement diagram would be similar to Figure 4.53d.

Consider again the von Mises truss with a spring from Figure 4.48, but assume now that the bars are so slender that they may buckle before the horizontal position is reached (Fig. 4.54). Assume the bars to be perfect, and the critical axial force of each bar to be P_{E}. Let P_{B} be the axial compressive force in the bars. For $P_B < P_{\mathrm{E}}$, the stiffness of each bar is EA/l where A = cross-section area of the bar and $l = L/\cos\alpha$ = bar length. For $P \geq P_{\mathrm{E}}$, the axial force in the bar depends approximately linearly on the bar shortening u, but the incremental stiffness drops to the value $C_b = P_{\mathrm{E}}/2l$, as established in Section 1.9.

After buckling ($P_B \geq P_{\mathrm{E}}$), the strain energy of the bar, Π_b, is equal to the shaded area under the axial load-shortening diagram in Figure 4.54b; $\Pi_b = \Pi_c^0 + P_{\mathrm{E}}u + \frac{1}{2}C_b u^2$, where Π_c^0 = constant, u = elongation of the bar (we choose $u > 0$ to mean that the bar shortens), $u = L/\cos\alpha - L/\cos q_1 \simeq \frac{1}{2}L(\alpha^2 - q_1^2)$. Combining this with the potential energy of the spring and of the load (Equation 4.8.1), and assuming angles α and q_1 to be small, we obtain instead of Equation

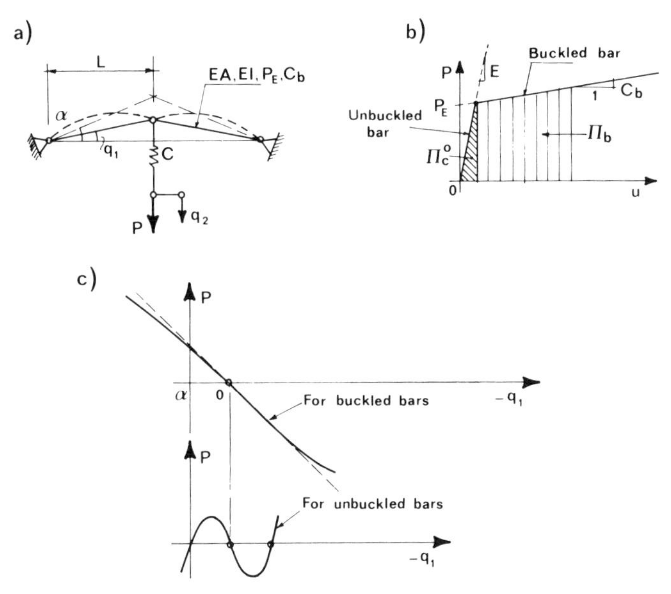

Figure 4.54 (a) Spring-loaded von Mises truss with slender bars; diagrams of (b) load versus axial shortening of bars, and (c) responses for buckled and unbuckled bars.

4.8.1 the potential-energy expression:

$$\Pi \simeq 2\Pi_c^0 + P_{\mathrm{E}}L(\alpha^2 - q_1^2) + \frac{C_b}{4}L^2(\alpha^2 - q_1^2)^2 + \frac{C}{2}[q_2 - L(\alpha - q_1)]^2 - Pq_2 \qquad (4.8.24)$$

Setting $\partial\Pi/\partial q_1 = 0$ and $\partial\Pi/\partial q_2 = 0$, we get a system of two equations for q_1 and q_2, and eliminating q_2 we obtain for the equilibrium path at $P_B \geq P_{\mathrm{E}}$ the equation:

$$P = 2P_{\mathrm{E}}q_1 + C_bL(\alpha^2 - q_1^2)q_1 \qquad (4.8.25)$$

This relation is plotted in Figure 4.54c; together with the path for the case when the bars do not buckle. The transition from one path to the other occurs at the point where they cross; this is a bifurcation point.

As a further approximation, we can introduce in Equation 4.8.25 $P \simeq 2P_{\mathrm{E}}q_1$ because C_b is much smaller than the prebuckling axial stiffness of the bar, EA/l. The solution may be illustrated as in Figure 4.55, in which the curved equilibrium diagrams correspond to our previous solution for unbuckled bars.

First consider the case of an infinitely stiff spring ($C \to \infty$). If the bars are sufficiently slender, that is, have a sufficiently small Euler load P_{E}, the bars will buckle before the first maximum point is reached (point 1 in Fig. 4.55a). Beyond that point, $P \simeq 2P_{\mathrm{E}}q_1$; this gives the line $\overline{135}$, which is approximately straight. The segment $\overline{01}$ is stable, and the segments $\overline{135}$ are unstable under load control. So the system will exhibit a snapthrough from point 1 to point 6. We see that in this case the maximum load is reduced due to buckling of the bars.

If the spring is sufficiently soft, the equilibrium load-deflection diagram for the case of unbuckled bars as well as the path after the buckling of the bars may exhibit a reversal (snapback), as shown in Figure 4.55b. In this case, not only the

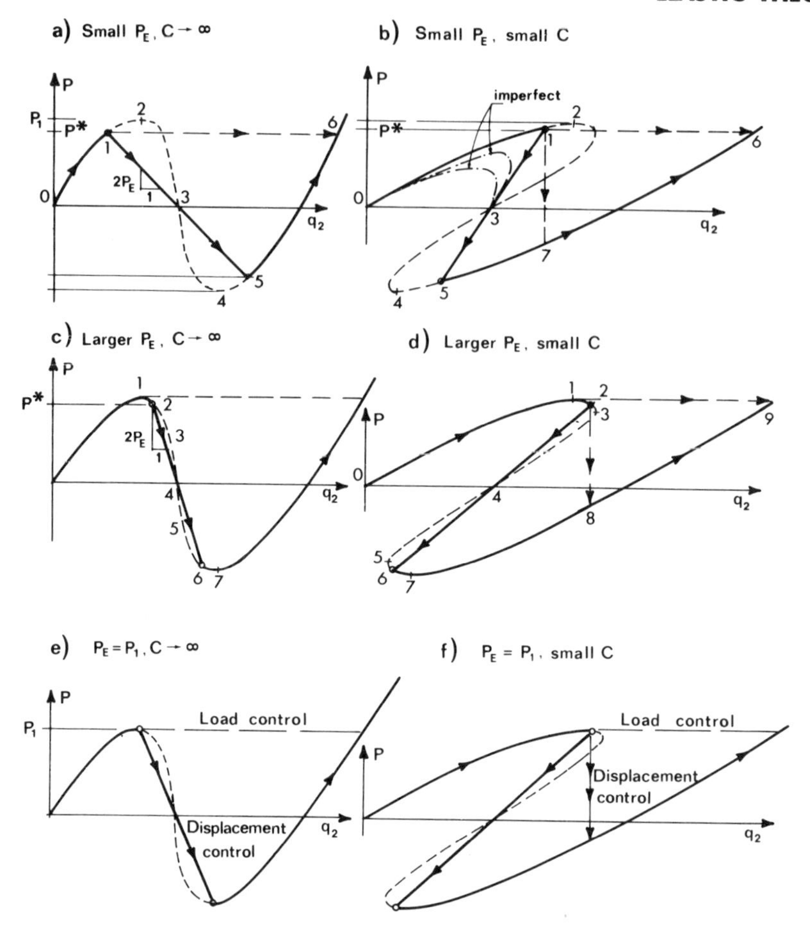

Figure 4.55 Load-displacement diagrams for various choices of parameters and types of control.

snapthrough occurs at a reduced load, but also the snapdown under displacement control occurs at a smaller displacement than it does for the case of no bar buckling.

For bars with a large P_E, the buckling of the bars may occur after the maximum point of the path for the unbuckled bars, as shown in Figure 4.55c and d. If C is sufficiently large, this has no effect on either the snapthrough load or the snapdown displacement. However, if the spring is soft, the displacement at snapdown may get reduced, as shown in Figure 4.55d.

In the construction of the diagrams in Figure 4.55 we already assumed that after bifurcation the path corresponding to buckled bars is stable, and the path corresponding to unbuckled bars is unstable. If we did not know this, we could alternatively also obtain this conclusion by considering Π as a function of three displacements q_1, q_2, q_3, the last one representing the maximum lateral deflection of the bars.

One purpose of the last example was to illustrate more complicated equilibrium paths that a structural system may take. In particular we see in Figure

4.55b that bifurcation may occur in the reverse direction (segment 135). Such behavior, which is often encountered in shells, shows extreme imperfection sensitivity. Indeed, if the bars in Figure 4.48 were considered imperfect, for example, with a small initial curvature, one would obtain in Figure 4.55b the dash-dot equilibrium paths. For such paths, the maximum load can be greatly reduced compared to the bifurcation load, even if the imperfection is very small.

For a certain slenderness of the bars, that is, for a certain value of P_E, it is possible that the bifurcation due to buckling of the bars occurs exactly at the maximum point of the equilibrium path for the unbuckled bars. Thus, two instabilities can occur simultaneously (Fig. 4.55e, f). Under load control, snapthrough would prevail over the bifurcation instability, while under displacement control the bifurcation instability would prevail except if the spring is sufficiently soft, in which case a snapdown can occur.

Problems

4.8.1 For the structures in Figure 4.56a, b, determine the snapdown load on the basis of the diagram $P(q_1)$ of load versus load-point displacement of the structure without the spring.

4.8.2 For the structures in Figure 4.56c, d, determine the snapdown load.

4.8.3 Do the same for the structures in Figure 4.57a, b, c, d, e, f, g.

4.8.4 Do the same for the structures in Figure 4.58a, b, c, d, e, f, g.

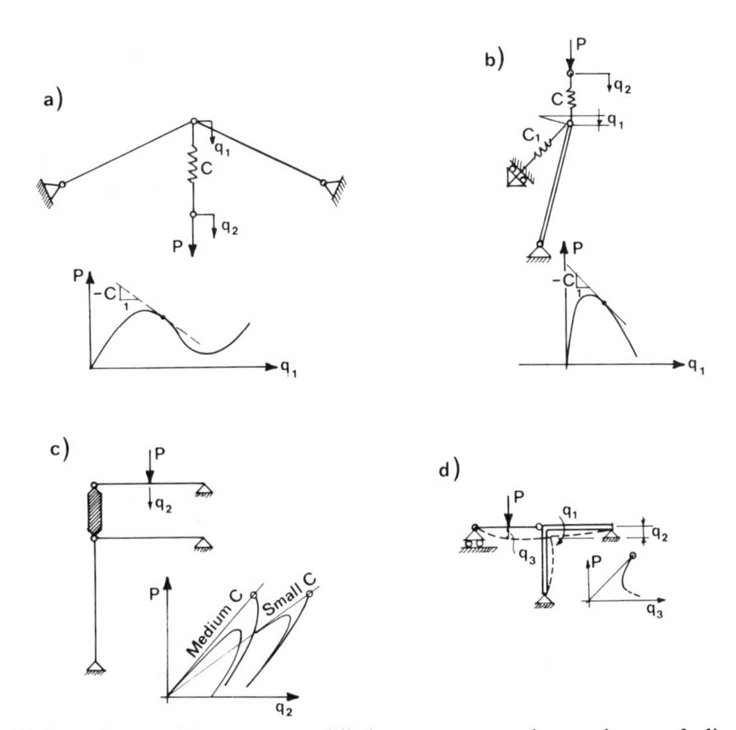

Figure 4.56 Exercise problems on equilibrium curves and snapdown of displacement-controlled systems.

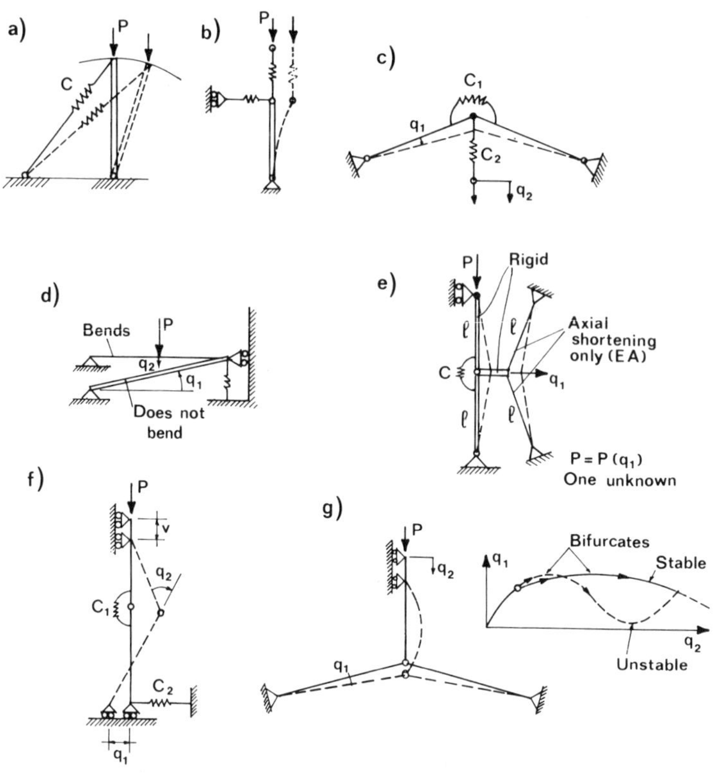

Figure 4.57 Further exercise problems on equilibrium curves and snapdown of displacement-controlled systems.

4.8.5 Calculate the response curves for the imperfect structure in Figure 4.58h.

4.8.6 Do the same for the structure in Figure 4.58i. [The resulting diagrams $P(q_2)$, $P(q_1)$ are plotted for this case.]

4.8.7 As a generalization of Bergan's truss (Prob. 4.6.8, Fig. 4.41) consider that the vertical load P is not applied directly but through a third spring of stiffness C_3. (Snapback can occur then, too.)

4.8.8 For the structure in Figure 4.48 (Eq. 4.8.1), obtain the eigenvectors and express Π as a sum of squares of orthogonal coordinates y_k according to Equation 4.1.10.

4.9 INCREMENTAL WORK CRITERION AT EQUILIBRIUM DISPLACEMENTS

The criterion of minimum potential energy is based on comparing the potential-energy value at the equilibrium state q_i under consideration with the potential-energy values at all adjacent nonequilibrium states $q_i + \delta q_i$ with the same load. In many applications, however, it is more convenient, albeit equivalent, to make energy comparisons with adjacent *equilibrium* states for a slightly different load. This can be done if the stability criterion is reformulated in terms of the work of equilibrium reactions on displacements (Bažant, 1985).

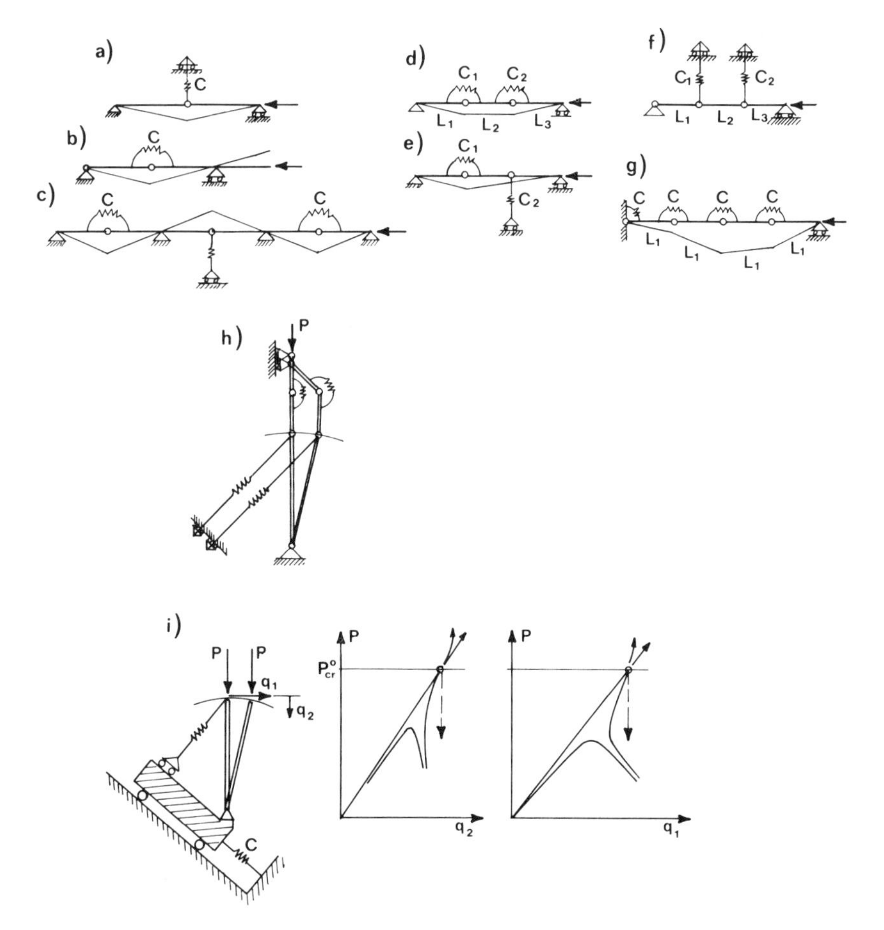

Figure 4.58 Further exercise problems on equilibrium curves and snapdown of displacement-controlled systems.

Stability Criterion

For infinitesimal variations δq_i of generalized displacements q_i, the energy increment from the equilibrium state q_i is, according to Equation 4.2.2, if $\Pi(q_i)$ is differentiable at least twice,

$$\Delta\Pi = \delta^2\Pi = \frac{1}{2}\sum_{i=1}^{N}\sum_{j=1}^{N} K_{ij}\,\delta q_i\,\delta q_j \qquad K_{ij} = \frac{\partial^2\Pi}{\partial q_i\,\partial q_j} \tag{4.9.1}$$

in which K_{ij} represents the incremental (tangential) stiffness matrix of the structure. In Equation 4.9.1 we take note of the fact that $\delta\Pi = 0$ if the initial state is an equilibrium state. In the special case that $\delta^2\Pi = 0$, higher-order variations $\delta^n\Pi$ $(n > 2)$ must be considered to decide stability, but for $\delta^2\Pi \neq 0$ they are irrelevant. (Equation 4.9.1 may be extended to structures for which Π does not exist and K_{ij} is not unique; see Chap. 10. Then all possible K_{ij} matrices must, of course, be considered. This occurs, e.g., when the unloading and loading stiffnesses differ.)

The potential-energy derivatives $\partial\Pi/\partial q_i = f_i$ represent the generalized forces associated with the generalized displacements q_i. Instead of considering deviations δq_i from equilibrium at no change of applied loads, we now consider that the applied loads are changed simultaneously with δq_i so that they are equal to the reactions, that is, equilibrium is maintained at the adjacent state $q_i + \delta q_i$. Increments of the applied forces that must be applied to preserve equilibrium are obviously expressed as

$$\delta f_i = \sum_{j=1}^{N} K_{ij}\, \delta q_j \tag{4.9.2}$$

The incremental work that must be done on the system to effect displacement variations δq_i may now be calculated, with second-order accuracy, as follows (see Fig. 4.59):

$$\Delta \mathcal{W} = \delta^2 \mathcal{W} = \sum_{i=1}^{N} \left(f_i + \frac{1}{2}\,\delta f_i\right) \delta q_i = \frac{1}{2} \sum_{i=1}^{N} \delta f_i\, \delta q_i = \frac{1}{2} \sum_i \sum_j K_{ij}\, \delta q_i\, \delta q_j = \delta^2 \Pi \tag{4.9.3}$$

In this equation we took note of the fact that, because of equilibrium in the initial state q_i, $\sum_i f_i\, \delta q_i = \sum_i (\partial\Pi/\partial q_i)\, \delta q_i = \delta\Pi = 0$. We see that, same as for the variation of the potential energy, the second-order terms decide, that is, $\Delta\mathcal{W}$ is a second-order expression (provided the structure behaves linearly at least near the state q_i).

We may now conclude that the incremental second-order work that must be done on the system in order to produce displacement variations δq_i in an equilibrium manner is exactly equal to the increment of the potential energy when passing from the initial equilibrium state q_i to the state $q_i + \delta q_i$ in a nonequilibrium manner, while the load is kept constant. This means that the stability criterion may be restated in terms of the incremental work at equilibrium displacements as follows:

$$\begin{aligned} &\text{If} \quad \delta^2\mathcal{W} > 0 \quad \text{for } \textit{all} \text{ kinematically admissible } \delta q_i\text{—stable} \\ &\text{If} \quad \delta^2\mathcal{W} = 0 \quad \text{for } \textit{some} \text{ kinematically admissible } \delta q_i\text{—critical} \\ &\text{If} \quad \delta^2\mathcal{W} < 0 \quad \text{for } \textit{some} \text{ kinematically admissible } \delta q_i\text{—unstable} \end{aligned} \tag{4.9.4}$$

In other words, if the incremental work expression $\delta^2\mathcal{W}$ at equilibrium displacements is positive definite then the structure is stable. This is the criterion of incremental work at equilibrium displacements. Although it deals only with equilibrium states of the system it is equivalent to the criterion of minimum

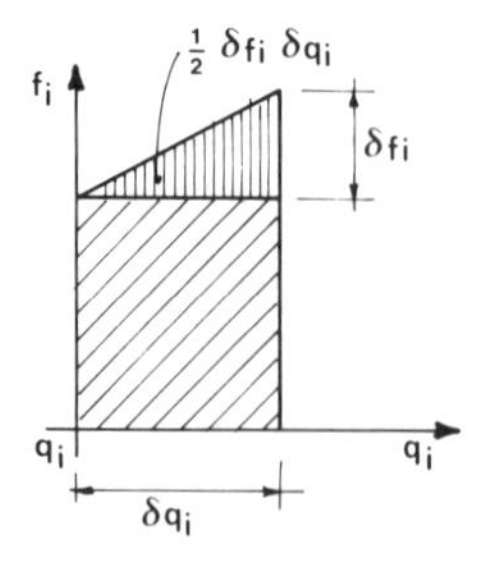

Figure 4.59 Incremental work.

potential energy, which refers to nonequilibrium changes since it compares the equilibrium state with adjacent nonequilibrium states. This criterion is also equivalent to the incremental stiffness criterion, which requires that, for stability, the incremental (tangential) stiffness matrix K_{ij} must be positive definite. (When K_{ij} is not unique, all possible K_{ij} must be positive definite to ensure stability.)

When $\delta^2\mathcal{W}=0$ and $\delta^3\mathcal{W}\neq 0$, the critical state is unstable. When $\delta^2\mathcal{W}=\delta^3\mathcal{W}=0$, stability of the critical state is decided by the sign of $\delta^4\mathcal{W}$.

Example 1: Hinged column. For the sake of illustration, let us analyze again the stability of a perfect simply supported column having length l, bending rigidity EI, and axial compression force P. We examine the stability of the initial equilibrium state $w(x)=0$. For this purpose we consider the variation of deflections $\delta w(x)$. To maintain equilibrium after this variation, one must apply the lateral distributed load:

$$\delta p(x) = EI\,\delta w^{\mathrm{IV}}(x) + P\,\delta w''(x) \tag{4.9.5}$$

The incremental work may be calculated as

$$\Delta\mathcal{W} = \int_0^l \tfrac{1}{2}\,\delta p(x)\,\delta w(x)\,dx \tag{4.9.6}$$

It is now convenient to introduce the Fourier series expansions:

$$\delta p(x) = \sum_{n=1}^{\infty} \delta f_n \sin\frac{n\pi x}{l} \qquad \delta w = \sum_{n=1}^{\infty} \delta q_n \sin\frac{n\pi x}{l} \tag{4.9.7}$$

Substituting then into Equation 4.9.5 we get

$$\sum_{n=1}^{\infty}\left[\left(EI\frac{n^4\pi^4}{l^4} - P\frac{n^2\pi^2}{l^2}\right)\delta q_n - \delta f_n\right]\sin\frac{n\pi x}{l} = 0 \tag{4.9.8}$$

which can be satisfied identically for all x only if the bracketed expression vanishes. This yields the equilibrium relation between the displacement parameters δq_n and the associated distributed load parameters δf_n:

$$\delta q_n = \frac{l^2}{n^2\pi^2(P_{\mathrm{cr}_n} - P)}\,\delta f_n \tag{4.9.9}$$

Substituting this into Equation 4.9.6, we obtain

$$\Delta\mathcal{W} = \sum_{n=1}^{\infty}\frac{1}{2}\,\delta f_n\,\delta q_n = \sum_{n=1}^{\infty}\frac{n^2\pi^2}{2l^2}(P_{\mathrm{cr}_n} - P)\,\delta q_n^2 \tag{4.9.10}$$

The column is stable if this quadradic form is positive definite. This obviously requires that $P_{\mathrm{cr}_n} - P > 0$ for all n. Thus, the stability condition is found to be $P < P_{\mathrm{cr}_1}$, as we already showed before on the basis of the potential energy (Eq. 4.3.5). Note that this stability analysis made no use of the potential-energy expression for the column; it used only the equilibrium relations and the work of loads. Note also that, due to the linearity of Equation 4.9.5, $\Delta\mathcal{W} = \delta^2\mathcal{W}$, that is, $\Delta\mathcal{W}$ contains no terms of order higher than 2.

Example 2: Rigid-bar column with two degrees of freedom. Consider the column consisting of two rigid bars connected by hinges and held upright by rotational springs of stiffness C, a problem we solved before in Section 4.3; see Figure 4.60. The bars have equal lengths l and are loaded by a vertical dead load P; the column has initial imperfections consisting of initial inclinations of the bars α_1 and α_2 (Fig. 4.60a). As generalized displacements, we choose the inclination angles q_1 and q_2 of the lower and upper bars. Based on our preceding considerations, we must introduce disturbing forces δf_1 and δf_2 associated with q_1 and q_2; they are represented by moments δf_1, δf_2 applied on the lower bar and the upper bar at the hinges, as shown in Figure 4.60a.

To formulate the conditions of equilibrium, we may consider, for example, each bar as a separate free body, as shown in Figure 4.60b. First we consider equilibrium at the initial state for which the axial load is P and the disturbing loads $\delta f_1 = \delta f_2 = 0$. If the deflections of the column are assumed to be small, the moment conditions of equilibrium for the lower bar and the upper bar then yield the equations

$$\begin{aligned} C(q_1 - \alpha_1) - C(q_2 - q_1 - \alpha_2 + \alpha_1) - Pl(q_1 - \alpha_1) &= 0 \\ C(q_2 - q_1 - \alpha_2 + \alpha_1) + Plq_2 &= 0 \end{aligned} \tag{4.9.11}$$

After disturbing loads δf_1 and δf_2 are applied, the column will find a new equilibrium condition characterized by generalized displacements $q_1 + \delta q_1$ and $q_2 + \delta q_2$. The moment conditions of equilibrium of the lower and upper bars then yield

$$\begin{aligned} C(q_1 + \delta q_1 - \alpha_1) - C(q_2 + \delta q_2 - q_1 - \delta q_1 - \alpha_2 + \alpha_1) - Pl(q_1 + \delta q_1 - \alpha_1) - \delta f_1 &= 0 \\ C(q_2 + \delta q_2 - q_1 - \delta q_1 - \alpha_2 + \alpha_1) + Pl(q_2 + \delta q_2) - \delta f_2 &= 0 \end{aligned} \tag{4.9.12}$$

Subtracting Equations 4.9.11 from Equations 4.9.12 we obtain incremental equilibrium conditions which may be cast in the following matrix form:

$$\begin{Bmatrix} \delta f_1 \\ \delta f_2 \end{Bmatrix} = \begin{bmatrix} 2C - Pl & -C \\ -C & C + Pl \end{bmatrix} \begin{Bmatrix} \delta q_1 \\ \delta q_2 \end{Bmatrix} \tag{4.9.13}$$

The square matrix in this equation represents the incremental stiffness matrix of the structure subjected to axial load P. For stability, this matrix must be positive definite. We may now note that this matrix is the same as the matrix of the

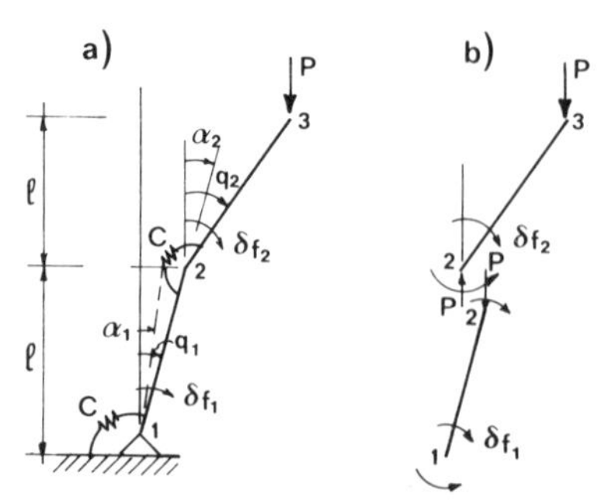

Figure 4.60 Free-standing two-bar column.

quadratic form $\Pi_{,ij}$ that we obtained previously (Sec. 4.3, Eqs. 4.3.16) from the principle of minimum potential energy. Therefore, the rest of the stability analysis is the same, and so are the results.

Alternatively, we may base the stability analysis on the incremental work expression at equilibrium displacements:

$$\Delta \mathcal{W} = \tfrac{1}{2}(\delta f_1\, \delta q_1 + \delta f_2\, \delta q_2) = \tfrac{1}{2}(K_{11}\, \delta q_1^2 + 2K_{12}\, \delta q_1\, \delta q_2 + K_{22}\, \delta q_2^2) \quad (4.9.14)$$

Stability requires that the quadratic form in this equation be positive definite. Obviously this yields again the same stability condition as before.

As in Example 1, note again that in this approach to stability we have used only equilibrium conditions and work of loads, and made no use of the expression for potential energy. Also note that this approach can be extended to rigid columns with springs that do not possess potential energy (e.g., elastoplastic springs). We will deal with such problems later in Chapters 8 and 10.

If the column is perfect ($\alpha_1 = \alpha_2 = 0$), the moment conditions of equilibrium of the lower and the upper bar directly yield Equation 4.9.13. We see that in this problem, under the assumption of small deflections, the stability regions of the perfect and imperfect columns are the same. This is, of course, not true when angles q_1 and q_2 are large, but the present approach can easily be generalized to this case (which was already solved on the basis of potential energy in Sec. 4.3).

Example 3: Spring-loaded von Mises truss. Consider again the von Mises truss loaded through a spring (Fig. 4.61)—a two-degree-of-freedom system, which we analyzed in Section 4.8. If the spring is soft enough, the system exhibits an equilibrium path $P(q_2)$ with a snapback and fails by snapdown instability. On the snapback segment, the equilibrium curve has a positive slope, which in general is necessary but not sufficient for stability (except for single-degree-of-freedom systems). Imagine now applying small incremental moments δf_1 at the support hinges in the direction of angle q_1 (Fig. 4.61a). If the system is in the snapback regime, the slope of the curve $P(q_1)$ is negative (Fig. 4.61b), and it is easy to see that this implies the slope of the curve $f_1(q_1)$ to be also negative, that is, the incremental stiffness for the disturbing moment δf_1 is negative. Thus, the incremental work $\Delta \mathcal{W} = \frac{1}{2}\, \delta f_1\, \delta q_1$ is negative for this particular disturbing load. This proves that the states on the snapback segment of the load-displacement curve are, despite its positive slope, unstable. Even though the incremental work approach is equivalent to potential-energy minimization, it is nevertheless often simpler, as this example illustrates.

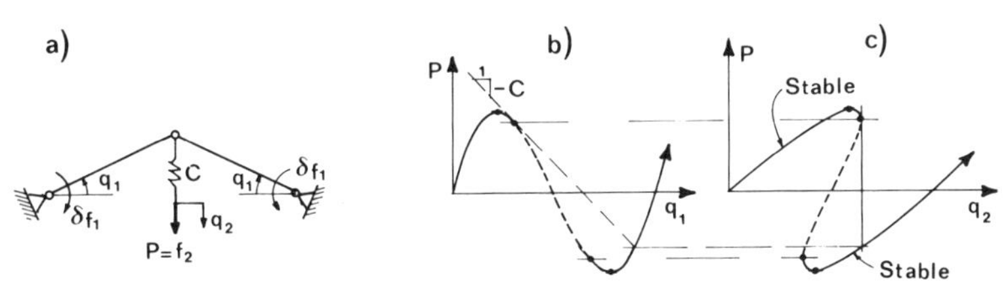

Figure 4.61 Spring-loaded von Mises truss and its stability limits under displacement control.

Possibility of Generalization to Inelastic Systems

The incremental work criterion at equilibrium displacements is more general than the criterion of minimum potential energy because it can be extended to problems with dissipative forces, which cannot be determined from potential energy. We will derive such an extension from the laws of thermodynamics in Chapter 10 where we will deal with inelastic structures, which exhibit dissipative processes such as plasticity, fracture, microcracking, void nucleation and growth, and other kinds of damage.

The incremental work criterion at equilibrium displacements is equivalent to the most fundamental thermodynamic condition that the internally produced entropy of the system (Sec. 10.1) cannot decrease, as required by the second law of thermodynamics. The negativeness of $\Delta \mathscr{W}$ means that the system can spontaneously release energy, which must then be dissipated as heat. A process that dissipates heat increases the internal entropy of the system (precisely by $-\Delta \mathscr{W}/T$ where $T =$ absolute temperature). Therefore, such a process *must* take place. This means that the system cannot remain in the initial equilibrium state, and so the initial state is unstable. On the other hand, if $\Delta \mathscr{W}$ is positive and no work is actually done on the structure by δf_i, the change for which $\Delta \mathscr{W}$ was calculated cannot happen because it would cause the internally produced entropy of the system to decrease by $-\Delta \mathscr{W}/T$, and so the initial state is stable. Obviously, these conditions do not require that a potential energy exist. For a more rigorous analysis, see Section 10.1.

Problems

4.9.1 Consider a hinged column with an axially sliding support and with initial curvature $z_0(x) = \sum a_n \sin (n\pi x/l)$. Find the stability condition under an axial load P using the concept of incremental work.

4.9.2 Use the criterion of incremental work to find the stability condition for the systems in Figure 4.62a, b, c.

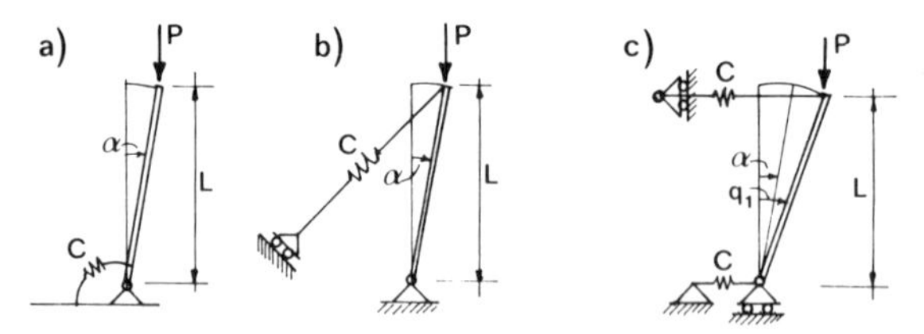

Figure 4.62 Exercise problems on stability of rigid-bar systems.

4.9.3 Solve the problems of Sections 4.3, 4.4, and 4.5 by the incremental work approach.

References and Bibliography

Andronov, A. A., and Pontryagin, L. S. (1937), "Coarse Systems," *Dokl. Akad. Nauk. SSSR,* 14:247; also in Andronov, A. A. (1956), "Sobraniye Trudov," *Izd. Akad. Nauk. SSSR,* p. 181 (Sec. 4.7).

Argyris, G. H., and Kelsey, S. (1960), *Energy Theorems and Structural Analysis,* Butterworths, London (Sec. 4.2).

Arnold, V. I. (1963), "Small Denominators and Problems of Stability of Motion in Classical and Celestial Mechanics," *Russian Mathematical Surveys,* 18(6):85 (Sec. 4.7).

Arnold, V. I. (1972), "Lectures on Bifurcations in Versal Families," *Russian Mathematical Surveys,* 27(54):54–123 (Sec. 4.7).

Augusti, G. (1961), "Sulla definizione di carico critico delle strutture elastiche" (in Italian), *Giornale del Genio Civile,* 99(12):946 (Sec. 4.5).

Augusti, G. (1964), "Stabilità di strutture elastiche elementari in presenza di grandi spostamenti" (in Italian), *Atti Accademia Scienze Fis. Mat., Napoli,* Vol. 4, Series 3, No. 5 (Sec. 4.6).

Bažant, Z. P. (1960), "Mechanical Analysis and New Designs of Safety Ski Binding" (in Czech), *Theory and Practice of Physical Education,* 8:562 (Sec. 4.4).

Bažant, Z. P. (1985), *Lectures on Stability of Structures,* Course 720–D24, Northwestern University, Evanston, Ill. (Sec. 4.1).

Bažant, Z. P., and Cedolin, L. (1989), "Koiter's Power Law for Imperfection Sensitivity in Buckling: Direct Derivation and Applications," Dept. Civil Eng. Report No. 89–9/C616k, Northwestern University, Evanston, Ill (Sec. 4.6).

Bellini, P. X. (1972), "The Concept of Snap-Buckling Illustrated by a Simple Model," *Int. J. NonLinear Mechanics,* 7:643–50.

Bellini, P. X., and Sritalapat, P. (1988), "Post-Buckling Behavior Illustrated by Two-DOF Model," *J. Eng. Mech.* (ASCE), 114(2):314–27.

Bergan, P. G. (1979), "Solution Algorithms for Non-Linear Structural Problems," *Proceedings of the International Conference on Engineering Application of the Finite Element Method,* Computas, Norway (Sec. 4.6).

Bergan, P. G. (1982), "Automated Incremental-Iterative Solution Methods in Structural Mechanics," in *Recent Advances in Non-Linear Computational Mechanics,* ed. by E. Hinton, D. R. J. Owen, and C. Taylor, Pineridge Press Ltd., Swansea, U.K., pp. 41–62 (Sec. 4.6).

Biezeno, C. B. (1938), "Das Durchschlagen eines schwach gekrümmten Stabes," *Zeitschrift Angew. Math. und Mech.* (*ZAMM*) 18:21 (Sec. 4.4).

Budiansky, B. (1974), "Theory of Buckling and Post-Buckling of Elastic Structures," in *Advances in Applied Mechanics,* vol. 14, ed. by C. S. Yih, Academic Press, New York, pp. 1–65 (Sec. 4.6).

Byskov, E., and Hutchinson, J. W. (1977), "Mode Interaction in Axially Stiffened Shells," *AIAA Journal,* 15:941–48 (Sec. 4.7).

Chilver, A. H. (1967), "Coupled Modes of Elastic Buckling," *J. Mech. Phys. Solids,* 15:15–28; *Appl. Mech. Rev.,* 20:Rev. 7755 (Sec. 4.6).

Courant, R., and Hilbert, D. (1962), *Methods of Mathematical Analysis,* Vol. 1, Interscience Publ. (John Wiley & Sons), New York (transl. from German, Springer Verlag 1937) (Sec. 4.1).

Croll, J. G. A., and Walker, A. C. (1972), *Elements of Structural Stability,* Macmillan, London (Sec. 4.6).

Dowling, P. J. (1975), "Strength of Box-Girder Bridges," *J. Struct. Eng.* (ASCE), 101(9):1929–46 (Sec. 4.6).

Faddeev, D. K., and Faddeeva, V. N. (1963), *Computational Methods of Linear Algebra,* Freeman, San Francisco–London.

Franciosi, V., Augusti, G., and Sparacio, R. (1964), "Collapse of Arches under Repeated Loading," *J. Struct. Eng.* (ASCE), 90(1):165 (Sec. 4.4).

Franklin, J. N. (1968), *Matrix Theory,* Prentice-Hall, Englewood Cliffs, N.J., (Sec. 4.1).

Fung, Y. C., and Kaplan, A. (1952), "Buckling of Low Arches of Curved Beams of Small Curvature," NACA TN 2840 (Sec. 4.4).

Gantmacher, F. R. (1959), *The Theory of Matrices I, II,* Chelsea (transl. from Russian, Gostekhizdat, Moscow, 1953) (Sec. 4.1).

Gjelsvik, A., and Bodner, S. R. (1962), "The Energy Criterion and Snap-Buckling of Arches," *J. Eng. Mech.* (ASCE), 88(5):87–134.

Guggenheim, E. A. (1980), "Thermodynamics," in *Encyclopedia Britannica*; also, "Thermodynamics, Classical and Statistical," in *Encyclopedia of Physics,* vol. III/2, ed. by S. Flügge, Springer-Verlag, Berlin, 1959, pp. 1–118 (Sec. 4.4).

Hansen, J. S. (1977), "Some Two-Mode Buckling Problems and Their Relation to Catastrophe Theory," *AIAA Journal,* 15:1638–44 (Sec. 4.7).

Ho, D. (1972), "The Influence of Imperfections on Systems with Coincident Buckling Loads," *Int. J. Nonlinear Mech.,* 7:311–21.

Ho, D. (1974), "Buckling Load of Nonlinear Systems with Multiple Eigenvalue," *Int. J. Solids and Structures,* 10:1315–30.

Hoff, N. J., and Bruce, V. G. (1953), "Dynamic Analysis of the Buckling of Laterally Loaded Flat Arches," *J. Math. and Phys.,* XXXII(4):276–88 (Sec. 4.4).

Hohn, F. E. (1958), *Elementary Matrix Algebra,* Macmillan, New York (Sec. 4.1).

Huddleston, J. V. (1968), "Finite Deflections and Snap-Through of High Circular Arches," *J. Appl. Mech.* (ASME), 35:763–69.

Hui, D. (1984), "Effects of Mode Interaction on Collapse of Short, Imperfect, Thin-Walled Columns," *J. Appl. Mech.* (ASME), 51:556–73 (Sec. 4.7).

Hui, D. (1985), "Amplitude Modulation Theory and Its Application to Two-Mode Buckling Problems," *Zeitschrift für Angew. Math. und Phys.* (*ZAMP*), 35(6):789–802.

Hui, D. (1986), "Viscoelastic Response of Floating Ice Plates under Distributed or Concentrated Loads," *J. Strain Analysis* 21(3):135–43. (Sec. 4.7).

Hui, D., and Hansen, J. S. (1980), "The Swallowtail and Butterfly Cuspoids and Their Application in the Initial Post-Buckling of Single-Mode Structural Systems," *Quart. Appl. Math.,* 38:17–36 (Sec. 4.7).

Hui, D., and Hansen, J. S. (1981), "The Parabolic Umbilic Catastrophe and Its Application in the Theory of Elastic Stability," *Quart. Appl. Math.,* 39:201–220 (Sec. 4.7).

Hunt, G. W. (1983), "Elastic Stability: in Structural Mechanics and Applied Mathematics; Collapse: the Buckling of Structures in Theory and Practice," *IUTAM Symposium held at University College, London,* pp. 125–47, Cambridge University Press, London (Sec. 4.7).

Hutchinson, J. W. (1967), "Imperfection-Sensitivity of Externally Pressurised Spherical Shells," *J. Appl. Mech.* (ASME), 34:49–55 (Sec. 4.6).

Hutchinson, J. W., and Koiter, W. T. (1970), "Post-Buckling Theory," *Appl. Mech. Rev.,* 13:1353–66 (Sec. 4.6).

Iooss, G., and Joseph, D. D. (1980), *Elementary Stability and Bifurcation Theory,* Springer-Verlag, New York, p. 286.

Johns, K. C., and Chilver, A. H. (1971), "Multiple Path Generation at Coincident Branching Points," *Int. J. Mech. Sci.,* 13:899–910.

Koiter, W. T. (1945), "Over de stabiliteit van het elastische evenwicht," Dissertation, Delft, Holland. Translation: "On the Stability of Elastic Equilibrium," NASA TT-F-10833, 1967 and AFFDL-TR-70-25, 1970 (Sec. 4.6).

Koiter, W. T. (1963), "Elastic Stability and Post-Buckling Behavior," in *Nonlinear Problems,* ed. by R. E. Langer, University of Wisconsin Press, Madison.

Koiter, W. T. (1967), "Post-Buckling Analysis of Simple Two-Bar Frame," in *Recent Progress in Applied Mechanics,* ed. by B. Broberg et al. (Folke Odqvist Volume), Almqvist and Wiksell, Sweden, p. 337 (Sec. 4.5).

Koiter, W. T., and Kuiken, G. D. C. (1971), "The Interaction between Local Buckling and Overall Buckling on the Behavior of Built-Up Columns," Report No. 447, Laboratory of Engineering Mechanics, Delft (Sec. 4.6).

Koiter, W. T., and Škaloud, M. (1963), "Interventions, comportement postcritique des plaques utilisées en construction métalliques," *Mém. Soc. Sci. Liège,* 5:64–68 and 103–104 (Sec. 4.7).

Korn, G. A., and Korn, T. M. (1968), *Mathematical Handbook for Scientists and Engineers,* 2nd ed., McGraw-Hill, New York (Sec. 4.1).

Langhaar, H. L. (1962), *Energy Methods in Applied Mechanics,* John Wiley & Sons, New York.

Lee, C. S., and Lardner, T. J. (1989), "Buckling of Three-Dimensional Rigid-Link Models," *J. Eng. Mech.* (ASCE), 115(1):163–77.

Magnus, R. J., and Poston, T. (1977), "On the Full Unfolding of the von Kármán Equations at a Double Eigenvalue," Math. Report No. 109, Battelle Advanced Studies Center, Geneva, Switz. (Sec. 4.7).

Marguerre, K. (1938), "Die Durchschlagskraft eines schwach gekrümmten Balken," *Sitz. Berlin Math. Ges.,* 37:92 (Sec. 4.4).

Masur, E. F., and Lo, D. L. C. (1972), "The Shallow Arch—General Buckling, Post-Buckling and Imperfection Analysis," *J. Struct. Mech.,* 1(1):91 (Sec. 4.4).

Oran, C. (1968), "Complementary Energy Method for Buckling of Arches," *J. Eng. Mech.* (ASCE), 94(2):639–51.

Pearson, C. E., ed. (1974), *Handbook of Applied Mathematics,* Van Nostrand Rheinhold Co. (Sec. 4.1).

Pecknold, D. A., Ghaboussi, J., and Healey, T. J. (1985), "Snap-Through and Bifurcation in a Simple Structure," *J. Eng. Mech.* (ASCE), 111(7):909–22. See also discussion, 113(12):1977–80.

Poston, T., and Stewart, I. (1978), *Catastrophe Theory and Its Applications,* Pitman, London (Sec. 4.7).

Rektorys, K. (1969), *Survey of Applicable Mathematics,* Iliffe Books, London, (Sec. 4.1).

Roorda, J. (1965), "Stability of Structures with Small Imperfections," *J. Eng. Mech.* (ASCE), 91(1):87–106 (Sec. 4.4).

Roorda, J. (1971), "An Experience in Equilibrium and Stability," Tech. Note No. 3, Solid Mech. Div., University of Waterloo, Canada (Sec. 4.5).

Schreyer, H. L., and Masur, E. F. (1966), "Buckling of Shallow Arches," *J. Eng. Mech.* (ASCE) 92(4):1–19.

Sewell, M. J. (1970), "On the Branching of Equilibrium Paths," *Proceedings of the Royal Society,* London, A315:499–518.

Shilov, G. E. (1977), *Linear Algebra,* Dover Publ., New York (transl. from Russian) (Sec. 4.1).

Simitses, G. J. (1973), "Snapping of Low Pinned Arches on an Elastic Foundation," *J. Appl. Mech.* (ASME), 40(3):741 (Sec. 4.4).

Simitses, G. J. (1976), *An Introduction to the Elastic Stability of Structures,* Prentice-Hall, Englewood Cliffs, N.J. (Sec. 4.1).

Simitses, G. J., and Rapp, I. H. (1977), "Snapping of Low Arches with Non-Uniform Stiffness," *J. Eng. Mech.* (ASCE), 103(1):51–65 (Sec. 4.4).

Smale, S. (1967), "Differentiable Dynamic Systems, " *Bull. Ann. Math. Soc.,* 73:747 (Sec. 4.7).

Smirnov, V. (1970), *Cours de mathématique supèrieure,* vol. 3, Mir Publishers, Moscow (Sec. 4.1).

Sridharan, S. (1983), "Doubly Symmetric Interactive Buckling of Plate Structures," *Int. J. Solids Struct.,* 19:625–41 (Sec. 4.7).

Stein, M. (1968), "Recent Advances in the Investigation of Shell Buckling," *AIAA Journal*, 6(12):2239–45 (Sec. 4.6).

Strang, G. (1980), *Linear Algebra and Its Applications*, Academic Press, New York (Sec. 4.1).

Strang, G. (1986), *Introduction to Applied Mathematics*, Wellesley-Cambridge Press, Wellesley, Mass. (Sec. 4.1).

Supple, W. J. (1967), "Coupled Branching Configurations in the Elastic Buckling of Symmetric Structural Systems," *Int. J. Mech. Sci.*, 9:97–112 (Sec. 4.6).

Szabo, J., Gáspár, Z., and Tarnai, T. (1986), *Post-Buckling of Elastic Structures*, Elsevier Science Publ., Netherlands.

Theocaris, P. S. (1984), "Instability of Cantilever Beams with Non-Linear Elements: Butterfly Catastrophe," *Int. J. Mech. Sci.*, 26(4):265–75.

Thom, R. (1975), *Structural Stability and Morphogenesis*, translated from French by D. H. Fowler, Benjamin, Reading (Sec. 4.7).

Thompson, J. M. T. (1969), "A General Theory for the Equilibrium and Stability of Discrete Conservative Systems," *Zeitschrift für Angew. Math. Phys. (ZAMP)*, 20:797 (Sec. 4.7).

Thompson, J. M. T. (1970), "Basic Theorems of Elastic Stability," *Int. J. Eng. Sci.*, 8:307 (Sec. 4.7).

Thompson, J. M. T. (1982), *Instabilities and Catastrophes in Science and Engineering*, John Wiley and Sons, Chichester–New York, 240 pp. (Sec. 4.7).

Thompson, J. M. T., and Hunt, G. W. (1973), *A General Theory of Elastic Stability*, John Wiley & Sons, London (Sec. 4.6).

Thompson, J. M. T., and Hunt, G. W. (1974) "Dangers of Structural Optimization," *Engineering Optimization*, 1:99–110.

Thompson, J. M. T., and Hunt, G. W. (1977), "A Bifurcation Theory for the Instabilities of Optimization and Design," *Synthese*, 36:315 (Sec. 4.7).

Thompson, J. M. T., and Hunt, G. W. (1984), *Elastic Instability Phenomena*, John Wiley & Sons, Chichester (Sec. 4.2).

Thompson, J. M. T., and Supple, W. J. (1973), "Erosion of Optimum Designs by Compound Branching Phenomena," *J. Mech. Phys. Solids*, 21:135 (Sec. 4.6).

Timoshenko, S. P. (1935), "Buckling of Curved Bars with Small Curvature," *J. Appl. Mech.* (ASME), 2(1):17 (Sec. 4.4).

Timoshenko, S. P., and Gere, J. M. (1961), *Theory of Elastic Stability*, McGraw-Hill, New York (Sec. 4.4).

Tvergaard, V. (1973), "Imperfection Sensitivity of a Wide Integrally Stiffened Panel under Compression," *Int. J. Solids Struct.*, 9:177–92 (sec. 4.7).

Tvergaard, V., and Needleman, A. (1980), "On the Localization of Buckling Patterns," *J. Appl. Mech.* (ASME), 47:613–9.

Van der Neut, A. (1968), 'The Interaction of Local Buckling and Column Failure of Thin-Walled Compression Members," *Proceedings of XII Internat. Cong. Appl. Mech.*, Stanford, pp. 389–99 (Springer-Verlag).

Washizu, K. (1975), *Variational Methods in Elasticity and Plasticity*, 2nd ed., Pergamon Press, Oxford–New York, 412 pp. (Sec. 4.2).

Wicks, P. J. (1987), "Compound Buckling of Elastically Supported Struts," *J. Eng. Mech.* (ASCE), 113(12):1861–69 (Sec. 4.7).

Zeeman, E. C. (1977), *Catastrophe Theory: Selected Papers 1972–1977*, Addison-Wesley, London (Sec. 4.7).

Zurmühl, R. (1958), *Matrizen*, Springer-Verlag, Berlin (Sec. 4.1).

5

Energy Analysis of Continuous Structures and Approximate Methods

The energy criterion of stability, as stated in the Lagrange–Dirichlet theorem, is the most effective way to analyze stability of conservative structural systems. The use of the energy criterion for stability analysis is simpler than the fundamental dynamic approach. However, this is not the only use of the energy criterion.

The energy approach is also valuable for obtaining approximate values of the critical loads of more complex conservative structural systems. For example, the exact solution of the critical load of a column of nonuniform stiffness, or a column with variable axial force, or a column whose buckling is resisted by a weak lateral spring placed between the column ends is, in principle, straightforward, however, computationally much more laborious. Quite accurate, yet simple approximations of the critical load can be obtained with the energy approach by using estimated approximate shapes of the deflection curve. These shapes may be chosen either on the basis of experience and experimental observations, or they may be taken the same as the exact solution of a simplified problem, for example, a column of uniform rather than nonuniform bending rigidity.

It turns out that the approximate values of critical loads obtained from the energy criterion always represent upper bounds on the exact values. For structural design, however, it would be preferable to have a principle that would yield a lower-bound approximation to the critical load. Obviously, it would better serve the safety of design. Unfortunately, though, determination of close, generally applicable lower bounds that would be as simple to calculate as the Rayleigh quotient P_R (Sec. 5.3) has proven to be an elusive goal and in general the quest has not yet succeeded, although a good lower bound of the first critical load can be obtained on the basis of successive approximations, provided that a not too poor lower bound for the second critical load is known (Sec. 5.8). The situation in stability is unlike that in plasticity, where both upper and lower bounds that give good approximations for the collapse load are available and are both equally easy to calculate.

When the approximate deflection curve can be guessed so that there is only one unknown parameter, a good upper-bound approximation to the critical load can be obtained simply by evaluation of an explicit expression, known as the Rayleigh quotient. When the deflection shape cannot be guessed with sufficient accuracy, one can obtain an upper-bound approximation for the lowest critical load by minimization of the potential energy with respect to several parameters of the assumed deflection shape. This is the essence of the Rayleigh-Ritz method. Similar results can be obtained on the basis of the differential equation using the Galerkin method.

In the present chapter we will develop these energy approximations in detail, and amply illustrate them by examples. At the beginning of the chapter we will need to clarify the relationship between the potential-energy expression and the differential equation of the problem with its boundary conditions. This relationship is provided by the calculus of variations, and for the reader's convenience we will present a brief overview of this branch of mathematics. Its understanding is essential for approximate energy-based methods as well as for the stability analysis of continuous structures in general, which represents the unifying theme of the present chapter.

5.1 INDIRECT VARIATIONAL METHOD AND EULER EQUATION

The stability analysis illustrated in the preceding chapter, in which the structure is discrete or discretized and the expression for the second variation of potential energy is reduced to a quadratic form, is called the direct variational approach. In contrast to this approach, continuous structures can be analyzed also by an indirect variational method in which the structure is not discretized but differential equations are obtained from the minimizing condition for the potential energy. In general, the potential energy is a function of a function, that is, $\Pi = \Pi[w(x)]$, which is called a functional. The conditions of a minimum (or maximum) of a functional are studied in the calculus of variations, and for the reader's convenience we now give a brief overview of this mathematical theory (e.g., Elsgol'ts, 1963; Fung, 1965; Courant and Hilbert, 1962; Stakgold, 1967).

Review of the Calculus of Variations

We are interested in determining the conditions of a minimum of functional Π of one function w of one variable x, that is,

$$\Pi[w(x)] = \int_0^l \phi(x, w, w', w'')\, dx \tag{5.1.1}$$

Function $w(x)$ is assumed to be continuous and have continuous first four derivatives in the interval $0 < x < l$, and, at the boundaries $x = 0$ and $x = l$, the function is subjected to the appropriate boundary conditions. The problem is to find the function $w(x)$ that makes the functional in Equation 5.1.1 a minimum.

Let $w(x)$ be the exact solution we seek (Fig. 5.1a), and consider all possible functions that are close to $w(x)$. These functions may be written as $w(x) + \delta w(x)$ where $\delta w(x) = \varepsilon \bar{w}(x) =$ variation of function w, $\varepsilon =$ variable parameter, and $\bar{w}(x) =$ any chosen (fixed) function that has the same continuity properties as $w(x)$ and is such that the sum $w + \varepsilon \bar{w}$ satisfies the same kinematic boundary conditions as w. Function $\bar{w}(x)$ is then said to be kinematically admissible; see Figure 5.1a. Replacing $w(x)$ by $w(x) + \varepsilon \bar{w}(x)$ in Equation 5.1.1, the functional Π

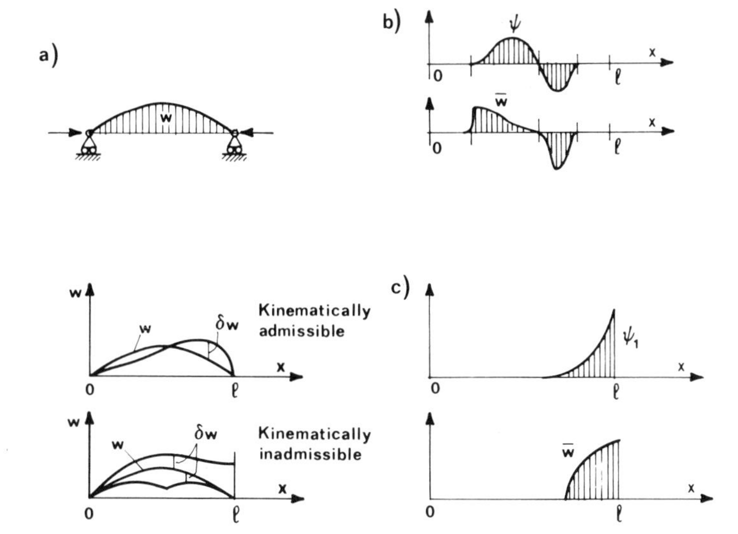

Figure 5.1 Variations of function w which represents the exact solution.

becomes a function of parameter ε alone because functions $w(x)$ and $\bar{w}(x)$ are fixed. So we have

$$\Pi[w(x)+\varepsilon\bar{w}(x)]=f(\varepsilon)=\int_0^l \phi(x,\, w+\varepsilon\bar{w},\, w'+\varepsilon\bar{w}',\, w''+\varepsilon\bar{w}'')\,dx \quad (5.1.2)$$

A necessary, albeit not sufficient, condition of a minimum of function $f(\varepsilon)$ is that $df/d\varepsilon=0$ at $\varepsilon=0$. Recalling the rules for differentiation of an integral and of an implicit function, we then obtain from Equation 5.1.2:

$$\frac{d\Pi}{d\varepsilon}=\frac{df(\varepsilon)}{d\varepsilon}=\int_0^l \frac{\partial\phi}{\partial\varepsilon}\,dx=\int_0^l [\phi_{,w}(x,\, w+\varepsilon\bar{w},\, w'+\varepsilon\bar{w}',\, w''+\varepsilon\bar{w}'')\bar{w}$$

$$+\,\phi_{,w'}(x,\, w+\varepsilon\bar{w},\, \ldots)\bar{w}'+\phi_{,w''}(x+w+\varepsilon\bar{w}'',\, \ldots)\bar{w}'']\,dx \quad (5.1.3)$$

in which we introduce the notations $\phi_{,w}=\partial\phi/\partial w$, $\phi_{,w'}=\partial\phi/\partial w'$, and $\phi_{,w''}=\partial\phi/\partial w''$. After setting $\varepsilon=0$, it is now customary to multiply this equation by ε, which yields

$$\delta\Pi=\delta\Pi[w(x)]=\varepsilon\left[\frac{d\Pi}{d\varepsilon}\right]_{\varepsilon=0}=\int_0^l [\phi_{,w}(x,\, w,\, w',\, w'')\,\delta w(x)$$

$$+\,\phi_{,w'}(x,\, w,\, w',\, w'')\,\delta w'(x)+\phi_{,w''}(x,\, w,\, w',\, w'')\,\delta w''(x)]\,dx=0 \quad (5.1.4)$$

in which $\delta\Pi$ is called the first variation of functional Π, and

$$\delta w(x)=\varepsilon\bar{w}(x) \qquad \delta w'(x)=\varepsilon\bar{w}'(x) \qquad \delta w''(x)=\varepsilon\bar{w}''(x) \quad (5.1.5)$$

These expressions define the first variation of the function $w(x)$ and its derivatives.

It is now useful to eliminate the derivatives of $\delta w(x)$ from Equation 5.1.4. This may be achieved if the second term in the integrand is integrated by parts once, and the third term by parts twice. The result is

$$\delta\Pi = [\psi_1\, \delta w]_0^l + [\psi_2\, \delta w']_0^l + \int_0^l \psi(x)\, \delta w(x)\, dx = 0 \tag{5.1.6}$$

with the notation

$$\psi(x) = \phi_{,w} - \frac{d}{dx}\phi_{,w'} + \frac{d^2}{dx^2}\phi_{,w''} \qquad \psi_1 = \phi_{,w'} - \frac{d}{dx}\phi_{,w''} \qquad \psi_2 = \phi_{,w''} \tag{5.1.7}$$

where the arguments x, w, w', w'' are omitted for brevity.

Equation 5.1.6 must be satisfied for *any* choice of function $\bar{w}(x)$ that is kinematically admissible, that is, for any shape of the curve of variation $\delta w(x)$. The word "any" is crucial. It has the consequence that (1) $\psi(x) = 0$ for all points x within the interval $(0, l)$, and (2) at each boundary ($x = 0$ and $x = l$) either $\delta w = 0$ or $\psi_1 = 0$, and either $\delta w' = 0$ or $\psi_2 = 0$. This consequence is known as the *fundamental lemma* of the calculus of variations (Fung, 1965).

This lemma may be proven by demonstrating the impossibility of the opposite. Thus, consider that the foregoing conclusion is not true, that is, that $\psi(x) \neq 0$, at least in some parts of the interval; see Figure 5.1b. Due to the continuity properties, function $\psi(x)$ must then be nonzero in some finite intervals. However, one can then choose some function $\bar{w}$ that is nonzero in the same intervals and has always the same sign as $\psi(x)$. Then, if the terms in front of the integral are made to vanish, $\delta\Pi$ is positive and thus cannot be zero. But this violates the necessary condition of the minimum. Therefore, $\psi(x) = 0$ is the only possibility. Similarly, if $\delta w \neq 0$ at the boundary point ($x = 0$ or $x = l$), and if we assume that ψ_1 may also be nonzero at that point (Fig. 5.1c), then the expression in Equation 5.1.6 can always be made nonzero by choosing the function $\bar{w}(x)$ to have some finite value at the boundary point. A similar argument can be made for the second boundary term (at $x = l$) in front of the integral in Equation 5.1.6. Thus, $\psi_1 = 0$ (or $\psi_2 = 0$) is the only possibility if δw (or $\delta w'$) can be nonzero at the boundary point according to the given boundary conditions.

Therefore, the condition of a stationary value of functional Π, which represents the necessary, albeit not sufficient, conditions of the minimum of Π, are as follows:

$$\frac{\partial\phi}{\partial w} - \frac{d}{dx}\frac{\partial\phi}{\partial w'} + \frac{d^2}{dx^2}\frac{\partial\phi}{\partial w''} = 0 \qquad \text{for } 0 \leq x \leq l \tag{5.1.8}$$

and for $x = 0$ and $x = l$:

$$\text{Either} \quad w = 0 \quad \text{or} \quad \phi_{w'} - \frac{d}{dx}\phi_{w''} = 0 \tag{5.1.9a}$$

and

$$\text{Either} \quad w' = 0 \quad \text{or} \quad \phi_{w''} = 0 \tag{5.1.9b}$$

As an example consider the total potential energy for an axially loaded inextensible column. It can be obtained by adding the strain energy of bending of the beam (with bending rigidity EI) and the work of the axial load and lateral load:

$$\Pi = \int_0^l \left[\tfrac{1}{2}EIw''^2 - P(\tfrac{1}{2}w'^2) - pw\right] dx \tag{5.1.10}$$

Evaluating the derivatives of the integrand, we have $\phi_{,w} = -p$, $\phi_{,w'} = -Pw'$, $\phi_{,w''} = EIw''$. Substituting this into Equation 5.1.8, we then get the well-known differential equation of equilibrium:

$$(EIw'')'' + (Pw')' = p \tag{5.1.11}$$

while from Equations 5.1.9 for the boundary points we get

$$\text{Either} \qquad w = 0 \qquad \text{or} \qquad (EIw'')' - Pw' = V = 0 \tag{5.1.12a}$$

and

$$\text{Either} \qquad w' = 0 \qquad \text{or} \qquad EIw'' = M = 0 \tag{5.1.12b}$$

which we recognize to be the well-known boundary conditions for a beam.

Equation 5.1.8 is called the Euler equation of the variational problem, first derived by Euler (who obtained it as the limiting case of finite difference approximations). The boundary conditions on the left of Equations 5.1.9a, b are called in mathematics the essential boundary conditions, and in mechanics the kinematic boundary conditions, while the boundary conditions on the right of Equations 5.1.9a, b are called in mathematics the natural boundary conditions, and in mechanics the static boundary conditions.

Application to Structures Possessing a Potential Energy

As one basic observation, we see that not every boundary condition that may randomly come to mind is admissible as a boundary condition if the problem is known to possess potential energy. The condition that the potential energy must be a strict minimum yields not only the differential equation of the problem but also the admissible forms of the boundary conditions. This observation sheds further light on the discussion of nonconservative loads in Section 3.2. In particular, this observation explains that the boundary conditions for the follower force are incompatible with the existence of a potential energy (since they disagree with the natural boundary conditions according to Eqs. 5.1.12a, b). This discrepancy further implies that the follower load is nonconservative, as we established before by other means.

The energy functional, of course, does not have to be quadratic. For example, the potential-energy expression for large deflections of a simply supported inextensible column can be written as $\Pi = \int_0^l [\frac{1}{2}EI\theta'^2 - P(1 - \cos\theta)]\, ds$ where θ = slope of a deflected beam, which is a function of the length coordinate s measured along the arc of the deflected beam. Denoting the integrand as ϕ, we have $\phi_{,\theta} = -P \sin\theta$, $\phi_{,\theta'} = EI\theta'$, and so the Euler equation yields $(EI\theta')' + P\sin\theta = 0$, which is the same as obtained before in Section 1.9 (Eq. 1.9.1).

The Euler equation as well as the associated kinematic and static admissible boundary conditions are easily generalized to problems where the energy density function ϕ depends on higher derivatives of w, to problems where it depends on several functions of one variable x, and to problems where it depends on one or generalizations when we analyze stability of thin-walled beams and of plates.

The Euler equation obtained in the foregoing mathematical derivation, of course, ensures only that the functional has a stationary value, but does not prove it achieves a minimum. However, when dealing with functionals such as the potential energy we are sure, on physical grounds, that the stable states correspond to a minimum rather than a maximum or no extreme at all [inflection point of $f(\varepsilon)$, Eq. 5.1.2]. (This is also clear from the fact that the operators involved in the potential-energy expression for the differential equation must be symmetric and self-adjoint in the stable states at very small loads.)

The Euler equation with its proper boundary conditions, which represents the condition that $\delta\Pi = 0$, is in mechanics generally equivalent to the equilibrium condition. However, in the frequent special case that $w(x) = 0$ is an equilibrium state, the Euler equation (with its proper boundary conditions) is also equivalent to the Trefftz condition for the critical load, $\delta(\delta^2\Pi) = 0$; see Equation 4.2.6. The reason is that, due to the fact that $\delta w(x) = w(x)$, we have in this special case $\delta^2\Pi = \Pi$ for linear problems, and $\delta^2\Pi \simeq \Pi$ with second-order accuracy for nonlinear problems.

Review of Positive-Definite and Self-Adjoint Operators

Before turning to practical applications in the next section, it might be useful to review from a course of mathematics some basic properties of positive-definite operators. The problem of critical load for a linear system is a special case of an eigenvalue problem, which is generally given by the linear differential equation

$$\underset{\sim}{M}w - P\underset{\sim}{N}w = 0 \qquad (5.1.13)$$

with homogeneous boundary conditions; $\underset{\sim}{M}$ and $\underset{\sim}{N}$ are linear differential operators and P is the eigenvalue to be found. In the special case of column buckling we have (cf. Eq. 5.1.11) $\underset{\sim}{M} = (d^2/dx^2)[EI(x)\,d^2/dx^2]$ and $\underset{\sim}{N} = -d^2/dx^2$. We may define a scalar product (or inner product) of two functions as $(u, w) = \int_0^l u(x)w(x)\,dx$, $(0, l)$ being the given domain. Operator $\underset{\sim}{M}$ is called *positive definite* if $(\underset{\sim}{M}w, w) > 0$ for any admissible function w (except $w = 0$). If $(\underset{\sim}{M}u, v) = (u, \underset{\sim}{M}v)$, the operator is called *symmetric,* and if it is bounded, it is also called *self-adjoint.* Since we will consider only bounded (continuous) operators, these two terms become synonymous.

Theorem 5.1.1 If $\underset{\sim}{M}$ and $\underset{\sim}{N}$ are positive-definite operators, then all eigenvalues of Equation 5.1.13 are positive.

Proof Taking a scalar product of Equation 5.1.13 with w, we have $((\underset{\sim}{M}w - P\underset{\sim}{N}w), w) = 0$ or $(\underset{\sim}{M}w, w) - P(\underset{\sim}{N}w, w) = 0$. Because $(\underset{\sim}{M}w, w) > 0$ and $(\underset{\sim}{N}w, w) > 0$, P must be positive.

Theorem 5.1.2 If $\underset{\sim}{M}$ and $\underset{\sim}{N}$ are symmetric, the eigenfunctions w_m and w_n associated with two different eigenvalues P_{cr_m} and P_{cr_n} are *orthogonal* in the sense that $(\underset{\sim}{M}w_m, w_n) = 0$ as well as $(\underset{\sim}{N}w_m, w_n) = 0$. In the special case that $\underset{\sim}{N}$ is an identity operator, we have $(w_m, w_n) = 0$.

Proof (Collatz, 1963; Courant and Hilbert, 1962) By definition $\underset{\sim}{M}w_m - P_{\mathrm{cr}_m}\underset{\sim}{N}w_m = 0$. Taking a scalar product of these equations with w_n and w_m, respectively, and subtracting the resulting equations, we have $(\underset{\sim}{M}w_m, w_n) - (\underset{\sim}{M}w_n, w_m) - P_{\mathrm{cr}_m}(\underset{\sim}{N}w_m, w_n) + P_{\mathrm{cr}_n}(\underset{\sim}{N}w_n, w_m) = 0$. Obviously, $(\underset{\sim}{N}w_n, w_m) = (\underset{\sim}{N}w_m, w_n)$ and $(\underset{\sim}{M}w_m, w_n) = (\underset{\sim}{M}w_n, w_m)$ because of symmetry. Thus $(P_{\mathrm{cr}_m} - P_{\mathrm{cr}_n})(\underset{\sim}{N}w_m, w_n) = 0$ and when $P_{\mathrm{cr}_m} \neq P_{\mathrm{cr}_n}$, we finally obtain $(\underset{\sim}{N}w_m, w_n) = 0$. Furthermore, taking again a scalar product of the equation $(\underset{\sim}{M}w_m - P_{\mathrm{cr}_m}\underset{\sim}{N}w_m)$ with w_n, we have $(\underset{\sim}{M}w_m, w_n) = P_{\mathrm{cr}_m}(\underset{\sim}{N}w_m, w_n)$ or $(\underset{\sim}{M}w_m, w_n) = 0$.

The differential equation for column buckling (Eq. 5.1.11) may now be considered as a special case of the equation $\underset{\sim}{L}w = p$, with $\underset{\sim}{L} = \underset{\sim}{M} - P\underset{\sim}{N}$. If $\underset{\sim}{M}$ and $\underset{\sim}{N}$ are symmetric, $\underset{\sim}{L}$ is also symmetric, as one can readily verify.

Theorem 5.1.3 (e.g., Dym and Shames, 1973, p. 156) If the equation $\underset{\sim}{L}w = p$ has a solution and if $\underset{\sim}{L}$ is a *self-adjoint positive-definite* operator, then the function which minimizes the functional

$$I(w) = \tfrac{1}{2}(\underset{\sim}{L}w, w) - (w, p) \tag{5.1.14}$$

is a solution of the differential equation and, conversely, the solution of the differential equation (with homogeneous boundary conditions) minimizes the functional.

Proof Consider a family of functions $w + \delta w$ such that $\delta w = \varepsilon\bar{w}$ where ε is an arbitrary parameter and $\bar{w}$ is any chosen function that is admissible for the field of definition of $\underset{\sim}{L}$ (which also implies that $\bar{w}$ satisfies homogeneous boundary conditions). Consider that w is the function that minimizes $I(w)$, that is, $I(w + \varepsilon\bar{w}) - I(w) > 0$ for $\varepsilon \neq 0$. Substituting from Equation 5.1.14, we obtain

$$\tfrac{1}{2}(\underset{\sim}{L}(w + \varepsilon\bar{w}), w + \varepsilon\bar{w}) - (w + \varepsilon\bar{w}, p) - \tfrac{1}{2}\underset{\sim}{L}(w, w) + (w, p) > 0 \tag{5.1.15}$$

This may be rearranged as

$$\tfrac{1}{2}[(\underset{\sim}{L}w, w) + \varepsilon(\underset{\sim}{L}w, \bar{w}) + \varepsilon(\underset{\sim}{L}\bar{w}, w) + \varepsilon^2(\underset{\sim}{L}\bar{w}, \bar{w})] - (w, p) - \varepsilon(\bar{w}, p) - \tfrac{1}{2}\underset{\sim}{L}(w, w) + (w, p) > 0 \tag{5.1.16}$$

or

$$\tfrac{1}{2}[(\underset{\sim}{L}w, \bar{w}) + (\underset{\sim}{L}\bar{w}, w) - 2(\bar{w}, p)]\varepsilon + \tfrac{1}{2}(\underset{\sim}{L}\bar{w}, \bar{w})\varepsilon^2 > 0 \qquad (\varepsilon \neq 0) \tag{5.1.17}$$

Since $\underset{\sim}{L}$ is assumed to be a self-adjoint operator, we have $(\underset{\sim}{L}w, \bar{w}) = (\underset{\sim}{L}\bar{w}, w)$, and this provides

$$[(\underset{\sim}{L}w, \bar{w}) - (\bar{w}, p)]\varepsilon + \tfrac{1}{2}(\underset{\sim}{L}\bar{w}, \bar{w})\varepsilon^2 > 0 \qquad (\varepsilon \neq 0) \tag{5.1.18}$$

This is a quadratic expression of the form $a_1\varepsilon + a_2\varepsilon^2$. Clearly such an expression can be positive for all $\varepsilon \neq 0$ only if $a_1 = 0$. Thus $(\underset{\sim}{L}w, \bar{w}) - (\bar{w}, p) = 0$, which may be rewritten as

$$((\underset{\sim}{L}w - p), \bar{w}) = 0 \tag{5.1.19}$$

Because this equation must hold for *any* (admissible) $\bar{w}$, it follows (by virtue of the basic lemma of the calculus of variations) that $\underset{\sim}{L}w - p = 0$, that is, w satisfies the differential equation. Furthermore, to prove the converse of the theorem, we consider that w is a solution of the differential equations with its homogeneous boundary conditions and we try to prove that w minimizes $I(w)$. To this end we express $I(w)$ by substituting $\underset{\sim}{L}w$ for p. This gives

$$I(w+\varepsilon\bar{w}) = \tfrac{1}{2}(\underset{\sim}{L}(w+\varepsilon\bar{w}),\ w+\varepsilon\bar{w}) - (\underset{\sim}{L}w,\ w+\varepsilon\bar{w})$$
$$= -\tfrac{1}{2}(\underset{\sim}{L}w,\ w) + \tfrac{1}{2}[(\underset{\sim}{L}\bar{w},\ w) + (\underset{\sim}{L}w,\ \bar{w})$$
$$- 2(\underset{\sim}{L}w,\ \bar{w})]\varepsilon + \tfrac{1}{2}\underset{\sim}{L}(\bar{w},\ \bar{w})\varepsilon^2.$$

Using the self-adjoint property, $(\underset{\sim}{L}\bar{w}, w) = (\underset{\sim}{L}w, \bar{w})$, and so we have

$$I(w+\varepsilon\bar{w}) = -\tfrac{1}{2}(\underset{\sim}{L}w,\ w) + \tfrac{1}{2}(\underset{\sim}{L}\bar{w},\ \bar{w})\varepsilon^2 \tag{5.1.20}$$

For I to be a minimum at $\varepsilon = 0$, it is necessary that $(\underset{\sim}{L}\bar{w}, \bar{w}) > 0$, that is, operator $\underset{\sim}{L}$ must be positive definite. Also, if $\underset{\sim}{L}$ is positive definite, I achieves a minimum at $\varepsilon = 0$.

Note that the use of the self-adjoint property in this proof has replaced the use of integration by parts. The second term on the right-hand side of Equation 5.1.20 is called the second variation of the functional, $\delta^2 I = \tfrac{1}{2}(\underset{\sim}{L}\bar{w}, \bar{w})\varepsilon^2$. The condition that the extreme is a minimum may be stated as $\delta^2 I > 0$, which is similar to Equation 4.2.5 for discrete systems.

As an example, let us consider a beam-column with homogeneous boundary conditions and verify that $I(w)$, defined by Equation 5.1.14, represents its potential energy, as expected. We have

$$I(w) = \frac{1}{2}\int_0^l w(EIw'')''\,dx - \frac{1}{2}P\int_0^l w(-w'')\,dx - \int_0^l pw\,dx \tag{5.1.21}$$

Integrating by parts, we get

$$I(w) = \int_0^l \left[-\frac{1}{2}w'(EIw'')' - \frac{P}{2}w'^2 - pw\right]dx + \frac{1}{2}[w\{(EIw'')' + Pww'\}]_0^l$$
$$= \int_0^l \left(\frac{1}{2}EIw''^2 - \frac{P}{2}w'^2 - pw\right)dx - \frac{1}{2}[Vw]_0^l - \frac{1}{2}[Mw']_0^l \tag{5.1.22}$$

where we used $-(EIw'')' - Pw' = -M' - Pw' = V =$ shear force. The boundary terms $[\ \ ]_0^l$ vanish due to the fact that the boundary conditions must be either kinematic ($w = 0$, $w' = 0$) or static ($V = 0$, $M = 0$). The remaining expression is indeed the potential energy of a beam-column.

Finally, note the analogy with matrix eigenvalue problems (Sec. 4.1), for example, the analogy between orthogonality of eigenvectors and of eigenfunctions. This is not surprising, since an eigenfunction can be regarded as the limiting case of an infinite-dimensional eigenvector.

Problems

5.1.1 Derive the Euler differential equation and boundary conditions for the energy integrand that also contains w''' and w^{IV}.

5.1.2 Assume an elastic but nonlinear moment-curvature relation $M = EI(1 - kw''^2)w''$ where EI, $k =$ constants, formulate the potential energy, and use the calculus of variations to derive the Euler differential equation for $w(x)$ with the boundary conditions. Check it from the equilibrium conditions for beam element dx.

5.1.3 Solve the same problem as above but $M = EIw''$ ($EI = \text{const.}$) and deflections are large. Use the approximation $M = EI/\rho = EIw''(1 - 3w'^2/2)$ (Sec. 1.1) and, for the work of load P, express $(ds - dx)/dx$ up to terms that are of the next higher order in w than the approximation $\frac{1}{2}\int Pw'^2\,dx$. Consider that $P = P_0 + P_1$ where P_0 = initial load and P_1 = increment during buckling, and assume the column as inextensible.

5.1.4 Do the same but assume $M = EI(w'' + kw^{\text{IV}})$, with EI, k = constants, while the deflections are small and the behavior is linear. (This expression for M appears in some recent nonlocal continuum theories for heterogeneous materials, e.g., Bažant, 1984).

5.1.5 Do the same, but assume $M = EI(w'' + w''')$.

5.1.6 Using the operator approach, prove that, for $P < P_{\text{cr}}$, the potential energy Π for a pin-ended beam-column has a minimum with respect to parameter ε for any chosen $\bar{w}(x)$ (and not a maximum or an inflection point).

5.1.7 Adding the term $\int \frac{1}{2}EAu'^2\,dx$ to the expression for Π, where u = axial displacement, show that variational calculus yields for $u(x)$ a separate differential equation that reads $\Delta P = -EAu'$ (provided the deflections are small and the behavior is linear). Discuss the result with respect to buckling of a column whose ends during buckling either slide freely or are restrained against sliding in the axial direction (Fig. 1.2).

5.1.8 Formulate Π for the free-standing column loaded on top by an elastic stretched cable anchored at the base. (a) Follow the procedure of variational calculus to obtain the boundary conditions on top, and check them by equilibrium considerations. (b) Do the same when the tendon is anchored at a distance a below or above the base. (c) Can you do the same if P is not applied by a cable but is a follower load (tangential to the column on top)? (d) Deduce from Π the boundary conditions of the columns in Figures 1.11 and 1.13.

5.1.9 Prove from the potential energy expression that a prestressed column (whose tendon has no lateral free play) cannot buckle.

5.1.10 Write the potential energy for a pipe column filled with water and loaded by force P applied on a piston in the pipe at its end (Fig. 1.22). Derive from it the differential equation with boundary conditions and the critical load.

5.1.11 (a) Express the potential energy of a beam with shear in general, and of a sandwich beam in particular. (b) Derive from it the differential equations and boundary conditions for $w(x)$ and $\psi(x)$ given in Section 1.7 (Eqs. 1.7.8).

5.1.12 Using the potential-energy expression for a beam with shear from Problem 5.1.11, assume a suitable deflection shape $w(x)$, $\psi(x)$ for a free cantilever column (Fig. 5.2) and determine the critical height h for loading by own weight (this is an estimate of maximum height of a tall building, wind and earthquake disregarded).

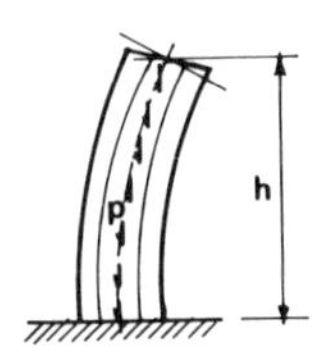

Figure 5.2 Exercise problem on cantilever column under own weight.

5.1.13 Formulate the potential energy for three-dimensional buckling of a doubly symmetric I-beam under axial load and torque, using all the assumptions from Section 1.10. Derive from it the differential equations for $v(x)$ and $w(x)$ from Section 1.10.

5.2 BEAM ON ELASTIC FOUNDATION

To illustrate applications of the calculus of variations, let us study beams on elastic foundations, which are widely encountered in engineering practice. They may be used for approximate description of the behavior of foundation beams or pavements resting on deformable subsoil (Fig. 5.3a, b). Ice sheets floating on water behave as beams or plates on elastic foundation. The beam on elastic foundation is also extensively used as a simplified model for the buckling of piles embedded in soil. Another type of problem that may be approximately treated as a beam on elastic foundation is the lateral buckling of the compression belt of a truss bridge of U-shaped cross section (belt $\overline{12}$ in Fig. 5.3c), provided the elastic reactions due to deformation of the U-shaped cross section are approximately treated as uniformly distributed. Axisymmetric deformations of axially compressed cylindrical shells (Fig. 5.3d) are also equivalent to the problem of a beam on elastic foundation.

Potential Energy and Differential Equations

The elastic foundation is usually considered as the limiting case of an infinitely dense distribution of a row of springs. Denoting the spring stiffness per unit length of the beam, called the foundation modulus, as c, we may express the distributed reaction p_r of the foundation against the beam as $p_r(x) = -cw(x)$ (Fig. 5.4a). In this formulation, called the Winkler foundation (Winkler, 1867), one

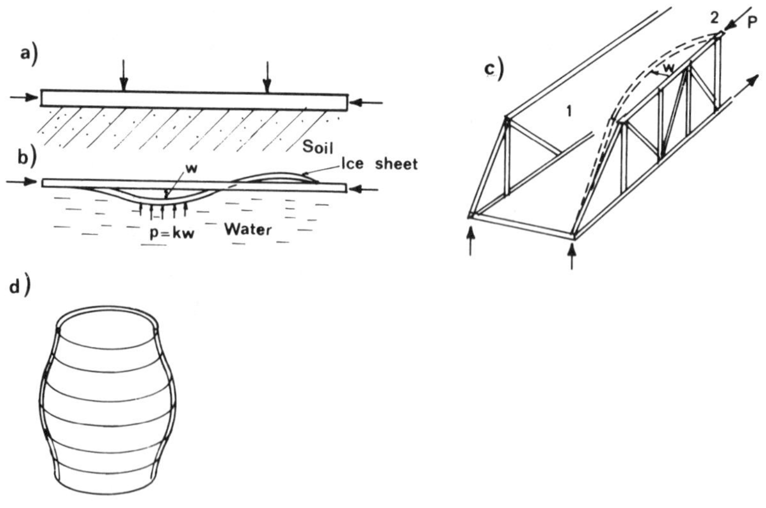

Figure 5.3 (a) Beam on an elastic foundation; (b) ice sheet floating on water; (c) compression belt of a truss bridge; (d) axisymmetric deformation of a cylindrical shell.

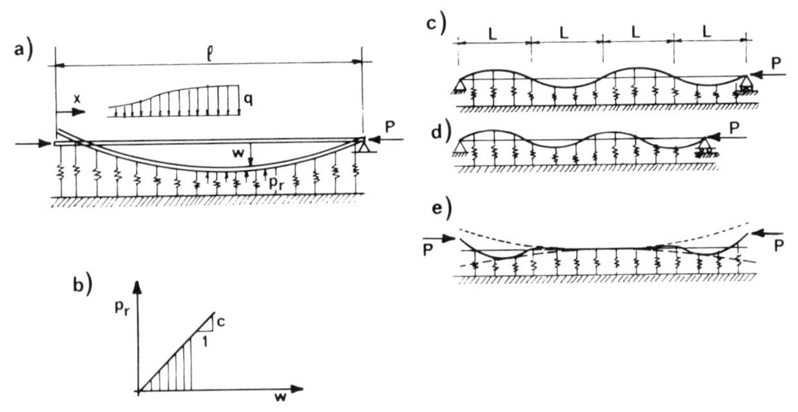

Figure 5.4 (a) Beam on Winkler foundation; (b) foundation modulus c; (c–e) buckling of beams with various end restraints.

assumes that the reaction p_r at point x depends only on deflection w at point x and is independent of deflection w at adjacent points. This assumption is satisfied exactly for ice sheets floating on water, for which $c = p_w$ = unit weight of water, as follows from Archimedes' law. For other applications, such as beams resting on foundation soil or on an elastic body, this assumption is not exactly satisfied, since $p_r(x)$ depends also on the deflections $w(x')$ in adjacent points x'. Various assumptions, which introduce such dependence, have been proposed; for example, Wieghardt's foundation, for which $p_r(x) = \int_{-\infty}^{\infty} c_0 e^{-c_1|x-x'|} w(x')\,dx'$ where c_0 = constant. However, these generalized foundation models have not found much application, since they do not seem to give significantly better results for various practical applications. Particularly, for beams on soils, the nonlinearity of foundation behavior causes the simple Winkler foundation to be usually a better model than the generalized foundations with dependence of the reaction on the adjacent deflections.

The strain energy of the foundation, represented by the area under the $p_r - w$ diagram in Figure 5.4b, is $(-p_r)w/2 = cw^2/2$. Together with the strain energy of bending of the beam (with bending rigidity EI) and the work of axial compression force P and of transverse distributed load $p(x)$, the expression for the total potential energy of the beam-foundation system is

$$\Pi = \int_0^l \left[\frac{1}{2}(EIw''^2 + cw^2 - Pw'^2) - pw \right] dx \tag{5.2.1}$$

To apply the Euler equation, we evaluate the derivatives of the integrand: $\phi_{,w} = cw - p$, $\phi_{,w'} = -Pw'$, $\phi_{,w''} = EIw''$. Thus, according to the Euler equation (Eq. 5.1.8), the differential equation for a beam on elastic foundation is

$$(EIw'')'' + (Pw')' + cw = p \tag{5.2.2}$$

The boundary conditions that are compatible with the existence of potential energy are, according to Equations 5.1.9,

$$\text{Either} \qquad w = 0 \qquad \text{or} \qquad -(EIw'')' - Pw' = V = 0 \tag{5.2.3}$$

and

$$\text{Either} \quad w' = 0 \quad \text{or} \quad EIw'' = M = 0 \tag{5.2.4}$$

The boundary conditions on the left are the essential ones (kinematic), and those on the right are the natural ones (static). We see that the foundation modulus c appears only in the differential equation but not in the boundary conditions. Boundary conditions of other forms, for example, $(EIw'')' = -Q = 0$ (Q defined in Sec. 1.3), are not compatible with the existence of potential energy and cause the structure to be nonconservative.

The same differential equation and boundary conditions can be easily obtained from equilibrium conditions of an infinitesimal element of the beam, as in Section 1.3. One can also obtain the differential equation from the one for beam columns (Eq. 5.1.11) by replacing p with the net resulting distributed load $p - cw$.

Solution for Different Boundary Conditions

Let us now solve a perfect beam on an elastic foundation ($p = 0$), with constant EI, P, and c. The solution may be sought in the form $w(x) = e^{\lambda x}$, and substitution into Equation 5.2.2 yields for λ the characteristic equation

$$EI\lambda^4 + P\lambda^2 + c = 0 \tag{5.2.5}$$

This is a biquadratic equation, that is, a quadratic equation for λ^2. It has four roots, given by $\lambda = i\kappa_1, -i\kappa_1, i\kappa_2, -i\kappa_2$ where

$$\begin{Bmatrix} \kappa_1 \\ \kappa_2 \end{Bmatrix} = \alpha\sqrt{\gamma \pm \sqrt{\gamma^2 - 1}} \qquad \gamma = \frac{P}{2\sqrt{cEI}} \qquad \alpha = \left(\frac{c}{EI}\right)^{1/4} \tag{5.2.6}$$

Consider now periodic solutions $w(x)$. They are possible only if λ is imaginary, which occurs when $\gamma \geq 1$. From Equations 5.2.6, the critical loads may be written as $P = P_{cr} = 2\gamma\sqrt{cEI}$. Clearly, the smallest P_{cr} occurs for the smallest γ that gives real values for κ_1 and κ_2, which is $\gamma = 1$. Thus, the smallest critical load for a periodic solution is

$$P_{cr_{min}} = 2\sqrt{cEI} \tag{5.2.7}$$

In this case $\lambda = i\alpha, i\alpha, -i\alpha, -i\alpha$, that is, we have two double roots. So the corresponding general solution is

$$w(x) = A \sin \alpha x + Bx \sin \alpha x + C \cos \alpha x + Dx \cos \alpha x \tag{5.2.8}$$

where A, B, C, D = arbitrary constants. The half-wavelength of the periodic terms is

$$L = \frac{\pi}{\alpha} = \pi\left(\frac{EI}{c}\right)^{1/4} \tag{5.2.9}$$

The solution in Equation 5.2.8 can satisfy the boundary conditions when the beam has pin supports ($w = w'' = 0$) at both ends (Fig. 5.4c), provided that the beam length is $l = nL$ where n = positive integer (in this case $B = D = 0$ since otherwise w would not be periodic). By placing the origin $x = 0$ into one end, the periodic solution can be written as $w = A \sin(\pi x/L)$, with $C = 0$ (because $w = 0$ at $x = 0$).

Note that for a beam on an elastic foundation the lowest critical load does not correspond to the longest wavelength permitted by boundary conditions. This property is different from beams and is similar to plates and shells.

The periodic solution obviously applies also for an infinitely long beam ($l \to \infty$, $n \to \infty$). So Equation 5.2.7 can be called the critical load of an infinitely long beam on an elastic foundation.

The lengths of actual beams are often not compatible with the wavelength L (Fig. 5.4d) and then γ is higher than 1, with the consequence that the critical load is higher than Equation 5.2.7 indicates. However, if we consider the beam on an elastic foundation to be getting infinitely long, a wavelength $L = l/n$ compatible with the end supports is getting infinitely close to the value π/α and so the critical load is approaching for $l \to \infty$ the value indicated by Equation 5.2.7.

The solution just obtained is applicable to ice sheets floating on water, and it determines the upper bound on the force than an ice sheet can exert on an obstacle; for example, the force of a river ice sheet on a bridge pier, or the force of a sea ice sheet on a fixed object such as an oil-drilling platform. However, these problems are in reality more complicated because the ice sheet is two-dimensional, and a simplification of the ice sheet to a strip of unit width is much too conservative. The in-plane compression force carried by ice sheets also influences their capability of carrying vertical loads because it reduces the effective bending stiffness, just like the axial load reduces the bending stiffness of a beam-column. River ice as well as ice sheets in the arctic are often under large in-plane forces, and these need to be taken into account in determining their carrying capacity (to decide, e.g., whether an aircraft of a certain weight can land on the ice sheet safely).

Let us now outline the general solution. For the case $\gamma > 1$, there are four distinct imaginary roots, and so the general solution has the form

$$w(x) = A \sin \kappa_1 x + B \cos \kappa_1 x + C \sin \kappa_2 x + D \cos \kappa_2 x \qquad (5.2.10)$$

where A, B, C, D are arbitrary constants to be found from boundary conditions. For $\gamma = 1$, the solution has already been given (Eq. 5.2.8). Finally, for $\gamma < 1$, the roots λ are complex: $\lambda = \rho + ir$, $\rho - ir$, $-\rho + ir$, $-\rho - ir$, where r and ρ are such that $2r^2 = \alpha^2(1+\gamma)$, and $2\rho^2 = \alpha^2(1-\gamma)$. This fact can be checked easily: $\lambda^2 = (\pm\rho \pm ir)^2 = -\alpha^2\gamma \pm i\alpha^2\sqrt{1-\gamma^2}$, which is obviously identical to Equation 5.2.6. Thus, the general solution is

$$w(x) = A \sin rx \sinh \rho x + B \cos rx \sinh \rho x + C \sin rx \cosh \rho x + D \cos rx \cosh \rho x \qquad (5.2.11)$$

For the case of a beam on an elastic foundation with both ends fixed, it can be shown (Hetényi, 1946) that the boundary conditions can be satisfied only with $\gamma > 1$, and using the solution in Equation 5.2.10 one finds the characteristic equation to be $(\kappa_1 l) \tan(\kappa_1 l) = (\kappa_2 l) \tan(\kappa_2 l)$ for symmetric buckling, and $(\kappa_1 l) \cot(\kappa_1 l) = (\kappa_2 l) \cot(\kappa_2 l)$ for antisymmetric buckling. The roots of these equations can be approximately found graphically and with great accuracy by Newton iteration. This then yields the critical load for a beam of given length. Since in this case the solution (Eq. 5.2.10) is periodic, it is not surprising that the critical load approaches the value given in Equation 5.2.7 as the beam length l approaches infinity. For finite lengths $l \neq nL$ where L is given by Equation 5.2.9, P_{cr} must be, of course, larger than Equation 5.2.7, since we proved that for periodic modes the maximum P_{cr} occurs at $\gamma = 1$.

When both ends of the beam are free, the boundary conditions can be satisfied only if $\gamma < 1$. Then, using Equation 5.2.11 one finds (as shown by Hetényi, 1946; see also Simitses, 1976) the characteristic equation $(3\rho^2 - r^2)r \sinh 2\rho l = \pm(\rho^2 - 3r^2)\rho \sin 2rl$ in which the plus sign applies for symmetric buckling and the minus sign for antisymmetric buckling. If the beam length l tends to infinity, $\sinh 2\rho l \rightarrow \infty$, while $\sin 2rl$ remains bounded; therefore, $3\rho^2 - r^2$ must approach zero, which indicates that $\gamma = \frac{1}{2}$. So the critical load of an infinitely long beam on elastic foundation with free ends is

$$P_{cr_{min}} = \sqrt{cEI} \qquad \text{(free ends)} \tag{5.2.12}$$

The deflection curve (Fig. 5.4e) is a damped sine curve whose amplitude decays away from the ends. Therefore, the buckling is essentially confined to the end portions of a very long beam. To make a very long beam buckle along its entire length, the axial load would have to be doubled to the value given by Equation 5.2.7.

The wavelength of the damped sine curve is (from Eq. 5.2.11) $2L = 2\pi/r = 2\pi\sqrt{2}/\alpha\sqrt{1+\gamma} = 4\pi/\sqrt{3}\,\alpha$, in which we have used $\gamma = \frac{1}{2}$. The rate of decay can be quantified by considering the ratio of the sine amplitude reductions at a distance of one wavelength, that is, $e^{\rho x}/e^{\rho(x+2L)} = e^{-\rho 2L} = e^{-2\pi/\sqrt{3}} = 1/37.6$. So the sine curve is damped very strongly.

As a further insight into the difference between the infinitely long beams with a supported end and an unsupported (free) end, we may consider the nature of solution decay. The solution can be either periodic (no real exponentials present) or consist of exponentially modulated oscillations. The exponential functions grow either toward the left or toward the right. Thus, if deflections within the central portion of an infinitely long beam are nonzero, the solution must be periodic, or else the deflections at the end would be infinite. The exponentials can be present only if the deflections vanish everywhere except within boundary segments whose length depends on the rate of decay (controlled by coefficient λ in $e^{\pm\lambda x}$) and occupies an infinitely short fraction of the infinite beam length. It is found that the critical load for the periodic solution is higher than the critical load for the exponentially modulated solution. So the latter solution must correspond to a free end, and the periodic solutions cannot therefore correspond to a free end. By exclusion, it must then correspond to the remaining possibilities of supported ends (fixed or hinged). It also follows that, for both types of support (hinged or fixed) in the central part of the beam, the periodic solution is the same, and so the critical loads are also the same. The foregoing considerations illustrate how to obtain critical loads without fully solving the problem (as is often done for shells).

The buckling mode that involves nonzero deflections over the entire beam (except for periodic nodal points) is an example of global buckling, and it occurs for the periodic solution with supported ends. The buckling mode for the free end, for which the deflections are nonzero only within limited segments near the ends, is an example of local buckling and, in particular, of boundary buckling. Similar distinctions can be introduced for shells.

A detailed exposition of the buckling of beams on an elastic foundation was presented by Hetényi (1946). The stiffness matrix of beam-columns on an elastic (Winkler) foundation is presented in Tuma's (1988) handbook—not only for the planar case we analyzed, but also for the spatial case of a beam with torsion (simple torsion, Chap. 6) encased in an elastic foundation resisting not merely deflections but also twisting rotations. Tuma also gives the transfer (or transport) matrices and the loading terms of the matrix stiffness equation of equilibrium for many typical transverse load cases, as well as the modified stiffness matrices for various end supports.

Fiber on Elastic Foundation

An interesting limiting case of a beam on elastic foundation is a fiber on elastic foundation (Fig. 5.5). The fiber represents the limiting case of the beam on an elastic foundation for $EI \rightarrow 0$, and thus Equation 5.2.7 yields $P_{cr} = 0$. This means that a fiber on a Winkler-type foundation cannot carry any compressive force (a fiber embedded in an elastic space can, but this means that the Winkler-type foundation is an inadequate model for it). Nevertheless, the fiber–foundation system has a useful analogy for columns as we will see in the next section. The potential energy of this system is

$$\Pi_f = \int_0^l \tfrac{1}{2} c w^2 \, dx + \int_0^l \tfrac{1}{2} N w'^2 \, dx \tag{5.2.13}$$

N denotes the force in the fiber, positive if tensile, and c is the foundation modulus. From the condition $\delta\Pi = 0$ we get, upon integrating by parts (for $P_f = \text{const.}$),

$$\delta\Pi_f = \int_0^l (cw\,\delta w + Nw'\,\delta w')\,dx = \int_0^l (cw - Nw'')\,\delta w\,dx + [Nw'\,\delta w]_0^l \tag{5.2.14}$$

Since $\delta\Pi_f$ must vanish for any continuous $\delta w(x)$, the following differential equation of equilibrium applies:

$$Nw'' - cw = 0 \tag{5.2.15}$$

with the admissible boundary conditions: either $w = 0$ (kinematic) or $Nw' = V = 0$ (static). The first boundary condition corresponds to a supported end of fiber, and the second one to a free end (V = shear force in the fiber). Note that the

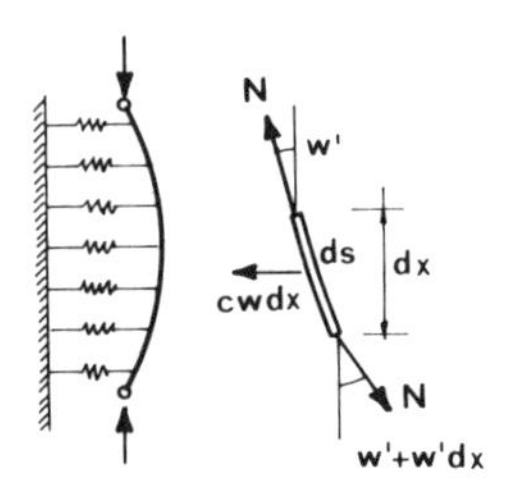

Figure 5.5 A fiber on an elastic foundation.

same differential equation results from the force diagram for element dx in Figure 5.5. Indeed, the resultant of transverse forces is $N(w' + w'' dx) - Nw' - cw\,dx = 0$ which yields Equation 5.2.15.

The differential equation $Nw'' - cw = 0$ can be made mathematically equivalent to the equation $w'' - (P/EI)w = 0$, which we solved in Section 1.2 (for the same boundary conditions $w = 0$ at $x = 0$ and $x = l$), if we set either $EI = 1/c$, $P = -1/N$ or $EI = -1/c$, $P = 1/N$. So, in analogy to the equation $P_{cr} = EIn^2\pi^2/l^2$, we have $(1/N)_{cr} = -(1/c)n^2\pi^2/l^2$ or $N_{cr} = -cl^2/n^2\pi^2$. The smallest $|N_{cr}|$ results for $n \to \infty$; $N_{cr} = 0$.

Problems

5.2.1 Consider periodic solutions $w = q \sin(\pi x/l)$. Substituting this into Equation 5.2.2, obtain P_{cr} as function $f(l)$ of l. Setting $df/dl = 0$, obtain $P_{cr_{min}}$ and check that it agrees with Equation 5.2.7.

5.2.2 Using the Euler equation of variational calculus, determine the differential equation for a beam on elastic foundation for which (a) c is variable, $c = kx$, and $EI = $ const.; (b) same but $c = k \sin ax$; (c) $c = $ const. but $EI = Rx$; (d) $c = $ const. but $EI = R \sin ax$, where k, R, $a = $ given constants; (e) do the same as (a), (b), (c), (d), but $c = $ const., $EI = $ const., and the axial load, rather than being applied at the ends, is introduced as a distributed axial load p ($p = $ const.) (Figure 5.6a).

5.2.3 Consider the continuous beam with n equal spans l, n being large, and simple supports (Fig. 5.6b). The supports rest on springs of equal spring constants C. A beam on elastic foundation with foundation modulus $c = C/l$ can approximate this continuous beam closely if the buckling half-wavelength is $L \gg l$, say $L > 4l$. (a) Find the condition for C and l to permit analysis as a beam on elastic foundation. Give the solution of P_{cr} and $w(x)$ (b) for $n \to \infty$; (c) for $n = 10$. (d) Using stability functions s and c (Chap. 2) formulate the algebraic equation system that gives the exact solution of this continuous beam for any given n. (e) Imagine the continuous beam to represent one compression belt of a truss bridge in Figure 5.6c, in which C is the spring stiffness of the cross-section U-frame, both compression belts buckling symmetrically or antisymmetrically, and calculate C from the data on the U-frame given in Figure 5.6c.

5.2.4 Solve P_{cr} and $w(x)$ for the semi-infinite beam on elastic foundation shown in Figure 5.6d, which has at the end $x = 0$ a transversely sliding restraint, that is, $w' = 0$, $V = 0$ at $x = 0$ (P, EI, $c = $ constants). Plot $w(x)$.

5.2.5 Do the same as Problem 5.2.4, but the end of the beam is loaded through a short rigid pin-ended link of length L_1 (Fig. 5.6e). Also discuss the dependence of the solution on L_1/L.

5.2.6 Solve $w(x)$ for an infinite beam on elastic foundation that is initially imperfect, having initial curvature $z_0(x) = a \sin(\pi x/b)$ where a, b are given constants (Fig. 5.6f). First obtain the differential equation from the Euler equation of variational calculus. Also plot w_{max} versus P.

5.2.7 Write the potential energy for a beam on an elastic but nonlinear foundation such that the foundation reaction (per unit length) is $p_r = (c_0 - c_1 w)w$ where c_0, $c_1 = $ constants. Using variational calculus obtain the differential equation for $w(x)$ (which must be nonlinear) and check it by

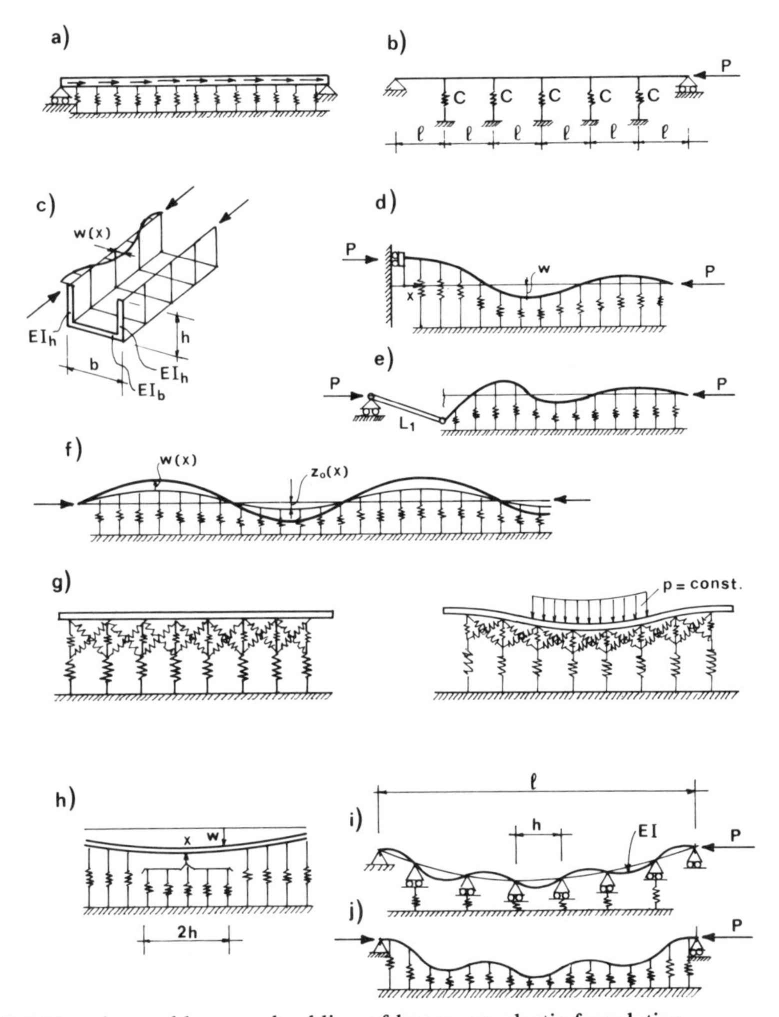

Figure 5.6 Exercise problems on buckling of beams on elastic foundation.

considering equilibrium of a beam element dx. Comment on possible solutions.

5.2.8 Do the same as in Problem 5.2.7, but the foundation is linear ($c = \text{const.}$) while the deflections are large. Use the approximations $M = EI/\rho \simeq EIw''(1 - 3w'^2/2)$ and the work of load per unit length is $P(ds - dx)/dx$ where $(ds - dx)/dx$ is expressed up to terms that are of the next higher order in w than the approximation $Pw'^2/2$ (cf. Prob. 5.1.3).

5.2.9 Do the same as Problem 5.1.4 but add an elastic foundation.

5.2.10 As a generalized approximate foundation model suitable for foundation piles (cf. Bažant and Masopust, 1986), Pasternak (1926) proposed that beam rotation w' is also resisted by shear stresses from the foundation acting along the beam surface; see also Kerr (1964), Vlasov and Leontiev (1966), and Soldini (1965). The distributed moment m that these stresses apply on the beam is $k_p w'$ where $k_p = \text{constant}$. The additional potential energy per unit beam length then is $\frac{1}{2}mw'$ or $\frac{1}{2}k_p w'^2$, and so the potential energy of an axially loaded beam on a Pasternak foundation (Fig. 5.6g) is $\Pi = \int_0^l (\frac{1}{2}EIw''^2 - \frac{1}{2}Pw'^2 + \frac{1}{2}kw^2 + \frac{1}{2}k_p w'^2)\, dx$. Derive from this the differential equation and admissible boundary contitions and discuss stability (noting that replacement of P with $P - k_p$ yields the ordinary foundation we solved already).

5.2.11 As an alternative foundation model, one may assume that the distributed reaction $p_f(x)$ at point x (Fig. 5.6h) represents the average of reactions $kw(x+s)$ from all points $x+s$ over a beam segment of length $2h$, that is, $p_f(x) = \int_{x-h}^{x+h} kw(x+s)\, ds$. Expanding $w(x+s)$ in a Taylor series about point x and truncating, show that the result is approximately equivalent to the equation for a Pasternak foundation.

5.2.12 Consider a long continuous beam with spans h (Fig. 5.6i) and supports resting on springs of stiffness C. Assume $l \gg h$. For the global buckling replace the discrete elastic supports by a continuous Winkler foundation of modulus $k = C/h$ (Fig. 5.6j). Considering a local buckling mode $w_1 = a \sin(\pi x/h)$ and a global buckling mode $w_2 = A \sin(\pi x/l)$, use the energy method to determine the condition for these two modes to have the same critical load. *Note*: In that case both modes occur simultaneously, that is, $w = w_1 + w_2$. The interaction of modes causes a postcritical imperfection sensitivity characteristic of the so-called double-cusp catastrophy (which does not belong to the seven elementary catastrophes of Thom; Table 4.7.1, last row). Its analysis requires considering nonlinear large deflections and determining the energy function up to the fourth-order term in a and A, and applying Koiter's postbifurcation theory; cf. Hansen (1977), Hui (1986), or Wicks (1987).

5.2.13 Including inertia forces in the differential equation for deflections, calculate the natural vibration frequency of a beam on elastic foundation with simply supported ends, subjected to constant axial force P. (This is a generalization of Sec. 3.1.)

5.2.14 Do the same as above, but $P(t)$ is pulsating. Using the energy method from Section 3.3, analyze parametric resonance. (This is again a generalization of Sec. 3.3.)

5.2.15 Generalizing Problem 5.1.13 formulate the potential energy for three-dimensional buckling of a beam that is subjected to axial force and torque and rests on an elastic foundation with the same foundation modulus k for y- and z-directions. Derive from it the differential equation form for $v(x)$ and $w(x)$.

5.2.16 Derive from potential energy the differential equation of a beam-column with torsion encased in an elastic foundation that resists also transverse rotation θ, such that the traverse distributed moment on the beam is $m = k_t\theta$ (simple torsion, Chap. 6). Also obtain kinematic and static boundary conditions.

5.2.17 For the above case, derive the stiffness matrix and the transfer matrix (cf. Tuma, 1988, p. 305).

5.2.18 Derive the load terms of the matrix stiffness equation of equilibrium for the planar cases: (a) transverse concentrated load at third point, (b) uniform load over a half-span, (c) triangular or distributed load over a quarter-span, and (d) concentrated moment applied at a half-span. Then do it for the spatial case of a transverse concentrated moment applied at a quarter-span (all results in Tuma, 1988).

5.2.19 Derive the modified stiffness matrix of a beam-column on elastic foundation having one end (a) hinged, or (b) free, all the reactions at the other end being nonzero (cf. Tuma, 1988).

5.2.20 Analyze the heavy rotating column with centrifugal forces, Problem 3.4.3, on the basis of potential energy. Discuss analogy with the beams on elastic foundation.

5.3 RAYLEIGH QUOTIENT

So far we have used the energy approach to determine the exact conditions of equilibrium and stability. However, the energy methods are also valuable as a means of obtaining bounds and approximate solutions, which are sometimes much simpler than exact solutions.

We now restrict consideration to linearized bifurcation problems, such that $w(x) = 0$ (or $q_1 = \cdots = q_n = 0$) is the equilibrium position whose stability is to be examined. Then the potential energy is $\Pi = \delta^2\Pi$ = quadratic functional or quadratic form. Assuming that either there is a single load P or all the loads are proportional to one load parameter, P, we can always express Π in the form:

$$\Pi = \delta^2\Pi = U - P\bar{W} = (P_R - P)\bar{W} \tag{5.3.1}$$

in which U is a positive-definite quadratic strain energy expression, independent of the load (or load parameter) P; $\bar{W}$ is a positive-definite quadratic expression defining the work per unit load; and P_R is defined as

$$P_R = \frac{U}{\bar{W}} \tag{5.3.2}$$

This quotient is called the Rayleigh quotient, after the famous physicist who introduced it in the general context of linear eigenvalue problems (Rayleigh, 1894, p. 110). Same as U and $\bar{W}$, P_R represents a functional of function $w(x)$, the deflection shape, that is, $P_R = P_R[w(x)]$.

Upper-Bound Property of Rayleigh Quotient

Since $\bar{W}$ is always positive, Equation 5.3.1 allows us to restate the stability criterion [i.e., Eq. 4.2.5 or the Lagrange–Dirichlet theorem (Theorem 3.6.1)] as follows: The structure is

1. Stable — If $P < P_R$ for *all* admissible $w(x)$ (5.3.3a)
2. Critical — If $P \leq P_R$ for *all* admissible $w(x)$ while $P = P_R$ for *some* admissible $w(x)$ (5.3.3b)
3. Unstable — If $P > P_R$ for *some* admissible $w(x)$ (5.3.3c)

where the admissible functions $w(x)$ are those that are continuous, with continuous slopes $w'(x)$, and satisfy all the given kinematic boundary conditions. Since the limit of stability is the first critical load P_{cr_1}, it follows that

$$P_{cr_1} = \min P_R \tag{5.3.4}$$

This is an important property. It means that the Rayleigh quotient represents an upper-bound approximation of P_{cr_1} and is equal to P_{cr_1} if and only if the exact equilibrium curve $w(x)$ is used to calculate U and $\bar{W}$.

Equation 5.3.4 implies that $\delta P_R = 0$ for the exact equilibrium curve $w(x)$ that corresponds to the first critical load. This fact may also be demonstrated as follows:

$$\bar{W}\,\delta P_R = \bar{W}\,\delta(U/\bar{W}) = (\bar{W}\,\delta U - U\,\delta\bar{W})\bar{W}^{-1} = \delta U - \delta\bar{W}\frac{U}{\bar{W}} = \delta U - P_R\,\delta\bar{W} = \delta\Pi \tag{5.3.5}$$

which shows that the conditions $\delta\Pi = 0$ [or $\delta(\delta^2\Pi) = 0$] and $\delta P_R = 0$ are equivalent.

The argument that we just used to prove the minimum property of P_R (Eq. 5.3.4) is physical. For a mathematical proof see, for example, Collatz (1963, sec. 8.1) and also the text which follows.

Can the Rayleigh quotient be used to determine the higher critical loads, for example, P_{cr_2}? It can, but only if the choice of the deflection curves $w(x)$ is restricted to those curves that represent linear combinations of the second and higher modes, excluding the first mode (Collatz, 1963). Obviously, such a choice of the deflection curve can be made only if the first buckling mode has already been solved [we will see that this requires restricting the choice of $w(x)$ to those deflection functions that are orthogonal to the first buckling mode].

If the equilibrium position to be examined is not characterized by $w(x) = 0$ (or $q_1 = \cdots = q_n = 0$), then Π does not coincide with the second variation $\delta^2\Pi = \delta^2 U - P\,\delta^2\bar{W}$, in which $\delta^2 U$ and $\delta^2\bar{W}$ are different from U and $\bar{W}$. By a similar argument as before, one can show that

$$P_{cr_1} = \min\frac{\delta^2 U}{\delta^2\bar{W}} \tag{5.3.6}$$

When Π is a nonlinear function of load P, one may consider a Taylor series expansion of Π as a function of the difference $P - P_{cr_1}$. Then, truncating all terms with higher than first powers of $P - P_{cr_1}$, one may obtain an energy expression of the form of Equation 5.3.1, from which again a Rayleigh quotient may be formulated.

Application to Beam-Columns

For beam-columns, for which Π is given by Equation 5.1.10 with $p = 0$, Equation 5.3.2 yields

$$P_R = \frac{\int_0^l \frac{1}{2}EIw''^2\,dx}{\int_0^l \frac{1}{2}w'^2\,dx} \tag{5.3.7}$$

Since, for the exact deflection shape $w(x)$, the Rayleigh quotient (Eq. 5.3.7) gives the exact critical load, one may naturally expect that for a deflection shape $w(x)$ that is very close to the critical state the Rayleigh quotient P_R will be very close to the exact critical load. This must be so because P_R is a continuous functional of $w(x)$. This fact is useful for approximate calculations of critical loads, and, according to Equations 5.3.2 and 5.3.3, such approximate calculations yield always upper bounds.

The stationary property of the Rayleigh quotient expression for a beam, in Equation 5.3.7, can be again demonstrated more directly. We showed that $\bar{W}\,\delta P_R = \delta U - P_R\,\delta\bar{W}$ (Eq. 5.3.5). Thus, calculating δU and $\delta\bar{W}$ from the numerator and denominator of Equation 5.3.7, and setting $P_R = P$ for the exact solution, we obtain

$$\bar{W}\,\delta P_R = \int_0^l EIw''\,\delta w''\,dx - \int_0^l Pw'\,\delta w'\,dx = \int_0^l [(EIw'')'' + (Pw')']\,\delta w\,dx \quad (5.3.8)$$

in which we transformed the integrals using integrations by parts along with the boundary conditions. We may now recognize that $\bar{W}\,\delta P_R$ must vanish because the first expression represents the statement of the principle of virtual work [since $EIw'' = M$ and $w'\,\delta w' = \delta(\frac{1}{2}w'^2) = \delta\varepsilon$], and also because $(EIw'')'' + (Pw')' = p = 0$. This proves again that, for the exact $w(x)$, P_R attains a stationary value with regard to all kinematically admissible variations of $w(x)$, but it, of course, does not prove the minimum property.

It should be emphasized that the chosen approximate deflection shape $w(x)$ (i.e., the trial function) must always be kinematically admissible, that is, it must satisfy all the kinematic boundary conditions (e.g., for a hinged column, $w = 0$ at both $x = 0$ and $x = l$). The trial function does not have to satisfy the static (or natural) boundary conditions, because the static boundary conditions as well as the differential condition of equilibrium are a consequence of the minimization of P_R (or Π) itself. However, if the static boundary conditions are satisfied, the shape of $w(x)$ is obviously much closer to the exact buckling shape, and so the resulting P_R approximates P_{cr_1} more accurately, usually much more accurately, as we will see later.

As an example, consider the simply supported Euler column (Fig. 5.7), and substitute the deflection shape $w(x) = A\sin(\pi x/l)$, which we know to be the exact buckling shape. Indeed, evaluation of Equation 5.3.7 yields $P_{cr_1} = EI\pi^2/l^2$ = the exact value of the first critical load.

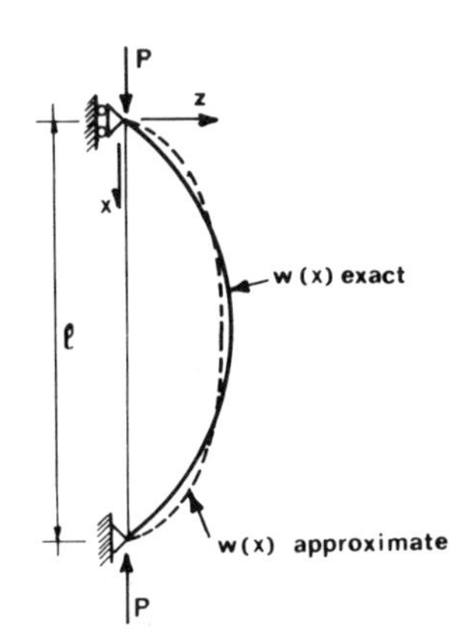

Figure 5.7 Exact and approximate deflection of a Euler column.

Now pretend we do not know the exact $w(x)$ and consider a parabolic deflection curve

$$w(x) = Ax(l - x) \tag{5.3.9}$$

which obviously satisfies the kinematic boundary conditions $w = 0$ at $x = 0$ and $x = l$. Substituting this into Equation 5.3.7 and integrating, we obtain

$$P_R = \frac{12EI}{l^2} = 1.215P_{cr_1} \tag{5.3.10}$$

This result is indeed larger than P_{cr_1} and exceeds P_{cr_1} by 21.5 percent. It would be a poor approximation to P_{cr_1}. Obviously, the choice of a parabola for the deflection curve is not very good, since the curvature of the deflection curve is constant, while it should vanish at column ends where the bending moment is zero.

To improve the result, we might consider a better function, which is not only kinematically admissible (i.e., $w = 0$ at $x = 0$ and $x = l$) but also satisfies the static boundary conditions $M = 0$ at $x = 0$ and $x = l$. This may be achieved by a parabolic distribution of curvature:

$$w''(x) = Ax(l - x) \tag{5.3.11}$$

(A = arbitrary constant). Integrating twice we have $w(x) = A(2lx^3 - x^4 + Cx + D)/12$. The integration constants C and D can be determined from the boundary conditions $w = 0$ at $x = 0$ and $x = l$, which yield $C = -l^3$, $D = 0$. So $w(x) = Ax(2lx^2 - x^3 - l^3)/12$. Substituting this and Equation 5.3.11 into Equation 5.3.7 and integrating, we obtain

$$P_R = \left(\frac{168}{17}\right)\left(\frac{EI}{l^2}\right) = 1.0013P_{cr_1} \tag{5.3.12}$$

Again the upper-bound property of P_R is verified, but the result is a far better approximation to P_{cr_1}, with an error of only 0.13 percent. This drastic improvement is a consequence of the fact that the convergence of P_R to P_{cr_1} is quadratic, as will be shown later. It also shows that one way to obtain a very good estimate for the deflection shape is to try to satisfy not only the kinematic boundary conditions but also the static boundary conditions.

Relation to Differential Equation

Now that we have illustrated the minimizing property of the Rayleigh quotient by examples, we will examine the minimization property more rigorously. Let us consider a self-adjoint linear eigenvalue problem defined as the boundary value problem with the homogeneous linear differential equation $\underline{L}w = \underline{M}w - P\underline{N}w = 0$ and homogeneous boundary conditions; $\underline{L}$, $\underline{M}$, $\underline{N}$ are linear differential operators, $\underline{M}$ and $\underline{N}$ being positive definite (see Sec. 5.1). Stable states are considered to be those for which operator $\underline{L}$ is positive definite. According to Theorem 5.1.3, this condition means that state $w(x) = 0$ is stable if $(w, \underline{L}w) = (w, \underline{M}w) - P(w, \underline{N}w) > 0$ for all admissible $w(x)$, critical if $(w, \underline{L}w) \geq 0$ for all admissible $w(x)$, and $(w, \underline{L}w) = 0$ for some admissible $w(x)$ and unstable if $(w, \underline{L}w) < 0$ for some admissible $w(x)$. Noting that $(w, \underset{\sim}{N}w) > 0$ because $\underset{\sim}{N}$ is positive definite, we may express P from these inequalities and find that the system is stable if, for all admissible $w(x)$, $P < P_R$ where

$$P_R = \frac{(w, \underset{\sim}{M}w)}{(w, \underset{\sim}{N}w)} = \frac{\int_0^l w(x)\underset{\sim}{M}[w(x)]\,dx}{\int_0^l w(x)\underset{\sim}{N}[w(x)]\,dx} \tag{5.3.13}$$

This definition of the Rayleigh quotient, which is normally used in mathematics, is more general than our definition in Equation 5.3.2, for it is applicable even if the differential operators $\underset{\sim}{M}$ and $\underset{\sim}{N}$ are not associated (according to variational calculus) with the strain energy and work.

From the condition that $(w, \underset{\sim}{L}w) > 0$ for all admissible $w(x)$ it follows that $P < (w, \underset{\sim}{M}m)/(w, \underset{\sim}{N}w)$ for all admissible $w(x)$ that do not satisfy the differential equation with its natural boundary conditions. This implies (by the same argument as used to derive Eq. 5.3.4) that min P_R is P_{cr_1} also for the definition of P_R according to Equation 5.3.13.

For the special case of a beam-column ($\underset{\sim}{M} = d^2/dx^2[EI(x)\,d/dx^2]$, $N = -d^2/dx^2$, see Eq. 5.1.11), Equation 5.3.13 becomes

$$P_R = \frac{\int_0^l w(EIw'')''\,dx}{-\int_0^l ww''\,dx} \tag{5.3.14}$$

Is this the same as Equation 5.3.7? It is, for by integrating by parts and taking into account the homogeneous essential or natural boundary conditions of the column, Equation 5.3.14 can be transformed into Equation 5.3.7. One may easily check that if the trial functions used in the previous examples (Eqs. 5.3.9 and 5.3.11) are substituted, one gets from Equation 5.3.14 the same results as those obtained before from Equation 5.3.7. Also, substitution of the exact deflection shape yields the exact first (lowest) critical load.

Proof of Upper-Bound Property and Convergence

The fact that the first critical mode minimizes the Rayleigh quotient can be alternatively, and perhaps also more conspicuously, proven on the basis of the eigenfunctions (buckling modes) $w_n(x)$, $n = 1, 2, 3, \ldots$. Since the eigenfunctions are linearly independent and mutually orthogonal, any admissible function $w(x)$ (i.e., any kinematically admissible deflection curve) may be expressed (exactly) by the eigenfunction expansion:

$$w(x) = \sum_{n=1}^{\infty} q_n w_n(x) \tag{5.3.15}$$

We may now substitute this expansion into Equation 5.3.13. This yields

$$P_R = \frac{\Sigma_n \Sigma_m (w_n, Mw_m) q_n q_m}{\Sigma_n \Sigma_m (w_n, Nw_m) q_n q_m} \tag{5.3.16}$$

According to the differential equation $\underline{M}w - P\underline{N}w = 0$ we may now substitute $\underline{M}w_m = P_{cr_m}\underline{N}w_m$ and assume that the eigenfunctions w_n are normalized in such a manner that $(w_n, \underline{N}w_m) = \delta_{nm}$ = Kronecker delta (= 1 if $n = m$ and 0 if $n \neq m$). Thus we can reduce Equation 5.3.16 to the form

$$P_R = \frac{\sum_{n=1}^{\infty} P_{cr_n} q_n^2}{\sum_{n=1}^{\infty} q_n^2} = P_{cr_1} \frac{q_1^2 + \dfrac{P_{cr_2}}{P_{cr_1}} q_2^2 + \dfrac{P_{cr_3}}{P_{cr_1}} q_3^2 + \dfrac{P_{cr_4}}{P_{cr_1}} q_4^2 + \cdots}{q_1^2 + q_2^2 + q_3^2 + q_4^2 + \cdots} \geq P_{cr_1} \quad (5.3.17)$$

The inequality $P_R \geq P_{cr_1}$ follows from the fact that each term in the denominator of the last fraction is equal to or less than the corresponding term in the numerator, that is, $q_n^2 \leq q_n^2 P_{cr_n}/P_{cr_1}$, since, by definition, P_{cr_1} is the lowest critical load (i.e., $P_{cr_1} < P_{cr_n}$ for all $n > 1$). The case $P_R = P_{cr_1}$ is obtained only if $q_2 = q_3 = q_4 = 0$, that is, if $w(x) = q_1 w_1(x)$ = first critical mode. Otherwise, $P_R > P_{cr_1}$. Thus we prove that $\min P_R = P_{cr_1}$.

If we require $w(x)$ to be orthogonal to $w_1(x)$, that is, restrict $w(x)$ to such functions that $(w, \underline{N}w_1) = 0$, then by a similar argument we can show that $\min P_R = P_{cr_2}$.

Noting that, according to Equation 5.3.17,

$$P_R - P_{cr_1} = \frac{\sum_{n=2}^{\infty} (P_{cr_n} - P_{cr_1}) q_n^2}{\sum_{n=1}^{\infty} q_n^2} \quad (5.3.18)$$

it is also apparent that the difference $P_R - P_{cr_1}$ decreases as the squares of the deviations from the first critical mode. This is an essential property; it indicates that even an approximate shape of the deflection curve should yield an excellent estimate of the first eigenvalue provided it is not very different from the first eigenfunction.

The last observation means that the convergence of the Rayleigh quotient to the first eigenvalue should be quadratic. This may be proven more directly as follows. Equation 5.3.18 shows that $P_R = P_{cr_1}$ when $w(x) = w_1(x)$ = first eigenfunction. For a general admissible deflection curve we may write $w(x) = w_1(x) + \varepsilon\bar{w}(x)$ in which $\bar{w}(x)$ is any kinematically admissible function satisfying the given kinematic boundary conditions (which are homogeneous) and ε is an arbitrary parameter. Thus, by definition (Eq. 5.3.13),

$$\begin{aligned} P_R[w(x)] &= \frac{(w_1 + \varepsilon\bar{w}, \underline{M}w_1 + \varepsilon\underline{M}\bar{w})}{(w_1 + \varepsilon\bar{w}, \underline{N}w_1 + \varepsilon\underline{N}\bar{w})} \\ &= \frac{(w_1, \underline{M}w_1) + \varepsilon(\bar{w}, \underline{M}w_1) + \varepsilon(w_1, \underline{M}\bar{w}) + \varepsilon^2(\bar{w}, \underline{M}\bar{w})}{(w_1, \underline{N}w_1) + \varepsilon(\bar{w}, \underline{N}w_1) + \varepsilon(w_1, \underline{N}\bar{w}) + \varepsilon^2(\bar{w}, \underline{N}\bar{w})} \end{aligned} \quad (5.3.19)$$

In the numerator we have $\varepsilon(w_1, \underline{M}\bar{w}) = \varepsilon(\bar{w}, \underline{M}w_1)$ because of the self-adjoint property of operator $\underline{M}$. Furthermore, because w_1 is the exact eigenfunction, we may substitute $\underline{M}w_1 = P_{cr_1}\underline{N}w_1$. In the denominator, $\varepsilon(w_1, \underline{N}\bar{w}) = \varepsilon(\bar{w}, \underline{N}w_1)$ because of the self-adjoint property of operator $\underline{N}$. Thus, Equation 5.3.19 yields

$$\begin{aligned} P_R[w(x)] &= \frac{P_{cr_1}(w_1, \underline{N}w_1) + 2\varepsilon P_{cr_1}(\bar{w}, \underline{N}w_1) + \varepsilon^2(\bar{w}, \underline{M}\bar{w})}{(w_1, \underline{N}w_1) + 2\varepsilon(\bar{w}, \underline{N}w_1) + \varepsilon^2(\bar{w}, \underline{N}\bar{w})} \\ &= P_{cr_1} + \varepsilon^2\left[\frac{(\bar{w}, \underline{M}\bar{w}) - P_{cr_1}(\bar{w}, \underline{N}\bar{w})}{(w_1, \underline{N}w_1) + 2\varepsilon(\bar{w}, \underline{N}w_1) + \varepsilon^2(\bar{w}, \underline{N}\bar{w})}\right] \end{aligned} \quad (5.3.20)$$

The term that multiplies ε^2 is not constant. However, it varies continuously and is nonzero when ε is zero. So this term is nearly constant within the range of infinitely small values of ε. For a sufficiently small ε, the denominator is positive. Because $\underline{L} = \underline{M} - P_{cr_1}\underline{N}$ is a positive-semidefinite operator, the numerator is nonnegative, that is, it is zero for some $\bar{w}(x)$, and positive for other $\bar{w}(x)$ [this would not be true if $w_1(x)$ were replaced by another $w_n(x)$]. Thus we conclude again that with diminishing ε the smallest eigenvalue is approached from above, that is, P_R is an upper bound. At the same time, we also verify that the error in the eigenvalue is of the order of ε^2 if the error in the eigenfunction is of the order of ε. This proves that as the deflection curve $w(x)$ approaches the exact first eigenfunction, the convergence is quadratic. This further implies that even a relatively crude estimate of the first eigenfunction should give a relatively accurate estimate of P_{cr_1}. See also, for example, Dym and Shames (1973), Collatz (1963), Courant and Hilbert (1962), and Morse and Feshbach (1953).

Extension to Free Vibration

Let us now generalize the Rayleigh quotient for free vibrations of elastic columns under a conservative axial load. Similar to Section 3.1, we assume that $w(x, t) = e^{i\omega t}v(x)$. Then, according to the D'Alembert principle and the principle of virtual work, the condition of dynamic equilibrium leads to the differential equation $(EIv)'' + (Pv')' - \mu\omega^2 v = 0$, with homogeneous boundary conditions. The system is stable if the differential operator of this differential equation (which is self-adjoint) is positive definite, which requires that

$$\int_0^l [(EIv'')'' + (Pv')' - \mu\omega^2 v]v \, dx > 0 \tag{5.3.21}$$

Then, integrating by parts and using the boundary conditions, which are assumed to be homogeneous, we may transform Equation 5.3.21 to the equation:

$$\int_0^l (\tfrac{1}{2}EIv''^2 - \tfrac{1}{2}Pv'^2 - \tfrac{1}{2}\mu\omega^2 v^2) \, dx > 0 \tag{5.3.22}$$

This immediately leads to the following stability conditions in terms of the Rayleigh quotients for either the circular frequency ω or the axial load P:

$$\omega^2 < \frac{\int_0^l \tfrac{1}{2}(EIv''^2 - Pv'^2) \, dx}{\int_0^l \tfrac{1}{2}\mu v^2 \, dx} \tag{5.3.23a}$$

$$P < \frac{\int_0^l \tfrac{1}{2}(EIv''^2 - \mu\omega^2 v) \, dx}{\int_0^l \tfrac{1}{2}v'^2 \, dx} \tag{5.3.23b}$$

In the plane (P, ω^2), either one of these inequalities defines an exterior approximation for the stable domain (Fig. 5.8). It consists of a straight line segment connecting the point of the Rayleigh quotient ω_R (Eq. 5.3.23a for $P = 0$)

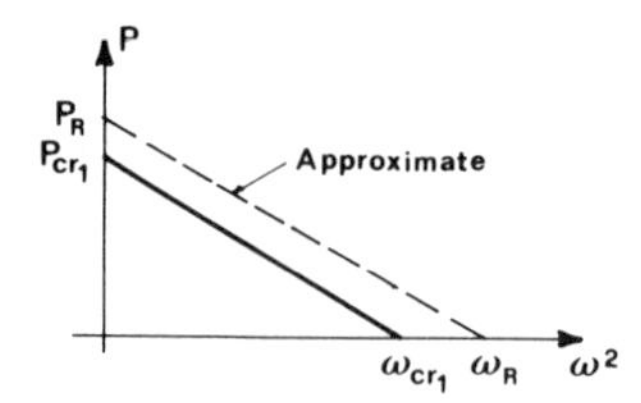

Figure 5.8 Variation of exact and approximate circular frequencies of free vibrations of an Euler column with axial load.

with the point of the Rayleigh quotient P_R (Eq. 5.3.23b for $\omega = 0$). In the first (positive) quadrant of Figure 5.8, this straight line segment lies entirely outside the straight line segment that connects the points of ω_{cr_1} and P_{cr_1} (Fig. 5.8) and was already obtained by the exact solution in Section 3.1 (Fig. 3.2a).

Problems

5.3.1 Consider a free-standing column (Fig. 5.9a) and transform Equation 5.3.14 into Equation 5.3.7.

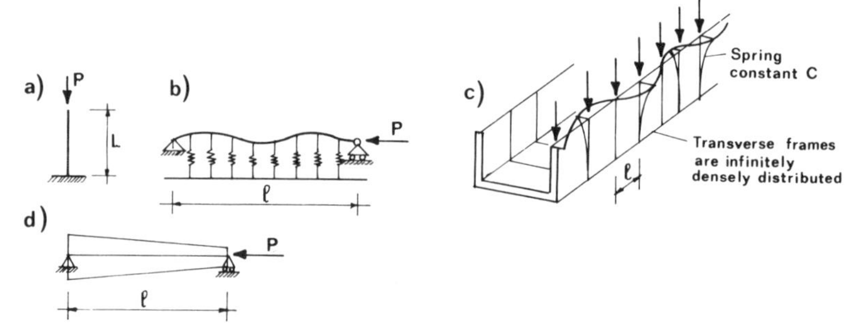

Figure 5.9 Exercise problems on Rayleigh quotient for columns and beams on elastic foundation.

5.3.2 Consider a beam on an elastic foundation (Fig. 5.9b) with hinged ends and length l. Solve it by the Rayleigh quotient assuming $w(x) = A \sin(n\pi x/l)$. (*Note*: $n = 1$ does not in general give the lowest critical load, and so one must minimize with respect to n.)

5.3.3 Do the same, but $l \to \infty$ and $w(x) = \sin(\pi x/L)$. (*Note*: Now one must minimize with respect to L, and this must give the exact solution, Sec. 5.2.) Also, check that $w(x) = [\sin(\pi x/l)]^m$, for example, for $m = 3$, yields an upper bound.

5.3.4 Soive the problem of the lateral buckling of the compression belt of a truss bridge (Fig. 5.9c, also Fig. 5.5c).

5.3.5 Solve the problem of a simply supported beam with variable cross section (Fig. 5.9d) by the Rayleigh quotient. Use (a) $I = I_0 + I_1 x/l$ or (b) $I = I_0 + 4I_1 x(l - x)/l^2$. (c) Repeat for the case of a cantilever beam and for a fixed-hinged beam.

5.3.6 Calculate the frequencies of axially loaded simply supported columns: (a) of variable cross section (see Prob. 5.3.5), (b) loaded by own weight. If $P = 0.95P_{cr_1}$, how does the frequency change with respect to $P = 0.90P_{cr_1}$?

5.3.7 Generalize the Rayleigh quotient for columns with shear (analyzed by equilibrium in Sec. 1.7).

5.3.8 Formulate the Rayleigh quotient for three-dimensional buckling of a column under axial force and torque (Sec. 1.10).

5.4 TIMOSHENKO QUOTIENT AND RELATIONS BETWEEN VARIOUS BOUNDS

Derivation

For structures that are statically determinate, there exists another upper-bound method that yields closer approximations. In this method, due to Timoshenko (Timoshenko and Gere, 1961, p. 90), we calculate the bending energy from the bending moment M, which is $M = -Pw$, for the pin-ended column (Fig. 5.10a), rather than from the curvature of the assumed approximate deflection function (trial function). Imposing the condition that the energy variation at the critical state must be zero, we have

$$\Phi = \int_0^l \frac{(Pw)^2}{2EI}\,dx - \int_0^l \frac{P}{2}\left(\frac{dw}{dx}\right)^2 dx = P^2 U_1 - P\bar{W} = 0 \tag{5.4.1}$$

where U_1 = strain energy based on M but calculated from $w(x)$, and $\bar{W}$ = work of load, same as used for P_R. Now we may solve for P and denote the expression as

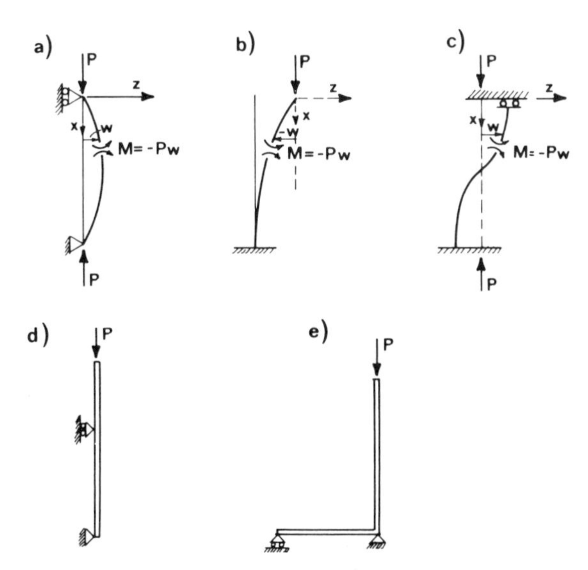

Figure 5.10 Examples of structures in which the bending energy can be calculated from the bending moment (expressed through deflection).

P_T:

$$P_T = \frac{\bar{W}}{U_1} = \frac{\int_0^l \frac{1}{2} w'^2 \, dx}{\int_0^l \frac{1}{2}[w^2/EI(x)] \, dx} \tag{5.4.2}$$

This expression is known in buckling theory as the Timoshenko quotient. It exists only for statically determinate structures. We will see that if P_T exists it always provides a closer upper bound than the Rayleigh quotient defined by Equation 5.3.7 or 5.3.14.

Equation 5.4.2 is applicable not only to pin-ended columns, but also more generally to a free-standing column provided that $w(x)$ is measured from the moving axis of load P rather than the fixed axis (Fig. 5.10b). This equation is also applicable to a column with one sliding rotational restraint and one fixed end (Fig. 5.10c), provided that w is measured from the vertical line passing through the inflection point at which $M = 0$. For multispan continuous beam-columns, one could define $w(x)$ differently in each span in order to make the expression $M = -Pw$ applicable. But in such cases, for example, the simply-supported beam-column with an overhang (Fig. 5.10d) or a free-standing column with an L-shaped base (Fig. 5.10e), it is preferable to write first the expression for energy Φ and then solve P from it.

A necessary, although not sufficient, condition for P_T to attain a minimum for the exact deflected equilibrium shape $w(x)$ is that $\delta P_T = 0$. This property may be easily checked from the relation $\delta P_T = \delta(\bar{W}/U_1) = (U_1 \, \delta\bar{W} - \bar{W} \, \delta U_1)U_1^{-2} = (\delta\bar{W} - P \, \delta U_1)/U_1$. By a procedure similar to Equation 5.3.8, we have

$$U_1 \, \delta P_T = \int_0^l w' \, \delta w' \, dx - P \int_0^l \frac{w}{EI} \, \delta w \, dx = -\int_0^l \left(w'' + \frac{Pw}{EI}\right) \delta w \, dx + [w' \, \delta w]_0^l \tag{5.4.3}$$

in which the first integral has been transformed by integration by parts and the bracketed expression vanishes since either $w = 0$ or $w' = 0$ at the boundary points. If equilibrium exists not only for $w(x) = 0$ but also for nonzero $w(x)$, the last integral in this equation vanishes because the integrand vanishes. This demonstrates that P_T is stationary if there is equilibrium at nonzero $w(x)$, but, of course, not that P_T is minimum.

Examples

Let us check how close an approximation we get from the Timoshenko quotient for the pin-ended column. First we can verify that by substituting $w(x) = \sin(\pi x/l)$ into Equation 5.4.2, we obtain the exact first critical load $P_{\text{cr}_1} = EI\pi^2/l^2$. Then we pretend again we do not know the exact shape and choose a parabola, Equation 5.3.9. Substitution into Equation 5.4.2 and integration then yields

$$P_T = \frac{10EI}{l^2} = 1.013 P_{\text{cr}_1} \tag{5.4.4}$$

This is a far closer upper bound than the value $1.215P_{\mathrm{cr}_1}$ we obtained from the Rayleigh quotient (Eq. 5.3.10).

Now adopt a parabola for the curvature distribution, as written in Equation 5.3.11, and substitute the associated expressions for $w(x)$ and $w'(x)$ that we determined before into Equation 5.4.2. This yields

$$P_T = 9.87096\frac{EI}{l^2} = 1.00014P_{\mathrm{cr}_1} \tag{5.4.5}$$

This is again closer to P_{cr_1} than the upper bound $1.0013P_{\mathrm{cr}_1}$, which we obtained from the Rayleigh quotient (Eq. 5.3.12). Later we will show in general that, if P_T exists, it always gives a closer upper bound than P_R (that is, $P_R \geq P_T$).

If we realize that the Timoshenko quotient involves lower-order derivatives than the Rayleigh quotient, it is not surprising that it gives closer bounds. In mathematics it is generally known that if a certain function has an error, then its derivative has a larger error. Therefore, it is generally desirable to make approximations on the basis of derivatives of the lowest order possible.

Relation to Differential Equation and Proof of Upper-Bound Property

From the fact that the Timoshenko quotient can be defined only for statically determinate structures we may have suspected it must be related to the integrated (second-order) differential equation of beam-columns, $w'' + wP/EI = 0$, with which we started in Section 1.2. This equation is of the form $\underline{L}^*w = \underline{M}^*w - P\underline{N}^*w = 0$, in which the differential operators are $\underline{M}^* = -d^2/dx^2$, $\underline{N}^* = 1/EI$. In mathematics, the general rule for forming the Rayleigh quotient is Equation 5.3.13, and substitution of these operators then yields, for $w(0) = 0$,

$$P = \frac{\int_0^l w\underline{M}^*w\,dx}{\int_0^l w\underline{N}^*w\,dx} = \frac{\int_0^l (-ww'')\,dx}{\int_0^l \frac{w^2}{EI}\,dx} = \frac{\int_0^l w'^2\,dx}{\int_0^l \frac{w^2}{EI}\,dx} = P_T \tag{5.4.6}$$

in which the last expression has been obtained through integration by parts. The boundary terms arising from the integration by parts vanish because at the boundaries $x = 0$ and $x = l$ we have either $w = 0$ or $w' = 0$. Note the boundary terms also vanish for the free-standing column if $w(x)$ is measured from the load axis (Fig. 5.10b).

Note that the last expression in Equation 5.4.6 is identical to the Timoshenko quotient (Eq. 5.4.2), at which we initially arrived by physical arguments. Consequently, even though P_T is not the Rayleigh quotient in the physical sense of $P_R = U/\bar{W}$ (= ratio of strain energy to work per unit load), it mathematically represents nothing else but the Rayleigh quotient for a twice-integrated differential equation of equilibrium of a beam-column. From this observation it immediately follows that P_T is an upper bound and attains a minimum for the exact first buckling mode, that is,

$$\min P = P_T \qquad \text{for all admissible } w(x) \tag{5.4.7}$$

The proof of the minimum property that we showed in Equations 5.3.19 and 5.3.20 applies also to P_T, and so does the demonstration of quadratic convergence (Eq. 5.3.20).

Note that the formation of the Rayleigh quotient according to Equation 5.4.6 is contingent upon the fact that operator $L^* = M^* - PN^*$ is self-adjoint. This property may be easily verified by integration by parts. Also, one can verify that operators M^* and N^* are positive definite. Furthermore, the conditions of positive definiteness of operator L^* require that functional Φ be negative definite. This may be verified by integrating by parts as follows:

$$\Phi^*[w(x)] = \frac{1}{2}(L^*w, w) = \frac{1}{2}(w, M^*w) - \frac{P}{2}(w, N^*w)$$
$$= \frac{1}{2}\int_0^l w(-w'')\,dx - \frac{P}{2}\int_0^l \frac{w^2}{EI}\,dx$$
$$= \frac{1}{P}\left[\int_0^l \frac{P}{2}w'^2\,dx - \int_0^l \frac{(Pw)^2}{2EI}\,dx\right] = -\frac{1}{P}\Phi[w(x)] \qquad (5.4.8)$$

which shows that $\Phi[w(x)] < 0$ for all admissible $w(x)$ if and only if $\Phi^*[w(x)] > 0$ for all admissible $w(x)$.

Relation to Rayleigh Quotient and Inequalities

From the examples we saw that P_T was a better upper bound than P_R, that is, $P_T \leq P_R$. Is this valid in general? We want to prove that

$$\frac{\displaystyle\int_0^l w'^2\,dx}{\displaystyle\int_0^l w^2\,dx/EI} \leq \frac{\displaystyle\int_0^l EIw''^2\,dx}{\displaystyle\int_0^l w'^2\,dx} \qquad (5.4.9)$$

or

$$\int_0^l EIw''^2\,dx \cdot \int_0^l \frac{w^2}{EI}\,dx \geq \left(\int_0^l w'^2\,dx\right)^2 \qquad (5.4.10)$$

Integration by parts and boundary conditions $w = 0$ yield

$$\int_0^l w'^2\,dx = -\int_0^l ww''\,dx \qquad (5.4.11)$$

Denoting $w''\sqrt{EI} = f(x)$ and $w/\sqrt{EI} = g(x)$, we have

$$\int_0^l f^2(x)\,dx \cdot \int_0^l g^2(x)\,dx \geq \left[\int_0^l f(x)g(x)\,dx\right]^2 \qquad (5.4.12)$$

Now this last inequality may be recognized to represent the well-known Cauchy-Schwarz inequality (e.g., Rektorys, 1969, p. 571). (This inequality may in turn be proven by approximating the integrals with sums and considering a limit of the algebraic inequality in Eq. 5.4.22.) So we conclude that if P_T exists then

$P_T \leq P_R$. Together with our previous result we thus have (Bažant, 1985)

$$P_R \geq P_T \geq P_{\mathrm{cr}_1} \tag{5.4.13}$$

The cases of equality occur here only if $w(x)$ is the exact solution.

The inequalities in Equation 5.4.13 also can be proven by considering again the expansion of a general deflection curve into the eigenfunctions that we already introduced in Equation 5.3.15; see, for example, Schreyer and Shih (1973). Consideration must now be restricted to statically determinate columns, for which the differential equation is (see Eq. 5.1.11) $\underset{\sim}{M}^*w - P\underset{\sim}{N}^*w = -w'' - (P/EI)w = 0$, where EI can be variable. Let us assume that the eigenfunctions $w^*(x)$ are normalized in such a way that

$$(w_m^*, \underset{\sim}{N}^* w_n^*) = \int_0^l w_m^*(x) w_n^*(x) \frac{dx}{EI} = \delta_{mn} \tag{5.4.14}$$

in which $\delta_{mn} = 1$ if $m = n$ and otherwise zero. (Note that δ_{mn} is not nondimensional but has the dimension of length/force.) Because the eigenfunctions satisfy the differential equation, we may substitute $w_n^* = -EIw_n^{*}{''}/P_{\mathrm{cr}_n}$ into Equation 5.4.14. This furnishes

$$\int_0^l EI w_m^{*}{''}(x) w_n^{*}{''}(x)\, dx = \delta_{mn} P_{\mathrm{cr}_n}^2 \tag{5.4.15}$$

Furthermore, we may substitute $w_n^* = -EIw_n^{*}{''}/P_{\mathrm{cr}_n}$ into Equation 5.4.14, and this yields $-\int_0^l w_m^{*}{''} w_n^*\, dx = \delta_{mn} P_{\mathrm{cr}_n}$. If we integrate by parts and use the boundary conditions $w_n^* = 0$ at $x = 0$ and $x = l$, we obtain

$$\int_0^l w_m^{*}{'}(x) w_n^{*}{'}(x)\, dx = \delta_{mn} P_{\mathrm{cr}_n} \tag{5.4.16}$$

We may now use Equations 5.4.14 to 5.4.16 to evaluate

$$\int_0^l [w(x)]^2 \frac{dx}{EI} = \int_0^l \left[\sum_n q_n^* w_n^*(x)\right]^2 \frac{dx}{EI} = \int_0^l \sum_m \sum_n \delta_{mn} q_m^* q_n^* w_m^*(x) w_n^*(x) \frac{dx}{EI} = \sum_{n=1}^{\infty} q_n^{*2} \tag{5.4.17}$$

$$\int_0^l [w'(x)]^2\, dx = \cdots = \sum_{n=1}^{\infty} P_{\mathrm{cr}_n} q_n^{*2} \tag{5.4.18}$$

$$\int_0^l EI[w''(x)]^2\, dx = \cdots = \sum_{n=1}^{\infty} P_{\mathrm{cr}_n}^2 q_n^{*2} \tag{5.4.19}$$

Substituting this into Equation 5.4.2, which defines the Timoshenko quotient, we find that

$$P_T = \frac{\sum_n P_{\mathrm{cr}_n} q_n^{*2}}{\sum_n q^{*2}} \geq \frac{P_{\mathrm{cr}_1} \sum_n q_n^{*2}}{\sum_n q_n^{*2}} = P_{\mathrm{cr}_1} \tag{5.4.20}$$

where the inequality ensues from the fact that $P_{\mathrm{cr}_n} > P_{\mathrm{cr}_1}$. This proves again that the Timoshenko quotient is an upper bound for the lowest critical load.

Equation 5.4.20 is identical to Equation 5.3.17. This observation is not surprising since Equation 5.3.17 is valid for any operator $\underset{\sim}{N}$, provided that the eigenfunctions are normalized with respect to the same operator, that is, $\underset{\sim}{N}$. If one substitutes Equations 5.4.18 and 5.4.19 into Equation 5.3.7 for the Rayleigh quotient, one obtains

$$P_R = \frac{\sum_n P_{\mathrm{cr}_n}^2 q_n^{*2}}{\sum_n P_{\mathrm{cr}_n} q_n^{*2}} \tag{5.4.21}$$

Let us now prove that $P_R \geq P_T$ always if P_T exists. Substituting into this inequality the expressions in Equations 5.4.20 and 5.4.21, multiplying by the expressions in the denominators (which are positive), and denoting $P_{\mathrm{cr}_n} q_n^* = a_n$, we reduce the inequality $P_R \geq P_T$ to the inequality

$$\sum_n a_n^2 \sum_n q_n^{*2} \geq \left(\sum_n a_n q_n^* \right)^2 \tag{5.4.22}$$

This inequality may be proven as follows. Consider the hyperspace with coordinates $q_1^*, q_2^*, \ldots$ and denote the vectors $\mathbf{q} = (q_1^*, q_2^*, \ldots)$ and $\mathbf{a} = (a_1, a_2, \ldots)$. Then the sums on the left-hand side of Equation 5.4.22 represent the squares of the magnitudes of vectors $\mathbf{a}$ and $\mathbf{q}$, and the right-hand side represents the square of the scalar product of these vectors. This means that inequality 5.4.22 is equivalent to $|\mathbf{a}|\,|\mathbf{q}| \geq |\mathbf{a} \cdot \mathbf{q}|$. The last inequality, as well as that in Equation 5.4.22, is known as the Cauchy–Schwarz inequality (in algebraic form). Its geometric proof is simple: see Figure 5.11. We divide the last inequality by $|\mathbf{a}|$ and note that $\mathbf{a}/|\mathbf{a}| = \mathbf{e}_a =$ unit direction vector of vector $\mathbf{a}$. Thus, the last inequality is equivalent to $|\mathbf{q}| \geq |\mathbf{e}_a \cdot \mathbf{q}|$, and this inequality states an obvious fact: The length of vector $\mathbf{q}$ in Figure 5.11 is not shorter than its projection onto the direction of vector $\mathbf{a}$. This completes the proof.

It also may be noted that for $n = 2$, the Cauchy–Schwarz inequality reads $(a_1^2 + a_2^2)(q_1^2 + q_2^2) \geq (a_1 q_1 + a_2 q_2)^2$; this can be reduced to $a_2^2 q_1^2 + a_1^2 q_2^2 \geq 2 a_1 a_2 q_1 q_2$ or $(a_2 q_1 - a_1 q_2)^2 \geq 0$, which again proves Equation 5.4.22 for $n = 2$.

The preceding proof can be also generalized to other boundary conditions (as shown by Schreyer and Shih, 1973). To this end, one needs to note that generally $EIM'' + PM = 0$, and so what we did before using eigenfunctions $w_n(x)$ can be repeated for moment eigenfunctions $M_n(x)$ with only slight modifications. One modification is to write $\int_0^l M_m' M_n' \, dx = P_{\mathrm{cr}_n} \delta_{mn}$ for the case of a free-standing column or a fixed-sliding column. For other types of supports, slightly different but similar relations may be written. In more detail, see Schreyer and Shih (1973).

The Rayleigh and Timoshenko quotients also can be defined for discrete

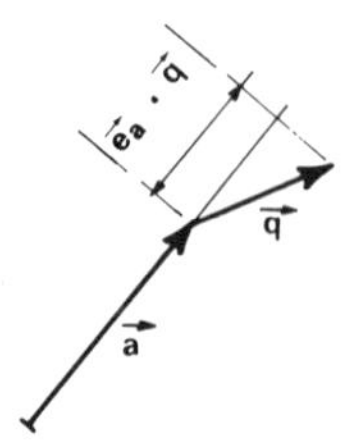

Figure 5.11 Projection of vector $\mathbf{q}$ onto the direction of vector $\mathbf{a}$.

systems. They may then be used for obtaining approximate solutions and bounds using a reduced number of kinematic variables q_n. The foregoing proofs can be directly translated from continuous to discrete systems, using an expansion in terms of discrete eigenvectors rather than continuous eigenfunctions.

To derive the inequality for P_T, one might be tempted to assume that the work expression $\int \frac{1}{2}(Pw)^2\,dx/EI - P\int \frac{1}{2}w'^2\,dx = P(P-P_T)\int \frac{1}{2}w^2\,dx/EI$ must be positive for stable states. But, curiously, this cannot be right because solving for P from the positiveness condition incorrectly indicates that stable states would occur, for all admissible $w(x)$, when $P > P_T$, which would imply that P_T should be a lower rather than an upper bound on P_{cr_1}—a consequence contradicted by our examples. So where is the error in this argument? The error lies in having tacitly assumed that the aforementioned work expression is positive definite. This expression may be recognized to coincide with Φ as defined before (Eq. 5.4.1), and we already showed (below Eq. 5.4.8) that in fact $\Phi < 0$. Φ is not a potential energy, since the equilibrium condition $M = -Pw$ has been introduced. Therefore $\Phi > 0$ is not required for stability (in contrast to the condition $\Pi = \delta^2\Pi > 0$ we used in Eq. 5.3.2).

An inequality for P_T can also be proven if one can find some structural system for which the expression in Equation 5.4.1 represents the potential energy. Such a system indeed exists. It is the fiber on an elastic foundation, whose potential energy we have already found to be $\Pi_f = \int \frac{1}{2}(cw^2 + Nw'^2)\,dx$ (Eq. 5.2.13). Comparing Equation 5.4.1 with Equation 5.2.13 we see that mathematical equivalence can occur if either $\Phi = P^2\Pi_f$, $c = 1/EI$, $N = -1/P$ or $\Phi = -P^2\Pi_f$, $c = -1/EI$, $N = 1/P$. Only the second choice, however, is acceptable, for two reasons: (1) an increase of stiffness EI must cause an increase in potential energy and (2) an increase of load P must cause a decrease in potential energy. Now from the stability condition $\Pi_f > 0$ it follows that the stability condition associated with Equation 5.4.1 is $P^2\Pi_f = -\Phi > 0$ or $\Phi < 0$ (and not $\Phi > 0$!). This means that

$$\Phi = P(P - P_T)\int_0^l \frac{w^2}{2EI}\,dx < 0 \tag{5.4.23}$$

that is, Φ is negative definite for stability (this also follows from Eq. 5.4.8). So we conclude that the column is stable if and only if $P < P_T$ for all admissible $w(x)$, which implies that $P_{\text{cr}_1} = \min P_T$.

Inapplicability to Dynamics

As we have seen, the Rayleigh quotient has an analogy in structural dynamics where it is used to approximate the first free vibration frequency. The Timoshenko quotient, however, has no analogy in dynamics. The reason is that the differential equation for free vibrations of a beam-column, which we derived in Section 3.1 (Eq. 3.1.2) contains the term $\mu\ddot{w}$, which is a zeroth-order derivative of w with respect to x, so that the equation cannot be integrated to reduce the highest derivative with respect to x from the fourth to the second order. This agrees with the fact that the bending moment in a vibrating pin-ended or free-standing column cannot be expressed in terms of the load and deflection alone.

The Question of Lower Bounds

From the viewpoint of safety, it would be comforting to designers to have an approximate method that would give a lower rather than upper bound. However, the problem is more difficult than the upper-bound problem. One type of lower bound, P_L, may be obtained on the basis of the Rayleigh and Timoshenko quotients, using the equation (Schreyer and Shih, 1973)

$$P_L = P_T\left(1 - \sqrt{\frac{P_R}{P_T} - 1}\right) \tag{5.4.24}$$

Unfortunately, this bound, as well as some other lower bounds discovered so far, is quite poor, that is, not sufficiently close to P_{cr_1}. Another type of a lower bound, which can be quite close, nevertheless exists; see Section 5.8.

Problems

5.4.1 Use the Timoshenko quotient to estimate P_{cr_1} of the columns in Figure 5.12a, b, c. Also solve these problems by the Rayleigh quotient and compare the results.

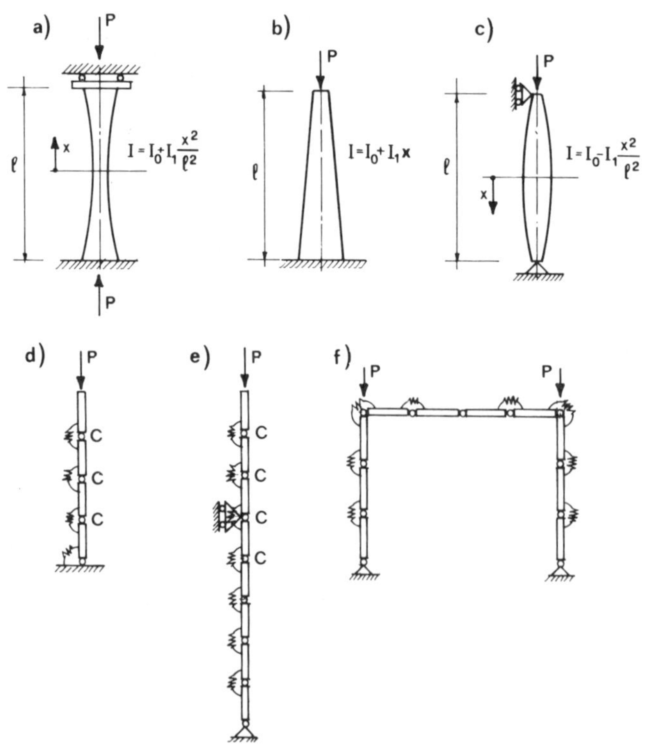

Figure 5.12 Exercise problems on Timoshenko quotient for columns and rigid-bar systems with elastic springs.

5.4.2 Can the Timoshenko quotient be formulated for (a) a simply supported beam on an elastic foundation? (b) A two-span continuous simply supported beam with a hinge in the middle of one span?

5.4.3 Solve, approximately, P_{cr_1} for the discrete systems made of rigid-body links and springs in Figure 5.12d, e, f.

5.4.4 Formulate and demonstrate the Timoshenko quotient for three-dimensional buckling of a statically determinate column under axial force and torque (Sec. 1.10).

5.5 BOUND APPROXIMATIONS FOR COLUMNS, FRAMES, AND HIGH ARCHES

We will now illustrate the practical calculation of upper-bound approximations for critical loads using the Rayleigh and Timoshenko quotients. This approach is often quicker and easier than the exact solution of critical loads.

Columns

Consider a column that is loaded by its own distributed weight $p(x)$ and by a concentrated force P_0 at the top (Fig. 5.13). The condition of zero energy

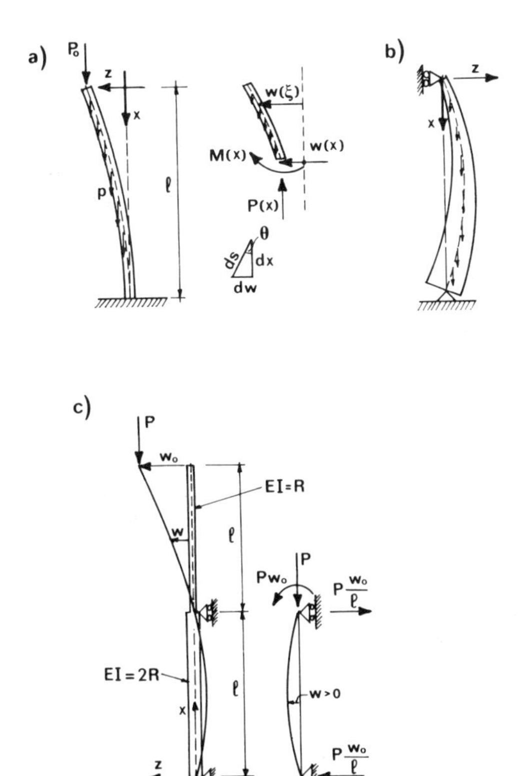

Figure 5.13 (a) Free-standing column loaded by a distributed weight; (b) column with variable cross section loaded by a distributed weight; (c) column with overhang.

variation of the critical load becomes

$$U - W = \int_0^l \tfrac{1}{2}EI(w'')^2\,dx - \int_0^l p(x)u(x)\,dx - P_0u(0) = 0 \qquad (5.5.1)$$

where $w(x)$ is an approximate solution and $u(x)$ is the vertical displacement due to deflection w. Integrating by parts the second integral, and using the boundary conditions $u(l)=0$ and $P(0)=P_0$, we obtain

$$\int_0^l \tfrac{1}{2}EI(w'')^2\,dx + \int_0^l P(x)u'(x)\,dx = 0 \qquad (5.5.2)$$

in which $P(x) = P_0 + \int_0^x p(\xi)\,d\xi$ is the variable axial force. The second integral in Equation 5.5.2 now represents the work of the axial force $-P(x)$ on the second-order axial strain component $u'(x) \simeq -w'^2(x)/2$ (which results upon imposing the condition $\varepsilon = 0$ in Eq. 4.3.7). Introducing this last expression in Equation 5.5.2, and expressing $P(x) = p_0 f(x)$, in which p_0 is a multiplier of the distributed load, one obtains the Rayleigh quotient

$$p_R^0 = \frac{\int_0^l \tfrac{1}{2}EIw''^2\,dx}{\int_0^l f(x)(w'^2/2)\,dx} \qquad (5.5.3)$$

Consider now a free-standing column of constant cross section (Fig. 5.13a), with no load at top, that is, $P_0 = 0$. We can then set $p = p_0 =$ constant and $f(x) = x$. Let us choose for the approximate buckling shape $w = A[1 - \sin(\pi x/2l)]$ where A is an undetermined constant. This choice has the advantage that it satisfies not only the kinematic boundary conditions ($w = w' = 0$ at base, $x = l$), but also one of the static boundary conditions on top of the column, $w'' = 0$ at $x = 0$. Substituting in Equation 5.5.3 one gets

$$p_R^0 = \left(\frac{\pi^4}{\pi^2 - 4}\right)\left(\frac{EI}{2l^3}\right) = \frac{8.29EI}{l^3} \qquad (5.5.4)$$

The exact solution can be obtained by integrating the differential equation $M' + Pw' = (EIw'')' + P(x)w' = 0$, which can be done with the help of Bessel functions. The result is $p_{cr}^0 = 7.837EI/l^3$. This means that the error in Equation 5.5.4 is 5.9 percent.

A better result may be obtained from the Timoshenko quotient, which can be calculated since the column is statically determinate. The relation analogous to Equation 5.4.1 reads

$$\Phi = \int_0^l \frac{M^2}{2EI}\,dx - \int_0^l p(x)u(x)\,dx = \int_0^l \frac{p_0^2}{2EI}\left(\frac{M}{p_0}\right)^2 dx - \int p_0 f(x)\frac{w'^2}{2}\,dx \qquad (5.5.5)$$

The expression of the Timoshenko quotient becomes

$$p_{0_T} = \frac{\int_0^l f(x)(w'^2/2)\,dx}{\int_0^l (1/2EI)(M/p_0)^2\,dx} \qquad (5.5.6)$$

For the previously considered case of the cantilever in Figure 5.13a we have $M(x) = \int_0^x p_0[w(\xi) - w(x)]\, d\xi$ and substituting the same approximate buckling shape as before, we obtain

$$p_{0_T} = \frac{7.888EI}{l^3} \tag{5.5.7}$$

which differs from the exact solution only by 0.65 percent and is much better than the result from the Rayleigh quotient, as expected.

Better approximations may be obtained by assuming the moment distribution to be parabolic, $M = M_0 x^2/l^2$, and by obtaining the corresponding deflection distribution by integration with the use of the boundary conditions $w = 0$ and $w' = 0$ for $x = l$. Substitution of the solution $w = (M_0/12EI)(x^4/l^2 - 4lx + 3l^2)$ into Equation 5.5.3 yields $p_R = 8EI/l^3$.

The approximate energy methods are also useful for estimating the critical loads of columns with variable cross sections (see Probs. 5.3.5 and 5.4.1). For the case of a beam of variable cross section loaded by its own weight and axially supported at its base (Fig. 5.13b), the expressions for p_{0_R} and p_{0_T} in Equations 5.5.3 and 5.5.6 (if they exist) remain valid (for various boundary condtions) provided that the origin of coordinate x is placed at the top (see Prob. 5.5.2).

The upper-bound approximations are not limited to simple columns. They can be obtained for any structure for which the buckling shape can be reasonably estimated either on the basis of similar known solutions or on the basis of experimental observations. As an example, consider the column with an overhang, loaded axially as shown in Figure 5.13c. Since this beam is statically determinate, the bending moment due to load P can be calculated for every point of the beam. To do this for points between the supports, it is, of course, necessary to first calculate the reactions (see Fig. 5.13c). One gets (1) for $0 < x < l$, $M = P(w_0 x/l - w)$ and (2) for $l < x < 2l$, $M = P(w_0 - w)$. So the equation which gives the Timoshenko quotient (Eq. 5.4.1) becomes

$$P_T = \frac{\displaystyle\int_0^{2l} \tfrac{1}{2} w'^2\, dx}{\displaystyle\int_0^{2l} (M/P)^2 (1/2EI)\, dx} \tag{5.5.8}$$

Assuming $w = w_0(x/2l^2)(x - l)$ one gets from Equation 5.3.7 $P_R = 2.57R/l^2$, and from Equation 5.5.8 $P_T = 1.83R/l^2$. Again one can see that the Timoshenko quotient gives a far better result. (The exact result is $P_{\text{cr}_1} = 1.789R/l^2$; see Prob. 2.2.11.)

Frames

Consider now the two-bar frame in Figure 5.14a. Assume distributions of deflections given by $w_1(x) = \theta_0(h/\pi)\sin(\pi x/h)$ for the vertical bar, and by $w_2(x) = \theta_0(x/l)(x - l)$ for the horizontal bar. One can verify that they satisfy compatibility at the joint, the rotation of which is denoted as θ_0. The Rayleigh

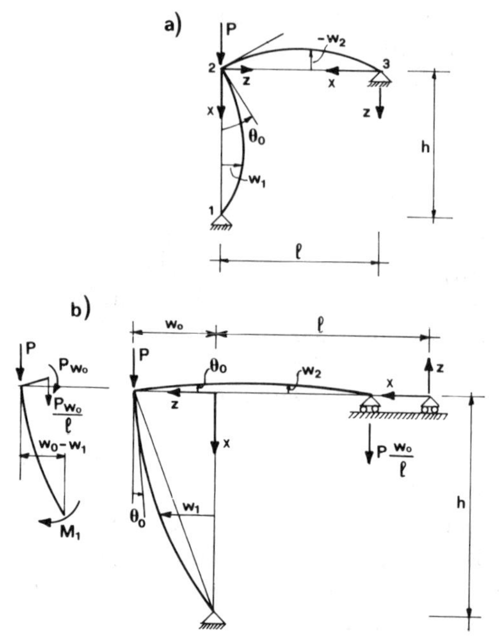

Figure 5.14 Two-bar frame with (a) pin support and (b) sliding pin support.

quotient

$$P_R = \frac{\displaystyle\int_0^h \tfrac{1}{2}EIw_1''^2\,dx + \int_0^l \tfrac{1}{2}EIw_2''^2\,dx}{\displaystyle\int_0^h \tfrac{1}{2}w_1'^2\,dx} \tag{5.5.9}$$

gives $P_R = (2EI/h)(\pi^2/2h + 4/l)$. For $h = l$ one gets $P_R = 1.81\pi^2 EI/l^2$. The exact value is $P_{\text{cr}_1} = 1.406\pi^2 EI/l^2$ (see Prob. 2.2.15).

Consider now the two-bar frame in Figure 5.14b, and let θ_0 and w_0 be the rotation and displacement of the joint of the two bars under an assumed deflection distribution given by $w_1 = w_0(1 - x/h) + [(w_0 - h\theta_0)/\pi]\sin(\pi x/h)$ for the vertical bar, and by $w_2 = \theta_0(x/l)(l - x)$ for the horizontal bar. One can verify that these distributions satisfy slope compatibility at the joint of the two bars. The Rayleigh quotient P_R is expressed again by Equation 5.5.9, but in this case it is a function of two variables, θ_0 and w_0. One may now divide both the numerator and the denominator by θ_0^2, obtaining P_R as a function of the ratio $X = w_0/\theta_0$. For any chosen ratio X, P_R yields an upper bound for P_{cr_1}. To get the best upper bound, P_R must be minimized with respect to ratio X. The necessary minimum condition $dP/dX = 0$ yields a quadratic equation for X. Its proper root, corresponding to the minimum $\bar{X}$, may then be substituted into the expression for P_R, giving $P_R = 0.201\pi^2 EI/l^2$. (The exact result for $h = l$ is $P_{\text{cr}_1} = 0.144\pi^2 EI/l^2$; see Prob. 2.2.15.) It may be noted that minimization of P_R with respect to X is equivalent to the Rayleigh-Ritz method for problems with more than one unknown (see Sec. 5.6).

In the foregoing example, however, it is also possible to calculate the

Timoshenko quotient, since the structure is statically determinate. One must first calculate the bending moment, which is given by $M_1 = -Pw_0 + P(w_0 - w_1) = -Pw_1$ for the vertical bar and by $M_2 = -Pw_0 x/l$ for the horizontal bar. The Timoshenko quotient is expressed by

$$P_T = \frac{\int_0^h \frac{1}{2} w_1'^2 \, dx}{\int_0^h (1/2EI)(M_1/P)^2 \, dx + \int_0^l (1/2EI)(M_2/P)^2 \, dx} \tag{5.5.10}$$

Again one finds that P_T needs to be minimized with respect to the ratio $X = w_0/\theta_0$. This leads to a quadratic equation for X and its proper root gives $P_T = 0.145\pi^2 EI/l^2$—a very close bound.

Elastically Supported Beams

Consider now a column laterally supported at midlength by a spring of stiffness C, as shown in Figure 5.15a. Let w_0 be the displacement at midlength. The symmetry and compatibility conditions at $x = 0$ may be satisfied by choosing $w = w_0 + a(x/l)^2 + b(x/l)^4$. Determining constants a and b so that both the kinematic and static boundary conditions are satisfied at $x = l/2$, that is, $w(l/2) = 0$, $w''(l/2) = 0$, one gets $w = w_0[1 - \frac{24}{5}(x/l)^2 + \frac{16}{5}(x/l)^4]$. The Rayleigh quotient is given by

$$P_R = \frac{\int_0^l \frac{1}{2} EI w''^2 \, dx + \frac{1}{2} C w_0^2}{\int_0^l \frac{1}{2} w'^2 \, dx} \tag{5.5.11}$$

Performing the calculations for $C = 20EI/l^3$, one obtains $P_R = 13.903EI/l^2 = 1.4087P_E$, which is very close to the exact value, $P_{cr_1} = 1.4076P_E$ (see Prob. 2.2.12). Again we see that if the assumed displacement distribution satisfies the static boundary conditions, the solution is very accurate.

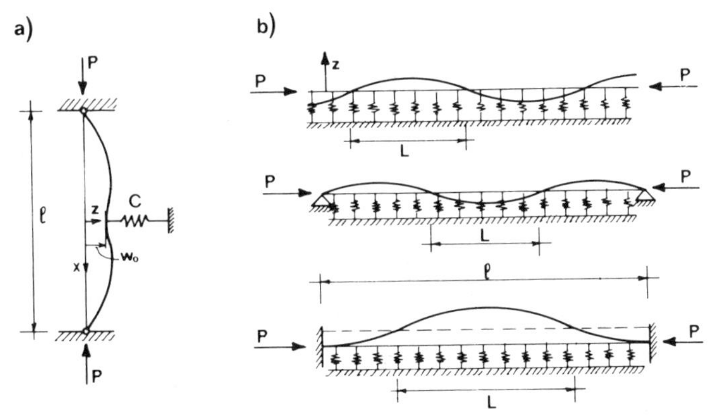

Figure 5.15 (a) Elastically supported beam; (b) beams on elastic foundations.

Since the spring reaction at the midlength support (which makes the structure redundant) may be expressed as a function of w_0, the bending moment becomes a known function of w_0 (for the assumed displacement shape) and consequently the Timoshenko quotient may be calculated from the equation

$$\int_0^l \frac{1}{2EI}\left[-P_T w + \frac{Cw_0}{2}\left(\frac{l}{2} - x\right)\right]^2 dx + \frac{1}{2} Cw_0^2 - P_T \int \frac{1}{2} w'^2 \, dx = 0 \quad (5.5.12)$$

Performing the calculation for $C = 20EI/l^3$, one obtains $P_T = 1.4078P_E$. Compared to the exact value $1.4076P_E$, this seems too good to be true but it is true.

Consider now an infinite beam on an elastic foundation, of modulus c, and assume that $w = w_0 \sin(\pi x/2L)$ (Fig. 5.15b). The expression for the Rayleigh quotient becomes

$$P_R = \frac{\int_0^L \frac{1}{2}EIw''^2 \, dx + \int_0^L \frac{1}{2}cw^2 \, dx}{\int_0^L \frac{1}{2}w'^2 \, dx} = \frac{EI\pi^2}{4L^2} + c\frac{4L^2}{\pi^2} \quad (5.5.13)$$

Minimizing with respect to L, we get $P_{R_{\min}} = 2\sqrt{EIc}$, which is the same as in Equation 5.2.7.

Note that the sinusoidal $w(x)$ can also be used for finite-length beams with hinged ends. Then it would be necessary to minimize P_R only with respect to the discrete values of L which are compatible with the hinged ends. For a beam of length l, $L = l/n$, $n = 1, 2, \ldots$ (Fig. 5.15b).

High Arches

Consider now high arches, that is, arches in which the shortening of the neutral axis has a negligible effect, which we studied in Section 2.8. Let us first treat the case of circular arches under a uniformly distributed radial load p whose direction remains constant during deflection (dead pressure). For this case we derived the differential equation (Eq. 2.8.11). This sixth-order equation may be rewritten as $\underline{M}u - k^2\underline{N}u = 0$, with $\underline{M} = d^6/d\phi^6 + 2d^4/d\phi^4 + d^2/d\phi^2$, $\underline{N} = -(d^4/d\phi^4 + 2d^2/d\phi^2 + 1)$, $k^2 = pR^3/EI$. Now, forming the Rayleigh quotient according to Equation 5.3.13, one gets

$$p_{\text{cr}}R < p_R R = \frac{EI}{R^2}\frac{\int_0^L u(u^{\text{IV}} + 2u'' + u)'' \, d\phi}{-\int_0^L u(u^{\text{IV}} + 2u'' + u) \, d\phi} = \frac{EI}{R^2}\frac{\int_0^L [(d^2w/ds^2) + (w/R^2)]^2 \, ds}{\int_0^L (1/R^2)[(dw/ds) + (u/R)]^2 \, ds} \quad (5.5.14)$$

where the primes denote differentiation with regard to the angular coordinate ϕ. The last expression in Equation 5.5.14 has been obtained by integration by parts, use of the boundary conditions for fixed and hinged ends, and consideration of the inextensibility condition $u' = w$. Equation 5.5.14 can be rewritten in the form

$$\frac{1}{2}EI \int_0^L \left(\frac{d^2w}{ds^2} + \frac{w}{R^2}\right)^2 ds + (-pR)\int_0^L \frac{1}{2}\left(\frac{dw}{ds} + \frac{u}{R}\right)^2 ds = 0 \quad (5.5.15)$$

Here the first term represents the strain energy due to bending, since $d^2w/ds^2 + w/R^2$ is the curvature change; see Equation 2.8.2. The second term represents the strain energy due to axial extension, since $-pR$ is the axial (normal) force, positive for tension, and $dw/ds + u/R$ represents the rotation (which is analogous to the slope dw/ds for a straight beam, for which we derived the expression of the potential energy in Eq. 4.3.8). We see in this way that Equation 5.5.14 could be obtained directly from the expression for the variation of potential energy.

We can use this energy approach to find the expression of the Rayleigh quotient for the case that the pressure remains normal to the deflected arch (i.e., for the case of normal pressure or hydrostatic pressure of a fluid). For this kind of loading, which is of follower type but is conservative, we need to subtract from the second variation of the strain energy the work $\Delta W = \int_0^L \frac{1}{2} p_t u \, ds$ done by the tangential component p_t of the load (Fig. 2.43c) during deflection. From Section 2.8 we have $p_t = -p\theta$, with $\theta = (u + w')/R =$ rotation of the arch cross section. Then, by integration by parts, use of the boundary conditions for hinged or fixed ends, and consideration of the inextensibility condition, one obtains

$$\Delta W = \frac{1}{2} pR \int_0^L \left(\frac{w^2}{R^2} - \frac{u^2}{R^2} \right) ds \tag{5.5.16}$$

Subtracting this term from the left-hand side of Equation 5.5.15 and integrating by parts, one obtains

$$p_{\mathrm{cr}} R < p_R^n R = \frac{EI}{R^2} \frac{\displaystyle\int_0^L [(d^2w/ds^2) + (w/R^2)]^2 \, ds}{\displaystyle\int_0^L (1/R^2)[(dw/ds)^2 - (w/R)^2] \, ds} \tag{5.5.17}$$

As an application, consider a semicircular arch with hinged ends, which we solved exactly in Section 2.8. Assume $w = w_0 \sin 2\phi$ for the radial displacement (this displacement shape represents an antisymmetric mode and satisfies the boundary conditions $w = 0$ for $\phi = \pm\pi/2$). The inextensibility condition $u' = w$ and the boundary conditions $u = 0$ for $\phi = \pm\pi/2$ suggest the assumption $u = -w_0(1 + \cos 2\phi)/2$. Substituting this into Equation 5.5.14, we find

$$p_R R = \frac{36EI}{11R^2} = \frac{\pi^2 EI}{[\beta(\pi R/2)]^2} \qquad \text{with } \beta = 1.1055 \tag{5.5.18}$$

This value is very close to the exact solution, for which we found $\beta = 1.1078$ (Sec. 2.8, below Eq. 2.8.17). The upper-bound character of the Rayleigh quotient is, of course, verified, since the lower the effective length, the higher is the critical load.

For the case of normal (hydrostatic) pressure, we substitute the same curve $w(\phi)$ as assumed previously for dead pressure into Equation 5.5.17. We find

$$p_R^n R = 3\frac{EI}{R^2} = \frac{\pi^2 EI}{[\beta(\pi R/2)]^2} \qquad \text{with } \beta = 1.1547 \tag{5.5.19}$$

which is the same value as calculated exactly from the differential equation (see Prob. 2.8.6). This can happen only if the trial function is the exact solution, and this is in fact what happened here (see Prob. 2.8.6).

It may now be interesting to construct the Rayleigh quotient starting from the approximate second-order differential equation given in Section 2.8, Equation 2.8.5. This equation can be written in the form $\underline{M}^*w - k^{*2}\underline{N}^*w = 0$, with $\underline{M}^* = -(d^2/ds^2 + 1/R^2)$, $\underline{N}^* = 1$, and $k^{*2} = pR/EI$. The Rayleigh quotient becomes

$$p_R^a R = \frac{EI}{R^2} \frac{\int_0^L [-w(d^2w/ds^2) - (w^2/R^2)]\, ds}{\int_0^L (1/R^2)w^2\, ds} = \frac{EI}{R^2} \frac{\int_0^L [(dw/ds)^2 - (w/R)^2]\, ds}{\int_0^L (1/R^2)w^2\, ds} \tag{5.5.20}$$

Substituting the same assumed function $w(\phi)$ as before, one finds

$$p_R^a R = 3\frac{EI}{R^2} = \frac{\pi^2 EI}{[\beta(\pi R/2)]^2} \qquad \text{with } \beta = 1.1547 \tag{5.5.21}$$

This is the same solution as that which we found for p_R^n. Again, this is not surprising because the assumed solution satisfies exactly Equation 2.8.5 if $pR = 3EI/R^2$, as can be verified. Since, however, Equation 2.8.5 is only approximate, the value given by Equation 5.5.21 cannot be assumed as an upper bound for the critical load.

Problems

5.5.1 For the free-standing column of constant cross section in Figure 5.13a determine the Rayleigh and Timoshenko bounds for the self-weight assuming for the deflection the expression $w = A(1 - x^2/l^2)$. How do the results compare with those we obtained before for the same example but with a different assumption for w?

5.5.2 For the hinged-sliding column in Figure 5.13b, determine the Rayleigh and Timoshenko quotients, assuming for the self-weight the distribution $p = p_0(1 + x/l)$, and for the deflection the expressions (a) $w = A \sin(\pi x/l)$, (b) $w = Ax^4 + Bx^3 + Cx^2 + Dx + E$ in which A, B, C, D, E can be determined (through the kinematic and boundary conditions) as functions of only one unknown parameter. Does a horizontal shear force V develop during buckling?

5.5.3 Do the same as Problem 5.5.2, but consider a fixed end at the bottom of the column. Find suitable expressions for the buckling shape. Can you calculate the Timoshenko quotient?

5.5.4 For the column with a lateral spring at midlength, as shown in Figure 5.15a, determine the Rayleigh and Timoshenko quotients, assuming either (a) $w = A \sin(\pi x/l)$ or (b) $w = Ax(l - x)$. Using both these approximations and the one used in our previous example, determine the effect of the spring constant, assuming $Cl^3/EI = 0.01$, 0.1, 0.5, 1, 2, 10, 100.

5.5.5 Use the Rayleigh quotient and, if possible, also the Timoshenko quotient to estimate P_{cr_1} of the columns in Figure 5.16a–k. Compare the results of the two methods.

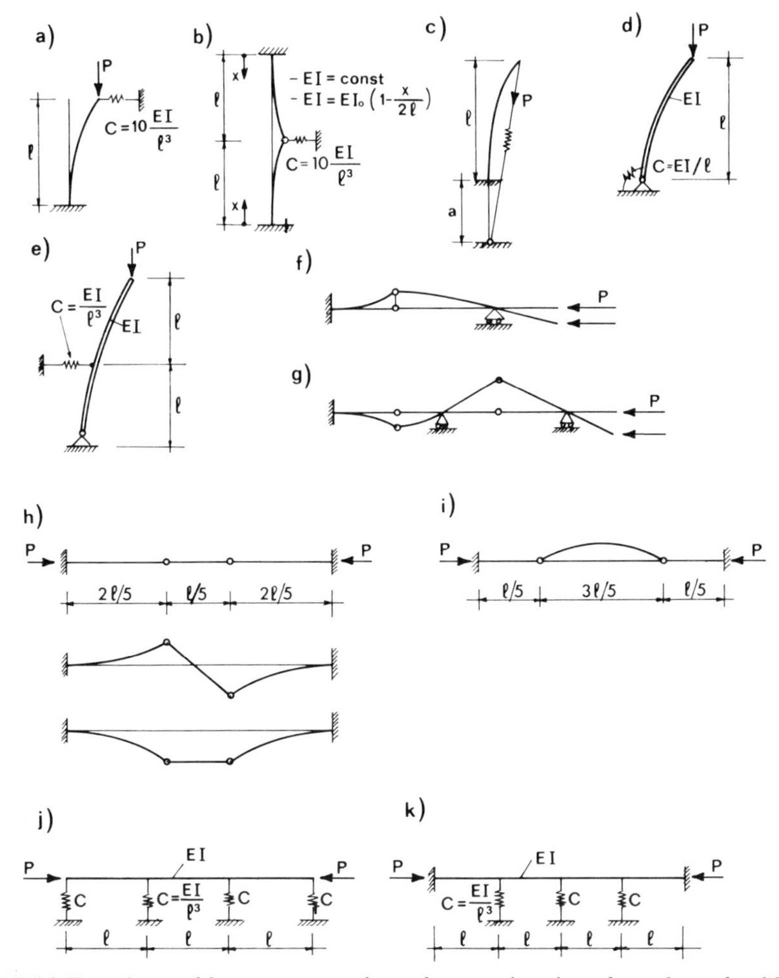

Figure 5.16 Exercise problems on upper-bound approximations for column buckling.

5.5.6 For the columns with L-shaped base in Figure 5.17a–d, calculate the Raleigh and Timoshenko quotients.

5.5.7 For the frames in Figure 5.17e–o, find an estimate of P_{cr_1} using Rayleigh, and for statically determinate frames, also Timoshenko quotients.

5.5.8 For the frames in Figure 5.17p, q, estimate P_{cr_1} treating the springs as an elastic foundation of foundation modulus $c = 3EI/l^4$.

5.5.9 Using Equations 5.5.14 and 5.5.17 for the Rayleigh quotient, solve the critical values of dead and normal pressures of a clamped circular arch. Compare the result with the exact solution given in Section 2.8 and with the value obtained from Equation 5.5.20. (Note that the assumed solution must now satisfy the boundary conditions $w = 0$ and $w' = 0$ at $\phi = \pm\pi/2$.)

5.5.10 Solve the approximate critical distributed load p for the circular three-hinge arch of constant cross section shown in Figure 5.18a. Also do the same for the arches in Figure 5.18b, c, d, for which $I = I_0 - I_1 s^2$.

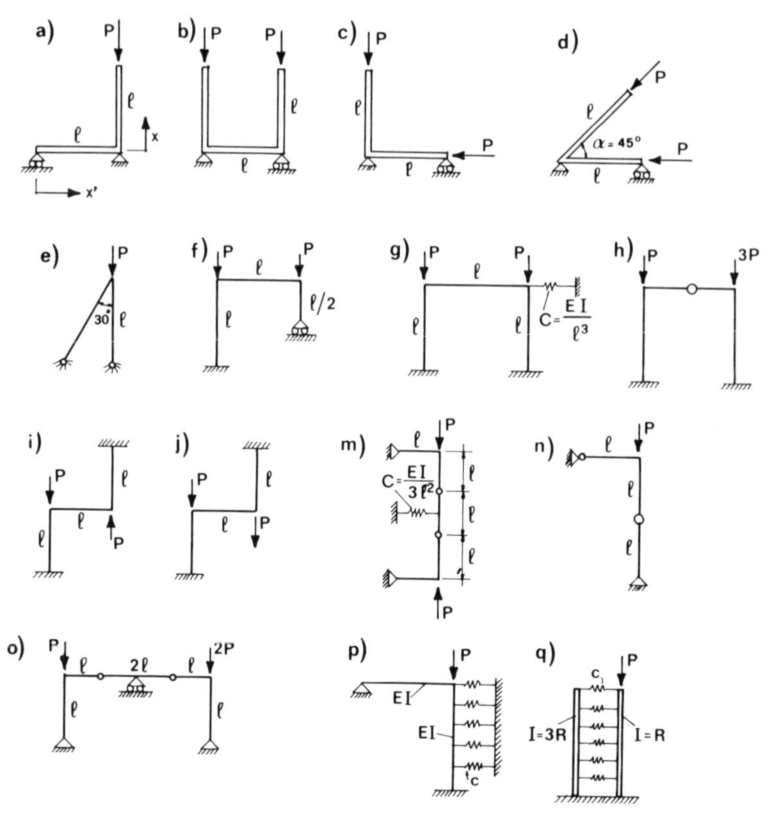

Figure 5.17 Exercise problems on upper-bound approximations for frame buckling.

5.5.11 Solve the approximate critical load for lateral (spatial) buckling of the pipe arch in Figure 5.18e, with lateral bending rigidity EI_y and torsional rigidity GI_t. Assume a constant circular thin cross section of thickness h and radius r. The arch is circular, of radius R, and has a slope α at its base. Solve it both by Rayleigh and Timoshenko quotients and compare the solutions (a) for distributed vertical load p, and (b) for concentrated load P. *Hint:* Assume lateral displacements $v = Af(y)$ and calculate strain energy $U = 2\int_0^l [(M_y^2/2EI_y) + (M_t^2/2GI_t)]\,ds$ where M_t = torque and M_y = lateral bending moment.

5.5.12 Solve the same as Problem 5.5.11, but for lateral buckling of the portal frame in Figure 5.18f.

5.5.13 Solve P_{cr} for the lateral buckling of the frames of pipe cross section in Figure 5.18g, h.

5.6 RAYLEIGH-RITZ VARIATIONAL METHOD

In many problems it is not possible to guess a function that provides a close enough approximation to the exact deflection curve. Then, of course, the upper-bound approximation obtained from the Rayleigh quotient is not very

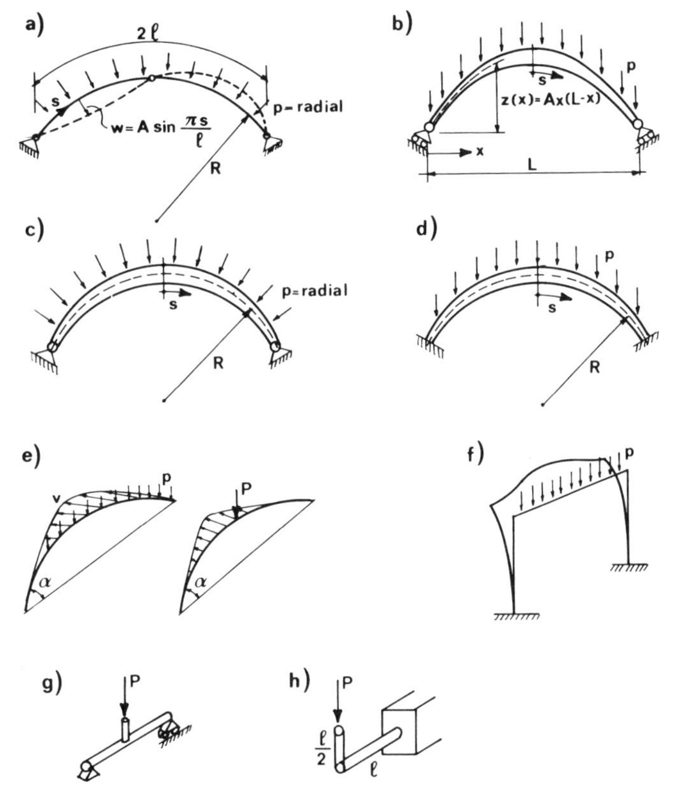

Figure 5.18 Exercise problems on upper-bound approximations for in-plane and lateral buckling of arches and frames.

close. An improvement in accuracy can be achieved by considering the deflection curve to be a linear combination of several assumed functions. The solution is then obtained by minimizing either the Rayleigh quotient, or more directly the potential energy itself, with respect to the unknown coefficients q_k of this linear combination. An approach of this kind is known as the Ritz method or the Rayleigh-Ritz method (it was presented in a rigorous general form by Ritz, a Swiss mathematician at the beginning of this century).

We begin with the example of a column of nonuniform bending rigidity $EI(x)$. Although for certain functions $I(x)$ it is possible to get an exact solution of the critical load in terms of some special functions [e.g., Bessel functions if $I(x)$ is linear], it is more effective to solve this problem by an approximate variational method.

Let us assume that $I(x) =$ given function of x, and let the column be subjected to a small lateral disturbing load $p = p_0 \cos(\pi x/l)$ (see Fig. 5.19a). It is obvious that the tapering of the column toward the ends will cause a decrease of curvature at midspan and an increase of curvature near the ends. Such changes of curvature can be described by superimposing on the function $\cos(\pi x/l)$ the function $-\cos(3\pi x/l)$. These functions are chosen so as to satisfy both the kinematic and the static boundary conditions at the column ends (the origin of the coordinates is

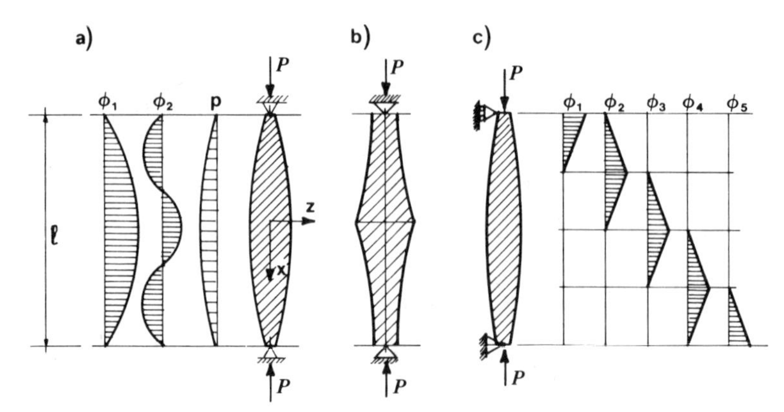

Figure 5.19 (a, b) Trial functions for columns of variable cross section; (c) piece-wise linear "hill" functions.

placed at midspan). So we introduce the linear combination

$$w = q_1\phi_1(x) + q_2\phi_2(x) \qquad \phi_1(x) = \cos\frac{\pi x}{l} \qquad \phi_2(x) = \cos\frac{3\pi x}{l} \tag{5.6.1}$$

where q_1 and q_2 are unknown coefficients (generalized displacements or kinematic variables).

Substituting this into the potential-energy expression we obtain

$$\begin{aligned}
\Pi &= \frac{1}{2}\int_{-l/2}^{l/2} (EIw''^2 - Pw'^2)\, dx - \int_{-l/2}^{l/2} pw\, dx \\
&= \frac{1}{2}\int_{-l/2}^{l/2} [EI(q_1\phi_1'' + q_2\phi_2'')^2 - P(q_1\phi_1' + q_2\phi_2')^2]\, dx \\
&\quad - \int_{-l/2}^{l/2} p_0 \cos\frac{\pi x}{l}(q_1\phi_1 + q_2\phi_2)\, dx \\
&= \tfrac{1}{2}(A_{11}q_1^2 + 2A_{12}q_1q_2 + A_{22}q_2^2) - \tfrac{1}{2}P(B_{11}q_1^2 \\
&\quad + 2B_{12}q_1q_2 + B_{22}q_2^2) - C_1q_1 - C_2q_2
\end{aligned} \tag{5.6.2}$$

in which we introduced the notations:

$$\begin{aligned}
A_{11} &= \int_{-l/2}^{l/2} EI(x)[\phi_1''(x)]^2\, dx = \int_{-l/2}^{l/2} EI\left(\frac{\pi^4}{l^4}\right)\left(\cos\frac{\pi x}{l}\right)^2 dx \\
A_{12} &= \int_{-l/2}^{l/2} EI(x)\phi_1''(x)\phi_2''(x)\, dx = \int_{-l/2}^{l/2} EI\left(\frac{9\pi^4}{l^4}\right)\left(\cos\frac{\pi x}{l}\right)\left(\cos\frac{3\pi x}{l}\right) dx \\
A_{22} &= \int_{-l/2}^{l/2} EI(x)[\phi_2''(x)]^2\, dx = \int_{-l/2}^{l/2} EI\left(\frac{81\pi^4}{l^4}\right)\left(\cos\frac{3\pi x}{l}\right)^2 dx \\
B_{11} &= \int_{-l/2}^{l/2} [\phi_1'(x)]^2\, dx = \int_{-l/2}^{l/2} \frac{\pi^2}{l^2}\left(\sin\frac{\pi x}{l}\right)^2 dx
\end{aligned} \tag{5.6.3}$$

$$B_{12} = \int_{-l/2}^{l/2} \phi_1'(x)\phi_2'(x)\,dx = \int_{-l/2}^{l/2} \frac{3\pi^2}{l^2}\left(\sin\frac{\pi x}{l}\right)\left(\sin\frac{3\pi x}{l}\right)dx$$

$$B_{22} = \int_{-l/2}^{l/2} [\phi_2']^2\,dx = \int_{-l/2}^{l/2} \frac{9\pi^2}{l^2}\left(\sin\frac{3\pi x}{l}\right)^2 dx$$

$$C_1 = \int_{-l/2}^{l/2} p_0\left(\cos\frac{\pi x}{l}\right)\phi_1(x)\,dx = \int_{-l/2}^{l/2} p_0\left(\cos\frac{\pi x}{l}\right)^2 dx$$

$$C_2 = \int_{-l/2}^{l/2} p_0\left(\cos\frac{\pi x}{l}\right)\phi_2(x)\,dx = \int_{-l/2}^{l/2} p_0\left(\cos\frac{\pi x}{l}\right)\left(\cos\frac{3\pi x}{l}\right)dx$$

All these coefficients are obviously positive. The condition of equilibrium, as a necessary condition of min Π, requires that

$$\begin{Bmatrix} \partial\Pi/\partial q_1 \\ \partial\Pi/\partial q_2 \end{Bmatrix} = \begin{bmatrix} A_{11} - PB_{11} & A_{12} - PB_{12} \\ A_{12} - PB_{12} & A_{22} - PB_{22} \end{bmatrix}\begin{Bmatrix} q_1 \\ q_2 \end{Bmatrix} = \begin{Bmatrix} C_1 \\ C_2 \end{Bmatrix} = 0 \tag{5.6.4}$$

Two kinds of problems may now be distinguished. If the lateral disturbing load p is zero, then $C_1 = C_2 = 0$, in which case Equation 5.6.4 represents a linear matrix eigenvalue problem in a nonstandard, that is, generalized, form. The solution is possible only if the matrix equation $(\mathbf{A} - P\mathbf{B})\mathbf{q} = \mathbf{0}$ or $(\mathbf{B}^{-1}\mathbf{A} - P\mathbf{I})\mathbf{q} = \mathbf{0}$ has a nonzero solution $\mathbf{q}$, that is, if

$$\text{Det}\,(\mathbf{A} - P\mathbf{B}) = 0 \qquad \text{or} \qquad \text{Det}\,(\mathbf{B}^{-1}\mathbf{A} - P\mathbf{I}) = 0 \tag{5.6.5}$$

in which $\mathbf{A}$ and $\mathbf{B}$ are 2×2 square matrices formed from A_{11}, $A_{12}, \ldots, B_{11}, \ldots, B_{12}$, and $\mathbf{I}$ is the identity matrix. The latter equation reduces the problem to a standard eigenvalue problem. Assuming $I(x) = I_0[1 - k(x/l)^2]$ (Fig. 5.19a), substituting into Equations 5.6.3, and expressing the critical load as $P_{\text{cr}_1} = c\pi^2 EI_0/l^2$, one gets for $k = 1$ the value $c = 0.9655$ and for $k = 2$ the value $c = 0.9268$. Assuming $I(x) = I_0/(1 + k\,|x/l|)$ (Fig. 5.19b) one gets for $k = 1$ the value $c = 0.8699$ and for $k = 2$ the value $c = 0.7689$. (Note that for $k = 2$ in both cases $I = 0.5I_0$ at the ends, but in the second case I decreases more rapidly away from the midspan.)

In the second kind of problem, p, C_1, and C_2 are nonzero. The solution then tends to infinity as the value of P approaches a critical value, which solves Equations 5.6.5. In either case, the salient property of the solution is the value of the critical load P, which follows from Equations 5.6.5. The lowest eigenvalue obtained from Equations 5.6.5 is an approximation for the first critical load P_{cr_1}. It is always an upper-bound approximation. The approximation is generally closer than that obtained with only one function $\phi_1(x) = \cos(\pi x/l)$. In that case the result of the aforementioned procedure would be the same as that obtained from the Rayleigh quotient.

Further improvements of accuracy can be obtained by including more functions in the linear combination, for example, $\phi_3(x) = \cos(5\pi x/l)$, $\phi_4(x) = \cos(7\pi x/l)$, etc. So generally one may use

$$w(x) = \sum_{k=1}^{N} q_k \phi_k(x) \tag{5.6.6}$$

This may be substituted into the potential-energy expression (Eq. 5.6.2), which may be written in the form $\Pi = U - P\bar{W} - W_p$ in which W_p denotes the work of disturbing loads, such as the lateral load $p(x)$ on a column. Then one may impose the conditions $\partial\Pi/\partial q_k = 0$ for all $k = 1, 2, \ldots, N$, which represent the equilibrium conditions if disturbing loads are present. The conditions $\partial\Pi/\partial q_k = 0$, however, should not be regarded as equilibrium conditions if such loads are absent, that is, in bifurcation problems (problems with adjacent equilibrium); rather, they represent the Trefftz conditions of critical state (Sec. 4.2). In this manner one obtains the equation

$$\frac{\partial U}{\partial q_k} - P\frac{\partial \bar{W}}{\partial q_k} - \frac{\partial W_p}{\partial q_k} = 0 \qquad (k = 1, 2, \ldots, N) \tag{5.6.7}$$

This equation yields again, for the critical load (at $W_p = 0$), the matrix equations 5.6.5. The coefficients of the matrices generally are similar to Equations 5.6.3:

$$\begin{aligned} A_{km} &= (EI\phi''_k, \phi''_m) = \int_0^l EI(x)\phi''_k(x)\phi''_m(x)\,dx \\ B_{km} &= (P\phi'_k, \phi'_m) = \int_0^l P\phi'_k(x)\phi'_m(x)\,dx \end{aligned} \tag{5.6.8}$$

in which the parenthesis notation for scalar products of two functions is used.

The foregoing method obviously can be generalized to the minimization of any functional (of one or more functions), as well as to two or three dimensions (see, e.g., plate buckling, Eq. 7.3.14).

The Ritz method is not restricted to quadratic functionals associated with linear problems. It may be applied to all nonlinear problems for which a minimizing functional Π exists, as we shall illustrate in Section 5.9 for the problem of large deflections of columns. The functional Π, of course, has a minimum only for $P < P_{cr_1}$, in which case the matrix $\mathbf{A} - P\mathbf{B}$ or $\mathbf{B}^{-1}\mathbf{A} - P\mathbf{I}$ in Equations 5.6.5 is positive definite. For $P > P_{cr_1}$, Equation 5.6.7 still applies but represents merely stationarity conditions since Π does not have a minimum.

Underlying the Rayleigh-Ritz method is the following *theorem of Ritz*:

Theorem 5.6.1 The limit of the approximations $w_N(x) = \sum_{k=1}^{N} q_k\phi_k(x)$ for $N \to \infty$ is the exact solution $w(x)$ if the system of the chosen functions $\phi_k(x)$ satisfies the following conditions:

1. Functions $\phi_k(x)$ are linearly independent (i.e., none of these functions can be expressed as a linear combination of the others).
2. Functions $\phi_k(x)$ form a complete system of functions.
3. Functions $\phi_k(x)$ satisfy the essential (kinematic) boundary conditions.

The chosen basis functions $\phi_k(x)$ do not need to satisfy the natural (static) boundary conditions; however, if they do, the approximations are better, usually substantially better. A sequence of functions ϕ_k $(k = 1, 2, 3, \ldots)$ is complete with respect to operator $\underset{\sim}{A}$ if any function $w(x)$ in a linear set M (more precisely, in a linear manifold) can be approximated by a linear combination $\sum_{k=1}^{N} a_k\phi_k$ so that for any given positive number ε $(\varepsilon > 0)$ one can achieve for sufficiently large N that

$$\left(\underset{\sim}{A}\left(w - \sum_{k=1}^{N} a_k\phi_k\right), w - \sum_{k=1}^{N} a_k\phi_k\right) < \varepsilon \tag{5.6.9}$$

An equivalent condition is: If $(\underset{\sim}{A}w, \phi_k) = 0$ for all k (i.e., up to ∞), and if ϕ_k is a complete system, then $w = 0$ everywhere. A linear set (linear manifold) is a set of functions such that if the set contains functions $w_1, w_2, \ldots, w_n$, then it also contains any linear combination $w = c_1 w_1 + c_2 w_2 + \cdots + c_n w_n$.

The theorem of Ritz applies only to the space L_2 of real square-integrable functions w on the given interval I (or region Ω). The approximations converge to the exact solution in the mean. More precisely, the convergence is such that $\|w_k - w_{\text{exact}}\| \leq [\Pi(w_k) - \Pi_{\text{exact}}]/k$ provided that operators $\mathbf{A}$ and $\mathbf{B}$ are positive definite in the sense that $(\mathbf{A}w, w) \geq k\,\|w\|$ where k is a positive number and $\|\cdot\cdot\cdot\|$ denotes a quadratic norm. If the problem is linear (i.e., if the functional is quadratic), the convergence is quadratic and is always from above. This signifies that the first critical load obtained from the Rayleigh-Ritz method is guaranteed to be larger than the exact critical load. We have an upper-bound approximation.

The Rayleigh-Ritz method does not require that functions $\phi_k(x)$ form an orthogonal system of functions [an orthogonal system is such that the scalar products of any two different functions from the system is zero, i.e., $(\phi_k, \phi_m) = 0$ for $k \neq m$]. However, for accuracy it is generally advantageous to choose an orthogonal system of functions. It is easy to show that if functions $\phi_k(x)$ represent the exact eigenmodes, the Ritz method yields the exact solution (eigenmodes are always orthogonal). In that case it is usually computationally convenient if the orthogonal system of functions is normalized, which is done by introducing in $\phi_k(x)$ multipliers such that $(\phi_k, \phi_k) = 1$.

For a column of constant cross section, trigonometric functions represent the exact eigenmodes and constitute the best complete orthogonal system of functions to choose. The solution of a linear problem for a column represents a Fourier trigonometric series. For a column with $I(x) = \text{const.} \cdot \text{x}$, the exact eigenmodes are the Bessel functions. For some other types of variation of $I(x)$, special functions that represent exact eigenmodes exist. These special functions represent systems of functions that are orthogonal with a weight (weighting function), a generalization of the concept of orthogonal functions. For example, if $I(x) = \text{const} \cdot \text{x}$, the orthogonality condition is $(x\phi_k, \phi_m) = 0$ for $k \neq m$, rather than $(\phi_k, \phi_m) = 0$, where x represents the weight.

Polynomials of different degrees represent linearly independent functions, and particularly the functions $1, x, x^2, x^3, x^4, \ldots$ represent a complete system of linearly independent functions. These functions, however, are not orthogonal. They can be orthogonalized on the basis of the conditions $(\phi_k, \phi_m) = 0$ for $k \neq m$, and the result of the orthogonalization are the Legendre polynomials: 1, x, $\frac{3}{2}x^2 - \frac{1}{2}$, $\frac{5}{3}x^3 - \frac{3}{2}x$, $\frac{35}{8}x^4 - \frac{15}{4}x^2 + \frac{3}{8}, \ldots$, which are orthogonal in the interval $(-1, 1)$.

A very important, nonclassical special case of the Rayleigh-Ritz method is the finite element method. This method is equivalent to using for functions $\phi_k(x)$, for example, the linear "hill" functions sketched in Figure 5.19c. These piecewise linear functions are obviously linearly independent and form a complete system, although they are not orthogonal. In the finite element context, by contrast, their orthogonalization is not computationally advantageous. The properties of the Rayleigh method that we described, of course, apply also to the finite element method.

Thus far our formulation of the Ritz method was based on potential energy. It turns out, however, that minimization of the Rayleigh quotient P_R is equivalent to minimization of potential energy Π, for two reasons: (1) both Π and P_R possess a minimum for the stable equilibrium states (i.e., for $P < P_{cr_1}$), and (2) the condition $\delta P_R = 0$ is equivalent to the condition $\delta\Pi = 0$, which is easily proven (as we already showed in Equation 5.3.5) by the following transformations:

$$\delta P_R = \delta(U/\bar{W}) = \bar{W}^{-2}(\bar{W}\,\delta U - U\,\delta\bar{W})$$
$$= \bar{W}^{-1}(\delta U - P_R\,\delta\bar{W}) = \bar{W}^{-1}\,\delta\Pi \tag{5.6.10}$$

Hence the name Rayleigh-Ritz method.

In a certain sense, though, minimization of the Rayleigh quotient is more generally applicable than minimization of the potential energy (Collatz, 1963). This is because the Rayleigh quotient can be formulated even when, for some boundary-value problem, no potential energy exists. This is because the Rayleigh quotient can be expressed in terms of the differential operators $\underset{\sim}{M}$ and $\underset{\sim}{N}$ from the differential equation (as stated already in Eq. 5.3.13 for the case when $\underset{\sim}{M}$ and $\underset{\sim}{N}$ are linear). Formulation of the Rayleigh quotient from operators $\underset{\sim}{M}$ and $\underset{\sim}{N}$ is equivalent to a variant of the Rayleigh-Ritz method, called the Galerkin method, which we outline next.

Problems

5.6.1 For the hinged column loaded by self-weight, shown in Figure 5.20a, determine the upper-bound approximation of the critical distributed load. Use the Rayleigh-Ritz method, assuming $w = q_1 \sin(\pi x/l) + q_2 \sin(2\pi x/l)$. Compare the result with the Rayleigh quotient value obtained for $w = w_0 \sin(\pi x/l)$.

5.6.2 For the symmetric portal frame in Figure 5.20b, find the upper-bound approximation of the first critical load using the Rayleigh-Ritz method. (Impose antisymmetric deflection and express this deflection through parameters $q_1 = \theta$ and $q_2 = \Delta$.)

5.6.3 Do the same, but for the two-bar frame in Figure 5.20c.

5.6.4 For the free-standing column (Fig. 5.20d) of $I(x) = I_0 + I_1(1 - x^2/l^2)$ find an upper bound on the critical load using the Rayleigh-Ritz method. Assume $w = q_1[1 - \cos(\pi x/2l)] + q_2[1 - \cos(\pi x/l)]$.

5.6.5 For the hinged column (Fig. 5.20e) with $I(x) = I_0 + I_1(1 - x/l)$, find an upper bound on the critical load using the Rayleigh-Ritz method. Assume $w = q_1 \sin(\pi x/l) + q_2 \sin(2\pi x/l)$.

5.6.6 Using the same functions as in Problem 5.6.4, find the upper bound on the critical distributed load for the free-standing column of Figure 5.20f using the Rayleigh-Ritz method.

5.6.7 For the beam supported laterally by a spring at midspan (Fig. 5.20g), find an upper bound on the critical load using the Rayleigh-Ritz method. Assume $w = q_1 \sin(\pi x/l) + q_2 \sin(3\pi x/l)$. Compare the result with the results for Figure 5.15a and for Problem 5.5.4 with $C = EI/l^3$.

5.6.8 For the free-standing column supported on an elastic foundation of modulus c (Fig. 5.20h), find an upper bound on the critical load using the Rayleigh-Ritz method.

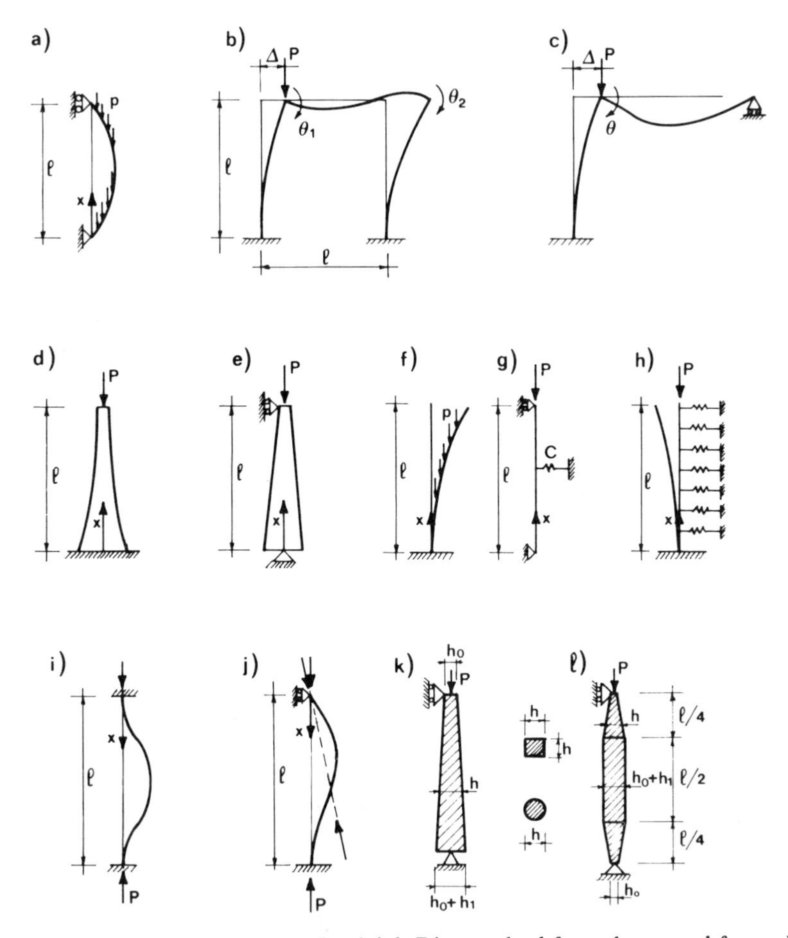

Figure 5.20 Exercise problems on Rayleigh-Ritz method for column and frame buckling.

5.6.9 Use the Rayleigh-Ritz method to find the relation between the load and the load-point displacement in large deflections for the fixed–fixed beam in Figure 5.20i. Assume $\theta = q_1 \sin(\pi x/l) + q_2 \sin(3\pi x/l)$.

5.6.10 Do the same as above, but for the fixed–hinged column in Figure 5.20j. Assume $\theta = q_1[\sin(\pi x/L_1) - k_1 x] + q_2[\sin(\pi x/L_2) - k_2 x]$. First determine for the parameters L_1, k_1, L_2, k_2 the conditions that the assumed functions satisfy the kinematic boundary conditions.

5.6.11 Using a linear combination of three or more trigonometric functions that satisfy the kinematic boundary conditions, find a close upper-bound approximation of the first critical load of the tapered column of Figure 5.20k. The cross-section dimensions vary linearly with height, $h = h_0 + h_1(1 - x/l)$. For $h_0 = h_1$ the exact result is $P_{cr_1} = 4.10 P_E^0$, $P_E^0 = \pi^2 E I_0 / l^2$ (Tuma's handbook, 1988, p. 225). (In the same handbook exact values are given for various ratios h_0/h_1, and also for different cross sections and end conditions.)

5.6.12 Do the same as Problem 5.6.11, but for the symmetrically tapered column in Figure 5.20l.

5.7 GALERKIN VARIATIONAL METHOD

This method is based on the differential equation rather than the potential energy, and therefore it is also applicable to problems for which no potential energy exists. The basic idea of the method, proposed by Galerkin, a Russian mathematician, early in this century, is to exploit the fact that the functional $\psi(x)$ is identically zero within the interval $(0, l)$ if the condition $\int_0^l \psi(x)\phi_k(x) = 0$ is satisfied for all the square-integrable functions $\phi_k(x)$ that are infinite in number ($k = 1, 2, \ldots$) and form a complete set of linearly independent functions. Now, if one deals with the differential equation $\underline{L}(w) = 0$ where $\underline{L}$ is a certain differential operator, one may set $\psi(x) = \underline{L}[w(x)] = 0$, and then by imposing the condition

$$\int_0^l \underline{L}[w(x)]\phi_k(x)\, dx = 0 \tag{5.7.1}$$

for all functions $\phi_k(x)$, infinite in number, one ensures that the differential equation is satisfied identically. This suggests that an approximation for the differential equation can be obtained with a finite set of square-integrable functions $\phi_k(x)$, yielding a finite set of equations. More precisely, the Galerkin method may be formulated by the following theorem:

Theorem 5.7.1 The approximation $w_N(x) = \sum_{k=1}^N q_k\phi_k(x)$ has as its limit for $N \to \infty$ the exact solution of the boundary-value problem if the coefficients q_k are solved from Equation 5.7.1 (after substituting $w = \sum_k q_k\phi_k$) and if the assumed functions $\phi_k(x)$ are such that

1. They are linearly independent.
2. They form a complete system of functions within the given domain.
3. Each of these functions satisfies the given essential (kinematic) boundary conditions.

Again, although these functions do not need to satisfy the static boundary conditions, the approximation is usually much better if they do. The convergence of the solution is in the mean (e.g., Rektorys, 1969, p. 699). For a proof of the theorem, see, for example, Collatz (1963), Rektorys (1969), Courant and Hilbert (1962), or Churchill (1963).

Applying integration by parts to Equation 5.7.1, one can easily show that if an associated minimizing functional (the potential energy) exists, then the Galerkin method is equivalent to the Ritz (or Rayleigh-Ritz) method and gives, for the same set of assumed functions $\phi_k(x)$, exactly the same solution.

For the sake of illustration consider the differential equation of beam-columns, $L[w(x)] = (EIw'')'' + (Pw')' - p = 0$. One may solve again the example of pin-ended columns of variable stiffness considered before (Fig. 5.19a), using a linear combination of two functions as defined in Equations 5.6.1. Substituting this linear combination into the differential equation, then the differential equation into Equation 5.7.1, and integrating, one obtains again Equation 5.6.4, with the same expressions as before (Eqs. 5.6.3). The rest of the solution is identical. Ultimately, instead of substituting the trigonometric functions for $\phi_k(x)$, one can keep the general form of functions $\phi_k(x)$ and integrate Equation 5.7.1 by parts, twice. This then leads to the general expressions in Equations 5.6.3.

We will show further examples of application of the Galerkin method, as well as the Rayleigh-Ritz method, when we deal with the buckling of thin-walled beams and plates (Chaps. 6 and 7).

As already mentioned for the Rayleigh-Ritz method, no particular advantage [e.g., for arbitrary $I(x)$] is generally derived for the Galerkin method from using an orthogonal set of functions that does not represent the exact solution of the eigenvalue problem. This is because the orthogonal functions available are defined with a certain weighting function (e.g., 1 for trigonometric functions, x for Bessel functions, x^2 for spherical harmonics), but often orthogonality with respect to a different weighting function may be required for the given problem. For example, if the weight is $I(x)^{-1} = a + bx + cx^2$, no orthogonal set of functions is found in mathematical handbooks and one would have to construct it by solving the buckling eigenmodes exactly.

Using the piecewise linear "hill" functions from Figure 5.19b as the system of linearly independent functions, the Galerkin method reduces to the finite element method. For boundary-value problems that do not possess a potential, the Galerkin method serves as the most general basis for the finite element method.

The Galerkin method is equivalent to the principle of virtual work. Indeed, after the substitution of the differential operator of beam-columns and integration by parts, Equation 5.7.1 yields $\int M\phi_k''\,dx - \int P\phi_k'\,dx - \int p\phi_k\,dx = 0$. According to Equation 2.38, this represents the virtual work equation that requires that the work of unbalanced transverse distributed loads on virtual deflections $\phi_k(x)$ be zero for all kinematically admissible deflection distributions of the system.

The Galerkin method, just as the principle of virtual work, is a more limited concept than the Rayleigh-Ritz method, carrying more limited information. The Rayleigh-Ritz method implies both equilibrium and stability conditions (minimizing conditions), while the Galerkin method implies only the equilibrium conditions. The fact that this method yields upper bounds and quadratic convergence for linear boundary-value problems that possess a potential is not due to the method per se, but to its equivalence to the Rayleigh-Ritz method for such problems.

Like the Rayleigh-Ritz method, the Galerkin method can be generalized to problems with more than one unknown function, as well as to two- or three-dimensional problems. We will show this when we deal with thin-walled beams and plates (Chaps. 6 and 7).

Another variational method for approximate solutions is the method of least squares (Rektorys, 1969). In this case the solution to the differential equation $\underline{M}w - P\underline{N}w = 0$ with homogeneous boundary conditions is obtained by minimizing the functional

$$\Phi = \int_0^l (\underline{M}w - P\underline{N}w)^2\,dx \tag{5.7.2}$$

after substitution of the approximation $w = \sum_{n=1}^{N} a_n\phi_n$. The minimizing conditions are $\partial\Phi/\partial a_m = 0$ ($m = 1, \ldots, N$). Obviously, this method is more generally applicable than the Ritz method; it can be applied even if the potential energy does not exist, for example, when operator $\underline{M}$ or $\underline{N}$ is not symmetric (i.e., not self-adjoint).

For multidimensional problems, still another possibility is the Trefftz variational method (Rektorys, 1969). It requires a system of exact solutions of the differential equation and minimizes the error in the boundary conditions. The boundary element method is a recent important variant to this idea.

Problem

5.7.1 Using the Galerkin method, repeat Problems 5.6.1, 5.6.4, 5.6.5, 5.6.6, 5.6.7, 5.6.8, 5.6.9, and 5.6.10 using Figure 5.20a, d, e, f, g, h, i, j.

5.8 METHOD OF SUCCESSIVE APPROXIMATIONS AND LOWER BOUNDS

We have already seen in our examples that the estimate of the deflection shape can be improved when the bending moment distribution is based on an initial estimate of the deflection shape and is then used to get an improved estimate by integrating the corresponding curvature distribution (see Sec. 5.5). In a similar spirit, one can generally improve the estimates of eigenvalues, or the critical loads, by the method of successive approximations. With a sufficient number of approximations, any desired accuracy can be achieved.

Formulation of the Method

Let us describe the method for the differential equation $\underset{\sim}{M}(w) - P\underset{\sim}{N}(w) = 0$. We start by choosing any kinematically admissible function $w = w^{(1)}(x)$ as the first iterate, $n = 1$. Then we proceed as follows:

$$
\begin{aligned}
&1.\ \text{Evaluate } p^{(1)}(x) = \underset{\sim}{N}[w^{(1)}(x)] \text{ and solve } \underset{\sim}{M}[w^{(2)}(x)] = p^{(1)}(x).\\
&2.\ \text{Evaluate } p^{(2)}(x) = \underset{\sim}{N}[w^{(2)}(x)] \text{ and solve } \underset{\sim}{M}[w^{(3)}(x)] = p^{(2)}(x).\\
&\quad\vdots\\
&n.\ \text{Evaluate } p^{(n)}(x) = \underset{\sim}{N}[w^{(n)}(x)] \text{ and solve } \underset{\sim}{M}[w^{(n+1)}(x)] = p^{(n)}(x).
\end{aligned}
\tag{5.8.1}
$$

Note that, if $w^{(1)}(x)$ were actually an eigenfunction, $w^{(2)}(x)$ would differ from $w^{(1)}(x)$ only by the factor P, since the equation $\underset{\sim}{M}w^{(2)} - \underset{\sim}{N}w^{(1)} = 0$ would hold. Note also that the functions $w^{(n)}(x)$ must all satisfy the kinematic boundary conditions, and so the integration constants in solving $w^{(n)}(x)$ must be determined from the boundary conditions.

The foregoing represents a sequence of boundary-value problems, in each of which $p^{(n)}$ are known functions of x (in the case of the beam-column differential equation, they represent the distributions of transverse load needed to balance the beam). Because $\underset{\sim}{M}(w_n)/\underset{\sim}{N}(w_n) = P_{\text{cr}_n}$ (if w_n is an eigenfunction) we expect that ratio $\underset{\sim}{M}(w^{(n)})/\underset{\sim}{N}(w^{(n)})$ should approach, for $n \to \infty$, the eigenvalue P, and function $w^{(n)}$ should approach the eigenfunction. However, this ratio depends on x, and in order to remove the dependence on x we may integrate from 0 to l with the factor $w^{(n)}(x)$. Thus, the nth approximation to the eigenvalue is given by the quotient:

$$P^{(n)} = \frac{\int_0^l w^{(n)}(x)\underset{\sim}{M}[w^{(n)}(x)]\,dx}{\int_0^l w^{(n)}(x)\underset{\sim}{N}[w^{(n)}(x)]\,dx} = \frac{\int_0^l w^{(n)}(x)\underset{\sim}{N}[w^{(n-1)}(x)]\,dx}{\int_0^l w^{(n)}(x)\underset{\sim}{N}[w^{(n)}(x)]\,dx} = \frac{a_{n-1}}{a_n} \quad (5.8.2)$$

where the last two equalities are of course valid only for $n > 1$. In mathematics, this is called the Schwarz quotient (Collatz, 1963, eq. 12.8) and the a_n are called the Schwarz constants. Note that the first approximation ($n = 1$) represents the Rayleigh quotient (Eq. 5.3.13).

As a check, consider a pin-ended column of constant cross section under axial load P, for which the solution is known. The differential equation is characterized by $\underset{\sim}{N} = 1/EI$, $\underset{\sim}{M} = -d^2/dx^2$. Considering as the first approximation $w^{(1)} = A\sin(\pi x/l)$, one obtains $p^{(1)}(x) = (A/EI)\sin(\pi x/l)$, and solving for $w^{(2)}$ gives $w^{(2)} = (Al^2/EI\pi^2)\sin(\pi x/l)$. From this, $w^{(1)}/w^{(2)} = \pi^2 EI/l^2$, which illustrates that the ratio of two successive terms in the sequence of approximation should tend to the eigenvalue (in this case it is equal to it since $w^{(1)}$ is exact). The same is true of the ratio $\underset{\sim}{M}(w^{(2)})/\underset{\sim}{N}(w^{(2)})$.

Newmark (1943) showed very effective practical solutions of column buckling loads by successive approximations describing functions $w^{(n)}(x)$ by a discrete set of values and by using finite difference expressions; see also Chen and Lui (1987, p. 443).

For a general eigenvalue problem, the differential equations that define the successive approximations may be deduced by subjecting the differential equations and the boundary conditions to the following operation (Collatz, 1963): (1) in all the terms containing eigenvalue λ replace w by $w^{(n-1)}$ and delete λ; and (2) in all the terms not containing λ replace w by $w^{(n)}$.

Example

Consider now the case of the tapered column solved before (Sec. 5.6, Fig. 5.19b), and assume $I(x) = I_0/(1 + k\,|x/l|)$ with either $k = 1$ or $k = 2$. We have $\underset{\sim}{N} = 1/EI(x)$ and $\underset{\sim}{M} = -d^2/dx^2$. As the first approximation, we choose $w^{(1)} = A\cos(\pi x/l)$, which is not the exact solution but satisfies the boundary conditions. Substitution into Equation 5.8.2 gives, as the first approximation to the eigenvalue, $P_{cr}^{(1)} = c(\pi^2 EI_0/l^2)$, with $c = 0.870565$ for $k = 1$ and $c = 0.770797$ for $k = 2$, values that are slightly higher than the approximations that we found for the same case with the Ritz method in Section 5.6 ($c = 0.8699$ and $c = 0.7689$, respectively). Note that the first approximation in this case coincides with the Timoshenko quotient, Equation 5.4.6, rather than with the Rayleigh quotient (because we use the second-order rather than the fourth-order differential equation of beam-columns).

We can then calculate $p^{(1)} = (A/EI_0)(1 + k\,|x/l|)\cos(\pi x/l)$, and solving $\underset{\sim}{M}w^{(2)}(x) = p^{(1)}(x)$ with the boundary conditions $w^{(2)}(l/2) = 0$ and $w^{(2)\prime}(0) = 0$, we obtain

$$w^{(2)} = \frac{Al^2}{\pi^2 EI_0}\left[\left(1 + k\frac{x}{l}\right)\cos\frac{\pi x}{l} - 2\frac{kl}{\pi}\sin\frac{\pi x}{l} + k\left(\frac{x}{l} + \frac{\alpha}{2}\right)\right] \quad (5.8.3)$$

with $\alpha = 4/\pi - 1$. Equation 5.8.2 then gives for the second approximation: $c = 0.869719$ if $k = 1$ and $c = 0.768444$ if $k = 2$.

To calculate the third approximation, we evaluate $p^{(2)}(x) = \underline{N}w^{(2)}(x)$ and solve $\underline{M}w^{(3)}(x) = p^{(2)}(x)$ with the boundary conditions $w^{(3)}(l/2) = 0$ and $w^{(3)\prime}(0) = 0$. Thus, we obtain

$$w^{(3)}(x) = A\left(\frac{l^2}{\pi^2 EI_0}\right)^2 [g(x) + C] \tag{5.8.4}$$

where

$$g(x) = \left(1 + \frac{2kx}{l} + \frac{k^2x^2}{l^2} - \frac{10k^2}{\pi^2}\right)\cos\frac{\pi x}{l} - \frac{6kl}{\pi}\left(1 + \frac{kx}{l}\right)\sin\frac{\pi x}{l}$$
$$- \frac{\pi^2 k^2}{12}\left(\frac{x^4}{l^4}\right) - \frac{8\pi^2 k}{3}\left(1 + \frac{\alpha k}{2}\right)\left(\frac{x^3}{l^3}\right) - \frac{\pi^2 \alpha k}{4}\left(\frac{x^2}{l^2}\right) + \frac{4kx}{l} \tag{5.8.5}$$

$$c = k\left[\frac{6}{\pi} - 2 + \frac{\pi^2}{16}\alpha + \frac{\pi^2}{48} + k\left(\frac{3}{\pi} + \frac{\pi^2}{96}\alpha + \frac{\pi^2}{192}\right)\right] \tag{5.8.6}$$

For $P_{\mathrm{cr}}^{(3)}$ one gets, from Equation 5.8.2, $c = 0.869708$ if $k = 1$ and $c = 0.768411$ if $k = 2$. Since the rate of convergence of this method is known to be quadratic (and monotonic from above; Collatz, 1963), and since the third approximation is very close to the second, we may expect the third approximation to be nearly exact.

From the foregoing procedure it appears that the successive approximations lead to rather complicated expressions. This makes the method unattractive, except perhaps if Newmark's (1943) finite difference version is used. However, an advantage of the method is that it allows lower bounds, and that these bounds are often surprisingly close.

Lower Bound

The way to use successive approximations to obtain lower bounds on P_{cr_1} was shown by Temple (1929) (see also Collatz, 1963, eq. 12.19). Temple's lower bound, which is often close, is given by

$$P_{\mathrm{cr}_1} \geq P^{(n)} - \frac{P^{(n-1)} - P^{(n)}}{(P_{\mathrm{cr}_2}^{-}/P^{(n)}) - 1} \tag{5.8.7}$$

where $P_{\mathrm{cr}_2}^{-}$ is some lower bound on the second eigenvalue P_{cr_2} such that $P_{\mathrm{cr}_2}^{-} > P^{(n)}$. $P_{\mathrm{cr}_2}^{-}$ can be conveniently obtained as the second critical load of a column of constant cross section equal to the minimum cross section of the actual column. Equation 5.8.7 is valid if the operator is self-adjoint and positive definite, if the first eigenvalue λ is not a multiple root, and if λ does not appear in the boundary conditions.

For the previously considered example with $k = 2$, Equation 5.8.7 yields $P_{\mathrm{cr}_1}^{-} l^2/\pi^2 EI_0 = 0.766976$ for $n = 2$ and 0.768390 for $n = 3$. The corresponding upper bounds are $c^{(2)} = 0.768444$ *and* $c^{(3)} = 0.768411$. As we see, these bounds are extremely close, and their spread is only 0.19 percent for $n = 2$ and 0.0027 percent for $n = 3$.

Problems

5.8.1 (a) Use the method of successive approximations to calculate the critical load of a hinged column of constant cross section. Although the exact solution is a sine curve, deliberately use a parabolic curve for $w^{(1)}$ so as to be able to check how good the results are. (b) Obtain also the lower bounds and compare them to the exact solution.

5.8.2 (a) Repeat the calculation of the critical load of the tapered column from the example in the text assuming a parabolic function for $w^{(1)}$. (b) Obtain also the lower bounds and compare them to the solution in the text. (The upper and lower bounds for $k=2$ are $c^+=0.768446$ *and* $c^-=0.766247$ for $n=2$, and $c^+=0.768411$ and $c^-=0.768389$ for $n=3$.)

5.8.3 Solve the same example as in the text, but with $k=-0.5$ and $k=-1$. Also obtain the lower bounds.

5.8.4 For the columns shown in Figure 5.20a and e (considered before in Probs. 5.6.1 and 5.6.5), assume $w^{(1)}=A\sin(\pi x/l)$. (a) Obtain second improved estimates of the critical load by successive approximations. (b) Also obtain a lower bound on P_{cr_2} (replacing p by axial load $P=pl$ at column top for the column shown in Fig. 5.20a), and then a lower bound on P_{cr_1}.

5.8.5 Solve the same problem as above, but characterize $w^{(1)}$ by values $w_i^{(1)}$ $(i=1,2,\ldots,6)$ at six points along the column. Approximating the derivatives and integrals by finite difference expressions and the Simpson rule, obtain better upper bounds on P_{cr_1} (this is the approach used by Newmark, 1943). Also obtain lower bounds. (Since the finite difference equations have an error, the solution does not necessarily represent bounds for exact solution of the actual column, but it does for the exact solution of the finite difference equations.)

5.9 NONLINEAR PROBLEMS: LARGE DEFLECTIONS OF COLUMNS

Let us now illustrate (following Bažant, 1985) how a nonlinear problem can be approximately solved by minimization of potential energy. The energy approach makes it possible to determine the approximate initial large-deflection behavior of columns in a simpler and more general manner than was done by equilibrium analysis in Section 1.9. We study large deflections of an inextensible column of length l. As a generalization compared to Section 1.9, we now allow the bending rigidity to be variable, that is, $EI(s)$ where $s=$ length coordinate measured along the deflection curve, with $s=0$ placed at midlength (see Fig. 5.21a), and the column, to be imperfect with initial imperfection, defines a curve with slopes $\theta_0(s)$. The exact potential-energy expression valid for arbitrarily large deflections is $\Pi=U-P\bar{W}$ where

$$U=\int_l\left[\frac{EI}{2}(\theta'-\theta_0')^2\right]ds,\qquad \bar{W}=\int_l(\cos\theta_0-\cos\theta)\,ds \tag{5.9.1}$$

where $\theta'-\theta_0'=$ change of curvature produced by load, and $(\ \ )'=d/ds$ (see Sec. 5.1). To discretize the problem, we assume the slope of the deflection curve to be

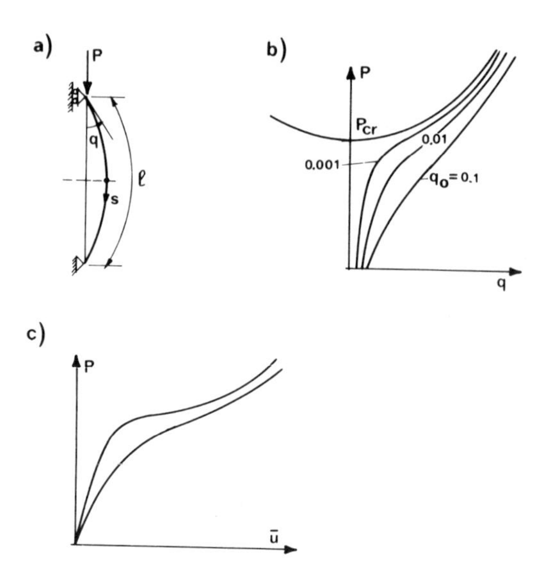

Figure 5.21 (a) Column with large deflections, (b) load-deflection curves, and (c) load-displacement curves.

approximately $\theta(s) = qf(s)$ where function $f(s)$ is given or suitably chosen. For the initial imperfection we assume $\theta_0(s) = q_0 f(s)$. Then

$$U(q) = (q - q_0)^2 \int_l \frac{EI}{2} f'^2 \, ds \qquad \bar{W}(q) = \int_l [\cos(q_0 f) - \cos(qf)] \, ds \quad (5.9.2)$$

The equilibrium condition is $\partial \Pi / \partial q = 0$ or $\partial U / \partial q - P \, \partial \bar{W} / \partial q = 0$, which yields

$$P = \frac{\partial U / \partial q}{\partial \bar{W} / \partial q} = \frac{(q - q_0) \int_l EIf'^2 \, ds}{\int_l f \sin(qf) \, ds} = P(q) \qquad (5.9.3)$$

Evaluating this integral yields the equilibrium curve $P(q)$ for any imperfection q_0. The column is stable if also $\partial^2 U / \partial q^2 > 0$ or $\partial^2 U / \partial q^2 - P \, \partial^2 \bar{W} / \partial q^2 > 0$ (with all derivatives taken at constant P). This leads to the condition:

$$\int_l EIf'^2 \, ds - P(q) \int_l f^2 \cos(qf) \, ds > 0 \qquad \text{(stable)} \qquad (5.9.4)$$

For the special case of a perfect column ($q_0 = 0$) and the start of buckling ($q \to 0$), Equation 5.9.4 gives

$$P = P_{cr} = \frac{\int_l EIf'^2 \, ds}{\int_l f^2 \, ds} \qquad (5.9.5)$$

which we recognize to be the Rayleigh quotient. If $f(s)$ is the exact initial slope of the column curve [e.g., $\sin(\pi s / l)$ for the case that $EI = \text{const.}$], Equation 5.9.5

must give the exact critical load; otherwise, only an approximation. For example, for a pin-ended column with $EI = \text{const.}$, function $f(s) = \sin(\pi s/l)$ furnishes $P_{cr} = EI\pi^2/l^2 = P_E =$ exact solution. If the first integral is expressed from Equation 5.9.5 and substituted into Equation 5.9.4, the stability condition of a perfect column (for which $q \to 0$, $\cos qf \to 1$) takes the form:

$$(P_{cr} - P) \int_l f^2 \, ds > 0 \qquad (\text{stable}, \ q = q_0 = 0) \tag{5.9.6}$$

This inequality must hold true for any kinematically admissible function $f(s)$, including the one that gives the exact buckling shape and exact P_{cr}. Equation 5.9.6 means that a perfect column is stable for $P < P_{cr}$, where P_{cr} is any critical load that can be obtained from the Rayleigh quotient in Equation 5.9.5. Since we know the minimum of this quotient to be the exact P_{cr}, we thus prove that the column is stable if P is less than the exact P_{cr} and unstable if larger—a result we have already proven by dynamic analysis (Sec. 3.1), as well as by eigenfunction explansion of $w(x)$ (Eq. 5.3.17), but could not prove from equilibrium analysis in Chapter 1.

For imperfect columns ($q_0 > 0$) and finite q, one can easily check that the stability condition in Equation 5.9.4 (in which the first integral is expressed from Eq. 5.9.3) is equivalent to the condition $\partial P/\partial q > 0$ (provided that $\int f \sin qf \, ds > 0$). This demonstrates that curves $P(q)$, which were already shown in Figure 1.25, are stable as long as they have a positive slope (for $q > 0$).

Let us now analyze a pin-ended column (Fig. 5.21a), choosing again $f(s) = \sin(\pi s/l)$. While in a pin-ended column there are no kinematic boundary conditions to be satisfied by $f(s)$, that is, by $\theta(s)$, this function satisfies the static boundary conditions $\theta' = 0$ at $s = -l/2$ and $s = l/2$. It also represents the exact buckling shape asymptotically, for $q \to 0$ (if $EI = \text{const.}$). Substitution into Equation 5.9.3 yields

$$P = \frac{\dfrac{\pi^2}{l^2}(q - q_0) \displaystyle\int_{-l/2}^{l/2} EI(s)\left(\cos\frac{\pi s}{l}\right)^2 ds}{\displaystyle\int_{-l/2}^{l/2} \left(\sin\frac{\pi s}{l}\right)\left[\sin\left(q \sin\frac{\pi s}{l}\right)\right] ds} \tag{5.9.7}$$

(Note that if $q_0 = 0$ and $EI = \text{const.}$, the limit of this expression for $q \to 0$ is $P_{cr} = EI\pi^2/l^2$.) We may now introduce in the denominator the expansion $\sin\alpha = \alpha - \frac{1}{6}\alpha^3 + \frac{1}{120}\alpha^5 - \cdots$ where $\alpha = q \sin(\pi s/l)$. For $EI = \text{const.}$ we may then evaluate all integrals in Equation 5.9.4 and get

$$P = \frac{(EI\pi^2/l^2)(q - q_0)(l/2)}{\displaystyle\int_{-l/2}^{l/2} \left[q\left(\sin\frac{\pi s}{l}\right)^2 - \frac{q^3}{6}\left(\sin\frac{\pi s}{l}\right)^4 + \frac{q^5}{120}\left(\sin\frac{\pi s}{l}\right)^6 - \cdots\right] ds}$$

or

$$\frac{P}{P_{cr}} = \frac{q - q_0}{\dfrac{2}{l}\left[q\dfrac{l}{2} - \dfrac{q^3}{6}\left(\dfrac{3l}{8}\right) + \dfrac{q^5}{120}\left(\dfrac{5l}{16}\right) - \cdots\right]} = \frac{q - q_0}{q(1 - \frac{1}{8}q^2 + \frac{1}{192}q^4 - \cdots)} \tag{5.9.8}$$

Since $f(s) = \sin(\pi s/l)$ is not exact, this expression is only approximate.

For small enough q, in Equation 5.9.8 we may neglect all the terms beginning with $q^4/192$ and also introduce $(1-\frac{1}{8}q^2)^{-1} \simeq 1+\frac{1}{8}q^2$. Thus we get the approximation:

$$\frac{P}{P_{\text{cr}}} = \frac{q-q_0}{q}\left(1+\frac{q^2}{8}\right) \tag{5.9.9}$$

(see the curves in Fig. 5.21b for various q_0). For $q_0 \to 0$, this equation describes the initial postcritical behavior and exactly agrees with Equation 1.9.7 that we got in Section 1.9 from the exact solution in terms of elliptic integrals. This shows that the term $1+q^2/8$ is the exact approximation up to the second order in q [the reason is that $f(s) = \sin(\pi s/l)$ gives the slope of the exact deflection shape for $q \to 0$ and $q_0 \to 0$].

Based on Equation 5.9.8 we can further deduce the initial dependence of P on maximum deflection w and/or the axial displacement u, generalizing the results obtained in Section 1.9 for $q_0 = 0$. We may approximately consider the deflection curve ordinates to be $z = -a\cos(\pi s/l)$ where a = midspan deflection ordinate. From this, $\theta \simeq z' = (\pi a/l)\sin(\pi s/l)$, and so $q = \pi a/l$. Thus, Equation 5.9.9 provides

$$\frac{P}{P_{\text{cr}}} = \left(1-\frac{a_0}{a}\right)\left(1+\frac{\pi^2}{8l^2}a^2\right) \tag{5.9.10}$$

where $a_0 = q_0 l/\pi$ = ordinate of initial imperfection at midspan (*note:* deflection at midspan $= a - a_0 = w_{\max}$). Note that for small a, Equation 5.9.10 becomes $P/P_{\text{cr}} = 1 - a_0/a$, which yields $a = a_0/(1-P/P_{\text{cr}})$—the same as derived in Chapter 1 (Eq. 1.5.10).

To calculate the dependence of P on the axial displacement u due to transverse deflections, we need to note the approximation $u \simeq \frac{1}{2}\int (z'^2 - z_0'^2)\,ds = \pi^2(a^2-a_0^2)/4l = (q^2-q_0^2)l/4$, from which $q^2 - q_0^2 = 4u/l$. Thus, Equation 5.9.9 yields

$$\frac{P}{P_{\text{cr}}} = \left[1 - q_0\left(q_0^2 + \frac{4u}{l}\right)^{-1/2}\right]\left(1+\frac{q_0^2}{8}+\frac{u}{2l}\right) \tag{5.9.11}$$

where $q_0 = \pi a_0/l$. Equations 5.9.10 and 5.9.11, describing the initial large-deflection behavior, generalize Equations 1.9.12 and 1.9.14 deduced for $q_0 = 0$.

The value of u in Equation 5.9.11 does not include the effect of axial elastic strains. To take them into account, we must introduce into Equation 5.9.11, an expression of u in terms of the total axial displacement $\bar{u}$, that is, substitute $u = \bar{u} - Pl/EA$ = axial displacement due to both the transverse deflection and the axial elastic shortening (EA = axial elastic stiffness). This yields the following implicit relation for P as a function of $\bar{u}$:

$$P = P_{\text{cr}}[1 - q_0(q_0^2 + 4\lambda)^{-1/2}]\left(1+\frac{1}{8}q_0^2+\frac{1}{2}\lambda\right) \qquad \lambda = \frac{\bar{u}}{l} - \frac{P}{EA} \tag{5.9.12}$$

which generalizes Equation 1.9.15. The graph of P versus $\bar{u}$ can be easily constructed by choosing successively increasing values of λ and calculating first the corresponding P and then $\bar{u}$ from Equations 5.9.12; $\bar{u} = l(\lambda + P/EA)$ (see Fig.

5.21c; Fig. 1.26d). Further discussion of these results would be analogous to that already made in Section 1.9.

For large deflections, better accuracy may be achieved by introducing into Equation 5.9.1 the expression $f(s) = \sum_n q_n f_n(s)$ where $f_n(s)$ are linearly independent chosen functions satisfying at least the kinematic boundary conditions for slope θ; $n = 1, 2, \ldots, N$. The equilibrium states then result from the conditions $\partial\Pi/\partial q_n = 0$ (all n), which represent a system of nonlinear algebraic equations. This approach is similar to the Ritz method for linear problems, but, unlike the Ritz method, convergence for $N \to \infty$ cannot be guaranteed, due to nonlinearity of the problem.

Nonlinear finite element analysis is today, of course, preferable to solutions of this type if a larger N is needed. But in the past, this was an important method.

Problems

5.9.1 Derive results equivalent to Equations 5.9.9 to 5.9.11 assuming that $f(s)$ is a polynomial such that $f(0) = f'(\pm l/2) = 0$ (which gives $M = 0$ at the ends). Alternatively, is $f(s) = s$ admissible and is an upper bound guaranteed?

5.9.2 Consider the same column (Fig. 5.21a) with $EI(s) = EI_0(1 - ks^2)$; I_0, k = constants. Calculate and plot the diagrams $P(q)$ for $q = 0$, 0.001, 0.01, 0.1. (It is easier to evaluate the integrals in Eq. 5.9.3 numerically on a calculator.) Alternatively, consider $EI(s) = EI_0 \exp(-ks^2)$ or $EI_0 \cos ks$.

5.9.3 Based on the above, obtain also the equivalents of Equations 5.9.9 to 5.9.11.

5.9.4 Do the same analysis as in Equations 5.9.1 to 5.9.11 but for a free-standing column (EI = const.).

5.9.5 Do the same, but a fixed–fixed column, using (a) $f(s) = \sin(2\pi s/l)$, (b) $f(s)$ = appropriate polynomial.

5.9.6 Do the same, but for a fixed–hinged column, using for $f(s)$ the exact solution of $w'(x)$ from Section 1.4.

5.9.7 Do the same but for a pinned column with rotational springs $c = 10EI/l$ at ends.

5.9.8 Do the same but for symmetric buckling of a portal frame (braced) with hinges at the base, using suitable polynomials for $f(s)$ in the columns and the beam.

5.9.9 Do the same as above, but with the frame legs fixed at the base.

5.9.10 Do the same as Problem 5.9.2 but EI = const. and P is variable due to its own weight; $P = P_0(1 - gs)$, where P_0, g = const.

5.9.11 Do the same type of analysis as in Equations 5.9.1 to 5.9.11 but based on $w(x)$ instead of $\theta(s)$. To express Π use $M = EIw''(1 - \frac{3}{2}w'^2)$, $1 - \cos\theta = \frac{1}{2}w'^2 - \frac{1}{24}w'^4$, $1 - \cos\theta_0 = \frac{1}{2}w_0'^2 - \frac{1}{24}w_0'^4$.

5.9.12 Do the same as Problem 5.9.2 but EI and P are constant while the moment curvature relation is nonlinearly elastic; $M = EI\theta'(1 - k\theta'^3)$.

5.9.13 For the same problem as in the text, assume that $f(s) = q_1 \sin(\pi s/l) + q_2 \sin(3\pi s/l)$, express Π, formulate the equilibrium conditions from $\partial\Pi/\partial q_1 = 0$, $\partial\Pi/\partial q_2 = 0$, and, using numerical integration for the integrals that arise, calculate the diagram of P versus w_{max} for (a) constant EI, or (b) variable $EI(s) = EI_0(1 - ks^2)$. Compare the result to the curves obtained before for $f(s) = \sin(\pi s/l)$. Are these new curves higher or lower?

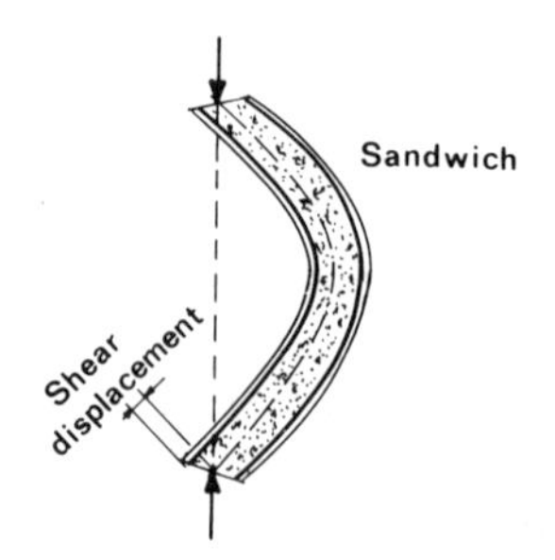

Figure 5.22 Exercise problems on buckling of sandwich column.

5.9.14 Extending the linearized small-deflection formulation for sandwich beams from Section 1.7 (Eqs. 1.7.1–1.7.2), use the energy method to calculate large deflections of a simply supported sandwich beam with very thin faces ($f \ll c$) (Fig. 5.22). Determine the approximate diagrams $P(w_{max})$ and $P(u)$.

References and Bibliography

Al-Sarraf, S. Z. (1979), "Upper and Lower Bounds of Elastic Critical Loads," *The Structural Engineer,* 57A(12):415–20.

Atanackovic, T. M., and Milisavljevic, B. M. (1981), "Some Estimates for a Buckling Problem," *Acta Mech.,* 41(1–2):63–71.

Bažant, Z. (1943), "Stability of Upper-Compressed Belt of a Bridge Girder without Top Transverse Stiffeners," *Technický obzor* 51, L(23–24).

Bažant, Z. P. (1984), "Imbricate Continuum and its Variational Derivation," *J. Eng. Mech.* (ASCE), 110(12):1693–712 (Sec. 5.1).

Bažant, Z. P. (1985), *Lectures on Stability of Structures,* Course 720-D24, Northwestern University, Evanston, Ill. (Sec. 5.9).

Bažant, Z., and Masopust, J. (1986), "Horizontal Loading on Large-Diameter Piles in Layered Foundation" (in Czech), *Inženýrské stavby* (Prague), No. 7, pp. 371–77 (Sec. 5.2).

Boley, B. A. (1963), "On the Accuracy of the Bernoulli-Euler Theory for Beams of Variable Section," *J. Appl. Mech.* (ASME), 30:373–78.

Chen, W. F., and Lui, E. M. (1987), *Structural Stability—Theory and Implementation,* Elsevier, New York (Sec. 5.8).

Chen, Y. Z., Cheung, Y. K., and Xie, J. R. (1989), "Buckling Loads of Columns with Varying Cross Sections," *J. Eng. Mech.* (ASCE), 115(3):662–67.

Chini, S. A., and Wolde-Tinsae, A. M. (1988), "Effect of Prestressing on Elastica Arches," *J. Eng. Mech.* (ASCE), 114(10):1791–800.

Chugh, A. K., and Biggers, S. B. (1976), "Stiffness Matrix for a Non-Prismatic Beam-Column Element," *Int. J. Numer. Methods Eng.,* 10(5):1125–42.

Churchill, R. V. (1963), *Fourier Series and Boundary Value Problems,* 2nd ed., McGraw-Hill, New York (Sec. 5.7).

Collatz, L. (1963), "Eigenwertaufgaben mit technischen Anwendungen (Eigenvalue Problems with Technical Applications)," 2nd ed., Akademische Verlagsgesellschaft, Geest & Portig, Leipzig (Sec. 5.1).

Courant, R., and Hilbert, D. (1962), *Methods of Mathematical Physics,* Interscience Publishers, New York (Sec. 5.1).

Culver, C. G., and Preg, S. M., Jr. (1968), "Elastic Stability of Tapered Beam-Columns," *J. Struct. Eng.* (ASCE), 94(2):455–70.

Dinnik, A. N. (1929), "Designs of Columns of Varying Cross Section," *Trans. ASME,* 51:105–14; see also 54:165–71 (1932).

Dym, C. L., and Shames, I. H. (1973), *Solid Mechanics—A Variational Approach,* McGraw-Hill, New York (Sec. 5.1).

Elishakoff, I., and Bert, C. W. (1988), "Comparison of Rayleigh's Non-Integer-Power Method with Rayleigh-Ritz Method," *Comput. Methods Appl. Mech. Eng.,* 67(3):297–309.

Elishakoff, I., and Pellegrini, F. (1987), "Exact and Effective Approximate Solutions of Some Divergence-Type Non-Conservative Problems," *J. Sound and Vibration,* 114(1):143–47.

Elsgol'ts, L. E. (1963), *Calculus of Variations,* Pergamon Press, Oxford-London-New York (Sec. 5.1).

Fogel, C. M., and Ketter, R. L. (1962), "Elastic Strength of Tapered Columns," *J. Struct. Eng.* (ASCE), 88(5):67–106.

Fung, Y. C. (1965), *Foundation of Solid Mechanics,* Prentice-Hall, Englewood Cliffs, N.J. (Sec. 5.1).

Gere, J. M., and Carter, W. D. (1962), "Critical Buckling Loads for Tapered Columns," *J. Struct. Eng.* (ASCE), 88(1):1–11.

Hanna, S. Y., and Michalopoulos, C. D. (1979), "Improved Lower Bounds for Buckling Loads and Fundamental Frequencies of Beams," *J. Appl. Mech.* (ASME), 46(3):696–98.

Hansen, J. S. (1977), "Some Two-Mode Buckling Problems and Their Relation to Catastrophe Theory," *AIAA Journal,* 15:1638–44 (Sec. 5.2).

Hetényi, M. (1946), *Beams on Elastic Foundation,* The University of Michigan Press, Ann Arbor (Sec. 5.2).

Hui, D. (1986), "Imperfection Sensitivity of Elastically-Supported Beams and Its Relations to the Double-Cusp Instability Model," *Proc. Royal Soc.* (London), 405:143–58 (Sec. 5.2).

Hui, D. (1988), "Postbuckling Behavior of Infinite Beams on Elastic Foundations Using Koiter's Improved Theory," *Int. J. Non-Linear Mech.,* 23(2):113–23.

Huseyin, K., and Plaut, R. H. (1974), "Extremum Properties of the Generalized Rayleigh Quotient Associated with Flutter Instability," *Quart. Appl. Math.,* 32(2):189–201.

Johnson, D. A. (1985), "Rayleigh Revisited," *J. Appl. Mech.* (ASME), 52(1):220–22.

Kantorovich, L. V., and Krylov, V. I. (1958), *Approximate Methods of Higher Analysis,* 4th ed. (translated from Russian), Interscience Publishers, New York.

Kerr, A. D. (1964), "Elastic and Viscoelastic Foundation Models," *J. Appl. Mech.* (ASME), 31:491–98 (Sec. 5.2).

Komatsu, S., Nishimura, N., and Nakata, N. (1985), "Simplified Formulae for Overall Lateral Buckling Load of Braced Twin Girder Bridges," *Trans. Jpn. Soc. Civ. Eng.,* 15:226–27.

Kounadis, A. N. (1980), "Buckling and Postbuckling of a Symmetrically-Loaded Three-Hinged Portal Frame," *J. Struct. Mech.,* 8(4):423–34.

Ku, A. B. (1977a), "Upper and Lower Bound Eigenvalues of a Conservative Discrete System," *J. Sound and Vibration,* 53(2):183–87.

Ku, A. B. (1977b), "Upper and Lower Bounds for Fundamental Natural Frequency of Beams," *J. Sound and Vibration,* 54(3):311–16.

Leighton, W. (1971), "Upper and Lower Bounds for Eigenvalues," *J. Math. Anal. Applic.,* 35:381–88.

Levine, H. A., and Protter, M. H. (1985), "Unrestricted Lower Bounds for Eigenvalues for Classes of Elliptic Equations and Systems of Equations with Applications to Problems in Elasticity," *Math. Methods Appl. Sci.*, 7(2):210–22.

Lind, N. C. (1975), "Newmark's Numerical Method," Special Publication, Solid Mechanics Division, University of Waterloo, Ontario, Canada.

Ly, B. L. (1986), "Simply Supported Column under a Tangential Follower Force as a Self-Adjoint System," *J. Appl. Mech.* (ASME), 53(2):466–67.

Ly, B. L. (1987a), "Lower Bound for Buckling Loads," *Ind. Math.*, 37(1):1–9.

Ly, B. L. (1987b), "Upper and a Lower Bound for Natural Frequencies," *J. Sound and Vibration*, 112(3):547–49.

Masur, E. F., and Popelar, C. H. (1976), "On the Use of the Complementary Energy in the Solution of Buckling Problems," *Int. J. Solids Struct.*, 12:203–16.

Mau, S. T. (1989), "Buckling and Post-Buckling Analyses of Struts with Discrete Supports," *J. Eng. Mech.* (ASCE), 115(4):721–39.

Morse, P. M., and Feshbach, H. (1953), *Methods of Theoretical Physics*, McGraw-Hill, New York (2 Vol.) (Sec. 5.3).

Murthy, G. K. (1973), "Buckling of Beams Supported by Pasternak Foundation," *J. Eng. Mech.* (ASCE), 99(3):565–79.

Newmark, N. M. (1943), "A Numerical Procedure for Computing Deflections, Moments, and Buckling Loads," *Trans. ASCE,* 108:1161–88 (Sec. 5.8).

Oran, C. (1967), "Complementary Energy Method for Buckling," *J. Eng. Mech.* (ASCE), 93(1):57–75.

Pasternak, P. (1926), "Die baustatische Theorie biegefester Balken und Platten auf elastischer Bettung," *Beton und Eisen,* 25:163–71, 178–86 (Also *Osnovi novago metoda raschota fundamentov na uprugom osnovanii pri pomoshchi dvukh koefficientov posteli,* Stroyizdat, Moscow, 1954) (Sec. 5.2).

Popelar, C. H. (1974), "Assured Upper Bounds via Complementary Energy," *J. Eng. Mech.* (ASCE), 100(4):623–33.

Raju, K. K., and Rao, G. V. (1988), "Free Vibration Behavior of Tapered Beam-Columns," *J. Eng. Mech.* (ASCE), 114(5):889–92.

Ramsey, H. (1985), "A Rayleigh Quotient for the Instability of a Rectangular Plate with Free Edges Twisted by Corner Forces," *J. Mech. Theor. Appl.*, 4(2):243–56.

Rayleigh, J. W. S. (1894), *Theory of Sound,* vol. 1, Macmillan, also Dover Publications, 1945 (Sec. 5.3).

Razaqpur, A. G. (1989), "Beam-Column Element on Weak Winkler Foundation," *J. Eng. Mech.* (ASCE), 115(8):1798–817.

Rektorys, K. (1969), *Survey of Applicable Mathematics,* Iliffe Books, London (Sec. 5.4).

Rutenberg, A., Leviathan, I., and Decalo, M. (1968), "Stability of Shear-Wall Structures," *J. Struct. Eng.* (ASCE), 114(3):707–16.

Schmidt, R. (1981), "Variant of the Rayleigh-Ritz Method," *Ind. Math.*, 31(1):37–46.

Schmidt, R. (1986), "Initial Postbuckling of Columns by the Rayleigh-Ritz Method," *Ind. Math.*, 36(1):67–76.

Schmidt, R. (1989), "Lower Bounds for Eigenvalues via Rayleigh's Method," *J. Eng. Mech.* (ASCE), 115(6):1365–70.

Schreyer, H. L., and Shih, P. Y. (1973), "Lower Bounds to Column Buckling Loads," *J. Eng. Mech.* (ASCE), 99(5):1011–22. See also discussion, 100(5):1053–55 (Sec. 5.4).

See, T., and McConnel, R. E. (1986), "Large Displacement Elastic Buckling of Space Structures," *J. Struct. Eng.* (ASCE), 112(5):1052–69.

Shih, P-Y., and Schreyer, H. L. (1978), "Lower Bounds to Fundamental Frequencies and Buckling Loads of Columns and Plates," *Int. J. Solids Struct.*, 14(12):1013–26.

Simitses, G. J. (1976), *An Introduction to the Elastic Stability of Structures,* Prentice-Hall, Englewood Cliffs, N.J. (Sec. 5.2).

Soldini, M. (1965), *Contribution a l'étude théorique et expérimentale des déformations d'un sol horizontal élastique à l'aide d'une loi de seconde approximation,* Leemann, Zürich (Sec. 5.2).

Stakgold, I. (1967), *Boundary Value Problems of Mathematical Physics,* The Macmillan Co., London (Sec. 5.1).

Stoman, S. H. (1988), "Stability Criteria for X-Bracing Systems," *J. Eng. Mech.* (ASCE), 4(8):1426–34.

Temple, G. (1929), "The Computation of Characteristic Numbers and Characteristic Functions," *Proc. London Math. Soc.*, 29(2):257–80 (Sec. 5.8).

Temple, G., and Bickley, W. G. (1933), *Rayleigh's Principle and its Applications to Engineering,* Cambridge University Press, London; also Dover Publications, New York, 1956.

Timoshenko, S. P., and Gere, J. M. (1961), *Theory of Elastic Stability,* McGraw-Hill, New York (Sec. 5.4).

Tuma, J. (1988), *Handbook of Structural and Mechanical Matrices,* McGraw-Hill, New York (Sec. 5.2).

Vlasov, V. Z., and Leontiev, N. N. (1966), "Beams, Plates and Shells on Elastic Foundation," Israel Program of Scientific Translations, Jerusalem (Sec. 5.2).

Vrouwenvelder, A., and Witteveen J. (1975), "Lower Bound Approximation for Elastic Buckling Loads," *Heron,* 20(4):1–27.

Wang, C. M., and Ang, K. K. (1988), "Buckling Capacities of Braced Heavy Columns under an Axial Load," *Comput. Struct.*, 28(5):563–71.

Wicks, P. J. (1987), "Compound Buckling of Elastically-Supported Struts," *J. Eng. Mech.* (ASCE), 113(12):1861–79 (Sec. 5.2).

Williams, F. W., and Aston, G. (1989), "Exact or Lower Bound Tapered Column Buckling Loads," *J. Struct. Eng.* (ASCE), 115(5):1088–99.

Williams, F. W., and Banerjee, J. R. (1985), "Flexural Vibration of Axially-Loaded Beams with Linear or Parabolic Taper," *J. Sound and Vibration,* 99(1):121–38.

Winkler, E. (1867), *Die Lehre von der Elastizität und Festigkeit,* H. Dominicus, Prague, Czechoslovakia, pp. 182–4 (Sec. 5.2.).

Wolde-Tinsae, A. M., and Foadian, H. (1989), "Asymmetrical Buckling of Prestressed Tapered Arches," *J. Eng. Mech.* (ASCE), 115(9):2020–33.

6

Thin-Walled Beams

As the term "beam" suggests, thin-walled beams are structures that can be treated as one-dimensional, characterized by cross-section deformation variables and cross-section internal forces that depend only on the longitudinal coordinate x. The basic forms of beam theory are the theory of bending and the theory of torsion. In the theory of bending, the shear deformations are neglected, and in the elementary theory of torsion of circular shafts, the longitudinal normal strains (as well as stresses) are zero. In the theory of thin-walled beams, by contrast, torsional loads also produce normal strains and stresses while longitudinal or transverse loads induce torsion and the thin-walled beam deformation involves both normal and shear strains. Torsion and bending become coupled.

The reason for this coupling is the fact that the cross section of a thin-walled beam can no longer be assumed to remain plane. The longitudinal displacements do not conform to a linear distribution over the cross-section plane and the cross section exhibits out-of-plane warping in response to torsion. The warping is generally not the same in adjacent cross sections, which obviously gives rise to longitudinal normal strains and stresses.

The loss of stability of thin-walled beams often occurs through a combination of bending and torsion, even if the loading consists of transverse and axial loads in one plane. The basic types of such instability are the lateral buckling of beams and the axial-torsional buckling of columns.

For the sake of easy comprehension, we will begin by analyzing warping torsion and the associated instabilities in I-beams. Then we generalize our formulation to thin-walled beams of arbitrary open cross section. We will make frequent use of the energy methods presented in the previous chapters, which are particularly efficacious for the treatment of thin-walled beams. Finally, we will briefly outline an analogous formulation for thin-walled beams of closed cross section, whose stability analysis is important for the safety of box–girder bridges.

A detailed exposition of the theory of thin-walled beams has been given in the works of Vlasov (1959), Bleich (1952), Timoshenko and Gere (1961), Murray (1984), and without stability also in Oden and Ripperger (1981). The basic developments took place during the 1930s and 1940s (see, e.g., Bleich, 1952, p. 105, or Nowinski, 1959). The early contributors included Wagner (1936), Vlasov (1959), Timoshenko (1945), Ostenfeld (1931), Bleich and Bleich (1936), Kappus (1937), Lundquist and Fligg (1937), Goodier (1941), and others.

6.1 POTENTIAL ENERGY AND DIFFERENTIAL EQUATIONS

The reason for warping of cross sections due to torsion can be most easily explained by considering first the behavior of I-beams. When the I-beam is subjected to a uniform torsional moment (or torque) M_t, the specific angle of twist θ' is distributed uniformly along the beam, as illustrated by the plan view of the beam in Figure 6.1a. The beam is said to undergo uniform torsion ($\theta' = \text{constant}$). However, for general boundary conditions, nonuniform M_t, or variable cross section, the torsion is nonuniform.

An easily understandable case is the torsion of an I-beam whose cross section is symmetric with regard to both the vertical and horizontal axes, and whose flanges and web are both fixed (clamped, built-in) at one end as shown in Figure 6.1c. The opposite end of the beam is free. As far as the horizontal displacements are concerned, each flange may be regarded as an independent beam. If the flanges at both ends are free to undergo axial displacements and the beam is subjected to a uniformly distributed torsional moment M_t (torque), the plan view of the deformation of the beam is as shown in Figure 6.1a. The cross sections of the two flanges rotate about the vertical axis by equal angles at all points along the beam. So the flanges are subjected to no bending and remain free of axial normal strains and stresses. The torque produces only shear stresses. This is the case of simple (St. Venant) torsion, which generally arises in beams of constant cross section if the torque is uniform. The cross sections of the I-beam nevertheless do not remain plane but warp out of their plane, since the cross sections of the opposite flanges rotate about the vertical axis in opposite sense.

If the flanges at one end are clamped (Fig. 6.1b), the rotation shown in Figure 6.1a cannot be accommodated at the clamped end. Thus, each flange undergoes bending in the horizontal plane. The top and bottom flanges bend to the opposite sides, and this again causes out-of-plane warping of the cross sections. In the case of nonuniform torsion, however, the warping is accompanied by axial strains and stresses associated with the bending of the flanges in the horizontal planes. The same is true when torque M_t varies along the beam.

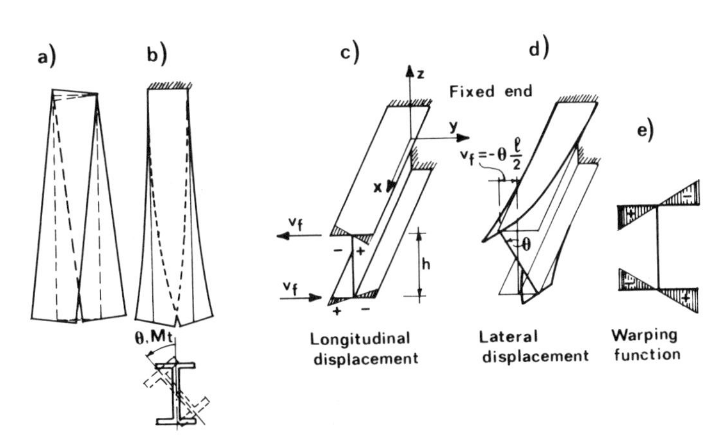

Figure 6.1 (a, b) I-beam subjected to torsional moment, (c, d) flange deformations, and (e) warping function.

Deformation of the Cross Section

Let us now formulate the behavior just described mathematically. We introduce a right-handed system of Cartesian coordinates x, y, z as shown in Figure 6.1c, x being the longitudinal coordinate, and y, z being the horizontal and vertical transverse coordinates. The x axis is the beam axis, and points (y, z) characterize the middle of the wall, which is assumed to be thin. The axial, transverse horizontal, and transverse vertical displacements of the points on the beam axis are denoted as u, v, w. Denoting as h the height of the I-beam and as θ the angle of twist as shown in Figure 6.1d (positive if clockwise when looking in he direction of x), the transverse horizontal displacements v_f of the top and bottom flanges and their longitudinal normal strains e_x are related to the angle of twist θ as follows:

$$\begin{aligned} &\text{Top flange:} && v_f = -\theta\frac{h}{2} && u_1 = -v_f'y = \theta'\frac{h}{2}y && e_x = -v_f''y = \theta''\frac{h}{2}y \\ &\text{Bottom flange:} && v_f = \theta\frac{h}{2} && u_1 = -v_f'y = -\theta'\frac{h}{2}y && e_x = -v_f''y = -\theta''\frac{h}{2}y \end{aligned} \tag{6.1.1}$$

in which u_1 represents the longitudinal displacement of points on the middle surface (Fig. 6.2a) and expressions for e_x are based on bending theory applied separately to each flange.

The difference in signs in Equations 6.1.1 between the top and bottom plates reflects the fact that the cross section of the I-beam ceases to be plane, that is, warps out of the plane as illustrated in Figure 6.1c. To preserve a planar form of the cross section after deformation, the signs for the top and bottom flanges would have to be the same.

It is now convenient, especially in view of further generalizations in Section 6.4, to describe the entire cross section by one equation. This can be done by writing

$$u_1 = -\theta'\omega \qquad e_x = -\theta''\omega \tag{6.1.2}$$

in which ω is a cross-section variable that is called the warping function and is defined as (Fig. 6.1e)

$$\begin{aligned} &\text{Top flange:} && \omega = -\frac{h}{2}y \\ &\text{Bottom flange} && \omega = \frac{h}{2}y \\ &\text{Web} && \omega = 0 \end{aligned} \tag{6.1.3}$$

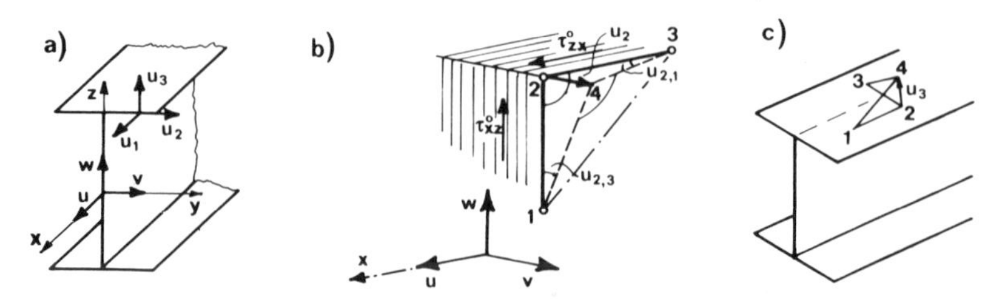

Figure 6.2 (a) Reference system for displacements of center line of cross section; (b, c) shear deformation due to transverse displacements.

The expression for e_x in Equations 6.1.2 defines the warping strains in the cross section.

Torsion is usually combined with bending and axial deformation, and then, according to the theory of bending, Equations 6.1.2 must be enlarged as

$$u_1 = u - v'y - w'z - \theta'\omega$$
$$e_x = u' - v''y - w''z - \theta''\omega \tag{6.1.4}$$

in which e_x is the axial normal strain at any point of the cross section and u, v, and w represent the axial, lateral, and vertical displacements (in directions x, y, and z) at the center 0 of the cross section (Fig. 6.2a). This center, which lies on the x axis of the beam, is chosen to coincide with the centroid of the I-beam cross section.

In the theory of warping torsion, the cross section is assumed to move in its own plane as a rigid body, that is, the in-plane deformations of the cross section are neglected while out-of-plane warping is not. Therefore, the lateral displacements u_2 and the vertical displacement u_3 of a point of the middle surface (Fig. 6.2a) may be calculated as

$$u_2 = v - \theta z \qquad u_3 = w + \theta y \tag{6.1.5}$$

Note that, although the section is no longer plane, we have for the web $\gamma_{xz}^{(1)} = \partial u_1/\partial z + \partial u_3/\partial x = 0$ and for the flanges $\gamma_{xy}^{(1)} = \partial u_1/\partial y + \partial u_2/\partial x = 0$, where $\gamma_{xz}^{(1)}$ and $\gamma_{xy}^{(1)}$ are the linear parts of the shear strain. These conditions, which can be verified immediately by substitutions of Equations 6.1.3, 6.1.4, and 6.1.5, will be generalized in Section 6.4.

The simplest approach to derive the governing differential equations of a thin-walled beam is the energy approach. Our energy analysis will be analogous to that by which we analyzed the beam-column under imposed axial displacement (Sec. 4.3). We are aiming at a linearized formulation that is correct for infinitely small incremental displacements u, v, θ but takes into account the effect of finite initial stresses σ^0 and τ^0 that exist prior to the displacements u, v, and θ. The strain energy is then a quantity that is second-order small in terms of displacement derivatives (gradients) u', v', and θ'. However, our expression for the axial strain e_x in Equations 6.1.4 is accurate only up to terms that are first-order small in terms of the displacement derivatives. Therefore, the axial strain expression must be enhanced by all second-order terms of the finite strain expression, particularly by the second-order axial strain $\varepsilon^{(2)}$ that is due to the transverse displacements v and w. For the same reason, we will have to calculate the second-order terms also for the shear strains, even though we have shown that the first-order parts of the shear strains are zero.

From the calculation of $\varepsilon^{(2)}$ from our analysis of columns (below Eq. 4.3.7) we recall that $\varepsilon^{(2)} = (ds_1 - dx)/dx$ in which ds_1 is the length of the longitudinal fiber after deformation due to displacements u_2 and u_3. As we calculated it for columns (Sec. 4.3), we have $ds_1 = (1 + u_2'^2 + u_3'^2)^{1/2}\,dx \simeq (1 + u_2'^2/2 + u_3'^2/2)$, from which $\varepsilon^{(2)} \simeq u_2'^2/2 + u_3'^2/2$. This is exact up to terms second-order small. Now the axial (normal) finite strain at a point undergoing displacements u_2, u_3 is expressed as $\varepsilon_x = e_x + \frac{1}{2}u_2'^2 + \frac{1}{2}u_3'^2$. Introducing the small (linearized) strain expression according to Equation 6.1.4, we thus obtain the result:

$$\varepsilon_x = u' - v''y - w''z - \theta''\omega + \tfrac{1}{2}(v' - \theta' z)^2 + \tfrac{1}{2}(w' + \theta' y)^2 \tag{6.1.6}$$

Consider now the shear angle γ_{xz} represented in Figure 6.2b by the change of the initially right angle $\sphericalangle 123$. Due to bending in the vertical plane associated with $w(x)$, this angle does not change because, for bending, the cross sections are assumed to remain normal to the deflected beam axis. In fact, we have seen (below Eqs. 6.1.5) that according to the assumed distribution of u_1, u_2, and u_3 (Eqs. 6.1.4–6.1.5), no first-order (linear) shear strain is produced. However, the lateral displacements u_2 produce a second-order small change γ_{xz} of angle $\sphericalangle 123$ (Fig. 6.2b), which causes the shear force V to work. The reason is that angle $\sphericalangle 143$ is less than 90° because plane 143 is not normal to the y axis (a plane normal to y is 123). The angle change may be calculated from angles $\sphericalangle 234 = u_{2,1}$ and $\sphericalangle 214 = u_{2,3}$ where $u_{2,1} = \partial u_2/\partial x$ and $u_{2,3} = \partial u_2/\partial z$. We will do a similar calculation for plates (see Sec. 7.2 and Fig. 7.3), and the result shows that the second-order shear angle is $\gamma_{xz}^{(2)} = \sphericalangle 123 - \sphericalangle 143 = u_{2,1}u_{2,3}$. Note that shear angle $\sphericalangle 123 - \sphericalangle 143$ is in the sense of γ_{xz}, and $\sphericalangle 143 - \sphericalangle 123$ would be against it. (The expression for γ_{xz} may also be deduced from the finite strain expression $\gamma_{13} = 2\varepsilon_{13} = u_{1,3} + u_{3,1} + u_{k,1}u_{k,3} \simeq u_{1,3} + u_{3,1} + u_{2,1}u_{2,3}$; see Sec. 7.2.) Because the cross section is rigid in its own plane, we may substitute $u_2 = v - \theta z$ (Eqs. 6.1.5) at any point of the cross section. So we finally get

$$\gamma_{xz} = \left(\frac{\partial u_2}{\partial x}\right)\left(\frac{\partial u_2}{\partial z}\right) = -v'\theta + \theta\theta' z$$
$$\gamma_{xy} = \left(\frac{\partial u_3}{\partial x}\right)\left(\frac{\partial u_3}{\partial y}\right) = w'\theta + \theta\theta' y \tag{6.1.7}$$

in which the second equation has been deduced by analogy (see also Fig. 6.2c). The shear strains given by Equations 6.1.7 are assumed to be constant through the small thickness of the wall.

Potential Energy

We will now restrict our attention to a beam under initial axial force P, initial lateral load p_z, initial bending moment M_z^0, and initial shear force V_z^0 in the xy plane. The strain energy due to incremental displacements u, v, and θ of the initially stressed I-beam may be calculated as

$$U = \int_0^l \int_A (\sigma^0 \varepsilon_x + \tfrac{1}{2}E e_x^2 + \tau_{xy}^0 \gamma_{xy} + \tau_{xz}^0 \gamma_{xz})\, dA\, dx + \int_0^l \tfrac{1}{2} GJ\theta'^2\, dx \tag{6.1.8}$$

in which l is the length of the I-beam, A is the cross-section area, σ^0 is the initial normal stress in the cross section, τ_{xy}^0 and τ_{xz}^0 are the initial shear stresses (constant throughout the thickness of wall), γ_{xy}, γ_{xz} are the associated shear angles, G is the elastic shear modulus, and GJ is the torsional stiffness for simple torsion. We include no strain energy associated with the stresses σ_{yy}, σ_{zz}, τ_{yz} in the plane of the cross section, which are assumed (and normally are) negligible.

The term $\sigma^0\varepsilon_x + \frac{1}{2}Ee_x^2$ in Equation 6.1.8 represents the strain energy due to axial normal strains. We use the finite strain ε_x to calculate the work $\sigma^0\varepsilon_x$ of the initial stresses σ^0, while at the same time we use the linearized, small-strain expression e_x to calculate the incremental strain energy. Inconsistent as this might first seem, our expression is nevertheless correct up to terms second-order small, which is all that we need for a linearized formulation at small incremental deformations. The term $\frac{1}{2}Ee_x^2$ is second-order small, and the further terms in the expression $\frac{1}{2}E\varepsilon_x^2$ with finite strain ε_x would be higher-order small, and would have to be truncated anyway for the purpose of a linearized formation. It is also obvious that the expression for $\sigma^0\varepsilon_x$ (in which σ^0 is not small) must be exact up to the second-order small terms because $\frac{1}{2}Ee_x^2$ is second-order small.

The work of initial shear stresses is $\tau_{xy}^0\gamma_{xy} + \tau_{xz}^0\gamma_{xz}$. As we already mentioned, the first-order components of γ_{xy} and y_{xz} are zero, and so this expression is of the second order. For this reason we have also omitted in Equation 6.1.8 the term $\int_0^l \int_A \frac{1}{2}G(\gamma_{xz}^2 + \gamma_{xy}^2)\, dA\, dx$; it would give only terms of higher than second order.

The last term in the expression of U (Eq. 6.1.8) represents the incremental strain energy due to shear stresses τ^* that results from the so-called Saint-Venant torsion, also called the simple torsion or uniform torsion (this energy was ignored by the foregoing assumption of constant strain distribution across the thickness). The Saint-Venant torsion is a torsional mode with $\theta' = \text{const}$. In this case torsion causes no longitudinal normal stresses, only shear stresses τ^* that are zero at the middle surface and vary approximately linearly across the wall thickness (Fig. 6.3b). In textbooks of mechanics of materials (e.g., Timoshenko and Goodier, 1973) it is shown that for an open cross section that consists of a set of thin plates of thicknesses t_j and lengths b_j, the torsional stiffness for Saint-Venant torsion, which results from these shear stresses, may be approximately calculated from the equation

$$GJ = G \sum_{j=1}^{n} \tfrac{1}{3} b_j t_j^3 \tag{6.1.9}$$

where n is the number of plates forming the cross section. This expression is derived as the limit for the torsional stiffness of a rectangular cross section and is exact asymptotically for a very small plate thickness, that is, $t_j/b_j \to 0$. According to Equation 6.1.9, the torsional stiffness of the cross section is the sum of the torsional stiffnesses of all its plates. This is true for open cross sections, but not for closed cross sections such as those of box girders.

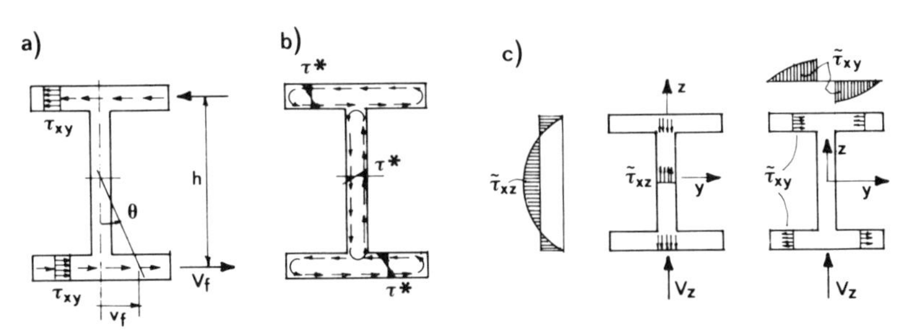

Figure 6.3 (a) Shear stress distributions due to warping torsion, (b) Saint-Venant torsion, and (c) shear force.

An interesting point is that the coefficient in Equation 6.1.9 is $\frac{1}{3}$ and not $\frac{1}{12}$ as the summation of the elementary torques from all the wall elements (by the shear stresses parallel to the wall surface) would seem to indicate. The reason is that additional torques arise from the shear fluxes across the wall thickness at the ends of the flanges.

The initial axial stresses σ^0 according to the bending theory and the initial shear stresses τ^0_{xz}, τ^0_{xy} may be expressed as

$$\sigma^0 = -\frac{P}{A} - \frac{M^0_z z}{I_z}$$
$$\tau^0_{xz} = \frac{V^0_z}{A} + \tilde{\tau}_{xz} \qquad \tau^0_{xy} = \tilde{\tau}_{xy} \qquad (V^0_y = 0) \tag{6.1.10}$$

in which the axial force P is positive for compression but stress σ^0 is positive for tension, $A =$ cross-section area, and $I_z = \int z^2\,dA =$ centroidal moment of inertia of the cross section about the y axis. The stresses $\tilde{\tau}_{xz}$ and $\tilde{\tau}_{xy}$ represent the deviations of the total shear stresses from the average shear stresses V^0_z/A and V^0_y/A, whose distribution may be approximately determined from the axial equilibrium condition; see, for example, Oden and Ripperger (1981, p. 120). $\tilde{\tau}_{xz}$ and $\tilde{\tau}_{xy}$ need not be calculated since their work is zero for a doubly symmetric cross section, as we will see later. Consequently, the work of the initial shear stresses can be expressed as the work of their average V^0_z/A.

It is certainly interesting that the initial shear stresses due to the shear force do nonnegligible work even though the deformed cross sections (except for their warping) are assumed to remain normal to the deflected beam axis. The need to include the work of shear stresses in the energy expression has been overlooked in most studies of stability of thin-wall beams, but was included in the formulation by Powell and Klingner (1970) and Pignataro, Rizzi, and Luongo (1983); Masur (1975) deduced its expression indirectly from the differential equations of equilibrium and the boundary conditions.

The increment ΔU of the strain energy of the beam compared to the initial equilibrium, which coincides with the total energy U if the zero energy value is associated with the initial state, may be calculated by substituting Equations 6.1.10 as well as Equations 6.1.4, 6.1.6, and 6.1.7 into Equation 6.1.8:

$$U = \int_0^l \int_A \Big[\Big(-\frac{P}{A}\Big)\Big(u' - v''y - w''z - \theta''\omega + \frac{1}{2}v'^2 - v'\theta'z + \frac{1}{2}\theta'^2 z^2$$
$$+ \frac{1}{2}w'^2 + w'\theta'y + \frac{1}{2}\theta'^2 y^2\Big)$$
$$+ \Big(-\frac{M^0_z}{I_z}\Big)\Big(u'z - v''yz - w''z^2 - \theta''\omega z + \frac{1}{2}v'^2 z - v'\theta'z^2$$
$$+ \frac{1}{2}\theta'^2 z^3 + \frac{1}{2}w'^2 z + w'\theta'yz + \frac{1}{2}\theta'^2 y^2 z\Big)$$
$$+ \Big(\frac{V^0_z}{A} + \tilde{\tau}_{xz}\Big)(-v'\theta + \theta\theta'z) + \tilde{\tau}_{xy}(w'\theta + \theta\theta'y)$$

$$+\frac{E}{2}(u'^2 + v''^2y^2 + w''^2z^2 + \theta''^2\omega^2 - 2u'v''y - 2u'w''z - 2u'\theta''\omega$$

$$+ 2v''w''yz + 2v''\theta''y\omega + 2w''\theta''z\omega)\Big]\, dA\, dx + \int_0^l \frac{GJ}{2}\theta'^2\, dx \qquad (6.1.11)$$

In this integral, the integration over the cross-section area A involves only the variables y, z, ω and their powers, while the variables u', v'', θ'', etc. depend only on the axial coordinate x and are constant during integration within each cross section. For doubly symmetric cross sections, such as that of an I-beam, the following integrals over the cross section vanish:

$$\int_A y\, dA = \int_A z\, dA = \int_A yz\, dA = \int_A \omega\, dA = \int_A y\omega\, dA$$

$$= \int_A z\omega\, dA = \int_A y^2 z\, dA = \int_A z^3\, dA = 0 \qquad (6.1.12)$$

provided the centroid of the cross section lies on the x axis. These relations ensue from the fact that the integrands have equal magnitudes and opposite signs at locations that are symmetric with respect to the z axis, the y axis, or center 0. Further we may introduce the notations:

$$A = \int_A dA \qquad I_y = \int_A y^2\, dA \qquad I_z = \int_A z^2\, dA \qquad I_\omega = \int_A \omega^2\, dA \quad (6.1.13)$$

The quantity I_ω (often also denoted as Γ) is called the warping moment of inertia of the cross section (alternatively, the term "sectorial moment of inertia" has also been used, for reasons that will be clear when we discuss arbitrary open cross sections). While the dimensions of the moment of inertia I_z and I_y are $(\text{length})^4$, the dimension of the warping moment of inertia I_ω is $(\text{length})^6$, as is clear if ω is substituted according to Equations 6.1.3. In this sense, I_ω is a higher-order cross-section parameter. For a doubly symmetric I-beam with negligible web area ($t_w/t_f \to 0$), integration according to Equations 6.1.3 and 6.1.13 yields $I_\omega = I_f h^2/2 = I_y h^2/4$, which $I_f = \frac{1}{12}t_f b_f^3$ = centroidal moment of inertia of one flange considered separately; t_f = thickness of the flange, and b_f = width (or depth) of the flange (Fig. 6.4a).

Bringing the integral over A into each term of Equation 6.1.11, we obtain for

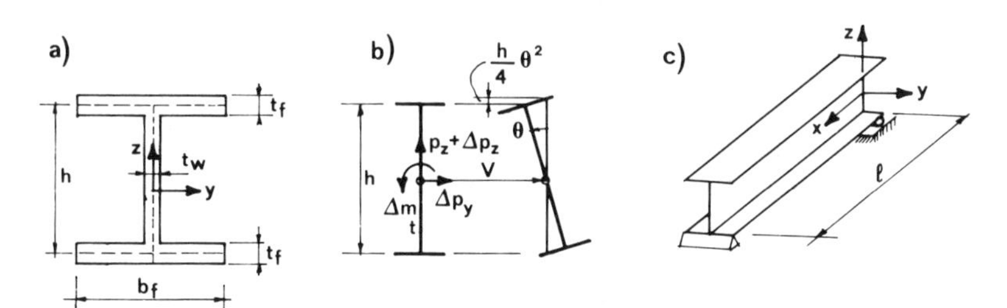

Figure 6.4 (a) Geometry of cross section; (b) applied loads at beam axis; (c) simply supported beam.

the beam the strain energy expression:

$$U=\int_0^l \left[-\frac{P}{A}\left(u'A+\frac{v'^2}{2}A+\frac{\theta'^2}{2}I_z+\frac{w'^2}{2}A+\frac{\theta'^2}{2}I_y\right)-\frac{M_z^0}{I_z}(-w''I_z-v'\theta'I_z)\right.$$

$$\left.+M_z^{0\prime}v'\theta+\frac{E}{2}(u'^2A+v''^2I_y+w''^2I_z+\theta''^2I_\omega)+\frac{GJ}{2}\theta'^2\right]dx \qquad (6.1.14)$$

In writing Equation 6.1.14 we took into account the fact that, due to the double symmetry of the cross section, the work of $\tilde{\tau}_{xz}$ and $\tilde{\tau}_{xy}$ (which have a zero resultant) is zero (see also Fig. 6.3c). Also, since we regard the deflection z^0 in the prebuckled state as negligibly small, we have put $V_z^0 \simeq -M_z^{0\prime}$ in the foregoing equation.

With the approximations just introduced, the terms in Equation 6.1.14 that involve both v and θ are $U_s=\int_0^l (M_z^0 v'\theta' + M_z^{0\prime}v'\theta)\,dx$. These terms may be simplified by integrating the first term by parts, which yields $U_s=\int_0^l(-M_z^{0\prime}v'\theta - M_z^0 v''\theta + M_z^{0\prime}v'\theta)\,dx + [M_z^0 v'\theta]_0^l$. The term $[\cdots]_0^l$ vanishes for many realistic boundary conditions at ends $x=0$ and $x=l$. So we conclude that $U_s = \int -M_z^0 v''\theta\,dx$. This form of the work term due to M_z^0 has been introduced in some classical works (see, e.g., the examples solved by Timoshenko and Gere, 1961, p. 266).

Regrouping the terms in Equation 6.1.14, and introducing the notation $(I_z+I_y)/A=r_P^2$, we finally obtain for the potential energy $\Pi=U-W$ the expression

$$\Pi=\int_0^l \frac{1}{2}(EI_y v''^2+EI_z w''^2+EI_\omega\theta''^2+GJ\theta'^2+2M_z^0 w''-2M_z^0 v''\theta)\,dx$$

$$-\int_0^l \frac{P}{2}(v'^2+w'^2+\theta'^2 r_P^2)\,dx+\int_0^l(\tfrac{1}{2}EAu'^2-Pu')\,dx - W \qquad (6.1.15)$$

The work of loads may be expressed as

$$W=\int_0^l[(p_z+\Delta p_z)w+\Delta p_y v+\Delta m_t\theta]\,dx+[\Delta M_z w']_0^l+[\Delta M_t\theta]_0^l-[(P+\Delta P)u]_0^l$$

$$(6.1.16)$$

in which Δp_y and Δp_z are small transversal distributed disturbing loads (added to the initial load p_z), Δm_t is a small perturbation representing an applied distributed moment about the beam axis, ΔM_y and ΔM_t are small moments about the z and x axes applied at the beam ends, and ΔP is a small disturbing increment in the axial compressive load. Keep in mind this is an incremental theory, which is accurate only if Δp_y, Δp_z, Δm_t, ΔM_y, ΔM_t, and ΔP represent increments that are small compared to p_z, M_z^0, and P.

The point of application of p_z in Equation 6.1.16 is assumed to be on the beam axis (i.e., at the beam centroid). If p_z is applied on the top flange, then the integrand in Equation 6.1.16 must be augmented by the term $-p_z(\frac{1}{4}h\theta^2)$ because rotation θ causes additional negative displacement $\frac{1}{4}h\theta^2$ as shown in Figure 6.4b (h = depth of the I-beam).

Differential Equations and Boundary Conditions

The differential equations of equilibrium in terms of u, w, v, and θ may now be obtained as the Euler variational equations for the sum of the integrands of Equations 6.1.15 and 6.1.16. The result is

$$(EAu')' = 0 \tag{6.1.17a}$$

$$(EI_z w'')'' + (Pw')' = \Delta p_z \tag{6.1.17b}$$

$$(EI_y v'')'' - (M_z^0 \theta)'' + (Pv')' = \Delta p_y \tag{6.1.18a}$$

$$(EI_\omega \theta'')'' - [(GJ - r_P^2 P)\theta']' - M_z^0 v'' = \Delta m_t \tag{6.1.18b}$$

In deriving Equation 6.1.17b we took into account the fact that $M_z^{0''} \simeq p_z$ because the initial state is an equilibrium state. The approximation $M_z^{0''} \simeq p_z$ results if the moment Pz^0 is neglected in calculating M_z^0.

Equation 6.1.18b is written for the case when load p_z is applied at the beam axis. If it is applied at the center of the top flange of the I-beam, the term $-p_z(h/2)\theta$ must be added to the right-hand side (because the term $-p_z\theta^2 h/4$ needs to be added to the integrand of Eq. 6.1.16, already mentioned).

The first two relations (Eqs. 6.1.17a, b) represent the well-known equations that govern axial extension and bending in the vertical plane. These equations are uncoupled from each other. By contrast, the differential equations for transverse rotation θ and for lateral displacement v (Eqs. 6.1.18a, b) are coupled. They represent a system of two linear simultaneous ordinary differential equations for the unknown functions $\theta(x)$ and $v(x)$.

The boundary conditions that are compatible with the existence of potential energy can be obtained from Equations 6.1.15 and 6.1.16 by the method of variational calculus (see Sec. 5.1). Alternatively, the boundary conditions can also be set up directly from the conditions of the stresses and strains at the end of the beam, as we will see later.

Equations 6.1.18a, b for $\Delta p_y = \Delta m_t = 0$ represent the basic differential equations that govern the lateral buckling of beams as well as the axial-torsional buckling of columns. Lateral buckling is the instability caused by bending moment M_z^0 when the axial force $P = 0$, and axial-torsional buckling is the instability due to the axial load P when the bending moment $M_z^0 = 0$. Naturally, a combination of lateral buckling and axial-torsional buckling is also possible.

If the boundary conditions are homogeneous and $\Delta p_y = \Delta m_t = 0$, Equations 6.1.18a, b define a linear eigenvalue problem for a beam that may be called perfect. When an applied distributed moment m_t about the x axis, or an applied lateral distributed load p_y in the direction of the y axis is considered, or if p_z has an eccentricity, then Equations 6.1.18a, b have nonzero right-hand sides, that is, are nonhomogeneous. In that case, the problem is to solve the rotations θ and lateral deflections v caused by these disturbing loads. Obviously, when an eigenstate of the associated homogeneous problem is approached, these rotations or deflections tend to infinity.

The preceding analysis suffices for formulating the boundary conditions for a fixed end of the beam as well as a simple support. If the end of the beam is fixed (clamped, built-in) over the entire cross section, we have the boundary conditions:

$$v = 0 \qquad \theta = 0 \qquad v' = 0 \qquad \theta' = 0 \qquad (\text{at } x = 0, x = l) \qquad (6.1.19)$$

The last condition follows from the fact that $u_1 = -\theta'\omega = 0$.

Another basic type of boundary condition is the simple support illustrated in Figure 6.4c. The boundary conditions then are

$$v = 0 \qquad \theta = 0 \qquad v'' = 0 \qquad \theta'' = 0 \qquad (6.1.20)$$

in which the last condition follows from the fact that $\sigma_x = Ee_x = -E\theta''\omega = 0$, but can nevertheless be obtained easily by the energy method.

The torque M_t is the sum of the Saint-Venant torque $GJ\theta'$ and torque $V_f h$ that is due to the horizontal shear forces V_f in the flanges (Fig. 6.3a) associated with the bending of flanges in the horizontal planes. Torque $GJ\theta'$ is due to the parts of the shear stress τ^* that are nonuniform (linear) across the wall thickness (Fig. 6.3b) and are zero at midthickness, while torque $V_f h$ is due to the shear stresses τ at midthickness of the wall, which represent the shear stress average over the thickness. According to the theory of bending, $V_f = -(EI_f v_f'')' = -[EI_f(h\theta/2)'']'$, and for $h =$ constant, $V_f h = -[EI_f(h\theta/2)'']'h = -EI_\omega\theta'''$ (because $I_f h^2/2 = I_\omega$). Therefore, the torque for a beam of uniform cross section is

$$M_t = -EI_\omega\theta''' + GJ\theta' \qquad (6.1.21)$$

The boundary condition of a free end is $M_t =$ applied torque, and $M_t = 0$ if no torque is applied. Note that the boundary condition is not $\theta' = 0$ (except when $I_\omega/Jl^2 \rightarrow 0$).

Problems

6.1.1 Derive the differential equations (Eqs. 6.1.18) directly by the variational procedure from the work expression (Eqs. 6.1.15–6.1.16) without using the Euler equation.

6.1.2 Extend the preceding assignment to determine by the variational procedure the kinematic (essential) and static (natural) boundary conditions for a (a) fixed end (Eqs. 6.1.19), (b) simple support (Eqs. 6.1.20), and (c) free end (Eq. 6.1.21).

6.1.3 What effect does a neglect of the work of shear stresses on the second-order shear strains have on the differential equations of equilibrium? For which practical situations is such neglect admissible?

6.1.4 How do the differential equations simplify when $t_w/t_f \rightarrow 0$ (ideal I-beam) and at the same time $t_f/h \rightarrow 0$ (very thin flanges)?

6.1.5 How do the differential equations simplify when $b_f = 0$ (i.e., the I–cross section degenerates into a rectangular cross section of width b_f and height h)?

6.1.6 A pile in the ground may be approximately treated as a thin-wall I-beam embedded in an elastic foundation that resists not only deflections but also rotation θ about beam axis. Derive the differential equation and boundary condition by minimizing Π (for $I_\omega = 0$ this must coincide with Prob. 5.2.16).

6.2 AXIAL-TORSIONAL BUCKLING OF COLUMNS

One characteristic property of thin-walled columns is that they may twist as they buckle due to axial load. This phenomenon can considerably reduce the critical load. It is called axial-torsional buckling and can be analyzed on the basis of Equations 6.1.18a, b, in which we now assume the cross section to be constant, $M_z^0 = 0$ and $P > 0$ (Fig. 6.5a). Then the coupling term between the transverse rotation θ and the lateral displacement v disappears. For the case of constant cross section, Equations 6.1.17b and 6.1.18a, which govern bending in the xz and xy plane, become

$$EI_z w^{\mathrm{IV}} + Pw'' = 0 \qquad EI_y v^{\mathrm{IV}} + Pv'' = 0 \tag{6.2.1}$$

These are the well-known bending differential equations that, along with the appropriate homogeneous boundary conditions, give the flexural (bending) critical loads. Equation 6.1.18b becomes

$$\theta^{\mathrm{IV}} + \frac{Pr_P^2 - GJ}{EI_\omega}\theta'' = 0 \tag{6.2.2}$$

and it is seen to be analogous in form to the bending differential equations (Eqs. 6.2.1). Also analogous are the boundary conditions for fixed or simple supports at the ends (Eqs. 6.1.19 and 6.1.20). We then have (with $n = 1, 2, 3, \ldots$):

Simple (torsional) supports (Fig. 6.5b): $$\frac{P_{\mathrm{cr}_n} r_P^2 - GJ}{EI_\omega} = \frac{n^2\pi^2}{l^2} \tag{6.2.3a}$$

Fixed (torsional) supports (Fig. 6.5c): $$\frac{P_{\mathrm{cr}_n} r_P^2 - GJ}{EI_\omega} = \frac{4n^2\pi^2}{l^2} \tag{6.2.3b}$$

As we now see, there exists for the I–cross section a critical load P_{cr_θ} for which the buckling mode represents pure torsion. The lowest critical load, which occurs

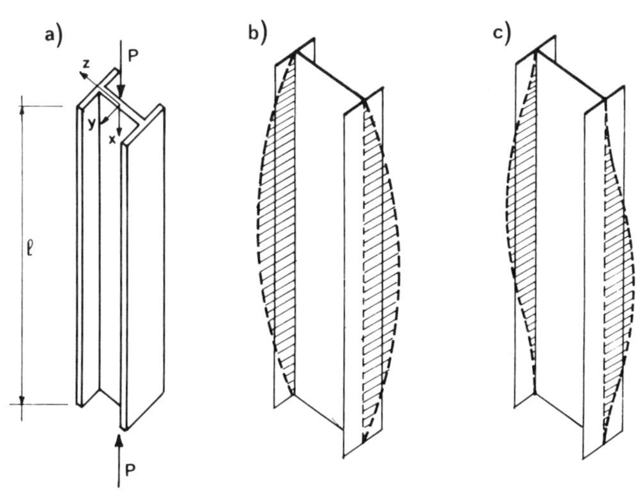

Figure 6.5 Axial-torsional buckling of I-beams with different end restraints.

for $n = 1$, is

Simple (torsional) supports (Fig. 6.5b):
$$P_{\text{cr}_\theta} = \frac{GJ + EI_\omega \dfrac{\pi^2}{l^2}}{r_P^2} \tag{6.2.4a}$$

Fixed (torsional) supports (Fig. 6.5c):
$$P_{\text{cr}_\theta} = \frac{GJ + EI_\omega \dfrac{4\pi^2}{l^2}}{r_P^2} \tag{6.2.4b}$$

There obviously exist also critical loads P_{cr_z} and P_{cr_y} for bending in the xz and xy plane. So the column will buckle for the lowest of the three loads P_{cr_y}, P_{cr_z}, P_{cr_θ} (for I-beam $P_{\text{cr}_z} > P_{\text{cr}_y}$ always). If imperfections are considered, all these modes are excited simultaneously to various extents. Note that the column length l has no effect when $I_\omega \to 0$. This has made the cruciform column an attractive test for determining certain material properties (cf. Eq. 8.1.25 and Fig. 8.10b).

The ratio of bending to torsional critical loads is, according to Equation 6.2.4a,

$$\frac{P_{\text{cr}_\theta}}{P_{\text{cr}_y}} = \frac{I_\omega}{r_P^2 I_y} + \frac{G}{\pi^2 E}\left(\frac{l}{r_P}\right)^2\left(\frac{J}{I_y}\right) \simeq \frac{1}{1 + \frac{1}{3}(b_f/h)^2} + \frac{4G}{\pi^2 E}\left(\frac{t_f}{b_f}\right)^2\left(\frac{l}{r_P}\right)^2 \tag{6.2.5}$$

where the last expression is valid for ideal doubly symmetric I-beams, $(t_w/t_f \to 0)$, for which $I_\omega = t_f b_f^3 h^2/24$, $I_y = t_f b_f^3/6$, and $r_P^2 = b_f^2/12 + h^2/4$. We now see that flexural buckling prevails when the column is sufficiently slender (l/r_P is high), while the flange thickness t_f is not too small. This condition often applies in practice. For example, cold-formed steel cross sections are more prone to torsional buckling than hot-rolled ones because they are thinner. Wide-flange I-beams are more prone to torsional buckling than regular ones, and stubby columns more than slender ones. However, for stubby columns, the failure due to reaching the strength or yield limit of the material often dominates over any type of buckling unless the walls are sufficiently thin. Obviously, torsional buckling is more important for high-strength materials.

Even when the flexural critical load is less than the torsional one, imperfections may induce significant torsional deformations below the critical load.

The conclusions we have drawn so far apply to doubly symmetric cross sections. But they also apply more generally to cross sections in which the shear center (see Sec. 6.4) coincides with the centroid (center of mass).

Finally, Figure 6.6 (Northwestern University, 1969) demonstrates the two possible buckling modes of cruciform columns: (1) axial-torsional buckling (left), which occurs when $P_{\text{cr}_\theta} < P_{\text{cr}_y}$ $(=P_{\text{cr}_z})$, and (2) planar flexural buckling (right), which occurs when $P_{\text{cr}_y} (= P_{\text{cr}_z}) < P_{\text{cr}_\theta}$.

Problems

6.2.1 For the beam that is simply supported (e.g., rests on an edge supported at one end and two spherical balls at the other end) and has a rectangular cross section (Fig. 6.7a), calculate the torsional buckling load. How does it depend on length L? How does it depend on the type of end supports?

6.2.2 For the I-beam in Figure 6.7c, calculate the torsional and flexural critical loads assuming $H = b_f$, $L = 20H$, and $H/t_f = 5$, 10, 20, 50, 100. Plot their ratio as a function of H/t_f.

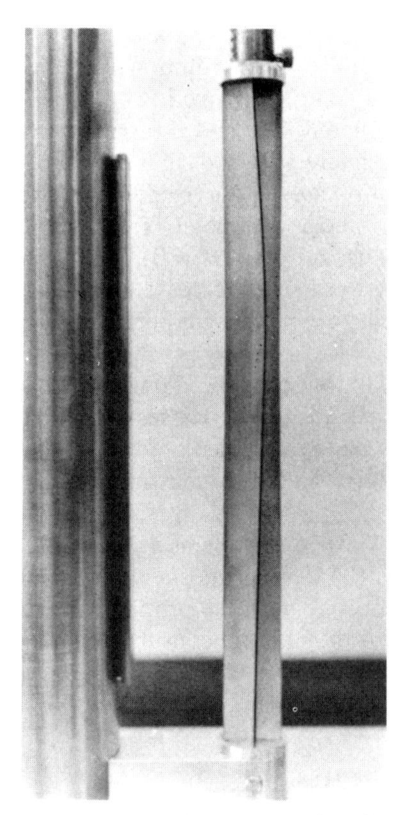

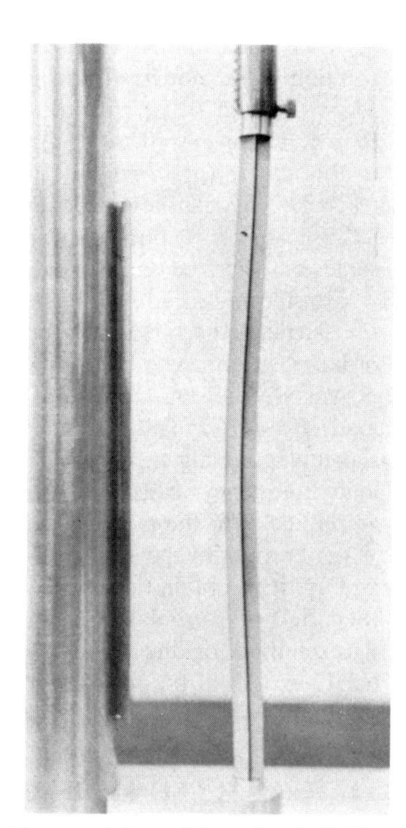

Figure 6.6 Northwestern University (1969) teaching models: axial-torsional (left) and planar (right) buckling of cruciform columns.

6.2.3 For the ideal I-beam, as well as for the beam of thin cruciform cross section shown in Figure 6.7b (for which $I_\omega/r^2J \rightarrow 0$), calculate the torsional and flexural critical loads for the following boundary conditions: (a) simply supported [so that $v = w = \theta = 0$ and $\sigma_x(y, z) \equiv 0$ at ends]; (b) same but two balls are replaced by one spherical ball, allowing free rotation θ at the end; (c) both ends fixed against all rotations ($v' = \theta' = 0$) but not against warping [$\sigma_x(y, z) = 0$]; (d) both ends completely fixed [$v' = w' = \theta' = 0$, $u(y, z) \equiv 0$];

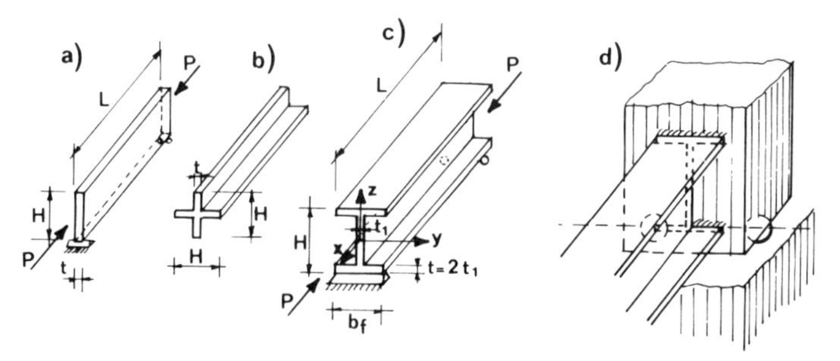

Figure 6.7 Exercise problems on axial-torsional buckling of columns.

(e) both ends rigidly stiffened against warping ($\theta = \theta' = 0$) and fixed against rotation θ but not fixed against rotations w' and v' (i.e., rotation v' is free, $M_z = 0$, Fig. 6.7d); (f) cantilever beam with a completely fixed (built-in) end $[\theta = \theta' = v' = w' = 0,\ u(y, z) \equiv 0]$ and a completely free end v (assume $J = 0$ in this case); (g) same but the free end is rigidly stiffened against warping ($\theta' = 0$); (h) cantilever beam with a support preventing slope and twist ($\theta' = v' = w' = 0$) but not warping (free θ'), and a completely free end; (i) same, but the free end is rigidly stiffened against warping ($\theta' = 0$).

6.2.4 Considering an ideal I-beam and including the inertia forces due to $\ddot{\theta}$ and $\ddot{\theta}'$, calculate the torsional vibration frequency for a simply supported column of length l subject to constant axial forces P.

6.2.5 Analyze the same I-beam as in Problem 6.2.4, but consider a pulsating axial load, $P = P_0 + P_t \sin \Omega t$. Note that bending of the flanges causes their axial shortening. Using the energy method from Section 3.3, calculate the smallest load frequency causing parametric resonance and the maximum stable amplitude P_t in the presence of damping.

6.2.6 (a) Formulate the Rayleigh quotient (Sec. 5.3) for Problems 6.2.1 to 6.2.3 and use it to obtain the upper bounds for P_{cr_1}. (b) Can a Timoshenko quotient (Sec. 5.4) be formulated for these problems? (c) Solve these problems by the Ritz method or the Galerkin method. (It is helpful to consult first Eqs. 6.3.12–6.3.15 of the next scction.)

6.3 LATERAL BUCKLING OF BEAMS AND ARCHES

Thin-walled beams can buckle by lateral twisting. This phenomenon, called lateral (or lateral-torsional) buckling, can again be analyzed on the basis of Equations 6.1.18a, b. To illustrate it, consider the example of a simply supported beam shown in Figure 6.8a. The cross section of the beam is constant and the axial force in the beam is P. For the sake of simplicity we consider first a uniformly distributed bending moment M_z^0 in the vertical plane, applied at the ends of the beam. This has the advantage that the coefficients of the governing differential equations are constant. In practical problems, of course, the bending moment causing instability is variable along the beam, and then these coefficients

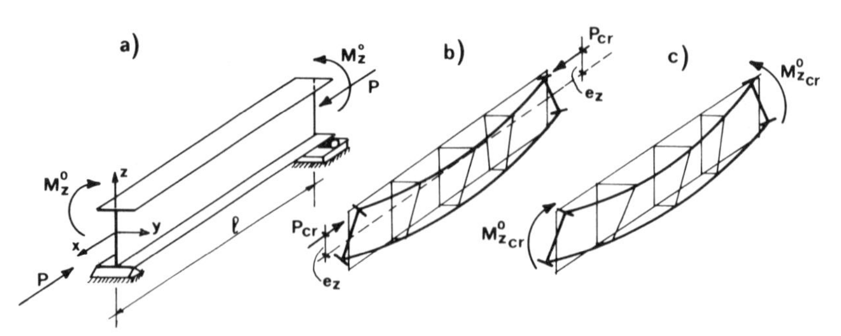

Figure 6.8 (a) Buckling of I-beam, subjected to (b) eccentric axial force or (c) uniform bending moment.

are not constant, which complicates the solution. The boundary conditions of simple supports at both ends (Eqs 6.1.20) can be satisfied, term by term, by the following Fourier series representations:

$$v = \sum_{n=1}^{\infty} a_n \sin \frac{n\pi x}{l} \qquad \theta = \sum_{n=1}^{\infty} b_n \sin \frac{n\pi x}{l} \tag{6.3.1}$$

in which a_n and b_n are constants. Substituting these series into Equations 6.1.18a, b, we obtain

$$\begin{gathered}\sum_{n=1}^{\infty} \left(EI_y a_n \frac{n^4\pi^4}{l^4} + M_z^0 b_n \frac{n^2\pi^2}{l^2} - P a_n \frac{n^2\pi^2}{l^2} \right) \sin \frac{n\pi x}{l} = 0 \\ \sum_{n=1}^{\infty} \left[EI_\omega b_n \frac{n^4\pi^4}{l^4} + (GJ - r_P^2 P) b_n \frac{n^2\pi^2}{l^2} + M_z^0 a_n \frac{n^2\pi^2}{l^2} \right] \sin \frac{n\pi x}{l} = 0\end{gathered} \tag{6.3.2}$$

To satisfy these equations identically for any x, it is necessary that for each n either the expressions in the parentheses vanish or $a_n, b_n = 0$. This condition (which is exact as long as all $n = 1, \ldots, \infty$ are considered) yields, for each n, an independent system of two linear algebraic homogeneous equations for a_n and b_n

$$\begin{bmatrix} P_{\mathrm{cr}_y}^{(n)} - P & M_z^0 \\ M_z^0 & (P_{\mathrm{cr}_\theta}^{(n)} - P) r_p^2 \end{bmatrix} \begin{Bmatrix} a_n \\ b_n \end{Bmatrix} = \begin{Bmatrix} 0 \\ 0 \end{Bmatrix} \tag{6.3.3}$$

where $P_{\mathrm{cr}_y}^{(n)} = n^2\pi^2 EI_y / l^2$ and $P_{\mathrm{cr}_\theta} = (GJ + EI_\omega n^2\pi^2/l^2)/r_p^2$ are the critical loads for purely flexural and purely torsional buckling defined in Section 6.2.

Axial-Torsional Buckling due to Eccentric Axial Force

Let us consider first the case of eccentric compression, that is, the case where the initial bending moment M_z^0 is caused by the eccentricity e_z of the axial load, $M_z^0 = Pe_z$. As we know, the column can deform in a planar bending mode ($w \neq 0$ at $v = \theta = 0$), which is governed solely by Equation 6.1.17b. We focus our attention exclusively on a combined mode that involves both v and θ, satisfies Equation 6.3.3, and is uncoupled. This mode arises when the determinant of Equation 6.3.3 vanishes. This condition gives for the first critical value ($n = 1$) $P = P_{\mathrm{cr}}$ the equation

$$XY = \left(\frac{e_z}{r_p} \right)^2 \qquad \text{with } X = \frac{P_{\mathrm{cr}_y}}{P_{\mathrm{cr}}} - 1, \qquad Y = \frac{P_{\mathrm{cr}_\theta}}{P_{\mathrm{cr}}} - 1 \tag{6.3.4}$$

which can be rearranged to the quadratic equation

$$\left(1 - \frac{e_z^2}{r_p^2} \right) P_{\mathrm{cr}}^2 - (P_{\mathrm{cr}_y} + P_{\mathrm{cr}_\theta}) P_{\mathrm{cr}} + P_{\mathrm{cr}_y} P_{\mathrm{cr}_\theta} = 0 \tag{6.3.5}$$

from which P_{cr} can be easily solved.

The left-hand side of Equations 6.3.5 is a function $f(P_{\mathrm{cr}})$. It is represented by the parabolas shown in Figure 6.9a. Their intersections with the P axis are the roots, that is, P_{cr}. If $r_p > |e_z| > 0$, the parabola is convex, and Figure 6.9a illustrates that the smaller root is positive and less than both P_{cr_y} and P_{cr_θ}. If $|e_z| > r_p$, the parabola is concave and the smaller root P_{cr} is negative, that is, a

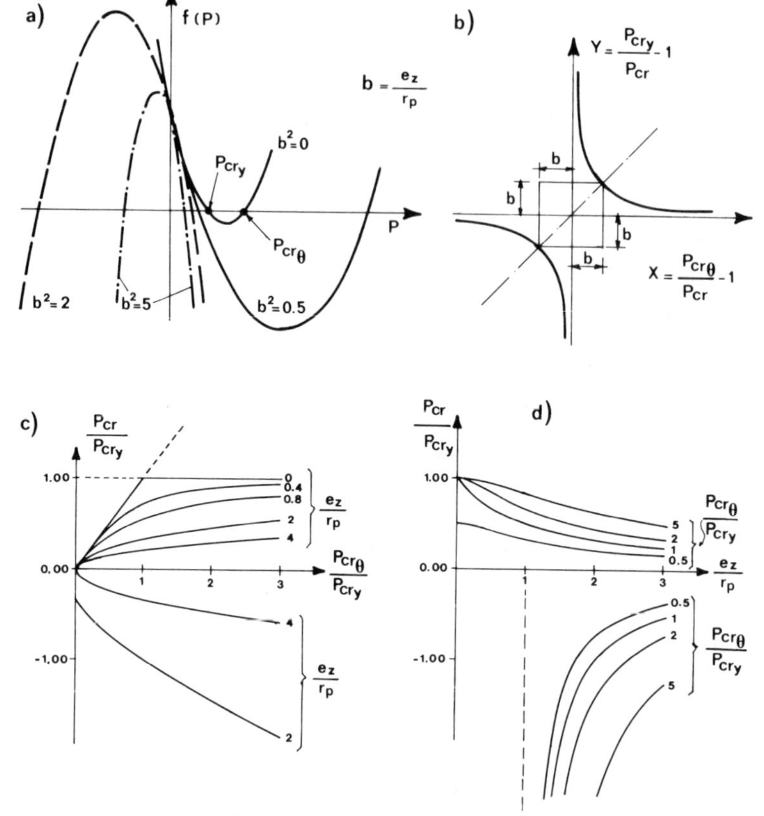

Figure 6.9 (a) Plot of the left-hand side of Equation 6.3.5; (b, c, d) graphic solutions of Equations 6.3.4 and 6.3.5.

tensile load causes the beam to buckle laterally, although the same beam would also buckle for a compressive load (the second, larger root), which is smaller in magnitude. For the case of compression ($P_{cr} > 0$) and for any value of e_z, Figure 6.9a shows that $P_{cr} < P_{cr_y}$ and $P_{cr} < P_{cr_\theta}$ simultaneously, that is, $P_{cr} < \min(P_{cr_y}, P_{cr_\theta})$. For tension ($P_{cr} < 0$) and for $|e_z| = r_p$, $P_{cr} \to -\infty$; so we see that buckling in tension is practically possible only if the eccentricity is substantially larger than r_p.

The solutions of Equations 6.3.5 are plotted in Figure 6.9c, d.

The basic properties of P_{cr} can also be deduced more directly from the plot of Equation 6.3.4 shown in Figure 6.9b, which makes it possible to calculate P_{cr_θ} if P_{cr} and P_{cry} is chosen. Because for compression ($P_{cr} > 0$) we always have $P_{cr} < P_{cr_y}$ and $P_{cr} < P_{cr_\theta}$, the relations among P_{cr}, P_{cr_y}, and P_{cr_θ} are represented by the positive branches $X > 0$, $Y > 0$. If $P_{cr_\theta} \gg P_{cr_y}$, then $Y \gg X$, and then $X \to 0$ or $P_{cr} \simeq P_{cr_y}$. If $P_{cr_\theta} \ll P_{cr_y}$, then $X \gg Y$, and then $Y \to 0$ or $P_{cr} \simeq P_{cr_\theta}$.

Now assume that $P_{cr_y} < P_{cr_\theta}$, which means that $X \ll Y$ (in design, typically one chooses $P_{cr_y} \ll P_{cr_\theta}$). We want to know when P_{cr}/P_{cr_y} is minimum, that is, P_{cr_y}/P_{cr} is maximum. This obviously happens for the maximum X value satisfying the condition $X \ll Y$. The point that satisfies this condition is the apex of the hyperbola in Figure 6.9b, which corresponds to the case $P_{cr_y} = P_{cr_\theta}$. Similarly, if we assume that $P_{cr_y} \geq P_{cr_\theta}$, that is, $X \geq Y$, we find that the minimum of P_{cr}/P_{cr_θ} occurs also for $P_{cr_y} = P_{cr_\theta}$. Therefore, the lowest relative depression of P_{cr} below P_{cr_y} and P_{cr_θ} occurs for the beam design such that $P_{cr_y} = P_{cr_\theta}$.

We see that coincidence of the critical loads for two modes causes the greatest relative reduction of the actual critical load (which corresponds to a combined mode). This is a general phenomenon, not restricted to this problem (see, e.g., Sec. 4.6).

Lateral Buckling due to Bending Moment

Let us consider now the case that $P = 0$, and the bending moments M_z^0 are applied at the two ends of the beam. Buckling is possible only if the determinant of equation system 6.3.3 vanishes. This yields for the first critical value ($n = 1$) of the initial bending moment the following result (which is exact):

$$M^0_{z_{cr_1}} = \pm\frac{\pi}{l}\left\{EI_yGJ\left[1+\frac{EI_\omega}{GJ}\left(\frac{\pi^2}{l^2}\right)\right]\right\}^{1/2} \tag{6.3.6}$$

This simple result may be used as a safe but crude approximation in design even if the bending moment is nonuniform; obviously, if the maximum bending moment does not exceed the value given by Equation 6.3.6, a more accurate solution is required only if the beam needs to be designed for a larger maximum bending moment. The first lateral buckling mode is sketched in Figure 6.8c.

Consider now some limiting cases. If the walls are extremely thin ($t \to 0$), J is negligible compared to I_ω, because J is proportional to t^3 while I_ω is proportional to t. Equation 6.3.6 then reduces to

$$M^0_{z_{cr_1}} = \pm\frac{\pi^2}{l^2}E\sqrt{I_yI_\omega} \tag{6.3.7}$$

On the other hand, if the walls are relatively thick, I_ω is negligible compared to J, and then Equation 6.3.6 reduces to

$$M^0_{z_{cr_1}} = \pm\frac{\pi}{l}\sqrt{EG}\,\sqrt{I_yJ} \tag{6.3.8}$$

It is interesting to note from Equation 6.3.6 or Equations 6.3.7 and 6.3.8 that in the former case (very thin walls) the critical bending moment is proportional to $1/l^2$, while in the latter case, it is proportional to $1/l$. Thus we see that for sufficiently short beams lateral buckling is dominated by the warping torsion, while for sufficiently long beams it is dominated by the simple torsion (Saint-Venant torsion).

Alternatively, the critical bending moment can, of course, be solved by direct minimization of the potential energy. In this approach, Equations 6.3.1 are directly substituted into Equation 6.1.15, and the conditions $\partial\Pi/\partial a_n = 0$ and $\partial\Pi/\partial b_n = 0$ are then found to yield again Equation 6.3.3 (with $P = 0$). The result is the same. This analysis however provides further information: it proves that the beam is stable for $M_z^0 < M^0_{z_{cr_1}}$ and unstable for $M_z^0 > M^0_{z_{cr_1}}$.

Lateral buckling can also be caused by a bending moment combined with an axial load. In this case Equation 6.3.3 gives the critical bending moment

$M^P_{z_{cr_1}} = r_P[(P_{cr_y} - P)(P_{cr_\theta} - P)]^{1/2}$. This can be cast in the form

$$M^P_{z_{cr_1}} = M^0_{z_{cr_1}}\left[\left(1 - \frac{P}{P_{cr_y}}\right)\left(1 - \frac{P}{P_{cr_\theta}}\right)\right]^{1/2} \tag{6.3.9}$$

where $M^0_{z_{cr_1}}$ is the critical moment at $P = 0$, given by Equation 6.3.6.

Lateral buckling of a beam with uniform M^0_z is also easily solved when the ends are fixed (the boundary conditions are then expressed by Eqs. 6.1.19). In that case the first critical mode can be described by the relations

$$v = a\left(1 - \cos\frac{2\pi x}{l}\right) \qquad \theta = b\left(1 - \cos\frac{2\pi x}{l}\right) \tag{6.3.10}$$

and by a similar procedure as before one obtains for the first critical bending moment the solution (cf., e.g., Chajes, 1974, p. 222)

$$M^0_{z_{cr}} = \pm\frac{2\pi}{l}\left[EI_y\left(GJ + 4EI_\omega\frac{\pi^2}{l^2}\right)\right]^{1/2} \tag{6.3.11}$$

This solution is exact. Its limiting behavior for a very thin short beam or a thicker very long beam is similar to Equations 6.3.7 and 6.3.8.

Approximate Solution for Variable M^0_z

For variable M^0_z, relatively simple solutions are possible in an approximate form, based on the Rayleigh quotient (Sec. 5.3) or the Rayleigh-Ritz method (Sec. 5.6). Let the bending moment distribution be described as $M^0_z(x) = \bar{M}_z f(x)$, where $f(x)$ is a fixed function and $\bar{M}_z$ is a moment parameter (of the dimension of a moment) whose critical value is to be found. As we did for columns, we may obtain the Rayleigh quotient from the stability condition $\Pi = \delta^2\Pi > 0$ with Π given by Equation 6.1.15 if the vertical load is applied at the centroidal axis rather than at the top flange. Solving Equation 6.1.15 for $\bar{M}_z$, we obtain for lateral buckling ($P = 0$, $u' = 0$, $w = 0$) the stability condition

$$\bar{M}_z < \bar{M}_{z_{cr}} = \frac{\displaystyle\int_0^l \tfrac{1}{2}(EI_y v''^2 + EI_\omega\theta''^2 + GJ\theta'^2)\,dx}{\displaystyle\int_0^l v''\theta f(x)\,dx} \tag{6.3.12}$$

For a simple upper-bound approximation we may assume

$$v = q_1 \sin\frac{\pi x}{l} \qquad \theta = q_2 \sin\frac{\pi x}{l} \tag{6.3.13}$$

Upon substitution into the Rayleigh quotient in Equation 6.3.12 we obtain $\bar{M}_{z_{cr}}$ as a function of the ratio $\xi = q_1/q_2$. The necessary condition of minimum of the Rayleigh quotient is $\partial\bar{M}_{z_{cr}}/\partial\xi = 0$. Minimization of the quotient then yields

$$\bar{M}_{z_{cr}} = \frac{1}{2Cl}[EI_y\pi^2(EI_\omega\pi^2 + GJl^2)]^{1/2} \qquad C = \int_0^l f(x)\sin^2\frac{\pi x}{l}\,dx \tag{6.3.14}$$

If, for example, we have a simply supported beam of constant cross section loaded by a uniform distribution load p_z applied at the axis of the beam (Fig. 6.10a) we may put $\bar{M}_z = p_z l^2/8 =$ value of M_z^0 at midspan, and $f(x) = 4(x - x^2/l)/l$. Substitution into Equations 6.3.14 leads to $C = l(3+\pi^2)/3\pi^2$ and $\bar{M}_{z_{cr}} = 1.150 M^0_{z_{cr_1}}$, with $M^0_{z_{cr_1}}$ given by Equation 6.3.6. The corresponding critical value of p_z is $p_{z_{cr}} = 8\bar{M}_{z_{cr}}/l^2$.

If we choose the Rayleigh-Ritz method, we substitute the expressions for v and θ (see Eqs. 6.3.13) into the potential-energy expression in Equation 6.1.15 (with $W = 0$, $P = 0$, $u' = 0$). Carrying out the integration and calculating then the derivatives with respect to q_1 and q_2, we obtain the following necessary equilibrium conditions:

$$\begin{aligned} \frac{\partial \Pi}{\partial q_1} &= \frac{1}{2} EI_y \frac{\pi^4}{l^3} q_1 - \bar{M}_z C \frac{\pi^2}{l^2} q_2 = 0 \\ \frac{\partial \Pi}{\partial q_2} &= \frac{1}{2}\left(EI_\omega \frac{\pi^4}{l^3} + GJ\frac{\pi^2}{l}\right) q_2 - \bar{M}_z C \frac{\pi^2}{l^2} q_1 = 0 \end{aligned} \tag{6.3.15}$$

This represents a system of two linear algebraic homogeneous equations for q_1 and q_2 and the critical state is obtained from the condition that the determinant of these equations must vanish. This yields exactly the same result as the minimization of the Rayleigh quotient (Eqs. 6.3.14), as expected. This value is, of course, an upper bound on the exact solution.

It may be checked that the foregoing solution based on the Rayleigh quotient or the Rayleigh-Ritz method yields the exact result for the case of a uniform initial bending moment distribution, for which $f(x) = 1$.

An accurate result for general $f(x)$ can be obtained by the Rayleigh-Ritz method if Equations 6.3.13 are replaced by a truncated Fourier series with sufficiently many terms.

The same procedure can be used for a beam with simply supported ends, subjected to a concentrated load F at midspan (Fig. 6.10b). Adopting for v and θ the same approximation as given in Equations 6.3.13 and denoting by $\bar{M}_z$ the value of M_z^0 at midspan, one finds

$$\bar{M}_{z_{cr}} = \frac{2\pi^2}{(\pi^2+4)l^2}[EI_y\pi^2(EI_\omega\pi^2 + GJl^2)]^{1/2} = 1.423 M^0_{z_{cr_1}} \tag{6.3.16}$$

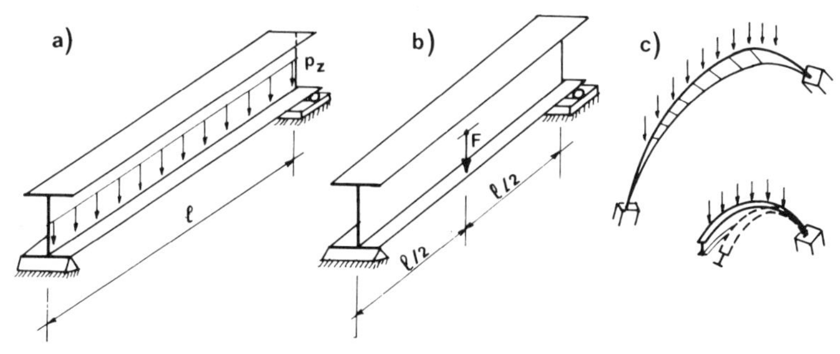

Figure 6.10 (a) Beam subjected to distributed transverse load; (b) beam subjected to concentrated transverse load; (c) lateral buckling of arches.

The corresponding critical value of F is $F_{cr} = 4\bar{M}^0_{z_{cr}}/l$. Tabulated values of F_{cr} for this case as well as for a cantilever beam with a concentrated load applied at the end can be found in Timoshenko and Gere (1961, pp. 264 and 259).

The lateral buckling of a cantilever subjected to a concentrated load at the free end is demonstrated in Figure 6.11, which portrays a teaching model developed at Northwestern University (1969).

Bimoment

It is customary to describe the stresses due to warping torsion in terms of the so-called bimoment B. The normal strains and stresses due to θ'' alone are $\varepsilon = -\theta''\omega$ and $\sigma = -E\theta''\omega$. In analogy to the bending moment definition, we may define the quantity

$$B = -\int_A \sigma\omega \, dA = E\theta'' \int_A \omega^2 \, dA = EI_\omega \theta'' \tag{6.3.17}$$

and call it the bimoment. The reason for this term is that, for a symmetric I-beam, $B = M_2 h/2 - M_1 h/2$, where M_1 and $M_2 = -M_1$ are bending moments in the upper and lower flange (in their planes), respectively. In terms of bimoment we now have $\sigma = -E\omega\theta'' = -E\omega(B/EI_\omega) = -B\omega/I_\omega$. Adding the normal stresses due to the axial load and bending moments, we may write, in general,

$$\sigma = -\frac{P}{A} - \frac{M_z z}{I_z} - \frac{M_y x}{I_y} - \frac{B\omega}{I_\omega} \tag{6.3.18}$$

where the last term is easily remembered as an analog to the bending stress formula. Note, however, that (in contrast to M_z, M_y, P, and M_t) B is not a static

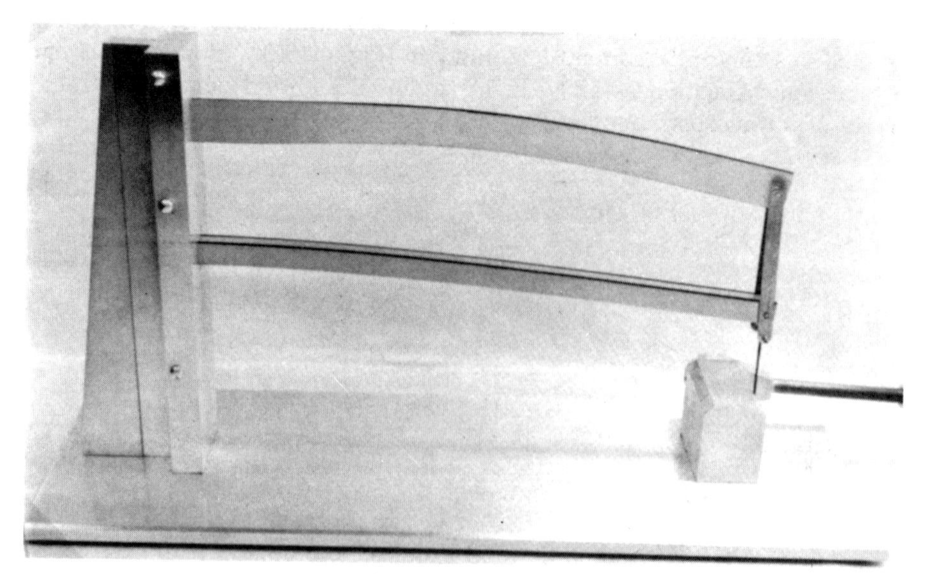

Figure 6.11 Norwhwestern University (1969) teaching models: lateral buckling of cantilever.

resultant of the stresses due to warping torsion; in fact this resultant is always zero for a doubly symmetric cross section.

The static boundary condition $\theta''=0$, used before, may alternatively be replaced by $B=0$. Except for load P applied at the centroid (free end, simply supported end), the static boundary conditions (in the absence of axial loads or restraints) are $\sigma=0$ and $M_z=M_y=0$, and so they may be written as $M_z=0$, $M_y=0$, and $B=0$.

Lateral Buckling of Arches

A practically important application of the thin-wall beam theory is the lateral buckling of arches (Fig. 6.10c). To solve the arch problem, the foregoing formulation for straight thin-wall beams must be generalized to curved thin-wall beams. The modification consists in replacing v'', w'', and θ' by suitable expressions for the bending curvatures and the specific angle of twist (torsional curvature) that take into account the curvature radius of the bar. Using such generalized curvature expressions (see, e.g., Vlasov, 1959, or Bažant, 1965), one can obtain an analogous expression for the potential-energy function (see Sec. 5.1) from which the differential equations of the problem follow. For detailed solutions see Timoshenko and Gere (1961, p. 317), Vlasov (1959, sec. 12.4), and for beams of variable cross section, Bažant (1965) and Bažant and El Nimeiri (1975).

Buckling formulas for lateral buckling of arches under various loadings are reported in Galambos (1988, pp. 594–605).

Problems

6.3.1 Find the value of the first critical load of the I-beam in Figure 6.8 under eccentric compression. Assume $h=b_f$; $l=20h$; $h/t_w=10$, 50, 100; $t_f/t_w=2$ (symbols defined in Fig. 6.4a); and $e_z=\frac{1}{2}h$, $2h$. What is the minimum eccentricity e_z for which buckling in tension becomes possible?

6.3.2 Do the same as Problem 6.3.1, but the ends are fixed ($v'=w'=\theta'=0$).

6.3.3 Find the exact value of the critical uniform bending moment for a fixed–fixed beam (Eq. 6.3.11), with each end fixed against warping and all rotations.

6.3.4 For the beam in Figure 6.10a: (a) find an approximation of the critical value of p_z for the case analyzed in the text; (b) do the same but consider p_z to be applied at a distance e_z above the axis of the beam (*hint:* how would Eq. 6.3.12 have to be modified?); (c) do the same for the problem in the text, but the end cross sections are fixed against warping ($\theta'=0$).

6.3.5 For the beam in Figure 6.10a: (a) solve the same problem as in the text, but consider both ends to be completely fixed against rotation as well as warping; (b) do the same, but consider the ends fixed only against slope and not against warping; (c) do the same for the case of a simple support at one end and fixed at the other end.

6.3.6 Solve the critical value of F in Figure 6.10b for the case of (a) completely fixed supports at both ends; (b) end cross sections fixed only against warping but not against rotation; (c) end cross sections fixed only against rotation but not against warping.

6.3.7 Consider a cantilever beam of I–cross section and find the critical loads for lateral buckling for (a) a distributed load p_z applied at a distance $H/2$ above the beam axis (Fig. 6.12a); (b) a concentrated load at the end (Fig. 6.12b). The beam is simply supported at both ends, with free rotations as well as warping. (c) Then solve the same problem but for both ends completely fixed; (d) then the same problem but the ends fixed only against rotation but not against warping.

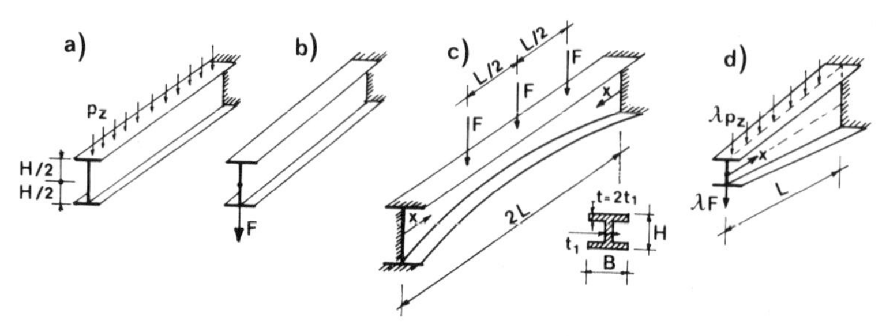

Figure 6.12 Exercise problem on lateral buckling of beams.

6.3.8 For the beam that has both ends completely fixed ($\theta = \theta' = w = w' = v = v' = 0$) (Fig. 6.12c) and has a variable cross section with $H = B[1 - 2x(L - x)/L^2]$ ($B = L/10$, $t = B/10$), find an approximate value of F for lateral buckling.

6.3.9 The cantilever beam in Figure 6.12d has the cross-section variation $H = B(1 + x/L)$ but is loaded by a combination of distributed and concentrated loads. Find the critical value of the load multiplier λ for lateral buckling.

6.3.10 A beam with simple supports ($w = w'' = v = v'' = \theta = \theta'' = 0$) is subjected to lateral load $p_z = f/L$ and axial load $P = 2f$. Find the approximate critical value of multiplier f. How does the solution change if $P = f$, $P = 4f$, $P = f/2$?

6.3.11 Do the same as (a) Problem 6.2.4 and (b) Problem 6.2.5, but consider lateral vibrations with torsion for a simply supported beam loaded by a uniform bending moment M_z^0 or by a vertical distributed load p_z.

6.4 BEAMS OF ARBITRARY OPEN CROSS SECTION

General Theory of Warping Torsion

Although our attention has so far been restricted to I-beams, a similar solution is possible for beams of arbitrary open cross section (Fig. 6.13a–j). The theory of thin-walled beams rests on the following two basic simplifying assumptions:

1. The cross section is perfectly rigid in its own plane.
2. The shear strains in the middle surface of the wall are negligible (this is known as Wagner's assumption).

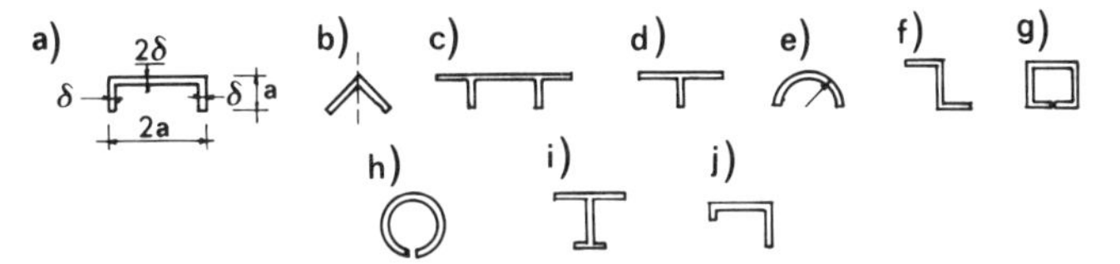

Figure 6.13 Typical open cross sections

Furthermore, the transverse normal stresses in the walls as well as the bending moments of normal stresses about any axes tangent to the midsurface of the wall (like the bending moments in shells) are assumed to be negligible. Also, the tangential normal stresses in the walls are considered to be small, so that they can be neglected in the relation between the longitudinal normal stress and strain. This assumption, of course, does not strictly conform to assumption 1, but (unless the cross section is stiffened by closely spaced diaphragms, see Gjelsvik, 1981, p. 17) appears to be closer to reality than the assumption of negligible strains in the tangential direction, which would be implied by a strict interpretation of assumption 1. (For Poisson ratio $\nu = 0$, however, this discrepancy disappears.)

These assumptions transform the three-dimensional problem to a one-dimensional problem. Assumption 2 obviously includes the bending theory of beams because a negligible value of the shear strain implies, in the case of bending, that the cross section remains plane and normal to the middle surface. All these assumptions, and especially Wagner's assumption, have been extensively verified by experiments and were found to be applicable to thin-walled beams that are sufficiently slender, approximately such that the length-to-width ratio exceeds 10.

Let s denote the length coordinate of the middle surface along the cross section, measured from some suitably chosen origin $0'$ (Fig. 6.14a, b). In the case of branched cross sections, such as T-beams or I-beams (Fig. 6.13d, i), coordinate

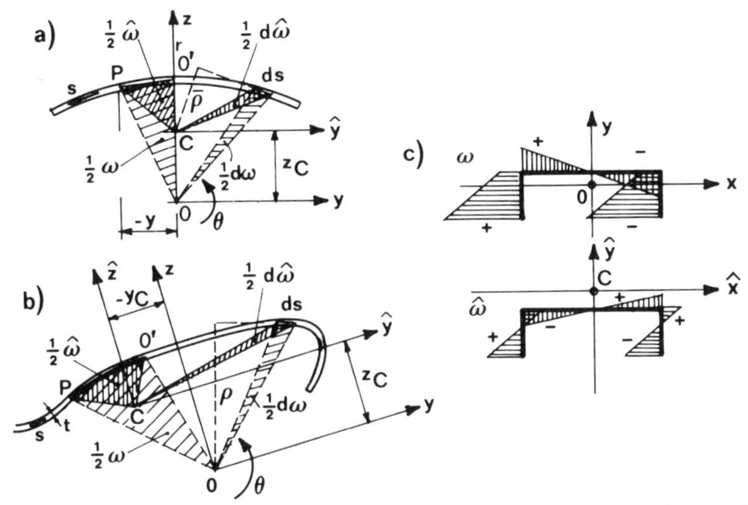

Figure 6.14 Geometry of (a) monosymmetric and (b) arbitrary cross sections and physical meaning of sectorial radius and sectorial coordinate for warping function; and (c) plots of warping functions for channel cross section.

s is defined separately for each flange but must be such that s is continuous through the branching points. The shear strain in the middle surface of the wall may be expressed as $\gamma_{xs} = (\partial u_1/\partial s) + (\partial u_s/\partial x)$ where u_1 = longitudinal displacement and u_s = transverse displacement in the direction of s. Wagner's assumption means that $\gamma_{xs} = 0$ or $\partial u_1/\partial s = -\partial u_s/\partial x$. By integration, $u_1 = \int_0^s (-\partial u_s/\partial x)\, ds + u_a$ where u_a is the axial displacement at the origin of coordinate s (point 0′). (For a further discussion of u_a, see Gjelsvik, 1981.)

Consider now the effect of rotation θ about the center 0 of coordinates y, z (Fig. 6.14a, b). Assumption 1 (rigidity of the cross-section shape) allows us to write $u_s = \theta\rho(s)$, where ρ represents the distance of the tangent to the middle surface at point s from point 0 and is taken positive if the vector $\mathbf{ds}$ turns counterclockwise (Fig. 6.14a, b) and negative if clockwise (so that it would give the correct sign for the component of u_s). The equation for u_1 now becomes $u_1 - u_a = -\int_0^s (\partial u_s/\partial x)\, ds = -\int_0^s (\partial \theta\rho/\partial x)\, ds = -\theta'\omega$, in which we introduce the notation

$$\omega = \int_0^s \rho(s)\, ds = \int_0^s (y\, dz - z\, dy) \tag{6.4.1}$$

Function $\omega(s)$ is called the sectorial coordinate (or the warping function), and $\frac{1}{2}\omega$ represents the area of the sector shown in Figure 6.14a, b. It is easy to check that the definition of ω in Equation 6.4.1 includes the expressions for an I-beam that we wrote in Equations 6.1.3. This documents that Wagner's assumption is the proper way to generalize the previously used idea that the flanges in an I-beam bend as independent beams.

Next we add to $u_1 - u_a = -\theta'\omega$ the axial and bending deformations (which also satisfy Wagner's assumption as already mentioned), including the axial displacement of point 0′ relative to point 0 due to bending. Thus we get

$$u_1 = u - v'y - w'z - \theta'\omega \tag{6.4.2}$$

in which u represents the axial displacement at the centroid (not at 0′). For thin-walled cross sections it can, of course, happen that there is no material point at the centroid. In that case u is an imagined displacement representing the average of the axial displacements at all material points of the cross section.

Equation 6.4.2 is of the same form as our previous Equations 6.1.4 for an I-beam. The expressions for the transverse displacement components in Equations 6.1.5 also apply in general. Consequently, the potential-energy expression that results from Equation 6.1.8 will be of the same form as before. So the derivation of the differential equations of the problem is also the same.

For a general choice of coordinates x and y, however, certain integrals over the cross section, which vanished before, are no longer negligible, in general. It is clear that substitution of Equation 6.4.2 into the potential-energy expression will lead to integrals over the cross section of the functions 1, y, z, ω, as well as of their squares and products. By choosing x and y as the principal centroidal axis of the cross section, we can achieve that the following cross-section integrals vanish: $\int y\, dA = \int z\, dA = \int yz\, dA = 0$.

By changing the pole of the sectorial coordinate ω to a different location it is possible to make further cross-section integrals vanish. To this end, we need to consider how the sectorial coordinate changes when the pole is moved to point C

of coordinates y_C, z_C (Fig. 6.14a, b). Let $\hat{y}$ and $\hat{z}$ be Cartesian coordinates that have their origin at point C and are parallel to the original coordinates y and z. The coordinate transformation is $y = y_C + \hat{y}$, $z = z_C + \hat{z}$. Substituting this into Equation 6.4.1 for ω, we obtain the transformation relation

$$\hat{\omega} = \omega + z_C y - y_C z + \omega_0 \tag{6.4.3}$$

in which $\hat{\omega} = \int \hat{\rho}\, ds = \int (\hat{y}\, d\hat{z} - \hat{z}\, d\hat{y})$ = sectorial coordinate at pole C, and $\omega_0 = z_C y_0 - y_C z_0$ is a constant that depends on the coordinates of pole C and of origin $0'$ of the curvilinear coordinate s. The relation between $\hat{\omega}$ and ω is illustrated in Figure 6.14c for a channel cross section.

In view of Equation 6.4.3, the mixed cross-sectional moments can be expressed as follows:

$$\begin{aligned} I_{\hat{\omega}y} &= \int \hat{\omega} y\, dA = \int \omega y\, dA + z_C \int y^2\, dA - y_C \int yz\, dA + \omega_0 \int y\, dA \\ I_{\hat{\omega}z} &= \int \hat{\omega} z\, dA = \int \omega z\, dA - y_C \int z^2\, dA + z_C \int zy\, dA + \omega_0 \int z\, dA \end{aligned} \tag{6.4.4}$$

Since y, z are the principal centroidal coordinates, the last two integrals in each of the two preceding equations vanish. We thus find that, in order to make $I_{\hat{\omega}y}$ and $I_{\hat{\omega}z}$ vanish, the following coordinates need to be chosen for the new pole C:

$$z_C = -\frac{\int \omega y\, dA}{\int y^2\, dA} = -\frac{I_{\omega y}}{I_y} \qquad y_C = \frac{\int \omega z\, dA}{\int z^2\, dA} = \frac{I_{\omega z}}{I_z} \tag{6.4.5}$$

The point C determined by Equations 6.4.5 coincides with the shear center of the cross section. The shear center is defined by the condition that for the transverse load resultants passing through point C the response consists of bending only (i.e., no twisting); see, for example, Oden and Ripperger (1981, p. 120).

Consider now a similar transformation of Equation 6.4.2. According to the assumption of rigidity of the cross-section form, we have $v = \hat{v} + z_C\theta$ and $w = \hat{w} - y_C\theta$. Substituting this into Equation 6.4.2, along with an expression for ω according to Equation 6.4.3, we find that the terms with z_C and y_C cancel out and the following expression results:

$$u_1 = u - \hat{v}'y - \hat{w}'z - \theta'\hat{\omega} + \theta'\omega_0 \tag{6.4.6}$$

By a suitable choice of origin $0'$ (Fig. 6.14a, b) of coordinate s it is possible to achieve that $\int \hat{\omega}\, dA$ vanishes. For this choice, however, ω_0 is generally nonzero. There is nevertheless a special case in which $\omega_0 = 0$; it is the case of monosymmetric cross sections as shown in Figure 6.14a. In that case, by choosing the origin of coordinate s to lie on the axis of symmetry (z axis), we achieve that both $\int \hat{\omega}\, dA$ and constant ω_0 vanish, as it appears from the value found for ω_0 in Equation 6.4.3. Indeed, Equation 6.4.3 in this case reduces to the simple expression $\omega = \hat{\omega} - z_C y$, the validity of which is geometrically obvious from an inspection of Figure 6.14a. In that case, Equation 6.4.6 takes the same form as Equation 6.4.2 as well as Equations 6.1.4 used before for symmetric I-beams. In the developments that follow, we will consider a cross section of general shape and choose $0'$ such that $\int \hat{\omega}\, dA = 0$.

Stresses and Bimoment in General Theory

The bimoment associated with the shear center is defined as $\hat{B} = -\int \sigma\hat{\omega}\, dA$. Substituting $\sigma = Eu_1'$ and expressing u_1' from Equation 6.4.6, we obtain

$$\hat{B} = -\int E(u' - \hat{v}''y - \hat{w}''z - \theta''\hat{\omega} + \theta''\omega_0)\hat{\omega}\, dA = EI_{\hat{\omega}}\theta'' \tag{6.4.7}$$

We see that normal stresses due to axial force and bending moment do not contribute to $\hat{B}$. Using again the same expression of u_1' and substituting it into the definitions $N_x = -P = \int_A \sigma\, dA$, $M_y = \int_A \sigma y\, dA$, $M_z = \int_A \sigma z\, dA$, we further obtain $u' + \theta''\omega_0 = -P/EA$, $\hat{v}'' = M_y/EI_y$, $\hat{w}'' = M/EI_z$. Finally, substituting these expressions together with $\theta'' = \hat{B}/EI_{\hat{\omega}}$ into the expression for $\sigma = Eu_1'$ according to Equation 6.4.6, we obtain the expression (Bažant, 1965):

$$\sigma = -\frac{P}{A} - \frac{M_y y}{I_y} - \frac{M_z z}{I_z} - \frac{\hat{B}\hat{\omega}}{I_{\hat{\omega}}} \tag{6.4.8}$$

Further, it is interesting to calculate the derivative of the bimoment:

$$\hat{B}' = \frac{d}{dx}\int \sigma\hat{\omega}t\, ds = \int \frac{\partial(\sigma t)}{\partial x}\hat{\omega}\, ds = -\int \frac{\partial(\tau_{xs}t)}{\partial s}\hat{\omega}\, ds \tag{6.4.9}$$

We utilized here the differential equation of equilibrium for stresses in the wall: $\partial(\sigma t)/\partial x = -\partial(\tau_{xs}t)/\partial s$, where τ_{xs} is the shear stress in the middle surface of the wall. The last integral in Equation 6.4.9 may be integrated by parts, and noting that $d\hat{\omega}/ds = \hat{\rho}$ (because $d\hat{\omega} = \hat{\rho}\, ds$), we have

$$\hat{B}' = -\int \tau_{xs}t\frac{d\hat{\omega}}{ds}\, ds + [\tau_{xs}t\omega]_{s_1}^{s_2} = -\int \tau_{xs}\hat{\rho}t\, ds = -M_{t_{\hat{\omega}}} \tag{6.4.10}$$

The boundary terms corresponding to the boundary coordinates s_1 and s_2 must vanish since $\tau_{xs} = 0$ at the boundary of the wall. The last expression in Equation 6.4.10, by definition, represents the torque due to stresses τ_{xs} in the middle surface of the wall. These stresses are due strictly to the warping torsion. The simple torsion (Saint-Venant's torsion) produces no shear stresses in the midsurface of the walls.

The last relation makes it possible to calculate the total torque transmitted in the cross section. The shear stresses τ acting at any point of the wall may be decomposed into two components: the shear stress τ_{xs} in the middle surface of the wall, and component τ^* that is zero at the middle surface and varies linearly across the wall thickness. The total torque M_t in the cross section may be calculated as

$$M_t = \int \tau\hat{\rho}_\tau\, dA = \int \tau_{xs}\hat{\rho}_\tau t\, ds + \int \tau^*\hat{\rho}_\tau\, dA \tag{6.4.11}$$

in which $\hat{\rho}_\tau$ denotes the arm from pole C to the vector of shear stress τ for any point within the wall thickness. The last integral in Equation 6.4.11 represents the torque due to simple (Saint-Venant's) torsion, denoted as M_{t_s}. According to Equations 6.4.11, 6.4.10, and 6.4.7, the total torque may now be expressed as

$$M_t = M_{t_{\hat{\omega}}} + M_{t_s} = -\hat{B}' + M_{t_s} = -(EI_{\hat{\omega}}\theta'')' + GJ\theta' \tag{6.4.12}$$

This agrees with our previous Equation 6.1.21 for I-beams.

The boundary condition for a free end of the beam is $M_t = 0$ (note that this does not imply $\theta' = 0$). Generally both the torque from simple torsion M_{t_s} and the warping torque $M_{t_{\hat{\omega}}}$ are nonzero at the free end but they exactly offset each other.

The presence of warping torque $-\hat{B}'$ means that there must exist shear stresses, τ_{sx}, which equilibrate this torque. They may be calculated by integrating the differential equation of equilibrium $\partial(\tau_{sx}t)/\partial x = -\partial(\sigma t)/\partial s$ (see, e.g., Oden and Ripperger, 1981) in the same way the shear stresses are calculated from σ in the bending theory. τ_{sx} are found to be proportional to $\hat{B}'$. The existence of τ_{sx}, of course, conflicts with the starting assumption that $\gamma_{sx} = 0$ in the midsurface of walls. This conflict is an inevitable paradox of the theory of warping torsion, in which the shear strains are implied to be zero due to the assumption that plane cross sections remain plane and normal to the deflection curve while the corresponding shear stresses must be nonzero to balance the shear force. For slender beams the error due to neglect of γ_{sx} is small, but not so for relatively deep or wide beams. In that case θ' needs to be augmented by an additional (secondary) specific twist $\theta'_{(2)}$ due to shear stresses associated with B'' (which may be approximately estimated from the condition of energy equivalence with the work of τ_{sx}).

Potential Energy and Differential Equations

Consider now the most general case of initial stresses σ^0 and τ^0 associated with the initial axial force P^0, initial bending moments M_y^0 and M_z^0, initial shear forces V_y^0 and V_z^0, initial bimoment $\hat{B}^0$, and initial Saint-Venant torque $M_{t_s}^0$. According to Equation 6.4.8 we have

$$\sigma^0 = -\frac{P}{A} - \frac{M_y^0 y}{I_y} - \frac{M_z^0 z}{I_z} - \frac{\hat{B}^0}{I_{\hat{\omega}}}\hat{\omega} \tag{6.4.13}$$

and, in analogy to Equations 6.1.10,

$$\tau_{xy}^0 = \frac{V_y^0}{A} + \bar{\tau}_{xy} \qquad \tau_{xz}^0 = \frac{V_z^0}{A} + \bar{\tau}_{xz} \tag{6.4.14}$$

where $\bar{\tau}_{xy}$ and $\bar{\tau}_{xz}$ now comprise also the secondary shear stresses τ_{sz} due to warping torsion. These stresses, representing deviations from the mean value of the shear stresses, have zero resultant. We will not give an expression for the stresses τ^* due to Saint-Venant torsion, since we will express their work directly through the resultant torque $M_{t_s}^0$.

The first-order, linearized parts of normal strains (small strains) are

$$e_x = \frac{\partial u_1}{\partial x} = u' - \hat{v}''y - \hat{w}''z - \theta''\hat{\omega} + \theta''\omega_0 \tag{6.4.15}$$

while the second-order, geometrically nonlinear parts of strains can be obtained by substituting the transformations $v = \hat{v} + z_c\theta$ and $w = \hat{w} - y_c\theta$ into Equations 6.1.5, and these equations then into Equations 6.1.6 and 6.1.7. The result is

$$\begin{aligned}\varepsilon_x - e_x &= \tfrac{1}{2}(\hat{v}'^2 + \theta'^2 z^2 - 2\hat{v}'\theta' z + z_c^2\theta'^2 + 2\hat{v}' z_c\theta' - 2\theta'^2 z z_c) \\ &\quad + \tfrac{1}{2}(\hat{w}'^2 + \theta'^2 y^2 + 2\hat{w}'\theta' y + y_c^2\theta'^2 - 2y_c\theta'\hat{w}' - 2\theta'^2 y y_c)\end{aligned} \tag{6.4.16}$$

$$\gamma_{xy} = \hat{w}'\theta - y_c\theta'\theta - y\theta'\theta \qquad \gamma_{xz} = -\hat{v}'\theta - z_c\theta'\theta - z\theta'\theta \tag{6.4.17}$$

The first-order, linearized (small) parts of γ_{xy} and γ_{xz} are zero as a consequence of the basic hypotheses of the theory. [This is obvious for the bending deformation. For the warping deformation, characterized by $u_1 = -\theta'\omega$, one can check it easily by substituting this equation, as well as the relations $u_y = -\theta z$, $u_z = \theta y$, $\partial\omega/\partial y = -z$, $\partial\omega/\partial z = y$, into the expressions ($\partial u_1/\partial y + \partial u_y/\partial x$ and $\partial u_1/\partial z + \partial u_z/\partial x$) for the first-order parts of the shear strains γ_{xy} and γ_{xz}.]

In analogy to Section 6.1, we can now calculate the strain energy:

$$U = \int_0^l \int_A \sigma^0 e_x \, dA \, dx + \int_0^l \int_A \sigma^0(\varepsilon_x - e_x) \, dA \, dx + \int_o^l \int_A (\tau^0_{xy}\gamma_{xy} + \tau^0_{xz}\gamma_{xz}) \, dA \, dx$$
$$+ \int_0^l \int_A \tfrac{1}{2} E e_x^2 \, dA \, dx + \int_0^l M^0_{t_s}\theta' \, dx + \int_0^l \tfrac{1}{2} GJ\theta'^2 \, dx \tag{6.4.18}$$

Note that in Equation 6.4.18 the term $\int_0^l \int_A \frac{1}{2} G(\gamma_{xz}^2 + \gamma_{xy}^2) \, dA \, dx$ is omitted because it is a higher-order term. Also negligible is the work of Saint-Venant torsional shear stresses τ^* on the second-order shear strains. The reason is that the average of τ^* is zero, so that the work of positive τ^* cancels the work of negative τ^* (and γ_{xz}, γ_{xy} = constants through the thickness of the wall).

Evaluating the integrals over cross-section area A, one obtains

$$\int_A \sigma^0 e_x \, dA = -P(u' + \theta''\omega_0) + M^0_y \hat{v}'' + M^0_z \hat{w}'' + \hat{B}^0\theta''$$

$$\int_A \sigma^0(\varepsilon_x - e_x) \, dA = -P(\tfrac{1}{2}\hat{v}'^2 + \tfrac{1}{2}\hat{w}'^2 + \tfrac{1}{2}r_c^2\theta'^2 + z_c\theta'\hat{v}' - y_c\theta'\hat{w}')$$
$$- M^0_y(\tfrac{1}{2}\beta_y\theta'^2 + \hat{w}'\theta') - M^0_z(\tfrac{1}{2}\beta_z\theta'^2 - \hat{v}'\theta') - \hat{B}^0(\tfrac{1}{2}\beta_\omega\theta'^2) \tag{6.4.19a}$$

$$\int_A (\tau^0_{xy}\gamma_{xy} + \tau^0_{xz}\gamma_{xz}) \, dA = V^0_y \hat{w}'\theta - V^0_z \hat{v}'\theta - V^0_y y_c\theta'\theta - V^0_z z_c\theta'\theta$$

$$\int_A \frac{1}{2} E e_x^2 \, dA = \frac{E}{2}(u'^2 A + \hat{v}''^2 I_y + \hat{w}''^2 I_z + \theta''^2 I_{\hat{\omega}} + \theta''^2 A\omega_0^2 + 2u'\theta'' A\omega_0)$$

in which

$$r_c^2 = \left(\frac{I_z}{A} + \frac{I_y}{A} + z_c^2 + y_c^2\right) \qquad \beta_y = \frac{1}{I_y}\int_A y(y^2 + z^2) \, dA - 2y_c$$
$$\beta_z = \frac{1}{I_z}\int_A z(y^2 + z^2) \, dA - 2z_c \qquad \beta_\omega = \frac{1}{I_{\hat{\omega}}}\int_A \hat{\omega}(y^2 + z^2) \, dA \tag{6.4.19b}$$

In the calculation of the work of shear stresses, the contributions of the self-equilibrated stresses $\tilde{\tau}_{xy}$ and $\tilde{\tau}_{xz}$ have been neglected since they appear to be indeed negligible at least for some common cross-section shapes; see Powell and Klingner (1970) or Kitipornchai and Chan (1987).

Further we need to calcualte the work of loads. We consider them to be distributed over the cross section (see Fig. 6.15a), having components $\tilde{p}_y(x, s) + \Delta\tilde{p}_y(x, s)$ and $\tilde{p}_z(x, s) + \Delta\tilde{p}_z(x, s)$. $\tilde{p}_y$ and $\tilde{p}_z$ denote the loads in the initial equilibrium state and the symbol Δ indicates small incremental quantities. The work of the distributed load is

$$W = \int_0^l \int_A [(\tilde{p}_y + \Delta\tilde{p}_y)u_2 + (\tilde{p}_z + \Delta\tilde{p}_z)u_3 + \tilde{p}_y \, \Delta u_2 + \tilde{p}_z \, \Delta u_3] \, dA \, dx \tag{6.4.20}$$

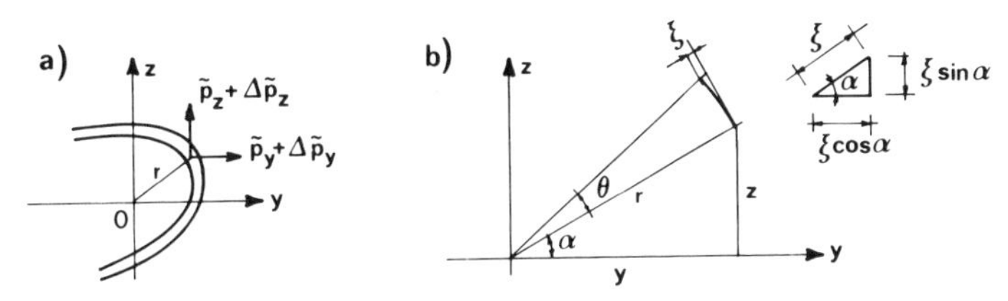

Figure 6.15 (a) Distributed loads over cross sections and (b) second-order components of radial displacement.

in which Δu_2 and Δu_3 are the second-order components of the displacements of the cross section. Noting that this second-order radial displacement is given by (Fig. 6.15b) $\xi = (r/\cos\theta) - r \simeq \frac{1}{2}\theta^2 r$, we obtain $\Delta u_2 = -\frac{1}{2}\theta^2 y$ and $\Delta u_3 = -\frac{1}{2}\theta^2 z$. Substituting these relations into Equation 6.4.20, together with Equations 6.1.5 and the transformation formulas between quantities referred to the centroid and to the shear center, we obtain

$$
\begin{aligned}
&W = \int_0^l [(p_y + \Delta p_y)\hat{v} + (p_z + \Delta p_z)\hat{w} + (\hat{m}_t + \Delta\hat{m}_t)\theta - \tfrac{1}{2}m_t\theta^2]\,dx \\
&p_y = \int_A \tilde{p}_y\,dA \qquad p_z = \int_A \tilde{p}_z\,dA \qquad \hat{m}_t = \int_A [\tilde{p}_z(y-y_c) - \tilde{p}_y(z-z_c)]\,dA \qquad (6.4.21) \\
&m_t = \int_A (\tilde{p}_y y + \tilde{p}_z z)\,dA
\end{aligned}
$$

To this work we must add (as in Eq. 6.1.16) the work of the incremental loads directly applied to the beam ends.

The total potential energy is given by $\Pi = U - W$. It can be obtained from Equations 6.4.18, 6.4.19, and 6.4.21. The expression for the potential energy will be used in dealing with finite element analysis for large deflections in Section 6.5. It is instructive, however, to derive also the Euler equations, according to Equation 5.1.8. To do so, we will introduce in the expression for U the approximations $V_y^0 \simeq -M_y^{0\prime}$, $V_z^0 \simeq -M_z^{0\prime}$ (in which the effect of initial deflection z^0 is neglected), and we will apply also the transformation $\int_0^l (-V_y^0)y_c\theta'\theta \simeq \int_0^l M_y^{0\prime} y_c\theta'\theta\,dx = y_c[M_y^{0\prime}\frac{1}{2}\theta^2]_0^l - \int_0^l \frac{1}{2}y_c M_y^{0\prime\prime}\theta^2\,dx$, and a similar transformation for $\int_0^l (-V_z^0)z_c\theta'\theta\,dx$. The Euler equations then become

$$
\begin{aligned}
&-EAu'' - EA\omega_0\theta''' = 0 \\
&EI_y\hat{v}^{\mathrm{IV}} - (M_z^0\theta)'' + P(\hat{v}'' + z_c\theta'') = \Delta p_y \\
&EI_z\hat{w}^{\mathrm{IV}} + (M_y^0\theta)'' + P(\hat{w}'' - y_c\theta'') = \Delta p_z \qquad (6.4.22) \\
&(EI_{\hat{\omega}} + EA\omega_0^2)\theta^{\mathrm{IV}} - (GJ - Pr_c^2)\theta'' + (M_y - Py_c)\hat{w}'' \\
&\quad + (Pz_c - M_z^0)\hat{v}'' - (M_y^{0\prime\prime}y_c + M_z^{0\prime\prime}z_c)\theta \\
&\quad + \beta_y(M_y^0\theta')' + \beta_z(M_z^0\theta')' + \beta_\omega(\hat{B}^0\theta')' + EA\omega_0 u''' = \Delta\hat{m}_t - m_t\theta
\end{aligned}
$$

The formulation we just described has been extended to beams of smoothly variable cross section and to beams with curved axis (Bažant, 1965). The formulation is also limited by the hypothesis of rigid cross section (i.e., the cross section keeps its shape). However, if the walls are very thin (as in cold-form steel profiles), or if stiffening ribs or diaphragms are missing, significant in-plane distortions of cross section can occur and cause interaction with local buckling (see, e.g., Sridharan and Ali, 1988; Pignataro and Luongo, 1987).

Monosymmetric Cross Section

Let us now consider the practically frequent special case of a monosymmetric cross section with z as the axis of symmetry. We assume the initial loading to consist only of axial force P of eccentricity e_z (Fig. 6.16). In this case we have $\omega_0 = 0$, $y_c = 0$, $M_y^0 = 0$, $M_z^0 = Pe_z$, and $\hat{B}^0 = 0$. The third of Equations 6.4.22 becomes uncoupled (same as it did for the doubly symmetric cross section treated in Sec. 4.3) and furnishes the flexural critical load P_{cr_z} (of the corresponding perfect beam). The second and fourth equations remain coupled and can be written as

$$
\begin{aligned}
EI_y\hat{v}^{\mathrm{IV}} + P\hat{v}'' + P(z_c - e_z)\theta'' &= 0 \\
EI_{\hat{\omega}}\theta^{\mathrm{IV}} - (GJ - Pr_c^2 - \beta_z Pe_z)\theta'' + P(z_c - e_z)\hat{v}'' &= 0
\end{aligned}
\tag{6.4.23}
$$

Substituting in Equations 6.4.23 the Fourier series expansion of Equations 6.3.1, one gets

$$
\begin{aligned}
&\sum_{n=1}^{\infty}\left[EI_y a_n\frac{n^4\pi^4}{l^4} - Pa_n\frac{n^2\pi^2}{l^2} - P(z_c - e_z)b_n\frac{n^2\pi^2}{l^2}\right]\sin\frac{n\pi x}{l} = 0 \\
&\sum_{n=1}^{\infty}\left[EI_{\hat{\omega}} b_n\frac{n^4\pi^4}{l^4} + (GJ - Pr_c^2 - P\beta_z e_z)b_n\frac{n^2\pi^2}{l^2} - P(z_c - e_n)a_n\frac{n^2\pi^2}{l^2}\right]\sin\frac{n\pi x}{l} = 0
\end{aligned}
\tag{6.4.24}
$$

To satisfy these equations identically for any x (with a_n and b_n in general nonzero), it is necessary that for each n the expressions in the brackets vanish, that is,

$$
\begin{bmatrix}
EI_y\dfrac{n^2\pi^2}{l^2} - P & -P(z_c - e_z) \\
-P(z_c - e_z) & EI_{\hat{\omega}}\dfrac{n^2\pi^2}{l^2} + GJ - P(r_c^2 + \beta_z e_z)
\end{bmatrix}
\begin{Bmatrix} a_n \\ b_n \end{Bmatrix} = \begin{Bmatrix} 0 \\ 0 \end{Bmatrix}
\tag{6.4.25}
$$

The critical load is obtained from the condition that the determinant of Equation 6.4.25 vanishes. For the case that the axial force is applied at the shear center ($e_z = z_c$) for which the differential equations 6.4.23 uncouple, we have a purely torsional mode, for which we find $P_{\mathrm{cr}_\theta}^{(n)} = (EI_{\hat{\omega}} n^2\pi^2/l^2 + GJ)/(r_c^2 + \beta_z z_c)$. (A purely torsional mode would result even for general cross sections.) For the case $e_z = z_c$ (and $n = 1$) we have the condition

$$
(P_{\mathrm{cr}_y} - P)(P_{\mathrm{cr}_\theta} - P) - P^2\alpha^2 = 0 \tag{6.4.26}
$$

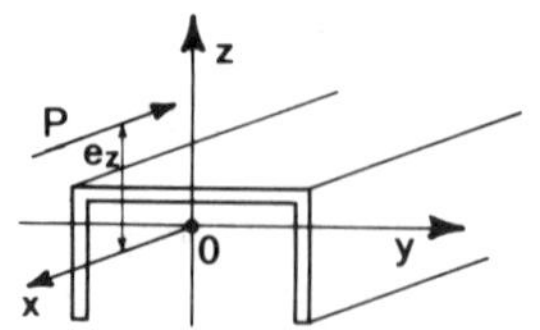

Figure 6.16 Monosymmetric channel cross section.

where $\alpha^2 = (z_c - e_z)^2/(r_c^2 + \beta_z e_z)$. Although here we have a more general expression for P_{cr_θ} than we did in Sections 6.2 and 6.3 (and $\alpha = e_z/r_p$ no longer), Equation 6.4.26 is formally identical to Equation 6.3.4. Therefore we do not need to repeat the discussion here. The only difference is that now $z_c \neq 0$, and consequently a centric axial load produces combined axial-torsional buckling rather than pure torsional and bending modes. The most important observation is that coupling of bending and torsional modes lowers the first critical load.

Problems

6.4.1 Calculate I_ω, $I_{\omega y}$, and the shear center location for the cross sections shown in Figure 6.13a–j.

6.4.2 Consider a nonsymmetric beam with simple or fixed supports. Prove that, for an axial load applied at the shear center, the flexural and torsional buckling modes uncouple. Calculate the critical loads.

6.4.3 Consider a nonsymmetric beam with simple or fixed supports loaded by an axial force at the centroid. Prove that the critical load for the combined mode is smaller than the critical loads for the uncoupled modes. Can this beam buckle due to tensile force?

6.4.4 Solve Problems 6.3.1 to 6.3.5 for (a) nonsymmetric beams; (b) for a T-beam expressing the critical load in terms of cross-section characteristics.

6.4.5 Formulate the solution to Problems 6.4.4 assuming the beam height varies so that $h = h_0 - h_1 x(L - x)$.

6.4.6 Generalize Problems 6.2.4, 6.2.5, and 6.3.11 to arbitrary open cross sections. *Hint:* Calculate the axial shortening of all longitudinal fibers and then take the work done on these shortenings by the initial normal stresses due to P^0 and M_z^0.

6.4.7 Calculate some of the elements of the transfer matrix that relates the column matrix $(v, \theta, v', \theta', V_y, M_t, M_y, B)^T$ for one cross section to that for another cross section (see also Prob. 2.1.11). (This is an initial-value rather than boundary-value problem. The result is in, e.g., Vlasov, 1959.)

6.4.8 Generalize the above to a beam on an elastic foundation that resists both deflection and rotation about the beam axis (cf. Probs. 6.1.6 and 5.2.16).

6.4.9 Do the same but for the stiffness (rather than the transfer) matrix.

6.5 LARGE DEFLECTIONS

In this section, which is based on a paper by Bažant and El Nimeiri (1973), we consider the most general case of finite deflections of thin-wall beams or thin-wall structures. This, of course, includes as a special case large deflections of ordinary beams or frames considered in Chapters 1 and 2. On the present-day computational scene, the proper approach is a geometrically nonlinear finite element analysis with incremental loading. To handle the geometric nonlinearity, the updated Lagrangian approach is appropriate. In this approach, the geometry of the beam is updated after each loading increment, and the small subsequent increments of deformations are solved on the basis of incrementally linearized equilibrium equations. This means that, even if the beam is initially straight, its updated shapes in the subsequent loading increments are curved, with a general spatial curvature.

The basic information needed for the analysis is the stiffness matrix of a thin-wall beam element. As the kinematic variables that characterize the deformation in a cross section, we may use the variables, u, $\hat{v}$, $\hat{w}$, θ, $\hat{w}'$, $\hat{v}'$, and θ', in which we mix variables referred to the centroid with variables referred to the shear center, since this brings about a simplification of the formulation. Labeling the end cross sections of a beam element by subscripts i and j (Fig. 6.17a, b), the column matrices of element displacements and the associated forces may be defined as

$$\mathbf{q} = (u_i, \hat{v}_i, \hat{w}_i, \theta_i, \hat{w}_i', \hat{v}_i', \theta_i', u_j, \hat{v}_j, \hat{w}_j, \theta_j, \hat{w}_j', \hat{v}_j', \theta_j')^T \tag{6.5.1}$$

$$\mathbf{F} = (-P_i, V_{y_i}, V_{z_i}, M_{t_i}, M_{y_i}, M_{z_i}, \hat{B}_i, -P_j, V_{y_j}, V_{z_j}, M_{t_j}, M_{y_j}, M_{z_j}, \hat{B}_j)^T \tag{6.5.2}$$

Note that these displacement and force variables are associated, that is, $\mathbf{F}^T\delta\mathbf{q}$ is the correct work expression for the element. It is for this reason that we put the minus sign in front of P in Equation 6.5.2. The fact that $\hat{B}_i\theta_i'$ is the correct work expression for bimoments at the ends is justified by noting from Equations 6.4.19a that the first-order work of B per unit length is $\hat{B}\theta''$, and then integrating by parts over the element length $\int_0^l \hat{B}\theta''\,dx = (\hat{B}_j\theta_j' - \hat{B}_i\theta_i') - \int_0^l \hat{B}'\theta'\,dx$.

From the viewpoint of vector notation, some authors used $(-w')$ rather than w' as the nodal value of rotation. This has the advantage that the axial vector of moment M_z is in the sense of y. With our notation, it is opposite to y. We prefer, however, the notation in Figure 6.17a, b because it makes it possible to treat bending in the xy and xz planes in an analogous manner (same shape functions and same expressions for bending in terms of the nodal values).

The distribution of displacements and rotations along the beam element may be introduced in the form:

$$\begin{aligned} u &= a_1 + b_1 x \qquad \hat{v} = a_2 + b_2 x + c_2 x^2 + d_2 x^3 \\ \hat{w} &= a_3 + b_3 x + c_3 x^2 + d_3 x^3 \qquad \theta = a_4 + b_4 x + c_4 x^2 + d_4 x^3 \end{aligned} \tag{6.5.3}$$

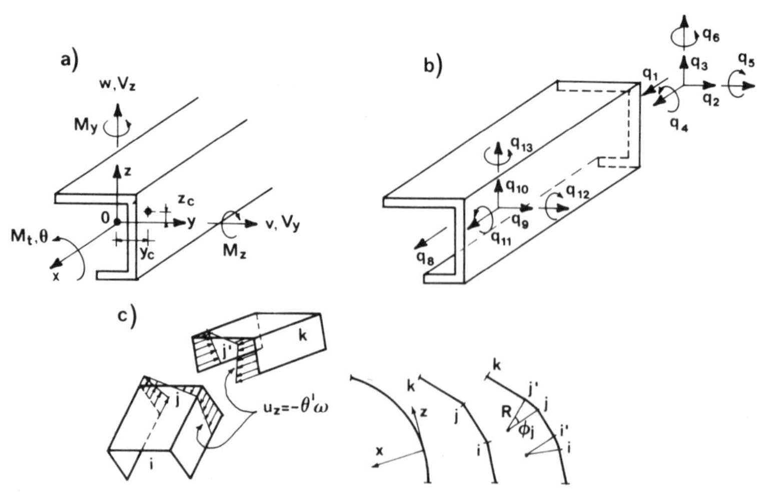

Figure 6.17 (a) Coordinate axes, displacements, and internal forces; (b) displacements of the beam element; (c) incompatibility of assumed warping displacement distributions when two elements meet at an angle and approximation of a curved beam.

in which $a_1, b_1, a_2, \ldots, d_4$ are arbitrary constants. Altogether, we have 14 arbitrary constants, which matches the number of element displacements in Equation 6.5.1. So these constants may be expressed in terms of the 14 components of q, which leads to a relation of the form

$$(u, \hat{v}, \hat{w}, \theta)^T = \mathbf{N}(x)\mathbf{q} \tag{6.5.4}$$

$\mathbf{N}(x)$ is a 4×14 matrix of the shape functions (distribution functions) whose coefficients are cubic polynomials in x. The choice of cubic polynomials in Equations 6.5.3 for transverse displacements agrees with what has already been shown in Section 2.3. The use of cubic polynomials for transverse displacement and rotation is necessary because the distribution functions must be able to describe a uniform distribution of the shear forces and the warping torque, which are proportional to the third derivatives of the transverse displacement and of rotation. To ensure convergence, the distribution functions in Equations 6.5.3 must also be able to describe rigid-body rotations without straining, as well as uniform distributions of the axial strain, curvatures, and the first and second derivatives of rotation θ. It is easily checked that these conditions are satisfied.

Substituting Equation 6.5.4 into the incremental strain energy expression in Equation 6.4.18, in which the integrals over the cross section are given by Equations 6.4.19a, one obtains a quadratic form in $\mathbf{q}$. In performing the integration over the length of the beam, the initial bending moments M_y^0, M_z^0 and the bimoment $\hat{B}^0$ may be assumed to be constant and equal to the averages $\frac{1}{2}(M_{y_j}^0 - M_{y_i}^0), \ldots$ of the nodal values, at the cost of some errors that tend to zero as the element length decreases. The elements of the incremental (tangential) stiffness matrix $\mathbf{K}$, 14×14 in size, can then be obtained as

$$K_{ij} = \frac{\partial^2 U}{\partial q_i \, \partial q_j} \tag{6.5.5}$$

The expressions for all elements of this stiffness matrix were presented by Bažant and El Nimeiri (1973) for a monosymmetric cross section and for loading in the vertical plane (however, without the terms that correspond to the work of the shear forces V_y^0, V_z^0 and that arise from the nonlinear part of shear strain expression). The resulting stiffness matrix for a general cross-section shape may be written in the form:

$$\mathbf{K} = \mathbf{K}_0 + P\mathbf{K}_1 + M_y^0\mathbf{K}_2 + M_z^0\mathbf{K}_3 + B^0\mathbf{K}_4 + V_y^0\mathbf{K}_5 + V_z^0\mathbf{K}_6 \tag{6.5.6}$$

in which $\mathbf{K}_0, \ldots, \mathbf{K}_6$ are constant 14×14 matrices; $\mathbf{K}_0$ depends on the elastic moduli and is called the elastic stiffness matrix, while matrices $\mathbf{K}_1, \ldots, \mathbf{K}_6$ are independent of the elastic moduli; $P\mathbf{K}_1, M_y^0\mathbf{K}_2, \ldots, V_z^0\mathbf{K}_6$ are called the geometric stiffness matrices. All these matrices are symmetric, and their elements are as follows:

Elements of matrix $\mathbf{K}_0$:

$$K_{1,1} = K_{8,8} = -K_{1,8} = \frac{EA}{L} \qquad K_{2,2} = K_{9,9} = -K_{2,9} = 12\frac{k_y}{L^3}$$

$$K_{6,6} = K_{13,13} = \frac{1}{2}K_{6,13} = 4\frac{k_y}{L} \qquad K_{2,6} = K_{2,13} = -K_{6,9} = -K_{9,13} = 6\frac{k_y}{L^2}$$

$$K_{3,3}=K_{10,10}=-K_{3,10}=12\frac{k_z}{L^3}\qquad K_{5,5}=K_{12,12}=\frac{1}{2}K_{5,12}=4\frac{k_z}{L}$$

$$K_{3,5}=K_{3,12}=-K_{5,10}=-K_{10,12}=6\frac{k_z}{L^2}\qquad K_{4,4}=K_{11,11}=-K_{4,11}=12\frac{k_\omega}{L^3}+\frac{6}{5}\frac{C}{L}$$

$$K_{7,7}=K_{14,14}=4\frac{k_\omega}{L}+\frac{2}{15}CL\qquad K_{7,14}=2\frac{k_\omega}{L}-\frac{1}{30}CL$$

$$K_{4,7}=K_{4,14}=-K_{7,11}=-K_{11,14}=6\frac{k_\omega}{L^2}+\frac{1}{10}C$$

$$K_{1,7}=-K_{1,14}=-K_{7,8}=K_{8,14}=\frac{EA\omega_0}{L}\tag{6.5.7}$$

in which $k_y=EI_y$, $k_z=EI_z$, $k_\omega=E(I_{\hat{\omega}}+A\omega_0^2)$, and $C=GJ$.

Elements of matrix $P\mathbf{K}_1+M_y^0\mathbf{K}_2+M^0\mathbf{K}_3+B^0\mathbf{K}_4$:

$$K_{4,4}=K_{11,11}=-K_{4,11}=-\frac{12}{10}\frac{1}{L}S_1\qquad K_{7,7}=K_{14,14}=-\frac{2}{15}LS_1$$

$$K_{7,14}=\frac{1}{30}LS_1\qquad K_{4,7}=K_{4,14}=-K_{7,11}=-K_{11,14}=-\frac{1}{10}S_1$$

$$K_{3,4}=-K_{3,11}=-K_{4,10}=K_{10,11}=\frac{3}{5}\frac{1}{L}S_2$$

$$K_{3,7}=K_{3,14}=K_{4,5}=-K_{5,11}=-K_{7,10}=-K_{10,14}=K_{4,12}=-K_{11,12}=\frac{1}{20}S_2$$

$$K_{5,7}=K_{12,14}=\frac{1}{15}LS_2\qquad K_{5,14}=K_{7,12}=-\frac{1}{60}LS_2$$

$$K_{2,4}=-K_{2,11}=-K_{4,9}=K_{9,11}=-\frac{3}{5}\frac{1}{L}S_3\tag{6.5.8}$$

$$K_{2,7}=K_{2,14}=K_{4,6}=-K_{6,11}=-K_{7,9}=-K_{9,14}=K_{4,13}=-K_{11,13}=-\frac{1}{20}S_3$$

$$K_{6,7}=K_{13,14}=-\frac{1}{15}LS_3\qquad K_{6,14}=K_{7,13}=\frac{1}{60}LS_3$$

$$K_{2,2}=K_{9,9}=-K_{2,9}=K_{3,3}=K_{10,10}=-K_{3,10}=-\frac{6}{5}\frac{P}{L}$$

$$K_{2,6}=K_{2,13}=-K_{6,9}=-K_{9,13}=K_{3,5}=K_{3,12}=-K_{5,10}=-K_{10,12}=-\frac{P}{10}$$

$$K_{6,6}=K_{13,13}=K_{5,5}=K_{12,12}=-\frac{2}{15}PL\qquad K_{6,13}=K_{5,12}=\frac{1}{30}PL$$

in which $S_1=Pr_c^2+\beta_yM_y^0+\beta_zM_z^0+\beta_\omega B^0$, $S_2=Py_c-2M_y^0$, and $S_3=Pz_c-2M_z^0$.

Elements of matrix $V_y^0\mathbf{K}_5 + V_z^0\mathbf{K}_6$:

$$K_{2,4} = K_{2,11} = -K_{4,9} = -K_{9,11} = \tfrac{1}{2}V_z^0 \qquad -K_{3,4} = -K_{3,11} = K_{4,10} = K_{10,11} = \tfrac{1}{2}V_y^0$$

$$K_{2,7} = -K_{2,14} = -K_{4,6} = K_{6,11} = -K_{7,9} = K_{9,14} = K_{4,13} = -K_{11,13} = \frac{L}{10}V_z^0$$

$$-K_{3,7} = K_{3,14} = K_{4,5} = -K_{5,11} = K_{7,10} = -K_{10,14} = -K_{4,12} = K_{11,12} = \frac{L}{10}V_y^0$$

$$K_{5,14} = -K_{7,12} = \frac{L^2}{60}V_y^0 \qquad -K_{6,14} = K_{7,13} = \frac{L^2}{60}V_z^0 \tag{6.5.9}$$

Note that the term in Equation 6.4.18 that represents the work of the initial Saint-Venant torque does not give rise to a geometric stiffness matrix since it does not contain any second-order terms. Note also that the stiffness matrix for short elements of a beam-column given in Equation 2.3.6 is a special case of Equation 6.5.5.

The shape functions in Equation 6.5.4 need to be also used to calculate according to the principle of virtual work the nodal loads that are equivalent to distributed loads over the length of the beam.

In the calculation of large deflections, the beam elements may be assumed, for the sake of simplicity, to be straight between their end points, even though the actual beam becomes curved. This approximation was proved to be satisfactory if the beam elements are short enough. Due to curvature of the beam, adjacent finite elements of the beam meet at their ends at an angle. A question then arises with regard to the transfer condition between the warping parameters θ' in the matrix of generalized displacements (Eq. 6.5.1, Fig. 6.17c). Detailed analysis (Bažant and El Nimeiri, 1973) showed that if the angle between the adjacent elements is small, the proper assumption is that the values of θ' for the ends of two adjacent elements meeting at an angle must be equal. This condition is easy to implement in the assembly of the global stiffness matrix, in which the remaining generalized displacements in Equation 6.5.1, representing displacements and rotations, are transformed to global coordinates in the usual manner.

Although the displacements are allowed to be large, the present formulation is nevertheless restricted to small strains. Nonlinear geometric properties in the present formulation arise exclusively from large rotations at small strains (see Sec. 11.1). This limitation of scope is adequate for most practical problems of thin-wall beams.

An incremental solution procedure with step-by-step loading, in which the locations of the nodes are updated after each loading step and the solution for each step is iterated to improve accuracy, was described in detail by Bažant and El Nimeiri (1973). Some of the results of their solution, which belongs to the class of updated Lagrangian solutions for large deformations, are exhibited in Figures 6.18, 6.19 and 6.20. Figure 6.18 shows a solution of large lateral buckling deflections of an I-beam for three different magnitudes of imperfections represented by small applied lateral distributed moments. We see that a significant deflection increase takes place near the critical load, and that lateral buckling exhibits a postcritical reserve rather than a postcritical loss of capacity. Thus, the lateral buckling of beams is not imperfection sensitive, which is similar to the

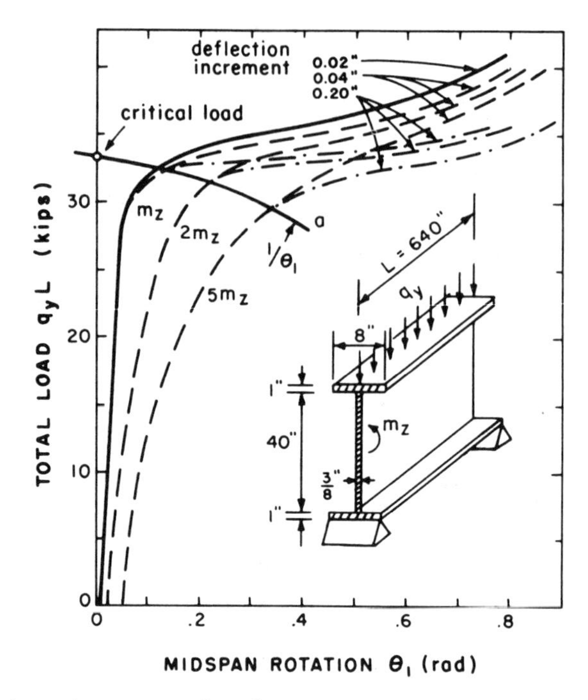

Figure 6.18 Load-rotation curves for three values of constant distributed disturbing torsional moments. (*After Bažant and El Nimeiri,* 1973.)

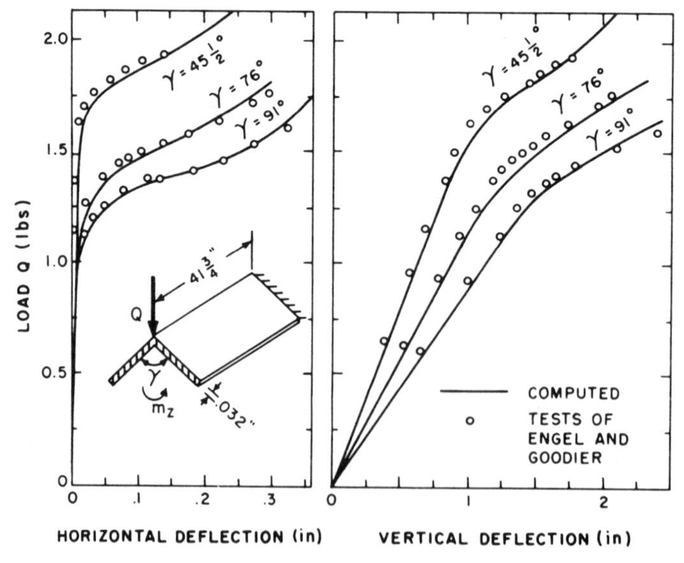

Figure 6.19 Comparison of computed and experimental load-deflection curves of angle cross section. (*After Bažant and El Nimeiri, 1973.*)

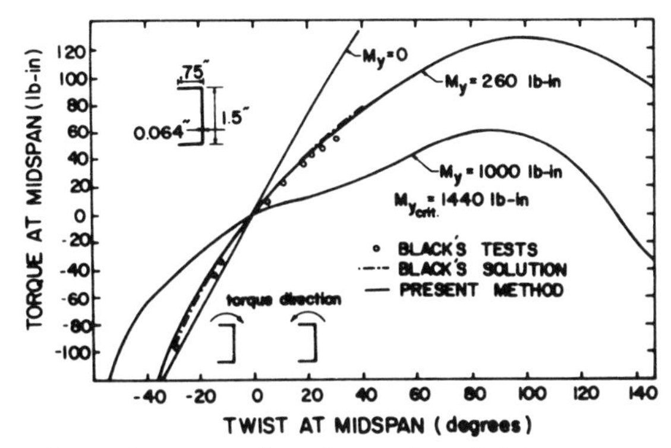

Figure 6.20 Comparison of computed and experimental torque versus rotation curves. (*After Bažant and El Nimeiri, 1973.*)

behavior of columns. Figures 6.19 and 6.20 show comparisons of Bažant and El Nimeiri's finite element incremental solutions with large lateral buckling deflections measured by Engel and Goodier (1953) and by Black (1967). The computational results in Figure 6.20, however, show that the load-deflection diagram of channel-section beam may exhibit at large deflections a limit point (maximum), leading to snapthrough instability. It is also seen from Figure 6.20 that, due to a symmetry of the cross section, the response diagrams for twisting to the right and twisting to the left are different. For more detail see Bažant and El Nimeiri (1973).

Interesting results were more recently presented by Kitipornchai and Chan (1987), who utilized a similar approach for the tracing of load-deflection curves of angle beam-columns under eccentric load. Their numerical results, shown in Figure 6.21, indicate the importance of updating the nodal coordinates of the

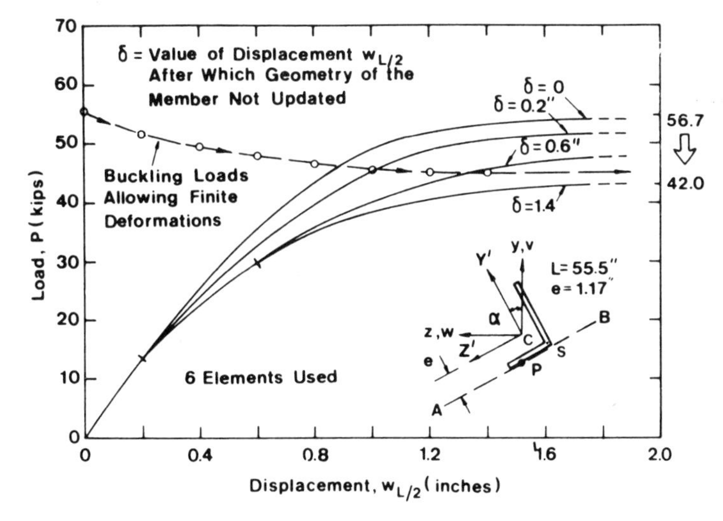

Figure 6.21 Influence of updating nodal coordinates after each loading step. (*After Kitipornchai and Chan, 1987.*)

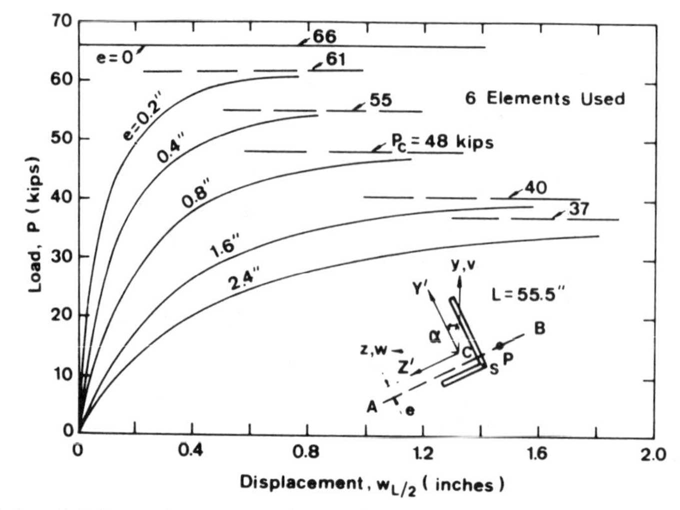

Figure 6.22 Load-deformation curves for various eccentricities. (*After Kitipornchai and Chan, 1987.*)

beam element after each loading step, and their results in Figure 6.22 illustrate how the maximum load decreases with increasing eccentricity.

Problems

6.5.1 Calculate at least some elements of the stiffness matrix in Equation 6.5.5 by using the generalized nodal displacements defined in Equation 6.5.1 and the shape functions defined in Equations 6.5.3 and 6.5.4. How would the results change if the nodal values of the rotations were all assumed to turn clockwise when looking in the directions of the x, y, and z axes?

6.5.2 Calculate the elements of the geometric stiffness matrices $M_y^0\mathbf{K}_2$, $M_z^0\mathbf{K}_3$, $B^0\mathbf{K}_4$ assuming a linear variation of M_y^0, M_z^0, and B^0 over the beam length, that is, $M_y^0 = -M_{y_i} + (M_{y_i} + M_{y_j})x/L$; $M_z^0 = -M_{z_i} + (M_{z_i} + M_{z_j})x/L$; $\hat{B}^0 = -\hat{B}_i + (\hat{B}_i + \hat{B}_j)x/L$.

6.6 BOX GIRDERS

Another stability problem, which can be approximately solved as a one-dimensional problem, is the buckling of slender box girders. Similarly to thin-wall beams of open cross section, box girders also exhibit warping of their cross sections. In contrast to thin-wall beams, however, box girders exhibit two further important modes of deformation: (1) shear strain of the middle surface of the walls (which is neglected in the formulation for open cross sections) and (2) distortion of the cross-section shape, unless the distortion is prevented by closely spaced rigid diaphragms.

The fact that a box girder must deform by shear straining in its middle surface becomes clear if we consider a longitudinal slit in the box girder, which transforms the closed cross section to an open cross section. In the latter case the shear strains in the middle surface are negligible, and according to the sectorial coordinate ω there must be a mismatch Δu of the longitudinal displacements across the slit (Fig. 6.23a). Obviously, to restore continuity across the slit, the walls must be made to deform by shear. This partially reduces the out-of-plane warping of the cross section, and so the behavior of box beams is influenced by warping less profoundly than the behavior of open cross section beams.

Deformation Modes and Postcritical Energy

Similar to our previous formulation for open cross sections, we assume again the deformation of the cross section to be approximately described by a linear combination of several basic modes:

$$u = \sum_{k=1}^{4} U_k(x)\phi_k(s) \qquad v = \sum_{k=1}^{4} V_k(x)\psi_k(s) \qquad w = \sum_{k=1}^{4} V_k(x)\chi_k(s) \quad (6.6.1)$$

in which u = longitudinal displacement at any point of the cross section, v and w = transverse displacements (i.e., displacements within the cross-section plane) of points on the middle surface in directions tangent and normal to the wall (Fig. 6.24a), $\phi_k(s)$ = longitudinal displacement distribution and warping mode, $\psi_k(s)$ and $\chi_k(s)$ = functions defining transverse displacements, rotations, and distortion of the cross section shape; s = length coordinate of the midsurface of the wall; $U_k(x)$, $V_k(x)$ = generalized displacements serving as parameters of the deformation modes; U_1 = axial displacement, U_2, U_3 = rotations about horizontal and vertical axes, $V_1 = \theta$ = rotation about beam axis; V_2, V_3 = transverse horizontal and vertical displacements, U_4 = out-of-plane warping parameter, V_4 = in-plane distortion parameter. Equations 6.6.1 were introduced by Bažant (1968), and their special case earlier by Umanskiy (1939); see also Křístek (1979).

The deformation modes that describe the warping and distortion of the cross section are ϕ_4, ψ_4, χ_4 and are graphically illustrated in Figure 6.24b. Together with the remaining deformation modes representing translations and rotations of the cross section, these functions may be described, for a rectangular cross section of width a and depth b, by the following expressions (Bažant, 1968; and Bažant

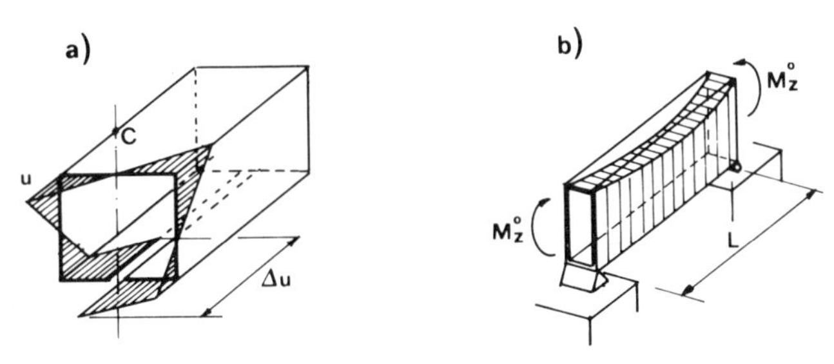

Figure 6.23 (a) Torsional behavior of an open cross section; (b) simply supported box girder subjected to constant bending moment.

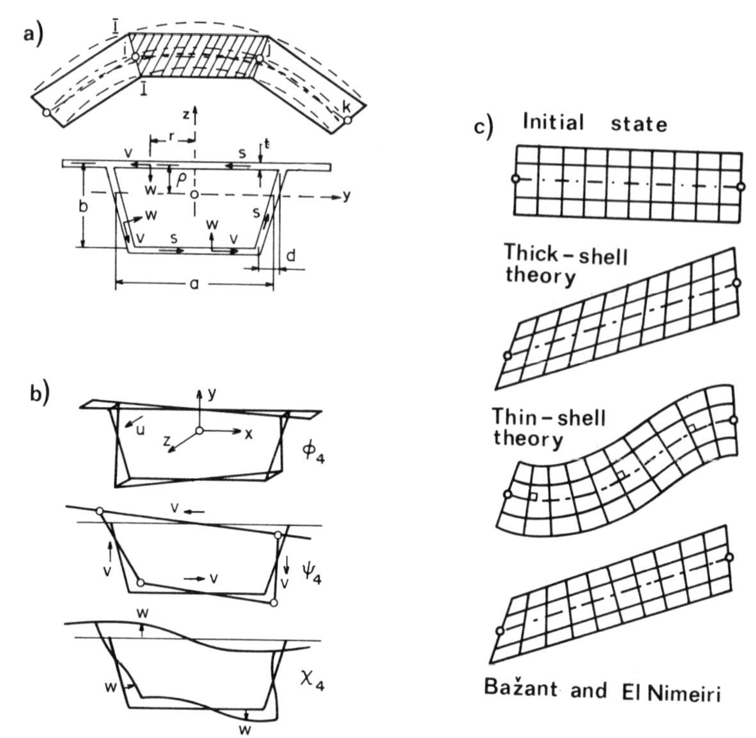

Figure 6.24 (a) Subdivision of a curved beam into finite elements and a box cross section; (b) deformation modes assumed for cross section; (c) assumed transverse displacements of cross section. (*After Bažant and El Nimeiri, 1974.*)

and El Nimeiri, 1974):

$$
\begin{aligned}
&\phi_1 = 1 \qquad \phi_2 = \frac{a}{2}v \qquad \phi_3 = \frac{b}{2}w \qquad \phi_4 = \frac{ab}{4}vw \\
&\psi_1 = \rho(s) \qquad \psi_2 = \frac{a}{2}v_{,s} \qquad \psi_3 = \frac{b}{2}w_{,s} \qquad \psi_4 = \frac{ab}{4}(vw_{,s} + wv_{,s}) \\
&\chi_1 = r(s) \qquad \chi_2 = \frac{b}{2}w_{,s} \qquad \chi_3 = \frac{a}{2}v_{,s} \\
&\chi_4 = \frac{vw}{4(at_1^3 + bt_2^3)}[a^3t_1^3(3 - v^2)v_{,s} + b^3t_2^3(3 - w^2)w_{,s}]
\end{aligned}
\tag{6.6.2}
$$

$\rho(s)$ = distance of cross-sectional wall tangent from the beam axis and $r(s)$ = distance of the transverse normal displacement vector from the beam axis (both being of positive sign when turning positive); t_1 and t_2 = thicknesses of the horizontal and vertical flanges of the box; s = length coordinate of the midsurface of the wall. The expression for function χ_4 represents the solution of the deformation shape of a rectangular frame due to an enforced change of length of its diagonal.

The potential-energy expression for the box girder is $\Pi = U - W$ where

W = work of applied loads and U = strain energy:

$$U=\int_0^l\int_A\left[\sigma^0\varepsilon_x+\frac{1}{2}E\left(\frac{\partial u}{\partial x}\right)^2+\tau_{xs}^0\gamma_{xs}+\frac{1}{2}G\left(\frac{\partial u}{\partial s}+\frac{\partial v}{\partial x}\right)^2\right]dA\,dx \tag{6.6.3}$$

in which ε_x and γ_{xs} are the finite strain expressions:

$$\varepsilon_x=\frac{\partial u}{\partial x}+\frac{1}{2}\left(\frac{\partial v}{\partial x}\right)^2+\frac{1}{2}\left(\frac{\partial w}{\partial x}\right)^2 \qquad \gamma_{xs}=\frac{\partial u}{\partial s}+\frac{\partial v}{\partial x}+\frac{\partial v}{\partial s}\left(\frac{\partial v}{\partial x}\right)+\frac{\partial w}{\partial s}\left(\frac{\partial w}{\partial x}\right) \tag{6.6.4}$$

In contrast to our previous expression in Equation 6.1.8 for thin-wall beams of open cross section, we must now include the strain energy due to the linear part of the shear strains in the middle surface of the wall (which was negligible for thin-wall beams of open cross section). Furthermore, we do not need a separate term for strain energy due to simple torsion (given by the last integral in Eq. 6.1.8). The reason is that this strain energy is included in the term with shear modulus G in Equation 6.6.3.

The differential equations for the unknown functions $U_k(x)$ and $V_k(x)$ and the boundary conditions may now be obtained by substituting Equations 6.6.1 into Equation 6.6.3 with Equations 6.6.4 and minimizing the potential-energy expression according to variational calculus. The resulting Euler equations are the differential equations of the problem.

Examples

For the sake of a simple illustration, consider a simply supported box girder of constant cross section and length L (Fig. 6.23b). Initially the girder is subjected to a uniform bending moment caused by applying initial bending moments M_z^0 at the ends. We assume that the cross section is braced so that the distortion V_4 is negligible. Minimizing the potential-energy expression, and neglecting the bending shear strains, we obtain the following set of three differential equations (see Bažant and El Nimeiri, 1973):

$$EI_zV_2^{\mathrm{IV}}+M_z^0V_1''=0 \tag{6.6.5a}$$

$$a_0U_4''-b_1U_4-b_2V_1'=0 \tag{6.6.5b}$$

$$b_1V_1''-M_x^0V_2''=0 \tag{6.6.5c}$$

in which the coefficients are

$$a_0=\int_A\phi_4^2\,dA \qquad b_1=\int_A\left(\frac{d\phi_4}{ds}\right)^2dA \qquad b_2=\int_A\frac{d\phi_4}{ds}\left(\frac{d\phi_1}{ds}\right)dA \tag{6.6.6}$$

To simplify the solution, we may set $U_4=f'(x)$. Then by integration of Equation 6.6.5b, we have $V_1=(a_0f''-b_1f)/b_2$. After substitution into Equations 6.6.5a, c, Equations 6.6.5 reduce to a single fourth-order differential equation for function $f(x)$:

$$f^{\mathrm{IV}}-2\alpha f''-\beta f=0 \tag{6.6.7}$$

in which $2\alpha = (b_1^2 - b_2^2)/a_0 b_1 - M_z^{0^2}/b_1 EI_z$, $\beta = M_z^{0^2}/a_0 EI_z$. The boundary conditions of simple support at $x = 0$ and $x = L$ are $V_1 = U_4' = f = f'' = 0$. Taking the solution in the form $f(x) = c_1 \sin mx + c_2 \cos mx + c_3 e^{nx} + c_4 e^{-nx}$ we find that a nonzero solution of the differential equations exists if $(\alpha^2 + \beta)^{1/2} - \alpha = \pi^2/L^2$. This yields for the critical value of the applied bending moment the expression, derived by Bažant and El Nimeiri (1974)

$$M_z^{\mathrm{cr}} = \frac{\pi}{L}\left[EI_z GJ\left(1 + \frac{\pi^2 a_0}{L^2 GJ}\right)\left(1 + \frac{\pi^2 a_0}{L^2 b_1}\right)^{-1}\right]^{1/2} \tag{6.6.8}$$

As another example, one can also solve the critical bending moment for lateral buckling of an arch of box cross section (Bažant and El Nimeiri, 1974, p. 2088), Figure 6.25.

Finite Element Solution

For beams of complex boundary conditions, beams of variable cross section, or curved beams it is preferable to solve the problem by the finite element method. Similar to our previous discussion of finite elements for beams of open cross section, it is sufficient (for beam elements that are sufficiently short), to consider straight box girder elements even if the beam is curved. A convenient way to formulate the stiffness matrix is to consider the finite element of a box girder of arbitrary shape to be a mapped image of a unit element whose length as well as box height and width equal unity. The transformation equations that describe this mapping are trilinear in Cartesian coordinates of the unit parent element. The stiffness matrix is formulated first for this unit element and then its transformation due to the mapping is carried out.

The distribution functions (shape functions) may be considered as linear for V_k $(k = 1, 2, 3, 4)$ and U_1, and quadratic for U_k $(k = 2, 3, 4)$. The shape functions are associated with 19 generalized displacements, represented by eight displacements $(V_1, \ldots, V_4, U_1, \ldots, U_4)$ at each end, and three displacements (U_2, U_3, U_4) at

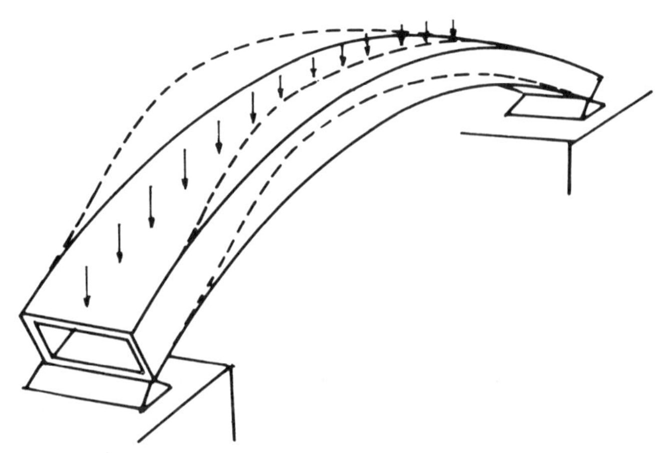

Figure 6.25 Lateral buckling of an arch of box cross section (*studied by Bažant and El Nimeiri, 1974*).

element midlength. The formulation leads to a 19×19 stiffness matrix for each finite element of the box girder. The stiffness matrix, which depends on the initial stress distribution in the box girder cross sections, is fully listed in Bažant and El Nimeiri (1974).

A sensitive aspect of finite elements for beams, and box girders in particular, is the spurious shear stiffness, which can occur not only in bending but also in torsion. It was found that the aforementioned finite element is free of spurious shear stiffness problems. In this formulation, the transverse displacements of the cross sections are assumed to vary linearly along the element (Fig. 6.24c), whereas the transverse rotations and the axial displacements, including those due to bending as well as warping, are taken as quadratic as is clear from the aforementioned distributions of V_k and U_k. This is a departure from the classical approach to bending of beams or torsion of thin-wall beams, in which a cubic variation of transverse displacements is normally introduced in order to achieve continuity of the transverse rotations expressed as derivatives of the transverse displacements. In the formulation just described, the rotations are considered as independent of the displacements. Consequently, the transverse displacements need not satisfy slope continuity conditions, which makes it possible to consider them linearly distributed (Fig. 6.24c). For a more detailed discussion of the avoidance of spurious shear stiffness with this finite element, see Bažant and El Nimeiri (1975).

Examples of finite element analysis according to the formulation just described have indicated good agreement with the tests of Křístek (1979) and of Aneja and Roll (1971), as well as with analytical solutions of certain special cases. Subsequent applications of this finite element approach to the calculation of deflections of prestressed concrete box girder bridges revealed that the method is sufficiently accurate only for relatively narrow and long box girder bridges, while for a higher width-to-span ratio, which frequently occurs in highway construction, the errors can be significant. These errors are due to local deformations and more complex transverse distributions of displacements than one can describe with the presently assumed transverse distribution modes (Van Zyl and Scordelis, 1979).

Interaction with Local Buckling

A very important aspect of stability of box birders, which is beyond the scope of thin-wall beam theory, is the local buckling of walls and the interaction of local buckling with global buckling. The local buckling of box girder walls and the effect of stiffeners are analyzed in detail by Křístek (1979). Interaction of various buckling modes can profoundly alter postcritical behavior. In particular, it may cause softening of the response after the maximum load is reached. This type of response, which is illustrated by some simple examples in Chapter 4, is found to occur for stiffened plate structures such as box girders; in more detail see Thompson and Hunt (1984, sec. 8.5). Insufficient knowledge of the changes in postcritical behavior due to interactive local–global buckling have contributed to some collapses of large box girder bridges.

Problems

6.6.1 For the simply supported box girder in Figure 6.23b, work out the solution in full detail and check the result.

6.6.2 Analyze the same girder as above, but with completely built ends.

6.6.3 The transfer matrix relates the (14×1) column matrix $(u, \hat{v}, \hat{w}, \theta, \hat{w}', \hat{v}', \theta', -P, V_y, V_z, M_t, M_y, M_z, B)^T$ for one cross section to that for another cross section. Calculate it by algebraic rearrangements of the stiffness matrix. [For thin-wall multispan beams or beams of variable cross section, the use of a transfer (or transport) matrix requires less computational work than the use of a stiffness matrix. This explains popularity of transfer matrices in the early computer days; however, for today's powerful computers it does not matter.]

References and Bibliography

Aida, T. (1986), "Applications of Extended Galerkin's Method to Non-Conservative Stability Problems of the Columns with Thin-Walled Open Cross Sections," *Comp. Meth. in Appl. Mech. Eng.*, 54(1):1–20.

Aida, T. (1989), "Dynamic Stability of Thin-Walled Structural Members under Periodic Axial Torque," *J. Eng. Mech.* (ASCE), 115(1):71–88.

Akhtar, M. N. (1987), "Element Stiffness of Circular Member," *J. Struct. Eng.* (ASCE), 113(4):867–72.

Ali, M. A., and Sridharan, S. (1988), "Versatile Model for Interactive Buckling of Columns and Beam-Columns," *Int. J. Solids Struct.*, 24(5):481–96.

Aneja, I., and Roll, F. (1971), "An Experimental and Analytical Investigation of a Horizontally Curved Box-Beam Highway Bridge Model," A.C.I. Special Publ. SP-26, pp. 379–410 (Sec. 6.6).

Attard, M. M. (1986), "Nonlinear Theory of Non-Uniform Torsion of Thin-Walled Open Beams," *Thin-Walled Struct.*, 4(2):101–34.

Attard, M. M. (1986), "Lateral Buckling Analysis of Beams by the FEM," *Comput. Struct.*, 23(2):217–31.

Barsoum, R. S., and Gallagher, R. H. (1970), "Finite Element Analysis of Torsional and Torsional-Flexural Stability Problems," *Int. J. Numer. Methods Eng.*, 2:335–52.

Bažant, Z. P. (1965), "Nonuniform Torsion of Thin-Walled Bars of Variable Cross Section," *International Association for Bridge and Structural Engineering* (IABSE), 25:245–67; see also Inženýrské Stavby, Prague, Czechoslovakia, Vol. 15, 1967, pp. 222–28 (Sec. 6.3).

Bažant, Z. P. (1968), "Pièces longues a voiles épais et calcul des poutres a section déformable," *Annales des Ponts et Chaussées,* No. 3, pp. 155–69; also Stavebnícky Časopis SAV, Vol. 15 (1967), pp. 541–55 (Sec. 6.6).

Bažant, Z. P., and El Nimeiri, M. (1973), "Large-Deflection Spatial Buckling of Thin-Walled Beams and Frames," *J. Eng. Mech.* (ASCE), 99(6):1259–81 (Sec. 6.5).

Bažant, Z. P., and El Nimeiri, M. (1974), "Stiffness Method for Curved Box Girders at Initial Stress," *J. Struct. Eng.* (ASCE) 100(10):2071–90 (Sec. 6.6).

Bažant, Z. P., and El Nimeiri, M. (1975), "Finite Element for Buckling of Curved Beams and Shells with Shear," *J. Struct. Eng.* (ASCE), 101(9):1997–2004 (Sec. 6.3).

Black, M. (1967), "Nonlinear Behavior of Thin-Walled Unsymmetrical Beam-Sections Subjected to Bending and Torsion," in *Thin-Walled Structures,* ed. by A. H. Chilver, John Wiley & Sons, New York, pp. 87–102 (Sec. 6.5).

Blandford, G. E. (1988), "Static Analysis of Flexibly Connected Thin-Walled Plane Frames," *Comput. Struct.*, 28(1):105–13.

Bleich, F. (1952), *Buckling Strengh of Metal Structures,* McGraw-Hill, New York (Sec. 6.0).

Bleich, F., and Bleich, H. (1936), "Bending Torsion and Buckling of Bars Composed of Thin Walls," Prelim. Pub. 2nd Cong. Intern. Assoc. Bridge and Struct. Eng. (IABSE), English ed., p. 871 (Sec. 6.0).

Bradford, M. A. (1989), "Buckling Strength of Partially Restrained I-Beams," *J. Struct. Eng.* (ASCE), 115(5): 1272–77.

Bradford, M. A., and Cuk, P. E. (1988), "Elastic Buckling of Tapered Monosymmetric I-Beams," *J. Struct. Eng.* (ASCE), 114(5): 977–96.

Bradford, M. A., Cuk, P. E., Gizejowski, M. A., and Trahair, N. S. (1987), "Inelastic Lateral Buckling of Beam-Columns," *J. Struct. Eng.* (ASCE), 113(11): 2259–77.

Chajes, A. (1974), *Principles of Structural Stability Theory,* Prentice-Hall, Englewood Cliffs, N.J. (Sec. 6.3).

Cheng, J.-J. R., Yura, J. A., and Johnson, C. P. (1988), "Lateral Buckling of Coped Steel Beams," *J. Struct. Eng.* (ASCE), 114(1): 1–15.

Chilver, A. H. (1967), *Thin-Walled Structures,* Chatto and Windus, London.

Clift, C. D., and Austin, W. J. (1989), "Lateral Buckling in Curtain Wall Systems," *J. Struct. Eng.* (ASCE), 115(10): 2481–95.

Davids, A. J., and Hancock, G. J. (1987), "Nonlinear Elastic Response of Locally Buckled Thin-Walled Beam-Columns," *Thin-Walled Struct.*, 5(3): 211–26.

Engel, H. L., and Goodier, J. N. (1953), "Measurements of Torsional Stiffness Changes and Instability due to Tension, Compression and Bending," *J. Appl. Mech.* (ASME), 20: 553–60 (Sec. 6.5).

Galambos, T. V. (ed.) (1988), *Guide to Stability Design Criteria for Metal Structures,* John Wiley & Sons, New York (Sec. 6.3).

Gjelsvik, A. (1981), *The Theory of Thin-Walled Bars,* John Wiley & Sons, New York (Sec. 6.4).

Goltermann, P., and Svensson, S. E. (1988), "Lateral-Distorsional Buckling: Predicting Elastic Critical Stress," *J. Struct. Eng.* (ASCE), 114(7): 1606–25.

Goodier, J. N. (1941), "The Buckling of Compressed Bars by Torsion and Flexure," Cornell Univ. Eng. Expt. Sta. Bull. No. 27 (Sec. 6.0).

Grimaldi, A., and Pignataro, M. (1975), "Post-Buckling Behaviour of Thin-Walled Open Cross Section Compression Members," Ist. Scienza delle Costruzioni, Report 2–180, University of Rome, Italy.

Grimaldi, A., and Pignataro, M. (1979), "Postbuckling Behaviour of Thin-Walled Open Cross-Sections Compression Members," *J. Struct. Mech.*, 7(2): 143–59.

Heins, C. P., and Potocko, R. A. (1979), "Torsional Stiffening of I-Girder Webs," *J. Struct. Eng.* (ASCE), 105(8): 1689–1700.

Hjelmstad, K. D. (1987), "Warping Effects in Transverse Bending of Thin-Walled Beams," *J. Eng. Mech.* (ASCE), 113(6): 907–24.

Hui, D. (1984), "Effects of Mode Interaction on Collapse of Short, Imperfect Thin-Walled Columns," *J. Appl. Mech.* (ASME), 50: 566–73.

Ings, N. L., and Trahair, N. S. (1987), "Beam and Column Buckling under Directed Loading," *J. Struct. Eng.* (ASCE), 113(6): 1251–63.

Jakubowski, S. (1988), "Buckling of Thin-Walled Girders Under Compound Load," *Thin-Walled Struct.*, 6(2): 129–50.

Kani, I. M., and McConnel, R. E. (1987), "Collapse of Shallow Lattice Domes," *J. Struct. Eng.* (ASCE), 113(8): 1806–19.

Kappus, R. (1937), "Drillknicken zentrisch gedrückter Stäbe mit offenem Profile im elastischen Bereich," *Luftfahrt-Forschung,* 14(9): 444–57; English translation, "Twisting Failure of Centrally Loaded Open-Sections Columns in the Elastic Range," NACA Tech. Mem. No. 851, 1938 (Sec. 6.0).

Kitipornchai, S., and Chan, S. L. (1987), "Nonlinear Finite Element Analysis of Angle and Tee Beam-Columns," *J. Struct. Eng.* (ASCE), 113(4): 721–39 (Sec. 6.4).

Kitipornchai, S., and Wang, C. M. (1986), "Lateral Buckling of Tee Beams Under Moment Gradient," *Comput. Struct.*, 23(1):69–76.

Kitipornchai, S., and Wang, C. M. (1988), "Out-of-Plane Buckling Formulas for Beam-Columns/Tie-Beams," *J. Struct. Eng.* (ASCE), 114(12):2773–89.

Kitipornchai, S., Wang, C. M., and Thevendran, V. (1986), "Buckling of Braced Monosymmetric Cantilevers: Timoshenko Energy Approach," Report No. CE74, Dep. Civ. Engng., Univ. of Queensland, St. Lucia, Aust., 32 p.

Kitipornchai, S., Wang, C. M., and Trahair, N. S. (1986), "Buckling of Monosymmetric I-Beams Under Moment Gradient," *J. Struct. Eng.* (ASCE), 112(4):781–99. See also discussion and closure, 113(6):1387–95.

Kolakowski, Z. (1989), "Some Thoughts on Mode Interaction in Thin-Walled Columns under Uniform Compression," *Thin-Walled Struct*, 7:23–35.

Krayterman, B. L., and Krayterman, A. B. (1987), "Generalized Nonuniform Torsion of Beams and Frames," *J. Struct. Eng.* (ASCE), 113(8):1772–87.

Křístek, V. (1979), *Theory of Box Girders,* John Wiley & Sons, New York (Sec. 6.6).

Kubo, M., and Fukumoto, Y. (1988), "Lateral-Torsional Buckling of Thin-Walled I-Beams," *J. Struct. Eng.* (ASCE), 114(4):841–55.

Lau, S. C. W., and Hancock, G. J. (1987), "Distortional Buckling Formulas for Channel Columns," *J. Struct. Eng.* (ASCE), 113(5):1063–78.

Lundquist, E. E., and Fligg, C. M. (1937), "A Theory for Primary Failure of Straight Centrally Loaded Columns," NACA Tech. Report No. 582 (Sec. 6.0).

Masur, E. F. (1975), "Discussion of the paper 'Large-Deflection Spatial Buckling of Thin-Walled Beams and Frames' by Z. P. Bažant and M. El Nimeiri," *J. Eng. Mech.* (ASCE), 101:82–6 (Sec. 6.1).

Milisavljevic, B. M. (1988), "Lateral Buckling of a Cantilever Beam with Imperfections," *Acta Mech.*, 74(1–4):123–37.

Mottershead, J. E. (1988), "Geometric Stiffness of Thin-Walled Open Section Beams Using a Semiloof Beam Formulation," *Int. J. Numer. Methods Eng.*, 26(10):2267–78.

Murray, N. W. (1984), *Introduction to the Theory of Thin-Walled Structures,* Oxford University Press, New York (Sec. 6.0).

Nethercot, D. A. (1973), "The Effective Lengths of Cantilevers as Governed by Lateral Buckling," *The Struct. Engineer,* 51(5):161–8.

Nethercot, D. A., and Rockey, K. C. (1971), "Finite Element Solutions for the Buckling of Columns and Beams," *Int. J. Mech. Sci.*, 13:945–9.

Nethercot, D. A., and Trahair, N. S. (1976), "Lateral Buckling Approximations for Elastic Beams," *The Struct. Engineer,* 54(6):197–204.

Northwestern University (1969), "Report on 'Models for Demonstrating Buckling of Structures'," prepared by R. A. Parmelee, G. Hermann, J. F. Fleming, and J. Schmidt under NSF Grant GE-1705 (Sec. 6.2).

Nowinski, J. (1959), "Theory of Thin-Walled Bars," *Applied Mech. Rev.*, 12(4):219–27; updated in *Applied Mechanics Surveys,* ed. by M. N. Abramson, et al., Spartan Books, Washington, D.C., 1966, pp. 325–38 (Sec. 6.0).

Oden, J. T., and Ripperger, E. A. (1981), *Mechanics of Elastic Structures,* McGraw-Hill, New York (Sec. 6.0).

Ojalvo, M. (1989), "The Buckling of Thin-Walled Open-Profile Bars," *J. Appl. Mech.* (ASME), 56:633–38.

Ojalvo, M., and Chambers, R. S. (1977), "Effect of Warping Restraints on I-Beam Buckling," *J. Struct. Eng.* (ASCE), 103(12):2351–60.

Osgood, W. R. (1943), "The Center of Shear Again," *Trans. ASME,* 10:A–62.

Ostenfeld, A. (1931), *Politecknisk Laereanstalts Laboratorium for Bygningsstatik,* Meddelelse No. 5, Copenhagen (Sec. 6.0).

Papangelis, J. P., and Trahair, N. S. (1987), "Flexural-Torsional Buckling of Arches," *J. Struct. Eng.*, 113(4):889–906.

Papangelis, J. P., and Trahair, N. S. (1987), "Flexural-Torsional Buckling Tests on Arches," *J. Struct. Eng.* (ASCE), 113(7):1433–43.

Pignataro, M., and Luongo, A. (1987), "Asymmetric Interactive Buckling of Thin-Walled Columns with Initial Imperfections," *Thin-Walled Struct.*, 5:365–86 (Sec. 6.4).

Pignataro, M., Rizzi, N., and Luongo, A. (1983), *Stabilità, biforcazione e comportamento postcritico delle strutture elastiche* (in Italian), ESA, Roma, (Sec. 6.1).

Powell, G., and Klingner, R. (1970), "Elastic Lateral Buckling of Steel Beams," *J. Struct. Eng.* (ASCE), 96(9):1919–32. See also discussion by E. F. Masur, 97(3):1008–1011 (Sec. 6.1).

Roberts, T. M., and Azizian, Z. M. (1984), "Instability of Monosymmetric I-Beams," *J. Struct. Eng.* (ASCE), 110(6):1415–19.

Sherbourne, A. N., and Pandey, M. D. (1989), "Elastic, Lateral-Torsional Stability of Beams: Moment Modification Factor," *J. Const. Steel Res.*, 13(4) 337–56.

Sridharan, S., and Ali, M. A. (1985), "Interactive Buckling in Thin-Walled Beam-Columns," *J. Eng. Mech.* (ASCE), 111(12):1470–86.

Sridharan, S., and Ali, M. A. (1988), "Behavior and Design of Thin-Walled Columns," *J. Struct. Eng.* (ASCE), 114(1):103–20 (Sec. 6.4).

Szewczak, R. M., Smith, E. A., and De Wolf, J. T. (1983), "Beams with Torsional Stiffeners," *J. Struct. Eng.* (ASCE), 109(7):1635–47.

Teng, J.-G., and Rotter, J. M. (1988), "Buckling of Restrained Monosymmetric Rings," *J. Eng. Mech.* (ASCE), 114(10):1651–71.

Thompson, J. M., and Hunt, G. W. (1984), *Elastic Instability Phenomena,* John Wiley & Sons, New York (Sec. 6.6).

Timoshenko, S. P. (1945), "Theory of Bending, Torsion and Buckling of Thin-Walled Members of Open Cross Section," *J. Franklin Inst.* (Philadelphia), 239(3):201; (4):248; (5):343 (Sec. 6.0).

Timoshenko, S. P., and Gere, J. M. (1961), *Theory of Elastic Stability,* McGraw-Hill, New York (Sec. 6.0).

Timoshenko, S. P., and Goodier, J. N. (1973), *Theory of Elasticity,* 3rd ed., McGraw-Hill, New York (Sec. 6.1).

Toneff, J. D., Stiemer, S. F., and Osterrieder, P. (1987), "Local and Overall Buckling in Thin-Walled Beams and Columns," *J. Struct. Eng.* (ASCE), 113(4):769–86.

Trahair, N. S., and Papangelis, J. P. (1987), "Flexural-Torsional Buckling of Monosymmetric Arches," *J. Struct. Eng.* (ASCE), 113(10):2271–88.

Tsai, W. T. (1979), "Shear Centers," *J. Eng. Mech.* (ASCE), 105(5):893–95.

Umanskiy, A. A. (1939), *Bending and Torsion of Thin-Walled Aricraft Structures* (in Russian), Oborongiz, Moscow (Sec. 6.6).

Van Zyl, S. F., and Scordelis, A. C. (1979), "Analysis of Curved Prestressed Segmental Bridges," *J. Struct. Eng.* (ASCE), 105(11):2399–417 (Sec. 6.6).

Vlasov, V. Z. (1959), *Thin-Walled Elastic Bars* (in Russian), 2nd ed., Fizmatgiz, Moscow; also, English Translation, Israel Program for Scientific Translations, Jerusalem, 1961 (Sec. 6.0).

Wagner, H. (1936), "Verdrehung und Knickung von offenen Profilen," 25th Anniversary Publication, Technische Hochschule Danzig, 1904–1929; English translation, NACA Tech. Mem. No. 807, 1936 (Sec. 6.0).

Wagner, H., and Pretschner, W. (1936), "Torsion and Buckling of Open Sections," NACA Tech. Mem. No. 784.

Wang, C. M., and Kitipornchai, S. (1986), "Buckling Capacities of Monosymmetric I-Beams," *J. Struct. Eng.* (ASCE), 112(11):2373–91.

Wang, C. M., and Kitipornchai, S. (1986),, "On Stability of Monosymmetric Cantilevers," *Eng. Struct.*, 8(3):169–80.

Wang, C. M., and Kitipornchai, S. (1989), "New Set of Buckling Parameters for Monosymmetric Beam-Columns/Tie-Beams," *J. Struct. Eng.* (ASCE), 115(6):1497–513.

Wang, C. M., Kitipornchai, S., and Thevendran, V. (1987), "Buckling of Braced Monosymmetric Cantilevers," *Int. J. Mech. Sci.*, 29(5):321–37.

Wekezer, J. M. (1985), "Instability of Thin-Walled Bars," *J. Eng. Mech.* (ASCE), 111(7):923–35.

Wen, R. K., and Medallah, K. (1986), "Elastic Stability of Deck-Type Arch Bridges," *J. Struct. Eng.*, 113(4):757–68.

Yang, Y.-B., and Kuo, S.-R. (1986), "Static Stability of Curved Thin-Walled Beams," *J. Eng. Mech.* (ASCE), 112(8):821–41. See also discussion, 114(5):915–18 (1988).

Yang, Y.-B., and Kuo, S.-R. (1987), "Effect of Curvature on Stability of Curved Beams," *J. Struct. Eng.* (ASCE), 113(6):1185–1202.

Yang, Y.-B., and Yau, J.-D. (1987), "Stability of Beams with Tapered I-Sections," *J. Eng. Mech.* (ASCE), 113(9):1337–57.

Yoo, C. H., and Pfeiffer, P. A. (1983), "Elastic Stability of Curved Members," *J. Struct. Eng.* (ASCE), 109(12):2922–40.

Yoshida, H., and Maegawa, K. (1984), "Lateral Instability of I-Beams with Imperfections," *J. Struct. Eng.* (ASCE), 110(8):1875–92.

Zahn, J. J. (1988), "Combined-Load Stability Criterion for Wood Beam-Columns," *J. Struct. Eng.* (ASCE), 114(11):2612–28.

7

Plates and Shells

The design of plates, and especially shells, is usually dominated by stability. In the preceding chapter we dealt with one type of shells, called thin-walled beams, or, alternatively, long shells. The buckling of such shells can be reduced to a one-dimensional problem, a simplifying feature made possible when the shell is sufficiently long, the cross section is sufficiently stiff or stiffened against distortions in its own plane, and local plate buckling does not intervene. In this chapter we will turn our attention to stability of plates and shells in general, which must be mathematically approached as a two-dimensional problem.

In some respects, buckling of plates and shells is analogous to buckling of columns and frames, while in other respects it is completely different. The similarities are bifurcation-type buckling with similar behavior near the critical loads, the possibility of solving the critical loads from linear eigenvalue problems, categorization into perfect and imperfect plates or shells, and qualitatively (not quantitatively) similar imperfection effects in the case of plates. The dissimilarities consist chiefly in postcritical reserve in plates, postcritical softening and imperfection sensitivity of shells, and the importance on nonlinear geometric effects at relatively small deflections.

Our analysis will begin with critical loads of plates. For this purpose we will derive the expression for the potential energy of an elastic plate, and by applying the calculus of variations we will obtain from it the governing differential equation and boundary conditions. Then we will solve the critical loads of various rectangular plates, the type of plates most frequently encountered in practice. Aside from exact solutions, we will also utilize approximate variational methods, particularly the Ritz and Galerkin methods, which were already presented in Chapter 5.

Subsequently we will discuss the postcritical behavior of plates and emphasize their postcritical reserve, which is usually much greater than that of columns, with loads becoming much larger than the critical loads already at relatively small deflections. This large postcritical reserve, which is obviously advantageous for design, is due to the capacity of a plate to redistribute the initial compressive in-plane forces, transferring them from the middle of the plate into strips along the edges, or from compressed diagonal strips into tensioned diagonal strips.

Next we will turn to stability of shells. This is a complex problem with a vast literature and is important for practical applications. In a book that is devoted to structural stability problems in general and explanation of the fundamental concepts in particular, it is impossible to offer a detailed and exhaustive treatment of the stability problems of shells. We will include only a compact exposition of the fundamentals. Fortunately, excellent detailed books and broad review articles on the subject are available (see, e.g., Brush and Almroth, 1975; Popov and Medwadowski, 1981; Calladine, 1983; Kollár and Dulácksa, 1984; Bushnell, 1985). In similarity to plates, and in contrast to columns, frames, arches, and thin-walled beams, the critical loads of perfect elastic shells have only a limited value for the designer since the collapse takes place at very different loads (with the exception of a few problems such as cylindrical shells under external pressure or torsion). However, in contrast to columns as well as plates, shells usually fail at loads that are only a small fraction (typically 20 to 35 percent) of the critical load of the perfect elastic shell. Whercas plates are imperfection-insensitive, most shells are imperfection-sensitive, and to a high degree. Mathematically, the types of imperfection sensitivity are the same as those we already illustrated for frames or asymmetrically supported columns; however, the imperfections have a much stronger influence, and even small imperfections that are inevitable even in the most careful casting, fabrication, and loading, tremendously decrease the failure load compared to the critical load of the perfect shell.

Calculation of the postcritical behavior and the collapse loads of plates and shells is generally quite complicated, and simple analytical solutions and closed-form expressions generally do not exist. Historically, some of these highly nonlinear problems were first solved by various series expansions and the classical Ritz or Galerkin methods. As important as these solutions were until recently, on the present-day design scene they lost importance since all these problems can now be solved more effectively by geometrically nonlinear finite element programs for shells. Therefore, it is proper to concentrate here on the description of fundamental behavior and its explanation, leaving the detailed calculations to finite element specialists. The finite element approach is beyond the scope of this chapter and this book.

7.1 CLASSICAL PLATE THEORY

The classical theory of thin plates is based on Kirchhoff's assumption (1850) that the normals to the middle surface of a plate remain (1) straight and (2) normal to the deflected middle surface. In addition, the transverse normal stresses σ_z are assumed to be zero. These assumptions are approximate, but they are exact in the limiting sense for an infinitely thin plate, as has been shown by determining the asymptotic form of the exact three-dimensional solutions of elasticity. When the deflections and rotations are small, Kirchhoff's assumption leads to the kinematic (geometric) relations:

$$\varepsilon_{xx} = -w_{,xx}z \qquad \varepsilon_{yy} = -w_{,yy}z \qquad \gamma_{xy} = -w_{,xy}z \tag{7.1.1}$$

where x, y = in-plane rectangular coordinates, z = transverse coordinate, w = transverse deflection (Fig. 7.1a); the subscripts preceded by a comma denote partial derivatives (e.g., $w_{,yy} = \partial^2 w/\partial y^2$); ε_{xx} and ε_{yy} = normal strains, and γ_{xy} = in-plane shear angle.

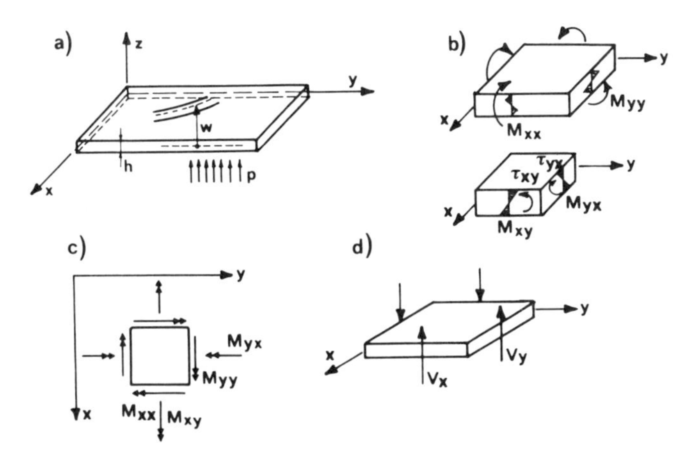

Figure 7.1 (a) Thin-plate deflection and definition of (b, c) positive bending, twisting moments, and (d) shear forces (per unit width).

There also exist higher-order plate theories in which Equations 7.1.1 are enhanced by nonlinear distributions depending on higher derivatives of w. But we will not consider such theories.

The internal forces, representing the stress resultants over the plate thickness and per unit width of the plate, are defined (Fig. 7.1b) as

$$M_{xx} = -\int_{-h/2}^{h/2} \sigma_{xx} z\, dz \qquad M_{yy} = -\int_{-h/2}^{h/2} \sigma_{yy} z\, dz \qquad M_{xy} = -\int_{-h/2}^{h/2} \tau_{xy} z\, dz \tag{7.1.2}$$

where h = thickness of the plate; σ_{xx}, σ_{yy} = in-plane normal stresses; τ_{xy} = in-plane shear stress; M_{xx}, M_{yy} = bending moments; M_{xy} = twisting moments; and $M_{yx} = M_{xy}$. Graphically, these moments are shown by their axial vectors with double arrows as in Figure 7.1c.

For the general case of orthotropic materials, the elastic stress-strain relations may be considered in the form:

$$\sigma_{xx} = E_{xx}\varepsilon_{xx} + E_{xy}\varepsilon_{yy} \qquad \sigma_{yy} = E_{xy}\varepsilon_{xx} + E_{yy}\varepsilon_{yy} \qquad \tau_{xy} = G_{xy}\gamma_{xy} \tag{7.1.3}$$

where E_{xx}, E_{yy}, E_{xy}, G_{xy} = elastic constants. Substituting Equations 7.1.1 into Equations 7.1.3 and then Equations 7.1.3 into Equations 7.1.2, we get the expressions for M_{xx}, M_{yy}, M_{xy} in terms of the deflection derivatives. For the special case of isotropic materials, for which $E_{xx} = E_{yy} = E/(1-\nu^2)$, $E_{xy} = \nu E_{xx}$, $G_{xy} = \frac{1}{2}E/(1+\nu)$, with E being Young's elastic modulus and ν Poisson's ratio, these expressions become

$$M_{xx} = D(w_{,xx} + \nu w_{,yy}) \qquad M_{yy} = D(w_{,yy} + \nu w_{,xx}) \qquad M_{xy} = M_{yx} = D(1-\nu)w_{,xy} \tag{7.1.4}$$

where $D = Eh^3/12(1-\nu^2)$ = cylindrical bending stiffness.

As in the theory of bending, the shear forces (always given per unit width of the plate; Fig. 7.1d) are determined from the following equilibrium relations:

$$V_x = -(M_{xx,x} + M_{yx,y}) \qquad V_y = -(M_{yy,y} + M_{xy,x}) \tag{7.1.5}$$

Substitution of Equations 7.1.5 into the condition of equilibrium with the transverse distributed load, $V_{x,x} + V_{y,y} = -p$, gives the equation

$$M_{xx,xx} + 2M_{xy,xy} + M_{yy,yy} = p \tag{7.1.6}$$

Using the expressions for M_{xx}, M_{yy}, M_{xy} given in Equations 7.1.4, one can obtain the differential equation that governs the behavior of thin plates in the form

$$D\nabla^4 w = D\left(\frac{\partial^4 w}{\partial x^4} + 2\frac{\partial^2 w}{\partial x^2\,\partial y^2} + \frac{\partial^4 w}{\partial y^4}\right) = p \tag{7.1.7}$$

in which $\nabla^4 = \nabla^2\nabla^2$, and ∇^2 = Laplacian operator. Equations 7.1.5 to 7.1.7 are valid only if no significant in-plane forces are present. Their influence will be investigated next.

Problems

7.1.1 Given the distribution of bending and twisting moments in a plate, derive the bending and twisting moments acting on the face of an element whose normal subtends an angle β with the axis x (Fig. 7.2a). Find the values of β for which the bending moment $M_{\beta\beta}$ is minimum or maximum (they are called the principal bending moments). What is the value of the corresponding twisting moment $M_{\beta t}$?

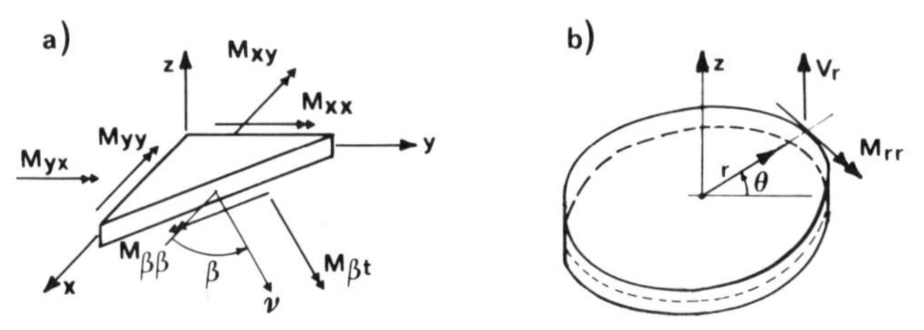

Figure 7.2 Internal forces acting on faces of variable orientation.

7.1.2 $M_{xx} + M_{yy}$ is invariant at coordinate rotation. So is $w_{,xx} + w_{,yy}$. Prove it. Also, M_{ij} and $w_{,ij}$ are tensors. Construct for them the Mohr circle.

7.1.3 Derive the differential equation for the deflections of a plate that is orthotropic.

7.1.4 Consider a circular plate under axisymmetric loading on its boundary. Assume a polar system of coordinates (Fig. 7.2b) and derive the differential equation for the deflection using a similar procedure as in the text.

7.2 DIFFERENTIAL EQUATION AND STRAIN ENERGY

To obtain the incremental potential energy of a plate under initial in-plane forces, we need to express the in-plane strains produced at the middle surface by transverse deflection $w(x, y)$. These forces must be expressed exactly up to terms of the second order in the derivatives of w.

Strains

The in-plane strains due to w may be obtained from the three-dimensional finite strain tensor ε_{ij}. Let us briefly indicate its derivation. We use a system of Cartesian coordinates $x_1 = x$, $x_2 = y$, $x_3 = z$, and we refer to these coordinates by letting lower-case subscripts take numerical values $i = 1, 2, 3$. The deformation causes the points of initial coordinates x_k to move to points of new coordinates $x'_k = x_k + u_k$. A line segment dx_k whose initial squared length is $(ds)^2 = dx_k\, dx_k$ transforms into a line segment dx'_k whose squared length is $(ds')^2 = dx'_k\, dx'_k$ (repeated subscripts imply summation from 1 to 3). Denoting partial derivatives by subscripts preceded by a comma (for example, $u_{i,j} = \partial u_i/\partial x_j$), we may write $dx'_k = dx_k + du_k = dx_k + u_{k,j}\, dx_j = (\delta_{kj} + u_{k,j})\, dx_j$, where δ_{kj} = Kronecker delta = 1 if $k = j$ and 0 if $k \neq j$. Thus we obtain

$$\begin{aligned}(ds')^2 - (ds)^2 &= (\delta_{ki} + u_{k,i})\, dx_i(\delta_{kj} + u_{k,j})\, dx_j - dx_k\, dx_k \\ &= \tfrac{1}{2}(u_{i,j} + u_{j,i} + u_{k,i}u_{k,j})2\, dx_i\, dx_j = 2\varepsilon_{ij}\, dx_i\, dx_j \end{aligned} \tag{7.2.1}$$

where we denote

$$\varepsilon_{ij} = \tfrac{1}{2}(u_{i,j} + u_{j,i} + u_{k,i}u_{k,j}) \tag{7.2.2}$$

Now we note that (1) for $\varepsilon_{ij} = 0$ there is no change in length and therefore no deformation, (2) ε_{ij} is a tensor, and (3) the linear terms $\frac{1}{2}(u_{i,j} + u_{j,i})$ coincide with the well-known linearized strain tensor e_{ij} that gives the correct expression for the work of stresses σ_{ij} on e_{ij}, namely, $\sigma_{ij}e_{ij}$. In view of these three properties, ε_{ij} may be used as a measure of finite strain. This measure is called the *finite* strain tensor (for more details, see Sec. 11.1). To emphasize the fact that this tensor is expressed in terms of the initial (or Lagrangian) coordinates of material points, rather than the current (final) coordinates (Eulerian coordinates), this tensor is also called the Lagrangian strain tensor (another name is Green's strain tensor).

For small transverse deflections of plates we set $x_1 = x$, $x_2 = y$, $u_3 = w$ and we neglect in Equation 7.2.2 all the terms that are not due to w. This furnishes the following second-order in-plane strains due to w:

$$\varepsilon_{11} = \varepsilon_{xx} = \frac{1}{2}\left(\frac{\partial w}{\partial x}\right)^2 \qquad \varepsilon_{22} = \varepsilon_{yy} = \frac{1}{2}\left(\frac{\partial w}{\partial y}\right)^2 \qquad \varepsilon_{12} = \frac{1}{2}\gamma_{xy} = \frac{1}{2}\left(\frac{\partial w}{\partial x}\right)\left(\frac{\partial w}{\partial y}\right) \tag{7.2.3}$$

These expressions may be written summarily in the form

$$\varepsilon_{ij} = \tfrac{1}{2}w_{,i}w_{,j} \qquad (i, j = 1, 2) \tag{7.2.4}$$

where subscripts i and j now run from 1 to only 2, not 3.

It is instructive to show also a simple geometric derivation of the second-order strains due to w (see also Timoshenko and Gere, 1961, p. 338). Referring to segment dx in Figure 7.3a,

$$\begin{aligned}\varepsilon_{11} &= (dx' - dx)/dx = (dx'/dx) - 1 = [\sqrt{(dx)^2 + (w_{,x}\, dx)^2}/dx] - 1 \\ &= \sqrt{1 + w_{,x}^2} - 1 \simeq 1 + \tfrac{1}{2}w_{,x}^2 - 1 = \tfrac{1}{2}w_{,x}^2.\end{aligned}$$

This agrees with ε_{11} as given in Equations 7.2.3. Furthermore, consider two orthogonal segments dx and dy, which transform into segments dx' and dy'. Referring to Figure 7.3b we consider plane $\overline{245}$, which is vertical and normal to segment $dy = \overline{02}$. Plane $\overline{024}$, however, is not normal to $dx' = \overline{03}$. The plane normal to $\overline{03}$ is $\overline{025}$, and this plane intersects plane $\overline{245}$ in line $\overline{25}$, which forms

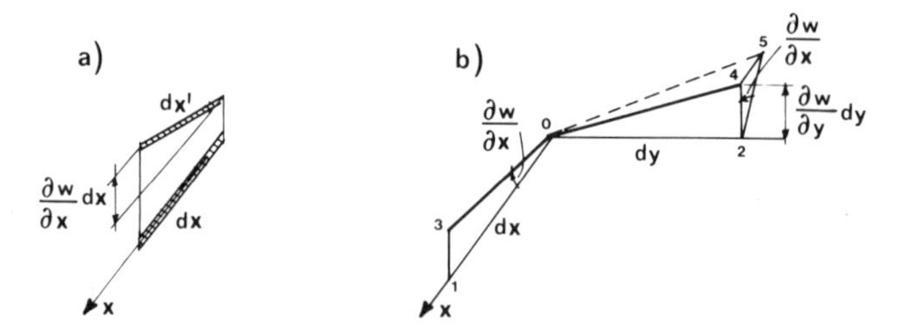

Figure 7.3 Second-order strains due to lateral deflection w: (a) extensional component, (b) shear component.

angle $w_{,x}$ with the vertical line $\overline{24}$. Also, $\overline{24} = dy\, w_{,y}$, and from triangle 245 we have $\overline{45} = (\overline{24})w_{,x} = (dy\, w_{,y})w_{,x}$. The change of the right angle initially formed by segments dx and dy is given by the angle between lines $\overline{04}$ and $\overline{05}$. This angle, which represents the shear angle γ_{xy}, is $\gamma_{xy} = (\overline{45})/dy = (dy\, w_{,y})w_{,x}/dy$ or $\frac{1}{2}\gamma_{xy} = \frac{1}{2}w_{,x}w_{,y} = \varepsilon_{12}$. Thus, ε_{12} as given by Equations 7.2.3 represents one-half of the shear angle caused by deflection w, as expected.

Potential Energy

Consider now a plate that is initially in equilibrium under in-plane forces N_{ij} at $w(x, y) = 0$. Let $\Pi = 0$ at this state. If the potential energy due to in-plane displacements u and v at the middle surface is excluded, the potential energy Π of the plate, which is due solely to deflections $w(x, y)$, may be written as

$$\Pi = U_1 + U_2 - W \tag{7.2.5}$$

where

$$U_1 = \iint_A \tfrac{1}{2}M_{ij}w_{,ij}\, dA = \iint_A D_{ijkm}w_{,ij}w_{,km}\, dA \qquad (i, j = 1, 2)$$

$$U_2 = \iint_A \tfrac{1}{2}N_{ij}w_{,i}w_{,j}\, dA \tag{7.2.6}$$

$$W = \iint_A pw\, dA$$

where A = area of plate; i, j = subscripts referring to in-plane Cartesian coordinates $x_1 = x$, $x_2 = y$ $(i, j = 1, 2)$; repeated subscripts imply summation from 1 to 2; w_i = deflections, which coincide with δw_i because $w_i = 0$ in the initial state; U_1 = strain energy of bending (twisting included), which is due to curvatures; U_2 = strain energy of in-plane deformations due to deflections, which is due to the work of initial in-plane forces N_{ij} on the second-order strain components; and W = work of transverse distributed loads p (the concentrated loads are included if p is allowed to be the Dirac delta function). Finally, D_{ijkm} are constants representing the plate-bending stiffnesses, which are defined by writing the moment-curvature relations in the form

$$M_{km} = D_{ijkm}w_{,ij} \tag{7.2.7}$$

Always $D_{ijkm} = D_{kmij} = D_{jikm} = D_{ijmk}$. For the special case of isotropic materials, we have, according to Equations 7.1.4, $D_{1111} = D_{2222} = D$, $D_{1122} = \nu D$, $D_{1212} = (1-\nu)D$, with all other D_{ijkm} being zero. M_{ij} and $w_{,ij}$ are second-order tensors in two-dimensional space (x_1, x_2), and D_{ijkm} is a fourth-order tensor in this space.

In component form, U_1 and U_2 (Eqs. 7.2.6) become

$$U_1 = \iint_A \tfrac{1}{2}(M_{xx}w_{,xx} + M_{yy}w_{,yy} + 2M_{xy}w_{,xy})\,dx\,dy \tag{7.2.8a}$$

$$U_2 = \iint_A \left(N_{xx}\frac{w_{,x}^2}{2} + N_{yy}\frac{w_{,y}^2}{2} + N_{xy}w_{,x}w_{,y}\right)dx\,dy \tag{7.2.8b}$$

and for isotropic plates substitution of Equations 7.1.4 gives

$$\begin{aligned} U_1 &= \iint_A \frac{D}{2}[w_{,xx}^2 + w_{,yy}^2 + 2\nu w_{,xx}w_{,yy} + 2(1-\nu)w_{,xy}^2]\,dx\,dy \\ &= \iint_A \frac{D}{2}[(w_{,xx} + w_{,yy})^2 + 2(1-\nu)(w_{,xy}^2 - w_{,xx}w_{,yy})]\,dx\,dy \end{aligned} \tag{7.2.9}$$

Differential Equations of Equilibrium

The equilibrium conditions may be expressed as $\delta\Pi = 0$. For a perfect plate, for which $p = 0$ and $w = 0$ at the initial state, this equilibrium condition at the same time represents the Trefftz criterion $\delta(\delta^2 V) = 0$ for the critical state (Eq. 4.2.6) because $\delta w = w$ and $\delta^2\Pi = \Pi$. Taking into account the symmetries $D_{ijkm} = D_{kmij}$ and $N_{ij} = N_{ji}$, we may calculate

$$\begin{aligned} \delta U_1 &= \iint_A D_{ijkm}w_{,ij}\delta w_{,km}\,dA \\ &= \iint_A (D_{ijkm}w_{,ij}\delta w_{,k})_{,m}\,dA - \iint_A (D_{ijkm}w_{,ij})_{,m}\delta w_{,k}\,dA \\ &= \iint_A (D_{ijkm}w_{,ij}\delta w_{,k})_{,m}\,dA - \iint_A [(D_{ijkm}w_{,ij})_{,m}\delta w]_{,k}\,dA \\ &\quad + \iint_A (D_{ijkm}w_{,ij})_{,km}\delta w\,dA \\ &= \int_s \nu_m D_{ijkm}w_{,ij}\delta w_{,k}\,ds - \int_s \nu_k D_{ijkm}w_{,ijm}\delta w\,ds + \iint_A D_{ijkm}w_{,ijkm}\delta w\,dA \end{aligned} \tag{7.2.10}$$

$$\begin{aligned} \delta U_2 &= \frac{1}{2}\iint_A (N_{ij}w_{,i}\delta w_{,j} + N_{ij}w_{,j}\delta w_{,i})\,dA \\ &= \iint_A N_{ij}w_{,i}\delta w_{,j}\,dA = \iint_A (N_{ij}w_{,i}\delta w)_{,j}\,dA - \iint_A (N_{ij}w_{,i})_{,j}\delta w\,dA \\ &= \iint_A (N_{ij}w_{,i}\delta w)_{,j}\,dA - \iint_A N_{ij,j}w_{,i}\delta w\,dA - \iint_A N_{ij}w_{,ij}\delta w\,dA \\ &= \int_s \nu_j N_{ij}w_{,i}\delta w\,ds + \iint_A F_i w_{,i}\delta w\,dA - \iint_A N_{ij}w_{,ij}\delta w\,dA \end{aligned} \tag{7.2.11}$$

$$\delta W = \iint_A p\,\delta w\,dA \tag{7.2.12}$$

where we applied Gauss' integral theorem in two dimensions to transform area integrals into integrals along the boundary curve s (*recall*: Gauss' theorem generally states that $\iint_A T_{ijk\ldots,r}\,dA = \int_s T_{ijk\ldots}\nu_r\,ds$ where $T_{ijk\ldots}$ = tensor of any order); ν_r = cosines of the angle formed by normal $\boldsymbol{\nu}$ with axis x_r (Fig. 7.4a); F_i = in-plane distributed loads in the initial state that may be expressed as $F_i = -N_{ij,j}$ because the initial state is an equilibrium state.

The equilibrium condition $\delta U_1 + \delta U_2 - \delta W = 0$ yields

$$\iint_A (D_{ijkm}w_{,ijkm} - N_{ij}w_{,ij} + F_i w_{,i} - p)\delta w\,dA + \int_s (V_\nu^N + V_\nu^M)\delta w\,ds + \int_s M_k \delta w_{,k}\,ds = 0 \tag{7.2.13}$$

with

$$V_\nu^N = \nu_j N_{ij} w_{,i} \qquad V_\nu^M = -\nu_k D_{ijkm} w_{,ijm} \qquad M_k = \nu_m D_{ijkm} w_{,ij} = \nu_m M_{km}$$

Since this variational condition must hold for any variation δw, the integrands

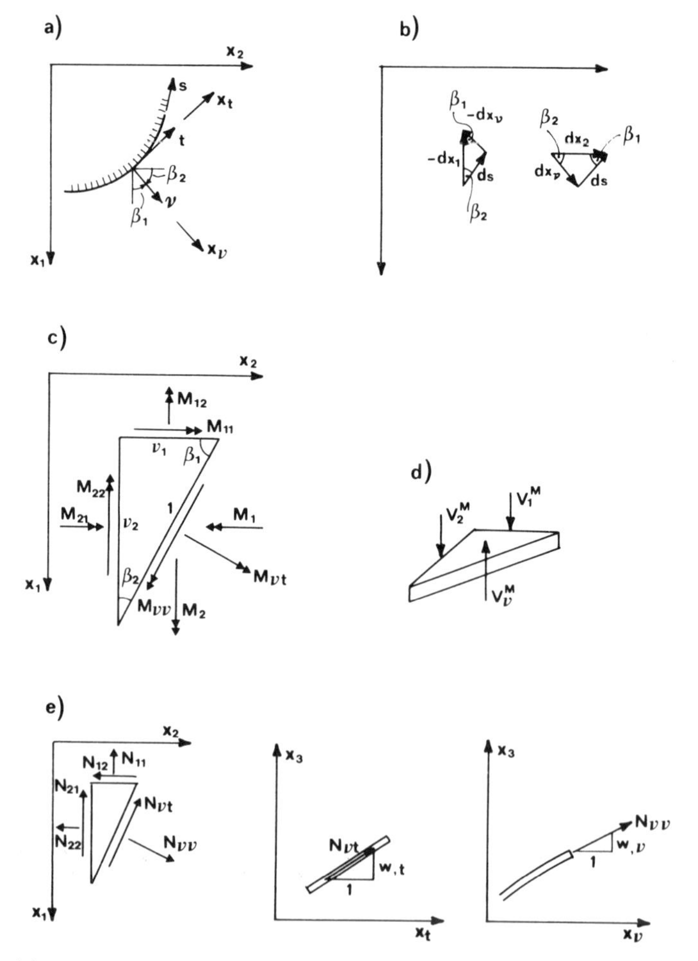

Figure 7.4 (a) Local coordinate system at boundary and (b–e) relations between components in local and global systems.

must vanish at each point of the plate. For isotropic plates, $D_{ijkm}w_{,ijkm} = D\nabla^4 w$, and so we get for the plate the differential equation

$$D\nabla^4 w = p + N_{ij}w_{,ij} - F_i w_{,i} \qquad (i, j = 1, 2) \tag{7.2.14}$$

In statics, the in-plane loads F_i (such as the weight of the plate, its inertia forces, or electromagnetic forces) are usually negligible, and then we may write

$$D\nabla^4 w = p + p_N \qquad \text{with } p_N = N_{xx}w_{,xx} + N_{yy}w_{,yy} + 2N_{xy}w_{,xy} \tag{7.2.15a}$$

in which p_N is an apparent transverse distributed load due to the initial inplane forces N_{xx}, N_{yy}, N_{xy} that satisfy the differential equilibrium equations;

$$N_{xx,x} + N_{xy,y} = 0 \qquad N_{xy,x} + N_{yy,y} = 0 \tag{7.2.15b}$$

Equations 7.2.15b can be derived by the same variational approach if the energy of in-plane deformation is included in the potential-energy expression.

Equations 7.2.15a, b represent a system of three simultaneous linear partial differential equations with four unknowns w, N_{xx}, N_{yy}, and N_{xy}. The in-plane membrane forces can, however, be expressed as functions of the middle surface strains (see Sec. 7.4), which in turn are functions (Eq. 7.2.2) of the transverse displacement w and the in-plane displacements u and v. So Equations 7.2.15a, b can be reduced to a system of three simultaneous differential equations in u, v, and w.

If, however, the deflection w is small (approximately $w \leq 0.2h$, $h =$ plate thickness, as indicated by Donnell, 1976, p. 175) experiments indicate that the effect of the deflection $w(x, y)$ on N_{ij} can be neglected, that is, N_{ij} in Equation 7.2.15a can be assumed to be constant and obtained by a separate solution of the in-plane problem given by Equations 7.2.15b. The small-deflection range, in which N_{ij} is independent of w, will be studied in Section 7.3 and the large-deflection range, in which N_{ij} depends on w, in Section 7.4.

Boundary Conditions

The boundary curve integrals in Equation 7.2.13 involve the variations δw and $\delta w_{,k}$ on the boundary curve s, which are not completely independent, since for a given δw the variation derivative $\delta w_{,t}$ in the direction tangent to s is also specified. It is then convenient to express $\delta w_{,k}$ as a function of $\delta w_{,t}$, as well as $\delta w_{,\nu}$, the latter representing the variation derivative along the normal $\mathbf{\nu}$ to the boundary curve. We have $\delta w_{,k} = \delta w_{,\nu}(dx_\nu/dx_k) + \delta w_{,t}(dx_t/dx_k)$ where x_ν, x_t represent a system of local Cartesian coordinates at the boundary point oriented in the direction of normal $\mathbf{\nu}$ and tangent $\mathbf{t}$, respectively (Fig. 7.4a). Since (according to Fig. 7.4b) $dx_\nu/dx_1 = \nu_1$, $dx_\nu/dx_2 = \nu_2$, $dx_t/dx_1 = ds/dx_1 = -\nu_2$, and $dx_t/dx_2 = ds/dx_2 = \nu_1$, we may write

$$\begin{aligned} M_k\delta w_{,k} &= M_k\delta w_{,\nu}\frac{dx_\nu}{dx_k} + M_k\delta w_{,s}\frac{dx_s}{dx_k} \\ &= (M_1\nu_1 + M_2\nu_2)\delta w_{,n} + (M_2\nu_1 - M_1\nu_2)\delta w_{,s} = M_{\nu\nu}\delta w_{,\nu} + M_{\nu t}\delta w_{,t} \end{aligned} \tag{7.2.16}$$

in which, as obtained from the moment equilibrium condition for a triangular plate element at the boundary (Fig. 7.4c), $M_k = \nu_m M_{km} =$ the moment component at the boundary whose axial vector is normal to the x_k axis; $M_{\nu\nu}$ and $M_{\nu t}$ are the bending and twisting moments acting on the plate boundary. We can now find for V_ν^M in Equation 7.2.13 the following expression:

$$V_\nu^M = -\nu_k M_{km,m} = -\nu_1(M_{11,1} + M_{12,2}) - \nu_2(M_{21,1} + M_{22,2}) = \nu_1 V_1^M + \nu_2 V_2^M \tag{7.2.17}$$

in which V_1^M and V_2^M are the shear forces (per unit length) that are required for moment equilibrium of an element of the plate (Eqs. 7.1.5), and V_ν^M is the shear force on the boundary that equilibrates V_1^M and V_2^M (Fig. 7.4d). The superscript M has been attached in order to indicate that this is the part of the shear force that is due exclusively to the bending and twisting moments.

Let us now find the meaning of V_ν^N in Equation 7.2.13. Considering the membrane components $N_{\nu\nu}$ and $N_{\nu t}$ in the local Cartesian system (x_ν, x_t) at the boundary (Fig. 7.4c) we have

$$\begin{aligned} V_\nu^N &= \nu_j N_{ij} w_{,\nu} \frac{dx_\nu}{dx_i} + \nu_j N_{ij} w_{,t} \frac{dx_s}{dx_i} \\ &= (\nu_1^2 N_{11} + 2\nu_1\nu_2 N_{12} + \nu_2^2 N_{22}) w_{,\nu} + [(N_{22} - N_{11})\nu_1\nu_2 + N_{12}(\nu_1^2 - \nu_2^2)] w_{,t} \\ &= N_{\nu\nu} w_{,\nu} + N_{\nu t} w_{,t} \end{aligned} \tag{7.2.18}$$

We recognize in V_ν^N a second-order shear force that arises if in-plane (membrane) forces are present and the deformed configuration is taken into account. We now see that the total shear force according to the second-order theory is $V_\nu = V_\nu^M + V_\nu^N$ (and, by inspection of Eqs. 7.2.17 and 7.2.18, we also notice an analogy with the relation $V = -M' - Pw'$ obtained in Sec. 1.3 for beam-columns in which P is positive for compression).

Since we have already ensured through Equation 7.2.14 that the integrand in the first parenthesis in Equation 7.2.13 will vanish, this last equation becomes (upon substitution of Eq. 7.2.16)

$$\int_s \left(V_\nu^N + V_\nu^M - \frac{\partial M_{\nu t}}{\partial t} \right) \delta w \, ds + [M_{\nu t} \delta w]_0^s + \int_s M_{\nu\nu} \delta w_{,\nu} ds = 0 \tag{7.2.19}$$

Here we integrated by parts the term $M_{\nu t}\delta w_{,t}$ in Equation 7.2.16. The term in brackets is zero for any smooth boundary curve, while at sharp corners it is required that $[M_{\nu t}\delta w] = 0$.

In order that Equation 7.2.19 be satisfied for any variations δw and $\delta w_{,\nu}$, the integrands in Equation 7.2.19 must vanish. This yields the boundary conditions

$$\begin{aligned} &\text{Either} \quad w = 0 \quad \text{or} \quad \frac{\partial M_{\nu t}}{\partial t} = V_\nu^N + V_\nu^M \\ &\text{Either} \quad w_{,\nu} = 0 \quad \text{or} \quad M_{\nu\nu} = 0 \end{aligned} \tag{7.2.20}$$

Note that the sign conventions for $M_{\nu t}$, V_ν, $N_{\nu t}$ on the boundary (Fig. 7.4) may but need not coincide with those adopted for the corresponding internal stress resultants on cross sections normal to the x or y axes (Figs. 7.1 and 7.5).

The conditions $w = 0$ or $w_{,\nu} = 0$ represent the kinematic boundary conditions, the latter one representing a fixed (clamped) edge, while the remaining ones represent the associated static boundary conditions. The condition for a free edge

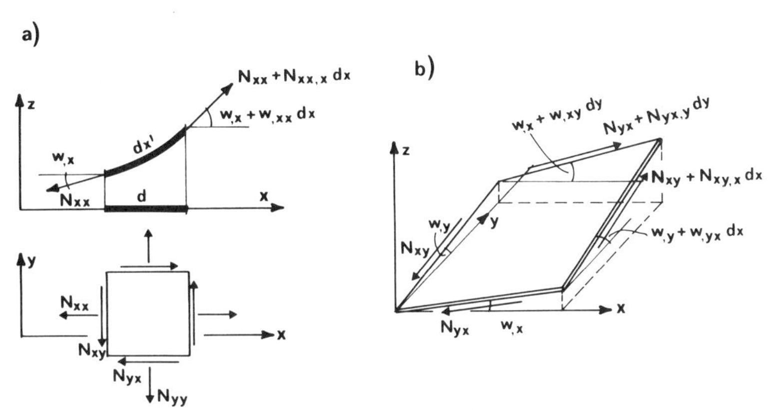

Figure 7.5 Equilibrium of deformed element: (a) effect of normal forces, (b) effect of shear in-plane forces.

($w \neq 0$) is the well-known Kirchhoff's boundary condition, which was much discussed when plate theory was evolving early in the nineteenth century. The notable aspect of this boundary condition is that one cannot prescribe at the boundary the transverse shear force and the twisting moment separately, but only a combination of the shear force and a derivative of the twisting moment, a fact conclusively demonstrated by Kirchhoff.

The boundary conditions pertinent to the in-plane displacements can be found by the same variational procedure upon including in the energy functional the in-plane deformation. This procedure also yields the in-plane differential equations of equilibrium.

Direct Derivation of Transverse Resultant of In-Plane Forces

It is instructive to show an elementary geometric derivation of p_N as given by Equation 7.2.15a (Timoshenko and Gere, 1961). We examine vertical equilibrium of a small rectangular element $dx\,dy$ that is transformed into a slightly inclined curved element $dx'\,dy'$ in Figure 7.5a. Let us consider first the normal forces acting in vertical cross section on the sides normal to x (Fig. 7.5a). The vertical resultant of these forces is

$$\begin{aligned} p_N^{(1)}\,dx\,dy &= -N_{xx}\,dy\,w_{,x} + (N_{xx} + N_{xx,x}\,dx)(w_{,x} + w_{,xx}\,dx)\,dy \\ &= N_{xx}w_{,xx}\,dx\,dy + N_{xx,x}w_{,x}\,dx\,dy + N_{xx,x}w_{,xx}\,dx^2\,dy \end{aligned} \tag{7.2.21}$$

An analogous expression is obtained for the vertical resultant of the normal forces acting on the sides normal to y:

$$p_N^{(2)}\,dx\,dy = N_{yy}w_{,yy}\,dy\,dx + N_{yy,y}w_{,y}\,dy\,dx + N_{yy,y}w_{,yy}\,dy^2\,dx \tag{7.2.22}$$

Furthermore, consider the shear forces on all four sides, shown in Figure 7.5b. The vertical resultant of these forces is

$$\begin{aligned} p_N^{(3)}\,dx\,dy = &-N_{xy}\,dy\,w_{,y} + (N_{xy} + N_{xy,x}\,dx)\,dy\,(w_{,y} + w_{,xy}\,dx) \\ &- N_{yx}\,dx\,w_{,x} + (N_{yx} + N_{yx,y}\,dy)\,dx\,(w_{,x} + w_{,xy}\,dy) \end{aligned} \tag{7.2.23}$$

Summing the expressions 7.2.21, 7.2.22, and 7.2.23 and neglecting higher-order terms such as $dx\,dy^2$, we obtain

$$\begin{aligned} p_N\,dx\,dy &= (p_N^{(1)} + p_N^{(2)} + p_N^{(3)})\,dx\,dy \\ &= [N_{xx}w_{,xx} + N_{yy}w_{,yy} + 2N_{xy}w_{,xy} + (N_{xx,x} + N_{xy,y})w_{,x} \\ &\quad + (N_{yy,y} + N_{xy,x})w_{,y}]\,dx\,dy \end{aligned} \tag{7.2.24}$$

in which $p_N\,dx\,dy$ is the apparent additional load per area $dx\,dy$, due to in-plane forces N_{ij}. Noting that $N_{xx,x} + N_{xy,y} = -F_x$ and $N_{yy,y} + N_{xy,x} = -F_y$, because of the differential equations of equilibrium for the initial state, we see that p_N is identical to the expression that we obtained before in Equation 7.2.15a.

Discussion and Summary

Equation 7.2.14 or 7.2.15a is a linear fourth-order partial differential equation for the unknown displacement function $w(x, y)$. Together with the appropriate boundary conditions, this differential equation defines a linear boundary-value problem. The solution of this problem yields, for the given boundary conditions, the deflections $w(x, y)$ as a function of the given load distribution $p(x, y)$.

When no transverse distributed load is applied ($p = 0$) and the boundary conditions are homogeneous, that is, when there are no loads other than the in-plane loads, and no enforced displacements, Equation 7.2.14 or 7.2.15a together with the boundary conditions define an eigenvalue problem. This problem has a nonzero solution $w(x, y)$ only for certain values of in-plane forces N_{ij}, which then represent the critical loads, at which a symmetric bifurcation occurs. For the case of nonzero lateral distributed load p, which represents the case of an imperfect plate, the solution yields deflections that tend to infinity if the critical values of the in-plane forces N_{ij} are approached.

To sum up, we have seen that the presence of significant in-plane forces N_{ij} in a plate contributes an additional second-order term to the potential-energy expression, as well as an additional apparent transverse distributed load to the differential equation of the plate. These additional terms are analogous to those for columns and become identical to them for the case of cylindrical bending, for which $N_{xx} = -P$, $D \rightarrow EI$, $w_{,y} = w_{,xy} = w_{,yy} = 0$, $N_{yy} = N_{xy} = 0$. We have also seen that the potential-energy expression and the differential equation with its boundary conditions each suffice to fully describe the problem; they are equivalent to each other.

For columns we used two different approaches to obtain the term due to the work of the axial load (Sec. 4.3). One approach was to assume the column center line to be inextensible and calculate the displacement Δl at the end of the column, for example, the simply supported column. The other approach, which we later showed for a hinged column whose ends are not actually sliding, was to calculate the normal strain of the column center line due to transverse deflection. In the case of plates, only the latter approach, which we have exemplified in this section, makes sense. For general behavior it is not possible to assume that the middle surface of the plate would be undeformable and calculate the corresponding displacements at the boundary (as we did for columns). This is one notable conceptual difference between the calculation of potential energy of plates and of columns.

Problems

7.2.1 Check that for a plate strip of width 1, with $w(x, y) = w(x)$, that is, no dependence on y, and $-N_{xx} = P$, the plate differential equation reduces to the differential equation for a beam-column. Also check that the boundary conditions reduce to those for columns, considering the fixed end, simply supported end, and free end.

7.2.2 Include in the derivation of potential energy the strain energy due to in-plane displacements u and v of the middle surface, and derive corresponding Euler equations. Does the equation for deflection remain the same? In which cases can the resulting equation system be solved as uncoupled equations?

7.2.3 Using the Gauss theorem in reverse order, starting from the differential equation of the plate, derive the potential-energy expression.

7.2.4 Find the potential-energy expression for an orthotropic plate and derive the differential equation.

7.2.5 Consider a simply supported circular plate (Fig. 7.6) under axisymmetric axial loading $N_{rr} = -\bar{N}$ (compressive). Assuming a polar system of coordinates, transform the potential-energy expression given in the text and find the Euler equation as well as the boundary conditions.

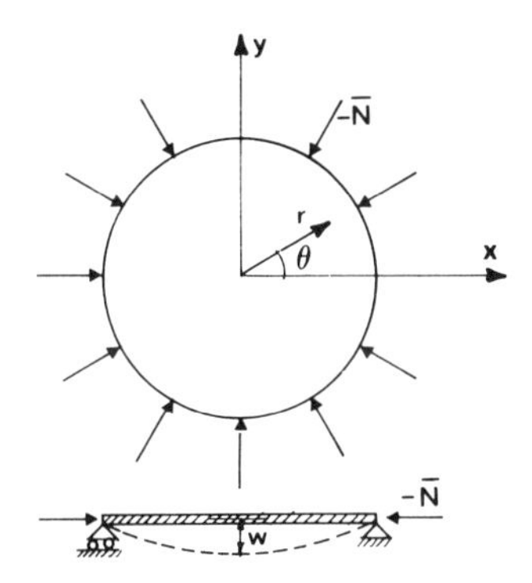

Figure 7.6 Circular plate under axisymmetric axial loading.

7.3 BUCKLING OF RECTANGULAR PLATES

Plate buckling governs the design of many types of structures, for example, the thickness of the walls used in thin-wall beams. The most efficient designs, used for large spans, usually employ stiffened plates. For the purpose of stability analysis, the wall plates between stiffening ribs may normally be analyzed approximately as isolated rectangular plates. Their analysis is relatively straightforward, and we will demonstrate it now.

While for columns the initial axial force is usually uniform throughout the column, for plates the initial in-plane forces N_{ij} are often nonuniform throughout the plate. This complicates the solution. First we take the easier case of uniform N_{ij}.

Buckling of Simply Supported Plates

Consider a rectangular plate of sides a and b such that $a \geq b$ (Fig. 7.7a). The plate is compressed in the direction of the longer side a by a uniformly distributed in-plane N_{xx} (<0, negative for compression); $N_{yy} = N_{xy} = 0$. According to Equation 7.2.15a, the differential equation of this problem is

$$w_{,xxxx} + 2w_{,xxyy} + w_{,yyyy} = \frac{N_{xx}}{D} w_{,xx} \tag{7.3.1}$$

This equation represents the condition of neutral equilibrium (or existence of an adjacent equilibrium state), and it implements Trefftz's criterion of critical state. We will seek the solution in the form of a Fourier series

$$w = \sum_{m=1}^{\infty} \sum_{n=1}^{\infty} q_{mn} \sin \frac{m\pi x}{a} \sin \frac{n\pi y}{b} \tag{7.3.2}$$

with unknown coefficients q_{mn} representing generalized displacements. Each term of this series satisfies the kinematic boundary conditions $w = 0$ on the boundaries, and it also satisfies the static boundary condition $M_{yy} = 0$ at the left and right boundaries and $M_{xx} = 0$ at the bottom and top boundaries (Fig. 7.7a). Because of Equations 7.1.4, and because $w_{,xx} = 0$ at the left and right boundaries and $w_{,yy} = 0$ at the top and bottom boundaries, these conditions reduce to $w_{,yy} = 0$

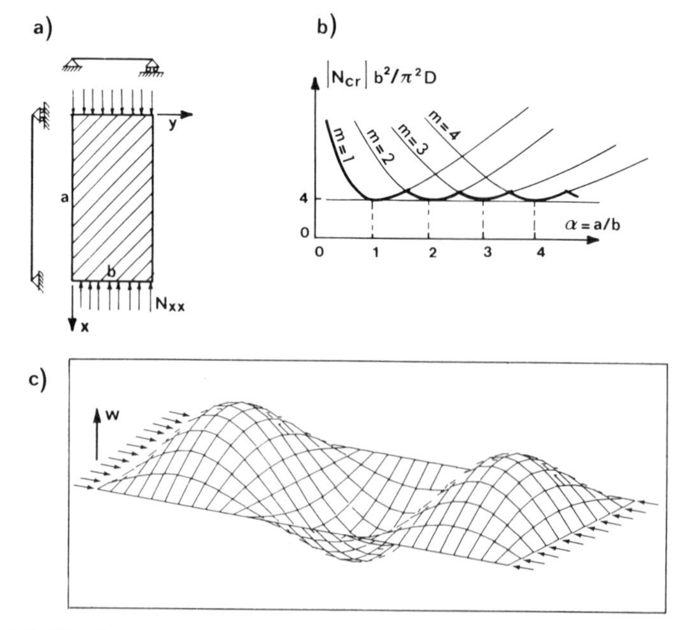

Figure 7.7 (a) Simply supported plate subjected to uniformly distributed normal forces N_{xx}, (b) critical value of normal force, and (c) buckling mode.

at the left and right boundaries and $w_{,xx}=0$ at the bottom and top boundaries. Substitution into Equation 7.3.1 yields

$$\sum_{m=1}^{\infty}\sum_{n=1}^{\infty} q_{mn}\left[\pi^4\left(\frac{m^4}{a^4}+\frac{2m^2n^2}{a^2b^2}+\frac{n^4}{b^4}\right)+\frac{N_{xx}}{D}\left(\frac{m^2\pi^2}{a^2}\right)\right]\sin\frac{m\pi x}{a}\sin\frac{n\pi y}{b}=0 \quad (7.3.3)$$

If this equation should be satisfied for all x and y at nonzero deflection w, it is necessary that, for at least one term of the sum, the bracketed expression vanishes (while q_{mn} can be zero for all the other terms). This condition yields the critical loads

$$-N_{xx}=\frac{\pi^2a^2D}{m^2}\left(\frac{m^2}{a^2}+\frac{n^2}{b^2}\right)^2 \quad (7.3.4)$$

Stability needs to be analyzed on the basis of the potential-energy expression in Equation 7.2.5. Substituting Equations 7.2.9 and 7.2.8 for U_1 and U_2, we find that

$$\Pi=\frac{\pi^2b}{8a}\sum_{m=1}^{\infty}\sum_{n=1}^{\infty} q_{mn}^2m^2\left[\frac{\pi^2a^2D}{m^2}\left(\frac{m^2}{a^2}+\frac{n^2}{b^2}\right)^2+N_{xx}\right] \quad (7.3.5)$$

If this expression should be positive definite, the bracketed terms must be nonnegative for all terms of the sum (i.e., all m and n), and positive for at least one term (i.e. one m–n combination). The critical state is obtained if the bracketed term vanishes for at least one term of the sum (while for all other terms q_{mn} can be zero). This again yields Equation 7.3.4.

By analogy with columns one might think that the lowest critical load magnitude would occur for $m=n=1$. This is not true, however. With regard to n, it is clear that the smallest $|N_{cr}|$ occurs when $n=1$; but $m=1$ does not necessarily give the smallest value of $|N_{cr}|$. Setting $a/b=\alpha$ and $n=1$, Equation 7.3.4 may be written as

$$|N_{cr}|=\frac{\pi^2D}{b^2}\left(\frac{m}{\alpha}+\frac{\alpha}{m}\right)^2 \quad (7.3.6)$$

We see that $|N_{cr}|$ as a function of α always has a positive curvature and the minimum of $|N_{cr}|$ occurs when $dN_{cr}/d\alpha=0$. This condition yields $\alpha=m$, and Equation 7.3.6 then yields

$$\min|N_{cr}|=\frac{4\pi^2D}{b^2} \qquad (\text{for } \alpha=m) \quad (7.3.7)$$

This value is independent of the number, m, of the half-waves, as well as of the ratio a/b. Plotting Equation 7.3.6 as a function of α for $m=1, 2, 3, \ldots,$ we obtain the curves shown in Figure 7.7b. From this figure, one can readily see which value of m yields the smallest $|N_{cr}|$ for a given a/b ratio; the line of smallest $|N_{cr}|$ is drawn as a heavy solid line. Note the tendency of the number of half-waves in the direction of compression to be such that the half-wavelength be as close to the width of b as possible (Fig. 7.7c).

Similar solutions are possible when the plate is under biaxial in-plane normal forces N_{xx} and N_{yy}. In this case the problem is to find for the given in-plane forces the value of a common multiplier, μ, which causes the plate to reach the lowest critical state. This value then represents the safety factor.

Rectangular Plate with Arbitrary Boundary Conditions

A different approach is needed for the rectangular plate shown in Figure 7.8a whose top and bottom edges are simply supported while the left and right edges can have different boundary conditions, including a clamped edge and a free edge. In this problem it is convenient to introduce a one-dimensional Fourier series in x, leaving the distribution of deflections in the y-direction as unknown, characterized by function $f(y)$. The boundary conditions at $x=0$ and $x=a$ are automatically satisfied by setting

$$w = f(y) \sum_{m=1}^{\infty} q_m \sin \frac{m\pi x}{a} \tag{7.3.8}$$

Substituting this into Equation 7.3.1, and imposing the condition that at least one of the terms multiplying $\sin(m\pi x/a)$ must vanish (while $q_m = 0$ for all the other terms), we find for $f(y)$ the following linear ordinary differential equation:

$$f_{,yyyy} - \frac{2m^2\pi^2}{a^2} f_{,yy} + \left[\frac{m^2\pi^4}{a^4} + \frac{N_{xx}}{D}\left(\frac{m^2\pi^2}{a^2}\right)\right] f = 0 \tag{7.3.9}$$

This equation can be solved easily when N_{xx} is constant. Then, assuming that $(-N_{xx})/D > m^2\pi^2/a^2$, the general solution is

$$f(y) = A \sinh \alpha y + B \sin \beta y + C \cosh \alpha y + D \cos \beta y \tag{7.3.10}$$

in which

$$\alpha, \beta = \left[\pm \frac{m^2\pi^2}{a} + \sqrt{\frac{-N_{xx}}{D}\left(\frac{m^2\pi^2}{a^2}\right)}\right]^{1/2} \tag{7.3.11}$$

The boundary conditions on the sides $y=0$ and $y=b$ may now be used to find the unknown constants A, B, C, and D.

Consider now that the side $y=0$ is simply supported and the side $y=b$ is a free edge. Then we have the boundary conditions:

At $y=0$: $\quad w=0 \quad w_{,yy}=0 \qquad$ (7.3.12a, b)

At $y=b$: $\quad w_{,yy} + \nu w_{,xx} = 0 \quad w_{,yyy} + (2-\nu) w_{,xxy} = 0 \qquad$ (7.3.12c, d)

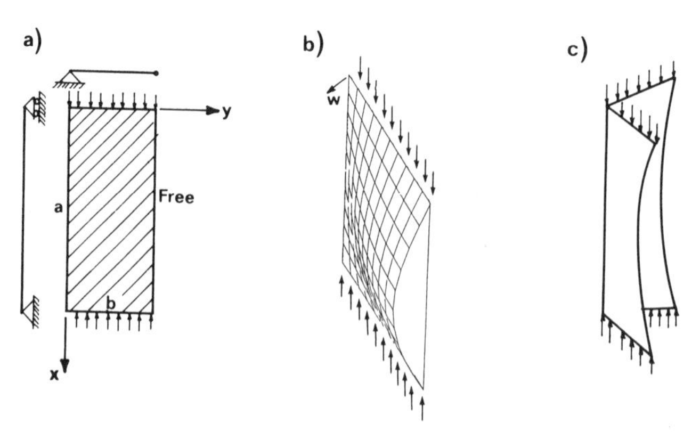

Figure 7.8 (a) Plate with three edges simply supported and one edge free, (b) buckling mode, and (c) analogy with buckling mode of L–cross section.

Equations 7.3.12c, d for the free edge mean that $M_{yy} = 0$ and $V_\nu - M_{\nu t,t} = V_\nu^M + V_\nu^N - M_{\nu t,t} = V_y^M - M_{yx,x} = 0$ (the minus sign in front of $M_{yx,x}$ results from the fact that for this boundary $M_{\nu t} = -M_{yx}$ and $dt = ds = -dx$; see also the comment below Eqs 7.2.20). Equations 7.3.12a, b yield $C = D = 0$, and Equations 7.3.12c, d provide two algebraic equations for A and B whose determinant must vanish;

$$\begin{vmatrix} \left(\alpha^2 - \nu\dfrac{m^2\pi^2}{a^2}\right)\sinh \alpha b & -\left(\beta^2 + \nu\dfrac{m^2\pi^2}{a}\right)\sin \beta b \\ \left[\alpha^3 - \alpha(2-\nu)\dfrac{m^2\pi^2}{a^2}\right]\cosh \alpha b & -\left[\beta^3 + \beta(2-\nu)\dfrac{m^2\pi^2}{a^2}\right]\cos \beta b \end{vmatrix} = 0 \quad (7.3.13)$$

The critical load is found to be $(-N_{xx})_{cr} = k\pi^2 D/b^2$, in which k is a coefficient that depends on the ratio a/b, and it occurs for $m = 1$. For long plates and for $\nu = 0.25$, one has $k \simeq 0.456 + b^2/a^2$. (In more detail see Timoshenko and Gere, 1961, p. 362). The deflection surface according to Eq. 7.3.8 is plotted in Fig. 7.8b.

The boundary conditions in Eqs 7.3.12a–d characterize the buckling of an axially compressed beam of an L–cross section with equal legs. Indeed, if the legs are equal and both legs buckle in concert, the joint of the legs (the corner point) rotates freely as if it were simply supported; see Figure 7.8c.

A similar solution is possible when the longitudinal edges include various combinations of fixed and simply supported edges or fixed and free edges (see, e.g., Timoshenko and Gere, 1961).

Buckling of Plate Subjected to Shear

When a rectangular plate is subjected to shear, N_{xy}, the sine and cosine functions cannot satisfy the differential equation, and thus an exact analytical solution in terms of trigonometric functions is impossible. An approximate solution, suitable for hand calculation, may be based on minimizing the potential energy with respect to an assumed class of functions. As an example, consider that the plate is simply supported and the in-plane shear force N_{xy} is uniform (Fig. 7.9a). We will assume that the deflection is adequately approximated as

$$w = q_1 \sin\frac{\pi x}{a}\sin\frac{\pi y}{b} + q_2 \sin\frac{2\pi x}{a}\sin\frac{2\pi y}{b} \quad (7.3.14)$$

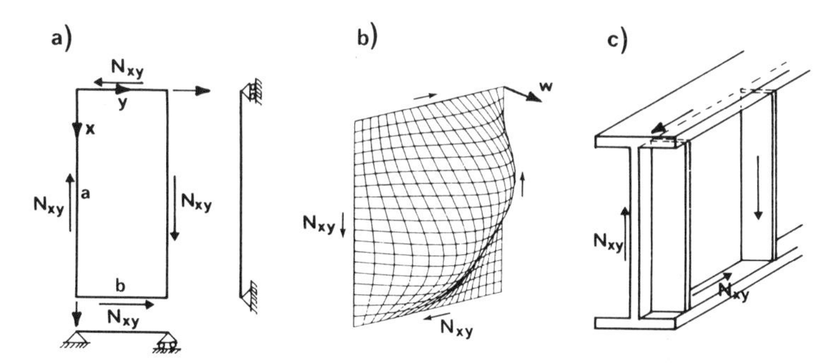

Figure 7.9 (a) Simply supported plate subjected to shear, (b) buckling mode, and (c) analogy with panel of beam web.

where q_1 and q_2 are unknown. Substituting this into the potential-energy expression $\Pi = U_1 + U_2$, with U_1 and U_2 given by Equations 7.2.9 and 7.2.8b, we obtain

$$U_1 = \frac{\pi^4}{8} Dab\left(\frac{1}{a^2}+\frac{1}{b^2}\right)^2 (q_1^2 + 16q_2^2) \tag{7.3.15}$$

$$U_2 = -\tfrac{32}{9} N_{xy} q_1 q_2 \tag{7.3.16}$$

The conditions of neutral equilibrium, which actually represent Trefftz's criterion for the critical state (Eq. 4.2.6), are $\partial\Pi/\partial q_1 = 0$ and $\partial\Pi/\partial q_2 = 0$. This yields

$$\begin{aligned} &\frac{\pi^4}{4} Dab\left(\frac{1}{a^2}+\frac{1}{b^2}\right)^2 q_1 - \frac{32}{9} N_{xy} q_2 = 0 \\ &-\frac{32}{9} N_{xy} q_1 + 4\pi^4 Dab\left(\frac{1}{a^2}+\frac{1}{b^2}\right)^2 q_2 = 0 \end{aligned} \tag{7.3.17}$$

A critical state occurs when these equations admit a nonzero solution, that is, when their determinant is zero. This yields the condition $\pi^8 D^2 a^2 b^2 (a^{-2} + b^{-2})^4 - (32N_{xy}/9)^2 = 0$, which gives the following approximation for the critical shear force:

$$N_{xy_{cr}} = \pm \frac{9}{32} \pi^4 Dab\left(\frac{1}{a^2}+\frac{1}{b^2}\right)^2 \tag{7.3.18}$$

This result is larger by 15 percent or more than the exact solution (Timoshenko and Gere, 1961, p. 381). The approximate deflection surface according to Eq. 7.3.14 is plotted in Fig. 7.9b.

The foregoing solution procedure in fact represents an application of the Ritz variational method, which we explained in Section 5.6 for one-dimensional problems. However, the principle of this method is also applicable to multidimensional problems, in which case the chosen system of linearly independent functions must be complete for the given multidimensional domain. A double trigonometric Fourier series represents such a complete system for a rectangular domain. From the properties of the Ritz method, we know that by increasing the number of terms in Equation 7.3.14 the solution converges to the exact solution. We also know, according to what has been said about the Ritz method, that the solution we obtained (Eq. 7.3.18) is an upper-bound approximation.

The problem can alternatively be solved by the Galerkin method on the basis of the differential equation. Since the present problem is linear and the potential energy exists, the result of the Galerkin solution must be exactly the same.

A simply supported rectangular plate under shear may be used as an approximation to the plate panel between two adjacent vertical stiffeners in the end regions of the web of a thin-wall beam, for example, an I-beam (Fig. 7.9c).

Nonuniform In-Plane Forces

A typical problem for the design of steel beams is the buckling of a rectangular plate that is subjected at its left and right boundaries to a linearly variable normal force N_{xx}. This loading is obtained when the plate forms a panel between two vertical stiffeners in the web of a beam, and the initial normal stress distribution in the beam is calculated according to the bending theory; see Figure 7.10a, b. Due to variation of the coefficient of the differential equation, this problem is best solved by an approximate method such as the Ritz or Galerkin method. By including a sufficient number of terms of a double trigonometric series, any desired accuracy can be achieved (see Timoshenko and Gere, 1961, p. 373; Gerard and Becker, 1957a; Flügge, 1962).

Solutions by Other Variational Methods

In multidimensional problems it is often difficult to choose the trial functions for the Ritz method or the Galerkin method in such a manner that only very few terms would yield an accurate result. This problem may be circumvented by a variant of the Ritz method or the Galerkin method, which is usually called the Kantorovich method.

In this method one chooses the distribution of an unknown function $w(x, y)$ in only one direction, say x, and in the other direction, that is, y, the distribution is characterized by an unknown function $f(y)$ in a manner that we already illustrated in our previous exact solution for a rectangular plate with three edges simply supported and one edge free (Eq. 7.3.8). Thus, an approximate solution of a rectangular simply supported plate, loaded by a linearly variable compression force $N_{xx} = -N_0 y$ (Fig. 7.10b), may be obtained by introducing either the one-term approximation:

$$w(x, y) = f(y) \sin \frac{\pi x}{a} \tag{7.3.19}$$

or the two-term approximation:

$$w(x, y) = f_1(y) \sin \frac{\pi x}{a} + f_2(y) \sin \frac{2\pi x}{a} \tag{7.3.20}$$

Substituting this approximation into the expression for the potential energy of the plate, one obtains a functional of function $f(y)$, or of functions $f_1(y)$ and $f_2(y)$. This functional must now be minimized with respect to these functions of y.

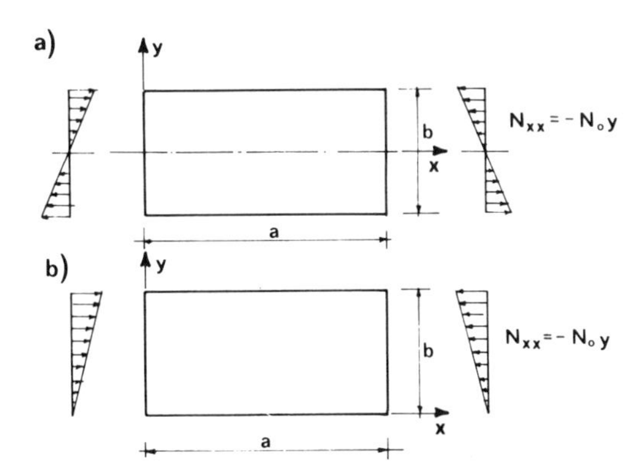

Figure 7.10 Simply supported plate under linearly varying normal forces.

One possible method of minimization is to apply the calculus of variations to obtain from this potential-energy expression either a single ordinary differential equation (Euler equation) for the unknown function $f(y)$, or a system of ordinary differential equations for the unknown functions $f_1(y)$ and $f_2(y)$. These differential equations are then solved either exactly, if possible, or approximately, for which purpose again a one-dimensional Galerkin method can be used. Alternatively, the one-dimensional Ritz method can be applied directly to the functional of function $f(y)$, or of functions $f_1(y)$ and $f_2(y)$. As still another alternative, one may introduce Equations 7.3.19 or 7.3.20 into the two-dimensional variational equation of the Galerkin method, and the result again is an ordinary differential equation for $f(y)$ or a system of two ordinary differential equations for $f_1(y)$ and $f_2(y)$.

Kantorovich's variational method, as we see, is not a fundamentally different type of variational method. Rather, it is an adaptation of the Ritz method or the Galerkin method for multidimensional problems, in which the dimensionality of the problem is reduced (in our examples, from two to one dimension).

In this sense, many theories of structural mechanics can be considered as a form of the Kantorovich method. For example, the bending theory may be regarded as an application of the Kantorovich method that reduces a two- or three-dimensional problem to a one-dimensional problem. This is accomplished by choosing functions $f(y)$ in the transverse directions of the beam to be linear, thus forcing the plane cross sections to remain planar. Similarly, the bending theory of plates can be regarded as an application of the Kantorovich method, which reduces the three-dimensional elasticity problem for a plate to a two-dimensional problem. This is achieved by choosing a linear transverse distribution of displacements. Higher-order theories for the bending of beams and plates, in which the transverse displacement distributions are described as a superposition of several chosen basis functions, have been developed in the past.

Other, fundamentally different, direct variational methods, such as the Trefftz method or the collocation method, have been applied to buckling of plates (Collatz, 1963). However, they are generally not as effective as the methods we have described.

Finally, we must stress that, on the contemporary scene, the approximate variational solutions for the buckling of plates make sense only as long as they are very simple and useful results can be obtained with one or two terms. When this is insufficient, it is better to use the finite element method for plate buckling, rather than to seek solutions with many trigonometric terms, as has been done in the past. However, even when the buckling problem of a plate is solved by finite elements, the simple analytical solutions we have just demonstrated are valuable due to the understanding and insight that they convey, and the checks that they provide.

Problems

7.3.1 (a) Consider the simply supported square plate in Figure 7.11a, loaded by a longitudinal in-plane compressive force N_{xx} (<0) and a transverse in-plane force N_{yy} that is either compressive or tensile. Let $N_{xx}/N_{yy} = \text{const}$. Find the critical value of the load multiplier μ of these in-plane forces. Discuss the

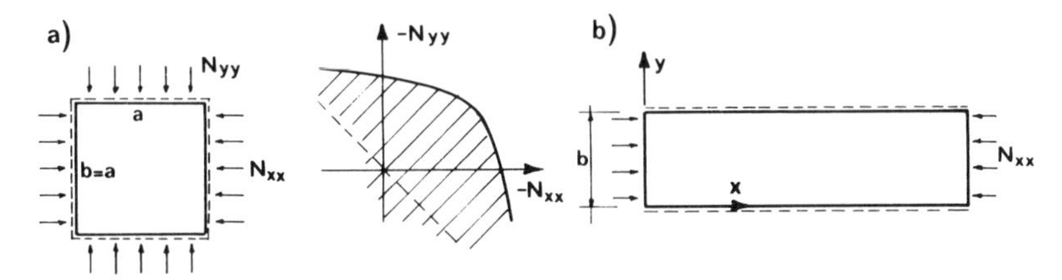

Figure 7.11 Exercise problems on buckling of plates.

effect of the ratio N_{xx}/N_{yy}, and particularly the effect of a change of sign of this ratio (note that $N_{xx}/N_{yy} = -1$ corresponds to pure shear). Construct the limit curve (interaction diagram) of the critical states in the plane (N_{xx}, N_{yy}). (b) Do the same for a rectangular plate with $b = 2a$.

7.3.2 Consider an infinite simply supported plate strip $0 \le y \le b$ (Fig. 7.11b), compressed by initial force N_{xx}. Find its lowest critical load. What is the corresponding half-wavelength L in the x-direction? (*Hint*: Consider L unknown and minimize the critical load with respect to L.)

7.3.3 Solve the same plate strip as above, but the plate carries forces N_{xx} and N_{yy} in both directions.

7.3.4 Consider the plate in Figure 7.8a, whose sides $y = 0$ and $y = b$ are clamped (i.e., fixed against rotation of the edge, $w_{,yy} = 0$). Using the same method as in the text, find the critical load.

7.3.5 Do the same as above, but the plate is a strip infinite in the x-direction.

7.3.6 Consider a square plate clamped on all edges and subjected to equal compression forces in two perpendicular directions. Find the approximate critical value of these forces using the Rayleigh quotient. Assume $w = q_1[1 - \cos(2\pi x/a)][1 - \cos(2\pi y/a)]$. Is this an upper bound? Does this function satisfy the differential equation?

7.3.7 Do the same as above, but with $a \ne b$, $w = q_1[1 - \cos(2\pi x/a)][1 - \cos(2\pi y/b)] + q_2[1 - \cos(4\pi x/a)][1 - \cos(4\pi x/b)]$, and use the Ritz method.

7.3.8 Consider again a simply supported plate subjected to an in-plane shear force (Fig. 7.9a), and solve it by the Galerkin method. Is the result the same as before? Is it an upper bound?

7.3.9 Consider the simply supported plate in Figure 7.10b, with $b = a$, and assume a two-term approximation $w = q_1 \sin(\pi x/a) \sin(\pi y/b) + q_2 \sin(\pi x/a) \sin(2\pi y/b)$. Find the critical value of N_0 using (a) the Ritz method; (b) the Galerkin method.

7.3.10 Do the same as above, but assume $w(x, y) = f(y) \sin(\pi x/a)$ and formulate a differential equation for $f(y)$ using the Kantorovich method.

7.3.11 An infinite simply supported plate strip $0 \le y \le b$ is subjected to shear force N_{xy}. Solve approximate $(N_{xy})_{cr}$.

7.3.12 Do the same as above but the critical forces are $N_{xx} = 2N_{xy}$.

7.3.13 A rectangular plate is simply supported on edges $x = 0$ and $y = 0$, and clamped on edges $x = a$ and $y = b$. Solve the approximate critical value of N_{xx} by the Ritz method.

7.3.14 A rectangular plate has simple supports on edges $x = 0$ and $x = a$, and on edges $y = 0$ and $y = b$ it is supported by an elastic Winkler foundation of foundation modulus c. Solve $(N_{xx})_{cr}$ and discuss the limiting case $c \to \infty$ and $c = 0$.

7.3.15 Do the same as above but the plate is infinitely long in the x-direction.

7.3.16 A rectangular plate is simply supported on edges $x = 0$ and $x = a$. On edges $y = 0$ and $y = b$ it is elastically supported on beams of bending stiffness EI each of which is simply supported at plate corners (the beams represent elastic stiffeners). Solve $(N_{xx})_{cr}$ by a Fourier series expansion.

7.3.17 For the circular plate in Problem 7.2.5 assume that it is simply supported and that the deflection is given by $w = q_1 + q_2(r/R)^2$. Find an approximate value of the critical load by the Ritz method. Is it an upper bound?

7.3.18 Including inertia forces in the partial differential equation for plate deflection, calculate the lateral vibration frequency of a simply supported square plate subject to constant N_{xx}, N_{yy}.

7.3.19 Analyze the same problem as above, but consider that $N_{xx}(t)$ and $N_{yy}(t)$ are pulsating. Using the energy method from Section 3.3, analyze parametric resonance. *Hint:* Determine the second-order in-plane strain, due to $w(x, y, t)$, calculate the work of $N_{xx}(t)$, $N_{yy}(t)$ on those strains, and compare it to the energy dissipated by damping that is proportional to $\dot{w}(x, y, t)$.

7.3.20 Thermal stress buckling. A square elastic simply supported plate whose edges cannot slide is uniformly heated by ΔT. Show that the critical values of N_{xx}, N_{yy} caused by ΔT are the same as if they were applied externally (although the postcritical behavior is very different). Express the stability limit in terms of ΔT. See Problem 4.3.13, note that $\sigma_{ij} = E_{ijkm}(\varepsilon_{km} - \varepsilon''_{km})$ where $\varepsilon''_{km} = \delta_{km}\alpha\,\Delta T$ = thermal strains, and substitute $\varepsilon_{ij} = -z\,\partial^2 w/\partial x_i\,\partial x_j$; $W = 0$, $\Pi = U = \int_V \frac{1}{2}(\varepsilon_{ij} - \varepsilon''_{ij})E_{ijkm}(\varepsilon_{km} - \varepsilon''_{km})\,dV$. {In general, the condition of stability in the presence of thermal stresses is of course again the positive definitiveness of $U[w(x, y)]$.}

7.4 LARGE DEFLECTIONS AND POSTCRITICAL RESERVE OF PLATES

Plates, like columns and some types of frames, and unlike typical shells, possess a postcritical reserve, which permits them to resist loads that are higher than the critical load. As we will see, however, this postcritical reserve of plates is usually much larger than it is for columns. It manifests itself already for moderately large deflections, due to a redistribution of in-plane (membrane) internal forces, which is not possible in beams. The moderately large deflections have no analogy for columns because there is no redistribution of internal forces.

In beam analysis we distinguished only between small deflections (linearized theory), for which the curvature expressions are linearized, and large or finite deflections (nonlinear theory), for which the curvature expressions must be kept in their nonlinear form. For plates (as well as shells), however, it is useful to distinguish:

1. A *linearized* (small-deflection) theory
2. A *nonlinear* theory for *moderately large* deflections, for which linearized expressions for the curvature are still valid while the in-plane forces redistribute due to deflections

3. A *nonlinear* theory for *very large* deflections

The final collapse of plates usually occurs at loads that are much larger than the lowest critical load (frequently 50–100 percent larger). We nevertheless assume that the material behaves linearly, which is true if the plate is sufficiently thin. Thicker plates or shells, however, develop plastic strains or damage before failure (see Section 8.6).

Von Kármán–Föppl Differential Equations

A classical approach to the calculation of the postcritical behavior of plates for moderately large deflections is based on the use of the Airy stress function $F(x, y)$ for in-plane stresses. The in-plane differential equations of equilibrium $N_{xx,x} + N_{xy,y} = 0$ and $N_{xy,x} + N_{yy,y} = 0$ (Eq. 7.2.15b) may be identically satisfied by setting

$$N_{xx} = F_{,yy} \qquad N_{yy} = F_{,xx} \qquad N_{xy} = -F_{,xy} \tag{7.4.1}$$

In this chapter we generally assume that even for very large deflections the strains are small enough for the material to behave linearly. The geometrically nonlinear effects then arise only from the large rotations that accompany large deflections (cf. Sec. 11.1). The strain expressions for large rotations but small in-plane strains are $\varepsilon_{xx} = u_{,x} + \frac{1}{2}w_{,x}^2$, $\varepsilon_{yy} = v_{,y} + \frac{1}{2}w_{,y}^2$, and $\gamma_{xy} = u_{,y} + v_{,x} + w_{,x}w_{,y}$. They satisfy the compatibility condition:

$$\varepsilon_{xx,yy} + \varepsilon_{yy,xx} - \gamma_{xy,xy} = w_{,xy}^2 - w_{,xx}w_{,yy} \tag{7.4.2}$$

The left-hand side represents the linear terms and the right-hand side the nonlinear terms. Note that for linear elasticity the right-hand side must vanish, and this yields the well-known compatibility condition of linear plane elasticity.

According to Hooke's law:

$$\varepsilon_{xx} = \frac{N_{xx} - \nu N_{yy}}{hE} \qquad \varepsilon_{yy} = \frac{N_{yy} - \nu N_{xx}}{hE} \qquad \gamma_{xy} = \frac{2(1+\nu)N_{xy}}{hE} \tag{7.4.3}$$

Substituting Equations 7.4.1 into 7.4.3 and Equations 7.4.3 into 7.4.2, we obtain

$$\nabla^4 F = Eh(w_{,xy}^2 - w_{,xx}w_{,yy}) \tag{7.4.4}$$

Furthermore, substituting Equations 7.4.1 into Equation 7.2.15a for lateral deflections, we also have

$$D\nabla^4 w = p + (F_{,yy}w_{,xx} + F_{,xx}w_{,yy} - 2F_{,xy}w_{,xy}) \tag{7.4.5}$$

Equations 7.4.4 and 7.4.5 represent a system of two coupled nonlinear fourth-order partial differential equations for functions $w(x, y)$ and $F(x, y)$, called von Kármán–Föppl equations. Föppl (1907) introduced the use of stress function F, and von Kármán (1910) obtained the final form of these equations. These equations are rather difficult to solve analytically, and they have been solved only approximately. The earliest results on the postcritical reserve in plate buckling were obtained on the basis of these differential equations.

Solution by Minimization of Potential Energy

Approximate solutions may be based on minimization of the potential energy (as long as the material of the plate is elastic). For deflections that are of the order of the thickness of the plate, experience shows that first-order approximations for the curvatures can still be used, and so the strain energy U_1 due to bending is still given by Equation 7.2.9. As the second part of the potential-energy expression, the strain energy due to the in-plane forces may be calculated as

$$U_2 = h\int_{-a}^{a}\int_{-b}^{b}\tfrac{1}{2}(\sigma_{xx}\varepsilon_{xx} + \sigma_{yy}\varepsilon_{yy} + \tau_{xy}\gamma_{xy})\,dx\,dy$$
$$= \frac{Eh}{2(1-\nu^2)}\int_{-a}^{a}\int_{-b}^{b}\left(\varepsilon_{xx}^2 + \varepsilon_{yy}^2 + 2\nu\varepsilon_{xx}\varepsilon_{yy} + \frac{1-\nu}{2}\gamma_{xy}^2\right)dx\,dy \qquad (7.4.6)$$

in which σ_{xx}, σ_{yy}, τ_{xy} represent the stresses in the middle surface of the plate due to in-plane forces N_{xx}, N_{yy}, N_{xy} (that is, $N_{xx} = h\sigma_{xx}$, $N_{yy} = h\sigma_{yy}$, $N_{xy} = h\sigma_{xy}$); ε_{xx}, ε_{yy}, and γ_{xy} represent the finite strains of the middle surface, which take into account deflection w; the material is assumed to be linear and isotropic; $2a$ and $2b$ are the length and width of the plate, and h is the thickness.

Aside from deflections $w(x, y)$, we must also solve the unknown in-plane displacements $u(x, y)$ and $v(x, y)$, which figure in the expressions for the finite strains ε_{xx}, ε_{yy}, and γ_{xy}. Assuming suitable distributions for $w(x, y)$, $u(x, y)$ and $v(x, y)$ in the sense of the Ritz method, one may then obtain a system of equilibrium equations by minimizing the potential energy with respect to the unknown parameters of these assumed distributions. This results in a system of nonlinear algebraic equations.

Let us demonstrate this approach with the example (Fig. 7.12a) of a square plate supported on all its sides and subjected to imposed uniform displacement $v = -2ae$ along the side $y = a$; e is a parameter that characterizes the loading. The constraints on the boundaries $x = -a$ and $x = a$ prevent any lateral displacements (sliding boundary). We assume the plate to be imperfect, with the initial warping defined by the ordinates:

$$z_0 = q_0 \cos\frac{\pi x}{2a}\cos\frac{\pi y}{2a} \qquad (7.4.7)$$

where q_0 is a given coefficient of the imperfections. The unknown distributions of displacements are assumed in the form:

$$w = q_1 \cos\frac{\pi x}{2a}\cos\frac{\pi y}{2a} \qquad u = q_2 \sin\frac{\pi x}{a}\cos\frac{\pi y}{2a} \qquad v = q_2 \cos\frac{\pi x}{2a}\sin\frac{\pi y}{a} - e(y+a) \qquad (7.4.8)$$

where q_1 and q_2 are unknown parameters. This choice satisfies all the kinematic boundary conditions and is supported by deflection surfaces observed in experiments. The deflection surface according to Equations 7.4.7 and 7.4.8 is plotted in Figure 7.12c.

The bending energy expression can be obtained from Equation 7.2.9 by

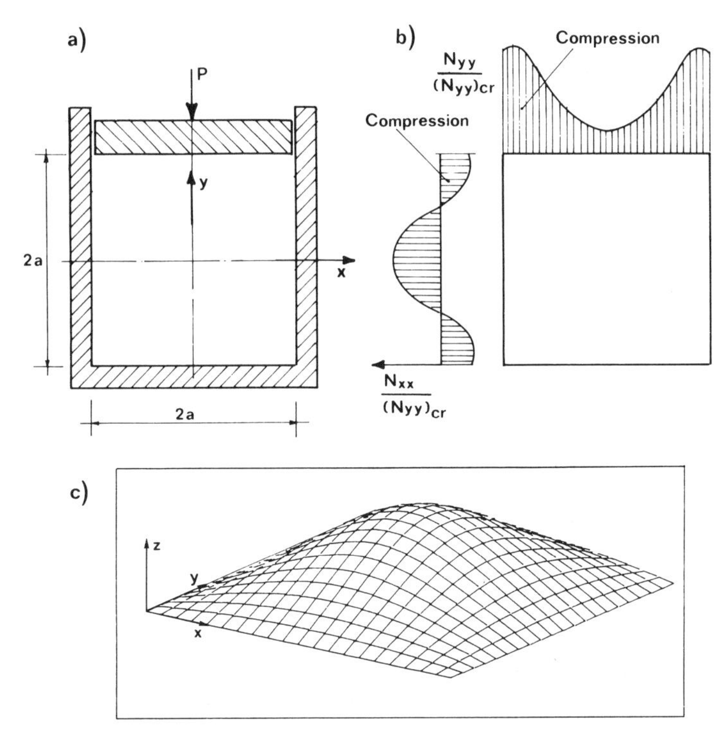

Figure 7.12 (a) Simply supported square plate subjected to uniform edge in-plane displacement, (b) normal force distribution at large displacements, and (c) deflection surface.

substituting for w the elastic part of the deflection, $w - z_0$:

$$U_1 = \frac{D}{2}\int_{-a}^{a}\int_{-a}^{a}\{[(w - z_0)_{,xx} + (w - z_0)_{,yy}]^2 + 2(1 - \nu)[(w - z_0)^2_{,xy}$$

$$- (w - z_0)_{,xx}(w - z_0)_{,yy}]\}\,dx\,dy$$

$$= \frac{Eh^3}{12(1 - \nu^2)}\left[\frac{\pi^4(q_1 - q_0)^2}{8a^2}\right] \tag{7.4.9}$$

in which the last expression has been obtained by substituting the expressions for w and $-z_0$ given in Equation 7.4.8 and 7.4.7.

The expressions for the finite strains in the middle surface of the plate are

$$\begin{aligned} \varepsilon_{xx} &= u_{,x} + \tfrac{1}{2}w^2_{,x} - \tfrac{1}{2}z^2_{0,x} \\ \varepsilon_{yy} &= v_{,y} + \tfrac{1}{2}w^2_{,y} - \tfrac{1}{2}z^2_{0,y} \\ \gamma_{xy} &= v_{,x} + u_{,y} + w_{,x}w_{,y} - z_{0,x}z_{0,y} \end{aligned} \tag{7.4.10}$$

Substituting these equations into Equations 7.4.6 and using the expressions in Equations 7.4.7 and 7.4.8, one obtains for the strain energy due to in-plane forces:

$$U_2 = \frac{Eh}{2(1-\nu^2)}\left[4a^2e^2 - q_1^2 e\frac{\pi^2(1+\nu)}{4} + q_1^4\frac{5\pi^4}{256a^2} - q_1^2 q_2\frac{\pi^2(5-3\nu)}{6a}\right.$$
$$+ q_2^2\frac{64+81\pi^2-\nu(9\pi^2-64)}{36} + q_0^4\frac{9\pi^4}{512a^2} - q_1^2q_0^2\frac{5\pi^4}{128a^2}$$
$$\left. + q_0^2 e\frac{\pi(1+\nu)}{4} + q_2 q_0^2\frac{\pi^2(5-3\nu)}{6a}\right] \tag{7.4.11}$$

The conditions of neutral equilibrium (Trefftz's condition of critical state) are obtained by imposing $\partial(U_1+U_2)/\partial q_1 = 0$ and $\partial(U_1+U_2)/\partial q_2 = 0$. Assuming $\nu = 0.3$, we obtain from the latter condition: $q_2 = 0.1418(q_1^2 - q_0^2)/a$. Substituting this into the former condition, we get

$$q_1[5.698(q_1^2 - q_0^2) + 4.059h^2 - 6.415ea^2] = 4.059h^2 q_0 \tag{7.4.12}$$

First we consider that the plate is perfect ($q_0 = 0$). Equation 7.4.12 reduces to $q_1(5.698q_1^2 + 4.059h^2 - 6.415ea^2) = 0$. This equation has the solutions:

$$\text{Either} \qquad q_1 = 0 \qquad \text{or} \qquad q_1 = \frac{(6.415ea^2 - 4.059h^2)^{1/2}}{5.6974} \tag{7.4.13}$$

which show that a real value of q_1 that is different from zero exists only if $e > e_{cr} = 0.6327h^2/a^2$ (as obtained for this problem by Timoshenko and Gere, 1961, p. 411). The stress resultant corresponding to this critical value of e (for which $q_1 = q_2 = 0$) is

$$-(N_{yy})_{cr} = \frac{Eh}{1-\nu^2}e_{cr} = \frac{0.6327Eh^3}{(1-\nu^2)a^2} \tag{7.4.14}$$

This represents the critical in-plane force N_{yy}. Since the lateral expansion is prevented by the sliding boundaries, we have $N_{xx} = \nu(N_{yy})_{cr}$ for the critical state.

Second, consider that the plate is imperfect ($q_0 \neq 0$), and calculate the value of N_{yy}. From Hooke's law, $N_{yy} = Eh(\varepsilon_{yy} + \nu\varepsilon_{xx})/(1-\nu^2)$. We are interested in the value of N_{yy} at $y = a$, for which $\varepsilon_{xx} = 0$. Calculating ε_{yy} from Equations 7.4.10 we obtain

$$(N_{yy})_{y=a} = -\frac{Eh}{1-\nu^2}\left[e + \frac{q_1^2 - q_0^2}{a^2}\left(0.4455 - 1.234\cos\frac{\pi x}{2a}\right)\cos\frac{\pi x}{2a}\right] \tag{7.4.15}$$

For a given value of e, we can calculate q_1 from Equation 7.4.12. For example, for $e = 10e_{cr}$ we obtain the distribution of N_{yy} plotted in Figure 7.12b, which also shows the distribution of N_{xx}, obtained in a similar way. The most interesting property of these results is the strong redistribution of N_{yy}, in which most of the axial in-plane force gets transferred from the middle strip of the plate into the edge strips along the sliding boundaries.

It is also useful to calculate the average value $\bar{N}_{yy} = (\int_{-a}^{a} N_{yy}\,dx)/2a$, which can be expressed as a function of q_1 by substituting into Equation 7.4.15 the value of e derived from Equation 7.4.12. One obtains

$$(\bar{N}_{yy})_{y=a} = -\frac{Eh}{1-\nu^2}\left(\frac{h^2}{a^2}\right)\left[0.6327\left(1 - \frac{q_0}{q_1}\right) + 0.5549\left(\frac{q_1^2}{h^2} - \frac{q_0^2}{h^2}\right)\right] \tag{7.4.16}$$

The corresponding diagram of average load versus the maximum deflection is shown in Figure 7.13 for various values of the initial imperfection q_0. As one can see, a large postcritical reserve develops even for moderately large values of the deflection q_1.

For comparison, we also plot on the same diagram the response of a unit plate strip considered as a hinged column with free sliding ends; its critical load is given by $N_{cr} = \pi^2 Eh^3/4a^2 = 0.2056Eh(h/a)^2$, and its nonlinear behavior is governed by Equation 5.9.10. Not only the critical load is lower, due to the lack of longitudinal supports $x = -a$ and $x = a$, but also the curvature of the postcritical response curve $N(q_1)$ is of the order of hundred times smaller.

The nonlinear behavior our example just illustrated fundamentally differs from the nonlinear behavior of columns. For columns, large-deflection energy analysis (Sec. 5.9) necessitates that the potential energy be expressed in terms of a nonlinear expression for curvature, which includes terms of at least the second order [that is, $1/\rho = d\theta/ds$ exactly, or $1/\rho \simeq w''(1 - 1.5w'^2)$ accurately up to second-order terms]. By contrast in our example we calculated the potential energy of the plate from the linearized curvatures, that is, $w_{,xx}$, $w_{,yy}$, and $w_{,xy}$, and the nonlinearity was due solely to the redistribution of the in-plane forces N_{xx}, N_{yy}, N_{xy}, which cannot occur in columns. This redistribution engenders an early stiffening of the response (postcritical reserve) that is encountered already at moderately large deflections, at which the nonlinear curvature effect is still secondary in importance.

Note that the use of the second-order strain term $\frac{1}{2}w'^2$ in the potential-energy expression for a column does not in itself represent a nonlinear effect. The reason is that the potential energy must always be represented exactly up to second-order small terms, even in a linear theory. What makes the effect of $\frac{1}{2}w_{,x}^2$ (or $w_{,x}w_{,y}$) nonlinear in plates is that the terms such that $N_{xx}w_{,x}^2$ also include a second-order change of N_{xx} due to $\frac{1}{2}w_{,x}^2$, making the term $N_{xx}w_{,x}^2$ accurate up to quantities that are fourth-order small in deflections.

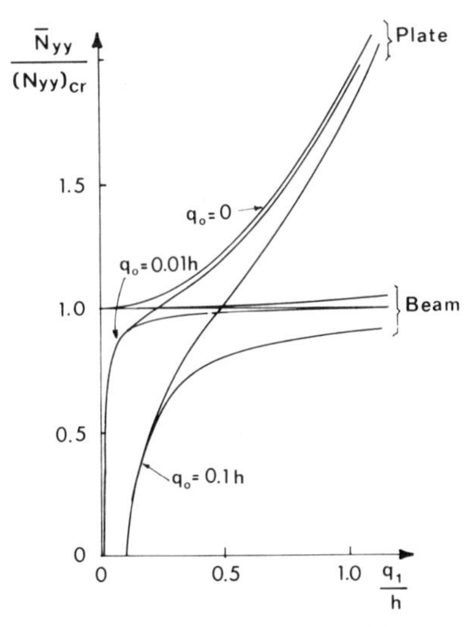

Figure 7.13 Force-displacement relations for perfect and imperfect plates and beams.

Further note that the principal cause of nonlinear stiffening of plates at moderately large deflection is not the dependence of N_{xx} on second-order strain $\frac{1}{2}w_{,x}^2$ (or $w_{,x}w_{,y}$), per se. Rather it is the redistribution of the in-plane forces in the plane of the plate. In a simply supported column whose ends are prevented from sliding axially, the axial force P, too, changes as a function of $\frac{1}{2}w'^2$. However, there can be no redistribution, such as the transfer of compression from the center of a plate toward its edge strips, which greatly helps the carrying capacity of a plate.

Not all the plates stiffen due to redistribution of the in-plane forces. For example, a rectangular plate simply supported on edges $x = 0$, $x = a$ and free on edges $y = 0$, $y = b$, compressed by N_{xx}, buckles with a cylindrical surface, and thus it behaves as a column in which there is no redistribution of N_{xx}. If very weak stiffeners are added to this plate along the edges $y = 0$ and $y = b$, there will be redistribution of in-plane forces, but only to a very small extent, with only a very small contribution to the postcritical reserve.

Large Deflections and Ultimate Strength

For very large deflections of plates, nonlinear expressions for the curvatures must, of course, be used also, same as for columns. In practice, however, it is preferable to analyze very large plate deflections by the finite element method with step-by-step loading, in which the nodal coordinates are updated after each loading step. In this approach, called the updated Lagrangian method, it is not necessary to use large-deflection theory with nonlinear curvature expressions provided that the loading steps are sufficiently small so that the deflection increments remain only moderately large compared to the initial state for each current loading step.

Unless the plate is very thin, nonlinear material properties may, of course, significantly affect the ultimate collapse mode. In general, the analysis of large-deflection postcritical collapse of plates may require not only a geometrically nonlinear analysis but also analysis for nonlinear inelastic material behavior. For such a general problem, the finite element method is ideally suited.

A simple approximate formula for the ultimate buckling strength of rectangular plates loaded by compression force N_{xx} was derived and experimentally verified by von Kármán, Sechler, and Donnell (1932). As the axial compression resultant P is gradually increased above the critical value, the distribution of N_{xx} becomes progressively more nonlinear as shown by curves 1, 2, and 3 in Figure 7.14a until at the ultimate state (curve 3) the N_{xx}-distribution becomes nearly rectangular, most of the axial force being transmitted by edge strips of a certain width c while the central portion of the plate becomes almost load-free. Because the edge strips are buckled for all the postcritical states, the average distributed force N_{xx} that is carried by the edge strips may be estimated as the critical value of a uniform N_{xx} for a simply supported rectangular plate of length a and width $2c$, which is imagined to be the width of the statically equivalent rectangular stress distributions in Figure 7.14a. Since the deflections are always small in the close vicinity of the edge, the local behavior near the edge may be approximately described by the classical formula for the critical load N_{xx}^{cr} (bifurcation load) even if the overall behavior of the plate is postcritical, with large deflections.

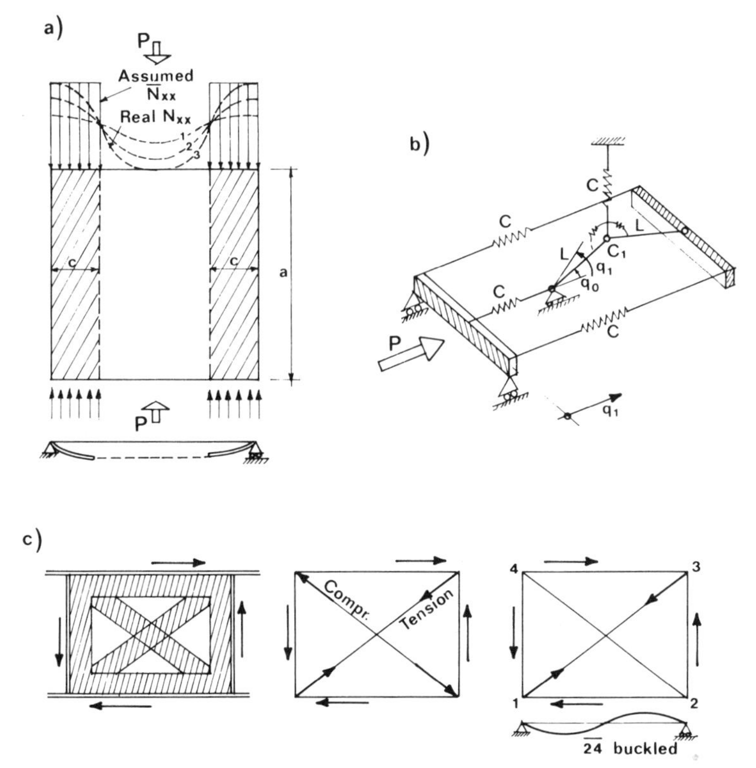

Figure 7.14 (a) Effect of nonlinear material behavior on normal force distribution in the postcritical state; (b) bar–spring model; (c) postcritical behavior of panel under shear.

According to Equation 7.3.7 the critical force on the edge strip of width $2c$ may be approximately expressed as $|\bar{N}_{xx}| = 4\pi^2 D/(2c)^2 = \pi^2 D/c^2$ (because Eq. 7.3.7 gives the critical N_{xx} for $b \ll a$, and we usually have $2c \ll a$). Because the edge strip is deflected for all $|N_{xx}|$ exceeding the elastic critical force of the entire plate, the last equation must be satisfied continuously for all load values. Therefore $c^2 = \pi^2 D/|\bar{N}_{xx}|$ and so an increase of $|\bar{N}_{xx}|$ due to increasing axial resultant P must produce a decrease of the effective width c of the edge strips. Failure occurs when c is reduced to a value such that $|\bar{N}_{xx}| = \pi^2 D/c^2 = hf_y$, where f_y = yield stress of the material and h = plate thickness. From this we get von Kármán's formula $c = \pi(D/hf_y)^{1/2}$ for the effective width at failure. The total ultimate force carried by the plate is $P_{\text{ult}} = 2chf_y$, and substitution for c, along with $D = Eh^3/12(1-\nu^2)$, yields von Kármán's ultimate load formula:

$$P_{\text{ult}} = k_u h^2 \sqrt{Ef_y} \tag{7.4.17}$$

where $k_u = \pi[3(1-\nu^2)]^{-1/2} = 1.9$ (for $\nu = 0.3$). It is noteworthy that P_{ult} is approximately independent of the plate dimensions a and b.

Based on more extensive test results, Winter (1947) proposed the empirical coefficient $k_u = 1.9[1 - 0.475(h/b)(E/f_y)^{1/2}]$ where b = total width of the plate. This correction is explained by the fact that the two strips of width c do not behave exactly as a simply supported plate of width $2c$. Another possible correction of coefficient k_u was suggested by Donnell (1976, p. 235) who also considered the case where the inner edge of the strip c is treated as a free edge.

If the edges of the plate are supported by flexible stiffening ribs of direction x, then the design must also satisfy Equation 7.4.17 in which f_y is replaced by the critical stress $\sigma^r_{cr} = |\bar{N}_{xx}|/h$ that causes the buckling of the ribs.

An excellent comprehensive survey of various ultimate load formulas for many situations is found in Johnston (1966, 1976) and Galambos (1988). For a discussion of Equation 7.4.17, see also Timoshenko and Gere (1961, p. 418).

A simply supported rectangular panel subjected to shear owes its large postcritical reserve to a different mechanism. As the buckling deflections become large, a large tension force is developed in that plane along one diagonal strip of the square plate, while the compression force that was originally produced by shear along the other diagonal strip is reduced to almost zero (Fig. 7.14c). Thus the panel, which initially carries the shear like a rectangular truss with two diagonals, one in tension and one in compression, acts after large deflections as a square truss with only one diagonal, the tension one, as if the compression diagonal failed.

Due to their large postcritical reserve, plates can in principle be designed with a smaller safety factor than columns or frames. The use of smaller safety factors, however, might be inadmissible if there are large repeated loads that could cause repeated large buckling deflections, or if there is danger of simultaneous buckling of stiffeners.

Measurement of Critical Loads

Due to their more pronounced nonlinearity, the Southwell plot (Eq. 1.5.13) is insufficient for plates. The critical load P_{cr_1} of a plate may be accurately identified by fitting a sufficiently accurate nonlinear solution to the measured curve of maximum deflection w vs. load (or load parameter) P. According to Williams and Aalami (1979), the $w(P)$ relation may be written as

$$w^2 - w_0^2 = h^2(C_1\psi + C_3\psi^3 + \cdots)^2 \tag{7.4.18}$$

where $\psi = (w_0/w - \mu^{-1})^{1/2}$, $\mu^{-1} = 1 - P/P_{cr_1}$ (Eq. 1.5.10), w_0 = initial imperfection; $C_1, C_3, \ldots$ = constants that take into account the boundary conditions and load configuration. According to Walker (1969) and Williams and Walker (1977), the second and higher terms of the expansion in Equation 7.4.18 are usually negligible. Using an optimization subroutine, one may fit Equation 7.4.18 to the measured $w(P)$ curve, which yields the optimum values of P_{cr_1}, w_0 and C_1. (See also Fok, 1984; Fok and Yuen, 1981; Spencer and Walker, 1975; Rhodes and Harvey, 1977; Dawson and Walker, 1972).

Problems

7.4.1 Derive Equations 7.4.9 and 7.4.11 for strain energies U_1 and U_2 and Equation 7.4.16 for the average in-plane (membrane) strain resultant $\bar{N}_{yy}$. Then derive the expressions for N_{xx} and $\bar{N}_{xx}$.

7.4.2 Do the same as above, but for a rectangular plate of length $2a$ and width $2b$. Evaluate the results for $b/a = 1$, 1.25, 1.50.

7.4.3 Consider the bar–spring model in Figure 7.14b and assume $C_1 = \frac{1}{2}CL^2$. By the energy approach, find the critical load and the postbuckling behavior, considering both the perfect ($q_0 = 0$) and imperfect ($q_0 \neq 0$) system. Show that the behavior of this model is analogous to the behavior of plates.

7.4.4 (a) Without referring to the text derive the von Kármán formulas for effective width and P_{ult} of a rectangular plate subjected to N_{xx}. (b) Then, using the same calculation procedure, generalize the formulas for the case of both N_{xx} and N_{yy} with a given ratio $k = N_{yy}/N_{xx}$ (to this end obtain first the critical value of N_{xx} for an infinite plate strip of width $2c$).

7.4.5 Calculate the value of k_u considering the infinite plate strip of width c to have a free edge opposite to the simply supported edge (this was done by Donnell, 1976).

7.5 AXISYMMETRIC BUCKLING OF CYLINDRICAL SHELLS

Buckling of shells is a vast subject whose comprehensive treatment requires an entire book. Various excellent specialized books and comprehensive articles devoted to shell buckling exist (Gerard and Becker, 1957b; Timoshenko and Woinosky-Krieger, 1959; Vol'mir, 1963; Brush and Almroth, 1975; Donnell, 1976; Calladine, 1983; Kollár and Dulácska, 1984; Popov and Medwadowski, 1981; and Bushnell, 1985). We must content ourselves with an explanation of only the basic methods and an analysis of only some typical cases. We begin with the axisymmetric buckling of axially compressed cylindrical shells.

Consider a circular cylindrical shell of constant thickness h and middle surface radius R, subjected to axial distributed in-plane force N_{xx} (positive for tension). Let x be the axial coordinate (see Fig. 7.15a), y the circumferential coordinate (length of arc), and z the outward normal coordinate. A strip of the shell of width

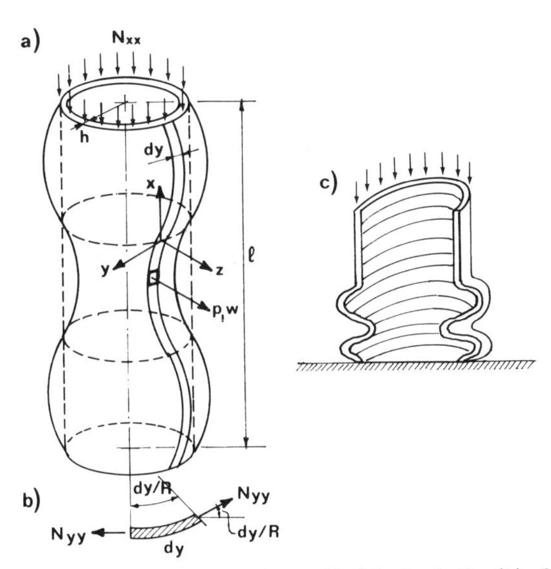

Figure 7.15 (a) Axially compressed circular cylindrical shell, (b) forces acting on a circumferential element, and (c) axisymmetric buckling mode.

dy behaves longitudinally as a beam or plate strip of bending stiffness $D = \frac{1}{12}Eh^3/(1-\nu^2)$ = cylindrical stiffness of the wall. So (according to Eq. 7.2.14 with $F_i = 0$) the normal outward force on the shell element $dx\,dy$ must be $p^*\,dx\,dy = (Dw_{,xxxx} - N_{xx}w_{,xx})\,dx\,dy$, in which p^* is the normal outward resultant per unit area and w = outward deflection.

An essential fact to note is that if we assume the shell deflection to be axisymmetric, deflection w changes the circle of radius R to a circle of radius $R + w$, and thus circumferential normal strain $\varepsilon_{yy} = [2\pi(R+w) - 2\pi R]/2\pi R = w/R$. So the deflection produces circumferential in-plane force $N_{yy} = Ehw/R$. From the sketch of the small element dy in Figure 7.15b we see that the forces N_{yy} on this element have a normal inward resultant $dx\,N_{yy}\,dy/R$. Therefore, $p^*\,dx\,dy = p\,dx\,dy - (Ehw/R)\,dx\,dy/R$, which yields

$$Dw_{,xxxx} - N_{xx}w_{,xx} + \frac{Eh}{R^2}w = p \tag{7.5.1}$$

where p is the applied transverse load. For buckling, we set $p = 0$.

The foregoing fourth-order differential equation is of the same form as our previous equation for the buckling of a beam on elastic foundation (Eq. 5.2.2). The equivalent foundation modulus is $c = Eh/R^2$. So the same solutions as those presented before (Sec. 5.2) may be applied again.

We consider here only the periodic solution of the type $w = w_0 \sin \alpha x$, which can occur either if the shell ends are simply supported and the shell length l is a multiple of the half-wavelength $L = \pi/\alpha$, or if the shell is very long and the boundary segments are not considered. The substitution of $w = w_0 \sin \alpha x$ into Equation 7.5.1 leads to the critical load:

$$-N_{xx}^{\text{cr}} = D\left[\alpha^2 + \frac{Eh}{R^2D}\left(\frac{1}{\alpha^2}\right)\right] \tag{7.5.2}$$

which depends on α. Setting the derivative of this equation with respect to α to zero, we find $\alpha = \alpha_1 = (Eh/R^2D)^{1/4}$, and substitution into Equation 7.5.2 then yields the minimum possible critical load:

$$-N_{xx_{\min}}^{\text{cr}} = \frac{1}{\sqrt{3(1-\nu^2)}}\left(\frac{Eh^2}{R}\right) \tag{7.5.3}$$

This solution was obtained by Lorenz (1908, 1911), Timoshenko (1910), and Southwell (1914).

If the length l of the shell is not compatible with the half-wavelength π/α_1, the critical load and the number of half-waves can be determined by plotting Equation 7.5.2 as shown in Figure 7.16 for various numbers n of half-wavelengths, with $\alpha = n\pi/l$. Obviously, if $\alpha \neq n\alpha_1$, that is, if the boundary conditions are not compatible with the half-wavelength π/α_1, the critical load is higher than Equation 7.5.3 indicates.

The potential energy of the shell may be calculated as the potential energy of all the beam strips of widths dy:

$$\Pi = 2\pi RhE \int_0^l \frac{1}{2}\left[Dw_{,xx}^2 + Eh\left(\frac{w}{R}\right)^2 - (-N_{xx})w_{,x}^2\right] dx \tag{7.5.4}$$

The same expression results from analogy with the beam on an elastic foundation.

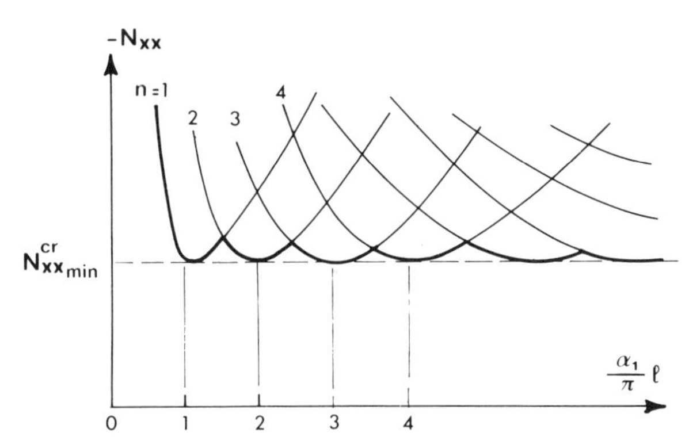

Figure 7.16 Critical values of normal compressive force.

The differential equation 7.5.1 may be derived from this expression by variational calculus as the corresponding Euler equation. Alternatively, substituting $w = w_0 \sin \alpha x$ into Equation 7.5.4, and setting $\partial\Pi/\partial w_0 = 0$, one obtains Equation 7.5.2 for the critical load directly.

Unfortunately, the foregoing simple solution applies only to the initial deformation of relatively thick and short tubes (Fig. 7.15c). The majority of cylindrical shells do not buckle axisymmetrically (and even the buckling mode of thick and short tubes later changes to nonaxisymmetric). That in itself would not be a major drawback for the foregoing solution since the critical load for the nonaxisymmetric buckling loads is the same or nearly the same. A more serious problem is that, due to the combined effects of nonlinearities and imperfections, the failure loads of axially compressed thin cylindrical shells are usually far smaller than the critical stress of the perfect shell (Eq. 7.5.3), as we will explain later.

Nevertheless, the result that the critical stress is approximately proportional to Eh^2/R is correct, and, of course, Equation 7.5.3 represents an upper bound on the actual collapse load. That is where the usefulness of Equation 7.5.3 lies.

Even the foregoing simple analysis suffices to bring to light one essential property of shells that distinguishes them from plates: The term Ehw/R^2 in Equation 7.5.1 is due to in-plane stretching of the shell, and thus it introduces a coupling between the bending and stretching modes of deformation. For plates such a coupling exists, too, but the coupling terms are of the second order, and therefore negligible for small deflections, that is, for the calculation of critical loads. Not so for shells. The differential equations for shells generally exhibit first-order bend–stretch coupling terms that are exemplified by Equation 7.5.1.

It is interesting that Equation 7.5.3 also gives the minimum possible critical value of the internal in-plane (membrane) force of a perfect spherical shell subjected to external pressure p. The critical value of p ensues from the critical value of N_{xx} (Eq. 7.5.3) according to the transverse equilibrium condition of the shell element: $N_{xx} = pR/2$ (Brush and Almroth, 1975; Timoshenko and Gere, 1961). Again however, due to imperfections and nonlinear effects, the actual critical load of a spherical shell under external pressure is much smaller than this classical result.

Problems

7.5.1 Check that Equation 7.5.3 can be derived from Equation 5.2.7 valid for beams on elastic foundation if the circumferential restraint of shell stiffness is directly interpreted as the foundation modulus of the beam.

7.5.2 Derive Equation 7.5.1 as the Euler equation using the expression of potential energy in Equation 7.5.4.

7.5.3 Construct the plot in Figure 7.16.

7.5.4 Solve the approximate critical value of liquid mass density ρ at which the conical tank shown in Figure 7.17(a) buckles axisymmetrically in the mode shown in Figure 7.17b. (Some water tower tanks have, in fact, collapsed in this manner.) Assume simple support at the base. *Hint:* First determine the membrane stresses required by equilibrium conditions; N_{xx} follows by equilibrium of the shell segment (x, l) with the water column above it, Figure 7.17c; N_{yy} follows by equilibrium of a ring ds for horizontal forces; see Figure 7.17c. Then calculate the potential energy Π due to bending in the mode shown in Figure 7.17b, taking into account the work of membrane forces due to deflection. Then assume various suitable approximate functions $w(x)$ and calculate the solution by minimizing Π. Alternatively, use the variational calculus to obtain a differential equation for $w(x)$ with boundary condition; its coefficients, however, are variable, which complicates analytical solution.

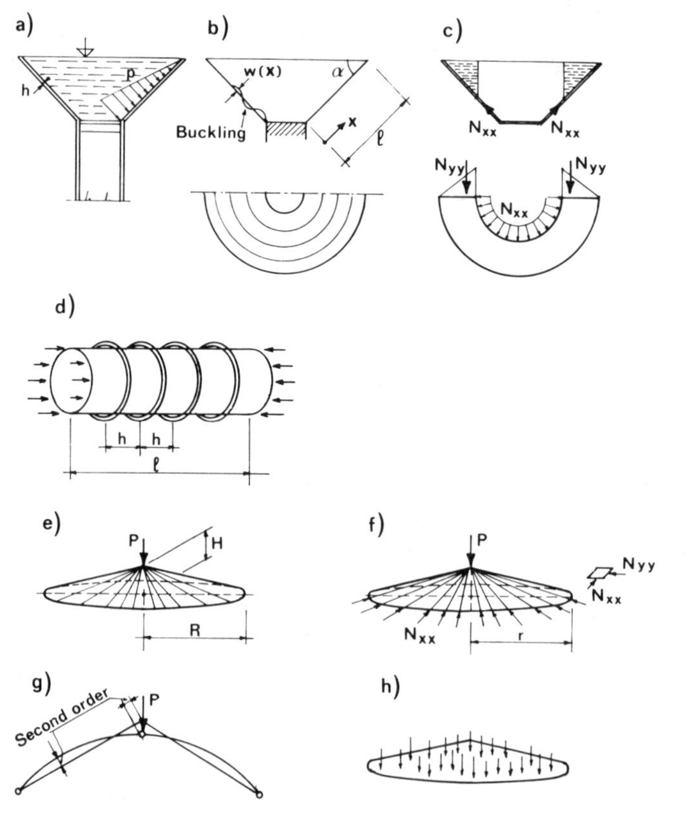

Figure 7.17 Exercise problems on buckling of thin shells.

7.5.5 Find under which conditions the critical loads for the global buckling mode $w_1 = A \sin(\pi x/l)$ and the local buckling mode $w_2 = a \sin(\pi x/h)$ of the ring-stiffened axially compressed cylindrical shell (Fig. 7.17d) are equal. Note that the problem is analogous to Problem 5.2.12 and the same comments apply.

7.5.6 Including inertia forces in the differential equation for axisymmetric deflection of a long tube under axial stress σ^0_{xx}, calculate its lowest natural vibration frequency.

7.5.7 Do the same as above, but the applied axial stress $\sigma^0_{xx}(t)$ is pulsating. Use the energy method from Section 3.3 to analyze parametric resonance. (This is analogous to Prob. 5.2.14 for a beam on elastic foundation.)

7.5.8 (a) Consider a flat conical shell (Fig. 7.17e), very thin (negligible bending stiffness), in a membrane state. Calculate the first-order membrane forces N_{xx}, N_{yy} (Fig. 7.17f) and the first-order in-plane strains as well as the transverse deflections of each originally straight line on the surface (see some text on membrane shell analysis, e.g., Flügge, 1973). From these calculate the second-order load-point displacement by integrating along the originally straight line on the shell (Fig. 7.17g). Evaluate and plot the diagram of P versus load-point displacement u and analyze stability. (b) Do the same as (a), but consider distributed pressure p (Fig. 7.17h). Define $\bar{u}$ as the average displacement $\bar{u} = \int u\, dA/S$ so that $p\, d\bar{u}$ = work of p on du (S = area of shell surface and dA = its element). Calculate and plot $p(\bar{u})$, and discuss stability.

7.6 SHALLOW OR QUASI-SHALLOW SHELLS

The general equations for shell buckling are rather complicated, however, a simplification that is valid in many situations is possible when the shell is shallow or quasi-shallow. A shallow shell is one whose rise with regard to any chord is small. A quasi-shallow shell is one which buckles in such a manner that each buckle alone represents a shallow shell. This means that the half-wavelength of buckling is short compared to the curvature radius R. Cylindrical, spherical, and other shells usually fail by forming many relatively small buckles (Fig. 7.18) even when the shell as a whole is not shallow. Thus, they may then treated as shallow shells.

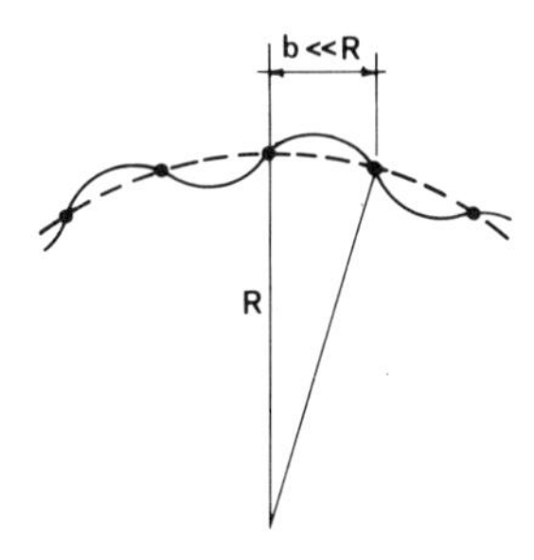

Figure 7.18 Buckling mode of quasi-shallow shell.

Basic Relations for Cylindrical Shells

We will now derive the basic differential equations for shallow cylindrical shells. First, we will consider the general nonlinear behavior and then carry out incremental linearization.

Similar to Kirchhoff's assumption for plates, thin shells may be analyzed under the assumption that the normals of the middle surface remain straight and normal to the middle surface, and that the transverse normal stress is negligible. This assumption, due to Love (1888), implies that the strains due to bending at any point of coordinate z measured from the middle surface are calculated as $\varepsilon_{xx} = -z\kappa_{xx}$, $\varepsilon_{yy} = -z\kappa_{yy}$, $\varepsilon_{xy} = -z\kappa_{xy}$, in which κ_{xx}, κ_{yy}, and κ_{xy} represent the additional curvatures due to bending and twisting of the shell wall. By a similar procedure as for plates, the same moment-curvature relations are obtained:

$$M_{xx} = D(\kappa_{xx} + \nu\kappa_{yy}) \qquad M_{yy} = D(\kappa_{yy} + \nu\kappa_{xx}) \qquad M_{xy} = D(1-\nu)\kappa_{xy} \quad (7.6.1)$$

where M_{xx}, M_{yy}, and M_{xy} are the bending moment and the twisting moment, and $D = Eh^3/12(1-\nu^2)$. For small displacements u, v, and w in the x-, y-, and z-directions, the curvature changes may be calculated as

$$\kappa_{xx} = w_{,xx} \qquad \kappa_{yy} = w_{,yy} + \left\{\frac{w}{R^2}\right\} \qquad \kappa_{xy} = w_{,xy} = \left\{\frac{1}{2}\frac{v_{,x}}{R}\right\} \quad (7.6.2)$$

The terms in braces are different from plates, and represent small corrections that can often be neglected unless the curvature radius R is too small. The term w/R^2 is the same as already derived for arches (Eq. 2.8.2). The term $v_{,x}/2R$ is due to the fact that a displcement, v, in the direction tangent to the middle surface must be accompanied by a rotation of the shell element: see Figure 7.19, which shows that the contribution of v to twisting curvature is $2\Delta\kappa_{xy} = d\phi/dx = (dv/R)/dx = v_{,x}/R$. For shallow shells, the additional curvature due to R, as written in braces in Equations 7.6.2, can be neglected.

The in-plane deformations at the middle surface are described, according to Hooke's law, by the relations:

$$N_{xx} = C(\varepsilon_{xx} + \nu\varepsilon_{yy}) \qquad N_{yy} = C(\varepsilon_{yy} + \nu\varepsilon_{xx}) \qquad N_{xy} = \frac{1-\nu}{2}C\gamma_{xy} \quad (7.6.3)$$

in which $\gamma_{xy} = 2\varepsilon_{xy}$ = shear angle, and $C = Eh/(1-\nu^2)$. The in-plane strains at the middle surface may be calculated the same way as for plates:

$$\varepsilon_{xx} = u_{,x} + [\tfrac{1}{2}w_{,x}^2] \qquad \varepsilon_{yy} = v_{,y} + \frac{w}{R} + [\tfrac{1}{2}w_{,y}^2] \qquad \gamma_{xy} = u_{,y} + v_{,x} + [w_{,x}w_{,y}] \quad (7.6.4)$$

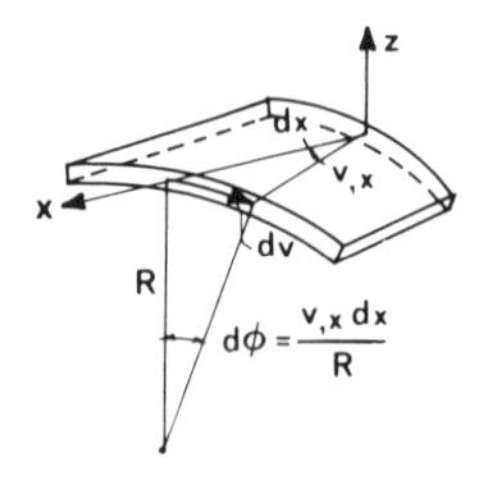

Figure 7.19 Twisting curvature generated by tangential displacement.

in which the terms in brackets are nonlinear and second-order small. They have already been derived in Section 7.2. Similar to plates, these terms must be considered in the potential-energy expression because the potential energy must be expressed accurately up to all the terms of second order. These terms, however, may normally be neglected in formulating the equilibrium equations for critical loads, since these equations should yield a linear eigenvalue problem, just like for plates.

The bending and twisting moments and the transverse shear forces Q_x and Q_y (which rotate with the cross section, see Sec. 1.7) are related, according to the moment equilibrium conditions, which read $M_{xx,x} + M_{xy,y} = Q_x$, $M_{yy,y} + M_{xy,x} = Q_y$. The shear forces Q_x and Q_y are assumed to rotate with the normal of the middle surface and thus they are analogous to the shear force Q that we introduced in Section 1.7 for columns with shear. They differ from the fixed direction shear forces V_x and V_y that are analogous to shear force V in beam-columns (Sec. 1.3). According to the condition of equilibrium of the forces normal to the shell, we also have $Q_{x,x} + Q_{y,y} + p + p_N = 0$ in which p is the applied normal distributed load and p_N is the transverse resultant (per unit area of shell) due to the membrane action, that is, due to the out-of-plane components of the in-plane forces in the curved middle surface of the shell. Combination of the foregoing equilibrium relations yields the differential equilibrium condition:

$$M_{xx,xx} + 2M_{xy,xy} + M_{yy,yy} = p + p_N \tag{7.6.5}$$

The transverse distributed force due to the membrane action has two components: one is due to the initial curvature, $1/R$, which equals $-N_{yy}/R$ according to Section 7.5 (Fig. 7.15b); and another one, which is due to the additional curvatures w_{xx}, w_{xy}, w_{yy}, is the same as previously derived for plates (Eq. 7.2.15a or 7.2.24). Therefore we have

$$p_N = -\frac{N_{yy}}{R} + (N_{xx}w_{,xx} + 2N_{xy}w_{,xy} + N_{yy}w_{,yy}) \tag{7.6.6}$$

The terms in the parentheses are a source of nonlinearity since the in-plane forces $N_{xx}, \ldots$ are in general unknown and variable. However, unless R is very large, these terms are higher-order small for the initial deflections for which $w_{,xx}$, $w_{,yy}$, and $w_{,xy}$ are infinitesimal.

The in-plane differential conditions of equilibrium may approximately be written in the usual form:

$$N_{xx,x} + N_{xy,y} = 0 \qquad N_{yy,y} + N_{xy,x} = 0 \tag{7.6.7}$$

since the term Q_y/R in the second of Equations 7.6.7 may be neglected due to the hypothesis that the shell is shallow.

Equations 7.6.1 to 7.6.7 represent all the necessary basic equations. The shell surface geometry and boundary conditions always have imperfections that induce initial bending deflections, and consequently there is no perfect initial loading path for which the transverse deflections would remain zero up to some critical load. Rather, the deflections increase continuously from the start of loading. Thus $N_{xx}, \ldots$ as well as w, u, v vary simultaneously, and so the resulting system of equations is nonlinear.

The solution can be simplified by the introduction of the Airy stress function $F(x, y)$, such that $N_{xx} = F_{,yy}$, $N_{yy} = F_{,xx}$, $N_{xy} = -F_{,xy}$. Substitution into the in-plane differential equilibrium conditions (Eqs. 7.6.7) then shows them to be automatically satisfied. The in-plane compatibility condition for strains ε_{xx}, ε_{yy}, γ_{xy}, which was already used (with $R \to \infty$) for deriving the von Kármán–Föppl equations (Eqs. 7.4.4–7.4.5), reads

$$\varepsilon_{xx,yy} + \varepsilon_{yy,xx} - \gamma_{xy,xy} = w^2_{,xy} - w_{,xx}w_{,yy} + \frac{1}{R} w_{,xx} \tag{7.6.8}$$

Expressing first ε_{xx}, ε_{yy}, and γ_{xy} from Hooke's law (Eqs. 7.6.3) and then N_{xx}, N_{yy}, N_{xy} in terms of F, we may obtain from Equation 7.6.8

$$\frac{1}{Eh} \nabla^4 F = w^2_{,xy} - w_{,xx}w_{,yy} + \frac{1}{R} w_{,xx} \tag{7.6.9a}$$

A second differential equation results by substituting Equations 7.6.2 (in which the terms in braces are neglected) into Equations 7.6.1, then Equations 7.6.1 into Equation 7.6.5, and finally Equation 7.6.6 (expressed in terms of Airy stress function) into Equation 7.6.5;

$$D\nabla^4 w = p + F_{,yy}w_{,xx} - 2F_{,xy}w_{,xy} + F_{,xx}w_{,yy} - \frac{1}{R} F_{,xx} \tag{7.6.9b}$$

Equations 7.6.9a, b were derived by Donnell (1934). Note that their special case for $R \to \infty$ is the von Kármán–Föppl equations for nonlinear buckling of plates.

Donnell's Equation

Let us now examine the effect of small displacement increments starting from the initial state F^0, w^0. We replace w in Equation 7.6.9b with $w^0 + w$ and F with $F^0 + F$ where w and F now denote small increments. Then we subtract the same equation written for the initial state w^0 and F^0 with p considered as constant and we neglect the higher-order terms. In this manner we obtain

$$D\nabla^4 w = F^0_{,yy}w_{,xx} + F_{,yy}w^0_{,xx} + F^0_{,xx}w_{,yy} + F_{,xx}w^0_{,yy} - 2F^0_{,xy}w_{,xy} - 2F_{,xy}w^0_{,xy} - \frac{1}{R} F_{,xx} \tag{7.6.10a}$$

In a similar manner, we obtain from Equation 7.6.9a

$$\frac{1}{Eh} \nabla^4 F = 2w^0_{,xy}w_{,xy} - w^0_{,xx}w_{,yy} - w^0_{,yy}w_{,xx} + \frac{1}{R} w_{,xx} \tag{7.6.10b}$$

Now, assuming that the initial state is a membrane state for which the shell is a perfect cylinder, we may substitute $w^0_{,xx} = w^0_{,yy} = w^0_{,yy} = 0$ (i.e., neglect initial curvatures). This reduces Equations 7.6.10a, b to

$$D\nabla^4 w = F^0_{,yy}w_{,xx} + F^0_{,xx}w_{,yy} - 2F^0_{,xy}w_{,xy} - \frac{1}{R} F_{,xx} \tag{7.6.11a}$$

$$\frac{1}{Eh} \nabla^4 F = \frac{1}{R} w_{,xx} \tag{7.6.11b}$$

Then, applying the operator ∇^4 to Equation 7.6.11a and noting that, according to Equation 7.6.11b, $\nabla^4 F_{,xx}/R = w_{,xxxx}Eh/R^2$, we obtain

$$D\nabla^8 w - \nabla^4(N^0_{xx}w_{,xx} + 2N^0_{xy}w_{,xy} + N^0_{yy}w_{,yy}) + \frac{Eh}{R^2}w_{,xxxx} = 0 \qquad (\text{if } w^0_{ij} = 0) \tag{7.6.12}$$

in which $\nabla^8 = (\nabla^4)^2 = (\nabla^2)^4$ where ∇^2 = the Laplacian. After solving the incremental deformations from Equation 7.6.12, in which N^0_{ij} are considered constant, one can solve F from Equation 7.6.11b, then solve strains from Equations 7.6.3 and finally calculate the displacements from strains.

Equation 7.6.12 is the famous linearized Donnell equation for shallow shells (1934). Its derivation, which we presented here (Bažant, 1985), is considerably simpler than the direct derivation previously presented in textbooks. It shows that derivation of linear differential equations by linearization of the nonlinear ones may be simpler than deducing the linearized equations directly.

Equation 7.6.12, which is valid only incrementally when $w^0_{,ij} = 0$, can be used to solve critical loads. It should be noted that, for general shells, buckling from an undeflected initial state ($w^0 = 0$) does not exist. Therefore, Equation 7.6.12 may be used only if the prebuckling curvature changes are negligible.

Axially Compressed Cylindrical Shell

Let us now apply Donnell's theory to calculate the critical load of an axially compressed cylindrical shell. We assume that the shell is initially in a perfect unbuckled membrane state, carrying an axial distributed force N^0_{xx} with $N^0_{yy} = N^0_{xy} = 0$, and $p = 0$. Then the shell buckles in a doubly periodic surface of alternating buckles assumed to have the form:

$$w = w_0 \cos \alpha x \cos \beta y \tag{7.6.13}$$

in which w_0, α, β = arbitrary constants, with $\alpha = \pi/L$, $\beta = \pi/b$, L and b being the half-wavelength of the buckles in the longitudinal and circumferential direction, respectively. Substitution of Equation 7.6.13 into Donnell's equation (Eq. 7.6.12) indicates that this equation is satisfied for any w_0 when

$$-N^0_{xx} = -N^{cr}_{xx} = \frac{Eh}{\pi^2 R^2 L^2}\left(\frac{1}{L^2} + \frac{1}{b^2}\right)^{-2} + \pi^2 L^2 D\left(\frac{1}{L^2} + \frac{1}{b^2}\right)^2 \tag{7.6.14}$$

According to the assumption of the quasi-shallow shell theory, this result is, of course, correct only if L and b are sufficiently smaller than the radius R.

One can also verify that if the expressions $u = u_0 \sin \alpha x \cos \beta y$ and $v = v_0 \cos \alpha x \sin \beta y$ are substituted into Equations 7.6.4 (the terms in brackets being neglected), and if $N_{xx}, \ldots$ are calculated from Equations 7.6.3, the trigonometric functions in Equations 7.6.7 cancel out.

For $\beta \to 0$ ($b \to \infty$) or $\cos \beta y \to 1$, deflections $w(x, y)$ according to Equation 7.6.13 are axisymmetric. Indeed Equation 7.6.14 reduces for this case to Equation 7.5.2 that we obtained earlier.

For doubly periodic modes one needs to find the minimum of Equation 7.6.14 with respect to variables L and b, subjected to the condition that suitable boundary conditions are satisfied. The simplest boundary conditions are simple supports at the ends of the tube. Obviously, they require an integer number of half-waves of length L over the length l of the shell. Along the circumference, there must also be an integer number of half-waves of length b. Minimization of Equation 7.6.14 under these constraints is not a straightforward matter; for details consult, for example, Calladine (1983, p. 485) or Chajes (1974, p. 312).

For a long tube and a high number of buckles along the circumference, unconstrained minimization of $-N_{xx}^{cr}$ (Eq. 7.6.14) with respect to L and b yields, curiously, the same critical value as Equation 7.5.3 for axisymmetric buckling, that is,

$$-N_{xx_{\min}}^{cr} = \frac{1}{\sqrt{3(1-\nu^2)}}\left(\frac{Eh^2}{R}\right) \qquad \text{if } Z \geq 2.85 \text{ where } Z = \frac{l^2}{Rh}\sqrt{1-\nu^2} \qquad (7.6.15)$$

Parameter Z, called the Batdorf parameter (1947), indicates the range over which this equation is a good approximation for the minimum critical load obtained by proper constrained optimization. For short cylinders, particularly for $Z < 2.85$, the approximate minimum is (Chajes, 1974, p. 314)

$$-N_{xx_{\min}}^{cr} \simeq \left(1 + \frac{12Z^2}{\pi^4}\right)\frac{\pi^2 D}{l^2} \qquad (7.6.16)$$

For this case one finds that there is only one half-wavelength over the entire circumference, that is, $b = 2\pi R$. This means that the assumptions of quasi-shallow shell theory are not satisfied, and so the solution cannot be accurate.

Further it may be noted that for a very short cylindrical shell ($Z \to 0$), the critical load expression approaches that for an infinitely wide rectangular plate, as may have been expected (this is not true of the axisymmetric solution, Eq. 7.5.3).

The most remarkable fact about the foregoing results is that the axially symmetric mode (Sec. 7.5) and the doubly periodic mode yield the same or nearly the same critical load. Furthermore, among the doubly periodic modes, one can usually find a number of combinations of half-wavelengths L and b that yield nearly the same critical load, with critical stresses differing by not over 2 percent of minimum N_{xx}^{cr} (Calladine, 1983, p. 493). This fact, which is a typical characteristic of shells and distinguishes them from other buckling problems we have discussed so far, has a major effect on the postcritical behavior.

We have already seen from some simple examples in Section 4.8 that the postcritical response is profoundly altered when two different buckling modes are associated with the same or nearly the same critical load. We saw (Fig. 4.55) that such a combination of modes may conspire to produce a softening response after the critical state. This is precisely what happens for shells and explains their strong sensitivity to imperfections. Thus we have an explanation why experiments generally yield failure loads that are much smaller than the critical load (except when Batdorf parameter Z is small). This phenomenon, to which we will return later (in Sec. 7.7), is illustrated in Figure 7.20a, which compares theoretical and experimental results (reported by Gerard and Becker, 1957b) for cylindrical shells subjected to axial compression. The theoretical results (taken from Brush and Almroth, 1975, p. 183) refer to cylinders with clamped ends, but the experimental results do not generally correspond to these boundary conditions.

It should be noted that our solution is inapplicable to very long shells, which buckle like a column, the circular cross sections remaining almost undeformed.

Such overall buckling cannot be predicted by Donnell's theory. This theory applies only to local buckling, that is, to situations where the half-wavelength of the buckles is small compared with the dimensions of the shell.

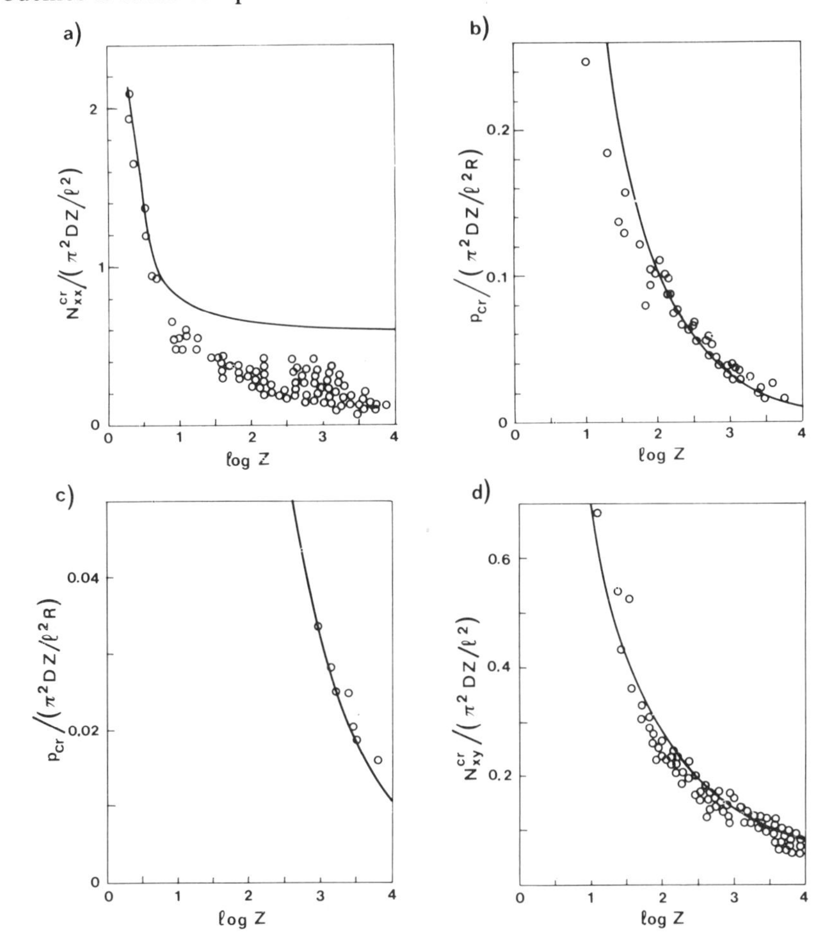

Figure 7.20 Comparison of theoretical and experimental results for cylindrical shells: (a) axial compression; (b) hydrostatic pressure; (c) lateral pressure; (d) torsion. (Adapted from Brush and Almroth, 1975.)

Effect of Lateral Pressure on Cylindrical Shells

The remarkable property of existence of many modes with nearly the same critical stress is lost when a lateral load p is superimposed (Fig. 7.21), producing nonzero N_{yy}. We will now examine the case in which the cylindrical shell is assumed to have initially a perfectly cylindrical form and be initially in a membrane state carrying initial in-plane forces N_{xx} and N_{yy}. An important special case is the loading by hydrostatic pressure, p, for which $N_{xx} = \frac{1}{2}N_{yy}$ and $N_{yy} = -pR$. We may assume for this loading the same doubly periodic displacements as we did in Equation 7.6.13. Substitution into Donnell's equation (Eq. 7.6.12) yields an algebraic relation from which one can solve the critical value of

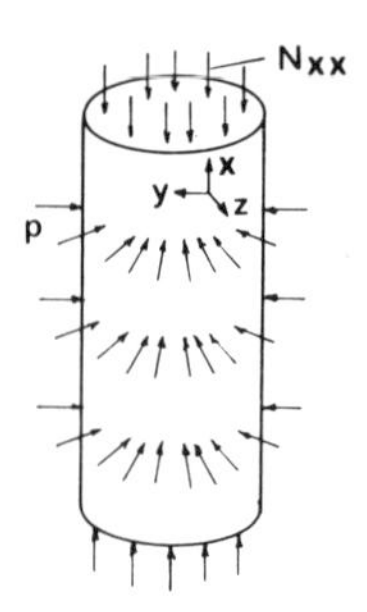

Figure 7.21 Combined normal axial forces and lateral pressure on cylindrical shell.

the transverse (hydrostatic) pressure:

$$p_{cr} = \frac{1}{R}\left[\frac{(k^2+n^2)^4(D/R^2)+k^4Eh}{(k^2+n^2)^2(n^2+\rho k^2)}\right] \qquad \left(\rho = \frac{N_{xx}}{N_{yy}},\ k = \alpha R = \frac{m\pi R}{l},\ n = \beta R\right) \tag{7.6.17}$$

in which $m = l/L$ = number of *half-wavelengths* along the length and $n = \pi R/b$ = number of *wavelengths* along the circumference.

For the case of hydrostatic pressure ($\rho = \frac{1}{2}$), Figure 7.20b shows the plot of $\bar{p}/Z$ as a function of the Batdorf parameter Z; $\bar{p}$ is the critical value of the nondimensional parameter $\bar{p} = l^2Rp/\pi^2D$. The same diagram also shows the experimental results reported by Gerard and Becker (1957b).

For the special case when only transverse pressure p is applied ($\rho = 0$), the smallest value of p_{cr} for any small n is found to correspond to $m = 1$, that is, a single half-wave along the length of the shell. Equation 7.6.17 may be rewritten for this case in the nondimensional form:

$$\bar{p}_{cr} = \frac{(1+\bar{n}^2)^2}{\bar{n}^2} + \frac{1}{\bar{n}^2(1+\bar{n}^2)^2}\left(\frac{12}{\pi^4}\right)Z^2 \tag{7.6.18}$$

in which we introduce a further nondimensional parameter, $\bar{n} = nl/\pi R$. Although n is not continuous, an approximate minimum can nevertheless be obtained by analytical minimization with respect to $\bar{n}$. This yields $\bar{p}_{cr} = f(Z)$ = a certain function of parameter Z. The values of $\bar{p}_{cr}/Z$ are plotted in Figure 7.20c, together with experimental results taken from Gerard and Becker (1957b).

The buckling modes produced by an internal vacuum in cylindrical shells of various dimensions are illustrated by the teaching models in Figure 7.22 (Northwestern University, 1969).

For lateral pressure different buckling modes have very different critical loads, as already mentioned. In keeping with the previous remarks, this means that the postcritical behavior is not expected to involve a steep softening, and experience confirms that. Therefore, the cylindrical shell is not highly imperfection sensitive for this type of loading. This is confirmed by the experimental data for transversely loaded cylindrical shells. Indeed, they yield failure loads that are quite close to the present theoretical critical loads (Eq. 7.6.18), in sharp contrast with the data for axial loading. Thus, for transverse pressure loading the linearized Donnell's theory is directly applicable.

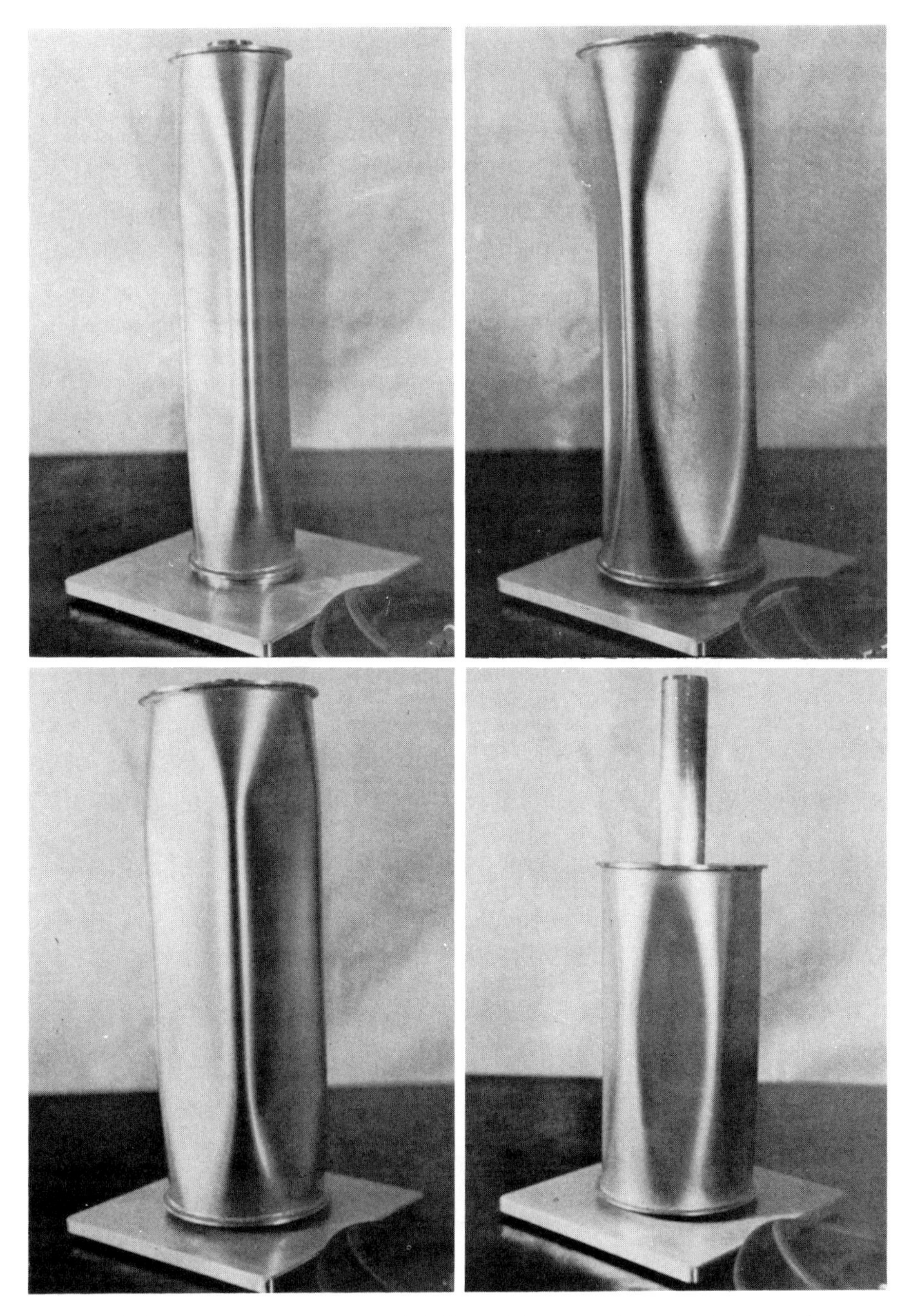

Figure 7.22 Northwestern University (1969) teaching models for buckling produced by an internal vacuum in cylindrical shells of various dimensions.

It may also be noted that the foregoing conclusion agrees with what is known about the buckling of high arches (Sec. 2.8). In fact, when the cylindrical shell is very long or infinite, the present solution for transverse pressure must approach the solution for a circumferential strip of the shell analyzed as an arch or ring.

Cylindrical Shell Subjected to Torsion

Another case where the linearized Donnell's theory yields results that are in good agreement with tests of failure loads is the case of torsional buckling of a cylindrical shell (Fig. 7.23). For this case, the displacements may be assumed to be of the form

$$w = w_0 \sin(\beta y - \alpha x) \tag{7.6.19}$$

in which w_0, α, β = constants. This deformation mode represents buckle strips that form a helix on the cylinder. Substituting Equation 7.6.19 into Donnell's equation (Eq. 7.6.12), we find that the trigonometric functions cancel out, and the solution of the characteristic equation then yields

$$N_{xy}^{\mathrm{cr}} = \frac{Dn^2}{2R^2k}(1+k^2)^2 + \frac{Ehk^3}{2n^2(1+k^2)^2} \tag{7.6.20}$$

Here $n = \beta R$ = number of wavelengths along the circumference and $k = \alpha/\beta$ = slope of the helix relative to the generating circle of the cylindrical surface.

Our initially assumed deflection $w = w_0 \sin(\beta y - \alpha x)$ does not satisfy any simple boundary conditions at the shell end, and so we must assume that our assumed deflection surface pertains to the central portion of a long shell that is little affected by the end conditions. In such a case the slope k is arbitrary. It may be shown that the minimum of N_{xy} occurs for a relatively small slope k, and so $1 + k^2$ may be approximately replaced with 1. This yields $N_{xy}^{\mathrm{cr}} = [(Dn^2/R^2k) + (Ehk^3/n^2)]/2$, and minimization of this expression with respect to k then yields

$$N_{xy_{\min}}^{\mathrm{cr}} = 0.136nEh\left(\frac{h}{R}\right)^{3/2}(1-\nu^2)^{-3/4} \tag{7.6.21}$$

The minimum corresponds to $k = n(h/6R)^{1/2}$. From this we can now see that for a thin shell the value of k^2 is indeed small compared to 1.

The minimum of Equation 7.6.21 occurs for $n = 1$, that is, for a single buckle along the circumference. However, a single buckle is usually unacceptable. This is realized upon considering the in-plane displacements, which are given by the

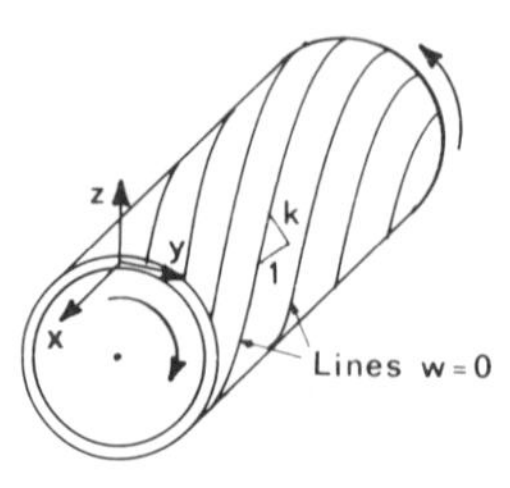

Figure 7.23 Torsional buckling of a cylindrical shell.

expressions $u = u_0 \cos(\beta y - \alpha x)$ and $v = v_0 \cos(\beta y - \alpha x)$ (it may be verified that these expressions, together with Eq. 7.6.19, satisfy the in-plane differential equilibrium equations; see Eqs. 7.6.26 given later). Now we may note that these expressions imply an antisymmetric distribution of the axial displacements, u, which can occur only if the end cross sections of the shells are permitted to rotate about an axis normal to the shell axis. The usual boundary conditions prevent this, and then the smallest possible number of buckles along the circumference is $n = 2$, which is the value to be substituted in Equation 7.6.21.

Near the ends of the shell, the value $n = 2$ is also too small, since long buckles along the circumference cannot form for the usual boundary conditions of a simple support or a fixed edge. Many short buckles arise near the ends of the shell.

It may be also noted that Equation 7.6.21 for $n = 2$ gives a value that is $\sqrt{\frac{4}{3}}$ times larger than the value given by Timoshenko and Gere (1961). This small discrepancy is due to our neglect of the effect of circumferential bending, which is appreciable for small n.

The values of the nondimensional critical shear force are plotted in Figure 7.20d as a function of the Batdorf parameter Z. Also shown in the figure are the test results according to the data reported by Gerard and Becker (1957b). As we see, the prediction of the linearized Donnell's theory is in this case in acceptable agreement with the measured failure loads of tubes subjected to torsion.

Variational Derivation from Potential Energy

The differential equations for shallow shells may, of course, be also obtained by variational calculus from the principle of stationary potential energy. Similar to Equation 7.2.5 for plates, the potential energy of a shallow shell may be expressed as $\Pi = U_m + U_b - W$ in which $W = -\iint_A pw\,dx\,dy$ = work of loads, U_m = strain energy due to in-plane stretching, that is, membrane action, and U_b = strain energy due to bending of the shell. For isotropic materials we then have

$$U_m = \frac{Eh}{2(1-\nu^2)} \iint_A [\varepsilon_{xx}^2 + \varepsilon_{yy}^2 + 2\nu\varepsilon_{xx}\varepsilon_{yy} + 2(1-\nu)\varepsilon_{xy}^2]\,dx\,dy \qquad (\varepsilon_{xy} = \tfrac{1}{2}\gamma_{xy}) \tag{7.6.22}$$

$$U_b = \frac{D}{2} \iint_A [\kappa_{xx}^2 + \kappa_{yy}^2 + 2\nu\kappa_{xx}\kappa_{yy} + 2(1-\nu)\kappa_{xy}^2]\,dx\,dy \tag{7.6.23}$$

One may now substitute Equations 7.6.4 and 7.6.2 for the in-plane strains and bending curvatures to obtain the potential-energy expression in terms of u, v, w. Note that, in contrast to our previous derivation of the differential equations of equilibrium, one may not neglect in this expression the second-order small term in the brackets of Equations 7.6.4. The reason is that the dominant terms in the potential-energy expression are already second-order small.

The nonlinear differential equations of equilibrium may be obtained as the Euler equations from the integrand of the potential-energy expression in Equations 7.6.22 and 7.6.23 (see Brush and Almroth, 1975, p. 149). The linearized differential equations of equilibrium at constant in-plane forces can also be directly obtained from the potential energy if the work of $N_{xx}, \ldots$ on the in-plane strains due to transverse deflection is included in Equation 7.6.22.

Cylindrical Shell Panels

The foregoing analysis of complete cylindrical shells may be readily extended to cylindrical shell panels limited by lines $x = \text{const.}$ and $y = \text{const.}$ The difference is that the number of half-waves in the circumferential direction must match the arc length B of the panel (Fig. 7.24) rather than the circumference $2\pi R$.

General Quasi-Shallow Shells

For shells of general shape, derivation of the differential equations from the potential-energy expression is particularly convenient. The shape of a general shell may be characterized by introducing lines-of-curvature coordinates x and y and specifying the distances along these coordinate lines by the relations $ds_x = A\,dx$, $ds_y = B\,dy$ (Fig. 7.25) in which A, B are called the Lamé coefficients. The shell surface may then be characterized by global Cartesian coordinates $X(x, y)$, $Y(x, y)$, $Z(x, y)$. From the relation $ds_x^2 = (X_{,x}\,dx)^2 + (Y_{,x}\,dx)^2 + (Z_{,z}\,dx)^2$ and a similar relation for ds_y it follows that (e.g., Novozhilov, 1953)

$$A = (X_{,x}^2 + Y_{,x}^2 + Z_{,x}^2)^{1/2} \qquad B = (X_{,y}^2 + Y_{,y}^2 + Z_{,y}^2)^{1/2} \tag{7.6.24}$$

The in-plane strains, rotations ϕ_x and ϕ_y of the normals, and the bending and twisting curvatures of shallow shells may then be generally calculated as (e.g., Brush and Almroth, 1975, p. 195)

$$\begin{gathered}
\varepsilon_{xx} = \frac{u_{,x}}{A} + \frac{A_{,y}v}{AB} + \frac{w}{R_x} \qquad \varepsilon_{yy} = \frac{v_{,y}}{B} + \frac{B_{,x}u}{AB} + \frac{w}{R_y} \\
\varepsilon_{xy} = \frac{v_{,x}}{A} + \frac{u_{,y}}{B} - \frac{B_{,x}v + A_{,y}w}{AB} + \phi_x\phi_y \\
\phi_x = -\frac{w_{,x}}{A} + \frac{u}{R_x} \qquad \phi_y = -\frac{w_{,y}}{B} + \frac{v}{R_y} \\
\kappa_{xx} = \frac{\phi_{x,x}}{A} + \frac{A_{,y}\phi_y}{AB} \qquad \kappa_{yy} = \frac{\phi_{y,y}}{B} + \frac{B_{,x}\phi_x}{AB} \\
2\kappa_{xy} = \frac{\phi_{y,x}}{A} + \frac{\phi_{x,y}}{B} - \frac{A_{,y}\phi_x + B_{,x}\phi_y}{AB}
\end{gathered} \tag{7.6.25}$$

in which R_x and R_y are the principal radii of curvature.

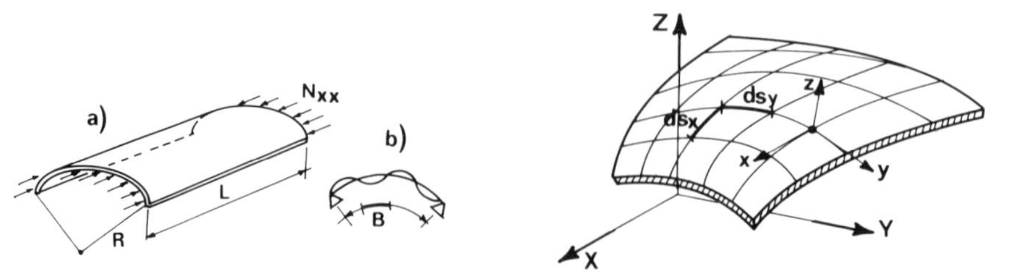

Figure 7.24 Buckling of a cylindrical panel.

Figure 7.25 General shell element.

Calculating the potential energy from these expressions, and proceeding according to the calculus of variations, one can then obtain the differential equations of equilibrium for a general shallow or quasi-shallow shell.

Incremental linearization of Equations 7.6.25 and quadratic approximation of the second variation $\Delta^2\Pi$ according to Equations 7.6.22 and 7.6.23 leads to linear differential equations for general shallow shells, which are usually called the Donnell-Mushtari-Vlasov theory (Donnell, 1934; Mushtari, 1938; Vlasov, 1949; see also Brush and Almroth, 1975, p. 199). A special case of such equations are the linearized Donnell's differential equations for shallow cylindrical shells that we derived before.

Problems

7.6.1 Derive in full detail the expression for the critical axial stress resultant N_{xx} in Equation 7.6.14.

7.6.2 Consider a simply supported elastic cylindrical shell of $R/h = 200$. Calculate the exact critical axial compression force N_{xx} for the case $l/R = 4$, using Equation 7.6.14. Compare the result with the expression in Equation 7.6.15.

7.6.3 For the same cylinder as in Problem 7.6.2, but with $l/R = 0.1$, calculate the critical axial force N_{xx} using Equation 7.6.14. Compare with the expression in Equation 7.6.16.

7.6.4 Solve $(N_{xx})_{cr}$ for a simply supported cylindrical shell panel (Fig. 7.24a) of length L and arc length $B = R/3$.

7.6.5 For the same cylinder as in Problem 7.6.2, calculate the critical value of (a) external transverse pressure and (b) hydrostatic pressure.

7.6.6 Derive in full detail the expression for the critical value of the external transverse pressure in Equation 7.6.17.

7.6.7 For the same cylinder as in Problem 7.6.2, calculate the critical value of the in-plane shear force N_{xy} under torsional loading.

7.6.8 Without referring to the text, check that for a sufficiently long shell $(l/R \to \infty)$ the critical value of transverse pressure p is the same as the critical radial distributed force of a ring of thickness h and a unit width.

7.6.9 Check that if l/R as well as h/R is sufficiently small, and the shell ends are simply supported, then (a) $(N_{xx})_{cr}$ can be obtained from Euler's formula for the critical load of a column; (b) $(N_{xy})_{cr}$ due to torsion agrees with the critical value of N_{xy} for a rectangular plate of sides l and $2\pi R$ (here we have only an approximate N_{xy} to go by); (c) the critical value of transverse pressure p tends to infinity; and (d) the critical value of hydrostatic pressure is $p_{cr} = 2\pi R N_{xx}/\pi R^2 = 2N_{xx}/R$ where N_{xx} is the same as for case (a).

7.6.10 Calculate the critical values of N_{xx} for various values of the ratio $\gamma = hp/N_{xx}$ where p is either (a) transverse pressure or (b) hydrostatic pressure. Plot the envelope of critical states (interaction diagram) for $0 \le \gamma \le 2$.

7.6.11 Starting from Equations 7.6.1 to 7.6.7, derive three differential equations of equilibrium for the x-, y-, z-direction in terms of displacements. Proceed as follows: Neglect the higher-order terms in the brackets and braces in Equations 7.6.2 and 7.6.4, then substitute Equations 7.6.4 into Equations 7.6.3 and Equations 7.6.3 into Equations 7.6.7, and also Equations 7.6.2 into

7.6.1, 7.6.1 into 7.6.5 and 7.6.6 also into 7.6.5. The result is

$$R[2u_{,xx} + (1 - \nu)u_{,yy} + (1 + \nu)v_{,xy}] + 2\nu w_{,x} = 0 \tag{7.6.26a}$$

$$R[2v_{,yy} + (1 - \nu)v_{,xx} + (1 + \nu)u_{,xy}] + 2w_{,y} = 0 \tag{7.6.26b}$$

$$D\nabla^4 w - N_{xx}w_{,xx} - N_{yy}(w_{,yy} - R^{-1}) - 2N_{xy}w_{,xy} = p \tag{7.6.26c}$$

Note: Problems in which the boundary conditions need to be specified in terms of u and v must, of course, be solved on the basis of these equations rather than Donnell's equation.

7.6.12 Derive Donnell's equation in the following manner (which is the procedure used in the previous works, e.g., Brush and Almroth, 1975; or Chajes, 1974). Apply operator $\partial^2/\partial x\, \partial y$ to Equation 7.6.26b and operators $\partial^2/\partial x^2$ and $\partial^2/\partial y^2$ (separately) to Equation 7.6.26a. Then algebraically rearrange these equations to the form:

$$R\nabla^4 w = \nu w_{,xxx} - w_{,xyy} \tag{7.6.27}$$

Apply now the operator $\partial^2/\partial x\, \partial y$ to Equation 7.6.26a and the operators $\partial^2/\partial x^2$ and $\partial^2/\partial y^2$ to Equation 7.6.26b. Then algebraically rearrange the resulting three equations to yield

$$R\nabla^4 v = (\nu + 2)w_{,xxy} + w_{,yyy} \tag{7.6.28}$$

Apply the operator ∇^4 to Equation 7.6.26c, and then substitute into the resulting equation the derivative of Equation 7.6.26a with respect to x and the derivative of Eq. 7.6.26b with respect to y. This finally gives Donnell's equation.

7.6.13 Check that if Equation 7.6.13 and the expressions $u = u_0 \sin \alpha x \cos \beta y$ and $v = v_0 \cos \alpha x \sin \beta y$ are substituted into Equations 7.6.26a, b, c, then the trigonometric functions cancel out. Check also that by setting the determinant of the linear equations for the unknowns u_0, v_0, w_0 equal to zero one obtains the same expression as in Equation 7.6.14 for the critical axial force N_{xx}.

7.6.14 Using Equations 7.6.22 and 7.6.23 for the potential energy, derive Equations 7.6.26a, b, c as the corresponding Euler equations.

7.6.15 Consider a shallow hyperbolic paraboloid shell given by the equation $z = (x/a)^2 - (y/b)^2$. Substituting this into Equations 7.6.24, find the values of the Lamé parameters A and B.

7.7 NONLINEAR ANALYSIS OF SHELL BUCKLING AND IMPERFECTIONS

As already emphasized, the classical critical load formulas obtained by linear shell analysis are in good agreement with the experimentally observed collapse loads only for some types of failure, for example, the buckling of cylindrical shells under external pressure or under torsion. For many shell problems, for example, the buckling of cylindrical shells under axial compression or bending, as well as the buckling of spherical shells, the observed collapse loads are much smaller, typically about 30 percent of the classical critical load, and sometimes as low as 10 percent of that load. Beginning with the experimental results of Robertson

(1928), Flügge (1932), and Donnell (1934), the inadequacy of the classical linearized shell theory has been demonstrated by numerous experiments (see Gerard and Becker's 1957b handbook, or the 1971 handbook of the Column Research Committe of Japan).

Reduction Factors for Classical Critical Loads

Large as it is, the error of the classical critical load formulas as estimates of the collapse load nevertheless does not render them useless, for several reasons: (1) The classical critical loads provide a useful upper bound on the collapse load. (2) They are also valuable as test cases for finite element programs (although it is quite difficult to predict these critical loads by finite elements correctly). (3) The classical critical states further yield good estimates of the length of the buckles on the shell surface, which may then be used for an approximate nonlinear analysis. (4) Finally, due to their simplicity, the classical critical load formulas serve as the basis of design specifications, in which an empirical correction based on extensive test results is introduced by a reduction factor ϕ.

As an example (e.g. Seide, 1981), the classical formula for the critical value of the axial distributed force in a cylindrical shell (Eq. 7.6.15) is modified for failure prediction as follows:

$$N_{xx}^{\mathrm{cr}} \simeq \phi \frac{1}{\sqrt{3(1-\nu^2)}}\left(\frac{Eh^2}{R}\right) \tag{7.7.1}$$

in which

$$\phi = 1 - 0.9(1 - e^{-\sqrt{R/h}/16}) \qquad \text{for } 0.5 \leq \frac{l}{R} \leq 5, \quad 100 \leq \frac{R}{h} \leq 3000 \tag{7.7.2}$$

This empirical formula for the reduction factor ϕ (also called the correction factor or the "knockdown" factor) has been determined as a lower-bound envelope of many test results. Equation 7.7.1 also applies to the critical value of the maximum normal force N_{xx} caused in the cross section by the bending moment; however, the reduction factor is, according to numerous tests, expressed as $\phi = 1 - 0.73[1 - \exp(-\frac{1}{16}\sqrt{R/h})]$.

For a combined loading by axial compression and bending, the critical value of the maximum N_{xx} in the cross section may be approximately and safely estimated, according to test results, from a linear interaction diagram:

$$\frac{N_{xx}^a}{N_{xx_{\mathrm{cr}}}^a} + \frac{N_{xx}^b}{N_{xx_{\mathrm{cr}}}^b} = 1 \tag{7.7.3}$$

in which N_{xx}^a, N_{xx}^b are maximum internal normal forces due to axial compression alone and to bending alone, and $N_{xx_{\mathrm{cr}}}^a$, $N_{xx_{\mathrm{cr}}}^b$ are their critical values. Based on tests, linear interaction diagrams also apply to other shell buckling problems, for example, the combination of axial compression and lateral pressure on a cylindrical shell.

Numerous other formulas, which are similar to Equation 7.7.1 and involve empirical reduction of the classical critical loads calculated by a linear theory, are given in various handbooks, as well as in the comprehensive review by Kollár and Dulácska (1984).

Physical Source of Postcritical Load Drop and High Imperfection Sensitivity

The reason that many shells fail at a much smaller load than the classical critical load consists in the combined effect of nonlinearity and imperfections. This was discovered by von Kármán with Dunn and Tsien (1940) (see also von Kármán and Tsien, 1941). In their revolutionizing paper, they demonstrated by approximate nonlinear analysis that, after reaching the critical states, the load can rapidly decrease at increasing deflection, that is, the structure undergoes softening. Further they demonstrated that even a small disturbance can cause the shell to jump to a postbuckling state at which the carrying capacity is greatly reduced. Shortly afterward, Koiter (in his dissertation in Dutch in 1945, which became known in the English speaking world only much later), recognized the possibility of developing a general theory of stability that asymptotically describes the immediate postcritical behavior, makes it possible to describe the effect of imperfections in general terms, and characterizes various types of postcritical imperfection sensitivity exhibited by various structures. Koiter introduced the idea of a multidimensional space whose coordinates are the load and the amplitudes of various possible buckling modes. He showed how different types of equilibrium paths can be obtained from the potential-energy expression by taking its first partial derivatives. He further showed how the stability of the subsequent equilibrium states on such paths can be decided on the basis of the second partial derivatives of the potential-energy expression.

Analysis of the role of imperfections in shell buckling was pioneered by Donnell and Wan (1950). An analysis of the effect of random imperfection near the critical state explained why, even in rather careful experiments, the measured collapse modes exhibit tremendous scatter. The classical mathematical result that a truly perfect axially compressed shell should not fail until the classical critical load based on linearized theory (Eq. 7.6.15) is reached had not been experimentally confirmed until Almroth, Holmes, and Brush (1964); Evensen (1964); Tennyson (1969); Tennyson and Muggeridge (1969); and Tennyson, Muggeridge, and Caswell (1971) did it almost half a century after the critical loads had been calculated by Lorenz (1908, 1911), Timoshenko (1910), and Southwell (1914). Tennyson's tests were made on epoxy resins shells manufactured with extraordinary precision.

The high imperfection sensitivity of shell buckling problems is caused by the fact that the polynomial expansion of the potential-energy expression, that is,

$$\Pi = a_0 + a_1 q + a_2 q^2 + a_3 q^3 + a_4 q^4 + \cdots \tag{7.7.4}$$

contains a cubic term; q = amplitude of the dominant buckling mode associated with the critical load, and $a_0, a_1, a_2, \ldots$ = coefficients that depend on the load P (or load parameter λ), as well as the type and magnitude of the imperfections. The cubic term is absent from the potential-energy expression of structures that are imperfection-insensitive and exhibit symmetric bifurcation, such as columns or plates. For shells, though, this term is nonzero and can be particularly large. In a linear stability analysis, the potential-energy expression is, of course, always quadratic. Thus the cubic (and higher-order) terms can be obtained only by nonlinear analysis.

The way the higher-order terms can influence stability was explained by von Kármán et al. (1940) by a simple and instructive example of a column restrained by a nonlinear spring; see Fig. 7.26. He assumed that the relationship of the force F and the displacement w of the spring is a quadratic curve shown in Figure 7.26b and described as $F = C_1 w - C_2 w^2$, in which C_1 and C_2 = positive constants. The potential energy of the system of two rigid bars and spring shown in Figure 7.26a may then be calculated as $\Pi = \int F\,dw - 2PL(\cos q_0 - \cos q)$ in which $\cos q_0 - \cos q \simeq (q^2 - q_0^2)/2$ and $w \simeq L(q - q_0)$. Integration yields the approximation:

$$\Pi = \tfrac{1}{2}C_1 L^2 (q - q_0)^2 - \tfrac{1}{3}C_2 L^3 (q - q_0)^3 - PL(q^2 - q_0^2) \qquad (7.7.5)$$

which is third-order accurate. The condition $\partial\Pi/\partial q = 0$ furnishes for the equilibrium paths the equation:

$$P = \left[\frac{C_1 L}{2} - \frac{C_2 L^2}{2}(q - q_0)\right]\frac{q - q_0}{q} \qquad (7.7.6)$$

From this equation it is now clear that the presence of the quadratic coefficient C_2 in the force-displacement relation of the spring causes the perfect system ($q_0 = 0$) to soften in its postcritical response [i.e., the curve $P(q)$ has a negative slope]. It is also clear from Equation 7.7.6 that the larger the imperfection q_0 the smaller is the maximum load of the perfect system (Fig. 7.26c).

As for the analogy with shells, we may now note that when a suitable buckling mode of amplitude q is introduced into the potential-energy expression in Equations 7.6.22 and 7.6.23 in which the proper nonlinear strain and curvature expressions are used, the resulting expression for the potential energy of the shell appears to be, in the simplest approximation, a cubic polynomial (analogous to Eq. 7.7.5). Since the potential-energy expression totally characterizes the response, the equilibrium path for the shell must be of the kind indicated in Figure 7.26c, as shown by von Kármán et al. (1940). Moreover, it is found that for shells the coefficient C_2 can be very large, causing the postcritical downward slope in Figure 7.26 to be rather steep or exhibit a snapback. Consequently the reduction of the carrying capacity of the shell is quite significant already at practically small deflections.

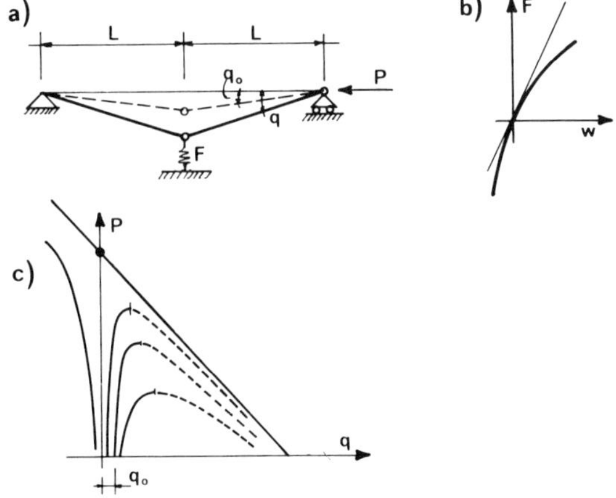

Figure 7.26 (a) Rigid-bar model with (b) nonlinear spring and (c) load-rotation curves (used by von Kármán et al., 1940).

Koiter's Laws of Imperfection Sensitivity

Koiter (1945) reached some simple, yet important, general conclusions regarding the initial asymptotic postcritical behavior. The equilibrium path of a perfect system, resulting from the condition $\partial\Pi/\partial q = 0$, may be described by the power series expansion:

$$\frac{\lambda}{\lambda_{cr}} = 1 - c_1 q - c_2 q^2 \tag{7.7.7}$$

in which q represents the amplitude of the buckling mode associated with the critical state, such that $q = 0$ at the critical state; $c_1 = c_2 =$ constants characterizing the given structure; $\lambda =$ load parameter (e.g., $\lambda = N_{xx}$ or $\lambda = P$); and $\lambda_{cr} =$ critical value of λ. Depending on the vanishing or nonvanishing of coefficient c_1, and on the sign of c_2, three basic types of behavior near the critical load can be distinguished, as shown in Figure 7.27. Koiter showed that the behavior of the perfect system near the critical state, as described by Equation 7.7.7 (or the associated potential-energy expression) completely determines the type of imperfection sensitivity.

For the type III response (Fig. 7.27c, $c_1 > 0$), Koiter showed that the maximum load parameter λ_{max} of an imperfect system is asymptotically (for small q_0) described by the formula:

$$\frac{\lambda_{max}}{\lambda_{cr}} = 1 - 2(q_0 \rho c_1)^{1/2} \tag{7.7.8}$$

in which ρ is a coefficient depending on the imperfection shape, and $q_0 =$ imperfection amplitude. This type of imperfection sensitivity is typical of spherical shells under external pressure or cylindrical shells under axial load or bending. The imperfection-sensitivity diagram according to Equation 7.7.8 is illustrated in Figure 7.27d, and it is noteworthy that its postcritical descent starts with a vertical tangent; hence, the ratio of the maximum load reduction to the magnitude of the imperfection tends to infinity as $q_0 \to 0$.

For type II behavior (Fig. 7.27b, $c_1 = 0$, $c_2 > 0$), Koiter showed the initial imperfection sensitivity to be, in general, described by the relation:

$$\frac{\lambda_{max}}{\lambda_{cr}} = 1 - 3\left[q_0\left(\frac{\rho}{2}\right)\sqrt{c_2}\right]^{2/3} \tag{7.7.9}$$

For type I behavior (Fig. 7.27a, $c_1 = 0$, $c_2 < 0$), Koiter showed the structure to be imperfection-insensitive.

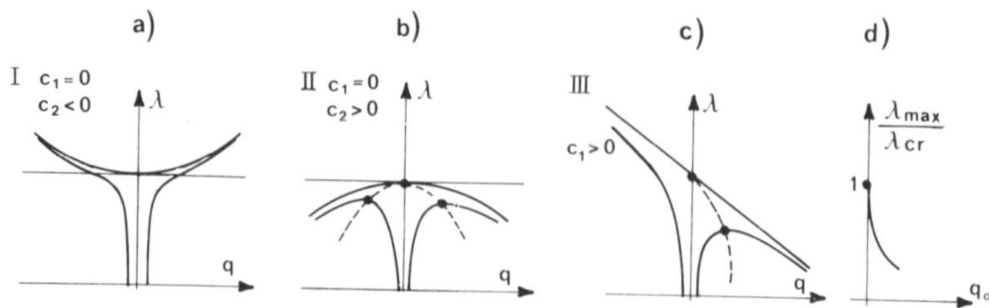

Figure 7.27 Basic types of postcritical behavior (as shown by Koiter, 1945).

Equations of the form of Equations 7.7.8 and 7.7.9 were already illustrated for simple systems in Chapter 4 (Eqs. 4.5.26 and 4.5.16). The power of Koiter's rules for imperfection sensitivity lies in their generality, that is, the fact that they are not restricted to the examples in Chapter 4. Koiter's general results were later also extended (Tvergaard, 1976) to the initial asymptotic behavior near a critical state that is associated with several buckling modes (eigenmodes).

Koiter's analysis, of course, gives only an asymptotic description of the initial postcritical behavior, and does not apply to large postcritical deflections. For this purpose, the higher-order terms of the potential-energy expression or of the equilibrium path in Equation 7.7.7 must be calculated. The typical postcritical response of imperfection-sensitive shells at large deflections is shown in Figure 7.28 for various magnitudes of the initial imperfection, characterized by parameter $\beta R/h$ where β is a certain nondimensional measure of the initial unevenness (see Popov and Medwadowski, 1981, p. 18).

Lest it be thought that a discrepancy between the measured collapse modes and the linear critical loads of shells is due solely to the effect of imperfections of the geometric shape, one should realize that even if they were zero one would still need to design for a greatly reduced maximum load. The reason is that very small dynamic disturbances, load misalignments, etc. can also cause the structure to jump over the peak of the equilibrium load-deflection curve if the postcritical softening is very steep. The ultimate cause of the high imperfection sensitivity of shells is the strong nonlinearity of their behavior.

Buckling Modes and Their Interaction

The nonlinearity of shell behavior, in turn, appears to be a consequence of the fact that, for imperfection-sensitive shells, there exist many different buckling loads associated with the same or nearly the same critical load. Consequently, the different buckling modes interact, and this interaction of modes is the culprit.

For example, in our linearized analysis of the cylindrical shell under axial compression, we have seen that the axisymmetric and many doubly periodic modes are associated with the same critical load. This means that these modes

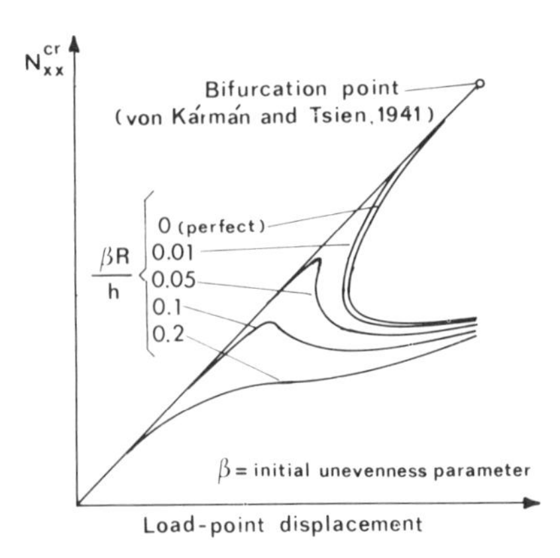

Figure 7.28 Postcritical response of axially compressed cylinders (after von Kármán and Tsien, 1941).

will interact as the critical load is approached. The way an interaction of two different buckling modes can affect the postcritical behavior was illustrated in Section 4.8. A simple and easily understandable example was the von Mises truss (arch), in which the individual bars buckle simultaneously with the snapthrough of the whole arch due to elastic shortening of the bars.

The fact that at the beginning of the postcritical response various buckling modes interact causes the buckling mode to change at larger deflections. For axially compressed cylindrical shells it is found that the initial square pattern of the buckles on the shell surface (Fig. 7.29a, Eq. 7.6.13) later changes to a diamond pattern that has ridges in two inclined directions and furrows in the direction normal to the cylinder axis. This pattern, called diamond buckles or the Yoshimura (1955) pattern (Fig. 7.29b), has one interesting geometric property: it represents an almost inextensional mapping of the shell surface, such that the combined length of all the furrows in each cross section is approximately equal to the length of the original circle. This means that, after the formation of the Yoshimura pattern, the membrane potential energy becomes negligibly small and most of the potential energy is due to bending at the ridges and the furrows.

While the Yoshimura final deformation pattern of an axially compressed cylindrical shell is almost inextensible, the shell cannot transit to this pattern from its initial shape in an extensible manner. Rather the shell must first pass through states with a relatively large membrane deformation. This observation explains why the load must first rise and then decrease in order to reach the final Yoshimura pattern.

In large postcritical deformations of axially compressed cylindrical shells, it is further observed that the buckles show a preference for growing inward rather than outward (see Fig. 7.30, taken from Liu and Lam, 1988). This is also intuitively obvious; whereas an inward deflection of each buckle on the shell surface can be accommodated through bending, an outward deflection obviously

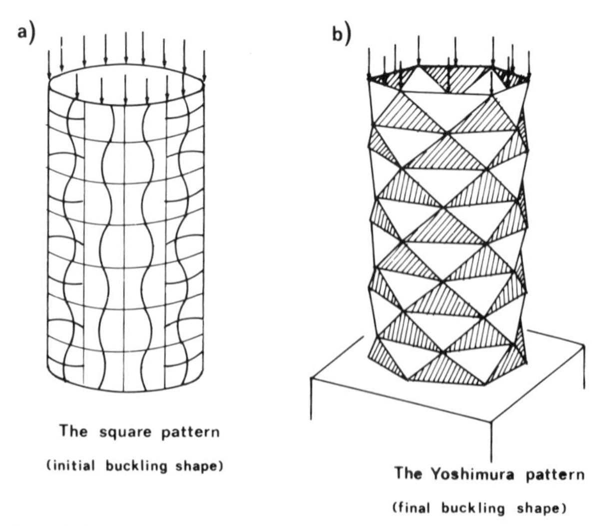

Figure 7.29 (a) Initial (small deflection) and (b) final (large deflection) buckling shapes of axially compressed cylinders.

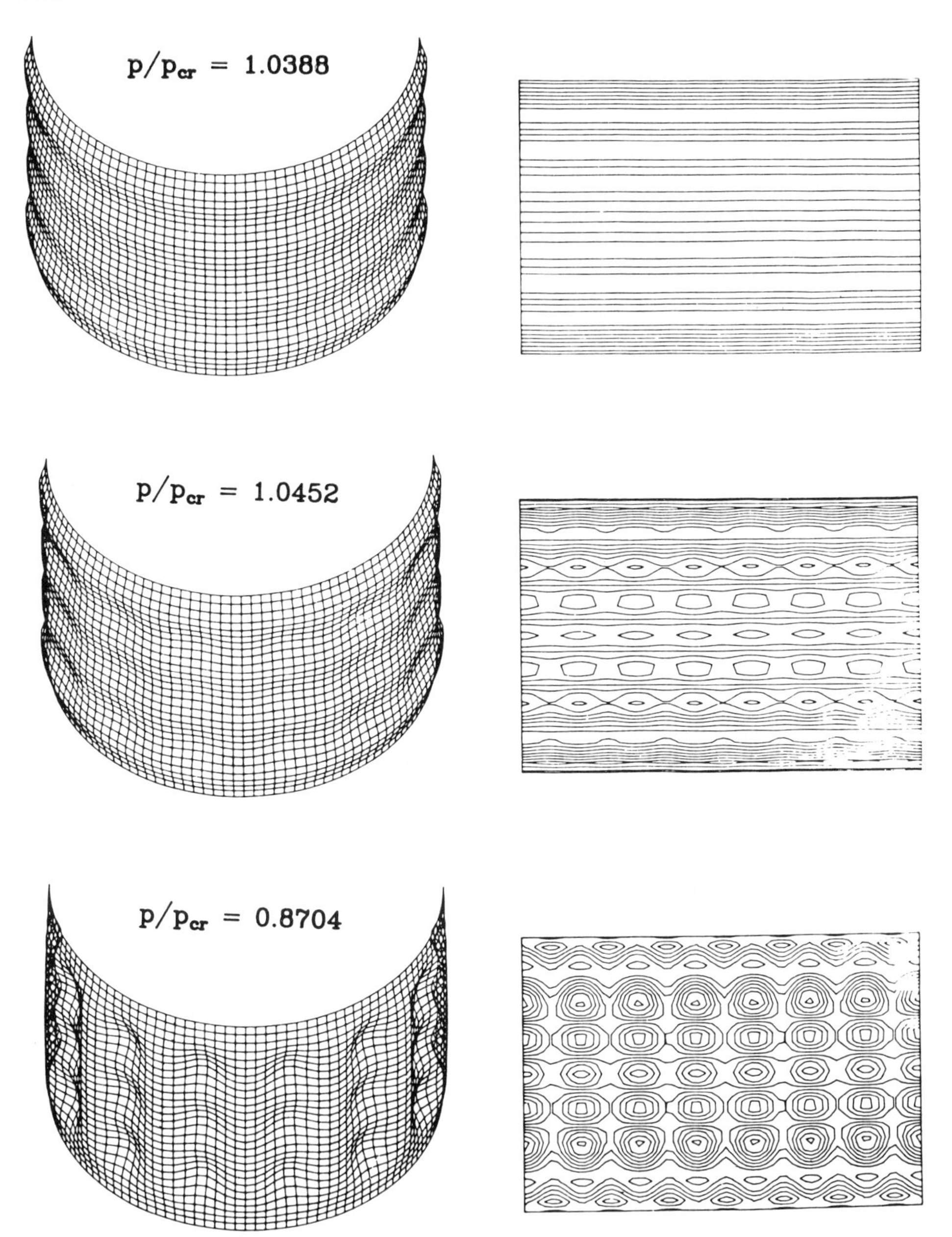

Figure 7.30 Postbuckling deformation of cylindrical shell (after Liu and Lam, 1988).

requires large membrane stretching. In a similar vein, it has been also observed that the inward buckles of the shell surface reduce their share of the transmitted axial compressive force and transfer that force to the outward buckles. Thus the share of axial compression carried by the outward buckles increases at large deflections.

Summary

Nonlinear buckling of imperfection-sensitive shells is a complex problem, but one that is presently well understood. Although the future of shell buckling analysis no doubt lies in finite element techniques, the design at present still needs to be based on the classical critical loads obtained by linear analysis and corrected on the basis of extensive experimental data. The need to use these empirical corrections is nevertheless deplorable for the theorists since it makes generalizations outside of the range of the existing experiments, and especially to different types of shells, uncertain.

Problems

7.7.1 Describe the stages of development of buckles on the surface of an axially compressed shell.

7.7.2 What is the reduction factor (Eq. 7.7.2) for the case of Problem 7.6.2?

7.7.3 Plot the interaction diagram based on Equation 7.7.3 and compare it to the interaction diagram obtained by linear analysis of the critical loads (see also Prob. 7.3.1).

7.7.4 Correlate the coefficients of Equations 7.7.8 and 7.7.9 to the examples of postcritical behavior in Section 4.5 (Figs. 4.21, 4.22, 4.23, and 4.24).

7.7.5 Try to find some simple elastic system other than the example of von Kármán that also exhibits a cubic term in the potential-energy expression, and express coefficients a_3 (Eq. 7.7.4) and c_1, c_2 for this system.

7.7.6 Without referring to the text, describe and illustrate by sketches the collapse process of an axially compressed cylindrical shell, with the formation of a Yoshimura pattern.

7.8 SANDWICH PLATES AND SHELLS

Sandwich plates or shells are layered plates or shells that have a very soft core of thickness c bonded to stiff but thin face sheets (skins) of thickness f; see Figure 1.21 in Chapter 1. Such structures, in which the skin is usually made of metal or fiber composite and the soft core is a hardened foam or has honeycomb or corrugated construction, are widely used in the aerospace industry and also find increasing applications in structural engineering.

Basic Relations for a Sandwich Plate Element

Usually one can assume that the skins carry all the bending and twisting moments and the core carries the shear forces. The principal difference from the regular plates we have analyzed so far is that the normals do not remain normal to the deflected middle surface. Thus, $u = -\phi(x, y)z$ and $v = -\psi(x, y)z$ where $\phi(x, y)$ and $\psi(x, y)$ are rotations, which, in contrast to plates, are independent of $w(x, y)$.

The normal stresses, for the case of orthotropy, are $\sigma_x = E_{xx}u_{,x} + E_{xy}u_{,y}$, $\sigma_y = E_{yy}u_{,y} + E_{yx}u_{,x}$, and the in-plane shear stresses are $\tau_{xy} = G_{xy}(u_{,y} + v_{,x})$ (cf. Eqs. 7.1.3). Substituting for u and v and integrating over the face sheet thicknesses f according to Equations 7.1.2, one gets

$$M_{xx} = D_{xx}\phi_{,x} + D_{xy}\psi_{,y} \qquad M_{yy} = D_{yx}\phi_{,x} + D_{yy}\psi_{,y} \tag{7.8.1}$$

$$M_{xy} = M_{yx} = H_{xy}(\phi_{,y} + \psi_{,x}) \tag{7.8.2}$$

where D_{xx}, D_{yy}, $D_{xy} = D_{yx}$, and H_{xy} are four constants characterizing the plate. If we assume that $f \ll c$, the rotations of the normals cause out-of-plane shear stresses $\tau_{xz} = G_{xz}\gamma_{xz} = G_{xz}(u_{,z} + w_{,x}) = G_{xz}(w_{,x} - \phi)$ and $\tau_{yz} = G_{yz}\gamma_{yz} = G_{yz}(v_{,z} + w_{,y}) = G_{yz}(w_{,y} - \psi)$. Multiplying them by thickness c yields the shear forces

$$Q_x = C_x(w_{,x} - \phi) \qquad Q_y = C_y(w_{,y} - \psi) \tag{7.8.3}$$

where C_x, C_y = constants. These shear forces rotate remaining normal to the deflection surface. Note that Equations 1.7.2 and 1.7.1, which are valid for sandwich beams if the faces are thin, are the special case of Equations 7.8.1 and 7.8.3 for cylindrical bending (i.e., for $\psi = \phi_{,y} = w_{,y} = 0$). A basic requirement is that Q_x and Q_y must vanish if the normal to the deflection surface remains orthogonal, that is, if $\phi = w_{,x}$ and $\psi = w_{,y}$; indeed Equations 7.8.3 satisfy this condition.

For isotropic plates, in analogy to Equations 7.1.4, we have $D_{xx} = D_{yy} = D$, $D_{xy} = D_{yx} = \nu D$, $H_{xy} = D(1 - \nu)$, and also $C_x = C_y = C$, where, in similarity to Equations 1.7.9 and 1.7.10 for cylindrical bending, $D = \frac{1}{2}f(c + f)^2 E_f/(1 - \nu_f^2)$ (if $f \ll c$), $C = G_c(f + c)$; $E_f = E_f$ = elastic Young's modulus of the face (skin), G_c = shear modulus of the core.

The differential equilibrium equations are the same as for any plate:

$$Q_x = -M_{xx,x} - M_{xy,y} \qquad Q_y = -M_{yy,y} - M_{xy,x} \tag{7.8.4}$$

$$-Q_{x,x} - Q_{y,y} = p + p_N \tag{7.8.5}$$

where p_N depends on $w(x, y)$ according to Equation 7.2.15a. Substituting Equations 7.8.1 to 7.8.3 we obtain a system of three differential equations for three unknown functions $w(x, y)$, $\phi(x, y)$, and $\psi(x, y)$.

A Rectangular Sandwich Plate and Other Problems

As an example consider a rectangular simply supported sandwich plate subjected to uniform N_{xx} (same as Fig. 7.7a), and $p = 0$. The solution may be sought in the form

$$w = Z \sin\frac{m\pi x}{a} \sin\frac{n\pi y}{b} \qquad \phi = X \cos\frac{m\pi x}{a} \sin\frac{n\pi y}{b} \qquad \psi = Y \sin\frac{m\pi x}{a} \cos\frac{n\pi y}{b} \tag{7.8.6}$$

where m and n are the numbers of half-waves in the x- and y-directions. Now we may note that with this choice we satisfy identically for all x and y the three differential equations in Equations 7.8.4 and 7.8.5 (after Eqs. 7.8.1–7.8.3 are substituted in them), and at the same time we fulfill the boundary conditions of simple supports, for example, $w = M_{yy} = 0$ for the edges x = const. Substitution of Equations 7.8.6 into these three differential equations yields a system of three homogeneous algebraic equations for three unknowns X, Y, Z. The determinant of this system can be considerably simplified and the condition of a vanishing determinant yields the critical value of N_{xx} as a function of m and n; see Plantema (1966). The numbers of half-waves are determined as we did it in Section 7.3. The results for the first critical value of N_{xx} are plotted in Figure 7.31, in which the curve for $1/s = 0$ corresponds to the case of infinite shear stiffness of the core. In that case, the normals remain normal. So this curve represents the solution of an ordinary plate (see Fig. 7.7b). Figure 7.31 is taken from the book by Plantema (1966). This book, as well as Allen (1969) or Kovařík and Šlapák (1973), also solves many other problems of sandwich plates and shells.

An important problem for sandwich plates and shells is the local buckling of the face sheet. The problem is similar to the beam on elastic foundation; for details see, for example, Plantema (1966). Another important problem is delamination of the faces; see, for example, Sallam and Simitses (1985, 1987).

The sandwich plate theory is the simplest and most important example of the so-called higher-order plate theories, in which the distribution of u and v is approximated by linear combinations of suitable assumed functions of z.

The formulation that we presented for shells in Section 7.6 can also be generalized for sandwich shells. In the curvature expressions, Equations 7.6.2, $w_{,xx}$, $w_{,yy}$, and $w_{,xy}$ are replaced with $\phi_{,x}$, $\psi_{,y}$, and $\phi_{,y} + \psi_{,x}$, respectively, and Equations 7.8.3 are introduced for Q_x and Q_y. Some problems, such as the critical loads for axisymmetric buckling or for cylindrical panels with a rectangular base, can be solved analytically without much difficulty (Plantema, 1966).

Problems

7.8.1 Solve the critical value of uniaxial compression N_{xx} for a square simply supported sandwich plate of side a (assuming $m = n = 1$).

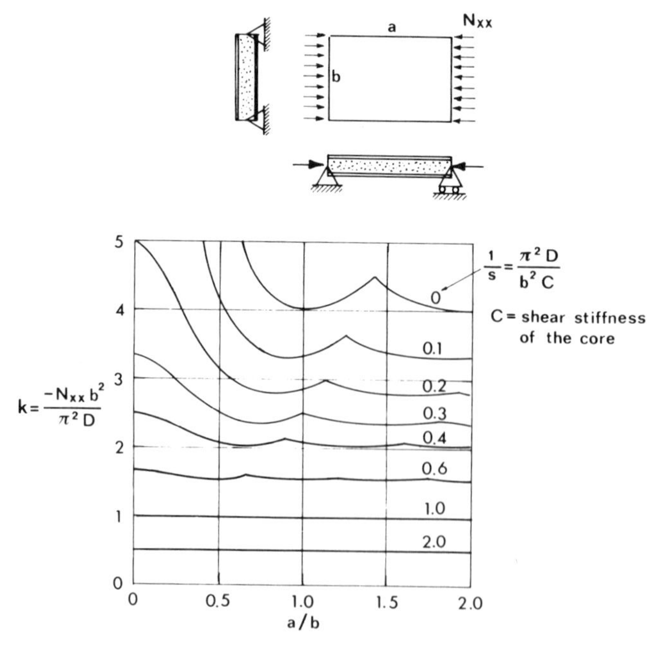

Figure 7.31 Critical value of normal compressive force for a sandwich plate (after Plantema, 1966).

7.8.2 Solve the same plate for biaxial compression with $N_{yy} = N_{xx}$.

7.8.3 Solve the same as Problem 7.8.1 but for an infinitely long plate of width b ($a \to \infty$).

7.8.4 Try to reduce the number of governing equations by eliminating some variables from Equations 7.8.1 to 7.8.5 (this is the usual approach, see, e.g., Plantema, 1966).

7.8.5 Using Equations 7.8.1 to 7.8.3, formulate the potential-energy expression for a sandwich plate and use it to solve the critical load of a square simply supported panel subjected to N_{xy} (proceed similarly to the solution for an ordinary plate, Eq. 7.3.18).

7.8.6 Solve the critical axial compression N_{xx} for cylindrical buckling of a sandwich plate on elastic foundation. Then solve the analogous problem of axisymmetric buckling of a cylindrical sandwich shell, axially compressed.

References and Bibliography

Abdel-Sayed, G. (1969), "Effective Width of Thin Plates in Compression," *J. Struct. Eng.* (ASCE), 95(10):2183–204.

Akkas, N. (1972), "On the Buckling and Initial Post-Buckling Behavior of Shallow Spherical and Conical Sandwich Shells," *J. Appl. Mech.* (ASME), 39(1):163–171.

Allen, H. G. (1969), *Analysis and Design of Structural Sandwich Panels,* Pergamon Press, Oxford (Sec. 7.8).

Almroth, B. O., Holmes, A. M. C., and Brush, D. O. (1964), "An Experimental Study of the Bulking of Cylinders under Axial Compression," *Exp. Mech.*, 4:263 (*Sec.* 7.7).

Babcock, C. D. (1983), "Shell Stability," *J. Appl. Mech.* (ASME), 50:935–40 (review, with 72 refs.).

Baginski, F. E. (1988), "Axisymmetric and Nonaxisymmetric Buckled States of a Shallow Spherical Cap," *Quart. Appl. Math.*, 46(2):331–51.

Batdorf, S. B. (1947), "A Simplified Method of Elastic Stability Analysis for Thin Cylindrical Shells," NACA Tech. Report No. 874 (Sec. 7.6).

Batdorf, S. B., and Stein, M. (1947), "Critical Combinations of Shear and Direct Stress for Simply Supported Rectangular Flat Plates," NACA Tech. Note No. 1223.

Batdorf, S. B., Schildcrout, M., and Stein, M. (1947a), "Critical Stress of Thin-Walled Cylinders in Axial Compression," NACA Tech. Note No. 1343.

Batdorf, S. B., Schildcrout, M., and Stein, M. (1947b), "Critical Stress of Thin-Walled Cylinders in Torsion," NACA Tech. Note No. 1344.

Bažant, Z. P. (1985), *Lectures on Stability of Structures,* Course 720–D24, Northwestern University, Evanston, Ill. (Sec. 7.6).

Ben-Heim, Y., and Elishakoff, I. (1989), "Non-Probabilistic Models of Uncertainty in the Nonlinear Buckling of Shells With General Imperfections: Theoretical Estimates of the Knockdown Factor," *J. Appl. Mech.* (ASME), 56:403–10.

Bijlaard, P. P. (1950), "Stability of Sandwich Plates in Combined Shear and Compression," *J. Aero. Sci.*, 17(1):63.

Bodner, S. R. (1957), "General Instability of Ring-Stiffened Circular Cylindrical Shell under Hydrostatic Pressure," *J. Appl. Mech.* (ASME), 24(2):269–77.

Brodland, G. W., and Cohen, H. (1987), "Deflection and Snapping of Spherical Caps," *Int. J. Solids Struct.*, 23(10):1341–56.

Brown, C. J., Yettram, A. L., and Burnett, M. (1987), "Stability of Plates with Rectangular Holes," *J. Struct. Eng.* (ASCE), 113(5):111–16.

Brush, D. O., and Almroth, B. O. (1975), *Buckling of Bars, Plates and Shells,* McGraw-Hill, New York (Sec. 7.0).

Bushnell, D. (1984), "Computerized Analysis of Shells—Governing Equations," *J. Comput. Struct.* 18(3):471–536.

Bushnell, D. (1985), "Static Collapse: A Survey of Methods and Modes of Behavior," *Finite Elements in Analysis and Design,* 1:165–205 (Sec. 7.0).

Bushnell, D. (1987), "Nonlinear Equilibrium of Imperfect, Locally Deformed Stringer-Stiffened Panels under Combined In-Plane Loads," *Comput. Struct.,* 27(4):519–39.

Byskov, E. (1982), "A Review of the Paper: Mode Interaction of Axially Stiffened Cylindrical Shells; Effects of Stringer Axial Stiffness, Torsional Rigidity and Eccentricity," *Appl. Mech. Rev.,* 35(8):1224.

Byskov, E., and Hansen, J. C. (1980), "Post-Buckling and Imperfection Sensitivity Analysis of Axially Stiffened Cylindrical Shells with Mode Interaction," *J. Struct. Mech.,* 8:205–24.

Byskov, E., and Hutchinson, J. W. (1977), "Mode Interaction in Axially Stiffened Cylindrical Shells," *AIAA Journal,* 15:941–48.

Cagan, J., and Taber, L. A. (1986), "Large Deflection Stability of Spherical Shells with Ring Loads," *J. Appl. Mech.* (ASME), 53:897–901.

Calladine, C. R. (1983), *Theory of Shell Structures,* Cambridge University Press, London (Sec. 7.0).

Carnoy, E. G. (1981), "Asymptotic Study of the Elastic Postbuckling Behavior of Structures by the Finite Element Method," *Comput. Methods Appl. Mech. Eng.,* 29(2):147–73.

Chajes, A. (1974), *Principles of Structural Stability Theory,* Prentice-Hall, Englewood Cliffs, N.J. (Sec. 7.6).

Chen, L.-W., and Chen, L.-Y. (1987), "Thermal Buckling of Laminated Cylindrical Plates," *Compos. Struct.,* 8(3):189–205.

Chen, L.-W., and Doong, J.-L. (1983), "Postbuckling Behavior of a Thick Plate," *AIAA Journal,* 21(8):1157–61.

Chen, Z. Q., and Simitses, G. J. (1988), "Delamination Buckling of Pressure-Loaded, Cross-Ply, Laminated Cylindrical Shells," *Z. für Angew. Math. und Mech.* (*ZAMM*), 68(10):491–501.

Chwalla, E. (1936), "Beitrag zur Stabilitätstheorie des Stegbleches vollwandiger Träger," *Stahlbau,* 9:81.

Collatz, L. (1963), *Eigenwertaufgaben mit technischen Anwendungen* (*Eigenvalue Problems with Technical Applications*), 2nd ed., Akademische Verlagsgesellschaft, Geest & Portig, Leipzig (Sec. 7.3).

Column Research Committee of Japan (1971), *Handbook of Structural Stability* , Corona Publishing Co., Ltd., Tokyo (Sec. 7.7).

Croll, J. G. A., and Batista, R. C. (1981), "Explicit Lower Bounds for the Buckling of Axially Loaded Cylinders," *Int. J. Mech. Sci.,* 23(6):331–43.

Croll, J. G. A., and Ellinas, C. P. (1983), "Reduced Stiffness Axial Load Buckling of Cylinders," *Int. J. Solids Struct.,* 19(5):461–77.

Davies, J. M. (1987), "Axially Loaded Sandwich Panels," *J. Struct. Eng.* (ASCE), 113(1):2212–30.

Dawe, D. J. (1985), "Buckling and Vibration of Plate Structures Including Shear Deformation and Related Effects," in *Aspects of the Analysis of Plate Structures, a Volume in Honour of W. H. Wittrick,* Clarendon Press, Oxford, England.

Dawe, D. J., and Roufaeil, O. L. (1982), "Buckling of Rectangular Mindlin Plates," *Comput. Struct.,* 15(4):461–71.

Dawson, R. G., and Walker, A. C. (1972), "Post-Buckling of Geometrically Imperfect Plates," *J. Struct. Eng.* (ASCE), 98(1):75–94 (Sec. 7.4).

Dickinson, S. M., and Di Blasio, A. (1986), "On the Use of Orthogonal Polynomials in the Rayleigh-Ritz Method for the Study of the Flexural Vibration and Buckling of Isotropic and Orthotropic Rectangular Plates," *J. Sound and Vibration,* 108(1):51–62.

Donnell, L. H. (1933), "Stability of Thin-Walled Tubes under Torsion," NACA Report No. 479.

Donnell, L. H. (1934), "A New Theory for the Buckling of Thin Cylinders under Axial Compression and Bending," *Trans. ASME,* 56:795–806 (Sec. 7.6).

Donnell, L. H. (1956), "Effect of Imperfections on Buckling of Thin Cylinders under External Pressure," *J. Appl. Mech.* (ASME), 23(4):569.

Donnell, L. H. (1976), *Beams, Plates and Shells,* McGraw-Hill, New York (Sec. 7.2).

Donnell, L. H., and Wan, C. C. (1950), "Effect of Imperfections on Buckling of Thin Cylinders and Columns under Axial Compression," *J. Appl. Mech.* (ASME), 17(1):73–88 (Sec. 7.7).

Dundrová, V., Kovařík, V., and Šlapák, P. (1970), *Biegungstheorie der Sandwich-Platten,* Academia, Prague; see also Kovařík, V., Dundrová, V., and Šlapák, P. (1965), *Stress and Deformation of Layered Plates Based on Geometrically Nonlinear Theory* (in Czech), Publishing House of ČSAV (Czechoslovak Academy of Sciences), Prague.

Eringen, A. C. (1951), "Buckling of a Sandwich Cylinder under Uniform Axial Compressive Load," *J. Appl. Mech.* (ASME), 18(2):195–202.

Evensen, D. A. (1964), "High Speed Photographic Observation of the Buckling of Thin Cylinders," *Exp. Mech.,* 4:110 (*Sec.* 7.7).

Flügge, W. (1932), "Die Stabilität der Kreiszylinderschale," *Ing. Arch.,* 3(5):436–506 (Sec. 7.7).

Flügge, W. (1962), *Handbook of Engineering Mechanics,* McGraw-Hill, New York (Sec. 7.3).

Flügge, W. (1973), *Stresses in Shells,* 2nd ed., Springer-Verlag, New York (Sec. 7.5).

Fok, W. C. (1984), "Evaluation of Experimental Data of Plate Buckling," *J. Eng. Mech.* (ASCE), 110(4):577–88 (Sec. 7.4).

Fok, W. C., and Yuen, M. F. (1981), "Modified Pivotal Plot for Critical Load Calculation of Rectangular Plate under Edge Compression," *J. Mech. Eng. Sci.,* 23(4):167–70 (Sec. 7.4).

Föppl, A. (1907), *Vorlesungen über technische Mechanik* (in six volumes), vol. 5, p. 132 Druck und Verlag Von B. G. Teubner, Leipzig (Sec. 7.4).

Galambos, T. V. (1988), *Guide to Stability Design Criteria for Metal Structures,* 4th ed., John Wiley & Sons, New York, (Sec. 7.4).

Gerard, G., and Becker, H. (1957a), "Handbook of Structural Stability: Part I, Buckling of Flat Plates," NACA Tech. Note No. 3781 (Sec. 7.3).

Gerard, G., and Becker, H. (1957b), "Handbook of Structural Stability: Part III, Buckling of Curved Plates and Shells," NACA Tech. Note No. 3783 (Sec. 7.5).

Goodier, J. N., and Hsu, C. S. (1954), "Nonsinusoidal Buckling Modes of Sandwich Plates," *J. Aero. Sci.,* 21:525–32.

Habip, L. M. (1965), "A Survey of Modern Development in the Analysis of Sandwich Structures," *Appl. Mech. Rev.,* 18(Feb.):93–98.

Horne, M. R., and Narayanan, R. (1977), "Design of Axially Loaded Stiffened Plates," *J. Struct. Eng.* (ASCE), 103(11):2243–57.

Hu, P. C., Lundquist, E. E., and Batdorf, S. B. (1946), "Effect of Small Deviations from Flatness on Effective Width and Buckling of Plates in Compression," NACA Tech. Note No. 1124.

Huang, N. C. (1972), "Dynamic Buckling of Some Elastic Shallow Structures Subject to Periodic Loading with High Frequency," *Int. J. Solids Struct.,* 8:315–26.

Hughes, T. J. R., and Liu, W. K. (1981a), "Nonlinear Finite Element Analysis of Shells: Part I, Three-Dimensional Shells," *Comput. Methods Appl. Mech. Eng.*, 26:331–62.

Hughes, T. J. R., and Liu, W. K. (1981b), "Nonlinear Finite Element Analysis of Shells: Part II, Two-Dimensional Shells," *Comput. Methods Appl. Mech. Eng.*, 27:167–82.

Hui, D. (1985), "Asymmetric Post-Buckling of Symmetrically Laminated Cross Ply, Short Cylindrical Panels under Compression," *Compos. Struct.*, 3:81–95.

Hui, D. (1986), "Imperfection Sensitivity of Axially Compressed Laminated Flat Plates Due to Bending-Stretching Coupling," *Int. J. Solids Struct.*, 22(1):13–22.

Hui, D., and Chen, Y. H. (1987), "Imperfection-Sensitivity of Cylindrical Panels under Compression Using Koiter's Improved Postbuckling Theory," *Int. J. Solids Struct.*, 23(7):969–82.

Hui, D., and Du, I. H. Y. (1987), "Imperfection Sensitivity of Long Antisymmetric Cross-Ply Cylindrical Panels under Shear Loads," *J. Appl. Mech.* (ASME), 54:1–7.

Hui, D., and Hansen, J. S. (1980), "Two Mode Buckling of an Elastically Supported Plate and its Relation to Catastrophe Theory," *J. Appl. Mech.* (ASME), 47:607–12.

Hui, R. C., Tennyson, R. C., and Hansen, J. S. (1981), "Mode Interaction in Axially Stiffened Cylindrical Shells: Effects of Stringer Axial Stiffness, Torsional Rigidity and Eccentricity," *J. Appl. Mech.* (ASME), 48:915–22.

Hutchinson, J. W. (1967), "Initial Post-Buckling Behavior of Toroidal Shell Segments," *Int. J. Solids Struct.*, 3:97–115.

Huttelmeier, H-P., and Epstein, M. (1989), "Multilayered Finite Element Formulation for Vibration and Stability Analysis of Plates," *J. Eng. Mech.* (ASCE), 115(2): 315–25.

Ignaccolo, S., Cousin, M., Jullien, J. F., and Waeckel, N. (1988), "Interaction of Mechanical and Thermal Stresses on the Instability of Cylindrical Shells," *Res Mech.*, 24(1):25–33.

Ilanko, S., and Dickinson, S. M. (1987), "Vibration and Postbuckling of Geometrically Imperfect, Simply Supported, Rectangular Plates under Uni-Axial Loading, Part I: Theoretical Approach; Part II: Experimental Investigation," *J. Sound and Vibration*, 118(2):313–36, 337–51.

Johnston, B. G. ed. (1966, 1976), *Guide to Stability Design Criteria for Metal Structures*, John Wiley & Sons, New York (Sec. 7.4).

Kao, R., and Perrone, N. (1978), "Dynamic Buckling of Axisymmetric Spherical Caps with Initial Imperfections," *Comput. Struct.*, 9:463–73.

Kapania, R. K., and Yang, T. V. (1987), "Buckling, Postbuckling and Nonlinear Vibrations of Imperfect Plates," *AIAA Journal*, 25(10):1338–46.

Khdeir, A. A. (1989), "Stability of Antisymmetric Angle-Ply Laminated Plates," *J. Eng. Mech.* (ASCE), 115(5):952–62.

Kirchhoff, G. R. (1850), *J. Math.* (Crelle), Vol. 40 (Sec. 7.1).

Kirkpatrick, S. W., and Holmes, B. S. (1989), "Effect of Initial Imperfections on Dynamic Buckling of Shells," *J. Eng. Mech.* (ASCE), 115(5):1075–91.

Koiter, W. T. (1945), *Over de Stabiliteit van het elastiche Evenwicht*, Dissertation, Delft, Holland. Translation: "On the Stability of Elastic Equilibrium," NASA TT F-10833, 1967, and AFFDL-TR-70-25 (1970) (Sec. 7.7).

Kollár, L., and Dulácska, E. (1984), *Buckling of Shells for Engineers*, John Wiley & Sons, Chichester, England (Sec. 7.0).

Kovařík, V., and Šlapák, P. (1973), *Stability and Vibrations of Sandwich Plates* (in Czech), Academia, Prague (Sec. 7.8).

Lange, C. G., and Newell, A. C. (1973), "Spherical Shells Like Hexagons; Cylinders Prefer Diamonds," *J. Appl. Mech.* (ASME), 40:575–81.

Leipholz, H. H. E. (1981), "On a Generalization of the Lower Bound Theorem for Elastic Rods and Plates Subjected to Compressive Follower Forces," *Comput. Methods Appl. Mech. Eng.*, 27(1): 101–20.

Leissa, A. W., and Ayoub, E. F. (1988), "Vibration and Buckling of a Simply Supported Rectangular Plate Subjected to a Pair of In-Plane Concentrated Forces," *J. Sound and Vibration,* 127(1): 155–71.

Libai, A. and Simmonds, J. G. (1988), *The Nonlinear Theory of Shells,* Academic Press, New York.

Liu, W. K., and Lam, D. (1988), "Numerical Analysis of Diamond Buckles," *Finite Elements in Analysis and Design,* 4: 291–302 (Sec. 7.7).

Lorenz, R. (1908), "Achsensymmetrische Verzerrungen in dünnwandingen Hohlzylindern," *Zeitschrift des Vereines Deutscher Ingenieure,* 52(43): 1706–13 (Sec. 7.5).

Lorenz, R. (1911), "Die nichtachsensymmetrische Knickung dünnwanger Hohlzylinder," *Physik. Z.*, 13: 241–60 (Sec. 7.5).

Love, A. E. H. (1888), "The Small Free Vibrations and Deformation of a Thin Elastic Shell," *Phil. Trans. Roy. Soc.* London, Vol. 179 Series A, p. 491 (Sec. 7.6).

Lukasiewicz, S. (1973), "Geometrical Analysis of the Post-Buckling Behavior of the Cylindrical Shell," *Int. Assoc. Shell and Spat. Struct.*, Kielce, Poland, pp. 305–22.

Maewal, A., and Nachbar, W. (1977), "Stable Post-Buckling Equilibria of Axially Compressed, Elastic Circular Cylindrical Shells: A Finite Element Analysis and Comparison with Experimental Results," *J. Appl. Mech.* (ASME), 44: 475–81.

Magnucki, K., Wegner, T., and Szye, W. (1988), "On Buckling of Ellipsoidal Cups under Internal Pressure," *Ing. Arch.*, 58(5): 339–42.

May, M., and Ganaba, T. H. (1988), "Elastic Stability of Plates with and without Openings," *Eng. Comput.*, 5: 50–52.

Mushtari, K. M. (1938), "Nekotorye obobshcheniya teorii tonkikh obolochek s prilozheniyami k zadache ustoichivosti uprugogo ravnovesiya (Some generalizations of the theory ot thin shells with application to the stability problem of elastic equilibrium)," Izv. Kaz. fiz-mat. o-va., seriya 3, Vol. XI (Sec. 7.6).

Navaneethakrishnan, P. V. (1988), "Buckling of Nonuniform Plates: Spline Method," *J. Eng. Mech.* (ASCE), 114(5): 893–98.

Ng, S. S. F., and Das, B. (1986), "Free Vibration and Buckling Analysis of Clamped Skew Sandwich Plates by the Galerkin Method," *J. Sound and Vibration,* 107(1): 97–106.

Nicholls, R., and Karbhari, V. (1989), "Nondestructive Load Predictions of Concrete Shell Buckling," *J. Struct. Eng.* (ASCE), 115(5): 1191–1211.

Niordson, F. I. (1985), *Shell Theory* (Ch. 15), North Holland, Amsterdam.

Noor, A. K., and Whiteworth, S. L. (1987), "Model-Size Reduction for Buckling and Vibration Analyses of Anisotropic Panels," *J. Eng. Mech.* (ASCE), 113(2): 170–85.

Northwestern University (1969), "Report on 'Models for Demonstrating Buckling of Structures'," prepared by R. A. Parmelee, G. Herrmann, J. F. Fleming, and J. Schmidt under NSF Grant GE-1705 (Sec. 7.6).

Novozhilov, V. V. (1953), *Foundations of the Nonlinear Theory of Elasticity,* Greylock Press, Rochester, N.Y. (Sec. 7.6).

Palazzotto, A. N., and Straw, A. D. (1987), "Shear Buckling of Cylindrical Composite Panels," *Comput. Struct.*, 27(5): 689–92.

Paliwal, D. N., and Rai, R. N. (1989), "Stability of Spherical Shells on Elastic and Viscoelastic Foundations," *J. Eng. Mech.* (ASCE), 115(5): 1121–28.

Plantema, F. J. (1966), *Sandwich Construction: The Bending and Buckling of Sandwich Beams, Plates and Shells,* John Wiley and Sons, New York (Sec. 7.8).

Popov, E. P., and Medwadowski, S. J. eds. (1981), *Concrete Shell Buckling,* ACI Special Publication SP-67 (Sec. 7.0).

Prevost, D. P. et al. (1984), "Buckling of a Spherical Dome in a Centrifuge," *Exp. Mech.*, 24(3):203–207.

Rajasekaran, S., and Weimar, K. (1989), "Buckling Analysis of Segmented Conical Concrete Shell Roof," *J. Struct. Eng.* (ASCE), 115(6):1514–24.

Raju, K. K., and Rao, G. V. (1984), "Thermal Post-Buckling of Circular Plates," *Comput. Struct.*, 18(6):1179–82.

Ramsey, H. (1985), "A Rayleigh Quotient for the Instability of a Rectangular Plate with Free Edges Twisted by Corner Forces," *J. Mech. Theor. Appl.*, 4(2):243–56.

Reddy, J. N. (1987), "A Small Strain and Moderate Rotation Theory of Elastic Anisotropic Plates," *J. of Appl. Mech.* (ASME), 54:623.

Rhodes, J., and Harvey, J. M. (1977), "Examination of Plate Postbuckling Behavior," *J. Eng. Mech.* (ASCE), 103(3):461–78 (Sec. 7.4).

Robertson, A. (1928), "The Strength of Tubular Struts," *Proc. Roy. Soc.* London, Series A, 121:558–85 (Sec. 7.7).

Roman, V. G., and Elwi, A. E. (1988), "Post-Buckling Shear Capacity of Thin Steel Tubes," *J. Struct. Eng.* (ASCE), 114(11):2511–23.

Roorda, J., and Hansen, J. S. (1971), "Random Buckling Behavior in Axially Loaded Cylindrical Shells with Axisymmetric Imperfections," Univ. of Waterloo, Solids Mech. Div., Report No. 80.

Rotter, J. M., and Teng, J.-G. (1989a), "Elastic Stability of Lap-Jointed Cylinders," *J. Struct. Eng.* (ASCE), 115(3):683–97.

Rotter, J. M., and Teng, J.-G. (1989b), "Elastic Stability of Cylindrical Shells with Weld Depressions," *J. Struct. Eng.* (ASCE), 115(5):1244–63.

Sabag, M., Stavsky, Y., and Greenberg, J. B. (1989), "Buckling of Edge Damaged, Cylindrical Composite Shells," *J. Appl. Mech.* (ASME), 56:121–6.

Sallam, S., and Simitses, G. J. (1985), "Delamination Buckling and Growth of Flat, Cross-Ply Laminated," *Compos. Struct.*, 4:361–81 (Sec. 7.8).

Sallam, S., and Simitses, G. J. (1987), "Delamination Buckling of Cylindrical Shells under Axial Compression," *Compos. Struct.*, 7:83–101 (Sec. 7.8).

Sang, Z. T., Chang, K. C., and Lee, G. C. (1988), "Simple Method for Measuring Local Buckling of Thin Plates," *Exp. Mech.*, 28(1):20–23.

Scheidl, R., and Troger, H. (1987), "Comparison on the Post-Buckling Behavior of Plates and Shells," *Comput. Struct.*, 27(1):157–63.

Schweizerhof, K., and Ramm, E. (1987), "Follower Force Effects on Stability of Shells under Hydrostatic Loads," *J. Eng. Mech.* (ASCE), 113(1):72–88.

Seide, P. (1981), "Stability of Cylindrical Reinforced Concrete Shells," in *Concrete Shell Buckling*, ed. by E. P. Popov and S. J. Medwadowski, Special Publication SP-67, American Concrete Inst., Detroit (Sec. 7.7).

Seide, P., and Weingarten, V. I. (1961), "On the Buckling of Circular Cylindrical Shells under Pure Bending," *J. of Appl. Mech.* (ASME), 28(1):112–16.

Shanmugam, N. E. (1988), "Effective Widths of Orthotropic Plates Loaded Uniaxially," *Comput. Struct.*, 29(4):705–13.

Sharifi, P., and Popov, E. P. (1973), "Nonlinear Finite Element Analysis of Sandwich Shells of Revolution," *AIAA Journal*, 11(5):715–22.

Sheinman, I. (1989), "Cylindrical Buckling Load of Laminated Columns," *J. Eng. Mech.* (ASCE), 115(3):659–61.

Sheinman, I., and Frostig, Y. (1988), "Post-Buckling Analysis of Stiffened Laminated Panel," *J. Appl. Mech.* (ASME), 55:635–40.

Simitses, G. J. (1986), "Buckling and Post-Buckling of Imperfect Cylindrical Shells," *Appl. Mech. Rev.*, 39:1517–24.

Simitses, G. J., and Chen, Z. (1988), "Buckling of Delaminated Long, Cylindrical Panels under Pressure," *Comput. Struct.*, 28(2):173–84.

Southwell, R. V. (1914), "On the General Theory of Elastic Stability," *Phil. Trans. Roy. Soc.* London, Series A, 213: 187 (Sec. 7.5).

Southwell, R. V. (1915), "On the Collapse of Tubes by External Pressure," *Philos. Mag.*, Part I, 25 (1913): 687–98; Part II, 26(1913): 502–11; Part III, 29(1915): 66–67.

Spencer, H. J., and Walker, A. C. (1975), "Critique of Southwell Plots with Proposals for Alternative Methods," *Exp. Mech.*, 15(8): 303–10 (Sec. 7.4).

Stein, M. (1968), "Some Recent Advances in the Investigation of Shell Buckling," *AIAA Journal*, 6(12): 2339–45.

Subbiah, J., and Natarajan, R. (1981), "Stability Analysis of Ring-Stiffened Shells of Revolution," *Comput. Struct.*, 13(4): 497–503.

Svalbonas, V., and Kalnins, A. (1977), "Dynamic Buckling of Shells: Evaluation of Various Methods," *Nucl. Eng. Des.*, 44: 331–56.

Tennyson, R. C. (1969), "Buckling Modes of Circular Cylindrical Shells under Axial Compression," *AIAA Journal*, 7: 1481 (Sec. 7.7).

Tennyson, R. C. and Hansen, J. S. (1983), "Optimum Design for Buckling of Laminated Cylinders," in *Collapse. The Buckling of Structures in Theory and Practice, Proc. of IUTAM Symp.*, ed. by *J. M. T. Thompson and G. W. Hunt*, Cambridge University Press, London pp. 409–29.

Tennyson, R. C., and Muggeridge, D. B. (1969), "Buckling of Axisymmetric Imperfect Circular Cylindrical Shells under Axial Compression," *AIAA Journal*, 7(11): 2127–31 (Sec. 7.7).

Tennyson, R. C., Muggeridge, D. B., and Caswell, R. D. (1971), "Buckling of Circular Cylindrical Shells Having Axisymmetric Imperfection Distribution," *AIAA Journal*, 9(5): 924–30 (Sec. 7.7).

Timoshenko, S. P. (1910), "Einige Stabilitätsprobleme der Elastizitätstheorie," *Zeitschrift für angew. Math. Phys. (ZAMP)*, 58(4): 337–85 (Sec. 7.5).

Timoshenko, S. P., and Gere, J. M. (1961), *Theory of Elastic Stability*, McGraw-Hill, New York (Sec. 7.2).

Timenshenko, S. P., and Woinowsky-Krieger, S. (1959), *Theory of Plates and Shells*, McGraw-Hill, New York (Sec. 7.5).

Tsang, S. K., and Harding, J. E. (1987), "Ring-Stiffened Cylinders under Interactive Loading," *J. Struct. Eng.* (ASCE), 113(9): 1977–93.

Tvergaard, V. (1973), "Imperfection-Sensitivity of a Wide Integrally Stiffened Panel under Compression," *Int. J. Solids Struct.*, 9: 177–92.

Tvergaard, V. (1976), "Buckling Behavior of Plate and Shell Structures," in *Proc. 14th Int. Congr. on Theoretical and Applied Mechanics*, ed. by W. T. Koiter, North-Holland, Amsterdam, pp. 233–47 (Sec. 7.7).

Tylikowski, A. (1989), "Dynamic Stability of Nonlinear Antisymmetrically-Laminated Cross-Ply Rectangular Plates," *J. Appl. Mech.* (ASME), 56: 375–81.

Vizzini, A. J., and Lagace, P. A. (1987), "Buckling of a Delaminated Sublaminate on an Elastic Foundation," *J. Compos Mater.*, 21(12): 1106–17.

Vlasov, V. Z. (1949), *Obshchaya teoriya obolochek i ee prilozheniya v tekhnike* (*The general theory of shells and its industrial applications*), Gostekhizdat, Moscow; see also *Allgemeine Schalentheorie und ihre Anwendung in der Technik*, Akademie-Verlag, Berlin, 1958 (Sec. 7.6).

Vol'mir, A. S. (1963), *Stability of Elastic Systems* (in Russian), Fizmatgiz, Moscow; English Translation, Wright-Patterson Air Force Base, FTD-MT (1964) and NASA, AD 628 508 (1965) (Sec. 7.5).

von Kármán, T. (1910), *Enzyklopädie der mathematischen Wissenschaften*, Vol. 4, p. 349 (Sec. 7.4).

von Kármán, T., and Tsien, H. S. (1941), "The Buckling of Thin Cylindrical Shells under Axial Compression," *J. Aero. Sci.*, 8(6): 303–312 (Sec. 7.7).

von Kármán, T., Dunn, L. G., and Tsien, H. S. (1940), "The Influence of Curvature on the Buckling Characteristics of Stuctures," *J. Aero. Sci.,* 7(7):276–89 (Sec. 7.7).

von Kármán, T., Sechler, E. E., and Donnell, L. H. (1932), "The Strength of Thin Plates in Compression," *Trans. ASME,* 54:53–58 (Sec. 7.4).

Walker, A. C. (1969), "The Post-Buckled Behavior of Simply Supported Square Plates," *Aeronaut. Q.,* 20(3):203–22 (Sec. 7.4).

Wang, S. S., Zahlan, N. M., and Suemasu, H. (1985), "Compressive Stability of Delaminated Random Short-Fiber Composites, Part I: Modeling and Methods of Analysis," *J. Compos. Mater,* 19(4):296–316.

Waszczyszyn, A., and Cichoń, C. (1988), "Nonlinear Stability Analysis of Structures under Multiple Parameter Loads," *Eng. Comput.,* 5:10–14.

Williams, D. G., and Aalami, B. (1979), *Thin Plate Design for In-Plane Loading,* J. Wiley and Sons, New York; see also V. Brown (1990), "Linearized Least-Squares Technique for Evaluating Plate Buckling Loads," *J. Eng. Mech.* (ASCE), 116(5):1050–57 (Sec. 7.4).

Williams, D. G., and Walker, A. C. (1977), "Explicit Solutions for Plate Buckling Analysis," *J. Eng. Mech.* (ASCE), 103(4), 549–59 (Sec. 7.4).

Winter, G. (1947), "Strength of Thin Steel Compression Flanges," *Trans. ASCE* vol. 112:527–76. (Sec. 7.4).

Wittrick, W. H., and Horsington, R. W. (1984), "Buckling and Vibration of Composite Folded-Plate Structures of Finite Length in Combined Shear and Compression," *Proc. Royal Soc.,* London Series A, 392(1802):107–44.

Yamada, S., and Croll, J. G. A. (1989), "Buckling Behavior of Pressure Loaded Cylindrical Panels," *J. Eng. Mech.* (ASCE), 115(2):327–44.

Yamaki, N., Otomo, K., and Matsuda, K. (1975), "Experiments on the Post-Buckling Behavior of Circular Cylindrical Shells under Compression," *Exp. Mech.* 15:23–28.

Yao, J. C. (1962), "Buckling of Sandwich Sphere under Normal Pressure," *J. Aerospace Sci.,* 29(3):264–68.

Yoshimura, Y. (1955), "On the Mechanism of Buckling of a Circular Shell under Axial Compression," NACA Tech. Note. No. 1390 (Sec. 7.7).

Yu, T. X., and Zhang, L. C. (1986), "Elastic Wrinkling of an Annular Plate under Uniform Tension on Its Inner Edge," *Int. J. Mech. Sci.,* 28(11):729–37.

Zintilis, G. M., and Croll, J. G. A. (1983), "Combined Axial and Pressure Buckling of End Supported Shells of Revolution," *Eng. Struct.,* 5(3):199–206.

II

INELASTIC, DAMAGE, AND FRACTURE THEORIES

8

Elastoplastic Buckling

As pointed out in the Introduction, structures can fail either due to material failure or to instability. But they can also fail due to a combination of both. The material failure is normally preceded by inelastic phenomena, which generally have a destabilizing influence on structures, and must therefore be taken into consideration. Even for structures that are elastic under service loads, achievement of a uniform safety margin requires the consideration of overloads, and overloads inevitably involve inelastic deformations.

Except for some dynamic solutions with energy dissipation due to damping (Chap. 3), all our analysis has so far been concerned with elastic behavior. To some extent, the Lagrange–Dirichlet theorem (Sec. 3.6) admits the presence of nonelastic phenomena because it admits the presence of dissipative forces. However, it does require the existence of the potential-energy function $\Pi(q_1, \ldots, q_n)$ since all the forces other than dissipative must be conservative. Dissipative phenomena such as velocity-dependent damping or friction do not violate this requirement. However, the potential-energy function does not exist, in general, for time-independent inelastic behavior, such as plasticity or fracturing, and time-dependent inelastic behavior, such as viscoelasticity or viscoplasticity. This makes the energy methods based on the Lagrange–Dirichlet theorem inapplicable. Whereas the dissipative phenomena that do not destroy the existence of potential energy (such as velocity-dependent damping, Chap. 3) cannot destabilize a structure, those which do violate it have generally a destabilizing effect, as we will see. Plasticity, as well as viscoelasticity, viscoplasticity, fracturing, and other damage, are such dissipative phenomena.

In this chapter we will be concerned with inelastic behavior of time-independent type, including perfectly plastic behavior as well as hardening elastoplastic behavior, but excluding fracturing, yield limit, degradation, and any softening damage. Our principal concern is the behavior of columns whose tangential bending stiffness varies continuously as the load is increased. Such behavior is obviously characteristic of columns made of aluminum alloys and high-strength alloy steels. However, it is also characteristic of structural steel columns, despite the fact that the stress-strain diagram of mild steel is not smooth but essentially elastic-perfectly plastic. The reason is that the residual stresses from hot-rolling, as well as those from welding or other heat treatment, cause a continuous variation of the tangential bending stiffness. When the structure is perfect, buckling of such columns is described by Shanley's theory, which generalizes the concept of bifurcation for the inelastic range. The main effect of plastic deformations is a reduction of the limit load at which a perfect column can start to deflect. This reduction is closely related to a reduction of the maximum load and can be very large.

Another important adverse influence on buckling is exerted by inevitable imperfections. While for elastic columns we found that imperfections merely increase deflections but do not lower the maximum load (if the buckling analysis is linearized), for elastic-plastic columns imperfections are found to decrease the maximum load, sometimes quite strongly. After analyzing small-deflection buckling of perfect and imperfect columns and the effect of residual stresses, we will finally be ready to explain and critically examine the basic design specifications in the current codes for metallic and concrete structures.

At very large deflections at which the load-deflection curve is descending, the critical cross section in a steel column becomes fully plastic and a plastic hinge forms. As we will see, the solutions of this behavior are particularly simple. Interpolation between them and the initial elastoplastic deflections provides useful approximations. Various simple estimates and bounds can be obtained in this manner for buckling and second-order effects in dynamic impact and blast problems.

Apart from the familiar instability in compression we will also briefly illustrate some instabilities that can be observed in other types of loading, for example, large extensions of a bar. We will demonstrate that even when the stress-strain diagram has a positive slope, a bar can lose stability in tension due to decrease of the cross-section area caused by Poisson's effect. This is a type of instability that has no counterpart in small strain theory.

Throughout this chapter we will concentrate on buckling analysis conducted on the basis of equilibrium conditions, as in Chapters 1 and 2. Since the principle of minimum potential energy is inapplicable to inelastic structures, the stability analysis must be based on more general minimum principles, which are furnished by thermodynamics. We prefer, however, to postpone the discussion of thermodynamic stability criteria to Chapter 10.

8.1 PERFECT COLUMNS OR STRUCTURES AND SHANLEY'S BIFURCATION

The ultimate failure of every column or structure is, of course, always caused by inelastic behavior, involving plastic yield or fracture, or both. Depending on structure geometry, especially column slenderness, the inelastic behavior is either irrelevant or important for buckling. Inelastic deformation always causes that for a given deflection the axial load becomes smaller than predicted by elasticity (Fig. 8.1). If the column is very slender, the load reduction due to inelastic behavior occurs at very large deflections, which are of no interest (curve a or b in Fig. 8.1). For such columns, inelastic behavior is obviously irrelevant for design. Inelastic behavior, however, becomes important for columns or structures that are not too slender (curve c or d in Fig. 8.1), for which a significant reduction of the

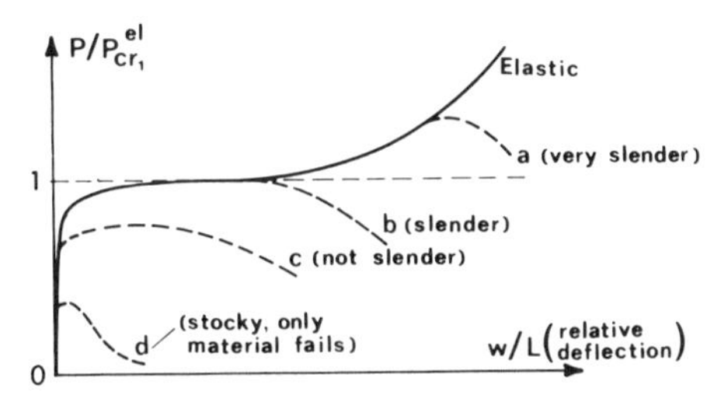

Figure 8.1 Effect of inelastic behavior and material failure on columns of various slendernesses.

maximum load occurs at relatively small deflections and must, therefore, be taken into account in design.

When the column is not very slender, it is possible to exceed the initial yield limit before the column buckles. After this happens, the behavior of the column is elastoplastic, with two important characteristics: (1) The tangent modulus E_t for continued loading is smaller than the initial elastic modulus E, and (2) the unloading modulus E_u is larger than E_t. For elastic-plastic materials that undergo no damage, one has $E_u = E$; however, in the presence of damage (e.g., microcracking or void growth) one has $E_u < E$.

Inelastic buckling is of interest for hardening elastic-plastic materials. They include high-strength alloy steels and aluminum alloys, for which the uniaxial stress-strain diagram above the plastic limit σ_p is smoothly curved, with a continuously decreasing slope, as shown in Figure 8.2a. As a simplification, one may sometimes consider the stress-strain diagram to be bilinear, with a constant tangent modulus E_t, as shown in Figure 8.2b. The limiting case of the bilinear diagram, which approximates mild steel, is the elastic–perfectly plastic material, for which $E_t \to 0$ (Fig. 8.2c). Mild steels (including structural carbon steels and high-strength low-alloy steels) approximately behave in such a manner.

For an elastic–perfectly plastic column for which the elastic critical load is larger than the axial yield load, it might seem that failure is completely controlled by the yield criterion and has nothing to do with buckling. In reality, however, the inevitable presence of residual stresses causes structural columns made of mild steel (even perfect columns) to behave like columns made of an elastic-hardening plastic material (see Sec. 8.3). It is for this reason that the present section applies to typical structural steel columns as well.

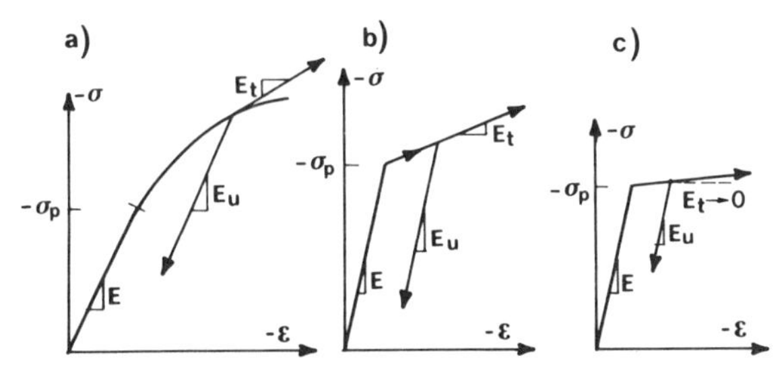

Figure 8.2 Idealized stress-strain curves of hardening plastic, bilinear, and almost perfectly plastic materials.

Reduced Modulus Load

Consider now a perfect pin-ended column of a rectangular cross section (Fig. 8.3a, b). In similarity to the results of linearized elastic analysis of buckling (Chap. 1), we will now assume that the column buckles at constant axial force P. Prior to buckling ($w = 0$), the stress distribution is uniform; $\sigma^0 = -P/A$, where A = cross-section area (Fig. 8.3c). If the column is not too short, we may assume again that plane cross sections remain plane and normal to the deflected center line of the column. This assumption is not as good as it is for elastic buckling, which takes place for columns that are more slender. It nevertheless usually is still acceptable, as verified by numerous experimental observations.

At the start of buckling, the concave face of the column (i.e., the one toward the center of curvature) undergoes further shortening, that is, loading, and the convex face undergoes extension, that is, unloading. (Note that if both faces of the column underwent shortening, the axial load could not remain constant.) Somewhere within the cross section there is a neutral axis at which the axial strain does not change; its distances from the convex and concave faces of the column are denoted as h_1 and h_2 (Fig. 8.3d), and $h_1 + h_2 = h$ = height of the cross section. Since the incremental moduli for loading and unloading are different (E_t and E_u, Fig. 8.2a), buckling causes a bilinear stress distribution within the cross section, as shown in Figure 8.3d. The incremental stresses at the convex and concave faces of the column are, according to the assumption of plane cross-section, $-E_u h_1/\rho$ and $E_t h_2/\rho$, where $1/\rho$ = curvature due to buckling. The condition of constant axial load that we imposed requires that the resultants F_1 and F_2 (Fig. 8.3d) of the incremental normal stresses be of opposite signs and equal magnitudes, and so $bh_1 E_u h_1/2\rho = bh_2 E_t h_2/2\rho$, in which b = width of the cross section. Substituting $h_2 = h - h_1$, we have $E_u h_1^2 = E_t(h - h_1)^2$, from which

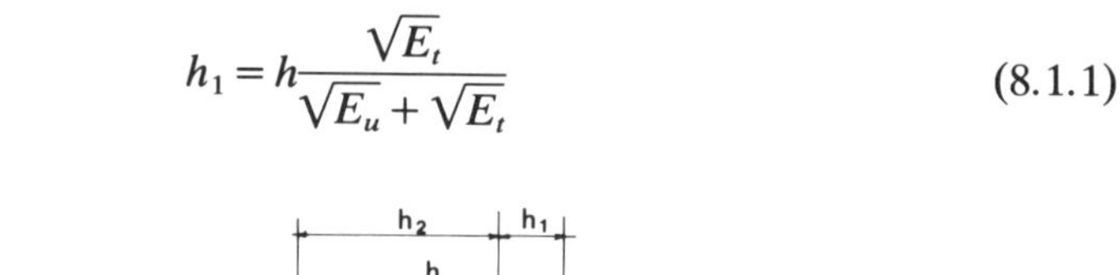

$$h_1 = h\frac{\sqrt{E_t}}{\sqrt{E_u} + \sqrt{E_t}} \tag{8.1.1}$$

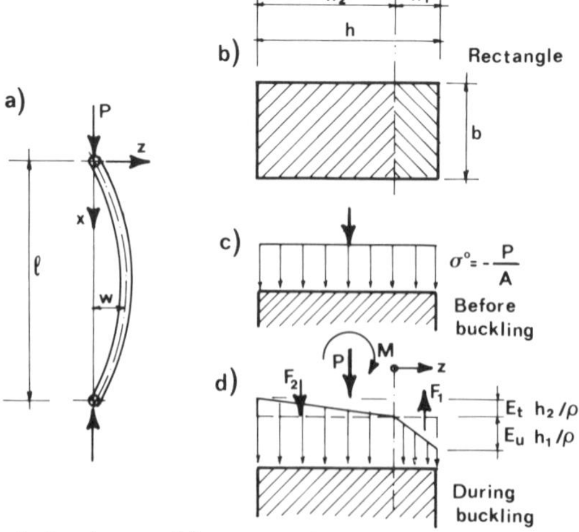

Figure 8.3 (a) Pin-ended column, (b) rectangular cross section, and stress distribution (c) before and (d) during buckling.

The bending moment representing the resultant of the positive and negative normal stress increments in the cross section may now be calculated as follows:

$$M = -F_1\left(\frac{2}{3}h_1 + \frac{2}{3}h_2\right) = -F_1\left(\frac{2}{3}h\right) = bh_1\left(\frac{E_u h_1}{2\rho}\right)\left(\frac{2}{3}h\right)$$
$$= \frac{bh^3}{(4)(3)}\left[\frac{4E_uE_t}{(\sqrt{E_u}+\sqrt{E_t})^2}\right]\left(\frac{1}{\rho}\right) \tag{8.1.2}$$

This may be rewritten as

$$M = \frac{E_r I}{\rho}, \qquad E_r = \left[\frac{1}{2}(E_u^{-1/2} + E_t^{-1/2})\right]^{-2} \tag{8.1.3}$$

in which $I = bh^3/12$ and E_r is called the reduced modulus (or double modulus). We see that for the start of buckling from the perfectly straight position, the bending moment is proportional to the curvature, same as for the linearized elastic theory; however, the bending stiffness is reduced.

According to the equilibrium condition for a pin-ended column, $M = -Pw$. Equations 8.1.3 yield the differential equation $E_r I w'' + Pw = 0$ (since $1/\rho \simeq w''$). The boundary conditions are $w = 0$ at $x = 0$ and $x = l$ (Fig. 8.3a). So the differential equation with the boundary conditions is the same as in Section 1.2, except that EI is replaced by $E_r I$. Therefore, the first critical load is

$$P_{\text{cr}} = P_r = \frac{\pi^2}{l^2} E_r I \tag{8.1.4}$$

Later (Sec. 10.1) we will see that this load, which characterizes a state of neutral equilibrium, represents the limit of stable equilibrium states of the columns when the load is controlled.

For a smoothly curved stress-strain diagram (Fig. 8.2a), $E_r = f(\sigma^0) =$ function of the uniform normal stress that exists just before buckling, $\sigma^0 = -P_r/A$. So Equation 8.1.4 represents an implicit nonlinear equation, $P_r = f(-P_r/A)I\pi^2/l^2$, from which P_r may be solved iteratively, for example by the Newton method. If the stress-strain diagram is bilinear ($E_t =$ const., Fig. 8.2b), then P_r is given by Equation 8.1.4 explicitly. However, it may happen (depending on the value of l) that $P_r < A\sigma_p =$ load at the yield limit, in which case P_r does not exist and equilibrium is lost upon reaching the load $P = A\sigma_p$ (or load $P = P_{\text{E}}$ if $P_{\text{E}} < A\sigma_p$) suddenly.

If the same analysis is carried out for the idealized, fully symmetric I cross section (i.e., an I-beam with a massless web), then the critical load is again given by Equation 8.1.4, however, with

$$E_r = [\tfrac{1}{2}(E_u^{-1} + E_t^{-1})]^{-1} \tag{8.1.5}$$

(this is the harmonic mean of E_u and E_t). Generally, in contrast to the elastic buckling theory, the flexural stiffness $E_r I$ depends not only on I but also on the shape of the cross section.

It is important to note that the reduced modulus values for buckling to the left and to the right are different if the cross section is nonsymmetric. Consider, for example, the ideal I-section with $t \ll h$, $t \ll b_1$, and $t \ll b_2$ ($t =$ plate thickness, $h =$ height, b_1, $b_2 =$ width of the cross section, Fig. 8.4). For buckling with the

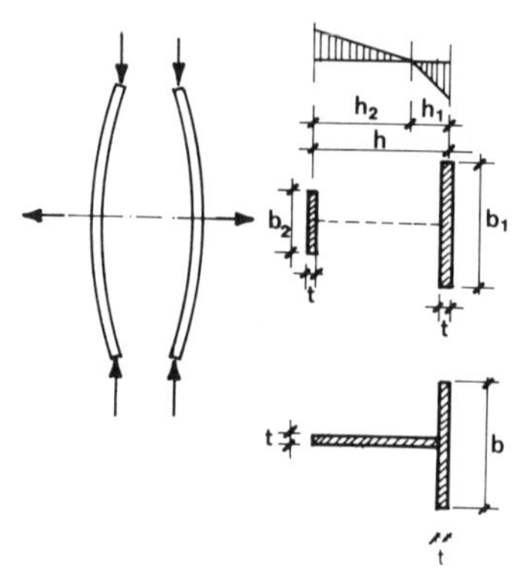

Figure 8.4 Buckling to the left or to the right of columns with nonsymmetric cross sections.

convex face to the right (Fig. 8.4), the axial equilibrium condition is $tb_1E_uh_1/\rho = tb_2E_th_2/\rho$, and the moment equilibrium condition is $M = th_1b_1E_uh_1/\rho + th_2b_2E_th_2/\rho$ where $h_2 = h - h_1$. Eliminating h_1 and h_2, we get $M = E_r^RI/\rho$ where $I = th^2(b_1^{-1} + b_2^{-1})^{-1}$ and

$$E_r^R = (b_1^{-1} + b_2^{-1})[(b_1E_u)^{-1} + (b_2E_t)^{-1}]^{-1} \qquad (8.1.6)$$

This is a weighted harmonic mean of E_t and E_u, with weights b_1 and b_2. For buckling with convex face to the left, E_t and E_u are interchanged, that is, $M = E_r^LI/\rho$ where

$$E_r^L = (b_1^{-1} + b_2^{-1})[(b_1E_t)^{-1} + (b_2E_u^{-1})]^{-1} \qquad (8.1.7)$$

Now we note that generally $E_r^L \neq E_r^R$ (except when $b_1 = b_2$, in which case Eqs. 8.1.6 and 8.1.7 reduce to Eq. 8.1.5). If $b_1 > b_2$, then always $E_r^R < E_r^L$, that is, the column will buckle with its convex, i.e., unloading face at the heavier flange.

Closed-form expressions for E_r^R and E_r^L can also be obtained for the T–cross section (Fig. 8.4). It is found that such a column will buckle with its flange at the convex face. For this case, the axial equilibrium condition (Fig. 8.4) is $h_2E_th_2t/2\rho = h_1E_uh_1t/2\rho + bh_1E_ut/\rho$ and the moment equilibrium equation is $M = t[(h_2^2E_t/2\rho)(2h_2/3) + (h_1^2E_u/2\rho)(2h_1/3) + (bh_1E_u/\rho)h_1]$ where $h_2 = h - h_1$. The first condition is a quadratic equation for h_1. Solving it and setting $M = E_rI/\rho$, we obtain

$$E_r^R = \frac{t}{3I}[E_t(h - h_1)^3 + E_uh_1^3 + 3E_ubh_1^2],$$
$$h_1 = -\frac{hE_t + bE_u}{E_u - E_t} + \left[\left(\frac{hE_t + bE_u}{E_u - E_t}\right)^2 + \frac{E_t}{E_u - E_t}h^2\right]^{1/2} \qquad (8.1.8)$$

The difference between E_r^R and E_r^L has important consequences for the effect of imperfections in monosymmetric columns (Sec. 8.2).

The reduced modulus theory, also called the double modulus theory, was in principle proposed by Considère (1891) and was later developed in detail by Engesser (1895) as his second theory of elastic-plastic buckling. It was further substantiated theoretically and experimentally by von Kármán (1910).

Tangent Modulus Load

Although the early test results agreed with the reduced modulus theory well, some experimental studies conducted after the publication of von Kármán's work (1910) revealed that columns can fail by buckling at loads that are significantly lower than the reduced modulus critical load (Eq. 8.1.4). The reason for this discrepancy with Engesser's second theory was explained by Shanley (1947). He showed that in a normal practical situation the column does not buckle at a constant load, as assumed in our previous calculations (Engesser's second theory). Rather, a perfect column may, and in fact must (see Sec. 10.3), start to deflect at a load that may lie significantly below the reduced modulus critical load. The buckling deflections occur simultaneously with a further increase of the axial load P. In this manner, it is possible that the tensile strain increments caused by the deflection may be compensated for by the axial shortening increment due to the increase of the axial load, so that there is no unloading anywhere in the cross section (Fig. 8.5a).

Consequently, the incremental modulus for the first stress increments at the initiation of buckling, for which the stress distribution is still uniform, is equal to the tangent modulus E_t for all points of the cross section. This is true for both the acual curved stress-strain diagram and the idealized bilinear stress-strain diagram (Fig. 8.2a, b). Thus it appears that, at an increasing axial load, a pin-ended column of length l starts to buckle at the tangent modulus load:

$$P_t = \frac{\pi^2}{l^2} E_t I \tag{8.1.9}$$

This is known as the tangent modulus theory of Shanley, and it coincides with Engesser's first theory, which Engesser introduced in 1889 but later discarded. Note that $P_t < P_r$, except when $E_t = E_u = E$ (elastic behavior), in which case

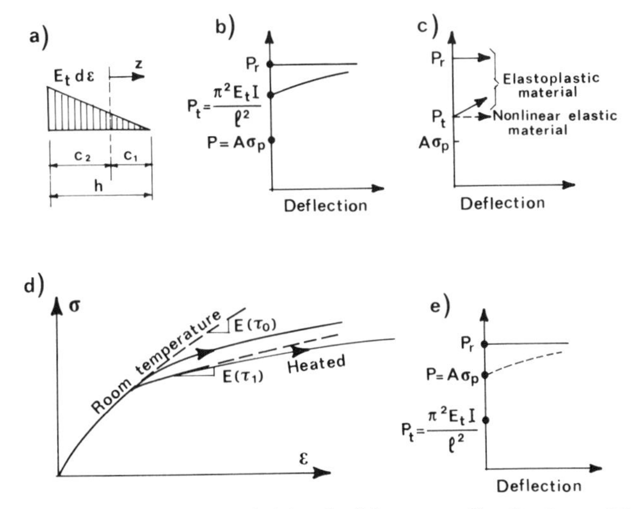

Figure 8.5 Buckling at increasing axial load: (a) stress distribution with no unloading; (b, c, e) load-deflection curves; (d) Young's modulus drop due to sudden heating.

$P_t = P_r > 0$, or when $E_t \to 0$, in which case the yield limit governs the start of deflection.

It may be noted that Equation 8.1.9 gives also the critical load of a column made of a nonlinear elastic material, for which, by definition, $E_u = E_t$.

For a smoothly curved stress-strain diagram (Fig. 8.2a) one has $E_t = F(\sigma^0) =$ function of the uniform normal stress $\sigma^0 = -P_t/A$ that exists just before buckling. Thus, Equation 8.1.9 represents an implicit nonlinear equation, $P_t = F(-P_t/A)I\pi^2/l^2$, from which P_t may be solved by Newton iteration.

For the special case of a bilinear stress-strain diagram (Fig. 8.2b), E_t is constant beyond the proportionality limit, σ_p, and Equation 8.1.9 then gives P_t explicitly. Two cases must be distinguished:

1. If the column is not too slender so that $P_t \geq A\sigma_p$ (Fig. 8.5b), then the value of P_t is admissible.
2. If the column is so slender that $P_t < A\sigma_p$ (Fig. 8.5e), then there is no solution for P_t. If also $A\sigma_p < P_{\mathrm{E}}$, the load can be raised in a stable manner until $P = A\sigma_p$ is reached. After that, can the load P be maintained, rather than dropping to the value of P_t corresponding to the value of E_t beyond the yield limit? The answer is yes, provided that $P < P_r$, which is true for the load-controlled mode of loading, whereas for the displacement-controlled mode of loading the condition is $P < P_{\mathrm{cr}}^D$ where P_{cr}^D is a certain critical value higher than P_r (see Eq. 10.3.16). This is evidenced by the fact, proven later in Section 10.3, that under the aforementioned condition the column is stable. Further this is justified by the fact that for any load $P < P_r$ (stability limit for load control) the load will subsequently increase rather than decrease as the deflection is increased from its initial zero value, as will be shown later (Eqs. 8.1.12 and 8.1.18). If, under this condition, the column were unstable, or if the load decreased with increasing deflection, a dynamic failure would occur upon reaching load $A\sigma_p$, and the column would have to be designed for load P_t ($<A\sigma_p$) rather than $A\sigma_p$ or a higher safety factor would have to be used. But fortunately this is not the case, that is, the column capacity is safely obtained as the yield limit $A\sigma_p$ if $P_t < A\sigma_p$.

A load $P_0 > P_t$ can also be reached if E_t drops instantly due to sudden heating (Fig. 8.5d), change of moisture content, damage due to a hysteretic loop of unloading, chemical damage, irradiation damage, etc.

If $P = P_r$ while the column is straight (undeflected), one has a state of neutral equilibrium (we have in mind only second-order accurate solutions in deflections; higher-order terms can make this state stable or unstable—see Sec. 4.5). This means that adjacent deflected states of equilibrium exist for the same load value (Fig. 8.5c). The tangent modulus load, on the other hand, is the load for which the column may start to deflect laterally under an increasing load (Fig. 8.5c). In this case, adjacent equilibrium positions at the same load do not exist, that is, there is equilibrium but not neutral equilibrium. When P_t is reached, there is more than one equilibrium path, but every path corresponds to an increasing load. Like perfect elastic columns, we have a bifurcation, but unlike them, the bifurcated path has a positive rather than zero slope. As we will prove in Section 10.3, the bifurcated path (branch) is the one that the column must actually follow, that is, a perfect column must start to buckle at $P = P_t$ (even though the undeflected states for $P_t \leq P < P_r$ are stable!). [As we saw, an exception is, of course, the case when $P_t < A\sigma_p$, in which case the column starts to buckle at $P = A\sigma_p$, and does so again with an increasing load. For nonlinear elastic materials, however, the load P_t would correspond to neutral equilibrium and represent a limit of stable states (Fig. 8.5c).]

The situation here is partially similar to the elastic asymmetric bifurcation we elucidated in Section 4.5 for linear elastic structures. In that case there were three branches after bifurcation but none with a zero slope, which means there was no neutral equilibrium at bifurcation (not even according to a second-order accurate solution). For the tangent modulus load (in contrast to the reduced modulus load), there is also no neutral equilibrium and three branches of nonzero slope after bifurcation exist (since the perfect column can deflect either left or right). However, in contrast to elastic asymmetric bifurcation, no branch has a negative slope (and therefore Shanley's bifurcation per se is not a cause of imperfection sensitivity of the maximum load although inelastic behavior is, for other reasons we will see later in Sec. 10.3).

It may be noted that, in contrast to symmetric as well as asymmetric bifurcations of elastic columns (Sec. 4.5), there is no neutral equilibrium at the bifurcation state for the tangent modulus load (according to a second-order accurate solution). Unlike the elastic behavior, a lateral force encounters a positive stiffness (yet, the determinant of the tangent stiffness matrix $\mathbf{K}$ vanishes at the bifurcation point). The reason is that the corresponding eigenvector does not satisfy the unloading criterion, which is absent for elastic structures. For further discussion, see Section 10.4 on bifurcation.

In the literature, one finds statements that $P_t \leq P_{cr} \leq P_r$ where P_{cr} = critical load = limit of stable states. This is not true, however, as we already remarked ($P_{cr} = P_r$, see Sec. 10.3). It is, however, true that, according to the small-deflection (linearized) theory, the maximum load, P_{max}, which a deflected perfect column can reach (according to the small-deflection theory), satisfies the bounds:

$$P_t < P_{max} < P_r \tag{8.1.10}$$

The experimentally observed P_{max} values are often much closer to P_t than to P_r. Thus P_t is not only a safe lower bound for design but often also a good estimate of the carrying capacity of a column.

It can be shown that the replacement of E in the elastic solution by the tangent modulus E_t generally yields safe lower bounds for the critical loads for all types of uniaxially stressed structures, including columns with arbitrary end restraints, frames, and arches (see Prob. 8.1.7). This fact greatly simplifies analysis. It means that the critical loads of real structures, which are usually inelastic, can be obtained by elastic analysis.

Column Strength Curve

For design, it is convenient to describe buckling in terms of the so-called column strength curve, which describes the dependence of the critical stress of bifurcation σ_u on the slenderness L/r of the column ($r = \sqrt{I/A}$ = radius of inertia of cross section). For linearly elastic behavior, σ_u coincides with σ_E (Sec. 1.2) and so the

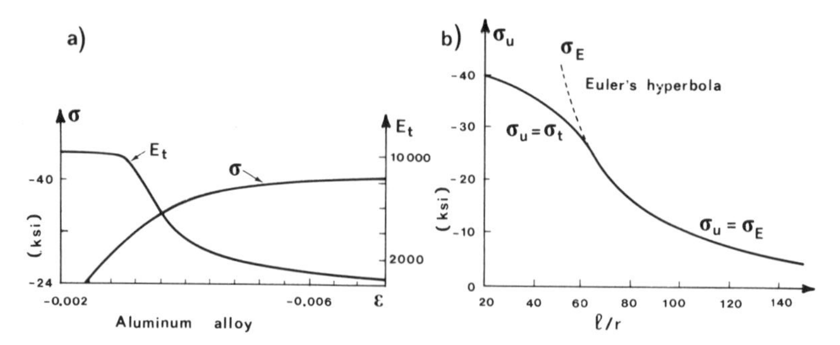

Figure 8.6 (a) Stress-strain diagram of aluminum alloy and (b) corresponding column strength curve. (*Adapted from Batterman and Johnston, 1967.*)

column strength curve coincides at high L/r with the Euler hyperbola (Fig. 8.6b). This hyperbola is valid up to stresses that do not exceed the proportionality limit of the stress-strain diagram. When this limit is exceeded, the column strength according to the tangent modulus theory is

$$-\sigma_t = \frac{P_t}{A} = \frac{\pi^2 E_t(\sigma_t) I}{L^2 A} = \frac{\pi^2 E_t(\sigma_t)}{(L/r)^2} \tag{8.1.11}$$

where E_t is the tangent modulus at the stress level $\sigma = \sigma_t$, L is the effective length of the column. In the literature, σ_t is often called (improperly) the critical stress, but in reality σ_t does not represent the critical stress, defined as the limit of stable states (see Sec. 10.3).

Equation 8.1.11 has the same form as Equation 1.2.8 for linearly elastic columns. However, unless $E_t = \text{const.}$ (i.e., the stress-strain diagram is bilinear), Equation 8.1.11 does not permit explicit calculation for σ_t; rather, it represents an implicit nonlinear equation because $E_t = E_t(\sigma_t) = \text{function of } \sigma_t$. The plot of σ_t versus L/r according to Equation 8.1.11 can, however, be easily obtained by choosing a sequence of σ_t values, evaluating E_t for each of them, and then calculating $L/r = \pi[E_t/(-\sigma_t)]^{1/2}$. As an example, Figure 8.6a shows the stress-strain diagram for an aluminum alloy (grade ASTM 6061–T6) and the corresponding variation of E_t. From these diagrams, one obtains the column strength curve in Figure 8.6b (Batterman and Johnston, 1967).

A caveat is in order. Due to the nonlinearity of Equation 8.1.11, its solution need not exist for all L/r (as already mentioned with regard to the equivalent equation 8.1.4). For example, if the stress–strain diagram is bilinear, there is no solution if $\pi\sqrt{E_p/\sigma_p} < L/r < \pi\sqrt{E/\sigma_p}$ where σ_p = proportionality limit and E_p = slope of the hardening branch beyond this point. For the same reason, if the slope E_t of the stress-strain curve becomes zero, as exhibited by an elastic–perfectly plastic material. Equations 8.1.8 can have no solution in the inelastic range. In such a case, Shanley's theory cannot be applied.

Postbifurcation Load-Deflection Diagram

The initial slope of the load-deflection diagram at $P = P_t$ is of interest for estimating the reserve capacity. Let $d\bar{w}$ be the first infinitesimal deflection increment at midspan from the perfect state. Assuming the deflection curve to be sinusoidal, $w = \bar{w} \sin (\pi x/l)$, we have at midheight the curvature increment $dw'' = -d\bar{w}\pi^2/l^2$. At $P = P_t$, the corresponding strain distribution is triangular (Fig. 8.5a), with $d\varepsilon = 0$ at the extreme point $z = c_1$ of the convex side where c_1 and c_2 are measured from the centroid. Therefore, $d\varepsilon = (c_1 - z)\, dw''$ (Fig. 8.5a). The corresponding increment dP of the axial stress resultant is obtained as $dP = -\int_A E_t\, d\varepsilon\, dA$ where A = cross-section area. Substituting for $d\varepsilon$ and then for dw'', and setting $E_t\pi^2/l^2 = P_t/I$, we obtain (for $P = P_t$)

$$\left[\frac{dP}{d\bar{w}}\right]_{\bar{w}=0} = \frac{P_t}{I}\int_A (c_1 - z)\, dA = \frac{P_t A c_1}{I} = \frac{P_t c_1}{r^2} \tag{8.1.12}$$

because $\int_A z\, dA = 0$. For an ideal I-beam, this equation yields $[dP/d\bar{w}]_0 = 2P_t/h$; for a rectangular cross section $[dP/d\bar{w}]_0 = 6P_t/h$.

Although computer analysis by finite elements is easy, an analytical solution of the postbifurcation behavior other than the initial slope is complicated, much more so than the preceding calculations. The main reason is that the boundary between loading and unloading in the cross section moves during buckling and its locations are not the same for various cross sections of the column.

To avoid these difficulties, we will study from now on the idealized pin-ended column considered by Shanley (1947); see Figure 8.7a. It consists of two rigid bars of length $l/2$, connected at midspan by a very short elastoplastic link (point hinge) of length $h \ll l$. This segment has the cross section of an ideal I-beam whose height we choose to be also h; then $I = Ah^2/4$ where A = cross-section area. (Note that changing the length of the plastic link from h to some value $h' \neq h$ is equivalent to replacing A with $A' = Ah'/h$.)

Let q_1 be the midspan deflection and q_2 the axial displacement on top of the column (positive for shortening). (It is important to realize that, in contrast to a column with a hinge, q_2 is kinematically independent of q_1 since the segment h

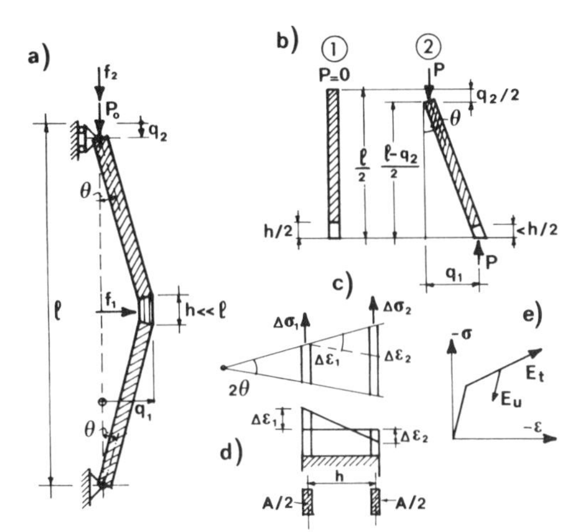

Figure 8.7 (a) Rigid-bar column with elastoplastic link (Shanley's column) (b) geometry of deformation, (c, d) strain and stress increments in link, and (e) loading and unloading moduli.

can shorten axially; moreover, in contrast to elastic behavior (cf. Sec. 4.3), this shortening is important.) The initial equilibrium of the undeflected column at some load $P = P_0$ is disturbed by applying at midheight a small lateral disturbing load f_1 while the axial load is raised by a small increment f_2 to $P = P_0 + f_2$. On the concave side, the deformable segment follows the tangent modulus, while on the convex side it follows modulus ξE_t where $\xi = 1$ if there is loading and $\xi = E_u/E_t > 1$ if there is unloading.

The compatibility of strains $\Delta\varepsilon_1$ and $\Delta\varepsilon_2$ at the concave and convex sides of a column with the displacements q_1 and q_2 requires that (see Fig. 8.7) $2 \tan \theta = (\Delta\varepsilon_2 - \Delta\varepsilon_1)h/h$ and $q_2 = -(\Delta\varepsilon_1 + \Delta\varepsilon_2)h/2 + l(1 - \cos\theta)$ where $\tan\theta = 2q_1/(l - q_2)$. Solving for $\Delta\varepsilon_1$ and $\Delta\varepsilon_2$, we have

$$\Delta\varepsilon_1 = \frac{l}{h}(1 - \cos\theta) - \frac{q_2}{h} + \frac{2q_1}{l - q_2} \tag{8.1.13a}$$

$$\Delta\varepsilon_2 = \frac{l}{h}(1 - \cos\theta) - \frac{q_2}{h} + \frac{2q_1}{l - q_2} \tag{8.1.13b}$$

where

$$\cos\theta = \left[1 + \left(\frac{2q_1}{l - q_2}\right)^2\right]^{-1/2}. \tag{8.1.13c}$$

The force and moment equilibrium conditions are

$$P - P_0 = f_2 = -\tfrac{1}{2}A(E_t\,\Delta\varepsilon_1 + \xi E_t\,\Delta\varepsilon_2) \tag{8.1.14}$$

$$Pq_1 + \tfrac{1}{4}f_1(l - q_2) = \tfrac{1}{4}Ah(\xi E_t\,\Delta\varepsilon_2 - E_t\,\Delta\varepsilon_1) \tag{8.1.15}$$

Shanley's (1947) original analysis corresponds to simplifying the solution by replacing $\cos\theta$ with 1 and $l - q_2$ with l in Equations 8.1.13 to 8.1.15. This simplification is possible if $q_2 \ll q_1$, which is true for small deflections. Then, after substituting Equations 8.1.13a and 8.1.13b into Equations 8.1.14 and 8.1.15, one gets

$$\frac{P - P_0}{P_t} = \frac{(\xi + 1)l}{2h^2}q_2 - \frac{\xi - 1}{h}q_1 \tag{8.1.16}$$

$$\frac{P}{P_t}q_1 + \frac{f_1 l}{4P_t} = \frac{\xi + 1}{2}q_1 - \frac{\xi - 1}{4}\left(\frac{l}{h}\right)q_2 \tag{8.1.17}$$

in which $P_t = AE_t h/l$ = tangent modulus load (this can be immediately verified by setting $f_1 = 0$ and $\xi = 1$ in Eq. 8.1.15, which expresses the moment equilibrium conditions, and substituting Eqs. 8.1.13a, b with $\cos\theta = 1$ and $l - q_2 = l$). Since q_2 now appears in Equations 8.1.16 and 8.1.17 linearly, it can be easily eliminated. After some rearrangement, this furnishes Shanley's formula:

$$P = \frac{P_0(\xi - 1)h + 4P_t\xi q_1}{(\xi - 1)h + 2(\xi + 1)q_1} \qquad (\text{for } f_1 = 0,\ q_1 \geq 0,\ q_2 \ll q_1). \tag{8.1.18}$$

If ξ = const., which happens for the idealized case of a bilinear stress-strain diagram, and if $P_t \geq \sigma_p A$ (σ_p = elastic limit, Fig. 8.2b), the curve $P(q_1)$ may be explicitly calculated from Equation 8.1.18. Now an important fact to note is that,

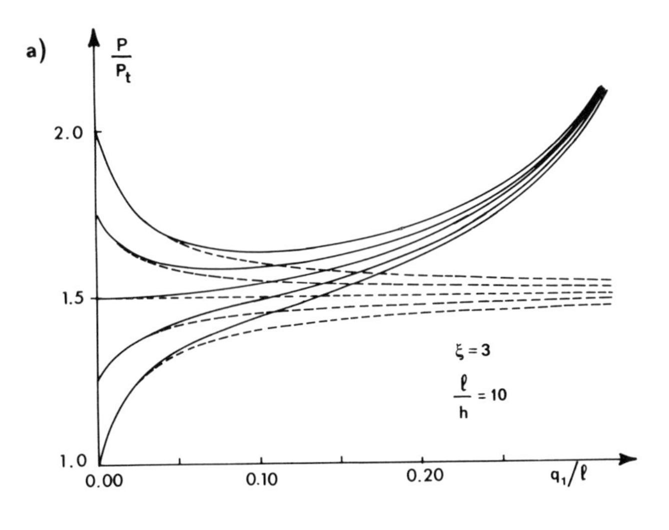

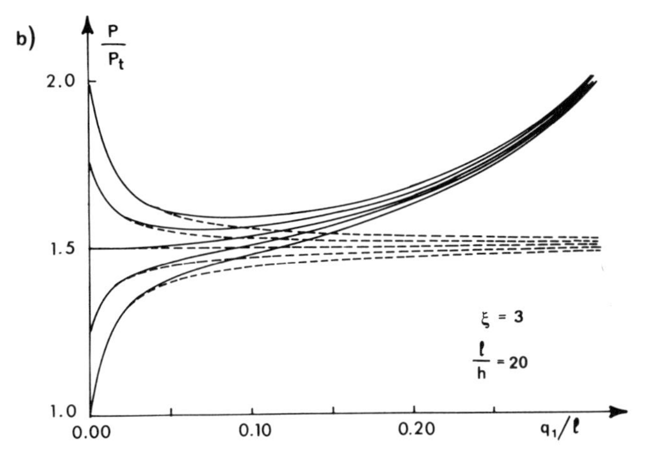

Figure 8.8 Load-deflection curves for Shanley's column: Shanley's approximate solution (dashed) and exact solution (solid).

for each initial value $P_0 \geq P_t$, there exists a distinct equilibrium path $P(q_1)$. These paths are plotted as dashed lines in Figure 8.8a, b for $\xi = 3$, and $l/h = 10$ and 20. The vertical line $q_1 = 0$ is the main equilibrium path. Thus we see that, beyond the point P_t, the main path consists of a continuous sequence of bifurcation points. This is a phenomenon not found in elastic stability theory. Later, on the basis of imperfection analysis (Sec. 8.2) as well as stability analysis (Sec. 10.3), we will see that the column must follow the path that bifurcates to the side at P_t and cannot reach any point above this path (unless the column is temporarily restrained). An exception is the case when a sudden drop in the slope of the σ-ε diagram causes $P_t < A\sigma_p$; then the state $P_0 = A\sigma_p > P_t$ and the corresponding path $P(q_1)$ can actually be reached by the column in a continuous loading process.

According to Equation 8.1.18, the curves $P(q_1)$ for all the initial values P_0

have a common horizontal asymptote (Fig. 8.1), that is,

$$\lim_{q_1 \to \infty} P = \frac{2\xi}{\xi + 1} P_t = P_r \qquad \left(\xi = \frac{E_u}{E_t} = \text{constant} \right) \tag{8.1.19}$$

where P_r denotes the reduced modulus load. That this must be so follows from the fact that P_r is defined as the value of P at which deflection q_1 can increase at constant load, and the asymptote is characterized by constancy of load ($dP/dq_1 = 0$). That P_r is given by Equation 8.1.19 can easily be verified by setting $f_1 = 0$, $f_2 = P - P_0$, and $\xi = E_u/E_t$ in Equations 8.1.14 and 8.1.15 and substituting Equations 8.1.13a, b with $\cos\theta = 1$, $l - q_2 = l$. (It should also be noted that the relation between P_r and P_t depends in general on the ratio E_r/E_t, and consequently Eq. 8.1.19 is valid for an idealized I-section.)

Differentiating Equation 8.1.18, one gets the slope of the equilibrium path that, at the onset of deflection ($q_1 = 0$), is given by $dP/dq_1 = [4\xi P_t - 2(\xi + 1)P_0]/h(\xi - 1)$, and, for $P_0 = P_t$, $dP/dq_1 = 2P_t/h$. If $P_0 = P_t$ the first infinitesimal increment of deflection occurs without unloading (Fig. 8.9a), because the resultant of the tensile stresses produced in the convex side flange by the bending moment, that is, $dM/h = P_0\, dq_1/h = P_t\, dq_1/h$, is exactly offset by the compressive resultant due to the increment of axial load, $dP/2 = P_t\, dq_1/h$. If $P_0 > P_t$, the first increment of deflection produces a larger tensile resultant, since P_0 is larger, and a smaller increment of the compressive resultant, since dP/dq_1 decreases with increasing P_0 (consequently, unloading occurs at the onset of deflection); see Figure 8.9a. Unloading then intensifies with increasing deflection along the equilibrium path (Fig. 8.9a).

For a curved stress-strain diagram, the value of ξ in Equation 8.1.18 decreases with increasing deflection. The dependence of ξ on ε_2 causes Equation 8.1.18 to become coupled with Equation 8.1.13b, with the result that the equilibrium deflection curve deviates downward from the curve for $\xi = \text{const.}$ and has no longer any asymptote but reaches at finite deflection a maximum value $P_{\max}$ that is less, and often much less, than the P_r value for the initial value of ξ at P_t; see Figure 8.9b. If E_t decreases rapidly with ε_2, which is often the case, $P_{\max}$ might be only slightly higher than P_t. The value of $P_{\max}$ is influenced also by the shape of the cross section. (When imperfections are considered, they too can further depress the $P_{\max}$ value; see Sec. 8.2.)

Among all the bifurcation states P_0, none represents a state of neutral

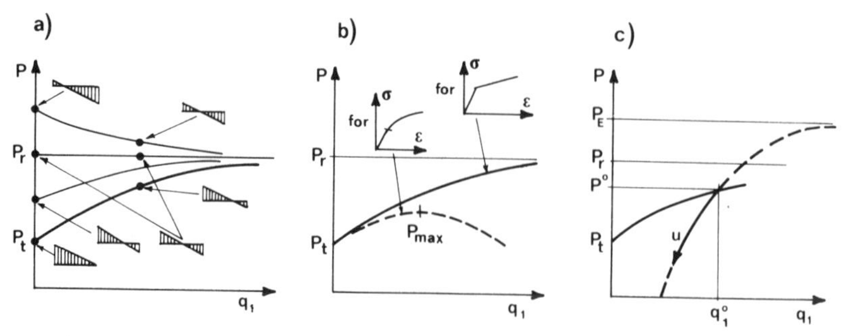

Figure 8.9 (a) Stress-increment distributions at various points on a load-deflection curve; (b) effect of curved stress-strain diagram; (c) column unloading.

equilibrium except $P = P_r$. However, even for deflections starting at $P = P_r$, the axial displacement q_2, unlike that in elasticity, increases with increasing deflections. Differentiating Equation 8.1.16 and substituting $dP/dq_1 = 0$, one gets:

$$dq_2 = \frac{\xi - 1}{\xi + 1}\left(\frac{2h}{l}\right) dq_1 \qquad \text{for } P = P_r \tag{8.1.20}$$

that is, dq_2 is of the same order as dq_1. In linear elasticity, by contrast, dq_2 at $P = P_{cr}$ is small of higher order than dq_1, that is, $dq_2/dq_1 \to 0$ at $P \to P_{cr}$ (Sec. 1.9).

An exact solution of the curves $P(q_1)$ can be obtained according to Equations 8.1.13a to 8.1.15. In general, for a curved stress-strain diagram, ξ is a function of ε_2, that is, $\xi = \xi(\varepsilon_2)$ where $\varepsilon_2 = \varepsilon_0 + \Delta\varepsilon_2$, ε_0 is the axial strain at $P = P_0$, $q_1 = 0$. By substituting Equations 8.1.13a–c into Equations 8.1.14 and 8.1.15 with $\xi = \xi(\varepsilon_2)$, one obtains a system of two nonlinear equations for q_1 and q_2. With a computer, the curve $P(q_1)$ (for $f_1 = 0$ or given f_1) can be solved easily. First one eliminates P from Equations 8.1.14 and 8.1.15 and obtains one nonlinear equation for q_1 and q_2. Then one chooses a sequence of q_1/l values, and for each of them one solves this equation for q_2/l and then calculates P/P_t.

The curves obtained by such calculations are shown for $\xi = \text{const.} = 3$ and $L/h = 10$, 20 as the solid curves in Figure 8.8a,b. It may be noted that for small q_1, these exact curves (solid curves) are very close to Shanley's approximate solutions (dashed curves). For larger q_1, these exact curves lie higher than the approximate solution (see Fig. 8.8a, b).

The most interesting property of the load-deflection curves in Figure 8.8a, b, approximately described by Equation 8.1.18, is the fact that there are infinitely many equilibrium bifurcation loads and infinitely many equilibrium paths, all of them infinitely close to each other. This property has no parallel in elastic stability theory. It is caused by the irreversible nature of inelastic strains, particularly the fact that, after unloading occurs, different stresses can correspond to the same strain (Fig. 8.2).

However, contrary to some statements found in the literature, the existence of a continuous series of bifurcation points does not mean that the critical load P_{cr}, defined as the limit of stable states, would have to be at P_t or somewhere below P_r; see Section 10.3. The bifurcation load is not necessarily the critical load for stability.

If the column has started to move along one equilibrium path $P(q_1)$, its state cannot shift to some other adjacent path. This is true, however, only as long as P increases monotonically, that is, the load-deflection curves in Figure 8.9a are valid only for increasing q_1. If q_1 starts to decrease (unloading of the column), the column behaves elastically, tracing backward the deflection curve for an elastic imperfect column; see curve u in Figure 8.9c and also Problem 8.1.14.

For arbitrary cross sections and arbitrary columns, the calculation of the postbifurcation load-deflection diagram is tedious and necessitates a computer. As a crude approximation, however, the formula for load P as a function of maximum deflection $\bar{w}$ may be assumed to have the same general form as Equation 8.1.18, that is, $P(\bar{w}) = (aP_0 + P_r\bar{w})/(a + \bar{w})$. Constant a may be determined from the initial slope $k = (dP/d\bar{w})$ at $\bar{w} = 0$, which is easy to calculate exactly for any column (see Eq. 8.1.12). Calculating the derivative, we find

$dP/dq_1 = (P_r - P_0)/a$ at $q_1 = 0$, from which $a = (P_r - P_0)/k$. So we obtain (Bažant, 1985)

$$P \simeq \frac{(P_r - P_0)P_0 + P_r k\bar{w}}{P_r - P_0 + k\bar{w}} \qquad (\text{for } E_t/E_u = \text{const.}) \tag{8.1.21}$$

Normally, one would substitute $P_0 = P_t$, but if $P_t < A\sigma_p$, then $P_0 = A\sigma_p$ (except when $A\sigma_p > P_r$, in which case $P_0 = P_r$).

Bifurcation in Plastic Structures with Multiaxial Stress

For multiaxially stressed structures, calculation of the tangent modulus load is complicated by problems of constitutive modeling. Consider first Hencky's deformation (total-strain) theory based on von Mises yield criterion (e.g., Fung, 1965). In this theory, the plastic hardening stress-strain relation for loading is

$$\varepsilon = \frac{\sigma}{3K} \qquad e_{ij} = \frac{s_{ij}}{2G_s} \qquad \frac{1}{G_s} = F(J_2) \tag{8.1.22}$$

where $\sigma = \sigma_{kk}/3$ = volumetric stress, $\varepsilon = \varepsilon_{kk}/3$ = volumetric strain, $s_{ij} = \sigma_{ij} - \sigma\delta_{ij}$ = deviatoric stresses, $e_{ij} = \varepsilon_{ij} - \varepsilon\delta_{ij}$ = deviatoric strains (subscripts i, j refer to Cartesian coordinates x_i, $i = 1, 2, 3$), K = bulk elastic modulus (constant), F = increasing function, G_s = secant shear modulus which depends on the second deviatoric stress invariant $J_2 = \frac{1}{2}s_{km}s_{km}$. The plastic deviatoric strain is $e^p_{ij} = e_{ij} - s_{ij}/2G$, where G = elastic shear modulus (constant). For the uniaxial test ($\sigma_{11} \neq 0$), $\sigma = \sigma_{11}/3$, $s_{11} = 2\sigma_{11}/3$, $s_{22} = s_{33} = -\sigma_{11}/3$ (all other $s_{ij} = 0$); Equation 8.1.22 yields $\varepsilon_{11} = (1/3K)(\sigma_{11}/3) + (1/2G_s)(2\sigma_{11}/3)$; since $\sigma_{11}/\varepsilon_{11} = E_s$ = secant (Young's) modulus (Fig. 8.10a), we get $1/G_s = 3/E_s - 1/3K$. To get the general tangent stress-strain relation, we note that $dJ_2 = s_{km}\, ds_{km}$, and upon differentiating Equation 8.1.22 and denoting $F' = d(1/G_s)/dJ_2 = -(dG_s/dJ_2)/G_s^2$, we get

$$d\varepsilon = \frac{d\sigma}{3K} \qquad de_{ij} = \frac{ds_{ij}}{2G_s} + \frac{F'}{2} s_{ij}\, dJ_2 \qquad dJ_2 = s_{km}\, d\varepsilon_{km} \tag{8.1.23}$$

Next consider the incremental theory of plasticity with associated flow rule

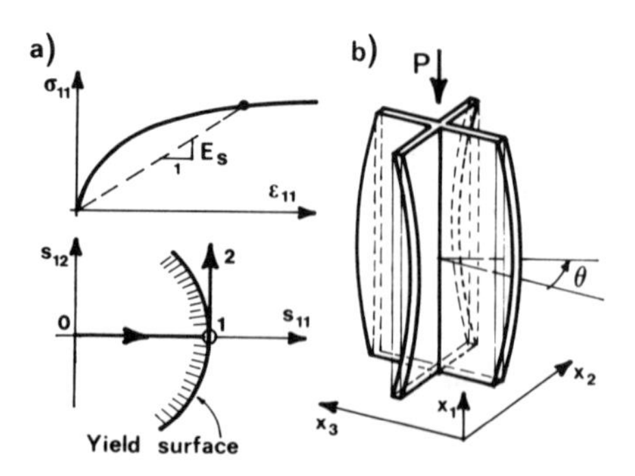

Figure 8.10 (a) Hardening plastic material and plastic stress increment tangential to the loading surface; (b) torsional buckling of cruciform column tested by Gerard and Becker (1957).

and normality (e.g., Fung, 1965 or Lin, 1968; cf. Eqs. 10.6.5–10.6.10). Assuming again the von Mises loading surface, we have the incremental stress-strain relation for loading:

$$d\varepsilon = \frac{d\sigma}{3K} \qquad de_{ij} = \frac{ds_{ij}}{2G} + de^p_{ij} \qquad de^p_{ij} = \frac{s_{ij}\, d\lambda}{\bar{\tau}} \qquad d\lambda = \frac{d\bar{\tau}}{2H} = \frac{s_{km}\, ds_{km}}{4H\bar{\tau}} \tag{8.1.24}$$

where $\bar{\tau} = \sqrt{J_2}$ = stress intensity, H = plastic hardening modulus for shear = function of J_2 or $\|\varepsilon^p_{11}\|$. For the uniaxial test, $J_2 = \sigma^2_{11}/3$ and Equation 8.1.24 yields $d\varepsilon^p_{11} = de^p_{11} = d\sigma_{11}/3H$ and $d\varepsilon_{11} = [(1/E) + (1/3H)]\, d\sigma_{11}$, from which $H^{-1} = 3(E_t^{-1} - E^{-1})$ where $E_t = d\sigma_{11}/d\varepsilon_{11}$ = tangent modulus in uniaxial test.

These two theories yield different plastic bifurcation loads. The difference is most blatant for a structure that is initially under uniaxial stress (primary path) but buckles due to shear (secondary path). This is the case for torsional buckling of a thin-wall column of cruciform cross section (Fig. 8.10b), for which $I_\omega = 0$ and, according to Equation 6.2.4, $P_{\text{cr}} = GJA/2I$ if the behavior is elastic ($J = 4bh^3/3$ = stiffness of cross section in simple torsion, $I = 2hb^3/3$, h = wall thickness, b = width of each of the four flanges). The plastic first bifurcation load P_t is obtained by replacing G with the tangent shear modulus G_t. For incremental shear deformation $ds_{12} \neq 0$ (with $ds_{11} = ds_{22} = ds_{13} = ds_{23} = 0$) we have $s_{km}\, ds_{km} = 0$ (because $s_{12} = 0$), and so $de_{12} = ds_{12}/2G_s$, that is, $G_t = G_s$ for this buckling mode. So, the first bifurcation stress is

$$\sigma_t = \frac{P_t}{A} = \frac{JG_s}{2I} = \left(\frac{h}{b}\right)^2 G_s = \left(\frac{h}{b}\right)^2 \left(\frac{3}{E_s} - \frac{1}{3K}\right)^{-1} \tag{8.1.25}$$

On the other hand, noting that $d\bar{\tau} = 0$, we get from Equation 8.1.24 $de_{ij} = ds_{ij}/2G$, that is, $G_t = G$ or $\sigma_t = G(h/b)^2$ where $G = [(3/E) - (1/3K)]^{-1}$. This means that the bifurcation load should be unaffected by plasticity—an untenable conclusion.

According to the famous experiments of Gerard and Becker (1957), Equation 8.1.25 is the correct one. It agrees with the tests very well. This noteworthy result has played an important role in the development of plasticity, revealing a fundamental deficiency of the classical incremental theory—the response to loading increments that are tangential to the loading surface (i.e., for which $s_{km}\, ds_{km} = 0$, for von Mises materials—see Fig. 8.10a) is mispredicted as elastic; in reality it is inelastic. Yet the deformation theory has other shortcomings: (1) when plastic loading resumes after unloading, ε^p_{ij} is in general predicted to have a discontinuous jump, and (2) the path independence implied by Equation 8.1.22 cannot be exactly true. A suitable remedy may be to endow the incremental theory of plasticity with a vertex (whose useful form has been formulated by J. Hutchinson at Harvard; also Bažant, 1987), or even better—use multisurface plasticity (cf. Secs. 10.6–10.7).

Finally consider a plate or shell in a uniform initial in-plane stress state $\sigma_{11} = N_{11}/h$, $\sigma_{22} = N_{22}/h$, with $\sigma_{12} = 0$. We have $3\sigma = \sigma_{11} + \sigma_{22}$, $3s_{11} = 2\sigma_{11} - \sigma_{22}$, $3s_{22} = 2\sigma_{22} - \sigma_{11}$, and $3dJ_2 = (2\sigma_{11} - \sigma_{22})\, d\sigma_{11} + (2\sigma_{22} - \sigma_{11})\, d\sigma_{22}$. Substituting this into Equation 8.1.23, we get the tangential stress-strain relation:

$$\begin{Bmatrix} d\varepsilon_{11} \\ d\varepsilon_{22} \\ d\varepsilon_{12} \end{Bmatrix} = \frac{1}{3} \begin{bmatrix} \frac{1}{3}K^{-1} + G_s^{-1} + \frac{1}{6}F'(2\sigma_{11} - \sigma_{22})^2 & \frac{1}{3}K^{-1} - \frac{1}{2}G_s^{-1} + \frac{1}{6}F'(2\sigma_{11} - \sigma_{22})(2\sigma_{22} - \sigma_{11}) & 0 \\ & \frac{1}{3}K^{-1} + G_s^{-1} + \frac{1}{6}F'(2\sigma_{22} - \sigma_{11})^2 & 0 \\ \text{sym} & & \frac{3}{2}G_s^{-1} \end{bmatrix} \times \begin{Bmatrix} d\sigma_{11} \\ d\sigma_{22} \\ d\sigma_{12} \end{Bmatrix} \quad (8.1.26)$$

The tangential compliance matrix seen in this equation is then used in quasi-elastic analysis of the first bifurcation load. For the incremental theory, a different matrix results from Equation 8.1.23. Better agreement with plastic buckling tests of plates is again obtained for the deformation theory (e.g., Bleich, 1952) (although according to Gjelsvik and Lin, 1987, the discrepancy of the incremental theory with experiments could be explained by edge friction).

For concrete structures, the tangential stiffness matrix needs to be based on damage theory (Chap. 13), and localization as well as creep needs to be taken into account.

Conclusion

To sum up, it is appropriate to quote von Kármán. In his discussion published with Shanley's (1947) celebrated paper, he epitomized Shanley's theory by stating that it determines "what is the smallest value of the axial load at which bifurcation of the equilibrium positions can occur, regardless of whether or not the transition to the bent position requires an increase of the axial load." The tangent modulus load obviously provides a *safe lower bound* on the critical load of an elastoplastic structure. This is the principal merit of Shanley's work.

Unfortunately, Shanley's theory does not indicate at which load a *perfect* column will actually start to deflect. This question can be theoretically settled for perfect columns only by stability analysis, which has been presented only recently (Bažant, 1988) and will be carried out in Section 10.3 on the basis of thermodynamics. We will see that, surprisingly, all the undeflected states for $P_t \leq P < P_r$ are stable though they cannot be reached in a continuous, stable loading path except when $P_t > A\sigma_p$. With this exception, the stable loading path is the deflected equilibrium path that emanates from P_t, and P_t is the limit of undeflected states that can be reached in a continuous loading process.

The fact that a perfect column must start to deflect at P_t can also be theoretically proven by considering the perfect column to be the limiting case of an imperfect column. This will be shown in the next section.

The first bifurcation load may be solved under the assumption that there is plastic loading everywhere, which leads to quasi-static analysis (see also Hill's method of linear comparison solid, Sec. 10.4). For structures with multiaxial stress, one needs to determine for this purpose the tangential compliance or stiffness matrix of the material, which is a difficult problem of constitutive modeling. These matrices differ for various plasticity theories if the buckling direction in the stress space ($\overrightarrow{12}$ in Fig. 8.10a) differs from the loading direction ($\overrightarrow{01}$) before buckling (they differ the most when the buckling direction is tangential to the loading surface; Fig. 8.10a). Hencky's deformation (total-strain) theory of plasticity generally gives results that agree with experiment much better than those of the incremental theory of plasticity.

Problems

8.1.1 Calculate the reduced modulus load for (a) an idealized I-beam, (b) a real I-beam, (c) a box beam, and (d) a tube.

8.1.2 Calculate the E_r values for buckling in the left and right directions of the z axis in Figure 8.11a, b, c, d, e.

8.1.3 Calculate E_r for buckling in the direction of the z'' axis of any orientation angle α'' in Figure 8.11f, g. Then find $\bar{\alpha}''$ for which E_r is minimum (this is the actual direction of buckling). Also calculate the direction angle α' for buckling at $P = P_t$ and compare α' with $\bar{\alpha}''$ (in general, they are not the same). *Hint:* Does z' coincide with the principal axis of inertia?

8.1.4 Considering an elastoplastic material that obeys the Ramberg–Osgood stress-strain law $\varepsilon/\varepsilon_0 = \sigma/\sigma_0 + m(\sigma/\sigma_0)^n$, with $m = \frac{3}{7}$, $n = 10$, $\sigma_0 = 315\ \text{N/mm}^2$, $E = 70{,}000\ \text{N/mm}^2$, and $\varepsilon_0 = \sigma_0/(0.7E)$, calculate the column strength curve in the inelastic range.

8.1.5 Assume that $E_t = E[1 - (\sigma/\sigma_\infty)^n]$ where n, σ_∞ are constants (σ_∞ = final asymptotic stress value). For $n = 1$ this yields the stress-strain curve $\sigma = \sigma_\infty[1 - \exp(-E\varepsilon/\sigma_\infty)]$, and for $n = 2$ the curve $\sigma = \sigma_\infty \tanh(E\varepsilon/\sigma_\infty)$ (which was proposed by Prager). Show that for both $n = 1$ and $n = 2$ an explicit formula σ_t as a function of L/r (column strength curve, Eq. 8.1.11) can be obtained.

8.1.6 Equations 8.1.4 and 8.1.9 represent implicit algebraic equations for P_r and P_t. Solve them, assuming the same Ramberg–Osgood material as in Problem

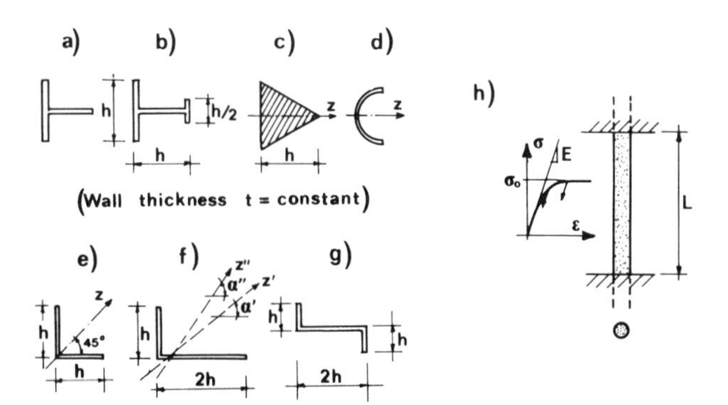

Figure 8.11 (a–g) Cross-section shapes of elastoplastic columns; (h) elastoplastic pipe under temperature and pressure loading.

8.1.4. Consider (a) an ideal I-beam, (b) a rectangular cross section, (c) a solid circular cross section, (d) a tubular cross section, and (e) a symmetric triangular cross section.

8.1.7 Consider a ring of radius R subjected to radial load p per unit length. The ring buckles (by deforming into an ellipse) for $p_{cr} = 3EI/R^3$ (Bresse formula, cf. Timoshenko and Gere, 1961, p. 291, and Sec. 2.8). (a) Showing that the initial axial stress in the ring is $\sigma = pR/A$ (A = cross-section area) and replacing E with $E_t = E[1 - (\sigma/\sigma_\infty)^n]$, $n = 1$, derive an explicit formula for p_{cr} as a function of R/r where $r^2 = I/A$ (the formula was given by Southwell, 1915). (b) Generalize this to a two-hinge arch (cf. Sec. 2.8).

8.1.8 For the system in Figure 8.12a consisting of Shanley's column loaded through a spring of spring constant C, find the dependence of the axial load P on deflection q_1 at midlength, assuming a bilinear stress-strain relation for the elastoplastic joint. *Hint:* Superimpose the effect of the spring on the diagram for $C \to \infty$.

8.1.9 Solve P_r, P_t, and the load-deflection diagrams of the inelastic systems in Fig. 8.12b, c, d, e, f, g, h, i, j, k, l. Assume for the elastic springs C = const. (>0) and for the elastoplastic links a bilinear stress-strain relation.

8.1.10 Calculate for a pin-ended column of length l the slope dP/dq_1 of the

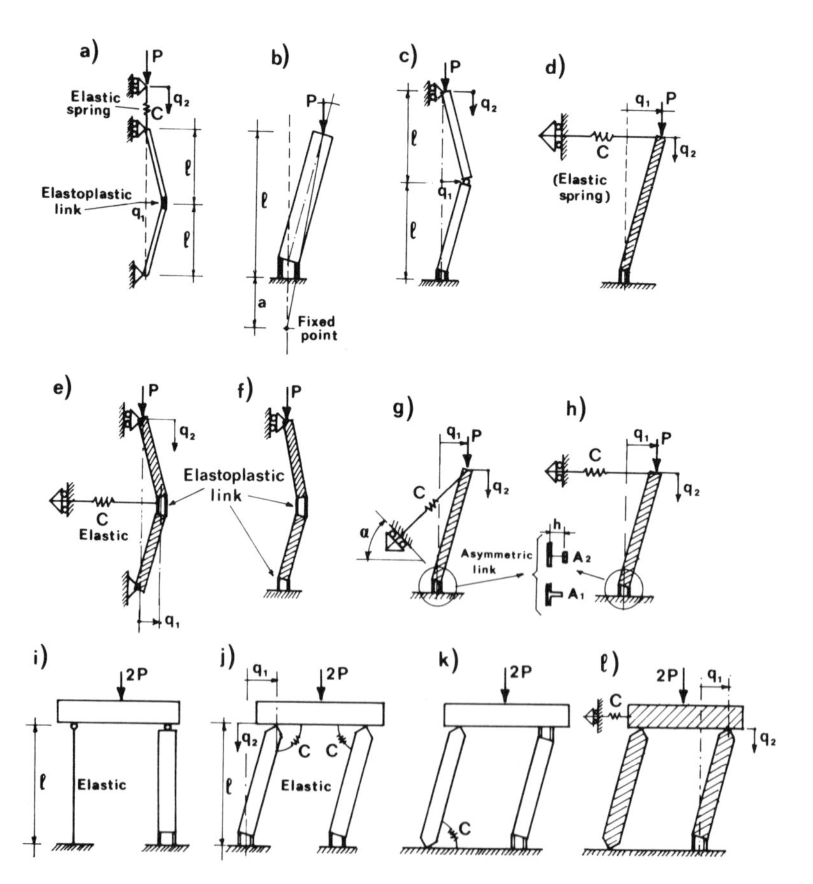

Figure 8.12 Exercise problems on buckling of elastoplastic systems.

load-deflection diagram $P(q_1)$ that emanates from Shanley's bifurcation point $P = P_t$, considering a sinusoidal deflection $w = q_1 \sin(\pi x/l)$. Assume (a) an ideal I-beam, (b) a rectangular cross section, (c) a tubular cross section, (d) a solid circle, and (e) a triangle. (*Hint:* Express dq_1 as well as dP as a function of curvature increment dq_1'', considering a triangular strain increment distribution according to Fig. 8.5a at P_t.) Determine approximately the q_1/L value for which $P = (P_t + P_r)/2$. Assume $\xi = 5 = \text{const.}$ (bilinear σ-ε diagram). Repeat for a fixed–fixed column and a fixed–hinged column of a thin tubular cross section and thin-wall rectangular section.

8.1.11 Calculate the ratio of the axial displacement increment to the maximum deflection increment, considering (a) the free-standing column in Figure 8.12b, (b) sinusoidal bending of a pin-ended column of a solid rectangular cross section, or (c) other cross sections such as the solid circular cross section, tubular cross section, and box cross section.

8.1.12 Using c_1 for buckling to the right and c_2 for buckling to the left, and considering a T–cross section, use Equation 8.1.12 to calculate the initial slopes $dP/d\bar{w}$ for buckling to the right and buckling to the left, starting at $P = P_t$. Note that they are not the same. (The curve that must be followed is that of smaller slope: see Sec. 10.2.)

8.1.13 Calculate the initial slope dP/dq_1 at $q_1 = 0$ for the case when $P_t < P_0 = A\sigma_p$ (which can happen for a bilinear σ-ε diagram). Consider cross sections: (a) ideal I, (b) rectangular, (c) tubular. (*Note:* The calculation is similar to that of P_r but simpler since the condition $\Delta P = 0$ is not present. Compare this slope with dP/dq_1 for $P_0 = P_t$ given below Eq. 8.1.19; it should always be smaller.)

8.1.14 Discuss the unloading behavior after a perfect column has previously deflected at monotonically increasing P. Realizing that at least initially the unloading must be elastic, calculate the unloading path from point (q_1^0, P^0) located on the loading curve $P(q_1)$ that emanates from P_t (Fig. 8.9c), considering Shanley's rigid-bar column in Figure 8.7. The result is sketched in Figure 8.9c. (*Hint:* The unloading curve is the same as the loading curve of an elastic imperfect column, but only until reverse yield begins.) What happens if the column is then reloaded to the previous maximum load, and beyond it?

8.1.15 (a) Solve P_r, P_t, and the load-deflection diagrams of the structure in Figure 8.12b assuming for the elastoplastic link $\sigma(\varepsilon) = E\varepsilon/(1 + k\varepsilon)$, $k = \text{constant}$. (b) Do the same for the other structures in Figure 8.12c–l.

8.1.16 (a) Solve the problem in Figure 8.12e with a nonlinear spring, $C = C_0(1 - kq_1)$ where C_0, $k = \text{constants}$. (When the link is elastic, this structure serves as a model for cylindrical shell buckling in compression; see Fig. 7.26.) The present structure is a model for elastoplastic shell buckling. (b) Do the same for the structure in Figure 8.12l.

8.1.17 Consider Eq. 8.1.25 for a cruciform column. (a) Assuming that each of the four flanges (Fig. 8.10b) has $b = 10h$, calculate the bifurcation stress P_t/A from Equation 8.1.25 considering the exponential uniaxial stress-strain relation (cf. Problem 8.1.5). (b) Repeat for the incremental theory of plasticity and compare the results. (c) Generalize the results for concrete assuming that $1/G_s = F(J_2, \sigma)$.

8.1.18 Consider a rectangular simply supported plate of sides a, b; $a = 2b$. (a) Let $\sigma_{11} = 2\sigma_{22}$, in which case the tangential compliance matrix in Equation

8.1.26 is orthotropic. Generalizing the solution from Equations 7.3.1 to 7.3.7 to an orthotropic material, derive the expression for the bifurcation load. (b) For the special case $\sigma_{11} = \sigma_{22}$, the tangent compliance matrix in Equation 8.1.26 is isotropic. Equating its elements E_{ij} to those of the isotropic elastic matrix with E_t, ν_t instead of E, ν, that is, $E_{11} = E_{22} = E_t/(1 - \nu_t^2)$, $E_{12} = \nu_t E_t/(1 - \nu_t^2)$, and $E_{33} = E_t/(1 + \nu_t)$, express the tangent E_t, ν_t in terms of σ_{11}, E_s, E, ν, and then, replacing D with $D_t = E_t h^3/12(1 - \nu_t^2)$, adapt Equation 7.3.7 to plastic buckling. (c) Assuming $\sigma_{11} \neq \sigma_{22}$ and $\sigma_{12} \neq 0$, generalize the matrix in Equation 8.1.26 and show that it is generally anisotropic. (d) Use the incremental theory of plasticity (Eq. 8.1.24) to derive the tangential compliance matrix analogous to Equation 8.1.26, and discuss the differences. (e) Repeat all for Tresca, Mohr–Coulomb, and Drucker-Prager yield surfaces. (f) Generalize to nonassociated theories (cf. Secs. 10.6–10.7).

8.1.19 The circular thin-wall pipe (thickness h, radius R) with built-in ends (Fig. 8.11h) is heated from temperature T_0, at which it is stress-free, to temperature T. Assume the uniaxial stress-strain relation $\sigma_{11} = E\varepsilon_{11}/(1 + E\varepsilon_{11}/\sigma_\infty)$ where E, σ_∞ = constants. Let α denote the thermal expansion coefficient. Let x_1, x_2 be the circumferential and axial directions. Determine the tangential compliance matrix from Equation 8.1.26 with $d\varepsilon_{22} = 0$ (as required for circumferential bending), and express the tangential cylindrical stiffness D_t. Then (a) calculate critical temperature T_{cr} at which the pipe starts to buckle, assuming pressure p in the pipe to be negligible. (b) With reference to Section 1.8, how does T_{cr} change if the pressure in the pipe is p? (c) In reality, the yield limit decreases with increasing temperature. How does T_{cr} change if $\sigma_\infty = \sigma_p(1 - kT)$ where σ_p, k = positive constants (and $p = 0$, E = const.)?

8.2 IMPERFECT COLUMNS AND STRUCTURES

Imperfections are inevitable and significantly influence the response of elastoplastic columns. Imperfections break symmetry. This eliminates path bifurcations and thus makes stability analysis of bifurcations unnecessary. The lack of symmetry, however, usually also makes the solution much harder. Since a perfect column must behave as the limiting case of an imperfect column with vanishing imperfections, analysis of very small imperfections makes it possible to decide which path is followed by a perfect column after bifurcation. Thus, analysis of imperfections can replace or supplement path stability analysis (Sec. 10.3), although it is normally more tedious.

Shanley's Rigid-Bar Column: Exact Solution

To illustrate the effect of imperfections, consider again Shanley's rigid-bar column (Figs. 8.7 and 8.13), for which the effect of imperfections is the easiest to solve. The calculations are particularly simple for an imperfection in the form of a small lateral load f_1 that is applied at midheight only after the axial load P has been raised to $P = P_t$. The column is initially straight and the axial load P is raised up to $P = P_t$ at zero deflection, $q_1 = 0$. At $P = P_t$, the disturbing load f_1 is suddenly applied and is then kept constant while the axial load P is increased further. For this type of lateral loading we have $\xi > 1$ as soon as the deflections start. If we

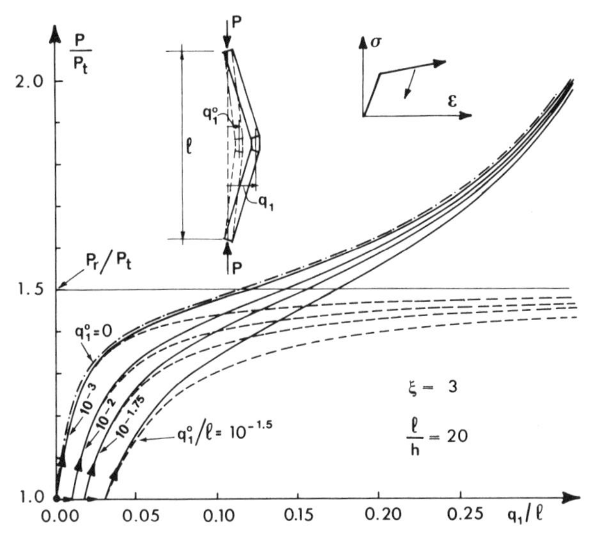

Figure 8.13 Load-deflection curves for an imperfect Shanley's column (bilinear material): exact solution (solid curves) and approximate solution (dashed curves)..

also assume the σ-ε diagram to be bilinear, that is, $\xi = \text{const.}$, then the exact equations 8.1.13 to 8.1.15 are valid through the entire deflection process, and ξ is known [i.e., need not be determined from a simultaneous nonlinear equation, $\xi = \xi(\varepsilon_2)$]. Eliminating P from Equations 8.1.14 and 8.1.15 and substituting Equations 8.1.13, one gets a nonlinear equation for q_1 and q_2. So, for each chosen value of q_1 one can calculate q_2 and then solve P from Equation 8.1.14. This solution is exact, even for finite deflections.

Figure 8.13 shows the load-deflection diagrams calculated in this manner for $\xi = 3 = \text{const.}$, $l/h = 20$, and various constant values of f_1 that were chosen so as to produce at $P = P_t$ the deflections $q_1^0/l = 10^{-1.5}$, $10^{-1.75}$, 10^{-2}, and 10^{-3}. Figure 8.13 also shows, for $q_1^0/l = 0$, the bifurcated load-deflection curve for the perfect column.

Now an important property that is demonstrated by Figure 8.13 is that, in the limit for $f_1 \to 0$, the load-deflection diagram $P(q_1)$ tends to the bifurcated branch of the solution for the perfect column that starts at $P_0 = P_t$ ($f_1 = 0$). It tends neither to the main branch, which corresponds to $q_1 = 0$, nor to some of the bifurcated branches that start at $P_0 > P_t$. Since a sufficiently small disturbing force f_1 must be considered to be inevitable, we now have a proof that the only solution of a perfect column that is actually possible is the load-deflection curve for which the deflection increase begins at P_t.

The foregoing argument reveals that stability analysis (Sec. 10.3) may be avoided by introducing an imperfection that breaks the symmetry of the system. However, that does not render stability analysis useless. Often the solution of the equilibrium path of an imperfect system is much more complicated than stability analysis of the perfect system, especially when arbitrary imperfections are considered.

An approximate solution of the foregoing problem may be obtained from Equations 8.1.16 and 8.1.17 (where $P_0 = P_t$). Eliminating q_2 from these two equations, we get (Bažant, 1985)

$$P = \frac{P_t(\xi - 1)h - \frac{1}{2}f_1 l(\xi + 1) + 4P_t\xi q_1}{(\xi - 1)h + 2(\xi + 1)q_1}. \tag{8.2.1}$$

The curves $P(q_1)$ plotted according to this simple formula are shown as the dashed lines in Figure 8.13. (Equation 8.2.1 for $f_1 \to 0$, of course, reduces to Shanley's formula in Eq. 8.1.18 for $P = P_t$.)

Arbitrary Imperfect Columns: Approximate Solution

From the design viewpoint, the most important effect of imperfections is that, in contrast to elastic columns, they normally decrease the maximum load P_{max} that the column can reach. The reason why such an effect has not been indicated by the preceding solutions (e.g., Eq. 8.2.1) is that ξ has been considered constant. In reality, ξ usually increases as the deflections grow. Then, however, analysis of Shanley's column ceases to be simpler than analysis of a more realistic deformable beam-column.

Therefore, to analyze deflections at increasing ξ, let us now consider a pin-ended beam-column of length l (Fig. 8.14a), with an imperfection in the form of a sinusoidal initial curvature given by the ordinates $z_0(x) = q_0 \sin(\pi x/l)$. We seek the deflection also as sinusoidal, given by the ordinates $z(x) = q \sin(\pi x/l)$. For the curvatures at midlength we have

$$w'' = z'' - z_0'' = -(q - q_0)\frac{\pi^2}{l^2}. \tag{8.2.2}$$

As a further simplification we assume the stress distribution between the concave and convex faces to be linear. This means that the portion of the stress-strain diagram between the strain values ε_2 and ε_1 for the concave and convex faces is replaced by a straight line. This line represents a chord of the segment of the actual stress-strain curve that corresponds to the strain values throughout the cross section; see Figure 8.14b. For an ideal I-beam, this assumption does not introduce any error since the stresses are needed only at two points.

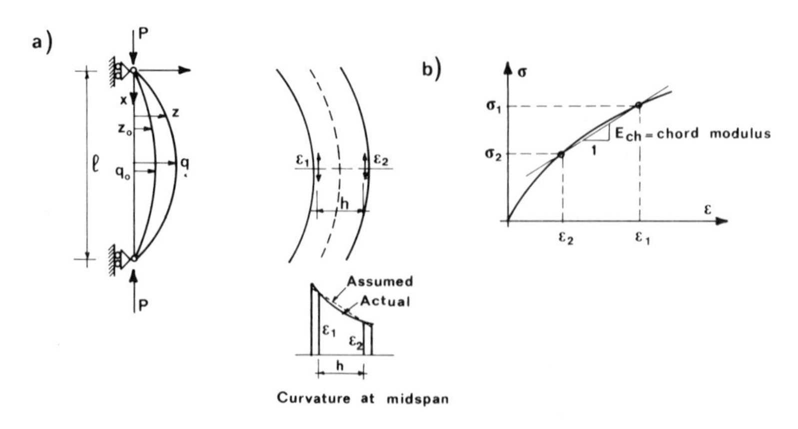

Figure 8.14 (a) Pin-ended column and (b) definition of chord modulus.

It may be noted that, for a general cross section, these two points would better be located so that, with a linear stress distribution, they would give rise to approximately the same bending moment as does the actual stress distribution. However, the locations of these two points cannot be given precisely since they would have to be varied with the stress level.

In view of the linearization of the stress distribution, the moment condition of equilibrium yields

$$w'' = \frac{M}{E_{ch}I} = \frac{-Pq}{E_{ch}I} \tag{8.2.3}$$

in which E_{ch} = chord modulus corresponding to the segment of the stress-strain curve between the strain values ε_2 and ε_1. Equating the last two expressions for the midlength curvature, w'', and solving for E_{ch}, we find

$$E_{ch} = \frac{Pl^2}{\pi^2 I}\left(\frac{q}{q - q_0}\right) \tag{8.2.4}$$

For the chord approximation of the stress-strain curve (Fig. 8.14b), we further have $\sigma_2 - \sigma_1 = E_{ch}(\varepsilon_2 - \varepsilon_1)$, in which σ_1, σ_2, ε_1, ε_2 are the stresses and strains at the concave and convex sides of the column. For the stress-strain curves we introduce the notations $\varepsilon_1 = \phi_1(\sigma_1)$ and $\varepsilon_2 = \phi_2(\sigma_2)$ where $\phi_1(\sigma_1)$ and $\phi_2(\sigma_2)$ are monotonically increasing functions describing the stress-strain relations for the concave and convex faces of the column. As long as the entire cross section undergoes loading, both functions are identical, that is, $\phi_1 = \phi_2 = \phi$, where $\varepsilon = \phi(\sigma)$ represents the stress-strain diagram for virgin loading. For large deflections, however, it normally happens that the convex face suffers strain reversal. In that case, function ϕ_2 must describe the unloading stress-strain diagram, which is different from $\phi_1(\varepsilon_1)$.

Expressing the stresses according to the bending theory, we thus have

$$\sigma_2 - \sigma_1 = E_{ch}[\phi_2(\sigma_2) - \phi_1(\sigma_1)] \tag{8.2.5}$$

$$\sigma_1 = -\frac{P}{A} - \frac{Pqh}{2I} \qquad \sigma_2 = -\frac{P}{A} + \frac{Pqh}{2I} \qquad \sigma_2 - \sigma_1 = \frac{Pqh}{I} \tag{8.2.6}$$

Finally, substituting Equation 8.2.4, we obtain the following nonlinear algebraic equation relating the axial load P and the maximum (midspan) deflection q:

$$\frac{\pi^2 h}{l^2}(q - q_0) = \phi_2\left(-\frac{P}{A} + \frac{Pqh}{2I}\right) - \phi_1\left(-\frac{P}{A} - \frac{Pqh}{2I}\right) \tag{8.2.7}$$

To solve this equation, one starts with $q = q_0$ = initial imperfection, and increments the value of q in small intervals. For each q, Equation 8.2.7 is solved by Newton iteration, using as the initial value the solution obtained for the previous q-value. Unless the initial curvature of the column is so high that the axial force resultant would be outside the core of the cross section, the entire cross section at first undergoes loading, that is, $\phi_1 = \phi_2 = \phi$. At a certain eccentricity, however, the convex face experiences strain reversal and afterward it is unloading. At the moment of reversal, function ϕ_2 must be reset so as to represent the unloading stress-strain diagram from the point of reversal.

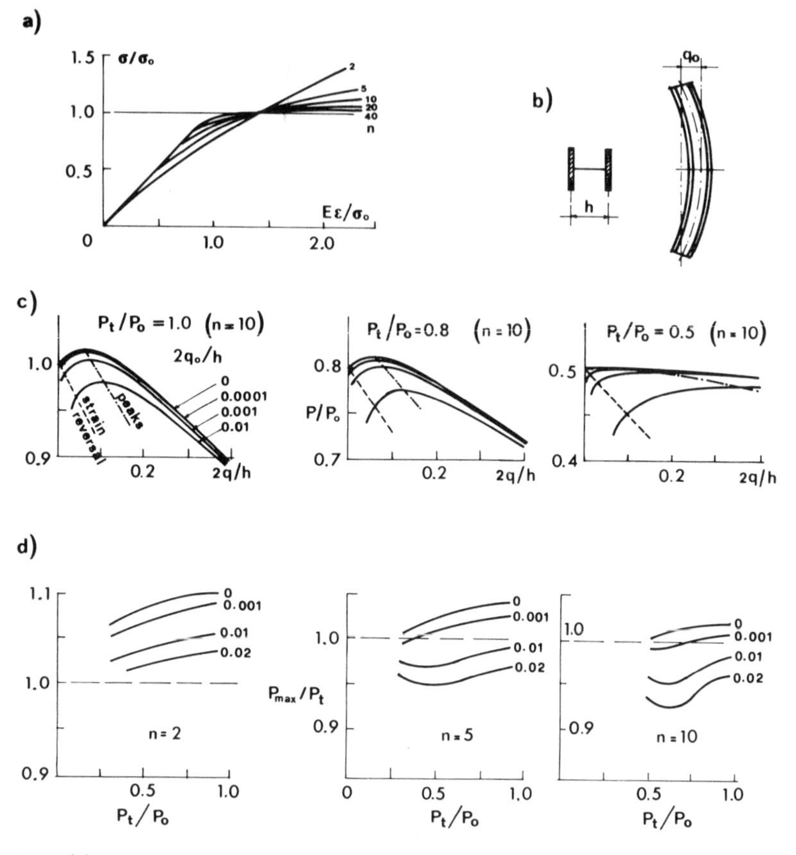

Figure 8.15 (a) Ramberg-Osgood stress-strain relation for various n values, (b) I-beam cross section, (c) load-deflection curves, and (d) curves of P_{max}. (*After Wilder, Brooks, and Mathauser, 1953.*)

Equation 8.2.7 was solved by Wilder, Brooks, and Mathauser (1953) for the stress-strain relation of Ramberg and Osgood (1943), which reads $\varepsilon/\varepsilon_0 = \sigma/\sigma_0 + m(\sigma/\sigma_0)^n$; $m = \frac{3}{7}$ and σ_0, ε_0 are the stress and strain associated with the secant drawn from the origin at a slope of $0.7E$ where E is the initial elastic modulus. The plot of this stress-strain curve for different values of the constant n is shown in Figure 8.15a, from which it appears that increasing values of n give a sharper "knee" of the curve. Values of n in the range 3 to 5 correspond to the typical behavior of high-strength high-alloy steels, values around 10 correspond to aluminum alloys, and values around 30 to 40 correspond to mild steel (but residual stresses are ignored).

Some of the results of these calculations, performed for an I-beam cross section with a massless web (Fig. 8.15b), are shown in Figure 8.15c, d. Figure 8.15c shows the load-deflection curves for $n = 10$ for different values of the initial deflection q_0 and for columns of different slendernesses, characterized by the values of the ratio between the tangent modulus load P_t and the squash load $P_0 = A\sigma_0$.

As the column becomes more slender, inelastic behavior disappears, and thus the maximum load approaches the Euler load. Figure 8.15d shows, for different values of n and of the initial imperfection, the values of P_{max} in relation to P_t. Obviously, imperfections can indeed significantly reduce P_{max}. From Figure 8.15c, d it is evident that, if the initial imperfections are large and if the stress-strain diagram is nearly elastic-perfectly plastic (larger n values), the maximum load of an imperfect column can even be smaller than the tangent modulus load of the perfect column.

The solution just demonstrated, in which the equilibrium is satisfied exactly only at two points of the cross section, represents what is generally known as the collocation method. Alternatively, the equilibrium conditions could be approximately satisfied in a global manner, for example, by using the principle of virtual work. This approach, however, yields more complicated equations.

Exact solutions, which were first obtained by Ježek (1934) (in terms of elliptic integrals) for an elastic-perfectly plastic σ-ε diagram and by Chwalla (1934, 1935) for a curved σ-ε diagram (but ignoring residual stresses), must take into account the fact that the deflection shape is not sinusoidal but varies during deflection. Accurate solutions are today best obtained by finite elements.

Effect of Cross-Section Nonsymmetry

An interesting behavior is found when the cross section is not symmetric (Fig. 8.16). Nonsymmetry does not influence the tangent modulus load (bifurcation load), but can strongly depress the maximum load P_{max} and the initial slope of the load-deflection curve.

The reason becomes clear by recalling that the reduced modulus loads P_r^R and P_r^L for buckling to the left and right are different (Eqs. 8.1.6 and 8.1.7). The load-deflection curves that emanate from the tangent modulus load (Fig. 8.16a) approach P_r^R and P_r^L asymptotically if $E_t/E_u = \text{const.}$, and so one curve must attain higher values than the other; see curves 1. Curves 2 for real columns, for which E_t/E_u is decreasing, deviate from curves 1 downward, and as is clear from Figure 8.16a, their peaks may be expected to occur at different heights. Calculations confirm that; see Figure 8.16b showing the numerical results of

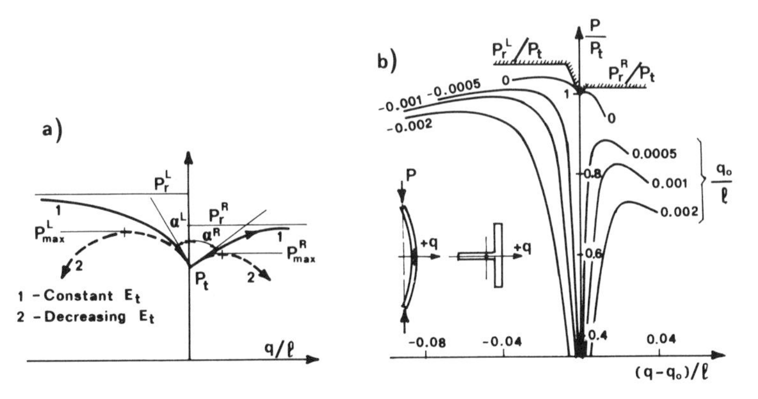

Figure 8.16 (a) Effect of cross-section asymmetry on load-deflection curve; (b) numerical results of Hariri. (*After Johnston, 1976, p. 47.*)

Hariri (1967) for some typical pin-ended columns of a T–cross section, reported also by Johnston (1976, p. 47).

It is important to note that the initial slopes α for buckling to the left and to the right are not the same (cf., Eq. 8.1.12 and Prob. 8.1.12) because they are proportional to the static (first) moment of the cross section about an axis tangent to the cross-section boundary at the extreme point to the right and to the left. The deflection curve that is actually followed from the bifurcation point at $P = P_t$ is that of smaller slope (this is proven in Sec. 10.3).

Problems

8.2.1 Solve Equation 8.2.7 explicitly for P as function of q using for the stress-strain relation the following formulas: (a) $\varepsilon = k \ln[1 + (\sigma/Ek)]$; (b) $\varepsilon = k \ln F(\sigma) + C_1$, with either $F(\sigma) = (1 + c\sigma)/(1 + b\sigma)$ or $F(\sigma) = 1 + c\sigma + b\sigma^2$; (c) $\varepsilon = E\sigma(1 - b\sigma)$; and (d) $\varepsilon = E\sigma/(1 + b\sigma)$, where k, b, c = constants. Obviously, the solution will be valid only until the point where strain ε_2 on the convex side reverses. Find this point ($d\varepsilon_2/dq = 0$).

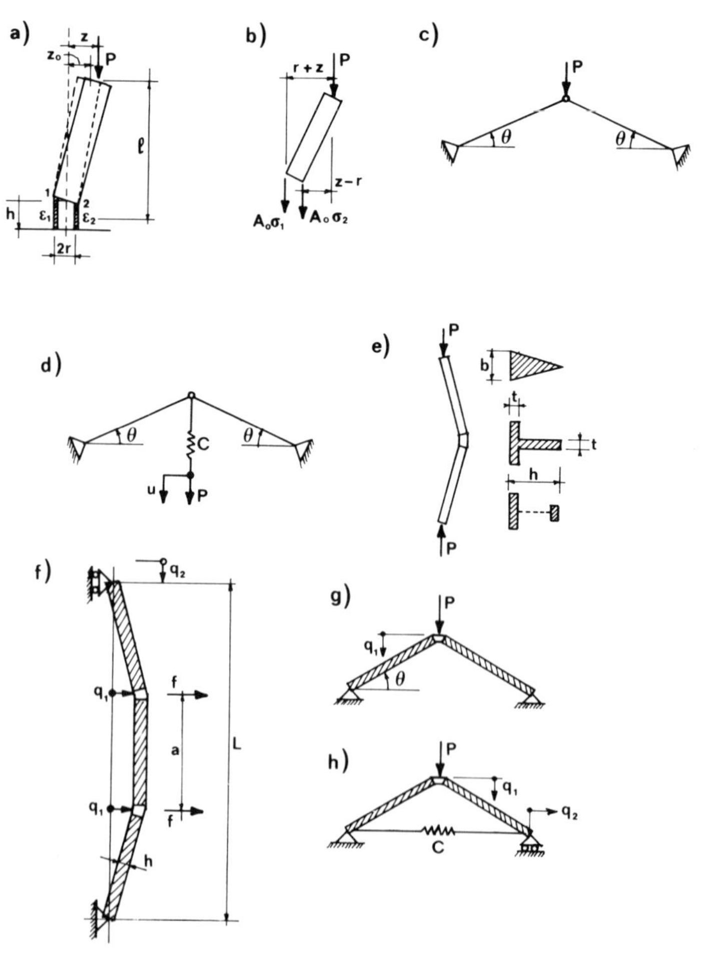

Figure 8.17 Exercise problems on buckling of elastoplastic systems.

8.2.2 Solve the load-deflection curve for the column supported by two elastoplastic links shown in Figure 8.17a. Let z_0 = initial imperfection. Assume $\varepsilon = k \ln(1 + \sigma/Ek)$ for the stress-strain curve of the links; $k = 0.01$, $l/r = 100$, $h = r$. Plot the results for various values of z_0/r. *Hint:* Express σ_1 and σ_2 from the equilibrium conditions (Fig. 8.17b) $P(r + z) + A_0\sigma_2 \cdot 2r = 0$, $P(r - z) - A_0\sigma_1 2r = 0$, and use the geometric relation $(\varepsilon_1 - \varepsilon_2)h/2r = (z - z_0)/l$.

8.2.3 Do the same, but consider Shanley's column, without or with an added axial spring (Fig. 8.12a). The initial imperfection is characterized by a given finite value $q_1 = q_1^0$ at $P = 0$.

8.2.4 Consider the same problem as solved in the text. According to the principle of virtual work, global equilibrium requires that $\delta W = \int_l \int_A f(\varepsilon)\, \delta\varepsilon\, dA\, dx - \int P[\delta\lambda + \delta(z'^2/2)]\, dx = 0$ where $f(\varepsilon) = \sigma$. Substituting $z = a \sin(\pi x/l)$, $z_0 = a_0 \sin(\pi x/l)$, and $\varepsilon = (z'' - z_0'')\zeta$ where ζ = transverse coordinate measured from the deflected centroidal axis, and λ = strain at centroid, express P and discuss calculation of the curve P versus a by incremental loading.

8.2.5 Calculate the curve $P(\theta)$ for the elastoplastic von Mises truss in Figure 8.17c, which exhibits an inelastic version of snapthrough. The bars do not buckle, they only shorten (elastoplastically). Consider the stress-strain relations: (a) $\sigma = E(\varepsilon - k\varepsilon^2)$; (b) $\sigma = E\varepsilon/(1 + k\varepsilon)$; (c) $\varepsilon = \sigma/E + k\sigma^n$; and (d) $\sigma = E\varepsilon(1 + k\varepsilon)/(1 + c\varepsilon)$ where k, c = constants ($k \leq c$). Find the critical load and critical θ. Compare the results with the elastic analysis in Section 4.4.

8.2.6 Calculate the load-displacement curve $P(u)$ for the elastoplastic von Mises truss in Figure 8.17d. Except for the added spring, everything is the same as in Problem 8.2.5. Compare the results with the elastic analysis in Section 4.4.

8.2.7 Consider again the elastoplastic von Mises trusses in Problems 8.2.5 and 8.2.6, and analyze their responses during unloading and reloading.

8.2.8 Consider Shanley's column (Fig. 8.7) in which, however, the deformable link at midspan does not have the cross section of an ideal I-beam but rather a nonsymmetric cross section: (a) asymmetric ideal (webless) I-section, (b) thin-wall T-section ($t \ll h$), (c) triangular cross section (Fig. 8.17e). Solve the load-deflection curves assuming various initial imperfections (out-of-straightness). Calculate the deflection curves both to the right and to the left. (*Note:* They will be different.) Comment on P_{max}.

8.2.9 Solve the response of the column in Figure 8.17f, perfect or imperfect. Consider only symmetric deflection modes. Analyze the effect of lateral forces f as an imperfection.

8.2.10 (a) Consider the von Mises truss made of rigid bars and a deformable inelastic link (Fig. 8.17g). Assuming $P = 0$ at $\theta = \theta_0 > 0$, and $\sigma = E\varepsilon/(1 + k\varepsilon)$, discuss the load-deflection path. Is there any bifurcation? (b) Same problem for the two-degrees-of-freedom system in Figure 8.17h.

8.3 EFFECT OF RESIDUAL STRESSES

The analysis presented so far in this chapter does not seem to apply to hot-rolled structural steels for which E_t (and thus also P_t) suddenly drops to zero. However, the picture is completely changed by the presence of large residual stresses. They make Shanley's theory applicable, too. We will discuss it in this section.

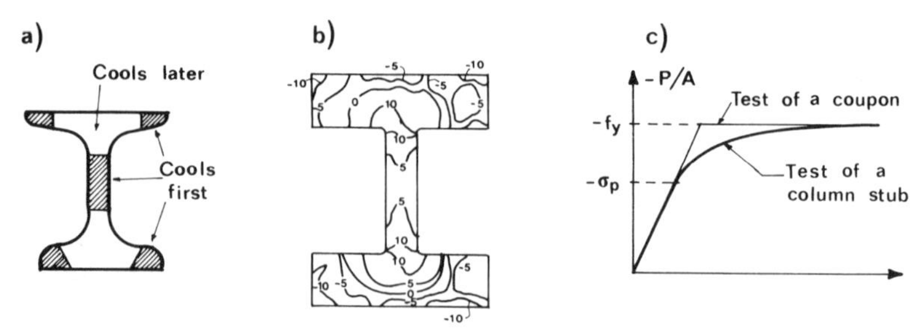

Figure 8.18 (a) Cooling sequence of hot-rolled beams; (b) residual stress contours (*adapted from Galambos, 1988, p. 36*); (c) relation between average stress and strain for a column cross section.

The disagreement between Euler's critical load and the experiments made in the early 1900s was attributed at first to the neglect of elastoplastic behavior and imperfections. After the theory had been modified to take these two phenomena into account, as already described in Section 8.2, good agreement between theory and experiment was achieved for certain elastoplastic materials, such as aluminum alloys. This was of considerable importance for aircraft design.

Serious discrepancies nevertheless persisted for tests of hot-rolled steel beams used in structural engineering. Subsequent research led to the recognition that the chief source of the discrepancy between theory and experiment was not the imperfections but residual stresses (this was suggested already in 1908 by Howard and in 1921 by Salmon.) Systematic studies (see, e.g., Osgood, 1951; Yang, Beedle, and Johnston, 1952; Beedle and Tall, 1960) in the 1950s and 1960s have established that the residual stresses have in fact a major effect on the load capacity of short hot-rolled steel columns while the effect of imperfections is secondary in comparison.

The residual stresses arise from nonuniform cooling of hot-rolled beams during their production. The rate of cooling is roughly inversely proportional to the square of their thickness. The tips of the flanges of an I-beam cool the fastest, and the flange web intersections the slowest (Fig. 8.18a). Thus, when the white regions in Fig. 8.18a contract, the shaded regions have already cooled and hardened, and therefore resist the contraction. This causes residual stresses that are compressive at the tips of flanges and tensile in the central parts of flanges. As an example, Figure 8.18b (adapted from Galambos, 1988, p. 36) shows the residual stress contours measured in a heavy hot-rolled shape.

Residual stresses are also caused by welding. After cooling, the region of the weld is normally in tension, and the rest of the cross section in compression. The plastic deformation due to the punching of holes for rivets or bolts or for cutting also causes residual stresses. Stress concentrations due to holes have a similar effect as the residual stresses, since they cause a part of the cross section to yield before the average stress reaches the yield limit.

Calculation of the Effect of Residual Stresses

As a consequence of the residual stresses, the yield limit is reached earlier when a compressive force is applied on the column. Even though the stress-strain curve $\sigma(\varepsilon)$ of mild structural steel is essentially elastic-perfectly plastic, one obtains for the column as a whole an average stress-strain curve of a gradually decreasing slope, as shown in Figure 8.18c. The curve of average stress versus strain for a cross section can be either determined experimentally or calculated on the basis of knowledge of the actual stress-strain diagram of the material and of the residual stress distribution. In terms of the average stress, the column behaves as if its apparent yield limit σ_p were much lower than the actual yield limit f_y; $\sigma_p = f_y - \sigma_r$ where σ_r is the maximum residual stress. When this apparent yield limit is exceeded, the parts of the cross section that have large residual compressive stresses yield plastically and thus reduce the average tangent modulus. Obviously, this engenders a reduction of Shanley's tangent modulus load, P_t.

Let us now show how to calculate the effect of given axial residual stresses $\sigma^0(y, z)$ on P_t. We assume $\sigma^0(y, z)$ to depend only on the cross-section coordinates y, z and to be independent of the axial coordinate x. The column is assumed to be perfect, and so there is no deflection until the tangent modulus load P_t is reached. The material stress-strain curve $\sigma(\varepsilon)$ is given, and its slope is $E_t(\varepsilon) = d\sigma(\varepsilon)/d\varepsilon$. If axial strain ε_1 is produced in the column by the applied load, the total strain is $\varepsilon = \varepsilon_1 + \varepsilon^0(y, z)$, where ε_1 is the load-produced strain and $\varepsilon^0(x, z) = \sigma^0(y, z)/E =$ the residual strain. When the column begins to buckle from its initial perfectly straight state, the bending moment (about the y axis) is $M = \int (E_t \kappa z) z \, dA = R_t \kappa$. Here $\kappa =$ curvature and $R_t =$ initial incremental (tangential) bending rigidity, which may be calculated as

$$R_t(\varepsilon_1) = \int_A E_t(\varepsilon) z^2 \, dy \, dz \qquad \varepsilon = \varepsilon_1 + \varepsilon^0(y, z) \tag{8.3.1}$$

where $A =$ cross-section area. Since Shanley's tangent modulus load must be equal to the resultant of axial stresses σ, we have the condition:

$$P_t(\varepsilon_1) = \frac{\pi^2}{L^2} R_t(\varepsilon_1) = -\int_A \sigma(\varepsilon) \, dy \, dz \qquad \varepsilon = \varepsilon_1 + \varepsilon^0(y, z) \tag{8.3.2}$$

where $L =$ effective length (Sec. 1.4) of the column. Substituting Equation 8.3.1 for R_t, we have a nonlinear algebraic equation for the axial strain ε_1 at which load P_t is reached. Strain ε_1 may be easily solved, for example, by the iterative Newton method. Then Shanley's load P_t is evaluated as $P_t(\varepsilon_1)$.

For the special case of an elastic–perfectly plastic material (mild steel), $E_t = 0$ in the yielded zone, and $E_t = E =$ elastic modulus outside that zone. So $R_t(\varepsilon_1) = \int_{A_e} E z^2 \, dy \, dz = E I_e$ and $P_t(\varepsilon_1) = -A_e E \varepsilon_1 - f_y(A - A_e)$, in which $f_y =$ yield stress (>0), A_e is the elastic portion of the cross section, and I_e is the moment of inertia of this portion of the cross section.

Examples

As an example, consider the wide-flange I-beam shown in Figure 8.19b. We consider the bending about the weak axis y, for which the effect of the residual stresses, concentrated at the flange tips, is more pronounced than for bending about the strong axis z. The effect of the web on bending rigidity is, for the weak axis bending, negligible. We assume the residual stress to vary monotonically from the middle of the flange to its tip (Fig. 8.19a). Denoting by ζ the length of

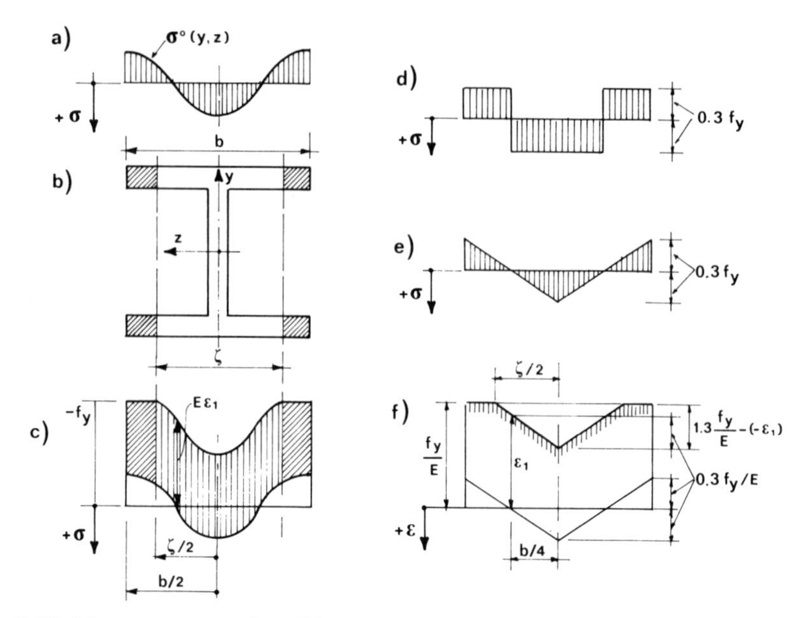

Figure 8.19 I-beam cross section (b); assumed residual stress distributions (a, d, e); and stress distributions at tangent modulus load (c, f).

the elastic segment of the flange (Fig. 8.19c), we have $R_t(\varepsilon) = Et\zeta^3/6$, in which t = flange thickness. Thus, Equation 8.3.2 takes the form:

$$P(\varepsilon_1) = -2E\varepsilon_1(\zeta)t\zeta + 2f_y t(b - \zeta) - 4t\int_{\zeta/2}^{b/2} (-\sigma^0)\, dz$$

$$= \frac{\pi^2}{L^2}\left(\frac{Et}{6}\right)\zeta^3 = \frac{\pi^2}{L^2} R_t(\varepsilon_1) \tag{8.3.3}$$

in which b = flange width, and the load-produced strain ε_1 can be considered as a function of ζ if the shape of the distribution of the residual stress $\sigma^0(y, z)$ is specified. The minus sign appears because P is positive for compression while the stress and strain are negative for compression. The value of P is equal to the shaded area in Figure 8.19c. Solving ζ from Equation 8.3.3, we obtain the tangent modulus load, P_t. Consider now two special cases:

Case 1. The residual stresses are distributed rectangularly, as shown in Figure 8.19d, and have the magnitude $0.3f_y$. In this case, it is necessary that $\zeta = b/2$, and Equation 8.3.3 yields the tangent modulus load

$$P_t = \frac{\pi^2 Etb^3}{48L^2} \tag{8.3.4}$$

Note that in this case the diagram of average stress $-P/A$ versus strain ε_1 is trilinear, the third segment being horizontal. Due to the discontinuity of slope of this diagram, however, P_t might not exist because Equation 8.3.4 might yield a value that is less than the P value at the first yield (see Sec. 8.1).

Case 2. A more realistic assumption is a triangular distribution of σ_0 (Fig. 8.19e). From the linear strain distribution in Figure 8.19f we figure out $\varepsilon_1 = -(1.3 - 0.6\zeta/b)f_y/E$ for $\varepsilon_1 \leq -0.7f_y/E$. Substituting this into Equation 8.3.3 and solving for ζ we can calculate the tangent modulus load from its expression as a function of the nonyielding part of the cross section, $P_t = \pi^2 EI_e/L^2 = \pi^2 Et\zeta^3/6L^2$. Alternatively, we could calculate the average stress-strain curve by substituting the expression for ζ as a function of ε_1 into the equation that gives the average stress $\sigma_{av} = -P_t(\zeta)/A = -f_y(1 - 0.3\zeta^2/b^2)$. The tangent modulus then becomes $E_t = d\sigma_{av}/d\varepsilon_1 = (d\sigma_{av}/d\zeta)(d\zeta/d\varepsilon_1) = E\zeta/b$, and P_t is obtained from Equation 8.1.11.

According to the latest studies at Lehigh University (cf. Chen and Lui, 1987, p. 122), the actual residual stress distributions usually are intermediate between the triangular and parabolic ones.

Problems

8.3.1 Derive the column strength curves for bending about the weak axis of the wide-flange beam shown in Figure 8.19b, for the rectangular and triangular residual stress distribution shown in Figure 8.19d, e.

8.3.2 Solve the same as Problem 8.3.1, but consider a parabolic residual stress distribution of maximum residual stress $0.3f_y$.

8.3.3 Solve the same as Problems 8.3.1 and 8.3.2 but for strong-axis bending.

8.3.4 Considering the linear residual stress distribution in Figure 8.19d and calculating $P(\varepsilon_1) = P(\zeta)$ according to Equation 8.3.3, check that $-dP/d\varepsilon_1 = E$ for $\xi = b$ and $dP/d\varepsilon_1 = 0$ for $\xi = 0$.

8.4 METAL COLUMNS AND STRUCTURES: DESIGN AND CODE SPECIFICATIONS

With various finite element or finite difference computer programs that have recently come into existence, it is now feasible to design columns and frames on the basis of the maximum load P_{max} (i.e., the load capacity). Such computer solutions are in close agreement with test results provided that they take into account: (1) the finite deflections and the equilibrium conditions on the deflected structure (which is, of course, the essence of buckling analysis); (2) the actual nonlinear stress-strain diagram; (3) the residual stresses from hot-rolling or cold-forming, as well as stress concentrations around holes, etc.; (4) the imperfections (inevitable initial crookedness); (5) the actual end restraints, with the elastic as well as inelastic and frictional properties of the end joints, and in the case of frames, the buckling interaction among all members; (6) possibly also load repetitions prior to the final overload up to P_{max} (which produce, in the case of inelastic behavior or friction, further residual stresses and residual bending moments in indeterminate structures); and (7) for statically indeterminate structures, the possibility that in some members the axial force-displacement curve may already descend while in others it still rises. Why then, despite the availability of these computer programs, do we still need design codes?

We still need them because the aforementioned completely rational analysis is quite involved and requires various extensive data that are not readily available to the design engineer. Cookbooks as they are, codes nevertheless fill four important functions: (1) The procedures, or "recipes," enacted by codes, provide reliable guidance to the designer who neither has the time and will to study the theory in depth and make his own choice among several possible procedures, nor is able to collect input data on crookedness, end joint friction, residual stresses, etc.; (2) by sanctioning certain procedures, the codes lend the designer legal protection (an engineer can get sued for lack of wisdom, but a code-writing committee has not been sued yet); (3) by favoring simple approximate formulas and procedures, the codes make the designer's work easier, more efficient; (4) at the time the code specifications are instituted, they speed up progress by forcing the designers to use better methods.

The benefit of the last function, however, gets reversed with the passage of time. It is patently difficult to change a code. Thus, all too often, code specifications tend to impede progress when they get old, when better methods become available. For this reason, and also because ripening of the theory in general obviates the need for codes, the future trend will probably be, and ought to be, away from codes.

Centrically Loaded Columns

For centrically loaded structural steel columns that are not too slender, tests show that the maximum load P_{max} is usually higher, but only slightly higher, than Shanley's tangent modulus load P_t. Moreover, the deflections at P_{max} are relatively small for short columns. For these reasons it is logical that the contemporary codes base design on P_t. As already emphasized (Sec. 8.3), the effect that makes Shanley's theory applicable to hot-rolled columns (in spite of the fact that mild steels, including structural carbon steels and high-strength low-alloy steels, are essentially elastic-perfectly plastic) is the effect of residual stresses.

Beginning about 1950, the residual stress magnitudes and distributions have been determined for typical steel beams. This was done either by direct measurement of the extension of a small coupon after it is cut out longitudinally from the beam or, indirectly, by calculation from the measured load-deformation curve of a short stub of the beam (comprising the whole cross section). Extensive studies of this type have been carried out at Lehigh University and elsewhere for many different cross-section shapes, effective column lengths, and residual stress magnitudes and distributions. An example of the predicted column strength, which Salmon and Johnson (1980) adapted from Johnston's guide (1976), is shown in Figure 8.20a.

According to the classical recommendation of Column Research Council (CRC) (cf. Johnston, 1976; or Galambos, 1988), a single curve of approximate column strength σ_u (ultimate stress) versus column slenderness $\lambda = L/r$ is used, approximating the mean trend of the curves for various types of columns, such as those shown in Figure 8.20a. The CRC curve, plotted as a function of the slenderness parameter $\lambda_c = \lambda(f_y/\pi^2 E)^{1/2}$ in Figure 8.20b (adapted from Chen and Lui, 1987), follows Euler hyperbola (Eq. 1.2.8) for $\sigma_u \leq 0.5 f_y$ (f_y = yield stress), and for $\sigma_u > 0.5 f_y$, it is assumed to have the shape of a parabola (Bleich, 1952)

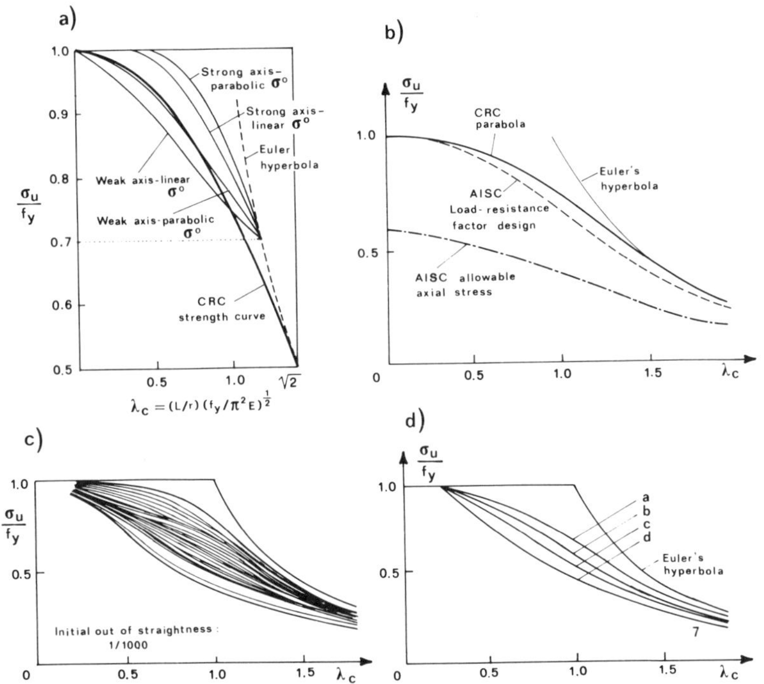

Figure 8.20 Column strength curves: (a) weak and strong axis bending (*adapted from Johnston's Guide, 1976*); (b) AISC curves (*adapted from Chen and Lui, 1987*); (c) effect of imperfections (*adapted from Johnston's guide, 1976*); (d) multiple column strength curves (*ECCS, European Convention of Constructional Steel Works, 1976*).

that has a horizontal tangent at $L/r \rightarrow 0$ and is tangent to the Euler hyperbola. The tangent point is found to be at $\sigma_u = 0.5f_y$, and the CRC curve is given by the equations:

$$\text{For } \frac{L}{r} < c: \qquad \sigma_u = f_y\left[1 - \frac{1}{2c^2}\left(\frac{L^2}{r^2}\right)\right] \tag{8.4.1a}$$

$$\text{For } \frac{L}{r} \geq c: \qquad \sigma_u = \sigma_{cr} = \frac{\pi^2 E}{(L/r)^2} \qquad \text{with } c = \pi\sqrt{\frac{2E}{f_y}} \tag{8.4.1b}$$

This curve (the solid curve in Fig. 8.20b) was used until 1986 by the American Institute of Steel Construction (AISC) for buildings, and has also been used by the American Association of State Highway and Transportation Officials (AASHTO) for bridges. Column strength in design equations usually is denoted as σ_{cr} instead of σ_u, but we prefer to avoid this notation since σ_u does not represent a critical stress, defined as the limit of stable states (see Sec. 10.3).

Length L is not the actual column length but the effective length (Sec. 1.4), which is normally calculated from the elastic critical load P_{cr}. The factor $K = L/l$, called the K-factor, is tabulated by approximate formulas in handbooks or textbooks for a variety of situations. In the elastic stability calculations of P_{cr} used for determining L (Chaps. 1 and 2), the columns and beams in a frame are in practice usually assumed to have the same elastic modulus E. However, as Yura

(1971) pointed out, for small slenderness $\lambda < c$ it is more realistic to use E only for the beams, and for the columns use the tangent modulus E_t. This is, of course, more tedious since the value of E_t depends on L (through the value of P_t) and thus needs to be determined iteratively.

Because friction at the end supports is inevitable, it is admissible, if a pin support is used, to take the end restraint stiffness not as $C = 0$ but as $C = 0.1C_0$, where C_0 is the end rotation stiffness of the column at zero axial load. In the case of continuous beams and frames, in which there are members adjoining the column to be designed, the stiffness of the adjacent beams or columns can be adjusted for calculations. For example, if there is supposed to be a hinge at the far end of the next member framed into the column, the flexibility of this member can be decreased by a factor of 2, on account of friction. On the other hand, if the far end is fixed, this flexibility is recommended to be increased by a factor of 2 since a perfect fixity is never achieved by the usual steel connection details.

Further recommendations on adjusting the theoretic elastic value of the K-factor are indicated in Figure 8.21. More details and further recommendations on trusses, lateral bracing, and connections are given in Galambos's guide (1988, pp. 51–57); see also Salmon and Johnson (1980, p. 265).

Equations 8.4.1a, b are used to determine the allowable axial stress of a column, $\sigma_{all} = \sigma_u/S$, where S is the safety factor. The AASHTO specifications (1983) have simply used $S = 2.12$ for any slenderness. However, this does not provide a uniform safety margin since the statistical scatter of test results is larger when the column is more slender (which is explained by an increased influence of random imperfections). Therefore, AISC defined the safety factor, for use in conjunction with Equations 8.4.1a, b, as follows:

$$\text{For } \lambda \le c: \quad S = \frac{5}{3} + \frac{3\lambda}{8c} - \frac{1}{8}\left(\frac{\lambda}{c}\right)^2$$
$$\text{For } \lambda \ge c: \quad S = \tfrac{23}{12} \qquad (8.4.2)$$

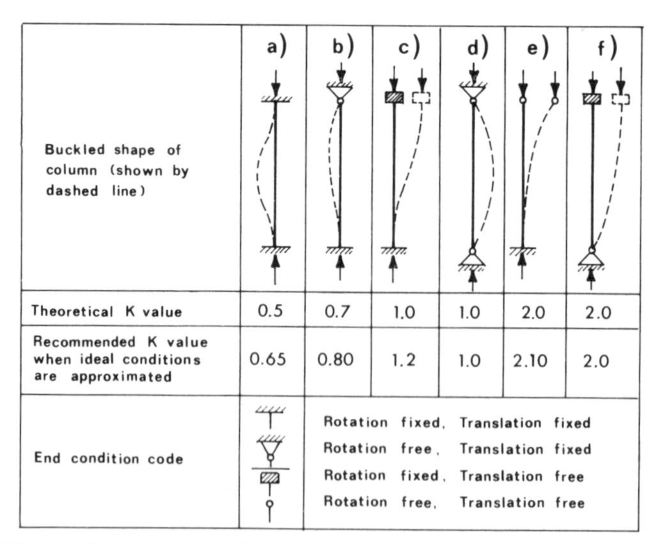

Figure 8.21 Effective-length factors for various end restraints.

This safety factor, which yields (according to Eqs. 8.4.1a, b and 8.4.2) the dash-dot curve in Fig. 8.4.1b, lumps the effects of statistical variabilities of the loads and the column strength. The statistical variability of column strength is largely due to randomness of residual stresses and the randomness of imperfections such as initial crookedness. According to Equations 8.4.2, the longer the column (larger L/r), the larger is its safety factor. This is supported by experiments that show the statistical scatter to be larger when the column is more slender, and is explained by the fact that imperfections have a larger effect for longer columns. Thus, even though no critical crookedness is assumed in calculations according to Equations 8.4.1a, b and 8.4.2, we can say that the imperfections are in some average sense taken into account by the safety factor. They indeed should be reflected in the safety factor because they are random.

On the other hand, to bury the effect of imperfections in the safety factor alone is, admittedly, a rather crude approach. In reality the effect of imperfections is influenced by many factors that can be brought under consideration only if the calculations assume the column to be initially crooked. Because these factors do not appear in Equations 8.4.1a, b, the CRC curve per se (without the safety factor) implies the design to be based on an initially straight (perfect) column, for which the depression of the strength-slenderness curve below the Euler hyperbola is due solely to the residual stresses, whose presence makes Shanley's theory applicable. Without the residual stresses, the strength-slenderness curve of a perfect column would have to follow the Euler hyperbola all the way up to the yield limit f_y, as long as the structural steel is elastic-perfectly plastic.

The preceding analysis emphasizes the viewpoint of AISC and AASHTO (rather than that of Eurocode No. 3, 1989), in which the column design is based on a single strength-slenderness curve (Eqs. 8.4.1a, b and Fig. 8.20a). This is, of course, an approximation, and in theory a crude one. Theoretically, the effect of residual stresses depends on the shape of the cross section (Sec. 8.3), and thus affects P_t and σ_{max}. Furthermore, the residual stresses themselves are different for different cross-section shapes, for example, when regular I-beams and wide-flange I-beams (called in Europe H-beams) are compared. The residual stresses also differ greatly among hot-rolled beams, cold-formed beams, round bars, beams welded from plates, etc. All these factors further cause the ratio P_{max}/P_t (which is generally close to 1) not to be constant but to vary.

The range of column strength curves that have been calculated for many different cross sections of columns used in practice is shown in Figure 8.20c. Since these curves fill a rather broad range, it is in principle desirable to abandon the use of a single strength-slenderness curve, and to introduce different curves for different types of beams. This has been recognized by the European Convention of Constructional Steelworks (ECCS), which in 1976 introduced for various typical cross sections the strength-slenderness curves shown in Figure 8.20d. Multiple column strength curves have also been recomended by the Structural Stability Research Council (SSRC, formerly CRC) (cf. Galambos, 1988) on the basis of a proposal by Bjorhovde and Tall (1971), and are also used in Eurocode No. 3 (1989); see also Ballio and Mazzolani, 1983.

The evidence in favor of multiple strength-slenderness curves, however, is incomplete at present and debate is continuing. The AISC has been studying the question in detail but so far adheres to a single curve (Eqs. 8.4.1a, b), not just for reasons of simplicity but also because other, possibly equally important, effects are still neglected by the curves in Figure 8.20d. Although in principle the use of multiple curves is doubtless superior under ideal conditions, it is unclear whether there is a significant gain in accuracy for *real* columns since other effects mentioned before are still neglected (even with multiple curves), for example, the uncertainties of end restraints and their inelastic (or frictional) behavior. Furthermore, the concept of effective length L, which is calculated by elastic analysis (Chaps. 1 and 2), is valid only for the tangential modulus load P_t and not for P_{max}. This means that the benefit of basing the strength-slenderness curves on the theoretical P_{max} values of an imperfect column might be lost, as these curves are valid only for one idealized type of end support. One might need to have also different curves for different end supports, different thicknesses of the flanges (because they affect the cooling rate, and thus the residual stresses), etc. Moreover, there are still considerable uncertainties about the residual stress distributions. When it comes to safety factors, it may be expected that properly they should be assigned different values for different types of columns, but this has not yet been sufficiently researched.

As part of a new code format—the load and resistance factor design—the AISC recently adopted for the strength-slenderness curve a new approximate formula (AISC, 1986):

For $\lambda_c < 1.5$:
$$\frac{\sigma_u}{f_y} = 0.658^{\lambda_c^2} \qquad \text{with } \lambda_c = \frac{L}{r}\left(\frac{1}{\pi}\sqrt{\frac{f_y}{E}}\right) \tag{8.4.3a}$$

For $\lambda_c \geq 1.5$:
$$\frac{\sigma_u}{f_y} = \frac{0.877}{\lambda_c^2} = \frac{0.877}{f_y}\frac{\pi^2 E}{(L/r)^2} \tag{8.4.3b}$$

The curve according to Equations 8.4.3a, b is plotted in Figure 8.20b as the dashed curve.

The curve of Equations 8.4.3 has been determined so as to be optimal under the following assumptions: (1) the end restraint stiffness is $C = 0.1C_0$ (as defined before); (2) the load is applied at the end centrically, and (3) the initial crookedness has a sinusoidal shape and amplitude $L/1500$ at midheight. As the last assumption reveals, a certain column imperfection has been considered in determining this curve. However, only a single design curve, not distinguishing among various cross sections, continues to be used by the AISC.

Load and Resistance Factor Design and Probabilistic Aspects

When the safety factor is used, the design condition is $\sigma \leq \sigma_u/S$. This simple classical approach (AISC, 1978), however, cannot distinguish between the differences in statistical scatter of the loads and strength. From the probabilistic viewpoint, a more realistic design condition that is used in the so-called Load and Resistance Factor Design (LRFD) should be written as (AISC, 1986)

$$\sum_i \gamma_i Q_i \leq \phi_u R \tag{8.4.4}$$

in which R = resistance (or capacity), for example, the column strength σ_u, Q_i = the forces or stresses (such as σ in the column) produced by various loads on

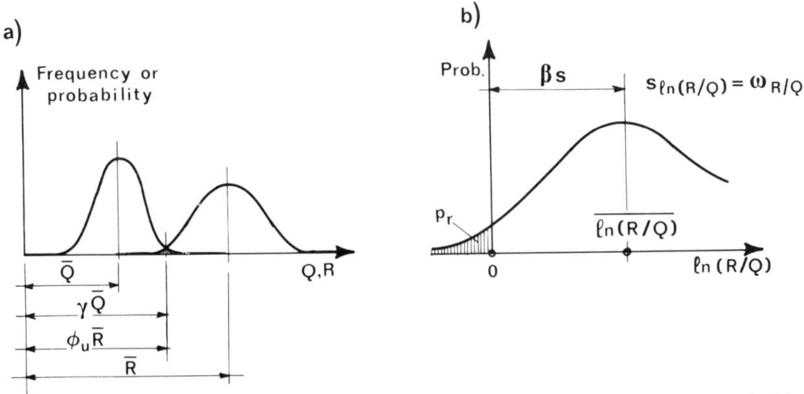

Figure 8.22 (a) Probabilistic distributions of load and structural resistance and (b) failure probability.

the structure ($i = 1, 2, \ldots, n$), ϕ_u = resistance factor (strength reduction factor, understrength factor, for example, $\phi_u = 0.85$ for a column) that characterizes the statistical scatter of the strength of the member and represents the combined effect of the statistical scatter of material properties, residual stresses, initial crookedness, etc.; and γ_i = load factor characterizing the statistical scatter of load i ($i = 1, 2, \ldots, n$), for example, 1.6 for the live loads, 1.2 for the dead loads acting simultaneously with the live loads, and 1.4 for the dead loads alone, according to the AISC.

For the case of one load, the LRFD condition is illustrated in Figure 8.22a; here $\bar{Q}$ and $\bar{R}$ are the statistical means of the load Q and the resistance R: $\gamma\bar{Q}$ is the load value that represents on the probability density distribution curve of the load a certain upper confidence limit (a probability cut-off) that is exceeded with a certain specified very small probability p_0 (e.g, $p_0 = 0.001$); and similarly $\phi_u\bar{R}$ is the resistance value that represents on the probability density distribution curve of the resistance a certain lower confidence limit that is exceeded with the same very small probability p_0. The probability distributions of Q and R are estimated empirically (usually as Gaussian distributions characterized by standard deviations), and from these the factors γ and ϕ_u are then deduced.

The LRFD approach, which is superior from the viewpoint of probabilistic theory of structural safety, is allowed to be used (AISC, 1986) as an alternative to the previous specifications (AISC, 1978), but is not mandatory.

For probabilistic treatment, it is convenient to consider the probability density distribution of $\ln(R/Q)$, which may be assumed to be Gaussian, that is, normal (which means that the distribution of R/Q is log-normal); see Figure 8.22b. Note that $\ln(R/Q)$ is preferable over R/Q since the Gaussian distribution extends to $-\infty$, which means that negative values can occur; those for $\ln(R/Q)$ are admissible but those for R/Q would not be since R and Q cannot be negative. The failure probability p_r is represented by the cross-hatched area in Figure 8.22b (for which $R < Q$). For any given p_r, one can determine from the probability distribution of $\ln(R/Q)$ the corresponding ratio $\beta = \overline{\ln(R/Q)}/s$ where $\overline{\ln(R/Q)}$ is the mean of $\ln(R/Q)$ and s is the standard deviation of $\ln(R/Q)$ (Fig. 8.22b), which is approximately equal (for small ω) to the coefficient of variation ω of R/Q [reason: $\delta(\ln z) = \delta(\ln z - \ln \bar{z}) = \delta \ln(z/\bar{z}) = \delta \ln(1 + \Delta z/\bar{z}) \simeq \delta(\Delta z/\bar{z}) = \delta(\Delta z)/\bar{z} = \delta z/\bar{z}$, where δ denotes a variation or error, $z = R/Q$, and $\Delta z =$

$z - \bar{z} = \text{small}]$. Ratio β is called the reliability index, and its meaning is that the load-design combinations of the same β have approximately the same reliability, that is, the same probability of nonfailure (as p_r is a function of β). As an approximation, $\overline{\ln(R/Q)} \simeq \ln(\bar{R}/\bar{Q})$, and also (since R and Q are independent random variables) $\omega \simeq (\omega_R^2 + \omega_Q^2)^{1/2}$ where $\omega_R = s_R/\bar{R}$ and $\omega_Q = s_Q/\bar{Q}$ are the coefficients of variation of R and Q (this is an application of Gauss approximation formulae; see e.g. Blom, 1989, p. 123, and Problem 8.4.5). Thus the reliability index can be approximately evaluated as

$$\beta = (\omega_R^2 + \omega_Q^2)^{-1/2} \ln \frac{\bar{R}}{\bar{Q}} \tag{8.4.5}$$

From the foregoing remarks it is clear that probabilistic analysis of structural stability is very important. This is still a relatively little explored subject; see, for example, Bolotin (1969, chap. 4), which deals with probability of imperfections, distribution of critical loads, and statistical methods for buckling of shells, or Elishakoff (1983, Secs. 5.6 and 11.7), which illustrate probabilistic analysis of static imperfections sensitivity and dynamic stability.

Beam-columns

Beam-columns are columns subjected to significant bending moments. For columns that are so short that stability considerations (second-order effects) are irrelevant, the ultimate load can be obtained by plastic limit analysis. As an approximation, one may use linear interaction equations that have, for the most general case of biaxial bending, the form (see, e.g., Galambos, 1988, p. 308):

$$\frac{P}{P_u} + 0.85\frac{M_y}{M_{y_p}} + 0.6\frac{M_z}{M_{z_p}} \le 1 \qquad \frac{M_y}{M_{y_p}} + \frac{M_z}{M_{z_p}} \le 1 \tag{8.4.6}$$

Here P = actual axial force; P_u = axial yield force at no bending moment; M_y, M_z = maximum magnitudes of the bending moments for the y- and z-directions (with z-direction corresponding to bending about weak y axis); and M_{y_p}, M_{z_p} = maximum plastic bending moments at no axial force. The special case of bending in one plane is obtained by setting $M_z = 0$, provided that the bending is about the strong axis. If it is about the weak axis, the interaction equation reads

$$\frac{P}{P_u} + 0.6\frac{M_z}{M_{z_p}} \le 1 \qquad \frac{M_z}{M_{z_p}} \le 1 \tag{8.4.7}$$

When the axial force P is not negligible compared to the elastic critical loads, then the interaction equations (Eqs. 8.4.6) need to be generalized by replacing the applied (first-order) moments M_y and M_z by the approximate magnified (second-order) moments M_y^*, M_z^*, as derived in Section 1.6, and P_u needs to be determined as the ultimate load capacity of a perfect-inelastic column, as we just described it. So the interaction equation for the most general case of biaxial bending becomes (AISC, 1978; Galambos, 1988)

$$\frac{P}{P_u} + \frac{M_y^*}{M_{y_p}} + \frac{M_z^*}{M_{z_p}} \le 1 \qquad M_y^* = \frac{M_y^0 C_{m_y}}{1 - P/P_{\mathrm{cr}_y}} \qquad M_z^* = \frac{M_z^0 C_{m_z}}{1 - P/P_{\mathrm{cr}_z}} \tag{8.4.8}$$

Here P_{cr_x}, P_{cr_y} = elastic critical loads for buckling of perfect column in x- and y-directions; M_y^0, M_z^0 = maximum applied (first-order) moments; and C_{m_y}, C_{m_z} =

correction coefficients that take into account the distribution of bending moments and the action of lateral forces in the y- and z-directions, determined by elastic analysis as explained in Section 1.6. While P_u is obtained by inelastic stability analysis based on Shanley's concept, the critical loads P_{cr_y}, P_{cr_z} are obtained by elastic stability analysis (Sec. 1.6). As a special case, Equations 8.4.8 cover also bending in one direction only ($M_{z_p}=0$). When axial-torsional buckling takes place, M_{y_p} and M_{z_p} must be calculated as the maximum bending moments when the axial load is zero (no second-order effects), taking into account axial-torsional buckling as well as plasticity and residual stresses.

Equations 8.4.6 to 8.4.8 do not suffice for design because they give no information on the (nonuniform) safety factor to be used. When second-order effects are negligible, the AISC conservatively omits the coefficients 0.85 and 0.6 from Equations 8.4.6, and after scaling this equation down with the appropriate safety factors one obtains the following criterion given in the AISC code (for biaxial bending):

$$\frac{\sigma_a}{0.6f_y}+\frac{\sigma_{b_y}}{F_{b_y}}+\frac{\sigma_{b_z}}{F_{b_z}}\leq 1 \tag{8.4.9}$$

Applying the same procedure to Equations 8.4.8 one obtains the following formula given in the AISC code (for the case in which the second-order effects must be taken into account);

$$\frac{\sigma_a}{F_a}+\frac{\sigma^*_{b_y}}{F_{b_y}}+\frac{\sigma^*_{b_z}}{F_{b_z}}\leq 1 \tag{8.4.10}$$

in which

$$\sigma^*_{b_y}=\frac{\sigma^0_{b_y}C_{m_y}}{1-\sigma_a/(\sigma_{cr_y}/S)} \qquad \sigma^*_{b_z}=\frac{\sigma^0_{b_z}C_{m_z}}{1-\sigma_a/(\sigma_{cr_z}/S)} \tag{8.4.11}$$

$\sigma_a=P/A=$ actual stress due to axial load, $F_a=P_u/AS=\sigma_u/S=$ allowable stress for the axial load (corresponding to safety factor S according to Eqs. 8.4.2), $\sigma_{b_y}=M_y y/I$, $\sigma^0_{b_y}=M^0_y y/I$, $\sigma_{cr_y}=P_{cr_y}/A$, and $F_{b_y}=$ allowable bending stress for bending in the y-direction; σ_{b_z}, $\sigma^0_{b_z}$, σ_{cr_z} and F_{b_z} are defined similarly ($A=$ cross-section area, $I=$ moment of inertia, $y=$ distance from centroid to column face). Note that F_{b_y}, F_{b_z} must take into account the possibility of lateral buckling (see AISC, 1978 and Sec. 6.3). Note also that the value of the safety factor used to scale down the critical stresses σ_{cr_y}, σ_{cr_z} in Equation 8.4.11 is $S=\frac{23}{12}$ (see Eq. 8.4.2). For details, see, for example, Salmon and Johnson (1980).

The derivation of Equation 8.4.10 from Equation 8.4.8 is as follows: One first sets $P=P_f$, $M^0_y=M^0_{yf}$, $M^0_z=M^0_{zf}$, with $P_f=$ axial failure load and M^0_{yf}, $M^0_{zf}=$ primary bending moments at failure, in which case Equation 8.4.8 becomes an equality. Then one divides both the numerator and denominator of each fraction by the safety factor S, and replaces P_f/S with P, M^0_{yf}/S with M^0_y, and M^0_{zf}/S with M^0_z, which means the equality is now changed back to $\leq$. Finally, one divides P by A, etc., to get the corresponding stress quantities.

The approximate semiempirical interaction Equations 8.4.6 to 8.4.11 have been justified by experiments as well as by second-order analysis taking into account plasticity and residual stresses. Many such analyses were made, in a simplified manner, already before the computer era. The easiest calculations of the interaction diagram begins by determination of the moment-curvature

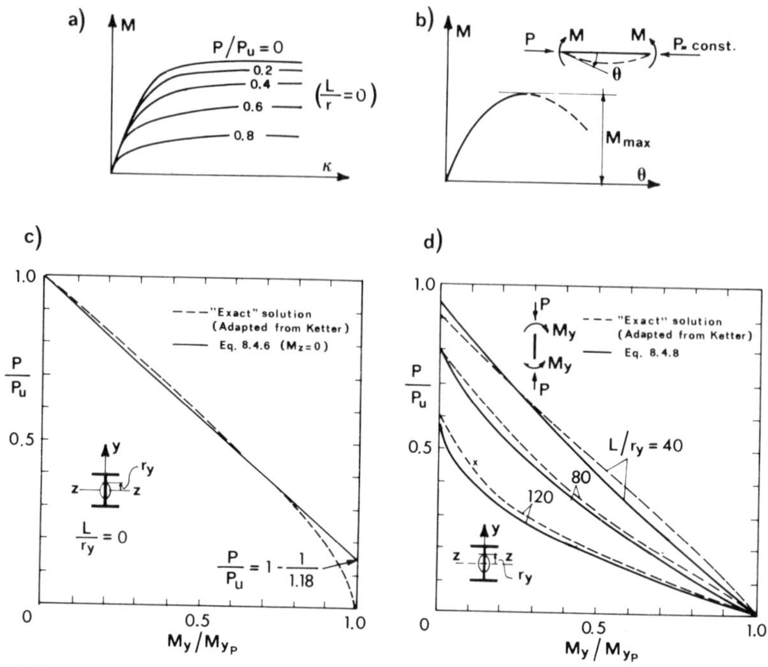

Figure 8.23 (a) Moment-curvature diagram; (b) maximum bending moment; (c, d) interaction diagrams (*adapted from Salmon and Johnson, 1980*).

relations for various fixed values of the axial force in a cross section of a given shape; see Figure 8.23a. Then one chooses a certain axial load value P and a certain effective buckling length L, and by considering an increasing sequence of the values of the end rotation of the column or maximum deflection, one calculates the corresponding column-deflection curve and the maximum bending moment using the moment-curvature diagrams in Figure 8.23a. An iterative procedure due to Newmark (1943) was often used for this purpose, but today the deflection curve or the moment distribution would, of course, be solved by finite elements with an incremental loading procedure. This calculation yields the maximum bending moment (Fig. 8.23b) for each chosen value of P and L. The result is an interaction diagram, which is exemplified by the dashed curves in Figure 8.23c, d (which are based on Salmon and Johnson, 1980). For comparison, Figure 8.23c, d also shows the interaction diagrams according to Equations 8.4.6 and 8.4.8. We see that the error of the approximate design formulas is acceptable.

Figure 8.24 shows an example of complete ultimate strength interaction curves for a wide-flange I-beam (W8 × 31, $f_y = 33{,}000$ psi) calculated under the assumption that the residual stress along the flange is distributed bilinearly, with maxima and minima of magnitude $0.3f_y$, as in Figure 8.19e. The figures are presented for the various initial bending moment distributions; they were obtained under the assumption that the column ends do not move, that is, the frame member is braced. The error of the interaction equation (Eqs. 8.4.8) compared to the interaction diagrams in Figure 8.24 has been found to be acceptable.

With the increasing power of computers, the second-order calculation of the

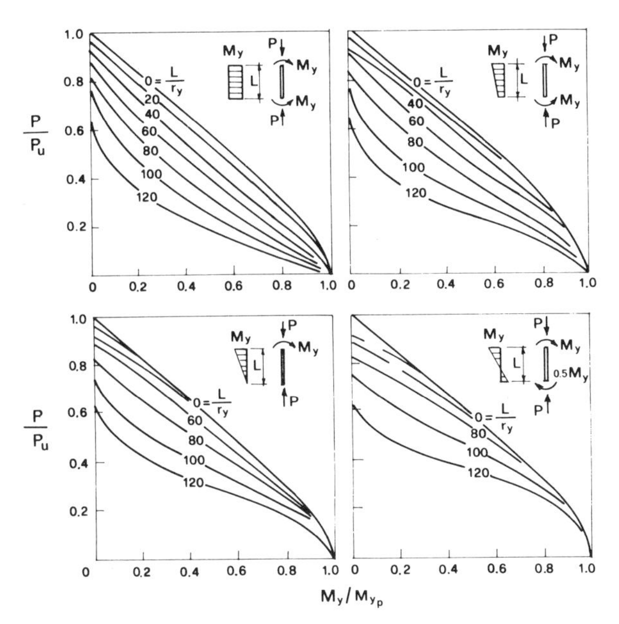

Figure 8.24 Interaction diagrams for wide-flange I-beams. (*Adapted from Salmon and Johnson, 1980.*)

ultimate strength interaction diagrams such as those in Figure 8.23d is becoming easier, and in the future the design practice may be expected to be increasingly based on finite element analysis.

In the latest specifications of AISC (1986), which are allowed to be used as an alternative to its 1978 specifications (AISC, 1978), a refinement is recommended for determining C_m for unbraced rectangular frames with lateral loads. Instead of the formula given in Section 1.6 and Equations 8.4.8 (AISC, 1978), the magnified moment ($M_y^* = M^*$) is obtained as

$$M^* = B_1 M_{nt} + B_2 M_{lt} \tag{8.4.12}$$

where M_{nt} = magnified moment due to loads that cause no lateral translation of column ends, defined as in Section 1.6; M_{lt} = additional moment due to loads causing lateral translations (sway, drift); $B_1 = C_m/(1 - P/P_{cr})$; and

$$B_2 = \left(1 - \frac{\Delta \sum P_u}{\sum HL}\right) \quad \text{or} \quad \left(1 - \frac{\sum P_u}{\sum P_{cr}}\right) \tag{8.4.13}$$

The summation $\sum$ runs over all the columns of the same floor, B_2 being common to all these columns; Δ = lateral translation of the floor (drift); and H = applied horizontal forces producing Δ. For details and justification see AISC (1986) with commentary (AISC, 1986, sec. H1).

It should be noted that buckling of some frames with lateral sway of columns can also involve torsion in which the entire floor rotates about a vertical axis (Wynhoven and Adams, 1972).

Plates, Shells, and Other Structures

In principle, it is safe to design any structures such as plates or shells for a critical load calculated on the basis of the tangent modulus E_t. However, this procedure is easy only if E_t is known in advance, that is, if the stress-strain diagram is bilinear. Otherwise, the determination of E_t is coupled with the calculation of the critical load and requires a nonlinear structural analysis that takes the plasticity of the material as well as the residual stresses into account.

There exists nevertheless a simple approximate method—the method of equivalent column slenderness (e.g., Bleich, 1952; Johnston, 1976; Herber, 1966), which is generally conservative (although sometimes too conservative). The reduction of the elastic critical stress σ_{cr}^{el} is assumed to be approximately the same as in a column of a certain equivalent slenderness $L/r = \lambda_{eq}$. Thus, one sets $\sigma_{cr}^{el} = \pi^2 E/\lambda_{eq}^2$ (Eq. 1.2.8 or 8.4.1b), from which

$$\lambda_{eq} = \pi\left(\frac{E}{\sigma_{cr}^{el}}\right)^{1/2} \tag{8.4.14}$$

Then one reduces the elastic critical stress σ_{cr}^{el} in the same ratio as required for a column of slenderness λ_{eq}.

We may illustrate it considering an infinite plate strip of width b and thickness h, subjected to longitudinal stress σ. The critical value of σ (Sec. 7.3) is given by $\sigma_{cr}^{el} = k\pi^2 Eh^2/[12(1-\nu^2)b^2]$ where k = coefficient depending on the boundary conditions. From Equation 8.4.14,

$$\lambda_{eq} = \frac{b}{h}\left[\frac{12}{k}(1-\nu^2)\right]^{1/2} \tag{8.4.15}$$

This method is often used for inelastic lateral buckling of beams, for which it works quite well, as well as tubes, shells, etc. (Johnston, 1976).

The local buckling of flanges in compression members is often simply handled by requiring the flange thickness to be so large that the buckling load will be higher than the load that causes the flange to yield. In this manner, elastoplastic stability analysis is avoided.

For structures such as plates, for which there exists a large postcritical reserve, calculation of the critical stress from the elastic solution is unnecessarily conservative if E is replaced with the tangent modulus E_t. Bleich (1952) proposed E to be replaced with $E_{ef} = (EE_t)^{1/2}$, which gives a larger critical stress; see Galambos (1988).

A detailed discussion and many other semiempirical formulas for various inelastic buckling problems are presented in Galambos's guide (1988) as well as other handbooks and textbooks (e.g., Salmon and Johnson, 1980). The future lies, however, in nonlinear finite element analysis.

Design Examples

Example 8.4.1 Consider a pin-ended steel column of length $l = 16$ ft (4.87 m), which carries an axial load $P = 180$ kips (799 kN) (Fig. 8.25a). Its cross

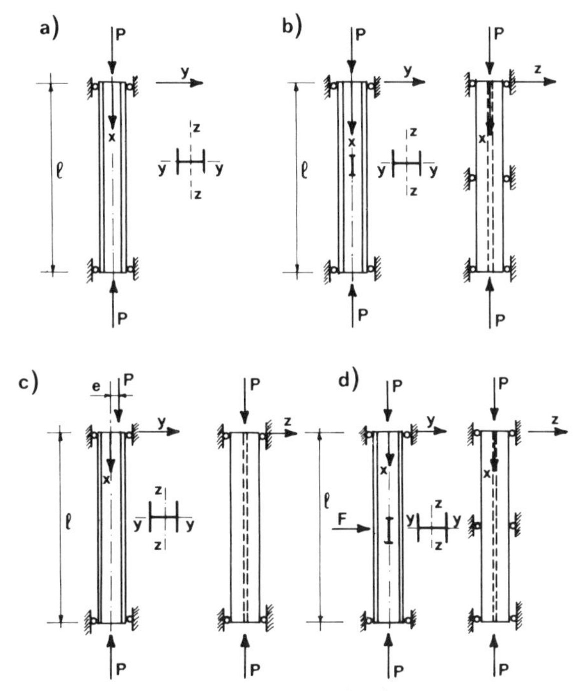

Figure 8.25 Design examples for buckling of steel columns.

section is a W8 × 48 shape, with area $A = 14.1\ \text{in}^2$ ($9096\ \text{mm}^2$), radii of gyration $r_z = 2.08$ in. (52.8 mm) in the weak direction of the z axis (i.e., about the y axis) and $r_y = 3.61$ in. (91.69 mm) in the strong direction of the y axis (i.e., about the z axis). Assuming $f_y = 36$ ksi ($247.7\ \text{N/mm}^2$), $E = 29{,}000$ ksi ($199.5\ \text{kN/mm}^2$), check the design according to the AISC Allowable Stress Design (Eqs. 8.4.1 and 8.4.2). *Hint:* In which direction will the column buckle?

Solution. A36 steel, $E = 29{,}000$ ksi, $f_y = 36$ ksi (1 ksi = 1000 psi).

$$c = \sqrt{\frac{2\pi^2 E}{f_y}} = 126.1 \qquad \lambda = \frac{Kl}{r_z} = \frac{16(12)}{2.08} = 92.3$$

$$\lambda_c = \lambda\sqrt{\frac{f_y}{\pi^2 E}} = \frac{\lambda\sqrt{2}}{c} = \frac{92.3\sqrt{2}}{126.1} = 1.035$$

$$\sigma_u = f_y\left(1 - \frac{\lambda^2}{2c^2}\right) = f_y\left(1 - \frac{\lambda_c^2}{4}\right) = 26.36\ \text{ksi}$$

$$S = \frac{5}{3} + \frac{3}{8}\left(\frac{1.035}{\sqrt{2}}\right) - \frac{1}{8}\left(\frac{1.035}{\sqrt{2}}\right)^3 = 1.89 \qquad F_a = \frac{\sigma_u}{S} = 13.93\ \text{ksi}$$

$$\sigma_a = \frac{P}{A} = \frac{180}{14.1} = 12.77\ \text{ksi} < F_a$$

Example 8.4.2 Same column as in Example 8.4.1, assuming that $D =$ 180 kips (799 kN) is the dead load (load factor 1.4). Check the design according to the AISC Load and Resistance Factor Design (Eqs. 8.4.3 and 8.4.4 with $\phi_u = 0.85$). Compare the results with Example 8.4.1.

Solution. $P = 1.4D$, $1.4D < 0.85R$, $D < R/1.65$.

$$\sigma_u = f_y(0.658^{\lambda_c^2}) = 36(0.658^{1.035^2}) = 22.99 \text{ ksi}$$

$$\frac{22.99}{1.65} = 13.93 \text{ ksi} = \text{same result}$$

Example 8.4.3 A hinged steel column of length $l = 28$ ft (9.75 m) has an intermediate support in the xz plane and carries an axial load of $P = 170$ kips (754.7 kN) (Fig. 8.25b). Its cross section is a W10 × 39 shape, characterized by $A = 11.5 \text{ in}^2$ (7419 mm^2), $r_z = 1.98$ in. (50.3 mm), $r_y = 4.27$ in. (109 mm). Assuming $f_y = 36$ ksi (247.7 N/mm^2), $E = 29{,}000$ ksi (199.5 kN/mm^2), check the design according to the AISC Allowable Stress Design (Eqs. 8.4.1 and 8.4.2).

Solution.

$$\lambda_z = \frac{Kl}{r_z} = \frac{0.5(28)(12)}{1.98} = 84.85$$

$$\lambda_y = \frac{Kl}{r_y} = \frac{28(12)}{4.27} = 78.69 \qquad \lambda_z > \lambda_y \quad \text{(weak axis governs).}$$

$$\lambda_c = \frac{\sqrt{2}\,\lambda_z}{c} = \frac{\sqrt{2}\,(84.85)}{126.1} = 0.952$$

$$\sigma_u = f_y\left(1 - \frac{\lambda_c^2}{4}\right) = 36\left(1 - \frac{0.952^2}{4}\right) = 27.85 \text{ ksi}$$

$$S = \frac{5}{3} + \frac{3}{8}\left(\frac{0.952}{\sqrt{2}}\right) - \frac{1}{8}\left(\frac{0.952}{\sqrt{2}}\right)^3 = 1.881$$

$$F_a = \frac{\sigma_u}{S} = 14.806 \text{ ksi} \qquad \sigma_a = \frac{P}{A} = \frac{170}{11.5} = 14.78 < F_a$$

Example 8.4.4 A steel column of length $l = 18$ ft (5.48 m) is hinged at the bottom and at the top, and loaded by an axial load $P = 150$ kips (666 kN) that has an eccentricity $e = 5$ in. (127 mm) at the top in the xy plane (Fig. 8.25c). Its cross section is W12 × 53, $A = 15.6 \text{ in}^2$ ($10{,}064 \text{ mm}^2$), $r_z = 2.48$ in. (63 mm), $r_y = 5.23$ in. (133 mm), S_y = cross-section modulus = 70.6 in^3 ($1.15 \times 10^6 \text{ mm}^3$). Assuming $f_y =$ 36 ksi (247.7 N/mm^2), $E = 29{,}000$ ksi (199.5 kN/mm^2), check the column design according to the AISC Allowable Stress Design (Eqs. 8.4.10 and 8.4.11). (*Hint:* Assume $F_b = 0.6f_y$ as the allowable stress in bending, which takes into account the danger of lateral buckling for the given section and length, and check also the cross section at the top for yielding.)

Solution. $r_y > r_z$, $\lambda_z > \lambda_y$, $\lambda_z = Kl/r_z = 18(12)/2.48 = 87.1$.

$$\lambda_c = \frac{\sqrt{2}\,\lambda_z}{c} = \frac{\sqrt{2}(87.1)}{126.1} = 0.977$$

$$\sigma_u = f_y\left(1 - \frac{\lambda_c^2}{4}\right) = 27.41 \text{ ksi}$$

$$S = \frac{5}{3} + \frac{3}{8}\left(\frac{0.977}{\sqrt{2}}\right) - \frac{1}{8}\left(\frac{0.977}{\sqrt{2}}\right)^3 = 1.885 \qquad F_a = \frac{\sigma_u}{S} = 14.54 \text{ ksi}$$

Assume $F_b = 0.60f_y = 21.6 \simeq 22$ ksi (from the AISC tables).

$$\lambda_y = \frac{Kl}{r_y} = \frac{18(12)}{5.23} = 41.3$$

$$F'_e = \frac{\sigma_{\text{cr}_y}}{S} = \frac{\pi^2 E}{\lambda_y^2(\frac{23}{12})} = 87.55 \text{ ksi}$$

$$\sigma_a = \frac{P}{A} = \frac{150}{15.6} = 9.61 \text{ ksi} \qquad \sigma_b = \frac{M}{S_y} = \frac{150(5)}{70.6} = 10.62 \text{ ksi}$$

$$\frac{\sigma_a}{F_a} + \frac{C_m \sigma_b}{F_b}\left(\frac{1}{1 - \sigma_a/F'_e}\right) = \frac{9.61}{14.54} + \frac{0.6(10.62)}{22}(1.123) = 0.986 < 1$$

$$\text{Yielding at the top: } \frac{\sigma_a}{0.6f_y} + \frac{\sigma_b}{F_b} = \frac{9.615}{0.6(36)} + \frac{10.62}{22} = 0.928 < 1$$

Example 8.4.5 A steel column of length $l = 16$ ft (4.88 m) is hinged at the bottom and at the top, and has an intermediate support at midlength in the xz plane (Fig. 8.25d). The axial load is $P = 100$ kips (443.9 kN) and the lateral load at midlength in the xy plane is $F = 8$ kips (35.51 kN). The cross section is a W10 × 39 shape, $A = 11.5\ \text{in}^2$ ($7.42 \times 10^3\ \text{mm}^2$), $r_z = 1.98$ in. (50.3 mm), $r_y = 4.27$ in. (108.4 mm), $S_y = 42.1\ \text{in}^3$ ($0.690 \times 10^6\ \text{mm}^3$). Assuming $f_y = 36$ ksi ($247.7\ \text{N/mm}^2$), $E = 29{,}000$ ksi ($199.5\ \text{kN/mm}^2$), check the column design according to the AISC Allowable Stress Design (Eqs. 8.4.10 and 8.4.11). (*Hint:* Assume $F_b = 0.66f_y$ as the allowable stress in bending, since for the given unbraced length and shape there is no danger of lateral buckling.)

Solution.

$$\lambda_z = \frac{Kl}{r_z} = \frac{8(12)}{1.98} = 48.5$$

$$\lambda_y = \frac{Kl}{r_y} = \frac{16(12)}{4.27} = 44.9 \qquad \lambda_z > \lambda_y$$

$$\lambda_c = \frac{\sqrt{2}\,\lambda_z}{c} = \frac{\sqrt{2}(48.5)}{126.1} = 0.544$$

$$\sigma_u = f_y\left(1 - \frac{\lambda_c^2}{4}\right) = 36\left(1 - \frac{0.544^2}{4}\right) = 33.34$$

$$S = \frac{5}{3} + \frac{3}{8}\left(\frac{0.544}{\sqrt{2}}\right) - \frac{1}{8}\left(\frac{0.544}{\sqrt{2}}\right)^3 = 1.804$$

$$F_a = \frac{\sigma_u}{S} = 18.48 \qquad \sigma_a = \frac{P}{A} = \frac{100}{11.5} = 8.70 \text{ ksi}$$

$$\frac{\sigma_a}{F_a} = \frac{8.70}{18.48} = 0.471 \qquad \sigma_b = \frac{M}{S_y} = \frac{\frac{8}{2}(8)(12)}{42.1} = 9.12 \text{ ksi}$$

$$F_b = 0.66f_y = 23.76 \text{ ksi} \qquad \frac{\sigma_b}{F_b} = \frac{9.12}{23.76} = 0.384$$

$$F'_e = \frac{\sigma_{\text{cr}_y}}{S} = \frac{\pi^2 E}{\lambda_y^2(\frac{23}{12})} = 74.1 \qquad \frac{1}{1 - \sigma_a/F'_e} = 1.133$$

$$C_m = 1 - \frac{0.2\sigma_a}{F'_e} = 0.977$$

$$\frac{\sigma_a}{F_a} + \frac{C_m \sigma_b}{F_b}\left(\frac{1}{1 - \sigma_a/F'_e}\right) = 0.471 + \frac{0.977(9.12)}{23.76}(1.133) = 0.896 < 1$$

Problems

8.4.1 Derive the coefficients of Equation 8.4.1a knowing the condition of tangency with the Euler hyperbola.

8.4.2 Describe the Load and Resistance Factor Design and its differences from the Allowable Stress Design.

8.4.3 Recalling that the interaction diagram of P and magnified moment is assumed to be linear, derive the interaction equation.

8.4.4 Solve (a) Examples 8.4.1 and 8.4.2 for $l = 17$ ft instead of 16 ft; (b) Example 8.4.3 for $l = 30$ ft instead of 28 ft and $P = 160$ kips instead of 170 kips; (c) Example 8.4.4 for $l = 20$ ft instead of 18 ft and $P = 115$ kips instead of 150 kips; (d) Example 8.4.5 for $l = 15$ ft instead of 16 ft and $F = 10$ kips instead of 8 kips.

8.4.5 Derive Equation 8.4.5 as a special case of the following derivation of Gauss approximation formulae for the mean and variance s_f^2 (s_f = standard deviation of f) of a function $f(x, y)$ of independent random variables x, y. (1) Mean: Expand $f(x, y) \simeq f(\bar{x}, \bar{y}) + (x - \bar{x})f_{,x} + (y - \bar{y})f_{,y}$ where $f_{,x} = f_{,x}(\bar{x}, \bar{y})$, $f_{,y} = f_{,y}(\bar{x}, \bar{y})$. From this, $\bar{f}(x, y) \simeq f(\bar{x}, \bar{y})$ since $\overline{(x - \bar{x})} = \overline{(y - \bar{y})} = 0$. (2) Variance:

$$s_f^2 = \overline{(f - \bar{f})^2} \simeq \overline{[f(x, y) - f(\bar{x}, \bar{y})]^2} \simeq \overline{[(x - \bar{x})f_{,x} + (y - \bar{y})f_{,y}]^2} = \overline{(x - \bar{x})^2}f_{,x}^2 + \overline{(y - \bar{y})^2}f_{,y}^2 + 2\overline{(x - \bar{x})(y - \bar{y})}f_{,x}f_{,y} = s_x^2 f_{,x}^2 + s_y^2 f_{,y}^2$$

since for independent variables $\overline{(x - \bar{x})(y - \bar{y})} = 0$.

8.5 CONCRETE COLUMNS AND STRUCTURES: DESIGN AND CODE SPECIFICATIONS

Reinforced concrete columns are composite structures whose load capacity depends on concrete, reinforcing steel bars, and their bond. Like steel, the uniaxial stress-strain diagram of concrete, whose typical form is sketched in Figure 8.26a, is also highly nonlinear. Unlike steel and other metals, the compressive and tensile strengths, f'_c and f'_t, are very different. Their ratio is about 10:1 (more precisely, $f'_t \simeq 6\sqrt{f'_c}\sqrt{\text{psi}}$, psi = 6895 Pa). The tensile resistance

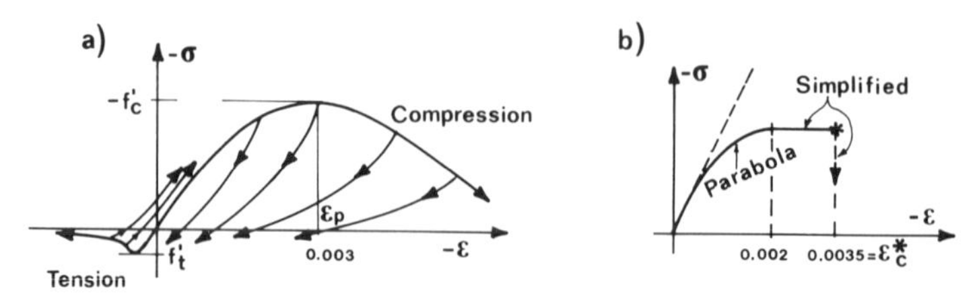

Figure 8.26 (a) Uniaxial stress-strain diagram of concrete and (b) simplified form with sudden stress drop adopted by CEB.

of concrete is usually neglected in the design of columns. This no-tension assumption is on the safe side.

The nonlinearity of the $\sigma(\varepsilon)$ diagram of concrete in compression or tension begins at a rather low stress level, roughly at $0.3f'_c$ or $0.3f'_t$. There is no plastic (yield) plateau. Rather, the stress begins to decline with increasing strain right after reaching the peak stress. The mathematical treatment of this phenomenon, which is called strain-softening and is due to distributed cracking, is difficult in general (Chap. 13), but not for the bending of columns.

The use of the smooth descending portion of the $\sigma(\varepsilon)$ diagram, of course, implies the hypothesis that strain-localization instabilities (Sec. 13.2) do not occur. In most cases of bending of not too large columns this is probably a reasonable hypothesis as long as the diagram of load versus load-point deflection is rising. This is normally the case nearly up to the peak point of the load-deflection diagram (but not completely up to the peak). On the other hand, when the diagram of load P versus the axial displacement u_1 under the load at the column end is descending, the possibility of localization instabilities, such as those in Section 13.2, is much greater. We accept the aforementioned hypothesis because the use of a smooth descending portion of the $\sigma(\varepsilon)$ diagram for both compression and tension has yielded good comparisons with experimentally measured deflections of reinforced concrete beams (Bažant and Oh, 1984) as well as partially prestressed beams (Chern, You, and Bažant, 1989).

In contrast to beams, strain-softening behavior is difficult to measure in uniaxial tests since strain localization occurs right after the peak (Chap. 13). In view of these difficulties, the practice has been to assume for concrete a stress-strain diagram terminating with a sudden stress drop (e.g., CEB, 1978, Fig. 8.26b). This is a simplification that is probably not very realistic for bending but is usually on the safe side, except for the use of a horizontal plateau at the end of the diagram (in reality the stress declines right after maximum σ is attained). However, the CEB Draft (1988) of the *Comité Euro-International du Béton* (CEB) Model Code recommends a smooth descending stress-strain diagram.

Unlike steel columns, the case of centric loading is not considered for concrete columns. This case is circumvented in codes by stipulating that every column must be designed for a certain minimum eccentricity, e_{min}.

Interaction Diagram (Failure Envelope)

As explained in the texts on reinforced concrete (e.g., Wang and Salmon, 1979; Park and Paulay, 1975), the ultimate compressive force P and the ultimate bending moment M for a concrete column cross section (Fig. 8.27a) are related by an interaction diagram (also called the failure envelope or failure surface). The cross-section failure may be due either to brittle failure of concrete in compression or to ductile failure of steel in tension. The interaction diagram has the characteristic "knee" shape shown in Figure 8.27b. Its slope is not continuously downward, but reverses sign. This is due to tensile cracking of concrete. If P/M is small, the main effect of increasing P is to suppress tensile cracking; hence M increases. On the other hand, if P/M is large, no cracks can form and the main effect of increasing P is to cause the cross section to reach the compression strength earlier; hence M decreases.

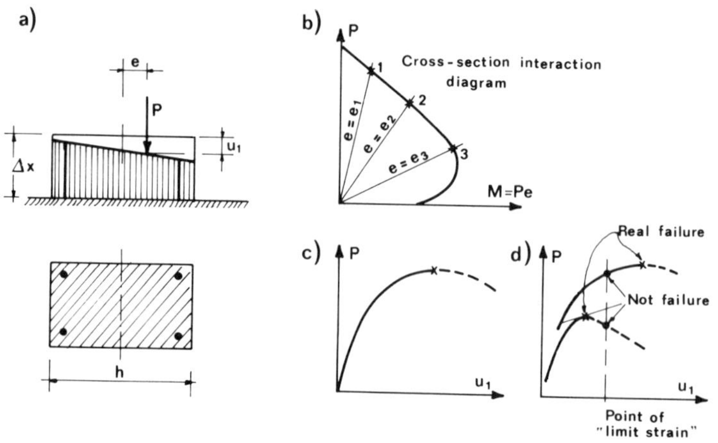

Figure 8.27 (a) Concrete column cross section, (b) cross-section interaction diagram, (c, d) load-displacement curves.

With regard to our further analysis, it may be helpful to discuss how the cross-section interaction diagram is defined. One considers a small element Δx of the beam (in which there are no second-order effects), one end cross section being fixed and the other one subjected to axial force P applied at constant eccentricity e (Fig. 8.27a). The load-point displacement u_1 is increased in small steps, and the corresponding values of P are calculated from equilibrium conditions and the stress-strain laws of concrete and steel, assuming the cross section to remain plane. If the curve $P(u_1)$ is rising, the beam element is stable, that is, no failure can take place (see also Sec. 10.1). If this curve is descending, the beam element is, under load control, unstable, (Sec. 10.1). The critical point of stability, that is, the failure point, is obtained as the peak-load point (max P) (see Fig. 8.27c). Thus, in a consistent mechanics analysis, the interaction diagram (at controlled load, as it is normally understood either directly or by implication) should be defined and calculated as the collection of the peak-load point of all the curves $P(u_1)$ obtained for all the eccentricities e.

In the practical engineering literature, this theoretically consistent definition of failure has normally not been adhered to. Failure has been assumed to occur, for example, when the maximum strain of concrete or steel reaches a certain specified limit, which is selected empirically. However, if calculations beyond this limit (with a realistic constitutive law, of course) would indicate a further increase of load (Fig. 8.27d), this is not really a failure state, and is not a state lying on the interaction diagram. Vice versa, if calculation would indicate that the load at this limit is already decreasing (Fig. 8.27d), then again this is not a failure state, since failure must have occurred earlier.

Consider now a hinged beam-column under the action of axial load P. It is usually reasonable to assume that load P is applied in such a manner that its eccentricity e remains constant (Fig. 8.28d) as the load increases. The path followed by the axial load and the bending moment acting on the cross section at column midlength can be shown in the same figure (Fig. 8.28e). In the absence of second-order effects (as in very short columns), the cross section would undergo proportional loading, and its state would follow the radial ray $\overline{01}$ (of slope

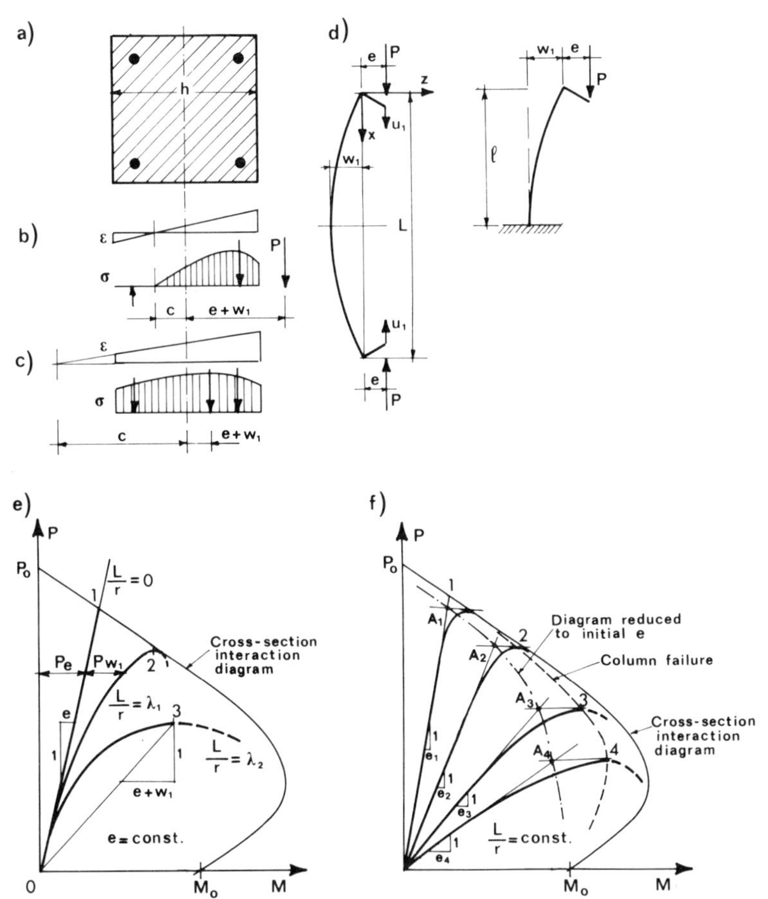

Figure 8.28 (a) Concrete column cross section; (b, c) strain and stress distributions; (d) columns subjected to axial load of constant eccentricity; loading paths for columns (e) of different slendernesses, or (f) of different eccentricities of axial load.

$e = M/P$) shown in Figure 8.28e, reaching a maximum at point 1 of the cross-section interaction diagram. For slender columns, however, the midlength deflection w_1 causes the path to deviate from the radial ray downward (path $\overline{02}$ in Fig. 8.28e). The larger the column slenderness λ ($\lambda = L/r$, where r = radius of gyration), the greater are the deviations.

For not too slender columns, the failure (i.e., the peak point) occurs at points rather close to the cross-section interaction diagram (path $\overline{02}$ in Fig. 8.28e). Such behavior corresponds to what has been called (according to the ACI) the cross-section failure. For very slender columns, on the other hand, the failure occurs well within the cross-section interaction diagram (path $\overline{03}$ in Fig. 8.28e), because of pronounced second-order effects. This corresponds to what had been called (according to the ACI) the stability failure. The type of instability is the snapthrough (limit point; Sec. 4.4).

Figure 8.28f compares the $P(M)$ path at constant e for columns of the same length L (or slenderness) and different end eccentricities e_i. Connecting the failure points (peak points) such as 1, 2, 3 in Figure 8.28f yields the interaction

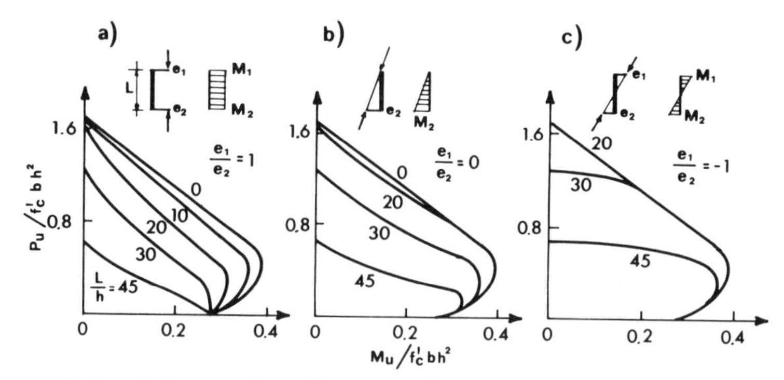

Figure 8.29 Reduced interaction diagrams. (*Adapted from MacGregor, Breen, and Pfrang, 1970.*)

diagram $P(M)$ of the column, which is modified for second-order effects. Projecting each failure point (maximum point) horizontally onto point A_i on the radial line of slope e_i (Fig. 8.28f), one obtains the column interaction diagram in terms of the ultimate primary bending moments $M = M_u^I = Pe$ (and ultimate load $P = P_u$). This diagram is called the reduced interaction diagram (a term used, e.g., in Italy) or the slender column interaction diagram (a term used in the United States). Figure 8.29 (adapted from MacGregor, Breen, and Pfrang, 1970) shows the reduced interaction diagrams for three different ratios of the end moments and for various values of the ratio L/h (where h = height of rectangular cross section, Fig. 8.27a). These diagrams, as well as the actual column interaction diagrams (curve 234 in Fig. 8.28f) are different for different column slendernesses.

Deflections and Interaction Diagram

To obtain approximate column deflections, we may assume, for the sake of simplicity, that the column deflection curve is sinusoidal, $w = -w_1 \sin(\pi x/L)$ where L = effective length of the column (Fig. 8.28d), equal to the column length if the supports are hinges. To simplify the solution, the equilibrium condition and the moment-curvature relation are matched at only the midlength of the column, in the spirit of the collocation method. The curvature at midlength is $\kappa = \pi^2 w_1/L^2$, from which $w_1 = L^2\kappa/\pi^2$. By equilibrium, the second-order bending moment is $M_{\text{II}} = Pw_1$, or

$$M_{\text{II}} = Pk_{\text{II}}\kappa \qquad k_{\text{II}} = \frac{L^2}{\pi^2} \tag{8.5.1}$$

(In reality, of course, coefficient k_{II} must be, for the same w_1, less than L^2/π^2 because nonlinear behavior tends to produce at midspan a sharper, more pointed curve.) The total moment at midlength is $M = M_{\text{I}} + M_{\text{II}}$ where $M_{\text{I}} = Pe$ = first-order (primary) bending moment, which is due to eccentricity e of load P at both column ends.

The maximum of the response curve $P(M)$ at constant e represents the failure points if load P is controlled. Consequently, the collection of all these maxima for various e determines the failure envelope of the column. This can be proven as follows.

The failure point under load control (snapthrough or limit-point instability) is obviously characterized by the condition $dP/du_1 = 0$ where u_1 is the axial displacement of the load point at the column end (the basic reason is that $P\,du_1$ is the correct expression for the work of load P; see Chap. 10). For a sinusoidal column shape, the rotations of the column ends are $\theta_1 = w_1[d \sin(\pi x/L)/dx] = w_1\pi/L$. So the load-point displacement is $u_1 = \theta_1 e + \frac{1}{2}\int \frac{1}{2} w'^2\, dx = \pi e w_1/L + \pi^2 w_1^2/8L$, in which the axial shortening of the column axis is neglected. Also $w_1 = L^2\kappa/\pi^2$. From this, approximately,

$$u_1 = \frac{eL}{\pi}\kappa + \frac{L^3}{8\pi^2}\kappa^2 \tag{8.5.2}$$

The column under load control is stable if $dP/du_1 > 0$ (see Sec. 10.1). It fails (loses stability) when the response first satisfies the condition $dP/du_1 \le 0$ or $\delta^2 W = \frac{1}{2}\,\delta P\,\delta u_1 \le 0$ (see Sec. 10.1), i.e. that is

$$\frac{dP}{du_1} = \frac{dP}{dM}\frac{dM}{d\kappa}\frac{d\kappa}{du_1} \le 0 \tag{8.5.3}$$

If the slope dP/du_1 varies continuously, then, of course, the failure condition is $dP/du_1 = 0$. According to Equation 8.5.2, $d\kappa/du_1 > 0$ always (since $\kappa \ge 0$, $e > 0$). Now it is possible that $dM/d\kappa$ at failure is either nonpositive or positive. If $dM/d\kappa \le 0$, the cross section fails even without the second-order (slenderness)

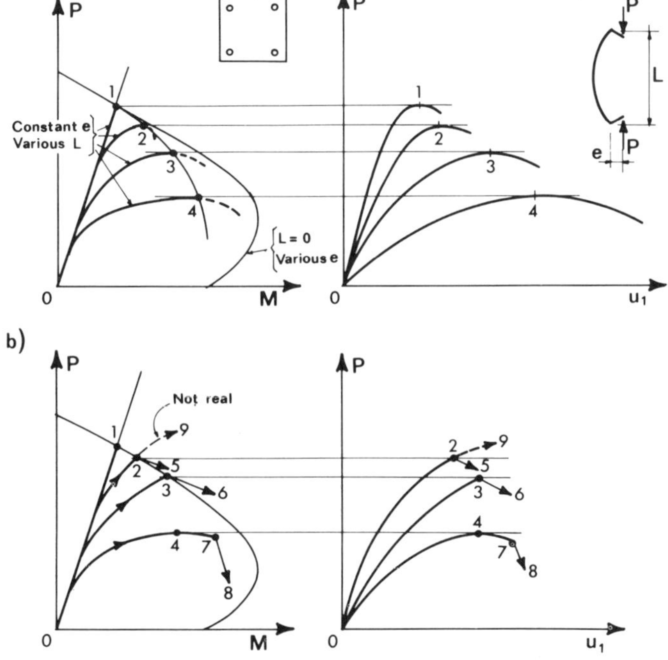

Figure 8.30 Column failure conditions (a) for smoothly varying P–M diagram, and (b) for discontinuous slope change.

effect. So this type of failure, which occurs at point 1 of Fig. 8.30a (the end of the straight path $\overline{01}$) is obtained for zero slenderness ($L=0$) and in this case we do not have $dP/dM=0$ for the points on the cross-section interaction diagram. For $L>0$ we must have $dM/d\kappa>0$. So the column fails (i.e., reaches a point on the column interaction diagram) as soon as the following condition is reached:

$$\frac{dP}{dM} \leq 0 \qquad (\text{if } L>0 \text{ and } e=\text{const.}) \tag{8.5.4}$$

This means that a peak point of the $P(M)$ diagram at constant e represents failure.

If the slope dP/dM at constant e varies continuously, then the column fails (loses stability) when $dP/dM=0$. In that case, the response curves $\overline{02}$, $\overline{03}$, and $\overline{04}$ in Figure 8.30a must intersect the column interaction diagram $P(M)$ horizontally.

If the slope changes discontinuously, then the intersection with the column interaction diagram (see Fig. 8.30b) is the first point where the slope dP/dM is nonnegative on the left and nonpositive on the right of the intersection. This situation may arise if the column fails when the steel bars begin to yield provided that the steel is assumed to behave as elastic–perfectly plastic and the steel area at each side is considered to be concentrated in its centroid.

Numerical Algorithm for Calculating Deflections and Interaction Diagram

For numerical calculation, it is convenient to subdivide the critical cross section at midlength of the column into many thin layers. Some of the layers are used to represent the steel reinforcement. Having the values of curvature κ and of the distance c from the beam axis to the neutral axis (Fig. 8.28b, c), one has $\varepsilon=-\kappa(z+c)$, and so one can use the given stress-strain diagram $\sigma(\varepsilon)$ of concrete and of steel for either loading or unloading to evaluate the stress at the center point of every layer (Fig. 8.28b, c). From these stress values one obtains the resultants $P=P(\kappa, c)$ and $M=M(\kappa, c)$. The $\sigma(\varepsilon)$ diagram of concrete should take into account the effect of elastic lateral confinement due to stirrups (or spirals).

To calculate the column response curve $P(M)$ at increasing deflection and constant eccentricity e at the column ends, the following algorithm may be used. One chooses an increasing sequence of κ values. For each κ value, one has $w_1=k_{\text{II}}\kappa$. So one needs to solve c from the equilibrium equation:

$$M(\kappa, c)-(e+w_1)P(\kappa, c)=0 \tag{8.5.5}$$

in which M and P are calculated as the resultants of the stresses in all the layers corresponding to strains $\varepsilon=-\kappa(z+c)$ (Eq. 8.5.5 ensures equilibrium to be satisfied only at column midlength). Equation 8.5.5 is a nonlinear equation that is quite easy to solve with a computer library subroutine such as that for the Marquardt-Levenberg algorithm. The convergence is very fast if the solution of c for the preceding κ value is used as the initial estimate of c for the new κ value. From the c value obtained, one may then evaluate $P(\kappa, c)$ and $M(\kappa, c)$, which define parametrically the curves $P(M)$ at constant e.

In the course of the foregoing calculation, one should compare the strain in each layer with the strain value at the previous load level, and if unloading is detected, then the proper $\sigma(\varepsilon)$ curve for unloading should be considered for that layer.

The foregoing new algorithm based on Equation 8.5.5 (Bažant, 1985, and Bažant, Cedolin, and Tabbara, 1989) is easy to program, is cheap to run (suitable for a small microcomputer), and is recommended as the best choice for a design office, among all the methods described in this section. This algorithm was used to calculate the response curves $P(M)$ at various constant values of e for concrete columns with the square cross section of side h shown in Figure 8.28. Various slendernesses L/r (with $r = h/\sqrt{12}$) were considered. The concrete was assumed to have no stiffness in tension ($\sigma_c = 0$ at $\varepsilon_c \geq 0$) and in compression to obey the formula recommended in the Draft (1988) of the 1990 CEB Model Code, which gives a smooth curve with peak value (strength) $f'_c = 5000$ psi and postpeak strain softening. For steel, up to yield limit $f_y = 60{,}000$ psi, $\sigma_s = E_s\varepsilon_s$ for both compression and tension, with $E_s = 29 \times 10^6$ psi. To achieve accurate results, the cross section was subdivided into 100 layers.

The calculated response curves $P(M)$ at constant e (Fig. 8.31) show that the

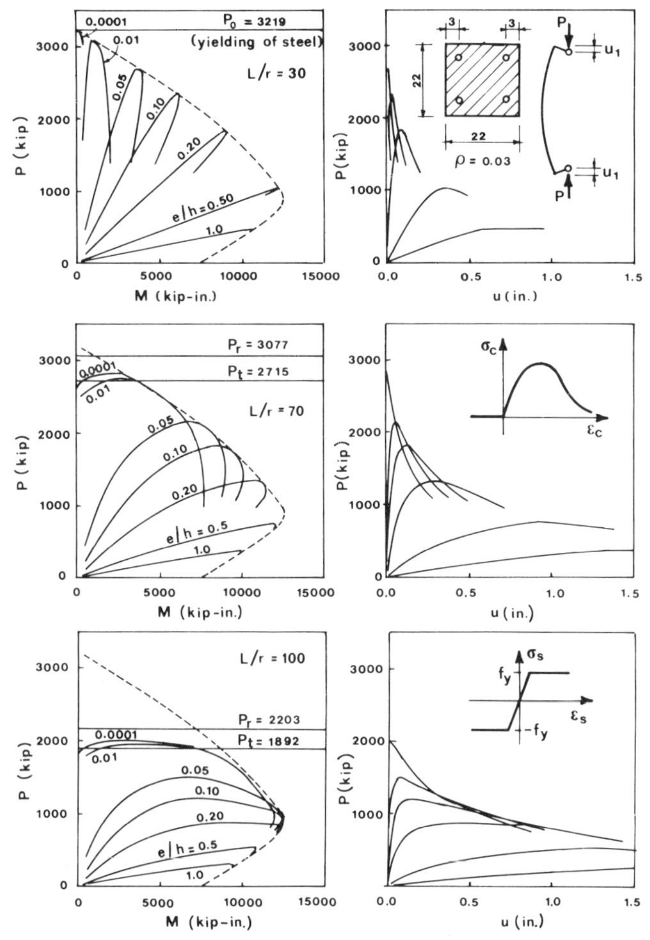

Figure 8.31 Loading paths and load-displacement curves for different eccentricities at the same column slenderness. (*After Bažant, Cedolin, and Tabbara, 1989.*)

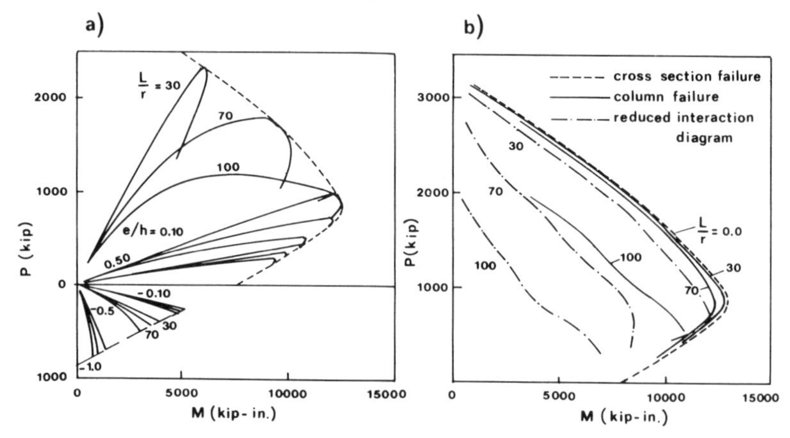

Figure 8.32 (a) Loading paths for different column slendernesses and various eccentricities, (b) failure envelopes of column for zero slenderness (dashed), and for actual slenderness (solid), and reduced failure envelopes in terms of primary bending moment (dash-dot). (*After Bažant, Cedolin, and Tabbara, 1989*.)

peak-load point always occurs within the cross-section interaction diagram, not on it. In textbooks it has been widely assumed that for small e (and not to large slendernesses) the response curve intersects the cross-section interaction diagram with a positive slope; but this is not the case. For columns of medium slenderness ($L/r = 70$), the peaks of the $P(M)$ curves (i.e., the failure points) lie rather close to the cross-section interaction diagram. This is so for small as well as large eccentricities e. On the other hand, for very slender columns ($L/r = 100$), these peaks are quite remote from the interaction diagram for all eccentricities except the very large ones ($e > 0.3h$). The corresponding diagrams $P(u_1)$, which have the same peak values of P determining the stability limits, are shown on the right. Figure 8.32a shows the effect of varying the column slenderness at various eccentricities. Note that the trend of the interaction envelope continues smoothly into the tensile side ($P < 0$). This means that tension stiffens the column, as it does in the elastic domain (Sec. 2.1). Also note that the second-order effect deflects the loading path again below (but to the left of) the straight radial path for zero slenderness. Figure 8.32b shows the failure envelope of columns of various slendernesses, calculated as the maximum value of P for various e/h. The same figure shows the reduced failure envelope in terms of primary bending moments. This envelope is close to both the column and the cross-section failure envelopes for low slendernesses ($L/r \simeq 30$), but it moves farther apart for higher slendernesses.

The responses in Figure 8.31 are calculated and plotted also for very small eccentricities e, for which they must approach Shanley's theory. Shanley's tangent modulus load P_t of the column has been calculated from the equations

$$P_t = \frac{\pi^2}{L^2}[E_c(\varepsilon)I_c + E_s I_s] \qquad P_t = -A_c\sigma_c(\varepsilon) - A_s E_s \varepsilon \tag{8.5.6}$$

where A_c, I_c, A_s, I_s = cross-sectional areas and inertia moments of the concrete and steel parts of the (uncracked) cross section, $\sigma_c(\varepsilon)$ is the given stress-strain diagram of concrete, and $E_c(\varepsilon) = d\sigma_c(\varepsilon)/d\varepsilon$ = tangent modulus of concrete. Setting both expressions for P_t equal, one gets a nonlinear algebraic equation for ε, from which ε can be solved iteratively by the Newton method, upon which P_t may be evaluated. If, however, $-E_s\varepsilon > f_y$, one must replace $-A_sE_s\varepsilon$ in Equation 8.5.6 by A_sf_y. The maximum load, P_{max}, of a perfect column ($e = 0$) is always higher than P_t. It is noteworthy, however, that P_{max} is nearly equal to P_t if the eccentricity e is small, but not very small, for example, $e = 0.01h$ in Figure 8.31 (which corresponds to $e/L = 1.15 \times 10^{-3}$ for $L/r = 30$, $e/L = 0.49 \times 10^{-3}$ for $\lambda = 70$, and $e/L = 0.34 \times 10^{-3}$ for $\lambda = 100$). This observation reveals that by requiring the eccentricity in column design to be at least $e_{min} \simeq 0.01L$ (which corresponds to $e/h = 0.09$ for $\lambda = 30$, $e/h = 0.20$ for $\lambda = 70$, and $e/h = 0.29$ for $\lambda = 100$), the need for using Shanley's theory is essentially circumvented.

The requirement for minimum eccentricity, e_{min}, as used in the existing codes, has previously been justified purely empirically. As we just showed, however, it has to some extent a logical explanation in Shanley's theory. This theory, however, at the same time points at a need for further refinement in which e_{min} is made dependent on the column properties, instead of being given as an empirical constant.

By a similar simple iterative calculation (see Prob. 8.5.10) one can calculate the value of the reduced modulus P_r, which is shown by horizontal lines in Figure 8.31. Note that P_r always represents an upper bound on P_{max}. This condition is satisfied as long as the steel has not yielded; otherwise, P_{max} admits only an upper bound P_0 calculated by setting $\varepsilon_c = \varepsilon_s$ = strain at the start of yielding.

Column Response for Unsmooth Stress-Strain Diagrams

As already mentioned, the change of slope $d\sigma_s/d\varepsilon_s$ due to transition of steel from elastic to perfectly plastic response causes a sudden change in the slope dP/dM of the response curve at constant e. It may then happen that the peak point of this curve, representing the failure point (point on the column interaction diagram), lies exactly on the cross-section interaction diagram; see Figure 8.30b. A sudden change in the slope $d\sigma/d\varepsilon$ of the bilinear $\sigma(\varepsilon)$ diagram of concrete, however, cannot cause a sudden change of slope dP/dM at constant e because sudden changes of $d\sigma/d\varepsilon$ can be simultaneous (for $\kappa > 0$) only in a vanishingly small area of the cross section (if $\kappa \neq 0$).

The stress-strain diagram of concrete has been, in practical concrete literature, assumed to terminate at a certain critical strain ε_c^* at which the stress drops suddenly to zero (Fig. 8.26b). This type of behavior does change the slope dP/dM at constant e. As a result, the maximum point of the response curve $P(M)$ may be without a horizontal tangent and can lie on the cross-section interaction diagram.

It should be noted, however, that the stress-strain diagrams with a sudden slope change or a sudden stress drop are not simplifications but complications because the discontinuities impair convergence of the numerical solution of the $P(M)$ diagram. Also, the assumption of a sudden stress drop is not realistic for column bending because strain-softening is stabilized by the preservation of planeness of cross section. [The assumption of a sudden stress drop at the start of strain-softening is nevertheless justified for sufficiently long tensile specimens, in which the strain-softening zone becomes unstable and leads to sudden dynamic failure (Sec. 13.2).]

According to concrete textbooks and codes (e.g., the ACI), the points of the cross-section failure envelope (interaction diagram) have been approximately determined under the assumptions (1) that the $\sigma(\varepsilon)$ diagram for compression is replaced by an equivalent rectangular stress block and (2) that the failure occurs at a certain value of the maximum compression strain in the cross section, namely, $\varepsilon_c^{max} = -0.0030$ (ACI) or -0.0035 (CEB). In that case it is possible that (for $L > 0$) the calculated response curve $P(M)$ for constant e (unlike the real response curve) intersects the actual cross-section interaction diagram with a positive slope (29 in Fig. 8.30b). This intersection, which occurs for small e, has then been considered as the failure point, even though the curve $P(M)$ may rise further after the intersection. In such a case, the practice has been to call this the cross-section failure, although this is not necessarily a failure state, as we already emphasized. Only for large e this method yields a peak ($dP/dM = 0$) before the cross-section interaction diagram is reached (point 4 in Fig. 8.30b). In that case, the practice has been to call this the stability failure.

Keeping in mind, though, that the distinction between the cross-section failure and the stability failure is but a figment, which does not exist in reality and merely results from the assumption that there is a *sudden stress drop* at the end of the $\sigma(\varepsilon)$ diagram. In reality, every column (with $L > 0$) loaded at constant e fails (if $L > 0$) at the peak of the response curve $P(M)$, and this peak must lie within, not on, the cross-section interaction diagram (even if only slightly within), regardless of slenderness. It is impossible for a real column to fail while the $P(M)$ curve has a positive slope to the right of the failure point.

Design Recommendations and the ACI Code

Design on the basis of a nonlinear calculation of the load-deflection relation and construction of the column interaction diagram is recommended by the ACI Code (ACI, 1977). This code, however, also permits the use of simple approximate formulas, which were shown to approximately agree with the interaction diagrams of the type shown in Figure 8.29 (e.g., MacGregor, Breen, and Pfrang, 1970). Similar to AISC formulas for steel beam-columns (Sec. 8.4), the ACI formulas are based on the magnification factor (Sec. 1.6). The magnification factor, however, is the only concept that is used to take the buckling phenomena into account. The strength-slenderness diagram used by the AISC for centric loading (Eqs. 8.4.1 and 8.4.3) is not used by the ACI. To avoid the need for specifications for centric loading, the ACI requires every column to be designed for a certain minimum eccentricity, namely $e_{min} = 0.03h + 0.6$ in. (h = cross-section height, Fig. 8.27a), even if the calculated initial load eccentricity is zero or less than ths value; the CEB requires $e_{min} = L/200$ or 2 cm. Based on numerous studies (MacGregor, Breen, and Pfrang, 1970; Blomeier and Breen, 1975; and Colville, 1975), the ACI code recommends the following formula, which uses the magnification factor derived in Sec. 1.6 (and is also recommended by the AISC for steel, see Sec. 8.4):

$$M_u^* = \frac{C_m M_u}{1 - P_u/P_{cr_1}^*} \tag{8.5.7}$$

in which $P_{cr_1}^* = \phi_u P_{cr_1}$; ϕ_u = ACI strength reduction factor (0.7 for tied columns and 0.75 for spiral columns) which takes into account the random statistical variability of column properties; $M_u^* =$ ultimate design moment magnified for second-order effects; M_u, $P_u =$ ultimate design moment and ultimate axial load, obtained by first-order analysis from the design loads multiplied by their safety factors (without the second-order effects); $P_{\mathrm{cr}_1} = \overline{EI}\pi^2/L^2$ where $L =$ effective column length, $\overline{EI} =$ effective bending stiffness of the column cross section; and $C_m =$ correction coefficient that takes into account the initial bending moment distribution along the column and the type of supports. The reduction factor ϕ_u is applied to P_{cr_1} because, if there is a random decrease of material strength, the effective (secant) E-value must also be lower (as the strain at peak stress is not lower for a weaker concrete). C_m is based on elastic analysis as already explained in Section 1.6, and is the same as that used for steel (see Sec. 8.4). Recall that $C_m = 0.6 + 0.4M_1/M_2$, $C_m \leq 1$, $C_m \geq 0.4$ (Eq. 1.6.7) for the columns without lateral sway, as in braced frames, and without lateral loads between the column ends; and $C_m = 1$ for all other cases. The design condition is that the point (M_u^*, P_u) must not lie outside the scaled cross-section interaction diagram ($\phi_u M$, $\phi_u P$), or the point (M_u^*/ϕ_u, P_u/ϕ_u) must not lie outside the actual cross-section interaction diagram (M, P) (Fig. 8.33a). The ACI Code allows the analysis of buckling (slenderness effects) to be skipped for short columns such that $L/r < 34 - 12M_1/M_2$ if the column is braced against sway, and $L/r < 22$ if not. [The CEB (1978) takes this limit as $L/r < 25$, with or without sway.]

For columns in a frame, the ACI Standard uses formulas analogous to Equations 8.4.12 and 8.4.13 for steel, in which the moments due to loads that do and do not cause lateral sway of the floor are magnified differently.

For the case of long-time loading, the ACI recommends the axial force P obtained from the interaction diagram, such as that shown in Figure 8.29, to be reduced by the factor $f(\beta_d)$. This factor takes into account the effect of creep and will be explained in Section 9.6. β_d is the ratio of the dead load to the live load.

The ACI code equation (Eq. 8.5.7 and also Eq. 1.6.1 with Eq. 1.6.7) does not consider how the initial moments and the axial loads are supposed to increase during the loading process. The values of C_m in Equation 8.5.7 were calculated (see Sec. 1.6) assuming the initial moments to be constant while P is increased (which means that the eccentricity M/P decreases with loading). A more severe assumption is that the initial moments increase in proportion to the axial load, which is equivalent to a column loaded solely by an eccentric axial load. For the special case of uniform eccentricity, e, this case is solved by Equation 1.5.11. That equation is equivalent to Equation 8.5.7 (or Eqs. 1.5.9–1.5.10) if one sets

$$C_m = \left(1 - \frac{P_u}{P^*_{\mathrm{cr}_1}}\right) \sec\left(\frac{\pi}{2}\sqrt{\frac{P_u}{P^*_{\mathrm{cr}_1}}}\right) \tag{8.5.8}$$

This expression yields $C_m = 1.024$ for $P_u = 0.1P^*_{\mathrm{cr}_1}$, 1.126 for $P_u = 0.5P^*_{\mathrm{cr}_1}$, and $4/\pi$ (Eq. 1.5.12) or 1.273 for $P_u \to P^*_{\mathrm{cr}_1}$. So the ACI (as well as the AISC) formula in which $C_m = 1$ for this case has an error on the unsafe side for such loading.

To eliminate this error, the code of the Swiss Engineers and Architects Association (SIA, 1976) introduces the approximate formula:

$$M_u^* = P_u e + P_u \frac{\Delta_1}{1 - P_u/P^*_{\mathrm{cr}_1}} \tag{8.5.9}$$

in which Δ_1 is the first-order deflection due to the applied end moments (see also MacGregor, 1986). For example, if the end moments are equal, $M_1 = M_2 = P_u e$, one has $\Delta_1 = P_u e l^2/8EI = \pi^2 P_u e/8P^*_{\text{cr}_1} = 1.234 P_u e/P^*_{\text{cr}_1}$ (which can be calculated, e.g., by the moment-area theorem). Substituting this into Equation 8.5.9, one finds that

$$M_u^* = P_u e \frac{1 + 0.234 P_u/P^*_{\text{cr}_1}}{1 - P_u/P^*_{\text{cr}_1}} \tag{8.5.10}$$

This formula is equivalent to the ACI equation (Eq. 8.5.7) if one uses $C_m = 1.023$ for $P_u = 0.1P^*_{\text{cr}_1}$, 1.117 for $P_u = 0.5P^*_{\text{cr}_1}$, and 1.234 for $P_u \rightarrow P^*_{\text{cr}_1}$. These values are indeed quite close to the exact ones listed below Equation 8.5.8. The advantage of Equation 8.5.9 is that it can be applied for columns with unequal initial end moments (i.e., nonuniform eccentricity), without needing coefficients C_m.

The effective bending rigidity for calculating P_{cr_1} in Equation 8.5.7 is allowed by the ACI code to be conservatively estimated from the following empirical equation due to MacGregor, Breen, and Pfrang (1970):

$$\overline{EI} = \frac{E'_c I_g + E_s I_s}{f(\beta_d)} \qquad E'_c = 0.2E_c \tag{8.5.11}$$

in which $f(\beta_d) = 1 + \beta_d$ where β_d = ratio of the maximum design dead load (moment) to the maximum design total load (moment) in the column, always taken as positive; E_s, E_c = elastic moduli of steel and concrete; I_s = moment of inertia of steel reinforcement; and I_g = gross cross-section area of the column (i.e., area including both concrete and reinforcement). The factor 0.2, which reduces the elastic modulus of concrete in Equation 8.5.11, is imagined to take into account the effect of nonlinearity of the moment-curvature diagram due mainly to concrete cracking (and also the effect of short-time creep). The value of $\overline{EI}$ roughly corresponds to the secant slope for the peak-moment point; see Figure 8.33b. Factor $f(\beta_d)$ is intended to introduce in a crude manner the effect of long-time creep. (For the effect of creep on the interaction diagram, see Secs. 9.5 and 9.6, with Fig. 9.16.)

A formula somewhat less conservative than Equation 8.5.11 was proposed by Medland and Taylor (1971), and another one by MacGregor, Oelhafen, and Hage (1975) who introduced $\overline{EI} = E'_c I_g + 1.2\rho_t E_s I_g$ where ρ_t = ratio of the total area of steel to the area of concrete and I_g = inertia moment of the gross section area.

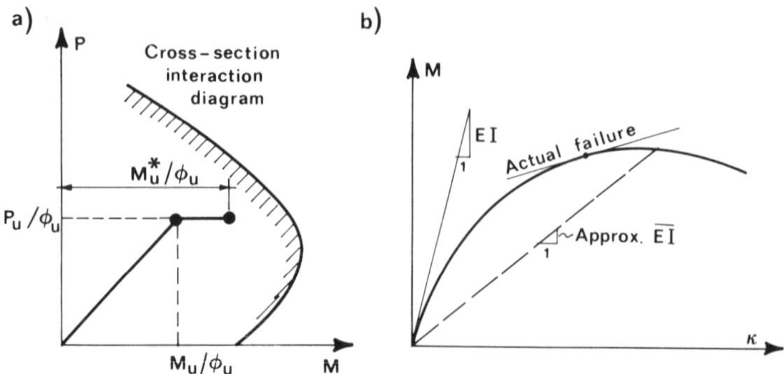

Figure 8.33 (a) Design condition according to ACI Code; (b) Effective bending rigidity adopted by the ACI Code for calculating P_{cr_1}.

It is difficult (and probably futile) to develop a simple formula that gives a close approximation for $\overline{EI}$, and the practice varies considerably. Menn (1974) proposed $\overline{EI} = M_f/\kappa_f$ where M_f and κ_f are the moment and curvature at failure calculated approximately under the assumption that the strains of tensile as well as compression steels are at the yield limit. The dependence of these $\overline{EI}$ values on M and P was studied by Wood and Shaw (1979). The Australian code [Australian Standards Association (ASA), 1984] uses $\overline{EI} = M_b/\kappa_b$ where M_b and κ_b correspond to the balanced condition of the cross section, that is, the state where the tensile steel strain is at the yield limit and simultaneously the compressive strain of concrete at the face is 0.003 (this yields $\overline{EI} \simeq 200M_bd$, in inches and psi, where d = depth to tensile reinforcement); Smith and Bridge's (1984) results indicate this approximation to be on the safe side. The British code [British Standards Institution (BSI), 1972] determines M at failure from the magnified eccentricity (as used by the CEB) and the curvature at failure. This curvature is calculated under the assumption that the maximum tensile strain in the cross section is $\varepsilon_y = f_y/E$ and the maximum compressive strain is $0.003\beta_c$ where factor β_c allows for creep ($\beta_c = 1.25$). The value thus obtained is reduced for very long columns according to an empirical formula.

The value of $\overline{EI}$ for buckling is no doubt influenced by initial cracking due to prior shrinkage, thermal strain, and differential creep. But these influences still remain to be systematically researched.

CEB Design Recommendations

Instead of using the $P(M)$ interaction diagrams for various constant values of e, the CEB recommends predicting the column response from the moment-curvature relations for various constant values of the axial force. In general, the CEB method requires calculation of the complete deflection curves of the column at increasing load. Finite difference calculations have been introduced for that purpose (CEB, 1978).

As a second, simplified alternative, the 1978 CEB Model Code permits the use of the so-called "model column method." This method is based on the approximate equation 8.5.1, which applies to a free-standing column of length $l = L/2$ or other equivalent columns such as the pin-ended column (Fig. 8.34a).

The CEB model column method considers the plots of the applied moment M and of the resisting moment M versus curvature κ at a constant P as sketched in Figure 8.34. The plot of the applied total moment $M = M_{\mathrm{I}} + M_{\mathrm{II}}$ versus $|\kappa|$ is, according to Equation 8.5.1, an inclined straight line of slope Pk_{II}; see Figure 8.34b. This line may intersect the associated resisting $M(\kappa)$ diagram at two points, or one point, or none. If there is no intersection, there exists no equilibrium state and the column fails dynamically. If there are two intersection points such as points 5 and 6 in Figure 8.34b, then point 5 is stable (because at that point the resisting M increases faster than the applied M), while point 6 is unstable. The maximum P for which a stable state exists occurs when the inclined straight line of applied M is tangent to its associated $M(\kappa)$ diagram; see line $\overline{124}$.

The failure envelope for a column of given length (or slenderness) is determined according to the CEB by selecting a number of constant P values. For

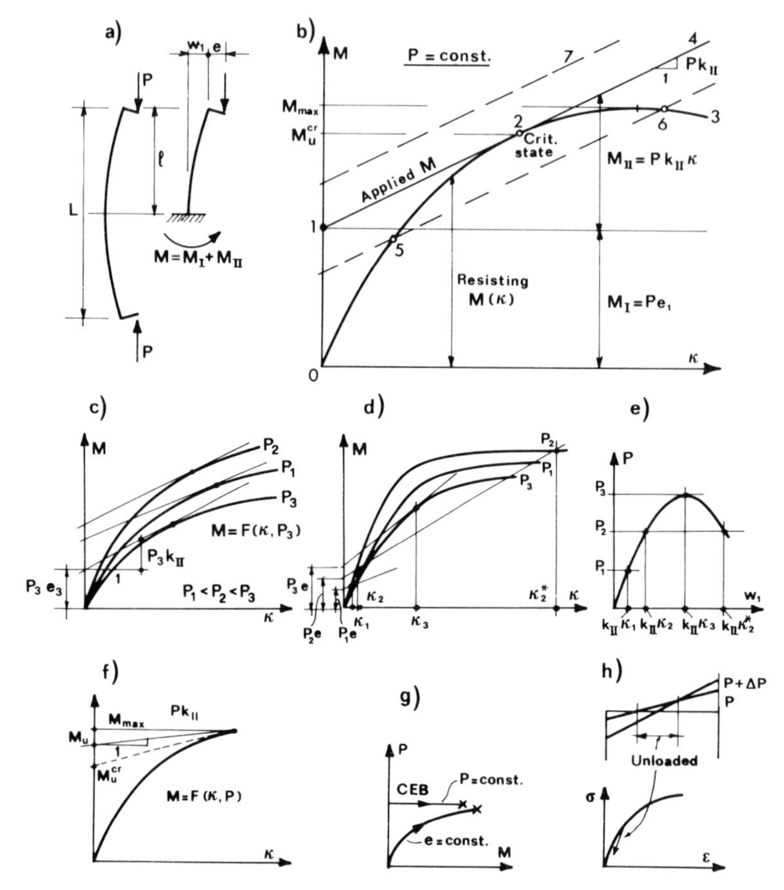

Figure 8.34 CEB model column method: (a) pin-ended and equivalent free-standing column; (b) critical state determination from moment-curvature diagram and total applied moment diagram; (c) cross-section moment-curvature diagrams for various axial loads; (d, e) construction of load-deflection curve; (f) failure condition coinciding with cross-section strength; (g) comparison of the CEB method and constant e method (Bažant, Cedolin, and Tabbara 1989); (h) loading-unloading reversal.

each of them, one calculates the $M(\kappa)$ diagram 0–2–3 (Fig. 8.34b). The tangent line 14 having slope Pk_{II} is then determined either graphically or by solving the κ value for the tangent point from the nonlinear equation $Pk_{II} = \partial M(\kappa)/\partial\kappa$. The tangent point (point 2) yields the ultimate bending moment for the critical state, M_u^{cr}. The point (M_u^{cr}, P) is shown in the $P(M)$ diagram as point 3 in Figure 8.28f. The collection of all such points obtained for various P values yields the failure envelope $\overline{234}$ (in Fig. 8.28f) for the total M $(M = M_I + M_{II})$ at failure. For design, however, it is more convenient to determine from Figure 8.34b the value M_I of the primary moment corresponding to the tangent point 2 (M_I = segment $\overline{01}$ in Fig. 8.34b). In the $P(M)$ diagram, the point (M_I, P) is shown as point A_i (Fig. 8.28f), and it represents the horizontal projection of the failure point onto the radial ray of slope $P/M_I = 1/e$.

The determination of the case for which the applied and resisting $M(\kappa)$ curves are tangent is equivalent to solving P and κ from two simultaneous nonlinear equations:

$$M_{\rm I} + Pk_{\rm II}\kappa = F(\kappa, P) \qquad Pk_{\rm II} = \frac{\partial F(\kappa, P)}{\partial \kappa} \tag{8.5.12}$$

where function $F(\kappa, P)$ represents the resisting moment-curvature diagram $M(\kappa)$ for any constant value P, as sketched in Figure 8.34c. The solution of Equations 8.5.12 with a standard computer library subroutine is a trivial matter once function $F(\kappa, P)$ has been formulated. Although it is more convenient to choose various constant values of P and solve for the corresponding values of $M_{\rm I}$, one may sometimes prefer to solve $P = P_{cr}$ for a given e.

The load-deflection curve for a fixed value of the eccentricity e at column ends, whose direct calculation we already explained (Eq. 8.5.1–8.5.2), can also be constructed on the basis of the CEB model column method (Migliacci and Mola, 1985), provided that the loading path dependence due to possible unloading is neglected. To this end one needs to determine the intersections of the curves $M = F(k, P)$ with the straight lines $M = Pe + k_{\rm II}P\kappa$ representing the equilibrium values of the applied moment (Fig. 8.34d), and then calculate the corresponding deflection $w_1 = k_{\rm II}\kappa$. Connecting these intersection points yields the load-deflection curve (Fig. 8.34e).

Note that the CEB model column method is applicable also for the case in which the failure is assumed to occur if the compression strain magnitude reaches 0.003 while at the same time $dP/dM > 0$ (Fig. 8.34f). For a smaller value of the slope of the applied moment line, one finds a larger M_u, and as this slope approaches zero, M_u approaches the peak moment that represents the cross-section strength.

The assumptions of a sinusoidal deflection curve and of enforcement of equilibrium only at midlength of the column (Eq. 8.5.1) are the same as those employed by the solution algorithm based on Equation 8.5.5. The results then must be exactly the same if unloading is neglected, provided that the numerical algorithms converged (in practice there are small differences, due to numerical errors). Unloading typically occurs in concrete during loading at constant P as the neutral axis moves into the previously compressed portion of cross section (Fig. 8.34h). The CEB model column method, however, cannot reproduce unloading in a meaningful way, since the (M, κ) diagram is calculated at constant P (and variable $e = M/P$) and thus does not represent the actual path followed by columns, which is closer to loading at constant e (Fig. 8.34g). But the effect of unloading is usually found to be small (Bažant, Cedolin, and Tabbara, 1989), because the loading-unloading reversal occurs near the neutral axis (with little effect on M) and at stress levels at which nonlinear effects are negligible (Fig. 8.34h). In any case, the constant e method described before is simpler and easier to use than the model column method, because it does not require the construction of (M, κ) curves and the numerical determination of the tangent to these curves.

The CEB model column method also can take into account the effect of constant lateral (horizontal) loads (Fig. 8.35a), which contribute to the primary moment. It can also be generalized to arbitrary columns, as long as one can determine a reasonably accurate relationship between $M_{\rm II}$ and κ for a certain

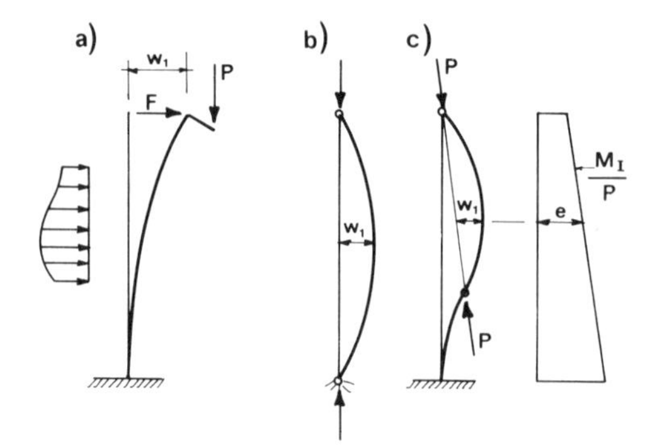

Figure 8.35 Generalizations of the CEB method to different load or end-constraints conditions.

critical cross section of the column (Eq. 8.5.1). This can always be done if the deflections can be assumed to have a sinusoidal shape, increasing proportionally along the column, and if the effective length L is known. For example, for a hinged–fixed column of length l, $k_{II} = (0.699l/\pi)^2$. Note that in writing $M_{II} = Pw_1$, one must consider w_1 to represent, in general, not the maximum deflection (Fig. 8.35b) but the displacement of the critical cross section from the line of the axial resultant (this resultant moves laterally and can also be inclined; cf. Figs. 1.7a and 8.35c). The critical cross section is not that for which w_1 is maximum but that for which $e + w_1$ is maximum ($e = M_I/P$ = initial eccentricity).

The effect of creep is introduced in the CEB model column method by scaling the κ-axis in the $M(\kappa)$ diagram with the factor $1 + \alpha\beta\phi$ where α is the ratio of sustained to total axial loads, β is the ratio of corresponding moments and ϕ is the creep coefficient (Sec. 9.4). But, according to MacGregor (1986, p. 19) this seems to underestimate the effect of creep.

It may be noted that all the aforementioned effects can also be introduced into the solution algorithm based on Equation 8.5.5.

As a third, still simpler alternative, the CEB (1978) also permits using the so-called second-order eccentricity method, which was adopted for the British Code (BSI, 1972). The cross section of a hinged column bent in single curvature is designed for the moment Pe_t with $e_t = e + e_a + e_2$. Here e_t = total eccentricity, e = first-order eccentricity due to initial moments, e_a = empirically specified additional eccentricity to account for geometric imperfections, and e_2 = second-order eccentricity due to column deflection, which is given by $e_2 = \kappa L^2/10$ (this is nearly the same as $\kappa L^2/\pi^2$, which corresponds to a sine curve). There is, however, a major problem with realistic determination of the curvature κ at failure, as already discussed. Note that if one substitutes $\kappa = Pe_t/\overline{EI}$ into the CEB relation $e_2 = e_t - e = \kappa L^2/\pi^2$, one gets $e_t = e/(1 - P/P_{cr})$, which is equivalent to the ACI moment magnifier equation (Eq. 8.5.7).

The second-order eccentricity method is quite convenient for introducing the effect of creep (CEB, 1978). This is done in terms of additional creep eccentricity e_c, which is defined as the residual eccentricity increase that is left when a column is suddenly unloaded after a long period of creep under the (factored) dead load, P_D. Under subsequent rapid overload to failure, the column behaves about the same way as a column with initial eccentricity $e + e_c$ instead of e (as discussed before Eq. 9.5.1). For the determination of e_c, Warner and Kordina (1975) and Warner, Rangan, and Hall (1976) proposed a formula based on Dischinger's method. This formula corresponds to Equation 9.4.11 derived in the next chapter, which gives the load eccentricity (ordinate) at time t as $z(t) = ef(t)$, where e = initial eccentricity due to initial moments. From this, e_c is obtained by dividing $z(t)$ by the magnification factor $1/(1 - P_D/P_{cr_t})$ and subtracting e, that is,

$$e_c = e\left[f(t)\left(1 - \frac{P_D}{P_{cr_t}}\right) - 1\right] \tag{8.5.13}$$

where P_{cr_t} is the elastic critical load corresponding to the elastic modulus $E(t)$ at time t (see also the discussion before Eq. 9.5.1). The value of $z(t)$ or $f(t)$ can be calculated by various methods for creep (Sec. 9.4). It appears that Equation 9.4.18, based on the age-adjusted effective modulus method, is both simpler and more accurate than the formula (Eq. 9.4.11) based on Dischinger's method.

Comparisons of Codes and Shortcomings

A questionable aspect of the ACI moment magnifier method is the value of $\overline{EI}$. Since failure of the cross section per se is determined by the peak point of the cross-section interaction diagram, the $\overline{EI}$ value in Equation 8.5.11 is assumed to correspond approximately to the secant slope of the $M(\kappa)$ diagram for the peak point (Fig. 8.33). This value of $\overline{EI}$ seems to be generally on the low (safe) side.

For small eccentricities, however, the ACI's value of $\overline{EI}$ seems to be less than for higher eccentricities. Below Equations 8.5.6, we observed (from Fig. 8.31) that Shanley's theory is indeed approached for small eccentricities such as $e = 0.01h$. But Shanley's theory would require calculating the bending stiffness from the tangent modulus (Eqs. 8.5.6), which gives a smaller bending stiffness than the approximate secant value $\overline{EI}$ used by the ACI.

Another error on the safe side is due to the neglect of redistribution of the initial moments that is caused by nonlinear behavior of a column in a frame. Indeed, if the largest M occurs at midlength, the bending stiffness at midlength is smaller than near the ends of the column, and this tends to transfer the initial bending moments toward the column ends.

As a third error, which tends to be on the safe side, the nonlinear behavior causes the effective length of the column in a frame to become less than that obtained by elastic analysis, since the inflection points move toward the midlength as the deflection curve becomes more pointed due to inelastic behavior.

The aforementioned effects further depend on the ratio of the initial end moments and on whether or not the column is braced against sidesway. The various sources of error might, but need not, offset each other. There is obviously much room for improvement in the quest for a design method that yields the same safety margin for all situations.

For the CEB model column method, likewise, several kinds of limitations or disadvantages need to be pointed out. The assumption of a sinusoidal shape of the deflection curve, along with the corresponding effective length L and the value of w_1, is exact for the initial deflection increment of a perfect column at the tangent modulus load, but is probably not too good at larger deflections for which the simplified CEB model column method is applied. Obviously, due to nonlinearity, the curvatures near the peak of the deflection curve become larger than for a sine curve. This effect would change the applied moment curves in Figure 8.34b from straight to curved. The sinusoidal shape is, of course, also implied in the ACI's use of the magnification factor. Same as the ACI, the CEB model column method also neglects the initial moment redistributions and changes of the effective length due to adjacent frame members.

Furthermore, the CEB model column method implies the value of M for a given κ depends only on the current value of P, but not on the path in the $P(M)$ response curve; that is, the method implies the assumption of path-independence, although some path-dependence might occur in reality if unloading takes place. (The ACI's use of the interaction diagram, of course, implies path-independence, too.)

The simplified CEB model column method is more tedious to implement (in hand calculatons) or to program, and it requires more information than the ACI method based on the magnification factor. Instead of one interaction diagram, it necessitates a set of moment-curvature diagrams for various P values.

Excellent practical appraisals of various code methods for concrete columns have recently been presented by MacGregor (1986) and Quast (1987). MacGregor points out several serious shortcomings: (1) the problem of determining a realistic $\overline{EI}$ value, which we have already discussed; (2) determination of the effective length L, taking into account the nonlinearities and overload (safety) factors; (3) lack of a good, simple approach for column failures involving biaxial bending or torsion, or both; and (4) lack of consensus on how to introduce the creep effects. MacGregor further indicates that accuracy of the simplified code methods is seriously reduced due to extensions (1) from uniform to nonuniform initial moments, (2) from isolated columns to columns in frames, and generally (3) from elastic behavior to nonlinear behavior with cracking and creep. As important recent developments, MacGregor (1986) calls attention to the following:

1. Calculation of interaction among columns that is involved in buckling with lateral sway (also called the lateral drift). Since all the columns of the floor have to buckle simultaneously, the failure depends not on the individual column loads but on $\sum P_i \Delta/L_i$ where P_i and L_i are the axial force and effective length of column number i, and Δ is the lateral sway displacement (cf. Eq. 8.4.13).
2. Formulation of a procedure (by Butler, 1977, and by Wood and Shaw, 1979) that uses a variable $\overline{EI}$, defined by a simple formula. This approach was shown to agree well with Cranston's (1972) computer simulations of column failures.
3. Recognition of fundamental differences between the initial column moments due to applied loads and those due to imposed deformations in frames (Thürlimann, 1984; Favre et al., 1984). The latter are not magnified as strongly.

To sum up, the simplified design procedure for reinforced concrete columns that is permitted by the ACI code as a replacement for the full second-order analysis is entirely based on the moment magnification factor and a comparison of the magnified moment with the cross-section interaction diagram of axial load versus bending moment. This approach is the same as that used by the AISC for steel columns (which is an advantage for engineers who do not specialize in concrete only). Nonlinear behavior, cracking, and creep are taken into account by the ACI rather approximately—by making a simple adjustment of the bending rigidity. The residual stresses and cracking due to prior shrinkage, thermal strains, and differential creep have so far been ignored, although for metal columns the residual stresses are known to be important. The need for a formula for centrically loaded columns is circumvented by prescribing a certain minimum load eccentricity.

Although the simplified code methods provide valuable insight, it appears increasingly preferable to obtain accurate values of the failure loads of columns and frames by finite elements considering geometric as well as material nonlinearities. Initial residual stresses and cracking due to prior shrinkage, thermal strains, and creep may have to be considered also. This should lead to more uniform true safety margins, and thus allow reduction of the safety factors, whose values for concrete columns are still quite large.

Prestressed Concrete Columns

In Section 1.8 we proved that an axial prestress (by a bonded tendon) does not alter the stability limit P_{cr} of a column. However, this is true only for columns that fail in the elastic range.

Consider two identical columns, one with prestressing force F and one without it, both under axial load P. If the column behaves nonlinearly, the effect of prestress is to change the resisting moment-curvature diagram from that corresponding to $P = P_2 = \text{const.}$ to that corresponding to $P_3 = P_2 + F$; see Fig. 8.36 [the construction of the resisting $M(\kappa)$ diagrams is explained in prestressed

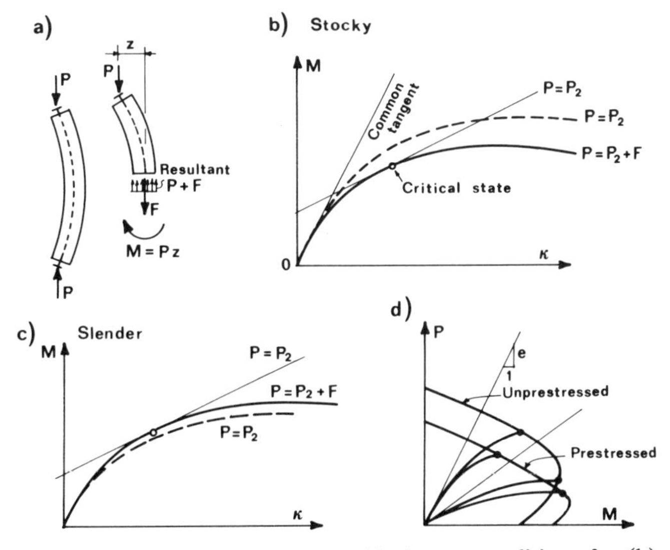

Figure 8.36 (a) Prestressed concrete column, critical state conditions for (b) stocky or (c) slender columns, and (d) comparison of column failure envelopes.

concrete textbooks; see Nilson, 1987; Lin and Burns, 1981]. (Note that the effect of prestress is similar to that of residual stresses.) At the same time, the straight-line diagram of the applied moment versus curvature remains unchanged because (as shown in Sec. 1.8) the axial force resultant in the cross section is not affected by F [the compressive prestress in concrete is balanced by the tension in steel (Fig. 8.36a)]. So the critical state does not occur when the line for P_2 that represents the applied moment becomes tangent to the diagram of the moment-curvature relation for $P = P_2$. Rather, the critical state (at constant axial force) occurs when the line for P_2 becomes tangent to the diagram for $P_3 = P_2 + F$; see Figure 8.36b. If the nonlinearity of concrete in compression dominates the shape of the resisting $M(\kappa)$ diagram, which is true for stocky columns, then the diagram for $P = P_2 + F$ lies below that for $P = P_2$ (Fig. 8.36b). So this procedure yields for a prestressed column a smaller critical load than for a nonprestressed column. If, however, the nonlinearity is due mainly to tensile cracking, which is true for slender columns, then the diagram for $P = P_2 + F$ lies above that for $P = P_2$ (Fig. 8.34c), and so a larger critical load is obtained than for a nonprestressed column. This is the case where the benefit from prestress is the greatest.

The effect of prestress on the critical load is also apparent in the ACI method. As shown in prestressed concrete textbooks, the effect of prestress is to change the ultimate state interaction diagram of P versus M. As shown in Figure 8.36d, this depresses the value of P_u at the terminal point of the $P(M)$ path if the column is not too slender (i.e., if it fails at a low ratio of $e = M/P$) but can increase P_u if the column is slender (i.e., if it fails at a high ratio of $e = M/P$).

The design of a prestressed column may also be done approximately on the basis of the magnification factor. One may then use Equation 8.5.7, the same as for nonprestressed columns:

$$M_u^* = \frac{C_m M_u}{1 - P_u/P_{cr_1}^*} \tag{8.5.14}$$

The prestress is taken into account in determining P_u from the interaction diagram of the cross section. The initial moment M_u should include not only the (factored) moments due to lateral loads with the moments and forces applied at the ends of the column, but also the (factored) initial moment $P(e + w_F)$ due to the axial force P where eccentricity of load P on the undeflected column and deflection due to prestress w_F are included. However, no moment due to axial prestressing force F should be included, as explained in Section 1.8.

Counterintuitive though it might seem, prestress is nevertheless used not only for slender columns (Fig. 8.36c) but also for some not-too-slender columns. This brings about various benefits. Prestress reduces cracking, which is generally advantageous. Prestress makes it possible for a lighter column to better withstand handling in transport. In concrete piles, prestress helps the pile to withstand the dynamic effects during driving.

Shells and Other Structures

In contradistinction to steel structures, there are no special code specifications for concrete structures other than columns, even though for some structures the analysis of stability is very important, for example, for shells. Generally, stability of shells is analyzed elastically using the tangent modulus E_t instead of the elastic modulus; see for example, Popov and Medwadowski (1981). This contrasts with the current practice for concrete columns where Shanley's tangent modulus theory is not used. The value of E_t is further reduced by the factor $1 + \phi$ where ϕ is the creep coefficient; see Section 9.4.

Stress-Strain Relations for Strain Softening

It needs to be emphasized that the softening (descending) portion of the effective stress-strain relation used in the analysis depends on many factors not normally considered in design. The reason is the localization of softening (see Chap. 13). Depending on the volume fraction of localized softening zones, the post-peak average slope can be mild or steep (or even a snapback). The aforementioned volume fraction must be expected to depend on the axial stiffness of reinforcement, the spacing of stirrups, the type of transverse reinforcement (ties, spirals), the shape of cross section, the shape of ties, the stiffness of ties, the aggregate size, the stiffness of matrix of concrete versus the stiffness of aggregate, the distribution of the axial bars, etc. Thus all these factors must be expected to influence considerably the steepness of the post-peak softening stress-strain relation. As long as these factors are neglected, it really makes no sense to argue whether one or another formula for the $\sigma(\varepsilon)$ curve is better. The fact that a laboratory test of a standard specimen gives a certain $\sigma(\varepsilon)$ curve is not too relevant. One would need, among other things, to actually calculate the localization of softening zone in the column in order to achieve more dependable results and be able to profit from a sophisticated stress-strain relation.

In this light the CEB formula for the $\sigma(\varepsilon)$ curve seems to be unnecessarily and unjustifiably complicated. A short formula such as $\sigma = E\varepsilon \exp(-k\varepsilon^2)$ is probably just as good, or just as poor.

Design Examples

Example 8.5.1 A column of length $l = 16$ ft (4877 mm) (Fig. 8.37a), braced against sidesway, is subjected to dead load $D = 105$ kips (467.2 kN) and live load

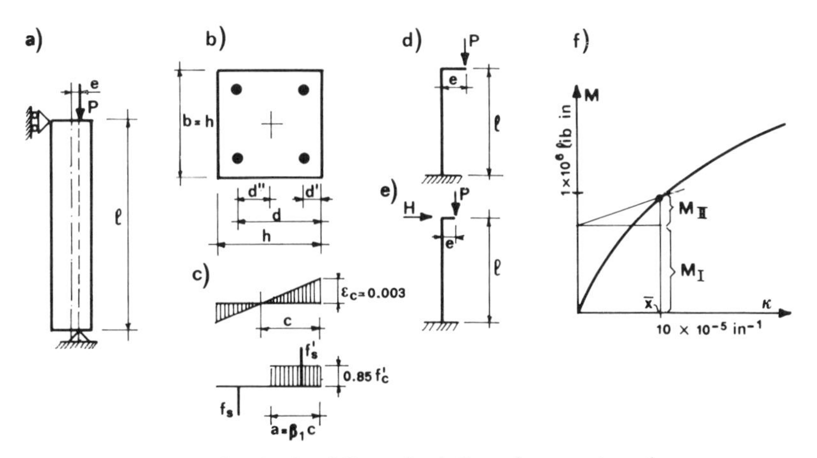

Figure 8.37 Design examples for buckling of reinforced concrete columns.

$L = 90$ kips (400.4 kN) acting with an eccentricity $e = 3$ in. (76.2 mm). The square cross section (Fig. 8.37b) of depth $h = 15$ in. (381.0 mm) is reinforced with a steel bar at each corner (area $= 0.875\ \text{in}^2 = 564.5\ \text{mm}^2$) whose centroid lies at $d' = 2.5$ in. (63.5 mm) from the edge. Assuming $f'_c = 3000$ psi ($20.6\ \text{N/mm}^2$), $f_y = 40 \times 10^3$ psi ($275.2\ \text{N/mm}^2$), $E_s = 30 \times 10^6$ psi ($206.4\ \text{kN/mm}^2$), $E_c = 57{,}000\sqrt{f'_c}$ (all in psi), check the column design according to the moment magnifier method. [*Hint:* The factored load is $P_u = 1.4D + 1.7L$; see Sec. 9.3 of ACI (1977) Code or Sec. 9.2.1 of updated ACI 1989 Code.] Calculate first the ultimate moment M_u for the cross section under P_u, writing the force and moment equilibrium conditions of the cross section and assuming a rectangular block for concrete stresses (Fig. 8.37c). Compare M_u with that resulting from Equations 8.5.7 and 8.5.10.

Solution. $P_u = 0.85f'_c ab + A'_s f_y - A_s f_s$, $\quad f_s = \varepsilon_s E_s = 0.003E_s(d-c)/c = 0.003E_s(\beta_1 d - a)/a$, $\quad M_u = 0.85f'_c ab(d - d'' - 0.5a) + A'_s f_y(d - d' - d'') + A_s f_s d''$, $\beta_1 = 0.85$.

$$f_s = \frac{0.003[0.85(12.5) - a]}{a}(30 \times 10^6) = \frac{90{,}000(10.625 - a)}{a}$$

$$300{,}000 = 0.85(3000a)(15) + 1.75(40{,}000) - \frac{1.75(90{,}000)(10.625 - a)}{a} \qquad a = 7.63\ \text{in.}$$

$$M_u = 0.85(3000)(7.63)(15)[12.5 - 5 - 0.5(7.63)] + 1.75(40{,}000)(12.5 - 2.5 - 5) + 1.75(90{,}000)\left(\frac{10.625 - 7.63}{7.63}\right)(5) = 1.734 \times 10^6\ \text{lb-in.}$$

$$E_c = 57{,}000\sqrt{3000} = 3.122 \times 10^6 \quad \text{(all in psi)} \qquad I_g = \tfrac{1}{12}(15)(15^3) = 4218\ \text{in}^4$$

$$E_s = 30 \times 10^6\ \text{psi} \qquad I_s = 4(0.875)(5^2) = 87.5\ \text{in}^2$$

$$\beta_d = \frac{1.4D}{1.4D + 1.7L} = \frac{1.4(105)}{1.4(105) + 1.7(90)} = 0.49$$

$$EI = \frac{0.2E_c I_g + E_s I_s}{1 + \beta_d} = 3.529 \times 10^9 \qquad C_m = 0.6 + 0.4 = 1$$

$$P_c = \frac{\pi^2 EI}{l^2} = \frac{\pi^2(3.529 \times 10^9)}{(192)^2} = 945{,}000$$

$$\delta = \frac{1}{1 - 300{,}000/0.7(945{,}000)} = 1.830$$

$$M_c = \delta P_u e = 1.83(300{,}000)(3) = 1.647 \times 10^6\ \text{lb-in.}$$

Example 8.5.2 Consider a free-standing column (Fig. 8.37d) of length $l = 13$ ft (3960 mm) subjected to an axial load of 250 kips (1110 kN) acting with an eccentricity $e = 3$ in. (76.2 mm). The cross-section area has depth $h = 15$ in. (381 mm) and is reinforced with a steel bar at each corner (Fig. 8.37b) (of area $0.875\ \text{in}^2 = 564.5\ \text{mm}^2$) whose centroid lies at $d' = 0.1d = 1.5$ in. $= 38.1$ mm from the edge. Assuming for concrete a "characteristic" strength $f_{ck} = 4800$ psi ($33\ \text{N/mm}^2$), $f_y = 40{,}000$ psi ($275.2\ \text{N/mm}^2$) for steel, relative axial load $\nu = 0.4$

and mechanical reinforcement ratio $w = 0.1$, the moment-curvature relation can be assumed (CEB Bulletin No. 123) to be given in tabular form by: For $\kappa/(10^{-5}\,\text{in}^{-1}) = 5$, 10, 15, 20, 25, 30, 35, 40, 45, the corresponding moment is $M/(10^6\,\text{lb-in.}) = 0.608$, 0.973, 1.22, 1.41, 1.56, 1.68, 1.76, 1.80, 1.81. Check stability according to the CEB model column method.

Solution. (cf. CEB, 1978): $f_c = 0.85 f_{ck}/1.5 = 0.85(4800)/1.5 = 2720$ psi.

$$f_s = \frac{f_y}{1.15} = \frac{40{,}000}{1.15} = 34{,}782 \text{ psi} \qquad P_c = f_c A_c = 2720(15^2) = 612{,}000 \text{ lb}$$

$$w = \frac{A_s f_s}{P_c} = \frac{2(0.875)(34{,}782)}{612{,}000} = 0.099 \simeq 0.1$$

$$v = \frac{P}{P_c} = \frac{250{,}000}{612{,}000} = 0.408 \simeq 0.4$$

$$M_{\text{I}} = 250(3) = 750 \text{ kips-in.} \qquad (1 \text{ kip} = 1000 \text{ lb})$$

$$k_{\text{II}} = \frac{4l^2}{\pi^2} = \frac{4[(13)(12)]^2}{\pi^2} = 9860 \text{ in}^2$$

$$M_{\text{II}} = P k_{\text{II}} \bar{\kappa} = 250{,}000(9860)(10.2 \times 10^{-5}) = 252 \text{ kips-in.} \qquad \text{(Fig. 8.32f)}$$

Problems

8.5.1 Describe the (a) ACI method and (b) CEB method for determining the load capacity of reinforced concrete columns.

8.5.2 Do the same as Example 8.5.2, but find the limit value of the eccentricity e.

8.5.3 So the same as Example 8.5.2, but find the limit value of the length l.

8.5.4 Do the same as Example 8.5.2, but with $e = 1$ in. (25.4 mm) and a horizontal force on top (Fig. 8.37c) given by $H = 2000$ kips (8.88 kN).

8.5.5 Solve (a) Example 8.5.1 for $h = 14$ in. instead of 15 in.; (b) Example 8.5.2 for $l = 14$ ft instead of 13 ft.

8.5.6 Program the step-by-step iterative solution of Equation 8.5.5 and solve the interaction diagram of the same column as in Figure 8.37 but with $\rho_s = 0.004$.

8.5.7 Calculate the stability limit of a prestressed concrete column that is the same as in Figure 8.37 but has uniform axial prestress $\sigma_c^p = -1500$ psi.

8.5.8 Repeat Problem 8.5.5 for a prestressed column with $\sigma_c^p = -1500$ psi.

8.5.9 The interaction diagrams in Figure 8.31, covering the top-right quadrant of the (P, M) plane, describe buckling to the left ($M > 0$). Buckling to the right is described by similar diagrams in the top-left quadrant. They are symmetric if the cross section is symmetric, but not if the cross section is nonsymmetric (cf. Fig. 8.16). Construct a portion of such interaction diagram assuming that the area of the left-side bars in Figure 8.31 is doubled.

8.5.10 Show that the reduced modulus load P_r of the column with the cross section in Figure 8.28a is given by $P_r = (\pi^2/L^2)(\Phi_c E_c^0 I_c + \Phi_s E_s^0 I_s)$ and $P_r = \sigma_c A_c + \sigma_s A_s$ where $\Phi_c E_c^0 I_c = (E_c^t h_L^3 + E_c^0 h_U^3)b/3$, $\Phi_s E_s^0 I_s = (A_s/2)[E_s^t(h_L - d)^2 + E_s^0(h_U - d)^2]$, $h_U = h - h_L$, and h_L (representing the portion of h that undergoes loading) is to be solved from the quadratic equation $bh_L^2(E_c^t - E_c^0) + h_L(2bhE_c^0 + A_sE_s^t + A_sE_s^0) - bE_c^0h^2 - A_sE_s^0h + dA_s(E_s^0 - E_s^t) = 0$, where

d = thickness of the concrete cover of steel bars. Solve P_r by iterations for the same data as used to calculate Figure 8.31 (see also Bažant, Cedolin, and Tabbara, 1989).

8.6 PERFECTLY PLASTIC LARGE-DEFLECTION BUCKLING, IMPACT, AND BLAST

The preceding sections dealt with stocky columns for which the initial buckling is inelastic. Now we focus attention on very large deflections of columns that are (1) so slender that their initial buckling is elastic, governed by the Euler hyperbola portion of the column strength curve, and (2) made of an elastic-plastic material whose stress-strain diagram terminates (both for compression and tension, Fig. 8.38d, e) with a long horizontal plateau at which the response becomes perfectly plastic. If the column is slender [and the hardening range of the $\sigma(\varepsilon)$ diagram is short or absent], the transition from an elastic state to a perfectly plastic state occurs relatively fast (i.e., within a small portion of the deflection range). At

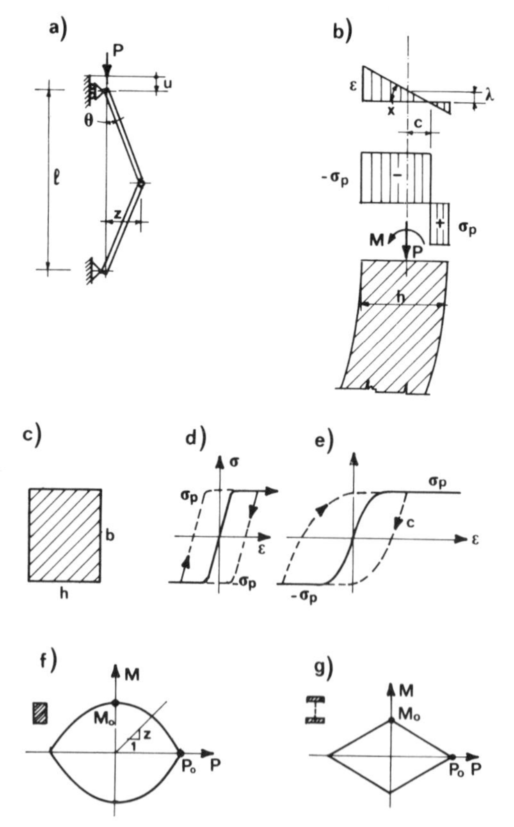

Figure 8.38 (a) Simply supported column with plastic hinge, (b) strain and stress distributions at plastic hinge, (c) column cross section, (d, e) stress-strain relations, (f, g) yield surface of rectangular and ideal-I cross sections.

sufficiently large deflections, nearly all of the cross section of maximum bending moment reaches the yield stress σ_p for the yield plateau. This makes it possible to obtain simple solutions, as we will see.

Since very large deflections are not covered by design codes, we were able to deal with the codes before addressing perfectly plastic behavior. Large buckling deflections are of interest mainly for determining the energy absorption capability of a structure, which is important for the behavior under dynamic impact or blast. They are also of interest for determining internal stress redistribution and the corresponding strength reserve in redundant structures during collapse, particularly for deciding whether the entire frame will collapse if one member buckles.

Perfectly plastic solutions of column buckling are of some interest also for stocky columns or frames, for which the initial buckling is well approximated by Shanley's theory. If the stress-strain diagram terminates with a long plateau for both compression and tension, then the cross section must eventually approach (along curve c in Fig. 8.38e) a state at which the unloading, too, reaches the yield plateau and every point is at yield in either tension or compression. In such a case, the perfectly plastic solution must be approached asymptotically at large deflections.

Load-Deflection Curve of Perfectly Plastic Columns

For perfectly plastic behavior, we may assume the plastic deformations in beams to be concentrated in plastic hinges. A simply supported column develops a plastic hinge (yield hinge) at midheight, as shown in Figure 8.38a. In the classical analysis of carrying capacity of beams and frames, the yield bending moment in the hinge may be assumed to be constant; however, in buckling one must take into account the influence of the axial load P on the yield moment M, that is, the dependence $M = f(P)$.

Function f is characterized by the geometry of the cross section. For example, for a rectangular cross section of height h and width b (Fig. 8.38c), the bending moment and axial force equilibrating the perfectly plastic stress distribution shown in Figure 8.38b may be calculated as

$$M = \sigma_p\left(\frac{b}{4}\right)(h^2) - \sigma_p bc^2 \qquad P = P_0 - 2\sigma_p b\left(\frac{h}{2} - c\right) \tag{8.6.1}$$

in which $P_0 = \sigma_p bh$ = centric axial yield force (also called the squash load), σ_p = yield limit of the material (Fig. 8.38d), and c = the z-coordinate of the neutral axis (point of $\varepsilon = 0$); see Figure 8.38b. Eliminating c from Eqs. 8.6.1 we get for a column of rectangular cross section

$$M = (P_0 - P)\left(\frac{h}{2} - \frac{P_0 - P}{4b\sigma_p}\right) = f(P) \qquad z = \frac{M}{P} \tag{8.6.2}$$

where the deflection ordinate z at column midheight is obtained from the equilibrium condition $|M| = Pz$. The deflection then is $w = z - z_0$ where z_0 is the initial imperfection, if any. The curve $M = f(P)$ defined by Equations 8.6.2 represents the yield surface of the cross section, and we see that for a rectangular cross section it is parabolic (Fig. 8.38f). It should be noted that, due to the variation of c during loading, some points of the cross section change their stress

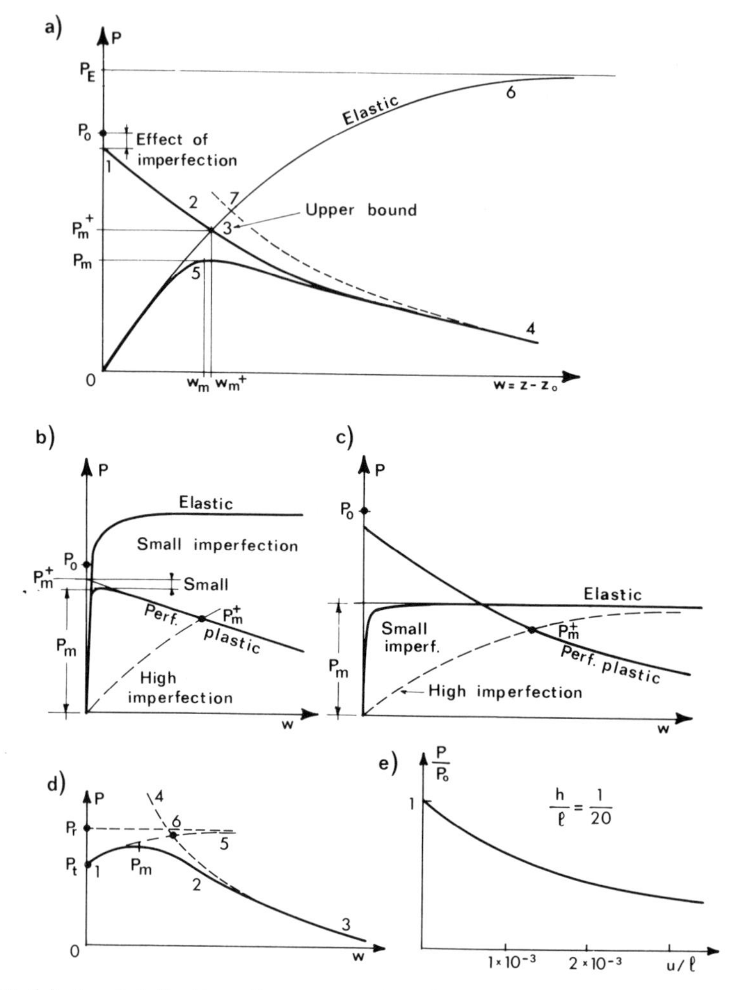

Figure 8.39 (a) Load-deflection diagrams under different assumptions and bounds for maximum load; (b, c) cases of short and slender columns; (d) transition to perfectly plastic response of Shanley's column; (e) load-displacement diagram.

from σ_p to $-\sigma_p$. Such behavior represents hysteresis, and it increases the plastic energy dissipation. If the axial load P has become small due to large deflections, and if the slope dM/dP of the cross-section yield surface at $P = 0$ is zero (i.e., the tangent is horizontal as in Fig. 8.38f), then the bending moment M can be assumed to be approximately constant and equal to the yield hinge moment at no axial load, that is, $f(P) = \text{const.} = M_0$. For this case, the column path is the dashed curve $\overline{74}$ in Figure 8.39a, which is an easily calculated upper bound on P. This curve is approached by the actual $P(w)$ curve $\overline{34}$ asymptotically for small w values.

If there are no imperfections ($z_0 = 0$), the load-deflection diagram of the column is given by the path $\overline{01234}$ in Figure 8.39a (see also Horne, 1971, p. 148). However, in practice there are always imperfections. They may cause the maximum load to be much less than P_0 (point 5 in Fig. 8.39a). If the column is imperfect, it will, of course, start deflecting before the yield stresses are reached, following initially a curve such as $\overline{036}$ in Figure 8.39a (this curve may be obtained from the magnification factor, Sec. 1.5). Now, as plastic strains start to develop, the actual load should be smaller than that for the elastic imperfect column, and so the actual behavior should follow some curve such as curve $\overline{05}$ in Figure 8.39a. At the same time, because the yield moment calculated according to the perfectly plastic stress-strain diagram in Figure 8.38d (Eqs. 8.6.1) is reached only at very large rotations in the hinge, and is larger than the bending moment for smaller rotations, the actual behavior must follow a curve that lies below the perfectly plastic response curve $\overline{1234}$ in Figure 8.39a and approaches this curve asymptotically for very large deflections. Thus, the actual behavior should follow a curve such as $\overline{54}$ in Figure 8.39a.

It is now obvious that the actual maximum load P_m (which represents the critical state, i.e., the limit of stability if the load is controlled) is always below the intersection point 3 of the perfect hinge response curve and perfectly elastic response curve of imperfect columns. Thus, the load P_m^+ at the intersection of these curves, which can be calculated easily, represents an upper bound on the collapse load or the actual maximum load. However, the easily calculated value P_m^+ is often not very useful, because the actual maximum load P_m can be substantially smaller.

Some approximate ways of estimating the peak point 5 (P_m) on the basis of P_E and P_0 have been formulated (see, e.g., Horne, 1971; and eq. 1.39 of Horne and Merchant, 1965). Unfortunately, they are not very accurate, in general. P_{max} may best be estimated on the basis of the intersection point 3.

If the imperfection is small (Fig. 8.39b) and the column is short, P_m is only slightly smaller than the squash load P_0. Then P_0 gives a good upper bound for the maximum load that the column can attain. If the column is slender, plasticity begins only at large deflections (Fig. 8.39c). The consequence is that the beam is prevented from displaying the postbuckling reserve that is calculated when the large deflections are elastic (Sec. 1.9). In this case, the maximum load P_m is close to the elastic critical load $P_{cr_1}^{el}$ if the initial imperfection is small.

To obtain the axial displacement u at column top (Fig. 8.38a), we need to calculate the plastic axial shortening displacement u_p in the plastic hinge at the beam axis. From Figure 8.38b we have $\lambda = \kappa c =$ axial strain at beam axis and $u_p = h_p \lambda = 2\theta c$ where $\kappa =$ curvature, $2\theta = \kappa h_p =$ total rotation in the hinge (Fig. 8.38a), and $h_p =$ length of the plastic hinge ($h_p \ll l$), whose value, however, is irrelevant. From Equations 8.6.1 we have $c = \frac{1}{2}h - (P_0 - P)/2\sigma_p b$. The total axial displacement under the load is $u = l(1 - \cos\theta) + u_p$, and $\theta = 2z/l$. So the curve $u(P)$ can be obtained as

$$u = l(1 - \cos\theta) + \left(h - \frac{P_0 - P}{\sigma_p b}\right)\theta \qquad \text{with } \theta = \frac{2M}{Pl} \tag{8.6.3}$$

where M is calculated from Equations 8.6.2. This equation, whose plot is shown in Figure 8.39e, makes it possible to calculate the work that is done by load P.

It may be noted that our calculation of u_p satisfies the so-called normality rule of plasticity. Indeed, differentiation of Equation 8.6.1 with respect to c yields $dM/dP = (dM/dc)/(dP/dc) = -c = -\lambda/\kappa = -(\lambda h_p)/(\kappa h_p) = -u_p/2\theta$, and so the vectors (u_p, θ) and $(-dP, dM)$ are orthogonal since their scalar product vanishes, that is, $u_p(-dP) + (2\theta)\, dM = 0$.

Equations 8.6.1 to 8.6.3 apply to rectangular cross sections. For the case of an ideal I-beam, the yield surface (failure envelope) $M(P)$ consists of straight lines as shown in Fig. 8.38g. Obviously, for each different cross section a different shape of the load-deflection curve is obtained.

Buckling of Perfectly Plastic Frames

The solution that we have shown for columns can be generalized to plastic buckling of frames and trusses. Let us illustrate it for the portal frame shown in Figure 8.40a. We will consider only the collapse mechanism shown in Figure 8.40b. In contrast to classical analysis, the load is not constant but decreases with increasing deflection. For an infinitely small increment of rotation from θ to $\theta + d\theta$, the condition that the incremental internal work and external work must be equal yields (Bažant, 1985)

$$\sum_k M_k\, d\theta_k = \sum_j P_j\, dw_j^{(1)} + \sum_i P_i\, du_i^{(2)} \tag{8.6.4}$$

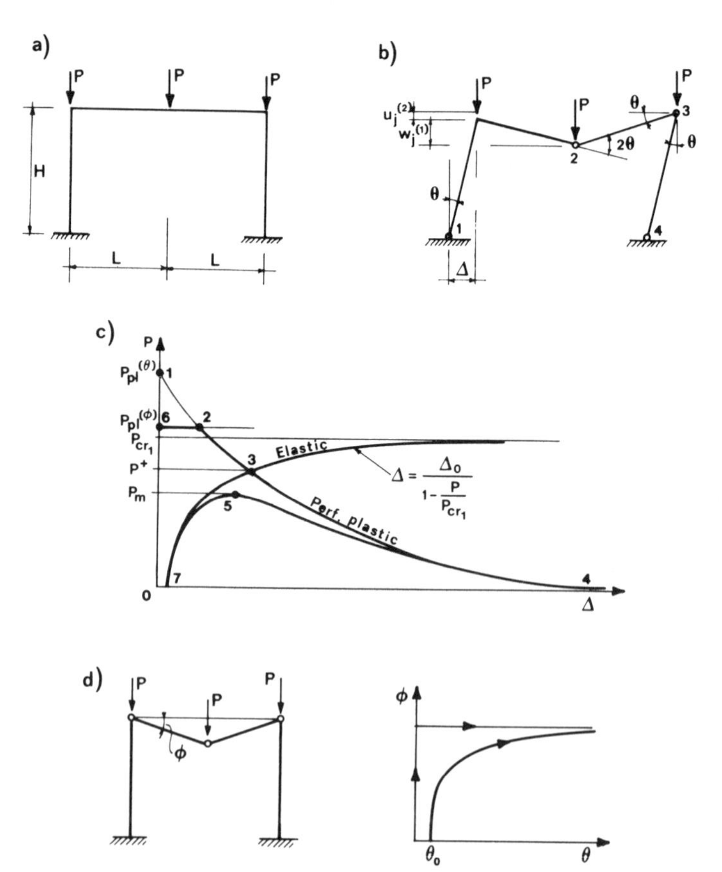

Figure 8.40 (a) Portal frame, (b) sway collapse mechanism, (c) load-deflection curves, and (d) beam collapse mechanism.

in which θ_k are the rotations in plastic hinges $k = 1, 2, 3, \ldots$; P_j or P_i are the concentrated loads; M_k are the bending moments in the hinges; $w_j^{(1)}$ are the deflections under the loads calculated according to the classical first-order theory; and $du_i^{(2)}$ are the additional displacements under the loads due to second-order analysis that takes into account a finite rotation θ. The last term containing $du_i^{(2)}$ represents the only difference from the classical, first-order analysis of plastic frames. For $\theta = 0$, $du_i^{(2)} = 0$.

In Equation 8.6.4, we neglect for the sake of simplicity the work of the axial forces on the axial plastic shortenings u_p in the hinges. This is acceptable only for slender frames. We took shortenings u_p into account in Equation 8.6.3, and they could also be introduced into Equation 8.6.4.

For the frame in Figure 8.40a and the collapse mechanism in Fig. 8.40b, with the notation shown, we have $w^{(1)} = L \sin \theta$, $u^{(2)} = H(1 - \cos \theta)$. At small θ we have the approximations $w^{(1)} \simeq L\theta$, $u^{(2)} \simeq \frac{1}{2}H\theta^2$ so that $dw^{(1)} = L\, d\theta$, $du^{(2)} = H\theta\, d\theta$. Now substituting this into Equations 8.6.4 we obtain the equation:

$$M_1 \cdot d\theta + M_2 \cdot 2\, d\theta + M_3 \cdot 2\, d\theta + M_4 \cdot d\theta = P \cdot L\, d\theta + 3P \cdot H\theta\, d\theta \quad (8.6.5)$$

from which $d\theta$ cancels out. For a slender frame, which is capable of very large deflections and rotations, the hinge moments for any cross-section shape may be supposed to be approximately equal to M_0 because the axial forces must be small. At $M \simeq M_0 = \text{const.}$ (which may be assumed after P has become small), Equation 8.6.5 yields $P = 6M_0/(L + 3\Delta)$, in which $\Delta = H\theta =$ lateral deflection of the frame (Fig. 8.40b). This expression yields a curve that represents, for small values of P, an asymptotic approximation to the curve $\overline{134}$ (Fig. 8.40c) of the actual perfectly plastic solution for which M is variable. The value $P_{pl}^{(\theta)}$ represents, for the mechanism considered, the plastic collapse load from the usual first-order plastic analysis of the frame. For a frame that is not slender, the dependence of M_k on the axial force in each plastic hinge must also be considered, and an iterative solution is then required.

It should be noted that the load P of the frame in Figure 8.40a has also an upper bound $P_{pl}^{(\phi)}$ (curve $\overline{62}$ in Fig. 8.40c) that corresponds to the beam collapse mechanism having hinges at the frame corners and under the load at midspan (Fig. 8.40d), for which there are no second-order effects. If the influence of the axial load on the yield moment is neglected and $M = M_0$ is equal for all plastic hinges, $P_{pl}^{(\phi)} = 4M_0/L$. The $P_{pl}^{(\phi)}$-value is always lower than the $P_{pl}^{(\theta)}$-value for the sway collapse mechanism considered previously; for example, under the same assumptions as mentioned before, $P_{pl}^{(\theta)} = 6M_0/L$. Then the P-value calculated from Equation 8.6.5 becomes valid only at large enough Δ at which it becomes less than $P_{pl}^{(\phi)}$ (curve $\overline{234}$ in Fig. 8.40c). If one considers the path in the (θ, ϕ) plane (Fig. 8.40d) where ϕ is the rotation angle of the left corner hinge in the beam collapse mechanism, the initial path direction is vertical ($\delta\theta = 0$, $\delta\phi > 0$), while (if the frame is slender) the path for large enough deflection tends to become horizontal ($\delta\theta > 0$, $\delta\phi = 0$). For an ideal frame the transition from vertical to horizontal is sudden, but for real (imperfect) frames it is smooth, with $\delta\theta$ and $\delta\phi$ increasing simultaneously in the transition stage. (For other examples of how different mechanisms may govern the postcollapse behavior of frames see, e.g., Corradi, 1978a.)

If the frame is imperfect, with imperfection characterized by a small initial angle $\theta_0 = \Delta_0/L$, then $u^{(2)} = \frac{1}{2}H(\theta^2 - \theta_0^2)$; this yields again $du^{(2)} = H\theta\, d\theta$, that is Equation 8.6.5 remains valid. So the initial imperfection does not influence the solution once the plastic hinges have been fully plasticized. Similar to Figure 8.39a, we plot in Figure 8.40c also the load-deflection diagram of the imperfect elastic frame; see curve $\overline{73}$ in Figure 8.40c. This diagram may be obtained by multiplying the initial imperfection Δ_0 by the elastic magnification factor $1/(1 - P/P_{cr_1})$ for the frame. The intersection point 3 in Figure 8.40c represents an upper bound on the actual maximum load. If the imperfection of the frame becomes smaller, the intersection point 3 moves upward and to the left. Again, some approximate considerations could be made to estimate the actual response diagram $\overline{754}$ (Fig. 8.40c). On the contemporary computational scene, however, this diagram is best obtained by finite elements.

Plastic Redistribution and Reserve Capacity of Structures

The foregoing type of solution describes plastic redistribution of bending moments in structures. This redistribution can endow the structure with reserve load capacity. The phenomenon is most easily illustrated for a truss. Consider the simple pin-jointed cross-braced square truss shown in Figure 8.41a (same as Fig. 2.24c). We assume the compressed frame bar $\overline{34}$ to be so strong that it never buckles, and then the only member susceptible to buckling is the diagonal brace $\overline{23}$. Since for slender members the axial yield force $\sigma_p A$ can be much higher than the maximum force in compression buckling, we assume the tensile diagonal $\overline{14}$ as well as all the other members to behave elastically while diagonal $\overline{23}$ buckles plastically (Fig. 8.41d).

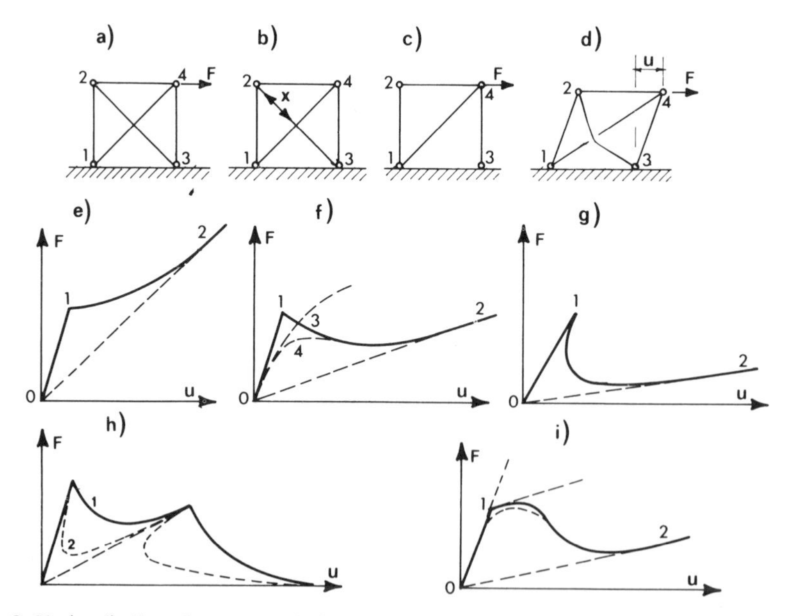

Figure 8.41 (a–d) Step-by-step solution of redundant truss; and load-deflection curves under different assumptions (e–h) regarding stiffness of diagonals and (i) regarding type of initial buckling.

The solution may be obtained incrementally with step-by-step loading. The axial compression force X in member $\overline{23}$ may be chosen as the redundant, and the condition of compatibility of the displacement increments at diagonal $\overline{23}$ may then be written, according to the principle of virtual work, as follows (Bažant, 1985):

$$\frac{du_X}{dX}\,\Delta X + \sum_i \bar{f}_i\,\Delta\delta_i = 0 \qquad \Delta\delta_i = \frac{\bar{f}_i\,\Delta X + f_i^L\,\Delta F}{EA_i}\,l_i \tag{8.6.6}$$

where du_X/dX is calculated from Equation 8.6.3 with Equations 8.6.2 ($X = P$) on the basis of the initial value of X for each loading step; u_X is the relative axial displacement between the ends of bar $\overline{23}$, associated with X; $\Delta\delta_i$ is the extension increment of member number i; $\bar{f}_i$, f_i^L is the axial force in the ith member caused either by compression force $X = 1$ at no load ($F = 0$, Fig. 8.41b), or by load F at $X = 0$ (Fig. 8.41c), respectively; A_i, l_i are the cross-section area and the length of member number i.

If the values of X have been solved up to a certain load level F, then the value of ΔX for a further small load increment ΔF can be solved from Equation 8.6.6. The corresponding load-point deflection is then obtained, according to the principle of virtual work, as $\Delta u = \sum_i f_i^L\,\Delta\delta_i$. In this manner one can construct the load-deflection curve $F(u)$. If the moment-curvature diagrams of the members are simplified as elastic–perfectly plastic and the members have no imperfections, the curves that can be obtained in this manner are as sketched in Figure 8.41e, f, g. The rising curve in Figure 8.41e is obtained if the diagonal $\overline{14}$ is rather stiff, and the descending curve $\overline{012}$ in Figure 8.41f is obtained if this diagonal is rather soft. The curve with a snapback in Figure 8.41g is obtained if members $\overline{24}$ and $\overline{14}$ are very soft. If both members $\overline{23}$ and $\overline{34}$ are not very strong and can buckle plastically, then the curve in Figure 8.41h may be obtained. If the members have imperfections, the response curve changes in the manner illustrated by curve $\overline{032}$ in Figure 8.41f. If, in addition, the moment-curvature diagram is smoothly curved before reaching a plateau (which is the reality, described by Eqs. 8.6.2–8.6.3), the response changes to curve $\overline{042}$ shown in Figure 8.41f.

All of our analysis so far dealt with slender columns for which the initial buckling is elastic and later there is relatively rapid transition to perfectly plastic behavior. The initial portions of the response curve become modified for stocky columns for which the initial buckling is elastoplastic. If the diagonal $\overline{23}$ is perfect and initially buckles at Shanley's tangent modulus load, then the response curve represents a transition from the rising initial Shanley's bifurcated path to the descending perfectly plastic response at large deflections, as sketched by curve $\overline{012}$ in Figure 8.41i. A slightly imperfect column does not exhibit the bifurcation, but its response is close to curve $\overline{012}$ in Figure 8.41i.

For the case of a perfect column (Fig. 8.38a) that is stocky rather than slender and buckles first at Shanley's tangent modulus load, one also observes a transition to a perfectly plastic response at very large deflections. This is illustrated by curve $\overline{0123}$ in Figure 8.39d. The intersection point 6 of Shanley's bifurcated curve (Eq. 8.1.18) with the perfectly plastic response curve $\overline{463}$ represents an upper bound on P_{max}.

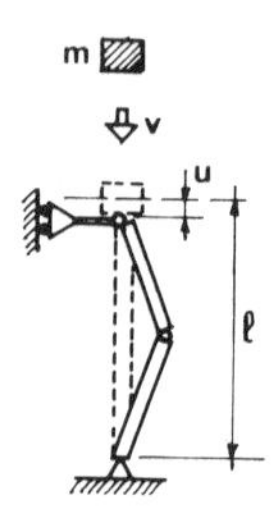

Figure 8.42 Simply supported column subjected to impact.

Dynamic Impact

Aside from redistribution and reserve capacity, the perfectly plastic analysis of buckling is useful in providing very simple bounds on deflections due to impact. To illustrate it, consider that an object of mass m impacts the top of a simply supported column (Fig. 8.42). An upper bound on the maximum column shortening u_1 may be obtained by assuming all the kinetic energy of mass m, that is, $\frac{1}{2}mv^2$, to be dissipated by the work W_p of the plastic deformation in the column. We have $W_p = \int P\,du$ where u is defined by Equations 8.6.3 and 8.6.2 as a function of P. The initial value of P is P_0 (see Fig. 8.39a). So we have the energy balance condition:

$$\frac{1}{2}mv^2 = W_p = \int_{P_0}^{P_1} P\frac{du}{dP}\,dP = (P_1 - P_0)u + \int_{P_1}^{P_0} u\,dP \qquad (8.6.7)$$

where integration by parts has been used to transform the integral, and P_1 is the axial force at maximum deflection u_1 ($P_1 < P_0$). After substituting Equations 8.6.3 and 8.6.2 for u one can solve for the value of P_1 from Equation 8.6.7, and then evaluate the upper bound u_1 from Equations 8.6.3 and 8.6.2.

Perfectly Plastic Buckling of Thick Plates

Although the load-deflection diagram of thin plates (in biaxial bending) has a very large postcritical reserve and thin plates remain elastic into relatively large deflections, a perfectly plastic analysis of plate buckling can give a useful approximation for the terminal stage of decreasing load at very large deflections of a thick plate (Fig. 8.43a). If the plate is thick, plasticity develops in the deflection range in which a geometrically linear formulation is applicable. Then a simple solution is possible by a second-order generalization of the yield line theory. If the plate is thin, however, plastic deflections cannot be analyzed in a geometrically linearized manner.

Assuming some familiarity with the yield line theory for upper bounds of plastic limit loads of plates, we consider for the sake of illustration the square simply supported plate of side a, which is shown in Figure 8.43b and is loaded by in-plane distributed normal forces N_{xx} and N_{yy} (<0). We assume the yield hinges to form in the pattern shown by the wavy lines in Figures 8.43b, c and to transmit per unit length the yield moment M_p (for the yield criterion of Tresca, $M_p = \sigma_p h^2/4$, and for that of von Mises, $M_p = \sigma_p h^2\sqrt{3}/2$ where h = plate thickness). We suppose M_p to be constant during deflection; this is, of course, an

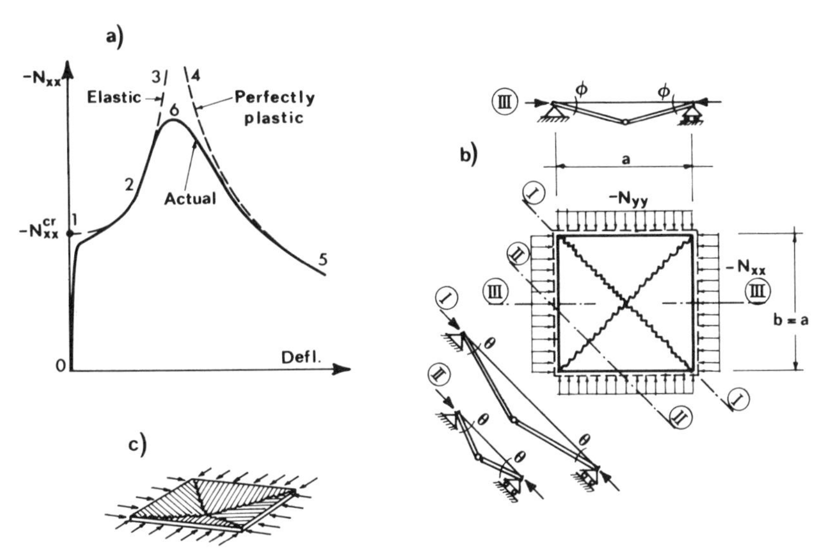

Figure 8.43 (a) Load-deflection diagram of thick plate at very large deflections and (b, c) yield line pattern.

approximation, which gets worse the thicker the plate, same as we saw it for a column (in reality M_p, of course, depends on N_{xx}, N_{yy}). The stress state of the plate pieces between the hinges is constant, and so they behave as rigid during the plastic buckling. Therefore, the deformation of the plate is characterized by one parameter, for example, the slope angle θ shown in Figure 8.43b.

Now an important point to note is that the relative shortening due to θ at any cross section such as I–I or II–II normal to the yield line is $1 - \cos\theta$, or approximately $\theta^2/2$, as the lowest-order approximation. Thus the boundaries of the plate move inward as if the plate underwent a uniform in-plane strain $\theta^2/2$ in any direction. So the work of the loads is $\Delta W_l = (-N_{xx}a)(a\theta^2/2) + (-N_{yy}a)(a\theta^2/2)$. The work dissipated in the hinges is $\Delta W_p = 2a\sqrt{2}\, M_p(2\theta)$. According to the principle of conservation of energy, we must have $\Delta W_l = \Delta W_p$, and solving θ from this condition we obtain for the maximum deflection $w = \theta a/\sqrt{2}$, which occurs at the plate center, the expression:

$$w = \frac{8M_p}{-N_{xx} - N_{yy}} \tag{8.6.8}$$

This gives the dashed curve $\overline{45}$ in Figure 8.43a.

Solutions of this type are again useful for determining crude deflection bounds in problems of dynamic impact or blast.

Transverse Impact or Blast on Plates or Columns with In-Plane or Axial Loads

When a plate (slab, panel) or column is subjected to impact or blast in the transverse direction and is at the same time under significant static in-plane or axial forces, then the second-order effects of these forces must be taken into

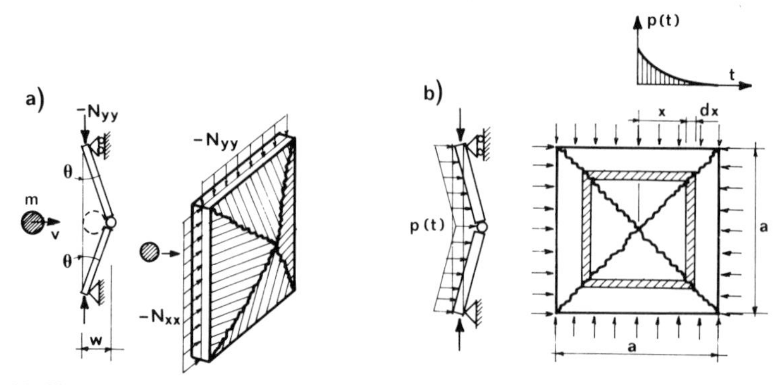

Figure 8.44 Simply supported plate subjected (a) to impact and (b) to blast.

account since they tend to increase deflections. The solution is again easy if the behavior is plastic. To demonstrate it, consider the same square panel as before. An object of mass m, traveling at velocity v, impacts the center of the plate in the normal direction (Fig. 8.44a). An upper bound on the transverse deflection w is obtained if all the kinetic energy $\frac{1}{2}mv^2$ together with the work of loads N_{xx} and N_{yy} is assumed to be converted into plastic work in the yield hinges, that is,

$$\frac{1}{2}mv^2 - (N_{xx} + N_{yy})\frac{a^2\theta^2}{2} = 4\sqrt{2}\, aM_p\theta \tag{8.6.9}$$

provided N_{xx} and N_{yy} remain constant during the deflection. Calculating the positive solution θ from this quadratic equation, one obtains the deflection bound $w = \theta a/\sqrt{2}$.

Consider now the effect of a blast that produces a given pressure history $p(t)$ on the surface of the plate (Fig. 8.44b). If energy dissipation outside the yield hinges is neglected, then the work of $p(t)$ and of N_{xx}, N_{yy} during time increment dt must be equal to the increment of the kinetic energy of the plate plus the work dissipated in the plastic hinges. This yields the condition:

$$p(t)\frac{a^2}{3}\,dw - (N_{xx} + N_{yy})a^2\theta\, d\theta = \int_0^{a/2}\left[\frac{1}{2}\mu(v + dv)^2 - \frac{1}{2}\mu v^2\right](8x\, dx) + 4\sqrt{2}\, aM_p\, d\theta \tag{8.6.10}$$

in which w = deflection at plate center, v = velocity at any point of the plate, μ = specific mass of the plate per unit area, and $(8x, dx)$ = area element of the plate (Fig. 8.44b) in which the velocity is $v = \dot{w}2x/a$. Evaluating the integral and substituting $\theta = w\sqrt{2}/a$, one obtains for the center deflection the differential equation:

$$\mu\frac{d^2w}{dt^2} + \frac{4}{a^2}(N_{xx} + N_{yy})w = \frac{2}{3}p(t) - 16\frac{M_p}{a^2} \tag{8.6.11}$$

Integration of this equation in time t yields the deflection history $w(t)$ due to the blast.

Problems

8.6.1 Derive the load-deflection diagram of a perfectly plastic column, replacing the equilibrium conditions with the condition that the work increment of the applied load $dW_l = P\,du$ must be equal to the internal work $dW_p = h_p \int \sigma\, d\varepsilon = M(2\,d\theta) + P\,du_p =$ work in the plastic hinge; $M =$ moment in the hinge; 2θ, $u_p =$ rotation and axial displacement in the hinge.

8.6.2 Derive equations corresponding to Equations 8.6.1 to 8.6.3 for the case of (a) an ideal I-beam, (b) an I-beam, (c) a solid circular cross section, (d) a tube, (e) a nonsymmetric ideal I-beam, (f) a T-beam, and in cases (e–f) also discuss the effect of nonsymmetry of the cross section.

8.6.3 Consider the frames in Figure 8.45a–i and solve them in the manner of Equations 8.6.4 and 8.6.5.

8.6.4 (a) Repeat Problem 8.6.2 for a fixed–hinged column (Fig. 8.45c). Consider a rectangular or any other cross section used in Problem 8.6.2. (*Note:* Find the distance of the plastic hinge from the base from the condition that the load P for a given axial shortening u would be minimized. This condition is indicated by the path stability analysis in Sec. 10.2.) (b) Repeat the above using a similar approach to solve the columns in Figure 8.45j—m, having a variable P or variable cross section (use the same condition to find the hinge location).

8.6.5 A fixed column obviously forms three hinges. Equations 8.6.2 for M applies to each hinge. However, Equation 8.6.3 and Equations 8.6.2 for z require modification. Derive it and plot the diagram $P = P(u)$. Consider (a) an ideal I-bar cross section and (b) a rectangular cross section.

8.6.6 Solve the same as Problem 8.6.5 but for a fixed–hinged column.

8.6.7 Consider the trusses in Figure 8.46 and solve them in a manner similar to Equation 8.6.6. Then determine the safety factor S_p due to plastic redistribution, defined as the ratio of the maximum load (at large deflection) to the load

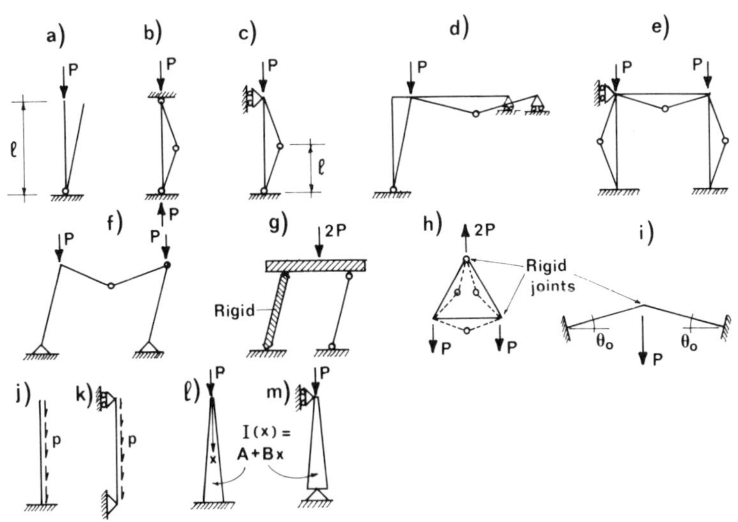

Figure 8.45 Exercise problems on buckling of columns and frames which develop perfectly plastic hinges.

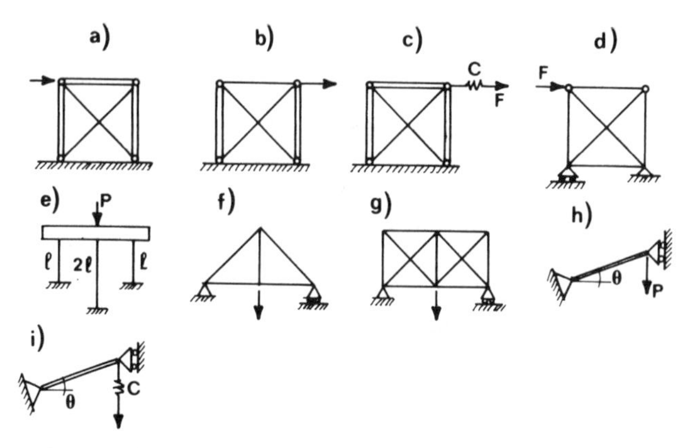

Figure 8.46 Exercise problems on buckling of trusses with perfectly plastic bars.

that causes the initial plastic deflection. (The problem is meaningful only if $S_p \geq 1$.)

8.6.8 (a) Recalling the derivation of Equation 8.6.3, formulate the axial shortenings u_p done in the hinges of the frame in Figure 8.40, and introduce the work of axial loads on du_p into Equations 8.6.4 and 8.6.5. (Such a solution would be valid even for frames that are not slender.) (b) Repeat for the L-frame, Figure 8.45d. (*Note:* Here the axial shortening causes asymmetry of deflections to the left and to the right, similar to that encountered in Sec. 2.6, Fig. 2.25.)

8.6.9 (a) Calculate the work that is done if the distance between the ends of the hinged column in Figure 8.38a is shortened to $0.5l$. (b) Repeat for the columns in Figure 8.45b, c.

8.6.10 A projectile of mass m impacts axially the top of a column (Fig. 8.42). Calculate the lower bounds on the impact velocity v that would cause the column initial length l to change to chord length $l_1 = 0.9l$, $0.5l$, $0.1l$. Consider both (1) an ideal I-beam and (2) a square cross section, and (a) a simply supported column, (b) a free-standing column, and (c) a fixed column. Assume $L/h = 20$. How would the results change if 30 percent of the kinetic energy is dissipated by shattering of the projectiles and friction on the column, 15 percent of the kinetic energy is transferred to the fragments flying away, and 5 percent is dissipated by hysteresis in the column at points away from the plastic hinge?

8.6.11 Solve the same as Problem 8.6.10, but determine an upper bound on axial displacement u if the mass m is dropped from a given height H above the column top.

8.6.12 Calculate the deflection of the plate in Figure 8.43b assuming its edges are fixed rather than simply supported.

8.6.13 Assume that the plate in Figure 8.44 is supported only on the top and bottom sides and that only one horizontal hinge forms at plate midheight. Calculate an upper bound for deflection w due to impact of mass m. Discuss the effect of N_{xx}, N_{yy}.

8.6.14 Solve the same as the impact problem in the text for Figure 8.44, but the boundary of the plate is fixed rather than simply supported.

8.6.15 Solve Equation 8.6.11 for blast loading $p(t) = p_0 e^{-t/\tau}$ for $t \geq 0$ and $p(t) = 0$ for $t < 0$, where $\tau = \text{constant}$.

8.6.16 Solve the same as Problem 8.6.13, but with blast loading as in Problem 8.6.15. Discuss the effect of N_{xx}, N_{yy}.

8.6.17 How would the preceding solutions change if N_{xx}, N_{yy} decreased during deflection? (This happens if there is a restraint against free in-plane displacements at the boundaries.)

8.7 GEOMETRIC TENSILE INSTABILITY, LOCALIZATION, AND NECKING

We will now shift attention to a different problem, considering structures that are subjected to tension. For small inelastic deformations, such structures are always stable. Not so, however, when large deformations take place. In that case, the Poisson effect may cause a significant reduction of the cross-section area. Since the load must be resisted over a reduced area, an increase of stress and strain results, and thus the Poisson effect may ultimately lead to instability. The material behavior may still be purely elastic, although a similar instability arises when the reduction of the cross-section area is caused by material plasticity or damage. (We will touch this aspect in Chap. 13.)

Role of Transverse Contraction and Finite Strain

To illustrate this type of stability loss, consider a rubber rod of initial length l_0 and initial cross section A_0, subjected to uniaxial tensile stress produced by axial tensile load F; see Figure 8.47a. Although this chapter is devoted to plastic materials, for the sake of clarity let us consider first a material that is perfectly elastic (reversible), highly deformable but incompressible in volume. The

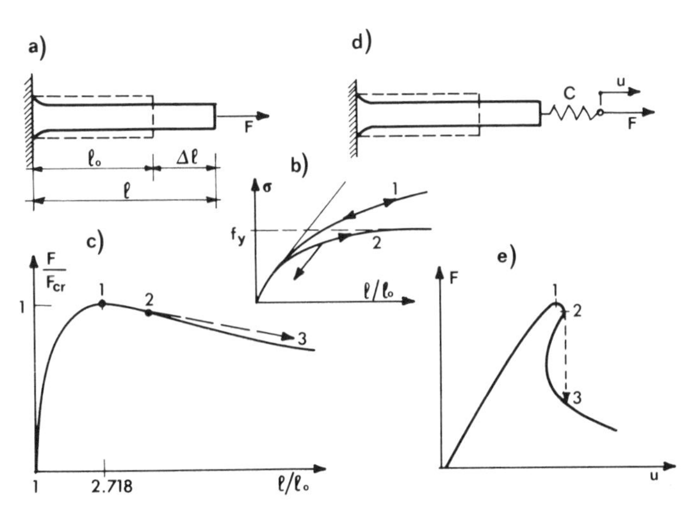

Figure 8.47 (a) Rubber rod subjected to tensile load; (b) stress-strain relation for large strains; (c) load-elongation curve; (d) loading through a spring; (e) snapdown instability.

condition of constant volume of the material may be written as $Al = A_0 l_0$ in which A = the current cross-section area, l = length of the rod (Fig. 8.47a), and A_0, l_0 = their initial values at $P = 0$. Substituting $A = A_0 l_0 / l$ into the equilibrium relation $F = \sigma A$ where σ is the true stress, that is, the force per unit area of the deformed rather than undeformed material, we find that $\sigma A_0 l_0 = Fl$ or

$$\sigma = \frac{F}{A_0}\left(\frac{l}{l_0}\right) \tag{8.7.1}$$

The uniaxial stress-strain relation for large strains is assumed to have the form

$$d\sigma = \frac{E\,dl}{l} \tag{8.7.2}$$

where E = constant. The strain increment $d\lambda = dl/l$ represents an increment of the so-called logarithmic strain λ (cf. Sec. 11.1), since $\lambda = \ln(l/l_0)$. Therefore, $\sigma = E\lambda$; see curve 1 in Figure 8.47b. Substituting this into Equation 8.7.1 we get

$$F = EA_0 \frac{\ln(l/l_0)}{l/l_0} = EA_0 \frac{\lambda}{e^{\lambda}} \tag{8.7.3}$$

The diagram of this load-elongation curve is plotted in Figure 8.47c. We see that the curve reaches a maximum after which it continuously declines and approaches asymptotically a zero load. Based on our previous discussion of stability of equilibrium curves that exhibit a limit point (Sec. 4.4), it is clear that the maximum point represents a critical state, after which the rod is unstable. Setting for the maximum point $dF/dl = 0$, we find that instability occurs when $l = l_{cr} = 2.718 l_0$ and the corresponding critical load is $F_{cr} = EA_0 l_0 / l_{cr}$. The instability is of the snapthrough type (see Sec. 4.4). But, in contrast to the snapthrough of arches, the structure cannot regain stability after snapthrough, since the equilibrium path has no second rising segment.

An interesting modification of this example is to attach a spring of stiffness C at the end of the rod (Fig. 8.47d) and load the rod through the spring. The behavior is then rather similar to the example of a von Mises truss with a spring attached at its apex (Fig. 4.48), which was discussed in Sec. 4.8. It is found that for a sufficiently soft spring such that its stiffness C is less than the magnitude of the steepest descending slope k_{min} in Figure 8.47c, the load-elongation diagram exhibits snapback, as shown in Figure 8.47e (from Eq. 8.7.3, $k_{min} = -EA_0 e^{-3}/2l_0$). This type of behavior exhibits instability even under displacement control, and snapdown instability may occur as shown by path $\overline{23}$ in Figure 8.47e. Without the attached spring, the tension rod is stable under displacement control for any value of the displacement, provided we assume the strain to remain uniform.

Similar tensile instabilities due to Poisson effect can occur in pressurized spherical shells (balloons) or tubes.

Consider now an elastoplastic material defined by

$$\sigma = \frac{F}{A} = \frac{E\lambda}{1 + E\lambda/f_y} \qquad \left(\lambda = \ln\frac{l}{l_0}\right) \tag{8.7.4}$$

where $\sigma = F/A$ = true stress (rather than the force per unit initial area); see curve

2 in Figure 8.47b. For the sake of simplicity let us replace λ with the linear strain $\varepsilon = (l - l_0)/l_0$ and suppose all deformations to happen at constant volume, in which case $\sigma = F/A = F(1+\varepsilon)/A_0$. Substituting this into Equation 8.7.4 with $\lambda = \varepsilon$, we get

$$F \simeq \frac{A_0 E\varepsilon}{(1+\varepsilon)(1+E\varepsilon/f_y)} \qquad (\varepsilon = l/l_0 - 1) \tag{8.7.5}$$

We find that, despite the absence of a peak in the $\sigma(\varepsilon)$ relation (Eq. 8.7.4), the curve $F(\varepsilon)$ does have a peak, given by $dF/d\varepsilon = 0$. For the peak-force point, representing the limit of stability (i.e., the point of snapthrough), this condition furnishes

$$\varepsilon_{cr} = \sqrt{\frac{f_y}{E}} \qquad \sigma_{cr} = f_y \frac{\sqrt{E}}{\sqrt{E} + \sqrt{f_y}} \tag{8.7.6}$$

The assumptions that $\lambda \simeq \varepsilon$ and that the volume is constant for the elastic part of deformation are, of course, not quite realistic; large (finite) strains and Equations 8.7.6 are acceptable only if ε_{cr} is small (for mild steel, Eqs. 8.7.6 yield $\varepsilon_{cr} \simeq 0.04$ and $\sigma_{cr} = 0.96 f_y$, which may be considered as small, although the assumptions we made do not really apply to mild steel). For finite strains, we need to characterize the elastic deformation by the relative length change $\varepsilon_e = (l_e/l_0) - 1$ and assume the plastic deformation to be defined by the relation $l/l_0 = (l_e/l_0)(l/l_e)$ where $(l/l_e) - 1 = \varepsilon_p$ = linear plastic strain. (This decomposition of l/l_0 is merely a reasonable assumption, not a law.) So we have $\lambda = \ln(l/l_0) = \lambda_e + \lambda_p$ where $\lambda_e = \ln(l_e/l_0) = \ln(1+\varepsilon_e)$ and $\lambda_p = \ln(l/l_e) = \ln(1+\varepsilon_p)$. The transverse linear contraction due to elastic axial strain may be considered as $\nu\lambda_e$ where ν = Poisson ratio. This ratio may be defined by writing, for the change of volume, $\ln(A_0/A) = \lambda_p + 2\nu\lambda_e$. This expression is justified by the fact that for the case of elastically incompressible material, $\nu = 0.5$, it reduces to $\ln(A_0/A) = \lambda_p + \lambda_e = \lambda = \ln(l/l_0)$, that is, $A_0/A = l/l_0$ or $Al = A_0 l_0$, which agrees with our previous assumption of volume incompressibility. Finally, considering the elastic strain to be given by $\lambda_e = F/AE$ (which is also merely a simplifying assumption, not a law), we get

$$A_0 = Ae^{\lambda_p + 2\nu\lambda_e} = Ae^{\lambda - (1-2\nu)F/AE} \tag{8.7.7}$$

Expressing A from Equation 8.7.4 and substituting it here, one gets a transcendental equation relating F and λ from which the curve $F(\lambda)$ can be numerically calculated. Again, this curve has a peak that represents a point of snapthrough instability.

We see that finite strains of an elastoplastic material pose new problems, which arise with regard to the manner of decomposition of strain into its elastic and plastic parts, and need to be decided on the basis of experiments. These problems are interesting, especially in three dimensions, and have been intensely debated. For the recent thinking, see, for example, Simo and Ortiz (1985) and Simo (1986).

Strain Localization

Consider again the first example of a rubber rod. In the preceding simple solution (Eqs. 8.7.1–8.7.3) we tacitly assumed the strain to remain uniform along the rod. Before reaching the peak load point this assumption is no doubt correct since only one value of strain corresponds to any given force. After the peak, however, the strain does not have to remain uniform since only a segment of the rod of some initial length h_0 ($h_0 < l_0$) (Fig. 8.48a, b) can increase its strain (Fig. 8.48d) while the rest of the rod length can unload at decreasing strain (Fig. 8.48e), maintaining uniform normal force F along the rod (Fig. 8.48c). Denoting $\lambda = \ln(l/l_0) =$ strain, we have in Equation 8.7.3 the function $F = f(\lambda)$. Its inverse is $\lambda = \phi_a(F)$ for post-peak response and $\lambda = \phi_b(F)$ for the pre-peak response. These relations are valid for each segment of the rod that has a uniform strain. Thus, the post-peak strains in the loading and unloading segments are $\lambda_a = \phi_a(F)$ and $\lambda_b = \phi_b(F)$. The corresponding lengths of the deformed segments are $l_a = h_0 e^{\lambda_a}$ and $l_b = (l_0 - h_0)e^{\lambda_b}$ (Fig. 8. 48d, e), and so the total post-peak length of the rod under force F is $l = l_a + l_b$, that is,

$$l = h_0 e^{\phi_a(F)} + (l_0 - h_0)e^{\phi_b(F)} \tag{8.7.8}$$

This defines the post-peak curve of F versus l, which can be graphically constructed as shown in Figure 8.48d–f.

Now we see that the peak point is a bifurcation point. After the peak we have two equilibrium paths as shown in Figure 8.48, one with uniform strain (dashed curve), and one with localized strain (solid curve). Which one actually occurs? This question can be answered only by thermodynamic stability analysis that we will carry out in Chapter 10. We will see that in a problem with a single controlled

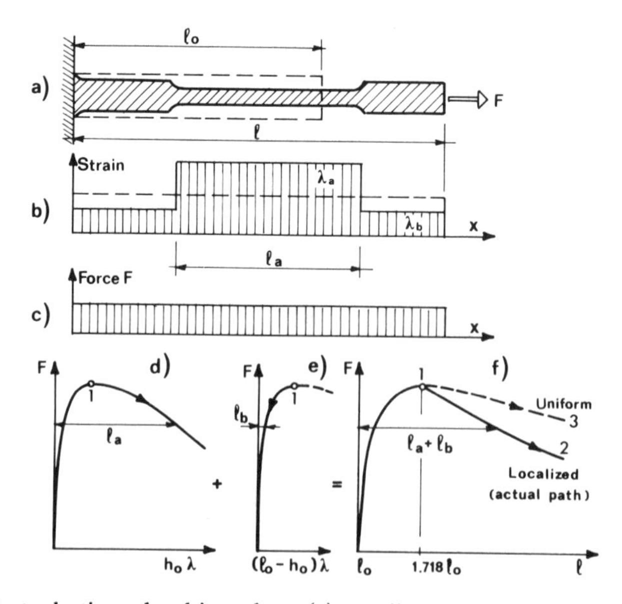

Figure 8.48 Elastoplastic rod subjected to (a) tensile force, (b) strain localization, (c) uniform distribution of axial force; force-elongation curves of (d, e) loading and unloading parts and (f) entire rod.

variable, such as l, the path that will actually take place is the path that descends steeper in the diagram of F versus l (because it gives a larger increase of internally produced entropy). Thus the strain will localize. A uniform strain after the peak cannot be obtained except if the rod were artificially restrained along its length.

Further we note that the post-peak path in Figure 8.48f depends on the length h_0 of the localization segment. For different values of h_0, different post-peak paths are obtained. Obviously, there are infinitely many descending equilibrium paths emanating from the peak point. The smaller the value of h_0, the steeper the descent of the path. The path that actually occurs is the steepest one possible. So it seems that the length h_0 of the localization segment will be vanishingly small, which would produce a sudden drop of force F.

In reality, however, length h_0 must have a certain minimum value $h_{\min}$ that is approximately proportional to the thickness b of the rod. The simple uniaxial analysis that we demonstrated is valid only if $h_0 \gg b$. For h_0 values that are of the same order as b, one must analyze the multiaxial stresses near the points of transition from one rod thickness to another, and such analysis imposes a certain minimum length of the transition zone. This problem is related to plastic necking, upon which we touch next, as well as to localization due to softening damage, which we will discuss in Chapter 13.

The influence of the h_0 value is obviously a source of a size effect. For the same material and the same cross section of the rod, a longer rod will exhibit a steeper descending curve in the plot of F vs l/l_0. For a very short rod of length $l = h_{\min}$, there would be no localization, that is, the strain will remain uniform. Our preceding analysis of snapback at displacement control is valid only for such a short rod. Otherwise the snapback analysis must be based on the response curve for a rod with localized strain.

If the material is inelastic, the principal difference is that function $\phi_b(\lambda)$ must describe the unloading curve starting from the peak point, which differs from the initial pre-peak loading curve (see Prob. 8.7.3).

Necking

An important example of tensile instability is plastic necking, which involves very large plastic deformations in the neck region (Fig. 8.49a). Necking is the typical mode of failure of elastic-plastic metals in tension. The problem has been studied in detail by finite elements (e.g. Needleman, 1982; Tvergaard, 1982; Becker and Needleman, 1986). It appears that there are two basic geometrically nonlinear

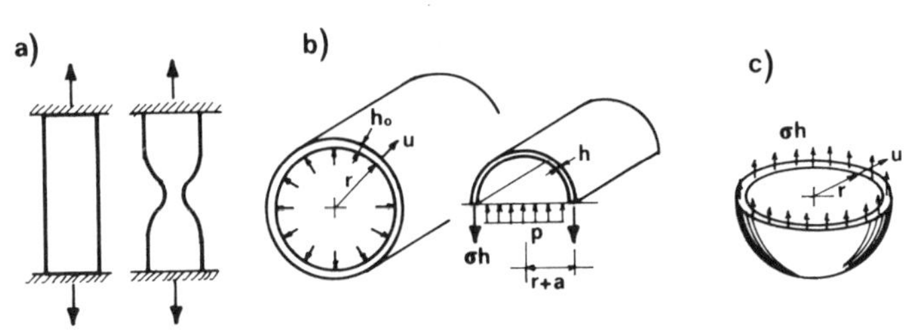

Figure 8.49 (a) Plastic necking; (b) pressurized cylindrical tube; (c) pressurized balloon.

causes of the necking instability: (1) a decrease of the cross-section area of the bar, causing a stress increase for the same load value, in the same manner as in our preceding example; and (2) formation of microscopic voids in the plastic metal. A cross section of the bar then intersects a certain number of the voids, with the result that the net resisting area of the material is less than the area of the cross section (see also Chap. 13). Approximately, the true stress to which the material between the voids is subjected is $\sigma_t = \sigma/(1 - v)$ where $\sigma = F/A =$ macroscopic stress (i.e., the stress in the usual, Eulerian definition), and $v =$ volume fraction of voids, which increases with the deformation during necking.

The elastic-plastic constitutive law of the metal may be approximately written as the relation between the true stress and the strain (finite strain must be considered). For the purpose of finite element analysis, this stress-strain relation must be converted to macroscopic stresses, which is usually done in Lagrangian formulations in terms of the strains and stresses that are referred to the original undeformed geometry (cf. Chap. 11).

The formulation of the three-dimensional elastoplastic constitutive relation at finite strain, as well as the evolution equation for the void fraction, is a difficult problem still debated at present. A further complicating factor is that the evolution of voids causes macroscopic strain-softening of the material, which causes the finite element analysis not to exhibit the correct structural size effect and to show spurious mesh-size sensitivity. It is now known that this sensitivity can be avoided by nonlocal continuum formulations. Among these, the most effective approach appears to be to consider the damage, represented here by the void fraction, as a function of mean strains and stresses from a certain representative volume of the material rather than a function of the local strains and stresses (Bažant and Pijaudier-Cabot, 1987).

Problems

8.7.1 Analyze tensile instability of the rod in Figure 8.47a assuming a "locking" elastic material, with the stress-strain law $\sigma = E\varepsilon(1 + k\varepsilon)$ where $k =$ positive constant. Assume the same definition of strain ε as before, as well as the condition of constant volume.

8.7.2 Analyze the snapdown behavior for the rod in Figure 8.47c, including the load-elongation diagram and the instability point.

8.7.3 (a) Repeat Problems 8.7.1 and 8.7.2 for an elastoplastic material described for monotonic uniaxial loading by the relation $\sigma = f_y \tanh(E\varepsilon/f_y)$ and for unloading by a straight σ-ε diagram of slope E. Again assume constant volume. (b) Do the same but for $\sigma = (2f_y/\pi)\tan^{-1}(\pi E\varepsilon/2f_y)$. (c) Do the same but for the Ramberg-Osgood stress-strain relation.

8.7.4 Solve tensile instability of the pressurized cylindrical tube in Figure 8.49b, assuming the strain to remain uniform and the relation between the membrane resultant N and the linear strain , $\varepsilon = (l - l_0)/l_0$, to be of the form $N = h_0 E\varepsilon(1 + k\varepsilon)$.

8.7.5 Solve the same as in Problem 8.7.4, but for a pressurized balloon (spherical shell, Fig. 8.49c).

References and Bibliography

AASHTO (1983), *Standard Specifications for Highway Bridges,* 13th ed. (Sec. 8.4).

Abo-Hamd, M. (1988), "Slender Composite Beam-Columns," *J. Struct. Eng.* (ASCE), 114(10):2254–67.

ACI (1977), *Building Code Requirements for Reinforced Concrete (ACI Standard 318–377),* Detroit; revised ed. 1983; also updated 1989, Code ACI 318–89. (Sec. 8.5).

Ackroyd, M. H., and Gerstle, K. H. (1983), "Strength of Flexibly-Connected Steel Frames," *J. Eng. Struct.,* 5:31–37.

Adams, P. F. (1972), "Discussion of 'The effective Length of Columns in Unbraced Frames' by J. A. Yura," *Engineering J.* (AISC), 9(1):40–41.

AISC (1978), *Specification for the Design, Fabrication and Erection of Structural Steel for Buildings,* Chicago (Sec. 8.4).

AISC (1986), *Load and Resistance Factor Design, Specifications for Structural Steel Buildings,* Chicago (Sec. 8.4).

Aluminum Association, The (1969), *Aluminum Construction Manual—Specifications for Aluminum Structures,* and also (1971), *Commentary on Specifications for Aluminum Structures.*

Aluminum Company of America (1960), *Alcoa Structural Handbook.*

American Society Civil Engineers (1971), *Plastic Design in Steel—a Guide and Commentary,* ASCE Manual No. 41, New York.

Arnold, R. R., and Parekh, J. C. (1987), "Buckling, Postbuckling and Failure of Stiffened Panels under Shear and Compression," *J. Aircr.,* 24(11):803–11.

ASA (1984), *Draft Australian Standard, Unified Concrete Structures Code,* BD 12/84-1 (Sec. 8.5).

Augusti, G. (1965a), "Studio dinamico di un'asta inelastica compressa" (in Italian), *Giornale del Genio Civile,* Fascicolo 8, pp. 396–405.

Augusti, G. (1965b), "Sulla stabilità dell'equilibrio di un'asta elasto-plastica," *Rend. Acc. Sc. Fis. e Mat.,* Ser. 4, Vol. XXXII, pp. 98–107.

Augusti, G. (1984), "Svergolamento e collasso di elementi strutturali in campo inelastico" (in Italian), *Ingegneria Civile,* 11:3–17.

Augusti, G., and Baratta, A. (1971), "Théorie probabiliste de la résistence des barres comprimées" (Probabilistic Theory of the Strength of Compressed Struts), *Construction Métallique,* No. 2.

Ballio, G., and Campanini, C. (1981), "Equivalent Bending Moments for Beam-Columns," *J. Constr. Steel Res.,* 1(3):13–23.

Ballio, G., and Mazzolani, F. M. (1983), *Theory and Design of Steel Structures,* Chapman and Hall, London (Sec. 8.4).

Ballio, G., Petrini, V., and Urbano, C. (1973), "The Effect of the Loading Process and Imperfections on the Load Bearing Capacity of Beam Columns," *Meccanica,* 8(1):56–67.

Batterman, R. H., and Johnston, B. G. (1967), "Behavior and Maximum Strength of Metal Columns," *J. Struct. Eng.* (ASCE), 93(2):205–30 (Sec. 8.1).

Bažant, Z. P. (1985), Lectures on "Stability of Structures," Course 720–D24, Northwestern University, Evanston, Ill. (Sec. 8.1).

Bažant, Z. P. (1987), Lectures on "Material Modeling Principles," Course 720–D30, Northwestern University, Evanston, Ill. (Sec. 8.1).

Bažant, Z. P. (1988), "Stable States and Paths of Structures with Plasticity and Damage," *J. Eng. Mech.* (ASCE), 114(12):2013–34 (Sec. 8.1).

Bažant, Z. P., and Oh, B. H. (1984), "Deformation of Progressively Cracking Reinforced Concrete Beams," *ACI Journal,* 81(3):268–78 (Sec. 8.5).

Bažant, Z. P., and Pijaudier-Cabot, G. (1987), "Nonlocal Damage Theory," *J. Eng. Mech.* (ASCE), 113(10):1512–33 (Sec. 8.7).

Bažant, Z. P., Cedolin, L., and Tabbara, M. T. (1989), "A New Method of Analysis for Slender Concrete Columns," Report, Center for Advanced Cement-Based Materials, Northwestern University, Evanston, Ill.; *ACI Structural Journal,* in press (Sec. 8.5).

Bazhenov, V. G., and Igonicheva, E. V. (1987), "Nonlinear Analysis of Nonaxisymmetric Buckling of Cylindrical and Conical Shells in Axial Impact," *Sov. Appl. Mech.,* 23(5):418–24.

Becker, R., and Needleman, A. (1986), "Effect of Yield Surface Curvature on Necking and Failure in Porous Plastic Solids," *J. Appl. Mech.* (ASME), 53:491–99 (Sec. 8.7).

Beedle, L. S., and Tall, L. (1960), "Basic Column Strength," *J. Struct. Eng.* (ASCE), 86(7):139–73 (Sec. 8.3).

Bell, J. F. (1988), "Dynamic Buckling of Rods at Large Plastic Strain," *Acta Mech.,* 74(1–4):51–67.

Berry, D. T. (1987), "Beyond Buckling a Nonlinear FE Analysis," *J. Mech. Eng.,* 109(8):40–44.

Bild, S., and Trahair, N. S. (1988), "Steel Columns Strength Models," *J. Constr. Steel Res.,* 11(1):13–26.

Bjorhovde, R. (1984), "Effect of End Restraint on Column Strength: Practical Applications," *J. Am. Inst. Steel Constr.,* 21(1):1–13.

Bjorhovde, R., and Tall, L. (1971), "Maximum Column Strength and the Multiple Column Curve Concept," Fritz Laboratory Report No. 337–29, Lehigh University (Sec. 8.4).

Bleich, F. (1952), *Buckling Strength of Metal Structures,* McGraw Hill, New York (Sec. 8.1).

Bljuger, F. (1987), "Probabilistic Analysis of Reinforced Concrete Columns and Walls in Buckling," *ACI Struct. J.,* 84(2):124–31.

Blom, G. (1989), *Probability and Statistics—Theory and Applications,* Springer-Verlag, New York (Sec. 8.4).

Blomeier, G. A., and Breen, J. E. (1975), "Effect of Yielding of Restraints on Slender Concrete Columns with Sidesway Prevented," Reinforced Concrete Columns, ACI Publication SP-50, Detroit, pp. 41–65 (Sec. 8.5).

Bolotin, V. V. (1969), *Statistical Methods in Structural Mechanics* (trans. from Russian), Holden-Day, San Francisco (Sec. 8.4).

Bradford, M. A., and Trahair, N. S. (1986), "Inelastic Buckling Tests on Beam-Columns," *J. Struct. Eng.* (ASCE), 112(3):538–49.

Bradford, M. A., Cuk, P. E., Gizejowski, A., and Trahair, N. (1987), "Inelastic Lateral Buckling of Beam-Columns," *J. Struct. Eng.,* (ASCE) 113(11):2259–77.

Broms, B., and Viest, M. (1961), "Ultimate Strength of Hinged Columns," *Transactions* (ASCE), Paper No. 3155, 126(II):309–97.

BSI (1972), *Structural Use of Concrete,* CP100, Part 1 (Sec. 8.5).

Budiansky, B., and Hutchinson, J. W. (1964), "Dynamic Buckling of Imperfection-Sensitive Structures," *Proc. Int. Cong. Appl. Mech.,* 12th, Munich, pp. 636–51.

Butler, D. J. (1977), "Strength of Restrained Reinforced Concrete Columns—A New Approach," *Magazine of Concrete Research* (London), 29(100):113–22 (Sec. 8.5).

Calladine, C. R. (1973), "Inelastic Buckling of Columns: the Effect of Imperfections," *Int. J. Mech. Sci.,* 15:593–604.

Canadian Standards Association (1974), *Steel Structures for Buildings—Limit State Design,* CSA Standard 516.1.

CEB (1978), *CEB-FIP Manual of Buckling and Instability,* Bulletin d'Information No. 123, The Construction Press, Lancaster, England (Sec. 8.5).

CEB (1988), *CEB-FIB Model Code 1990— First Predraft 1988,* Bulletin d'Information No. 190a, Lausanne (Sec. 8.5).

Chan, S. L., and Kitipornchai, S. (1988), "Inelastic Post-Buckling Behavior of Tabular Struts," *J. Struct. Eng.* (ASCE), 114(5):1091–1105.

Chen, W. F. (1970), "General Solution of Inelastic Beam-Column Problem," *J. Eng. Mech. Div.* (ASCE), 96(4):421–42.

Chen, W. F. and Atsuta, T. (1972a), "Column Curvature Curve Method for Analysis of Beam-Columns," *The Structural Engineer* 50(6):233–40.

Chen, W. F., and Atsuta, T. (1972b), "Interaction Equations for Biaxially Loaded Sections," *J. Struct. Div.* (ASCE), 98(5):1035–52.

Chen, W. F., and Atsuta, T. (1972c), "Simple Interaction Equations for Beam-Columns," *J. Struct. Eng.* (ASCE), 98(7):1413–26.

Chen, W. F., and Atsuta, T. (1973), "Ultimate Strength of Biaxially Loaded Steel H-Columns," *J. Struct. Eng.* (ASCE), 100(3):469–89.

Chen, W. F., and Atsuta, T. (1976), *Theory of Beam-Columns,* Vols. 1 and 2, McGraw-Hill, New York.

Chen, W. F., and Lui, E. M. (1985), "Columns with End Restraint and Bending in Load and Resistance Factor Design," *Eng. J. Am. Inst. Steel Constr.*, Third Quarter, pp. 105–132.

Chen, W. F., and Lui, E. M. (1987), *Structural Stability—Theory and Implementation,* Elsevier, New York (Sec. 8.3).

Chen, W. F., and Zhou, S. P. (1987), "Inelastic Analysis of Steel Braced Frames with Flexible Joints," *J. Solids Struct.*, 23(5):631–650.

Chen, Y. Z. (1986), "A Numerical Procedure for Evaluating the Plastic Limit Load of Circular Plate Using Mises Yield Criterion," *J. Comput. Struct.*, 24(1):821–22.

Chen, Z.-S., and Mang, H. A. (1988), "Buckling of Multi-Lamellae Compression Flanges of Welded I-Beams: A Unilateral Elastoplastic Plate-Stability Problem," *Int. J. Numer. Methods Eng.*, 26(6):1403–31.

Chern, J. C., You, C.-M., and Bažant, Z. P. (1989), "Deflection of Progressively Cracking Partially Prestressed Concrete Beams," Rep. No. 88–6/C606d, Center for Concrete and Geomaterials, Northwestern University, Evanston (Sec. 8.5); *PCI Journal,* in press.

Chiew, S.-P., Lee, S.-L., and Shanmugam, N. E. (1987), "Experimental Study of Thin-Walled Steel Box Columns," *J. Struct. Eng.* (ASCE), 113(10):2208–20.

Chuenmei, G. (1984), "Elastoplastic Buckling of Single Angle Columns," *J. Struct. Eng.*, 110(6):1391–95.

Chung, B. T., and Lee, G. C. (1971), "Buckling Strength of Columns Based on Random Parameters," *J. Struct. Div.* (ASCE), 97(7):1927–44.

Chwalla, E. (1934), "Über die experimentelle Untersuchung des Tragverhaltens gedrückter Stäbe aus Baustahl," *Der Stahlbau,* 7:17 (Sec. 8.2).

Chwalla, E. (1935), "Der Einfluss der Querschnittsform auf das Tragvermögen aussermittig gedrückter Baustahlstäbe," *Der Stahlbau,* 8:193 (Sec. 8.2).

Chwalla, E. (1937), "Aussermittig gedrückte Baustahlstäbe mit elastisch eingespannten Enden und verschieden grossen Angriffshebeln," *Der Stahlbau,* 10:49–52 and 57–60.

Clark, J. W., and Rolf, R. L. (1966). "Buckling of Aluminium Columns, Plates and Beams," *J. Struct. Eng.* (ASCE), 92(3):17–37.

Colville, J. (1975), "Slenderness Effects in Reinforced Concrete Square Columns," Reinforced Concrete Columns, ACI Publication SP-50, Detroit, pp. 149–64 (Sec. 8.5).

Considère, A. (1891), "Résistance des pièces comprimées," Congrès international des procédés des construction, Paris, Vol. 3 (Sec. 8.1).

Cook, N. E., and Gerstle, K. H. (1985), "Load History Effects on Structural Members," *J. Struct. Eng.* (ASCE), 111(3):628–40.

Cook, N. E., and Gerstle, K. H. (1987), "Safety of Type 2 Steel Frames under Load Cycles," *J. Struct. Eng.* (ASCE), 113(7):1444–55.

Corradi, L. (1977), "On a Stability Condition for Elastic Plastic Structures," *Meccanica,* 12:24–37.

Corradi, L. (1978a), *Instabilità delle strutture,* Clup, Milano (Sec. 8.6).

Corradi, L. (1978b), "Stability of Discrete Elastic Plastic Structures with Associated Flow Laws," *Solid Mechanics Archives,* 3(3):201–59.

Cranston, W. B. (1972), "Analysis and Design of Reinforced Concrete Columns," Research Report No. 20, Cement and Concrete Association, London (Sec. 8.5).

Dawe, J. L., and Kulak, G. L. (1986), "Local Buckling Behavior of Beam-Columns," *J. Struct. Eng.* (ASCE), 112(11):2447–61.

Diamantidis, D. I. (1988), "Limit-State Analysis of Submerged Concrete Cylinders," *ACI Struct. J.,* 85(1):53–60.

Disque, R. O. (1973), "Inelastic K-Factor for Column Design," *Engineering J.,* (AISC), 10(2):33–35.

Duan, L., and Chen, W.-F. (1988), "Design Rules of Built-Up Members in Load and Resistance Factor Design," *J. Struct. Eng.* (ASCE), 114(11):2544–54.

Duberg, J. E. (1962), "Inelastic Buckling," chapter 52 of *Handbook of Engineering Mechanics,* ed. by W. Flügge, McGraw-Hill, New York.

Duberg, J. E., and Wilder, T. W. III (1950a), "Column Behavior in the Plastic Stress Range," *J. Aeronautical Sci.,* 17(6):323–27.

Duberg, J. E., and Wilder, T. W. III (1950b), "Inelastic Column Behavior," NACA Tech. Note No. 2267, pp. 287–302.

ECCS (1976), *Manual on the Stability of Steel Structures,* 2nd ed., Paris (Sec. 8.4).

Elishakoff, I. (1978), "Impact Buckling of Thin Bar via Monte Carlo Method," *J. Appl. Mech.* (ASME), 45:586–90.

Elishakoff, I. (1983), *Probabilistic Methods in the Theory of Structures,* John Wiley & Sons, New York (Sec. 8.4).

El-Zanaty, M. H., and Murray, D. W. (1983), "Nonlinear Finite Element Analysis of Steel Frames," *J. Struct. Eng.* (ASCE), 109(2):353–68.

Engesser, F. (1895), "Über Knickfragen," *Schweizerische Bauzeitung,* 26:24–26. (Sec. 8.1).

Engesser, F. (1889), "Über die Knickfestigkeit gerader Stäbe," *Zeitschrift für Architektur und Ingenieurwesen,* Vol. 35 (Sec. 8.1).

Eurocode No. 3 (1989), *Design of Steel Structures,* Commission of the European Communities, Bruxelles (Sec. 8.4).

Favre, R., Najdanovic, D., Suter, R., and Thürlimann, C. (1984), "New Design Concept for Reinforced Concrete Columns in Buildings," Final Report, 12th Congress of Int. Ass. for Bridge and Struct. Eng. (IABSE), Vancouver, pp. 879–86 (Sec. 8.5).

Ferguson, P. M., and Breen, J. E. (1966), "Investigation of the Long Column in a Frame Subject to Lateral Loads," Symposium on Reinforced Concrete Columns, ACI Publication SP-13, pp. 75–119.

Ferreira, C. M. C., and Rondal, J. (1988), "Influence of Flexural Residual Stresses on the Stability of Compressed Angles," *J. Constr. Steel Res.,* 9(3):169–77.

Fulton, R. E. (1965), "Dynamic Axisymmetric Buckling of Shallow Conical Shells Subjected to Impulsive Loads," *J. Appl. Mech.* (ASME), 32(1):129–34.

Fung, Y. C. (1965), *Foundations of Solid Mechanics,* Prentice-Hall, Englewood Cliffs, N.J. (Sec. 8.1).

Furlong, R. W. (1976), "Guidelines for Analyzing Column Slenderness by a Rational Analysis of an Elastic Frame," *ACI Journal,* 73:138–40.

Furlong, R. W. (1983), "Comparison of AISC, SSLC and ACI Specifications for Composite Columns," *J. Struct Div.* (ASCE), 109(9):1784–803.

Furlong, R. W., and Ferguson, P. M. (1966), "Tests of Frames with Columns in Single

Curvature," Symposium on Reinforced Concrete Columns, ACI Publication SP-13, Detroit, pp. 55–74.

Furnes, O. (1981), "Buckling of Reinforced and Prestressed Concrete Members—Simplified Calculation Methods," *J. Prestressed Concr. Inst.*, 26(4):86–115.

Galambos, T. V. (1988), *Guide to Stability Design Criteria for Metal Structures,* (Column Research Council, 4th ed.), John Wiley & Sons, New York (Sec. 8.3).

Galambos, T. V., and Ketter, R. L. (1959), "Columns under Combined Bending and Thrust," *J. Eng. Mech.* (ASCE), 85(2):1–30.

Galletly, G. D., Blachut, J., and Kruzelecki, J. (1987), "Plastic Buckling of Imperfect Hemispherical Shells Subjected to External Pressure," *Proc. Inst. Mech. Eng.* Part. C, 201(3):153–70.

Genna, F., and Symonds, P. S. (1988), "Dynamic Plastic Instabilities in Response to Short-Pulse Excitation: Effects of Slenderness Ratio and Damping," *Proc. Royal Soc. London,* 417(1852):31–44.

Gerard, G., and Becker, H. (1952), "Column Behavior under Conditions of Impact," *J. Aero. Sci.,* 19:58–65.

Gerard, G., and Becker, H. (1957), "Handbook of Structural Stability: Part I, Buckling of Flat Plates," NACA Tech. Note No. 3781 (Sec. 8.1).

Gjelsvik, A., and Lin, G. S. (1987), "Plastic Buckling of Plates with Edge Frictional Shear Effects," *J. Eng. Mech.* (ASCE), 113(7):953–64 (Sec. 8.1).

Hariri, R. (1967), "Post-Buckling Behavior of Tee-Shaped Aluminum Columns," Ph.D. Dissertation, Univ. of Michigan (Sec. 8.2).

Hellesland, J., and Scordelis, A. C. (1981), "Analysis of RC Bridge Columns under Imposed Deformations," Advanced Mechanics of Reinforced Concrete, IABSE Report. No. 34.

Herber, K. H. (1966), "Vorschlag von Berechnung für Beul-und Traglasten von Schalen," *Stahlbau,* 35:142–51 (Sec. 8.4).

Hill, C. D., Blandford, G. E., and Wang, S. T. (1989), "Post-Buckling Analysis of Steel Space Trusses," *J. Struct. Eng.* (ASCE), 115(4):900–19.

Hognestad, E. (1951), "A Study of Combined Bending and Axial Load in Reinforced Concrete Members" (Bull. No. 399), Univ. of Illinois Eng. Exp. Station.

Horne, M. R. (1971), *Plastic Theory of Structures,* Nelson, London (Sec. 8.6).

Horne, M. R., and Merchant, W. (1965), *The Stability of Frames,* Pergamon Press, Oxford (Sec. 8.6).

Housner, J. M., and Knight, N. F., Jr. (1983), "The Dynamic Collapse of a Column Impacting a Rigid Surface," *AIAA Journal,* 21(8):1187–95.

Howard, J. E. (1908), "Some Results of the Tests of Steel Columns," *Proc. ASTM,* 8:336 (Sec. 8.3).

Hutchinson, J. W. (1972), "On the Post-Buckling Behavior of Imperfection-Sensitive Structures in the Plastic Range," *J. Appl. Mech.* (ASME), 39:155–62.

Hutchinson, J. W. (1974), "Plastic Buckling," in *Advances in Applied Mechanics,* ed. by C.-S. Yih, vol. 14, Academic Press, New York, pp. 67–144.

Iori, I., and Mirabella, G. B. (1987), "Analisi del comportamento di elementi snelli in cemento armato soggetti a carichi ripetuti di breve durata" (in Italian), *L'Industria Italiana del Cemento,* No. 609, pp. 214–20.

Ježek, K. (1934), "Die Tragfähigkeit des exzentrisch beanspruchten und des querbelasteten Druckstabes aus einem ideal plastischen Material," *Sitzungsberichte der Akademie der Wissenschaften in Wien,* Abt. IIa, Vol. 143 (Sec. 8.2).

Ježek, K. (1935), "Näherungsberechnung der Tragkraft exzentrisch gedrückter Stahlstäbe," *Der Stahlbau,* 8:89.

Ježek, K. (1936), "Die Tragfähigkeit axial gedrückter und auf Biegung beanspruchter Stahlstäbe," *Der Stahlbau,* 9:12.

Jingping, L., Zaitian, G., and Shaofan, C. (1988), "Buckling of Transversely Loaded I-Beam Columns," *J. Struct. Eng.* (ASCE), 114(9):2109–17.

Johnston, B. G. (1963), "Buckling Behavior Above the Tangent Modulus Load," *Transactions* (ASCE), 128(I):819–48.

Johnston, B. G. (1964), "Inelastic Buckling Gradient," *J. Eng. Mech. Div.* (ASCE), 90(6):31–47; see also discussion, 91(4):199–203.

Johnston, B. G. (1976), *Guide to Stability Design Criteria for Metal Structures* (Column Research Council, 3rd ed.), John Wiley & Sons, New York (Sec. 8.2).

Kabanov, V. V., and Kurtsevich, G. I. (1985), "Stability of a Circular Cylindrical Shell with Axisymmetric Initial Deflections in Compression Beyond the Elastic Limit," *Soviet Appl. Mech.*, 21(2):138–44.

Kamiya, F. (1988), "Buckling of Sheathed Walls: Nonlinear Analysis," *J. Struct. Eng.* (ASCE), 114(3):625–41.

Ketter, R. L. (1958), "The Influence of Residual Stresses on the Strength of Structural Members," Welding Research Council Bulletin No. 44.

Key, P. W., Hasan, S. W., and Hancock, G. J. (1988), "Column Behavior of Cold-Formed Hollow Sections," *J. Struct. Eng.* (ASCE), 114(2):390–407.

Kitipornchai, S., and Wong-Chung, A. D. (1987), "Inelastic Buckling of Welded Monosymmetric I-Beams," *J. Struct. Eng.* (ASCE), 113(4):740–56.

Kollbrunner, C. F. (1938), "Zentrischer un exzentrischer Druck von an beiden Enden gelenkig gelagerten Rechteckstäben aus Avional M und Baustahl," *Der Stahlbau*, 11:25.

Korn, A., Galambos, T. V. (1968), "Behavior of Elastic-Plastic Frames," *J. Struct. Eng.* (ASCE), 94(5):1119–42.

Kupfer, H., and Feix, J. (1988), "Geschlossene Lösung für die Ermittlung der zulässigen Schlankheit von Druckgliedern aus Stahlbeton (Direct Determination of the Admissible Slenderness of Reinforced Concrete Columns)," *Bauingenieur*, 63(1):43–45.

Lai, S.-M. A., and MacGregor, J. G. (1983), "Geometric Nonlinearities in Multi-Story Frames," *J. Struct. Eng.* (ASCE), 109(11):2528–45.

Leckie, F. A. (1969), "Plastic Instability of a Spherical Shell," in *Theory of Thin Shells*, ed. by F. I. Niordson, Springer-Verlag, Berlin and New York.

Lee, S.-L., Shanmugam, N. E., and Chiew, S.-P. (1988), "Thin-Walled Box Columns under Arbitrary End Loads," *J. Struct. Eng.* (ASCE), 114(6):1390–401.

Liew, J. Y. R., Shanmugam, N. E., and Lee, S. L. (1989), "Tapered Box Columns under Biaxial Loading," *J. Struct. Eng.* (ASCE), 115(7):1697–710.

Lin, T. H. (1968), *Theory of Inelastic Structures*, John Wiley and Sons, New York (Sec. 8.1).

Lin, T. Y., and Burns, N. H. (1981), *Design of Prestressed Concrete Structures*, John Wiley & Sons, New York (Sec. 8.5).

Lindberg, H. E., Rubin, M. B., and Schwer, L. E. (1987), "Dynamic Buckling of Cylindrical Shells from Oscillating Waves Following Axial Impact," *Int. J. of Solids Struct.*, 23(6):669–92.

Lui, E. M., and Chen, W. F. (1984), "Simplified Approach to the Analysis and Design of Columns with Imperfections," *Eng. J.* (AISC), 21:99–117.

MacGregor, J. G. (1986), "Stability of Reinforced Concrete Building Frames," in *Concrete Framed Structures*, ed. by R. Narayanan, Elsevier, London, pp. 1–41 (Sec. 8.5).

MacGregor, J. G., and Barter, S. L. (1966), "Long Eccentrically Loaded Concrete Columns Bent in Double Curvature," Symposium on Reinforced Concrete Columns, ACI Publication SP-13, pp. 139–56.

MacGregor, J. G., and Hage, S. E. (1977), "Stability Analysis and Design of Concrete Frames," *J. Struct. Eng.* (ASCE), 103(10):1953–70.

MacGregor, J. G., Breen, J. E., and Pfrang, E. O. (1970), "Design of Slender Columns," *ACI Journal,* 67(1):6–28 (Sec. 8.5).

MacGregor, J. G., Oelhafen, U. H., and Hage, S. E. (1975), "A Re-Examination of the EI Value for Slender Columns," Reinforced Concrete Columns, ACI Publication SP-50, Detroit, pp. 1–40 (Sec. 8.5).

Maier, G. (1971), "Incremental Plastic Analysis in the Presence of Large Displacements and Physical Instabilizing Effects," *Int. J. Solids Struct.,* 7:345–72.

Manuel, R. F., and MacGregor, J. G. (1967), "Analysis of Restrained Reinforced Concrete Columns under Sustained Load," *ACI Journal,* 64(1):12–24.

Massonnet, C. (1959), "Stability Considerations in the Design of Steel Columns," *J. Struct. Eng.* (ASCE), 85(7):75–111.

Massonnet, C. (1976), "Forty Years of Research on Beam-Columns," *Solid Mech. Arch.,* 1(2):27–157.

Mau, S. T., and El-Mabsout, M. (1989), "Inelastic Buckling of Reinforcing Bars," *J. Eng. Mech.,* 115(1):1–17.

Mazzolani, F. M., and Frey, F. (1983), "ECCS Stability Code for Aluminium-Alloy Members: Present State and Work in Progress," 3rd Int. Colloq. Stab. Met. Struct., Prelim. Rep., Paris.

Medland, I. C., and Taylor, D. A. (1971), "Flexural Rigidity of Concrete Column Sections," *J. Struct. Eng.* (ASCE), 97(11):573–86 (Sec. 8.5).

Menegotto, P., and Pinto, P. E. (1977), "Slender RC Compressed Members in Biaxial Bending," *J. Struct. Eng.* (ASCE), 103:587–605.

Menn, C. (1974), "Simple Method for Determining the Ultimate Load of Slender Compression Members" (in German), Preliminary Report, Symposium on the Design and Safety of Reinforced Concrete Compression Members, Int. Ass. for Bridge and Struct. Eng. (IABSE), Vol. 16, pp. 136–44 (Sec. 8.5).

Migliacci, A., and Mola, F. (1985), *Progetto agli stati limite di strutture in cemento armato* (in Italian), Vol. 2, Masson Italia Editori, Milano (Sec. 8.5).

Mirza, S. A., Lee, P. M., and Morgan, D. L. (1987), "ACI Stability Resistance Factor for RC Columns," *J. Struct. Eng.,* 113(9):1963–76.

Mola, F., Malerba, P. G., and Pisani, M. A. (1984), "Effetti della precompressione sulla capacità portante ultima di sezioni ed elementi snelli in cemento armato" (in Italian), *Studi e Ricerche*, Vol. 6, Politecnico di Milano, Milan, pp. 231–56.

Mortelhand, I. M. (1987), "Heron's Fountain. What Shoulder, and What Art," *Heron,* no. 32, pp. 115–21.

Nathan, N. D. (1985), "Rational Analysis and Design of Prestressed Concrete Beam Columns and Wall Panels," *J. Prestressed Concrete Institute,* 30(3):82–133.

Neale, K. W. (1975), "Effect of Imperfections on the Plastic Buckling of Rectangular Plates," *J. Appl. Mech.* (ASME), 97:115.

Needleman, A. (1982), "A Numerical Study of Necking in Circular Cylindrical Bars," *J. Mech. Phys. Solids,* 20:111–20 (Sec. 8.7).

Newmark, N. M. (1943), "Numerical Procedure for Computing Deflections, Moments and Buckling Loads," *Trans. ASCE,* 108:1161–88 (Sec. 8.4).

Nicholls, R., and Freeman, J. (1988), "Buckling of Fabric-Reinforced Concrete Shells," *J. Struct. Eng.* (ASCE) 114(4):765–82.

Nilson, A. H. (1987), *Design of Prestressed Concrete,* John Wiley & Sons, New York (Sec. 8.5).

Onat, E. T., and Drucker, D. C. (1953), "Inelastic Instability and Incremental Theories of Plasticity," *J. Aero. Sci.,* 20:181.

Osgood, W. R. (1947), "Beam-Columns," *J. Aero. Sci.,* 14:167.

Osgood, W. R. (1951), "The Effect of Residual Stress on Column Strength," Proc. 1st U.S. Nat. Cong. Appl. Mech., p. 415 (Sec. 8.3).

Papadrakakis, M., and Loukakis, K. (1988), "Inelastic Cyclic Response of Restrained Imperfect Columns," *J. Eng. Mech.* (ASCE), 114(2):295–313.

Papia, M., and Russo, G. (1989), "Compressive Concrete Strain at Buckling of Longitudinal Reinforcement," *J. Struct. Eng.*, 115(2):382–97.

Park, R., and Paulay, T. (1975), *Reinforced Concrete Structures*, John Wiley & Sons, New York (Sec. 8.5).

Pearson, C. E. (1950), "Bifurcation Criterion and Plastic Buckling of Plates and Columns," *J. Aero. Sci.*, 17:417–24.

Pfrang, E. O., and Siess, C. P. (1964), "Behavior of Restrained Reinforced Concrete Columns," *J. Struct. Eng.* (ASCE), 5:113–286.

Phillips, A. (1972), "Solution of a Plastic Buckling Paradox," *AIAA Journal*, 10:951.

Popov, E. P., and Medwadowski, S. J. (1981), "Stability of Reinforced Concrete Shells: State-of-the-Art Overview," Concrete Shell Buckling, ACI Publication SP-67, pp. 1–42 (Sec. 8.5).

Prestressed Concrete Institute (1975), *Design Handbook*, Chicago.

Quast, U. (1987), "Stability of Compression Members," in *Concrete Framed Structures*, ed. by R. Narayanan, Elsevier, London, pp. 43–69 (Sec. 8.5).

Ramberg, W., and Osgood, W. R. (1943), "Description of Stress-Strain Curves by Three Parameters," NACA Tech. Note No. 902 (Sec. 8.2).

Roorda, J. (1988), "Buckling Behavior of Thin-Walled Columns," *Can. J. Civ. Eng.*, 15(1):107–16.

Salmon, C. G. and Johnson, J. E. (1980), *Steel Structures*, Harper and Row, New York (Sec. 8.4).

Salmon, E. H. (1921), *Columns*, Oxford Technical Publication, London (Sec. 8.3).

Scordelis, A. C. (1981), "Stability of Reinforced Concrete Domes and Hyperbolic Paraboloid Shells," Concrete Shell Buckling, ACI Publication SP-67, pp. 63–110.

Scribner, C. F. (1986), "Reinforcement Buckling in Reinforced Concrete Flexural Members," *J. Am. Concr. Inst.*, 83(6):966–73.

Sewell, M. J. (1963), "A General Theory of Elastic and Inelastic Plate Failure: Part I," *J. Mech. Phys. Solids*, 11:377–93. See also "Part II," Vol 12, 1964, pp. 279–97.

Sewell, M. J. (1972), "A Survey of Plastic Buckling," chapter 5 in *Stability*, ed. by H. Leipholz, Univ. of Waterloo Press, Ontario, pp. 85–197.

Sfintesco, D. (1970), "Experimental Basis for the European Column Curves," *Constr. Met.*, 7(3):5–12.

Shanley, F. R. (1947), "Inelastic Column Theory," *J. Aero. Sci.*, 14(5):261–8 (Sec. 8.1).

SIA (1976), "Ultimate Load Design for Compression Members, Revisions to Articles 3.08, 3.09, 3.24 of the SIA Code 162 (1968)" in German, Zürich (Sec. 8.5).

Simitses, G. J. (1984), "Suddenly-Loaded Structural Configurations," *J. Eng. Mech.* (ASCE), 110:1320–34.

Simitses, G. J., Kounadis, A. N., and Girl, J. (1979), "Dynamic Buckling of Simple Frames under a Step Load," *J. Eng. Mech.* (ASCE), 105:896–900.

Simo, J. C. (1986), "On the Computational Significance of the Intermediate Configuration and Hyperelastic Relations in Finite Deformation Elastoplasticity," *Mechanics of Materials*, 4:439–51 (Sec. 8.7).

Simo, J. C. and Ortiz, N. (1985), "A Unified Approach to Finite Deformation Elastoplasticity Based on the Use of Hyperelastic Constitutive Equations," *Comp. Meth. Appl. Mech. Eng.*, 49:221–35 (Sec. 8.7).

Smith, C. V., Jr (1976), "On Inelastic Column Buckling," *Engineering J.* (AISC), 13(3):86–88.

Smith, R. G., and Bridge, R. Q. (1984), "Design of Concrete Columns," Top-Tier Design Methods in the Draft Concrete Code, School of Civil and Mining Eng., University of Sydney, Australia, pp. 2.1–2.95 (Sec. 8.5).

Sohal, I. S., and Chen, W.-F. (1988), "Local and Post-Buckling Behavior of Tubular

Beam-Columns," *J. Struct. Eng.* (ASCE), 114(5):1073–90.

Southwell, R. V. (1915), "On the Collapse of Tubes by External Pressure," *Phil. Mag.*, 29:67–77 (Sec. 8.1).

Structural Stability Research Council (SSRC), European Convention for Constructional Steelwork (ECCS), Column Research Committee of Japan (CRCJ), and Council of Mutual Economic Assistance (CMEA) (1982), *Stability of Metal Structures—A World View,* Am. Inst. Steel Constr., Chicago.

Templin, R. L., Sturm, R. G., Hartmann, E. C., and Holt, M. (1983), "Column Strength of Various Aluminum Alloys," Aluminum Research Labs., Tech. Paper 1, Aluminum Company of America, Pittsburgh.

Thürlimann, C. B. (1984), "Design of Reinforced Concrete Columns Subjected to Imposed End Deformations," Int. Ass. for Bridge and Struct. Eng. (IABSE), Zurich, pp. 89–108 (Sec. 8.5).

Timoshenko, S. P., and Gere, J. M. (1961), *Theory of Elastic Stability,* McGraw-Hill, New York (Sec. 8.1).

Tvergaard, V. (1982), "On Localization in Ductile Materials Containing Spherical Voids," *Int. J. Fracture,* 18:237–52 (Sec. 8.7).

Van Manen, S. E. (1982), "Plastic Design of Braced Frames Allowing Plastic Hinges in the Columns," *Heron,* 27(2).

Vielsack, P. (1987), "Simple Analysis for the Blowup Phenomenon of Inelastic Members in Compression," *J. Appl. Mech.* (ASME), 54(2):459–60.

Vinnakota, S., and Aystö, P. (1974), "Inelastic Spatial Stability of Restrained Beam-Columns," *J. Struct. Eng.* (ASCE), 100(11):2235–54.

von Kármán, T. (1910), "Untersuchungen über Knickfestigkeit," Mitteilungen über Forschungsarbeiten auf dem Gebiete des Ingenieurwesens, Berlin, No. 81 (Sec. 8.1).

von Kármán, T. (1947), "Discussion of Inelastic Column Theory," *J. Aero. Sci.*, 14:267–8.

Wang, C. K., and Salmon, C. G. (1979), *Reinforced Concrete Design,* Harper and Row, New York (Sec. 8.5).

Warner, R. F., and Kordina, K. (1975), "Influence of Creep on the Deflections of Slender Reinforced Concrete Columns," *Deutscher Ausschuss für Stahlbeton,* Vol. 250, pp. 39–156 (Sec. 8.5).

Warner, R. F., Rangan, B. V., and Hall, A. S. (1976), *Reinforced Concrete,* Pitman, Carlton, Australia (Sec. 8.5).

Westergaard, H. M., and Osgood, W. R. (1928), "Strength of Steel Columns," *Trans. ASME,* 49(50):65.

Wilder, T. W., Brooks, W. A., and Mathauser, E. E. (1953), "The Effect of Initial Curvature on the Strength of an Inelastic Column," NACA Tech. Note No. 2872 (Sec. 8.2).

Wood, R. H., and Shaw, M. R. (1979), "Developments in the Variable-Stiffness Approach to Reinforced Concrete Design," *Mag. of Concr. Res.* (London), 31(108):127–41 (Sec. 8.5).

Wynhoven, J. H., and Adams, P. F. (1972), "Analysis of Three-Dimensional Structures," *J. Struct. Eng.* (ASCE), 98(7):1361–76 (Sec. 8.4).

Yamamoto, K., Akiyama, N., and Okumura, T. (1988), "Buckling Strengths of Gusseted Truss Joints," *J. Struct. Eng.* (ASCE), 114(3):575–90.

Yang, C. H., Beedle, L. S., and Johnston, B. G. (1952), "Residual Stress and the Yield Strength of Steel Beams," *Welding Journal* (Supplement), 31(4):205–29 (Sec. 8.3).

Yang, T. Y., and Liaw, D. G. (1988), "Elastic-Plastic Dynamic Buckling of Thin-Shell Finite Elements with Asymmetric Imperfections," *AIAA Journal,* 26(4):479–86.

Yura, J. A. (1971), "The Effective Length of Columns in Unbraced Frames," *Eng. J. Am. Inst. Steel Constr.*, 8(2):37–42 (Sec. 8.4).

Zeris, C. A., and Mahin, S. A. (1988), "Analysis of Reinforced Concrete Beam-Columns under Uniaxial Excitation," *J. Struct. Eng.* (ASCE), 114(4):804–20.

9

Creep Buckling

The inelastic behavior of some materials, such as concrete and polymers, and at high temperature also metals, is strongly time-dependent and may be characterized as viscoelastic or viscoplastic. In viscoelastic behavior, typical of polymers and to some extent of concrete within the service stress range, the linearity of the stress–strain relation—an essential aspect of all the analysis in Chapters 1 to 7—is preserved. By virtue of this fact, most of the elastic solutions for stability can be transplanted to the viscoelastic domain, using the so-called elastic–viscoelastic analogy. If inevitable imperfections are taken into account, it is found that viscoelastic structures may lose stability slowly, over a long period of time. This phenomenon may be characterized either in terms of the long-time critical load, which may be much smaller than the short-time (elastic) critical load, or in terms of the deflection at the end of the projected life span.

An entirely new type of behavior is encountered when the time-dependent behavior is nonlinear, that is, viscoplastic. In that case, as it turns out, columns under sustained loads are generally unstable for any value of the load, if the usual definition of stability of solutions is adopted. However, this does not mean that such columns could not be used. It appears that the loss of stability may occur long after the end of the projected service life of the structure. Then it is more reasonable to characterize stability in terms of the critical time (life time), defined as the load duration for which the ratio of the buckling deflection to the initial imperfection becomes infinite. The existence of a critical time is the salient feature of viscoplastic (or nonlinear) creep buckling. For linear viscoelastic behavior, the critical time is infinite and therefore useless as a distinction between stable and unstable behavior.

In design applications to concrete structures, the viscoelastic and viscoplastic analysis of buckling of reinforced columns must further take into account the variation of creep properties of the material with age, a phenomenon called aging. We will briefly discuss some recent developments in this regard, and illustrate the analysis of long-time carrying capacity of concrete columns.

With regard to the thermodynamic analysis of stability, which follows in the next chapter, we must note that no state with a nonzero creep rate can represent stable thermodynamic equilibrium since this implies a positive energy dissipation rate (and thus a positive rate of internally produced entropy; Sec. 10.1). When we use the word "equilibrium" in this chapter, we will mean merely static (mechanical) equilibrium (i.e., absence of inertia forces), unless thermodynamic equilibrium is explicitly mentioned.

9.1 VISCOELASTIC STRESS–STRAIN RELATIONS

Some materials, such as polymers or concrete, continue to increase their deformation while the stress is kept constant. This phenomenon is now generally called creep. When a number of specimens are loaded to various stress levels at time $t = t_0$, and their strains are recorded at various specified times $t_0, t_1, t_2, \ldots,$ one can obtain the creep isochrones as the stress–strain curves that connect the points reached at the same time (in Greek "chronos" means time, and the prefix "iso" means equal); see Figure 9.1a. For some materials, such as polymers or reinforced polymers, the isochrones are initially straight, that is, linear, up to a certain proportionality limit. This is the special case of linear creep, also called viscoelasticity (or linear viscoelasticity). If the isochrones do not depend on the age of the material at the moment of loading, we speak of nonaging viscoelasticity, in contrast to aging viscoelasticity that is a typical property of concrete. The behavior above the proportionality limit of the isochrones represents nonlinear creep (or nonlinear viscoelasticity). For some materials, such as metals at high temperatures or clays, there is no linear range, that is, the creep isochrones have a significant curvature beginning with zero stress (Fig. 9.1b). In this section we will consider only nonaging linear viscoelastic behavior.

Compliance Function and Integral-Type Creep Law

A typical shape of the creep curve $\varepsilon(t)$ at constant stress σ is shown in Figure 9.2a. Due to the second law of thermodynamics, the creep curve must always have a nonnegative slope, and for materials that undergo no damage the slope of the creep curve must not increase with time. Due to the proportionality property, the creep curves of viscoelastic materials are fully characterized by the creep curve for unit constant stress, called the compliance function, $J(t - t_0)$, where t is the current time and t_0 is the time at which the unit stress was applied. For any stress within the linear range, the creep curve is then obtained as $\varepsilon(t) = \sigma J(t - t_0)$.

Another typical type of response is the stress variation when the material is deformed at time t_0 and subsequently the strain ε is kept constant (Fig. 9.2b). The ensuing stress response is called stress relaxation. According to the second law of thermodynamics the stress–time relaxation curve may not increase. For non-damaging materials the magnitude of the slope cannot increase.

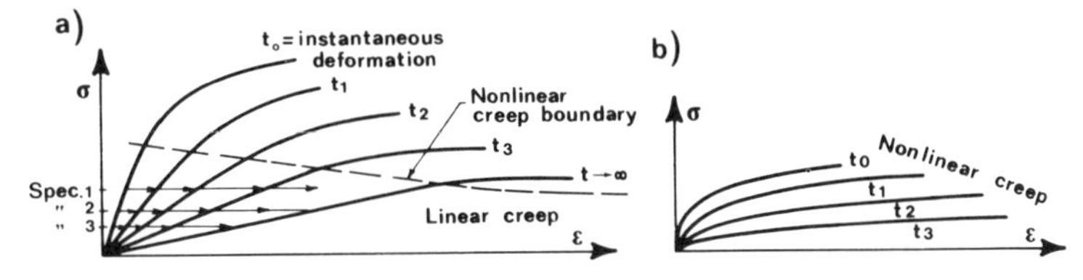

Figure 9.1 Creep isochrones: (a) initially linear creep, (b) nonlinear creep.

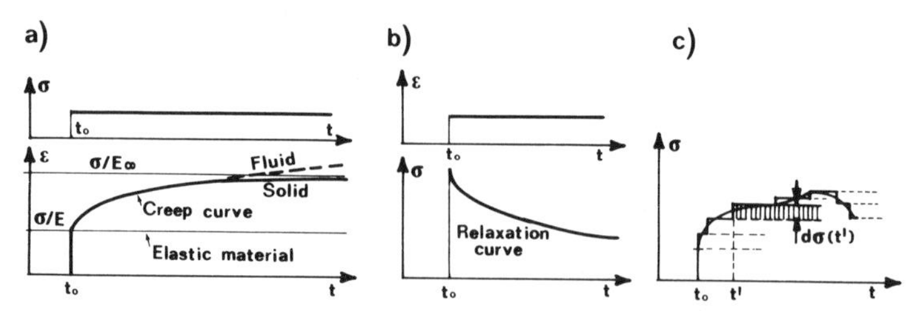

Figure 9.2 (a) Creep curve; (b) relaxation curve; (c) decomposition of stress history into infinitesimal stress increments.

For the analysis of creep buckling, we need a stress–strain relation applicable at time–variable stress, $\sigma(t)$. It appears that if the material should be described as linear, in the sense that the general stress–strain relation is characterized by linear differential (or integral) equations in time, it is necessary to satisfy the principle of superposition in time. This principle, which is not implied by linearity of the isochrones (see Bažant, Tsubaki, and Celep, 1983), requires that if two stress histories are superimposed, then the strain response to the combined stress history is a sum of the strain responses to each of the stress histories. An arbitrary stress history $\sigma(t)$ can be imagined to be a sum of infinitesimal stress increments $d\sigma(t')$, as illustrated by the horizontal strips in Figure 9.2c. According to the principle of superposition, the strain response should then be obtained as the sum of the responses to the individual steps $d\sigma(t')$, that is,

$$\varepsilon(t) = \int_0^t J(t - t')\, d\sigma(t') \tag{9.1.1}$$

(Note that the lower limit could be t_0 instead of 0 because $\sigma = 0$ for $t < t_0$). This equation, expressing the principle of superposition, agrees reasonably well (albeit not exactly) with the test results for many viscoelastic materials, particularly polymers, in their linear (low-stress) range. The integral in Equation 9.1.1 is the Stieltjes integral, which is defined even if $\sigma(t)$ varies discontinuously (piecewise continuously). If $\sigma(t)$ is continuous and differentiable, then $d\sigma(t') = \dot{\sigma}(t')\, dt'$ where $\dot{\sigma}(t') = d\sigma(t')/dt'$, and Equation 9.1.1 thus becomes the usual (Riemann) integral. For a stress history $\sigma(t)$ that is continuous and differentiable for $t > t_0$ but begins with a jump from 0 to $\sigma(t_0)$, Equation 9.1.1 implies that $\varepsilon(t) = J(t - t_0)\sigma(t_0) + \int_{t_0}^{t} J(t - t')\dot{\sigma}(t')\, dt'$.

If the stress history is unknown in advance, Equation 9.1.1 represents an integral equation for $\sigma(t)$. An integral equation of this type is called the Volterra integral equation. It is a special case of the integral equations of the second kind.

Differential-Type Creep Law and Rheologic Models

For the purpose of structural analysis, it is often more convenient to formulate the stress–strain relation as a differential equation in time. This can be accomplished by describing the material with a rheologic model consisting of springs and dashpots. Such models are purely phenomenologic and have nothing

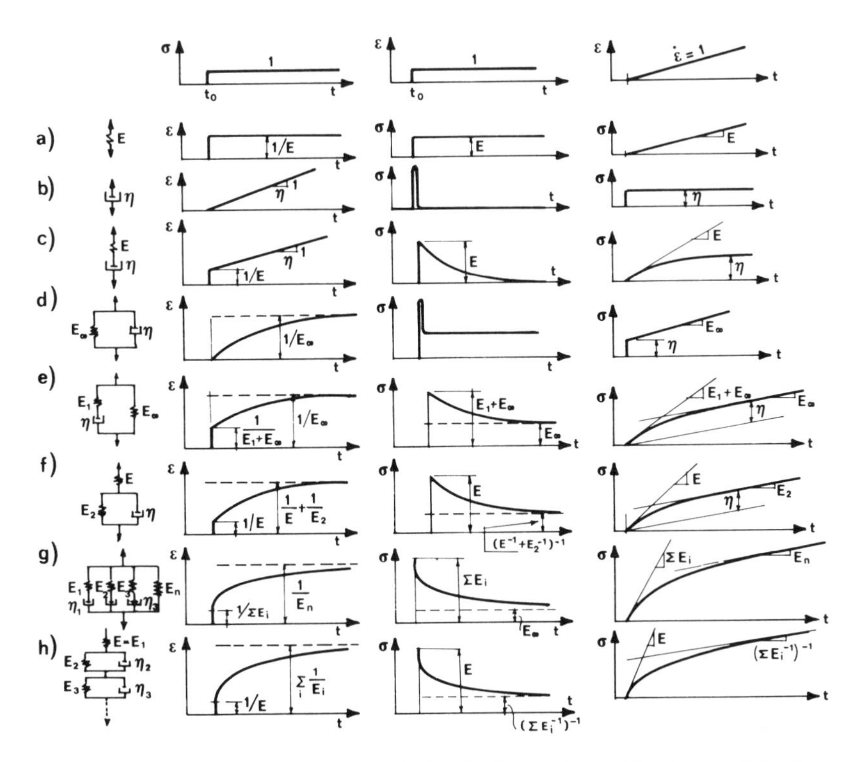

Figure 9.3 Responses of various types of rheologic models: (a) elastic; (b) viscous; (c) Maxwell; (d) Kelvin (Voigt); (e) standard solid (Maxwell type); (s) standard solid (Kelvin type); (g) Maxwell chain; (h) Kelvin chain.

to do with the actual physical mechanism of viscoelastic behavior. It has been shown (e.g., Roscoe, 1950) that any type of viscoelastic behavior can be approximated by a spring-dashpot model with any desired accuracy.

Various linear rheologic models are illustrated in Figure 9.3. The force applied on the model represents the stress, σ, and the relative displacements between the end points of the model represent the strain, ε. The model is characterized by its spring moduli E and dashpot viscosities η. Since the material is assumed to be nonaging, E and η are independent of time t, that is, constant. Figure 9.3 also shows the typical responses to constant unit stress or to constant unit strain (Heaviside step functions), as well as to constant strain rate. The spring is described by the relation $\sigma = E\varepsilon$, and the dashpot by the relation $\sigma = \eta\dot{\varepsilon}$ (superimposed dots denote time derivatives, that is, $\dot{\varepsilon} = d\varepsilon/dt$).

The Maxwell unit, describing material behavior that was first considered by J. C. Maxwell in England in the late nineteenth century, consists of a spring and dashpot coupled in series (Fig. 9.3c). This model exhibits constant-rate creep (at a constant stress), and stress relaxation ($\varepsilon = \text{const.}$) of a decaying rate, approaching zero stress at $t \to \infty$. Another simple model is the Kelvin unit (or Kelvin–Voigt unit) shown in Figure 9.3d. It describes material behavior first considered by Lord Kelvin in England and Voigt in Germany, and consists of a spring and dashpot coupled in parallel. This model exhibits creep at a decaying rate, but no instantaneous elastic deformation, in contrast to the Maxwell unit. The simplest rheologic model that gives both instantaneous elastic deformation and subsequent creep at a decaying rate is the standard solid model, for which there exist two types: the Maxwell type, which consists of a Maxwell unit and a spring coupled in parallel (Fig. 9.3e), and the Kelvin type, which consists of a spring and a Kelvin unit coupled in series (Fig. 9.3f).

The most general linear viscoelastic behavior can be approximated as closely as desired (Roscoe, 1950) by the Maxwell chain model (Fig. 9.3g) or the Kelvin chain model (Fig. 9.3h), which consists of either a sequence of Maxwell units coupled in parallel or a sequence of Kelvin units coupled in series. With these models it is possible to capture the fact that, as is true for many real materials, the creep or relaxation curve begins with an almost vertical slope, and its slope decays over many orders of magnitude of load duration, much slower than an exponential curve (exponential decay is typical, as one can check, for the standard solid models). Since the Maxwell and Kelvin chain models can describe material behavior as closely as desired, there is no need for other arrangements of springs and dashpots than those shown in Figure 9.3.

In viscoelastic behavior we may distinguish fluids and solids. Fluids are those materials for which the creep curve at constant stress is unbounded, as is true for the viscous unit (Fig. 9.3b) and the Maxwell unit (Fig. 9.3c). When the creep curve at constant stress is bounded, the material is called a solid. The Kelvin unit and the standard solid models (Fig. 9.3d, e, f) are solids. The Maxwell chain (Fig. 9.3g) represents a solid if the last Maxwell unit ($i = n$) in the chain has no dashpot (that is, $\eta_n \to \infty$). The Kelvin chain (Fig. 9.3h) always represents a solid; however, it exhibits instantaneous elastic strain only if the first Kelvin unit in the chain has no dashpot (that is, $\eta_1 \to 0$). The viscoelastic strain of a solid that remains after subtracting the instantaneous elastic strain is called delayed elastic strain.

It should be kept in mind, however, that the distinction between solids and fluids in viscoelasticity is blurred and for materials exhibiting creep of very long durations becomes an academic question. For concrete, for example, it is known that the creep curve does not approach a horizontal asymptote up to load durations of about 50 years, but whether an asymptote is approached for longer durations is irrelevant for design. The characterization as a solid is meaningful only for those materials that attain the final asymptotic value of the creep curve at load durations that are much smaller than the lifetime of the structure.

Let us now illustrate a stress–strain relation in the form of a differential equation, considering the standard solid model of Kelvin type (Fig. 9.3f). Let E_2 = spring constant and σ_2 = stress in the spring of the Kelvin unit. The total deformation of the model is

$$\varepsilon = \frac{\sigma}{E} + \frac{\sigma_2}{E_2} \tag{9.1.2}$$

The stress in the dashpot is, due to the equilibrium condition, $\sigma - \sigma_2$, and so the total strain rate of the model may be expressed as

$$\dot{\varepsilon} = \frac{\dot{\sigma}}{E} + \frac{\sigma - \sigma_2}{\eta} \tag{9.1.3}$$

Multiplying Equations 9.1.2 and 9.1.3 by E_2 and η, and adding them, we obtain the first-order linear differential equation:

$$\dot{\varepsilon}\eta + \varepsilon E_2 = \dot{\sigma}\frac{\eta}{E} + \sigma\left(1 + \frac{E_2}{E}\right) \tag{9.1.4}$$

With the notation $\tau_r = \eta / E_2$, $E_\infty = (E^{-1} + E_2^{-1})^{-1}$, this equation may be rewritten as

$$\tau_r \dot{\varepsilon} + \varepsilon = \frac{\tau_r}{E}\dot{\sigma} + \frac{1}{E_\infty}\sigma \tag{9.1.5}$$

in which the material constants E, E_∞, and τ_r represent the elastic modulus for instantaneous deformation, the elastic modulus for long-time deformation, and a constant with the dimension of time, called the retardation time.

The creep curve for a constant stress σ applied at time $t = 0$ may be obtained by integrating Equation 9.1.5 for $\dot{\sigma} = 0$. The initial condition is $\varepsilon = \sigma / E$ at $t = 0$, and integration yields for the creep curve (Fig. 9.3f) the expression:

$$\varepsilon = \sigma\left[\frac{1}{E_\infty} - \left(\frac{1}{E_\infty} - \frac{1}{E}\right)e^{-t/\tau_r}\right] \tag{9.1.6}$$

By optimum fitting of this formula to a measured creep curve for a given material, one can identify the material constants E, E_∞, and τ_r, from which one can then calculate also E_2 and η.

The stress relaxation curve may be obtained by integrating Equation 9.1.5 for a constant strain ε ($\dot{\varepsilon} = 0$) imposed at time $t = 0$. The initial condition is $\sigma = E\varepsilon$ at $t = 0$, and the integration yields the stress relaxation curve (Fig. 9.3f):

$$\sigma = \varepsilon[E_\infty + (E - E_\infty)e^{-t/(\tau_r E_\infty / E)}] \tag{9.1.7}$$

Elastic–Viscoelastic Analogy

The differential stress–strain relation for the standard solid, as well as any other linear rheologic model, may be generally written in the form

$$\sigma = \underset{\sim}{E}\varepsilon \qquad \text{or} \qquad \varepsilon = \underset{\sim}{E}^{-1}\sigma \tag{9.1.8}$$

$\underset{\sim}{E}$ and $\underset{\sim}{E}^{-1}$ represent differential operators that are inverse to each other; $\underset{\sim}{E}$ is called the relaxation operator, and $\underset{\sim}{E}^{-1}$ is called the creep operator. For the standard solid the relaxation and creep operators may be written in the form

$$\underset{\sim}{E} = \frac{1 + \tau_r(\partial/\partial t)}{(1/E_\infty) + (\tau_r/E)(\partial/\partial t)} \qquad \underset{\sim}{E}^{-1} = \frac{(1/E_\infty) + (\tau_r/E)(\partial/\partial t)}{1 + \tau_r(\partial/\partial t)} \tag{9.1.9}$$

It can be shown that for a general viscoelastic material, the relaxation operator may be expressed as

$$\underset{\sim}{E} = \frac{E + a_1(\partial/\partial t) + a_2(\partial^2/\partial t^2) + \cdots}{1 + b_1(\partial/\partial t) + b_2(\partial^2/\partial t^2) + \cdots} \tag{9.1.10}$$

where $a_1, a_2, b_1, b_2, \ldots =$ material constants. With certain minor restrictions the differential operators can be manipulated according to the same rules as those of linear algebra. The principal exception is that $(\partial/\partial t)/(\partial/\partial t)$ cannot be replaced by 1 because the equation $[(\partial/\partial t)/(\partial/\partial t)]y = z$ is equivalent to $\dot{y} = \dot{z}$ and has the solution $y = z + \text{const.}$, not $y = z$. A rigorous formulation requires the use of the Laplace transform, but for the sake of simplicity we refrain from it here. As long as the rules of linear algebra are applicable, we may obtain the governing differential equation for a certain problem simply by replacing E or $1/E$ in the associated elastic formulation with the relaxation operator $\underset{\sim}{E}$ or the creep operator $\underset{\sim}{E}^{-1}$, respectively. The possibility of replacing the elastic constants by the corresponding creep or relaxation operators and manipulating them according to the rules of linear algebra, is called the elastic–viscoelastic analogy (also called the correspondence principle by Biot, 1965, p. 359). This analogy was stated by Alfrey (1944) for nonaging viscoelastic materials (although for concrete as an aging viscoelastic material, this analogy was introduced previously by McHenry, 1943).

Problems

9.1.1 Determine the expression of the creep curve and the relaxation curve for the (a) Maxwell model (Fig. 9.3c); (b) Kelvin–Voigt model (Fig. 9.3d); (c) standard solid of Maxwell type.

9.1.2 For the same models, determine the σ-ε relations for $\dot{\sigma} = \text{const.}$ and for $\dot{\varepsilon} = \text{const.}$ (i.e., the stress-controlled or displacement-controlled tests at constant loading rate).

9.1.3 For those acquainted with the Laplace transform: Obtain the transform of the stress–strain relation for the standard solid of Kelvin type, which has the form of an elastic stress–strain relation. Use it to solve the load–displacement relation for the structures in Figure 9.4 under either constant load or constant displacement. Then, by inversion of the Laplace transform obtain the displacement or force as a function of time.

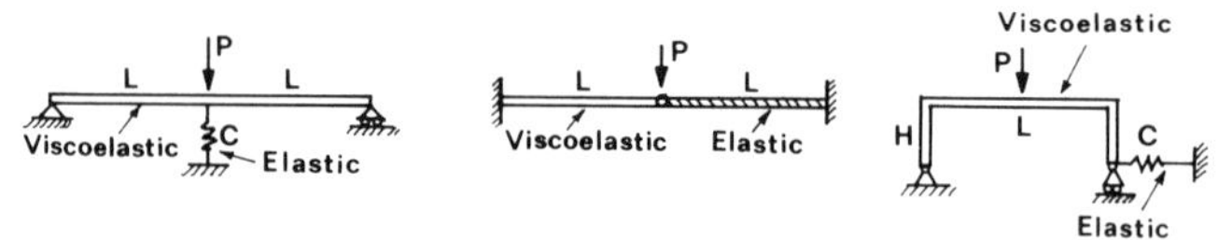

Figure 9.4 Exercise problems on viscoelasticity.

9.2 VISCOELASTIC BUCKLING

Viscoelastic buckling consists of a slow growth of deflection with time. Depending on the load magnitude, deflections may either asymptotically approach a finite value, which represents a stable response, or grow at $t \to \infty$ beyond any bounds, which is considered to be a long-time instability. If the problem is linear, the deflection cannot become unbounded at a finite time after load application, but if nonlinearities are present it can. For viscoelastic buckling it is essential to consider imperfections (Freudenthal, 1950; Libove, 1952; Kempner, 1954; Hoff, 1954, 1958; Lin, 1956; Rabotnov and Shesterikov, 1957; Huang, 1967). Without

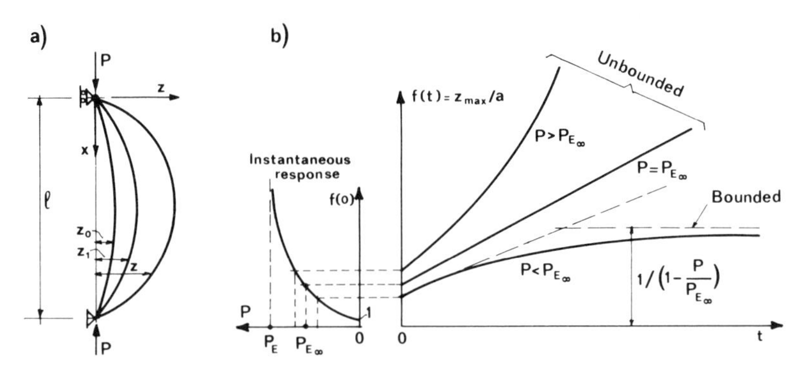

Figure 9.5 (a) Imperfect viscoelastic column subjected to axial load and (b) deflection histories.

the consideration of imperfections, the static form of the problem becomes meaningless.

Let us now analyze the buckling of a viscoelastic column, using the integrated differential equation for the deflections of a simply supported beam-column. The column is considered to be imperfect, with an initial curvature defined by function $z_0(x)$. Let $z(x, t)$ = deflection ordinates (Fig. 9.5a). According to Section 1.2, the elastic behavior is described by the differential equation $EI(z'' - z_0'') + Pz = 0$, in which derivatives with respect to spatial coordinate x are denoted by primes, P is the axial load, and I is the centroidal moment of inertia. Replacing E with the relaxation operator $\underline{E}$ we have

$$\underline{E}I(z'' - z_0'') + Pz = 0 \tag{9.2.1}$$

and upon substitution of the operator for the standard solid (Eq. 9.1.9):

$$\frac{1 + \tau_r(\partial/\partial t)}{(1/E_\infty) + (\tau_r/E)(\partial/\partial t)}(z'' - z_0'') + \frac{P}{I}z = 0 \tag{9.2.2}$$

In view of the analogy with the rules of linear algebra, we now multiply Equation 9.2.2 by the operator in the denominator:

$$\left(1 + \tau_r\frac{\partial}{\partial t}\right)(z'' - z_0'') - \frac{P}{I}\left[\frac{1}{E_\infty} + \frac{\tau_r}{E}\left(\frac{\partial}{\partial t}\right)\right]z = 0 \tag{9.2.3}$$

or

$$z'' - z_0'' + \tau_r\dot{z}'' + \frac{P}{E_\infty I}z + \frac{P}{EI}\tau_r\dot{z} = 0 \tag{9.2.4}$$

This is a partial differential equation in space and time, which governs the time-dependent buckling of a viscoelastic pin-ended imperfect column. The boundary conditions are $z = 0$ at $x = 0$ and $x = l$. The initial condition is given by the instantaneous elastic deformation calculated on the basis of elastic modulus E at time $t = 0$.

Deflection History and Long-Time Critical Load

Although the solution can be obtained in general for a Fourier series expansion of the initial imperfection $z_0(x)$, we consider for the sake of simplicity only a sinusoidal initial curvature, $z_0 = a \sin(\pi x/l)$ with $a = \text{constant}$, and seek the solution in the form:

$$z = f(t) a \sin \frac{\pi x}{l} \tag{9.2.5}$$

which agrees with the initial imperfection when $f = 1$. This form of the solution obviously satisfies the boundary conditions at the simply supported ends of the column. Substitution into Equation 9.2.4 yields

$$\tau_r\left(1 - \frac{P}{P_{\mathrm{E}}}\right)\frac{df}{dt} + \left(1 - \frac{P}{P_{\mathrm{E}_\infty}}\right)f = 1 \tag{9.2.6}$$

in which we introduced the notations:

$$P_{\mathrm{E}} = \frac{\pi^2}{l^2} EI \qquad P_{\mathrm{E}_\infty} = \frac{\pi^2}{l^2} E_\infty I \tag{9.2.7}$$

P_{E} is the elastic Euler load based on modulus E, and P_{E_∞} may be considered as the Euler load corresponding to modulus E_∞.

The initial elastic deflection for a sinusoidally curved column is obtained by multiplying the imperfection ordinates with a magnification factor from Equation 1.5.10, and so the initial condition is

$$\text{For } t = 0: \qquad f = \frac{1}{1 - P/P_{\mathrm{E}}} \tag{9.2.8}$$

Equation 9.2.6 is a linear ordinary differential equation with constant coefficients. Its solution for the given initial condition is

$$f(t) = f_\infty(1 - e^{-t/\tau_b}) + \frac{1}{1 - P/P_{\mathrm{E}}} e^{-t/\tau_b} \tag{9.2.9}$$

in which we introduced the notation:

$$\tau_b = \tau_r \frac{1 - P/P_{\mathrm{E}}}{1 - P/P_{\mathrm{E}_\infty}} \qquad f_\infty = \frac{1}{1 - P/P_{\mathrm{E}_\infty}} \tag{9.2.10}$$

Let us now discuss this result. For $P < P_{\mathrm{E}_\infty}$, the solution according to Equation 9.2.9 indicates a gradual increase of the deflection ratio $f(t)$ that approaches for $t \to \infty$ a final value f_∞ (Fig. 9.5b). If the initial imperfection is very small, that is, a is very small, then the final deflection at midspan, af_∞, is also very small. According to the general notion of stability, the column therefore must be considered stable.

When $P = P_{\mathrm{E}_\infty}$, we have $f_\infty \to \infty$, as well as $\tau_b \to \infty$. We therefore must calculate the limit of the solution in Eq. 9.2.9. Since $1/\tau_b \to 0$, the exponent t/τ_b of the exponential is infinitely small, and so we may either set $e^{-t/\tau_b} \simeq 1 - t/\tau_b$, and calculate the limit as follows:

$$\lim_{P \to P_{E_\infty}} f(t) = \lim_{P \to P_{E_\infty}} \left(f_\infty \frac{t}{\tau_b} + \frac{1}{1 - P/P_E} e^{-t/\tau_b} \right)$$

$$= \lim_{P \to P_{E_\infty}} \left[\frac{(1 - P/P_{E_\infty})t}{(1 - P/P_{E_\infty})^2 \tau_r} + \frac{1}{1 - P/P_E} \right] = \frac{1}{1 - P/P_E} \left(1 + \frac{t}{\tau_r} \right) \quad (9.2.11)$$

or get the same result using the rule of L'Hôpital. We also get the same result by setting $P = P_{E_\infty}$ in Eq. 9.2.6.

We see (Fig. 9.5b) that the deflection increases linearly with time. After a sufficient load duration, the deflection obviously will exceed any chosen bound no matter how small the initial imperfection a is ($a > 0$). According to the general notion of stability, this behavior must be considered as unstable.

When $P_{E_\infty} < P < P_E$ we have $f_\infty < 0$, $\tau_b < 0$, and $1 - e^{-t/\tau_b} < 0$. Thus, the exponent of the exponential functions in Equation 9.2.9 is positive, and so the exponential grows beyond any specified bound (Fig. 9.5b) no matter how small the initial imperfection a is ($a > 0$). This behavior is obviously also unstable.

When $P \geq P_E$, the column loses stability instantly, at $t = 0$, as we already know (Chaps. 1–3), and so it makes no sense to investigate long-time deflections.

P_{E_∞} is called the long-time critical load of the column, while P_E, calculated for the instantaneous loading at $t = 0$, is now called the instantaneous critical load. The fact that the long-time critical load of a viscoelastic column is always less than the instantaneous (elastic) critical load may be intuitively understood by realizing that the mechanism of viscoelastic buckling has a feedback due to the increasing arm of the axial force. As the deflection increases, the bending moment due to axial load increases even if the axial load is constant; this in turn causes an increase of creep, which causes a further increase of deflection, which causes an increase of bending moment, etc.

The same result can be obtained if the initial imperfection is considered as an inevitable eccentricity at the column end, or as a small lateral disturbing load $p(x)$ applied on a perfect column. Similar results are obtained for all types of columns as well as other structures. See Freudenthal (1950), Rosenthal and Baer (1951), Hilton (1952), Lin (1953, 1956), Kempner (1954), and others.

The Concept of Stability for Viscoelastic Structures

As a special case of the foregoing solution, we may note that for a perfect column ($a = 0$) no deflections are possible (that is, $z = 0$) except when $P = P_E$. This means that neutral equilibrium does not exist for $P < P_E$, that is, the long-time critical load cannot be obtained by analysis of neutral equilibrium (that is, adjacent equilibrium states or bifurcation of equilibrium path); see, for example, Kempner (1962). This is generally true for all kinds of viscoelastic structures.

In the foregoing analysis we have applied the general definition of stability in a somewhat generalized sense. As stated in Chapter 3 (Sec. 3.5), the fundamental dynamic definition of stability according to Liapunov considers a change in the solution caused by a change in the initial conditions. Instead of that we have compared the solution for an imperfect column with that for a perfect column and considered as unstable the situations where even for an arbitrarily small (nonzero) imperfection the difference between these two solutions for $t \to \infty$ can become finite (larger than any given positive number δ). This represents a slight generalization of Liapunov's stability definition, in which we consider as unstable all the situations in which an arbitrarily small change in some parameters of the problem, such as the imperfection amplitude a, can cause a finite change in the response. This is called the parametric concept of stability (see Leipholz, 1970).

Can the critical load be obtained from the general dynamic definition of stability according to Liapunov (Sec. 3.5)? Yes, it can, and the dynamic approach is the only one where creep buckling analysis can be limited to perfect columns only. Unlike the dynamic approach, the static approach does not work for perfect columns. The reason is that creep buckling deflections cannot occur unless there already is a bending moment. A bending moment may be created either by an imperfection or by a dynamic disturbance. For the purpose of dynamic analysis of a perfect column, one may consider imparting to the column at $t = 0$ a certain initial velocity (characterized by $\dot{a}$); see, for example, Dost and Glockner (1982). This induces oscillations (if $P < P_E$). As soon as the deflections become nonzero, creep buckling deflections due to bending moment from the axial load are superposed on these oscillations. Eventually, after all the kinetic energy dissipates due to creep and the oscillations cease, a finite creep buckling deflection has already been accumulated. This residual creep deflection plays the role of an initial imperfection (Hoff, 1958) in that it gives rise to a bending moment due to P, which causes the deflection to grow further, as a result of creep. Thus we can see that, after the motion that follows a dynamic disturbance of a perfect column comes to a standstill, the problem becomes the same as creep buckling of an imperfect column. So the general dynamic concept of stability leads to the same conclusion as our imperfection approach. The dynamic analysis of a perfect column, however, is much more complicated than the static analysis of an imperfect column.

Stability of a column in its asymptotic equilibrium state at $t \to \infty$ and $P < P_{E_\infty}$ can be proven on the basis of the Lagrange–Dirichlet theorem (Sec. 3.6), which applies even in the presence of dissipative forces if the potential energy exists. The energy dissipation rate, which is $D = \int_V \sum_i \eta_i \dot{\varepsilon}_i^2 \, dV$ (where η_i, ε_i = viscosity and strain in dashpot i), vanishes for infinitely slow ($\dot{\varepsilon}_i \to 0$) deviations from the final equilibrium state at $t \to \infty$. For infinitely slow deviations, there exists a potential energy given by Equation 4.3.8 in which E is replaced by E_∞. This energy is positive definite if $P < P_{E_\infty}$, and nonpositive definite if $P \geq P_{E_\infty}$, which means that $P < P_{E_\infty}$ is the condition of stability of the final asymptotic state. This final asymptotic state is a state of thermodynamic equilibrium, and so the question of stability of this state can also be answered from thermodynamics using the condition that there exists no small deviation from the final asymptotic state, even a very slow one, for which the (internally produced) entropy would increase (cf. Sec. 10.1). If some such infinitely slow deviation exists, then such a deviation must occur spontaneously, which means the final state is unstable (even though the motion away from this state is not necessarily dynamic, but infinitely slow).

If the column is imperfect, the deflection rate due to creep is nonzero at any finite time t. Thermodynamics (i.e., the classical equilibrium thermodynamics) does not apply to such states, which are in the realm of irreversible thermodynamics since the energy dissipation rate is finite. Stability of the entire solution $P(t)$ is proven, as already mentioned, on the basis of the general dynamic definition of Liapunov.

If, however, the column is perfect and axially undeformable (i.e., deforms only by bending), then $w(x, t) = 0$ (for all t and all x) is a possible solution, and so a constant displacement state exists at any t. The infinitely slow deviations from such states are the same as elastic deviations for modulus E_∞ instead of E. Thus it follows from the Lagrange–Dirichlet theorem that a perfect axially undeformable column is in a stable equilibrium state at any time t if $P < P_{E_\infty}$ (and in an unstable state if $P \geq P_{E_\infty}$).

Extensions and Ramifications

It can be shown that the stability condition $P < P_{E_\infty}$ applies generally for any viscoelastic material that represents a solid, characterized by long-time elastic modulus E_∞ (e.g., Distéfano, 1961, 1965). This may be verified, for example, by carrying out the solution (e.g. in terms of the Laplace transform) for the Maxwell chain or the Kelvin chain (Fig. 9.3g, h).

A solution based on the integral-type stress–strain relation (Eq. 9.1.1) utilizing the compliance function $J(t - t')$, is completely general. The governing equation for the buckling of a hinged column may then be obtained most easily by recalling that the elastic buckling equation can be written in the form $z'' + (P/I)E^{-1}z = z_0''$, and then replacing E^{-1} by the integral-type creep operator $\underset{\sim}{E}^{-1}$ defined by writing Equation 9.1.1 in the form $\varepsilon(t) = \underset{\sim}{E}^{-1}\sigma(t)$. This replacement, which is a manifestation of the elastic–viscoelastic analogy, yields

$$\frac{\partial^2 z(x, t)}{\partial x^2} + \frac{P}{I}\int_0^t J(t - t')\frac{dz(x, t')}{dt'}\,dt' = \frac{d^2 z_0(x)}{\partial x^2} \tag{9.2.12}$$

This is a Volterra-type integro-differential equation. It can be solved numerically in small finite time steps. To be able to integrate to infinity and decide whether the asymptotic deflection is finite, one may substitute a new variable such as $\tau = t/(\lambda + t)$, where λ = positive constant, and then use time steps $\Delta\tau$ instead of Δt. Alternatively, recognizing that the integral in Equation 9.2.12 is a convolution integral, one may apply the Laplace transform to reduce Equation 9.2.12 to an ordinary differential equation in x. After solving this equation for the appropriate boundary conditions, one may obtain the solution by inverse Laplace transformation. The Laplace transform can also be applied to viscoelastic buckling of structures other than columns.

This nature of the stability problem is different when the viscoelastic material is a fluid. In that case the compliance function $J(t - t')$ is unbounded (and the spring-dashpot model possesses no continuous connection from one end to the other exclusively through springs). Then the long-time asymptotic deflection of an imperfect column is found to be infinite for any $P > 0$, and the long-time critical load is $P_{cr_\infty} = 0$. However, the asymptotic critical load for $t \rightarrow \infty$ is not a meaningful measure of column capacity if the load durations that yield unacceptably large creep deflections exceed the practical lifetime of the structure, such as 50 years. For such behavior, typical, for example, of concrete columns, one needs to base the analysis of long-time carrying capacity either on the magnitude of deflections at the end of the lifetime (Bažant, 1965, 1968a) or consider an overload and carry out a nonlinear solution (Sec. 9.6).

Viscoelastic buckling becomes more complicated when the column or structure is inhomogeneous in its creep properties. For example, in a reinforced or composite column, one part of the cross section may exhibit creep while the other does not. In these problems there is a redistribution of internal stresses throughout the cross section, occurring simultaneously with the creep deflections. Some elementary problems of this type can be solved exactly (see Sec. 9.4), while generally a computer analysis by finite elements is preferable.

As a simple practical conclusion, we note that the stability limits for long-time behavior of columns made of viscoelastic solids are the same as the elastic stability limits when the elastic modulus is replaced by the long-time modulus E_∞ (or, more generally, the effective modulus; see Sec. 9.4). This conclusion applies generally, not just for the pin-ended column, but also for columns with any supports, frames, arches, thin-walled beams, plates, shells, etc. The method of replacement of E with E_∞ is also applicable to columns of composite cross section, for example, reinforced columns.

Long-time stability is one reason why the building code for concrete prescribes for the determination of carrying capacity of columns a much smaller bending stiffness EI than the true short-time stiffness. For example, American Concrete Institute (ACI) Standard 318 divides the short-time stiffness by factor $(1+\beta_d)$ where β_d accounts in a rather approximate manner for the creep effect (see Secs. 8.4 and 9.4).

Viscoelastic buckling is important for determining the force exerted by a floating ocean ice sheet moving against an obstacle, such as a stationary oil platform, or a river ice sheet moving against a bridge pier. See, for example, Hui (1986), who simplified ice as a linearly viscoelastic Maxwell solid, which enabled him to generalize an elastic buckling solution according to the elastic–viscoelastic analogy (but, in reality, the creep of ice is strongly nonlinear).

Problems

9.2.1 Solve the long-time critical load of the rigid bar hinged at the bottom and supported at the top by a Kelvin (Voigt) unit as depicted in Figure 9.6a. Let α be the initial imperfection. *Hint*: Relate the strain in the link to the rotation θ and write the moment equilibrium condition.

9.2.2 Solve the same as Problem 9.2.1, but with a support that behaves according to (a) the Maxwell unit, or (b) the standard solid model (Fig. 9.6b).

9.2.3 Solve the deflection and long-time critical loads for columns in Figure 9.6c–h in which the springs are elastic unless labeled viscoelastic. In that case, assume the standard solid model.

9.2.4 Using the elastic–viscoelastic analogy, calculate the deflection histories for (a) axial-torsional buckling, (b) lateral buckling of viscoelastic beams, and analyze long-time stability.

9.2.5 Do the same as Problem 9.2.4 for (a) a viscoelastic beam on an elastic foundation, (b) an elastic beam on a viscoelastic foundation, (c) a von Mises truss loaded through a spring, (d) spatial twist buckling of a shaft as in Section 1.10, (e) a simply supported square plate compressed uniformly in two directions, and (f) a portal frame as in Section 2.2.

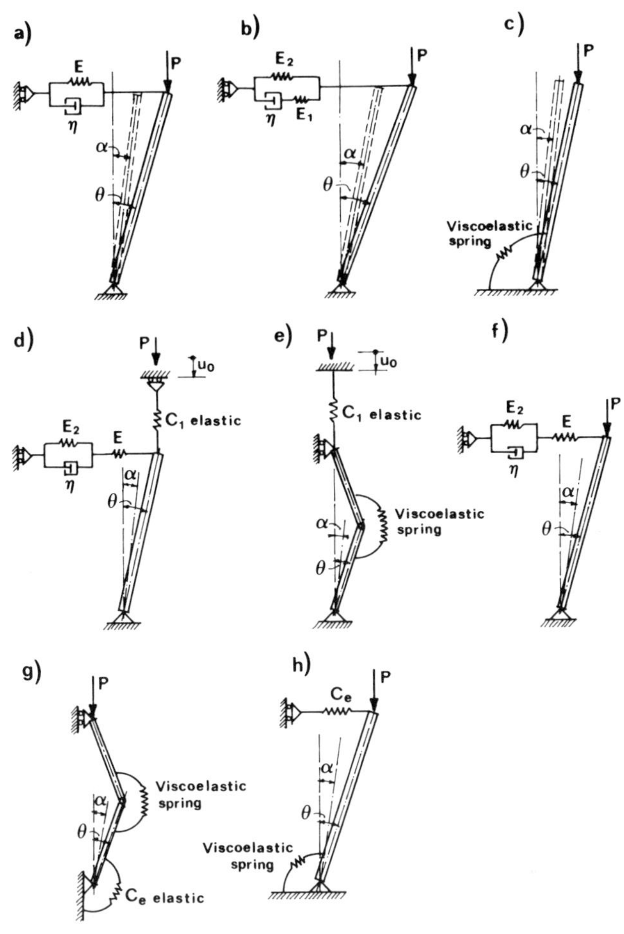

Figure 9.6 Exercise problems on viscoelastic buckling.

9.2.6 Use the Laplace transform to solve $z(x, t)$ from Equation 9.2.12 with z according to Equation 9.2.5. Consider (a) $J(t-t') = E^{-1} + c(t-t')^{1/2}$, or (b) $J(t-t') = E^{-1} + ct/(\lambda + t)$; c, λ = constants.

9.2.7 Express the potential energy of a perfect free-standing viscoelastic column for infinitely slow small deviations from its undeflected state, assuming that $E_\infty > 0$. Discuss stability.

9.2.8 Generalize equations 9.2.5–9.2.10 considering arbitrary $z_0(x)$ given as a Fourier series.

9.3 VISCOPLASTIC BUCKLING

When viscoplastic materials are loaded to high stress levels, their stress–strain relation ceases to obey the principle of superposition and becomes markedly nonlinear. For other materials, such as metals at high temperature, the stress–strain relation is strongly nonlinear for the entire stress range. We then speak of viscoplastic behavior, also called nonlinear creep. The behavior of

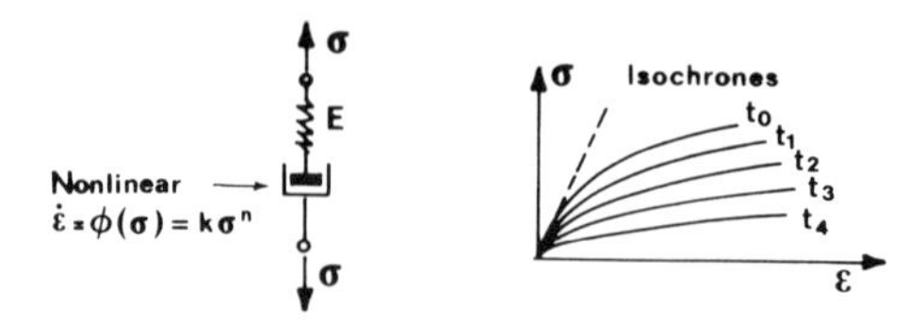

Figure 9.7 Nonlinear creep model.

various metals can be reasonably well approximated by the uniaxial viscoplastic stress–strain relation

$$\dot{\varepsilon} = \frac{\dot{\sigma}}{E} + \phi(\sigma) \tag{9.3.1}$$

in which $\phi(\sigma)$ is a positive monotonic nondimensional function of σ whose slope increases as the magnitude of σ increases; for example, $\phi(\sigma) = k\sigma^n$. The creep isochrones for this stress–strain relation are shown in Figure 9.7. As we will see, the nonlinearity of the creep law can greatly accelerate the growth of deflections of an imperfect column and cause the deflection to tend to infinity at a finite load duration. This phenomenon cannot occur in the small-deflection theory of linear viscoelastic buckling.

Note that Equation 9.3.1 can be visualized by the Maxwell model (Fig. 9.3c or Fig. 9.7) for which the spring is linearly elastic but the dashpot is nonlinearly viscous. More complicated arrangements of springs and nonlinear dashpots can be used to obtain other viscoplastic stress–strain relations.

Viscoplastic buckling was analyzed by Kempner (1954), Hoff (1954) (who developed the imperfection approach), Libove (1952, 1953), Higgins (1952), Lin (1956), Gerard and Papinno (who in 1963 proposed the critical deflection approach), and Rabotnov (1957, 1969) (who proposed the dynamic approach); see also the survey by Hoff (1958). Creep buckling in snapthrough problems was discussed by Boyle and Spence (1983), and so was creep buckling of thin-wall structures (Gerdeen and Sazawal, 1974).

Rigid-Bar Model Column

As a simple illustration of nonlinear creep buckling, consider the column consisting of a rigid bar of length l supported on a hinge at the bottom and loaded by a vertical load on top, as shown in Figure 9.8a. The column is held upright by a horizontal link of length a and cross section area A, located at height b above the hinge. The material behavior of the link is given by Equation 9.3.1. The column is imperfect, with an imperfection consisting of an initial inclination α. The rotation of the column, θ, as well as angle α, is assumed to be very small. From geometry, the strain rate in the horizontal link is $\dot{\varepsilon} = \dot{\theta}b/a$, and from equilibrium the stress in the horizontal link is $\sigma = F/A = PL\theta/bA$. Substituting these expressions into the stress–strain relation in Equation 9.3.1 and solving the

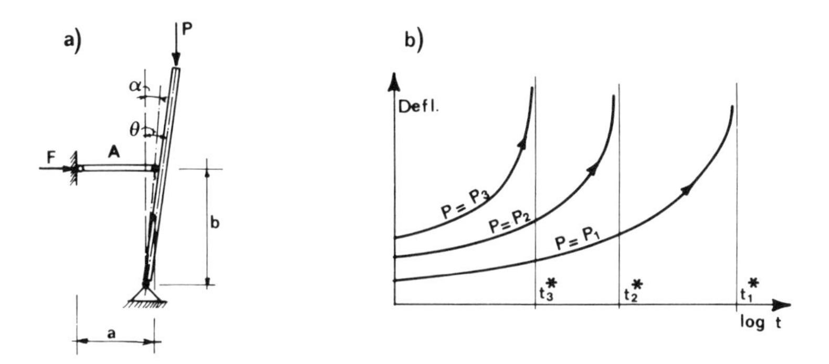

Figure 9.8 (a) Rigid-bar column with horizontal viscoplastic link and (b) deflection histories with vertical asymptotes (critical times).

resulting equation for $\dot{\theta}$, we obtain

$$\dot{\theta} = \frac{\phi(PL\theta/bA)}{(b/a) - (PL/bEA)} = \frac{a}{b}\left\{\frac{\phi[(PL/bA)\theta]}{1 - P/P_{cr}}\right\} \tag{9.3.2}$$

in which $P_{cr} = EAb^2/aL$ = instantaneous critical load of the column. Equation 9.3.2 is an ordinary nonlinear differential equation that can be easily solved by separation of variables. The initial condition is given by the instantaneous deflection of the imperfect column, which may be calculated elastically;

$$\text{For } t = 0: \qquad \theta = \theta_0 = \frac{\alpha}{1 - P/P_{cr}} \tag{9.3.3}$$

Critical Time and Stability Concept

Consider now that $\sigma > 0$ in the link, $\phi(\sigma) = k\sigma^n$, where k and n are material constants, $n \geq 1$ (coefficient k for metals is strongly temperature-dependent). Denoting $c = (ak/b)(PL/bA)^n(1 - P/P_{cr})^{-1}$ = constant, Equation 9.3.2 may be written as $d\theta/\theta^n = c\,dt$. Integration yields the inclination angle θ as a function of time t:

$$\theta = \left[\frac{1}{\theta_0^{n-1}} - (n-1)ct\right]^{-1/(n-1)} \tag{9.3.4}$$

We see that, for $n > 1$, the deflection θ tends to infinity as $t \rightarrow t^*$, where

$$t^* = \frac{1}{c(n-1)\alpha^{n-1}}\left(1 - \frac{P}{P_{cr}}\right)^{n-1} \tag{9.3.5}$$

This value is called the critical time. The curves of deflection versus time are plotted for various values of the vertical load in Figure 9.8b where the critical times correspond to the vertical asymptotes. Note that the higher the load P, the shorter the critical time. Also, the higher the exponent n of the creep law (i.e., the stronger the nonlinearity), the shorter is the critical time. On the other hand, for $n \to 1$, we have $t^* \to \infty$. This is the case of linear viscoelasticity, for which we have already demonstrated, in the previous section, that the deflections can approach infinity only at infinite time. For the linear case ($n = 1$), the solution of the differential equation is $\theta = \alpha e^{ct}(1 - P/P_{cr})^{-1}$. We see that the deflection tends to infinity at $t \to \infty$ for any positive value of load P. This is because the stress-strain relation in Equation 9.3.1 describes a fluid. (There is no bound on strain ε.) This again confirms that a column made of a linearly viscoelastic material that is a fluid has a zero long-time critical load.

For nonlinear creep buckling, the design of a column is usually based on the critical time. The critical time obviously must be sufficiently larger than the design lifespan of the structure.

For normal environmental temperatures, steel does not creep, but it does creep at high temperatures. Therefore, viscoplastic buckling of metals is primarily of interest for machines operating at high temperatures. It is also important for predicting performance of a steel building structure in a fire. When a steel column is sufficiently heated by the fire, it may creep, and its critical time must obviously be sufficiently larger than the maximum possible duration of the fire. The critical time can be made sufficiently large by slowing the rise of temperatures through thermal insulation.

Now how does the general definition of stability of solutions apply to viscoplastic buckling? When $\alpha = 0$ (perfect column) and $P < P_{cr}$, the only solution of Equation 9.3.2 is $\theta = 0$. But when parameter α is positive, no matter how small (imperfect column), θ becomes infinite at a certain time $t^* > 0$ (if $P > 0$). Thus, an arbitrarily small change in a parameter of the problem causes an infinitely large change in the response. Therefore, in accordance with the general definition of stability of solutions from Section 3.5, generalized for parametric instability (as explained in Sec. 9.2), we must conclude that a viscoplastic column is unstable for any load. Obviously, we face here a situation for which the general stability definition that considers the solutions up to infinite times does not yield practically useful information. However, modifying the stability definition by stipulating that the effect of α on the deflections is considered only for times that are less than the design lifetime t_L of the structure, we find that the deflection becomes infinite only if the load exceeds a certain finite value. In practice it is further necessary to require that the deflections at the end of the lifetime remain within certain reasonable limits. So buckling of viscoplastic structures should, in general, be approached through deformation limitations and lifetime requirements, rather than strictly according to the general stability definition.

Real Columns

An approximate solution can also be easily obtained for a flexible viscoplastic column. We assume that the pin-ended column shown in Figure 9.9a has an initial sinusoidal imperfection $z_0(x, t) = a \sin(\pi x/l)$ where a = small positive constant. We seek a solution of the form

$$z(x, t) = f(t) a \sin \frac{\pi x}{l} \tag{9.3.6}$$

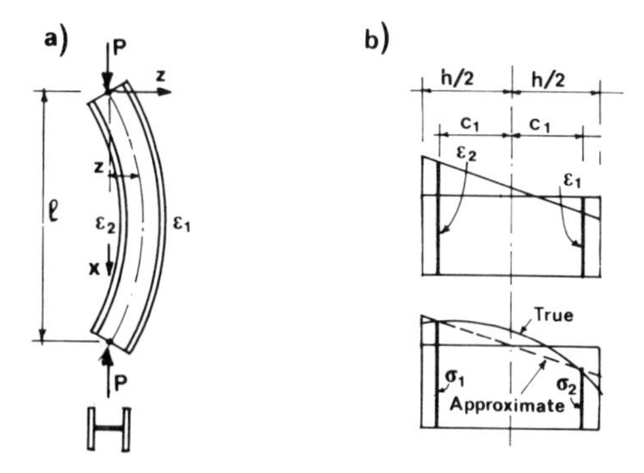

Figure 9.9 (a) Pin-ended column and (b) assumed strain and stress distributions over cross section.

where $f(t)$ is an unknown function to be found. Equation 9.3.6 satisfies the boundary conditions of a pin-ended column. The stress–strain relation is described by Equation 9.3.1. Evidently, the stress distribution throughout the cross section is generally nonlinear; however, we approximate it by a linear distribution characterized by stress values σ_1 and σ_2 at distances c_1 from the center line of the column; see Figure 9.9b. For the cross section at the midheight of the column, these stresses may be calculated as

$$\left.\begin{matrix}\sigma_1\\ \sigma_2\end{matrix}\right\} = -\frac{P}{A} \pm \frac{Paf}{I} c_1 \tag{9.3.7}$$

Substituting these values into the viscoplastic stress–strain relation, Equation 9.3.1, we may calculate the difference in strain rates:

$$\dot{\varepsilon}_1 - \dot{\varepsilon}_2 = \frac{P2c_1}{EI} a\dot{f} + \phi\left(-\frac{P}{A} + \frac{Pc_1}{I} af\right) - \phi\left(-\frac{P}{A} - \frac{Pc_1}{I} af\right) \tag{9.3.8}$$

Now, the difference can also be expressed from geometry of the deformation. For small deflections, and according to the assumption of plane cross sections, we have $\dot{\varepsilon}_1 - \dot{\varepsilon}_2 = 2c_1 \dot{z}_{,xx} = 2c_1 a\dot{f}\pi^2/l^2$. Substituting this into Equation 9.3.8, we obtain the following differential equation for the deflection ratio $f(t)$:

$$\dot{f} = \frac{l^2}{\pi^2 a 2c_1}\left(1 - \frac{P}{P_E}\right)^{-1}\left[\phi\left(-\frac{P}{A} + \frac{Pc_1 a}{I} f\right) - \phi\left(-\frac{P}{A} - \frac{Pc_1 a}{I} f\right)\right] \tag{9.3.9}$$

The initial condition for the instantaneous elastic deflection at $t = 0$ is $f = f_0 = (1 - P/P_E)^{-1}$. Equation 9.3.9 is an ordinary nonlinear first-order differential equation of the general form $\dot{f} = F(f; P)$, which is the standard form. Such a differential equation can generally be solved step-by-step, for example, by the Euler method or, more accurately, the Runge–Kutta method. Integration yields the curves of deflection versus time for various values of load P; they have similar shapes as those in Figure 9.8b. Their characteristic feature is again the existence of a critical time t^* depending on P, for which the deflection becomes infinite.

The foregoing approximate solution is an example of the collocation method. The differential equation is not satisfied for all x, but only for some discrete values of x, in the present case only for $x = l/2$. The stress distribution is collocated at two points, $z = \mp c_1$. For the special case of the idealized I-cross section (massless web), one may set $c_1 = h/2$, and the stress distribution is then represented exactly. The substitution $c_1 = h/2$ also can be made for other cross sections, but then the solution, which is only approximate, lacks optimum accuracy. A better result is achieved if c_1 is selected so that the moments of positive and negative areas between the true curved stress distribution and the assumed straight-line distribution are approximately equal, as sketched in Figure 9.9b.

More accurate solutions can, of course, be obtained by the finite element method. This approach is needed when the cross section is nonhomogeneous or composite, or when viscoplastic buckling of more complicated structures (frames, plates, shells, etc.) is to be analyzed. Nevertheless the simple solutions that we demonstrated provide valuable insight, not easily gained by finite element analysis.

Problems

9.3.1 Solve the problem defined by Figure 9.8a for the following expressions of function ϕ: (a) $\phi(\sigma) = \exp(k\sigma - 1)$ and (b) $\phi(\sigma) = \sinh k\sigma$.

9.3.2 Solve the problem in Figure 9.6a, assuming for the dashpot a nonlinear response, given by $\dot{\varepsilon} = k\sigma^n$. (*Note*: This case possesses a finite long-time critical load, and only above the long-time critical load one gets a critical time.)

9.3.3 Do the same as Problem 9.3.2, but for Figure 9.2b.

9.3.4 Solve the problem of the wide-flange column in Fig. 9.9a (Eq. 9.3.9) by forward integration for several time steps, using the Euler method or the Runge–Kutta method.

9.3.5 Assuming an ideal I-cross section, analyze (a) a free-standing column and (b) a fixed-end column.

9.3.6 Do the problems in Figure 9.10(a, b).

9.3.7 Do the problems in Figure 9.10(c, d, e), assuming an ideal I-beam cross section. Assume a suitable one-parameter deflection shape.

9.3.8 Analyze viscoplastic buckling of a square plate uniformly compressed in two directions. (*Hint*: Same as for the wide-flange beam, collocate the differential equation only at the central point of the plate, and across the thickness collocate the stress–strain law at points $\pm c_1$.)

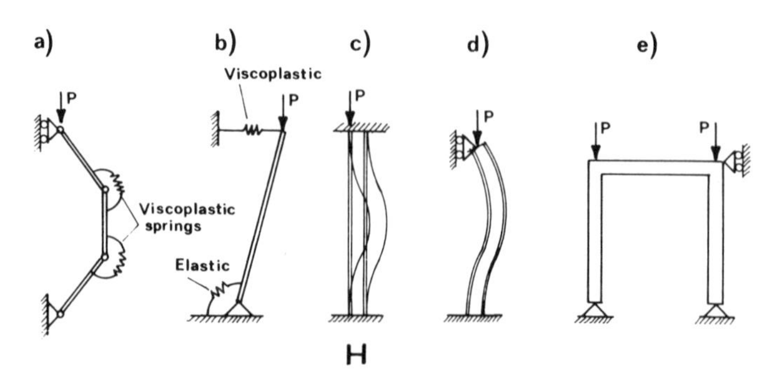

Figure 9.10 Exercise problems on viscoplastic buckling.

9.4 BUCKLING OF AGING VISCOELASTIC STRUCTURES

In the context of viscoelasticity, aging is understood as the age of dependence of material properties, such as the elastic moduli and viscosities of the springs and dashpots in a rheologic model. Aging is a characteristic property of the creep of concrete. It is caused by cement hydration, which becomes, after the initial hardening period, a very slow process advancing over many years. This process changes creep by an order of magnitude during a typical lifetime of a concrete structure, and so the aging phenomenon must be taken into account.

It is proper to distinguish a small and a large ratio of the live load to the dead load. The case of a small dead-to-live load ratio is the case when the stresses in the column under the dead load multiplied by its safety factor are within the service stress range. In this range, the creep law of concrete is approximately linear, which brings about a great simplification of the analysis. The failure of the column occurs in this case due to application of a sudden overload representing the live load multiplied by its safety factor. The effect of creep is to increase the deflections due to initial imperfections prior to the application of the live load. During the failure process caused by sudden overload (live load), creep plays no role and the analysis may be carried out in the usual time-independent manner, although the nonlinear behavior of concrete under the overload needs to be taken into account. The purpose of creep buckling analysis is to provide the initial conditions for the analysis of buckling due to the rapid overload. The increase of the column deflection due to creep prior to the overload must be taken into account in the analysis. Also, one may have to consider changes in the nonlinear properties caused by previous creep.

The case of large dead-to-live load ratio is the case when dead load multiplied by its safety factor puts the material of the column into the range of nonlinear creep. The creep analysis in this case must be nonlinear, which will be discussed in the next section.

Aging Maxwell Solid (Dischinger-Type Methods)

The simplest stress–strain relation to describe linear aging creep is the so-called rate-of-creep method, introduced by Glanville (1933) and Whitney (1932), in Germany also called the Dischinger formulation, after Dischinger (1937, 1939) who was first to use this approach extensively in structural calculations. In this case, the material is assumed to be characterized by an aging Maxwell model (see Fig. 9.3c), for which the stress–strain relation is

$$\dot{\varepsilon} = \frac{\dot{\sigma}}{E(t)} + \frac{\sigma}{\eta(t)} \tag{9.4.1}$$

$E(t)$ and $\eta(t)$ are the age-dependent elastic modulus and viscosity of the material. The compliance function $J(t, t_1)$, which represents the strain at age t caused by a unit constant stress applied at age t_1 (Sec. 9.1), is obtained by integrating Equation 9.4.1 for constant stress $\sigma = 1$, applied at age t_1:

$$J(t, t_1) = \varepsilon(t) = \frac{1}{E(t_1)} + \int_{t_1}^{t} \frac{dt'}{\eta(t')} = \frac{1 + \psi(t)}{E_1} \tag{9.4.2}$$

Here we introduced the notations $E_1 = E(t_1)$ and

$$\psi(t) = E(t_1) \int_{t_1}^{t} \frac{dt'}{\eta(t')} \tag{9.4.3}$$

Note that the compliance function $J(t, t_1)$ is not merely a function of a single variable, the time lag $(t - t_1)$ (Eq. 9.1.1), but depends now on two variables, the current age, t, and the age at loading, t_1. Function $\psi(t, t_1)$ indicates the ratio of the creep deformation to the initial elastic deformation, and is called the creep coefficient. The age dependence of the viscosity of the Maxwell model may be expressed as $\eta(t) = E(t_1)/\dot{\psi}(t)$, which becomes very large near the end of the lifetime (which makes the use of the Maxwell model possible even though in principle it represents a fluid). The age dependence of the elastic modulus is $E(t) = 1/J(t, t)$. According to Equation 9.4.2, $\psi(t) = E(t_1)J(t, t_1) - 1$. By virtue of Equation 9.4.3, Equation 9.4.1 may be written as $\dot{\varepsilon} = \dot{\sigma}/E(t) + [\sigma/E(t_1)]\dot{\psi}$. This may also be rewritten as

$$\frac{\partial \varepsilon}{\partial \psi} = \frac{1}{E(t)} \frac{\partial \sigma}{\partial \psi} + \frac{\sigma}{E_1} \tag{9.4.4}$$

because function $\psi(t, t_1)$ is monotonically increasing and may be taken as the independent variable, serving as a kind of reduced time.

The typical compliance function of concrete is shown in Figure 9.11a by the solid curves. We see that the creep coefficient function $\psi(t, t_1)$ (Eq. 9.4.2) can be fully determined on the basis of the specified or measured creep curve $J(t, t_1)$ where t_1 is the age at first loading, representing the age at which the load is applied on the column. From the definition of function $J(t, t_1)$ in Equation 9.4.2 we can further note that for other ages at loading, such as t_2 and t_3 (Fig. 9.11a), the creep curves $J(t, t_2)$, $J(t, t_3)$ are simply obtained by shifting the creep curve $J(t, t_1)$ vertically downward; see Figure 9.11a, in which these curves are shown as the dashed curves 5–9 and 10–11. It is important to note that the creep curves

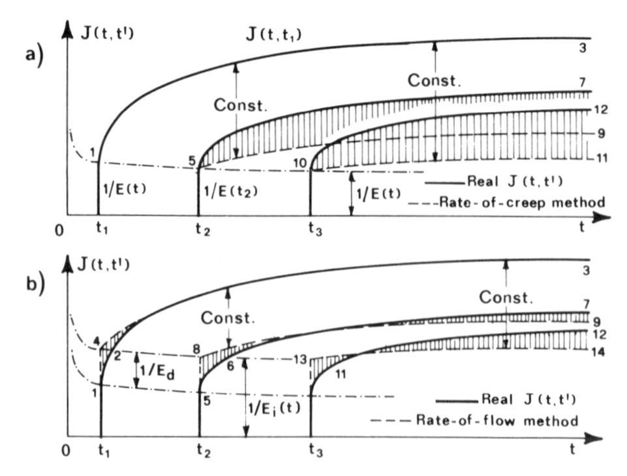

Figure 9.11 Creep curves for aging material (concrete): (a) rate-of-creep approximation (Dischinger formulation); (b) rate-of-flow approxmation (improved Dischinger formulation).

$J(t, t_2)$, $J(t, t_3)$ obtained in this manner substantially underestimate the creep for loads applied at ages much higher than t_1 (see the cross-hatched areas in Fig. 9.11a). This is an inevitable error in the rate-of-creep method.

The error of the rate-of-creep method can be substantially reduced by deliberately increasing the value of the instantaneous (elastic) deformation at the time of loading so that the aforementioned error is made smaller for longer ages at loading. Such a modification was in effect proposed by England and Illston (1965) under the name rate-of-flow method. This proposal was further improved by Nielsen (1970), and also by Rüsch, Jungwirth, and Hilsdorf (1973) under the name improved Dischinger method, which is currently embodied in the Comité Euro-International du Béton–Fédération Internationale de la Précontrainte (CEB–FIP, 1978) design recommendations. These authors assumed that (1) the additional initial instantaneous deformation represents the final value of the delayed elastic portion of creep (reversible creep) and (2) the final value of this delayed elastic deformation is age-independent; see deformation $1/E_d$ in Figure 9.11b. In the light of recent experimental evidence, however, the age-independence assumption for E_d does not appear to be very good (Bažant and Osman, 1975; Bažant, 1982; RILEM TC69, 1986). Nevertheless, as Figure 9.11 illustrates, by making the initial deformation $1/E_i(t)$ much larger than the actual initial elastic deformation, one considerably reduces the overall error, as shown by the cross-hatched areas in Figure 9.11b.

To formulate the differental equation for column buckling, it is convenient to apply the operator method. Equation 9.4.4 may be written as $\sigma = \underline{E}\varepsilon$, in which $\underline{E}$ is the following linear time-dependent differential operator:

$$\underline{E} = \frac{\partial / \partial \psi}{\dfrac{1}{E(t)} \dfrac{\partial}{\partial \psi} + \dfrac{1}{E_1}} \tag{9.4.5}$$

This type of operator obeys the same rules as linear algebra, except for the commutative law. The differential equation for aging creep may be obtained from the elastic differential equation of the beam-column by replacing the elastic modulus E with operator $\underline{E}$.

Deflections According to Aging Maxwell Model

Consider now a reinforced concrete column with a symmetric cross section (Fig. 9.12). Let I and I_s be the centroidal moments of inertia of the concrete and the steel bars, and let E_s be the elastic modulus of steel. The steel reinforcement does not creep (we exclude the case of very high temperatures, and we do not consider prestressed bars). The column is hinged and has an initial imperfection given by the initial shape $z_0(x)$. The elastic differential equation is $(E_s I_s + EI)(z'' - z_0'') + Pz = 0$. Now replacing E with the operator in Equation 9.4.5 we obtain

$$\left[E_s I_s + I \frac{\partial / \partial \psi}{\dfrac{1}{E(t)} \dfrac{\partial}{\partial \psi} + \dfrac{1}{E_1}} \right] \left(\frac{\partial^2 z}{\partial x^2} - \frac{\partial^2 z_0}{\partial x^2} \right) + Pz = 0 \tag{9.4.6}$$

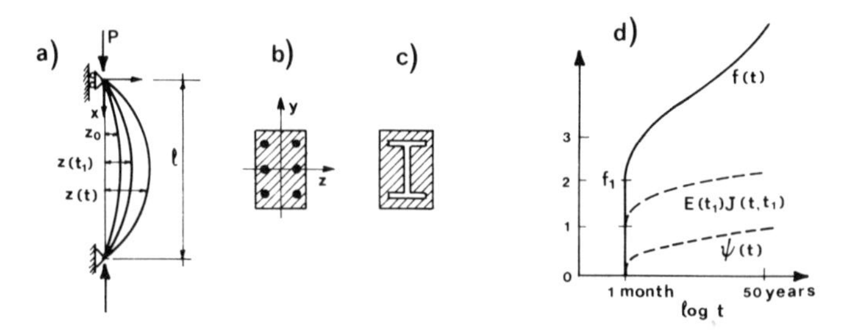

Figure 9.12 (a) Pin-ended imperfect column, (b) reinforced concrete or (c) composite cross sections, and (d) deflection history. (*After Bažant and Najjar, 1973; Bažant, 1968a.*)

Multiplying this equation with the differential operator in the denominator yields for the deflection ordinates $z(x, t)$ the following linear partial differential equation:

$$\left[\frac{1}{E(t)}\frac{\partial}{\partial\psi}+\frac{1}{E_1}\right]\left[E_sI_s\left(\frac{\partial^2 z}{\partial x^2}-\frac{\partial^2 z_0}{\partial x^2}\right)+Pz\right]+I\frac{\partial}{\partial\psi}\left(\frac{\partial^2 z}{\partial x^2}-\frac{\partial^2 z_0}{\partial x^2}\right)=0 \quad (9.4.7)$$

Although the solution can be obtained for arbitrary initial imperfection $z_0(x)$ expressed as a Fourier series, consider for the sake of simplicity that the initial curvature is sinusoidal. The sinusoidal shape may then be assumed also for the deflection ordinates, $z(x, t)$, and so

$$z_0(x)=a\sin\frac{\pi x}{l} \qquad z(x, t)=f(t)a\sin\frac{\pi x}{l} \quad (9.4.8)$$

in which $a=$ constant, $f(t)$ is a function of time t to be solved, and $l=$ length of the pin-ended column (Fig. 9.12). Substituting these equations into Equation 9.4.7, and introducing the notations $E(t)I\pi^2/l^2=P_{E_t}$, $E_sI_s\pi^2/l^2=P_{E_s}$, we get the following ordinary nonhomogeneous linear differential equation for function $f(t)$:

$$\frac{E_0}{E(t)}(P_{cr_t}-P)\frac{df}{d\psi}-(P-P_{E_s})f=P_{E_s} \quad (9.4.9)$$

in which $P_{cr_t}=P_{E_s}+P_{E_t}=$ instantaneous elastic critical load at time t.

Consider that the load P is applied at age t_1 and is constant afterward. Equation 9.4.9 has time-dependent coefficients due to the time variation of elastic modulus $E(t)$. However, this time variation is relatively small, much less than the variation of creep properties with age. For this reason, and also because for concrete the aging Maxwell model (rate-of-creep theory, Dischinger theory) introduces a relatively large error regardless of $E(t)$, we may replace $E(t)$ by a constant equal to the average value of $E(t)$ over the time interval of interest, (t_1, t), that is, we set $E_t=\frac{1}{2}[E_1+E(t)]$, and also $P_{E_t}=E_tI\pi^2/l^2$. The integral of Equation 9.4.9 then has the form $f=Ae^{-\lambda\psi}+f_p$ where f_p is the particular solution for which one finds $f_p=P_{E_s}/(P-P_{E_s})$; A, $\lambda=$ constants; λ is determined by substituting $f=e^{-\lambda\psi}$ into the homogeneous part of Equation 9.4.9 and A is found from the initial condition. The initial condition is represented by the instantaneous elastic deflection at the moment of application of the load P, which is given by $f(t_1)=f_1=(1-P/P_{cr_1})^{-1}$ in which $P_{cr_1}=P_{E_1}+P_{E_s}=$ instantaneous critical load of reinforced column at age t_1, and $P_{E_1} = E_1 I\pi^2/l^2$. Thus the initial condition ($t = t_1$, $\psi = 0$) requires that $f_1 = A + f_p$, and this then leads finally to the solution (Bažant, 1966):

$$f(t) = \left(\frac{P_{cr_1}}{P_{cr_1} - P} - \frac{P_{E_s}}{P - P_{E_s}}\right) \exp\left[\frac{P - P_{E_s}}{P_{cr_t} - P}\left(\frac{E_t}{E_1}\right)\psi(t)\right] + \frac{P_{E_s}}{P - P_{E_s}} \tag{9.4.10}$$

For weakly reinforced concrete columns, one can approximately neglect the steel, that is, set $E_s I_s / EI \to 0$, and then Equation 9.4.10 simplifies to

$$f(t) \simeq \frac{P_{E_1}}{P_{E_1} - P} \exp \frac{\psi(t)}{(P_{E_1}/P) - (E_1/E_t)} \tag{9.4.11}$$

Similar solutions were given by Warner and Kordina (1975) and Warner, Rangan, and Hall (1976).

If we would assume that $\psi \to \infty$, Equation 9.4.10 would indicate the long-time critical load for $t \to \infty$ (according to the linearized small-deflection theory) to be zero (even though the creep rate of concrete decays with time quite rapidly, no bound on the creep deformation is known to exist for concrete). However, within any reasonable lifetime such as 50 years, function $\psi(t)$ is bounded (and not larger than about 6), and so, according to Equation 9.4.10, the long-time critical load, defined as the load for which f becomes infinite within the lifetime, is equal to the initial instantaneous critical load of the entire cross section, P_{cr_1}. The fact that the Dischinger formulation gives no reduction of the critical load for the design lifetime is an unrealistic aspect of this formulation. It is inherent to the simplification in the creep model itself.

The bending moment M_s transmitted by the reinforcing steel is obtained from the curvature, which is $z_0''(f - 1)$. So the moment in steel, the total moment, and the moment in concrete are

$$M_s = E_s I_s z_0''[f(t) - 1] \qquad M = Pz = Pz_0 f(t) \qquad M_c = M - M_s \tag{9.4.12}$$

It is now interesting to note that M_s and M_c do now grow in proportion to M or $z(x, t)$. The portions of the total moment Pz carried by steel and by concrete, that is, the ratios M_s/M and M_c/M, vary in time. This redistribution of internal forces due to creep is a typical phenomenon in nonhomogeneous structures. Generally, the portion of the bending moment increases in the cross-section part that creeps less or does not creep at all, which is the steel reinforcement.

A typical solution of the deflection multiplier $f(t)$ versus the logarithm of age is plotted in Figure 9.12d (Bažant and Najjar, 1973; Bažant, 1968a).

Unfortunately, the aging Maxwell model on which the foregoing calculation is based is inherently incapable of describing the material behavior accurately. The aging Maxwell model is fully calibrated by matching one unit creep curve $J(t, t_0)$ for a single (chosen) age at loading t_0; but then the given or experimental curves $J(t, t')$ for $t' \neq t_0$ cannot be matched and are not even closely approximated. In the original Dischinger's version, in which the Maxwell model is calibrated according to the creep curve $J(t, t_1)$ for the age at loading t_1 equal to the age when the long-time load is applied on the structure, the creep due to the stress increments arising after t_1 is always underpredicted, and in consequence the long-time magnification factor f according to the original Dischinger's formulation (Eq. 9.4.11, and also Eq. 9.5.1) has generally an error on the low (unsafe) side.

In the so-called improved Dischinger version (England and Illston, 1965; Nielsen, 1970; Rüsch, Jungwirth, and Hilsdorf, 1973), in which the calibration of the aging Maxwell model in effect represents a compromise between the calibrations based on the creep curves for load applications at age t_0 and at a high age, the errors of the long-time prediction can be on either the low or the high side and are generally smaller. Nevertheless, they are still quite large (as is clear from Fig. 9.11b), and generally much larger than the errors of the age-adjusted effective modulus method. This method, which we will explain later, is also simpler to use. Thus, after the emergence of this method in the early 1970s, the foregoing solutions based on the aging Maxwell model lost most of their practical significance.

Deflection According to More Realistic Rheologic Models

A long-time critical load P_{cr_∞} that is less than P_{E_1} can be obtained by adopting more sophisticated constitutive relations. For example, a solution has been obtained (Distéfano, 1965; Bažant, 1965, 1966, 1968a, b) according to the standard solid rheologic model (Fig. 9.3e or f) in which both spring moduli E and E_2 as well as viscosity η depend on age t. The solution leads to a linear ordinary differential equation with a strongly time-dependent coefficient, which can be solved by means of the incomplete gamma function (Bažant, 1965, 1968a, b). This solution is more realistic than the preceding solution based on the aging Maxwell solid, but still the error can be substantial. This is because an accurate representation of concrete creep requires a Kelvin or Maxwell chain models (Figs. 9.3g, h) in which all the elastic moduli and viscosities depend on age. Such solutions can be obtained with great accuracy by step-by-step computer integration of the time history.

Deflection According to Effective Modulus

Another approximate solution method, whose error is generally not larger than that of the classical rate-of-creep method (aging Maxwell solid, with $E_1 = E$), is the effective modulus method. In this method, the creep strain is calculated assuming the stress to be constant from the beginning and equal to the current final stress $\sigma(t)$. Such an assumption would be totally unacceptable for a material whose creep curve is approximated by a power curve $(t - t')^n$ with large exponent n (about $n > 0.5$) but is not too bad for materials for which n is small, as is the case for concrete (for which $n \simeq 0.1$). The reason is that, for small n, a long portion of the stress history curve before time t is almost horizontal and the initial rise of stress toward this horizontal line is rather steep (Fig. 9.13). The stress–strain relation for the effective modulus method is assumed to be quasi-elastic, $\sigma(t) = E_{\text{eff}}\varepsilon(t)$, with $E_{\text{eff}} = 1/J(t, t_1) =$ effective modulus. The deflection multiplier for time t is then obtained according to the elastic magnification factor:

$$f(t) = \frac{1}{1 - P/P_{cr_{\text{eff}}}(t, t_1)} \qquad P_{cr_{\text{eff}}}(t, t_1) = (E_{\text{eff}}I + E_s I_s)\frac{\pi^2}{l^2} \tag{9.4.13}$$

in which $P_{cr_{\text{eff}}} =$ critical load of reinforced column based on the effective modulus

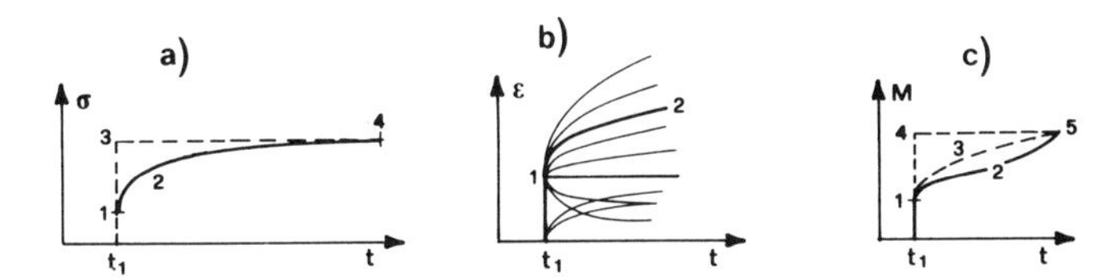

Figure 9.13 (a) Typical stress history and constant stress assumption of effective modulus method; (b) strain histories for which the age-adjusted effective modulus method is exact (Bažant, 1972a, b); (c) histories of bending moment for different values of load P.

for concrete. We see that according to this approximate solution, the long-time critical load is $P = P_{\mathrm{cr_{eff}}}$, a value which decreases with time.

The solution according to the effective modulus method is on the safe side. This is important since this method is widely used for concrete structures (it is implied in ACI Equations 8.5.7 and 8.5.11, in which factor $1 + \beta_d$ in effect gives the effective modulus). The solution according to the classical rate-of-creep method (with $E_1 = E$), on the other hand, always assumes less creep than in reality, as already mentioned, and so it represents a lower bound. Generally, z from Equation 9.4.10 for $E_1 = E$ is less than the actual z, which again is less than z from Equation 9.4.13.

Deflection According to Age-Adjusted Effective Modulus

There exists another, more recent method that gives a surprisingly accurate solution in most cases, yet is about as simple as the effective modulus method. It is the age-adjusted effective modulus method (Bažant, 1972b), which represents a refinement of Trost's method (1967) and is recently often called the Trost–Bažant method (e.g., Neville, Dilger, and Brooks, 1983, Ghali and Favre, 1986). This method is based on the following quasi-elastic stress–strain relation written for the entire time interval from the time of loading t_1 to the current time t:

$$\Delta\varepsilon = \frac{\Delta\sigma}{E^*} + \Delta\varepsilon^c \qquad \Delta\varepsilon^c = \frac{\sigma(t_1)}{E(t_1)}\,\phi \tag{9.4.14}$$

in which $\Delta\varepsilon = \varepsilon(t) - \varepsilon(t_1)$, $\Delta\sigma = \sigma(t) - \sigma(t_1)$, and

$$\phi = E(t_1)J(t, t_1) - 1 \qquad E^* = \frac{E(t_1) - R(t, t_1)}{\phi} \tag{9.4.15}$$

(for a proof see Appendix II to this section); $R(t, t_1)$ is the relaxation function of the material, which represents the stress at age t caused by a unit strain imposed at age t_1. This function can be calculated accurately and easily by a step-by-step numerical solution of the integral equation 9.4.19 (Bažant, 1972a, 1975). It can also be estimated with good accuracy from an approximate semiempirical formula (Eq. 9.4.24) developed by Bažant and Kim (1979). $\Delta\varepsilon^c$ plays the role of inelastic strain and represents the creep strain based on the initial stress $\sigma(t_1)$; E^* is called the age-adjusted effective modulus, and $\phi = \phi(t, t_1)$ = creep coefficient based on time t_1 of first loading = ratio of creep strain to the initial elastic strain [for the Maxwell model we denoted the creep coefficient as ψ since, due to the limitation of that model, it could be made equal to the actual creep coefficient $\phi(t, t_1)$ only for one value of t_1 used to calibrate that model].

The quasi-elastic stress–strain relation in Equation 9.4.14 is exact, according to the linear stress–strain relation based on compliance function $J(t, t')$, under the condition that the strains in the structure vary as a linear function of $J(t, t_1)$ [or $\phi(t, t_1)$]; this follows from Bažant's (1972b) theorem (see also RILEM TC69 State-of-Art Report, 1986, chap. 2; and Appendix II to this section). Such strain histories usually approximate quite closely the actual strain histories in structures, including the buckling of columns. They encompass all the stress histories sketched in Figure 9.13b. By constrast, the effective modulus method and the aging Maxwell model, either in the form of the rate-of-creep method or in the form of the improved Dischinger method, are exact only for one stress–strain history, namely that which corresponds to constant stress (curve 1–2 in Fig. 9.13b). This is the reason why Equation 9.4.14 is generally much more accurate, compared to the exact solutions according to linear aging viscoelasticity (this is not necessarily true in comparison to measurements because of various nonlinear effects, moisture effects, and temperature effects in real structures, which may sometimes offset the errors of the aging Maxwell model or the rate-of-creep method).

The age-adjusted effective modulus is frequently written in the form $E^* = E(t_1)/(1 + \chi\phi)$ in which $\chi = \chi(t, t_1) = [1 - R(t, t_1)/E(t_1)]^{-1} - \phi(t, t_1)^{-1}$; χ is called the aging coefficient (Bažant, 1972b) because in the absence of aging its value is almost exactly 1 for the typical form of $J(t, t')$ for concrete [the reason for this is that $J(t, t') \sim (t - t')^n$ where $n \ll 1$; it would not be true for n close to 1]. Always $\chi \leq 1$, and usually $\chi = 0.7$ to 0.9. As a crude mean estimate, one may use $\chi \simeq 0.8$, but for an accurate solution χ needs to be considered to be a function of t and t_1 (Bažant, 1972b; RILEM TC69 State-of-Art Report, chap. 2, 1986).

According to the quasi-elastic stress–strain relation in Equation 9.4.14, the change of the bending moment M_c in concrete from the time of loading t_1 to time t is $\Delta M_c = E^*I(\Delta z'' - \kappa_1\phi)$ in which $\kappa_1\phi = \Delta\kappa^c$ = change of curvature κ due to creep calculated as if M_c were constant and equal to the initial value $M_c(t_1)$; κ_1 is the initial elastic curvature change at time t_1 caused by instantaneously applied load P; $\kappa_1 = z_0''(f_1 - 1)$ in which $f_1 = f(t_1) = (1 - P/P_{\mathrm{cr}_1})^{-1}$, $P_{\mathrm{cr}_1} = P_{\mathrm{E}_1} + P_{\mathrm{E}_s}$ = initial elastic critical load of the reinforced column. Substituting these expressions into the equilibrium condition $\Delta M = \Delta M_c + E_sI_s\,\Delta z'' = -P\,\Delta z$ for a hinged column, we obtain the relation

$$\Delta M = E^*I\left(\Delta z'' - \frac{P}{P_{\mathrm{cr}_1} - P} z_0''\phi\right) + E_sI_s\,\Delta z'' = -P\,\Delta z \qquad (9.4.16)$$

Assuming the initial curvature to be sinusoidal, that is, $z_0 = a \sin(\pi x/l)$, we have $z = f(t)z_0$, $\Delta z = z(t) - z(t_1) = (f - f_1)z_0$, and so we obtain

$$E^*I\left[-\frac{\pi^2}{l^2}(f - f_1) + \frac{\pi^2}{l^2}\left(\frac{P}{P_{\mathrm{cr}_1} - P}\right)\phi\right] - E_sI_s\left(\frac{\pi^2}{l^2}\right)(f - f_1) = -P(f - f_1) \qquad (9.4.17)$$

Solving for f and substituting $f_1 = (1 - P/P_{\mathrm{cr}_1})^{-1}$, we finally get the result

$$f(t) = \frac{P_{\mathrm{cr}_1}}{P_{\mathrm{cr}_1} - P}\left[1 + \frac{P_{\mathrm{E}}^*\phi}{P_{\mathrm{cr}}^* - P}\left(\frac{P}{P_{\mathrm{cr}_1}}\right)\right] \qquad (9.4.18)$$

where $P_{\mathrm{E}}^* = E^*I\pi^2/l^2$ = critical load of the concrete part if it has modulus E^*, and $P_{\mathrm{cr}}^* = P_{\mathrm{E}}^* + P_{\mathrm{E}_s}$ = critical load of the reinforced column when concrete has modulus E^*.

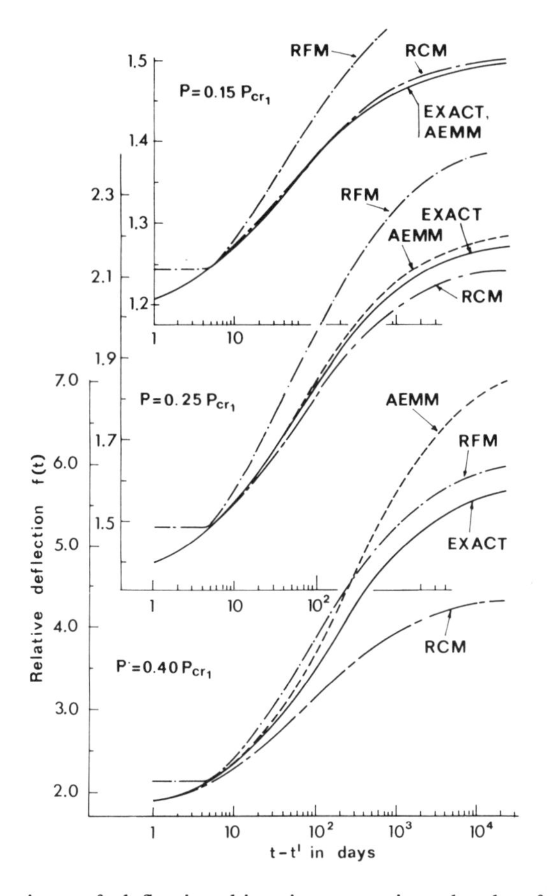

Figure 9.14 Comparison of deflection histories at various levels of axial load. (*After Bažant and Najjar, 1973.*) (AEMM = age-adjusted effective modulus method; RCM = rate-of-creep method, RFM = rate-of-flow method)

Note that this solution gives infinite deflection when $P = P^*_{cr}$, which is a value that is less than the initial instantaneous critical load P_{cr_1} but is larger than the critical load $P_{E_{eff}}$ based on the effective modulus. This is a realistic feature of this solution. For P close to P^*_{cr}, the solution has a large error, predicting deflections to be much higher than the actual ones. This is because, for high loads P, the history of bending moment M reverses curvature as shown by curve 1-2-5 in Figure 9.13c; this curve can be considerably different from curve 1-3-5 that is linearly dependent on the compliance function, as required for achieving good accuracy of this method. On the other hand, for small P such as $P \leq 0.3P_{cr_1}$, this solution is highly accurate compared to the exact solution according to aging linear viscoelasticity; this has been demonstrated by Bažant and Najjar (1973); see Figure 9.14.

An important advantage of the foregoing solution according to the age-adjusted effective modulus method is that the solution in time is algebraic, same as for the classical effective modulus method. No differential or integral equation in time needs to be solved. This makes it easy to apply this method for any type of buckling problem for which an elastic solution is available, including arbitrary columns, continuous beams and frames, lateral buckling, plate and shell buckling, etc. The long-time buckling behavior of these structures is similar to that we expounded for columns.

Deflection According to Integral-Type Stress–Strain Relation

Let us now outline the exact solution according to aging linear viscoelasticity. By definition of the compliance function, the strain history caused by stress increment $d\sigma(t')$ applied at time t' and held constant afterward is $J(t, t')\,d\sigma(t')$. Linearity implies the principle of superposition, which states that the strain history is a sum of the contributions of all the previous stress increments. In the limit, the summation becomes an integral. So the aging linear viscoelastic stress–strain relation for variable stress in general reads

$$\varepsilon(t) = \int_{t_1}^{t} J(t, t')\,d\sigma(t') = J(t, t_1)\sigma(t_1) + \int_{t_1^+}^{t} J(t, t')\frac{d\sigma(t')}{dt'}\,dt'. \qquad (9.4.19)$$

Here t_1 is the time when the first stress is applied. The first integral is the Stieltjes integral, which is applicable for continuous as well as discontinuous stress histories. The second integral is the usual Riemann integral, applicable only for continuous and differentiable histories $\sigma(t)$. The second form in Equation 9.4.19 is written for the case when the stress history has a sudden jump at the time of first loading, t_1, and varies continuously afterwards.

The equation governing creep buckling may again be obtained according to the elastic–viscoelastic analogy. Equation 9.4.19 may be written in the form $\varepsilon = \underset{\sim}{E}^{-1}\sigma$ where $\underset{\sim}{E}^{-1}$ is a Volterra integral operator defined by Equation 9.4.19. Such operators again can be manipulated according to the rules of linear algebra, except for the lack of commutativity. Considering the elastic relation for a pin-ended column: $M = (EI + E_s I_s)(z'' - z_0'') = -Pz$, multiplying it by E^{-1} and replacing E^{-1} with the operator $\underset{\sim}{E}^{-1}$, we obtain the following integro-differential equation of the buckling problem:

$$I(z'' - z_0'') + \underset{\sim}{E}^{-1}[E_s I_s(z'' - z_0'') + Pz] = 0 \qquad (9.4.20)$$

Considering a pin-ended column with $z_0 = a \sin(\pi x/l)$, $z = f(t)z_0$, Equation 9.4.20 reduces to the integral equation:

$$\underset{\sim}{E}^{-1}\left[\frac{l^2}{\pi^2}Pf - E_s I_s(f - 1)\right] = I(f - 1) \qquad (9.4.21)$$

in which $\underset{\sim}{E}^{-1}$ is an integral operator. At the time of loading, t_1, Pf has a jump from 0 to Pf_1, after which P is constant and f varies continuously. Function $(f - 1)$ has at time t_1 a jump from 0 to the value $f_1 - 1$. Thus, in similarity to the last form in Equation 9.4.19, Equation 9.4.21 may be written as follows:

$$J(t, t_1)\left[\frac{l^2}{\pi^2}Pf_1 - E_s I_s(f_1 - 1)\right] + \left(\frac{l^2}{\pi^2}P - E_s I_s\right)\int_{t_1^+}^{t} J(t, t')\,df(t') = I[f(t) - 1] \qquad (9.4.22)$$

This is a linear integral equation of Volterra type for function $f(t)$. This equation may be solved relatively easily by approximating the integral with a finite sum, which leads to a system of linear algebraic equations. It may be verified that, for $t = t_1$, Equation 9.4.22 yields the expression $f_1 = (1 - P/P_{cr_1})^{-1}$, as expected.

A more general approach is to replace the integral by a finite sum already in the stress–strain relation (Eq. 9.4.19). This leads to an incremental quasi-elastic stress-strain relation $\Delta\sigma = E^0(\Delta\varepsilon - \Delta\varepsilon^0)$, in which E^0 is a certain quasi-elastic modulus expressed in terms of $J(t, t')$; $\Delta\varepsilon^0$ is calculated by a summation involving the previous stress increments and can be evaluated if the solution up to the current time is known. This reduces the problem to a sequence of elastic buckling problems with initial strains $\Delta\varepsilon^0$. Each of these may be solved in the same manner as shown before for the age-adjusted effective modulus method (Eqs. 9.4.14, 9.4.16–9.4.18); see Bažant and Najjar (1973).

Appendix I—Compliance Function and Relaxation Function of Concrete

A simple expression is the double power law, $J(t, t') = \{1 + \phi_1[(t')^{-m} + \alpha] \times (t - t')^n\}/E_0$ where n, m, α, ϕ_1, E_0 = empirical constants. [Typically $n = \frac{1}{8}$, $m = \frac{1}{3}$, $\alpha = 0.05$ if t, t' are in days, $\phi_1 = 2$ to 6 (Bažant, 1975; Bažant and Osman, 1975).] A more realistic expression for the compliance function (Bažant and Chern, 1985), called the log-double-power law, is

$$J(t, t') = \frac{1}{E_0} + \frac{\psi_0}{E_0} \ln \{1 + \psi_1[(t')^{-m} + \alpha](t - t')^n\} \tag{9.4.23}$$

in which E_0, ψ_0, ψ_1, α, m, and n are empirical constants; their typical values are $n = 0.25$, $m = 0.5$, $\psi_0 = \psi_1 = 1$, $\alpha = 0.02$ if t' is in days. The conventional elastic modulus is expressed as $E(t) = 1/J(t + \Delta, t)$, in which $\Delta \simeq 0.1$ day (= duration of a typical short-time loading test of a structure). A typical semilogarithmic plot of the compliance function is shown in Figure 9.15. Recently, a still more realistic expression has been developed by Bažant and Prasannan (1989a, b); it is based on the analysis of the solidification process (Bažant, 1977) and has two important advantages: (1) All the unknown material parameters can be identified from test data by linear regression (Bažant and Kim, 1989); and (2) the elastic moduli and viscosities of the spring-dashpot model are constant (since aging is introduced by two transformations of time) and can be determined from the compliance function by explicit formulas.

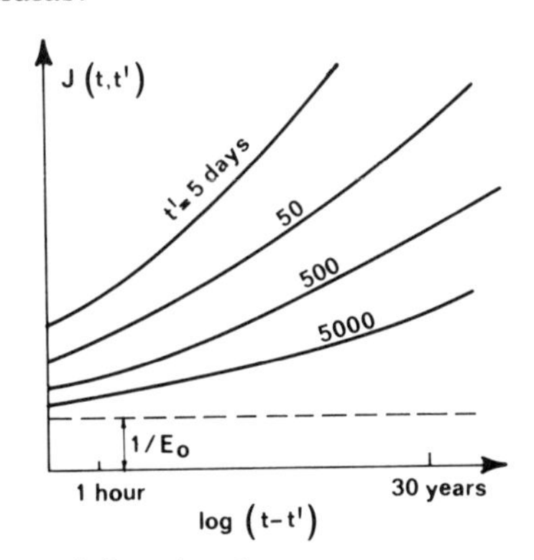

Figure 9.15 Compliance (creep) function for concrete.

Bažant and Kim's (1979) semiempirical formula for the relaxation function is

$$R(t, t') = \frac{0.992}{J(t, t')} - \frac{0.115}{J(t, t-1)}\left[\frac{J(t-\Delta, t')}{J(t, t'+\Delta)} - 1\right] \qquad \Delta = \frac{t-t'}{2} \tag{9.4.24}$$

This formula is approximate, but its error compared to the exact solution from the integral equation (Eq. 9.4.19) is normally within 1 percent, for any realistic form of $J(t, t')$ for concrete.

Appendix II—Proof of Age-Adjusted Effective Modulus Method

The method is based on the following theorem (Bažant, 1972b) for $t \geq t_1$:

Theorem 9.4.1 If, for constant α and β,

$$\varepsilon(t) = \alpha + \beta J(t, t_1) \tag{9.4.25}$$

then

$$\sigma(t) = \beta + \alpha R(t, t_1) \tag{9.4.26}$$

provided that $\varepsilon(t) = 0$ for $t < t_1$.

Proof Equation 9.4.25 may be rewritten as $\varepsilon(t) = \alpha H(t - t_1) + \beta J(t, t_1)$ where $H(t - t_1)$ = Heaviside step function ($=1$ for $t > t_1$, 0 for $t < t_1$). Then, multiplying this by the relaxation operator $\underset{\sim}{E}$, and noting that: $\underset{\sim}{E}\varepsilon(t) = \sigma(t)$ (by definition of $\underset{\sim}{E}$), $\underset{\sim}{E}H(t - t_1) = R(t, t_1)$ = relaxation function, and $\underset{\sim}{E}J(t, t_1) = H(t - t_1) = 1$ for $t > t_1$ (by definition), we get Equation 9.4.26. This theorem was later generalized by Bažant (1987) to triaxial stress states and to arbitrary internal force and deformation matrices, by Lazić and Lazić (1984) to internal forces in composite beams, and by Khazanovich (1989) to time-variable loads, such as thermal loads.

Denote now $\Delta\sigma(t) = \sigma(t) - \sigma(t_1)$ and $\Delta\varepsilon(t) = \varepsilon(t) - \varepsilon(t_1)$. From Equations 9.4.25 and 9.4.26, $\Delta\varepsilon = \beta[J(t, t_1) - 1/E(t_1)] = \beta\phi(t, t_1)/E(t_1)$, and $\Delta\sigma = \alpha[R(t, t_1) - E(t_1)]$. Solving for α and β and substituting into Equation 9.4.26 for $t = t_1$, that is, $\varepsilon(t_1) = \alpha + \beta/E(t_1)$, we get $\varepsilon(t_1) = \Delta\sigma(t)/[R(t, t_1) - E(t_1)] + \Delta\varepsilon(t)/\phi(t, t_1)$. Multiplying this equation by $\phi(t, t_1)$, and setting $\varepsilon(t_1) = \sigma(t_1)/E(t_1)$, we get

$$\Delta\sigma = \frac{E(t_1) - R(t, t_1)}{\phi(t, t_1)}\left[\Delta\varepsilon - \frac{\sigma(t_1)}{E(t_1)}\phi(t, t_1)\right] \tag{9.4.27}$$

which proves Equations 9.1.14 and 9.4.15 (Bažant, 1972b).

Problems

9.4.1 Consider (1) a free-standing, (2) a fixed unreinforced concrete column of length l, and solve aging creep buckling deflections using (a) the aging Maxwell model, (b) the effective modulus model, and (c) the age-adjusted effective modulus model.

9.4.2 Do the same, but for a reinforced column, symmetric cross section.

9.4.3 Show that the foregoing solutions are equivalent to those for a pin-ended column of length (1) $L = 2l$; (2) $L = \frac{1}{2}l$. Knowing this, write the deflection solution for (a) a pinned–fixed column, (b) a free-standing column.

9.5 EFFECT OF CREEP DEFLECTION ON CONCRETE COLUMN STRENGTH

For concrete columns, one cannot base their design on the long-time critical load P_{cr_∞}, for which the deflection tends to infinity at $t \to \infty$. One reason is that concrete creep does not appear to reach a final asymptotic value within the typical lifetimes of structures. Neither can the design be based on some critical lifetime, since the linearity hypothesis does not allow the deflection to become infinite within a finite and nonzero time. Therefore, the design must be based on the effect that the creep deflection has on the strength when the live load is subsequently superimposed.

Equation 9.4.18 or 9.4.13 indicates, for a given permanent load $P = P_D$ (dead load), the value of the ratio $f(t)$ of the deflection ordinates $z(x, t)$ at time t to the initial ones, $z_0(x)$, before load P_D is applied. Consider now that at time t the load is rapidly increased from P_D to $P_D + P_L$ where P_L is the live load. For this overload, the deflection ordinates $z(x, y) = z_0(x)f(t)$ play the role of initial imperfection. To figure out how they are increased due to P_L, we must imagine the column to be suddenly unloaded from P_D to 0 and then reloaded to $P_D + P_L$. Dividing by the magnification factor (Sec. 1.5), we figure that after unloading the ordinates become $\bar{z} = z/[1 - P_D/P_{cr}(t)]^{-1}$ in which $P_{cr}(t)$ = instantaneous elastic critical load at time t. [This corresponds to $\bar{z} = e + e_c$ as introduced in Sec. 8.5, e_c = creep eccentricity; see CEB (1978), Warner and Kordina (1975), Warner, Rangan, and Hall (1976).] Then, due to reloading, the ordinates become $z_u = \bar{z}[1 - (P_D + P_L)/P_{cr}(t)]^{-1}$ (now we multiply by the magnification factor). So, for load $P_D + P_L$ we have the ordinates $z_u(x, t) = z_0(x)f_u(t)$ where

$$f_u(t) = f(t)\frac{1 - P_D/P_{cr}(t)}{1 - (P_D + P_L)/P_{cr}(t)} \tag{9.5.1}$$

The bending moment under load $P_D + P_L$ at time t then is $M(x, t) = (P_D + P_L)z_0(x)f_u(t)$ (for the case of hinged columns).

How should we now use the calculated f_u or M_u in the ultimate strength design of columns? The effect of creep is to magnify the arm on which the axial force acts, which is the same as the effect of the elastic deformation under the axial load. So we may adopt the same design philosophy as in the ACI design method for short-time loads (Eq. 8.5.7). In particular, the safety factors may be incorporated and the interaction diagram used in the same manner.

Therefore, to take the statistical uncertainties of the column properties and the loads into account, we may proceed similarly as in Equation 8.5.7. First, the value used in calculating $f(t)$ and $f_u(t)$ must be the factored dead and live loads P_D^* and P_L^*, not their actual design values P_D and P_L (by the ACI, $P_D^* = 1.4P_D$ and $P_L^* = 1.7P_L$). Second, in the evaluation of Equation 9.5.1 we need to apply factor ϕ_u to the critical load $P_{cr}(t)$ for the same reasons as stated below Equation 8.5.7 (ϕ_u = ACI strength reduction factor, which is 0.70 for tied columns and 0.75 for spiral columns).

According to Equation 9.5.1, and in similarity to Equation 8.5.7, the magnification factor for the ultimate state becomes

$$f_u^*(t) = f^*(t)\frac{1 - P_D^*/\phi_u P_{cr}(t)}{1 - (P_D^* + P_L^*)/\phi_u P_{cr}(t)} \tag{9.5.2}$$

where $f^*(t)$ is calculated on the basis of $P = P_D^*$, not $P = P_D$. The bending moment to be used in the failure criterion then is

$$M^*(x, t) = (P_D^* + P_L^*) f_u^*(t) z_0(x) \tag{9.5.3}$$

Let the equation $M = F(P)$ describe the interaction diagram of the column cross section in the absence of second-order effects (no buckling). Then the design requirement may be expressed as

$$\frac{M_u^*}{\phi_u} \le F\left(\frac{P_{u_D}^* + P_{u_L}^*}{\phi_u}\right) \tag{9.5.4}$$

that is, the point $[M_u^*/\phi_u, (P_{u_D}^* + P_{u_L}^*)/\phi_u]$ must not fall outside the cross-section interaction diagram (Fig. 8.27).

The design approach is illustrated by Figure 9.16a. The curve 0–1 describes the rapid initial loading at $t = t_0$, the curvature being due to second-order effects. The segment 1–2 represents the growth of deflection (and thus M) due to creep at constant $P = P_D^*/\phi_u$. The curve 2–3 represents the rapid application of the live load, P_L^*/ϕ_u. It terminates at the failure point 3 on the interaction diagram (failure envelope).

The foregoing method ignores the fact that the strength for a live load applied after a period of creep under constant load P_D is not equal to the short-time strength. Usually it is higher (provided that the long-time stress is within the service stress range); see Section 9.6. Consequently, the appropriate interaction diagram looks roughly as shown by the dashed curve in Fig. 9.16b. Accordingly, the failure point is point 4.

The combined load $P_{u_D} + P_{u_L}$ is inevitably in the range of nonlinear material behavior but the foregoing calculation is based on the hypothesis of linearity. However, the same is true of the ACI method for short-time loads (Eq. 8.5.1).

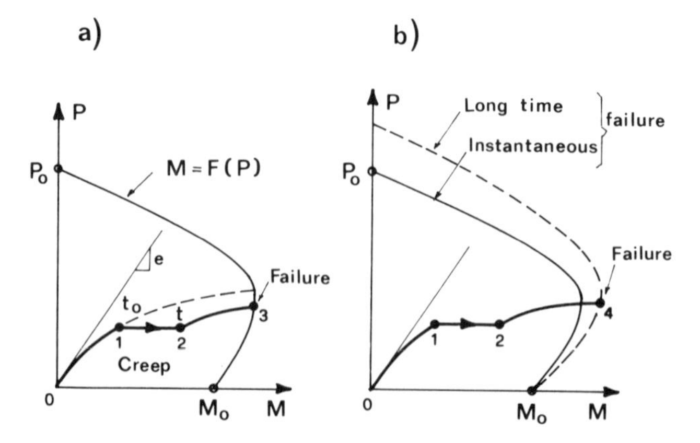

Figure 9.16 (a) Loading path to failure; (b) modification of interaction diagram for live load application.

The fact that this method has been calibrated so as to agree reasonably well with experiments lends credence to using the same approach for long-time loads.

It should also be noted that the response to sudden overload after creep is not completely equivalent to instantaneous loading of a column with the imperfection given by Equation 9.4.13 or 9.4.18. The reason is that the long-time creep due to P_D produces in the cross section self-equilibrated residual stresses that to some extent affect the response to overload P_L (as explained in Sec. 8.3). Further residual stresses are caused by shrinkage, and cracking may result from all these stresses.

There are also alternative design approaches. One can calculate the value of an effective elastic modulus of concrete for which a short-time load equal to $P_D^* + P_L^*$ gives the same magnification factor as Equation 9.4.18 or 9.4.13, and then use this modulus to design the column as if all the loading were short-time. Or one can calculate an effective load for which the short-time magnification factor is the same as that given by Equation 9.4.18 or 9.4.13, and then use this load to design the column like for a short-time load (Bažant, 1965, 1966, 1968a). Whether these alternative approaches adequately reflect the reality, however, has not been determined.

A more realistic calculation of the column response to rapid overload (live load) can be carried out in small loading steps, subdividing the length of the column into layered finite elements. The calculation must take into account the yielding of reinforcement, compression nonlinearity of concrete, and tensile cracking or softening (e.g., Bažant and Tsubaki, 1980).

With regard to the CEB method of column design for instantaneous loads (Sec. 8.5, Fig. 8.34g), the role of creep deflection can be explained by means of Figure 9.17. The curve $\overline{08}$ is the resisting instantaneous diagram of bending moment M versus curvature κ at time t_1; this curve is already scaled down by the capacity reduction factor ϕ_u and is taken, for the sake of simplification, at

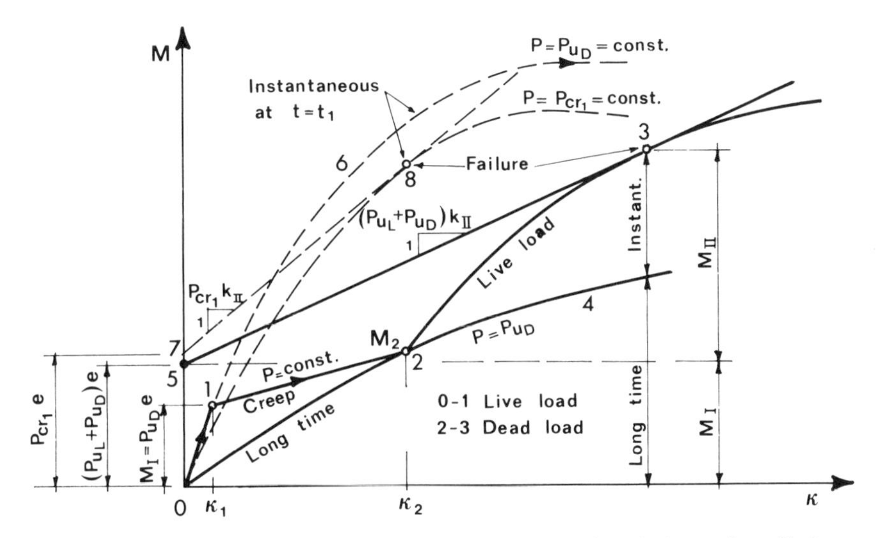

Figure 9.17 Moment-curvature relation for cross section and variations of applied moment according to assumptions of the CEB method.

$P = \text{const.}$ as M increases. The line $\overline{78}$ is the instantaneous applied moment, including the second-order moment M_{II}, and point 8 is the instantaneous failure point, as explained in Section 8.5.

Now, if the load is sustained, the short-time $M(\kappa)$ diagram $\overline{016}$ for $P = P_{u_D} = \text{const.}$ (which differs from that for $P = P_{\text{cr}_1}$) will change, at time t, to some long-time $M(\kappa)$ diagram $\overline{024}$ for $P = P_{u_D}$ const., which is also scaled down by factor ϕ_u. As the (factored) dead load P_{u_D} is applied at time t_1, the segment $\overline{01}$ is followed. Creep at constant load P_{u_D}, which happens at increasing M, causes the column state to move from point 1 to point 2, at which the long-time (isochronous) $M(\kappa)$ diagram $\overline{024}$ for $P = P_{u_D}$ (and fixed time t) is reached. The moment at point 2 can be determined as $M_2 = M_1 f(t)/f(t_1)$ where $M_1 = \text{moment}$ at point 1 and $f(t) = \text{factor}$ given by Equation 9.4.18 or 9.4.13. The bending curvature κ_2 at point 2 is obtained from $z_2'' = \kappa_2 + z_0'' = (\kappa_1 + z_0'')f(t)/f(t_1)$ where $\kappa_1 = \text{bending curvature}$ at point 1 and $z_0'' = \text{initial curvature}$ at no load.

Instantaneous superposition of the (factored) live load P_{u_L} at time t causes the column state to move from point 2 to point 3. The line $\overline{53}$ is the line of the applied moment, including the second-order effect that is caused by the constant long-time load P. The static failure occurs if line $\overline{53}$ is tangent to curve $\overline{23}$, as shown in Figure 9.17. If there is no intersection, dynamic failure will occur, and if there are two intersections, the column will not fail for that load value (Fig. 8.34).

For long-time loading, the slope of the applied moment line, which is equal to Pk_{II}, is obviously less than the slope for short-time loading (lines $\overline{78}$ and $\overline{53}$ in Fig. 9.17). Therefore, the column capacity is also less.

Problems

9.5.1 Consider $I_s = 0.05I$, $E_s = 7E(t_1)$, $J(t, t')$ according to Equation 9.4.23 with the typical parameter values; $t_1 = 28$ days, $P_D = P_L = 0.3P_{\text{cr}_1}$. The cross-section interaction diagram is approximated as bilinear with $P^0 = 2P_{\text{E}_1}$, $M^0 = 0.05IP_u^0$ (Fig. 9.18). The initial imperfection is sinusoidal with $a = 0.005L$. Calculate the deflection at $t = 40$ years for P_D as well as $P_D + P_L$. Determine the load factor μ such that the loads μP_D and $1.21\mu P_L$ put the column state on the interaction diagram (*note*: the ratio of the load factors for P_D and P_L is $1.7/1.4 = 1.21$).

9.5.2 The long-term column design may be based on the residual ordinate z (called creep eccentricity) that is obtained after sudden unloading at time t.

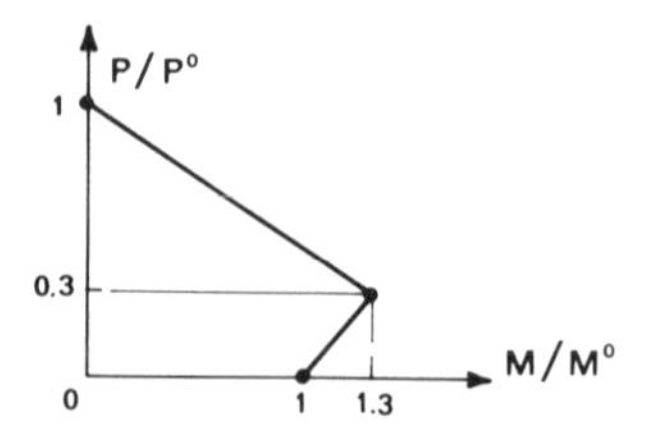

Figure 9.18 Approximate cross-section interaction diagram.

Calculate it for the Dischinger method, using Equation 9.4.10 or 9.4.11. (The result is particularly simple if one uses $E_t = E_1$ and $P_{E_t} = P_{E_1}$, which was done by Warner and Kordina, 1975, and Warner, Rangan, and Hall 1976.)

9.5.3 Use the age-adjusted effective modulus to analyze creep buckling deflections of (a) an infinite concrete beam on an elastic (not viscoelastic) foundation, (b) an axisymmetric buckling of an axially compressed thin circular concrete shell [this is not equivalent to case (a) if the foundation does not creep], (c) a thin square plate compressed by N_{xx} only, (d) a braced portal frame, (e) a snapthrough of a von Mises truss, and (f) a lateral buckling of a simply supported I-beam, either without or with reinforcement.

9.6 NONLINEAR CREEP AND LONG-TIME STRENGTH OF CONCRETE STRUCTURES

In Section 9.4 (as well as 9.5) the aging creep law of concrete was assumed to be linear. However, linearity is valid only for stresses less than about 0.5 of the strength, which happen to fall within the service stress range of structures. For higher stresses, the creep of concrete is strongly nonlinear. The principal consequence is that (as we already showed in Sec. 9.3 for viscoplastic behavior in general) the deflection calculated from a linearized small-deflection theory may tend to infinity within a certain finite time, called the critical time. Thus, a load magnitude that does not cause collapse under short-time loading may cause collapse if the load is sustained.

The increased specific creep (per unit stress), which is observed at high stresses and is often called flow, is not the only type of nonlinearity exhibited by concrete creep. Another type of nonlinearity, called adaptation, is observed in the service stress range. A low sustained compressive stress causes strengthening of concrete, diminishing the creep for subsequent load increments as well as increasing the strength. By modifying the integral-type stress–strain relation from Equation 9.4.19, Bažant and Kim (1979) described both the flow and the adaptation nonlinearities of concrete, and their constitutive relation was then used by Bažant and Tsubaki (1980) in a step-by-step incremental analysis of columns; see Figures 9.19 to 9.21 (taken from Bažant and Tsubaki).

Figure 9.19a shows the deflection histories of a column with assumed sinusoidal initial curvature. Two different reinforcement ratios ρ are considered. The deflection histories are plotted for various magnitudes of a constant sustained load, relative to Af'_c where A = cross-section area of concrete and f'_c = standard compression strength. These results illustrate that, depending on the load magnitude, the deflection after a certain period of time starts to increase rapidly and quickly leads to collapse. The larger the load, the shorter is the time to collapse. However, for loads that are less than about 50 percent of the short-time strength, collapse does not occur within a 30-year lifetime.

As explained in the preceding section, one needs to distinguish the dead load P_D and the live load P_L (i.e., the long-time and short-time loads). When the ratio P_D/P_L is small, the stresses caused by P_D alone are inevitably within the linear

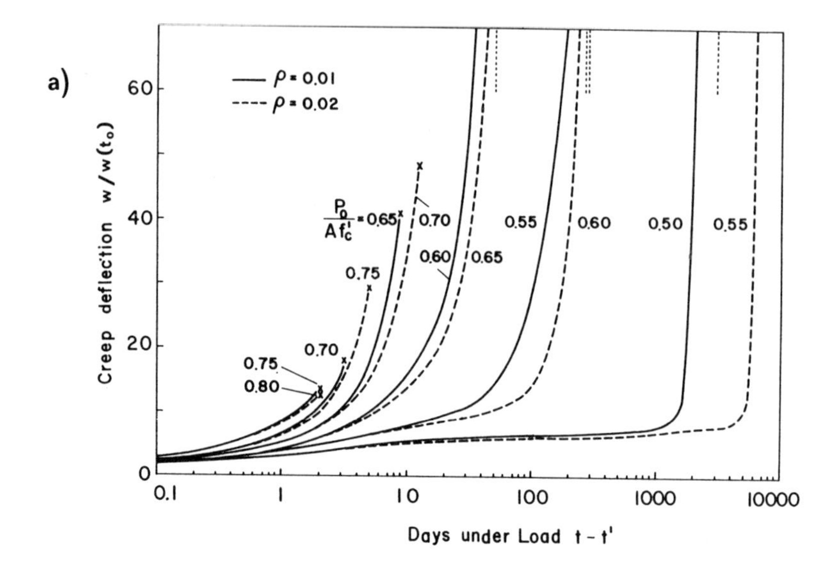

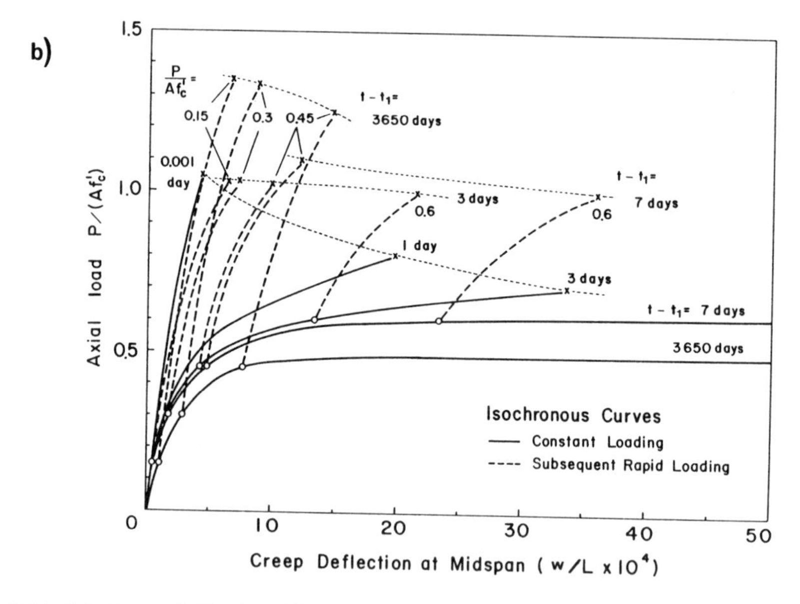

Figure 9.19 (a) Creep deflection history and (b) axial load variation. (*After Bažant and Tsubaki, 1980.*)

range, and then the creep part of the analysis can be based on a linear aging creep law, which was the subject of the preceding section. When the ratio P_D/P_L is not small, the creep is nonlinear and should be analyzed nonlinearly.

Some typical results of Bažant and Tsubaki's (1980) nonlinear creep analysis of a typical concrete column with a high P_D/P_L are shown in Figures 9.19b and 9.20. Figure 9.19b shows the isochrones of the nondimensionalized axial load P versus the nondimensionalized midspan deflection. (Remember from Sec. 9.1 that each isochrone is determined by collecting, for the same time, the results

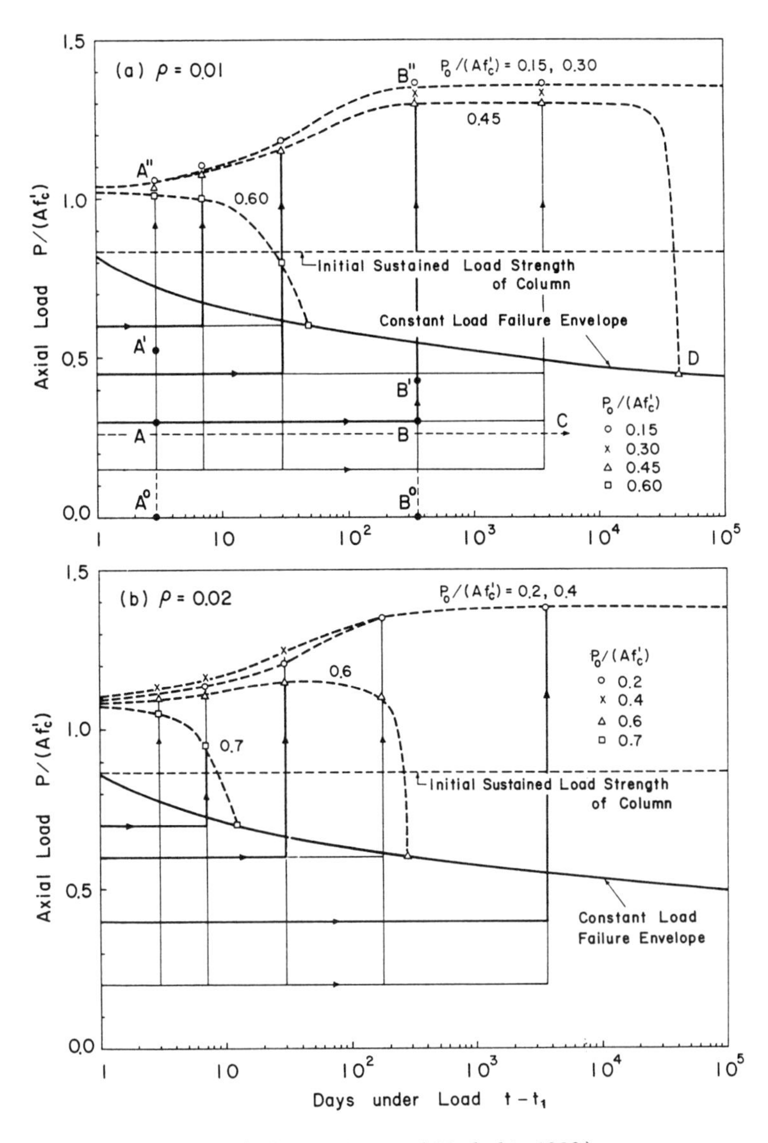

Figure 9.20 Axial load history. (*After Bažant and Tsubaki, 1980.*)

obtained for many specimens, each loaded with a different constant axial load.) The isochrones terminate at failure points marked by crosses. The solid curves correspond to the sustained load (dead load) before the live load is applied. As the live load is applied, the response is instantaneous and therefore can be represented in the same plot with the isochrones, as shown by the dashed curves for various instants of application of the live load. The dotted lines represent failure envelopes connecting the failure points for the same value of the sustained load.

Figure 9.20 shows the results in the plot of the axial load versus the logarithm of load duration $(t - t_1)$. The horizontal lines represent the sustained loads, and the vertical lines the instantaneous loading that terminates at the failure point. The failure points calculated for various values of the sustained load are connected by dashed curves, and the failure points for no live load (constant load up to failure) are connected by the solid curve. We see that for small sustained loads the points of failure caused by the subsequent sudden overload indicate a substantial increase in strength. This phenomenon, however, appears almost nonexistent in the tests of Drysdale and Huggins (1971), perhaps because of drying that prevents the growth of concrete strength due to hydration and was not considered in the calculations. The strength increase should nevertheless be observed in columns that are sealed or are subjected to moist environments. For larger values of the sustained load, the calculation results in Figure 9.20 indicate a substantial decrease of column strength if the duration of the preceding sustained load is sufficiently long.

Figure 9.21a shows the plot of P_u^0/P_u^t versus β_d where $\beta_d = P_D/(P_D + P_L) =$ ratio of dead load to total load, $P_u^0 =$ short-time failure load, and $P_u^t =$ long-time failure load. Figure 9.21b shows a similar plot from the test results of Drysdale and Huggins (1971) and of Kordina (1975). As already mentioned, the test results do not exhibit the column strengthening obtained in calculations. On the other hand, calculations show for some cases also a rather large reduction of capacity, which appears to be absent from the test results. This may be due to the fact that these large reductions of carrying capacity were calculated for very long load durations (30 years), while the sustained loads used in the tests were of much shorter durations (under 1 year). The solid and dashed curves represent the reductions of column strength due to sustained loading according to the following formulas specified in the ACI code and proposed in Bažant and Tsubaki (1980), respectively:

$$\frac{P_u^0}{P_u^t} = \begin{cases} 1 + \beta_d & \text{(ACI code)} \\ 1 + 1.3\beta_d^2 & \text{(Bažant \& Tsubaki)} \end{cases} \tag{9.6.1}$$

The strength of concrete columns under sustained loads and in combination with live loads was also studied by Bridge (1979); Fouré (1978); Hughes and Ash (1970); Manuel and MacGregor (1967); Mauch and Holley (1963); Mauch (1966);

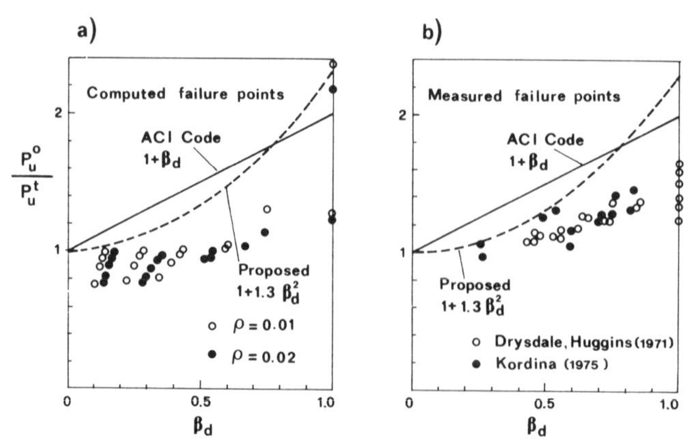

Figure 9.21 Computed and measured failure points. (*After Bažant and Tsubaki, 1980*).

McClure, Gerstle, and Tulin (1973); Wilhelm and Zia (1970); and Wu and Huggins (1977).

In general, analysis of creep buckling of concrete structures up to failure is a rather complicated problem that requires a realistic, and therefore sophisticated, constitutive law for nonlinear creep of concrete, as well as short-time nonlinear behavior including cracking. Effects of drying, shrinkage stresses, and temperature should also be included. Such calculations require a computer and can be accomplished by finite elements. On the other hand, random scatter in the behavior of actual columns, which is quite large, weakens the usefulness of very sophisticated calculations unless they are carried out probabilistically. For this reason, design practice has been limited to a rather simple semiempirical calculation method only partly supported by test results. The price paid for this simplicity is that the safety factors for concrete columns must be rather large.

Problems

9.6.1 The aging linear creep laws from Section 9.4 can be generalized for the nonlinear range. Consider the so-called "time-hardening," which corresponds to the aging Maxwell model with a linear spring and nonlinear aging dashpot, for which $\dot{\varepsilon} = (\dot{\sigma}/E) + f(t)\sigma^n$, $f(t) =$ given function of age t such as $f(t) = kt^{-m}$ $(0 < m < 1,\ n > 1)$. Analyze small buckling deflections due to P_D and P_L for (a) Shanley's column and (b) a deformable ideal I-beam.

9.6.2 Another useful approximation for nonlinear creep of concrete is $\varepsilon = f(\sigma, t) = (\sigma/E) + g(t)\sigma^n$ (this is a nonlinear generalization of the effective modulus approach). Use it for Problem 9.6.1.

9.7 CREEP BUCKLING AT FINITE DEFLECTIONS

Similar to elastic buckling, the finite deflection analysis of viscoelastic buckling must yield finite deflections for all times and all load values. Therefore, the long-time critical load P_{cr}^{∞} introduced in Section 9.2 cannot be characterized by an infinite deflection ordinate z at $t \to \infty$. Rather the infinite deflection values, which we used in the small-deflection theory (Sec. 9.2) as the indicator of instability, must be interpreted in relative terms, namely in relation to the magnitude of initial imperfection. Thus, within the framework of finite deflection theory, the long-time critical load P_{cr}^{∞} needs to be defined by the following property:

$$\lim_{z_0 \to 0} \frac{z(\infty)}{z_0} = \begin{cases} < \infty & \text{for } P < P_{cr}^{\infty} \\ = \infty & \text{for } P \geq P_{cr}^{\infty} \end{cases} \tag{9.7.1}$$

where $z(\infty) =$ deflection ordinate at $t \to \infty$, $z_0 =$ initial imperfection.

Example of Imperfection-Sensitive Rigid-Bar Column

Despite the boundedness of deflections, the deflection rate of a linearly viscoelastic column at large deflection may reach an infinite value at a finite time, provided the elastic characteristics of the structure are imperfection sensitive

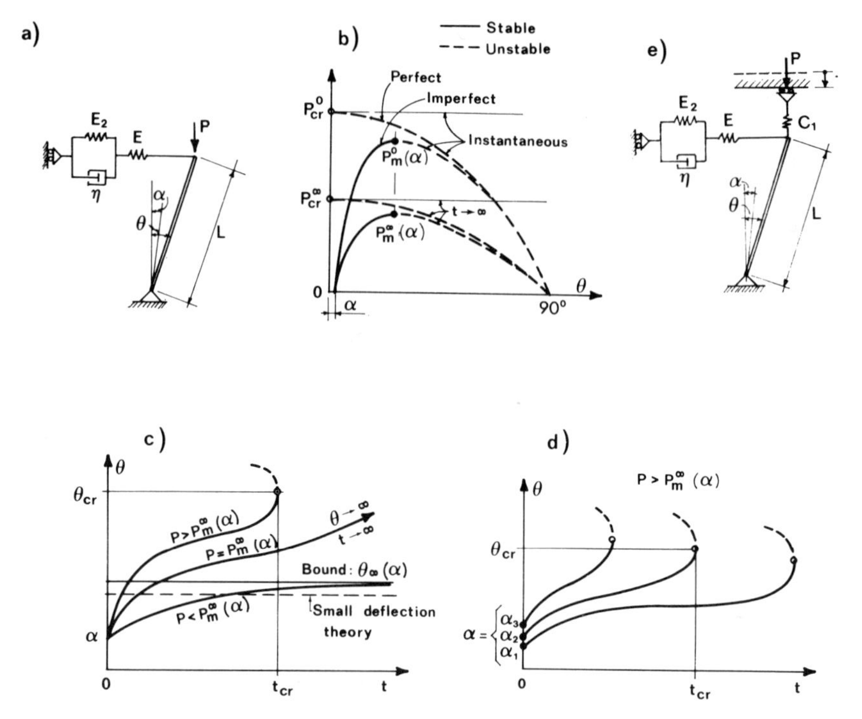

Figure 9.22 (a) Rigid-bar column with horizontal Kelvin-type support; (b) load-rotation curves; (c, d) rotation histories; and (e) load applied through elastic spring.

(i.e., P decreases at increasing deflection). This was discovered by Szyszkowski and Glockner (1985). We will demonstrate it by considering a rigid free-standing column with a hinge at the base and a viscoelastic lateral support on top. The support, characterized by a standard solid model of Kelvin type (Fig. 9.3f), slides vertically, always remaining horizontal (Fig. 9.22a). From Section 4.5 we know that the elastic behavior of a perfect column of this type is characterized by a decrease of P at increasing θ. Let L = height of the column; b, c = length and cross-section area of the lateral support; ε, σ = strain and stress in the support. From geometry we have $\varepsilon = L(\sin\theta - \sin\alpha)/b$ where α = initial inclination of the column (i.e., an imperfection), and from moment equilibrium we have $\sigma = (PL\sin\theta)/(cL\cos\theta) = (P\tan\theta)/c$. Substituting the foregoing expressions for ε and σ into the equation of the standard solid, $\tau_r\dot{\varepsilon} = (\sigma/E_\infty) - \varepsilon - \dot{\sigma}(\tau_r/E)$ (Eq. 9.1.5), we acquire the differential equation

$$\tau_r\left(\frac{L}{b}\cos\theta - \frac{P}{Ec\cos^2\theta}\right)\frac{d\theta}{dt} = \frac{P}{E_\infty c}\tan\theta - \frac{L}{b}(\sin\theta - \sin\alpha) + \frac{\tau_r}{Ec}\tan\theta\frac{dP}{dt} \quad (9.7.2)$$

When P is constant ($\dot{P} = dP/dt = 0$), this equation has the form $f(\theta)\,d\theta/dt = g(\theta)$. Integration may then be easily accomplished by separation of variables, which yields $t = t(\theta) = \int [f(\theta)/g(\theta)]\, d\theta$. Inverting function $t(\theta)$ one gets the deflection history $\theta(t)$.

Consider that at $t = 0$ the load is instantly raised to the value P and then is kept constant ($\dot{P} = 0$ at $t > 0$). The response may or may not reach a final state at rest (constant θ). If it does, the final state is determined by Equation 9.7.2 with $\dot{\theta} = \dot{P} = 0$. This equation coincides with the equation for the elastic behavior of a column whose lateral support is elastic, characterized by modulus E_∞. As shown in Section 4.5, for each imperfection angle α there exists a maximum load $P_m^0(\alpha)$; see Figure 9.22b. As a manifestation of imperfection sensitivity of this column, $P_m^\infty(\alpha)$ is less than the long-time critical load P_{cr}^∞; P_{cr}^∞ = critical load of a perfect-elastic column with modulus E_∞; $P_m^\infty(\alpha) = P_m^0(\alpha)E_\infty/E$ where $P_m^0(\alpha) =$ max P for instantaneous elastic loading. Furthermore, $P_{cr}^\infty = P_{cr}^0 E_\infty/E < P_{cr}^0$, where P_{cr}^0 is the instantaneous critical load based on modulus E.

For $P > P_m^\infty(\alpha)$, no final state at rest can exist. The lateral support cannot impose any bound on θ. Angle θ, therefore, cannot stop growing and must eventually reach some value $\theta = \theta_{cr}$ such that the expression in parenthesis on the left-hand side of Equation 9.7.2 vanishes, that is,

$$\cos\theta_{cr} = \left(\frac{P}{P_{cr}^0}\right)^{1/3} \qquad (P < P_{cr}^0) \tag{9.7.3}$$

where $P_{cr}^0 = ELc/b$ = instantaneous (elastic) critical load. Consequently, $d\theta/dt \to \infty$ at $\theta = \theta_{cr}$ (the right-hand side of Eq. 9.7.2 being nonzero); see Figure 9.22c, d, which also shows the effects of the values of P and α. The time t_{cr}, which corresponds to $\theta = \theta_{cr}$ and may be obtained from the aforementioned solution $t(\theta)$, is finite if $P > P_m^\infty(\alpha)$. This can be shown from Equation 9.7.2 and has been demonstrated by Szyszkowski and Glockner (1985). For $P = P_m^\infty(\alpha)$, one has $\theta_{cr} \to \infty$ (Fig. 9.22). Since at t_{cr} the left-hand side of Equation 9.7.2 changes sign, the curve $\theta(t)$ is indicated to turn back after t_{cr} as shown in Figure 9.22c, d. This is, however, impossible because the time would have to decrease. So there can be no equilibrium after t_{cr} (if P = const.), and dynamic analysis is required.

Broader Implications and Ramifications

Similar behavior is found for buckling to the right of the column in Figure 9.23 whose elastic behavior exhibits asymmetric bifurcation and is strongly imperfec-

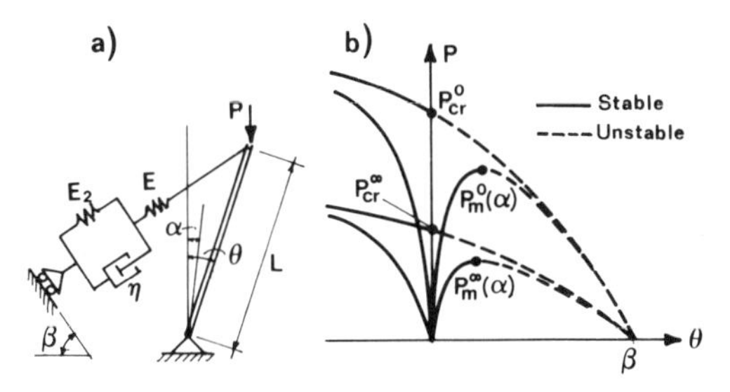

Figure 9.23 (a) Rigid-bar column with inclined Kelvin-type support and (b) load-rotation curves.

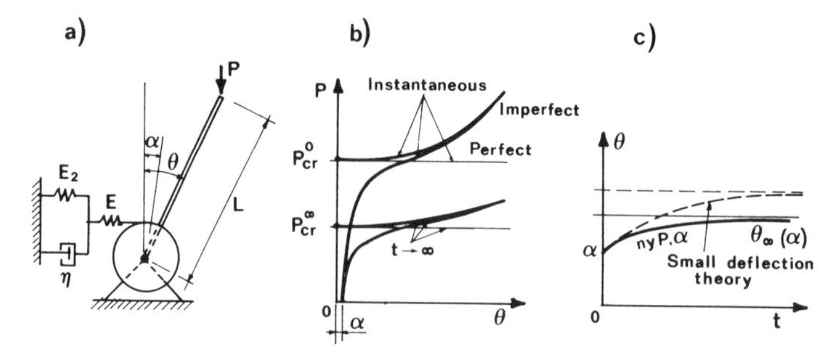

Figure 9.24 (a) Rigid-bar column with rotational viscoelastic spring, (b) load-rotation curves, and (c) rotation history.

tion sensitive (Sec. 4.5). However, the depression of $P_m^0(\alpha)$ compared to P_{cr}^{∞} is much stronger, causing a critical time to be encountered for smaller loads and smaller deflections (Fig. 9.23). For buckling to the left of this column, the maximum point $P_m^{\infty}(\alpha)$ does not exist, and neither does θ_{cr}. The elastic behavior of the column in Figure 9.24 exhibits stable symmetric bifurcation and is imperfection insensitive. In that case, $\theta(t)$ for $t \rightarrow \infty$ (and $P < P_{cr}^{\infty}$) approaches an asymptote located below that for the small-deflection theory (Fig. 9.24).

In Section 9.3 we saw that the existence of a critical time is typical of nonlinear creep buckling. Now we conclude that critical time can also exist for imperfection-sensitive columns with linear creep, provided that the deflection becomes so large that geometric nonlinearity matters. The most important conclusion is that for imperfection-sensitive structures the safe load limit is not the long-time critical load P_{cr}^{∞} for small-deflection theory, but load $P_m^{\infty}(\alpha)$ representing the maximum load for large-deflection elastic response of an elastic column with modulus E_{∞} instead of E. However, this is true only for large enough imperfection α.

For sufficiently small α, time t_{cr} is longer than the design lifetime of the structure, and then the safe load limit is still P_{cr}^{∞} rather than $P_m^{\infty}(\alpha)$. This usually is the case for concrete columns and frames in engineering practice, but not for thin plates and shells.

The value of the present analysis lies mainly in that it illustrates the type of behavior that can be expected for plates and shells. Although the creep buckling of such structures would today be analyzed by finite elements, it is important to realize what to look for, what kind of responses to expect.

The fact that the stability limit $P_m^{\infty}(\alpha)$ is decided by the long-time elastic modulus indicates that the analysis of the effect of large deflections on the stability limit of structures (within the range of linear viscoelastic behavior) may be conducted elastically, replacing E with E_{∞}. To capture the effect of a finite lifetime, one may use instead of E_{∞} the effective modulus and, in the case of aging, the age-adjusted effective modulus (Sec. 9.4).

Variable load

Equation 9.7.2, as written, is valid also for time-variable load P. As one special case, it can be used to solve the relaxation of force P when θ is held constant after $t = 0$ (Szyszkowski and Glockner, 1985). In that case, the left-hand side of Equation 9.7.2 is identically zero, and the right-hand side (with $\theta = \text{const.}$) represents a differential equation for $P(t)$. Its solution is a decaying exponential; $\theta(t)$ always decreases, and there is no critical time.

For a column that forms part of a larger redundant structure, the axial force P is generally variable even if the loading of the structure is constant. This is due to internal force redistributions caused by creep in the structure. As a simple example, we may consider the load to be applied on the column in Figure 9.22e through a spring of stiffness C_1. The spring is suddenly compressed at time $t = 0$ to produce a given force P_0, and a subsequently the top of the spring is prevented from moving vertically. The vertical force on the column must then satisfy the compatibility relation $P_0 - P(t) = C_1 L(\cos\alpha - \cos\theta)$. This yields $\dot{P} = C_1 L \dot{\theta} \sin\theta$, which may be substituted into Equation 9.7.2. The result is that the force $P(t)$ initially decreases with increasing deflection. The deflection rate $\dot{\theta}$ may again become unbounded at some critical time. The minimum value of P_0 above which this happens is generally less than $P_m^\infty(\alpha)$. The influence of structural redundancy is stronger if the instantaneous response exhibits asymmetric bifurcation than if it exhibits symmetric unstable bifurcation.

The geometric nonlinearity obviously affects also the creep buckling of viscoplastic structures if the deflections become large.

Problems

9.7.1 Analyze large-deflection creep buckling for the columns in Figure 9.6a, b, c, d, e, f, g, h. In particular, determine θ_{cr}, t_{cr}, and the condition for which these exist. Relate the behavior to elastic imperfection sensitivity, bifurcation stability, and bifurcation symmetry.

References and Bibliography

Alfrey, T. (1944), "Nonhomogeneous Stress in Viscoelastic Media," *Quart. Appl. Math.*, 2(2): 113–19 (Sec. 9.1).

American Society of Civil Engineers (ASCE) Task Committee on Finite Element Analysis of Reinforced Concrete Structures, Subcommittee 7 (1981), "Time-Dependent Effects," chap. 6 in *Report, Finite Element Analysis of Reinforced Concrete Structures,* ASCE, New York.

Arnold, S. M., Robinson, D. N., and Saleeb, A. F. (1989), "Creep Buckling of Cylindrical Shell under Variable Loading," *J. Eng. Mech.* (ASCE), 115(5): 1054–74.

Bažant, Z. P. (1964), "Die Berechnung des Kriechens und Schwindens nichthomogener Betonkonstruktionen," Proc. 7th Congr. Int. Ass. for Bridge and Struct. Eng. (IABSE), Rio de Janeiro, pp. 887–96.

Bažant, Z. P. (1965), "Effect of Creep on Long-Time Stability and Buckling Strength of Concrete Columns" (in Czech), Proc. 5th Nat. Conf. on Prestr. Concr., held in Žilina, publ. by ČS-VTS (Czech. Sci. Techn. Soc.), Bratislava, part I, pp. 115–32 (Sec. 9.2).

Bažant, Z. P. (1966), *Creep of Concrete in Structural Analysis* (in Czech), State Publishers of Technical Literature (SNTL), Prague; based on Ph.D. Dissertation defended at Czechoslovak Academy of Sciences (ČSAV) in Prague in April 1963 (Sec. 9.4).

Bažant, Z. P. (1968a), "Creep Stability and Buckling of Concrete Columns," *Mag. Concr. Res.*, 20:85–94 (Sec. 9.2).

Bažant, Z. P. (1968b), "Long-Time Stability and Buckling Strength of Concrete Columns" (in Czech), *Inženýrské Stavby*, 16:171–9 (Sec. 9.4).

Bažant, Z. P. (1972a), "Numerical Determination of Long-Range Stress History from Strain History in Concrete," *Materials and Struct.* (RILEM, Paris), 5(27):135–41 (Sec. 9.4).

Bažant, Z. P. (1972b). "Prediction of Concrete Creep Effects Using Age-Adjusted Effective Modulus Method," *J. Am. Concr. Inst. J.*, 69:212–17 (Sec. 9.4).

Bažant, Z. P. (1975), "Theory of Creep and Shrinkage in Concrete Structures: a Précis of Recent Developments," in *Mechanics Today*, vol. 2, Pergamon Press, New York, pp. 1–93 (Sec. 9.4).

Bažant, Z. P. (1977), "Viscoelasticity of Solidifying Porous Material—Concrete," *J. Eng. Mech.* (ASCE), 103(6):1049–67 (Sec. 9.4).

Bažant, Z. P. (1982), "Mathematical Models for Creep and Shrinkage of Concrete," in *Creep and Shrinkage in Concrete Structures*, ed by Z. P. Bažant and F. H. Wittmann, John Wiley & Sons, London, pp. 163–256 (Sec. 9.4).

Bažant, Z. P. (1987), "Matrix Force Displacement Relations in Aging Viscoelasticity," *J. Eng. Mech.* (ASCE), 113(8):1239–43 (Sec. 9.4).

Bažant, Z. P., ed. (1988), *Mathematical Modeling of Creep and Shrinkage of Concrete*, John Wiley & Sons, Chichester and New York.

Bažant, Z. P., and Chern, J. C. (1985), "Log-Double Power Law for Concrete Creep," *Am. Concr. Inst. J.*, 82:665–75 (Sec. 9.4).

Bažant, Z. P., and Kim, S. S. (1979), "Approximate Relaxation Function for Concrete," *J. Struct. Eng.* (ASCE), 105(12):1695–705 (Sec. 9.4).

Bažant, Z. P., and Kim, J.-K. (1989), "Improved Prediction Model for Time-Dependent Deformations of Concrete," Internal Report, Northwestern University; also Materials and Structures (RILEM, Paris), in press (Sec. 9.4).

Bažant, Z. P., and Najjar, J. (1973), "Comparison of Approximate Linear Methods for Concrete Creep," *J. Struct. Eng.* (ASCE), 99(9):1851–74 (Sec. 9.4).

Bažant, Z. P., and Osman, E. (1975), "On the Choice of Creep Function for Standard Recommendations on Practical Analysis of Structures," *Cem. Concr. Res.*, 5:631–41; see also discussion (1976), 7:111–30; (1978), 8:129–30 (Sec. 9.4).

Bažant, Z. P., and Prasannan, S. (1989a), "Solidification Theory For Concrete Creep. I: Formulation," *J. Eng. Mech.* (ASCE), 115(8):1691–703 (Sec. 9.4).

Bažant, Z. P., and Prasannan, S. (1989b), "Solidification Theory for Concrete Creep. II: Verification and Application," *J. Eng. Mech.* (ASCE), 115(8):1704–25 (Sec. 9.4).

Bažant, Z. P., and Tsubaki, T. (1980), "Nonlinear Creep Buckling of Reinforced Concrete Columns," *J. Struct. Eng.* (ASCE), 106(11):2235–57 (Sec. 9.5).

Bažant, Z. P., Tsubaki, T., and Celep, Z. (1983), "Singular History Integral for Creep Rate of Concrete," *J. Eng. Mech.* (ASCE), 109(3):866–84 (Sec. 9.1).

Behan, J. E., and O'Connor, C. (1982), "Creep Buckling of Reinforced Concrete Columns," *J. Struct. Eng.* (ASCE), 108(12):2799–818.

Biot, M. A. (1965), *Mechanics of Incremental Deformation*, John Wiley & Sons, New York (Sec. 9.1).

Boyle, J. T., and Spence, J. (1983), *Stress Analysis for Creep*, Butterworth, London (Sec. 9.3).

Bridge, R. Q. (1979), "Composite Columns under Sustained Load," *J. Struct. Eng.* (ASCE), 105(3):563–76 (Sec. 9.6).

Bukowski, R., and Wojewodzki, W. (1984), "Dynamic Buckling of Viscoplastic Spherical Shell," *Int. J. Solids Struct.*, 20(8):761–76.

CEB (1978), "CEB-FIP Model Code for Concrete Structures," Bulletin No. 124/125-E,

Comité Eurointernational du Béton, Paris (Sec. 9.4).
CEB-FIP (1980), "Design Manual—Structural Effects of Time-Dependent Behavior of Concrete," Bulletin d'Information No. 136, Paris.
Chang, W. P. (1986), "Creep Buckling of Nonlinear Viscoelastic Columns," *Acta Mech.*, 60(3–4):199–215.
de Jongh, A. W. (1980), "Simplified Approach to Creep Buckling and Creep Instability of Steel Concrete Columns," *Heron,* 25(4).
Dischinger, F. (1937), "Untersuchungen über die Knicksicherheit, die elastische Verformung und das Kriechen des Betons bei Bogenbrücken," *Der Bauingenieur,* 18:487–520, 539–62, 595–621 (Sec. 9.4).
Dischinger, F. (1939), "Elastische und plastische Verformungen bei Eisenbetontragwerke," *Der Bauingenieur,* 20:53–63, 286–94, 426–37, 563–72 (Sec. 9.4).
Distéfano, J. N. (1961), "Creep Deflections in Concrete and Reinforced Concrete Columns," *Int. Ass. Bridge Struct. Eng.* (IABSE), 21:37–47 (Sec. 9.2).
Distéfano, J. N. (1965), "Creep Buckling of Slender Columns," *J. Struct. Eng.* (ASCE), 91(3):127–50 (Sec. 9.2).
Distéfano, J. N. (1966), "Creep Buckling of Viscoelastic Structures under Stochastic Loads," RILEM Int. Symp. on Effects of Repeated Loading of Materials and Structures, Universidad de Buenos Aires.
Distéfano, J. N., and Sackman, J. L. (1968), "On the Stability of an Initially Imperfect, Nonlinearly Viscoelastic Column," *Int. J. Solids Struct.*, 4:341–54.
Dost, S., and Glockner, P. G. (1982), "On the Dynamic Stability of Viscoelastic Perfect Columns," *Int. J. Solids Struct.*, 18(7):587–96 (Sec. 9.2).
Dost, S., and Glockner, P. G. (1983), "On the Behavior of Viscoelastic Column with Imperfection," *Trans. Can. Soc. Mech. Eng.*, 7(4):198–202.
Drozdov, A. D., Kolmanovskii, V. B., and Potapov, V. D. (1984), "Stability of Rods of Nonuniform Aging Viscoelastic Material," *Mech. Solids,* 19(2):176–86.
Drysdale, R. G., and Huggins, M. W. (1971), "Sustained Biaxial Load on Slender Concrete Columns," *J. Struct. Eng.* (ASCE), 97(5):1423–43 (Sec. 9.6).
Dulacska, E. (1981), "Buckling of Reinforced Concrete Shells," *J. Struct. Eng.* (ASCE), 107(12):2381–401.
England, G. L., and Illston, J. M. (1965), "Methods of Computing Stress in Concrete from a History of Measured Strain," *Civ. Eng. Publ. Works Rev.*, pp. 513–17, 692–4, 845–7 (Sec. 9.4).
Faessel, P. (1970a), "Comparison des résultats des essais du flambement sous charges soutenues de M. M. Thürlimann, Baumann, Grenacher et Ramu avec les charges critiques calculées," IABSE Symposium, Madrid.
Faessel, P. (1970b), "Influence du fluage sur les phénomenes d'instabilité," IABSE Symposium, Madrid.
Fouré, B. (1978), "Le flambement des poteaux compte tenu du fluage du béton," Annales de l'Institut Technique du Bâtiment et des Travaux Publics (Théories et Méthodes de Calcul No. 214), Paris, No. 359, pp. 4–58 (Sec. 9.6).
Freudenthal, A. M. (1950), *The Inelastic Behavior of Engineering Materials and Structures,* John Wiley & Sons, New York (Sec. 9.2).
Gerdeen, J. C., and Sazawal, V. K. (1974), "A Review of Creep Instability in High Temperature Piping and Pressure Vessels," Bulletin No. 195, Welding Research Council, pp. 33–56 (Sec. 9.3).
Ghali, A., and Favre, R. (1986), *Concrete Structures: Stresses and Deformations,* Chapman and Hall, London (Sec. 9.4).
Gilbert, R. I., and Mickleborough, N. C. (1989), "Creep Effects in Slender Reinforced and Prestressed Concrete Columns," ACI Annual Convention, Atlanta, Georgia, Feb. 19–24.

Glanville, W. H. (1933), "Creep of Concrete Under Load," *The Structural Engineer,* 11(2):54–73 (Sec. 9.4).

Hewitt, J. S., and Mazumdar, J. (1977), "Buckling of Viscoelastic Plates," *AIAA Journal,* 15(4): 451–2.

Higgins, T. P. (1952), "Effect of Creep on Column Deflection," in *Weight-Strength Analysis of Aircraft Structures,* ed. by F. R. Shanley, McGraw-Hill, New York (Sec. 9.3).

Hilton, H. H. (1952), "Creep Collapse of Viscoelastic Columns with Initial Curvatures," *J. Aero. Sci.,* 19:844–6 (Sec. 9.2).

Hoff, N. J. (1954), "Buckling and Stability," *J. Roy. Aero. Soc.,* 58:1–52 (Sec. 9.2).

Hoff, N. J. (1958), "A Survey of the Theories of Creep Buckling," Proc. 3rd U.S. Nat. Congr. Appl. Mech., pp. 29–49, Providence, R.I. (Sec. 9.2).

Huang, N. C. (1967), "Nonlinear Creep Buckling of Some Simple Structures," *J. Appl. Mech.* (ASME), 34:651–8 (Sec. 9.2).

Hughes, B. P., and Ash, J. E. (1970), "Some Factors Influencing the Long-Term Strength of Concrete," *Materials and Structures* (RILEM, Paris), 3(14):81–84 (Sec. 9.6).

Hui, D. (1986), "Viscoelastic Response of Floating Ice Plates under Distributed or Concentrated Loads," *J. Strain Analysis,* 21(3):135–43 (Sec. 9.2).

Hui, D., and de Oliveria, J. G. (1986), "Dynamic Plastic Analysis of Impulsively Loaded Viscoplastic Rectangular Plates with Finite Deflections," *J. Appl. Mech* (ASME), 53:667–74.

Iori, I., and Mirabella-Roberti, G. (1987), "Analisi del comportamento di elementi snelli in cemento armato soggetti a carichi ripetuti di breve durata," L'Industria Italiana del Cemento, No. 609, pp. 214–20.

Kempner, J. (1954), "Creep Bending and Buckling of Nonlinear Viscoelastic Columns," NACA Tech. Note No. 3137 (Sec. 9.2).

Kempner, J. (1962), "Viscoelastic Buckling," in *Handbook of Engineering Mechanics,* ed. by W. Flügge, McGraw-Hill, New York (Sec. 9.2).

Khazanovich, L. (1989), "Age-Adjusted Effective Modulus Method for Time-Variable Loads," private communication to Z. P. Bažant; also *J. Eng. Mech.* (ASCE), in press (Sec. 9.4).

Kim, C-G., and Hong, C-S. (1988), "Viscoelastic Sandwich Plates with Crossply Faces," *J. Struct. Eng.* (ASCE), 114(1): 150–64.

Klyushnikov, V. D., and Tan, T. V. (1986), "Creep Stability: A Variant of the Theory and an Experiment," *Mech. Solids,* 21(3):89–98.

Koth, C. G. Kelly, J. M. (1989), "Viscoelastic Stability Model for Elastomeric Isolation Bearings," *J. Struct. Eng.* (ASCE), 115(2):285–301.

Kordina, K. (1975), "Langzeitversuche an Stahlbetonstützen," in *Deutscher Ausschuss für Stahlbeton,* Heft 250, W. Ernst und Sohn, Berlin, pp. 1–36 (Sec. 9.6).

Lazić, J. D., and Lazić, V. B. (1984), "Generalized Age-Adjusted Effective Modulus Method for Creep in Composite Beam Structures. Part I. Theory," *Cem. Concr. Res.,* 14:819–932, See also "Part II" (1985), 15:1–12 (Sec. 9.4).

Leipholz, H. (1970), *Stability Theory,* Academic Press, New York. See also 2nd. ed., John Wiley & Sons, Chichester (1987) (Sec. 9.2).

Leu, L.-J., and Yang, Y. B. (1989), "Viscoelastic Stability of Columns on Continuous Support," *J. Eng. Mech.* (ASCE), 115(7):1488–99.

Libove, C. (1952), "Creep Buckling of Columns," *J. Aero. Sci.,* 19:459–67 (Sec. 9.2).

Libove, C. (1953), "Creep Buckling Analysis of Rectangular-Section Columns," NACA Tech. Note No. 2956 (Sec. 9.3).

Lin, T. H. (1953), "Stresses in Columns with Time Dependent Elasticity," Proc. 1st Midwestern Conf. Solid Mech., pp. 196–99, Urbana, Illinois (Sec. 9.2).

Lin, T. H. (1956), "Creep Stresses and Deflections of Columns," *J. Appl. Mech.* (ASME), 23:214–18 (Sec. 9.2).

Maier, A. (1986), Berechnung der Momente des dauerbelasteten, schlanken Stahlbetondruckglieders nach der Theorie II. Ordnung," *VDI Zeitschrift,* 81(10):263–7.

Manuel, R. F., and MacGregor, J. G. (1967), "Analysis of Prestressed Reinforced Concrete Columns under Sustained Load," *ACI Journal,* 64:12–23 (Sec. 9.6).

Massonnet, C. (1974), "Buckling Behavior of Imperfect Elastic and Linearly Viscoelastic Structures," *Int. J. Solids Struct.,* 10(7):755–84.

Mauch, S. P. (1966), "Effect of Creep and Shrinkage on the Capacity of Concrete Columns," Symp. on Reinforced Concrete Columns, ACI Publication SP-50, Detroit, pp. 299–324 (Sec. 9.6).

Mauch, S., and Holley, M. J. (1963), "Creep Buckling of Reinforced Concrete Columns," *J. Struct. Eng.* (ASCE), 89(4):451–81 (Sec. 9.6).

McClure, G. S., Jr., Gerstle, K. H., and Tulin, L. G. (1973), "Sustained and Cyclic Loading of Concrete Beams," *J. Struct. Eng.* (ASCE), 99(2):243–57 (Sec. 9.6).

McHenry, D. (1943), "A New Aspect of Creep in Concrete and Its Application to Design," *Proc. ASTM,* 43:1069–86 (Sec. 9.1).

Mignot, F., and Puel, J. P. (1984), "Buckling of a Viscoelastic Rod," *Arch. Ration. Mech. Anal.,* 85(3):251–77.

Miyazaki, N. (1986), "On the Finite Element Formulation of Bifurcation Mode of Creep Buckling of Axisymmetric Shells," *Comput. Struct.,* 23(3):357–63.

Miyazaki, N. (1988), "Creep Buckling Analysis of Circular Cylindrical Shell under Both Axial Compression and Internal or External Pressure," *Comput. Struct.,* 28(4):437–41.

Miyazaki, N., Hagihara, S., and Munakata, T. (1988), "Application of the Finite Element Method to Elastic Plastic Creep Buckling Analysis of Partial Spherical Shell," *Nippon Kikai Gakkai Ronbunshu A Hen,* 54(500):794–8.

Mola, F. (1978), "Effetti della viscosità negli elementi pressonflessi in cemento armato (Creep Effects in Reinforced Concrete Elements Subjected to Bending Moment and Axial Load)," Atti del Congresso CTE, Siena, Italy.

Mola, F. (1983), "Stato limite di instabilità—Effetti della Viscosità," in *Progetto delle Strutture in Cemento Armato con il Metodo agli Stati Limite* (Mola, F., editor) Clup, Milano.

Mola, F., and Creazza, G. (1980), "General and Approximate Method for the Analysis of Linear Viscoelastic Structures Sensitive to Second Order Effects," Symposium on Fundamental Research on Creep and Shrinkage of Concrete, Lausanne.

Mola, F., Iori, I. (1979), "Influenza della viscosità sulla deformabilità di sezioni in cemento armato soggette a pressoflessione," in *Studi e Ricerche,* vol. 1, Politecnico di Milano, Milano.

Neville, A. M., Dilger, W. H., and Brooks, J. J. (1983), *Creep of Plain and Structural Concrete,* Construction Press–Longman, London, New York (Sec. 9.4).

Nielsen, L. F. (1970), "Kriechen und Relaxation des Betons," *Beton- und Stahlbetonbau,* 65:272–5 (Sec. 9.4).

Poshivalov, V. P. (1987), "Buckling of Stiffened Cylindrical Shells with Creep According to the Reinforcement Theory," *Mech. Solids,* 22(1):149–54.

Rabotnov, G. N., and Shesterikov, S. A. (1957), "Creep Stability of Columns and Plates," *J. Mech. Phys. Solids,* 6:27–34 (Sec. 9.2).

Rabotnov, Y. N. (1969), *Creep Problems in Structural Members,* North-Holland, Amsterdam (Sec. 9.3).

RILEM (Réunion Int. des Laboratoires d'Essais et de Recherches sur les Matériaux et les Constructions) Committee TC69 (1986), *Mathematical Modeling of Creep and Shrinkage of Concrete,* State-of-Art Report, Preprints, 4th International Symposium on Creep and Shrinkage of Concrete, pp. 41–455, ed. by Z. P. Bažant,

Evanston; also, Proceedings, ed. by Z. P. Bažant, John Wiley & Sons, Chichester and New York (1988) (Sec. 9.4).

Romanov, P. P. (1981), "Critical Time of Compression of Metal Rods under Creep Due to the Joint Effect of Constant Loads and Elevated Temperatures," *Sov. Appl. Mech.*, 17(8):770–4.

Roscoe, R. (1950), "Mechanical Models for the Representation of Viscoelastic Properties," *Br. J. Appl. Phys.* 1:171–3 (Sec. 9.1).

Rosenthal, D., and Baer, H. W. (1951), "An Elementary Theory of Creep Buckling of Columns," Proc. 1st U.S. Nat. Congr. Appl. Mech., pp. 603–11, Chicago (Sec. 9.2).

Rüsch, H., Jungwirth, D., and Hilsdorf, H. (1973), "Kritische Sichtung der Verfahren zur Berücksichtigung der Einflüsse von Kriechen und Schwinden des Betons auf das Verhalten der Tragwerke," *Beton- und Stahlbetonbau,* 63(3):49–60; 63(4):76–86; 63(5):152–8; see also discussion, 69(1974)(6):150–1 (Sec. 9.4).

Sjolind, S-G. (1985), "Viscoelastic Buckling Analysis of Floating Ice Sheets," *Cold. Reg. Sci. Technol.*, 11(3):241–6.

Solomentsev, Yu. E. (1985), "Thermal Stability of Nonuniformly Aging Viscoelastic Rods with Small Deformation," *Mech. Solids,* 20(5):185–8.

Szyszkowski, W., and Glockner, P. G. (1985). "Finite Deformation Analysis of Linearly Viscoelastic Simple Structures," *Int. J. Nonlinear Mech.*, 20(3):153–75 (Sec. 9.7).

Szyszkowski, W., and Glockner, P. G. (1987), "Further Results on the Stability of Viscoelastic Columns," *Trans. Can. Soc. Mech. Eng.*, 11(3):179–94.

Ting, E. C., and Chen, W. F. (1982), "Creep and Viscoelastic Buckling of Beams on Foundations," Purdue Univ., Sch. Civ. Eng. Struct., Tech. Rep. CE-STR No. 82-16.

Trost, H. (1967), "Auswirkungen des Superpositionsprinzips auf Kriech- und Relaxationsprobleme bei Beton und Spannbeton," *Beton und Stahlbetonbau,* 62(10):230–8; 62(11):261–9 (Sec. 9.4).

Vinogradov, A. M. (1985), "Nonlinear Effects in Creep Buckling Analysis of Columns," *J. Eng. Mech.* (ASCE), 111(6):757–67

Vinogradov, A. M. (1987), "Buckling of Viscoelastic Beam-Columns," *AIAA Journal,* 25(3):479–83.

Vinogradov, A. M., and Glockner, P. G. (1980), "Buckling of Spherical Viscoelastic Shells," *J. Struct. Eng.* (ASCE), 106(1):59–67.

Wang, S-C., and Meinecke, E. A. (1985a). "Buckling of Viscoelastic Columns, Part I: Constant Load Buckling," *Rubber Chem. Technol.*, 58(1):154–63.

Wang, S-C., and Meinecke, E. A. (1985b), "Buckling of Viscoelastic Columns: Part II: Constant Deformation Rate Buckling," *Rubber Chem. Technol.*, 58(1):164–75.

Warner, R. F., and Kordina, K. (1975), "Influence of Creep on the Deflections of Slender Reinforced Concrete Columns," in *Deutscher Ausschuss für Stahlbeton,* No. 250, pp. 39–156 (Sec. 9.4).

Warner, R. F., Rangan, B. V., and Hall, A. S. (1976), *Reinforced Concrete,* Pittman, Carlton, Australia (Sec. 9.4).

Whitney, G. S. (1932), "Plain and Reinforced Concrete Arches," *ACI Journal,* 28(7):479–519; see also discussion, 29:87–100 (Sec. 9.4).

Wilhelm, W. J., and Zia, P. (1970), "Effect of Creep and Shrinkage on Prestressed Concrete Columns," *J. Struct. Eng.* (ASCE), 96(10):2103–23 (Sec. 9.6).

Wilson, D. W., and Vinson, J. R. (1984), "Viscoelastic Analysis of Laminated Plate Buckling," *AIAA Journal,* 22(7):982–8.

Wojewodzki, W. (1973), "Buckling of Short Viscoplastic Cylindrical Shells Subjected to Radial Impulse," *Int. J. Non-Linear Mech.*, 8(4):325–43.

Wu, H., and Huggins, M. W. (1977), "Size and Sustained Load Effects in Concrete Columns," *J. Struct. Eng.* (ASCE), 103(3):493–506 (Sec. 9.6).

Zhang, W., and Ling, F. H. (1986), "Dynamic Stability of the Rotating Shaft Made of Boltzmann Viscoelastic Solid," *J. Appl. Mech.* (ASME), 53:424–9.

10

Stability of Inelastic Structures, Bifurcation and Thermodynamic Basis

Although in Chapter 8 (and partly also in Chap. 9) we succeeded to solve the equilibrium states, paths, and bifurcations for various practically important inelastic structures, we have not yet addressed the problem of stability. Neither have we discussed the general properties of path bifurcations and the general methods for their determination. We will now tackle these problems in this chapter.

First we will use the second law of thermodynamics to develop various useful criteria for stability of equilibrium states of inelastic structures. Then, on the basis of the second law, we will determine which equilibrium branch is actually followed after bifurcation and derive general criteria for the bifurcation states. We will illustrate our results using the example of an elastoplastic column. We will conclude by considering the stability aspects of small loading-unloading cycles and some stability problems of frictional materials.

10.1 THERMODYNAMIC CRITERIA OF STABLE STATE

As stated in the Lagrange-Dirichlet theorem (Theorem 3.6.1), the minimization of potential energy Π (Sec. 4.2) represents the fundamental criterion for stability of equilibrium of structures with conservative and dissipative forces. The proof in Section 3.6, however, is predicated on the existence of an elastic potential that is path-independent and is such that a given value of the potential corresponds to a unique contour in the space of generalized displacements. Closed contours around the origin indicate positive definiteness and guarantee stability.

In Chapters 4 to 7 we applied the potential-energy criterion to elastic structures. Although the Lagrange-Dirichlet theorem permits dissipative forces (such as damping) that do not preclude the existence of elastic potential, the potential-energy criterion is not applicable to inelastic structures, which do not possess a potential as their behavior is path-dependent. This behavior includes plasticity, as well as damage, fracture, and friction.

Stability of equilibrium of inelastic structures can in principle be analyzed on the basis of the dynamic definition of Poincaré and Liapunov (Sec. 3.5). In practice, however, this approach necessitates integrating the nonlinear equations of motion of the structure, which is actually quite complicated. Such studies have often been inconclusive.

A simpler approach is to use the second law of thermodynamics, as we will show now. The second law, which is not derivable from the laws of mechanics (i.e., Newton's laws), is in fact a more general and more fundamental approach to stability of equilibrium states and paths than are the laws of mechanics. Keep in mind, though, that since the second law of thermodynamics applies only to macroscopic systems that are in equilibrium or very close to equilibrium, the theromodynamic approach does not apply to systems that lose stability in a dynamic manner, which as we saw (Chap. 3) can happen under nonconservative loads.

The treatment of thermodynamics in the existing works on continuum or structural mechanics has generally been limited to thermodynamics of constitutive equations and field equations, while the thermodynamic aspects of structural stability have not received proper attention. We will, therefore, attempt a rather thorough exposition of this subject, following Bažant (1987a; 1988; 1989a, b).

First and Second Laws of Thermodynamics

The incremental work done on the structure by the loads is defined by

$$\Delta W = \sum_i P_i \, \Delta q_i = \mathbf{P}^T \, \Delta \mathbf{q} \leftrightarrow \int_S p_k \, \Delta u_k \, dS + \int_V F_k \, \Delta u_k \, dV \qquad (10.1.1)$$

Superscript T denotes a transpose, $\mathbf{q} = n$-dimensional column matrix (vector) of displacements or discrete deformation parameters q_i $(i = 1, \ldots, n)$ that characterize the state of the structure, $\mathbf{P}$ = column matrix (vector) of the given associated loads P_i such that $\mathbf{P}^T \, d\mathbf{q}$ represents the correct expression for work (note that some q_i may have to be introduced even if $P_i = 0$; see the example in Sec. 10.3); S, V = surface and volume of the structure; p_k, F_k = distributed surface and volume forces (loads) associated with displacements u_k that depend on the position vector $\mathbf{x}$ $(k = 1, 2, 3)$.

For the sake of brevity, we will be writing from now on simply $\Delta W = \mathbf{P}^T \, \Delta \boldsymbol{q}$, assuming this expression to be interchangeable with the integrals in Eq. 10.1.1. This is permissible because the concentrated loads can be regarded as the limit case of distributed loads having the form of Dirac delta functions, and vice versa the distributed loads can be regarded as the limiting case of infinitely small concentrated loads that are infinitely densely distributed. Note also that for fluids (the object of interest in the basic textbooks of thermodynamics) Equation 10.1.1 reduces to $\Delta W = -p \, \Delta V$ where p = pressure and V = volume of fluid. (We use the sign convention of solid mechanics; in fluid mechanics, ΔW is usually defined as $p \, \Delta V$, i.e., the work is considered positive if done *by* the system rather than *on* the system.)

According to the first law of thermodynamics, which expresses the law of conservation of energy, the total energy U of a structure (sometimes also called the internal energy) is incrementally defined by

$$\Delta U = \Delta W + \Delta Q \qquad (10.1.2)$$

Here ΔQ is the heat that has flowed into the structure from its surroundings.

To be able to apply the second law of thermodynamics, the system we consider must be in, or close to, thermodynamic equilibrium. Thus, we are allowed to consider only structures whose temperature is uniform or nearly so. The increment of (total) entropy of the structure is defined by

$$\Delta S = \frac{\Delta Q}{T} + (\Delta S)_{\text{in}} \tag{10.1.3}$$

in which T = absolute temperature, $(\Delta S)_{\text{in}}$ = internally produced increment of the entropy of the structure, and $\Delta Q/T$ = externally produced entropy increment, due to the influx of heat into the structure.

The second law of thermodynamics (e.g., Guggenheim, 1959; Denbigh, 1968; de Groot and Mazur, 1962; Brophy, Rose, and Wulff, 1964; Fermi, 1937; Planck, 1945) can be stated as follows: A change of state of a structure such that

$$(\Delta S)_{\text{in}} = \Delta S - \frac{\Delta Q}{T} \begin{cases} < 0 & \text{cannot occur} \\ = 0 & \text{can occur} \\ > 0 & \text{must occur} \end{cases} \tag{10.1.4}$$

Here, the case $(\Delta S)_{\text{in}} = 0$ is a change maintaining thermodynamic equilibrium, which is by definition reversible, and the case $(\Delta S)_{\text{in}} > 0$ is an irreversible change. In the following text, when we use the term "equilibrium" we will mean only mechanical equilibrium (i.e., equilibrium of forces) unless we explicitly state "thermodynamic equilibrium" or "reversible process."

An equilibrium state of a system is stable if no deviation from this state can take place by itself, that is, without any change in loads or boundary displacements. Therefore, the structure is

$$\begin{array}{lll} \text{Stable} & \text{if } (\Delta S)_{\text{in}} < 0 & \text{for all vectors } \delta\mathbf{q} \\ \text{Critical} & \text{if } (\Delta S)_{\text{in}} = 0 & \multirow{2}{*}{for some vector $\delta\mathbf{q}$} \\ \text{Unstable} & \text{if } (\Delta S)_{\text{in}} > 0 & \end{array} \tag{10.1.5}$$

This is the most fundamental criterion for stability of equilibrium of any physical system, whose concept was stated in 1875 by Willard Gibbs in the context of his work on chemical systems.

Note that for a system that is thermally isolated, that is, adiabatic ($\Delta Q = 0$), $(\Delta S)_{\text{in}}$ in Equation 10.1.5 may be replaced by ΔS.

After a structure (i.e. mechanical system) becomes unstable, the energy $-T(\Delta S)_{\text{in}}$ becomes the kinetic energy of the structure, and the structure is set into motion. Eventually the kinetic energy gets converted into heat due to dissipative processes such as viscosity, plasticity, friction, damage, and fracture. This is the practical significance of all the expressions for $-T\,(\Delta S)_{\text{in}}$ that we are going to present in this section.

Tangentially Equivalent Elastic Structure

The incremental mechanical properties of the structure are characterized by the relations $d\mathbf{f} = \mathbf{K}\,d\mathbf{q}$ or more precisely

$$d\mathbf{f}_T = \mathbf{K}_T(\mathbf{v})\,d\mathbf{q} \qquad \text{(isothermal, } dT = 0) \tag{10.1.6}$$

$$d\mathbf{f}_S = \mathbf{K}_S(\mathbf{v})\,d\mathbf{q} \qquad \text{(isentropic, } dS = 0) \tag{10.1.7}$$

which describe changes that can occur in mechanical equilibrium; $\mathbf{f}_T$ and $\mathbf{f}_S$ are the column matrices of the equilibrium forces f_{T_i} and f_{S_i} (or reactions) associated with $\mathbf{q}$, which depend on q_i so as to maintain mechanical equilibrium. Subscripts T and S label the isothermal and isentropic values of the tangential stiffness matrix $\mathbf{K}$, which can in general be history dependent and nonsymmetric. $\mathbf{K}_T$ and $\mathbf{K}_S$ are to be evaluated from the isothermal or isentropic tangential moduli of the material, which are not the same (see, e.g., Fung, 1965). $\mathbf{v}$ is the vector of direction cosines of the vector $d\mathbf{q}$ in the n-dimensional space of $q_1, \ldots, q_n$; $\mathbf{v} = d\mathbf{q}/(d\mathbf{q}^T d\mathbf{q})^{1/2}$. For inelastic materials, the value of $\mathbf{K}$ generally depends on $\mathbf{v}$; however, for many types of materials (e.g., plastic-hardening metals), there exist radial sectors (fans, cones) of directions $\mathbf{v}$ in the space of $d\mathbf{q}$ for which the value of $\mathbf{K}$ is constant.

To be able to analyze stability for the given inelastic structure, we need its equation of state. In thermodynamics, the equation of state describes the reversible properties of a system or structure (e.g., ideal gas equation $pV = RT$). It might seem that the equation of state is represented by the constitutive relation of the material or the corresponding force-displacement relation of the structure (i.e., Eqs. 10.1.6 and 10.1.7). The inelastic constitutive relation, however, is *not* an equation of state because it is irreversible, as the material is inelastic.

On the other hand, an elastic stress-strain relation of a material, or an elastic force-displacement relation of a structure, is an equation of state. This observation provides us a clue. As we know, the response of an inelastic structure can be solved in small loading steps by a series of quasi-elastic incremental analyses. The exact solution is the limiting case of infinitely small steps, which corresponds to approximating the given inelastic stress-strain or force-displacement relation in each loading step by its tangential approximation (Eq. 10.1.6 or 10.1.7) which has the form of an elastic stress-strain or force-displacement relation. In stability analysis, we need to deal only with infinitely small loading increments. Therefore, we may replace the given inelastic structure by a tangentially equivalent elastic structure whose elastic stiffness matrix is equal to $\mathbf{K}(\mathbf{v})(= \mathbf{K}_T$ or $\mathbf{K}_S)$.

Of course, there may exist many such structures since there may be many matrices $\mathbf{K}$ depending on the loading direction $\mathbf{v}$ (there are at least two matrices $\mathbf{K}$—one for loading and one for unloading). So one has to try various tangentially equivalent elastic structures, each corresponding to a different matrix $\mathbf{K}$. Among these, only those for which the solution $\delta\mathbf{q}$ is within the assumed sector of directions are valid.

To sum up, when the force-displacement relations in Equations 10.1.6 and 10.1.7 are used in the sense of the tangentially equivalent elastic structure, they do represent an *equation of state* because elastic deformation is a reversible process that preserves thermodynamic equilibrium. Certain hypotheses implied in this approach will be discussed in the subsection "Hypothesis Implied in Present Thermodynamic Approach" in Section 10.3. Introduction of the tangentially equivalent elastic structure is inevitable if the question of stability of equilibrium should be answered. Without this concept, we would have no state of thermodynamic equilibrium to analyze, and so the question of stability could not even be posed.

Total Energy U and Helmholtz Free Energy F

In practice, it is often convenient to express the stability criterion in Equation 10.1.5 in terms of other thermodynamic state functions. Introducing $dQ =$

$T[dS - (dS)_{in}]$ into Equation 10.1.2 (and replacing increments Δ by differentials d), we get

$$dU = \mathbf{P}^T d\mathbf{q} + T\,dS - T(dS)_{in}. \tag{10.1.8}$$

In situations in which the temperature is controlled, it is convenient to use the transformation $F = U - TS$, called the Legendre transformation, to introduce a new thermodynamic state function, F, called the Helmholtz free energy. Differentiating, we have $dF = dU - d(TS) = dU - S\,dT - T\,dS$, and substituting for dU from Equation 10.1.6, the terms $T\,dS$ cancel and we get

$$dF = \mathbf{P}^T d\mathbf{q} - S\,dT - T(dS)_{in} \tag{10.1.9}$$

For changes during which thermodynamic equilibrium is maintained, we have $(dS)_{in} = 0$. For such changes, U as well as F are reversible and path-independent, and so they represent thermodynamic potentials (Table 10.1.1). Their total differentials are $dU = \mathbf{P}^T d\mathbf{q} + T\,dS$ and $dF = \mathbf{P}^T d\mathbf{q} - S\,dT$, and consequently $U = U(\mathbf{q}, S)$ and $F = F(\mathbf{q}, T)$. Since dF involves dT, function F is convenient for isothermal conditions ($dT = 0$), and since dU involves dS, function U is convenient for isentropic conditions ($dS = 0$). Equations 10.1.8 and 10.1.9 mean that for irreversible processes $dU < \mathbf{P}^T d\mathbf{q} + T\,dS$ and $dF < \mathbf{P}^T d\mathbf{q} - S\,dT$ (see, e.g., eq. 11.9 and p. 8 in Guggenheim, 1959).

For isothermal ($dT = 0$) or isentropic ($dS = 0$) deformation increments at which mechanical equilibrium is maintained, we may write

$$dF = \mathbf{f}_T^T d\mathbf{q} \qquad \text{for } dT = 0 \tag{10.1.10}$$

$$dU = \mathbf{f}_S^T d\mathbf{q} \qquad \text{for } dS = 0 \tag{10.1.11}$$

in which superscript T denotes a transpose and the equilibrium forces (reactions). $\mathbf{f}_T$ and $\mathbf{f}_S$ are to be determined according to the isothermal or isentropic material properties (Eqs. 10.1.6–10.1.7). These forces must be distinguished from the applied loads P_i. In the initial equilibrium state we have $f_{T_i} = f^0_{T_i} = P_i$ or $f_{S_i} = f^0_{S_i} = P_i$ ($i = 1, \ldots, n$), but after the incremental deformation, f_{T_i} or f_{S_i} generally differ from P_i because the applied loads obey their own law such as gravity (or fluid pressure, centrifugal force, electrostatic force, electromagnetic force, aerodynamic force, hydraulic force, etc.).

Although Equations 10.1.10 and 10.1.11, having the form of incremental work, might be intuitively clear, they in fact follow from Equations 10.1.8 and 10.1.9 in which dS or dT vanishes and the term $-T\,(dS)_{in}$ must be omitted. The reason for omitting $(dS)_{in}$ is that, instead of the given inelastic structure, we consider the tangentially equivalent elastic structure whose deformation is reversible, as we already said.

Equations 10.1.8 to 10.1.11 imply that

$$f_{T_i} = \frac{\partial F}{\partial q_i} \qquad f_{S_i} = \frac{\partial U}{\partial q_i} \tag{10.1.12}$$

Because of the principle of conservation of energy (the first law of thermodynamics), we have for an equilibrium deformation path

$$F = \int_V \int_\varepsilon \boldsymbol{\sigma}_T : d\boldsymbol{\varepsilon}\,dV \qquad U = \int_V \int_\varepsilon \boldsymbol{\sigma}_S : d\boldsymbol{\varepsilon}\,dV \tag{10.1.13}$$

where V = volume of the body, $\delta\boldsymbol{\varepsilon}$ = strain tensor increment, $\boldsymbol{\sigma}_T$, $\boldsymbol{\sigma}_S$ = current stress tensor calculated (from current $\boldsymbol{\varepsilon}$ and its history) on the basis of isothermal or isentropic material properties, and (:) denotes a tensor product contracted on two indices. Note that when the problem is geometrically nonlinear, the equality of the work of stresses (Eqs. 10.1.13) to the work of reactions (Eqs. 10.1.10 and 10.1.11) is achieved only for some appropriate types of stress and strain tensors (e.g., the second Piola-Kirchhoff stress tensor and the Lagrangian finite strain tensor); see Chapter 11 (also, e.g., Hill, 1958). If the problem is geometrically linear, the equivalence of Equations 10.1.13 and Equation 10.1.10 or 10.1.11 can also be proven from the principle of virtual work, because $\boldsymbol{\sigma}$ is in equilibrium with $\mathbf{f}$ and $\boldsymbol{\varepsilon}$ is compatible with $\mathbf{q}$.

When the structure is loaded through some elastic device (a spring, a testing frame), it is convenient to consider this device as part of the structure. Its potential energy is then included in F or U, and its flexibility or stiffness is included in the calculation of structural stiffness (e.g., in Eq. 10.1.25 below).

Functions F and U represent the Helmholtz free energy and the total energy of the structure alone, without the loads. If the loads are conservative, it is more useful for stability analysis to introduce the Helmholtz free energy, $\mathscr{F}$, and the total energy, $\mathscr{U}$, of the structure-load system. Then, because the energy of the loads is $-W$ (see Sec. 4.1),

$$\mathscr{F} = F - W \tag{10.1.14}$$

$$\mathscr{U} = U - W \tag{10.1.15}$$

Obviously, these definitions represent generalizations of the concept of potential energy Π (Eq. 4.1.1). According to Equations 10.1.9 and 10.1.8 we have

$$d\mathscr{F} = -S\,dT - T\,(dS)_{\text{in}} \tag{10.1.16}$$

$$d\mathscr{U} = T\,dS - T\,(dS)_{\text{in}} \tag{10.1.17}$$

for increments preserving mechanical equilibrium ($\mathbf{f}_T = \mathbf{P}$, $\mathbf{f}_S = \mathbf{P}$) (but not necessarily thermal equilibrium). These equations have the advantage that, compared to Equations 10.1.9 and 10.1.8, $\mathbf{P}^T d\mathbf{q}$ does not appear in them. If thermodynamic equilibrium is disturbed *only* mechanically ($\mathbf{f}_T \neq \mathbf{P}$ or $\mathbf{f}_S \neq \mathbf{P}$) (but not thermally) according to Equations 10.1.10, 10.1.11, and 10.1.1 we may also write

$$d\mathscr{F} = \mathbf{f}_T^T\,d\mathbf{q} - \mathbf{P}^T\,d\mathbf{q} \qquad \text{at } T = \text{const.} \tag{10.1.18}$$

$$d\mathscr{U} = \mathbf{f}_S^T\,d\mathbf{q} - \mathbf{P}^T\,d\mathbf{q} \qquad \text{at } S = \text{const.} \tag{10.1.19}$$

$\mathscr{F}$ represents the isothermal potential energy and $\mathscr{U}$ the isentropic potential energy.

When Equations 10.1.16–10.1.17 are applied to an elastic (or tangentially equivalent elastic) structure, $(dS)_{\text{in}}$ of course does not represent an entropy produced within the structure *per se* because the deformations of an elastic structure are reversible. Rather, $(dS)_{\text{in}}$ represents the entropy produced in the structure-load *system* due to disequilibrium *between* the structure and the load.

Second Variation of $\mathscr{F}$ or $\mathscr{U}$

Consider now a change from the initial equilibrium state $\mathbf{q} = \mathbf{q}^0$ to an adjacent state $\mathbf{q} = \mathbf{q}^0 + \delta\mathbf{q}$, which may, but generally need not, be an equilibrium state.

By introducing the tangentially equivalent elastic structure, we suppose the inelastic response to be path-independent in the small (i.e., infinitesimally). We expand $\mathcal{F}$ into multiple Taylor series, writing $\Delta\mathcal{F} = \delta\mathcal{F} + \delta^2\mathcal{F} + \delta^3\mathcal{F} + \cdots$ where $\delta^n\mathcal{F} = (1/n!)\sum_i \cdots \sum_r (\partial^n\mathcal{F}/\partial q_i\,\partial q_j \cdots \partial q_r)\,\delta q_i\,\delta q_j \cdots \delta q_r = (1/n!)\sum_i \cdots \sum_r [\partial^{n-1}(\partial\mathcal{F}/\partial q_i)/\partial q_j \cdots \partial q_r]\,\delta q_i\,\delta q_j \cdots \delta q_r$, and substitute $\partial\mathcal{F}/\partial q_i = f_{T_i} - P_i$ (according to Eq. 10.1.18). Also, we assume that the applied loads P_i are constant, that is, dead loads. (However, if there are variable conservative loads, they may be incorporated into the structural system so that the remaining loads are constant.) For the change of the Helmholtz free energy of the structure-load system, we thus obtain the expression

$$\Delta\mathcal{F} = \int_{\mathbf{q}^0}^{\mathbf{q}^0+\delta\mathbf{q}} [\mathbf{f}_T^T(\mathbf{q})\,d\mathbf{q} - \mathbf{P}^T\,d\mathbf{q}] = \sum_i \int_{q_i^0}^{q_i^0+\delta q_i} \left\{ f_{T_i}^0 + \frac{1}{2!}\sum_j \left[\frac{df_{T_i}}{dq_j}\right]^0 (q_j - q_j^0) + \frac{1}{3!}\sum_j\sum_k \left[\frac{d^2 f_{T_i}}{dq_j\,dq_k}\right]^0 (q_j - q_j^0)(q_k - q_k^0) + \cdots - P_i \right\} dq_i \quad (10.1.20)$$

in which the initial state q_i^0 is labeled by superscript 0. Since the initial state is an equilibrium state, we have $f_{T_i}^0 = P_i$ (but generally $f_{T_{i,j}}^0 \neq P_{i,j}$ and $f_{T_{i,jm}}^0 \neq P_{i,jm}$). For the case of dead loads we further have $P_i = \text{const.}$ (and $P_{i,j} = 0$). Consequently, the first-order work terms due to $f_{T_i}^0$ are canceled in Equation 10.1.20 by the work of P_i. Integrating, neglecting the terms of higher than second order, and noting that $\sum_j [\partial f_{T_i}/\partial q_j]^0\,\delta q_j = \delta f_{T_i} =$ first-order equilibrium change of reactions, we obtain

$$\Delta\mathcal{F} = \delta^2\mathcal{F} = \tfrac{1}{2}\delta\mathbf{f}_T^T\,\delta\mathbf{q} \quad (10.1.21)$$

where $\delta^2\mathcal{F} = \delta^2 F =$ second-order change of the Helmholtz free energy of the structure, which is equal to the second-order work of isothermal equilibrium reactions due to δq_i (see the triangular area in Fig. 10.1b, in which the first-order variation $\delta\mathcal{F} = \delta F - \delta W$ is represented by the rectangular areas δF and δW).

By a procedure analogous to Equations 10.1.20 and 10.1.21, in which $\mathcal{F}$, F, and subscripts T are replaced by $\mathcal{U}$, U, and subscripts S, one finds that

$$\Delta\mathcal{U} = \delta^2\mathcal{U} = \tfrac{1}{2}\delta\mathbf{f}_S^T\,\delta\mathbf{q} \quad (10.1.22)$$

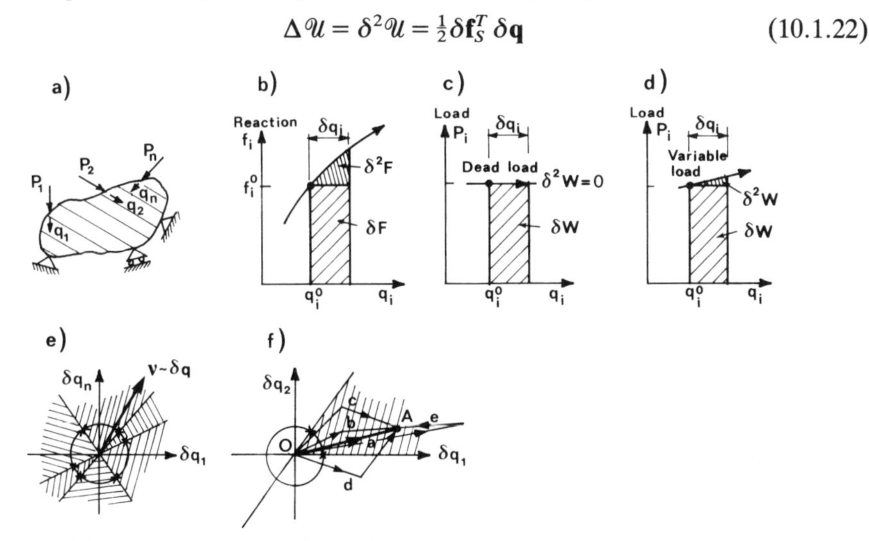

Figure 10.1 (a) Inelastic structure; (b, c, d) first- and second-order variations of Helmholtz free energy; (e, f) path independence limited to a certain range and path dependence.

The increments of isothermal or isentropic reactions occurring at mechanical equilibrium can be expressed as $\delta\mathbf{f}_T = \mathbf{K}_T(\mathbf{v})\,\delta\mathbf{q}$, $\delta\mathbf{f}_S = \mathbf{K}_S(\mathbf{v})\,\delta\mathbf{q}$ (Eqs. 10.1.6–10.1.7), and substitution into Equations 10.1.21 and 10.1.22 yields

$$\Delta\mathcal{F} = \delta^2\mathcal{F} = \tfrac{1}{2}\delta\mathbf{q}^T\mathbf{K}_T(\mathbf{v})\,\delta\mathbf{q} \tag{10.1.23}$$

$$\Delta\mathcal{U} = \delta^2\mathcal{U} = \tfrac{1}{2}\delta\mathbf{q}^T\mathbf{K}_S(\mathbf{v})\,\delta\mathbf{q} \tag{10.1.24}$$

According to Equations 10.1.21 and 10.1.22,

$$K_{T_{ij}} = \frac{\partial f_{T_i}}{\partial q_j} \qquad K_{S_{ij}} = \frac{\partial f_{S_i}}{\partial q_j} \tag{10.1.25}$$

Here $K_{T_{ij}}$ and $K_{S_{ij}}$ are the components of the tangential (incremental) isothermal ($dT = 0$) or isentropic ($dS = 0$) stiffness matrices $\mathbf{K}_T(\mathbf{v})$ and $\mathbf{K}_S(\mathbf{v})$ of the structure that are associated with q_i and must be evaluated on the basis of isothermal or isentropic tangential moduli (stiffnesses) of the material. These moduli characterize the total strain increments representing the sum of elastic and inelastic (irreversible) strain increments. The isothermal and isentropic moduli are not the same (see, e.g., Fung, 1965). For a discussion of the differences see Biot (1965, p. 286).

An important feature (which, as we will see, spoils path-independence) is that the tangential stiffness matrix $\mathbf{K}_T$ or $\mathbf{K}_S$ in general depends on the vector $\mathbf{v}$ of the direction cosines (Fig. 10.1e) of the displacements $\delta\mathbf{q}$ in the n-dimensional space of $q_1, \ldots, q_n$; $\mathbf{v} = \delta\mathbf{q}/(\delta\mathbf{q}^T\,\delta\mathbf{q})^{1/2}$. As we will see from the example in Section 10.3, the tangential stiffness matrix for the directions that represent loading-only everywhere in the structure differs from the tangential stiffness matrices which represent unloading-only or some of the combinations of loading in one part of the structure and unloading in another part of the structure.

Note that the tangential stiffnesses $\mathbf{K}_T$ or $\mathbf{K}_S$ include not only the stiffness for the reversible (elastic) part of the deformation but also the stiffness for the irreversible part of the deformation due to plasticity, damage, or fracture. The fact that the irreversible part of the response is entirely covered by these matrices makes it possible to introduce the tangentially equivalent elastic structures. This in turn justifies the omission of the term $-T(\Delta S)_{\text{in}}$ from Equations 10.1.18 and 10.1.19 as well as Equations 10.1.21 and 10.1.22, on which we already commented.

The meaning of isentropic deformations deserves some discussion. The term $-T(\Delta S)_{\text{in}}$ does not and should not appear in Equations 10.1.18 and 10.1.19 [that is, $(\Delta S)_{\text{in}} = 0$ must be used for these equations] because the tangential stiffnesses $\mathbf{K}_T$, $\mathbf{K}_S$ already comprise all incremental irreversible responses (plastic, damage, fracture). Consequently $dQ = T\,dS$ (Eq. 10.1.3) in conjunction with these equations. It follows that the isentropic conditions, $dS = 0$, are equivalent to adiabatic conditions, $dQ = 0$, if the tangentially equivalent elastic structure is considered. So the isentropic incremental stiffness matrix $\mathbf{K}_S$ is then equivalent to the adiabatic incremental stiffness matrix, which is calculated on the basis of adiabatic tangential elastic moduli.

The adiabatic deformations are approximated by very fast deformations, since heat transfer to material elements requires some time and is negligible for short times. The isothermal deformations are approximated by very slow deformations, which allow enough time for the temperature to equilibrate. Although the difference between isothermal and adiabatic material properties is often neglected, it may be quite appreciable.

Path Dependence and Incremental Potentials

Internal friction, damage, and fracture present potent physical arguments against symmetry of the tangential stiffness matrix $\mathbf{K}_T$ or $\mathbf{K}_S$ (see Bažant, 1980 and 1984; Bažant and Prat, 1988; Mandel, 1964). Therefore, the thermodynamic state functions $F(\mathbf{q})$, $U(\mathbf{q})$, $\mathcal{F}(\mathbf{q})$, $\mathcal{U}(\mathbf{q})$ do not have to be path-independent potentials. However, under certain frequently used simplifying hypotheses about the constitutive law that exclude friction, damage, and fracturing (e.g., Drucker's postulate in plasticity), the tangential stiffness matrix must be symmetric (Hill and Rice, 1973). Then Equations 10.1.6 and 10.1.7 imply that

$$K_{T_{ij}}(\mathbf{v}) = \frac{\partial f_{T_i}}{\partial q_j} = \frac{\partial f_{T_j}}{\partial q_i} = \frac{\partial^2 F}{\partial q_i\, \partial q_j} \tag{10.1.26}$$

$$K_{S_{ij}}(\mathbf{v}) = \frac{\partial f_{S_i}}{\partial q_j} = \frac{\partial f_{S_j}}{\partial q_i} = \frac{\partial^2 U}{\partial q_i\, \partial q_j} \tag{10.1.27}$$

This means that the tangential stiffness matrix represents the Hessian of function F or U with respect to the displacement vector $\mathbf{q}$. Furthermore, if P_i is independent of q_i (see Eqs. 10.1.38 and 10.1.39), we also have

$$K_{T_{ij}}(\mathbf{v}) = \frac{\partial^2 \mathcal{F}}{\partial q_i\, \partial q_j} \qquad K_{S_{ij}}(\mathbf{v}) = \frac{\partial^2 \mathcal{U}}{\partial q_i\, \partial q_j} \qquad \text{(dead loads)} \tag{10.1.28}$$

For such symmetric stiffness matrices, functions $\mathcal{F}$ and $\mathcal{U}$, or F and U, might be expected to be path-independent, and thus to represent incremental potentials. Not so, however.

These functions would be path-independent only if $\mathbf{K}_T$ or $\mathbf{K}_S$ were independent of the path direction $\mathbf{v}$ in the space of $\delta\mathbf{q}$. The consequence of the dependence of $\mathbf{K}_T$ and $\mathbf{K}_S$ on $\mathbf{v}$ may be explained by means of Figure 10.1f. We assume $\mathbf{K}_T(\mathbf{v})$ has the value $\mathbf{K}_T^a$ for $\mathbf{v}$-directions in the cross-hatched sector, and the value $\mathbf{K}_T^b$ for directions outside this sector. We consider five infinitesimally different paths a, b, c, d, e leading to the same point A. Path a is radial. For paths b, c, d, e, the infinitesimal increment $\delta\mathbf{q} = \overrightarrow{OA}$ is a sum of two infinitesimal subincrements $\delta\mathbf{q}_{\text{I}}$ and $\delta\mathbf{q}_{\text{II}}$ corresponding to the two straight path segments shown. We must now associate direction $\mathbf{v}$ with each of these two path segments rather than with the chord vector $\overrightarrow{OA}$.

Path b leads at point A to the same value of $\mathcal{F}$ as path a because at all times the direction of this path belongs to the same sector of $\mathbf{v}$-directions, particularly to the cross-hatched sector for which $\mathbf{K}_T(\mathbf{v}) = \mathbf{K}_T^a$. Path c, however, leads at point A to a different value of $\mathcal{F}$ because the direction of the second straight segment of this path belongs outside the cross-hatched sector. Path d, which goes out and into the sector, obviously also leads to a different value of $\mathcal{F}$. So does path e, which lies entirely in the sector, because the direction of its second segment (which involves unloading) lies outside the sector of $\mathbf{v}$-directions.

Consequently, if $\mathbf{K}_T$ or $\mathbf{K}_S$ depends on direction $\mathbf{v}$, then the thermodynamic state functions F, U, $\mathcal{F}$, and $\mathcal{U}$ are not in general path-independent, that is, they do not represent potentials in the mathematical sense. If the dependence of $\mathbf{K}_T$ or

$\mathbf{K}_S$ on the direction $\mathbf{v}$ is piecewise constant—a behavior typical of plasticity—then F, U, $\mathcal{F}$, and $\mathcal{U}$ are path-independent for a certain finite range of paths (such as paths a and b in Fig. 10.1f) and do represent potentials restricted to this range. This range includes all the radial and nonradial paths such that the path direction always belongs to the range of directions for the sector in which the point on the path lies. This means that the boundary between two sectors of constant $\mathbf{v}$ cannot be crossed (path d in Fig. 10.1f), that is, the path must lie entirely within one sector, and also no segment of the path must represent unloading (path e).

If the dependence on $\mathbf{v}$ is not piecewise constant but continuous—a behavior exhibited, for example, by endochronic theory—then F, U, $\mathcal{F}$, and $\mathcal{U}$ are never path-independent and do not represent potentials, not even within a limited range.

The lack of path independence caused by $\mathbf{v}$ has some important consequences. In Section 10.3 we will see that it may cause the equilibrium states on all the postbifurcation branches to be stable.

Second-Order Work of Stresses and Geometric Stiffness

According to the principle of conservation of energy (the first law of thermodynamics), the second-order work of reactions must be equal to the second-order work of stresses. Therefore,

$$\delta^2\mathcal{F} = \tfrac{1}{2}\delta\mathbf{f}_T^T\,\delta\mathbf{q} = \int_{V_0} \tfrac{1}{2}\delta\boldsymbol{\sigma}_T : \delta\mathbf{e}\,dV_0 + \delta^2\mathcal{W}_\sigma = \int_{V_0} \tfrac{1}{2}\delta\mathbf{e} : \mathbf{E}_T : \delta\mathbf{e}\,dV_0 + \delta^2\mathcal{W}_\sigma \quad (10.1.29)$$

$$\delta^2\mathcal{U} = \tfrac{1}{2}\delta\mathbf{f}_S^T\,\delta\mathbf{q} = \int_{V_0} \tfrac{1}{2}\delta\boldsymbol{\sigma}_S : \delta\mathbf{e}\,dV_0 + \delta^2\mathcal{W}_\sigma = \int_{V_0} \tfrac{1}{2}\delta\mathbf{e} : \mathbf{E}_S : \delta\mathbf{e}\,dV_0 + \delta^2\mathcal{W}_\sigma \quad (10.1.30)$$

Here V_0 = initial volume of the body (before displacements $\delta\mathbf{u}$); the colon denotes a product of tensors contracted on two indices (for example, $\boldsymbol{\sigma} : \boldsymbol{\varepsilon} = \sigma_{ij}\varepsilon_{ij}$ where repeated subscripts imply summation over $i = 1, 2, 3$); $\mathbf{E}_T$, $\mathbf{E}_S$ = fourth-order tensors of isothermal or isentropic (adiabatic) tangential moduli, which have different values for loading and unloading, $\delta\mathbf{e}$ = field of linearized (small) strain increments associated with displacements $\delta\mathbf{q}$, and $\delta^2\mathcal{W}_\sigma$ = work of the initial stresses $\boldsymbol{\sigma}_0$ on the second-order parts $\delta\boldsymbol{\varepsilon}^{(2)}$ of the finite strain increments $\delta\boldsymbol{\varepsilon}$ corresponding to $\delta\mathbf{q}$, that is,

$$\delta^2\mathcal{W}_\sigma = \int_{V_0} \boldsymbol{\sigma}_0 : \delta\boldsymbol{\varepsilon}^{(2)}\,dV_0 \quad (10.1.31)$$

The reason that the work $\boldsymbol{\sigma}_0 : \delta\boldsymbol{\varepsilon}^{(2)}$ must be included is that it is also second-order small, same as $\frac{1}{2}\delta\boldsymbol{\sigma} : \delta\mathbf{e}$. On the other hand, the reason that we can write $\frac{1}{2}\delta\boldsymbol{\sigma} : \delta\mathbf{e}$ instead of $\frac{1}{2}\delta\boldsymbol{\sigma} : \delta\boldsymbol{\varepsilon}$ is that $\delta\boldsymbol{\sigma}$ is small, so that the second-order part of $\delta\boldsymbol{\varepsilon}$ would contribute only to third-order work. An illustration and explanation of the work of $\delta\boldsymbol{\varepsilon}^{(2)}$ was given in Sections 4.3 and 6.1. For general three-dimensional formulation, see Chapter 11.

In the absence of thermal and inelastic effects, Equation 10.1.29 or 10.1.30 must reduce to the potential-energy expressions given in Chapter 4, and indeed it does (e.g., Eq. 4.3.10).

When the strains $\delta\boldsymbol{\varepsilon}$ and $\delta\boldsymbol{\varepsilon}^{(2)}$ are expressed in terms of $\delta\mathbf{q}$, $\delta^2\mathcal{W}_\sigma$ must be quadratic in $\delta\mathbf{q}$. So we may further write $\delta^2\mathcal{W}_\sigma = \frac{1}{2}\delta\mathbf{q}^T\mathbf{K}^\sigma\,\delta\mathbf{q}$ where $\mathbf{K}^\sigma$ is the geometric stiffness matrix, which depends only on the initial stresses and the geometry of the structure. It is independent of the tangential moduli of the material, and for this reason it is also independent of the loading direction $\mathbf{v}$. Its components are

$$K_{ij}^\sigma = \frac{\partial^2(\delta^2\mathcal{W}_\sigma)}{\partial q_i\,\partial q_j} \tag{10.1.32}$$

because $\delta^2\mathcal{W}_\sigma = \frac{1}{2}\sum_i\sum_j K_{ij}^\sigma(q_i - q_i^0)(q_j - q_j^0)$ with $q_i - q_i^0 = \delta q_i$.

Note that in stability analysis the calculation of $\delta\mathbf{f}_T$ (or $\delta\mathbf{f}_S$) from $\delta\mathbf{q}$ must be based on the equilibrium conditions for the deflected structure and must include the second-order effects of the initial forces $\mathbf{f}^0$ or the initial stresses $\boldsymbol{\sigma}^0$ (we will illustrate that in Sec. 10.3). In view of the split of the work expressions in Equations 10.1.29 and 10.1.30, we may also write

$$\delta\mathbf{f}_T = [\mathbf{K}_T^0(\mathbf{v}) + \mathbf{K}^\sigma]\,\delta\mathbf{q} \qquad \delta\mathbf{f}_S = [\mathbf{K}_S^0(\mathbf{v}) + \mathbf{K}^\sigma]\,\delta\mathbf{q} \tag{10.1.33}$$

$$\delta^2\mathcal{F} = \tfrac{1}{2}\,\delta\mathbf{q}^T[\mathbf{K}_T^0(\mathbf{v}) + \mathbf{K}^\sigma]\,\delta\mathbf{q} \qquad \delta^2\mathcal{U} = \tfrac{1}{2}\,\delta\mathbf{q}^T[\mathbf{K}_S^0(\mathbf{v}) + \mathbf{K}^\sigma]\,\delta\mathbf{q} \tag{10.1.34}$$

in which $\mathbf{K}_T^0(\mathbf{v})$ and $\mathbf{K}_S^0(\mathbf{v})$ are the stiffness matrices calculated from the isothermal or isentropic tangential moduli neglecting the second-order geometric effect of the initial stresses (or initial loads). (For details, see Eq. 11.8.8.)

In the case of a single load P, we may also write $\delta^2\mathcal{W}_\sigma = P\,\delta u^{(2)}$ where $\delta u^{(2)}$ is the second-order part of the load-point displacement due to $\delta\mathbf{q}$ [for example, $\delta u^{(2)} = \Delta l$ = shortening of a column due to the deflections $w(x)$; see Sec. 4.3].

If the special case that $\boldsymbol{\sigma}_0 : \delta\boldsymbol{\varepsilon}^{(2)}$ is negligible compared to $\delta\boldsymbol{\sigma} : \delta\mathbf{e}$, we may neglect $\delta^2\mathcal{W}_\sigma$. The resulting equation can alternatively be justified also by the principle of virtual work, because $\delta\mathbf{f}_T$ (or $\delta\mathbf{f}_S$) are in equilibrium with $\delta\boldsymbol{\sigma}_T$ (or $\delta\boldsymbol{\sigma}_S$), while $\delta\mathbf{q}$ is compatible with $\delta\boldsymbol{\varepsilon}$. However, the principle of virtual work in this form is inapplicable when the second-order geometric effects of finite strain are not negligible (and a correction due to the geometric stiffness matrix is required; see, e.g., Maier, 1971, and Maier and Drucker, 1973).

Criterion of Stable State for the Case of Dead Loads

Assume that all the loads are specified as constants, that is, dead loads. Many of the loads P_i can, of course, be zero since this is a special case of dead loads. It is also admissible that some displacements are prescribed, in which case $\delta q_i = 0$ for these displacements. From Equations 10.1.14, 10.1.15, 10.1.19, and 10.1.20 we obtain

1. for isothermal conditions ($dT = 0$):

$$-T(\Delta S)_{\text{in}} = \Delta\mathcal{F} = \tfrac{1}{2}\delta\mathbf{f}_T^T\,\delta\mathbf{q} \tag{10.1.35}$$

2. For isentropic conditions ($dS = 0$):

$$-T(\Delta S)_{\text{in}} = \Delta\mathcal{U} = \tfrac{1}{2}\delta\mathbf{f}_S^T\,\delta\mathbf{q} \tag{10.1.36}$$

Substituting this into Equation 10.1.5, the criterion of stable state is expressed in terms of the second-order work of the equilibrium reactions. This criterion may be restated as follows. An equilibrium state of a structure is

$$\begin{array}{lll} \text{Stable} & \text{if } -T(\Delta S)_{\text{in}} = \Delta \mathscr{W} = \tfrac{1}{2}\delta \mathbf{f}^T\, \delta \mathbf{q} = \tfrac{1}{2}\delta \mathbf{q}^T \mathbf{K}(\mathbf{v})\, \delta \mathbf{q} > 0 & \text{for all vectors } \delta \mathbf{q} \\ \text{Critical} & \text{if } -T(\Delta S)_{\text{in}} = \Delta \mathscr{W} = \tfrac{1}{2}\delta \mathbf{f}^T\, \delta \mathbf{q} = \tfrac{1}{2}\delta \mathbf{q}^T \mathbf{K}(\mathbf{v})\, \delta \mathbf{q} = 0 & \multirow{2}{*}{for some vector $\delta \mathbf{q}$} \\ \text{Unstable} & \text{if } -T(\Delta S)_{\text{in}} = \Delta \mathscr{W} = \tfrac{1}{2}\delta \mathbf{f}^T\, \delta \mathbf{q} = \tfrac{1}{2}\delta \mathbf{q}^T \mathbf{K}(\mathbf{v})\, \delta \mathbf{q} < 0 & \end{array} \tag{10.1.37}$$

in which we introduce the tangential stiffness matrix from Equation 10.1.21 or 10.1.22; for isothermal conditions $\delta \mathbf{f} = \delta \mathbf{f}_T$, $\mathbf{K} = \mathbf{K}_T$, $\Delta \mathscr{W} = \Delta \mathscr{F}$, while for isentropic conditions $\delta \mathbf{f} = \delta \mathbf{f}_S$, $\mathbf{K} = \mathbf{K}_S$, and $\Delta \mathscr{W} = \Delta \mathscr{U}$. Thus, the state is stable if the second-order work of the reaction increments is positive definite.

Note that the criterion of incremental work, which was stated for elastic structures in Section 4.9, is a special case of Equation 10.1.37.

Consider now the special case of an elastic structure with a single dead load P, associated displacement q, and stiffness K. According to Equation 10.1.50, the entropy change of the structure-load system at a small deviation δq away from the equilibrium state $q = P/K$ is $(\Delta S)_{\text{in}} = -\frac{1}{2}K\, \delta q^2/T$. One might at first have doubts since no entropy can be produced in an elastic structure. True, but we have a structure-load system. Since $(\Delta S)_{\text{in}} = 0$ characterizes changes that maintain thermodynamic equilibrium, $(\Delta S)_{\text{in}}$ must be nonzero for changes away (or toward) the equilibrium state, even if the structure is elastic.

The present stability criterion in terms of the work of equilibrium reactions was stated and rigorously proven in Bažant (1988, 1989a, b). A vague justification with application to uniaxial strain softening was indicated in Bažant (1976, 1985). Also, the stability condition $\Delta \mathscr{W} > 0$ was used by Maier (1971) for general discretized continuous solids, and by Maier, Zavelani, and Dotreppe (1973) for a beam with softening hinges. For the special case of elastoplastic structures and boundary conditions of fixed displacements, the foregoing stability criterion was given (without derivation) by Hill (1958). In his famous classical paper, he formulated this criterion in terms of incremental strains and stresses rather than displacements and reactions, and considered finite strains, which produce geometrically nonlinear effects. The stability criterion in terms of the work of the reaction increments on δq_i (Eq. 10.1.37) is, of course, valid even if geometrical nonlinearity is present.

It should be noted that in practical calculations, especially in finite element codes, the work of the equilibrium reactions can usually be calculated easier and more efficiently than the work of incremental stresses.

Extensions to Variable Loads

Consider now that the loads P_i are not constant (dead loads) but variable. An important point is that the variation of load is specified independently of the structure. For example, if the load (central force) is produced by attraction or repulsion between two electric charges or between two magnets, we have $P_i = a_i/(r_i - q_i)^2$ where a_i, r_i are constants, or if the load is produced by hydrostatic pressure of a heavy liquid, we have $P_i = a_i(r_i - q_i)$. The derivation of the stability criterion then proceeds similarly as before (Eqs. 10.1.14–10.1.20) except that a Taylor series expansion of the function $\mathbf{P}(\mathbf{q})$ must be introduced in Equation 10.1.20, that is, P_i must be replaced by $P_i^0 + \delta P_i$ with $\delta P_i = \sum_j [\partial P_i/\partial q_j]^0 (q_j - q_j^0) + \cdots$. Instead of Equations 10.1.21 and 10.1.22, integration of Equation 10.1.20 (or a similar equation for $\Delta\mathcal{U}$) then yields

$$\Delta\mathcal{F} = \delta^2\mathcal{F} = \tfrac{1}{2}\delta\mathbf{f}_T^T\,\delta\mathbf{q} - \tfrac{1}{2}\delta\mathbf{P}^T\,\delta\mathbf{q} \tag{10.1.38}$$

$$\Delta\mathcal{U} = \delta^2\mathcal{U} = \tfrac{1}{2}\delta\mathbf{f}_S^T\,\delta\mathbf{q} - \tfrac{1}{2}\delta\mathbf{P}^T\,\delta\mathbf{q} \tag{10.1.39}$$

The equilibrium state is stable if $\Delta\mathcal{F}$ (for isothermal conditions) or $\Delta\mathcal{U}$ (for isentropic conditions), as given by these equations, is positive for all $\delta\mathbf{q}$.

Stability at Critical State

At the first critical state, $\delta^2\mathcal{F}$ or $\delta^2\mathcal{U}$ is zero for some vector $\delta\mathbf{q}$ and positive for all other vectors $\delta\mathbf{q}$. To decide whether this critical state is stable or unstable, one needs to keep the third-order, fourth-order, and possibly higher-order terms of the Taylor series expansion of $\mathbf{f}_T(\mathbf{q})$ or $\mathbf{f}_S(\mathbf{q})$ in Equation 10.1.20, and the same for $\mathbf{Q}(\mathbf{P})$ in Equation 10.1.47. This yields expressions for $\delta^3\mathcal{F}$ and $\delta^4\mathcal{F}$ (or $\delta^3\mathcal{U}$ and $\delta^4\mathcal{U}$). Stability is decided by the positive definiteness of these expressions. If $\delta^3\mathcal{F}$ (or $\delta^3\mathcal{U}$) is nonzero, the state cannot be stable (because $\delta^3\mathcal{F}$ is always negative for some variations). The state is stable if $\delta^3\mathcal{F} = 0$ (or $\delta^3\mathcal{U} = 0$) and for all $\delta\mathbf{q}$, $\delta^4\mathcal{F} > 0$ (or $\delta^4\mathcal{U} > 0$). If $\delta^3\mathcal{F} = \delta^4\mathcal{F} = 0$, then the same could be said of $\delta^5\mathcal{F}$ and $\delta^6\mathcal{F}$, etc.

Gibbs Free Energy and Enthalpy

Gibbs free energy G and enthalpy H are thermodynamic generalizations of the complementary energy (see Sec. 4.2). For a structure, they may be defined as

$$G = F - \mathbf{P}^T\mathbf{q} = U - TS - \mathbf{P}^T\mathbf{q} \tag{10.1.40}$$

$$H = U - \mathbf{P}^T\mathbf{q} \tag{10.1.41}$$

in which the terms $\mathbf{P}^T\mathbf{q}$ implement Legendre transformations of the work terms (see Table 10.1.1). Differentiation and substitution of Equations 10.1.8 and 10.1.9 yield

$$dG = -dW^* - S\,dT - T\,(dS)_{\text{in}} \tag{10.1.42}$$

$$dH = -dW^* + T\,dS - T\,(dS)_{\text{in}} \tag{10.1.43}$$

in which

$$dW^* = \mathbf{q}^T\,d\mathbf{P} \tag{10.1.44}$$

Table 10.1.1. Total Differentials of Basic Thermodynamic Functions of a Structure at Equilibrium Changes (boxed) and Legendre Transformations (circled)

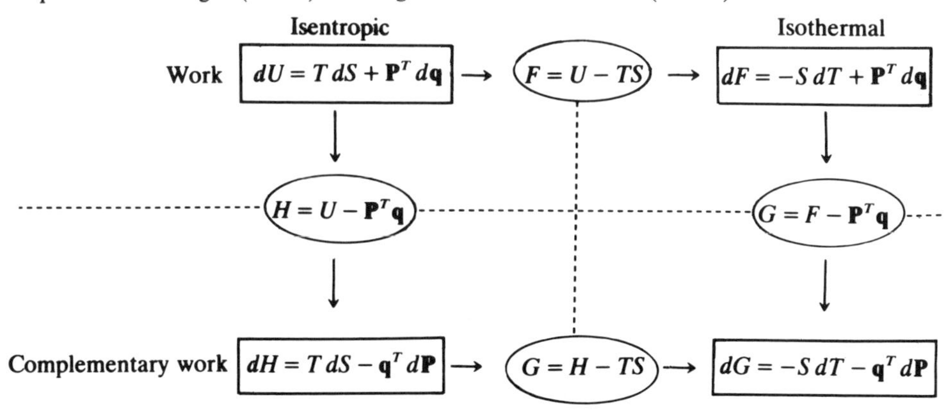

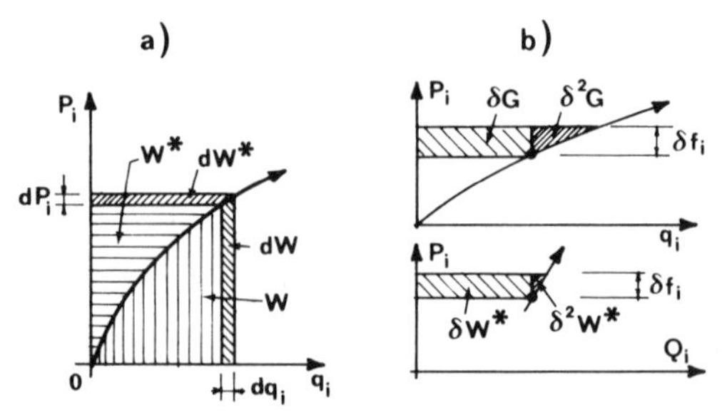

Figure 10.2 (a) Work and complementary work; (b) first and second variation of Gibbs free energy and of complementary work.

Here $W^* = \int \boldsymbol{q}^T d\mathbf{P}$ = complementary work (cf. Sec. 4.2); see Fig. 10.2*a*. For equilibrium changes we have $(dS)_{\text{in}} = 0$, and then G and F represent thermodynamic potentials. Their total (exact) differentials are $dG = -\mathbf{q}^T d\mathbf{P} - S\,dT$, $dH = -\mathbf{q}^T d\mathbf{P} + T\,dS$, and consequently $G = G(\mathbf{P}, T)$, $H = H(\mathbf{P}, S)$. Note also that, for fluids, $dW^* = -V\,dp$ where p = hydrostatic pressure and V = volume of fluid.

Let $\mathbf{q}_T$, $\mathbf{q}_S$ denote the displacements corresponding to $\mathbf{P}$ that maintain equilibrium of the structure and are calculated on the basis of the isothermal or isentropic material properties; and, if the function $\mathbf{P}(\mathbf{q})$ can be inverted, let $\mathbf{Q} = \mathbf{Q}(\mathbf{P})$ denote the displacements corresponding to $\mathbf{P}$ according to the given law of applied forces (e.g. electromagnetic force). Similar to Equations 10.1.14 and 10.1.15, it is again convenient to define the Gibbs free energy $\mathscr{G} = G + W^*$ and enthalpy $\mathscr{H} = H + W^*$ of the structure-load system. Then, for increments preserving mechanical equilibrium ($\mathbf{q}_T = \mathbf{Q}$, $\mathbf{q}_S = \mathbf{Q}$) (but not necessarily thermal equilibrium), $d\mathscr{G} = dG + dW^* = -\mathbf{q}^T d\mathbf{P} - S\,dT - T(dS)_{\text{in}} + \mathbf{q}^T d\mathbf{P}$, that is, $d\mathscr{G} = -S\,dT - T(dS)_{\text{in}}$. Similarly, $d\mathscr{H} = T\,dS - T(dS)_{\text{in}}$. When thermodynamic equilibrium is disturbed *only* mechanically ($\mathbf{q}_T \neq \mathbf{Q}$, $\mathbf{q}_S \neq \mathbf{Q}$) (but not thermally), we may write, according to Equations 10.1.42 and 10.1.43, $d\mathscr{G} = dG + dW^* = -\mathbf{q}^T d\mathbf{P} - S\,dT + \mathbf{Q}^T d\mathbf{P}$ and $d\mathscr{H} = \ldots$, that is,

$$d\mathscr{G} = \mathbf{Q}^T d\mathbf{P} - \mathbf{q}_T^T d\mathbf{P} \qquad \text{at } T = \text{const.} \tag{10.1.45}$$

$$d\mathscr{H} = \mathbf{Q}^T d\mathbf{P} - \mathbf{q}_S^T d\mathbf{P} \qquad \text{at } S = \text{const.} \tag{10.1.46}$$

These equations are analogous to Equations 10.1.18 and 10.1.19.

Stability Criteria Based on Complementary Work

From Equations 10.1.45 and 10.1.46 we conclude that the important case of dead loads must be excluded from considerations of stability of equilibrium state based on $\mathscr{G}$ or $\mathscr{H}$. The reason that this case cannot be handled in terms of the complementary work is that if $\mathbf{P}$ does not depend on $\mathbf{Q}$, the function $\mathbf{P}(\mathbf{Q})$ cannot be inverted. For the same reason, all the displacements q_j for which there are no loads (P_j = const. = 0) must be eliminated in advance by using equilibrium conditions. However, if $\det(d\mathbf{P}/d\mathbf{Q}) \neq 0$ (Fig. 10.2b) then the inverse functions $\mathbf{Q} = \mathbf{Q}(\mathbf{P})$ do exist. In the equations that follow, there must be a variable load associated with every displacement.

Consider now a change from the initial equilibrium state $\mathbf{P}^0$ to an adjacent state $\mathbf{P}^0 + \delta\mathbf{P}$, which may but generally is not a state of mechanical equilibrium (but thermal equilibrium is preserved). By Taylor series expansion, we obtain similarly to Equation 10.1.20

$$\Delta\mathcal{G} = \int_{\mathbf{P}^0}^{\mathbf{P}^0+\delta\mathbf{P}} [\mathbf{Q}^T\, d\mathbf{P} - \mathbf{q}_T^T(\mathbf{P})\, d\mathbf{P}] = \sum_i \int_{P_i^0}^{P_i^0+\delta P_i} \left\{ Q_i^0 + \frac{1}{2}\sum_j \frac{\partial Q_i^0}{\partial P_j}(P_j - P_j^0) + \cdots \right.$$
$$\left. - q_{T_i}^0 - \frac{1}{2}\sum_j \left[\frac{\partial q_{T_i}}{\partial P_j}\right]^0 (P_j - P_j^0) - \cdots \right\} dP_i \tag{10.1.47}$$

Integrating, neglecting all the terms of higher than second order, denoting $\sum_j [\partial q_{T_i}/\partial P_j]^0 (P_j - P_j^0) = \delta P_i$, observing that $q_{T_i}^0 = Q_i^0$ if the initial state is a compatible equilibrium state (Fig. 10.2b), and noting that, according to Equation 10.1.42, $\Delta\mathcal{G} = -T(\Delta S)_{\text{in}}$ for isothermal conditions ($dT = 0$), we conclude (in analogy to Equation 10.1.37) that the structure is

$$\begin{array}{lll} \text{Stable} & \text{if } -T(\Delta S)_{\text{in}} = \Delta\mathcal{G} = \delta^2\mathcal{G} = \frac{1}{2}\delta\mathbf{P}^T\,\delta\mathbf{Q} - \frac{1}{2}\delta\mathbf{P}^T\,\delta\mathbf{q}_T > 0 & \text{for all } \mathbf{q} \\ \text{Critical} & \text{if } -T(\Delta S)_{\text{in}} = \Delta\mathcal{G} = \delta^2\mathcal{G} = \frac{1}{2}\delta\mathbf{P}^T\,\delta\mathbf{Q} - \frac{1}{2}\delta\mathbf{P}^T\,\delta\mathbf{q}_T = 0 & \multirow{2}{*}{\text{for some } \mathbf{q}} \\ \text{Unstable} & \text{if } -T(\Delta S)_{\text{in}} = \Delta\mathcal{G} = \delta^2\mathcal{G} = \frac{1}{2}\delta\mathbf{P}^T\,\delta\mathbf{Q} - \frac{1}{2}\delta\mathbf{P}^T\,\delta\mathbf{q}_T < 0 & \end{array} \tag{10.1.48}$$

where

$$\tfrac{1}{2}\delta\mathbf{P}^T\,\delta\mathbf{q}_T = \tfrac{1}{2}\delta\mathbf{P}^T\mathbf{C}_T(\mathbf{v})\,\delta\mathbf{P} \tag{10.1.49}$$

Here $\mathbf{C}_T$ = isothermal tangential compliance matrix of the structure, which depends on direction $\mathbf{v}$ of loading.

Now one difficulty with the use of $\mathcal{G}$ becomes apparent. Because of the nature of the loading-unloading criteria for materials, direction $\mathbf{v}$ is the direction of the displacement vector δq_{T_i}, and not the direction of δP_i that is used in Equation 10.1.49 as the independent variable. Also note that for the cases of $P_i = 0$ or $P_i =$ const. (dead loads), which we had to exclude from our analysis, we would have to consider $\delta Q_i \to \infty$ for any $\delta P_i \neq 0$ in Equation 10.1.48 [because $\lim (dP_i/dQ_i) = 0$ for deal loads]. (Other difficulties with the use of compliance matrices were discussed in Sec. 4.2.)

Under isentropic conditions, a similar theorem holds for $\delta^2\mathcal{H}$. Its derivatiion proceeds similarly to Equations 10.1.45 to 10.1.49, with $\mathcal{G}$ and subscript T being replaced by $\mathcal{H}$ and subscript S.

Due to the limitation excluding $P_i =$ const., the Gibbs free energy and enthalpy (which are widely used in chemical engineering and heat machines) are of very limited usefulness for stability analysis of structures and have not been used so far. In the next section, however, we will find them useful to deal with the stability of paths.

Structures with a Single Load or a Single Controlled Displacement

Stability of structures with a single load f or a single controlled displacement q can be decided on the basis of tangential stiffness $K = dP/dq$ or compliance $C = 1/K$. They have the value K_t or C_t for loading of the structure ($dq > 0$) and the value K_u or C_u for unloading of the structure ($dq \leq 0$). (Note: some authors prefer to define loading as $dq \geq 0$, although practically the difference is inconsequential.) The case $dq \leq 0$ does not imply the material to be unloading everywhere in the structure; there can be a combination of loading and unloading of the material in different parts of the structure.

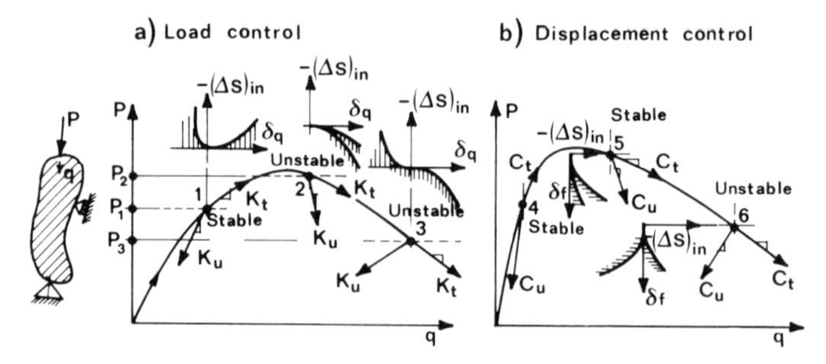

Figure 10.3 Stability of structures with (a) a single load and (b) a single controlled displacement.

Let us assume isothermal conditions, in which case K_u and K_t have the isothermal values. If the load is controlled, q is the independent variable and the governing thermodynamic function is the Helmholtz free energy, $\mathscr{F}$. Thus we have, according to Eq. 10.1.35,

$$-T(\Delta S)_{\text{in}} = \Delta\mathscr{F} = \begin{cases} \frac{1}{2}K_t\,\delta q^2 & \text{if } \delta q \geq 0 \quad \text{(loading)} \\ \frac{1}{2}K_u\,\delta q^2 & \text{if } \delta q < 0 \quad \text{(unloading)} \end{cases} \tag{10.1.50}$$

Stability is assured if this is positive definite, that is, if both K_t and K_u are positive. It follows that, for the normal unloading behavior (see points 1, 2, and 3 in Fig. 10.3a), inelastic structures under a single controlled load are stable in the prepeak hardening regime and unstable in the postpeak softening regime. Note that the actual movement of the state point (q, P) is along the horizontal dashed line through point 1, 2, or 3 in Figure 10.3a, which represents a path away from equilibrium. The lines of slopes K_t and K_u characterize the equilibrium responses that cannot happen under the constraint $P = \text{const}$.

If the displacement is controlled, one must use load P (in fact, a reaction) as the independent variable. Therefore, the governing thermodynamic function is the Gibbs free energy, $\mathscr{G}$. Thus we have, according to Equation 10.1.48,

$$-T(\Delta S)_{\text{in}} = \Delta\mathscr{G} = \begin{cases} -\frac{1}{2}C_t\,\delta P^2 & \text{(without loading)} \\ -\frac{1}{2}C_u\,\delta P^2 & \text{(with unloading)} \end{cases} \tag{10.1.51}$$

This equation is applicable only in the postpeak softening regime (points 5 and 6 in Fig. 10.3b) and indicates that the structure is stable if both C_t and C_u are negative as illustrated at point 5 in Figure 10.3b, and unstable if C_u is positive, as illustrated at point 6 in Figure 10.3b. The latter case (positive C_u) represents snapback instability (Sec. 4.8).

The reason that Equation 10.1.51 cannot be applied to the prepeak hardening regime is that a structure under displacement control has one degree of freedom less than the same structure under load control; see Section 4.8. Therefore, instability can occur only due to internal degrees of freedom, which in fact are the cause of the difference between C_t and C_u. At softening states (points 5 and 6), the structure does have an internal degree of freedom, manifested by the existence of two slopes C_t and C_u. But at hardening states (point 4), the structure has no degree of freedom if the displacement is controlled. (This is because internal equilibrium of the structure is implied in the use of C_t and C_u.) Hence, the state remains fixed (a change $\Delta\mathscr{G}$ cannot occur), which means the structure is stable. We will clarify this behavior when we analyze a strain-softening bar taking strain localization due to internal degrees of freedom actually into consideration (Sec. 13.2).

The foregoing conclusions also apply to structures with many loads P_i and many associated displacements q_i $(i = 1, \ldots, n)$ if all the loads depend on a single load parameter λ and all the displacements depend on a single displacement parameter μ.

Summary

1. Thermodynamic stability analysis requires the given inelastic structure to be replaced by tangentially equivalent elastic structures applicable to various sectors or cones of loading directions.
2. An equilibrium state is stable if the internally produced entropy increment of the structure-load system is negative for all possible deviations.
3. At isothermal (isentropic) conditions, a state is stable if the second variation of the Helmholtz free energy (total energy) of the structure-load system is positive for all possible deviations. For dead loads, this second variation is equal to the (isothermal or isentropic) work of the equilibrium reaction increments on the generalized displacement increments.
4. The tangential (isothermal or isentropic) stiffness matrix may be, but need not be, symmetric. But even if it is symmetric it does not guarantee incremental path independence and existence of an incremental potential. The reason: dependence of this matrix on the direction of the vector of generalized displacement increments.
5. Inelastic structures under a single controlled load are stable if the load-deflection curve has a positive slope (hardening), and unstable if it has a negative slope (softening). Under displacement control, they are stable except if a snapback path exists.

Problems

10.1.1 Consider a slow displacement δq of a spring of isothermal stiffness $\mathbf{K}_T$ away from an equilibrium state under a dead load, and show that for the spring-load system $(\Delta S)_{\text{in}} = -K_T(\delta q)^2/2T$ (see second paragraph below Eq. 10.1.37).

10.1.2 Considering elastic structures, derive the principle of minimum potential energy from the second law of thermodynamics and distinguish between isothermal and isentropic (adiabatic) elastic constants.

10.1.3 Prove the stability criterion for variable loads (Eq. 10.1.38 or 10.1.39) in full detail.

10.1.4 Mechanical (force) equilibrium is a special case of thermodynamic equilibrium. Therefore, it must be possible to derive the principle of virtual work from the second law of thermodynamics. Show this derivation. (*Note*: The condition of min $\mathcal{F}$ or min $\mathcal{U}$ is equivalent to the second law. One must distinguish isothermal and isentropic conditions.)

10.1.5 The points of the structure at which there is a fixed support need not be included in the second-order work sums. However, the points at which there is no applied load must be included, because a nonzero reaction may be

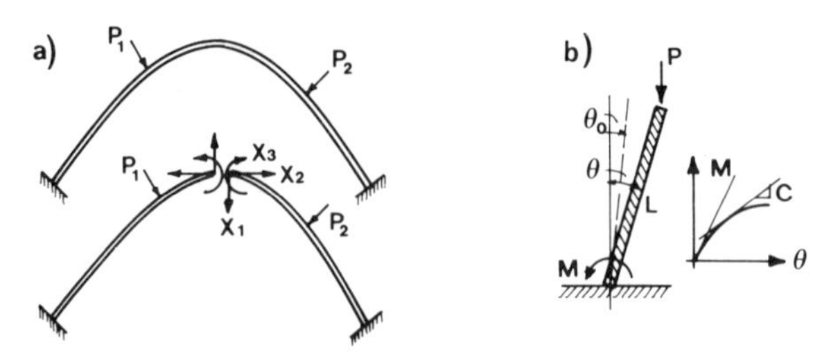

Figure 10.4 (a) Redundant internal forces; (b) rigid-bar column with elastoplastic hinge.

required to equilibrate some vector $\delta \mathbf{q}_i$. Explain it by an example of some structure (one example is δf_1 for Shanley's column in Sec. 10.3).

10.1.6 The redundant internal forces X_k $(k = 1, \ldots, m)$ in a structure (Fig. 10.4a) can be determined from the condition of stationary value of the complementary energy $\mathbf{W}^*$, that is, $\delta \mathbf{W}^*$ (where $\mathbf{W}^*$ represents either $\mathscr{G}$ or $\mathscr{H}$). However, the second-order work of δX_k (Eq. 10.1.49) has nothing to do with stability of the given structure. Explain why. (Note that the work is done on the relative displacements associated with X_k in the primary structure; these displacements are zero for the actually given structure, which represents the conditions of compatibility. However, the second-order work of δX_k would determine stability of the primary structure, if that were the question.)

10.1.7 Apply the present stability criterion to an imperfect rigid-bar column (Fig. 10.4b) with an elastoplastic hinge assumed to be characterized by the equation $dM = C(\theta)\, d\theta$ where M, θ = moment and rotation in the hinge, $C(\theta) = C_0 e^{-k(\theta - \theta_0)}$ where C_0, k = constants, and $\theta = \theta_0$ = initial inclination at $P = 0$. Determine the range of P for which the column is stable.

10.1.8 Derive the stability criterion for the critical state assuming that $\delta^2 \mathscr{F} = \delta^3 \mathscr{F} = 0$ and $\delta^4 \mathscr{F} \neq 0$.

10.2 THERMODYNAMIC CRITERIA OF STABLE PATH

In contrast to elasticity, for inelastic structures it can happen that, after a bifurcation of the equilibrium path, the states on all the postbifurcation branches are stable. So is the choice of the branch left to chance? That would be philosophically unacceptable. "The god does not play dice," (except on the subatomic scale), as Albert Einstein once quipped. There must be some fundamental law that determines the path that will actually be followed. Such a law is provided by thermodynamics. In this section (following Bažant, 1987a, 1988, 1989a, b) we will present the derivation of the thermodynamic criterion of stable path. Its application will be illustrated in Section 10.3.

One necessary condition for a stable path is that it must consist entirely of stable states. There exists a second necessary condition that is provided by the second law of thermodynamics. However, the form of the second law of thermodynamics stated in Equation 10.1.4 or 10.1.5 is not quite pertinent. We will need another form of the second law (Guggenheim, 1959), which was stated in 1875 (in the context of chemical thermodynamics) by Willard Gibbs: Every system approaches equilibrium in such a manner that

$$(\Delta S)_{\text{in}} = \Delta S - \frac{\Delta Q}{T} = \max \tag{10.2.1}$$

among all reachable states. For adiabatic conditions ($\Delta Q = 0$), this reduces to the condition $\Delta S = \max$.

Equation 10.2.1, which is applicable only to states infinitely close to thermodynamic equilibrium, should not be confused with the principle of maximum entropy production. That principle is used in irreversible (nonequilibrium) thermodynamics for processes at which the state is not close to equilibrium. It is a different principle; see also Section 10.3.

Path Stability for Basic Types of Control

An equilibrium path of a structure represents a series of infinitesimal deviations from equilibrium and restorations of equilibrium. Let us consider for an arbitrary (irreversible) inelastic structure a small (infinitesimal) loading step along the equilibrium path α ($\alpha = 1$ or 2) that starts at the bifurcation state A (characterized by $\mathbf{f}^0$, $\mathbf{q}^0$) and ends at another state C on path α (Fig. 10.5). To be able to apply Equation 10.2.1, we decompose this step into two substeps, the first one (I) *away* from the initial equilibrium state A and ending at some intermediate nonequilibrium state B, and the second one (II) *toward* a new equilibrium state, ending on one of the equilibrium paths at state C; see Figure 10.5 (this decomposition was introduced in Bažant, 1985). Since these two substeps are intended as approximations for the actual equilibrium path, matrix $\mathbf{K}(\mathbf{v})$ must be evaluated in both substeps on the basis of the actual direction $\mathbf{v}^{(\alpha)}$ of the equilibrium path, and not on the basis of the directions of the two substeps.

The displacements or forces that are controlled are denoted as q_m, f_m ($m = 1, 2, \ldots, N$; $N \leq n$). If q_m is controlled (which is called displacement control), we consider δq_m to be changed in the first substep (Fig. 10.5a) while f_m is kept constant, which, of course, destroys equilibrium. Displacements δq_m are kept constant during the second substep in which δf_m is allowed to change so as to regain equilibrium, that is, reaches a state on one of the equilibrium paths. If f_m is controlled (which is called load control), we consider δf_m to be changed in the first substep (Fig. 10.5b) while q_m is frozen (constant), which destroys equilibrium. Forces δf_m are kept constant during the second substep in which δq_m is

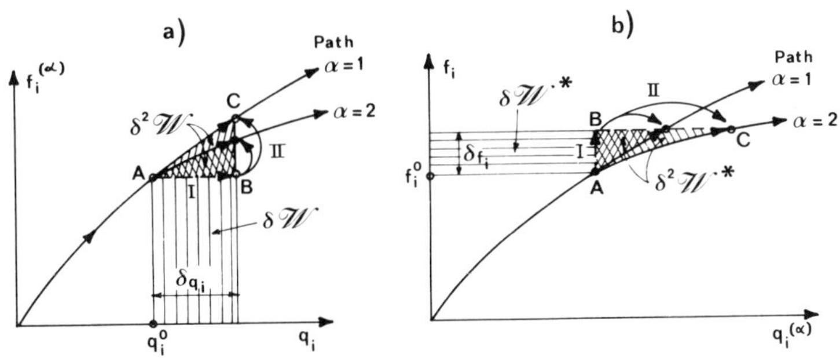

Figure 10.5 Decomposition of a loading step along an equilibrium path after bifurcation: (a) for displacement control, (b) for load control.

allowed to change so as to restore equilibrium. We will separately consider four basic controls:

1a. First we consider displacement control (that is, $\delta \mathbf{q}$ prescribed, Fig. 10.5a) and isothermal conditions ($dT = 0$). Then $\delta \mathbf{q}$ is the same for all the paths α but the equilibrium force increments $\delta \mathbf{f}^{(\alpha)T}$ are different. The increment of the Helmholtz free energy of the structure over the entire step happens at mechanical equilibrium and $dT = 0$, and so (according to Eq. 10.1.10) we have, up to second-order terms,

$$\Delta F = (\mathbf{f}^0 + \tfrac{1}{2}\delta \mathbf{f}^{(\alpha)})^T \,\delta \mathbf{q} = \Delta F_{\mathrm{I}} + \Delta F_{\mathrm{II}}^{(\alpha)} \tag{10.2.2}$$

Here $\delta \mathbf{f}^{(\alpha)}$ is the equilibrium reaction change; $\Delta F_{\mathrm{I}} = \mathbf{f}^{0^T} \delta \mathbf{q}$ = increment of F over the first substep, which is the same for both paths $\alpha = 1, 2$, and $\Delta F_{\mathrm{II}}^{(\alpha)} = \frac{1}{2}\delta \mathbf{f}^{(\alpha)^T} \delta \mathbf{q} = \delta^2 F^{(\alpha)} = \delta^2 \mathscr{W}^{(\alpha)}$ = increment of F in the second substep in which the q_m-values are constant, while the forces f_m change by $\delta f_m^{(\alpha)}$ to find new equilibrium on path α; $\delta^2 \mathscr{W}^{(\alpha)}$ is the second-order work along path α. According to Equation 10.1.9 with $dT = 0$, we have for the second substep (in which $\delta \mathbf{q} = \mathbf{0}$) $\Delta F_{\mathrm{II}}^{(\alpha)} = -T(\Delta S)_{\mathrm{in}}$. The second law of thermodynamics (Eq. 10.2.1) indicates that the structure will approach the equilibrium state that maximizes ΔS_{in}, that is, minimizes $\Delta F_{\mathrm{II}}^{(\alpha)}$. Hence, the path α that actually occurs (stable path) is that for which

$$-T(\Delta S)_{\mathrm{in}} = \delta^2 \mathscr{W}^{(\alpha)} = \tfrac{1}{2}\delta \mathbf{f}^{(\alpha)T} \,\delta \mathbf{q} = \tfrac{1}{2}\delta \mathbf{q}^T \mathbf{K}^{(\alpha)} \,\delta \mathbf{q} = \min_{(\alpha)} \tag{10.2.3}$$

(if $\mathbf{q}$ is controlled). $\mathbf{K}^{(\alpha)}$ is the tangential stiffness matrix $\mathbf{K}_T$ for path α, which must be based on isothermal material properties. It is again important to keep in mind that $\mathbf{K}^{(\alpha)}$ is considered to represent the stiffness matrix of a tangentially equivalent elastic structure (recall our discussion below Eqs. 10.1.6–10.1.7).

1b. Next consider displacement control (δq_m prescribed, Fig. 10.5a) and isentropic conditions ($dS = 0$). The increment of the total energy of the structure over the entire step occurs while mechanical equilibrium is maintained, and so (according to Eq. 10.1.11) we have

$$\Delta U = (\mathbf{f}^0 + \tfrac{1}{2}\delta \mathbf{f}^{(\alpha)})^T \,\delta \mathbf{q} = \Delta U_{\mathrm{I}} + \Delta U_{\mathrm{II}}^{(\alpha)} \tag{10.2.4}$$

Here $\Delta U_I = \mathbf{f}^{0^T} \delta \mathbf{q}$, which is the same for both paths α, and $\Delta U_{\mathrm{II}}^{(\alpha)} = \frac{1}{2}\delta \mathbf{f}^{(\alpha)T} \delta \mathbf{q} = \delta^2 \mathscr{W}^{(\alpha)}$ = increment of U in the second substep in which the q_m-values are constant while the f_m-values change by $\delta f_m^{(\alpha)}$ to find new equilibrium on path α. According to Equation 10.1.8 with $dS = 0$, we have for the second substep (in which $\delta \mathbf{q} = \mathbf{0}$) $\Delta U_{\mathrm{II}}^{(\alpha)} = -T(\Delta S)_{\mathrm{in}}$. The second law of thermodynamics indicates that, on approach to equilibrium, $\Delta U_{\mathrm{II}}^{(\alpha)}$ must be minimized. Hence, the path that occurs (stable path) is again determined by Equation 10.2.3, in which however $\mathbf{K}^{(\alpha)}$ now represents the tangential stiffness matrix $\mathbf{K}_S$ that must be based on isentropic rather than isothermal material properties.

2a. Futhermore, consider now load control (Fig. 10.5b) and isothermal conditions. The proper thermodynamic function now is Gibbs' free energy, which is defined by Equation 10.1.40. The increment of G of the structure over the entire step occurs while mechanical equilibrium is maintained, and so (according to Eq. 10.1.45 without loads) we have, up to second-order terms and for both substeps combined,

$$\Delta G = -(\mathbf{q} + \tfrac{1}{2}\delta \mathbf{q}^{(\alpha)})^T \,\delta \mathbf{f} = \Delta G_{\mathrm{I}} + \Delta G_{\mathrm{II}}^{(\alpha)} \tag{10.2.5}$$

Here $\Delta G_{\text{I}} = -\sum q_m \,\delta f_m$ = increment of G over the first substep, which is the same for both paths $\alpha = 1, 2$, and $\Delta G_{\text{II}}^{(\alpha)} = \frac{1}{2}\mathbf{q}^{(\alpha)^T}\,\delta\mathbf{f} = -\delta^2\mathcal{W}^{*(\alpha)}$ = increment of G over the second substep in which the f_m-values are constant while the q_m-values are allowed to change by $\delta q_m^{(\alpha)}$ so as to restore equilibrium by reaching path α; $\delta^2\mathcal{W}^{*(\alpha)}$ is the second-order complementary work along path α. According to Equation 10.1.42 with $dT = 0$, we have for the second substep (in which $\delta\mathbf{f} = 0$) $\Delta G_{\text{II}}^{(\alpha)} = -T(\Delta S)_{\text{in}}$. Based on the second law of thermodynamics (Eq. 10.2.1), the approach to the new equilibrium state must maximize $(\Delta S)_{\text{in}}$, that is, minimize $\Delta G_{\text{II}}^{(\alpha)}$. Hence, the path α that actually occurs (stable path) is that for which

$$T(\Delta S)_{\text{in}} = \delta^2\mathcal{W}^{*(\alpha)} = \tfrac{1}{2}\delta\mathbf{q}^{(\alpha)^T}\,d\mathbf{f} = \tfrac{1}{2}\delta\mathbf{f}^T\,\delta\mathbf{q}^{(\alpha)} = \tfrac{1}{2}\delta\mathbf{f}^T\mathbf{C}^{(\alpha)}\,\delta\mathbf{f} = \underset{(\alpha)}{\text{Max}} \qquad (10.2.6)$$

(provided that f_m is controlled). $\mathbf{C}^{(\alpha)}$ is the $n \times n$ tangential compliance matrix of the structure for path α, which must be based on isothermal material properties. Note that, in contrast to Equation 10.2.3, the path label (α) now appears with $\delta\mathbf{q}$ rather than $\delta\mathbf{f}$.

2b. Finally, consider load control (Fig. 10.5b) and isentropic conditions ($dS = 0$). The proper thermodynamic function is now the enthalpy, H, which is defined by Equation 10.1.41. The increment of H of the structure over the entire step occurs while mechanical equilibrium is maintained, and so (according to Eq. 10.1.46 without loads) we have, up to second-order terms and for both substeps combined,

$$\Delta H = -(\mathbf{q} + \tfrac{1}{2}\delta\mathbf{q}^{(\alpha)})^T\,\delta\mathbf{f} = \Delta H_{\text{I}} + \Delta H_{\text{II}}^{(\alpha)} \qquad (10.2.7)$$

Here $\Delta H_{\text{I}} = -\mathbf{q}^T\,\delta\mathbf{f}$ = increment of enthalpy H over the first substep, which is the same for both paths $\alpha = 1, 2$; and $\Delta H_{\text{II}}^{(\alpha)} = -\frac{1}{2}\delta\mathbf{q}^{(\alpha)T}\,\delta\mathbf{f} = -\delta^2\mathcal{W}^{*(\alpha)}$ = increment of H over the second substep in which the f_m-values are constant (Fig. 10.5b) while q_m are allowed to change so as to restore equilibrium. According to Equation 10.1.43 with $dS = 0$, we have for the second substep (in which $\delta\mathbf{f} = \mathbf{0}$) $\Delta H_{\text{II}}^{(\alpha)} = -T(\Delta S)_{\text{in}}$. In view of the second law of thermodynamics (Eq. 10.2.1), the approach to new equilbrium must maximize $(\Delta S)_{\text{in}}$, that is, minimize $\Delta H_{\text{II}}^{(\alpha)}$. Hence, the path that occurs (stable path) is again that indicated by Equation 10.2.6, in which however $\mathbf{C}^{(\alpha)}$ must be based on isentropic rather than isothermal material properties.

Mixed Controls of Loads and Displacements

In general, it is possible that the load and displacement controls are mixed, that is, some loads and some displacements are controlled. A more general path stability condition is then needed. Consider that displacements $\mathbf{q}$ and (non-associated) loads $\mathbf{f}$ are controlled. The force and displacement responses, which are different for various paths α, are $\mathbf{f}^{(\alpha)}$ and $\mathbf{q}^{(\alpha)}$. Note that in the present notation column matrix $\mathbf{f}$ does not contain all the f_m components of the structure, nor does $\mathbf{q}$ contain all the q_m components; the components of $\mathbf{q}$ and $\mathbf{q}^{(\alpha)}$ are different, and so are the components of $\mathbf{f}$ and $\mathbf{f}^{(\alpha)}$.

The simplest way to treat this case is to imagine that for mixed loading each step $\delta\mathbf{q}$ and $\delta\mathbf{f}$ consists of two simple infinitesimal steps; in the first step, only the components of $\delta\mathbf{q}$ are changed, and in the second step, only the components of

$\delta\mathbf{f}$ are changed. During the first step, according to Equation 10.2.3, we have

$$T(\Delta S)_{\text{in}} = -\tfrac{1}{2}\delta\mathbf{f}^{(\alpha)T}\,\delta\mathbf{q} \tag{10.2.8}$$

and in the second step, according to Equation 10.2.6,

$$T(\Delta S)_{\text{in}} = \tfrac{1}{2}\delta\mathbf{f}^{T}\,\delta\mathbf{q}^{(\alpha)} \tag{10.2.9}$$

For both steps combined into a single infinitesimal step, the $T(\Delta S)_{\text{in}}$ values from Equations 10.2.8 and 10.2.9 must be summed. Therefore, in view of Equation 10.2.1, the path α that is stable under mixed controls is that for which

$$T(\Delta S)_{\text{in}} = \tfrac{1}{2}\delta\mathbf{q}^{(\alpha)T}\,\delta\mathbf{f} - \tfrac{1}{2}\delta\mathbf{f}^{(\alpha)T}\,\delta\mathbf{q} = \max_{(\alpha)} \tag{10.2.10}$$

This result can also be derived by a direct thermodynamic argument. We consider isothermal conditions and introduce a semicomplementary thermodynamic function Z that involves the complementary work for $\mathbf{f}^{(\alpha)}$ but the actual work for $\mathbf{q}^{(\alpha)}$. The function is obtained by Legendre transformation of $\mathbf{f}$: $Z = F - \mathbf{f}^T\mathbf{q}^{(\alpha)}$. If only the displacements are controlled, there is no term with $\mathbf{q}^{(\alpha)}$, and then $Z = F =$ Helmholtz free energy. If only the loads are controlled, $\mathbf{q}^{(\alpha)}$ comprises all the loads, and then $Z = G =$ Gibbs' free energy. Differentiating, $dZ = dF - \mathbf{f}^T\,d\mathbf{q}^{(\alpha)} - \mathbf{q}^{(\alpha)T}\,d\mathbf{f}$, substituting $dF = -S\,dT + dW$, $dW = \mathbf{f}^{(\alpha)T}\,d\mathbf{q} + \mathbf{f}^T\,d\mathbf{q}^{(\alpha)}$ (work of all reactions and loads), and integrating from $\mathbf{f}^0$, $\mathbf{q}^0$ to $\mathbf{f}^0 + \delta\mathbf{f}$, $\mathbf{q}^0 + \delta\mathbf{q}$ we obtain

$$\Delta Z = -\int S^{(\alpha)}\,dT + \int \mathbf{f}^{(\alpha)T}\,d\mathbf{q} - \int \mathbf{q}^{(\alpha)T}\,d\mathbf{f} \tag{10.2.11}$$

where $\mathbf{f}^{(\alpha)}$ and $\mathbf{q}^{(\alpha)}$ vary to maintain equilibrium, $\mathbf{f}^{(\alpha)} = \mathbf{f}_T^{(\alpha)}$, $\mathbf{q}^{(\alpha)} = \mathbf{q}_T^{(\alpha)}$. Expanding $\mathbf{f}^{(\alpha)}$ and $\mathbf{q}^{(\alpha)}$ into a Taylor series about $\mathbf{f}^0$ and $\mathbf{q}^0$, integrating, and neglecting the terms of higher than second order, we get

$$\Delta Z = -\int S^{(\alpha)}\,dT + \Delta Z_{\text{I}} + \Delta Z_{\text{II}} \tag{10.2.12}$$

in which

$$\Delta Z_{\text{II}} = \mathbf{f}^{0^T}\,\delta\mathbf{q} - \mathbf{q}^{0^T}\,\delta\mathbf{f} \tag{10.2.13}$$

$$\Delta Z_{\text{II}} = \tfrac{1}{2}\delta\mathbf{f}^{(\alpha)T}\,\delta\mathbf{q} - \tfrac{1}{2}\delta\mathbf{q}^{(\alpha)T}\,\delta\mathbf{f} \tag{10.2.14}$$

On the other hand, if the same changes of controlled variables $\delta\mathbf{q}$ and $\delta\mathbf{f}$ are carried out at constant $\mathbf{f} = \mathbf{f}^0$ (dead loads) and constant displacements $\mathbf{q} = \mathbf{q}^0$, we have a nonequilibrium change, for which the term $-T(\Delta S)_{\text{in}}$ must be added to dF and dZ in the preceding derivation (cf. Eq. 10.1.16). Hence,

$$\Delta Z = -\int S^{(\alpha)}\,dT + \mathbf{f}^{0^T}\,\delta\mathbf{q} - \mathbf{q}^{0^T}\,\delta\mathbf{f} - T(\Delta S)_{\text{in}} \tag{10.2.15}$$

Here $\mathbf{f}^0$ and $\mathbf{q}^0$ are the same for all paths α because we consider increments from a common bifurcation point. Subtracting Equation 10.2.15 from Equation 10.2.12 with Equations 10.2.13 and 10.2.14, we finally obtain $T(\Delta S)_{\text{in}} = -\Delta Z_{\text{II}}$, which yields Equation 10.2.10.

The linear terms in Equations 10.2.12 to 10.2.14 and Equation 10.2.15 are identical. Therefore, the internal entropy change during the first substep, in which the controlled variables q_i and f_j are changed while their responses f_i and q_j

are kept constant (frozen), is the same for each path α. So the differences in internal entropy among paths $\alpha = 1, 2, \ldots$ arise only during the second substep in which the controlled variables q_i and f_j are kept constant while $f_i^{(\alpha)}$ and $q_j^{(\alpha)}$ change to find an equilibrium state on one of the paths α. Therefore, the equation $T(\Delta S)_{\text{in}} = -\Delta Z_{\text{II}}$, which corresponds to the second substep, represents the internal entropy change on approach to an equilibrium state. According to the second law of thermodynamics (Eq. 10.2.1) this entropy change must be maximized. This leads to the general path stability condition in Equation 10.2.10.

A similar argument shows that this is also true when the total entropy S is constant (isentropic conditions). To derive this result one needs to introduce, instead of Z, the semicomplementary thermodynamic function $Z^* = U - \mathbf{f}^T\mathbf{q}^{(\alpha)}$ that differs from Z only by the thermal term.

Note that the previously stated path stability conditions for the cases when only the displacements are controlled or only the loads are controlled result as special cases of Equation 10.2.10.

The general criteria of stable path in Equations 10.2.3, 10.2.6, and 10.2.10 were derived from thermodynamics by Bažant (1988, 1989a, b) (based on a 1987 report). A sketch of the proof for displacement control appeared in Bažant (1985). The special criterion for displacement control in Equation 10.2.3 has been used as a postulate without thermodynamic proof by Maier and Zavelani (1970), Maier, Zavelani, and Dotreppe (1973), Bažant (1976), Nguyen (1984, 1987), Petryk (1985a, b), Nguyen and Stolz (1986), Stolz (1989); and a simplified justification based on the second law of thermodynamics was also given by Bažant (1976), Nemat-Nasser (1979), Nemat-Nasser, Sumi, and Keer (1980) and others.

The Case of Equal $(\Delta S)_{\text{in}}$ for Two Branches

In the special case that the $T(\Delta S)_{\text{in}}$ values at bifurcation load P_t are equal for both bifurcation branches, there are two ways to select the correct path: (1) Either one must calculate the higher-order term in the work terms, or (2) one considers a bifurcation load $P_t + \Delta P$ where ΔP is infinitesimal, and if $T(\Delta S)_{\text{in}}$ is larger for one of the branches, this is the branch that occurs for $\Delta P \to 0$ (this actually happens for the Shanley column in Sec. 10.3). However, in the case of symmetry, the branch that is followed is decided by random imperfections (e.g., whether the column in Sec. 10.3 deflects right or left).

Second-Order Work of Stresses along the Path

In view of Equations 10.1.29 and 10.1.30 representing the principle of conservation of energy, we may alternatively express the second-order work along path α as follows:

$$\delta^2\mathcal{W}^{(\alpha)} = \tfrac{1}{2}\delta\mathbf{f}^{(\alpha)^T}\,\delta\mathbf{q} = \int_V \tfrac{1}{2}\delta\boldsymbol{\sigma}^{(\alpha)} : \delta\mathbf{e}^{(\alpha)}\,dV + \tfrac{1}{2}\delta\mathbf{q}^T\mathbf{K}^\sigma\,\delta\mathbf{q} \qquad (10.2.16)$$

in which $\mathbf{K}^\sigma$ = the geometric stiffness matrix introduced in Equation 10.1.32 (for more detail, see Chapter 11); $\delta\mathbf{f}^{(\alpha)} = \delta\mathbf{f}_T^{(\alpha)}$ for isothermal changes, $\delta\mathbf{f}^{(\alpha)} = \delta\mathbf{f}_S^{(\alpha)}$ for isentropic (adiabatic) changes; $\delta\boldsymbol{\sigma}^{(\alpha)}$, $\delta\boldsymbol{\varepsilon}^{(\alpha)}$ = stress and small strain increments along path α. However, a similar expression for the complementary second-order work $\delta^2\mathscr{W}^{*(\alpha)}$ generally does not exist because $\mathscr{W}^*$ must be expressed in terms of the compliance tensor but the compliance corresponding to $\mathbf{K}^\sigma$ need not exist, that is, $\mathbf{K}^\sigma$ may be, and typically is, singular and cannot be inverted. In the special case that nonlinear geometric effects are absent, $\mathbf{K}^\sigma = \mathbf{0}$, and then

$$\delta^2\mathscr{W}^{*(\alpha)} = \tfrac{1}{2}\delta\mathbf{f}^T\,\delta\mathbf{q}^{(\alpha)} = \int_V \tfrac{1}{2}\delta\boldsymbol{\sigma}^{(\alpha)}:\delta\mathbf{e}^{(\alpha)}\,dV \tag{10.2.17}$$

Also in that case, Equation 10.2.17, lacking the nonlinear geometric term, can alternatively be obtained from the principle of virtual work because $\delta\mathbf{f}$ is in equilibrium with $\delta\boldsymbol{\sigma}(\mathbf{x})$ and $\delta\mathbf{e}(\mathbf{x})$ is compatible with $\delta\mathbf{q}$.

Note that in Equations 10.2.16 and 10.2.17, the path label (α) appears with both $\delta\boldsymbol{\sigma}$ and $\delta\mathbf{e}$, but only with one of the variables $\delta\mathbf{q}$ and $\delta\mathbf{f}$. In finite element programs, it is generally much more efficient to calculate $\delta\mathbf{f}^T\,\delta\mathbf{q}$ than the volume integrals.

Structures with a Single Load or a Single Controlled Displacement

In this case the structural response is characterized by tangential stiffness $K = df/dq$ (whose isothermal value $K = K_T$ should be distinguished from its isentropic value $K = K_S$). Expressing the second-order work and second-order complementary work, we have for the stable path the conditions

$$\delta^2\mathscr{W}^{(\alpha)} = \tfrac{1}{2}\delta f^{(\alpha)}\,\delta q = \frac{K^{(\alpha)}}{2}\,\delta q^2 = \min_{(\alpha)} \tag{10.2.18}$$

$$\delta^2\mathscr{W}^{*(\alpha)} = \tfrac{1}{2}\delta f\,\delta q^{(\alpha)} = \frac{1}{2K^{(\alpha)}}\,\delta f^2 = \max_{(\alpha)} \tag{10.2.19}$$

From this the following theorem ensues:

Theorem 10.2.1 If the bifurcation state is stable, the stable path is that for which the tangential stiffness $K^{(\alpha)}$ is minimum regardless of whether the displacement or the load is controlled.

This theorem holds not only for $K^{(\alpha)} > 0$ (hardening) but also for $K^{(\alpha)} \to 0$ (perfect plasticity) and $K^{(\alpha)} < 0$ (softening). Note that for $K^{(\alpha)} < 0$ the load control is excluded because such an initial state is unstable. For example, the stable paths, as indicated by this theorem, are the lower ones in Figure 10.5, for both displacement control (left) and load control (right).

For uniaxial test specimens, this theorem implies that the strain must start to localize right after the peak stress point, even though nonlocalized (uniform) strain states in the softening range may be stable way beyond the peak (as proven in Bažant, 1976). For more details, see Section 13.2.

Stable States on Postbifurcation Branches

At the outset we alluded to the following theorem (Bažant, 1987a):

Theorem 10.2.2 Let the main (primary, symmetric) path be along the direction of axis q_n. If the structure is elastic, no secondary postbifurcation branches that are not initially orthogonal to axis q_n can consist of stable states.

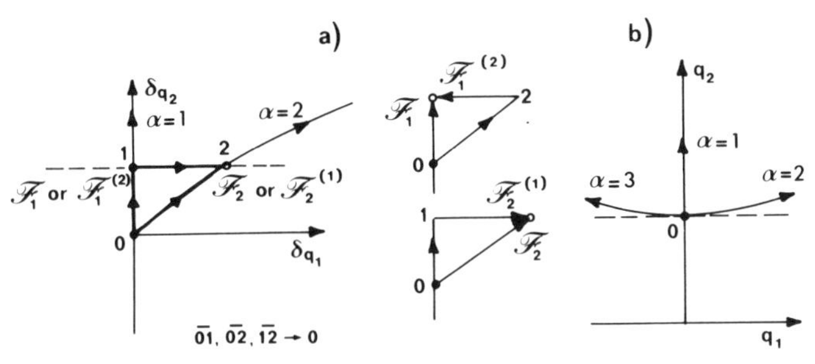

Figure 10.6 Path dependence of post bifurcation equilibrium states.

But if the structure is inelastic, both such branches can consist of stable states. This difference in behavior is due to the dependence of the tangential stiffness matrix on the direction of $\delta\mathbf{q}$.

Proof Consider isothermal conditions (in the case of isentropic conditions, simply replace $\mathcal{F}$ by $\mathcal{U}$ in all that follows). At point 0, at which the Helmholtz free energy is $\mathcal{F} = \mathcal{F}_0$, the equilibrium path in the space of $q_1, \ldots, q_n$ bifurcates into paths $\alpha = 1$ and 2 leading toward adjacent (infinitely close) points 1 and 2 corresponding to the same load; see Figure 10.6. Moving along these paths, the values of $\mathcal{F}$ at points 1 and 2 are $\mathcal{F}_1$ and $\mathcal{F}_2$. If point 1 is reached along path $\overline{021}$, the value of $\mathcal{F}$ at point 1 is $\mathcal{F}_1^{(2)}$. If point 2 is reached along path $\overline{012}$, the value of $\mathcal{F}$ at point 2 is $\mathcal{F}_2^{(1)}$. Stability at point 1 requires that $\Delta\mathcal{F}_{12} = \mathcal{F}_2^{(1)} - \mathcal{F}_1 > 0$. Stability at point 2 requires that $\Delta\mathcal{F}_{21} = \mathcal{F}_1^{(2)} - \mathcal{F}_2 > 0$.

If the structure is elastic, function $\mathcal{F}$ is *path-independent,* and so $\mathcal{F}_1^{(2)} = \mathcal{F}_1$ and $\mathcal{F}_2^{(1)} = \mathcal{F}_1$. Then $\Delta\mathcal{F}_{12} = \mathcal{F}_2 - \mathcal{F}_1$ and $\Delta\mathcal{F}_{21} = \mathcal{F}_1 - \mathcal{F}_2$. So $\Delta\mathcal{F}_{12} = -\Delta\mathcal{F}_{21}$. Obviously, either $\Delta\mathcal{F}_{12}$ or $\Delta\mathcal{F}_{21}$ must be nonpositive, and so either state 1 or state 2 must be unstable.

If, however, the structure is not elastic, $\mathcal{F}$ is *path-dependent* (see our discussion of path dependence in Sec. 10.1 in relation to Fig. 10.1f). Therefore, $\mathcal{F}_2^{(1)}$ need not be equal to $\mathcal{F}_2$, and $\mathcal{F}_1^{(2)}$ need not be equal to $\mathcal{F}_1$. Consequently, $\Delta\mathcal{F}_{12}$ need not be equal to $-\Delta\mathcal{F}_{21}$. Clearly, both $\Delta\mathcal{F}_{12}$ and $\Delta\mathcal{F}_{21}$ can now be positive. This means that states 1 and 2 in Fig. 10.6 can both be stable (this will be illustrated by an example in Sec. 10.3, particularly paths $\overline{13}$ and $\overline{123}$ in Fig. 10.8c).

The foregoing argument, however, is invalid if state 2 of an elastic structure lies on a branch that, as illustrated in Figure 10.6b, starts in a direction orthogonal to the main path (axis q_2) at the bifurcation point (this happens, e.g., in symmetric bifurcation of perfect columns, Chaps. 4 and 8). Then line $\overline{02}$ of Fig. 10.6a tends to a horizontal, and so point 2 is no longer infinitely close to point 1; then $\Delta\mathcal{F}_{12}$ corresponds to a finite segment $\overline{12}$ (even if $\overline{01}$ is infinitesimal) and so it need not be positive if state 1 is stable; similarly $\Delta\mathcal{F}_{21}$ need not be positive if state 2 is stable. Therefore, this case must be excepted. Q.E.D.

The catastrophy theory (Sec. 4.7), as general as it is purported to be, is nevertheless rather limited. It deals only with path-independent systems for which the behavior near the bifurcation point is characterized by a single potential surface. In its present form, the catastrophe theory does not apply to structures that are inelastic (i.e., path-dependent).

Further Comments and Conclusion

There is some philosophical question whether the present path stability criterion must always be exactly followed. An analogous example from physics is the condensation of a vapor on cooling (described, e.g., by van der Waals' equation for gases). According to the condition $(\Delta S)_{in} = \max$, condensation into a liquid would have to begin right after the critical point of vapor-liquid equilibrium is passed. However, as is well known, the start of condensation requires nucleation of liquid droplets, and if nucleating inhomogeneities are not present, it is in fact possible to get a supercooled vapor for which the Gibbs free energy per unit mass is lower than that for a mixture of vapor and liquid [that is, $(\Delta S)_{in}$ per unit mass is higher]. In this light, we can see that it might be possible for a structure to start along some path for which $(\Delta S)_{in}$ is not maximum. However, such a behavior would, of course, be metastable (since a path toward a higher entropy would exist nearby), and could not be expected to last in the presence of small imperfections or any type of small disturbances (which are analogous to nucleation of liquid droplets).

To sum up, path stability is decided by the second-order work along the equilibrium path. For displacement control, this work must be minimized, while for load control it must be maximized.

We will relegate further discussion of the path stability concept and bifurcations until an example will have been presented in the next section.

Problems

10.2.1 If the loads are not dead loads, the condition of stability of state acquires an additional second-order term due to the load variation (Eq. 10.1.38). Consider that the load control is affected by changing the attraction force between two electromagnets and determine whether the condition of path stability must also be modified.

10.2.2 How is a stable path to be decided if there exist two paths along which $\delta^2 \mathscr{W} = 0$ (all the states being stable)?

10.3 APPLICATION TO ELASTOPLASTIC COLUMNS AND BROADER IMPLICATIONS

To illustrate the criteria of stable states established in the previous section, let us now study (following Bažant, 1987a, 1988, 1989a, b) again the idealized column from Figure 8.7 that was considered in the epoch-making paper by Shanley (1947). Equilibrium of this column was analyzed in Section 8.1. The column (Fig. 10.7) is pin-ended and consists of two rigid bars of lengths $l/2$, which are connected by a very short elastoplastic link (point hinge) of length $h \ll l$ and width h, having an ideal I-beam cross section of area A. Before loading, the column is perfectly straight. The lateral deflection and the axial displacement at the load point (positive if shortening) are denoted as q_1 and q_2, respectively. The rotation of the rigid bars, assumed to be small, is $\delta\theta = 2\delta q_1/l$. The column is loaded by an axial centric load P (positive if compressive). The initial equilibrium

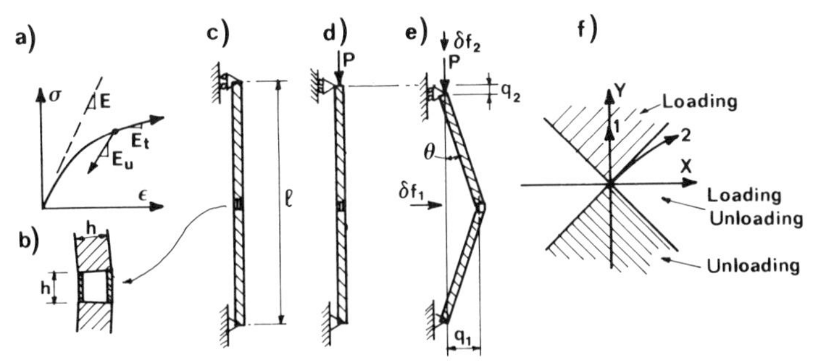

Figure 10.7 Shanley's elastoplastic column.

under load $P = P_0$ at zero lateral deflection is imagined to be disturbed by raising the axial load to $P = P_0 + \delta f_2$ and applying a small lateral load δf_1.

Loading-Unloading Combinations and Equilibrium Paths

The incremental moduli for loading and unloading are denoted as E_t and E_u (Fig. 10.7a). These moduli have different values for isothermal and isentropic (adiabatic) deformations. E_t is called the tangential modulus (although E_u has the meaning of tangential modulus for unloading). Always $E_t < E_u$, except when the material is elastic, in which case $E_t = E_u$. E_t is a given function of the initial uniform stress $\sigma = \sigma_1 = \sigma_2 = -P/A$. It is convenient to express the moduli at the left and right faces as $E_1 = \eta E_t$ and $E_2 = \eta \xi E_t$ and define the nondimensional displacements $X = \delta q_1/l$ and $Y = \delta q_2/2h$. The force variables associated with X and Y by work are $f_X = l\,\delta f_1$ and $f_Y = 2h\,\delta f_2$. Using the expressions for small (linearized) strains at the left and right flanges:

$$\delta e_1 = -\delta\theta - \frac{\delta q_2}{h} = -2(X+Y) \qquad \delta e_2 = \delta\theta - \frac{\delta q_2}{h} = 2(X-Y) \quad (10.3.1)$$

one may obtain for buckling to the right ($X > 0$) the following loading-unloading criteria (Fig. 10.7e):

$$\begin{aligned} &1)\ \text{For } Y > X \text{ (loading only):} && \xi = 1 && \eta = 1 \\ &2)\ \text{for } -X \le Y \le X \text{ (loading-unloading):} && \xi = \xi_u && \eta = 1 \\ &3)\ \text{For } Y < -X \text{ (unloading only):} && \xi = 1 && \eta = \xi_u \end{aligned} \quad (10.3.2)$$

where $\xi_u = E_u/E_t$. The values of ξ and η characterize the ranges of directions $\mathbf{v}$ introduced in Section 10.1. The dependence on ξ and η is a dependence on $\mathbf{v}$.

Based on the incremental stresses $\delta\sigma_1 = E_1\,\delta e_1$ and $\delta\sigma_2 = E_2\,\delta e_2$ at the left and right flanges, the moment and axial conditions of equilibrium at the midspan lead, for buckling to the right, to the equations:

$$\begin{Bmatrix} l\,\delta f_1 \\ 2h\,\delta f_2 \end{Bmatrix} = 2\eta P_t l \begin{bmatrix} 1 + \xi - \dfrac{2P_0}{\eta P_t} & 1 - \xi \\ 1 - \xi & 1 + \xi \end{bmatrix} \begin{Bmatrix} X \\ Y \end{Bmatrix} \quad (10.3.3)$$

in which $P_t = E_t Ah/l$ = Shanley's tangent modulus (see Sec. 8.1). (Note that for loading only, $\xi = \eta = 1$, the determinant of this equation vanishes if $P_0 = P_t$.) If

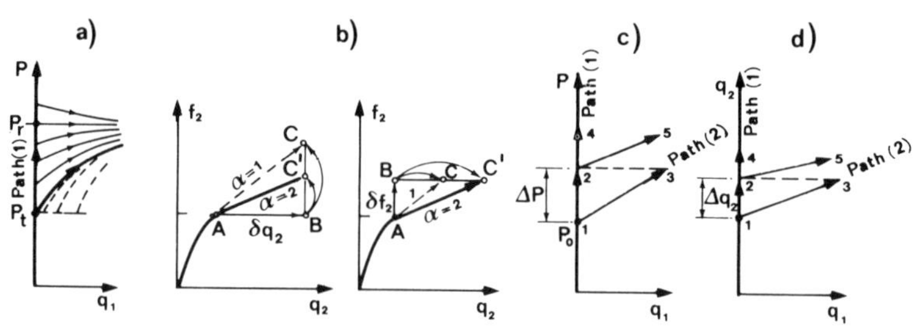

Figure 10.8 Bifurcations of equilibrium paths.

$\delta f_2 = 0$ ($P = \text{const.}$) and $\delta f_1 = 0$ (i.e., if Eqs. 10.3.3 are homogeneous), then the only nonzero solution of Equations 10.3.3 is $P_0 = 2\xi_u P_t/(\xi_u + 1) = P_r$. Load P_r represents the reduced modulus load of Engesser and von Kármán (Sec. 8.1), at which there is neutral equilibrium (Fig. 10.8a).

For a straight segment of the σ-ε diagram, E_t, P_t, and P_r are constant. When E_t depends on σ (curved σ-ε diagram), P_t and P_r depend on $\sigma_0 = -P/A$ through $E_t(\sigma_0)$.

As shown in Section 8.1, Equations 10.3.3 for $\delta f_1 = 0$ permit more than one equilibrium path. One path, called the primary (or main) path or path 1 (Fig. 10.8), is characterized by zero deflection and $\xi = \eta = 1$:

$$X^{(1)} = 0 \qquad \delta f_2^{(1)} = 2P_t\left(\frac{l}{h}\right)Y^{(1)} \tag{10.3.4}$$

For $P_0 \to P_t$, however, $X^{(1)}$ is arbitrary since the matrix of Equations 10.3.3 becomes singular (but $X^{(1)} < Y^{(1)}$ because $\xi = 1$). Another path, called the secondary path or path 2 (Fig. 10.8), is characterized by positive deflection (at $\delta f_1 = 0$), $\eta = 1$ and $\xi > 1$. For this path Equations 10.3.3 yield

$$X^{(2)} = \frac{\xi_u - 1}{\xi_u + 1 - 2P_0/P_t}\,Y^{(2)} \qquad \delta f_2^{(2)} = \frac{P_t l}{h}\left[\frac{4\xi_u P_t - 2(\xi_u + 1)P_0}{(\xi_u + 1)P_t - 2P_0}\right]Y^{(2)} \tag{10.3.5}$$

The superscripts (1) and (2) refer to paths 1 and 2. Path 2 according to Equations 10.3.5 is possible only if $\xi > 1$, and $\xi > 1$ is possible only if $P_0 \geq P_t$ (Sec. 8.1). Since the solutions in Equations 10.3.4 and 10.3.5 exist for each point $P_0 \geq P_t$ (solid curves in Fig. 10.8a), the primary path for $P \geq P_t$ represents a continuous sequence of bifurcation points. The first bifurcation occurs at $P_0 = P_t$, as shown by Shanley (Fig. 10.8a). Note that for $P_0 = P_t$ we have $Y^{(2)}/X^{(2)} = 1$, that is, the secondary path starts along the boundary of the loading-only sector in the (X, Y) plane.

Second-Order Work

According to the preceding section (Sec. 10.1), analysis of stability necessitates that we calculate the second-order incremental work $\delta^2\mathscr{W}$ of small equilibrium forces δf_1 and δf_2 on arbitrary small incremental displacements δq_1 and δq_2. We have $\delta^2\mathscr{W} = \frac{1}{2}(\delta f_1\,\delta q_1 + \delta f_2\,\delta q_2)$, and substitution of Equations 10.3.3 provides, after some algebraic rearrangements,

$$\delta^2\mathscr{W}(X, Y) = \frac{P_t l\eta}{\xi + 1}\left\{[(\xi + 1)Y - (\xi - 1)\,|X|]^2 + 4\xi\left[1 - \frac{(\xi + 1)P_0}{2\xi P_t \eta}\right]X^2\right\} \tag{10.3.6}$$

The absolute value $|X|$ is introduced here in order to make Equation 10.3.6 valid for buckling to both right and left.

As indicated in Section 10.1, $\delta^2\mathscr{W}(X, Y)$ can also be calculated as the work of stresses instead of reactions (which is the procedure we favored in Chaps. 4–7). According to Equation 10.1.29 or 10.1.30, we have

$$\delta^2\mathscr{W} = \frac{h}{2}\left(\frac{A}{2}\right)(\delta\sigma_1\,\delta e_1 + \delta\sigma_2\,\delta e_2) + \delta^2\mathscr{W}_\sigma \qquad \delta^2\mathscr{W}_\sigma = -P_0\,\Delta l \qquad (10.3.7)$$

in which

$$\delta\sigma_1 = -2\eta E_t(X+Y) \qquad \delta\sigma_2 = 2\eta\xi E_t(X-Y) \qquad (10.3.8)$$

Here σ_1, σ_2, e_1, e_2 = stresses and small (linearized) strain in the left and right flanges of the elastoplastic link, and $\Delta l = l(1-\cos\delta\theta) \simeq l(\delta\theta)^2/2 = 2lX^2$ = axial length change of the column due to lateral deflection, which is second-order small. It may be checked that Equation 10.3.7 yields for $\delta^2\mathscr{W}$ the same expression as Equation 10.3.6. Also note that since $\delta^2\mathscr{W}_\sigma = -2P_0lX^2$, the geometric stiffness matrix $\mathbf{K}^\sigma$ with respect to variables X and Y has the components $K_{11}^\sigma = -4P_0l$, $K_{12}^\sigma = K_{21}^\sigma = K_{22}^\sigma = 0$, and is independent of ξ and η. The stiffness matrix for zero initial stress, $\mathbf{K}^0$, is then obtained by subtracting $\mathbf{K}^\sigma$ from the total tangential stiffness matrix in Equations 10.3.3.

Equation 10.3.6 applies in general if both the axial load and the column length can change during the incremental deformation. Under a displacement-controlled mode of loading, we have $\delta q_2 = 0$ during buckling, and then $\delta^2\mathscr{W} = \frac{1}{2}\delta f_1\,\delta q_1$, that is, (for $Y \geq 0$):

$$\delta^2\mathscr{W} = [\delta^2\mathscr{W}]_{Y=\text{const.}} = \delta^2\mathscr{W}(X) = \eta P_t l\left(\xi + 1 - \frac{2P_0}{\eta P_t}\right)X^2 \qquad (10.3.9)$$

On the other hand, under a load-controlled mode of loading ($\delta f_2 = 0$):

$$\delta^2\mathscr{W} = [\delta^2\mathscr{W}]_{P=\text{const.}} = \delta^2\mathscr{W}(X) = \eta P_t l\left(\frac{4\xi}{\xi+1} - \frac{2P_0}{\eta P_t}\right)X^2 \qquad (10.3.10)$$

To identify the stable path, we will also need the work $\delta^2\mathscr{W}$ done along each equilibrium path (for $\delta f_1 = 0$). For path 2, $\delta^2\mathscr{W}^{(2)} = \frac{1}{2}\delta f_2^{(2)}\,\delta q_2^{(2)}$. Substituting Equations 10.3.5 we get, in terms of Y,

$$\delta^2\mathscr{W}^{(2)} = \frac{2\xi_u P_t - (\xi_u + 1)P_0}{(\xi_u+1)P_t - 2P_0}(2P_t l Y^2) \qquad (10.3.11)$$

and in terms of δf_2

$$\delta^2\mathscr{W}^{(2)} = \frac{(\xi_u+1)P_t - 2P_0}{2\xi_u P_t - (\xi_u+1)P_0}\left(\frac{h^2}{2P_t l}\right)\delta f_2^2 \qquad (10.3.12)$$

For path 1, $\delta^2\mathscr{W}^{(1)} = \frac{1}{2}\delta f_2^{(1)}\,\delta q_2^{(1)} = 2P_t l Y^2$ in terms of Y, or $\delta^2\mathscr{W}^{(1)} = h^2\,\delta f_2^2/2P_t l$ in terms of δf_2, according to Equations 10.3.4. After some algebra, the difference is found to be, when Y is the same for both paths,

$$\delta^2\mathscr{W}^{(1)} - \delta^2\mathscr{W}^{(2)} = \frac{(P_0 - P_t)(\xi_u - 1)P_t l}{[1 + (\xi_u - 1)^2/4\xi_u]P_r - P_0}Y^2 \qquad (10.3.13)$$

and when δf_2 is the same for both paths,

$$\delta^2 \mathscr{W}^{(1)} - \delta^2 \mathscr{W}^{(2)} = -\frac{(P_0 - P_t)(\xi_u - 1)h^2}{2(P_r - P_0)(\xi_u + 1)P_t l} \delta f_2^2 \tag{10.3.14}$$

Equations 10.3.13 and 10.3.14 characterize the differences between the substeps considered in Section 10.2, which are for the present column represented by segments $\overline{BC}$ (substep toward the primary path) and $\overline{BC'}$ (substep toward the secondary path) and are marked by arrow arcs in Figure 10.8b.

Stable Equilibrium States of Elastoplastic Column

According to Equation 10.1.37, the condition of stability is that the expressions in Equation 10.3.10 (for a prescribed axial load) or Equation 10.3.9 (for a prescribed axial displacement) must be positive for all real X and Y (i.e., positive definite). From this we conclude that, under load control, Shanley's column is stable if $P_0 < P_{cr}^L$ and unstable if $P_0 > P_{cr}^L$ where

$$P_{cr}^L = P_r = \frac{2\xi_u}{\xi_u + 1} P_t \tag{10.3.15}$$

Under displacement control, the column is stable if $P_0 < P_{cr}^D$ and unstable if $P_0 > P_{cr}^D$ where

$$P_{cr}^D = \frac{\xi_u + 1}{2} P_t = \left[1 + \frac{(\xi_u - 1)^2}{4\xi_u}\right] P_r > P_r \tag{10.3.16}$$

For elastic columns, by contrast, $P_{cr}^L = P_{cr}^D$. The physical reason for P_{cr}^D to be higher than P_{cr}^L is that lateral deflection of an elastoplastic column at constant P is accompanied by axial displacement (Eqs. 10.3.5). Note that P_{cr}^D is the P_0 value for which Equations 10.3.3 with $\delta f_1 = Y = 0$ have a nonzero solution (at $\delta f_2 \neq 0$), while P_{cr}^L is the P_0 value for which Equations 10.3.3 with $\delta f_1 = \delta f_2 = 0$ ($P = \text{const.}$) have a nonzero solution.

Also note that the value of P_{cr}^L can be obtained from Equations 10.3.5 as the value of P_0 for which the slope $\delta f_2 / Y$ ceases to be positive, and the value of P_{cr}^D from the condition that Y/X ceases to be positive.

The main aspects of the present stability problem can be illustrated by the surfaces in Figure 10.9. This figure shows (for $E_u = 3E_t$) three-dimensional views of the surfaces of $T(\Delta S)_{in} = -\delta^2 \mathscr{W}$ given by Equation 10.3.6 as functions of X and Y. The equilibrium state is characterized by $\partial(\delta^2 \mathscr{W})/\partial X = 0$ and $\partial(\delta^2 \mathscr{W})/\partial Y = 0$. Accordingly, all the surfaces shown have zero slopes at the origin, for any direction. These surfaces show the values of $\delta^2 \mathscr{W}$ that are reached along radial paths from the origin (i.e., paths for which $\delta q_2 / \delta q_1 = \text{const.}$). For nonradial paths for which the path direction at all points of the path belongs to the same sector (cone) of $\mathbf{v}$-directions in the space of $\delta\mathbf{q}$, the behavior is path-independent, that is, the same $\delta^2 \mathscr{W}$ is obtained. For other paths, however, there is path dependence, that is, different $\delta^2 \mathscr{W}$ is obtained.

Equation 10.3.6, which can be written as $\delta^2 \mathscr{W} = \sum \frac{1}{2} K_{jk} \, \delta q_j \, \delta q_k$, appears to be a quadratic form but is not, because ξ and η depend on the ratio $\delta q_2 / \delta q_1$. The surfaces in Figure 10.9 consist of quadratic portions separated by the lines $X = \pm Y$ at which ξ or η changes discontinuously. These are lines of curvature

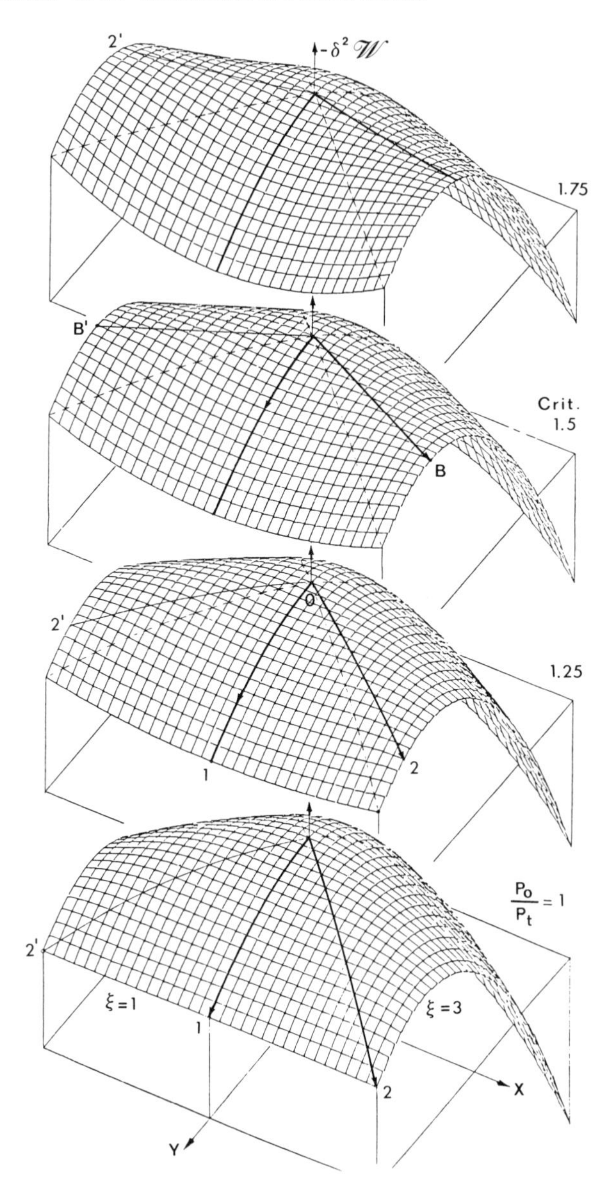

Figure 10.9 Surfaces of internally produced entropy increment $(\Delta S)_{\text{in}} = -\delta^2\mathcal{W}/T$ for Shanley's column at various load levels.

discontinuity. Therefore, in contrast to the potential-energy surfaces for elastic stability problems (Chap. 4), the surfaces of $\delta^2\mathcal{W}$ are not smooth.

We have constructed the surface $\delta^2\mathcal{W}(X, Y)$ for radial paths. Because the tangential stiffness matrix K_{ij} is symmetric (Eqs. 10.3.3), and because the values of K_{ij} are constant for finite ranges (cones) of directions characterized by the values of ξ and η, the same surface $\delta^2\mathcal{W}$ also applies to a set of infinitesimal nonradial paths such that the direction of the path at any point of the path belongs to the same sector (cone) of the q-space in which the point lies (see our discussion in Sec. 10.1). For this particular set of paths, the values of $\delta^2\mathcal{W}$ are path-independent,

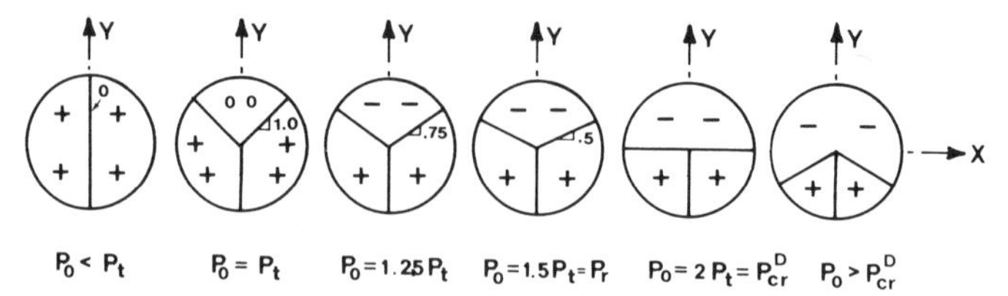

Figure 10.10 Evolution of surfaces $\delta^2\mathcal{W}(X, Y)$ for Shanley's column at increasing load P.

$\delta^2\mathcal{W}$ does represent an incremental potential, and the surface of $\delta^2\mathcal{W}$ is continuous (as seen in Fig. 10.9). For other paths, the behavior is path-dependent.

The surfaces of $\delta^2\mathcal{W}$ have continuous slopes, because the gradient represents the equilibrium forces ($f_k = \partial\mathcal{F}/\partial q_k$ or $\partial\mathcal{U}/\partial q_k$), which must change continuously. The curvatures, however, must be discontinuous because they represent the tangential stiffnesses K_{ij} of the structure, which change discontinuously, for example, due to a change from E_t to E_u.

The stack of the surfaces for $P_0/P_t = 1$, 1.25, 1.5 (crit.), 1.75 (Fig. 10.9) shows how these surfaces evolve as the load is increased. The limit of stable states ($P_0 = 1.5P_t = P_{cr}$) is manifested on these surfaces by the existence of a horizontal path emanating from the origin 0 (line B or B' in the figures). Instability is characterized by the existence of a path for which $\delta^2\mathcal{W}$ descends [or $T(\Delta S)_{in} = -\delta^2\mathcal{W}$ rises] while moving away from the origin. Absence of any such path ensures stability. Note that for $P_0/P_t = 1.25$ a portion of the surface in Figure 10.9 for $Y > |X|$ is a hyperbolic paraboloid, even though the state is stable. Figure 10.10 shows the evolution of the $\delta^2\mathcal{W}$ surfaces as seen from the top.

While stability under load control (dead load P) is determined by the positive definiteness of the entire two-dimensional surface in Figure 10.9, stability at displacement control (Y fixed) is determined only by the positive definiteness of the cross section $Y = 0$; this cross section is one-dimensional, that is, a curve. The critical load P_{cr}^D is obtained when the path $Y = 0$ away from the origin is horizontal, which occurs for $P = 2P_t$; see Figure 10.10 (not shown in Fig. 10.9).

Stable Equilibrium Path of an Elastoplastic Column

Inspecting Equations 10.3.13 and 10.3.14 for Shanley's column, we find that under displacement control (same Y) we always have $\delta^2\mathcal{W}^{(2)} < \delta^2\mathcal{W}^{(1)}$ if $P_0 > P_t$, and under load control (same δf_2), we always have $\delta^2\mathcal{W}^{(2)} > \delta^2\mathcal{W}^{(1)}$ if $P_0 > P_t$. This means that, for $P_0 > P_t$, path 2 must occur and is, therefore, stable, while path 1 cannot occur and is, therefore, unstable. So the column must deflect for $P_0 > P_t$. Shanley's load P_t represents the maximum load of an undeflected column that can be achieved in a continuous loading process, provided E_t varies continuously.

What is then the meaning of the stable states of a perfect column for $P_0 > P_t$? They can be reached if temporary restraints are placed at the sides of the column to prevent it from buckling. The load may then be raised up to some value $P_0 > P_t$. If $P_0 < P_{cr}^D$ at axial displacement control, or if $P_0 < P_{cr}^L$ at axial load control, this column will not deflect when the lateral restraint is removed (provided that the column is perfect, of course). So the column is stable at such a load because the initial state does not change. Deflection occurs only if the load is increased further.

If E_t decreases discontinuously, which happens, for example, if the σ-ε diagram is bilinear or if the temperature suddenly increases, then an undeflected equilibrium state $P_0 > P_t$ can be reached by a continuous loading process (without any lateral restraints).

The equilibrium paths leading away from the origin are marked in Figure 10.9 as 1, 2, and 2′. In the plots of $\delta^2\mathscr{W}$, the structure follows the path that descends steeper with respect to Y (since X is controlled), and less steeply for the plots of $-\delta^2\mathscr{W}$. The limit of stability of the primary path is characterized by the fact that points 1 and 2 (of equal Y) are at equal altitude (Fig. 10.9). This happens on the surfaces for $P_0/P_t = 1$. Instability of the primary path is characterized by the fact that point 2 in the plots of $-\delta^2\mathscr{W}$ lies at a higher altitude than point 1 (and a lower altitude for the plots of $\delta^2\mathscr{W}$). (Note also that the states on the primary path $\overline{01}$ in Fig. 10.9 cannot be called metastable because for $P_0 > P_t$ it is not possible to move from point 1 to point 2.)

The static structural stability studies in the literature, even those conducted in the most general sense of catastrophy theory (Sec. 4.7), have mostly been confined to elastic structures that possess a potential. For inelastic structures, we have two crucial differences: (1) While the surface of the elastic potential energy is always smooth, the surface of $-\delta^2\mathscr{W}$ that determines stability of inelastic structures is unsmooth, exhibiting lines of curvature discontinuity. (2) While the elastic potential surface characterizes a path-independent response, the surface of $\delta^2\mathscr{W}$ applies only to *radial* outward paths (for example, $\delta q_2/\delta q_1 = \text{const.}$). Nonradial loading paths, such as $\overline{012}$ in Figure 10.6a, produce changes $-T(\Delta S)_{\text{in}}$ that do not lie on this surface, in contrast to elastic behavior.

As we have seen in general in Section 10.2 and illustrated for Shanley's column, irreversible systems have the following striking properties: The first bifurcation point on the equilibrium path of an inelastic structure does not have to represent the limit of stability, that is, the states on all the branches emanating from the bifurcation point can be stable (which cannot occur in elasticity); yet at the same time, the stable states on one branch beyond the first bifurcation point cannot be reached by a continuous loading process.

As proven in Section 10.2, the ultimate cause for this behavior lies in the irreversibility of inelastic deformation. Let us describe it in more detail considering the path of Shanley's column in the (q_1, P) and (q_1, q_2) planes. After bifurcation at point 1 in Figure 10.8c, d, a subsequent prescribed increment of either axial load P or axial displacement q_2 can occur along two distinct equilibrium paths leading to points 2 and 3 (actually, if buckling to the left is also considered, there is also a third path $\overline{13'}$, but it need not be analyzed since it is symmetric to path $\overline{13}$). This is similar to elastic bifurcation. However, contrary to elastic bifurcation, the structure cannot move along path $\overline{23}$, not even in a nonequilibrium (dynamic) manner, and cannot reach at point 3 the same values of q_1, q_2, and P. The cause is evidently the irreversibility (path dependence) of plastic strain, which prohibits reaching the same values of q_1, q_2, and P as those reached along path $\overline{12}$.

An elastic structure, though, can move along path $\overline{23}$ in a nonequilibrium manner, and it does reach at point 3 the *same* values of q_1, q_2, and P as does path $\overline{12}$. This is dictated by path independence of elastic deformation. If the structure were elastic (reversible, path-independent) then admissibility of path $\overline{23}$ would cause the potential energy at point 2 to be non-positive definite. For inelastic structures, on the other hand, the state (q_1, q_2, P) at point 3 (Fig. 10.8c, d) cannot manifest itself in the incremental work expression at point 2 since path $\overline{23}$ is kinematically inadmissible. It is for this reason that point 2 in Figure 10.8c, d can be stable for inelastic structures but never for elastic structures.

For the same reason, point 2, even if it is infinitely close to point 1, can be a bifurcation state itself, permitting as the subsequent equilibrium paths both path $\overline{24}$ and path $\overline{25}$ in Figure 10.8c, d. Bifurcation states infinitely close to each other, occupying a continuous path (such as $\overline{124}$), are impossible for elastic structures (reversible systems).

The foregoing observations justify our broadening (in Sec. 10.2) of the general concept of stability by distinguishing between (1) a stable equilibrium *state,* and (2) a stable equilibrium *path.* The stable path is that which (1) consists entirely of stable states and (2) maximizes $(\Delta S)_{\text{in}}$ compared to all the other paths (which satisfy the given constraints of load or displacement control). So the stable path is a narrower concept than a stable state. For elastic (reversible) structures, both concepts are equivalent, and so this distinction does not exist. For irreversible systems, however, an equilibrium state can be stable while the equilibrium path on which it lies may be unstable. This stable state cannot, in reality, be reached. For such systems, examination of stable states is obviously insufficient.

Note also that stability of the state is decided on the basis of deviations *away* from equlibrium, while stability of the path is decided on the basis of approaches *toward* equilibrium.

The concept of a stable path does not quite fit the general definition of stability of solutions, as stated in the dynamic definition of stability in the sense of Poincaré and Liapunov (Sec. 3.5). If an infinitely small disturbance (such as lateral load f_1) is introduced at the bifurcation point (point 1 or 2 of Fig. 10.8c, d), it does not change path $\overline{124}$ to path $\overline{13}$ or $\overline{125}$; rather it excludes path $\overline{124}$ from paths $\overline{124}$, $\overline{13}$, and $\overline{125}$ that are all possible in the absence of any disturbance. Thus, instability of a path is not manifested by the creation of a second, distinct path, as a consequence of an infinitely small disturbance. It is manifested by the opposite, namely by the *exclusion* of one of several possible paths.

Breakdown of Symmetry

The fact that real columns must start to deflect at P_t can be independently proven by analyzing the effect of imperfections; see Sec. 8.2 (Eq. 8.2.1 and Fig. 8.13). Imperfections generally are an effect that breaks symmetry of the structure.

The bifurcations that we have illustrated correspond to a breakdown of symmetry. The perfect column has the symmetric choice of deflecting either left of right, but once it has deflected to the right it no longer has the choice of deflecting to the left, that is, its symmetry has broken down. Structures without symmetry (or at least some hidden symmetry) do no exhibit equilibrium path bifurcations. This is, for example, the case for our column if it is perturbed by a lateral load (Eq. 8.2.1).

Symmetry of any system can be eliminated by introducing suitable imperfections. Does this render the preceding stability analysis useless? Hardly. Imperfect systems are in general harder to solve than the perfect (symmetric) systems. A particular difficulty is that, in principle, all the possible imperfections have to be considered (in Eq. 8.2.1 we considered only one type of imperfection). Analysis of the perfect system is required to get a complete picture of the behavior near the bifurcation point.

Hypothesis Implied in Present Thermodynamic Approach

A general thermodynamic description of inelastic (irreversible) systems requires the use of internal variables that characterize the dissipative process in the material. In that approach, the equilibrium state is the terminal state of the dissipative processes, and so none of the states along the (static) equilibrium path of a structure can be considered to be an equilibrium state. But then it is also impossible to consider stability. Yet, from a practical viewpoint it certainly does make sense to consider stability of an inelastic structure. As already pointed out in our discussion of elastic structures in Sec. 10.1, the way out of this conceptual roadblock is to define stability in terms of the tangentially equivalent elastic structure, and deliberately avoid the theory of internal variables for inelastic materials.

There are nevertheless some fine points to worry about. A tangentially equivalent *elastic* structure exists only if the response to infinitesimal loading increments is path-independent (e.g., if application of de_{11} followed by de_{12} produces the same response as a simultaneous proportional application of both de_{11} and de_{12}). Most inelastic constitutive models satisfy this condition of incremental path independence, but some do not; for example, the nonassociated Drucker-Prager material, which violates the normality rule. In those cases, the approximation $d\mathbf{f} = \mathbf{K}(\mathbf{v})\, d\mathbf{q}$ is still possible but matrix $\mathbf{K}$ is not symmetric (and thus does not correspond to elastic behavior). In uniaxial behavior of the material, these problems, of course, do not arise.

Our approach to stability of state as well as path has been based on maximization of entropy. This might seem to be similar to the principle of maximum entropy production that is widely used in irreversible thermodynamics. Unlike the second law of thermodynamics, that principle is not a natural law and can be violated. If we use the tangentially equivalent elastic structure, however, we are not in the realm of irreversible thermodynamics but in the realm of equilibrium (classical) thermodynamics, which deals only with processes infinitely close to an equilibrium state. In that sense, the maximization of entropy is a natural law and does not represent application of the principle of maximum entropy production from irreversible thermodynamics. It should also be noted that plasticity, unlike viscoplasticity, is not a well-defined process of irreversible thermodynamics because the dissipation rate is undefined, as time plays the role of an arbitrary increasing parameter. From this viewpoint, plasticity is more like elasticity than viscoplasticity.

In view of the aforementioned conceptual difficulties, it is not surprising that Hill (1958) and others chose to state the stability criterion of positive-definite second-order work as a mathematical postulate (hypothesis), without any thermodynamic justification. However, by the reverse of the present argument, this postulate of Hill in effect implies the approximation of the actual structure by the tangentially equivalent elastic structure. It appears that this approximation cannot be avoided if stability should be investigated.

In the case of path stability analysis, an additional conceptual difficulty can arise from the need to compare the entropy increases for various postbifurcation paths. It may happen that paths $\alpha = 1$ and $\alpha = 2$ correspond to different tangential stiffness matrices $\mathbf{K}$ (e.g., one purely for loading and another one for a combination of loading and unloading, which happens for Shanley's column bifurcations at $P > P_t$). In that case $(\Delta S)_{\text{in}}^{(1)}$ and $(\Delta S)_{\text{in}}^{(2)}$ correspond to *different* tangentially equivalent elastic structures. We have tacitly assumed that the maximum $(\Delta S)_{\text{in}}$ still decides the path that actually happens.

Summary

1. While stability of a state is decided by considering deviations *away* from equilibrium, stability of a path is decided by considering approaches *toward* equilibrium.
2. Among all the equilibrium paths emanating from a bifurcation point, the internal entropy of the structure is maximized for the stable path. It follows that, among all the equilibrium paths, the stable path is that which consists of stable states and either minimizes the second-order work along the path if the displacements are controlled, or maximizes it if the loads are controlled, as compared to all other equilibrium paths.
3. The undeflected states of Shanley's perfect elastoplastic column are stable up to the reduced modulus load P_r if the axial load is controlled, and up to an even higher load if the displacements are controlled. However, the stable undeflected states for loads P above the tangent modulus load P_t are not reachable in a continuous loading process, except when E_t decreases discontinuously. The stable equilibrium path is such that the deflection becomes nonzero as soon as P exceeds P_t.
4. If there is only a single load, the stable path is that for which the tangential stiffness is minimum, provided the initial state is stable.

Problems

10.3.1 Analyze stable states and the stable path for the symmetric buckling mode of a perfect rigid-bar column in Figure 10.11a that has two elastoplastic links and is characterized by two displacements q_1 and q_2.

10.3.2 Analyze stable states and the stable path for the free-standing perfect rigid-bar column in Figure 10.11b whose link is elastic perfectly plastic (yield stress f_y) and is restrained against rotation by a spring of stiffness C.

10.3.3 Find the stability limit P_{cr}^{D} at controlled axial displacement for a perfect hinged continuously deformable elastoplastic column of constant rectangular cross section (Fig. 10.11c). Assume f_1 to be the amplitude of a sinusoidally distributed lateral load.

10.3.4 Calculate surface $\delta^2 \mathscr{W}$ for the column in Problem 10.3.3, and discuss stability of states and paths. (*Note*: For the initial deflection, the deflection shape is sinusoidal.)

10.3.5 Do the same as above, but for (a) a continuous two-span column, (b) a fixed column (Figs. 10.11d, e).

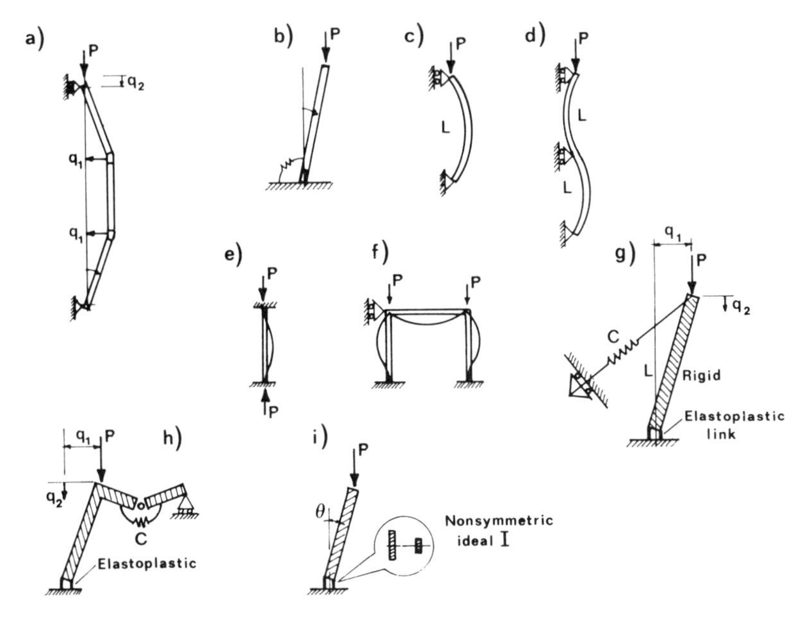

Figure 10.11 Exercise problems on buckling of elastoplastic systems.

10.3.6 Do the same as above but for a perfect portal frame (Fig. 10.11f): (a) with no sway, (b) in the sway mode. Note that since the differential equation for the initial deflections is linear, the s and c stability functions from Chapter 2 can be used for all the calculations. The variables of $\delta^2\mathcal{W}$ are the amplitude of the first buckling mode and the axial shortening of the column.

10.3.7 Calculations of $\delta^2\mathcal{W}$ from the work of reactions (Eq. 10.1.21 or 10.1.22) and from the work of stresses (Eq. 10.1.29 or 10.1.30) must yield the same result. Prove it for (a) Shanley's column, Figure 10.7, (b) the columns in Figure 10.11a, b, c, d, e.

10.3.8 Verify path dependence of surface $\delta^2\mathcal{W}(X, Y)$ for Shanley's column (Fig. 10.7). We already calculated this surface when point (X, Y) is reached radially (Fig. 10.12a). Consider that any point (X, Y) is reached by moving (1) first along axis Y (at $X = \text{const.}$), then in the direction of X ($Y = \text{const.}$), Fig. 10.12b; (2) first along the X axis (at $Y = \text{const.}$), then in the direction of Y ($X = \text{const.}$), Fig. 10.12c; (3) first in arbitrary direction ϕ (Fig. 10.12d), then in the direction of X. Show that each case yields a different surface $\delta^2\mathcal{W}(X, Y)$, and that there in fact exist infinitely many possible surfaces

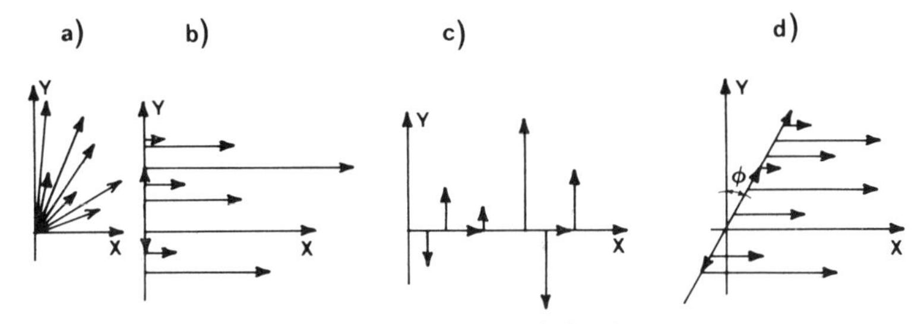

Figure 10.12 Different paths for Shanley's column deflection.

$\delta^2\mathscr{W}(X, Y)$ if nonradial paths are considered. Also show that for elastic behavior ($\xi_u = 1$, $E_t = E_u$), all these surfaces are the same.

10.3.9 Do the same as Problem 10.3.4 but for the structures in Figure 10.11g, h, i.

10.3.10 Check that Equation 10.3.7 indeed yields the same $\delta^2\mathscr{W}$ as given in Equation 10.3.6.

10.4 CRITICAL STATES OF STABILITY AND BIFURCATION

According to Sections 10.1 to 10.3, the critical states of stability and bifurcation can be found by examining certain quadratic forms as demonstrated for elastoplastic columns in Section 10.3. For more complex structures, however, more general criteria are needed.

The analysis that follows first ignores the dependence of the stiffness matrix on direction $\mathbf{v}$ in the space of $\delta\mathbf{q}$, and later brings it under consideration.

Critical State for Structures with a Symmetric Stiffness Matrix

The critical state of stability represents the limit of stable states and the onset of instability. Stable states are those for which $-T(\Delta S)_{\text{in}} = \delta^2\mathscr{W} = \frac{1}{2}\delta\mathbf{q}^T\mathbf{K}\,\delta\mathbf{q} > 0$ for all $\delta\mathbf{q}$ (i.e., $\delta^2\mathscr{W}$ is positive definite); here $\delta\mathbf{q}$ = column matrix of displacements $q_1, \ldots, q_n$, and $\mathbf{K} = [K_{ij}(\mathbf{v})]$ = tangential stiffness matrix (either isothermal or isentropic) as introduced in Section 10.1. Thus, the critical state of stability is the state at which the quadratic form $\delta^2\mathscr{W}$ vanishes for some vector $\delta\mathbf{q} = \delta\mathbf{q}^*$, that is, it becomes positive semidefinite (see also Langhaar, 1962, p. 326). Assuming $\mathbf{K}$ to be symmetric, a quadratic form is of this type (Korn and Korn, 1968; Strang, 1976) if and only if among the eigenvalues λ_k of matrix $\mathbf{K}$ ($k = 1, 2, \ldots, n$) at least one is zero and the remaining ones are positive. Therefore, and also because $\det\mathbf{K} = \lambda_1\lambda_2\cdots\lambda_n$ (Vieta's rule, Sec. 4.1), the critical state of stability (for symmetric $\mathbf{K}$) occurs if

$$\det\mathbf{K} = 0 \tag{10.4.1}$$

In this case the matrix equation $\mathbf{K}\,\delta\mathbf{q} = 0$ has a nonzero solution vector $\delta\mathbf{q} = \delta\mathbf{q}^*$ and consequently the critical state of stability represents neutral equilibrium. The vector $\delta\mathbf{q}^*$ is the eigenvector associated with a zero eigenvalue of $\mathbf{K}$ and describes the mode of instability (critical mode). For $\delta\mathbf{q} = \delta\mathbf{q}^*$ we obviously have $\delta^2\mathscr{W} = \frac{1}{2}\delta\mathbf{q}^T(\mathbf{K}\,\delta\mathbf{q}) = \frac{1}{2}\delta\mathbf{q}^T\cdot\mathbf{0} = 0$.

According to Sylvester's criterion (Sec. 4.1), positive definiteness is lost not only when $\det\mathbf{K} = 0$ but also when $\det_m\mathbf{K} = 0$ where $\det_m\mathbf{K}$ is the mth principal subdeterminant (a minor) of matrix $\mathbf{K}$. However when $\det_m\mathbf{K}$ for subvector $(\delta q_1, \delta q_2, \ldots, \delta q_m)$ with $m < n$ becomes zero while at the same time $\det\mathbf{K} \neq 0$, one does not have a critical state, in other words, the quadratic form of matrix $\mathbf{K}$ is not semidefinite but indefinite (because if $\det\mathbf{K} \neq 0$ it cannot be semidefinite and if $\det_m\mathbf{K} = 0$ it cannot be positive definite) and there is no neutral equilibrium, that is, $\mathbf{K}\,\delta\mathbf{q} = \delta\mathbf{f} \neq 0$ for all vectors $\delta\mathbf{q}$. Furthermore (under the hypothesis that the condition $\det\mathbf{K} = 0$ corresponds to a change of sign of determinant from positive to negative) one can prove that (Langhaar, 1962, p.

326) in the loading process, the determinant itself vanishes before any of its principal minors.

Critical States for Structures with a Nonsymmetric Stiffness Matrix

In certain situations, for example, if there is internal friction in the structure (Sec. 10.6), the tangential stiffness matrix $\mathbf{K}$ may be nonsymmetric. In that case

$$-T(\Delta S)_{\text{in}} = \delta^2 \mathcal{W} = \frac{1}{2}\,\delta\mathbf{q}^T\mathbf{K}\,\delta\mathbf{q} = \frac{1}{2}\,\delta\mathbf{q}^T\left(\frac{\mathbf{K}+\mathbf{K}^T}{2}\right)\delta\mathbf{q} = \frac{1}{2}\,\delta\mathbf{q}^T\hat{\mathbf{K}}\,\delta\mathbf{q} \quad (10.4.2)$$

where $(\mathbf{K}+\mathbf{K}^T)/2 = \hat{\mathbf{K}} =$ symmetric part of $\mathbf{K}$, and superscript ^{T} denotes the transpose. This is proven by noting that $2\delta\mathbf{q}^T\hat{\mathbf{K}}\,\delta\mathbf{q} = \delta\mathbf{q}^T\mathbf{K}\,\delta\mathbf{q} + (\delta\mathbf{q}^T\mathbf{K}^T\,\delta\mathbf{q})^T = 2\delta\mathbf{q}^T\mathbf{K}\,\delta\mathbf{q}$. So we must conclude that stability is decided solely by the symmetric part of the tangential stiffness matrix.

For nonsymmetric matrices, the condition of critical state (stability limit), $\delta\mathbf{q}^T\mathbf{K}\,\delta\mathbf{q} = 0$, can be satisfied not only when $\mathbf{K}\,\delta\mathbf{q} = 0$, which is the case of neutral equilibrium, but also when the vector $\mathbf{K}\,\delta\mathbf{q} = \delta\mathbf{f}$ is orthogonal to vector $\delta\mathbf{q}$, that is, when $\delta\mathbf{f} \neq 0$ but

$$\delta\mathbf{q}^T\,\delta\mathbf{f} = 0 \quad (10.4.3)$$

Therefore (in contrast to the case of symmetric $\mathbf{K}$), the critical state of (stability limit) of a tangentially nonsymmetric structure (in the sense that $\delta^2\mathcal{W} = 0$ at nonzero $\mathbf{q}$) can also be obtained when $\det\mathbf{K} \neq 0$, that is, when no neutral equilibrium exists ($\mathbf{K}\,\delta\mathbf{q} \neq 0$) and all the eigenvalues of $\mathbf{K}$ are nonzero. This critical state is characterized by

$$\det\hat{\mathbf{K}} = 0 \qquad \text{or} \qquad \det(\mathbf{K}+\mathbf{K}^T) = 0 \quad (10.4.4)$$

The corresponding smallest critical load represents the limit of stability, such that $\delta^2\mathcal{W}$ is positive definite for loads below this limit. This limit is typically lower than the lowest critical load for the state of neutral equilibrium ($\det\mathbf{K} = 0$). The positiveness of all the eigenvalues $\hat{\lambda}_i$ of matrix $\hat{\mathbf{K}}$, but not of matrix $\mathbf{K}$, is a necessary and sufficient condition of stability of a structure with a nonsymmetric tangential stiffness matrix (e.g., de Borst, 1987a, b).

Theorem 10.4.1 (Bromwich, 1906; see also Mirsky, 1955, p. 389). Every eigenvalue λ of a nonsymmetric matrix $\mathbf{K}$ satisfies the inequalities (Bromwich bounds)

$$\hat{\lambda}_1 \le \operatorname{Re}\lambda \le \hat{\lambda}_n \qquad \tilde{\lambda}_1 \le \operatorname{Im}\lambda \le \tilde{\lambda}_n \quad (10.4.5)$$

where $\hat{\lambda}_1$ and $\hat{\lambda}_n$ are the smallest and largest eigenvalues of the symmetric matrix $\hat{\mathbf{K}} = \frac{1}{2}(\mathbf{K}+\mathbf{K}^T)$, and $\tilde{\lambda}_1$ and $\tilde{\lambda}_n$ are the smallest and largest eigenvalues of the antisymmetric (Hermitian) matrix $\tilde{\mathbf{K}} = (\mathbf{K}-\mathbf{K}^T)/2i$. (Thus the real part of any eigenvalue of $\mathbf{K}$ lies within the spectrum of $\hat{\mathbf{K}}$.)

Proof The definition of eigenvalues λ is $\mathbf{q}\lambda = \mathbf{K}\mathbf{q}$ where λ and $\mathbf{q}$ may be complex if $\mathbf{K}$ is nonsymmetric. Let the eigenvectors $\mathbf{q}$ be normalized so that $\bar{\mathbf{q}}^T\mathbf{q} = 1$ where the overbar denotes the complex conjugate. Then $\lambda = \bar{\mathbf{q}}^T\mathbf{q}\lambda = \bar{\mathbf{q}}^T\mathbf{K}\mathbf{q}$ and $\bar{\lambda} = \mathbf{q}^T\bar{\mathbf{K}}\bar{\mathbf{q}} = (\bar{\mathbf{q}}^T\bar{\mathbf{K}}^T\mathbf{q})^T = \bar{\mathbf{q}}^T\bar{\mathbf{K}}^T\mathbf{q} = \bar{\mathbf{q}}^T\mathbf{K}^T\mathbf{q}$ because $\bar{\lambda}^T = \bar{\lambda}$ (λ is a number, not a matrix) and $\mathbf{K}$ is assumed to be real, $\bar{\mathbf{K}} = \mathbf{K}$ [recall the rules $\overline{\mathbf{AB}} = \bar{\mathbf{A}}\bar{\mathbf{B}}$, $(\mathbf{AB})^T = \mathbf{B}^T\mathbf{A}^T$]. Therefore, $\operatorname{Re}\lambda = \frac{1}{2}(\lambda+\bar{\lambda}) = \bar{\mathbf{q}}^T\frac{1}{2}(K+K^T)\mathbf{q} = \bar{\mathbf{q}}^T\hat{\mathbf{K}}\mathbf{q}$ and $\operatorname{Im}\lambda =$

$(\lambda - \bar{\lambda})/2i = \bar{q}^T(\mathbf{K} - \mathbf{K}^T)\mathbf{q}/2i = \bar{\mathbf{q}}^T\tilde{\mathbf{K}}\mathbf{q}$. Further consider the linear substitution $\mathbf{q} = \mathbf{Q}\mathbf{y}$ that transforms $\hat{\mathbf{K}}$ to a diagonal form. The columns of the square matrix $\mathbf{Q}$ (which is real) are the normalized eigenvectors of $\hat{\mathbf{K}}$, and $\mathbf{Q}^{-1} = \mathbf{Q}^T$ (Sec. 4.1). Now, obviously, $\hat{\lambda}_1(y_1^2 + y_1^2 + \cdots + y_n^2) \le \hat{\lambda}_1 y_1^2 + \hat{\lambda}_2 y_2^2 + \cdots + \hat{\lambda}_n y_n^2 \le \hat{\lambda}_n(y_1^2 + y_2^2 + \cdots + y_n^2)$, which may be concisely written as $\hat{\lambda}_1 \bar{\mathbf{y}}^T\mathbf{y} \le \bar{\mathbf{y}}^T\hat{\mathbf{\Lambda}}\mathbf{y} \le \hat{\lambda}_n \bar{\mathbf{y}}^T\mathbf{y}$ where $\hat{\mathbf{\Lambda}}$ is the diagonal matrix of all the eigenvalues $\hat{\lambda}_1, \ldots, \hat{\lambda}_n$ of $\hat{\mathbf{K}}$ (written in Eq. 4.1.9). Here $\bar{\mathbf{y}}^T\mathbf{y} = \bar{\mathbf{q}}^T\mathbf{Q}\mathbf{Q}^T\mathbf{q} = \bar{\mathbf{q}}^T\mathbf{q} = 1$. Also, $\bar{\mathbf{q}}^T\hat{\mathbf{K}}\mathbf{q} = \bar{\mathbf{y}}^T\mathbf{Q}^T\hat{\mathbf{K}}\mathbf{Q}\mathbf{y} = \bar{\mathbf{y}}^T\hat{\mathbf{\Lambda}}\mathbf{y}$ (according to Theorem 4.1.5). So the inequality becomes $\hat{\lambda}_1 \le \bar{\mathbf{q}}^T\hat{\mathbf{K}}\mathbf{q} \le \hat{\lambda}_n$, that is, $\hat{\lambda}_1 \le \operatorname{Re}\lambda \le \hat{\lambda}_n$. The inequality for $\operatorname{Im}\lambda$ is proven similarly. In more detail, see Mirsky (1955, p. 389); cf. also Householder (1964). Q.E.D.

To sum up, structures with a nonsymmetric stiffness matrix have two kinds of critical states: (1) the critical state of *neutral equilibrium* ($\det\mathbf{K} = 0$), and (2) the critical state of *stability limit* ($\det\hat{\mathbf{K}} = 0$). The latter one is generally more severe. Positive definiteness of all the eigenvalues of the nonsymmetric tangential stiffness matrix $\mathbf{K}$ is a necessary but not sufficient condition for stability. Also, $\det\mathbf{K} = 0$ is not a condition of the critical state of stability if $\mathbf{K}$ is nonsymmetric. Since neutral equilibrium is an unstable state, it also follows that the critical load of neutral equilibrium cannot be less than the critical load of stability.

Example of a Nonsymmetric Stiffness Matrix

Let us now illustrate the foregoing conclusion by an example similar to Problem 4.1.5. Consider a structure with (tangential) stiffness matrix

$$\mathbf{K} = \begin{bmatrix} 5 - 2P & 2 - 2P - \sqrt{5} \\ 2 - 2P + \sqrt{5} & 4 - 4P \end{bmatrix} \tag{10.4.6}$$

Neutral equilibrium occurs when the equations $\sum_j K_{ij}q_j = 0$ have a nonzero solution q_i, that is, when $\det K_{ij} = 0$. This condition yields the quadratic equation $4P^2 - 20P + 21 = 0$, which has the two roots $P_{\text{cr}_1} = 1.5$ and $P_{\text{cr}_2} = 3.5$. The symmetric part of $\mathbf{K}$ is

$$\hat{\mathbf{K}} = \begin{bmatrix} 5 - 2P & 2 - 2P \\ 2 - 2P & 4\text{-}4P \end{bmatrix} \tag{10.4.7}$$

This matrix becomes singular (i.e., $\det\hat{\mathbf{K}} = 0$) when $4P^2 - 20P + 16 = 0$, which has the roots: $\hat{P}_{\text{cr}_1} = 1$ and $\hat{P}_{\text{cr}_2} = 4$. Note that the critical P-values of $\mathbf{K}$ (i.e., 1.5 and 3.5) lie between the critical P-values of $\hat{\mathbf{K}}$ (i.e., between 1 and 4), as required by Bromwich bounds, Equation 10.4.5 (the critical P-values represent the generalized eigenvalues of matrix $\mathbf{K}$ or $\hat{\mathbf{K}}$). Also note that the eigenvalues $\hat{\lambda} = \hat{\lambda}_i$ of $\hat{\mathbf{K}}$ follow from the equation

$$\det\hat{\mathbf{K}} = \begin{bmatrix} 5 - 2P - \lambda & 2 - 2P \\ 2 - 2P & 4 - 4P - \lambda \end{bmatrix} = 0 \tag{10.4.8}$$

which, for $P = 1$, reads $(\hat{\lambda} - 3)\hat{\lambda} = 0$, yielding $\hat{\lambda}_1 = 0$ and $\hat{\lambda}_2 = 3$ (so the lowest eigenvalue of $\hat{\mathbf{K}}$, but not of $\mathbf{K}$, vanishes at $P = 1$). Further one can verify that

$$2\delta^2\mathcal{W} = \sum_i \sum_j K_{ij}q_iq_j = (1 - P)(q_1 + 2q_2)^2 + (4 - P)q_1^2 \tag{10.4.9}$$

This quadratic form (which can be derived from the eigenvectors of $\hat{\mathbf{K}}$ similarly to Eqs. 4.3.18) is positive definite if and only if $P<1$, that is, $P<\hat{P}_{cr_1}$. So the limit of stability is $P=\hat{P}_{cr_1}=1$, which is less than the critical load of neutral equilibrium, $P_{cr_1}=1.5$. Note that for $P=\hat{P}_{cr_1}$ there is no neutral equilibrium; indeed, substituting $P=1$, the equilibrium equations $\sum_j K_{ij}q_j=0$ read $3q_1-\sqrt{5}\,q_2=f_1$ and $\sqrt{5}\,q_1=f_2$, which have no nonzero solution if $f_1=f_2=0$. The eigenvector of $\hat{\mathbf{K}}$ associated with $P=1$ is $q_1^{(1)}=0$, $q_2^{(1)}=1$, and the corresponding forces are $f_1^{(1)}=K_{11}q_1^{(1)}+K_{12}q_2^{(1)}=-\sqrt{5}$ and $f_2^{(1)}=K_{21}q_1^{(1)}+K_{22}^{(1)}=0$. Therefore, $2\delta^2\mathcal{W}=f_1q_1+f_2q_2=-\sqrt{5}\cdot 0+0\cdot 1=0$ at $P=1$. So $\delta^2\mathcal{W}$ vanishes not because $f_1=f_2=0$, as in the case of neutral equilibrium, but because vectors $\mathbf{f}^{(1)}=(f_1^{(1)},f_2^{(1)})$ and $\mathbf{q}^{(1)}=(q_1^{(1)},q_2^{(1)})$ are orthogonal at $P=1$, that is, $\delta^2\mathcal{W}=\frac{1}{2}\mathbf{f}^{(1)^T}\mathbf{q}^{(1)}=0$ (Eq. 10.4.3).

Symmetric and Asymmetric Bifurcations at the Critical State

The quadratic forms $\delta^2\mathcal{W}=\delta^2\mathcal{F}$ (or $\delta^2\mathcal{W}=\delta^2\mathcal{U}$) determine only stability. At the critical state of stability $\delta\mathcal{F}=\delta^2\mathcal{F}=0$ for some $\delta\mathbf{q}$, and $\delta^2\mathcal{F}\neq 0$ for all $\delta\mathbf{q}$ near the critical state. In similarity to our analysis in Sections 4.5 and 4.6 based on potential energy, the behavior near the critical state depends on the third and fourth variations of $\mathcal{F}$ (for isothermal conditions) or $\mathcal{U}$ (for isentropic conditions). Keeping the higher-order terms in Equation 10.1.20, and considering the possibility that the loads P_i are not dead loads but can vary, one may use a procedure similar to Equations 10.1.20 and 10.1.21 to obtain, for the states near the critical state, $\Delta\mathcal{F}=\delta^2\mathcal{F}+\delta^3\mathcal{F}+\delta^4\mathcal{F}$, in which

$$\delta^3\mathcal{F}=\sum_{i,j,k}\frac{1}{3!}\left[\frac{\partial^2 f_{T_i}}{\partial q_j\,\partial q_k}\right]_0\delta q_i\,\delta q_j\,\delta q_k$$

$$\delta^4\mathcal{F}=\sum_{i,j,k,m}\frac{1}{4!}\left[\frac{\partial^3 f_{T_i}}{\partial q_j\,\partial q_k\,\partial q_m}\right]_0\delta q_i\,\delta q_j\,\delta q_k\,\delta q_m \tag{10.4.10}$$

Now, for the same reasons as in Sections 4.5 and 4.6, the behavior near the critical state is a symmetric bifurcation if $\delta^3\mathcal{F}=0$ for $\delta\mathbf{q}=\delta\mathbf{q}^*$ and $\delta^4\mathcal{F}\neq 0$ for $\delta\mathbf{q}^*$. The behavior is an asymmetric bifurcation if $\delta^3\mathcal{F}\neq 0$ for some $\delta\mathbf{q}$. The critical state at asymmetric bifurcation is always unstable and strongly imperfection-sensitive. The critical state at symmetric bifurcation can be stable (if $\delta^4\mathcal{F}>0$ for all $\delta\mathbf{q}^*$) and imperfection-insensitive, or unstable (if $\delta^4\mathcal{F}<0$ for some $\delta\mathbf{q}^*$) and imperfection-sensitive. If $\delta^4\mathcal{F}=0$ for some $\delta\mathbf{q}^*$ and $\delta^4\mathcal{F}>0$ for all other δq, and also $\delta^2\mathcal{F}=\delta^3\mathcal{F}=0$ for all $\delta\mathbf{q}$, one needs to consider $\delta^5\mathcal{F}$ and $\delta^6\mathcal{F}$.

Uniqueness

The critical state of bifurcation implies nonuniqueness, and so the critical state condition can also be derived as the condition of loss of uniqueness. For small steps along the equilibrium path, we have $\mathbf{K}\,\delta\mathbf{q}=\delta\mathbf{f}$. There are now two possibilities. If the loads are such that $\delta\mathbf{f}=\mathbf{0}$ at this state, we have the limit point (maximum point, snapthrough point) of the load-deflection path. At that point we must obviously have $\det\mathbf{K}=0$. If $\delta\mathbf{f}\neq 0$ at the bifurcation point, we must have at the same time another vector $\delta\mathbf{q}'$ for which $\mathbf{K}\,\delta\mathbf{q}'=\delta\mathbf{f}$ also. Since both equations must hold simultaneously, we may subtract them, yielding $\mathbf{K}(\delta\mathbf{q} - \delta\mathbf{q}') = \mathbf{0}$, where $\delta\mathbf{q} - \delta\mathbf{q}'$ is nonzero. This, or course, implies that $\det \mathbf{K} = 0$, for symmetric as well as nonsymmetric matrix $\mathbf{K}$. So, a necessary condition of the loss of uniqueness and bifurcation is that $\det \mathbf{K} = 0$. This condition, however, also characterizes the limit point.

The condition $\det \mathbf{K} = 0$, as we showed, coincides with the condition of critical state of stability only if $\mathbf{K}$ is symmetric. If $\mathbf{K}$ is nonsymmetric and if $\det \mathbf{K} = 0$ while at the same time $\det(\mathbf{K} + \mathbf{K}^T) \neq 0$, there is loss of uniqueness (bifurcation) but no critical state of stability.

However, as we know from Section 10.3, bifurcation (loss of uniqueness) can occur without loss of stability (i.e., in a stable manner) and without neutral equilibrium even if $\mathbf{K}$ is symmetric. This happens (and can happen only) for inelastic structures for which $\mathbf{K}$ depends on direction $\mathbf{v}$ of vector $\delta\mathbf{q}$ due to the distinction between loading and unloading, which we have ignored so far. From now on we take this important phenomenon into consideration.

Bifurcation for Inelastic Structures and Hill's Linear Comparison Solid

We consider again the example of the Shanley column (Fig. 10.7). Let L be the loading-only sector of direction $\mathbf{v}$ in the space of $\delta\mathbf{q}$, and U any one of the adjacent loading-unloading sectors (Fig. 10.13b). Let $\mathbf{K}^L$, $\mathbf{K}^U$ be the corresponding tangential stiffness matrices $\mathbf{K}$ (either $\mathbf{K}_T$ or $\mathbf{K}_S$).

If the column has a choice of two paths under load control, then $\mathbf{K}^L\,\delta\mathbf{q}^{(1)} = \delta\mathbf{f}$ and $\mathbf{K}^U\,\delta\mathbf{q}^{(2)} = \delta\mathbf{f}$ where $\delta\mathbf{f}$ is given. The direction $\mathbf{v}^{(1)}$ of $\delta\mathbf{q}^{(1)}$ always lies in sector L. Prior to the first bifurcation, the direction $\mathbf{v}^{(2)}$ of $\delta\mathbf{q}^{(2)}$ lies outside the corresponding sector U for all possible sectors U, that is, no path 2 exists. After the first bifurcation, along the primary (main) path, the direction $\mathbf{v}^{(2)}$ lies within the corresponding sector U for at least one sector U, and then path 2 exists.

Suppose now that the tangential stiffnesses K_{ij} vary continuously along the loading path. Then the direction $\mathbf{v}^{(2)}$ should also vary continuously. So, at the first bifurcation, the direction $\mathbf{v}^{(2)}$ must coincide with the boundary of sector L (as illustrated by Shanley's column, Sec. 10.3). But then we must have not only $\mathbf{K}^U\,\delta\mathbf{q}^{(2)} = \delta\mathbf{f}$ but also $\mathbf{K}^L\,\delta\mathbf{q}^{(2)} = \delta\mathbf{f}$. Subtracting this from $\mathbf{K}^L\,\delta\mathbf{q}^{(1)} = \delta\mathbf{f}$ we get $\mathbf{K}^L(\delta\mathbf{q}^{(2)} - \delta\mathbf{q}^{(1)}) = \mathbf{0}$ where $\delta\mathbf{q}^{(2)} \neq \delta\mathbf{q}^{(1)}$. Consequently, the first bifurcation is indicated by singularity of matrix $\mathbf{K}^L$, that is, by the fact that

$$\det \mathbf{K}^L = 0 \tag{10.4.11}$$

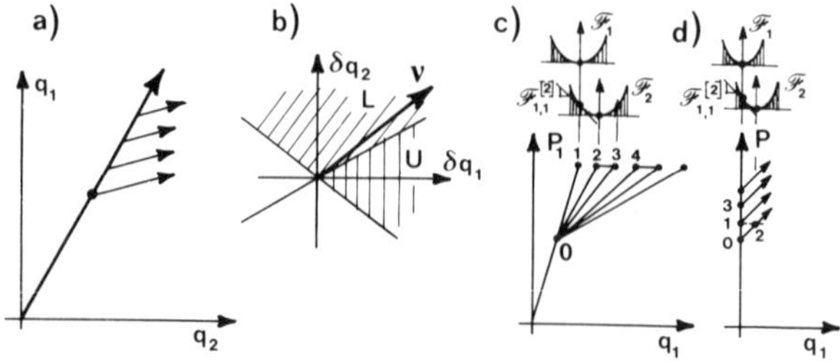

Figure 10.13 Various postbifurcation paths for inelastic structures.

or that the smallest eigenvalue λ_1 of matrix $\mathbf{K}^L$ vanishes (our example in Sec. 10.3 illustrates that). This is the well-known bifurcation criterion of Hill (1961, 1962c). He derived this criterion on the basis of consideration of uniqueness rather than stability. The solid corresponding to matrix $\mathbf{K}^L$ (for which one finds multiple solutions) is called Hill's linear comparison solid. It is a solid in which there is loading at all points of the structure.

If, however, the tangential stiffnesses change along the loading path discontinuously (see the case of bilinear material in Sec. 8.1), then the first bifurcation may occur when the value of λ_1 jumps from positive to negative, without $\mathbf{K}^L$ ever becoming singular (i.e., without the vanishing of $\det \mathbf{K}^L$). In this case, matrix $\mathbf{K}$ along the initial bifurcated path corresponds to a combination of loading and unloading (rather than loading only, as assumed for the linear comparison solid).

The eigenvector $\delta\mathbf{q}^*$ of the singular matrix $\mathbf{K}^L$ at the first bifurcation can lie either inside or outside sector L. If $\delta\mathbf{q}^*$ lies inside L, then there exists path 2 such that $\mathbf{K}^L\,\delta\mathbf{q}^{(2)} = \mathbf{0}$ where $\delta\mathbf{q}^{(2)} = \delta\mathbf{q}^*$. This means that there is neutral equilibrium, which represents the limit point instability (or snapthrough).

If $\delta\mathbf{q}^*$ lies outside sector L (which is the case for Shanley's column), then $\delta\mathbf{q}^{(2)}$ cannot coincide with $\delta\mathbf{q}^*$ but must lie at the boundary of sector L; then $\mathbf{K}^L\,\delta\mathbf{q}^{(2)} = \delta\mathbf{f}$ where $\delta\mathbf{f}$ is nonzero. This means that the secondary path at the first bifurcation occurs at increasing load, which represents the Shanley-type bifurcation.

If λ_1 is an eigenvalue of $\mathbf{K}^L$, we have $(\mathbf{K}^L - \lambda_1\mathbf{I})\,\delta\mathbf{q}^* = 0$ where $\mathbf{I}$ is the identity matrix. It follows that, for $\delta\mathbf{q} = \delta\mathbf{q}^*$, $\delta^2\mathcal{W} \sim \delta\mathbf{q}^{*T}\mathbf{K}^L\,\delta\mathbf{q} = \delta\mathbf{q}^{*T}\lambda_1\mathbf{I}\,\delta\mathbf{q} = \lambda_1\,\delta\mathbf{q}^{*T}\,\delta\mathbf{q}^* < 0$ if $\lambda_1 < 0$. But this does not imply instability of state if the associated eigenvector $\mathbf{q}^*$ lies outside L (which has been ignored in some papers). However, the existence of negative smallest eigenvalue λ_1 of tangential stiffness matrix $\mathbf{K}^t$ always means that a bifurcation point must have been passed and that the state cannot lie on a stable path. Alternatively, this situation is indicated by the existence of a negative pivot in the Gaussian elimination process of solving the system of tangential stiffness equations (cf. Theorem 4.1.11).

So the computational strategy may be to check always for negativeness of the smallest eigenvalue λ_1 of $\mathbf{K}^t$ (or the existence of a negative pivot). If a negative λ_1 is detected, one must determine the other postbifurcation paths, and the path that is actually followed is that for which all the eigenvalues of $\mathbf{K}^t$ are nonnegative (i.e., all the pivots are nonnegative). Proceeding along a path with negative λ_1 (or with a negative pivot) is incorrect.

Consider now displacement control. Let δq_n be controlled and $\delta f_1 = \cdots = \delta f_{n-1} = 0$. One may now take the foregoing case of load control for which $\mathbf{K}^L$ is singular at the first bifurcation point, and then scale δf_n and δq_n by a common factor so as to make δq_n for both paths equal. Since such a scaling does not change the eigenvalues of $\mathbf{K}^L$, the condition $\det \mathbf{K} = 0$ must also characterize the first bifurcation point under displacement control (provided that the tangential stiffnesses vary continuously). The checks for bifurcation are analogous.

The typical variations of $\det \mathbf{K}^L$ for loading only and $\det \mathbf{K}^U$ for combined loading and unloading are illustrated in Figure 10.14 for an elastoplastic column.

In regard to bifurcations in structures with damage, many other recent works are also of interest; for example, Billardon and Doghri (1989), Sulem and

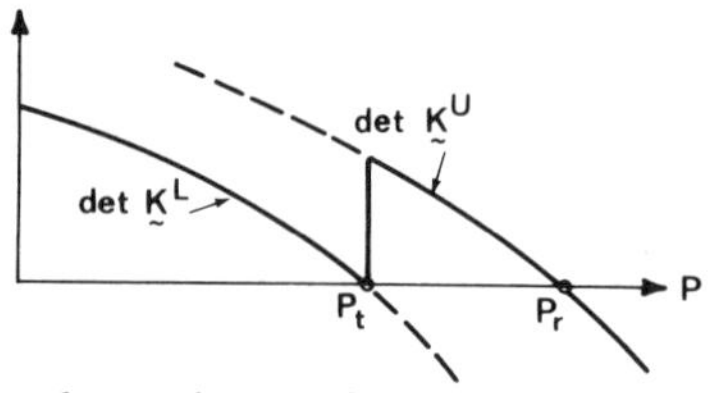

Figure 10.14 Variation of the determinant of the tangential stiffness matrix.

Vardoulakis (1989), Stolz (1989), Borré and Maier (1989), Runesson, Larsson and Sture (1989), de Borst (1989), and Leroy and Ortiz (1989).

Distribution of Bifurcation Points and Postbifurcation Branches

In Section 10.2 (Fig. 10.6) we proved Theorem 10.2.2 stating that all postbifurcation branches can consist of stable states if the structure is inelastic. We now give two related theorems (Bazant, 1987b).

Theorem 10.4.2 For inelastic (irreversible) structures, there can be two or infinitely many postbifurcation branches that consist of stable states and have directions that are infinitely close to each other (Fig. 10.13c). But for elastic (reversible) structures this is impossible.

Proof Consider the space $(q_1, \ldots, q_n)$. Let $\mathscr{F}_{1,1}$, $\mathscr{F}_{2,1}$, and $\mathscr{F}_{3,1}$ be the partial derivatives of $\mathscr{F}$ in the direction of coordinate q_1 at points 1, 2, 3 that correspond to the same load and lie at a finite distance from point 0 (Fig. 10.13c). If state 2 is an equilibrium state, then $\mathscr{F}_{2,1} = 0$, and if it is stable then $\mathscr{F}^{[2]}_{3,1} > 0$ and $\mathscr{F}^{[2]}_{1,1} < 0$ at the adjacent points 1 and 3 where superscript $^{[2]}$ means that the points are accessed by traveling to them from point 2 (Fig. 10.13c). At point 1 the derivative $\mathscr{F}_{1,1}$ is taken in the direction of q_1, while the derivative $\mathscr{F}^{[2]}_{1,1}$ is taken in the direction opposite to q_1. Due to path-dependence, the values of $\mathscr{F}_{1,1}$ and $\mathscr{F}^{[2]}_{1,1}$ at point 1 can be different for an inelastic structure, which means that $\mathscr{F}_{1,1}$ can be zero even though $\mathscr{F}^{[2]}_{1,1}$ is negative (Fig. 10.13c). So there can be an equilibrium branch also at point 1 infinitely close to point 2 (such behavior is typical for strain-softening and fracture problems; see Sec. 13.2 and Fig. 13.10). However, if the structure is elastic, that is, $\mathscr{F}$ is path-independent, then we must have $\mathscr{F}_{1,1} = \mathscr{F}^{[2]}_{1,1}$; consequently $\mathscr{F}_{1,1}$ cannot be zero, that is, branch $\overline{01}$ cannot be an equilibrium path. Q.E.D.

Theorem 10.4.3 For inelastic structures, a state [2] on a stable equilibrium path that is adjacent (infinitely close) to a bifurcation state can also be a bifurcation state, so that the equilibrium path can consist of infinitely many bifurcation points that are spaced infinitely closely (Fig. 10.13a, d). But for elastic structures this is impossible.

Proof Consider now infinitely close points 0, 1, 3 shown in Fig. 10.13d. If point 1 is a bifurcation state on an equilibrium path, we must have $\mathscr{F}_{1,1} = 0$, and if it is stable, we must have $\mathscr{F}^{[1]}_{2,1} > 0$ when point 2 is reached from point 1. If point 2 is an equilibrium state, we must have $\mathscr{F}_{2,1} = 0$. When the structure is elastic we must have $\mathscr{F}_{2,1} = \mathscr{F}^{[1]}_{2,1}$ (due to path-independence), and so point 2 cannot be an equilibrium state. But then point 0, which is infinitely close, cannot be a bifurcation point (we except the case where the angle 102 is vanishing since then segment $\overline{01}$ would not be infinitesimal). On the other hand, when the structure is inelastic we can have $\mathscr{F}_{2,1} \neq \mathscr{F}_{2,1}^{[1]}$ since path-dependence can occur (either because of the lack of symmetry of the stiffness matrix or its dependence on the displacement direction; see Sec. 10.1); this means that 02 can be an equilibrium path, and so point 0 can be a bifurcation point. Repeating for points 1, 3 the same argument as for points 0, 1 etc., we conclude that for an inelastic structure 0, 1, 3 can be bifurcation points. We will see that such behavior is typical for strain-softening structures and interacting crack systems (Secs. 13.2 and 13.8). Q.E.D.

Numerical Finite Element Analysis

In nonlinear finite element analysis, the given loads and prescribed displacements are increased at small loading steps. In each step, the tangential stiffness matrix **K** ($\mathbf{K} = \mathbf{K}_T$ or $\mathbf{K}_S$) may first be estimated on the basis of the values of forces, stresses, strains, and displacements, and iterations of the loading step may be used to improve this matrix so as to correspond to the state in the middle of the loading step (in which case the numerical error is second-order small). Stability checks can be made if the signs of the eigenvalues $\lambda_1, \ldots, \lambda_n$ of matrix **K** are determined at each loading step and each iteration.

Due the finiteness of the chosen loading steps as well as round-off errors, the lowest eigenvalue λ_1 of **K** is never exactly zero. It is not convenient to vary the size of the loading step to make λ_1 exactly zero. So one does not try to obtain the critical state exactly. The sign of λ_1 is monitored, and when a negative λ_1 is encountered while all other λ_i ($i = 2, \ldots, n$) are positive, it means that either a bifurcation point or a critical state has been passed. (Alternatively, the negativeness of λ_1 is signaled by the occurrence of a negative pivot in the Gaussian elimination process.)

If the value of $-T(\Delta S)_{\text{in}}$ obtained from matrix **K** for further loading (Eq. 10.1.23 or 10.1.24) is positive for this loading step, it means that a bifurcation point has been passed (and the state is stable). In the case of a single load or a single controlled displacement, this is the case where λ_1 becomes negative while the load is rising.

The foregoing numerical approach has initially been developed for buckling of elastic structures (e.g., Riks, 1979), for which **K** is unique, independent of loading direction **v**. For inelastic structures, in principle, the values of $-T(\Delta S)_{\text{in}}$ should be checked for matrices **K**(**v**) associated with all the sectors of **v** (Sec. 10.1). However, if **K** varies continuously along the loading path, only matrix **K** for further loading needs to be considered because it determines the first bifurcation (see Hill's criterion).

A negative value of λ_1 can also mean that the limit point (snapthrough point, maximum load point) has been passed. In this case the load is descending [and the value of $-T(\Delta S)_{\text{in}}$ is negative for some **v** if the controlled displacement is considered as a free variable, i.e., is included among the degrees of freedom associated with **K**]. This state is stable (despite negative λ_1) if the displacement is controlled. If the second lowest eigenvalue λ_2 becomes negative while the load is descending, it means that either a bifurcation point on the descending branch or a snapback point has been passed.

After bifurcation, $T(\Delta S)_{\text{in}}$ should, in principle, be calculated for the steps along each branch, and the branch for which $T(\Delta S)_{in} = \max$ should then be chosen. In practice, if there are two postbifurcation equilibrium branches, the primary one preserving symmetry and the secondary one breaking symmetry, the latter branch needs to be followed. Continuation along the symmetry-breaking branch can be enforced for the first iteration by adding the eigenmode $\mathbf{q}^*$ that is associated with vanishing λ_1 to the vector of incremental displacements $\Delta\mathbf{q}^{(1)}$ along the primary (symmetry-preserving) path (see Riks, 1979; de Borst, 1986, 1987a, b), that is,

$$\Delta\mathbf{q} = \Delta\mathbf{q}^{(1)} + k\mathbf{q}^* \tag{10.4.12}$$

where k is an unknown parameter. To obtain k, one may assume for the first iteration that the trial vector $\Delta\mathbf{q}$ is orthogonal to the primary path, that is, $\Delta\mathbf{q}^T \Delta\mathbf{q}^{(1)} = 0$, or $(\Delta\mathbf{q}^{(1)} + k\mathbf{q}^*)^T \Delta\mathbf{q}^{(1)} = 0$, from which

$$k = -\frac{\Delta\mathbf{q}^{(1)T} \Delta\mathbf{q}^{(1)}}{\Delta\mathbf{q}^{(1)T}\mathbf{q}^*} \tag{10.4.13}$$

This procedure, however, does not work when $\Delta\mathbf{q}^{(1)T}\mathbf{q}^* = 0$, that is, when the eigenmode happens to be orthogonal to the primary path. A simple remedy (de Borst, 1986, 1987a, b) is to use such k that $\Delta\mathbf{q}^T \Delta\mathbf{q} = \Delta\mathbf{q}^{(1)T} \Delta\mathbf{q}^{(1)}$ in such iterations. In the subsequent iterations of the step, the trial vector generally will not remain orthogonal to $\Delta\mathbf{q}^{(1)}$, but the orthogonality assumption in Equation 10.4.13 will maximize the chance that $\Delta\mathbf{q}$ would converge to the symmetry-breaking (secondary, bifurcated) path rather than to the symmetry-preserving (primary) path.

When several secondary branches emanate from the bifurcation point, this procedure will not necessarily converge to the branch for which $T(\Delta S)_{in} = \max$. In previous works, such a situation was assumed to be revealed by a check for a negative eigenvalue of the tangential stiffness matrix $\mathbf{K}^t$ for the points adjacent to the bifurcation point. This check is not necessarily sufficient, however. From Sections 10.2 and 10.3 we know that the states on all the postbifurcation branches can be stable, and if they are stable then either $\mathbf{K}^t$ has only positive eigenvalues or $\mathbf{K}^t$ has a negative eigenvalue but the associated eigenvector is outside the sector of directions to which $\mathbf{K}^t$ applies.

Summary

1. In the case that the tangential stiffness matrix $\mathbf{K}$ does not depend on the direction of the vector of generalized displacement increments, a structure is stable if and only if all the eigenvalues of the symmetric part of this matrix are positive, and critical if and only if at least one of these eigenvalues vanishes. The vanishing of at least one eigenvalue of a nonsymmetric $\mathbf{K}$ implies either a limit point or bifurcation (loss of uniqueness), but not a critical state of stability.
2. If $\mathbf{K}$ depends on the direction $\mathbf{v}$ of the generalized displacement increment vector, the vanishing of at least one eigenvalue of $\mathbf{K}$ implies (Fig. 10.13a, c, d) either a limit point (with or without bifurcation) or a bifurcation of Shanley type, which occurs at increasing load. The former

case happens if the associated eigenvector lies in the sector of directions $\mathbf{v}$ for which $\mathbf{K}$ is valid. The latter case happens if the eigenvector lies outside.

3. If $\mathbf{K}$ varies discontinuously along the loading path, the first bifurcation occurs when the smallest eigenvalue of $\mathbf{K}$ becomes zero or negative. If it jumps from a positive value to a negative value, then the initial postbifurcation direction of the secondary path does not involve loading-only at all points of the structure (i.e., is not governed by Hill's linear comparison solid).
4. For inelastic structures, bifurcation points can be continuously distributed along the equilibrium path, and infinitely many branches with a continuous angular distribution can emanate from each bifurcation point (Fig. 10.13a, c, d). For elastic structures, this is impossible.
5. Calculation of $(\Delta S)_{in}$ according to Section 10.2 may be used to decide which branch will be followed after bifurcation. Hill's bifurcation criterion gives no information in this regard. At a point on a stable path, all the eigenvalues of the tangential stiffness matrix must be positive. (So, if at least one is not positive, the point does not lie on a stable path.)
6. Calculation of $(\Delta S)_{in}$ from $\delta \mathbf{q}^{(\alpha)}$ and $\delta \mathbf{f}^{(\alpha)}$ (Sec. 10.2) makes it possible to detect bifurcations even without actually calculating the tangential stiffness matrix $\mathbf{K}$. This enables nonlinear finite element analysis to utilize step-by-step algorithms in which iterations are not based on the tangential stiffness but on the elastic stiffness or on the secant stiffness (see, e.g., Owen and Hinton, 1980). For such algorithms, Hill's method of linear comparison solid cannot be implemented. A complicating factor in this approach is that all the branching paths α must first be detected before $(\Delta S)_{in}$ can be calculated from $\delta \mathbf{q}^{(\alpha)}$ and $\delta \mathbf{f}^{(\alpha)}$.

Remark. In some finite element studies (e.g., Droz and Bažant, 1989; Bažant, 1989a, b) the calculation of the lowest eigenvalue of $\mathbf{K}$ was omitted and the secondary path was obtained by introducing slight symmetry-breaking imperfections in the nodal coordinates or small disturbing loads. Subsequently, the values of $(\Delta S)_{in}$ could be compared without calculating the tangential $\mathbf{K}$.

Problems

10.4.1 Use the condition $\det \mathbf{K} = 0$ for loading only ($\xi = \eta = 1$) to determine the tangent modulus load of Shanley's column. Determining the eigenvector direction, prove that there is no neutral equilibrium at this load. Discuss the variation of $\det \mathbf{K}$ during loading if the stress-strain diagram is nonlinear.

10.4.2 Do the same as above but for the columns in Figure 10.11a, b, c, d, e and the frame in Figure 10.11f.

10.4.3 Analyze again the structures in Figure 8.11g (Prob. 8.1.13) and in Figure 8.11e, l (Prob. 8.1.14) from the energy viewpoint. Plot surface $\delta^2 \mathscr{W}(X, Y)$. Analyze stable states and stable paths. Check bifurcation from (a) the second-order work criterion and (2) Hill's criterion.

10.4.4 In the example treated in this section (Eq. 10.4.6), loads P were real, but for a nonsymmetric stiffness matrix they can also be complex. Show that this occurs for $K_{11} = 1 - P$, $K_{22} = 4 - P$, $K_{12} = 3$, $K_{21} = -2$. Prove that the limit of stability is $P = \hat{P}_{cr} = (10 - \sqrt{85})/4$, and that there is no state of neutral equilibrium, because the P values for which $\mathbf{Kq} = \mathbf{0}$ are complex. (See also Prob. 4.1.6.)

10.4.5 Prove Bromwich bounds for the generalized eigenvalue problem $\mathbf{Nq}\lambda = \mathbf{Mq}$ (Eq. 4.1.6) where $\mathbf{N}$ and $\mathbf{M}$ are nonsymmetric real square matrices.

10.5 STABILITY AT INFINITESIMAL LOADING CYCLES

For some applications it is useful to consider stability from the viewpoint of infinitesimal loading cycles, in which either a load increment $\delta\mathbf{f}$ or a displacement increment $\delta\mathbf{q}$ is applied and subsequently removed. In the case of elastic structures under conservative loads, such a cycle reveals nothing because the net work over the cycle is zero. For inelastic structures, though, this work can be positive or negative, and this has some interesting consequences for material modeling.

Internal Entropy Changes for Cycles in Shanley's Column

First consider again Shanley's perfect column (Figs. 8.7 and 10.7). The column is initially in equilibrium under load P_0 at no deflection ($q_1 = 0$), and we assume the axial displacement q_2 to be controlled. As indicated by the principle of virtual work, the first-order work $\Delta W^{(1)}$ during application as well as removal of δf_1 vanishes because the column is initially in equilibrium. So we need to consider only the second-order work $\delta^2\mathscr{W}$.

Imagine a load-unload cycle in which the lateral load δf_1 is applied and then removed while the axial displacement is kept constant ($\delta q_2 = 0$). For $\delta q_2 = 0$ we have, from Equation 10.3.3, $\delta f_1 = 2\eta P_t[1 + \xi - (2P_0/\eta P_t)]\,\delta q_1/l$. During the application of δf_1, Equation 10.3.3 for δf_1 applies, with $\eta = 1$, $\xi = \xi_u$ where $\xi_u = E_u/E_t$ (Fig. 10.15c). After some algebraic manipulations, one finds

$$\delta^2\mathscr{W}_{\mathrm{I}} = \frac{1}{2}\delta f_1\,\delta q_1 = \frac{l\,\delta f_1^2}{4P_t C_{\mathrm{I}}} \qquad C_{\mathrm{I}} = 2\frac{P_{cr}^D - P_0}{P_t} \tag{10.5.1}$$

where P_{cr}^D is the critical load at controlled axial displacement (Eq. 10.3.16). During the removal of δf_1, Equations 10.3.3 for δf_1 can be applied again, but with $\eta = \xi_u$ and $\xi = 1$ (one side of the cross section unloads and the other one reloads, Fig. 10.15d). After some algebra, one gets

$$\delta^2\mathscr{W}_{\mathrm{II}} = \frac{1}{2}(-\delta f_1)\,\delta q_1 = -\frac{l\,\delta f_1^2}{4P_t C_{\mathrm{II}}} \qquad C_{\mathrm{II}} = 2\frac{P_{cr}^{el} - P_0}{P_t} \qquad P_{cr}^{el} = \xi_u P_t \tag{10.5.2}$$

where P_{cr}^{el} is the critical load if the column is elastic, with modulus E_u.

Over the entire load cycle, the net change of q_1 is nonzero because the inelastic behavior is irreversible. If the conditions are isothermal, the thermodynamic state function that depends on displacements is the Helmholtz free energy $\mathscr{F}$. From Section 10.1 we know that $\Delta\mathscr{F} = -T(\Delta S)_{\mathrm{in}}$. For isothermal deformations at mechanical equilibrium, the change $\Delta\mathscr{F}$ is given by the net work

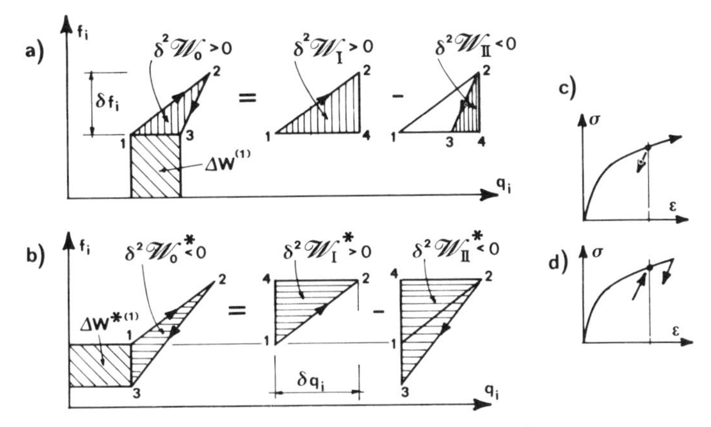

Figure 10.15 Work on infinitesimal load or displacement cycle.

$\delta^2\mathcal{W}_0$ done during the cycle, that is,

$$T(\Delta S)_{\text{in}} = -\Delta\mathcal{F} = -\delta^2\mathcal{W}_0 = -\delta^2\mathcal{W}_I - \delta^2\mathcal{W}_{II} = \frac{l\,\delta f_1^2}{4P_t}\left(\frac{C_I - C_{II}}{C_I C_{II}}\right) \quad (10.5.3)$$

The values of $\delta^2\mathcal{W}_0$, $\delta^2\mathcal{W}_I$, and $\delta^2\mathcal{W}_{II}$ are graphically represented by the cross-hatched areas in Figure 10.15a. Note that the first-order work $\Delta W^{(1)}$ is canceled by the work of applied (actual) loads (due to the principle of virtual work) and thus has no effect on stability.

Second, consider a cycle in which the lateral displacement δq_1 is applied and removed at constant axial displacement ($\delta q_2 = 0$). During the application of δq_1, Equations 10.3.3 for δf_1 apply, with $\eta = 1$, $\xi = \xi_u$. During the removal of δq_1, the same equation applies with $\eta = \xi_u$, $\xi = 1$. After some algebra, one obtains

$$\delta^2\mathcal{W}_I^* = \frac{1}{2}\,\delta f_1\,\delta q_1 = \frac{P_t\,\delta q_1^2}{l} C_I \quad (10.5.4)$$

$$\delta^2\mathcal{W}_{II}^* = \frac{1}{2}\,\delta f_1(-\delta q_1) = -\frac{P_t\,\delta q_1^2}{l} C_{II} \quad (10.5.5)$$

where the stiffnesses C_I and C_{II} are the same as before. Note that $\delta^2\mathcal{W}_I^*$ and $\delta^2\mathcal{W}_{II}^*$ must be interpreted as the complementary work (while $\delta^2\mathcal{W}_I$ and $\delta^2\mathcal{W}_{II}$ represent the work); see the cross-hatched areas in Figure 10.15b. Also note that the first-order complementary work $\Delta W^{*(1)}$ can have no effect on stability (Sec. 10.1).

Over the entire displacement cycle, the net change of force f_1 is nonzero. If the conditions are isothermal, the thermodynamic state function that depends on forces is the Gibbs free energy $\mathcal{G}$. Thus, according to the definition of $\mathcal{G}$ (Eq. 10.1.42), we have $\Delta\mathcal{G} = -\delta^2\mathcal{W}_0^*$ for isothermal deformations at mechanical equilibrium. From Sec. 10.1 we also recall that $\Delta\mathcal{G} = -T(\Delta S)_{\text{in}}$. For the change over the entire cycle we get

$$T(\Delta S)_{\text{in}} = -\Delta\mathcal{G} = \delta^2\mathcal{W}_0^* = \delta^2\mathcal{W}_I^* + \delta^2\mathcal{W}_{II}^* = \frac{2P_t\,\delta q_1^2}{l}(C_I - C_{II}) \quad (10.5.6)$$

Note that in the preceding analysis the physical meaning of $\delta^2 \mathscr{W}_0^*$ as minus the change of Gibbs (rather than Helmholtz) free energy is crucial. Without realizing that, one would get an incorrect sign for $\delta^2 \mathscr{W}_0^*$. It is also important to realize that, for a displacement cycle, the potential is the Gibbs free energy rather than the Helmholtz free energy, even though the controlled variable is the displacement.

Stability

Now consider stability. If $(\Delta S)_{\text{in}} < 0$, the cycle that would create such a deviation from the initial state cannot happen, and so the structure is stable. If $(\Delta S)_{\text{in}} > 0$, the cycle can (and will) happen, and so the structure is unstable.

Stability with respect to one-way deviations requires that $P_0 < P_{\text{cr}}^D$, as Equation 10.5.1 confirms. Then we have $C_{\text{I}} > 0$, $C_{\text{II}} > 0$ (if $E_t > 0$). Furthermore,

$$C_{\text{I}} - C_{\text{II}} = \frac{2}{P_t}\left(\frac{P_{\text{cr}}^D - P_{\text{cr}}^{el}}{P_t}\right) = \frac{1 - \xi_u}{P_t} \tag{10.5.7}$$

Because $\xi_u > 1$, we conclude that always $T(\Delta S)_{\text{in}} < 0$, that is, an unstable cycle cannot occur if the column is stable. So the consideration of a cycle does not impose any further stability restriction.

In view of this conclusion, the consideration of loading cycles might seem useless. It is nevertheless useful from the viewpoint of material modeling.

For example, can a stress-strain diagram of the material be assumed to have an unloading slope that is less than the loading slope, that is, $0 < E_u < E_t$? The stability conditions for one-way deviation from equilibrium, $\delta^2 \mathscr{W}_{\text{I}} > 0$ and $\delta^2 \mathscr{W}_{\text{I}}^* < 0$ (Eqs. 10.5.1 and 10.5.4), do not prohibit E_u to be less than E_t, since they still yield a positive stability limit for P. However, the stability condition for a cycle does prohibit E_u to be less than E_t. Indeed, $\xi_u < 1$ yields, according to Equation 10.5.7, $(\Delta S)_{\text{in}} > 0$ for a cycle at any load, even an arbitrarily small load P. So a material model for which $E_u < E_t$ is physically impossible.

To sum up, the consideration of load cycles may pose material model restrictions that are not obtained from the consideration of one-way deformations.

Implications of load cycles for material models are not particular to the consideration of columns. We will discuss them in a broader sense in the next section.

Further conceptual differences with regard to the standard concept of stability, which is based on one-way, monotonic deviations, are illustrated in Figure 10.16. To actually obtain (for $E_u < E_t$) a cycle that produces positive $(\Delta S)_{\text{in}}$, the structure must first deflect along a monotonic path segment that is stable (segment $\overline{01}$ in Fig. 10.16b, c). Therefore an initial positive energy input, $\Delta \mathscr{E}$, is required to effect the cycle ($\overline{013}$ or $\overline{015}$ in Fig. 10.16b, c) even if the cycle as a whole makes a net release of energy ($\delta^2 \mathscr{W}_0 < 0$). On the other hand, if the equilibrium state is unstable, the structure requires no positive energy input to get moving along the unstable path (path $\overline{02}$ in Fig. 10.16a).

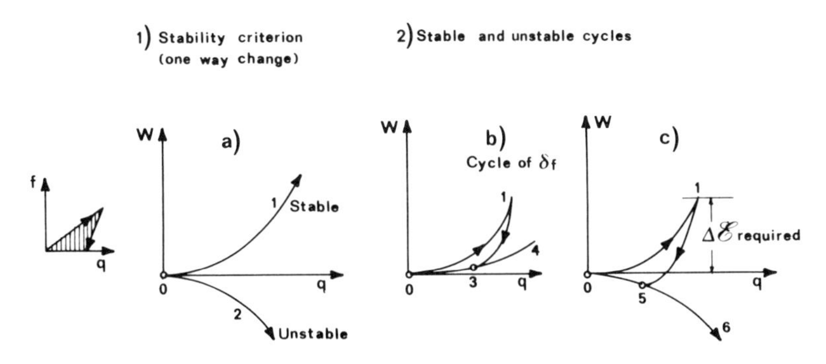

Figure 10.16 Conceptual difference between stability of a structure and stability of a cycle.

Structures with a Single Cyclic Load or Displacement

In this case, we generally have $-T(\Delta S)_{\text{in}} = \delta^2\mathscr{W}_0$ where $\delta^2\mathscr{W}_0$ is the second-order work done during the cycle, which is represented by the area (in the diagram of load versus displacement) that is enclosed during the cycle; see Figure 10.17. The stable cycles are those for which $\delta^2\mathscr{W}_0 > 0$, and the unstable cycles are those for which $\delta^2\mathscr{W}_0 < 0$. Various types of stable and unstable cycles are illustrated in Figure 10.17.

Incremental Collapse

Under a cyclic load, the structure may settle after a few cycles (i.e. "shake down") into an elastic response mode—a phenomenon called shakedown. Shanley's column behaves in this manner. But for certain other structures it is possible that plastic deformations accumulate from cycle to cycle and grow beyond any bounds with an increasing number of cycles, as illustrated in Figure 10.18b and d. This well-known phenomenon, in which shakedown is not obtained, is called incremental collapse. This may represent a cyclic instability because a cyclic load of arbitrarily small amplitude may produce after a sufficient number of cycles any given deflection. If shakedown does not occur, it is also possible that cycles of plastic deformations of opposite signs continue indefinitely. Such alternating plasticity leads to material failure.

Problems

10.5.1 If, for Shanley's column, the axial load P rather than the axial displacement is controlled, that is, $\delta f_2 = 0$ (and $\delta q_2 =$ arbitrary), the work of the axial force on δq_2 is of the first order ($P\,\delta q_2$) because $\delta f_2 = 0$, and the second-order work of axial load, $\frac{1}{2}\delta f_2\,\delta q_2$, vanishes (same as for axial displacement control). However, the expression for δf_1 is different: $\delta f_1 = \{[4\eta P_t\xi/(\xi+1)] - 2P_0\}\,2\delta q_1/l$. Proceeding similarly as Equations 10.5.1 to 10.5.6, show that, under axial load control, also lateral load or displacement cycles of Shanley's column are stable if $\xi_u > 1$ and unstable if $\xi_u < 1$.

10.5.2 Calculate $\delta^2\mathscr{W}_0$ for load cycles as well as displacement cycles of the rigid-bar column with two elastoplastic links shown in Figure 10.11a, and discuss stability of the cycle for $E_u > E_t$ as well as $E_u = E_t$ and $E_u < E_t$.

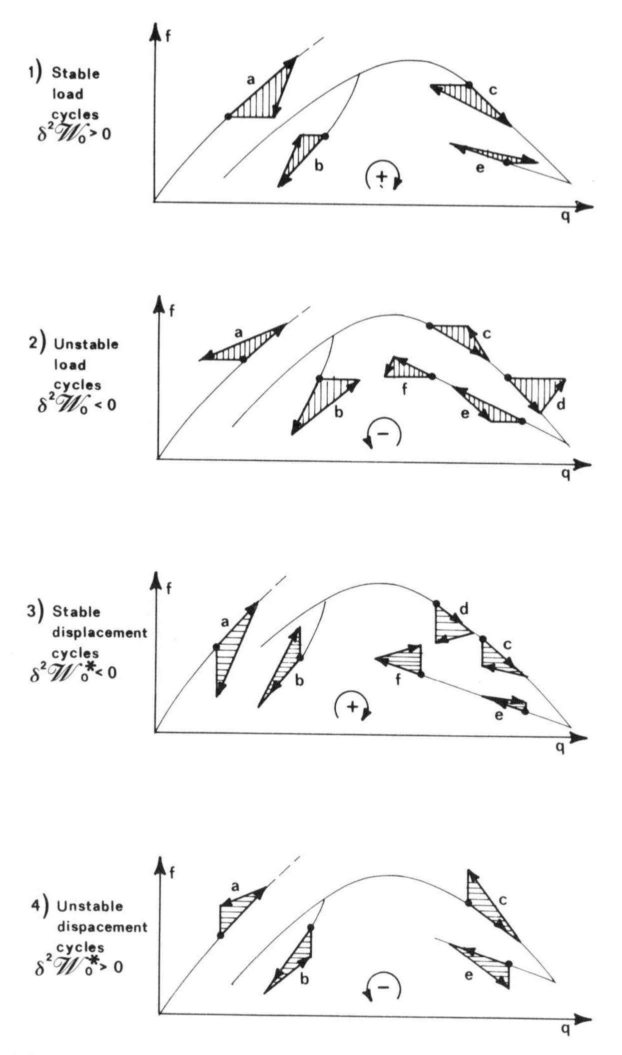

Figure 10.17 Stable and unstable cycles of load or displacement for the case of a single load.

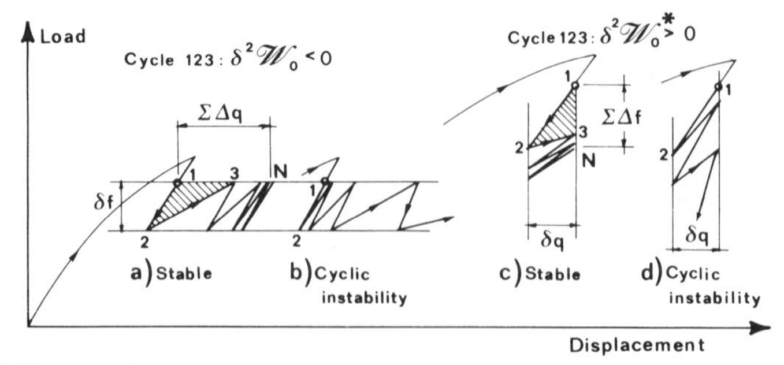

Figure 10.18 For unstable cycles the structure can be stable or unstable.

10.5.3 Do the same as above, but for the rigid-bar column in Figure 10.11b.

10.5.4 Do the same, but for the continuously deformable column in Figure 10.11c. (Assume f_1 to be the amplitude of a sinusoidally distributed lateral load.)

10.6 DRUCKER'S AND IL'YUSHIN'S POSTULATES FOR STABLE MATERIALS

The preceding section brought us to a point at which we can discuss certain stability postulates that are important for the theory of inelastic constitutive equations and are intimately related to normality and symmetry (Rice, 1971).

Drucker's Postulate

Consider a uniformly strained body that is subjected to a small loading-unloading cycle in which a small uniform stress $\delta\boldsymbol{\sigma}$ is applied and removed (Fig. 10.19). The residual strain that remains after the cycle is, by definition, the inelastic strain, $\delta\boldsymbol{\varepsilon}''$. The first-order work over the cycle, represented by the rectangular area 1365 in Figure 10.19a, determines the initial equilibrium but is irrelevant for stability.

The second-order work (per unit volume) done on applying $\delta\boldsymbol{\sigma}$ is $\Delta\bar{W}_l = \delta^2 W_l = \frac{1}{2}\delta\boldsymbol{\sigma}:\delta\boldsymbol{\varepsilon}$ (area 124), and the second-order work done on removing $\delta\boldsymbol{\sigma}$ is $\Delta\bar{W}_u = \delta^2\bar{W}_u = \frac{1}{2}\delta\boldsymbol{\sigma}:\sigma\boldsymbol{\varepsilon}^e$ (area 423) where $\delta\boldsymbol{\varepsilon}^e = \mathbf{C}:\delta\boldsymbol{\sigma}$ = elastic strain, determined by the current values of the elastic compliance tensor $\mathbf{C}$ of the material ($\boldsymbol{\sigma}$ and $\boldsymbol{\varepsilon}$ are second-order tensors and $d\boldsymbol{\sigma}:d\boldsymbol{\varepsilon} = d\sigma_{ij}\,d\varepsilon_{ij}$). The second-order work dissipated per unit volume over the cycle 123 is $\Delta W_c = \delta^2 W_c = \Delta W_l - \Delta W_u = \frac{1}{2}\delta\boldsymbol{\sigma}:(\delta\boldsymbol{\varepsilon} - \delta\boldsymbol{\varepsilon}^e) = \frac{1}{2}\delta\boldsymbol{\sigma}:\delta\boldsymbol{\varepsilon}^p$ in which $\delta\boldsymbol{\varepsilon}^p = \delta\boldsymbol{\varepsilon} - \delta\boldsymbol{\varepsilon}^e$ = inelastic strain increment, by definition. For isothermal changes we have $\Delta\bar{W}_c V = \Delta\mathcal{F}$ (Helmholtz free

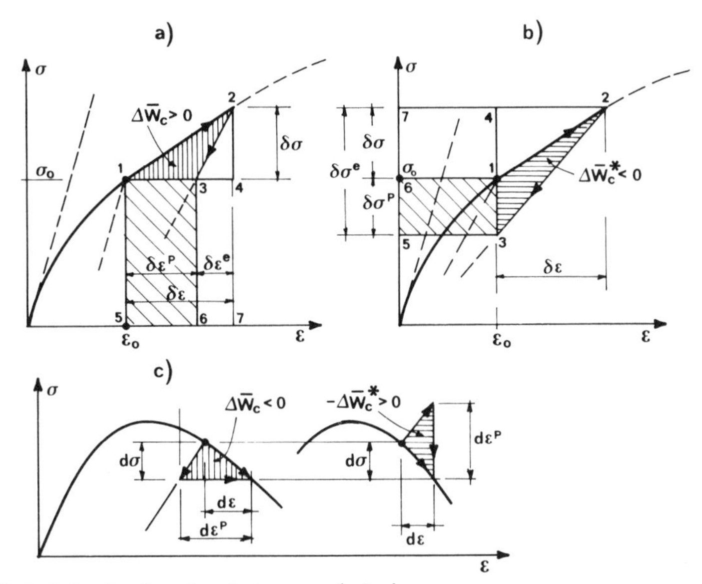

Figure 10.19 Infinitesimal cycle of stress and strain.

energy), and for isentropic changes $\Delta \bar{W}_c V = \Delta \mathcal{U}$ (total energy); V = volume of the uniformly strained body.

The second-order work given by $\Delta \bar{W}_c$, which is represented by the shaded triangle 123 in Figure 10.19a, is related to stability, as is clear from the preceding section. Based on Section 10.1, $T(\Delta S)_{\text{in}} = -\Delta \bar{W}_c V$ and $(\Delta S)_{\text{in}}$ = the internally produced entropy change over the cycle. Since stability requires that $(\Delta S)_{\text{in}} < 0$, we may follow Drucker (1950, 1959) to define a "stable" material as a material for which, for any tensor $\delta\boldsymbol{\sigma}$,

$$-\frac{T}{V}(\Delta S)_{\text{in}} = \Delta \bar{W}_c = \tfrac{1}{2}\delta\boldsymbol{\sigma} : \delta\boldsymbol{\varepsilon}^p \begin{cases} >0 & \text{for loading} \\ =0 & \text{for unloading or neutral loading} \end{cases} \tag{10.6.1}$$

This definition is called Drucker's postulate (Drucker 1950, 1954, 1959, 1964). [Note that the strict inequality (>), which is used here to conform with Equations 4.2.5, 4.9.4, and 10.1.37, excludes the limiting case of perfect plasticity from the definition of stable materials; often, however, Equation 10.6.1 has been written, as a matter of definition, with the sign $\geq$, in order that perfect plasticity be included among "stable" materials.]

The term "stable material," however, is somewhat misleading—a negative $\Delta \bar{W}_c$ for a cycle does not necessarily imply instability of the body, as demonstrated in the preceding section. Moreover, if we consider bodies that are not uniformly strained, nonnegativeness of $\Delta \bar{W}_c$ for some part of the body does not necessarily imply nonpositiveness of $(\Delta S)_{\text{in}}$ over the cycle of the body as a whole (see Chapter 11). So-called "unstable" states of the material for which $\Delta \bar{W}_c < 0$ for some $\delta\boldsymbol{\sigma}$ seem to exist in real structures in a stable manner (e.g., states of localized softening damage or frictional plastic states).

It may be noted that, in applications to plasticity, Drucker's postulate is equivalent to Hill's (1948) postulate of maximum plastic work (see also Simo, 1989). It has been thought by some that Drucker's postulate might be a consequence of the second law of thermodynamics, but this is not true. As Drucker himself pointed out, his postulate can be and is violated by real materials, for example, due to friction. Unfortunately this causes considerable mathematical complications (e.g., Bažant, 1980).

Il'yushin's Postulate

Instead of cycles of stress $\delta\boldsymbol{\sigma}(\mathbf{x})$ (where $\delta\boldsymbol{\sigma}$ depends on coordinate vector $\mathbf{x}$), one may consider as well a uniformly strained body subjected to cycles of strain $\delta\boldsymbol{\varepsilon}(\mathbf{x})$; see Figure 10.19b. The first-order complementary work, given by the rectangular area 1356 in Fig. 10.19b, is irrelevant for stability. The second-order complementary work (per unit volume) done in a uniformly strained body on applying $\delta\boldsymbol{\varepsilon}$ is $\Delta \bar{W}_l^* = \frac{1}{2}\delta\boldsymbol{\sigma} : \delta\boldsymbol{\varepsilon}$ (area 124 in Fig. 10.19b). The second-order complementary work done on removing $\delta\boldsymbol{\varepsilon}$ is $\Delta \bar{W}_u^* = \frac{1}{2}\delta\boldsymbol{\sigma}^e : \delta\boldsymbol{\varepsilon}$ (area 423) where $\delta\boldsymbol{\sigma}^e = \mathbf{E} : \delta\boldsymbol{\varepsilon}$ = elastic stress, determined by the current values of elastic moduli tensor $\mathbf{E}$ of the material. The second-order complementary work (per unit volume) dissipated over cycle 123 is $\Delta \bar{W}_c^* = \Delta \bar{W}_l^* - \Delta \bar{W}_u^* = \frac{1}{2}(\delta\boldsymbol{\sigma} - \delta\boldsymbol{\sigma}^e) : \delta\boldsymbol{\varepsilon} = -\frac{1}{2}\delta\boldsymbol{\sigma}^p : \delta\boldsymbol{\varepsilon}$ where $\delta\boldsymbol{\sigma}^p = \delta\boldsymbol{\sigma}^e - \delta\boldsymbol{\sigma}$ = inelastic stress decrement tensor, by the customary definition. For isothermal changes we have $\Delta \bar{W}_c^* V = -\Delta \mathcal{G}$ (Gibbs free energy), and for isentropic changes $\Delta \bar{W}_c^* V = -\Delta \mathcal{H}$ (enthalpy).

According to Sections 10.5 and 10.1, $T(\Delta S)_{\text{in}} = -\Delta \bar{W}_c^* V$. Therefore, following Il'yushin (1961), one may alternatively define a "stable" material as a material for which, for any tensor $\delta\boldsymbol{\varepsilon}$,

$$-\frac{T}{V}(\Delta S)_{\text{in}} = -\Delta \bar{W}_c^* = \tfrac{1}{2}\delta\boldsymbol{\sigma}^p : \delta\boldsymbol{\varepsilon} \begin{cases} >0 & \text{for loading} \\ =0 & \text{for unloading or neutral loading} \end{cases} \tag{10.6.2}$$

This definition is called Il'yushin's postulate. It is not equivalent to Drucker's postulate. For example, states of strain-softening damage violate Drucker's postulate ($\Delta \bar{W}_c^* < 0$) but not Il'yushin's postulate ($-\Delta \bar{W}_c^* > 0$); see Figure 10.19c.

For the same reason as Drucker's postulate, violation of Il'yushin's postulate, too, does not necessarily imply instability of the body, and if the strain state of the body is nonuniform, it does not even imply $(\Delta S)_{\text{in}}$ over the cycle of the body as a whole to be nonnegative. "Unstable" material states, in the sense of Il'yushin's postulate, can exist in a stable manner.

Nonuniformly Strained Bodies

The fact that the stability implications of Drucker's and Il'yushin's postulates are limited to uniformly strained bodies has often been overlooked. It is usually considered that, in any structure, some external agency applies stresses $\delta\boldsymbol{\sigma}(\mathbf{x})$ and then removes them. However, it would have to be a supernatural agency to actually carry that out under general conditions. Stresses cannot in general be applied; only loads (or displacements) can.

In reality, stresses $\delta\boldsymbol{\sigma}(\mathbf{x})$, once introduced in an inelastic body by means of some loads $\delta\mathbf{f}$, cannot be removed, except for some special cases such as uniformly strained bodies. Rather, residual stresses $\delta\boldsymbol{\sigma}^r(\mathbf{x})$ will remain after forces $\delta\mathbf{f}$ are removed (unloaded) (regarding some stability implications of residual stresses, see, e.g., Maier, 1966a, 1967). Therefore, the net strain change over a cycle of applying and removing $\delta\mathbf{f}$ in an elastic body is $\delta\boldsymbol{\varepsilon}^c = \delta\boldsymbol{\varepsilon} - \delta\boldsymbol{\varepsilon}^u = \delta\boldsymbol{\varepsilon} - \delta\boldsymbol{\varepsilon}^e - \delta\boldsymbol{\varepsilon}^r$ where $\delta\boldsymbol{\varepsilon} = \mathbf{C}^t : \delta\boldsymbol{\sigma}$ = strains at loading, $\delta\boldsymbol{\varepsilon}^e = \mathbf{C} : \delta\boldsymbol{\sigma}$ = elastic strains due to $\delta\boldsymbol{\sigma}$, $\delta\boldsymbol{\varepsilon}^r = \mathbf{C} : \delta\boldsymbol{\sigma}^r$ = residual strains, $\mathbf{C}^t$ = tangential compliances for loading, and $\mathbf{C}$ = elastic compliances for unloading (fourth-order tensors). The reason is that the field of strains $\delta\boldsymbol{\varepsilon}^e = \mathbf{C} : \mathbf{E}^t : \delta\boldsymbol{\varepsilon}$ (where $\mathbf{E}^t$ = tangential moduli tensor, inverse to $\mathbf{C}^t$) generally do not satisfy the compatibility conditions if $\delta\boldsymbol{\varepsilon}$ does, unless $\mathbf{C}$ and $\mathbf{E}^t$ are constant over the body.

For the same reason, a cycle of $\delta\boldsymbol{\varepsilon}$ will in general also leave residual stresses at the end of the cycle, except for uniformly strained bodies. (The problem, however, becomes more involved when one recognizes that in real materials $\boldsymbol{\sigma}$ and $\boldsymbol{\varepsilon}$ are volume averages of microstresses and microstrains that are always nonuniform; see, e.g., Maier, 1966a, 1967.)

The second-order work in a structure during a cycle that consists of loading with stresses $\delta\boldsymbol{\sigma}(\mathbf{x})$ and unloading is

$$\Delta W_0 = \left[\int_V (\boldsymbol{\sigma} + \tfrac{1}{2}\delta\boldsymbol{\sigma}) : \delta\boldsymbol{\epsilon}\, dV - \mathbf{P}^T \delta\mathbf{q}\right] - \left[\int_V (\boldsymbol{\sigma} + \tfrac{1}{2}\delta\boldsymbol{\sigma}) : \delta\boldsymbol{\varepsilon}^u\, dV - \mathbf{P}^T \delta\mathbf{q}^u\right]$$

$$= \int_V \tfrac{1}{2}\delta\boldsymbol{\sigma} : (\delta\boldsymbol{\varepsilon} - \delta\boldsymbol{\varepsilon}^u)\, dV + \left[\int_V \boldsymbol{\sigma} : \delta\boldsymbol{\varepsilon}\, dV - \mathbf{P}^T \delta\mathbf{q}\right] - \left[\int_V \boldsymbol{\sigma} : \delta\boldsymbol{\varepsilon}^u\, dV - \mathbf{P}^T \delta\mathbf{q}^u\right]$$

$$= \int_V \tfrac{1}{2}\delta\boldsymbol{\sigma} : (\delta\boldsymbol{\varepsilon} - \delta\boldsymbol{\varepsilon}^u)\, dV \qquad (10.6.3)$$

Here $\delta\boldsymbol{\varepsilon}$, $\delta\mathbf{q}$ are the strain and displacement changes corresponding to the introduction of $\delta\boldsymbol{\sigma}$, and $\delta\boldsymbol{\varepsilon}^u$, $\delta\mathbf{q}^u$ are those during the unloading part of the cycle ($\delta\boldsymbol{\varepsilon}^u = \delta\boldsymbol{\varepsilon}^e + \delta\boldsymbol{\varepsilon}^r$). The last two expressions in the square brackets vanish due to the principle of virtual work because $\boldsymbol{\sigma}$ is in equilibrium with $\mathbf{P}$, and because $\delta\boldsymbol{\varepsilon}$ is compatible with $\delta\mathbf{q}$ and $\delta\boldsymbol{\varepsilon}^e$ is compatible with $\delta\mathbf{q}^e$. Note that here we do not need to assume equilibrium to exist after the loading part of the cycle and after the entire cycle. Also note that the geometrically nonlinear effects are assumed here to be negligible.

For uniformly strained bodies we have $\delta\boldsymbol{\varepsilon}^u = \delta\boldsymbol{\varepsilon}^e$ ($\delta\boldsymbol{\varepsilon}^r = 0$), and then Equation 10.6.3 yields $\Delta W_0 = \int \frac{1}{2}\delta\boldsymbol{\sigma} : \delta\boldsymbol{\varepsilon}^p\, dV = \frac{1}{2}\delta\boldsymbol{\sigma} : \delta\boldsymbol{\varepsilon}^p V$ for the body as a whole. Therefore, validity of Drucker's postulate implies nonnegativeness of ΔW_0 for the whole body. However, this is not true for a nonuniformly strained body because $\delta\boldsymbol{\varepsilon}^u \neq \delta\boldsymbol{\varepsilon}^e$.

Similarly, the complementary second-order work done in a nonuniformly strained body during a cycle that consists of loading with strains $\delta\boldsymbol{\varepsilon}$ and unloading is

$$\Delta W_0^* = \int_V (\boldsymbol{\varepsilon} + \tfrac{1}{2}\delta\boldsymbol{\varepsilon}) : \delta\boldsymbol{\sigma}\, dV - \int_V (\boldsymbol{\varepsilon} + \tfrac{1}{2}\delta\boldsymbol{\varepsilon}) : \delta\boldsymbol{\sigma}^u\, dV$$

$$= -\int_V (\delta\boldsymbol{\sigma}^u - \delta\boldsymbol{\sigma}) : \delta\boldsymbol{\varepsilon}\, dV + \int_V \boldsymbol{\varepsilon} : \delta\boldsymbol{\sigma}\, dV - \int_V \boldsymbol{\varepsilon} : \delta\boldsymbol{\sigma}^u\, dV$$

$$= -\int_V (\delta\boldsymbol{\sigma}^u - \delta\boldsymbol{\sigma}) : \delta\boldsymbol{\varepsilon}\, dV \qquad (10.6.4)$$

where $\delta\boldsymbol{\sigma}^u$ = residual stresses after the cycle. The integrals of $\boldsymbol{\varepsilon} : \delta\boldsymbol{\sigma}$ and $\boldsymbol{\varepsilon} : \delta\boldsymbol{\sigma}^u$ vanish because of the principle of virtual work, since $\delta\boldsymbol{\sigma}$ and $\delta\boldsymbol{\sigma}^u$ are equilibrium stress fields and $\boldsymbol{\varepsilon}$ = compatible stress field. If the body is uniformly strained, then $\delta\boldsymbol{\sigma}^u = \delta\boldsymbol{\sigma}^e$ and $\delta\boldsymbol{\sigma}^u - \delta\boldsymbol{\sigma} = \delta\boldsymbol{\sigma}^p$ = inelastic stresses, in which case positiveness of ΔW_0 (and thus also stability of the structure for the cycle) follows from Il'yushin's postulate. But for nonuniformly strained bodies this does not follow.

To conclude, the material stability postulates are neither necessary nor sufficient for stability of the initial equilibrium state of the structure in general. One reason is that stability is defined with respect to one-way (monotonic) deviations from the initial state (see the discussion in Sec. 10.5). Another reason is that each of these postulates suffices to prevent only some kind of unstable deformations. It ensures neither stability for arbitrary cycles (not even in the absence of nonlinear geometric effects), nor stability for one-way (monotonic) deviations from the critical equilibrium state (see Fig. 10.20). It is basically a

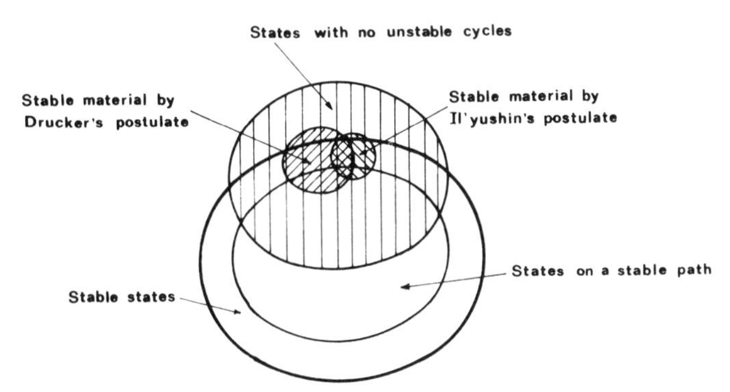

Figure 10.20 Static stability concept for inelastic structures.

convenience hypothesis with a vital but partial relationship to stability (Bažant, 1980). Nevertheless, the important role that Drucker's postulate played in the development of plasticity theory and formulation of bound theorems must be recognized.

Normality Rule for Plasticity

Although there exist questions about its applicability to real materials, Drucker's stability postulate brings in four important practical advantages: (1) it implies a certain special form of elastoplastic constitutive relation that usually leads to a good, albeit imperfect, description of experimental reality; (2) it greatly reduces the number of unknown functions and parameters that need to be identified from test results; (3) the resulting formulation is efficient and well-behaved in numerical applications; and (4) formulation of certain useful bounding principles is rendered possible. We will now briefly discuss the implications of Drucker's postulate.

To satisfy tensorial invariance restrictions such as isotropy, the loading-unloading boundary in the stress space (a space with σ_{ij} as coordinates) must be characterized in terms of a scalar loading function $f(\boldsymbol{\sigma})$ that depends on the proper invariants of the stress tensor $\boldsymbol{\sigma}$. In the case of isotropy (same properties after any rotation of coordinate axes), function f can depend only on the three basic invariants of $\boldsymbol{\sigma}$ or their functions. Function f may also depend on further scalar parameters κ_k that introduce the influence of loading history. Thus, the current plastic boundary (yield locus, or yield surface) in the stress space is described by the equation $f(\boldsymbol{\sigma}, \kappa_k) = 0$. The states for which $f(\boldsymbol{\sigma}, \kappa_k) < 0$ are elastic, representing either the initial elastic loading or unloaded states after previous plastic yielding. The case $f(\boldsymbol{\sigma}, \kappa_k) > 0$ is by definition impossible (increase of the yield limit is handled by parameter κ_k). The simplest example is the von Mises loading surface $f(\boldsymbol{\sigma}, \kappa_1) = \bar{\tau} - \kappa_1 = 0$ where $\bar{\tau} = (\frac{1}{2}s_{ij}s_{1j})^{1/2}$ = stress intensity, s_{ij} = deviatoric stresses, and κ_1 = yield limit in pure shear, which may vary as a result of previous plastic straining.

During plastic loading, parameters κ_k are changing, and so the surfaces $f = 0$ at constant κ_k evolve; see Figure 10.21a showing the subsequent loading surfaces at times t_1 and t_2. Since the material remains plastic, we must have $df = 0$, that is,

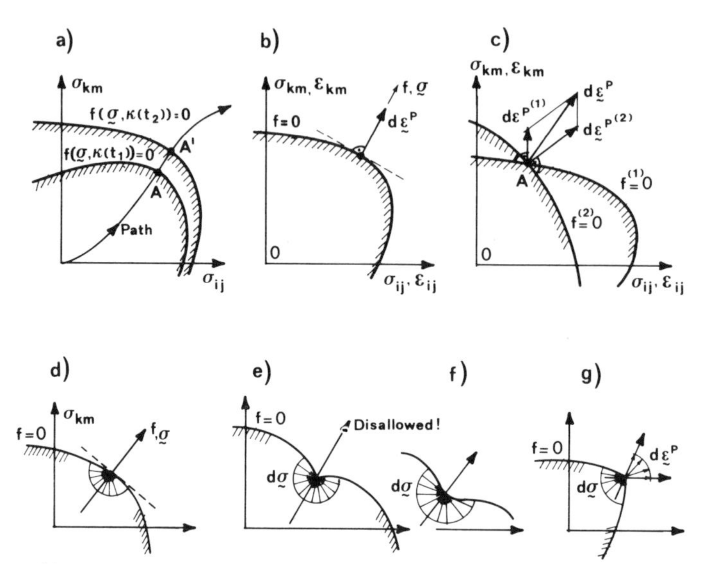

Figure 10.21 (a) Subsequent loading surfaces; (b) normality rule; (c) multisurface plasticity; (d) geometric interpretation of Drucker's postulate; (e, f) inadmissible and (g) admissible forms of loading surface.

$df = f_{,\boldsymbol{\sigma}}:d\boldsymbol{\sigma} + \sum_k f_{,\kappa_k}\, d\kappa_k = 0$, where $f_{,\boldsymbol{\sigma}} = \partial f/\partial\boldsymbol{\sigma}$ = tensor of components $\partial f/\partial\sigma_{ij}$ and $f_{,\boldsymbol{\sigma}}:d\boldsymbol{\sigma} = (\partial f/\partial\sigma_{ij})\, d\sigma_{ij}$ (summation implied by repeated subscripts). This sign of f can always be chosen in such a manner that $\sum f_{,x_k}\, d\kappa_k < 0$ for loading. With this choice, the loading criterion is formulated as

$$f_{,\boldsymbol{\sigma}}:d\boldsymbol{\sigma}\begin{cases} >0 & \text{plastic loading} \\ =0 & \text{neutral loading} \end{cases} \tag{10.6.5}$$

while $f_{,\boldsymbol{\sigma}}:d\boldsymbol{\sigma} < 0$ represents unloading, which is elastic. The neutral loading is the limit case of plastic loading at which $d\kappa_k = 0$. (In the limit case of perfect plasticity, though, one always has $f_{,\boldsymbol{\sigma}}:d\boldsymbol{\sigma} = 0$ and loading occurs unless $d\lambda = 0$, based on Eq. 10.6.8 below.)

Now, adopting Drucker's postulate, both Equations 10.6.1 and 10.6.5 must be positive, and so we may set

$$\frac{d\boldsymbol{\varepsilon}^p : d\boldsymbol{\sigma}}{f_{,\boldsymbol{\sigma}}:d\boldsymbol{\sigma}} = d\lambda > 0 \qquad \text{(for plastic loading)} \tag{10.6.6}$$

Because $f_{,\boldsymbol{\sigma}}:d\boldsymbol{\sigma} > 0$ for plastic loading, we may multiply this inequality by the denominator, and we get for plastic loading (Bažant, 1980):

$$d\boldsymbol{\varepsilon}^s : d\boldsymbol{\sigma} = 0 \qquad \text{where } d\boldsymbol{\varepsilon}^s = d\boldsymbol{\varepsilon}^p - f_{,\boldsymbol{\sigma}}\, d\lambda \tag{10.6.7}$$

This condition must hold for all possible increments $d\boldsymbol{\sigma}$ that represent plastic loading. One way to satisfy this condition is to set $d\boldsymbol{\varepsilon}^s = \mathbf{0}$. Then it follows that

$$d\boldsymbol{\varepsilon}^p = f_{,\boldsymbol{\sigma}}\, d\lambda \tag{10.6.8}$$

This is the famous plastic flow rule of Prandtl (1924) and Reuss (1930); see also Fung (1965) and Malvern (1969). It is also called the normality rule (Prager,

1949); the reason is that, in a nine-dimensional space (not six-simensional—unless a 6×1 column strain matrix with shear angles $\gamma_{12} = 2\varepsilon_{12}, \ldots$ is used), in which coordinates $d\varepsilon_{ij}^p$ are superimposed over coordinates σ_{ij}, the vector of $f_{,\boldsymbol{\sigma}}$ is normal to the surface $f(\boldsymbol{\sigma}, \kappa_k) = 0$ at $\kappa_k = \text{const.}$; see Figure 10.21b. This is proven by differentiation: $df = f_{,\boldsymbol{\sigma}} : d\boldsymbol{\sigma}$ at $\kappa_k = \text{const.}$; the vector of $d\boldsymbol{\sigma}$ must be tangential to the loading surface if $d\kappa_k = 0$, and the vector of $f_{,\boldsymbol{\sigma}}$ is normal to $d\boldsymbol{\sigma}$ because $f_{,\boldsymbol{\sigma}} : d\boldsymbol{\sigma}$ represents a scalar product of two nine-dimensional vectors.

Drucker's postulate, $d\boldsymbol{\sigma} : d\boldsymbol{\varepsilon}^p \geq 0$, may be geometrically interpreted as a condition that the projection of the vector of $d\boldsymbol{\sigma}$ onto the vector $d\boldsymbol{\varepsilon}_p$, that is, onto the normal vector $f_{,\boldsymbol{\sigma}}$, be nonnegative for any vector $d\boldsymbol{\sigma}$ that does not point inside the loading surface $f = 0$; see Figure 10.21d. Consequently, a reentrant corner on a yield surface is not allowed by Drucker's stability postulate; see Figure 10.21e. Furthermore, if Drucker's postulate is extended from infinitesimal to small but finite cycles of $\Delta\boldsymbol{\sigma}$, (Drucker's postulate "in the large") a concave form (Fig. 10.21f) of the loading surface is also inadmissible. Hence, the loading surface is required to be convex.

It is admissible that the loading surface has a corner (vertex) of less than 180°. Then the projection of the vector of $d\boldsymbol{\sigma}$ for loading onto the vector of $d\boldsymbol{\varepsilon}^p$ is positive for all the vectors shown in Figure 10.21g. Hence, Drucker's postulate allows $d\boldsymbol{\varepsilon}^p$ to be anywhere between the normals of the loading surface on the sides of a corner (which is called the generalized normality rule). The direction of $d\boldsymbol{\varepsilon}^p$ is in this case indeterminate, and all the possible directions fill the fan (or cone) of directions between these normals; see Figure 10.21g.

Since the material continues to be plastic after a loading increment, we must have $df = 0$. Consequently, $df = f_{,\boldsymbol{\sigma}} : d\boldsymbol{\sigma} + \sum_k f_{,\kappa_k} (d\kappa_k / d\lambda)\, d\lambda = 0$, which represents Prager's (1949) continuity relation. Denoting $H = -\sum_k f_{,\kappa_k}\, d\kappa_k / d\lambda$, we obtain

$$d\lambda = \frac{1}{H} f_{,\boldsymbol{\sigma}} : d\boldsymbol{\sigma} \tag{10.6.9}$$

provided that $f_{,\boldsymbol{\sigma}}\, d\boldsymbol{\sigma} \geq 0$. Coefficient H, to be determined from experiments, is called the plastic hardening modulus. Equation 10.6.9 is due to Melan (1938).

The total strain tensor $\boldsymbol{\varepsilon}$ is a sum of elastic and plastic strain tensors. that is, $d\boldsymbol{\varepsilon} = \mathbf{C} : d\boldsymbol{\sigma} + d\boldsymbol{\varepsilon}^p$ where $\mathbf{C}$ = fourth-order elastic compliance tensor. Substituting here Equation 10.6.8 with $d\lambda$ according to Equation 10.6.9, and factoring out $d\lambda$, we obtain

$$d\boldsymbol{\varepsilon} = \mathbf{C}^t : d\boldsymbol{\sigma} \qquad \mathbf{C}^t = \mathbf{C} + \frac{1}{H} f_{,\boldsymbol{\sigma}} f_{,\boldsymbol{\sigma}} \tag{10.6.10}$$

for loading $(f_{,\boldsymbol{\sigma}} : d\boldsymbol{\sigma} > 0)$, while for unloading $(f_{,\boldsymbol{\sigma}} : d\boldsymbol{\sigma} \leq 0)$ $d\boldsymbol{\varepsilon} = \mathbf{C} : d\boldsymbol{\sigma}$. Here $\mathbf{C}^t$ = tangential compliance tensor. Note that this tensor is symmetric, which is a consequence of the normality rule. If, on the other hand, the material does not obey the normality rule, $\mathbf{C}^t$ is nonsymmetric. Thus, symmetry is in plasticity synonymous with normality and is implied by Drucker's postulate. (But in continuum damage mechanics $\mathbf{C}^t$ can be nonsymmetric, even if there is normality; see Maier and Hueckel, 1979.)

Nonsymmetry is allowed in the so-called nonassociated plasticity, in which one assumes $d\boldsymbol{\varepsilon}^p = g_{,\boldsymbol{\sigma}}\, d\lambda$ where $g = g(\boldsymbol{\sigma}, \kappa_k)$ = plastic potential (or flow potential) that is different from the yield function $f(\boldsymbol{\sigma}, \kappa_k)$.

To obtain the tangential moduli tensor $\mathbf{E}^t$, the expression $d\boldsymbol{\sigma} = \mathbf{E}:(d\boldsymbol{\varepsilon} - d\boldsymbol{\varepsilon}^p) = \mathbf{E}:(d\boldsymbol{\varepsilon} - f_{,\boldsymbol{\sigma}}\, d\lambda)$ where $\mathbf{E}$ = fourth-order elastic moduli tensor may be substituted into Equation 10.6.9 and the resulting equation solved to get an expression for $d\lambda$. After substituting this expression again into the relation $d\boldsymbol{\sigma} = \boldsymbol{E}:(d\boldsymbol{\varepsilon} - f_{,\boldsymbol{\sigma}}\, d\lambda)$, one obtains

$$d\boldsymbol{\sigma} = \mathbf{E}^t\, d\boldsymbol{\varepsilon} \qquad \mathbf{E}^t = \mathbf{E} - \frac{\mathbf{E}:f_{,\boldsymbol{\sigma}} \otimes f_{,\boldsymbol{\sigma}}:\mathbf{E}}{\mathrm{f}_{,\boldsymbol{\sigma}}:\mathbf{E}:f_{,\boldsymbol{\sigma}} + H} \tag{10.6.11}$$

for loading. The symbol $\otimes$ denotes the tensor product, which is not contracted on any index that is, $f_{,\boldsymbol{\sigma}} \otimes f_{,\boldsymbol{\sigma}}$ is the fourth-order tensor of $f_{,\sigma_{ij}} f_{,\sigma_{km}}$). Note again that tensor $\mathbf{E}^t$ is fully symmetric ($E^t_{ijkm} = E^t_{ijmk} = E^t_{jikm} = E^t_{kmij}$) for associated plasticity.

The formulation that we just presented may be generalized, assuming that at each state the plastic strain consists of several components, $d\boldsymbol{\varepsilon}^p = \sum_i d\boldsymbol{\varepsilon}^{p(i)}$, each governed by a different yield (loading) surface $f^{(i)}(\boldsymbol{\sigma}, \kappa_k) = 0$ $(i = 1, \ldots, n_p)$. The same logic then leads to the following generalizations of Equations 10.6.8 and 10.6.9 (Koiter, 1953):

$$d\boldsymbol{\varepsilon}^{p(i)} = f^{(i)}_{,\boldsymbol{\sigma}}\, d\lambda^{(i)} \qquad d\lambda^{(i)} = \frac{1}{H_i} f^{(i)}_{,\boldsymbol{\sigma}} : d\boldsymbol{\sigma} \tag{10.6.12}$$

This formulation, illustrated in Figure 10.21c, is called multisurface plasticity. Although in principle much more realistic in the description of material behavior, it is not easy to apply because identification of the material parameters from test data appears to be much more difficult. (In this regard note that, for example, the Drucker-Prager "cup" surface endowed with a "cap" does not represent multisurface plasticity because two simultaneous yield surfaces do not intersect and two plastic strain increment vectors are not superposed at every stress point. Rather, it is a single loading surface with a vertex, at which two parts of the surface, defined by different equations, meet. In practice, though, the intersection of cap and cone has been treated according to multisurface plasticity.)

There is another way to satisfy Equation 10.6.7: We may require the vector of $d\boldsymbol{\varepsilon}^s$ in a nine-dimensional space to be normal to the vector of $d\boldsymbol{\sigma}$. Then $d\boldsymbol{\epsilon}^p = d\boldsymbol{\varepsilon}^n + d\boldsymbol{\varepsilon}^t$, in which $d\boldsymbol{\varepsilon}^n = f_{,\boldsymbol{\sigma}}\, d\lambda$ = normal plastic strain and $d\boldsymbol{\varepsilon}^t$ = tangential plastic strain that is normal to the vector of $f_{,\boldsymbol{\sigma}}$, that is, is tangential to the loading surface (for detail, see Bažant, 1980). A special form of such a formulation has been proposed by Rudnicki and Rice (1975). The attractive feature of using $d\boldsymbol{\varepsilon}^t$ is that actual materials do show deviations from the normality rule, with plastic strain components parallel to the loading surface. In numerical applications, however, the use of $d\boldsymbol{\varepsilon}^t$ leads to convergence difficulties. Rather than using $d\boldsymbol{\varepsilon}^t$, it appears to be numerically preferable as well as more realistic to use multisurface plasticity, which makes it possible to obtain various directions of vector $d\boldsymbol{\varepsilon}^p$. (As another way to introduce deviation from single-surface normality, flow rules in which $d\boldsymbol{\varepsilon}^p$ depends on $d\boldsymbol{\sigma}$ have been proposed; cf. Maier, 1969.)

Based on the work $\Delta\bar{W} = \frac{1}{2} d\boldsymbol{\sigma} : d\boldsymbol{\varepsilon}^p$ in an infinitesimal cycle, one can define the tangential compliance magnitude $C_p = 2\Delta\bar{W}/\|d\boldsymbol{\sigma}\|$ where $\|d\boldsymbol{\sigma}\| = d\sigma_{ij}\, d\sigma_{ij}$. It can be shown that for plasticity $C_p = k \cos^2\theta$ for loading and $C_p = 0$ for unloading, in which θ = angle between the vector of $d\boldsymbol{\sigma}$ and the normal to the loading surface, and k = coefficient independent of $d\boldsymbol{\sigma}$. The dependence of C_p on θ appears to be a simple basic characteristic of inelastic constitutive theory. For theories other than the single-surface associated plasticity, for example, the deformation theory (also called the total strain theory), multisurface plasticity, or endochronic theory, the dependence of C_p on θ is rather different (see Bažant, 1978, 1980).

Il'yushin's postulate, too, can be used as the basis of constructing inelastic constitutive relations. The loading surfaces are considered in the strain space rather than the stress space, and a normality rule in the strain space, analogous to Equation 10.6.8, may be derived. Whether this is more realistic depends on experiment. The inelastic behavior can be characterized in this formulation in terms of either inelastic strain increments (e.g., Naghdi and Trapp, 1975) or inelastic stress decrements (Dougill, 1975, 1976; Bažant and Kim, 1979; Bažant, 1980). The use of the strain space is particularly suitable for the description of the strain-softening (see Prob. 10.6.7). However, additional complexities arise if the inelastic stress decrements are associated with degradation of elastic moduli. It is also possible and apparently advantageous to combine in the same constitutive law plastic strains based on a loading surface in the stress space with inelastic stress increments based on loading surfaces in the strain space (Bažant and Kim, 1979).

Problems

10.6.1 Describe the relation of Drucker's postulate to the stability of (a) a uniformly stressed body and (b) a nonuniformly stressed body.

10.6.2 Show that for nonassociated plasticity there exist cycles of applying and removing $d\boldsymbol{\sigma}$ for which the work done is negative.

10.6.3 Using Il'yushin's postulate, derive the normality rule for strain-space plasticity.

10.6.4 Derive the tangential compliance and stiffness matrices, given that (a) $f(\boldsymbol{\sigma}, \kappa) = J_2 + kI_1$ and (b) $f(\boldsymbol{\sigma}, \kappa) = J_3 + kI_1$, where $I_1 = \sigma_{kk}$, $J_2 = \frac{1}{2}s_{ij}s_{ij}$, $J_3 = \frac{1}{3}s_{ij}s_{jk}s_{ki}$. Assume normality.

10.6.5 Rewrite all the equations of this section in tensor component notation, for example, $f_{,\boldsymbol{\sigma}} : d\boldsymbol{\sigma} = (\partial f / \partial \sigma_{ij})\, d\sigma_{ij}$.

10.6.6 Arranging the components of stress and strain tensors as 6×1 column matrices, $\boldsymbol{\sigma}$ and $\boldsymbol{\varepsilon}$, rewrite all the equations of this section in matrix notation. (*Caution*: The column matrix for strains must involve shear angles $\gamma_{ij} = 2\varepsilon_{ij}$ so that $\boldsymbol{\sigma}^T d\boldsymbol{\varepsilon}$ be the correct expression for work.)

10.6.7 Show that the loss of positive definiteness of the tangential stiffness matrix $\mathbf{E}^t$ of the material (called strain-softening, Chap. 13) violates Drucker's postulate but not Il'yushin's postulate.

10.7 STABILITY OF FRICTIONAL MATERIALS AND STRUCTURES

Frictional phenomena have a profound influence on stability. It is of particular interest to extend our preceding discussion of Drucker's postulate to frictional materials. As Drucker (1950, 1959) pointed out, internal friction in the material can cause $\frac{1}{2}d\boldsymbol{\sigma} : d\boldsymbol{\varepsilon}^p$ to be negative yet the material can be stable.

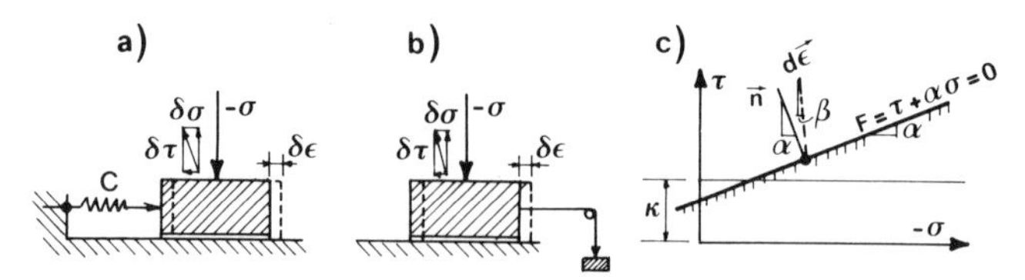

Figure 10.22 Frictional block at the point of sliding loaded by (a) a spring or (b) a constant force.

Frictional Block Preloaded by a Spring

To illustrate this phenomenon, consider first the simple example in Figure 10.22a given by Mandel (1964) (and further discussed by Maier, 1971). A block that slides on a rough surface is preloaded through a horizontal spring of stiffness C. The slip $\delta\varepsilon$ may be imagined to correspond to plastic shear angle γ^p, the horizontal applied force to shear stress τ, and the vertical applied force to normal stress σ. The roughness of the surface causes a small slip $d\gamma^p$ to be accompanied by a certain vertical displacement $d\varepsilon^p = \beta\,|d\gamma^p|$ simulating dilatancy; β is the given dilatancy factor.

In the absence of the spring, the slip condition of the block may be written as $f(\sigma, \tau) = \tau + \alpha\sigma - \kappa = 0$ where α = given friction coefficient and κ = current slip limit, representing a hardening parameter. The slip condition is graphically represented by the line in Figure 10.22c. The normal to this slip surface has the inclination $1/\alpha$. The vector $d\vec{\varepsilon}$ of slip displacement and vertical displacement, plotted in the same diagram, gives a line of inclination $1/\beta$. Obviously, normality exists only for $\beta = \alpha$, but the value of β is independent of α and, in particular, β may be zero. So the sliding of the block violates the normality rule.

The presence of the spring causes a change in the limit value of $\tau + \alpha\sigma$, which represents hardening. The current cohesion limit is $\kappa = \tau_0 + C\gamma^p$, where τ_0 = given initial cohesion limit. Assume now that the spring force $f(\sigma, \tau)$ is such that sliding of the block is imminent. Consequently,

$$f(\sigma, \tau) = \tau + \alpha\sigma - \tau_0 - C\gamma^p = 0 \qquad (10.7.1)$$

We apply load increments $d\tau$ and $d\sigma$ such that $d\tau$ is opposite to the spring force and $|d\tau| < k\,|d\sigma|$ (Fig. 10.22a). Now we realize that this causes the block to slide to the right by $d\gamma^p$. The reason is that $d\sigma$ reduces the friction capacity more than $d\tau$ increases it by relieving the spring. Equilibrium after sliding requires that $f + df = (\tau + d\tau) + \alpha(\sigma + d\sigma) - \tau_0 - C(\gamma^p + d\gamma^p) = 0$. Subtracting from this Equation 10.7.1, we get the condition of continuing equilibrium $df = d\tau + \alpha\,d\sigma - C\,d\gamma^p = 0$, from which $d\gamma^p = (d\tau + \alpha\,d\sigma)/C$. Also, slip $d\gamma^p$ causes the block to rise by a distance $\beta\,d\gamma^p$. So the second-order work done by forces $d\tau$ and $d\sigma$ on the inelastic displacements $d\gamma^p$ and $d\varepsilon^p$ will be

$$\Delta W = \frac{1}{2}(d\tau\,d\gamma^p + d\sigma\,d\varepsilon^p) = \frac{1}{2C}(d\tau + \beta\,d\sigma)(d\tau + \alpha\,d\sigma) \qquad (10.7.2)$$

This work is obviously equivalent to the work in Drucker's postulate (Eq. 10.7.1).

If $\beta = \alpha$, which is the case of normality (Fig. 10.22c), the expression for ΔW is symmetric and we always have $\Delta W > 0$. However, if $\beta = 0$ (flat surface, no dilatancy) and if we choose $d\tau < 0$ and $d\sigma > -d\tau/\alpha$ we get $\Delta W < 0$, that is, energy is released by the block. The block is nevertheless stable because infinitely small loads $d\tau$ and $d\sigma$ cause an infinitely small slip $\delta\gamma^p$. More generally, if $\alpha > \beta > 0$ and if we choose $d\sigma > 0$ and $-\alpha\, d\sigma < d\tau < -\beta\, d\sigma$ (<0), we always get $\Delta W < 0$, that is, the block is still stable.

Note, however, that if the spring force were replaced with a constant force (e.g., the pull of a weight, Fig. 10.22b), no new equilibrium would exist, that is, the system would be unstable. Thus, the stability of the block is obviously due to the fact that the driving force decreases with increasing displacements, as is true for the release of elastic energy.

From the finding that $\Delta W < 0$ is not an unstable situation in these cases we may conclude that a release of frictionally blocked elastic energy is harmless for stability. We have seen that this can occur only if $\beta \neq \alpha$ (lack of normality). Thus, it is expedient to rewrite Equation 10.7.2 in the form

$$\Delta W = \Delta W_n + \Delta W_f \tag{10.7.3}$$

in which

$$\Delta W_n = \frac{1}{2C}(d\tau + \beta\, d\sigma)^2 \qquad \Delta W_f = \frac{\alpha - \beta}{2C}\, d\sigma(d\tau + \beta\, d\sigma) \tag{10.7.4}$$

Here ΔW_n is always positive. It is solely ΔW_f that may cause ΔW to become negative.

Generalization to Frictional Continuum

Following Bažant (1980), we now establish a continuum analogy to the preceding example. We need to (1) express ΔW and f by means of differentials of the same variables; (2) express ΔW in terms of the stress invariants because f must be given in terms of stress invariants; (3) express ΔW by means of only two stress variables and two strain variables, and (4) express ΔW such that no cross products be present, just as neither $d\sigma\, d\gamma^p$ nor $d\tau\, d\varepsilon^p$ is present in Equation 10.7.2. This last condition is the salient property that defines friction. Generally a friction-producing force (such as σ in Fig. 10.22a) is any force that does *no work* on some displacement (on $d\gamma^p$ in Fig. 10.22a) yet at the same time does affect this displacement.

The foregoing conditions can be met by writing

$$\Delta W = \tfrac{1}{2} d\sigma\, d\varepsilon^p_{kk} + \tfrac{1}{2} ds_{ij}\, d\gamma^p_{ij} = \tfrac{1}{2} d\sigma(3\, d\varepsilon^p) + \tfrac{1}{2} p\, d\bar{\tau}\, d\hat{\gamma}^p \tag{10.7.5}$$

in which

$$p = p_{ij} q_{ij} \qquad p_{ij} = \frac{ds_{ij}}{d\bar{\tau}} \qquad q_{ij} = \frac{d\gamma^p_{ij}}{d\hat{\gamma}^p} \qquad \bar{\tau} = (\tfrac{1}{2} s_{ij} s_{ij})^{1/2} \qquad d\hat{\gamma}^p = (\tfrac{1}{2} d\gamma^p_{ij}\, d\gamma^p_{ij})^{1/2} \tag{10.7.6}$$

and $s_{ij} = \sigma_{ij} - \delta_{ij}\sigma$ = deviatoric stresses, $\sigma = \sigma_{kk}/3$ = volumetric stress, $\gamma_{ij} = \varepsilon_{ij} - \delta_{ij}\varepsilon$ = deviatoric strains, $\varepsilon = \varepsilon_{kk}/3$ = volumetric strain, $\bar{\tau}$ = stress intensity, $\hat{\gamma}^p$ = length of plastic strain path, $d\bar{\tau} = s_{ij}\, ds_{ij}/2\bar{\tau}$, and $\sigma = \sigma_{kk}/3$. Variables p_{ij} and q_{ij} characterize the directions of vectors ds_{ij} and $d\gamma^p_{ij}$ in the stress space, and coefficient p is a function of the angle between these two vectors and of the angle between ds_{ij} and s_{ij}. Here we chose to normalize p_{ij} and q_{ij} in different ways. Instead of the plastic path length $\hat{\gamma}^p$ (Odquist's hardening parameter) one might think of using $\bar{\gamma}^p = (\gamma^p_{ij}\gamma^p_{ij}/2)^{1/2}$ and write $q_{ij} = d\gamma^p_{ij}/d\bar{\gamma}^p$ (where $d\bar{\gamma}^p = \gamma_{km}\, d\gamma_{km}/2\bar{\gamma}^p$); but this would be inconvenient since it is $\hat{\gamma}^p$, rather than $\bar{\gamma}^p$, which is suitable as a hardening parameter in the loading function. Alternatively, one might think of introducing $p_{ij} = ds_{ij}/d\hat{\tau}$ where $d\hat{\tau} = (ds_{ij}ds_{ij}/2)^{1/2}$ = stress path length; but this would again be inconvenient because the loading function depends on $\bar{\tau}$ rather than $\hat{\tau}$.

Comparison of Equation 10.7.5 with Equation 10.7.2 indicates that the variables $d\sigma$, $d\tau$, $d\varepsilon^p$, and $d\gamma^p$ for the block correspond to continuum variables $d\sigma$, $d\bar{\tau}$, $3d\varepsilon^p$, and $p\, d\hat{\gamma}^p$, respectively. A general loading function for isotropic materials may be considered in the form

$$f(\sigma, \bar{\tau}, J_3, \varepsilon^p, \hat{\gamma}^p, \kappa_k) = 0 \tag{10.7.7}$$

where $J_3 = s_{ij}s_{km}s_{mi}/3$ = third invariant of s_{ij}; κ_k are possible further hardening parameters in addition to ε^p and $\hat{\gamma}^p$. Differentiating f, we get

$$df = \frac{\partial f}{\partial \sigma}\, d\sigma + \frac{Df}{D\bar{\tau}}\, d\bar{\tau} + \frac{Df}{D\hat{\gamma}^p}\left(\frac{p\, d\hat{\gamma}^p}{p}\right) = 0 \tag{10.7.8}$$

where

$$\frac{Df}{D\bar{\tau}} = \frac{\partial f}{\partial \bar{\tau}} + \frac{\partial J_3}{\partial \bar{\tau}}\left(\frac{\partial f}{\partial J_3}\right) \qquad \frac{Df}{D\hat{\gamma}^p} = \frac{\partial f}{\partial \hat{\gamma}^p} + \frac{2\beta}{3}\frac{\partial f}{\partial \varepsilon^p} + \frac{\partial f}{\partial \kappa_k}\left(\frac{\partial \kappa_k}{\partial \hat{\gamma}^p}\right) \tag{10.7.9}$$

$$\beta = \frac{3d\varepsilon^p}{2d\hat{\gamma}^p} \tag{10.7.10}$$

The last expression is chosen to define the dilatancy factor f, in which $2d\hat{\gamma}^p$ is used because, for pure shear, it represents $2de^p_{12}$ = plastic shear angle increment. Note that if f depends on J_3, then $Df/D\bar{\tau}$ depends on the direction of vector ds_{ij} and if $\beta \neq 0$ or $\partial f/\partial \kappa_k \neq 0$, then $Df/D\gamma^p$ depends on the direction of vector de^p_{ij}. Dividing by $Df/D\bar{\tau}$ and keeping in mind the proper correspondence of variables with the frictional block, comparison of Equation 10.7.8 with $df = d\tau + \alpha\, d\sigma - C\, d\gamma^p = 0$ (Eq. 10.7.1) yields

$$C = -\frac{1}{p}\left(\frac{Df/D\hat{\gamma}^p}{Df/D\bar{\tau}}\right) \qquad \alpha = \frac{\partial f/\partial \sigma}{Df/d\bar{\tau}} \qquad (\text{if } C, \alpha \geq 0) \tag{10.7.11}$$

The dilatancy factor for the block, $d\varepsilon^p/d\gamma^p$, corresponds according to definition (10.7.1) to the ratio $3d\varepsilon^p/p\, d\hat{\gamma}^p$ that equals $2\beta/p$ where β is given by Equation 10.7.10. Thus, according to Equations 10.7.4, the frictionally blocked second-order elastic energy may be expressed as

$$\Delta W_f = \frac{\beta' - \beta^*}{2C}\, d\sigma(d\bar{\tau} + \beta^*\, d\sigma) \qquad \text{in which } \beta^* = \frac{2}{p}\beta \tag{10.7.12}$$

This expression is general, applicable to any loading function. Note that the equivalent stiffness C for the frictionally blocked elastic energy as well as dilatancy factor β^* depends on the directions of ds_{ij} and de^p_{ij}. So does the friction coefficient α if $\partial\alpha/\partial J_3 \neq 0$.

Obviously we must have $C \geq 0$ and $\alpha \geq 0$. Not only the derivatives of f, but also p must be checked for this purpose. Normally $(Df/D\hat{\gamma}^p)/(Df/D\bar{\tau}) < 0$, and then p must be positive; this is so if $ds_{ij}\, d\gamma^p_{ij} > 0$.

Stability Condition of Frictional Materials

Let us now introduce the work expression

$$\Delta \bar{W} = \Delta W - \Delta W_f = \tfrac{1}{2} d\sigma_{ij}\, d\varepsilon_{ij}^{p} - \Delta W_f \tag{10.7.13}$$

Now, although ΔW can become negative due to release of the frictionally blocked elastic energy, $\Delta \bar{W}$ will still remain positive. So the state is stable. On the other hand if ΔW becomes negative for other reasons ($\Delta W_f = 0$), so will $\Delta \bar{W}$. Thus the following proposition, which gives for bodies under uniform strain a less restrictive (more general) sufficient condition for material stability than does Drucker's postulate, appears to be true for isotropic materials under controlled stress or strain (Bažant, 1980):

$$\textit{If either} \quad \Delta W > 0 \quad \textit{or} \quad \Delta \bar{W} > 0, \quad \textit{the material is stable} \tag{10.7.14}$$

Note that we cannot discard the condition $\Delta W > 0$ because ΔW_f can be negative when $\beta\, d\sigma < -d\bar{\tau}$ even if $\Delta W > 0$.

In the stress space $(\sigma, \bar{\tau})$, the domain of $d\sigma_{ij}$-vectors that give $\Delta W > 0$ occupies the half-plane a in Figure 10.23b, and the domain of those that give $\Delta \bar{W} > 0$ occupies a certain other half-plane b. The combined domain of vectors of applied stress increments $d\sigma_{ij}$ for which the response is inelastic and is assured to be stable (for bodies under uniform strain) occupies the union of these two half-planes, that is, the reentrant wedge (Fig. 10.23b), one side of which is tangent to the loading surface.

Condition 10.7.14 may equivalently be stated as follows (Bažant, 1980):

$$\textit{If} \quad \Delta W - \chi\, \Delta W_f > 0 \quad \textit{for any} \quad \chi \in (0, 1), \quad \textit{the material is stable} \tag{10.7.15}$$

Since $\Delta W - \chi\, \Delta W_f$ is a linear function of χ, the extremes can occur only at $\chi = 0$ and $\chi = 1$, and so condition (10.7.13) follows from condition (10.7.14) and vice versa.

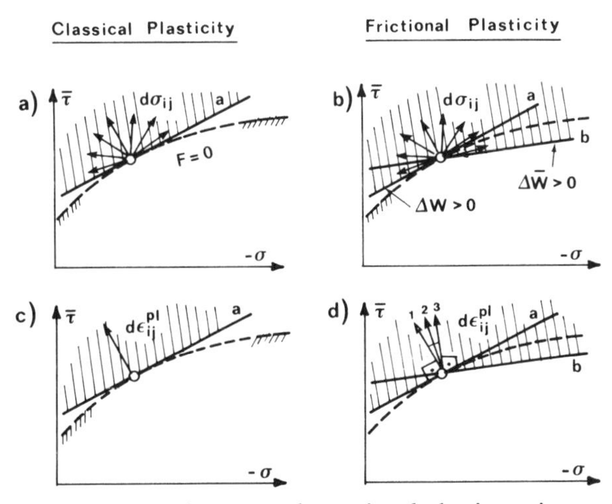

Figure 10.23 Stable increments of stress and associated plastic strain.

Plastic Strain Increment for Frictional Materials

In view of the fact that the flow rule can be derived from $\Delta W > 0$, it is interesting to see what follows by the same line of reasoning from the more general condition $\Delta W - \chi\, \Delta W_f > 0$. We have

$$\Delta W - \chi\, \Delta W_f = \frac{1}{2} ds_{ij}\, d\gamma_{ij}^p + \frac{3}{2} d\sigma\, d\varepsilon^p - \chi \frac{\alpha - \beta^*}{2C} d\sigma(d\bar{\tau} + \beta^*\, d\sigma) > 0 \quad (10.7.16)$$

The loading criterion may be written as $df = f_{,s_{ij}}\, ds_{ij} + f_{,\sigma}\, d\sigma > 0$. The ratio of this expression to that in Equation 10.7.16 obviously must be positive, and denoting it as $d\mu/2$ ($d\mu > 0$), we get

$$\left(\frac{\partial f}{\partial \sigma_{ij}} d\mu - d\gamma_{ij}^p\right) ds_{ij} + \left[\frac{\partial f}{\partial \sigma} d\mu - 3d\varepsilon^p + \chi \frac{\alpha - \beta^*}{C}(d\bar{\tau} + \beta^*\, d\sigma)\right] d\sigma = 0. \quad (10.7.17)$$

This equation must hold for any ds_{ij} and $d\sigma$. Pursuing the same line of argument as we did in developing the normality rule of classical plasticity, we note that this is possible only if the bracketed expressions vanish, that is, if (Bažant, 1980)

$$d\gamma_{ij}^p = \frac{\partial f}{\partial s_{ij}} d\mu \qquad 3d\varepsilon^p = \frac{\partial f}{\partial \sigma} d\mu - \chi \frac{\alpha - \beta^*}{C}(d\bar{\tau} + \beta^*\, d\sigma)$$

$$= \frac{\partial f}{\partial \sigma} d\mu - \frac{\chi}{C}\left(\alpha - \frac{3\, d\varepsilon^p}{p\hat{f}\, d\mu}\right)\left(d\bar{\tau} + \frac{3\, d\varepsilon^p}{p\hat{f}\, d\mu} d\sigma\right) \quad (10.7.18)$$

where we used $\beta^* = (2/p)3d\varepsilon^p/2d\bar{\gamma}^p$ and substituted $d\hat{\gamma}^p = \hat{f}\, d\mu$ where $\hat{f} = [(\partial f/\partial s_{ij})(\partial f/\partial s_{ij})/2]^{1/2}$, which follows from the above expression for $d\gamma_{ij}^f$.

Equation 10.7.18 governs the ratio of the $d\varepsilon_{ij}^p$ components, that is, the direction of the vector $d\varepsilon_{ij}^p$. Using the same logic as in classical plasticity, we could further consider the magnitude of $d\varepsilon_{ij}^p$ to be proportional to $\Delta W - \chi\, \Delta W_f$.

Except for $\chi = 0$, Equation 10.7.18 is nonlinear with regard to $d\varepsilon_{ij}^p/d\mu$. Moreover, C, α, and p depend on the direction of vectors $d\sigma_{ij}$ and $d\varepsilon_{ij}^p$. These aspects complicate applications. Equation 10.7.18, however, is instructive.

What we should observe is that, by pursuing basically the same line of reasoning as we did in classical plasticity to derive the flow rule, we now obtain no unique direction of the vector of $d\varepsilon_{ij}^p$ but a continuous set of infinitely many possible directions characterized by an arbitrary parameter $\chi \in (0, 1)$. In the volumetric section of stress space, all the possible directions of $d\varepsilon_{ij}^p$ fill a continuous fan of finite angle (1-2-3 in Fig. 10.23d). One boundary direction of the fan is the normal to the loading surface ($\chi = 0$). The other boundary direction (3 in Fig. 10.23d, $\chi = 1$) can be thought to be normal to some other surface (*b* in Fig. 10.23d).

This situation resembles that encountered in classical plasticity at the corner of the loading surface. It is also similar to what is assumed in nonassociated plasticity; however, the direction of the fan boundary ($\chi = 1$) is not unique and is not known in advance as it is not uniquely determined by the current loading surface in the stress space. Moreover, material stability (in the sense of Drucker's postulate) is assured for all loading directions $d\sigma_{ij}$ within the fan, while in nonassociated plasticity the stability is not assured.

Inverse Material Friction

A friction-producing force, in a generalized sense, is a force that does no work on some displacement yet does affect that displacement. From Equation 10.7.5 we saw that $d\sigma$ does no work on $d\hat{\gamma}^p$ but affects it if the loading function depends on both σ and $\hat{\gamma}^p$ (Eq. 10.7.7). However, we may now notice from Equation 10.7.5 that, conversely, $d\bar{\tau}$ does no work on $d\varepsilon^p$ yet can affect $d\varepsilon^p$ if the loading function depends on ε^p, as is the case for geomaterials. So, $d\bar{\tau}$ alternatively may be regarded as a friction-producing force and may be assumed to correspond to $d\sigma$ for the block (Fig. 10.22a), while $d\sigma$, $p\,d\hat{\gamma}^p$ and $3d\varepsilon^p$ are assumed to correspond to $d\tau$, $d\varepsilon^p$, and $d\gamma^p$ for the block. This phenomenon was called the inverse friction (Bažant, 1980). Instead of Equation 10.7.8, we may now write the differential of Equation 10.7.7 in the form

$$df = \frac{Df}{D\bar{\tau}}\,D\bar{\tau} + \frac{\partial f}{\partial \sigma}\,d\sigma + \frac{1}{3}\frac{Df}{D\varepsilon^p}(3d\varepsilon^p) = 0 \qquad (10.7.19)$$

where

$$\frac{Df}{D\varepsilon^p} = \frac{\partial f}{\partial \varepsilon^p} + \frac{3}{2\beta}\frac{\partial f}{\partial \bar{\gamma}^p} \qquad (10.7.20)$$

while $Df/D\bar{\tau}$ and β are given by Equations 10.7.9 and 10.7.10. Comparison of Equation 10.7.19 with the equation $df = d\tau + \alpha\,d\sigma - C\,d\gamma^p$ stated below Equation 10.7.1 now yields

$$C = -\frac{1}{3}\left(\frac{Df/D\varepsilon^p}{\partial f/\partial \sigma}\right) \qquad \alpha = \frac{Df/D\bar{\tau}}{\partial f/\partial \sigma} \qquad (10.7.21)$$

where α may now be called the "inverse friction" coefficient.

The expression for ΔW_f has again the form of Equation 10.7.12 in which, though, $\beta^* = p\,d\gamma^p/3d\varepsilon^p = p/2\beta$ where $\beta = 3d\varepsilon^p/2d\hat{\gamma}^p$. Note that this expression for ΔW_f cannot be reduced to the previous one (Eqs. 10.7.11–10.7.12).

The material stability condition (Eq. 10.7.14) may now be broadened. We may define $\Delta\tilde{W} = \Delta W - \Delta\tilde{W}_f$ where $\Delta\tilde{W}_f = \Delta W_f$ as given by Equations 10.7.21 and 10.7.12 with $\beta^* = p/2\beta$, while $\Delta\bar{W}$ remains to be given by Equations 10.7.11 and 10.7.13. Then:

$$\textit{If} \quad \Delta W > 0 \quad \textit{or} \quad \Delta\bar{W} > 0 \quad \textit{or} \quad \Delta\tilde{W} > 0, \quad \textit{the material is stable} \qquad (10.7.22)$$

(Bažant, 1980). Instead of a reentrant wedge, the domain of $d\sigma_{ij}$ vectors that produce inelastic strain and stable response now becomes a reentrant pyramid, one side of which is tangent to the loading surface. Equivalently, we can state that

$$\textit{If} \quad \Delta W - \chi\,\Delta W_f - \psi\,\Delta\tilde{W}_f > 0 \quad \textit{for any} \quad \chi \in (0, 1) \quad \textit{and} \quad \psi \in (0, 1), \ \textit{the material is stable} \qquad (10.7.23)$$

To make a distinction, $\Delta\tilde{W}_f$ may be called the frictionally blocked volumetric elastic energy, while the previously introduced ΔW_f is the frictionally blocked deviatoric elastic energy.

The line of reasoning that is used in classical plasticity to deduce the normality rule would now generalize Equation 10.7.18 to a form that contains two arbitrary parameters χ and ψ and indicates that all stable plastic strain increment vectors fill a cone (hypercone) rather than just a fan. The normal to f lies on one side of this cone.

It may be instructive to illustrate the meaning of coefficient p from Equation 10.7.5. Consider the special case when the medium principal axes of ds_{ij} and γ^p_{ij} coincide and let them lie on the x_2 axis. Also assume that the medium principal values of ds_{ij} and $d\varepsilon^p_{ij}$ are zero, that is, $ds_{22} = d\gamma^p_{22} = 0$. For a suitable choice of the x_1 and x_2 axes, the stress state in the x_1, x_2 plane can be represented as hydrostatic stress $d\sigma$ superimposed on a pure shear stress of magnitude $d\hat{\tau}$. Likewise, in some other x_1' and x_2' axes the strain state in the x_1, x_2 plane can be represented as volumetric strain $d\varepsilon^p$ superimposed on a pure shear strain of magnitude $d\hat{\gamma}^p$. Since $ds_{22} = -ds_{11}$, we have $ds_{11} = \pm d\hat{\tau}/\sqrt{2}$. Furthermore, working in the principal axes of ds_{ij}, we have $ds_{12} = 0$, and because of $d\gamma^p_{22}$ we have $d\gamma^p_{11} = \pm d\hat{\gamma}^p \cos 2\omega/\sqrt{2}$ where ω = angle between the maximum principal directions of $d\sigma_{ij}$ and $d\varepsilon^p_{ij}$. Thus, from $\Delta W = \frac{3}{2}\, d\sigma\, d\varepsilon^p + \frac{1}{2} ds_{ij}\, d\gamma^p_{ij}$ we obtain

$$\Delta W = \tfrac{3}{2} d\sigma\, d\varepsilon^p + \tfrac{1}{2} d\hat{\tau}(2 d\hat{\gamma}^p \cos 2\omega) \tag{10.7.24}$$

So, coefficient p from Equation 10.7.5 is simply equal to $2 \cos 2\omega$, and we see that it may vary between 2 and -2. When $d\sigma_{ij}$ and $d\varepsilon^p_{ij}$ are coaxial (as in a cubic triaxial test) $p = \cos 2\omega = 2$ and we have a one-to-one correspondence with the friction block example, without introduction of any further arbitrary factor.

Frictional Phenomena in Other Constitutive Theories

An analogous frictional formulation can be developed for stress-strain relations based on loading surfaces in the strain space. This has been done (Bažant, 1980) for the so-called fracturing material models in which the inelastic deviatoric stress decrements depend on the volumetric strain.

Problems

10.7.1 Consider the frictional block in Figure 10.22b, which is loaded by a weight over a pulley rather than by a spring. Show that if slipping is imminent, any infinitesimal disturbance $(d\sigma, d\tau)$ for which $\Delta W = \frac{1}{2}(d\tau\, d\gamma^p + d\sigma\, d\varepsilon^p) < 0$ causes the block to slip a finite distance. The equilibrium is therefore unstable.

10.7.2 Can the block be stable if it is preloaded by a nonlinear spring?

References and Bibliography

Argyris, J. H., and Kelsey, J. (1960), *Energy Theorems and Structural Analysis*, Butterworth, London; reprinted from *Aircraft Engineering*, series of articles between October 1954 and May 1955.

Bažant, Z. P. (1976), "Instability, Ductility and Size Effect in Strain-Softening Concrete," *J. Eng. Mech.* (ASCE), 102(2):331–44; "Closure," 103:357–8 and 775–7 (Sec. 10.1).

Bažant, Z. P. (1978), "Endochronic Inelasticity and Incremental Plasticity," *Int. J. Solids and Structures*, 14:691–714 (Sec. 10.6).

Bažant, Z. P. (1980), "Work Inequalities for Plastic-Fracturing Material," *Int. J. Solids and Structures*, 16:873–901 (Sec. 10.1).

Bažant, Z. P. (1984), "Microplane Model for Strain-Controlled Inelastic Behavior," chapter 3 in *Mechanics of Engineering Materials,* ed. by C. S. Desai and R. H. Gallagher, John Wiley and Sons, London (Sec. 10.1).

Bažant, Z. P. (1985), "Distributed Cracking and Nonlocal Continuum," in *Finite Element Methods for Nonlinear Problems,* ed. by P. Bergan et al., Springer Verlag, Berlin, pp. 77–102 (Proc. of Europe-U.S. Symp. held in Trondheim) (Sec. 10.1).

Bažant, Z. P. (1987a), "Stable States and Paths of Inelastic Structures: Completion of Shanley's Theory," Report No. 87–10/606s, Dept. of Civil Engineering, Northwestern University, Evanston, Ill., October 1987, p. 30 (Sec. 10.1).

Bažant, Z. P. (1987b), "Lectures on Material Modeling Principles," Course 720-D30, Northwestern University, Evanston, Ill. (Sec. 10.4).

Bažant, Z. P. (1988), "Stable States and Paths of Structures with Plasticity or Damage," *J. Eng. Mech.* (ASCE), 114(12):2013–34 (Sec. 10.1).

Bažant, Z. P. (1989a), "Stable States and Stable Paths of Propagation of Damage Zones and Interactive Fractures," in *Cracking and Damage—Strain Localization and Size Effect,* ed. by J. Mazars and Z. P. Bažant, Elsevier, London and New York, pp. 183–206 (Proc. of France–U.S. Workshop held at ENS, Cachan, September 1988) (Sec. 10.1).

Bažant, Z. P. (1989b), "Bifurcation and Thermodynamic Criteria of Stable Paths of Structures Exhibiting Plasticity and Damage Propagation," in *Computational Plasticity,* ed. by D. R. J. Owen, E. Hinton, and E. Oñate (Proc. 2nd Int. Conf. held in Barcelona, September 1989), Pineridge Press, Swansea, U.K., pp. 1–26 (Sec. 10.1).

Bažant, Z. P., and Kim, S. S. (1979), "Plastic-Fracturing Theory for Concrete," *J. Eng. Mech.* (ASCE), 105:407–28, with "Errata," in Vol. 106, 1980, p. 421 (Sec. 10.6).

Bažant, Z. P., and Prat, P. C. (1988), "Microplane Model for Brittle Plastic Material: I. Theory," *J. Eng. Mech.* (ASCE), 114(10):1672–88. See also Part II, "Verification," 114(10):1689–1702 (Sec. 10.1).

Billardon, R., and Doghri, I. (1989), "Localization Bifurcation Analysis for Damage Softening Elasto-Plastic Materials," in *Cracking and Damage—Strain Localization and Size Effect,* ed. by J. Mazars and Z. P. Bažant, Elsevier, London and New York, pp. 295–307 (Proc. of France–U.S. Workshop held at ENS, Cachan, September 1988) (Sec. 10.4).

Biot, M. A. (1965), *Mechanics of Incremental Deformation,* John Wiley and Sons, New York, p. 504 (Sec. 10.1).

Borrè, G., and Maier, G. (1989), "On Linear Versus Nonlinear Flow Rules in Strain Localization Analysis," *Meccanica,* 24(1):36–41 (Sec. 10.4).

Bromwich, J. T. l'A. (1906), "On the Roots of the Characteristic Equation of a Linear Substitution," *Acta Mathematica,* 30:297–304 (*Sec.* 10.4).

Brophy, J. H., Rose, R. M., and Wulff, J. (1964), *The Structure and Properties of Materials,* Vol. II, *Thermodynamics of Structure,* John Wiley and Sons, New York (Sec. 10.1).

Bruhns, O. T. (1984), "Bifurcation Problems in Plasticity," in *Stability in the Mechanics of Continua,* ed. by T. Lehmann, Springer-Verlag, Berlin, pp. 46–56.

Castigliano, A. (1879), "Théorie de l'équilibre des systèmes élastiques," Thesis, Turin Polytechnical Institute, Turin.

Corradi, L. (1978), "Stability of Discrete Elastic Plastic Structures with Associated Flow Laws," *Solid Mechanics Archives,* 3(3):201–59.

Crisfield, M. A. (1981), "A Fast Incremental/Iterative Procedure That Handles Snap-Through," *Computers and Structures,* 13:55–62.

Crisfield, M. A. (1986), "Snapthrough and Snapback Response in Concrete Structures and the Dangers of Underintegration," *Int. J. Numer. Methods Eng.,* 22:751–68.

de Borst, R. (1986), "Nonlinear Analysis of Frictional Materials," Doctoral Dissertation, Delft University of Technology, Delft, Netherlands (Sec. 10.4).

de Borst, R. (1987a), "Stability and Uniqueness in Numerical Modeling of Concrete Structures," Proc. IABSE Symposium on Computational Mechanics of Concrete, held at Delft University of Technology, Delft, Netherlands, publ. by IABSE, Zürich, pp. 167–86 (Sec. 10.4).

de Borst, R. (1987b), "Computation of Post-Bifurcation and Post-Failure Behavior of Strain-Softening Solids," *Computers and Structures,* 25:211–24 (Sec. 10.4).

de Borst, R. (1989), "Numerical Methods for Bifurcation Analysis in Geomechanics," *Ingenieur-Archiv,* 59:160–74 (Sec. 10.4).

de Groot, S. R., and Mazur, P. (1962), *Non-Equilibrium Thermodynamics,* North-Holland Publishing Co., Amsterdam; also Dover Publications, 1984 (Sec. 10.1).

Denbigh, K. (1968), *The Principles of Chemical Equilibrium,* Cambridge University Press, Cambridge, U.K. (Sec. 10.1).

Dimaggio, F. L. (1960), "Principle of Virtual Displacements in Structural Analysis," *J. Struct. Eng.* (ASCE), 86(11):65–78.

Dougill, J. W. (1975), "Some Remarks on Path Independence in the Small in Plasticity," *Quart. Appl. Math.* 32:233–43 (Sec. 10.6).

Dougill, J. W. (1976), "On Stable Progressively Fracturing Solids," *Zeits. Angew. Math. Phys. (ZAMP),* 27(4):423–37 (Sec. 10.6).

Droz, P., and Bažant, Z. P. (1989), "Nonlocal Analysis of Stable States and Stable Paths of Propagation of Damage Shear Bands," in *Cracking and Damage—Strain Localization and Size Effect,* ed. by J. Mazars and Z. P. Bažant, Elsevier, London and New York, 415–23 (Proc. of France–U.S. Workshop held at ENS, Cachan, September 1988) (Sec. 10.4).

Drucker, D. C. (1950), "Some Implications of Work Hardening and Ideal Plasticity," *Quart. Appl. Math.,* 7:411–18 (Sec. 10.6)

Drucker, D. C. (1954), "Coulomb Friction, Plasticity and Limit Loads," *J. Appl. Mech.* (ASME), 21:71–74 (Sec. 10.6).

Drucker, D. C. (1959), "A Definition of Stable Inelastic Material," *J. Appl. Mech.* (ASME), 26:101–106 (Sec. 10.6).

Drucker, D. C. (1962), *Stress-Strain Time Relations and Irreversible Thermodynamics,* Pergamon Press, New York pp. 331–51 (Proc. of IUTAM Symp. on Second-Order Effects in Elasticity, Plasticity and Fluid Dynamics).

Drucker, D. C. (1964), "On the Postulate of Stability of Material in the Mechanics of Continua," *J. Mécanique,* 3:235–49 (Sec. 10.6).

Fermi, E. (1937), *Thermodynamics,* Prentice-Hall, New York; see also Dover, 1956, New York (Sec. 10.1).

Fung, Y. C. (1965), *Foundations of Solid Mechanics,* Prentice-Hall, Englewood Cliffs, N.J. (Sec. 10.1).

Germain, P., Nguyen, Q. S., and Suquet, P. (1983), "Continuum Thermodynamics," *J. Appl. Mech.* (ASME), 50:1010–1020.

Gibbs, J. W. (1875), "On the Equilibrium of Heterogeneous Substances," *Transactions of the Connecticut Academy,* III; see also *Collected Works of J. Willard Gibbs,* Vol. 1, *Thermodynamics,* Yale University Press, New Haven, Conn. 1957, pp. 55–353 (Sec. 10.1).

Guggenheim, E. A. (1959), "Thermodynamics, Classical and Statistical," in *Encyclopedia of Physics,* ed. by S. Flügge, Vol. III/2, Springer Verlag, Berlin, pp. 1–118. (Sec. 10.1).

Hill, R. (1948), "A Variational Principle of Maximum Plastic Work in Classical Plasticity," *J. Quart. Mech. Math.,* 1:18–28 (Sec. 10.6).

Hill, R. (1957), "On the Problem of Uniqueness in the Theory of a Rigid–Plastic Solid—III," *J. Mech. Phys. Solids,* 5:153–61.

Hill, R. (1958), "A General Theory of Uniqueness and Stability in Elastic-Plastic Solids," *J. Mech. Phys. Solids,* 6:236–49 (Sec. 10.1).

Hill, R. (1961), "Bifurcation and Uniqueness in Non-Linear Mechanics of Continua," in *Problems of Continuum Mechanics,* Soc. of Industr. & Appl. Math., Philadelphia, Pa., pp. 155–64 (Sec. 10.4).

Hill, R. (1962a), "Acceleration Waves in Solids," *J. Mech. Phys. Solids,* 10:1–16.

Hill, R. (1962b), "Constitutive Law and Waves in Rigid/Plastic Solids," *J. Mech. Phys. Solids,* 10:89–98.

Hill, R. (1962c), "Uniqueness Criteria and Extremum Principles in Self-Adjoint Problems of Continuum Mechanics," *J. Mech. Phys. Solids,* 10:185–94 (Sec. 10.4).

Hill, R. (1967), "Eigenmodal Deformations in Elastic-Plastic Continua," *J. Mech. Phys. Solids,* 15:371–86.

Hill, R., and Rice, J. R. (1973), "Elastic Potentials and the Structure of Inelastic Constitutive Laws," *J. Appl. Math.* (SIAM), 25(3):448–61 (Sec. 10.1).

Householder, A. S. (1964), *The Theory of Matrices in Numerical Analysis,* Blaisdell, New York (Sec. 10.4).

Hutchinson, J. W. (1974), "Plastic Buckling," *Adv. Appl. Mech.,* 14:67–144.

Il'yushin, A. A. (1961), "On the Postulate of Plasticity," *Appl. Math. Mech.,* 25:746–52 (Sec. 10.6).

Keenan, J. H. K., Matsopoulos, G. N., and Gyftopoulos, E. P. (1980), "Thermodynamics," in *Encyclopedia Britannica,* 15th ed., Vol. 18 (Macropedia), pp. 290–315.

Kestin, J. (1966), *On the Application of the Principles of Thermodynamics to Strained Solid Materials,* ed. by H. Parkus and L. Sedov, Springer, New York, pp. 177–212 (Proc. of the IUTAM Symposium, Vienna).

Koiter, W. T. (1953), "Stress-Strain Relations, Uniqueness and Variational Theorems for Elastic-Plastic Material with a Singular Yield Surface," *Quart. Appl. Math.,* 11(3):29–53 (Sec. 10.6).

Korn, G. A., and Korn, T. M. (1968), *Mathematical Handbook for Scientists and Engineers,* 2nd ed., McGraw-Hill, New York (Sec. 10.4).

Lade, P., Nelson, R., and Ito, M. (1987), "Non-Associated Flow and Stability of Granular Materials," *J. Eng. Mech.* (ASCE), 113(9):1302–18; see also discussion, 115(8):1842–47.

Langhaar, H. L. (1962), *Energy Methods in Mechanics,* John Wiley and Sons, New York (Sec. 10.4).

Leroy, Y., and Ortiz, M. (1989), "Finite Element Analysis of Strain Localization in Frictional Materials," *Int. J. of Num. Anal. Methods in Geomech.,* 13:53–74 (Sec. 10.4).

Lin, T. H. (1968), *Theory of Inelastic Structures,* John Wiley and Sons, New York.

Maier, G. (1966a), "Behavior of Elastic-Plastic Trusses with Unstable Bars," *J. Eng. Mech.* (ASCE), 92(3):67–91 (Sec. 10.6).

Maier, G. (1966b). "Sui legami associati tra sforzi e deformazioni incrementali in elastoplasticità" (in Italian), *Istituto Lombardo di Scienze e Lettere,* 100(A):809–38.

Maier, G. (1967), "On Elastic-Plastic Structures with Associated Stress-Strain Relations Allowing for Work Softening," *Meccanica,* 2(1):55–64 (Sec. 10.6).

Maier, G. (1969), "Linear Flow Laws of Elasto-Plasticity: A Unified Approach," *Accademia Nazionale dei Lincei, Serie* VIII, XLVII(5):266–76 (Sec. 10.6).

Maier, G. (1971), "Incremental Plastic Analysis in the Presence of Large Displacements and Physical Instabilizing Effects," *Int. J. Solids Structures,* 7:345–72 (Sec. 10.1).

Maier, G., and Drucker, D. C. (1973), "Effects of Geometry Change on Essential Features of Inelastic Behavior," *J. Eng. Mech.* (ASCE), 99(4):819–34 (Sec. 10.1).

Maier, G., and Hueckel, T. (1979), "Non-Associated and Coupled Flow Rules of Elastoplasticity for Rock-Like Materials," *Int. J. Rock Mech. Min. Sci. Geomech. Abstr.*, 16:77–92 (Sec. 10.6).

Maier, G., and Zavelani, A. (1970), "Sul comportamento di aste metalliche compresse eccentricamente," *Costruzioni Metalliche,* No. 4, pp. 282–97 (Sec. 10.2).

Maier, G., Zavelani, A., and Dotreppe, J. C. (1973), "Equilibrium Branching Due to Flexural Softening," *J. Eng. Mech.* (ASCE), 99(4):897–901 (Sec. 10.1).

Malvern, L. E. (1969), *Introduction to the Mechanics of a Continuous Medium,* Prentice-Hall, Englewood Cliffs, N.J. (Sec. 10.6).

Mandel, J. (1964), "Conditions de stabilité et postulat de Drucker," in *Rheology and Soil Mechanics,* ed. by J. Kravtchenko and P. M. Sirieyes, Springer-Verlag, Berlin, 1966, pp. 56–68 (IUTAM Symp. held in Grenoble) (Sec. 10.1).

Martin, J. B. (1975), *Plasticity: Fundamentals and General Results,* MIT Press, Cambridge, Mass.

Melan, E. (1938), "Zur Plastizität des räumlichen Kontinuums," *Ing.-Arch.*, 9:116–26 (Sec. 10.6).

Mirsky, L. (1955), *An Introduction to Linear Algebra,* Oxford University Press, London (also Dover Publ., New York, 1983) (Sec. 10.4).

Mróz, Z. (1963), "Non-Associated Flow Laws in Plasticity," *J. de Mécanique,* 2:21–42

Mróz, Z. (1966), "On Forms of Constitutive Laws in Elastic-Plastic Solids," *Arch. Mech. Stos.* (Warsaw), 18(1):3–15.

Naghdi, P. M., and Trapp, J. A. (1975), "The Significance of Formulating the Plasticity Theory with Reference to Loading Surface in Strain Space," *Int. J. Eng. Sci.* 13:785–97 (*Sec.* 10.6).

Needleman, A., and Tvergaard V. (1982), "Aspects of Plastic Post-Buckling Behaviour," in *Mechanics of Solids,* ed. by H. G. Hopkins and M. J. Sewell, Pergamon Press, Oxford, pp. 453–98.

Nemat-Nasser, S. (1979), "The Second Law of Thermodynamics and Noncollinear Crack Growth," in *Proceedings of the Third ASCE/EMD Specialty Conference,* September 17–19, Austin, Texas, pp. 449–52 (Sec. 10.2).

Nemat-Nasser, S., Sumi, Y., and Keer, L. M. (1980), "Unstable Growth of Tension Cracks in Brittle Solids," *Int. J. Solids Structures,* 16:1017–35 (Sec. 10.2).

Nguyen, Q. S. (1984), "Bifurcation et stabilité des systèmes irréversibles obéissant au principe de dissipation maximale," *Journal de mécanique-théorique et appliquée,* 3(1):41–61 (Sec. 10.2).

Nguyen, Q. S. (1987), "Bifurcation and Postbifurcation Analysis in Plasticity and Brittle Fracture," *J. Mech. Phys. Solids,* 35(3):303–24 (Sec. 10.2).

Nguyen, Q. S., and Stolz, C. (1986), "Energy Methods in Fracture Mechanics: Bifurcation and Second Variations," IUTAM-ISIMM Symposium, Application of Multiple Scaling in Mechanics, Paris (Sec. 10.2).

Owen, D. R. J., and Hinton, E. (1980), *Finite Elements in Plasticity—Theory and Practice,* Pineridge Press, Swansea, U.K. (Sec. 10.4).

Palmer, A. C., Maier, G., and Drucker, D. C. (1967), "Normality Relations and Convexity of Yield Surfaces for Unstable Materials or Structural Elements," *J. Appl. Mech.* (ASME), 34:464–70.

Petryk, H. (1985a), "On Energy Criteria of Plastic Instability," in *Plastic Instability,* publication by Ecole Nat. des Ponts et Chaussées, Paris, pp. 215–26 (Proc. of Considère Memorial) (Sec. 10.2).

Petryk, H. (1985b), "On Stability and Symmetry Condition in Time-Dependent Plasticity,"*Arch. Mech. Stos.* (Warszaw), 37(4–5):503–20 (Sec. 10.2).

Planck, M. (1945), *Treatise on Thermodynamics,* Dover, New York (Sec. 10.1).

Prager, W. (1949), "Recent Developments in the Mathematical Theory of Plasticity," *J. Appl. Physics,* 20:235–41 (Sec. 10.6).

Prandtl, L. (1924), Proc. 1st Int. Cong. Appl. Mech., Delft, p. 43 (Sec. 10.6).

Reuss, A. (1930), "Berücksichtigung der elastischen Formänderung in der Plastizitätstheorie," *Zeits. Ang. Math. Mech. (ZAMM),* 10:266 (Sec. 10.6).

Rice, J. R. (1971) "Inelastic Constitutive Relation for Solids: an Internal Variable Theory and its Application to Metal Plasticity," *J. Mech. Phys. Solids,* 19:433–55 (Sec. 10.6).

Riks, E. (1979), "An Incremental Approach to the Solution of Snapping and Buckling Problems," *Int. J. Solids Structures,* 15:529–51 (Sec. 10.4).

Rudnicki, J. W., and Rice, J. R. (1975), "Conditions for the Localization of Deformation in Pressure-Sensitive Dilatant Materials," *J. Mech. Phys. Solids,* 23:371–94 (Sec. 10.6).

Runnesson, K., and Sture, S. (1989), "Stability of Frictional Materials," *J. Eng. Mech.* (ASCE), 115(8):1828–33.

Runesson, K., Larsson, R., and Sture, S. (1989), "Characteristics and Computational Procedure in Softening Plasticity," *J. Eng. Mech.* (ASCE), 115:1628–46 (Sec. 10.4).

Sandler, I. S. (1978), "On the Uniqueness and Stability of Endochronic Theories of Material Behavior," *J. Applied Mech.* (ASME), 45:263–6.

Schapery, R. A. (1968), *On a Thermodynamic Constitutive Theory and its Applications to Various Nonlinear Materials,* ed. by B. A. Boley, Springer-Verlag, New York (Proc. IUTAM Symp. East Kilbride, June).

Shanley, F. R. (1947), "Inelastic Column Theory," *J. Aero. Sci.,* 14:261–8 (Sec. 10.3).

Simo, J. C. (1989), "Strain Softening and Dissipation: a Unification of Approaches," in *Cracking and Damage—Strain Localization and Size Effect,* ed. by J. Mazars and Z. P. Bažant, Elsevier, London and New York, pp. 440–61 (Proc. of France–U.S. Workshop held at ENS, Cachan, September 1988) (Sec. 10.6).

Stolz, C. (1989), "On Some Aspect of Stability and Bifurcation in Fracture and Damage Mechanics," in *Cracking and Damage—Strain Localization and Size Effect,* ed. by J. Mazars and Z. P. Bažant, Elsevier, London and New York, pp. 207–16 (Proc. of France–U.S. Workshop held at ENS, Cachan, September 1988) (Sec. 10.2).

Strang, G. (1976), *Linear Algebra and Its Applications,* Academic Press, New York (Sec. 10.4).

Sulem, J., and Vardoulakis, I. (1989), "Bifurcation Analysis of the Triaxial Test on Rock Specimens," in *Cracking and Damage—Strain Localization and Size Effect,* ed. by J. Mazars and Z. P. Bažant, Elsevier, London and New York, pp. 308–22 (Proc. of France–U.S. Workshop held at ENS, Cachan, September 1988) (Sec. 10.4).

ter Haar, D., and Wergeland, H. (1966), *Elements of Thermodynamics,* Addison-Wesley Publ. Co., Reading Mass.

11

Three-Dimensional Continuum Instabilities and Effects of Finite Strain Tensor

The exposition of elasticity in textbooks normally begins by a general three-dimensional formulation, which is subsequently simplified to solve one- and two-dimensional problems. Why haven't we followed this route?—because the general theory of stability for multidimensional continuous bodies is considerably more difficult than the theory of beams, plates, and shells. The difficulty stems from the geometrically nonlinear nature of the finite strain tensor.

As we have seen in Chapters 4 and 10, stability depends on the second-order incremental work. One contribution to this work comes from $\frac{1}{2}\sigma_{ij}e_{ij}$ where e_{ij} can be taken as the small (linearized) strain increment since the stress increment σ_{ij} is also small. However, there is another contribution from $S_{ij}^0\varepsilon_{ij}$ where S_{ij}^0 is the stress in the initial state. Now since S_{ij}^0 is not small, the incremental strain ε_{ij} must be expressed accurately up to terms second-order small in $u_{i,j}$, that is, one must use finite strain. This complicates the concept of stress.

In the usual (Cauchy) definition, the stress (true stress) can be defined by forces acting on a small elemental cube cut from the material. Due to the finiteness of strain, this cube would have to be cut after deformation ε_{ij}, but then the material properties cannot be related to the initial state and the body surface would have a more complicated shape. For these reasons it is necessary to write the equilibrium conditions in terms of stresses T_{ij} that act on a deformed element (which was a small elemental cube in the initial state) and are referred to the initial areas of the faces of the cube. But these stresses do not give the correct work expression in terms of ε_{ij} and are nonsymmetric, while the stresses used in the stress–strain relation must be symmetric. To remedy it, still another symmetric kind of stress, σ_{ij}, must be introduced in order to write the stress–strain relation. In this regard, one must further realize that there is not one but infinitely many equally justified expressions for finite strain, ε_{ij}, and the meaning of σ_{ij} depends on the choice of this expression. Hence the incremental moduli C_{ijkm} must also depend on the choice of ε_{ij}.

It is important to use only a formulation that involves a conjugate group of ε_{ij}, σ_{ij}, and C_{ijkm}. Unfortunately, many formulations presented in the literature mixed nonconjugated variables. Biezeno and Hencky (1928) were apparently the first to present one correct statement of stability that happens to meet the conjugacy requirements, although these requirements were not clearly stated until much later.

For most problems of thin bodies (beams, plates, shells), the aforementioned difficulties due to conjugacy of variables are avoided because it suffices to consider that only the deflections and rotations are large, while the material strains remain small. This has been the case for all the problems considered so far in this book. In this chapter we finally focus on three-dimensional massive bodies in which the finite strains of the material must be taken into account.

The critical states of stability of three-dimensional bodies that are not thin can generally occur, as we will see, only if the compression stress is of the same order of magnitude as the tangential shear modulus or transverse modulus of the material. Consequently, this chapter is of practical interest only for

1. Highly anisotropic materials, such as fiber composites
2. Composite structures having very soft components, such as sandwich plates
3. Continuum approximations of latticed structures
4. Materials that undergo a drastic reduction of tangential stiffness due to plasticity or damage (see Chap. 13).

The stiffness reduction we have in mind is not necessarily instantaneous. In the sense of the effective modulus treatment of long-time creep (Chap. 9), the large stiffness reduction can come about as a result of long-time creep (viscoelastic or viscoplastic). Thus, for example, the folding of rock strata, a basic problem in geology, may be regarded as a long-time three-dimensional instability, even though the instantaneous stiffness of rock remains very high.

While the finite strain theory is usually approached by means of coordinate transformations and tensorial invariance arguments, we will find it simpler to derive the entire theory solely by means of energy and variational arguments (Bažant, 1967, 1971). Moreover, the latter approach will also reveal the stress conjugacy restrictions that cannot be detected by the former. Then we will proceed to show some applications that can be solved by hand and, therefore, understood easily—buckling of a thick column or plate with shear, surface buckling of an orthotropic half-space, bulging of a compressed orthotropic wall or cylinder, and fiber buckling in composites.

All these applications are inherently multidimensional continuum problems. They illustrate that finite strain (as opposed to finite rotation) and a two- or three-dimensional form of buckling are important only when some of the stresses are of the same order of magnitude as the tangential moduli of the material. If the body is not thin, such a situation can arise only for highly anisotropic incremental material moduli or for stress states that are close to the peak of the stress–strain diagram. The anisotropy need not be natural but can be induced by previous inelastic straining or damage.

11.1 FINITE STRAIN

As illustrated in Chapters 4 to 7, in stability problems the work of initial stresses must be calculated on the basis of finite strain tensor components that are accurate up to terms second-order small in displacement gradients $u_{i,j}$. The reason is that the work of small stress increments is second-order small in $u_{i,j}$. We have already introduced finite strain in Sections 4.3, 6.1, and 7.2, however, only for thin bodies (beams, plates, and shells). For such bodies, as we will see (Eq. 11.1.19), the second-order part of strain depends principally on material rotations ω_{ij}. To treat bodies that are not thin, we need to introduce a more rigorous measure of finite strain than we did before.

Notations and Basic Requirements

For the sake of simplicity we will use only Cartesian tensors in component notation, that is, $x_1 = x$, $x_2 = y$, $x_3 = z$, and refer to the coordinates by italic lowercase subscripts running from 1 to 3. The current (or final) spatial coordinates x_i ($i = 1, 2, 3$) of material points, also called Eulerian coordinates, must now be distinguished from the initial coordinates X_i of material points, also called the material coordinates, reference coordinates, or Lagrangian coordinates. The initial (or reference) state in stability problems is generally a stressed state, and we measure the finite strain with respect to this state, not with respect to the natural stress-free state of the material. The deformation of the structure represents a mapping, $\mathbf{x} = \mathbf{x}(\mathbf{X})$. Locally this mapping is characterized by the deformation gradient tensor $x_{i,j} = \partial x_i / \partial X_j$ (the subscripts preceded by a comma will always denote partial derivatives with respect to X_i, not x_i). In terms of displacements $u_i = x_i - X_i$, obviously $x_{i,j} = \delta_{ij} + u_{i,j}$ where $\delta_{ij} = 1$ for $i = j$ and 0 for $i \neq j$ (Kronecker delta). The displacement gradient $u_{i,j}$ is in general a nonsymmetric tensor and may be decomposed as

$$u_{i,j} = e_{ij} + \omega_{ij} \qquad e_{ij} = \tfrac{1}{2}(u_{i,j} + u_{j,i}) \qquad \omega_{ij} = \tfrac{1}{2}(u_{i,j} - u_{j,i}) \tag{11.1.1}$$

where e_{ij} is called the small (or linearized) strain tensor, and ω_{ij} is the small (or linearized) rotation tensor of the material; e_{ij} and ω_{ij} depend on $u_{i,j}$ linearly and describe strain and rotation accurately only up to first-order terms in $u_{i,j}$.

The definition of finite strain ε_{ij} must satisfy four requirements:

I. ε_{ij} must be a second-order tensor.
II. ε_{ij} must be symmetric.
III. ε_{ij} must vanish for all rigid-body motions.
IV. ε_{ij} must depend on $u_{i,j}$ in a continuous, continuously differentiable, and monotonic manner.

Requirement I follows from the fact that the work increment per unit volume is $dW = \sigma_{ij}\, d\varepsilon_{ij}$ where σ_{ij} is a certain type of stress tensor. If ε_{ij} were not a tensor, then either σ_{ij} could not be a tensor or W could not be a scalar, each of which is inadmissible. Furthermore, since σ_{ij} is symmetric (see Sec. 11.2), a nonsymmetric part of $d\varepsilon_{ij}$ would do no work and would, therefore, be arbitrary, indeterminate. This is the reason for requirement II. Requirement III ensures that the work of stresses on rigid-body rotation is zero (this condition is violated by e_{ij}, the error being of the second order in $u_{i,j}$). Requirement IV implies that if there is no rotation ($\omega_{ij} = 0$), the dependence of ε_{ij} on $u_{i,j}$ must be unique and invertible (i.e., monotonicity holds). Monotonicity in requirement IV means that if the length of any line segment $d\mathbf{X}$ increases, the resolved normal component of ε_{ij} in the direction of $d\mathbf{X}$ must increase, too. For the sake of convenience, we further choose to impose a fifth requirement:

V. The first-order (linear) part of ε_{ij} must coincide with e_{ij} (Eq. 11.1.1).

Lagrangian (Green's) Finite Strain Tensor

To deduce tensor ε_{ij} satisfying the foregoing requirements, we consider a small line segment dX_i of length $|d\mathbf{X}|$ that gets transformed to line segment dx_i of length $|d\mathbf{x}|$ (Fig. 11.1a). Partially duplicating our argument in Section 7.2, we may set

$$2\varepsilon_{ij}\, dX_i\, dX_j = |d\mathbf{x}|^2 - |d\mathbf{X}|^2. \tag{11.1.2}$$

Repetition of italic lowercase subscripts implies summation over 1, 2, 3 (Einstein's summation rule). Equation 11.1.2 automatically satisfies (1) requirement III, because $|d\mathbf{X}| = |d\mathbf{x}|$ for rigid-body rotations; (2) requirement I, because the dyad $dX_i\, dX_j$ is a second-order tensor; and (3) requirement II, because this tensor is symmetric. Substituting $|d\mathbf{X}|^2 = dX_k\, dX_k$, $|d\mathbf{x}|^2 = dx_k\, dx_k$, and noting that $dx_k = x_{k,i}\, dX_i$, we get

$$\begin{aligned}
2\varepsilon_{ij}\, dX_i\, dX_j &= x_{k,i}\, dX_i\, x_{k,j}\, dX_j - dX_k\, dX_k = (x_{k,i}x_{k,j} - \delta_{ij})\, dX_i\, dX_j \\
&= [(X_k + u_k)_{,i}(X_k + u_k)_{,j} - \delta_{ij}]\, dX_i\, dX_j \\
&= [(\delta_{ki} + u_{k,i})(\delta_{kj} + u_{k,j}) - \delta_{ij}]\, dX_i\, dX_j \\
&= (u_{i,j} + u_{j,i} + u_{k,i}u_{k,j})\, dX_i\, dX_j.
\end{aligned}$$

Hence (see also Eq. 7.2.2):

$$\varepsilon_{ij} = \tfrac{1}{2}(x_{k,i}x_{k,j} - \delta_{ij}) = \tfrac{1}{2}(u_{i,j} + u_{j,i} + u_{k,i}u_{k,j}) = e_{ij} + \tfrac{1}{2}u_{k,i}u_{k,j} \tag{11.1.3}$$

Equation 11.1.3 is called the Lagrangian (finite) strain tensor (because it is based on the Lagrangian coordinates X_i, not because Lagrange would have invented it) or Green's strain tensor (Green, 1839). The expression $C_{ij} = x_{k,i}x_{k,j}$ is called Green's or Cauchy-Green's deformation tensor (Cauchy, 1828). In view of Equation 11.1.2, Equation 11.1.3 obviously satisfies requirement IV.

The factor 2 is introduced in Equations 11.1.2 and 11.1.3 in order to satisfy the convenience requirement V (this is clear upon noting that $u_{k,i}u_{k,j}$ in Eq. 11.1.3 is second-order small).

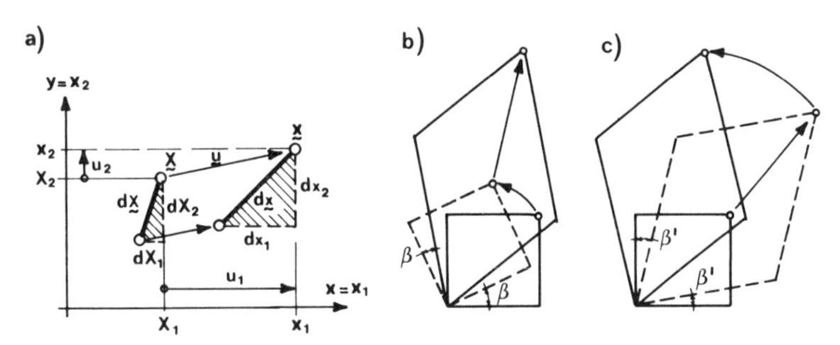

Figure 11.1 (a) Large deformation of line segment and material element; (b, c) polar decompositions of deformation, corresponding to left and right stretch tensors.

Note that while $\frac{1}{2}(u_{i,j}+u_{j,i})$ is not the exact expression for finite strain, the analogous expression $\dot{e}_{ij}=\frac{1}{2}[(\partial v_i/\partial x_j)+(\partial v_j/\partial x_i)]$ gives nevertheless the strain rate exactly ($v_i=\dot{x}_i=\dot{u}_i=$ velocity of material point, and a superior dot denotes a time derivative). For proof, we consider u_i to be the displacements from any deformed state taken as the initial state, and set $u_i=v_i\,\Delta t$ ($t=$ time); then $\dot{\varepsilon}_{ij}=\lim[\varepsilon_{ij}(\Delta t)/\Delta t]=\lim\frac{1}{2}(v_{i,j}\,\Delta t+v_{j,i}\,\Delta t+v_{k,i}v_{k,j}\,\Delta t^2)/\Delta t$ for $\Delta t\to 0$. The term with Δt^2 vanishes, and so $\lim\dot{\varepsilon}_{ij}=\dot{e}_{ij}$ at the initial deformed state, that is, for $u_i\to 0$, $x_i\to X_i$, and $\partial/\partial x_i\to\partial/\partial X_i$.

Biot's Finite Strain Tensor

Equation 11.1.3 is at present a generally adopted finite strain measure. However, it is not the only possible way to define finite strain in Lagrangian coordinates X_i. In fact, other definitions were either used or tacitly implied in most original contributions to stability theory of multidimensional continuous bodies. This generated unnecessary controversies in the past and prevented acceptance of some important original results. We will now show other possible definitions.

Biot (1934, 1939, 1965) introduced an alternative physical definition of finite strain that we denote as ε^b_{ij}. We imagine that the material element is first subjected to a symmetric deformation ($\omega_{ij}=0$) with a displacement gradient equal to ε^b_{ij} (Fig. 11.1c), and then the deformed element is rotated as a rigid body to its final position characterized by coordinates $x_i(\mathbf{X})$. Let x'_i be the coordinates of material points X_i after the symmetric deformation. As the subsequent rotation does not change the length of an infinitesimal line segment, we have $|d\mathbf{x}'|^2=|d\mathbf{x}|^2$ or $dx'_k\,dx'_k=dx_k\,dx_k$. Since $dx'_k\,dx'_k=x'_{k,i}\,dX_i\,x'_{k,j}\,dX_j=(\delta_{ki}+\varepsilon^b_{ki})(\delta_{kj}+\varepsilon^b_{kj})\,dX_i\,dX_j$ and $dx_k\,dx_k=(\delta_{ki}+u_{k,i})(\delta_{kj}+u_{k,j})\,dX_i\,dX_j$ we have $2\varepsilon^b_{ij}+\varepsilon^b_{ki}\varepsilon^b_{kj}=u_{i,j}+u_{j,i}+u_{k,i}u_{k,j}$, that is,

$$\varepsilon^b_{ij}+\tfrac{1}{2}\varepsilon^b_{ki}\varepsilon^b_{kj}=\varepsilon_{ij} \tag{11.1.4}$$

Finite strain ε^b_{ij}, called the Biot strain (Biot 1934, 1939, 1965), is a solution of this nonlinear equation (which represents a system of six quadratic equations). Similarly as before, one can check that ε^b_{ij} fulfills requirements I, II, III, and V, and fulfillment of requirement IV is physically clear from the decomposition of deformation into pure strain and rotation. A convenient feature of Biot's strain is that $\varepsilon^b_{ij}=u_{i,j}=u_{j,i}=e_{ij}$ if there is no rotation ($\omega_{ij}=0$). (On the other hand, $\varepsilon_{ij}\neq e_{ij}$ even for no rotation.)

The fact that Equation 11.1.4 defines ε^b_{ij} implicitly is inconvenient. However, for stability analysis we need ε^b_{ij} with only second-order accuracy. So we may approximate $\varepsilon^b_{ki}\varepsilon^b_{kj}$ as $e_{ki}e_{kj}$, and thus we get the second-order approximation:

$$\varepsilon^b_{ij}\simeq\varepsilon_{ij}-\tfrac{1}{2}e_{ki}e_{kj} \tag{11.1.5}$$

To obtain a third-order approximation to ε^b_{ij}, we may substitute the expression in Equation 11.1.5 into the term $\frac{1}{2}\varepsilon^b_{ki}\varepsilon^b_{kj}$ in Equation 11.1.4. This yields

$$\varepsilon^b_{ij}\simeq\varepsilon_{ij}-\tfrac{1}{2}(\varepsilon_{ki}-\tfrac{1}{2}e_{nk}e_{ni})(\varepsilon_{kj}-\tfrac{1}{2}e_{nk}e_{nj}) \tag{11.1.6}$$

Second-Order Approximations of Other Finite Strain Tensors

Comparing Equation 11.1.5 for ε_{ij}^b with Equation 11.1.3 for ε_{ij}, Bažant (1971) proposed that the tensors

$$\varepsilon_{ij}^{(m)} = \varepsilon_{ij} - \alpha e_{ki} e_{kj} \qquad \alpha = 1 - \frac{m}{2} \tag{11.1.7}$$

where coefficient m (or α) can have any value, could be used as acceptable second-order approximations to finite strain (the meaning of m will become apparent after Eq. 11.1.13). That Equation 11.1.7 as well as Equation 11.1.5 is an admissible second-order approximation of finite strain follows from the fact that this expression violates requirement III only by terms of higher than second order in $u_{i,j}$, and satisfies requirement IV for sufficiently small $u_{i,j}$. The latter fact means that the range of monotonicity of Equation 11.1.7 as well as Equation 11.1.5 is limited, while that of ε_{ij} is unbounded.

For $m = 2$ ($\alpha = 0$), Equation 11.1.7 (as well as the foregoing deformation sequence) yields the second-order Lagrangian strain ε_{ij}, that is, $\varepsilon_{ij}^{(2)} = \varepsilon_{ij}$ [the lack of superscript (m) will signify $m = 2$]. For $m = 1$ ($\alpha = \frac{1}{2}$), Equation 11.1.7 (as well as the foregoing sequence) yields the second-order Biot strain, that is, $\varepsilon_{ij}^{(1)} = \varepsilon_{ij}^b$. For $m = 0$ ($\alpha = 1$), Equation 11.1.7 gives $\varepsilon_{ij}^{(0)} = \varepsilon_{ij} - e_{ki}e_{kj} \simeq \varepsilon_{ij}^{\log}$; this represents the second-order approximation to what is called the logarithmic strain or Hencky strain. This strain, often favored for representing large strain test data, is associated (see Sec. 11.3) with the differential equations of equilibrium of Biezeno and Hencky (1928), as well as with the Jaumann stress rate (Prager, 1961; Truesdell and Noll, 1965; Masur, 1965). For $m = -2$ ($\alpha = 2$), Equation 11.1.7 yields a finite strain that is associated with Timoshenko's differential equilibrium equations for beam-columns with shear (see Sec. 11.6). And for $m = -1$ ($\alpha = \frac{3}{2}$), Equation 11.1.7 yields a finite strain that is associated (Sec. 11.3) with the so-called convected stress rate of Cotter and Rivlin (Prager, 1961; Masur, 1965).

The reason for calling $\varepsilon_{ij}^{(0)}$ with $m = 0$ the logarithmic strain (Bažant 1971) is that the principal strain (i.e., the normal strain ε_{11} at $\varepsilon_{12} = \varepsilon_{13} = 0$) according to Equation 11.1.7 is $\varepsilon_{11}^{(0)} = e_{11} + \frac{1}{2}e_{11}^2 - e_{11}^2 = e_{11} - \frac{1}{2}e_{11}^2$. The logarithmic normal strain is defined as $\varepsilon_{11}^{\log} = \int de_{11}/(1 + e_{11}) = \ln(1 + e_{11})$, the expansion of which is $\varepsilon_{11}^{\log} \simeq e_{11} - \frac{1}{2}e_{11}^2 + \frac{1}{3}e_{11}^3 - \cdots$. This coincides with Equation 11.1.7 up to the second-order terms.

The most general second-order approximation that satisfies requirement III except for terms of higher than second order, as well as requirement IV for sufficiently small $u_{i,j}$, has the form (Bažant, 1971):

$$\varepsilon_{ij}^* \simeq \varepsilon_{ij} - \alpha e_{ki} e_{kj} + a\delta_{ij} e_{kk} e_{nn} + b e_{ij} e_{kk} + c\delta_{ij} e_{km} e_{km} \tag{11.1.8}$$

where α, a, b, c are any constants.

Further Measures of Finite Strain

Let $\mathbf{F}$ denote the transformation tensor $\mathbf{F} = \nabla \mathbf{x}$ (where ∇ = gradient vector). The components of $\mathbf{F}$ are $F_{ij} = \partial x_i / \partial X_j = x_{i,j}$. The transformation $\mathbf{F}$ can be decomposed either as a strain followed by a rotation (Fig. 11.1c) or a rotation followed by a strain (Fig. 11.1b). This is described by the polar decomposition theorem (e.g., Ogden, 1984; Marsden and Hughes, 1983; Truesdell and Noll, 1965):

Theorem 11.1.1 For any nonsingular tensor $\mathbf{F}$ there exist unique positive-definite symmetric second-order tensors $\mathbf{U}$ and $\mathbf{V}$ such that

$$\mathbf{F} = \mathbf{RU} = \mathbf{VR} \tag{11.1.9}$$

in which $\mathbf{R}$ is a (finite) rotation tensor, which is orthogonal, that is, $\mathbf{R}^T\mathbf{R} = \mathbf{I} =$ unit tensor. (Product $\mathbf{RU}$ denotes a tensor product contracted on one index, also written as $\mathbf{R} \cdot \mathbf{U}$; the component form of Equation 11.1.9 is $F_{ij} = R_{ik}U_{kj} = V_{ik}R_{kj}$.)

To prove it, one obtains from Equation 11.1.9

$$\mathbf{C} = \mathbf{F}^T\mathbf{F} = \mathbf{U}^T\mathbf{R}^T\mathbf{R}\mathbf{U} = \mathbf{U}^T\mathbf{U} = \mathbf{U}^2 \tag{11.1.10}$$

$$\mathbf{D} = \mathbf{F}\mathbf{F}^T = \mathbf{V}\mathbf{R}\mathbf{R}^T\mathbf{V}^T = \mathbf{V}\mathbf{V}^T = \mathbf{V}^2 \tag{11.1.11}$$

So $\mathbf{U} = (\mathbf{F}^T\mathbf{F})^{1/2}$ and $\mathbf{V} = (\mathbf{F}\mathbf{F}^T)^{1/2}$. Such square roots are known to exist and be unique since tensors $\mathbf{F}^T\mathbf{F}$ and $\mathbf{F}\mathbf{F}^T$ are symmetric and positive definite. Indeed, the principal values of $\mathbf{U}$ may be calculated as $\lambda_{(i)} = \sqrt{\mu_{(i)}}$ $(i = 1, 2, 3)$ where $\mu_{(i)}$ are the principal values of the symmetric tensor $\mathbf{F}^T\mathbf{F}$, and the unit principal direction vectors $n_j^{(i)}$ of tensors $\mathbf{U}$ and $\mathbf{F}^T\mathbf{F}$ are the same. Obviously, $\mu_{(i)}$ are positive if $\mathbf{F}^T\mathbf{F}$ is positive definite. The spectral representation of $\mathbf{U}$ (e.g. Ogden, 1984) is

$$U_{jk} = \sum_{i=1}^{3} \sqrt{\mu_{(i)}}\, n_j^{(i)} n_k^{(i)} \tag{11.1.12}$$

The uniqueness may be proven by showing the impossibility of two different tensors $\mathbf{U}$ and $\mathbf{U}'$.

Tensor $\mathbf{C}$ (Eq. 11.1.10) represents Green's (or Cauchy-Green's) deformation tensor (Cauchy, 1828); $\mathbf{U}$ is called the right stretch tensor (Fig. 11.1c) and $\mathbf{V}$ is called the left stretch tensor (Fig. 11.1b) ("right" and "left" refer to the positions in Eq. 11.1.9). The Lagrangian (Green's) strain is obtained as

$$\boldsymbol{\varepsilon} = \tfrac{1}{2}(\mathbf{C} - \mathbf{I}) = \tfrac{1}{2}(\mathbf{F}^T\mathbf{F} - \mathbf{I}) = \tfrac{1}{2}(\mathbf{U}^2 - \mathbf{I}) \tag{11.1.13}$$

and from this, $\mathbf{U} = (\mathbf{I} + 2\boldsymbol{\varepsilon})^{1/2}$ ($\mathbf{I}$ is the unit tensor, which has components δ_{ij}).

The relation $\mathbf{F} = \mathbf{RU}$ describes exactly the sequence of transformations that are used to obtain the Biot strain tensor $\boldsymbol{\varepsilon}^b$ (Eq. 11.1.4) (and $U_{ij} = x'_{i,j} = \delta_{ij} + \varepsilon^b_{ij}$). Therefore

$$\boldsymbol{\varepsilon}^b = \mathbf{U} - \mathbf{I} \tag{11.1.14}$$

The last two equations suggest defining generalized Lagrangian finite strains:

$$\boldsymbol{\varepsilon}^{(m)} = \frac{1}{m}(\mathbf{U}^m - \mathbf{I}) \quad \text{for } m \neq 0$$
$$\boldsymbol{\varepsilon}^{(m)} = \ln \mathbf{U} \quad \text{for } m = 0 \tag{11.1.15}$$

as proposed by Doyle and Ericksen (1956). These tensors all represent acceptable strain measures, satisfying requirements I to V. Tensors $\boldsymbol{\varepsilon}^{(m)}$ have the principal values

$$\varepsilon_{(i)}^{(m)} = \frac{1}{m}(\lambda_{(i)}^m - 1) \quad \text{for } m \neq 0$$
$$\varepsilon_{(i)}^{(m)} = \ln \lambda_{(i)} \quad \text{for } m = 0 \tag{11.1.16}$$

Here $\lambda_{(i)} = \sqrt{\mu_{(i)}}$ = principal values of $\mathbf{U}$, called the principal stretches ($i = 1, 2, 3$), $\lambda_{(i)}$ geometrically represents the stretch ratio $|d\mathbf{x}|/|d\mathbf{X}|$ if vector $d\mathbf{X}$ lies in the principal direction $\mathbf{n}^{(i)}$ associated with $\lambda_{(i)}$ (e.g., Ogden, 1984). [Note that $\lim (\lambda^m - 1)/m$ for $m \to 0$ is $\ln \lambda$.] When m is not an integer, $\mathbf{U}^m$ (the mth power of tensor $\mathbf{U}$) is defined by the aforementioned principal values $\varepsilon_{(i)}^{(m)}$ and the unit principal direction vectors $\mathbf{n}^{(i)}$ of tensor $\mathbf{U}$ (or $\mathbf{C}$, $\boldsymbol{\varepsilon}$):

$$[\mathbf{U}^m]_{jk} = \sum_{i=1}^{3} \lambda_{(i)}^m n_j^{(i)} n_k^{(i)} \tag{11.1.17}$$

which constitutes the spectral representation of tensor $\mathbf{U}^m$. When m is an integer, $\mathbf{U}^2 = \mathbf{UU} = U_{ik}U_{kj}$, $\mathbf{U}^3 = \mathbf{U}(\mathbf{UU}) = U_{ik}U_{km}U_{mj}$, etc.; $\mathbf{U}^{1/n}$ is the solution $\mathbf{Y}$ of the equation $\mathbf{Y}^n = \mathbf{U}$. $\ln \mathbf{U}$ may also be defined by the power series expansion of the logarithm, and $\lambda_{(i)}^m$ in Equation 11.1.17 must in this case be replaced by $\ln \lambda_{(i)}^m$. Equation 11.1.17 shows that all the tensors $\mathbf{U}^{(m)}$, $\boldsymbol{\varepsilon}^{(m)}$ are coaxial (i.e., have the same principal directions). Thus, they also are coaxial with $\mathbf{C}$ and $\boldsymbol{\varepsilon}$.

If the X_1 axis is rotated to coincide with the principal direction $i = 1$, then $\lambda_1 = 1 + e_{11}$ (for finite strain). Thus $\varepsilon_{11}^{(m)} = (1/m)[(1 + e_{11})^m - 1]$ if $m \neq 0$ and $\varepsilon_{11}^{(m)} = \ln(1 + e_{11})$ if $m = 0$. Noting that $(1 + e_{11})^m = 1 + me_{11} + (m/2)(m-1)e_{11}^2 + \cdots$ and $\ln(1 + e_{11}) = e_{11} - \frac{1}{2}e_{11}^2 + \cdots$, we obtain the second-order approximation

$$\varepsilon_{11}^{(m)} = e_{11} + \tfrac{1}{2}(m-1)e_{11}^2 \tag{11.1.18}$$

Now, for comparison, Equation 11.1.7 yields the second-order approximation $\varepsilon_{11}^{(m)} = \varepsilon_{11} - \alpha e_{11}^2 = e_{11} + \frac{1}{2}(m-1)e_{11}^2$. On this basis it has been realized that the tensors $\varepsilon_{ij}^{(m)}$ in Equation 11.1.7 proposed by Bažant (1971) for the purpose of stability analysis represent second-order approximations to the finite strain tensors $\varepsilon_{ij}^{(m)}$ in Equations 11.1.15, proposed by Doyle and Ericksen (1956) for hyperelastic materials.

The finite strain tensor can also be defined with reference to the Eulerian coordinates x_i. A counterpart of Green's strain tensor ε_{ij} is then the Almansi (1911) (or Eulerian) strain tensor defined as $\alpha_{ij} = \frac{1}{2}(\mathbf{I} - \mathbf{BB}^t) = \frac{1}{2}(\mathbf{I} - \mathbf{V}^{-2})$ where $\mathbf{B}$ is the tensor of $\partial X_i/\partial x_j$. Analogs to all the tensors given in this section can be formulated. For the analysis of solids, however, the Lagrangian coordinates X_i of material points are usually more convenient. The Eulerian coordinates, on the other hand, are more suitable for fluids.

To sum up, infinitely many equally justifiable definitions of the finite strain tensor $\boldsymbol{\varepsilon}$ are possible. Therefore, it would be purely by chance if the stress–strain relation $\boldsymbol{\sigma}(\boldsymbol{\varepsilon})$ in finite strain were linear. The tangential moduli $\mathbf{C}^t = d\boldsymbol{\sigma}/d\boldsymbol{\varepsilon}$ are obviously different if different definitions of $\boldsymbol{\varepsilon}$ are used for the same material. The choice of $\boldsymbol{\varepsilon}$ depends only on convenience. In general, the Lagrangian strain tensor is the simplest to calculate at large strain and should perhaps be preferred for the sake of standardization.

The Special Case of Thin Bodies

Except for very large deflections, the deformations of thin plates and shells normally satisfy the following three hypotheses:

(1) Strains e_{ij} are negligible compared to out-of-plane rotations ω_{13}, ω_{23}, that is, $\max |e_{ij}| \ll \max(|\omega_{13}|, |\omega_{23}|)$.

(2) The in-plane rotation ω_{12} is of the same order of magnitude as e_{ij} and thus negligible compared to ω_{13} and ω_{23}.
(3) The out-of-plane shear and normal strains e_{13}, e_{23}, e_{33} are negligible compared to the in-plane strains e_{11}, e_{22}, e_{12}.

Because of hypothesis (1), $e_{ki}e_{kj}$ is negligible compared to $\omega_{ki}\omega_{kj}$, and so, from Equations 11.1.7 and 11.1.3, $\varepsilon_{ij}^{(m)} \simeq \varepsilon_{ij} \simeq e_{ij} + \frac{1}{2}(\omega_{ki}\omega_{kj} + e_{ki}\omega_{kj} + e_{kj}\omega_{ki})$. Now, from hypothesis (2) we find the terms $e_{12}\omega_{12}$ and $e_{21}\omega_{21}$ to be of the same order of magnitude as $e_{ki}e_{kj}$ and thus negligible compared to $\omega_{ki}\omega_{kj}$; and from hypothesis (3) we find $e_{31}\omega_{31}$ and $e_{32}\omega_{32}$ as well as $e_{31}\omega_{32}$, $e_{32}\omega_{31}$ to be negligible compared to $\omega_{ki}\omega_{kj}$. Thus, for the in-plane finite strain components in plates and shells we may write

$$\varepsilon_{ij} \simeq \varepsilon_{ij}^{(m)} \simeq e_{ij} + \tfrac{1}{2}\omega_{ki}\omega_{kj} \qquad (i = 1, 2; j = 1, 2; \text{all } m) \tag{11.1.19}$$

Decomposition of Strain into Elastic and Inelastic Parts

In large strains, one no longer can justify the customary summation decomposition $\varepsilon_{ij} = \varepsilon_{ij}^{e} + \varepsilon_{ij}^{p}$ where ε_{ij}^{e} = elastic strain and ε_{ij}^{p} = plastic (or other inelastic) strain. Assuming that the plastic strain increment occurs first (since it is treated as the initial strain) and the elastic one afterward, the transformation tensor $\mathbf{F}$ of the total displacement gradients $F_{ij} = u_{i,j}$ is properly decomposed in the form of polar decomposition $\mathbf{F} = \mathbf{F}^e\mathbf{F}^p$ where $\mathbf{F}^e$ and $\mathbf{F}^p$ are transformation tensors for the elastic strain and the plastic strain. The elastic and plastic strains may then be determined on the basis of $\mathbf{F}^e$ and $\mathbf{F}^p$. See, for example, Lee (1969), and some of the recent discussions in Simo and Ortiz (1985) and Simo (1986).

Problems

11.1.1 Why is $\log(\delta_{ij} + \varepsilon_{ij})$ unacceptable as the components of a strain measure? *Hint:* Is it a tensor? What about $\exp(\delta_{ij} + \varepsilon_{ij})$?

11.1.2 The tensor $\bar{\varepsilon}_{ij} = \varepsilon_{ij} - \varepsilon_{ik}\varepsilon_{kj}$ satisfies all the requirements for finite strain except that requirement IV is satisfied only in a limited range of strains. Find the maximum uniaxial strain ε_{11} beyond which the dependence of $\bar{\varepsilon}_{ij}$ on e_{ij} at $\omega_{ij} = 0$ is not invertible.

11.1.3 Let the x_3 axis be identified with the axis of rigid-body rotation $\omega = \omega_{12}$. Then $x_1 = X_1 \cos\omega - X_2 \sin\omega$, $x_2 = -X_1 \sin\omega + X_3 \cos\omega$. Calculate $u_1 = x_1 - X_1$, $u_2 = x_2 - X_2$ and set $\sin\omega \simeq \omega - \frac{1}{3}\omega^3$, $\cos\omega \simeq 1 - \frac{1}{2}\omega^2 + \frac{1}{24}\omega^4$. Check that Equation 11.1.5 is accurate only up to second order, and express the error up to the fourth order.

11.1.4 Explain why Equation 11.1.3 cannot be written in the standard matrix notation for strains? *Hint:* Can $u_{k,i}$ be written as a 6×1 column matrix?

11.1.5 Show that, for thin bodies, Equation 11.1.19 is a correct approximation of any tensor $\varepsilon_{ij}^{(m)}$, for any value α. (Thus, the distinctions between various types of finite strain tensors disappear for thin bodies.)

11.1.6 Given that $u_{1,1} = -0.3$, $u_{1,2} = 0.2$, $u_{2,1} = -0.1$, $u_{2,2} = 0.1$, all other $u_{i,j} = 0$, calculate all the principal values and unit principal direction vectors of tensors e_{ij}, ε_{ij}, $\mathbf{F}^T\mathbf{F}$ $(= \mathbf{U}^2)$, $\mathbf{C}$, $\mathbf{U}$, $\mathbf{R}$ $(= \mathbf{U}^{-1}\mathbf{F})$, $\mathbf{U}^{(m)}$, $\boldsymbol{\varepsilon}^{(m)}$ (for $m = 2, 1, 0, -2$). Then calculate the second-order approximations to $\boldsymbol{\varepsilon}^{(m)}$ (Eq. 11.1.5) as well as the third-order approximations. Then calculate for all these tensors the components with subscripts (1, 1), (1, 2), (2, 1), (1, 1) using the spectral representation of a tensor when convenient.

11.1.7 Do the same, but (a) for $u_{1,1}=-3$, $u_{1,2}=2$, $u_{2,1}=-1$, $u_{2,2}=1$; (b) for $u_{1,1}=-0.025$, $u_{1,2}=0.02$, $u_{2,1}=-0.01$, $u_{2,2}=0.01$ (in this case the second-order approximations will be very good).

11.1.8 By expressing $\mathbf{U}^2=[U_{jk}U_{km}]$ according to Equation 11.1.12 and noting that $n_k^{(i)}n_k^{(j)}=\delta_{ij}$, prove that $\mathbf{U}^2=\mathbf{F}^T\mathbf{F}$.

11.1.9 Prove Equation 11.1.13.

11.2 STRESSES, WORK, AND EQUILIBRIUM AT FINITE STRAIN

The finiteness of strain makes it necessary to distinguish between stresses referred to the initial and final configurations of the body, as well as between the true stresses and the stresses that are associated by work with various finite strain tensors.

Virtual Work Relations and Equilibrium

The initial state and the final (current) state of the body after incremental displacements u_i are assumed to be equilibrium states. The condition of equilibrium in the final state may be expressed in terms of the principle of virtual work, which can be written in three ways:

$$\int_{V'} S_{ij}\frac{\partial \delta u_i}{\partial x_j}\,dV' - \int_{V'} \rho' f_i \delta u_i\,dV' - \int_{S'} p_i' \delta u_i\,dS' = 0 \tag{11.2.1}$$

$$\int_V T_{ij}\frac{\partial \delta u_i}{\partial X_j}\,dV - \int_V \rho f_i \delta u_i\,dV - \int_S p_i \delta u_i\,dS = 0 \tag{11.2.2}$$

$$\int_V \Sigma_{ij}^{(m)}\delta\varepsilon_{ij}^{(m)}\,dV - \int_V \rho f_i \delta u_i\,dV - \int_S p_i \delta u_i\,dS = 0 \tag{11.2.3}$$

Here δu_i = arbitrary kinematically admissible variations of u_i, $\delta\varepsilon_{ij}$ = associated finite strain variations; V, S, V', S' = volume and surface of the body in its initial (stressed) configuration and final (current) configuration; ρ, ρ' = mass densities in the initial and final configurations; f_i = prescribed body forces per unit mass; p_i, p_i' = prescribed distributed surface loads (or, tractions) in the final state that are referred to the initial or final surface areas, respectively; S_{ij}, T_{ij}, and $\Sigma_{ij}^{(m)}$ = various types of stresses in the final state, which we are going to discuss in detail.

The first equation expresses the work on the basis of the final (current) configuration. The second and third equations express the work on the basis of the initial configuration. In the last equation the work is expressed in terms of the (symmetric) finite strain tensor, while in the first and second equations the work is expressed in terms of the (nonsymmetric) displacement gradients $\partial u_i/\partial x_j$ and $\partial u_i/\partial X_j$ with respect to the final coordinates x_i and the initial coordinates X_i.

The integrand of the first integral in Equation 11.2.2 or 11.2.3 represents the work per unit initial volume, that is, $\delta\mathscr{W}=T_{ij}\delta u_{i,j}=\Sigma_{ij}^{(m)}\delta\varepsilon_{ij}^{(m)}$. According to

Section 10.1, $\bar{\mathscr{W}}$ is the Helmholtz free energy F per unit initial volume if the conditions are isothermal, and the total energy U per unit volume if the conditions are isentropic (adiabatic). It follows that

$$T_{ij} = \frac{\partial \bar{\mathscr{W}}}{\partial u_{i,j}} \qquad \Sigma_{ij}^{(m)} = \frac{\partial \bar{\mathscr{W}}}{\partial \varepsilon_{ij}^{(m)}} \tag{11.2.4}$$

Note, however, that $\bar{\mathscr{W}}$ is not path-independent (see Sec. 10.1); it does not represent a potential, except if the material is elastic in finite strain (i.e., hyperelastic).

Applying the Gauss integral theorem to the first volume integral in Equation 11.2.1 as well as Equation 11.2.2, we may get rid of the derivatives of δu_i in these equations and obtain

$$\int_{S'} (p'_i - n_j S_{ij}) \delta u_i \, dS' + \int_{V'} \left(\frac{\partial S_{ij}}{\partial x_j} + \rho' f_i \right) \delta u_i \, dV' = 0 \tag{11.2.5}$$

$$\int_{S} (p_i - \nu_j T_{ij}) \delta u_i \, dS + \int_{V} \left(\frac{\partial T_{ij}}{\partial X_j} + \rho f_i \right) \delta u_i \, dV = 0 \tag{11.2.6}$$

in which ν_j and n_j are the unit outward normals of the surfaces S and S'. Since these variational equations must be satisfied for any variation δu_i, it is necessary (according to the fundamental lemma of the calculus of variations, see Sec. 5.1) that

$$\frac{\partial S_{ij}}{\partial x_j} + \rho' f_i = 0 \quad \text{in } V' \qquad S_{ij} n_j = p'_i \quad \text{on } S' \tag{11.2.7}$$

$$\frac{\partial T_{ij}}{\partial X_j} + \rho f_i = 0 \quad \text{in } V \qquad T_{ij} \nu_j = p_i \quad \text{on } S \tag{11.2.8}$$

These are the differential equations of equilibrium and the static boundary conditions of the final state based on the final and initial configurations.

True (Cauchy) Stress

Tensor S_{ij} is the actual stress tensor that is referred to the current (final) configuration and works on the strain rate; see Equation 11.2.11 and Figure 11.2. It represents the forces in the x_i-directions on a small unit cube that is cut out from the body in its final configuration (i.e., after the incremental deformation). S_{ij} is called the true stress or Cauchy stress (Cauchy, 1828; Malvern, 1969; Ogden, 1984); it has also been called the Eulerian stress (Prager 1961), since the x_i coordinates are called the Eulerian coordinates. Tensor S_{ij} must be symmetric (this follows from the equilibrium conditions of a small material element, as well as the fact that asymmetric S_{ij} would, in Eq. 11.2.1, yield nonzero work for rigid-body displacements). So we have $S_{ij}\, \partial(\delta u_i / \partial x_j) = S_{ij} \delta d_{ij}$ were

$$\delta d_{ij} = \frac{1}{2} \left(\frac{\partial \delta u_i}{\partial x_j} + \frac{\partial \delta u_j}{\partial x_i} \right) \tag{11.2.9}$$

d_{ij} represents a symmetric strain tensor that is linear in displacement gradients,

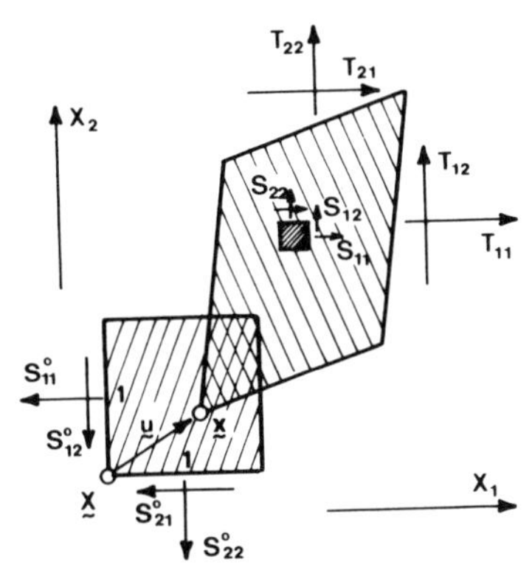

Figure 11.2 Two-dimensional representation of Cauchy stress tensor S_{ij} and Piola–Kirchhoff stress tensor T_{ij}.

same as e_{ij}. It is referred to the current (final) configuration. By contrast, $e_{ij} = \frac{1}{2}(\partial u_i/\partial X_j + \partial u_j/\partial X_i)$ is referred to the initial configuration. Equation 11.2.1 may also be written as

$$\int_{V'} S_{ij}\delta d_{ij}\, dV' - \int_{V'} \rho' f_i \delta u_i\, dV - \int_{S'} p_i \delta u_i\, dS' = 0 \tag{11.2.10}$$

Even though the first virtual work relation in Equation 11.2.1 and the equivalent relation in Equation 11.2.10 are the simplest ones and involve a symmetric stress tensor, they are not convenient to treat the finite strain of solids. The reason is that the tensor δd_{ij} in Equation 11.2.10 does not take into account the initial configuration, which is important for solids.

The foregoing equations can be divided by δt where $t =$ time-like parameter, not necessarily the real time. From δu_i one gets $\delta u_i/\delta t = \dot{u}_i = v_i =$ velocities of material points; $\partial \delta u_i/\partial x_j$, $\partial \delta u_i/\delta X_j$, and $\delta \varepsilon_{ij}$ become $\partial \dot{u}_i/\partial x_j$, $\partial \dot{u}_i/\partial X_j$, and $\dot{\varepsilon}_{ij}$. Strain d_{ij} yields the strain-rate tensor

$$\frac{\delta d_{ij}}{\delta t} = \frac{1}{2}\left(\frac{\partial v_i}{\partial x_j} + \frac{\partial v_j}{\partial x_i}\right) \tag{11.2.11}$$

Stress Referred to Initial Configuration and Working on Displacement Gradient

As is clear from Equation 11.2.2, stress T_{ij} is referred to the initial configuration and is associated by work with the displacement gradient. According to Equations 11.2.2 and 11.2.6, the stress tensor T_{ij} represents the forces acting in the final configuration in X_i-directions on a deformed material element that was in the initial configuration a unit cube; see Figure 11.2. This tensor is nonsymmetric. Indeed there is no need for T_{ij} to be symmetric since $\partial \delta u_i/\partial X_j$ is a nonsymmetric tensor. Tensor T_{ij} is usually called the first Piola–Kirchhoff stress tensor (Piola, 1835; Kirchhoff, 1852; Truesdell and Noll, 1965; Malvern, 1969) or the nominal stress tensor (Ogden, 1984); it has also been called the Lagrangian stress tensor (Prager, 1961) or Boussinesq's stress tensor (Mandel, 1966), or the mixed stress tensor ("mixed," because T_{ij} characterizes the forces that act on a deformed cubic element in the final configuration of the x_i coordinates but are referred to the facet areas of an undeformed element in the initial configuration of the X_i coordinates). In most texts, the initial configuration to which T_{ij} is referred is considered unstressed, but to deal with stability we must consider a stressed initial configuration.

To establish the relationship of T_{ij} and S_{ij}, we transform the integral over V in Equation 11.2.2 to an integral over V', that is,

$$\int_V T_{ik} \frac{\partial \delta u_i}{\partial X_k} dV = \int_{V'} T_{ik} \frac{\partial \delta u_i}{\partial x_j} \left(\frac{\partial x_j}{\partial X_k} \right) J^{-1} dV' \tag{11.2.12}$$

in which $J = dV/dV'$ = Jacobian of the transformation;

$$J = \det \frac{\partial x_i}{\partial X_j} \qquad J^{-1} = \det \frac{\partial X_i}{\partial x_j} \tag{11.2.13}$$

Now comparing Equation 11.2.12 to the first volume integral in Equation 11.2.1 (and noting that $x_{j,k} = \delta_{jk} + u_{j,k}$) we conclude that

$$S_{ij} = \frac{1}{J} \frac{\partial x_j}{\partial X_k} T_{ik} = \frac{1}{J} (T_{ij} + T_{ik} u_{j,k}) \tag{11.2.14}$$

Multiplying Equation 11.2.14 by $\partial X_m / \partial x_j$ and noting that $(\partial X_m / \partial x_j)(\partial x_j / \partial X_k) = \delta_{mk}$, we obtain the inverse relation

$$T_{im} = \frac{\partial X_m}{\partial x_j} J S_{ij} = \frac{\partial X_m}{\partial x_j} \bar{S}_{ij} \qquad \text{or} \qquad \mathbf{T} = J\mathbf{S}\mathbf{F}^{-T} = \bar{\mathbf{S}}\mathbf{F}^{-T} \tag{11.2.15}$$

in which $\partial X_m / \partial x_j = \partial (x_m - u_m)/\partial x_j = \delta_{jm} - \partial u_m / \partial x_j = (\mathbf{F}^{-1})_{mj}$, $\mathbf{F}^{-T}$ denotes $(\mathbf{F}^{-1})^T$, and $\bar{S}_{ij} = J S_{ij}$; $\bar{S}_{ij}$ is called the Kirchhoff stress. The same equation results when the integral over V' in Equation 11.2.1 is transformed (similarly to Eq. 11.2.12) to an integral over V and is compared to the first integral in Equation 11.2.2.

Note that the relationship between T_{ij} and S_{ij} is independent of the choice of the finite strain tensor, which we discussed in the previous section.

Stress Referred to Initial Configuration and Working on Finite Strain

Consider now stress $\Sigma_{ij}^{(m)}$, which is referred to the initial configuration and is conjugated by work with the finite strain tensor $\varepsilon_{ij}^{(m)}$. To derive its relationship to T_{ij}, we may observe that the integrands of the first integrals in Equations 11.2.2 and 11.2.3 must be identical, that is,

$$T_{kj} \delta u_{k,j} = \Sigma_{jn}^{(m)} \delta \varepsilon_{jn}^{(m)} \tag{11.2.16}$$

For the Lagrangian strain ($m = 2$), we drop the labels (m) and we have

$$\Sigma_{jn} \delta \varepsilon_{jn} = \Sigma_{jn} \tfrac{1}{2} \delta (x_{k,n} x_{k,j} - \delta_{jn}) = \Sigma_{jn} \tfrac{1}{2} (x_{k,n} \delta x_{k,j} + x_{k,j} \delta x_{k,n})$$
$$= \Sigma_{jn} x_{k,n} \delta x_{k,j} = \Sigma_{jn} x_{k,n} \delta u_{k,j}$$

because tensor Σ_{jn} is symmetric (symmetry of Σ_{ij} is required because, in Eq. 11.2.3, any nonsymmetric part of Σ_{ij} would always contribute no work, i.e., would be irrelevant, indeterminate). Also, $\delta x_{k,j} = \delta(X_k + u_k)_{,j} = \delta(\delta_{kj} + u_{k,j}) = \delta u_{k,j}$. Thus Equation 11.2.16 becomes $(T_{kj} - \Sigma_{jn} x_{k,n})\delta u_{k,j} = 0$. Since this equation must hold for any variation $\delta u_{k,j}$, it follows that

$$T_{kj} = \frac{\partial x_k}{\partial X_n} \Sigma_{jn} \qquad \text{or} \qquad \mathbf{T} = \mathbf{F}\boldsymbol{\Sigma} \qquad (m = 2) \tag{11.2.17}$$

where $\partial x_k / \partial X_n = x_{k,n} = F_{kn}$. Furthermore, multiplying this equation by $\partial X_i / \partial x_k$ and noting that $\Sigma_{jn}(\partial X_i / \partial x_k)(\partial x_k / \partial X_n) = \Sigma_{jn}\delta_{ni} = \Sigma_{ij}$, we obtain the inverse relation

$$\Sigma_{ij} = \frac{\partial X_i}{\partial x_k} T_{kj} \qquad \text{or} \qquad \boldsymbol{\Sigma} = \mathbf{F}^{-1}\mathbf{T} \qquad (m = 2) \tag{11.2.18}$$

where $\mathbf{F}^{-1}$ is a tensor of components $\partial X_i / \partial x_j$. Tensor $\Sigma_{ij} = \Sigma_{ij}^{(2)}$ is called the second Piola–Kirchhoff stress (Piola, 1835; Kirchhoff, 1852; Truesdell and Noll, 1965) or simply the Piola–Kirchhoff stress (Ogden, 1984). Substitution of Equation 11.2.17 into Equation 11.2.14 provides

$$S_{ij} = \frac{1}{J} \frac{\partial x_i}{\partial X_k} \frac{\partial x_j}{\partial X_m} \Sigma_{km} \qquad \text{or} \qquad \mathbf{S} = \frac{1}{J} \mathbf{F}\boldsymbol{\Sigma}\mathbf{F}^T \qquad (m = 2) \tag{11.2.19}$$

(see also, e.g., Ogden, 1984; Hill, 1968; Truesdell and Noll, 1965). Using $x_{i,j} = \delta_{ij} + u_{i,j}$, one further obtains $S_{ik} = J^{-1}(\Sigma_{ik} + u_{i,j}\Sigma_{jk} + u_{k,j}\Sigma_{ji} + u_{i,j}u_{k,n}\Sigma_{jn})$.

Substitution of Equation 11.2.15 into Equation 11.2.18 provides the inverse relation

$$\Sigma_{ij} = \frac{\partial X_i}{\partial x_k} \frac{\partial X_j}{\partial x_m} J S_{km} \qquad \text{or} \qquad \boldsymbol{\Sigma} = \mathbf{F}^{-1}\bar{\mathbf{S}}\mathbf{F}^{-T} \qquad (m = 2) \tag{11.2.20}$$

To obtain the stress tensor $\Sigma_{ij}^{(1)}$ that is work-conjugate to the Biot strain $\varepsilon_{ij}^b = \varepsilon_{ij}^{(1)}$, we note from Equation 11.1.13 that the Green's (Lagrangian) strain variation may be expressed as $\delta\boldsymbol{\varepsilon} = \frac{1}{2}\delta(\mathbf{U}^2)$ where $\mathbf{U} = (\mathbf{F}^T\mathbf{F})^{1/2}$ = right stretch tensor (Eq. 11.1.9) and $\mathbf{F} = \partial\mathbf{x}/\partial\mathbf{X}$ or $F_{ij} = x_{i,j}$. Consequently, $\delta\boldsymbol{\varepsilon} = \frac{1}{2}(\mathbf{U}\delta\mathbf{U} + \delta\mathbf{U}\mathbf{U})$, and $\delta\bar{\mathscr{W}} = \Sigma_{kj}\delta\varepsilon_{kj} = \frac{1}{2}(\Sigma_{kj}U_{ki}\delta U_{ij} + \Sigma_{kj}\delta U_{ki}U_{ij}) = \frac{1}{2}(U_{ik}\Sigma_{kj} + \Sigma_{ik}U_{kj})\delta U_{ij}$ where $\Sigma_{ij} = \Sigma_{ij}^{(2)}$. From Equation 11.1.14, $\delta U_{ij} = \delta\varepsilon_{ij}^b$. The stress tensor $\Sigma_{ij}^{(1)}$, called the Biot stress, is defined by $\delta\bar{\mathscr{W}} = \Sigma_{ij}^{(1)}\delta\varepsilon_{ij}^b = \Sigma_{ij}^{(1)}\delta U_{ij}$. Since the foregoing two expressions for $\delta\bar{W}$ must be equivalent for any δU_{ij}, it follows that

$$\Sigma_{ij}^{(1)} = \tfrac{1}{2}(U_{ik}\Sigma_{kj} + \Sigma_{ik}U_{kj}) \qquad \text{or} \qquad \boldsymbol{\Sigma}^{(1)} = \tfrac{1}{2}(\mathbf{U}\boldsymbol{\Sigma} + \boldsymbol{\Sigma}\mathbf{U}) \qquad (m = 1) \tag{11.2.21}$$

For any m, the stress tensors $\Sigma_{ij}^{(m)}$ that are conjugate to $\varepsilon_{ij}^{(m)}$ are related to Σ_{ij} and U_{ij} by more complicated implicit equations (Ogden, 1984, p. 158). However, their second-order approximations are calculated easily (Bažant, 1971), as we show in the next section.

Finally, it should be emphasized that the stress referred to the initial configuration and the finite strain involved in the constitutive relation must represent a *conjugate* pair. If the Lagrangian strain ε_{ij} is chosen, the stress–strain relations must be written in terms of Σ_{ij}, and not T_{ij}, S_{ij}, or $\Sigma_{ij}^{(1)}$. If Biot strain $\varepsilon_{ij}^{(1)}$ is chosen, they must be written in terms of $\Sigma_{ij}^{(1)}$, not Σ_{ij}, T_{ij}, or S_{ij}. Furthermore, if the material is elastic in finite strain (i.e., hyperelastic), only the use of a conjugate pair of stress $\Sigma_{ij}^{(m)}$ and finite strain $\varepsilon_{ij}^{(m)}$ satisfies the condition that $\Sigma_{ij}^{(m)} = \partial\bar{\mathscr{W}}/\partial\varepsilon_{ij}^{(m)}$ where $\bar{\mathscr{W}}$ = strain energy per unit initial volume.

Problems

11.2.1 Using tensor **F** whose components are $F_{ij} = x_{i,j}$, rewrite all the relations of this section and reproduce the derivations in tensor notation.

11.2.2 Deduce the second-order accurate relationships between $\Sigma_{ij}^{(1)}$ and T_{ij} and between $\Sigma_{ij}^{(1)}$ and S_{ij} when the second-order accurate Biot strain ε_{ij}^{b} (Eq. 11.1.5) is used.

11.2.3 Do the same for logarithmic (Hencky) strain $\varepsilon_{ij}^{(0)}$ ($m = 0$ or $\alpha = 1$).

11.2.4 Assume that $S_{11}^0 = -5000\ \text{Pa}$, $S_{12}^0 = S_{21}^0 = 1000\ \text{Pa}$, $S_{22}^0 = -1000\ \text{Pa}$, all other $S_{ij}^0 = 0$, and use the results of Problems 11.1.6 to 11.1.8. Calculate the components of $\Sigma_{ij}^{(2)}$ and $\Sigma_{ij}^{(1)}$, their principal values, and the unit principal direction vectors. (Use the spectral representations for **U**.)

11.3 INCREMENTAL EQUILIBRIUM AND OBJECTIVE STRESS RATES

For calculating the critical states and analyzing stability, the finite strain tensors need to be only second-order accurate in $u_{i,j}$. Then the Lagrangian (Green's) strain tensor no longer has any particular advantage in simplicity and other finite strain measures discussed in Section 8.1 can be used just as well without causing any increase in complexity. In this section we will use the second-order finite strain approximations to formulate the incremental equilibrium conditions of three-dimensional initially stressed bodies and then will proceed to determine the corresponding objective stress increments and rates.

Incremental Equilibrium Conditions

The stress increments s_{ij}, τ_{ij}, and $\sigma_{ij}^{(m)}$ with respect to the Cauchy stress S_{ij}^0 (true stress) in the initial state may be defined by the relations:

$$S_{ij} = S_{ij}^0 + s_{ij} \qquad T_{ij} = S_{ij}^0 + \tau_{ij} \qquad \Sigma_{ij}^{(m)} = S_{ij}^0 + \sigma_{ij}^{(m)} \tag{11.3.1}$$

The fact that the initial state is an equilibrium state is expressed by the virtual work relation

$$\int_V S_{ij}^0 \frac{\partial \delta u_i}{\partial X_j}\, dV - \int_V \rho f_i^0 \delta u_i\, dV - \int_S p_i^0 \delta u_i\, dS = 0 \tag{11.3.2}$$

Subtracting this equation from Equation 11.2.2, we obtain

$$\int_V \tau_{ij} \delta u_{i,j}\, dV - \int_V \rho \bar{f}_i \delta u_i\, dV - \int_S \bar{p}_i \delta u_i\, dS = 0 \tag{11.3.3}$$

in which $\delta u_{i,j} = \partial \delta u_i / \partial X_j$; $\bar{f}_i = f_i - f_i^0 =$ increment of body forces per unit mass, and $\bar{p}_i = p_i - p_i^0 =$ increment of given surface forces (loads, tractions) per unit initial area. Writing $\tau_{ij}\delta u_{i,j} = (\tau_{ij}\delta u_i)_{,j} - \tau_{ij,j}\delta u_i$ and applying the Gauss integral theorem to the volume integral of the first term, we get a variational equation from which the following incremental differential equilibrium conditions and incremental boundary conditions result:

$$\begin{aligned} &\tau_{ij,j} + \rho \bar{f}_i = 0 && \text{(in volume } V\text{)} \\ &\tau_{ij} \nu_j = \bar{p}_i && \text{(on the stress boundary part of } S\text{)} \end{aligned} \tag{11.3.4}$$

Note that the incremental equations are valid irrespective of the material properties and of the choice of the finite strain measure.

Increments of Cauchy (True) Stresses

If the displacement gradients are small, the calculation of $J = \det x_{i,j} = \det(\delta_{ij} + u_{i,j})$ and neglection of all the second- and third-order terms yields the approximations

$$J \simeq 1 + u_{k,k} \qquad J^{-1} \simeq 1 - u_{k,k} \tag{11.3.5}$$

where $u_{k,k} = e_{kk} = dV/dV_0$ = relative volume change of material element. We substitute this into Equation 11.2.14, set $T_{ik} = S^0_{ik} + \tau_{ik}$, $S_{ij} = S^0_{ij} + s_{ij}$, $\partial x_j/\partial X_k = \delta_{jk} + u_{j,k}$, and neglect terms of higher than first order in $u_{i,j}$. Considering τ_{ij} to be linearly related to $u_{i,j}$, τ_{ij} are small if $u_{i,j}$ are small. Thus we get for the true stress increments

$$s_{ij} = \tau_{ij} + S^0_{ik}u_{j,k} - S^0_{ij}u_{k,k} \tag{11.3.6}$$

which is accurate up to the first-order terms in $u_{i,j}$ (s_{ij} and τ_{ij} are small compared to S^0_{ij}). This relation is, of course, independent of the choice of finite strain measure.

Objective Stress Increments Conjugate to Strain Increments

To relate σ_{ij} (or Σ_{ij}) to τ_{ij} and T_{ij}, we consider the two different expressions (in Eqs. 11.2.2 and 11.2.3) for the work variation $\delta\bar{\mathscr{W}}$ (per unit initial volume):

$$\delta\bar{\mathscr{W}} = T_{ij}\delta u_{i,j} = (S^0_{ij} + \tau_{ij})\delta u_{i,j} \tag{11.3.7}$$

$$\delta\bar{\mathscr{W}} = \Sigma^{(m)}_{ij}\delta\varepsilon^{(m)}_{ij} = (S^0_{ij} + \sigma^{(m)}_{ij})\delta\varepsilon^{(m)}_{ij} \tag{11.3.8}$$

where we admit the possibility of various finite strain measures $\varepsilon^{(m)}_{ij}$ and attach superscript m to $\sigma^{(m)}_{ij}$ because the meaning of σ_{ij} depends on the choice of $\varepsilon^{(m)}_{ij}$, as we will see. Tensors $\varepsilon^{(m)}_{ij}$ are symmetric, and so is S^0_{ij}. Therefore, the tensors $\Sigma^{(m)}_{ij}$ and $\sigma^{(m)}_{ij}$, too, must be symmetric (their nonsymmetric part would be arbitrary, doing no work). Tensor $\delta u_{i,j}$ is nonsymmetric, and so there is no reason for τ_{ij} to be symmetric.

The tensor s_{ij} characterizes the force increments on a material element cut from the deformed material in the final configuration, which is a different material element than that cut in the initial configuration and subsequently deformed. Therefore s_{ij} cannot be used to calculate work on the initially cut material element as it deforms. On the other hand, both τ_{ij} and σ_{ij} characterize the force increments acting in the final configuration on the deformed material element that was cut in the initial configuration. Of these two, however, only σ_{ij} is symmetric. Therefore it is only tensor $\sigma^{(m)}_{ij}$, rather than τ_{ij} or s_{ij}, whose product with strain gives the correct work expression. The value of $\sigma^{(m)}_{ij}$ is objective in the sense that it is invariant with respect to the coordinate transformations associated with the deformation (called observer transformations; see, e.g., Malvern, 1969).

To derive the relationship between $\sigma^{(m)}_{ij}$ and τ_{ij}, we subtract Equation 11.3.7 from Equation 11.3.8. This yields $S^0_{ij}(\delta\varepsilon^{(m)}_{ij} - \delta u_{i,j}) + \sigma^{(m)}_{ij}\delta\varepsilon^{(m)}_{ij} - \tau_{ij}\delta u_{i,j} = 0$ where we now admit the possibility of any finite strain measure $\varepsilon^{(m)}_{ij}$, not just the

Lagrangian strain ε_{ij}. Noting that $S^0_{ij}\delta u_{i,j} = S^0_{ij}\delta e_{ij}$ (as S^0_{ij} is symmetric), and that $\sigma^{(m)}_{ij}\delta\varepsilon^{(m)}_{ij} \simeq \sigma^{(m)}_{ij}\delta e_{ij} = \sigma^{(m)}_{ij}\delta u_{i,j}$ (as σ_{ij}, ε_{ij}, $u_{i,j}$ are small, $\sigma^{(m)}_{ij}\delta\varepsilon^{(m)}_{ij}$ is second-order small, and only second-order accuracy is required), we obtain

$$\left[\sigma^{(m)}_{ij} + S^0_{pq}\frac{\partial(\varepsilon^{(m)}_{pq} - e_{pq})}{\partial u_{i,j}} - \tau_{ij}\right]\delta u_{i,j} = 0 \tag{11.3.9}$$

This must hold for any variation $\delta u_{i,j}$, and so we obtain the following general relation (Bažant, 1967, 1971):

$$\tau_{ij} = \sigma^{(m)}_{ij} + S^0_{pq}\frac{\partial(\varepsilon^{(m)}_{pq} - e_{pq})}{\partial u_{i,j}} \tag{11.3.10}$$

For small incremental deformations, $u_{i,j}$, $\sigma^{(m)}_{ij}$, and τ_{ij} are first-order small while S^0_{ij} is finite. Therefore, it suffices for the finite strain expression in Equation 11.3.10 to be accurate only up to the second-order terms in $u_{i,j}$.

An important property to notice is that, unlike the relationship between s_{ij} and τ_{ij}, the relationship between the objective stress increment and the incremental first Piola–Kirchhoff (mixed) stress τ_{ij} does depend on the choice of strain tensor, $\varepsilon^{(m)}_{ij}$, for which infinitely many possibilities exist, as we know from Section 11.1. Let us now consider some of these possibilities.

Substituting $\varepsilon^{(m)}_{ij} = \varepsilon_{ij}$ = Lagrangian (Lagrange–Green) finite strain tensor ($m = 2$, Eq. 11.1.3), and noting that $\partial(u_{m,p}u_{m,q})/\partial u_{i,j} = \delta_{mi}\delta_{pj}u_{m,q} + \delta_{mi}\delta_{qj}u_{m,p} = \delta_{pj}u_{i,q} + \delta_{qj}u_{i,p}$, $\partial e_{mp}/\partial u_{i,j} = \frac{1}{2}(\partial u_{m,p}/\partial u_{i,j} + \partial u_{p,m}/\partial u_{i,j}) = \delta_{mi}\delta_{pj} + \delta_{pi}\delta_{mj}$, etc., we obtain from Equation 11.3.10

$$\tau_{ij} = \sigma_{ij} + S^0_{kj}u_{i,k} \tag{11.3.11}$$

Here the lack of superscript (m) at σ_{ij} labels the objective stress increment associated with the Lagrangian strain ($m = 2$) according to Equation 11.1.3 (i.e., $\sigma_{ij} = \sigma^{(2)}_{ij}$).

Substituting Biot's finite strain ($\varepsilon^{(1)}_{ij} = \varepsilon^b_{ij}$) (Eq. 11.1.5 or 11.1.7 with $m = 1$), referred to by superscript (1), Equation 11.3.10 yields

$$\tau_{ij} = \sigma^{(1)}_{ij} + S^0_{kj}u_{i,k} - \tfrac{1}{4}S^0_{pq}\,\partial(e_{kp}e_{kq})/\partial u_{i,j} = \sigma^{(0)}_{ij} + S^0_{kj}u_{i,k} - \tfrac{1}{2}S^0_{pq}e_{pq}(\delta_{ki}\delta_{pj} + \delta_{pi}\delta_{kj}),$$

that is,

$$\tau_{ij} = \sigma^{(1)}_{ij} + S^0_{kj}u_{i,k} - \tfrac{1}{2}(S^0_{ik}e_{kj} + S^0_{jk}e_{ki}) \tag{11.3.12}$$

Another case of interest is to substitute $\varepsilon^{(0)}_{ij}$ (Eq. 11.1.7 for $m = 0$ or $\alpha = 1$). This corresponds to the logarithmic strain, as already remarked. We have $\varepsilon^{(0)}_{ij} = \varepsilon_{ij} - e_{ki}e_{kj}$, and for this choice Equation 11.3.10 yields

$$\begin{aligned}\tau_{ij} &= \sigma^{(0)}_{ij} + S^0_{kj}u_{i,k} - \tfrac{1}{2}S^0_{pq}\,\partial(e_{kp}e_{kq})/\partial u_{i,j}\\ &= \sigma^{(0)}_{ij} + S^{(0)}_{kj}u_{i,k} - S^{(0)}_{pq}e_{pq}(\delta_{ki}\delta_{pj} + \delta_{pi}\delta_{kj})\\ &= \sigma^{(0)}_{ij} + S^0_{kj}u_{i,k} - S^0_{jq}e_{iq} - S^0_{iq}e_{jq},\end{aligned}$$

that is,

$$\tau_{ij} = \sigma^{(0)}_{ij} + S^0_{kj}\omega_{ik} - S^0_{ik}e_{kj} \tag{11.3.13}$$

Substitution of Equations 11.3.11 and 11.3.12 into Equation 11.3.4 yields the incremental differential equations of equilibrium that were proposed, on the basis of various geometric considerations, by Biezeno and Hencky (1928). They were apparently the first to present one of the correct statements of these equations.

Southwell (1914) and Neuber (1943, 1952) presented incremental equilibrium conditions corresponding to

$$\tau_{ij} = \sigma_{ij}^{(0)} + S_{kj}^0 \omega_{ik} - S_{ik}^0 e_{kj} + S_{ij}^0 e_{kk} \tag{11.3.14}$$

in which the last term differs from Equation 11.3.13. For the case that S_{ij}^0 is uniform, Equation 11.3.13 corresponds to the differential equations proposed by Neuber (1943, 1952). Biot (1965), too, used Equation 11.3.13 (aside from Eq. 11.3.2). Except if the material is incompressible, the last term of Eq. 11.3.14 is incorrect since it does not give correct work in the expression $\delta^2 \bar{\mathscr{W}} = \frac{1}{2}\tau_{ij}\delta u_{i,j}$ (for any definition of $\sigma_{ij}^{(0)} = C_{ijkm}\delta u_{k,m}$; see Sec. 11.4).

If we substitute the general expression, $\varepsilon_{pq}^{(m)} = \varepsilon_{pq} - \alpha e_{kp} e_{kq}$ (Eq. 11.1.7) in Equation 11.3.10, we need to calculate

$$\begin{aligned}\frac{\partial(e_{kp}e_{kq})}{\partial u_{i,j}} &= \frac{\partial(e_{kp}e_{kq})}{\partial e_{mn}}\left[\frac{\frac{1}{2}\partial(u_{m,n}+u_{n,m})}{\partial u_{i,j}}\right] \\ &= \tfrac{1}{2}(\delta_{km}\delta_{pn}e_{kq} + \delta_{km}\delta_{qn}e_{kp})(\delta_{mi}\delta_{nj} + \delta_{ni}\delta_{mj}) \\ &= \tfrac{1}{2}(e_{ip}\delta_{jq} + e_{iq}\delta_{jp} + e_{jp}\delta_{iq} + e_{jq}\delta_{ip})\end{aligned} \tag{11.3.15}$$

(which represents the symmetric part of the fourth-order tensor $\mathbf{e} \otimes \boldsymbol{\delta}$). Equation 11.3.10 then yields the general relation

$$\tau_{ij} = \sigma_{ij}^{(m)} + S_{kj}^0 u_{i,k} - \left(1 - \frac{m}{2}\right)(S_{ik}^0 e_{kj} + S_{jk}^0 e_{ki}) \tag{11.3.16}$$

from which Equations 11.3.12 and 11.3.13 result by setting $m = 1$ and $m = 0$.

Objective Stress Rates

If the deformation increment is associated with time interval δt, then $\lim(\sigma_{ij}^{(m)}/\delta t)$ for $\delta t \to 0$ represents an objective stress rate, denoted as $\hat{S}_{ij}^{(m)}$. On the other hand, $\lim(s_{ij}/\delta t) = \dot{S}_{ij}$ = material rate of stress, which represents the rate of change of the Cauchy stress in a material element as it deforms (in Lagrangian coordinates X_i, the material time derivative is simply $\partial/\partial t$, while in Eulerian coordinates it is $\partial/\partial t + v_k\, \partial/\partial x_k$).

The strain rate associated with $\hat{S}_{ij}^{(m)}$ is $\lim(\varepsilon_{ij}^{(m)}/\delta t)$. However, in the limit $\delta t \to 0$, all the higher-order terms in the strain increment vanish, for example, $\lim(u_{k,i}u_{k,j}/\delta t) = \lim(v_{k,i}\delta t v_{k,j}\delta t/\delta t) = \lim(v_{k,i}v_{k,j}\delta t) = 0$. So the strain rate associated with $\hat{S}_{ij}^{(m)}$ is simply $\dot{e}_{ij} = \frac{1}{2}(v_{i,j} + v_{j,i})$, regardless of which finite strain measure is used in the derivation of $\hat{S}_{ij}^{(m)}$. Therefore, the second-order work increment that governs stability (per unit initial volume) is

$$\delta^2 \bar{\mathscr{W}} = \tfrac{1}{2}\hat{S}_{ij}^{(m)}\dot{e}_{ij}\delta t^2 \tag{11.3.17}$$

The expressions $\frac{1}{2}\dot{S}_{ij}\dot{e}_{ij}\delta t^2$ or $\frac{1}{2}\dot{T}_{ij}\dot{e}_{ij}\delta t^2$ do not represent the second-order work increments.

Dividing Equations 11.3.6 and 11.3.10 by δt, and considering the limit $\delta t \to 0$, one gets for the objective stress rate $\hat{S}_{ij}^{(m)}$ associated with $\varepsilon_{ij}^{(m)}$ the general expressions

$$\hat{S}_{ij}^{(m)} = \dot{T}_{ij} - S_{pq}\frac{\partial^2(\varepsilon_{pq}^{(m)} - e_{pq})}{\partial u_{i,j}\,\partial t} \qquad \dot{T}_{ij} = \dot{S}_{ij} - S_{ik}v_{j,k} + S_{ij}v_{k,k} \tag{11.3.18}$$

in which one must express $\varepsilon_{pq}^{(m)}$ and e_{pq} on the basis of $u_{i,j} = v_{i,j}\,dt$. For $\varepsilon_{pq}^{(m)}$ one can use any admissible finite strain tensor. Using the tensor given by Equation 11.1.7, and noting Equations 11.3.18, one obtains

$$\hat{S}_{ij}^{(m)} = \hat{S}_{ij} + \left(1 - \frac{m}{2}\right)(S_{ik}\dot{e}_{kj} + S_{jk}\dot{e}_{ki}) \tag{11.3.19}$$

In particular, Equations 11.3.11 to 11.3.13, which correspond to Equation 11.3.10 for $m = 2$, 1, and 0, and also the equation that corresponds to Equation 11.3.10 for $m = -1$, yield the following expressions for the objective stress rates:

$$\hat{S}_{ij}^{(2)} = \hat{S}_{ij} = \dot{S}_{ij} - S_{kj}v_{i,k} - S_{ki}v_{j,k} + S_{ij}v_{k,k} \tag{11.3.20a}$$

$$\hat{S}_{ij}^{(1)} = \dot{S}_{ij} - S_{kj}v_{i,k} - S_{ki}v_{j,k} + \tfrac{1}{2}(S_{ik}\dot{e}_{kj} + S_{jk}\dot{e}_{ki}) + S_{ij}v_{k,k} \tag{11.3.20b}$$

$$\hat{S}_{ij}^{(0)} = \dot{S}_{ij} - S_{kj}\dot{\omega}_{ik} + S_{ik}\dot{\omega}_{kj} + S_{ij}v_{k,k} \tag{11.3.20c}$$

$$\hat{S}_{ij}^{(-1)} = \dot{S}_{ij} + S_{ik}v_{k,j} + S_{jk}v_{k,i} + S_{ij}v_{k,k} \tag{11.3.20d}$$

in which $\dot{\omega}_{jk} = \frac{1}{2}(v_{i,j} - v_{j,i})$ = rotation rate, $\dot{e}_{ij} = \frac{1}{2}(v_{i,j} + v_{j,i})$ = strain rate, and $v_i = \dot{u}_i$ = velocity of material point. Same as $\hat{S}_{ij}$, $\hat{S}_{ij}^{(m)}$ must be symmetric. The reason is that the second-order work is $\delta^2\mathscr{W} = \frac{1}{2}\hat{S}_{ij}^{(m)}\dot{e}_{ij}\delta t^2$, and a nonsymmetric part of $\hat{S}_{ij}^{(m)}$ would do no work as $\dot{e}_{ij}$ is symmetric.

$\hat{S}_{ij}$ is known in continuum mechanics as Truesdell's stress rate (Truesdell, 1953, 1955; Truesdell and Toupin, 1960). The rate $\hat{S}_{ij}^{(1)}$, which corresponds to Biot strain, may be called Biot's objective stress rate.

Another widely used expression is the Jaumann's stress rate (also called the corotational stress rate or Jaumann derivative of Cauchy stress tensor; Jaumann, 1911; Prager, 1961):

$$\hat{S}_{ij}^{J} = \dot{S}_{ij} - S_{kj}\dot{\omega}_{ik} + S_{ik}\dot{\omega}_{kj} \tag{11.3.21}$$

It differs from Equation 11.3.20c only by the missing term $S_{ij}v_{k,k}$, which is negligible for incompressible materials. However, if this term is not negligible it is disturbing that Jaumann's stress rate cannot be obtained by substituting the most general second-order finite strain approximation (Eq. 11.1.8 into Eqs. 11.3.18). Thus, as pointed out by Bažant (1971), the Jaumann's rate does not appear to be associated by work with any admissible finite strain expression. This has the consequence that $\frac{1}{2}\hat{S}_{ij}^{J}\dot{e}_{ij}\delta t^2$ is *not* the correct expression for the second-order work of stress increments, whose value is the basic indicator of stability (cf. Sec. 10.1).

Another well-known objective stress rate is that of Cotter and Rivlin (1955) (see Prager, 1961):

$$\hat{S}_{ij}^{R} = \dot{S}_{ij} + S_{ik}v_{k,j} + S_{jk}v_{k,i} \tag{11.3.22}$$

It coincides with Equation 11.3.20d (for $m = -1$) except that the term $S_{ij}v_{k,k}$ is missing, same as for Jaumann's rate. Consequently, $\frac{1}{2}\hat{S}_{ij}^{R}\dot{e}_{ij}\delta t^2$ is also an incorrect expression for the second-order work of stress increments except when the material is incompressible.

The same comment can be made regarding the objective stress rates of Green (1956) and of Oldroyd (1950) (given e.g. in Eringen, 1962) and the rate that corresponds to Biot's (1965) "incremental" stress. They coincide with $\hat{S}_{ij}^{(2)}$ and $\hat{S}_{ij}^{(0)}$, respectively, except that the term $S_{ij}v_{k,k}$ is again missing.

The aforementioned deficiency of Jaumann's stress rate may be remedied by applying the Jaumann (corotational) rate to the Kirchhoff stress $\bar{S}_{ij} = JS_{ij}$ (Eq. 11.2.15) rather than to the Cauchy (true) stress S_{ij}. This yields the rate

$$\hat{S}_{ij}^{(0)} = (JS_{ij})^{\cdot} - (JS_{kj})\dot{\omega}_{ik} + (JS_{ik})\dot{\omega}_{kj} \tag{11.3.23}$$

that is called the Jaumann rate of Kirchhoff stress and has recently been popular in finite-strain plasticity. Substituting $J = 1 + u_{k,k}$, noting that $(JS_{ij})^{\cdot} = J\dot{S}_{ij} + \dot{J}S_{ij} \simeq \dot{S}_{ij} + S_{ij}v_{k,k}$, and neglecting higher-order small terms, one finds that this rate is equal to $\hat{S}_{ij}^{(0)}$ as given by Equation 11.3.20c. Replacing $\mathbf{S}$ with $\bar{\mathbf{S}}$, a similar remedy can be obtained for the rates of Cotter and Rivlin, Oldroyd, Green, etc.

Oldroyd's rate represents what is known in mathematics as the Lie derivative, $L_{\mathbf{v}}\mathbf{S}$, of tensor $\mathbf{S}$ with respect to vector field $\mathbf{v}(\mathbf{x}, t)$; $L_{\mathbf{v}}\mathbf{S} = \dot{S}_{ij} - S_{kj}v_{i,k} - S_{ki}v_{j,k}$ (Guo, 1963; Marsden and Hughes, 1983, p. 100). By missing the term $S_{ij}v_{k,k}$, $L_{\mathbf{v}}\mathbf{S}$ suffers the same problem as the Jaumann's rate of $\mathbf{S}$. Correctly, however, the Lie derivative must be applied to Kirchhoff stress, that is, $L_{\mathbf{v}}\bar{\mathbf{S}}$, which coincides with Truesdell's rate (and is free of the aforementioned problem). Other objective stress rates can be obtained as the Lie derivatives of various transformations of tensor $\bar{\mathbf{S}}$.

It is interesting to note that for $\delta t \rightarrow 0$

$$\hat{\mathbf{S}}^{(2)} = \lim_{\delta t \rightarrow 0} \frac{\mathbf{F}^{-1}\bar{\mathbf{S}}\mathbf{F}^{-T} - \mathbf{S}^0}{\delta t} = L_v\bar{\mathbf{S}} \tag{11.3.24}$$

$$\hat{\mathbf{S}}^{(0)} = \lim_{\delta t \rightarrow 0} \frac{\mathbf{R}^T\bar{\mathbf{S}}\mathbf{R} - \mathbf{S}^0}{\delta t} \tag{11.3.25}$$

$$\hat{\mathbf{S}}^{(-1)} = \lim_{\delta t \rightarrow 0} \frac{\mathbf{F}^T\bar{\mathbf{S}}\mathbf{F} - \mathbf{S}^0}{\delta t} \tag{11.3.26}$$

in which $\bar{\mathbf{S}} = J\mathbf{S} =$ Kirchhoff stress. To prove the last relation, we write

$$\begin{aligned} S_{km}F_{mj} &= S_{km}(\delta_{mj} + u_{m,j}) = S_{kj} + S_{km}u_{m,j} \\ S_{ij}^* &= JF_{ki}S_{km}F_{mj} = J(\delta_{ki} + u_{k,i})(S_{kj} + S_{km}u_{m,j}) \\ &\simeq (1 + u_{k,k})(S_{ij} + S_{ik}u_{k,j} + S_{kj}u_{k,i}) \\ &\simeq S_{ij} + S_{ik}u_{k,j} + S_{kj}u_{k,i} + S_{ij}u_{k,k}, \end{aligned}$$

and then $\lim (S_{ij}^* - S_{ij}^0)/\delta t$ (with $S_{ij} \simeq S_{ij}^0$) yields Equation 11.3.26. Equations 11.3.24 and 11.3.25 can be proven similarly, noting that tensor $\mathbf{F}^{-1}$ has components $\partial X_i/\partial x_j \simeq \delta_{ij} - u_{i,j}$, and tensor $\mathbf{R}$ has components $(\delta_{ij} + \omega_{ij})$ (only small displacements and rotations need to be considered).

In concluding, we must emphasize that all the different finite-strain formulations and the associated objective stress rates must be physically equivalent. Therefore, different objective stress rates must be used in conjunction with different characterizations of incremental material properties, such that the results would indeed be physically equivalent (Bažant, 1971). How to do that, we discuss next.

Problems

11.3.1 Derive Equation 11.3.6 directly from the work equivalence, $\delta \mathscr{W} = (S_{ij}^0 + \tau_{ij})\delta u_{i,k} = (S_{ij}^0 + \sigma_{ij})\delta \varepsilon_{ij}$, rather than from Equation 11.2.14.
11.3.2 Work out in detail the derivation of Equations 11.3.11 to 11.3.13.
11.3.3 Work out in detail the derivation of Equations 11.3.20a–c.
11.3.4 Derive $\hat{S}_{ij}^{(m)}$ for $m = -2$ $(\alpha = 2)$.
11.3.5 Rewrite the equations of this section in (a) tensor symbolism and (b) matrix symbolism.
11.3.6 Substituting Equation 11.1.8 into Equations 11.3.18, derive the most general possible expression for an objective stress rate.
11.3.7 Prove Equations 11.3.24 to 11.3.26 in detail.

11.4 TANGENTIAL MODULI AT LARGE INITIAL STRESS

In the preceding sections we developed the finite-strain formulations for many rather than just one finite-strain measure. Vain as this exercise might have seemed, it is nevertheless rather revealing with regard to the meaning of the tangential moduli C_{ijkm} of three-dimensional bodies at initial stress. It brings to light a certain degree of arbitrariness in the determination of C_{ijkm}, and at the same time it shows that this arbitrariness merely represents different ways of looking at the same material properties.

In Section 10.6 we have seen that plasticity theory leads to incremental stress–strain relations that are linear for a certain sector or cone of directions in the space of strain increments. Such linearity in fact characterizes most constitutive models. Thus we will assume in this section that the material properties are adequately described by a set of tangential moduli C_{ijkm} relating the stress and strain increments (or rates).

Based on the foregoing considerations it is clear that the tangential moduli cannot relate to the strain rate the time derivative of the true stress, $\partial S_{ij}/\partial t$, because the material elements on which the stresses S_{ij} are defined at two subsequent times are not the same, that is, do not involve the same piece of material (in the Eulerian coordinate description the material is moving through a fixed elemental volume in space). So the requirement to describe the deformation process on the same small piece of the material excludes the increments s_{ij} of Cauchy stress S_{ij} and permits only the increments τ_{ij} and $\sigma_{ij}^{(m)}$. However, τ_{ij} must be excluded for a different reason—namely that $(S_{ij}^0 + \tau_{ij})\delta\varepsilon_{ij}^{(m)}$ is not the correct expression for work in terms of $\varepsilon_{ij}^{(m)}$. Rather, the correct expression is $(S_{ij}^0 + \sigma_{ij}^{(m)})\delta\varepsilon_{ij}^{(m)}$. Therefore the tangential moduli must refer to the increments $\sigma_{ij}^{(m)}$, or to the corresponding objective stress rates $\hat{S}_{ij}^{(m)}$. It follows that the incremental stress–strain relation must have the form $\sigma_{ij}^{(m)} = C_{ijkm}e_{km}$ or $\hat{S}_{ij}^{(m)} = C_{ijkm}\dot{e}_{km}$.

Since the expression for $\hat{S}_{ij}^{(m)}$ is a matter of choice as far as the m value is concerned, the corresponding moduli $C_{ijkm}^{(m)}$ must be different, so as to always represent the same material properties. Considering finite strain tensors $\varepsilon_{ij}^{\alpha} = \varepsilon_{ij} - \alpha e_{ki}e_{kj}$ $(\alpha = 1 - \frac{1}{2}m$, Eq. 11.1.7), we will label the corresponding quantities $\sigma_{ij}^{(m)}$, $\hat{S}_{ij}^{(m)}$, and $C_{ijkm}^{(m)}$. For $m = 2$ (Lagrangian strain) we have σ_{ij}, $\hat{S}_{ij}$, C_{ijkm}; for

$m = 1$ (Biot's strain) we have $\sigma_{ij}^{(1)}$, $\hat{S}_{ij}^{(1)}$, $C_{ijkm}^{(1)}$; for $m = 0$ (logarithmic strain) we have $\sigma_{ij}^{(0)}$, $\hat{S}_{ij}^{(0)}$, $C_{ijkm}^{(0)}$, etc. The tangential stress–strain relations must now be written as (Bažant, 1971)

$$\hat{S}_{ij}^{(m)} = C_{ijkm}^{(m)} \dot{e}_{km} \qquad \text{or} \qquad \sigma_{ij}^{(m)} = C_{ijkm}^{(m)} e_{km} \tag{11.4.1}$$

The uses of different tangential moduli $C_{ijkm}^{(m)}$ must be physically equivalent. This is so if they yield the same second-order work $\delta^2 \bar{\mathscr{W}}$ per unit initial volume. Since $\delta^2 \bar{\mathscr{W}} = \frac{1}{2} \dot{T}_{ij} v_{i,j} \delta t^2 = \frac{1}{2} \tau_{ij} \delta u_{i,j}$, we see that $\dot{T}_{ij}$ or τ_{ij} must be the same for any $C_{ijkm}^{(m)}$. We express τ_{ij} both from Equation 11.3.10 ($m \neq 2$) and Equation 11.3.11 ($m = 2$), and set them to be equal:

$$C_{ijkm}^{(m)} e_{km} + S_{pq}^0 \frac{\partial(\varepsilon_{pq}^{(m)} - e_{pq})}{\partial u_{i,j}} = C_{ijkm} e_{km} + S_{mj}^0 u_{i,m} \tag{11.4.2}$$

where we may replace S_{ij}^0 with S_{ij} since we consider infinitesimal increments only. Because $C_{ijkm}^{(m)} e_{km} = C_{ijkm}^{(m)} u_{k,m}$ and $\varepsilon_{pq}^{(m)} - e_{pq} \simeq [\partial(\varepsilon_{pq}^{(m)} - e_{pq})/\partial u_{k,m}] u_{k,m}$ we may rewrite the last equation as

$$\left[C_{ijkm}^{(m)} + S_{pq} \frac{\partial^2(\varepsilon_{pq}^{(m)} - e_{pq})}{\partial u_{i,j} \, \partial u_{k,m}} - C_{ijkm} - S_{jm} \delta_{ik} \right] u_{k,m} = 0 \tag{11.4.3}$$

This equation must hold true for any $u_{k,m}$. It follows that

$$C_{ijkm}^{(m)} = C_{ijkm} + S_{jm} \delta_{ik} - S_{pq} \frac{\partial^2(\varepsilon_{pq}^{(m)} - e_{pq})}{\partial u_{i,j} \, \partial u_{k,m}} \tag{11.4.4}$$

If we now substitute $\varepsilon_{pq}^{(m)} = \varepsilon_{pq} - \alpha e_{kp} e_{kq} = \frac{1}{2}(u_{p,q} + u_{q,p} + u_{k,p} u_{k,q}) - (\alpha/4)(u_{k,p} + u_{p,k})(u_{k,q} + u_{q,k})$, $\alpha = 1 - \frac{1}{2}m$, and differentiate with respect to $u_{i,j}$ and $u_{k,m}$ (for example, $\partial u_{p,q}/\partial u_{i,j} = \delta_{pi} \delta_{jq}$), we obtain (Bažant, 1971)

$$C_{ijkm}^{(m)} = C_{ijkm} + \tfrac{1}{4}(2 - m)(S_{ik} \delta_{jm} + S_{jk} \delta_{im} + S_{im} \delta_{jk} + S_{jm} \delta_{ik}) \tag{11.4.5}$$

where $C_{ijkm} = C_{ijkm}^{(2)}$ = tangential moduli associated with Lagrangian strain. Note that these relations preserve all the symmetries, $C_{ijkm}^{(m)} = C_{jikm}^{(m)} = C_{ijmk}^{(m)} = C_{kmij}^{(m)}$. So if these symmetries hold for C_{ijkm}, they hold also for other $C_{ijkm}^{(m)}$, which is, of course, required from the physical point of view.

Substituting for $\varepsilon_{ij}^{(m)}$ other finite strain tensors, for example, Equation 11.1.8, the corresponding $C_{ijkm}^{(m)}$ can be obtained from Equation 11.4.4.

To examine the Jaumann rate of Cauchy stress (Eq. 11.3.21), we may express $\dot{T}_{ij}$ from both Equations 11.3.20a and 11.3.21 and we can verify that $\hat{S}_{ij}^J = C_{ijkm}^J \dot{e}_{km}$ and $\hat{S}_{ij}^{(0)} = C_{ijkm}^{(0)} \dot{e}_{km}$ yield the same $\dot{T}_{ij}$ if and only if (Bažant, 1971)

$$C_{ijkm}^J = C_{ijkm}^{(0)} - \delta_{km} S_{ij} \tag{11.4.6}$$

Note that this relation does not preserve the symmetry of the 6×6 matrix of tangential moduli, that is, $C_{ijkm}^J \neq C_{kmij}^J$ if $C_{ijkm} = C_{kmij}$. This is a questionable feature (Bažant, 1971), which however disappears if the material is incompressible (because in that case $\hat{S}_{ij}^J \simeq \hat{S}_{ij}^{(0)}$). (The same deficiency occurs for the rates of Cotter and Rivlin and of Oldroyd.) The problem is eliminated by using the Jaumann rate of Kirchhoff stress (Eq. 11.3.20c).

The differences between the various types of tangential moduli need to be explained in relation to experiment. Assuming that a uniform strain field in a test specimen can be achieved, one can determine the stresses from the forces measured or applied at the specimen surface. These forces yield stresses T_{ij} (or τ_{ij}), not $\Sigma_{ij}^{(m)}$ or S_{ij} ($\sigma_{ij}^{(m)}$ or t_{ij}). To determine the tangential moduli for each stress level T_{ij}^0, one cannot avoid first using Equation 11.3.10 to determine the objective stress increments $\sigma_{ij}^{(m)}$, and then $C_{ijkm}^{(m)}$ can be determined on the basis of $\sigma_{ij}^{(m)}$ and e_{ij} (many types of tests are, of course, needed for that). Now the important point is that before one can do this, one must obviously first choose the type of finite-strain measure to be used in Equation 11.3.10. For example, if the Lagrangian strain is chosen, one uses the special form of Equation 11.3.10 indicated in Equation 11.3.11. So it is clear that the tangential moduli measured are inevitably associated with a certain type of finite strain tensor, and different moduli are obtained from the same test results if different finite strain tensors are used or implied in the evaluation of tests. The arbitrariness lies neither in the material nor in the measurement method, but in the chosen method of description.

If the tangential moduli happen to be constant (independent of S_{ij}) in one formulation, they will not be constant in another formulation, as is revealed by Equation 11.4.4 or 11.4.5. Since there are infinitely many possible finite-strain measures, and none of them is physically better justified than the others, we realize that the case of constant tangential moduli (independent of S_{ij}) must be purely speculative and could happen only exceptionally, by chance. Equation 11.4.5 also confirms that incremental isotropy of the material in finite strain is but a figment not usually met in practice.

In practical calculations it is important to employ the same type of finite-strain formulation as that used in evaluating experimental data. In the absence of tests, one must at least be consistent enough to use only equations corresponding to one and the same finite-strain formulation. Unfortunately, it has been a widespread practice to use, for example, incremental equilibrium conditions that correspond to the Lagrangian strain ($\tau_{ij,j} = 0$, with $\tau_{ij} = \sigma_{i,j} + S_{kj}^0 u_{i,k}$; Eq. 11.3.11) but at the same time adopt for the objective stress rate the Jaumann rate of Cauchy stress (Eq. 11.3.21) or of Kirchhoff stress (Eq. 11.3.20c). Even if we discount the aforementioned limitation of the Jaumann rate of Cauchy stress to incompressibility, such a practice is incorrect. It implies an incorrect expression for the second-order work, and thus leads to incorrect predictions of stability limits and path bifurcations. The only objective stress rate that can be used in this case is Truesdell's rate (Eq. 11.3.20a). Any other objective rate yields a wrong value for the second-order work, although the condition of objectivity (invariance at all observer transformations) is satisfied. Objectivity does not guarantee a correct value of work.

The aforementioned incorrect practices can have serious consequences only if the strains are truly large. For the case of thin bodies in which only the deflections and rotations are large while the strains may remain small, the differences between various finite-strain measures are negligible, and so are the differences between the associated tangential moduli values and incremental equilibrium relations. In such a case one can freely mix any objective stress rate with any finite-strain measure and any incremental equilibrium relation (the cases of large $|m|$-values excepted). At very large deflections, however, even thin bodies can be

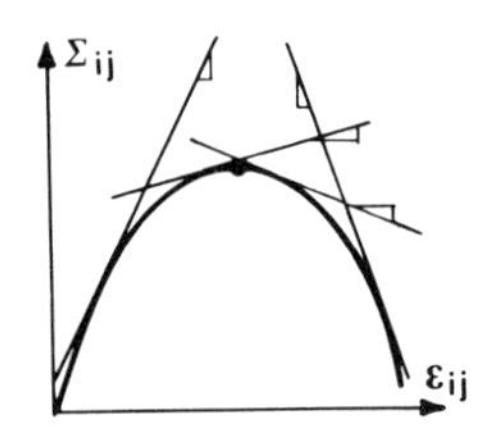

Figure 11.3 Reduction of tangential modulus in inelastic materials.

subjected to large strains. Then, of course, the foregoing comment does not apply.

The differences between the tangential moduli associated with various finite-strain formulations (Eq. 11.4.5) are important only if the stresses reach the same order of magnitude as the tangential moduli. Structural materials (steel, concrete, aluminum) are so stiff that stresses in beams, plates, and shells are much smaller than the elastic modulus. However, for stiff structural materials, the differences between various finite-strain formulations can become important if the material is inelastic and the tangential modulus is greatly reduced, which happens near the peak of the stress–strain diagram (Fig. 11.3). Therefore, the finite-strain formulation of stability problems for stiff structural materials can be important only for material states that are close to perfect plasticity. If $\|S_{ij}\| \ll \|C_{ijkm}\|$, the differences among various objective stress rates and among various finite-strain measures can be ignored.

On the other hand, for soft materials such as rubber, the finite-strain formulation is important in most situations.

At this point it may be helpful to summarize the correlations among various formulations of finite strain, objective stress rates, incremental equilibrium equations, etc.; see Table 11.4.1 (taken from Bažant, 1971). The last column of this table, for $\alpha = 2$, and the lines for stability criterion and for columns with shear, will be explained in Sections 11.5 and 11.6.

Problems

11.4.1 Express the difference between second-order works based on $C^{(-1)}_{ijkm}$ and on $\hat{S}^{(2)}_{ij}$.

11.4.2 Derive C_{ijkm} associated with the Cotter–Rivlin objective stress rate and show that they are nonsymmetric unless the material is incompressible.

11.4.3 Express τ_{ij} from Equations 11.3.11 to 11.3.13, and substituting Equation 11.4.5 for $C^{(1)}_{ijkm}$ $(\alpha = \frac{1}{2})$ and $C^{(0)}_{ijkm}$ $(\alpha = 1)$ verify that indeed the same expressions for τ_{ij} are obtained.

11.4.4 Carry out the detailed derivation of Equation 11.4.5 from Equation 11.4.4.

11.4.5 Consider plane strain and assume that $S^0_{11} = -5000$, $S^0_{12} = S^0_{21} = 1000$, $S^0_{22} = -1000$, all other $S^0_{ij} = 0$; $v_{1,1} = 5$, $v_{1,2} = 2$, $v_{2,1} = 1$, $v_{2,2} = -1$; $C^0_{1111} = 12{,}000$, $C^0_{2222} = 2000$, $C^0_{1212} = 3000$, $C^0_{1122} = C^0_{2211} = 1000$, $C^0_{1112} = C^0_{1211} = 200$, all other $C_{ijkm} = 0$. Calculate $\hat{S}^{(2)}_{ij}$, $\hat{S}^{(1)}_{ij}$, $\hat{S}^{(0)}_{ij}$, $\hat{S}^{(-2)}_{ij}$, $\hat{S}^{J}_{ij}$, $\hat{S}^{R}_{ij}$.

Table 11.4.1 Bažant's (1971) Correlations of Various Stability Formulations.

Form	(a) $m=2$ $(\alpha=0)$	(b) $m=1$ $(\alpha=\frac{1}{2})$	(c) $m=0$ $(\alpha=1)$	(d) $m=0$ $(\alpha=1)$	(e) $m=-1$ $(\alpha=\frac{3}{2})$	(f) $m=-2$ $(\alpha=2)$
1. Finite strain tensor	ε_{ij} $\varepsilon_{11}=e_{11}+\frac{1}{2}e_{11}^2$ Eq. 11.1.3 (Green's or Lagrangian)	$\varepsilon_{ij}^{(1)}=\varepsilon_{ij}-\frac{1}{2}e_{ki}e_{kj}$ $\varepsilon_{11}^{(1)}=e_{11}$ Eq. 11.1.5 (Biot's pure deformation part of displacement gradient)	$\varepsilon_{ij}^{(0)}=\varepsilon_{ij}-e^{ki}e^{kj}$ $\varepsilon_{11}^{(0)}=e_{11}-\frac{1}{2}e_{11}^2\approx$ $\ln(1+\varepsilon_{11})$ Eq. 11.1.7 (logarithmic or Hencky's strain)	$\varepsilon_{ij}^{d}=\varepsilon_{ij}^{(0)}$	$\varepsilon_{ij}^{(-1)}=\varepsilon_{ij}-\frac{3}{2}e_{ki}e_{kj}$ $\varepsilon_{11}^{(-1)}=e_{11}-e_{11}^2$ Eq. 11.1.7	$\varepsilon_{ij}^{(-2)}=\varepsilon_{ij}-2e_{ki}e_{kj}$ $\varepsilon_{11}^{(-2)}=e_{11}-\frac{3}{2}e_{11}^2$ Eq. 11.6.12
2. Incremental equilibrium equation (Eqs. 11.3.4)	Eq. 11.3.11 (Trefftz)	Eq. 11.3.12 (Biot)	Eq. 11.3.13 (Biezeno and Hencky)	Eq. 11.3.14 (Biot, Neuber, Southwell)	—	Eq. 11.3.9
3. Objective material stress tensor	Eq. 11.2.18, Eq. 11.3.11 (second Piola-Kirchhoff tensor)	Eq. 11.3.12 (Biot's alternative stress)	Eq. 11.3.13	Eq. 11.3.14 (Biot's incremental stress)	—	Eq. 11.3.10

4. Incremental moduli if a potential exists	$C^{(2)}_{ijkm}$ (symmetric)	$C^{(1)}_{ijkm}$ (symmetric)	$C^{(0)}_{ijkm}$ (symmetric)	C^{J}_{ijkm} ($C^{J}_{ijkm} = C^{J}_{kmij}$)	$C^{(-1)}_{ijkm}$	$C^{(-2)}_{ijkm}$ Eq. 11.4.5
5. Objective stress rate	$\hat{S}^{(2)}_{ij}$, Eq. 11.3.20a (Truesdell's rate) (if incompressible, also Oldroyd's and Green's rate*)	$\hat{S}^{(1)}_{ij}$, Eq. 11.1.30b	$\hat{S}^{(0)}_{ij}$, Eq. 11.3.20c (Jaumann rate of Kirchhoff stress)	$\hat{S}^{J}_{ij}$, Eq. 11.3.21 [Jaumann (corotational) rate of Cauchy stress*]	$\hat{S}^{(-1)}_{ij}$, Eq. 11.3.20d [Lie derivative of Kirchhoff stress (if incompressible, also convected rate of Cotter and Rivlin*)]	
6. Stability criterion	Eq. 11.5.4 (Hadamard, Trefftz, Pearson, Hill)	Eq. 11.5.5 (Biot)	Eq. 11.5.1	—	—	Eq. 11.5.1
7. Buckling of column with shear	$E^{(2)}/G^{(2)}$ in Fig. 11.5 Eq. 11.6.6 (Engesser)	$E^{(1)}/G^{(1)}$ in Fig. 11.5	$E^{(0)}/G^{(0)}$ in Fig. 11.5	—	—	Eq. 11.6.11 (Haringx's shear column)
8. Surface buckling	Fig. 11.9 (Bažant)	Eq. 11.7.9 (Biot)				

* Give incorrect $\delta^2 \mathscr{W}$ unless incompressible.

11.5 STABLE STATES AND PATHS FOR MULTIDIMENSIONAL CONTINUOUS BODIES

Now that we have formulated strains, stresses, and tangential moduli in the presence of geometric nonlinearity due to finite strain, we are ready to discuss stability of equilibrium states and equilibrium paths. First let us start from the point where we ended in Chapter 10, particularly Equations 10.1.29 and 10.1.30, which show that $-T(\Delta S)_{\text{in}} = \int_V \frac{1}{2}\sigma_{ij}^{(m)}\varepsilon_{ij}^{(m)}\,dV + \delta^2\mathscr{W}_\sigma$ where T = absolute temperature, $(\Delta S)_{\text{in}}$ = internally produced increment of entropy of the structure, $\sigma_{ij}^{(m)}$ = stress increments caused by strain increments $\varepsilon_{ij}^{(m)}$, and $\delta^2\mathscr{W}_\sigma$ = additional work due to geometric nonlinearity, representing the work of initial stresses S_{ij}^0 on the second-order part of strains. In the case of finite strains, we must now be more precise and write the foregoing integral in terms of volume V_0 in the initial configuration. The second-order work of the initial stresses is $\delta^2\mathscr{W}_\sigma = \int_V S_{ij}^0(\varepsilon_{ij}^{(m)} - e_{ij})\,dV$. The stress increments $\sigma_{ij}^{(m)}$ must be understood as the objective stress increments $\sigma_{ij}^{(m)}$ associated with $\varepsilon_{ij}^{(m)}$ (Sec. 11.3), and $\sigma_{ij}^{(m)} = C_{ijkm}^{(m)}e_{km}$ where $C_{ijkm}^{(m)}$ are the tangential moduli as discussed in Section 11.4; but in view of Section 10.1 we must distinguish between isothermal and isentropic (adiabatic) tangential moduli, $C_{Tijkm}^{(m)}$ and $C_{Sijkm}^{(m)}$.

Based on Section 10.1, equilibrium of a body with boundary conditions of prescribed displacements and dead loads is stable if $\delta^2\mathscr{W}$ is positive definite (with respect to all kinematically admissible fields $\delta\varepsilon_{ij}^{(m)}$), that is, if

$$-T(\Delta S)_{\text{in}} = \delta^2\mathscr{W} = \int_V \left[S_{ij}^0(\varepsilon_{ij}^{(m)} - e_{ij}) + \frac{1}{2}e_{ij}C_{ijkm}^{(m)}e_{km} \right] dV > 0 \qquad (11.5.1)$$

for all admissible fields $\varepsilon_{ij}^{(m)}$ (Bažant, 1967). The state is unstable if this energy expression is indefinite.

Based on Section 10.2, under controlled displacements the stable equilibrium branch after a bifurcation point is that which minimizes this integral among all the emanating equilibrium branches.

Since the principles on which this criterion is based have been developed in an abstract sense by thermodynamic considerations, it might be helpful to also derive Equation 11.5.1 in a simpler, albeit less general, manner. We consider a deviation from the initial equilibrium state characterized by displacement field $u_i(\mathbf{x})$ with finite strains $\varepsilon_{ij}^{(m)}(\mathbf{x})$, and load-point displacements q_k under given dead loads P_k ($k = 1, \ldots, n$). The total work that must be done on the structure-load system to produce this deviation is $\Delta\mathscr{W} = \sum_k \int f_k\,dq_k - \sum_k P_k\,dq_k$ where f_k are the equilibrium reactions at the load points depending on q_k (see Eqs. 10.1.6–10.1.7). If $\Delta\mathscr{W}$ is positive for all possible deviations q_k, $\varepsilon_{ij}^{(m)}$, the system is stable.

For small enough deviations q_k we have $\int f_k\,dq_k \simeq \sum_k (f_k^0 + \frac{1}{2}\tilde{f}_k)q_k$ where f_k^0 = initial values of f_k and $\tilde{f}_k$ are their increments. According to the principle of conservation of energy, $\sum_k f_k^0 q_k = \int_V S_{ij}^0\varepsilon_{ij}^{(m)}\,dV$ and $\sum_k \frac{1}{2}\tilde{f}_k q_k = \int_V \frac{1}{2}\sigma_{ij}^{(m)}e_{ij}\,dV$. Here, in order to calculate the work of objective stress increments $\sigma_{ij}^{(m)}$, we can take small strains e_{ij} because we need only second-order accuracy; but for the work of the initial Cauchy (true) stresses S_{ij}^0 we need to take second-order accurate finite strain $\varepsilon_{ij}^{(m)}$ because S_{ij}^0 is not small. Thus we have (Bažant, 1967)

$$\Delta \mathcal{W} = \int_V (S_{ij}^0 \varepsilon_{ij}^{(m)} + \tfrac{1}{2}\sigma_{ij}^{(m)} e_{ij})\, dV - \sum_k P_k q_k \tag{11.5.2}$$

in which $\sigma_{ij}^{(m)} = C_{ijkm}^{(m)} e_{km}$ and $C_{ijkm}^{(m)}$ represents either the isothermal moduli, $C_{Tijkm}^{(m)}$, or the isentropic moduli, $C_{Sijkm}^{(m)}$.

Since the initial state is an equilibrium state, the principle of virtual work (Eq. 11.2.3) requires that

$$\int_V S_{ij}^0 e_{ij}\, dV - \sum_k P_k q_k = 0 \tag{11.5.3}$$

because the field of S_{ij}^0 is in equilibrium with loads P_k and strains e_{ij} are compatible with q_k. Subtracting the last equation from Equation 11.5.2, we obtain for $\Delta \mathcal{W}$ the expression in Equation 11.5.1 and we rename it $\delta^2 \mathcal{W}$ because it includes only second-order work terms. The fact that $\delta^2 \mathcal{W} = -T(\Delta S)_{\text{in}}$ was established in Section 10.1.

The diversity of admissible forms of finite-strain formulations projects itself into the stability criterion for three-dimensional continuous bodies (attempted perhaps first by Bryan, 1888). Using $\varepsilon_{ij}^{(m)} = \varepsilon_{ij} =$ Lagrangian strain ($m = 2$) with the corresponding $C_{ijkm}^{(m)} = C_{ijkm}$ ($m = 2$), Equation 11.5.1 can be reduced to the stable equilibrium conditions:

$$-T(\Delta S)_{\text{in}} = \delta^2 \mathcal{W} = \int_V (C_{ijkm} + \delta_{ik} S_{jm}^0) u_{i,j} u_{k,m}\, dV > 0 \tag{11.5.4}$$

in which $u_i(\mathbf{x})$ can be any kinematically admissible displacement field. This condition was presented by Trefftz (1933). In a special sense, an identical criterion was obtained by Hadamard (1903) from the condition that in a stable body the wave propagation velocities must be real. Equation 11.5.4 is also equivalent to the criterion given by Goodier and Plass (1952), Pearson (1956), Hill (1957), Prager (1961), and Truesdell and Noll (1965).

For Biot's second-order strain ($m = 1$), Equation 11.5.1 yields the stability criterion

$$-T(\Delta S)_{\text{in}} = \delta^2 \mathcal{W} = \int_V \tfrac{1}{2}[S_{ij}^0 (u_{ki} u_{kj} - e_{ki} e_{kj}) + u_{i,j} C_{ijkm}^{(1)} u_{k,m}]\, dV > 0 \tag{11.5.5}$$

Other forms of the stability criterion would result by using $\varepsilon_{ij}^{(m)}$ for other m values (Bažant, 1971). Of course, all these forms are physically equivalent.

Problems

11.5.1 Assume the special case of uniaxial stress and apply criterion 11.5.4 to the Euler column.

11.5.2 Consider the finite strain tensor for a thin body at small strains but finite rotations, and apply Equation 11.5.4 to a fixed column.

11.5.3 Do the same as above but use Biot's strain tensor ($m = 1$).

11.5.4 Use the calculus of variations to derive, by minimization of the functional in Equation 11.5.4, the differential equations of equilibrium and the boundary conditions in terms of $\sigma_{ij}^{(2)}$ (for details, see Bažant, 1967).

11.5.5 Do the same as above but for the more general functional in Equation 11.5.1.

11.6 COLUMN OR PLATE WITH SHEAR: FINITE-STRAIN EFFECT

The effect of shear on the buckling of columns or plates, which is important for built-up (reticulated, latticed) columns, composite materials, and sandwich plates, was already studied in Sections 1.7 and 7.8. However, we did not, and could not, observe that the form of the differential equations of equilibrium in the presence of shear depends on the choice of finite-strain measure, and that the elastic moduli cannot be considered constant but must depend on this choice, too. But in the light of this chapter we may sense a problem. We know that in thin bodies there is no difference among the uses of various finite-strain measures because the second-order work depends only on rotations (Eq. 11.1.19) in which there is no ambiguity. In beams or plates with shear, though, there are two different rotations at each cross section: the rotation of the beam axis, $w_{,x}$, and the rotation of the cross section, $\psi(x)$. So which rotation counts? Let us revisit this subject.

Differential Equations

We consider a perfect hinged column of length L and a constant cross section of area A and centroidal moment of inertia I. The material is elastic, and isotropic or orthotropic. The column is initially straight and carries uniform uniaxial stress S_{11}^0. Let $x_1 = x =$ longitudinal axis and $x_3 = z =$ deflection direction. The deflection curve is $w(x)$. The cross sections are assumed to remain plane but not normal to the deflection curve. The cross-section rotation is $\psi(x)$ (Fig. 11.4a), and the longitudinal displacements are $u_1 = -z\psi(x)$. According to Equation 11.5.1, the condition of stable equilibrium of the column is that

$$-T(\Delta S)_{\text{in}} = \delta^2 \mathscr{W} = \int_0^L \int_A [S^0(\varepsilon_{11}^{(m)} - e_{11}) + \tfrac{1}{2}E^{(m)}e_{11}^2 + \tfrac{1}{2}G^{(m)}\gamma^2]\, dA\, dx > 0 \quad (11.6.1)$$

in which $S^0 = S_{11}^0$; $A =$ cross-section area; $E^{(m)}$, $G^{(m)} =$ (tangential) Young's modulus and shear modulus; and label (m) indicates the choice of finite strain tensor $\boldsymbol{\varepsilon}^{(m)}$ to which these moduli refer. The critical state is obtained when the first variation of the integral in Equation 11.6.1 vanishes, that is,

$$\int_0^L \int_A [S^0\delta(\varepsilon_{11}^{(m)} - e_{11}) + E^{(m)}e_{11}\delta e_{11} + G^{(m)}\gamma\delta\gamma]\, dA\, dx = 0 \quad (11.6.2)$$

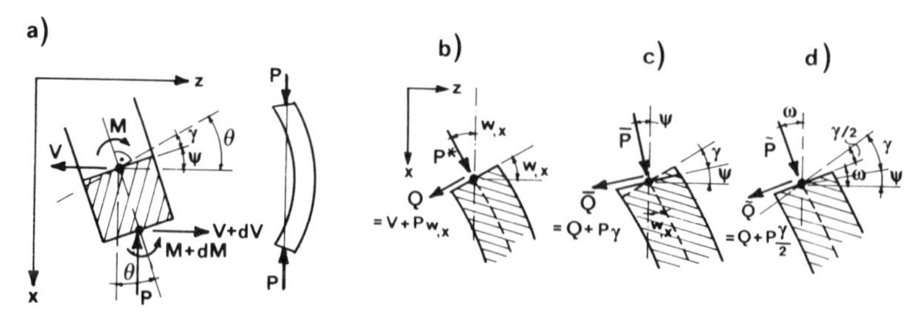

Figure 11.4 Beam with shear deformation and various choices of the direction of the shear force component.

Substituting $\varepsilon_{11}^{(m)} - e_{11} = \frac{1}{2}u_{1,1}u_{1,1} + \frac{1}{2}u_{3,1}u_{3,1} - \alpha e_{11}e_{11} - \alpha e_{31}e_{31}$ $(\alpha = 1 - \frac{1}{2}m)$, $u_{1,1} = e_{11} = -z\psi_{,x}$, $u_{3,1} = w_{,x}$, and $\gamma = 2e_{13} = u_{1,3} + u_{3,1} = w_{,x} - \psi$ (shear angle), we have

$$\int_0^L \int_A \{S^0\delta\left[\left(\frac{1}{2} - \alpha\right)z^2\psi_{,x}^2 + \frac{1}{2}w_{,x}^2 - \frac{\alpha}{4}(w_{,x} - \psi)^2\right] + E^{(m)}z^2\psi_{,x}\delta\psi_{,x}$$
$$+ G^{(m)}(w_{,x} - \psi)(\delta w_{,x} - \delta\psi)\Big\}\, dA\, dx = 0 \quad (11.6.3)$$

and noting that $\int dA = A$, $\int z^2\, dA = I$, we obtain

$$\int_0^L \Big\{S^0\left[(1 - 2\alpha)I\psi_{,x}\delta\psi_{,x} + Aw_{,x}\delta w_{,x} - \frac{\alpha}{2}A(w_{,x} - \psi)(\delta w_{,x} - \delta\psi)\right]$$
$$+ E^{(m)}I\psi_{,x}\delta\psi_{,x} + G^{(m)}A(w_{,x} - \psi)(\delta w_{,x} - \delta\psi)\Big\}\, dx = 0 \quad (11.6.4)$$

Now we integrate by parts the terms containing $\delta\psi_{,x}$, take into account the boundary conditions ($w = \psi_{,x} = 0$), and group all the terms with $\delta\psi$ as well as those with $\delta w_{,x}$, that is,

$$\int_0^L \Big\{\left[(E^{(m)}I\psi_{,x} + S^0(1 - 2\alpha)I\psi_{,x})_{,x} - \frac{\alpha}{2}S^0A(w_{,x} - \psi) + G^{(m)}A(w_{,x} - \psi)\right]\delta\psi$$
$$- \left[S^0Aw_{,x} - \frac{\alpha}{2}S^0A(w_{,x} - \psi) + G^{(m)}A(w_{,x} - \psi)\right]\delta w_{,x}\Big\}\, dx = 0 \quad (11.6.5)$$

If this should hold true for any admissible variations $\delta\psi$ and $\delta w_{,x}$, the expressions in the square brackets must vanish. This yields two differential equations. By adding the second equation to the first one and then integrating the first equation, Bažant (1971) obtained the following system of two homogeneous linear differential equations for $\psi(x)$ and $w(x)$:

$$\begin{aligned} [E^{(m)} + (m - 1)S^0]I\psi_{,x} - S^0Aw &= 0 \\ [G^{(m)} - \tfrac{1}{4}(2 - m)S^0](w_{,x} - \psi) + S^0w_{,x} &= 0 \end{aligned} \quad (11.6.6)$$

Substituting $S^0 = -P/A$, differentiating the second equation, and eliminating $\psi_{,x}$ from both equations, one obtains a single second-order linear differential equation $w_{,xx} + k^2w = 0$, in which

$$k^2 = \frac{P\left[1 + \dfrac{2 - m}{4}\left(\dfrac{P}{G^{(m)}A}\right)\right]}{E^{(m)}I\left[1 - \dfrac{2 + m}{4}\left(\dfrac{P}{G^{(m)}A}\right)\right]\left[1 - (m - 1)\dfrac{P}{E^{(m)}A}\right]} \quad (11.6.7)$$

This differential equation is of the same form as Equation 1.7.6. The solution may be sought again in the form $w = a \sin kx$, and from the boundary conditions $w = 0$ at $x = 0$, we get $k = n\pi/L$ where $n = 1, 2, 3, \ldots$, same as in Section 1.7. Substituting this into Equation 11.6.7, we obtain for the critical loads a quadratic equation.

The shear is important in columns only if $G^{(m)} \ll E^{(m)}$. The axial stress P/A cannot have a higher order of magnitude than $G^{(m)}$, and so $P/A \ll E^{(m)}$.

Therefore, $(m-1)P/E^{(m)}A \ll 1$. With this simplification, Equation 11.6.7 becomes:

For $m=2$:

$$k^2 = \frac{P}{E^{(2)}I[1-(P/G^{(2)}A)]} \tag{11.6.8}$$

For $m=-2$:

$$k^2 = \frac{P}{E^{(-2)}I}\left(1+\frac{P}{G^{(-2)}A}\right) \tag{11.6.9}$$

Theories of Engesser and Haringx

Substituting $k^2 = \pi^2/L^2$, introducing the notation $P_E^{(m)} = \pi^2 E^{(m)} I/L^2$, and solving for P, Equation 11.6.8 yields for a hinged column Engesser's (1889, 1891) formula

$$P_{\mathrm{cr}_1} = \frac{P_E^{(2)}}{1+(P_E^{(2)}/G^{(2)}A)} \tag{11.6.10}$$

already presented as Equation 1.7.7, while Equation 11.6.9 yields the formula

$$P_{\mathrm{cr}_1} = \frac{G^{(-2)}A}{2}\left[\left(1+4\frac{P_{\mathrm{E}}^{(-2)}}{G^{-2)}A}\right)^{1/2}-1\right] \tag{11.6.11}$$

proposed, albeit without the labels (-2), by Haringx (1942) (and also given by Timoshenko and Gere, 1961, p. 143). Haringx initially proposed this formula for buckling of a helical spring treated approximately as a column, but later Haringx's formula was used also for piles of elastomeric bridge bearings or for rubber bearings, as well as for laced and battened structural members; see Chan and Kelly (1970), Lin, Glauser, and Johnston (1970), Johnston (1976, p. 359), Nänni (1971), etc.

For not too slender columns, the results of both formulas are quite close, much closer to each other than to the Euler formula. However, in the limit of vanishing slenderness, that is, $P_{\mathrm{E}} \to \infty$, Haringx's formula gives $P_{\mathrm{cr}_1} \to \infty$ while Engesser's formula (Eq. 11.6.10 or 1.1.7) yields $P_{\mathrm{cr}_1} \to G^{(2)}A$, which is a very different result. For very small slenderness L/r (i.e., very small P_{E}), Haringx found his formula to agree with the results of the tests on springs much better than Engesser's formula, which had previously been proposed for springs by Biezeno and Koch (1925).

The dependence of $\sigma_{\mathrm{cr}_1} = -P_{\mathrm{cr}_1}/A$ on column slenderness L/r ($r^2 = I/A$) is plotted in Figure 11.5a for five different orthotropic materials characterized by $E^{(2)}/G^{(2)} = 25$, $E^{(1)}/G^{(1)} = 25$, $E^{(0)}/G^{(0)} = 25$, $E^{(-1)}/G^{(-1)} = 25$, and $E^{(-2)}/G^{(-2)} = 25$. The solid curves are the exact (and theoretically fully consistent) solutions of Equation 11.6.7, while the dashed curves are the approximate solutions when $(m-1)P/E^{(m)}A$ is neglected compared to 1, which includes Engesser's and Haringx's formulas. For $L \to 0$, all the curves except those for $m=-2$ tend to a finite value of P_{cr_1}. The corresponding curves for $E^{(m)}/G^{(m)} = 3$ are plotted in Figure 11.5b.

Haringx's equation is associated with the finite strain tensor $\varepsilon_{ij}^{(m)}$

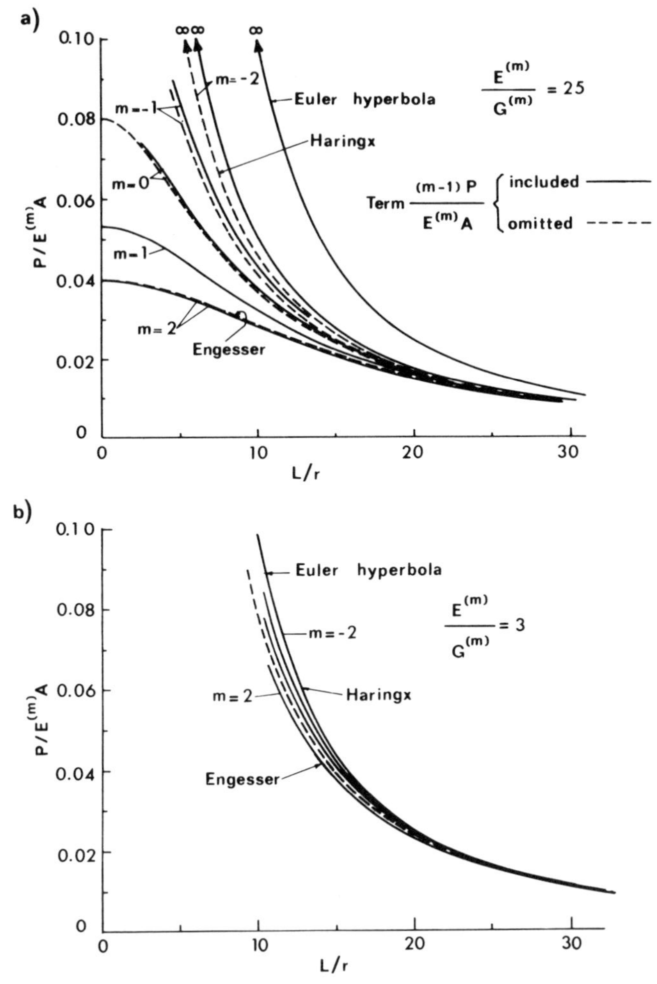

Figure 11.5 Bažant's (1971) solution of critical stress for thick columns with shear for different definitions of finite strain.

corresponding to $m = -2$ ($\alpha = 2$) (Bažant, 1971), that is,

$$\varepsilon_{ij}^{(-2)} = \varepsilon_{ij} - 2e_{ki}e_{kj} \tag{11.6.12}$$

This tensor has not been used in other types of problems but is perfectly admissible, just as the Lagrangian strain tensor.

Regarding the differences between Engesser's and Haringx's formulas, much confusion persists in the literature, even though the reason for the differences was explained by Bažant (1971). Contrary to frequent claims, these formulas do not contradict each other, but are in fact equivalent. The only reason for the differences is different definitions of incremental elastic moduli of a stressed material, corresponding to different finite-strain formulations. This is described by Equation 11.4.5, which yields, for the case of uniaxial stress ($S_{11}^0 = S^0 = -P/A$)

and under the assumption that $C_{1133}^{(m)} = 0$, the following relations:

$$E^{(2)} = E^{(1)} - S^0 = E^{(0)} + 2S^0 = E^{(-2)} - 4S^0 \tag{11.6.13}$$

$$G^{(2)} = G^{(1)} + \tfrac{1}{4}S^0 = G^{(0)} - \tfrac{1}{2}S^0 = G^{(-2)} - S^0 \tag{11.6.14}$$

The transformation in Equation 11.6.14 is essential but that in Equation 11.6.13 can be neglected since shear in columns is important only when $|S^0| \ll E^{(2)}$. Using Equation 11.6.14, one can for example transform Engesser's formula (Eq. 11.6.10) into Haringx's formula (Eq. 11.6.11).

The implication is that the effective incremental shear stiffness of any column with shear cannot really be determined without taking into account the initial stresses and their work on the geometrically nonlinear part of finite deformation of a beam element. This is important particularly for shear columns with very small slendernesses. The fact that the buckling tests of springs are apparently described by Haringx's equation with constant $G^{(-2)}$ better than by Engesser's equation with constant $G^{(2)}$ means that the transformation in Equation 11.6.14 happens to compensate for some geometric nonlinearity of springs that causes $G^{(2)}$ to depend on S^0.

Engesser's equations 1.7.8 for buckling of columns with shear were originally derived from the equilibrium conditions written under the assumption that the shear force that produces shear angle $\gamma = w_{,x} - \psi$ is given by $Q = -Pw_{,x}$. This shear force represents the tangential force component caused by the initial axial force P ($P = -S^0A$) on a cross section that is normal to the deflected beam axis (Fig. 11.4b). On the other hand, Haringx's equation is derived assuming the shear force that produces γ to be given by $\bar{Q} = P\psi$, which represents the tangential force component produced by the axial load P on a cross section that was normal to the beam axis in the initial state, after this cross section has rotated through angle ψ (Fig. 11.4c).

In the case of solid columns, whether we take the shear force on a cross section inclined by angle $w_{,x}$ (Fig. 11.4b) or by angle ψ (Fig. 11.4c) or by a weighted average of these two angles is entirely a matter of choice. But for each choice one needs to define and measure the tangential stiffness for finite deformation of a solid column in a different manner, obtaining for them different values; see Equations 11.6.13 and 11.6.14. (Conversely, a good way of measuring $E^{(m)}$ and $G^{(m)}$ under initial uniaxial stress experimentally might be to make buckling tests and calculate $E^{(m)}$ and $G^{(m)}$ from the measured values of P_{cr_1}.)

Correlation to Built-Up Columns

In the light of the foregoing discussion, it is interesting to examine the meaning of $G^{(m)}$ for built-up columns. The material in built-up columns remains in small strain, and so in contrast to solid shear columns, there can be no ambiguity in the values of the elastic moduli of the material. Therefore, the value of the effective shear stiffness must be fully determined by the small strain, Young's modulus E, and by the geometry of the cells of the column, and the transformation in Equation 11.6.14 must be entirely explicable by geometric effects.

To illustrate it, consider a battened column (Fig. 11.6a). In Sec. 2.7 we analyzed its equilibrium considering an undeformed H-cell (Fig. 2.34), but correctly the equilibrium should be analyzed considering a deformed cell as

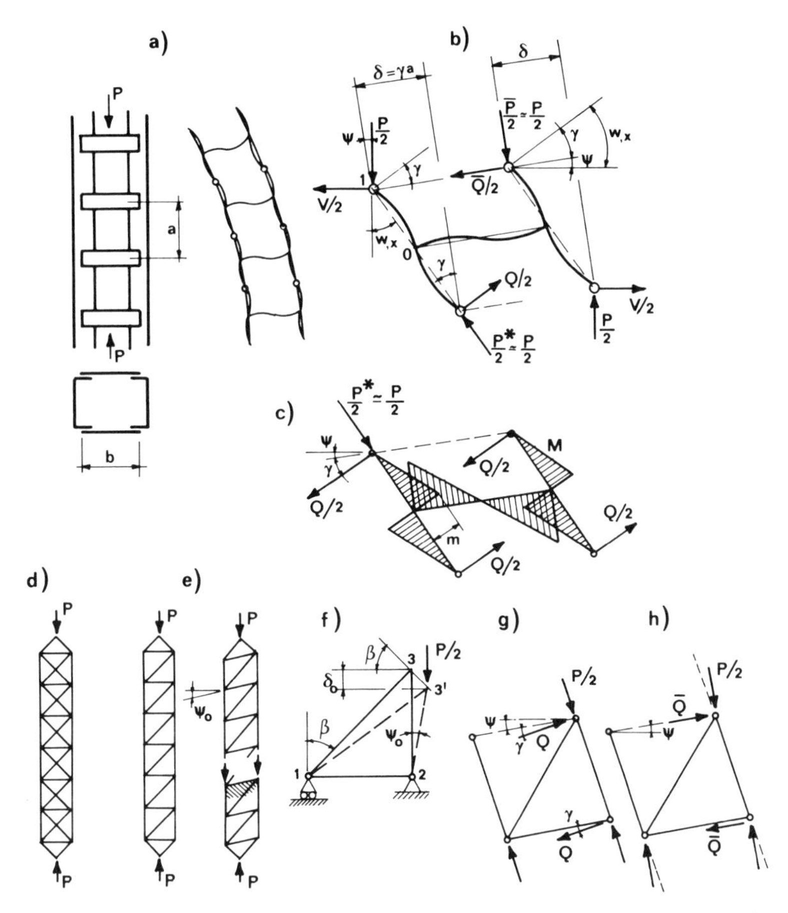

Figure 11.6 Deformed cell configurations for (a–c) battened columns and (d–h) laced columns with symmetric and asymmetric lattices.

shown in Figure 11.6b. The internal forces P and V in the coordinate directions can be replaced by statically equivalent forces P^* and Q whose inclinations are equal to slope $w_{,x}$, or by forces $\bar{P}$ and $\bar{Q}$ whose inclinations are equal to cross-section rotation ψ (Fig. 11.6b). Because angles $w_{,x}$ and γ are small and because $|V| \ll |P|$, we have $P^* = P \cos w_{,x} - V \sin w_{,x} \simeq P$, $Q = P \sin w_{,x} + V \cos w_{,x} \simeq V + Pw_{,x}$, $\bar{P} = P \cos \psi - V \sin \psi \simeq P$, $\bar{Q} = P \sin \psi + V \cos \psi \simeq V + P\psi$. Now, the relative shear displacement δ of the cell ($\delta = \gamma a$) is proportional to bending moment m in the flange near the joint (Fig. 11.6c). Among the shear forces V, $\bar{Q}$, and Q, only Q determines m (because P^* does not contribute to γ while P and $\bar{P}$ do—see Fig. 11.6b, c). Therefore, the correct shear angle expression is $\gamma = Q/GA_0$ (and not $\bar{Q}/GA_0$ or V/GA_0), which is what we used in Section 2.7 (GA_0 = effective shear stiffness and $A_0 = GA_0/G$).

Knowing that $Q = GA_0\gamma$ for battened columns, we can also deduce the transformation in Equation 11.6.14 purely geometrically. Static equivalence (according to Figs 11.4b, c and 11.6b) requires that $\bar{Q} = Q \cos \gamma - P^* \sin \gamma \simeq Q - P\gamma$. Substituting $Q = GA_0\gamma$ we have $\bar{Q} = (GA_0 - P)\gamma$, which may be written as $\bar{Q} = \bar{G}A_0\gamma$ where

$$\bar{G} = G + S^0 \qquad \left(S^0 = -\frac{P}{A_0}\right) \tag{11.6.15}$$

Since $\bar{G} = G^{(-2)}$ and $G = G^{(2)}$, we have verified Equation 11.6.14, originally deduced from finite-strain theory.

Similar results are obtained for laced columns with symmetric lattice, such as that in Figure 11.6d, or regular building frames treated as columns (Sec. 2.9). The behavior is more complicated for laced columns with unsymmetric lattice, for example, that in Figure 11.6e. The cells of this lattice, assumed to be pin-jointed, undergo shearing by angle $\psi_0 = (\delta_0/a)\cot\beta$ (with $\delta_0 = Pa/2EA_b$) already before any lateral deflection (see Fig. 11.6f; A_b = area of the flange, β = inclination of the diagonal); but the shear force Q in this undeflected state is zero (while $\bar{Q} = P\psi_0 \neq 0$). Considering then the deformed cell in Figure 11.6g, h, the additional shear angle γ must be due entirely to force Q normal to the flange (Fig. 11.6g), which is associated with force P parallel to the flange. An inclined force $\bar{P}$ associated with $\bar{Q}$, as shown in Figure 11.6h, would produce additional shear displacement in the cell. Therefore, we have again $Q = GA_0\gamma$ (and not $\bar{Q} = GA_0\gamma$) where GA = effective shear stiffness of the cell in Figure 11.6f (calculated on p. 137 of Timoshenko and Gere, 1961).

In the light of the foregoing arguments, we may conclude: The effective shear stiffness GA_0 calculated by first-order theory from a cell of a built-up (battened or laced) column whose material is under small strain represents the shear stiffness in Engesser's theory ($m = 2$) (and not in Haringx's theory, $m = -2$, or any other theory with $m \neq 2$), and in general corresponds to a formulation based on the Lagrangian strain.

Summary

While in no-shear columns only the axial stresses work, in shear columns, second-order work is done also by the shear stresses. This is the reason why the finite strain tensor must be considered for shear columns, while the geometric nonlinearity of the deflection curve suffices for no-shear columns.

Problems

11.6.1 Generalize the present variational analyses to an imperfect column that is initially straight but carries small constant lateral load $p(x)$.

11.6.2 In the formulation of Equations 11.6.2 to 11.6.11, all the points of the cross section contribute to both shear and axial stress transfer. Generalize the solution for sandwich beams and plates in cylindrical bending using the assumptions and idealizations of Section 1.7 and develop for them a finite-strain stability formulation.

11.6.3 Derive the differential equations for $\psi(x)$ and $w(x)$ that are associated with (a) Biot strain, $m = 1$; (*b*) logarithmic strain, $m = 0$ (and the Jaumann rate of Kirchhoff stress).

11.6.4 Solve exact P_{cr_1} according to Equation 11.6.7 with $k = \pi/L$ for any m, and show that for large P_{E} it asymptotically approaches the Euler formula.

11.6.5 Use Equation 11.6.14 or 11.6.15 to transform Engesser's formula into Haringx's formula.

11.6.6 The material rotation in a shear column is neither $w_{,x}$ nor ψ, but $\omega = (w_{,x} + \psi)/2 = w_{,x} - \gamma/2$. Let $\tilde{Q}$ be the shear force on a cross section rotated by ω (which is inclined by $\gamma/2$ with respect to the normal; Fig. 11.4d). Show that the differential equations obtained upon assuming that $\tilde{Q} = \tilde{G}A\gamma$ (where $\tilde{G}$ = material constant) correspond to $m = 0$, that is, $\tilde{G} = G^{(0)}$, and calculate P_{cr_1} for a hinged column.

11.6.7 Do the same, but $G''A\gamma$ (G'' = shear modulus) acts on a cross section rotated by $(w_{,x} + \omega)/2$.

11.6.8 Use small strain theory to calculate the effective shear stiffness GA_0 of the latticed column in Figure 11.6e and prove that it must be used in Engesser's rather than Haringx's formula (cf. eq. 2-62 in Timoshenko and Gere, 1961).

11.6.9 (a) A column consisting of two rigid bars is supported on two small elastic cubes of sides h as shown in Figure 11.7a. Taking into account not only the axial forces but also the shear forces on these small cubes, considering finite normal and shear strains of these cubes, and assuming these cubes to deform homogeneously, calculate the critical load and plot it as a function of L/b. (b) Do the same for the column in Figure 11.7b.

11.6.10 Consider a cruciform column (see Fig. 8.10b) and generalize Equation 8.1.25 for critical stress σ_t to the case of large initial axial strain ε_{11}. Start by expressing $\delta^2\mathscr{W}$ directly from rotation $\theta(x)$, similar to Equation 11.6.1. Discuss the differences between various m, and the possibility of using σ_t for measuring $G^{(m)}$.

11.7 SURFACE BUCKLING AND INTERNAL BUCKLING OF ANISOTROPIC SOLIDS

Three-dimensional instabilities are important for solids with a high degree of incremental anisotropy, which can be either natural, as is the case for many fiber composites and laminates, or stress-induced, as is the case for highly damaged

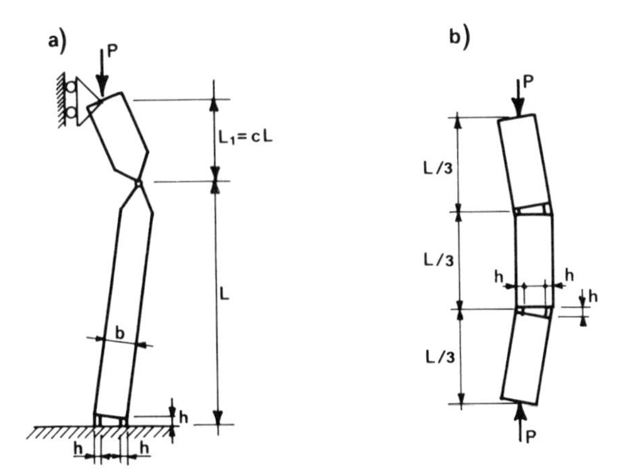

Figure 11.7 Exercise problems on rigid-bar columns with deformable cubes under finite strain.

states of materials. The typical three-dimensional instabilities are the surface buckling and internal buckling, as well as bulging and strata folding.

Basic Relations for Incompressible Orthotropic Solids

Let us consider an orthotropic material in plane strain, whose incremental stress-strain relations (Eqs. 11.4.1 for $m=2$) reduce to the form:

$$\sigma_{11}^{(m)} = C_{1111}^{(m)} u_{1,1} + C_{1122}^{(m)} u_{2,2} \qquad \sigma_{22}^{(m)} = C_{2211}^{(m)} u_{1,1} + C_{2222}^{(m)} u_{2,2}$$
$$\sigma_{12}^{(m)} = C_{1212}^{(m)}(u_{1,2} + u_{2,1}) \tag{11.7.1}$$

The solution is easier if we assume incompressibility, $u_{1,1} = -u_{2,2}$. This assumption is often quite realistic if the shear modulus is much smaller than the longitudinal moduli, which is the case when the three-dimensional instabilities are of most practical interest. Equations 11.7.1 become

$$\sigma_{11}^{(m)} - \sigma^{(m)} = 2N^{(m)} u_{1,1} \qquad \sigma_{22}^{(m)} - \sigma^{(m)} = 2N^{(m)} u_{2,2} \qquad \sigma_{12}^{(m)} = Q^{(m)}(u_{1.2} + u_{2,1}) \tag{11.7.2}$$

in which $\sigma^{(m)} = (\sigma_{11}^{(m)} + \sigma_{22}^{(m)})/2$ and

$$N^{(m)} = \tfrac{1}{4}(C_{1111}^{(m)} + C_{2222}^{(m)} - C_{1122}^{(m)} - C_{2211}^{(m)}) \qquad Q^{(m)} = C_{1212}^{(m)} \tag{11.7.3}$$

So we have only two independent elastic constants, $Q^{(m)}$ and $N^{(m)}$. Since for isotropic materials $Q^{(m)} = N^{(m)}$, the ratio $N^{(m)}/Q^{(m)}$ is a measure of the degree of orthotropy.

Consider now that S_{11}^0 and S_{22}^0 are the only nonzero initial stresses. Then, according to Equation 11.4.5,

$$N^{(1)} = N^{(2)} + \tfrac{1}{4}(S_{11}^0 + S_{22}^0) \qquad N^{(0)} = N^{(2)} + \tfrac{1}{2}(S_{11}^0 + S_{22}^0) \tag{11.7.4}$$

where superscripts (2), (1), and (0) refer to Lagrangian (Green's) strain, Biot strain, and logarithmic strain. According to Equation 11.3.11 (for $\alpha = 0$) we have

$$\tau_{11} = \sigma_{11}^{(2)} + S_{11}^0 u_{1,1} \qquad \tau_{12} = \sigma_{12}^{(2)} + S_{22}^0 u_{1,2}$$
$$\tau_{21} = \sigma_{21}^{(2)} + S_{11}^0 u_{2,1} \qquad \tau_{22} = \sigma_{22}^{(2)} + S_{22}^0 u_{2,2} \tag{11.7.5}$$

(note that $\sigma_{12}^{(m)} = \sigma_{21}^{(m)}$ but $\tau_{12} \neq \tau_{21}$). The differential equations of equilibrium are $\tau_{11,1} + \tau_{12,2} = 0$ and $\tau_{21,1} + \tau_{22,2} = 0$, which yields

$$\sigma_{11,1}^{(2)} + S_{11}^0 u_{1,11} + \sigma_{12,2}^{(2)} + S_{22}^0 u_{1,22} = 0$$
$$\sigma_{12,1}^{(2)} + S_{11}^0 u_{2,11} + \sigma_{22,2}^{(2)} + S_{22}^0 u_{2,22} = 0 \tag{11.7.6}$$

Surface Buckling of an Orthotropic Half-Space

Let us now discuss the surface buckling of a homogeneous orthotropic half-space, $x_2 < 0$ (in two dimensions). The boundary conditions on the half-space surface require that $\tau_{22} = \tau_{21} = 0$, that is,

$$\sigma_{22}^{(2)} + S_{22}^0 u_{2,2} = 0 \qquad \sigma_{12}^{(2)} + S_{11}^0 u_{2,1} = 0 \qquad (\text{at } x_2 = 0) \tag{11.7.7}$$

(note that $\sigma_{22}^{(2)} \neq 0$, $\sigma_{12}^{(2)} \neq 0$ at the half-space surface). The differential equations for functions $u_1(x_1, x_2)$ and $u_2(x_1, x_2)$ result by substituting Equations 11.7.1 or 11.7.2 into Equations 11.7.6 and 11.7.7.

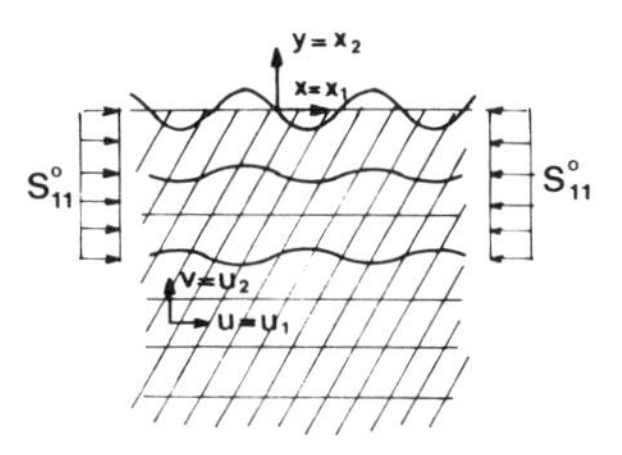

Figure 11.8 Surface buckling for homogeneous orthotropic half-space.

The solution for a half-space ($x_2 < 0$, Fig. 11.8) as well as some other bodies, may be sought in the form

$$u_1 = -f'(x_2) \sin \omega x_1 \qquad u_2 = kf(x_2) \cos \omega x_1 \tag{11.7.8}$$

where k = constant. Substituting this into the differential equations and boundary conditions one finds them to be identically satisfied if $k = 1/\omega$, and one obtains a homogeneous fourth-order linear ordinary differential equation with constant coefficients for the unknown function $f(x_2)$, along with a set of homogeneous linear boundary conditions at $x_2 = 0$. This represents an eigenvalue problem whose solutions are of the type $f(x_2) = A \exp(\beta x_2)$; β is a constant for which one obtains a quadratic equation whose roots are real or complex. The roots having positive real parts must be discarded since $\lim f(x_2)$ for $x_2 \to \infty$ must be finite. This means the displacements decay away from the surface exponentially, either as an exponential curve or an exponentially modulated sine curve. Thus, buckling is confined to a layer near the surface.

In full detail the solution was given by Biot (1963a, b). Using exclusively the elastic moduli $N^{(1)}$, $Q^{(1)}$ associated with Biot finite strain $\varepsilon_{ij}^{(1)}$, he showed that the critical states are given by the relation

$$\frac{N^{(1)}}{Q^{(1)}} = \frac{1}{2} S\left[\left(\frac{1+\zeta}{1-\zeta}\right)^{1/2} - 1\right] \qquad \zeta = \frac{S_{22}^0 - S_{11}^0}{2Q^{(1)}} \tag{11.7.9}$$

which is plotted as curve b in Figure 11.9. Replacing the elastic moduli according

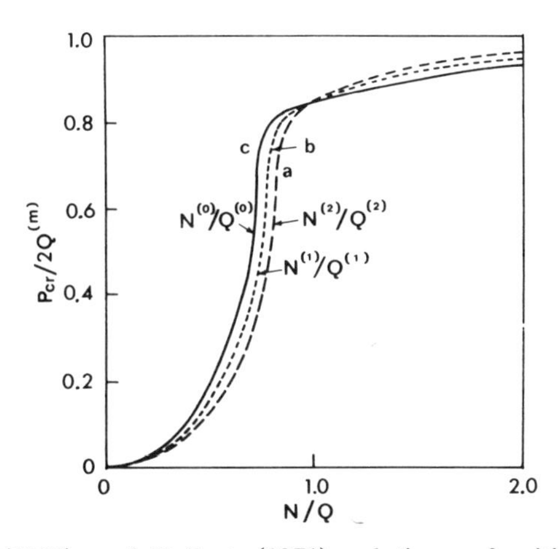

Figure 11.9 Biot's (1965) and Bažant (1971) solutions of critical stresses for surface instability of orthotropic half-space ($P = S_{22}^0 = -S_{11}^0$).

to Equation 11.4.5, one can generalize Biot's solution for the elastic moduli associated with other finite strain tensors $\varepsilon_{ij}^{(m)}$ (Bažant, 1971); see curves a and c for $m = 2$ (Lagrangian strain) and $m = 0$ (logarithmic strain). It is noteworthy that for $N^{(m)}/Q^{(m)} = 0.75$ these curves differ by as much as 2:1, which indicates an enormous influence of the choice of finite-strain measure. The differences disappear when $N^{(2)} = Q^{(2)}$, which corresponds to material isotropy.

For $S_{22}^0 - S_{11}^0 \rightarrow 0$, the ratio of the values $N^{(2)}/Q^{(2)}$, $N^{(1)}/Q^{(1)}$, and $N^{(0)}/Q^{(0)}$ approaches 1:1:1 asymptotically. This means that, for a strongly orthotropic material, the choice of the finite strain tensor becomes irrelevant. The reason is that buckling occurs at stresses that are much smaller than the elastic moduli, in which case only material rotations are large while the strains are small, just like in thin bodies.

A solution for orthotropic compressible materials has also been presented (Bažant, 1967).

Note from Figure 11.9 that the critical stress can be of a smaller order of magnitude than the tangential elastic moduli only if the material is strongly orthotropic, with $N^{(m)} \ll Q^{(m)}$.

Internal Buckling and Other Instabilities

Using a formulation corresponding to Equations 11.7.2 to 11.7.7 with $\alpha = \frac{1}{2}$, Biot (1965) considered the instability modes shown in Figure 11.10f, g, h, called the internal instabilities. They are described by the displacement field

$$u_1 = a\omega_2 \cos \omega_1 x_1 \sin \omega_2 x_2 \qquad u_2 = a\omega_1 \sin \omega_1 x_1 \cos \omega_2 x_2 \qquad (11.7.10)$$

where a, ω_1, ω_2 = real constants. This field can satisfy the boundary conditions of a rigid perfectly lubricated rectangular boundary (Fig. 11.10f, g, h). Biot showed that for one kind of instability the critical initial stresses S_{11}^0, S_{22}^0 are given by the

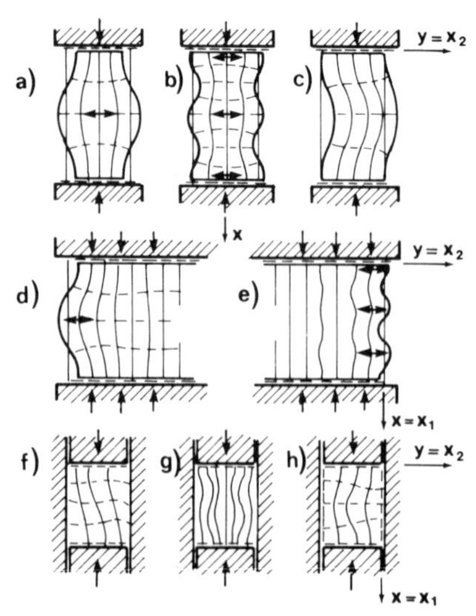

Figure 11.10 Surface and internal instability modes. (*After Bažant, 1967.*)

condition (Biot, 1965, eq. 3.23):

$$(2Q^{(1)} + P)\xi^4 + 4(2N^{(1)} - Q^{(1)})\xi^2 + 2Q^{(1)} - P = 0 \qquad \xi = \omega_2/\omega_1 \quad (11.7.11)$$

where $P = S^0_{22} - S^0_{11}$, and $Q^{(1)}$, $N^{(1)}$ $(m = 1)$ represent the formulation associated with Biot finite strain. From Equation 11.7.1, $\min P = 2Q^{(1)}$ or

$$\min (S^0_{22} - S^0_{11}) = 2C^{(1)}_{1212} \quad (11.7.12)$$

In view of the analysis in Bažant (1971) (Eqs. 11.7.4), we should note that the equivalent condition in terms of the modulus associated with the Lagrangian finite strain is

$$\min (S_{22} - S^0_{11}) = 2C^{(2)}_{1212} + \tfrac{1}{2}(S^0_{11} + S^0_{22}) \quad (11.7.13)$$

We see that the hydrostatic pressure component is important in determining the incremental shear modulus $C^{(m)}_{1212}$.

Biot (1965, eq. 3.47) also showed that another kind of internal instability occurs for $P^2 = 16N^{(1)}(Q^{(1)} - N^{(1)})$.

To solve the three-dimensional buckling of a thick compressed rectangular solid in two dimensions (Fig. 11.10a, b, c; Fig. 11.11), Biot considered the displacement field given by Equations 11.7.8 and showed that the solutions that can satisfy the boundary conditions of free surfaces of the layer have the forms (Biot, 1965, eqs. 5.4, 5.12):

$$f(x_2) = C_1 \cosh \omega_1 x_2 + C_2 \cosh \omega_2 x_2 \quad \text{(antisymmetric)} \quad (11.7.14)$$

$$f(x_2) = C_1 \sinh \omega_1 x_2 + C_2 \sinh \omega_2 x_2 \quad \text{(symmetric)} \quad (11.7.15)$$

where ω_1, ω_2, C_1, C_2 = real constants.

Figure 11.11 shows the plot of load versus slenderness obtained by Biot (1963e, 1965). He assumed that the material exhibits rubberlike finite elasticity whose incremental form corresponds to $\sigma^{(1)}_{12} = C^{(1)}_{1212} e_{12}$ $(C^{(1)}_{1212} = Q^{(1)})$. The solution is compared in the figure to Euler's buckling formula, and we see that this formula deviates appreciably from Biot's two-dimensional solution only for $\pi b/2L < 0.3$, that is, for slenderness <17. Such comparisons, of course, depend on the assumed material properties in finite strain.

The same comment applies for doubly periodic modes called internal instability shown in Figure 11.10f, g, h (Biot, 1965).

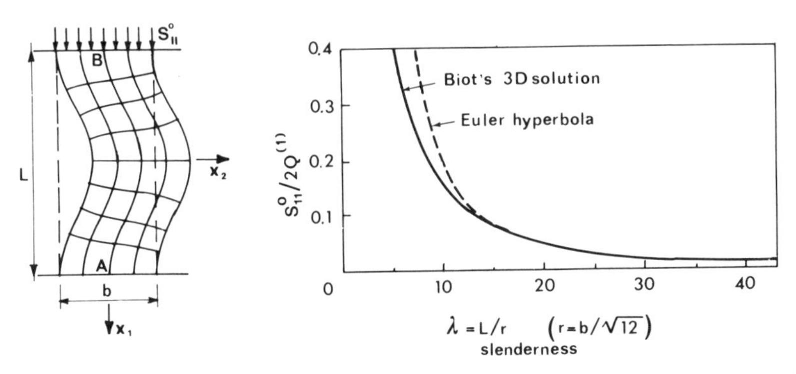

Figure 11.11 Biot's (1963e, 1965) solution of critical stress of a thick rectangular solid.

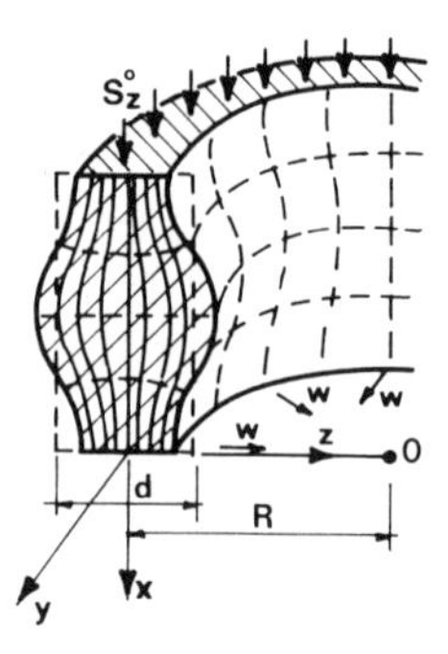

Figure 11.12 Thick-wall bulging instability considered by Bažant (1967) to explain his measurements of the compression failure stresses.

Biot (1965) also presented solutions for buckling of thick layers embedded in an elastic half-space—a problem of interest in geology for the folding of strata. He has also generalized the foregoing solutions for viscoelastic behavior.

The three-dimensional buckling modes described in this section no doubt play some role in the final phase of compression failures. For example, Bažant (1967) showed that a formula based on the thick-wall buckling mode shown in Figure 11.12 agrees with his measurements of the effects of the radius-to-wall thickness ratio on the compressive failure stress of fiber-glass laminate tubes. On the other hand, other physical mechanisms, particularly the propagation of fractures or damage bands, are no doubt more important for the theory of compression failure. The reason is two-fold: (1) The calculated critical states for the three-dimensional instabilities require some of the tangential moduli to be reduced to the same order of magnitude as some of the applied stress components, which can occur only in the final stage of the failure process; and (2) the body at this stage might no longer be adequately treated as a homogeneous continuum.

By generalization of Biot's solution for a thick layer, one can solve the periodic buckling of a compressed laminated plate (or a laminated half-space) in which the wavelength of the layers is not much larger than their thickness. The solution can be effectively obtained by means of transfer matrices that relate the integration constants of the solution for one layer to those for the next layer (these transfer matrices are obtained from the matrix that relates the interface stresses and displacements to the integration constants). See Bufler (1965), Mühlhaus (1985), Mühlhaus and Vardoulakis (1986), Vardoulakis (1984), Horii and Nemat–Nasser (1982), and a similar solution for a layer on a half-space by Dorris and Nemat-Nasser (1980).

General Solution

Finally we consider general incremental anisotropy. Substituting $\sigma_{ij}^{(2)} = C_{ijkm}^{(2)} u_{k,m}$ into the differential equations of equilibrium $\tau_{ij,j} = 0$ (Eqs. 11.3.4, $\bar{f}_i = 0$), we obtain for functions u_i the differential equations

$$(C_{ijkm}^{(2)} u_{k,m} + S_{mj}^{0} u_{i,m})_{,j} = 0 \tag{11.7.16}$$

The three-dimensional instabilities may be considered to have the general form

$$u_k = a_k e^{\omega_1 x_1} e^{\omega_2 x_2} e^{\omega_3 x_3} = a_k \exp\left(\Sigma \omega_r x_r\right) \tag{11.7.17}$$

where a_k and ω_r are complex constants. Substituting Equation 11.7.17 into Equation 11.7.16 and assuming $C^{(2)}_{ijkm}$ and S^0_{mj} to be independent of x_i, one finds this equation to be satisfied identically if

$$A_{ik} a_k = 0 \qquad \text{with } A_{ik} = \left(C^{(2)}_{ijkm} + \delta_{ik} S^0_{jm}\right) \omega_j \omega_m \tag{11.7.18}$$

This is a system of three homogeneous linear algebraic equations for amplitudes a_i. Nonzero solutions are possible if $\det A_{ij} = 0$. The roots of this algebraic equation are the eigenvalues ω_i, to each of which there corresponds a solution.

There are many such solutions, whose linear combination can satisfy the boundary conditions for various typical instabilities. These include the surface instability (Fig. 11.10d, e), as well as the doubly periodic instabilities called internal buckling (Fig. 11.10f, g, h), the bulging instabilities (Fig. 11.10a, b) of compressed prismatic specimens, the buckling of thick compressed layers (Fig. 11.10c), the buckling of layers confined in an infinite space, the buckling of layered space, and the interfacial buckling. These solutions are of interest in geology for the theory of folding of strata in the earth crust. Many problems of this type were solved by Biot (1957, 1963a, b, c, d, 1964a, b, 1965), who used moduli $C^{(1)}_{ijkm}$ rather than $C^{(2)}_{ijkm}$ and considered incompressible orthotropic materials or incrementally isotropic materials. Some further solutions for orthotropic materials were given in terms of the Lagrangian strain formulation in Bažant (1967). Stability of thick layered or sandwich plates was analyzed by Bufler (1965), and in terms of the aforementioned Neuber's differential equilibrium equations also by Neuber (1952, 1965). Buckling of rectangular thick bodies was also examined by Kerr and Tang (1967), and Wu and Widera (1969).

An impediment to practical applications of the solutions of three-dimensional instabilities is the fact that they necessitate the full anisotropic matrix of the tangential moduli C_{ijkm} (for all loading directions). These moduli need to be deduced from the finite-strain inelastic constitutive relations, which are generally not known at present. The research community is still struggling with the constitutive relations for small strain plasticity and damage.

Problems

11.7.1 Check that Equations 11.7.8 satisfy Equations 11.7.6 and 11.7.7 with Equations 11.7.2, and derive the differential equation for $f(x_2)$ with the boundary conditions. Discuss the solution of the above differential equation for (a) surface buckling, (b) internal buckling.

11.7.2 The condition of incompressibility in plane strain reads $u_{1,1} + u_{2,2} = 0$. To satisfy it identically in plane strain, Biot introduced function $\Phi(x_1, x_2)$ such that $u_1 = \partial\Phi/\partial x_2$, $u_2 = -\partial\Phi/\partial x_1$. Instead of Equations 11.7.8, one may then introduce $\Phi(x_1, x_2) = \phi(x_2) \sin \omega x_1$. Use Equations 11.7.5 and 11.7.6 to obtain the differential equation and boundary conditions for function $\Phi(x_2)$, and derive from it the solution for surface instability.

11.8 CONSISTENT GEOMETRIC STIFFNESS MATRIX OF FINITE ELEMENTS

For problems more complex than those analyzed in the preceding sections, analytical solutions become either ineffective or impossible, and one has no choice but to resort to the finite element method. Although treatment of the numerical aspects of this method is outside the scope of this text, let us at least examine how the geometric nonlinearity due to finite strain is properly introduced.

The structure is analyzed in small loading increments. In each of them, we need the incremental stiffness matrix **K** of each finite element. To this end, we substitute $\Sigma_{ij}^{(m)} = S_{ij}^0 + \sigma_{ij}^{(m)}$ (Eqs. 11.3.1) into the virtual work equation 11.2.3 for the final state of the loading increment, and subtract from it the virtual work equation $\int_V S_{ij}^0 \delta e_{ij}\, dV - \int_V \rho f_i^0 \delta u_i\, dV - \int_S p_i^0 \delta u_i\, dS = 0$ expressing the equilibrium condition for the initial state at the start of the loading increment, at which the load values are f_i^0 and p_i^0. This yields the fundamental incremental virtual work relation

$$\delta \mathscr{W} = \int_V [\sigma_{ij}^{(m)} \delta e_{ij} + S_{ij}^0(\delta \varepsilon_{ij}^{(m)} - \delta e_{ij})]\, dV - \int_V \rho \tilde{f}_i \delta u_i\, dV - \int_S \tilde{p}_i \delta u_i\, dS = 0 \tag{11.8.1}$$

$\tilde{f}_i = f_i - f_i^0$ and $\tilde{p}_i = p_i - p_i^0$ are the increments of the applied volume and surface forces over the loading step.

Let $\mathbf{q} = (n \times 1)$ column matrix of the nodal displacements of the finite element. The basic hypothesis of the finite element method is that the displacements $u_i(\mathbf{x})$ at any point $\mathbf{x}$ of the finite element can be approximated as $u_i = \mathbf{H}_i^T \mathbf{q}$ $(i = 1, 2, 3)$ where $\mathbf{H}_i = \mathbf{H}_i(\mathbf{x}) = (n \times 1)$ column matrices of the interpolation polynomials of the finite element (e.g., Zienkiewicz, 1977). Recalling that the linearized (small) strains are $e_{ij} = (u_{i,j} + u_{j,i})/2$, we obtain by differentiation:

$$u_{k,i} = \mathbf{H}_{k,i}^T \mathbf{q} \qquad e_{ij} = \tfrac{1}{2}(\mathbf{H}_{i,j} + \mathbf{H}_{j,i})^T \mathbf{q} = \mathbf{R}_{ij}^T \mathbf{q} \tag{11.8.2}$$

where $\mathbf{H}_{k,i}$ and $\mathbf{R}_{ij}$ are $(n \times 1)$ column matrices depending on coordinate vector $\mathbf{x}$. Further we recall from Equation 11.1.7 that $\varepsilon_{ij}^{(m)} - e_{ij} = \frac{1}{2} u_{k,i} u_{k,j} - \alpha e_{ki} e_{kj}$ (only second-order accuracy is needed for small increments), substitute this into Equation 11.8.1, and note that $\mathbf{H}_{k,i}^T \mathbf{q} = \mathbf{q}^T \mathbf{H}_{k,i}$, $\mathbf{R}_{ki}^T \mathbf{q} = \mathbf{q}^T \mathbf{R}_{ki}$. Thus we acquire the basic variational relation:

$$\int_V [\sigma_{ij}^{(m)} \delta e_{ij} + S_{ij}^0 \delta(\tfrac{1}{2}\mathbf{q}^T \mathbf{H}_{k,i} \mathbf{H}_{k,j}^T \mathbf{q} - \alpha \mathbf{q}^T \mathbf{R}_{ki} \mathbf{R}_{kj}^T \mathbf{q})]\, dV - \int \cdots\, dS = 0 \tag{11.8.3}$$

It is now advantageous to assume that the nodal coordinates of the finite element are updated after each small loading increment so that $\mathbf{q}$, u_i, and e_{ij} represent the increments of nodal displacements, continuum displacements, and small strains from the initial state of the loading step, in which the stress is S_{ij}^0. Then, from Equations 11.4.1, $\sigma_{ij}^{(m)} = C_{ijkm}^{(m)} e_{km} = C_{ijkm}^{(m)} \mathbf{R}_{km}^T \mathbf{q}$ where $C_{ijkm}^{(m)}$ are the tangential moduli for the initial state or, more accurately, for the midstep. Substituting this and $\delta e_{ij} = \mathbf{R}_{ij}^T \delta \mathbf{q} = \delta \mathbf{q}^T \mathbf{R}_{ij}$ into Equation 11.8.3 and taking the variations, we have

$$\int_V \{\delta\mathbf{q}^T\mathbf{R}_{ij}C^{(m)}_{ijkm}\mathbf{R}^T_{km}\mathbf{q} + S^0_{ij}[\tfrac{1}{2}(\delta\mathbf{q}^T\mathbf{H}_{k,i}\mathbf{H}^T_{k,j}\mathbf{q} + \mathbf{q}^T\mathbf{H}_{k,i}\mathbf{H}^T_{k,j}\delta\mathbf{q})$$

$$-\alpha(\delta\mathbf{q}^T\mathbf{R}_{ki}\mathbf{R}^T_{kj}\mathbf{q} + \mathbf{q}^T\mathbf{R}_{ki}\mathbf{R}^T_{kj}\delta\mathbf{q})]\}\, dV - \int \cdots dS = 0 \quad (11.8.4)$$

Now, if we note that $\mathbf{q}^T\mathbf{H}_{k,i}\mathbf{H}^T_{k,j}\delta\mathbf{q} = \delta\mathbf{q}^T\mathbf{H}_{k,j}\mathbf{H}^T_{k,i}\mathbf{q}$ and factor out $\delta\mathbf{q}^T$ as well as $\mathbf{q}$, we obtain the variational relation

$$\delta\mathbf{q}^T(\mathbf{K}\mathbf{q} - \mathbf{F}) = 0 \qquad \mathbf{F} = \int_V \mathbf{H}_i\tilde{f}_i\rho\, dV + \int_S \mathbf{H}_i\tilde{p}_i\, dS \quad (11.8.5)$$

in which

$$\mathbf{K} = \mathbf{K}^{0(m)} + \mathbf{K}^{\sigma(m)} \quad (11.8.6)$$

$$\mathbf{K}^{0(m)} = \int_V \mathbf{R}_{ij}C^{(m)}_{ijkm}\mathbf{R}^T_{km}\, dV \quad (11.8.7)$$

$$\mathbf{K}^{\sigma(m)} = \int_V S^0_{ij}[\tfrac{1}{2}(\mathbf{H}_{k,i}\mathbf{H}^T_{k,j} + \mathbf{H}_{k,j}\mathbf{H}^T_{k,i}) - \alpha(\mathbf{R}_{ki}\mathbf{R}^T_{kj} + \mathbf{R}_{kj}\mathbf{R}^T_{ki})]\, dV \quad (11.8.8)$$

with $\alpha = 1 - (m/2)$. Summations over repeated tensorial subscripts i, j, k are implied in Equations 11.8.5 to 11.8.8; $\mathbf{K}$ is the incremental stiffness matrix of the finite element; $\mathbf{K}^{0(m)}$ represents the usual, small-strain incremental stiffness matrix, and $\mathbf{K}^{\sigma(m)}$ represents the incremental geometric stiffness matrix (which is referred to the initial state at the start of the loading step and was already introduced in Equations 10.1.32–10.1.34). From the fact that Equation 11.8.5 must hold for arbitrary variations $\delta\mathbf{q}^T$, it follows that $\mathbf{K}\mathbf{q} = \mathbf{F}$, and so $\mathbf{F}$ represents the $(n \times 1)$ column matrix of the incremental nodal forces resulting from the prescribed load increments. The evaluation of the integrals such as those in Equations 11.8.5 to 11.8.8 by numerical integration is standard in finite element courses.

Now it is important to realize that both $\mathbf{K}^{0(m)}$ and $\mathbf{K}^{\sigma(m)}$ depend on m, that is, on the choice of the finite strain tensor for the incremental displacements. On the other hand, for physical reasons, the total incremental stiffness matrix $\mathbf{K}$ must, of course, be independent of the choice of m. So the differences in $\mathbf{K}^{\sigma(m)}$ due to various choices of the finite strain tensor are exactly compensated by the corresponding differences in $\mathbf{K}^{0(m)}$, resulting from Equation 11.4.5 for $C^{(m)}_{ijkm}$, provided that the values of $\mathbf{K}^{(\sigma)}$ and C_{ijkm} are mutually consistent (same m). This consistency condition for the geometric stiffness has often been overlooked.

The geometric stiffness matrix has its simplest form for the case of the Lagrangian (Green's) strain tensor ($m = 2$, $\alpha = 0$), for which the term with α in the integrand of Equation 11.8.8 vanishes.

Problems

11.8.1 Is $\mathbf{K}^{\sigma(m)}$ fully symmetric?

11.8.2 Use Equation 11.4.5 to show that $\mathbf{K}$, according to Equations 11.8.6 to 11.8.8, is independent of the choice of m.

11.8.3 In finite element courses, the $(1 \times n)$ vectors $\mathbf{R}_{ij}^T$ for all the six ij-combinations are usually grouped into one $(6 \times n)$ rectangular matrix $\mathbf{B} = (\mathbf{R}_{11}^T, \ldots, \mathbf{R}_{31}^T)$, called the geometric matrix, such that $\mathbf{e} = \mathbf{Bq}$. Another rectangular matrix $\mathbf{H}$ can be set up from matrices $\mathbf{H}_{i,j}^T$. Derive the matrix form of Equations 11.8.7 and 11.8.8 that has no subscripts. (Our derivation in terms of $\mathbf{R}_{ij}$ and $\mathbf{H}_{i,j}$ was more transparent than that in terms of $\mathbf{B}$ and $\mathbf{H}$.)

11.8.4 A planar four-node quadrilateral element is subjected to various uniform deformation fields $u_{i,j}$. Using $\mathbf{K}^{0^{(m)}}$ and $\mathbf{K}^{\sigma^{(m)}}$, calculate the corresponding nodal forces. Supposing that one can deform a rectangular specimen in the laboratory in the same way (this, is, of course, difficult), discuss how the force resultants measured on the sides of the specimen could be used to determine the tangential stiffnesses $C_{ijkm}^{(m)}$.

11.8.5 Considering a beam element with a cubic interpolation polynomial for w, prove that its geometric stiffness matrix (Eq. 2.3.6) is a special case of Equation 11.8.8.

11.9 BUCKLING OF CURVED FIBERS IN COMPOSITES

As we saw in Section 11.7, orthotropic composites that have a very high stiffness in one direction and a small shear stiffness may suffer three-dimensional instabilities such as internal buckling or surface buckling. These instabilities, which involve buckling of stiff fibers (glass, carbon, metal) restrained by a relatively soft matrix (polymer), are analogous to the buckling of perfect columns. When the fibers are initially curved, one may expect behavior analogous to the buckling of imperfect columns. In particular, the initial curvature of fibers causes fiber buckling, which reduces the stiffness of the composite. It also gives rise to transverse tensions, which may promote delamination failure.

For the sake of simplicity, we will assume the fibers to be sinusoidally undulated. Following Bažant (1968a), we will now show a simple solution without taking recourse to the general three-dimensional stability theory expounded in the previous sections. This solution is valid only for small incremental deformations of a composite with strong fiber imperfections and cannot yield the critical loads and instability as the limiting case.

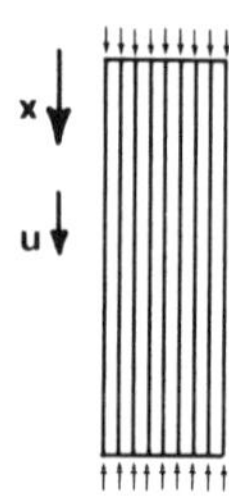

Figure 11.13 Orthotropic composite with straight parallel fibers analyzed by Bažant (1965).

Macroscopic Elastic Stress–Strain Relations

Consider first the case of plane strain in Cartesian coordinates x and z. Assuming the material to be elastic and orthotropic, one may write the small-strain stress–strain relations in the form

$$\sigma_x = E_{xx}u_{,x} + E_{xz}w_{,z} \qquad \sigma_z = E_{xz}u_{,x} + E_{zz}w_{,z} \qquad \tau_{xz} = G_{xz}(u_{,z} + w_{,x}) \quad (11.9.1)$$

where σ_x, σ_z = normal stresses, τ_{xz} = shear stress, and u, w = displacements in x- and z-directions. If all the fibers have the direction of the x axis (Fig. 11.13), one can use the approximate estimates (Bažant, 1965):

$$\begin{aligned}
&E_x = E_m V_m + E_f V_f && G_{xz} = (G^{-1}V_m + G_f^{-1}V_f)^{-1} \\
&E_{zz} = (E_m'^{-1}V_m + E_f'^{-1}V_f)^{-1} && E_x^{-1}\nu_x = E_z^{-1}\nu_z = E_x^{-1}(\nu_m V_m + \nu_f V_f) \\
&E_{xz} = E_{zz}E_x E_z^{-1}\nu_z && E_{xx} = E_x + E_{zz}^{-1}E_{xz}^2 = E_x + E_{zz}(\nu_m V_m + \nu_x V_f)^2 \\
&E_z = E_{zz} - E_{xx}^{-1}E_{xz}^2 && (11.9.2)
\end{aligned}$$

in which E_f, E_m, G_f, G_m, ν_f, ν_m are the longitudinal elastic moduli, shear moduli, and Poisson ratios of the fibers and the matrix; $E_m' = E_m/(1-\nu_m^2)$, $E_f' = E_f/(1-\nu_f^2)$; V_f and $V_m = 1 - V_f$ are the volume fractions of the fibers and the matrix. The foregoing formulas are based on the following assumptions: (1) The fibers are sufficiently densely distributed, so that macroscopically the composite may be treated as a continuum; (2) the fibers are and remain nearly straight and their curvature radius is much larger than the fiber diameter; (3) $E_f \gg E_m$.

Decrease of Elastic Moduli Due to Fiber Undulation

Case A. Consider now a composite in which the fibers are not initially straight but undulated in the form of parallel sine curves (Fig. 11.14a) described by ordinates $\zeta = C\cos(\pi x/l)$ where C, l = constants, C = initial amplitude and l = half-wavelength; $C \ll l$. If the fibers carry axial stress σ_x^f, they transmit into the composite the radial forces (per unit volume)

$$f_z = \sigma_x^f V_f \zeta_{,xx} = -f_0 \cos\frac{\pi x}{l} \quad (11.9.3)$$

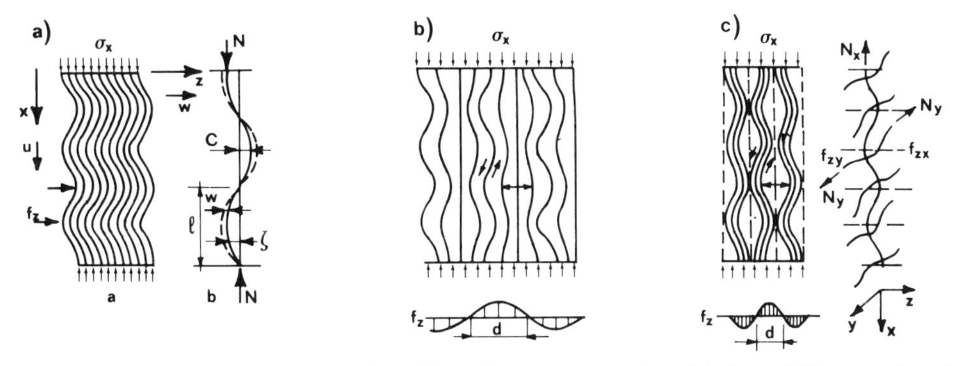

Figure 11.14 Orthotropic composite with various patterns of undulation of fibers analyzed by Bažant (1968a).

in which $f_0 = C\sigma_x^f V_f \pi^2/l^2$. The differential equations of equilibrium for the composite are $\sigma_{x,x} + \tau_{xz,z} = 0$, $\tau_{xz,x} + \sigma_{z,z} + f_z = 0$, and upon substitution of Equations 11.9.1 this yields

$$\begin{aligned}(E_{xx}u_{,x} + E_{xz}w_{,z})_{,x} + [G_{xz}(u_{,z} + w_{,x})]_{,z} &= 0\\ [G_{xz}(u_{,z} + w_{,x})]_{,x} + (E_{zz}w_{,z} + E_{xz}u_{,x})_{,z} + f_z &= 0\end{aligned} \tag{11.9.4}$$

Instead of boundary conditions we assume translational symmetry of the solution in the z-direction, and symmetry with respect to all the planes of symmetry of the sine curve undulations. The solution has the form:

$$u = \varepsilon_x x \qquad w = -c\cos\frac{\pi x}{l} \tag{11.9.5}$$

where ε_x = constant (strain) and $c = f_0 l^2/\pi^2 G_{xz} = \sigma_x^f C V_f/G_{xz}$. We note that the axial stress magnifies the initial undulation, and the magnification depends, among the moduli, only on G_{xz}.

Similar to our analysis of columns (Sec. 4.3), the contribution of the magnification of fiber undulations to the axial deformation of composite per unit length is described by

$$\begin{aligned}\varepsilon_x l &= \varepsilon_x^f l - \frac{1}{2}\int_0^l [(\zeta + w)_{,x}^2 - \zeta_{,x}^2]\,dx\\ &= \varepsilon_x^f l + \left(Cc - \frac{1}{2}c^2\right)\frac{\pi^2}{2l} \simeq \varepsilon_x^f l + \frac{\pi^2 Cc}{2l}\end{aligned} \tag{11.9.6}$$

where ε_x^f = axial strain of fibers. So the effect of the initial undulation is to change the composite modulus E_x to the effective value $\bar{E}_x = E_m V_m + \bar{E}_f V_f$ in which $\bar{E}_f = \sigma_x^f/\varepsilon_x$, that is,

$$\bar{E}_f = \frac{E_f}{1 + kV_f(C^2/l^2)(E_f/G_{xz})} \tag{11.9.7}$$

with $k = \pi^2/2$.

Case B. Consider now the initial fiber undulation that is variable as shown in Figure 11.14b and is described by $\zeta = C\cos(\pi x/l)\sin(\pi z/d)$. In this case $f_z = -f_0\cos(\pi x/l)\sin(\pi z/d)$. The solution of Equations 11.9.4 has the form

$$\begin{aligned}u &= \varepsilon_x x - a\sin\frac{\pi x}{l}\cos\frac{\pi z}{d}\\ w &= -c\cos\frac{\pi x}{l}\sin\frac{\pi z}{d}\end{aligned} \tag{11.9.8}$$

Since the shortenings of various fibers are now different, we must express the effective stiffness on the basis of the potential energy

$$\Pi = \int_0^l \left(\tfrac{1}{2}E_{zz}w_{,z}^2 + \tfrac{1}{2}G_{xz}w_{,x}^2 - f_z w\right)dx\,dz \tag{11.9.9}$$

From the minimizing condition $\partial\Pi/\partial c = 0$ we get

$$c = \frac{f_0/\pi^2}{(G_{xz}/l^2) + (E_{zz}/d^2)} \tag{11.9.10}$$

To obtain $\varepsilon_x l$, we consider now the average fiber shortening, that is,

$$\varepsilon_x l = \varepsilon_x^f l - \frac{1}{2d}\int_{x=0}^{l}\int_{z=0}^{d} [(\zeta + w)_{,x}^2 - \zeta_{,x}^2]\, dx\, dz \tag{11.9.11}$$

Again we obtain Equation 11.9.7, in which now

$$k = \frac{\pi^2}{4[1 + (E_{zz}/G_{xz})(l^2/d^2)]} \tag{11.9.12}$$

Case C. In a laminate reinforced by fabric, all the waves of fibers have about the same amplitudes, and the fiber shape may be idealized as shown in Figure 11.14c, for which $\zeta = \pm C \cos(\pi x/l)$. We may consider this to be approximately equivalent to the previous case provided that the sum of the radial forces from the fibers is the same (per length l), that is,

$$\int_0^d f_z\, dz = -V_f C\left(\frac{\pi^2}{l^2}\right)\sigma_x^f d \cos\frac{\pi x}{l} \tag{11.9.13}$$

in which we may assume $f_z = -\bar{f}_0 \cos(\pi x/l)\sin(\pi z/d)$. From Equation 11.9.13 it follows that $\bar{f}_0 = f_0\pi/2$. Replacing f_0 by $\bar{f}_0$ in Equation 11.9.10, we get

$$c = \frac{\bar{f}_0/\pi^2}{(G_{xz}/l^2) + (E_{zz}/d^2)} = \frac{f_0 l^2}{\pi^2 \bar{G}_{xz}} \qquad \bar{G}_{xz} = \frac{2}{\pi}\left(G_{xz} + \frac{l^2}{d^2}E_{zz}\right) \tag{11.9.14}$$

The shortening due to undulation of variable amplitude is again determined by the integral in Equation 11.9.6. This yields $\bar{E}_x = E_m V_m + \bar{E}_f V_f$, with $\bar{E}_f$ given by Equation 11.9.7 in which (Bažant, 1968a)

$$k = \frac{\pi^3}{4[1 + (E_{zz}/G_{xz})\,(l^2/d^2)]} \tag{11.9.15}$$

For $E_m/E_f \to 0$ this simplifies as

$$\lim \bar{E}_x = E_m V_m + G_{xz} V_f \frac{l^2}{kC^2} \tag{11.9.16}$$

Generalization to Three Dimensions

Case D. Consider now composites reinforced by a fabric woven from fibers of two directions x and y, whose volumes are V_f^1 and V_f^2 ($V_f^1 + V_f^2 = V_f$). We assume that (1) the mean normal stresses of x-direction in the matrix and in the fibers of orientation y are equal; (2) the shear stresses τ_{xz}, τ_{xy}, τ_{yz} in fibers and matrix are also equal; (3) the axial strains of fibers and matrix are equal. Then, approximately (Bažant, 1965),

$$\begin{aligned} G_{xy} = G_{xz} = G_{yz} &= (G_m^{-1} V_m + G_f^{-1} V_f)^{-1} \\ E_x &= E_m^1 V_m + E_f V_f^1 \qquad \text{etc.} \end{aligned} \tag{11.9.17}$$

in which $V_m^1 = V_m + V_f^2$, $E_m^1 V_m^1 = (E_m^{-1} V_m + E_f^{-1} V_f^2)^{-1}$. We assume the fiber shape in the woven fabric is, approximately,

$$\begin{aligned} \zeta_1 &= \pm C_1 \cos\frac{\pi x}{l_1} \qquad \text{for } \cos\frac{\pi y}{l_2} \gtrless 0 \\ \zeta_2 &= \mp C_2 \cos\frac{\pi y}{l_2} \qquad \text{for } \cos\frac{\pi x}{l_1} \gtrless 0 \end{aligned} \tag{11.9.18}$$

The radial distributed forces from the fibers on the matrix are $f_z = -\tilde{f}_0 \cos(\pi x/l_1)\cos(\pi y/l_2)$. We introduce an averaging condition similar to Equation 11.9.13:

$$\int_0^{l_2} f_2\, dy = -l_2 \sigma_x^f V_f^1 C_1 \frac{\pi^2}{l_1^2} \cos\frac{\pi x}{l_1} \tag{11.9.19}$$

with $\tilde{f}_0 = f_0 \pi/2$, $f_0 = l_1 \sigma_x^f V_f^1 \pi^3 / l_1^2$. A three-dimensional generalization of the differential equations 11.9.4 can be solved by

$$u = v = \varepsilon_x x = 0 \qquad w = -c\cos\frac{\pi x}{l_1}\cos\frac{\pi x}{l_2} \tag{11.9.20}$$

with

$$c = \frac{\tilde{f}_0/\pi_2}{(G_{xz}/l_1^2) + (G_{yz}/l_2^2)} = \frac{f_0 l_1^2}{\pi^2 \tilde{G}_{xz}} \qquad \tilde{G}_{xz} = \left(G_{xz} + \frac{G_{yz} l_1}{l_2}\right)^2 \frac{2}{\pi} \tag{11.9.21}$$

Next we proceed in the same manner as from Equation 11.9.6 to 11.9.7, gaining the result $\tilde{E}_x = E_m^1 V_m^1 + \tilde{E}_f^1 V_f^1$, in which (Bažant, 1968a)

$$\tilde{E}_f^1 = \frac{E_f}{1 + kV_f^1 (C_1^2/l_1^2)(E_f/G_{xz})} \qquad \text{with} \qquad k = \frac{\pi^3/4}{1 + (G_{yz}/G_{xz})(l_1^2/l_2^2)} \tag{11.9.22}$$

The undulation of fibers also affects the Poisson ratio in the xy plane. Extending the technique illustrated here, Bažant (1968a, eq. 25) obtained an expression that shows the Poisson ratio increases in proportion to the undulation amplitude c. Furthermore, he calculated estimates for the effective moduli for some other three-dimensional arrangements. Some extensions and experimental verifications of Bažant's results were presented by Tarnopol'skii and Roze (1969) and others.

Stresses Due to Fiber Undulation

Fiber undulation causes transverse shear stresses τ_{xz} and normal stresses σ_z. The shear stresses are obtained as $\tau_{xz} = G_{xz}(u_{,z} + w_{,x})$. For case A, substitution of Equations 11.9.5 yields $\tau_{xz} = (\pi/l) G_{xz} c \sin(\pi x/l)$ and $\max \tau_{xz} = \pi G_{xz} c/l$. After substituting $c = f_0 l^2 / \pi^2 G_{xz}$, $f_0 = C\sigma_x^f V_f \pi^2 / l^2$, one gets

$$\max \tau_{xz} = \frac{\pi}{l} C \sigma_x^f V_f \tag{11.9.23}$$

So the maximum shear stress is proportional to the amplitude of fiber undulation. Similar expressions can be obtained for cases B, C, and D. Analogously, one can calculate $\sigma_z = E_{zz} w_{,z}$.

In closing this subject, it needs to be pointed out that there are many factors influencing the stiffness and strength of fibrous composites. The initial curvature of fibers is not the most important one. Nevertheless, as we saw, this factor can be taken into account easily.

Problem

11.9.1 Generalize the solution for case A to three dimensions, assuming that the initial shapes of the fibers running in the x- and y-directions are $\zeta_1 = \zeta_2 = C \cos(\pi x/l_1) \cos(\pi y/l_2)$.

References and Bibliography

Almansi, E. (1911), "Sulle deformazioni finite dei solidi elastici isotropi," I and III, *Rend. Accad. Lincei (5A),* 20(1):705–44; 20(2):289–96 (Sec. 11.1).

Bardet, J. P. (1989), "Numerical Modeling of Rockburst," in *Numerical Models in Geomechnics (NUMOG 3),* ed. by S. Pietruszczak and G. N. Pande, Elsevier, New York, pp. 311–18.

Bažant, A. P. (1965), "Mathematische Ermittlung der rheologischen Eigenschaften von glasfaserverstärkten Plasten für die Berechnung von Konstruktionen" (Mathematical Formulation of Rheological Properties of Fiber-Reinforced Polymers), *Plaste und Kautschuk* (Leipzig, Germany), 12:592–9 (Sec. 11.9).

Bažant, Z. P. (1967), "L'instabilité d'un milieu continu et la résistence en compression" (Continuum Instability and Compression Strength), Bulletin RILEM (Paris), No. 35, pp. 99–112 (Sec. 11.0).

Bažant, Z. P. (1968a) "Effect of Folding of Reinforcing Fibres on the Elastic Moduli and Strength of Composite Materials" (in Russian), *Mekhanika Polimerov* (Riga), 4:314–21 (Sec. 11.9), see also English translation in *Mechanics of Polimers.*

Bažant, Z. P. (1968b), "Conditions of Deformation Instability of a Continuum and Their Application for Thick Slabs and a Half Space" (in Czech), *Stavebnícky Časopis SAV* (Bratislava), 16(1):48–64; see also "On Instability of Three-Dimensional Bodies," *Acta Polytechnica (Prague),* 1967(3):5–17.

Bažant, Z. P. (1971), "Correlation Study of Formulations of Incremental Deformations and Stability of Continuous Bodies," *J. Appl. Mech.* (ASME), 38:919–28 (Sec. 11.0).

Benedetti, D., Brebbia, C., and Cedolin, L. (1972), "Geometrically Nonlinear Analysis of Structures by Finite Elements," *Meccanica,* 7(1):1–10.

Biezeno, C. B., and Hencky, H. (1928), "On the General Theory of Elastic Stability," Koninklijke Akademie van Wettenschappen te Amsterdam, Proceedings of the Section of Sciences, Vol. 31, 1928, pp. 569–92 (Meetings December 1927, January 1928), and vol. 32, 1929, pp. 444–56 (Meeting April 1929); cf. also Biezeno, C. B., and Grammel, R. (1939), *Technische Dynamik,* chap. 1, p. 85, Springer, Berlin (English translation: Blackie, London) (Sec. 11.0).

Biezeno, C. B. and Koch, J. J. (1925), "Die Knickung von Schraubenfedern," *Z. Angew. Math. Mech. (ZAMM),* 5:279 (Sec. 11.6).

Biot, M. A. (1934), "Sur la stabilité de l'équilibre élastique. Equations de l'élasticité d'un milieu soumis à tension initiale," *Annales de la Société Scientifique de Bruxelles,* Vol. 54, Series B, Part I, pp. 18–21 (Sec. 11.1).

Biot, M. A. (1939), "Nonlinear Theory of Elasticity and the Linearized Case for a Body under Initial Stress," *Philosophical Magazine,* 27(7):468–89 (Sec. 11.1).

Biot, M. A. (1940), "Elastizitätstheorie zweiter Ordnung mit Anwendungen," *Z. Angew. Math. Mech. (ZAMM)*, 20(2):89–99.

Biot, M. A. (1957), "Folding Instability of a Layered Viscoelastic Medium under Compression," *Proc. Roy. Soc.*, Series A, 242:444–54 (Sec. 11.7).

Biot, M. A. (1959), "On the Instability and Folding Deformation of a Layered Viscoelastic Medium in Compression," *J. Appl. Mech.* (ASME), 26:393–400.

Biot, M. A. (1963a), "Internal Buckling under Initial Stress in Finite Elasticity," *Proc. Roy. Soc.*, Series A, 273:306–28 (Sec. 11.7).

Biot, M. A. (1963b), "Surface Instability in Finite Anisotropic Elasticity under Initial Stress," *Proc. Royal Soc. A*, 273:329–39 (Sec. 11.7).

Biot, M. A. (1963c), "Theory of Stability of Multilayered Continua in Finite Anisotropic Elasticity," *J. Franklin Institute*, 276(2):128–53 (Sec. 11.7).

Biot, M. A. (1963d), "Stability of Multilayered Continua Including the Effect of Gravity and of Viscoelasticity," *J. Franklin Institute*, 276(3):231–52 (Sec. 11.7).

Biot, M. A. (1963e), "Exact Theory of Buckling of a Thick Slab," *Applied Scientific Research A*, 12:182–98 (Sec. 11.7).

Biot, M. A. (1964a), "Continuum Theory of Stability of an Embedded Layer in Finite Elasticity under Initial Stress," *Quart. Mech. and Appl. Math.*, 27(I):19–22 (Sec. 11.7).

Biot, M. A. (1964b), "Theory of Stability of Periodic Multilayered Continua," *Quart. Mech. and Appl. Math.*, 17(II):217–24 (Sec. 11.7).

Biot, M. A. (1965), *Mechanics of Incremental Deformation*, John Wiley & Sons, New York, p. 504 (Sec. 11.1).

Biot, M. A., and Odé, H. (1962), "On the Folding of a Viscoelastic Medium with Adhering Layer under Compressive Initial Stress," *Quart. Appl. Math.*, 19(4):351–5.

Bryan, G. H. (1888), "On the Stability of Elastic Systems," *Proc. of the Cambridge Philos. Soc.*, 6:199–210 (Spec. Gen. Meeting on Feb. 27) (Sec. 11.5).

Bufler, H. (1965), "Compression Stability of Rectangular Composite Plates," in German, *Ingenieur Archiv*, 34:104–28 (Sec. 11.7).

Cauchy, A. (1828), "Sur les équations qui expriment les conditions d'équilibre ou les lois du mouvement intérieur d'un corps solide, élastique ou non-élastique," *Exercises de Mathématiques*, 3:160–87 [Ouevres (2), 8, pp. 253–77] (Sec. 11.1).

Chan, G. K., and Kelly, J. (1970), "Viscoelastic Stability Model for Elastomeric Isolation Bearings," *J. Struct. Eng.* (ASCE), 96(7):1377 (Sec. 11.6).

Cotter, B. A., and Rivlin, R. S. (1955), "Tensors Associated with Time-Dependent Stress," *Quart. Appl. Math.*, 13:177–82 (Sec. 11.3).

Dorris, J. P., and Nemat-Nasser, S. (1980), "Instability of a Layer on a Halfspace," *J. Appl. Mech.* (ASME), 47:304–12 (Sec. 11.7).

Doyle, T. C., and Ericksen, J. L. (1956), "Nonlinear Elasticity," *Advances Appl. Mech.*, 4:53–115 (Sec. 11.1).

Duhem, P. (1906), *Recherches sur l'élasticité, IIIe Partie, La stabilité des milieux élastiques*, Gauthier-Villars, Paris.

Engesser, F. (1889), "Die Knickfestigkeit gerader Stäbe," *Z. Architekten und Ing. Vereins zu Hannover*, 35:455 (Sec. 11.6).

Engesser, F. (1891), "Die Knickfestigkeit gerader Stäbe," *Zentralblatt der Bauverwaltung*, 11:483 (Sec. 11.6).

Eringen, A. C. (1962), *Nonlinear Theory of Continuous Media*, McGraw-Hill, New York (Sec. 11.3).

Goodier, J. N., and Plass, H. J. (1952), "Energy Theorems and Critical Load Approximations in the General Theory of Elastic Stability," *Quart. Appl. Math.*, 9(4):371–380 (Sec. 11.5).

Goto, Y., and Chen, W-F. (1987), "Second-Order Elastic Analysis for Frame Design," *J. Struct. Eng.* (ASCE), 113(7):1501–19; see also "Discussion," 115(2):500–506.

Green, A. E. (1956), "Hypo-Elasticity and Plasticity," *Proc. Roy. Soc.*, Series A, 234:46–59 (Sec. 11.3).

Green, A. E., and Adkins, J. E. (1960), *Large Elastic Deformations and Nonlinear Continuum Mechanics,* Clarendon Press, Oxford, chap. IX.

Green, G. (1839), "On the Laws of Reflection and Refraction of Light in Crystallized Media,"*Trans., Cambridge Philos. Soc.*, 7 (1838–1842):121–40 (Sec. 11.1).

Guo, Z.-H. (1963), "Time Derivatives of Tensor Fields in Nonlinear Continuum Mechanics," *Arch. mech. stosowanej* (Warszaw), 1 (15):131–63 (Sec. 11.3).

Hadamard, J. (1903), *Leçons sur la propagation des ondes,* Hermann, Paris, chap. VI (Sec. 11.5).

Haringx, J. A. (1942), "On the Buckling and Lateral Rigidity of Helical Springs," *Proc. Konink. Ned. Akad. Wet.*, 45:533; see also Timoshenko and Gere (1961), p. 142 (Sec. 11.6).

Hill, R. (1957), "On Uniqueness and Stability in the Theory of Finite Elastic Strain," *J. Mech. and Phys. of Solids,* 5:229–41 (Sec. 11.5).

Hill, R. (1968), "On Constitutive Inequalities for Simple Materials," *J. Mech. and Phys. of Solids,* 16:229–42, 315–22 (Sec. 11.2).

Hill, R. (1972), "On Constitutive Macro-Variables for Heterogeneous Solids at Finite Strain," *Proc. Roy. Soc.*, Series A, 326:131–47.

Hill, R. (1978), "Aspects of Invariance in Solid Mechanics," *Advances in Appl. Mech.*, 18:1–75.

Horii, H., and Nemat–Nasser, S. (1982), "Instability of Halfspace with Frictional Materials," *Z. Angew. Math. Phys.* (*ZAMP*), 33:1–15 (Sec. 11.7).

Hutchinson, J. W., and Tvergaard, V. (1980), "Surface Instabilities on Statically Strained Plastic Solids," *Int. J. Mech. Sci.*, 22:339–54.

Jaumann, G. (1911), "Geschlossenes System physikalischer und chemischer Differentialgesetze," *Sitzungsberichte Akad. Wiss. Wien.* (*IIa*), 120:385–530 (Sec. 11.3).

Johnston, B. G. (1976), *Guide to Stability Design Criteria for Metal Structures,* 3rd ed., Column Research Council, John Wiley and Sons, New York (Sec. 11.6).

Kerr, A. D., and Tang, S. (1967), "The Instability of a Rectangular Solid," *Acta Mechanica,* 4:43–63 (Sec. 11.7).

Koiter, W. T. (1965), "The Energy Criterion of Stability for Continuous Bodies," Proc., Konink. Ned. Akad. Wet., Amsterdam, Serie B, *Phys. Sciences,* 68:178–202.

Kirchhoff, G. (1852), "Über die Gleichungen des Gleichgewichts eines elastischen Körpers bei nicht unendlich kleinen Verschiebungen seiner Teile," *Sitzungsberichte Akad. Wiss. Wien,* 9:762–73 (Sec. 11.2).

Lee, E. H. (1969), "Elastic-Plastic Deformations in Finite Strains," *J. Appl. Mech.*, (ASME), 36:1–60 (Sec. 11.1).

Lin, F. J., Glauser, E. C., and Johnston, B. G. (1970), "Behavior of Laced and Battened Structural Members," *J. Struct. Eng.* (ASCE), 96(7):1377 (Sec. 11.6).

Malvern, L. E. (1969), *Introduction to the Mechanics of a Continuous Medium,* Prentice-Hall, Englewood Cliffs, N.J. (Sec. 11.2).

Mandel, J. (1966), *Cours de mécanique,* Dunod, Paris (Annexe XIV) (Sec. 11.2).

Marsden, J. E., and Hughes, T. J. R. (1983), *Mathematical Foundations of Elasticity,* Prentice-Hall, Englewood Cliffs, N.J. (Sec. 11.1).

Masur, E. F. (1965), "On Tensor Rates in Continuum Mechanics," *Z. Angew. Math. Phys.* (*ZAMP*), Birkhäuser, Basel, 16:191–201 (Sec. 11.1).

Mühlhaus, H. B. (1985), "Surface Instability of Layered Halfspace with Bending Stiffness," *Ingenieur Archiv,* 55:388–95 (Sec. 11.7).

Mühlhaus, H. B. (1988), "Delamination Phenomena in Prestressed Rock," *Proc. 2nd Int. Symp. on Rockbursts and Seismicity in Mines,* University of Minnesota.

Mühlhaus, H. B., and Vardoulakis, I. (1986), "Axially Symmetric Buckling of the Surface of a Laminated Halfspace with Bending Stiffness," *Mechanics of Materials,* 5:109–20 (Sec. 11.7).

Nänni, J. (1971), "Das Eulersche Knickproblem unter Berücksichtigung der Querkräfte," *Z. Angew. Math. Physik (ZAMP),* 22:156 (Sec. 11.6).

Neuber, H. (1943), "Die Grundgleichungen der elastischen Stabilität in allgemeinen Koordinaten und ihre Integration, *Z. Angew. Math. Mech. (ZAMM),* 23(6):321–30 (Sec. 11.3).

Neuber, H. (1952), "Theorie der Druckstabilität der Sandwich-Platte," *Z. Angew. Math. Mech. (ZAMM),* 32(11/12):325; 33(1953):10 (Sec. 11.3).

Neuber, H. (1965), "Theorie der elastischen Stabilität bei nichtlinearer Vorverformung," *Acta Mechanica,* Springer, Wien, No. 3, p. 285 (Sec. 11.7).

Novozhilov, V. V. (1958), *Teoria uprugosti,* Sudpromgiz, Leningrad, translated: *Theory of Elasticity,* Pergamon Press, New York, 1961; see also, *Osnovy nelineinoi teorii uprugosti,* Techteorizdat, Moscow, 1948, translated: *Foundations of the Nonlinear Theory of Elasticity,* Graylock Press, Rochester, 1953.

Ogden, R. W. (1984), *Non-linear Elastic Deformations,* Ellis Horwood, Ltd., Chichester, U.K., and John Wiley & Sons, Chichester, U.K. (Sec. 11.1).

Oldroyd, J. G. (1950), "On the Formulation of Rheological Equations of State," *Proc. Roy. Soc.* (London), Series A, 200:523–41 (Sec. 11.3).

Pearson, C. E. (1956), "General Theory of Elastic Stability," *Quart. Appl. Math.,* 14(2):133–44 (Sec. 11.5).

Piola, G. (1835), "Nuova analisi per tutte le questioni della meccanica molecolare," *Mem. Mat. Fis. Soc. Ital. Modena,* 21:155–231 (Sec. 11.2).

Prager, W. (1947), "The General Variational Principle of the Theory of Structural Stability," *Quart. Appl. Math.,* 4(4):378–84.

Prager, W. (1961), *Introduction to Mechanics of Continuous Media,* Ginn, Boston, 1961; German ed., *Einführung in die Kontinuumsmechanik,* Birkhäuser, Basel (Sec. 11.1).

Simo, J. C. (1986), "On the Computational Significance of the Intermediate Configuration and Hyperelastic Constitutive Equations," *Mechanics of Materials,* 4:439–51 (Sec. 11.1).

Simo, J. C., and Ortiz, M. (1985), "A Unified Approach to Finite Deformation Elastoplasticity Based on the Use of Hyperelastic Constitutive Equations," *Comp. Meth. Appl. Mech. and Eng.,* 49:177–208 (Sec. 11.1).

Southwell, R. V. (1914), "On the General Theory of Elastic Stability," *Phil. Trans. Roy. Soc.* (London), Series A, 213:187–244 (Sec. 11.3).

Sulem, J., and Vardoulakis, I. (1988), "A New Approach to Borehole Stability Based on Bifurcation Theory," *Proc. 6th Int. Conf. Num. Meth. in Geomech.,* Innsbruck, 3:1929–35.

Tarnopol'skii, Yu. M., and Roze, A. V. (1969), "Osobennosti rascheta detalei iz armirovannykh plastikov," in *Analysis of Reinforced Plastics,* Zinatne, Riga, Latvia, pp. 47–48 (Sec. 11.9).

Timoshenko, S. P. (1913), "Sur la stabilité des systèmes élastiques," *Annales des Ponts et Chaussées,* No. 3, p. 496, No. 4, p. 73, No. 5, p. 372.

Timoshendo, S. P., and Gere, J. M. (1961), *Theory of Elastic Stability,* 2nd ed., McGraw-Hill, New York. (Sec. 11.6).

Trefftz, E. (1933), "Zur Theorie der Stabilität des elastischen Gleichgewichts," *Z. Angew. Math. Mech. (ZAMM)*, 12(2):160–5; see also: "Über die Ableitung der Stabilitätskriterien des elastischen Gleichgewichts aus Elastizitätstheorien der endlichen Deformationen," *Proceeding of the Third International Congress on Technical Mechanics,* Stockholm 1930, Vol. 3, 1931, pp. 44–50 (Sec. 11.5).

Truesdell, C. (1953), Corrections and Additions to "The Mechanical Foundations of Elasticity and Fluid Dynamics," *J. Rat. Mech. Analysis,* 2:505–616 (eq. 55b2) (Sec. 11.3).

Truesdell, C. (1955), "The Simplest Rate Theory of Pure Elasticity," *Comm. Pure Appl. Math.,* 8:123–32 (Sec. 11.3).

Truesdell, C., and Noll, W. (1965), "The Non-Linear Field Theories of Mechanics," in *Encyclopedia of Physics,* ed. by S. Flügge, vol. III/3, chaps. 68–70, Springer-Verlag, Berlin (Sec. 11.1).

Truesdell, C., and Toupin, R. (1960), "The Classical Field Theories," in *Handbuch der Physik* III/1, ed. by S. Flügge, Springer-Verlag, Berlin (Sec. 11.3).

Vardoulakis, I. (1984), "Rock Bursting as a Surface Instability Phenomenon," *Int. J. Rock Mech. Min. Sci. and Geomech. Abstr.,* 21(3):137–44 (Sec. 11.7).

Vardoulakis, I., and Mühlhaus, H. B. (1986), "Local Rock Surface Instabilities," *Int. J. Rock Mech. Min. Sci. and Geomech. Abstr.,* 23:379–83.

Vardoulakis, I. G., and Papanastasiou, P. C. (1988), "Bifurcation Analysis of Deep Boreholes: I. Surfaces instabilities," *Int. J. Numer. Anal. Methods Geomech.,* 12(4):379–99.

Vardoulakis, I., Sulem, J., and Guenot, A. (1988), "Borehole Instabilities as Bifurcation Phenomena," *Int. J. Rock Mech. & Min. Sci. and Geomech. Abstr.,* 25:159–70.

Wan, R. G., Chan, D. H., and Morgenstern, N. R. (1989), "The Numerical Modeling of the Development of Shear Bands in Geomechanics," in *Numerical Models in Geomechanics (NUMOG 3),* ed. by S. Pietruszczak and G. N. Pande, Elsevier, New York, pp. 319–29.

Wu, C. H., and Widera, O. E. (1969), "Stability of a Thick Rubber Solid Subject to Pressure Loads," *Int. J. Solids Structures,* 5:1107–17 (Sec. 11.7).

Zienkiewicz, O. C. (1977), *The Finite Element Method,* McGraw-Hill, New York (Sec. 11.8).

12

Fracture as a Stability Problem

As we saw in Chapters 8 and 10, material plasticity has a great influence on the stability of a structure. But it does not in itself cause instability. The destabilizing cause is solely the nonlinear geometric effect. Without geometric nonlinearity there are no instabilities in hardening elastoplastic structures.

Fracture, by contrast, has a profound destabilizing influence. Instabilities arise even if nonlinear geometric effects are absent. Fracture alone can be a cause of instability.

Before we can focus on stability problems of fracture, we first need to explain the elementary concepts of linear and nonlinear fracture mechanics. After that we will apply the general thermodynamic stability criteria to structures exhibiting fracture and analyze single, isolated cracks as well as systems of interacting cracks. We will discuss in more detail some important types of instabilities such as the snapback instability in the terminal phase of tearing of a cross section. Finally, we will briefly examine the fracture instabilities that govern the spacing and width of cracks in parallel crack systems.

12.1 LINEAR ELASTIC FRACTURE MECHANICS

The basic theory of fracture, originated by Griffith (1921, 1924), is linear elastic fracture mechanics (LEFM). This theory deals with sharp cracks and its central assumption is that all of the fracture process takes place at one point—the crack tip. Thus, if the material is linearly elastic, the entire volume of the structure is in a linearly elastic state, and so the methods of linear elasticity may be applied. The basic problem is to determine when an existing crack will grow.

Creation of a crack in an elastic body under uniform uniaxial tension disrupts the trajectories of the maximum principal stress in the manner shown in Figure 12.1. This reveals that stress concentrations arise near the crack tip. They were first calculated by Inglis (1913) as the limit case of his solution for an elliptical hole.

Stress Singularity and Fracture Energy

From Inglis's solution, Griffith noted that the strength criterion cannot be applied because the stress at the tip of a sharp crack is infinite no matter how small the load (Fig. 12.2a). He proposed that the formation of a crack requires a certain energy per unit length and unit width of fracture surface, which is a material

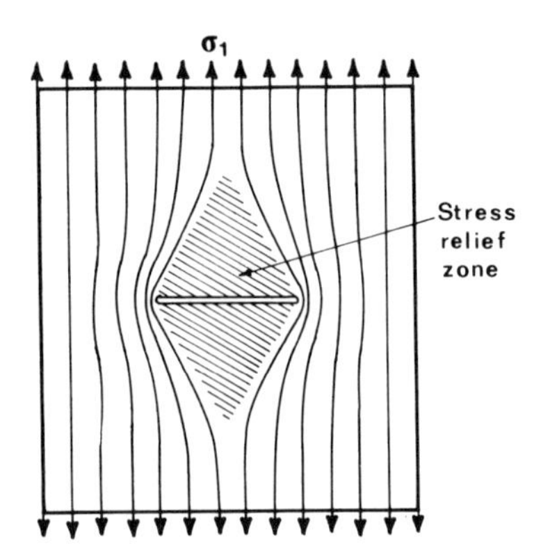

Figure 12.1 Maximum principal stress trajectories in a cracked plate under tension.

property and is called the fracture energy G_f. Crack propagation is possible if

$$\mathcal{G} = G_f \qquad \text{where} \qquad b\mathcal{G} = -\left[\frac{\partial \Pi}{\partial a}\right]_u = \left[\frac{\partial \Pi^*}{\partial a}\right]_P \tag{12.1.1}$$

a = crack length, b = thickness of the structure, $\mathcal{G}$ represents the energy release of the structure (per unit length of crack and unit width of crack front), u = load-point displacement, $\Pi = \Pi(a, u)$ = potential energy of the structure-load system (Helmholtz free energy in the case of isothermal conditions, or total energy in the case of isentropic conditions); and $\Pi^* = \Pi^*(a, P) = Pu(P) - \Pi(a, u(P))$ = complementary energy of the structure-load system (Gibbs free energy in the case of isothermal conditions or enthalpy in the case of isentropic conditions; see Eqs. 10.1.40–10.1.41). Label u or P in Equation 12.1.1 means that the derivative is to be calculated at constant u or constant P. Note that the minus sign in Equation 12.1.1 appears with $\partial\Pi/\partial u$ but not with $\partial\Pi^*/\partial u$. By

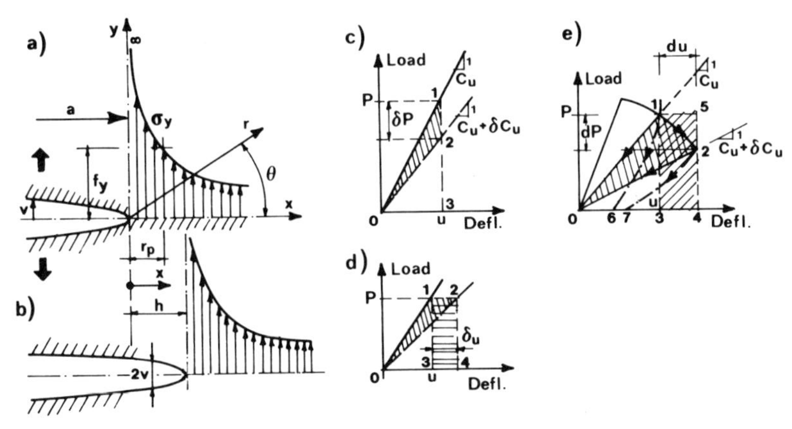

Figure 12.2 Elastic stress distributions at the crack tip (a) before and (b) after crack advance; and load-displacement curves for unloading after crack length increments (c) at deflection control, (d) at load control, (e) at arbitrary control.

analogy with the relation $f = -\partial\Pi/\partial u$ (f = force, u = displacement), and because $b\mathcal{G}$ has the dimension of a force Irwin also called $\mathcal{G}$ the crack driving force.

The most direct way to calculate $\mathcal{G}$ by finite elements is to evaluate the potential energies Π_1 and Π_2 of the specimen or structure, corresponding to crack lengths $a - \Delta a/2$ and $a + \Delta a/2$. Then $b\mathcal{G} = -\partial\Pi(a)/\partial a = -(\Pi_2 - \Pi_1)/\Delta a$.

The energy release rate may be determined from the stress and displacement fields near the crack tip. If we consider a sufficiently small region around the crack tip, the boundaries of the body become infinitely remote compared to the size of this small region, and so the shape of the boundaries of the body can have no effect on the stress and displacements fields in such a relatively small region. This physical deduction is indeed verified by the solutions of the stresses for various body geometries. Furthermore, if we consider a microregion around the crack tip that is still much smaller, the stress and displacement fields in it must be similar. So the near-tip field must be such that the transformation $r \rightarrow kr$ (k = constant, r = radial coordinate from the crack tip) would not change the angular distribution. It follows that all the asymptotic near-tip stress and displacement distributions must be separable in the form $F(r)f(\theta)$, where F and f are some functions of polar coordinates r and θ centered at the crack tip (Fig. 12.2a). It is found that $F(r) \rightarrow \infty$ as $r \rightarrow 0$, and that (see, e.g., Kanninen and Popelar, 1985; Broek, 1977 and 1988; Knott, 1981; Owen and Fawkes, 1983; Hellan, 1984), regardless of the body shape and for arbitrary loading, the near-tip asymptotic field of stresses σ_{ij} and displacements u_i in an isotropic material always has the form

$$\begin{aligned} \sigma_{ij} &= [K_{\mathrm{I}} f_{ij}(\theta) + K_{\mathrm{II}} g_{ij}(\theta) + K_{\mathrm{III}} h_{ij}(\theta)] r^{\lambda-1} \\ u_i &= [K_{\mathrm{I}} \phi_i(\theta) + K_{\mathrm{II}} \psi_i(\theta) + K_{\mathrm{III}} \chi_i(\theta)] r^{\lambda} \end{aligned} \tag{12.1.2}$$

where $\lambda = \frac{1}{2}$, K_{I}, K_{II}, K_{III} are parameters called the stress intensity factors, and f_{ij}, g_{ij}, h_{ij}, ϕ_i, ψ_i, χ_i are certain functions of polar angle θ. For example (see, e.g. Broek, 1978), if the crack lies in axis $x = x_1$,

$$\begin{aligned} \left.\begin{matrix} f_{11}(\theta) \\ f_{22}(\theta) \end{matrix}\right\} &= \frac{1}{\sqrt{2\pi}} \cos\frac{\theta}{2} \left(1 \mp \sin\frac{\theta}{2} \sin\frac{3\theta}{2}\right) \\ f_{12}(\theta) &= \frac{1}{\sqrt{2\pi}} \cos\frac{\theta}{2} \sin\frac{\theta}{2} \cos\frac{3\theta}{2} \end{aligned} \tag{12.1.3a}$$

$$\begin{aligned} \phi_1(\theta) &= \sqrt{\frac{2}{\pi}} \left(\frac{1+\nu}{E}\right) \cos\frac{\theta}{2} \left(1 - 2\bar{\nu} + \sin^2\frac{\theta}{2}\right) \\ \phi_2(\theta) &= \sqrt{\frac{2}{\pi}} \left(\frac{1+\nu}{E}\right) \sin\frac{\theta}{2} \left(2 - 2\bar{\nu} - \cos^2\frac{\theta}{2}\right) \end{aligned} \tag{12.1.3b}$$

where $\bar{\nu} = \nu$ for plane strain and $\bar{\nu} = \nu/(1+\nu)$ for plane stress, E = Young's elastic modulus, and ν = Poisson's ratio. Similar functions are associated with K_{II} and K_{III}. The three terms in Equation 12.1.2 represent deformation modes that are symmetric, antisymmetric in a plane, and antisymmetric out-of-plane. These terms correspond to the three basic fracture modes shown in Figure 12.3, which are mutually orthogonal. They are called mode I (opening), mode II (in-plane shear), and mode III (antiplane shear).

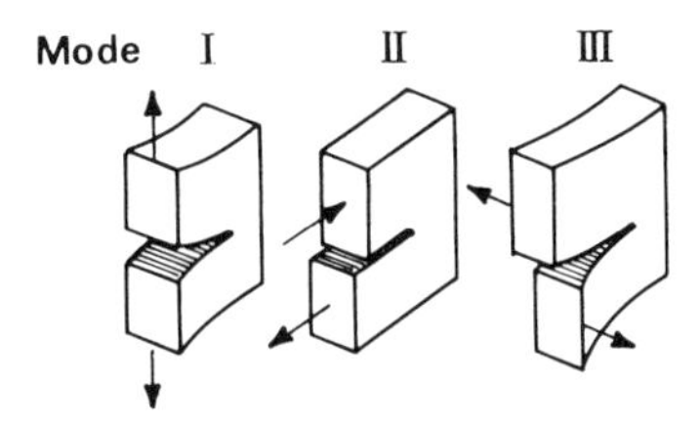

Figure 12.3 Basic fracture modes: I—opening, II—in-plane shear, and III—antiplane shear.

The easiest way to derive Equations 12.1.3b is to substitute $u_i = K\phi_i(\theta)r^{\lambda}$ (where K = constant) into the differential equations of equilibrium $\sigma_{ij,j} = 0$ after setting $\sigma_{ij} = C_{ijkm}u_{k,m}$. This yields for functions $\phi_1(\theta)$ and $\phi_2(\theta)$ an eigenvalue problem defined by a system of two homogeneous ordinary differential equations with homogeneous boundary conditions ($\sigma_{22} = 0$) at $\theta = -\pi$ and π. It is found that the smallest λ value for which this eigenvalue problem has a nonzero solution (such that u_i is finite at $r \to 0$) is $\lambda = \frac{1}{2}$, and the corresponding eigenfunctions are those appearing in Equations 12.1.3b. This method can be generalized to determine the asymptotic fields at singularities for anisotropic materials, sharp corners, three-dimensional situations, etc. (see Williams, 1952; Karp and Karal, 1962; Bažant, 1974; Bažant and Estenssoro, 1979).

For points on the crack extension line ($\theta = 0$, $r = x$), Equations 12.1.3a yield $\sigma_y = K_I/\sqrt{2\pi x}$, and so $\sigma_y\sqrt{2\pi x} = K_I$ for the near-tip asymptotic field. Similar relations hold for τ_{xy} and τ_{xz}. Therefore, the stress intensity factors for arbitrary bodies under any loading are defined as

$$K_I = \lim_{x \to 0^+} \sigma_y\sqrt{2\pi x} \qquad K_{II} = \lim_{x \to 0^+} \tau_{xy}\sqrt{2\pi x} \qquad K_{III} = \lim_{x \to 0^+} \tau_{xz}\sqrt{2\pi x} \quad (12.1.4)$$

Energy Release Rate

Because the asymptotic stress field is unique, and because the rate of energy flow into the crack tip must depend only on this field, there must exist a unique relationship between the energy release rate, $\mathcal{G}$, and the stress intensity factors. We consider mode I and imagine the crack tip to move by an infinitesimal distance h in the direction of axis $x_1 = x$ (Fig. 12.2b). After that happens the new near-tip displacements behind the new crack tip location are obtained from Equations 12.1.2 and 12.1.3b by setting $r = h - x$, $\theta = \pi$, which yields ($v = u_2$)

$$v = \sqrt{\frac{8}{\pi}}\left(\frac{K_I}{E'}\right)\sqrt{h - x} \qquad (12.1.5)$$

in which $E' = E/(1 - \nu^2)$ for plane strain and $E' = E$ for plane stress. (Thus, according to linear elastic fracture mechanics, the shape of every opened crack near the tip is a parabola.) The original stresses ahead of the original crack tip are obtained from Equations 12.1.3a by setting $\theta = 0$, which yields ($\sigma_y = \sigma_{22}$)

$$\sigma_y = \frac{K_I}{\sqrt{2\pi x}} \qquad (12.1.6)$$

These stresses are reduced to zero as the crack tip advances. So the energy release rate may be obtained as the work of σ_y on v, that is,

$$\mathscr{G} = 2 \lim_{h \to 0} \frac{1}{h} \int_0^h \frac{1}{2} \sigma_y v \, dx \tag{12.1.7}$$

where the factor $\frac{1}{2}$ is due to the fact that, for small h, the stresses must reduce to zero linearly and the factor 2 to the fact that the crack opening width is $2v$. Substituting Equations 12.1.5 and 12.1.6 into Equation 12.1.7 and integrating (by means of the substitution $x = h \sin^2 \omega$), one gets for mode I

$$\mathscr{G} = \frac{K_\text{I}^2}{E'} \qquad \text{(mode I)} \tag{12.1.8}$$

(Irwin, 1957). Doing a similar calculation for a combination of modes I, II, and III, one gets, in general,

$$\mathscr{G} = \frac{K_\text{I}^2}{E'} + \frac{K_\text{II}^2}{E'} + \frac{(1+\nu)K_\text{III}^2}{E} \tag{12.1.9}$$

According to Equations 12.1.1 and 12.1.8, a mode I crack can propagate if

$$K_\text{I} = K_c \qquad K_c = \sqrt{E' G_f} \tag{12.1.10}$$

where K_c = critical value of stress intensity factor K_I, called fracture toughness. Similar critical values may be introduced for K_II and K_III.

Equation 12.1.7 makes it possible to prove most easily that the singularity exponent λ must indeed be $\frac{1}{2}$. If exponents $\frac{1}{2}$ in the expressions for v and σ_y were replaced by $\lambda \neq \frac{1}{2}$, then evaluation of the integral in Equation 12.1.7 would indicate that $\mathscr{G}$ is 0 if $\lambda > \frac{1}{2}$ or ∞ if $\lambda < \frac{1}{2}$, either of which is impossible. Replacing r^λ by other functions such as $\ln \lambda$ a similar conclusion ensues.

Another way to calculate the energy release rate is to choose around the crack tip some closed path Γ that is moving as a rigid body as the crack tip advances by Δa. Since no energy is dissipated in the material (which is elastic), the energy $\Delta \mathscr{E}$ that has flowed into the crack tip must be equal to the energy that has flowed into the contour Γ as it moves with the crack tip regardless of the shape of the contour. This energy is the sum of the energies $\bar{U} \, \Delta a \, dy$ contained in all the elemental areas $\Delta a \, dy$ that moved into contour Γ, cross-hatched in Figure 12.4a

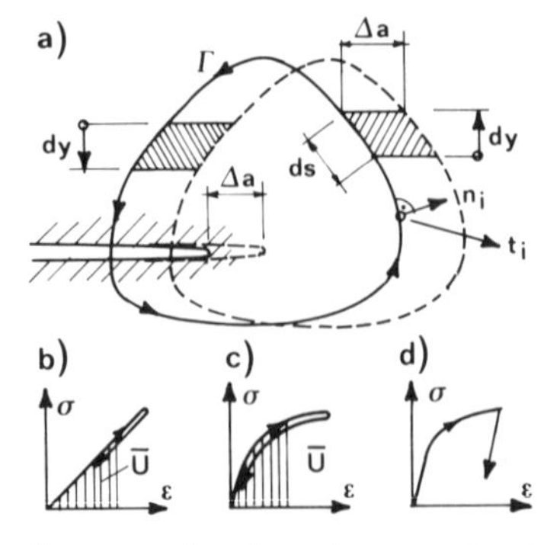

Figure 12.4 (a) Closed path around advancing crack tip; (b) linearly elastic; (c) nonlinearly elastic, and (d) inelastic stress–strain relations.

(where $\bar{U} = \frac{1}{2}\sigma_{ij}\varepsilon_{ij}$ = strain energy density in the material), minus all the work of the surface tractions acting on this contour, $t_i = \sigma_{ij}n_j$, done on the displacement increments $\Delta u_i = (\partial u_i/\partial x)\,\Delta a$ that occur at a point of contour Γ as it moves by distance Δa; n_i is the unit outward normal of contour Γ and $x = x_1$. (*Note:* The fact that the cross-hatched region on the left in Fig. 12.4a is moving out of the contour is taken into account by the negative value of dy.) In consequence, $\Delta\mathscr{E} = \oint_\Gamma \bar{U}\,\Delta a\,dy - \oint_\Gamma t_i\,\Delta u_i\,ds$. So the rate of energy flow into the crack tip is $\mathscr{G} = J = \Delta\mathscr{E}/\Delta a$ where

$$J = \oint_\Gamma \left(\bar{U}\,dy - \sigma_{ij}n_j \frac{\partial u_i}{\partial x}\,ds\right) = \oint_\Gamma P_{1j}n_j\,ds \tag{12.1.11}$$

in which $P_{ij} = \bar{U}\delta_{ij} - \sigma_{jk}u_{k,i}$ (which follows by setting $dy = n_1\,ds = \delta_{1j}n_j\,ds$); P_{ij} is called Eshelby's energy momentum tensor. Equation 12.1.11, introduced into fracture mechanics by Rice (1968), is called the J-integral. From the fact that energy is dissipated only at the crack tip and nowhere else within the contour, it is physically clear that the J-integral must be path-independent. For the same reasons, the path-independence of J is also true for nonlinearly elastic materials, for which the unloading stress–strain diagram coincides with the loading stress–strain diagram (Fig. 12.4c); in that case $\bar{U} = \int \sigma_{ij}\,d\varepsilon_{ij}$, not $\frac{1}{2}\sigma_{ij}\varepsilon_{ij}$. But J is not path-independent for inelastic materials, which dissipate energy throughout their volume because the unloading and loading diagrams (Fig. 12.4d) differ (often, though, path-independence is a good approximation). Nevertheless, J is path-independent for such paths that do not go through an inelastic region. The path-independence of the J-integral has also been proven mathematically.

If Γ is taken, for example, as a circle of radius $r = R$ around the crack tip, one may use Equation 12.1.11 to calculate $\mathscr{G} = J$. The result is the same as before (Eq. 12.1.8). The fact that $\lambda = \frac{1}{2}$ can be derived even without exact calculation of the J-integral (cf. Bažant and Estenssoro, 1979, p. 411). Near the crack tip we have $u_i \sim r^\lambda$ where $\sim$ denotes proportionality. Hence $\partial u_i/\partial x \sim r^{\lambda-1}$, $\varepsilon_{ij} \sim \sigma_{ij} \sim r^{\lambda-1}$, $\sigma_{ij}\,\partial u_i/\partial x \sim r^{2\lambda-2}$, $\bar{U} \sim r^{2\lambda-2}$, and $J \sim R^{2\lambda-2}R = R^{2\lambda-1}$. Consequently, for J to be bounded and nonzero as $R \to \infty$ it is necessary that $\mathrm{Re}\,(2\lambda - 1) = 0$, that is, $\mathrm{Re}\,(\lambda) = \frac{1}{2}$. (That λ is a real number can be shown from the field equation.)

Determination of $\mathscr{G}$ and G_f from Compliance Changes

The load-point displacement may be expressed as $u = C_u(a)P(a)$ where $C_u(a)$ = unloading compliance of the structure or specimen at crack length a. The complementary energy of the structure is $\Pi^* = \frac{1}{2}Pu = \frac{1}{2}C_u(a)P^2$. Therefore, the energy release rate $\mathscr{G} = (\partial\Pi^*/\partial a)/b$ (Eq. 12.1.1) is given by

$$\mathscr{G}(a) = \frac{P^2}{2b}\frac{dC_u(a)}{da} \tag{12.1.12}$$

If the crack is propagating, then, of course, $G_f = \mathscr{G}(a)$. Equation 12.1.12 serves as the basis of measurement of the fracture energy.

Equation 12.1.12 can also be derived from the potential energy Π. Since $\Pi = P(u)u - \Pi^*(a, P(u)) = u^2/C_u - \frac{1}{2}Pu = u^2/C_u - u^2/2C_u = u^2/2C_u(a)$, we have $b\mathscr{G} - \partial\Pi/\partial a = -(-u^2/2C_u^2)\,dC_u/da$, and substituting $u = C_uP$, we obtain again Equations 12.1.12.

It is instructive to derive Equation 12.1.12 by elementary energy balance considerations. When the crack is assumed to advance by δa while the loading point is fixed (fixed grip, $du = 0$), load P does no work, $dW = 0$. We have $U = \frac{1}{2}Pu$ = area 0130 in Figure 12.2c, and $u = C_u P$. Therefore, $-b\mathscr{G}\,da = d\Pi = dU = d(\frac{1}{2}Pu) = d(\frac{1}{2}u^2/C_u) = -\frac{1}{2}(u/C_u)^2\,dC_u = -\frac{1}{2}P^2\,dC_u$ (= area 0120 in Fig. 12.2c), which yields Equations 12.1.12.

When the crack is assumed to advance while the load is fixed (dead load, $dP = 0$), load P does work. We have $-b\mathscr{G}\,da = d\Pi = dU - dW$, which is the difference between areas 0120 and 12431 in Fig. 12.2d; we calculate $d\Pi = d(\frac{1}{2}Pu) - P\,du = d(\frac{1}{2}C_u P^2) - P(P\,dC_u) = \frac{1}{2}P^2\,dC_u - P^2\,dC_u = -\frac{1}{2}P^2 C_u$. So the result for dead load is the same as for the case of fixed grip (Eq. 12.1.12), even though the strain energy for the case of dead load increases while for the case of fixed grip it decreases.

In real testing, du is normally accompanied by load drop dP. In that case $dU = d(\frac{1}{2}Pu) = \frac{1}{2}P\,du + \frac{1}{2}u\,dP$ = area 0240 − area 0130 in Figure 12.2e; $d\Pi = dU - dW = dU$ − area 15431 $= dU - P\,du = -\frac{1}{2}(P\,du - u\,dP)$. At the same time, $dC_u = d(u/P) = (P\,du - u\,dP)/P^2$. Therefore, $dC_u/d\Pi = -dC_u/b\mathscr{G}\,da = -2/P^2$, which again yields Equation 12.1.12. Note that the result is independent of the softening slope dP/du, and that $-d\Pi$ = area 0120 in Figure 12.2e.

Equation 12.1.12 serves as the basis of the compliance method for measuring the fracture energy. This method, however, becomes questionable if the fracture process zone is so large that the value of equivalent crack length, a, is uncertain (Sec. 12.2), or if a substantial amount of energy is dissipated by plasticity and friction outside the fracture region. The latter problem is manifested by the fact that the experimental unloading curves are not straight and do not return to the origin; see curves 16 and 27 in Figure 12.2e. The residual displacements 06 and 07 are due to plasticity, and various attempts to compensate for them have been made.

Some Simple Elastic Solutions[1]

For a line crack of length $2a$ in an infinite elastic plane (Fig. 12.5a) subjected at infinity to a uniform uniaxial normal stress σ in the direction orthogonal to the crack plane

$$K_{\mathrm{I}} = \sigma\sqrt{\pi a} \tag{12.1.13}$$

A crack of length $2a$ subjected to uniform remote stress σ becomes critical when $K_{\mathrm{I}} = \sigma\sqrt{\pi a} = K_c$ (according to Eq. 12.1.13). This occurs when the crack length reaches the critical value $a_{\mathrm{cr}} = (K_c/\sigma)^2/\pi$, which represents the limit of stability (Sec. 12.3). This expression may be used as an estimate of a_{cr} even for other situations (provided that $a_{\mathrm{cr}}\,|\mathrm{grad}\,\sigma_{ij}^0| \ll |\sigma_{ij}^0|$ for all i, j where σ_{ij}^0 is the stress field for no crack).

For a normal surface crack (Fig. 12.5b) of lenght a in a half-plane subjected to

1. See, e.g., Tada, Paris, and Irwin (1985), Murakami (1987), or Kanninen and Popelar (1985).

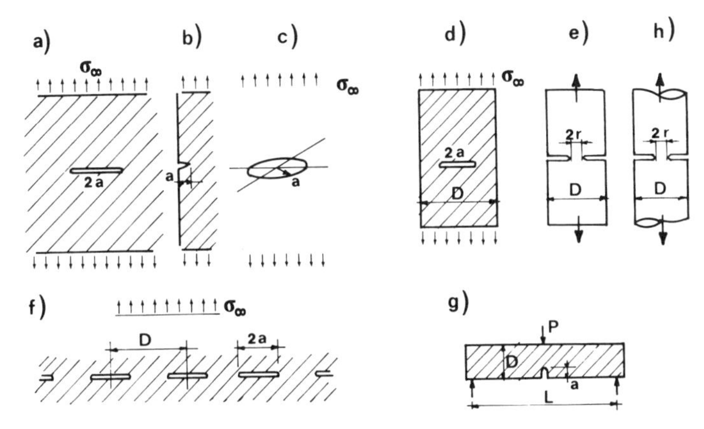

Figure 12.5 Various crack geometries: (a) line crack in infinite elastic plane; (b) edge crack in half-plane; (c) circular (penny-shaped) crack; (d) central crack in infinite strip; (e) edge crack in infinite strip; (f) collinear equidistant cracks; (g) edge crack in beam (three-point-bend specimen); (h) circumferential edge crack in circular bar with circular ligament.

stress σ parallel to the surface,

$$K_{\mathrm{I}} = 1.12\sigma\sqrt{\pi a} \tag{12.1.14}$$

For a circular (penny-shaped) crack (Fig. 12.5c) of radius a in an infinite elastic space subjected at infinity to a uniform uniaxial normal stress σ in the direction orthogonal to the crack plane

$$K_{\mathrm{I}} = \frac{2}{\pi}\sigma\sqrt{\pi a} \tag{12.1.15}$$

For finite two-dimensional bodies, the formulas for K_{I} generally have the form

$$K_{\mathrm{I}} = \frac{P}{b\sqrt{D}}k(\alpha) \qquad \alpha = \frac{a}{D} \tag{12.1.16}$$

where P = load, b = thickness, D = characteristic dimension, a = crack length, and $k(\alpha)$ = function of α depending on the shape but not the size of the structure. For example, for a long planar strip (Fig. 12.5d) of width D containing a centric transverse crack of length $2a$ and subjected to axial tensile stress $\sigma = P/D$ at remote cross sections,

$$K_{\mathrm{I}} \simeq \sigma\sqrt{\pi a}\left(\cos\frac{\pi a}{D}\right)^{-1/2} f_1 \qquad \alpha = \frac{a}{D} \tag{12.1.17}$$

where approximately $f_1 = 1$, but more accurately $f_1 = 1 - 0.1\alpha^2 + 0.96\alpha^4$.

For a planar stip (Fig. 12.5e) of width D and thickness b that is axially loaded by tensile force P at infinity and is weakened by a transverse crack such that only a small ligament of length $2r \ll D$ remains at the center,

$$K_{\mathrm{I}} = \frac{P}{b\sqrt{\pi r}} \qquad k(\alpha) = \sqrt{\frac{D}{\pi r}} \qquad \alpha = \frac{1}{2} - \frac{r}{D} \tag{12.1.18}$$

For a row of collinear equidistant cracks (Fig. 12.5f) of length $2a$ and center-to-center spacing D in an infinite plane loaded at infinity by stress σ normal to the crack plane,

$$K_I = \sigma\left(D \tan\frac{\pi a}{D}\right)^{1/2} \qquad (12.1.19)$$

This has the form of Equation 12.1.16 with $P = \sigma D b$ = load per crack, and $k(\alpha) = [\tan(\pi a/D)]^{1/2}$.

For a three-point bend specimen of depth D and span $L = 4D$, with one crack of length a (Fig. 12.5g), Equation 12.1.16 holds, with (Srawley 1976)

$$k(\alpha) = \frac{3L}{2D}\sqrt{\alpha}\left[\frac{1.99 - \alpha(1-\alpha)(2.15 - 3.93\alpha + 2.7\alpha^2)}{(1+2\alpha)(1-\alpha)^{3/2}}\right] \qquad \alpha = \frac{a}{D} \qquad (12.1.20)$$

For finite three-dimensional bodies, the formulas for K_I generally have the form

$$K_I = PD^{-3/2}k(\alpha) \qquad \alpha = \frac{A}{D} \qquad (12.1.21)$$

where function $k(\alpha)$ again depends on the shape but not the size of the structure. For example, for a circular bar (Fig. 12.5h) of radius $D/2$ that is axially loaded by tensile force P and is weakened by a transverse crack such that only a small ligament of radius $r \ll D/2$ remains,

$$K_I = \frac{P}{2\sqrt{\pi r^3}} \qquad k(\alpha) = \frac{1}{2\sqrt{\pi}}\left(\frac{D}{r}\right)^{3/2} \qquad (12.1.22)$$

Approximation by Stress Relief Zone

The principal stress trajectories in Figure 12.1 reveal that the formation of a crack causes stress relief in the shaded triangular regions next to the crack. As an approximation one may assume the stress relief region to be limited by lines of some constant slope k, (Fig. 12.6), called the "stress diffusion" lines, and further assume the stresses inside the stress relief region to drop to zero while remaining unchanged outside. Based on this assumption, the total loss of strain energy (per unit thickness) due to the formation of a crack of length $2a$ at controlled (fixed) boundary displacements is $\Delta\Pi = -2ka^2(\sigma^2/2E)$ where $\sigma^2/2E$ = initial strain energy density; therefore, the energy release rate per crack tip is $\mathcal{G} = -\partial(\Delta\Pi)/\partial a = 2ka\sigma^2/E$. Setting $\mathcal{G} = K_I^2/E$, one obtains

$$K_I = \sigma\sqrt{2ka} \qquad (12.1.23)$$

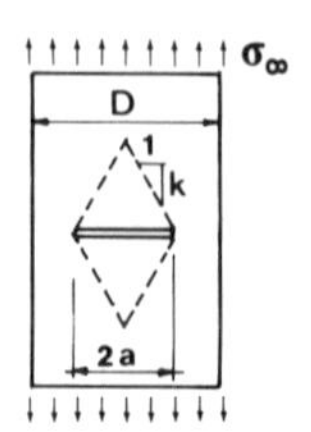

Figure 12.6 Stress relief zone.

This is in exact agreement with Equation 12.1.13 if one assumes that $k = \pi/2 = 1.571$. (If the stress relief zone is assumed to be a circle of radius a passing through the crack tips, the exact K_I results.)

For the penny-shaped crack (Fig. 12.5c), the stress relief region consists of two cones of base πa^2 and height ka. Therefore, $\Delta\Pi = -2\frac{1}{3}\pi a^2 ka(\sigma^2/2E)$. Also $\mathscr{G} = -(\partial\,\Delta\Pi/\partial a)/2\pi a$ (per unit length of crack perimeter), that is, $\mathscr{G} = k\sigma^2 a/2E = K_\text{I}^2/E$. From this

$$K_\text{I} = \sigma\sqrt{\frac{ka}{2}} \tag{12.1.24}$$

This is in exact agreement with Equation 12.1.15 if one assumes that $k = 8/\pi$. (If the stress relief zone is assumed to be a rotational prolate ellipsoid of minor axis a and major axis $2a$, the exact K_I results.)

The approximate method of stress relief zones can be applied in diverse situations for a quick estimate of $\mathscr{G}$ or K_I. The value of k depends on geometry and its order of magnitude is 1. The error of intuitive estimations of k can be substantial; however, the form of the equation obtained for K_I is correct.

Examples Solvable by Bending Theory

If the beams or structures shown in Figure 12.7 are slender, the energy release ratio may be solved by means of the theory of bending. The solutions are exact asymptotically, as the slenderness tends to infinity.

If the double cantilever specimen in Figure 12.7a is slender, the cantilevers may approximately be treated as beams having a fixed end at the cross section with the crack tip. The load-point relative displacement u due to load P is, according to the principle of virtual work, $u = 2\int M\bar{M}\,dx/EI = 8Pa^3/Ebh^3$ (where $I = bh^3/12$, b = thickness, $\bar{M} = x$, $M = Px$). From this, $P = Ebh^3u/8a^3$. If the relative displacement u is controlled, the potential energy is $\Pi = Pu/2 = Ebh^3u^2/16a^3$. Hence, $b\mathscr{G} = -\partial\Pi/\partial a = 3Ebh^3u^2/16a^4$. From $\mathscr{G} = G_f$, the fracture propagation condition is

$$u = u_{\text{cr}} = 4\left(\frac{G_f a^3}{3Eh^3}\right)^{1/2}\sqrt{a} \qquad P = P_u = \frac{b}{2}\left(EG_f\frac{h^3}{3a^3}\right)^{1/2}\sqrt{a} \tag{12.1.25}$$

The stress intensity factor is $K_\text{I} = \sqrt{E\mathscr{G}}$ or

$$K_\text{I} = \left(3\frac{h^3}{a^3}\right)^{1/2}\left(\frac{Eu}{4\sqrt{a}}\right) = \frac{2P}{b\sqrt{a}}\left(3\frac{a^3}{h^3}\right)^{1/2} \tag{12.1.26}$$

As another example, consider a wall with a longitudinal crack of length $2a$ at a small depth h below the surface, subjected to compressive stress $\sigma_0 < 0$ (Fig. 12.7f). Sufficient compression causes the layer above the crack to buckle as a slender column whose ends at the crack tip are assumed to remain fixed. The buckling stress is $\sigma_{\text{cr}} = -E'I\pi^2/a^2h$ where $I = h^3/12$, $E' = E/(1-\nu^2)$. Without buckling, the layer may be brought to some stress $|\sigma_0| > |\sigma_{\text{cr}}|$, and due to buckling the axial stress in the layer drops to σ_{cr}. This contributes to fracture the energy loss $U = -\Delta\Pi_0$ corresponding to the cross-hatched triangle in Fig. 12.7f, $-\Delta\Pi_0 = 2ah(\sigma_0 - \sigma_{\text{cr}})^2/2E'$ (per unit thickness). If the bending energy of the layer is neglected, the crack will propagate when $-\partial(\Delta\Pi_0)/\partial a = 2G_f$. From this

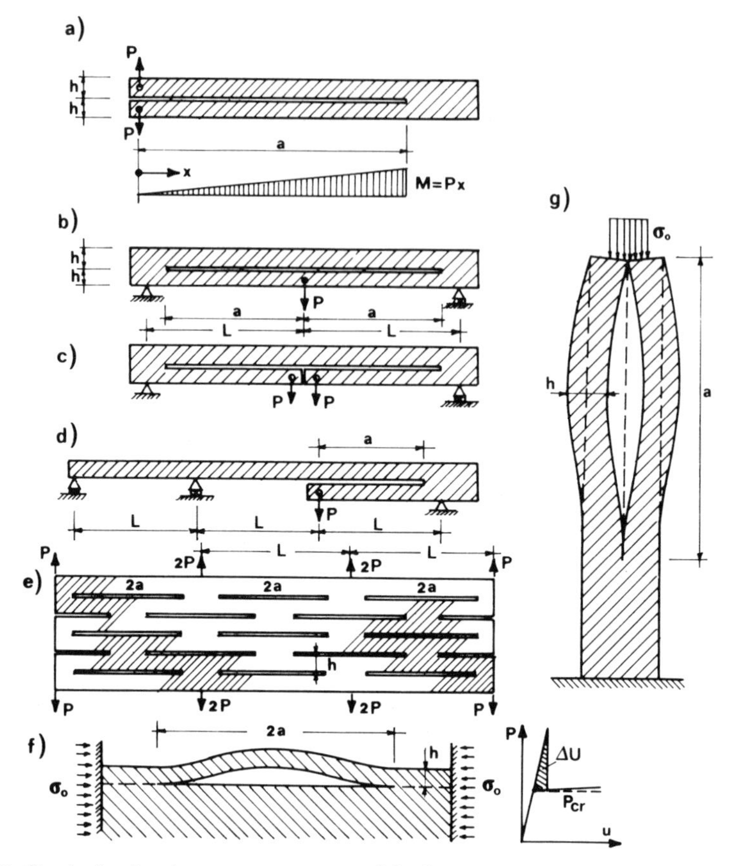

Figure 12.7 Cracks in slender structures treated by beam theory.

we obtain the condition

$$\sigma_0^2 - 2\sigma_0 E'I\pi^2 a^{-2}h^{-1} - 3(E'I)^2\pi^4 a^{-4}h^{-2} = \frac{2E'G_f}{h} \tag{12.1.27}$$

from which one can solve for the stress σ_0 that would cause the crack to propagate.

The foregoing calculation captures some but not all of the important aspects of the problem of delamination in layered composites (Sallam and Simitses, 1985, 1987; Yin, Sallam, and Simitses, 1986).

Herrmann's Method to Obtain Approximate K_I by Beam Theory

A remarkably simple method for close approximation of K_I in notched beams was recently discovered by Kienzler and Herrmann (1986) and Herrmann and Sosa (1986). The method was derived from a certain unproven hypothesis (postulate) regarding the energy release when the thickness of the fracture band is increased. We show a different derivation of this method (Bažant, 1988) that is simpler and at the same time indicates that the hypothesis used by Herrmann et al. might not be exact but merely a good approximation. Also, Herrmann's method relies on

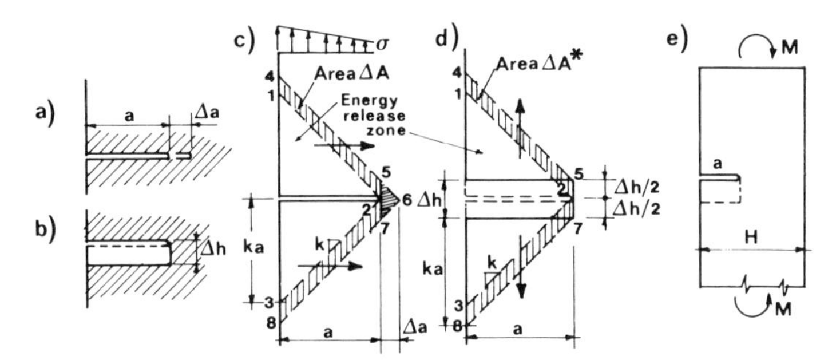

Figure 12.8 (a) Line crack advance; (b) crack band widening; (c, d) corresponding stress relief zones; (e) notched beam under constant bending moment.

more sophisticated concepts (material forces) that seem more complicated than necessary to obtain the result.

Let us estimate the energy release rate $\mathcal{G}$ by considering the expansion of the stress relief zone at crack extension Δa and employing the simplified "stress diffusion lines" of empirical slope k; see Figure 12.8a, c. Consider now that instead of crack extension Δa the crack is widened into a crack band of width Δh; see Figure 12.8b, d. This widens the stress relief zone from area 1231 to area 45784 as shown by arrows in Figure 12.8d. Since the triangular area 56725 in Figure 12.8c is second-order small if Δa is small, the increments of the stress relief zones 123876541 and 12387541 in Figure 12.8c and d are identical provided that $\Delta h/2 = k\,\Delta a$. Therefore (at fixed boundary displacements, for which $\Pi = U$),

$$-b\mathcal{G} = \frac{\partial U}{\partial a} = 2k\frac{\partial U}{\partial h} \tag{12.1.28}$$

(Bažant, 1988); b = thickness of body. The case $k = 1$ coincides with the hypothesis (postulate) on p. 41 of Kienzler and Herrmann (1986). However, there seems no reason to assume that $k = 1$, and numerical results in Bažant (1988) show that more accurate results can be obtained if the empirical constant k is allowed to differ from 1.

The advantage of Equation 12.1.28 is that it can be used even in problems where the initial stress field before the appearance of the crack is nonuniform. The reason is that, approximately, the widening of the crack into a band has merely the effect of shifting the stress field as a rigid body along with the shift of the triangular stress relief zones as indicated by arrows in Figure 12.8d, while the strain energy in the band can be easily estimated.

Following Kienzler and Herrmann (1986), we illustrate their method by considering the beam in Figure 12.8e subjected to bending moment M. The bending stiffness of the beam is EI_1, and the notched cross section has bending stiffness EI_2, where $I_1 = bH^3/12$ and $I_2 = b(H-a)^3/12$ (moments of inertia, b = thickness). The energy release due to widening of the notch to thickness (length) Δh is $\Delta U = M^2(1/EI_1 - 1/EI_2)\,\Delta hb/2$. From Equation 12.1.28, $\mathcal{G} = (-\Delta U/\Delta h)(2k/b)$ or

$$\mathcal{G} = \frac{k}{b}\left(\frac{1}{EI_2} - \frac{1}{EI_1}\right)M^2 \tag{12.1.29}$$

from which $K_I = (E\mathscr{G})^{1/2}$. According to Kienzler and Herrmann (1986, figs. 3 and 4) this compares (for $k = 1$) very well with the accurate solutions from handbooks (but it appears the agreement would be even better for some value $k \neq 1$).

Problems

12.1.1 Show that Equation 12.1.7 may be obtained by applying the J-integral to a path of two straight segments at $y = 0$ that run along the surfaces of the crack extension.

12.1.2 Assume that the near-tip displacement field is separable as $u = r^\lambda \phi(\theta)$ and $v = r^\lambda \psi(\theta)$. Substitute this into differential equations of equilibrium and crack surface boundary conditions written in terms of u and v. Variable r cancels out and one obtains an eigenvalue problem for λ consisting of two linear homogeneous differential equations for $\phi(\theta)$ and $\psi(\theta)$ and associated homogeneous boundary conditions. Show that the physically admissible eigenvalues are $\lambda = n/2$ $(n = 1, 2, 3, \ldots)$, and obtain the corresponding $\phi(\theta)$, $\psi(\theta)$. By differentiation, get the stresses and strains. Also use this method for the case of a corner of a finite angle, for which $\lambda > \frac{1}{2}$ (for details see Karp and Karal, 1962).

12.1.3 $\mathscr{G}$, K_I, and the crack propagation criteria for the structures in Figure 12.7b, c, d, e are solvable by the theory of bending. Carry out these solutions. Also solve Fig. 12.7a with P replaced by opposite applied moments M.

12.1.4 Figure 12.7g illustrates compression failure due to pressure applied over one-half of the top surface. If there is an axial crack as shown, the load applied on each column strip is eccentric and causes the strip to buckle. Calculate the stress σ_0 that makes the crack of length a grow.

12.1.5 What is the critical crack length a_{cr} for Figure 12.5f (Eq. 12.1.19)?

12.1.6 As already remarked, Equation 12.1.8 can be obtained by calculating the J-integral along a circle of radius r using Equations 12.1.3a and $\partial v / \partial x = v/h$. Show it.

12.1.7 The regular field near a sharp corner of a finite angle is characterized by $u = r^\lambda \phi(\theta)$, $v = r^\lambda \psi(\theta)$ with $\lambda > \frac{1}{2}$. Can a nonzero finite energy flow into the corner point? Can an energy criterion be used for the start of crack propagation from the corner? The answers are no, but why?

12.2 NONLINEAR FRACTURE MECHANICS AND SIZE EFFECT

If all the fracture process happens at one point—the crack tip, the whole body is elastic and linear elasticity may be used. If the fracture process happens either on a finite line segment or in a finite zone, part of the body behaves in a nonlinear manner. There exist many brittle materials in which the fracture process zone is so large that it cannot be approximated as a point. Fracture behaviour of this type is generally desirable; it increases the material resistance to fracture growth. We will now briefly outline the typical methods to deal with this type of fracture. Then we will consider the consequences for the size effect. But the examination of the stability aspect of size effect will be postponed to Chapter 13.

Inelastic Zone and Equivalent Elastic Crack

Although linear elastic fracture mechanics assumes the size of the inelastic zone at the crack tip to be zero, in reality this zone must have some finite size, r_p (Fig. 12.9a). This size was estimated by Irwin (1958). For a crude estimate, we note that the point on the crack extension line where $\sigma_y = f_t'$ ($=$ tensile strength of the material) lies at the distance $x = r_p$ from the crack tip given by $\sigma_y = K_c(2\pi r_p)^{-1/2} = f_t'$. This yields, for mode I (see Fig. 12.9a),

$$r_p = \frac{1}{2\pi}\left(\frac{K_c}{f_t'}\right)^2 = \frac{1}{2\pi}\left(\frac{E'G_f}{f_t'^2}\right) \tag{12.2.1}$$

In reality, the nonlinear zone must be larger than this because the stress resultant from an inelastic zone of size r_p would be less than the stress resultant of the elastically calculated stresses $\sigma_y = K_c(2\pi r)^{-1/2}$. If these resultants were made

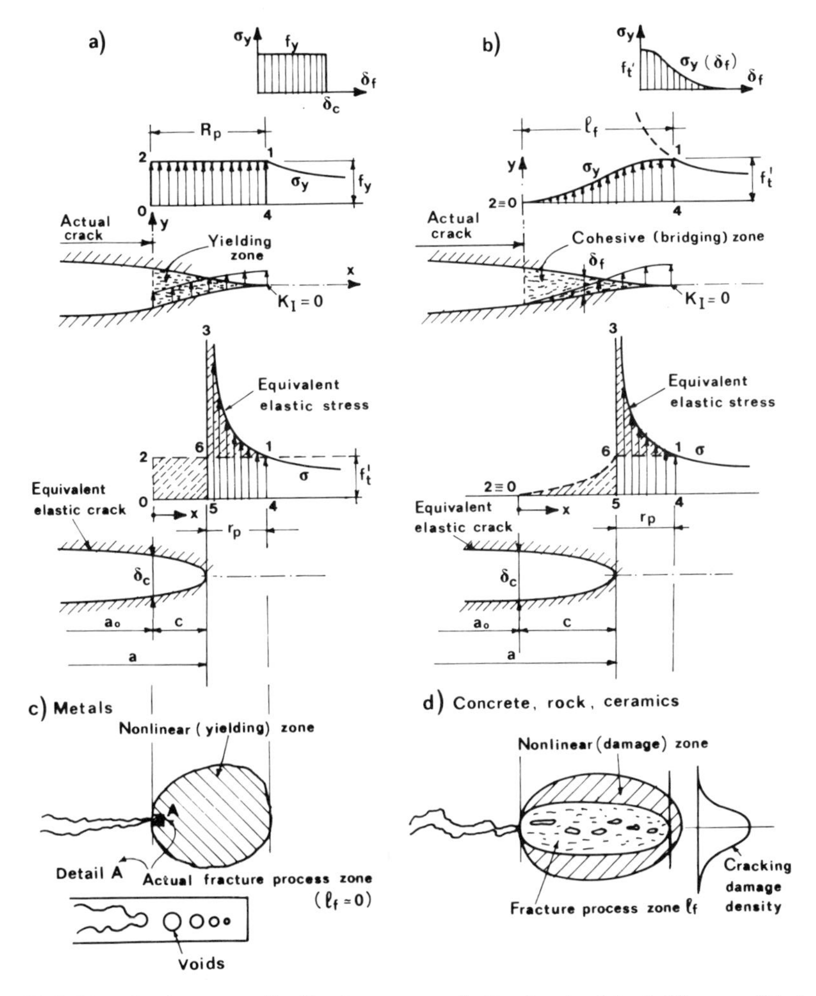

Figure 12.9 Nonelastic stress distributions at crack tip due to (a) yielding or (b) damage, and (c, d) corresponding fracture process zones.

coincident and equal in magnitude, then the stress field farther from the nonlinear process zone would be unaffected, due to Saint-Venant's principle. Equal magnitude of these resultants may be achieved by assuming the tip of an equivalent elastic crack to be shifted forward by distance $R_p - r_p$ ahead of the true crack tip, where R_p is an improved estimate of the size of the nonlinear zone. Graphically, this means that the two cross-hatched areas (area 04120 and 54135) in Figure 12.9a must be equal (this means that area 1361 is replaced by equal area 05620). If we also assume the material to be plastic (stress $= f_y$) over the entire length R_p (Fig. 12.9a) (which implies all the fracture processes to happen at point 0 at which the stress suddenly drops to zero), then the condition of equal magnitudes of the stress resultants over length R_p (i.e., equality of areas 04120 and 54135 in Fig. 12.9a) reads

$$\int_{R_p - r_p}^{R_p} K_c[2\pi(x - R_p - r_p)]^{-1/2}\, dx = R_p f_y \tag{12.2.2}$$

where f_y = yield limit and r_p is given by Equation 12.2.1 with $f_t' = f_y$. Solving for R_p from this equation, Irwin (1958), obtained for the length of the yielding zone the improved estimate $R_p = 2r_p = k_p K_c^2/f_y^2$, $k_p = 1/\pi$. However, the moments of the stress distributions (i.e., of areas 04120 and 13541) about point 0 are still not equal, which means the lines of the stress resultants do not coincide. More seriously, the stress intensity factor K_{I} calculated for point 4 is nonzero, which is impossible because the stress must be finite. Therefore Dugdale (1960) calculated length R_p from the condition that the value of K_{I} caused by the plastic stresses in a body with a large crack up to point 4 must vanish, and he obtained

$$R_p = k_p\left(\frac{K_c}{f_y}\right)^2 = k_p \frac{E' G_f}{f_y^2} \qquad k_p = \frac{\pi}{8} \tag{12.2.3}$$

The hypothesis that stress σ_y is constant is justified for ductile metals, in which typically there is a large nonlinear (plastic) zone ahead of the crack tip but only a very small fracture process zone embedded in it (Fig. 12.9c). In materials such as concrete, rocks, and toughened ceramics, the nonlinear zone is also large but most of it constitutes the fracture process zone (Fig. 12.9d), in which the material undergoes progressive distributed damage such as microcracking. For such materials, σ_y declines gradually along the nonlinear zone (Fig. 12.9b). To describe this behavior one may assume that σ_y is a decreasing function of the crack opening δ. Since a change from a uniform to a descending stress profile shifts the stress resultant forward, one has $R_p = k_p K_{\text{I}}^2/f_y^2$ where k_p is larger than Irwin's estimate $1/\pi$, $k_p \geq 1/\pi$. Coefficient k_p depends on the stress profile throughout the inelastic zone, which in turn depends on the relationship between σ_y and the relative displacement δ_f across the width of the nonlinear zone.

The opening displacement δ_c at the front of the actual crack may be assumed to be approximately the same as the opening displacement of the equivalent elastic crack with length c_0 at that point. We have $\delta_c = \delta(x)$ for $x = 0$, where according to Equations 12.1.3b for $\theta = \pi$, $\delta(x) = 2u_2(x) = (4K_{\text{I}}/E)(1+\nu)(1-\bar{\nu}) \times [2(R_p - r_p - x)/\pi]^{1/2}$. Setting $c_0 = R_p - r_p$ this yields (for $K_{\text{I}} = K_c$) the estimate

$$\delta_c = k_0 \frac{K_c^2}{E f_y} = k_0 \frac{G_f}{f_y} \tag{12.2.4}$$

where $k_0 = \text{constant} = 4(1+\nu)(1-\bar{\nu})(f_y/K_c)(2c_0/\pi)^{1/2}$. For Dugdale's model, for which $\mathscr{G} = \delta_c f_y$, one gets $k_0 = 1$ along with $c_0 = \pi k_c^2/32 f_y^2$ for plane stress (while Planas and Elices, 1987, by a certain asymptotic analysis, obtained $c_0 = \pi K_c^2/24 f_y^2$ for Dugdale's model).

Fracture Models with a Nonlinear Zone

Equivalent linear elastic cracks do not suffice for describing materials in which the nonlinear zone is not very small compared to the cross-section dimensions. In this case, the inelastic deformations accumulated across the width of the nonlinear zone may be lumped into relative displacements δ_f across the center line of the crack, and all the material on the sides of the centerline may be approximately treated as elastic. Various approaches have been devised.

One approach, introduced by Dugdale (1960) and Barrenblatt (1959), is to consider a fictitious crack that extends all the way to the front of the inelastic zone, in which case the stress intensity factor for the tip of this crack must be zero ($K_\text{I} = 0$). The action of the inelastic zone at the crack front is represented by stresses σ_y applied on the fictitious crack surfaces, which tend to close the crack and are distributed over a certain length. Depending on the material properties, there are two possibilities: (1) The material (Fig. 12.9a) is plastic and the stress σ_y drops suddenly to zero when the fictitious crack opening δ_f reaches a certain critical value δ_c (Dugdale, 1960; Barrenblatt, 1959). (2) The material (Fig. 12.9b) exhibits progressive softening, in which case one needs to specify $\sigma_y(\delta_f)$ as a decreasing function (Knauss, 1973; Wnuk, 1974; Kfouri and Rice, 1977; Hillerborg, Modéer, and Petersson, 1976); here the area under the curve $\sigma_y(\delta_f)$ may be approximately interpreted as the fracture energy:

$$G_f = \int_0^\infty \sigma_y(\delta_f)\, d\delta_f \tag{12.2.5}$$

For finite element analysis, the crack is modeled in this approach as an interelement line crack and the stresses $\sigma_y(\delta_f)$ are modeled as internodal forces (Fig. 12.10a). Equation 12.2.5 is exact only if the fracture process zone is assumed to be a line and all the behavior outside the crack is reversible.

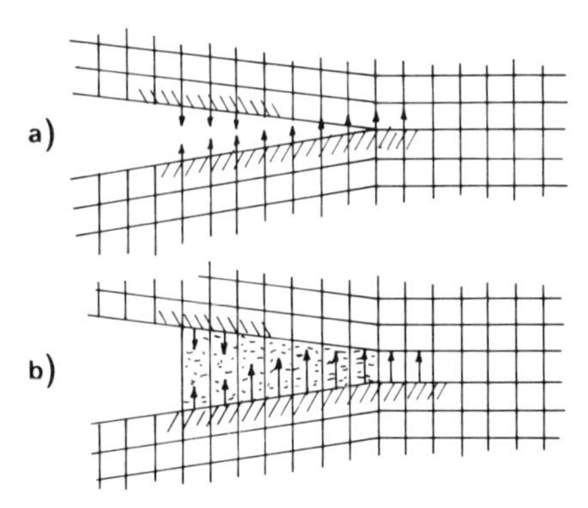

Figure 12.10 Finite element representation of (a) a line crack and (b) a crack band.

The $\sigma_y(\delta)$ relation for mortar and, consequently, also G_f) was estimated from laser-interferometric measurements by Cedolin, Dei Poli, and Iori (1983, 1987) along with the extension of the fracture process zone.

To deal with materials exhibiting a large zone of microcracking at the fracture front (e.g., toughened ceramics), a sharp line crack, characterized by stress intensity factor K_I', is considered to exist inside the microcracking zone. The effect of this zone is to reduce the value of K_I' compared to the stress intensity factor K_I that characterizes the elastic field around the microcracking zone. The reduction from K_I to K_I', called crack tip shielding, is approximately described according to various simplified models.

Instead of line cracks, a convenient alternative for finite element analysis is to assume the nonlinear deformation near the crack front to be uniformly smeared over a certain band, called the crack band, whose width w_c is approximately equal to the width of the nonlinear softening zone. The crack band is represented by a row of finite elements (Fig. 12.10b) and the material in the crack band is characterized by a softening stress–strain relation, whose component transverse to the crack is specified as a function $\sigma_y(\bar{\varepsilon}_y)$, where $\bar{\varepsilon}_y$ is the mean strain across the crack band. The width w_c must be considered to be a material property, and the fracture energy may be approximately expressed as

$$G_f = w_c \int_0^{\infty} \sigma_y(\bar{\varepsilon}_y)\, d\bar{\varepsilon}_y \tag{12.2.6}$$

This approach (Bažant, 1976b; Bažant and Cedolin, 1979, 1980, 1983; Bažant, 1982; Bažant and Oh, 1983a) is essentially equivalent to the line crack approach if $\varepsilon_y = \delta_f / w_c$.

A refined version of this approach is the nonlocal formulation (Sec. 13.10) in which the crack band width need not be specified as a material property but is obtained by solving the problem (in which case it depends on geometry, albeit not strongly). In the nonlocal formulation the fracturing strain at any point $\mathbf{x}$ (strain-softening or damage) is assumed to be a function of the spatial average of the strains from a neighbourhood of point $\mathbf{x}$ whose size is determined by the characteristic length l, a property of the material. The nonlocal approach is free of mesh bias and allows crack band propagation in arbitrary oblique directions through the mesh (see Sec. 13.10).

Size Effect

The most important consequence of fracture mechanics is the size effect on failure loads. It may be defined by considering geometrically similar specimens or structures of different sizes, with geometrically similar notches or initial cracks. We describe it in terms of the nominal strength (or nominal stress at failure):

$$\sigma_N = \frac{c_n P_u}{bD} \quad \text{for 2D similarity}$$
$$\sigma_N = \frac{c_n P_u}{D^2} \quad \text{for 3D similarity} \tag{12.2.7}$$

where P_u = maximum load (ultimate load); b = thickness of a two-dimensional structure, the same for all structure sizes; D = characteristic dimension of the structure or specimen; and c_n = coefficient introduced for convenience. For example, if D is the depth of the beam, the elastic formula for the maximum bending stress in a simply supported beam of span L is $\sigma_N = 1.5P_uL/bD^2 = c_nP_u/bD$ with $c_n = 1.5L/D$ (=constant if the structures are geometrically similar). One can also use the plastic bending stress formula, $\sigma_N = P_uL/bD^2 = c_nP_u/bD$, for which $c_n = L/D$ (= constant). Alternatively, one may define $\sigma_N = P_u/bL = c_nP_u/bD$ in which $c_n = D/L$ (= constant). This illustrates that any stress formula can be brought to the form of Equation 12.2.7.

It is known that plastic limit analysis, as well as elastic analysis with an allowable stress criterion or any failure criterion based on stress, exhibits no size effect (i.e., structures of different sizes fail at the same σ_N). Not fracture mechanics, though.

Due to similarity of elastic stress fields in similar two-dimensional structures of different size, the strain energy of the structure must have the form $U = V(\sigma_N^2/2E)f(\alpha)$ where $V = c_0bD^2$ = volume of the structure (c_0 is some constant) and $f(\alpha)$ is a decreasing function of the relative crack length $\alpha = a/D$, which depends on the shape of the structure. Therefore $b\mathscr{G} = -\partial U/\partial a = -(\partial U/\partial\alpha)/D = -c_0bD(\sigma_N^2/2E)f'(\alpha)$, from which, for two-dimensional problems,

$$\mathscr{G} = \frac{P_u^2 g(\alpha)}{E'b^2D} \qquad K_{\mathrm{I}} = \sqrt{\mathscr{G}E'} = \frac{P_u k(\alpha)}{b\sqrt{D}} \qquad \text{(2D)} \qquad (12.2.8)$$

where $E' = E/(1-\nu^2)$ for plane strain, $E' = E$ for plane stress and for 3D, $g(\alpha) = -f'(\alpha)c_0c_n^2/2$ = certain function of α depending on the structure shape, and $k(\alpha) = \sqrt{g(\alpha)}$. For basic specimen geometries, formulas for function $k(\alpha)$ can be found in handbooks (Tada, Paris, and Irwin, 1985; Murakami, 1987), and the values of $k(\alpha)$ can also be obtained by linear elastic finite element analysis. It may be checked from handbooks that all the formulas for K_{I} have the form of Equation 12.2.8 (check Eqs. 12.1.18 and 12.1.26).

For three-dimensional similarity, we have $U = V(\sigma_N^2/2E)f(\alpha)$ where $V = c_0D^3$ and $\mathscr{G}p(\alpha)D = -\partial U/\partial a = -(\partial U/\partial\alpha)/D = -c_0D^2(\sigma_N^2/2E)f'(\alpha)$, where $p(\alpha)D$ = length of the perimeter of the fracture front and $\mathscr{G}$ = average energy release rate per unit length of the perimeter. Therefore

$$\mathscr{G} = \frac{P_u^2 g(\alpha)}{E'D^3} \qquad K_{\mathrm{I}} = \sqrt{\mathscr{G}E'} = \frac{P_u k(\alpha)}{\sqrt{D^3}} \qquad \text{(3D)} \qquad (12.2.9)$$

where $g(\alpha) = -f'(\alpha)c_0c_n^2/2p(\alpha)$ and $k(\alpha) = \sqrt{g(\alpha)}$.

When $g'(\alpha) > 0$, which applies to most practical situations, linear fracture mechanics indicates that the maximum load occurs at infinitely small crack extensions. Then α at failure is the same for all the sizes of specimens of similar geometry. Setting $\mathscr{G} = G_f$ or $K_{\mathrm{I}} = K_c$, we recognize from Equations 12.2.8 as well as 12.2.9 that the size effect of linear elastic fracture mechanics generally is,

$$\sigma_N = \frac{\text{const.}}{\sqrt{D}} \qquad \text{(2D and 3D)} \qquad (12.2.10)$$

Based on our preceding discussion of nonlinear fracture mechanics the maximum loads for various structure sizes may be assumed to occur when the tip of the equivalent elastic crack lies at a certain distance c ahead of the tip of the initial crack or notch, that is, $a = a_0 + c$ or $\alpha = \alpha_0 + (c/D)$ where $\alpha_0 = a_0/D$, a_0 = initial crack or notch length. Let c_f denote the value of c for $D \to \infty$, for which $\mathcal{G} = G_f$ at failure.

At the beginning of loading, the fracture process zone starts with a zero length. Subsequently its length (characterized by c) grows as the load is increased, remaining attached to the notch tip provided that the specimen geometry is such that $g'(\alpha) > 0$ (which usually is the case). At maximum load P_u, the zone reaches approximately its maximum length. In the postpeak regime the zone detaches itself from the notch tip and travels forward while keeping its length approximately constant (Bažant, Gettu, and Kazemi, 1989). The $\mathcal{G}$-value required for fracture growth is basically determined by the length of the fracture process zone, and since this length determines the value of c, the $\mathcal{G}$ value at $P = P_u$ depends on c. At the same time the magnitude of c at failure fully determines the value of $g(\alpha)$ at failure, and so the ratio $\mathcal{G}/g(\alpha)$ at maximum load must be approximately equal to $G_f/g(\alpha_f)$ for a specimen of the same size. Therefore $\mathcal{G} \simeq G_f g'(\alpha)/g(\alpha_f)$. If we substitute this expression into Equations 12.2.8, make the approximation $g(\alpha_f) \simeq g(\alpha_0) + g'(\alpha_0)(c_f/D)$ (based on the Taylor series expansion) where $\alpha_f = c_f/D$ and $g'(\alpha_0) \simeq dg(\alpha_0)/d\alpha$, and if we also set $P_u^2 = (\sigma_N bD/c_n)^2$ and solve the resulting equation for σ_N, we get for the size effect law (Bažant, 1984) the following expression

$$\sigma_N = c_n \left[\frac{E'G_f}{g'(\alpha_0)c_f + g(\alpha_0)D} \right]^{1/2} = \frac{Bf_u}{\sqrt{1+\beta}} \qquad \beta = \frac{D}{D_0} \tag{12.2.11}$$

(see also Sec. 13.9). Here we introduce the tensile strength $f_u = f_t'$ and denote $B = [E'G_f/g'(\alpha_0)c_f]^{1/2} c_n/f_u$, $D_0 = c_f g'(\alpha_0)/g(\alpha_0)$. Equation 12.2.11 may be rewritten as (Bažant and Kazemi, 1988)

$$\tau_N = \left(\frac{E'G_f}{c_f + \bar{D}} \right)^{1/2} \tag{12.2.12}$$

where $\tau_N = \sqrt{g'(\alpha_0)}\, P_u/bD$, $\bar{D} = Dg(\alpha_0)/g'(\alpha_0)$; τ_N = intrinsic (shape-independent) nominal strength (nominal stress at failure), and $\bar{D}$ = intrinsic (shape-independent) size of the structure. Equations 12.2.11 and 12.2.12 are, of course, valid only if $g'(\alpha_0) > 0$, which comprises most practical situations.

For three-dimensional similarity, the size effect is again found to be approximately described by Equation 12.2.11 or 12.2.12 (Bažant, 1987a).

Parameter β, which is called the brittleness number (Bažant, 1987a, Bažant and Pfeiffer, 1987), may also be expressed as

$$\beta = \frac{D}{c_f} \left[\frac{g(\alpha_0)}{g'(\alpha_0)} \right] = \frac{\bar{D}}{c_f} \tag{12.2.13}$$

An alternative expression for β may be obtained (Bažant, 1987a) by calculating D_0 as the value of D at the intersection of the strength asymptote $\sigma_N = Bf_t'$ with the inclined asymptote $\sigma_N = c_n[G_f E'/g(\alpha_0)D]^{1/2}$ based on linear elastic fracture mechanics (Fig. 12.11). Equating these two expressions yields D_0, and then

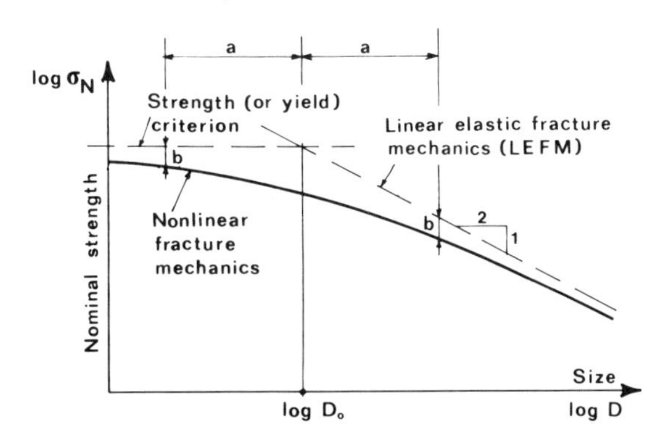

Figure 12.11 Size effect law proposed by Bažant (1984).

$\beta = D/D_0$ or

$$\beta = \frac{B^2 f_u^2 g(\alpha_0)}{c_n^2 E' G_f} D \tag{12.2.14}$$

The value of B characterizes σ_N for very small structures and can be determined on the basis of plastic limit analysis from the relation $B = \sigma_N / f_u$. Thus, Equation 12.2.14 is suitable for small β. On the other hand, Equation 12.2.13 gives β solely on the basis of the fracture mechanics function g, without taking any plastic solution into account. Thus Equation 12.2.13 is preferable for large β. Since Equations 12.2.13 and 12.2.14 are based on different asymptotic approximations ($D \to \infty$ and $D \to 0$), the fitting of test data may give slightly different B values.

The size effect law in Equation 12.2.11, giving the approximate relation of σ_N (or τ_N) to D, is plotted in Figure 12.11. For large β ($\beta = D/D_0$) such that $\beta > 10$, Equation 12.2.11 gives (with an error under 5% of σ_N) the approximation $\sigma_N \sim D^{-1/2}$, which is the size effect of linear elastic fracture mechanics (Eq. 12.2.10). So in this range, linear elastic fracture mechanics may be used. For small β such that $\beta < 0.1$, Equation 12.2.11 gives (again with an error under 5 percent of σ_N) $\sigma_N = Bf_u = \text{const.}$ or $\tau_N = (EG_f/c_f)^{1/2} = \text{const.}$, that is, there is no size effect and the failure load is proportional to the strength of the material. In this case, plastic limit analysis may be used (and the value of B can be deduced from such analysis). For $0.1 < \beta < 10$, the size effect is transitional between linear elastic fracture mechanics and plastic limit analysis. In this range, nonlinear fracture mechanics must be used.

Since the size effect vanishes for $\beta \to 0$ and $\sigma_N \to Bf_u$, the value of Bf_u or B can be obtained by solving the failure load according to plastic limit analysis. On the other hand, since $\sigma_N \simeq Bf_u \sqrt{D_0}/\sqrt{D}$ for $\beta \to \infty$, the value of $Bf_u \sqrt{D_0}$ can be obtained by solving the failure load according to linear elastic fracture mechanics. From these two results, D_0 can be identified.

The brittleness number β, proposed by Bažant (1987a), is capable of characterizing the type of failure regardless of structure geometry. The effect of geometry is captured by the ratio $g(\alpha_0)/g'(\alpha_0)$. This is, of course, only approximate and fails when $g'(\alpha_0) \leq 0$. However, structural geometries for which $g'(\alpha_0) \leq 0$ are rare (one is a panel with a short enough centric crack loaded by concentrated forces at the middle of the crack). Usually $g'(\alpha_0) > 0$ (which was called type G geometries by Jenq and Shah, 1985, and positive geometries by Planas and Elices, 1988a).

The size effect law in Equation 12.2.11 has also been derived by dimensional analysis and similitude arguments. This derivation (Bažant, 1984) was based on the hypothesis that the total energy ΔU released by the crack formation is a function of both the crack length a and the length c_f of the fracture process zone, which is assumed to be a material property. An alternative hypothesis that leads to the same law is that there is a crack band of frontal width w_c that is a material property (i.e., same for any size) and the total energy ΔU released by the crack band formation depends both on the length a and the area aw_c of the crack band (Bažant, 1984, 1985a, 1987a). If ΔU is assumed to depend only on a but not on c_f, the size effect of linear elastic fracture mechanics (that is, $\sigma_N \sim D^{-1/2}$) results. So it does if ΔU is assumed to depend only on the crack band's length a but not on its area aw_c.

The size effect law proposed by Bažant has been shown to agree well with fracture tests of very different geometries (Bažant and Pfeiffer, 1987), not only for Mode I but also for Mode II (Bažant and Pfeiffer, 1986) and Mode III (Bažant and Prat, 1987). The agreement was demonstrated for concretes, as well as rocks, certain toughened ceramics, and tough aluminum alloys (Bažant and Kazemi, 1988, 1989; Bažant, Lee, and Pfeiffer, 1987).

Fundamentally, there is no reason why the derivation of the size effect law should be based on a Taylor series expansion of $g(\alpha)$. Alternatively, one can expand function $k(\alpha) = \sqrt{g(\alpha)}$ or function $g(\alpha)^m$ for any value of m. Thus one gets the generalized size effect law $\sigma_N = Bf_t'(1 + \beta^r)^{-1/2r}$ with additional parameter r (Bažant, 1985a, 1987a). For $r = 0.44$, this formula gave very close agreement with some size effect predictions of Hillerborg's model (Bažant, 1985b, 1987a); but the fitting of tests of specimens of various geometries indicated the optimum value to be $r \simeq 1$ (Bažant and Pfeiffer, 1987).

The size effect law can be used to determine the fracture energy, G_f, and the length of fracture process zone, c_f, using only the maximum load P_u measured on similar specimens of significantly different sizes (of ratio at least 1:4). In fact, the size effect may be used to provide the following unambiguous definition of G_f and c_f as fundamental material properties, independent of specimen size and shape (Bažant, 1987a):

G_f and c_f are the energy required for crack growth and the elastically equivalent (or effective) fracture process zone length in an *infinitely large* specimen.

Mathematically, this definition can be stated as $G_f = \lim \mathscr{G}_0 = \lim (K_{I0}^2/E')$ for $D \to \infty$ where $\mathscr{G}_0$ or K_{I0} can be determined as the value of $\mathscr{G}$ or K_I at the measured maximum load P_u calculated for the initial crack length a_0 (notch length). (Alternatively, $\mathscr{G}_0$ or K_{I0} could be evaluated for the tip location of the equivalent elastic crack.) For this limiting procedure, however, it is necessary to restrict consideration only to those geometries for which $g(\alpha)$ increases with increasing crack length because otherwise the value of $\alpha = a/D$ would not approach $\alpha_0 = a_0/D$ as $D \to \infty$ (Planas and Elices, 1988a, 1989).

The foregoing definition would provide the exact values of G_f and c_f if the size effect law were known exactly. Equation 12.2.11, however, is only approximate (valid for a size range of only up to about 1:20). Therefore, the values of G_f and

c_f obtained from this definition are also approximate. But this is not really a serious drawback because the size range 1:20 is sufficient for most practical purposes. The obtained values of G_f and c_f then do not correspond to the true extrapolation to infinity, but that does not really matter as long as they correctly describe the energetics of fracture in the size range needed.

The reason that G_f and c_f in an infinitely large specimen are independent of the specimen shape is as follows: In an infinitely large specimen the fracture process zone occupies a vanishingly small volume fraction of the body, so that all of the body is elastic. Consequently the stress and displacement fields surrounding the fracture process zone are the asymptotic elastic fields. They are known to be the same for any specimen geometry (Eqs. 12.1.2–12.1.4), and so no influence of the boundary's shape can be transmitted to the fracture process zone.

Based on the foregoing definitions of G_f and c_f, Equation 12.2.11 can be used for their experimental determination. A simple formula for G_f (Bažant, 1987a; Bažant and Pfeiffer, 1987) can be obtained by taking the limit of Equations 12.2.8 in which $P_u = bD\sigma_N/c_n$ and σ_N is expressed from Equation 12.2.11 (and by noting that $\lim \alpha = \alpha_0$ for $D \to \infty$);

$$G_f = \lim_{D\to\infty} \frac{P_u^2 g(\alpha)}{E'b^2 D} = \frac{B^2 f_u^2}{c_n^2 E'} g(\alpha_0) \lim_{D\to\infty} \frac{D}{1 + D/D_0} = \frac{B^2 f_u^2}{c_n^2 E'} D_0 g(\alpha_0) \quad (12.2.15)$$

Furthermore, from Equation 12.2.13 (where $\beta = D/D_0$),

$$c_f = \frac{D_0 g(\alpha_0)}{g'(\alpha_0)}. \quad (12.2.16)$$

So, in order to determine G_f and c_f, one needs only to calculate $g(\alpha_0)$ and $g'(\alpha_0)$ according to linear elastic fracture mechanics and then find Bf_u and D_0 by fitting Equation 12.2.11 to the σ_N data (Bažant, 1987a; Bažant and Pfeiffer, 1987). This may be accomplished by linear regression since Equation 12.2.11 can be transformed to the linear plot $Y = AX + C$ where $X = D$, $Y = (f_u/\sigma_N)^2$, $B = C^{-1/2}$, and $D_0 = C/A$. Linear regression formulas also yield the coefficients of variation of A, C, G_f, and c_f.

Fracture parameters G_f and c_f can be related to the critical crack tip opening displacement δ_c, which occurs at some distance $k_f c_f$ behind the tip of the elastically equivalent crack; k_f is an empirical coefficient (approximately 1), which depends on the shape of the softening stress-displacement diagrams (Planas and Elices, 1988a). From Eq. 12.1.5, $\delta_c = 2v = 2(8k_f c_f/\pi)^{1/2} K_c/E$ (for plane stress) in which $K_c^2 = EG_f$. Denoting $k_c = 32k_f/\pi$ (an empirical coefficient), one gets (Bažant, Gettu, and Kazemi, 1989)

$$\delta_c = \left(\frac{k_c c_f G_f}{E'}\right)^{1/2} \quad (12.2.17)$$

It is now obvious that instead of characterizing nonlinear fracture properties by G_f and c_f, one can equivalently characterize them by G_f and δ_c, or by K_c and δ_c. Several models of this type have been proposed, for example, Jenq and Shah's two-parameter model for concrete. We see that the parameters of these models can also be obtained solely on the basis of the measured effect of size on σ_N.

The size effect law in Equation 12.2.11 has also been shown to describe well the existing test data on the size effect in various brittle failures of concrete structures, in particular; (1) the diagonal shear failure of reinforced concrete beams (unprestressed or prestressed, without or with stirrups), (2) the torsional failure of concrete beams, (3) the punching shear failure of reinforced concrete slabs, (4) the pull-out failure of steel bars embedded in concrete, and (5) the ring and beam failures of plain concrete pipes (Bažant and Kim, 1984; Bažant and Cao, 1986a, b; and 1987; Bažant and Prat, 1988b; Bažant and Sener, 1987, 1988; Bažant and Kazemi, 1988, 1989). Since, in contrast to fracture specimens, concrete structures have no notches, these applications rest on two additonal hypotheses, which are normally correct for concrete structures: (1) the failure (maximum load) does not occur at crack initiation but only after a relatively long crack or crack band has developed; and (2) the shape of this crack or crack band is about the same for structures of different sizes.

It may also be noted that describing the size effect in (unnotched) structures by Equation 12.2.11 is in conflict with the classical Weibull statistical theory of size effect. However, this theory needs to be made nonlocal (cf. Sec. 13.10), and then this classical theory is found to apply only asymptotically to very small structures, while for large structures there is a transition to the LEFM size effect, similar to Equation 12.2.11 (Bažant and Xi, 1989).

Problems

12.2.1 Let $\sigma_y = F(\delta) =$ given descending function of opening δ of the equivalent elastic crack shown in Figure 12.9b. Formulate the condition that the resultant of the equivalent elastic stresses σ_y over length r_p (Fig. 12.9b) be equal to the resultant of stresses $\sigma_y = F(\delta)$ over length l_f (Fig. 12.9b). (From this condition, one can estimate the ratio of l_f to $E'G_f/f_{y^*}^2$.)

12.2.2 Rearrange Equation 12.2.11 algebraically so that Bf_t' and D_0 can be obtained from σ_N data by linear regression (Bažant, 1984).

12.3 CRACK STABILITY CRITERION AND *R* CURVE

The propagation of a crack is a problem of equilibrium and stability governed by the same laws as those for inelastic structures in general. This section will discuss the criterion of stability of a crack and explain a simple approach to handle in an equivalent elastic manner the nonlinearity of fracture caused by the existence of a nonlinear zone at the crack tip.

R Curve and Fracture Equilibrium Condition

After a crack starts from a smooth surface or a notch, the size of the fracture process grows as the crack advances. The consequence is that the energy release rate R required for crack propagation (also called the crack resistance to propagation) increases and may be considered to be a function of the distance c

of the advance of the tip of the equivalent elastic crack ($c = a - a_0$ where a = current crack length and a_0 = initial crack length or notch length). If the function $R(c)$ is known, the crack propagation may be approximately analyzed by methods of linear elastic fracture mechanics, in which constant G_f is replaced by the function $R(c)$, called the R curve (Fig. 12.12a).

To some extent, the function $R(c)$ may be approximately considered to be a fixed material property, as proposed by Irwin (1960) and Krafft, Sullivan, and Boyle (1961). It has been found, however, that the shape of the R curve depends considerably on the shape of the specimen or structure. The R curve may be assumed to be unique only for a narrow range of specimen or structure geometries. Thus, it is necessary to determine the R curve for the given geometry prior to fracture analysis. A simple method to do that, utilizing the size effect law, has been proposed by Bažant, Kim, and Pfeiffer (1986) and refined by Bažant and Kazemi (1988), and by Bažant, Gettu, and Kazemi (1989). In the analysis that follows we will assume that the R curve for the given geometry is known.

The energy that must be supplied to an elastic structure under isothermal conditions in order to produce a crack of length a is

$$\mathscr{F} = \int_0^c bR(c')\,dc' + \Pi(a) \qquad c = a - a_0 \tag{12.3.1}$$

where b = thickness of the structure and a_0 = initial crack length or notch length. $\mathscr{F}$ represents the Helmholtz free energy and Π is the total potential energy of the structure due to formation of a crack of length $c = a - a_0$. An equlibrium state of fracture occurs when $\delta\mathscr{F} = 0$, in which case neither energy needs to be supplied, nor energy is released if the crack length changes from a to $a + \delta a$. Since $\delta\mathscr{F} = (\partial\mathscr{F}/\partial a)\delta a = 0$ and (from Eq. 12.3.1) $\partial\mathscr{F}/\partial a = bR(c) + \partial\Pi/\partial a = 0$

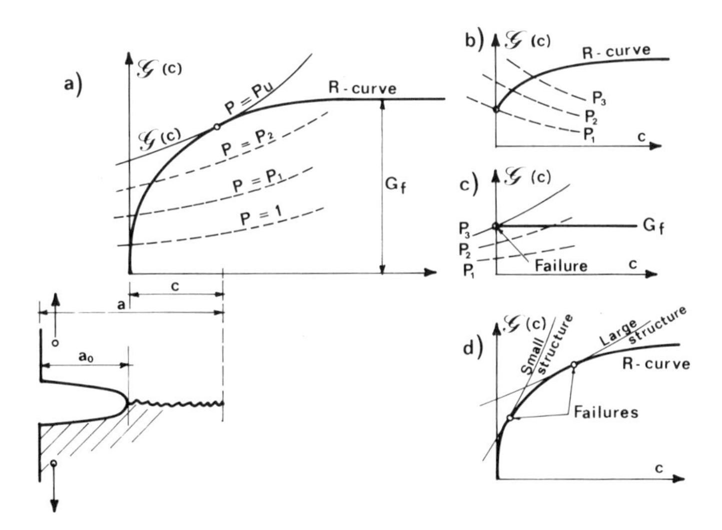

Figure 12.12 Curves of crack resistance to crack propagation (R curves) and of energy release rate: (a) $\mathscr{G}' > 0$; (b) $\mathscr{G}' < 0$; (c) R = const. = G_f; (d) critical stress for structures of different sizes. (*After Bažant and Cedolin, 1984.*)

(where $\partial\Pi/\partial a = \partial\Pi/\partial c$), it follows that a growing crack is in equilibrium if

$$\mathcal{G}(a) = R(c) \qquad \text{with } \mathcal{G}(a) = -\frac{\Pi'(a)}{b} \qquad c = a - a_0 \tag{12.3.2}$$

Here $\mathcal{G}(a)$ = energy release rate of the structure when the crack length is a, $\Pi'(a) = \partial\Pi(a)/\partial a$. In the special case of linear elastic fracture mechanics we have $R(c) = G_f$ = constant, and then Equation 12.3.2 becomes $-\partial\Pi/\partial a = bG_f$, as already indicated in Equation 12.1.1, where $\mathcal{G}$ = energy release rate of structure.

If $0 < \mathcal{G} < R(c)$, the crack can neither grow nor shorten, and so it is in equilibrium as a stationary crack.

If $\mathcal{G} = 0$, the crack can start closing near the tip, as the crack tip opening displacement δ_c is zero if $\mathcal{G} = 0$ or $K_\mathrm{I} = 0$ (see Eqs. 12.1.5 and 12.2.4); this case represents the equilibrium state of crack shortening ($\delta a < 0$) because $\delta\mathcal{F} = (\partial\mathcal{F}/\partial a)\delta a = -b\mathcal{G}\delta a = 0$, that is, the energy required for crack closing is zero [as if $R(c) = 0$].

Under isentropic conditions, the only the change needed to be made in the foregoing equations is to replace $\mathcal{F}$ with total energy of the structure, $\mathcal{U}$. However, the values of $R(c)$ and G_f for isentropic conditions are different (larger) than for isothermal conditions.

Fracture Stability Criterion and Critical State

If the fracture equilibrium state is stable, the crack cannot propagate by itself, that is, without any change of loading (applied force or prescribed displacement). If the fracture equilibrium state is unstable, the crack will propagate by itself, without any change of applied load or boundary displacement. The fracture equilibrium state is stable if the second variation $\delta^2\mathcal{F}$ is positive (same as in Sec. 10.1). Since $\delta^2\mathcal{F} = \frac{1}{2}(\partial^2\mathcal{F}/\partial a^2)\delta a^2$ and $\partial^2\mathcal{F}/\partial a^2 = b(dR/dc) + \partial^2\Pi/\partial a^2$, we conclude that the equilibrium state of a growing crack satisfying Equation 12.3.2 is

$$\begin{array}{lll} \text{Stable if} & R'(c) - \mathcal{G}'(a) > 0 & \\ \text{Critical if} & R'(c) - \mathcal{G}'(a) = 0 & [\text{if } \mathcal{G} = R(c)] \\ \text{Unstable if} & R'(c) - \mathcal{G}'(a) < 0 & \end{array} \tag{12.3.3}$$

where $R'(c) = dR(c)/dc$ and $\mathcal{G}'(a) = \partial\mathcal{G}/\partial a = -(1/b)\partial^2\Pi(a)/\partial a^2$.

If $0 < \mathcal{G} < R(c)$, the crack is stable regardless of the sign of $R'(c) - \mathcal{G}'(a)$ because it can neither extend nor shorten.

If $\mathcal{G} = 0$ (or $K_\mathrm{I} = 0$), we have an equilibrium state of crack shortening ($\delta a < 0$). It is

$$\begin{array}{lll} \text{Stable if} & \mathcal{G}'(a) < 0 & \\ \text{Critical if} & \mathcal{G}'(a) = 0 & (\text{if } \mathcal{G} = 0) \\ \text{Unstable if} & \mathcal{G}'(a) > 0 & \end{array} \tag{12.3.4}$$

If the equilibrium state of crack growth (or crack shortening) is unstable, the crack starts to propagate (or shorten) dynamically, and inertia forces must then be taken into account.

In the special case of linear elastic fracture mechanics, for which $R(c) = G_f =$ constant, we have $dR(c)/dc = 0$. In that case, conditions 12.3.3 for stability of a growing crack become identical to conditions 12.3.4.

For a structure with a single load P (or a system of loads with a single parameter P), Π is proportional to P^2. According to Equations 12.2.8, $\mathscr{G}(a) = P_u^2 g(\alpha)/Eb^2D$. For most structures and fracture specimens, $g(\alpha)$, and thus also $\mathscr{G}(a)$, are increasing functions. The plots of functions $\mathscr{G}(a)$ for a succession of increasing values $P = P_1, P_2, P_3, \ldots$ then look as shown by the dashed curve in Figure 12.12a (Bažant and Cedolin, 1984). According to Equation 12.3.2, the equilibrium states of crack propagation for various load values are the intersections of these dashed curves with the R curve. According to Equations 12.3.3, these equilibrium states are stable if $R'(c) > \mathscr{G}'(a)$ at the intersection point (Fig. 12.12a). As the load increases and the crack grows, the difference between the slopes $R'(c)$ and $\mathscr{G}'(a)$ gradually diminishes until, at a certain point, the slopes become equal (i.e. the curves become tangent); this is then the critical state, at which the load is maximum and the structure fails if the load is controlled. Beyond this point the crack extension is, under load control, unstable and occurs dynamically; the excess energy $\mathscr{G}(a) - R(c)$ goes into kinetic energy. The portion of the R curve before the critical state represents a stable crack growth (also called the "slow crack growth," to indicate the growth is not dynamic).

In the case that $g'(\alpha) < 0$ or $\mathscr{G}'(a) < 0$ (Fig. 12.12b), stability is guaranteed because $R'(c) > 0$. This case occurs for the double cantilever fracture specimen with a relatively short crack and for a rectangular specimen with a small centric crack loaded on the crack (as well as for specimens with chevron notches). For most other geometries, though, $g'(\alpha) > 0$ or $\mathscr{G}'(a) > 0$.

In the case that $R(c) = G_f = \text{const.}$ (Fig. 12.12c), stability under controlled load requires that $\mathscr{G}'(a) < 0$. So there can be no stable crack growth in linear elastic fracture mechanics except when $\mathscr{G}'(a) < 0$. Conversely, if a stable growth is observed and $\mathscr{G}'(a) > 0$, it means that the fracture law must be nonlinear.

Comparing structures that are geometrically similar (with similar notches) but of different sizes, the curves of $\mathscr{G}(c)$ are of similar shape (Fig. 12.12d), while the R curve remains the same. Consequently, the larger the structure, the larger is the crack length c at the maximum load (critical state under load control).

Determination of Geometry-Dependent *R* Curve from Size Effect Law

The foregoing quasi-elastic analysis of equilibrium propagation and stability can be carried out only if a realistic R curve is known. The R curve can be determined experimentally, but that is quite demanding and fraught with possible ambiguities. The main trouble, however, is that the experimentally measured R curve is valid only for specimens of similar geometry. For different geometries, the R curves are rather different (Bažant and Cedolin, 1984; Bažant, Kim, and Pfeiffer, 1986). This fact narrows the applicability of R curves, despite the attractive simplicity of this approach.

A different twist, however, has recently appeared, making the R curve approach much more versatile (Bažant and Kazemi, 1988). It has been found that the size effect law proposed by Bažant is much more broadly applicable than a single R curve. One and the same size effect law, based on the same G_f and c_f (Eq. 12.2.11) applies for specimens of very different geometries while their R curves are very different (see fig. 8 in Bažant, Kim, and Pfeiffer, 1986). So the point is how to determine the R curve from the size effect law.

Consider now that the maximum load P_u has been measured for a set of geometrically similar specimens of different sizes D. For each size one has $\mathscr{G}(a) = P_u^2 g(\alpha)/Eb^2D$, where function $g(\alpha)$ is the same for all sizes D (Eqs. 12.2.8). On each curve $\mathscr{G}(a)$, there is normally one and only one point $a = a_1$ that represents the failure point (critical state). At this point, the $\mathscr{G}(a)$ curve must be tangent to the R curve. Consequently, the R curve is the envelope of the family of all the fracture equilibrium curves $\mathscr{G}(a)$ for different sizes, as shown in Figure 12.13.

To describe the envelope mathematically, we write the condition of equilibrium fracture propagation $f(c, D) = G(\alpha) - R(c) = 0$ where $\alpha = a/D = \alpha_0 + c/D$. If we slightly change size D to $D + \delta D$ but keep the geometric shape (that is, $\alpha_0 = \text{const.}$), failure (max P) occurs at a slightly different crack length $c + \delta c$. Since $f(c, D)$ must vanish both for D and $D + \delta D$, we must have $\partial f(c, D)/\partial D = 0$. Geometrically, the condition $\partial f(c, D)/\partial D = 0$ together with $f(c, D) = 0$ means that the R curve is the envelope of the family of fracture equilibrium curves $f(c, D) = 0$ for various D values (Bažant, Kim, and Pfeiffer, 1986). Because the $R(c)$ curve is a size-independent property, $\partial R(c)/\partial D = 0$. Therefore, the envelope condition is

$$\frac{\partial \mathscr{G}(\alpha)}{\partial D} = 0 \tag{12.3.5}$$

We have $P_u^2 = (\sigma_N bD/c_n)^2 = (Bf_u bD/c_n)^2/(1 + D/D_0)$ where $(Bf_u)^2 = c_n^2 E' G_f / D_0 g(\alpha_0)$ (according to Eq. 12.2.15). If we substitute this into $\mathscr{G}(\alpha) = P_u^2 g(\alpha)/E'b^2D$ (Eqs. 12.2.8), we obtain for the critical states

$$\mathscr{G}(\alpha) = G_f \frac{g(\alpha)}{g(\alpha_0)} \left(\frac{D}{D + D_0} \right) \tag{12.3.6}$$

Substituting this into the condition for the envelope, $\partial \mathscr{G}(\alpha)/\partial D = 0$ (Eq. 12.3.5),

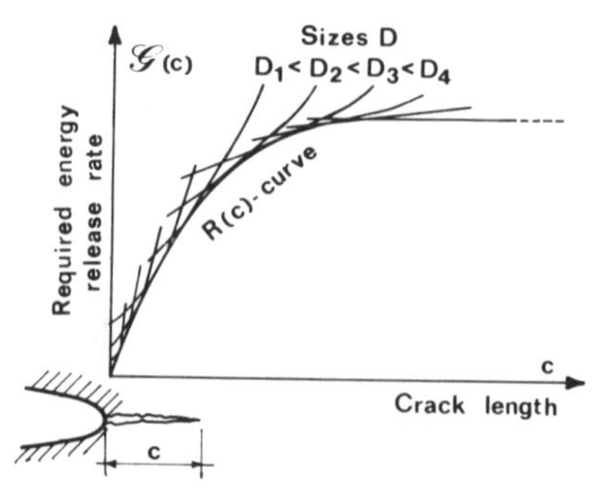

Figure 12.13 R curve as an envelope of fracture equilibrium curves. (*After Bažant, Kim, and Pfeiffer, 1986.*)

differentiating and noting that $\partial\alpha/\partial D = \partial\alpha_0/\partial D + \partial(c/D)/\partial D = -c/D^2 = -(\alpha - \alpha_0)/D$ (because $\alpha_0 = \text{const.}$ or $\partial\alpha_0/\partial D = 0$ for geometrically similar structures), we get

$$D + D_0 = \frac{D_0 g(\alpha)}{(\alpha - \alpha_0)g'(\alpha)} \tag{12.3.7}$$

Furthermore, substituting this, along with the relations $(\alpha - \alpha_0)D = c$ and $D_0 = c_f g'(\alpha_0)/g(\alpha_0)$ (from Eq. 12.2.16), into Equation 12.3.6, and setting $\mathcal{G}(\alpha) = R(c)$, we obtain the following result (Bažant and Kazemi, 1988, 1989):

$$R(c) = G_f \frac{g'(\alpha)}{g'(\alpha_0)}\left(\frac{c}{c_f}\right) \qquad \left(\alpha = \alpha_0 + \frac{c}{D}\right) \tag{12.3.8}$$

Equations 12.3.8 and 12.3.7 define the R curve parametrically. To calculate the R curve, we must first obtain G_f and D_0 from the size effect law (Eq. 12.2.11). Then we choose a series of α values. For each of them we evaluate D from Equation 12.3.7, get $c = (\alpha - \alpha_0)D$, and calculate R from Equation 12.3.8. When c is specified, then R needs to be solved by Newton iterations.

For different geometric shapes, functions $g(\alpha)$ are different, and so Equation 12.3.8 gives different R curves for different geometries. The R curves obtained in this manner, as well as the load-deflection diagrams calculated from such R curves, have been found to be in good agreement with various data on concrete and rocks, as well as aluminum alloys.

The foregoing derivation presumed the fracture process zone to remain attached to the tip of the notch or initial crack. This ceases to be true after passing the peak load; the fracture process zone becomes detached from the tip and subsequently its size remains approximately constant. Therefore it dissipates roughly the same amount of energy per unit crack extension. Consequently, the values of $\mathcal{G}$ after the peak load must be kept constant and equal to the value that $R(c)$ attained at the peak load (Bažant, Gettu, and Kazemi, 1989).

Determination of the R curve from the size effect does not work in all circumstances. It obviously fails when $g'(\alpha_0) = 0$, and does not work when $g'(\alpha_0) < 0$ or $g'(\alpha) \leq 0$, because $g'(\alpha) > 0$ ($D_0 > 0$) was implied in the derivation of Equation 12.3.8. It also fails when $g'(\alpha_0)$ or $g'(\alpha)$ is too small, because of the scatter of test results. So this method must be limited to specimen geometries for which $g'(\alpha_0) > 0$ and $g'(\alpha) > 0$. This nevertheless comprises most practical situations.

Crack Propagation Direction

Another stability problem in crack propagation is that of propagation direction. Equilibrium modes of propagation can generally be found for many directions emanating from the tip of a crack or notch but only one will occur in reality. There exist several theories to decide the actual direction.

One theory assumes that the crack propagates in the direction normal to the maximum principal stress. If the trajectory is smoothly curved, this implies the propagation to occur in such a direction that the crack tip field would be of mode I. Another theory, due to Sih (1974), assumes the crack to propagate in the direction of minimum strain energy density. A third theory (Wu, 1978) holds that

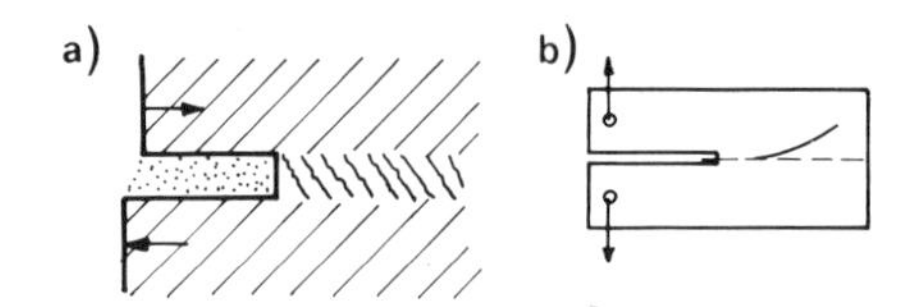

Figure 12.14 (a) Shear crack propagation in brittle materials; (b) crack kinking.

the crack should propagate in the direction that maximizes the energy release rate of the structure or specimen.

The last theory appears to be most reasonable since it in a certain sense maximizes the internally produced entropy increment at a deviation from the initial equilibrium state. The prediction of the last theory is often very close to that of the maximum principal stress theory. However, cracks in concrete have been observed to propagate under certain conditions in shear following the mode II direction (Bažant and Pfeiffer, 1987) or the mode III direction (Bažant, Prat, and Tabbara, 1989). The possibility of shear crack propagation seems to be typical of brittle materials with a coarse microstructure, in which the shear fracture propagates as a band of tensile (mode I) microcracks that are inclined with regard to the direction of propagation (Fig. 12.14a) and later coalesce into a continuous shear crack.

Kinking of Cracks and Three-Dimensional Instability of Front Edge

In some specimens with a symmetric (mode I) loading, the crack does not propagate straight, along the symmetry line, but deviates to the side. This phenomenon, called kinking, occurs, for example, in double cantilever specimens (Fig. 12.14b). Rice and Cotterel (1980) analyzed the kinking as a stability problem and showed that the straight-line propagation along the symmetry line is stable only if there is a normal compressive stress σ_x of sufficient magnitude in the propagation direction; see also Sumi, Nemat-Nasser, and Keer (1985).

Recently Rice also studied the condition when a propagating circular crack in an axisymmetric situation ceases to be circular. He showed that the crack front edge can develop a wavelike shape superimposed on the basic circular shape.

Problems

12.3.1 Supposing that $R(c) = G_f c/(k + c)$ where $k = \text{constant}$, calculate the critical crack length and the stability region for the cracks in Figure 12.5a–h (Eqs. 12.1.13–12.1.22).

12.3.2 Do the same for $R(c) = G_f(1 - e^{-kc})$.

12.4 SNAPBACK INSTABILITY OF A CRACK AND LIGAMENT TEARING

In the preceding section, we formulated the stability criterion in terms of crack length increments that cause deviations from the fracture equilibrium state. From Section 10.1 we recall, however, that in the case of a single load or displacement (or load parameter, displacement parameter) stability can be determined from the load-displacement diagram $P(u)$. With this aim in mind, we will now show how to calculate this diagram. Then we will apply the procedure to the terminal phase of the fracture process in which the distance between two crack tips or between a crack tip and a surface, called the ligament, is being reduced to zero.

General Procedure for Load-Displacement Relation at Growing Crack

Handbooks such as Tada, Paris, and Irwin (1985) or Murakami (1987) contain the solutions of many elastic crack problems. In most cases, though, they list only the stress intensity factor K_I (or K_{II}, K_{III}) as a function of the crack length a and the load P, but the displacement is not given because it does not directly figure in many analytical methods of elastic analysis. Yet, if the entire structure is elastic, u can be calculated from $K_I(a)$ quite easily as follows (Bažant, 1987b).

Consider a body with a single load P (or a single loading parameter P). Instead of the actual process of equilibrium crack growth at $\mathscr{G} = R(c)$, the current state with load P, load-point displacement u, and crack length a may be imagined to be reached by two other loading processes at which mechanical equilibrium is maintained [but the condition of $\mathscr{G} = R(c)$ is violated].

Process I. Load P is applied first on an uncracked specimen (Fig. 12.15a) and then, while P is kept constant, a crack grows from length 0 to length a (Fig. 12.15b), which causes additional displacement u_f. The energy release rate is $\mathscr{G} = K_I^2/E'$ (Eq. 12.1.8) where $K_I = Pk(\alpha)/b\sqrt{D}$ for two-dimensional similarity (2D, Eqs. 12.2.8) or $K_I = Pk(\alpha)D^{-3/2}$ for three-dimensional similarity (3D, Eqs. 12.2.9), and $E' = E/(1 - \nu^2)$ for plane strain, $E' = E$ for plane stress and for 3D. The energy dissipated by the crack tip is

For 2D:

$$W_f = b \int \mathscr{G} D \, d\alpha = \frac{P^2 \phi(\alpha)}{E'b} \qquad \phi(\alpha) = \int_0^{\alpha} [k(\alpha')]^2 \, d\alpha' \qquad (12.4.1a)$$

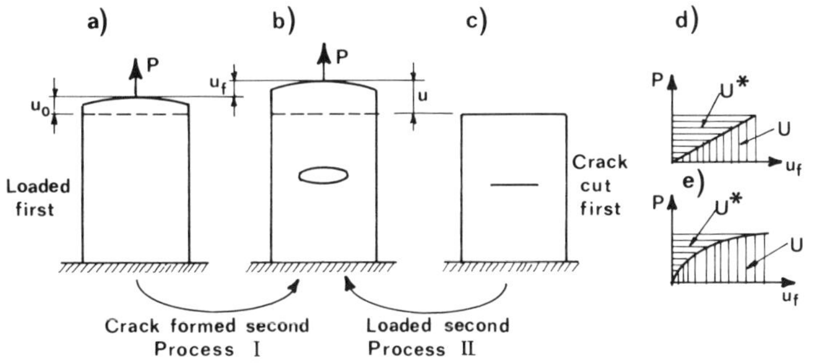

Figure 12,15 (a, b, c) Two loading and cracking sequences leading to the same final state of an elastic solid; strain energy and complementary strain energy for (d) linear and (e) nonlinear elasticity.

For axisymmetric or other 3D situations:

$$W_f = \int p(\alpha) D \mathscr{G} D \, d\alpha = \frac{P^2 \psi(\alpha)}{E'D} \qquad \psi(\alpha) = \int_0^\alpha p(\alpha')[k(\alpha')]^2 \, d\alpha' \tag{12.4.1b}$$

in which $D\,d\alpha = da$ and $p(\alpha)D$ = perimeter of the crack front edge in three-dimensional situations. For example, $p(\alpha) = 2\pi\alpha$ for a circular crack of radius a in a bar of diameter D.

The change of potential energy needed to produce the crack is $\Delta\mathscr{F} = \Pi + W_f$ where $\Pi = U - W = U - \int P\,du_f = U - Pu_f$ = potential-energy change without regard to fracture, W = work of load, U = strain energy change (at constant temperature), and $\Delta\mathscr{F}$ represents the Helmholtz free energy of the system (cf. Eq. 12.5.1). Since there is no external energy input other than load P, we have $\Delta\mathscr{F} = 0$, from which

$$W_f = Pu_f - U \tag{12.4.2}$$

(alternatively we could have written this relation directly on the basis of energy conservation requirements). Now we notice that the expression $U^* = Pu_f - U$ coincides with the definition of the complementary strain energy (Fig. 12.15d). Moreover, for constant load ($dP = 0$), the complementary energy of the structure-load system is $\Pi_f^* = U^* - W^* = U^*$ because $W^* = \int u_f\,dP = 0$ (complementary work of the load at $dP = 0$). So we conclude that

$$\Pi_f^* = W_f \tag{12.4.3}$$

The total complementary energy of the cracked structure is $\Pi^* = \Pi_0^* + \Pi_f^*$ where $\Pi_0^* = \Pi_0^*(P)$ = complementary strain energy of the structure if there is no crack.

Process II. The crack of length a is imagined to be cut prior to loading (in an unstressed body, Fig. 12.15c), and then the load is increased from 0 to P while the crack length a is kept constant. Because the body is elastic, the principle of conservation of energy must apply, and so the complementary energy at the end of this process must be the same, that is, $\Pi^* = \Pi_0^* + \Pi_f^*$ where Π_0^* = change of complementary energy calculated for a body with no crack.

Process II has the advantage that we may apply Castigliano's theorem to calculate the displacements; $u = \partial\Pi^*/\partial P$. Therefore,

$$u(P) = \frac{\partial\Pi_0^*}{\partial P} + \frac{\partial\Pi_f^*}{\partial P} = u_0(P) + u_f(P) \tag{12.4.4}$$

in which $u_0(P) = \partial\Pi_0^*(P)/\partial P$ = displacement calculated for a body with no crack (Fig. 12.15a), and $u_f(P) = \partial\Pi_f^*(P)$ where $u_f(P)$ represents the additional load-point displacement due to crack, which must be equal to the displacement caused in process I by creating the entire crack at constant load P (Fig. 12.15b). The additivity of the displacement due to crack, stated by Equation 12.4.4, is a basic simple principle for calculating deformations of cracked elastic bodies.

According to $u_f = \partial\Pi_f^*/\partial P$, Equations 12.4.1 yields

For 2D:

$$u_f = \frac{\partial\Pi_f^*}{\partial P} = \frac{2P}{E'b}\phi(\alpha) \tag{12.4.5a}$$

For axisymmetric or other 3D situations:

$$u_f = \frac{\partial \Pi_f^*}{\partial P} = \frac{2P}{E'D}\psi(\alpha) \tag{12.4.5b}$$

Now consider the actual loading process, in which the crack grows gradually as the load is increased. During the actual crack growth (in contrast to processes I and II) we must have $\mathscr{G} = R(c)$ or $K_I = K_I^R(c)$ satisfied all the time; $K_I^R(c) = [E'R(c)]^{1/2}$ = critical stress intensity factor depending on the crack extension $c = \alpha D - a_0$ according to the R curve (which must be calculated in advance as we showed in Sec. 12.3). Consequently, from Equations 12.2.8 and 12.2.9,

For 2D:

$$P = \left[\frac{E'b^2D}{g(\alpha)}R(c)\right]^{1/2} = \frac{b\sqrt{D}}{k(\alpha)}K_I^R(c) \tag{12.4.6a}$$

For axisymmetric or other 3D situations:

$$P = \left[\frac{E'D^3}{g(\alpha)}R(c)\right]^{1/2} = \frac{D^{3/2}}{k(\alpha)}K_I^R(c) \tag{12.4.6b}$$

Analogous equations, in which K_I^R is replaced by K_{II}^R or K_{III}^R, apply to mode II or mode III fracture.

Equations 12.4.5a, b and 12.4.6a, b describe the load-deflection curve $P(u)$ at advancing crack in a parametric way, with a as the parameter. For any value of a, one may calculate P from Equations 12.4.6a, b and u_f from Equations 12.4.5a, b. Adding $u_0(P)$, one obtains u.

A similar derivation can be made when the boundary conditions consist of a specified remote stress σ_∞ instead of load P.

Snapback Instability at Crack Coalescence in Two Dimensions

To demonstrate stability analysis based on the foregoing procedure, consider a periodic array of collinear cracks of length $2a$ and center-to-center spacing D in an infinite space subjected at infinity to tensile stress σ normal to the cracks (Ortiz, 1987; Horii, Hasegawa, and Nishino, 1987; Bažant, 1987b) (Fig. 12.16a). This problem is of interest for micromechanics of the fracture process zone in brittle heterogeneous materials, such as concrete or modern toughened ceramics. According to Equation 12.1.19 (Tada, Paris, and Irwin, 1985) we have

$$\frac{1}{b}\frac{\partial \Pi_f^*}{\partial a} = 2\mathscr{G} = \frac{2K_I^2}{E'} = \frac{2\sigma^2}{E'}D\tan\frac{\pi a}{D} \tag{12.4.7}$$

with $E' = E/(1-\nu^2)$, for both crack tips combined. By integration, the strain energy released by symmetric crack formation is

$$\Pi_f^* = -\frac{2\sigma^2 D^2 b}{\pi E'}\ln\left(\cos\frac{\pi a}{D}\right) \tag{12.4.8}$$

The relativc displacement due to cracks, measured between two planes remote

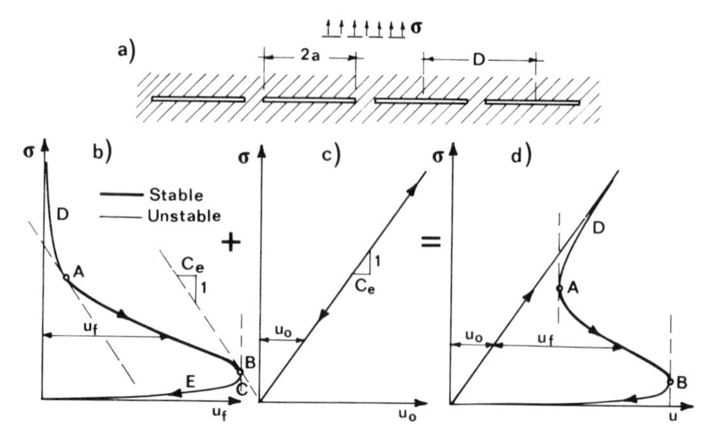

Figure 12.16 (a) Array of collinear cracks, (b, c, d) determination of snapback instability and graphical construction of stress–displacement diagram.

from the crack plane, is (by Castigliano's theorem)

$$u_f = \frac{\partial \Pi_f^*}{\partial P} = \frac{1}{bD} \frac{\partial \Pi_f^*}{\partial \sigma} = -\frac{4\sigma D}{\pi E'} \ln \left(\cos \frac{\pi a}{D} \right) \tag{12.4.9}$$

where $P = \sigma D b$ = force per crack. Although linear elastic fracture mechanics does not apply to macroscopic fracture of concrete or toughened ceramics, it may probably be assumed to apply to microcracks, and so we may use $R(c) = G_f =$ const. Setting $\mathscr{G} = R(c)$, we also have

$$\sigma = \left(\frac{E' G_f}{D \tan (\pi a/D)} \right)^{1/2} \tag{12.4.10}$$

Equations 12.4.9 and 12.4.10 define the relation $\sigma(u_f)$ in a parametric way, with a as a parameter. We may assume the R curve in the form $R(a) = G_f[1 - (1 - a/c_m)^n]$ for $a \leq c_m$ and $R(a) = G_f$ for $a \geq c_m$ where G_f, c_m, and n are material constants and c_m has the dimension of length (Bažant, Kim, and Pfeiffer, 1986). Then, if we choose various values of a, we can calculate the corresponding values of $\sigma(a)$ and $u_f(a)$. The resulting curve $\sigma(u_f)$ is plotted in Figure 12.16b (for $n = 2.8$, $c_m = 0.1$ in.). An interesting property is that, after a maximum displacement u_f, this curve exhibits snapback.

The stress σ, in reality, is not applied or controlled at infinity but at some remote planes at distance L, parallel to the cracks. To judge stability one needs the total displacement $u(\sigma) = u_f(\sigma) + u_0(\sigma)$ where $u_0(\sigma) = C_e \sigma$, $C_e = L/E' =$ relative displacements between these planes if no cracks existed. Based on this relation, one may construct the $u(\sigma)$ diagram graphically by adding the abscissas for the same value of σ as shown in Figure 12.16b, c, d. Clearly, the resulting diagram $\sigma(u)$ (Fig. 12.16d) also exhibits snapback.

Since the diagram $\sigma(u)$ is descending, the states of growing cracks cannot be stable under load control. They can be stable only if displacement u is controlled, but never after the snapback point. The stability condition is $du/dP < 0$ (see Sec. 10.1). The critical state is obtained for $du/dP = 0$, which means drawing vertical tangents as shown in Figure 12.16d. Stability (under displacement control) exists only between the tangent points. Since $u = u_f + C_e\sigma$, one finds that the state of growing cracks under displacement control is

$$\begin{aligned} &\text{Stable if} && \frac{du_f}{d\sigma} < -C_e \\ &\text{Critical if} && \frac{du_f}{d\sigma} = -C_e \\ &\text{Unstable if} && \frac{du_f}{d\sigma} > -C_e \end{aligned} \tag{12.4.11}$$

This stability condition is illustrated in Figure 12.16b where dashed straight lines of slope $-1/C_e$ are drawn as tangents to the calculated $u_f(\sigma)$ diagram. There are two tangent points A and B representing the critical states. Between these critical states (segment AB) the growing crack is stable, and the remaining points on the $u_f(\sigma)$ curve (branches DA and BE) are unstable. The shorter length L is, the steeper is the slope $-1/C_e$, which causes the stability region to increase. However, no matter how steep this slope is, the critical state cannot be pushed beyond point C with a vertical tangent.

Snapback Instability at Tearing of Circular Ligament

For micromechanics of the fracture process zone in concrete or toughened ceramics, a more realistic model for the terminal process of ligament tearing is shrinking circular ligaments of spacing D, connecting two elastic half-spaces in three dimensions. As stated in Equation 12.1.22, $K_\mathrm{I} = \frac{1}{2}P(\pi r^3)^{-1/2}$ (Tada, Paris, and Irwin, 1985) where $r =$ ligament radius and $P =$ transverse force transmitted by one ligament. From this, $-\partial\Pi_f^*/\partial r = \partial\Pi_f^*/\partial a = 2\pi r\mathscr{G} = 2\pi r K_\mathrm{I}^2/E' = P^2/2E'r^2$ where $E' = E/(1-\nu^2)$. By integration,

$$\Pi_f^* = \frac{P^2}{2E'}\left(\frac{1}{r} - \frac{\kappa}{D}\right) \qquad r = \frac{D}{2} - a \tag{12.4.12}$$

where $\kappa =$ integration constant. By Castigliano's second theorem,

$$u_f = \frac{\partial\Pi_f^*}{\partial P} = \frac{P}{E'}\left(\frac{1}{r} - \frac{\kappa}{D}\right) \tag{12.4.13}$$

Also, setting $\mathscr{G} = R(c) = G_f$ we obtain $r = (P^2/4\pi E'G_f)^{1/3}$, and Equation 12.4.13 then becomes

$$u_f = \frac{P}{E'}\left[\left(\frac{4\pi E'G_f}{P^2}\right)^{1/3} - \frac{\kappa}{D}\right] \tag{12.4.14}$$

Constant κ can be determined from some known state at finite r. Requiring Equation 12.4.14 to pass through the snapback point obtained from a more realistic model, Bažant (1987b) found $\kappa = 2.299$.

The plot of Equation 12.4.14 is shown in Figure 12.17b as the lower dashed curve. Obviously, Equation 12.4.14 represents snapback, and we could repeat a similar discussion as before. Figure 12.17a, b also exhibits the solutions obtained by Bažant (1987b) for the two idealized three-dimensional situations shown,

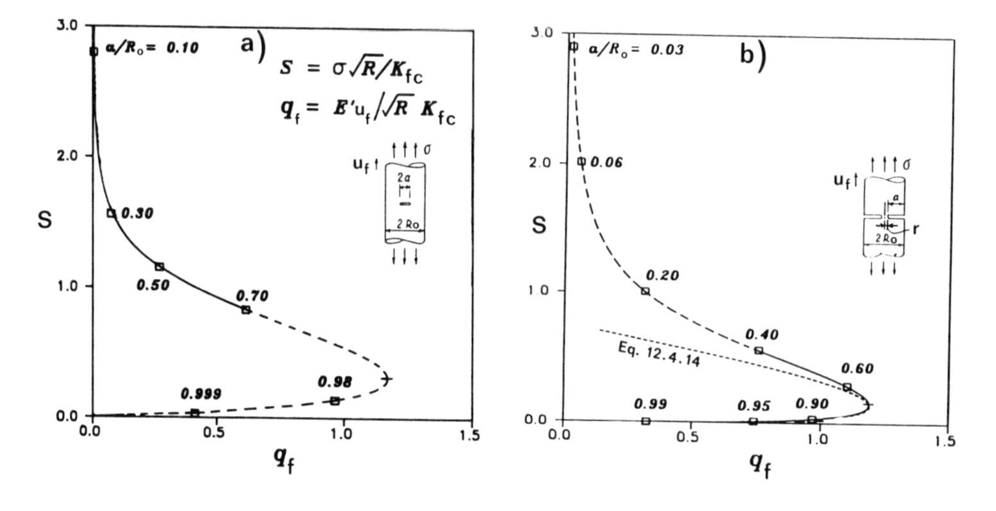

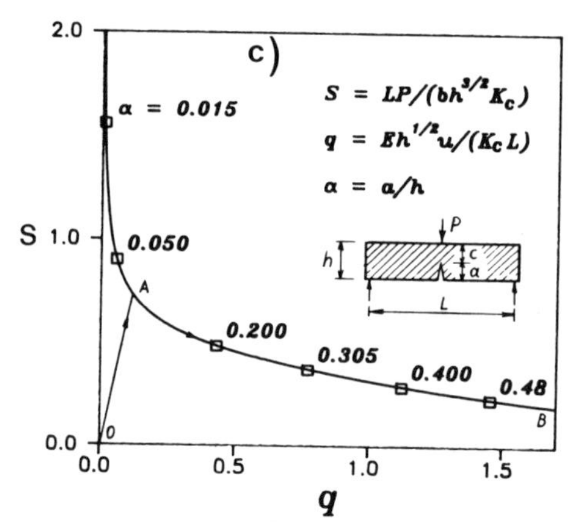

Figure 12.17 Stress–displacement diagrams for (a) circular (penny-shaped) crack, (b) circular ligament, (c) three-point bent beam. (*After Bažant, 1987b.*)

which model the evolution of fracture in the process zone from small circular cracks to small circular ligaments.

With regard to the stress–displacement (or stress–strain) relation for the fracture process zone, the foregoing results suggest that it should be considered to possess a certain maximum displacement (or strain). Conceivably, though, various other inelastic phenomena could spoil this picture; see the discussion in Bažant (1987b).

General Condition for Snapback at Ligament Tearing

Does the snapback always take place in the terminal phase of fracture? It does not. For example, using Srawley's expression for K_I, Bažant (1987b) used the technique we just demonstrated to calculate the $P(u_f)$ diagram for a three-point bend fracture specimen; see Figure 12.17c; it exhibits no snapback instability. So what is the property that causes snapback?

Let us consider the ligament size to be infinitely small compared to any cross-section dimension of the structure. We assume the subsequent ligament shapes to remain similar as the ligament shrinks. Let P and M be the internal force of any direction and the internal moment about any axis transmitted across the ligament (Fig. 12.18). (The special cases of P include a normal force or a shear force, and of M a bending moment or a twisting moment.) According to Saint-Venant's principle, P or M can produce significant stresses and significant strain energy density only in a three-dimensional region whose size (L_1 and L_2 in Fig. 12.18) is of the same order of magnitude as the ligament size $2r$. The strain energy produced in this region by P or M is

$$\Pi_1^* = \frac{P^2}{2EA} k_1 r = \frac{P^2}{2Ek_3 r} \qquad \Pi_2^* = \frac{M^2}{2EI} k_2 r = \frac{M^2}{2Ek_4 r^3} \tag{12.4.15}$$

where $A = k_5 r^2$ = cross-section area of the ligament, $I = k_6 r^4$ = moment of inertia of the cross section of the ligament, and $k_1, k_2, \ldots, k_6$ = constants. The remote displacement u_f and rotation θ associated with P and M, respectively, are

$$u_f = \frac{\partial \Pi_1^*}{\partial P} = \frac{P}{Ek_3 r} \qquad \theta = \frac{\partial \Pi_2^*}{\partial M} = \frac{M}{Ek_4 r^3} \tag{12.4.16}$$

The energy release rates due to P and M per unit circumference of the ligament cross section are

$$\mathscr{G}_1 = -\frac{1}{k_7 r}\frac{\partial \Pi_1^*}{\partial r} = \frac{P^2}{2Ek_3 k_7 r^3} \qquad \mathscr{G}_2 = -\frac{1}{k_8 r}\frac{\partial \Pi_2^*}{\partial r} = \frac{3M^2}{2Ek_4 k_8 r^5} \tag{12.4.17}$$

Setting $\mathscr{G}_1 = G_f$ or $\mathscr{G}_2 = G_f$, we have

$$\begin{aligned} &\text{For } P: \qquad r = \left(\frac{P^2}{2Ek_3 k_7 G_f}\right)^{1/3} \\ &\text{For } M: \qquad r = \left(\frac{3M^2}{2Ek_4 k_8 G_f}\right)^{1/5} \end{aligned} \tag{12.4.18}$$

Substituting this into Equations 12.4.16 we get, for very small ligament size r, the

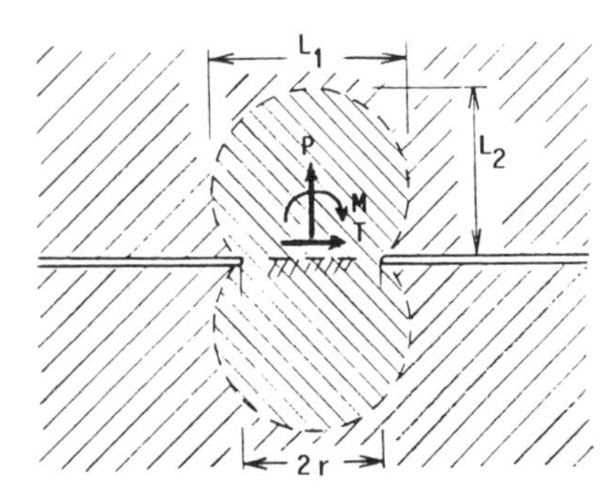

Figure 12.18 Tearing of ligament joining two half-spaces or half-planes. (*After Bažant, 1987b.*)

asymptotic approximations:

$$\begin{aligned} &\text{For } P: \qquad && v = c_1 P^{1/3} \\ &\text{For } M: \qquad && \theta = c_2 M^{-1/5} \end{aligned} \tag{12.4.19}$$

where $c_1, c_2 =$ constants. Note that Equation 12.4.19 for P agrees with the asymptotic form of Equation 12.4.14, which is $u_f^3 = 4\pi G_f P / E'^2$ for $P \to 0$.

From Equations 12.4.19 we conclude that if the ligament is loaded by a force, the curve $u_f(P)$ must return to the origin ($u_f = P = 0$) as $P \to 0$ ($r \to 0$). This implies that there must be snapback at some finite P value.

On the other hand, if the ligament is loaded by a moment, the curve $\theta(M)$ tends to infinity as $M \to 0$ ($r \to 0$). So there can be no snapback.

For two-dimensional problems a similar asymptotic analysis is possible, but only for the moment loading. We have $I = k_6 b r^3$ where $b =$ thickness of the body, and instead of Equations 12.4.15 to 12.4.17 we obtain

$$\Pi_2^* = \frac{M^2}{2EI} k_2 r = \frac{M^2}{2Ek_4 b r^2} \tag{12.4.20}$$

$$\theta = \frac{\partial \Pi_2^*}{\partial M} = \frac{M}{Ek_4 b r^2} \tag{12.4.21}$$

$$\mathcal{G}_2 = -\frac{\partial \Pi_2^*}{b\, \partial r} = \frac{M^2}{Ek_4 b^2 r^3} \tag{12.4.22}$$

Setting $\mathcal{G} = G_f$, we have $r = (M^2 / G_f E k_4 b^2)^{1/3}$, and substituting this into Equation 12.4.21 we get, for small r,

$$\theta = c_2 M^{-1/3} \tag{12.4.23}$$

So for moment loading in two dimensions there cannot be any snapback either.

For two-dimensional problems in which the ligament is loaded by a force, the foregoing approach fails because, as it turns out, the curve $u_f(P)$ is not of a power type as $P \to 0$. For a sufficiently short ligament, the stress field must be the same as that near a ligament joining two elastic half-planes. For that problem it is known that $K_I = (P/b)(\pi r)^{-1/2}$ where $P =$ normal (centric) force and $r =$ half-length of the ligament (Fig. 12.18). Therefore $-\partial \Pi_f^* / \partial r = b\mathcal{G} = bK_I^2 E' = P^2 / \pi E' b r$, and by integration the total strain energy release is

$$\Pi_f^* = -\frac{P^2}{\pi E' b} \ln \frac{r}{r_0} \tag{11.4.24}$$

where $r_0 =$ integration constant. Furthermore,

$$u_f = \frac{\partial \Pi_f^*}{\partial P} = -\frac{2P}{\pi E' b} \ln \frac{r}{r_0} \tag{12.4.25}$$

Setting $K_I = K_{Ic} =$ critical value of K_I, we also get $r = P^2 / \pi b^2 K_{Ic}^2$, and substitution into Equation 12.4.25 yields

$$u_f = -\frac{2P}{\pi E' b} \ln \frac{P^2}{\pi b^2 K_I^{R^2} r_0} \tag{12.4.26}$$

The curve $u_f(P)$ described by this equation is not of a power type, which explains why the type of approach used in Equations 12.4.15 to 12.4.23 would fail.

The curve $u_f(P)$ given by Equation 12.4.26 obviously exhibits a snapback since, for $P \to 0$, $\lim u_f = 0$. The critical state is characterized by the condition $\partial u_f / \partial P = 0$. u_f yields for the snapback instability the critical load $P_{cr} = K_I^R b\sqrt{\pi r_0}/7.389$; max u_f follows from Equation 12.4.26.

Note: Coalescence of adjacent circular voids in a plastic material is a related stability problem. It is of interest for micromechanics of fracture propagation in metals.

Alternative Calculation of Displacement from Compliance Variation

Instead of using Castigliano's theorem and complementary energy, we can alternatively calculate the load-point displacement u by integrating the changes of compliance C. We again consider a process in which load P is applied first on an uncracked structure and then, while P is kept constant, the crack grows from length 0 to length a (see process I in Fig. 12.15a, b). According to Equation 12.1.1, the energy release rate during this process is, for 2D, $\mathcal{G} = (\partial \Pi^* / \partial a)/b = b^{-1} d[\frac{1}{2}C(a)P^2]/da = (P^2/2b)\, dC(a)/da$ (Eq. 12.1.12), and for axisymmetric or other 3D situations $\mathcal{G} = (\partial \Pi^* / \partial a)/p(\alpha)D$. Substituting $\mathcal{G} = K_I^2/E'$, we obtain

For 2D:

$$\frac{dC(a)}{da} = \frac{2bK_I^2(a)}{P^2E'} = \frac{2[k(\alpha)]^2}{bDE'} \tag{12.4.27a}$$

For axisymmetric or other 3D situations:

$$\frac{dC(a)}{da} = \frac{2p(\alpha)DK_I^2(a)}{P^2E'} = \frac{2p(\alpha)[k(\alpha)]^2}{D^2E'} \tag{12.4.27b}$$

where $\alpha = a/D$. The initial value of C for $a = 0$ is the compliance C_0 of the same structure with no crack. Thus, setting $da = D\,d\alpha$ and integrating Equations 12.4.1a, b at $P =$ const. from $C(0) = C_0$ to $C(a)$, we obtain

$$C(a) = C_0 + C_f(a) \tag{12.4.28}$$

where

For 2D:

$$C_f(a) = \frac{2\phi(\alpha)}{bE'} \qquad \phi(\alpha) = \int_0^{\alpha} [k(\alpha')]^2\, d\alpha' \tag{12.4.29a}$$

For axisymmetric or other 3D situations:

$$C_f(a) = \frac{2\psi(\alpha)}{DE'} \qquad \psi(\alpha) = \int_0^{\alpha} p(\alpha')[k(\alpha')]^2\, d\alpha' \tag{12.4.29b}$$

$C_f(a)$ represents the additional compliance due to the crack. Equation 12.4.28 proves that the compliance of the uncracked structure and the compliance due to the crack are additive. According to Equation 12.4.28, the load-point displacement is $u(P) = u_0(P) + u_f(P)$ where $u(P) = C_0P$ and $u_f(P) = C_f(a)P$. Substituting Equations 12.4.29a, b, we get for u_f the same expression as in Equations 12.4.5a, b.

Problems

12.4.1 Derive Equation 12.4.25 as the asymptotic approximation of Equations 12.4.9 and 12.4.10 for a very small ligament, such that $2r/D \ll 1$, $2r = D - 2a$. {Note that $\cos(\pi a/D) \simeq \sin[\frac{1}{2}\pi(1 - 2a/D)] \simeq \pi r/D$.}

12.4.2 Using the same procedure as that in Equations 12.4.24 to 12.4.26, calculate $u_f(P)$ for a tensioned infinite planar strip with a centric normal crack, for which K_I is given by Equation 12.1.17. Find $u_{f,\max}$ for snapback instability.

12.4.3 Derive the asymptotic $u_f(P)$ associated with Equation 12.1.22 for $r \to 0$ (tensioned circular bar with a circular crack).

12.4.4 Analyze possible snapback instability in a very slender double cantilever specimen, using Equation 12.1.26.

12.4.5 Referring to Section 10.1, formulate the second variation of the Helmholtz free energy, $\delta^2\mathcal{F}$, for the crack problem described by Equations 12.4.24 to 12.4.26, and then discuss stability directly on the basis of $\delta^2\mathcal{F}$.

12.4.6 Generalize Equations 12.4.8 to 12.4.10 assuming that $R(c)$ is not a constant but $R(c) = G_f c/(k + c)$, $c = a - a_0$. Discuss the effect of variable $R(c)$ on stability.

12.4.7 Generalize Equations 12.4.1 to 12.4.4 to a nonlinear elastic structure, for which the definition of complementary energy is illustrated in Figure 12.15e.

12.4.8 All the problems in this section can alternatively be solved on the basis of Equation 12.4.28, without any use of Castigliano's theorem. Do that.

12.5 STABLE STATES AND STABLE PATHS OF INTERACTING CRACKS

So far we have investigated only the stability of a structure with a single (active) crack. Various applications, however, call for the analysis of a structure with a system of interacting cracks. This is a more intricate problem. The equilibrium path of a crack system can exhibit bifurcations of various types. There exist bifurcations in which progress along each branch requires an increase of load, similar to the behavior of Shanley's column.

Conditions of Equilibrium and Stability in Terms of Crack Length

Consider a two-dimensional brittle elastic structure (Fig. 12.19a) that contains a crack system with N crack tips and crack lengths a_i ($i = 1, 2, \ldots, N$). The cracks propagate in mode I along known paths (so that the question of propagation direction need not be considered). Stability may be analyzed on the basis of the energy $\mathcal{F}$ that has to be supplied to the body in order to produce the cracks;

$$\mathcal{F} = \Pi(a_1, \ldots, a_N; \lambda) + \sum_{i=1}^{N} \int R_i(a_i)\, da_i \tag{12.5.1}$$

in which λ is a loading parameter, for example, a parameter controlling the applied forces, or enforced boundary displacements, or temperature distribution; $R_i(a_i) = R(c_i)$ ($c_i = a_i - a_{0i}$) is the energy required for the growth of crack a_i (i.e.,

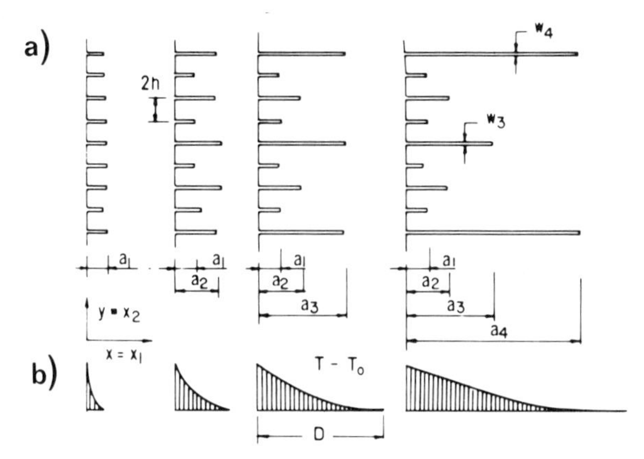

Figure 12.19 System of parallel cracks in a half-space. (*After Bažant, Ohtsubo, and Aoh, 1979.*)

the R curve, treated as a given property), and $\Pi = U - W$ = potential energy of the structure (U = strain energy and W = work of applied loads). In the special case of fixed boundary displacements, $W = 0$ and $\Pi = U$. We assume the conditions to be isothermal, and then, from the thermodynamic viewpoint, $\mathcal{F}$ represents the Helmholtz free energy of the structure (Sec. 10.1) and the values of R_i must correspond to isothermal fracture. (If the conditions are isentropic, all our analysis to come is the same except that $\mathcal{F}$ needs to be replaced by the total energy, $\mathcal{U}$, and the R_i values for adiabatic conditions need to be used.) In the special case of linear elastic fracture mechanics, $R_i(a_i) = G_f$ = const.

Consider now that the crack tips numbered $i = 1, \ldots, m$ advance ($\delta a_i > 0$), the crack tips numbered $i = m + 1, \ldots, n$ close near the tip and retreat ($\delta a_i < 0$), and the crack tips numbered $n + 1, \ldots, N$ remain stationary ($\delta a_i = 0$); obviously $0 \le m \le n \le N$. The special case $m = n$ means that no crack closes, and $n = N$ means that no crack remains immobile. The work, $\Delta\mathcal{F}$, that would have to be supplied in order to change the crack lengths by δa_i at constant λ (no change of loading) may be expanded as $\Delta\mathcal{F} = \delta\mathcal{F} + \delta^2\mathcal{F} + \cdots$ where (for a body of unit thickness)

$$\delta\mathcal{F} = \sum_{i=1}^{m} \left(\frac{\partial \Pi}{\partial a_i} + R_i \right) \delta a_i + \sum_{j=m+1}^{n} \frac{\partial \Pi}{\partial a_j} \delta a_j \tag{12.5.2}$$

$$\delta^2\mathcal{F} = \frac{1}{2} \sum_{i=1}^{n} \sum_{j=1}^{n} \frac{\partial^2 \Pi}{\partial a_i \, \partial a_j} \delta a_i \delta a_j + \frac{1}{2} \sum_{i=1}^{m} \frac{\partial R_i}{\partial a_i} (\delta a_i)^2$$

$$= \frac{1}{2} \sum_{i=1}^{n} \sum_{j=1}^{n} \mathcal{F}_{,ij} \, \delta a_i \delta a_j \qquad (i, j = 1, \ldots, n) \tag{12.5.3}$$

in which

$$\mathcal{F}_{,ij} = \mathcal{F}_{,ji} = \frac{\partial^2 \Pi}{\partial a_i \, \partial a_j} + \frac{\partial R_i}{\partial a_i} \delta_{ij} H(\delta a_i) \qquad i, j = 1, \ldots, n. \tag{12.5.4}$$

δ_{ij} = Kronecker delta and H = Heaviside function, that is, $H(\delta a_i) = 1$ if $\delta a_i > 0$ and 0 if $\delta a_i < 0$. In the case of a homogeneous body obeying linear elastic fracture mechanics, $\partial R_i / \partial a_i = 0$ and $\mathcal{F}_{,ij} = \Pi_{,ij}$. (Note that in the above all the partial derivatives with respect to a_i and a_j are calculated at constant λ, and so they correspond in general to deviations from equilibrium.)

For the cracks to change their lengths in an equilibrium manner, $\delta\mathcal{F}$ must vanish for any δa_i. One must distinguish whether a crack extends ($\delta a_i > 0$) or closes ($\delta a_i < 0$). According to Equation 12.5.2, $\delta\mathcal{F} = 0$ occurs if and only if

$$\begin{aligned}
&\text{For } \delta a_i > 0: && -\frac{\partial \Pi}{\partial a_i} = R_i \\
&\text{For } \delta a_i = 0: && 0 \le -\frac{\partial \Pi}{\partial a_i} \le R_i \\
&\text{For } \delta a_i < 0: && -\frac{\partial \Pi}{\partial a_i} = 0
\end{aligned} \tag{12.5.5}$$

This includes the Griffith criterion, and $-\partial\Pi/\partial a_i$ is the energy release rate. An equivalent form of this equation can be given in terms of the stress intensity factors K_i, defined by $K_i^2 = -(\partial\Pi/\partial a_i)E'$ where $E' = E/(1-\nu^2)$ for plane strain and $E' = E$ for plane stress;

$$\begin{aligned}
&\text{For } \delta a_i > 0: && K_i = K_{c_i} \\
&\text{For } \delta a_i = 0: && 0 \le K_i \le K_{c_i} \\
&\text{For } \delta a_i < 0: && K_i = 0
\end{aligned} \tag{12.5.6}$$

in which $K_{c_i} = (R_i E')^{1/2}$ = critical value of K_i (fracture toughness). From the foregoing expression for K_i^2 and the symmetry property $\Pi_{,ij} = \Pi_{,ji}$ for growing cracks it follows that

$$K_i \frac{\partial K_i}{\partial a_j} = K_j \frac{\partial K_j}{\partial a_i} \tag{12.5.7}$$

An equilibrium state of fracture is stable if and only if no a_i can change without a change in the loading (λ = const.). Stability is assured if the work $\Delta\mathcal{F}$ done on any admissible δa_i is positive, for δa_i cannot occur if this work is not done on the body. On the other hand, if $\Delta\mathcal{F} < 0$ for some δa_i, energy is released, and when a release of energy is possible, δa_i will occur spontaneously, $\Delta\mathcal{F}$ being transformed into kinetic energy and ultimately dissipated as heat (which follows from the second law of thermodynamics).

An unstable situation obviously arises when $-\partial\Pi/\partial a_i > R_i$ or $K_i > K_{c_i}$ for $\delta a_i > 0$. Indeed, $\delta\mathcal{F} < 0$ for $\delta a_i > 0$, and so $K_i > K_{c_i}$ cannot be a stable equilibrium state. Similarly, the case $-\partial\Pi/\partial a_i < 0$ or $K_i < 0$ is also unstable. Therefore, stability requires that $0 \le -\partial\Pi/\partial a_i \le R_i$ or $0 \le K_i \le K_{c_i}$ at all times. Combining this with Equations 12.5.6, we see that only the following crack length variations are admissible:

$$\begin{aligned}
&\text{For } K_i = K_{c_i}: && \delta a_i \ge 0 \\
&\text{For } 0 < K_i < K_{c_i}: && \delta a_i = 0 \\
&\text{For } K_i = 0: && \delta a_i \le 0
\end{aligned} \tag{12.5.8}$$

Now stability of an equlibrium state of growing cracks ($\delta\mathscr{F}=0$) will be ensured if (Bažant and Ohtsubo, 1977)

$$\delta^2\mathscr{F}=\frac{1}{2}\sum_{i=1}^{n}\sum_{j=1}^{n}\mathscr{F}_{,ij}\,\delta a_i\,\delta a_j>0 \qquad \text{for any admissible } \delta a_i \tag{12.5.9}$$

(i.e., if $\delta^2\mathscr{F}$ is positive definite). Instability occurs if $\delta^2\mathscr{F}<0$ for some admissible δa_i. The critical state, for which $\delta^2\mathscr{F}=0$ for some admissible δa_i, can be stable or unstable depending on the higher variations of $\mathscr{F}$.

These conditions are the same as those for any structure (Sec. 10.1). But the variations are taken here in the crack lengths rather than displacements q_i, and an important restriction on the admissibility of δa_i must be imposed here. If matrix $\mathscr{F}_{,ij}$ is positive definite, stability is obviously ensured. But stability can exist even if matrix $\mathscr{F}_{,ij}$ is not positive definite provided that negative $\delta^2\mathscr{F}$ occurs only for inadmissible δa_i.

Stability of Parallel Cooling or Shrinkage Cracks

Consider the two-dimensional problem of a homogeneous isotropic elastic half-space in which a system of parallel equidistant cracks normal to the surface is produced by cooling or drying shrinkage. This problem arises in many applications. It was studied in detail with respect to a proposed hot-dry-rock geothermal energy scheme; see Bažant and Ohtsubo (1977), Bažant, Ohtsubo, and Aoh (1979), Bažant and Wahab (1979), Nemat-Nasser, Keer, and Parihar (1978), Bažant (1976a), Keer, Nemat-Nasser, and Oranratnachai (1979), Sumi, Nemat-Nasser, and Keer (1980) (and also, with respect to crack width in beams, see Bažant and Wahab, 1980). In this scheme, a large primary crack is created by forcing water under high pressure into a bore several kilometers deep (hydraulic fracturing). Heat is extracted from the rock by circulating water through the crack. However, this cools the rock adjacent to the crack wall, and so the scheme can be viable only if further cracks are produced by the cooling water. Another application is the shrinkage cracking of concrete (Bažant and Raftshol, 1982), as well as shrinkage cracks in dried lake beds or in lava flows, which are of interest to geologists.

Let the half-space be initially at uniform temperature $T=T_0$ and at time $t=0$ the temperature of the surface $x=0$ is suddenly changed to T_1. The temperature field is assumed to have the form $T-T_0=f(\xi)(T_1-T_0)$ where $\xi=\sqrt{3}x/\lambda$, $x=$ coordinate normal to the surface, and $\lambda=\lambda(t)=$ penetration depth of cooling. If all the heat is transferred by conduction in the solid, the solution is well known (Carslaw and Jaeger, 1959) and is given by $f(\xi)=\operatorname{erfc}\xi=2\int_\xi^\infty\exp(-\eta^2)\,d\eta/\sqrt{\pi}$, $\lambda=\sqrt{12ct}$, $c=$ heat diffusivity of the material. The cooling profiles $T(x,t)-T_0$ advance into the half-space as shown in Figure 12.19b. T approaches T_0 asymptotically for $t\to\infty$. The depth λ is defined as the depth of an equivalent parabolic profile that gives the same heat flux at the surface.

Because T is constant along lines parallel to the surface, one may expect a possible solution to be a system of equidistant equally long cracks. However, there also exist solutions with unequal crack lengths, for example, with the crack lengths alternating from a_1 to a_2 (Fig. 12.19). Thus we suspect that in the space

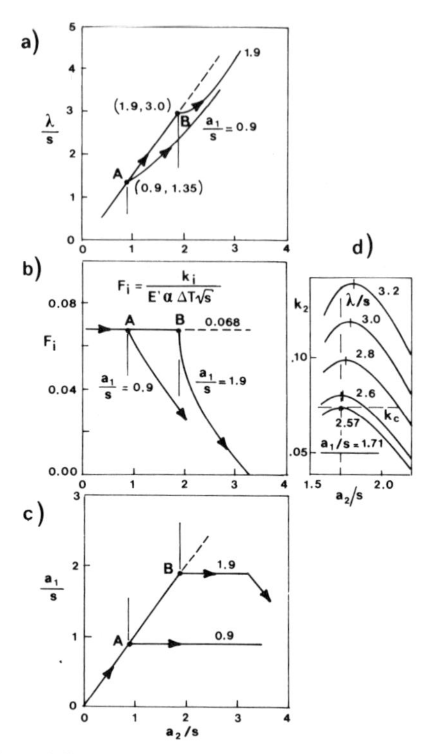

Figure 12.20 System of equidistant parallel cooling or shrinkage cracks in a half-plane, exhibiting path bifurcation in which every other crack gets arrested. (*After Bažant and Tabbara, 1989.*)

(a_1, a_2), the path representing the solution bifurcates (Fig. 12.20). The bifurcation may or may not be the limit of stable states.

First, let us examine bifurcation at the limit of stable states for an initial state in which all the cracks are equally long ($a_2 = a_1$) and critical ($K_i = K_c$) (Bažant and Ohtsubo, 1977). One simple possibility is that only every other crack extends by δa_2 while the intermediate cracks remain stationary ($\delta a_1 = 0$). (Fig. 12.20a). Then we have $m = 1$, $n = N = 2$, and $\det \mathcal{F}_{,ij} = \mathcal{F}_{,22} = \partial^2 \mathcal{F} / \partial a_2^2$, $\delta^2 \mathcal{F} = \frac{1}{2} \mathcal{F}_{,22} (\delta a_2)^2$. Assuming the material to obey linear elastic fracture mechanics, we have $\mathcal{F}_{,22} = \Pi_{,22} = U_{,22} = -\partial G / \partial a_2$ where $G = -\partial U / \partial a_2 = K_2^2 / E'$ = energy release rate for a body of unit thickness. According to Equation 12.5.9, the state is

$$\text{Stable if} \qquad \frac{\partial K_2}{\partial a_2} < 0$$

$$\text{Critical if} \qquad \frac{\partial K_2}{\partial a_2} = 0 \tag{12.5.10}$$

$$\text{Unstable if} \qquad \frac{\partial K_2}{\partial a_2} > 0$$

The instability mode at critical state consists of a sudden advance of every other crack by $\delta a_2 > 0$, occurring at $\delta a_1 = 0$ and at constant λ, that is, at no change in temperature distribution (Fig. 12.20b). This represents a state of neutral equilibrium. Finite element calculations indicate that K_1 decreases during the instability, that is, the stationary cracks cease to be critical, while, of course, K_2 remains equal to K_c.

The bifurcation with stability loss and the equilibrium postbifurcation path calculated by finite elements are plotted in Figure 12.20a, b, c, in which s = crack spacing. The calculation has been made (by Bažant an Ohtsubo, 1977, and Bažant, Ohtsubo, and Aoh, 1979) for $T_0 - T_1 = 100°\text{C}$ and the typical properties of Westerly granite: thermal expansion coefficient $\alpha = 8 \times 10^{-6}/°C$, $E = 37{,}600\ \text{MPa}$, $\nu = 0.305$, $c = 1.08 \times 10^{-6}\ \text{m}^2/\text{s}$, and $G_f = 208\ \text{J/m}^2$. (Similar results have been presented by Nemat-Nasser, Keer, and Parihar, 1978.) The instability is found to occur when $a_2/s \simeq 1.8$, $\lambda/s \simeq 2.5$. The plots of K_2 versus a_2/s from which the condition $\partial K_2/\partial a_2 = 0$ is identified are shown in Figure 12.20d. After bifurcation, cracks a_2 grow as the cooling front advances ($K_2 = K_c$), while cracks a_1 are arrested (a_1 = const.) and their K_1 decreases (Fig. 12.20c). Eventually (at $\lambda/s \simeq 3.3$), K_1 becomes zero. As the cooling front advances further, cracks a_1 are closing. The solution now becomes approximately equivalent to that at the start—a system of equally long cracks of length a_2, but the spacing is now doubled to $2s$ (Fig. 12.19). When $a_2/2s = 1.8$ or $a_2/s = 3.6$, the same type of bifurcation repeats itself. After that every other remaining opened crack is again arrested and eventually closes, thus causing the spacing of the open cracks to double again, etc.

The practical significance of this behavior is that the crack opening δ_c at the crack mouth is approximately $\delta_c = \varepsilon^0 s^0$ where $\varepsilon^0 = \alpha(T_0 - T_1)$ and $s^0 = s$, $2s, 4s, 8s, \ldots$ = spacing of the opened cracks. So the crack width increases with the advance of cooling (a common experience in observing, e.g. a drying lake bed). For the aforementioned geothermal energy system this means that, due to increasing crack width, more cooling water can circulate through these widely spaced new cracks—a desirable behavior. However, circulation of water through the newly formed cracks alters the cooling profile, making it steeper at the cooling front. It appears that the profile of temperature has a very large effect on the occurrence of bifurcation instability (Fig. 12.21). The steeper the profile at the front, the larger the ratio λ/s at which the bifurcation occurs. If the profile is sufficiently steep at the front, bifurcation never occurs, and the parallel, equally long cracks can propagate indefinitely, at constant spacing, without any instability (see Fig. 12.21). For details, see Bažant and Wahab, 1979.

A question remains: Can an instability mode at constant λ (i.e., at neutral equilibrium) occur with both δa_1 and δa_2 being nonzero? It cannot. In that case one has $m = n = N = 2$, and a sufficient condition of stability is the positive definiteness of the matrix $\mathscr{F}_{,ij} = \Pi_{,ij} = U_{,ij} = \partial^2 U/\partial a_i\, \partial a_j$, which requires that

$$U_{,22} = U_{,11} > 0 \qquad \text{and} \qquad \begin{vmatrix} U_{,11} & U_{,12} \\ U_{,21} & U_{,22} \end{vmatrix} > 0 \tag{12.5.11}$$

Now, finite element calculations indicate that, for a certain range of G_f values, the determinant condition is violated before the condition $\partial K_2/\partial a_2 < 0$ that we examined before. Does it mean that an instability mode associated with the determinant condition occurs earlier? It does not.

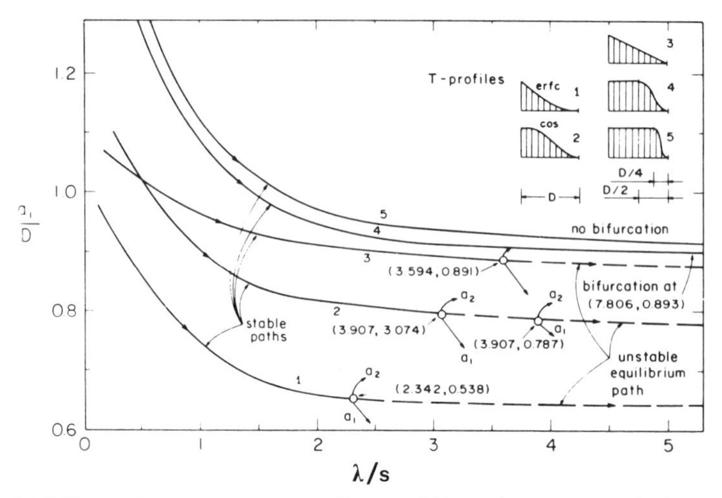

Figure 12.21 Effect of temperature profile on bifurcation points. (*After Bažant and Wahab, 1979.*)

A bifurcation of the basic equilibrium path $a_1 = a_2$ at neutral equilibrium would occur if $\delta^2 \mathscr{F} = \frac{1}{2}\Sigma_i(\Sigma_j U_{,ij}\delta a_j)\delta a_i = 0$ for some admissible δa_i. This condition is satisfied if $\Sigma_j U_{,ij}\delta a_j = 0$ or

$$\begin{aligned} U_{,11}\delta a_1 + U_{,12}\delta a_2 &= 0 \\ U_{,21}\delta a_1 + U_{,22}\delta a_2 &= 0 \end{aligned} \tag{12.5.12}$$

in which $U_{,12} = U_{,21}$. Since $U_{,11} = U_{,22}$ for $a_1 = a_2$, the condition of vanishing determinant is $U^2_{,22} - U^2_{,12} = 0$ or $U_{,22} = \pm U_{,12} = \pm U_{,21}$. Thus Equations 12.5.12 indicate that

$$\frac{\delta a_1}{\delta a_2} = \pm 1 \tag{12.5.13}$$

at the point of bifurcation with neutral equilibrium. The plus sign refers to the main equilibrium path of the crack system ($\delta a_1 = \delta a_2$). The minus sign refers to the secondary path and indicates that the cracks a_1 would have to shorten ($\delta a_1 < 0$) for this type of instability to occur. But shortening of these cracks would violate Equations 12.5.8 since we have assumed that $K_2 = K_1 = K_c$. So we must conclude that an instability mode at constant λ in which both δa_1 and δa_2 are nonzero is impossible (Bažant and Ohtsubo, 1977; Nemat-Nasser, Keer, and Parihar, 1978).

Stable Path and Bifurcation at Advancing Cooling Front

The instability mode corresponding to $\partial K_2/\partial a_2 = 0$ occurs at constant loading parameter. With regard to our previous analysis of plastic buckling (Sec. 10.3), this is analogous to the instability at the reduced modulus load, which happens at constant load. Is there a situation analogous to that at Shanley's tangent modulus load? The answer is yes.

Along the equilibrium path of a crack system, the condition $\mathscr{F}_{,i}=0$ (equivalent to $K_i = K_{ic}$) must be satisfied at every point, and $\mathscr{F}_{,i} = U_{,i} + G_f$ if we assume K_{ic} to be constant. Therefore $d\mathscr{F}_{,i}/d\lambda=0$ or $dU_{,i}/d\lambda=0$ where $U=U(a_1, \ldots, a_N; \lambda)$; λ is a loading parameter (which may represent a boundary displacement, an applied force, the depth of cooling front, etc.). Since a_i are functions of loading parameter λ, this means that $\sum_j U_{,ij}a'_j + U'_{,i}=0$ where the primes denote partial derivatives with respect to λ. At a bifurcation point, this equation must hold for both postbifurcation branches, that is,

$$\sum_j U_{,ij}a_j'^{(1)} + U'_{,i}=0$$
$$\sum_j U_{,ij}a_j'^{(2)} + U'_{,i}=0 \tag{12.5.14}$$

where superscript (1) labels the main path ($\delta a_1 = \delta a_2$) and superscript (2) the secondary path ($\delta a_1 \neq \delta a_2$). Subtraction of the equations now yields

$$\sum_{j=1} U_{,ij}a_j^* = 0 \qquad \text{where } a_j^* = a_j'^{(2)} - a_j'^{(1)} \tag{12.5.15}$$

Since this is a system of homogeneous linear algebraic equations for a_j^*, it follows that bifurcation (i.e., the case $a_j'^{(2)} \neq a_j'^{(1)}$) occurs if and only if

$$\det U_{,ij}=0 \tag{12.5.16}$$

This agrees with the second condition in Equations 12.5.11.

Let us now return to the system of parallel cooling cracks in a half-space. As already mentioned, finite element analysis indicates that the condition of vanishing $\det U_{,ij}$ may occur before the first instability point ($\partial K_2/\partial a_2=0$). Therefore, bifurcation may occur in a stable manner during an advance of the cooling front, without any instability. As we already showed (Eq. 12.5.13), the eigenvector a_j^* associated with $\det U_{,ij}=0$ is such that $a_1^* = -a_2^*$, that is, $a_1'^{(2)} - a_1'^{(1)} = -(a_2'^{(2)} - a_2'^{(1)})$. Because for the main path $a_1'^{(1)} = a_2'^{(1)}$, we get $a_1'^{(2)} + a_2'^{(2)} = 2a_2'^{(1)}$. For the main path we have $a_1'^{(1)} = a_2'^{(1)} = a'^{(1)} > 0$. For the secondary path with $a_2'^{(2)}>0$ it is necessary that $a_1'^{(2)}=0$; the reason is that if both a_1 and a_2 propagate we must have $K_1 = K_2 = K_c$, but the stress intensity factors cannot be equal if $a_1 \neq a_2$ while $a_1'^{(2)} \neq 0$ would lead to $a_1 \neq a_2$. Therefore, the first bifurcation is characterized by $a_2'^{(2)} = 2a_2'^{(1)}$ or

$$\frac{1}{2}\frac{\partial a_2^{(2)}}{\partial \lambda} = \frac{\partial a_2^{(1)}}{\partial \lambda} = \frac{\partial a_1^{(1)}}{\partial \lambda} \qquad \frac{\partial a_1^{(2)}}{\partial \lambda}=0 \tag{12.5.17}$$

The behavior at first and second bifurcation states and the corresponding postbifurcation branches are schematically drawn in Figure 12.20a, b, c.

The foregoing analysis shows that bifurcation is possible but does not show whether the secondary path must actually be followed by the crack system. The first bifurcation path can be decided from the condition that $\delta^2\mathscr{F}$ or $\delta^2 U$ must be minimized for the actual (stable) path. It turns out that bifurcation must occur, that is, every other crack must get arrested, as soon as the condition $\det \mathscr{F}_{,ij}>0$ (or $\det U_{,ij}>0$, Eqs. 12.5.11) becomes violated. Yet, as shown below Equations 12.5.11, the states on the main path at and after this point are stable. In fact, both postbifurcation branches consist of stable states (and so analysis of stability of equilibrium cannot decide which branch must be followed). The situation is analogous to Shanley bifurcation at the tangent modulus load of a column (Sec. 10.3).

By dimensional analysis one may conclude that the stability behavior of the parallel crack system can depend only on the nondimensional spacing

$$\bar{s} = \frac{sE'}{G_f} = \frac{sE'^2}{K_c^2} \tag{12.5.18}$$

Numerical results of Bažant and Tabbara (1989) show that, for any $\bar{s}$, the bifurcation at $\delta\lambda > 0$ (that is, $\det U_{,ij} = 0$) is reached before the bifurcation at $\delta\lambda = 0$ (that is, $\partial K_2/\partial a_2 = 0$).

It may be noted that, in the foregoing analysis, the spacing of the parallel cracks was left indeterminate. We will discuss its determination in Section 12.6.

The critical state of stability limit (neutral equilibrium, Eq. 12.5.10) was, for a system of parallel cooling cracks, found by Bažant (1976a) and Bažant and Ohtsubo (1977), and analyzed in detail by Bažant, Ohtsubo, and Aoh (1979); Bažant and Wahab (1979, 1980); Nemat-Nasser, Keer, and Parihar (1978); Keer, Nemat-Nasser, and Oranratnachai (1979); Sumi, Nemat-Nasser, and Keer (1980); and Bažant and Tabbara (1989). The fact that a system of parallel cooling cracks exhibits a critical state with $\det U_{,ij} = 0$ (Eq. 12.5.16) and that this critical state precedes the stability limit was noted by Bažant (1976a) on the basis of finite element results of Ohtsubo. The paths at and near this critical state were determined by Sumi, Nemat-Nasser, and Keer (1980), and more generally by Bažant and Tabbara (1989).

Three-Dimensional Pattern of Cooling or Shrinkage Cracks

Analyzing the cooling or shrinkage cracks in a half-space as two-dimensional is only a simplified approximation. In three dimensions, the pattern of the cracks in the planes parallel to the surface is hexagonal; see Figure 12.22d.

To show that it ought to be so, we need to note that the actual crack pattern (under isothermal conditions) should minimize the average Helmholtz free energy $\bar{\mathcal{F}}$ per unit cell of the crack pattern (see Sec. 10.1), that is,

$$\bar{\mathcal{F}} = \bar{U}(a_1, \ldots, a_N; \lambda) + G_f \bar{A}_c = \min \tag{12.5.19}$$

where we assume that $R = G_f = \text{const.}$, and denote as A_c the area of the cracks per unit cell. For the sake of simplification we may probably assume that the strain energy $\bar{U}$ per unit volume is approximately the same for all the patterns. Thus it is necessary to minimize $\bar{A}_c$ for the same cell volume V_c. Parallel planar cracks (Fig. 12.22a) must be rejected because they do not relieve the normal stress in all directions. The regular patterns that relieve the stress in all directions are the square, triangular, and hexagonal patterns (Fig. 12.22b, c, d). The volumes of cells of unit depth are $V_c = s^2$, $s^2\sqrt{3}/4$, and $s^2\sqrt{27}/2$, and so the sides are $s = \sqrt{V_c}$, $(4V_c/\sqrt{3})^{1/2}$, and $(2V_c/\sqrt{27})^{1/2}$, respectively. Noting that by repeating the pattern of boldly drawn sides of the cross-hatched elements in Figure 12.22 one can generate the entire mesh, we find that the crack areas per cell of the same volume are $2\sqrt{V_c}$, $\frac{3}{2}(4V_c/\sqrt{3})^{1/2}$, and $3(2V_c/\sqrt{27})^{1/2}$, that is, $2\sqrt{V_c}$, $2.280\sqrt{V_c}$,

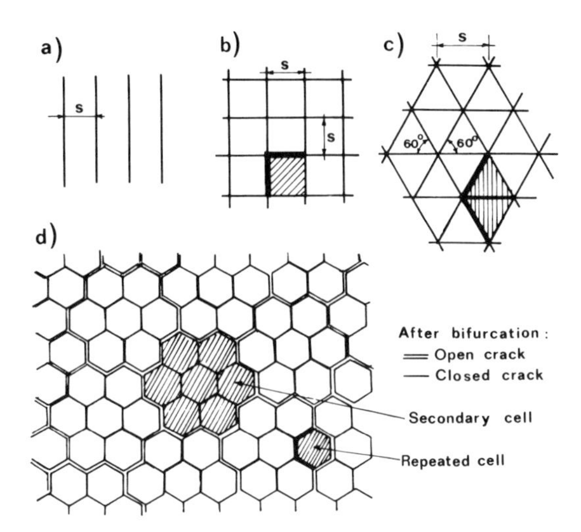

Figure 12.22 Possible patterns of shrinkage cracks in 3D, dominant hexagonal pattern and formation of larger quasi-hexagons.

and $1.861\sqrt{V_c}$, respectively. The last value is the smallest, which indicates the pattern should indeed be hexagonal. Although in nature the crack pattern is disturbed by random inhomogeneities, observations (e.g., dried lake beds) confirm that such a pattern indeed dominates (Lachenbruch, 1961; Lister, 1974). Rigorous proof, however, would require also calculating the values of $\bar{U}$, which could be done by finite elements.

As cooling or drying advances into the half-space, a similar phenomenon happens as we showed for two dimensions. At a certain moment some hexagon sides cease to grow into the half-space. The remaining sides which grow further, increase their opening width (see the double lines in Fig. 12.22d) and transform themselves into rough hexagons whose size is three times larger. The triplication of the hexagons is repeated as cooling penetrates deeper, etc.

Shrinkage cracks on the surface of concrete seem to behave similarly, but their precise pattern is complicated by the effect of reinforcement as well as geometry of the structure.

Stability of Parallel Cracks in Reinforced Concrete

Reinforcement can greatly alter the evolution of parallel shrinkage or cracking stress in concrete. It can either postpone or entirely suppress the occurrence of bifurcation with instability in which the growth of every other crack is arrested. The behavior is very sensitive to the nature of the bond between steel and concrete. Bond slip must take place and cannot be ignored, for if there were no bond slip at the points where a steel bar crosses the crack, then the crack could not open at all unless the bar broke.

Two-dimensional stability analysis of various problems of this type was conducted by Bažant and Wahab (1980), who modeled the frictional bond slip by means of an equivalent free bond slip length as introduced by Bažant and Cedolin (1980). Some of the results are plotted in Figure 12.23a, b. It is seen that, despite

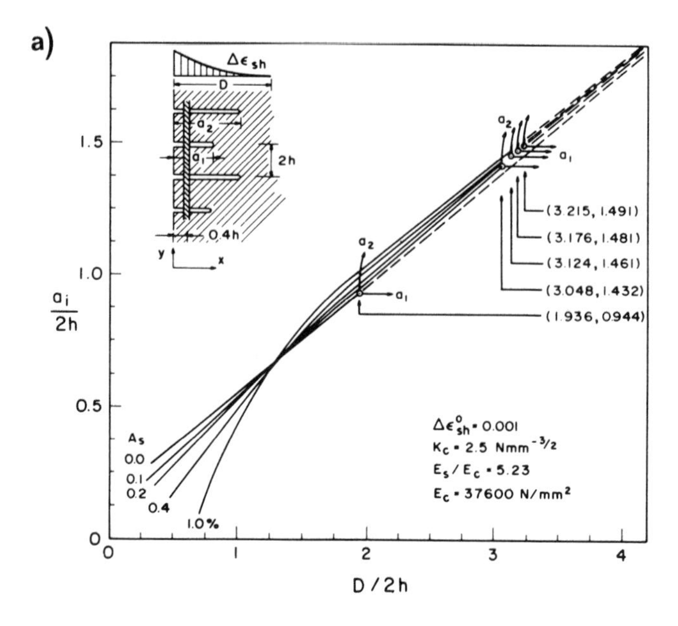

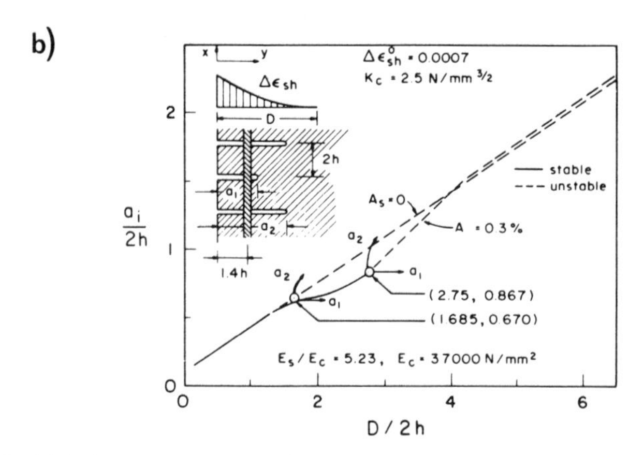

Figure 12.23 Critical states and growth of shrinkage or cooling cracks for reinforcement (a) close to the surface, (b) deeper within the solid. (*After Bažant and Wahab, 1980*).

reinforcement, instability at which every other crack gets arrested does occur. Even a very small reinforcement area, such as 0.1 percent of crack spacing s (times unit thickness), causes the point of instability to be pushed from depth $a_1 \simeq s$ to $a_1 \simeq 1.5s$, which is located well behind the reinforcing bar. A further increase in reinforcement, however, has only a marginal effect.

This type of behavior is not restricted to shrinkage or cooling cracks. The arrest of every other crack seems to occur generally when parallel cracks are propagating toward a zone of concrete without tension—for example, in a reinforced concrete beam subjected to bending, as demonstrated in Figure 12.24 (after Bažant and Wahab, 1980). It is seen that a reinforcement percentage of 3 percent pushes the point of instability from $a_1 = 0.9s$ to $a_1 = 1.8s$.

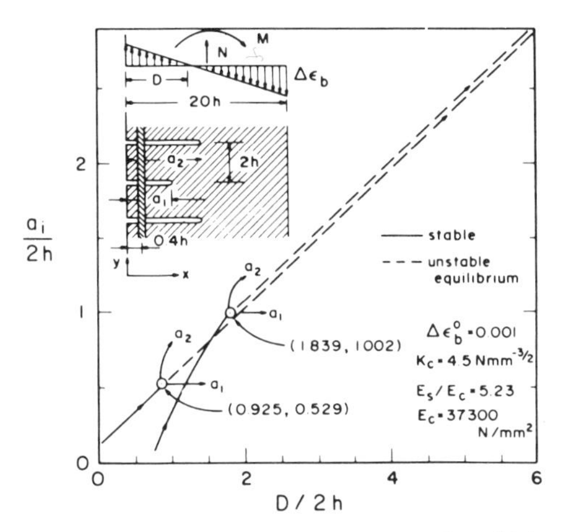

Figure 12.24 Critical states and growth of bending cracks. (*After Bažant and Wahab, 1980.*)

The results just presented give only the limit of stability, that is, the state where every other crack suddenly jumps ahead at no change of loading. No doubt a stable bifurcation at an increase of loading (without instability) occurs earlier, but numerical results are unavailable to confirm it. It should also be kept in mind that bifurcation and stability are not the only phenomena that decide the width of opened cracks in reinforced concrete.

Stability Analysis in Terms of Displacements

Instead of analyzing stable states and paths in terms of crack length variations δa_i, one may equally well carry out the analysis on the basis of the tangential stiffness matrices associated with structural displacements q_i. In some problems, this approach is more convenient.

All that has been said about this approach in Sections 10.1 and 10.2 is applicable to structures with propagating cracks. An important point, though, is that the tangential stiffness matrices $\mathbf{K}^t$ must now be calculated for cracks that advance simultaneously with loading (in an equilibrium manner, $K_i = K_{ic}$). In theory, these matrices must be calculated for all possible combinations of advancing or stationary tips for those cracks for which $K_i = K_{ic}$, and receding or stationary tips for those cracks for which $K_i = 0$. According to Equation 10.1.37, the structure with moving cracks is stable if $-T(\Delta S)_{\text{in}} = \delta^2\mathscr{F} = \frac{1}{2}\delta\mathbf{q}^T\mathbf{K}^t\mathbf{q} > 0$ for all vectors $\delta\mathbf{q}$ with the associated matrices $\mathbf{K}^t$. Bifurcation states, which need not represent a limit of stability, are found from the condition $\det \mathbf{K}^t = 0$. Here, for the first bifurcation, matrix $\mathbf{K}^t$ needs to be evaluated only for loading, which means continuing crack advance (this is analogous to Hill's method of the linear comparison solid for plasticity; cf. Sec. 10.4). Determination of the stable postbifurcation branch necessitates calculating from $\mathbf{K}^t$ the values of $\delta^2\mathscr{F}^{(\alpha)}$ along each path α.

To describe the procedure in more detail, consider a structure whose state is characterized by displacements q_i ($i = 1, \ldots, n$). Let all the cracks be initially at the state of propagation; the crack tips with coordinates $a_1, a_2, \ldots, a_m$ are growing ($m \leq N$) and those with $a_{m+1}, \ldots, a_N$ are unloading. The condition that the initial state is a state of propagation may be written as $\partial\Pi/\partial a_k = -R_k(a_k)$ where $R_k(a_k)$ is the given R curve of crack a_k ($R_k = G_f =$ constant for linear elastic fracture mechanics). The condition that the cracks remain propagating are $\delta(\partial\Pi/\partial a_k) = -\delta R_k(a_k)$ where δ is a total variation. This yields

$$\sum_{l=1}^{m} \frac{\partial}{\partial a_l}\left(\frac{\partial \Pi}{\partial a_k}\right)\delta a_l + \sum_{j=1}^{n} \frac{\partial}{\partial q_j}\left(\frac{\partial \Pi}{\partial a_k}\right)\delta q_j = -\frac{dR_k}{da_k}\,\delta a_k$$

or

$$\sum_{l=1}^{m} \Phi_{kl}\,\delta a_l = -\sum_{j=1}^{n} \frac{\partial^2 \Pi}{\partial a_k\,\partial q_j}\,\delta q_j \tag{12.5.20}$$

in which

$$\Phi_{kl} = \Phi_{lk} = \frac{\partial^2 \Pi}{\partial a_k\,\partial a_l} + \frac{dR_l}{da_l}\,\delta_{kl} \tag{12.5.21}$$

Assuming that the inverse, Ψ_{ij}, of matrix Φ_{ij} exists, Equation 12.5.20 may be solved for δa_l:

$$\delta a_l = -\sum_{k=1}^{m} \Psi_{kj} \sum_{j=1}^{n} \frac{\partial^2 \Pi}{\partial a_k\,\partial q_j}\,\delta q_j \qquad \text{with} \qquad [\Psi_{kl}] = [\Phi_{ij}]^{-1} \tag{15.5.22}$$

The total force changes at propagating cracks are

$$\delta f_i = \sum_{j=1}^{n} \frac{\partial f_i}{\partial q_j}\,\delta q_j + \sum_{l=1}^{m} \frac{\partial f_i}{\partial a_l}\,\delta a_l = \sum_{j=1}^{n} K_{ij}^{s}\delta q_j + \sum_{l=1}^{m} \frac{\partial^2 \Pi}{\partial q_i\,\partial a_l}\,\delta a_l \tag{12.5.23}$$

where $\partial f_i/\partial q_j = K_{ij}^s = \partial^2\Pi/\partial q_i\,\partial q_j =$ current secant elastic stiffness matrix of the structure damaged by cracks, and $\partial\Pi/\partial q_i = f_i =$ force associated with q_i. Substituting now Equation 12.5.22, we obtain a force-displacement relation that may be written in the form $\delta f_i = \sum_j K_{ij}^t\,\delta q_j$, in which

$$K_{ij}^t = K_{ij}^s - \sum_{k=1}^{m}\sum_{l=1}^{m} \Psi_{kl} \frac{\partial^2 \Pi}{\partial a_l\,\partial q_i}\frac{\partial^2 \Pi}{\partial a_k\,\partial q_j} = K_{ij}^s - \sum_k \sum_l \Psi_{kl} \frac{\partial f_i}{\partial a_l}\frac{\partial f_j}{\partial a_k} \tag{12.5.24}$$

This is the tangential stiffness matrix at growing cracks. As we see, it is symmetric. The second-order work is $-T(\Delta S)_{\text{in}} = \delta^2 W = \sum_i \sum_j \frac{1}{2} K_{ij}^t\,\delta q_i\,\delta q_j$, from which stability and the postbifurcation path may be determined.

Equation 12.5.24 gives admissible values only for such vectors of displacements δq_i for which (1) $\delta a_k \geq 0$ for all $k = 1, \ldots, m$ (as calculated from Eq. 12.5.22) and (2) $\delta(\partial\Pi/\partial a_k) = \sum_i (\partial^2\Pi/\partial a_k\,\partial q_i)\,\delta q_i \geq 0$ for all $k = m+1, \ldots, N$; that is, the stress intensity factor of the critical nongrowing cracks may not increase. If these conditions are not satisfied, one must analyze other choices of the cracks assumed to propagate, either with the same total number m (but the cracks renumbered), or a different number m. To characterize the entire surface $\delta^2\mathscr{F}$ as a function of δq_i, one must try successively $m = N, N-1, \ldots, 1, 0$, and check also all the possible crack numberings for each number m. The second derivatives of Π at a_i and q_i in Equations 12.5.21 and 12.5.24 can be approximated by finite differences after Π is calculated by finite elements for $a_i + \Delta a_i$, $a_i - \Delta a_i$, $q_i + \Delta q_i$, $q_i - \Delta q_i$. On the basis of $\delta^2\mathscr{F}$ or K_{ij}^t, stability and critical states can be analyzed by the methods described in Chapter 10.

Consider now a path bifurcation, choosing as q_i only those displacements that are controlled. Then δq_i are the same for paths 1 and 2 but $\delta a_i = \delta a_i^{(1)}$ or $\delta a_i = \delta a_i^{(2)}$ are not. Writing Equation 12.5.20 for these two paths and subtracting the equations, we get $\sum_l \Phi_{kl}\, \delta a_l^* = 0$ where $\delta a_l^* = \delta a_l^{(2)} - \delta a_l^{(1)}$. This shows (in accordance with Eq. 12.5.16) that Φ_{kl} at bifurcation must be singular ($\det \Phi_{kl} = 0$), provided that the derivatives $\partial^2\Pi/\partial a_k\, \partial a_l$ are evaluated for the case that only the controlled displacements are kept constant while other displacements vary. This was the case for the system of parallel cooling cracks, for which λ may be interpreted as q_1.

If, on the other hand, we include among q_i some displacements that are associated with prescribed forces and are not controlled, then such δq_i are not the same for paths 1 and 2. So one would have to substitute $\delta q_i = \delta q_i^{(1)}$ or $\delta q_i^{(2)}$ in Equation 12.5.20. Then, of course, $\sum_l \Phi_{kl}\, \delta a_l^* \neq 0$, and so matrix Φ_{kl} (which equals $\delta^2\Pi/\partial a_k \partial a_l$, with derivatives evaluated while *all* q_i are kept constant) is not singular at the first bifurcation. But in terms of K_{ij}^t we must simultaneously have $\sum_j K_{ij}^t\, \delta q_j^{(1)} = 0$ and $\sum_j K_{ij}^t\, \delta q_j^{(2)} = 0$. By subtraction, $\sum_j K_{ij}^t\, \delta q_j^* = 0$ where $\delta q_j^* = \delta q_j^{(2)} - \delta q_j^{(1)}$. Since the vector of δq_j^* is nonzero, K_{ij}^t must be singular at bifurcation ($\det K_{ij}^t = 0$), provided, of course, that $\delta q_j^{(1)}$ and $\delta q_j^{(2)}$ belong to the same sector of the loading directions, with the same set of cracks that are growing. As shown in general in Section 10.4, this case must occur for the first bifurcation, if all the properties of the system change continuously along the loading path.

The method and results can be illustrated by examples analyzed by Bažant, Tabbara, and Kazemi (1989). These examples, in which linear elastic fracture mechanics is used ($R = G_f = \text{const.}$), consider bodies with two symmetric crack tips that are simultaneously critical ($K_1 = K_2 = K_c$). Bifurcation into three branches is possible at every state: (1) for the main path, both crack tips advance; (2) for the secondary path, only one crack tip (K_1) advances while the other crack tip unloads, its stress intensity factor K_2 decreasing below the critical value K_c, $\delta K_2 < 0$ (actually this comprises two paths—either tip 1 advances or tip 2 advances, but for symmetric situations both are equivalent); (3) for the third path, both crack tips unload, $\delta K_1 = \delta K_2 < 0$.

Figure 12.25 shows the results for a long strip containing a transverse crack with tips at distances a_1, a_2 from the axis of symmetry. Initially, the state is symmetric, $a_1 = a_2$, and propagation with $a_1 = a_2$, normally assumed in practice, represents the main path. As explained in Section 12.1, the problem can be completely solved from the values of the stress intensity factors that are known for this problem (see Murakami, 1987). The calculated plots of average axial stress σ versus the additional displacement q_c due to the cracks are descending, which means the specimen can be stable only under displacement control at the ends provided the snapback point has not been reached or exceeded. Every point of the main path is a bifurcation point. Since the secondary path is always steeper, it is stable. So the specimen cannot follow the main path, that is, the crack cannot grow in a symmetric manner. The actual stable path is the secondary path emanating from the initial state with a symmetric notch (the solid curve). After this path reaches the snapback point, the shorter ligament tears suddenly in a dynamic manner (Sec. 12.3) subsequently a single one-sided edge crack is gradually loaded up to its critical state, after which it grows in a stable manner until complete failure of the specimen.

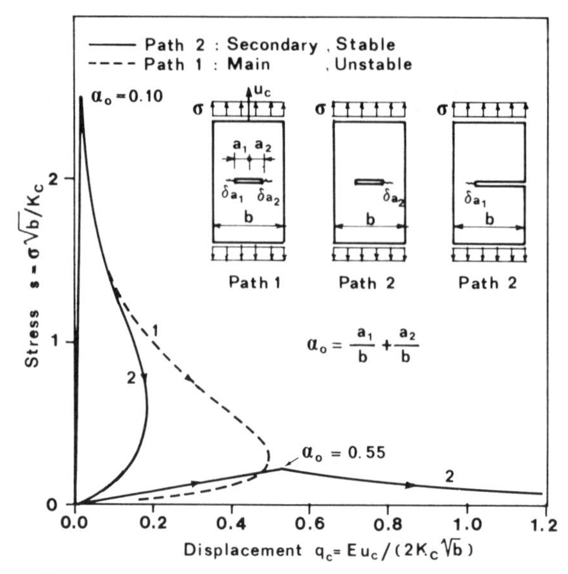

Figure 12.25 Stable (solid curve) and unstable (dashed curve) paths of center-cracked panel. (*After Bažant, Tabbara, and Kazemi, 1989.*)

A similar problem is a long strip with two transverse coplanar edge cracks of lengths a_1 and a_2. The results obtained by finite elements (Bažant, Tabbara, and Kazemi, 1989) are plotted in Figure 12.26. Again, every point of the main path ($a_1 = a_2$, the dashed curve in Fig. 12.26a, b) is a bifurcation point, and since the secondary path descends more steeply, the main (symmetric) path cannot be actually followed. The stable (actual) path is the solid curve corresponding to the growth of only one crack (the solid curve in Fig. 12.26a).

Figure 12.26c shows the surfaces of $\delta^2\mathscr{F}$ $[= -T(\Delta S)_{\text{in}}]$ for the state $a_1 = a_2 = 0.2b$, which is stable under axial displacement control, and the state $a_1 = a_2 = 0.76b$, which is unstable (b = strip width). The coordinates for these surfaces are the axial average displacement δu and the rotation $\delta\theta$ at the end cross section. For convenience of calculations, the controlled variables in the finite element program were the crack lengths, and δu and $\delta\theta$ were obtained as the output. The values of $\delta^2\mathscr{F}$ for various values of δu and $\delta\theta$ were calculated as $\delta^2\mathscr{F} = \frac{1}{2}\delta f\,\delta u + \frac{1}{2}\delta m\,\delta\theta$ where δf, δm are the axial stress resultant and the moment at the end cross section. The main (symmetric) path ($\alpha = 1$) corresponds to $\delta\theta = 0$. The solid curves labeled 1 and 2 represent the main (symmetric) path, which corresponds to $\delta\theta = 0$, and the secondary (bifurcated, nonsymmetric) path.

Stability of path 2 (only one crack grows) is revealed by the fact that, for a given δu, the point on path 2 lies lower than the point on path 1 (i.e., minimizes $\delta^2\mathscr{F}$). Stability of the state is revealed by the cross section of the surface for $\delta u = 0$ (controlled variable); if it curves upward (at the left in Fig. 12.26a), the state is stable (that is, $\delta^2\mathscr{F}$ has a minimum); if it curves downward (at the right), the state is unstable (that is, $\delta^2\mathscr{F}$ does not have a minimum). The surfaces of $\delta^2\mathscr{F}$ represent a patch-up of these quadratic surfaces corresponding to the tangential stiffnesses for (1) both cracks growing, (2) just one growing, (3) none growing.

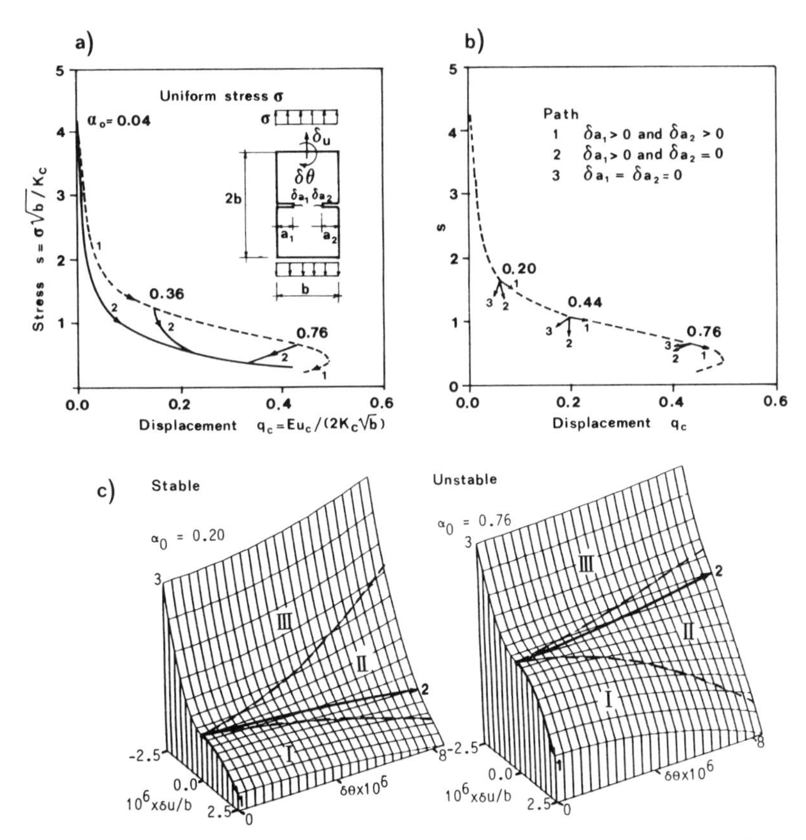

Figure 12.26 Stable and unstable paths of edge-cracked panel and surfaces of $\delta^2\mathcal{F}$ (δu, $\delta\theta$ = axial displacement and rotation at specimen ends). (*After Bažant, Tabbara, and Kazemi, 1989.*)

The dividing lines between these surfaces are shown as the dashed lines. The surface is always continuous and has a continuous slope across these dividing lines. It is interesting to notice the similarities and differences of these surfaces compared to those for plastic column buckling shown in Section 10.3.

The lack of a maximum for the present surfaces, in contrast to some of those in Section 10.3, is due to the fact that the state is stable only under displacement control (while the state of the Shanley column in Sec. 10.3 was stable for both load and displacement control when $P < P_t$). It needs to be also pointed out that the curves and surfaces shown in Figure 12.26 have been calculated solely from the compliance matrix due to crack growth, neglecting the additional compliance $\mathbf{C}_0$ due to elastic deformation of the uncracked specimen. This does not affect the choice of postbifurcation path, but does affect the limit of stability. The present stability limits apply only if the uncracked specimen is assumed to have infinite stiffness. Nevertheless, Figure 12.26 serves the purpose of simple illustration.

In general it appears that symmetric crack systems usually do not lie on a stable path. A stable path usually is that for which the fracture process localizes into a single crack tip, while all other crack tips unload. For the specimen in Fig. 12.26 this was experimentally demonstrated by Cedolin, Dei Poli and Iori (1983),

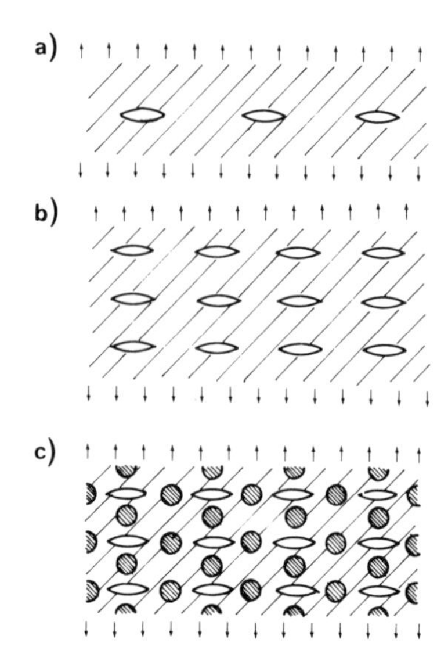

Figure 12.27 Regular crack systems.

and Labuz, Shah and Dowding (1985). Other examples are the systems in Figure 12.27. Therefore, a homogeneous elastic solid with a periodic crack array such as that in Figure 12.27a, b is not a realistic model for the fracture process zone in concrete or similar material. A realistic model may be a periodic array of interacting cracks and inclusions (Fig. 12.27c). Pijaudier-Cabot, Bažant, and Berthaud (1989) showed that the presence of cracks makes simultaneous propagation of many cracks stable.

The results in Figure 12.26 cannot be applied to for materials such as concrete, rock, or ceramics because these materials exhibit pronounced R-curve behavior, that is, the energy required for crack propagation is not constant but varies as a function of the effective crack length (Sec. 12.3). Introduction of an R curve into the calculations is necessary to achieve agreement with test data, such as those of Labuz, Shah, and Dowding (1985); see the data points in Figure 12.28. Using the specimens shown in Figure 12.28c, these investigators achieved stability by controlling the average of the crack mouth opening displacements (CMOD) measured by two clip gauges mounted over both notches as shown in Figure 12.28c. The load-point displacements were measured with two LVDT displacement transducers mounted on the face of these specimens across the notches, and the average of these two displacements was recorded. The load-displacement curves for paths 1 and 2 shown in Figure 12.28a were calculated by Bažant and Tabbara (1989). They used the R curve given by Equation 12.3.8 with Equation 12.3.7 ($G_f = 0.165\,\mathrm{N/m}$, $c_f = 27.9\,\mathrm{mm}$). Observe that: (1) The load-deflection curve now bifurcates already on the rising branch; in fact the first bifurcation occurs at the origin, which is due to assuming $R(0)$ to be zero according to Equations 12.5.10. (2) The fit of the LVDT-measured displacements (Fig. 12.28a) is good and verifies that the nonsymmetric path 2 is the stable path. (3) Due to snapback in the curve of the load versus load-point displacement, this test specimen would have become unstable shortly after the peak load if the load-point displacement was controlled instead of CMOD. The

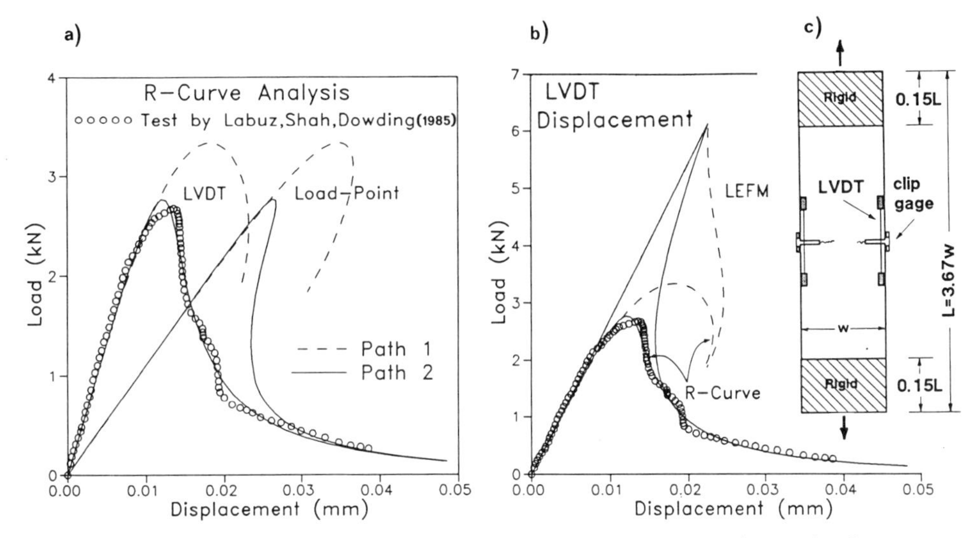

Figure 12.28 Load-displacement curves predicted solely from maximum load measurements for specimens of various sizes by means of R curve and by means of LEFM. (a) Load-point displacement. (b) Relative displacement over LVDT gauge length shown. (c) Edge-cracked strip specimen used. Data points = test results. (*After Bažant and Tabbara, 1989; see also Bažant, Gettu, and Kazemi, 1989.*)

test was also analyzed according to LEFM; the results shown in Figure 12.28b prove that LEFM cannot give a good fit, except at the end of the declining branch.

For further analysis of the tensile test specimen from the viewpoint of damage, see Sections 13.2 and 13.8.

Problems

12.5.1 Generalize Equations 12.5.11 assuming an R curve (variable R).

12.5.2 Assuming that initially $a_2 = a_1$, and $R = G_f =$ constant, show that when the cracks in Figure 12.29a become critical only one of them advances (the load is applied through a rigid cable over a pulley; displacement q is controlled). Assuming that $h \ll a_1$, nearly all of the strain energy is due to bending of the cantilever arms and may be calculated by bending theory. Solve the problem in terms of a_1, a_2. Repeat for the structure in Figure 12.29b.

12.5.3 Do the same as above but solve the problem in terms of displacements q_1 and q_2 at the ends of the cantilever arms. Also consider the case when $q_1 = q_2 = q$ (i.e., no pulley).

12.5.4 Do the same as Problem 12.5.1 but consider $R = G_f c/(k + c)$.

12.5.5 Denoting as q_1 and q_2 the displacements at the cantilever ends, calculate the surface $\delta^2\mathscr{F}(q_1, q_2)$ and plot it.

12.5.6 Using the approximate method for calculating stress intensity factors, as explained in Section 12.1 (Eqs. 12.1.28–12.1.29), solve the bifurcation and stability in beams with two growing cracks sketched in Figure 12.29c, d, e, f.

12.5.7 Discuss generalization of Equations 12.4.1 and 12.4.4 for load-point deflection to a body with a system of (a) noninteracting cracks, (b) interacting cracks.

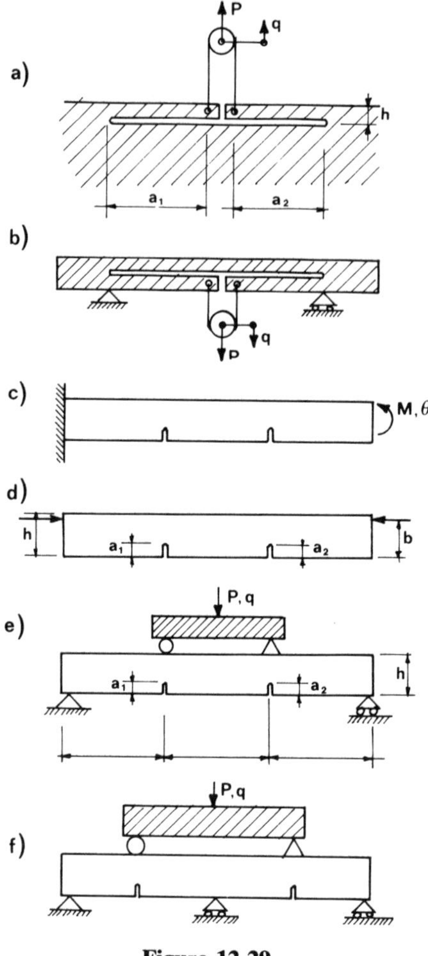

Figure 12.29

12.6 CRACK SPACING

In concrete structures as well as in some geotechnical or geological problems, the spacing s of parallel cracks is an important factor, mainly because it determines the crack width. Very narrow cracks are bridged and thus they still transmit some tensile stresses while widely opened cracks do not. Due to surface roughness of cracks (or interlock of asperities, aggregate pieces), narrow cracks transmit shear stresses. Very narrow cracks, for example, less than about 0.1 mm in concrete, are not continuous, and they do not serve as conduits for moisture or corrosive agents (Bažant and Raftshol, 1982). Diffusivity of a densely cracked material is a function of crack width and thus of crack spacing (Bažant, Sener, and Kim, 1987).

The spacing of cracks depends on various factors such as structure geometry or reinforcement outlay. Stability and energy balance are important considerations, and we will now discuss them briefly.

Spacing of Parallel Initial Drying or Cooling Cracks: Energy Balance

In the preceding section we showed how the spacing of opened cracks in a parallel crack system (Fig. 12.19) evolves but we did not determine the initial crack spacing, s. One limiting factor is the heterogeneity of the material, for example, the (macroscopic) shrinkage cracks (which matter for continuum analysis) cannot be closer than the size of the maximum aggregate pieces in concrete or the size of the largest grains in rock. Normally, though, the minimum possible spacing is larger. This minimum depends on the balance of energy during the initiation of cracks from a (macroscopically) smooth surface.

Consider first parallel planar cracks growing into a half-space (Fig. 12.19). Let x, y, z be cartesian coordinates, x being normal to the half-space surface. Let $\varepsilon^0(x)$ be the profile of free (unrestrained) shrinkage or thermal strain corresponding to a certain depth $x = \lambda$ of the drying or cooling front. The stresses produced before cracking by $\varepsilon^0(x)$ may be solved from the condition $(\sigma_y^0 - \nu\sigma_z^0)/E_{\text{eff}} - \varepsilon^0 = 0$ where $\sigma_y^0 = \sigma_z^0$, ν = Poisson ratio, and $E_{\text{eff}} = E/(1+\phi)$ = effective modulus for elastic deformation plus creep; ϕ = creep coefficient (Sec. 9.4) corresponding to the direction of drying or cooling. Consideration of creep is important for concrete since it significantly reduces the shrinkage stresses. For a material such as granite, creep may be neglected, that is, $\phi = 0$ and $E_{\text{eff}} = E$. Solving the foregoing equation, we get the stresses before cracking:

$$\sigma_y^0 = \sigma_z^0 = \frac{E_{\text{eff}}}{1-\nu}\,\varepsilon^0(x) \tag{12.6.1}$$

At a certain moment, the initial cracks normal to the y axis suddenly form. This reduces the stresses σ_y^0 to much smaller values and the stress becomes nonuniformly distributed. Let $\Delta\sigma_z$ be the change of σ_z, and $\Delta\varepsilon_y^m$ be the change of strain ε_y of the material between the cracks (while $\Delta\varepsilon_z^m = 0$). From Hooke's law $\Delta\varepsilon_y^m = (\Delta\sigma_y - \nu\,\Delta\sigma_z)/E$ and $\Delta\varepsilon_z = (\Delta\sigma_z - \nu\,\Delta\sigma_y)/E$. For the sake of simplicity, we assume the stresses σ_y^0 to be reduced uniformly to 0, that is, the stress change is $\Delta\sigma_y = -\sigma_y^0$ everywhere. The solution then is $\Delta\sigma_z^0 = -\nu\sigma_y^0$ and $\Delta\varepsilon_y^m = -(1-\nu^2)\sigma_y^0/E$. Now the loss of strain energy due to cracking per unit area of half-space surface is

$$\Delta U_1 = \int_0^\lambda \left(\sigma_y^0 + \frac{1}{2}\Delta\sigma_y\right)\Delta\varepsilon_y^m\,dx = \int_0^\lambda \frac{\sigma_y^0}{2}\left[\frac{(1-\nu^2)\sigma_y^0}{E}\right]dx$$

$$= \left(\frac{E_{\text{eff}}}{1-\nu}\right)^2\left(\frac{1-\nu^2}{2E}\right)\int_0^\lambda \varepsilon^{0^2}(x)\,dx = \frac{\lambda E}{10}\left(\frac{\varepsilon_s^0}{1+\phi}\right)^2\left(\frac{1+\nu}{1-\nu}\right) \tag{12.6.2}$$

where we assumed the profile of $\varepsilon^0(x)$ to be parabolic, expressed as $\varepsilon^0(x) = (1-x/\lambda)^2\varepsilon_s^0$ for $x \le \lambda$ and $\varepsilon^0(x) = 0$ for $x \ge \lambda$; ε_s^0 = constant = free shrinkage on thermal strain at half-space surface, and λ = penctration depth of drying or cooling.

Stresses σ_y^0 actually are not reduced completely to zero, and so the loss of strain energy will be $r\,\Delta U_1$ where $0 < r < 1$ and probably r is close to 1. The balance of energy during crack formation requires that $r\,\Delta U_1 s = G_f a$ where

a = crack length. This yields the following estimate for the spacing of parallel planar cracks (Bažant and Ohtsubo, 1977):

$$s = \frac{10}{r}\left(\frac{1-\nu}{1+\nu}\right)\left(\frac{G_f}{E}\right)\left(\frac{1+\phi}{\varepsilon_s^0}\right)^2\left(\frac{a}{\lambda}\right) \tag{12.6.3}$$

The order of magnitude of ratio a/λ is 1. An accurate value of λ/a as well as r can be solved by elastic finite element analysis from the condition that the crack length a should be such that $K_I = K_c$ for the given temperature or free shrinkage profile. Such analysis provides $\lambda/a \simeq 1.5$ (and $a \simeq 2s$); see Bažant and Wahab, 1979. This gives the initial crack length as $a \simeq 2\lambda/3$.

Size of Initial Hexagonal Cracking Cells from Energy Balance

The planar cracks normal to one axis, x, can occur only if cracking normal to the other axis, z, is somehow inhibited, for example, by reinforcing bars laid in the z-direction. In the absence of such an inhibitor, the true crack pattern on the half-space surface consists of a hexagonal mesh, of side s, as shown in Section 12.5. For that case, both σ_y and σ_z are reduced about equally and may be assumed to drop to 0. One finds that

$$\Delta U_1 = \int_0^{\lambda} \frac{E}{2}\left(\frac{\varepsilon^0}{1+\phi}\right)^2 dx = \frac{\lambda E}{10}\left(\frac{\varepsilon_s^0}{1+\phi}\right)^2 \tag{12.6.4}$$

Noting that the area of one hexagonal cell is $s^2\sqrt{27}/2$ and the crack area per cell is $3sa$ (see Fig. 12.22), we have the energy balance condition $r\,\Delta U_1\, s^2\sqrt{27}/2 = 3saG_f$, from which we find for the side of the hexagonal cells of the initial cracking pattern the estimate:

$$s = \frac{20}{r\sqrt{3}}\left(\frac{G_f}{E}\right)\left(\frac{1+\phi}{\varepsilon_s^0}\right)^2\left(\frac{a}{\lambda}\right) \tag{12.6.5}$$

Snapthrough Formation of Cracks According to LEFM

Now, from the stability viewpoint, what is the nature of the formation of the initial cracks? First of all, linear elastic fracture mechanics does not permit cracks to start from a smooth surface. Furthermore, if we imagine the cracks growing from the surface to their final length a corresponding to Equation 12.6.3 (or 12.6.5) while the profile of $\varepsilon^0(x)$ is fixed, the stress intensity factor K_I varies with x approximately as shown in Figure 12.30 by curve 01234. The energy release per unit length is K_I^2/E'. Therefore, the strain energy release ΔU_1 given by Equation 12.6.2 must be equal to $\int (K_I^2/E')\,da'$, which is represented by the cross-hatched area 0123450 in Figure 12.30. According to the energy balance condition that we used to derive Equation 12.6.3, this area must be equal to area 06750 under the horizontal line representing K_c^2/E. From Figure 12.30 it is clear that the final state obtained from the overall energy balance condition has $K_I < K_c$, which means it is not a critical state for fracture propagation. So the drying or cooling front needs to advance further before the K_I value at the tip of the initial crack can reach K_c and start to propagate further. During the first phase of crack

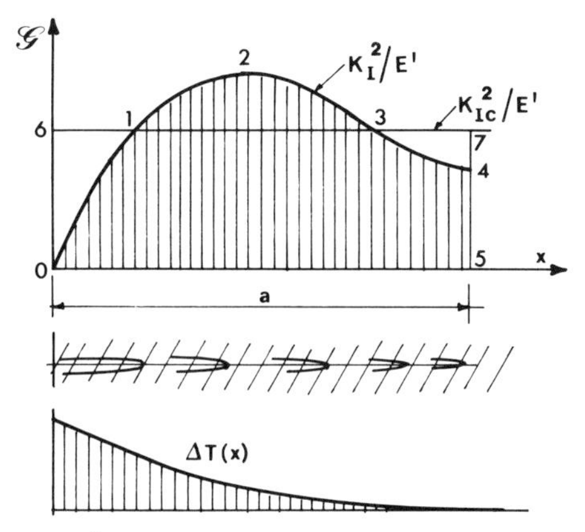

Figure 12.30 Variation of K_I^2/E' with thermal crack growth.

formation, area 0610 represents the energy deficiency that must be overcome by some dynamic impulse of imperfection. Subsequently, area 1231 represents excess energy which goes into kinetic energy. Finally, area 3743 represents again energy deficiency, which reduces the existing kinetic energy. Thus, the process of crack formation implied in our analysis represents a snapthrough process, analogous to that illustrated for elastic systems in Chapter 4. Path 6137 is the static equivalent of the dynamic snapthrough instability, with the same overall energy balance.

The dynamic snapthrough formation of cracks is, of course, true only in theory, according to linear elastic fracture mechanics. According to nonlinear fracture mechanics with a finite-size fracture process zone, the formation of the initial crack can happen as an equilibrium process. However, nonlinear fracture mechanics does not yield a result as simple as Equation 12.6.5 or 12.6.3.

One useful aspect of nonlinear fracture mechanics is that it can take into account the tensile strength, f_t'. However, a separate strength consideration can supplement Equation 12.6.5 or 12.6.3 and provide a lower bound on the value of the surface free shrinkage or cooling strain, ε_s^0, at which the initial cracks can form (see Bažant, 1985a).

Crack Spacing in Loaded Reinforced Concrete Beams

The overall energy balance condition of the type just illustrated, which in theory implies dynamic crack formation by snapthrough instability, is also useful for simple estimates of crack spacing and width of reinforced concrete beams subjected to external loads. Such fracture mechanics estimates can supplement and perhaps even supplant the existing provisions for crack width in concrete design codes, which are empirical.

To avoid using nonlinear fracture mechanics, which embodies both fracture and tensile strength, one may separately impose the strength criterion and the energy balance criterion for the formation of cracks. The former governs the crack initiation in the form of discontinuous microcracks, while the latter governs the formation of continuous (macroscopic) cracks.

The energy balance condition for the sudden formation of the complete cracks (by a dynamic snapthrough) must take into account the fact, already mentioned in Section 12.5, that fractures that have a nonzero width at the points of crossing of reinforcing bars must be accompanied by bond slip between the bars and concrete. Thus, the complete energy balance relation should read $\Delta U = G_f A_c + \Delta W_b$, in which ΔU = total release of strain energy due to fracture, $G_f A_c$ = energy needed to produce cracks of area A_c, and ΔW_b = energy dissipated by the bond slip that is casued by fracture.

Consider now a round concrete rod of diameter b, which is subjected to tension and has a single steel bar in the middle (Fig. 12.31). (This rod approximately simulates the behavior of the concrete zone in a beam surrounding one of the bars in a tensioned beam or panel with many parallel reinforcing bars.) We distinguish two cases.

(a) Cracks without bond slip. As an approximation, we can imagine that the sudden formation of a complete crack relieves the stress from the triangular area 012, cross-hatched in Figure 12.31. This area is limited by a "stress diffusion" line of slope k, already explained just before Equation 12.1.23. The volume of the region obtained by rotating this area about the bar axis is $V_1 = \pi b^3/12k$, provided the cracks are assumed to be spaced so far apart that their stress relief zones do not overlap (for the case they do overlap, see Bažant and Oh, 1983c). The strain energy release is $\Delta U = V_1 \sigma_1^2/2E_c$ where $\sigma_1 = E_c \varepsilon_s$ = initial axial stress in concrete prior to cracking, E_c = elastic modulus of concrete (which should be replaced by $E_{\text{eff}} = E_c/(1+\phi)$ in the case of long-time loading), and ε_s = axial strain in the steel bar. Substituting into the energy balance equation $\Delta U = G_f A_c$ and solving for ε_s, we find that complete transverse cracks (which have a zero width at bar crossing) can form if

$$\varepsilon_s \geq \left(\frac{6kG_f}{E_c b}\right)^{1/2} \tag{12.6.6}$$

(b) Cracks with bond slip. In this case, instead of $\sigma_1 = E_c \varepsilon_s$ the condition of force equilibrium with the bond stresses over length $s/2$ furnishes the relation $\sigma_1 \pi b^2/4 = U_b' s/2$ where s = crack spacing and U_b' = average value of bond strength. All the other equations are the same as in case (a). Neglecting the

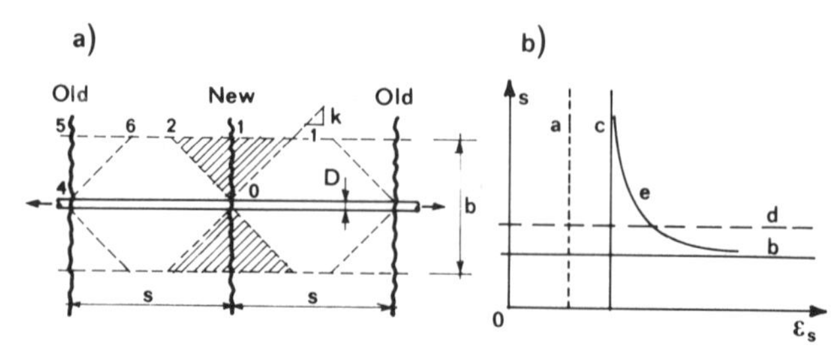

Figure 12.31 Parallel cracks normal to a reinforcing bar, and (b) limits on crack spacing and strain at cracking (a—strength criterion, no slip; b—strength criterion, slip; c—energy criterion, no slip; d—energy criterion, slip; e—curve for more accurate solution). (*After Bažant and Oh, 1983c.*)

energy of bond slip ($\Delta W_b \simeq 0$), one obtains

$$s \geq 2\pi\left(3kb^3 \frac{E_c G_f}{8U_b'}\right)^{1/2} \tag{12.6.7}$$

The foregoing results were derived in Bažant and Oh (1983c) where much more detail as well as comparisons with test data can be found. The main practical usefulness of calculating s is that it allows estimation of the crack width.

Snapthrough Crack Formation in a Drying Tube

After thin tubular specimens of hardened cement paste (Fig. 12.32) are exposed to a drying environment for some time, a long longitudinal crack may suddenly appear in the wall. To explain it, consider the energy required to produce this crack. It must come from the bending energy of the tube walls due to drying shrinkage. Assume the tube dries only from the outside, and consider the distribution of circumferential normal stress σ to be approximately parabolic (Fig. 12.32c) when the drying front just reaches the inside face. The bending moment of this distribution about the midthickness of the wall is $M = \Delta\sigma h^2/12$ where h = wall thickness and $\Delta\sigma = (1+\nu)E'\,\Delta\varepsilon_{sh}$, $E' = E/(1-\nu^2)$. The formation of the longitudinal crack relieves the entire energy of circumferential bending of the tube, and so (per unit length of tube) $U = 2\pi r M^2/(2E'I)$, $I = h^3/12$, r = radius of the midsurface of the wall. This must be equal to the fracture energy, that is, $U = G_f h$, which yields (Bažant and Raftshol, 1982)

$$\Delta\varepsilon_{sh}^{cr} = \left[\left(\frac{1-\nu}{1+\nu}\right)\left(\frac{12G_f}{\pi r E}\right)\right]^{1/2} \tag{12.6.8}$$

as the critical shrinkage strain that is capable of causing the crack. The process of formation of these cracks is a snapthrough instability satisfying the overall energy balance but not the incremental balance of energy that would be required for a static growth of the crack.

Problems

12.6.1 Work out the derivation of Equation 12.6.4 in complete detail.

12.6.2 A square pattern of cracks can form if the material is macroscopically orthrotropic with a weaker strength in the x- and y-directions. Such weakening can arise in composites or concrete by an orthogonal layout of fibers or reinforcing bars. Using the same procedure as that to get Equation 12.6.3 or 12.6.5, derive the minimum possible size s of a cell in a square crack pattern.

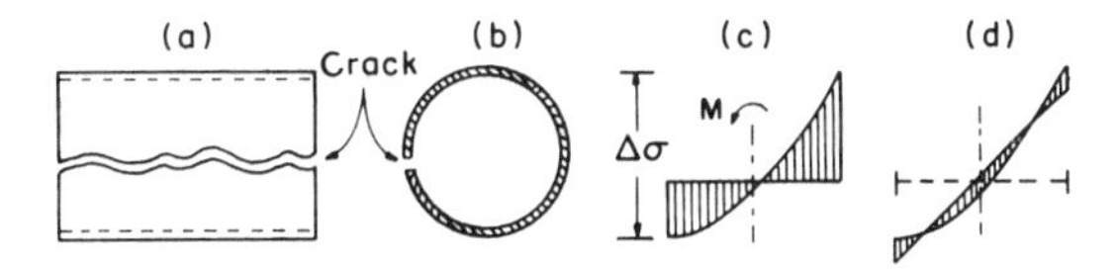

Figure 12.32 Longitudinal cracking of thin-walled tubular drying specimen.

12.6.3 Derive formulas analogous to Equations 12.6.6 and 12.6.7 assuming that the initial cracks are so close that the cones of stress relief zone V_1 overlap before they reach the surface of the round reinforcing rod (the derivation is given in Bažant and Oh, 1983b).

References and Bibliography

Barrenblatt, G. I. (1959), "The Formation of Equilibrium Cracks during Brittle Fracture—General Ideas and Hypothesis, Axially Symmetric Cracks," *Prikl. Math. Mekh.*, 23(3):434–44 (Sec. 12.2).

Bažant, Z. P. (1974), "Three-Dimensional Harmonic Functions near Termination or Intersection of Singularity Lines: A General Numerical Method," *Int. J. Eng. Science,* 12:221–43 (Sec. 12.1).

Bažant, Z. P. (1976a), "Growth and Stability of Thermally Induced Cracks in Brittle Solids," draft manuscript communicated to L. M. Keer, S. Nemat-Nasser, and K. S. Parihar of Northwestern University (Sec. 12.5).

Bažant, Z. P. (1976b), "Instability, Ductility and Size Effect in Strain-Softening Concrete," *J. Eng. Mech.* (ASCE), 102(2):331–44; see also discussion, 103:357–8, 775–6, 104:501–2 (based on Struct. Engng. Report No. 74-8/640, Northwestern University, August 1974) (Sec. 12.2).

Bažant, Z. P. (1980), "Instability of Parallel Thermal Cracks and Its Consequences for Geothermal Energy," Proc., Int. Conf. on Thermal Stresses in Severe Environments, held in March at VPI, Blacksburg, Virginia, pp. 169–81.

Bažant, Z. P. (1982), "Crack Band Model for Fracture of Geomaterials," Proc. 4th Int. Conf. on Numer. Methods in Geomech., Edmonton, Canada, Vol. 3, pp. 1137–52 (Sec. 12.2).

Bažant, Z. P. (1983), "Fracture in Concrete and Reinforced Concrete," Preprints, *IUTAM Prager Symposium on Mechanics of Geomaterials: Rocks, Concrete, Soils,* ed. by Z. P. Bažant, Northwestern University, pp. 281–316.

Bažant, Z. P. (1984), "Size Effect in Blunt Fracture: Concrete, Rock, Metal," *J. Eng. Mech.* (ASCE), 110(4):518–35; also Preprints, 7th Int. Conf. on Struct. Mech. in Reactor Technology (SMIRT 7), held in Chicago, 1983 (Sec. 12.2).

Bažant, Z. P. (1985a), "Fracture Mechanics and Strain-Softening in Concrete," Preprints, US.–Japan Seminar on Finite Element Analysis of Reinforced Concrete Structures, Tokyo, Vol. 1, pp. 47–69 (Sec. 12.2).

Bažant, Z. P. (1985b), "Comment on Hillerborg's Comparison of Size Effect Law with Fictitious Crack Model," Dei Poli Anniversary Volume, Politecnico di Milano, Italy, pp. 335–38 (Sec. 12.2).

Bažant, Z. P. (1987a), "Fracture Energy of Heterogeneous Material and Similitude," Preprints, SEM-RILEM Int. Conf. on Fracture of Concrete and Rock, Houston, Texas, publ. by SEM (Soc. for Exper. Mech.), pp. 390–402 (Sec. 12.2).

Bažant, Z. P. (1987b), "Snapback Instability at Crack Ligament Tearing and Its Implication for Fracture Micromechanics," *Cement and Concrete Research,* 17:951–67 (Sec. 12.4).

Bažant, Z. P. (1988), "Justification and Improvement of Kienzler and Herrmann's Approximate Method for Stress Intensity Factors of Cracks in Beams," Report No. 88–12/498j, Center for Concrete and Geomaterials, Northwestern University; also *Engng. Fracture Mech.*, 36(3):523–25 (1990) (Sec. 12.1).

Bažant, Z. P. (1989), "Smeared-Tip Superposition Method for Nonlinear and Time-Dependent Fracture," Report, ACBM Center, Northwestern University; also *Mech. Research Communications,* 17 (1990), in press.

Bažant, Z. P., and Cao, Z. (1986a), "Size Effect in Shear Failure of Prestressed Concrete Beams," *ACI Journal,* 83(2):260–8 (Sec. 12.2).

Bažant, Z. P., and Cao, Z. (1986b), "Size Effect in Brittle Failure of Unreinforced Pipes," *ACI Journal,* 83(3):369–73 (Sec. 12.2).

Bažant, Z. P., and Cao, Z. (1987), "Size Effect in Punching Shear Failure of Slabs," *ACI Journal,* 84(1):44–53 (Sec. 12.2).

Bažant, Z. P., and Cedolin, L. (1979), "Blunt Crack Band Propagation in Finite Element Analysis," *J. Eng. Mech.* (ASCE), 105(2):297–315 (Sec. 12.2).

Bažant, Z. P., and Cedolin, L. (1980), "Fracture Mechanics of Reinforced Concrete," *J. Eng. Mech.* (ASCE), 106(6):1287–1306; see also discussion, 108:464–71 (Sec. 12.2).

Bažant, Z. P., and Cedolin, L. (1983), "Finite Element Modeling of Crack Band Propagation," *J. Struct. Eng.* (ASCE), 109(1):69–92 (Sec. 12.2).

Bažant, Z. P., and Cedolin, L. (1984), "Approximate Linear Analysis of Concrete Fracture by *R*-Curves," *J. Struct. Eng.* (ASCE), 110(6):1336–55 (Sec. 12.3).

Bažant, Z. P., and Estenssoro, L. F. (1979), "Surface Singularity and Crack Propagation," *Int. J. Solids and Structures,* 15:405–26; "Addendum," 16:479–81 (Sec. 12.1).

Bažant, Z. P., Gettu, R., and Kazemi, M. T. (1989), "Identification of Nonlinear Fracture Properties from Size Effect Tests and Structural Analysis Based on Geometry-Dependent *R*-Curves," Rep. No. 89-3/498 p, Center for Advanced Cement-Based Materials, Northwestern University, Evanston, Illinois; *Int. J. Rock Mech. and Min. Sci.,* in press (Sec. 12.2).

Bažant, Z. P., and Kazemi, M. T. (1988), "Determination of Fracture Energy, Process Zone Length and Brittleness Number from Size Effect, with Application to Rock and Concrete," Report, ACBM Center, Northwestern University; also *Int. J. of Fracture,* 44:111–31 (1990); summary in "Brittleness and Size Effect in Concrete Structures," Preprints, Engng. Found. Conf. on "*Advances in Cement Manufacture and Use,*" held at Trout Lodge, Potosi, MO, 1988, Paper No. 5 (Sec. 12.2).

Bažant, Z. P., and Kazemi, M. T. (1989), "Size Effect in Fracture of Ceramics and Its Use to Determine Fracture Energy and Effective Process Zone Length," Internal Report, Center for Advanced Cement-Based Materials, Northwestern University, Evanston, Illinois; also *J. Amer. Ceramic Soc.* (1990), 73(7):1841–53 (1990) (Sec. 12.2).

Bažant, Z. P., and Kim, J.-K., (1984), "Size Effect in Shear Failure of Longitudinally Reinforced Beams," *ACI Journal,* 81(5):456–68 (Sec. 12.2).

Bažant, Z. P., Kim, J.-K., and Pfeiffer, P. A. (1986), "Nonlinear Fracture Properties from Size Effect Tests," *J. Struct. Eng.* (ASCE), 112(2):289–307 (Sec. 12.3).

Bažant, Z. P., Lee, S.-G., and Pfeiffer, P. A. (1987), "Size Effect Tests and Fracture Characteristics of Aluminum," *Eng. Fracture Mechanics,* 26(1):45–57 (Sec. 12.2).

Bažant, Z. P., and Oh, B.-H. (1983a), "Crack Band Theory for Fracture of Concrete," *Materials and Structures* (RILEM, Paris), 16:155–7 (Sec. 12.2).

Bažant, Z. P., and Oh, B.-H. (1983b), "Crack Spacing in Reinforced Concrete: Approximate Solution," *J. Struct. Eng.* (ASCE), 109:2207–12.

Bažant, Z. P., and Oh, B. E. (1983c), "Spacing of Cracks in Reinforced Concrete," *J. Struct. Eng.* (ASCE), 109(9):2066–85 (Sec. 12.6).

Bažant, Z. P., and Ohtsubo, H. (1977), "Stability Conditions for Propagation of a System of Cracks in a Brittle Solid," *Mechanics Research Communications,* 4(5):353–66 (Sec. 12.5).

Bažant, Z. P., Ohtsubo, H., and Aoh, K. (1979), "Stability and Post-Critical Growth of a System of Cooling or Shrinkage Cracks," *Int. J. Fracture,* 15(5):443–56 (Sec. 12.5).

Bažant, Z. P., and Pfeiffer, P. A. (1986), "Shear Fracture Tests of Concrete," *Materials and Structures* (RILEM, Paris), 19(110):111–21 (Sec. 12.2).

Bažant, Z. P., and Pfeiffer, P. A. (1987), "Determination of Fracture Energy from Size Effect and Brittleness Number," *ACI Materials Journal,* 84(6):463–80 (Sec. 12.2).

Bažant, Z. P., and Prat, P. C. (1988a), "Effect of Temperature and Humidity on Fracture Energy of Concrete," *ACI Materials Journal,* 84:262–71.

Bažant, Z. P., and Prat, P. C. (1988b), "Measurement of Mode III Fracture Energy of Concrete," *Nuclear Engineering and Design,* 106:1–8 (Sec. 12.2).

Bažant, Z. P., Prat, P. C., and Tabbara, M. R. (1989), "Antiplane Shear Fracture Tests (Mode III)," *ACI Materials Journal,* in press (Sec. 12.3).

Bažant, Z. P., and Raftshol, W. J. (1982), "Effect of Cracking in Drying and Shrinkage Specimens," *Cement and Concrete Research,* 12:209–26; see also discussion, 12:797–8 (Sec. 12.6).

Bažant, Z. P., and Sener, S. (1987), "Size Effect in Torsional Failure of Longitudinally Reinforced Concrete Beams," *J. Struct. Eng.* (ASCE), 113(10):2125–36 (Sec. 12.2).

Bažant, Z. P., and Sener, S. (1988), "Size Effect in Pullout Tests," *ACI Materials Journal,* 85(5):347–51 (Sec. 12.2).

Bažant, Z. P., Sener, S., and Kim, J.-K. (1987), "Effect of Cracking on Drying Permeability and Diffusivity of Concrete," *ACI Materials J.,* 84:351–7 (Sec. 12.6).

Bažant, Z. P., and Tabbara, M. (1989), "Stability, Bifurcation and Localization in Structures with Propagating Interacting Cracks," Report No. 89–7/428s, Center for Advanced Cement-Based Materials, Northwestern University; Submitted to *Int. J. of Fracture* (Sec. 12.5).

Bažant, Z. P., Tabbara, M. R., and Kazemi, M. T. (1989), "Stable Path of Interacting Crack Systems and Micromechanics of Damage," in *Advances in Fracture Research,* Proceedings of the 7th Int. Conf. on Fracture (ICF7), Houston, Texas, Vol. 3, pp. 2141–52 (Sec. 12.5).

Bažant, Z. P., and Wahab, A. B. (1979), "Instability and Spacing of Cooling or Shrinkage Cracks," *J. Eng. Mech.* (ASCE), 105:873–89 (Sec. 12.5).

Bažant, Z. P., and Wahab, A. B. (1980), "Stability of Parallel Cracks in Solids Reinforced by Bars," *Int. J. Solids Structures,* 16:97–105 (Sec. 12.5).

Bažant, Z. P., and Xi, Y. (1989), "Statistical Size Effect in Concrete Structures: Nonlocal Theory," Internal Report, ACBM Center, Northwestern University, Evanston, Ill. (Sec. 12.2).

Belytschko, T., Fish, J., and Engelmann, B. E. (1988), "A Finite Element with Embedded Localization Zones," *Comp. Meth. Appl. Mech. Eng.,* 70(1):59–89.

Billardon, R., and Moret-Bailly, L. (1987), "Fully Coupled Strain and Damage Finite Element Analysis of Ductile Rupture," *Nucl. Eng. Design,* 105:43–49.

Broek, D. (1978), *Elementary Engineering Fracture Mechanics,* Sijthoff & Noordhoff, The Netherlands; also 4th revised edition, Martinus Nijhoff, Dordrecht–Boston (1986) (Sec. 12.1).

Broek, D. (1988), *The Practical Use of Fracture Mechanics,* Kluwer Academic Publishers, Dordrecht, Boston (Sec. 12.1).

Carpinteri, A. (1988), "Snap-Back and Hyperstrength in Lightly Reinforced Concrete Beams," *Mag. Concrete Research,* 40:209–15.

Carslaw, H. S., and Jaeger, J. C. (1959), *Conduction of Heat in Solids,* 2nd ed., Clarendon Press, Oxford, pp. 102–5, 198–201 (Sec. 12.5).

Cedolin, L., Dei Poli, S., and Iori, I. (1983), "Experimental Determination of the Fracture Process Zone in Concrete," *Cement and Concrete Research,* 13:557–67 (Sec. 12.2).

Cedolin, L., Dei Poli, S., and Iori, I. (1987), "Tensile Behavior of Concrete," *J. Eng. Mech.* (ASCE), 113(3):431–49 (Sec. 12.2).

Collins, W. D. (1962), "Some Axially Symmetric Stress Distribution in Elastic Solids Containing Penny-Shaped Cracks," Proc. R. Soc., Serial A266 (1324), pp. 359–86.

Cornelissen, H. A. W., Hordijk, D. A., and Reinhardt, H. W. (1985), "Experiments and Theory for the Application of Fracture Mechanics to Normal and Light-Weight Concrete," Preprints, Int. Conf. on Fracture Mech. of Concrete, Lausanne, Switzerland, pp. 423–46.

de Borst, R., and Nauta, P. (1984), "Smeared Crack Analysis of Reinforced Concrete Beams and Slabs Failing in Shear," in *Computer-Aided Analysis and Design of Concrete Structures,* (Proc. Int. Conf. held in Split, Yugoslavia, Sept. 1984), ed. by F. Damjanić and E. Hinton, Pineridge Press, Swansea, U.K., pp. 261–74.

de Borst, R., and Nauta, P. (1985), "Non-Orthogonal Cracks in a Smeared Finite Element Model," *Eng. Computation,* 2:35–46.

Dugdale, D. S. (1960), "Yielding of Steel Sheets Containing Slits," *J. Mech. Phys. Solids,* 8:100–108 (Sec. 12.2).

Erdogan, F. (1983), "Stress Intensity Factors," *J. Appl. Mech.* (ASME), 50:992–1001.

Fairhurst, C., and Cornet, F. H. (1981), "Rock Fracture and Fragmentation," *Rock Mechanics from Research to Applications,* 22nd U.S. Symp. Rock Mech., Massachussetts Institute of Technology.

Gao, H., and Rice, J. R. (1987), "Nearly Circular Connections of Elastic Half Spaces," *J. Appl. Mech.* (ASME), 54:627–31.

Grady, D. (1985), "The Mechanics of Fracture Under High-Rate Stress Loading," in *Mechanics of Geomaterials, Rocks, Concrete, Soils,* ed. by Z. P. Bažant, John Wiley and Sons, Chichester, pp. 129–56.

Griffith, A. A. (1921), "The Phenomena of Rupture and Flow in Solids," *Phil. Trans. Roy. Soc. of London,* A221:163–97 (Sec. 12.1).

Griffith, A. A. (1924), "The Theory of Rupture," Proc. 1st Int. Congress Appl. Mech., pp. 55–63. Also, Biezeno and Burgers (ed.), Waltman, London (Sec. 12.1).

Hellan, K. (1984), *Introduction to Fracture Mechanics,* McGraw-Hill, New York (Sec. 12.1).

Herrmann, G., and Sosa, H. (1986), "On Bars with Cracks," *Eng. Fracture Mechanics,* 24(6):889–94 (*Sec.* 12.1).

Hillerborg, A. (1988), "Existing Methods to Determine and Evaluate Fracture Toughness of Aggregative Materials," RILEM Recommendation on Concrete, Preprints, Int. Workshop of Fracture Toughness and Fracture Energy Test Methods for Concrete and Rock, Tohoku University, Sendai, Japan.

Hillerborg, A., Modéer, M., and Petersson, P. E. (1976), "Analysis of Crack Formation and Crack Growth in Concrete by Means of Fracture Mechanics and Finite Elements," *Cement and Concrete Research,* 6:773–82 (Sec. 12.2).

Horii, H., Hasegawa, A., and Hishino, F. (1987), "Process Zone Model and Influencing Factors in Fracture of Concrete," Preprints, SEM-RILEM Int. Conf. on Fracture of Concrete and Rock, Houston, Texas; published by SEM (Soc. for Exper. Mechanics), pp. 299–307 (Sec. 12.4).

Hutchinson, J. W. (1983), "Fundamentals of the Phenomenological Theory of Nonlinear Fracture Mechanics," *J. Appl. Mech.* (ASME), 50:1042–51.

Inglis, C. E. (1913), "Stresses in a Plate Due to the Presence of Cracks and Sharp Corners," *Trans. Inst. Naval Architects,* 55:219–41 (Sec. 12.1).

Irwin, G. R. (1957), "Analysis of Stresses and Strains Near the End of a Crack Transversing a Plate," *J. Appl. Mech.* (ASME), 24:361–64 (Sec. 12.1).

Irwin, G. R. (1958), "Fracture," in *Handbuch der Physik, VI,* ed. by W. Flügge, Springer, Berlin, pp. 551–90, (Sec. 12.2).

Irwin, G. R. (1960), "Fracture Testing of High Strength Sheet Material," *ASTM Bulletin,* p. 29 (Sec. 12.3).

Jenq, Y. S., and Shah, S. P. (1985), "Two Parameter Fracture Model," *J. Eng. Mech.* (ASCE), 111(10):1227–41 (Sec. 12.2).

Kanninen, M. F., and Popelar, C. H. (1985), *Advanced Fracture Mechanics,* Oxford University Press, New York (Sec. 12.1).

Karp, S. M., and Karal, F. C. (1962), "The Elastic Field Behavior in the Neighborhood of a Crack of Arbitrary Angle," *Communications in Pure and Appl. Math.,* 15:413–21 (Sec. 12.1).

Keer, L. M., Nemat-Nasser, S., and Oranratnachai, A. (1979), "Unstable Growth of Thermally Induced Interacting Cracks in Brittle Solids: Further Results," *Int. J. Solids and Structures,* 15:111–126 (Sec. 12.5).

Kfouri, A. P., and Rice, J. R. (1977), "Elastic–Plastic Separation Energy Rate for Crack Advance in Finite Growth Steps," in *Fracture 1977,* ed. by D. M. R. Taplin, Proc. 4th Int. Conf. on Fracture, Univ. of Waterloo, Canada, Vol. 1, pp. 43–59 (Sec. 12.2).

Kienzler, R., and Herrmann, G. (1986), "An Elementary Theory of Defective Beams," *Acta Mechanica,* 62:37–46 (Sec. 12.1).

Kirby, G. C., and Mazurs, C. J. (1985), "Fracture Toughness Testing of Coal," *Research & Engineering Applications in Rock Masses,* 26th U.S. Symp. Rock Mech., Published by A. A. Balkema, Boston.

Knauss, W. C. (1973), "On the Steady Propagation of a Crack in a Viscoelastic Sheet—Experiments and Analysis," in *The Deformation in Fracture of High Polymers,* ed. by H. H. Kausch, Plenum, New York, pp. 501–41 (Sec. 12.2).

Knott, I. F. (1981), *Fundamentals of Fracture Mechanics,* 2nd ed., Butterworths, London (Sec. 12.1).

Krafft, J. M., Sullivan, A. M., and Boyle, R. W. (1961), "Effect of Dimensions on Fast Fracture Instability of Notched Sheets," *Cranfield Symposium,* 1: 8–28 (Sec. 12.3).

Kuszmaul, J. S. (1987), "A Technique for Predicting Fragmentation and Fragment Sizes Resulting from Rock Blasting," *Rock Mechanics,* Proc. 28th U.S. Symp., published by A. A. Balkema, Boston.

Labuz, J. F., Shah, S. P., and Dowding, C. M. (1985), "Experimental Analysis of Crack Propagation in Granite," *Int. J. Rock Mech. Min. Sci. and Geomech. Abstracts,* 22(2):85–98 (Sec. 12.5).

Lachenbruch, A. H. (1961), "Depth and Spacing of Tension Cracks," *J. Geophysical Research,* 66:4273 (Sec. 12.5).

Lister, C. R. B. (1974), "On the Penetration of Water into Hot Rock," *Geophysics J. Royal Astronomical Society* (London), 39:465–509 (Sec. 12.5).

Murakami, Y. (ed.) (1987), *Stress Intensity Factors Handbook,* Pergamon Press, Oxford–New York (Sec. 12.1).

Nguyen, Q. S., and Stolz, C. (1985), "Sur le problème en vitesses de propagation de fissure et de déplacement en rupture fragile ou ductile," *C.R. Acad. des Sciences* (Paris), 301:661–4.

Nguyen, Q. S., and Stolz, C. (1986), "Energy Methods in Fracture Mechanics: Bifurcation and Second Variations," IUTAM–ISIMM Symposium on Appl. of Multiple Scaling in Mechanics, Paris.

Nemat–Nasser, S. (1979), "The Second Law of Thermodynamics and Noncollinear Crack Growth," Proc. 3rd ASCE (Eng. Mec. Div.) Specialty Conference, University of Texas, Austin.

Nemat-Nasser, S., Keer, L. M., and Parihar, K. S. (1978), "Unstable Growth of Thermally Induced Interacting Cracks in Brittle Solids," *Int. J. Solids and Structures,* 14:409–30 (Sec. 12.5).

Nemat-Nasser, S., Sumi, Y., and Keer, L. M. (1980), "Unstable Growth of Tensile Crack in Brittle Solids," *Int. J. Solids & Structures,* 16:1017–35.

Ortiz, M., (1987), "Microcrack Coalescence and Macroscopic Crack Growth Initiation in Brittle Solids," manuscript privately communicated to Bažant in June 1987; also *Int. J. Solids and Structures,* 24(3):231–50 (1988) (Sec. 12.4).

Ouchterlony, F. (1982), "A Simple *R*-Curve Approach to Fracture Toughness Testing of Rock Core Specimens," *Issues in Rock Mech.*, 23rd U.S. Symp. Rock Mech., Soc. of Mining Engrs.

Owen, D. R. J., and Fawkes, A. J. (1983), *Engineering Fracture Mechanics; Numerical Methods and Application,* Pineridge Press Ltd., Swansea, U.K. (Sec. 12.1).

Pijaudier-Cabot, G., Bažant, Z. P., and Berthaud, Y. (1989), "Interacting Crack Systems in Particulate or Fiber Reinforced Composites," Report, Northwestern University; also Proc., 5th Int. Conf. on Numerical Methods in Fracture Mechanics, Freiburg, Germany, ed. by A. R. Luxmoore, Pineridge Press, Swansea, U.K.; and a paper submitted to *J. Eng. Mech.* (ASCE) (Sec. 12.5).

Planas, J., and Elices, M. (1987), "Asymptotic Analysis of the Development of a Cohesive Crack Zone in Mode I Loading for Arbitrary Softening Curves," Preprints, SEM-RILEM Int. Conf. on Fracture of Concrete and Rock, Houston, Texas, publ. by SEM (Soc. for Exper. Mech.), p. 384 (Sec. 12.2).

Planas, J., and Elices, M. (1988a). "Conceptual and Experimental Problems in the Determination of the Fracture Energy of Concrete," presented at Int. Workshop on Fracture Testing and Fracture Energy Test Methods for Concrete and Rock, Tohoku University, Sendai, Japan, Paper 2A-3 (Sec. 12.2).

Planas, J., and Elices, M. (1988b). "Fracture Criteria for Concrete: Mathematical Approximations and Experimental Validation," Proc. Int. Conf. on Fracture and Damage of Concrete and Rock, Vienna, Austria, July 4–6.

Planas, J., and Elices, M. (1989), "Size Effect in Concrete Structures: Mathematical Approximations and Experimental Validation," in *Cracking and Damage—Strain Localization and Size Effect,* ed. by J. Mazars and Z. P. Bažant, Elsevier, London and New York, pp. 462–76; see also Preprints of France–U.S. Workshop at ENS, Cachan, France, in September 1988 (Sec. 12.2).

Rice, J. R. (1968), "A Path Independent Integral and the Approximate Analysis of Strain Concentration by Notches and Cracks," *J. Appl. Mech.* (ASME), 35:379–86; see also "Mathematical Analysis in the Mechanics of Fracture," in *Fracture, an Advanced Treatise,* ed. by H. Liebowitz, Academic Press, New York, 1968 (Sec. 12.1).

Rice, J. R., and Cotterel, B. (1980), "Slightly Curved or Kinked Cracks," *Int. J. Fracture,* 16:155–69 (Sec. 12.3).

Rots, J. G. (1988), "Computational Modeling of Concrete Fracture," Dissertation, Delft University of Technology, Delft, September.

Rots, J. G., and de Borst, R. (1987), "Analysis of Mixed-Mode Fracture in Concrete," *J. Eng. Mech.* (ASCE), 113:1739–58.

Rots, J. G., Hordijk, D. A., and de Borst, R. (1987), "Numerical Simulation of Concrete Fracture in Direct Tension," in *Numerical Methods in Fracture Mechanics,* Proc. 4th Int. Conf. held in San Antonio, Tex.; A. R. Luxmoore et al. (eds.), Pineridge Press, Swansea, U.K., pp. 457–471.

Rots, J. G., Nauta, P., Kusters, G. M. A., and Blaauwendraad, J. (1985), "Smeared Crack Approach and Fracture Localization in Concrete," *HERON,* 30(1): 5–48.

Sakay, M., Yoshimura, J.-I., Goto, Y., and Inagaki, M. (1988), "*R*-Curve Behavior of a Polycrystalline Graphite: Microcracking and Grain Bridging in the Wake Region," *J. Am. Ceram. Soc.*, 71(8):609–16.

Sallam, S., and Simitses, G. J. (1985), "Delamination Buckling and Growth of Flat, Cross-Ply Laminates," *Composites and Structures,* 4:361–81 (Sec. 12.1).

Sallam, S., and Simitses, G. J. (1987), "Delamination Buckling of Cylindrical Shells under Axial Compression," *Composite and Structures,* 7:83–101 (Sec. 12.1).

Sih, G. C. (1974), "Strain Energy Density Factor Applied to Mixed Mode Crack Problems," *Int. J. Fracture,* 10:305–32 (Sec. 12.3).

Srawley, J. E. (1976), "Wide Range Stress Intensity Factor Expression for ASTM E399 Standard Fracture Toughness Specimen," *Int. J. Fracture,* 12:475 (Sec. 12.1).

Sumi, Y., Nemat-Nasser, S., and Keer, L. M. (1980), "A New Combined Analytical and Finite Element Solution Method for Stability Analysis of the Growth of Interacting Tension Cracks in Brittle Solids," *Int. J. Eng. Science,* 18(1):211–24 (Sec. 12.5).

Sumi, Y., Nemat-Nasser, S., and Keer, L. M. (1985), "On Crack Path Stability in Finite Body," *Eng. Fracture Mech.,* 22(5):759–71 (Sec. 12.3).

Tada, H., Paris, P. C., and Irwin, J. K. (1985), *The Stress Analysis of Cracks Handbook,* 2nd ed., Paris Production Inc., St. Louis, Mo. (Sec. 12.1).

Triantafyllidis, N., and Aifantis, E. (1986), "A Gradient Approach to Localization of Deformation. I. Hyperelastic Materials," *J. Elasticity,* 16:225–36.

Willam, K. J., Bicanic, N., and Sture, S. (1985), "Experimental Constitutive and Computational Aspects of Concrete Failure," Preprints, U.S.–Japan Seminar on Finite Element Analysis of Reinforced Concrete Structures, Tokyo, pp. 149–72.

Willam, K., Bicanic, N., Pramono, E., and Sture, S. (1986), "Composite Fracture Model for Strain Softening Computations of Concrete," in *Fracture Toughness and Fracture Energy of Concrete,* ed. by F. H. Wittmann, Elsevier, New York, pp. 149–62.

Williams, M. L. (1952), "Stress Singularities Resulting from Various Boundary Conditions in Angular Corners of Plates in Extension," *J. Appl. Mech.* (ASME), 19:526–8 (Sec. 12.1).

Wnuk, M. P. (1974), "Quasi-Static Extension of a Tensile Crack Contained in Viscoelastic Plastic Solid," *J. Appl. Mech.* (ASME), 41(1):234–48 (Sec. 12.2).

Wu, C. H. (1978), "Fracture Under Combined Loads by Maximum Energy Release Rate Criterion," *J. Appl. Mech.* (ASME), 45:553–8 (Sec. 12.3).

Yin, W-L., Sallam, S., and Simitses, G. J. (1986), "Ultimate Axial Load Capacity of a Delaminated Beam-Plate," *AIAA Journal,* 24(1):123–8 (Sec. 12.1).

13

Damage and Localization Instabilities

To complete our exposition of stability theory of nonelastic structures begun in Chapter 8, in this last chapter we will tackle the instabilities due to damage—a problem that recently attained prominence in research, and in which rapid advances are taking place. Damage is a broad term describing a decrease of material stiffness or strength, or both. The physical cause of damage is chiefly microcracking and void formation. Damage may be regarded as continuously distributed (smeared) fractures. Vice versa, fracture may be regarded as damage localized into a line or surface, in the sense of the Dirac delta function.

Damage, just like fracture, has a strong destabilizing influence and can cause instability of a structure even without any geometric nonlinearity. Not all damage causes instability. Whether it does depends on the tangential stiffness matrix. If this matrix is positive definite, in which case the stress–strain curve is rising and the material is said to exhibit strain hardening, damage causes no instability. When this matrix ceases to be positive definite, in which case the stress–strain curve is descending and the material is said to exhibit strain softening, instabilities and bifurcations arise. They generally consist of localization of damage into a zone of the minimum possible size permitted by the continuum model. As we will see, this creates a new twist to the problem of constitutive modeling—it makes no sense to formulate stress–strain relations for strain softening without at the same time coping with the problem of localization instabilities (Bažant, 1976).

As pointed out by Hadamard (1903), the lack of positive definiteness of the elastic moduli matrix makes the wave propagation speed imaginary. This observation, which is also true of the tangential moduli matrix of an inelastic material, causes the differential equation of motion to change its type from hyperbolic to elliptic. This in turn causes the initial-boundary-value problem of dynamics to cease being well-posed (i.e., become ill-posed, which means that an infinitely small change in the initial or boundary conditions can cause a finite change in response). However, this is not as problematic as thought until recently because the unloading wave speed is real as the unloading modulus remains positive even in the strain-softening range (Bažant, 1976; Reinhardt and Cornelissen, 1984).

Due to the imaginary wave speed, strain softening has been considered an inadmissible property of a continuum and some scholars have argued that strain softening simply does not exist (Read and Hegemier, 1984; Sandler 1984). True, but only on the microscale. Strain softening does not exist in the heterogeneous microstructure at sufficient resolution. It is merely an abstraction, a necessary expedient to model the material on the macroscale. It would be impossible for an analyst to take the vast number of microcracks into account individually.

Similar phenomena, which cause ill-posedness of the initial-boundary-value problem, are known and accepted as real in other branches of physics. For example, the liquid–vapor phase transition described by van der Waals' equation of state involves a region where pressure increases at increasing volume. This behavior, which must be classified as volumetric strain softening, characterizes metastable superheated liquids or supersaturated vapors. In astrophysics, a continuum equation of state of stellar matter that exhibits what we might call volumetric strain softening is a well-established concept, which explains the gravitational collapse of a white dwarf into a neutron star or the gravothermal collapse (Thompson, 1982; Misner and Zapolsky, 1964; Misner, Thorne, and Wheeler, 1973; Harrison et al., 1965). Ill-posedness arises in certain boundary-layer problems of viscous fluids, and so forth.

The central problem of this chapter is the stability aspect of strain softening. We will begin by clarifying the phenomenon of localization and ill-posedness in dynamics as well as statics. Then, observing that, due to the finiteness of the energy dissipation per unit volume, localization of damage to a zone of zero volume would unrealistically imply the structural failure to occur at zero energy dissipation, we will show that the usual (local) finite element solutions are unobjective with regard to the analyst's choice of the mesh and exhibit spurious convergence. We will see that the remedy is to introduce into the material model a mathematical device that limits the localization of damage to a zone of a certain minimum thickness that is a material property. Then we will discuss the crack band model that limits localization in the crudest but simplest manner. Furthermore, we will analyze some multidimensional localization instabilities due to strain softening that can be solved exactly, and we will also briefly discuss localization instabilities due to frictional phenomena. Finally we will briefly examine more general material models that limit localization by nonlocal continuum concepts. We will close by pointing out some consequences for the structural size effect and the constitutive modeling. With this we will complete a fairly comprehensive picture of contemporary structural stability theory that has been attempted in this book.

13.1 WAVES IN STRAIN-SOFTENING MATERIALS

To clarify the mathematical difficulties caused by strain-softening damage, we present the exact solution of a one-dimensional wave propagation recently obtained by Bažant and Belytschko (1985).

The material of a bar has the stress–strain diagram *OPF* shown in Figure 13.1, which exhibits elastic behavior with Young's modulus E up to strain ε_p at peak stress f_t' (strength), followed by a strain-softening curve $F(\varepsilon)$—a positive monotonic continuous function that has a negative slope $F'(\varepsilon)$ but, otherwise, an arbitrary shape, and that attains zero stress either at some finite strain or

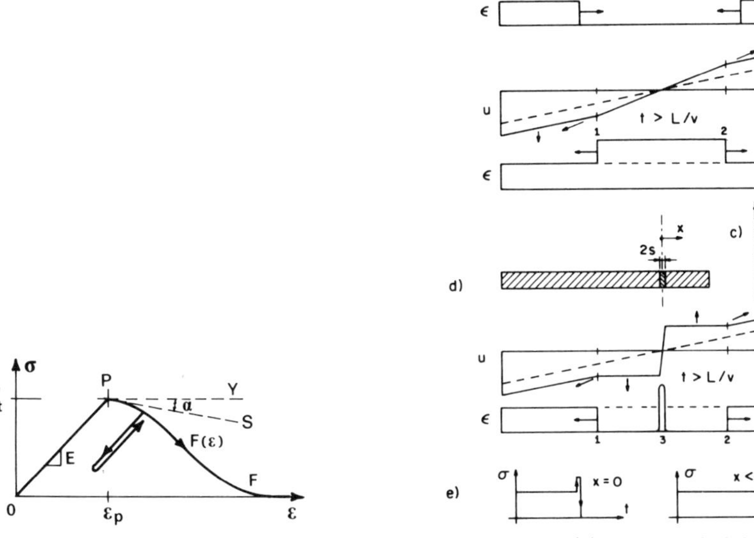

Figure 13.1. Stress–strain diagram with strain softening.

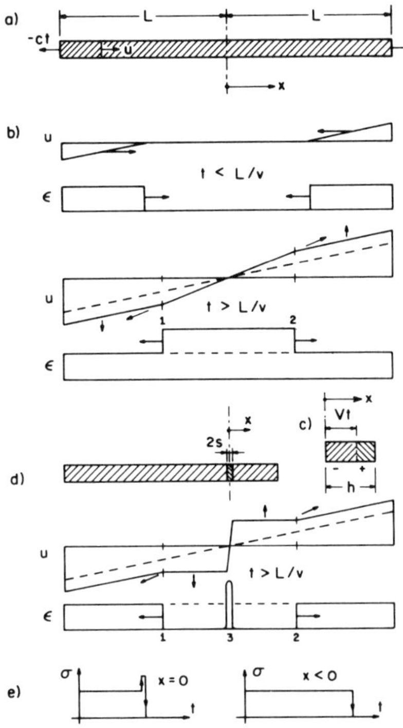

Figure 13.2. (a) Bar loaded by imposed motions of constant velocity at opposite ends; (b) elastic solution; (c,d,e) solution with strain softening. (*After Bažant and Belytschko, 1985.*)

asymptotically for $\varepsilon \to \infty$. The unloading ($\dot{\varepsilon} < 0$) and reloading ($\dot{\varepsilon} \geq 0$), up to the last previous maximum strain, is elastic with modulus E and, if the strain increases beyond this maximum, the (virgin) strain-softening diagram is followed.

Consider a bar of length $2L$, with a unit cross section and mass ρ per unit length (Fig. 13.2a). Let the bar be loaded by forcing both ends to move simultaneously outward, with constant opposite velocities of magnitude c. The boundary conditions are

$$\begin{aligned} &\text{For } x = -L: && u = -ct \\ &\text{For } x = L: && u = ct \end{aligned} \Bigg\} \quad (\text{for } t \geq 0) \qquad (13.1.1)$$

in which x = length coordinate measured from the midlength; t = time; and $u(x, t)$ = displacement in the x-direction. Initially (at $t = 0$), the bar is undeformed and at rest ($u = \dot{u} = 0$). Due to symmetry, the problem is equivalent to a bar fixed at $x = 0$.

Exact Solution of Strain-Softening Bar

Suppose first that strains exceeding ε_p are never produced, that is, the bar remains linearly elastic. The differential equation of motion is hyperbolic and reads

$$v^2\frac{\partial^2 u}{\partial x^2}=\frac{\partial^2 u}{\partial t^2} \qquad \text{with } v^2=\frac{E}{\rho} \tag{13.1.2}$$

For the given boundary and initial conditions, the solution is

$$u=-c\left\langle t-\frac{x+L}{v}\right\rangle+c\left\langle t+\frac{x-L}{v}\right\rangle \qquad \left(\text{for } t\leq\frac{2L}{v}\right) \tag{13.1.3}$$

in which the symbol $\langle\ \ \rangle$, called Macaulay brackets, is defined as $\langle A\rangle=A$ if $A>0$ and $\langle A\rangle=0$ if $A\leq 0$, and $v=$ wave velocity. The strain is

$$\varepsilon=\frac{\partial u}{\partial x}=\frac{c}{v}\left[H\left(t-\frac{x+L}{v}\right)+H\left(t+\frac{x-L}{v}\right)\right] \tag{13.1.4}$$

in which H denotes Heaviside step function. The stress is $\sigma=E\varepsilon$. The strain consists of two tensile step waves of magnitude c/v, emanating from the ends of the bar and converging on the center. After the waves meet at the midpoint, the strain is doubled (Fig. 13.2b). Obviously, if $c/v\leq\varepsilon_p/2$, the assumption of elastic behavior holds for $x\leq 2Lv$, that is, until the time each wavefront runs the entire length of the bar. If, however, $\varepsilon_p/2<c/v\leq\varepsilon_p$, the previous solution applies only for $t<L/v$ and the midpoint cross section ($x=0$) enters the strain-softening regime at $t=L/v$, that is, when the wavefronts meet at the midpoint.

It is interesting to look at the behavior in a very small neighborhood of the interface between the elastic and the strain-softening region, imagining a segment (control volume) of length h (Fig. 13.2c) that is fixed within the material and contains the interface at the distance Vt from the left end of the segment (placed at $x=0$), where V denotes velocity of the interface. The displacements at points x just to the right of the interface are $u^+=U+\varepsilon^+(x-Vt)$, and those just to the left of it are $u^-=U+\varepsilon^-(x-Vt)$, in which $U=$ displacement at the interface, and ε^+, $\varepsilon^-=$ strains just to the right and to the left of the interface. Differentiating, we get the material velocities (material derivatives of u) $\dot{u}^+=\dot{U}-V\varepsilon^+$ and $\dot{u}^-=\dot{U}-V\varepsilon^-$ just to the right and to the left of the interface. The equation of motion of the small element h, fixed within the material, may be stated as follows: the rate of the linear momentum of element h equals the total force applied on this element. Thus, we have

$$\frac{\partial}{\partial t}\left[\int_0^{Vt}\rho(\dot{U}-V\varepsilon^-)\,dx+\int_{Vt}^{h}\rho(\dot{U}-V\varepsilon^+)\,dx\right]=\sigma^+-\sigma^- \tag{13.1.5}$$

from which we get the jump relation, relating the jumps in stress and strain:

$$\sigma^+-\sigma^-=\rho V^2(\varepsilon^+-\varepsilon^-) \tag{13.1.6}$$

Now consider again the whole bar and assume that $\varepsilon_p/2<c/v<\varepsilon_p$. Then the waves are elastic until they meet at time $t_1=L/v$. Strain softening begins immediately at time t_1 at $x=0$ (midpoint). Due to symmetry, a centrally located strain-softening segment of length $2s$ with initial value $s=0$ is created in the middle of the bar at $t=t_1$ (Fig. 13.2d).

It might be useful to mention what happens if the step waves are considered as the limiting cases of strain waves with a wavefront that rises very sharply but continuously. Then the stress at $x=0$ would rise to $\sigma=f_t'$ before strain softening begins, and so an elastic stress wave of magnitude $f_t' - E\,c/v$, superimposed on the incoming wave of magnitude Ec/v, would pass through the midpoint ($x = 0$) from right to left. However, this passing elastic wave would become in the limit a finite magnitude stress pulse of an infinitely short duration, because the strain softening is reached instantly in the limit. The strain energy of such a pulse is zero; thus it can have no effect on the bar and need not be considered in our solution.

The differential equation of motion within the strain-softening segment $2s$ has the form $\partial\sigma/\partial x = \rho\,\partial^2 u/\partial t^2$. Since $\partial\sigma/\partial x = (\partial\sigma/\partial\varepsilon)\,\partial\varepsilon/\partial x = F'(\varepsilon)\,\partial^2 u/\partial x^2$, we have

$$\alpha^2 \frac{\partial^2 u}{\partial x^2} + \frac{\partial^2 u}{\partial t^2} = 0 \qquad \text{with } \alpha^2 = \frac{-F'(\varepsilon)}{\rho} \tag{13.1.7}$$

Because $F'(\varepsilon) < 0$, this equation is *elliptic,* which means that interaction over finite distances is immediate; however, that need not be questionable as long as s remains infinitely small.

One possible solution of Equation 13.1.7 is

$$u = [a(t - t_1) + \varepsilon_p]x \qquad (\text{for } -s \leq x \leq s,\ t > t_1) \tag{13.1.8}$$

in which a is some constant. This solution, implying a uniform strain distribution, can adequately describe the strain-softening segment as long as s is infinitely small. Note that Equation 13.1.8 satisfies the symmetry condition, and also the condition $\partial u/\partial x = \varepsilon_p$ at $t = t_1$.

Check that, for $-v(t - t_1) \leq x < -s$ (i.e., outside of segment s, on its left), Equation 13.1.2 is solved by the following expression:

$$u = c\left(\frac{x + L}{v} - t\right) + f(\xi) = c\left(\frac{2x}{v} - \xi\right) + f(\xi) \qquad \xi = t - \frac{L - x}{v} \tag{13.1.9}$$

in which f is an arbitrary function describing a wave propagating toward the left. By differentiation of Equation 13.1.9

$$\varepsilon = \frac{\partial u}{\partial x} = \frac{1}{v}[c + f'(\xi)] \qquad \text{for } -v\left(t - \frac{L}{v}\right) < x < -s \tag{13.1.10}$$

in which $f'(\xi) = df(\xi)/d\xi$.

Now we need to formulate the interface conditions for displacements and stresses at $x = -s$. Continuity of displacement requires, for $t > (L + s)/v$, that

$$f(\xi_1) - c\left(\xi_1 + \frac{2s}{v}\right) = -[a(t - t_1) + \varepsilon_p]s \qquad \xi_1 = t - \frac{L + s}{v} \tag{13.1.11}$$

The interface stresses must satisfy the jump condition in Equation 13.1.6. The interface at the left end of segment s can be either stationary (constant s, $s \to 0$) or it can move to the left at velocity $V = -\dot{s}$. It cannot move to the right, since then the softening segment would not exist, and there would be no strain softening.

Suppose that the interface moves to the left. Then the material points are entering the strain-softening regime as the interface moves through them. Therefore, σ^- on the left of the interface must be equal to the strength f_t'. From this we know that $\sigma^- \geq \sigma^+$. At the same time, $\varepsilon^+ > \varepsilon^-$ because the strain must be larger than ε_p inside the strain-softening segment and must not be larger than ε_p outside it. Thus, we see from Equation 13.1.6 that $V^2 < 0$ if the interface moves. This, however, is impossible, since V must not be imaginary. The only remaining possibility is that the interface does not move ($V = 0$) and s remains infinitesimal. Therefore, according to Equation 13.1.6

$$\sigma^+ = \sigma^- \qquad (\text{at } x = s) \tag{13.1.12}$$

that is, the stress must be continuous at $x = s$ (strain-softening boundary) for $t > t_1 + s/v$, although the strain is discontinuous. It follows that $E\varepsilon^- = F(\varepsilon^+)$. Substituting here Equation 13.1.10 and calculating ε^+ from Equation 13.1.8 as $\partial u/\partial x$, we get (for $t > t_1 + s/v$)

$$\frac{E}{v}[c + f'(\xi_1)] = F(\varepsilon^+) \qquad \varepsilon^+ = a(t - t_1) + \varepsilon_p \tag{13.1.13}$$

Eliminating a from Equations 13.1.11 and 13.1.13, we find

$$\varepsilon^+ = \frac{1}{s}\left[c\left(\xi_1 + \frac{2s}{v}\right) - f(\xi_1)\right] \tag{13.1.14}$$

Now, since the strain-softening segment must remain infinitely short, that is, $s \to 0$, we have $\varepsilon^+ \to \infty$, and consequently $F(\varepsilon^+) \to 0$. Therefore, $c + f'(\xi_1) = 0$, that is, $f(\xi_1) = -c\xi_1$ or $f(\xi) = -c\xi$. Hence

$$f(\xi) = -c\left\langle t - \frac{L - x}{v}\right\rangle \tag{13.1.15}$$

Consequently, the complete solution for $0 \leq t \leq 2L/v$ and $x < 0$ is

$$u = -c\left\langle t - \frac{x + L}{v}\right\rangle - c\left\langle t - \frac{L - x}{v}\right\rangle \tag{13.1.16}$$

and

$$\varepsilon = \frac{c}{v}H\left(t - \frac{x + L}{v}\right) - \frac{c}{v}H\left(t - \frac{L - x}{v}\right) \tag{13.1.17}$$

For the right half of the bar, $x > 0$, a symmetric solution applies. For $x \to 0^-$ the displacements are $u = -2c\langle t - L/v\rangle$, and for $x \to 0^+$ they are $u = 2c\langle t - L/v\rangle$. So, after time $t_1 = L/v$, the displacements develop a discontinuity at $x = 0$, with a jump of magnitude $4c\langle t - L/v\rangle$. Therefore, the strain near $x = 0$ is $\varepsilon = 4c\langle t - L/v\rangle\delta(x)$ in which $\delta(x) =$ the Dirac delta function. This expression for ε satisfies the condition that $\int_{-s}^{s} \varepsilon\, dx = 4c\langle t - L/v\rangle$ for $s \to 0$.

The complete strain field for $x \leq 0$ and $0 \leq t \leq 2L/v$ is

$$\varepsilon = \frac{c}{v}\left[H\left(t - \frac{x + L}{v}\right) - H\left(t - \frac{L - x}{v}\right) + 4\langle vt - L\rangle\delta(x)\right] \tag{13.1.18}$$

This solution is sketched in Figure 13.2d, e. It may be checked that no unloading (i.e., strain reversal) occurs within the strain-softened material (at $x = 0$), as supposed.

A subtle question in regard to unloading still remains. We tacitly implied no unloading and showed that a solution exists. Can we find another solution if unloading is assumed to occur after the start of strain softening? Apparently we cannot. Indeed, for unloading, the tangent modulus becomes positive; thus the equation of motion in segment $2s$ becomes hyperbolic, which means that the wave arriving from the right is transmitted to the left across the midpoint. But then the superposiiton of the converging waves yields an increase of strain, which contradicts the assumption of unloading. Thus, our solution for $c < v\varepsilon_p/2$ (Eq. 13.1.16) appears to be unique.

Stability Aspects and Unrealistic Properties of Solution of a Bar

Some properties are rather interesting and unrealistic. The solution does not depend on the shape of the strain-softening diagram; the result is the same for strain-softening stress–strain diagrams of very mild slope or very steep slope, and even for a vertical stress drop in the σ-ε diagram. The latter case is equivalent to having a bar that splits in the middle at time $t = L/v$, with each half having a free end at $x = 0$; indeed, this behavior leads to the same solution.

Consider the dependence of the solution on the boundary conditions. When c is just slightly smaller than $v\varepsilon_p/2$, the solution is given by Equation 13.1.3, and when c is just slightly larger than $v\varepsilon_p/2$, the solution is given by Equation 13.1.16, which differs from Equation 13.1.3 by a finite amount for $t > L/v$. Thus, an infinitely small change in the boundary condition can lead to a finite change in the response. Hence, the solution of the dynamic problem for a strain-softening material does not depend continuously on the boundary conditions, that is, the response may be termed unstable (for $c = v\varepsilon_p/2$), similarly to the definition in Section 3.5.

The solution also exhibits a discontinuous dependence on the parameters of the stress–strain diagram. Compare the solutions for stress–strain diagram OPS in Figure 13.1 for which the magnitude of the downward slope PS is as small as desired but nonzero, and for the stress–strain diagram OPY in Figure 13.1, for which the straight line PY is horizontal and represents plastic yielding. For the former, the present solution applies. For the latter, plastic case, Equation 13.1.9 also applies but must be subjected to the boundary condition $E\,\partial u/\partial x = f_t'$ at $x = -s$. The resulting solution is well known and is entirely different from the present solution. Thus the response is discontinuous in regard to the strain-softening slope E_t as $E_t \to 0$.

As we see from our solution, strain softening cannot happen within a finite segment of the bar. It is confined to a single cross-sectional plane of the bar, at which the strains become infinite within an instant (i.e., an infinitely short time interval after the start of softening), regardless of the shape and slope of the strain-softening part of the σ-ε diagram. Strain softening cannot happen as a field in the usual classical type of a deformable continuum as considered here. The parameters of the strain-softening diagram, such as the slope $F'(\varepsilon)$, cannot be considered as characteristic properties of a classical continuum, since they have no effect on the solution.

This conclusion agrees with that drawn by Bažant (1976) on the basis of static stability analysis of strain localization. To circumvent the phenomenon of localization into a single cross-sectional plane, one would have to postulate some special type of continuum in which the strains cannot localize into a single cross-sectional plane. A nonlocal continuum can ensure that, as we will see.

Strain softening consumes or dissipates no energy because the volume of the strain-softening region is zero while the energy density in the strain-softening domain is finite, not infinite. This is also confirmed by the fact that, for $t > L/v$, the present solution is the same as the elastic solution for a bar that is split initially in the middle into two bars, for which the mechanical energy (potential plus kinetic) must be conserved.

Note that strain softening can be produced in the interior of a body, and not just at the boundaries, as has been recently suggested by some. To produce strain softening, it is also not necessary to have a symmetric problem; for example, we may enforce velocity $c = c_1 = -0.6\varepsilon_p/v$ at the left end, and $c = c_2 = 0.9\varepsilon_p/v$ at the right end, and strain softening would again be produced at $x = 0$.

To sum up, strain softening in a classical (local) continuum is not a mathematically meaningless concept. A solution exists for given initial and boundary conditions. However, the hypothesis that strain softening may occur in a classical (local) continuum is not representative of known strain-softening materials (concrete, geomaterials, some composites), in which strain softening consumes finite energy and strain-softening regions of finite size are observed experimentally.

It has been demonstrated that numerical step-by-step solutions by finite elements converge to the present solution (Bažant, Belytschko, and Chang, 1984).

Bar with Rehardening or Residual Yield Strength

Some materials, such as fiber-reinforced concrete, exhibit rehardening that follows a strain-softening drop of stress (Fig. 13.3a). Other materials, for example, some soils in compression, exhibit a residual yield strength (Fig. 13.3b). We will now present the solution for the same bar as before (Fig. 13.3c), which was obtained by Belytschko et al. (1987).

We assume again that $Ec/v > \sigma_a/2$ where $\sigma_a(=f'_t) =$ peak stress so that when the opposite waves meet, strain softening begins. Based on our preceding experience, we will assume the strain softening to be limited to a single point $x = s(t)$, and the strain at that point to increase instantaneously to the strain ε_b at which the stress attains the postpeak minimum, so after strain softening $\varepsilon \geq \varepsilon_b$ (Fig. 13.3a).

Rehardening (Fig. 13.3a). We will see that a solution can be found if, after softening, the stress assumes any value σ_s such that $\sigma_b \leq \sigma_s \leq \sigma_a$. Figure 13.3d, e illustrates the form of the solution we seek. In domain 1, behind the initial elastic wavefront, $\dot{u}_1 = -c$, $\varepsilon_1 = c/v$, $\sigma_1 = Ec/v$. We will see that after the opposite waves meet, the strain-softening point moves, but slower than the elastic unloading wave.

At the unloading wavefront of velocity v (interface between domains 1 and 2 in Fig. 13.3e), the following jump conditions must be satisfied (cf. Eq. 13.1.6):

$$\dot{u}_2 - \dot{u}_1 = v(\varepsilon_2 - \varepsilon_1) \qquad \sigma_2 - \sigma_1 = \rho v^2(\varepsilon_2 - \varepsilon_1) \tag{13.1.19}$$

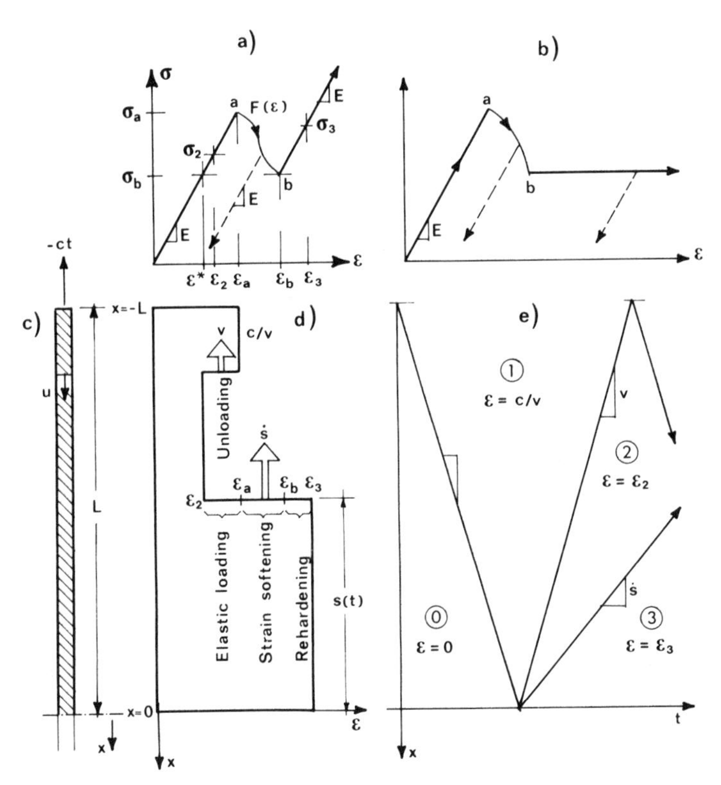

Figure 13.3. Bar loaded by imposed motions of constant velocity at opposite ends (c) and solutions (d, e) for (a) stress–strain diagram with rehardening or (b) residual yield strength. (*After Belytschko, Wang, Bažant, and Hyun, 1987.*)

where subscripts 1 and 2 refer to domains 1 and 2 in Figure 13.3e. Substituting $\dot{u}_1 = -c$ and $\varepsilon_1 = c/v$, we have

$$\dot{u}_2 = v\left(\varepsilon_2 - \frac{2c}{v}\right) \tag{13.1.20}$$

The displacement field in domains 1 and 2 then is

$$u(x, t) = -c\left\langle \xi - \frac{2x}{v} \right\rangle + (v\varepsilon_2 - c)\langle \xi \rangle \qquad \xi = t - \frac{L - x}{v} \tag{13.1.21}$$

Hence $u(0, t) = (v\varepsilon_2 - 2c)\langle t - L/v \rangle$.

If the displacements are to remain continuous at $x = 0$, another wave, of speed $V = -\dot{s}$ $(V < v)$, must emanate from that point, except when $\varepsilon_2 = 2c/v$. This wave is represented by the interface between domains 2 and 3, Figure 13.3e. The velocity jump condition yields

$$\dot{s}(\varepsilon_3 - \varepsilon_2) = \dot{u}_3 - \dot{u}_2 = 2c - v\varepsilon_2 \tag{13.1.22}$$

where it has been noted that $\dot{u}_3 = 0$ and Equation 13.1.20 has been substituted. The stress-jump condition (Eq. 13.1.6) gives

$$\sigma_3 - \sigma_2 = \rho\dot{s}^2(\varepsilon_3 - \varepsilon_2) \tag{13.1.23}$$

and the stress–strain law for rehardening reads

$$\sigma_3 - \sigma_b = E(\varepsilon_3 - \varepsilon_b) \tag{13.1.24}$$

Subtracting the stress changes corresponding to $\varepsilon_3 - \varepsilon_b$ and $\varepsilon_2 - \varepsilon^*$ according to Figure 13.3a, we have $\sigma_3 - \sigma_2 = \rho v^2(\varepsilon_3 - \varepsilon_b - \varepsilon_2 + \varepsilon^*)$ where $\rho v^2 = E$ and $\varepsilon^* = \sigma_b/E$ (Fig. 13.3a). Substituting this into Equation 13.1.23 we obtain

$$\dot{s}^2(\varepsilon_3 - \varepsilon_2) = v^2(\varepsilon_3 - \varepsilon_b - \varepsilon_2 + \varepsilon^*) \tag{13.1.25}$$

Expressing $\dot{s}$ from Equation 13.1.22 and substituting it here, we gain for ε_3 a quadratic equation whose solution is

$$\varepsilon_3 = \varepsilon_2 + \frac{\varepsilon_b - \varepsilon^*}{2} + \left[\left(\frac{\varepsilon_b - \varepsilon^*}{2}\right)^2 + \left(\frac{2c}{v} - \varepsilon_2\right)^2\right]^{1/2} \tag{13.1.26}$$

This is a family of solutions with ε_2 as an arbitrary parameter such that $\sigma_b/E \le \varepsilon_2 \le \sigma_a/E$. A second solution with a minus in front of the radical has been rejected since it is necessary that $\varepsilon_3 > \varepsilon_b$.

Solving ε_3 from Equation 13.1.22, substituting it into Equation 13.1.26, and rearranging, we obtain the equation $(\dot{s}/c)^2 + 2A(\dot{s}/c) - 1 = 0$ where

$$A = \frac{v(\varepsilon_b - \varepsilon^*)}{2(2c - v\varepsilon_2)} \tag{13.1.27}$$

and solving this quadratic equation for $\dot{s}/c$ we get

$$\dot{s} = c(-A + \sqrt{1 + A^2}) \tag{13.1.28}$$

It follows that if $A \ge 0$, then $\dot{s} \le c$. This condition is satisfied if $c/v > \varepsilon_2/2$, which is necessary to initiate strain softening at $x = 0$.

An interesting special case, to which numerical solutions have converged (Belytschko et al., 1987), arises for $\varepsilon_2 = \sigma_b/E = \varepsilon^*$. Substituting this into Equation 13.1.26, one can show that

$$\varepsilon_2 + \frac{\varepsilon_b + \varepsilon^*}{2} \le \varepsilon_3 \le \frac{2c}{v} + \varepsilon_b - \varepsilon^* \qquad (\text{for } \varepsilon_2 = \varepsilon^*) \tag{13.1.29}$$

Note that if $c/v \to \varepsilon^*/2$, Equation 13.1.27 shows that $A \to \infty$, and from Equation 13.1.28, $\dot{s} \to 0$. The solution for $-L \le x < -s$ then is

$$u(x, t) = -c\left\langle \xi - \frac{2x}{c} \right\rangle + (v\varepsilon^* - c)\langle \xi \rangle \tag{13.1.30}$$

$$\varepsilon = \frac{c}{v} H\left(\xi - \frac{2x}{v}\right) + \left(\varepsilon^* - \frac{c}{v}\right) H(\xi) \tag{13.1.31}$$

and for $-s \le x \le 0$ the solution is $u = \varepsilon_3 x H(\xi)$ and $\varepsilon = \varepsilon_3 H(\xi)$.

The character of the solution is shown in Figure 13.3d, e. The most interesting point is that now, in contrast to Fig. 13.2, the strain-softening point moves. The strain at this point also involves a postsoftening elastic increase of strain, such that the overall stress change (from σ_2 to σ_3 in Figure 13.3a) is an increase. This means that the interface $s(t)$ is actually an interface between two elastic domains with different uniform strains. (Further this means that, according to

Equation 13.1.22, $\dot{s}>0$, because $\varepsilon_3>\varepsilon_2$.) Since the strain-softening point moves, the energy dissipation is, in contrast to the previous problem, nonzero. The shape of the strain-softening portion *ab* of the stress–strain diagram (Fig. 13.3a) has no effect on the solution.

Another interesting property is that, at the moving strain-softening points $x=\pm s(t)$, the stress takes the values σ in the range $\sigma_a<\sigma<\sigma_b$ three times at the same time (albeit at different strains). This clouds the question of differentiability of $\partial\sigma/\partial x$ in the equation of motion. In contrast to the preceding problem (Fig. 13.2), no displacement discontinuity is caused by strain softening. But again the problem is not a well-posed one, since at $\varepsilon_a=Ev/2\sigma_a$ an infinitely small change in prescribed boundary velocity c causes a finite change in the solution.

Residual yield strength (Fig. 13.3b). The stress at the strain-softening point becomes σ_b after the jump in strain, and so the elastic solution in domains 1 and 2 becomes

$$\varepsilon_2=\varepsilon^* \qquad \sigma_2=\sigma_b \qquad u(x,t)=-c\left\langle \xi-\frac{2x}{v}\right\rangle+(v\varepsilon^*-2c)\langle\xi\rangle \quad (13.1.32)$$

In particular, at $x=0$ $(s=0)$, we have $u(0,t)=(v\varepsilon^*-2c)\langle t-L/c\rangle$. If strain softening has taken place, $v\varepsilon^*-2c<0$. So, since the stresses in both the elastic and strain-softening domains are σ_b, there is no way a wave eliminating the displacement discontinuity could develop. The only way to satisfy the boundary conditions is to allow a discontinuity in the displacement at $x=0$, same as in the first problem with softening of stress to zero. Hence, Equations 13.1.32 must hold for $x\le 0$, with $s=0$, and the magnitude of displacement discontinuity for $t\ge L/v$ is $2(v\varepsilon^*-2c)$. Noting that $\partial\langle\xi\rangle/\partial\xi=H(\xi)$ and differentiating Equation 13.1.32, we get for the strain field

$$\varepsilon(x,t)=\frac{c}{v}H\left(t-\frac{L+x}{c}\right)-\left(\frac{c}{v}-\varepsilon^*\right)H\left(t-\frac{L-x}{v}\right)+(2v\varepsilon^*-4c)\,\delta(x)H\left(t-\frac{L}{v}\right)$$

$$(\text{for } x\le 0) \quad (13.1.33)$$

where the term with the Dirac delta function $\delta(x)$ is added to represent the discontinuity at $x=0$. In contrast to the first problem (Fig. 13.2), this function causes the energy dissipation W to be finite (occurring in the region $-s\le x\le s$ where $s\to 0$); $W=2\sigma_b(v\varepsilon^*-2c)$. Note that when $\sigma_b\to 0$, W vanishes, which agrees with the first solution. Thus, energy dissipation is due solely to the residual yield strength σ_b. It is independent of the magnitude of the stress peak σ_a and of the shape of the strain-softening curve.

Belytschko et al. (1987) showed that step-by-step finite element solutions converge to the foregoing analytical solutions, although slowly and with much noise near the wavefronts.

Cylindrically or Spherically Converging Waves

Waves that propagate inward along the radii of a cylindrical or spherical coordinate system exhibit further strange phenomena when the material is strain softening. Such waves, with a step wavefront, can be produced by a sudden increase of pressure along the surface of a cylinder or sphere (caused, e.g., by an explosion). For the case of elastic material behavior, there exists a simple closed-form formula for the solution (e.g., Achenbach, 1973). This elastic solution is plotted in Figure 13.4b. As can be seen, the magnitude of the elastic wavefront grows, and it approaches ∞ as the wavefront approaches the center.

If the material exhibits strain softening after the elastic limit is exceeded, the wavefront will grow (Fig. 13.4b, right) to eventually reach the strain-softening limit, regardless of the magnitude of the surface pressure applied. Once strain softening is initiated, the strain suddenly increases and the stress suddenly drops, as suggested by the experience with uniaxial waves. This is borne out by the numerical step-by-step finite element solutions, which are reproduced in Figure 13.5. These plots, showing the radial profiles of radial displacement and volumetric strain at various times, were calculated for a material that has unlimited elasticity in shear and a triangular strain-softening diagram in volumetric response (Belytschko et al. 1986; Bažant, 1986a). The pertinent plots are those (on the left) marked "local." For the sake of comparison, the figure also shows (on the right) results marked "nonlocal," although these are pertinent to Section 13.10.

An interesting point revealed by the solutions in Figure 13.5 is that the sudden stress drop that occurs after the elastic limit is first reached is not the end of strain softening. A strange thing happens: A part of the wavefront, with an uncertain magnitude, creeps under the strain-softening limit and propagates further inward. Due to the radially converging nature of the wave, the wavefront grows again, and reaches again the strain-softening limit. At that moment the stress drops suddenly, but again some part of the wavefront, whose magnitude is uncertain, gets through and grows further. In this manner, strain softening is produced at many singular points, all of them located beyond the point where the elastic limit is reached first. The numerical solution beyond the elastic domain does not appear to converge as the mesh is refined, and the locations of the subsequent softening points vary randomly, giving an impression of a chaotic response. It nevertheless seems that the locations of the strain-softening points remain sparse, so that the total energy dissipation in all the softening finite elements converges to zero as the mesh is refined to zero. Again this is an unrealistic feature of the strain-softening concept for an ordinary (local) continuum.

General Three-Dimensional Condition for Waves to Exist

Consider now a homogeneous space of tangential moduli D^t_{ijkm} and mass density ρ. The differential equations of motion are $\sigma_{ij,j} = \rho \ddot{u}_i$ where u_i = displacement increments, σ_{ij} = stress increments and the subscripts refer to cartesian coordinates x_i ($i = 1, 2, 3$). The tangential stress–strain relation may be written as $\sigma_{ij} = D^t_{ijkm} u_{k,m}$, and so the equation of motion becomes

$$D^t_{ijkm} u_{k,jm} = \rho \ddot{u}_i \tag{13.1.34}$$

which is a system of three hyperbolic partial differential equations for $u_i(x_k, t)$. The solutions may be sought in the form:

$$u_k = A_k \exp\left[i\omega(x_r n_r v^{-1} - t)\right] \tag{13.1.35}$$

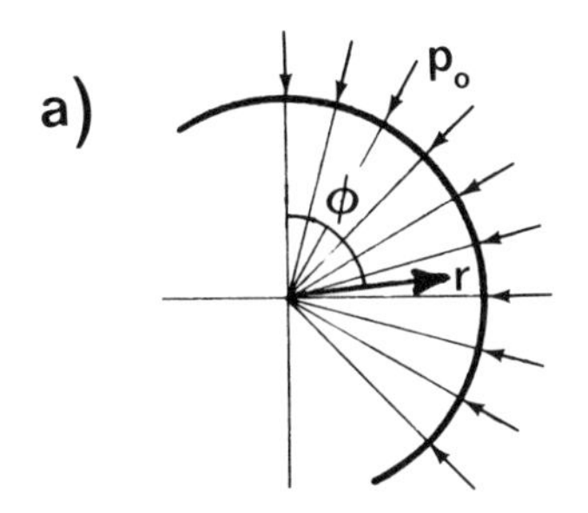

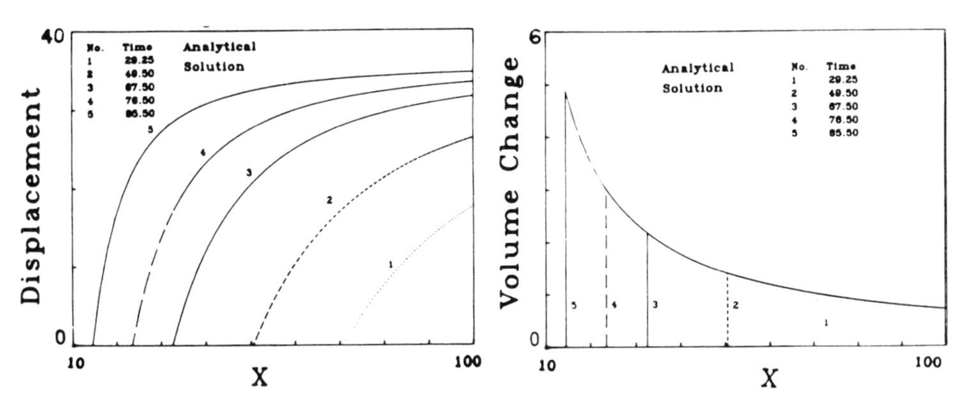

Figure 13.4. (a) Inward propagating wave in a sphere, and (b) elastic solution. (*After Belytschko, Bažant, Hyun, and Chang, 1986.*)

where $i^2 = -1$; A_k, ω, v = constants; n_r = unit vector indicating the direction of propagation; and v = wave velocity. Substitution into Equation 13.1.34 yields a system of three homogeneous linear equations (Synge, 1957; Achenbach, 1973):

$$(Z_{ik} - \lambda\delta_{ik})A_k = 0 \quad \text{with} \quad Z_{ik} = D^t_{ijkm} n_j n_m, \quad \lambda = \rho v^2 \qquad (13.1.36)$$

Waves of any direction can propagate only if v is real, i.e. if all the eigenvalues λ of Z_{ik} are positive. Thus the (3×3) matrix Z_{ik} (called the Christoffel stiffness) must be positive definite, that is, $Z_{ik} q_i q_k > 0$ for all nonzero q_i. Substituting for Z_{ik} and denoting $q_i n_j = \xi_{ij}$, which is a symmetric second-order tensor, we conclude that Z_{ij} is positive for all propagation directions n_i if and only if

$$D^t_{ijkm} \xi_{ij} \xi_{km} > 0 \quad \text{for any nonzero } \xi_{ij} \qquad (13.1.37)$$

This means that the 6×6 matrix of D^t_{ijkm} must be positive definite (Hadamard, 1903). Otherwise the material cannot propagate loading waves in at least one direction n_i, and the type of the partial differential equation (Eq. 13.1.34) changes from hyperbolic to elliptic.

Since u_k according to Equation 13.1.35 can have either sign, the use of D^t_{ijkm}

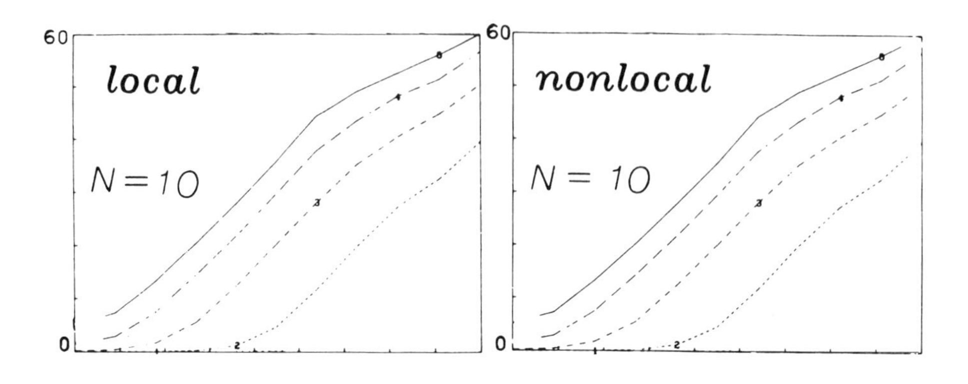

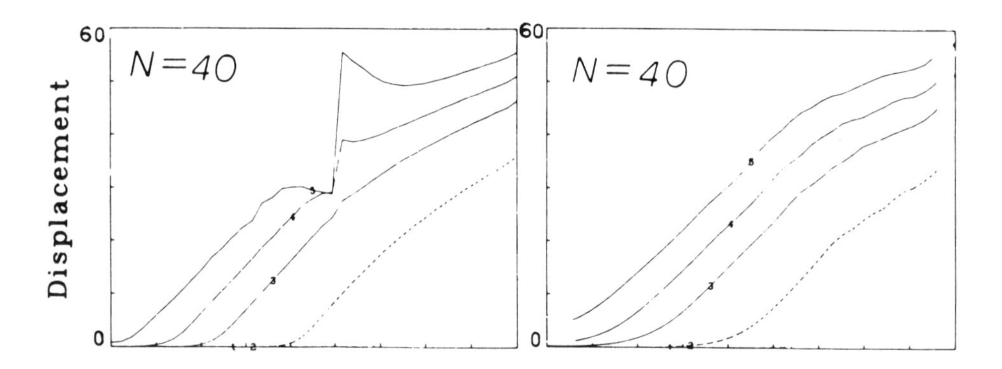

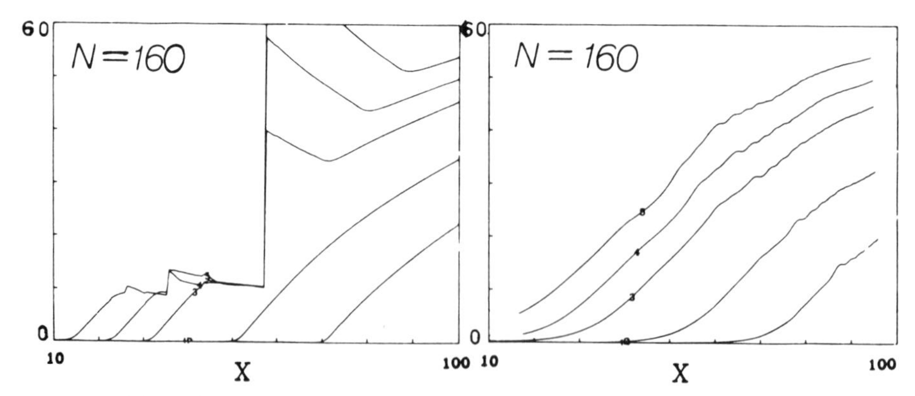

Figure 13.5 (a) Radial displacement and (b) volumetric strain for local and nonlocal materials (spherical geometry). (*After Belytschko, Bažant, Hyun, and Chang, 1986.*)

b)

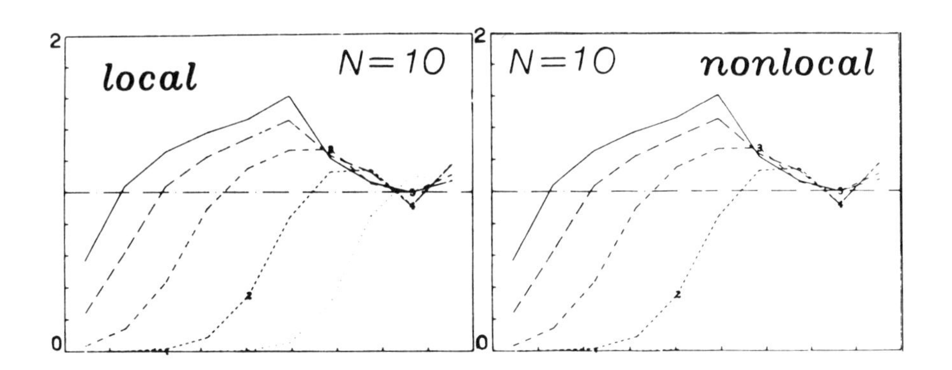

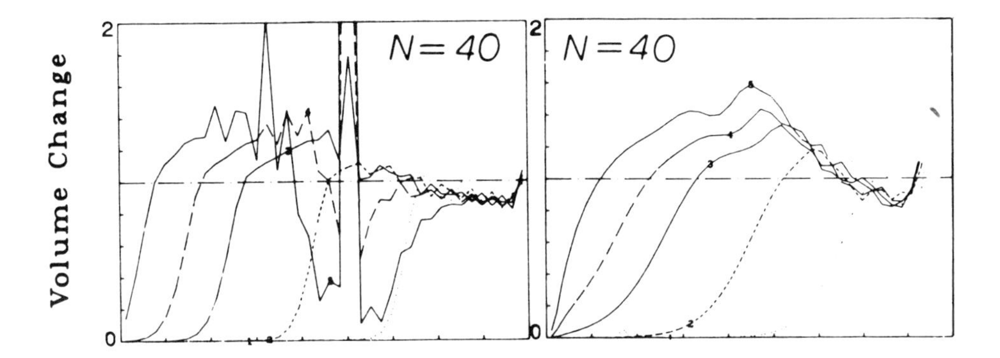

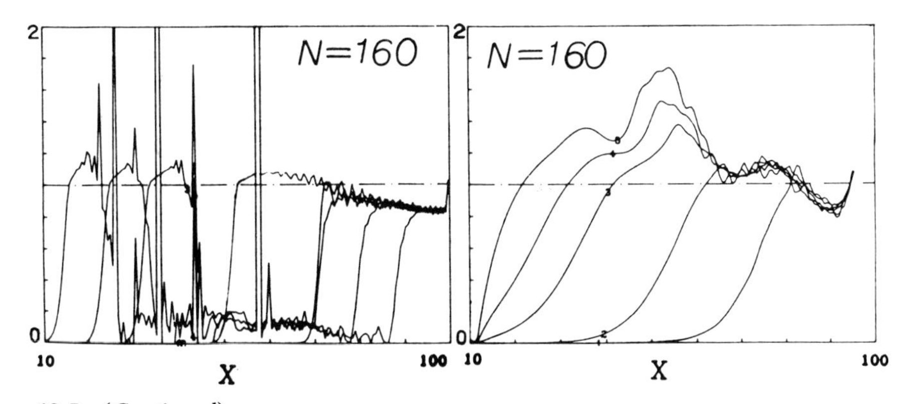

Figure 13.5—(*Continued*)

for loading might seem inconsistent. But it is not, for two reasons: (1) a sum of many terms u_k given by Equation 13.1.35 can approximate any wave profile, and (2) the same result ensues by considering a loading step wave whose wavefront has the form of a discontinuity line with stress σ_{ij}^+ ahead of the front and σ_{ij}^- behind the front. For a discontinuity line whose unit normal is n_i, Equation 13.1.6 may be generalized as $[\sigma_{ij}]n_j = -\rho v[\dot{u}_i]$, where [ ] is the standard notation for a jump, that is, $[\dot{u}_i] = \dot{u}_i^+ - \dot{u}_i^-$. The material rates of displacements are $\dot{u}_i = -vu_{i,r}n_r$, both ahead and behind the discontinuity. Hence $[\dot{u}_i] = -v[u_{i,r}]n_r$ (Achenbach, 1973, p. 142), which gives

$$[\sigma_{ij}]n_j = \rho v^2[u_{i,r}]n_r \tag{13.1.38}$$

Now we may seek solutions $[u_{i,r}]$ represented as a sum of terms of the form A_in_r. By virtue of the superposition principle, we may consider each one of them separately, that is, set $[u_{i,r}] = A_in_r$ where A_i are some constants. Multiplying this by n_r, we get $[u_{i,r}]n_r = A_i$, which we substitute into Equation 13.1.38. From the tangential stress–strain relation we have $[\sigma_{i,j}] = D^t_{ijkm}[\varepsilon_{km}] = D^t_{ijkm}[u_{k,m}] = D^t_{ijkm}A_kn_m$, which we also substitute into Equation 13.1.38. Thus we get for A_i the equation system $D^t_{ijkm}A_kn_jn_m = \rho v^2A_i$. But this is identical to Equation 13.1.36, and so the conclusion is the same.

Summary

Exact dynamic solutions to some strain-softening problems can be found, and numerical finite element solutions converge to these exact solutions (and in fact do so quite rapidly). The typical feature of these solutions is that strain softening remains isolated at singular points. Consequently, the energy dissipation due to strain softening is zero, which is an unacceptable feature. In the numerical solutions this is manifested as spurious mesh sensitivity, a phenomenon which will be discussed in Section 13.5.

Remarks. Before closing, it is interesting to add that strain softening may apparently lead to chaos (Sec. 3.9). This was discovered by G. Maier (discussion recorded by Roelfstra, 1989) when he studied forced vibrations of a beam with a softening hinge (of the type discussed in Sec. 13.6). A chaotic structure was indicated by the Poincaré diagram in the velocity-displacement plane.

Valuable studies illuminating the ill-posedness and other aspects of strain softening in a continuum (local, Sec. 13.10) were presented by Sandler and Wright (1983), Hegemier and Read (1983), Read and Hegemier (1984), Wu and Freund (1984), Sandler (1984), Needleman (1987), and others.

Problems

13.1.1 Beginning at $t = 0$, the left end of the bar considered before (Fig. 13.1) is moved at velocity $(vf'_t/E) - \delta$ to the left, and the right end is moved at velocity $2v\delta/E$ to the right; δ is infinitely small ($\delta \to 0$). Obviously, both waves are elastic until they meet at midlength, at which time strain softening is produced. The solution is obviously not symmetric. Derive it, assuming that strain softening goes all the way down to zero stress.

13.1.2 Do the same as above, but the material exhibits rehardening, as in Figure 13.3a.

13.1.3 Do the same, but the material exhibits residual yield, as in Figure 13.3b.

13.2 SERIES-COUPLING MODEL FOR LOCALIZATION DUE TO SOFTENING

As transpired from the preceding dynamic analysis, strain-softening damage leads to localization of strain and energy dissipation. In statics, the consequence of strain softening is that a uniform state of strain and damage that satisfies all the field equations and boundary conditions may become an impossible solution that cannot occur in reality. The strain and damage may become nonuniform and localize into a relatively small zone, representing the failure zone. The localization is driven by the release of strain energy from an unloading region outside the localization zone. In statics, the localization presents itself as a problem of bifurcation of equilibrium path, in which the main (primary) path, which preserves the uniform strain field, becomes impossible and the secondary path, which produces localization, must occur. The localization after the bifurcation state can represent either an instability, in which case it happens at constant load, or a stable bifurcation at increasing load, analogous to Shanley bifurcation in plastic columns (Sec. 10.3).

In this section, we first analyze a simple series-coupling model and then focus attention on localization in bars or specimens under uniaxial stress. Discussion of multidimensional localization problems is left for the subsequent sections.

Stable States

A simple prototype of localization behavior, which typifies the behavior of many structures, is a series-coupling model (Fig. 13.6a) in which one part of the

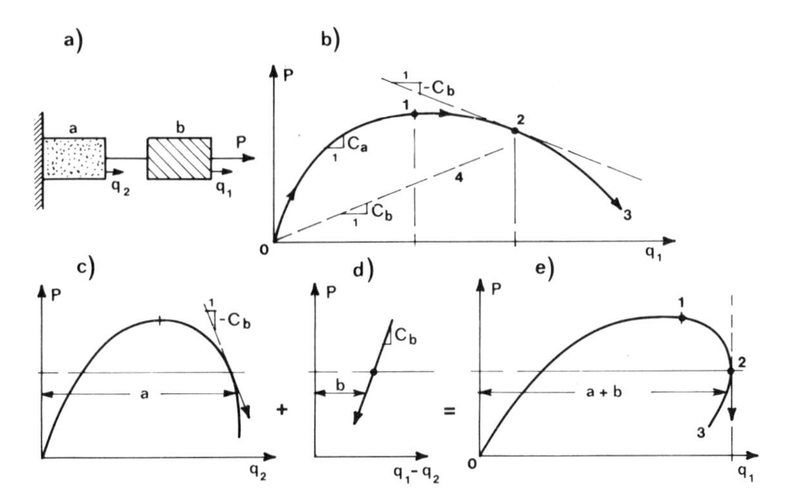

Figure 13.6 (a) Series-coupling model; (b) load-displacement diagram of softening part; (c, d, e) graphic construction of load-displacement curve.

structure (the localization part) undergoes further loading (suffering further softening damage) while the remaining part of the structure that is coupled in series unloads (and suffers no more damage). The strain energy released from the unloading part helps to drive the localization with further softening damage in the other part. Let C_a and C_b be the tangential stiffnesses of parts a and b; $C_a > 0$ if part a is unloading and $C_b > 0$ if part b is unloading; $C_a < 0$ if part a is loading and softening and $C_b < 0$ if part b is loading and softening. If we denote as δq_1 the load-point displacement increment (under load P) and δq_2 the displacement increment at the interface between parts a and b (Fig. 13.6), then the force increments in parts a and b are $\delta F_a = C_a\,\delta q_2$ and $\delta F_b = C_b(\delta q_1 - \delta q_2)$. So the second-order work is $\delta^2 \mathscr{W} = \frac{1}{2}\delta F_a \delta q_2 + \frac{1}{2}\delta F_b(\delta q_1 - \delta q_2)$ or

$$\delta^2 \mathscr{W} = \tfrac{1}{2} C_a\,\delta q_2^2 + \tfrac{1}{2} C_b(\delta q_1 - \delta q_2)^2 \tag{13.2.1}$$

Note that according to Section 10.1, $\delta^2 \mathscr{W} = -T(\Delta S)_{\text{in}}$ where $T =$ absolute temperature and $(\Delta S)_{\text{in}} =$ internally produced entropy increment of the tangentially equivalent elastic structure. $\delta^2 \mathscr{W}$ represents the second variation of the Helmholtz free energy of the structure-load system if the conditions are isothermal or the enthalpy if the conditions are isentropic. The first-order work, that is, $\delta \mathscr{W} = F_a\,\delta q_2 + F_b(\delta q_1 - \delta q_2) - P\delta q_1$, must vanish (according to the principle of virtual work) if the initial state is an equilibrium state (this is indeed true since in equilibrium $P = F_a = F_b$).

Based on Section 10.1, the initial state is stable if $\delta^2 \mathscr{W} > 0$ [or $(\Delta S)_{\text{in}} < 0$] for all admissible displacements, and unstable if $\delta^2 \mathscr{W} < 0$ for some admissible displacement. When the load is controlled, both δq_1 and δq_2 can be nonzero, and so the structure is stable only if $C_a > 0$ and $C_b > 0$, that is, if there is no softening in any part. If the load-point displacement is controlled, we have $\delta q_1 = 0$ for any possible excursion from the initial state, and then Equation 13.2.1 yields

$$\delta^2 \mathscr{W} = \tfrac{1}{2}(C_a + C_b)\,\delta q_2^2 \tag{13.2.2}$$

From the condition $\delta^2 \mathscr{W} > 0$ it now follows that the state is

$$\begin{aligned} &\text{Stable if} && -C_a < C_b \\ &\text{Unstable if} && -C_a > C_b \end{aligned} \tag{13.2.3}$$

The critical state $-C_a = C_b$ can be stable or unstable depending on higher variations of $\mathscr{W}$.

An elementary derivation of Equation 13.2.3 (Bažant, 1976) can alternatively be made as follows. The force that would have to be applied at the interface between parts a and b in order to preserve equilibrium is $\delta f_2 = (C_a + C_b)\,\delta q_2$. Its work is $\delta^2 \mathscr{W} = \frac{1}{2}\delta f_2\,\delta q_2$. When this work is positive, no change of state can happen if force δf_2 is not applied. So $C_a + C_b > 0$ implies stability. If $\delta^2 \mathscr{W} < 0$, then δq_2 leads to a spontaneous release of energy, which must go first into kinetic energy but must eventually be dissipated as heat, due to inevitable damping. Whenever a spontaneous release of heat is possible, it must happen, according to the second law of thermodynamics, and so the initial state cannot be maintained. Hence $\delta^2 \mathscr{W} < 0$, or $C_a + C_b < 0$, implies instability.

Equations 13.2.3 have a simple graphical meaning shown in Figure 13.6b. Curve 0123 represents the response of part a, which is softening for states located beyond the peak point 1. The limit of stability is obtained by passing a tangent line of slope $-C_b$ (a similar construction was used for elastic softening structures in Sec. 4.8). Tangent point 2 is the stability limit (point 2).

If $C_a < 0$ and $C_b > 0$, the stability condition, $-C_a < C_b$, may also be written as $-1/C_a > 1/C_b$ or $C_a^{-1} + C_b^{-1} < 0$. So the structure is

$$\text{Stable if} \qquad C = (C_a^{-1} + C_b^{-1})^{-1} < 0 \tag{13.2.4}$$

where C represents the tangential stiffness of the whole structure. Hence, a structure with a softening part is stable (under displacement control) as long as the stiffness of the entire structure associated with the controlled displacement is negative. When this stiffness becomes positive ($C < 0$), the structure is unstable (even under displacement control). Furthermore, the critical state condition $C_a \to -C_b$ may be written as $C_a^{-1} \to -C_b^{-1}$ or $C_a^{-1} + C_b^{-1} \to 0$ or $C = (C_a^{-1} + C_b^{-1})^{-1} \to \infty$.

Therefore, snapback in the response of the entire structure represents instability. The stability limit (critical state) is represented by the snapback point, that is, point 2 in Figure 13.6e (in which curve 0123 describes the response of the whole structure). This response curve can be graphically obtained from the response of softening part a simply by adding to each ordinate a the ordinate b of part b for the same force P, as illustrated in Figures 13.6c, d, e.

The foregoing stability conditions are similar to those shown in Section 4.8 for elastic structures whose parts exhibit snapback. The difference is that now one needs to check various combinations of loading and unloading stiffnesses for stability. If both parts a and b have softening properties, then localization that destabilizes the structure is not decided by a combination of loading (softening) stiffnesses in both parts but by a combination of the loading (negative) stiffness of one part (a or b) and the unloading (positive) stiffness of the other part.

The stability condition can also be interpreted in terms of energy exchange, for which only the second-order work needs to be considered. The release of energy due to unloading of part b is $(\delta P)^2/2C_b$ and the energy influx into part a required for further softening is $-(\delta P)^2/2C_a$. If $(\delta P)^2/2C_b < -(\delta P)^2/2C_a$, the energy released does not suffice to drive the localization, and so instability cannot happen. This yields the stability condition $C_b^{-1} + C_a^{-1} < 0$, same as before. If $(\delta P)^2/2C_b > -(\delta P)^2/2C_a$, there is an excess release of energy that can manifest itself as nothing else but kinetic energy, and so equilibrium can no longer exist.

Surface of Second-Order Work

The surface of second-order work (or internally produced entropy of structure-load system) given by Equation 13.2.1 is plotted in Figure 13.7 (Bažant, 1988c) for the cases of a mild softening with $-C_a = C_b/6$, medium softening with $-C_a = C_b/3$, and steep softening with $-C_a = 2C_b/3$. The surface is a patch-up of quadratic surfaces in various radial sectors of the plane of displacements $(\delta q_1, \delta q_2)$. For the same reasons as explained in Section 10.3, the quadratic surfaces must be joined continuously and with a continuous slope, but with a discontinuity in curvature. Paths $\overline{01}$, $\overline{02}$, $\overline{03}$, and $\overline{04}$ represent the infinitesimal equilibrium path increments for (1) loading in both parts, (2) loading in part a and unloading in part b, (3) unloading in part a and loading in part b, and (4) unloading in both parts, as shown.

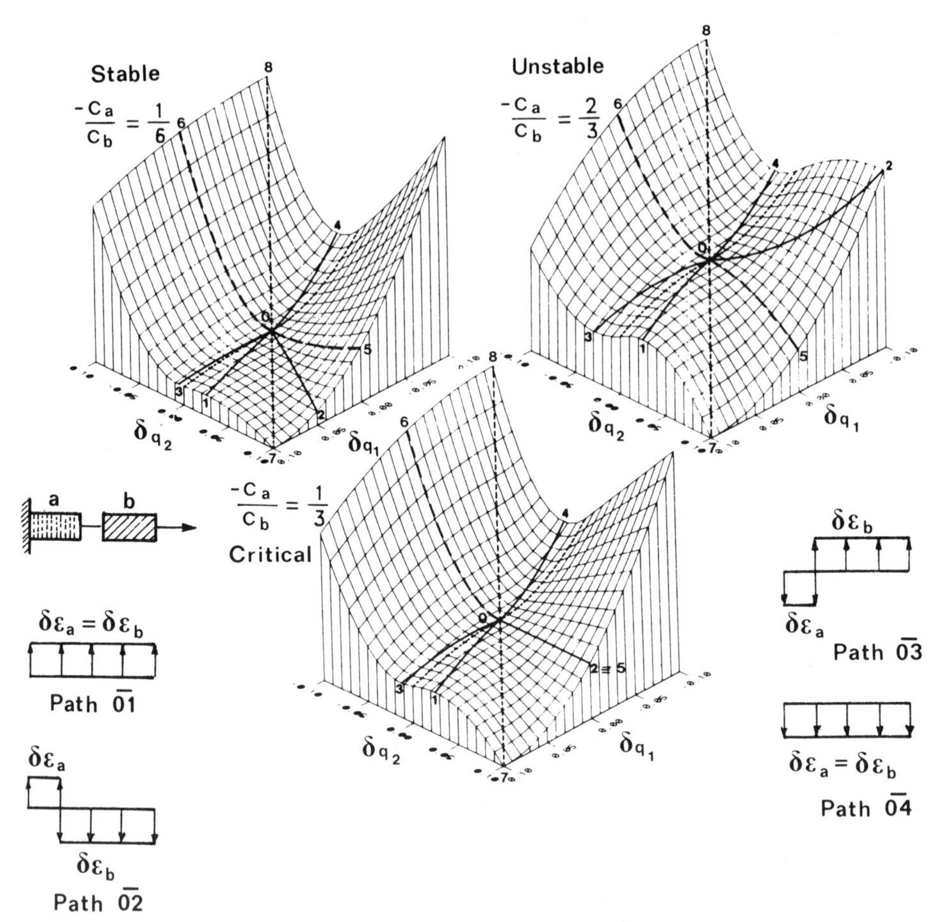

Figure 13.7 Surfaces of second-order work $\delta^2 \mathcal{W}$ for uniaxial specimen. (*After Bažant, 1988c.*)

Application to Uniaxially Stressed Bars or Specimens

As a simple but important practical application, consider a uniaxially stressed bar or specimen of length L, uniform cross section of area A, and uniform material properties (Fig. 13.8a). One end of the specimen is fixed and the other is loaded through a spring of stiffness C_s that, for example, simulates the behavior of the frame of the testing machine. The applied axial force P is either tensile or compressive. The stress–strain diagram exhibits strain softening; see Figure 13.8b.

Experimental observations of uniaxial softening response of concrete as well as rock in tension were reported in many studies (in tension: Rüsch and Hilsdorf, 1963; Hughes and Chapman, 1966; Evans and Marathé, 1968; Petersson, 1981; Reinhardt and Cornelissen, 1984; Labuz, Shah, and Dowding, 1985; in compression: van Mier, 1984, 1986; see also review by Bažant, 1986b).

We assume the specimen to be initially in a uniform state of strain $\varepsilon = \varepsilon_0$ that lies in the strain-softening (postpeak) range. We restrict consideration only to

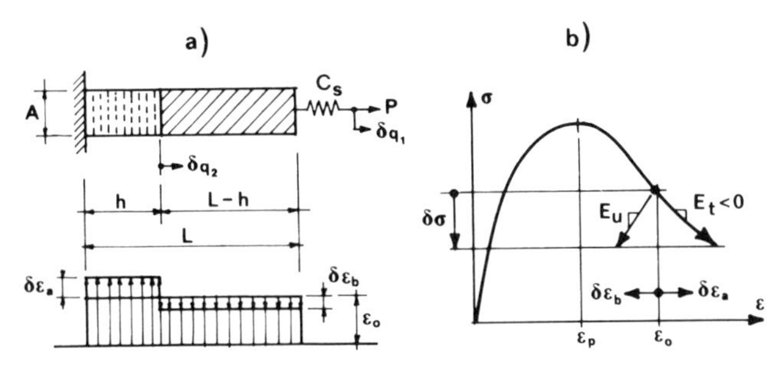

Figure 13.8 (a) Uniaxially stressed specimen and (b) softening stress--strain diagram.

such infinitesimal deviations from this state for which the strains $\delta\varepsilon_a$ and $\delta\varepsilon_b$, shown in Fig. 13.8b, can be different but uniform within segments of lengths h and $L-h$. Without any loss of generality, we may assume $\delta\varepsilon_a$ to represent loading, characterized by a negative tangent modulus E_t, and $\delta\varepsilon_b$ to represent unloading, characterized by a positive unloading modulus E_u. Such a deformation mode represents localization of strain into segment h.

The behavior is obviously equivalent to a series-coupling model in which segment h corresponds to part a of stiffness C_a, and segment $L-h$ together with the spring corresponds to part b of stiffness C_b. The compliance of segment $L-h$ alone is $(L-h)/E_uA$, and the compliance of segment $L-h$ together with the spring is $C_b^{-1}=C_s^{-1}+(L-h)/E_uA$. Also, $C_a=E_tA/h$. Substituting for C_a and C_b in the stability condition $C_a+C_b>0$ (Eqs 13.2.3), we conclude, after rearrangements, that the specimen is (Bažant, 1976)

$$\text{Stable if} \quad -\frac{E_t}{E_u}<-\frac{E_t^{\text{cr}}}{E_u}=\frac{1}{(L/h)-1+(AE_u/C_sh)} \tag{13.2.5}$$

where E_t^{cr} is the strain-softening (negative) tangent modulus at the stability limit (critical state), at which the specimen is in a state of neutral equilibrium. If the inequality sign is reversed, the specimen is unstable.

In Figure 13.8a, the localization segment is shown to develop at one end of the specimen. In a uniaxial model, however, the location of this segment is irrelevant and indeterminate. It can develop anywhere within the specimen length. In fact, the segment can even be subdivided in discontinuous subsegments and only their combined length matters.

Effects of Size and Support Stiffness

Equation 13.2.5 can explain the effects of specimen length and testing frame stiffness on the failure strain $\varepsilon_0=\varepsilon_{\text{cr}}$ of a uniformly strained specimen. We assume the postpeak $\sigma(\varepsilon)$ diagram to be concave (that is, $d^2\sigma/d\varepsilon^2<0$), which is true [for smooth $\sigma(\varepsilon)$ diagrams] for at least some distance beyond the peak-stress point. Then, the smaller the magnitude $|E_t^{\text{cr}}|$ of the tangent modulus ($|E_t^{\text{cr}}|=-E_t^{\text{cr}}$), the smaller is ε_{cr}. From Equation 13.2.5 it is now clear that an increase of L (at constant h) causes a decrease of $|E_t^{\text{cr}}|$ at the critical state. Therefore, the longer the specimen, the smaller is ε_{cr}. Furthermore, it is clear that an increase in C_s causes an increase of $|E_t^{\text{cr}}|$ at the critical state. Therefore, the stiffer the testing frame, the larger is ε_{cr}. Since a softer testing frame has a larger stored energy (for the same P), it also follows that the larger the stored energy of the structure, the smaller is ε_{cr}.

To be able to observe the softening response at large strains, the test specimen must be short enough, and the testing frame must be stiff enough. The importance of testing frame stiffness has been known to experimentalists for a long time and was highlighted for rock especially by the researchers at the Chamber of Mines in Johannesburg in the 1950s and 1960s (e.g., Cook, Bieniawski, Fairhurst), and for concrete by Glucklich and Ishai, as well as others. In this classical literature, however, the localization within the specimen length was not considered. Even without it, one can show that the stiffer the testing frame, the larger is ε_{cr}, but the effect of specimen length is not revealed.

Specimen Ductility as a Function of Length and Loading Frame Stiffness

The foregoing conclusions are illustrated in Figure 13.9 (from Bažant, 1976) that shows for concrete the diagrams of tensile ductility as a function of the relative specimen length L/nd_a, for various values of the relative stiffness C_sL/AE of the testing frame; here d_a = maximum size of the aggregate in concrete, n = empirical

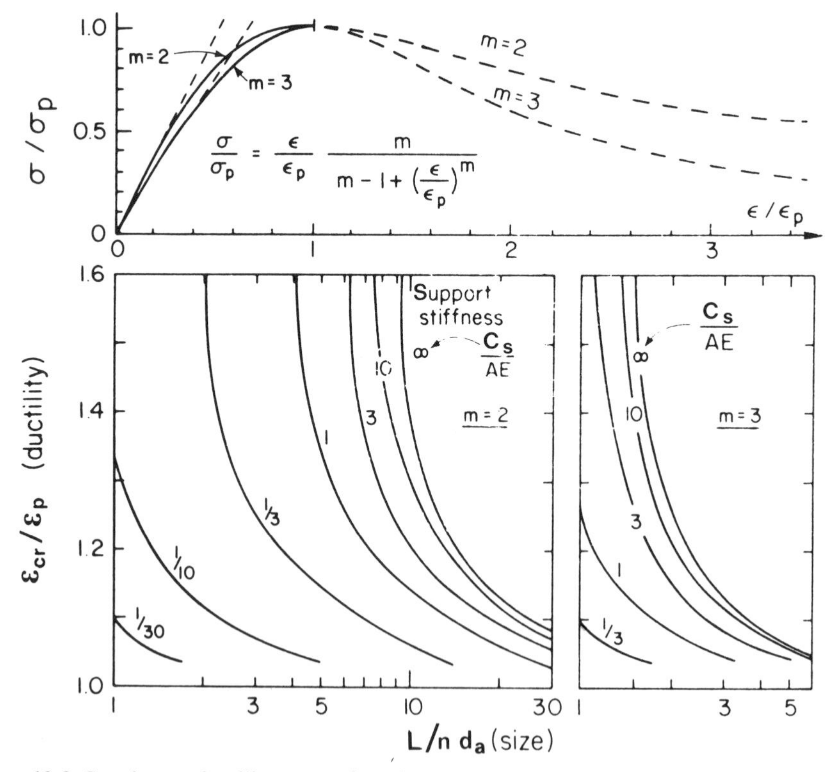

Figure 13.9 Specimen ductility as a function of relative specimen length and relative stiffness of testing frame.

factor taken as $n = 3$ (see the crack band model, Sec. 13.10). The ductility is defined as the ratio $\varepsilon_{cr}/\varepsilon_p$ where ε_p = strain at peak stress ($\sigma = f'_t$). The diagrams were calculated assuming the stress–strain diagram to be given by Popovics' formula stated in the figure, with two values of empirical coefficient m ($m = 2$ and 3, which applies to lower- and higher-strength concretes).

Inadmissibility of Arbitrarily Small Size of Localization Region

Now consider the effect of h, the size (length) of the localization (softening) region. For the concave part of the postpeak $\sigma(\varepsilon)$ diagram, Equation 13.2.5 shows that the smaller the value of h, the smaller is ε_{cr}. Especially,

$$\lim_{h \to 0} E_t^{cr} = 0 \qquad \lim_{h \to 0} \varepsilon_{cr} = \varepsilon_p \tag{13.2.6}$$

where ε_p = strain at peak stress for which $E_t = 0$. It must be emphasized that the diagrams in Figure 13.9 (bottom) show the limits of stable states of uniform strain, corresponding to snapback instability with neutral equilibrium. As we will see, localization develops in a stable manner before these limits are reached, and the stability loss (snapback) occurs at a smaller average strain than these diagrams indicate. But the solution shown in Figure 13.9 has the advantage of simplicity. Since instability occurs when $\delta^2 \mathscr{W} < 0$ becomes possible at any h, it follows that, for the first instability,

$$h \to 0 \tag{13.2.7}$$

To examine the energy consequence, assume further that the $\sigma(\varepsilon)$ diagram before the peak is straight, that is, elastic. In that case, the unloading retraces the elastic loading, and so the energy dissipation in the unloading segment $L - h$ is zero. In the loading (softening) segment the energy dissipation per unit length of the bar is nonzero and finite, but because $h \to 0$, the total energy dissipation during strain localization in the specimen is zero. Since strain localization is the mechanism of failure, we have the same unrealistic property as obtained in the preceding section on wave propagation in strain-softening materials.

From the fact that softening does exist in experiments and that energy dissipation during failures due to strain localization must be finite, Bažant (1976) concludes that a realistic macroscopic continuum model must be endowed with some property that limits localization to a region of a certain minimum (finite) size. The simplest way to do that is to require that the size h of the softening region cannot be less than a certain characteristic length l that is a material property. A model based on this condition, called the crack band model (Bažant and Oh, 1983) was shown to yield results that are in good agreement with experiments. A more general and fundamental way to prevent localization of softening into a region of vanishing size is to replace the ordinary (local) continuum model with a nonlocal model (see Sec. 13.10).

Bifurcation and Stable Path

The bifurcations we analyzed so far represent instabilities and occur at neutral equilibrium. However, bifurcations can also occur without neutral equilibrium at increasing displacement and no loss of stability. After the peak-stress state, two different strain increments $\delta\varepsilon = \delta\sigma/E_t$ and $\delta\varepsilon = \delta\sigma/E_u$ are possible for the same stress decrement $\delta\sigma$ in the specimen. Therefore, each state after the peak-stress state is a bifurcation state; there is a continuous sequence of infinitely many bifurcation states (Fig. 13.10c, d; see also Theorems 10.4.2 and 10.4.3). If h is fixed, then just two paths emanate from each bifurcation point. However, if the subdivision of specimen length into softening and unloading segments h and $L-h$ is arbitrary, a fan of infinitely many possible paths emanates from each bifuraction point (Fig. 13.10g).

To determine the stable path, we need to compare the second-order work $\delta^2\mathscr{W}$ done along the equilibrium path. This work is calculated easily as $\delta^2\mathscr{W} = \frac{1}{2}\delta P \delta q_1 = \frac{1}{2}C\delta q_1^2$ where δP = equilibrium reaction change due to imposed δq_1, and $C = (C_a^{-1} + C_b^{-1})^{-1}$ = overall structural stiffness. According to Section 10.2, the path that occurs must minimize $\delta^2\mathscr{W}$, and so it must minimize C. For path 1, which preserves uniform strain, we have $C = [C_s^{-1} + (h/E_tA) + (L-h)/E_tA]^{-1}$. For path 2, which involves localization, we have $C = [C_s^{-1} + (h/E_tA) + (L-h)/E_uA]^{-1}$. Path 2 must occur if the latter expression for

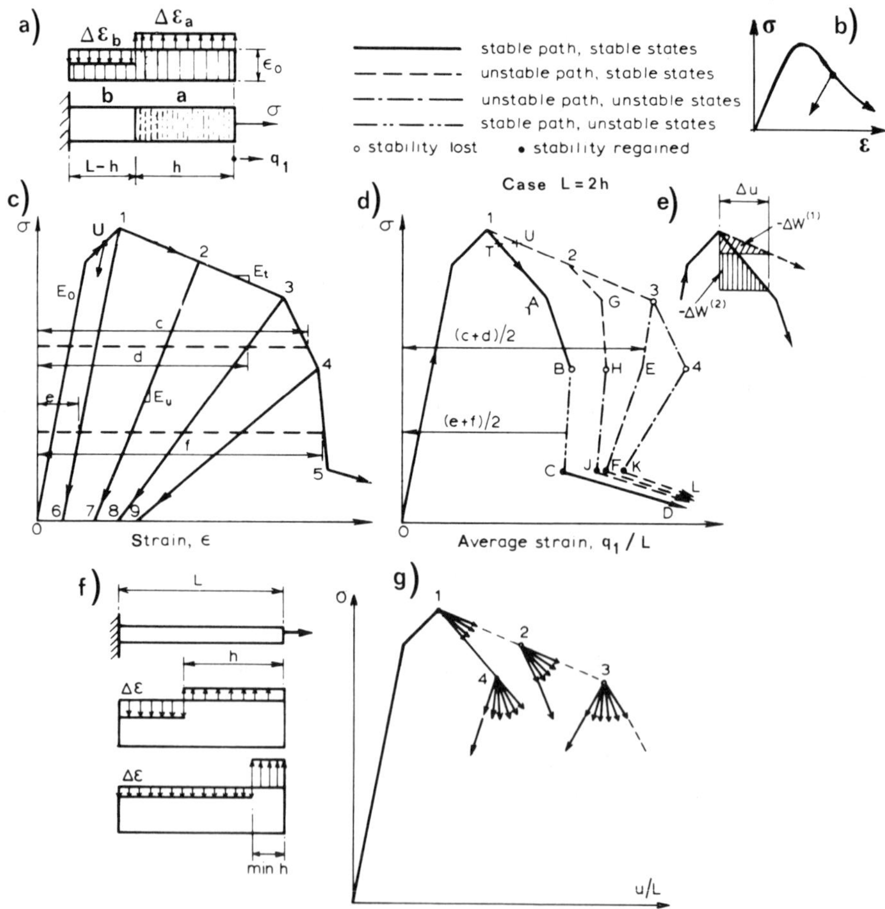

Figure 13.10 Bifurcation of equilibrium path in strain-softening uniaxial test specimen. (*After Bažant, 1988c.*)

C is less than the former. But this is always true because $E_u > 0$ and $E_t < 0$. Hence, the strain must localize as soon as a localization path exists, which is at the peak-stress point. Furthermore, if we compare the values of C for various h, we see that the smaller the value of h, the smaller is C. Therefore, the strain localization segment will be the shortest possible. In view of our previous argument below Equation 13.2.7, it is now again necessary to introduce into the material model some material property that limits localization (e.g. the characteristic length of nonlocal continuum, see Sec. 13.10).

The foregoing conclusions can be instructively derived from the diagrams in Figure 13.10. We assume a polygonal $\sigma(\varepsilon)$ diagram, as shown in Figure 13.10c, with unloading defined by lines $\overline{16}$, $\overline{27}$, $\overline{38}$, and $\overline{49}$. For the sake of illustration suppose that $h = L/2$, and that there is no spring (that is, $C_s \to \infty$). By equilibrium, stress σ in the softening and unloading segments must be the same. The response diagrams of stress versus the average strain $\bar{\varepsilon} = q_1/L$, shown in Figure 13.10d, can be constructed graphically by considering various stress values, taking for each of them the corresponding abscissa c for the softening segment and abscissa d for the unloading segment, and finally plotting the averages $(c + d)/2$ as the abscissae in the $\sigma(\bar{\varepsilon})$ diagram. (In the general case, this would be weighted averages, with the weights determined by the values of h and C_s, and varying h would produce a fan of slopes from points 1, 2, 3, 4.) Assuming the localization to begin at various states 1, 2, 3, one obtains various paths shown as $\overline{1ABCD}$, $\overline{2GHJ}$, $\overline{3EF}$, $\overline{4KL}$, etc., starting at bifurcation points 1, 2, 3, 4.

For structures with a single controlled displacement, the limit of stable states is the snapback point. Therefore, the stability limit for uniform strain is point 4 in Figure 13.10d. For the various localized paths, the limits of stable states are points B, H, and E in Figure 13.10d, and not the bifurcation points 1, 2, 3. At the bifurcation points the equilibrium (under displacement control) is stable (note the similarity with Shanley's tangent modulus load, see Sec. 10.3).

The stable postbifurcation path of a softening structure with a single controlled displacement is that of the steepest descent possible (Sec. 10.2). This follows from the values of the second-order work for the same δq_1, which are represented by the areas of the cross-hatched triangles in Figure 13.10e (they are both negative, and so the area that is larger in magnitude indicates the correct path). Therefore, at every bifurcation point such as 1, 2, or 3 (Fig. 13.10d), the steeper path shown is followed. In a continuous loading process, the localization must begin at the peak-stress state (point 1), and the specimen must follow the path $\overline{01ABCD}$, which leads to snapdown failure at point D.

Can the stable softening states of uniform strain, such as point 2 or 3, ever be reached? They can, but only by some other process, for example, if E_t were made negative by heating or irradiation at constant q_1, or if the specimen were temporarily forced to deform uniformly by gluing it to a parallel stiff plate, extending it jointly with the plate, and the glue were dissolved as the ends are kept fixed. The specimen would then remain stable in a state such as point 2, and upon further extension would start to localize along path $\overline{2G}$.

Now we must realize that the ductility diagrams in Figure 13.9, based on snapback instability of uniform strain, give only upper bounds on the true ductility. Snapback takes place at a smaller $\varepsilon_{cr}/\varepsilon_p$ because the strain distribution must have already localized (in a stable manner) before. However, this solution of ductility, which is not as simple as that for uniform strain, is still not perfect since it neglects the three-dimensional behavior during failure.

Some further interesting viewpoints of the uniaxial localization instability were presented in Ottosen (1986) and a subsequent discussion by Borrè and Maier (1988).

Alternative: Imperfection Approach to Bifurcation

The fact that softening must localize right at the peak-stress state can also be proven, independently of thermodynamics, by considering imperfections. Suppose that, for example, the strength (peak-stress σ_p) in segment h is slightly smaller than the strength in segment $L-h$. Consequently, when segment h reaches the peak stress, segment $L-h$ has not quite reached it yet (point U in Fig. 13.10c). After that both stresses decrease. Segment h can undergo strain softening along line $\overline{12}$ (Fig. 13.10c), but segment $L-h$ cannot pass over the peak-stress point into the softening range, and so it can only unload.

Further suppose that the value of strength σ_p is a random variable of Weibull distribution, which has a lower bound. Then the probability that the minimum value of σ_p within the specimen could be reached in more than one cross section is zero. This means that only one cross section can reach strain softening, that is, $h \to 0$ and $L-h \to L$. But this implies that a specimen that is elastic up to the peak stress would fail with zero energy dissipation, which is physically impossible. The remedy is to introduce some spatial averaging, that is, adopt a nonlocal formulation (Sec. 13.10).

Identification of Softening Stress-Strain Relations from Tests

The strain localization complicates experimental determination of strain-softening material properties from tests. Previous evaluations of test data from small tensile or compression specimens have generally assumed the strain to be uniform even after the peak stress. However, this can be assumed only for small enough specimens whose size does not exceed the minimum possible length l of the localization segment (which was introduced below Equation 13.2.7 and is roughly equal to the characteristic length l, Section 13.10).

To simplify the material identification and avoid complicated finite element analysis of test specimens, one may assume a series-coupling model, in which a loading region of volume fraction f undergoes uniform strain softening and an unloading region of volume fraction $1-f$ unloads. The observed mean strain is $\bar{\varepsilon} = f\varepsilon + (1-f)\varepsilon_u$ where ε = true postpeak strain in the loading (softening) region and ε_u = strain in the unloading region, which follows the unloading branch from the peak-stress point. From this (Bažant, 1989a):

$$\varepsilon = \frac{1}{f}\bar{\varepsilon} + \left(\frac{1}{f} - 1\right)\varepsilon_u \tag{13.2.8}$$

For tensile specimens, one may assume that $f = l/L$. Obviously, ε can be determined only if l is known (see Prob. 13.2.8). For specimens that exhibit postpeak softening in compression, shear, or other complex modes, the value of l might depend on the size and shape of the cross section.

Equation 13.2.8 may be demonstrated using recent uniaxial compression data for concrete reported by van Mier (1984, 1986). His measured diagrams of average stress σ versus average strains $\bar{\varepsilon}$ for specimens of the same concrete, same cross section, and lengths $L = 5$, 10, and 20 cm are plotted as the data points in Figure 13.11. From the present uniaxial series-coupling model we have $\bar{\varepsilon} = f\varepsilon + (1 - f)\varepsilon_u$ where $f = l/L$ and ε_u = unloading strain assumed as shown in the figure. Assuming $l = 5$ cm, we use the curve for $L = 5$ cm as the material $\sigma(\varepsilon)$ curve, and the curves for $L = 10$ cm and 20 cm calculated on this basis are plotted in the figure (after Bažant, 1989a). The agreement with the data is seen to be very close. This confirms that the series-coupling model is an acceptable approximation, despite the fact that the compression failure mechanism is three-dimensional.

The figure also shows the calculated curve for $L = 40$ cm; it exhibits a snapback, and so a test of a specimen of this length could not have succeeded.

Some related but more sophisticated and highly interesting ideas for identification of softening constitutive relations have been advanced and developed for shear bands in three dimensions by Ortiz (1989).

Relation of Strain Softening to Fracture Energy

Strain softening in brittle heterogeneous materials such as concrete, rocks, or ceramics is the macroscopic consequence of distributed microcracking. The combined energy dissipated by the microcracks over length h of the softening zone per unit cross-section area is the fracture energy G_f. Consider a finite drop of stress from the peak value $\sigma = f'_t$ (= tensile strength) all the way to zero. Setting $A = 1$, the work done on the softening segment of length h may be expressed as $\Delta W = -\frac{1}{2}f'^2_t h/\bar{E}_t$ or as $\Delta W = G_f - \Delta U$ where $\Delta U = \frac{1}{2}f'^2_t h/\bar{E}_u$ = energy released from the material between the microcracks in the softening region, in which $\bar{E}_t$ and $\bar{E}_u$ now represent the average softening and unloading moduli for the stress drop from the peak point all the way down to zero (Fig. 13.12). Equating these two expressions for ΔW, we get

$$G_f = \frac{f'^2_t h}{2}\left(\frac{1}{\bar{E}_u} - \frac{1}{\bar{E}_t}\right) = h\left(\int_{\varepsilon_1}^{\varepsilon_p} \sigma\, d\varepsilon + \int_{\varepsilon_p}^{\infty} \sigma\, d\varepsilon\right) \tag{13.2.9}$$

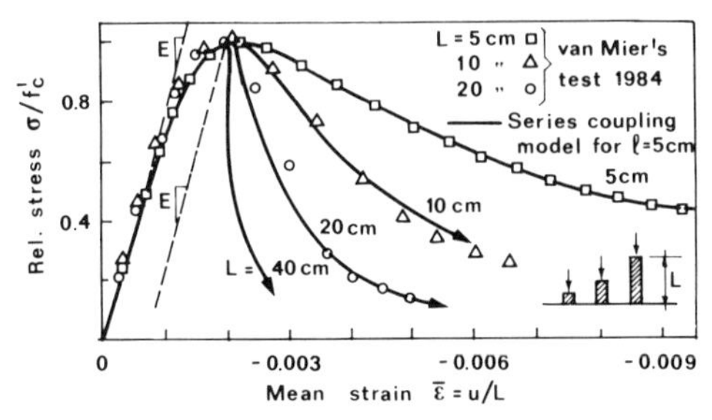

Figure 13.11 Bažant's (1989a) fit of uniaxial compression stress–strain diagrams measured by van Mier (1984, 1986).

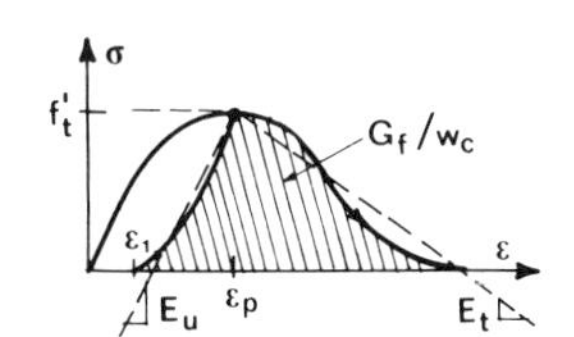

Figure 13.12 Release of strain energy density due to localization. (*After Bažant, 1982a.*)

where ε_1 is the strain at full unloading. So we conclude that the fracture energy is equal to the area under the curves of loading and unloading emanating from the peak-stress point; see Figure 13.12 (Bažant, 1982a).

Summary

The series coupling of a softening element with an elastic element becomes unstable when the load-deflection diagram reaches a snapback point. This occurs at a certain finite negative tangential stiffness of the softening element. In a uniaxially stressed specimen of a strain-softening material, the strain localizes into a segment whose length must be determined by material properties. In a continuous loading process the localization starts at the peak-stress point, but snapback instability may occur later, at a certain finite value of negative tangent modulus. The localization must be taken into account in identification of material properties from test data. The series-coupling model provides a simple description of the effects of size (or length) and of support stiffness on the localization instability.

Problems

13.2.1 Assume the $\sigma(\varepsilon)$ diagram to be a parabola, $\sigma = E\varepsilon(1 - \varepsilon/a)$ where $a =$ constant. Calculate the limit of stability of uniform strain assuming that (a) $h = L/4$, $C_s \to \infty$, (b) $h = L/4$, $C_s = EA/L$, (c) $h = L/8$, $C_s \to \infty$.

13.2.2 Do the same, but calculate the postpeak curve of σ versus the average strain $\bar{\varepsilon}$ assuming localization to begin at the peak stress. Then find the failure state as the snapback point of this curve.

13.2.3 Some experimentalists, for example, Evans and Marathé (1968), managed to measure tensile softening response of concrete specimens in a soft testing machine. The trick was to stabilize the specimen by coupling to its ends very stiff parallel steel bars of stiffness C_p, as shown in Figure 13.13a where C_m represents the stiffness of the testing machine. Segment $L - h$ together with springs C_m and C_p represents part b; calculate its stiffness and then analyze the effect of C_p on stability. [Bažant and Panula (1978), showed that this model agrees with test results.]

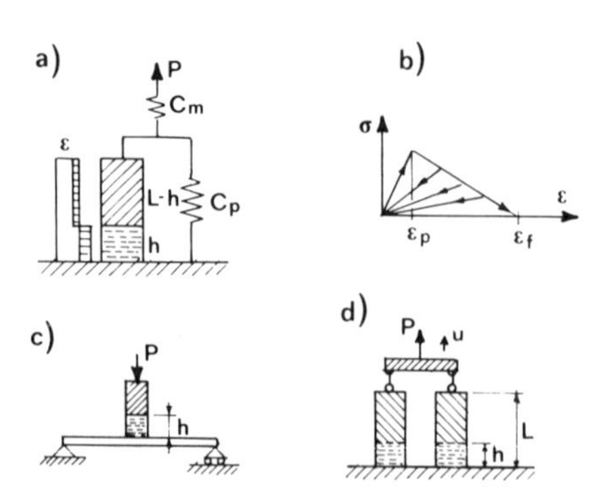

Figure 13.13

13.2.4 Damage causes a decrease of unloading modulus. Consider the triangular $\sigma(\varepsilon)$ diagram in Figure 13.13b with $\varepsilon_f = 5\varepsilon_p$, and assume the unloading diagrams to coincide with the secants. Calculate the limit of stability of uniform strain for the same cases as in Problem 13.2.1.

13.2.5 The series-coupling model can describe the effect of structure size on ductility, but not on its maximum nominal stress at failure (cf. size effect law in Sec. 12.2). However, a system of localizing softening bars coupled in parallel, with a statistical distribution of their strength values, can describe a kind of effect, as shown by Bažant and Panula (1978). Illustrate it in a simple manner, considering only two parallel bars having the same properties as in Problem 13.2.4 but one of them with a strength 25 percent larger than the other. Consider that $h = L/4$ for each bar.

13.2.6 Analyze stability and bifurcation for a localizing softening specimen that rests on a simply supported beam of span l and bending stiffness EI (Fig. 13.13c).

13.2.7 Analyze stability and bifurcation in the system of two identical softening specimens loaded as shown in Figure 13.13d. First consider the balance beam to be rigid. (A similar problem is solved in Sec. 13.7.) Then consider the balance beam to be flexible.

13.2.8 Assuming the series-coupling model for material test specimens, consider the compatibility conditions $l\varepsilon(\sigma) + (L_i - l)\varepsilon_u(\sigma) = L_i\bar{\varepsilon}_i(\sigma)$ where L_i = various lengths of specimens of the same cross section and $\bar{\varepsilon}_i(\sigma)$ = corresponding mean strains measured. It might seem that by writing these conditions for two different lengths L_i, one could solve from these equations both $X = l$ and $Y = l\varepsilon(\sigma)$ and thus identify the value of l from size effect data. Show that, unfortunately, this equation system has no solution because it has a zero determinant. (This proves that l cannot be identified solely from test data on the effect of specimen length; different l-values give equally good fits.)

13.3 LOCALIZATION OF SOFTENING DAMAGE INTO PLANAR BANDS

A uniform state of strain softening in an infinite space or in a layer can localize, in a stable or unstable manner, into an infinite planar band. This localization problem, which is of considerable interest to the mechanics of earthquakes as well as shear banding in metals, is one-dimensional and similar to the uniaxial softening that we just analyzed, except that the triaxiality of stress needs to be taken into account. These localizations are rather sensitive to the form of the nonlinear triaxial constitutive relation.

Localization into planar bands is the simplest mode of instability or bifurcation in a uniformly stressed space. Some of the basic principles ensue from Hadamard's work (1903), which was extended to inelastic behavior by Thomas (1961), Hill (1962a), and Mandel (1966). An in-depth discussion was given in a review by Rice (1976). Thorough studies of bifurcation and neutral equilibrium states were made by Rudnicki and Rice (1975) and Rice (1976). These studies were focused primarily on localization caused by the geometrically nonlinear effects of finite strain before the peak of the stress–strain diagram (i.e., in the plastic-hardening range), although critical states for negative values of the plastic-hardening modulus were also identified. Although these studies represented an important advance, they did not actually address the stability conditions but were confined to neutral equilibrium conditions for the critical state. They did not consider a general incremental stress–strain relation but were limited to von Mises plasticity (Rudnicki and Rice, 1975) or Drucker–Prager plasticity (Rudnicki, 1977) (in some cases enhanced with a vertex-hardening term). They also did not consider bodies of finite dimensions, for which the size of the localization region usually has a major influence on the critical state and, in the case of planar localization bands, did not consider the effect of unloading outside the localization band, which is important for finite bodies. A more general analysis which takes these conditions into account was presented in Bažant (1988a). The present section, based on this article, explains exact analytical solutions for some multidimensional localization problems with softening, using the method expounded in the previous section. The presentation emphasizes the work inequalities that describe the stability condition and determine the postbifurcation path, while the previous studies have dealt merely with the equations for neutral equilibrium or bifurcation.

Stability Condition for the Softening Band within a Layer or Infinite Solid

Let us analyze the stability of a uniform state of softening against localization in an infinite layer, which is called the softening band (or localization band) and forms inside an infinite layer of thickness L $(L \geq h)$; see Figure 13.14. The minimum possible thickness h of the band is assumed to be a material property, proportional to the characteristic length l. The layer is initially in equilibrium under a uniform (homogeneous) state of strain ε_{ij}^0 and stress σ_{ij}^0 assumed to be in the strain-softening range. Lowercase subscripts refer to Cartesian coordinates x_i $(i = 1, 2, 3)$ of material points in the initial state.

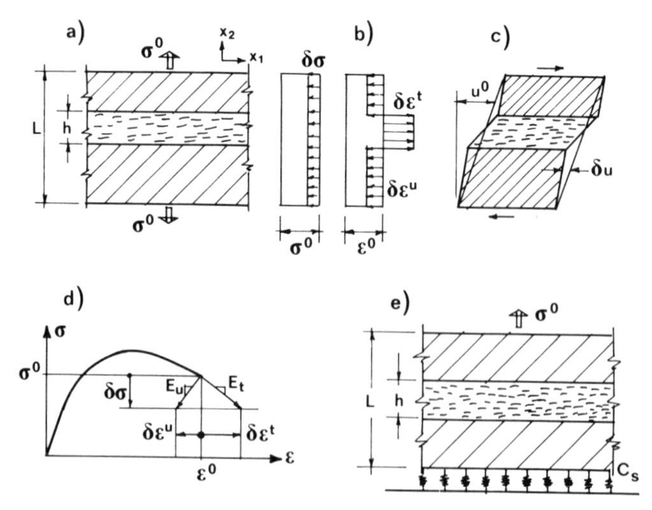

Figure 13.14 (a, b) Planar localization band in a layer; (c) pure shear; (d) stress–strain diagram with softening and unloading; (e) layer supported on an elastic foundation.

The initial equilibrium is considered to be disturbed by small incremental displacements δu_i whose gradients $\delta u_{i,j}$ are uniform both inside and outside the band. The values of $\delta u_{i,j}$ inside and outside the band are denoted as $\delta u^t_{i,j}$ and $\delta u^u_{i,j}$ and are assumed to represent further loading and unloading, respectively. So the increments $\delta u_{i,j}$ represent strain localization. We choose the x_2 axis to be normal to the layer (Fig. 13.14). As the boundary conditions, we assume that the surface points of the layer are fixed during the incremental deformation, that is, $\delta u_i = 0$ at the surfaces $x_2 = 0$ and $x_2 = L$ of the layer. If the strains inside and outside the band are homogeneous, compatibility of displacements at the band surfaces requires that

$$\delta u^u_{i,1} = \delta u^t_{i,1} = 0 \qquad \delta u^u_{i,3} = \delta u^t_{i,3} = 0 \tag{13.3.1}$$

$$h\,\delta u^t_{i,2} + (L-h)\,\delta u^u_{i,2} = 0 \qquad (i = 1, 2, 3) \tag{13.3.2}$$

Assume the incremental material properties to be characterized by incremental moduli tensors D^t_{ijpq} and D^u_{ijpq} for loading and unloading. These two tensors may either be prescribed by the given constitutive law directly as functions of ε^0_{ij} and possibly other state variables, or they may be implied indirectly. The latter case occurs, for example, for continuum damage mechanics. The crucial fact is that D^u_{ijpq} differs from D^t_{ijpq} and is always positive definite, even if there is strain softening.

Noting that $\varepsilon_{ij} = (u_{i,j} + u_{j,i})/2$ and assuming symmetry $D_{ijkm} = D_{jimk}$, we may write the incremental stress–strain relations as follows:

$$\delta\sigma^t_{ji} = D^t_{jikm}\,\delta\varepsilon^t_{km} = D^t_{jikm}\,\delta u^t_{k,m} \qquad \text{for loading} \tag{13.3.3}$$

$$\delta\sigma^u_{ji} = D^u_{jikm}\,\delta\varepsilon^u_{km} = D^u_{jikm}\,\delta u^u_{k,m} \qquad \text{for unloading} \tag{13.3.4}$$

Stability of the initial equilibrium state (Sec. 10.1) may be decided on the basis of the second-order work $\delta^2\mathscr{W}$ that must be done on the layer per unit area in the x_1x_3 plane in order to produce the increments δu_i. This work represents the Helmholtz free energy under isothermal conditions and the total energy under isentropic conditions. Using Equations 13.3.3, 13.3.4, and 13.3.1, as well as the relation $\delta u^u_{i,2} = -\delta u^t_{i,2}h/(L-h)$, which follows from Equation 13.3.2, we get

$$\begin{aligned}\delta^2\mathscr{W} &= \frac{h}{2}\,\delta\sigma^t_{2i}\,\delta u^t_{i,2} + \frac{L-h}{2}\,\delta\sigma^u_{2i}\,\delta u^u_{i,2} = \frac{h}{2}(\delta\sigma^t_{2i} - \delta\sigma^u_{2i})\,\delta u^t_{i,2}\\ &= \frac{h}{2}(D^t_{2ij2}\,\delta u^t_{j,2} - D^u_{2ij2}\,\delta u^u_{j,2})\,\delta u^t_{i,2}\end{aligned} \tag{13.3.5}$$

$$\delta^2\mathscr{W} = \frac{h}{2}\left(D^t_{2ij2} + \frac{h}{L-h}D^u_{2ij2}\right)\delta u^t_{j,2}\,\delta u^t_{i,2} \tag{13.3.6}$$

We may now denote the 3×3 matrix:

$$Z_{ij} = D^t_{2ij2} + \frac{h}{L-h}D^u_{2ij2} \tag{13.3.7}$$

$$Z = [Z_{ij}] = \begin{bmatrix} D^t_{2112} & D^t_{2122} & D^t_{2132}\\ D^t_{2212} & D^t_{2222} & D^t_{2232}\\ D^t_{2312} & D^t_{2322} & D^t_{2332}\end{bmatrix} + \frac{h}{L-h}\begin{bmatrix} D^u_{2112} & D^u_{2122} & D^u_{2132}\\ D^u_{2212} & D^u_{2222} & D^u_{2232}\\ D^u_{2312} & D^u_{2322} & D^u_{2332}\end{bmatrix} \tag{13.3.7a}$$

This matrix is symmetric ($Z_{ij} = Z_{ji}$) if and only if $D^t_{2ij2} = D^t_{2ji2}$ and $D^u_{2ij2} = D^u_{2ji2}$. This is assured if D^t_{kijm} as well as D^u_{kijm} has a symmetric matrix. Our derivation, however, is valid in general for nonsymmetric Z_{ij} or D^t_{kijm}.

The necessary stability condition (Bažant, 1988a) may be stated according to Equation 13.3.6 as follows:

$$\delta^2 \mathscr{W} = \frac{h}{2}\, \delta u^t_{i,2} Z_{ij}\, \delta u^t_{j,2} > 0 \qquad \text{for any } \delta u^t_{i,2} \tag{13.3.8}$$

Positive definiteness of $\delta^2 \mathscr{W}$ (or the 3×3 matrix Z_{ij}) is a necessary condition of stability. (We cannot claim this to be sufficient for stability since we have not analyzed all the possible localization modes; however, changing $>$ to $<$ yields a sufficient condition for instability.)

Positive definiteness of the 6×6 matrix $\mathbf{Z}^* = \mathbf{D}_t + \mathbf{D}_u h/(L-h)$ (where $\mathbf{D}_t$ and $\mathbf{D}_u$ are the 6×6 matrices of incremental moduli for loading and unloading) of course implies stability. However, the body can be stable even if $\mathbf{Z}^*$ is not positive definite.

Discussion of Various Cases

If the softening band is infinitely thin ($h/L \to 0$), or if the layer is infinitely thick ($L/h \to \infty$), we have $Z_{ij} = D^t_{2ij2}$, and so matrix $\mathbf{Z}$ loses positive definiteness when the (3×3) matrix of D^t_{2ij2} ceases to be positive definite. This condition, whose special case for von Mises plasticity was obtained by Rudnicki and Rice (1975) and Rice (1976), indicates that instability may occur right at the peak of the stress–strain diagram. However, for a continuum approximation of a heterogeneous material, for which h must be finite, the loss of stability can occur only after the strain undergoes a finite increment beyond the peak of the stress–strain diagram. For $L \to h$, the softening band (under displacement control) is always stable.

The strain-localization instability in a uniaxially stressed bar, which we analyzed in the previous section, may be obtained from the present three-dimensional solution as the special case for which the softening material is incrementally orthotropic, with $D^t_{2222} = E_t$ (<0) and $D^u_{2222} = E_u$ (>0) as the only nonzero incremental moduli. Equations 13.3.7 and 13.3.8 then yield the stability condition in Equation 13.2.5 for $C_s = 0$ (Bažant, 1976). This simple condition clearly illustrates that for finite L/h the localization instability can occur only at a finite slope $|E_t|$, that is, some finite distance beyond the peak of the stress–strain diagram. If the end of the bar at $x = L$ is not fixed but has an elastic support with spring constant C_s, one may obtain the solution by imagining the bar length to be augmented to length L', the additional length being chosen so that it has the same stiffness as the spring; that is, $L' - L = C_s/E_u$. Therefore, the stability condition is $-E_t/E_u < h(L' - h)$, which yields the same condition as given in Equation 13.2.5.

The stability condition for a layer whose surface points are supported by an elastic foundation (Fig. 13.14e) may be treated similarly, that is, by adding to the layer of thickness L another layer of thickness $L' - L$ such that its stiffness is equivalent to the given foundation modulus.

The case when the outer surfaces of the layer are kept at constant load $p_i^0 = \sigma_{ij}^0 n_j$ during localization is equivalent to adding a layer of infinite thickness $L' - L$. Therefore, the necessary stability condition is that the 3×3 matrix of D_{2ij2}^t be positive definite, that is, no softening can occur.

Another simple type of localization may be caused by softening in pure shear, Figure 13.14c. In this case, $\delta u_{1,2} \neq 0$, $\delta u_{2,2} = \delta u_{3,2} = 0$, and $D_{2112}^t = G_t$, $D_{2112}^u = G_u$ are the shear moduli. According to Equations 13.3.7 and 13.3.8, the necessary stability condition is

$$-\frac{G_t}{G_u} < \frac{h}{L-h} \tag{13.3.9}$$

When the normal to the layer and the localization band has an arbitrary orientation with respect to the material coordinates defined by direction cosines n_i, then the components σ_{2i} and $u_{i,2}$ must be replaced by $\sigma_{ij}n_j$ and $u_{i,j}n_j$. This may be done in every step of the preceding derivation, and the conclusion is that the state of uniform strain is stable against localization into a band of any orientation n_i if the 3×3 matrix

$$Z_{ij}(\mathbf{n}) = n_k\left(D_{kijm}^t + \frac{h}{L-h} D_{kijm}^u\right)n_m \tag{13.3.10}$$

is positive definite. If it is indefinite the state is unstable, and if it is positive semidefinite (the critical state), the state can be stable or unstable depending on higher variations. Matrix $\mathbf{Z}$, of components Z_{ij}, is called the localization matrix. For the special case of a localization band in an infinite space, Equation 13.3.10 becomes

$$Z_{ij}(\mathbf{n}) = n_k D_{kijm}^t n_m \qquad (L \to \infty) \tag{13.3.11}$$

This special localization matrix can be derived in a variety of ways, for example, directly from the conditions of uniqueness or bifurcation, or from the condition that the wave speed should be real (e.g., Ortiz, 1987).

While in a layer an infinitely long softening band must be parallel to the layer surface, for an infinite solid the softening band can have any orientation. Since $L \to \infty$, we have $Z_{ij} = D_{2ij2}^t$, and if the band is normal to the x_2 axis, stability requires that $\delta^2 \mathscr{W} = \delta u_{i,2}^t D_{2ij2}^t \, \delta u_{j,2}^t h/2 =$ positive definite. To generalize this condition to a band of arbitrary orientation, we may carry out an arbitrary rotation transformation of coordinates from x_i to x_i'. The transformation relations are $x_i = c_{ij}x_j'$ where c_{ij} are the direction cosines of the old coordinate base vectors in the new coordinates. According to the rules of transformation of tensors, we now have $2\delta^2\mathscr{W} = h(c_{ik}c_{2m}\,\delta u'_{k,m})(c_{2p}c_{iq}c_{jr}c_{2s}D_{pqrs}^{t'})(c_{ju}c_{2v}u'_{u,v})$ where the primes refer to the new coordinates x_j'. Noting that $c_{ik}c_{iq} = \delta_{kq}$, $c_{jr}c_{ju} = \delta_{ru}$, and that $D_{ijkm}^{t'}$ must be symmetric, we obtain

$$2\delta^2\mathscr{W} = h\,\delta a_{pq} D_{pqsr}^{t'}\,\delta a_{sr} \qquad \text{with } \delta a_{pq} = \tfrac{1}{2}(c_{2p}\,\delta u_{q,m}\,c_{2m} + c_{2p}\,\delta u_{m,q}\,c_{2m}) \tag{13.3.12}$$

In the space of six strain components, δa_{pq} is a 6×1 column matrix. For arbitrary rotations, δa_{pq} can have any values. Thus the 6×6 matrix of moduli $D_{pqsr}^{t'}$, and also D_{ijkm}^t, must be positive definite in order to ensure that the strain cannot localize into an infinite planar band of any orientation.

The same requirement was stated by Hadamard (1903), who derived it from the condition that the wave speed would not become imaginary (cf. Eq. 13.1.37). Hadamard's analysis, however, implied that $\mathbf{D}^u = \mathbf{D}^t$.

Finally, consider another case of boundary conditions: The case when the plane surfaces of an infinite layer slide freely over rigid bodies during the localization. The boundary conditions at $x_2 = 0$ and $x_2 = L$ now are $\delta\sigma_{21} = \delta\sigma_{23} = 0$. Similar to Equation 13.3.6, we now have

$$\delta^2 \mathcal{W} = \frac{h}{2}\left(D^t_{2222} + \frac{h}{L-h} D^u_{22222}\right) \delta u^t_{2,2}\, \delta u^t_{2,2} + \tfrac{1}{2}(\delta N_{11}\, \delta u_{1,1} + \delta N_{33}\, \delta u_{3,3} + 2\delta N_{13}\, \delta u_{1,3}) \quad (13.3.13)$$

in which δN_{11}, δN_{33}, and δN_{13} are homogeneously distributed in-plane incremental normal and shear force resultants over the whole thickness of the layer. Overall equilibrium now requires that $\delta N_{11} = \delta N_{33} = \delta N_{13} = 0$. Hence, the necessary condition of stability against localization is

$$D^t_{2222} + \frac{h}{L-h} D^u_{2222} > 0 \quad (13.3.14)$$

Numerical Examples

The limit of stable states according to Equation 13.3.7 was analyzed in Bažant and Lin (1989) for the case of associated and nonassociated Drucker–Prager elastoplastic materials, for which the loading function and the plastic potential function are

$$f(\boldsymbol{\sigma}, \kappa) = \bar{\tau} + \Psi(\sigma^v) - \kappa = 0 \qquad g(\boldsymbol{\sigma}, \kappa) = \bar{\tau} + \Phi(\sigma^v) - \kappa = 0 \quad (13.3.15)$$

in which $\bar{\tau} = \sqrt{J_2} = (\frac{1}{2}s_{ij}s_{ij})^{1/2}$ = stress intensity, $\sigma^v = \sigma_{kk}/3$ = volumetric stress, κ = plastic-hardening parameter, and Φ, Ψ = empirical monotonically increasing functions. As usual, we introduce the notations

$$H = \frac{\partial \kappa}{\partial \lambda} \qquad \beta = \frac{\partial \Phi}{\partial \sigma^v} \qquad \beta' = \frac{\partial \Psi}{\partial \sigma^v} \quad (13.3.16)$$

where λ is the proportionality parameter in the flow rule, $d\varepsilon^p_{ij} = (\partial g/\partial \sigma_{ij})\, d\lambda$; H = plastic modulus (which is positive for strain hardening and negative for strain softening), β' = internal friction, and β = dialatancy ratio of the material. Using the flow rule and eliminating $d\lambda$ (as shown in general in Sec. 10.6), one can show that, for loading,

$$D^t_{ijkm} = D^e_{ijkm} - \frac{\left(\frac{G}{\bar{\tau}} s_{ij} + K\beta\delta_{ij}\right)\left(\frac{G}{\bar{\tau}} s_{km} + K\beta'\delta_{km}\right)}{G + K\beta\beta' + H} \quad (13.3.17)$$

in which D^e_{ijkm} = isotropic tensor of elastic moduli; $D^e_{ijkm} = (K - \frac{2}{3}G)\delta_{ij}\delta_{km} + 2G\delta_{ik}\delta_{jm}$ and G, K = shear and bulk elastic moduli. Note that for nonassociated plasticity ($\beta' \neq \beta$), D^t_{ijkm} is nonsymmetric, and for associated plasticity ($\beta' = \beta$), it is symmetric (with respect to interchanging ij with km).

The condition of stability limit here depends on four nondimensional material parameters: ν, H/G, β, and β'. In addition, it depends on the relative size L/h and on the ratios $s_{ij}/\bar{\tau}$ characterizing the initial stress state. The influence of these parameters was studied numerically by Bažant and Lin (1989). The eigenvalues of the localization matrix **Z** (Eq. 13.3.7) were calculated with a computer for various combinations of the parameter values. Using Newton's method, the material parameter combinations for which the smallest eigenvalue vanishes were found. Some of the results are plotted in Figure 13.15 for initial states representing uniaxial tension and pure shear. The states below each curve are stable, and those above it unstable.

As expected, these curves show that the thicker the layer, the lower the magnitude of the plastic tangential modulus H at which the layer becomes unstable. However, the asymptotic values of the curves for $L \to \infty$ show a surprise. For an infinitely thick layer (infinite space), the limit of stability against localization does not occur at $H \to 0$, but at some finite magnitude of H. By contrast, the stability limit of uniaxial localization (Sec. 13.2) at $L \to 0$ occurs at $H = 0$. This is also true for the present triaxial solution when the incremental moduli are assumed to be isotropic (as will be shown in Fig. 13.21, curves 2 and 3).

When D^t_{ijkm} is nonsymmetric, so is matrix Z_{ij} (Eq. 13.3.10). In that case, the states for which Z_{ij} is singular ($\det Z_{ij} = 0$) represent only states of path bifurcation (or loss of uniqueness, cf. Sec. 10.4) but not states at the limit of stability (i.e., the limit of positive definiteness of Z_{ij}). Since, for any matrix Z_{ij}, $\sum_i \sum_j Z_{ij} x_i x_j = \sum_i \sum_j \hat{Z}_{ij} x_i x_j$ where x_i = arbitrary column matrix and $\hat{Z}_{ij} = (Z_{ij} + Z_{ji})/2$ = symmetric part of matrix Z_{ij}, the limit of stable states (i.e., of positive definiteness of Z_{ij}) is characterized by the singularity of matrix $\hat{Z}_{ij}$ (which is not the same as the singularity of Z_{ij}; see Prob. 4.1.5 and Sec. 10.4). So, in the case of a nonsymmetric matrix, one needs to distinguish between (1) the critical states of path bifurcation, for which $\det(Z_{ij}) = 0$ (i.e., one eigenvalue of Z_{ij} is zero), and (2) the critical states of stability limit, for which $\det(\hat{Z}_{ij}) = 0$ (i.e., the smallest eigenvalue of $\hat{Z}_{ij}$ is zero). The second condition is more stringent (cf. Bromwhich bounds, Equation 10.4.5; for an illustration, see Prob. 4.1.5). The curves shown in Figure 13.15 (taken from Bažant and Lin, 1989) indicate the states of path bifurcation (which represents neutral equilibrium if the corresponding eigenvector does not represent unloading). However, the differences between these curves and those for the stability limit are small for the present problem (and graphically distinguishable only for $\beta' < 0.5$).

Generalization for Geometrically Nonlinear Effects

Strain localization in an infinite planar band can be solved easily even when the geometrically nonlinear effects due to finite strains and finite rotations are taken into account. The deviations δu_i and $\delta \varepsilon_{ij}$ from the initial state with homogeneous stresses σ^0_{ij} again are considered to be infinitely small, and the energy expression $\delta^2 \mathscr{W}$ that governs stability is second-order small. The contribution to $\delta^2 W$ that arises from the geometrically nonlinear finite-strain expression, is also second-order small, and so it may not be neglected. Then the material stress increment $\delta\sigma_{ij}$ in the work expression needs to be replaced by the mixed (first) Piola–Kirchhoff stress increment $\delta\tau_{ij}$, which is referred to the initial state and is nonsymmetric (Sec. 11.2). As is well known, $\delta\tau_{ij} = \delta s_{ij} - \sigma^0_{ik}\, \delta u_{j,k} + \sigma^0_{ij}\, \delta u_{k,k}$ where δs_{ij} = material increment of the true (Cauchy) stress. Since neither $\delta\sigma_{ij}$ nor $\delta\tau_{ij}$ is invariant at coordinate rotations, one must use in the incremental

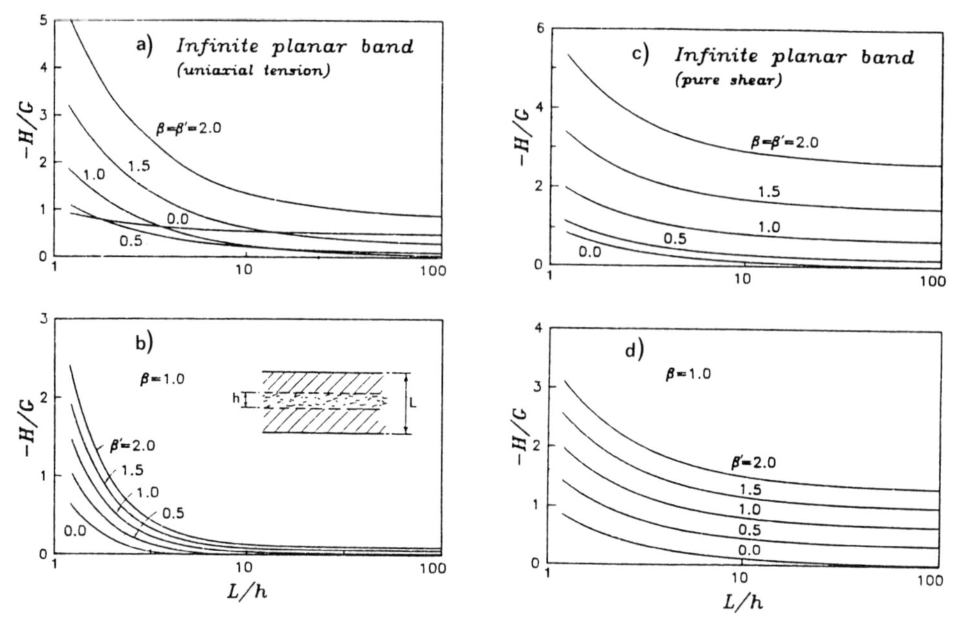

Figure 13.15 Plastic modulus H at the limit of stability against localization into planar bands in (a, b) pure tension and (c, d) pure shear, as a function of ratio L/h of thicknesses of layer and planar band. (*After Bažant and Lin, 1989.*)

stress–strain relation the objective stress increment $\delta\sigma_{ij}$ (representing the objective stress rate times the increment of time); $\delta\sigma_{ij}$ is symmetric. The relationship between $\delta\sigma_{ij}$ and $\delta\tau_{ij}$ (Sec. 11.2) may be written in the general form

$$\delta\tau_{ij} = \delta\sigma_{ij} + R_{ijkmrs}\sigma^0_{rs}\,\delta u_{k,m} = H_{ijkm}\,\delta u_{k,m} \tag{13.3.18}$$

with

$$H_{ijkm} = D_{ijkm} + R_{ijkmrs}\sigma^0_{rs} \tag{13.3.19}$$

in which we substituted $\delta\sigma_{ij} = D_{ijkm}\,\delta u_{k,m}$. Coefficients R_{ijpqrs} are certain constants that take into account the geometric nonlinearity of finite strain. The values of R_{ijkmpq} are different for various possible choices of the objective stress rate and the associated type of the finite-strain tensor, as shown in Chapter 11. The expressions that are admissible according to the requirements of tensorial invariance and objectivity are (Bažant, 1971)

$$R_{ijpqrs}\sigma^0_{rs} = \delta_{ip}\sigma^0_{qj} - \alpha(\delta_{ip}\sigma^0_{qj} + \delta_{iq}\sigma^0_{pj} + \delta_{jp}\sigma^0_{qi} + \delta_{jq}\sigma^0_{pi}) \tag{13.3.20}$$

where α can be an arbitrary constant (Chap. 11). For each different α value, different values of incremental moduli D_{ijpq} must be used so as to obtain physically equivalent results; generally (cf. Bažant, 1971, or Eq. 11.4.5)

$$D_{ijpq} = [D_{ijpq}]_{\alpha=0} + \alpha(\delta_{ip}\sigma^0_{qj} + \delta_{iq}\sigma^0_{pj} + \delta_{jp}\sigma^0_{qi} + \delta_{jq}\sigma^0_{pi}) \tag{13.3.21}$$

From Equation 13.3.5, in which $\delta\sigma_{ij}$ must now be replaced by $\delta\tau_{ij}$, the necessary stability condition is (Bažant, 1988a)

$$\begin{aligned}2\delta^2\mathcal{W} &= h\,\delta\tau^t_{2i}\,\delta u^t_{i,2} + (L-h)\,\delta\tau^u_{2i}\,\delta u^u_{i,2} \\ &= h(\delta\tau^t_{2i} - \delta\tau^u_{2i})u^t_{i,2} = h(H^t_{2ij2}\,\delta u^t_{j,2} - H^u_{2ij2}\,\delta u^u_{j,2})\,\delta u^t_{i,2} \\ &= h\,\delta u^t_{i,2} Z_{ij}\,\delta u^t_{j,2} > 0 \qquad \text{for any } \delta u^t_{i,2}\end{aligned} \tag{13.3.22}$$

in which

$$Z_{ij} = H^t_{2ij2} + \frac{h}{L-h} H^u_{2ij2} = D^t_{2ij2} + \frac{h}{L-h} D^u_{2ij2} + \frac{L}{L-h} R_{2ij2rs}\sigma^0_{rs} \tag{13.3.23}$$

In particular, if we use the Lagrangian (Green's) finite strain that is associated with Truesdell's objective stress rate ($\alpha = 0$), we have

$$Z_{ij} = \left[D^t_{2ij2} + \frac{h}{L-h} D^u_{2ij2} \right]_{\alpha=0} + \frac{L}{L-h}\,\sigma^0_{i2}\delta_{2j} \tag{13.3.24}$$

The condition $\det(Z_{ij}) = 0$ obviously represents the critical state of path bifurcation with strain localization (tantamount to stability loss if Z_{ij} is symmetric). Rudnicki and Rice (1975) derived a special case of this condition for which (1) the constitutive law consists of von Mises plasticity, possibly enhanced by vertex hardening; (2) $L/h \to \infty$ (infinite space); (3) $\alpha = \frac{1}{2}$ (Jaumann's rate); and (4) the initial stress σ^0_{ij} is pure shear σ^0_{12}. Their analysis was concerned only with the critical state of neutral equilibrium or bifurcation, rather than with stability. The values of the unloading moduli D^u_{ijkm} and the fact that they are different from D^t_{ijkm} and positive definite were irrelevant for their analysis.

Rudnicki and Rice (1975) showed that, due to geometric nonlinearity, the critical state of strain localization can develop in plastic materials while the matrix of D^t_{ijkm} is still positive definite, that is, before the final yield plateau of the stress-strain diagram is reached. This may explain the formation of shear bands in plastic (nonsoftening) materials. The destabilizing effect is then due exclusively to geometric nonlinearity. According to a geometrically linear analysis (small-strain theory), strain localization could develop in plastic (nonsoftening) materials only upon reaching the yield plateau but not earlier.

From Equation 13.3.23, it appears that the geometric nonlinearity can have a significant effect on strain localization only if the incremental moduli for loading are of the same order of magnitude as the initial stresses σ^0_{ij}; precisely, if $\max|D^t_{ijkm}|$ and $\max|\sigma^0_{ij}|$ are of the same order of magnitude. (The unloading moduli D^u_{ijkm} of structural materials are always several orders of magnitude larger.) Thus the importance of geometric nonlinearity depends on D^t_{ijkm}. For strain-softening types of localization, the geometric nonlinearity can be important only if the instability develops at a very small negative slope of the stress–strain diagram, which occurs very close to the peak-stress point. This can occur only if the layer thickness L is much larger than l. If $L - h \ll l$, instability occurs when the downward slope of the stress–strain diagram (Fig. 13.14d) is of the same order of magnitude as the initial elastic modulus E, and this is inevitably orders of magnitude larger than the initial stresses σ_{ij}^0. Therefore, we must conclude that the role of geometric nonlinearity in localization due to strain softening can be significant only when the localization occurs very near the peak point of the stress–strain diagram (or possibly another point of very small slope).

Bifurcation and Stable Path

For the special case when only one of the three stress components σ_{2i} at the surface is nonzero and does work, the conclusion from Section 10.2 for systems with a single load is applicable. It immediately follows (same as in Sec. 13.1) that localization must begin right at the peak-stress states.

When more than one of the components σ_{2i} at the surface works, the peak-stress state is not unambiguously defined. If there is bifurcation, the postbifurcation path is determined by comparing the work $\delta^2\mathscr{W}^{(2)}$ along the localization path with the work $\delta^2\mathscr{W}^{(1)}$ along the path preserving uniform strain (main path). Since, in contrast to stability analysis, we now need to consider only equilibrium states with uniform stress, $\sigma_{2i}^u = \sigma_{2i}^t = \sigma_{2i}$, it is more convenient to use the compliance tensors C_{ijkm}^t and C_{ijkm}^u whose 6×6 matrices are the inverses of those of D_{ijkm}^t and D_{ijkm}^u. We have $\delta u_{i,2}^t = C_{2ij2}\,\delta\sigma_{2j}$ and $\delta u_{i,2}^u = C_{2ij2}^u\,\delta\sigma_{2j}$, and so the relative displacements between the opposite surfaces of the layer are

$$\delta v_i = [hC_{2ij2}^t + (L-h)C_{2ij2}^u]\,\delta\sigma_{2j} \tag{13.3.25}$$

The second-order work done on the layer along the equilibrium path is $\delta^2\mathscr{W} = \frac{1}{2}\delta\sigma_{2i}\,\delta v_i$. Solving $\delta\sigma_{2i}$ from Equation 13.3.25, we have, along the path of localization (path 2),

$$\delta^2\mathscr{W}^{(2)} = \tfrac{1}{2}[h\mathbf{C}^t + (L-h)\mathbf{C}^u]_{ij}^{-1}\,\delta v_i\,\delta v_j \tag{13.3.26}$$

where $\mathbf{C}^t$ and $\mathbf{C}^u$ are the 3×3 matrices of C_{2ij2}^t and C_{2ij2}^u. For the path that preserves uniform strain (path 1), we obtain $\delta^2\mathscr{W}^{(1)}$ simply by replacing $\mathbf{C}^u$ with $\mathbf{C}^t$, which yields

$$\delta^2\mathscr{W}^{(1)} = \tfrac{1}{2}[L\mathbf{C}^t]_{ij}^{-1}\,\delta v_i\,\delta v_j \tag{13.3.27}$$

Taking the difference,

$$\delta^2\mathscr{W}^{(1)} - \delta^2\mathscr{W}^{(2)} = \tfrac{1}{2}\{[L\mathbf{C}^t]^{-1} - [h\mathbf{C}^t + (L-h)\mathbf{C}^u]^{-1}\}_{ij}\,\delta v_i\,\delta v_j \tag{13.3.28}$$

Since softening states are unstable under load control, we need to consider only the case of displacement control. The postbifurcation path that occurs is that for which $\delta^2\mathscr{W}$ is smaller. Therefore, the localization path must occur (that is, is the stable path) if the matrix of the quadratic form in Equation 13.2.28 is positive definite. Calculations indicate that if localization is possible, the localized path will occur. If the stability limit has not yet been reached, then the localization happens as a bifurcation at increasing load (i.e., similar to Shanley's bifurcation in plastic columns).

If the tangential moduli or compliances vary continuously, then the first bifurcation must happen when the stiffness matrix for loading everywhere becomes singular, as shown in general in Section 10.4. The layer with moduli D^t_{ijkm} applicable everywhere represents Hill's comparison solid. The condition of first bifurcation in a layer normal to x_2 is that the 3×3 matrix of D^t_{2ij2} becomes singular. For a layer of any orientation, the condition is that the localization matrix given by Equation 13.3.11 becomes singular. For an infinite space, it suffices if this happens for one direction vector n_i.

If moduli D^t_{ijkm} vary discontinuously, then the appearance of a negative eigenvalue of the 3×3 matrix of D^t_{2ij2} or of $Z_{ij}(\mathbf{n})$ in Equation 13.3.11 implies that a bifurcation state has just been passed and that localization must have started.

Although nonlocalized stable states beyond the first bifurcation state cannot be reached in a continuous loading process, they can be reached if the material properties change discontinuously, or if the loss of positive definiteness of D^t_{2ij2} is caused by heating of the material, etc.

Localization into Shear Bands Due to Nonassociatedness in Frictional Materials

Frictional plastic materials are usually considered as those in which the deviatoric yield limit depends on the hydrostatic pressures as described by the Drucker and Prager (1952) yield surface (Eqs. 13.3.15–13.3.17) (a more general concept of friction in materials was discussed in Sec. 10.7). Because of the physical nature of friction, the flow rule for such materials should violate normality, that is, be nonassociated. Examples of localization in a nonassociated frictional material are presented in Figure 13.15. In these examples, however, the instabilities are still caused primarily by strain softening, although they are affected by the lack of associatedness (violations of normality).

The nonassociatedness, however, can by itself cause instabilities due to localization into shear bands, even if there is no strain softening. These instabilities can be detected on the basis of the criteria already stated—see Equations 13.3.7, 13.3.10 and 13.3.11. A detailed study of localization due to nonassociatedness, based on the condition of positive definiteness of the localization matrix Z_{ij} in Equation 13.3.11, has been carried out by Leroy and Ortiz (1989) and Bažant and Lin (1989). Other pertinent studies were presented by de Borst (1988a); de Borst (1986); Vermeer and de Borst (1984); Raniecki and Bruhns (1981); and some numerical problems in finite element simulation were addressed by Ortiz, Leroy, and Needleman (1987) (in a local context).

Sand Liquefaction as a Localization Instability

Cyclic shear deformations of loose sand (of undercritical density, underconsolidated) due to earthquake or other dynamic loads may produce volume compaction. If the sand is saturated by water, as is often the case in foundations, the compaction causes the pore pressure p to rise from p_0 to $p_0 + \Delta p_c$. The value of Δp_c depends also on how fast the water can flow out of the densified zone (diffusion problem). The frictional forces between sand grains are proportional to the so-called effective stresses $\sigma'_{ij} = \sigma_{ij} + \delta_{ij} p$. Although the initial principal effective stresses $\sigma_{ij} + \delta_{ij} p_0$ are all negative (compressive), the change of p by Δp_c can cause the effective stress $\sigma_{ij} + \delta_{ij}(p_0 + \Delta p_c)$ to become nonnegative (tensile or zero). Thus the sand loses its capability to transmit shear stresses and flows as a liquid (until water flows out), which is manifested in the matrix D^t_{ijkm} (e.g., Bažant, and Krizek, 1975). The foundation may then collapse (e.g., Blazquez, Bažant and Krizek, 1980; Bažant, Ansal, and Krizek, 1982; Ansal, Krizek, and Bažant, 1982). This collapse may again be regarded as a strain-softening instability, in which the strain softening is caused by the loss of shear stiffness of sand. Usually it is assumed that the foundation collapses as soon as the maximum principal value σ'_{I} or σ'_{ij} becomes zero. In reality this need not yet represent collapse; collapse occurs in the form of strain-localization instability, of which the simplest to analyze is localization into a planar band, as just discussed (the liquefaction band must have a certain minimum thickness h, due to the nonlocal material property). The foregoing stability criteria can be used to decide liquefaction failure as a localization instability. (The analysis of the next section is also applicable.)

Summary

Localization of strain softening into planar bands in three dimensions is a problem similar to the uniaxial localization treated in the previous section. However, the triaxial material properties, particularly the tangent moduli tensors for loading with strain softening and unloading, affect the localization instability as well as the bifurcation of the equilibrium path. While in infinite space the localization instability and bifurcation depend only on the material properties, localization into a band inside a layer of finite thickness depends also on the ratio of the band thickness to the layer thickness. The band thickness must be nonzero and a material property.

Problems

13.3.1 Derive in detail the matrices $Z_{ij}(\mathbf{n})$ in Equations 13.3.10 and 13.3.11, introducing from the outset the components $n_j \delta\sigma_{ij}$ and $n_j \delta u_{j,i}$ instead of $\delta\sigma_{2i}$ and $\delta u_{2,i}$.

13.3.2 For infinite space, the bifurcation condition that $\det(D^t_{2ij2}) = 0$ can be derived most directly as follows. We assume $\mathbf{D}^t$ to be applicable everywhere and subtracting the stress-strain relations for the inside and outside of the band, labeled by superscripts (i) and (o), we have $\delta\sigma^{(i)}_{2i} - \delta\sigma^{(o)}_{2i} = D^t_{2ij2}(\delta u^{(i)}_{2,j} - \delta u^{(o)}_{2,j})$. Since the stresses must be continuous, we must have $\delta\sigma^{(i)}_{2i} = \delta\sigma^{(o)}_{2i}$, and since $\delta u^{(i)}_{2,j} \neq \delta u^{(o)}_{2,j}$ if there is localization, it follows that the 3×3 matrix of D^t_{2ij2} must be singular. Furthermore, since this is not an instability mode, it follows that the localization must happen during loading. Generalize this argument (a) to a band with any normal vector n_i (see, e.g., Ortiz, 1987), and (b) to a band within a layer. (This approach is simpler than that in the text but tells nothing about stability of state and path.)

13.4 LOCALIZATION OF SOFTENING DAMAGE INTO ELLIPSOIDAL REGIONS

The strain-localization solutions in the preceding section dealt with unidirectional localization of strain into an infinite planar band. If the body is finite, localization into such a band does not represent an exact solution for certain boundary conditions because, for example, the fixed boundary conditions cannot be satisfied at the location where the localization band intersects the boundary. In this section (which closely follows Bažant, 1988b), we will seek exact solutions for multidirectional localization due to strain softening in finite regions.

In particular, we will study localization into ellipsoidal regions for which analytical solutions can be found. Except for the special case of cylindrical and spherical localization regions, these solutions are available only for an infinite solid, and generally cannot satisfy the boundary conditions for a finite body. However, in contrast to the infinite localization band, they can at least satisfy the boundary conditions approximately, provided the body is sufficiently large compared to the size of the localization ellipsoid. This is due to the fact that the stresses, strains, and displacements in the analytical solution for the ellipsoidal localization region in an infinite solid decay rapidly with the distance from the ellipsoid, thus becoming negligible at a certain sufficient distance from the ellipsoid. If the boundary lies beyond that distance, the solution is nearly vanishing at the boundary and can, therefore, be used as an approximate solution for a finite body.

Eshelby's Theorem

Localization in ellipsoidal regions can be solved by application of Eshelby's (1957) theorem for ellipsoidal inclusions with uniform eigenstrain. Consider an ellipsoidal hole (Fig. 13.16a) in a homogeneous isotropic infinite medium that is elastic and is characterized by elastic moduli matrix $\mathbf{D}_u$. We imagine fitting and glueing into this hole an ellipsoidal plug of the same material (Fig. 13.16a), which must first be deformed by uniform strain $\boldsymbol{\varepsilon}^{\sigma}$ (the eigenstrain) in order to fit into the hole perfectly (note that a uniform strain always changes an ellipsoid into another ellipsoid). Then the strain in the plug is unfrozen, which causes the plug to deform with the surrounding medium to attain a new equilibrium state. The famous discovery of Eshelby (1957) was that if the plug is ellipsoidal and the elastic medium is homogeneous and infinite, the strain increment $\boldsymbol{\varepsilon}^e$ in the plug that occurs during this deformation is *uniform* and is expressed as

$$\varepsilon_{ij}^{e} = S_{ijkm}\varepsilon_{km}^{0} \tag{13.4.1}$$

where S_{ijkm} are components of a fourth-order tensor that depend only on the ratios a_1/a_3 and a_2/a_3 of the principal axes of the ellipsoid and for the special case

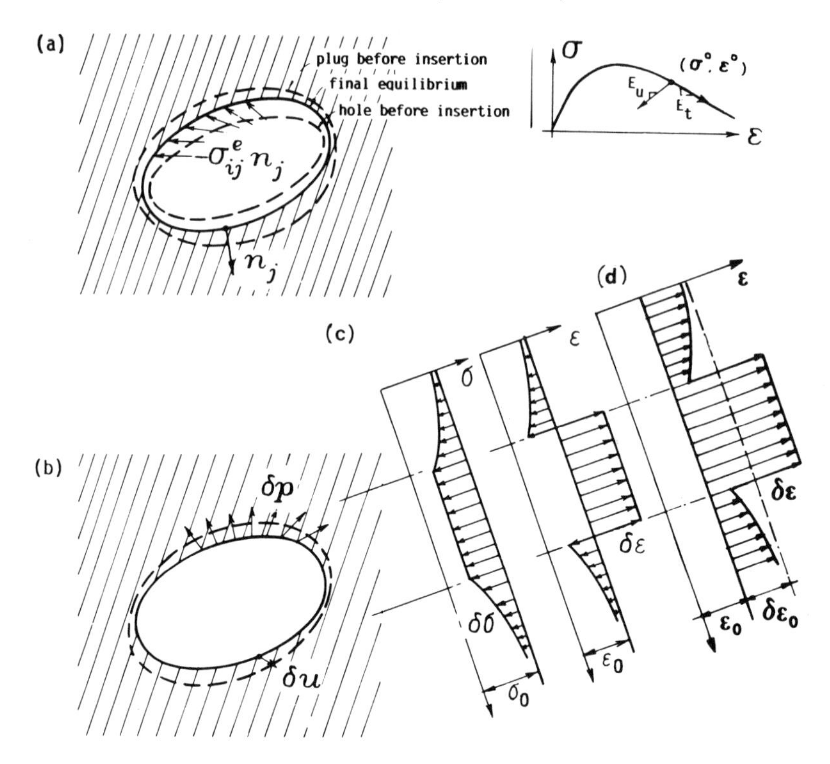

Figure 13.16 (a) Ellipsoidal plug (inclusion) inserted into infinite elastic solid; (b) interface tractions and displacements, (c) distribution of stress and strain increments during bifurcation at localization instability; (d) distribution of strain increments during stable bifurcation with strain localization.

of isotropic materials, on Poisson ratio ν_u; see, for example, Mura (1982) and Christensen (1979). Due to symmetry of ε_{ij}^0 and ε_{ij}^e, $S_{ijkm} = S_{jikm} = S_{ijmk}$, but, in general, $S_{ijkm} \neq S_{kmij}$. Coefficients S_{ijkm} are, in general, expressed by elliptic integrals; see also Mura (1982). Extension of Eshelby's theorem to generally anisotropic materials was later accomplished by Kinoshita and Mura (1971) and Lin and Mura (1973).

It will be convenient to rewrite Equation 13.4.1 in a matrix form

$$\boldsymbol{\varepsilon}^e = \mathbf{Q}_u \boldsymbol{\varepsilon}^0 \tag{13.4.2}$$

or

$$\begin{matrix} \varepsilon_{11}^e \\ \varepsilon_{22}^e \\ \varepsilon_{33}^e \\ 2\varepsilon_{12}^e \\ 2\varepsilon_{23}^e \\ 2\varepsilon_{31}^e \end{matrix} = \begin{matrix} S_{1111} & S_{1122} & S_{1133} & S_{1112} & S_{1123} & S_{1131} \\ S_{2211} & S_{2222} & S_{2233} & S_{2212} & S_{2223} & S_{2231} \\ S_{3311} & S_{3322} & S_{3333} & S_{3312} & S_{3323} & S_{3331} \\ 2S_{1211} & 2S_{1222} & 2S_{1233} & 2S_{1212} & 2S_{1223} & 2S_{1231} \\ 2S_{2311} & 2S_{2322} & 2S_{2333} & 2S_{2312} & 2S_{2323} & 2S_{2331} \\ 2S_{3111} & 2S_{3122} & 2S_{3133} & 2S_{3112} & 2S_{3123} & 2S_{3131} \end{matrix} \begin{matrix} \varepsilon_{11}^0 \\ \varepsilon_{22}^0 \\ \varepsilon_{33}^0 \\ 2\varepsilon_{12}^0 \\ 2\varepsilon_{23}^0 \\ 2\varepsilon_{31}^0 \end{matrix} \tag{13.4.3}$$

in which $\boldsymbol{\varepsilon} = (\varepsilon_{11}, \varepsilon_{22}, \varepsilon_{33}, 2\varepsilon_{12}, 2\varepsilon_{23}, 2\varepsilon_{31})^T$ and superscript T denotes the transpose of a matrix.

For isotropic materials, the only nonzero elements of matrix $\mathbf{Q}_u$ are those between the dashed lines marked in Equation 13.4.3, which is the same as for the stiffness matrix. The factors 2 in matrix $\mathbf{Q}_u$ in Equation 13.4.3 are due to the fact that the column matrix of strains is 6×1 rather than 9×1 and, therefore, must involve shear angles $2\varepsilon_{12}$, $2\varepsilon_{23}$, and $2\varepsilon_{31}$ rather than tensorial shear strain components ε_{12}, ε_{23}, and ε_{31} or else $\boldsymbol{\sigma}^T \delta\boldsymbol{\varepsilon}$ where $\boldsymbol{\sigma}^T = (\sigma_{11}, \sigma_{22}, \sigma_{33}, \sigma_{12}, \sigma_{23}, \sigma_{31})$ would not be a correct work expression. (The work expression $\sigma_{ij}\delta\varepsilon_{ij}$, as well as the sum implied in Equation 13.4.1 for each fixed i, j, has nine terms in the sum, not six.) For example, writing out the terms of Equation 13.4.1, we have

$$\varepsilon_{11}^e = \cdots + (S_{1112}\varepsilon_{12}^0 + S_{1121}\varepsilon_{21}^0) + \cdots = \cdots + S_{1112}(2\varepsilon_{12}^0) + \cdots \quad (13.4.4)$$

while the factors 2 arise as follows:

$$\begin{aligned} 2\varepsilon_{12}^e &= 2[\cdots + S_{1233}\varepsilon_{33}^0 + (S_{1112}\varepsilon_{12}^0 + S_{1121}\varepsilon_{21}^0) + (S_{1223}\varepsilon_{23}^0 + S_{1232}\varepsilon_{32}^0) + \cdots] \\ &= \cdots (2S_{1233})\varepsilon_{33}^0 + (2S_{1212})2\varepsilon_{12}^0 + (2S_{1223})2\varepsilon_{23}^0 + \cdots \end{aligned} \quad (13.4.5)$$

The stress in the ellipsoidal plug $\boldsymbol{\sigma}^e$ (which is uniform), may be expressed according to Hooke's law as

$$\boldsymbol{\sigma}^e = \mathbf{D}_u(\boldsymbol{\varepsilon}^e - \boldsymbol{\varepsilon}^0) \quad (13.4.6)$$

After substituting $\boldsymbol{\varepsilon}^0 = \mathbf{Q}_u^{-1}\boldsymbol{\varepsilon}^e$, according to Equation 13.4.2, we get $\boldsymbol{\sigma}^e = \mathbf{D}_u(\boldsymbol{\varepsilon}^e - \mathbf{Q}_u^{-1}\boldsymbol{\varepsilon}^e)$ or

$$\boldsymbol{\sigma}^e = \mathbf{D}_u(\mathbf{I} - \mathbf{Q}_u^{-1})\boldsymbol{\varepsilon}^e \quad (13.4.7)$$

where $\mathbf{I}$ is a unit 6×6 matrix. The surface tractions that the ellipsoidal plug exerts upon the surrounding infinite medium, Figure 13.16a, are $p_i^e = \sigma_{ij}^e n_j$, in which n_j denotes the components of a unit normal $\mathbf{n}$ of the ellipsoidal surface (pointed from the ellipsoid outward).

Stability of Uniform Strain against Ellipsoidal Localization

Consider now infinitesimal variations $\delta\mathbf{u}$, $\delta\boldsymbol{\varepsilon}$, $\delta\boldsymbol{\sigma}$ from the initial equilibrium state of uniform strain $\boldsymbol{\varepsilon}^0$ in an infinite homogeneous anisotropic solid (without any hole). The matrices of incremental moduli corresponding to $\boldsymbol{\varepsilon}^0$ are $\mathbf{D}_t$ for further loading and $\mathbf{D}_u$ for unloading, $\mathbf{D}_u$ being positive definite. We imagine that the initial equilibrium state is disturbed by applying surface tractions δp_i over the surface of the ellipsoid with axes a_1, a_2, a_3, Figure 13.16b. We expect δp_i to produce loading inside the ellipsoid and unloading outside. We try to calculate the displacements δu_i produced by tractions δp_i at all loading points on the ellipsoid surface.

Let $\delta\varepsilon_{ij}^e$, δu_i^e be the strain and displacement variations produced (by tractions δp_i) in the ellipsoid, and denote the net tractions acting on the softening ellipsoid as δp_i^t, and those on the rest of the infinite body, that is, on the exterior of the ellipsoid, as δp_i^e. As for the distributions of δp_i^t and δp_i^e over the ellipsoid surface, we assume them to be such that $\delta p_i^t = \delta\sigma_{ij}^t n_j$ and $\delta p_i^e = \delta\sigma_{ij}^e n_j$ where $\delta\sigma_{ij}^t$ and $\delta\sigma_{ij}^e$ are arbitrary constants; $\delta\sigma_{ij}^t$ is the stress within the softening ellipsoidal region, which is uniform (and represents an equilibrium field), and $\delta\sigma_{ij}^e$ is a fictitious uniform stress in this region that would equilibrate δp_i^e.

Equilibrium requires that $\delta p_i = \delta p_i^t - \delta p_i^e$. The first-order work $\delta \mathscr{W}$ done by σ_{ij}^0 must vanish if the initial state is an equilibrium state. The second-order work done by δp_i may be calculated as

$$\delta^2 \mathscr{W} = \int_S \frac{1}{2} \delta p_i \, \delta u_i^e \, dS = \frac{1}{2} \int_S \delta p_i^t \, \delta u_i^e \, dS - \frac{1}{2} \int_S \delta p_i^e \, \delta u_i^e \, dS$$
$$= \frac{1}{2} \delta \sigma_{ij}^t \int_S n_j \, \delta u_i^e \, dS - \frac{1}{2} \delta \sigma_{ij}^e \int_S n_j \, \delta u_i^e \, dS \tag{13.4.8}$$

where S = surface of the softening ellipsoidal region. Note that $\delta \sigma_{ij}^e$ are not the actual stresses in the solid but merely serve the purpose of characterizing the surface tractions δp_i^e. Applying Gauss' integral theorem and exploiting the symmetry of tensors $\delta \sigma_{ij}^t$ and $\delta \sigma_{ij}^e$, we further obtain

$$\delta^2 \mathscr{W} = \frac{1}{2} (\delta \sigma_{ij}^t - \delta \sigma_{ij}^e) \int_V \delta u_{i,j}^e \, dV = \int_V \frac{1}{2} (\delta \sigma_{ij}^t - \delta \sigma_{ij}^e) \frac{1}{2} (\delta u_{i,j}^e + \delta u_{j,i}^e) \, dV$$
$$= \int_V \frac{1}{2} \delta \boldsymbol{\epsilon}^{e^T} (\delta \boldsymbol{\sigma}^t - \delta \boldsymbol{\sigma}^e) \, dV \tag{13.4.9}$$

where V = volume of the softening ellipsoidal region, and subscripts preceded by a comma denote partial derivatives. We changed here to matrix notation and also recognized that $\frac{1}{2}(\delta u_{i,j}^e + \delta u_{j,i}^e) = \delta \varepsilon_{ij}^e$. Now we may substitute

$$\delta \boldsymbol{\sigma}^t = \mathbf{D}_t \, \delta \boldsymbol{\varepsilon}^e \tag{13.4.10}$$

According to Equation 13.4.7, we also have, as a key step,

$$\delta \boldsymbol{\sigma}^e = \mathbf{D}_u (\mathbf{I} - \mathbf{Q}_u^{-1}) \delta \boldsymbol{\varepsilon}^e \tag{13.4.11}$$

In contrast to our previous consideration of the elastic ellipsoidal plug made of the same elastic material (Eqs. 13.4.2–13.4.7), the sole meaning of $\delta \boldsymbol{\sigma}^e$ now is to characterize the tractions δp_i^e acting on the ellipsoidal surface of the infinite medium lying outside the ellipsoid. Noting that the integrand in Equation 13.4.9 is constant, we thus obtain

$$\delta^2 \mathscr{W} = \tfrac{1}{2} \delta \boldsymbol{\varepsilon}^{e^T} \mathbf{Z} \, \delta \boldsymbol{\varepsilon}^e V = \tfrac{1}{2} Z_{ijkm} \, \delta \varepsilon_{ij}^e \, \delta \varepsilon_{km}^e V \tag{13.4.12}$$

in which Z_{ijkm} are the tensor components corresponding to the following 6×6 matrix $\mathbf{Z}$:

$$\mathbf{Z} = \mathbf{D}_t + \mathbf{D}_u (\mathbf{Q}_u^{-1} - \mathbf{I}) \tag{13.4.13}$$

Equation 13.4.12 defines a quadratic form. If the initial uniform strain $\boldsymbol{\varepsilon}^0$ is such that the associated $\mathbf{D}_t$ and $\mathbf{D}_u$ give $\delta^2 \mathscr{W} > 0$ for all possible $\delta \varepsilon_{ij}^e$, then no localization in an ellipsoidal region can begin from the initial state of uniform strain $\boldsymbol{\varepsilon}^0$ spontaneously, that is, without applying loads δp. If, however, $\delta^2 \mathscr{W}$ is negative for some $\delta \varepsilon_{ij}^e$, the localization leads to a release of energy, which is first manifested as kinetic energy and is ultimately dissipated as heat. Such a localization obviously increases entropy of the system, and so it will occur, as required by the second law of thermodynamics. Therefore, the necessary condition of stability of a uniform strain field in an infinite solid is that matrix $\mathbf{Z}$ given by Equation 13.4.13 must be *positive definite*.

To check whether the result in Equation 13.4.13 is unique, we may note that instead of the equilibrium relation $\delta p_i^e = \delta\sigma_{ij}^e n_j$ with which we started one can most generally write $\delta p_i^e = [\delta\sigma_{ij}^e + f_{ij}(\mathbf{n})]n_j$ where $f_{ij}(\mathbf{n}) = (m_i k_j + m_j k_i)\phi(\mathbf{n})$; $\phi(\mathbf{n})$ = arbitrary scalar function of orientation $\mathbf{n}$; m_i, k_i = components of mutually orthogonal unit vectors, each being normal to $\mathbf{n}$, that is, $m_j n_j = 0$, $k_j n_j = 0$. Tracing the effect of functions $f_{ij}(\mathbf{n})$ through the entire derivation, one finds that the term $-\frac{1}{2}\int_V f_{ij}(\mathbf{n})\delta\varepsilon_{ij}^e\,dV$ needs to be added to the right-hand side of Equation 13.4.12. Considering now the limiting case of neutral equilibrium, $\delta^2\mathscr{W} = 0$, we conclude that $\frac{1}{2}[Z_{ijkm}\,\delta\varepsilon_{ij}^e - f_{ij}(\mathbf{n})]\,\delta\varepsilon_{km}^e = 0$ for any $\delta\varepsilon_{km}^e$, from which $f_{ij}(\mathbf{n}) = Z_{ijkm}\,\delta\varepsilon_{ij}^e$. This relation can be satisfied for arbitrary $\delta\varepsilon_{ij}^e$ only if $f_{ij}(\mathbf{n}) = 0$, as tacitly assumed at the outset.

The expressions for Eshelby's coefficients S_{ijkm} from which matrix $\mathbf{Q}_u$ is formed (see, e.g., Mura, 1982) depend on the ratios a_1/a_3, a_2/a_3 of the axes of the ellipsoidal localization region. They also depend on the ratios of the unloading moduli D_{ijkm}^u. If, for example, the unloading behavior is assumed to be isentropic, they depend on the unloading Poisson ratio ν_u. Matrix $\mathbf{D}_u$ is determined by ν_u and unloading Young's modulus E_u. If, just for the sake of illustration, the loading behavior is assumed to be also isotropic (which is, of course, not realistic for inelastic materials, see Chaps. 10 and 11), matrix $\mathbf{D}_t$ is determined by ν_t and E_t (Poisson's ratio and Young's modulus for loading); E_t, ν_t, E_u, ν_u, in turn, depend on the strain ε_{ij}^0 at the start of localization. Since a division of $\mathbf{Z}$ by E_u does not affect positive definiteness, only the ratio E_t/E_u matters. Thus, $\mathbf{Z}$ is a function of the form

$$\mathbf{Z} = E_u\tilde{\mathbf{Z}}\left(\frac{a_1}{a_3}, \frac{a_2}{a_3}, \nu_u, \nu_t, \frac{E_t}{E_u}\right) = E_u\bar{\mathbf{Z}}\left(\frac{a_1}{a_3}, \frac{a_2}{a_3}, \varepsilon_{ij}^0\right) \tag{13.4.14}$$

where $\tilde{\mathbf{Z}}$ and $\bar{\mathbf{Z}}$ are nondimensional matrix functions.

Note that matrix $\mathbf{Z}$, which decides the localization instability, is independent of the size of the ellipsoidal localization region. (This is the same conclusion as already made in Sec. 13.3 for a planar localization band in an infinite solid.) No doubt, the size of the localization ellipsoid would matter for finite-size solids, same as it does for localization bands in layers.

The previously obtained solution for a planar localization band in an infinite solid must be a special case of the present solution for an ellipsoid with $a_2/a_1 \to 0$ and $a_2/a_3 \to 0$ (Fig. 13.17). Localization in line cracks also must be obtained as a special case for $a_2 \to 0$; however, the present solution is not realistic for this case since energy dissipation due to strain softening is finite per unit volume and, therefore, vanishes for a crack (the volume of which is zero). For this case, it would be necessary to include the fracture energy (surface energy) in the energy criterion of stability, same as in fracture mechanics. In the present approach, though, we take the view that, due to material heterogeneity, it makes no sense to apply a continuum analysis to localization regions whose width is less than a certain length h proportional to the maximum size of material inhomogeneities.

Numerical Examples of Stability Limits and Discussion

Figure 13.17 shows some numerical results (from Bažant, 1988b) for localization of strain into ellipsoidal domains in infinite space. The results were calculated for

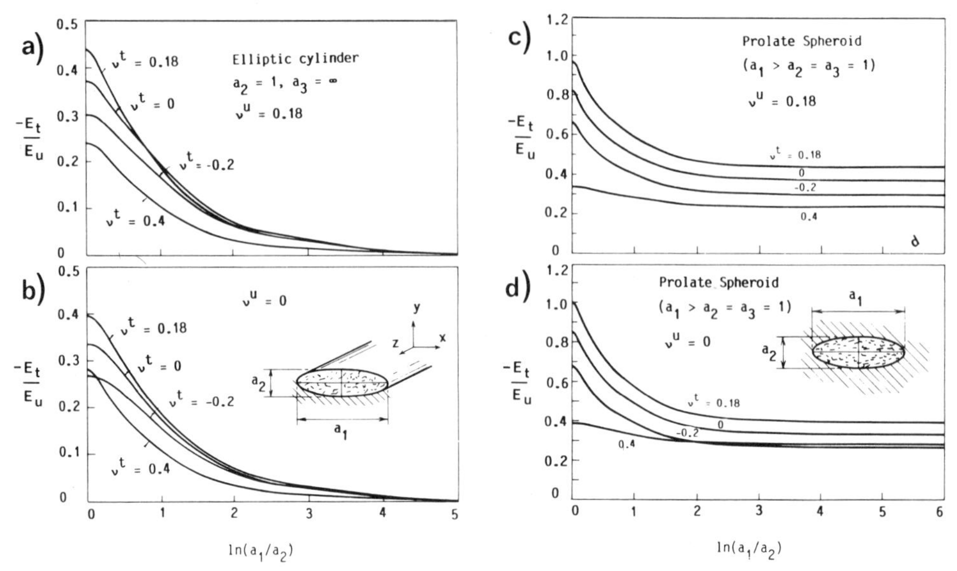

Figure 13.17 Tangential modulus E_t at the limit of stability against localization into ellipsoidal region as a function of ratio a_1/a_2 of principal axes of ellipsoid. (*After Bažant, 1988b.*)

domains in the shape of infinitely long elliptic cylinders ($a_3 \rightarrow \infty$ and various a_1/a_2) as well as prolate spheroids ($a_1 > a_2 = a_3$, various ratios a_1/a_2). The material was assumed to be incrementally isotropic, with matrices $\mathbf{D}_t$ and $\mathbf{D}_u$ characterized by Young's moduli E_t and E_u, and Poisson ratios $\nu_u = 0.18$ with various values of ν_t. (The assumption of incremental isotropy is here made for the sake of simplicity; in reality, the incremental moduli at strain softening must be expected to be anisotropic, except when the initial state is a purely volumetric strain; cf. Chap. 11.)

Matrix $\mathbf{Z}$, Equation 13.4.13, was evaluated by computer on the basis of S_{ijkm} taken from Mura (1982, Eqs. 11.22 and 11.29). The smallest eigenvalue of the symmetric part of matrix $\mathbf{Z}$ was calculated by a computer library subroutine. Iterative solution by the Newton method was used to find the value of E_t/E_u for which the smallest eigenvalue is zero and is about to become negative, which indicates loss of positive definiteness.

The results are plotted in Figure 13.17a–d. For the infinite cylinder (Fig. 13.17a, b), as a_1/a_2 increases, the localization instability occurs at smaller $|E_t/E_u|$. The case $a_1/a_2 \rightarrow \infty$ corresponds to an infinite planar band, and the results are identical to those given for this case in Section 13.3. In particular, $|E_t|$ tends to 0 as $a_1/a_2 \rightarrow \infty$; that is, instability occurs right at the peak of the stress–strain diagram.

For the prolate spheroid, Figure 13.17c, d, the instability also occurs at decreasing $|E_t/E_u|$ as a_1/a_2 increases, but for $|a_1| \rightarrow \infty$, which corresponds to an infinite circular tube, a finite value of $|E_t|$, depending on Poisson's ratio, is still required for instability. This limiting case is equivalent to two-dimensional localization in a circular region (Sec. 13.5). On the other hand, the case $a_1/a_2 = 1$, Figure 13.17c, d, is equivalent to localization into a spherical region (Sec. 13.5).

The results show that a localization instability in the form of a planar band always develops at a smaller $|E_t|$, and thus at a smaller initial strain, than the localization instability in the form of an ellipsoidal softening region. That does not mean, however, that the planar band would always occur in practice. A planar localization band cannot accommodate the boundary conditions of a finite solid restrained on its boundary, and a localization region similar to an ellipsoid may then be expected to form. It is remarkable how slowly the slope $|E_t|$ at instability decreases as a function of a_1/a_2. The value of the aspect ratio that is required to reduce $|E_t|$ at instability from about 0.4 to about 0.04 of E_u is $a_1/a_2 = e^3 \simeq 20$. This means that if a very long planar softening band cannot be accommodated within a given solid, the deformation required for softening instability is considerably increased.

Figure 13.18 shows some of the results obtained by Bažant and Lin (1989) for the Drucker–Prager elastoplastic material (Eq. 13.3.17). The limiting case $a_3/a_2 \to \infty$ is equivalent to localization in a planar band, and the results indeed coincide with those from Figure 13.15. It is again noteworthy that the magnitude of H at the stability limit is not zero for the limit case of a band ($a_3/a_2 \to \infty$) except when $\beta' = 0$.

As already pointed out below Equation 13.4.13, matrix $\mathbf{D}_t$, and thus also matrix $\mathbf{Z}$, can be nonsymmetric for certain materials with internal friction or damage. As explained, it is then necessary to distinguish between (1) the critical state of bifurcation, for which $\mathbf{Z}\delta\boldsymbol{\varepsilon}^e = \mathbf{0}$ or $\det \mathbf{Z} = 0$, that is, one eigenvalue of $\mathbf{Z}$ is zero (this represents neutral equilibrium if the associated eigenvector $\delta\boldsymbol{\varepsilon}^e$ does not correspond to unloading); and (2) the critical state of stability limit, at which $\delta^2\mathscr{W}$ ceases to be positive definite. The latter critical state depends only on the

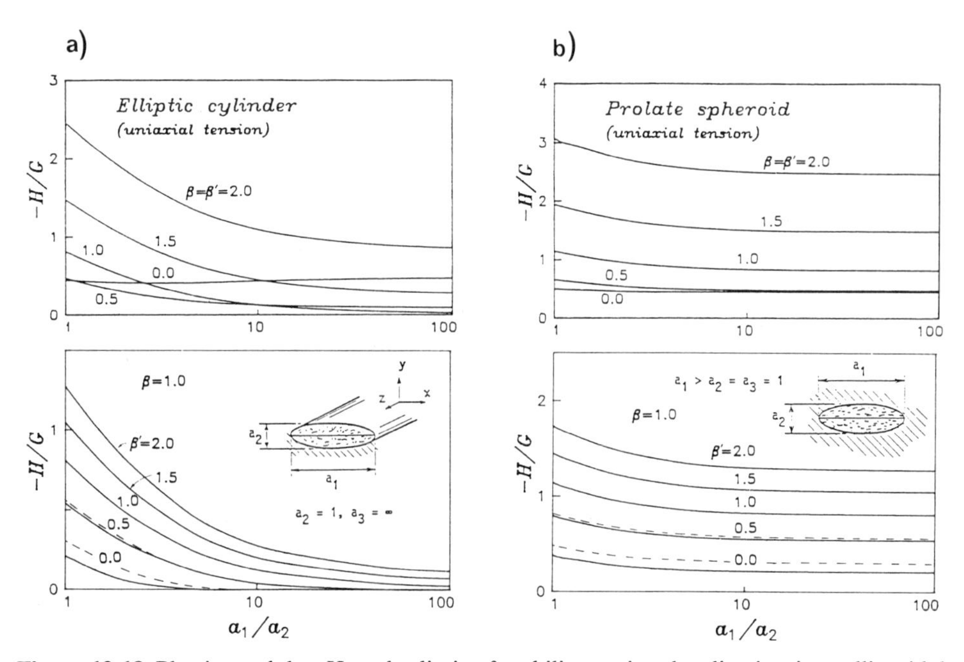

Figure 13.18 Plastic modulus H at the limit of stability against localization into ellipsoidal regions as a function of ratio a_1/a_2 of principal axes of ellipsoid. (*After Bažant and Lin, 1989.*)

symmetric part $\hat{\mathbf{Z}}$ of matrix $\mathbf{Z}$, that is, $\hat{\mathbf{Z}} = (\mathbf{Z} + \mathbf{Z}^T)/2$, and is characterized by $\det \hat{\mathbf{Z}} = 0$ (i.e., the smallest eigenvalue of $\hat{\mathbf{Z}}$ is zero). The latter condition is more stringent (cf. Bromwich bounds, Eq. 10.4.5 and Prob. 4.1.5); however, the numerical results for ellipsoidal localization (Bažant and Lin, 1989) indicate that the differences are usually very small except for some cases of very strong nonsymmetry of $\mathbf{D}_t$. Figures 13.18 and 13.19 show by the solid curves the critical

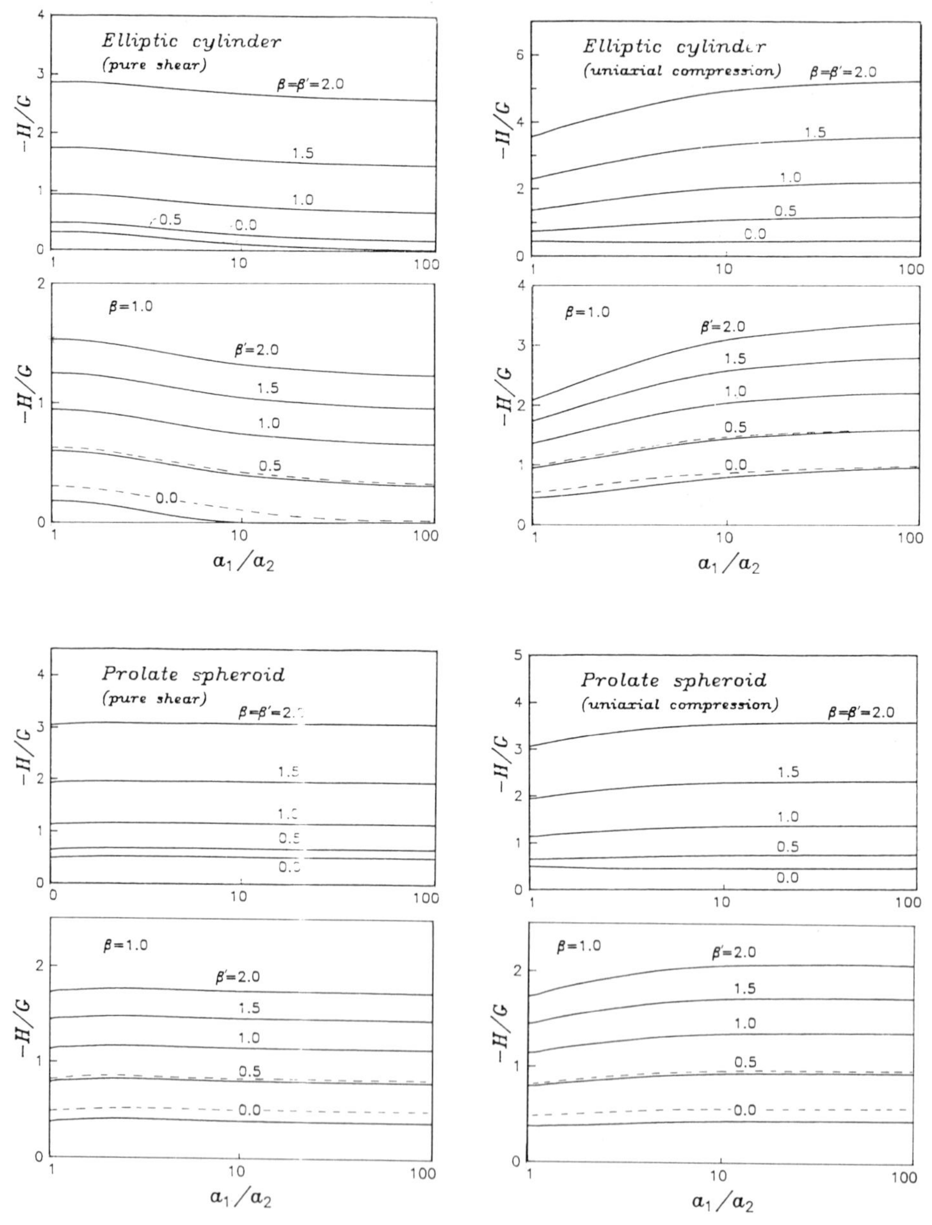

Figure 13.19 Plastic modulus H at the limit of stability against localization into ellipsoidal regions as a function of ratio a_1/a_2 of principal axes of ellipsoid. (*After Bažant and Lin, 1989.*)

states of stability limit ($\det \hat{\mathbf{Z}} = 0$) and by the dashed curves the critical states of bifurcation ($\det \mathbf{Z} = 0$).

Bifurcation and Stable Path of Ellipsoidal Localization

As we know in general from Sections 10.2 and 10.4, and illustrated for uniaxial localization in Section 13.2, the condition of stable equilibrium does not necessarily indicate which path will be followed by an inelastic structure after a bifurcation point. The states on the equilibrium branches emanating from a bifurcation point can all be stable, yet only one branch is stable, that is, can be followed by the structure. Such behavior is again exhibited by ellipsoidal localization.

The loss of stability due to strain localization is considered to occur while the remote displacements, strains, and stresses are constant (Fig. 13.16). Bifurcation of the equilibrium path, on the other hand, can occur while the remote displacements, strains, and stresses increase. In this manner, it can happen that the strains increase everywhere, but most in the ellipsoidal region so that the strain localizes in a stable manner simultaneously with the progress of loading (increasing ε at infinity). In the mode of instability (Fig. 13.16c, neutral equilibrium), matrix $\mathbf{D}_t$ of the loading moduli applies for the interior of the ellipse, and matrix $\mathbf{D}_u$ of the unloading moduli applies for the exterior, but in the stable bifurcation mode (in Fig. 13.16d), matrix $\mathbf{D}_t$ applies for both the interior and the exterior of the ellipse.

By examining the derivation of Equation 13.4.13, we readily find that the second-order work along the equilibrium path when loading takes place both inside and outside the ellipsoid is $\delta^2 \mathscr{W} = \frac{1}{2} \delta\boldsymbol{\varepsilon}^T \mathbf{Z}_t\, \delta\boldsymbol{\varepsilon}$ where $\mathbf{Z}_t$ is obtained from $\mathbf{Z}$ by replacing $\mathbf{D}_u$ with $\mathbf{D}_t$. Therefore, $\mathbf{Z}_t = \mathbf{D}_t(\mathbf{I} + \mathbf{Q}_t^{-1} - \mathbf{I})$ or

$$\mathbf{Z}_t = \mathbf{D}_t \mathbf{Q}_t^{-1} \tag{13.4.15}$$

Assuming $\mathbf{D}_t$ to vary during the loading process continuously, the first bifurcation of the equilibrium path is obtained when $\delta^2 \mathscr{W} = \frac{1}{2} \delta\boldsymbol{\varepsilon}^T \mathbf{Z}_t\, \delta\boldsymbol{\varepsilon} = 0$ for some nonzero vector $\delta\boldsymbol{\varepsilon}$ (Bažant, 1987b, 1988c). This case is obtained when the matrix equation

$$\mathbf{Z}_t\, \delta\boldsymbol{\varepsilon} = 0 \tag{13.4.16}$$

admits a nonzero solution $\delta\boldsymbol{\varepsilon}$. This means that matrix $\mathbf{Z}_t$ must be singular at the first bifurcation.

It is now useful to realize the basic physical meaning of matrix $\mathbf{Q}_t$. We consider an infinite homogeneous elastic body that has elastic moduli $\mathbf{D}_t$ and contains an ellipsoidal hole. An ellipsoidal plug of different shape is deformed by a uniform eigenstrain $\delta\boldsymbol{\varepsilon}^0$ so that it would fit exactly and could be glued into the hole. Then the eigenstrain is unfrozen and the infinite body with the plug finds its new equilibrium state. As already mentioned, Eshelby's discovery was that this state involves a uniform strain $\delta\boldsymbol{\varepsilon}$ within the plug, and that this strain is given by the equation $\delta\boldsymbol{\varepsilon} = \mathbf{Q}_t\, \delta\boldsymbol{\varepsilon}^0$. Conversely, the eigenstrain necessary to obtain $\delta\boldsymbol{\varepsilon}$ is $\delta\boldsymbol{\varepsilon}^0 = \mathbf{Q}_t^{-1}\, \delta\boldsymbol{\varepsilon}$. From this physical meaning it is now clear that if $\mathbf{D}_t$ is positive definite, then a finite $\delta\boldsymbol{\varepsilon}$ can be produced only by a finite $\delta\boldsymbol{\varepsilon}^0$, and if $\mathbf{D}_t$ is nearly singular (a state near the peak of the stress–strain diagram) then a finite $\delta\boldsymbol{\varepsilon}$

occurs even for a vanishingly small $\delta\boldsymbol{\varepsilon}^0$. Therefore, if $\mathbf{D}_t$ is positive definite, so must be $\mathbf{Q}_t^{-1}$, and if $\mathbf{D}_t$ is singular, so must be $\mathbf{Q}_t^{-1}$. Hence, in view of Equation 13.4.15, singularity of matrix $\mathbf{Z}_t$ implies that the tangential moduli matrix $\mathbf{D}_t$ must be singular, that is,

$$\det \mathbf{D}_t = 0 \tag{13.4.17}$$

(Bažant, 1988b). This is equivalent to the condition for the peak point of the stress–strain diagram and represents the condition that separates the strain-hardening regime of material ($\mathbf{D}_t$ positive definite) from the strain-softening regime ($\mathbf{D}_t$ indefinite).

So we must conclude that, during a loading process in which the displacements are continuously increased at infinity, localization of homogeneous strain into an ellipsoidal region must begin as soon as $\mathbf{D}_t$ loses positive definiteness, that is, as soon as strain softening begins.

Let us now discuss the type of bifurcation. In view of Equation 13.4.15, Equation 13.4.16 for eigenvector $\delta\boldsymbol{\varepsilon}^*$ of matrix $\mathbf{Z}$ may be written as

$$\mathbf{D}_t\mathbf{x}^* = \mathbf{0} \tag{13.4.18}$$

in which $\mathbf{x}^* = \mathbf{Q}_t^{-1}\,\delta\boldsymbol{\varepsilon}^*$ or

$$\delta\boldsymbol{\varepsilon}^* = \mathbf{Q}_t\mathbf{x}^* \tag{13.4.19}$$

and $\mathbf{x}^*$ is the eigenvector of matrix $\mathbf{D}_t$ corresponding to its zero eigenvalue.

Two cases may now be distinguished: (1) Either the eigenvector $\delta\boldsymbol{\varepsilon}^*$ lies in the sector of loading or (2) it does not.

If it does, then the bifurcation state would be a state of neutral equilibrium, which represents the limit of stability. However, we have shown that, except for the limit case of an infinite layer (which is equivalent to an ellipsoid for which the two ratios of its axes tend to infinity), the limit of stability does not occur when $\det \mathbf{D}_t = 0$ but only later, when matrix $\mathbf{D}_t$ becomes indefinite (i.e., at a certain finite distance after the peak of the stress–strain diagram). It follows that the eigenvector $\delta\boldsymbol{\varepsilon}^*$ must lie outside the loading sector. So the actual increment $\delta\boldsymbol{\varepsilon}^{eq}$ along the equilibrium path cannot coincide with $\delta\boldsymbol{\varepsilon}^*$ because it would imply unloading, which we have ruled out. Therefore, the actual $\delta\boldsymbol{\varepsilon}^{eq}$ must lie at the boundary of the loading sector (cf. Sec. 10.4) and must differ from the eigenvector $\delta\boldsymbol{\varepsilon}^*$. Hence, $\mathbf{Z}_t\,\delta\boldsymbol{\varepsilon}^{eq} \neq 0$ (that is, $\mathbf{D}_t\mathbf{x}^{eq} \neq 0$ where $\mathbf{x}^{eq} = \mathbf{Q}_t^{-1}\,\delta\boldsymbol{\varepsilon}^{eq}$). This means that the increment $\delta\boldsymbol{\varepsilon}^{eq}$ along the equilibrium path must be happening at increasing boundary displacements (or increasing strain) at very remote points. This is similar to the Shanley bifurcation in plastic columns. The bifurcation state and all the immediate postbifurcation states are stable.

We have seen that the loss of stable equilibrium with localization into an ellipsoidal domain occurs only when matrix $\mathbf{D}_t$ becomes indefinite, that is, the material is in a strain-softening state. However, now we find that along the equilibrium path the localization into an ellipsoidal domain occurs already when matrix $\mathbf{D}_t$ becomes semidefinite, that is, at the peak-stress state, which always precedes the state of stability loss.

The classical bifurcation condition of Hill (1962c), which serves as the basis of the method of linear comparison solid (Sec. 10.4), also indicates that singularity of matrix $\mathbf{D}_t$ is the condition of first bifurcation; see also Rudnicki and Rice

(1975), Rice (1976), Leroy and Ortiz (1989); and de Borst (1988a). Note, however, that Hill's condition applies only to localization into an infinite band in an infinite space. The boundary conditions of such a localization mode cannot be accommodated for finite bodies. The present analysis (Bažant, 1988b) proves that Hill's bifurcation condition (that is, $\det \mathbf{D}_t = 0$) is also correct for localization into ellipsoidal regions, although the mode of localization is different.

The foregoing analysis that led to the bifurcation condition $\det \mathbf{D}_t = 0$ shows only when localization can occur. To show that it must occur, the path-stability criterion from Bažant (1987b, 1988c) requires it to prove that, for the conditions of prescribed displacements at infinity, the value of $\delta^2 \mathscr{W}$ is smaller for the localizing path than for the nonlocalized path (for which the strain field remains uniform). The calculation of $\delta^2 \mathscr{W}$ along the equilibrium path can be done numerically.

The preceding analysis shows that the case of stability loss with localization into an ellispoidal domain can never occur in a continuous loading process in which the displacements at infinity are controlled. Localization along the loading path, without stability loss, always precedes such instability. So may the solution of stability loss have in this case any practical application? It may—either if the tangential moduli matrix $\mathbf{D}_t$ changes suddenly during the loading process (which happens, e.g., for bilinear stress–strain diagrams; cf. Secs. 8.1 and 10.3, or if the uniform strain state in the strain-softening range is reached by some other type of process). Examples are processes in which some displacements inside the body are controlled, or processes in which a finite sudden change of $\mathbf{D}_t$ is caused by a change of temperature, change in moisture content or pore pressure, crystallographic conversion (as in certain ceramics), chemical conversion, irradiation, etc., or processes in which $\mathbf{D}_t$ is changed due to hysteretic cycles.

Simpler Derivation of Bifurcation Condition

It is instructive to show a simpler direct derivation of the bifurcation condition $\det \mathbf{Z} = 0$. According to Equation 13.4.7, the (fictitious) stress variations inside the ellipsoidal region that correspond to Eshelby's solution for the outside are $\delta\boldsymbol{\sigma}^e = \mathbf{D}_u(\mathbf{I} - \mathbf{Q}_u^{-1})\,\delta\boldsymbol{\varepsilon}^e$. The components of surface traction vector that must be transmitted from the ellipsoidal region to the outside in order to provide the correct boundary conditions for Eshelby's solution for the outside is $\delta p_i^e = -\delta\sigma_{ij}^e n_j$ where n_j is the unit vector of the normals to the ellipsoid, pointed outward from the ellipsoid. Therefore

$$\delta p_i^e = -\{\mathbf{D}_u(\mathbf{I} - \mathbf{Q}_u^{-1})\,\delta\boldsymbol{\epsilon}^e\}_{ij} n_j \qquad (13.4.20)$$

were $\{\ \}_{ij}$ denotes the tensorial components extracted from a 6×1 column matrix. At the same time, the stress variations inside the ellipsoidal region may be expressed as $\delta\boldsymbol{\sigma}^e = \mathbf{D}_t \delta\boldsymbol{\varepsilon}^e$. Therefore, the vector of surface tractions acting on the surface of the ellipsoid is $\{\mathbf{D}_t \delta\boldsymbol{\varepsilon}^e\}_{ij} n_j$, and the vector of surface tractions applied from the ellipsoidal region on the outside is $\delta p_i^e = -\{\mathbf{D}_t \delta\boldsymbol{\varepsilon}^e\}_{ij} n_j$. Substituting this into Equation 13.4.20, one obtains for the strain variations $\delta\boldsymbol{\varepsilon}^e$ in the ellipsoidal localization region the condition:

$$\{\delta\mathbf{X}\}_{ij} n_j = 0 \qquad \text{with} \qquad \delta\mathbf{X} = \mathbf{Z}\,\delta\boldsymbol{\varepsilon}^e \qquad (13.4.21)$$

in which $\mathbf{Z}$ is defined by Equation 13.4.13.

Column matrix $\delta\mathbf{X}$ is the same for all the points of the ellipsoidal surface because the strain $\delta\boldsymbol{\varepsilon}^e$ in the ellipsoidal region is homogeneous, both according to Eshelby's solution for the outside and the assumed strain-softening deformation inside. Equation 13.4.21 represents a system of three homogeneous linear algebraic equations for the three components n_j. These equations must be satisfied not just for one vector n_j, but for infinitely many vectors of all the normals n_j of the ellipsoidal surface. This is possible if and only if $\delta\mathbf{X} = \mathbf{0}$. Therefore,

$$\mathbf{Z}\,\delta\boldsymbol{\varepsilon}^e = \mathbf{0} \tag{13.4.22}$$

This is a system of six homogeneous linear algebraic equations for the six components of the column matrix $\delta\boldsymbol{\varepsilon}^e$. A nonzero solution is possible if and only if the determinant of this equation system vanishes, that is, $\det \mathbf{Z} = 0$.

Summary

Localization of strain softening into ellipsoidal regions in an infinite homogeneous body is a three-dimensional localization problem with a finite-size localization region for which an exact analytical solution can be found. The solution is based on Eshelby's theorem from elasticity theory, whose applicability is due to the fact that the infinite region outside the ellipsoid is unloading, and thus behaves elastically. Bifurcation of localization type begins in a stable manner (at increasing load) when the tangential moduli matrix becomes singular (which corresponds to the peak of the stress–strain diagram), but localization instabilities arise later. Also the localization instability occurs later than in an infinite band. The more elongated the ellipsoidal region, the earlier the instability develops.

Problems

13.4.1 Without referring to the text, derive Equation 13.4.7.

13.4.2 Consider that further loading (strain softening) occurs outside the ellipsoid, while the ellipsoidal region begins to unload. Adapting the present solution, formulate the condition of stable state. Note that in this case the Eshelby coefficients must be calculated for an anisotropic material, except when the D^t_{ijkm} are assumed to be isotropic.

13.5 LOCALIZATION OF SOFTENING DAMAGE INTO SPHERICAL OR CIRCULAR REGIONS

Localization of softening damage into spherical or circular regions is a special case of the preceding solution for ellipsoidal regions. That solution, however, was limited to infinite space. Following Bažant (1988b), we will now show that for spherical or circular regions the solution can be easily obtained even for finite size bodies (spheres and circular disks or cylinders).

Localization Instability for Spherical Geometry

Consider a spherical hole of radius a inside a sphere of radius R (Fig. 13.20a). We assume polar symmetry of the deformation field and restrict our attention to

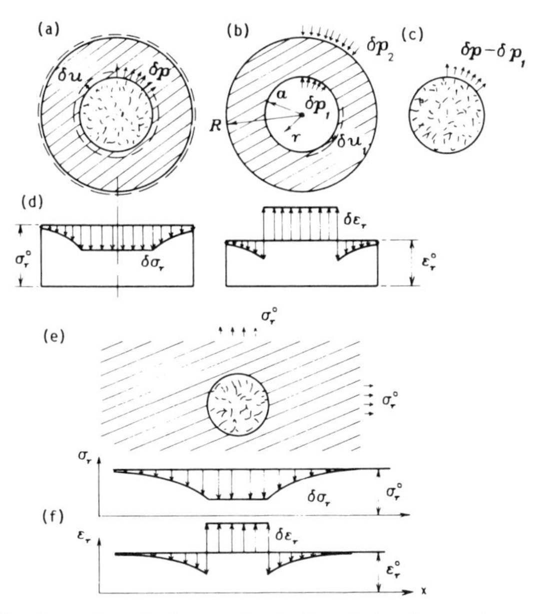

Figure 13.20 Localization of strain into spherical and circular regions.

materials that are isotropic for unloading. As shown by Lamé (1852) (see, e.g., Timoshenko and Goodier, 1970, p. 395), the elastic solution for the radial displacements and the radial normal stresses at a point of radial coordinate r is

$$u = Ar + Dr^{-2} \qquad \sigma_r = E_u(\bar{A} - 2\bar{D}r^{-3}) \tag{13.5.1}$$

where $\bar{A} = A/(1 - 2\nu_u)$; $\bar{D} = D/(1 + \nu_u)$; E_u, ν_u = Young's modulus and Poisson's ratio of the sphere; and A, D = arbitrary constants to be found from the boundary conditions.

We now consider a solid sphere of radius R that is initially under uniform hydrostatic stress σ^0 and strain ε^0 ($\sigma^0 = \sigma^0_{kk}/3$, $\varepsilon^0 = \varepsilon^0_{kk}/3$). We seek the conditions for which the initial strain may localize in an unstable manner into a spherical region of radius a. Such localization may be produced by applying on the solid sphere at $r = a$ radial outward tractions δp (i.e., pressure) uniformly distributed over the spherical surface of radius a, Figure 13.20a. To determine the second-order work of δp, we need to calculate the radial outward displacement δu_1 at $r = a$. We will distinguish several types of boundary conditions on the outer surface $r = R$.

Outer surface kept under constant load. As the boundary condition during localization, we assume that the initial radial pressure p^0_2 applied at outer surface $r = R$ is held constant, that is, $\delta p_2 = 0$. For $\delta\sigma_r = -\delta p_1$ at $r = a$ and $\delta\sigma_r = -\delta p_2 = 0$ at $r = R$, Equations 13.5.1 may be solved to yield $\bar{A} = a^3\,\delta p_1/E(R^3 - a^3)$, $\bar{D} = \bar{A}R^3/2$, and then

$$\delta u_1 = \frac{\delta p_1}{C_u} \qquad \frac{1}{C_u} = \frac{a}{E_u(R^3 - a^3)}\left[(1 - 2\nu_u)a^3 + \frac{1 + \nu_u}{2}R^3\right] \tag{13.5.2}$$

The inner spherical softening region of radius a, Figure 13.20c, is assumed to remain in a state of uniform hydrostatic stress and strain, and its strain-softening properties to be isotropic, characterized by E_t and ν_t. Thus the strains for $r<a$ are $\delta\varepsilon = \delta u_1/a$, the stresses are $\delta\sigma = 3K_t\delta u_1/a = C_t\delta u_1$ with $C_t = 3K_t/a$; $K_t =$ bulk modulus for further loading (softening); $3K_t = E_t/(1-2\nu_t)$, with $E_t<0$ and $K_t<0$ for softening. The surface tractions acting on the softening region, Figure 13.20c, are equal to $C_t\delta u_1$ where $C_t = 3K_t/a$. Hence, by equilibrium, the total distributed traction at surface $r=a$ must be $\delta p = C_u\delta u_1 + C_t\delta u_1$ and the second-order work done by δp is $\delta^2\mathcal{W} = 4\pi a^2(\frac{1}{2}\delta p\delta u_1) = 2\pi a^2(C_u + C_t)\delta u_1^2$. Thus the necessary condition of stability of the initial uniform strain ε^0 in the solid sphere is $C_u + C_t > 0$, which yields the necessary condition for stability

$$-\frac{E_t}{E_u} < \frac{2(1-2\nu_t)(R^3-a^3)}{2(1-2\nu_u)a^3 + (1+\nu_u)R^3} \tag{13.5.3}$$

Changing $<$ to $>$, we obtain the sufficient condition for strain-localization instability.

Outer surface kept fixed. In this case we assume that during localization $\delta u = 0$ at $r=R$, and $\delta\sigma_r = -\delta p_1$ at $r=a$. From Equations 13.5.1, we may then solve $D = -AR^3$ and

$$A = -\frac{\delta p_1}{E_u}\left[\frac{1}{1-2\nu_u} + \frac{2R^3}{(1+\nu_u)a^3}\right]^{-1} \tag{13.5.4}$$

which yields

$$\delta p_1 = C_u\,\delta u_1 \qquad C_u = \frac{E_u a^2}{R^3-a^3}\left[\frac{1}{1-2\nu_u} + \frac{2R^3}{(1+\nu_u)a^3}\right] \tag{13.5.5}$$

The second-order work done on the solid sphere by tractions $\delta p = C_u\delta u_1 + C_t\delta u_1$ applied at surface $r=a$ is $\delta^2\mathcal{W} = 2\pi a^2(C_u + C_t)\delta u_1^2$ where $C_t = E_t/(1-2\nu_t)a$, as before. Thus the necessary stability condition is $C_u + C_t > 0$, which can now be reduced to the condition

$$-\frac{E_t}{E_u} < \frac{1-2\nu_t}{R^3-a^3}\left(\frac{a^3}{1-2\nu_u} + \frac{2R^3}{1+\nu_u}\right) \tag{13.5.6}$$

Assuming that $|E_t|$ increases continuously after the peak of the stress–strain diagram as ε_0 is increased, instability develops at the value $a = a_{cr}$ that minimizes $|E_t|$ under the restriction $h/2 \le a \le R$ where h is the given minimum admissible size of the strain-softening region, representing a material property. For the case of prescribed pressure at the boundary $r=R$, we find from Equation 13.5.3 that $a_{cr} = R$, which corresponds to $E_t = 0$. So the sphere becomes unstable right at the start of strain softening, that is, no strain softening can be observed when the boundary is not fixed. For the case of a fixed (restrained) boundary at $r=R$, Equation 13.5.6, one can verify that $\min|E_t|$ is finite and occurs at $a_{cr} = \min a = h/2$ (provided that $\nu_u \ge 0$).

For $R/a \to \infty$, Equation 13.5.6 yields the stability condition for the case of an infinite solid fixed at infinity (Fig. 13.20e, f)

$$-\frac{E_t}{E_u} < \frac{2(1-2\nu_t)}{1+\nu_u} \tag{13.5.7}$$

It is interesting that for $R/a \to \infty$, Equation 13.5.3 yields the same condition, but this limit case is of questionable significance since we found that $a_{cr} = R$ when the boundary is not fixed. Note that the stability condition in Equation 13.5.7 does not depend on the radius a of the softening region; yet, unlike the softening in a layer in an infinite solid, solved before, instability does not begin at the peak of the stress–strain diagram (where $E_t = 0$) but begins only at a certain finite negative slope of the stress–strain diagram. This slope can in fact be rather steep ($-E_t = 2E_u$ for $\nu_t = \nu_u = 0$).

Localization Instability for Circular or Cylindrical Geometry

Working in two dimensions, consider a circular hole of radius a inside a homogeneous isotropic circular disk of radius R, Figure 13.20a, b. We assume a plane-stress state, and then, according to Lamé's solution, the radial displacement u and the radial normal stresses are (see, e.g., Flügge, 1962, p. 37-13; or Timoshenko and Goodier, 1970, p. 70)

$$u = \frac{1-\nu_u}{E_u}\left(\frac{a^2 p_1 - R^2 p_2}{R^2 - a^2}\right) r + \frac{1+\nu_u}{E_u}\left[\frac{R^2 a^2 (p_1 - p_2)}{(R^2 - a^2) r}\right] \tag{13.5.8}$$

$$\sigma_r = \frac{R^2 a^2 (p_2 - p_1)}{(R^2 - a^2) r^2} + \frac{a^2 p_1 - R^2 p_2}{R^2 - a^2} \qquad \varepsilon_z = \frac{2\nu_u (p_2 R^2 - p_1 a^2)}{E_u (R^2 - a^2)} \tag{13.5.9}$$

p_1 and p_2 are the pressures applied along the hole perimeter and along the outer perimeter of the disk, respectively, and ε_z is the transverse strain in the plate, which is independent of r. Depending on the boundary conditions, we distinguish two cases.

Outer boundary kept under constant load. As the boundary condition during the strain-localization instability, we now assume that the initial radial pressure p_2 applied at the outer boundary $r = R$ of the disk is held constant, that is, $\delta p_2 = 0$. Equation 13.5.8 then yields for $\delta u_1 = \delta u$ at $r = a$, Figure 13.20b, the relation:

$$\delta u_1 = \frac{\delta p_1}{C_u} \qquad \frac{1}{C_u} = \frac{a}{E_u}\left(\nu_u + \frac{R^2 + a^2}{R^2 - a^2}\right) \tag{13.5.10}$$

The inner circular softening region of radius a, Figure 13.20c, is assumed to remain in a uniform state of stress and strain. Thus the strains for $r < a$ are $\delta\varepsilon_r = \delta u_1 / a$. Assuming the plate to be thin compared to radius a, we may assume the strain-softening region to be also in a plane-stress state, and then $\delta\varepsilon_r = \delta\sigma_r (1 - \nu_t)/E_t$. Hence,

$$\delta\sigma_r = C_r\, \delta u_1 \qquad C_t = \frac{1}{a}\left(\frac{E_t}{1 - \nu_t}\right) \tag{13.5.11}$$

Now, we consider uniformly distributed outward tractions δp to be applied along the circle $r = a$ on the solid disk (without the hole). By equilibrium, $\delta p = C_u\, \delta u_1 + C_t\, \delta u_1$ and the work done by δp is $\delta^2 \mathscr{W} = 2\pi a(\frac{1}{2}\delta p\, \delta u_1) = \pi a (C_u + C_t)\, \delta u_1^2$. Thus the necessary condition of stability of the initial state of uniform strain is $C_u + C_t > 0$. According to Equations 13.5.10 and 13.5.11, the stability condition for plane stress becomes

$$-\frac{E_t}{E_u} < \frac{1-\nu_t}{\nu_u + (R^2+a^2)/(R^2-a^2)} \qquad \text{(thin plate)} \qquad (13.5.12)$$

As another limiting case, we may consider a long cylinder of radius R (and length $\gg R$), in which case the strain in the softening region along the cylinder axis is forced to be equal to the strain ε_z in the unloading region. However, the softening region is not in a plane-strain state either. Assuming the planes normal to the cylinder axis to remain plane, consider now that, unlike before, the axial stresses σ_z are nonzero. We must impose the equilibrium condition that the resultants of σ_z^t in the softening region and of σ_z^u in the unloading region cancel each other, that is, $\pi(R^2-a^2)\sigma_z^u = -\pi a^2 \sigma_z^t$. We leave it to a possible user to work out the solution in detail and we now restrict our attention to the case $a \ll R$, for which (according to Eq. 13.5.9 with $\delta p_2 = 0$), we have $\delta\varepsilon_z = 2\nu_u\,\delta p_1 a^2/(a^2-R^2)E_u = 0$ (for $r \geq a$), while also $\delta\sigma_z = 0$. Therefore, we may assume for the incremental deformation in the strain-softening region a state of plane strain. The solution may then be obtained simply by replacing E_t with $E_t' = E_t/(1-\nu_t^2)$, and ν_t with $\nu_t' = \nu_t/(1-\nu_t)$ (the unloading region remains in plane stress in this case). Thus Equation 13.5.12 transforms to

$$-\frac{E_t}{E_u} < \frac{(1+\nu_t)(1-2\nu_t)}{\nu_u + (R^2+a^2)/(R^2-a^2)} \qquad \text{(long cylinder, } a \ll R) \qquad (13.5.13)$$

When a/R is not very small, the solution may be expected to lie between Equations 13.5.12 and 13.5.13.

Outer boundary kept fixed. In this case we have during localization $\delta u = 0$ at $r = R$ and $\delta\sigma_r = -\delta p_1$ at $r = a$. Taking the variations of Equations 13.5.8 at $r = R$ and $r = a$, we get

$$\frac{1-\nu_u}{E_u}\left(\frac{a^2\,\delta p_1 - R^2\,\delta p_2}{R^2-a^2}\right)R + \frac{1+\nu_u}{E_u}\left[\frac{R^2a^2(\delta p_1 - dp_2)}{(R^2-a^2)R}\right] = 0 \qquad (13.5.14)$$

$$\delta u_1 = \frac{1-\nu_u}{E_u}\left(\frac{a^2\,\delta p_1 - R^2\,\delta p_2}{R^2-a^2}\right)a + \frac{1+\nu_u}{E_u}\left[\frac{R^2a^2(\delta p_1 - \delta p_2)}{(R^2-a^2)a}\right] \qquad (13.5.15)$$

Eliminating δp_2 from these two equations, we get the relation $\delta p_1 = C_u\,\delta u_1$ with

$$\frac{1}{C_u} = \frac{1}{E_u(R^2-a^2)}\left\{a[(1-\nu_u)a^2 + (1+\nu_u)R^2] - \frac{4R^2a^3}{(1-\nu_u)R^2 + (1+\nu_u)a^2}\right\} \qquad (13.5.16)$$

By the same reasoning as before, the necessary condition for the stability of the initial uniform strain ε^0 is $C_u + C_t > 0$ where C_t is again given by Equation 13.5.11. This condition yields

$$-\frac{E_t}{E_u} < \frac{(1-\nu_t)(R^2-a^2)}{R^2 + a^2 + \nu_u(R^2-a^2) - (4R^2a^2)/((1-\nu_u)R^2 + (1+\nu_u)a^2)} \qquad \text{(thin plate)} \qquad (13.5.17)$$

For the case of a long cylinder of length $\gg R$ and with $a \ll R$, we may obtain the solution again by replacing E_t, ν_t with E_t', ν_t'. This yields

$$-\frac{E_t}{E_u} < \frac{(1+\nu_t)(1-2\nu_t)(R^2-a^2)}{R^2+a^2+\nu_u(R^2-a^2)-(4R^2a^2)/((1-\nu_u)R^2+(1+\nu_u)a^2)}$$

(long cylinder, $a \ll R$) (13.5.18)

Instability develops at the value $a = a_{cr}$ that minimizes $|E_t|$ under the restriction that $h/2 \leq a \leq R$. For the case of a prescribed load at the outer boundary we find from Equation 13.5.12 or 13.5.13 that $a_{cr} = R$, which corresponds to $E_t = 0$. Thus the disk becomes unstable right at the start of strain softening, that is, no strain softening can be observed. For the case of a fixed (restrained) boundary at $r = R$, we find that, for $a \to R$, $\lim(-E_t/E_u) = \infty$ (to verify it, one needs to substitute $a = R - \delta$ and consider $\delta \to 0$); consequently $\min |E_t|$ is finite, and it is found to occur at $a_{cr} = \min a = h/2$.

For $R/a \to \infty$, Equation 13.5.17 or 13.5.18 yields the stability condition for the case of an infinite plate fixed at infinity, Figure 13.5.20e, f,

$$-\frac{E_t}{E_u} < \frac{1-\nu_t}{1+\nu_u} \qquad \text{for a thin plate} \qquad (13.5.19)$$

$$-\frac{E_t}{E_u} < \frac{(1+\nu_t)(1-2\nu_t)}{1+\nu_u} \qquad \text{for massive solid} \qquad (13.5.20)$$

It is interesting that for $R/a \to \infty$, Equations 13.5.12 and 13.5.13 yield the same conditions, but these limits are of questionable significance since we found that $a_{cr} = R$ when the boundary is not fixed. Note that the stability conditions in Equations 13.5.19 and 13.5.20 are independent of the size of the localization region, same as we found it for spherical localization regions and layers.

Numerical Examples

Figure 13.21 shows the plots of $|E_t|/E_u$ at the limit of stable states as a function of R/a for spherical geometry (curve 1) and for circular geometry (curve 4), both for fixed displacement at the outer boundary. For comparison, the figure also shows the solutions for localization in planar bands under transverse uniaxial stress (curve 2) and under shear stress (curve 3). We see that the value of E_t at instability depends strongly on the relative size $R(a)$ of the body as well as well as Poisson's ratios. For infinite body size ($R/a \to \infty$), instability of planar bands occurs at $E_t = 0$, that is, at the peak of the stress–strain diagram. The same happens under load-controlled conditions for the cases of spherical or circular regions with $R/a \to 1$ (i.e., the smallest possible body size for which the body remains at homogeneous strain). For $R/a > 1$, the spherical or circular regions generally require a finite slope $|E_t|$ to produce localization instability, provided the boundary is under prescribed displacement during the localization.

In the preceding analysis of ellipsoidal softening regions, we solved only the case of infinite solids and were unable to examine the effect of the boundary conditions at infinity. Now, from the fact that spherical and circular softening regions are special cases of ellipsoidal ones, we must conclude that our solution for ellipsoidal region is applicable only if the ellipsoidal region is of finite size,

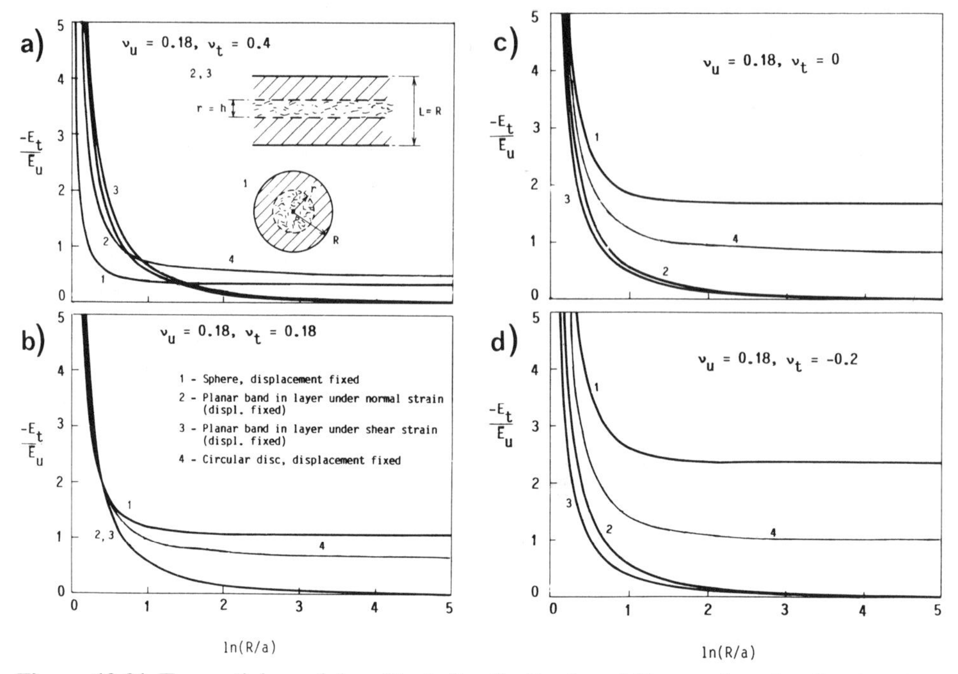

Figure 13.21 Tangential modulus E at the limit of stability against localization into spherical and circular regions of radius r within a sphere or disk of radius R. (*After Bažant, 1988b.*)

which is guaranteed only if the infinite body is fixed at infinity rather than having prescribed loads at infinity. Otherwise the limit cases $a_{cr} = R$ for spherical or circular softening regions discussed after Equations 13.5.6 and 13.5.18 would not be satisfied.

The present solutions represent upper bounds on $|E_t|$ at actual localization. When the stability condition for some of the previously considered softening regions is violated, instability with such a region is possible and must occur since it leads to an increase of entropy. However, it is possible that localization instability with some other form of localization region that we could not solve would occur earlier, at a smaller strain. For this reason, the present stability conditions are only necessary rather than sufficient. However, the opposite inequalities (that is, $<$ changed to $>$) represent sufficient conditions for instability.

Bifurcation and Stable Path

The preceding analysis has dealt only with stability of an initially uniform strain state against spherical or circular localization. A bifurcation with localization may, of course, take place in a stable manner, at increasing load. Proceeding similarly as for the ellipsoidal regions, the present approach may be adapted to show that bifurcation with stable localization occurs at smaller $|E_t|/E_u$.

Summary

Circular or spherical bodies with a circular or spherical localization region represent a softening localization problem for which an exact analytical solution can be found when both the localization region and the body have a finite size. The localization instability occurs later than in an elongated ellipsoidal region or in a layer. The smaller the ratio of body size to the size of the localization region, the later the localization instability takes place.

Problems

13.5.1 A ceramic circular disk of thickness b_1, which is in a strain-softening state, is perfectly bonded to a circular disk of the same radius R and thickness b_1, which is in an elastic state. Generalize Equation 13.5.13 to obtain the stability condition in terms of $|E_t|$ for the ceramic material.

13.5.2 Assuming that spherical localization happens with the same tangent modulus E_t both inside and outside a sphere of radius $r = a$, derive the condition of bifurcation. Then calculate the ratio $\delta p_2/\delta u_2$ (at $\delta p = 0$), show that $\delta p_2 \neq 0$ (which means this cannot be a state of neutral equilibrium), and calculate $\delta^2 \mathscr{W}$ done by δp_2. Then show that the localization path is the stable path.

13.6 LOCALIZATION IN BEAMS AND SOFTENING HINGES

Plastic limit analysis of beams and frames is valid only if the moments in all the simultaneously active plastic hinges (yield hinges) of the collapse mechanism carry their maximum moment simultaneously at least at one instant of the collapse process. This condition is guaranteed only if the plastic hinges exhibit no softening, that is, no decrease of moment at increasing rotation. Often this condition is met by practical designs. For reinforced concrete beams with no axial compression, absence of softening is achieved if the reinforcement ratio is sufficiently less than the so-called balanced steel ratio. Such a design is called the underreinforced cross section and is required by design codes. In many important cases, however, absence of softening cannot be guaranteed—for example, in reinforced concrete beams carrying a sufficiently large axial force, such as prestressed beams, columns, or beams in frames, a pronounced softening is seen in experiments (e.g., Darvall, 1983, 1984, 1985; Darvall and Mendis, 1985; Mendis and Darvall, 1988; Warner, 1984; Nylander and Sahlin, 1955; Cranston, 1965). In steel beams, likewise, softening in a plastic hinge region can arise if fracture develops in the plastic hinge, or if a stiffener of a flange or web locally buckles during plastic deformation, or if the cross-section shape suffers large distortion (Maier and Zavelani, 1970). The diagram of buckling moment M versus hinge rotation θ then has the form shown in Figure 13.22a. For different axial forces, the curves are different.

In reinforced concrete beams, softening of the hinge is caused by strain softening of concrete (which itself is a macroscopic manifestation of microfracturing). Based on softening material properties one may calculate the diagram of

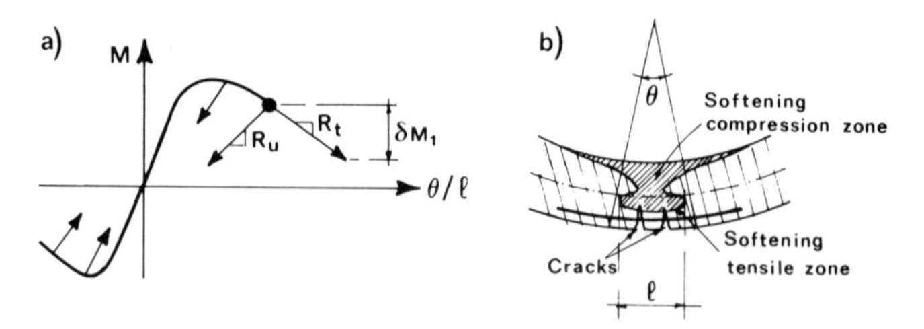

Figure 13.22 (a) Moment-rotation diagram; (b) plastic hinge.

bending moment M versus beam curvature κ. The $M(\theta)$ diagram may then be obtained by setting $\kappa = \theta/l$ where l is the effective length of the plastic hinge (Fig. 13.22b). A discussion of the value of l will better be postponed. If M is variable throughout the length l, as is usually the case, the $M(\theta)$ relation refers to the average of M over length l.

Stability Limit and Snapback

In view of the preceding sections, we must expect softening of plastic hinges to cause bifurcation and instability. Following Bažant (1976) and Bažant, Pijaudier-Cabot, and Pan (1987), we will now analyze stability of a structure with a single softening hinge—for example, the beam of span L loaded at midspan by load P

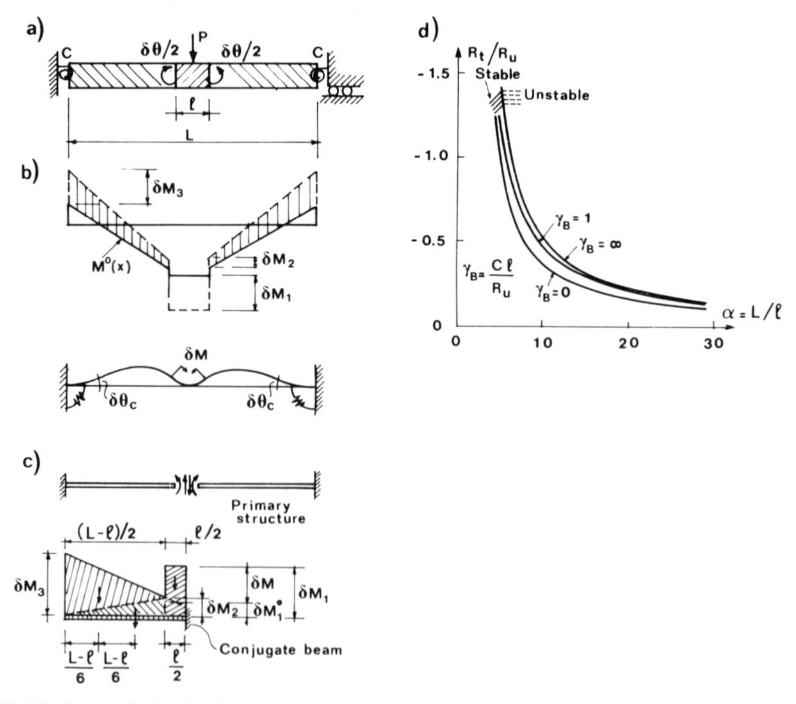

Figure 13.23 (a, b, c) Definition of curvature localization zone; (d) influence of relative length and spring constant on stability limit. (*After Bažant, Pijaudier-Cabot, and Pan, 1987.*)

and restrained at the ends by rotational springs of stiffness C (Fig. 13.23a). The beam is loaded by controlling the load-point (midspan) displacement w. For $C \to 0$, we have a simply supported beam, and for $C \to \infty$ we have a fixed beam. For the sake of simplicity, we assume the incremental bending rigidities R_t for further loading and R_u for unloading to be distributed uniformly. Except when the $M(\kappa)$ diagram is triangular, R_t and R_u in fact vary along the beam, but we content ourselves with using their average values for the loading and unloading parts of the beam.

To analyze stability, we consider infinitesimal variations $\delta M(x)$ and $\delta\kappa(x)$ of the bending moment and curvature, superimposed on a certain initial equilibrium state of beam, which is characterized by the initial bending moments $M^0(x)$ and initial curvatures $\kappa^0(x)$. The beam is loaded in a displacement-controlled mode, and so the load-point deflection, w, is fixed during the variation ($\delta w = 0$), and the load P, representing a reaction, can change arbitrarily. We assume that small variations $\delta\theta/2$ of rotation are enforced at the ends of the curvature localization segment of length l (Fig. 13.23a). We try to calculate the applied moment reactions δM at these points and the work done by these moment reactions. The bending moment variations are characterized by variation δM_1 within segment l, δM_2 just outside this segment, and δM_3 at the beam ends. From δM_2 to δM_3, the bending moment varies linearly. At the ends of segment l, we must assume for the calculation of curvatures that there are moments of magnitude $\delta M_1 = \delta M + \delta M_1^0$, $\delta M_1^0 = \delta M_2 + (\delta M_2 - \delta M_3)l/2(L - l)$, where δM_1^0 represents the average moment in segment l for $\delta M = 0$ (see Fig. 13.23c).

The conditions of compatibility require that, if a primary structure is created by a cut at midspan, the variations of rotation, $\delta\phi$, and of deflection, δw, at the midspan are equal to zero. We may calculate them applying the moment-area theorem to the cantilever represented by the left half of the beam, that is, calculating the vertical reaction and the support moment on the conjugate cantilever due to the curvature diagram sketched in Figure 13.23c (or by applying the virtual work principle to the primary structure). Thus we get

$$\delta\phi = \int_0^{L/2} \delta\kappa(x)\, dx + \delta\theta_c = \frac{\delta\theta}{2} + \frac{L-l}{2}\left(\frac{\delta M_2 + \delta M_3}{2R_u}\right) + \frac{\delta M_3}{C} = 0 \quad (13.6.1)$$

$$\delta w = \int_0^{L/2} \delta\kappa(x)\left(\frac{L}{2} - x\right) dx + \delta\theta_c \frac{L}{2} = \frac{\delta\theta}{2}\left(\frac{l}{4}\right) + \frac{L-l}{2}\left(\frac{L}{2} - \frac{L-l}{6}\right)\left(\frac{\delta M_3}{2R_u}\right)$$
$$+ \frac{L-l}{2}\left(\frac{L}{2} - \frac{L-l}{3}\right)\left(\frac{\delta M_2}{2R_u}\right) + \frac{\delta M_3}{C}\left(\frac{L}{2}\right) = 0 \quad (13.6.2)$$

in which the x coordinate is meaured from the left end, and $\delta\theta_c = \delta M_3/C =$ rotation in the spring. Now we substitute $\delta\theta = \delta M_1 l/R_t$ along with the expression previously found for δM_1. Thus we reduce Equations 13.6.1 and 13.6.2 to a single relation, $\delta M = k\,\delta\theta$, in which the incremental stiffness k is given as $k = \bar{k}R_u/l$, with

$$\bar{k} = \frac{R_t}{R_u} + \frac{1}{\alpha - 1}\left(\chi + \frac{\chi + \psi}{2(\alpha - 1)}\right) \quad (13.6.3)$$

in which we have set $\alpha = L/l$, $\psi = (2\alpha + 1)/(\alpha - 1 + \beta\alpha\gamma)$, $\chi = \psi/2 + 3(2\alpha - 1)/2(\alpha - 1)$, $\gamma = R_u/CL$. The variation of the load during the instability mode may be calculated as $\delta P = 2(\delta M_2 - \delta M_3)/(L - l)$. The second-order work done on the structure by enforcing the rotations $\delta\theta/2$ is $\delta^2\mathcal{W} = \delta M\,\delta\theta/2 = k(\delta\theta)^2/2$. If $\delta^2\mathcal{W} > 0$, the structure is stable, that is, the deformation increments will not occur if work $\delta^2\mathcal{W}$ is not supplied. However, if $\delta^2\mathcal{W} < 0$, the structure is unstable because a kinematically admissible deflection variation releases energy. It follows that $\bar{k} > 0$ signifies a stable state, $\bar{k} < 0$ an unstable state, and $\bar{k} = 0$ a critical state.

Numerical evaluations of Equation 13.6.3 have been used by Bažant, Pijaudier-Cabot, and Pan (1987) to obtain the stability limits. The results, plotted in Figure 13.23d (with some corrections), show the lines of critical $R_t L R_u$ values as functions of the relative beam size for various fixed values of the relative stiffness Cl/R_u of the elastic restraints at the beam ends. The states below these lines are stable, and the states above these lines are unstable. We see that with an increasing size of the beam or with a decreasing stiffness of the end restraint, the magnitude of R_t/R_u for the critical state decreases (which means that instability occurs closer to the peak point of the moment-curvature diagram). The diagrams in Figure 13.23d permit an approximate assessment of instability of our model beam (approximate because R_t and R_u are assumed to be uniformly distributed within segments of the beam).

By analogy with plastic limit analysis, the softening in previous works was assumed to be localized in a point hinge (Baker and Amarakone, 1964; Barnard, 1965; Darvall, 1983, 1984; Ghosh and Cohn, 1972; Maier, 1967a, b, 1971; Mróz, 1985). Let us examine whether this simplifying assumption makes a significant difference. Consider the beam in Figure 13.24 and assume that, at midspan, there is a softening hinge such that its rotation $\Delta\theta$ equals the additional rotation difference (between the ends of the segment of length $l \simeq h$) due to strain softening, which is obtained as the total rotation difference (between these ends) caused by moments M applied at the ends of the segment, minus the rotation difference (between the ends) corresponding to elastic deformation only. This yields for the hinge the moment-rotation relation:

$$\delta\theta = l\frac{M_p - M}{\bar{R}_t} \tag{13.6.4}$$

$$\frac{1}{\bar{R}_t} = \frac{1}{R_t} - \frac{1}{R_u} \tag{13.6.5}$$

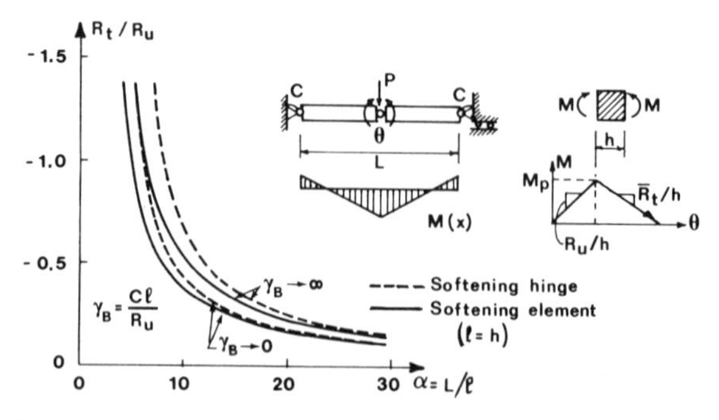

Figure 13.24 Stability limits for softening hinge analysis, and comparison with softening element analysis. (*After Bažant, Pijaudier-Cabot, and Pan, 1987*).

(where $l > 0$). It is important to note that it would be incorrect to replace $\bar{R}_t$ with R_t.

We may now repeat the same analysis as before, except that the segment of length l is replaced by a point hinge characterized by Equations 13.6.4 and 13.6.5. The result must be equal to the limit of Equation 13.6.3 for $l \to 0$. Instead of Equation 13.6.3 one gets

$$k = \frac{\bar{R}_t}{l} + \frac{R_u}{L}\left[\frac{4(1+6\gamma)}{1+8\gamma}\right] \tag{13.6.6}$$

The stability limits according to this equation are plotted in Figure 13.24 (for various C) as the solid lines.

For comparison, Figure 13.24 also shows as the dashed lines the stability limits for a strain-softening segment that has length $l = h$ and the same secant stiffness at uniform moment distribution as the hinge with the elastic segments of length $l/2$ adjacent to it.

If the beam is very long compared to length l of the softening region, the agreement must be close. However, even for normal slenderness values (L/l around 20), Figure 13.24 shows R_t/R_u values that are distinctly lower (more conservative) than the classical results for a hinge, independently of C. When the beam slenderness is very small, the difference of the present calculation from the classical analysis (i.e., a softening point hinge) becomes large. Thus, the accuracy of failure analysis based on a softening hinge is normally poor, and for deep beams very poor.

On the load-deflection diagram, the instability we obtained corresponds to the snapback point at which the descending slope of the load-deflection diagram becomes vertical (point 3 or 4 in Fig. 13.25). (That this must be so is clear if we note that the instability involves load variation δP at constant midspan deflection w.) If the beam is not sufficiently long, or the end restraints are not sufficiently weak, the load-deflection diagram may exhibit no snapback stability (curve 01 or 02 in Fig. 13.25). Otherwise, snapback instability occurs, and the load-deflection diagram after the point of instability descends with a positive slope, at which the equilibrium is unstable. Under displacement-controlled conditions, the structure snaps dynamically from 3 to 5 (Fig. 13.25). Nevertheless, the knowledge of the equilibrium path $\overline{049}$ (Fig. 13.25) is important; the area under the curve (cross-hatched) represents the energy dissipated by the structure. [The fact that

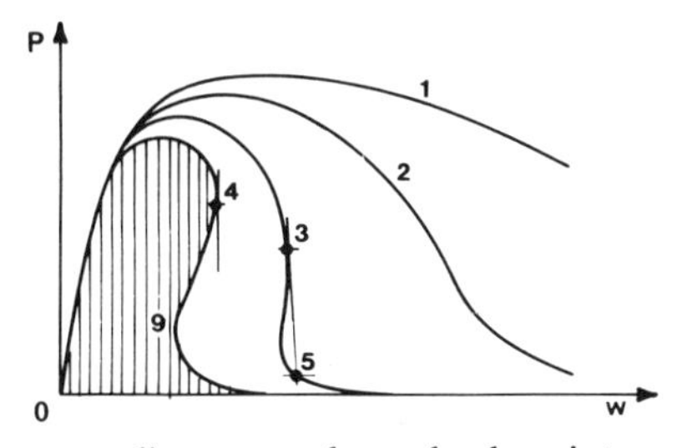

Figure 13.25 Load-displacement diagram and snapback point.

strain softening may cause snapback in beams was shown by Mróz (1985), Bažant (1976), Maier, Zavelani, and Dotreppe (1973), and others.]

Note that without imagining the curvature to depend on the average bending moment within segment l (which is accomplished by assuming M to be uniform over l) one could not explain how strain softening could spread from the midspan point into a segment of finite length. This is known as Wood's (1968) paradox.

Rotation Capacity or Ductility of Hinges in Concrete Beams

From the diagrams in Figure 13.23 we see that with an increasing slenderness of the beam, or with a decreasing relative stiffness of the end restraint, the magnitude of R_t/R_u for the stability limit decreases. Since l is proportional to beam depth h, it follows that this stability limit also decreases with increasing depth of the beam.

The ratio of hinge rotation $\theta = \theta_{cr}$ at the stability limit to the elastic rotation at $M = M_p$ characterizes the rotation capacity of the hinge (also called the ductility of a softening hinge). Because normally the moment-curvature diagram is concave $(d^2M/d\kappa^2 < 0)$, it further follows that the rotation capacity decreases with an increasing depth of the beam, with an increasing slenderness of the beam, and with a decreasing relative stiffness of the beam.

If the formulas for the $M(\kappa)$ diagram are specified, then one can calculate the function defining R_t, that is, $R_t = dM/d\kappa = f(\kappa)$. In terms of the initial hinge rotation, $\kappa = \theta_{cr}/l$. Then $\theta = g(R_t)l$ where g is the inverse function of f (in the postpeak range). Furthermore, the diagrams in Figure 13.23 may be described by the function $R_t/R_u = \phi(L/l, Cl/R_u)$. Therefore, we have

$$\theta_{cr} = g\left[R_u \phi\left(\frac{L}{l}, \frac{Cl}{R_u}\right)\right] l \qquad (13.6.7)$$

which represents an implicit function. Furthermore, since l is proportional to beam depth h, this equation may be rewritten in the form:

$$\theta_{cr} = F(h, L, C) \qquad (13.6.8)$$

where F is a certain function describing how the rotation capacity decreases with increasing beam depth h, increasing beam length L, and decreasing support stiffness C. It seems that, for typical concrete properties, function F has roughly the form

$$\theta_{cr} = \frac{k}{h} \qquad (13.6.9)$$

where $k \simeq$ constant if L and C are kept constant.

The fact that the rotation capacity θ_{cr} depends on beam depth h as well as span L was derived in the above manner by Bažant (1976). Recently Hillerborg (1988) analyzed the problem of rotation capacity in a somewhat different although analogous manner. He assumed the softening to be due to transverse line fracture in compression occurring in the compression zone of a concrete beam that he treated as a uniaxially stressed compressed bar assuming its length to be proportional to its depth (or to the beam depth). Based on his assumptions he concluded that the rotation capacity should be inversely proportional to the beam depth (Eq. 13.6.9) and supported this conclusion by some test data of Corley (1966). However, the assumption that the compression zone behaves as a uniaxially compressed beam seems to be an oversimplification. So does a uniaxial treatment of the compression fracture at bending, since this type of fracture is inherently a three-dimensional phenomenon.

Length of the Softening Region

In the preceding analysis we assumed the length l of the softening region to be given. The value of l is a central question of the analysis. Can it be arbitrary?

From Equation 13.6.3 we may notice that for $l = 0$ we always have $\bar{k} < 0$ if $R_t < 0$. This implies the first instability to occur at the peak moment point, which further implies that, for $l = 0$, the softening behavior of a frame could never be observed experimentally. But it can be. Hence, the softening length cannot be zero. So there must exist a certain lower bound $l_{\min}$ on the value of l.

The same conclusion may be reached from the value of the postbifurcation stiffness, $K = \partial P / \partial u$, where u = load-point displacement. It may be shown that for $l \rightarrow 0$ it is equal to the elastic stiffness (this may be readily verified by Equation 13.6.11). The positiveness of K means that there is always snapback instability when $l = 0$. Again, it follows that softening could not be observed, but it can.

Moreover, the fact that $K^{(2)}$ equals the elastic stiffness means that all the elastic energy has been recovered, and so the energy dissipation at softening failure vanishes if $l = 0$. This is a general property due to the fact that by introducing a moment-curvature diagram to describe softening we implied the energy dissipation due to softening, per unit length of the beam, to be zero. So if $l = 0$, the dissipation of energy must be zero, which is physically inadmissible. Therefore l must be finite.

From these arguments (advanced by Bažant, 1976), it follows that strain softening cannot localize into a segment that is shorter than a certain characteristic value of l. Based on the experience with softening in continua (Secs. 13.1–13.5), l must be bounded by the size of inhomogeneities in the material and cannot, for example, be less than several aggregate sizes in concrete. It seems, however, that this bound is too small for bending. The value of l must also be related to the depth h of the cross section. This is due to the fact that the bending theory that we are using is not valid for large deformations occurring within a length that is less than approximately the size of the cross section. As an approximation we may assume l to be equal to the depth h of the cross section.

Therefore, in our subsequent analysis we will assume that the minimum possible length of the softening region l is nonzero. But, for the sake of convenience of calculations, we will assume that the hinge rotation is concentrated into a point.

Bifurcation Due to Interaction of Softening Hinges

When more than one plastic hinge enters the softening range, their interaction produces a multitude of equilibrium paths. To illustrate it, let us analyze the simply supported beam of span L (Fig. 13.26) that receives two loads $P/2$ at the

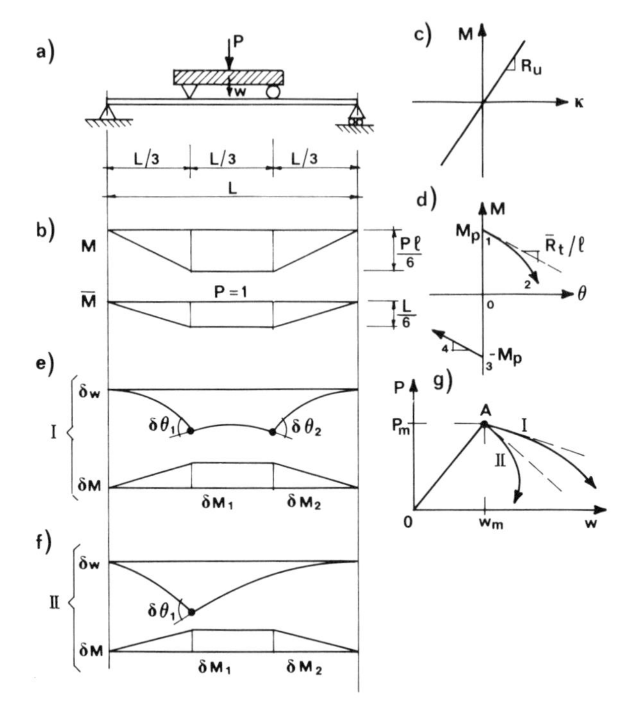

Figure 13.26 (a) Simply supported beam under third-point loads; (b) bending moment distribution; (c) initial linear moment-curvature relation; (d) softening moment-rotation diagram; (e, f) incremental deflection shapes; (g) equilibrium curves according to localized (II) and unlocalized (I) paths.

one-third points as reactions from a shorter, perfectly rigid beam. From the bending moment distributions in Figure 13.26b we see that two softening hinges may form simultaneously at the one-third points of the beam (they may also form anywhere between the one-third points, but we do not analyze that case, although the method will be obvious). Now we assume, for the sake of simplicity, that the hinges are point hinges and that the $M(\kappa)$ diagram is linear up to moment M_p at which softening begins with the initial slope $dM/d\theta = \bar{R}_t/l$ ($\bar{R}_t < 0$).

The load-point deflection w (at the center of the rigid beam) is $w = w_e + w_s$ where w_e = deflection due to the elastic curvature M/R_u and w_s = deflection due to rotations θ_1 and θ_2 in the left and right hinges (R_u = elastic bending rigidity = EI where I = cross-section moment of inertia). From the principle of virtual work, $w_e = \int \bar{M}M\,dx/R_u$ where $\bar{M}$ is the moment distribution due to $P = 1$. The distributions of $\bar{M}$ and M shown in Figure 13.26 yield $w_e = 5PL^3/324R_u$ and $w_s = (\theta_1 + \theta_2)L/6$. After the state $M = M_p$ is reached at the one-third points, either both hinges can soften simultaneously, which represents the nonlocalized path (Fig. 13.26e), or only one may soften ($\delta\theta_1 > 0$) while the other one unloads ($\delta M < 0$ at $\delta\theta_2 = 0$), which represents the localized path (Fig. 13.26f).

For the nonlocalized path, $\delta\theta_1 = \delta\theta_2 = \delta M_1 l/\bar{R}_t$ ($\bar{R}_t < 0$) where $\delta M_1 = \delta M_2 = \delta PL/6$ (by equilibrium), and so $\delta w_s = \delta\theta_1 L/3 = \delta PL^2 l/18\bar{R}_t$. Summing δw_s and

δw_e, we get $\delta w = \delta P/K^{(1)}$ where

$$\frac{1}{K^{(1)}} = \frac{L^2}{18}\left(\frac{l}{\bar{R}_t} + \frac{5L}{18R_u}\right) \tag{13.6.10}$$

For the localized path, $\delta\theta_1 = \delta M_1 l/\bar{R}_t$ (>0) and $\delta\theta_2 = 0$ where $\delta M_1 = \delta M_2 = \delta PL/6$ (by equilibrium). Therefore, $\delta w_s = (L/2)\delta\theta_1/3 = \delta PL^2 l/36\bar{R}_t$. Summing δw_s and δw_e, we get $\delta w = \delta P/K^{(2)}$ where

$$\frac{1}{K^{(2)}} = \frac{L^2}{18}\left(\frac{l}{2\bar{R}_t} + \frac{5L}{18R_u}\right) \tag{13.6.11}$$

We see that for the nonlocalized and localized paths we get postbifurcation branches of different slopes (see Fig. 13.26g where I labels the nonlocalized path and II the localized path).

From our analysis of structures with a single load in Section 10.2, we know that instability at controlled displacement is indicated by snapback, that is, by positiveness of K. Therefore, path stability needs to be analyzed only for bifurcation states for which $K < 0$ for both paths.

According to Section 10.2, the stable path at controlled δw is the branch for which $\delta^2\mathscr{W}$ is smaller. We have $\delta^2\mathscr{W} = \frac{1}{2}\delta P \delta w = \frac{1}{2}K\delta w^2$. Therefore, the localized path is the stable path if $K^{(2)} < K^{(1)}$. Substituting Equations 13.6.10 and 13.6.11 and noting that we need to consider only the cases for which $K^{(1)} < 0$, $K^{(2)} < 0$, we find that this inequality is always satisfied. This means that softening will localize in only one of the two possible softening hinges and the beam response will be nonsymmetric.

A similar example of bifurcation in a beam with softening hinges has been solved by Maier, Zavelani, and Dotreppe (1973) (see also Maier, 1970, 1971; and Maier and Zavelani, 1970). They identified the correct path assuming that $\delta^2 W$ should be minimized.

Imperfection Approach

The correct path after bifurcation, of course, can also be determined by considering inevitable very small imperfections. For example, in the beam in Figure 13.26 the elastic limit M_p in one hinge can be slightly smaller than in the other hinge. Then softening rotation starts in this hinge. But this causes unloading in the other hinge, and so this hinge can never reach the elastic limit and can only unload. This is the same conclusion we reached from the condition of stable path.

Bifurcation and Localization in Redundant Structures

In a statically determinate structure, the load starts to decrease as soon as a softening hinge forms. In redundant structures, though, more than one hinge must rotate to produce collapse, and the load may continue to increase after the first softening hinge forms. As a simple example, consider the singly redundant beam in Figure 13.27a that has one end fixed and one simply supported. Load P is applied at distance a from the fixed end. The $M(\kappa)$ diagram is again linear up to M_p. The bending moment distribution is characterized by moments M_1 at the

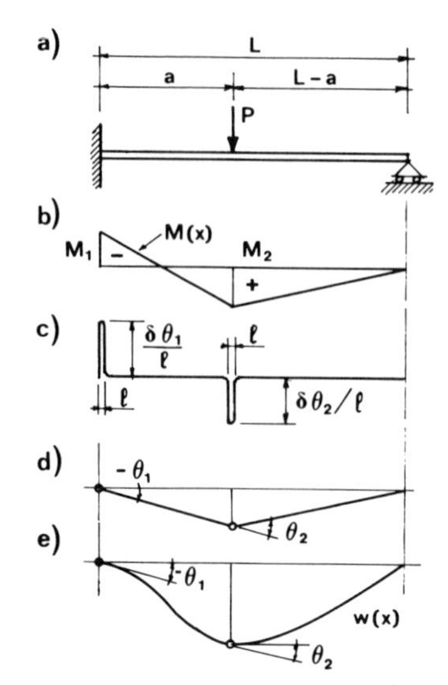

Figure 13.27 (a) Redundant beam; (b) bending moment distribution; (c, d) collapse mechanisms; (e) incremental deflection shapes.

fixed end and M_2 under the load (Fig. 13.27b). Due to redundancy, we now have a compatibility condition that requires that if the simple support is removed, the cantilever obtained must have a zero deflection. From the moment–area theorem, we may calculate the elastic part of this deflection as the moment of the M-distribution (Fig. 13.27b) about the right end of the beam, divided by R_u. The softening part of this deflection may be calculated as the moment of forces $\delta\theta_1 l/R_t$ and $\delta\theta_2 l/R_t$ about the right end. In this manner, we obtain the compatibility condition

$$\frac{L}{2}\left(\frac{\delta M_2}{R_u}\right)\left(\frac{L}{L-a}\right)\left(\frac{2L}{3}\right)-\frac{a}{2}\left[\frac{\delta M_2}{R_u}\left(\frac{L}{L-a}\right)-\frac{\delta M_1}{R_u}\right]\left(L-\frac{a}{3}\right)$$

$$+\,\delta\theta_1 L+\delta\theta_2(L-a)=0 \quad (13.6.12)$$

in which $\delta\theta_1=\xi_1\,\delta M_1 l/\bar{R}_t$ and $\delta\theta_2=\xi_2\,\delta M_2 l/\bar{R}_t$ where coefficients ξ_1, ξ_2 distinguish loading (with softening) from unloading; $\xi_1=1$ or $\xi_2=1$ for loading, and $\xi_1=0$ or $\xi_2=0$ for unloading in softening hinge 1 or 2. We expect $\delta M_1>0$, $\delta M_2<0$, $\delta\theta_1<0$, $\delta\theta_2>0$ ($R_t<0$).

Using a collapse mechanism with hinges (real hinges) inserted under the load and at the fixed support (Fig. 13.27d) in order to obtain the equilibrium condition, according to the principle of virtual work, we get $M_1(-\theta_1)+M_2\theta_2=P(-\theta_1)a$, where $\theta_2=-\theta_1 L/(L-a)$. From this we have

$$\delta M_2=-(1-\alpha)\,\delta M_1-\alpha(1-\alpha)L\delta P \qquad \left(\alpha=\frac{a}{L}\right) \quad (13.6.13)$$

Substituting this into Equation 13.6.12 and rearranging, one finds that $\delta M_1=$

$C_1\delta P$ and $\delta M_2 = C_2\delta P$ where

$$C_1 = \frac{\alpha L(1-\alpha)\left[2-3\alpha+\alpha^2+\xi_2\frac{6l}{L}\left(\frac{R_u}{\bar{R}_t}\right)(1-\alpha)^2\right]}{\alpha(1-\alpha)(3-\alpha)+\xi_1\frac{6l}{L}\left(\frac{R_u}{\bar{R}_t}\right)(1-\alpha) - (1-\alpha)\left[2-3\alpha+\alpha^2+\xi_2\frac{6l}{L}\left(\frac{R_u}{\bar{R}_t}\right)(1-\alpha)^2\right]} \tag{13.6.14}$$

$$C_2 = (1-\alpha)(-C_1-\alpha L)$$

The load-point deflection w may now be calculated as the deflection at the end of the cantilever segment of length a. By the moment–area theorem, one takes the moment of the $M(\kappa)$ distribution to the left of the load about the load point. Thus one gets $\delta w = \delta P/K$ in which

$$\frac{1}{K} = \frac{\delta w}{\delta P} = \left[\delta\theta_1 a + \frac{\delta M_1}{R_u}\left(\frac{a}{2}\right)\left(\frac{2a}{3}\right) + \frac{\delta M_2}{R_u}\left(\frac{a}{2}\right)\left(\frac{a}{3}\right)\right]\left(\frac{1}{\delta P}\right)$$
$$= \frac{L^3\alpha}{R_u}\left[\left(\frac{\xi_1 l R_u}{L\bar{R}_t}+\frac{\alpha}{3}\right)\frac{C_1}{L}+\frac{\alpha}{6}\left(\frac{C_2}{L}\right)\right] \tag{13.6.15}$$

K represents the tangential stiffness of the structure. The second-order work under controlled displacement δw can now be evaluated as $\delta^2\mathcal{W} = \frac{1}{2}\delta P\delta w = \frac{1}{2}K\,\delta w^2$.

As displacement w is increased, first one softening hinge forms, either at θ_1 or at θ_2 depending on the value of α. This corresponds to point A in Figure 13.28. Afterward this hinge rotates and softens (line 13 or 26 in Fig. 13.28b), which

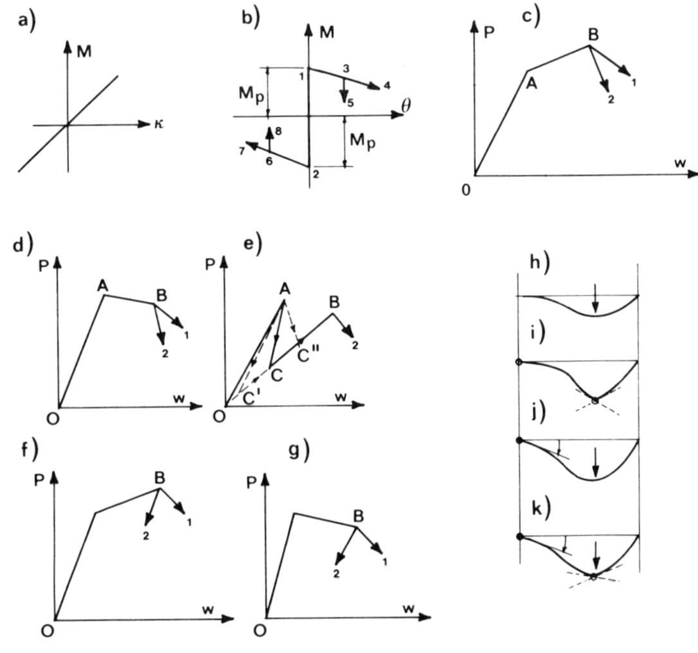

Figure 13.28 (a, b) Moment-curvature and moment-rotation diagrams; (c, g) equilibrium paths; (h–k) incremental deflection shapes.

causes the slope of the $P(w)$ curve to drop (lines AB in the figure). Later, at point B, the second softening hinge forms as its moment reaches M_p. Subsequently, there exist two equilibrium paths (branches, shown as lines $B1$ and $B2$): (1) both hinges rotate and soften (line 34 or 67 in Fig. 13.28b), or (2) the new hinge rotates and softens while the previously formed hinge starts to unload (line 35 or 68). (The case of the first hinge rotating and the new one unloading does not exist at point B since $|M|$ would exceed M_p.)

For path 1, $\xi_1 = \xi_2 = 1$. For path 2, either ξ_1 or ξ_2 is 1 and the other is 0. Using these combinations, one can calculate the corresponding stiffnesses $K = K^{(1)}$ and $K = K^{(2)}$ from Equation 13.6.15 with Equation 13.6.14. If at least one of them is positive, snapback instability takes place. If they are both negative, the structure is still stable, and in that case the stable path is that for which K is smaller (that is, $\delta^2 \mathscr{W}$ is smaller, for the same δw).

Depending on the values of α and R_t/R_u, the structure can exhibit various kinds of responses as shown in Figure 13.28c, d, e, f, g, among which cases c, d, e are stable states at the bifurcation point B and cases f and g are unstable states. The incremental deflection shapes for various combinations of active and nonactive hinges are shown in Figure 13.28h, i, j, k.

A path of the kind shown in Figure 13.28e is obtained when the first-formed softening hinge softens so rapidly (high $|R_t|$) that its moment is reduced to zero before the second softening hinge is activated. The moment at the first hinge becomes 0 at point C, and along path $\overline{CB}$ the structure behaves elastically, but with reduced stiffness (as if a real hinge were provided). If $|R_t| \rightarrow \infty$ for the first-formed hinge, which corresponds to a sudden moment drop in this hinge, then the response (dashed line AC') nearly retraces the initial elastic loading path $A0$. On the other hand, if the softening slope is sufficiently mild (small $|R_t|$), then there may be no snapback after point A (dashed line AC''). If the first hinge has softened to a zero moment before the second hinge is activated (i.e., before point B), there can be no bifurcation at point B.

Bifurcation at Simultaneous Formation of Several Softening Hinges

In the preceding examples we tacitly ignored the possibility the two hinges may form simultaneously. This case arises if the elastic moment distribution in the beam (for $\theta_1 = \theta_2 = 0$) has max M and min M of the same magnitude M_p. To find the α value for which this happens, we substitute $\delta M_1 = -M_p$, $\delta M_2 = M_p$, $\delta\theta_1 = \delta\theta_2 = 0$, and then the compatibility Equation 13.6.12 for this case reduces to the equation $\alpha^3 - 5\alpha^2 + 6\alpha - 2 = 0$. The root of this cubic equation that is practically useful is $\alpha = 0.5859$, which indicates the value of $\alpha = a/L$ such that the hinges at the fixed end and under the load would form simultaneously. If $\alpha < 0.5859$, then the softening hinge forms first at the fixed end, and if $\alpha > 0.5859$, then it forms first under the load. When the hinges become activated, there exist three different tangent stiffnesses $K = K^{(1)}$, $K^{(2)}$, $K^{(3)}$ with three combinations (paths): (1) $\xi_1 = 1$, $\xi_2 = 0$; (2) $\xi_1 = 0$, $\xi_2 = 1$; and (3) $\xi_1 = 1$, $\xi_2 = 1$, which may be evaluated from Equation 13.6.15. The correct postbifurcation path is that for which K is minimum. As an example, for $L = 12l$ and $R_u/R_t = 3$ (and $a = 0.5859l$), Equation 13.6.15 yields $K^{(1)} = 47.47R_u/L^3$, $K^{(2)} = 15.50R_u/L^3$, and $K^{(3)} = 23.44R_u/L^3$ for the three paths. Therefore, path 2 is the stable one. Only one hinge is loading along this path, and so again we see that the softening has a tendency to localize.

The typical responses that can be traced by the structure when three hinges form simultaneously are sketched in Figure 13.29a–f. Cases a, d, e are stable (no snapback), whereas cases b, c, f are unstable, due to snapback. Figure 13.29d, e, f illustrates what happens when the structure size, particularly the ratio L/l, is increased while the geometric shape of the structure (beam or frame) remains the same. The softening behavior becomes steeper (Fig. 13.29e) and for a sufficiently large size it leads to snapback instability (Fig. 13.29f).

A statically indeterminate (redundant) structure of a sufficiently large size exhibits in its $P(w)$ diagram multiple peaks (Fig. 13.29e), and for a still larger size it exhibits multiple snapback instabilities. The number of peaks or snapbacks is equal to the number of hinges in the plastic collapse mechanism (which is equal to the redundancy degree plus one), provided that no two hinges form simultaneously. For each pair of simultaneously forming hinges, the number of peaks and snapbacks is reduced by one.

If the structure is displacement controlled, it may be designed (with the proper safety factor) for the highest peak load (Fig. 13.29e), provided there is no snapback before it. If the structure is under load control, it must be designed for the first peak load, even if there is no snapback. Under mixed load control (loading through an elastic device), either case may apply. If there is snapback, the structure must be designed for the first snapback displacement.

If the structure is loaded through an elastic loading device of stiffness C_s

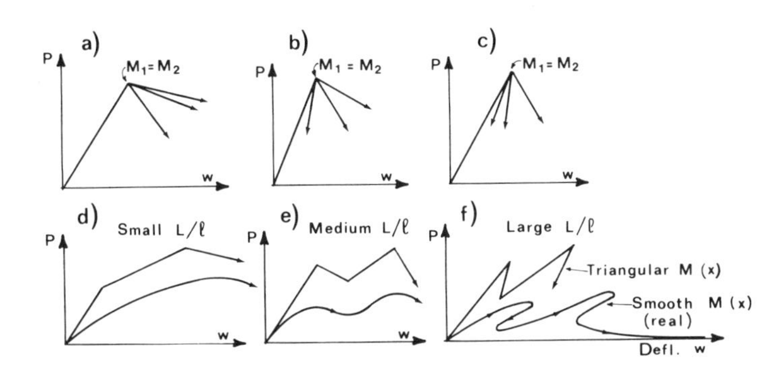

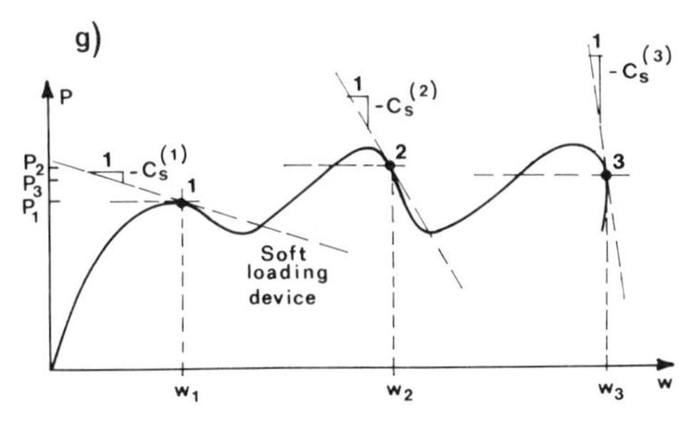

Figure 13.29 (a–f) Equilibrium paths for simultaneous formation of three hinges; (g) graphical construction of failure points for a structure loaded through an elastic device.

(coupled in series), then the load P and deflection w at which the structure fails (and for which it must be designed) can be determined by the graphic construction in Figure 13.29g, which is a generalization of the construction in Figure 13.6b already justified before. A line of slope $-C_s$ (dashed in Fig. 13.29g) is passed as the first possible tangent to the load-deflection curve $P(w)$. Depending on the value of stiffness C_s, this can produce failure states such as 1, 2, or 3 shown in Fig. 13.29g. (Note, for example, that the line of slope $C_s^{(2)}$ cannot be made tangent before the second load peak.)

It might be thought that, for $|R_t|\, l \ll R_u L$, the beams between the softening hinges could be considered as rigid during a small deflection increment after bifurcation. However, in contrast to plastic analysis of frames, this cannot be done. The structure would then be found to be always stable under displacement control (this can be verified, e.g., by the fact that the limit of Equation 13.6.3 for $R_t \to 0$ always yields $\bar{k} > 0$). Softening instabilities are driven by release of stored elastic energy, but a structure with rigid beams has no stored energy. So the calculation of elastic deflections is an essential part of stability analysis of softening beams or frames.

As for bifurcations, in structures with beams that are rigid between the hinges they exist only if there are more hinges than necessary to create a collapse mechanism (this is verified by the fact that Equations 13.6.10 and 13.6.11 give different $K^{(1)}$ and $K^{(2)}$ for $R_u \to \infty$). No bifurcations can be detected if the number of hinges just suffices to create a mechanism. This is illustrated by the fact that if the beam in Figure 13.27 is rigid and only one hinge can rotate (while the other one would have to unload), no deflection is possible.

Softening Frames and Trusses

The type of behavior we illustrated for beams is, of course, exhibited by continuous beams and frames. Examples of structures with such behavior are given in the exercise problems.

The axial response of bars can also exhibit softening. Then, whether or not the material softening is localized along the bar, the overall diagram $F(u)$ of axial force F versus the axial extension u of the bar exhibits softening (Fig. 13.30a). When this happens in a truss, the behavior is analogous to that described for beams.

For example, the triangular truss in Figure 13.30b under displacement control can be stable or unstable depending on the softening slope dF/du and the elastic properties. The truss in Figure 13.30c having the same member properties in tension and compression exhibits a bifurcation as soon as the elastic limit is reached because there exist three paths for which either both bars extend and

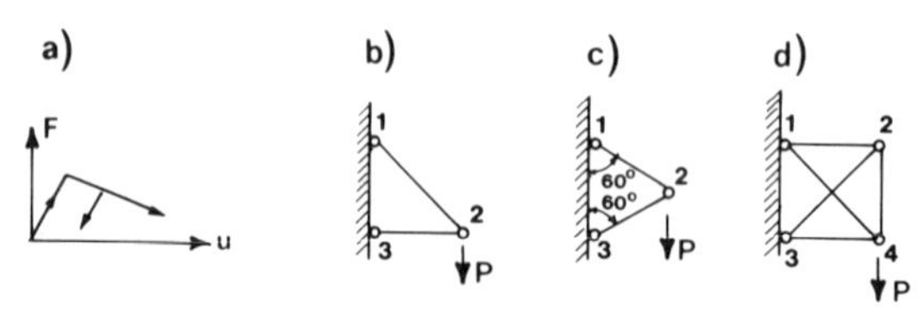

Figure 13.30 Trusses (b–d) with bars having softening response (a).

soften, or only one bar extends and the other one unloads. The stable path is that for which only one bar extends. The redundant truss in Figure 13.30d exhibits bifurcation similar to the redundant beam in Figure 13.27.

An important aspect of the inelastic behavior of frames and trusses is their capacity to redistribute their internal forces so as to carry higher loads than elastic analysis permits. Softening behavior reduces that capacity or entirely eliminates it. This has been illustrated by the beam in Figure 13.27. Determination of the extent to which redistribution of internal forces (moments) is possible is basically a problem of stable path (rather than stable state).

Softening in Metallic Structures

Development of softening hinges is not limited to reinforced concrete beams. Hinges with pronounced (steep) softening develop also in thin-wall metallic beams such as tubes or T-sections subjected to bending or eccentric compression. This in turn produces a softening response in the axial force-shortening diagram of a metallic structural member. The reason for softening in metallic structures is local buckling of the walls and deformation of the cross section, for example, the flattening of the circular cross section of a tube. This was experimentally demonstrated by Maier and Zavelani (1970); see Figure 13.31 (adapted from their figs. 2 and 3).

Summary

Softening damage in beams localizes in softening hinges and causes instability with snapback. The rotation capacity of softening hinges is limited and is not merely a property of the material. Rather, it is determined by structural stability considerations. It depends on the cross-section depth, the beam slenderness, and the support stiffness. Interaction of several softening hinges causes bifurcations of the equilibrium path, and the correct path may be identified by the path stability criterion (Sec. 10.2). In redundant softening structures, instabilities and bifurcations affect the maximum load, and successive formation of softening hinges may produce multiple load peaks, possibly also with snapback instabilities after the successive peaks. The structure must then be designed for the first peak load, or

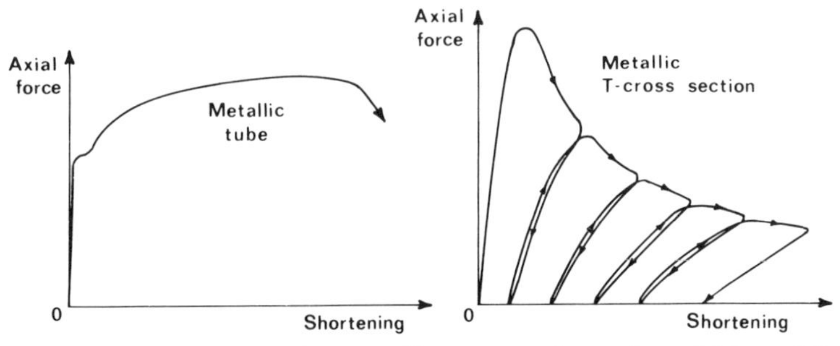

Figure 13.31 Softening response of a metallic structural member. (*Adapted from Maier and Zavelani, 1970.*)

the overall peak load, or the first snapback displacement, depending on the type of loading.

Problems

In all the problems consider a triangular $M(\kappa)$ or $M(\theta)$ diagram.

13.6.1 Verify that the limit cases of Equations 13.6.6 for $C \to 0$ and $C \to \infty$ give correct results, valid for a simply supported beam and a fixed-end beam.

13.6.2 Calculate the snapback instabilities for the cantilevers in Figure 13.32a, b assuming a softening region of finite length l. Discuss the size effects.

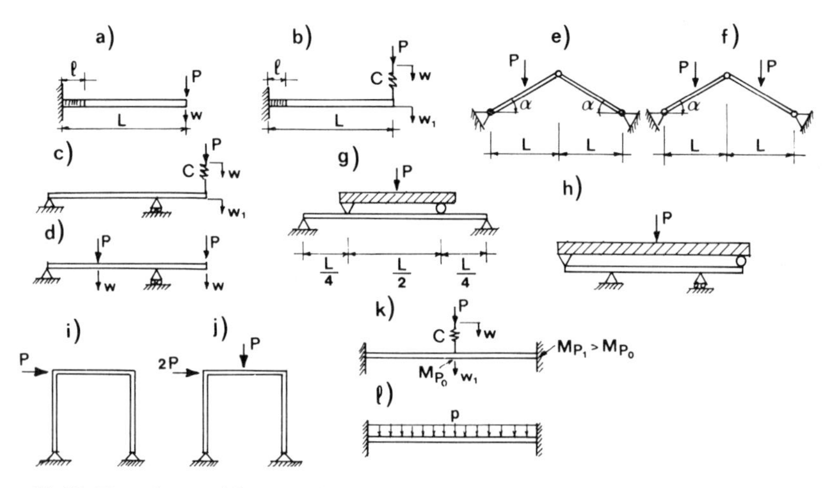

Figure 13.32 Exercise problems with softening hinges.

13.6.3 Calculate the maximum loads and snapback instabilities for the structures in Figure 13.32a, b, c, d, e, f assuming softening hinges equivalent to softening segments of length $l > 0$. Discuss the size effect.

13.6.4 Analyze path bifurcation and stable path for the statically determinate structures in Figure 13.32g, h. Discuss the size effect.

13.6.5 Analyze successive load peaks and successive snapbacks for the redundant structures in Figure 13.32i, j, k, l. Consider the effect of size on failure load.

13.7 FRICTION: STATIC AND DYNAMIC

Frictional phenomena have much in common with softening damage and cause similar instabilities. Due to the difference between static and dynamic friction coefficients k_s and k_d ($k_d < k_s$), a slip displacement causes the frictional force to drop, which represents a kind of softening damage to the rubbing surfaces. Therefore, systems of frictional bodies exhibit instabilities and bifurcations similar to those in softening beams and frames that we discussed in the preceding section.

Due to these similarities, we will touch the subject only superficially. Instabilities caused by frictional phenomena have been studied in depth by Ruina (1981), Rice and Ruina (1983), Kosterin and Kragel'skii (1960), Brockley and Ko (1970), Gu, Rice, and Ruina (1982), Schaeffer (1987), Schaeffer and Pitman, (1988), Pitman and Schaeffer (1987), and others.

Paradox in Sudden Friction Drop

Consider the rigid block of length $2L$ shown in Figure 13.33a, which is pulled by a spring of stiffness C over a rough surface. The static and dynamic frictional limits of the friction force F on the block are F_s and F_d. The block first slips when $P = F_s$, after which F drops instantly to $F = F_d$ (Fig. 13.33c).

Rigid bodies, however, are an idealization that does not exist in nature. They are the limiting case of elastic bodies whose stiffness tends to infinity. Therefore let us imagine, for instance, that in the middle of the block there exists a thin layer of stiffness C_1 joining two halves of the block (Fig. 13.33b). As $C_1 \to \infty$, the case of a single rigid block must be obtained.

Now, if $C_1 < \infty$, then the friction force on the left half must be zero ($F_1 = 0$) until the right half slips. The right half slips at $P = F_s/2$. Then, as the load-point displacement u increases, the slip q_1 of the right half increases and the force in the elastic layer grows as $C_1 q_1$. The left half then slips when $C_1 q_1 = F_d/2$, in which case the pulling force is $P = F_d$. For this case the $P(u)$ diagram is as shown in Figure 13.33d, which is sketched under the assumption that $F_d > F_s/2$. If $F_d < F_s/2$ (and $C_1 \to \infty$), then, after the first slip of the right half, the force in the elastic layer immediately rises to the value $F_s/2$ that makes the left half also slip immediately, after which force P must immediately drop to F_d. The $P(u)$ diagram then looks as shown in Figure 13.33e.

Now we cannot escape noting that the maximum forces needed to make the block move are different: $P_{max} = F_s$ for the rigid block and $P_{max} = F_d$ or $F_s/2$ for the elastic blocks with stiffness approaching infinity. This result is a paradox. (Its

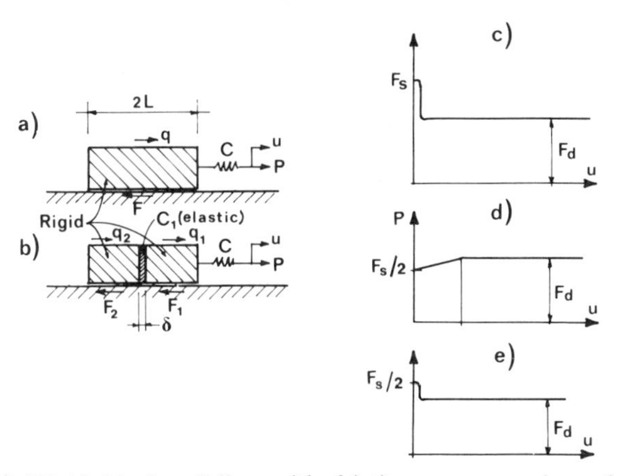

Figure 13.33 (a, b) Rigid blocks sliding with friction over rough surface; (c, d) load-displacement diagrams.

essence was pointed out by A. Ruina in his seminar at Northwestern University in 1988.)

So the assumption that the friction force drops from the static value to the dynamic value instantly is not quite reasonable. In reality, of course, this drop must occur gradually over a certain finite slip distance h (Fig. 13.34a) that is a property of the rubbing surfaces. This is explained by the fact that the microscopic asperities whose interlock causes friction are of finite size and the scraping of the peaks that causes the friction to drop requires a displacement at least equal to the size of these asperities.

Bifurcation, Stable Path, and Localization of Frictional Slip

Consider now the pair of two identical rigid blocks (Fig. 13.34b) that are attached by springs of stiffness C to a rigid balance rod and are pulled over a frictional surface by load P. The blocks are identical and have equal weights and the same frictional properties, described by static and dynamic frictional force limits F_s and F_d, such that $F_d < F_s$. The actual frictional forces on the blocks, denoted as F_1, must be equal due to the equilibrium condition on the balance rod. At the moment of the first slip of a block, its friction force equals F_s.

If F under each block is assumed to drop from F_s to F_d instantly, then the load-deflection diagram $P(u)$ looks as sketched in Figure 13.34c. Now consider that the frictional forces drop gradually (Fig. 13.34a). Let the initial slope of the diagram of friction force versus slip displacement be C_f ($C_f < 0$). There are two possibilities:

1. Both blocks start slipping when $P = 2F_s$, in which case $q_1 = q_2 = u - P/2C$ and $F_1 = F_2 = F_s + C_f q_1$ in each block. So the pulling force is $P = 2(F_s + C_f q_1) = 2F_s + 2C_f(u - P/2C)$, from which

$$K^{(1)} = \frac{dP}{du} = 2\left(\frac{1}{C} + \frac{1}{C_f}\right)^{-1} \tag{13.7.1}$$

2. One block starts slipping when $P = 2F_s$ while the other block does not move. So $q_1 = (u/2) - (P/2C)$, and the force in the slipping block is $F_1 = F_s + C_f q_1$. Due to equilibrium conditions, the force in the other block must be the same, and so $P = 2F_1 = 2F_s + 2C_f q_1 = 2F_s + 2C_f(u - P/C)/2$, from which

$$K^{(2)} = \frac{dP}{du} = \left(\frac{1}{C} + \frac{1}{C_f}\right)^{-1} = \frac{1}{2}K^{(1)} \tag{13.7.2}$$

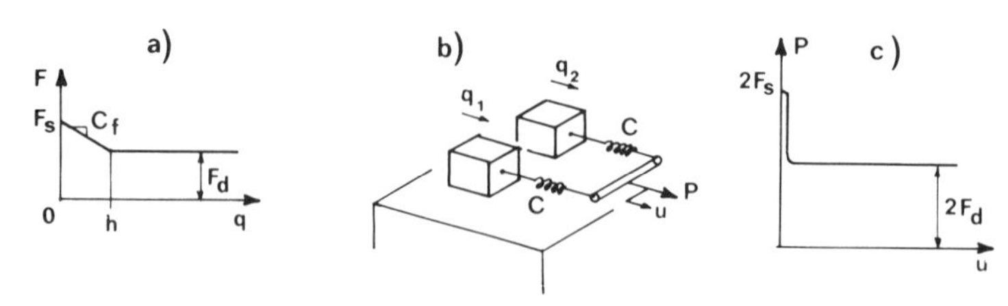

Figure 13.34 (a) Initial friction force drop; (b) identical blocks on rough surface; (c) load-displacement diagram.

So we see there is a bifurcation of equilibrium path. If $C > -C_f$, then $K^{(1)} > 0$ (which implies $K^{(2)} > 0$), and so the system exhibits snapback and is unstable. If $C = -C_f$, the system is critical, and if $C < -C_f$, it is stable. Path stability needs to be considered only for stable states, that is, for $C + C_f < 0$. Since this is a structure with a single load (Sec. 10.2), the stable path is that for which the tangential stiffness is less. We find that $K^{(2)} < K^{(1)}$ always. So we conclude that the path that is stable in path 2, for which the frictional slip localizes into one block.

For the case of a sudden drop of friction forces from F_s to F_d we have $K^{(1)} \to \infty$ and $K^{(2)} \to \infty$, and so on this basis one cannot decide which path will take place. However, on the basis of the limiting process $C_f \to \infty$, one must conclude that path 2 should occur.

It may be noted that the system of frictional blocks we analyzed (Fig. 13.34b) is mathematically equivalent to the system of two parallel softening specimens in Problem 13.2.7 and Figure 13.13d.

Frictional Supports in Columns

Friction also has considerable effect on the buckling of columns. Solutions to many problems in Chapters 1 to 4 are modified introducing friction. For example, consider the perfect columns in Figure 13.35, in which we assume for the sake of simplicity the dynamic and static friction to be equal, characterized by constant friction coefficient k. In the column (a), a frictional property of the support has no effect because the horizontal reaction H on top of the column is zero. For columns (b)–(e), however, H becomes nonzero as soon as deflection begins. After bifurcation, H is proportional to deflection, and so is the frictional reaction kH.

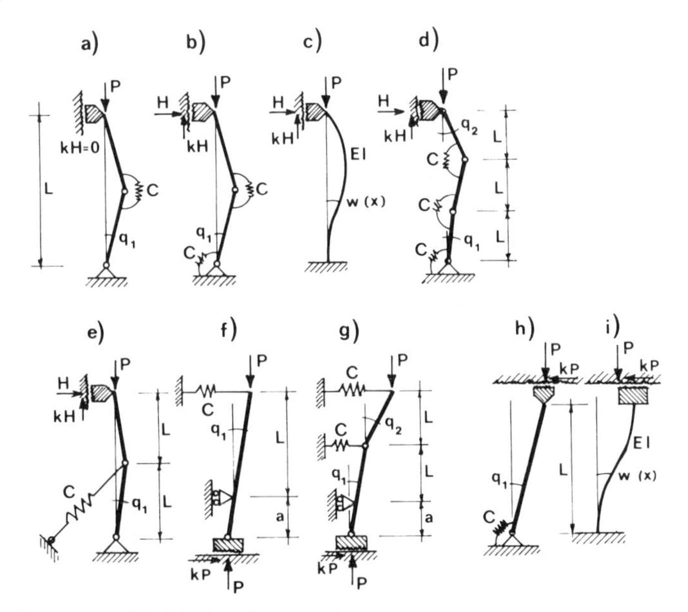

Figure 13.35 Columns with frictional supports.

For column (b), the moment equilibrium condition of all forces about the base joint yields $H = Cq_1/L$, and the moment equilibrium condition of the top half of the column about the middle joint reads $2Cq_1 - (P - kH)q_1L/2 + HL/2 = 0$, from which

$$P = P_{cr}\left(1 + \frac{kq_1}{5}\right) \qquad \text{with } P_{cr} = \frac{5C}{L} \tag{13.7.3}$$

We see that the bifurcation load is the same as that without friction, but friction causes P to increase after bifurcation. Thus, there is no neutral equilibrium and the bifurcation becomes of Shanley type, with all the immediate postbifurcation states being stable (this can be shown by calculating $\delta^2\mathcal{W}$).

Columns (c) and (d) behave similarly. Column (e) would, without friction, exhibit asymmetric bifurcation (Sec. 4.3), but the increase of P due to friction tends to offset the decrease of P associated with asymmetric bifurcation; if k is too small P still decreases after bifurcation for buckling to the right, but if k is sufficiently large, P increases for buckling both to the left and to the right.

In columns (f)–(i), friction is nonzero for any $P > 0$ and causes the equations of equilibrium of the column to become nonhomogeneous, with the right-hand sides containing P. Column (g) first buckles at no frictional slip ($q_1 = 0$, $q_2 > 0$), while column (f) requires $P \rightarrow \infty$ to start buckling, but P then declines rapidly with increasing q_1 (similar to the perfectly plastic columns considered in Sec. 8.6).

Structures with Stiffness Matrix Asymmetry

To demonstrate asymmetry of the stiffness matrix of the type seen in Problem 4.1.5 and the example in Section 10.4, we need to consider frictional-elastic structures with multidimensional action, for example, those in Figure 13.36. The blocks I, II or I, II, III are rigid. The frictional forces on the horizontal and vertical sliding surfaces are $F_1 = k_1N_1$ and $F_2 = k_2N_2$ where k_1, k_2 = given constant friction coefficients. A salient feature is that f_1 affects q_2 through F_2, and f_2 affects q_1 through f_1 even though f_1 does no work on q_2 and f_2 no work on q_1 (cf. Sec. 10.7). Let us assume that, on each frictional surface, the sliding is imminent, the structure is initially in equilibrium, and $dq_1 > 0$ and $dq_2 > 0$. Then, writing the horizontal equilibrium condition of blocks I and II and the vertical equilibrium condition of block II in Figure 13.36a, we get $C_1\,dq_1 + dF_1 = df_1$ and $C_2\,dq_2 +$

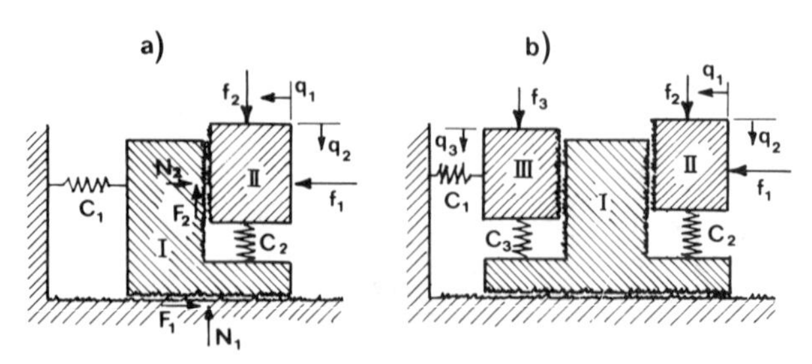

Figure 13.36 Multidimensional frictional-elastic systems.

$dF_2 = df_2$ where $dF_1 = k_1\, df_2$ and $dF_2 = k_2\, df_1$. Rearranging, we obtain

$$\begin{bmatrix} C_1 & \kappa_1 C_2 \\ \kappa_2 C_1 & C_2 \end{bmatrix} \begin{Bmatrix} dq_1 \\ dq_2 \end{Bmatrix} = \begin{Bmatrix} (1-\kappa_1\kappa_2)\, df_1 \\ (1-\kappa_1\kappa_2)\, df_2 \end{Bmatrix} \tag{13.7.4}$$

in which $\kappa_1 = k_1$, $\kappa_2 = k_2$. If $dq_1 < 0$, $dq_2 > 0$, we get the same result but with $\kappa_1 = -k_1$, $\kappa_2 = k_2$; if $dq_1 > 0$, $dq_2 < 0$: $\kappa_1 = k_1$, $\kappa_2 = -k_2$; and if $dq_1 < 0$, $dq_2 < 0$: $\kappa_1 = -k_1$, $\kappa_2 = -k_2$. Furthermore, when $dq_1 = 0$, we have a system with only one degree of freedom, provided that $-k_1 f_2 \leq F_1 \leq k_1 f_2$, and similarly when $dq_2 = 0$.

From Equation 13.7.4 it is now interesting to note that, in general, the stiffness matrix is nonsymmetric. The consequences of nonsymmetry for stability have already been discussed in Section 10.7, the example in Section 10.4, and Problem 4.1.5. The multidimensional elastic-frictional systems in Figure 13.36 are the most elementary prototypes of the salient aspects of stability of granular materials such as sand (Sec. 10.7).

Problems

13.7.1 Determine the stable path in the example in Fig. 13.34 by considering infinitely small imperfections in the values of F_s.

13.7.2 Consider that the springs in Figure 13.34 are replaced by a single spring transmitting load P into the rod and analyze the stable path.

13.7.3 Construct examples of frictional bodies mathematically equivalent to problems in Equations 10.4.6 to 10.4.7.

13.7.4 Write the equilibrium equations of the columns in Figure 13.35d, e, discuss the type of bifurcation and the postbifurcation response.

13.7.5 Do the same for columns in Figure 13.35f–i.

13.7.6 For the system in Figure 13.36a, determine the critical state of bifurcation and the stability limit.

13.7.7 Do the same for the system in Figure 13.36b.

13.7.8 Discuss problems 13.7.5 to 13.7.7 under the assumption that the dynamic friction coefficients k_{1_d}, k_{2_d} are smaller than the static ones, k_{1_s}, k_{2_s}.

13.8 BIFURCATIONS DUE TO INTERACTION OF SOFTENING DAMAGE ZONES

In Sections 13.2 to 13.7, we have seen how localization of softening damage into a small zone from an adjacent unloading zone can produce bifurcations and instabilities. The same effect can also be caused by interaction of nonadjacent zones of softening damage.

Interaction of Damage (Cracking) Fronts and Stable Paths

Distributed cracking is in finite element analysis often modeled as a sudden drop of the maximum principal stress to zero. If the element size at the cracking front is fixed as a material property (as is done in crack band theory), this approach is objective. Figures 13.37 and 13.38 show examples of application of this approach to a tensioned rectangular panel, calculated by Lin (1985) and Bažant

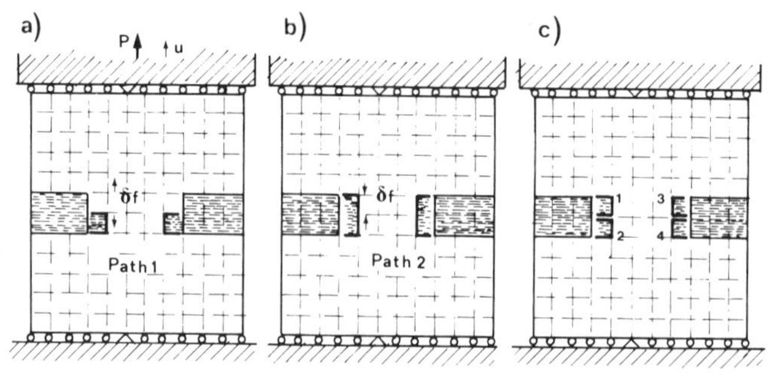

Figure 13.37 Various crack band advances in tensioned rectangular panel.

(1985b) (the mesh actually used has been much finer than that shown). One horizontal mesh line coincides with the axis of symmetry of the panel, and the pairs of elements at the left and right boundaries bordering on the symmetry axis are assumed to have a slightly smaller strength than the rest of the panel. The loading is displacement controlled, the displacement at the top and bottom being uniform. The material is assumed to obey the Tresca yield criterion. Softening is introduced by decreasing the yield limit linearly as a function of the maximum principal strain. For the sake of simplicity, the left and right sides of the panel are forced to respond symmetrically.

As the displacement is increased, a band of smeared cracking starts growing horizontally from each edge. Now, at every advance of the crack band there are two possible equilibrium paths which satisfy all the boundary conditions and compatibility conditions:

Path 1 (Fig. 13.37a), for which only one element cracks while its symmetric element unloads elastically, so that the crack band front becomes a single element wide and nonsymmetric

Path 2 (Fig. 13.37b) for which a symmetric pair of elements cracks simultaneously and the crack band front remains two elements wide.

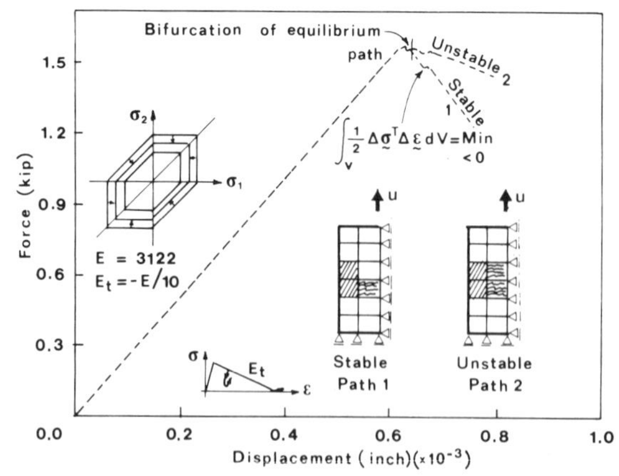

Figure 13.38 Bifurcation of loading path in strain-softening panel. (*After Bažant 1985b and Lin, 1985.*)

By applying infinitesimal force pairs δf as shown in Figure 13.37a, b, it has been checked that the states for both paths are stable. Once a crack band front with single-element width forms, forces δf cannot cause the element on the side of the frontal crack band element to crack, so as to make the cracking band two elements wide. The reason is that the element at the other side of the band has unloaded (elastically). On the other hand, once a band front of two-element width forms, forces δf cannot cause the width of the crack band front to decrease to one element since the cracking that occurred in one element cannot be reversed. More generally, $\delta^2 \mathscr{W}$ for the structure is found to be positive definite for the equilibrium states on path 1 as well as path 2.

So the evolution of cracking is decided by stability of path rather than stability of state, just as it is for Shanley's plastic column. The stable path may be determined by comparing $\delta^2 \mathscr{W}^{(1)}$ and $\delta^2 \mathscr{W}^{(2)}$ for paths 1 and 2. (Their values are best calculated from the increments of the boundary displacement and reaction force, although their calculation on the basis of stress and strain increments in all the elements yields the same result.) In our example, since the displacement is forced to be uniform at the top and bottom boundaries, one may even more simply exploit the fact that the stable equilibrium path is that of a steeper downward slope (Sec. 10.2) in the load-deflection diagram. Now the calculated paths plotted in Figure 13.38 show that the stable path is path 1. It means that the cracking front must localize to a single-element width.

Because the load-displacement diagrams for both cases are identical, the path stability criterion cannot decide whether the cracking in Figure 13.37 localizes into the top element or the bottom element. That is decided by random imperfections.

Convergence of Iterations Depends on Stability of State, not Path

It is interesting to observe that the existing general-purpose large finite element codes generally produce path 2, the incorrect path. Moreover they give no indication of trouble. The iterations of the loading step based on initial elastic stiffness matrix, as normally used in nonlinear finite element programs, converge well and about equally fast for increments along paths 1 and 2.

The problem is that the convergence of these iterations has nothing to do with path stability (Bažant 1988c, 1988b). It depends only on the stability of state. The reason is that, during the iterations of a loading step, the loading (prescribed displacement or load) does not change. Consequently, the errors in the iterated states represent deviations away from the equilibrium state. Path stability, on the other hand, can be checked only when approaches toward a new equilibrium change are involved, which requires consideration of responses to load changes.

Incorrect path can, however, be detected by checking the signs of eigenvalues (or pivots) of the tangential stiffness matrix $\mathbf{K}^t$ (de Borst, 1987). If one of them is negative, the state cannot lie on a stable path (see Sec. 10.4). One must return to a previous load step and find a loading path for which no eigenvalue (or pivot) of $\mathbf{K}^t$ is negative.

Multiple Interacting Crack Band Fronts

The two paths in Figure 13.37a, b, that we considered so far are not all the possible equilibrium paths in this example. In every loading step, the cracking front can expand into any of the elements 1, 2, 3, 4 marked in Figure 13.37c, in any combination. Altogether, the possible combinations consist of cracking in the following elements: 1, 2, 3, 4, 12, 23, 34, 41, 13, 24, 234, 341, 412, 123, and 1234, which represents 15 possible paths. This enumeration shows that, in general, if there are n elements into which the damage can spread, in any combination, the number of all the paths is

$$N = \binom{n}{0} + \binom{n}{1} + \binom{n}{2} + \cdots + \binom{n}{n-1} = 2^n - 1 \tag{13.8.1}$$

Another way to deduce this result is to note that the number of all the possible combinations is $2 \times 2 \times \cdots \times 2 = 2^n$. But one combination represents no cracking in all the elements, which must be excluded, yielding $N = 2^n - 1$.

From this conclusion we now realize that in larger finite element systems the number of all the possible paths is enormous. It is not feasible to check all of them. If the stable path from the initial stress-free state is to be determined, states on an incorrect path can be excluded by checking the sign of the smallest eigenvalue (or pivot) of $\mathbf{K}^t$, as already said. This does not work, however, for determining the correct path that begins from a given initial stable state that lies on an unstable path.

The random imperfection approach might be another way to circumvent the problem. By assigning small random deviations to the strength limits or other softening parameters in all the candidate elements, the symmetries are broken and the number of possible paths might be reduced to one. This approach nevertheless also presents difficulties: In theory, all the possible imperfections, infinite in number, should be checked. Moreover, the imperfections must be very small, but then the loading steps must also be sufficiently small. Otherwise, several elements might straddle their strength peaks simultaneously, in the same loading step.

The number of combinations can be reduced by assuming the cracking to localize into a single-element front. Indeed, this is often the case, but not always. Crack bands in the limit represent fractures, and we may recall that in the analogous problem of interacting crack tips (Sec. 12.5) it is possible that two interacting crack tips may propagate simultaneously. This happens, for example, in a center-cracked square specimen loaded at the middle of the crack. An analogous specimen with a crack band must behave similarly if the crack band is sufficiently narrow.

Interaction of Multiple Shear Bands

Problems of bifurcation and stable path also arise when multiple shear bands form and interact. This may be illustrated by the example in Figure 13.39 taken from Droz and Bažant (1989). The rectangular specimen, solved by finite elements, is subjected to uniform controlled displacement at the top boundary. The top and bottom boundaries are bonded to rigid blocks. The material is assumed to yield according to the Mohr–Coulomb criterion and follow the normality rule (Sec. 10.6). The yield limit decreases with increasing effective plastic strain, which is described by a negative value of the plastic-hardening modulus. A strip of elements in the center of the specimen (cross-hatched in

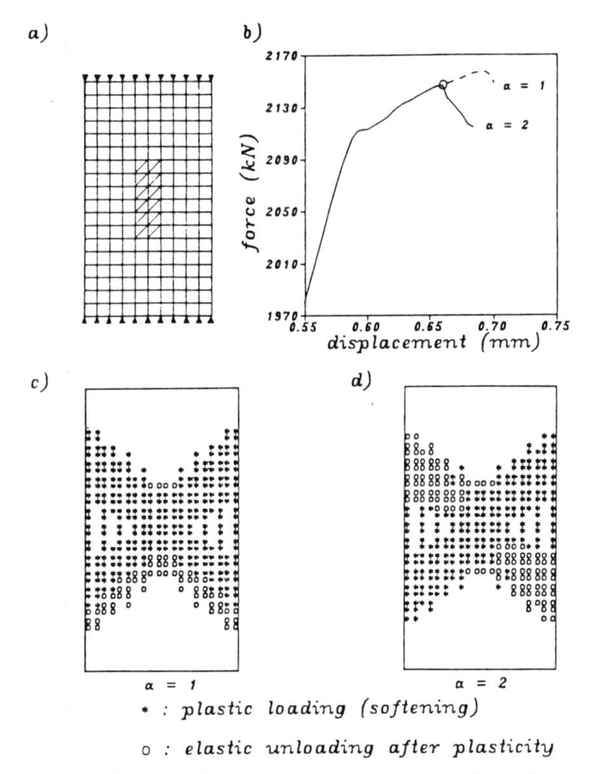

Figure 13.39 (a) Rectangular panel with central elements of smaller yield strength; (b) load-displacement curves; (c, d) failure patterns. (*After Droz and Bažant, 1988.*)

Figure 13.39a) is assumed to have a slightly smaller yield strength than the rest. The width of the shear band is forced to be at least several elements wide, which is accomplished by the nonlocal continuum formulation with local strain, which will be described in Section 13.10.

Two symmetric shear bands of opposite inclinations develop at first, as seen in Figure 13.39c. Symmetric shear bands, however, do not remain the only solution. At a certain moment, while the load is still rising, another solution becomes possible (Fig. 13.39d). This solution is asymmetric—one shear band continues to exhibit loading (with plastic softening) while the other shear band starts to unload after it has experienced some plastic softening. For this second path, the load decreases—see Figure 13.39b. According to Theorem 10.2.1 about structures with a single load, this second, unsymmetric path must be the stable path. The shear softening localizes into a single asymmetric shear band.

Various examples of periodic eigenmodal deformations indicating instabilities and bifurcations were found by de Borst (1989) to occur in finite element systems with material behavior exhibiting either strain softening or nonassociated frictional plasticity; see also de Borst (1987), (1988b), Vardoulakis (1983), Sulem and Vardoulakis (1989), and Vardoulakis, Sulem, and Guenot (1988). These eigenmodes, which trigger shear band formation, resemble the internal instabilities analyzed in Section 11.7 (however, in contrast to Sec. 11.7, de Borst's treatment did not consider finite strains).

Example: Buckling in Direct Tensile Test

In Section 12.5 we saw that a double-edge-notched fracture specimen does not respond symmetrically. Rather, due to interaction of two crack tips, it buckles to the side while only one crack extends and the other is arrested (Bažant, 1985e). The symmetric tensile failure of a specimen with distributed cracking may be modeled by propagation of two symmetric crack bands shown in Figure 13.40a; however, such symmetric growth of two crack bands does not represent a stable path and in reality only one crack band will grow, causing the specimen to buckle to the side. Rots and de Borst (1989) demonstrated such response both by experiment and by two-dimensional finite element analysis with smeared cracking. Pijaudier-Cabot and Akrib (1989) demonstrated a similar asymmetric response using layered beam finite elements with smeared cracking.

This kind of response may be demonstrated quite easily, without finite elements, by the tensile specimen in Figure 13.40c, consisting of two rigid bars of length L connected by a deformable link. The link, of width $2b$ and length $2h$, consists of two very thin flanges of cross-section areas A, which can exhibit strain softening. The structure is symmetric including the properties of flanges. The specimen is supported at the ends by hinges with springs of stiffness C. The deflection is characterized by angle θ. Assuming that initially $\theta = 0$, the

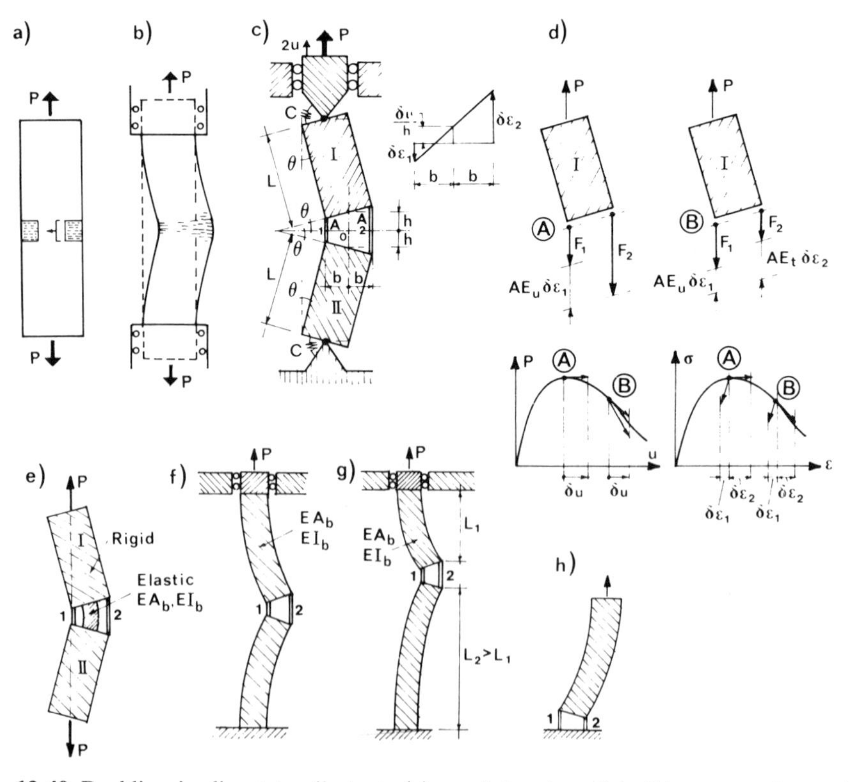

Figure 13.40 Buckling in direct tensile test: (a) crack band model; (b) test specimen; (c) rigid-bar model with deformable softening links; (d) force and strain increments; (e, f, g, h) exercise problems.

conditions of force and moment equilibrium in the link require that $\delta F_1 + \delta F_2 = \delta P$ and $b(\delta F_1 - \delta F_2) = (PL + C)\delta\theta$ where P = axial tensile load, F_1, F_2 = forces in the flanges of the link, and $\delta\theta = \theta$. The strains in the link are $\delta\varepsilon_1 = (\delta u/h) - b\,\delta\theta/h$, $\delta\varepsilon_2 = (\delta u/h) + b\,\delta\theta/h$, and, if one link is loading (modulus E_t) and the other unloading (modulus E_u), $\delta F_1 = E_u A\,\delta\varepsilon_1$ and $\delta F_2 = E_t A\,\delta\varepsilon_2$ where E_t is assumed to be negative (for strain softening) and E_u positive (unloading). We also assume there is no localization along the flanges. Eliminating $\delta\varepsilon_1$, $\delta\varepsilon_2$, δF_1, and δF_2 from these equations we obtain, with the notation $k = b^2A/(PL + C)h$,

$$\delta\theta = \frac{k(E_u - E_t)}{b[1 + k(E_u + E_t)]}\,\delta u \tag{13.8.2}$$

and, for $\delta\theta \neq 0$, $\delta P = K^{(2)}\,\delta u$ where

$$K^{(2)} = \frac{A(E_u + E_t + 4kE_uE_t)}{h[1 + k(E_u + E_t)]} \tag{13.8.3}$$

The condition that one flange unloads requires that $\delta\varepsilon_1 < 0$ or $\delta u < b\,\delta\theta$, which, by substitution of Equation 13.8.2, can be reduced to the condition

$$-2E_tAb^2 > (PL + C)h \tag{13.8.4}$$

Of course, there exists also a symmetric (primary) path, for which both flanges are loading and the lateral deflection is zero ($\theta = 0$, $\varepsilon_1 = \varepsilon_2$). For this path $\delta P = K^{(1)}\,\delta u$ where $K^{(1)} = 2E_tA/h$. Path 2, which represents buckling of the specimen, occurs when $K^{(2)} < K^{(1)}$. According to Equation 13.8.3, and if $E_u + E_t > 0$, this condition can be reduced to the same inequality 13.8.4.

The inequality 13.8.4 is satisfied if the length h of the link is sufficiently small, or if $|E_t|$ is sufficiently large. In such a case the direct tensile test in which the deformation is symmetric and all the points of the cross section have the same strain cannot be carried out for the complete strain-softening branch. The symmetric response path is unstable, bifurcation occurs, and buckling is inevitable. Note that the bifurcation with localization does not begin at the start of strain softening; see also Problem 13.8.3.

Problems

13.8.1 Consider small random imperfections (deviations in the strength limit) to argue that the localized path (path 1) is the path that must occur in Figure 13.37. (Take into account the fact that if one element cracks, the stress in the element on its side decreases rather than increases.)

13.8.2 Apply the same imperfection argument to the shear bands in Figure 13.39.

13.8.3 For the specimen in Figure 13.40c it has been shown in the text that buckling may occur if $K^{(2)} < K^{(1)}$. $K^{(1)} = 0$ at the start of strain softening and $K^{(2)} < 0$ [assuming the slope of the $\sigma(\varepsilon)$ diagram to be continuous]; since $K^{(2)} < K^{(1)}$ one might think that the bifurcation with localization would begin at the start of strain softening. Show that this is impossible. (*Hint*: At the start of strain softening the stress in the loading flange does not change while the stress in the unloading flange decreases. Is the sign of the resulting couple compatible with moment equilibrium in the deflected configuration? See Figure 13.40d and note that buckling can occur only when $F_2 < F_1$.)

13.8.4 Determine conditions under which the tensile specimens in Figure 13.40e, f, g, h cannot remain undeflected. Also, comparing (f) and (g), show that (f) is the case that occurs. This is no doubt the reason that direct tensile specimens tend to break usually at midlength (except for improper grips in which stress concentrations promote breakage at the grip).

13.9 SIZE EFFECT, MESH SENSITIVITY, AND ENERGY CRITERION FOR CRACK BANDS

Practically the most important property of the localization instabilities due to strain softening which were described in the preceding sections is that they cause size effect in structural failures. In finite element analysis in which strain softening is allowed to localize into bands of single-element width, the size effect in turn gets manifested as spurious mesh sensitivity that causes the results to become unobjective, depending on the analyst's subjective choice of mesh size and layout. In this section, we will first explain the size effect, then demonstrate the spurious mesh sensitivity, and finally outline one simple remedy based on an energy criterion for crack band propagation.

Localization as a Cause of Size Effect

In Section 12.2 we have already seen how the blunting of a crack front by a softening zone of finite length causes a size effect that is transitional between plastic limit analysis (which has no size effect) and linear elastic fracture mechanics. A similar size effect is due to fracture front blunting by a crack band (or softening band) of a finite frontal width w_c (Bažant, 1984c, 1986a, 1987a). Let us give a simple illustration, considering the rectangular panels in Figure 13.41, which are geometrically similar and are of different sizes, characterized by dimension D. Without any crack band, the stress is uniform and equal to the nominal stress $\sigma_N = P/bD$ (Eq. 12.2.7, $c_n = 1$) where P = load resultant and b = panel thickness. The strain energy density before cracking is $\sigma_N^2/2E$ (E =

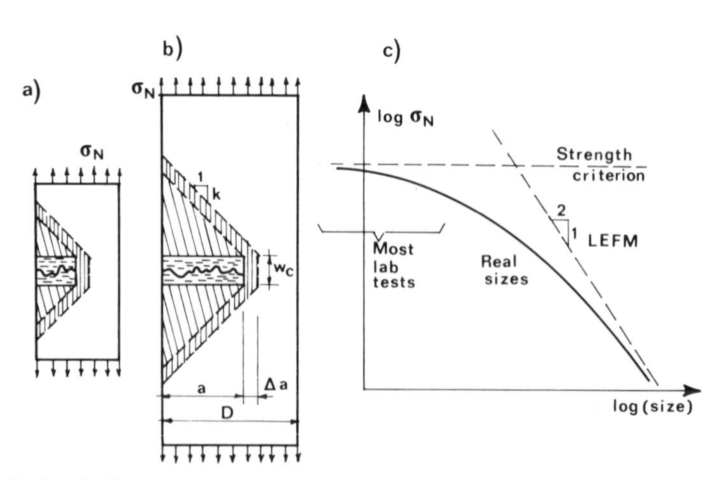

Figure 13.41 (a, b) Crack band advance in panels of different sizes; (c) size effect law. (*Bažant, 1984c.*)

Young's modulus). As argued in Sections 13.2 and 13.3 on the basis of localization instabilities, the frontal width w_c of the crack band has to be a material property, and thus has to be the same for any panel size D.

Creation of the crack band of length a may approximately be imagined to relieve the strain energy from (1) the area of the crack band and (2) the cross-hatched triangular areas limited by "stress-diffusion" lines of slope k. If the crack band advances by Δa, additional strain energy is released from the densely cross-hatched area $\Delta A_c = w_c\, \Delta a + 2ka\, \Delta a$. The energy release rate for $b = 1$ may be calculated as $G = -\partial U/\partial a = (\sigma_N^2/2E)\, \Delta A_c/\Delta a$ (assuming fixed boundary conditions, for which $\Pi = U$). If the crack band propagates, we must also have $G = G_f$ = fracture energy of the band, which may be approximately considered as a material property (i.e., a constant). From this we obtain for the nominal stress at failure the following approximate size effect law (Bažant 1984c):

$$\sigma_N = \frac{Bf_t'}{\sqrt{1+\beta}} \qquad \beta = \frac{D}{D_0} \tag{13.9.1}$$

in which $D_0 = (w_c/2k)D/a$ = constant (because D/a is constant for geometrically similar failure situations), and $Bf_t' = (2EG_f/w_c)^{1/2}$.

For small sizes, $\beta \ll 1$, and Equation 13.9.1 yields $\sigma_N \simeq$ const. For larger sizes, $\beta \gg 1$, and then 1 may be neglected compared to β in Equation 13.9.1, which gives $\sigma_N \simeq \text{const.}/\sqrt{D}$ or $\log \sigma_N = -\frac{1}{2}\log D + \text{const.}$ This is the size effect of linear elastic fracture mechanics (Eq. 12.2.10), which is represented in the plot of $\log \sigma_N$ versus $\log D$ by a straight line of slope $-\frac{1}{2}$; see Figure 13.41c. Equation 13.9.1 represents a smooth transition from the strength criterion to linear elastic fracture mechanics, as sketched in Figure 13.41c.

Equation 13.9.1 ensues by the foregoing type of approximate argument for various shapes of bodies with crack bands (Bažant, 1984c). Moreover, Equation 13.9.1 coincides with Equation 12.2.12 that we derived in Section 12.2 for line cracks with a cohesive zone whose maximum effective length c_f is a material property. Therefore, there must be some more general argument leading to this law.

Indeed, such an argument (Bažant, 1984c, 1986a, 1987a) can be based on dimensional analysis and similitude considerations, starting from the following fundamental hypothesis: The total potential-energy release caused by the fracture (or the crack band) is a function of both (1) the length a of the fracture (crack band) and (2) a length constant of the material, l_0. Here l_0 may represent the material constant combination $l_0 = E'G_f/f_t'^2$ (Irwin's characteristic size of nonlinear near-tip zone, Sec. 12.2), or the effective length c_f of fracture process zone in an infinitely large specimen, or the effective width w_c of this zone. [If only part (1) of this hypothesis were adopted, the size effect of linear elastic fracture mechanics would ensue, and if only part (2) were adopted, there would be no size effect, as in plasticity.)

We consider geometrically similar two-dimensional bodies of different sizes, and assume the ratio a/D at failure to be the same. The potential-energy release $-\Pi$ must depend on a and l_0 through some nondimensional parameters, which may be defined as $\theta_1 = a/D$ and $\theta_2 = l_0/D$ (this can be rigorously proven by applying Buckingham's theorem of dimensional analysis; see, e.g., Barenblatt, 1979). The energy release may then be written in the form $-\Pi = F(\theta_1, \theta_2)V(\sigma_N^2/2E)$, in which $\sigma_N = P/bD$ = nominal stress at failure, P = maximum load, $V = bD^2 =$

volume of the structure times some constant, b = thickness of the body, and F = function characteristic of the shape of the structure and the fracture. Energy balance at fracture propagation requires that $-\partial\Pi/\partial a = bG_f$. From this, noting that $\partial F/\partial a = F'/D$ where $F' = \partial F/\partial\theta_1$, we have $(F'/D)bD^2\sigma_N^2/2E = bG_f$, from which

$$\sigma_N = \left[\frac{2EG_f}{DF'(\theta_1, \theta_2)}\right]^{1/2} \tag{13.9.2}$$

For geometrically similar bodies, $\theta_1 = a/D$ = constant, and so F' depends only on θ_2. This dependence must be smooth and expandable into a Taylor series. Taking only the first two terms of the series we have $F'(\theta_1, \theta_2) \simeq F'_0 + F'_1\theta_2$ in which F'_0 and F'_1 are constants if geometrically similar bodies are considered. Now, with this approximation, Equation 13.9.2 takes the form of Equation 13.9.1 (or Eq. 12.2.12) in which $Bf'_t = (2EG_f/F'_1l_0)^{1/2}$ and $D_0 = w_cF'_1/F'_0$.

An analogous derivation has been given for structures similar in three dimensions (Bažant, 1987a).

The foregoing derivation indicates that the law in Equation 13.9.2 should have a very general applicability. Even though it is only approximate, this law has indeed been shown to describe quite closely the available test data for various types of brittle failure of concrete structures, including (1) diagonal shear failures of reinforced concrete beams, nonprestressed or prestressed, and without or with stirrups; (2) torsional failures of beams; (3) punching shear failures of slabs; (4) pullout failures of bars and anchors; and (5) beam and ring failures of pipes (Bažant and Kim, 1984; Bažant and Cao, 1986a,b and 1987; Bažant and Sener, 1987a,b). Applicability of the law in Equation 13.9.2 to these problems rests on two additional hypotheses: (1) the failure modes of structures of different sizes are geometrically similar and (2) the failure does not occur at initiation of cracking. These hypotheses are normally applicable to concrete structures. The second hypothesis represents a requirement of good design.

Further arguments and applications of the law in Equation 13.9.2 were presented by Planas and Elices (1989). This law is also applicable to fracture specimens of materials such as concrete or rock, as already indicated in Section 12.2.

Inobjectivity or Spurious Mesh Sensitivity

Strain softening, in the form of smeared cracking with a sudden stress drop, was introduced into finite element analysis of concrete structures by Rashid (1968) and a generalization to gradual strain softening was proposed by Scanlon (1971). This concept has found broad application in finite element analysis of concrete structures as well as geotechnical problems, and has been implemented in large general-purpose codes (e.g., NONSAP, ADINA, ABAQUS). Soon, however, it was discovered (Bažant, 1976) that the convergence properties of strain-softening models are incorrect and the calculation results are nonobjective with regard to the analyst's choice of the mesh (see also Bažant and Cedolin, 1979, 1980, 1983; Darwin, 1986; Rots et al., 1985; Bažant 1986a, de Borst, 1989).

The problem is illustrated by the finite element results in Figures 13.42 and 13.43 (taken from Bažant and coworkers). Figure 13.42 shows the results for a rectangular panel with a symmetric crack band, obtained for three different

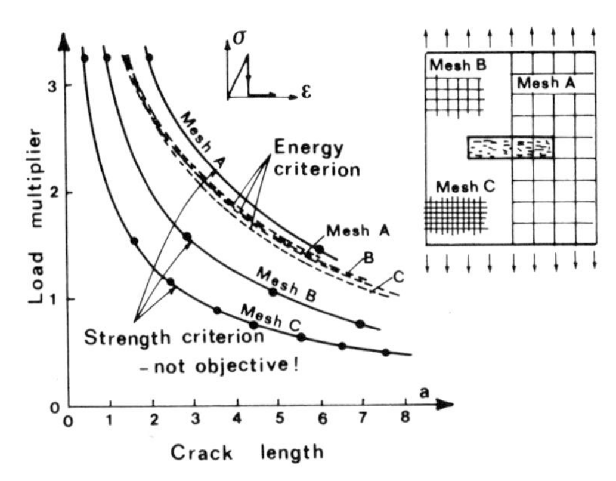

Figure 13.42 Spurious mesh sensitivity and incorrect convergence. (*After Bažant and Cedolin, 1979.*)

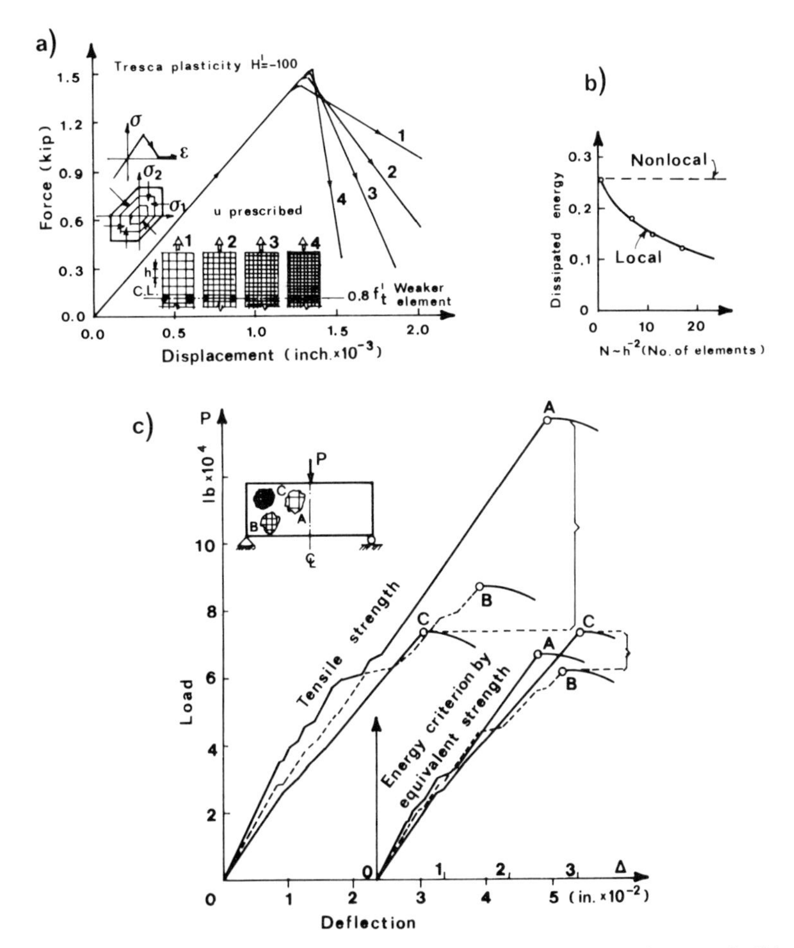

Figure 13.43 (a) Mesh sensitivity for softening stress–strain relations and (b) energy dissipated by failure (*After Bažant 1985b and Lin, 1985*); (c) mesh sensitivity of predictions of diagonal shear failure of reinforced concrete beams. (*After Bažant and Cedolin, 1983.*)

meshes of sizes 1:2:4. The material is assumed to crack when the maximum principal tensile stress reaches the strength limit. The calculated diagrams of the load needed for further propagation versus the crack band length reveal great differences. In the limit of vanishing element size, an arbitrarily small load will suffice to propagate the crack, which is physically incorrect. The reason for this behavior is that, according to the localization instabilities we analyzed before, the crack band will localize into a band of elements a single element wide. The stress in the element just in front of the crack band increases as the mesh size (element size) decreases, and for any given load it can exceed the strength limit if the mesh size is small enough. Figure 13.42, taken from Bažant and Cedolin (1980) was calculated for sudden stress drop at cracking, but similar behavior has been demonstrated for gradual strain softening (e.g., Bažant and Oh, 1983).

Figure 13.43a (which was calculated under the assumption of symmetric response by F.-B. Lin at Northwestern University) shows an example of the effect of the mesh size h on the postpeak softening slope of the load-deflection diagram. For mesh sizes smaller than shown, the postpeak response is a snapback. The great differences between various meshes are again caused by localization. Figure 13.43b presents the values of the energy dissipated by failure of the panel. As the mesh is refined, this energy is found to approach zero. This behavior is, of course, physically incorrect. Figure 13.43c, taken from Bažant and Cedolin (1983), shows that the mesh size can also have a great effect on the maximum load.

The foregoing properties are generally the consequence of bifurcation, which causes strain softening to localize into a band whose front is single element wide. In the limit of vanishing element size, the softening zone has a vanishing volume, and because stress–strain relations always give a finite energy dissipation density, the structure is incorrectly indicated to fail in the limit with a zero energy dissipation.

Note that stability failures of elastic structures (the subject of Chaps. 1–7) exhibit no size effect, same as failure loads when the second-order geometric effects are ignored. This can be verified by the formula for the critical stress in Euler columns: $\sigma_E \pi^2 E/(l/r)^2$. There is an effect of slenderness l/r, which characterizes the structure's shape, but no effect of size. For geometrically similar structures, l/r is constant; hence no size effect.

Energy Criterion for Crack Band and Stability

Correct strain-softening formulations must in some way circumvent the pathological behavior just demonstrated. The simplest remedy is to base the crack band propagation on an energy criterion (Bažant and Cedolin, 1979, 1980, 1983) instead of a strength criterion. This approach is applicable if a sudden stress drop is assumed, and in view of the R curve, if the structure is so large that the energy required to propagate a crack band by a unit length is approximately constant.

Generalizing Rice's (1968) energy analysis of the extension of a notch, Bažant and Cedolin (1979, 1980) calculated the aforementioned required energy for reinforced concrete assuming sudden unidirectional cracking. The crack band extension by length Δa into volume ΔV (of the next finite element, Fig. 13.44) may be decomposed for calculation purposes into two stages.

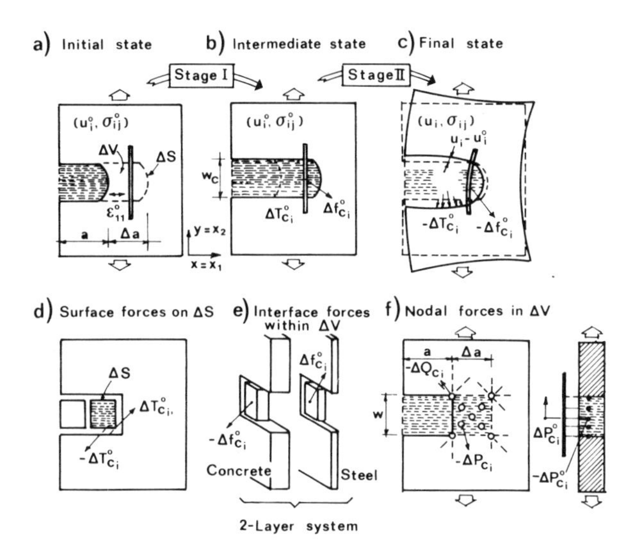

Figure 13.44 Assumed stages of extension of a crack band in reinforced concrete. (*After Bažant and Cedolin, 1980.*)

Stage I. Cracks are created in concrete inside volume ΔV of the element ahead of the crack in the direction of principal tensile stress (Fig. 13.44b), while, at the same time, the deformations and stresses in the rest of the body are imagined to remain fixed (frozen). This means that one must introduce surface tractions $\Delta T^0_{c_i}$ applied on the boundary ΔS of volume ΔV, and distributed forces $\Delta f^0_{c_i}$ applied at the concrete-steel interface, such that they replace the action of concrete that has cracked upon the remaining volume $V - \Delta V$ and upon the reinforcement within ΔV.

Stage II. Next, forces $\Delta T^0_{c_i}$ and $\Delta f^0_{c_i}$ (Fig. 13.44c) are released (unfrozen) by gradually applying the opposite forces $-\Delta T^0_{c_i}$ and $-\Delta f^0_{c_i}$, reaching in this way the final state.

Let u^0_i and ε^0_{ij} be the displacements and strains before the crack band advance, and let u_i and ε_{ij} be the same quantities after the crack band advance. For the purpose of analysis, the reinforcement may be imagined to be smeared in a separate parallel layer undergoing the same strains as concrete. The interface forces between steel and concrete, $\Delta f^0_{c_i}$, then appear as volume forces applied on the concrete layer.

Upon passing from the initial to the intermediate state (Stage I), the strains are kept unchanged, while cracking goes on. Thus, the corresponding stress changes in concrete in ΔV are given by $\Delta\sigma^c_{11} = \sigma^{c^0}_{11} - E'\varepsilon^0_{11} = (\varepsilon^0_{11} + \nu'\varepsilon^0_{22})E'/(1-\nu'^2) - E'\varepsilon^0_{11}$; $\Delta\sigma^c_{22} = \sigma^{c^0}_{22}$; $\Delta\sigma^c_{12} = 0$ (cracks are assumed to propagate in the principal stress direction). Here, $\sigma^{c^0}_{ij}$ denotes the stress fraction carried before cracking by concrete alone, defined as the force in concrete per unit area of the steel-concrete composite; E and ν are the Young's modulus and Poisson's ratio of concrete. The values $E' = E$ and $\nu' = \nu$ apply to plane stress and $E' = E/(1-\nu^2)$ and $\nu' = \nu/(1-\nu)$ to plane strain. The change in potential energy of the system during Stage I in Figure 13.44b is given by the elastic energy initially stored in ΔV

and released by cracking, that is,

$$\Delta U_1 = -\int_{\Delta V} \tfrac{1}{2}(\sigma_{ij}^{c^0} \varepsilon_{ij}^0 - E'\varepsilon_{11}^{0^2})\, dV \qquad (13.9.3)$$

The change in potential energy during Stage II is given by the work done by the forces $\Delta T_{c_i}^0$ and $\Delta f_{c_i}^0$ while they are being released, that is,

$$\Delta U_2 = \int_{\Delta S} \tfrac{1}{2}\, \Delta T_{c_i}^0 (u_i - u_i^0)\, dS + \int_{\Delta V} \tfrac{1}{2} \Delta f_{c_i}^0 (u_i - u_i^0)\, dV \qquad (13.9.4)$$

Coefficients $\frac{1}{2}$ must be used because forces T_{c_i} and f_{c_i} reduce to zero at the end of Stage II (and for small enough Δa they should reduce nearly linearly, similarly as in Eq. 12.1.7). If the concrete is reinforced, part of the energy is consumed by the bond slip of reinforcing bars during cracking within volume ΔV. This part may be expressed as $\Delta W_b = \int_s F_b \delta_b\, ds$, in which δ_b represents the relative tangential displacement between the bars and the concrete, F_b is the average bond force during displacement δ_b per unit length of the bar (force during the slip) and s is the length of the bar segment within the fracture process zone w_c (and not within volume ΔV since the energy consumed by bond slip would then depend on the chosen element size). The energy criterion for the crack band extension may now be expressed as

$$\delta^2 \mathscr{W} = G_f\, \Delta a - \Delta U_1 - \Delta U_2 - \Delta W_b \begin{cases} >0 & \text{stable} \\ =0 & \text{critical} \\ <0 & \text{unstable} \end{cases} \qquad (13.9.5)$$

where $\delta^2 \mathscr{W} = -T(dS)_{in}$ can be regarded as energy (second order work) that would have to be externally supplied to the structure if the crack band of width h should be extended by length Δa. In Bažant and Cedolin (1983), the foregoing formulation has been developed in detail and various aspects such as zigzag crack bands and bond slip have been dealt with (see also Bažant, 1985d).

The results obtained with the energy criterion, shown in Figures 13.42 and 13.43c, are seen to be sufficiently close to each other. The practical applicability of the energy criterion in Equations 13.9.5 is nevertheless limited. This criterion is easy to apply only if the stress may be assumed to drop suddenly after the strength limit has been reached. This is generally possible only if the structure is sufficiently large compared to the crack band width h (w_c or $h \geq 2EG_f/f_t'^2$; see Eq. 13.10.4). For many problems one must consider gradual strain softening, and also one cannot base the term $G_f\, \Delta a$ in Equation 13.9.5 on a constant fracture energy value G_f. Then one needs a more general and fundamental remedy of spurious mesh sensitivity, which requires the localization to be limited to bands of a certain minimum thickness that is a material property (or depends on material properties). This is achieved by nonlocal material models, which we discuss in the next section.

Problems

13.9.1 If the mesh size is enlarged in proportion to the size of the structure, a (local) solution based on stress–strain relations (as opposed to stress-displacement relations) must exhibit no size effect, that is, the failure must occur at the same nominal stress σ_N. On the other hand, if the structure is enlarged while the mesh size is kept constant, there must be a size effect and it should roughly follow Equation 3.9.1. Use these considerations to show that the effect of mesh size h on σ_N at fixed structure size D is approximately given by the relation

$$\sigma_N = Bf_t'\left(1 + \frac{D}{h}\right)^{-1/2} \tag{13.9.6}$$

which for very small h becomes $\sigma_N \simeq \text{const.} \times \sqrt{h}$ (note that $\sigma_N \to 0$ for $h \to 0$).

13.9.2 How does the energy criterion in Equation 13.9.5 simplify if all the stress components in ΔV are reduced to zero and reinforcement is absent? (Calculation of Rice, 1968.)

13.9.3 Consider the formulas for the critical loads of a thin-wall beam in lateral buckling, a square simply supported beam, and a cylindrical axially compressed shell. Examine the nominal stress at failure, and show that the stability failures of these elastic structures exhibit no size effect.

13.10 NONLOCAL CONTINUUM AND ITS STABILITY

Distributed damage such as cracking is conveniently described by stress–strain relations with strain softening on the macroscale. For general-purpose programs, this formulation seems computationally more efficient than the modeling of localized damage through stress-displacement relations for the fracture process zone of a line crack and is more realistic in those situations where damage does not localize. As we have seen, however, objective and properly convergent results require that the localization be artificially limited to a zone of a certain minimum thickness that is essentially a material property. This can, in general, be achieved by introducing a nonlocal continuum concept. In this section, we discuss various forms of this concept, illustrate some spurious instabilities that can be caused by nonlocality, and present some effective nonlocal formulations.

Crack Band Model

The simplest but also crudest way to limit localization is the crack band model that was formulated in detail in Bažant (1982a) and Bažant and Oh (1983, 1984) on the basis of the general idea from Bažant (1976) and of the studies by Bažant and Cedolin (1979, 1980). The basic idea is (1) to prescribe a certain minimum admissible width w_c of the crack band (or damage band) that represents a material property (Bažant, 1976), and (2) in case w_c as the finite element size would be too small, to adjust the softening stress–strain relation so as to achieve the correct energy dissipation by the crack band (Bažant and Cedolin, 1979, 1980) (Fig. 13.45a). It has been shown that this simple approach makes it possible to achieve good agreement with all the basic fracture test data for concrete as well as rock. By data fitting it was found that the optimal values of w_c for normal concretes are between $2d_a$ and $8d_a$, with $w_c \simeq 3d_a$ as the overall optimum for concrete and $w_c \simeq 5d_g$ for rocks where d_a, d_g = maximum size of aggregate in concrete or grain in rock. To avoid bias, it is best to use a square element mesh in the region near the fracture, although some small mesh irregularities are not a

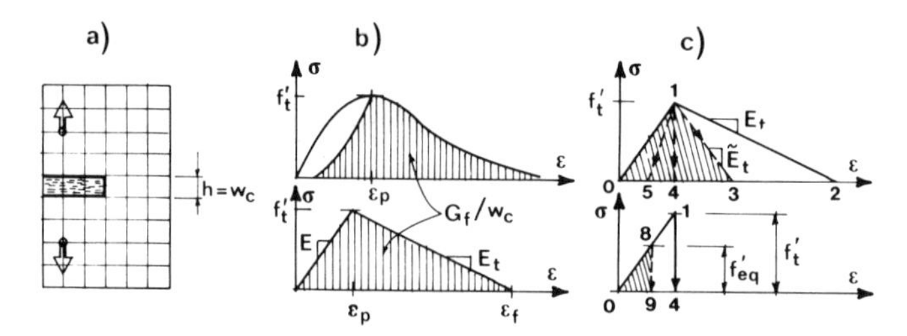

Figure 13.45 (a) Crack band advance, (b) softening stress–strain relations, and (c) adjustment for correct energy dissipation.

problem. The elements near the boundary of the fracture process zone are preferably of low order (e.g., three-node triangles or four-node quadrilaterals) since they can best simulate discontinuities in strains.

The uniaxial stress–strain relation in the crack band model can most simply be considered as linear (Fig. 13.45b), characterized by tensile strength f_t' and softening modulus E_t (<0). In that case, the fracture energy is

$$G_f = w_c\left(\frac{f_t'^2}{2}\right)\left(\frac{1}{E} - \frac{1}{E_t}\right) \tag{13.10.1}$$

In reality this relation is no doubt curved, and, for example, a decaying exponential curve seems to give better results (Darwin, 1986). In that case G_f/w_c is equal to the area under the softening curve and the unloading curve emanating from the peak-stress point (Fig. 13.45b). However, in view of other simplifications involved, the simple triangular stress–strain relations might not give inferior results.

The triaxial stress–strain relation for softening is obtained under the assumption that the total strain is decomposed as $\boldsymbol{\varepsilon} = \boldsymbol{\varepsilon}^e + \boldsymbol{\varepsilon}^{\text{fr}}$ where $\boldsymbol{\varepsilon}^e$ = elastic strain and $\boldsymbol{\varepsilon}^{\text{fr}}$ = fracturing strain. The fracturing strain is usually assumed to be a function of the normal strain ε_n in the direction normal to the cracks, in which case the stress–strain relation can be manipulated either into a form of an orthotropic total stress–strain relation (Bažant, 1985d; Bažant and Oh, 1983) or damage theory. Basically, there are two variants.

In the classical variant, the smeared cracks are assumed to start forming in the direction normal to the maximum prinicipal stress σ_{I} when the strength limit is first reached, and the crack orientation is assumed to remain fixed even if σ_{I} subsequently rotates. Then it is possible that the crack surfaces would later receive shear stresses. The stiffness of the cracked material in shear has usually been assumed to be βG where G = shear modulus and β = empirical shear retention factor ($0 < \beta < 1$), as proposed by Suidan and Schnobrich (1973) and Yuzugullu and Schnobrich (1973). Cedolin and Dei Poli (1977) proposed the use of a value of β decreasing with the crack width.

In a new variant, it is assumed that the cracks rotate so as to remain normal to σ_{I} and that no shear stresses arise on the crack surfaces. This "rotating (or swinging) crack" hypothesis (which really means that cracks of one direction close and cracks of another direction open) has been shown to agree relatively well with some experimental observations (Gupta and Akbar, 1984; Cope, 1984), perhaps better than the fixed crack direction hypothesis. The general stress–strain relation for this variant has been developed by Bažant and Lin (1988a). For loading it has the form

$$\varepsilon_{ij} = \left[C_{ijkm} + \frac{\omega}{(1-\omega)E'} n_i n_j n_k n_m \right] \sigma_{km} \tag{13.10.2}$$

where C_{ijkm} = isotropic tensor of elastic moduli, ω = damage function of maximum principal strain ε_I, and n_i = unit vector of maximum principal strain direction.

When w_c is too small and the finite element size h needs to be made larger than w_c, the softening stress–strain relation for the crack band model must be adjusted so as to ensure that the energy dissipation by a crack band of width h is about the same as the actual crack dissipation by a crack band of width w_c. The manner of adjustment is illustrated in Figure 13.45c, where 012 is the actual stress–strain curve associated with crack band width w_c. When h is increased beyond the value of w_c, one first adjusts the value of softening modulus from E_t to $\tilde{E}_t$ so that area 0130 times h would equal area 0120 times w_c. From this condition one finds

$$-\frac{1}{\tilde{E}_t} = \frac{w_c}{h}\left(\frac{1}{E} - \frac{1}{E_t}\right) - \frac{1}{E} = \frac{2G_f}{hf_t'^2} - \frac{1}{E} \qquad (\text{if } \tilde{E}_t < 0) \tag{13.10.3}$$

However, for a band wider than $h = 2l_0$ where

$$l_0 = \frac{EG_f}{f_t'^2} \tag{13.10.4}$$

the value of $\tilde{E}_t$ becomes positive, which means that one would need to use a stress–strain curve that exhibits snapback (curve 015 in Figure 13.45c). To make static calculations possible, one must then replace snapback curve 015 with curve 089 for which the stress drop is vertical and the strength is reduced from f_t' to a certain equivalent strength f'_{eq} such that area 0890 times h would equal G_f (or area 0140 times h_{cr}). This condition yields (Bažant and Cedolin, 1979, 1980; Bažant and Oh, 1983)

$$f'_{\mathrm{eq}} = \frac{\sqrt{2EG_f}}{\sqrt{h}} \qquad (\text{if } f'_{\mathrm{eq}} \le f'_t) \tag{13.10.5}$$

Note that the reduction of the equivalent strength is identical to that obtained from the size effect law in Equation 13.9.2 when $D_0 = \text{const.} \times h$ while D is fixed and is much larger than D_0.

When the elements are not exactly square in shape, some approximate adjustments are possible to preserve applicability of the model. The value of h may be defined as $h = A/\Delta a$ where A = area of finite element that undergoes cracking and Δa = length of advance of crack band corresponding to this element.

Nonlocal Continuum Concept

The crack band model has some less than satisfactory features. From the mathematical viewpoint it is disconcerting that convergence of the finite element solutions cannot be checked since the element size cannot be reduced below w_c. A more fundamental model should permit arbitrary subdivisions of the fracture process zone by finite elements. (Even though the material is on such a scale very heterogeneous, the calculations refer to a macroscopic smoothing continuum that describes the statistical mean of the scattered material response.) Another weakness is that the exact value of the effective width of the fracture process zone that gives the correct rate of energy dissipation is probably not exactly a constant but varies slightly due to interaction with boundaries and the type of loading. So what is the material constant governing localization and energy dissipation? It seems that a true and more fundamental constant is the characteristic length l of a nonlocal continuum model for the material.

Nonlocal continuum is a continuum for which the constitutive relation for a point involves some variables obtained by averaging over a neighborhood of this point. For example, the average (nonlocal) strain is defined as

$$\bar{\boldsymbol{\varepsilon}}(\mathbf{x}) = \frac{1}{V_r(x)} \int_V \alpha(\mathbf{x}-\mathbf{s})\boldsymbol{\varepsilon}(\mathbf{s})\, dV(\mathbf{s}) \tag{13.10.6}$$

with

$$V_r(\mathbf{x}) = \int_V \alpha(\mathbf{x}-\mathbf{s})\, dV(\mathbf{s}) \tag{13.10.7}$$

in which the integral may be considered to extend over the whole structure, $\boldsymbol{\varepsilon}(\mathbf{s})$ = the usual (local) strain at point $\mathbf{s}$, α = given weight function, and the overbar denotes the spatial averaging operator.

As the simplest form of the weight function, one may consider $\alpha = 1$ within a certain representative volume V_0 centered at point $\mathbf{x}$, and $\alpha = 0$ outside this volume. Convergence of numerical solutions, however, is better if α is a smooth function. An effective choice is the bell-shaped function (Fig. 13.46a):

$$\alpha = \left[1 - \left(\frac{r}{\rho_0 l}\right)^2\right]^2 \quad \text{if } |r| < \rho_0 l \qquad \alpha = 0 \quad \text{if } |r| \geq \rho_0 l \tag{13.10.8}$$

where $r = |\mathbf{x}-\mathbf{s}|$ = distance from point $\mathbf{x}$, l = characteristic length (material property), and ρ_0 = coefficient chosen in such a manner that the volume under function α from Equation 13.10.8 is equal to the volume under function $\alpha = 1$ for $r \leq l/2$ and $\alpha = 0$ for $r > l/2$ (which represents a line segment in 1D, a circle in 2D, and a sphere in 3D). From this requirement, $\rho_0 = \frac{15}{16} = 0.9375$ for one dimension, $\rho_0 = \sqrt[3]{\frac{3}{4}} = 0.9086$ for two dimensions, and $\rho_0 = (\frac{105}{192})^{1/3} = 0.8178$ for three dimensions. For points whose distance from all the boundaries is larger than $\rho_0 l$, we have $V_r(\mathbf{x}) = l$ in one dimension, $\pi l^2/4$ in two dimensions, and $\pi l^3/6$ in

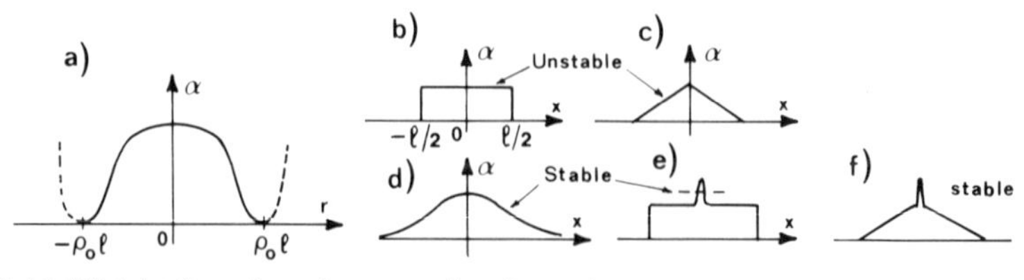

Figure 13.46 Weight functions for a nonlocal continuum.

three dimensions. For points closer to the boundary, the averaging volume protrudes outside the body, and then $V_r(\mathbf{x})$ is variable because the domain of averaging is not constant. The Gaussian (normal) distribution function (Fig. 13.46d) has also been found to be effective for function α.

The concept of a nonlocal continuum was conceived and for a long time studied for elastic materials with heterogeneous microstructure (Kröner, 1967; Krumhansl, 1968; Kunin, 1968; Beran and McCoy, 1970; Levin, 1971; Eringen and Edelen, 1972; Eringen and Ari, 1983). The macroscopic continuum stresses $\boldsymbol{\sigma}(\mathbf{x})$ and strains $\boldsymbol{\varepsilon}(\mathbf{x})$ are defined as statistical averages of the randomly scattered microstresses $\boldsymbol{\sigma}_M(\mathbf{x})$ and microstrains $\boldsymbol{\varepsilon}_M(\mathbf{x})$ (Fig. 13.47) taken over a suitable representative volume around point $\mathbf{x}$. The statistical theory of the microstructure developed in the 1960s showed that if $\boldsymbol{\varepsilon}(\mathbf{x})$ is nonuniform the macroscopic constitutive relation is not exactly of the form $\boldsymbol{\sigma}(\mathbf{x})$ = function of $\boldsymbol{\varepsilon}(x)$ but $\boldsymbol{\sigma}(x)$ also depends on the average macroscopic strain $\bar{\boldsymbol{\varepsilon}}(\mathbf{x})$ (even though $\boldsymbol{\varepsilon}$ itself is an average of $\boldsymbol{\varepsilon}_M$). This provided the original impetus for the development of nonlocal elasticity.

Periodic Instabilities Due to Nonlocal Concept

The nonlocal concept has for a long time been used in various analytical studies of elastic heterogeneous materials until finite element studies (Bažant, Belytschko, and Chang, 1984) revealed the existence of spurious instabilities due to spatial averaging of the elastic strain. Let us briefly summarize their analysis made in Bažant and Chang (1984). We consider a long bar with the nonlocal uniaxial elastic stress–strain relation:

$$\sigma(x) = E\bar{\varepsilon}(x) \qquad \bar{\varepsilon}(x) = \int_{-\infty}^{\infty} \varepsilon(x+r)\alpha(r)\,dr \tag{13.10.9}$$

in which $\int_{-\infty}^{\infty} \alpha(r)\,dr = 1$, and E = Young's modulus. Now, any theory of an elastic continuum must obviously satisfy the following two requirements: (1) If the stresses $\sigma(x)$ are everywhere zero, the strains in a stable material must also be zero, that is, no unresisted deformation (zero-energy deformation mode) may be permitted by the theory. (2) In a stable material, the wave propagation velocity v must be real. From these two requirements it follows that the Fourier transform

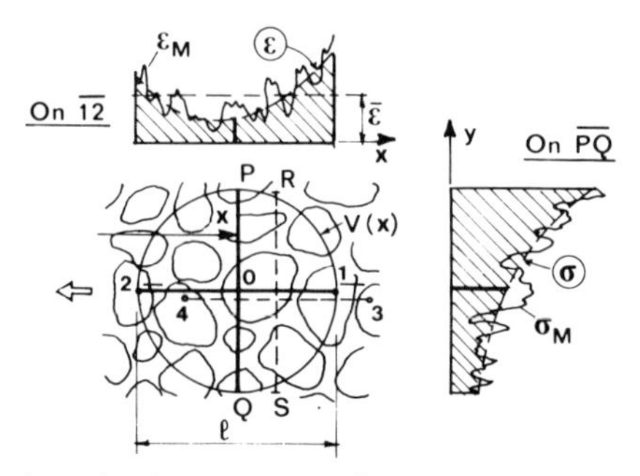

Figure 13.47 Stresses and strains in concrete microstructure.

of the weight function, $\alpha(r)$, must be positive for all real ω, that is,

$$\alpha^*(\omega) = \int_{-\infty}^{\infty} e^{-i\omega r}\alpha(r)\,dr > 0 \tag{13.10.10}$$

To prove this condition, consider Requirement 1 first. According to Equation 13.10.9, $\sigma(x) = 0$ occurs when

$$\int_{-\infty}^{\infty} \varepsilon(x+r)\alpha(r)\,dr = 0 \tag{13.10.11}$$

This condition must not have any nonzero solution (eigenstate). A general strain distribution may be approximated as $\varepsilon(x) = \sum_k a_k \exp(i\omega_k x)$, in which a_k and ω_k are some real numbers ($k = 1, 2, 3, \ldots$), and the actual strain is to be understood as the real part. No single term of this expansion, that is, $\varepsilon(x) = a \exp(i\omega x)$ with a real amplitude a_r and a real frequency ω_r, may satisfy Equation 13.10.11. Substituting this into Equation 13.10.11, we obtain the condition that the equation $\int_{-\infty}^{\infty} \alpha(r) \exp[i\omega(x+r)]\,dr = 0$ must not have any solution, and dividing this equation by $\exp(i\omega x)$, we conclude that $\alpha^*(\omega)$ must not be zero for any ω. Since $\alpha^*(\omega)$ must be continuous, it must be either positive everywhere or negative everywhere.

Second, consider Requirement 2. We restrict attention to small deformations, such that in one dimension $\varepsilon(x) = \partial u(x)/\partial x$, in which u = displacement. The equation of motion is $\partial\sigma/\partial x = \rho\,\partial^2 u/\partial t^2$, in which t = time and ρ = mass density. Equation 13.10.9 then yields

$$E\frac{\partial}{\partial x}\int_{-\infty}^{\infty} \frac{\partial u(r)}{\partial s}\,\alpha(x-s)\,ds = \rho\frac{\partial^2 u}{\partial t^2} \qquad (s = x + r) \tag{13.10.12}$$

Any wave may be decomposed into harmonic components of the type $u(x) = a\exp[i\omega(t - vt)]$ in which v = wave velocity; ω, a = real constants; and $\omega \neq 0$ (since for $\omega = 0$ there is no strain). Substituting this into Equation 13.10.12, multiplying the equation by $\exp(i\omega vt)/i\omega\rho a$, and setting $s = x + r$, $ds = dr$, we obtain

$$\frac{E}{\rho}\frac{\partial}{\partial x}\left[e^{i\omega x}\int_{-\infty}^{\infty} e^{i\omega r}\,\alpha(-r)\,dr\right] = v^2 i\omega e^{i\omega x} \tag{13.10.13}$$

Furthermore, substituting $r = -y$, and noting that $\alpha(-r) = \alpha(r)$, we find that the integral in this equation equals $\int_{-\infty}^{\infty} \exp(-i\omega y)\alpha(y)\,dy$, which represents $\alpha^*(\omega)$. Then, differentiating $\alpha^*(\omega)\exp(i\omega x)$ with respect to x, and dividing the equation by $i\omega\exp(i\omega x)$, we finally get

$$v^2 = \frac{E}{\rho}\,\alpha^*(\omega) \tag{13.10.14}$$

We see that v is real if and only if $\alpha^*(\omega)$ is positive for all ω except $\omega = 0$. The fact that $\alpha^*(\omega)$ must be positive also for $\omega = 0$ follows from Requirement 1, as already proven. Note that for the usual local continuum we have $\alpha(r) = \delta(r)$ = Dirac delta function, and so $\alpha^*(\omega) = 1$ for all ω.

For a uniform weighting function $\alpha(r) = 1/l$, the Fourier transform is $\alpha^*(\omega) = (2/\omega l)\sin(\omega l/2)$, which can be positive, negative, or zero. Therefore, a nonlocal continuum with this $\alpha(s)$ is unstable. The triangular weight function $\alpha(r) = (2/l)(1 - 2|r|/l)$ also produces instability because its Fourier transform is $\alpha^*(\omega) = 2(2/\omega l)^2(1 - \cos\frac{1}{2}\omega l)$, which can be zero. On the other hand, the Gaussian (normal) distribution function $\alpha(r) = (l\sqrt{2\pi})^{-1}\exp(-r^2/2l^2)$ produces no instability because $\alpha^*(\omega) = \exp(-\omega^2 l^2/2) > 0$. Neither does the bilateral exponential distribution, the Cauchy distribution, and the hyperbolic sine function (for details see Bažant and Chang, 1984). The uniform distribution may be stabilized by adding to it a spike of the Dirac delta function, $\alpha(r) = c\delta(r) + (1 - c)/l$, and from the Fourier transform it is then found that instabilities are prevented if $c > 0.17847$. The spiked triangular distribution, $\alpha(r) = c\delta(r) + 2(1 - c)(1 - 2|r|/l)/l$, is found to produce no instabilities if $c > 0$.

Nonlocal Continuum with Local Strain

After realizing that the nonlocal continuum concept can serve to limit localization of strain softening to a band of a certain minimum thickness, the first nonlocal finite element model was based on a spiked uniform distribution (Bažant, Belytschko, and Chang, 1984; Bažant, 1984a). As a manageable approach to programming, it was shown that this formulation can be considered as the limiting case of a system of elastic-softening finite elements that are imbricated (i.e., overlap each other in a regular manner) and are further overlaid by a local elastic layer. Although the imbricate model was found to indeed limit localization of softening and guarantee mesh insensitivity, especially in terms of the energy dissipated by failure, it was still quite complicated, due to the fact that the differential equations of equilibrium as well as the boundary conditions were of nonstandard form (involving either spatial integrals or higher-order derivatives); see Bažant (1984a). Moreover, the need for using an overlay with a local elastic continuum in order to suppress spurious zero-energy periodic modes of instability was an unrealistic artifice.

The property that gives rise to nonstandard differential equations of equilibrium and boundary conditions may be explained by considering their derivation from the virtual work relation:

$$\delta W = \int \sigma_{ij}(\bar{\boldsymbol{\varepsilon}})\, \delta\bar{\varepsilon}_{ij}\, dV - \int f_i\, \delta u_i\, dV - \int p_i\, \delta u_i\, dS = 0 \qquad (13.10.15)$$

where V, S = volume and surface of the structure; f_i, p_i = given volume and surface forces; u_i = displacements, and σ_{ij} = stresses. The fact that the strain in the variation $\delta\bar{\varepsilon}_{ij}$ is nonlocal, that is, expressed in terms of a spatial integral, poses difficulties. For a local continuum, by contrast, one can set $\delta\varepsilon_{ij} = \delta u_{i,j}$ and use the Gauss integral theorem to obtain the differential equations and boundary conditions of the problem (Chaps. 5 and 6). But this standard procedure is impossible if $\delta\bar{\varepsilon}_{ij}$ is expressed in terms of a spatial integral (even if the integral is approximated by a Taylor series). A much more complex procedure was found to be necessary, with a complicated result (Bažant, 1984a). This observation led to the idea of a partially nonlocal continuum in which $\delta\bar{\varepsilon}_{ij}$ is is replaced with $\delta\varepsilon_{ij}$ in Equation 13.10.15, but σ_{ij} is still determined from $\bar{\varepsilon}_{ij}$ while the elastic strains are local. After this modification, the Gauss integral theorem can be applied to the first integral in Equation 13.10.15, and then the differential equations of equilibrium and boundary conditions that arise are of the standard form.

Such a nonlocal model, called the nonlocal continuum with local strain (Bažant and Pijaudier-Cabot, 1987b; Bažant, Lin, and Pijaudier-Cabot, 1987; and Pijaudier-Cabot and Bažant, 1987), has proven to be quite simple and effective. In this approach the usual constitutive relation for strain softening is simply modified by calculating all the state variables that characterize strain softening from the nonlocal rather than local strains. All that is necessary to change in a local finite element program is to provide a subroutine (Bažant and Ožbolt, 1989a) that delivers at each integration point, each loading step, and each iteration the value of $\bar{\varepsilon}_{ij}$. The formulation can be of various kinds. In nonlocal damage theory, one needs to either calculate nonlocal damage $\bar{\omega}$ from $\bar{\varepsilon}_{ij}$ or replace ω with its spatial average $\bar{\omega}$. If strain softening is due to degradation of the yield limit τ_p (Bažant and Lin, 1988b), either τ_p must be replaced by $\bar{\tau}_p$ or the value of the effective plastic strain from which τ_p is calculated must be replaced by its spatial average. If strain softening is described through the fracturing strain $\boldsymbol{\varepsilon}^{fr}$, one needs to replace $\boldsymbol{\varepsilon}^{fr}$ with its spatial average $\bar{\boldsymbol{\varepsilon}}^{fr}$ or calculate it from $\bar{\varepsilon}$.

The essential property of all the variants is that the energy dissipation density rate due to damage must be nonlocal. Then it can be mathematically proven (Bažant and Pijaudier-Cabot, 1987b, 1988b) that the energy dissipation density rate cannot localize into a vanishing volume. Briefly, the proof goes as follows. The strains as well as ω, τ_p, or $\boldsymbol{\varepsilon}^{fr}$ have at least C_0-continuity (i.e., they could be Dirac delta functions); so the spatial averages $\bar{\varepsilon}$, $\bar{\omega}$, $\bar{\tau}_p$, $\bar{\varepsilon}^{fr}$ have at least C_1-continuity, and so they cannot be Dirac delta functions; but this means that the damage cannot localize into a zero volume (as numerical solutions confirm).

A micromechanics argument for the nonlocal continuum with local strain has also been given (Bažant, 1987c), albeit only for a rather simplified situation that consisted of a uniaxially stressed elastic continuum containing a cubic array of either circular (penny-shaped) cracks or circular ligaments that are small compared to their spacing. For this problem one can satisfy the homogeneization conditions for the micro–macro compatibility of displacements and equality of work exactly, and the result is a continuum damage formulation in which the elastic strains are local while the damage is nonlocal. Intuitively, the reason for the macroscopic nonlocality of fracturing strain (or damage) due to microcracks is that formation of a crack releases energy from a region of nonzero volume adjacent to the crack, and that the additional displacement due to formation of any crack is not manifested immediately at the crack but only at sufficient distance from the crack. Conversely, formation of a crack or other damage at a certain point cannot depend on the local deformation at that point but must be determined by the overall deformation of a certain region around that point. In Figure 13.47, picturing the microstructure of concrete, the damage on cross section *PQ* depends mainly on the relative displacement of points 1 and 2 and not on the local strains at point 0 (also note the requirement for symmetry; the relative displacement of points 3 and 4 would determine the damage on line *RS* rather than *PQ*).

Further mathematical arguments for a nonlocal continuum with local elastic strain and nonlocal energy dissipation were advanced by Simo (1989) (from the viewpoint of regularization). An argument based on Weibull-type statistical theory of strength, which also leads to a stochastic generalization of the size effect law in Equation 13.9.2, is given in Bažant and Xi (1989).

One-Dimensional Localization Instability

The characteristic length l of a nonlocal continuum is not equal to the width w_c of a localized crack band, but must be related to it. This relationship may be established by stability analysis on the basis of the given nonlocal constitutive model. Such an analysis has been carried out for one dimension by Bažant and Pijaudier-Cabot (1988b), and we will explain it now. The analysis is based on a uniaxial constitutive law of continuum damage mechanics

$$\sigma = (1 - \Omega)E\varepsilon \tag{13.10.16}$$

where Ω = nonlocal damage. The specific free energy per unit volume is $\rho\psi = \frac{1}{2}(1 - \Omega)E\varepsilon^2$, and the damage energy release rate (Lemaitre and Chaboche, 1985) is

$$Y = -\frac{\partial(\rho\psi)}{\partial\Omega} = \frac{E\varepsilon^2}{2} \tag{13.10.17}$$

The energy dissipation rate per unit volume is $\phi = -\partial(\rho\psi)/\partial t = -\dot{\Omega}\,\partial(\rho\psi)/\partial\Omega = \dot{\Omega}Y$. The local damage evolution is, in general, defined by an evolution equation of the type $\dot{\omega} = f(\varepsilon, \omega)$; but for the sake of simplicity we use an integrable special form of this relation:

$$\omega = g(Y) = 1 - [1 + b(Y - Y_1)]^{-n} \tag{13.10.18}$$

in which b, n, Y_1 = positive material constants; $n > 2$; and Y_1 = local damage threshold. The damage is assumed to grow only for loading. For unloading or reloading, $\dot{\omega}$ must vanish in the nonlocal (average) sense. This means the response is elastic. The loading criterion and the nonlocal damage Ω are defined as follows:

$$\begin{aligned} &\text{If} \quad F(\bar{\omega}) = 0 \quad \text{and} \quad \dot{F}(\bar{\omega}) = 0, \qquad \text{then} \quad \dot{\Omega} = \dot{\bar{\omega}} \\ &\text{If} \quad F(\bar{\omega}) < 0, \quad \text{or if} \quad F(\bar{\omega}) = 0 \quad \text{and} \quad F(\dot{\bar{\omega}}) < 0, \qquad \text{then} \quad \dot{\Omega} = 0 \end{aligned} \tag{13.10.19}$$

where the overbar denotes the spatial averaging operator (Eq. 13.10.6). Function $F(\bar{\omega})$ represents the loading function and is defined as $F(\bar{\omega}) = \bar{\omega} - k(\bar{\omega})$, where $k(\bar{\omega})$ is a softening parameter, which is set to be equal to the maximum value of $\bar{\omega}$ achieved up to the present. The initial value of $k(\bar{\omega})$ is zero. The formulation of the loading formulation automatically satisfies the dissipation inequality. The density of the energy dissipation rate due to damage is $\phi = \dot{\Omega}Y$, and since $\dot{\Omega} \geq 0$, we have $\phi \geq 0$. The damage expression in Equation 13.10.18 was found to approximate acceptably the behavior of concrete, provided that different local damage thresholds Y_1 are introduced for tension and compression.

The fact that Equation 13.10.16 uses nonlocal rather than local damage, and that the unloading condition is stated in terms of the nonlocal damage, represents all that is different from the classical (local) damage theory.

Consider now for the sake of simplicity the one-dimensional problem of a bar loaded through two springs of stiffness C (Fig. 13.48). This problem was used by Bažant (1976) to demonstrate the localization instability due to strain softening (see also Sec. 13.2). The length coordinate is x, the bar length is L, and the bar ends are at $x = 0$ and $x = L$. The bar is initially in a state of uniform strain ε_0 and stress σ_0, with uniform damage $\Omega_0 = \omega_0$. The initial state is in the

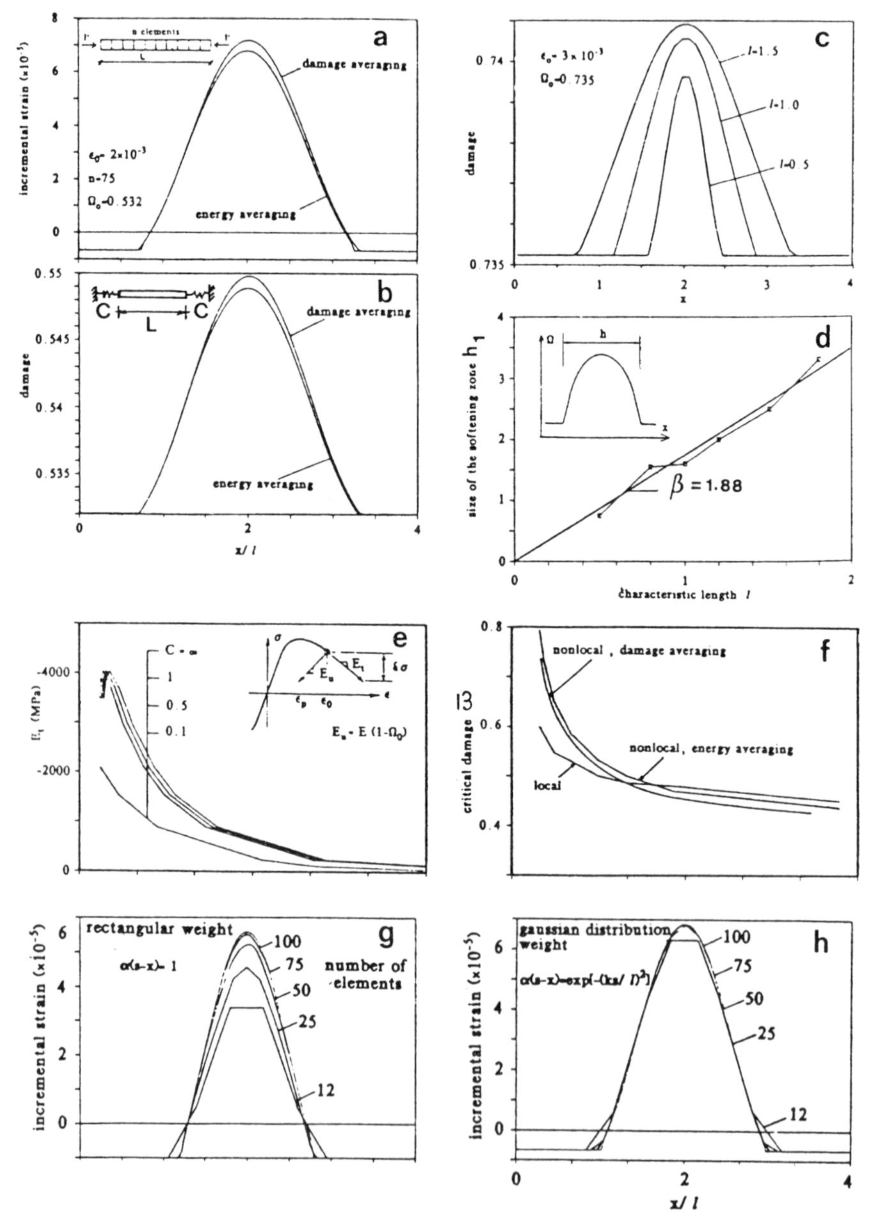

Figure 13.48 Bar loaded through elastic springs: (a, b, c) localization profiles of incremental strain; (d) length of softening zone; (e, f) critical tangent modulus; (g, h) profiles of incremental strains for various element subdivisions and weight functions.

strain-softening range and satisfies the relation $\sigma_0 = (1 - \Omega_0)E\varepsilon_0$. If the constitutive relation for damage is given in the form of Equation 13.10.18, we have $\omega_0 = 1 - [1 + b(\frac{1}{2}E\varepsilon_0^2 - Y_1)]^{-n}$.

We now consider a small deviation from the initial state caused by incremental variation of the load at the bar ends. Let $\delta\varepsilon(x) = \eta(x) =$ incremental strain and $\delta\sigma = \tau =$ incremental stress. To maintain equilibrium, it is necessary that $\tau =$ constant along the bar. The compatibility condition for fixed supports requires that

$$\int_0^L \eta(x)\,dx + \frac{2\tau}{C} = 0 \tag{13.10.20}$$

Taking the one-dimensional form of the constitutive law in Equation 13.10.16 with the averaging according to Equation 13.10.6 and the loading criterion according to Equation 13.10.19, we find

For $\int_0^L \alpha(s-x)\eta(s)\,ds > 0$:

$$\alpha_0 + \tau = \left[1 - \frac{1}{l'(x)}\int_0^L \alpha(s-x)\omega(s)\,ds\right]E[\varepsilon_0 + \eta(x)] \tag{13.10.21}$$

Otherwise: $\qquad \tau = (1 - \omega_0)E\eta(x)$

in which $l'(x) = \int_0^L \alpha(s-x)\,dx =$ effective averaging length for point x. For infinitely small increments $\eta(x)$, the equations can be simplified by incremental linearization. To this end, we may introduce the linear approximation $\omega(x) = \omega_0 + \omega_\varepsilon\eta(x)$, with $\omega_\varepsilon = \partial\omega/\partial\varepsilon$ for $\omega = \omega_0$. Substituting into Equation 13.10.21, neglecting the quadratic term $\eta(x)\eta(s)$, and subtracting the equation $\sigma_0 = E(1 - \Omega_0)\varepsilon_0$, we can reduce Equation 13.10.21 to the form

$$\tau = (1 - \omega_0)E\eta(x) - \frac{k}{l'(x)}\left\langle\int_0^L \alpha(s-x)\eta(s)\,ds\right\rangle \tag{13.10.22}$$

with $k = E\varepsilon_0\omega_\varepsilon$. The symbol $\langle\ \rangle$, which introduces the loading criterion, is defined as $\langle x\rangle = x$ if $x > 0$ and otherwise $\langle x\rangle = 0$. For the special damage constitutive law in Equation 13.10.18, we may evaluate $\omega_\varepsilon = E\varepsilon_0\omega_Y$, with $\omega_Y = bn(\frac{1}{2}E\varepsilon_0^2 - Y_1)^{n-1}[1 + b(Y - Y_1)]^{-2n}$.

Equation 13.10.22 has the form of a linear integral equation of the second kind. However, the problem is not simply that of solving an integral equation, because Equation 13.10.22 is an integral equation only for those x for which loading takes place. Outside the loading region, Equation 13.10.21 reduces to the usual linear elastic differential equation for displacements and the behavior is then local, elastic, and unaffected by the strain localization in another segment of the bar.

If we find one solution such that there exist elastic segments of finite length at both ends of the bar, then we can obtain another solution simply by shifting the softening segment along with the softening solution profile as a rigid body. This shift is arbitrary provided the entire original length of the softening region, along with a small neighborhood of points located just outside the softening zone, remains within the bar. Due to this fact, we can calculate the length of the softening region and the solution profile through it by analyzing any bar of a shorter length, provided the bar length exceeds the length of the softening region.

The actual boundary conditions and the compatibility condition may be disregarded in such analysis and satisfied afterward. Obviously, one can have infinitely many solutions. Arbitrary shifts of the softening region, however, do not affect the overall response of the bar. The length of the softening region and the solution profile through it is nevertheless unique. The localization profiles terminating at the boundary points are different from those at the interior and require a special analysis. For real materials, the actual location of the strain-softening segment that does not reach the boundary is decided by inevitable random fluctuations of material strength along the bar.

Equation 13.10.22, which is also obtained for various other types of nonlocal continuum with local strain (Bažant and Pijaudier-Cabot, 1988b), can be easily solved numerically. If we subdivide the bar length into N equal elements of length $\Delta x = L/N$, use for integration the trapezoidal rule, express τ from Equation 13.10.20, and substitute it into Equation 13.10.22, we may approximate the resulting equation as follows:

$$\sum_{j=1}^{N} K_{ij}\eta_j = 0 \qquad \text{with} \qquad K_{ij} = (1-\omega_0)E\delta_{ij} - \frac{k\,\Delta x}{l'_i} I_i \alpha(x_i - x_j) + \frac{C}{2} \qquad (13.10.23)$$

Subscripts i, j refer to centroids of elements number i or j, and $I_i = 1$ if $\int_0^L \alpha(x_i - s)\eta(s)\,ds \simeq \sum_i \alpha(x_j - x_i)\eta_j\,\Delta x \geq 0$; otherwise $I_i = 0$. Equation 13.10.23 represents a system of homogeneous linear algebraic equations for η_j. The critical state occurs at such ε_0 for which $\det(K_{ij}) = 0$. The loading segment, characterized by $I_i = 1$, is, of course, unknown in advance, and so it must be determined iteratively. To search for the critical state with the smallest ε_0 and the shortest damage localization segment, the values of ε_0 may be incremented in small loading steps and, for each ε_0, the following algorithm (by Bažant and Pijaudier-Cabot, 1988b) may be used:

1. In the first iteration, we assume that only one element i undergoes loading (the central element). In the subsequent iterations, we increase the number of elements that undergo loading by one (either on the left or on the right), unless the number of loading elements already equals N, in which case we start a new loading step with a larger ε_0.

2. For each assumed set of elements with loading, we calculate the lowest eigenvalue λ_1 of K_{ij} and the corresponding eigenvector $\eta_i^{(1)}$. If the positive values of $\eta_i^{(1)}$ in this eigenvector do not agree with the assumed set of loading elements, we return to step 1 and repeat the calculation for the same ε_0 and a larger number of softening elements. If they agree but $\lambda_1 > 0$, we also return to step 1 (after storing the λ_1 value) and commence the next load step for an incremented value of ε_0; otherwise ($\lambda_1 \leq 0$) the critical state $\varepsilon_0 = \varepsilon_0^{cr}$ has just been passed. In that case, we interpolate with regard to the λ_1 value in the preceding loading step, in order to estimate more closely the value of ε_0 for which $\lambda_1 = 0$. Then we print this ε_1 value, calculate the corresponding eigenvector $\eta_i^{(1)}$ and stop.

As we can see from Equation 13.10.23, the tangential stiffness matrix K_{ij} is nonsymmetric ($K_{ij} \neq K_{ji}$). The nonsymmetry is inevitable for the present formulation and is due to the loading–unloading criterion, brought about through parameter I_i, as well as to the variability of l'_i caused by overlap of the averaging region with the unloading segment or its protrusion outside the body. If I_i as well as l_i' had the same values for all elements, K_{ij} would be symmetric. Similarly, the part of matrix K_{ij} for which l_i and l_i' are constant is symmetric. Matrix K_{ij} also becomes symmetric for $\varepsilon_0 \rightarrow 0$ (elastic behavior). For small ε_0 it is weakly nonsymmetric. The fact that K_{ij} is nonsymmetric implies that the integral operator in Equation 13.10.22 is nonsymmetric, too, which means it is not self-adjoint. This further implies that a minimizing functional associated with Equation 13.10.23 does not exist.

Numerical results were obtained (Bažant and Pijaudier-Cabot, 1988b) for the bar shown in Figure 13.48 with $b = 20.5$, $Y_0 = 0.00854$, and $l = 0.25L$. This figure shows (a, b) the localization profiles of incremental strain ε and damage ω at the limit of stable states, for formulations in which either damage ω or the energy release rate is averaged; (c) profiles for various other characteristic lengths l; (d) ratio of the length h_1 of the softening zone to the characteristic length for various values of l; (e, f) the values of tangent modulus E_t and critical nonlocal damage $\bar{\omega}$ as a function of L/l; and (g, h) the profiles of incremental strain for subdivisions with various numbers of elements, N, and for rectangular or Gaussian weight functions. The results confirm the absence of localization into a vanishing volume and, for the Gaussian weight function, good convergence as the subdivision is refined.

Bifurcations leading to stable localization at increasing load have not yet been studied, but they are likely to occur before the instabilities just analyzed. The localization instabilities in a nonlocal continuum with local strain were further clarified by a dynamic analysis in Pijaudier-Cabot and Bažant (1988). An interesting uniaxial model for strain softening in which higher-order derivatives are used to approximately characterize nonlocal properties has been presented by Schreyer and Chen (1986) and Schreyer (1989). The use of higher-order derivatives has also been studied by Belytschko and Lasry (1989).

By contrast to the nonlocal continuum with local strain, the operators and stiffness matrices for the originally formulated imbricate nonlocal continuum (Bažant, Belytschko, and Chang, 1984; Bažant, 1984a) are symmetric. In fact, the requirement of symmetry was shown to lead to the imbricate nonlocal model (Bažant, 1984a). Convenient though the symmetry is for numerical solutions, this advantage is nevertheless more than offset by other disadvantages of the imbricate model. For uniaxial deformations or bending of bars, however, the localization instabilities of imbricate continuum can be solved analytically (Bažant and Zubelewicz, 1988).

Measurement of Characteristic Length of Nonlocal Continuum

The preceding stability analysis indicates one simple way of measuring the characteristic length l of the material. From Figure 13.48d we see that the length of the softening zone in a uniaxially tensioned bar is about $h_1 = \beta l$ where $\beta = \text{constant}$ ($\beta = 1.88$ for the model used), and the shape of the damage distribution throughout this zone is also constant. Replacing this distribution by a rectangle of equal area, one would get the width h_1/α where $\alpha = 1.93$. The fracture energy G_f represents the energy dissipation over the width of the localization band, and balance of energy requires that $G_f = (h_1/\alpha)W_s$ where W_s is the energy dissipation per unit volume of a uniformly strained specimen. Thus,

$l = h_1/\beta = \alpha G_f/\beta W_s = 1.02 G_f/W_s$, or approximately

$$l = \frac{G_f}{W_s} \tag{13.10.24}$$

Note that this fraction has indeed the dimension of length because the dimensions of G_f and W_s are J/m^2 and J/m^3.

So the characteristic length can be easily determined if a uniaxially tensioned specimen can be deformed macroscopically uniformly. As proposed and demonstrated by Bažant and Pijaudier-Cabot (1987a, 1988a, 1989), this can be achieved by gluing a set of parallel steel rods to the specimen sides with epoxy (Fig. 13.49). The rods must be considerably thinner than the maximum aggregate size, so as not to alter the nonlocal properties of the material. The specimen (Fig. 13.49) should be as thin as it can be cast, so as to avoid localization in the transverse direction. Replacement of thin rods by a steel sheet is not suitable, because the sheet would interfere with the Poisson effect.

The basic condition of the test is that the slope of the load-displacement curve of the specimen with the steel rods must always be positive, since this guarantees absence of localization, according to our stability analysis in Section 13.2. The characteristic shape of the load-displacement curve is shown in Figure 13.49 (curve 012). The dashed straight line 03 represents the response of the steel rods without concrete. The response of concrete is represented by the difference in the ordinates of curve 012 and line 03. After concrete has been completely damaged, the response of the specimen must approach the straight line 03, as sketched. The strain-softening response of concrete begins at the point at which the tangent is parallel to line 03.

Using this method, it was found that, for one typical concrete, $l = 2.7 d_a$ where d_a = maximum aggregate size (Bažant and Pijaudier-Cabot, 1987a, 1988a, 1989). Further refinements of this measurement method were made by Mazars and Berthaud (1989).

Example: Stability of a Tunnel

As an example of application in large-scale finite element analysis, one can mention a recent study of stability of a subway tunnel that is excavated in a strain-softening soil with large inhomogeneities (Bažant and Lin, 1988b). The soil

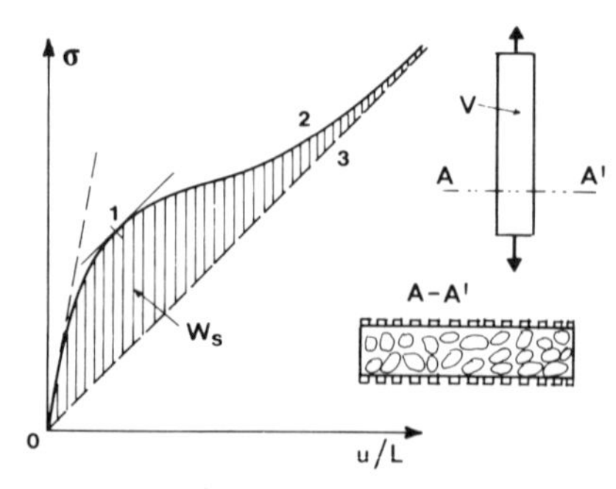

Figure 13.49 Load-displacement curves for restrained concrete specimen (solid curve) and for steel bars alone (dashed line.)

is a clayey sand with gravel, which is injected by cement grout from the surface in order to strengthen it so as to allow excavation by machines, without temporary supports (Z. J. Bažant, 1983). Zones of compressive strain-softening develop on the sides of the tunnel (Fig. 13.50). The characteristic length was assumed to be about three times the mean spacing betweeen highly and little cemented chunks of soil. The excavation process was simulated in two dimensions by gradually reducing the stresses acting on the outline of the tunnel to zero. If the displacements remain small and finite, the tunnel is stable, and if they become too large, the tunnel is unstable.

To assess convergence, the problem has been solved with meshes of increasing fineness, two of which are shown in Figure 13.50 (the fine mesh involves 3248 degrees of freedom). Interestingly, the nonlocal solution for the finest mesh ran (on a Cray II computer) faster than the corresponding local solution, which is no doubt due to the stabilizing effect of the nonlocal operators (which tends to make iterations converge faster). Figure 13.50 shows the boundaries of the strain-softening zones at the end of excavation, obtained for various meshes (Bažant and Lin, 1988b). Note that these boundaries mutually agree very well, while in local analysis they are quite different.

Gradient Approximation to Nonlocal Continuum

If the averaging zone is symmetric (which means it must not protrude outside the boundary), the averaging integral in Equation 13.10.6 may be approximated on the basis of the second strain gradient (see eq. 44 in Bažant, 1984a):

$$\bar{\boldsymbol{\varepsilon}}(\mathbf{x}) = (1 + \lambda^2 \nabla^2)\boldsymbol{\varepsilon}(\mathbf{x}) \tag{13.10.25}$$

where $\lambda^2 = k_0 l^2$. For $\alpha(\mathbf{x}) = 1$ within a sphere of diameter l (in 3D), $k_0 = \frac{1}{40}$; a circle of diameter l (in 2D), $k_0 = \frac{1}{32}$; and a segment of length l (in 1D), $k_0 = \frac{1}{24}$ (see eqs. 49 and 50 in Bažant, 1984a). Equation 13.10.25 can be derived by substituting into Equation 13.10.6 the truncated Taylor series expansion $\boldsymbol{\varepsilon}(\mathbf{s}) = \boldsymbol{\varepsilon}(\mathbf{x}) + \boldsymbol{\varepsilon}_{,k}(\mathbf{x})r_k + \frac{1}{2}\boldsymbol{\varepsilon}_{,km}(\mathbf{x})r_k r_m$ where $r_k = s_k - x_k$, integrating and noting that, for

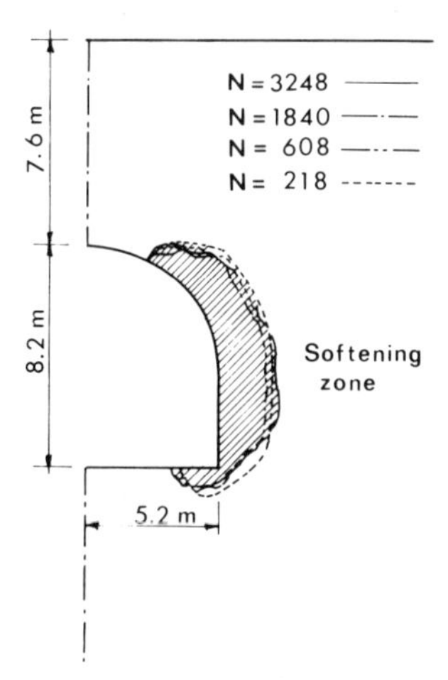

Figure 13.50 Softening zone in tunnel analysis. (*After Bažant and Lin, 1988b.*)

integrals over a circle or sphere of diameter l, $\int r_k r_m \, dV = A\delta_{km}$ where A is a certain constant.

Consider now again the localization instability for the uniaxial bar in Figure 13.51, with the same notations as before (Fig. 13.48). For the incremental stress–strain relation, we again assume that the elastic strain is local and only the inelastic damage strain is subjected to the nonlocal gradient operator, that is,

$$\delta\sigma = E_u \, \delta\varepsilon - (E_u - E_t)\langle \delta\bar{\varepsilon} \rangle \tag{13.10.26}$$

where $\langle \delta\bar{\varepsilon} \rangle = \delta\varepsilon$ for $\delta\bar{\varepsilon} \geq 0$ (loading) and $\langle \partial\bar{\varepsilon} \rangle = 0$ for $\delta\bar{\varepsilon} < 0$ (unloading) and E_t, E_u = moduli for loading and unloading ($E_t < 0$, $E_u > 0$). For the loading segment, $-h/2 < x < h/2$, substitution of Equation 13.10.25 with $\nabla^2 = d^2/dx^2$ yields the linear differential equation $E_t \, \delta u'(x) - (E_u - E_t)\lambda^2 \, \delta u'''(x) = \delta\sigma$ where $\delta\sigma$ = constant, $\delta u(x)$ = axial displacement, and $\delta u'(x) = \delta\varepsilon(x)$. Integration under the conditions of symmetric deformation of the bar yields

$$\delta u(x) = \left(\frac{1}{E_t} x + A \sin \mu x\right)\delta\sigma \qquad \mu = \frac{1}{\lambda}\left(1 - \frac{E_u}{E_t}\right)^{-1/2} \tag{13.10.27}$$

where A is an arbitrary constant. The coordinate $x = h/2$ at the end of the loading segment is given by the condition $\delta\bar{\varepsilon}(x) = 0$ or $\delta u'(x) + \lambda^2 \delta u'''(x) = 0$. Together with the compatibility condition $\delta u(h/2) + [(L - h)/2E_u + 1/C] \, \delta\sigma = 0$ (where L = bar length and C = stiffness of the springs at bar ends), this yields two equations from which A can be eliminated and $\delta\sigma$ cancels out. Thus one gets the equation

$$2 \tan \frac{\mu h}{2} = \mu(1 - \lambda^2 \mu^2)\left[h + (L - h)\frac{E_t}{E_u} + \frac{2E_t}{C}\right] \tag{13.10.28}$$

from which the width h of the softening region may be solved. For the limiting case $C \to 0$ or $L \to \infty$, one gets localization at constant stress ($\delta\sigma = 0$), and Equation 13.10.28 provides

$$h = \frac{\pi}{\mu} = \pi\lambda \sqrt{1 - \frac{E_u}{E_t}} \tag{13.10.29}$$

In this case the strain profile throughout the localization region is a half sine wave and the strain is continuous. For finite L and $C > 0$, the localization segment is shorter; there is a jump in $\delta\varepsilon$ between the localization segment and the unloading

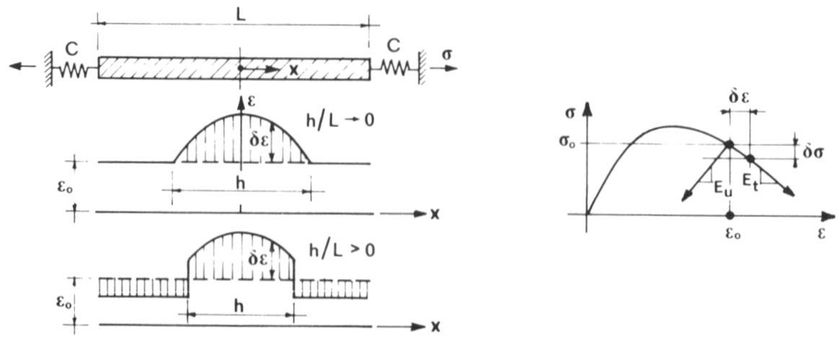

Figure 13.51 Softening zone in a bar loaded through elastic springs.

segment, and the strain profile is an incomplete half sine wave (Fig. 13.51). For $|E_t/E_u| \to \infty$ (vertical stress drop) one gets $h \to 0$.

For points close to the boundary for which the domain of the averaging integral in Equation 13.10.6 protrudes outside the boundary, one finds from the Taylor series expansion that grad $\boldsymbol{\varepsilon}$ should also appear in Equation 13.10.25, and in fact its effect may dominate over $\nabla^2\boldsymbol{\varepsilon}$.

Another approach, which also leads to a formulation with the gradient of strain, is the micropolar (or Cosserat) continuum (cf. Sec. 2.10); this was proposed by Sulem and Vardoulakis (1989).

Summary

Various materials exhibit distributed damage, which may be described in a smeared continuum manner as macroscopic strain softening. The continuum model, however, must be formulated in such a manner that instabilities and bifurcations in which the energy dissipation due to strain-softening damage would localize to a region of vanishing volume are prevented. The simplest, but also crudest, way to achieve it is the crack band model in which the element size is not allowed to become less than a certain length that is a material property. A more general way to achieve it is to make the continuum model nonlocal. The nonlocal operator, however, can produce zero-energy periodic modes of instability. These spurious instabilities can be prevented by making the continuum only partially nonlocal, so that the strain as a kinematic variable, and especially the elastic response, would be local and only the softening damage would be nonlocal. This brings about a further advantage in that the differential equations of equilibrium as well as the boundary (or interface) conditions have the standard form, the same as for a local continuum. Analysis of localization instabilities in such a continuum shows that the size of the strain-softening region is proportional to the characteristic length, and knowledge of their ratio can be exploited for measurement of the characteristic length. The nonlocal continuum with local strain is easily programmed in finite element codes and its efficiency has been verified in some large-scale computations.

Problems

13.10.1 Show that a nonlocal elastic continuum with a triangular weight function is unstable, and that the Gaussian function makes it stable.

13.10.2 Consider a nonlocal von Mises continuum with a degrading yield limit, and formulate for it the integral equation governing uniaxial localization instability.

13.11 CONSTITUTIVE EQUATIONS FOR STRAIN SOFTENING

Although there is no way to fit a thorough discussion of constitutive equations into this text, some closing comments are nevertheless in order because the triaxial nature of the constitutive law for strain softening can tremendously influence the localization instabilities (as confirmed by some results quoted in Secs. 13.3–13.5). It is now clear that not only the strain-softening stiffness but also accuracy in the representation of such multiaxial phenomena as internal friction, nonassociatedness, dilatancy, and vertices on the loading surfaces can be very important.

Inelastic constitutive equations can be subdivided into two broad categories: (1) phenomenological, and (2) micromechanical. In the phenomenological constitutive relations, the tensorial invariance restrictions are identically satisfied a priori by the use of proper stress and strain invariants. But since the invariants have no simple meaning one must find their proper combinations empirically and intuitively. The constitutive laws in this category include (for a detailed review see Bažant, 1986b):

1. Plasticity with yield limit degradation, either associated or nonassociated (i.e., with or without normality, Sec. 10.6), in which strain softening is achieved by gradual reduction of the yield limit (see, e.g., Lin, Bažant, Chern, and Marchertas, 1987; and the bounding surface model of Yang, Dafalias, and Herrmann, 1985).
2. Fracturing theory (Dougill 1975, 1976) and plastic-fracturing theory (Bažant and Kim, 1979; Chen and Mau, 1989), where strain softening is achieved by reduction of material stiffness, while, similarly to plasticity, loading surfaces with normality rule are used (although in the strain space or in both the stress and strain spaces).
3. Continuous damage mechanics (originally proposed by Kachanov, 1958, for creep but later adapted to strain softening), where strain softening is also achieved by reduction of material stiffness. But instead of loading surfaces the central concept is that of damage, which is often simplified as a scalar although properly it should be a fourth-order tensor; for concrete, see Mazars and Pijaudier-Cabot (1989), Bažant and Carol (1989), and Ortiz (1985).
4. Total strain models, which are based on the simplifying assumption of path independence (Bažant and Tsubaki, 1980).
5. Heuristic incremental models, such as the orthotropic models, which however have problems with the form-invariance requirement (Bažant, 1983).
6. Endochronic theory, an idea due to Valanis (1971) and Schapery (1968), which was adapted to strain-softening concrete by Bažant and Bhat (1976) (also Bažant and Shieh, 1980). This theory, which is particularly powerful for cyclic loading, is incrementally nonlinear while all the previous ones are piecewise linear in the sectors of loading directions in the stress or strain spaces.

In the micromechanical constitutive models, the tensorial invariance restrictions are not imposed a priori by means of stress and strain invariants, but are satisfied a posteriori only after the constitutive behavior has been described locally in the microstructure model. This brings about a great conceptual simplification and makes it possible to utilize various laws from physics (e.g., frictional slip, activation energy theory for load ruptures), but at the expense of greatly increased computational requirements.

One successful formulation has been the microplane model, which is based on an idea of Taylor (1938). In this model the stress–strain relation is specified independently on the planes of various orientations in the material, and the responses from all the orientations are then combined using some weak constraint such as the equivalence of virtual work. The original models for metals (known as the slip theory of plasticity of Batdorf and Budianski, 1949) used a static constraint in which the stresses on the microplanes are assumed to be the resolved components of the macroscopic stress tensor. To model strain softening, however, stability considerations (Bažant and Oh, 1984) dictate the use of either a kinematic constraint, in which the strain components on the microplane are assumed to be the resolved components of the macroscopic strain tensor (a combined kinematic-static constraint can also be used). This approach has allowed so far the best representation of the existing triaxial test data for concrete; see the model of Bažant and Prat (1988a, b), which achieves conceptual simplicity by assuming that on each microplane the response for monotonic loading is path independent and all the macroscopic path dependence (as well as friction and dilatancy) results from combinations of loading and unloading on the microplanes of all orientations.

The microplane model takes into account only interactions between various orientations in the microstructure, but neglects interaction at distance. Although those are partially reflected in the nonlocal concept (which can also be combined with the microplane model), a more realistic description of the interactions at distance is obtained by direct simulation of the random microstructure of concrete or other heterogeneous materials. Adapting an idea of Cundall for sand, Zubelewicz (1983), Zubelewicz and Bažant (1987), and Bažant, Tabbara, Kazemi, and Pijaudier-Cabot (1989) developed this approach for the fracturing of concrete. Concrete is represented as a system of randomly generated rigid particles whose interactions are described by interparticle force–displacement relations, which may involve both normal and shear interactions but can be simplified to normal interactions only. It has been demonstrated that this approach can describe the strain-softening test data very realistically and, unlike all the previously mentioned models, need not be enhanced by a nonlocal formulation. The nonlocality is an implied feature of the particle models. This is evidenced by the fact that they exhibit limited localization and a size effect that is in good agreement with the size effect law in Equation 13.9.2 and represents a transition from plasticity to linear elastic fracture mechanics. A serious disadvantage of particle simulation is that it poses extreme demands for computer time and storage, which are at present almost unsurmountable if the particle system is three-dimensional, as it should be.

There are many further possible avenues to explore in the micromechanics approach to constitutive modeling. This approach seems to be more promising than the phenomenologic constitutive equations, which have been studied intensely for a long time and probably leave little room for further improvements. Substantial improvements, however, will be needed in order to master predictions of instabilities and bifurcations caused by constitutive properties rather than geometric nonlinearities, or by both of these combined.

References and Bibliography

Achenbach, J. D. (1973), *Wave Propagation in Elastic Solids,* North Holland, Amsterdam (Sec. 13.1).

Ansal, A. M., Krizek, R. J., and Bažant, Z. P. (1982), "Seismic Analysis of an Earth Dam Based on Endochronic Theory," in *Numerical Methods in Geomechanics* (Proc. of Int. Symp. held in Zürich, September 1982), ed. by R. Dungar, Balkema, Rotterdam, pp. 559–76 (Sec. 13.3).

Baker, A. L. L., and Amarakone, A. M. N. (1964), "Inelastic Hyperstatic Frames Analysis," Proc. Int. Symp. on the Flexural Mech. of Reinforced Concrete, held in Miami, Fla.; see also American Concrete Institute Special Publication No. 12, pp. 85–142 (Sec. 13.6).

Barenblatt, G. I. (1979), *Similarity, Self-Similarity and Intermediate Asymptotics,* Consultants Bureau, New York (transl. from Russian, *Podobie, avtomodel'nost', promezhutochnaia asimptotika,* Gidrometeoizdat, Moscow, 1978) (Sec. 13.9).

Barnard, P. R. (1965), "The Collapse of Reinforced Concrete Beams," Proc. Int. Symp. Flexural Mech. of Reinforced Concrete, held in Miami, Fla.; see also American Concrete Institute Special Publication No. 12 (Sec. 13.6).

Batdorf, S. B., and Budianski, B. (1949), "A Mathematical Theory of Plasticity Based on the Concept of Slip," NACA Tech. Note TN 1871, April (Sec. 13.11).

Bažant, Z. J. (1983), *Methods of Foundation Engineering,* Elsevier, New York (Sec. 13.10).

Bažant, Z. P. (1971), "A Correlation Study of Incremental Deformations and Stability of Continuous Bodies," *J. Appl. Mech.* (ASME), 38:919–28 (Sec. 13.3).

Bažant, Z. P. (1976), "Instability Ductility and Size Effect in Strain-Softening Concrete," *J. Eng. Mech.* (ASCE), 102(2):331–44; see also discussion, 103:357–8, 775–6; 104:501–502 (based on Struct. Eng. Report No. 74-8/640, Northwestern University, August 1974) (Sec. 13.0).

Bažant, Z. P. (1982a), "Crack Band Model for Fracture of Geomaterials," Proc. 4th Intern. Conf. on Num. Meth. in Geomechanics, Edmonton, Canada, 3:1137–52 (Sec. 13.2).

Bažant, Z. P. (1982b), "Crack Band Propagation and Stress–Strain Relations for Fracture Process Zone of Geomaterials," in *Numerical Models in Geomechanics* (Proc., Int. Symp., held in Zürich, September 1982), ed. by R. Dungar et al., A. A. Balkema, Rotterdam, pp. 189–97.

Bažant, Z. P. (1983), "Comment on Orthotropic Models for Concrete and Geomaterials," *J. Eng. Mech.* (ASCE), 109:849–65 (Sec. 13.11).

Bažant, Z. P. (1984a), "Imbricate Continuum and Its Variational Derivation," *J. Eng. Mech.* (ASCE), 110:1693–712 (Sec. 13.10).

Bažant, Z. P. (1984b), "Imbricate Continuum and Progressive Fracturing of Concrete and Geomaterials," *Meccanica* (Italy) (Special Issue Commemorating the Centennial of A. Castigliano's Death), 19:86–93.

Bažant, Z. P. (1984c), "Size Effect in Blunt Fracture: Concrete, Rock, Metal," *J. Eng. Mech.* (ASCE), 110:518–35 (Sec. 13.9).

Bažant, Z. P. (1985a), *Comment on Hillerborg's Size Effect Law and Fictitious Crack Model,* ed. by P. Gambarova et al., Dei Poli Anniversary Volume, Politecnico di Milano, Italy, p. 335–8.

Bažant, Z. P. (1985b), "Distributed Cracking and Nonlocal Continuum," in *Finite Element Methods for Nonlinear Problems,* ed. by P. Bergan et al., Springer, Berlin, 1986, pp. 77–102; also Symp. Preprints, Trondheim, 1985, paper II-2 (Sec. 13.8).

Bažant, Z. P. (1985c), "Fracture in Concrete and Reinforced Concrete," chapter 13 in *Mechanics of Geomaterials*: *Rocks, Concretes, Soils* (Proc. IUTAM Prager Symposium held at Northwestern University, September 1984), ed. by Z. P. Bažant, John Wiley & Sons, London, pp. 259–303.

Bažant, Z. P. (1985d), "Mechanics of Fracture and Progressive Cracking in Concrete Structures," chapter 1 in *Fracture Mechanics of Concrete*: *Structural Application and*

Numerical Calculation, ed. by G. C. Sih and A. Di Tommaso, Martinus Nijhoff, Dordrecht and Boston (Sec. 13.9).

Bažant, Z. P. (1985e), Lectures on "Stability of Structures," Course 720-D24, Northwestern University, Evanston, Ill. (Sec. 13.8).

Bažant, Z. P. (1986a), "Fracture Mechanics and Strain-Softening of Concrete," in *Finite Element Analysis of Reinforced Concrete Structures,* ed. by C. Meyer and H. Okamura, ASCE, New York, pp. 121–50; see also Preprints, U.S.–Japan Seminar, held in Tokyo, Japan, May 1985, 1:47–69 (Sec. 13.1).

Bažant, Z. P. (1986b), "Mechanics of Distributed Cracking," *Appl. Mech. Rev.* 39(5):675–705 (*Sec.* 13.2).

Bažant, Z. P. (1987a), "Fracture Energy of Heterogeneous Materials and Similitude," Preprints SEM-RILEM Int. Conf. on Fracture of Concrete and Rock (held in Houston, Texas, June 1987), ed. by S. P. Shah and S. E. Swartz, publ. by SEM (Soc. for Exper. Mech.), pp. 390–402 (Sec. 13.9).

Bažant, Z. P. (1987b), "Stable States and Paths of Inelastic Structures: Completion of Shanley's Theory," Report 87-10/606s, Dept. of Civil Engineering, Northwestern University, Evanston, Ill. (Sec. 13.4).

Bažant, Z. P. (1987c), "Why Continuum Damage Is Nonlocal: Justification by Quasiperiodic Microcrack Array," *Mech. Res. Communications,* 14:407–19; also "Why Continuum Damage is Nonlocal: Energy Release by Microcracks," submitted (1990) to ASCE *J. Eng. Mech.* (Sec. 13.10).

Bažant, Z. P. (1987d), "Tests of Size Effect in Pullout of Reinforcing Bars from Concrete," Proc. IABSE Colloquium on Computational Mechanics of Concrete Structures (held in Delft, Netherlands, August 1987), pp. 261–84.

Bažant, Z. P. (1987e), Lectures on "Material Modeling Principles," Course 720-D30, Northwestern University, Evanston, Ill.

Bažant, Z. P. (1988a), "Softening Instability: Part I—Localization into a Planar Band," *J. Appl. Mech.* (ASME) 55:517–22 (Sec. 13.3).

Bažant, Z. P. (1988b), "Softening Instability: Part II—Localization into Ellipsoidal Regions," *J. Appl. Mech.* (ASME) 55:523–9 (Sec. 13.4).

Bažant, Z. P. (1988c), "Stable States and Paths of Structures with Plasticity or Damage," *J. Eng. Mech.* (ASCE), 114(12):2013–34 (Sec. 13.2).

Bažant, Z. P. (1989a), "Identification of Strain-Softening Constitutive Relation from Uniaxial Tests by Series Coupling Model for Localization," *Cement and Concrete Research,* 19(6):973–7 (Sec. 13.2).

Bažant, Z. P. (1989b), "Stable States and Stable Paths of Propagation of Damage Zones and Interactive Fractures," in *Cracking and Damage—Strain Localization and Size Effect,* ed. by J. Mazars and Z. P. Bažant, Elsevier, London and New York, pp. 183–206 (Proc. of France–U.S. Workshop held at ENS, Cachan, September 1988) (Sec. 13.8).

Bažant, Z. P. (1989c), "Recent Advances in Failure Localization and Nonlocal Models," Report, Northwestern University, Evanston, Ill.; also Proc., Int. Conf. on Micromechanics of Failure of Quasibrittle Materials, ed. by M.-L. Wang and S. P. Shah, held at Univ. of New Mexico, Albuquerque, June 1990, Elsevier, London.

Bažant, Z. P., Ansal, A. M., and Krizek, R. J. (1982), "Endochronic Models for Soils," in *Soil Mechanics—Transient and Cyclic Loads* (Proc. of Int. Conf. in Swansea, U.K., 1980), ed. by G. N. Pande and O. C. Zienkiewicz, John Wiley and Sons, London, pp. 419–38 (Sec. 13.3).

Bažant, Z. P., and Belytschko, T. B. (1985), "Wave Propagation in a Strain-Softening Bar," *J. Eng. Mech.* (ASCE), 111(3):381–9 (Sec. 13.1).

Bažant, Z. P., and Belytschko, T. B. (1987), "Strain-Softening Continuum Damage: Localization and Size Effect", in *Constitutive Laws of Engineering Materials:*

Theory and Applications (Proc. of 2nd Int. Conf. held in Tucson, Ariz., January 1985), ed. by C. S. Desai et al., Elsevier, New York, pp. 11–33.

Bažant, Z. P., Belytschko, T. B., and Chang, T. P. (1984), "Continuum Model for Strain-Softening," *J. Eng. Mech.* (ASCE), 110:1666–92 (Sec. 13.1).

Bažant, Z. P., and Bhat, P. (1976), "Endochronic Theory of Inelasticity and Failure of Concrete," *J. Eng. Mech.* (ASCE), 102:701–21 (Sec. 13.11).

Bažant, Z. P., and Cao, Z. (1986a), "Size Effect in Shear Failure of Prestressed Concrete Beams," *ACI Journal,* 83(2):260–8 (Sec. 13.9).

Bažant, Z. P., and Cao, Z. (1986b), "Size Effect in Brittle Failure of Unreinforced Pipes," *ACI Journal,* 83(3):369–73 (Sec. 13.9).

Bažant, Z. P., and Cao, Z. (1987), "Size Effect in Punching Shear Failure of Slabs," ACI *Journal,* 84(1):44–53 (Sec. 13.9).

Bažant, Z. P., and Carol, I. (1989), "Geometric Damage Tensor, Based on Microplane Model," Report, Northwestern University, Evanston, Ill.; also Proc., Southeastern Conference on Applied Mechanics, Georgia Inst. of Technology, Atlanta, March 1990 (Sec. 13.11).

Bažant, Z. P., and Cedolin, L. (1979), "Blunt Crack Band Propagation in Finite Element Analysis," *J. Eng. Mech.* (ASCE) 105(2):297–315 (Sec. 13.9).

Bažant, Z. P., and Cedolin, L. (1980), "Fracture Mechanics of Reinforced Concrete," *J. Eng. Mech.* (ASCE) 106:1287–1306; see also discussion, 108:464–71 (Sec. 13.9).

Bažant, Z. P., and Cedolin, L. (1983), "Finite Element Modeling of Crack Band Propagation," *J. Struct. Eng.* (ASCE), 109(1):69–92 (Sec. 13.9).

Bažant, Z. P., and Chang, T. P. (1984), "Instability of Nonlocal Continuum and Strain-Averaging," *J. Eng. Mech.* (ASCE), 110:1441–50 (Sec. 13.10).

Bažant, Z. P., and Chang, T. P. (1987), "Nonlocal Finite Element Analysis of Strain-Softening Solids," *J. Struct. Eng.* (ASCE), 113(1):89–105.

Bažant, Z. P., and Gambarova, P. (1984), "Crack Shear in Concrete: Crack Band Microplane Model," *J. Struct. Eng.* (ASCE), 110:2015–35.

Bažant, Z. P., and Kazemi, M. T. (1989a), "Size Dependence of Concrete Fracture Energy Determined by RILEM Work-of-Fracture Method," Report, Northwestern University, Evanston, Ill.; also *Int. J. of Fracture,* in press.

Bažant, Z. P., and Kazemi, M. T. (1989b), "Size Effect in Fracture of Ceramics and Its Use to Determine Fracture Energy and Effective Process Zone Length," Report No 89-6/498s, Center for Advanced Cement-Based Materials, Northwestern University, Evanston, Ill.; also *J. Amer. Ceramic Soc.*; 73(7):1841–53, 1990.

Bažant, Z. P., and Kim, J.-K. (1984), "Size Effect in Shear Failure of Longitudinally Reinforced Beams," *ACI Journal,* 81(5):456–68 (Sec. 13.9).

Bažant, Z. P., Kim, J.-K., and Pfeiffer, P. (1985), "Continuum Model for Progressive Cracking and Identification of Nonlinear Fracture Parameters," in *Applications of Fracture Mechanics to Cementitious Composites,* ed. by S. P. Shah, Martinus Nijhoff, Dordrecht–Boston, pp. 197–246.

Bažant, Z. P., Kim, J.-K., and Pfeiffer, P. (1986), "Nonlinear Fracture Properties from Size Effect Tests," *J. Struct. Eng.* (ASCE), 112(2):289–307.

Bažant, Z. P., and Kim, S. S. (1979), "Plastic-Fracturing Theory for Concrete," *J. Eng. Mech.* (ASCE), 105:407–28 (Sec. 13.11).

Bažant, Z. P., and Krizek, R. J. (1975), "Saturated Sand as an Inelastic Two-Phase Medium," *J. Eng. Mech.* (ASCE), 101:317–32 (Sec. 13.3).

Bažant, Z. P., and Lin, F.-B. (1988a), "Nonlocal Smeared Cracking Model for Concrete Fracture," *J. Struct. Eng.* (ASCE), 114(11):2493–510 (Sec. 13.10).

Bažant, Z. P., and Lin, F-B. (1988b), "Nonlocal Yield Limit Degradation," *Int. J. Numerical Methods Eng.,* 26:1805–23 (Sec. 13.10).

Bažant, Z. P., and Lin, F-B. (1988c), "Localization Instability for Softening in Ellipsoidal Regions and Bands," in *Mechanics of Composite Materials—1988,* AMD 92, ed. by

G. J. Dvorak and N. Laws, Am. Soc. of Mech. Engnrs., New York. (Joint ASME/SES Conference, Berkeley, Calif., June), pp. 79–85.

Bažant, Z. P., and Lin, F.-B. (1989), "Stability against Localization of Softening into Ellipsoids and Bands: Parameter Study," *Int. J. Solids and Structures,* 25(12): 1483–98 (Sec. 13.3).

Bažant, Z. P., Lin, F.-B., and Pijaudier-Cabot, G. (1987), "Yield Limit Degradation: Nonlocal Continuum with Local Strain," in *Computational Plasticity* (Proc. of Int. Conf. held in Barcelona, Spain, April 1987), ed. by D. R. J. Owen, E. Hinton, and E. Oñate, Pineridge Press, Swansea, U.K., pp. 1757–79 (Sec. 13.10).

Bažant, Z. P., and Oh, B. H. (1983), "Crack Band Theory for Fracture of Concrete," *Materials and Structures* (RILEM, Paris), 16: 155–77 (based on a 1981 Report) (Sec. 13.2).

Bažant, Z. P., and Oh, B. H. (1984), "Rock Fracture via Strain-Softening Finite Elements," *J. Eng. Mech.* (ASCE), 110: 1015–35 (Sec. 13.10).

Bažant, Z. P., and Ožbolt, J. (1989a), "Nonlocal Microplane Model for Fracture, Damage and Size Effect in Structures," Report, Northwestern University, Evanston, Ill.; also *J. Eng. Mech.* (ASCE), 116, 1990, in press (Sec. 13.10).

Bažant, Z. P., and Ožbolt, J. (1989b), "Nonlocal Microplane Model for Concrete Tensile, Compression and Splitting Fractures," in *Computer-Aided Design of Reinforced Concrete Structures* (Proc., 2nd Intern. Conf. held at Zell am See, Austria, in April 1990) ed. by H. Mang and N. Bicanic.

Bažant, Z. P., Pan, J.-Y., and Pijaudier-Cabot, G. (1987), "Softening in Reinforced Concrete Beams and Frames," *J. Struct. Eng.* (ASCE), 113(12): 2333–47.

Bažant, Z. P., and Panula, L. (1978), "Statistical Stability Effects in Concrete Failure," *J. Eng. Mech.* (ASCE), 104: 1195–212 (Sec. 13.2).

Bažant, Z. P., and Pfeiffer, P. (1986), "Shear Fracture Tests of Concrete," *Materials and Structures* (RILEM, Paris), 19: 111–21.

Bažant, Z. P., and Pfeiffer, P. (1987), "Determination of Fracture Energy from Size Effect and Brittleness Number," *ACI Materials J.*, 84: 463–80.

Bažant, Z. P., and Pijaudier-Cabot, G. (1987a), "Measurement of Characteristic Length of Nonlocal Continuum," Report No. 87-12/498m, Center for Concrete and Geomaterials, Northwestern University, Evanston, Ill. (Sec. 13.10).

Bažant, Z. P., and Pijaudier-Cabot, G. (1987b), "Modeling of Distributed Damage by Nonlocal Continuum with Local Strains," *Numerical Methods in Fracture Mechanics* (Proc. of 4th Int. Conf. held in San Antonio, Texas, March 1987), ed. by A. R. Luxmoore et al., Pineridge Press, Swansea, U.K., pp. 411–31 (Sec. 13.10).

Bažant, Z. P., and Pijaudier-Cabot, G. (1988a), "Nonlocal Continuum Damage and Measurement of Characteristic Length," in *Mechanics of Composite Materials*—1988, AMD Vol. 92, ed. by G. J. Dvorak and N. Laws, Am. Soc. Mech. Engrs., New York. (Joint ASME/SES Conference, Berkeley, Calif., June) (Sec. 13.10).

Bažant, Z. P., and Pijaudier-Cabot, G. (1988b), "Nonlocal Continuum Damage, Localization Instability and Convergence," *J. Appl. Mech.* (ASME), 55: 287–93 (Sec. 13.10).

Bažant, Z. P., and Pijaudier-Cabot, G. (1989), "Measurement of Characteristic Length of Nonlocal Continuum," *J. Eng. Mech.* (ASCE), 115(4): 755–67 (Sec. 13.10).

Bažant, Z. P., Pijaudier-Cabot, G., and Pan, J. (1987), "Ductility, Snapback, Size Effect, and Redistribution in Softening Beams or Frames," *J. Struct. Eng.* (ASCE), 113(12): 2348–64 (Sec. 13.6).

Bažant, Z. P., and Prat, P. C. (1988a), "Microplane Model for Brittle-Plastic Material: I. Theory," *J. Eng. Mech.* (ASCE), 114(10): 1672–88 (Sec. 13.11).

Bažant, Z. P., and Prat, P. C. (1988b), "Microplane Model for Brittle-Plastic Material: II. Verification," *J. Eng. Mech.* (ASCE), 114(10): 1689–1702 (Sec. 13.11).

Bažant, Z. P., and Sener, S. (1987a), "Size Effect in Torsional Failure of Longitudinally Reinforced Concrete Beams," *J. Struct. Eng.* (ASCE), 113(10):2125–36 (Sec. 13.9).

Bažant, Z. P., and Sener, S. (1987b), "Tests of Size Effect in Pullout of Reinforcing Bars from Concrete," Proc. IABSE Colloquium on Computational Mechanics of Concrete Structures (held in Delft, The Netherlands, August 1987), pp. 261–84.

Bažant, Z. P., and Sener, S. (1988), "Size Effect in Pullout Tests," *ACI Materials J.*, 85:347–51.

Bažant, Z. P., and Shieh, C. L. (1980), "Hysteretic Fracturing Endochronic Theory for Concrete," *J. Eng. Mech.* (ASCE), 106:929–50 (Sec. 13.11).

Bažant, Z. P., and Sun, H-H. (1987), "Size Effect in Diagonal Shear Failure: Influence of Aggregate Size and Stirrups," *ACI Materials J.*, 84(4):259–72.

Bažant, Z. P., Tabbara, M. T., Kazemi, M. T., and Pijaudier-Cabot, G. (1989), "Random Particle Model for Fracture of Aggregate or Fiber Composites," Report, Northwestern University, Evanston, Ill.; also *J. Eng. Mech.* (ASCE), 116(8):1686–1705 1990, in press; and also Proc., Int. Conf. on Numerical Methods in Engineering, held at Univ. of Wales, Swansea, U.K., January 1990 (Sec. 13.11).

Bažant, Z. P., and Tsubaki, T. (1980), "Local Strain Theory and Path-Dependence of Concrete," *J. Eng. Mech.*, (ASCE) 106:1151–73 (Sec. 13.11).

Bažant, Z. P., and Xi, Y. (1989), "Statistical Size Effect in Quasibrittle Structures: II Nonlocal Theory," Report, Northwestern University; submitted to *J. Eng. Mech.* (ASCE) (Sec. 13.10).

Bažant, Z. P., and Xu, K. (1989), "Size Effect in Fatigue Fracture," Report No. 89-12/498f, Center for Advanced Cement-Based Materials, Northwestern University, Evanston, Ill.; also *ACI Materials J.*, in press.

Bažant, Z. P., and Zubelewicz, A. (1988), "Strain-Softening Bar and Beam: Exact Nonlocal Solution," *Int. J. Solids and Structures*, 24(7):659–73 (Sec. 13.10).

Belytschko, T., and Lasry, D. (1989), "Localizations Limiters and Numerical Strategies for Strain Softening Materials," in *Cracking and Damage—Strain Localization and Size Effect*, ed. by J. Mazars and Z. P. Bažant, Elsevier, London and New York, pp. 269–94 (Proc. of France–U.S. Workshop held at ENS, Cachan, September 1988) (Sec. 13.10).

Belytschko, T. B., Bažant, Z. P., Hyun, Y. W., and Chang, T. P. (1986), "Strain-Softening Materials and Finite Element Solutions," *Computers and Structures*, 23(2):163–80 (Sec. 13.1).

Belytschko, T. B., Wang, X-J., Bažant, Z. P., and Hyun, Y. (1987), "Transient Solutions for One-Dimensional Problems with Strain-Softening," *J. Appl. Mech.* (ASME), 54(3):513–18 (Sec. 13.1).

Beran, M. J., and McCoy, J. J. (1970), "Mean Field Variations in a Statistical Sample of Heterogeneous Elastic Solids," *Int. J. Solids and Structures*, 6:1035–54 (Sec. 13.10).

Billardon, R., and Moret-Bailly, L. (1987), "Fully Coupled Strain and Damage Finite Element Analysis of Ductile Fracture," *Nuclear Eng. Des.*, 105:43–49.

Blázquez, R., Bažant, Z. P., and Krizek, R. J. (1980), "Site Factors Controlling Liquefaction,' *J. Geotech. Div.* (ASCE), 106:785–801 (Sec. 13.3).

Borré, G. and Maier, G. (1988), "Discussion of the Paper by Ottosen, 1986," *J. Eng. Mech.* (ASCE), 114(12):2207–09 (Sec. 13.2).

Brockley, C. A., and Ko, P. L. (1970), "Quasi-Harmonic Friction-Induced Vibration," *Trans. ASME, J. Lubrication Technology*, 92:550–6 (Sec. 13.7).

Cedolin, L., and Dei Poli, S. (1977), "Finite Element Studies of Shear-Critical R/C Beams," *J. Eng. Mech.* (ASCE), 103(3):395–410 (Sec. 13.10).

Chang, Y. W., and Asaro, R. J. (1981), "An Experimental Study of Shear Localization in Aluminium-Copper Single Crystals," *Acta Metall.*, 29:241–57.
Chen, B., and Mau, S. T. (1989), "Recalibration of Plastic-Fracturing Model for Concrete Confinement," *Cement and Concrete Research,* 19(1):143–54 (Sec. 13.11).
Christensen, R. M. (1979), *Mechanics of Composite Materials,* John Wiley and Sons, New York (Sec. 13.4).
Chudnovsky, A., and Kachanov, M. (1983), "Interaction of a Crack with a Field of Microcracks," *Int. J. Eng. Sci.*, 21(8):1009–18.
Cope, R. J. (1984), "Material Modeling of Real Reinforced Concrete Slabs," in *Computer Aided Analysis and Design of Concrete Structures* (Proc. of Int. Conf. held in Split, Yugoslavia, September 1984), ed. by F. Damjanić and E. Hinton, Pineridge Press, Swansea, U.K., pp. 85–118 (Sec. 13.10).
Corley, W. G. (1966), "Rotational Capacity of Reinforced Concrete Beams," *J. Struct. Eng.* (ASCE), 92(5):121–46 (Sec. 13.6).
Cranston, W. B. (1965), "Tests on Reinforced Concrete Frames. 1: Pinned Portal Frames," Tech. Rep. TRA/392, Cement and Concrete Assoc., London, England (Sec. 13.6).
Crisfield, M. A. (1989), "No-Tension Finite Element Analysis of Reinforced Concrete Beams Subject to Shear," in *Cracking Damage—Strain Localization and Size Effect,* ed. by J. Mazars and Z. P. Bažant, Elsevier, London and New York, pp. 323–34; see also Preprints of France—U.S. Workshop at ENS, Cachan, France, in September 1988.
Darvall, P. L. (1983), "Critical Softening of Hinges in Indeterminate Beams and Portal Frames," *Civ. Engng. Trans.* (I.E. Australia), CE 25(3):199–210 (Sec. 13.6).
Darvall, P. L. (1984), "Critical Softening of Hinges in Portal Frames," *J. Struct. Eng.* (ASCE), 110(1):157–62 (Sec. 13.6).
Darvall, P. L. (1985), "Stiffness Matrix for Elastic-Softening Beams," *J. Struct. Eng.* (ASCE), 111(2):469–73 (Sec. 13.6).
Darvall, P. L., and Mendis, P. A. (1985), "Elastic-Plastic-Softening Analysis of Plane Frames," *J. Struct. Eng.* (ASCE), 111(4):871–88 (Sec. 13.6).
Darwin, D. (1986), "Crack Propagation in Concrete: Study of Model Parameters", in *Finite Element Analysis of Reinforced Concrete Structures,* ed. by C. Meyer and H. Okamura, ASCE, New York, pp. 184–203; see also Preprints, U.S.–Japan Seminar, held in Tokyo, Japan, May 1985 (Sec. 13.1).
de Borst, R. (1986), "Nonlinear Analysis of Frictional Materials," Ph.D. Dissertation, Technische Hogeschool Delft, Netherlands (Sec. 13.3).
de Borst, R. (1987), "Computation of Post-Bifurcation and Post-Failure Behavior of Strain-Softening Solids," *Comp. and Structures,* 25:211–24 (Sec. 13.8).
de Borst, R. (1988a), "Bifurcations in Finite Element Models with a Nonassociated Flow Law," *Int. J. Numer. Analytical Meth. in Geomechanics,* 12:99–116 (Sec. 13.3).
de Borst, R. (1988b), "Numerical Methods for Bifurcation Analysis in Geomechanics," *Ingenieur-Archiv.*, Springer-Verlag, Berlin (in press) (Sec. 13.8).
de Borst, R. (1989), "Analysis of Spurious Kinematic Modes in Finite Element Analysis of Strain-Softening Solids," in *Cracking and Damage—Strain Localization and Size Effect,* ed. by J. Mazars and Z. P. Bažant, Elsevier, London and New York, pp. 335–48 (Proc. of France—U.S. Workshop held at ENS, Cachan, September 1988) (Sec. 13.8).
de Borst, R., and Nauta, P. (1985), "Non-Orthogonal Cracks in a Smeared Finite Element Model," *Eng. Comput.*, 2:35–46.
Dougill, J. W. (1975), "Some Remarks on Path Independence in the Small in Plasticity," *Quart. Appl. Math.*, 32:233–43 (Sec. 13.11).

Dougill, J. W. (1976), "On Stable Progressively Fracturing Solids," Zeit. Ang. Math. Phys. (*ZAMP*), 27(4):423–37 (Sec. 13.11).

Dougill, J. W. (1983), "Constitutive Relations for Concrete and Rock. Applications and Extension of Elasticity and Plasticity Theory," Preprints W. Prager Symp. on Mechanics of Geomaterials; Rocks, Concrete, Soils, Northwestern Univ., Evanston, Ill.

Dragon, A., and Mróz, Z. (1979), "A Continuum Model for Plastic Brittle Behavior of Rock and Concrete," *Int. J. Eng. Sci.,* 17:137–45.

Droz, P., and Bažant, Z. P. (1989), "Nonlocal Analysis of Stable States and Stable Paths of Propagation of Damage Shear Bands," in *Cracking and Damage—Strain Localization and Size Effect,* ed. by J. Mazars and Z. P. Bažant, Elsevier, London and New York, pp. 415–25 (Proc. of France—U.S. Workshop held at ENS, Cachan, September 1988) (Sec. 13.8).

Drucker, D. C., and Prager, W. (1952), "Soil Mechanics and Plastic Analysis or Limit Design," *Quart. J. Appl. Math.,* 10:157–65 (Sec. 13.3).

Eringen, A. C., and Ari, N. (1983), "Nonlocal Stress Field at Griffith Crack," *Crist. Latt. Amorph. Materials,* 10:33–38 (Sec. 13.10).

Eringen, A. C., and Edelen, D. G. B. (1972), "On Nonlocal Elasticity," *Int. J. Eng. Sci.,* 10:233–248 (Sec. 13.10).

Eshelby, J. D. (1957), "The Determination of the Elastic Field of an Ellipsoidal Inclusion and Related Problems," *Proc. of the Royal Soc. of London,* A 241:376 (Sec. 13.4).

Evans, R. H., and Marathé, M. S. (1968), "Microcracking and Stress–Strain Curves for Concrete in Tension," *Materials and Structures* (RILEM, Paris), 1:61–4 (Sec. 13.2).

Flügge, W. (1962), *Handbook of Engineering Mechanics,* McGraw-Hill, New York (Sec. 13.5).

Gettu, R., Bažant, Z. P., and Karr, M. E. (1989), "Fracture Properties of High Strength Concrete," Report No. 89-10/B627f, Center for Advanced Cement-Based Materials, Northwestern University, Evanston, Ill.; also *ACI Materials J.,* in press.

Ghosh, S. K., and Cohn, M. Z. (1972), "Nonlinear Analysis of Strain-Softening Structures," in *Inelasticity and Nonlinearity in Structural Concrete,* ed. by M. Z. Cohn, Study No. 8, University of Waterloo Press, Waterloo, Ontario, Canada, pp. 315–32 (Sec. 13.6).

Gu, J. C., Rice, J. R., and Ruina, A. L. (1982), "A Nonlinear Analysis of Motion in a Spring and Massless Block Model with a Rate and State Dependent Friction Law," Proc., 9th U.S. Nat. Congress of Appl. Mech., Cornell University, Ithaca, N.Y. (Sec. 13.7).

Gupta, A. K., and Akbar, H. (1984), "Cracking in Reinforced Concrete Analysis," *J. Struct. Eng.* (ASCE), 110(8):1735 (Sec. 13.10).

Gurson, A. L. (1977), "Continuum Theory of Ductile Rupture by Void Nucleation and Growth: Part I—Yield Criteria and Flow Rules for Ductile Media," *J. Eng. Mat. Tech.,* 99:2–15.

Hadamard, J. (1903), *Leçons sur la propagation des ondes,* chap. VI, Hermann et Cie, Paris (Sec. 13.0).

Harrison, B. K., Thorne, K. S., Wakano, M., and Wheeler, J. A. (1965), *Gravitation Theory and Gravitational Collapse,* The University Press, Chicago (Sec. 13.0).

Hegemier, G. A., and Read, H. E. (1983), "Some Comments on Strain-Softening" (manuscript, 19 pages) presented at the DARPA-NSF Workshop, Northwestern University, Evanston, Ill. (Sec. 13.1).

Hill, R. (1962a), "Acceleration Waves in Solids," *J. Mech. Phys. Solids,* 10:1–16 (Sec. 13.3).

Hill, R. (1962b), "Constitutive Laws and Waves in Rigid-Plastic Solids," *J. Mech. Phys. Solids,* 10:89–98.

Hill, R. (1962c), "Uniqueness Criteria and Extremum Principles in Self-Adjoint Problems of Continuum Mechanics," *J. Mech. Phys. Solids,* 10:185–94 (Sec. 13.4).

Hillerborg, A. (1988), "Fracture Mechanics Concepts Applied to Moment Capacity and Rotational Capacity of Reinforced Concrete Beams," Proc. of Int. Conf. on Fracture of Concrete and Rock, held in Vienna, July 1988 (Sec. 13.6).

Hordijk, D. A., Reinhardt, H. W., and Cornelissen, H. A. W. (1987), "Fracture Mechanics Parameters of Concrete from Uniaxial Tensile Tests as Influenced by Specimen Length", Preprints of the SEM/RILEM Int. Conf. on Fracture of Concrete and Rock, Houston, Tex., ed. by S. P. Shah and S. Swartz, SEM (Soc. of Exper. Mech.), pp. 138–49.

Hughes, B. P., and Chapman, B. P. (1966), "The Complete Stress–Strain Curve for Concrete in Direct Tension," RILEM, Paris, Bull. No. 30, pp. 95–97 (Sec. 13.2).

Kachanov, L. M. (1958), "Time of Rupture Process Under Creep Conditions," *Izv. Akad. Nauk. SSR, Otd. Tekh. Nauk.,* No. 8, pp. 26–31 (Sec. 13.11).

Kachanov, M. (1987), "Elastic Solids with Many Cracks: A Simple Method of Analysis," *Int. J. Solids Struct.,* 23(1):23–43.

Kinoshita, N., and Mura, T. (1971), "Elastic Fields of Inclusions in Anisotropic Media," *Phys. Stat. Sol. (a),* 5:759–68 (Sec. 13.4).

Kosterin, Y. I., and Kragel'skii, I. V. (1960), "Relaxation Oscillation in Elastic Friction Systems," in *Friction and Wear in Machinery,* ed. by M. Kruschchov, ASME, New York, pp. 111–34 (Sec. 13.7).

Krajcinovic, D. (1983), "Constitutive Equations for Damaging Material," *J. Appl. Mech.* (ASME), 50:355–60.

Krajcinovic, D. (1989), "A Mesomechanical Model of Concrete and Its Application to the Size Effect Phenomenon," in *Cracking Damage—Strain Localization and Size Effect,* ed. by J. Mazars and Z. P. Bažant, Elsevier, London and New York, pp. 164–82; see also Preprints of France—U.S. Workshop at ENS, Cachan, France, in September 1988.

Krajcinovic, P., and Fonseka, G. U. (1981), "The Continuous Damage Theory of Brittle Materials," *J. Appl. Mech.* (ASME), 48:809–15.

Kröner, E. (1967), "Elasticity Theory of Materials with Long-Range Cohesive Forces," *Int. J. Solids Structures,* 3:731–42 (Sec. 13.10).

Krumhansl, J. A., (1968), "Some Considerations on the Relations between Solid State Physics and Generalized Continuum Mechanics," in *Mechanics of Generalized Continua,* ed. by E. Kröner, Springer-Verlag, Heidelberg, Germany, pp. 298–331 (Sec. 13.10).

Kunin, I. A. (1968), "The Theory of Elastic Media with Microstructure and the Theory of Dislocations," in *Mechanics of Generalized Continua,* ed. by E. Kröner, Springer-Verlag, Heidelberg, Germany, pp. 321–8 (Sec. 13.10).

Labuz, J. F., Shah, S. P., and Dowding, C. M. (1985) "Experimental Analysis of Crack Propagation in Granite," *Int. J. Rock Mech. Min. Sci. and Geomech. Abstracts,* 22(2):85–98 (Sec. 13.2).

Lamé, G. (1852), *Leçons sur la théorie de l'elasticité,* Gauthier-Villars, Paris (Sec. 13.5).

Leckie, F. A., and Onat, E. T. (1981), "Tensorial Nature of Damage Measuring Internal Variables," in *Physical Nonlinearities in Structural Analysis,* ed. by J. Hult and J. Lemaitre, Proceedings of the May 1980, IUTAM Symp. held at Senlis, France, Springer-Verlag, Berlin, pp. 140–55.

Lemaitre, J., and Chaboche, J. L. (1985), *Méchanique des Matériaux Solides,* Dunod-Bordas, Paris (Sec. 13.10).

Leroy, Y., and Ortiz, M. (1989), "Finite Element Analysis of Strain Localization in

Frictional Materials," *Int. J. Numer. Analytical Meth. in Geomechanics,* 13:53–74 (Sec. 13.3).

Levin, V. M. (1971), "The Relation between Mathematical Expectation of Stress and Strain Tensors in Elastic Micro-Heterogeneous Media" (in Russian), *Prikl. Mat. Mekh.,* 35:694–701 (Sec. 13.10).

Lin, F.-B. (1985), Private Communication on Unpublished Doctoral Research at Northwestern University (Sec. 13.8).

Lin, F.-B., Bažant, Z. P., Chern, J.-C., and Marchertas, A. H. (1987), "Concrete Model with Normality and Sequential Identification," *Computers and Structures,* 26(6):1011–25 (Sec. 13.11).

Lin, S. C., and Mura, T. (1973), "Elastic Field of Inclusion in Anisotropic Media (II)," *Phys. Stat. Sol. (a),* 15:281–5 (Sec. 13.4).

Maier, G. (1967a), "On Elastic-Plastic Structures with Associated Stress–Strain Relations Allowing for Work Softening," *Meccanica,* 2(1):55–64 (Sec. 13.6).

Maier, G. (1967b), "Extremum Theorems for the Analysis of Elastic-Plastic Structures Containing Unstable Elements," *Meccanica,* 2(4):235–42 (Sec. 13.6).

Maier, G. (1970), "Unstable Flexural Behavior in Elastoplastic Beams" (in Italian), *Rendic. Ist. Lombardo (A),* 102:648–77 (Sec. 13.6).

Maier, G. (1971), "Instability Due to Strain Softening," in *Stability of Continuous System,* Int. Union of Theoretical and Applied Mechanics Symposium, held in Karlsruhe, 1969; publ. by Springer-Verlag, Berlin, pp. 411–17 (Sec. 13.6).

Maier, G., and Zavelani, A. (1970), "Sul comportamento di aste metalliche compresse eccentricamente. Indagine sperimentale e considerazioni teoriche," *Costruz. Metall.,* 22(4):282–97 (Sec. 13.6).

Maier, G., Zavelani, A., and Dotreppe, J. C. (1973), "Equilibrium Branching due to Flexural Softening," *J. Eng. Mech.* (ASCE) 99(4):897–901 (Sec. 13.6).

Mandel, J. (1966), "Conditions de stabilité et postulat de Drucker," in *Rheology and Soil Mechanics,* ed. by J. Kravtchenko and P. M. Sirieys, Springer-Verlag, Berlin, pp. 56–68 (Sec. 13.3).

Marquis, D., and Lemaitre, J. (1988), "Constitutive Equations for the Coupling between Elasto-Plasticity, Damage and Aging," *Revue Phys. Appl.,* 23:615–24.

Mazars, J. (1986), "A Description of Micro- and Macro-Scale Damage of Concrete Structures," *Eng. Fract. Mech.,* 25:729–37.

Mazars, J., and Berthaud, Y. (1989), *Report,* Lab. de Méchanique et Technologie, ENS, Cachan, France (Sec. 13.10).

Mazars, J., and Pijaudier-Cabot, G. (1989), "Continuum Damage Theory—Application to Concrete," *J. Eng. Mech.* (ASCE), 115(2):327–44 (Sec. 13.11).

Mendis, P. A., and Darvall, P. L. (1988), "Stability Analysis of Softening Frames," *J. Struct. Eng.* (ASCE), 114(5):1057–72 (Sec. 13.6).

Mirsky, L. (1955), *An Introduction to Linear Algebra,* Oxford–Clarendon Press, London (also Dover Publ., New York, 1983).

Misner, C. W., Thorne, K. S., and Wheeler, J. A. (1973), *Gravitation,* Freeman, San Francisco (Sec. 13.0).

Misner, C. W., and Zapolsky, H. S. (1964), "High-Density Behavior and Dynamical Stability of Neutron Star Models," in *Phys. Rev. Letters,* 12, p. 535 (Sec. 13.0).

Mróz, Z. (1985), "Current Problems and New Directions in Mechanics of Geomaterials," chapter 24 in *Mechanics of Geomaterials: Rocks, Concretes, Soils,* ed. by Z. P. Bažant, John Wiley and Sons, Chichester and New York, pp. 534–66 (Sec. 13.6).

Mura, T. (1982), *Micromechanics of Defects in Solids,* Martinus Nijhoff Publishers, The Hague (Sec. 13.4).

Murkami, S. (1987), "Anisotropic Damage Theory and its Application to Creep Crack Growth Analysis," in *Constitutive Laws for Engineering Materials: Theory and*

Applications, (Proc. 2nd Int. Conf. held in Tucson, Ariz., Jan. 1985) ed. by C. S. Desai et al., Elsevier, New York–Amsterdam–London, Vol. 1, pp. 187–94.

Needleman, A. (1987), "Material Rate Dependence and Mesh Sensitivity in Localization Problems," *Comp. Meth. Appl. Mech. Engng.,* 67:68–85 (Sec. 13.1).

Needleman, A., and Tvergaard, V. (1984), "An Analysis of Ductile Rupture in Notched Bars," *J. Mech. Phys. Solids,* 3:461–90.

Nylander, H., and Sahlin, S. (1955), "Investigation of Continuous Concrete Beams at Far Advanced Compressive Strains in Concrete", *Betong,* 40(3); Translation No. 66, Cement and Concrete Assoc., London, England (Sec. 13.6).

Ortiz, M. (1985), "A Constitutive Theory for the Inelastic Behavior of Concrete," *Mech. of Materials,* 4:67–93 (Sec. 13.11).

Ortiz, M. (1987), "An Analytical Study of the Localized Failure Modes in Concrete," *Mech. of Materials,* 6:159–74 (Sec. 13.3).

Ortiz, M. (1988), "Microcrack Coalescence and Macroscopic Crack Growth Initiation in Brittle Solids," *Int. J. Solids Struct.,* 24(3): 231–50.

Ortiz, M. (1989), "Extraction of Constitutive Data from Specimens Undergoing Strain Localization," *J. Eng. Mech.* (ASCE), 115(8):1748–60 (Sec. 13.2).

Oritz, M., Leroy, Y., and Needleman, A. (1987), "A Finite Element Method for Localized Failure Analysis," *Comp. Meth. Appl. Mech. Eng.,* 61:189–224 (Sec. 13.3).

Ottosen, N. S. (1986), "Thermodynamic Consequences of Strain Softening in Tension," *J. Eng. Mech.* (ASCE), 112(11):1152–64 (Sec. 13.2).

Pande, G. N., and Sharma, K. G. (1980), "A Micro-Structural Model for Soils under Cyclic Loading," Proc. Int. Symp. on Soil under Cyclic and Transient Loadings, Swansea, Balkema Press, Rotterdam, Vol. 1, pp. 451–62.

Pande, G. N., and Sharma, K. G. (1981), "Multi-Laminate Model of Clays—a Numerical Study of the Influence of Rotation of the Principal Stress Axes," in *Implementation of Computer Procedures and Stress–Strain Laws in Geotechnical Engineering,* ed. by C. S. Desai and S. K. Saxena, Vol. 2, Acorn Press, Durham, N.C., pp. 575–90.

Petersson, P. E. (1981), "Crack Growth and Development of Fracture Zones in Plain Concrete and Similar Materials," Doctoral Dissertation, Lund Institute of Technology, Lund, Sweden (Sec. 13.2).

Pietruszczak, S., and Mróz, Z. (1981), "Finite Element Analysis of Deformation of Strain-Softening Materials," *Int. J. Num. Meth. Eng.,* 17:327–34.

Pijaudier-Cabot, G., and Akrib, A. (1989), "Bifurcation et résponse postbifurcation de structures en béton," Preprint, Colloquium GRECO on Rhéologie des Géomateriaux, France (Sec. 13.8).

Pijaudier-Cabot, G., and Bažant, Z. P. (1987), "Nonlocal Damage Theory," *J. Eng. Mech.* (ASCE), 113(10):1512–33 (Sec. 13.10).

Pijaudier-Cabot, G., and Bažant, Z. P. (1988), "Dynamic Stability Analysis with Nonlocal Damage," *Computer and Structures,* 29(3):503–507 (Sec. 13.10).

Pijaudier-Cabot, G., Bažant, Z. P., and Tabbara, M. (1988), "Comparison of Various Models for Strain-Softening," *Engineering Computations,* 5:141–50.

Pitman, E. B., and Schaeffer, D. G. (1987), "Stability of Time Dependent Compressible Granular Flow in Two Dimensions," *Comm. on Pure and Appl. Math.,* 40:421–47 (Sec. 13.7).

Planas, J., and Elices, M. (1989), "Size-Effect in Concrete Structures: Mathematical Approximations and Experimental Validation," in *Cracking and Damage—Strain Localization and Size Effect,* ed. by J. Mazars and Z. P. Bažant, Elsevier, London and New York, pp. 462–76 (Proc. of France–U.S. Workshop held at ENS, Cachan, September 1988) (Sec. 13.9).

Raniecki, B., and Bruhns, D. T. (1981), "Bounds to Bifurcation Stresses in Solids with Nonassociated Plastic Flow Laws and Finite Strain," *J. Mech. Phys. Solids,* 29:153–72 (Sec. 13.3).

Rashid, Y. R. (1968), "Analysis of Prestressed Concrete Pressure Vessels," *Nucl. Eng. Des.,* 7(4):334–55 (Sec. 13.9).

Read, H. E., and Hegemier, G. P. (1984), "Strain-Softening of Rock, Soil and Concrete—A Review Article," *Mech. Mat.,* 3(4):271–94 (Sec. 13.0).

Reinhardt, H. W., and Cornelissen, H. A. W. (1984), "Post-Peak Cyclic Behavior of Concrete in Unixial and Alternating Tensile and Compressive Loading," *Cement and Concrete Research,* 14(2):263–70 (Sec. 13.0).

Rice, J. R. (1968), "A Path Independent Integral and the Approximate Analysis of Strain Concentration by Notches and Cracks," *J. Appl. Mech.* (ASME), 35:379–86; see also "Mathematical Analysis in the Mechanics of Fracture," in *Fracture, an Advanced Treatise,* ed. by H. Liebowitz, Academic Press, New York, 1968, Vol. 2, pp. 191–250 (Sec. 13.9).

Rice, J. R. (1976), "The Localization of Plastic Deformation," Congress of Theor. and Appl. Mech., Delft, the Netherlands, Vol. 1, pp. 207–20 (Sec. 13.3).

Rice, J. R., and Ruina, A. L. (1983), "Stability of Steady Frictional Slipping," *J. of Applied Mech.* (ASME), 50:343–9 (Sec. 13.7).

Roelfstra, P. E. (1989), "Nonlocal Models, Localization Limiters and Size Effect, Part 2 and General Discussion," in *Cracking and Damage—Strain Localization and Size Effect,* ed. by J. Mazars and Z. P. Bažant, Elsevier, London and New York, pp. 539–46 (Proc. of France—U.S. Workshop at ENS, Cachan, France, September 1988) (Sec. 13.1.)

Rots, J. G., Nauta, P., Kusters, G. M. A., and Blaauwenraad, J. (1985), "Smeared Crack Approach and Fracture Localization in Concrete," *Heron,* 30(1):5–48 (Sec. 13.9).

Rots, J. G., and de Borst, R. (1989), "Analysis of Concrete Fracture in Direct Tension," *Int. J. Solids Struct.,* 25(12):1381–94 (Sec. 13.8).

Rudnicki, J. W. (1977), "The Inception of Faulting in a Rock Mass with a Weakened Zone," *J. Geophysical Research,* 82(5):844–54 (Sec. 13.3).

Rudnicki, J. W., and Rice, J. R. (1975), "Conditions for the Localization of Deformation in Pressure-Sensitive Dilatant Materials," *J. Mech. Phys. Solids,* 23:371–94 (Sec. 13.3).

Ruina, A. L. (1981), "Friction Laws and Instabilities: A Quasistatic Analysis of Some Dry Frictional Behavior," Ph.D. Thesis, Brown University (Sec. 13.7).

Rüsch, H., and Hilsdorf, H. (1963), "Deformation Characteristics of Concrete under Axial Tension" (in German), *Voruntersuchungen,* Bericht 44, Munich (Sec. 13.2).

Sandler, I. S. (1984), "Strain-Softening for Static and Dynamic Problems," in Proc. Symp. on Constit. Equations, ASME, Winter Annual Meeting, New Orleans, pp. 217–31 (Sec. 13.0).

Sandler, I., and Wright, J. (1983), "Summary of Strain-Softening," in *Theoretical Foundations for Large-Scale Computations of Nonlinear Material Behavior,* ed. by S. Nemat-Nesser, DARPA-NSF Workshop, Northwestern University, Evanston, Ill, pp. 285–315 (Sec. 13.1).

Scanlon, A. (1971), "Time Dependent Deflections of Reinforced Concrete Slabs," Ph.D. Dissertation, Univ. of Alberta, Edmonton, Canada (Sec. 13.9).

Schaeffer, D. G. (1987), "Instability in the Evolution Equation Describing Incompressible Granular Flow," *J. Differential Equations,* 66(1):19–50 (Sec. 13.7).

Schaeffer, D. G., and Pitman, E. B. (1988), "Illposedness in Three-Dimensional Plastic Flow," Report, Dept. of Math., Duke University, Durham, N.C. (Sec. 13.7).

Schapery, R. A. (1968), "On a Thermodynamic Constitutive Theory and its Applications to Various Nonlinear Materials," *Proceedings of the International Union of*

Theoretical and Applied Mechanics Symposium, ed. by B. A. Boley, Springer-Verlag, New York (Sec. 13.11).

Schreyer, H. L. (1989), "Formulations for Nonlocal Softening in a Finite Zone with Anisotropic Damage," in *Cracking and Damage—Strain Localization and Size Effect,* ed. by J. Mazars and Z. P. Bažant, Elsevier, London and New York, pp. 426–39 (Proc. of France–U.S. Workshop held at ENS, Cachan, September 1988) (Sec. 13.10).

Schreyer, H. L., and Chen, Z. (1986), "The Effect of Localization on the Softening Behavior of Structural Members," *J. Appl. Mech.* (ASME), 53:791–97; also, Proc., Symposium on Constitutive Equations: Micro, Macro, and Computational Aspects, ASME Winter Annual Meeting, New Orleans, December 1984, ed. by K. Willam, ASME, New York, pp. 193–203 (Sec. 13.10).

Simo, J. C. (1989), "Strain Softening and Dissipation: A Unification of Approaches," in *Cracking and Damage—Strain Localization and Size Effect,* ed. by J. Mazars and Z. P. Bažant, Elsevier, London and New York, pp. 440–61 (Proc. of France—U.S. Workshop held at ENS, Cachan, September 1988) (Sec. 13.10).

Simo, J. C., and Ju, J. W. (1987a), "A Continuum Strain Based Damage Model. Part I," *Int. J. Solids Structures,* 23(7):821–40.

Simo, J. C., and Ju, J. W. (1987b), "A Continuum Strain Based Damage Model. Part II," *Int. J. Solids Structures,* 23(7):841–69.

Suidan, M., and Schnobrich, W. C. (1973), "Finite Element Analysis of Reinforced Concrete," *J. Struct. Eng.* (ASCE), 99(10):2109–22 (Sec. 13.10).

Sulem, J., and Vardoulakis, I. (1989), "Bifurcation Analysis of the Triaxial Test on Rock Specimen," in *Cracking and Damage—Strain Localization and Size Effect,* ed. by J. Mazars and Z. P. Bažant, Elsevier, London and New York, pp. 308–22 (Proc. of France–U.S. Workshop held at ENS, Cachan, September 1988) (Sec. 13.8).

Synge, J. L. (1957), *J. of Mathematics and Physics,* 35:323 (Sec. 13.1).

Taylor, G. I. (1938), "Plastic Strain in Metals," *J. Inst. Metals* (London), 62:307–24 (Sec. 13.11).

Thomas, T. Y. (1961), *Plastic Flow and Fracture of Solids,* Academic Press, New York (Sec. 13.3).

Thompson, J. M. T. (1982), *Instabilities and Catastrophes in Science and Engineering,* John Wiley and Sons, New York (Sec. 13.0).

Timoshenko, S. P., and Goodier, J. N. (1970), *Theory of Elasticity,* 3rd ed., McGraw-Hill, New York (Sec. 13.5).

Trent, B. C., Margolin, L. G., Cundall, P. A., and Gaffney, E. S. (1987), "The Micro-Mechanics of Cemented Granular Material," in *Constitutive Laws for Engineering Materials: Theory and Applications*, Proc. 2nd Int. Conf. held in Tucson, Ariz. Jan. 1985, ed. by C. S. Desai et al., Elsevier, New York, vol. 1, pp. 795–802.

Triantafyllidis, N., and Aifantis, E. (1986), "A Gradient Approach to Localization of Deformation: I. Hyperelastic Materials," *J. Elasticity,* 16:225–37.

Tvergaard, V., Needleman, A., and Lo, K. K. (1981), "Flow Localization in the Plane Strain Tensile Test," *J. Mech. Phys. Solids,* 29:115–42.

Valanis, K. C. (1971), "A Theory of Viscoplasticity without a Yield Surface," *Archiwum Mechaniki Stossowanej,* 23:517–51 (Sec. 13.11).

van Mier, J. G. M. (1984), "Strain-Softening of Concrete Under Multiaxial Loading Conditions," Ph.D. Dissertation, De Technische Hogeschool Eindhoven, The Netherlands (Sec. 13.2).

van Mier, J. G. M. (1986), "Multiaxial Strain-Softening of Concrete; Part I: Fracture, Part II: Load Histories," *Materials and Structures,* 111(19):179–200 (Sec. 13.2).

Vardoulakis, I. (1983), "Rigid Granular Plasticity Model and Bifurcation in the Triaxial Test," *Acta Mechanica,* 49:57–79 (Sec. 13.8).

Vardoulakis, I., Sulem, J., and Guenot, A. (1988), "Borehole Instabilities as Bifurcation Phenomena," *Int. J. Rock Mech. and Min. Sci. and Geomech. Abstracts,* 25:159–70.

Vermeer, P. A., and de Borst, R. (1984), "Nonassociated Plasticity for Soils, Concrete and Rock," *Heron,* No. 3, Delft, Netherlands (Sec. 13.3).

Warner, R. F. (1984), "Computer Simulation of the Collapse Behavior of Concrete Structures with Limited Ductility," in *Computer Aided Analysis and Design of Concrete Structures* (Proc. of Int. Conf. held in Split, Yugoslavia, September 1984), ed. by F. Damjanić and E. Hinton, Pineridge Press, Swansea, U.K., pp. 1257–70 (Sec. 13.6).

Wastiels, J. (1980), "A Softening Plasticity Model for Concrete," *Proc. Int. Conf. Numerical Methods for Nonlinear Problems*, Pineridge Press, Swansea, U.K., pp. 481–93.

Wood, R. H. (1968), "Some Controversial and Curious Developments in the Plastic Theory of Structures," in *Engineering Plasticity,* ed. by J. Heyman and F. A. Leckie, Cambridge Univ. Press, Cambridge, England, pp. 665–91 (Sec. 13.6).

Wu, F. H., and Freund, L. B. (1984), "Deformation Trapping Due to Thermoplastic Instability in One-Dimensional Wave Propagation," *J. Mech. Phys. Solids,* 32(2):119–32 (Sec. 13.1).

Yang, B. L., Dafalias, Y. F., and Herrmann, L. R. (1985), "A Bounding Surface Plasticity Model for Concrete," *J. Eng. Mech.* (ASCE) 111:359–80 (Sec. 13.11).

Yuzugullu, O., and Schnobrich, W. C. (1973), "A Numerical Procedure for the Determination of the Behavior of a Shear Wall Frame System," *ACI Journal,* 70:474–9 (Sec. 13.10).

Zubelewicz, A. (1983), "A Proposal for a New Structural Model of Concrete," *Archiwum Inzynierii Ladowej* (Poland), 24(4):417–39 (Sec. 13.11).

Zubelewicz, A., and Bažant, Z. P. (1987), "Constitutive Model with Rotating Active Plane and True Stress," *J. Eng. Mech.* (ASCE) 113(3):398–416 (Sec. 13.11).

"First my fear;
then my court'sy;
last my speech.
My fear is your displeasure;
my court'sy, my duty;
and my speech, to beg your pardons."

William Shakespeare
(*King Henry IV*)

The more time you spend on reporting on what you are doing, the less time you have for doing anything. *Stability* is achieved when you spend all the time doing nothing but reporting on the nothing you are doing (Cohn's law).

Glossary of Symbols

A	= cross-section area (1.1)
A_0	= A/m (m = shear correction coefficient) (1.7)
A_k	= constant (3.7)
B	= bimoment (6.3) or constant of size effect law (12.2)
$\mathbf{B}$	= transformation matrix (4.1)
$C, C_1, C_2, \ldots$	= spring stiffness (1.2)
C	= $Eh/(1-\nu^2)$ (7.6)
$\tilde{C}$	= variable spring stiffness (3.9)
C_a, C_b	= tangential stiffness (13.2)
C_m	= correction coefficient (1.6)
C_f	= slope of friction force diagram (13.7)
C_t	= tangential compliance for loading (10.1)
C_u	= tangential compliance for unloading (10.1)
$\mathbf{C}$	= compliance (flexibility) matrix (4.1) or elastic compliance tensor of the material, C_{ijkl} (10.6) or Green's (Cauchy–Green's) deformation tensor (11.1)
$\bar{\mathbf{C}}$	= flexibility matrix of primary structure (2.2)
$\mathbf{C}_S, \mathbf{C}_T$	= tangential compliance matrix of structure in isentropic or isothermal conditions (10.1)
$\mathbf{C}^{(\alpha)}$	= compliance matrix for path α (10.2)
D	= characteristic dimension (size) of structure (12.1) or cylindrical stiffness of plate or shell, $Eh^2/12(1-\nu^2)$ (7.1) or determinant (2.2) or energy dissipated by damping (3.3)
D_{ijkm}	= plate bending stiffnesses (7.2)
D^t_{ijkm}	= incremental elastic moduli (13.1) or incremental moduli for loading (tensor $\mathbf{D}_t$) (13.3)
D^u_{ijkm}	= incremental moduli for unloading (tensor $\mathbf{D}_u$) (13.3)
D_0	= constant of size effect law (12.2)
$D_1, D_2, \ldots$	= principal minors (4.1)
E	= Young's elastic modulus (1.1) or total energy (potential plus kinetic) (3.6)

	or instantaneous elastic modulus (9.1)
E_{ch}	= chord modulus for elastoplastic column (8.2)
E_r	= reduced modulus for elastoplastic column (8.1)
E_t	= tangential (incremental) modulus for loading (8.1)
E_u	= incremental modulus for unloading (8.1)
E_{xx}, E_{yy}, E_{xy}	= elastic moduli of orthotropic material (7.1)
E_∞	= long-time (quasi) elastic modulus (9.1)
$\underset{\sim}{E}$	= relaxation operator (9.1)
$\underset{\sim}{E}^{-1}$	= creep operator (9.1)
E'	= $E/(1-\nu^2)$ (12.1)
E^*	= age-adjusted effective modulus for creep of concrete (9.4)
$\overline{EI}$	= effective bending stiffness (8.5)
EI_ω	= warping stiffness of thin-walled beam cross section (6.4)
$\mathbf{E}_S, \mathbf{E}_T$	= isentropic and isothermal tensors of tangential moduli (10.1)
F	= prestressing force (1.8)
	or reference load (2.2)
	or stress function (7.4)
	or axial tensile force (8.7)
	or Helmholtz free energy of structure (10.1)
$\mathcal{F}$	= Helmholtz free energy of structure-load system (10.1)
F_i	= in-plane distributed loads (7.2)
F_a	= allowable compressive stress for axial load (8.4)
F_b	= allowable compressive stress for bending (8.4)
$\mathbf{F}$	= column matrix of forces (2.2)
	or transformation tensor (11.1)
$\mathbf{F}^L$	= column matrix of fixed-end forces due to loads (2.2)
G	= elastic shear modulus (1.7)
G	= Gibbs free energy of structure (10.1)
$\mathcal{G}$	= Gibbs free energy of structure-load system (10.1)
	or energy release rate (energy released per unit length and unit width of crack) (12.1)
G_f	= fracture energy (12.1)
G_{xy}	= elastic shear modulus of orthotropic material (7.1)
GJ	= torsional stiffness for simple torsion (6.1)
H	= height (4.5)
	or horizontal force (4.2)
	or Heaviside step function (9.4)
	or enthalpy of structure (10.1)
	or hardening (tangential) modulus (10.6)
$\mathcal{H}$	= enthalpy of structure-load system (10.1)
$\mathbf{H}$	= Hurwitz matrix (3.7)
I	= centroidal moment of inertia (1.1)
	or functional (5.1)
	or invariant of stress tensor (I_1, I_2, I_3) (10.6)
I_y	= moment of inertia in y direction (about z axis) (6.1)
I_z	= moment of inertia in z direction (about y axis) (6.1)
I_ω	= warping (sectorial) moment of inertia (6.1)
$I_{\omega y}, I_{\omega z}$	= mixed cross-sectional moments (6.4)

$\mathbf{I}$	= identity matrix (2.3)
J	= pure-torsion stiffness divided by G (Sec. 3.2) or compliance function (creep strain per unit constant stress) (9.1) or J-integral (12.1)
J_2	= second invariant of deviatoric part of stress tensor (10.6)
K	= stiffness coefficient (2.1) or bulk modulus (Sec. 8.1)
K_t	= tangential stiffness for loading (10.1)
K_u	= tangential stiffness for unloading (10.1)
$K_{\mathrm{I}}, K_{\mathrm{II}}, K_{\mathrm{III}}$	= stress intensity factors for modes I, II, III (12.1)
K_c	= critical value of stress intensity factor (12.1)
$\mathbf{K}$	= stiffness matrix (2.2)
$\hat{\mathbf{K}}$	= symmetric part of matrix $\mathbf{K}$ (4.1)
$\tilde{\mathbf{K}}$	= antisymmetric part of matrix $\mathbf{K}$ (10.4)
$\mathbf{K}_e$	= linear elastic stiffness matrix (2.3)
$\mathbf{K}_T$	= tangential stiffness matrix in isothermal conditions (10.1)
$\mathbf{K}_S$	= tangential stiffness matrix in isentropic conditions (10.1)
$\mathbf{K}^t$	= tangential stiffness matrix (12.5)
$\mathbf{K}^\sigma$	= geometric stiffness matrix (2.3)
L	= length (4.4) or effective length of column (1.4)
L_x, L_y	= length in x or y directions (2.7)
$\underset{\sim}{L}$	= self-adjoint positive definite operator (5.1)
M	= bending moment (1.1) or moment in general (4.5)
M_0	= bending moment for no axial load (1.2)
$M_{\max}$	= maximum bending moment (1.6)
M_t	= torsional moment (torque) (1.10)
M_u	= ultimate design moment (8.5)
M_y	= bending moment for y direction (about axis z) (1.10)
M_z	= bending moment for z direction (about axis y) (1.10)
M_y^*, M_z^*	= magnified bending moments (8.4)
M_{xx}, M_{yy}	= bending moments (per unit width) in a plate or shell (7.1)
$M_{xy} = M_{yx}$	= twisting moment (per unit width) (7.1)
M_{I}	= first-order moment (disregarding second-order geometric effects) (8.5)
M_{II}	= second-order moment (8.5)
$\mathbf{M}_t$	= torque vector (1.10)
M^T	= bending component of torque vector (1.10)
$\underset{\sim}{M}$	= linear differential operator (5.1)
N	= axial force (positive if tensile) (1.3)
N_{ij}	= in-plane forces per unit width of plate (7.2)
$\mathbf{N}$	= column matrix of interpolation functions (2.3)
$\underset{\sim}{N}$	= linear differential operator (5.1)
P	= axial load (positive for compression) (1.2)
P_D	= permanent axial load (9.5)
P_E	= Euler load (of pin-ended column) (1.2)

P_{E_∞} = long-time Euler load corresponding to modulus E_∞ (9.2)
P_L = lower bound for critical load (5.3)
or live load (axial) (9.5)
P_R = Rayleigh quotient (5.3)
P_T = Timoshenko quotient (5.3)
P_{cr} = critical value of axial load (1.2)
P_{cr_y}, P_{cr_z} = critical load for bending in the y or z direction (6.2)
P_{cr_θ} = critical load for pure torsional buckling (6.2)
P_{cr}^D = critical load under displacement control (8.1)
P_m = maximum axial load (8.6)
P_r = reduced modulus load (8.1)
P_t = tangent modulus loal (8.1)
P_u = axial force corresponding to yielding of the material (2.5)
= ultimate (design) axial force (8.5)
= ultimate load (12.2)
P_x, P_y = axial force in directions x, y (2.7)
P_0 = axial yield force (squash load) (Sec. 8.6)
$\mathbf{P}$ = force field of components P_x, P_y, P_z (3.2)
$\mathbf{P}$ = column matrix of loads (10.1)
Q = shear force normal to the deflected beam axis (1.3)
or heat flow into the structure (10.1)
Q_k = force component in direction k (2.3)
or constant (3.7)
Q_x, Q_y = (rotating) transverse shear forces normal to deflection surface, per unit width (7.6)
Q_{0_n} = Fourier coefficients (1.5)
$\mathbf{Q}$ = square matrix (4.1)
or column matrix of displacements (10.1)
$\mathbf{Q}_u$, $\mathbf{Q}_t$ = matrices of Eshelby coefficients for unloading and loading (13.4)
R = curvature radius (2.8)
or radius (7.5)
or resistance (8.4)
or energy release required for crack propagation (resistance to crack growth) (12.3)
R_p = inelastic zone length (12.2)
R_t = tangential (incremental) bending stiffness (8.3)
or tangential bending stiffness for loading (13.6)
R_u = incremental bending stiffness for unloading (13.6)
$\mathbf{R}$ = unitary matrix (4.1)
or rotation tensor (11.1)
S = safety factor (8.4)
or surface of structure (10.1)
or entropy (10.1)
$(\Delta S)_{\text{in}}$ = internally produced entropy increment of structure (10.1)
S_{ij}, $\mathbf{S}$ = true stress tensor (Cauchy stress, Eulerian stress) (11.2)
$\hat{S}_{ij}$ = Truesdell stress rate (11.3)
$\hat{S}_{ij}^R$ = Cotter–Rivlin stress rate (11.3)

$\hat{S}^J_{ij}$ = Jaumann stress rate (11.3)
S_{ijkm} = fourth-order tensor of Eshelby coefficients for ellipsoidal domain (13.4)
T = kinetic energy (3.6)
or period (3.9)
or absolute temperature (4.9)
T_{ij} = first Piola–Kirchhoff stress tensor (Lagrangian stress) (11.2)
U = strain energy (2.9)
or total internal energy of structure (10.1)
$\mathcal{U}$ = total internal energy of structure-load system (10.1)
U_k = generalized displacement parameters (6.6)
U_b = strain energy due to bending in shells (7.6)
U_m = strain energy due to membrane action in shells (7.6)
U_1 = strain energy of bending and twisting in plates (7.2)
U_2 = strain energy of in-plane deformation in plates (7.2)
$\mathbf{U}$ = right stretch tensor (11.1)
V = shear force (normal to undeflected beam axis) (1.3)
or volume of structure (4.3)
V_0 = initial volume of body (10.1)
V_k = generalized displacement parameters (6.6)
V_x, V_y = fixed direction shear forces per unit width (7.1)
$\mathbf{V}$ = left stretch tensor (11.1)
W = work of loads (3.2)
$\mathcal{W}$ = work at equilibrium displacements (4.9)
$\bar{\mathcal{W}}$ = work per unit initial volume (11.2)
W_{ext} = work of external forces (2.5)
W_f = energy dissipated at the crack tip (12.4)
W_{int} = work of internal forces (2.5)
W^* = complementary work (10.1)
$\bar{W}^*$ = complementary work per unit volume (10.6)
$\bar{W}$ = work per unit load (4.7)
or work per unit volume (10.6)
$\mathcal{W}_\sigma$ = work of initial stresses (10.1)
X_i, $\mathbf{X}$ = initial spatial coordinates of material point of continuum (11.1)
Z = Batdorf parameter for shells (7.6)
$\mathbf{Z}$, Z_{ij} = matrix deciding localization instability (13.3, 13.4)
a = distance (2.7)
or crack length (12.1)
b = distance (2.7)
or damping parameter (3.1)
or width (8.1)
or thickness (12.1)
c = sandwich core thickness (1.7)
or stability function (carry-over factor) of beam-column (2.1)
or foundation modulus (5.2)
or limit slenderness (8.4)
or z-coordinate of neutral axis (8.6)

	or length of (equivalent elastic) crack advance (12.2)
	or velocity of imposed displacement (13.1)
c_f	= effective length of fracture process zone for infinite-size structure (12.2)
$[c_{ij}]$	= damping matrix (3.5)
e	= eccentricity of axial load (1.4)
	or first-order normal strain (4.3)
e_x	= first-order normal strain (6.1)
$\mathbf{e}$, e_{ij}	= small (linearized) strain tensor (10.1)
f	= skin thickness of sandwich plate (1.7)
	or small (disturbing) load (2.2)
f_y	= yield limit (1.2)
f'_c	= compression strength (1.2)
f'_t	= tensile strength (12.2)
f_u	= ultimate stress (strength) (1.9)
$\mathbf{f}$	= column matrix of generalized forces (2.3)
$\mathbf{f}_S$, $\mathbf{f}_T$	= column matrix of equilibrium forces at isentropic or isothermal conditions (10.1)
$g(\alpha)$	= nondimensional energy release rate of a crack (function of α and structure shape) (12.2)
h	= thickness (2.4)
	or length or thickness of localization zone (13.2, 13.3)
k	= $\sqrt{P/EI}$ (1.2)
$k(\alpha)$	= function of structure shape (12.1)
l	= length of column (1.1)
l^*	= half wavelength of the deflection curve (2.1)
m	= shear correction coefficient (1.7)
	or small applied disturbing moment (2.9)
	or point mass (3.2)
m_{xz}, m_{yz}	= couple stresses in micropolar continuum (2.10)
$[m_{ij}]$	= mass matrix (3.5)
$\mathbf{n}$, n_i	= unit vector in principal strain direction (11.1)
	or unit normal vector (11.2)
p	= lateral distributed load (1.2)
p_{cr}	= critical value of hydrostatic pressure (1.8)
p_h	= hydrostatic pressure (1.8)
p_N	= transverse load due to in-plane forces (7.2)
p_n	= Fourier coefficients (4.3)
p_r	= radial distributed forces (1.8)
	or distributed reaction (5.2)
q	= generalized displacement (3.5)
q_n	= mode amplitude (1.2)
	or degree of freedom (3.6)
	or Fourier coefficient (1.5)
$q_i^{(k)}$	= displacements for mode k (eigenvector) (4.3)
$\mathbf{q}$	= column matrix of generalized displacements (3.5)
$\mathbf{q}^{(k)}$	= buckling mode (4.3)
r	= radius of inertia (1.2)

r_p	= Irwin's inelastic zone length (12.2)
r_p^2	$= (I_z + I_y)/A$ (6.1)
s	= length coordinate along deflected member centerline (1.8)
	or stability function of beam-column (2.1)
	or span (4.5)
	or standard deviation (8.4)
$\bar{s}$	$= s(1 + c)$ = stability function (2.1)
s^*	$= 2\bar{s} - \pi^2 P/P_E$ = stability function (2.1)
t	= time (3.1)
	or thickness (6.1)
t'	= time (9.1)
t^*	= critical time (9.3)
t_0, t_1	= age of concrete at loading (9.1)
$\mathbf{t}$	= unit tangent vector (7.2)
u	= displacement in direction x (1.9)
	or kl (1.4)
	or load-point displacement (4.4)
	or radial displacement (13.5)
u_i, $\mathbf{u}$	= displacements of material point of continuum in x_i direction (11.1)
u_s	= transverse displacement in direction s (6.4)
u_1, u_2, u_3	= displacements of wall middle surface in x, y, z directions (6.1)
$\mathbf{u}$	= column matrix of joint displacements (2.2)
v	= displacement in direction y (1.10)
	or wind velocity (3.2)
	or velocity (8.6)
	or wave velocity (13.1)
v_i	= velocity of material point in the x_i-direction (11.1)
w	= deflection (in direction z) (1.1)
w_c	= width of crack band (12.2)
w_k	= load-point displacement (4.3)
w_{max}	= maximum column deflection (1.1)
w_0	= deflection at no axial load (1.6)
x	= coordinate (along undeformed centerline for beams) (1.1)
x_i, $\mathbf{x}$	= current (final) spatial coordinates of material point of continuum (11.1)
y	= coordinate (in lateral direction for beams) (7.1)
	or distance from centroid to column face (8.4)
y_i	= transformed variable (4.1)
y_k	= amplitude of buckling mode (4.3)
z	= coordinate (transverse for beams, plates and shells) (1.1)
z_i	= transformed variable (4.1)
z_0	= coordinate of beam centerline at no axial load (1.5)
α	= nondimensional imperfection (2.6)
	or angle (1.8)
	or friction coefficient (10.7)
	or a/D = relative crack length (12.2)
	or weight function (13.10)

α_{ij} = Almansi (Eulerian) strain tensor (11.1)
β = $1 - P/GA_0$ (1.7)
or damping coefficient (3.1)
or dilatancy factor (10.7)
or brittleness number (12.2)
or shear retention factor (13.10)
γ = shear strain (1.7)
$\gamma^{(1)}$ = linear part of shear strain (6.1)
$\gamma^{(2)}$ = second-order part of shear strain (6.1)
γ^p = plastic slip (10.7)
δ = deflection (2.7)
or crack opening displacement (12.2)
or variation sign (4.2)
$\delta(x)$ = Dirac delta function of x (13.1)
δ_{ij} = Kronecker delta (= 1 for $i = j$ and 0 for $i \neq j$) (7.2)
δ_c = opening displacement at crack front (12.2)
δ_f = fictitious crack opening displacement (12.2)
ε = normal strain (1.1)
ε_1 = normal strain at any generic point of cross section (4.3)
or load-produced strain (8.3)
ε^0 = residual strain (8.3)
$\varepsilon^{(2)}$ = second-order component of normal strain (4.3)
ε_{ij}^b = Biot's finite strain (11.1)
$\varepsilon_{ij}^{(m)}$ = finite strain associated with parameter m (11.1)
$\boldsymbol{\varepsilon}$, ε_{ij} = strain tensor (4.3) (finite strain tensor in Ch. 11)
$\boldsymbol{\varepsilon}^p$ = inelastic (or plastic) strain tensor (10.6)
ξ = E_u/E_t (8.1)
ξ_u = E_u/E_t (10.1)
η = dashpot viscosity (9.1)
θ = slope angle (1.3)
or rotation (2.1)
or angle of twist (6.1)
κ = curvature (2.8)
or current slip limit (10.7)
κ_k = scalar parameter (10.6)
λ = $kl = \pi\sqrt{P/P_E}$ (2.1)
or load parameter (4.1)
or column slenderness (8.4)
or logarithmic strain (8.7)
$\lambda_1, \lambda_2, \ldots$ = eigenvalues (4.1)
or control parameters in catastrophe theory (4.7)
λ_c = column slenderness parameter (8.4)
$\lambda_{(i)}$ = principal values of tensor $\mathbf{U}$ (11.1)
μ = magnification factor (1.5)
or load multiplier (1.5)
or mass per unit length (3.1)
ν = Poisson ratio (1.7)
$\boldsymbol{\nu}$ = unit normal vector (7.2)

ρ	= curvature radius (1.1) or P/P_E (2.1) or distance (6.4) or mass per unit length (13.1)
σ	= normal stress (1.1)
σ_{cr}	= critical stress (1.7)
σ_E	= Euler stress (1.2)
σ_N	= nominal stress at maximum load (12.2)
σ_p	= stress at proportionality limit (8.1) or apparent yield limit (8.3) or yield limit of the material (8.5)
σ_r	= residual stress (8.3)
σ_t	= stress at tangent modulus load (8.1)
σ_u	= ultimate stress (strength) (8.4)
$\sigma_{xx}, \sigma_{xy}, \ldots$	= normal and shear stresses in cartesian coordinates x, y (Sec. 2.10)
$\boldsymbol{\sigma}, \sigma_{ij}$	= stress tensor (4.3)
τ	= shear stress (6.1)
τ^*	= shear stress due to Saint-Venant torsion (6.1)
τ_N	= intrinsic (shape-independent) nominal stress (12.2)
τ_r	= retardation time (9.1)
τ_{xs}	= shear stress in wall middle surface (6.4)
ϕ	= rotation (2.4) or collapse load reduction factor for shells (7.7)
$\phi(t)$	= creep coefficient (9.4)
ϕ_s	= stability function (2.1)
ϕ_u	= strength reduction factor (8.4)
χ	= aging coefficient for concrete creep (9.4)
ψ	= rotation of beam cross section (1.7)
$\psi(t)$	= creep coefficient (9.4)
ψ_s	= stability function (2.1)
ω	= angular velocity (3.4) or warping function (sectorial coordinate) (6.1) or coefficent of variation (8.4) or damage (13.10)
ω, ω_n	= undamped natural circular frequency (3.1)
ω', ω_n'	= damped natural circular frequency (3.1)
ω^0	= natural frequency of coiumn at no axial load (3.3)
ω_{ij}	= tensor of small (linearized) rotation of the material (11.1)
Δ	= deflection (1.4) or relative displacement (2.1) or increment sign (2.6)
$\Delta_1, \Delta_2, \ldots$	= principal minors (3.7)
Π	= potential energy (2.10)
Π^*	= complementary energy (4.2)
$\bar{\Pi}^*$	= complementary energy of primary structure (4.2)
$\Sigma_{ij}^{(m)}$	= second Piola–Kirchhoff stress tensor (11.2)
Φ, Φ^*	= energy functionals (3.8)

Φ	= potential energy of loads (4.3) or energy dissipation rate (13.10)
Ω	= forcing frequency (3.3) or nonlocal damage (13.10)

ADDITIONAL SYMBOLS

Superscripts

$'$	prime indicates derivative with respect to length coordinate
$\dot{}$	superimposed dot indicates derivative with respect to time
$\hat{}$	indicates symmetric part of a matrix
$\tilde{}$	indicates antisymmetric part of a matrix
T	superscript T indicates transpose of a matrix
0	superscript 0 indicates initial equilibrium state
(m)	superscript (m) refers to finite strain tensor definition (Sec. 11.1) with which the variable is associated

Subscripts

$,i$	a subscript preceded by a comma (e.g. i) indicates a partial derivative (with respect to coordinate x_i)

Other Symbols

$:$	denotes a tensor product contracted on two inner indices
$\cdot$	denotes a tensor product contracted on one inner index
$\otimes$	denotes a tensor product contracted on no index
$\langle A \rangle$	$= A$ if $A > 0$; $\langle A \rangle = 0$ if $A \leq 0$ (Macaulay brackets)

Author Index

Italic page numbers indicate reference citations.

Subject Index